Quanta, Matter, and Change

A molecular approach to physical chemistry

Peter Atkins
Professor of Chemistry,
University of Oxford,
and Fellow of Lincoln College, Oxford

Julio de Paula
Professor and Dean of the College of Arts and Sciences,
Lewis and Clark College,
Portland, Oregon

Ronald Friedman
Professor and Chair of Chemistry,
Indiana University Purdue University Fort Wayne,
Fort Wayne, Indiana

W. H. Freeman and Company
New York

Library of Congress Control Number: 2005936591

Quanta, Matter, and Change: A molecular approach to physical chemistry

ISBN: 0-7167-6117-3
ISBN: 978-0-7167-6117-4

Published in the United States and Canada by W. H. Freeman and Company
This edition has been authorized by Oxford University Press for sale in the United States and Canada only and not for export therefrom.

First printing 2009

Typeset by Graphicraft Ltd, Hong Kong
Printed and bound in China by C&C Offset Printing Co. Ltd

W. H. Freeman and Company
41 Madison Avenue
New York, NY 10010
www.whfreeman.com

General data and fundamental constants

Quantity	Symbol	Value	Power of ten	Units
Speed of light	c	2.997 925 58*	10^{8}	$\mathrm{m\ s^{-1}}$
Elementary charge	e	1.602 176	10^{-19}	C
Faraday's constant	$F = N_A e$	9.648 53	10^{4}	$\mathrm{C\ mol^{-1}}$
Boltzmann's constant	k	1.380 65	10^{-23}	$\mathrm{J\ K^{-1}}$
Gas constant	$R = N_A k$	8.314 47		$\mathrm{J\ K^{-1}\ mol^{-1}}$
		8.314 47	10^{-2}	$\mathrm{dm^3\ bar\ K^{-1}\ mol^{-1}}$
		8.205 74	10^{-2}	$\mathrm{dm^3\ atm\ K^{-1}\ mol^{-1}}$
		6.236 37	10^{1}	$\mathrm{dm^3\ Torr\ K^{-1}\ mol^{-1}}$
Planck constant	h	6.626 08	10^{-34}	J s
	$\hbar = h/2\pi$	1.054 57	10^{-34}	J s
Avogadro's constant	N_A	6.022 14	10^{23}	$\mathrm{mol^{-1}}$
Atomic mass constant	m_u	1.660 54	10^{-27}	kg
Mass				
electron	m_e	9.109 38	10^{-31}	kg
proton	m_p	1.672 62	10^{-27}	kg
neutron	m_n	1.674 93	10^{-27}	kg
Vacuum permittivity	$\varepsilon_0 = 1/c^2\mu_0$	8.854 19	10^{-12}	$\mathrm{J^{-1}\ C^2\ m^{-1}}$
	$4\pi\varepsilon_0$	1.112 65	10^{-10}	$\mathrm{J^{-1}\ C^2\ m^{-1}}$
Vacuum permeability	μ_0	4π	10^{-7}	$\mathrm{J\ s^2\ C^{-2}\ m^{-1}\ (= T^2\ J^{-1}\ m^3)}$
Magneton				
Bohr	$\mu_B = e\hbar/2m_e$	9.274 01	10^{-24}	$\mathrm{J\ T^{-1}}$
nuclear	$\mu_N = e\hbar/2m_p$	5.050 78	10^{-27}	$\mathrm{J\ T^{-1}}$
g value of the electron	g_e	2.002 32		
Bohr radius	$a_0 = 4\pi\varepsilon_0\hbar^2/m_e e^2$	5.291 77	10^{-11}	m
Fine-structure constant	$\alpha = \mu_0 e^2 c/2h$	7.297 35	10^{-3}	
	α^{-1}	1.370 36	10^{2}	
Second radiation constant	$c_2 = hc/k$	1.438 78	10^{-2}	m K
Stefan–Boltzmann constant	$\sigma = 2\pi^5 k^4/15h^3c^2$	5.670 51	10^{-8}	$\mathrm{W\ m^{-2}\ K^{-4}}$
Rydberg constant	$R = m_e e^4/8h^3 c\varepsilon_0^2$	1.097 37	10^{5}	$\mathrm{cm^{-1}}$
Standard acceleration of free fall	g	9.806 65*		$\mathrm{m\ s^{-2}}$
Gravitational constant	G	6.673	10^{-11}	$\mathrm{N\ m^2\ kg^{-2}}$

**Exact value*

The Greek alphabet

A, α	alpha	H, η	eta	N, ν	nu	Y, υ	upsilon
B, β	beta	Θ, θ	theta	Ξ, ξ	xi	Φ, φ	phi
Γ, γ	gamma	I, ι	iota	Π, π	pi	X, χ	chi
Δ, δ	delta	K, κ	kappa	P, ρ	rho	Ψ, ψ	psi
E, ε	epsilon	Λ, λ	lambda	Σ, σ	sigma	Ω, ω	omega
Z, ζ	zeta	M, μ	mu	T, τ	tau		

Quanta, Matter, and Change

A molecular approach to physical chemistry

About the book

Our *Physical Chemistry* has always started with thermodynamics, progressed on to quantum mechanics, and then brought these two great rivers together by considering statistical thermodynamics. We always took care to enrich the thermodynamics with molecular understanding, and wrote the text so that it could be used flexibly to suit the pedagogical inclinations of its users. There are many, though, who consider it more appropriate to build an understanding of the subject from a firm foundation of quantum theory and then to show how the concepts of thermodynamics emerge as the microscopic evolves into the macroscopic. This text is directed at them.

We have taken the cloth of *Physical Chemistry*, unravelled it, and woven a new cloth that begins from quantum theory, establishes the link with the macroscopic world by introducing statistical thermodynamics, and then shows how thermodynamics is used to describe bulk properties of matter. But this is no mere reordering of topics. As we planned the book and then progressed through its writing, we realized that we had to confront issues that required fundamentally new approaches even to very familiar material. In fact, we experienced a kind of intellectual liberation that comes from looking at a familiar subject from a new perspective. Therefore, although readers will see material that has appeared throughout the editions of *Physical Chemistry*, there is an abundance of new material, new approaches to familiar topics, and—we hope—a refreshing new insight into the familiar.

The text is divided into five parts and preceded by a *Fundamentals* section that reviews the material that we presume is already familiar to readers at this level but about which their memories might need a gentle prod. In *Part 1, Quantum theory*, we set out the foundations of quantum mechanics in terms of its postulates and then show how these principles are used to describe motion in one and more dimensions. We have acknowledged the present surge of interest in nanoscience, and have built our presentation around these exciting systems. In *Part 2, Atoms molecules, and assemblies*, we turn to the more traditional nanosystems of chemistry and work progressively through the building blocks of chemistry, ending with solids. We have paid particular attention to computational chemistry, which is, of course, of great practical significance throughout chemistry. We have confronted head on the sheer difficulty of presenting computational chemistry at this level by illustrating all the major techniques by focusing on an almost trivially simple system. Our aim in this important chapter was to give a sense of reality to this potentially recondite subject: we develop understanding and provide a launching platform for those who wish to specialize further. *Part 3, Molecular spectroscopy*, brings together all the major spectroscopic techniques, building on the principles of quantum mechanics introduced in Part 1.

Part 4, Molecular thermodynamics, was for us the most challenging—and therefore the most exciting—part to write, for here we had to make the awesome passage from the quantum theory of microscopic systems to the thermodynamic properties of bulk matter. The bridge is provided by that most extraordinary concept, the Boltzmann distribution. Once that concept has been established, it can be used to develop an understanding of the central thermodynamic properties of internal energy and entropy. We have trodden carefully through this material, trying to maintain the sense that thermodynamics is a self-contained subject dealing with phenomenological relations between properties but, at the same time, showing the illumination that comes from a molecular perspective. We hope this sensitivity to the subject is apparent and that the new insights that we ourselves have acquired in the course of developing this material will be found to be interesting and informative. There are parts of traditional thermodynamics (phase equilibria, among them), we have to admit, that are not open to this kind of elucidation or at least would be made unduly complicated, and we have not hesitated where our judgement persuaded us to set the molecular aside and present the material from a more straightforward classical viewpoint.

In *Part 5, Chemical dynamics*, we turn to another main stream of physical chemistry, the rates of reactions. Some of this material—the setting up of rate laws, for instance—can be expressed in a purely traditional manner, but there are aspects of the dynamics of chemical reactions that draw heavily on what has gone before.

The 'Using the book' section that follows gives details of the pedagogical apparatus in the book, but there is one feature that is so important that it must be mentioned in this Preface. The principal impediment to the 'quantum first' approach adopted by this text is the level of mathematics required, or at least the *perceived* level if not the actual level, for we have taken great pains to step carefully through derivations. The actual level of mathematics needed to understand the material is not great, but the thought that it exists can be daunting. To help overcome this barrier to understanding we have included a series of *Mathematical background* features between various chapters. These sections (there are eight) give background support to the mathematics that has been used in the preceding chapter and which will be drawn on in later chapters. We are aware that many chemists prefer the concrete to the abstract, and have illustrated the material with numerous examples.

We hope that you will enjoy using the book as much as we have enjoyed—and learned from—writing it and will appreciate that we have aimed to produce a book that illuminates physical chemistry from a new direction.

PWA
JdeP
RSF

Using the book

We have paid attention to the needs of the student, and have provided a lot of pedagogical features to make the learning process more enjoyable and effective. This section reviews these features. Paramount among them, though, is something that pervades the entire text: we have tried throughout to *interpret* the mathematical expressions, for mathematics is a language, and it is crucially important to be able to recognize what it is seeking to convey. We have paid particular attention to the level at which we introduce information, the possibility of progressively deepening one's understanding, and providing background information to support the development in the text. We have also been very alert to the demands associated with problem solving, and have provided a variety of helpful procedures.

Organizing the information

Checklist of key ideas

We have summarized the principal concepts introduced in each chapter as a checklist at the end of the chapter. We suggest checking off the box that precedes each entry when you feel confident about the topic.

Checklist of key ideas

☐ 1. A van der Waals interaction between closed-shell molecules is inversely proportional to the sixth power of their separation.

☐ 2. The permittivity is the quantity ε in the Coulomb potential energy, $V = Q_1 Q_2 / 4\pi\varepsilon r$.

☐ 3. A polar molecule is a molecule with a permanent electric dipole moment; the magnitude of a dipole moment is the product of the partial charge and the separation.

☐ 4. The potential energy of the dipole–dipole interaction between two fixed (non-rotating) molecules is proportional to $\mu_1\mu_2/r^3$ and that between molecules that are free to rotate

☐ 8. A hydrogen bond is an interacti where A and B are N, O, or F.

☐ 9. The Lennard-Jones (12,6) poten is a model of the total intermole

☐ 10. In real gases, molecular interact state; the true equation of state coefficients $B, C, \ldots : pV_m = RT$

☐ 11. The van der Waals equation of the true equation of state in whi represented by a parameter a ar by a parameter b: $p = nRT/(V -$

Impact sections

Where appropriate, we have separated the principles from their applications: the principles are constant; the applications come and go as the subject progresses. The *Impact* sections show how the principles developed in the chapter are currently being applied in a variety of modern contexts, especially biology and materials science.

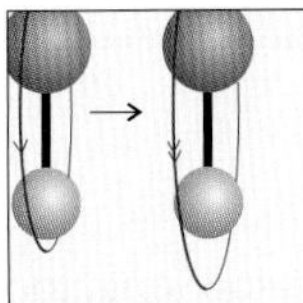

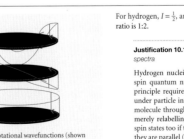

IMPACT ON BIOCHEMISTRY
I13.1 The helix–coil transition in polypeptides

The hydrogen bonds between amino acids of a polypeptide give rise to stable helical or sheet structures, which may collapse into a random coil when certain conditions are changed. The unwinding of a helix into a random coil is a *cooperative transition*, in which the polymer becomes increasingly more susceptible to structural changes once the process has begun. We examine here a model based on the principles of statistical thermodynamics that accounts for the cooperativity of the helix–coil transition in polypeptides.

To calculate the fraction of polypeptide molecules present as helix or coil we need to set up the partition function for the vari-

ward: we simply replace the upper li

$$\frac{q}{q_0} = \sum_{i=0}^{n} C(n,i) s^i$$

A cooperative transformation is modate, and depends on building a facilitate each other's conformational *model*, conversion from h to c is all cent to the one undergoing the conv Thus, the zipper model allows a tran $\ldots \rightarrow \ldots hhhcc \ldots$, but not a tran $\ldots \rightarrow \ldots hchch \ldots$. The only excep the very first conversion from h to

Notes on good practice

Science is a precise activity and its language should be used accurately. We have used this feature to help encourage the use of the language and procedures of science in conformity to international practice (as specified by IUPAC, the International Union of Pure and Applied Chemistry) and to help avoid common mistakes.

bsorbed by a molecule, the angular
 system is conserved. If the
me sense as the spin of the
reases by 1.

Self-test 10.4 Repeat the problem for $C^{35}ClH_3$ (see Self-test 10.2 for details).

[Lines of separation 0.944 cm⁻¹ (28.3 GHz)]

A note on good practice For the discussion of spectroscopic transitions, the upper state is written first. So $X \rightarrow Y$ is an emission and $X \leftarrow Y$ is an absorption, where X and Y specify the states in some way (such as by giving the value of the rotational quantum number J, as we have done above).

Justifications

On first reading it might be sufficient simply to appreciate the 'bottom line' rather than work through detailed development of a mathematical expression. However, mathematical development is an intrinsic part of physical chemistry, and to achieve full

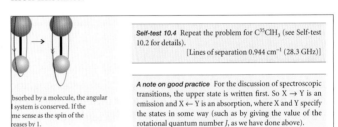

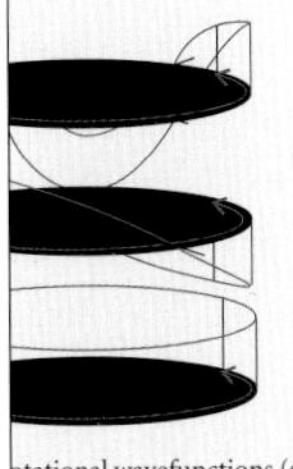

For hydrogen, $I = \frac{1}{2}$, and the ratio is 3:1. For N_2, with $I = 1$, the ratio is 1:2.

Justification 10.1 *The effect of nuclear statistics on rotational spectra*

Hydrogen nuclei are fermions (particles with half-integer spin quantum number; in their case $I = \frac{1}{2}$), so the Pauli principle requires the overall wavefunction to change sign under particle interchange. However, the rotation of an H_2 molecule through 180° has a more complicated effect than merely relabelling the nuclei, because it interchanges their spin states too if the nuclear spins are paired ($\uparrow\downarrow$) but not if they are parallel ($\uparrow\uparrow$).

For the overall wavefunction of the molecule to change

tational wavefunctions (shown
imensional rotor) under a rotation

understanding it is important to see how a particular expression is obtained. The *Justifications* let you adjust the level of detail that you require to your current needs, and make it easier to review material.

interActivities

You will find that many of the graphs in the text have an *interActivity* attached: this is a suggestion about how you can use the on-line-resources of the book's website to explore the consequences of changing various parameters or of carrying out a more elaborate investigation related to the material in the illustration.

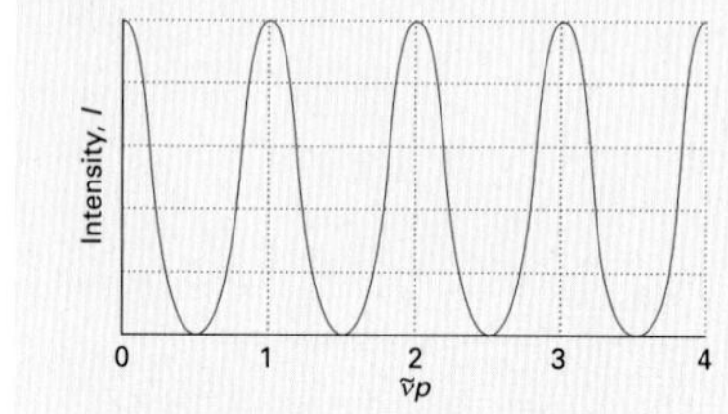

Fig. 10.25 An interferogram produced as the path length p is changed in the interferometer shown in Fig. 10.24. Only a single frequency component is present in the signal, so the graph is a plot of the function $I(p) = I_0(1 + \cos 2\pi\tilde{\nu}p)$, where I_0 is the intensity of the radiation.

interActivity Referring to Fig. 10.24, the mirror M_1 moves in finite distance increments, so the path difference p is also incremented in finite steps. Explore the effect of increasing the step size on the shape of the interferogram for a monochromatic beam of wavenumber $\tilde{\nu}$ and intensity I_0. That is, draw plots of $I(p)/I_0$ against $\tilde{\nu}p$, each with a different number of data points spanning the same total distance path taken by the movable mirror M_1.

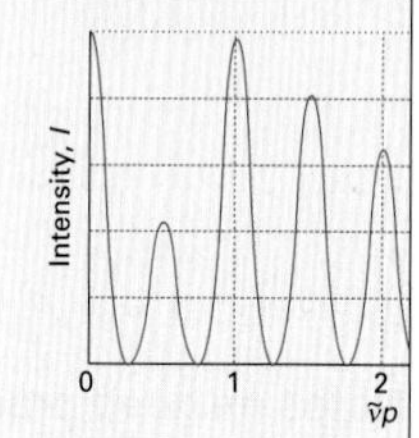

Fig. 10.26 An interferogram obtained three) frequencies are present in the

interActivity For a signal consis monochromatic beams, the in replaced by a sum over the finite nu this information to draw your own v go on to explore the effect of varying intensities of the three components o of the interferogram.

where $I(0)$ is given by eqn 10.23 wi carried out numerically in a comp

Further information

In some cases, we have judged that a derivation is too long, too detailed, or too different in level for it to be included in the text. In these cases, the derivations will be found less obtrusively at the end of the chapter.

Further information 13.2 *The partition functions of polyatomic rotors*

The energies of a symmetric rotor are

$$E_{J,K,M_J} = hc\tilde{B}J(J+1) + hc(\tilde{A} - \tilde{B})K^2$$

with $J = 0, 1, 2, \ldots$, $K = J, J-1, \ldots, -J$, and $M_J = J, J-1, \ldots, -J$. Instead of considering these ranges, we can cover the same values by allowing K to range from $-\infty$ to ∞, with J confined to $|K|, |K|+1, \ldots, \infty$ for each value of K (Fig. 13.23). Because the energy is independent of M_J, and there are $2J+1$ values of M_J for each value of J, each value of J is $(2J+1)$-fold degenerate. It follows that the partition function

$$q = \sum_{J=0}^{\infty}\sum_{K=-J}^{J}\sum_{M_J=-J}^{J} e^{-E_{J,K,M_J}/kT}$$

can be written equivalently as

$$q = \sum_{K=-\infty}^{\infty}\sum_{J=|K|}^{\infty}(2J+1)e^{-E_{J,K,M_J}/kT}$$

$$= \sum_{K=-\infty}^{\infty}\sum_{J=|K|}^{\infty}(2J+1)e^{-hc\{\tilde{B}J(J+1)+(\tilde{A}-\tilde{B})K^2\}/kT} \qquad (13.59)$$

Now we assume that the temperature is s occupied and that the sums may be appr

$$q = \int_{-\infty}^{\infty} e^{-\{hc(\tilde{A}-\tilde{B})/kT\}K^2}\int_{|K|}^{\infty}(2J+1)e^{-hc}$$

As before, the integral over J can be recog derivative of a function, which is the func

$$\int_{|K|}^{\infty}(2J+1)e^{-hc\tilde{B}J(J+1)/kT}dJ = \int_{|K|}^{\infty}\left(-\frac{kT}{hc}\right.$$

$$= \left(-\frac{kT}{hc\tilde{B}}\right)e$$

$$= \left(\frac{kT}{hc\tilde{B}}\right)e^{-h}$$

$$\approx \left(\frac{kT}{hc\tilde{B}}\right)e^{-h}$$

Synoptic tables and the Resource section

Long tables of data are helpful for assembling and solving exercises and problems, but can break up the flow of the text. The *Resource section* at the end of the text consists of a *Data section* with a lot of useful numerical information and a collection of other useful tables. Short extracts in the *Synoptic tables* in the text itself give an idea of the typical values of the physical quantities we are introducing.

Part 1 Data section

Physical properties of selected materials

	ρ/(g cm^{-3}) at 293 K†	T_f/K	T_b/K		ρ/(g cm^{-3}) at 293 K†
Elements				**Inorganic compounds**	
Aluminium(s)	2.698	933.5	2740	$CaCO_3$(s, calcite)	2.71
Argon(g)	1.381	83.8	87.3	$CuSO_4\cdot 5H_2O$(s)	2.284
Boron(s)	2.340	2573	3931	HBr(g)	2.77
Bromine(l)	3.123	265.9	331.9	HCl(g)	1.187
Carbon(s, gr)	2.260	3700s		HI(g)	2.85
Carbon(s, d)	3.513			H_2O(l)	0.997
Chlorine(g)	1.507	172.2	239.2	D_2O(l)	1.104
Copper(s)	8.960	1357	2840	NH_3(g)	0.817
Fluorine(g)	1.108	53.5	85.0	KBr(s)	2.750
Gold(s)	19.320	1338	3080	KCl(s)	1.984
Helium(g)	0.125		4.22	NaCl(s)	2.165
Hydrogen(g)	0.071	14.0	20.3	H_2SO_4(l)	1.841
Iodine(s)	4.930	386.7	457.5		
Iron(s)	7.874	1808	3023	**Organic compounds**	
Krypton(g)	2.413	116.6	120.8	Acetaldehyde, CH_3CHO(l)	0.788
Lead(s)	11.350	600.6	2013	Acetic acid, CH_3COOH(l)	1.049

Mathematics support

A brief comment

A topic often needs to draw on a mathematical procedure or a concept of physics; *A brief comment* is a quick reminder of the procedure or concept.

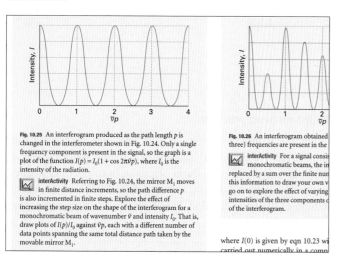

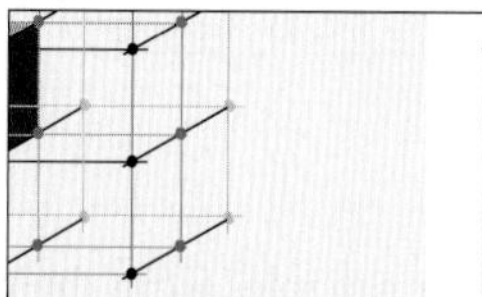

sided (but not necessarily
n the entire crystal structure can be
slations (not reflections, rotations,

A brief comment A *symmetry operation* is an action (such as a rotation, reflection, or inversion) that leaves an object looking the same after it has been carried out. There is a corresponding *symmetry element* for each symmetry operation, which is the point, line, or plane with respect to which the symmetry operation is performed. For instance, an *n-fold rotation* (the symmetry operation) about an *n-fold axis of symmetry* (the corresponding symmetry element) is a rotation through 360°/*n*. See Chapter 7 for a more detailed discussion of symmetry.

Mathematical background

It is often the case that you need a more full-bodied account of a mathematical concept, either because it is important to understand the procedure more fully or because you need to use a series of tools to develop an equation. The eight *Mathematical background* sections are located between chapters, primarily where they are first needed, and include many illustrations of how each concept is used.

MATHEMATICAL BACKGROUND 8: MULTIVA

MATHEMATICAL BACKGROUND 8

Multivariate calculus

A property of a system typically depends on a number of variables, such as the pressure depending on the amount, volume, and temperature according to an equation of state, $p = f(n,T,V)$. To understand how these properties vary with the conditions we need to understand how to manipulate their derivatives. This is the field of **multivariate calculus**, the calculus of several variables.

MB8.1 **Partial derivatives**

A **partial derivative** of a function of more than one variable, such as $f(x,y)$, is the slope of the function with respect to one of the variables, all the other variables being held constant (Fig. MB8.1). Although a partial derivative shows how a function changes when one variable changes, it may be used to determine how the function changes when more than one variable changes by an infinitesimal amount. Thus, if f is a function of x and y then, when x and y change by dx and dy, respectively, f changes by

$$df = \left(\frac{\partial f}{\partial x}\right)_y dx + \left(\frac{\partial f}{\partial y}\right)_x dy \qquad \text{(MB8.1)}$$

● A BRIEF ILLUSTRATION

Suppose that $f(x,y) = ax^3y +$ … Fig. MB8.1), then

$$\left(\frac{\partial f}{\partial x}\right)_y = 3ax^2y \qquad \left(\frac{\partial f}{\partial y}\right)_x = \ldots$$

Then, when x and y undergo inf… by

$$df = 3ax^2y\,dx + (ax^3 + 2by)d\ldots$$

To verify that the order of taking … is irrelevant, we form

$$\left(\frac{\partial}{\partial y}\left(\frac{\partial f}{\partial x}\right)_y\right)_x = \left(\frac{\partial(3ax^2y)}{\partial y}\right)\ldots$$

$$\left(\frac{\partial}{\partial x}\left(\frac{\partial f}{\partial y}\right)_x\right)_y = \left(\frac{\partial(ax^3 + 2b\ldots}{\partial x}\right.$$

Self test MB8.1 Evaluate df for … verify that the order of taking the … is irrelevant. [$df = 4x$…

Problem solving

A brief illustration

A brief illustration is a short example of how to use an equation that has just been introduced in the text. In particular, we show how to use data and how to manipulate units correctly.

● A BRIEF ILLUSTRATION

Consider a complex salt with three unpaired electrons per complex cation at 298 K, of mass density 3.24 g cm^{-3}, and molar mass 200 g mol^{-1}. First note that

$$\frac{N_A g_e^2 \mu_0 \mu_B^2}{3k} = 6.3001 \times 10^{-6}\ \text{m}^3\ \text{K}^{-1}\ \text{mol}^{-1}$$

Consequently,

$$\chi_m = 6.3001 \times 10^{-6} \times \frac{S(S+1)}{T/\text{K}}\ \text{m}^3\ \text{mol}^{-1}$$

Substitution of the data with $S = \frac{3}{2}$ gives $\chi_m = 7.9 \times 10^{-8}$ m^3 mol^{-1}. Note that the density is not needed at this stage. To obtain the volume magnetic susceptibility, the molar susceptibility is divided by the molar volume $V_m = M/\rho$, where ρ is the mass density. In this *illustration*, $V_m = 61.7$ cm^3 mol^{-1}, so $\chi = 1.3 \times 10^{-3}$. ●

(c) Induced magnetic moments

An applied magnetic field induces the ci… currents. These currents give rise to a ma… opposes the applied field, so the substan… few cases the induced field augments th… substance is then paramagnetic.

The great majority of molecules with… spins are diamagnetic. In these cases, the… rents occur within the orbitals of the mol… in its ground state. In the few cases in whi… magnetic despite having no unpaired … electron currents flow in the opposite … can make use of unoccupied orbitals that… in energy. This orbital paramagnetism… from spin paramagnetism by the fact that… pendent: this is why it is called **temperat… magnetism** (TIP).

We can summarize these remarks as

Worked examples

Each *Worked example* has a *Method section* to suggest how to set up the problem (another way might seem more natural: setting up problems is a highly personal business) and use or find the necessary data. Then there is the worked-out *Answer*, where we emphasize the importance of using units correctly.

Example 13.5 *Evaluating the rotational partition function explicitly*

Evaluate the rotational partition function of ^{1}H^{35}Cl at 25°C, given that $\tilde{B} = 10.591$ cm^{-1}.

Method We use eqn 13.19 and evaluate it term by term. Once again, we use $kT/hc = 207.224$ cm^{-1} at 298.15 K. The sum is readily evaluated by using mathematical software.

Answer To show how successive terms contribute, we draw up the following table by using $hc\tilde{B}/kT = 0.051\,11$ (Fig. 13.8):

J	0	1	2	3	4	⋯10
$(2J+1)e^{-0.05111J(J+1)}$	1	2.71	3.68	3.79	3.24	⋯0.08

The sum required by eqn 13.19 (the sum of the numbers in the second row of the table) is 19.9, hence $q^R = 19.9$ at this temperature. Taking J up to 50 gives $q^R = 19.902$. Notice that about ten J-levels are significantly populated but the number of populated *states* is larger on account of the $(2J+1)$-fold degeneracy of each level. We shall shortly encounter the approximation that $q^R \approx kT/hc\tilde{B}$, which in the present case gives $q^R = 19.6$, in good agreement with the exact value and with much less work.

CO_2 ν_1 ν_2 ν_3

* For more values, see Table 10. and use $hc/k = 1.439$ K cm.

At room temperature $kT/hc \approx 20$… stants of many molecules are close … 13.2) and often smaller (though the … which $\tilde{B} = 60.9$ cm^{-1}, is one excep… rotational levels are populated at n… this is the case, the partition functio…

$$\text{Linear rotors:} \quad q^R = \frac{kT}{hc\tilde{B}}$$

$$\text{Non-linear rotors:} \quad q^R = \left(\frac{kT}{hc}\right)^{3\ldots}$$

where $\tilde{A}$, $\tilde{B}$, and $\tilde{C}$ are the rotationa… expressed as wavenumbers. Howeve… sions, read on (to eqns 13.21 and 13…

Self-tests

Each *Worked example* has a *Self-test* with the answer provided as a check that the procedure has been mastered. There are also a number of free-standing *Self-tests* that are located where we thought it a good idea to provide a question to check your understanding. Think of *Self-tests* as in-chapter exercises designed to help you monitor your progress.

Discussion questions

The end-of-chapter material starts with a short set of questions that are intended to encourage reflection on the material and to view it in a broader context than is obtained by solving numerical problems.

Discussion questions

17.1 Explain how the mixing of reactants and products affects the position of chemical equilibrium.

17.2 Explain how a reaction that is not spontaneous may be driven forward by coupling to a spontaneous reaction.

17.3 Use concepts of statistical thermodynamics to describe the molecular features that determine the magnitudes of equilibrium constants and their variation with temperature.

17.4 Suggest how the thermodynamic equilibrium constant may respond differently to changes in pressure and temperature from the equilibrium constant expressed in terms of partial pressures.

17.5 Account for Le Chatelier's principle in terms of thermodynamic quantities. Can you think of a reason why the principle might fail?

17.6 State the limits to the generality of t… as in eqn 17.28.

17.7 Distinguish between galvanic, electr…

17.8 Explain why salt bridges are routine… measurements.

17.9 Discuss how the electrochemical ser… redox reaction is spontaneous.

17.10 Describe a method for the determi… of a redox couple.

17.11 Describe at least one non-calorime… determining a standard reaction enthalpy

Exercises and Problems

The core of testing understanding is the collection of end-of-chapter *Exercises* and *Problems*. The *Exercises* are straightforward numerical tests that give practice with manipulating numerical data. The *Problems* are more searching. They are divided into 'numerical', where the emphasis is on the manipulation of data, and 'theoretical', where the emphasis is on the manipulation of equations before (in some cases) using numerical data. At the end of the *Problems* are collections of problems that focus on practical applications of various kinds, including the material covered in the *Impact* sections. Although this text includes many of the hundreds of *Exercises* and *Problems* that are present in the 8th edition of *Physical chemistry*, well more than half of them are entirely new or have been modified.

Exercises

17.1(a) Write the expressions for the equilibrium constants of the following reactions in terms of (i) activities and (ii) where appropriate, the ratios $p/p^{\ominus}$ and the products $\gamma b/b^{\ominus}$:

(a) $CO(g) + Cl_2(g) \rightleftharpoons COCl(g) + Cl(g)$
(b) $2\,SO_2(g) + O_2(g) \rightleftharpoons 2\,SO_3(g)$
(c) $Fe(s) + PbSO_4(aq) \rightleftharpoons FeSO_4(aq) + Pb(s)$
(d) $Hg_2Cl_2(s) + H_2(g) \rightleftharpoons 2\,HCl(aq) + 2\,Hg(l)$
(e) $2\,CuCl(aq) \rightleftharpoons Cu(s) + CuCl_2(aq)$

17.1(b) Write the expressions for the equilibrium constants of the following reactions in terms of (i) activities and (ii) where appropriate, the ratios $p/p^{\ominus}$ and the products $\gamma b/b^{\ominus}$:

(a) $H_2(g) + Br_2(g) \rightleftharpoons 2\,HBr(g)$
(b) $2\,O_3(g) \rightleftharpoons 3\,O_2(g)$
(c) $2\,H_2(g) + O_2(g) \rightleftharpoons 2\,H_2O(l)$
(d) $H_2(g) + O_2(g) \rightleftharpoons H_2O_2(aq)$
(e) $H_2(g) + I_2(g) \rightleftharpoons 2\,HI(aq)$

17.2(a) Identify the stoichiometric numbers in the reaction $Hg_2Cl_2(s) + H_2(g) \rightarrow 2\,HCl(aq) + 2\,Hg(l)$.

17.2(b) Identify the stoichiometric numbers in the reaction $CH_4(g) + 2\,O_2(g) \rightarrow CO_2(g) + 2\,H_2O(l)$.

17.4(b) The equilibrium pressure of H_2 o
hydride, UH_3, at 500 K is 139 Pa. Calculat
formation of $UH_3(s)$ at 500 K.

17.5(a) From information in the *Data sec*
Gibbs energy and the equilibrium consta
for the reaction $PbO(s) + CO(g) \rightleftharpoons Pb(s)$
reaction enthalpy is independent of temp

17.5(b) From information in the *Data sec*
Gibbs energy and the equilibrium consta
the reaction $CH_4(g) + 3\,Cl_2(g) \rightleftharpoons CHCl_3($
reaction enthalpy is independent of temp

17.6(a) For $CaF_2(s) \rightleftharpoons Ca^{2+}(aq) + 2\,F^-(aq$
the standard Gibbs energy of formation o
Calculate the standard Gibbs energy of fo

17.6(b) For $PbI_2(s) \rightleftharpoons Pb^{2+}(aq) + 2\,I^-(aq)$
the standard Gibbs energy of formation o
Calculate the standard Gibbs energy of fo

17.7(a) In the gas-phase reaction 2 A + B
when 1.00 mol A, 2.00 mol B, and 1.00 m
come to equilibrium at 25°C, the resultin
at a total pressure of 1.00 bar. Calculate (

The Book Companion Site

The Book Companion Site to accompany *Quanta, Matter, and Change* provides teaching and learning resources to augment the printed book. It is free of charge, and provides additional material for download, which can be incorporated into a virtual learning environment.

The book companion site can be accessed by visiting

www.whfreeman.com/pchem.

Note that instructor resources are available only to registered adopters of the textbook. To register, simply visit www.whfreeman.com/pchem and follow the appropriate links. You will be given the opportunity to select your own username and password, which will be activated once your adoption has been verified.

Student resources are openly available to all, without registration.

The materials on the book companion site include:

Living graphs

A *Living graph* can be used to explore how a property changes as a variety of parameters are changed. To encourage the use of this resource (and the more extensive *Explorations in physical chemistry;* see below), we have included a suggested *interActivity* to many of the illustrations in the text.

Artwork

An instructor may wish to use the figures from this text in a lecture. Almost all the figures are available in PowerPoint® format and can be used for lectures without charge (but not for commercial purposes without specific permission).

Tables of data

All the tables of data that appear in the chapter text are available and may be used under the same conditions as the artwork.

Group theory tables

Comprehensive group theory tables are available for downloading.

Weblinks

There is a huge network of information available about physical chemistry, and it can be bewildering to find your way to it. Also, a piece of information may be needed that we have not included in the text. The website might suggest where to find the specific data or indicate where additional data can be found.

Other resources

Explorations in Physical Chemistry by Valerie Walters, Julio de Paula, and Peter Atkins

Explorations in Physical Chemistry consists of interactive Mathcad® worksheets and interactive Excel® workbooks, complete with thought-stimulating exercises. They motivate students to simulate physical, chemical, and biochemical phenomena with their personal computers. Harnessing the computational power of Mathcad® by Mathsoft, Inc. and Excel® by Microsoft Corporation, students can manipulate over 75 graphics, alter simulation parameters, and solve equations to gain deeper insight into physical chemistry.

Explorations in Physical Chemistry can be purchased at www.whfreeman.com/explorations; ISBN 0-7167-0841-8.

Solutions manuals

Two solutions manuals accompany this book; both are written by Charles Trapp, Marshall Cady, and Carmen Giunta.

A *Student's Solutions Manual* (ISBN 1-4292-2375-8) provides full solutions to the 'a' exercises, and the odd-numbered problems.

An *Instructor's Solutions Manual* (1-4292-2374-x) provides full solutions to the 'b' exercises, and the even-numbered problems.

About the authors

Peter Atkins is a fellow of Lincoln College in the University of Oxford and the author of more than sixty books for students and a general audience. His texts are market leaders around the globe. A frequent lecturer in the United States and throughout the world, he has held visiting professorships in France, Israel, Japan, China, and New Zealand. He was the founding chairman of the Committee on Chemistry Education of the International Union of Pure and Applied Chemistry and was a member of IUPAC's Physical and Biophysical Chemistry Division.

Julio de Paula is Professor of Chemistry and Dean of the College of Arts & Sciences at Lewis & Clark College. A native of Brazil, Professor de Paula received a B.A. degree in chemistry from Rutgers, The State University of New Jersey, and a Ph.D. in biophysical chemistry from Yale University. His research activities encompass the areas of molecular spectroscopy, biophysical chemistry, and nanoscience. He has taught courses in general chemistry, physical chemistry, biophysical chemistry, instrumental analysis, and writing.

Ronald Friedman is Professor and Chair of the Chemistry Department at Indiana University Purdue University Fort Wayne. He received a B.S. in chemistry from the University of Virginia, a Ph.D. in chemistry from Harvard University, and did postdoctoral work at the University of Minnesota. He teaches general chemistry and physical chemistry at IPFW and has also taught at the University of Michigan and at the Technion (Israel). His research interests are theories of reaction dynamics.

Acknowledgements

The authors have received a great deal of help during the preparation and production of this text and wish to thank all their colleagues who have made such thought-provoking and useful suggestions. In particular, we wish to record publicly our thanks to:

Temer S. Ahmadi, Villanova University
David L. Andrews, University of East Anglia
Martin Bates, University of York
Warren Beck, Michigan State University
Nikos Bentenitis, Southwestern University
Holly Bevsek, The Citadel
D. Neal Boehnke, Jacksonville University
Dale J. Brugh, Ohio Wesleyan University
William A. Burns, Arizona State University
Colleen Byron, Ripon College
Rosemarie C. Chinni, Alvernia College
Stephen Davis, University of Mississippi
Peter Derrick, University of Warwick
Paul A. DiMilla, Northeastern University
Robert A. Donnell, Auburn University
Michael Dorko, The Citadel
Mel Dutton, California State University-Bakersfield
Matthew Elrod, Oberlin College
James M. Farrar, University of Rochester
Stephen Fletcher, Loughborough University
Krishna L. Foster, California State University-Los Angeles
James Gimzewski, UCLA
Robert Glinski, Tennessee Tech University
Rebecca Goyan, Simon Fraser University
Alexander Grushow, Rider University
Dorothy A. Hanna, Kansas Wesleyan University
Gerard S. Harbison, University of Nebraska at Lincoln
Shizuka Hsieh, Smith College
Markus M. Hoffmann, SUNY-Brockport
William Hollingsworth, Carleton College
Karl James Jalkanen, Technical University of Denmark
Evguenii I. Kozliak, University of North Dakota
Jeffrey E. Lacy, Shippensburg College
Daniel B. Lawson, University Of Michigan-Dearborn
John W. Logan, San Jose State University
Arthur A. Low, Tarleton State University
Michael Lyons, Trinity College Dublin
Charles McCallum, Pacific College
Preston MacDougall, Middle Tennessee College
Roderick M. Macrae, Marian College
Jeffrey D. Madura, Duquesne University
Clyde Metz, College of Charleston
David A. Micha, University of Florida
Brian G. Moore, Augustana College
Dale Moore, Mercer University
Stephan P. A. Sauer, University of Copenhagen
Richard Schwenz, University of Northern Colorado-Greeley
Olle Soderman, Lund University
Jie Song, University of Michigan-Flint
David Smith, University of Bristol
Julian Talbot, Duquesne University
Earle Waghorne, University College Dublin
Michael Wedlock, Gettysburg College
Benjamin Whitaker, University of Leeds
Kurt Winkelmann, Florida Institute of Technology
Dawn C. Wiser, Lake Forest College

The publication of a book does not cease when the authors lay down their (these days, virtual) pens. We are particularly grateful to our copy editor Claire Eisenhandler and equally to Valerie Walters, who read through the proofs with meticulous attention to detail and caught in private what might have been a public grief. We are particularly grateful to Charles Trapp, Carmen Giunta, and Marshall Cady for their critical reading of the end of chapter Exercises and Problems and their recommendations for modifications.

Last, but by no means least, we wish to acknowledge the wholehearted and unstinting support of our two commissioning editors, Jonathan Crowe of Oxford University Press and Jessica Fiorillo of W.H. Freeman & Co., who—in other projects as well as this—have helped the authors to realize their vision and have done so in such an agreeable and professional a manner.

Summary of contents

Contents

List of Impact sections

Fundamentals

Chemistry is the science of matter and the changes it can undergo. **Physical chemistry** is the branch of chemistry that establishes and develops the principles of the subject in terms of the underlying concepts of physics and the language of mathematics. It provides the basis for developing new spectroscopic techniques and their interpretation, for understanding the structures of molecules and the details of their electron distributions, and for relating the bulk properties of matter to their constituent atoms. Physical chemistry also provides a window on to the world of chemical reactions, and allows us to understand in detail how they take place. In fact, the subject underpins the whole of chemistry, providing the principles in terms of which we understand structure and change and which form the basis of all techniques of investigation.

Throughout the text we shall draw on a number of concepts that should already be familiar from introductory chemistry. This section reviews them. In almost every case the following chapters will provide a deeper discussion, but we are presuming that we can refer to these concepts at any stage of the presentation. Because physical chemistry lies at the interface between physics and chemistry, we also need to review some of the concepts from elementary physics that we need to draw on in the text.

F.1 Atoms

Matter consists of atoms. The atom of an element is characterized by its **atomic number**, Z, which is the number of protons in its nucleus. The number of neutrons in a nucleus is variable to a small extent, and the **nucleon number** (which is also commonly called the *mass number*), A, is the total number of protons and neutrons, which are collectively called **nucleons**, in the nucleus. Atoms of the same atomic number but different nucleon number are the **isotopes** of the element.

According to the **nuclear model**, an atom of atomic number Z consists of a nucleus of charge $+Ze$ surrounded by Z electrons each of charge $-e$ (e is the fundamental charge: see inside the front cover for its value). These electrons occupy **atomic orbitals**, which are regions of space where they are most likely to be found, with no more than two electrons in any one orbital. The atomic orbitals are arranged in **shells** around the nucleus, each shell being characterized by the **principal quantum number**, $n = 1, 2, \ldots$. A shell consists of n^2 individual orbitals, which are grouped together into n **subshells**; these subshells, and the orbitals they contain, are denoted s, p, d, and f. For all neutral atoms other than hydrogen, the subshells of a given shell have slightly different energies.

The sequential occupation of the orbitals in successive shells results in periodic similarities in the **electronic configurations**, the specification of the occupied orbitals, of atoms when they are arranged in order of their atomic number, which leads to the formulation of the **periodic table** (a version is shown inside the back cover). The vertical columns of the periodic table are called **groups** and (in the modern convention) numbered from 1 to 18. Successive rows of the periodic table are called **periods**, the number of the period being equal to the principal quantum number of the **valence shell**, the outermost shell of the atom. The periodic table is divided into s, p, d, and f **blocks**, according to the subshell that is last to be occupied in the formulation of the electronic configuration of the atom. The members of the d block (specifically in Groups 3–12) are also known as the **transition metals**; those of the f block (which is not divided into numbered groups) are sometimes called the **inner transition metals**. The upper row of the f block (Period 6) consists of the **lanthanoids** (still commonly the 'lanthanides') and the lower row (Period 7) consists of the **actinoids** (still commonly the 'actinides'). Some of the groups also have familiar names: Group 1 consists of the **alkali metals**, Group 2 (more specifically, calcium, strontium, and barium) of the **alkaline earth metals**, Group 17 of the **halogens**, and Group 18 of the **noble gases**. Broadly speaking, the elements towards the left of the periodic table are **metals** and those towards the right are **nonmetals**; the two classes of substance meet at a diagonal line running from boron to polonium, which constitute the **metalloids**, with properties intermediate between those of metals and nonmetals.

An **ion** is an electrically charged atom. When an atom gains one or more electrons it becomes a negatively charged **anion**; when it loses them it becomes a positively charged **cation**. The charge number of an ion is called the **oxidation number** of the element in that state (thus, the oxidation number of magnesium in Mg^{2+} is +2 and that of oxygen in O^{2-} is −2). It is appropriate, but not always done, to distinguish between the oxidation number and the **oxidation state**, the latter being the physical state of the atom with a specified oxidation number. Thus, the oxidation number *of* magnesium is +2 when it is present as Mg^{2+}, and it is present *in* the oxidation state Mg^{2+}. The elements form ions that are characteristic of their location in the periodic table: metallic elements typically form cations by losing the electrons of their outermost shell and acquiring the electronic configuration of the preceding noble gas. Nonmetals typically form anions by gaining electrons and attaining the electronic configuration of the following noble gas.

F.2 Molecules

A **chemical bond** is the link between atoms. Compounds that contain a metallic element typically, but far from universally, form **ionic compounds** that consist of cations and anions in a crystalline array. The 'chemical bonds' in an ionic compound are due to the Coulombic interactions (Section F.6) between all the ions in the crystal, and it is inappropriate to refer to a bond between a specific pair of neighbouring ions. The smallest unit of an ionic compound is called a **formula unit**. Thus $NaNO_3$, consisting of a Na^+ cation and a NO_3^- anion, is the formula unit of sodium nitrate. Compounds that do not contain a metallic element typically form **covalent compounds** consisting of discrete molecules. In this case, the bonds between the atoms of a molecule are **covalent**, meaning that they consist of shared pairs of electrons.

The pattern of bonds between neighbouring atoms is displayed by drawing a **Lewis structure**, in which bonds are shown as lines and **lone pairs** of electrons, pairs of valence electrons that are not used in bonding, are shown as dots. Lewis structures are constructed by allowing each atom to share electrons until it has acquired an **octet** of eight electrons (for hydrogen, a *duplet* of two electrons). A shared pair of electrons is a **single bond**, two shared pairs constitute a **double bond**, and three shared pairs constitute a **triple bond**. Atoms of elements of Period 3 and later can accommodate more than eight electrons in their valence shell and 'expand their octet' to become **hypervalent**, that is, form more bonds than the octet rule would allow (for example, SF_6), or form more bonds to a small number of atoms (for example, a Lewis structure of SO_4^{2-} with double bonds). When more than one Lewis structure can be written for a given arrangement of atoms, it is supposed that **resonance**, a blending of the structures, may occur and distribute multiple-bond character over the molecule (for example, the two Kekulé structures of benzene). Examples of these aspects of Lewis structures are shown in Fig. F.1.

Except in the simplest cases, a Lewis structure does not express the three-dimensional structure of a molecule. The simplest approach to the prediction of molecular shape is **valence-shell electron pair repulsion theory** (VSEPR theory). In this approach, the regions of high electron density, as represented by bonds—

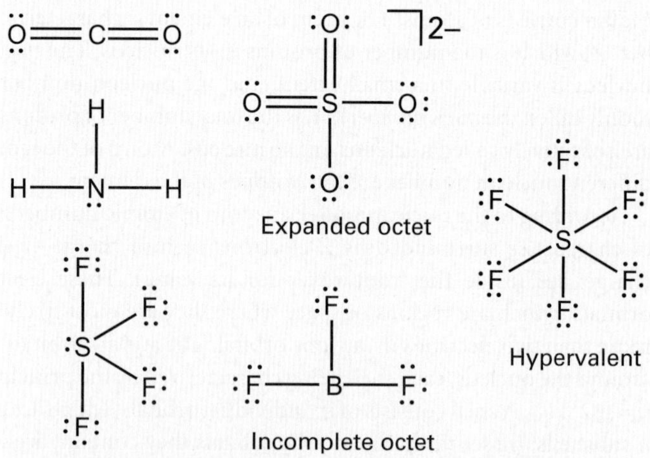

Fig. F.1 A collection of typical Lewis structures for simple molecules and ions. The structures show the bonding patterns and lone pairs and except in simple cases do not express the shape of the species.

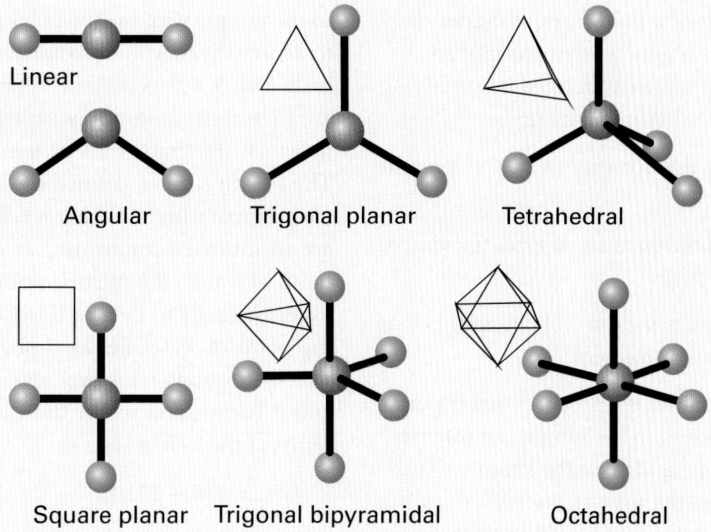

Fig. F.2 The names of the shapes of the geometrical figures used to describe symmetrical polyatomic molecules and ions.

whether single or multiple—and lone pairs, take up orientations around the central atom that maximize their separations. Then the position of the attached atoms (not the lone pairs) is noted and used to classify the shape of the molecule. Thus, four regions of electron density adopt a tetrahedral arrangement; if an atom is at each of these locations (as in CH_4), then the molecule is tetrahedral; if there is an atom at only three of these locations (as in NH_3), then the molecule is trigonal pyramidal, and so on. The names of the various shapes that are commonly found are shown in Fig. F.2. In a refinement of the theory, lone pairs are assumed to repel bonding pairs more strongly than bonding pairs repel each other, and the shape a molecule adopts, if it is not determined fully by symmetry, is such as to minimize repulsions from lone pairs. Thus, in SF_4 the lone pair adopts an equatorial position and the two axial S—F bonds bend away from it slightly, to give a bent see-saw shaped molecule (Fig. F.3).

Covalent bonds may be **polar**, or correspond to an unequal sharing of the electron pair, with the result that one atom has a partial positive charge (denoted δ+) and the other a partial negative charge (δ−). The ability of an atom to attract electrons to itself when part of a molecule is measured by the **electronegativity**, χ (chi), of the element. The juxtaposition of equal and opposite partial charges constitutes an **electric dipole.** If those charges are $+Q$ and $-Q$ and they are separated by a distance d, the magnitude of the **electric dipole moment** is $\mu = Qd$. Whether or not a molecule as a whole is polar depends on the arrangement of its bonds, for in highly symmetrical molecules there may be no net dipole. Thus, although the linear CO_2 molecule has polar CO bonds, their effects cancel and the OCO molecule as a whole is nonpolar.

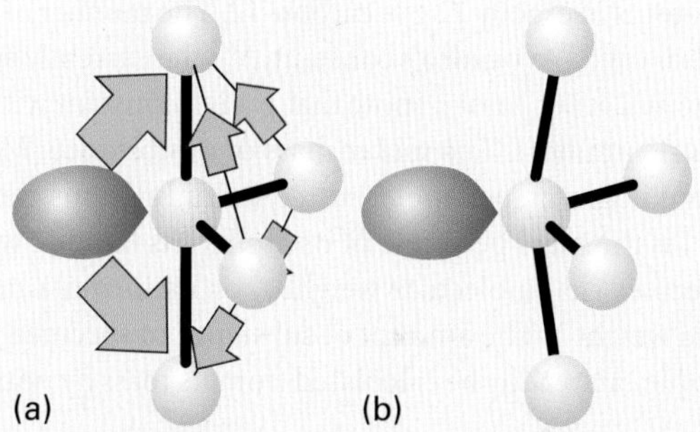

Fig. F.3 (a) The influences on the shape of the SF_4 molecule according to the VSEPR model. (b) As a result, the molecule adopts a bent see-saw shape.

F.3 Bulk matter

Bulk matter consists of large numbers of atoms, molecules, or ions. Its physical state may be solid, liquid, or gas:

A **solid** is a form of matter that adopts and maintains a shape that is independent of the container it occupies.

A **liquid** is a form of matter that adopts the shape of the part of the container it occupies (in a gravitational field, the lower part) and is separated from the unoccupied part of the container by a definite surface (for example, a meniscus).

A **gas** is a form of matter that immediately fills any container it occupies.

A liquid and a solid are examples of a **condensed state** of matter. A liquid and a gas are examples of a **fluid** form of matter: they flow in response to forces (such as gravity) that are applied.

The state of a bulk sample of matter is defined by specifying the values of various properties. Among them are:

The **mass**, m, a measure of the quantity of matter present (unit: kilogram, kg).

The **volume**, V, a measure of the quantity of space the sample occupies (unit: cubic metre, m^3).

The **amount of substance**, n, a measure of the number of specified entities present (unit: mole, mol).

An **extensive property** of bulk matter is a property that depends on the amount of substance present in the sample; an **intensive property** is a property that is independent of the amount of substance. The volume is extensive; the mass density, ρ (rho), the mass of a sample divided by its volume, $\rho = m/V$, is intensive.

The amount of substance, n (colloquially, 'the number of moles'), is a measure of the number of specified entities present in the sample. 'Amount of substance' is the official name of the quantity; it is commonly simplified to 'chemical amount' or simply 'amount'. The unit 1 mol is defined as the number of carbon atoms in exactly 12 g of carbon-12. The number of entities per mole is called **Avogadro's constant,** N_A; the currently accepted value is $6.022 \times 10^{23}\ \text{mol}^{-1}$ (note that N_A is a constant with units, not a pure number). The **molar mass of a substance,** M (units: formally kilograms per mole but commonly grams per mole, g mol^{-1}) is the mass per mole of its atoms (its 'atomic weight'), its molecules (its 'molecular weight'), or its formula units (its 'formula weight'). The amount of substance of specified entities in a sample can readily be calculated from its mass by noting that

$$n = \frac{m}{M} \qquad \text{(F.1)}$$

A sample of matter may be subjected to a **pressure,** p (unit: pascal, Pa; $1\ \text{Pa} = 1\ \text{kg m}^{-1}\ \text{s}^{-2}$), which is defined as the force, F, it is subjected to, divided by the area, A, to which that force is applied. A sample of gas exerts a pressure on the walls of its container because the molecules of gas are in ceaseless, random motion, and exert a force when they strike the walls. The frequency of the collisions is normally so great that the force, and therefore the pressure, is perceived as being steady. Although pascal is the SI unit of pressure (Section F.8), it is also common to express pressure in bars (1 bar = 10^5 Pa) or atmospheres (1 atm = 101 325 Pa exactly), both of which correspond to typical atmospheric pressure. The **standard pressure** is currently defined as $p^{\ominus} = 1$ bar exactly.

To specify the state of a sample fully it is also necessary to give its **temperature,** T. The temperature is formally a property that determines in which direction energy will flow as heat when two samples are placed in contact through thermally conducting walls: energy flows from the sample with the higher temperature to the sample with the lower temperature. The symbol T is used to denote the **thermodynamic temperature,** which is an absolute scale with $T = 0$ as the lowest point. Temperatures above $T = 0$ are then most commonly expressed by using the **Kelvin scale,** in which the gradations of temperature are called **kelvin** (K). The Kelvin scale is defined by setting the triple point of water (the temperature at which ice, liquid water, and water vapour are in mutual equilibrium) at exactly 273.16 K. The freezing point of water (the melting point of ice) at 1 atm is then found experimentally to lie 0.01 K below the triple point, so the freezing point of water lies at approximately 273.15 K. The Kelvin scale is unsuitable for everyday measurements of temperature, and it is common to use the **Celsius scale,** which is defined in terms of the Kelvin scale as

$$\theta/^{\circ}\text{C} = T/\text{K} - 273.15 \qquad \text{(F.2)}$$

Thus, the freezing point of water is 0°C and its boiling point (at 1 atm) is found to be 100°C. Note that in this text T invariably denotes the thermodynamic (absolute) temperature and that temperatures on the Celsius scale are denoted θ (theta).

The properties that define the state of a system are not in general independent of one another. The most important example of a relation between them is provided by the idealized fluid known as a **perfect gas** (also, commonly, an 'ideal gas'):

$$pV = nRT \qquad \text{(F.3)}$$

Here R is the **gas constant,** a universal constant (in the sense of being independent of the chemical identity of the gas) with the value $8.314\ \text{J K}^{-1}\ \text{mol}^{-1}$. Equation F.3, the **perfect gas law,** is a summary of three empirical conclusions, namely, Boyle's law ($p \propto 1/V$ at constant temperature and amount), Charles's law ($p \propto T$ at constant volume and amount), and Avogadro's principle ($V \propto n$ at constant temperature and pressure).

All gases obey the perfect gas law ever more closely as the pressure is reduced towards zero. That is, eqn F.3 is an example of a **limiting law,** a law that becomes increasingly valid in a particular limit, in this case as the pressure is reduced to zero. In practice, normal atmospheric pressure at sea level (about 1 atm) is already low enough for most gases to behave almost perfectly and, unless stated otherwise, we shall always assume in this text that the gases we encounter behave perfectly and obey eqn F.3.

A mixture of perfect gases behaves like a single perfect gas. According to **Dalton's law,** the total pressure of such a mixture is the sum of the **partial pressures** of the constituents, the pressure to which each gas would give rise if it occupied the container alone:

$$p = p_A + p_B + \cdots \qquad \text{(F.4)}$$

Each partial pressure, p_J, can be calculated from the perfect gas law in the form $p_J = n_J RT/V$.

F.4 Thermodynamic properties

The central unifying concept of physical chemistry is energy. The systematic discussion of the transfer and transformation of energy in bulk matter is called **thermodynamics**. This subtle subject is treated in detail in Part 4 of the text, but it will be familiar from introductory chemistry that there are two central concepts, the **internal energy**, U (unit: joule, J), and the **entropy**, S (unit: joules per kelvin, J K^{-1}).

The internal energy is the total energy of a system. The **first law of thermodynamics** states that the internal energy is constant in a system isolated from external influences. The internal energy of a sample of matter increases as its temperature is raised, and we write

$$\Delta U = C\Delta T \qquad \text{(F.5)}$$

where ΔU is the change in internal energy when the temperature of the sample is raised by ΔT. The constant C is called the **heat capacity** (units: joules per kelvin, J K^{-1}), of the sample. If the heat capacity is large, a small increase in temperature results in a large increase in internal energy. This remark can be expressed in a physically more significant way by inverting it: if the heat capacity is large, then even a large transfer of energy into the system leads to only a small rise in temperature. The heat capacity is an extensive property, and values for a substance are commonly reported as the **molar heat capacity**, $C_m = C/n$ (units: joules per kelvin per mole, J K^{-1} mol^{-1}) or the **specific heat capacity**, $C_s = C/m$ (units: joules per kelvin per gram, J K^{-1} g^{-1}), both of which are intensive properties.

Thermodynamic properties are often best discussed in terms of infinitesimal changes, in which case we would write eqn F.5 as $dU = CdT$. When this expression is written in the form

$$C = \frac{dU}{dT} \qquad \text{(F.6)}$$

we see that the heat capacity can be interpreted as the slope of the plot of the internal energy of a sample against the temperature.

As will also be familiar from introductory chemistry and will be explained in detail later, for systems maintained at constant pressure it is usually more convenient to modify the internal energy by adding to it the quantity pV, and introducing the **enthalpy**, H (unit: joule, J):

$$H = U + pV \qquad \text{(F.7)}$$

The enthalpy, an extensive property, greatly simplifies the discussion of chemical reactions, in part because changes in enthalpy can be identified with the energy transferred as heat from a system maintained at constant pressure (as in common laboratory experiments).

The **entropy**, S, is a measure of the *quality* of the energy of a system. If the energy is distributed over many modes of motion (for example, the rotational, vibrational, and translational motions for the particles which comprise the system), then the entropy is high. If the energy is distributed over only a small number of modes of motion, then the entropy is low. The **second law of thermodynamics** states that any spontaneous (that is, natural) change in an isolated system is accompanied by an increase in the entropy of the system. This tendency is commonly expressed by saying that the natural direction of change is accompanied by dispersal of energy from a localized region or to a less organized form.

The entropy of a system and its surroundings is of the greatest importance in chemistry because it enables us to identify the spontaneous direction of a chemical reaction and to identify the composition at which the reaction is at **equilibrium**. In a state of *dynamic* equilibrium, which is the character of all chemical equilibria, the forward and reverse reactions are occurring at the same rate and there is no net tendency to change in either direction. However, to use the entropy to identify this state we need to consider both the system and its surroundings. This task can be simplified if the reaction is taking place at constant temperature and pressure, for then it is possible to identify the state of equilibrium as the state when the **Gibbs energy**, G (unit: joule, J), of the system has reached a minimum. The Gibbs energy is defined as

$$G = H - TS \qquad \text{(F.8)}$$

and is of the greatest importance in chemical thermodynamics. The Gibbs energy, which informally is called the 'free energy', is a measure of the energy stored in a system that is free to do useful work, such as driving electrons through a circuit or causing a reaction to be driven in its unnatural (nonspontaneous) direction.

F.5 The relation between molecular and bulk properties

The energy of a molecule, atom, or subatomic particle that is confined to a region of space is **quantized**, or restricted to certain discrete values. These permitted energies are called **energy levels**. The values of the permitted energies depend on the characteristics of the particle (for instance, its mass) and the extent of the region to which it is confined. The quantization of energy is most important—in the sense that the allowed energies are widest apart—for particles of small mass confined to small regions of space. Consequently, quantization is very important for electrons in atoms and molecules, but usually unimportant for macroscopic bodies, for which the separation of translational energy levels of particles in containers of macroscopic dimensions is so small that for all practical purposes their translational motion is unquantized and can be varied virtually continuously. As we shall see in detail in Part 1, quantization becomes increasingly important as we change focus from rotational to vibrational and then to electronic motion. The separation of rotational

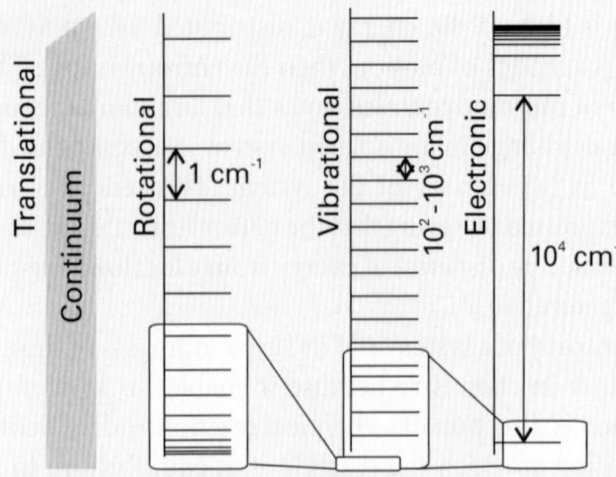

Fig. F.4 The energy level separations (expressed as wavenumbers, see p. 12) typical of four types of system.

energy levels (in small molecules, about 10^{-23} J or 0.01 zJ, corresponding to about 0.01 kJ mol^{-1}) is smaller than that of vibrational energy levels (about 10 kJ mol^{-1}), which itself is smaller than that of electronic energy levels (about 10^{-18} J or 1 aJ, where a is another uncommon but useful SI prefix (Section F.8), standing for atto, 10^{-18}, corresponding to about 10^3 kJ mol). Figure F.4 depicts these typical energy level separations.

(a) The Boltzmann distribution

The continuous thermal agitation that the molecules experience in a sample at $T > 0$ ensures that they are distributed over the available energy levels. One particular molecule may be in a state corresponding to a low energy level at one instant, and then be excited into a high energy state a moment later. Although we cannot keep track of the state of a single molecule, we can speak of the *average* numbers of molecules in each state; even though individual molecules may be changing their states as a result of collisions, the average number in each state is constant (provided the temperature remains the same).

The average number of molecules in a state is called the **population** of the state. Only the lowest energy state is occupied at $T = 0$. Raising the temperature excites some molecules into higher energy states, and more and more states become accessible as the temperature is raised further (Fig. F.5). The formula for calculating the relative populations of states of various energies is called the **Boltzmann distribution** and was derived by the Austrian scientist Ludwig Boltzmann towards the end of the nineteenth century. This formula gives the ratio of the numbers of particles in states with energies E_i and E_j as

$$\frac{N_i}{N_j} = e^{-(E_i - E_j)/kT} \qquad \text{(F.9)}$$

where k is **Boltzmann's constant**, a fundamental constant with the value $k = 1.381 \times 10^{-23}$ J K^{-1}. This constant occurs throughout

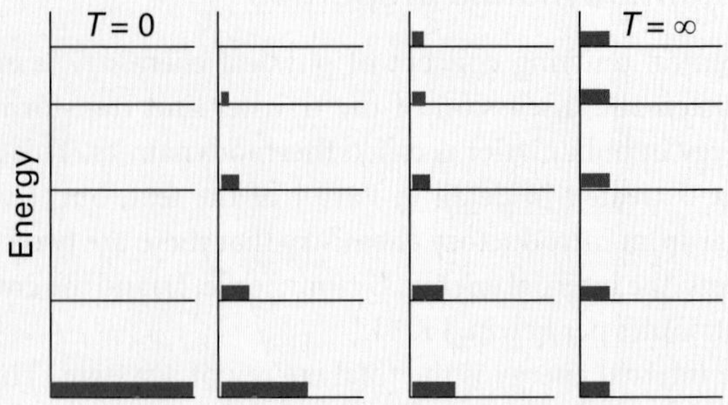

Fig. F.5 The Boltzmann distribution of populations for a system of five energy levels as the temperature is raised from zero to infinity.

physical chemistry, often in a disguised 'molar' form as the gas constant, for

$$R = N_A k \qquad \text{(F.10)}$$

where N_A is Avogadro's constant. We shall see in Part 4 that the Boltzmann distribution provides the crucial link for expressing the macroscopic properties of matter in terms of microscopic behaviour.

The important features of the Boltzmann distribution to bear in mind are that the distribution of populations is an exponential function of energy and temperature, and that more levels are significantly populated if they are close together in comparison with kT (like rotational and translational states) than if they are far apart (like vibrational and electronic states). Moreover, at a high temperature more energy levels are occupied than at a low temperature. Figure F.6 summarizes the form of the Boltzmann distribution for some typical sets of energy levels. The peculiar shape of the population of rotational levels stems from the fact that eqn F.9 applies to *individual states* and, for molecular rotation, the number of rotational states corresponding to a given energy level—broadly speaking, the number of planes of

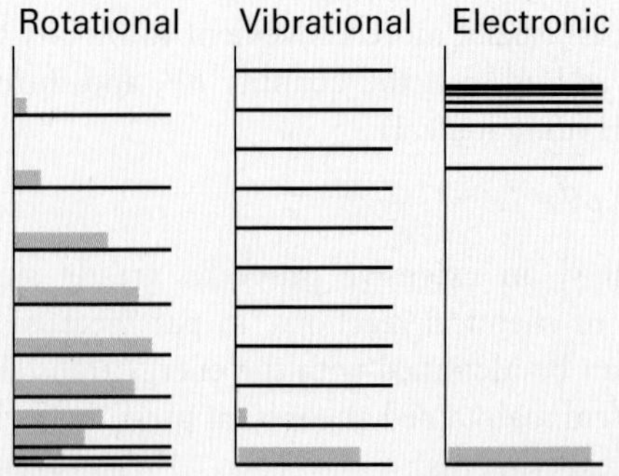

Fig. F.6 The Boltzmann distribution of populations for rotational, vibrational, and electronic energy levels at room temperature.

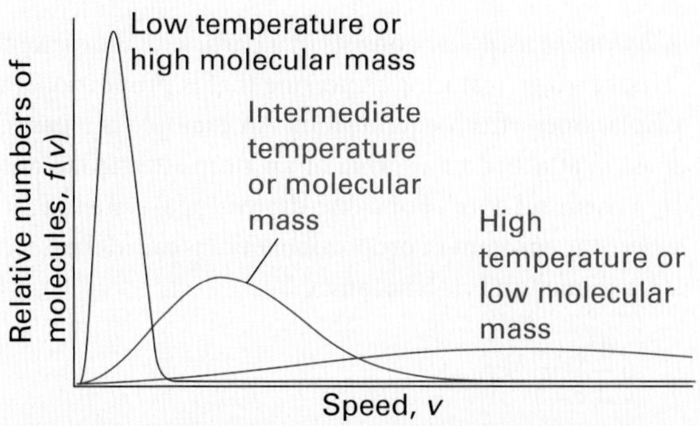

Fig. F.7 The distribution of molecular speeds with temperature and molar mass. Note that the most probable speed (corresponding to the peak of the distribution) increases with temperature and with decreasing molar mass, and simultaneously the distribution becomes broader.

interActivity (a) Plot different distributions by keeping the molar mass constant at 100 g mol^{-1} and varying the temperature of the sample between 200 K and 2000 K. (b) Use mathematical software or the *Living graph* applet from the text's web site to evaluate numerically the fraction of molecules with speeds in the range 100 m s^{-1} to 200 m s^{-1} at 300 K and 1000 K. (c) Based on your observations, provide a molecular interpretation of temperature.

rotation—increases with energy; therefore, although the population of each *state* decreases with energy, the population of the *levels* goes through a maximum.

One of the simplest examples of the relation between microscopic and bulk properties is provided by **kinetic molecular theory**, a model of a perfect gas. In this model, it is assumed that the molecules, imagined as particles of negligible size, are in ceaseless, random motion and do not interact except during their brief collisions. Different speeds correspond to different energies, so the Boltzmann formula can be used to predict the proportions of molecules having a specific speed at a particular temperature. The expression giving the fraction of molecules that have a particular speed is called the **Maxwell distribution**, and has the features summarized in Fig. F.7. The Maxwell distribution, which is discussed more fully in Chapter 18, can be used to show that the mean speed, $\bar{c}$, of the molecules depends on the temperature and their molar mass as

$$\bar{c} \propto \left(\frac{T}{M}\right)^{1/2} \qquad \text{(F.11)}$$

Thus, the average speed is high for light molecules at high temperatures. The distribution itself gives more information. For instance, the tail towards high speeds is longer at high temperatures than at low, which indicates that at high temperatures more molecules in a sample have speeds much higher than average.

(b) Equipartition

The Boltzmann distribution can be used to calculate the average energy associated with each mode of motion of a molecule. However, for modes of motion that can be described by classical mechanics (which in practice means translation of any molecule and the rotation of all except the lightest molecules) there is a short cut, called the **equipartition theorem**. This theorem states that, in a sample at a temperature T, *all quadratic contributions to the total energy have the same mean value, namely* $\frac{1}{2}kT$. A 'quadratic contribution' simply means a contribution that depends on the square of the position or the velocity (or momentum). For example, because the kinetic energy of a body of mass m free to undergo translation in three dimensions is $E_k = \frac{1}{2}mv_x^2 + \frac{1}{2}mv_y^2 + \frac{1}{2}mv_z^2$, there are three quadratic terms. The theorem implies that the average kinetic energy of motion parallel to the x-axis is the same as the average kinetic energy of motion parallel to the y-axis and to the z-axis. That is, in a normal sample (one at thermal equilibrium throughout), the total energy is equally 'partitioned' over all the available modes of motion. One mode of motion is not especially rich in energy at the expense of another. Because the average contribution of each mode is $\frac{1}{2}kT$, the average kinetic energy of a molecule free to move in three dimensions is $\frac{3}{2}kT$, as there are three quadratic contributions to the kinetic energy.

F.6 Particles

Molecules are built from atoms and atoms are built from subatomic particles. To understand their structures we need to know how these bodies move under the influence of the forces they experience. There are two approaches to the description of the motion of particles: **classical mechanics**, the description of matter formulated by Isaac Newton in the seventeenth century, and **quantum mechanics**, formulated in the twentieth century. Quantum mechanics is the more fundamental theory and must be used to describe small particles, such as electrons, atoms, and molecules. Classical mechanics is acceptable for the description of large particles. Quantum mechanics is dealt with at length in Part 1 of the text and is not considered here.

In this section we consider some elementary aspects of the classical mechanics of particles. This discussion is necessarily more mathematical than the foregoing. However, that is a virtue, for it introduces us to the origin of the power of physical chemistry: that it provides a *quantitative* foundation for chemistry. A great deal of the usefulness of physical chemistry stems from the fact that its central ideas can be expressed precisely in terms of mathematics and that the consequences of those ideas can be deduced precisely and logically—once again, in terms of mathematics. Mathematics is a central feature of physical chemistry and, provided we constantly keep in mind the *interpretation* of the equations that arise, through it we acquire a whole new language for the description of nature.

(a) Force and work

'Translation' is the motion of a particle through space. The **velocity**, v, of a particle is the rate of change of its position r:

$$v = \frac{dr}{dt} \qquad (F.12)$$

For motion confined to a single dimension, we would write $v_x = dx/dt$. The velocity and position are vectors, with both direction and magnitude. The magnitude of the velocity is the **speed**, v. The **linear momentum**, p, of a particle of mass m is related to its velocity, v, by

$$p = mv \qquad (F.13)$$

Like the velocity vector, the linear momentum vector points in the direction of travel of the particle (Fig. F.8); its magnitude is denoted p.

According to **Newton's second law of motion**, *the rate of change of momentum is equal to the force acting on the particle*:

$$\frac{dp}{dt} = F \qquad (F.14a)$$

For motion confined to one dimension, we would write $dp_x/dt = F_x$. Equation F.14a may be taken as the definition of force. The SI unit of force is the newton (N), with

$$1\,N = 1\,kg\,m\,s^{-2}$$

Because $p = m(dr/dt)$, it is sometimes more convenient to write eqn F.14a as

$$ma = F \qquad a = \frac{d^2r}{dt^2} \qquad (F.14b)$$

where a is the **acceleration** of the particle, its rate of change of velocity. It follows that, if we know the force acting everywhere and at all times, solving eqn F.14 will give the **trajectory**, the position and momentum of the particle at each instant.

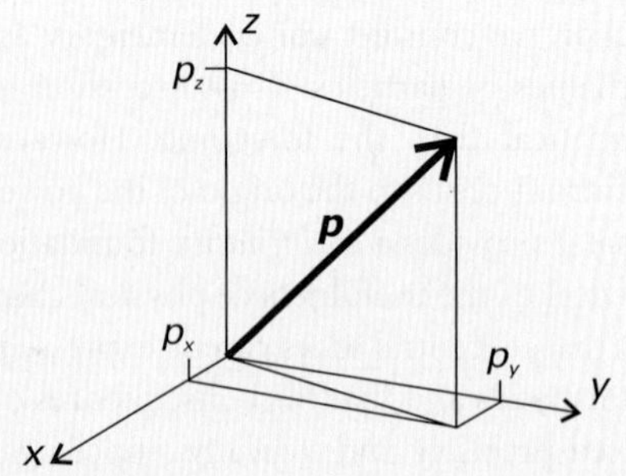

Fig. F.8 The linear momentum p is denoted by a vector of magnitude p and an orientation that corresponds to the direction of motion.

● A BRIEF ILLUSTRATION

A *harmonic oscillator* consists of a particle that experiences a 'Hooke's law' restoring force, one that is proportional to its displacement from equilibrium. An example is a particle of mass m attached to a spring or an atom attached to another by a chemical bond. For a one-dimensional system, $F_x = -kx$, where the constant of proportionality k is called the *force constant*. Equation F.14b becomes

$$m\frac{d^2x}{dt^2} = -kx$$

If $x = 0$ at $t = 0$, a solution (as may be verified by substitution) is

$$x(t) = A \sin 2\pi\nu t \qquad \nu = \frac{1}{2\pi}\left(\frac{k}{m}\right)^{1/2}$$

This solution shows that the position of the particle varies *harmonically* (that is, as a sine function) with a frequency ν, and that the frequency is high for light particles (m small) attached to stiff springs (k large). ●

The description of rotation is very similar to that of translation. The rotational motion of a particle about a central point is described by its **angular momentum**, J. The angular momentum is a vector: its magnitude gives the rate at which a particle circulates and its direction indicates the axis of rotation (Fig. F.9). The magnitude of the angular momentum, J, is

$$J = I\omega \qquad (F.15)$$

where ω is the **angular velocity** of the body, its rate of change of angular position (in radians per second), and I is the **moment of inertia**, a measure of its resistance to rotational acceleration. For a point particle of mass m moving in a circle of radius r, the moment of inertia about the axis of rotation is $I = mr^2$. To

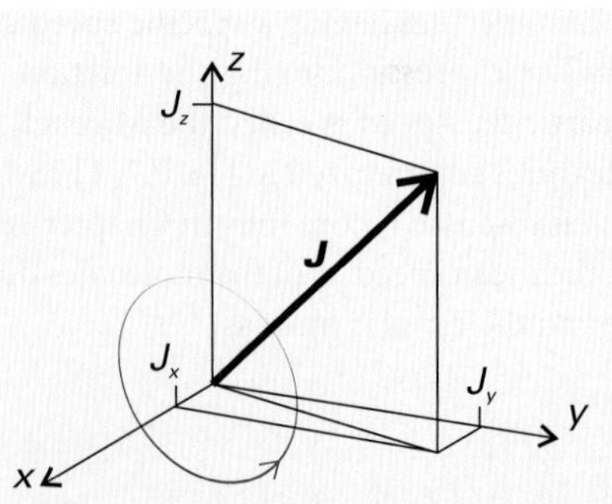

Fig. F.9 The angular momentum of a particle is represented by a vector along the axis of rotation and perpendicular to the plane of rotation. The length of the vector denotes the magnitude of the angular momentum. The direction of motion is clockwise to an observer looking in the direction of the vector.

Table F.1 Analogies between translation and rotation

Translation		Rotation	
Property	Significance	Property	Significance
Mass, m	Resistance to the effect of a force	Moment of inertia, I	Resistance to the effect of a torque
Speed, v	Rate of change of position	Angular velocity, ω	Rate of change of angle
Magnitude of linear momentum, p	$p = mv$	Magnitude of angular momentum, J	$J = I\omega$
Translational kinetic energy, E_k	$E_k = \frac{1}{2}mv^2 = p^2/2m$	Rotational kinetic energy, E_k	$E_k = \frac{1}{2}I\omega^2 = J^2/2I$
Equation of motion	$dp/dt = F$	Equation of motion	$dJ/dt = T$

accelerate a rotation it is necessary to apply a **torque**, T, a twisting force. Newton's equation is then

$$\frac{dJ}{dt} = T \tag{F.16}$$

The analogous roles of m and I, of v and ω, and of p and J in the translational and rotational cases, respectively, should be remembered because they provide a ready way of constructing and recalling equations. These analogies are summarized in Table F.1.

Work, w, is motion against an opposing force. For an infinitesimal motion through $d\boldsymbol{s}$ (a vector), the work done is

$$dw = -\boldsymbol{F}\cdot d\boldsymbol{s} \tag{F.17}$$

where $\boldsymbol{F}\cdot d\boldsymbol{s}$ is the scalar product of the vectors $\boldsymbol{F}$ and $d\boldsymbol{s}$. Generally speaking, the scalar product of two vectors $\boldsymbol{u}$ and $\boldsymbol{v}$ is written $\boldsymbol{u}\cdot\boldsymbol{v} = uv\cos\theta$, where θ is the angle between them. For motion in one dimension, we write $dw = -F_x dx$. The total work done along a path is the integral of this expression, allowing for the possibility that $\boldsymbol{F}$ changes in direction and magnitude at each point of the path. With force in newtons and distance in metres, the unit of work is the joule (J), with

$$1\,\text{J} = 1\,\text{N m} = 1\,\text{kg m}^2\,\text{s}^{-2}$$

● A BRIEF ILLUSTRATION

The work needed to stretch a chemical bond that behaves like a spring through an infinitesimal distance dx is

$$dw = -F_x dx = -(-kx)dx = kx dx$$

The total work needed to stretch the bond from zero displacement ($x = 0$) at its equilibrium length R_e to a length R, corresponding to a displacement $x = R - R_e$, is

$$w = \int_0^{R-R_e} kx\,dx = k\int_0^{R-R_e} x\,dx = \tfrac{1}{2}k(R - R_e)^2$$

We see that the work required increases as the square of the displacement: it takes four times as much work to stretch a bond through 20 pm as it does to stretch it through 10 pm. ●

(b) Energy

As we have indicated, much of chemistry is concerned with transfers and transformations of energy, and it is now appropriate to define this familiar quantity precisely: **energy** is the capacity to do work. The SI unit of energy is the same as that of work, namely, the joule. Calories (cal) and kilocalories (kcal) are still encountered in the chemical literature. Caution needs to be exercised as there are several different kinds of calorie. The thermochemical calorie, cal_{15}, is the energy required to raise the temperature of 1 g of water at 15°C by 1°C. The calorie is now defined in terms of the joule, with 1 cal = 4.184 J (exactly).

A particle may possess two kinds of energy, kinetic energy and potential energy. The **kinetic energy**, E_k, of a body is the energy the body possesses as a result of its motion. For a body of mass m travelling at a speed v,

$$E_k = \tfrac{1}{2}mv^2 \tag{F.18}$$

It follows from Newton's second law that if a particle of mass m is initially stationary and is subjected to a constant force F for a time τ, then its speed increases from zero to $F\tau/m$ and therefore its kinetic energy increases from zero to

$$E_k = \frac{F^2\tau^2}{2m} \tag{F.19}$$

The energy of the particle remains at this value after the force ceases to act. Because the magnitude of the applied force, F, and the time, τ, for which it acts may be varied at will, eqn F.19 implies that the energy of the particle may be increased to any value.

The **potential energy**, E_p or V, of a body is the energy it possesses as a result of its position. Because (in the absence of losses) the work that a particle can do when it is stationary in a given location is equal to the work that had to be done to bring it there, we can use the one-dimensional version of eqn F.17 to write $dV = -F_x\,dx$, and therefore

$$F_x = -\frac{dV}{dx} \tag{F.20}$$

No universal expression for the potential energy can be given because it depends on the type and magnitude of the force the body experiences. For a particle of mass m close to the surface of the Earth, the gravitational force (the 'weight' of the particle) is $-mg$ (with x measured outwards), where g is the **acceleration of free fall** ($g = 9.81\ \mathrm{m\ s^{-2}}$); so the potential energy of the particle at an altitude h is

$$V(h) - V(0) = -\int_0^h (-mg)\mathrm{d}x = mgh \qquad \text{(F.21)}$$

The zero of potential energy is arbitrary. For a particle close to the surface of the Earth, it is common to set $V(0) = 0$.

The **total energy** of a particle is the sum of its kinetic and potential energies:

$$E = E_\mathrm{k} + V \qquad \text{(F.22)}$$

We make use of the apparently universal law of nature that *energy is conserved*, that is, energy can neither be created nor destroyed. Although energy can be transferred from one location to another and transformed from one form to another, the total energy is constant. In terms of the linear momentum, the total energy of a particle is

$$E = \frac{p^2}{2m} + V \qquad \text{(F.23)}$$

This expression may be used in place of Newton's second law to calculate the trajectory of a particle.

● A BRIEF ILLUSTRATION

Consider an argon atom free to move in one direction (along the x-axis) in a region where $V = 0$ (so the energy is independent of position). Because $v = \mathrm{d}x/\mathrm{d}t$, it follows from eqns F.12 and F.18 that $\mathrm{d}x/\mathrm{d}t = (2E_\mathrm{k}/m)^{1/2}$. As may be verified by substitution, a solution of this differential equation is

$$x(t) = x(0) + \left(\frac{2E_\mathrm{k}}{m}\right)^{1/2} t$$

The linear momentum is

$$p(t) = mv(t) = m\frac{\mathrm{d}x}{\mathrm{d}t} = (2mE_\mathrm{k})^{1/2}$$

and is a constant. Hence, if we know the initial position and momentum, we can predict all later positions and momenta exactly. ●

(c) The Coulomb potential energy

One of the most important forms of potential energy in chemistry is the **Coulomb potential energy** between two electric charges. For a point charge Q_1 at a distance r in a vacuum from another point charge Q_2, their potential energy is

$$V = \frac{Q_1 Q_2}{4\pi\varepsilon_0 r} \qquad \text{(F.24)}$$

Charge is expressed in coulombs (C), often as a multiple of the fundamental charge, e. Thus, the charge of an electron is $-e$ and that of a proton is $+e$; the charge of an ion is ze, with z the **charge number** (positive for cations, negative for anions). The constant ε_0 (epsilon zero) is the **vacuum permittivity**, a fundamental constant with the value $8.854 \times 10^{-12}\ \mathrm{C^2\ J^{-1}\ m^{-1}}$. In a medium other than a vacuum, the potential energy of interaction between two charges is reduced, and the vacuum permittivity is replaced by the **permittivity**, ε, of the medium. The permittivity is commonly expressed as a multiple of the vacuum permittivity:

$$\varepsilon = \varepsilon_\mathrm{r}\varepsilon_0 \qquad \text{(F.25)}$$

with ε_r the dimensionless **relative permittivity** (formerly, the *dielectric constant*). The Coulomb potential energy is equal to the work that must be done to bring up a charge Q_1 from infinity to a distance r from a charge Q_2. It is conventional (as in eqn F.24) to set the potential energy equal to zero at infinite separation of charges. Then two opposite charges have a negative potential energy at finite separations, whereas two like charges have a positive potential energy.

Care should be taken to distinguish *potential energy* from *potential*. The potential energy of a charge Q_1 in the presence of another charge Q_2 can be expressed in terms of the **Coulomb potential**, ϕ (phi):

$$V = Q_1\phi \qquad \phi = \frac{Q_2}{4\pi\varepsilon_0 r} \qquad \text{(F.26)}$$

The units of potential are joules per coulomb, $\mathrm{J\ C^{-1}}$ so, when ϕ is multiplied by a charge in coulombs, the result is in joules. The combination joules per coulomb occurs widely and is called a volt (V):

$$1\ \mathrm{V} = 1\ \mathrm{J\ C^{-1}}$$

If there are several charges $Q_2, Q_3, \ldots$ present in the system, the total potential experienced by the charge Q_1 is the sum of the potential generated by each charge:

$$\phi = \phi_2 + \phi_3 + \cdots \qquad \text{(F.27)}$$

Just as the potential energy of a charge Q_1 can be written $V = Q_1\phi$, so the magnitude of the force on Q_1 can be written $F = Q_1\mathcal{E}$, where $\mathcal{E}$ is the magnitude of the **electric field strength** (units: volts per metre, $\mathrm{V\ m^{-1}}$) arising from Q_2 or from some more general charge distribution. The electric field strength (which, like the force, is actually a vector quantity) is the negative gradient of the electric potential. In one dimension, we write the magnitude of the electric field strength as

$$\mathcal{E} = -\frac{d\phi}{dx} \quad \text{(F.28)}$$

The language we have just developed also inspires an important alternative energy unit, the **electronvolt** (eV): 1 eV is defined as the kinetic energy acquired when an electron is accelerated from rest through a potential difference of 1 V. The relation between electronvolts and joules is

$$1 \text{ eV} = 1.602 \times 10^{-19} \text{ J}$$

Many processes in chemistry involve energies of a few electronvolts. For example, to remove an electron from a sodium atom requires about 5 eV.

(d) Power

The rate of supply of energy is called the power (P), and is expressed in watts (W):

$$1 \text{ W} = 1 \text{ J s}^{-1}$$

A particularly important way of supplying energy in chemistry (as in the everyday world) is by passing an electric current through a resistance. An **electric current** (I) is defined as the rate of supply of charge, $I = dQ/dt$, and is measured in *amperes* (A):

$$1 \text{ A} = 1 \text{ C s}^{-1}$$

If a charge Q is transferred from a region of potential ϕ_i, where its potential energy is $Q\phi_i$, to where the potential is ϕ_f and its potential energy is $Q\phi_f$, and therefore through a potential difference $\Delta\phi = \phi_f - \phi_i$, the change in potential energy is $Q\Delta\phi$. The rate at which the energy changes is $(dQ/dt)\Delta\phi$, or $I\Delta\phi$. The power is therefore

$$P = I\Delta\phi \quad \text{(F.29)}$$

With current in amperes and the potential difference in volts, the power is in watts. The total energy, E, supplied in an interval Δt is the power (the rate of energy supply) multiplied by the duration of the interval:

$$E = P\Delta t = I\Delta\phi\Delta t \quad \text{(F.30)}$$

The energy is obtained in joules with the current in amperes, the potential difference in volts, and the time in seconds.

F.7 Waves

Several important investigative techniques in physical chemistry, such as spectroscopy and X-ray diffraction, involve electromagnetic radiation, a wave-like electromagnetic disturbance. We shall also see that the properties of waves are central to the quantum mechanical description of electrons in atoms and molecules. To prepare for those discussions, we need to understand the mathematical description of waves.

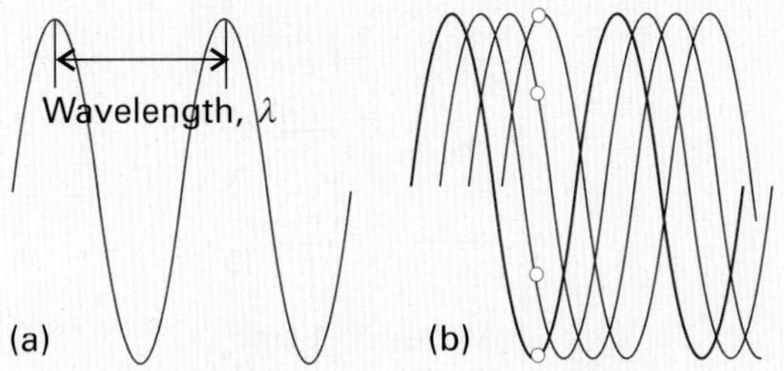

Fig. F.10 (a) The wavelength, λ, of a wave is the peak-to-peak distance. (b) The wave is shown travelling to the right at a speed c. At a given location, the instantaneous amplitude of the wave changes through a complete cycle (the four dots show half a cycle) as it passes a given point. The frequency, ν, is the number of cycles per second that occur at a given point. Wavelength and frequency are related by $\lambda\nu = c$.

A **wave** is an oscillatory disturbance that travels through space. Examples of such disturbances include the collective motion of water molecules in ocean waves and of gas particles in sound waves. A **harmonic wave** is a wave with a displacement that can be expressed as a sine or cosine function.

An **electromagnetic field** is an oscillating electric and magnetic disturbance that spreads as a harmonic wave through empty space, the vacuum. The wave travels at a constant speed called the *speed of light*, $c = 2.998 \times 10^8$ m s^{-1}. As its name suggests, an electromagnetic field has two components, an **electric field** that acts on charged particles (whether stationary or moving) and a **magnetic field** that acts only on moving charged particles. The electromagnetic field is characterized by a **wavelength**, λ (lambda), the distance between the neighbouring peaks of the wave, and its **frequency**, ν (nu), the number of times per second at which its displacement at a fixed point returns to its original value (Fig. F.10). The frequency is measured in *hertz*, where 1 Hz = 1 s^{-1}. The wavelength and frequency of an electromagnetic wave are related by

$$\lambda\nu = c \quad \text{(F.31)}$$

Therefore, the shorter the wavelength, the higher the frequency. The speed of light in a medium, c', is less than in a vacuum. The difference is expressed in terms of the **refractive index**, n_r, of the medium, where

$$n_r = \frac{c}{c'} \quad \text{(F.32)}$$

The refractive index depends on the frequency of the light, and typically increases with frequency. The characteristics of a wave

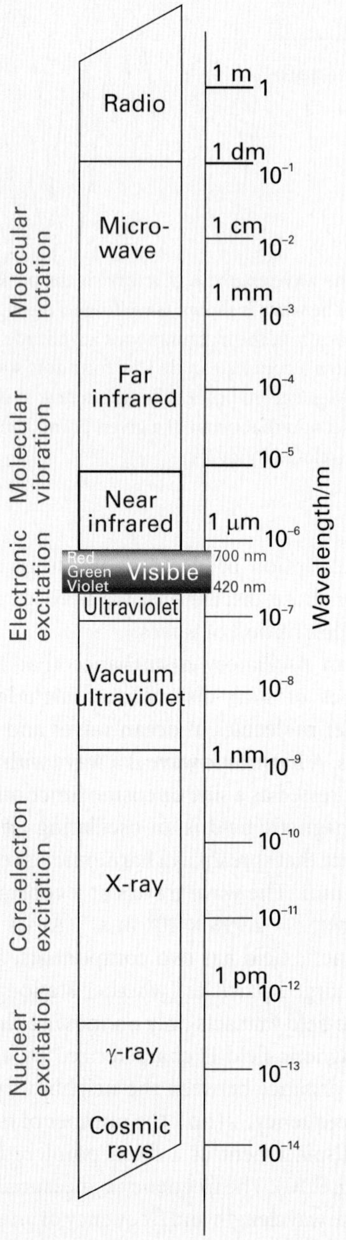

Fig. F.11 The electromagnetic spectrum and its classification into regions (the boundaries are not precise).

are also reported by giving the **wavenumber**, $\tilde{\nu}$ (nu tilde), of the radiation, where

$$\tilde{\nu} = \frac{\nu}{c} = \frac{1}{\lambda} \tag{F.33}$$

A wavenumber can be interpreted as the number of complete wavelengths in a given length. Wavenumbers are normally reported in reciprocal centimetres (cm^{-1}), so a wavenumber of 5 cm^{-1} indicates that there are 5 complete wavelengths in 1 cm.

The classification of the electromagnetic field according to its frequency and wavelength is summarized in Fig. F.11.

The functions that describe the oscillating electric field, $\mathcal{E}(x,t)$, and magnetic field, $\mathcal{B}(x,t)$, travelling along the x-direction with wavelength λ and frequency ν are

$$\mathcal{E}(x,t) = \mathcal{E}_0 \cos\{2\pi\nu t - (2\pi/\lambda)x + \phi\} \tag{F.34a}$$

$$\mathcal{B}(x,t) = \mathcal{B}_0 \cos\{2\pi\nu t - (2\pi/\lambda)x + \phi\} \tag{F.34b}$$

where $\mathcal{E}_0$ and $\mathcal{B}_0$ are the (vector) amplitudes of the electric and magnetic fields, respectively, and ϕ is the **phase** of the wave, which may lie between $-\pi$ and π and gives the relative location of the peaks of two waves (Fig. F.12). Note that the magnetic field is perpendicular to the electric field and both are perpendicular to the propagation direction.

Equation F.34 describes electromagnetic radiation that is **plane-polarized**; it is so called because the electric and magnetic fields each oscillate in a single plane (Fig. F.13). The plane of polarization may be orientated in any direction around the direction of propagation. An alternative mode of polarization is **circular polarization**, in which the electric and magnetic fields rotate around the direction of propagation in either a clockwise or a counter-clockwise sense but remain perpendicular to it and each other. A plane-polarized beam can be regarded as a superposition of two oppositely rotating circularly polarized components (and vice versa). By convention, in right-handed circularly polarized light the electric vector rotates clockwise as seen by an observer facing the oncoming beam (Fig. F.14).

If two waves, in the same region of space, with the same wavelength have phases that differ by $\pm\pi$ (so the peaks of one wave coincide with the troughs of the other), then the resultant wave, the sum of the two, will have a diminished amplitude. This effect is called **destructive interference**. If the phases of the two waves are the same (coincident peaks), the resultant has an enhanced amplitude. This effect is called **constructive interference**.

It follows by differentiating eqn F.34a or F.34b that

$$\frac{\partial^2 \psi(x,t)}{\partial x^2} = -\frac{4\pi^2}{\lambda^2}\psi(x,t) \qquad \frac{\partial^2 \psi(x,t)}{\partial t^2} = -4\pi^2\nu^2\psi(x,t)$$

where $\psi(x,t)$ is either $\mathcal{E}(x,t)$ or $\mathcal{B}(x,t)$ and the derivatives are 'partial derivatives', which will be explained in more detail in the text. Briefly, in the first expression the second derivative of $\psi(x,t)$ with respect to x is calculated with t treated as a constant. Likewise, in the second expression, the second derivative of $\psi(x,t)$ with respect to t is calculated with x treated as a constant. By comparing these two equations we find that

$$c^2 \frac{\partial^2 \psi(x,t)}{\partial x^2} = \frac{\partial^2 \psi(x,t)}{\partial t^2} \tag{F.35}$$

This 'partial differential equation' is known as the **wave equation**.

According to classical electromagnetic theory, the intensity of electromagnetic radiation is proportional to the square of the

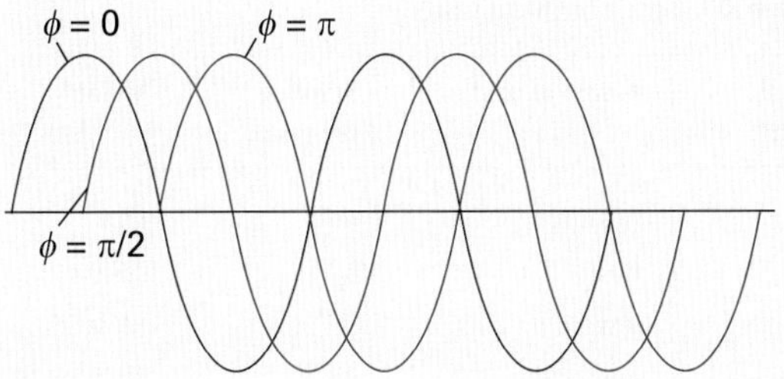

Fig. F.12 The phase (ϕ) of a wave specifies the relative location of its peaks.

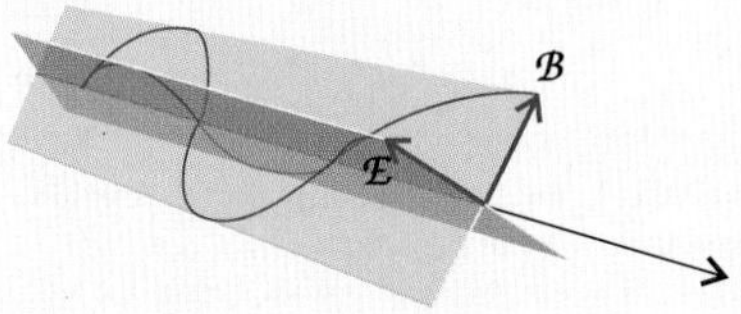

Fig. F.13 In a plane-polarized wave, the electric and magnetic fields oscillate in orthogonal planes and are perpendicular to the direction of propagation.

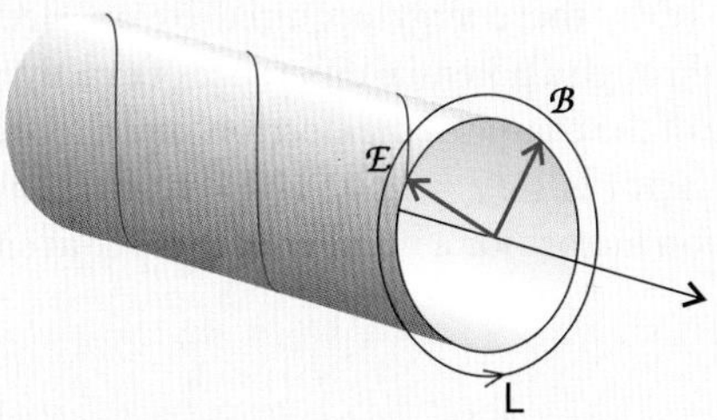

Fig. F.14 In a circularly polarized wave, the electric and magnetic fields rotate around the direction of propagation but remain perpendicular to one another. The illustration also defines 'right' and 'left (L)-handed' polarizations.

Table F.2 The SI base units

Physical quantity	Symbol for quantity	Base unit
Length	l	metre, m
Mass	m	kilogram, kg
Time	t	second, s
Electric current	I	ampere, A
Thermodynamic temperature	T	kelvin, K
Amount of substance	n	mole, mol
Luminous intensity	I_v	candela, cd

Table F.3 A selection of derived units

Physical quantity	Derived unit*	Name of derived unit
Force	1 kg m s^{-2}	newton, N
Pressure	1 kg m^{-1} s^{-2}	pascal, Pa
	1 N m^{-2}	
Energy	1 kg m^2 s^{-2}	joule, J
	1 N m	
	1 Pa m^3	
Power	1 kg m^2 s^{-3}	watt, W
	1 J s^{-1}	

* Equivalent definitions in terms of derived units are given following the definition in terms of base units.

amplitude of the wave. For example, the light detectors discussed in Chapter 11 are based on the interaction between the electric field of the incident radiation and the detecting element, so light intensities are proportional to $\mathcal{E}_0^2$.

F.8 Units

The measurement of a physical property is expressed as

Physical property = numerical value × unit

For example, a length (l) may be reported as l = 5.1 m, if it is found to be 5.1 times as great as a defined unit of length, namely 1 metre (1 m). Units are treated as algebraic quantities, and may be multiplied and divided. Thus, the same length could be reported as l/m = 5.1. The symbols for physical properties are always italic (sloping; thus V for volume, not V), including Greek symbols (thus, μ for electric dipole moment, not μ), but available typefaces are not always so obliging.

In the **International System** of units (SI, from the French *Système International d'Unités*), the units are formed from seven **base units** listed in Table F.2. All other physical quantities may be expressed as combinations of these physical quantities and reported in terms of **derived units**. Thus, volume is (length)^3 and may be reported as a multiple of 1 metre cubed (1 m^3), and density, which is mass/volume, may be reported as a multiple of 1 kilogram per metre cubed (1 kg m^{-3}).

A number of derived units have special names and symbols. The names of units derived from names of people are lower case (as in torr, joule, pascal, and kelvin), but their symbols are upper case (as in Torr, J, Pa, and K). Among the most important for our purposes are those listed in Table F.3.

In all cases (both for base and derived quantities), the units may be modified by a prefix that denotes a factor of a power of 10. The Greek prefixes of units are upright (as in μm, not μm).

Table F.4 Common SI prefixes

Prefix	z	a	f	p	n	μ	m	c	d
Name	zepto	atto	femto	pico	nano	micro	milli	centi	deci
Factor	10^{-21}	10^{-18}	10^{-15}	10^{-12}	10^{-9}	10^{-6}	10^{-3}	10^{-2}	10^{-1}
Prefix	k	M	G	T	P				
Name	kilo	mega	giga	tera	peta				
Factor	10^{3}	10^{6}	10^{9}	10^{12}	10^{15}				

Among the most common prefixes are those listed in Table F.4. Examples of the use of these prefixes are

$$1\ \text{nm} = 10^{-9}\ \text{m} \qquad 1\ \text{ps} = 10^{-12}\ \text{s} \qquad 1\ \mu\text{mol} = 10^{-6}\ \text{mol}$$

The kilogram (kg) is anomalous: although it is a base unit, it is interpreted as 10^3 g, and prefixes are attached to the gram (as in 1 mg = 10^{-3} g). Powers of units apply to the prefix as well as the unit they modify:

$$1\ \text{cm}^3 = 1\ (\text{cm})^3 = 1\ (10^{-2}\ \text{m})^3 = 10^{-6}\ \text{m}^3$$

Note that 1 cm^3 does not mean 1 $\text{c}(\text{m}^3)$. When carrying out numerical calculations, it is usually safest to write out the numerical value of an observable as a power of 10.

There are a number of units that are in wide use but are not a part of the International System. Some are exactly equal to multiples of SI units. These include the *litre* (L), which is exactly 10^3 cm^3 (or 1 dm^3) and the *atmosphere* (atm), which is exactly 101.325 kPa. Others rely on the values of fundamental constants, and hence are liable to change when the values of the fundamental constants are modified by more accurate or more precise measurements. Thus, the size of the energy unit *electronvolt* (eV), the energy acquired by an electron that is accelerated through a potential difference of exactly 1 V, depends on the value of the charge of the electron, and the present (2008) conversion factor is 1 eV = $1.602\,177 \times 10^{-19}$ J. Table F.5 gives the conversion factors for a number of these convenient units.

Table F.5 Some common units

Physical quantity	Name of unit	Symbol for unit	Value*
Time	minute	min	60 s
	hour	h	3600 s
	day	d	86 400 s
Length	ångström	Å	10^{-10} m
Volume	litre	L, l	1 dm^3
Mass	tonne	t	10^3 kg
Pressure	bar	bar	10^5 Pa
	atmosphere	atm	101.325 kPa
Energy	electronvolt	eV	$1.602\,177 \times 10^{-19}$ J
			96.485 31 kJ mol^{-1}

* All values in the final column are exact, except for the definition of 1 eV, which depends on the measured value of *e*.

Exercises

F.1 Atoms

F1.1(a) Summarize the nuclear model of the atom.

F1.1(b) Define the terms atomic number, nucleon number, mass number.

F1.2(a) Express the typical ground-state electron configuration of an atom of an element in (a) Group 2, (b) Group 7, (c) Group 15 of the periodic table.

F1.2(b) Express the typical ground-state electron configuration of an atom of an element in (a) Group 3, (b) Group 5, (c) Group 13 of the periodic table.

F1.3(a) Identify the oxidation numbers of the elements in (a) $MgCl_2$, (b) FeO, (c) Hg_2Cl_2.

F1.3(b) Identify the oxidation numbers of the elements in (a) CaH_2, (b) CaC_2, (c) LiN_3.

F1.4(a) Where in the periodic table are metals and nonmetals found?

F1.4(b) Where in the periodic table are transition metals, lanthanoids, and actinoids found?

F.2 Molecules

F2.1(a) Summarize what is meant by a single and multiple bond.

F2.1(b) Identify a molecule with (a) one, (b) two, (c) three lone pairs on the central atom.

F2.2(a) Draw the Lewis (electron dot) structures of (a) SO_3^{2-}, (b) XeF_4, (c) P_4.

F2.2(b) Draw the Lewis (electron dot) structures of (a) O_3, (b) ClF_3^+, (c) N_3^-.

F2.3(a) Summarize the principal concepts of the VSEPR theory of molecular shape.

F2.3(b) Identify four hypervalent compounds.

F2.4(a) Use VSEPR theory to predict the structures of (a) PCl_3, (b) PCl_5, (c) XeF_2, (d) XeF_4.

F2.4(b) Use VSEPR theory to predict the structures of (a) H_2O_2, (b) FSO_3^-, (c) KrF_2, (d) PCl_4^+.

F2.5(a) Identify the polarities (by attaching partial charges δ+ and δ−) of the bonds (a) C—Cl, (b) P—H, (c) N—O.

F2.5(b) Identify the polarities (by attaching partial charges $\delta+$ and $\delta-$) of the bonds (a) C—H, (b) P—S, (c) N—Cl.

F2.6(a) State whether you expect the following molecules to be polar or nonpolar: (a) CO_2, (b) SO_2, (c) N_2O, (d) SF_4.

F2.6(b) State whether you expect the following molecules to be polar or nonpolar: (a) O_3, (b) XeF_2, (c) NO_2, (d) C_6H_{14}.

F2.7(a) Arrange the molecules in Exercise F2.6a by increasing dipole moment.

F2.7(b) Arrange the molecules in Exercise F2.6b by increasing dipole moment.

F.3 Bulk matter

F3.1(a) Compare and contrast the properties of the solid, liquid, and gas states of matter.

F3.1(b) Compare and contrast the properties of the condensed and gaseous states of matter.

F3.2(a) Classify the following properties as extensive or intensive: (a) mass, (b) mass density, (c) temperature, (d) number density.

F3.2(b) Classify the following properties as extensive or intensive: (a) pressure, (b) specific heat capacity, (c) weight, (d) molality.

F3.3(a) Calculate (a) the amount of C_2H_5OH (in moles) and (b) the number of molecules present in 25.0 g of ethanol.

F3.3(b) Calculate (a) the amount of $C_6H_{12}O_6$ (in moles) and (b) the number of molecules present in 5.0 g of glucose.

F3.4(a) Calculate (a) the mass, (b) the weight on the surface of the Earth (where $g = 9.81\ \mathrm{m\ s^{-2}}$) of 10.0 mol $H_2O(l)$.

F3.4(b) Calculate (a) the mass, (b) the weight on the surface of Mars (where $g = 3.72\ \mathrm{m\ s^{-2}}$) of 10.0 mol $C_6H_6(l)$.

F3.5(a) Calculate the pressure exerted by a person of mass 65 kg standing (on the surface of the Earth) on shoes with soles of area 150 cm^2.

F3.5(b) Calculate the pressure exerted by a person of mass 60 kg standing (on the surface of the Earth) on shoes with stiletto heels of area 2 cm^2 (assume that the weight is entirely on the heels).

F3.6(a) Express the pressure calculated in Exercise F3.5a in atmospheres.

F3.6(b) Express the pressure calculated in Exercise F3.5b in atmospheres.

F3.7(a) Express a pressure of 1.45 atm in (a) pascal, (b) bar.

F3.7(b) Express a pressure of 222 atm in (a) pascal, (b) bar.

F3.8(a) Convert blood temperature, 37.0°C, to the Kelvin scale.

F3.8(b) Convert the boiling point of oxygen, 90.18 K, to the Celsius scale.

F3.9(a) Equation F.2 is a relation between the Kelvin and Celsius scales. Devise the corresponding equation relating the Fahrenheit and Celsius scales and use it to express the boiling point of ethanol (78.5°C) in degrees Fahrenheit.

F3.9(b) The Rankine scale is a version of the thermodynamic temperature scale in which the degrees (°R) are the same size as degrees Fahrenheit. Derive an expression relating the Rankine and Kelvin scales and express the freezing point of water in degrees Rankine.

F3.10(a) A sample of hydrogen gas was found to have a pressure of 110 kPa when the temperature was 20.0°C. What can its pressure be expected to be when the temperature is 7.0°C?

F3.10(b) A sample of 325 mg of neon occupies 2.00 dm^3 at 20.0°C. Use the perfect gas law to calculate the pressure of the gas.

F3.11(a) At 500°C and 93.2 kPa, the mass density of sulfur vapour is 3.710 $\mathrm{kg\ m^{-3}}$. What is the molecular formula of sulfur under these conditions?

F3.11(b) At 100°C and 16.0 kPa, the mass density of phosphorus vapour is 0.6388 $\mathrm{kg\ m^{-3}}$. What is the molecular formula of phosphorus under these conditions?

F3.12(a) Calculate the pressure exerted by 22 g of ethane behaving as a perfect gas when confined to 1000 cm^3 at 25.0°C.

F3.12(b) Calculate the pressure exerted by 7.05 g of oxygen behaving as a perfect gas when confined to 100 cm^3 at 100.0°C.

F3.13(a) A vessel of volume 10.0 dm^3 contains 2.0 mol H_2 and 1.0 mol N_2 at 5.0°C. Calculate the partial pressure of each component and their total pressure.

F3.13(b) A vessel of volume 100 cm^3 contains 0.25 mol O_2 and 0.034 mol CO_2 at 10.0°C. Calculate the partial pressure of each component and their total pressure.

F.4 Thermodynamic properties

F4.1(a) The heat capacity of a sample of iron was 3.67 $\mathrm{J\ K^{-1}}$. By how much would its temperature rise if 100 J of energy was transferred to it as heat?

F4.1(b) The heat capacity of a sample of water was 5.77 $\mathrm{J\ K^{-1}}$. By how much would its temperature rise if 50.0 kJ of energy was transferred to it as heat?

F4.2(a) The molar heat capacity of lead is 26.44 $\mathrm{J\ K^{-1}\ mol^{-1}}$. How much energy must be supplied (by heating) to 100 g of lead to increase its temperature by 10°C?

F4.2(b) The molar heat capacity of water is 75.2 $\mathrm{J\ K^{-1}\ mol^{-1}}$. How much energy must be supplied by heating to 10.0 g of water to increase its temperature by 10.0°C?

F4.3(a) The molar heat capacity of ethanol is 111.46 $\mathrm{J\ K^{-1}\ mol^{-1}}$. What is its specific heat capacity?

F4.3(b) The molar heat capacity of sodium is 28.24 $\mathrm{J\ K^{-1}\ mol^{-1}}$. What is its specific heat capacity?

F4.4(a) The specific heat capacity of water is 4.18 $\mathrm{J\ K^{-1}\ g^{-1}}$. What is its molar heat capacity?

F4.4(b) The specific heat capacity of copper is 0.384 $\mathrm{J\ K^{-1}\ g^{-1}}$. What is its molar heat capacity?

F4.5(a) By how much does the molar enthalpy of oxygen gas differ from its molar internal energy at 298 K? Assume perfect gas behaviour.

F4.5(b) By how much does the molar enthalpy of hydrogen gas differ from its molar internal energy at 1000°C? Assume perfect gas behaviour.

F4.6(a) The mass density of lead is 11.350 $\mathrm{g\ cm^{-3}}$. By how much does the molar enthalpy of lead differ from its molar internal energy at 1 bar?

F4.6(b) The mass density of water is 0.997 $\mathrm{g\ cm^{-3}}$. By how much does the molar enthalpy of water differ from its molar internal energy at 1 bar?

F4.7(a) Which do you expect to have the greater entropy at 298 K and 1 bar, liquid water or water vapour?

F4.7(b) Which do you expect to have the greater entropy at 0°C and 1 atm, liquid water or ice?

F4.8(a) Which do you expect to have the greater entropy, 100 g of iron at 300 K or 3000 K?

F4.8(b) Which do you expect to have the greater entropy, 100 g of water at 0°C or 100°C?

F4.9(a) State the second law of thermodynamics.

F4.9(b) Can the entropy of the system that is not isolated from its surroundings decrease during a spontaneous process?

F4.10(a) Give three examples of a system that is in dynamic equilibrium. What might happen when the equilibrium is disturbed?

F4.10(b) Give three examples of a system that is in static equilibrium. What might happen when the equilibrium is disturbed?

F.5 The relation between molecular and bulk properties

F5.1(a) What is meant by quantization of energy?

F5.1(b) In what circumstances are the effects of quantization most important for microscopic systems?

F5.2(a) Suppose two states differ in energy by 1.0 eV (electronvolts, see inside the front cover); what is the ratio of their populations at (a) 300 K, (b) 3000 K?

F5.2(b) Suppose two states differ in energy by 2.0 eV (electronvolts, see inside the front cover); what is the ratio of their populations at (a) 200 K, (b) 2000 K?

F5.3(a) Suppose two states differ in energy by 1.0 eV, what can be said about their populations when $T = 0$?

F5.3(b) Suppose two states differ in energy by 1.0 eV, what can be said about their populations when the temperature is infinite?

F5.4(a) A typical vibrational excitation energy of a molecule corresponds to a wavenumber of 2500 cm^{-1} (convert to an energy separation by multiplying by hc; see Section F.7). Would you expect to find molecules in excited vibrational states at room temperature (20°C)?

F5.4(b) A typical rotational excitation energy of a molecule corresponds to a frequency of about 10 GHz (convert to an energy separation by multiplying by h; see Section F.7). Would you expect to find gas-phase molecules in excited rotational states at room temperature (20°C)?

F5.5(a) What are the assumptions of the kinetic molecular theory?

F5.5(b) What are the main features of the Maxwell distribution of speeds?

F5.6(a) Suggest a reason why most molecules survive for long periods at room temperature.

F5.6(b) Suggest a reason why the rates of chemical reactions typically increase with increasing temperature.

F5.7(a) Calculate the relative mean speeds of N_2 molecules in air at 0°C and 40°C.

F5.7(b) Calculate the relative mean speeds of CO_2 molecules in air at 20°C and 30°C.

F5.8(a) Calculate the relative mean speeds of N_2 and CO_2 molecules in air.

F5.8(b) Calculate the relative mean speeds of Hg_2 and H_2 molecules in a gaseous mixture.

F5.9(a) Use the equipartition theorem to calculate the contribution of translational motion to the internal energy of 5.0 g of argon at 25°C.

F5.9(b) Use the equipartition theorem to calculate the contribution of translational motion to the internal energy of 10.0 g of helium at 30°C.

F5.10(a) Use the equipartition theorem to calculate the contribution to the total internal energy of a sample of 10.0 g of (a) carbon dioxide, (b) methane at 20°C; take into account translation and rotation but not vibration.

F5.10(b) Use the equipartition theorem to calculate the contribution to the total internal energy of a sample of 10.0 g of lead at 20°C, taking into account the vibrations of the atoms.

F5.11(a) Use the equipartition theorem to compute the molar heat capacity of argon.

F5.11(b) Use the equipartition theorem to compute the molar heat capacity of helium.

F5.12(a) Use the equipartition theorem to estimate the heat capacity of (a) carbon dioxide, (b) methane.

F5.12(b) Use the equipartition theorem to estimate the heat capacity of (a) water vapour, (b) lead.

F.6 Particles

F6.1(a) A particle of mass 1.0 g is released near the surface of the Earth, where the acceleration of free fall is $g = 9.81$ m s^{-2}. What will be its speed and kinetic energy after (a) 1.0 s, (b) 3.0 s? Ignore air resistance.

F6.1(b) The same particle is released near the surface of Mars, where the acceleration of free fall is $g = 3.72$ m s^{-2}. What will be its speed and kinetic energy after (a) 1.0 s, (b) 3.0 s? Ignore air resistance.

F6.2(a) An ion of charge ze moving through water is subject to an electric field of strength $\mathcal{E}$ which exerts a force $ze\mathcal{E}$, but it also experiences a frictional drag proportional to its speed s and equal to $6\pi\eta Rs$, where R is its radius and η (eta) is the viscosity of the medium. What will be its terminal velocity?

F6.2(b) A particle descending through a viscous medium experiences a frictional drag proportional to its speed s and equal to $6\pi\eta Rs$, where R is its radius and η (eta) is the viscosity of the medium. If the acceleration of free fall is denoted g, what will be the terminal velocity of a sphere of radius R and mass density ρ (rho)?

F6.3(a) Confirm that the general solution of the harmonic oscillator equation of motion ($md^2x/dt^2 = -kx$) is $x(t) = A \sin \omega t + B \cos \omega t$ with $\omega = (k/m)^{1/2}$.

F6.3(b) Given the general solution of a harmonic oscillator in Exercise F6.3a, how does the momentum of the oscillator vary with time?

F6.4(a) Consider a harmonic oscillator with $B = 0$ (in the notation of Exercise F6.3a); relate the total energy at any instant to its maximum displacement amplitude.

F6.4(b) Identify the *turning points* of a harmonic oscillator, the displacements at the ends of its swing.

F6.5(a) In an early ('semiclassical') picture of a hydrogen atom, an electron travels in a circular path of radius 53 pm at 2188 km s^{-1}. What is the magnitude of the average acceleration that the electron undergoes during one-quarter of a revolution?

F6.5(b) Given the acceleration calculated in Exercise F6.5a, what is the magnitude of the average force that the electron experiences in its orbit?

F6.6(a) Use the information in Exercise F6.5a to calculate the magnitude of the angular momentum of an electron in the semiclassical picture of the hydrogen atom. Go on to express your result as a multiple of $h/2\pi$, where h is Planck's constant (see inside front cover).

F6.6(b) In a continuation of the semiclassical picture, the electron is excited into an orbit of radius $4a_0$ but continues to travel at 2188 km s^{-1}. Calculate the magnitude of the angular momentum of the electron and express your result as a multiple of $h/2\pi$, where h is Planck's constant (see inside front cover).

F6.7(a) The force constant of a C—H bond is about 450 N m^{-1}. How much work is needed to stretch such bond by (a) 10 pm, (b) 20 pm?

F6.7(b) The force constant of the H—H bond is about 575 N m^{-1}. How much work is needed to stretch such bond by (a) 10 pm, (b) 20 pm?

F6.8(a) An electron is accelerated in an electron microscope from rest through a potential difference $\Delta\phi = 100$ kV and acquires an energy of $e\Delta\phi$. What is its final speed? What is its energy in electronvolts (eV)?

F6.8(b) A $C_6H_4^{2+}$ ion is accelerated in a mass spectrometer from rest through a potential difference $\Delta\phi = 20$ kV and acquires an energy of $2e\Delta\phi$. What is its final speed? What is its energy in electronvolts (eV)?

F6.9(a) Calculate the work that must be done in order to remove a Na^+ ion from 200 pm away from a Cl^- ion to infinity (in a vacuum). What work would be needed if the separation took place in water? ($\varepsilon_r = 78$)

F6.9(b) Calculate the work that must be done in order to remove an Mg^{2+} ion from 250 pm away from an O^{2-} ion to infinity (in a vacuum). What work would be needed if the separation took place in water? ($\varepsilon_r = 78$)

F6.10(a) Calculate the electric potential due to the nuclei at a point in a LiH molecule located at 200 pm from the Li nucleus and 150 pm from the H nucleus.

F6.10(b) Plot the electric potential due to the nuclei at a point in a Na^+Cl^- ion pair located on a line halfway between the nuclei (the internuclear separation is 283 pm) as the point approaches from infinity and ends at the midpoint between the nuclei.

F6.11(a) An electric heater is immersed in a flask containing 200 g of water, and a current of 2.23 A from a 15.0 V supply is passed for 12.0 minutes. How much energy is supplied to the water? Estimate the rise in temperature (for water, $C = 75.3$ J K^{-1} mol^{-1}).

F6.11(b) An electric heater is immersed in a flask containing 150 g of ethanol, and a current of 1.12 A from a 12.5 V supply is passed for 172 s. How much energy is supplied to the ethanol? Estimate the rise in temperature (for ethanol, $C = 111.5$ J K^{-1} mol^{-1}).

F.7 Waves

F7.1(a) Calculate the wavenumber and frequency of yellow light, of wavelength 590 nm.

F7.1(b) Calculate the wavenumber and frequency of red light, of wavelength 710 nm.

F7.2(a) What is the speed of light in water if the refractive index of water is 1.33?

F7.2(b) What is the speed of light in benzene if the refractive index of water is 1.52?

F7.3(a) The wavenumber of a typical vibrational transition of a hydrocarbon is 2500 cm^{-1}. Calculate the corresponding wavelength and frequency.

F7.3(b) The wavenumber of a typical vibrational transition of an O—H bond is 3600 cm^{-1}. Calculate the corresponding wavelength and frequency.

F7.4(a) Draw the graph of the wave $f(x) = \cos\{2\pi x/\lambda\}$ against x for $\lambda = 1$ cm, and then on the same graph the wave $f(x) = \cos\{(2\pi x/\lambda) + \phi\}$ with $\phi = \pi/3$.

F7.4(b) Draw the graph of the wave $f(x) = \cos\{2\pi x/\lambda\}$ against x for $\lambda = 1$ cm, and then on the same graph the wave $f(x) = \cos\{(2\pi x/\lambda) + \phi\}$ with $\phi = -\pi/3$.

F7.5(a) Confirm that $f(x,t) = \cos\{2\pi\nu t - (2\pi/\lambda)x\}$ satisfies the wave equation.

F7.5(b) Confirm that $f(x,t) = \cos\{2\pi\nu t - (2\pi/\lambda)x + \pi\}$ satisfies the wave equation.

F.8 Units

F8.1(a) Express a volume of 1.45 cm^3 in cubic metres.

F8.1(b) Express a volume of 1.45 dm^3 in cubic centimetres.

F8.2(a) Express a mass density of 11.2 g cm^{-3} in kilograms per cubic metre.

F8.2(b) Express a mass density of 1.12 g dm^{-3} in kilograms per cubic metre.

F8.3(a) Express pascal per joule in base units.

F8.3(b) Express (joule)^2 per (newton)^3 in base units.

F8.4(a) The expression kT/hc sometimes appears in physical chemistry. Evaluate this expression at 298 K in reciprocal centimetres (cm^{-1}).

F8.4(b) The expression kT/e sometimes appears in physical chemistry. Evaluate this expression at 298 K in millivolts (mV).

F8.5(a) Given that $R = 8.3144$ J K^{-1} mol^{-1}, express R in decimetre cubed atmospheres per kelvin per mole.

F8.5(b) Given that $R = 8.3144$ J K^{-1} mol^{-1}, express R in pascal centimetre cubed per kelvin per molecule.

F8.6(a) Convert 1 dm^3 atm into joules.

F8.6(b) Convert 1 J into litre-atmospheres.

F8.7(a) Determine the SI units of $e^2/\varepsilon_0 r^2$. Express them in (a) base units, (b) units containing newtons.

F8.7(b) Determine the SI units of $\mu_B^2/\mu_0 r^3$, where μ_B is the Bohr magneton ($\mu_B = e\hbar/2m_e$) and μ_0 is the vacuum permeability (see inside front cover). Express them in (a) base units, (b) units containing joules.

MATHEMATICAL BACKGROUND 1

Differentiation and integration

Two of the most important mathematical techniques in the physical sciences are differentiation and integration. They occur throughout the subject, and it is essential to be aware of the procedures involved. However, both techniques have much richer implications than these remarks might suggest, as will become apparent as the subject unfolds.

MB1.1 Differentiation: definitions

Differentiation is concerned with the slopes of functions, such as the rate of change of a variable with time. The formal definition of the **derivative**, $\mathrm{d}f/\mathrm{d}x$, of a function $f(x)$ is

$$\frac{\mathrm{d}f}{\mathrm{d}x} = \lim_{\delta x \to 0} \frac{f(x+\delta x) - f(x)}{\delta x} \tag{MB1.1}$$

As shown in Fig. MB1.1, the derivative can be interpreted as the slope of the tangent to the graph of $f(x)$. A positive first derivative indicates that the function slopes upwards (as x increases), and a negative first derivative indicates the opposite. It is sometimes convenient to denote the first derivative as $f'(x)$. The **second derivative**, $\mathrm{d}^2f/\mathrm{d}x^2$, of a function is the derivative of the first derivative:

$$\frac{\mathrm{d}^2f}{\mathrm{d}x^2} = \lim_{\delta x \to 0} \frac{f'(x+\delta x) - f'(x)}{\delta x} \tag{MB1.2}$$

It is sometimes convenient to denote the second derivative f''. As shown in Fig. MB1.1, the second derivative of a function can be interpreted as an indication of the sharpness of the curvature of the function. A positive second derivative indicates that the function is ⌣ shaped, and a negative second derivative indicates that it is ⌢ shaped.

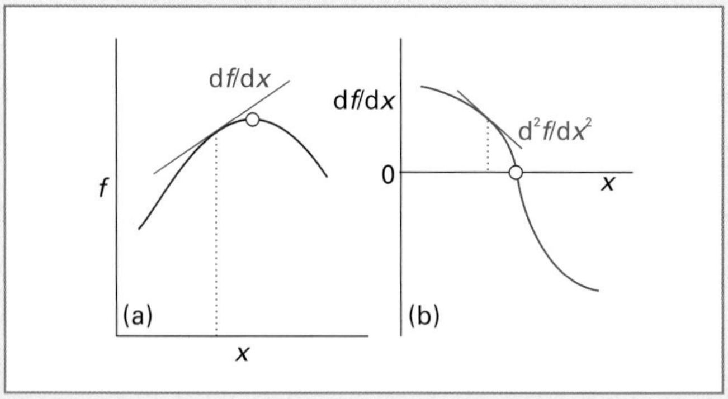

Fig. MB1.1 (a) The first derivative of a function is equal to the slope of the tangent to the graph of the function at that point. The small circle indicates the extremum (in this case, maximum) of the function, where the slope is zero. (b) The second derivative of the same function is the slope of the tangent to a graph of the first derivative of the function. It can be interpreted as an indication of the curvature of the function at that point.

The derivatives of some common functions are as follows:

$$\frac{\mathrm{d}}{\mathrm{d}x}x^n = nx^{n-1} \tag{MB1.3a}$$

$$\frac{\mathrm{d}}{\mathrm{d}x}e^{ax} = ae^{ax} \tag{MB1.3b}$$

$$\frac{\mathrm{d}}{\mathrm{d}x}\sin ax = a\cos ax \qquad \frac{\mathrm{d}}{\mathrm{d}x}\cos ax = -a\sin ax \tag{MB1.3c}$$

MB1.2 Differentiation: manipulations

It follows from the definition that a variety of combinations of functions can be differentiated by using the following rules:

$$\frac{\mathrm{d}}{\mathrm{d}x}(u+v) = \frac{\mathrm{d}u}{\mathrm{d}x} + \frac{\mathrm{d}v}{\mathrm{d}x} \tag{MB1.4a}$$

$$\frac{\mathrm{d}}{\mathrm{d}x}uv = u\frac{\mathrm{d}v}{\mathrm{d}x} + v\frac{\mathrm{d}u}{\mathrm{d}x} \tag{MB1.4b}$$

$$\frac{\mathrm{d}}{\mathrm{d}x}\frac{u}{v} = \frac{1}{v}\frac{\mathrm{d}u}{\mathrm{d}x} - \frac{u}{v^2}\frac{\mathrm{d}v}{\mathrm{d}x} \tag{MB1.4c}$$

The last of these three relations follows from the second.

● **A BRIEF ILLUSTRATION**

To differentiate the function $f = \{(\sin ax)/x\}^2$, we use eqn MB1.4 to write

$$\frac{\mathrm{d}}{\mathrm{d}x}\left(\frac{\sin ax}{x}\right)^2 = \frac{\mathrm{d}}{\mathrm{d}x}\left(\frac{\sin ax}{x}\right)\left(\frac{\sin ax}{x}\right) = 2\left(\frac{\sin ax}{x}\right)\frac{\mathrm{d}}{\mathrm{d}x}\left(\frac{\sin ax}{x}\right)$$

$$= 2\left(\frac{\sin ax}{x}\right)\left\{\frac{1}{x}\frac{\mathrm{d}}{\mathrm{d}x}\sin ax + \sin ax\frac{\mathrm{d}}{\mathrm{d}x}\frac{1}{x}\right\}$$

$$= 2\left\{\frac{a}{x^2}\sin ax\cos ax - \frac{\sin^2 ax}{x^3}\right\}$$

The function and this first derivative are plotted in Fig. MB1.2. ●

MB1.3 Partial derivatives

When a function depends on more than one variable, we need the concept of a **partial derivative**, $\partial f/\partial x$. Note the change from d to ∂: partial derivatives are dealt with at length in *Mathematical background 8*; all we need know at this stage is that they signify that all variables other than the stated variable are regarded as constant when evaluating the derivative.

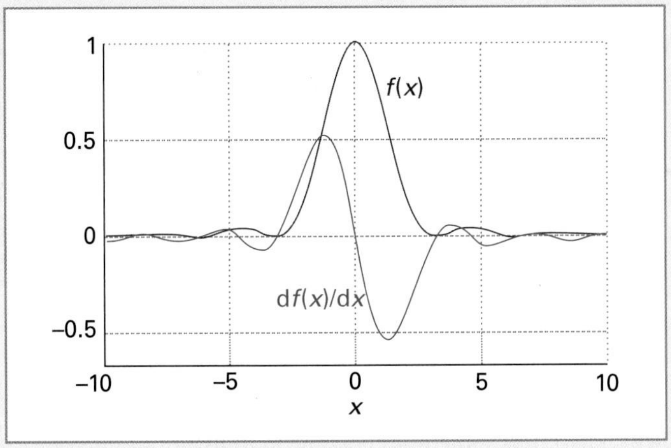

Fig. MB1.2 The function considered in the *illustration* and its first derivative.

● **A BRIEF ILLUSTRATION**

Suppose we are told that f is a function of two variables, and specifically $f = 4x^2y^3$. Then, to evaluate the partial derivative of f with respect to x, we regard y as a constant (like the 4), and obtain

$$\frac{\partial f}{\partial x} = \frac{\partial(4x^2y^3)}{\partial x} = 4y^3\frac{\partial x^2}{\partial x} = 8xy^3$$

Similarly, to evaluate the partial derivative of f with respect to y, we regard x as a constant (again, like the 4), and obtain

$$\frac{\partial f}{\partial y} = \frac{\partial(4x^2y^3)}{\partial y} = 4x^2\frac{\partial y^3}{\partial y} = 12x^2y^2 \text{ ●}$$

MB1.4 Series expansions

One application of differentiation is to the development of power series for functions. The **Taylor series** for a function $f(x)$ in the vicinity of $x = a$ is

$$f(x) = f(a) + \left(\frac{\mathrm{d}f}{\mathrm{d}x}\right)_a (x-a) + \frac{1}{2!}\left(\frac{\mathrm{d}^2 f}{\mathrm{d}x^2}\right)_a (x-a)^2$$

$$+ \cdots + \frac{1}{n!}\left(\frac{\mathrm{d}^n f}{\mathrm{d}x^n}\right)_a (x-a)^n + \cdots$$

$$= \sum_{n=0}^{\infty}\frac{1}{n!}\left(\frac{\mathrm{d}^n f}{\mathrm{d}x^n}\right)_a (x-a)^n \qquad \text{(MB1.5)}$$

where the notation $(\ldots)_a$ means that the derivative is evaluated at $x = a$ and $n!$ denotes a **factorial**, given by

$$n! = n(n-1)(n-2)\ldots 1 \qquad \text{(MB1.6)}$$

By definition $0! = 1$. The **Maclaurin series** for a function is a special case of the Taylor series in which $a = 0$.

● **A BRIEF ILLUSTRATION**

To evaluate the expansion of cos x around $x = 0$ we note that

$$\left(\frac{\mathrm{d}\cos x}{\mathrm{d}x}\right)_0 = (-\sin x)_0 = 0 \qquad \left(\frac{\mathrm{d}^2\cos x}{\mathrm{d}x^2}\right)_0 = (-\cos x)_0 = -1$$

and in general

$$\left(\frac{\mathrm{d}^n\cos x}{\mathrm{d}x^n}\right)_0 = \begin{cases} 0 \text{ for } n \text{ odd} \\ (-1)^{n/2} \text{ for } n \text{ even} \end{cases}$$

Therefore,

$$\cos x = \sum_{n \text{ even}}^{\infty}\frac{(-1)^{n/2}}{n!}x^n = 1 - \tfrac{1}{2}x^2 + \tfrac{1}{24}x^4 - \cdots \text{ ●}$$

The following Taylor series (specifically, Maclaurin series) are used at various stages in the text:

$$(1+x)^{-1} = 1 - x + x^2 - \cdots = \sum_{n=0}^{\infty}(-1)^n x^n \qquad \text{(MB1.7a)}$$

$$\mathrm{e}^x = 1 + x + \tfrac{1}{2}x^2 + \cdots = \sum_{n=0}^{\infty}\frac{x^n}{n!} \qquad \text{(MB1.7b)}$$

$$\ln(1+x) = x - \tfrac{1}{2}x^2 + \tfrac{1}{3}x^3 - \cdots = \sum_{n=1}^{\infty}(-1)^{n+1}\frac{x^n}{n} \qquad \text{(MB1.7c)}$$

Taylor series are used to simplify calculations, for when $x \ll 1$ it is possible, to a good approximation, to terminate the series after one or two terms. Thus, provided $x \ll 1$ we can write

$$(1+x)^{-1} \approx 1 - x \qquad \text{(MB1.8a)}$$

$$\mathrm{e}^x \approx 1 + x \qquad \text{(MB1.8b)}$$

$$\ln(1+x) \approx x \qquad \text{(MB1.8c)}$$

A series is said to **converge** if the sum approaches a finite, definite value as n approaches infinity. If the sum does not approach a finite, definite value, then the series is said to **diverge**. Thus, the series in eqn MB1.7a converges for $x < 1$ and diverges for $x \geq 1$.There are a variety of tests for convergence, which are explained in mathematics texts.

MB1.5 Integration: definitions

Integration (which formally is the inverse of differentiation) is concerned with the areas under curves. The **integral** of a function $f(x)$, which is denoted $\int f\,\mathrm{d}x$ (the symbol $\int$ is an elongated S denoting a sum), between the two values $x = a$ and $x = b$ is defined by imagining the x-axis as divided into strips of width δx and evaluating the following sum:

$$\int_a^b f(x)\mathrm{d}x = \lim_{\delta x \to 0}\sum_i f(x_i)\delta x \qquad \text{(MB1.9)}$$

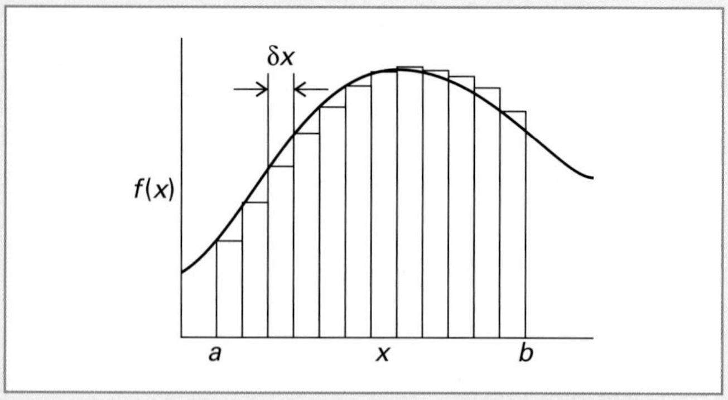

Fig. MB1.3 A definite integral is evaluated by forming the product of the value of the function at each point and the increment δx, with $\delta x \to 0$, and then summing the products $f(x)\delta x$ for all values of x between the limits a and b. It follows that the value of the integral is the area under the curve between the two limits.

As can be appreciated from Fig. MB1.3, the integral is the area under the curve between the limits a and b. The function to be integrated is called the **integrand.** It is an astonishing mathematical fact that the integral of a function is the inverse of the differential of that function in the sense that, if we differentiate f and then integrate the resulting function, then we obtain the original function f (to within a constant). The function in eqn MB1.9 with the limits specified is called a **definite integral.** If it is written without the limits specified, then we have an **indefinite integral.** If the result of carrying out an indefinite integration is $g(x)+C$, where C is a constant, the following notation is used to evaluate the corresponding definite integral:

$$I=\int_a^b f(x)\mathrm{d}x=\{g(x)+C\}\Big|_a^b=\{g(b)+C\}-\{g(a)+C\}$$
$$=g(b)-g(a) \qquad \text{(MB1.10)}$$

Note that the constant of integration disappears.

Some of the common indefinite integrals encountered in chemistry are as follows (with C a constant in each case):

$$\int x^n\,\mathrm{d}x=\frac{x^{n+1}}{n+1}+C \text{ (provided } n\neq-1) \qquad \text{(MB1.11a)}$$

$$\int \frac{1}{x}\mathrm{d}x=\ln x+C \qquad \text{(MB1.11b)}$$

$$\int \mathrm{e}^{ax}\,\mathrm{d}x=\frac{\mathrm{e}^{ax}}{a}+C \qquad \text{(MB1.11c)}$$

$$\int \ln ax\,\mathrm{d}x=x\ln ax-x+C \qquad \text{(MB1.11d)}$$

$$\int \sin ax\,\mathrm{d}x=-\frac{1}{a}\cos ax+C \qquad \text{(MB1.11e)}$$

$$\int \cos ax\,\mathrm{d}x=\frac{1}{a}\sin ax+C \qquad \text{(MB1.11f)}$$

As may be verified, these relations are the inverse of those in eqn MB1.3.

MB1.6 Integration: manipulations

When an indefinite integral is not in the form of one of eqns MB1.11a–f, it is sometimes possible to transform it into one of the forms by using integration techniques such as:

Substitution. Introduce a variable u related to the independent variable x (for example, an algebraic relation such as $u=x^2-1$ or a trigonometric relation such as $u=\sin x$). Express the differential dx in terms of du (for these substitutions, d$u=2x\,$dx and d$u=\cos x\,$dx, respectively). Then transform the original integral written in terms of x into an integral in terms of u upon which, in some cases, a standard form such as one of those above can be used.

A BRIEF ILLUSTRATION

To evaluate the indefinite integral $\int\cos^2 x\sin x\,\mathrm{d}x$, which occurs in the discussion of atomic structure, we make the substitution $u=\cos x$. It follows that $\mathrm{d}u/\mathrm{d}x=-\sin x$, and therefore that $\sin x\,\mathrm{d}x=-\mathrm{d}u$. The integral is therefore

$$\int\cos^2 x\sin x\,\mathrm{d}x=-\int u^2\,\mathrm{d}u=-\tfrac{1}{3}u^3+C=-\tfrac{1}{3}\cos^3 x+C$$

To evaluate the corresponding definite integral, we have to convert the limits on x into limits on u. Thus, if the limits are $x=0$ and $x=\pi$, the limits become $u=\cos 0=1$ and $u=\cos\pi=-1$:

$$\int_0^\pi\cos^2 x\sin x\,\mathrm{d}x=-\int_1^{-1}u^2\,\mathrm{d}u=-\{\tfrac{1}{3}u^3+C\}\Big|_1^{-1}=\tfrac{2}{3}$$

Integration by parts. For two functions $f(x)$ and $g(x)$:

$$\int f\frac{\mathrm{d}g}{\mathrm{d}x}\mathrm{d}x=fg-\int g\frac{\mathrm{d}f}{\mathrm{d}x}\mathrm{d}x \qquad \text{(MB1.12a)}$$

which may be abbreviated as:

$$\int f\,\mathrm{d}g=fg-\int g\,\mathrm{d}f \qquad \text{(MB1.12b)}$$

● **A BRIEF ILLUSTRATION**

Integrals over xe^{-ax} and their analogues occur commonly in the discussion of atomic structure and spectra. They may be integrated by parts, as in the following:

$$\int_0^\infty \overbrace{x}^{f}\overbrace{e^{-ax}}^{dg/dx}\,dx = \overbrace{x}^{f}\overbrace{\frac{e^{-ax}}{-a}}^{g}\bigg|_0^\infty - \int_0^\infty \overbrace{\frac{e^{-ax}}{-a}}^{g}\,dx \quad \left(\text{because } \frac{df}{dx}=1\right)$$

$$= -\frac{1}{a}xe^{-ax}\bigg|_0^\infty + \frac{1}{a}\int_0^\infty e^{-ax}\,dx$$

$$= 0 - \frac{e^{-ax}}{a^2}\bigg|_0^\infty = \frac{1}{a^2} \; ●$$

MB1.7 **Multiple integrals**

A function may depend on more than one variable, in which case we may need to integrate over both the variables:

$$I = \int_a^b\int_c^d f(x,y)\,dx\,dy \qquad \text{(MB1.13)}$$

The conventions vary, but we shall adopt the convention that a and b are the limits of the variable x and c and d are the limits for y. This procedure is simple if the function is a product of functions of each variable and of the form $f(x,y) = X(x)Y(y)$. In this case, the double integral is just a product of each integral:

$$I = \int_a^b\int_c^d X(x)Y(y)\,dx\,dy = \int_a^b X(x)\,dx\int_c^d Y(y)\,dy \qquad \text{(MB1.14)}$$

● **A BRIEF ILLUSTRATION**

Double integrals of the form

$$I = \int_0^{L_1}\int_0^{L_2} \sin^2(\pi x/L_1)\sin^2(\pi y/L_2)\,dx\,dy$$

occur in the discussion of the translational motion of a particle in two dimensions, where L_1 and L_2 are the maximum extents of travel along the x- and y-axes, respectively. To evaluate I we use eqn MB1.14 and the indefinite integral

$$\int \sin^2 ax\,dx = \frac{x}{2} - \frac{\sin 2ax}{4a} + C$$

to write

$$I = \int_0^{L_1} \sin^2(\pi x/L_1)\,dx\int_0^{L_2} \sin^2(\pi y/L_2)\,dy$$

$$= \left\{\frac{x}{2} - \frac{L_1}{4\pi}\sin(2\pi x/L_1) + C\right\}\bigg|_0^{L_1}$$

$$\times\left\{\frac{y}{2} - \frac{L_2}{4\pi}\sin(2\pi y/L_2) + C\right\}\bigg|_0^{L_2}$$

$$= \tfrac{1}{4}L_1L_2 \; ●$$

PART 1
Quantum theory

The phenomena of chemistry cannot be understood thoroughly without a firm understanding of the principal concepts of quantum mechanics, the most fundamental description of matter that we currently possess. The same is true of virtually all the spectroscopic techniques that are now so central to investigations of composition and structure. Present-day techniques for studying chemical reactions have progressed to the point at which the information is so detailed that quantum mechanics has to be used in its interpretation. And, of course, the very currency of chemistry—the electronic structures of atoms and molecules—cannot be discussed without making use of quantum mechanical concepts.

Part 1 of the text introduces the fundamental principles of quantum mechanics and its basic manipulation. We begin by describing the results of three crucial experiments that led to the conclusion that energy is quantized and that particles have wave-like dynamical properties that can be deduced from a wavefunction. Then we determine the wavefunctions corresponding to several basic types of motion: translation (in one and several dimensions); vibration; and rotation (in two and three dimensions). Along the way we discover phenomena and properties that are purely quantum mechanical in nature and have no classical analogues. In subsequent parts of the text we use these concepts to understand atomic structure, molecular structure, and chemical reactivity.

The principles of quantum theory

1

This chapter introduces some of the basic principles of quantum mechanics. First, it reviews the experimental results that called into question the fundamental assumptions of classical physics. These experiments led to the conclusion that particles cannot in general have an arbitrary energy and that the classical concepts of 'particle' and 'wave' blend together. The result was the formulation of quantum mechanics, which postulates that the properties of a system are expressed in terms of a wavefunction obtained by solving the Schrödinger equation. In this chapter, we interpret wavefunctions and see that they lead to the uncertainty principle, one of the most profound departures from classical physics. We also introduce some of the techniques of quantum mechanics by making use of mathematical constructs known as operators.

The first goal of our study of physical chemistry is to gain a firm understanding of atomic and molecular structure. It was once thought that the motion of atoms and subatomic particles could be expressed using **classical mechanics**, the laws of motion introduced in the seventeenth century by Isaac Newton, for these laws explained the motion of common objects and planets. However, towards the end of the nineteenth century, experimental evidence accumulated showing that classical mechanics failed when it was applied to particles as small as electrons, and the appropriate concepts and equations for describing them were not discovered until the 1920s. In this chapter, we introduce this new mechanics, which is called **quantum mechanics**.

The concepts of quantum mechanics are used by computational chemists in theoretical studies of molecular structure and reactivity. Quantum mechanical phenomena also form the basis of virtually all the modes of spectroscopy and microscopy that are now so central to investigations of composition and structure in both chemistry and biology. Present-day techniques for studying chemical reactions have progressed to the point where the information is so detailed that quantum mechanics has to be used in its interpretation. For these reasons, the language of quantum mechanics will be used throughout the text.

Three crucial experiments

The basic principles of classical mechanics are reviewed in *Fundamentals*. In brief, they show that classical physics (1) predicts a trajectory for particles, with precisely specified locations and momenta at each instant, and (2) allows the translational, rotational, and vibrational modes of motion to be excited to any energy simply by controlling the forces that are applied. These conclusions agree with everyday experience. Everyday experience, however, does not extend to individual atoms and molecules, and careful experiments of the type described below have shown that classical mechanics fails when applied to the transfers of very small energies and to objects of very small mass.

1.1 Quantization of energy

The most compelling evidence that particles can absorb or release energy only in discrete amounts (that is, that the energy of particles is quantized) comes from **spectroscopy**, the detection and analysis of the electromagnetic radiation absorbed, emitted, or scattered by a substance. The record of light intensity transmitted or scattered by a molecule as a function of frequency (ν), wavelength (λ), or wavenumber ($\tilde{\nu} = \nu/c$) is called its **spectrum** (from the Latin word for appearance).

A typical atomic spectrum is shown in Fig. 1.1 and a typical molecular spectrum is shown in Fig. 1.2. The obvious feature of both is that radiation is emitted or absorbed at a series of discrete frequencies. This observation can be understood if the energy of the atoms or molecules is also confined to discrete values, for then energy can be discarded or absorbed only in discrete amounts (Fig. 1.3). Then, if the energy of an atom decreases by ΔE, the energy is carried away as radiation of frequency ν, and an emission 'line', a sharply defined peak, appears in the spectrum. We say that a molecule undergoes a **spectroscopic transition**, a change of state, when the **Bohr frequency condition**

$$\Delta E = h\nu \qquad (1.1)$$

is fulfilled. The fundamental constant h, known as **Planck's constant**, has a value 6.626×10^{-34} J s.

We develop the principles and applications of atomic spectroscopy in Chapter 4 and of molecular spectroscopy in Chapters 10–12.

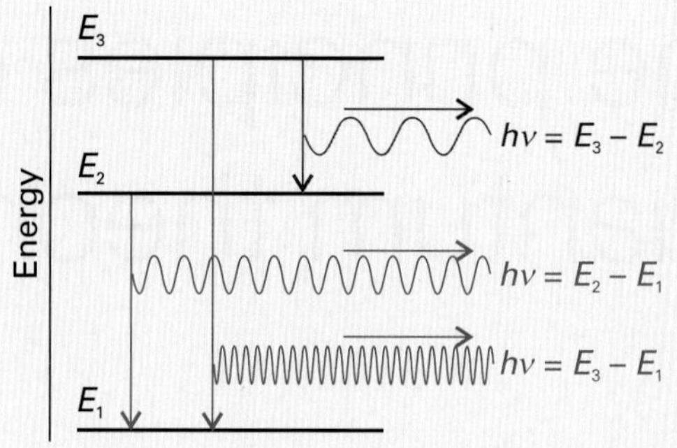

Fig. 1.3 Spectroscopic transitions, such as those shown above, can be accounted for if we assume that a molecule emits electromagnetic radiation as it changes between discrete energy levels. Note that high-frequency radiation is emitted when the energy change is large.

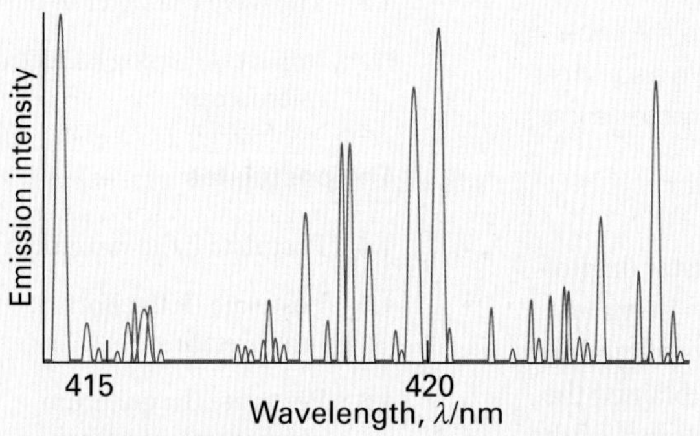

Fig. 1.1 A region of the spectrum of radiation emitted by excited iron atoms consists of radiation at a series of discrete wavelengths (or frequencies).

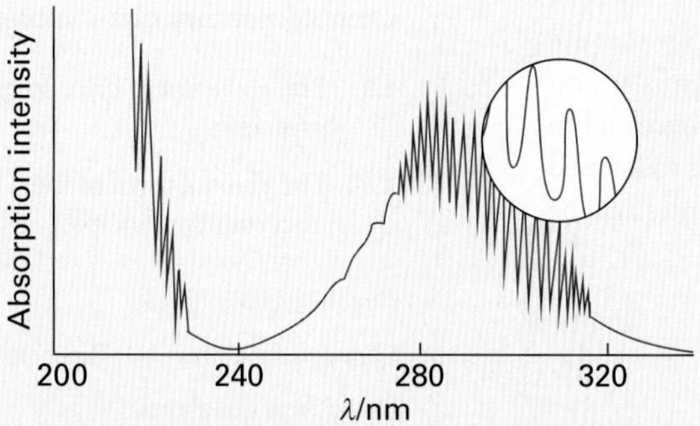

Fig. 1.2 When a molecule changes its state, it does so by absorbing radiation at definite frequencies. This spectrum is part of that due to the electronic, vibrational, and rotational excitation of sulfur dioxide (SO_2) molecules. This observation suggests that molecules can possess only discrete energies, not an arbitrary energy.

1.2 The particle character of electromagnetic radiation

Classical physics treats electromagnetic radiation as a wave. However, the **photoelectric effect**, the ejection of electrons from metals when they are exposed to ultraviolet radiation, suggests that radiation also has properties normally associated with particles. The experimental characteristics of the effect are as follows.

1. No electrons are ejected, regardless of the intensity of the radiation, unless the frequency of the radiation exceeds a threshold value characteristic of the metal.

2. The kinetic energy of the ejected electrons increases linearly with the frequency of the incident radiation but is independent of the intensity of the radiation.

3. Even at low light intensities, electrons are ejected if the frequency is above the threshold.

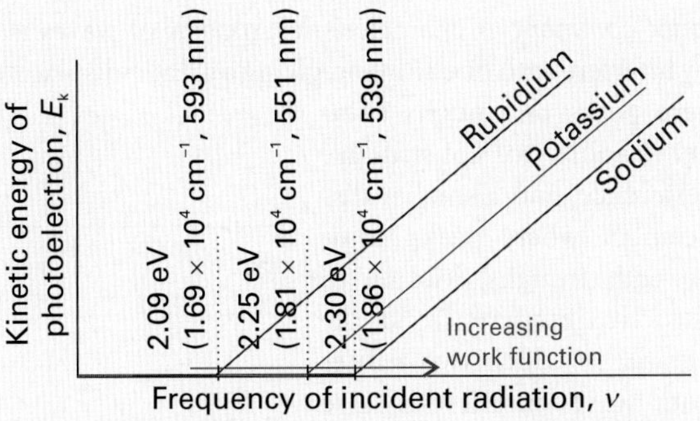

Fig. 1.4 In the photoelectric effect, it is found that no electrons are ejected when the incident radiation has a frequency below a value characteristic of the metal and, above that value, the kinetic energy of the photoelectrons varies linearly with the frequency of the incident radiation.

interActivity Calculate the value of Planck's constant given that the following kinetic energies were observed for photoejected electrons irradiated by radiation of the wavelengths noted.

λ_i/nm	320	330	345	360	385
E_k/eV	1.17	1.05	0.885	0.735	0.511

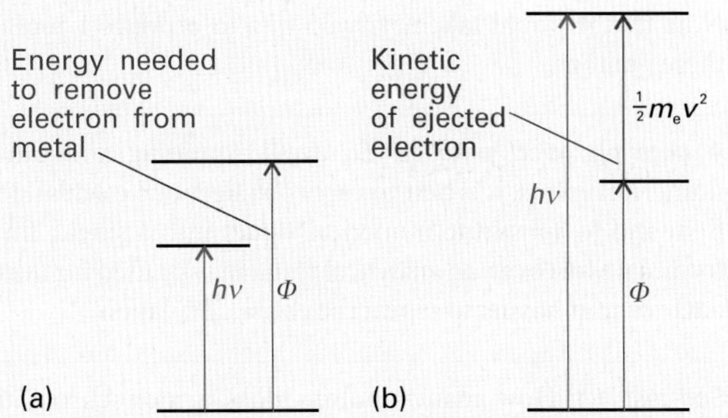

Fig. 1.5 The photoelectric effect can be explained if it is supposed that the incident radiation is composed of photons that have energy proportional to the frequency of the radiation. (a) The energy of the photon is insufficient to drive an electron out of the metal. (b) The energy of the photon is more than enough to eject an electron, and the excess energy is carried away as the kinetic energy of the photoelectron (the ejected electron).

Figure 1.4 illustrates the first and second characteristics.

Albert Einstein explained these observations by suggesting that the photoelectric effect depends on the ejection of an electron when it is involved in a collision with a particle-like projectile that carries enough energy to eject the electron from the metal. If we suppose that the projectile is a **photon**, a particle of electromagnetic radiation, of energy $h\nu$, where ν is the frequency of the radiation, then it follows from the conservation of energy that the kinetic energy of the ejected electron should be given by

$$\tfrac{1}{2}m_e v^2 = h\nu - \Phi \tag{1.2}$$

In this expression Φ (uppercase phi) is a characteristic of the metal called its **work function**, the energy required to remove an electron from the metal to infinity (Fig. 1.5), the analogue of the ionization energy of an individual atom or molecule. Photoejection cannot occur if $h\nu < \Phi$ because the photon brings insufficient energy: this conclusion accounts for observation (1). Equation 1.2 predicts that the kinetic energy of an ejected electron should increase linearly with frequency, in agreement with observation (2). When a photon collides with an electron, it gives up all its energy, so we should expect electrons to appear as soon as the collisions begin, provided the photons have sufficient energy: this conclusion agrees with observation (3). A practical application of eqn 1.2 is that it provides a technique for the determination of Planck's constant, for the slopes of the lines in Fig. 1.4 are all equal to h.

The revolutionary idea behind this interpretation of the photoelectric effect is the view that a beam of electromagnetic radiation is a collection of particles, the photons, each with energy $h\nu$. It follows that the total energy emitted by a source of radiation of frequency ν is $Nh\nu$, where N is the number of photons emitted.

Example 1.1 *Calculating the number of photons*

Calculate the number of photons emitted by a monochromatic (single frequency) 100 W sodium vapour lamp in 1.0 s. Take the wavelength as 589 nm and assume 100 per cent efficiency.

Method Each photon has an energy $h\nu$, so the total number of photons needed to account for an energy E is $E/h\nu$. To use this equation, we need to know the frequency of the radiation (from $\nu = c/\lambda$) and the total energy emitted by the lamp. The latter is given by the product of the power (P, in watts) and the time interval for which the lamp is turned on ($E = P\Delta t$).

Answer The number of photons is

$$N = \frac{E}{h\nu} = \frac{P\Delta t}{h(c/\lambda)} = \frac{\lambda P\Delta t}{hc}$$

Substitution of the data gives

$$N = \frac{(5.89\times10^{-7}\ \text{m})\times(100\ \text{J s}^{-1})\times(1.0\ \text{s})}{(6.626\times10^{-34}\ \text{J s})\times(2.998\times10^{8}\ \text{m s}^{-1})} = 3.0\times10^{20}$$

Note that it would take nearly 35 min to produce 1 mol of these photons.

A note on good practice To avoid rounding and other numerical errors, it is best to carry out algebraic calculations first, and to substitute numerical values into a single, final formula. Moreover, an analytical result may be used for other data without having to repeat the entire calculation.

Self-test 1.1 How many photons does a monochromatic infrared source of power 1 mW and wavelength 1000 nm emit in 0.1 s? [5×10^{14}]

A brief comment A characteristic property of waves is that they interfere with one another, giving a greater displacement where peaks or troughs coincide, leading to constructive interference, and a smaller displacement where peaks coincide with troughs, leading to destructive interference. See the diagram, in which two lighter colour waves interfere to give the darker one; (a) constructive, (b) destructive.

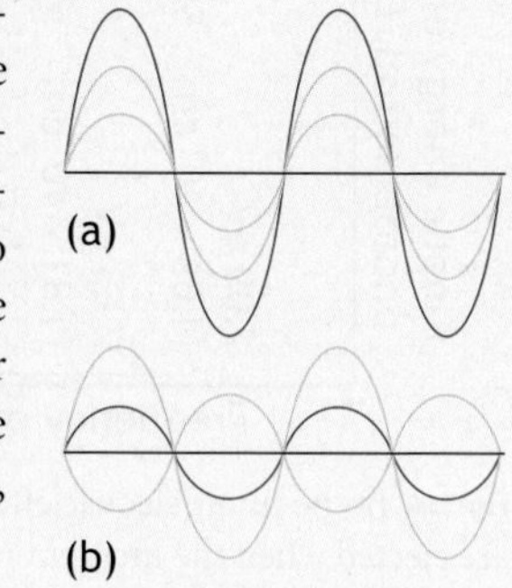

1.3 The wave character of particles

Classical physics treats electrons as particles, but experiments carried out in 1925 required consideration of the possibility that electrons, and matter in general, possessed wave-like properties. The crucial experiment was performed by the American physicists Clinton Davisson and Lester Germer, who observed the diffraction of electrons by a crystal (Fig. 1.6). **Diffraction** is the interference caused by an object in the path of waves. Depending on whether the interference is constructive or destructive, the result is a region of enhanced or diminished intensity of the wave. Davisson and Germer's success was a lucky accident, because a chance rise of temperature caused their polycrystalline sample to anneal, and the ordered planes of atoms then acted as a diffraction grating. At almost the same time, George Thomson, working in Scotland, showed that a beam of electrons was diffracted when passed through a thin gold foil. Electron diffraction is the basis for special techniques in microscopy used by biologists and materials scientists (*Impact I1.1* and Section 9.4).

The Davisson–Germer experiment, which has since been repeated with other particles (including α particles and molecular hydrogen), shows clearly that particles have wave-like properties, and the diffraction of neutrons is a well-established technique for investigating the structures and dynamics of condensed phases (Chapter 9). We have also seen that waves of electromagnetic radiation have particle-like properties. Thus we are brought to the heart of modern physics. When examined on an atomic scale, the classical concepts of particle and wave melt together, particles taking on the characteristics of waves, and waves the characteristics of particles.

Some progress towards coordinating these properties had already been made by the French physicist Louis de Broglie when, in 1924, he suggested that any particle, not only photons, travelling with a linear momentum p should have (in some sense) a wavelength given by the **de Broglie relation**:

$$\lambda = \frac{h}{p} \tag{1.3}$$

That is, a particle with a high linear momentum has a short wavelength (Fig. 1.7). Macroscopic bodies have such high momenta

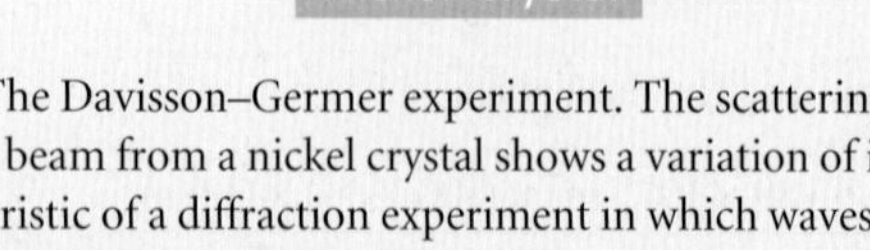

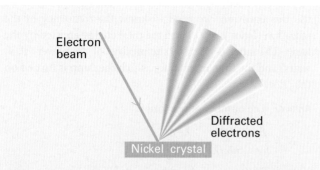

Fig. 1.6 The Davisson–Germer experiment. The scattering of an electron beam from a nickel crystal shows a variation of intensity characteristic of a diffraction experiment in which waves interfere constructively and destructively in different directions.

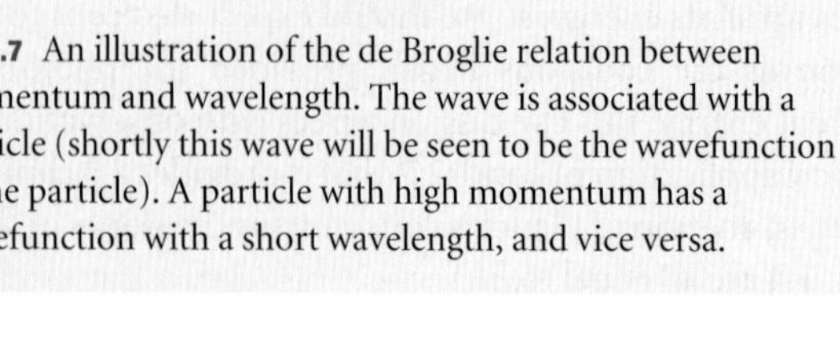

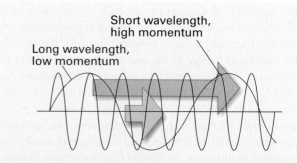

Fig. 1.7 An illustration of the de Broglie relation between momentum and wavelength. The wave is associated with a particle (shortly this wave will be seen to be the wavefunction of the particle). A particle with high momentum has a wavefunction with a short wavelength, and vice versa.

even when they are moving slowly (because their mass is so great), that their wavelengths are undetectably small, and the wave-like properties cannot be observed.

Example 1.2 *Estimating the de Broglie wavelength*

Estimate the wavelength of electrons that have been accelerated from rest through a potential difference of 40 kV.

Method To use the de Broglie relation, we need to know the linear momentum, p, of the electrons. To calculate the linear momentum, we note that the energy acquired by an electron accelerated through a potential difference V is eV, where e is the magnitude of its charge. At the end of the period of acceleration, all the acquired energy is in the form of kinetic energy, $E_k = \frac{1}{2}m_e v^2 = p^2/2m_e$, so we can determine p by setting $p^2/2m_e$ equal to eV. As before, carry through the calculation algebraically before substituting the data.

Answer The expression $p^2/2m_e = eV$ implies that $p = (2m_e eV)^{1/2}$; then, from the de Broglie relation $\lambda = h/p$,

$$\lambda = \frac{h}{(2m_e eV)^{1/2}}$$

Substitution of the data and the fundamental constants (from inside the front cover) gives

$$\lambda = \frac{6.626 \times 10^{-34}\ \text{J s}}{\{2 \times (9.109 \times 10^{-31}\ \text{kg}) \times (1.602 \times 10^{-19}\ \text{C}) \times (4.0 \times 10^4\ \text{V})\}^{1/2}}$$

$$= 6.1 \times 10^{-12}\ \text{m}$$

where we have used 1 V C = 1 J and 1 J = 1 kg m^2 s^{-2}. The wavelength of 6.1 pm is shorter than typical bond lengths in molecules (about 100 pm). Electrons accelerated in this way are used in the technique of electron diffraction for the determination of molecular structure (Section 9.4).

Self-test 1.2 Calculate (a) the wavelength of a neutron with a translational kinetic energy equal to kT at 300 K, (b) a tennis ball of mass 57 g travelling at 80 km h^{-1}.

[(a) 178 pm, (b) 5.2×10^{-34} m]

We now have to conclude that not only has electromagnetic radiation the character classically ascribed to particles, but electrons (and all other particles) have the characteristics classically ascribed to waves. This joint particle and wave character of matter and radiation is called **wave–particle duality**. Duality strikes at the heart of classical physics, where particles and waves are treated as entirely distinct entities. We have also seen that the energies of electromagnetic radiation and of matter cannot be varied continuously and that, for small objects, the discreteness of energy is highly significant. In classical mechanics, in contrast, energies could be varied continuously. Such total failure of classical physics for small objects implied that its basic concepts were false. A new mechanics had to be devised to take its place.

IMPACT ON BIOLOGY

I1.1 Electron microscopy

The basic approach of illuminating a small area of a sample and collecting light with a microscope has been used for many years to obtain magnified images of small specimens. However, the *resolution* of a microscope, the minimum distance between two objects that leads to two distinct images, is of the order of the wavelength of light used as a probe. Therefore, conventional microscopes employing visible light have resolutions in the micrometre range and cannot resolve features on a scale of nanometres.

There is great interest in the development of new experimental probes of very small specimens that cannot be studied by traditional light microscopy. For example, our understanding of biochemical processes, such as enzymatic catalysis, protein folding, and the insertion of DNA into the cell's nucleus, will be enhanced if it becomes possible to image individual biopolymers—with dimensions much smaller than visible wavelengths—at work. One technique that is often used to image nanometre-sized objects is *electron microscopy*, in which a beam of electrons with a well defined de Broglie wavelength replaces the lamp found in traditional light microscopes. Instead of glass or quartz lenses, magnetic fields are used to focus the beam. In *transmission electron microscopy* (TEM), the electron beam passes through the specimen and the image is collected on a screen. In *scanning electron microscopy* (SEM), electrons scattered back from a small irradiated area of the sample are detected and the electrical signal is sent to a video screen. An image of the surface is then obtained by scanning the electron beam across the sample.

As in traditional light microscopy, the wavelength of the incident beam and the ability to focus the beam—in this case a beam of electrons—govern the resolution. Electron wavelengths in typical electron microscopes can be as short as 10 pm, but it is not possible to focus electrons well with magnetic lenses so, in the end, typical resolutions of TEM and SEM instruments are about 2 nm and 50 nm, respectively. It follows that electron microscopes cannot resolve individual atoms (which have diameters of about 0.2 nm). Furthermore, only certain samples can be observed under certain conditions. The measurements must be conducted under high vacuum. For TEM observations, the samples must be very thin cross-sections of a specimen and SEM observations must be made on dry samples. A consequence of these requirements is that neither technique can be used to study living cells. In spite of these limitations, electron

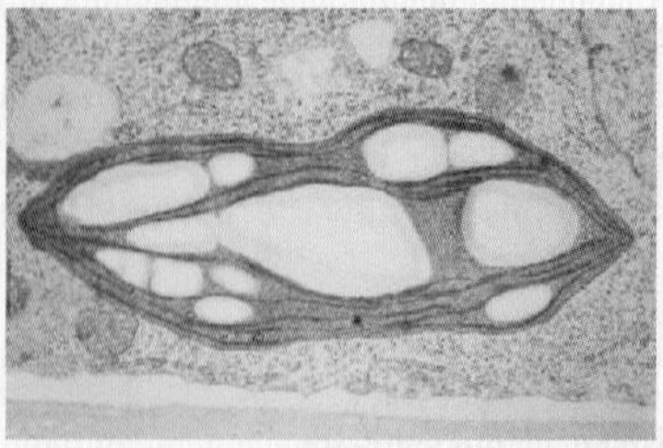

Fig. 1.8 A TEM image of a cross-section of a plant cell showing chloroplasts, organelles responsible for the reactions of photosynthesis (Chapter 19). Chloroplasts are typically 5 μm long. (Image supplied by Brian Bowes.)

microscopy is very useful in studies of the internal structure of cells (Fig. 1.8).

The postulates

We have seen that classical physics was unable to explain the results of several experiments involving electromagnetic radiation and particles as small as electrons and atoms. Although the work of Einstein and de Broglie successfully explained a number of these phenomena, it soon became clear that the development of a new theory of matter was needed to understand the behaviour of all known forms of matter, including electrons, atoms, and molecules. The new theory of matter that emerged is called **quantum mechanics.** In the system of mechanics we are about to present, it should not be too surprising that Planck's constant will play an important role given its appearance in the Bohr frequency condition (eqn 1.1), the photoelectric effect (eqn 1.2), and the de Broglie relation (eqn 1.3).

There are two approaches to the formal introduction of quantum mechanics. One is to see the theory gradually emerging from the work of Planck, Einstein, Heisenberg, Schrödinger, and Dirac, in which experiment and intuition together determined the form of the theory. The other approach is to stand at a point in time at which the theory has already been well-developed and look at its underlying structure. We adopt the latter approach here and see how quantum mechanics can be expressed in terms of and developed from a small set of underlying principles or postulates.

1.4 Postulate I: the wavefunction

Quantum mechanics acknowledges the wave–particle duality of matter by supposing that, rather than travelling along a definite path, a particle is distributed through space like a wave. This remark, which may seem mysterious, is interpreted and developed more fully below. The mathematical representation of the wave that in quantum mechanics replaces the classical concept of a trajectory is called a **wavefunction**, ψ (psi). A principal tenet of quantum mechanics is that the wavefunction contains information about all the properties of the system that are subject to experimental determination.

The wavefunction depends on the spatial coordinates ($\boldsymbol{r}_1$, $\boldsymbol{r}_2, \ldots$) of all the particles (1, 2, . . .) that constitute the system and, in general, the time t. The wavefunction $\Psi(\boldsymbol{r}_1, \boldsymbol{r}_2, \ldots; t)$ is called the **time-dependent wavefunction.** When we are not concerned with the evolution of the system over time we use the **time-independent wavefunction** $\psi(\boldsymbol{r}_1, \boldsymbol{r}_2, \ldots)$. Throughout this chapter we shall consider only time-independent wavefunctions and take up the question of their time dependence in Chapter 4. The wavefunction may also depend on the spin states of the particles but we ignore this property for now and return to it in Chapter 3.

This discussion is summarized by the first postulate of quantum mechanics.

> **Postulate I** The state of the system is described as fully as possible by the wavefunction $\psi(\boldsymbol{r}_1, \boldsymbol{r}_2, \ldots)$.

We need to know how to calculate the wavefunction of any system and extract information from it. We address the latter question first.

1.5 Postulate II: the Born interpretation

The wavefunction contains all the dynamical information about the system it describes. Here we concentrate on the information it carries about the location of the particles. For simplicity, we assume initially that the system is composed of a single particle and that the wavefunction is simply $\psi(\boldsymbol{r})$, or ψ for short.

The interpretation of the wavefunction is based on a suggestion made by Max Born. He made use of an analogy with the wave theory of light, in which the square of the amplitude of an electromagnetic wave in a region is interpreted as its intensity and therefore (in quantum terms) as a measure of the probability of finding a photon present in the region. The **Born interpretation** of the wavefunction focuses on the square of the wavefunction (or the square modulus, $|\psi|^2 = \psi^*\psi$, if ψ is complex):

> **Postulate II′** For a system described by the wavefunction $\psi(\boldsymbol{r})$, the probability of finding the particle in the volume element $d\tau$ at $\boldsymbol{r}$ is proportional to $|\psi|^2 d\tau$.

(Postulate II′, which is relevant to a system composed of a single particle, is a special case of the more general Postulate II presented below.) Thus, $|\psi|^2$ is the **probability density**, and to obtain the probability it must be multiplied by the volume of the infinitesimal region $d\tau$ (Fig. 1.9) The wavefunction ψ itself is called the **probability amplitude.** The prime on this postulate

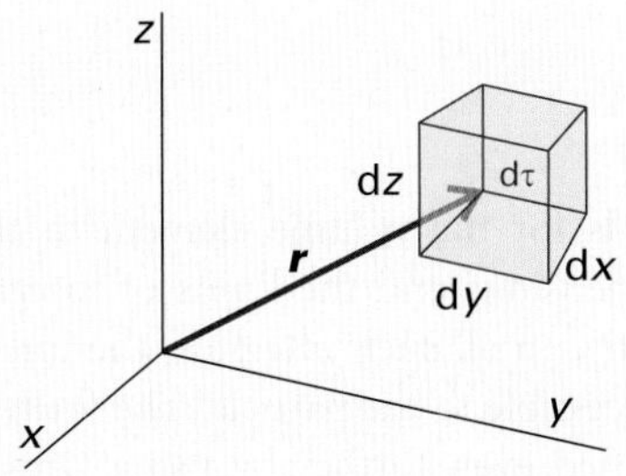

Fig. 1.9 The Born interpretation of the wavefunction in three-dimensional space implies that the probability of finding the particle in the volume element $d\tau = dx\,dy\,dz$ at some location $\boldsymbol{r}$ is proportional to the product of $d\tau$ and the value of $|\psi|^2$ at that location.

number will be discarded when we generalize it to more than one particle at the end of this section.

A brief comment Complex numbers and functions are discussed in *Mathematical background 3*. They have the form $z = x + iy$, where $i = (-1)^{1/2}$. To form the complex conjugate, ψ^*, of a complex function, replace i wherever it occurs by $-i$. For instance, the complex conjugate of e^{ikx} is e^{-ikx}. If the wavefunction is real, $|\psi|^2 = \psi^2$.

The Born interpretation does away with any worry about the significance of a negative (and, in general, complex) value of ψ because $|\psi|^2$ is real and never negative. There is no *direct* significance in the negative (or complex) value of a wavefunction: only the square modulus, a positive quantity, is directly physically significant, and both negative and positive regions of a wavefunction may correspond to a high probability of finding a particle in a region (Fig. 1.10). However, later we shall see that the presence of positive and negative regions of a wavefunction is of great *indirect* significance, because it gives rise to the possibility of constructive and destructive interference between different wavefunctions.

Example 1.3 *Interpreting a wavefunction*

We shall see in Chapter 4 that the wavefunction of an electron in the lowest energy state of a hydrogen atom is proportional to e^{-r/a_0}, with a_0 a constant and r the distance from the nucleus. (Notice that this wavefunction depends only on this distance, not the angular position relative to the nucleus.) Calculate the relative probabilities of finding the electron inside a region of volume 1.0 pm^3, which is small even on the scale of the atom, located at (a) the nucleus and at (b) a distance a_0 from the nucleus.

Method The region of interest is so small on the scale of the atom that we can ignore the variation of ψ within it and write the probability, P, as proportional to the probability density (ψ^2; note that ψ is real) evaluated at the point of interest multiplied by the volume of interest, δV. That is, $P \propto \psi^2 \delta V$, with $\psi^2 \propto e^{-2r/a_0}$.

Answer In each case $\delta V = 1.0\ \text{pm}^3$. (a) At the nucleus, $r = 0$, so

$$P \propto e^0 \times (1.0\ \text{pm}^3) = (1.0) \times (1.0\ \text{pm}^3)$$

(b) At a distance $r = a_0$ in an arbitrary direction,

$$P \propto e^{-2} \times (1.0\ \text{pm}^3) = (0.14) \times (1.0\ \text{pm}^3)$$

Therefore, the ratio of probabilities is 1.0/0.14 = 7.1. Note that it is more probable (by a factor of about 7) that the electron will be found at the nucleus than in a volume element of the same size located at a distance a_0 from the nucleus. The negatively charged electron is attracted to the positively charged nucleus, and is likely to be found close to it.

A note on good practice The square of a wavefunction is not a probability: it is a probability density, and (in three dimensions) has the dimensions of 1/length^3. It becomes a (unitless) probability when multiplied by a volume. (The probabilities in *Example* 1.3 appear to carry units because we deal with proportionalities, and the unshown constant of proportionality—see below—cancels the units.) In general, we have to take into account the variation of the amplitude of the wavefunction over the volume of interest, but here we are supposing that the volume is so small that the variation of ψ in the region can be ignored.

Self-test 1.3 The wavefunction for the electron in its lowest energy state in the ion He^+ is proportional to e^{-2r/a_0}. Repeat the calculation for this ion. Any comment?

[55; more compact wavefunction]

Postulate II′ refers to a *proportionality* between probability and $|\psi|^2 d\tau$. To determine the actual value of the probability we write $\psi' = N\psi$, where N is a (real) constant selected so that $|\psi'|^2 d\tau$ is *equal* to the probability that the particle is in the volume element $d\tau$. To determine this constant, we note that the total probability of finding the particle anywhere in space must be 1 (it must be somewhere). If the system is one-dimensional, the total probability of finding the particle is the sum (integral) of all the infinitesimal contributions $|\psi'|^2 d\tau$, and we can write

$$\int_{-\infty}^{\infty} (\psi')^* \psi' dx = 1$$

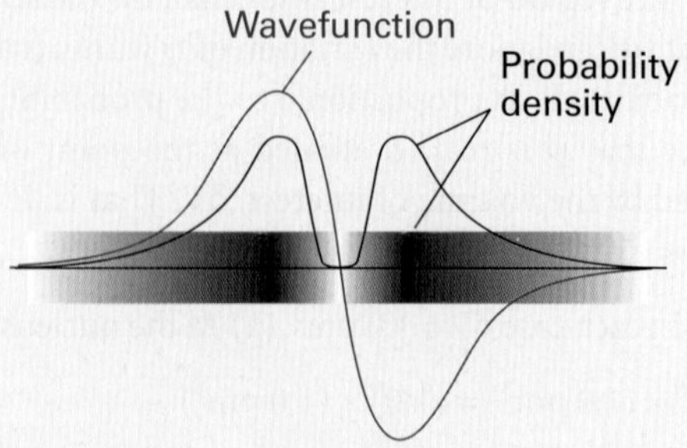

Fig. 1.10 The sign of a wavefunction has no direct physical significance: the positive and negative regions of this wavefunction both correspond to the same probability distribution (as given by the square modulus of ψ and depicted by the density of shading).

A wavefunction that satisfies this condition is said to be **normalized** (strictly, normalized to 1). In terms of the original wavefunction this equation becomes

$$N^2\int_{-\infty}^{\infty}\psi^*\psi\,dx = 1$$

It follows that the **normalization constant** N is given by

$$N = \frac{1}{\left(\int_{-\infty}^{\infty}\psi^*\psi\,dx\right)^{1/2}} \tag{1.4}$$

A brief comment Almost all wavefunctions go to zero at sufficiently great distances so there is rarely any difficulty with the evaluation of this integral. Wavefunctions for which the integral in eqn 1.4 exists (in the sense of having a finite value) are said to be 'square-integrable'.

From now on, unless we state otherwise, we always use wavefunctions that have been normalized to 1. That is, from now on we assume that ψ already includes a factor that ensures that (in one dimension)

$$\int_{-\infty}^{\infty}\psi^*\psi\,dx = 1 \tag{1.5a}$$

In three dimensions, the wavefunction is normalized if

$$\int_{-\infty}^{\infty}\int_{-\infty}^{\infty}\int_{-\infty}^{\infty}\psi^*\psi\,dx\,dy\,dz = 1 \tag{1.5b}$$

In general, the normalization condition may be written

$$\int\psi^*\psi\,d\tau = 1 \tag{1.6}$$

where $d\tau$ stands for the volume element in the appropriate number of dimensions and the limits of integrations are not written explicitly. In all such integrals, the integration is over all the space accessible to the particle. The form of this integral is often simplified even further by using Dirac notation; see *Further information 1.1.*

Example 1.4 *Normalizing a wavefunction*

Carbon nanotubes are thin hollow cylinders of carbon atoms that are excellent electrical conductors and can be used as wires in nanodevices (Section 9.9). The tubes have diameters between 1 and 2 nm and lengths of several micrometres. A long carbon nanotube can be modelled as a one-dimensional structure. According to a simple model introduced in Chapter 2, the electrons of the nanotube are described by the wavefunction $\sin \pi x/L$, where L is the length of the nanotube. Find the normalized wavefunction.

Method We need to carry out the integration specified in eqn 1.5a where the limits of integration are 0 and L. The wavefunction is real, so $\psi^* = \psi$. The following indefinite integral is required:

$$\int \sin^2 ax\,dx = \frac{x}{2} - \frac{\sin 2ax}{4a} + C$$

Answer We write the wavefunction as $\psi = N\sin \pi x/L$, where N is the normalization factor. It follows that

$$\int\psi^*\psi\,d\tau = N^2\overbrace{\int_0^L \sin^2\frac{\pi x}{L}\,dx}^{L/2} = \tfrac{1}{2}N^2L = 1$$

and

$$N = \left(\frac{2}{L}\right)^{1/2}$$

The normalized wavefunction is then

$$\psi = \left(\frac{2}{L}\right)^{1/2}\sin\frac{\pi x}{L}$$

Note that, because L is a length, the dimensions of ψ are $1/\text{length}^{1/2}$ and therefore those of ψ^2 are 1/length as is appropriate for a probability density.

Self-test 1.4 Normalize the wavefunction $\sin 2\pi x/L$ for a particle confined to a region of length L. $[N = (2/L)^{1/2}]$

The Born interpretation puts severe restrictions on the acceptability of wavefunctions. The principal constraint is that ψ must not be infinite over a finite region. If it were, the integral in eqn 1.4 would be infinite and the normalization constant would be zero. The normalized function would then be zero everywhere, except where it is infinite, which would be unacceptable. The Born interpretation also implies that a wavefunction cannot have more than one value because it would be absurd to have more than one probability that a particle is at the same point. This restriction is expressed by saying that the wavefunction must be *single-valued*, that is, have only one value at each point of space.

A brief comment Infinitely sharp spikes are acceptable provided they have zero width. In elementary quantum mechanics the simpler restriction, that ψ cannot be infinite anywhere, is sufficient.

Finally, we need to be aware of the interpretation of the wavefunction for a system with more than one particle. In this case, $\psi(\boldsymbol{r}_1, \boldsymbol{r}_2, \ldots)$ is used to calculate the *overall* probability of finding each particle in its own specific volume element.

Postulate II For a system described by the wavefunction $\psi(\boldsymbol{r}_1, \boldsymbol{r}_2, \ldots)$, the probability of finding particle 1 in the volume element $d\tau_1$ at $\boldsymbol{r}_1$, particle 2 in the volume element $d\tau_2$ at $\boldsymbol{r}_2$, etc. is proportional to $|\psi|^2 d\tau_1 d\tau_2 \ldots$

1.6 Postulate III: quantum mechanical operators

We shall now begin to see how to deduce the form of the wavefunction and in the process introduce two more postulates (there are five altogether).

In 1926, the Austrian physicist Erwin Schrödinger proposed a special second-order differential equation for finding the wavefunction of any system. (For an introduction to differential equations, see *Mathematical background 2*.) The **time-independent Schrödinger equation** for a particle of mass m moving in one dimension with energy E is

$$-\frac{\hbar^2}{2m}\frac{d^2\psi}{dx^2} + V\psi = E\psi \qquad (1.7)$$

where V is the potential energy of the particle and $\hbar = h/2\pi$ (which is read h-cross or h-bar) is a convenient modification of Planck's constant. Extensions of the Schrödinger equation to more than one dimension and its time-dependent form are shown in Table 1.1.[1] We could regard eqn 1.7 itself as a postulate, but it turns out to be far more fruitful to interpret it in a special way and to regard it as a consequence of deeper, more general postulates.

First, though, we should note that the fact that a wavefunction is a solution to a second-order differential equation (the Schrödinger equation) introduces two further restrictions on its acceptability in addition to the two implied by the Born interpretation. For eqn 1.7 to be meaningful, the second derivative of ψ must be well-defined everywhere. We can take the second derivative of a function only if its first derivative, its slope, is continuous (so there are no kinks in the function), and we can take the first derivative only if the function itself is continuous

Table 1.1 The Schrödinger equation

For one-dimensional systems:

$$-\frac{\hbar^2}{2m}\frac{d^2\psi}{dx^2} + V(x)\psi(x) = E\psi(x)$$

where $V(x)$ is the potential energy of the particle and E is its total energy.

For two-dimensional systems:

$$-\frac{\hbar^2}{2m}\left(\frac{\partial^2\psi}{\partial x^2} + \frac{\partial^2\psi}{\partial y^2}\right) + V(x,y)\psi(x,y) = E\psi(x,y)$$

For three-dimensional systems:

$$-\frac{\hbar^2}{2m}\nabla^2\psi + V\psi = E\psi$$

where V may depend on position and ∇^2 ('del squared') is

$$\nabla^2 = \frac{\partial^2}{\partial x^2} + \frac{\partial^2}{\partial y^2} + \frac{\partial^2}{\partial z^2}$$

In systems with spherical symmetry three equivalent forms are

$$\nabla^2 = \frac{1}{r}\frac{\partial^2}{\partial r^2}r + \frac{1}{r^2}\Lambda^2$$

$$= \frac{1}{r^2}\frac{\partial}{\partial r}r^2\frac{\partial}{\partial r} + \frac{1}{r^2}\Lambda^2$$

$$= \frac{\partial^2}{\partial r^2} + \frac{2}{r}\frac{\partial}{\partial r} + \frac{1}{r^2}\Lambda^2$$

where

$$\Lambda^2 = \frac{1}{\sin^2\theta}\frac{\partial^2}{\partial\phi^2} + \frac{1}{\sin\theta}\frac{\partial}{\partial\theta}\sin\theta\frac{\partial}{\partial\theta}$$

In the general case the Schrödinger equation is written

$$\hat{H}\psi = E\psi$$

where $\hat{H}$ is the hamiltonian operator for the system:

$$\hat{H} = -\frac{\hbar^2}{2m}\nabla^2 + V$$

For the evolution of a system with time, it is necessary to solve the time-dependent Schrödinger equation:

$$\hat{H}\Psi = i\hbar\frac{\partial\Psi}{\partial t}$$

[1] A detailed discussion of the form of these operators can be found in our *Molecular quantum mechanics*, Oxford University Press, Oxford (2005).

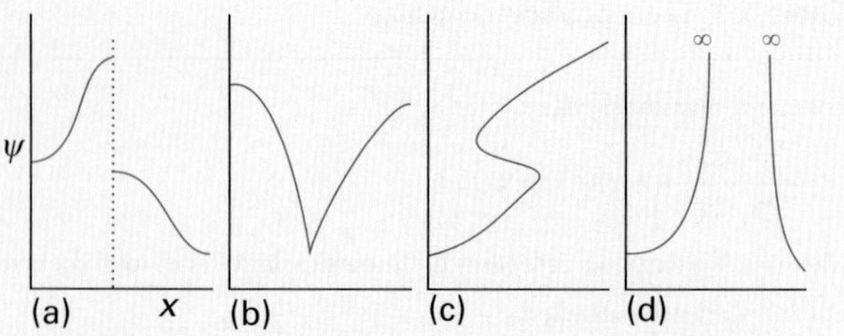

Fig. 1.11 The wavefunction must satisfy stringent conditions for it to be acceptable. (a) Unacceptable because it is not continuous; (b) unacceptable because its slope is discontinuous; (c) unacceptable because it is not single-valued; (d) unacceptable because it is infinite over a finite region.

(so there are no sharp steps in it). In summary, at this point we know that a wavefunction must:

1. be continuous;
2. have a continuous slope;
3. be single-valued;
4. be square-integrable.

These restrictions are summarized in Fig. 1.11. They are so severe that we shall find that, in general, acceptable solutions of the Schrödinger equation can be found only for certain values of the energy E. You should be able to sense that the quantization of energy is starting to emerge. We shall see in Chapter 2 specific examples of the origins of quantization. However, by adopting the more general route that we shall now describe, we shall discover that the quantization of energy is only one of the extraordinary consequences of quantum mechanics.

A brief comment There are cases, and we shall meet them, where acceptable wavefunctions have kinks. These cases arise when one of the terms in the Schrödinger equation (namely, the potential energy) has peculiar properties, such as rising abruptly to infinity. There are only two cases of this behaviour in elementary quantum mechanics, and the peculiarity will be mentioned when we meet them.

To start our journey of generalization, we note that eqn 1.7 and its three-dimensional counterparts in Table 1.1 may all be written in the succinct form

$$\hat{H}\psi = E\psi \tag{1.8a}$$

where in one dimension

$$\hat{H} = -\frac{\hbar^2}{2m}\frac{d^2}{dx^2} + V(x)\times \tag{1.8b}$$

(For reasons that will shortly become clear, we have noted that the potential energy of the particle depends on its position, x.)

The quantity $\hat{H}$ is an **operator**, something that carries out a mathematical operation on the function ψ. In this case, the operation is to take the second derivative of ψ and (after multiplication by $-\hbar^2/2m$) to add the result to the outcome of multiplying ψ by the value of V at the position x. The operator $\hat{H}$ plays a special role in quantum mechanics, and is called the **hamiltonian operator** after the nineteenth century mathematician William Hamilton, who developed a form of classical mechanics that, it subsequently turned out, is well suited to the formulation of quantum mechanics. We can infer from the form of eqn 1.8a that the hamiltonian operator is the operator corresponding to the total energy of the system, the sum of the kinetic and potential energies. It then follows that the first term in eqn 1.8b (the term proportional to the second derivative) must be the operator for the kinetic energy.

Equation 1.8a is highly suggestive of a more general formulation. First, it suggests that there might be other **observables**, or measurable properties, of a system that can be represented by other operators, and that the structure

$$[\text{energy operator (hamiltonian)}]\,\psi = [\text{value of energy}] \times \psi$$

is a special case of the more general form

$$[\text{operator for the observable } \Omega]\,\psi = [\text{value of the observable } \Omega] \times \psi$$

From now on, we represent the operator corresponding to the observable Ω (uppercase omega) by $\hat{\Omega}$, the value of the observable Ω by ω (lowercase omega), and write the last equation as

$$\hat{\Omega}\psi = \omega\psi \tag{1.9}$$

Our immediate problem is to discover how to formulate the operator corresponding to an observable. Once again, eqn 1.8 gives us a clue. In classical mechanics, the total energy of a particle in one dimension can be expressed in terms of the linear momentum p as

$$E = \frac{p^2}{2m} + V(x)$$

Comparison of this expression with eqn 1.8b strongly suggests that the operator for position, x, is just multiplication by position ($x\times$), because then the potential energy $V(x)$ is represented by the multiplicative operation $V(x)\times$. For instance, if the potential energy is proportional to x^2, as it is for a harmonic oscillator (see Chapter 2, where we see that $V = \frac{1}{2}kx^2$), then the potential energy operator becomes

$$\tfrac{1}{2}k(x\times)^2 = \tfrac{1}{2}kx\times x\times = \tfrac{1}{2}kx^2\times = V(x)\times$$

as required. The comparison also strongly suggests that the operator for the linear momentum is proportional to the operation of taking the derivative of a function. If we write $\hat{p} = (\hbar/i)d/dx$, then the operator corresponding to $p^2/2m$ becomes

$$\frac{\hat{p}^2}{2m} = \frac{1}{2m}\left(\frac{\hbar}{i}\frac{d}{dx}\right)\left(\frac{\hbar}{i}\frac{d}{dx}\right) = -\frac{\hbar^2}{2m}\frac{d^2}{dx^2}$$

which corresponds to the first term in the hamiltonian operator.

This discussion is summarized by the following postulate.

Postulate III For each observable property Ω of a system there is a corresponding operator $\hat{\Omega}$ built from the following position and linear momentum operators:

$$\hat{x} = x\times \qquad \hat{p}_x = \frac{\hbar}{i}\frac{d}{dx} \qquad [1.10]$$

The multiplication sign for multiplicative operators is not normally written, and from now on we shall omit it. However, it must always be remembered that x is the operation that multiplies any function on its right. When we are discussing one-dimensional systems, we discard the index x on p; for three-dimensional systems, we use subscripts x, y, and z to denote the components of the vector $\boldsymbol{p}$ along each direction, with the corresponding operators defined analogously.

A brief comment The rules summarized by eqn 1.10 apply to observables that depend on spatial variables; intrinsic properties, such as spin (see Section 3.5) are treated differently.

With Postulate III established, we can immediately write down the operator for the kinetic energy, E_k, of a particle in one dimension:

$$\text{Because } E_k = \frac{p^2}{2m}, \qquad \hat{E}_k = \frac{\hat{p}^2}{2m} = -\frac{\hbar^2}{2m}\frac{d^2}{dx^2} \qquad (1.11)$$

In mathematics, the second derivative of a function is a measure of its curvature: a large second derivative indicates a sharply curved function (Fig. 1.12). It follows that a sharply curved wavefunction is associated with a high kinetic energy, and one with a low curvature is associated with a low kinetic energy. This interpretation is consistent with the de Broglie relation, which predicts a short wavelength (a sharply curved wavefunction) when the linear momentum (and hence the kinetic energy) is high. However, it extends the interpretation to wavefunctions that do not spread through space and resemble those shown in Fig. 1.12. The curvature of a wavefunction in general varies from place to place. Wherever a wavefunction is sharply curved, its contribution to the total kinetic energy is large (Fig. 1.13). Wherever the wavefunction is not sharply curved, its contribution to the overall kinetic energy is low.

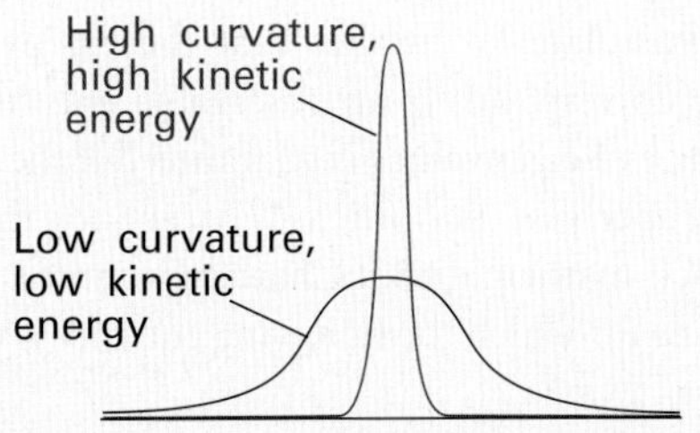

Fig. 1.12 Even if a wavefunction does not have the form of a periodic wave, it is still possible to infer from it the average kinetic energy of a particle by noting its average curvature. This figure shows two wavefunctions: the sharply curved function corresponds to a higher kinetic energy than the less sharply curved function.

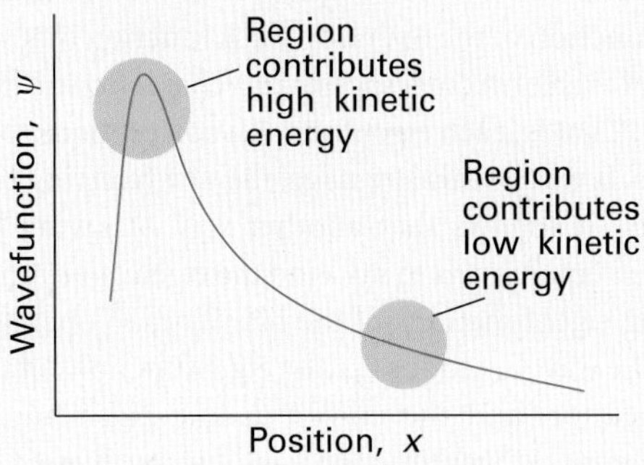

Fig. 1.13 The observed kinetic energy of a particle is an average of contributions from the entire space covered by the wavefunction. Sharply curved regions contribute a high kinetic energy to the average; slightly curved regions contribute only a small kinetic energy.

A brief comment We are using the term 'curvature' informally: the precise technical definition of the curvature of a function f is $(d^2f/dx^2)/\{1 + (df/dx)^2\}^{3/2}$.

A list of the operators more commonly encountered in quantum mechanics is collected in the *Resource section* at the end of the text.

1.7 Postulate IV: eigenvalues and eigenfunctions

We know how to construct operators for observables, we know (in principle) the solutions of the Schrödinger equation, so we know (in principle) the wavefunction for the system, and we know that the value of the observable of interest is the ω in eqn 1.9 ($\hat{\Omega}\psi = \omega\psi$). The interpretation of this equation must now be expressed more precisely.

First, we note that eqn 1.9 has the form

$$(\text{Operator})(\text{function}) = (\text{constant factor}) \times (\text{same function}) \qquad (1.12)$$

An equation of this form is called an **eigenvalue equation**; the constant factor is called the **eigenvalue** of the operator and the function that occurs on both sides of the equation is called the **eigenfunction** of the operator. Each function that satisfies an equation of this form corresponds to a characteristic multiplicative factor. That is, each eigenfunction corresponds to a specific eigenvalue. If we use this language, then by referring to eqn 1.8a ($\hat{H}\psi = E\psi$) it follows that another way of saying 'solve the Schrödinger equation' is to say 'find the eigenvalues and the corresponding eigenfunctions of the hamiltonian operator for the system'. The wavefunctions are the eigenfunctions of the hamiltonian operator, and the corresponding eigenvalues are the allowed energies. Eigenfunctions and eigenvalues of operators play a crucial role in quantum mechanics and we will encounter them throughout the text.

Example 1.5 *Identifying an eigenfunction*

Show that e^{ikx} is an eigenfunction of the linear momentum operator, and find the corresponding eigenvalue. Then show that the bell-shaped 'Gaussian function' e^{-ax^2} is not an eigenfunction of this operator.

Method We need to operate on the function with the operator and check whether the result is a constant factor times the original function. In each case we identify the operator $\hat{\Omega}$ with the linear momentum operator $\hat{p} = (\hbar/i)d/dx$.

Answer For $\psi = e^{ikx}$:

$$\hat{p}\psi = \frac{\hbar}{i}\frac{d}{dx}e^{ikx} = \frac{\hbar}{i}ike^{ikx} = k\hbar\psi$$

Therefore e^{ikx} is indeed an eigenfunction of $\hat{p}$, and its eigenvalue is $k\hbar$. For $\psi = e^{-ax^2}$,

$$\hat{p}\psi = \frac{\hbar}{i}\frac{d}{dx}e^{-ax^2} = \frac{\hbar}{i}(-2ax)e^{-ax^2} = 2ia\hbar x \times \psi$$

We have used $-2/i = 2i$. This is not an eigenvalue equation even though the same function ψ occurs on the right, because ψ is now multiplied by a variable factor ($2ia\hbar x$), not a constant factor. Alternatively, if the right-hand side is written $2ia\hbar(xe^{-ax^2})$, we see that it is a constant ($2ia\hbar$) times a *different* function.

Self-test 1.5 Is the function cos ax an eigenfunction of (a) the linear momentum operator, (b) the kinetic energy operator? [(a) No, (b) yes]

There are far-reaching consequences that result from the interpretation of eqn 1.9 (and the Schrödinger equation, eqn 1.8) as an eigenvalue equation. First, we need to note that the allowed value of an observable Ω (such as the energy E) is the eigenvalue of an eigenvalue equation. Therefore, once we have found all the permitted eigenvalues of an operator, we shall know the permitted values of that observable. Each eigenvalue corresponds to a particular eigenfunction (specifically, each energy corresponds to a particular wavefunction), and we have already seen that there are severe restrictions on the acceptability of wavefunctions. We can anticipate, therefore, that, because only certain wavefunctions are acceptable, only certain eigenvalues are allowed. In other words, in general, *an observable is quantized*. We summarize the discussion so far in the language introduced in this section with the following postulate:

Postulate IV If the system is described by a wavefunction ψ that is an eigenfunction of $\hat{\Omega}$ such that $\hat{\Omega}\psi = \omega\psi$, then the outcome of a measurement of Ω will be the eigenvalue ω.

● A BRIEF ILLUSTRATION

We saw in Example 1.5 that e^{ikx} is an eigenfunction of the linear momentum operator with eigenvalue $k\hbar$. Therefore, if the wavefunction of an electron accelerated in a linear accelerator to a certain energy is e^{ikx}, then we know from Postulate IV that, if we were to measure its linear momentum, we would find the value $p = +k\hbar$. Similarly, if the wavefunction is e^{-ikx}, then, because the eigenvalue is now $-k\hbar$ (note the change in sign), a measurement of the linear momentum would give the value $p = -k\hbar$. The magnitude of the linear momentum of the electron is the same in each case ($k\hbar$), but the signs are different: in (a) the electron is travelling to the right (positive x) but in (b) the linear accelerator is pointed in the opposite direction and the electron is travelling to the left (negative x). ●

Self-test 1.6 As a result of its acceleration in a linear accelerator, the wavefunction of a proton became cos kx. What is the kinetic energy of the proton? [$E_k = k^2\hbar^2/2m_p$]

Because the value of an observable is a real quantity (real, that is, in the mathematical sense of not involving the imaginary number i), Postulate IV implies that the eigenvalues of any operator that corresponds to an observable must themselves be real. The reality of eigenvalues is guaranteed if the operator has the special property of 'hermiticity' (named for the nineteenth century French mathematician Charles Hermite). A **hermitian operator** is one for which the following equality is true:

$$\text{Hermiticity:} \int f^*\hat{\Omega}g\,dx = \left\{\int g^*\hat{\Omega}f\,dx\right\}^* \qquad [1.13]$$

where f and g are any two wavefunctions. It is easy to confirm that the position operator ($x\times$) is hermitian because we are free to change the order of the factors in the integrand:

$$\int_{-\infty}^{\infty} f^* x g\,dx \overset{\text{change order}}{=} \int_{-\infty}^{\infty} g x f^*\,dx \overset{\text{use } g^{**}=g \text{ and } x^*=x}{=} \int_{-\infty}^{\infty} g^{**}x^*f^*\,dx = \left\{\int_{-\infty}^{\infty} g^* x f\,dx\right\}^*$$

The demonstration that the linear momentum operator is hermitian is more involved because we cannot just alter the order of functions we differentiate, but it is hermitian, as we show in the following *Justification*.

Justification 1.1 *The hermiticity of the linear momentum operator*

Our task is to show that

$$\int_{-\infty}^{\infty} f^*\hat{p}g\,dx = \left\{\int_{-\infty}^{\infty} g^*\hat{p}f\,dx\right\}^*$$

with $\hat{p}$ given in eqn 1.10. To do so, we use 'integration by parts', the relation

$$\int u\frac{dv}{dx}dx = uv - \int v\frac{du}{dx}dx$$

with $u = f^*$ and $v = g$. In the present case we write

$$\int_{-\infty}^{\infty} f^*\hat{p}g\,dx = \frac{\hbar}{i}\int_{-\infty}^{\infty} f^*\frac{dg}{dx}dx = \frac{\hbar}{i}f^*g\Big|_{-\infty}^{\infty} - \frac{\hbar}{i}\int_{-\infty}^{\infty} g\frac{df^*}{dx}dx$$

The first term on the right is zero, because all wavefunctions are zero at infinity in either direction (or, in special cases, have equal values at each infinity), so we are left with

$$\int_{-\infty}^{\infty} f^*\hat{p}g\,dx = -\frac{\hbar}{i}\int_{-\infty}^{\infty} g\frac{df^*}{dx}dx \overset{\text{use } i^*=-i}{=} \left\{\frac{\hbar}{i}\int_{-\infty}^{\infty} g^*\frac{df}{dx}dx\right\}^* = \left\{\int_{-\infty}^{\infty} g^*\hat{p}f\,dx\right\}^*$$

as we set out to prove.

Self-test 1.7 Confirm that the kinetic energy operator is hermitian.

Equation 1.13 might seem to be far removed from being equivalent to the statement that the eigenvalues of hermitian operators are real, but in fact the proof is quite straightforward, as the following *Justification* shows.

Justification 1.2 *The reality of eigenvalues of hermitian operators*

For a wavefunction ψ that is normalized to 1 and is an eigenfunction of an hermitian operator $\hat{\Omega}$ with eigenvalue ω, we can write

$$\int \psi^*\hat{\Omega}\psi\,d\tau \overset{\hat{\Omega}\psi=\omega\psi}{=} \int \psi^*\omega\psi\,d\tau = \omega\overbrace{\int \psi^*\psi\,d\tau}^{1} = \omega$$

However, by taking the complex conjugate we can write

$$\omega^* = \left\{\int \psi^*\hat{\Omega}\psi\,d\tau\right\}^* \overset{\text{eqn 1.13 with } f=g=\psi}{=} \int \psi^*\hat{\Omega}\psi\,d\tau = \omega$$

The conclusion that $\omega^* = \omega$ confirms that ω is real.

1.8 Postulate V: superpositions and expectation values

We have seen that, if the wavefunction of a particle is an eigenfunction of the operator corresponding to an observable, then it is easy to identify the value of that observable: we just pick out the corresponding eigenvalue. Suppose, though, that the wavefunction of the system is not an eigenfunction of the operator corresponding to the property of interest: what can we then say? For instance, suppose the wavefunction of a particle is cos kx. This is an eigenfunction of the kinetic energy operator (with eigenvalue $k^2\hbar^2/2m$) so we know that a measurement of the kinetic energy will certainly give that value. However, cos kx is not an eigenfunction of the linear momentum operator (because $d\cos kx/dx = -k\sin kx$), and so in this case we cannot use Postulate IV to predict the outcome of a measurement of the linear momentum.

The clue we need in order to make progress is to note that we can use Euler's formula, $e^{ix} = \cos x + i\sin x$ and $e^{-ix} = \cos x - i\sin x$, to write

$$\cos kx = \tfrac{1}{2}e^{ikx} + \tfrac{1}{2}e^{-ikx}$$

and each of the exponential functions is an eigenfunction of the linear momentum operator (as we saw in the previous *illustration*) with eigenvalues $+k\hbar$ and $-k\hbar$, respectively. We say that the actual wavefunction is a **linear combination** or **superposition** of e^{ikx} and e^{-ikx} and that the actual wavefunction is a superposition of more than one eigenfunction. Symbolically we can write the superposition as

$$\underset{\text{Particle with linear momentum } +k\hbar}{\psi = \psi_{\rightarrow}} + \underset{\text{Particle with linear momentum } -k\hbar}{\psi_{\leftarrow}}$$

A brief comment In general, a linear combination of two functions f and g is $c_1f + c_2g$, where c_1 and c_2 are numerical coefficients, so a linear combination is a more general term than 'sum'. In a sum, $c_1 = c_2 = 1$. A linear combination might have the form $0.567f + 1.234g$, for instance, so it is more general than the simple sum $f + g$. The function cos kx can be written as the linear combination $\frac{1}{2}e^{ikx} + \frac{1}{2}e^{-ikx}$ since $e^{ikx} = \cos kx + i \sin kx$.

The interpretation of this composite wavefunction is that, if the momentum of the particle is repeatedly measured in a long series of observations, then its *magnitude* will be found to be $k\hbar$ in all the measurements (because that is the value for each of the eigenfunctions). However, because the two eigenfunctions occur equally in the superposition (the same numerical coefficient occurs in the linear combination), half the measurements will show that the particle is moving to the right ($p = +k\hbar$), and half the measurements will show that it is moving to the left ($p = -k\hbar$). According to quantum mechanics, we cannot predict in which direction the particle will in fact be found to be travelling; all we can say is that, in a long series of observations, if the particle is described by this wavefunction, then there are equal probabilities of finding the particle travelling to the right and to the left. Furthermore, since half the measurements yield $p = +k\hbar$ and half yield $p = -k\hbar$, we expect the average value of a large number of measurements to be zero.

This discussion motivates the following generalization to the case when the system is known to be a superposition of many different eigenfunctions of the operator $\hat{\Omega}$ corresponding to the observable of interest and written as the linear combination

$$\psi = c_1\psi_1 + c_2\psi_2 + \cdots = \sum_k c_k\psi_k \tag{1.14}$$

where the c_k are numerical (and possibly complex) coefficients and the ψ_k correspond to different eigenfunctions of the operator, with $\hat{\Omega}\psi_k = \omega_k\psi_k$. Then:

Postulate V When the value of an observable Ω is measured for a system that is described by a linear combination of eigenfunctions of $\hat{\Omega}$, with coefficients c_k, each measurement gives one of the eigenvalues ω_k of $\hat{\Omega}$ with a probability proportional to $|c_k|^2$.

If the system is described by a wavefunction that is normalized, then the probability of obtaining the eigenvalue ω_k equals $|c_k|^2$.

● A BRIEF ILLUSTRATION

A linear accelerator does not accelerate particles to a precisely defined linear momentum, so the wavefunction of the particles is a superposition of functions corresponding to the range of momenta present in the beam. Suppose we model the wavefunction of the electrons in the beam as the (normalized) superposition

$$\psi = (\tfrac{3}{8})^{1/2}\psi_1 - \tfrac{1}{2}\psi_2 + i(\tfrac{3}{8})^{1/2}\psi_3$$

where ψ_1 corresponds to the linear momentum $+k_1\hbar$, ψ_2 to $+k_2\hbar$, and ψ_3 to $+k_3\hbar$. According to Postulate V, when we measure the linear momentum of the electrons in the beam we will get one of the values $+k_1\hbar$, $+k_2\hbar$, or $+k_3\hbar$, but which one we get is unpredictable. The probabilities that we measure each value will be $|(\frac{3}{8})^{1/2}|^2 = \frac{3}{8}$ for $+k_1\hbar$, $|-\frac{1}{2}|^2 = \frac{1}{4}$ for $+k_2\hbar$, and $|i(\frac{3}{8})^{1/2}|^2 = \frac{3}{8}$ for $+k_3\hbar$. ●

Self-test 1.8 What would be the result of measuring the kinetic energy of the electrons of the previous *illustration*?

[individual measurements: $k_1^2\hbar^2/2m_e$ (probability = 3/8); $k_2^2\hbar^2/2m_e$ (probability = 1/4); $k_3^2\hbar^2/2m_e$ (probability = 3/8)]

The mean (that is, average) value from measurement of the observable Ω is equal to the **expectation value** of the operator $\hat{\Omega}$, denoted $\langle\Omega\rangle$ and defined as

$$\langle\Omega\rangle = \frac{\int\psi^*\hat{\Omega}\psi\mathrm{d}\tau}{\int\psi^*\psi\mathrm{d}\tau} \qquad [1.15a]$$

This definition applies whether or not ψ is written as a linear combination of wavefunctions. For a normalized wavefunction, the denominator is 1 and this expression simplifies to

$$\langle\Omega\rangle = \int\psi^*\hat{\Omega}\psi\mathrm{d}\tau \tag{1.15b}$$

If ψ happens to be an eigenfunction of $\hat{\Omega}$, then every measurement of the observable will yield a single eigenvalue, say ω, and the average value is also ω. In this case, the expectation value of $\hat{\Omega}$ is simply

$$\langle\Omega\rangle = \int\psi^*\hat{\Omega}\psi\mathrm{d}\tau \overset{\hat{\Omega}\psi=\omega\psi}{=} \int\psi^*\omega\psi\mathrm{d}\tau = \omega\overbrace{\int\psi^*\psi\mathrm{d}\tau}^{1} = \omega$$

If ψ is not an eigenfunction of $\hat{\Omega}$, we have more work to do to justify Postulate V.

First, we need to note a further very special feature of hermitian operators other than the fact that their eigenvalues are real. We show in the *Justification* below that *eigenfunctions corresponding to different eigenvalues of the same hermitian operator are orthogonal.* To say that two different functions ψ_i and ψ_j are **orthogonal** means that the integral (over all space) of their product is zero:

$$\text{Orthogonality:} \quad \int\psi_i^*\psi_j\mathrm{d}\tau = 0 \qquad [1.16]$$

For example, the hamiltonian operator is hermitian (it corresponds to an observable, the energy). Therefore, if ψ_i corresponds to one energy, and ψ_j corresponds to a different energy, then we know at once that the two functions are orthogonal and that the integral of their product is zero. Being able to set integrals to zero in this way greatly simplifies the calculations that we shall do in the following chapters and also provides a foundation for the justification of Postulate V, as we shall see.

Justification 1.3 *The orthogonality of eigenfunctions*

Suppose we have two eigenfunctions of $\hat{\Omega}$ with unequal eigenvalues:

$$\hat{\Omega}\psi_i = \omega_i\psi_i \text{ and } \hat{\Omega}\psi_j = \omega_j\psi_j$$

with ω_i not equal to ω_j. Multiply the first of these eigenvalue equations on both sides by ψ_j^* and the second by ψ_i^*, and integrate over all space:

$$\int \psi_j{}^*\hat{\Omega}\psi_i\,\mathrm{d}\tau = \omega_i\int \psi_j{}^*\psi_i\,\mathrm{d}\tau$$

$$\int \psi_i{}^*\hat{\Omega}\psi_j\,\mathrm{d}\tau = \omega_j\int \psi_i{}^*\psi_j\,\mathrm{d}\tau$$

Now take the complex conjugate of the first of these two expressions (noting that, by the hermiticity of $\hat{\Omega}$, the eigenvalues are real):

$$\left\{\int \psi_j{}^*\hat{\Omega}\psi_i\,\mathrm{d}\tau\right\}^* = \omega_i\int \psi_j\psi_i{}^*\mathrm{d}\tau = \omega_i\int \psi_i{}^*\psi_j\,\mathrm{d}\tau$$

However, by hermiticity, the first term on the left is

$$\left\{\int \psi_j{}^*\hat{\Omega}\psi_i\,\mathrm{d}\tau\right\}^* = \int \psi_i{}^*\hat{\Omega}\psi_j\,\mathrm{d}\tau = \omega_j\int \psi_i{}^*\psi_j\,\mathrm{d}\tau$$

Subtraction of this line from the preceding line then gives

$$0 = (\omega_i - \omega_j)\int \psi_i{}^*\psi_j\,\mathrm{d}\tau$$

But we know that the two eigenvalues are not equal, so the integral must be zero, as we set out to prove.

Example 1.6 *Verifying orthogonality*

We shall see in Chapter 2 that two possible wavefunctions for an electron confined to a one-dimensional quantum dot (a collection of atoms with dimensions in the range of nanometres and of great interest in nanotechnology) are of the form sin x and sin $2x$. These two wavefunctions are eigenfunctions of the kinetic energy operator, which is hermitian, and correspond to different eigenvalues:

$$\hat{E}_\mathrm{k}\sin x = -\frac{\hbar^2}{2m_\mathrm{e}}\frac{\mathrm{d}^2\sin x}{\mathrm{d}x^2} = \frac{\hbar^2}{2m_\mathrm{e}}\sin x$$

$$\hat{E}_\mathrm{k}\sin 2x = -\frac{\hbar^2}{2m_\mathrm{e}}\frac{\mathrm{d}^2\sin 2x}{\mathrm{d}x^2} = \frac{2\hbar^2}{m_\mathrm{e}}\sin 2x$$

Verify that the two wavefunctions are mutually orthogonal.

Method To verify the orthogonality of two functions, we integrate their product, sin $2x$ sin x, over all space, which we may take to span from $x = 0$ to $x = 2\pi$, because both functions repeat themselves outside that range. Hence proving that the integral of their product is zero within that range implies that the integral over the whole of space is also zero (Fig. 1.14). We need the standard integral

$$\int \sin ax \sin bx\,\mathrm{d}x = \frac{\sin(a-b)x}{2(a-b)} - \frac{\sin(a+b)x}{2(a+b)} + \text{constant},$$

if $a^2 \neq b^2$

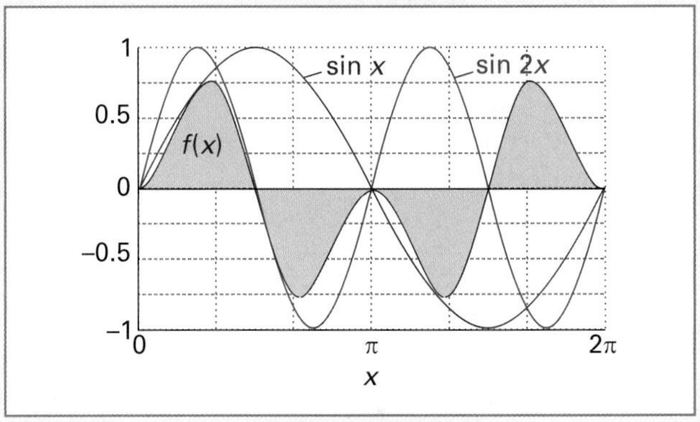

Fig. 1.14 The integral of the function $f(x) = \sin 2x \sin x$ is equal to the area (tinted) below the brown curve, and is zero, as can be inferred by symmetry. The function—and the value of the integral—repeats itself for all replications of the section between 0 and 2π, so the integral from $-\infty$ to ∞ is zero.

Answer It follows that, for $a = 2$ and $b = 1$, and given the fact that $\sin 0 = 0$, $\sin 2\pi = 0$, and $\sin 6\pi = 0$,

$$\int_0^{2\pi} \sin 2x \sin x\,\mathrm{d}x = \left.\frac{\sin x}{2}\right|_0^{2\pi} - \left.\frac{\sin 3x}{6}\right|_0^{2\pi} = 0$$

and the two functions are mutually orthogonal.

Self-test 1.9 When the electron is excited to higher energies, its wavefunction may become sin $3x$. Confirm that the functions sin x and sin $3x$ are mutually orthogonal.

$[\int_{-\infty}^{\infty} \sin x \sin 3x\,\mathrm{d}x = 0]$

Now we take the final step, and demonstrate in the following *Justification* that the expectation value of $\hat{\Omega}$ can be written in terms of the coefficients c_k and the individual eigenvalues ω_k as

$$\langle\Omega\rangle = \sum_k |c_k|^2 \omega_k \qquad (1.17)$$

This conclusion shows that the expectation value is a weighted average of the eigenvalues of $\hat{\Omega}$, with the weighting equal to the square modulus of the expansion coefficients c_k. This is the basis of Postulate V, for it strongly suggests that the measurement of the property Ω gives a series of values ω_k, with occurrences that are determined by the values of $|c_k|^2$.

● A BRIEF ILLUSTRATION

For the system described in the previous *illustration* that followed Postulate V, the mean value of the linear momentum, from eqn 1.17, is

$$\langle p\rangle = \tfrac{3}{8}k_1\hbar + \tfrac{1}{4}k_2\hbar + \tfrac{3}{8}k_3\hbar \bullet$$

Self-test 1.10 What is the mean value of the kinetic energy of the system of Self-test 1.8?

$[\langle E_\mathrm{k}\rangle = (\hbar^2/2m_\mathrm{e})(3k_1^2/8 + k_2^2/4 + 3k_3^2/8)]$

Justification 1.4 *The expectation value of an operator as a weighted average*

For simplicity, suppose the (normalized) wavefunction is the sum of two eigenfunctions (the general case can easily be developed). Then, from eqn 1.15b,

$$\langle\Omega\rangle = \int (c_1\psi_1 + c_2\psi_2)^*\hat{\Omega}(c_1\psi_1 + c_2\psi_2)\mathrm{d}\tau$$

$$= \int (c_1\psi_1 + c_2\psi_2)^*(c_1\hat{\Omega}\psi_1 + c_2\hat{\Omega}\psi_2)\mathrm{d}\tau$$

$$= \int (c_1\psi_1 + c_2\psi_2)^*(c_1\omega_1\psi_1 + c_2\omega_2\psi_2)\mathrm{d}\tau$$

$$= c_1{}^*c_1\omega_1\overbrace{\int \psi_1{}^*\psi_1\mathrm{d}\tau}^{1} + c_2{}^*c_2\omega_2\overbrace{\int \psi_2{}^*\psi_2\mathrm{d}\tau}^{1}$$

$$+ c_2{}^*c_1\omega_1\overbrace{\int \psi_2{}^*\psi_1\mathrm{d}\tau}^{0} + c_1{}^*c_2\omega_2\overbrace{\int \psi_1{}^*\psi_2\mathrm{d}\tau}^{0}$$

The first two integrals on the right are both equal to 1 because the eigenfunctions are individually normalized. Because ψ_1 and ψ_2 correspond to different eigenvalues of an hermitian operator, they are orthogonal, so the third and fourth integrals on the right are zero. We can conclude that

$$\langle\Omega\rangle = |c_1|^2\omega_1 + |c_2|^2\omega_2$$

in accord with eqn 1.17.

Example 1.7 *Calculating an expectation value*

Calculate the average value of the position of an electron in the carbon nanotube described in Example 1.4.

Method The average value is the expectation value of the operator corresponding to position, which is multiplication by x. To evaluate $\langle x\rangle$, we need to know the normalized wavefunction (from Example 1.4) and then evaluate the integral in eqn 1.15b. The following integral is required:

$$\int x \sin^2 ax\,\mathrm{d}x = \frac{x^2}{4} - \frac{x\sin 2ax}{4a} - \frac{\cos 2ax}{8a^2} + C$$

Answer The average value is given by the expectation value

$$\langle x\rangle = \int \psi^*\hat{x}\psi\,\mathrm{d}x \quad \text{with } \psi = \left(\frac{2}{L}\right)^{1/2}\sin\frac{\pi x}{L} \quad \text{and } \hat{x} = x\times$$

which evaluates to

$$\langle x\rangle = \frac{2}{L}\overbrace{\int_0^L x\sin^2\frac{\pi x}{L}\mathrm{d}x}^{L^2/4} = \tfrac{1}{2}L$$

This result means that, if a very large number of measurements of the position of the electron are made, then their mean value will be exactly one-half the length of the nanotube. However, each different observation will give a different and unpredictable individual result because the wavefunction is not an eigenfunction of the operator corresponding to x. (Note that we used eqn 1.15b, rather than eqn 1.17, because ψ was not expressed as a linear combination of eigenfunctions of $\hat{x}$.)

Self-test 1.11 Evaluate the root mean square position, $\langle x^2\rangle^{1/2}$, of the electron using the indefinite integral

$$\int x^2 \sin^2 ax\,\mathrm{d}x = \frac{x^3}{6} - \left(\frac{x^2}{4a} - \frac{1}{8a^3}\right)\sin 2ax - \frac{x\cos 2ax}{4a^2} + C$$

We shall see in Section 1.9 a relationship between the root mean square position and the uncertainty in the position of the particle. $[L\{1/3 - 1/(2\pi^2)\}^{1/2}]$

Table 1.2 The postulates of quantum mechanics

I. The state of the system is described as fully as possible by the wavefunction $\psi(\boldsymbol{r}_1, \boldsymbol{r}_2, \ldots)$ where $(\boldsymbol{r}_1, \boldsymbol{r}_2, \ldots)$ are the spatial coordinates of the particles $(1, 2, \ldots)$.

II. For a system described by the normalized wavefunction $\psi(\boldsymbol{r}_1, \boldsymbol{r}_2, \ldots)$, the probability of finding particle 1 in the volume element $d\tau_1$ at $\boldsymbol{r}_1$, particle 2 in the volume element $d\tau_2$ at $\boldsymbol{r}_2$, etc. is equal to $|\psi|^2 d\tau_1 d\tau_2 \ldots$.

III. For each observable property Ω of a system there is a corresponding hermitian operator $\hat{\Omega}$ built from the following position and linear momentum operators:

$$\hat{x} = x \times \qquad \hat{p}_x = \frac{\hbar}{i}\frac{d}{dx}$$

IV. Suppose the system is described by a wavefunction ψ that is an eigenfunction of $\hat{\Omega}$ with eigenvalue ω:

$$\hat{\Omega}\psi = \omega\psi$$

Then the outcome of a measurement of the observable property Ω will be ω.

V. Suppose the system is described by a normalized wavefunction ψ that is a linear combination of eigenfunctions of $\hat{\Omega}$:

$$\psi = c_1\psi_1 + c_2\psi_2 + \cdots = \sum_k c_k\psi_k \qquad \text{with } \hat{\Omega}\psi_k = \omega_k\psi_k$$

Then when the value of an observable Ω is measured, each measurement gives one of the eigenvalues ω_k with a probability equal to $|c_k|^2$. The mean (that is, average) value of the measurements is equal to the expectation value $\langle\Omega\rangle$.

$$\langle\Omega\rangle = \int \psi^*\hat{\Omega}\psi d\tau$$

We have reached an important point in our study of quantum mechanics. Having described the postulates, which are summarized in Table 1.2, we can now use them to understand atomic and molecular structure and chemical change. This task will occupy our attention for the remainder of the text.

Complementary observables

The fact that particles are described by wavefunctions and that the outcome of observations depends on the properties of operators and eigenvalues leads to profound differences between quantum mechanics and classical mechanics. We can begin to appreciate these differences by considering the wavefunction for a particle travelling in one dimension towards positive x with linear momentum $k\hbar$. As we saw in Section 1.7, the wavefunction is Ne^{ikx}, where N is the (real) normalization factor.

Where is the particle? To answer this question, we use Postulate II to calculate the probability density:

$$|\psi|^2 = (Ne^{ikx})^*(Ne^{ikx}) = N^2(e^{-ikx})(e^{ikx}) = N^2 \qquad (1.18)$$

This probability density is independent of x, so, wherever we look on the x-axis, there is an equal probability of finding the particle (Fig. 1.15). In other words, if the wavefunction of the particle is given by Ne^{ikx}, then we cannot predict where we will find the particle. The extraordinary conclusion is that, if we know the linear momentum precisely, then we can say nothing about the position. This **complementarity** of two observables, in this case linear momentum and position, pervades the whole of quantum mechanics.

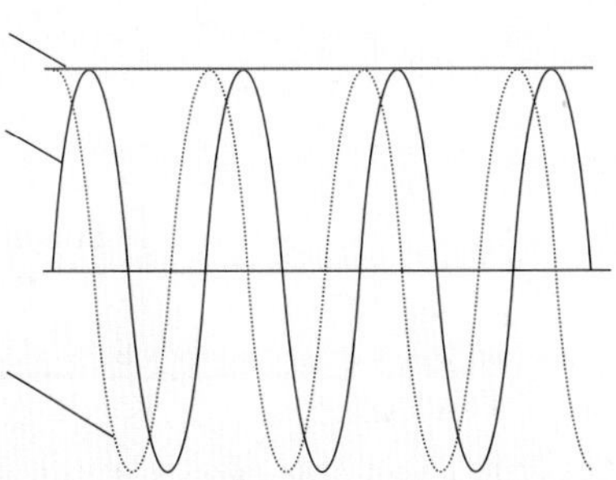

Fig. 1.15 The square modulus of a wavefunction corresponding to a definite state of linear momentum is a constant, so it corresponds to a uniform probability of finding the particle anywhere.

How do we recognize complementary observables, and what are their consequences? Can we specify the energy of a molecule at the same time, for instance, as its dipole moment or are they complementary too? First, we consider linear momentum and position in more detail, then generalize to other properties.

1.9 The Heisenberg uncertainty principle

The conclusion that, if the momentum is specified precisely, then it is impossible to predict the location of the particle is one conclusion that we can draw from the **Heisenberg uncertainty principle** proposed by Werner Heisenberg in 1927:

> It is impossible to specify simultaneously, with arbitrary precision, both the momentum and the position of a particle.

Before discussing the principle further, we must establish its other half: that if we know the position of a particle exactly then we can say nothing about its momentum. The argument draws on the idea of regarding a wavefunction as a superposition of eigenfunctions, and runs as follows.

If we know that the particle is at a definite location, its wavefunction must be large there and zero everywhere else (Fig. 1.16). Such a wavefunction can be created by superimposing a large number of harmonic (sine and cosine) functions, or, equivalently, a number of e^{ikx} functions. In other words, we can create a sharply localized wavefunction, called a **wavepacket**, by forming a linear combination of wavefunctions that correspond to many different linear momenta. The superposition of a few harmonic functions gives a wavefunction that spreads over a range

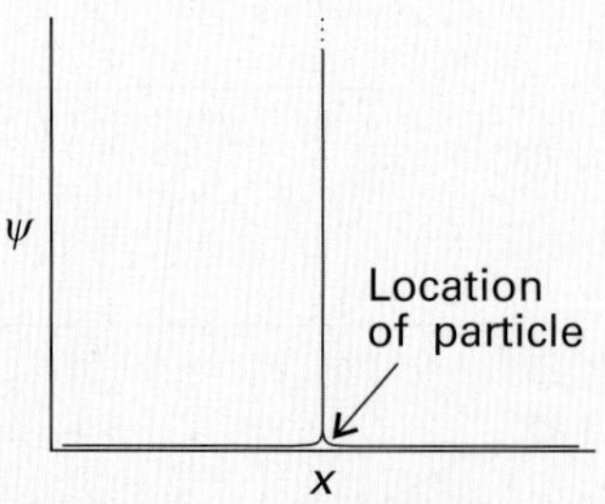

Fig. 1.16 The wavefunction for a particle at a well-defined location is a sharply spiked function which has zero amplitude everywhere except at the particle's position.

of locations (Fig. 1.17). However, as the number of wavefunctions in the superposition increases, the wavepacket becomes sharper on account of the more complete interference between the positive and negative regions of the individual waves. When an infinite number of components is used, the wavepacket is a sharp, infinitely narrow spike, which corresponds to perfect localization of the particle. Now the particle is perfectly localized.

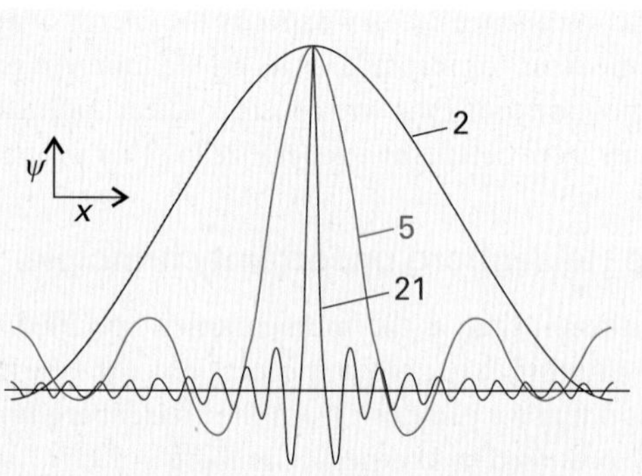

Fig. 1.17 The wavefunction for a particle with an ill-defined location can be regarded as the superposition of several wavefunctions of definite wavelength that interfere constructively in one place but destructively elsewhere. As more waves are used in the superposition (as given by the numbers attached to the curves), the location becomes more precise at the expense of uncertainty in the particle's momentum. An infinite number of waves is needed to construct the wavefunction of a perfectly localized particle.

interActivity Use mathematical software or an electronic spreadsheet to construct superpositions of cosine functions as $\psi(x) = \sum_{k=1}^{N} (1/N)\cos(k\pi x)$, where the constant $1/N$ is introduced to keep the superpositions with the same overall magnitude. Explore how the probability density $\psi^2(x)$ changes with the value of N.

However, we have lost all information about its momentum because, as we saw above, a measurement of the momentum will give a result corresponding to any one of the infinite number of waves in the superposition and which one it will give is unpredictable. Hence, if we know the location of the particle precisely (implying that its wavefunction is a superposition of an infinite number of momentum eigenfunctions), then its momentum is completely unpredictable.

The quantitative version of the Heisenberg uncertainty principle is

$$\Delta p \Delta q \geq \tfrac{1}{2}\hbar \qquad (1.19a)$$

In this expression Δp is the 'uncertainty' in the linear momentum parallel to the axis q, and Δq is the uncertainty in position along that axis. These 'uncertainties' are precisely defined, for they are the root mean square deviations of the properties from their mean values:

$$\Delta p = \{\langle p^2\rangle - \langle p\rangle^2\}^{1/2} \qquad \Delta q = \{\langle q^2\rangle - \langle q\rangle^2\}^{1/2} \qquad [1.19b]$$

If there is complete certainty about the position of the particle ($\Delta q = 0$), then the only way that eqn 1.19a can be satisfied is for $\Delta p = \infty$, which implies complete uncertainty about the momentum. Conversely, if the momentum parallel to an axis is known exactly ($\Delta p = 0$), then the position along that axis must be completely uncertain ($\Delta q = \infty$).

The p and q that appear in eqn 1.19 refer to the same direction in space. Therefore, whereas simultaneous specification of the position on the x-axis and momentum parallel to the x-axis is restricted by the uncertainty relation, simultaneous location of position on x and motion parallel to y or z is not restricted. Table 1.3 summarizes the restrictions that the uncertainty principle implies.

Table 1.3* Constraints of the uncertainty principle

	Variable 1					
Variable 2	x	y	z	p_x	p_y	p_z
x				■		
y					■	
z						■
p_x	■					
p_y		■				
p_z			■			

* Pairs of observables that cannot be determined simultaneously with arbitrary precision are marked with a blue rectangle; all others are unrestricted.

Example 1.8 *Using the uncertainty principle*

Suppose the speed of a projectile of mass 1.0 g is known to within 1 μm s^{-1}. Calculate the minimum uncertainty in its position.

Method Estimate Δp from $m\Delta v$, where Δv is the uncertainty in the speed; then use eqn 1.19a to estimate the minimum uncertainty in position, Δq.

Answer The minimum uncertainty in position is

$$\Delta q = \frac{\hbar}{2m\Delta v}$$

$$= \frac{1.055\times10^{-34}\ \mathrm{J\,s}}{2\times(1.0\times10^{-3}\ \mathrm{kg})\times(1\times10^{-6}\ \mathrm{m\,s^{-1}})} = 5\times10^{-26}\ \mathrm{m}$$

where we have used 1 J = 1 kg m^2 s^{-2}. The uncertainty is completely negligible for all practical purposes concerning macroscopic objects. However, if the mass is that of an electron, then the same uncertainty in speed implies an uncertainty in position far larger than the diameter of an atom (the analogous calculation gives Δq = 60 m), so the concept of a trajectory, the simultaneous possession of a precise position and momentum, is untenable.

Self-test 1.12 Estimate the minimum uncertainty in the speed of an electron in a one-dimensional region of length $2a_0$, where a_0 = 53 pm (the Bohr radius) [547 km s^{-1}]

1.10 The general form of the uncertainty principle

The Heisenberg uncertainty principle is a special case of a very general form of an uncertainty principle developed by H.P. Robertson in 1929. The latter applies to any pair of observables called **complementary observables**, which are defined in terms of the properties of their operators. Specifically, two observables Ω_1 and Ω_2 are complementary if

$$\hat{\Omega}_1\hat{\Omega}_2\psi \neq \hat{\Omega}_2\hat{\Omega}_1\psi \qquad (1.20)$$

When the effect of two operators depends on their order (as this equation implies), we say that they do not **commute**. The different outcomes of the effect of applying $\hat{\Omega}_1$ and $\hat{\Omega}_2$ in a different order are expressed by introducing the **commutator** of the two operators, which is defined as

$$[\hat{\Omega}_1,\hat{\Omega}_2] = \hat{\Omega}_1\hat{\Omega}_2 - \hat{\Omega}_2\hat{\Omega}_1 \qquad [1.21]$$

Example 1.9 *Demonstrating non-commutativity*

Show that the operators for position and momentum do not commute and therefore that these observables are complementary.

Method Consider the effect of $\hat{x}\hat{p}$ on a wavefunction ψ (that is, the effect of $\hat{p}$ followed by the effect on the outcome of multiplication by x), then the effect of $\hat{p}\hat{x}$ on the same function (that is, the effect of multiplication by x followed by the effect of $\hat{p}$ on the outcome). Finally, take the difference of the two results.

Answer The effect of $\hat{x}\hat{p}$ on ψ is

$$\hat{x}\hat{p}\psi = x\times\frac{\hbar}{\mathrm{i}}\frac{\mathrm{d}\psi}{\mathrm{d}x}$$

The effect of $\hat{p}\hat{x}$ on the same function is

$$\hat{p}\hat{x}\psi = \frac{\hbar}{\mathrm{i}}\frac{\mathrm{d}}{\mathrm{d}x}x\psi = \frac{\hbar}{\mathrm{i}}\left(\psi + x\frac{\mathrm{d}\psi}{\mathrm{d}x}\right)$$

(For this step we have used the standard rule about differentiating a product of functions.) The second expression is clearly different from the first, so the two operators do not commute. Their commutator is calculated by subtracting the second of these two equations from the first:

$$(\hat{x}\hat{p} - \hat{p}\hat{x})\psi = \frac{\hbar}{\mathrm{i}}\left(x\frac{\mathrm{d}\psi}{\mathrm{d}x}\right) - \frac{\hbar}{\mathrm{i}}\left(\psi + x\frac{\mathrm{d}\psi}{\mathrm{d}x}\right) = \mathrm{i}\hbar\psi$$

Because this relation is true for any ψ, it follows that $[\hat{x},\hat{p}] = \mathrm{i}\hbar$. This relation is valid for linear momentum along the x-axis (that is, when $\hat{p}$ is actually $\hat{p}_x$). The same kind of calculation can be used to deduce that $[\hat{x},\hat{p}_y] = 0$ and $[\hat{x},\hat{p}_z] = 0$.

Self-test 1.13 Do the operators for potential energy and kinetic energy commute?

$$\left[\text{No: } [\hat{V},\hat{E}_\mathrm{K}] = \frac{\hbar^2}{2m}\left(\frac{\mathrm{d}^2V}{\mathrm{d}x^2} + 2\frac{\mathrm{d}V}{\mathrm{d}x}\frac{\mathrm{d}}{\mathrm{d}x}\right)\right]$$

We can conclude from the calculation in Example 1.9 that the commutator of the operators for position and linear momentum parallel to the same axis is

$$[\hat{x},\hat{p}_x] = \mathrm{i}\hbar \qquad (1.22)$$

with similar expressions for other axes. This commutator is of such vital significance in quantum mechanics that it is taken as a fundamental distinction between classical mechanics and quantum mechanics. In fact, rather than giving the explicit forms of the operators for position and linear momentum in eqn 1.10, we could have stated Postulate III (Section 1.6) as:

For every observable property Ω of a system there is a corresponding hermitian operator $\hat{\Omega}$ built from position and linear momentum operators that satisfy the commutation relation stated in eqn 1.22.

With the concept of the commutator established, the most general form of the uncertainty principle can be given. For *any* two pairs of observables, Ω_1 and Ω_2, the uncertainties (to be precise, the root mean square deviations of their values from the mean) in simultaneous determinations are related by

$$\Delta\Omega_1\Delta\Omega_2 \geq \tfrac{1}{2}|\langle[\hat{\Omega}_1,\hat{\Omega}_2]\rangle| \qquad (1.23)$$

We obtain the special case of eqn 1.19a when we identify the observables with x and p_x and use eqn 1.22 for their commutator.

A brief comment The 'modulus' notation $|\ldots|$ means take the magnitude of the term the bars enclose: for a real quantity x, $|x|$ is the magnitude of x (its value without its sign); for an imaginary quantity $\mathrm{i}y$, $|\mathrm{i}y|$ is the magnitude of y; and—most generally—for a complex quantity $z = x + \mathrm{i}y$, $|z|$ is the value of $(z^*z)^{1/2}$. For example, $|-2| = 2$, $|3\mathrm{i}| = 3$, and $|-2+3\mathrm{i}| = \{(-2-3\mathrm{i})(-2+3\mathrm{i})\}^{1/2} = 13^{1/2}$. Physically, the modulus on the right of eqn 1.23 ensures that the product of uncertainties has a real, non-negative value.

Example 1.10 *Determining the complementarity of two observables*

Can the electric dipole moment and the energy of a molecule be specified simultaneously?

Method First, determine if the electric dipole moment and the energy are complementary observables by evaluating the commutator, eqn 1.21, of the operators corresponding to the observables. Then use the general form of the uncertainty principle, eqn 1.23, to determine if the observables can be specified simultaneously. That is, can the product of their uncertainties be identically zero?

Answer The electric dipole moment of a charge distribution of a proton at the origin and an electron at a position $\boldsymbol{r}$ is $\boldsymbol{\mu} = -e\boldsymbol{r}$; therefore the operator corresponding to the dipole moment is $\hat{\boldsymbol{\mu}} = -e\boldsymbol{r}\,\times$, which is normally written simply $\hat{\boldsymbol{\mu}} = -e\boldsymbol{r}$, with multiplication understood. For a one-dimensional system, the electric dipole moment operator is $-ex$ and the energy operator is the hamiltonian operator in eqn 1.8b. To evaluate the commutator of $-ex$ and $\hat{H}$, we need to see if x commutes with the potential energy operator and with the kinetic energy operator. The potential energy operator commutes with x because they are both multiplicative, and $xV(x) = V(x)x$. To decide whether the kinetic energy operator commutes with x we need to evaluate the following expression:

$$[\hat{x},\hat{E}_\mathrm{k}]\psi = -\frac{\hbar^2}{2m_\mathrm{e}}\left(x\frac{\mathrm{d}^2}{\mathrm{d}x^2} - \frac{\mathrm{d}^2}{\mathrm{d}x^2}x\right)\psi$$

To do so, we note that

$$\frac{\mathrm{d}^2}{\mathrm{d}x^2}x\psi = \frac{\mathrm{d}}{\mathrm{d}x}\left(\frac{\mathrm{d}}{\mathrm{d}x}x\psi\right) = \frac{\mathrm{d}}{\mathrm{d}x}\left(\psi + x\frac{\mathrm{d}\psi}{\mathrm{d}x}\right)$$

$$= \frac{\mathrm{d}\psi}{\mathrm{d}x} + \left(\frac{\mathrm{d}\psi}{\mathrm{d}x} + x\frac{\mathrm{d}^2\psi}{\mathrm{d}x^2}\right) = 2\frac{\mathrm{d}\psi}{\mathrm{d}x} + x\frac{\mathrm{d}^2\psi}{\mathrm{d}x^2}$$

It then follows that

$$[\hat{x},\hat{E}_\mathrm{k}]\psi = \frac{\hbar^2}{m_\mathrm{e}}\frac{\mathrm{d}\psi}{\mathrm{d}x} = \frac{\mathrm{i}\hbar}{m_\mathrm{e}}\hat{p}_x\psi$$

and therefore that

$$[\hat{x},\hat{E}_\mathrm{k}] = \frac{\mathrm{i}\hbar}{m_\mathrm{e}}\hat{p}_x$$

Hence,

$$[\hat{x},\hat{H}] = [\hat{x},\hat{E}_\mathrm{k} + \hat{V}] = [\hat{x},\hat{E}_\mathrm{k}] + [\hat{x},\hat{V}] = [\hat{x},\hat{E}_\mathrm{k}] = \frac{\mathrm{i}\hbar}{m_\mathrm{e}}\hat{p}_x$$

Because the electric dipole moment operator does not commute with the hamiltonian, in general, the electric dipole moment and the energy are complementary observables. However, eqn 1.23 tells us that the restriction on their simultaneous determination is

$$\Delta\mu\Delta E = -e\Delta x\Delta E \geq -\tfrac{1}{2}e\,|\langle[\hat{x},\hat{H}]\rangle|$$

$$\geq -\tfrac{1}{2}e\left|\left\langle\frac{\mathrm{i}\hbar}{m_\mathrm{e}}\hat{p}_x\right\rangle\right| = -\frac{\hbar e}{2m_\mathrm{e}}\langle\hat{p}_x\rangle$$

Therefore, provided the electron has no net linear momentum (so $\langle\hat{p}_x\rangle = 0$), there will be no restriction on the simultaneous determination of the electric dipole moment and the energy even though the corresponding operators are complementary.

Self-test 1.14 Can the potential and kinetic energies be simultaneously specified for an electron undergoing oscillatory motion in one dimension x with a potential energy proportional to x^2?

[No: $[\hat{x}^2,\hat{E}_\mathrm{k}] = \frac{\hbar}{m_\mathrm{e}}(\hbar + 2x\mathrm{i}\hat{p}_x)$]

Complementary observables are observables with noncommuting operators (that is, those with nonzero commutators). With the discovery that some pairs of observables are complementary (we meet more examples in the later chapters), we are at the heart of the difference between classical and quantum mechanics. Classical mechanics supposed, falsely as we now know, that the position and momentum of a particle could be specified simultaneously with arbitrary precision. However, quantum mechanics shows that position and momentum are complementary, and that we have to make a choice: we can specify position at the expense of momentum, or momentum at the expense of position.

The realization that some observables are complementary allows us to make considerable progress with the calculation of atomic and molecular properties, but it does away with some of the most cherished concepts of classical physics.

Checklist of key ideas

- ☐ 1. Spectroscopic transitions are changes in populations of quantized energy levels of a system involving the absorption, emission, or scattering of electromagnetic radiation, $\Delta E = h\nu$.
- ☐ 2. The photoelectric effect is the ejection of electrons from metals when they are exposed to ultraviolet radiation: $\frac{1}{2}m_e v^2 = h\nu - \Phi$, where Φ is the work function, the energy required to remove an electron from the metal to infinity.
- ☐ 3. The photoelectric effect and electron diffraction are phenomena that confirm wave–particle duality, the joint particle and wave character of matter and radiation.
- ☐ 4. The de Broglie relation, $\lambda = h/p$, relates the momentum of a particle with its wavelength.
- ☐ 5. A wavefunction is a mathematical function obtained by solving the Schrödinger equation and which contains all the dynamical information about a system.
- ☐ 6. The Born interpretation of the wavefunction states that the value of $|\psi|^2$, the probability density, at a point is proportional to the probability of finding the particle at that point.
- ☐ 7. Quantization is the confinement of a dynamical observable to discrete values.
- ☐ 8. An acceptable wavefunction must be continuous, have a continuous first derivative, be single-valued, and be square-integrable.
- ☐ 9. The time-independent Schrödinger equation in one dimension is $-(\hbar^2/2m)(\mathrm{d}^2\psi/\mathrm{d}x^2) + V(x)\psi = E\psi$.
- ☐ 10. An operator is something that carries out a mathematical operation on a function. The position and momentum operators are $\hat{x} = x\times$ and $\hat{p}_x = (\hbar/\mathrm{i})\mathrm{d}/\mathrm{d}x$, respectively.
- ☐ 11. The hamiltonian operator is the operator for the total energy of a system, $\hat{H}\psi = E\psi$, and is the sum of the operators for kinetic energy and potential energy.
- ☐ 12. An eigenvalue equation is an equation of the form $\hat{\Omega}\psi = \omega\psi$. The eigenvalue is the constant ω in the eigenvalue equation; the eigenfunction is the function ψ in the eigenvalue equation.
- ☐ 13. The expectation value of an operator is $\langle\Omega\rangle = \int\psi^*\hat{\Omega}\psi\,\mathrm{d}\tau$ for a normalized wavefunction.
- ☐ 14. A hermitian operator is one for which $\int\psi_i{}^*\hat{\Omega}\psi_j\,\mathrm{d}x = (\int\psi_j{}^*\hat{\Omega}\psi_i\,\mathrm{d}x)^*$. The eigenvalues of hermitian operators are real and correspond to observables, measurable properties of a system. The eigenfunctions of hermitian operators are orthogonal, meaning that $\int\psi_i^*\psi_j\,\mathrm{d}\tau = 0$.
- ☐ 15. The Heisenberg uncertainty principle states that it is impossible to specify simultaneously, with arbitrary precision, both the momentum and the position of a particle; $\Delta p\Delta q \geq \frac{1}{2}\hbar$.
- ☐ 16. Two operators commute when $[\hat{\Omega}_1,\hat{\Omega}_2] = \hat{\Omega}_1\hat{\Omega}_2 - \hat{\Omega}_2\hat{\Omega}_1 = 0$.
- ☐ 17. Complementary observables are observables corresponding to non-commuting operators.
- ☐ 18. The general form of the uncertainty principle is $\Delta\Omega_1\Delta\Omega_2 \geq \frac{1}{2}|\langle[\hat{\Omega}_1,\hat{\Omega}_2]\rangle|$.

Further information

Further information 1.1 *Dirac notation*

The normalization condition of eqn 1.6 is often written as (replacing ψ by ψ_n)

$$\langle n|n\rangle = 1$$

and the orthogonality relation of eqn 1.16 as

$$\langle n|n'\rangle = 0 \qquad (n' \neq n)$$

This **Dirac bracket notation** is much more succinct than writing out the integrals in full. It also introduces the words 'bra' and 'ket' into the language of quantum mechanics. Thus, the **bra** $\langle n|$ corresponds to ψ_n^* and the **ket** $|n'\rangle$ corresponds to the wavefunction $\psi_{n'}$. When the bra and ket are put together as in the above expressions, integration over all space is understood. The two expressions can be combined into one:

$$\langle n|n'\rangle = \delta_{nn'} \tag{1.24}$$

Here $\delta_{nn'}$, which is called the **Kronecker delta**, is 1 when $n' = n$ and 0 when $n' \neq n$.

Integrals of the form $\int \psi_n{}^*\hat{\Omega}\psi_m \mathrm{d}\tau$, which we first encounter in the definition of an hermitian operator (Section 1.7), and which are commonly called 'matrix elements', are incorporated into the bracket notation by writing

$$\langle n|\hat{\Omega}|m\rangle = \int \psi_n{}^*\hat{\Omega}\psi_m \mathrm{d}\tau \qquad [1.25]$$

Note how the operator stands between the bra and the ket (which may denote different states), in the place of the c in ⟨bra|c|ket⟩. An integration is implied whenever a complete bracket is written. In this notation, an expectation value for a normalized wavefunction (eqn 1.15b) is

$$\langle \Omega \rangle = \langle n|\hat{\Omega}|n\rangle \qquad (1.26)$$

with the bra and the ket corresponding to the same state (with wavefunction ψ_n). An operator is hermitian (eqn 1.13) if

$$\langle n|\hat{\Omega}|m\rangle = \langle m|\hat{\Omega}|n\rangle^* \qquad (1.27)$$

Discussion questions

1.1 Summarize the evidence that led to the introduction of quantum mechanics.

1.2 Describe how a wavefunction determines the dynamical properties of a system and how those properties may be predicted.

1.3 Account for the uncertainty relation between position and linear momentum in terms of the shape of the wavefunction.

1.4 Suggest how the general shape of a wavefunction can be predicted without solving the Schrödinger equation explicitly.

1.5 Explain the meaning and consequences of wave–particle duality.

1.6 Describe the relationship between operators and observables in quantum mechanics.

1.7 Compare the results of experimental measurements of an observable when the wavefunction is (a) an eigenfunction of the corresponding operator, (b) a superposition of eigenfunctions of that operator.

1.8 Describe the properties of wavepackets in terms of the Heisenberg uncertainty principle.

Exercises

1.1(a) Atomic sodium produces a yellow glow (as in some street lamps): what is the energy separation of the levels responsible for the radiation of wavelength 590 nm?

1.1(b) Neon lamps emit red radiation of wavelength 736 nm. What is the energy separation of the levels responsible for the emission?

1.2(a) Calculate the size of the quantum involved in the excitation of (a) a molecular vibration of period 20 fs, (b) a pendulum of period 2.0 s. Express the results in joules and kilojoules per mole.

1.2(b) Calculate the size of the quantum involved in the excitation of (a) a molecular vibration of period 3.2 fs, (b) a balance wheel of period 1.0 ms. Express the results in joules and kilojoules per mole.

1.3(a) The work function for metallic caesium is 2.14 eV. Calculate the kinetic energy and the speed of the electrons ejected by light of wavelength (a) 580 nm, (b) 250 nm.

1.3(b) The work function for metallic rubidium is 2.09 eV. Calculate the kinetic energy and the speed of the electrons ejected by light of wavelength (a) 520 nm, (b) 355 nm.

1.4(a) In an experiment to study the photoelectric effect, a photon of radiation of wavelength 465 nm was found to eject an electron from a metal with a kinetic energy 2.11 eV. What is the maximum wavelength capable of ejecting an electron from the metal?

1.4(b) A photon of radiation of wavelength 305 nm ejects an electron from a metal with a kinetic energy 1.77 eV. What is the maximum wavelength capable of ejecting an electron from the metal?

1.5(a) When light of wavelength 165 nm strikes a certain metal surface, electrons are ejected with a speed of 1.24×10^6 m s^{-1}. Calculate the speed of electrons ejected from the metal surface by light of wavelength 265 nm.

1.5(b) When light of wavelength 195 nm strikes a certain metal surface, electrons are ejected with a speed of 1.23×10^6 m s^{-1}. Calculate the speed of electrons ejected from the metal surface by light of wavelength 255 nm.

1.6(a) In an X-ray photoelectron experiment, a photon of wavelength 150 pm ejects an electron from the inner shell of an atom and it emerges with a speed of 2.14×10^7 m s^{-1}. Calculate the binding energy of the electron.

1.6(b) In an X-ray photoelectron experiment, a photon of wavelength 121 pm ejects an electron from the inner shell of an atom and it emerges with a speed of 5.69×10^7 m s^{-1}. Calculate the binding energy of the electron.

1.7(a) Calculate the energy per photon and the energy per mole of photons for radiation of wavelength (a) 620 nm (red), (b) 570 nm (yellow), (c) 380 nm (blue).

1.7(b) Calculate the energy per photon and the energy per mole of photons for radiation of wavelength (a) 188 nm (ultraviolet), (b) 125 pm (X-ray), (c) 1.00 cm (microwave).

1.8(a) A sodium lamp emits yellow light of wavelength 590 nm. How many photons does it emit each second if its power is (a) 10 W, (b) 250 W?

1.8(b) A laser used to read CDs emits red light of wavelength 700 nm. How many red photons does it emit each second if its power is (a) 0.25 W, (b) 1.5 mW?

1.9(a) Calculate the de Broglie wavelength of (a) a mass of 2 g travelling at 1 cm s^{-1}, (b) the same, travelling at 250 km s^{-1}, (c) a He atom travelling at 1000 m s^{-1} (a typical speed at room temperature).

1.9(b) Calculate the de Broglie wavelength of an electron accelerated from rest through a potential difference of (a) 100 V, (b) 15 kV, (c) 250 kV.

1.10(a) Electron diffraction makes use of electrons with wavelengths comparable to bond lengths. To what speed must an electron be accelerated for it to have a wavelength of 100 pm? What accelerating potential difference is needed?

1.10(b) Could *proton diffraction* be an interesting technique for the investigation of molecular structure? To what speed must a proton be accelerated for it to have a wavelength of 100 pm? What accelerating potential difference is needed?

1.11(a) The impact of photons on matter exerts a force that can move it, but the effect of a single photon is insignificant except when it strikes an atom or subatomic particle. Calculate the speed to which a stationary electron would be accelerated if it absorbed a photon of 150 nm radiation.

1.11(b) Similarly, calculate the speed to which a stationary H atom would be accelerated if it absorbed a photon of 100 nm radiation.

1.12(a) Consider a time-independent wavefunction of a particle moving in three-dimensional space. Identify the variables upon which the wavefunction depends.

1.12(b) Consider a time-dependent wavefunction of a particle moving in two-dimensional space. Identify the variables upon which the wavefunction depends.

1.13(a) Consider a time-independent wavefunction of a hydrogen atom. Identify the variables upon which the wavefunction depends.

1.13(b) Consider a time-dependent wavefunction of a helium atom. Identify the variables upon which the wavefunction depends.

1.14(a) Repeat Exercise 1.13a but use spherical polar coordinates.

1.14(b) Repeat Exercise 1.13b but use spherical polar coordinates.

1.15(a) An unnormalized wavefunction for a light atom rotating around a heavy atom to which it is bonded is $\psi(\phi) = e^{i\phi}$ with $0 \le \phi \le 2\pi$. Normalize this wavefunction.

1.15(b) An unnormalized wavefunction for an electron in a carbon nanotube of length L is $\sin(2\pi x/L)$. Normalize this wavefunction.

1.16(a) For the system described in Exercise 1.15a, what is the probability of finding the light atom in the volume element $d\phi$ at $\phi = \pi$?

1.16(b) For the system described in Exercise 1.15b, what is the probability of finding the electron in the range dx at $x = L/2$?

1.17(a) For the system described in Exercise 1.15a, what is the probability of finding the light atom between $\phi = \pi/2$ and $\phi = 3\pi/2$?

1.17(b) For the system described in Exercise 1.15b, what is the probability of finding the electron between $x = L/4$ and $x = L/2$?

1.18(a) Construct the operator for kinetic energy of a particle moving in two dimensions.

1.18(b) Construct the potential energy operator of a particle subjected to a Coulombic potential.

1.19(a) Complex functions of the form e^{ikx} can be used to model the wavefunctions of particles in a linear accelerator. Show that any linear combination of the complex functions e^{2ix} and e^{-2ix} is an eigenfunction of the operator d^2/dx^2 and identify its eigenvalue.

1.19(b) Functions of the form $\sin(nx)$ can be used to model the wavefunctions of electrons in a carbon nanotube. Show that any linear combination of the functions $\sin(3x)$ and $\cos(3x)$ is an eigenfunction of the operator d^2/dx^2 and identify its eigenvalue.

1.20(a) The momentum operator is proportional to d/dx. Which of the following functions are eigenfunctions of d/dx: (a) e^{ikx}, (b) e^{ax^2}, (c) x, (d) x^2, (e) $ax + b$, (f) $\sin(x + 3a)$? Give the corresponding eigenvalue where appropriate.

1.20(b) The kinetic energy operator is proportional to d^2/dx^2. Which of the following functions are eigenfunctions of d^2/dx^2: (a) e^{ax}, (b) e^{-ax^2}, (c) k, (d) kx^2, (e) $ax + b$, (f) $\cos(kx + 5)$? Give the corresponding eigenvalue where appropriate.

1.21(a) Functions of the form $\sin(n\pi x/L)$ can be used to model the wavefunctions of electrons in a carbon nanotube of length L. Show that the wavefunctions $\sin(n\pi x/L)$ and $\sin(m\pi x/L)$, where $n \ne m$, are orthogonal for a particle confined to the region $0 \le x \le L$.

1.21(b) Functions of the form $\cos(n\pi x/L)$ can be used to model the wavefunctions of electrons in metals. Show that the wavefunctions $\cos(n\pi x/L)$ and $\cos(m\pi x/L)$, where $n \ne m$, are orthogonal for a particle confined to the region $0 \le x \le L$.

1.22(a) A light atom rotating around a heavy atom to which it is bonded is described by a wavefunction of the form $\psi(\phi) = e^{im\phi}$ with $0 \le \phi \le 2\pi$ and m an integer. Show that the $m = +1$ and $m = +2$ wavefunctions are orthogonal.

1.22(b) Repeat Exercise 1.22a for the $m = +1$ and $m = -1$ wavefunctions.

1.23(a) An electron in a carbon nanotube of length L is described by the wavefunction $\psi(x) = \sin(2\pi x/L)$. Compute the expectation value of the position of the electron.

1.23(b) An electron in a carbon nanotube of length L is described by the wavefunction $\psi(x) = \sin(\pi x/L)$. Compute the expectation value of the kinetic energy of the electron.

1.24(a) An electron in a one-dimensional metal of length L is described by the wavefunction $\psi(x) = \sin(\pi x/L)$. Compute the expectation value of the momentum of the electron.

1.24(b) A light atom rotating around a heavy atom to which it is bonded is described by a wavefunction of the form $\psi(\phi) = e^{i\phi}$ with $0 \le \phi \le 2\pi$. If the operator corresponding to angular momentum is given by $(\hbar/i)d/d\phi$, compute the expectation value of the angular momentum of the light atom.

1.25(a) Confirm that the kinetic energy operator, $-(\hbar^2/2m)d^2/dx^2$, is hermitian.

1.25(b) When we discuss rotational motion, we shall see that the operator corresponding to the angular momentum of a particle is $(\hbar/i)d/d\phi$, where ϕ is an angle. Is this operator hermitian?

1.26(a) You might come across an operator of the form $\hat{x} + ia\hat{p}_x$, where a is a real constant, and wonder if it corresponds to an observable. Could it?

1.26(b) Likewise, you might wonder if $\hat{x}^2 - ia\hat{E}_k$, where a is a real constant, corresponds to an observable. Could it?

1.27(a) The speed of a certain proton is 6.1×10^6 m s^{-1}. If the uncertainty in its momentum is to be reduced to 0.0100 per cent, what uncertainty in its location must be tolerated?

1.27(b) The speed of a certain electron is 1000 km s^{-1}. If the uncertainty in its momentum is to be reduced to 0.0010 per cent, what uncertainty in its location must be tolerated?

1.28(a) Calculate the minimum uncertainty in the speed of a ball of mass 500 g that is known to be within 1.0 μm of a certain point on a bat. What is the minimum uncertainty in the position of a bullet of mass 5.0 g that is known to have a speed somewhere between 350.000 00 m s^{-1} and 350.000 01 m s^{-1}?

1.28(b) An electron in a nanoparticle is confined to a region of length 0.10 nm. What are the minimum uncertainties in (a) its speed, (b) its kinetic energy?

1.29(a) Evaluate the commutators (a) $[\hat{x}, \hat{y}]$, (b) $[\hat{p}_x, \hat{p}_y]$, (c) $[\hat{x}, \hat{p}_x]$, (d) $[\hat{x}^2, \hat{p}_x]$, (e) $[\hat{x}^n, \hat{p}_x]$.

1.29(b) Evaluate the commutators (a) $[(1/\hat{x}), \hat{p}_x]$, (b) $[(1/\hat{x}), \hat{p}_x^2]$, (c) $[\hat{x}\hat{p}_y - \hat{y}\hat{p}_x, \hat{y}\hat{p}_z - \hat{z}\hat{p}_y]$, (d) $[\hat{x}^2(\partial^2/\partial y^2), \hat{y}(\partial/\partial x)]$.

Problems*

Numerical problems

1.1 Suppose that the normalized wavefunction for an electron in a carbon nanotube of length $L = 10.0$ nm is $\psi = (2/L)^{1/2}\sin(\pi x/L)$. Calculate the probability that the electron is (a) between $x = 4.95$ nm and 5.05 nm, (b) between $x = 7.95$ nm and 9.05 nm, (c) between $x = 9.90$ nm and 10.00 nm, (d) in the left half of the box, (e) in the central third of the box.

1.2 The normalized wavefunction for the electron in a hydrogen atom is

$$\psi = \left(\frac{1}{\pi a_0^3}\right)^{1/2} e^{-r/a_0}$$

where $a_0 = 53$ pm (the Bohr radius). (a) Calculate the probability that the electron will be found somewhere within a small sphere of radius 1.0 pm centred on the nucleus. (b) Now suppose that the same sphere is located at $r = a_0$. What is the probability that the electron is inside it?

1.3 A hydrogen atom attached to a metallic surface is undergoing oscillatory motion so that the state of the atom is described by a wavefunction that is proportional to the square of the atom's displacement from the metallic surface. Assume that the motion of the H atom is constrained to one dimension between $x = 0$ and $x = \pi$ and that its state is described by the unnormalized wavefunction $\psi(x) = x^2$. If the probability of finding the atom between $x = 0$ and $x = a$ is $\frac{1}{2}$, what is the value of a?

1.4 A particle free to move along one dimension x (with $0 \le x < \infty$) is described by the unnormalized wavefunction $\psi(x) = e^{-ax}$ with $a = 2$ m^{-1}. What is the probability of finding the particle at a distance $x \ge 1$ m?

1.5 The rotation of a light atom around a heavier atom to which it is bonded can be described quantum mechanically. The unnormalized wavefunctions for a light atom confined to move on a circle (with a heavier atom at the circle's centre) are $\psi(\phi) = e^{-im\phi}$, where $m = 0, \pm 1, \pm 2, \pm 3, \ldots$ and $0 \le \phi \le 2\pi$. Determine $\langle \phi \rangle$.

1.6 Atoms in a chemical bond vibrate around the equilibrium bond length. An atom undergoing vibrational motion is described by the wavefunction $\psi(x) = Ne^{-x^2/(2a^2)}$, where a is a constant and $-\infty < x < \infty$. (a) Normalize this function. (b) Calculate the probability of finding the particle in the range $-a \le x \le a$. *Hint.* The integral encountered in part (b) is the error function. It is defined and tabulated in M. Abramowitz and I.A. Stegun, *Handbook of mathematical functions*, Dover (1965) and is provided in most mathematical software packages.

1.7 Suppose that the state of the vibrating atom in Problem 1.6 is described by the wavefunction $\psi(x) = Nxe^{-x^2/(2a^2)}$. Where is the most probable location of the particle?

1.8 An atom undergoing vibrational motion is described by the wavefunction $\psi(x) = (2a/\pi)^{1/4}e^{-ax^2}$, where a is a constant and $-\infty < x < \infty$. Verify that the value of the product $\Delta p \Delta x$ is consistent with the predictions from the uncertainty principle.

1.9 A particle freely moving in one dimension x with $0 \le x < \infty$ is in a state described by the wavefunction $\psi(x) = a^{1/2}e^{-ax/2}$, where a is a constant. Determine the expectation values of the position and momentum operators.

Theoretical problems

1.10 Normalize the following wavefunctions: (a) $\sin(n\pi x/L)$ in the range $0 \le x \le L$, where $n = 1, 2, 3, \ldots$ (this wavefunction can be used to describe delocalized electrons in a linear polyene), (b) a constant in the range $-L \le x \le L$, (c) $e^{-r/a}$ in three-dimensional space (this wavefunction can be used to describe the electron in the ion He^+), (d) $xe^{-r/2a}$ in three-dimensional space. *Hint.* The volume element in three dimensions is $d\tau = r^2 dr \sin\theta\, d\theta\, d\phi$, with $0 \le r < \infty$, $0 \le \theta \le \pi$, $0 \le \phi \le 2\pi$.

1.11 Write the time-independent Schrödinger equations for (a) an electron moving in one dimension about a stationary proton and subjected to a Coulombic potential, (b) a free particle, (c) a particle subjected to a constant, uniform force.

1.12 (a) Two (unnormalized) excited state wavefunctions of the H atom are

$$\text{(i) } \psi = \left(2 - \frac{r}{a_0}\right)e^{-r/a_0} \qquad \text{(ii) } \psi = r\sin\theta\cos\phi\, e^{-r/2a_0}$$

Normalize both functions to 1. (b) Confirm that these two functions are mutually orthogonal.

1.13 Confirm the following properties of commutators:

(i) $[\hat{A}, \hat{B}] = -[\hat{B}, \hat{A}]$;

(ii) $[\hat{A}, \hat{B} + \hat{C}] = [\hat{A}, \hat{B}] + [\hat{A}, \hat{C}]$ (we used this relation in Example 1.10);

(iii) $[\hat{A}^2, \hat{B}] = \hat{A}[\hat{A}, \hat{B}] + [\hat{A}, \hat{B}]\hat{A}$ (this relation was used in Self-test 1.14).

1.14 Evaluate the commutators (a) $[\hat{H}, \hat{p}_x]$ and (b) $[\hat{H}, \hat{x}]$, where $\hat{H} = \hat{p}_x^2/2m + \hat{V}(x)$. Choose (i) $V(x) = V$, a constant, (ii) $V(x) = \frac{1}{2}kx^2$.

1.15 Evaluate the limitation on the simultaneous specification of the following observables: (a) the position and momentum of a particle in one dimension, (b) the three components of linear momentum of a particle, (c) the kinetic energy and potential energy of a particle in one

* Problems denoted with the symbol ‡ were supplied by Charles Trapp, Carmen Giunta, and Marshall Cady.

dimension, (d) the electric dipole moment and the total energy of a one-dimensional system, (e) the kinetic energy and the position of a particle in one dimension.

1.16 Construct quantum mechanical operators for the following observables: (a) kinetic energy in one and in three dimensions, (b) the inverse separation, $1/x$, (c) electric dipole moment in one dimension, (d) the mean square deviations of the position and momentum of a particle in one dimension from the mean values.

1.17 Determine which of the following functions are eigenfunctions of the inversion operator $\hat{\imath}$ (which has the effect of making the replacement $x \to -x$): (a) $x^3 - kx$, (b) $\cos kx$, (c) $x^2 + 3x - 1$. State the eigenvalue of $\hat{\imath}$ when relevant.

1.18 The wavefunction of an electron in a linear accelerator is $\psi = (\cos\chi)e^{ikx} + (\sin\chi)e^{-ikx}$, where χ (chi) is a parameter. What is the probability that the electron will be found with a linear momentum (a) $+k\hbar$, (b) $-k\hbar$? What form would the wavefunction have if it were 90 per cent certain that the electron had linear momentum $+k\hbar$?

1.19 Evaluate the kinetic energy of the electron with wavefunction given in Problem 1.18.

1.20 Calculate the average linear momentum of a particle described by the following wavefunctions: (a) e^{ikx}, (b) $\cos kx$, (c) e^{-ax^2}, where in each one x ranges from $-\infty$ to $+\infty$.

1.21 Evaluate the expectation values of r and r^2 for a hydrogen atom with wavefunctions given in Problem 1.12.

1.22 Calculate (a) the mean potential energy and (b) the mean kinetic energy of an electron in the hydrogen atom whose state is described by the wavefunction given in Problem 1.2.

1.23 Show that the expectation value of an operator that can be written as the square of an hermitian operator is positive.

1.24 (a) Given that any operators used to represent observables must satisfy the commutation relation in eqn 1.22, what would be the operator for position if the choice had been made to represent linear momentum parallel to the x-axis by multiplication by the linear momentum? These different choices are all valid 'representations' of quantum mechanics. (b) With the identification of $\hat{x}$ in this representation, what would be the operator for $1/x$? *Hint.* Think of $1/x$ as x^{-1}.

1.25 Max Planck found that he could account for the experimental observations of a black-body radiator (an object capable of emitting and absorbing all frequencies of radiation uniformly) by proposing that the energy of each electromagnetic oscillator was quantized. He derived the expression

$$\rho(\lambda) = \frac{8\pi hc}{\lambda^5(e^{hc/\lambda kT} - 1)}$$

for the spectral energy density (so that $\rho(\lambda)d\lambda$ is the energy density of radiation in the range λ to $\lambda + d\lambda$, and $\rho(\lambda)Vd\lambda$ is the total energy in that wavelength range in a region of volume V). (a) Plot the Planck distribution for ρ as a function of wavelength (take $T = 298$ K). (b) Show mathematically that, as $\lambda \to 0$, $\rho \to 0$ and therefore that the 'ultraviolet catastrophe' (the infinite acccumulation of radiation energy at short wavelengths) is avoided. (c) Show that for long wavelengths ($hc/\lambda kT \ll 1$), the Planck distribution reduces to the *Rayleigh–Jeans law*, the classical expression $\rho(\lambda) = 8\pi kT/\lambda^4$.

1.26 Derive *Wien's law*, that $\lambda_{max}T$ is a constant, where λ_{max} is the wavelength corresponding to the maximum in the Planck distribution (Problem 1.25) at the temperature T, and deduce an expression for the constant as a multiple of the second radiation constant, $c_2 = hc/k$. Values of λ_{max} from a small pinhole in an electrically heated container were determined at a series of temperatures, and the results are given below. Deduce a value for Planck's constant using the values of c_2 and k.

θ/°C	1000	1500	2000	2500	3000	3500
λ_{max}/nm	2181	1600	1240	1035	878	763

Applications: to astrophysics and nanoscience

1.27‡ The temperature of the Sun's surface is approximately 5800 K. On the assumption that the human eye evolved to be most sensitive at the wavelength of light corresponding to the maximum in the Sun's radiant energy distribution, determine the colour of light to which the eye is the most sensitive. *Hint.* See Problem 1.26.

1.28 We saw in *Impact I1.1* that electron microscopes can obtain images with several hundredfold higher resolution than optical microscopes because of the short wavelength obtainable from a beam of electrons. For electrons moving at speeds close to c, the speed of light, the expression for the de Broglie wavelength (eqn 1.3) needs to be corrected for relativistic effects:

$$\lambda = \frac{h}{\left\{2m_e eV\left(1 + \dfrac{eV}{2m_e c^2}\right)\right\}^{1/2}}$$

where c is the speed of light in vacuum and V is the potential difference through which the electrons are accelerated. (a) Use the expression above to calculate the de Broglie wavelength of electrons accelerated through 50 kV. (b) Is the relativistic correction important?

1.29 Suppose that the wavefunction of an electron in a carbon nanotube is a linear combination of $\cos(nx)$ functions. Use mathematical software to construct superpositions of cosine functions and determine the probability that a given momentum will be observed. If you plot the superposition (which you should), set $x = 0$ at the centre of the screen and build the superposition there. Evaluate the root mean square location of the packet, $\langle x^2\rangle^{1/2}$.

MATHEMATICAL BACKGROUND 2

Differential equations

A **differential equation** is a relation between a function and its derivatives, as in

$$a\frac{d^2f}{dx^2}+b\frac{df}{dx}+cf=0 \qquad \text{(MB2.1)}$$

where *f* is a function of the variable *x* and the factors *a*, *b*, *c* may be either constants or functions of *x*. If the unknown function depends on only one variable, as in this example, the equation is called an **ordinary differential equation**; if it depends on more than one variable, as in

$$a\frac{\partial^2f}{\partial x^2}+b\frac{\partial^2f}{\partial y^2}+cf=0 \qquad \text{(MB2.2)}$$

it is called a **partial differential equation**. Here, *f* is a function of *x* and *y*, and the factors *a*, *b*, *c* may be either constants or functions of both variables. Note the change in symbol from d to ∂ to signify a 'partial derivative' (see *Mathematical background* 1).

MB2.1 The structure of differential equations

The **order** of the differential equation is the order of the highest derivative that occurs in it: both examples above are second-order equations. Only rarely in science is a differential equation of order higher than 2 encountered.

A **linear differential equation** is one for which if *f* is a solution then so is constant × *f*. Both examples above are linear. If the 0 on the right were replaced by a different number or a function other than *f*, then they would cease to be linear.

Solving a differential equation means something different from solving an algebraic equation. In the latter case, the solution is a value of the variable *x* (as in the solution $x=\pm2$ of the quadratic equation $x^2-4=0$). The solution of a differential equation is the entire function that satisfies the equation, as in

$$\frac{d^2f}{dx^2}+f=0 \quad \text{has the solution } f=A\sin x+B\cos x \qquad \text{(MB2.3)}$$

with *A* and *B* constants. The process of finding a solution of a differential equation is called **integrating** the equation. The solution in eqn MB2.3 is an example of a **general solution** of a differential equation, that is, it is the most general solution of the equation and is expressed in terms of a number of constants (*A* and *B* in this case). When the constants are chosen to accord with certain specified **initial conditions** (if one variable is the time) or certain **boundary conditions** (to fulfil certain spatial restrictions on the solutions), we obtain the **particular solution** of the equation. The particular solution of a first-order differential equation requires one such condition; a second-order differential equation requires two.

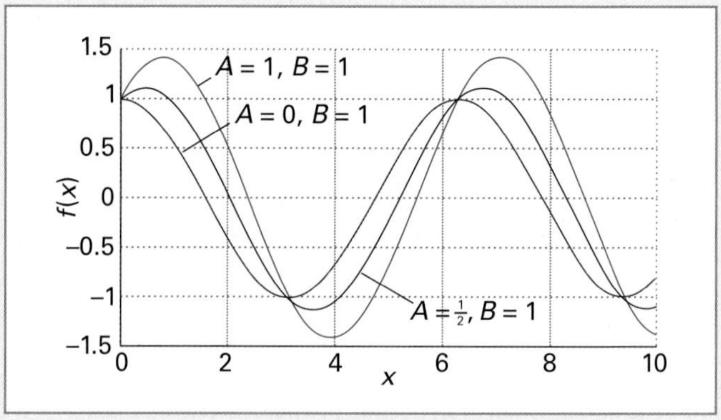

Fig. MB2.1 The solution of the differential equation in eqn MB2.3 with three different boundary conditions (as indicated by the resulting values of the constants *A* and *B*).

● **A BRIEF ILLUSTRATION**

If we are informed that $f(0)=0$, then because, from eqn MB2.3, it follows that $f(0)=B$, we can conclude that $B=0$. That still leaves *A* undetermined. If we are also told that $df/dx=2$ at $x=0$ (that is, $f'(0)=2$, where the prime denotes a first derivative), then, because the general solution (but with $B=0$) implies that $f'(x)=A\cos x$, we know that $f'(0)=A$, and therefore $A=2$. The particular solution is therefore $f(x)=2\sin x$. Figure MB2.1 shows a series of particular solutions corresponding to different boundary conditions. ●

MB2.2 The solution of ordinary differential equations

The first-order linear differential equation

$$\frac{df}{dx}+af=0 \qquad \text{(MB2.4a)}$$

with *a* a function of *x* or a constant can be solved by direct integration. To proceed, we use the fact that the quantities d*f* and d*x* (called *differentials*) can be treated algebraically like any quantity and rearrange the equation into

$$\frac{df}{f}=-a\,dx \qquad \text{(MB2.4b)}$$

and integrate both sides. For the left-hand side, we use the familiar result $\int dy/y=\ln y+\text{constant}$. After pooling all the constants into a single constant *A*, we obtain:

$$\ln f=-\int a\,dx+A \qquad \text{(MB2.4c)}$$

● A BRIEF ILLUSTRATION

Suppose that in eqn MB2.4a the factor $a = 2x$; then the general solution, eqn MB2.4c, is

$$\ln f = -2\int x\,\mathrm{d}x + A = -x^2 + A$$

(We have absorbed the constant of integration into the constant A.) Therefore

$$f = \mathrm{e}^{A}\mathrm{e}^{-x^2}$$

If we are told that $f(0) = 1$, then we can infer that $A = 0$ and therefore that $f = \mathrm{e}^{-x^2}$. ●

The solution of even first-order differential equations quickly becomes more complicated. A non-linear first-order equation of the form

$$\frac{\mathrm{d}f}{\mathrm{d}x} + af = b \qquad \text{(MB2.5a)}$$

with a and b functions of x (or constants) has a solution of the form

$$f\mathrm{e}^{\int a\mathrm{d}x} = \int \mathrm{e}^{\int a\mathrm{d}x} b\,\mathrm{d}x + A \qquad \text{(MB2.5b)}$$

as may be verified by differentiation. Mathematical software packages can often perform the required integrations.

Second-order differential equations are in general much more difficult to solve than first-order equations. One powerful approach commonly used to lay siege to second-order differential equations is to express the solution as a power series:

$$f(x) = \sum_{n=0}^{\infty} c_n x^n \qquad \text{(MB2.6)}$$

and then to use the differential equation to find a relation between the coefficients. This approach results, for instance, in the Hermite polynomials that form part of the solution of the Schrödinger equation for the harmonic oscillator (Section 2.5). Many of the second-order differential equations that occur in this text are tabulated in compilations of solutions or can be solved with mathematical software, and the specialized techniques that are needed to establish the form of the solutions may be found in mathematical texts.

MB2.3 The solution of partial differential equations

The only partial differential equations that we need to solve are those that can be separated into two or more ordinary differential equations by the technique known as **separation of variables.** To discover if the differential equation in eqn MB2.2 can be solved by this method we suppose that the full solution can be factored into functions that depend only on x or only on y, and write $f(x,y) = X(x)Y(y)$. At this stage there is no guarantee that the solution can be written in this way. Substituting this trial solution into the equation and recognizing that

$$\frac{\partial^2 XY}{\partial x^2} = Y\frac{\mathrm{d}^2X}{\mathrm{d}x^2} \qquad \frac{\partial^2 XY}{\partial y^2} = X\frac{\mathrm{d}^2Y}{\mathrm{d}y^2}$$

we obtain

$$aY\frac{\mathrm{d}^2X}{\mathrm{d}x^2} + bX\frac{\mathrm{d}^2Y}{\mathrm{d}y^2} + cXY = 0$$

We are using d instead of ∂ at this stage to denote differentials because each of the functions X and Y depend on one variable, x and y, respectively. Division through by XY turns this equation into

$$\frac{a}{X}\frac{\mathrm{d}^2X}{\mathrm{d}x^2} + \frac{b}{Y}\frac{\mathrm{d}^2Y}{\mathrm{d}y^2} + c = 0$$

Now suppose that a is a function only of x, b a function of y, and c a constant. (There are various other possibilities that permit the argument to continue.) Then the first term depends only on x and the second only on y. If x is varied, only the first term can change. But as the other two terms do not change and the sum of the three terms is a constant (0), even that first term must be a constant. The same is true of the second term. Therefore, because each term is equal to a constant, we can write

$$\frac{a}{X}\frac{\mathrm{d}^2X}{\mathrm{d}x^2} = c_1 \qquad \frac{b}{Y}\frac{\mathrm{d}^2Y}{\mathrm{d}y^2} = c_2 \qquad \text{with } c_1 + c_2 = -c$$

We now have two ordinary differential equations to solve by the techniques described in Section MB2.2. An example of this procedure is given in Section 3.1, for a particle in a two-dimensional region.

2 Nanosystems 1: motion in one dimension

To find the properties of systems according to quantum mechanics we need to solve the appropriate Schrödinger equation. This and the following chapter present the essentials of the solutions for three basic types of motion: translation, vibration, and rotation. We shall see that only certain wavefunctions and their corresponding energies are acceptable. Hence, quantization emerges as a natural consequence of the equation and the conditions imposed on it. The solutions bring to light a number of highly nonclassical, and therefore surprising, features of particles, especially their ability to tunnel into and through regions where classical physics would forbid them to be found. In this chapter, we consider translational and vibrational motion in one dimension and illustrate the postulates introduced in Chapter 1. By restricting our attention to motion in a single dimension, we keep the mathematical treatment relatively simple and focus on extracting physical insight from the Schrödinger equation that can be applied to problems in multiple dimensions as well.

Modern chemistry is increasingly interested in nanomaterials, materials composed of entities with dimensions up to about 100 nm. It always has been: from its earliest days, chemistry has been concerned with atoms, which have dimensions of the order of 0.1 nm, and molecules, which have dimensions of the order of 1 nm. Now, though, its attention is turning to aggregates of atoms and molecules that have dimensions of up to 100 nm and which—for quantum mechanical reasons—have properties that are often strikingly different from those of bulk matter.

This chapter will show how quantum mechanics is applied to a range of systems. For simplicity, we confine attention to one-dimensional systems initially. Although they are simple, one-dimensional systems illustrate many of the principles of quantum mechanics in an uncluttered way, and in some cases are directly relevant to the discussion of nanomaterials.

We shall build on the foundations laid in Chapter 1. Because, according to Postulate I, all the dynamical information about a system is contained in its wavefunction, we must determine the wavefunction $\psi(x)$ for the system of interest by solving the appropriate Schrödinger equation. We have to be aware, though, that an acceptable wavefunction of any system must satisfy the constraints specified in Section 1.6: it must be continuous, its slope must be continuous, it must be single-valued, and not become infinite. To extract the information in the wavefunction, we shall draw on the Born interpretation (Postulate II) and the properties of operators (Postulates III to V). We have to be prepared, though, for the concept of complementarity (Section 1.9) to restrict the type of question that, in quantum mechanics, it is meaningful to ask.

Translational motion

Translation (motion through space) is the most primitive form of motion and is where we begin. Despite being primitive, it is of great importance: it is the form of motion by which gas-phase molecules store energy, it is the type of motion by which electrons conduct electricity, and—when the particle is trapped into a small region of space—it accounts for some of the properties of nanomaterials.

2.1 Free motion

The Schrödinger equation for a particle of mass m (with $m = m_e$ for an electron, $m = m_p$ for a proton, and so on) free to move in one dimension was stated in eqn 1.7:

$$-\frac{\hbar^2}{2m}\frac{d^2\psi}{dx^2} + V\psi = E\psi \qquad (2.1a)$$

For a region of space where the potential energy is zero, we can set $V = 0$ and consider the equation

$$-\frac{\hbar^2}{2m}\frac{d^2\psi}{dx^2} = E\psi \qquad (2.1b)$$

The general solutions of this equation are

$$\psi_k = Ae^{ikx} + Be^{-ikx} \qquad E_k = \frac{k^2\hbar^2}{2m} \qquad (2.2)$$

with A and B constants, as may be verified by substitution. Note that we are now labelling both the wavefunctions and the energies (that is, the eigenfunctions and eigenvalues of $\hat{H}$) with the index k. These solutions are continuous, have continuous slope everywhere, are single-valued, and do not go to infinity, and so—in the absence of any other information—are acceptable for all values of k. Because the energy of the particle is proportional to k^2, all values of the energy are permitted. It follows that the translational energy of a free particle is not quantized.

The values of the constants A and B depend on how the state of motion of the particle was prepared. If it is shot towards positive x, then its linear momentum will be $+k\hbar$, as explained in Section 1.7, and its wavefunction will be proportional to e^{ikx}. In this case $B = 0$ and A will be a normalization factor. If the particle is shot in the opposite direction, towards negative x, then its linear momentum will be $-k\hbar$ and its wavefunction proportional to e^{-ikx}. In this case, $A = 0$ and B will be the normalization factor. The wavefunctions e^{ikx} and e^{-ikx} exemplify a general feature of quantum mechanics: a particle with net motion is described by a complex wavefunction. A real wavefunction (for instance, if $A = B = \frac{1}{2}$ in eqn 2.2 and $\psi = \cos kx$) corresponds to zero net motion.

A brief comment The wavefunction for a free particle, $e^{\pm ikx}$, is not square-integrable in a region of infinite length. To get round this problem, we assume temporarily that the particle is in a region of length L, and normalize the wavefunction. At the end of any subsequent calculation using the wavefunction, L is allowed to become infinite: see Problem 2.11.

If the particle is in either of the pure momentum states e^{ikx} or e^{-ikx}, its probability density $|\psi|^2$ is uniform. According to the Born interpretation (Postulate II, Section 1.5), nothing further can be said about the location of the particle. That conclusion is consistent with the uncertainty principle because, if the momentum is certain, then the position cannot be specified (the operators corresponding to x and p do not commute, Section 1.10).

2.2 A particle in a box

In this section, we consider a **particle in a box**, in which a particle of mass m is confined between two walls at $x = 0$ and $x = L$: the potential energy is zero inside the box but rises abruptly to infinity at the walls (Fig. 2.1). Although this problem is very elementary, there has been a resurgence of research interest in it now that nanostructures are used to trap electrons in cavities resembling square wells. The particle in a box model is also an idealization of the potential energy of a gas-phase molecule that is free to move in a one-dimensional container, and forms the basis of a primitive treatment of the electronic structure of metals (Chapter 9) and conjugated molecules such as butadiene. The particle in a box is also used in statistical thermodynamics in assessing the contribution of the translational motion of molecules to their thermodynamic properties (Chapter 13).

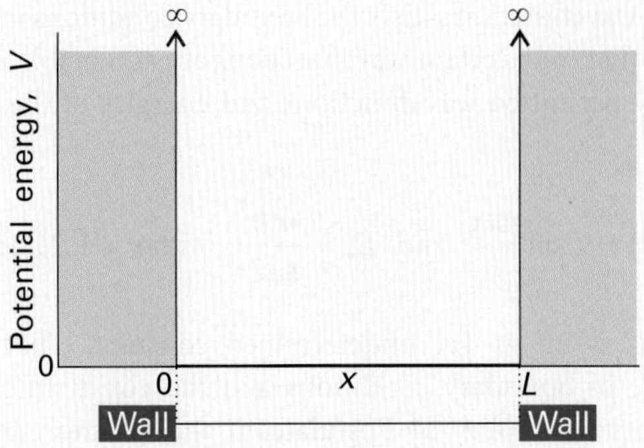

Fig. 2.1 A particle in a one-dimensional region with impenetrable walls. Its potential energy is zero between $x = 0$ and $x = L$, and rises abruptly to infinity as soon as it touches the walls.

(a) The acceptable solutions

The Schrödinger equation for a particle when it is between the walls (and where $V = 0$) is the same as for a free particle (eqn 2.1b), so the general solutions given in eqn 2.2 are also the same. However, it will prove convenient to use $e^{\pm ix} = \cos x \pm i \sin x$ to write

$$\psi_k = Ae^{ikx} + Be^{-ikx} = A(\cos kx + i \sin kx) + B(\cos kx - i \sin kx)$$
$$= (A + B)\cos kx + (A - B)i \sin kx$$

If we absorb all numerical factors into two new coefficients C and D, then the general solutions and their energies take the form

$$\text{For } 0 \le x \le L, \quad \psi_k(x) = C \sin kx + D \cos kx \quad E_k = \frac{k^2\hbar^2}{2m} \tag{2.3a}$$

When the particle touches the walls, its potential energy rises sharply to infinity. As a result, it is never found inside the material of the walls and its wavefunction is zero there. That is

$$\text{For } x < 0 \text{ and } x > L, \ \psi_k(x) = 0 \tag{2.3b}$$

The requirement of the continuity of the wavefunction (which stems from Postulates II and III) then implies that $\psi_k(x)$ as given by eqn 2.3a must also be zero at the walls, for it must match the wavefunction inside the material of the walls, where the functions meet. That is, the wavefunction must satisfy the following two **boundary conditions**, or constraints on the function at certain locations:

$$\psi_k(0) = 0 \text{ and } \psi_k(L) = 0 \tag{2.4}$$

In a more sophisticated treatment, the potential energy is supposed not to be infinite initially, and the Schrödinger equation is solved for all three regions. Then the potential energy is allowed to rise to infinity. The continuity of the wavefunction at the two walls then results in the same two boundary conditions.

As we show in the following *Justification*, the requirement that the wavefunction satisfy the boundary conditions in eqn 2.4 implies that only certain wavefunctions are acceptable and that the only permitted wavefunctions and energies of the particle are

$$\psi_n(x) = C \sin\frac{n\pi x}{L} \text{ and } E_n = \frac{n^2h^2}{8mL^2} \quad \text{with } n = 1, 2, \ldots \tag{2.5}$$

where C is an as yet undetermined constant. That is, the presence of boundary conditions and the constraints on the wavefunction implied by Postulates II and III imply that only certain wavefunctions are acceptable and hence that the energy is quantized. Note that the wavefunction and energy are now labelled with the dimensionless integer n instead of the quantity k.

Justification 2.1 *The energy levels and wavefunctions of a particle in a one-dimensional box*

For an informal demonstration of quantization, we consider each wavefunction to be a de Broglie wave that fits into the container in the sense that an integral number of half wavelengths (one bulge, two bulges, . . . , Fig. 2.2) is equal to the length of the box:

$$n \times \tfrac{1}{2}\lambda = L \qquad n = 1, 2, \ldots$$

and therefore

$$\lambda = \frac{2L}{n} \qquad \text{with } n = 1, 2, \ldots$$

According to the de Broglie relation, these wavelengths correspond to the momenta

$$p = \frac{h}{\lambda} = \frac{nh}{2L}$$

The particle has only kinetic energy inside the box (where $V = 0$), so the permitted energies are

$$E = \frac{p^2}{2m} = \frac{n^2h^2}{8mL^2} \qquad \text{with } n = 1, 2, \ldots$$

as in eqn 2.5.

A more formal and widely applicable approach is as follows. From the boundary condition $\psi_k(0) = 0$ and the fact that, from eqn 2.3a, $\psi_k(0) = D$ (because $\sin 0 = 0$ and $\cos 0 = 1$), we can conclude that $D = 0$. It follows that the wavefunction must be of the form

$$\psi_k(x) = C \sin kx$$

From the second boundary condition, $\psi_k(L) = 0$, we know that $\psi_k(L) = C \sin kL = 0$. We could take $C = 0$, but doing so would give $\psi_k(x) = 0$ for all x, which would conflict with Postulate II and the Born interpretation (the particle must be somewhere). The alternative is to require that kL be chosen so that $\sin kL = 0$. This condition is satisfied if

$$kL = n\pi \qquad n = 1, 2, \ldots$$

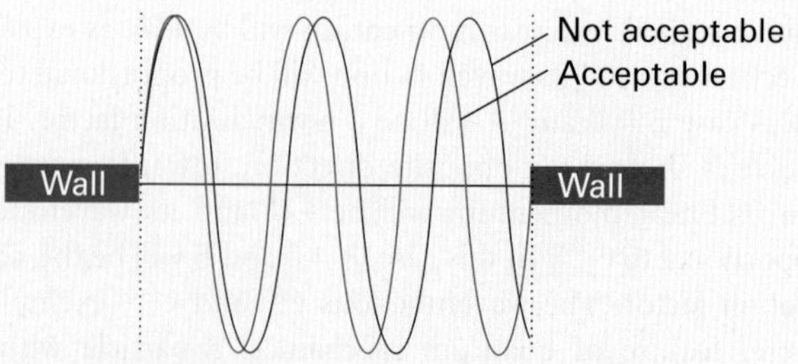

Fig. 2.2 An acceptable wavefunction must have a de Broglie wavelength such that the wave fits into the box.

The value $n=0$ is ruled out, because it implies $k=0$ and $\psi_k(x) = 0$ everywhere (because $\sin 0 = 0$), which is unacceptable. Negative values of n merely change the sign of $\sin kL$ (because $\sin(-x) = -\sin x$) and do not result in new solutions. The wavefunctions are therefore

$$\psi_n(x) = C \sin(n\pi x/L) \qquad n = 1, 2, \ldots$$

(This is the point where we start to label the solutions with the index n instead of k.) Because k and E_k are related by eqn 2.3a, and k and n are related by $kL = n\pi$, it follows that the energy of the particle is limited to $E_n = n^2h^2/8mL^2$, the values obtained by the informal procedure and stated in eqn 2.5.

We conclude that the energy of the particle in a one-dimensional box is quantized and that this quantization arises from the boundary conditions that ψ must satisfy. This is a general conclusion: *the need to satisfy boundary conditions in conjunction with the general requirements of Postulates II and III implies that only certain wavefunctions are acceptable, and hence restricts observables to discrete values.* So far, only energy has been quantized; shortly we shall see that other physical observables may also be quantized.

We need to determine the constant C in eqn 2.5. To do so, we normalize the wavefunction to 1. Because the wavefunction is zero outside the range $0 \le x \le L$, we use

$$\int_0^L \psi^2 dx = C^2 \int_0^L \sin^2 \frac{n\pi x}{L} = C^2 \times \frac{L}{2} = 1, \qquad \text{so } C = \left(\frac{2}{L}\right)^{1/2}$$

for all n. Therefore, the complete solution to the particle in a box problem is

$$E_n = \frac{n^2h^2}{8mL^2} \qquad n = 1, 2, \ldots \qquad (2.6a)$$

$$\psi_n(x) = \left(\frac{2}{L}\right)^{1/2} \sin\left(\frac{n\pi x}{L}\right) \qquad \text{for } 0 \le x \le L \qquad (2.6b)$$

It should be recalled from Section 1.8 that the hermiticity of the hamiltonian operator implies that wavefunctions corresponding to different energies are orthogonal. For the present system, orthogonality means that

$$\int_0^L \psi_{n'}(x)\psi_n(x)dx = 0 \qquad \text{for } n' \ne n \qquad (2.7)$$

This relation was confirmed in *Example 1.6* in Section 1.8 for the specific examples of wavefunctions with $n = 1$ and $n' = 2$.

Self-test 2.1 Provide the intermediate steps for the determination of the normalization constant C. *Hint.* Use the standard integral $\int \sin^2 ax\, dx = \frac{1}{2}x - (1/4a)\sin 2ax + \text{constant}$ and the fact that $\sin 2n\pi = 0$, with n an integer.

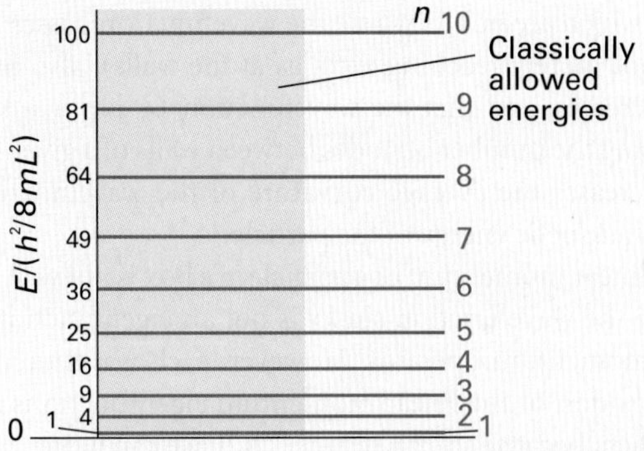

Fig. 2.3 The allowed energy levels for a particle in a box. Note that the energy levels increase as n^2, and that their separation increases as the quantum number increases.

The energies and wavefunctions are labelled with the 'quantum number' n. A **quantum number** is an integer (in some cases, as we shall see, a half-integer) that labels the state of the system. For a particle in a box there is an infinite number of acceptable solutions, and the quantum number n specifies the one of interest (Fig. 2.3). As well as acting as a label, a quantum number can often be used to calculate the energy corresponding to the state and to write down the wavefunction explicitly (in the present example, by using the relations in eqn 2.6).

(b) The properties of the solutions

Figure 2.4 shows some of the wavefunctions of a particle in a box: they are all sine functions with the same amplitude but different wavelengths. Shortening the wavelength results in a sharper average curvature of the wavefunction and therefore an increase in the kinetic energy of the particle (its only source of

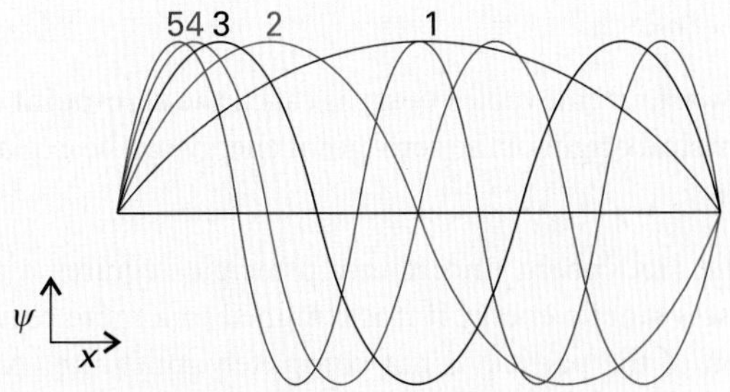

Fig. 2.4 The first five normalized wavefunctions of a particle in a box. Each wavefunction is a standing wave, and successive functions possess one more half wave and a correspondingly shorter wavelength.

interActivity Plot the probability density for a particle in a box with $n = 1, 2, \ldots 5$ and $n = 50$. How do your plots illustrate the correspondence principle?

energy because $V = 0$ inside the box). Note that the number of **nodes**, which are points where the wavefunction passes *through* zero (not merely reaching zero, as at the walls), also increases as n increases, and that the wavefunction ψ_n has $n - 1$ nodes. Increasing the number of nodes between walls of a given separation increases the average curvature of the wavefunction and hence the kinetic energy of the particle.

The linear momentum of a particle in a box is not well defined because the wavefunction sin kx is not an eigenfunction of the linear momentum operator. However, each wavefunction is a superposition of the linear momentum eigenfunctions e^{ikx} and e^{-ikx}. Then, because $\sin x = (e^{ix} - e^{-ix})/2i$, we can write

$$\psi_n = \left(\frac{2}{L}\right)^{1/2} \sin\frac{n\pi x}{L} = \frac{1}{2i}\left(\frac{2}{L}\right)^{1/2}(e^{ikx} - e^{-ikx}) \quad k = \frac{n\pi}{L} \qquad (2.8)$$

It follows (from Postulate V) that measurement of the linear momentum will give the value $+k\hbar$ for half the measurements of momentum and $-k\hbar$ for the other half. This detection of opposite directions of travel with equal probability is the quantum mechanical version of the classical picture that a particle in a box rattles from wall to wall and in any given period spends half its time travelling to the left and half travelling to the right.

Self-test 2.2 What is (a) the average value of the linear momentum of a particle in a box with quantum number n, (b) the average value of p^2?

[(a) $\langle p\rangle = 0$, (b) $\langle p^2\rangle = n^2h^2/4L^2$]

Because n cannot be zero, the lowest energy that the particle may possess is not zero (as would be allowed by classical mechanics, corresponding to a stationary particle) but

$$E_1 = \frac{h^2}{8mL^2} \qquad (2.9)$$

This lowest, irremovable energy is called the **zero-point energy**. The physical origin of the zero-point energy can be explained in two ways:

- The Heisenberg uncertainty principle requires a particle to possess kinetic energy if it is confined to a finite region: the location of the particle is not completely indefinite ($\Delta x \neq \infty$), so the uncertainty in its momentum cannot be precisely zero ($\Delta p \neq 0$). Because $\Delta p = (\langle p^2\rangle - \langle p\rangle^2)^{1/2} = \langle p^2\rangle^{1/2}$ in this case, $\Delta p \neq 0$ implies that $\langle p^2\rangle \neq 0$, which implies that the particle must always have nonzero kinetic energy.
- If the wavefunction is to be zero at the walls, but smooth, continuous, and not zero everywhere, then it must be curved, and curvature in a wavefunction implies the possession of kinetic energy.

The separation between adjacent energy levels with quantum numbers n and $n + 1$ is

$$E_{n+1} - E_n = \frac{(n+1)^2h^2}{8mL^2} - \frac{n^2h^2}{8mL^2} = (2n+1)\frac{h^2}{8mL^2} \qquad (2.10)$$

This separation decreases as the length of the container increases, and is very small when the container has macroscopic dimensions. The separation of adjacent levels becomes zero when the walls are infinitely far apart. Atoms and molecules free to move in normal laboratory-sized vessels may therefore be treated as though their translational energy is not quantized. We have already seen (Section 2.1) that the translational energy of completely free particles (those not confined by walls) is not quantized.

Example 2.1 *Estimating an absorption wavelength*

β-Carotene (**1**) is a linear polyene in which 10 single and 11 double bonds alternate along a chain of 22 carbon atoms.

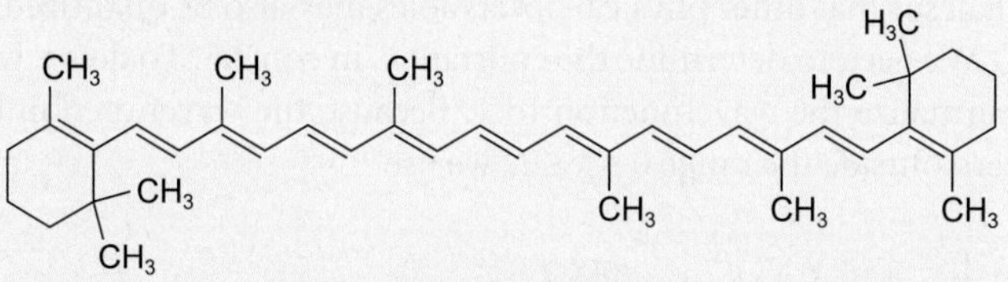

1 β-Carotene

If we take each CC bond length to be about 140 pm, then the length L of the molecular box in β-carotene is $L = 2.94$ nm. Estimate the wavelength of the light absorbed by this molecule from its ground state to the next higher excited state.

Method For reasons that will be familiar from introductory chemistry, each C atom contributes one p electron to the π orbitals. Use eqn 2.10 to calculate the energy separation between the highest occupied and the lowest unoccupied levels, and convert that energy to a wavelength by using the Bohr frequency relation (eqn 1.1).

Answer There are 22 C atoms in the conjugated chain; each contributes one p electron to the levels, so each level up to $n = 11$ is occupied by two electrons. The separation in energy between the ground state and the state in which one electron is promoted from $n = 11$ to $n = 12$ is

$$\Delta E = E_{12} - E_{11}$$

$$= (2 \times 11 + 1)\frac{(6.626 \times 10^{-34}\,\mathrm{J\,s})^2}{8 \times (9.109 \times 10^{-31}\,\mathrm{kg}) \times (2.94 \times 10^{-9}\,\mathrm{m})^2}$$

$$= 1.60 \times 10^{-19}\,\mathrm{J}$$

It follows from the Bohr frequency condition ($\Delta E = h\nu$) that the frequency of radiation required to cause this transition is

$$\nu = \frac{\Delta E}{h} = \frac{1.60 \times 10^{-19}\,\text{J}}{6.626 \times 10^{-34}\,\text{J s}} = 2.41 \times 10^{14}\,\text{s}^{-1}$$

or 241 THz (1 THz = 10^{12} Hz). The experimental value is 603 THz ($\lambda = 497$ nm), corresponding to radiation in the visible range of the electromagnetic spectrum. Considering the crudeness of the model we have adopted here, we should be encouraged that the computed and observed frequencies agree to within a factor of 2.5.

Self-test 2.3 Estimate a typical nuclear excitation energy in electronvolts (1 eV = 1.602×10^{-19} J; 1 GeV = 10^9 eV) by calculating the first excitation energy of a proton confined to a square well with a length equal to the diameter of a nucleus (approximately 1×10^{-15} m, or 1 fm). [0.6 GeV]

The probability density for a particle in a box is

$$\psi^2(x) = \frac{2}{L}\sin^2\frac{n\pi x}{L} \tag{2.11}$$

and varies with position. The nonuniformity is pronounced when n is small (Fig. 2.5), but—provided we ignore the fine detail of the increasingly rapid oscillations—$\psi^2(x)$ becomes more uniform as n increases. The probability density at high quantum numbers reflects the classical result that a particle bouncing between the walls spends, on the average, equal times at all points. That the quantum result corresponds to the classical prediction at high quantum numbers is an illustration of the **correspondence principle**, which states that classical mechanics emerges from quantum mechanics as high quantum numbers are reached.

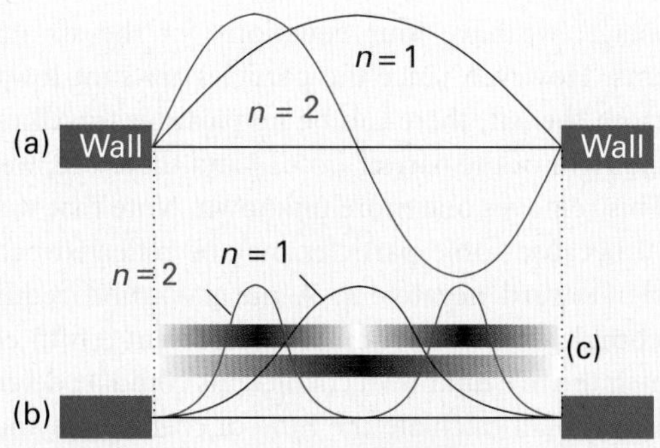

Fig. 2.5 (a) The first two wavefunctions, (b) the corresponding probability densities, and (c) a representation of the probability density in terms of the darkness of shading.

Example 2.2 *Using the particle-in-a-box solutions*

The wavefunctions of an electron in a conjugated polyene can be approximated by particle-in-a-box wavefunctions. What is the probability, P, of locating the electron between $x = 0$ (the left-hand end of a molecule) and $x = 0.2$ nm in its lowest energy state in a conjugated molecule of length 1.0 nm?

Method According to the Born interpretation, $\psi^2 dx$ is the probability of finding the particle in the small region dx located at x; therefore, the total probability of finding the electron in the specified region is the integral of $\psi^2 dx$ over that region. The wavefunction of the electron is given in eqn 2.6b with $n = 1$. A useful integral is provided in Self-test 2.1.

Answer The probability of finding the particle in a region between $x = 0$ and $x = l$ is

$$P = \int_0^l \psi_n^2 dx = \frac{2}{L}\int_0^l \sin^2\frac{n\pi x}{L}dx = \frac{l}{L} - \frac{1}{2n\pi}\sin\frac{2\pi nl}{L}$$

We then set $n = 1$ and $l = 0.2$ nm, which gives $P = 0.05$. The result corresponds to a chance of 1 in 20 of finding the electron in the region. As n becomes infinite, the sine term, which is multiplied by $1/n$, makes no contribution to P and the classical result, $P = l/L$, is obtained.

Self-test 2.4 Calculate the probability that an electron in the state with $n = 1$ will be found between $x = 0.25L$ and $x = 0.75L$ in a conjugated molecule of length L (with $x = 0$ at the left-hand end of the molecule). [0.82]

2.3 Tunnelling

If the potential energy of a particle does not rise to infinity when it is in the walls of the container, and $E < V$, the wavefunction does not decay abruptly to zero. If the walls are thin (so that the potential energy falls to zero again after a finite distance), then the wavefunction oscillates inside the box, varies smoothly inside the region representing the wall, and oscillates again on the other side of the wall outside the box (Fig. 2.6). Hence the particle might be found on the outside of a container whereas according to classical mechanics the particle has insufficient energy to escape and will be reflected off the wall. Such leakage by penetration through a classically forbidden region is called **tunnelling** and is a consequence of the wave character of matter. Tunnelling has very important implications for the electronic properties of materials, not only nanomaterials, for the rates of electron transfer reactions, the properties of acids and bases, and for the techniques currently used to study surfaces.

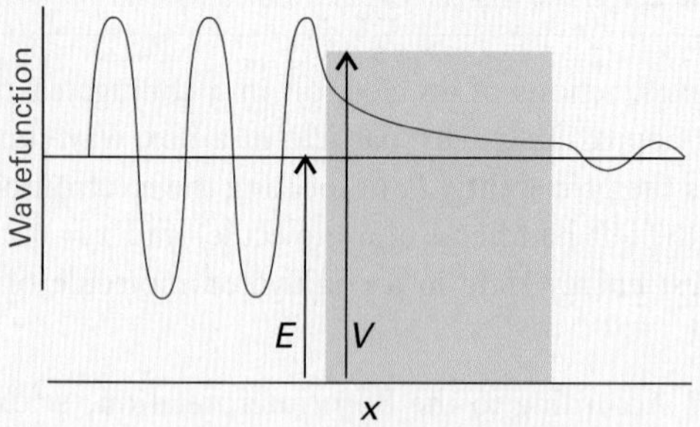

Fig. 2.6 A particle incident on a barrier from the left has an oscillating wavefunction, but inside the barrier there are no oscillations (for $E < V$). If the barrier is not too thick, the wavefunction is nonzero at its opposite face, and so oscillations begin again there. (Only the real component of the wavefunction is shown.)

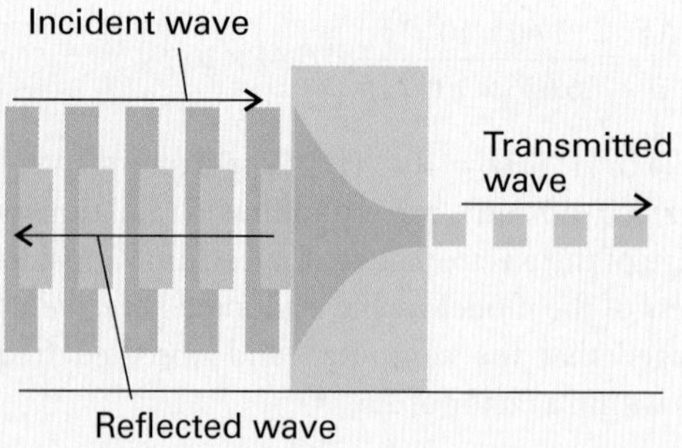

Fig. 2.7 When a particle is incident on a barrier from the left, the wavefunction consists of a wave representing linear momentum to the right, a reflected component representing momentum to the left, a varying but not oscillating component inside the barrier, and a (weak) wave representing motion to the right on the far side of the barrier.

The Schrödinger equation can be used to calculate the probability of tunnelling of a particle of mass m incident from the left on a barrier that extends from $x = 0$ to $x = L$. For simplicity, we suppose that, once inside the barrier, the particle has a constant potential energy V. On the left of the barrier ($x < 0$) the wavefunctions are those of a particle with $V = 0$, so from eqn 2.2 we can write

$$\psi = Ae^{ikx} + Be^{-ikx} \qquad k\hbar = (2mE)^{1/2} \tag{2.12}$$

The Schrödinger equation for the region representing the barrier ($0 \le x \le L$), where the potential energy is the constant V, is

$$-\frac{\hbar^2}{2m}\frac{d^2\psi}{dx^2} + V\psi = E\psi \tag{2.13}$$

We shall consider particles that have $E < V$ (so, according to classical physics, the particle has insufficient energy to pass through the barrier), and therefore $V - E$ is positive. The general solutions of this equation are

$$\psi = Ce^{\kappa x} + De^{-\kappa x} \qquad \kappa\hbar = \{2m(V - E)\}^{1/2} \tag{2.14}$$

as we can readily verify by differentiating ψ twice with respect to x. The important feature to note is that the two exponentials in eqn 2.14 are now real functions, as distinct from the complex, oscillating functions for the region where $V = 0$. (Oscillating functions would be obtained in the region $0 \le x \le L$ if we considered energies above the barrier height, $E > V$; see Problem 2.14.) To the right of the barrier ($x > L$), where $V = 0$ again, the wavefunctions are

$$\psi = A'e^{ikx} + B'e^{-ikx} \qquad k\hbar = (2mE)^{1/2} \tag{2.15}$$

We now investigate the relationships among the coefficients A, B, C, D, A', and B', which might be complex numbers. The complete wavefunction for a particle incident from the left consists of (Fig. 2.7):

- an incident wave (recall that Ae^{ikx} corresponds to positive momentum);
- a wave reflected from the barrier (recall that Be^{-ikx} corresponds to negative momentum, motion to the left);
- the exponentially changing amplitudes inside the barrier (eqn 2.14); and
- an oscillating wave (eqn 2.15) representing the propagation of the particle to the right after tunnelling through the barrier successfully.

The acceptable wavefunctions must obey the conditions set out in Section 1.6. In particular, they must be continuous at the edges of the barrier (at $x = 0$ and $x = L$, remembering that $e^0 = 1$):

$$A + B = C + D \qquad Ce^{\kappa L} + De^{-\kappa L} = A'e^{ikL} + B'e^{-ikL} \tag{2.16a}$$

Their slopes (their first derivatives) must also be continuous there (Fig. 2.8):

$$ikA - ikB = \kappa C - \kappa D$$
$$\kappa Ce^{\kappa L} - \kappa De^{-\kappa L} = ikA'e^{ikL} - ikB'e^{-ikL} \tag{2.16b}$$

At this stage, we have four equations for the six unknown coefficients. However, since the particles are shot towards the barrier from the left, there can be no particles travelling to the left on the right of the barrier ($x > L$) and therefore, we can set $B' = 0$. This removes one more unknown. Note that we cannot set $B = 0$ because some particles may be reflected back from the barrier toward negative x. A normalization requirement for the complete wavefunction would present a fifth equation from which the five unknown coefficients could be determined. However, we shall calculate the *ratio* of coefficients and therefore not need to invoke normalization.

The probability that a particle is travelling towards positive x (to the right) on the left of the barrier ($x < 0$) is proportional to $|A|^2$, and the probability that it is travelling to the right on the

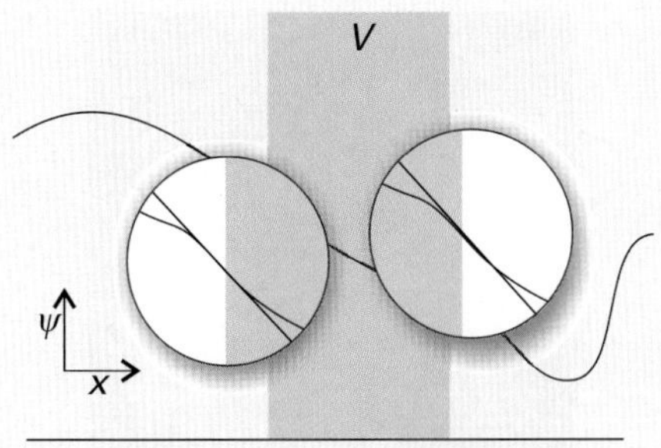

Fig. 2.8 The wavefunction and its slope must be continuous at the edges of the barrier. The conditions for continuity enable us to connect the wavefunctions in the three zones and hence to obtain relations between the coefficients that appear in the solutions of the Schrödinger equation.

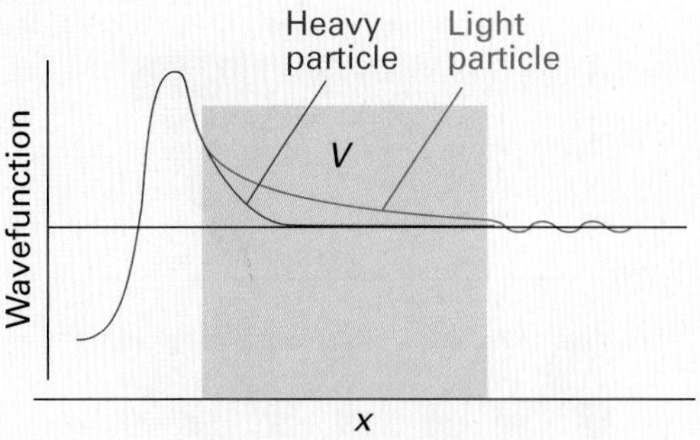

Fig. 2.10 The wavefunction of a heavy particle decays more rapidly inside a barrier than that of a light particle. Consequently, a light particle has a greater probability of tunnelling through the barrier.

right of the barrier ($x > L$) is $|A'|^2$. The ratio of these two probabilities is called the **transmission probability**, T. After some algebra (see Problem 2.13) we find

$$T = \left\{1 + \frac{(e^{\kappa L} - e^{-\kappa L})^2}{16\varepsilon(1-\varepsilon)}\right\}^{-1} \tag{2.17a}$$

where $\varepsilon = E/V$. This function is plotted in Fig. 2.9; the transmission coefficient for $E > V$ (which you are invited to calculate in Problem 2.14) is shown there too. For high, wide barriers (in the sense that $\kappa L \gg 1$), eqn 2.17a simplifies to

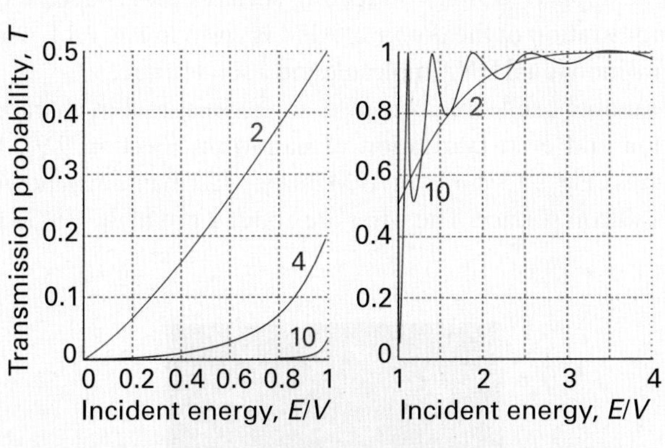

Fig. 2.9 The transmission probabilities for passage through a barrier. The horizontal axis is the energy of the incident particle expressed as a multiple of the barrier height. The curves are labelled with the value of $L(2mV)^{1/2}/\hbar$. The graph on the left is for $E < V$ and that on the right for $E > V$. Note that $T > 0$ for $E < V$ whereas classically T would be zero. However, $T < 1$ for $E > V$, whereas classically T would be 1.

interActivity Plot T against ε for a hydrogen molecule, a proton, and an electron.

$$T \approx 16\varepsilon(1-\varepsilon)e^{-2\kappa L} \tag{2.17b}$$

The transmission probability decreases exponentially with the thickness of the barrier and with $m^{1/2}$. It follows that particles of low mass are more able to tunnel through barriers than heavy ones (Fig. 2.10). Tunnelling is very important for electrons and muons ($m_\mu \approx 207m_e$), and moderately important for protons ($m_p \approx 1840m_e$); for heavier particles it is less important.

A number of effects in chemistry (for example, the isotope-dependence of some reaction rates) depend on the ability of the proton to tunnel more readily than the deuteron. The very rapid equilibration of proton transfer reactions is also a manifestation of the ability of protons to tunnel through barriers and transfer quickly from an acid to a base. Tunnelling of protons between acidic and basic groups is also an important feature of the mechanism of some enzyme-catalysed reactions.

IMPACT ON NANOSCIENCE
I2.1 Scanning probe microscopy

As we have indicated, *nanoscience* is the study of atomic and molecular assemblies with dimensions ranging from 1 nm to about 100 nm and *nanotechnology* is concerned with the incorporation of such assemblies into devices. The future economic impact of nanotechnology could be very significant. For example, increased demand for very small digital electronic devices has driven the design of ever smaller and more powerful microprocessors. However, there is an upper limit on the density of electronic circuits that can be incorporated into silicon-based chips with current fabrication technologies. As the ability to process data increases with the number of circuits in a chip, it follows that soon chips and the devices that use them will have to become bigger if processing power is to increase indefinitely. One way to circumvent this problem is to fabricate devices from nanometre-sized components.

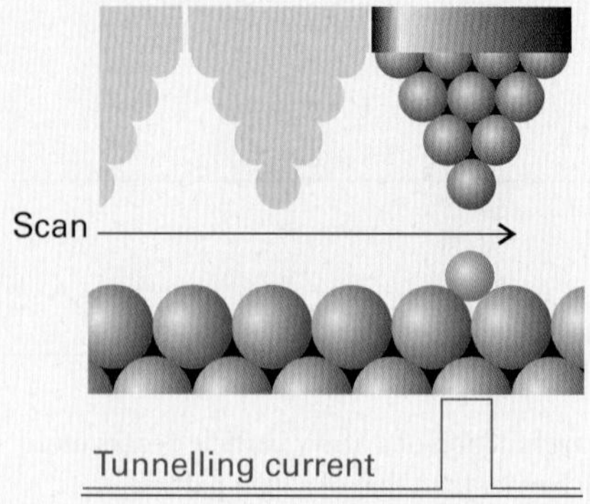

Fig. 2.11 A scanning tunnelling microscope makes use of the current of electrons that tunnel between the surface and the tip. That current is very sensitive to the distance of the tip above the surface.

We will explore several concepts of nanoscience throughout the text. We begin with the description of *scanning probe microscopy* (SPM), a collection of techniques that can be used to visualize and manipulate objects as small as atoms on surfaces. Consequently, SPM has far better resolution than electron microscopy (*Impact I1.1*).

One version of SPM is *scanning tunnelling microscopy* (STM), in which a platinum–rhodium or tungsten needle is scanned across the surface of a conducting solid. When the tip of the needle is brought very close to the surface, electrons tunnel across the intervening space (Fig. 2.11). In the 'constant-current mode' of operation, the stylus moves up and down corresponding to the form of the surface, and the topography of the surface, including any adsorbates, can be mapped on an atomic scale. The vertical motion of the stylus is achieved by fixing it to a piezoelectric cylinder, which contracts or expands according to the potential difference it experiences. In the 'constant-z mode', the vertical position of the stylus is held constant and the current is monitored. Because the tunnelling probability is very sensitive to the size of the gap, the microscope can detect tiny, atom-scale variations in the height of the surface.

Figure 2.12 shows an example of the kind of image obtained with a surface, in this case of gallium arsenide, that has been modified by addition of atoms, in this case caesium atoms. Each 'bump' on the surface corresponds to an atom. In a further variation of the STM technique, the tip may be used to nudge single atoms around on the surface, making possible the fabrication of complex and yet very tiny nanometre-sized structures.

In *atomic force microscopy* (AFM) a sharpened stylus attached to a cantilever is scanned across the surface. The force exerted by the surface and any bound species pushes or pulls on the stylus and deflects the cantilever (Fig. 2.13). The deflection is monitored either by interferometry or by using a laser beam. Because no current is needed between the sample and the probe, the technique can be applied to non-conducting surfaces too. A spectacular demonstration of the power of AFM is given in Fig. 2.14, which shows individual DNA molecules on a solid surface.

Scanning probe microscopy is an essential tool in characterization and even fabrication of nanowires (Section 9.9). For example, Fig. 2.15 shows an AFM image of germanium nanowires on a silicon surface. The wires are about 2 nm high, 10–32 nm

Fig. 2.12 An STM image of caesium atoms on a gallium arsenide surface.

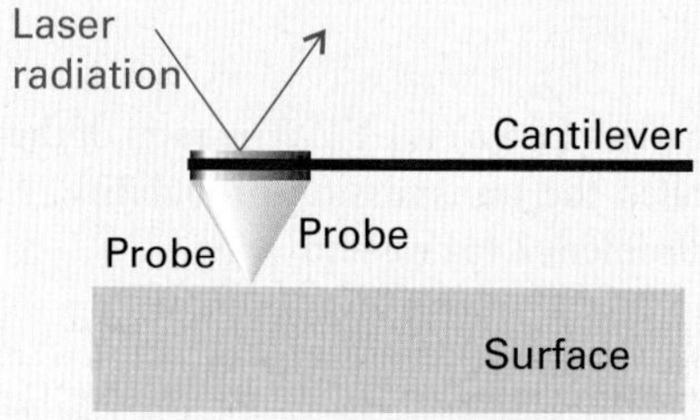

Fig. 2.13 In atomic force microscopy, a laser beam is used to monitor the tiny changes in the position of a probe as it is attracted to or repelled from atoms on a surface.

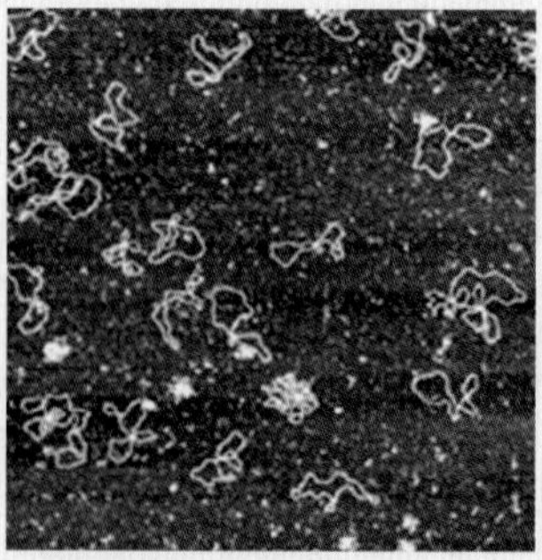

Fig. 2.14 An AFM image of bacterial DNA plasmids on a mica surface. (Courtesy of Veeco Instruments.)

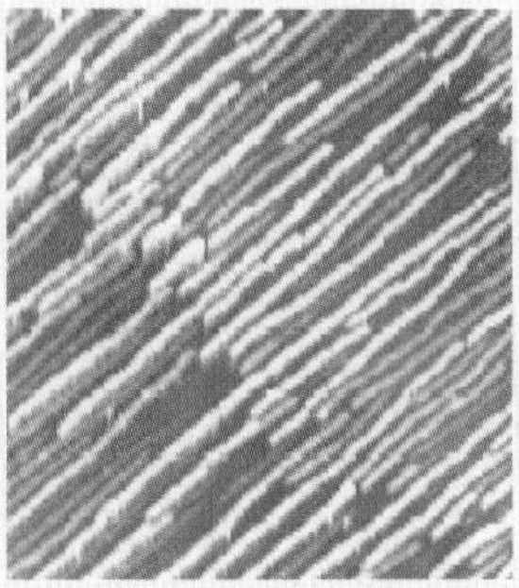

Fig. 2.15 Germanium nanowires fabricated on to a silicon surface by molecular beam epitaxy. (Reproduced with permission from T. Ogino *et al.*, *Acc. Chem. Res.* 32, 447 (1999).)

wide, and 10–600 nm long. Direct manipulation of atoms on a surface can lead to the formation of nanowires. The Coulomb attraction between an atom and the tip of an STM can be exploited to move atoms along a surface, arranging them into patterns, such as wires.

Example 2.3 *Exploring the origin of the current in scanning tunnelling microscopy*

To appreciate the distance dependence of the tunnelling current in STM, suppose that the wavefunction of the electron in the gap between sample and needle is given by $\psi = Be^{-\kappa x}$, where $\kappa = \{2m_e(V-E)/\hbar^2\}^{1/2}$; take $V - E = 2.0$ eV. By what factor would the current drop if the needle is moved from $L_1 = 0.50$ nm to $L_2 = 0.60$ nm from the surface?

Method We regard the tunnelling current to be proportional to the transmission probability *T*, so the ratio of the currents is equal to the ratio of the transmission probabilities. To choose between eqn 2.17a or 2.17b for the calculation of *T*, first calculate κL for the shortest distance L_1: if $\kappa L_1 > 1$, then use eqn 2.17b.

Answer When $L = L_1 = 0.50$ nm and $V - E = 2.0$ eV $= 3.20 \times 10^{-19}$ J the value of κL is

$$\kappa L_1 = \left\{\frac{2m_e(V-E)}{\hbar^2}\right\}^{1/2} L_1$$

$$= \left\{\frac{2\times(9.109\times10^{-31}\text{ kg})\times(3.20\times10^{-19}\text{ J})}{(1.055\times10^{-34}\text{ J s})^2}\right\}^{1/2} \times(5.0\times10^{-10}\text{ m})$$

$$= (7.24\times10^{9}\text{ m}^{-1})\times(5.0\times10^{-10}\text{ m}) = 3.6$$

Because $\kappa L_1 > 1$, we use eqn 2.17b to calculate the transmission probabilities at the two distances. It follows that

$$\frac{\text{current at } L_2}{\text{current at } L_1} = \frac{T(L_2)}{T(L_1)} = \frac{16\varepsilon(1-\varepsilon)e^{-2\kappa L_2}}{16\varepsilon(1-\varepsilon)e^{-2\kappa L_1}} = e^{-2\kappa(L_2-L_1)}$$

$$= e^{-2\times(7.24\times10^{9}\text{ m}^{-1})\times(1.0\times10^{-10}\text{ m})} = 0.24$$

We conclude that, at a distance of 0.60 nm between the surface and the needle, the current is 24 per cent of the value measured when the distance is 0.50 nm.

Self-test 2.5 The ability of a proton to tunnel through a barrier contributes to the rapidity of proton transfer reactions in solution and therefore to the properties of acids and bases. Estimate the relative probabilities that a proton and a deuteron ($m_d = 3.342 \times 10^{-27}$ kg) can tunnel through the same barrier of height 1.0 eV (1.6×10^{-19} J) and length 100 pm when their energy is 0.9 eV. Any comment?

[$T_H/T_D = 3.1 \times 10^2$; we expect proton transfer reactions to be much faster than deuteron transfer reactions.]

Vibrational motion

Vibrational motion is the second type of motion we consider. It is important in chemistry because atoms in molecules and solids vibrate around their mean positions as bonds stretch, compress, and bend. The detection and interpretation of vibrational frequencies is the basis of infrared spectroscopy, and we need to understand molecular vibration in order to interpret thermodynamic properties such as heat capacities. Molecular vibration also plays a role in the rates of chemical reactions, so we need this material when discussing the quantum mechanical aspects of chemical kinetics.

2.4 The energy levels

The only type of vibrational motion we need consider at this stage is 'harmonic motion' in one dimension. A particle undergoes **harmonic motion** if it experiences a restoring force proportional to its displacement:

$$F = -kx \tag{2.18}$$

where *k* is the **force constant**: the stiffer the 'spring', the greater the value of *k*. Because force is related to potential energy by $F = -dV/dx$ (see *Fundamentals Section F.6*), the force in eqn 2.18 corresponds to the particle having a potential energy

$$V(x) = \tfrac{1}{2}kx^2 \tag{2.19}$$

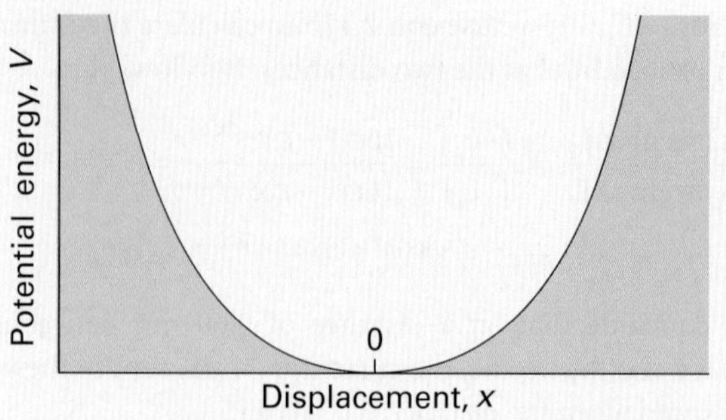

Fig. 2.16 The parabolic potential energy $V = \frac{1}{2}kx^2$ of a harmonic oscillator, where x is the displacement from equilibrium. The narrowness of the curve depends on the force constant k: the larger the value of k, the narrower the well.

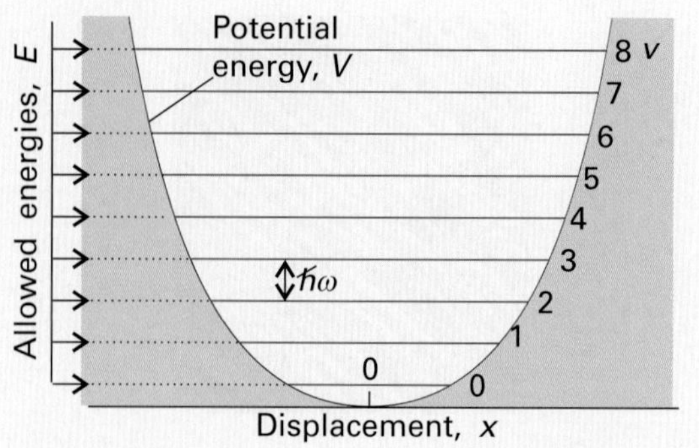

Fig. 2.17 The energy levels of a harmonic oscillator are evenly spaced with separation $\hbar\omega$, with $\omega = (k/m)^{1/2}$. Even in its lowest state, an oscillator has an energy greater than zero.

when it is displaced through a distance x. This expression, which is the equation of a parabola (Fig. 2.16), is the origin of the term 'parabolic potential energy' for the potential energy characteristic of a harmonic oscillator. The Schrödinger equation for the particle is therefore

$$-\frac{\hbar^2}{2m}\frac{d^2\psi}{dx^2} + \frac{1}{2}kx^2\psi = E\psi \qquad (2.20)$$

We can anticipate that the energy of an oscillator will be quantized because the wavefunction has to satisfy boundary conditions: it will not be found with very large extensions because its potential energy rises to infinity there. That is, when we impose the boundary conditions $\psi = 0$ at $x = \pm\infty$, we can expect to find that only certain wavefunctions and their corresponding energies are possible.

Equation 2.20 is a standard equation in the theory of differential equations and its solutions are well known to mathematicians.[1] The permitted energy levels are

$$E_v = (v + \tfrac{1}{2})\hbar\omega \qquad \omega = \left(\frac{k}{m}\right)^{1/2} \qquad v = 0, 1, 2, \ldots \qquad (2.21)$$

where v is another example of a quantum number. Note that ω (omega) increases with increasing force constant and decreasing mass. It follows that the separation between adjacent levels is

$$E_{v+1} - E_v = \hbar\omega \qquad (2.22)$$

which is the same for all v. Therefore, the energy levels form a uniform ladder of spacing $\hbar\omega$ (Fig. 2.17). The energy separation $\hbar\omega$ is negligibly small for macroscopic objects (with large mass), but is of great importance for objects with mass similar to that of atoms. As an example, which we describe in the illustration below, for a diatomic molecule X—H, where X is a heavy atom treated as a stationary anchor and the light hydrogen atom moving and vibrating as a harmonic oscillator, we can substitute m_H for m.

Because the smallest permitted value of v is 0, it follows from eqn 2.21 that a harmonic oscillator has a zero-point energy

$$E_0 = \tfrac{1}{2}\hbar\omega \qquad (2.23)$$

The mathematical reason for the zero-point energy is that v cannot take negative values, for if it did the wavefunction would be ill-behaved. The physical reason is the same as for the particle in a box: the particle is confined, its position is not completely uncertain, and therefore its momentum, and hence its kinetic energy, cannot be exactly zero. We can picture this zero-point state as one in which the particle fluctuates incessantly around its equilibrium position; classical mechanics would allow the particle to be perfectly still.

● A BRIEF ILLUSTRATION

Atoms vibrate relative to one another in molecules with the bond acting like a spring. Consider an X—H chemical bond, where a heavy X atom forms a stationary anchor for the very light H atom. That is, only the H atom moves, vibrating as a harmonic oscillator. Therefore, eqn 2.21 describes the allowed vibrational energy levels of an X—H bond. The force constant of a typical X—H chemical bond is around 500 N m^{-1}. For example k = 516.3 N m^{-1} for the $^{1}H^{35}Cl$ bond. Because the mass of a proton is about 1.7×10^{-27} kg, using k = 500 N m^{-1} in eqn 2.21 gives $\omega \approx 5.4 \times 10^{14}$ s^{-1} (540 THz). It follows from eqn 2.22 that the separation of adjacent levels is $\hbar\omega \approx 5.7 \times 10^{-20}$ J (57 zJ, about 0.36 eV). This energy separation corresponds to 34 kJ mol^{-1}, which is chemically significant. From eqn 2.23, the zero-point energy of this molecular oscillator is about 28 zJ, which corresponds to 0.18 eV, or 17 kJ mol^{-1}.

The excitation of the vibration of the bond from one level to the level immediately above requires 57 zJ. Therefore, if it is caused by a photon, the excitation requires radiation of

[1] For details, see our *Molecular quantum mechanics*, Oxford University Press, Oxford (2005).

frequency $\nu = \Delta E/h = 86$ THz and wavelength $\lambda = c/\nu = 3.5\ \mu$m. It follows that transitions between adjacent vibrational energy levels of molecules are stimulated by or emit infrared radiation (see Chapter 10). ●

2.5 The wavefunctions

It is helpful at the outset to identify the similarities between the harmonic oscillator and the particle in a box, for then we shall be able to anticipate the form of the oscillator wavefunctions without detailed calculation. Like the particle in a box, a particle undergoing harmonic motion is trapped in a symmetrical well in which the potential energy rises to large values (and ultimately to infinity) for sufficiently large displacements (compare Figs. 2.1 and 2.16). However, there are two important differences. First, because the potential energy climbs towards infinity only as x^2 and not abruptly, the wavefunction approaches zero more slowly at large displacements than for the particle in a box. Second, as the kinetic energy of the oscillator depends on the displacement in a more complex way (on account of the variation of the potential energy), the curvature of the wavefunction also varies in a more complex way.

(a) The form of the wavefunctions

The detailed solution of eqn 2.20 shows that the wavefunctions for a harmonic oscillator have the form

$$\psi(x) = N \times (\text{polynomial in } x) \times (\text{bell-shaped Gaussian function})$$

where N is a normalization constant. A Gaussian function is a bell-shaped function of the form e^{-x^2} (Fig. 2.18). The precise form of the wavefunctions is

$$\psi_v(x) = N_v H_v(y) e^{-y^2/2} \qquad y = \frac{x}{\alpha} \qquad \alpha = \left(\frac{\hbar^2}{mk}\right)^{1/4} \tag{2.24}$$

The factor $H_v(y)$ is a **Hermite polynomial**; their form and some of their properties are listed in Table 2.1.

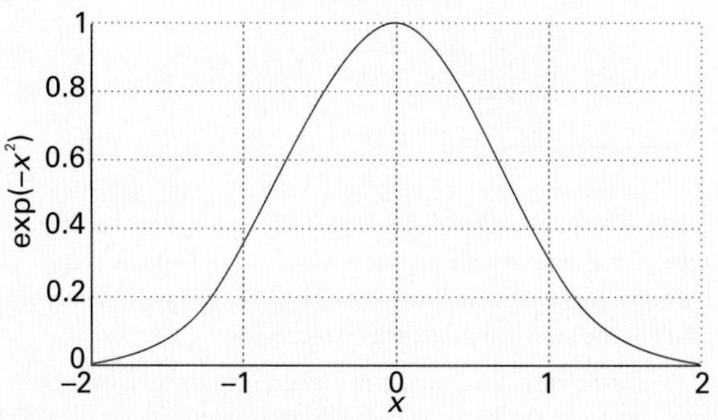

Fig. 2.18 The graph of the Gaussian function, $f(x) = e^{-x^2}$.

Table 2.1 The Hermite polynomials $H_v(y)$

v	$H_v(y)$
0	1
1	$2y$
2	$4y^2 - 2$
3	$8y^3 - 12y$
4	$16y^4 - 48y^2 + 12$
5	$32y^5 - 160y^3 + 120y$
6	$64y^6 - 480y^4 + 720y^2 - 120$

The Hermite polynomials are solutions of the differential equation

$$H_v'' - 2yH_v' + 2vH_v = 0$$

where primes denote differentiation. They satisfy the recursion relations

$$H_{v+1} - 2yH_v + 2vH_{v-1} = 0 \qquad \frac{dH_v}{dy} = 2vyH_{v-1}$$

An important integral is

$$\int_{-\infty}^{\infty} H_{v'}H_v e^{-y^2} dy = \begin{cases} 0 & \text{if } v' \neq v \\ \pi^{1/2}2^v v! & \text{if } v' = v \end{cases}$$

A brief comment Hermite polynomials are members of a class of functions called *orthogonal polynomials*. These polynomials have a wide range of important properties that allow a number of quantum mechanical calculations to be done with relative ease.

Because $H_0(y) = 1$, the wavefunction for the ground state (the lowest energy state, with $v = 0$) of the harmonic oscillator is

$$\psi_0(x) = N_0 e^{-y^2/2} = N_0 e^{-x^2/2\alpha^2} \tag{2.25a}$$

and the corresponding probability density is

$$\psi_0^2(x) = N_0^2 e^{-y^2} = N_0^2 e^{-x^2/\alpha^2} \tag{2.25b}$$

The wavefunction and the probability density are shown in Fig. 2.19. Both curves have their largest values at zero displacement (at $x = 0$), so they capture the classical picture of the zero-point energy as arising from the ceaseless fluctuation of the particle about its equilibrium position.

The wavefunction for the first excited state of the oscillator, the state with $v = 1$, is obtained by noting that $H_1(y) = 2y$ (note that some of the Hermite polynomials are very simple functions!):

$$\psi_1(x) = N_1 2ye^{-y^2/2} = \left(\frac{2N_1}{\alpha}\right) xe^{-x^2/2\alpha^2} \tag{2.26}$$

This function has a node at zero displacement ($x = 0$), and the probability density has maxima at $x = \pm\alpha$ (Fig. 2.20). The shapes of several wavefunctions are shown in Fig. 2.21. The shading

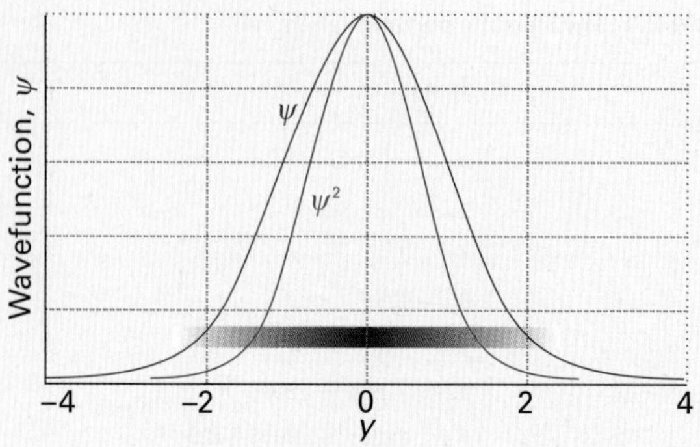

Fig. 2.19 The normalized wavefunction and probability density (shown also by shading) for the lowest energy state of a harmonic oscillator.

in Figs 2.19 and 2.20 that represents the probability density is based on the squares of these functions. At high quantum numbers, harmonic oscillator wavefunctions have their largest amplitudes near the turning points of the classical motion (the locations at which $V = E$, so the kinetic energy is zero). We see classical properties emerging in the correspondence limit of high quantum numbers, for a classical particle is most likely to be found at the turning points (where it travels most slowly) and is least likely to be found at zero displacement (where it travels most rapidly).

We shall pause frequently throughout the text to interpret various mathematical expressions. In the case of the harmonic oscillator wavefunctions in eqn 2.24, we should note the following:

• The Gaussian function goes very strongly to zero as the displacement increases (in either direction), so all the wavefunctions approach zero at large displacements.

• The exponent y^2 is proportional to $x^2 \times (mk)^{1/2}$, so the wavefunctions decay more rapidly for large masses and stiff springs.

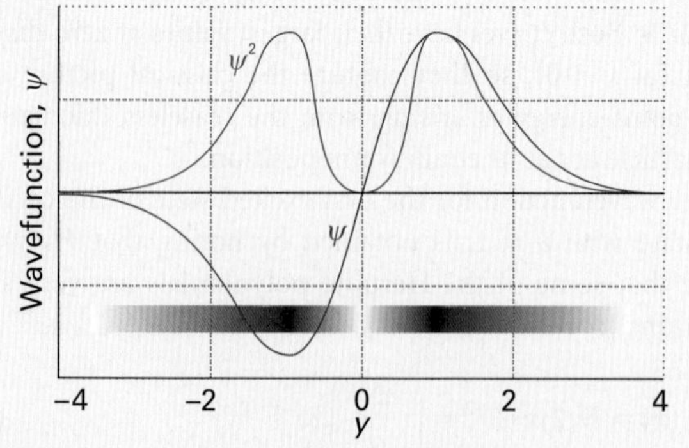

Fig. 2.20 The normalized wavefunction and probability density (shown also by shading) for the first excited state of a harmonic oscillator.

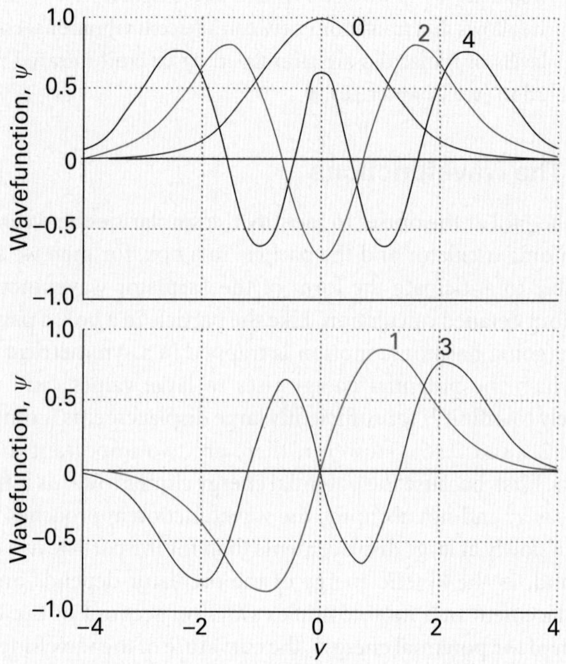

Fig. 2.21 The normalized wavefunctions for the first five states of a harmonic oscillator. Note that the number of nodes is equal to v and that alternate wavefunctions are symmetrical or antisymmetrical about $y = 0$ (zero displacement).

• As v increases, the Hermite polynomials become larger at large displacements (as x^v), so the wavefunctions grow large before the Gaussian function damps them down to zero: as a result, the wavefunctions spread over a wider range as v increases (Fig. 2.22).

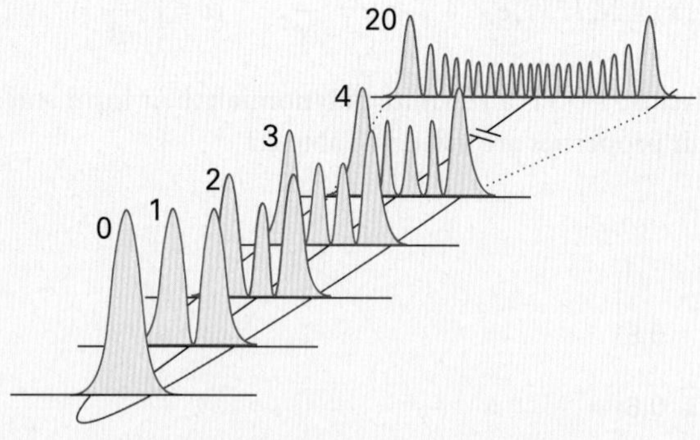

Fig. 2.22 The probability densities for the first five states of a harmonic oscillator and the state with $v = 20$. Note how the regions of highest probability density move towards the turning points of the classical motion as v increases.

interActivity To gain some insight into the origins of the nodes in the harmonic oscillator wavefunctions, plot the Hermite polynomials $H_v(y)$ for $v = 0$ through 5.

Example 2.4 *Normalizing a harmonic oscillator wavefunction*

Find the normalization constant for the harmonic oscillator wavefunctions.

Method Normalization is carried out by evaluating the integral of $|\psi|^2$ over all space and then finding the normalization factor from eqn 1.4. The normalized wavefunction is then equal to $N\psi$. In this one-dimensional problem, the volume element is dx and the integration is from $-\infty$ to $+\infty$. The wavefunctions are expressed in terms of the dimensionless variable $y = x/\alpha$, so begin by expressing the integral in terms of y by using $dx = \alpha dy$. The integrals required are given in Table 2.1.

Answer The unnormalized wavefunction is

$$\psi_v(x) = H_v(y)e^{-y^2/2}$$

It follows from the integrals given in Table 2.1 that

$$\int_{-\infty}^{\infty} \psi_v{}^*\psi_v dx = \alpha\int_{-\infty}^{\infty} \psi_v{}^*\psi_v dy = \alpha\int_{-\infty}^{\infty} H_v^2(y)e^{-y^2} dy$$
$$= \alpha\pi^{1/2}2^v v!$$

where $v! = v(v-1)(v-2)\ldots 1$. Therefore,

$$N_v = \left(\frac{1}{\alpha\pi^{1/2}2^v v!}\right)^{1/2}$$

Note that for a harmonic oscillator N_v is different for each value of v.

Self-test 2.6 Confirm, by explicit evaluation of the integral, that ψ_0 and ψ_1 are orthogonal.

[Evaluate the integral $\int_{-\infty}^{\infty}\psi_0{}^*\psi_1\,dx$ by using the information in Table 2.1]

Example 2.5 *Locating the nodes of a harmonic oscillator*

Consider the X—H chemical bond of the *illustration* on p. 62 where the X atom acts as a stationary anchor for the H atom. If the H atom is undergoing harmonic motion with $v = 2$, determine the X—H bond distances at which there is zero probability density of finding the proton. Take the mass of the proton to be 1.7×10^{-27} kg, the force constant to be 500 N m^{-1}, and the X—H equilibrium bond length to be 1.20×10^{-10} m.

Method The distances at which there is zero probability density of finding the H atom are the nodes of the $v = 2$ harmonic oscillator wavefunction. Since the Gaussian function does not pass through zero, the nodes will be those values of x at which the Hermite polynomial passes though zero.

Answer Because $H_2(y) = 4y^2 - 2$, we need to solve $4y^2 - 2 = 0$ which has solutions at $y = \pm 1/2^{1/2}$. The nodes are therefore at $x = \pm\alpha/2^{1/2}$. From eqn 2.24 and the values of m and k given, we find $x = \pm 7.6$ pm. Because x is the displacement from equilibrium, these displacements correspond to X—H bond distances of 112 pm and 128 pm. For higher values of v, it is best and often necessary to use numerical methods (for example, a root-extraction procedure of a mathematics package) to locate zeroes.

Self-test 2.7 Suppose the molecule is vibrationally excited to the state $v = 3$. Where will the proton not be found?

[$x = 0, \pm\alpha(3/2)^{1/2}$; 120 pm, 107 pm, 133 pm]

(b) The properties of oscillators

We saw in Section 1.8 that the average value of a property is calculated by evaluating the expectation value of the corresponding operator (eqn 1.15). Now that we know the wavefunctions of the harmonic oscillator, we can start to explore its properties by evaluating integrals of the type

$$\langle\Omega\rangle = \int_{-\infty}^{\infty} \psi_v{}^*\hat{\Omega}\psi_v dx \tag{2.27}$$

(Here and henceforth, the wavefunctions are all taken to be normalized to 1.) When the explicit wavefunctions are substituted, the integrals look fearsome, but the Hermite polynomials have many simplifying features. For instance, we show in the following example that the mean displacement, $\langle x\rangle$, and the mean square displacement, $\langle x^2\rangle$, of the oscillator when it is in the state with quantum number v are

$$\langle x\rangle = 0 \qquad \langle x^2\rangle = (v + \tfrac{1}{2})\frac{\hbar}{(mk)^{1/2}} \tag{2.28}$$

The result for $\langle x\rangle$ shows that the oscillator is equally likely to be found on either side of $x = 0$ (like a classical oscillator). The result for $\langle x^2\rangle$ shows that the mean square displacement increases with v. This increase is apparent from the probability densities in Fig. 2.22, and corresponds to the classical amplitude of swing increasing as the oscillator becomes more highly excited.

Example 2.6 *Calculating properties of a harmonic oscillator*

Consider the harmonic oscillator motion of the X—H molecule encountered in *Example* 2.5. Calculate the mean displacement of the oscillator when it is in a state with quantum number v.

Method Normalized wavefunctions must be used to calculate the expectation value. The operator for position along x is multiplication by the value of x (Section 1.6). The resulting integral can be evaluated either by inspection (the integrand is the product of an odd and an even function), or by explicit evaluation using the formulas in Table 2.1. To give practice in this type of calculation, we illustrate the latter procedure. We shall need the relation $x = \alpha y$, which implies that $dx = \alpha dy$.

A brief comment An even function is one for which $f(-x) = f(x)$; an odd function is one for which $f(-x) = -f(x)$. The product of an odd and even function is itself odd, and the integral of an odd function over a symmetrical range about $x = 0$ is zero.

Answer The integral we require is

$$\langle x\rangle = \int_{-\infty}^{\infty} \psi_v{}^* x \psi_v \mathrm{d}x = N_v^2 \int_{-\infty}^{\infty} (H_v \mathrm{e}^{-y^2/2}) x (H_v \mathrm{e}^{-y^2/2}) \mathrm{d}x$$

$$= \alpha^2 N_v^2 \int_{-\infty}^{\infty} (H_v \mathrm{e}^{-y^2/2}) y (H_v \mathrm{e}^{-y^2/2}) \mathrm{d}y$$

$$= \alpha^2 N_v^2 \int_{-\infty}^{\infty} H_v y H_v \mathrm{e}^{-y^2} \mathrm{d}y$$

Now use the recursion relation (Table 2.1) to form

$$yH_v = vH_{v-1} + \tfrac{1}{2}H_{v+1}$$

which turns the integral into

$$\int_{-\infty}^{\infty} H_v y H_v \mathrm{e}^{-y^2} \mathrm{d}y = v \int_{-\infty}^{\infty} H_{v-1} H_v \mathrm{e}^{-y^2} \mathrm{d}y + \tfrac{1}{2} \int_{-\infty}^{\infty} H_{v+1} H_v \mathrm{e}^{-y^2} \mathrm{d}y$$

Both integrals are zero, so $\langle x\rangle = 0$. As remarked in the text, the mean displacement is zero because the displacement occurs equally on either side of the equilibrium position. The following Self-test extends this calculation by examining the mean square displacement, which we can expect to be nonzero and to increase with increasing v.

Self-test 2.8 Calculate the mean square displacement $\langle x^2\rangle$ of the H atom (attached to the stationary X atom) from its equilibrium position. (Use the recursion relation twice.)

[eqn 2.28]

The mean potential energy of an oscillator, the expectation value of $V = \tfrac{1}{2}kx^2$, can now be calculated very easily:

$$\langle V\rangle = \langle \tfrac{1}{2}kx^2\rangle = \tfrac{1}{2}(v + \tfrac{1}{2})\hbar\left(\frac{k}{m}\right)^{1/2} = \tfrac{1}{2}(v + \tfrac{1}{2})\hbar\omega \qquad (2.29a)$$

Because the total energy in the state with quantum number v is $(v + \tfrac{1}{2})\hbar\omega$, it follows that

$$\langle V\rangle = \tfrac{1}{2}E_v \qquad (2.29b)$$

The total energy is the sum of the potential and kinetic energies, so it follows at once that the mean kinetic energy of the oscillator is (as could also be shown using the kinetic energy operator)

$$\langle E_k\rangle = \tfrac{1}{2}E_v \qquad (2.29c)$$

The result that the mean potential and kinetic energies of a harmonic oscillator are equal (and therefore that both are equal to half the total energy) is a special case of the **virial theorem**:

> If the potential energy of a particle has the form $V = ax^b$, then its mean potential and kinetic energies are related by
>
> $$2\langle E_k\rangle = b\langle V\rangle \qquad (2.30)$$

For a harmonic oscillator $b = 2$, so $\langle E_k\rangle = \langle V\rangle$, as we have found. The virial theorem is a short cut to the establishment of a number of useful results, and we shall use it again.

An oscillator may be found at extensions with $V > E$ that are forbidden by classical physics, for they correspond to negative kinetic energy. For example, it follows from the shape of the wavefunction (see the *Justification* below) that in its lowest energy state there is about an 8 per cent chance of finding an oscillator stretched beyond its classical limit and an 8 per cent chance of finding it with a classically forbidden compression. These tunnelling probabilities are independent of the force constant and mass of the oscillator. The probability of being found in classically forbidden regions decreases quickly with increasing v, and vanishes entirely as v approaches infinity, as we would expect from the correspondence principle. Macroscopic oscillators (such as pendulums) are in states with very high quantum numbers, so the probability that they will be found in a classically forbidden region is wholly negligible. Molecules, however, are normally in their vibrational ground states, and for them the probability is very significant.

Justification 2.2 *Tunnelling in the quantum mechanical harmonic oscillator*

According to classical mechanics, the turning point, x_{tp}, of an oscillator occurs when its kinetic energy is zero, which is when its potential energy $\tfrac{1}{2}kx^2$ is equal to its total energy E. This equality occurs when

$$x_{tp}^2 = \frac{2E}{k} \quad \text{or} \quad x_{tp} = \pm\left(\frac{2E}{k}\right)^{1/2}$$

with E given by eqn 2.21. The probability of finding the oscillator stretched beyond a displacement x_{tp} is the sum of the probabilities $\psi^2 dx$ of finding it in any of the intervals dx lying between x_{tp} and infinity:

$$P = \int_{x_{tp}}^{\infty} \psi_v^2 dx$$

The variable of integration is best expressed in terms of $y = x/\alpha$ with $\alpha = (\hbar^2/mk)^{1/4}$, and then the turning point on the right lies at

$$y_{tp} = \frac{x_{tp}}{\alpha} = \left\{\frac{2(v+\frac{1}{2})\hbar\omega}{\alpha^2 k}\right\}^{1/2} = (2v+1)^{1/2}$$

For the state of lowest energy ($v = 0$), $y_{tp} = 1$ and the probability is

$$P = \int_{x_{tp}}^{\infty} \psi_0^2 dx = \alpha N_0^2 \int_1^{\infty} e^{-y^2} dy$$

The integral is a special case of the *error function*, erf z, which is defined as follows:

$$\text{erf}\, z = 1 - \frac{2}{\pi^{1/2}} \int_z^{\infty} e^{-y^2} dy$$

The values of this function are tabulated and available in mathematical software packages, and a small selection of values is given in Table 2.2. In the present case (see Example 2.4 for N_0 and use $0! = 1$)

$$P = \tfrac{1}{2}(1 - \text{erf}\, 1) = \tfrac{1}{2}(1 - 0.843) = 0.079$$

It follows that in 7.9 per cent of a large number of observations, any oscillator in the state $v = 0$ will be found stretched to a classically forbidden extent. There is the same probability of finding the oscillator with a classically forbidden compression. The total probability of finding the oscillator tunnelled into a classically forbidden region (stretched or compressed) is about 16 per cent. A similar calculation for the state with $v = 6$ shows that the probability of finding the oscillator outside the classical turning points has fallen to about 7 per cent.

Table 2.2 The error function

z	erf z
0	0
0.01	0.0113
0.05	0.0564
0.10	0.1125
0.50	0.5205
1.00	0.8427
1.50	0.9661
2.00	0.9953

Techniques of approximation

All the applications treated so far have had exact solutions and we shall encounter in the following two chapters several more examples where the Schrödinger equation can be solved exactly. However, many problems—and almost all the problems of interest in chemistry—do not have exact solutions. For example, if we were modelling a nanometre-sized, electrically conducting metallic particle by a square-well potential (a 'quantum well'), it might be more realistic to allow for a small variation in potential energy across the material rather than supposing that $V = 0$ everywhere. Also, a molecular vibration is not exactly harmonic because a bond is not a perfect spring: at large extensions the bond breaks and at high compressions the potential energy rises much more rapidly than parabolically. To make progress with these problems we need to develop techniques of approximation.

2.6 An overview of approximation techniques

There are three major approaches to finding approximate solutions. The first is to try to guess the shape and the mathematical form of the wavefunction. *Variation theory* provides a criterion of success with such an approach and, as it is most commonly encountered in the context of molecular orbital theory, we consider it there (Chapter 5). The second approach, the *self-consistent field procedure*, is useful for trying to find solutions of the Schrödinger equation for systems of many particles. This iterative method will be described in Chapter 6.

The third approach takes the hamiltonian operator for the problem that cannot be solved exactly and separates it into two pieces: one piece represents a model hamiltonian for which the Schrödinger equation can be solved exactly and the other piece (which is the difference between the true and model hamiltonians) represents a 'perturbation'. *Perturbation theory*, which provides the mathematical tools for solving complex problems by this approach, comes in two flavours depending on whether or not the perturbation varies with time. We consider *time-independent perturbation theory* below and postpone a discussion of *time-dependent perturbation theory* until Chapter 4 where we consider the response of atoms and molecules to time-dependent electromagnetic fields.

2.7 Time-independent perturbation theory

In **perturbation theory**, we suppose that the hamiltonian for the problem we are trying to solve, $\hat{H}$, can be expressed as the sum of a simple hamiltonian, $\hat{H}^{(0)}$, which has known eigenvalues and eigenfunctions, and a contribution, $\hat{H}^{(1)}$, which represents the extent to which the true hamiltonian differs from the 'model' hamiltonian:

$$\hat{H} = \hat{H}^{(0)} + \hat{H}^{(1)} \tag{2.31}$$

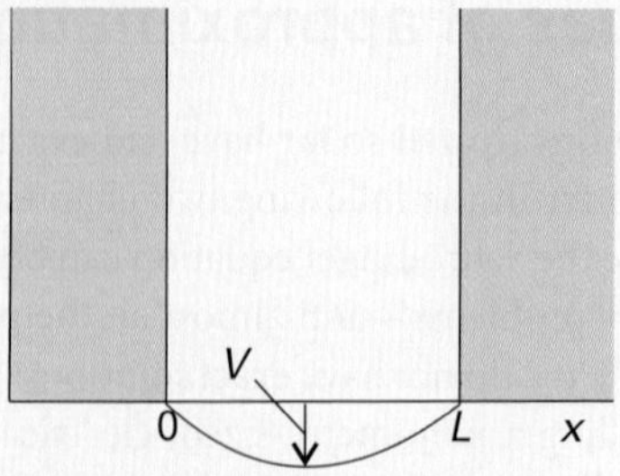

Fig. 2.23 The potential energy for a particle in a box with a potential that varies as V across the floor of the box. We can expect the particle to accumulate more in the centre of the box (in the ground state at least) than in the unperturbed box.

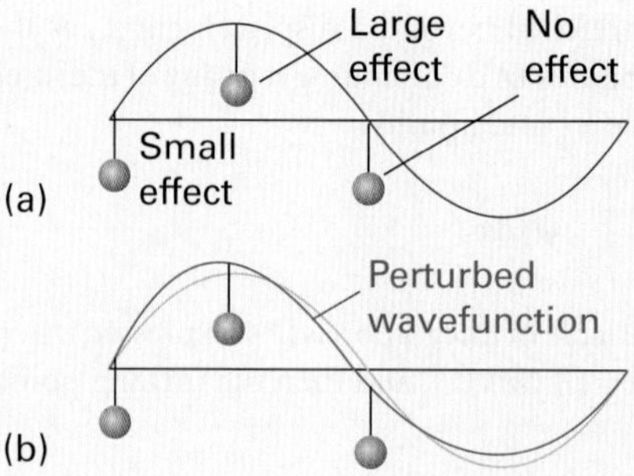

Fig. 2.24 (a) The first-order energy is an average of the perturbation (represented by the hanging weights) over the unperturbed wavefunction. (b) The second-order energy is a similar average, but over the distortion induced by the perturbation.

In **time-independent perturbation theory**, the perturbation is always present and unvarying. For example, if we were modelling a metallic nanoparticle, then $\hat{H}^{(0)}$ would be the hamiltonian for the particle in a box and the perturbation might represent a dip in the potential energy of a particle in a box in some region along the length of the box.

In time-independent perturbation theory, we suppose that the true energy of the system differs from the energy of the simple system, and that we can write

$$E = E^{(0)} + E^{(1)} + E^{(2)} + \cdots \tag{2.32}$$

where $E^{(1)}$ is the 'first-order correction' to the energy, a contribution proportional to $\hat{H}^{(1)}$, and $E^{(2)}$ is the 'second-order correction' to the energy, a contribution proportional to $\hat{H}^{(1)2}$, and so on. For instance, if the potential energy of an electron trapped in a metallic nanoparticle modelled as a square well is lowered by an amount V when it is near the centre of the nanoparticle (Fig. 2.23), then the first-order correction would be proportional to V, the second-order correction would be proportional to V^2, and so on. The true wavefunction also differs from the 'simple' wavefunction, and we write

$$\psi = \psi^{(0)} + \psi^{(1)} + \psi^{(2)} + \cdots \tag{2.33}$$

In practice, we need to consider only the 'first-order correction' to the wavefunction, $\psi^{(1)}$, the contribution to the wavefunction that depends on $\hat{H}^{(1)}$. In our metallic-nanoparticle model, we would say that the distortion of the unperturbed (particle-in-a-box) wavefunction was proportional to the depth of the central part of the well.

In most applications it is sufficient to know only the corrections to the energy. We show in *Further information 2.1* that the first-order correction to the energy of the ground state (with the wavefunction $\psi_0^{(0)}$) is

$$E^{(1)} = \int \psi_0^{(0)*}\hat{H}^{(1)}\psi_0^{(0)}\mathrm{d}\tau \tag{2.34}$$

The integral will be recognized as an expectation value: in this case it is the expectation value of the perturbation calculated using the unperturbed ground-state wavefunction. We can therefore interpret $E^{(1)}$ as the average value of the effect of the perturbation. An analogy is the shift in energy of vibration of a violin string when small weights are hung along its length. The weights hanging close to the nodes have little effect on its energy of vibration. Those hanging at the antinodes (locations of maximum amplitude), however, have a pronounced effect (Fig. 2.24a). The overall effect is the average of all the weights.

Example 2.7 *Using perturbation theory*

In this example we model the effect of a varying potential in a one-dimensional metallic nanoparticle by supposing that the potential energy of an electron trapped in a box of length L is not zero but $V(x) = -\varepsilon \sin(\pi x/L)$ (see Fig. 2.23). Evaluate the first-order correction to the energy of the ground state.

Method Identify the first-order perturbation hamiltonian and evaluate $E^{(1)}$ from eqn 2.34. We can expect a small lowering of the energy because the average potential energy of the particle is lower in the distorted box.

Solution The perturbation hamiltonian is $\hat{H}^{(1)} = -\varepsilon \sin(\pi x/L)$ and the unperturbed ground-state ($n = 1$) wavefunction in the notation of Section 2.2 is $\psi_1^{(0)}(x) = (2/L)^{1/2}\sin(\pi x/L)$. Therefore, the first-order correction to the energy is

$$E^{(1)} = \int_0^L \psi_1 \hat{H}^{(1)}\psi_1 \mathrm{d}x = -\frac{2\varepsilon}{L}\overbrace{\int_0^L \sin^3\frac{\pi x}{L}\mathrm{d}x}^{4L/3\pi} = -\frac{8\varepsilon}{3\pi}$$

Note that the energy is lowered by the perturbation, as would be expected for the shape shown in Fig. 2.23.

Self-test 2.9 Evaluate the first-order correction to the energy of the ground state if, in the same model, $V(x) = -\varepsilon \sin^2(\pi x/L)$.

[$E^{(1)} = -3\varepsilon/4$]

The second-order correction to the energy of the unperturbed ground state is rather more complicated, but has a straightforward interpretation:

$$E^{(2)} = \sum_{n\neq 0} \frac{|\int \psi_n^{(0)*} \hat{H}^{(1)} \psi_0^{(0)} \mathrm{d}\tau|^2}{E_0^{(0)} - E_n^{(0)}} \tag{2.35}$$

The sum is over all states of the system other than the ground state (the state for which we are calculating the correction to the energy). This correction also represents a similar average of the perturbation, but now it is an average taken over the *perturbed* wavefunctions. In terms of the violin analogy, the average is now taken over the distorted waveform of the vibrating string, in which the nodes and antinodes are slightly shifted (Fig. 2.24b).

We shall not need to carry out an explicit calculation of $E^{(2)}$ in this text, but it will be the basis of our discussion of the electric and magnetic properties of molecules. It will then be important to have in mind the following three features:

- Because $E_n^{(0)} > E_0^{(0)}$, all the terms in the denominator of eqn 2.35 are negative, and because the numerators are all positive, $E^{(2)}$ is negative. That is, the second-order energy correction, but not necessarily the first-order correction, always *lowers* the energy of the ground state.
- The perturbation appears (as its square) in the numerator; so the stronger the perturbation, the greater the lowering of the ground-state energy.
- If the energy levels of the system are widely spaced, all the denominators are large, so the sum is likely to be small. In this case the perturbation has little effect on the energy of the system: the system is 'stiff', and unresponsive to perturbations. The opposite is true when the energy levels lie close together.

Checklist of key ideas

☐ 1. The wavefunction of a free particle is $\psi_k = Ae^{ikx} + Be^{-ikx}$, $E_k = k^2\hbar^2/2m$.

☐ 2. The wavefunctions and energies of a particle in a one-dimensional box of length L are, respectively, $\psi_n(x) = (2/L)^{1/2}\sin(n\pi x/L)$ and $E_n = n^2h^2/8mL^2$, $n = 1,2,\ldots$. The zero-point energy, the lowest possible energy, is $E_1 = h^2/8mL^2$.

☐ 3. The correspondence principle states that classical mechanics emerges from quantum mechanics as high quantum numbers are reached.

☐ 4. The functions ψ_n and $\psi_{n'}$ are orthogonal if $\int \psi_n^* \psi_{n'} \mathrm{d}\tau = 0$; all wavefunctions corresponding to different energies of a system are orthogonal.

☐ 5. Tunnelling is the penetration into or through classically forbidden regions. The transmission probability is given by eqn 2.17a.

☐ 6. Harmonic motion is the motion in the presence of a restoring force proportional to the displacement, $F = -kx$, where k is the force constant. As a consequence, $V = \frac{1}{2}kx^2$.

☐ 7. The wavefunctions and energy levels of a quantum mechanical harmonic oscillator are given by eqns 2.24 and 2.21, respectively.

☐ 8. The virial theorem states that, if the potential energy of a particle has the form $V = ax^b$, then its mean potential and kinetic energies are related by $2\langle E_k\rangle = b\langle V\rangle$.

☐ 9. Perturbation theory is a technique that supplies approximate solutions to the Schrödinger equation and in which the hamiltonian for the problem is expressed as a sum of simpler hamiltonians.

☐ 10. In time-independent perturbation theory, the perturbation is always present and unvarying. The first- and second-order corrections to the energy are given by eqns 2.34 and 2.35, respectively.

Further information

Further information 2.1 *Time-independent perturbation theory*

To develop the expressions for the corrections to the wavefunction and energy of a system subjected to a time-independent perturbation, we write

$$\psi = \psi^{(0)} + \lambda\psi^{(1)} + \lambda^2\psi^{(2)} + \cdots$$

where λ is a dummy variable that will help us keep track of the order of the correction. At the end of the calculation, we discard it. Likewise, we write

$$\hat{H} = \hat{H}^{(0)} + \lambda\hat{H}^{(1)}$$

and

$$E = E^{(0)} + \lambda E^{(1)} + \lambda^2 E^{(2)} + \cdots$$

When these expressions are inserted into the Schrödinger equation, $\hat{H}\psi = E\psi$, we obtain

$$(\hat{H}^{(0)} + \lambda\hat{H}^{(1)})(\psi^{(0)} + \lambda\psi^{(1)} + \lambda^2\psi^{(2)} + \cdots)$$
$$= (E^{(0)} + \lambda E^{(1)} + \lambda^2 E^{(2)} + \cdots)(\psi^{(0)} + \lambda\psi^{(1)} + \lambda^2\psi^{(2)} + \cdots)$$

which we can rewrite as

$$\hat{H}^{(0)}\psi^{(0)} + \lambda(\hat{H}^{(1)}\psi^{(0)} + \hat{H}^{(0)}\psi^{(1)}) + \lambda^2(\hat{H}^{(0)}\psi^{(2)} + \hat{H}^{(1)}\psi^{(1)}) + \cdots$$
$$= E^{(0)}\psi^{(0)} + \lambda(E^{(0)}\psi^{(1)} + E^{(1)}\psi^{(0)}) + \lambda^2(E^{(2)}\psi^{(0)} + E^{(1)}\psi^{(1)} + E^{(0)}\psi^{(2)}) + \cdots$$

By comparing powers of λ, we find

Terms in λ^0: $\hat{H}^{(0)}\psi^{(0)} = E^{(0)}\psi^{(0)}$
Terms in λ: $\hat{H}^{(1)}\psi^{(0)} + \hat{H}^{(0)}\psi^{(1)} = E^{(0)}\psi^{(1)} + E^{(1)}\psi^{(0)}$
Terms in λ^2: $\hat{H}^{(0)}\psi^{(2)} + \hat{H}^{(1)}\psi^{(1)} = E^{(2)}\psi^{(0)} + E^{(1)}\psi^{(1)} + E^{(0)}\psi^{(2)}$

and so on. At this point λ has served its purpose, and can now be discarded.

The equations we have derived are applicable to any state of the system. From now on we shall consider only the ground state ψ_0 with energy E_0. The first equation, which we now write as

$$\hat{H}^{(0)}\psi_0^{(0)} = E_0^{(0)}\psi_0^{(0)}$$

is the Schrödinger equation for the ground state of the unperturbed system, which we assume we can solve (for instance, it might be the equation for the ground state of the particle in a box, with the solutions given in eqn 2.6). To solve the next equation, which is now written as

$$\hat{H}^{(1)}\psi_0^{(0)} + \hat{H}^{(0)}\psi_0^{(1)} = E_0^{(0)}\psi^{(1)} + E_0^{(1)}\psi_0^{(0)}$$

we suppose that the first-order correction to the wavefunction can be expressed as a linear combination of the wavefunctions of the unperturbed system, and write

$$\psi_0^{(1)} = \sum_n c_n\psi_n^{(0)} \tag{2.36}$$

Substitution of this expression gives

$$\hat{H}^{(1)}\psi_0^{(0)} + \sum_n c_n\hat{H}^{(0)}\psi_n^{(0)} = \sum_n c_n E_0^{(0)}\psi_n^{(0)} + E_0^{(1)}\psi_0^{(0)}$$

We can isolate the term in $E_0^{(1)}$ by making use of the fact that the $\psi_n^{(0)}$ form a complete orthogonal and normalized set in the sense that

$$\int \psi_0^{(0)*}\psi_n^{(0)}\mathrm{d}\tau = 0 \text{ if } n \neq 0, \text{ but } 1 \text{ if } n = 0$$

Therefore, when we multiply through by $\psi_0^{(0)*}$ and integrate over all space, we get

$$\int \psi_0^{(0)*}\hat{H}^{(1)}\psi_0^{(0)}\mathrm{d}\tau + \sum_n c_n \overbrace{\int \psi_0^{(0)*}\hat{H}^{(0)}\psi_n^{(0)}\mathrm{d}\tau}^{E_0^{(0)} \text{ if } n=0,\ 0 \text{ otherwise}}$$
$$= \sum_n c_n E_0^{(0)} \overbrace{\int \psi_0^{(0)*}\psi_n^{(0)}\mathrm{d}\tau}^{1 \text{ if } n=0,\ 0 \text{ otherwise}} + E_0^{(1)} \overbrace{\int \psi_0^{(0)*}\psi_0^{(0)}\mathrm{d}\tau}^{1}$$

That is,

$$\int \psi_0^{(0)*}\hat{H}^{(1)}\psi_0^{(0)}\mathrm{d}\tau = E_0^{(1)}$$

which is eqn 2.34.

To find the coefficients c_n, we multiply the same expression through by $\psi_k^{(0)*}$, where now $k \neq 0$, and integrate over all space, which gives

$$\int \psi_k^{(0)*}\hat{H}^{(1)}\psi_0^{(0)}\mathrm{d}\tau + \sum_n c_n \overbrace{\int \psi_k^{(0)*}\hat{H}^{(0)}\psi_n^{(0)}\mathrm{d}\tau}^{E_k^{(0)} \text{ if } n=k,\ 0 \text{ otherwise}}$$
$$= \sum_n c_n E_0^{(0)} \overbrace{\int \psi_k^{(0)*}\psi_n^{(0)}\mathrm{d}\tau}^{1 \text{ if } n=k,\ 0 \text{ otherwise}} + E_0^{(1)} \overbrace{\int \psi_k^{(0)*}\psi_0^{(0)}\mathrm{d}\tau}^{0}$$

Since both summations above reduce to a single term,

$$\int \psi_k^{(0)*}\hat{H}^{(1)}\psi_0^{(0)}\mathrm{d}\tau + c_k E_k^{(0)} = c_k E_0^{(0)}$$

which we can rearrange into

$$c_k = -\frac{\int \psi_k^{(0)*}\hat{H}^{(1)}\psi_0^{(0)}\mathrm{d}\tau}{E_k^{(0)} - E_0^{(0)}} = -\frac{H_{k0}^{(1)}}{E_k^{(0)} - E_0^{(0)}} \tag{2.37}$$

The second-order energy is obtained starting from the second-order expression, which for the ground state is

$$\hat{H}^{(0)}\psi_0^{(2)} + \hat{H}^{(1)}\psi_0^{(1)} = E_0^{(2)}\psi_0^{(0)} + E_0^{(1)}\psi_0^{(1)} + E_0^{(0)}\psi_0^{(2)}$$

To isolate the term $E_0^{(2)}$ we multiply both sides by $\psi_0^{(0)*}$, integrate over all space, and obtain

$$\overbrace{\int \psi_0^{(0)*}\hat{H}^{(0)}\psi_0^{(2)}\mathrm{d}\tau}^{E_0^{(0)}\int \psi_0^{(0)*}\psi_0^{(2)}\mathrm{d}\tau} + \int \psi_0^{(0)*}\hat{H}^{(1)}\psi_0^{(1)}\mathrm{d}\tau$$
$$= E_0^{(2)} \underbrace{\int \psi_0^{(0)*}\psi_0^{(0)}\mathrm{d}\tau}_{1} + E_0^{(1)}\int \psi_0^{(0)*}\psi_0^{(1)}\mathrm{d}\tau + E_0^{(0)}\int \psi_0^{(0)*}\psi_0^{(2)}\mathrm{d}\tau$$

In the first term, we have used the hermiticity of $\hat{H}^{(0)}$. The first and last terms cancel, and we are left with

$$E_0^{(2)} = \int \psi_0^{(0)*}\hat{H}^{(1)}\psi_0^{(1)}\mathrm{d}\tau - E_0^{(1)}\int \psi_0^{(0)*}\psi_0^{(1)}\mathrm{d}\tau$$

We have already found the first-order corrections to the energy and the wavefunction, so this expression could be regarded as an explicit expression for the second-order energy. However, we can go one step further by substituting eqn 2.36:

$$E_0^{(2)} = \sum_n c_n \overbrace{\int \psi_0^{(0)*}\hat{H}^{(1)}\psi_n^{(0)}\mathrm{d}\tau}^{H_{0n}^{(1)}} - \sum_n c_n E_0^{(1)} \overbrace{\int \psi_0^{(0)*}\psi_n^{(0)}\mathrm{d}\tau}^{1 \text{ if } n=0,\ 0 \text{ otherwise}}$$
$$= \sum_n c_n H_{0n}^{(1)} - c_0 E_0^{(1)}$$

where we use the notation $H_{ij} = \int \psi_i{}^*\hat{H}\psi_j\,\mathrm{d}\tau$. The final term cancels the term $c_0 H_{00}^{(1)}$ in the sum, and we are left with

$$E_0^{(2)} = \sum_{n\neq 0} c_n H_{0n}^{(1)}$$

Substitution of the expression for c_n in eqn 2.37 now produces the final result, eqn 2.35.

Discussion questions

2.1 Discuss the physical origin of quantization of energy for a particle confined to motion inside a one-dimensional box.

2.2 Discuss the correspondence principle and provide two examples.

2.3 Define, justify, and provide examples of zero-point energy.

2.4 Discuss the physical origins of quantum mechanical tunnelling. Identify chemical systems where tunnelling might play a role.

2.5 Describe the features that stem from nanometre-scale dimensions that are not found in macroscopic objects.

2.6 Describe three approximation techniques used in quantum mechanics and explain why they are useful.

2.7 Explain why the particle in a box and the harmonic oscillator are useful models for quantum mechanical systems: what chemically significant systems can they be used to represent?

Exercises

2.1(a) Determine the linear momentum and kinetic energy of a free electron described by the wavefunction e^{ikx} with $k = 3\ m^{-1}$.

2.1(b) Determine the linear momentum and kinetic energy of a free proton described by the wavefunction e^{-ikx} with $k = 5\ m^{-1}$.

2.2(a) Write the wavefunctions for an electron travelling to the right ($x > 0$) after being accelerated from rest through a potential difference of (a) 1.0 V, (b) 10 kV.

2.2(b) Write the wavefunction for a particle of mass 1.0 g travelling to the right at $10\ m\ s^{-1}$.

2.3(a) Calculate the energy separations in joules, kilojoules per mole, wavenumbers, and electronvolts between the levels (a) $n = 2$ and $n = 1$, (b) $n = 6$ and $n = 5$ of an electron in a one-dimensional nanoparticle modelled by a box of length 1.0 nm.

2.3(b) Calculate the energy separations in joules, kilojoules per mole, wavenumbers, and electronvolts between the levels (a) $n = 3$ and $n = 1$, (b) $n = 7$ and $n = 6$ of an electron in a one-dimensional nanoparticle modelled by a box of length 1.5 nm.

2.4(a) A conjugated polyene can be modelled by a particle in a one-dimensional box. Calculate the probability that an electron will be found between $0.49L$ and $0.51L$ in a box of length L when it has (a) $n = 1$, (b) $n = 2$. Take the wavefunction to be a constant in this narrow range.

2.4(b) A conjugated polyene can be modelled by a particle in a one-dimensional box. Calculate the probability that a particle will be found between $0.65L$ and $0.67L$ in a box of length L when it has (a) $n = 1$, (b) $n = 2$. Take the wavefunction to be a constant in this narrow range.

2.5(a) Calculate the expectation values of $\hat{p}$ and $\hat{p}^2$ for a particle in the state $n = 1$ in a square-well potential used to model a one-dimensional nanoparticle.

2.5(b) Calculate the expectation values of $\hat{p}$ and $\hat{p}^2$ for a particle in the state $n = 2$ in a square-well potential used to model a one-dimensional nanoparticle.

2.6(a) An electron is squeezed between two confining walls, one of which can be moved inwards. At what separation of the walls will the zero-point energy of the electron be equal to its rest mass energy, m_ec^2? Express your answer in terms of the parameter $\lambda_C = h/m_ec$, the 'Compton wavelength' of the electron.

2.6(b) Now replace the electron in Exercise 2.6a by a proton. At what separation of the walls will the zero-point energy of the proton be equal to its rest mass energy, m_pc^2?

2.7(a) What are the most likely locations of a particle in a box of length L in the state $n = 3$?

2.7(b) What are the most likely locations of a particle in a box of length L in the state $n = 4$?

2.8(a) Suppose that the junction between two semiconductors can be represented by a barrier of height 2.0 eV and length 100 pm. Calculate the (transmission) probability that an electron with energy 1.5 eV can tunnel through the barrier.

2.8(b) Suppose that a proton of an acidic hydrogen atom is confined to an acid that can be represented by a barrier of height 2.0 eV and length 100 pm. Calculate the (transmission) probability that a proton with energy 1.5 eV can escape from the acid.

2.9(a) Calculate the zero-point energy of a harmonic oscillator consisting of a proton attached to a metal surface by a bond of force constant $155\ N\ m^{-1}$.

2.9(b) Calculate the zero-point energy of a harmonic oscillator consisting of a rigid CO molecule adsorbed to a metal surface by a bond of force constant $285\ N\ m^{-1}$.

2.10(a) For a harmonic oscillator of effective mass 1.33×10^{-25} kg, the difference in adjacent energy levels is 4.82 zJ. Calculate the force constant of the oscillator.

2.10(b) For a harmonic oscillator of effective mass 2.88×10^{-25} kg, the difference in adjacent energy levels is 3.17 zJ. Calculate the force constant of the oscillator.

2.11(a) Suppose a hydrogen atom is adsorbed on the surface of a gold nanoparticle by a bond of force constant $855\ N\ m^{-1}$. Calculate the wavelength of a photon needed to excite a transition between its neighbouring vibrational energy levels.

2.11(b) Suppose an oxygen atom ($m = 15.9994m_u$) is adsorbed on the surface of a nickel nanoparticle by a bond of force constant $544\ N\ m^{-1}$. Calculate the wavelength of a photon needed to excite a transition between its neighbouring vibrational energy levels.

2.12(a) Refer to Exercise 2.11a and calculate the wavelength that would result from replacing hydrogen by deuterium.

2.12(b) Refer to Exercise 2.11b and calculate the wavelength that would result from replacing the oxygen atom by a rigid dioxygen molecule.

2.13(a) Confirm that the wavefunction for the ground state of a one-dimensional linear harmonic oscillator given in eqn 2.25a is a solution of the Schrödinger equation for the oscillator and that its energy is $\frac{1}{2}\hbar\omega$.

2.13(b) Confirm that the wavefunction for the first excited state of a one-dimensional linear harmonic oscillator given in equation 2.26 is a solution of the Schrödinger equation for the oscillator and that its energy is $\frac{3}{2}\hbar\omega$.

2.14(a) Locate the nodes of the harmonic oscillator wavefunction with $v = 4$.

2.14(b) Locate the nodes of the harmonic oscillator wavefunction with $v = 5$.

2.15(a) Calculate the normalization constant for an oscillator with $v = 2$ and confirm that its wavefunction is orthogonal to the wavefunction for the state $v = 4$.

2.15(b) Calculate the normalization constant for an oscillator with $v = 3$ and confirm that its wavefunction is orthogonal to the wavefunction for the state $v = 1$.

2.16(a) Assuming that the vibrations of a $^{35}Cl_2$ molecule are equivalent to those of a harmonic oscillator with a force constant $k = 329$ N m^{-1}, what is the wavenumber of the radiation needed to excite the molecule vibrationally? The mass of a ^{35}Cl atom is $34.9688m_u$; the mass to use in the expression for the vibrational frequency of a diatomic molecule is the 'effective mass' $\mu = m_A m_B/(m_A + m_B)$, where m_A and m_B are the masses of the individual atoms.

2.16(b) Assuming that the vibrations of a $^{14}N_2$ molecule are equivalent to those of a harmonic oscillator with a force constant $k = 2293.8$ N m^{-1}, what is the wavenumber of the radiation needed to excite the molecule vibrationally? The mass of a ^{14}N atom is $14.0031m_u$; see Exercise 2.16a.

2.17(a) Calculate the probability that an O—H bond treated as a harmonic oscillator will be found at a classically forbidden extension when $v = 1$.

2.17(b) Calculate the probability that an O—H bond treated as a harmonic oscillator will be found at a classically forbidden extension when $v = 2$.

2.18(a) What is the relation between the mean kinetic and potential energies for a particle if the potential is proportional to x^3?

2.18(b) What is the relation between mean kinetic and potential energies of an electron in a hydrogen atom?

2.19(a) Calculate the first-order correction to the energy of an electron in a one-dimensional nanoparticle modelled as a particle in a box when the perturbation is $V(x) = -\varepsilon \sin(2\pi x/L)$ and $n = 1$.

2.19(b) Calculate the first-order correction to the energy of an electron in a one-dimensional nanoparticle modelled as a particle in a box when the perturbation is $V(x) = -\varepsilon \sin(3\pi x/L)$ and $n = 1$.

2.20(a) Calculate the first-order correction to the energy of a particle in a box when the perturbation is $V(x) = -\varepsilon \cos(2\pi x/L)$ and $n = 1$.

2.20(b) Calculate the first-order correction to the energy of a particle in a box when the perturbation is $V(x) = -\varepsilon \cos(3\pi x/L)$ and $n = 1$.

2.21(a) Suppose that the 'floor' of a one-dimensional box slopes up from $x = 0$ to ε at $x = L$. Calculate the first-order effect on the energy of the state $n = 1$.

2.21(b) Suppose that the 'floor' of a one-dimensional box slopes up from $x = 0$ to ε at $x = L$. Calculate the first-order effect on the energy of the state $n = 2$.

2.22(a) Does the vibrational frequency of an O—H bond depend on whether it is horizontal or vertical at the surface of the Earth? Suppose that a harmonic oscillator of mass m is held vertically, so that it experiences a perturbation $V(x) = mgx$, where g is the acceleration of free fall. Calculate the first-order correction to the energy of the ground state.

2.22(b) Repeat the previous exercise to find the change in excitation energy from $v = 1$ to $v = 2$ in the presence of the perturbation.

2.23(a) Evaluate the second-order correction to the energy of the harmonic oscillator for the perturbation described in Exercise 2.22a. *Hint.* You will find that there is only one term that contributes to the sum in eqn 2.35.

2.23(b) Evaluate the second-order correction to the energy of a particle in a one-dimensional square well for the perturbation described in Exercise 2.20(a). *Hint.* The only term that contributes to the sum in eqn 2.35 in $n = 3$.

Problems*

Numerical problems

2.1 Calculate the separation between the two lowest translational energy levels of an O_2 molecule in a one-dimensional container of length 5.0 cm. At what value of n does the energy of the molecule reach $\frac{1}{2}kT$ at 300 K, and what is the separation of this level from the one immediately below?

2.2 The state of an electron in a one-dimensional cavity of length 1.0 nm in a semiconductor is described by the normalized wavefunction $\psi(x) = \frac{1}{2}\psi_1(x) + (1/2\text{i})\psi_2(x) - (\frac{1}{2})^{1/2}\psi_4(x)$, where $\psi_n(x)$ is given by eqn 2.6b. When the energy of the electron is measured, what is the outcome? What is the expectation value of the energy?

2.3 The mass to use in the expression for the vibrational frequency of a diatomic molecule is the effective mass $\mu = m_A m_B/(m_A + m_B)$, where m_A and m_B are the masses of the individual atoms. The following data on the infrared absorption wavenumbers ($\tilde{\nu} = 1/\lambda = \nu/c$) of molecules is taken from G. Herzberg, *Spectra of diatomic molecules*, van Nostrand (1950):

	$H^{35}Cl$	$H^{81}Br$	HI	CO	NO
$\tilde{\nu}$/cm^{-1}	2990	2650	2310	2170	1904

Calculate the force constants of the bonds and arrange them in order of increasing stiffness.

2.4 An electron confined to a metallic nanoparticle is modelled as a particle in a one-dimensional box of length L. If the electron is in the state $n = 1$, calculate the probability of finding it in the following regions: (a) $0 \le x \le \frac{1}{2}L$, (b) $0 \le x \le \frac{1}{4}L$, (c) $\frac{1}{2}L - \delta x \le x \le \frac{1}{2}L + \delta x$.

2.5 Repeat Problem 2.4 for a general value of n.

2.6 Suppose that the floor of a one-dimensional nanoparticle has an imperfection that can be represented by a small step in the potential

* Problems denoted with the symbol ‡ were supplied by Charles Trapp, Carmen Giunta, and Marshall Cady.

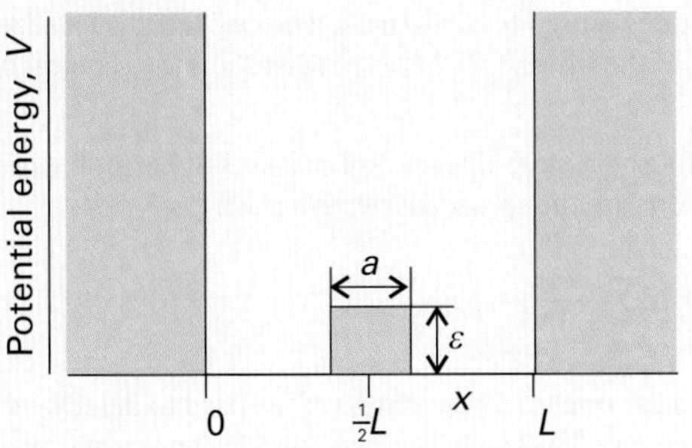

Fig. 2.25 The definition of the potential in Problem 2.6.

energy, as in Fig. 2.25. (a) Write a general expression for the first-order correction to the ground-state energy, $E_0^{(1)}$. (b) Evaluate the energy correction for $a = L/10$ (so the blip in the potential occupies the central 10 per cent of the well) with $n = 1$.

2.7 We normally think of the one-dimensional well as being horizontal. Suppose it is vertical; then the potential energy of the particle depends on *x* because of the presence of the gravitational field. Calculate the first-order correction to the zero-point energy, and evaluate it for an electron in a box on the surface of the Earth. Account for the result. *Hint.* The energy of the particle depends on its height as *mgh*, where $g = 9.81\ \text{m s}^{-2}$. Because *g* is so small, the energy correction is small; but it would be significant if the box were near a very massive star.

2.8 Calculate the second-order correction to the energy for the system described in Problem 2.7 and calculate the ground-state wavefunction. Account for the shape of the distortion caused by the perturbation. *Hint.* The following integrals are useful

$$\int x \sin ax \sin bx \, dx = -\frac{d}{da}\int \cos ax \sin bx \, dx$$

$$\int \cos ax \sin bx \, dx = \frac{\cos(a-b)x}{2(a-b)} - \frac{\cos(a+b)x}{2(a+b)} + \text{constant}$$

2.9 The vibrations of molecules are only approximately harmonic because the energy of a bond is not exactly parabolic. Calculate the first-order correction to the energy of the ground state of a harmonic oscillator subjected to an anharmonic potential of the form $ax^3 + bx^4$, where *a* and *b* are small (anharmonicity) constants. Consider the three cases in which the anharmonic perturbation is present (a) during bond expansion ($x \geq 0$) and compression ($x \leq 0$), (b) during expansion only, (c) during compression only.

2.10 To make progress with this problem you may wish to review concepts of chemical kinetics introduced in introductory chemistry. The *kinetic isotope effect* is the decrease in the rate constant of a chemical reaction upon replacement of one atom in a reactant by a heavier isotope. The effect arises from the change in activation energy that accompanies the replacement of an atom by a heavier isotope on account of changes in the zero-point vibrational energies. Consider a reaction in which a C—H bond is cleaved. If scission of this bond is the rate-determining step, then the reaction coordinate corresponds to the stretching of the C—H bond and the potential energy profile is shown in Fig. 2.26. On deuteration, the dominant change is the reduction of the zero-point energy of the bond (because the deuterium atom is heavier). The whole reaction profile is not lowered, however, because the relevant vibration in the activated complex has a very low force constant, so there is little zero-point energy associated with the reaction coordinate in either form of the activated complex. (a) Assume that the change in the activation energy E_a arises only from the change in zero-point energy of the stretching vibration and show that

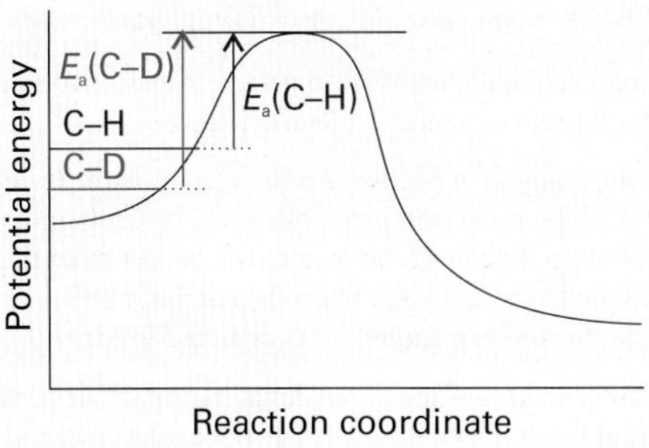

Fig. 2.26 Changes in the reaction profile when a C—H bond undergoing cleavage is deuterated. In this figure the C—H and C—D bonds are modelled as simple harmonic oscillators. The only significant change is in the zero-point energy of the reactants, which is lower for C—D than for C—H. As a result, the activation energy is greater for C—D cleavage than for C—H cleavage.

$$E_a(\text{C—D}) - E_a(\text{C—H}) = \tfrac{1}{2}N_A hc\tilde{\nu}(\text{C—H})\left\{1 - \left(\frac{\mu_{\text{CH}}}{\mu_{\text{CD}}}\right)^{1/2}\right\}$$

where $\tilde{\nu}$ is the relevant vibrational wavenumber and μ is the relevant effective mass. (b) Now consider the effect of deuteration on the rate constant, *k*, of the reaction. (i) Starting with the Arrhenius equation, $k = Ae^{-E_a/kT}$, and assuming that the pre-exponential factor *A* does not change upon deuteration, show that the rate constants for the two species should be in the ratio

$$\frac{k(\text{C—D})}{k(\text{C—H})} = e^{-\lambda} \qquad \text{with } \lambda = \frac{hc\tilde{\nu}(\text{C—H})}{2kT}\left\{1 - \left(\frac{\mu_{\text{CH}}}{\mu_{\text{CD}}}\right)^{1/2}\right\}$$

(ii) Does $k(\text{C—D})/k(\text{C—H})$ increase or decrease with decreasing temperature? (c) From infrared spectroscopy, the fundamental vibrational wavenumber for stretching of a C—H bond is about $3000\ \text{cm}^{-1}$. Predict the value of the ratio $k(\text{C—D})/k(\text{C—H})$ at 298 K. (d) In some cases, substitution of deuterium for hydrogen results in values of $k(\text{C—D})/k(\text{C—H})$ that are too low to be accounted for by the model described above. Explain this effect.

Theoretical problems

2.11 The wavefunction for a free particle e^{ikx} is not square-integrable and therefore cannot be normalized in a box of infinite length. However, to circumvent this problem, we suppose that the particle is in a region of finite length *L*, normalize the wavefunction, and then allow *L* to become infinite at the end of the calculations that use the wavefunction. Find the normalization constant for the wavefunction e^{ikx}, assuming that a free particle is in a region of length *L*.

2.12 Consider two different particles moving in one dimension *x*, one (particle 1) described by the (unnormalized) wavefunction $\psi_1(x) = e^{i(x/m)}$ and the second (particle 2) described by the (unnormalized) wavefunction $\psi_2(x) = \tfrac{1}{2}(e^{2i(x/m)} + e^{3i(x/m)} + e^{-2i(x/m)} + e^{-3i(x/m)})$. If the positions of the particles were measured, which would be found to be more localized in space (that is, which has a position known more precisely)? Explain your answer with a diagram.

2.13 Derive eqn 2.17a, the expression for the transmission probability and show that it reduces to eqn 2.17b when $\kappa L \gg 1$.

2.14 Repeat the analysis of Section 2.3 to determine the transmission coefficient T and the reflection probability, R, the probability that a particle incident on the left of the barrier will reflect from the barrier and be found moving to the left away from the barrier, for $E > V$. Suggest a physical reason for the variation of T as depicted in Fig. 2.9.

2.15 An electron inside a one-dimensional nanoparticle has a potential energy different from its potential energy once it has escaped through the confining barrier. Consider a particle moving in one dimension with $V = 0$ for $-\infty < x \le 0$, $V = V_2$ for $0 < x \le L$, and $V = V_3$ for $L \le x < \infty$ and incident from the left. The energy of the particle lies in the range $V_2 > E > V_3$. (a) Calculate the transmission coefficient, T. (b) Show that the general equation for T reduces to eqn 2.17a when $V_3 = 0$.

2.16 The wavefunction inside a long barrier of height V is $\psi = Ne^{-\kappa x}$. Calculate (a) the probability that the particle is inside the barrier and (b) the average penetration depth of the particle into the barrier.

2.17 Confirm that a function of the form $e^{-\kappa x^2}$ is a solution of the Schrödinger equation for the ground state of a harmonic oscillator and find an expression for κ in terms of the mass and force constant of the oscillator.

2.18 Calculate the mean kinetic energy of a harmonic oscillator by using the relations in Table 2.1.

2.19 Calculate the values of $\langle x^3 \rangle$ and $\langle x^4 \rangle$ for a harmonic oscillator by using the relations in Table 2.1.

2.20 Determine the values of $\Delta x = (\langle x^2 \rangle - \langle x \rangle^2)^{1/2}$ and $\Delta p = (\langle p^2 \rangle - \langle p \rangle^2)^{1/2}$ for the ground state of (a) a particle in a box of length L and (b) a harmonic oscillator. Discuss these quantities with reference to the uncertainty principle.

2.21 Repeat Problem 2.20 for (a) a particle in a box and (b) an harmonic oscillator in a general quantum state (n and v, respectively).

2.22 Show for a particle in a box that Δx approaches its classical value as $n \to \infty$. *Hint.* In the classical case the distribution is uniform across the box, and so in effect $\psi(x) = 1/L^{1/2}$.

2.23 We shall see in Chapter 4 that the intensity of spectroscopic transitions between the vibrational states of a molecule are proportional to the square of the integral $\int \psi_{v'} x \psi_v \mathrm{d}x$ over all space. Use the relations between Hermite polynomials given in Table 2.1 to show that the only permitted transitions are those for which $v' = v \pm 1$ and evaluate the integral in these cases.

2.24 The potential energy of the rotation of one CH_3 group relative to its neighbour in ethane can be expressed as $V(\phi) = V_0 \cos 3\phi$, where ϕ is the angle shown in 2. (a) Show that for small displacements the motion of the group is harmonic and calculate the (molar) energy of excitation from $v = 0$ to $v = 1$. (b) What is the force constant for these small-amplitude oscillations? (c) The energy of impacts with any surrounding molecules is typically kT, where k is Boltzmann's constant. Should you expect the oscillations to be excited? (d) What do you expect to happen to the energy levels and wavefunctions as the excitation increases?

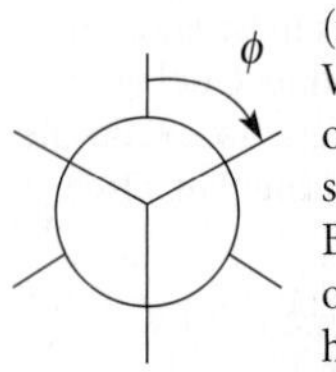

2

2.25 The motion of a pendulum can be thought of as representing the location of a wavepacket that migrates from one turning point to the other periodically. Show that, whatever superposition of harmonic oscillator states is used to construct a wavepacket, it is localized at the same place at the times 0, T, $2T$, . . . , where T is the classical period of the oscillator.

2.26 The 'most classical' linear combinations of harmonic oscillator wavefunctions are the so-called *coherent states*:

$$\psi_\alpha(x) = N \sum_{v=0}^{\infty} \frac{\alpha^v}{(v!)^{1/2}} \psi_v(x)$$

where α is a parameter. These states can be used to describe the radiation generated by lasers. (a) Show that the normalization constant is $N = e^{-|\alpha|^2/2}$. (b) Show that two coherent states ψ_α and ψ_β are not in general orthogonal. (c) Go on to show that a coherent state is the 'most classical' in the sense that the uncertainty relation for position and momentum for a particle it describes has its minimum value (that is, $\Delta x \Delta p = \frac{1}{2}\hbar$). *Hint.* Use a recursion relation in Table 2.1.

Applications: to biology and nanotechnology

2.27 When β-carotene is oxidized *in vivo*, it forms two molecules of retinal (vitamin A), a precursor to the pigment in the retina responsible for vision (*Impact I11.1*). The conjugated system of retinal consists of 11 C atoms and one O atom. In the ground state of retinal, each level up to $n = 6$ is occupied by two electrons. Assuming an average internuclear distance of 140 pm, calculate (a) the separation in energy between the ground state and the first excited state in which one electron occupies the state with $n = 7$, and (b) the frequency and wavelength of the radiation required to produce a transition between these two states. (c) Using your results, choose among the words in parentheses in the following sentence to generate a rule for the prediction of frequency shifts in the absorption spectra of linear polyenes:

> The absorption spectrum of a linear polyene shifts to (higher/lower) frequency as the number of conjugated atoms (increases/decreases).

2.28 Many biological electron transfer reactions, such as those associated with biological energy conversion, may be visualized as arising from electron tunnelling between protein-bound co-factors, such as cytochromes, quinones, flavins, and chlorophylls. This tunnelling occurs over distances that are often greater than 1.0 nm, with sections of protein separating electron donor from acceptor. For a specific combination of donor and acceptor, the rate of electron tunnelling is proportional to the transmission probability, with $\kappa \approx 7\ \mathrm{nm}^{-1}$ (eqn 2.17). By what factor does the rate of electron tunnelling between two co-factors increase as the distance between them changes from 2.0 nm to 1.0 nm?

2.29 Carbon monoxide binds strongly to the Fe^{2+} ion of the haem group of the protein myoglobin. Estimate the vibrational frequency of CO bound to myoglobin by using the data in Problem 2.3 and by making the following assumptions: the atom that binds to the haem group is immobilized, the protein is infinitely more massive than either the C or O atom, the C atom binds to the Fe^{2+} ion, and binding of CO to the protein does not alter the force constant of the C≡O bond.

2.30 Of the four assumptions made in Problem 2.29, the last two are questionable. Suppose that the first two assumptions are still reasonable and that you have at your disposal a supply of myoglobin, a suitable buffer in which to suspend the protein, $^{12}C^{16}O$, $^{13}C^{16}O$, $^{12}C^{18}O$, $^{13}C^{18}O$, and an infrared spectrometer. Assuming that isotopic substitution does not affect the force constant of the C≡O bond, describe a set of experiments that: (a) proves which atom, C or O, binds to the haem group of myoglobin, and (b) allows for the determination of the force constant of the C≡O bond for myoglobin-bound carbon monoxide.

2.31 When in Chapter 8 we come to study macromolecules, such as synthetic polymers, proteins, and nucleic acids, we shall see that one conformation is that of a random coil. For a one-dimensional random coil of N units, the restoring force at small displacements and at a temperature T is

$$F = -\frac{kT}{2l}\ln\left(\frac{N+n}{N-n}\right)$$

where l is the length of each monomer unit and nl is the distance between the ends of the chain. Show that for small extensions ($n \ll N$) the restoring force is proportional to n and therefore that the coil undergoes harmonic oscillation with force constant kT/Nl^2. Suppose that the mass to use for the vibrating chain is its total mass Nm, where m is the mass of one monomer unit, and deduce the root mean square separation of the ends of the chain due to quantum fluctuations in its vibrational ground state.

2.32 The forces measured by AFM (atomic force microscopy) arise primarily from interactions between electrons of the stylus and on the surface. To get an idea of the magnitudes of these forces, calculate the force acting between two electrons separated by 2.0 nm. To calculate the force between the electrons, use $F = -\mathrm{d}V/\mathrm{d}r$ where V is their mutual Coulombic potential energy and r is their separation.

MATHEMATICAL BACKGROUND 3

Complex numbers

We describe here general properties of complex numbers and functions, which are mathematical constructs frequently encountered in quantum mechanics.

MB3.1 Definitions

Complex numbers have the general form

$$z = x + \mathrm{i}y \qquad \text{(MB3.1)}$$

where $\mathrm{i} = (-1)^{1/2}$. The real numbers x and y are, respectively, the real and imaginary parts of z, denoted $\mathrm{Re}(z)$ and $\mathrm{Im}(z)$. When $y = 0$, $z = x$ is a real number; when $x = 0$, $z = \mathrm{i}y$ is a pure imaginary number. Two complex numbers $z_1 = x_1 + \mathrm{i}y_1$ and $z_2 = x_2 + \mathrm{i}y_2$ are equal when $x_1 = x_2$ and $y_1 = y_2$. Although the general form of the imaginary part of a complex number is written iy, a specific numerical value is typically written in the reverse order; for instance, as 3i.

The **complex conjugate** of z, denoted z^*, is formed by replacing i by –i:

$$z^* = x - \mathrm{i}y \qquad \text{(MB3.2)}$$

The product of z^* and z is denoted $|z|^2$ and is called the **square modulus** of z. From eqn MB3.1,

$$|z|^2 = (x + \mathrm{i}y)(x - \mathrm{i}y) = x^2 + y^2 \qquad \text{(MB3.3)}$$

since $\mathrm{i}^2 = -1$. The square modulus is a real number. The **absolute value** or **modulus** is itself denoted $|z|$ and is given by:

$$|z| = (z^*z)^{1/2} = (x^2 + y^2)^{1/2} \qquad \text{(MB3.4)}$$

Since $zz^* = |z|^2$ it follows that $z \times (z^*/|z|^2) = 1$, from which we can identify the (multiplicative) inverse of z (which exists for all nonzero complex numbers):

$$z^{-1} = \frac{z^*}{|z|^2} \qquad \text{(MB3.5)}$$

● A BRIEF ILLUSTRATION

Consider the complex number $z = 8 - 3\mathrm{i}$. Its square modulus is

$$|z|^2 = z^*z = (8 - 3\mathrm{i})^*(8 - 3\mathrm{i}) = (8 + 3\mathrm{i})(8 - 3\mathrm{i}) = 64 + 9 = 73$$

The modulus is therefore $|z| = 73^{1/2}$. From eqn MB3.5, the inverse of z is

$$z^{-1} = \frac{8 + 3\mathrm{i}}{73} = \tfrac{8}{73} + \tfrac{3}{73}\mathrm{i} \; ●$$

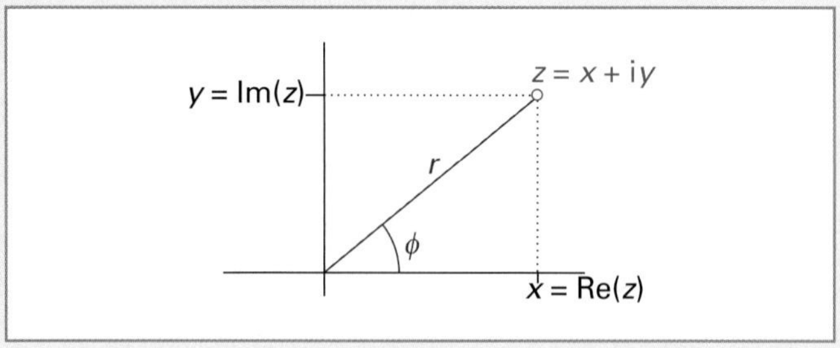

Fig. MB3.1 The representation of a complex number z as a point in the complex plane using cartesian coordinates (x, y) or polar coordinates (r, ϕ).

MB3.2 Polar representation

The complex number $z = x + \mathrm{i}y$ can be represented as a point in a plane, the **complex plane**, with $\mathrm{Re}(z)$ along the x-axis and $\mathrm{Im}(z)$ along the y-axis (Fig. MB3.1). If, as shown in the figure, r and ϕ denote the polar coordinates of the point, then since $x = r\cos\phi$ and $y = r\sin\phi$, we can express the complex number in **polar form** as

$$z = r(\cos\phi + \mathrm{i}\sin\phi) \qquad \text{(MB3.6)}$$

The angle ϕ, called the **argument** of z, is the angle that z makes with the x-axis. Because $y/x = \tan\phi$, it follows that the polar form can be constructed from

$$r = (x^2 + y^2)^{1/2} = |z| \qquad \phi = \arctan\frac{y}{x} \qquad \text{(MB3.7a)}$$

To convert from polar to Cartesian form, use

$$x = r\cos\phi \text{ and } y = r\sin\phi \text{ to form } z = x + \mathrm{i}y \qquad \text{(MB3.7b)}$$

One of the most useful relations involving complex numbers is **Euler's formula:**

$$\mathrm{e}^{\mathrm{i}\phi} = \cos\phi + \mathrm{i}\sin\phi \qquad \text{(MB3.8a)}$$

The simplest proof of this relation is to expand the exponential function as a power series and to collect real and imaginary terms. It follows that

$$\cos\phi = \tfrac{1}{2}(\mathrm{e}^{\mathrm{i}\phi} + \mathrm{e}^{-\mathrm{i}\phi}) \qquad \sin\phi = -\tfrac{1}{2}\mathrm{i}(\mathrm{e}^{\mathrm{i}\phi} - \mathrm{e}^{-\mathrm{i}\phi}) \qquad \text{(MB3.8b)}$$

The polar form in eqn MB3.6 then becomes

$$z = r\mathrm{e}^{\mathrm{i}\phi} \qquad \text{(MB3.9)}$$

● A BRIEF ILLUSTRATION

Consider the complex number $z = 8 - 3\mathrm{i}$. From the previous *illustration*, $r = |z| = 73^{1/2}$. The argument of z is

$$\phi = \arctan\frac{-3}{8} = -0.359 \text{ rad or } -20.6°$$

The polar form of the number is therefore

$$z = 73^{1/2}\mathrm{e}^{-0.359\mathrm{i}} \; ●$$

MB3.3 Operations

The following rules apply for arithmetic operations for the complex numbers $z_1 = x_1 + iy_1$ and $z_2 = x_2 + iy_2$.

1. Addition: $z_1 + z_2 = (x_1 + x_2) + i(y_1 + y_2)$ (MB3.10a)
2. Subtraction: $z_1 - z_2 = (x_1 - x_2) + i(y_1 - y_2)$ (MB3.10b)
3. Multiplication: $z_1z_2 = (x_1 + iy_1)(x_2 + iy_2)$
   $= (x_1x_2 - y_1y_2) + i(x_1y_2 + y_1x_2)$ (MB3.10c)
4. Division: We interpret z_1/z_2 as $z_1z_2^{-1}$ and use eqn MB3.5 for the inverse:

$$\frac{z_1}{z_2} = z_1z_2^{-1} = \frac{z_1z_2^{\star}}{|z_2|^2} \qquad \text{(MB3.10d)}$$

● **A BRIEF ILLUSTRATION**

Consider the complex numbers $z_1 = 6 + 2i$ and $z_2 = -4 - 3i$. Then

$$z_1 + z_2 = (6 - 4) + (2 - 3)i = 2 - i$$

$$z_1 - z_2 = 10 + 5i$$

$$z_1z_2 = \{6(-4) - 2(-3)\} + \{6(-3) + 2(-4)\}i = -18 - 26i$$

$$\frac{z_1}{z_2} = (6 + 2i)\left(\frac{-4 + 3i}{25}\right) = -\tfrac{6}{5} + \tfrac{2}{5}i \; ●$$

The polar form of a complex number is commonly used to perform arithmetical operations. For instance the product of two complex numbers in polar form is

$$z_1z_2 = (r_1e^{i\phi_1})(r_2e^{i\phi_2}) = r_1r_2e^{i(\phi_1+\phi_2)} \qquad \text{(MB3.11)}$$

This multiplication is depicted in the complex plane as shown in Fig. MB3.2. The nth power and the nth root of a complex number are

$$z^n = (re^{i\phi})^n = r^ne^{in\phi} \qquad z^{1/n} = (re^{i\phi})^{1/n} = r^{1/n}e^{i\phi/n} \qquad \text{(MB3.12)}$$

The depictions in the complex plane are shown in Fig. MB3.3.

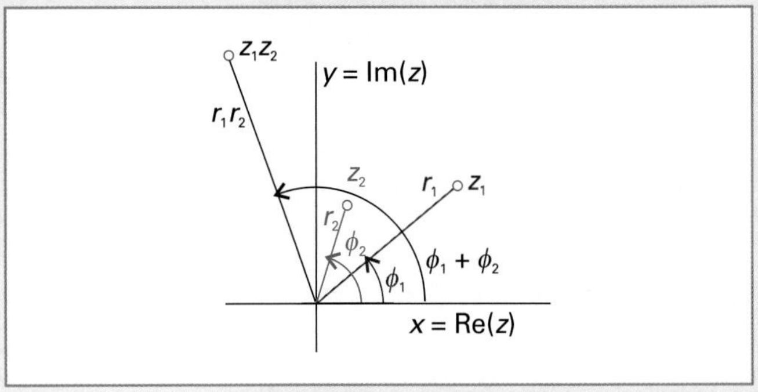

Fig. MB3.2 The multiplication of two complex numbers depicted in the complex plane.

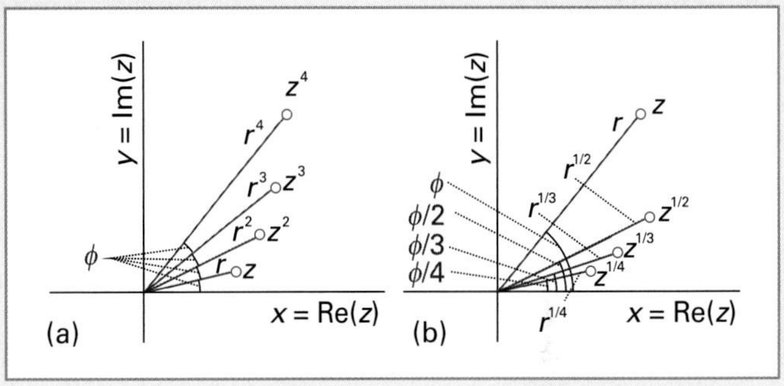

Fig. MB 3.3 (a) The nth powers and (b) the nth roots ($n = 1, 2, 3, 4$) of a complex number depicted in the complex plane.

● **A BRIEF ILLUSTRATION**

To determine the 5th root of $z = 8 - 3i$, we note that from the *illustration* in Section MB3.2 its polar form is

$$z = 73^{1/2}\, e^{-0.359i} = 8.544e^{-0.359i}$$

The 5th root is therefore

$$z^{1/5} = (8.544e^{-0.359i})^{1/5} = 8.544^{1/5}\, e^{-0.359i/5} = 1.536e^{-0.0718i}$$

It follows that $x = 1.536 \cos(-0.0718) = 1.532$ and $y = 1.536 \times \sin(-0.0718) = -0.110$ (note that we work in radians), so

$$(8 - 3i)^{1/5} = 1.532 - 0.110i \; ●$$

3 Nanosystems 2: motion in several dimensions

This chapter presents the solutions for translational and rotational motions in two and three dimensions. We use the technique of separation of variables to find acceptable wavefunctions and quantized energies for these basic types of motion. In the discussion of rotational motion, angular momentum plays an important role and introduces a property of the electron, its spin, that has no classical counterpart.

Real systems are three-dimensional. The clusters of atoms that constitute real metallic nanoparticles are three-dimensional, atoms are three-dimensional, and so are molecules. Moreover, in one dimension there is no such mode of motion as rotation, yet the rotations of molecules make an important contribution to their thermodynamic and spectroscopic properties. Clearly, we need to be able to extend the considerations so far to higher than one dimension.

That does not mean that the material in Chapter 2 is not useful. First, a number of real systems can be modelled by a one-dimensional analogue, and certain nanosystems—we have in mind electron corals, in which an electron is confined to a surface by a ring of atoms—are intermediate in dimension between 1 and 3 (that is, of dimension 2). We shall also see that the equations for three-dimensional systems can often be broken down into equations for one-dimensional systems, so we can draw on the conclusions in Chapter 2 to describe them.

Finally, some properties are not manifested in ordinary space but are intrinsic to the particle itself. In a sense, this property of 'spin' extends the number of dimensions we need to discuss to more than three if we are to understand the properties of electrons and other subatomic particles fully.

Translational motion

We presented translation in one dimension in Chapter 2 as an example of a primitive kind of motion. We saw that, if the particle was free to move without constraint, then its energy was not quantized. However, as soon as its motion is restricted to a finite region of space, its energy is quantized. The same is true in two and three dimensions, as for the molecules of a gas in a container. Our problem is to discover the allowed energy levels and wavefunctions for these more realistic systems. They will prove essential for understanding the thermodynamic properties of gases and the behaviour of electrons in metals and semiconductors.

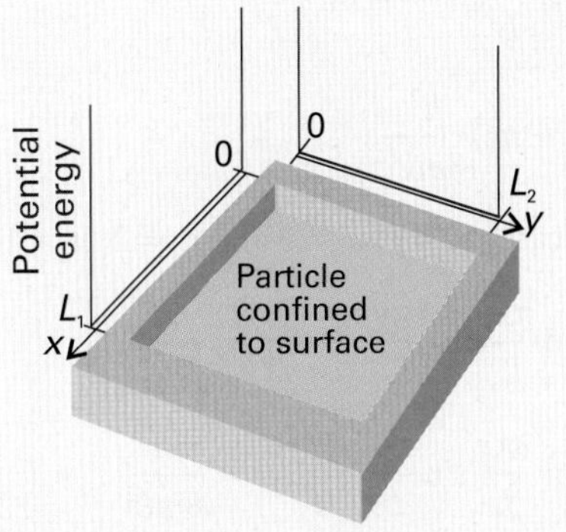

Fig. 3.1 A two-dimensional square well. The particle is confined to the plane bounded by impenetrable walls. As soon as it touches the walls, its potential energy rises to infinity.

3.1 Motion in two dimensions

We shall work up from one dimension to three in two steps: first we consider a two-dimensional system, and then generalize it to three. As we have indicated, even a two-dimensional system is now of practical significance, for it is a model of an electron confined to a small region of a surface by a ring of atoms.

To keep the discussion simple, we shall consider a rectangular two-dimensional region of a surface with length L_1 in the x-direction and L_2 in the y-direction; the potential energy is zero everywhere except at the walls, where it is infinite (Fig. 3.1). Because the particle has contributions to its kinetic energy from its motion in both the x- and y-directions, the Schrödinger equation has two kinetic energy terms, one for each axis, and for a particle of mass m in a region where the potential energy is zero the equation is

$$-\frac{\hbar^2}{2m}\left(\frac{\partial^2\psi}{\partial x^2}+\frac{\partial^2\psi}{\partial y^2}\right)=E\psi \qquad (3.1)$$

This is a *partial* differential equation (*Mathematical background 2*), a differential equation in more than one variable, and the resulting wavefunctions are functions of both x and y, denoted $\psi(x,y)$. All this means is that the wavefunction and the corresponding probability density depend on where we are in the plane, with each position specified by the coordinates x and y.

(a) Separation of variables

A partial differential equation of the form of eqn 3.1 can be simplified by the **separation of variables technique**, which divides the equation into two or more ordinary differential equations, one for each variable. To implement this approach, we explore whether a solution of eqn 3.1 can be found by writing the wavefunction as a product of functions, one depending only on x and the other only on y:

$$\psi(x,y)=X(x)Y(y)$$

With this substitution, we show in the *Justification* below that eqn 3.1 does indeed separate into two ordinary differential equations, one for each coordinate:

$$-\frac{\hbar^2}{2m}\frac{d^2X}{dx^2}=E_XX \qquad -\frac{\hbar^2}{2m}\frac{d^2Y}{dy^2}=E_YY \qquad E=E_X+E_Y \qquad (3.2)$$

The quantity E_X is the energy associated with the motion of the particle parallel to the x-axis, and likewise for E_Y and motion parallel to the y-axis.

Justification 3.1 *The separation of variables*

The first step to confirm that the Schrödinger equation is separable and the wavefunction factorizable into the product of two functions X and Y is to note that, because X is independent of y and Y is independent of x, we can write

$$\frac{\partial^2\psi}{\partial x^2}=\frac{\partial^2XY}{\partial x^2}=Y\frac{d^2X}{dx^2} \qquad \frac{\partial^2\psi}{\partial y^2}=\frac{\partial^2XY}{\partial y^2}=X\frac{d^2Y}{dy^2}$$

Then eqn 3.1 becomes

$$-\frac{\hbar^2}{2m}\left(Y\frac{d^2X}{dx^2}+X\frac{d^2Y}{dy^2}\right)=EXY$$

Next, we divide both sides by XY, and rearrange the resulting equation into

$$\frac{1}{X}\frac{d^2X}{dx^2}+\frac{1}{Y}\frac{d^2Y}{dy^2}=-\frac{2mE}{\hbar^2}$$

The first term on the left is independent of y, so if y is varied only the second term of the two on the left can change. But the sum of these two terms is a constant given by the right-hand side of the equation; therefore, if the second term did change then the right-hand side could not be constant. Consequently, even the second term cannot change when y is changed. In other words, the second term is a constant, which we write $-2mE_Y/\hbar^2$. By a similar argument, the first term is a constant when x changes, and we write it $-2mE_X/\hbar^2$, with $E=E_X+E_Y$. Therefore, we can write

$$\frac{1}{X}\frac{d^2X}{dx^2}=-\frac{2mE_X}{\hbar^2} \qquad \frac{1}{Y}\frac{d^2Y}{dy^2}=-\frac{2mE_Y}{\hbar^2}$$

which rearrange into the two ordinary (that is, single variable) differential equations in eqn 3.2.

Each of the two ordinary differential equations in eqn 3.2 is the same as the one-dimensional particle-in-a-box Schrödinger equation. The boundary conditions are also the same, apart from the detail of requiring $X(x)$ to be zero at $x=0$ and L_1, and

$Y(y)$ to be zero at $y = 0$ and L_2. We can therefore adapt the results obtained in Section 2.2 without further calculation:

$$X_{n_1}(x) = \left(\frac{2}{L_1}\right)^{1/2} \sin\frac{n_1\pi x}{L_1} \qquad Y_{n_2}(y) = \left(\frac{2}{L_2}\right)^{1/2} \sin\frac{n_2\pi y}{L_2}$$

Then, because $\psi = XY$, we obtain

$$\psi_{n_1,n_2}(x,y) = \frac{2}{(L_1L_2)^{1/2}}\sin\frac{n_1\pi x}{L_1}\sin\frac{n_2\pi y}{L_2} \tag{3.3a}$$

$$0 \le x \le L_1, 0 \le y \le L_2$$

Outside the box, the wavefunction is everywhere zero. Similarly, because $E = E_X + E_Y$, the energy of the particle is limited to the values

$$E_{n_1,n_2} = \left(\frac{n_1^2}{L_1^2} + \frac{n_2^2}{L_2^2}\right)\frac{h^2}{8m} \tag{3.3b}$$

with the two quantum numbers taking the values $n_1 = 1, 2, \ldots$ and $n_2 = 1, 2, \ldots$ independently.

Some of the wavefunctions are plotted as contours in Fig. 3.2. They are the two-dimensional versions of the wavefunctions shown in Fig. 2.4. Whereas in one dimension the wavefunctions resemble states of a vibrating string with ends fixed, in two dimensions the wavefunctions correspond to vibrations of a plate with fixed edges.

(b) Degeneracy

A special feature of the solutions arises when the box is not merely rectangular but square, with $L_1 = L_2 = L$. Then the wavefunctions and their energies become

$$\psi_{n_1,n_2}(x,y) = \frac{2}{L}\sin\frac{n_1\pi x}{L}\sin\frac{n_2\pi y}{L}$$

$$E_{n_1,n_2} = (n_1^2 + n_2^2)\frac{h^2}{8mL^2} \tag{3.4}$$

Consider the cases $n_1 = 1, n_2 = 2$ and $n_1 = 2, n_2 = 1$:

$$\psi_{1,2} = \frac{2}{L}\sin\frac{\pi x}{L}\sin\frac{2\pi y}{L} \qquad E_{1,2} = \frac{5h^2}{8mL^2}$$

$$\psi_{2,1} = \frac{2}{L}\sin\frac{2\pi x}{L}\sin\frac{\pi y}{L} \qquad E_{2,1} = \frac{5h^2}{8mL^2}$$

We see that, although the wavefunctions are different, they have the same energy. The technical term for different wavefunctions corresponding to the same energy is **degeneracy**, and we say that the state with energy $5h^2/8mL^2$ is 'doubly degenerate'.

The occurrence of degeneracy is related to the symmetry of the system. Figure 3.3 shows contour diagrams of the two degenerate functions $\psi_{1,2}$ and $\psi_{2,1}$. Because the box is square, we can convert one wavefunction into the other simply by rotating the plane by 90°. Interconversion by rotation through 90° is not possible when the plane is not square, and $\psi_{1,2}$ and $\psi_{2,1}$ are then not degenerate. We shall see many other examples of degeneracy in the pages that follow (for instance, in the hydrogen atom), and all of them can be traced to the symmetry properties of the system (see Section 7.4).

3.2 Motion in three dimensions

We are now ready to take the final step, to three dimensions. The system consists of a particle of mass m confined to a box of length L_1 in the x-direction, L_2 in the y-direction, and L_3 in the z-direction. Inside the box, the potential energy is zero and at the

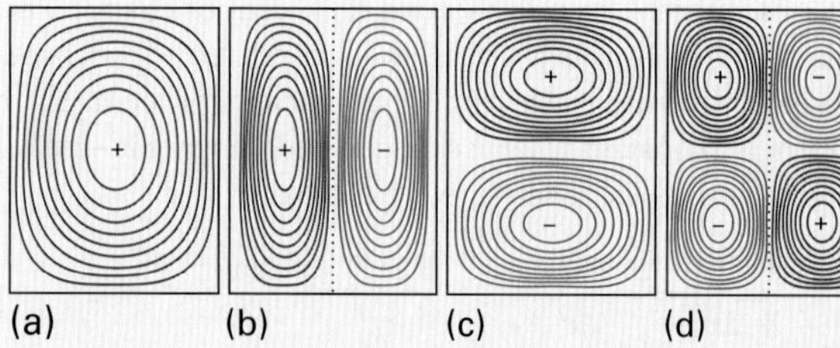

Fig. 3.2 The wavefunctions for a particle confined to a rectangular surface depicted as contours of equal amplitude. (a) $n_1 = 1, n_2 = 1$, the state of lowest energy, (b) $n_1 = 1, n_2 = 2$, (c) $n_1 = 2, n_2 = 1$, and (d) $n_1 = 2, n_2 = 2$.

interActivity Use mathematical software to generate three-dimensional plots of the functions in this illustration. Deduce a rule for the number of nodal lines in a wavefunction as a function of the values of n_1 and n_2.

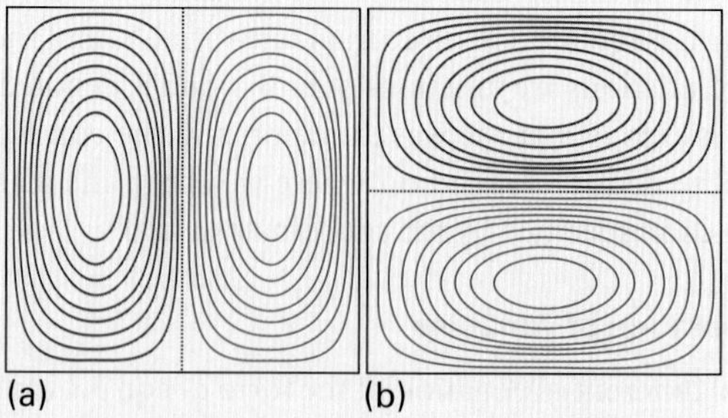

Fig. 3.3 The wavefunctions for a particle confined to a square surface. Note that one wavefunction can be converted into the other by a rotation of the box by 90°. The two functions correspond to the same energy. Degeneracy and symmetry are closely related.

walls it is infinite. This system could be a model for a quantum mechanical system of a gas in a container of macroscopic dimensions or for an electron confined to a small cavity in a solid.

The step is easy to take, for it should be obvious (and can be proved by the method of separation of variables) that the wavefunction simply has another factor:

$$\psi_{n_1,n_2,n_3}(x,y,z)$$

$$= \left(\frac{8}{L_1L_2L_3}\right)^{1/2} \sin\frac{n_1\pi x}{L_1}\sin\frac{n_2\pi y}{L_2}\sin\frac{n_3\pi z}{L_3} \qquad (3.5a)$$

Outside the box, the wavefunction is zero. Likewise, the energy has a third contribution from motion in the z-direction:

$$E_{n_1,n_2,n_3} = \left(\frac{n_1^2}{L_1^2}+\frac{n_2^2}{L_2^2}+\frac{n_3^2}{L_3^2}\right)\frac{h^2}{8m} \qquad (3.5b)$$

The quantum numbers n_1, n_2, and n_3 are all positive integers that can be varied independently. The system has a zero-point energy and, as for the two-dimensional system, there can be degeneracies.

Example 3.1 *Analysing transitions in a cubic box*

Solutions of metals in liquid ammonia are widely used as reducing agents in organic synthesis. For example, the addition of sodium to ammonia generates a solvated electron that is effectively trapped in a cavity 0.3 nm in diameter formed by ammonia molecules. The solvated electron can be modelled as a particle moving freely inside a cubic box with ammonia molecules on the surface of the cube. If the length of the box is taken to be 0.3 nm, what energy is required for the electron to undergo a transition from its lowest energy state to the state that is second-lowest in energy?

Method The quantized energies are given in eqn 3.5b. The lowest energy state corresponds to ($n_1 = 1, n_2 = 1, n_3 = 1$). The electron makes a transition to the state ($n_1 = 2, n_2 = 1, n_3 = 1$) which has the next lowest energy and is also degenerate with (1,2,1) and (1,1,2). Compute $\Delta E = E_{2,1,1} - E_{1,1,1}$.

Answer Using eqn 3.5b with $L_1 = L_2 = L_3 = L = 0.3 \times 10^{-9}$ m, we find

$$\Delta E = E_{2,1,1} - E_{1,1,1} = (2^2+1^2+1^2)\frac{h^2}{8mL^2} - (1^2+1^2+1^2)\frac{h^2}{8mL^2}$$

$$= \frac{3h^2}{8mL^2} = \frac{3\times(6.626\times10^{-34}\ \text{J s})^2}{8\times(9.109\times10^{-31}\ \text{kg})\times(0.3\times10^{-9}\ \text{m})^2}$$

$$= 2.01\times10^{-18}\ \text{J}$$

or 2.01 aJ. If this transition were caused by radiation, it would follow from the Bohr frequency condition that the transition frequency would be $\nu = \Delta E/h = 3.03 \times 10^{15}$ Hz (3.03 PHz).

Self-test 3.1 In dilute solutions of sodium in ammonia, there is an absorption of 0.20 PHz radiation by the solvated electrons which accounts for the blue colour of the solution. In the particle-in-a-box model of the solvated electron, what box length L would account for an absorption of this frequency? [$L = 1.2$ nm]

IMPACT ON NANOSCIENCE
I3.1 Quantum dots

In *Impact I2.1* we outlined some advantages of working in the nanometre regime. Another is the possibility of using quantum mechanical effects that render the properties of an assembly dependent on its size. Here we focus on the origins and consequences of these quantum mechanical effects.

Ordinary bulk metals conduct electricity because, in the presence of an electric field, electrons become mobile when they are easily excited by thermal motion. Ignoring the attraction of the electrons to the nuclei and the repulsions between them, we can treat these electrons as occupying the energy levels characteristic of a three-dimensional box. Because the box has macroscopic dimensions, we know from eqn 3.5b that the separation between neighbouring levels is so small that they form a continuum. Consequently, we are justified in neglecting quantum mechanical effects on the properties of the material. However, in a **nanocrystal**, a small cluster of atoms with dimensions in the nanometre scale, eqn 3.5b predicts that quantization of energy will be significant and will affect the properties of the sample. This quantum mechanical effect can be observed in 'boxes' of any shape. For example, you are invited to show in Problem 3.33 that the (spherically symmetrical, $l = 0$) energy levels of an electron in a spherical cavity of radius R are given by

$$E_n = \frac{n^2h^2}{8m_eR^2} \qquad (3.6)$$

The quantization of energy in nanocrystals has important technological implications when the material is a semiconductor, in which electrical conductivity increases with increasing temperature or upon excitation by light. That is, transfer of energy to a semiconductor increases the mobility of electrons in the material (see Chapter 9 for a more detailed discussion). Three-dimensional nanocrystals of semiconducting materials containing 10 to 10^5 atoms are called **quantum dots**. They can be made in solution or by depositing atoms on a surface, with the size of the nanocrystal being determined by the details of the synthesis (see, for example, *Impact I2.1*).

For every electron that moves to a different site in a semiconductor a unit of positive charge, called a *hole*, is left behind. The holes are also mobile, so a complete mathematical model of electrical conductivity in semiconductors should consider the movement of electron–hole pairs, not simply electrons. However, for the sake of gaining physical insight into the properties of quantum dots, it suffices to use eqn 3.6 as a guide. First, we see that the required energy to induce electronic transitions from lower to higher energy levels, thereby increasing the mobility of electrons and inducing electrical conductivity, depends on the size of the quantum dot. The electrical properties of large, macroscopic samples of semiconductors cannot be tuned in this way. Second, in many quantum dots, such as the nearly spherical nanocrystals of cadmium selenide (CdSe), mobile electrons can be generated by absorption of visible light. Therefore, we predict that, as the radius of the quantum dot decreases, the excitation wavelength decreases. That is, as the size of the quantum dot varies, so does the colour of the material. This phenomenon is indeed observed in suspensions of CdSe quantum dots of different sizes.

Because quantum dots are semiconductors with tunable electrical properties, there are many uses for these materials in the manufacture of transistors. But the special optical properties of quantum dots can also be exploited. Just as the generation of an electron–hole pair requires absorption of light of a specific wavelength, so does recombination of the pair result in the emission of light of a specific wavelength. This property forms the basis for the use of quantum dots in the visualization of biological cells at work. For example, a CdSe quantum dot can be modified by covalent attachment of an organic spacer to its surface. When the other end of the spacer reacts specifically with a cellular component, such as a protein, nucleic acid, or membrane, the cell becomes labelled with a light-emitting quantum dot. The spatial distribution of emission intensity and, consequently, of the labelled molecule can then be measured with a microscope. Though this technique has been used extensively with organic molecules as labels, quantum dots are more stable and are stronger light emitters.

Rotational motion

There is no such thing as rotation in one dimension, so we are about to embark on the description of a different type of system. Rotation is enormously important throughout chemistry. We have already remarked that molecular rotation has thermodynamic and spectroscopic consequences. However, even more fundamental to chemistry is the motion of electrons around nuclei in atoms, which gives rise to the concept of atomic orbitals that will be familiar from introductory chemistry and thence to the structure of the periodic table.

As in the discussion of translational motion, we shall work up in the number of dimensions. However, we have to start with two dimensions, and consider a particle circulating on a ring. Actually, that model applies much more widely than its artificiality might suggest. For instance, once we have solved that problem, we can stack together rings of different sizes to form a sphere, and use it to discuss rotation in three dimensions. Once we have got to that point, we shall have the basis for describing molecular rotation, the motion of electrons in atoms, and particles trapped in spherical cavities.

This discussion will draw on the concepts of angular momentum and moment of inertia, so it would be a good idea to review them in *Fundamentals F.6*.

3.3 Rotation in two dimensions: a particle on a ring

Just as *linear* momentum was the central concept for the discussion of linear motion, *angular* momentum is the central concept for the discussion of rotational motion. For a particle moving on a circle of radius r in the xy-plane and having a linear momentum of magnitude p at some instant, the angular momentum around the perpendicular z-axis is

$$J_z = \pm pr \tag{3.7}$$

A positive sign corresponds to clockwise motion (seen from below) and a negative sign corresponds to counter-clockwise motion. The kinetic energy of a particle of mass m and linear momentum p is $E_k = p^2/2m$; the potential energy for the particle, freely moving on the circle, is 0. Therefore, in terms of the angular momentum, the total energy $E = J_z^2/2mr^2$. However, the moment of inertia, I, of the particle around the centre of rotation is mr^2, so

$$E = \frac{J_z^2}{2I} \tag{3.8}$$

This expression applies to any body of moment of inertia I undergoing two-dimensional rotation, not just a point mass on a circle. For instance, it applies to a circular disc of mass M and radius R, with $I = \frac{1}{2}MR^2$, and to a diatomic molecule of bond length R composed of atoms of masses m_A and m_B, with the moment of inertia interpreted as $I = \mu R^2$, with $\mu = m_A m_B/(m_A + m_B)$.

(a) The qualitative origin of quantized rotation

The linear momentum that appears in eqn 3.7 can be expressed as a wavelength by using the de Broglie relation $p = h/\lambda$, which gives $J_z = \pm hr/\lambda$. Likewise, the energy in eqn 3.8 becomes $E = h^2r^2/2I\lambda^2$. Suppose for the moment that λ can take an arbitrary value. In that case, the wavefunction depends on the azimuthal angle ϕ as shown in Fig. 3.4a. When ϕ increases beyond 2π, the wavefunction continues to change, but for an arbitrary

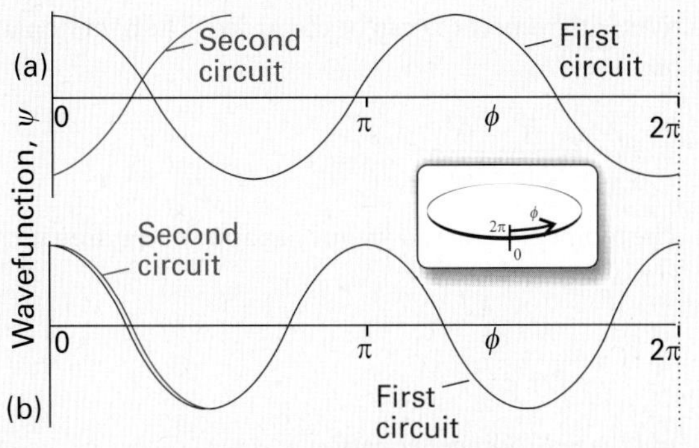

Fig. 3.4 Two solutions of the Schrödinger equation for a particle on a ring. The circumference has been opened out into a straight line; the points at $\phi = 0$ and 2π are identical. The solution in (a) is unacceptable because it is not single-valued. Moreover, on successive circuits it interferes destructively with itself, and does not survive. The solution in (b) is acceptable: it is single-valued, and on successive circuits it reproduces itself.

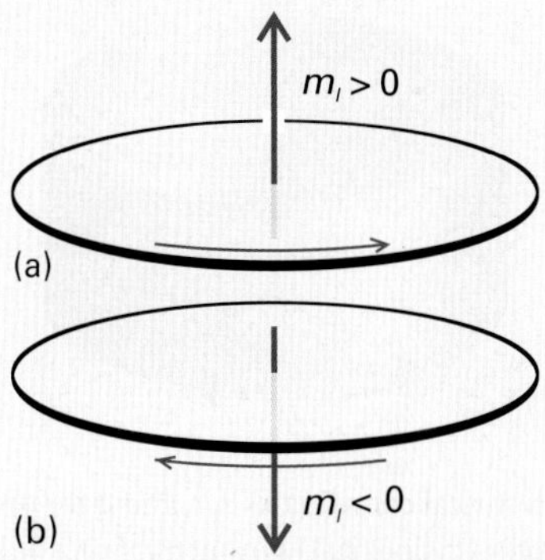

Fig. 3.5 The angular momentum of a particle confined to a plane can be represented by a vector of length $|m_l|$ units along the z-axis and with an orientation that indicates the direction of motion of the particle. The direction is given by the right-hand screw rule.

wavelength it gives rise to a different value at each point, which is unacceptable because a wavefunction must be single-valued. An acceptable solution is obtained only if the wavefunction reproduces itself on successive circuits, as in Fig. 3.4b. Because only some wavefunctions have this property, it follows that only some angular momenta are acceptable, and therefore that only certain rotational energies exist. Hence, the energy of the particle is quantized. Specifically, an integer number of wavelengths must fit the circumference of the ring (which is $2\pi r$):

$$n\lambda = 2\pi r \qquad n = 0, 1, 2, \ldots$$

The value $n = 0$ corresponds to $\lambda = \infty$; a 'wave' of infinite wavelength has a constant height at all values of ϕ. The angular momentum is therefore limited to the values

$$J_z = \pm\frac{hr}{\lambda} = \pm\frac{nhr}{2\pi r} = \frac{m_l h}{2\pi} \qquad \text{with } m_l = 0, \pm1, \pm2, \ldots$$

where we have allowed m_l (the conventional notation for this quantum number) to have positive or negative values. That is,

$$J_z = m_l\hbar \qquad m_l = 0, \pm1, \pm2, \ldots \tag{3.9}$$

Positive values of m_l correspond to rotation in a clockwise sense around the z-axis (as viewed in the direction of z, Fig. 3.5) and negative values of m_l correspond to counter-clockwise rotation around z. It then follows from eqn 3.8 that the energy is limited to the values

$$E_{m_l} = \frac{J_z^2}{2I} = \frac{m_l^2\hbar^2}{2I} \tag{3.10a}$$

We shall see shortly that the corresponding normalized wavefunctions are

$$\psi_{m_l}(\phi) = \frac{e^{im_l\phi}}{(2\pi)^{1/2}} \tag{3.10b}$$

The wavefunction with $m_l = 0$ is $\psi_0(\phi) = 1/(2\pi)^{1/2}$, and has the same value at all points on the circle and $E_{m_l} = 0$; there is no zero-point energy in this system.

We have arrived at a number of conclusions about rotational motion by combining some classical notions with the de Broglie relation. Such a procedure can be very useful for establishing the general form (and, as in this case, the exact energies) for a quantum mechanical system. However, to be sure that the correct solutions have been obtained, and to obtain practice for more complex problems where this less formal approach is inadequate, we need to solve the Schrödinger equation explicitly. The formal solution is described in the *Justification* that follows.

Justification 3.2 *The energies and wavefunctions of a particle on a ring*

The hamiltonian for a particle of mass m travelling on a circle in the xy-plane (with $V = 0$) is the same as that given in eqn 3.1:

$$\hat{H} = -\frac{\hbar^2}{2m}\left(\frac{\partial^2}{\partial x^2} + \frac{\partial^2}{\partial y^2}\right)$$

but with the constraint to a path of constant radius r. It is always a good idea to use coordinates that reflect the full symmetry of the system, especially when there is a restriction of the particle to a definite path corresponding to that symmetry, so we introduce the coordinates r and ϕ (Fig. 3.6),

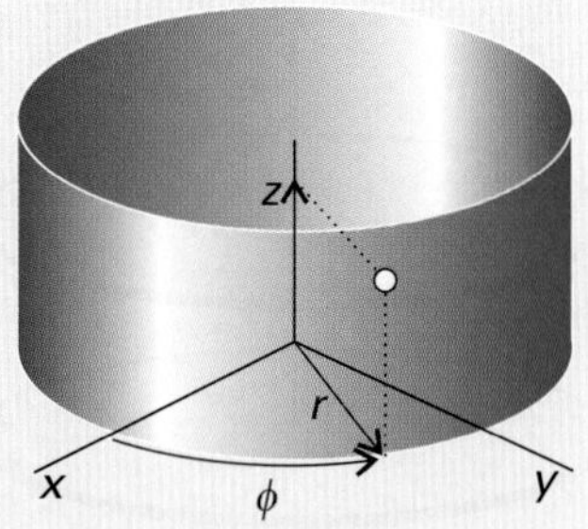

Fig. 3.6 The cylindrical coordinates z, r, and ϕ for discussing systems with axial (cylindrical) symmetry. For a particle confined to the xy-plane, only r and ϕ can change.

Table 3.1 Useful relations between Cartesian and other coordinate systems

Cylindrical coordinates (z, r, ϕ). See Fig. 3.6.

$$x = r\cos\phi$$

$$y = r\sin\phi$$

$$\mathrm{d}x\mathrm{d}y = r\mathrm{d}r\mathrm{d}\phi$$

$$\frac{\partial^2}{\partial x^2}+\frac{\partial^2}{\partial y^2}+\frac{\partial^2}{\partial z^2}=\frac{\partial^2}{\partial r^2}+\frac{1}{r}\frac{\partial}{\partial r}+\frac{1}{r^2}\frac{\partial^2}{\partial\phi^2}+\frac{\partial^2}{\partial z^2}$$

$$x\frac{\partial}{\partial y}-y\frac{\partial}{\partial x}=\frac{\partial}{\partial\phi}$$

Spherical polar coordinates (r, θ, ϕ). See Fig. 3.11.

$$x = r\sin\theta\cos\phi$$

$$y = r\sin\theta\sin\phi$$

$$z = r\cos\theta$$

$$\mathrm{d}x\mathrm{d}y\mathrm{d}z = r^2\sin\theta\,\mathrm{d}r\mathrm{d}\theta\mathrm{d}\phi$$

$$\frac{\partial^2}{\partial x^2}+\frac{\partial^2}{\partial y^2}+\frac{\partial^2}{\partial z^2}=\frac{1}{r}\frac{\partial^2}{\partial r^2}r+\frac{1}{r^2}\Lambda^2$$

$$=\frac{1}{r^2}\frac{\partial}{\partial r}r^2\frac{\partial}{\partial r}+\frac{1}{r^2}\Lambda^2$$

$$=\frac{\partial^2}{\partial r^2}+\frac{2}{r}\frac{\partial}{\partial r}+\frac{1}{r^2}\Lambda^2$$

where

$$\Lambda^2=\frac{1}{\sin^2\theta}\frac{\partial^2}{\partial\phi^2}+\frac{1}{\sin\theta}\frac{\partial}{\partial\theta}\sin\theta\frac{\partial}{\partial\theta}$$

where $x = r\cos\phi$ and $y = r\sin\phi$. By standard manipulations (see Table 3.1 for useful relations among coordinate systems) we can write

$$\frac{\partial^2}{\partial x^2}+\frac{\partial^2}{\partial y^2}=\frac{\partial^2}{\partial r^2}+\frac{1}{r}\frac{\partial}{\partial r}+\frac{1}{r^2}\frac{\partial^2}{\partial\phi^2} \tag{3.11}$$

However, because the radius of the path is fixed, the derivatives with respect to r can be discarded. The hamiltonian then becomes

$$\hat{H}=-\frac{\hbar^2}{2mr^2}\frac{\mathrm{d}^2}{\mathrm{d}\phi^2}$$

The moment of inertia $I = mr^2$ has appeared automatically, so $\hat{H}$ may be written

$$\hat{H}=-\frac{\hbar^2}{2I}\frac{\mathrm{d}^2}{\mathrm{d}\phi^2} \tag{3.12}$$

and the Schrödinger equation is

$$\frac{\mathrm{d}^2\psi}{\mathrm{d}\phi^2}=-\frac{2IE}{\hbar^2}\psi \tag{3.13}$$

The (unnormalized) general solutions of this equation are

$$\psi_{m_l}(\phi)=\mathrm{e}^{\mathrm{i}m_l\phi}\qquad m_l=\pm\frac{(2IE)^{1/2}}{\hbar}$$

as can be verified by substitution. The quantity m_l is just a dimensionless number at this stage.

We now select the acceptable solutions from among these general solutions by imposing the condition that the wavefunction should be single-valued. That is, the wavefunction ψ must satisfy a **cyclic boundary condition** and match at points separated by a complete revolution: $\psi(\phi+2\pi)=\psi(\phi)$. On substituting the general wavefunction into this condition, we find

$$\psi_{m_l}(\phi+2\pi)=\mathrm{e}^{\mathrm{i}m_l(\phi+2\pi)}=\mathrm{e}^{\mathrm{i}m_l\phi}\mathrm{e}^{2\pi\mathrm{i}m_l}=\psi_{m_l}(\phi)\mathrm{e}^{2\pi\mathrm{i}m_l}$$
$$=\psi_{m_l}(\phi)(\mathrm{e}^{\pi\mathrm{i}})^{2m_l}$$

As $\mathrm{e}^{\mathrm{i}\pi}=\cos\pi+\mathrm{i}\sin\pi=-1$, this relation is equivalent to

$$\psi_{m_l}(\phi+2\pi)=(-1)^{2m_l}\psi_{m_l}(\phi)$$

Because we require $(-1)^{2m_l}=1$, $2m_l$ must be a positive or a negative even integer (including 0), and therefore m_l must be an integer: $m_l = 0, \pm1, \pm2, \ldots$.

We now normalize the wavefunction by finding the normalization constant N given by eqn 1.4:

$$N=\frac{1}{\left(\displaystyle\int_0^{2\pi}\psi^*\psi\,\mathrm{d}\phi\right)^{1/2}}$$
$$=\frac{1}{\left(\displaystyle\int_0^{2\pi}\mathrm{e}^{-\mathrm{i}m_l\phi}\mathrm{e}^{\mathrm{i}m_l\phi}\mathrm{d}\phi\right)^{1/2}}=\frac{1}{(2\pi)^{1/2}} \tag{3.14}$$

and we obtain the normalized wavefunction of eqn 3.10b. The expression for the energies of the states is obtained by rearranging the relation $m_l=\pm(2IE)^{1/2}/\hbar$.

(b) Quantization of rotation

We can summarize the conclusions so far as follows. The energy for a particle on a ring is quantized and restricted to the values given in eqn 3.10a ($E_{m_l} = m_l^2\hbar^2/2I$). The occurrence of m_l as its square means that the energy of rotation is independent of the sense of rotation (the sign of m_l), as we expect physically. In other words, states with a given value of $|m_l|$ are doubly degenerate, except for $m_l = 0$, which is non-degenerate. The non-degeneracy of the ground state is consistent with the interpretation that, when $m_l = 0$, the particle has an infinite wavelength and is 'stationary'; the question of the direction of rotation does not arise.

We have also seen that the angular momentum is quantized and confined to the values given in eqn 3.9 ($J_z = m_l\hbar$). The increasing angular momentum is associated with the increasing number of nodes in the real ($\cos m_l\phi$) and imaginary ($\sin m_l\phi$) parts of the wavefunction $e^{im_l\phi}$ (the complex function does not have nodes but its real and imaginary components do): the wavelength decreases stepwise as $|m_l|$ increases, so the momentum with which the particle travels round the ring increases (Fig. 3.7). We can come to the same conclusion more formally by using the argument about the relation between eigenvalues and the values of observables established in Section 1.7 by Postulate IV. In the discussion of translational motion in one dimension, we saw that the opposite signs in the wavefunctions e^{ikx} and e^{-ikx} correspond to opposite directions of travel, and that the linear momentum is given by the eigenvalue of the linear momentum operator. The same conclusions can be drawn here, but now we need the eigenvalues of the angular momentum operator.

In classical mechanics the angular momentum l_z about the z-axis is defined as the 'moment of momentum':

$$l_z = xp_y - yp_x \tag{3.15}$$

where p_x is the component of linear momentum parallel to the x-axis and p_y is the component parallel to the y-axis. The operators for the linear momentum components are given in eqn 1.10, so the operator for angular momentum about the z-axis is

$$\hat{l}_z = \frac{\hbar}{i}\left(x\frac{\partial}{\partial y} - y\frac{\partial}{\partial x}\right) \tag{3.16a}$$

When expressed in terms of the coordinates r and ϕ, by standard manipulations (see Table 3.1) this equation becomes

$$\hat{l}_z = \frac{\hbar}{i}\frac{\partial}{\partial\phi} \tag{3.16b}$$

With the angular momentum operator available, we can test if the wavefunction in eqn 3.10b is an eigenfunction. Since the wavefunction depends only on the coordinate ϕ, the partial derivative in eqn 3.16b can be replaced by a complete derivative and we find

$$\hat{l}_z\psi_{m_l} = \frac{\hbar}{i}\frac{d\psi_{m_l}}{d\phi} = im_l\frac{\hbar}{i}\frac{e^{im_l\phi}}{(2\pi)^{1/2}} = m_l\hbar\psi_{m_l} \tag{3.17}$$

That is, ψ_{m_l} is an eigenfunction of $\hat{l}_z$, and corresponds to an angular momentum $m_l\hbar$. When m_l is positive, the angular momentum is positive (clockwise rotation when seen from below); when m_l is negative, the angular momentum is negative (counter-clockwise when seen from below).

These features are the origin of the vector representation of angular momentum, in which the magnitude is represented by the length of a vector and the direction of motion by its orientation (Fig. 3.8). This vector representation of angular momentum is also useful in classical physics but there is one crucial difference: in quantum mechanics the length of the vector is restricted to discrete values (corresponding to permitted values of m_l), whereas in classical physics the length is continuously variable.

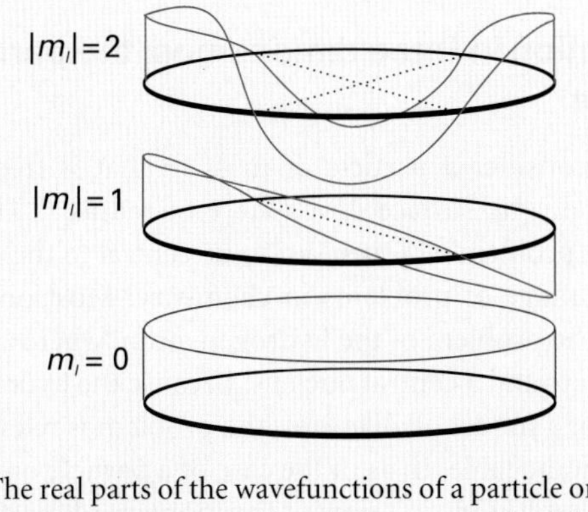

Fig. 3.7 The real parts of the wavefunctions of a particle on a ring. As shorter wavelengths are achieved, the magnitude of the angular momentum around the z-axis grows in steps of $\hbar$.

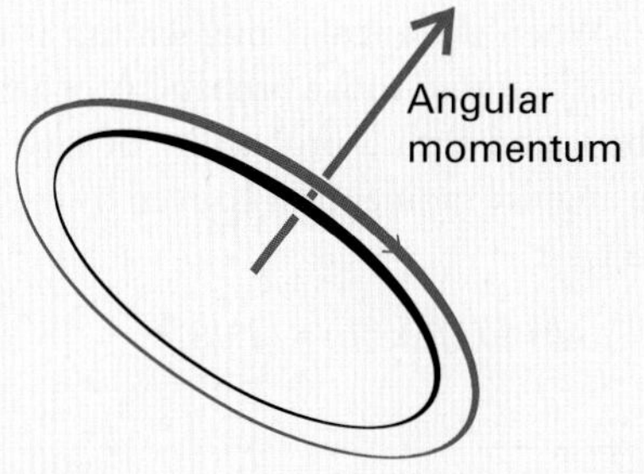

Fig. 3.8 The basic ideas of the vector representation of angular momentum: the magnitude of the angular momentum is represented by the length of the vector, and the orientation of the motion in space by the orientation of the vector (using the right-hand screw rule).

Example 3.2 *Finding expectation values of a particle-on-a-ring wavefunction*

A synchrotron, a type of cyclic particle accelerator, is used to accelerate protons along an effectively circular path. Since the synchrotron does not accelerate particles to a precisely defined angular momentum, the wavefunction of the protons is a superposition of functions. If, at some instant of time, the state of the proton is described by the wavefunction

$$\psi = \left(\frac{5}{12\pi}\right)^{1/2} e^{4i\phi} - i\left(\frac{1}{12\pi}\right)^{1/2} e^{3i\phi}$$

what value(s) of the angular momentum would be found? If more than one value is possible, what is the probability of obtaining each value and what is the average value?

Method Postulate IV tells us that, if ψ is an eigenfunction of the angular momentum operator $\hat{l}_z$, then only a single value of the angular momentum (the corresponding eigenvalue) will be found. Rather than explicitly testing if ψ is an eigenfunction by performing the operation $\hat{l}_z\psi$, we can use a short cut by recognizing that the wavefunction is written as a linear combination of eigenfunctions ψ_{m_l} of eqn 3.10b whose eigenvalues are given in eqn 3.9. However, we must be careful when using Postulate V to identify the coefficients c_k in the superposition.

Answer We rewrite the wavefunction as

$$\psi = \left(\frac{5}{6}\right)^{1/2} \times \left(\frac{1}{2\pi}\right)^{1/2} e^{4i\phi} + \frac{-i}{6^{1/2}} \times \left(\frac{1}{2\pi}\right)^{1/2} e^{3i\phi}$$

$$= \left(\frac{5}{6}\right)^{1/2} \psi_4 + \frac{-i}{6^{1/2}}\psi_3$$

The wavefunction is a linear combination of normalized eigenfunctions ψ_{m_l} (with eigenvalues $m_l\hbar$) and we can use Postulate V. When the angular momentum is measured, a value of $4\hbar$ will be found with a probability of 5/6 and a value of $3\hbar$ will be found with a probability of 1/6. The average value of the angular momentum is given by the expectation value in eqn 1.17:

$$\langle l_z \rangle = (5/6)(4\hbar) + (1/6)(3\hbar) = (23/6)\hbar$$

Self-test 3.2 Repeat the above problem for the energy of the proton. Assume that the synchrotron has an effective radius r.

[With $I = m_p r^2$, $\langle H \rangle = (5/6)(16\hbar^2/2I) + (1/6)(9\hbar^2/2I) = 89\hbar^2/12I$]

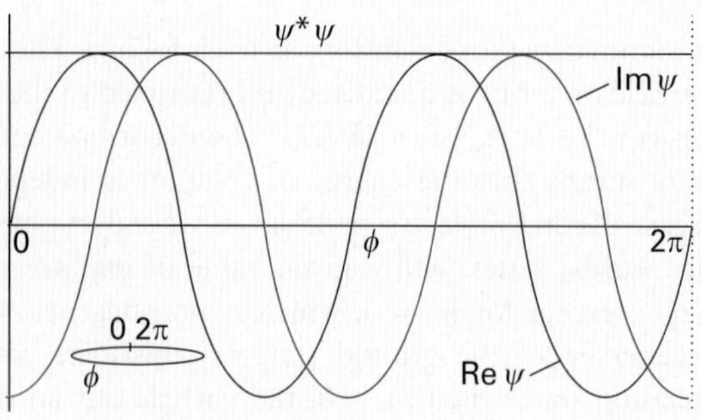

Fig. 3.9 The probability density for a particle in a definite state of angular momentum is uniform, so there is an equal probability of finding the particle anywhere on the ring.

To locate the particle given its wavefunction in eqn 3.10b, we form the probability density:

$$\psi_{m_l}{}^*\psi_{m_l} = \left(\frac{e^{im_l\phi}}{(2\pi)^{1/2}}\right)^* \left(\frac{e^{im_l\phi}}{(2\pi)^{1/2}}\right)$$

$$= \left(\frac{e^{-im_l\phi}}{(2\pi)^{1/2}}\right)\left(\frac{e^{im_l\phi}}{(2\pi)^{1/2}}\right) = \frac{1}{2\pi}$$

Because this probability density is independent of ϕ, the probability of locating the particle somewhere on the ring is also independent of ϕ (Fig. 3.9). Hence the location of the particle is completely indefinite (other than knowing that the particle must be somewhere on the ring), and knowing the angular momentum precisely eliminates the possibility of specifying the particle's location. Angular momentum and angle are a pair of complementary observables (in the sense defined in Section 1.10; see Problem 3.19), and the inability to specify them simultaneously with arbitrary precision is another example of the uncertainty principle.

3.4 Rotation in three dimensions: the particle on a sphere

We now consider a particle of mass m that is free to move anywhere on the surface of a sphere of radius r. This hugely important problem will turn out to be central to the discussion of a wide range of problems in chemistry. Most obviously, it is part of the problem of the hydrogen atom, where an electron circulates around a central nucleus. Because the hydrogen atom is a basis of a model of all atoms, the problem is relevant to the entire periodic table. If, as in the case of a particle on a ring, we express the solutions in terms of the moment of inertia $I = mr^2$, then we can use them to describe the three-dimensional rotation of any body of that moment of inertia, such as a solid ball

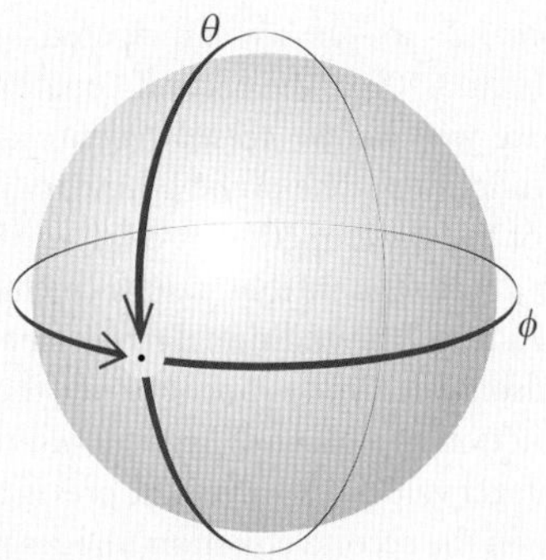

Fig. 3.10 The wavefunction of a particle on the surface of a sphere must satisfy two cyclic boundary conditions; this requirement leads to two quantum numbers for its state of angular momentum.

(which, if its mass is M and its radius is R, has $I = \frac{2}{5}MR^2$) or a molecule, such as CH_4 (which, if the C—H bond length is R, has $I = (8/3)m_H R^2$). The solutions we are about to find also occur in the description of particles trapped in spherical cavities, such as that used to model a three-dimensional quantum dot, a molecule in a helium droplet, or an electron in a metal–ammonia solution. Even more fundamentally, the discussion leads to an account of angular momentum, which is one of the most important problems in quantum mechanics, with applications that range from the classification of atomic states to many aspects of molecular spectroscopy.

We shall build on the work we have already done on the particle on a ring. This should not be surprising given that a sphere can be thought of as a three-dimensional stack of rings with the additional freedom for the particle to migrate from one ring to another. The cyclic boundary condition for the particle on a ring will once again lead to a quantum number, the same m_l that we have already encountered. However, the requirement that the wavefunction should match as a path is traced over the poles as well as round the equator of the sphere surrounding the central point introduces a second cyclic boundary condition and therefore a second quantum number (Fig. 3.10).

(a) The Schrödinger equation

The hamiltonian operator for motion in three dimensions (Table 1.1) is

$$\hat{H} = -\frac{\hbar^2}{2m}\nabla^2 + V \qquad \nabla^2 = \frac{\partial^2}{\partial x^2} + \frac{\partial^2}{\partial y^2} + \frac{\partial^2}{\partial z^2} \tag{3.18a}$$

As remarked in that table, the laplacian, ∇^2 (read 'del squared'), is a convenient abbreviation for the sum of the three second derivatives. For the particle confined to a spherical surface, $V = 0$ wherever it is free to travel and r is a constant. The wavefunction is therefore a function of the colatitude θ, and the azimuth ϕ

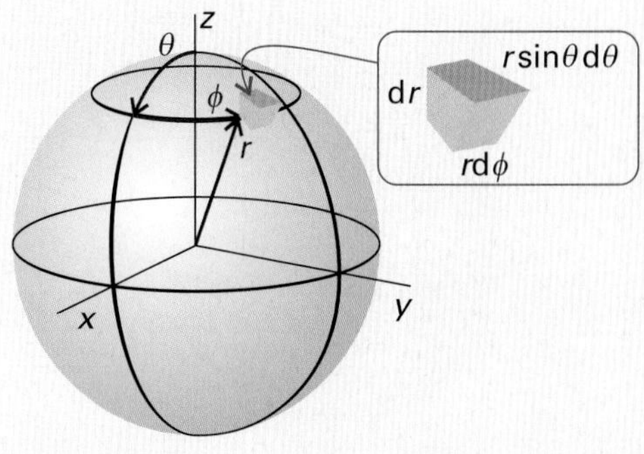

Fig. 3.11 The spherical polar coordinates used for discussing systems with spherical symmetry.

(Fig. 3.11), and we write it $\psi(\theta,\phi)$. The Schrödinger equation is therefore

$$-\frac{\hbar^2}{2m}\nabla^2\psi = E\psi \tag{3.18b}$$

Our remark that a sphere can be regarded as a stack of rings suggests that the solutions of eqn 3.18b can be written

$$\psi(\theta,\phi) = \Theta(\theta)\Phi(\phi) \tag{3.19}$$

where Θ is a function only of θ, and Φ, a function only of ϕ, is the solution for a particle on a ring. This separation of variables is confirmed in the following *Justification*, where we show that the solutions are specified by quantum numbers l and m_l restricted to the values

$$l = 0, 1, 2, \ldots \qquad m_l = l, l-1, \ldots, -l$$

The **orbital angular momentum quantum number** l is non-negative and, for a given value of l, there are $2l + 1$ permitted values of the **magnetic quantum number**, m_l.

Justification 3.3 *The separation of variables technique applied to the particle on a sphere*

To take advantage of the symmetry of the problem and the fact that r is a constant for a particle on a sphere, we use **spherical polar coordinates**, the radius r, the colatitude θ, and the azimuth ϕ (Fig. 3.12), with

$$x = r\sin\theta\cos\phi \qquad y = r\sin\theta\sin\phi \qquad z = r\cos\theta$$

We saw in Table 1.1 and Table 3.1 that the laplacian in spherical polar coordinates is

$$\nabla^2 = \frac{\partial^2}{\partial r^2} + \frac{2}{r}\frac{\partial}{\partial r} + \frac{1}{r^2}\Lambda^2 \tag{3.20a}$$

where the legendrian, Λ^2, is

$$\Lambda^2 = \frac{1}{\sin^2\theta}\frac{\partial^2}{\partial\phi^2} + \frac{1}{\sin\theta}\frac{\partial}{\partial\theta}\sin\theta\frac{\partial}{\partial\theta} \tag{3.20b}$$

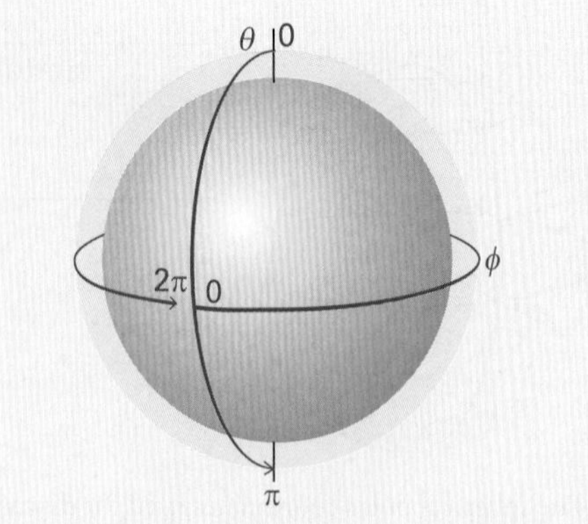

Fig. 3.12 The surface of a sphere is covered by allowing θ to range from 0 to π, and then sweeping that arc around a complete circle by allowing ϕ to range from 0 to 2π.

Because r is constant, we can discard the part of the laplacian that involves differentiation with respect to r, and so write the Schrödinger equation as

$$-\frac{\hbar^2}{2mr^2}\Lambda^2\psi = E\psi$$

The moment of inertia, $I = mr^2$, has appeared. This expression can be rearranged into

$$\Lambda^2\psi = -\varepsilon\psi \qquad \varepsilon = \frac{2IE}{\hbar^2}$$

To verify that this expression is separable, we try the substitution $\psi = \Theta\Phi$ with the form of the legendrian in eqn 3.20b:

$$\Lambda^2\Theta\Phi = \frac{1}{\sin^2\theta}\frac{\partial^2(\Theta\Phi)}{\partial\phi^2} + \frac{1}{\sin\theta}\frac{\partial}{\partial\theta}\sin\theta\frac{\partial(\Phi\Theta)}{\partial\theta} = -\varepsilon\Theta\Phi$$

We now use the fact that Θ and Φ are each functions of one variable, so the partial derivatives become complete derivatives:

$$\frac{\Theta}{\sin^2\theta}\frac{\mathrm{d}^2\Phi}{\mathrm{d}\phi^2} + \frac{\Phi}{\sin\theta}\frac{\mathrm{d}}{\mathrm{d}\theta}\sin\theta\frac{\mathrm{d}\Theta}{\mathrm{d}\theta} = -\varepsilon\Theta\Phi$$

Division through by $\Theta\Phi$, multiplication by $\sin^2\theta$, and minor rearrangement give

$$\frac{1}{\Phi}\frac{\mathrm{d}^2\Phi}{\mathrm{d}\phi^2} + \frac{\sin\theta}{\Theta}\frac{\mathrm{d}}{\mathrm{d}\theta}\sin\theta\frac{\mathrm{d}\Theta}{\mathrm{d}\theta} + \varepsilon\sin^2\theta = 0$$

The first term on the left depends only on ϕ and the remaining two terms depend only on θ. By the same argument used in *Justification 3.1*, each term is equal to a constant. Thus, if we set the first term equal to the numerical constant $-m_l^2$ (using a notation chosen with an eye to the future), the separated equations are

$$\frac{1}{\Phi}\frac{\mathrm{d}^2\Phi}{\mathrm{d}\phi^2} = -m_l^2 \qquad \frac{\sin\theta}{\Theta}\frac{\mathrm{d}}{\mathrm{d}\theta}\sin\theta\frac{\mathrm{d}\Theta}{\mathrm{d}\theta} + \varepsilon\sin^2\theta = m_l^2$$

The first of these two equations is the same as that in *Justification 3.2*, so—as we anticipated—it has the same solutions (eqn 3.10b). The second equation is new, but its solutions are well known to mathematicians as 'associated Legendre functions'. The cyclic boundary condition for the matching of the wavefunction at $\phi = 0$ and 2π restricts m_l to positive and negative integer values (including 0), as for a particle on a ring. The additional requirement that the wavefunction also match on a journey over the poles results in the introduction of a second quantum number, l, with non-negative integer values. However, the presence of the quantum number m_l in the second equation implies that the ranges of the two quantum numbers are linked, and it turns out that l can take the values 0, 1, 2, . . . and, for a given value of l, m_l ranges in integer steps from $-l$ to $+l$, as quoted in the text.

The normalized wavefunctions $\psi(\theta,\phi)$ for a given l and m_l are usually denoted $Y_{lm_l}(\theta,\phi)$ and are called the **spherical harmonics** (Table 3.2). They are as fundamental to the description of waves on spherical surfaces as the harmonic (sine and cosine) functions

Table 3.2 The spherical harmonics

l	m_l	$Y_{l,m_l}(\theta,\phi)$
0	0	$\left(\frac{1}{4\pi}\right)^{1/2}$
1	0	$\left(\frac{3}{4\pi}\right)^{1/2}\cos\theta$
	±1	$\mp\left(\frac{3}{8\pi}\right)^{1/2}\sin\theta\,\mathrm{e}^{\pm i\phi}$
2	0	$\left(\frac{5}{16\pi}\right)^{1/2}(3\cos^2\theta - 1)$
	±1	$\mp\left(\frac{15}{8\pi}\right)^{1/2}\cos\theta\sin\theta\,\mathrm{e}^{\pm i\phi}$
	±2	$\left(\frac{15}{32\pi}\right)^{1/2}\sin^2\theta\,\mathrm{e}^{\pm 2i\phi}$
3	0	$\left(\frac{7}{16\pi}\right)^{1/2}(5\cos^3\theta - 3\cos\theta)$
	±1	$\mp\left(\frac{21}{64\pi}\right)^{1/2}(5\cos^2\theta - 1)\sin\theta\,\mathrm{e}^{\pm i\phi}$
	±2	$\left(\frac{105}{32\pi}\right)^{1/2}\sin^2\theta\cos\theta\,\mathrm{e}^{\pm 2i\phi}$
	±3	$\mp\left(\frac{35}{64\pi}\right)^{1/2}\sin^3\theta\,\mathrm{e}^{\pm 3i\phi}$

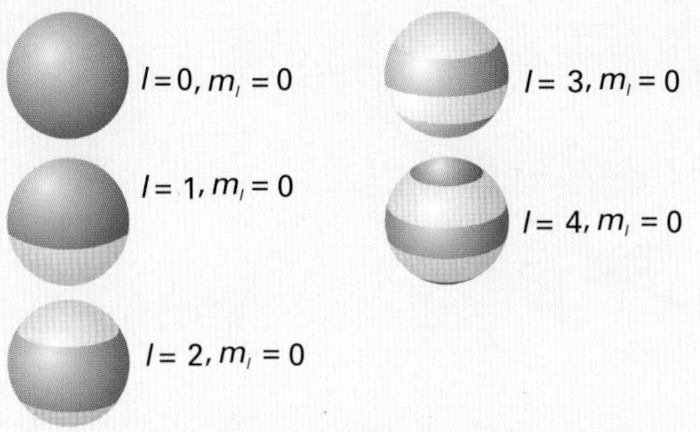

Fig. 3.13 A representation of the wavefunctions of a particle on the surface of a sphere that emphasizes the location of angular nodes: dark and light shading correspond to different signs of the wavefunction. Note that the number of nodes increases as the value of l increases. All these wavefunctions correspond to $m_l = 0$; a path round the vertical z-axis of the sphere does not cut through any nodes.

are to the description of waves on lines and planes. Figure 3.13 is a representation of the spherical harmonics for $l = 0$ to 4 and $m_l = 0$. There are no angular nodes (the positions at which the wavefunction passes through zero) around the z-axis for functions with $m_l = 0$. The spherical harmonic with $l = 0$, $m_l = 0$ has no nodes at all: it is a 'wave' of constant height at all positions of the surface. The spherical harmonic with $l = 1$, $m_l = 0$ has a single angular node at $\theta = \pi/2$; therefore, the equatorial plane is a nodal plane. The spherical harmonic with $l = 2$, $m_l = 0$ has two angular nodes (at colatitudes $\theta = 54.7°$ and $125.3°$; close to the latitudes of Los Angeles and Buenos Aires on an actual globe). In general, the number of angular nodes is equal to l. As the number of nodes increases, the wavefunctions become more buckled, and with this increasing curvature we can anticipate that the kinetic energy of the particle increases.

The increase in energy with increasing l is confirmed by the full solution of the Schrödinger equation, which shows that the energy of the particle is restricted to the values

$$E = l(l+1)\frac{\hbar^2}{2I} \qquad l = 0, 1, 2, \ldots \tag{3.21}$$

independent of the value of m_l. Because there are $2l + 1$ different wavefunctions (one for each value of m_l) that correspond to the same energy, it follows that a level with quantum number l is $(2l + 1)$-fold degenerate. Notice also that there is no zero-point energy because the ground-state ($l = 0$, $m_l = 0$) wavefunction has a constant value and all its derivatives are zero. The appearance of the factor $l(l + 1)$ in eqn 3.21 has its roots in the properties of the angular momentum operator, in particular the operator for the square of the magnitude of the angular momentum, $\hat{l}^2$, introduced in Section 3.4(d).

(b) Angular momentum

Just as important as the quantization of energy is the quantization of the particle's angular momentum. We can infer that the angular momentum is quantized with very little further calculation by noting that the energy of a rotating particle is related classically to its angular momentum J by $E = J^2/2I$ (see *Fundamentals* F.6). Therefore, by comparing this equation with eqn 3.21, we can deduce that the magnitude of the angular momentum is confined to the values

$$\text{Magnitude of angular momentum} = \{l(l+1)\}^{1/2}\hbar \qquad l = 0, 1, 2 \ldots \tag{3.22a}$$

We have already seen (in the context of rotation in a plane) that the angular momentum about the z-axis is quantized, and that for a given value of l it has the values

$$z\text{-Component of angular momentum} = m_l\hbar \qquad m_l = +l, l-1, \ldots, -l \tag{3.22b}$$

A note on good practice When quoting the value of m_l, always give the sign, even if m_l is positive. Thus, write $m_l = +2$, not $m_l = 2$.

The fact that the number of nodes in $Y_{lm_l}(\theta,\phi)$ increases with l reflects the fact that higher angular momentum implies higher kinetic energy, and therefore a more sharply buckled wavefunction.

● A BRIEF ILLUSTRATION

As a first approximation to the description of the rotation of a $^1H^{127}I$ molecule, we can imagine an ^1H atom orbiting a heavy, stationary ^{127}I atom at a distance $r = 160$ pm, the equilibrium bond distance. The moment of inertia of $^1H^{127}I$ is then $I = m_H r^2 = 4.284 \times 10^{-47}$ kg m^2. It follows that

$$\frac{\hbar^2}{2I} = \frac{(1.055 \times 10^{-34}\ \text{J s})^2}{2 \times (4.284 \times 10^{-47}\ \text{kg m}^2)} = 1.299 \times 10^{-22}\ \text{J}$$

or 0.1299 zJ. This energy corresponds to 78.23 J mol^{-1}. From eqn 3.21, the first few rotational energy levels are therefore 0 ($l = 0$), 0.2598 zJ ($l = 1$), 0.7794 zJ ($l = 2$), and 1.559 zJ ($l = 3$). The degeneracies of these levels are 1, 3, 5, and 7, respectively (from $2l + 1$), and the magnitudes of the angular momentum of the molecule are 0, $2^{1/2}\hbar$, $6^{1/2}\hbar$, and $(12)^{1/2}\hbar$ (from eqn 3.22a). It follows from our calculations that the $l = 0$ and $l = 1$ levels are separated by $\Delta E = 0.2598$ zJ. A transition between these two rotational levels of the molecule can be brought about by the emission or absorption of a photon with a frequency given by the Bohr frequency condition (eqn 1.1):

$$\nu = \frac{\Delta E}{h} = \frac{2.598 \times 10^{-22}\ \text{J}}{6.626 \times 10^{-34}\ \text{J s}} = 3.921 \times 10^{11}\ \text{Hz} = 392.1\ \text{GHz}$$

Radiation with this frequency belongs to the microwave region of the electromagnetic spectrum, so microwave spectroscopy is a convenient method for the study of molecular rotations. Because the transition energies depend on the moment of inertia, microwave spectroscopy is a very accurate technique for the determination of bond lengths. We discuss rotational spectra further in Chapter 10. ●

Self-test 3.3 Repeat the calculation for a $^{2}H^{127}I$ molecule (same bond length as $^{1}H^{127}I$).

[Energies are smaller by a factor of two; same angular momenta and numbers of components]

(c) Space quantization

The result that m_l is confined to the values $l, l-1, \ldots, -l$ for a given value of l means that the component of angular momentum about the z-axis—the contribution to the total angular momentum of rotation around that axis—may take only $2l+1$ values. If we represent the angular momentum by a vector of length $\{l(l+1)\}^{1/2}$, then it follows that this vector must be oriented so that its projection on the z-axis is m_l and that it can have only $2l+1$ orientations rather than the continuous range of orientations of a rotating classical body (Fig. 3.14). The remarkable implication is that *the orientation of a rotating body is quantized.*

The quantum mechanical result that a rotating body may not take up an arbitrary orientation with respect to some specified axis (for example, an axis defined by the direction of an externally applied electric or magnetic field) is called **space quantization**. It was confirmed by an experiment first performed by Otto Stern and Walther Gerlach in 1921, who shot a beam of silver atoms through an inhomogeneous magnetic field (Fig. 3.15). The idea behind the experiment was that a rotating, charged body (such as the $5s^1$ electron in the silver atom) behaves like a magnet and interacts with the applied field. According to classical mechanics, because the orientation of the angular momentum can take any value, the associated magnet can take any orientation. Because the direction in which the magnet is driven by the inhomogeneous field depends on the magnet's orientation, it follows that a broad band of atoms is expected to emerge from the region where the magnetic field acts. According to quantum mechanics, however, because the orientation of the angular momentum is quantized, the associated magnet lies in a number of discrete orientations, so several sharp bands of atoms are expected.

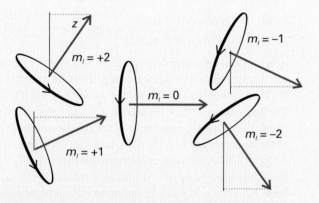

Fig. 3.14 The permitted orientations of angular momentum when $l = 2$. We shall see soon that this representation is too specific because the azimuthal orientation of the vector (its angle around z) is indeterminate.

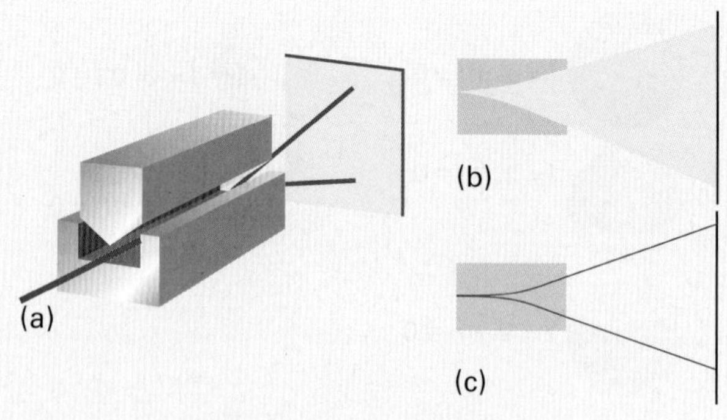

Fig. 3.15 (a) The experimental arrangement for the Stern–Gerlach experiment: the magnet provides an inhomogeneous field. (b) The classically expected result. (c) The observed outcome using silver atoms.

In their first experiment, Stern and Gerlach appeared to confirm the classical prediction. However, the experiment is difficult because collisions between the atoms in the beam blur the bands. When the experiment was repeated with a beam of very low intensity (so that collisions were less frequent), they observed discrete bands, and so confirmed the quantum prediction.

(d) The vector model

So far, we have discussed the magnitude of the angular momentum and its z-component. In classical physics, we would be able to specify the components about the x- and y-axes too and be able to represent the angular momentum by a vector with a definite orientation. According to quantum mechanics, we *already* have as complete a description of the angular momentum of a rotating object as it is possible to have, and can say nothing further about the orientation of the vector.

The reason for this restriction is that the components of angular momentum are complementary to each other (in the sense described in Section 1.10) and that, if we specify any one of them (the z-component, typically), then the other two components cannot be specified. To verify that the components are complementary, we need to examine the corresponding operators. Each one is like the operator in eqn 3.16a:

$$\hat{l}_x = \frac{\hbar}{\mathrm{i}}\left(y\frac{\partial}{\partial z} - z\frac{\partial}{\partial y}\right) \qquad \hat{l}_y = \frac{\hbar}{\mathrm{i}}\left(z\frac{\partial}{\partial x} - x\frac{\partial}{\partial z}\right)$$
$$\hat{l}_z = \frac{\hbar}{\mathrm{i}}\left(x\frac{\partial}{\partial y} - y\frac{\partial}{\partial x}\right) \qquad (3.23)$$

As you are invited to show in Problem 3.27, these three operators do not commute with one another:

$$[\hat{l}_x,\hat{l}_y] = \mathrm{i}\hbar\hat{l}_z \qquad [\hat{l}_y,\hat{l}_z] = \mathrm{i}\hbar\hat{l}_x \qquad [\hat{l}_z,\hat{l}_x] = \mathrm{i}\hbar\hat{l}_y \qquad (3.24)$$

Therefore, l_x, l_y, and l_z are complementary observables and we cannot specify more than one of them at a time (unless $l = 0$). On the other hand, the operator for the square of the magnitude of the angular momentum is

$$\hat{l}^2 = \hat{l}_x^2 + \hat{l}_y^2 + \hat{l}_z^2 \qquad (3.25)$$

This operator does commute with all three components (see Problem 3.29):

$$[\hat{l}^2,\hat{l}_q] = 0 \qquad q = x, y, \text{ and } z \qquad (3.26)$$

Therefore, because $\hat{l}^2$ and $\hat{l}_q$ commute, we may specify precisely and simultaneously the magnitude of the angular momentum and any *one* of the components of the angular momentum. It follows that the schematic in Fig. 3.14, which is summarized in Fig. 3.16a, gives a false impression of the state of the system, because it suggests definite values for the x- and y-components. A more accurate picture must reflect the impossibility of specifying l_x and l_y if l_z is known.

The **vector model** of angular momentum uses pictures like that in Fig. 3.16b. The cones are drawn with sides of length $\{l(l+1)\}^{1/2}$, and represent the magnitude of the angular momentum. Each

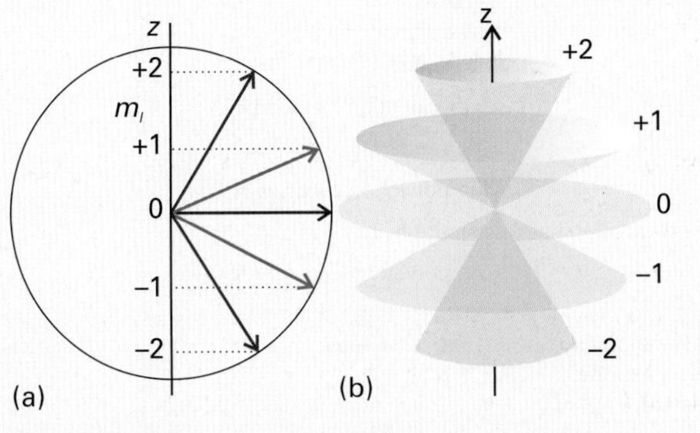

Fig. 3.16 (a) A summary of Fig. 3.14. However, because the azimuthal angle of the vector around the z-axis is indeterminate, a better representation is as in (b), where each vector lies at an unspecified azimuthal angle on its cone.

cone has a definite projection (of m_l) on the z-axis, representing the system's precise value of l_z. The l_x and l_y projections, however, are indefinite. The vector representing the state of angular momentum can be thought of as lying with its tip on any point on the mouth of the cone. At this stage it should not be thought of as sweeping round the cone; that aspect of the model will be added later when we allow the picture to convey more information.

3.5 Spin

Stern and Gerlach observed *two* bands of Ag atoms in their experiment. This observation seems to conflict with one of the predictions of quantum mechanics, because an angular momentum l gives rise to $2l + 1$ orientations, which is equal to 2 only if $l = \frac{1}{2}$, contrary to the conclusion that l must be an integer. The conflict was resolved by the suggestion that the angular momentum they were observing was not due to orbital angular momentum (the motion of an electron around the atomic nucleus) but arose instead from the motion of the electron about its own axis. This intrinsic angular momentum of the electron is called its **spin**. The explanation of the existence of spin emerged when Dirac combined quantum mechanics with special relativity and established the theory of relativistic quantum mechanics.

The spin of an electron about its own axis does not have to satisfy the same boundary conditions as those for a particle circulating around a central point, so the quantum number for spin angular momentum is subject to different restrictions. To distinguish this spin angular momentum from orbital angular momentum we use the **spin quantum number** s (in place of l; like l, s is a non-negative number) and m_s, the **spin magnetic quantum number**, for the projection on the z-axis. The magnitude of the spin angular momentum is $\{s(s+1)\}^{1/2}\hbar$ and the component $m_s\hbar$ is restricted to the $2s + 1$ values

$$m_s = s, s-1, \ldots, -s \qquad (3.27)$$

A note on good practice You will sometimes see the quantum number s used in place of m_s, and written $s = \pm\frac{1}{2}$. That is wrong: like l, s is never negative and denotes the magnitude of the spin angular momentum. For the z-component, use m_s.

The detailed analysis of the spin of a particle is sophisticated and shows that the property should not be taken to be an actual spinning motion. It is better to regard 'spin' as an intrinsic property like mass and charge. However, the picture of an actual spinning motion can be very useful when used with care. For an electron it turns out that only one value of s is allowed, namely $s = \frac{1}{2}$, corresponding to an angular momentum of magnitude $(\frac{3}{4})^{1/2}\hbar = 0.866\hbar$. This spin angular momentum is an intrinsic property of the electron, like its rest mass and its charge, and every electron has exactly the same value: the magnitude of the

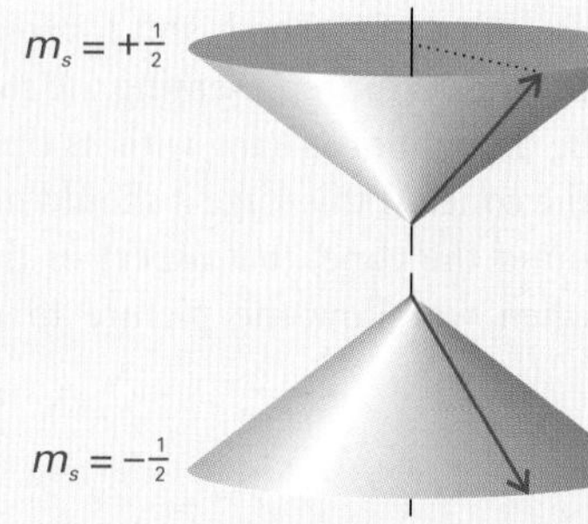

Fig. 3.17 An electron spin ($s = \frac{1}{2}$) can take only two orientations with respect to a specified axis. An α electron (top) is an electron with $m_s = +\frac{1}{2}$; a β electron (bottom) is an electron with $m_s = -\frac{1}{2}$. The vector representing the spin angular momentum lies at an angle of 55° to the z-axis (more precisely, the half-angle of the cones is $\arccos(m_s/\{s(s+1)\}^{1/2}) = \arccos(1/3^{1/2})$).

spin angular momentum of an electron cannot be changed. The spin may lie in $2s + 1 = 2$ different orientations (Fig. 3.17). One orientation corresponds to $m_s = +\frac{1}{2}$ (this state is often denoted α or ↑); the other orientation corresponds to $m_s = -\frac{1}{2}$ (this state is denoted β or ↓).

The outcome of the Stern–Gerlach experiment can now be explained if we suppose that each Ag atom possesses an angular momentum due to the spin of a single electron, because the two bands of atoms then correspond to the two spin orientations. Why the atoms behave like this is explained in Chapter 4 (but it is already probably familiar from introductory chemistry that the ground-state configuration of a silver atom is $[Kr]4d^{10}5s^1$, a single unpaired electron outside a closed shell).

Like the electron, other elementary particles have characteristic spin. For example, protons and neutrons are spin-$\frac{1}{2}$ particles (that is, $s = \frac{1}{2}$) and invariably spin with angular momentum $(\frac{3}{4})^{1/2}\hbar = 0.866\hbar$. Because the masses of a proton and a neutron are so much greater than the mass of an electron, yet they all have the same spin angular momentum, the classical picture would be of these two particles spinning much more slowly than an electron. Some elementary particles have $s = 1$, and so have an intrinsic angular momentum of magnitude $2^{1/2}\hbar$. Some mesons are spin-1 particles (as are some atomic nuclei), but for our purposes the most important spin-1 particle is the photon. We shall see the importance of photon spin in the next chapter and the importance of proton spin in Chapter 12 (magnetic resonance).

Particles with half-integral spin are called **fermions** and those with integral spin (including 0) are called **bosons**. Thus, electrons and protons are fermions and photons are bosons. It is a very deep feature of nature that all the elementary particles that constitute matter are fermions whereas the elementary particles that are responsible for the forces that bind fermions together are all bosons. Photons, for example, transmit the electromagnetic force that binds together electrically charged particles. Matter, therefore, is an assembly of fermions held together by forces conveyed by bosons.

The properties of angular momentum that we have developed are set out in Table 3.3. As mentioned there, when we use the quantum numbers l and m_l we shall mean orbital angular momentum (circulation in space). When we use s and m_s we shall mean spin angular momentum (intrinsic angular momentum). When we use j and m_j we shall mean either (or, in some contexts to be described in Chapter 4, a combination of orbital and spin momenta).

Table 3.3 Properties of the angular momentum of an electron

Quantum number	Symbol*	Values†	Specifies
Orbital angular momentum	l	$0, 1, 2, \ldots$	Magnitude, $\{l(l+1)\}^{1/2}\hbar$
Magnetic	m_l	$l, l-1, \ldots, -l$	Component on z-axis, $m_l\hbar$
Spin	s	$\frac{1}{2}$	Magnitude, $\{s(s+1)\}^{1/2}\hbar$
Spin magnetic	m_s	$\pm\frac{1}{2}$	Component on z-axis, $m_s\hbar$
Total‡	j	$l+s, l+s-1, \ldots, \lvert l-s\rvert$	Magnitude, $\{j(j+1)\}^{1/2}\hbar$
Total magnetic	m_j	$j, j-1, \ldots, -j$	Component on z-axis, $m_j\hbar$

*For many-electron systems, the quantum numbers are designated by upper-case letters (L, M_L, S, M_S, etc.).
†Note that the quantum numbers for magnitude (l, s, j, etc.) are never negative.
‡To combine two angular momenta, use the Clebsch–Gordan series (Section 4.5):

$$j = j_1 + j_2, j_1 + j_2 - 1, \ldots, |j_1 - j_2|$$

Checklist of key ideas

☐ 1. The wavefunctions and energies of a particle in a two-dimensional box are given by eqn 3.3.

☐ 2. Degenerate wavefunctions are different wavefunctions corresponding to the same energy. Degeneracy results from the symmetry of the system.

☐ 3. A particle in a three-dimensional box has wavefunctions and energies given by eqn 3.5.

☐ 4. Angular momentum is the moment of momentum around a point.

☐ 5. The wavefunctions and energies of a particle on a ring are, respectively, $\psi_{m_l}(\phi) = (1/2\pi)^{1/2}e^{im_l\phi}$ and $E_{m_l} = m_l^2\hbar^2/2I$, with $I = mr^2$ and $m_l = 0, \pm1, \pm2, \ldots$.

☐ 6. The wavefunctions of a particle on a sphere are the spherical harmonics, the functions $Y_{l,m_l}(\theta,\phi)$ (Table 3.2). The energies are $E = l(l+1)\hbar^2/2I$, $l = 0, 1, 2, \ldots$.

☐ 7. For a particle on a sphere, the magnitude of the angular momentum is $\{l(l+1)\}^{1/2}\hbar$ and the z-component of the angular momentum is $m_l\hbar$, $m_l = l, l-1, \ldots, -l$.

☐ 8. Space quantization is the restriction of the component of angular momentum around an axis to discrete values.

☐ 9. For three-dimensional rotation, because of the uncertainty principle, only the magnitude of the angular momentum and one of its components can be specified precisely and simultaneously.

☐ 10. Spin is an intrinsic angular momentum of a fundamental particle. A fermion is a particle with a half-integral spin quantum number; a boson is a particle with an integral spin quantum number.

☐ 11. For an electron, the spin quantum number is $s = \frac{1}{2}$.

☐ 12. The spin magnetic quantum number is $m_s = s, s-1, \ldots, -s$; for an electron, $m_s = +\frac{1}{2}, -\frac{1}{2}$.

Discussion questions

3.1 Discuss the physical origin of quantization of energy for a particle confined to motion around a ring.

3.2 Discuss the appearance or absence of zero-point energy for translational and rotational motions in 2 and 3 dimensions.

3.3 Distinguish between a fermion and a boson. Provide examples of each type of particle.

3.4 Describe the features of the solution of the particle in a one-dimensional box that appear in the solutions of the particle in two- and three-dimensional boxes. What concept applies to the latter but not to a one-dimensional box?

3.5 Describe the features of the solution of the particle on a ring that appear in the solution of the particle on a sphere. What concept applies to the latter but not to the former?

3.6 Describe the vector model of angular momentum in quantum mechanics. What features does it capture? What is its status as a model?

3.7 Compare and contrast the properties of spin angular momentum and the properties of angular momentum arising from rotational motion in two and three dimensions.

Exercises

3.1(a) Some nanostructures can be modelled as an electron confined to a two-dimensional region. Calculate the energy separations in joules, kilojoules per mole, wavenumbers, and electronvolts between the levels (a) $n_1 = n_2 = 2$ and $n_1 = n_2 = 1$, (b) $n_1 = n_2 = 6$ and $n_1 = n_2 = 5$ of an electron in a square box with sides of length 1.0 nm.

3.1(b) Some nanostructures can be modelled as an electron confined to a three-dimensional region. Calculate the energy separations in joules, kilojoules per mole, wavenumbers, and electronvolts between the levels (a) $n_1 = n_2 = n_3 = 2$ and $n_1 = n_2 = n_3 = 1$, (b) $n_1 = n_2 = n_3 = 6$ and $n_1 = n_2 = n_3 = 5$ of an electron in a cubic box with sides of length 1.0 nm.

3.2(a) Nanostructures commonly show physical properties that distinguish them from bulk materials. Calculate the wavelength and frequency of the radiation required to cause a transition between the levels in Exercise 3.1a.

3.2(b) Nanostructures commonly show physical properties that distinguish them from bulk materials. Calculate the wavelength and frequency of the radiation required to cause a transition between the levels in Exercise 3.1b.

3.3(a) Suppose a nanostructure is modelled by an electron confined to a rectangular region with sides of lengths $L_1 = 1.0$ nm and $L_2 = 2.0$ nm and is subjected to thermal motion with a typical energy equal to kT where k is Boltzmann's constant. How low should the temperature be for the thermal energy to be comparable to (a) the zero-point energy, (b) the first excitation energy of the electron?

3.3(b) Suppose a nanostructure is modelled by an electron confined to a three-dimensional region with sides of lengths $L_1 = 1.0$ nm, $L_2 = 2.0$ nm, and $L_3 = 1.5$ nm and is subjected to thermal motion with a typical energy equal to kT where k is Boltzmann's constant. How low should the temperature be for the thermal energy to be comparable to (a) the zero-point energy, (b) the first excitation energy of the electron?

3.4(a) For quantum mechanical reasons, particles confined to nanostructures are not distributed uniformly through them.

Calculate the probability that an electron confined to a square region with sides of length L will be found in the region $0.49L \le x \le 0.51L$ and $0.49L \le y \le 0.51L$ when it is in a state with (a) $n_1 = n_2 = 1$, (b) $n_1 = n_2 = 2$. Take the wavefunction to be a constant in this narrow range.

3.4(b) For quantum mechanical reasons, particles confined to nanostructures are not distributed uniformly through them. Calculate the probability that a hydrogen atom in a cubic cavity with sides of length L will be found in the region $0.49L \le x \le 0.51L$, $0.49L \le y \le 0.51L$, and $0.49L \le z \le 0.51L$ when it has (a) $n_1 = n_2 = n_3 = 1$, (b) $n_1 = n_2 = n_3 = 2$. Take the wavefunction to be a constant in this narrow range.

3.5(a) What are the most likely locations of an electron in a nanostructure modelled by a particle in a square box with sides of length L when it is in the state $n_1 = 2$, $n_2 = 3$?

3.5(b) What are the most likely locations of an electron in a nanostructure modelled by a particle in a cubic box with sides of length L when it is in the state $n_1 = 1$, $n_2 = 4$, $n_3 = 5$?

3.6(a) Locate the nodes of the wavefunction of an electron in a nanostructure modelled by a particle in a square region with sides of length L when it is in the state $n_1 = 2$, $n_2 = 3$.

3.6(b) Locate the nodes of the wavefunction of an electron in a nanostructure modelled by a particle in a cubic well with sides of length L when it is in the state $n_1 = 3$, $n_2 = 4$, $n_3 = 5$.

3.7(a) Also for quantum mechanical reasons, particles confined to nanostructures cannot be perfectly still even at $T = 0$. Calculate the expectation values of $\hat{p}_x$ and $\hat{p}^2$ for an electron in the ground state of a nanostructure modelled by a square box with sides of length L.

3.7(b) Also for quantum mechanical reasons, particles confined to nanostructures cannot be perfectly still even at $T = 0$. Calculate the expectation values of $\hat{p}_x$ and $\hat{p}^2$ for an electron in the ground state of a nanostructure modelled by a cubic box with sides of length L.

3.8(a) In Exercise 2.6a you were invited to explore whether compression could cause the zero-point energy of an electron to rise to equal its rest mass, m_ec^2, in one dimension. Repeat that calculation for a two-dimensional container. Express your answer in terms of the parameter $\lambda_C = h/m_ec$, the 'Compton wavelength' of the electron.

3.8(b) Repeat Exercise 3.8a for an electron squeezed inside a cubic box.

3.9(a) For a particle in a rectangular box with sides of length $L_1 = L$ and $L_2 = 2L$, find a state that is degenerate with the state $n_1 = n_2 = 2$. Degeneracy is normally associated with symmetry; why, then, are these two states degenerate?

3.9(b) For a particle in a rectangular box with sides of length $L_1 = L$ and $L_2 = 2L$, find a state that is degenerate with the state $n_1 = n_2 = 4$. Degeneracy is normally associated with symmetry; why, then, are these two states degenerate?

3.10(a) Consider a particle in a cubic box. What is the degeneracy of the level that has an energy three times that of the lowest level?

3.10(b) Consider a particle in a cubic box. What is the degeneracy of the level that has an energy $\frac{14}{3}$ times that of the lowest level?

3.11(a) Calculate the percentage change in a given energy level of a particle in a cubic box when the length of the side of the cube is decreased by 10 per cent in each direction.

3.11(b) Calculate the percentage change in a given energy level of a particle in a square box when the length of the side of the square is decreased by 10 per cent in each direction.

3.12(a) Should a gas be treated quantum mechanically? An oxygen molecule is confined in a cubic box of volume 2.00 m^3. Assuming that the molecule has an energy equal to $\frac{3}{2}kT$ at $T = 300$ K, what is the value of $n = (n_1^2 + n_2^2 + n_3^2)^{1/2}$, for this molecule? What is the energy separation between the levels n and $n + 1$? What is its de Broglie wavelength? Would it be appropriate to describe this particle as behaving classically?

3.12(b) Should a gas be treated quantum mechanically? A nitrogen molecule is confined in a cubic box of volume 1.00 m^3. Assuming that the molecule has an energy equal to $\frac{3}{2}kT$ at $T = 300$ K, what is the value of $n = (n_1^2 + n_2^2 + n_3^2)^{1/2}$, for this molecule? What is the energy separation between the levels n and $n + 1$? What is its de Broglie wavelength? Would it be appropriate to describe this particle as behaving classically?

3.13(a) Confirm that wavefunctions for a particle on a ring with different values of the quantum number m_l are mutually orthogonal.

3.13(b) Confirm that the wavefunction for a particle on a ring (eqn 3.10b) is normalized.

3.14(a) The rotation of a molecule can be represented by the motion of a point mass rotating on the surface of a sphere. Calculate the magnitude of its angular momentum when $l = 1$ and the possible components of the angular momentum on an arbitrary axis. Express your results as multiples of $\hbar$.

3.14(b) The rotation of a molecule can be represented by the motion of a point mass rotating on the surface of a sphere with angular momentum quantum number $l = 2$. Calculate the magnitude of its angular momentum and the possible components of the angular momentum on an arbitrary axis. Express your results as multiples of $\hbar$.

3.15(a) Draw scale vector diagrams to represent the states (a) $s = \frac{1}{2}$, $m_s = +\frac{1}{2}$, (b) $l = 1$, $m_l = +1$, (c) $l = 2$, $m_l = 0$. What is the angle that the vector makes to the z-axis?

3.15(b) Draw scale vector diagrams for all the permitted rotational states of a body with $l = 6$. What are the angles that the vectors make to the z-axis?

3.16(a) In later chapters we shall see that the number of states corresponding to a given energy plays a crucial role in atomic structure and thermodynamic properties. Determine the degeneracy of a body rotating with $l = 3$.

3.16(b) In later chapters we shall see that the number of states corresponding to a given energy plays a crucial role in atomic structure and thermodynamic properties. Determine the degeneracy of a body rotating with $l = 4$.

3.17(a) The moment of inertia of an $H^{35}Cl$ molecule is 2.73×10^{-47} kg m^2. What energy is needed to excite the molecule from its rotational ground state to the next higher level? What is the wavelength of electromagnetic radiation that would achieve that excitation? Where in the electromagnetic spectrum does that radiation lie?

3.17(b) The average moment of inertia of a benzene molecule is 1.5×10^{-45} kg m^2. What energy is needed to excite the molecule from its rotational ground state to the next higher level? What is the wavelength of electromagnetic radiation that would achieve that excitation? Where in the electromagnetic spectrum does that radiation lie?

3.18(a) The classical picture of an electron is that of a sphere of radius $r_e = 2.82$ fm. On the basis of this model, how fast is a point on the equator of the electron moving? Is this answer plausible?

3.18(b) A proton has a spin angular momentum with $I = \frac{1}{2}$. Suppose it is a sphere of radius 2 pm. On the basis of this model, how fast is a point on the equator of the proton moving?

Problems

Numerical problems

3.1 Calculate the separation between the two lowest translational energy levels for an O_2 molecule in a cubic box with sides of length 5.0 cm. At what value of $n = n_1 = n_2 = n_3$ does the energy of the molecule reach $\frac{3}{2}kT$ at 300 K, and what is the separation of this level from one of the degenerate levels immediately below?

3.2 The particle in a two-dimensional box is a useful model for the motion of electrons around the indole rings (**1**), the conjugated cyclic compound found in the side chain of the amino acid tryptophan. As a first approximation, we can model indole as a rectangle with sides of length 280 pm and 450 pm, with 10 electrons in the conjugated system (the N atom provides two from its lone pair). Assume that in the ground state of the molecule each of the lowest available energy levels is occupied by two electrons. (a) Calculate the energy of an electron in the highest occupied level. (b) Calculate the wavelength of the radiation that can induce a transition between the highest occupied and lowest unoccupied levels.

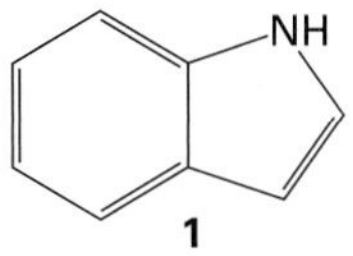

1

3.3 A very crude model of the buckminsterfullerene molecule (C_{60}) is to treat it as a collection of electrons in a cube with sides of length equal to the mean diameter of the molecule (0.7 nm). Suppose that only the π electrons of the carbon atoms contribute, and predict the wavelength of the first excitation of C_{60}. (The actual value is 730 nm.)

3.4 Now treat the buckminsterfullerene molecule as a sphere of radius $a = 0.35$ nm and predict the wavelength of the lowest energy transition of C_{60} resulting from excitation into an energy level not completely filled. You need to know that the energies are

$$E_{n,l} = \frac{F_{n,l}^2 h^2}{8ma^2}$$

with the factors F and degeneracies g as follows:

n,l	1,0	1,1	1,2	2,0	1,3	2,1	1,4	2,2
$F_{n,l}$	1	1.430	1.835	2	2.224	2.459	2.605	2.895
$g_{n,l}$	1	3	5	1	7	3	9	5

3.5 When alkali metals dissolve in liquid ammonia, they lose an electron and give rise to a deep blue solution that contains unpaired electrons occupying cavities in the solvent. These 'metal–ammonia solutions' have a maximum absorption at 1500 nm. Suppose that the absorption is due to the excitation of an electron in a spherical square well from its ground state to the next higher state (see the preceding problem for information), what is the radius of the cavity?

3.6 The detailed distribution of particles within nanostructures is of interest. Use mathematical software to draw contour maps of the wavefunctions and probability densities of a particle confined to a square surface with $n_1 = 4$ and $n_2 = 6$ (or other values of your choice). This problem is taken further in Problem 3.31. Go on—and this is a real challenge—to devise a way to depict the wavefunctions and probability densities of a cubic quantum dot in various states.

3.7 A synchrotron accelerates protons along a circular path of radius r. The state of the proton is described by the unnormalized wavefunction $\psi(\phi) = \psi_{-1}(\phi) + 3^{1/2}\mathrm{i}\psi_{+1}(\phi)$. (a) Normalize this wavefunction. If measurements are made to determine (b) the total angular momentum and (c) the total energy of the proton, what will be the outcome? (d) What are the expectation values of these quantities?

3.8 Can the electronic structures of aromatic molecules be treated as electrons on a ring? Use such a model to predict the π electronic structure of benzene, allowing two electrons to occupy each state and supposing a radius of 133 pm. What is the wavelength of the first absorption band that you would predict on the basis of this model? (The actual value is 185 nm.)

3.9 The rotation of an $^1H^{127}I$ molecule can be pictured as the orbital motion of an H atom at a distance 160 pm from a stationary I atom. (This picture is quite good; to be precise, both atoms rotate around their common centre of mass, which in this case is very close to the I nucleus.) Suppose that the molecule rotates only in a plane (a restriction removed later in the next problem). (a) Calculate the wavelength of electromagnetic radiation needed to excite the molecule into rotation. (b) What, apart from 0, is the minimum angular momentum of the molecule?

3.10 Modify Problem 3.9 so that the molecule is free to rotate in three dimensions, using for its moment of inertia $I = \mu R^2$, with $\mu = m_H m_I/(m_H + m_I)$ and $R = 160$ pm. Calculate the energies and degeneracies of the lowest four rotational levels, and predict the wavelength of electromagnetic radiation emitted in the $l = 1 \rightarrow 0$ transition. In which region of the electromagnetic spectrum does this wavelength appear?

3.11 A diatomic molecule with $\mu = 2.000 \times 10^{-26}$ kg and bond length 250.0 pm is rotating about its centre of mass in the xy-plane. The state of the molecule is described by the normalized wavefunction $\psi(\phi)$. When the total angular momentum of different molecules is measured, two possible results are obtained: a value of $3\hbar$ for 25 per cent of the time and a value of $-3\hbar$ for 75 per cent of the time. However, when the rotational energy of the molecules is measured, only a single result is obtained. (a) What is the expectation value of the angular momentum? (b) Write down an expression for the normalized wavefunction $\psi(\phi)$. (c) What is the result of measuring the energy?

3.12 A helium atom moving on the surface of a buckminsterfullerene molecule before it diffuses into the molecule's interior can be modelled as a free particle on the surface of a sphere of radius 0.35 nm. Suppose the state of the atom is described by a wavepacket of composition $\psi(\theta,\phi) = 2^{1/2}Y_{2,+1}(\theta,\phi) + 3\mathrm{i}Y_{2,+2}(\theta,\phi) + Y_{1,+1}(\theta,\phi)$. (a) Normalize this wavefunction. If (b) the total angular momentum, (c) the z-component of angular momentum, and (d) the total energy of the atom are measured, what results will be found? (e) What are the expectation values of these observables?

3.13 Use the properties of the spherical harmonics to identify the most probable angles a rotating linear molecule will make to an arbitrary axis when $l = 1$, 2, and 3.

Theoretical problems

3.14 The energy levels of an electron in a nanoparticle and confined to a geometrically square region are proportional to $n^2 = n_1^2 + n_2^2$. This is an equation for a circle of radius n in (n_1,n_2)-space (with meaningful values in one quadrant). (a) Produce an argument that uses this relation to predict the degeneracy of a level with a high value of n. (b) Extend this argument to three dimensions.

3.15 Use the separation of variables method to derive the wavefunctions of eqn 3.5a.

3.16 Can the location and momentum of an electron confined to two-dimensional motion in a nanostructure be determined precisely and simultaneously? Determine the values of $\Delta x = (\langle x^2\rangle - \langle x\rangle^2)^{1/2}$ and $\Delta p_x = (\langle p_x^2\rangle - \langle p_x\rangle^2)^{1/2}$ for a particle in a square box of length L in its lowest energy state. Go on to calculate $\Delta p_y = (\langle p_y^2\rangle - \langle p_y\rangle^2)^{1/2}$. Discuss these quantities with reference to the uncertainty principle.

3.17 Confirm by explicit differentiation that the wavefunction given in eqn 3.3a is a solution of the Schrödinger equation, eqn 3.1, for a particle in a two-dimensional box with energies given by eqn 3.3b.

3.18 What is the average angular position for a proton accelerated to a well-defined angular momentum in a synchrotron? Calculate the mean value of $\langle\phi\rangle$ for a particle on a ring described by the wavefunction in eqn 3.10b. Explain your answer.

3.19 The uncertainty principle takes on a different form for cyclic systems: $\Delta l_z \Delta \sin\phi \geq \frac{1}{2}\hbar|\langle\cos\phi\rangle|$, where $\Delta X = \{\langle X^2\rangle - \langle X\rangle^2\}^{1/2}$ in each case. Evaluate the quantities that appear in this expression for (a) a particle with angular momentum $+\hbar$, (b) a particle with wavefunction proportional to $\cos\phi$. Is the uncertainty principle satisfied in each case? Is there a difference between the two cases; if so, why?

3.20 Evaluate the z-component of the angular momentum and the kinetic energy of a proton in a synchroton of radius r whose state is described by the (unnormalized) wavefunctions (a) $e^{i\phi}$, (b) $e^{-2i\phi}$, (c) $\cos\phi$, and (d) $(\cos\chi)e^{i\phi} + (\sin\chi)e^{-i\phi}$.

3.21 If a proton were accelerated on an elliptical ring rather than a circular ring, how would solution of the relevant Schrödinger equation proceed? In particular, is the Schrödinger equation for a particle on an elliptical ring of semimajor axes a and b separable? *Hint.* Although r varies with angle ϕ, the two are related by $r^2 = a^2\sin^2\phi + b^2\cos^2\phi$.

3.22 Use mathematical software to construct a wavepacket of the form

$$\Psi(\phi,t) = \sum_{m_l=0}^{m_{l,\max}} c_{m_l} e^{i(m_l\phi - E_{m_l}t/\hbar)} \qquad E_{m_l} = m_l^2\hbar^2/2I$$

with coefficients c of your choice (for example, all equal). Explore how the wavepacket migrates on the ring but spreads with time.

3.23 Confirm that the spherical harmonics (a) $Y_{0,0}$, (b) $Y_{2,-1}$, and (c) $Y_{3,+3}$ satisfy the Schrödinger equation for a particle free to rotate in three dimensions, and find its energy and angular momentum in each case.

3.24 Confirm by explicit integration that $Y_{1,+1}$ and $Y_{2,0}$ are orthogonal. (The integration required is over the surface of a sphere.)

3.25 Confirm that $Y_{3,+3}$ is normalized to 1. (The integration required is over the surface of a sphere.)

3.26 Derive an expression in terms of l and m_l for the half-angle of the apex of the cone used to represent an angular momentum according to the vector model. Evaluate the expression for an α spin. Show that the minimum possible angle approaches 0 as $l \to \infty$.

3.27 Derive (in Cartesian coordinates) the quantum mechanical operators for the three components of angular momentum starting from the classical definition of angular momentum, $\boldsymbol{l} = \boldsymbol{r} \times \boldsymbol{p}$. Show that any two of the components do not mutually commute, and find their commutator.

3.28 Starting from the operator $\hat{l}_z = \hat{x}\hat{p}_y - \hat{y}\hat{p}_x$, prove that in spherical polar coordinates $\hat{l}_z = -i\hbar\partial/\partial\phi$.

3.29 Show that the commutator $[\hat{l}^2,\hat{l}_z] = 0$, and then, without further calculation, justify the remark that $[\hat{l}^2,\hat{l}_q] = 0$ for all $q = x, y$, and z.

Applications: to biology and nanotechnology

3.30 The particle on a ring is a useful model for the motion of electrons around the porphine ring (**2**), the conjugated macrocycle that forms the structural basis of the haem group and the chlorophylls. We may treat the group as a circular ring of radius 440 pm, with 22 electrons in the conjugated system moving along the perimeter of the ring. We assume that in the ground state of the molecule each level is occupied by two electrons. (a) Calculate the energy and angular momentum of an electron in the highest occupied level. (b) Calculate the frequency of radiation that can induce a transition between the highest occupied and lowest unoccupied levels.

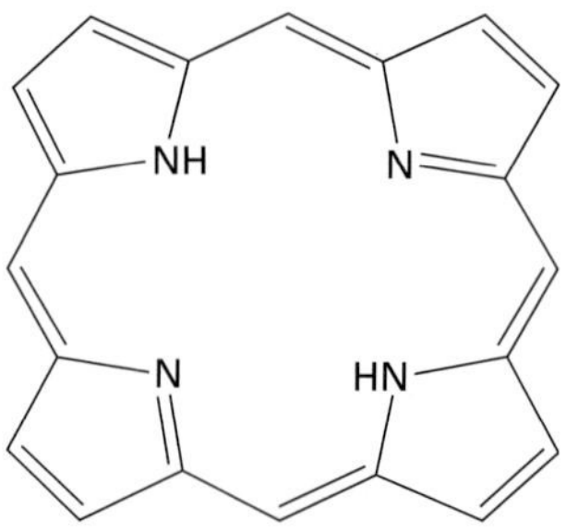

2

3.31 In Problem 3.6 you were invited to plot contours showing the amplitude of the wavefunction and the probability density for various states of a particle confined to a plane. Develop that visualization in relation to the porphine ring (Problem 3.30), treating it as a square. Plot contours of the highest occupied wavefunction and the corresponding probability density superimposed on a drawing of the molecule. Does your map bear any relation to reality?

3.32 Here we explore further the idea introduced in *Impact I3.1* that quantum mechanical effects need to be invoked in the description of the electronic properties of metallic nanocrystals, here modelled as three-dimensional boxes. (a) Set up the Schrödinger equation for a particle of mass m in a three-dimensional rectangular box with sides L_1, L_2, and L_3. Show that the Schrödinger equation is separable. (b) Show that the wavefunction and the energy are defined by three quantum numbers. (c) Specialize the result from part (b) to an electron moving in a cubic box of side $L = 5$ nm and draw an energy diagram showing the first 15 energy levels. Note that each energy level may consist of degenerate energy states. (d) Compare the energy level diagram from part (c) with the energy level diagram for an electron in a one-dimensional box of length $L = 5$ nm. Are the energy levels becoming more or less sparsely distributed in the cubic box than in the one-dimensional box?

3.33 We remarked in *Impact I3.1* that a particle confined to within a spherical cavity is a reasonable starting point for the discussion of the electronic properties of spherical metal nanoparticles. Here, we justify eqn 3.6, which shows that the energy of an electron in a sphere is quantized. (a) The hamiltonian for a particle free to move inside a sphere of radius a is

$$\hat{H} = -\frac{\hbar^2}{2m}\nabla^2$$

Show that the Schrödinger equation is separable into radial and angular components. That is, begin by writing $\psi(r,\theta,\phi) = R(r)Y(\theta,\phi)$, where $R(r)$ depends only on the distance of the particle away from the centre of the sphere, and $Y(\theta,\phi)$ is a spherical harmonic. Then show that the Schrödinger equation can be separated into two equations, one for $R(r)$, the radial equation, and the other for Y, the angular equation. You may wish to consult *Further information 4.1* for additional help. (b) Consider the case $l = 0$. Show by differentiation that the solution of the radial equation has the form

$$R(r) = (2\pi a)^{-1/2}\frac{\sin(n\pi r/a)}{r}$$

(c) Now go on to show (by acknowledging the appropriate boundary conditions) that the allowed energies are given by $E_n = n^2h^2/8ma^2$. With substitution of m_e for m and of R for a, this is eqn 3.6 for the energy.

PART 2
Atoms, molecules, and assemblies

In this part of the text we apply the principles of quantum mechanics to systems of great chemical significance. First, we draw on the foregoing material to describe the electronic structures of atoms. As we shall see, a few simple ideas based on the simplest atom of all, the hydrogen atom, can be developed to account for the structure of the periodic table and the periodicity of the properties of the elements. Then we allow the atoms to form bonds and see that, once again, a simple species, in this case the hydrogen molecule, provides a model on which the structures of even quite elaborate molecules can be built. In both of these chapters we focus on the concepts involved; in the chapter that follows we show how modern computational chemistry is brought to bear on the same issues and results in quantitative accounts of the properties of molecules. Here too we illustrate the varieties of computation that are used in contemporary quantum chemistry by focusing once again on the hydrogen molecule and demonstrating as explicitly as possible the underlying formulation of modern molecular structure computational software. One of the most powerful tools for setting up molecular structure calculations, formulating molecular orbitals, and assessing a molecule's properties and spectroscopic behaviour is group theory, the mathematical theory of symmetry, and we show how straightforward ideas about symmetry can be made quantitative. Then we turn to bulk systems. Here we need to know the interactions that result in the aggregation of molecules, either weakly as in gases and liquids or strongly as in molecular solids. Finally, we explore the properties of solids in which atoms, ions, and molecules lie in regular arrays. We see how to classify crystals, how atoms form different types of solids, and how these structures are established experimentally by X-ray diffraction. Solids have characteristic mechanical, electrical, optical, and magnetic properties, and we conclude this part by showing how the ideas developed in the preceding chapters are relevant to extended arrays of atoms.

Atomic structure and spectra

4

We now use the principles of quantum mechanics introduced in the earlier chapters to describe the internal structures of atoms. We see what experimental information is available from a study of the spectrum of atomic hydrogen. Then we set up the Schrödinger equation for an electron in an atom and separate it into angular and radial parts. The wavefunctions obtained are the 'atomic orbitals' of hydrogenic atoms. Next, we use these hydrogenic atomic orbitals to describe the structures of many-electron atoms. In conjunction with the Pauli exclusion principle, we account for the periodicity of atomic properties. The spectra of many-electron atoms are more complicated than those of hydrogen, but the same principles apply. We see in the closing sections of the chapter how such spectra are described by using term symbols, and the origin of the finer details of their appearance.

In this chapter we see how to use quantum mechanics to describe the **electronic structure** of an atom, the distribution of electrons around its nucleus. The concepts we meet are of central importance for understanding the structure and reactivity of atoms and molecules, and hence have extensive chemical applications. We need to distinguish between two types of atom. A **hydrogenic atom** is a one-electron atom or ion of general atomic number Z; examples of hydrogenic atoms are H, He^+, Li^{2+}, O^{7+}, and even U^{91+}. A **many-electron atom** (or *polyelectronic atom*) is an atom or ion with more than one electron; examples include all neutral atoms other than H. So even He, with only two electrons, is a many-electron atom. Hydrogenic atoms are important because their Schrödinger equations can be solved exactly. They also provide a set of concepts that are used to describe the structures of many-electron atoms and, as we shall see in the following chapter, the structures of molecules too.

Hydrogenic atoms

When an electric discharge is passed through gaseous hydrogen, the H_2 molecules are dissociated and the energetically excited H atoms that are generated emit light of discrete frequencies, producing a spectrum of a series of 'lines' (Fig. 4.1). The Swedish spectroscopist Johannes Rydberg noted (in 1890) that all of the lines are described by the expression

$$\tilde{\nu} = R_{\mathrm{H}}\left(\frac{1}{n_1^2} - \frac{1}{n_2^2}\right) \qquad R_{\mathrm{H}} = 109\,677\ \mathrm{cm}^{-1} \tag{4.1}$$

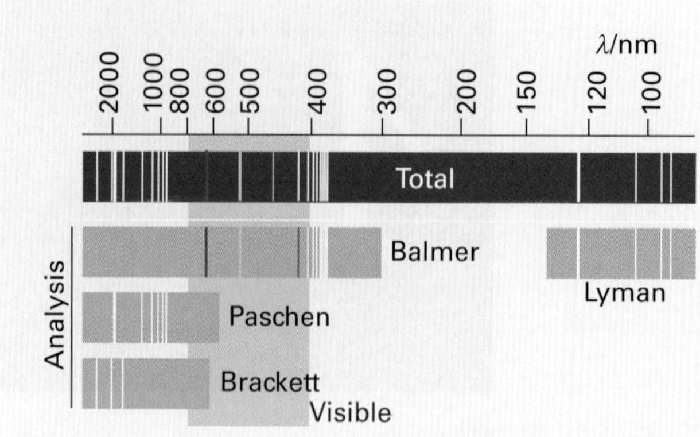

Fig. 4.1 The spectrum of atomic hydrogen. Both the observed spectrum and its resolution into overlapping series are shown. Note that the Balmer series lies in the visible region.

with $n_1 = 1$ (the *Lyman series*), 2 (the *Balmer series*), and 3 (the *Paschen series*), and that in each case $n_2 = n_1 + 1, n_1 + 2, \ldots$. The constant R_H is now called the **Rydberg constant** for the hydrogen atom.

The form of eqn 4.1 strongly suggests that each spectral line arises from a **transition**, a jump from one state to another, each with an energy proportional to R_H/n^2, with the difference in energy discarded as electromagnetic radiation of frequency (and wavenumber) given by the Bohr frequency condition (Section 1.1). Our tasks in the first part of this chapter are therefore to determine the origin of this energy quantization, to find the permitted energy levels, and to account for the value of R_H.

Example 4.1 *Calculating the shortest and longest wavelength lines in a series*

Determine the shortest and longest wavelength lines in the Balmer series.

Method Identify the value of n_1 for the Balmer series. The shortest wavelength line corresponds to the largest wavenumber; from eqn 4.1, recognize that this line will arise from $n_2 = \infty$. The longest wavelength corresponds to the smallest wavenumber, which will arise from $n_2 = n_1 + 1$. To convert from wavenumber to wavelength, use $\lambda = 1/\tilde{\nu}$.

Answer The Balmer series corresponds to $n_1 = 2$. The largest wavenumber (use $n_2 = \infty$), calculated from eqn 4.1, is 27 419 cm^{-1}, which corresponds to a wavelength of 365 nm. The smallest wavenumber (use $n_2 = 3$) is 15 233 cm^{-1}, which corresponds to a wavelength of 656 nm.

Self-test 4.1 Calculate the shortest and longest wavelength lines in the Paschen series. [821 nm, 1876 nm]

4.1 The structure of hydrogenic atoms

The Coulomb potential energy of an electron in a hydrogenic atom of atomic number Z (and nuclear charge Ze) is (see *Fundamentals F.6*)

$$V = -\frac{Ze^2}{4\pi\varepsilon_0 r} \tag{4.2}$$

where r is the distance of the electron from the nucleus and ε_0 is the vacuum permittivity. Because this expression depends only on the distance from a single point (here, the nucleus), it is an example of a *central potential*. The hamiltonian operator for the electron and a nucleus of mass m_N is therefore

$$\hat{H} = \hat{E}_{\text{k,electron}} + \hat{E}_{\text{k,nucleus}} + \hat{V}$$
$$= -\frac{\hbar^2}{2m_e}\nabla_e^2 - \frac{\hbar^2}{2m_N}\nabla_N^2 - \frac{Ze^2}{4\pi\varepsilon_0 r} \tag{4.3}$$

The subscripts on ∇^2 indicate differentiation with respect to the electronic or nuclear coordinates.

(a) The separation of variables

Physical intuition suggests that the full Schrödinger equation ought to separate into two equations, one for the motion of the atom as a whole through space and the other for the motion of the electron relative to the nucleus. We show in *Further information 4.1* how this separation is achieved, and that the Schrödinger equation for the internal motion of the electron relative to the nucleus is

$$-\frac{\hbar^2}{2\mu}\nabla^2\psi - \frac{Ze^2}{4\pi\varepsilon_0 r}\psi = E\psi \qquad \mu = \frac{m_e m_N}{m_e + m_N} \tag{4.4}$$

where differentiation is now with respect to the coordinates of the electron relative to the nucleus. The quantity μ is called the **reduced mass**. The reduced mass is very similar to the electron mass because m_N, the mass of the nucleus, is much larger than the mass of an electron, so $\mu \approx m_e$. In all except the most precise work, the reduced mass can be replaced by m_e.

Because we can think of the electron as free to move around the nucleus on a spherical surface—the 'particle-on-a-sphere' problem treated in Section 3.4—with the additional freedom to move between concentric surfaces of different radii, we can suspect that the problem will separate into angular and radial parts. Therefore, we write

$$\psi(r,\theta,\phi) = R(r)Y(\theta,\phi) \tag{4.5}$$

and examine whether the Schrödinger equation can be separated into two equations, one for R and the other for Y. To do so, we use the laplacian in three dimensions from Table 1.1 to write the Schrödinger equation in eqn 4.4 as

$$-\frac{\hbar^2}{2\mu}\left(\frac{\partial^2}{\partial r^2} + \frac{2}{r}\frac{\partial}{\partial r} + \frac{1}{r^2}\Lambda^2\right)RY + VRY = ERY$$

Because R depends only on r and Y depends only on the angular coordinates, this equation becomes

$$-\frac{\hbar^2}{2\mu}\left(Y\frac{\mathrm{d}^2R}{\mathrm{d}r^2}+\frac{2Y}{r}\frac{\mathrm{d}R}{\mathrm{d}r}+\frac{R}{r^2}\Lambda^2Y\right)+VRY=ERY$$

On multiplying through by r^2/RY, we obtain

$$-\frac{\hbar^2}{2\mu R}\left(r^2\frac{\mathrm{d}^2R}{\mathrm{d}r^2}+2r\frac{\mathrm{d}R}{\mathrm{d}r}\right)+Vr^2-\frac{\hbar^2}{2\mu Y}\Lambda^2Y=Er^2$$

At this point we employ the usual argument for the separation of variables (Section 3.1). The term in Y is the only one that depends on the angular variables, so it must be a constant and the differential equation separates into two equations:

$$-\frac{\hbar^2}{2\mu Y}\Lambda^2Y=\text{constant}$$

$$-\frac{\hbar^2}{2\mu R}\left(r^2\frac{\mathrm{d}^2R}{\mathrm{d}r^2}+2r\frac{\mathrm{d}R}{\mathrm{d}r}\right)+Vr^2-Er^2=-\text{constant}$$

When we write the constant as $\hbar^2 l(l+1)/2\mu$, we immediately obtain a familiar equation for Y:

$$\Lambda^2Y=-l(l+1)Y \tag{4.6}$$

This equation is the same as the Schrödinger equation for a 'particle on a sphere' (Section 3.4); its solutions are the spherical harmonics (Table 3.2), and they are labelled with the quantum numbers l and m_l that specify the angular momentum of the particle. We consider them in more detail shortly. The appearance of the second separated equation is simplified if we write $R=u(r)/r$, for it then becomes

$$-\frac{\hbar^2}{2\mu}\frac{\mathrm{d}^2u}{\mathrm{d}r^2}+V_{\text{eff}}u=Eu \tag{4.7a}$$

where

$$V_{\text{eff}}=-\frac{Ze^2}{4\pi\varepsilon_0 r}+\frac{l(l+1)\hbar^2}{2\mu r^2} \tag{4.7b}$$

Equation 4.7 is called the **radial wave equation** and describes the motion of a particle of mass μ in a one-dimensional region $0\le r<\infty$ where the potential energy is V_{eff}.

(b) The radial solutions

We can anticipate some features of the shapes of the radial wavefunctions by analysing the form of V_{eff}. The first term in eqn 4.7b is the Coulomb potential energy of the electron in the field of the nucleus. The second term, which depends on the angular momentum of the electron around the nucleus, stems from what in classical physics would be called the centrifugal effect. When $l=0$, the electron has no angular momentum, and the effective potential energy is purely Coulombic and attractive at all radii (Fig. 4.2). When $l\neq 0$, the centrifugal term gives a positive (repulsive) contribution to the effective potential energy. When the electron is close to the nucleus ($r\approx 0$), this repulsive term, which is proportional to $1/r^2$, dominates the attractive Coulombic component, which is proportional to $1/r$, and the net result is an effective repulsion of the electron from the nucleus. The two effective potential energies, the one for $l=0$ and the one for $l\neq 0$, are qualitatively very different close to the nucleus. However, they are similar at large distances because the centrifugal contribution tends to zero more rapidly (as $1/r^2$) than the Coulombic contribution (as $1/r$). Therefore, we can expect the solutions with $l=0$ and $l\neq 0$ to be quite different near the nucleus but similar far away from it. We show in the following *Justification* that:

1. Close to the nucleus the radial wavefunction is proportional to r^l.

2. Far from the nucleus all wavefunctions approach zero exponentially.

It follows that when $l\neq 0$ (and $r^l=0$ when $r=0$) there is a zero probability density for finding the electron at the nucleus, and the higher the orbital angular momentum, the less likely the electron is to be found near the nucleus (Fig. 4.3). However, when $l=0$ (and $r^l=1$ even when $r=0$) there is a nonzero probability density of finding the electron at the nucleus. The contrast

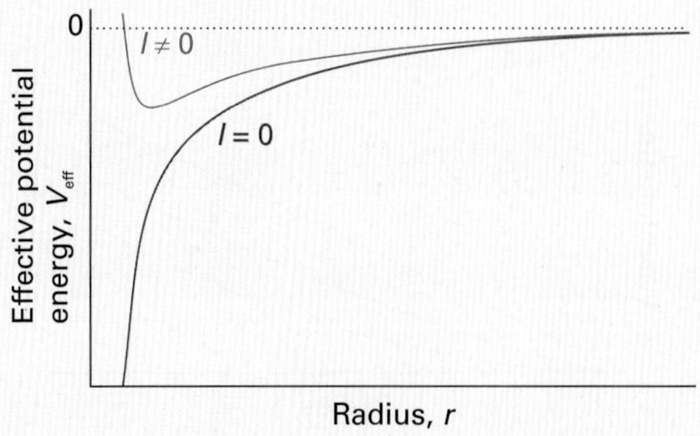

Fig. 4.2 The effective potential energy of an electron in the hydrogen atom. When the electron has zero orbital angular momentum, the effective potential energy is the Coulombic potential energy. When the electron has nonzero orbital angular momentum, the centrifugal effect gives rise to a positive contribution that is very large close to the nucleus. We can expect the $l=0$ and $l\neq 0$ wavefunctions to be very different near the nucleus.

interActivity Plot the effective potential energy against r for several nonzero values of the orbital angular momentum l. How does the location of the minimum in the effective potential energy vary with l?

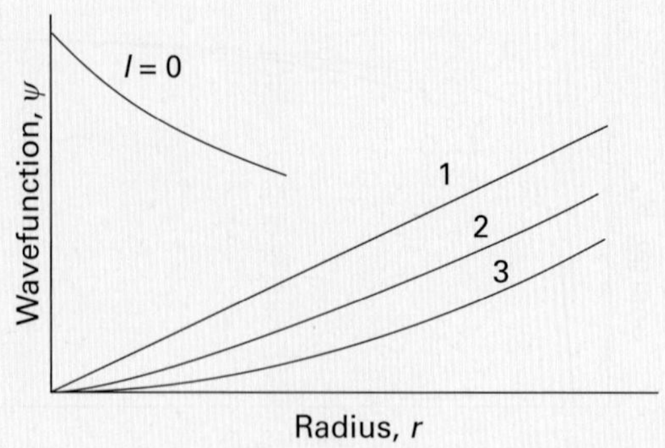

Fig. 4.3 Close to the nucleus, orbitals with $l = 1$ are proportional to r, orbitals with $l = 2$ are proportional to r^2, and orbitals with $l = 3$ are proportional to r^3. Electrons are progressively excluded from the neighbourhood of the nucleus as l increases. An orbital with $l = 0$ has a finite, nonzero value at the nucleus.

in behaviour has profound implications for chemistry, as we shall see, for it underlies the structure of the periodic table. The exponential decay of wavefunctions has a further important implication: it means that atoms can, to a good approximation, be represented by spheres with reasonably well defined radii. This feature is especially important when we come to discuss solids, which are commonly modelled as aggregates of spheres representing their atoms and ions.

Justification 4.1 *The shape of the radial wavefunction*

When r is very small (close to the nucleus), $u = rR \approx 0$, so the right-hand side of eqn 4.7a is zero; we can also ignore all but the largest terms (those depending on $1/r^2$) in eqn 4.7b and write

$$-\frac{d^2u}{dr^2} + \frac{l(l+1)}{r^2}u \approx 0$$

The solution of this equation (for $r \approx 0$) is

$$u \approx Ar^{l+1} + \frac{B}{r^l}$$

as may be verified by substitution of the solution into the differential equation. Because $R = u/r$, and R cannot be infinite anywhere and specifically at $r = 0$, we must set $B = 0$, and hence obtain $R \approx Ar^l$, as we wanted to show.

Far from the nucleus, when r is very large and we can ignore all terms in $1/r$ and $1/r^2$, eqn 4.7a becomes

$$-\frac{\hbar^2}{2\mu}\frac{d^2u}{dr^2} \simeq Eu$$

where $\simeq$ means 'asymptotically equal to'. Because

$$\frac{d^2u}{dr^2} = \frac{d^2(rR)}{dr^2} = r\frac{d^2R}{dr^2} + 2\frac{dR}{dr} \simeq r\frac{d^2R}{dr^2}$$

this equation has the form

$$-\frac{\hbar^2}{2\mu}\frac{d^2R}{dr^2} \simeq ER$$

The acceptable (finite) solution of this equation (for r large and $E < 0$) is

$$R \simeq e^{-(2\mu|E|/\hbar^2)^{1/2}r}$$

and the wavefunction decays exponentially towards zero as r increases.

We shall not go through the technical steps of solving the radial equation for the full range of radii, and see how the form r^l close to the nucleus blends into the exponentially decaying form at great distances.[1] It is sufficient to know that the two limits can be matched only for integral values of a new quantum number n, and that the allowed energies corresponding to the allowed solutions are

$$E_n = -\frac{Z^2\mu e^4}{32\pi^2\varepsilon_0^2\hbar^2n^2} \quad (4.8)$$

with $n = 1, 2, \ldots$. Likewise, it is found that the radial wavefunctions depend on the values of both n and l (but not on m_l) with l restricted to values $0, 1, \ldots, n-1$, and all radial wavefunctions have the form

$$R(r) = r^l \times (\text{polynomial in } r) \times (\text{decaying exponential in } r)$$

These functions are most simply written in terms of the dimensionless quantity ρ (rho), where (for simplicity, and introducing negligible error, replacing μ with m_e)

$$\rho = \frac{2Zr}{na_0} \qquad a_0 = \frac{4\pi\varepsilon_0\hbar^2}{m_e e^2} \quad (4.9)$$

The **Bohr radius**, a_0, has the value 52.9 pm; it is so called because the same quantity appeared in Bohr's early model of the hydrogen atom as the radius of the electron orbit of lowest energy. Specifically, the radial wavefunctions for an electron with quantum numbers n and l are the (real) functions

$$R_{n,l}(r) = N_{n,l}\rho^l L_{n+1}^{2l+1}(\rho)e^{-\rho/2} \quad (4.10)$$

where L is a polynomial in ρ called an *associated Laguerre polynomial*, a function commonly used in mathematical physics: it connects the $r \approx 0$ solutions on its left (corresponding to $R \propto \rho^l$) to the exponentially decaying function on its right. The notation might look fearsome, but the polynomials have quite simple forms, such as 1, ρ, and $2 - \rho$ (they can be picked out in Table 4.1, with $\rho \propto r$). We can interpret the components of this expression as follows:

1. The exponential factor ensures that the wavefunction approaches zero far from the nucleus.

[1] For details, see our *Molecular quantum mechanics*, Oxford University Press, Oxford (2005).

Table 4.1 Hydrogenic radial wavefunctions

Orbital	n	l	$R_{n,l}$
1s	1	0	$2\left(\dfrac{Z}{a}\right)^{3/2} e^{-\rho/2}$
2s	2	0	$\dfrac{1}{8^{1/2}}\left(\dfrac{Z}{a}\right)^{3/2} (2-\rho)e^{-\rho/2}$
2p	2	1	$\dfrac{1}{24^{1/2}}\left(\dfrac{Z}{a}\right)^{3/2} \rho e^{-\rho/2}$
3s	3	0	$\dfrac{1}{243^{1/2}}\left(\dfrac{Z}{a}\right)^{3/2} (6-6\rho+\rho^2)e^{-\rho/2}$
3p	3	1	$\dfrac{1}{486^{1/2}}\left(\dfrac{Z}{a}\right)^{3/2} (4-\rho)\rho e^{-\rho/2}$
3d	3	2	$\dfrac{1}{2430^{1/2}}\left(\dfrac{Z}{a}\right)^{3/2} \rho^2 e^{-\rho/2}$

$\rho = (2Z/na)r$ with $a = 4\pi\varepsilon_0\hbar^2/\mu e^2$. For an infinitely heavy nucleus (or one that may be assumed to be so), $\mu = m_e$ and $a = a_0$, the Bohr radius. The full wavefunction is obtained by multiplying R by the appropriate Y given in Table 3.2.

2. The factor ρ^l ensures that (provided $l > 0$) the wavefunction vanishes at the nucleus.

3. The associated Laguerre polynomial is a function that oscillates from positive to negative values and accounts for the presence of radial nodes.

Expressions for some radial wavefunctions are given in Table 4.1 and illustrated in Fig. 4.4. Note that, because r is never negative, the zero in the radial wavefunctions at $r = 0$ (for $l > 0$) is not a node: the wavefunction does not pass *through* zero there.

● A BRIEF ILLUSTRATION

To calculate the probability density at the nucleus for an electron with $n = 1$, $l = 0$, and $m_l = 0$, we begin by evaluating $\psi_{1,0,0}$ at $r = 0$:

$$\psi_{1,0,0}(0,\theta,\phi) = R_{1,0}(0)Y_{0,0}(\theta,\phi) = 2\left(\frac{Z}{a_0}\right)^{3/2}\left(\frac{1}{4\pi}\right)^{1/2}$$

The probability density is therefore

$$\psi_{1,0,0}(0,\theta,\phi)^2 = \frac{Z^3}{\pi a_0^3}$$

which evaluates to 2.15×10^{-6} pm^{-3} when $Z = 1$. ●

Self-test 4.2 Evaluate the probability density at the nucleus for an electron with $n = 2$, $l = 0$, $m_l = 0$. $[(Z/a_0)^3/8\pi]$

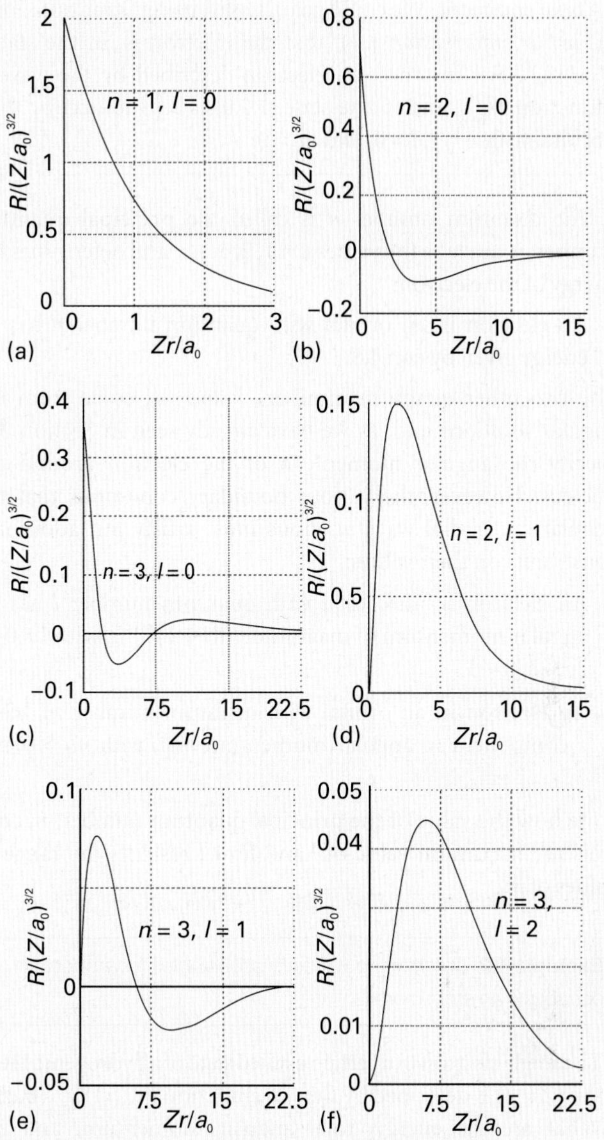

Fig. 4.4 The radial wavefunctions of the first few states of hydrogenic atoms of atomic number Z. Note that the orbitals with $l = 0$ have a nonzero and finite value at the nucleus. The horizontal scales are different in each case: orbitals with high principal quantum numbers are relatively distant from the nucleus.

interActivity Use mathematical software to find the locations of the radial nodes in hydrogenic wavefunctions with n up to 3.

4.2 Atomic orbitals and their energies

An **atomic orbital** is a one-electron wavefunction for an electron in an atom. Each hydrogenic atomic orbital (eqn 4.5) is defined by three quantum numbers, designated n, l, and m_l. When an electron is described by one of these wavefunctions, we say that it 'occupies' that orbital.

A brief comment We could go on to say, using Dirac notation (*Further information 1.1*), that the electron is 'in the state $|n,l,m_l\rangle$'. For instance, an electron described by the wavefunction $\psi_{1,0,0}$ and in the state $|1,0,0\rangle$ is said to occupy the orbital with $n=1$, $l=0$, and $m_l=0$.

The quantum number n is called the **principal quantum number**; it can take the values $n=1, 2, 3, \ldots$ and determines the energy of the electron:

An electron in an orbital with quantum number n has an energy given by eqn 4.8.

The two other quantum numbers, l and m_l, come from the angular solutions and, as we have already seen in Section 3.4, specify the angular momentum of the electron around the nucleus. However, the various boundary conditions that the angular and radial wavefunctions must satisfy put additional constraints on their values:

An electron in an orbital with quantum number l has an angular momentum of magnitude $\{l(l+1)\}^{1/2}\hbar$, with $l=0, 1, 2, \ldots, n-1$.

An electron in an orbital with quantum number m_l has a z-component of angular momentum $m_l\hbar$, with $m_l = 0, \pm1, \pm2, \ldots, \pm l$.

Note how the value of the principal quantum number, n, controls the maximum value of l and how l restricts the range of values of m_l.

Example 4.2 *Determining values of observables for an electron occupying an atomic orbital*

The single electron in a certain excited state of a hydrogenic He^+ ion ($Z=2$) is described by the wavefunction $R_{3,2}(r)Y_{2,-1}(\theta,\phi)$. What are the energy, total angular momentum, and z-component of the angular momentum of its electron?

Method Identify the quantum numbers n, l, and m_l by noting that radial wavefunctions are designated $R_{n,l}$ and angular wavefunctions are designated Y_{l,m_l}. Then use eqn 4.8 to calculate the energy from n and infer the values of the angular momentum from l and m_l. To a good approximation, the reduced mass in eqn 4.8 can be replaced by m_e. As remarked in the text, the energy of a hydrogenic atom depends on n but is independent of the values of l and m_l.

Answer We identify $n=3$, $l=2$, and $m_l=-1$. For the energy, eqn 4.8 gives:

$$E_3 = -\frac{2^2 \times (9.109\times10^{-31}\ \text{kg}) \times (1.602\times10^{-19}\ \text{C})^4}{32\pi^2 \times (8.854\times10^{-12}\ \text{J}^{-1}\ \text{C}^2\ \text{m}^{-1})^2 \times (1.055\times10^{-34}\ \text{J s})^2 \times 3^2}$$
$$= -9.676\times10^{-19}\ \text{J}$$

or -0.9676 aJ (a, for atto, is the prefix that denotes 10^{-18}). Because $l=2$, the magnitude of the angular momentum is $6^{1/2}\hbar$; because $m_l=-1$, the z-component is $-\hbar$. For greater accuracy, we should use the reduced mass of the electron and helium nucleus.

Self-test 4.3 Repeat the problem for the electron in an excited state of a Li^{2+} ion ($Z=3$) for which the wavefunction is known to be $R_{4,3}(r)Y_{3,-2}(\theta,\phi)$.
[-1.225 aJ (1 aJ $=10^{-18}$ J), $2(3)^{1/2}\hbar$, $-2\hbar$]

To define the state of an electron in a hydrogenic atom fully we need to specify not only the orbital it occupies but also its spin state. We saw in Section 3.5 that an electron possesses an intrinsic angular momentum, its 'spin', that is described by the two quantum numbers s and m_s (the analogues of l and m_l). The value of s is fixed at $\frac{1}{2}$ for an electron, so we do not need to consider it further at this stage. However, m_s may be either $+\frac{1}{2}$ or $-\frac{1}{2}$, and to specify the electron's state in a hydrogenic atom we need to specify which of these values describes it. It follows that to specify the state of an electron in a hydrogenic atom, we need to give the values of four quantum numbers, namely n, l, m_l, and m_s.

(a) The energy levels

The energy levels predicted by eqn 4.8 are depicted in Fig. 4.5. The energies, and also the separation of neighbouring levels, are proportional to Z^2, so the levels are four times as wide apart (and the ground state four times deeper in energy) in He^+ ($Z=2$) than in H ($Z=1$). All the energies given by eqn 4.8 are negative. They refer to the **bound states** of the atom, in which the energy of the atom is lower than that of the infinitely separated, stationary electron and nucleus (which corresponds to the zero of energy and $n=\infty$). There are also solutions of the Schrödinger equation with positive energies. These solutions correspond to **unbound states** of the electron, the states to which an electron is raised

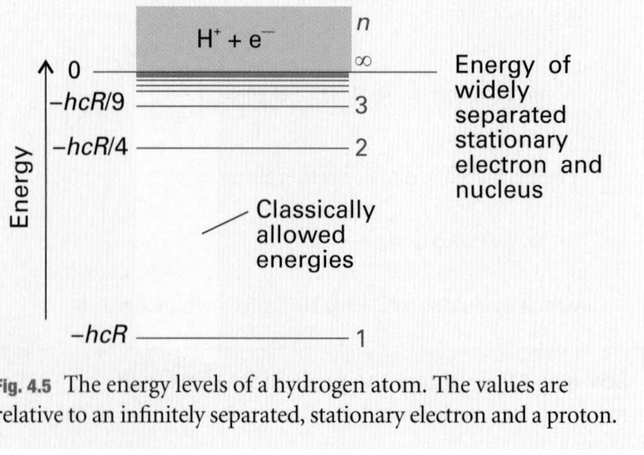

Fig. 4.5 The energy levels of a hydrogen atom. The values are relative to an infinitely separated, stationary electron and a proton.

when it is ejected from the atom by a high-energy collision or photon. The energies of the unbound electron are not quantized and form the **continuum states** of the atom.

Equation 4.8 is consistent with the spectroscopic result summarized by eqn 4.1, which suggested that the energy levels are proportional to $1/n^2$, and we can identify the Rydberg constant for hydrogen ($Z = 1$), after using $\tilde{\nu} = \nu/c$ and the Bohr frequency condition eqn 1.1, as

$$hcR_{\mathrm{H}} = \frac{\mu_{\mathrm{H}} e^4}{32\pi^2\varepsilon_0^2\hbar^2} \qquad [4.11]$$

where μ_{H} is the reduced mass for hydrogen. The **Rydberg constant** itself, R, is defined by the same expression except for the replacement of μ_{H} by the mass of an electron, m_e, corresponding to a nucleus of infinite mass:

$$R_{\mathrm{H}} = \frac{\mu_{\mathrm{H}}}{m_e}R \qquad R = \frac{m_e e^4}{8\varepsilon_0^2 h^3 c} \qquad [4.12]$$

Insertion of the values of the fundamental constants into the expression for R_{H} gives almost exact agreement with the experimental value. The only discrepancies arise from the neglect of relativistic corrections (in simple terms, the increase of mass with speed), which the non-relativistic Schrödinger equation ignores.

(b) Ionization energies

The **ionization energy**, I, of an element is the minimum energy required to remove an electron from the ground state, the state of lowest energy, of one of its atoms in the gas phase. Because the ground state of hydrogen is the state with $n = 1$, with energy $E_1 = -hcR_{\mathrm{H}}$, and the atom is ionized when the electron has been excited to the level corresponding to $n = \infty$ (see Fig. 4.5), the energy that must be supplied is

$$I = hcR_{\mathrm{H}} \qquad (4.13)$$

The value of I is 2.179 aJ, which corresponds to 13.60 eV.

Example 4.3 *Measuring an ionization energy spectroscopically*

The emission spectrum of atomic hydrogen shows lines at 82 259, 97 492, 102 824, 105 292, 106 632, and 107 440 $\mathrm{cm^{-1}}$, which correspond to transitions to the same lower state. Determine (a) the ionization energy of the lower state, (b) the value of the Rydberg constant.

Method The spectroscopic determination of ionization energies depends on the determination of the 'series limit', the wavenumber at which the series terminates and becomes a continuum. If the upper state lies at an energy $-hcR_{\mathrm{H}}/n^2$, then, when the atom makes a transition to E_{lower}, a photon of wavenumber

$$\tilde{\nu} = -\frac{R_{\mathrm{H}}}{n^2} - \frac{E_{\mathrm{lower}}}{hc}$$

is emitted. However, because $I = -E_{\mathrm{lower}}$, it follows that

$$\tilde{\nu} = \frac{I}{hc} - \frac{R_{\mathrm{H}}}{n^2}$$

A plot of the wavenumbers against $1/n^2$ should give a straight line of slope $-R_{\mathrm{H}}$ and intercept I/hc. Use mathematical software to make a least-squares fit of the data to get a result that reflects the precision of the data.

Answer The wavenumbers are plotted against $1/n^2$ in Fig. 4.6. The (least-squares) intercept lies at 109 679 $\mathrm{cm^{-1}}$, so the ionization energy is 2.1788 aJ (1312.1 kJ $\mathrm{mol^{-1}}$). The slope is, in this instance, numerically the same, so R_{H} = 109 679 $\mathrm{cm^{-1}}$. A similar extrapolation procedure can be used for many-electron atoms (see Section 4.4).

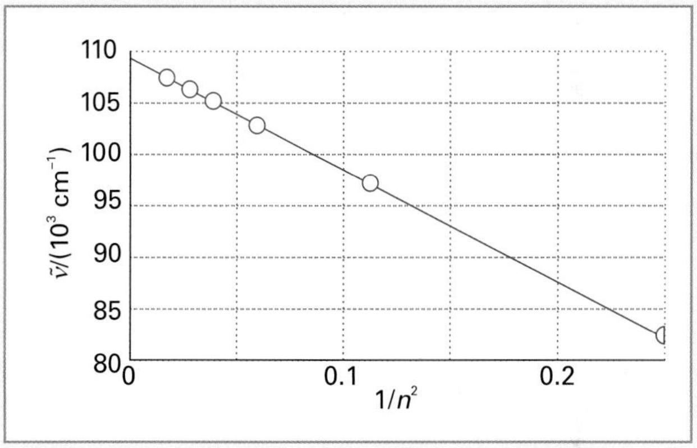

Fig. 4.6 The plot of the data in Example 4.3 used to determine the ionization energy of an atom (in this case, of H).

interActivity The initial value of n was not specified in Example 4.3. Show that the correct value can be determined by making several choices and selecting the one that leads to a straight line.

Self-test 4.4 The emission spectrum of atomic deuterium shows lines at 15 238, 20 571, 23 039, and 24 380 $\mathrm{cm^{-1}}$, which correspond to transitions to the same lower state. Determine (a) the ionization energy of the lower state, (b) the ionization energy of the ground state, (c) the mass of the deuteron (by expressing the Rydberg constant in terms of the reduced mass of the electron and the deuteron, and solving for the mass of the deuteron).

[(a) 328.1 kJ $\mathrm{mol^{-1}}$, (b) 1312.4 kJ $\mathrm{mol^{-1}}$, (c) 2.8×10^{-27} kg, a result very sensitive to R_{D}]

(c) Shells and subshells

All the orbitals of a given value of n are said to form a single **shell** of the atom. In a hydrogenic atom, all orbitals of given n, and therefore belonging to the same shell, have the same energy. It is common to refer to successive shells by upper-case letters:

$n =$	1	2	3	4...
	K	L	M	N...

Thus, all the orbitals of the shell with $n = 2$ form the L shell of the atom, and so on.

The orbitals with the same value of n but different values of l are said to form a **subshell** of a given shell. These subshells are generally referred to by lower-case letters:

$l =$	0	1	2	3	4	5	6...
	s	p	d	f	g	h	i...

The letters then run alphabetically omitting j (because some languages do not distinguish between i and j). Figure 4.7 is a version of Fig. 4.5 which shows the subshells explicitly. Because l can range from 0 to $n - 1$, giving n values in all, it follows that there are n subshells of a shell with principal quantum number n. Thus, when $n = 1$, there is only one subshell, the one with $l = 0$. When $n = 2$, there are two subshells, the 2s subshell (with $l = 0$) and the 2p subshell (with $l = 1$).

When $n = 1$ there is only one subshell, that with $l = 0$, and that subshell contains only one orbital, with $m_l = 0$ (the only value of m_l permitted). When $n = 2$, there are four orbitals, one in the s subshell with $l = 0$ and $m_l = 0$, and three in the $l = 1$ subshell with $m_l = +1, 0, -1$. When $n = 3$ there are nine orbitals (one with $l = 0$, three with $l = 1$, and five with $l = 2$). The organization of orbitals in the shells is summarized in Fig. 4.8. In general, the number of orbitals in a shell of principal quantum number n is n^2, so in a hydrogenic atom each energy level is n^2-fold degenerate.

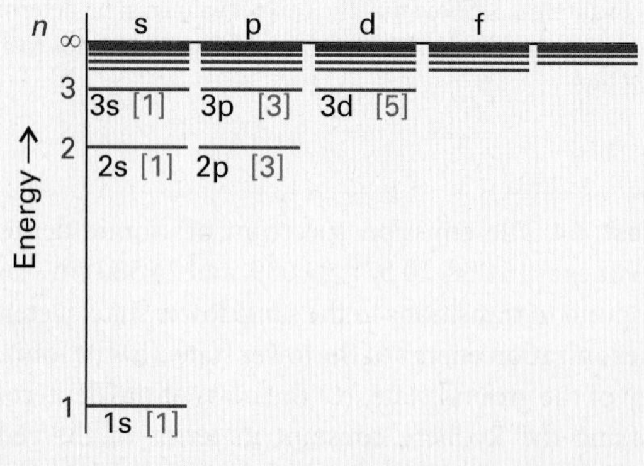

Fig. 4.7 The energy levels of the hydrogen atom showing the subshells and (in square brackets) the numbers of orbitals in each subshell. In hydrogenic atoms, all orbitals of a given shell have the same energy.

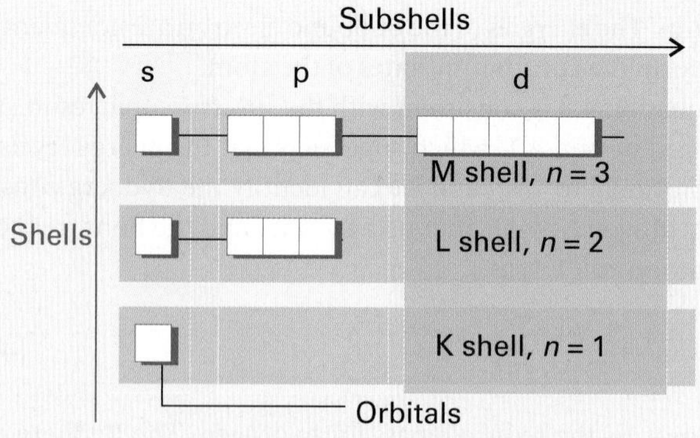

Fig. 4.8 The organization of orbitals (white squares) into subshells (characterized by l) and shells (characterized by n).

(d) Atomic orbitals

The orbital occupied in the ground state is the one with $n = 1$ (and therefore with $l = 0$ and $m_l = 0$, the only possible values of these quantum numbers when $n = 1$). From Tables 3.2 and 4.1 we can write (for $Z = 1$):

$$\psi = \frac{1}{(\pi a_0^3)^{1/2}} e^{-r/a_0} \tag{4.14}$$

This wavefunction is independent of angle and has the same value at all points of constant radius; that is, the 1s orbital is *spherically symmetrical.* The wavefunction decays exponentially from a maximum value of $1/(\pi a_0^3)^{1/2}$ at the nucleus (at $r = 0$). It follows that the most probable point at which the electron will be found is at the nucleus itself.

We can understand the general form of the ground-state wavefunction by considering the contributions of the potential and kinetic energies to the total energy of the atom. The closer the electron is to the nucleus on average, the lower its average potential energy. This dependence suggests that the lowest potential energy should be obtained with a sharply peaked wavefunction that has a large amplitude at the nucleus and is zero everywhere else (Fig. 4.9). However, this shape implies a high kinetic energy, because such a wavefunction has a very high average curvature. The electron would have very low kinetic energy if its wavefunction had only a very low average curvature. However, such a wavefunction spreads to great distances from the nucleus and the average potential energy of the electron will be correspondingly high (that is, less negative). The actual ground-state wavefunction is a compromise between these two extremes: the wavefunction spreads away from the nucleus (so the expectation value of the potential energy is not as low as in

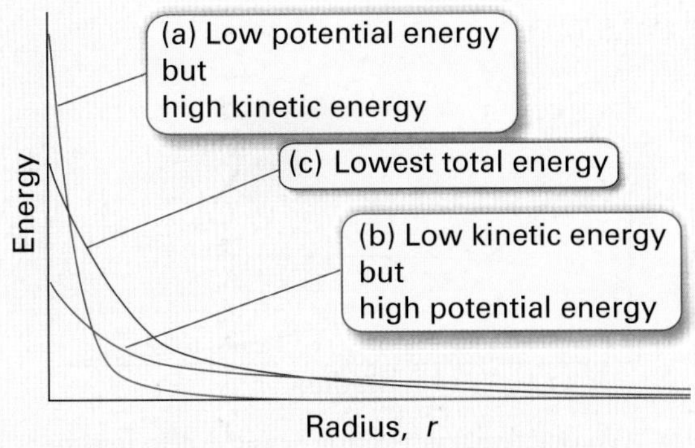

Fig. 4.9 The balance of kinetic and potential energies that accounts for the structure of the ground state of hydrogen (and similar atoms). (a) The sharply curved but localized orbital has high mean kinetic energy, but low mean potential energy; (b) the mean kinetic energy is low, but the potential energy is not very favourable; (c) the compromise of moderate kinetic energy and moderately favourable potential energy.

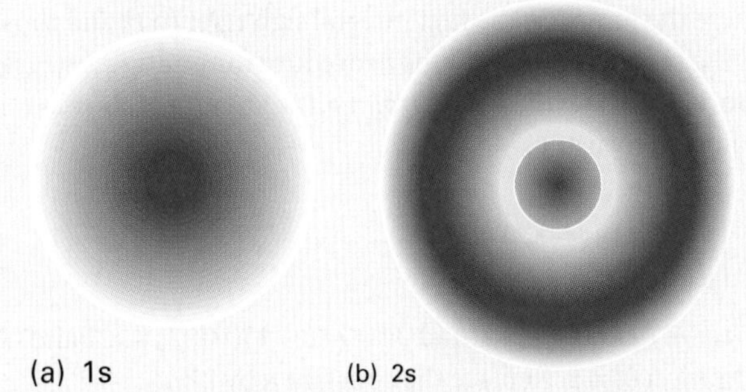

Fig. 4.10 Representations of cross-sections through the 1s and 2s hydrogenic atomic orbitals in terms of their electron probability densities (as represented by the density of shading).

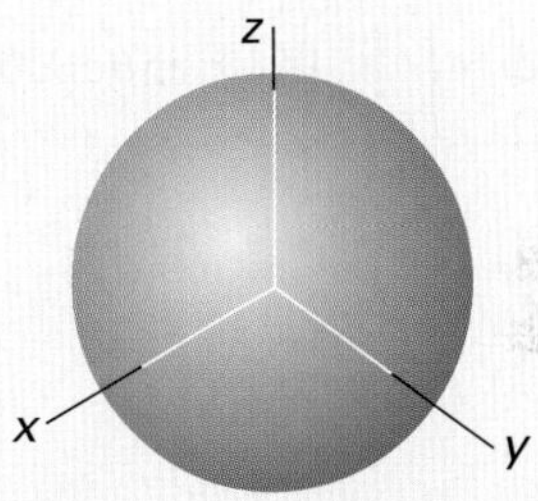

Fig. 4.11 The boundary surface of an s orbital, within which there is a 90 per cent probability of finding the electron.

the first example, but nor is it very high) and has a reasonably low average curvature (so the expectation value of the kinetic energy is not very low, but nor is it as high as in the first example).

A brief comment It is sometimes awkward to keep in mind what 'higher' and 'lower' potential energy means. When attractions are dominant, the potential energy is negative and a lower (more negative) potential energy is 'more stable' and a higher (less negative) potential energy is 'less stable'.

The energies of ns orbitals increase (become less negative; the electron becomes less tightly bound) as n increases because the average distance of the electron from the nucleus increases. By the virial theorem with $b=-1$ (eqn 2.30), $\langle E_k\rangle = -\frac{1}{2}\langle V\rangle$ so, even though the average kinetic energy decreases as n increases, the total energy is equal to $\frac{1}{2}\langle V\rangle$, which becomes less negative as n increases.

One way of depicting the probability density of the electron is to represent $|\psi|^2$ by the density of shading (Fig. 4.10). A simpler procedure is to show only the **boundary surface**, the surface that captures a high proportion (typically about 90 per cent) of the electron probability density. For the 1s orbital, the boundary surface is a sphere centred on the nucleus (Fig. 4.11).

All s orbitals are spherically symmetric, but differ in the number of radial nodes. For example, the 1s, 2s, and 3s orbitals have 0, 1, and 2 radial nodes, respectively (see Fig. 4.4 and note the number of zeroes in the radial wavefunction). In general, an ns orbital has $n-1$ radial nodes. (In general, an orbital with quantum numbers n and l has $n-l-1$ radial nodes.)

Self-test 4.5 Locate the nodes of (a) a 2s orbital, (b) a 3s orbital in a hydrogenic atom of atomic number Z.

[(a) $2a_0/Z$; (b) $1.90a_0/Z$ and $7.10a_0/Z$]

Example 4.4 *Calculating the mean radius of an orbital*

Use hydrogenic orbitals to calculate the mean radius of a 1s orbital.

Method The mean radius is the expectation value

$$\langle r\rangle = \int \psi^* r\psi \,d\tau = \int r|\psi|^2 \,d\tau$$

We therefore need to evaluate the integral using the wavefunctions given in Table 4.1; the volume element in spherical polar coordinates is $d\tau = r^2 dr \sin\theta \,d\theta d\phi$. The angular parts of the wavefunction are normalized in the sense that

$$\int_0^{\pi}\int_0^{2\pi} |Y_{l,m_l}|^2 \sin\theta \,d\theta d\phi = 1$$

where the limits on the first integral sign refer to θ, and those on the second to ϕ (recall the procedure for multiple integration in *Mathematical background 1*). The integral over r is evaluated by using

$$\int_0^\infty x^n e^{-ax} dx = \frac{n!}{a^{n+1}}$$

where $n!$ denotes a factorial: $n! = n(n-1)(n-2)\ldots 1$ and by definition $0! = 1$. Replace the constant a by $2Z/a_0$.

Answer With the wavefunction written in the form $\psi = RY$, the integration is

$$\langle r \rangle = \int_0^\infty \int_0^\pi \int_0^{2\pi} r R_{n,l}^2 |Y_{l,m_l}|^2 r^2\, dr \sin\theta d\theta d\phi$$

$$= \int_0^\infty r^3 R_{n,l}^2 dr \times \int_0^\pi \int_0^{2\pi} |Y_{l,m_l}|^2 \sin\theta d\theta d\phi$$

$$= \int_0^\infty r^3 R_{n,l}^2 dr$$

For a 1s orbital,

$$R_{1,0} = 2\left(\frac{Z}{a_0}\right)^{3/2} e^{-Zr/a_0}$$

Hence

$$\langle r \rangle = \frac{4Z^3}{a_0^3} \int_0^\infty r^3 e^{-2Zr/a_0} dr = \frac{3a_0}{2Z}$$

For H, $\langle r \rangle = 79$ pm; for He^+, with its higher nuclear charge, $\langle r \rangle = 40$ pm.

Self-test 4.6 Evaluate the mean radius of (a) a 3s orbital and (b) a 3p orbital by integration. [(a) $27a_0/2Z$; (b) $25a_0/2Z$]

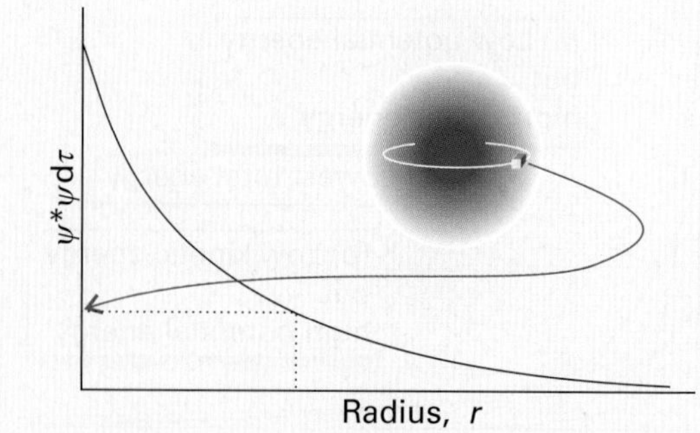

Fig. 4.12 A constant-volume electron-sensitive detector (the small cube) gives its greatest reading at the nucleus, and a smaller reading elsewhere. The same reading is obtained anywhere on a circle of given radius: the s orbital is spherically symmetrical.

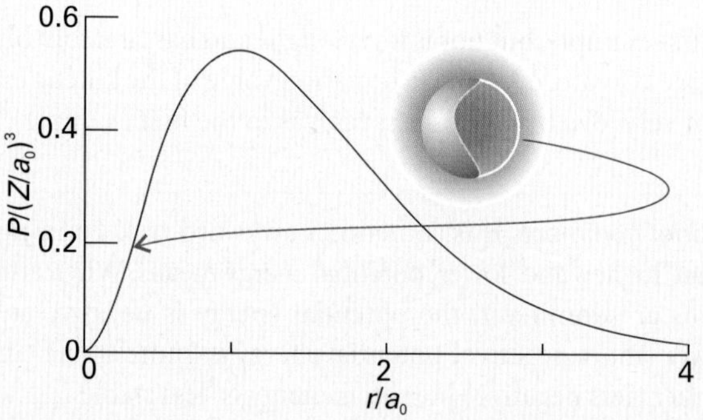

Fig. 4.13 The radial distribution function P gives the probability that the electron will be found anywhere in a shell of radius r. For a 1s electron in hydrogen, P is a maximum when r is equal to the Bohr radius a_0. The value of P is equivalent to the reading that a detector shaped like a spherical shell would give as its radius is varied.

(e) Radial distribution functions

The wavefunction tells us, through the value of $|\psi|^2$, the probability of finding an electron in any region in space. We can imagine a probe with a volume $d\tau$ that is sensitive to electrons and can move around near the nucleus of a hydrogen atom. Because the probability density in the ground state of the atom is $|\psi|^2 \propto e^{-2Zr/a_0}$, the reading from the probe decreases exponentially as the probe is moved out along any radius but is constant if the probe is moved on a circle of constant radius (Fig. 4.12).

Now consider the probability of finding the electron *anywhere* between the two walls of a spherical shell of thickness dr at a radius r. The sensitive volume of the probe is now the volume of the shell (Fig. 4.13), which is $4\pi r^2 dr$ (the product of its surface area, $4\pi r^2$, and its thickness, dr). The probability that the electron will be found between the inner and outer surfaces of this shell is the probability density at the radius r multiplied by the volume of the probe, or $|\psi|^2 \times 4\pi r^2 dr$. This expression has the form $P(r)dr$, where

$$P(r) = 4\pi r^2 |\psi|^2 \qquad (4.15a)$$

In the following *Justification* we show that the more general expression, which also applies to orbitals that are not spherically symmetrical, is

$$P(r) = r^2R(r)^2 \tag{4.15b}$$

where $R(r)$ is the radial wavefunction for the orbital in question.

Justification 4.2 *The general form of the radial distribution function*

The probability of finding an electron in a volume element $d\tau$ when its wavefunction is $\psi = RY$ is $R^2|Y|^2 d\tau$ with $d\tau = r^2 dr \sin\theta d\theta d\phi$ (recall R is real). The total probability of finding the electron at any angle at a constant radius is the integral of this probability over the surface of a sphere of radius r, and is written $P(r)dr$, so

$$P(r)dr = \int_0^{\pi}\int_0^{2\pi} \overbrace{R(r)^2|Y(\theta,\phi)|^2}^{|\psi|^2} \overbrace{r^2 dr \sin\theta d\theta d\phi}^{d\tau}$$

$$= r^2R(r)^2 dr \overbrace{\int_0^{\pi}\int_0^{2\pi} |Y(\theta,\phi)|^2 \sin\theta d\theta d\phi}^{1} = r^2R(r)^2 dr$$

The last equality follows from the fact that the spherical harmonics are normalized to 1 (see Example 4.4). It follows that $P(r) = r^2R(r)^2$, as stated in the text.

The **radial distribution function**, $P(r)$, is a probability density in the sense that, when it is multiplied by dr, it gives the probability of finding the electron anywhere between the two walls of a spherical shell of thickness dr at the radius r. For a 1s orbital,

$$P(r) = \frac{4Z^3}{a_0^3} r^2 e^{-2Zr/a_0} \tag{4.16}$$

Let's interpret this expression:

1. Because $r^2 = 0$ at the nucleus, at the nucleus $P(0) = 0$. Although the probability density itself is a maximum at the nucleus, the radial distribution function is zero at $r = 0$ on account of the r^2 factor.
2. As $r \to \infty$, $P(r) \to 0$ on account of the exponential term.
3. The increase in r^2 and the decrease in the exponential factor means that P passes through a maximum at an intermediate radius (see Fig. 4.13).

The maximum of $P(r)$, which can be found by differentiation, marks the most probable radius at which the electron will be found and, for a 1s orbital in hydrogen occurs at $r = a_0$, the Bohr radius. When we carry through the same calculation for the radial distribution function of the 2s orbital in hydrogen, we find that the most probable radius is $5.2a_0 = 275$ pm. This larger value reflects the expansion of the atom as its energy increases.

Example 4.5 *Calculating the most probable radius*

Calculate the most probable radius, r^*, at which an electron will be found when it occupies a 1s orbital of a hydrogenic atom of atomic number Z, and tabulate the values for the one-electron species from H to Ne^{9+}.

Method We find the radius at which the radial distribution function of the hydrogenic 1s orbital has a maximum value by solving $dP/dr = 0$. If there are several maxima, then we choose the one corresponding to the greatest amplitude (the largest value of P).

Answer The radial distribution function is given in eqn 4.16. It follows that

$$\frac{dP}{dr} = \frac{4Z^3}{a_0^3}\left(2r - \frac{2Zr^2}{a_0}\right)e^{-2Zr/a_0} = \frac{8Z^3}{a_0^3} r\left(1 - \frac{Zr}{a_0}\right)e^{-2Zr/a_0}$$

This function is zero where the term in parentheses is zero (ignore $r = 0$ which corresponds to a minimum in P), which is at

$$r^* = \frac{a_0}{Z}$$

Then, with $a_0 = 52.9$ pm, the most probable radius is at

	H	He^+	Li^{2+}	Be^{3+}	B^{4+}	C^{5+}	N^{6+}	O^{7+}	F^{8+}	Ne^{9+}
r^*/pm	52.9	26.5	17.6	13.2	10.6	8.82	7.56	6.61	5.88	5.29

Notice how the 1s orbital is drawn towards the nucleus as the nuclear charge increases. At uranium the most probable radius is only 0.58 pm, almost 100 times closer than for hydrogen. (On a scale where $r^* = 10$ cm for H, $r^* = 1$ mm for U^{91+}.) The electron then experiences strong accelerations and relativistic effects are important.

Self-test 4.7 Find the most probable distance of an electron from the nucleus in a hydrogenic atom when it occupies a 2s orbital. $[(3 + 5^{1/2})a_0/Z]$

(f) p Orbitals

The three 2p orbitals are distinguished by the three different values that m_l can take when $l = 1$. Because the quantum number m_l tells us the orbital angular momentum around an axis, these different values of m_l denote orbitals in which the electron has different orbital angular momenta around an arbitrary z-axis but the same magnitude of that momentum (because l is the same for all three). The orbital with $m_l = 0$, for instance, has zero angular momentum around the z-axis. Its angular variation is proportional to $\cos\theta$, so the probability density, which is

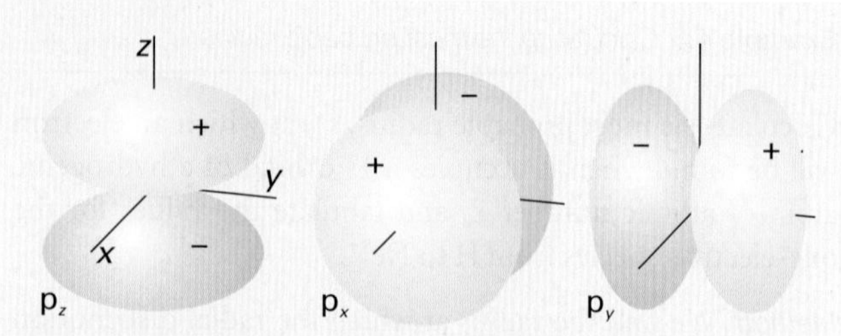

Fig. 4.14 The boundary surfaces of p orbitals. A nodal plane passes through the nucleus and separates the two lobes of each orbital. The dark and light areas denote regions of opposite sign of the wavefunction.

interActivity Use mathematical software to plot the boundary surfaces of the real parts of the spherical harmonics $Y_{1,m_l}(\theta,\phi)$. The resulting plots are not strictly the p orbital boundary surfaces, but sufficiently close to be reasonable representations of the shapes of hydrogenic orbitals.

proportional to $\cos^2\theta$, has its maximum value on either side of the nucleus along the z-axis (at $\theta = 0$ and 180°). The wavefunction of a 2p orbital with $m_l = 0$ is

$$\psi_{\mathrm{p}_0} = R_{2,1}(r)Y_{1,0}(\theta,\phi) = \frac{1}{4(2\pi)^{1/2}}\left(\frac{Z}{a_0}\right)^{5/2} r\cos\theta\, \mathrm{e}^{-Zr/2a_0}$$
$$= r\cos\theta\, f(r) \qquad (4.17a)$$

where $f(r)$ is a function only of r. Because in spherical polar coordinates $z = r\cos\theta$, this wavefunction may also be written

$$\psi_{\mathrm{p}_z} = zf(r) \qquad (4.17b)$$

All p orbitals with $m_l = 0$ have wavefunctions of this form regardless of the value of n (of course, the function f differs with n). This way of writing the orbital is the origin of the name 'p_z orbital': its boundary surface is shown in Fig. 4.14. The wavefunction is zero everywhere in the xy-plane, where $z = 0$, so the xy-plane is a **nodal plane** of the orbital and $\theta = \pi/2$ is an angular node: the wavefunction changes sign on going from one side of the nodal plane to the other.

The wavefunctions of 2p orbitals with $m_l = \pm 1$ have the following form:

$$\psi_{\mathrm{p}_{\pm 1}} = R_{2,1}(r)Y_{1,\pm 1}(\theta,\phi)$$
$$= \mp\frac{1}{8\pi^{1/2}}\left(\frac{Z}{a_0}\right)^{5/2} r\,\mathrm{e}^{-Zr/2a_0}\sin\theta\,\mathrm{e}^{\pm\mathrm{i}\phi} \qquad (4.18a)$$
$$= \mp\frac{1}{2^{1/2}} r\sin\theta\,\mathrm{e}^{\pm\mathrm{i}\phi} f(r)$$

We remarked in Section 2.1 that a moving particle is described by a complex wavefunction. In the present case, these functions correspond to nonzero angular momentum about the z-axis: $\mathrm{e}^{+\mathrm{i}\phi}$ corresponds to clockwise rotation when viewed from below, and $\mathrm{e}^{-\mathrm{i}\phi}$ corresponds to counter-clockwise rotation (from the same viewpoint). They have zero amplitude where $\theta = 0$ and 180° (along the z-axis) and maximum amplitude at 90°, which is in the xy-plane. To draw the functions it is usual to represent them as standing waves. To do so, we take the real linear combinations

$$\psi_{\mathrm{p}_x} = -\frac{1}{2^{1/2}}(\mathrm{p}_{+1} - \mathrm{p}_{-1}) = r\sin\theta\cos\phi\, f(r) = xf(r) \qquad (4.18b)$$

$$\psi_{\mathrm{p}_y} = \frac{\mathrm{i}}{2^{1/2}}(\mathrm{p}_{+1} + \mathrm{p}_{-1}) = r\sin\theta\sin\phi\, f(r) = yf(r) \qquad (4.18c)$$

These linear combinations are indeed standing waves with no net orbital angular momentum around the z-axis, as they are superpositions of states with equal and opposite values of m_l. The p_x orbital has the same shape as a p_z orbital, but it is directed along the x-axis (see Fig. 4.14); the p_y orbital is similarly directed along the y-axis. The wavefunction of any p orbital of a given shell can be written as a product of x, y, or z and the same radial function (which depends on the value of n). The following *Justification* explains why it is permissible to take linear combinations of degenerate orbitals when we want to indicate a particular point.

Justification 4.3 *The linear combination of degenerate wavefunctions*

The freedom to take linear combinations of degenerate functions rests on the fact that, whenever two or more wavefunctions correspond to the same energy (such as p_{+1} and p_{-1}), any linear combination of them (such as p_x or p_y) is an equally valid solution of the Schrödinger equation.

Suppose ψ_1 and ψ_2 are both solutions of the Schrödinger equation with energy E; then we know that

$$\hat{H}\psi_1 = E\psi_1 \qquad \hat{H}\psi_2 = E\psi_2$$

Now consider the linear combination

$$\psi = c_1\psi_1 + c_2\psi_2$$

where c_1 and c_2 are arbitrary coefficients. Then it follows that

$$\hat{H}\psi = \hat{H}(c_1\psi_1 + c_2\psi_2) = c_1\hat{H}\psi_1 + c_2\hat{H}\psi_2$$
$$= c_1E\psi_1 + c_2E\psi_2 = E(c_1\psi_1 + c_2\psi_2) = E\psi$$

Hence, the linear combination is also a solution corresponding to the same energy E. Furthermore, the result that a linear combination of eigenfunctions of an operator (all having the same eigenvalue) is also an eigenfunction of the operator (with the same eigenvalue) applies to all quantum mechanical operators, not just the hamiltonian.

(g) d Orbitals

When $n = 3$, l can be 0, 1, or 2. As a result, this shell consists of one 3s orbital, three 3p orbitals, and five 3d orbitals. The five d

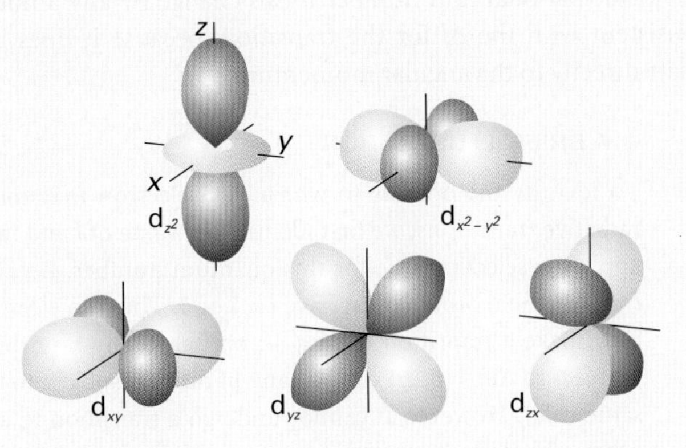

Fig. 4.15 The boundary surfaces of d orbitals. Two nodal planes in each orbital intersect at the nucleus and separate the lobes of each orbital. The dark and light areas denote regions of opposite sign of the wavefunction.

interActivity To gain insight into the shapes of the f orbitals, use mathematical software to plot the boundary surfaces of the spherical harmonics $Y_{3,m_l}(\theta,\phi)$.

orbitals have $m_l = +2, +1, 0, -1, -2$ and correspond to five different angular momenta around the z-axis (but the same *magnitude* of angular momentum, because $l = 2$ in each case). As for the p orbitals, d orbitals with opposite values of m_l (and hence opposite senses of motion around the z-axis) may be combined in pairs to give real standing waves and the boundary surfaces of the resulting shapes are shown in Fig. 4.15. The real combinations have the following forms:

$$d_{xy} = xyf(r) \qquad d_{yz} = yzf(r) \qquad d_{zx} = zxf(r)$$
$$d_{x^2-y^2} = \tfrac{1}{2}(x^2 - y^2)f(r) \qquad d_{z^2} = (\tfrac{1}{2}\sqrt{3})(3z^2 - r^2)f(r) \qquad (4.19)$$

4.3 Spectroscopic transitions

The energies of the hydrogenic atoms are given by eqn 4.8. When the electron undergoes a transition from an orbital with quantum numbers n_1, l_1, m_{l1} to another (lower energy) orbital with quantum numbers n_2, l_2, m_{l2}, it undergoes a change of energy ΔE and discards the excess energy as a photon of electromagnetic radiation with a frequency ν given by the Bohr frequency condition (eqn 1.1).

It is tempting to think that all possible transitions are permissible, and that a spectrum arises from the transition of an electron from any initial orbital to any other orbital. However, this is not so, because a photon has an intrinsic spin angular momentum corresponding to $s = 1$ (Section 3.5). The change in angular momentum of the electron must compensate for the angular momentum carried away by the photon. Thus, an electron in a d orbital ($l = 2$) cannot make a transition into an s orbital ($l = 0$) because the photon cannot carry away enough angular momentum. Similarly, an s electron cannot make a transition to another s orbital, because there would then be no change in the electron's angular momentum to make up for the angular momentum carried away by the photon. It follows that some spectroscopic transitions are **allowed**, meaning that they can occur, whereas others are **forbidden**, meaning that they cannot occur. A **selection rule** is a statement about which transitions are allowed. We describe below more completely the basis for selection rules.

(a) Selection rules and the transition dipole moment

According to classical physics, for an atom to absorb or emit a photon of frequency ν, it must possess, at least temporarily, an electric dipole that oscillates at that frequency. An electric dipole consists of two electric charges $+Q$ and $-Q$ separated by a vector R; the electric dipole moment is $\boldsymbol{\mu} = QR$, and can oscillate either because the charges change or because their vector separation changes (see *Mathematical background 4* for a discussion of vectors).

To develop a quantum mechanical view of absorption and emission from an initial state ψ_i to a final state ψ_f, we first need to write an expression for the dipole moment operator using Postulate III (Section 1.6). For a one-electron atom, the operator $\hat{\boldsymbol{\mu}}$ is multiplication by $-e\boldsymbol{r}$, where e is the fundamental charge and $\boldsymbol{r}$ is a vector from the origin (that is, the nucleus) to the electron with components x, y, and z. It follows that $\hat{\boldsymbol{\mu}}$ is a vector with components $\hat{\mu}_x = -ex$, $\hat{\mu}_y = -ey$, $\hat{\mu}_z = -ez$. The expression for the rate of a spectroscopic transition, and hence the intensity of absorption or emission of radiation, is derived by making use of **time-dependent perturbation theory** in which the perturbation arises from an oscillating electric field of strength $\mathcal{E}(t)$.

We begin by writing the hamiltonian for the system as

$$\hat{H} = \hat{H}^{(0)} + \hat{H}^{(1)}(t) \qquad (4.20)$$

where $\hat{H}^{(1)}(t)$ is the time-dependent perturbation and $\hat{H}^{(0)}$ is the hamiltonian in the absence of the perturbation. Because the perturbation arises from the effect of an oscillating electric field with the electric dipole moment, we write

$$\hat{H}^{(1)}(t) = -\hat{\boldsymbol{\mu}}\cdot\boldsymbol{\mathcal{E}}(t) \qquad (4.21a)$$

or for the z-component (we generalize to x and y below)

$$\hat{H}^{(1)}(t) = -\hat{\mu}_z\mathcal{E}\cos\omega t \qquad (4.21b)$$

where ω is the frequency of the field and $\mathcal{E}$ is its amplitude. We suppose that the perturbation is absent until $t = 0$, and then it is turned on.

We show in *Further information 4.2* that the rate of change of population of the state ψ_f due to transitions from state ψ_i, denoted $w_{f\leftarrow i}$, is proportional to the square modulus of the matrix element of the perturbation between the two states:

$$w_{f\leftarrow i} \propto |H_{fi}^{(1)}|^2 \qquad H_{fi}^{(1)} = \int \psi_f^*\hat{H}^{(1)}\psi_i\,d\tau \qquad (4.22)$$

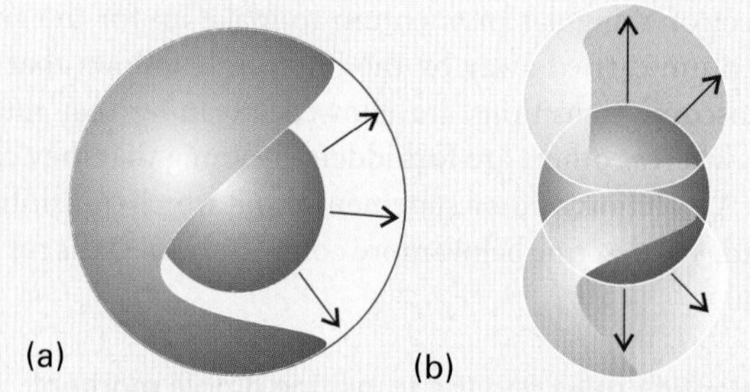

Fig. 4.16 (a) When a 1s electron becomes a 2s electron, there is a spherical migration of charge. Since there is no dipole moment associated with this migration of charge, this transition is electric-dipole forbidden. (b) In contrast, when a 1s electron becomes a 2p electron, there is a dipole associated with the charge migration; this transition is allowed. (There are subtle effects arising from the sign of the wavefunction that give the charge migration a dipolar character, which this diagram does not attempt to convey.)

Because in our case the perturbation is that of the interaction of the electric component of the electromagnetic field with an atom (eqn 4.21b), we conclude that

$$w_{\mathrm{f}\leftarrow\mathrm{i}} \propto |\mu_{z,\mathrm{fi}}|^2 \mathcal{E}^2 \tag{4.23}$$

where

$$\mu_{z,\mathrm{fi}} = \int \psi_\mathrm{f}^* \hat{\mu}_z \psi_\mathrm{i} \mathrm{d}\tau \qquad [4.24]$$

The x- and y-components are defined similarly. The transition rate is proportional to the square modulus of the **transition dipole moment**, $\boldsymbol{\mu}_\mathrm{fi}$, with $|\boldsymbol{\mu}_\mathrm{fi}|^2 = |\mu_{x,\mathrm{fi}}|^2 + |\mu_{y,\mathrm{fi}}|^2 + |\mu_{z,\mathrm{fi}}|^2$. The magnitude of the transition dipole can be regarded as a measure of the extent and form of the charge redistribution that accompanies a transition: a transition will be active (and generate or absorb photons) only if the accompanying charge redistribution is dipolar (Fig. 4.16). In other words, the transition is forbidden if all three components of the transition dipole moment are zero; it is allowed if any one component of the transition dipole moment is nonzero.

To identify the selection rules (for both atomic and molecular spectra), we must establish the conditions for which $\boldsymbol{\mu}_\mathrm{fi} \neq 0$. **Specific selection rules** express the allowed transitions in terms of the changes in particular quantities, most usually quantum numbers. For the atomic spectra of hydrogenic atoms, the specific selection rules are derived by identifying the transitions that conserve angular momentum when a photon is emitted or absorbed and are shown in the *Justification* below to be

$$\Delta l = \pm 1 \qquad \Delta m_l = 0, \pm 1 \tag{4.25}$$

The principal quantum number n can change by any amount consistent with the Δl for the transition, because it does not relate directly to the angular momentum.

● A BRIEF ILLUSTRATION

To identify the orbitals to which a 4d electron may make radiative transitions, we first identify the value of l and then apply the selection rule for this quantum number. Because $l = 2$, the final orbital must have $l = 1$ or 3. Thus, an electron may make a transition from a 4d orbital to any np orbital (subject to $\Delta m_l = 0, \pm 1$) and to any nf orbital (subject to the same rule). However, it cannot undergo a transition to any other orbital, so a transition to any ns orbital or to another nd orbital is forbidden. ●

Self-test 4.8 To what orbitals may an electron in a 4s orbital make electric-dipole allowed radiative transitions?

[to np orbitals only]

Justification 4.4 *The identification of selection rules*

To determine the selection rules for atoms, we need to identify the conditions for which the transition dipole moment, $\boldsymbol{\mu}_\mathrm{fi}$, connecting the final state ψ_f and the initial state ψ_i is nonzero:

$$\boldsymbol{\mu}_\mathrm{fi} = \int \psi_\mathrm{f}^* \hat{\boldsymbol{\mu}} \psi_\mathrm{i} \mathrm{d}\tau$$

We consider each component in turn. For example, for the z-component,

$$\mu_{z,\mathrm{fi}} = -e\int \psi_\mathrm{f}^* z \psi_\mathrm{i} \mathrm{d}\tau$$

To evaluate the integral, we note from Table 3.2 that $z = (4\pi/3)^{1/2} r Y_{1,0}$, so

$$\int \psi_\mathrm{f}^* z \psi_\mathrm{i} \mathrm{d}\tau =$$

$$\int_0^\infty \int_0^\pi \int_0^{2\pi} \overbrace{R_{n_\mathrm{f},l_\mathrm{f}} Y_{l_\mathrm{f},m_{l,\mathrm{f}}}}^{\psi_\mathrm{f}^*}{}^* \overbrace{\left(\frac{4\pi}{3}\right)^{1/2} r Y_{1,0}}^{z} \overbrace{R_{n_\mathrm{i},l_\mathrm{i}} Y_{l_\mathrm{i},m_{l,\mathrm{i}}}}^{\psi_\mathrm{i}} \overbrace{r^2 \mathrm{d}r \sin\theta \mathrm{d}\theta \mathrm{d}\phi}^{\mathrm{d}\tau}$$

This multiple integral is the product of three factors, an integral over r and two integrals over the angles, so the factors on the right can be grouped as follows:

$$\int \psi_\mathrm{f}^* z \psi_\mathrm{i} \mathrm{d}\tau =$$

$$\left(\frac{4\pi}{3}\right)^{1/2} \int_0^\infty R_{n_\mathrm{f},l_\mathrm{f}} r R_{n_\mathrm{i},l_\mathrm{i}} r^2 \mathrm{d}r \int_0^\pi \int_0^{2\pi} Y_{l_\mathrm{f},m_{l,\mathrm{f}}}{}^* Y_{1,0} Y_{l_\mathrm{i},m_{l,\mathrm{i}}} \sin\theta \mathrm{d}\theta \mathrm{d}\phi$$

We now use the property of the spherical harmonics that

$$\int_0^{\pi}\int_0^{2\pi} Y_{l'',m_l''}(\theta,\phi)^* Y_{l',m_l'}(\theta,\phi) Y_{l,m_l}(\theta,\phi)\sin\theta \mathrm{d}\theta \mathrm{d}\phi = 0$$

unless l, l', and l'' are integers that can form the sides of a triangle and $m_l + m_l' + m_l'' = 0$. It follows that the integral

$$\int_0^{\pi}\int_0^{2\pi} Y_{l_f,m_{l,f}}{}^* Y_{1,m} Y_{l_i,m_{l,i}} \sin\theta \mathrm{d}\theta \mathrm{d}\phi$$

is zero unless $l_f = l_i \pm 1$ and $m_{l,f} = m_{l,i} + m$. Because $m = 0$ in the present case, the angular integral, and hence the z-component of the transition dipole moment, is zero unless $\Delta l = \pm 1$ and $\Delta m_l = 0$, which is a part of the set of selection rules. The same procedure, but considering the x- and y-components, results in the complete set of rules.

The selection rules and the atomic energy levels jointly account for the structure of a **Grotrian diagram** (Fig. 4.17), which summarizes the energies of the states and the transitions between them. The thicknesses of the arrows depicting the transitions denote their relative intensities in the spectrum obtained by evaluating the transition dipole moments.

(b) Spectral linewidths

A number of effects contribute to the widths of spectroscopic lines. Some contributions to linewidths can be modified by changing the conditions, and to achieve high resolutions we need to know how to minimize these contributions.

One important broadening process in atomic gaseous samples is the **Doppler effect**, in which radiation is shifted in frequency when the source is moving towards or away from the observer: the transition frequency remains unchanged (because $\Delta E/h$ remains unchanged), but an observer *detects* different frequencies. When a source emitting electromagnetic radiation of frequency ν moves with a speed s relative to an observer, the observer detects radiation of frequency

$$\nu_{\text{receding}} = \nu\left(\frac{1-s/c}{1+s/c}\right)^{1/2} \qquad \nu_{\text{approaching}} = \nu\left(\frac{1+s/c}{1-s/c}\right)^{1/2} \tag{4.26a}$$

where c is the speed of light. For nonrelativistic speeds ($s \ll c$), these expressions simplify to

$$\nu_{\text{receding}} \approx \frac{\nu}{1+s/c} \qquad \nu_{\text{approaching}} \approx \frac{\nu}{1-s/c} \tag{4.26b}$$

Atoms reach high speeds in all directions in a gas, and a stationary observer detects the corresponding Doppler shifted range of frequencies. Some atoms approach the observer, some move away; some move quickly, others slowly. The detected spectral 'line' is the absorption or emission profile arising from all the resulting Doppler shifts. As shown in the following *Justification*, the profile reflects the distribution of atomic velocities parallel to the line of sight, which is a bell-shaped Gaussian curve. The Doppler line shape is therefore also a Gaussian (Fig. 4.18), and we show in the *Justification* that, when the temperature is T and the mass of the atom is m, then the observed width of the line at half-height (in terms of frequency or wavelength) is

$$\delta\nu_{\text{obs}} = \frac{2\nu}{c}\left(\frac{2kT\ln 2}{m}\right)^{1/2} \qquad \delta\lambda_{\text{obs}} = \frac{2\lambda}{c}\left(\frac{2kT\ln 2}{m}\right)^{1/2} \tag{4.27}$$

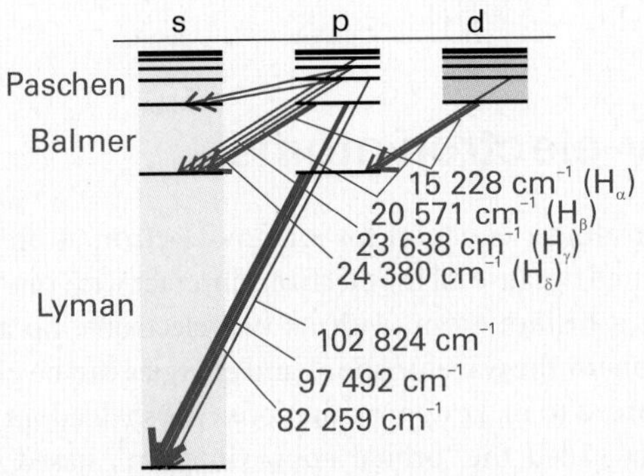

Fig. 4.17 A Grotrian diagram that summarizes the appearance and analysis of the spectrum of atomic hydrogen. The thicker the line, the more intense the transition.

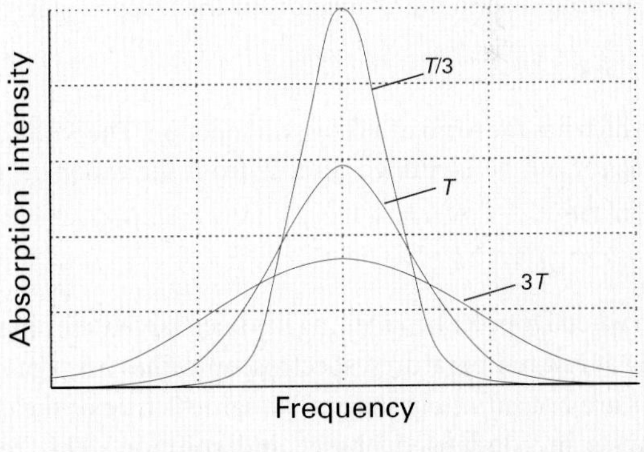

Fig. 4.18 The Gaussian shape of a Doppler-broadened spectral line reflects the Maxwell distribution of speeds in the sample at the temperature of the experiment. Notice that the line broadens as the temperature is increased.

interActivity In a spectrometer that makes use of *phase-sensitive detection* the output signal is proportional to the first derivative of the signal intensity, $\mathrm{d}I/\mathrm{d}\nu$. Plot the resulting line shape for various temperatures. How is the separation of the peaks related to the temperature?

A brief comment A Gaussian function of the general form $y(x) = ae^{-(x-b)^2/2\sigma^2}$, where a, b, and σ are constants, has a maximum $y(b) = a$ and a width at half-height $\delta x = 2\sigma(2\ln 2)^{1/2}$.

For atomic hydrogen at room temperature ($T \approx 300$ K), $\delta\nu/\nu \approx 1.2 \times 10^{-5}$. Doppler broadening increases with temperature because the atoms acquire a wider range of speeds. Therefore, to obtain spectra of maximum sharpness, it is best to work with cold samples.

Justification 4.5 *Doppler broadening*

We know from the Boltzmann distribution (*Fundamentals F.5*) that the probability that a gas atom or molecule of mass m and speed s in a sample with temperature T has kinetic energy $E_k = \frac{1}{2}ms^2$ is proportional to $e^{-ms^2/2kT}$. The observed frequencies, ν_{obs}, emitted or absorbed by the atom or molecule are related to its speed by eqn 4.26b:

$$\nu_{obs} = \nu\left(\frac{1}{1 \pm s/c}\right)$$

where ν is the unshifted frequency. When $s \ll c$, the Doppler shift in the frequency is

$$\nu_{obs} - \nu \approx \pm \nu s/c$$

which implies a symmetrical distribution of observed frequencies with respect to atomic or molecular speeds. More specifically, the intensity I of a transition at ν_{obs} is proportional to the probability of finding the atom or molecule that emits or absorbs at ν_{obs}, so it follows from the Boltzmann distribution and the expression for the Doppler shift that

$$I(\nu_{obs}) \propto e^{-mc^2(\nu_{obs}-\nu)^2/2\nu^2kT}$$

which has the form of a Gaussian function. The width at half-height can be identified directly from the exponent to give eqn 4.27.

Even when Doppler broadening has been largely eliminated by working at low temperatures, spectroscopic lines from gas-phase samples are not infinitely sharp. The same is true of the spectra of samples in condensed phases and solution. This residual broadening is due to quantum mechanical effects. Specifically, when the Schrödinger equation is solved for a system that is changing with time, it is found that it is impossible to specify the energy levels exactly. If on average a system survives in a state for a time τ (tau), the **lifetime** of the state, then its energy levels are blurred to an extent of order δE, where

$$\delta E \approx \frac{\hbar}{\tau} \qquad (4.28a)$$

This expression is reminiscent of the Heisenberg uncertainty principle (Section 1.9), and consequently this **lifetime broadening** is often called 'uncertainty broadening'. When the energy spread is expressed as a wavenumber through $\delta E = hc\delta\tilde{\nu}$, and the values of the fundamental constants introduced, this relation becomes

$$\delta\tilde{\nu} \approx \frac{5.3 \text{ cm}^{-1}}{\tau/\text{ps}} \qquad (4.28b)$$

No excited state has an infinite lifetime; therefore, all states are subject to some lifetime broadening, and the shorter the lifetimes of the states involved in a transition, the broader the corresponding spectral lines.

Two processes are responsible for the finite lifetimes of excited states. The dominant one for low frequency transitions is **collisional deactivation**, which arises from collisions between atoms or with the walls of the container. If the **collisional lifetime**, the mean time between collisions, is τ_{col}, the resulting collisional linewidth is $\delta E_{col} \approx \hbar/\tau_{col}$. Because τ_{col} is inversely proportional to the collision frequency and, as we shall see in Section 18.1, the collision frequency is proportional to the pressure, we see that the collisional linewidth is proportional to the pressure. The collisional linewidth can therefore be minimized by working at low pressures.

Excited states also emit radiation spontaneously (this process is treated quantitatively in *Further information 10.1*) and this rate of spontaneous emission cannot be changed. Hence it is a natural limit to the lifetime of an excited state, and the resulting lifetime broadening is the **natural linewidth** of the transition. The natural linewidth is an intrinsic property of the transition, and cannot be changed by modifying the conditions. Natural linewidths depend strongly on the transition frequency, so low frequency transitions have very small natural linewidths, and collisional and Doppler line-broadening processes are dominant. Electronic transitions occur at high frequencies and therefore have short natural lifetimes and large natural linewidths. For example, a typical electronic excited state natural lifetime is about 10^{-8} s (10 ns), corresponding to a natural width of about 5×10^{-4} cm^{-1} (15 MHz).

Many-electron atoms

The Schrödinger equation for a many-electron atom is highly complicated because all the electrons interact with one another. Even for a helium atom, with its two electrons, no analytical expression for the wavefunctions and energies can be given, and we are forced to make approximations. We shall adopt a simple approach, called the 'orbital approximation', based on what we already know about the structure of hydrogenic atoms and the energies of orbitals. In Chapter 6, we shall see the types of numerical computations that are used to obtain accurate wavefunctions and energies.

Atomic spectra rapidly become very complicated as the number of electrons increases, but there are some important and moderately simple features that make atomic spectroscopy useful in the study of the composition of samples as large and as complex as stars (*Impact I4.1*). The general idea is straightforward: lines in the spectrum (in either emission or absorption) occur when the atom undergoes a transition with a change of energy $|\Delta E|$, and emits or absorbs a photon of frequency $\nu = |\Delta E|/h$ and $\tilde{\nu} = |\Delta E|/hc$. Hence, we can expect the spectrum to give information about the energies of electrons in atoms. However, the actual energy levels are not given solely by the energies of the orbitals, because the electrons interact with one another in various ways, and it is necessary to consider contributions to the energy beyond those of the orbital approximation.

IMPACT ON ASTROPHYSICS
I4.1 The spectroscopy of stars

The bulk of stellar material consists of neutral and ionized forms of hydrogen and helium atoms, with helium being the product of 'hydrogen burning' by nuclear fusion. However, nuclear fusion also makes heavier elements and contributes to the process of nucleosynthesis. It is generally accepted that the outer layers of stars are composed of lighter elements, such as H, He, C, N, O, and Ne in both neutral and ionized forms. Heavier elements, including neutral and ionized forms of Si, Mg, Ca, S, and Ar, are found closer to the stellar core. The core itself contains the heaviest elements and ^{56}Fe is particularly abundant because it is a very stable nuclide. All of these elements are in the gas phase on account of the very high temperatures in stellar interiors. For example, the temperature is estimated to be 3.6 MK halfway to the centre of the Sun.

Astronomers use spectroscopic techniques to determine the chemical composition of stars because each element, and indeed each isotope of an element, has a characteristic spectral signature that is transmitted through space by the star's light. To understand the spectra of stars, we must first know why they shine. Nuclear reactions in the dense stellar interior generate radiation that travels to less dense outer layers. Absorption and re-emission of photons by the atoms and ions in the interior give rise to a quasi-continuum of radiation energy that is emitted into space by a thin layer of gas called the *photosphere*. To a good approximation, the distribution of energy emitted from a star's photosphere resembles the intensity distribution for a very hot body, which for the Sun corresponds to a temperature of 5.8 kK. Superimposed on the radiation continuum are sharp absorption and emission lines from neutral atoms and ions present in the photosphere. Analysis of stellar radiation with a spectrometer mounted on to a telescope yields the chemical composition of the star's photosphere by comparison with known spectra of the elements. The data can also reveal the presence of small molecules, such as CN, C_2, TiO, and ZrO, in certain 'cold' stars, which are stars with relatively low effective temperatures.

The two outermost layers of a star are the *chromosphere*, a region just above the photosphere, and the *corona*, a region above the chromosphere that can be seen (with proper care) during eclipses. The photosphere, chromosphere, and corona comprise a star's 'atmosphere'. Our Sun's chromosphere is much less dense than its photosphere and its temperature is much higher, rising to about 10 kK. The reasons for this increase in temperature are not fully understood. The temperature of our Sun's corona is very high, rising up to 1.5 MK, so black-body emission is strong from the X-ray to the radiofrequency region of the spectrum. The spectrum of the Sun's corona is dominated by emission lines from electronically excited species, such as neutral atoms and a number of highly ionized species. The most intense emission lines in the visible range are from the Fe^{13+} ion at 530.3 nm, the Fe^{9+} ion at 637.4 nm, and the Ca^{4+} ion at 569.4 nm.

Because only light from the photosphere reaches our telescopes, the overall chemical composition of a star must be inferred from theoretical models of its interior and from spectral analysis of its atmosphere. Data on the Sun indicate that it is 91 to 94 per cent hydrogen and 6 to 9 per cent helium by atom. The remaining mass is due to heavier elements, among which C, N, O, Ne, and Fe are the most abundant. More advanced analysis of spectra also permits the determination of other properties of stars, such as their relative speeds and their effective temperatures.

4.4 The orbital approximation

The wavefunction of a many-electron atom is a very complicated function of the coordinates of all the electrons, and we should write it $\psi(\boldsymbol{r}_1, \boldsymbol{r}_2, \ldots)$, where $\boldsymbol{r}_i$ is the vector from the nucleus to electron i. However, in the **orbital approximation** we suppose that a reasonable first approximation to this exact wavefunction is obtained by thinking of each electron as occupying its 'own' orbital, and write the product

$$\psi(\boldsymbol{r}_1, \boldsymbol{r}_2, \ldots) = \psi(\boldsymbol{r}_1)\psi(\boldsymbol{r}_2)\cdots \tag{4.29}$$

We can think of the individual orbitals as resembling the hydrogenic orbitals, but corresponding to nuclear charges modified by the presence of all the other electrons in the atom. This description is only approximate, as explained in the following *Justification*, but it is a useful model for discussing the chemical properties of atoms, and is the starting point for more sophisticated descriptions of atomic structure.

Justification 4.6 *The orbital approximation*

The orbital approximation would be exact if there were no interactions between electrons. To demonstrate the validity of this remark, we need to consider a system in which the hamiltonian for the energy is the sum of two contributions, one for electron 1 and the other for electron 2:

$$\hat{H} = \hat{H}_1 + \hat{H}_2$$

In an actual atom (such as helium), there is an additional term corresponding to the interaction of the two electrons, but we are ignoring that term. We shall now show that, if $\psi(r_1)$ is an eigenfunction of $\hat{H}_1$ with energy E_1, and $\psi(r_2)$ is an eigenfunction of $\hat{H}_2$ with energy E_2, then the product $\psi(r_1,r_2) = \psi(r_1)\psi(r_2)$ is an eigenfunction of the combined hamiltonian $\hat{H}$. To do so we write

$$\begin{aligned}\hat{H}\psi(r_1,r_2) &= (\hat{H}_1 + \hat{H}_2)\psi(r_1)\psi(r_2)\\ &= \hat{H}_1\psi(r_1)\psi(r_2) + \psi(r_1)\hat{H}_2\psi(r_2)\\ &= E_1\psi(r_1)\psi(r_2) + \psi(r_1)E_2\psi(r_2)\\ &= (E_1 + E_2)\psi(r_1)\psi(r_2)\\ &= E\psi(r_1,r_2)\end{aligned}$$

where $E = E_1 + E_2$. This is the result we need to prove. However, if the electrons interact (as they do in fact), then the proof fails; nevertheless, it remains a reasonable and almost universally used starting point for the discussion of atomic structure.

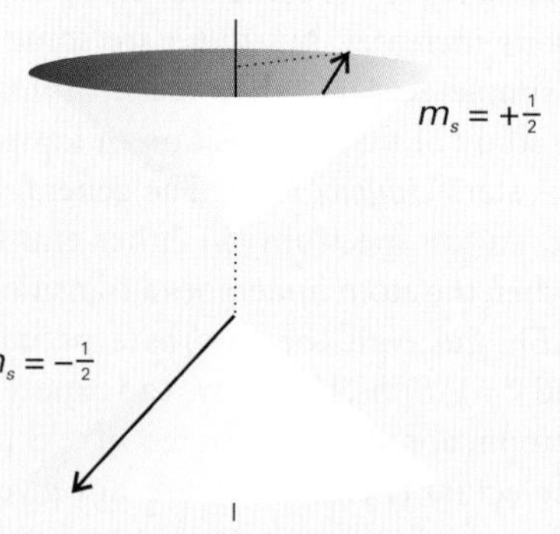

Fig. 4.19 Electrons with paired spins have zero resultant spin angular momentum. They can be represented by two vectors that lie at an indeterminate position on the cones shown here but, wherever one lies on its cone, the other points in the opposite direction; their resultant is zero.

(a) The helium atom

The orbital approximation allows us to express the electronic structure of an atom by reporting its **configuration**, the list of occupied orbitals (usually, but not necessarily, in its ground state). Thus, as the ground state of a hydrogenic atom consists of the single electron in a 1s orbital, we report its configuration as $1s^1$.

The He atom has two electrons. We can imagine forming the atom by adding the electrons in succession to the orbitals of the bare nucleus (of charge $2e$). The first electron occupies a 1s hydrogenic orbital, but because $Z = 2$ that orbital is more compact than in H itself. The second electron joins the first in the 1s orbital, so the electron configuration of the ground state of He is $1s^2$.

(b) The Pauli principle

Lithium, with $Z = 3$, has three electrons. The first two occupy a 1s orbital drawn even more closely than in He around the more highly charged nucleus. The third electron, however, does not join the first two in the 1s orbital because that configuration is forbidden by the **Pauli exclusion principle**:

> No more than two electrons may occupy any given orbital and, if two do occupy one orbital, then their spins must be paired.

Electrons with paired spins, denoted ↑↓, have zero net spin angular momentum because the spin of one electron is cancelled by the spin of the other. Specifically, one electron has $m_s = +\frac{1}{2}$, the other has $m_s = -\frac{1}{2}$ and they are orientated on their respective cones so that the resultant spin is zero (Fig. 4.19). The exclusion principle is the key to the structure of complex atoms, to chemical periodicity, and to molecular structure. It was proposed by Wolfgang Pauli in 1924 when he was trying to account for the absence of some lines in the spectrum of helium. Later he was able to derive a very general form of the principle from theoretical considerations.

The Pauli exclusion principle in fact applies to any pair of identical fermions (particles with half integral spin). Thus it applies to protons, neutrons, and ^{13}C nuclei (all of which have spin $\frac{1}{2}$) and to ^{35}Cl nuclei (which have spin $\frac{3}{2}$). It does not apply to identical bosons (particles with integral spin), which include photons (spin 1) and ^{12}C nuclei (spin 0). Any number of identical bosons may occupy the same state (that is, be described by the same wavefunction).

The Pauli exclusion principle is a special case of a general statement called the **Pauli principle**:

> When the labels of any two identical fermions are exchanged, the total wavefunction changes sign; when the labels of any two identical bosons are exchanged, the total wavefunction retains the same sign.

By 'total wavefunction' is meant the entire wavefunction, including the spin of the particles, that is, the total wavefunction must be a function of the positions as well as spins of the particles. To see that the Pauli principle implies the Pauli exclusion principle, we consider the (total) wavefunction for two electrons $\psi(1,2)$. The Pauli principle implies that it is a fact of nature (which has its roots in the theory of relativity) that the wavefunction must change sign if we interchange the labels 1 and 2 wherever they occur in the function:

$$\psi(2,1) = -\psi(1,2) \qquad (4.30)$$

Suppose the two electrons in an atom occupy an orbital ψ, a spatial wavefunction which depends on the coordinates in space of the electron; then in the orbital approximation the overall spatial wavefunction is $\psi(1)\psi(2)$. To apply the Pauli principle, we must consider the total wavefunction, the wavefunction including spin. There are several possibilities for two spins: both electrons can be in state α, denoted $\alpha(1)\alpha(2)$; both β, denoted $\beta(1)\beta(2)$; and one α the other β, denoted either $\alpha(1)\beta(2)$ or $\alpha(2)\beta(1)$. Because we cannot tell which electron is α and which is β, in the last case it is appropriate to express the spin states as the (normalized) linear combinations

$$\sigma_+(1,2) = (1/2^{1/2})\{\alpha(1)\beta(2) + \beta(1)\alpha(2)\} \quad (4.31a)$$

$$\sigma_-(1,2) = (1/2^{1/2})\{\alpha(1)\beta(2) - \beta(1)\alpha(2)\} \quad (4.31b)$$

because these combinations allow one spin to be α and the other β with equal probability. The total wavefunction of the system is therefore the product of the orbital part and one of the four spin states:

$$\psi(1)\psi(2)\alpha(1)\alpha(2) \qquad \psi(1)\psi(2)\beta(1)\beta(2)$$
$$\psi(1)\psi(2)\sigma_+(1,2) \qquad \psi(1)\psi(2)\sigma_-(1,2)$$

The notation for the first two functions can be made more compact by introducing the **spinorbital**, a combination of one orbital function and one spin function, such as $\psi^\alpha(1) = \psi(1)\alpha(1)$. More generally, for electron n with an orbital function $\psi(n)$, there are two possible spinorbitals, $\psi^\alpha(n) = \psi(n)\alpha(n)$ and $\psi^\beta(n) = \psi(n)\beta(n)$. The first two wavefunctions can then be denoted $\psi^\alpha(1)\psi^\alpha(2)$ and $\psi^\beta(1)\psi^\beta(2)$.

A brief comment A stronger justification for taking linear combinations in eqn 4.31 is that they correspond to eigenfunctions of the total spin operators $\hat{S}^2$ and $\hat{S}_z$, with $S = 1$, $M_S = 0$ for σ_+ and $S = 0$, $M_S = 0$ for σ_- (see Section 4.5). The linear combinations are also orthogonal in the sense that the integral of their product over spin 'coordinates' vanishes.

The Pauli principle says that, for a total wavefunction to be acceptable (for electrons), it must change sign when the electrons are exchanged. In each case, exchanging the labels 1 and 2 converts the factor $\psi(1)\psi(2)$ into $\psi(2)\psi(1)$, which is the same, because the order of multiplying the functions does not change the value of the product. The same is true of $\alpha(1)\alpha(2)$ and $\beta(1)\beta(2)$. Therefore, the two overall products (in the first row of the two rows listed above) are not allowed, because they do not change sign. The combination $\sigma_+(1,2)$ changes to

$$\sigma_+(2,1) = (1/2^{1/2})\{\alpha(2)\beta(1) + \beta(2)\alpha(1)\} = \sigma_+(1,2)$$

because it is simply the original function written in a different order. The third overall product is therefore also disallowed. Finally, consider $\sigma_-(1,2)$:

$$\sigma_-(2,1) = (1/2^{1/2})\{\alpha(2)\beta(1) - \beta(2)\alpha(1)\}$$
$$= -(1/2^{1/2})\{\alpha(1)\beta(2) - \beta(1)\alpha(2)\} = -\sigma_-(1,2)$$

This combination does change sign (it is 'antisymmetric'). The product $\psi(1)\psi(2)\sigma_-(1,2)$ also changes sign under particle exchange, and therefore it is acceptable.

Now we see that only one of the four possible states is allowed by the Pauli principle, and the one that survives has paired α and β spins. This is the content of the Pauli exclusion principle. The exclusion principle is irrelevant when the orbitals occupied by the electrons are different, and both electrons may then have (but need not have) the same spin state. Nevertheless, even then the overall wavefunction must still be antisymmetric, and must still satisfy the Pauli principle itself.

A final point in this connection is that the acceptable product wavefunction $\psi(1)\psi(2)\sigma_-(1,2)$ can be expressed as a determinant:

$$\frac{1}{2^{1/2}}\begin{vmatrix}\psi^\alpha(1) & \psi^\beta(1)\\ \psi^\alpha(2) & \psi^\beta(2)\end{vmatrix} = \frac{1}{2^{1/2}}\{\psi^\alpha(1)\psi^\beta(2) - \psi^\alpha(2)\psi^\beta(1)\}$$
$$= \frac{1}{2^{1/2}}\{\psi(1)\alpha(1)\psi(2)\beta(2) - \psi(2)\alpha(2)\psi(1)\beta(1)\}$$
$$= \psi(1)\psi(2)\sigma_-(1,2)$$

Any acceptable wavefunction for a **closed-shell** species (a species with a noble-gas configuration) can be expressed as a determinant, which in this context is called a **Slater determinant**. In general, for N electrons in orbitals $\psi_a, \psi_b, \ldots$, the total wavefunction Ψ (we use upper-case psi whenever the wavefunction includes both spatial and spin parts) is

$$\Psi(1,2,\ldots,N) = \frac{1}{(N!)^{1/2}}\begin{vmatrix}\psi_a^\alpha(1) & \psi_a^\beta(1) & \psi_b^\alpha(1) & \cdots & \psi_z^\beta(1)\\ \psi_a^\alpha(2) & \psi_a^\beta(2) & \psi_b^\alpha(2) & \cdots & \psi_z^\beta(2)\\ \psi_a^\alpha(3) & \psi_a^\beta(3) & \psi_b^\alpha(3) & \cdots & \psi_z^\beta(3)\\ \vdots & \vdots & \vdots & \vdots & \vdots\\ \psi_a^\alpha(N) & \psi_a^\beta(N) & \psi_b^\alpha(N) & \cdots & \psi_z^\beta(N)\end{vmatrix} \quad [4.32]$$

Writing a many-electron wavefunction in this way ensures that it is antisymmetric under the interchange of any pair of electrons, as is explored in Problem 4.25. Slater determinants will appear again in Chapter 6, as they are commonly used in calculations on atoms and, more importantly for chemists, molecules.

Now we can return to lithium. In Li ($Z = 3$), the third electron cannot enter the 1s orbital because that orbital is already full: we say the K shell is **complete** and that the two electrons form a closed shell. Because a similar closed shell is characteristic of the He atom, we denote it [He]. The third electron is excluded from the K shell and must occupy the next available orbital, which is one with $n = 2$ and hence belonging to the L shell. However, we now have to decide whether the next available orbital is the 2s orbital or a 2p orbital, and therefore whether the lowest energy configuration of the atom is [He]$2s^1$ or [He]$2p^1$.

(c) Penetration and shielding

In hydrogenic atoms all orbitals of a given shell are degenerate. In many-electron atoms, although orbitals of a given subshell remain degenerate, the subshells themselves have different

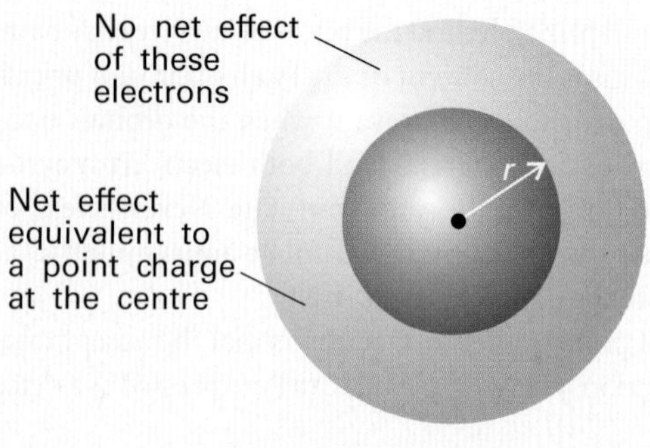

Fig. 4.20 An electron at a distance r from the nucleus experiences a Coulombic repulsion from all the electrons within a sphere of radius r and which is equivalent to a point negative charge located on the nucleus. The negative charge reduces the effective nuclear charge of the nucleus from Ze to $Z_{\text{eff}}e$.

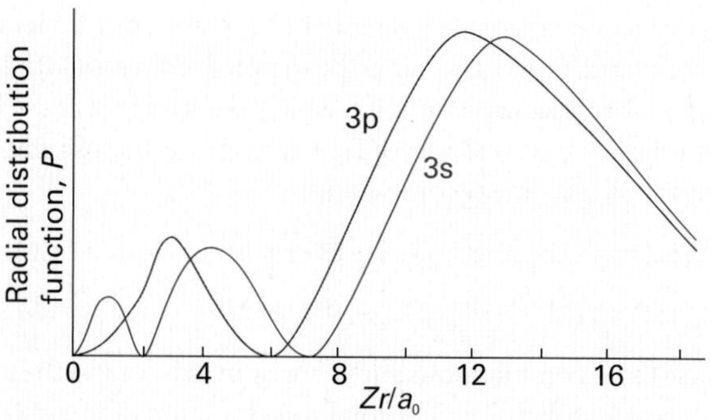

Fig. 4.21 An electron in an s orbital (here a 3s orbital) is more likely to be found close to the nucleus than an electron in a p orbital of the same shell (note the closeness of the innermost peak of the 3s orbital to the nucleus at $r = 0$). Hence an s electron experiences less shielding and is more tightly bound than a p electron.

interActivity Calculate and plot the graphs given above for $n = 4$.

energies. The difference can be traced to the fact that an electron in a many-electron atom experiences a Coulombic repulsion from all the other electrons present. If it is at a distance r from the nucleus, it experiences an average repulsion that can be represented by a point negative charge located at the nucleus and equal in magnitude to the total charge of the electrons within a sphere of radius r (Fig. 4.20). The effect of this point negative charge, when averaged over all the locations of the electron, is to reduce the full charge of the nucleus from Ze to $Z_{\text{eff}}e$, the **effective nuclear charge**. In everyday parlance, Z_{eff} itself is commonly referred to as the 'effective nuclear charge'. We say that the electron experiences a **shielded** nuclear charge, and the difference between Z and Z_{eff} is called the **shielding constant**, σ:

$$Z_{\text{eff}} = Z - \sigma \qquad [4.33]$$

The electrons do not actually 'block' the full Coulombic attraction of the nucleus: the shielding constant is simply a way of expressing the net outcome of the nuclear attraction and the electronic repulsions in terms of a single equivalent charge at the centre of the atom.

The shielding constant is different for s and p electrons because they have different radial distribution functions (Fig. 4.21). An s electron has a greater **penetration** through inner shells than a p electron, in the sense that it is more likely to be found close to the nucleus than a p electron of the same shell (the wavefunction of a p orbital, remember, is zero at the nucleus). Because only electrons inside the sphere defined by the location of the electron (in effect, the core electrons) contribute to shielding, an s electron experiences less shielding than a p electron. Consequently, by the combined effects of penetration and shielding, an s electron is more tightly bound than a p electron of the same shell. Similarly, a d electron penetrates less than a p electron of the same shell (recall that the wavefunctions of orbitals are proportional to r^l close to the nucleus and therefore that a d orbital varies as r^2 close to the nucleus, whereas a p orbital varies as r), and therefore experiences more shielding. The consequence of penetration and shielding is that the energies of subshells of a shell in a many-electron atom in general lie in the order s < p < d < f.

Shielding constants for different types of electrons in atoms have been calculated from their wavefunctions obtained by numerical solution of the Schrödinger equation (Table 4.2). We see that, in general, valence-shell s electrons do experience higher effective nuclear charges than p electrons, although there are some discrepancies. We return to this point shortly.

We can now complete the Li story. Because the shell with $n = 2$ consists of two non-degenerate subshells, with the 2s orbital lower in energy than the three 2p orbitals, the third electron occupies the 2s orbital. This occupation results in the ground-state configuration $1s^22s^1$, with the central nucleus surrounded by a complete helium-like shell of two 1s electrons, and around that a more diffuse 2s electron. The electrons in the outermost shell of an atom in its ground state are called the **valence electrons** because they are largely responsible for the chemical

Synoptic table 4.2* Effective nuclear charge, $Z_{\text{eff}} = Z - \sigma$

Element	Z	Orbital	Z_{eff}
He	2	1s	1.6875
C	6	1s	5.6727
		2s	3.2166
		2p	3.1358

* More values are given in the *Data section*.

bonds that the atom forms. Thus, the valence electron in Li is a 2s electron and its other two electrons belong to its core.

(d) The building-up principle

The extension of this argument is called the **building-up principle**, or the *Aufbau principle*, from the German word for building up, which will be familiar from introductory courses. In brief, we imagine the bare nucleus of atomic number Z, and then feed into the orbitals Z electrons in succession. The order of occupation is

1s 2s 2p 3s 3p 4s 3d 4p 5s 4d 5p 6s

and each orbital may accommodate up to two electrons. As an example, consider the carbon atom, for which $Z = 6$ and there are six electrons to accommodate. Two electrons enter and fill the 1s orbital, two enter and fill the 2s orbital, leaving two electrons to occupy the orbitals of the 2p subshell. Hence the ground-state configuration of C is $1s^2 2s^2 2p^2$ or, more succinctly, $[He]2s^2 2p^2$ with [He] the helium-like $1s^2$ core. However, we can be more precise: we can expect the last two electrons to occupy different 2p orbitals because they will then be further apart on average and repel each other less than if they were in the same orbital. Thus, one electron can be thought of as occupying the $2p_x$ orbital and the other the $2p_y$ orbital (the x, y, z designation is arbitrary, and it would be equally valid to use the complex forms of these orbitals), and the lowest energy configuration of the atom is $[He]2s^2 2p_x^1 2p_y^1$. The same rule applies whenever degenerate orbitals of a subshell are available for occupation. Thus, another rule of the building-up principle is:

> Electrons occupy different orbitals of a given subshell before doubly occupying any one of them.

For instance, nitrogen ($Z = 7$) has the configuration $[He]2s^2 2p_x^1 2p_y^1 2p_z^1$, and only when we get to oxygen ($Z = 8$) is a 2p orbital doubly occupied, giving $[He]2s^2 2p_x^2 2p_y^1 2p_z^1$. When electrons occupy orbitals singly we invoke **Hund's maximum multiplicity rule**:

> An atom in its ground state adopts a configuration with the greatest number of electrons with unpaired spins.

The explanation of Hund's rule is subtle, but it reflects the quantum mechanical property of **spin correlation**, that electrons with parallel spins behave as if they have a tendency to stay well apart (see the following *Justification*), and hence repel each other less. In essence, the effect of spin correlation is to allow the atom to shrink slightly, so the electron–nucleus interaction is improved when the spins are parallel. We can now conclude that in the ground state of the carbon atom, the two 2p electrons have the same spin, that all three 2p electrons in the N atom have the same spin, and that the two 2p electrons in different orbitals in the O atom have the same spin (the two in the $2p_x$ orbital are necessarily paired).

Justification 4.7 *Spin correlation*

Suppose electron 1 is described by a spatial wavefunction $\psi_a(\boldsymbol{r}_1)$ and electron 2 is described by a wavefunction $\psi_b(\boldsymbol{r}_2)$; then, in the orbital approximation, the joint wavefunction of the electrons is the product $\psi = \psi_a(\boldsymbol{r}_1)\psi_b(\boldsymbol{r}_2)$. However, this wavefunction is not acceptable, because it suggests that we know which electron is in which orbital, whereas we cannot keep track of electrons. According to quantum mechanics, the correct description is either of the two following wavefunctions:

$$\psi_\pm = (1/2^{1/2})\{\psi_a(\boldsymbol{r}_1)\psi_b(\boldsymbol{r}_2) \pm \psi_b(\boldsymbol{r}_1)\psi_a(\boldsymbol{r}_2)\}$$

According to the Pauli principle, because ψ_+ is symmetrical under particle interchange, it must be multiplied by an antisymmetric spin function (the one denoted σ_- in eqn 4.31b). That combination corresponds to a spin-paired state. Conversely, ψ_- is antisymmetric, so it must be multiplied by one of the three symmetric spin states. These three symmetric states correspond to electrons with parallel spins (as will be explained in Section 4.5).

Now consider the values of the two combinations $\psi_\pm$ when one electron approaches another, and eventually $r_1 = r_2$. We see that ψ_- vanishes, which means that there is zero probability of finding the two electrons at the same point in space when they have parallel spins. The decreasing probability that the electrons approach one another in the state ψ_- is called a **Fermi hole**. The other combination does not vanish when the two electrons are at the same point in space. Because the two electrons have different relative spatial distributions depending on whether their spins are parallel or not, it follows that their Coulombic interaction is different, and hence that the two states have different energies, with the state corresponding to parallel spins being lower in energy.

However, we have to be cautious with this explanation, for it supposes that the original wavefunctions are unchanged. Detailed numerical calculations have shown that in the specific case of a helium atom electrons with parallel spins are actually closer together than those with antiparallel spins. The explanation in this case is that spin correlation between electrons with parallel spins allows the entire atom to shrink. Therefore, although the average separation is reduced, the electrons are found closer to the nucleus, which lowers their potential energy.

Neon, with $Z = 10$, has the configuration $[He]2s^2 2p^6$, which completes the L shell. This closed-shell configuration is denoted [Ne], and acts as a core for subsequent elements. The next electron must enter the 3s orbital and begin a new shell, so an Na atom, with $Z = 11$, has the configuration $[Ne]3s^1$. Like lithium with the configuration $[He]2s^1$, sodium has a single s electron outside a complete core. This analysis has brought us to the origin of chemical periodicity. The L shell is completed by eight electrons, so the element with $Z = 3$ (Li) should have similar properties to the element with $Z = 11$ (Na). Likewise, Be ($Z = 4$) should be similar to $Z = 12$ (Mg), and so on, up to the noble gases He ($Z = 2$), Ne ($Z = 10$), and Ar ($Z = 18$).

Ten electrons can be accommodated in the five 3d orbitals, which accounts for the electron configurations of scandium to zinc. Calculations of the type discussed in Chapter 6 show that for these atoms the energies of the 3d orbitals are always lower than the energy of the 4s orbital. However, spectroscopic results show that Sc has the configuration [Ar]$3d^14s^2$, instead of [Ar]$3d^3$ or [Ar]$3d^24s^1$. To understand this observation, we have to consider the nature of electron–electron repulsions in 3d and 4s orbitals, where the effects are particularly finely balanced because there is only a small difference in energy between the orbitals. The most probable distance of a 3d electron (with no radial nodes) from the nucleus is less than that for a 4s electron (with three radial nodes), so two 3d electrons repel each other more strongly than two 4s electrons. As a result, Sc has the configuration [Ar]$3d^14s^2$ rather than the two alternatives, for then the strong electron–electron repulsions in the 3d orbitals are minimized. The total energy of the atom is least despite the cost of allowing electrons to populate the high energy 4s orbital (Fig. 4.22). The effect just described is generally true for scandium through zinc, so their electron configurations are of the form [Ar]$3d^n4s^2$, where $n = 1$ for scandium and $n = 10$ for zinc. Two notable exceptions, which are observed experimentally, are Cr, with electron configuration [Ar]$3d^54s^1$, and Cu, with electron configuration [Ar]$3d^{10}4s^1$. The theoretical basis of the exceptions represented by Cr and Cu is the additional energy lowering characteristic of half-filled and completely filled d subshells.

At gallium, the building-up principle is used in the same way as in preceding periods. Now the 4s and 4p subshells constitute the valence shell, and the period terminates with krypton.

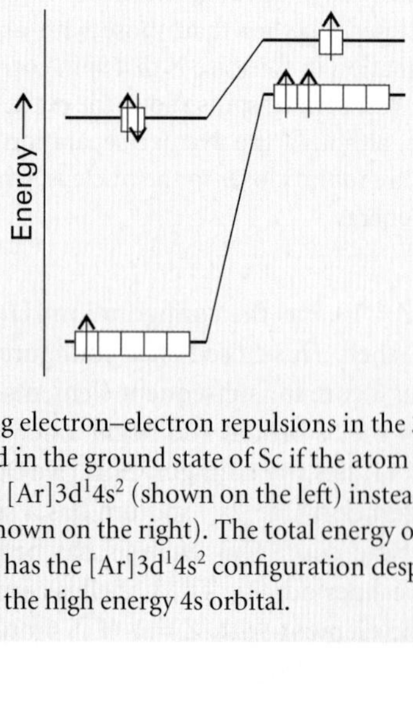

Fig. 4.22 Strong electron–electron repulsions in the 3d orbitals are minimized in the ground state of Sc if the atom has the configuration [Ar]$3d^14s^2$ (shown on the left) instead of [Ar]$3d^24s^1$ (shown on the right). The total energy of the atom is lower when it has the [Ar]$3d^14s^2$ configuration despite the cost of populating the high energy 4s orbital.

Because 18 electrons have intervened since argon, this period is the first 'long period' of the periodic table. The existence of the d-block elements (the 'transition metals') reflects the stepwise occupation of the 3d orbitals, and the subtle shades of energy differences and effects of electron–electron repulsion along this series gives rise to the rich complexity of inorganic d-metal chemistry. A similar intrusion of the f orbitals in Periods 6 and 7 accounts for the existence of the f block of the periodic table (the lanthanoids and actinoids).

We derive the configurations of cations of elements in the s, p, and d blocks of the periodic table by removing electrons from the ground-state configuration of the neutral atom in a specific order. First, we remove valence p electrons, then valence s electrons, and then as many d electrons as are necessary to achieve the specified charge. For instance, because the configuration of V is [Ar]$3d^34s^2$, the V^{2+} cation has the configuration [Ar]$3d^3$. It is reasonable that we remove the more energetic 4s electrons in order to form the cation, but it is not obvious why the [Ar]$3d^3$ configuration is preferred in V^{2+} over the [Ar]$3d^14s^2$ configuration, which is found in the isoelectronic Sc atom. Calculations show that the energy difference between [Ar]$3d^3$ and [Ar]$3d^14s^2$ depends on $Z_{\rm eff}$. As $Z_{\rm eff}$ increases, transfer of a 4s electron to a 3d orbital becomes more favourable because the electron–electron repulsions are compensated by attractive interactions between the nucleus and the electrons in the spatially compact 3d orbital. Indeed, calculations reveal that, for a sufficiently large $Z_{\rm eff}$, [Ar]$3d^3$ is lower in energy than [Ar]$3d^14s^2$. This conclusion explains why V^{2+} has an [Ar]$3d^3$ configuration and also accounts for the observed [Ar]$4s^03d^n$ configurations of the M^{2+} cations of Sc through Zn.

The configurations of anions of the p-block elements are derived by continuing the building-up procedure and adding electrons to the neutral atom until the configuration of the next noble gas has been reached. Thus, the configuration of the O^{2-} ion is achieved by adding two electrons to [He]$2s^22p^4$, giving [He]$2s^22p^6$, the same as the configuration of neon.

(e) Ionization energies and electron affinities

The minimum energy necessary to remove an electron from a many-electron atom in the gas phase is the **first ionization energy**, I_1, of the element. The **second ionization energy**, I_2, is the minimum energy needed to remove a second electron (from the singly charged cation). Some numerical values are given in Table 4.3. The **electron affinity**, $E_{\rm ea}$, is the energy released when an electron attaches to a gas-phase atom (Table 4.4). In a common, logical, but not universal convention (which we adopt), the electron affinity is positive if energy is released when the electron attaches to the atom.

As will be familiar from introductory chemistry, ionization energies and electron affinities show periodicities (Fig. 4.23). The former is more regular and we concentrate on it. Lithium has a low first ionization energy because its outermost electron

Synoptic table 4.3* First and second ionization energies

Element	I_1/(kJ mol^{-1})	I_2/(kJ mol^{-1})
H	1312	
He	2372	5250
Mg	738	1451
Na	496	4562

* More values are given in the *Data section.*

Synoptic table 4.4* Electron affinities, E_{ea}/(kJ mol^{-1})

Cl	349		
F	322		
H	73		
O	141	O^-	−844

* More values are given in the *Data section.*

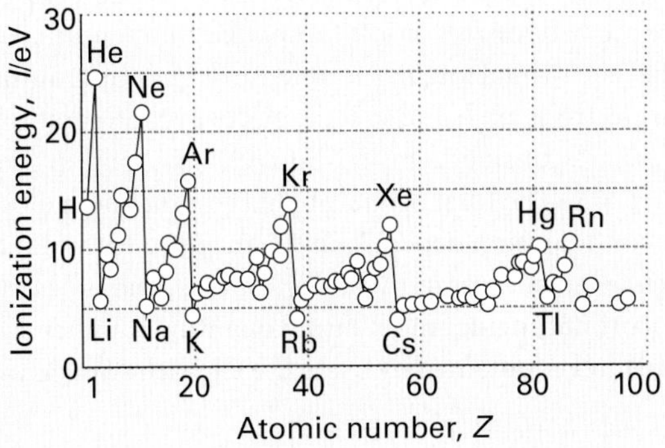

Fig. 4.23 The first ionization energies of the elements plotted against atomic number.

is well shielded from the nucleus by the core electrons ($Z_{eff} = 1.3$, compared with $Z = 3$). The ionization energy of beryllium ($Z = 4$) is greater but that of boron is lower because in the latter the outermost electron occupies a 2p orbital and is less strongly bound than if it had been a 2s electron. The ionization energy increases from boron to nitrogen on account of the increasing nuclear charge. However, the ionization energy of oxygen is less than would be expected by simple extrapolation. The explanation is that at oxygen a 2p orbital must become doubly occupied, and the electron–electron repulsions are increased above what would be expected by simple extrapolation along the row. In addition, the loss of a 2p electron results in a configuration with a half-filled subshell (like that of N), which is an arrangement of low energy, so the energy of $O^+ + e^-$ is lower than might be expected, and the ionization energy is correspondingly low too. (The kink is less pronounced in the next row, between phosphorus and sulfur because their orbitals are more diffuse.) The values for oxygen, fluorine, and neon fall roughly on the same line, the increase of their ionization energies reflecting the increasing attraction of the more highly charged nuclei for the outermost electrons.

The outermost electron in sodium is 3s. It is far from the nucleus, and the latter's charge is shielded by the compact, complete neon-like core. As a result, the ionization energy of sodium is substantially lower than that of neon. The periodic cycle starts again along this row, and the variation of the ionization energy can be traced to similar reasons.

Electron affinities are greatest close to fluorine, for the incoming electron enters a vacancy in a compact valence shell and can interact strongly with the nucleus. The attachment of an electron to an anion (as in the formation of O^{2-} from O^-) invariably requires the absorption of energy, so E_{ea} is negative. The incoming electron is repelled by the charge already present. Electron affinities are also small, and may be negative, when an electron enters an orbital that is far from the nucleus (as in the heavier alkali metal atoms) or is forced by the Pauli principle to occupy a new shell (as in the noble gas atoms).

The values of ionization energies and electron affinities can help us to understand a great deal of chemistry and, through chemistry, biology. For example, we can now begin to see why carbon is an essential building block of complex biological structures. Among the elements in Period 2, carbon has intermediate values of the ionization energy and electron affinity, so it can share electrons (that is, form covalent bonds) with many other elements, such as hydrogen, nitrogen, oxygen, sulfur, and, more importantly, other carbon atoms. As a consequence, such networks as long carbon–carbon chains (as in lipids) and chains of peptide links can form readily. Because the ionization energy and electron affinity of carbon are neither too high nor too low, the bonds in these covalent networks are neither too strong nor too weak. As a result, biological molecules are sufficiently stable to form viable organisms but are still susceptible to dissociation and rearrangement. In Chapter 5 we shall develop additional concepts that will complete this story about carbon.

(f) Self-consistent field calculations

The treatment we have given to the electronic configuration of many-electron species is only approximate because it is hopeless to expect to find exact solutions of a Schrödinger equation that take into account the interaction of all the electrons with one another. However, computational techniques are available that give very detailed and reliable approximate solutions for the wavefunctions and energies. The techniques were originally introduced by D. R. Hartree (before computers were available) and then modified by V. Fock to take into account the Pauli principle correctly. These techniques are of great interest to chemists when applied to molecules, and are explained in detail in Chapter 6; however, we should be aware of the general principles at this stage too. In broad outline, the **Hartree–Fock self-consistent field** (HF-SCF) procedure is as follows.

Imagine that we have an approximate idea of the structure of the atom. In the Ne atom, for instance, the orbital approximation suggests the configuration $1s^2 2s^2 2p^6$ with the orbitals approximated by hydrogenic atomic orbitals. Now consider one of the 2p electrons. A Schrödinger equation can be written for this electron by ascribing to it a potential energy due to the nuclear attraction and the repulsion from the other electrons. Although the equation is for the 2p orbital, it depends on the wavefunctions of all the other occupied orbitals in the atom. To solve the equation, we guess an approximate form of the wavefunctions of all the orbitals except 2p and then solve the Schrödinger equation for the 2p orbital. The procedure is then repeated for the 1s and 2s orbitals. This sequence of calculations gives the form of the 2p, 2s, and 1s orbitals, and in general they will differ from the set used initially to start the calculation. These improved orbitals can be used in another cycle of calculation, and a second improved set of orbitals and a better energy are obtained. The recycling continues until the orbitals and energies obtained are insignificantly different from those used at the start of the current cycle. The solutions are then self-consistent and accepted as solutions of the problem.

Figure 4.24 shows plots of some of the HF-SCF radial distribution functions for sodium. They show the grouping of electron density into shells, as was anticipated by the early chemists, and the differences of penetration as discussed above. These SCF calculations therefore support the qualitative discussions that are used to explain chemical periodicity. They also considerably extend that discussion by providing detailed wavefunctions and precise energies.

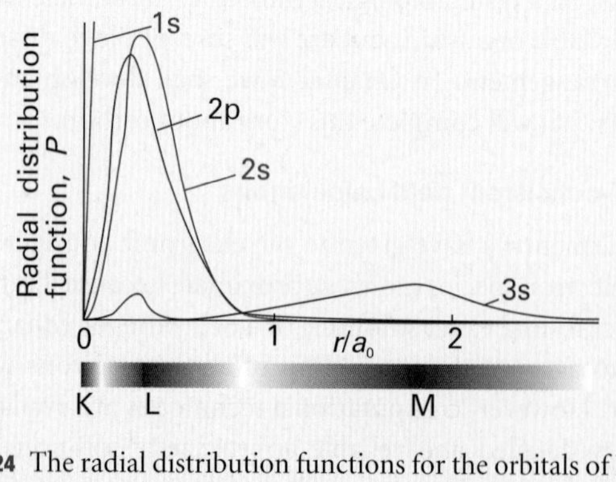

Fig. 4.24 The radial distribution functions for the orbitals of Na based on SCF calculations. Note the shell-like structure with the 3s orbital outside the inner K and L shells.

4.5 Term symbols

Now that we know how to account for the ground-state configurations of atoms it is appropriate to examine the states of these atoms in more detail, especially their excited states, and to consider the transitions between them. A complication we immediately encounter is that a single configuration of an atom, such as the excited configuration of He, $1s^1 2s^1$ for instance, or the ground state of C, $[He]2s^2 2p^2$, can give rise to a number of different individual states with various energies. Our first task is to identify these states and find a way to label them with a 'term symbol'. Atomic term symbols, and their molecular counterparts, play a crucial role in spectroscopy, in the discussion of magnetic properties, in the photochemistry of the atmosphere, and in the description of the operation of lasers.

The key to identifying the various states that can arise from a configuration and attaching a term symbol is the angular momentum of the electrons: that includes their orbital angular momentum, their spin, and their total angular momentum. Our first job is to identify the allowed values of these angular momenta for atoms with more than one electron.

(a) The total orbital angular momentum

Let's consider an atom with two electrons outside a closed core, so there are two sources of orbital angular momentum. We suppose that the orbital angular momentum quantum numbers of the two electrons are l_1 and l_2. If the electron configuration of the atoms we are considering is p^2, both electrons are in p orbitals and $l_1 = l_2 = 1$. The total orbital angular momenta that can arise from a configuration depends on the magnitudes and the relative orientation of these individual momenta, and is described by the **total orbital angular momentum quantum number**, L, a non-negative integer obtained by using the **Clebsch–Gordan series:**

$$L = l_1 + l_2, l_1 + l_2 - 1, \dots, |l_1 - l_2| \qquad (4.34)$$

For instance, if $l_1 = l_2 = 1$, then $L = 2$, 1, and 0. Once we know the value of the L we can calculate the magnitude of the total orbital angular momentum from $\{L(L+1)\}^{1/2}\hbar$. As for any angular momentum, the total orbital angular momentum has $2L+1$ orientations distinguished by the quantum number M_L, which can take the values $L, L-1, \dots, -L$.

Just as we use lower-case letters to tell us the value of l, so we use an upper-case letter to tell us the value of L. The code for converting the value of L into a letter is the same as for the s, p, d, f, ... designation of orbitals, but uses upper-case letters:

L:	0	1	2	3	4	5	6...
	S	P	D	F	G	H	I...

Thus a p^2 configuration can give rise to D, P, and S terms. A closed shell has zero orbital angular momentum because all the individual orbital angular momenta sum to zero. Therefore, when working out term symbols, we need consider only the electrons of the unfilled shell. In the case of a single electron outside a closed shell, the value of L is the same as the value of l, so the configuration [Ne]$3s^1$ has only an S term.

The terms that arise from a given configuration differ in energy due to the Coulombic interaction between the electrons. For example, to achieve a D ($L = 2$) term from a $2p^1 3p^1$ configuration, both electrons need to be circulating in the same direction around the nucleus but, to achieve an S ($L = 0$) term, they would need to be circulating in opposite directions. In the former arrangement, they do not meet; in the latter they do. On the basis of this classical picture, we can suspect that the repulsion between them will be higher if they meet, and therefore that the S term will lie higher in energy than the D term. The quantum mechanical analysis of the problem supports this interpretation.

Example 4.6 *Deriving the total orbital angular momentum of a configuration*

Find the terms that can arise from the configurations (a) d^2, (b) p^3.

Method Use the Clebsch–Gordan series and begin by finding the minimum value of L (so that we know where the series terminates). When there are more than two electrons to couple together, use two series in succession: first couple two electrons, and then couple the third to each combined state, and so on.

Answer (a) The minimum value is $|l_1 - l_2| = |2 - 2| = 0$. Therefore,

$$L = 2 + 2, 2 + 2 - 1, \ldots, 0 = 4, 3, 2, 1, 0$$

corresponding to G, F, D, P, S terms, respectively. (b) Coupling two electrons gives a minimum value of $|1 - 1| = 0$. Therefore,

$$L' = 1 + 1, 1 + 1 - 1, \ldots, 0 = 2, 1, 0$$

Now couple $l_3 = 1$ with $L' = 2$, to give $L = 3, 2, 1$; with $L' = 1$, to give $L = 2, 1, 0$; and with $L' = 0$, to give $L = 1$. The overall result is

$$L = 3, 2, 2, 1, 1, 1, 0$$

giving one F, two D, three P, and one S term.

Self-test 4.9 Repeat the question for the configurations (a) f^1d^1 and (b) d^3.

[(a) H, G, F, D, P; (b) I, 2H, 3G, 4F, 5D, 3P, S]

(b) The total spin angular momentum

The energy of a term also depends on the relative orientation of the electron spins. We have seen a glimmer of this dependence already, when we saw that spin correlation results in states with parallel spins having a lower energy than states with antiparallel spins (Section 4.4). Parallel (unpaired) and antiparallel (paired) spins differ in their overall spin angular momentum. In the paired case, the two spin momenta cancel each other, and there is zero net spin (as was depicted in Fig. 4.19), and so once again we are brought to a correlation between an angular momentum, in this case spin, and an energy.

When there are several electrons to be taken into account, we must assess their **total spin angular momentum quantum number**, S (a non-negative integer or half integer). To do so, we use the Clebsch–Gordan series, this time in the form

$$S = s_1 + s_2, s_1 + s_2 - 1, \ldots, |s_1 - s_2| \quad (4.35)$$

noting that each electron has $s = \frac{1}{2}$, which gives $S = 1, 0$ for two electrons (Fig. 4.25). If there are three electrons, the total spin angular momentum is obtained by coupling the third spin to each of the values of S for the first two spins, which results in $S = \frac{3}{2}$, and $S = \frac{1}{2}$. The value of S for a term is expressed by giving the **multiplicity** of a term, the value of $2S + 1$, as a left-superscript on the term symbol. Thus, ^{1}P is a 'singlet' term ($S = 0, 2S + 1 = 1$) and ^{3}P is a 'triplet' term ($S = 1, 2S + 1 = 3$). The multiplicity actually tells us the number of permitted values of $M_S = S, S - 1, \ldots, -S$ for the given value of S, and hence the number of orientations in space that the total spin can adopt. We shall see the importance of this information shortly.

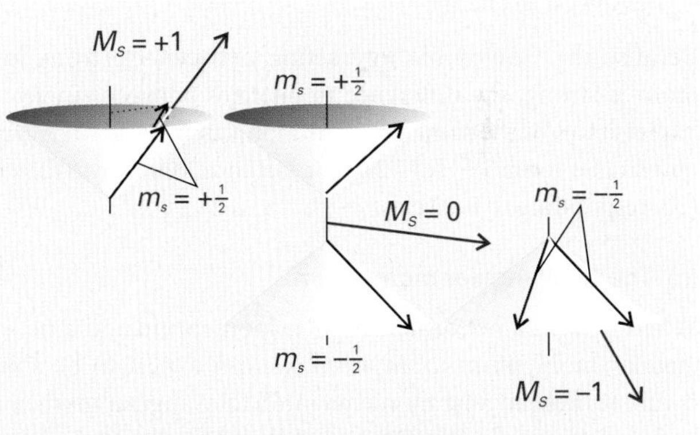

Fig. 4.25 When two electrons have parallel spins, they have a nonzero total spin angular momentum. There are three ways of achieving this resultant, which are shown by these vector representations. Note that, although we cannot know the orientation of the spin vectors on the cones, the angle between the vectors is the same in all three cases, for all three arrangements have the same total spin angular momentum (that is, the resultant of the two vectors has the same length in each case, but points in different directions). Compare this diagram with Fig. 4.19, which shows the antiparallel case. Note that, whereas two paired spins are precisely antiparallel, two 'parallel' spins are not strictly parallel.

A note on good practice Throughout our discussion of atomic spectroscopy, distinguish italic *S*, the total spin quantum number, from Roman S, the term label. Thus, ^{3}S is a triplet term with $S = 1$ (and $L = 0$). All state symbols are upright; all quantum numbers and physical observables are sloping.

When $S = 0$ (as for a closed shell, like $1s^2$), $M_S = 0$, the electron spins are all paired and there is no net spin: this arrangement gives a singlet term, ^{1}S. A single electron has $S = s = \frac{1}{2}$ ($M_S = m_s = \pm\frac{1}{2}$), so a configuration such as [Ne]$3s^1$ can give rise to a doublet term, ^{2}S. Likewise, the configuration [Ne]$3p^1$ is a doublet, ^{2}P. When there are two electrons with unpaired spins, $S = 1$ ($M_S = \pm 1, 0$), so $2S + 1 = 3$, giving a triplet term, such as ^{3}D. We saw in Section 4.4 that the energies of two states, one with paired spins and one with unpaired spins, will differ on account of the different effects of spin correlation. The fact that the parallel arrangement of spins, as in the $1s^12s^1$ configuration of the He atom, lies lower in energy than the antiparallel arrangement can now be expressed by saying that the triplet state of the $1s^12s^1$ configuration of He lies lower in energy than the singlet state. This is a general conclusion that applies to other atoms (and molecules) and, *for states arising from the same configuration, the triplet state generally lies lower in energy than the singlet state.* The latter is an example of Hund's rule of maximum multiplicity (Section 4.4) which can be restated as:

> For a given configuration, the term of greatest multiplicity lies lowest in energy.

Because the Coulombic interaction between electrons in an atom is strong, the difference in energies between singlet and triplet states of the same configuration can be large. The triplet and singlet terms of He$1s^12s^1$, for instance, differ by 6421 cm^{-1} (corresponding to 0.80 eV).

(c) The total angular momentum

When there is a net orbital angular momentum and a net spin angular momentum in an atom, we can expect to be able to combine these angular momenta into a total angular momentum and—perhaps—for the energy of the atom to depend on its value. The **total angular momentum quantum number**, *J* (a non-negative integer or half integer), takes the values

$$J = L + S, L + S - 1, \ldots, |L - S| \qquad (4.36)$$

and, as usual, we can calculate the magnitude of the total angular momentum from $\{J(J+1)\}^{1/2}\hbar$. The specific value of *J* is given as a right-subscript on the term symbol; for example, a ^{3}P term with $J = 2$ is fully dressed as ^{3}P$_2$.

If $S \leq L$, there are $2S + 1$ values of *J* for a given *L*, so the number of values of *J* is the same as the multiplicity of the term. Each possible value of *J* designates a **level** of a term so, provided $S \leq L$, the multiplicity tells us the number of levels. For example, the [Ne]$3p^1$ configuration of sodium (an excited state) has $L = 1$ and $S = \frac{1}{2}$ and a multiplicity of 2; the two levels are $J = \frac{3}{2}$ and $\frac{1}{2}$ and the ^{2}P term therefore has two levels, ^{2}P$_{3/2}$ and ^{2}P$_{1/2}$.

Before moving on, we should note that there is a hidden assumption in eqn 4.36. We have assumed that the orbital angular momenta of the electrons all combine to give a total, that their spins all combine to give a total spin, and that only then do these two totals combine to give the overall total angular momentum of the atom. This procedure is called **Russell–Saunders coupling**. An alternative is that the orbital and spin angular momenta of each electron combine separately into a resultant for each one (with quantum number *j*), and then those resultants combine to give an overall total. We shall not deal with this so-called ***jj*-coupling** case: Russell–Saunders coupling turns out to be reasonably accurate for light atoms.

Example 4.7 *Deriving term symbols*

Write the term symbols arising from the ground-state configurations of (a) Na and (b) F, and (c) the excited configuration $1s^22s^22p^13p^1$ of C.

Method Begin by writing the configurations, but ignore inner closed shells. Then couple the orbital momenta to find *L* and the spins to find *S*. Next, couple *L* and *S* to find *J*. Finally, express the term as $^{2S+1}\{L\}_J$, where $\{L\}$ is the appropriate letter. For F, for which the valence configuration is $2p^5$, treat the single gap in the closed-shell $2p^6$ configuration as a single particle.

Answer (a) For Na, the configuration is [Ne]$3s^1$, and we consider the single 3s electron. Because $L = l = 0$ and $S = s = \frac{1}{2}$, it is possible for $J = j = s = \frac{1}{2}$ only. Hence the term symbol is ^{2}S$_{1/2}$. (b) For F, the configuration is [He]$2s^22p^5$, which we can treat as [Ne]$2p^{-1}$ (where the notation $2p^{-1}$ signifies the absence of a 2p electron). Hence $L = 1$, and $S = s = \frac{1}{2}$. Two values of $J = j$ are allowed: $J = \frac{3}{2}, \frac{1}{2}$. Hence, the term symbols for the two levels are ^{2}P$_{3/2}$, ^{2}P$_{1/2}$. (c) We are treating an excited configuration of carbon because, in the ground configuration, $2p^2$, the Pauli principle forbids some terms, and deciding which survive (^{1}D, ^{3}P, ^{1}S, in fact) is quite complicated.[2] That is, there is a distinction between 'equivalent electrons', which are electrons that occupy the same orbitals, and 'inequivalent electrons', which are electrons that occupy different orbitals; we only consider the latter here. The excited configuration of C under consideration is effectively $2p^13p^1$. This is a two-electron problem, and $l_1 = l_2 = 1$, $s_1 = s_2 = \frac{1}{2}$. It follows that

[2] For details, see our *Inorganic chemistry*, Oxford University Press and W. H. Freeman & Co. (2006).

$L = 2, 1, 0$ and $S = 1, 0$. The terms are therefore ^{3}D and ^{1}D, ^{3}P and ^{1}P, and ^{3}S and ^{1}S. For ^{3}D, $L = 2$ and $S = 1$; hence $J = 3, 2, 1$ and the levels are $^3D_3$, $^3D_2$, and $^3D_1$. For ^{1}D, $L = 2$ and $S = 0$, so the single level is $^1D_2$. The triplet of levels of ^{3}P is $^3P_2$, $^3P_1$, and $^3P_0$, and the singlet is $^1P_1$. For the ^{3}S term there is only one level, $^3S_1$ (because $J = 1$ only), and the singlet term is $^1S_0$.

Self-test 4.10 Write down the terms arising from the configurations (a) $2s^12p^1$, (b) $2p^13d^1$.

[(a) $^3P_2$, $^3P_1$, $^3P_0$, $^1P_1$;
(b) $^3F_4$, $^3F_3$, $^3F_2$, $^1F_3$, $^3D_3$, $^3D_2$, $^3D_1$, $^1D_2$, $^3P_2$, $^3P_1$, $^3P_0$, $^1P_1$]

The different levels of a term, such as $^2P_{1/2}$ and $^2P_{3/2}$, have different energies due to **spin–orbit interaction**, a magnetic interaction between angular momenta. To see the origin of this coupling, we need to note that a circulating current gives rise to a magnetic moment (Fig. 4.26). The spin of an electron is one source of magnetic moment and its orbital angular momentum is another. Two magnetic dipole moments close to each other interact to an extent that depends on their relative orientation. However, the relative orientation of the two momenta also determines the electron's total angular momentum, so there is a correlation between the energy of interaction and the value of J (Fig. 4.27). Magnetic dipole moments that are antiparallel (that is, lie in opposite directions) will be lower in energy than when they are parallel; therefore, a lower energy is achieved when the orbital and spin angular momenta are antiparallel, corresponding to a lower value of J. In the case of the ^{2}P term, we predict that the $^2P_{1/2}$ level is lower in energy than the $^2P_{3/2}$ level. For a system with many electrons, a detailed analysis yields the following general statement:

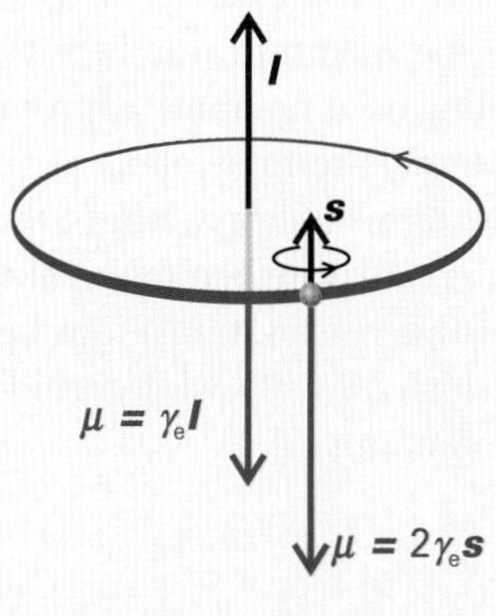

Fig. 4.26 Angular momentum gives rise to a magnetic moment ($\boldsymbol{\mu}$). For an electron, the magnetic moment is antiparallel to the orbital angular momentum, but proportional to it. For spin angular momentum, there is a factor 2, which increases the magnetic moment to twice its expected value. The constant of proportionality $\gamma_e = -e/2m_e$, is called the *magnetogyric ratio*.

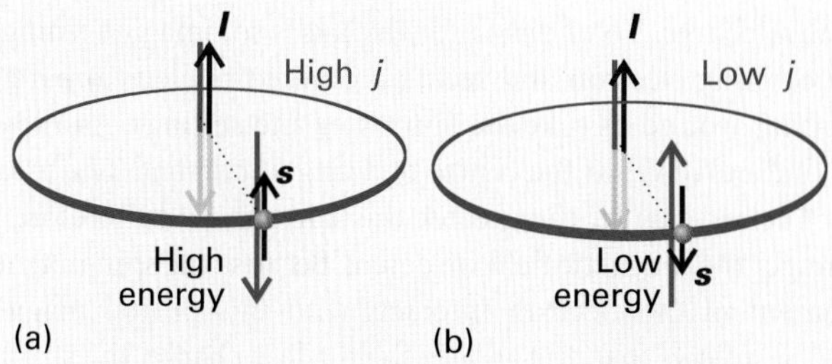

Fig. 4.27 Spin–orbit coupling is a magnetic interaction between spin and orbital magnetic moments. When the angular momenta are parallel, as in (a), the magnetic moments are aligned unfavourably; when they are opposed, as in (b), the interaction is favourable. This magnetic coupling is the cause of the splitting of a configuration into levels.

For atoms with less-than-half-full shells, the level with the smallest value of J will be lowest in energy; for atoms with more-than-half-full shells, the level with the largest value of J will be lowest in energy.

For a quantitative treatment of spin–orbit coupling, we need to include in the hamiltonian a term that depends on the relative orientation of the vectors that represent the spin and orbital angular momenta. The simplest procedure is to write the contribution as

$$\hat{H}_{so} = \lambda \boldsymbol{L}\cdot\boldsymbol{S} \qquad (4.37)$$

where $\boldsymbol{L}\cdot\boldsymbol{S}$ is the scalar product of the vectors $\boldsymbol{L}$ and $\boldsymbol{S}$ (as we see in *Mathematical background* 4, $\boldsymbol{L}\cdot\boldsymbol{S}$ is proportional to $\cos\theta$, where θ is the angle between the two vectors). To use this expression, we note that the total angular momentum is $\boldsymbol{J} = \boldsymbol{L} + \boldsymbol{S}$, so

$$\boldsymbol{J}\cdot\boldsymbol{J} = (\boldsymbol{L}+\boldsymbol{S})\cdot(\boldsymbol{L}+\boldsymbol{S}) = L^2 + S^2 + 2\boldsymbol{L}\cdot\boldsymbol{S}$$

and therefore

$$\lambda \boldsymbol{L}\cdot\boldsymbol{S} = \tfrac{1}{2}\lambda(J^2 - L^2 - S^2)$$

The eigenvalues of this expression are

$$E_{so} = \tfrac{1}{2}\lambda\hbar^2\{J(J+1) - L(L+1) - S(S+1)\} \qquad (4.38)$$

● A BRIEF ILLUSTRATION

When $L = 1$ and $S = \frac{1}{2}$, as in a ^{2}P term,

$$E_{so} = \tfrac{1}{2}\lambda\hbar^2\{J(J+1) - 2 - \tfrac{3}{4}\} = \tfrac{1}{2}\lambda\hbar^2\{J(J+1) - \tfrac{11}{4}\}$$

Therefore, for a level with $J = \frac{3}{2}$, $E_{so} = \frac{1}{2}\lambda\hbar^2$, and for a level with $J = \frac{1}{2}$ from the same configuration, $E_{so} = -\lambda\hbar^2$. The separation of the two levels is therefore $\Delta E_{so} = \frac{3}{2}\lambda\hbar^2$. ●

The strength of the spin–orbit coupling depends on the nuclear charge. To understand why this is so, imagine riding on the orbiting electron and seeing a charged nucleus apparently orbiting around us (like the Sun rising and setting). As a result, we find ourselves at the centre of a ring of current. The greater the nuclear charge, the greater this current, and therefore the stronger the magnetic field we detect. Because the spin magnetic moment of the electron interacts with this orbital magnetic field, it follows that, the greater the nuclear charge, the stronger the spin–orbit interaction. The coupling increases sharply with atomic number (as Z^4 in hydrogenic atoms). Whereas it is only small in H (giving rise to shifts of energy levels of no more than about 0.4 cm^{-1}), in heavy atoms like Pb it is very large (giving shifts of the order of thousands of reciprocal centimetres).

Two spectral lines are observed when the p electron of an electronically excited alkali metal atom undergoes a transition and falls into a lower s orbital. The higher frequency line is due to a transition starting in a $^2P_{3/2}$ level and the other line is due to a transition starting in the $^2P_{1/2}$ level of the same configuration. The presence of these two lines is an example of **fine structure**, the structure in a spectrum due to spin–orbit coupling. Fine structure can be clearly seen in the emission spectrum from sodium vapour excited by an electric discharge (for example, in one kind of street lighting). The yellow line at 589 nm (close to 17 000 cm^{-1}) is actually a doublet composed of one line at 589.76 nm (16 956.2 cm^{-1}) and another at 589.16 nm (16 973.4 cm^{-1}); the components of this doublet are the 'D lines' of the spectrum (Fig. 4.28). Therefore, in Na, the spin–orbit coupling affects the energies by about 17 cm^{-1}.

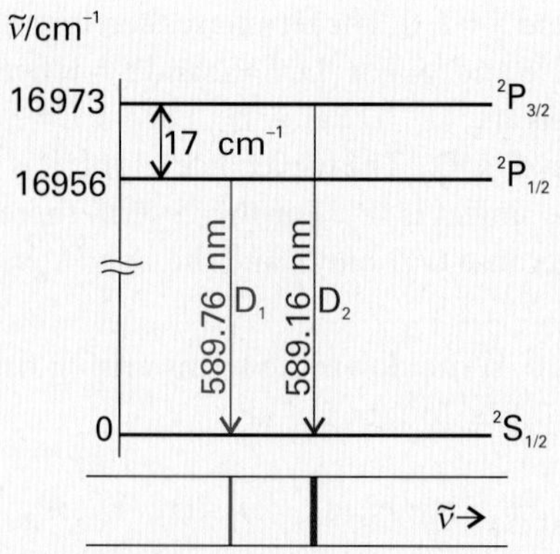

Fig. 4.28 The energy-level diagram for the formation of the sodium D lines. The splitting of the spectral lines (by 17 cm^{-1}) reflects the splitting of the levels of the ^2P term.

(d) Selection rules

Any state of the atom, and any spectral transition, can be specified by using term symbols. For example, the transitions giving rise to the yellow sodium doublet (which were shown in Fig. 4.28) are

$$3p^1\ ^2P_{3/2} \rightarrow 3s^1\ ^2S_{1/2} \qquad 3p^1\ ^2P_{1/2} \rightarrow 3s^1\ ^2S_{1/2}$$

By convention, the upper term precedes the lower. The corresponding absorptions are therefore denoted $^2P_{3/2} \leftarrow {}^2S_{1/2}$ and $^2P_{1/2} \leftarrow {}^2S_{1/2}$ (the configurations have been omitted).

We have seen that selection rules arise from the conservation of angular momentum during a transition and from the fact that a photon has a spin of 1. They can therefore be expressed in terms of the term symbols, because the latter carry information about angular momentum. A detailed analysis leads to the following rules:

$$\Delta S = 0 \qquad \Delta L = 0, \pm 1 \qquad \Delta l = \pm 1$$
$$\Delta J = 0, \pm 1, \text{ but } J = 0 \nleftrightarrow J = 0 \qquad (4.39)$$

where the symbol $\nleftrightarrow$ denotes a forbidden transition. The rule about ΔS (no change of overall spin) stems from the fact that the light does not affect the spin directly. The rules about ΔL and Δl express the fact that the orbital angular momentum of an individual electron must change (so $\Delta l = \pm 1$), but whether or not this results in an overall change of orbital momentum depends on the coupling.

The selection rules given above apply when Russell–Saunders coupling is valid (in light atoms). If we insist on labelling the terms of heavy atoms with symbols like ^3D, then we shall find that the selection rules progressively fail as the atomic number increases because the quantum numbers S and L become ill defined as *jj*-coupling becomes more appropriate. As explained above, Russell–Saunders term symbols are only a convenient way of labelling the terms of heavy atoms: they do not bear any direct relation to the actual angular momenta of the electrons in a heavy atom. For this reason, transitions between singlet and triplet states (for which $\Delta S = \pm 1$), while forbidden in light atoms, are allowed in heavy atoms.

Checklist of key ideas

☐ 1. A hydrogenic atom is a one-electron atom or ion of general atomic number Z. A many-electron (polyelectronic) atom is an atom or ion with more than one electron.

☐ 2. The Lyman, Balmer, and Paschen series in the spectrum of atomic hydrogen arise, respectively, from the transitions $n \rightarrow 1$, $n \rightarrow 2$, and $n \rightarrow 3$.

☐ 3. The wavenumbers of all the spectral lines of a hydrogen atom can be expressed as $\tilde{\nu} = R_H(1/n_1^2 - 1/n_2^2)$, where R_H is the Rydberg constant for hydrogen.

☐ 4. The wavefunction of a hydrogenic atom is the product of a radial wavefunction and an angular wavefunction (spherical harmonic) and is labelled by the quantum numbers n, l, and m_l: $\psi_{n,l,m_l}(r,\theta,\phi) = R_{n,l}(r)Y_{l,m_l}(\theta,\phi)$.

☐ 5. An atomic orbital is a one-electron wavefunction for an electron in an atom.

☐ 6. The energies of an electron in a hydrogenic atom are given by $E_n = -Z^2\mu e^4/32\pi^2\varepsilon_0^2\hbar^2 n^2$, where n is the principal quantum number, $n = 1, 2, \ldots$; the total orbital angular momentum is given by $\{l(l+1)\}^{1/2}\hbar$, where $l = 0, 1, 2, \ldots, n-1$; the z-component of angular momentum is given by $m_l\hbar$, where $m_l = 0, \pm1, \pm2, \ldots, \pm l$.

☐ 7. All the orbitals of a given value of n belong to a given shell; orbitals with the same value of n but different values of l belong to different subshells.

☐ 8. The radial distribution function is a probability density that, when it is multiplied by dr, gives the probability of finding the electron anywhere in a shell of thickness dr at the radius r; $P(r) = r^2R(r)^2$.

☐ 9. The rate of change of population of the state ψ_f due to transitions from state ψ_i is $w_{f\leftarrow i} \propto |\mu_{z,fi}|^2\mathcal{E}^2$, where $\mu_{z,fi} = \int \psi_f^* \hat{\mu}_z \psi_i d\tau$ is the transition dipole moment. The expression for the rate of change arises from time-dependent perturbation theory.

☐ 10. Doppler broadening and uncertainty broadening contribute to the widths of spectroscopic lines.

☐ 11. A selection rule is a statement about which spectroscopic transitions are allowed; a specific selection rule expresses the allowed transitions in terms of the changes in quantum numbers.

☐ 12. A Grotrian diagram is a diagram summarizing the energies of the states of the atom and the transitions between them.

☐ 13. In the orbital approximation it is supposed that each electron occupies its 'own' orbital, $\psi(r_1, r_2, \ldots) = \psi(r_1)\psi(r_2)\cdots$. The configuration is the list of occupied orbitals.

☐ 14. The Pauli exclusion principle states that no more than two electrons may occupy any given orbital and, if two do occupy one orbital, then their spins must be paired.

☐ 15. The Pauli principle states that, when the labels of any two identical fermions are exchanged, the total wavefunction changes sign; when the labels of any two identical bosons are exchanged, the total wavefunction retains the same sign.

☐ 16. The effective nuclear charge Z_{eff} is the net charge experienced by an electron allowing for electron–electron repulsions.

☐ 17. Shielding is the effective reduction in charge of a nucleus by surrounding electrons; the shielding constant σ is given by $Z_{eff} = Z - \sigma$.

☐ 18. Penetration is the ability of an electron to be found inside inner shells and close to the nucleus.

☐ 19. The building-up (*Aufbau*) principle is the procedure for filling atomic orbitals that leads to the ground-state configuration of an atom.

☐ 20. Hund's maximum multiplicity rule states that an atom in its ground state adopts a configuration with the greatest number of electrons with unpaired spins.

☐ 21. The first ionization energy I_1 is the minimum energy necessary to remove an electron from a many-electron atom in the gas phase; the second ionization energy I_2 is the minimum energy necessary to remove an electron from an ionized many-electron atom in the gas phase.

☐ 22. The electron affinity E_{ea} is the energy released when an electron attaches to a gas-phase atom.

☐ 23. The allowed values of the total orbital angular momentum L of a configuration are obtained by using the Clebsch–Gordan series $L = l_1 + l_2, l_1 + l_2 - 1, \ldots, |l_1 - l_2|$.

☐ 24. The allowed values of the total spin angular momentum S are obtained by using the Clebsch–Gordan series $S = s_1 + s_2, s_1 + s_2 - 1, \ldots, |s_1 - s_2|$.

☐ 25. Spin–orbit coupling is the interaction of the spin magnetic moment with the magnetic field arising from the orbital angular momentum.

☐ 26. Russell–Saunders coupling is a coupling scheme based on the view that, if spin–orbit coupling is weak, then it is effective only when all the orbital momenta are operating cooperatively.

☐ 27. The total angular momentum J, in the Russell–Saunders coupling scheme, has possible values $J = L + S,\ L + S - 1, \ldots, |L - S|$.

☐ 28. A level is a group of states with a common value of J.

☐ 29. The multiplicity of a term is the value of $2S + 1$; provided $L \geq S$, the multiplicity is the number of levels of the term.

☐ 30. A term symbol is a symbolic specification of the state of an atom, ${}^{2S+1}\{L\}_J$.

☐ 31. Fine structure is the structure in a spectrum due to spin–orbit coupling.

☐ 32. The selection rules for spectroscopic transitions in polyelectronic atoms are: $\Delta S = 0$, $\Delta L = 0, \pm1$, $\Delta l = \pm1$, $\Delta J = 0, \pm1$, but $J = 0 \nleftrightarrow J = 0$. These selection rules apply when Russell–Saunders coupling is valid.

Further information

Further information 4.1 *The separation of internal and external motion*

Consider a one-dimensional system in which the potential energy depends only on the separation of the two particles. The total energy is

$$E = \frac{p_1^2}{2m_1} + \frac{p_2^2}{2m_2} + V$$

where $p_1 = m_1\dot{x}_1$, $p_2 = m_2\dot{x}_2$, and the dot signifies differentiation with respect to time, as in $\dot{x} = dx/dt$. The centre of mass (Fig. 4.29) is located at

$$X = \frac{m_1}{m}x_1 + \frac{m_2}{m}x_2 \qquad m = m_1 + m_2$$

and the separation of the particles is $x = x_1 - x_2$. It follows that

$$x_1 = X + \frac{m_2}{m}x \qquad x_2 = X - \frac{m_1}{m}x$$

The linear momenta of the particles can be expressed in terms of the rates of change of x and X:

$$p_1 = m_1\dot{x}_1 = m_1\dot{X} + \frac{m_1m_2}{m}\dot{x} \qquad p_2 = m_2\dot{x}_2 = m_2\dot{X} - \frac{m_1m_2}{m}\dot{x}$$

Then it follows that

$$\frac{p_1^2}{2m_1} + \frac{p_2^2}{2m_2} = \tfrac{1}{2}m\dot{X}^2 + \tfrac{1}{2}\mu\dot{x}^2$$

where μ is given by $1/\mu = 1/m_1 + 1/m_2$. By writing $P = m\dot{X}$ for the linear momentum of the system as a whole and defining p as $\mu\dot{x}$, we find

$$E = \frac{P^2}{2m} + \frac{p^2}{2\mu} + V$$

The corresponding hamiltonian (generalized to three dimensions) is therefore

$$\hat{H} = -\frac{\hbar^2}{2m}\nabla^2_{\text{c.m.}} - \frac{\hbar^2}{2\mu}\nabla^2 + V$$

where the first term differentiates with respect to the centre of mass coordinates and the second with respect to the relative coordinates.

Now we write the overall wavefunction as the product $\psi_{\text{total}} = \psi_{\text{c.m.}}\psi$, where the first factor is a function of only the centre of mass coordinates and the second is a function of only the relative coordinates. The overall Schrödinger equation, $\hat{H}\psi_{\text{total}} = E_{\text{total}}\psi_{\text{total}}$, then separates by the argument that we have used in Sections 3.1 and 3.4, with $E_{\text{total}} = E_{\text{c.m.}} + E$.

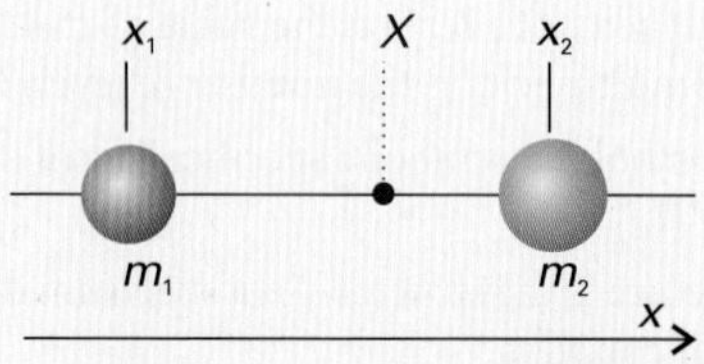

Fig. 4.29 The coordinates used for discussing the separation of the relative motion of two particles from the motion of the centre of mass.

Further information 4.2 *Time-dependent perturbation theory*

To cope with a perturbed wavefunction that evolves with time, we need to solve the time-dependent Schrödinger equation,

$$\hat{H}\Psi = i\hbar\frac{\partial\Psi}{\partial t} \tag{4.40}$$

If we write the first-order correction to the wavefunction as

$$\Psi_0^{(1)}(t) = \sum_n c_n(t)\Psi_n(t) = \sum_n c_n(t)\psi_n^{(0)}e^{-iE_n^{(0)}t/\hbar} \tag{4.41a}$$

then the coefficients in this expansion are given by

$$c_n(t) = \frac{1}{i\hbar}\int_0^t H_{n0}^{(1)}(t)e^{i\omega_{n0}t}dt \tag{4.41b}$$

The formal demonstration of eqn 4.41 is quite lengthy.[3] Here we shall show that, given eqn 4.41b, then a perturbation that is switched on very slowly to a constant value gives the same expression for the coefficients as we obtained for time-independent perturbation theory. For such a perturbation, we write

$$\hat{H}^{(1)}(t) = \hat{H}^{(1)}(1 - e^{-t/\tau})$$

and take the time constant τ to be very long (Fig. 4.30). Substitution of this expression into eqn 4.41b gives

$$\begin{aligned} c_n(t) &= \frac{1}{i\hbar}H_{n0}^{(1)}\int_0^t (1 - e^{-t/\tau})e^{i\omega_{n0}t}dt \\ &= \frac{1}{i\hbar}H_{n0}^{(1)}\left\{\frac{e^{i\omega_{n0}t} - 1}{i\omega_{n0}} - \frac{e^{(i\omega_{n0}-1/\tau)t} - 1}{i\omega_{n0} - 1/\tau}\right\} \end{aligned}$$

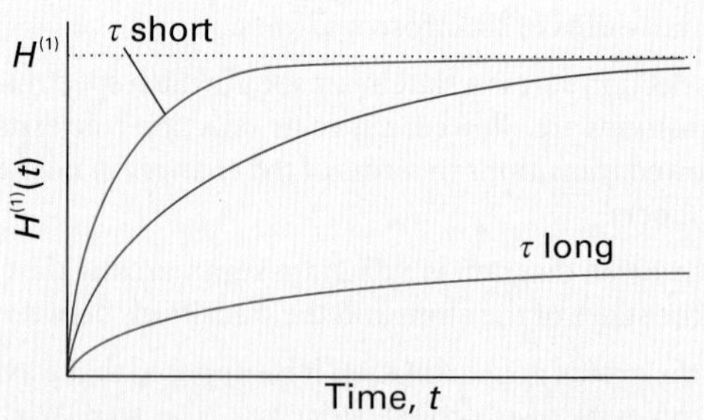

Fig. 4.30 The time-dependence of a slowly switched perturbation. A large value of τ corresponds to very slow switching.

[3] For details, see our *Molecular quantum mechanics*, Oxford University Press (2005).

At this point we suppose that the perturbation is switched slowly, in the sense that $\tau \gg 1/\omega_{n0}$ (so that the $1/\tau$ in the second denominator can be ignored). We also suppose that we are interested in the coefficients long after the perturbation has settled down into its final value, when $t \gg \tau$ (so that the exponential in the second numerator is close to zero and can be ignored). Under these conditions,

$$c_n(t) = -\frac{H_{n0}^{(0)}}{\hbar\omega_{n0}}\mathrm{e}^{\mathrm{i}\omega_{n0}t}$$

Now we recognize that $\hbar\omega_{n0} = E_n^{(0)} - E_0^{(0)}$, which gives

$$c_n(t) = -\frac{H_{n0}^{(0)}}{E_n^{(0)} - E_0^{(0)}}\mathrm{e}^{\mathrm{i}E_n^{(0)}t}\mathrm{e}^{-\mathrm{i}E_0^{(0)}t/\hbar}$$

When this expression is substituted into eqn 4.41a, we obtain the time-independent expression, eqn 2.37 (apart from an irrelevant overall phase factor).

In accord with the general rules for the interpretation of wavefunctions, the probability that the system will be found in the state n is proportional to the square modulus of the coefficient of the state, $|c_n(t)|^2$. Therefore, the rate of change of population of a final state ψ_f due to transitions from an initial state ψ_i is

$$w_{\mathrm{f}\leftarrow\mathrm{i}} = \frac{\mathrm{d}|c_\mathrm{f}|^2}{\mathrm{d}t}$$

Because the coefficient is proportional to the matrix elements of the perturbation, $w_{\mathrm{f}\leftarrow\mathrm{i}}$ is proportional to the square modulus of the matrix element of the perturbation between the two states:

$$w_{\mathrm{f}\leftarrow\mathrm{i}} \propto |H_{\mathrm{fi}}^{(1)}|^2$$

which is eqn 4.22.

Discussion questions

4.1 Discuss the origin of the series of lines in the emission spectra of hydrogen. What region of the electromagnetic spectrum is associated with each of the series shown in Fig. 4.1?

4.2 Discuss the separation of variables procedure as it is applied to simplify the description of a hydrogenic atom free to move through space.

4.3 List and discuss the significance of the quantum numbers needed to specify the internal state of a hydrogenic atom.

4.4 Describe how the presence of orbital angular momentum affects the shape of the atomic orbital.

4.5 Specify and account for the selection rules for transitions in hydrogenic atoms. Are they strictly valid?

4.6 Discuss the significance of (a) a boundary surface and (b) the radial distribution function for hydrogenic orbitals.

4.7 Discuss the relationship between the location of a many-electron atom in the periodic table and its electron configuration.

4.8 Describe and account for the variation of first ionization energies along Period 2 of the periodic table. Would you expect the same variation in Period 3?

4.9 Describe the orbital approximation for the wavefunction of a many-electron atom. What are the limitations of the approximation?

4.10 Describe why the Slater determinant provides a useful representation of electron configurations of many-electron atoms. Why is it an approximation to the true wavefunction?

4.11 Explain the origin of spin–orbit coupling and how it affects the appearance of a spectrum.

4.12 Describe the physical origins of linewidths in absorption and emission spectra. Do you expect the same contributions for species in condensed and gas phases?

Exercises

4.1(a) Determine the shortest and longest wavelength lines in the Lyman series.

4.1(b) The Pfund series has $n_1 = 5$. Determine the shortest and longest wavelength lines in the Pfund series.

4.2(a) Compute the wavelength, frequency, and wavenumber of the $n = 2 \rightarrow n = 1$ transition in He^+.

4.2(b) Compute the wavelength, frequency, and wavenumber of the $n = 5 \rightarrow n = 4$ transition in Li^{2+}.

4.3(a) What is the orbital angular momentum of an electron in the orbitals (a) 2s, (b) 3p, (c) 5f? Give the numbers of angular and radial nodes in each case.

4.3(b) What is the orbital angular momentum of an electron in the orbitals (a) 3d, (b) 4f, (c) 3s? Give the numbers of angular and radial nodes in each case.

4.4(a) Compute the ionization energy of the He^+ ion.

4.4(b) Compute the ionization energy of the Li^{2+} ion.

4.5(a) When ultraviolet radiation of wavelength 58.4 nm from a helium lamp is directed on to a sample of krypton, electrons are ejected with a speed of 1.59×10^6 m s^{-1}. Calculate the ionization energy of krypton.

4.5(b) When ultraviolet radiation of wavelength 58.4 nm from a helium lamp is directed on to a sample of xenon, electrons are ejected with a speed of 1.79×10^6 m s^{-1}. Calculate the ionization energy of xenon.

4.6(a) What is the degeneracy of an energy level in the L shell of a hydrogenic atom?

4.6(b) What is the degeneracy of an energy level in the M shell of a hydrogenic atom?

4.7(a) State the orbital degeneracy of the levels in a hydrogen atom that have energy (a) $-hcR_\mathrm{H}$, (b) $-\frac{1}{4}hcR_\mathrm{H}$, (c) $-\frac{1}{16}hcR_\mathrm{H}$.

4.7(b) State the orbital degeneracy of the levels in a hydrogenic atom (Z in parentheses) that have energy (a) $-hcR_{\text{atom}}$, (2), (b) $-\frac{1}{4}hcR_{\text{atom}}$ (4), and (c) $-\frac{25}{16}hcR_{\text{atom}}$ (5).

4.8(a) The wavefunction for the ground state of a hydrogen atom is Ne^{-r/a_0}. Determine the normalization constant N.

4.8(b) The wavefunction for the 2s orbital of a hydrogen atom is $N(2 - r/a_0)e^{-r/2a_0}$. Determine the normalization constant N.

4.9(a) By differentiation of the 2s radial wavefunction, show that it has two extrema in its amplitude and locate them.

4.9(b) By differentiation of the 3s radial wavefunction, show that it has three extrema in its amplitude and locate them.

4.10(a) Locate the radial node in the 2s orbital of an H atom.

4.10(b) Locate the radial node in the 3p orbital of an H atom.

4.11(a) Calculate the average kinetic and potential energies of an electron in the ground state of an He^+ ion.

4.11(b) Calculate the average kinetic and potential energies of a 3s electron in an H atom.

4.12(a) Compute the mean radius and the most probable radius for a 2s electron in a hydrogenic atom of atomic number Z.

4.12(b) Compute the mean radius and the most probable radius for a 2p electron in a hydrogenic atom of atomic number Z.

4.13(a) Write down the expression for the radial distribution function of a 3s electron in a hydrogenic atom and determine the radius at which the electron is most likely to be found.

4.13(b) Write down the expression for the radial distribution function of a 3p electron in a hydrogenic atom and determine the radius at which the electron is most likely to be found.

4.14(a) Locate the angular nodes and nodal planes of each of the 2p orbitals of a hydrogenic atom of atomic number Z. To locate the angular nodes, give the angle that the nodal plane makes with the z-axis.

4.14(b) Locate the angular nodes and nodal planes of each of the 3d orbitals of a hydrogenic atom of atomic number Z. To locate the angular nodes, give the angle that the nodal plane makes with the z-axis.

4.15(a) Which of the following transitions are allowed in the normal electronic emission spectrum of an atom: (a) $3s \rightarrow 1s$, (b) $3p \rightarrow 2s$, (c) $5d \rightarrow 2p$?

4.15(b) Which of the following transitions are allowed in the normal electronic emission spectrum of an atom: (a) $5d \rightarrow 3s$, (b) $5s \rightarrow 3p$, (c) $6f \rightarrow 4p$?

4.16(a) What is the Doppler-shifted wavelength of a red (680 nm) traffic light approached at 60 km h^{-1}?

4.16(b) At what speed of approach would a red (680 nm) traffic light appear green (530 nm)?

4.17(a) Estimate the lifetime of a state that gives rise to a line of width (a) 0.20 cm^{-1}, (b) 2.0 cm^{-1}.

4.17(b) Estimate the lifetime of a state that gives rise to a line of width (a) 200 MHz, (b) 2.45 cm^{-1}.

4.18(a) A molecule in a liquid undergoes about 1.0×10^{13} collisions in each second. Suppose that (a) every collision is effective in deactivating the molecule vibrationally and (b) that one collision in 100 is effective. Calculate the width (in cm^{-1}) of vibrational transitions in the molecule.

4.18(b) A molecule in a gas undergoes about 1.0×10^9 collisions in each second. Suppose that (a) every collision is effective in deactivating the molecule rotationally and (b) that one collision in 10 is effective. Calculate the width (in hertz) of rotational transitions in the molecule.

4.19(a) Write a Slater determinant for the ground state of a magnesium atom.

4.19(b) Write a Slater determinant for the ground state of a fluorine atom.

4.20(a) What are the values of the quantum numbers n, l, m_l, s, and m_s for each of the valence electrons in the ground state of a carbon atom?

4.20(b) What are the values of the quantum numbers n, l, m_l, s, and m_s for each of the valence electrons in the ground state of a nitrogen atom?

4.21(a) Write the ground-state electron configurations of the d-metals from scandium to zinc.

4.21(b) Write the ground-state electron configurations of the d-metals from yttrium to cadmium.

4.22(a) (a) Write the electron configuration of the Pd^{2+} ion. (b) What are the possible values of the total spin quantum numbers S and M_S for this ion?

4.22(b) (a) Write the electron configuration of the Nb^{2+} ion. (b) What are the possible values of the total spin quantum numbers S and M_S for this ion?

4.23(a) Calculate the permitted values of j for (a) a d electron, (b) an f electron.

4.23(b) Calculate the permitted values of j for (a) a p electron, (b) an h electron.

4.24(a) An electron in two different states of an atom is known to have $j = \frac{5}{2}$ and $\frac{1}{2}$. What are the possible orbital angular momentum quantum numbers in each case?

4.24(b) An electron in two different states of an atom is known to have $j = \frac{7}{2}$ and $\frac{3}{2}$. What are the possible orbital angular momentum quantum numbers in each case?

4.25(a) What are the allowed total angular momentum quantum numbers of a composite system in which $j_1 = 1$ and $j_2 = 2$?

4.25(b) What are the allowed total angular momentum quantum numbers of a composite system in which $j_1 = 4$ and $j_2 = 2$?

4.26(a) What information does the term symbol $^3P_2$ provide about the angular momentum of an atom?

4.26(b) What information does the term symbol $^2D_{3/2}$ provide about the angular momentum of an atom?

4.27(a) Suppose that an atom has (a) 2, (b) 3 electrons in different orbitals. What are the possible values of the total spin quantum number S? What is the multiplicity in each case?

4.27(b) Suppose that an atom has (a) 4, (b) 5, electrons in different orbitals. What are the possible values of the total spin quantum number S? What is the multiplicity in each case?

4.28(a) What atomic terms are possible for the electron configuration ns^1nd^1? Which term is likely to lie lowest in energy?

4.28(b) What atomic terms are possible for the electron configuration np^1nd^1? Which term is likely to lie lowest in energy?

4.29(a) What values of J may occur in the terms (a) ^3S, (b) ^2D, (c) ^1P? How many states (distinguished by the quantum number M_J) belong to each level?

4.29(b) What values of J may occur in the terms (a) ^3F, (b) ^4G, (c) ^2P? How many states (distinguished by the quantum number M_J) belong to each level?

4.30(a) Give the possible term symbols for (a) Na [Ne]$3s^1$, (b) K [Ar]$3d^1$.

4.30(b) Give the possible term symbols for (a) Y [Kr]$4d^15s^2$, (b) I [Kr]$4d^{10}5s^25p^5$.

4.31(a) Which of the following transitions between terms are allowed in the normal electronic emission spectrum of a many-electron atom: (a) $^3D_2 \rightarrow {}^3P_1$, (b) $^3P_2 \rightarrow {}^1S_0$, (c) $^3F_4 \rightarrow {}^3D_3$?

4.31(b) Which of the following transitions between terms are allowed in the normal electronic emission spectrum of a many-electron atom: (a) $^2P_{3/2} \rightarrow {}^2S_{1/2}$, (b) $^3P_0 \rightarrow {}^3S_1$, (c) $^3D_3 \rightarrow {}^1P_1$?

Problems*

Numerical problems

4.1 The *Humphreys series* is a group of lines in the spectrum of atomic hydrogen. It begins at 12 368 nm and has been traced to 3281.4 nm. What are the transitions involved? What are the wavelengths of the intermediate transitions?

4.2 A series of lines in the spectrum of atomic hydrogen lies at 656.46 nm, 486.27 nm, 434.17 nm, and 410.29 nm. What is the wavelength of the next line in the series? What is the ionization energy of the atom when it is in the lower state of the transitions?

4.3 The Li^{2+} ion is hydrogenic and has a Lyman series at 740 747 cm^{-1}, 877 924 cm^{-1}, 925 933 cm^{-1}, and beyond. Show that the energy levels are of the form $-hcR/n^2$ and find the value of R for this ion. Go on to predict the wavenumbers of the two longest-wavelength transitions of the Balmer series of the ion and find the ionization energy of the ion.

4.4 A series of lines in the spectrum of neutral Li atoms rise from combinations of $1s^22p^1\ {}^2P$ with $1s^2nd^1\ {}^2D$ and occur at 610.36 nm, 460.29 nm, and 413.23 nm. The d orbitals are hydrogenic. It is known that the ^2P term lies at 670.78 nm above the ground state, which is $1s^22s^1\ {}^2S$. Calculate the ionization energy of the ground-state atom.

4.5‡ Wijesundera *et al.* (*Phys. Rev.* A **51**, 278 (1995)) attempted to determine the electron configuration of the ground state of lawrencium, element 103. The two contending configurations are [Rn]$5f^{14}7s^27p^1$ and [Rn]$5f^{14}6d^17s^2$. Write down the term symbols for each of these configurations, and identify the lowest level within each configuration. Which level would be lowest according to a simple estimate of spin–orbit coupling?

4.6 The characteristic emission from K atoms when heated is purple and lies at 770 nm. On close inspection, the line is found to have two closely spaced components, one at 766.70 nm and the other at 770.11 nm. Account for this observation, and deduce what information you can.

4.7 Calculate the mass of the deuteron given that the first line in the Lyman series of H lies at 82 259.098 cm^{-1} whereas that of D lies at 82 281.476 cm^{-1}. Calculate the ratio of the ionization energies of H and D.

4.8 Positronium consists of an electron and a positron (same mass, opposite charge) orbiting round their common centre of mass. The broad features of the spectrum are therefore expected to be hydrogen-like, the differences arising largely from the mass differences. Predict the wavenumbers of the first three lines of the Balmer series of positronium. What is the binding energy of the ground state of positronium?

4.9 The *Zeeman effect* is the modification of an atomic spectrum by the application of a strong magnetic field. It arises from the interaction between applied magnetic fields and the magnetic moments due to orbital and spin angular momenta (recall the evidence provided for electron spin by the Stern–Gerlach experiment, Section 3.4). To gain some appreciation for the so-called *normal Zeeman effect*, which is observed in transitions involving singlet states, consider a p electron, with $l = 1$ and $m_l = 0, \pm 1$. In the absence of a magnetic field, these three states are degenerate. When a field of magnitude $\mathcal{B}$ is present, the degeneracy is removed and it is observed that the state with $m_l = +1$ moves up in energy by $\mu_B\mathcal{B}$, the state with $m_l = 0$ is unchanged, and the state with $m_l = -1$ moves down in energy by $\mu_B\mathcal{B}$, where $\mu_B = e\hbar/2m_e = 9.274 \times 10^{-24}$ J T^{-1} is known as the Bohr magneton. Therefore, a transition between a $^1S_0$ term and a $^1P_1$ term consists of three spectral lines in the presence of a magnetic field where, in the absence of the magnetic field, there is only one. (a) Calculate the splitting in reciprocal centimetres between the three spectral lines of a transition between a $^1S_0$ term and a $^1P_1$ term in the presence of a magnetic field of 2 T (where 1 T = 1 kg s^{-2} A^{-1}). (b) Compare the value you calculated in (a) with typical optical transition wavenumbers, such as those for the Balmer series of the H atom. Is the line splitting caused by the normal Zeeman effect relatively small or relatively large?

4.10 In 1976 it was mistakenly believed that the first of the 'superheavy' elements had been discovered in a sample of mica. Its atomic number was believed to be 126. What is the most probable distance of the innermost electrons from the nucleus of an atom of this element? (In such elements, relativistic effects are very important, but ignore them here.)

4.11 An electron in the ground-state He^+ ion undergoes a transition to a state described by the wavefunction $R_{4,1}(r)Y_{1,+1}(\theta,\phi)$. (a) Describe the transition using term symbols. (b) Compute the wavelength, frequency, and wavenumber of the transition. (c) By how much does the mean radius of the electron change due to the transition?

4.12 The electron in a Li^{2+} ion is prepared in a state that is the following superposition of hydrogenic atomic orbitals:

$$\psi(r,\theta,\phi) = -(\tfrac{1}{3})^{1/2}R_{4,2}(r)Y_{2,-1}(\theta,\phi) + \tfrac{2}{3}\mathrm{i}R_{3,2}(r)Y_{2,+1}(\theta,\phi) - (\tfrac{2}{9})^{1/2}R_{1,0}(r)Y_{0,0}(\theta,\phi)$$

(a) If the total energy of different Li^{2+} ions in this state is measured, what values will be found? If more than one value is found, what is the probability of obtaining each result and what is the average value? (b) After the energy is measured, the electron is in a state described by an eigenfunction of the hamiltonian. Are transitions to the ground state of Li^{2+} allowed and, if so, what are the frequency and wavenumber of the transition(s)?

4.13 (a) Calculate the probability of the electron being found anywhere within a sphere of radius 53 pm for a hydrogenic atom. (b) If the radius of the atom is defined as the radius of the sphere inside which there is a 90 per cent probability of finding the electron, what is the atom's radius?

* Problems denoted with the symbol ‡ were supplied by Charles Trapp, Carmen Giunta, and Marshall Cady.

4.14 The collision frequency z of a molecule of mass m in a gas at a pressure p is $z = 4\sigma(kT/\pi m)^{1/2}p/kT$, where σ is the collision cross-section. Find an expression for the collision-limited lifetime of an excited state assuming that every collision is effective. Estimate the width of a rotational transition in HCl ($\sigma = 0.30$ nm^2) at 25°C and 1.0 atm. To what value must the pressure of the gas be reduced in order to ensure that collision broadening is less important than Doppler broadening?

Theoretical problems

4.15 At what point in the hydrogen atom is there maximum probability of finding a (a) $2p_z$ electron, (b) $3p_z$ electron? How do these most probable points compare to the most probable radii for the locations of $2p_z$ and $3p_z$ electrons?

4.16 Show by explicit integration that hydrogenic (a) 1s and 2s orbitals are mutually orthogonal, (b) $2p_x$ and $2p_z$ orbitals are mutually orthogonal.

4.17‡ Explicit expressions for hydrogenic orbitals are given in Tables 4.1 and 3.2. (a) Verify that the $3p_x$ orbital is normalized and that $3p_x$ and $3d_{xy}$ are mutually orthogonal. (b) Determine the positions of both the radial nodes and nodal planes of the 3s, $3p_x$, and $3d_{xy}$ orbitals. (c) Determine the mean radius of the 3s orbital. (d) Draw a graph of the radial distribution function for the three orbitals (of part (b)) and discuss the significance of the graphs for interpreting the properties of many-electron atoms. (e) Create both xy-plane polar plots and boundary surface plots for these orbitals. Construct the boundary plots so that the distance from the origin to the surface is the absolute value of the angular part of the wavefunction. Compare the s, p, and d boundary surface plots with that of an f orbital; for example, $\psi_f \propto x(5z^2 - r^2) \propto \sin\theta\,(5\cos^2\theta - 1)\cos\phi$.

4.18 Show that d orbitals with opposite values of m_l may be combined in pairs to give real standing waves with boundary surfaces as shown in Fig. 4.15 and with forms that are given in eqn 4.19.

4.19 The 'size' of an atom is sometimes considered to be measured by the radius of a sphere that contains 90 per cent of the probability density of the electrons in the outermost occupied orbital. Calculate the 'size' of a hydrogen atom in its ground state according to this definition. Go on to explore how the 'size' varies as the definition is changed to other percentages, and plot your conclusion.

4.20 A quantity important in some branches of spectroscopy is the probability density of an electron being found at the same location as the nucleus. Evaluate this probability density for an electron in the 1s-, 2s-, and 3s-orbitals of a hydrogenic atom. What happens to the probability density when an orbital other than s is considered?

4.21 Some atomic properties depend on the average value of $1/r$ rather than the average value of r itself. Evaluate the expectation value of $1/r$ for (a) a hydrogen 1s orbital, (b) a hydrogenic 2s orbital, (c) a hydrogenic 2p orbital.

4.22 Atomic units of length and energy may be based on the properties of a particular atom. The usual choice is that of a hydrogen atom, with the unit of length being the Bohr radius, a_0, and the unit of energy being the (negative of the) energy of the 1s orbital. If the positronium atom (e^+,e^-) were used instead, with analogous definitions of units of length and energy, what would be the relation between these two sets of atomic units?

4.23 Some of the selection rules for hydrogenic atoms were derived in *Justification 4.4*. Complete the derivation by considering the x- and y-components of the electric dipole moment operator.

4.24‡ Stern–Gerlach splittings of atomic beams are small and require either large magnetic field gradients or long magnets for their observation. For a beam of atoms with zero orbital angular momentum, such as H or Ag, the deflection is given by $x = \pm(\mu_B L^2/4E_k)\mathrm{d}\mathcal{B}/\mathrm{d}z$, where μ_B is the Bohr magneton (Problem 4.9), L is the length of the magnet, E_k is the average kinetic energy of the atoms in the beam, and $\mathrm{d}\mathcal{B}/\mathrm{d}z$ is the magnetic field gradient across the beam. (a) Given that the average translational kinetic energy of the atoms emerging as a beam from a pinhole in an oven at temperature T is $\frac{1}{2}kT$, calculate the magnetic field gradient required to produce a splitting of 2.00 mm in a beam of Ag atoms from an oven at 1200 K with a magnet of length 80 cm.

4.25 The wavefunction of a many-electron closed-shell atom can be expressed as a Slater determinant (Section 4.4). A useful property of determinants is that interchanging any two rows or columns changes their sign and therefore that, if any two rows or columns are identical, then the determinant vanishes. Use this property to show that (a) the wavefunction is antisymmetric under particle exchange, (b) no two electrons can occupy the same orbital with the same spin.

Applications: to astrophysics and biochemistry

4.26 Hydrogen is the most abundant element in all stars. However, neither absorption nor emission lines due to neutral hydrogen are found in the spectra of stars with effective temperatures higher than 25 000 K. Account for this observation.

4.27 The distribution of isotopes of an element may yield clues about the nuclear reactions that occur in the interior of a star. Show that it is possible to use spectroscopy to confirm the presence of both $^4He^+$ and $^3He^+$ in a star by calculating the wavenumbers of the $n = 3 \rightarrow n = 2$ and of the $n = 2 \rightarrow n = 1$ transitions for each isotope.

4.28‡ Highly excited atoms have electrons with large principal quantum numbers. Such *Rydberg atoms* have unique properties and are of interest to astrophysicists. Derive a relation for the separation of energy levels for hydrogen atoms with large n. Calculate this separation for $n = 100$; also calculate the average radius, the geometric cross-section, and the ionization energy. Could a thermal collision with another hydrogen atom ionize this Rydberg atom? What minimum velocity of the second atom is required? Could a normal sized neutral H atom simply pass through the Rydberg atom leaving it undisturbed? What might the radial wavefunction for a 100s orbital be like?

4.29 The spectrum of a star is used to measure its *radial velocity* with respect to the Sun, the component of the star's velocity vector that is parallel to a vector connecting the star's centre to the centre of the Sun. The measurement relies on the Doppler effect. When a star emitting electromagnetic radiation of frequency ν moves with a speed s relative to an observer, the observer detects radiation of frequency $\nu_{\text{receding}} = \nu f$ or $\nu_{\text{approaching}} = \nu/f$, where $f = \{(1 - s/c)/(1 + s/c)\}^{1/2}$ and c is the speed of light. (a) Three Fe I lines of the star HDE 271 182, which belongs to the Large Magellanic Cloud, occur at 438.882 nm, 441.000 nm, and 442.020 nm. The same lines occur at 438.392 nm, 440.510 nm, and 441.510 nm in the spectrum of an Earth-bound iron arc. Determine whether HDE 271 182 is receding from or approaching the Earth and estimate the star's radial speed with respect to the Earth. (b) What additional information would you need to calculate the radial velocity of HDE 271 182 with respect to the Sun?

4.30 In Problem 4.29, we saw that Doppler shifts of atomic spectral lines are used to estimate the speed of recession or approach of a star. From the discussion in Section 4.3, it can be inferred that Doppler broadening of an atomic spectral line depends on the temperature of the star that emits the radiation. A spectral line of $^{48}Ti^{8+}$ (of mass $47.95m_u$) in a distant star was found to be shifted from 654.2 nm to 706.5 nm and to be broadened to 61.8 pm. What is the speed of recession and the surface temperature of the star?

4.31 The d-metals iron, copper, and manganese form cations with different oxidation states. For this reason, they are found in many oxidoreductases and in several proteins of oxidative phosphorylation and photosynthesis (*Impacts I17.3* and *I19.1*). Explain why many d-metals form cations with different oxidation states.

4.32 Thallium, a neurotoxin, is the heaviest member of Group 13 of the periodic table and is found most usually in the +1 oxidation state. Aluminium, which causes anaemia and dementia, is also a member of the group but its chemical properties are dominated by the +3 oxidation state. Examine this issue by plotting the first, second, and third ionization energies for the Group 13 elements against atomic number. Explain the trends you observe. *Hints*. The third ionization energy, I_3, is the minimum energy needed to remove an electron from the doubly charged cation: $E^{2+}(g) \rightarrow E^{3+}(g) + e^-(g)$, $I_3 = E(E^{3+}) - E(E^{2+})$. For data, see the links to databases of atomic properties provided in the text's web site.

MATHEMATICAL BACKGROUND 4

Vectors

A **scalar physical property** (such as temperature) in general varies through space and is represented by a single value at each point of space. A **vector physical property** (such as the electric field strength) also varies through space, but in general has a different direction as well as a different magnitude at each point.

MB4.1 Definitions

A **vector** v has the general form (in three dimensions):

$$v = v_x i + v_y j + v_z k \qquad \text{(MB4.1)}$$

where i, j, and k are **unit vectors**, vectors of magnitude 1, pointing along the positive directions on the x-, y-, and z-axes and v_x, v_y, and v_z are the **components** of the vector on each axis (Fig. MB4.1). The **magnitude** of the vector is denoted v or $|v|$ and is given by

$$v = (v_x^2 + v_y^2 + v_z^2)^{1/2} \qquad \text{(MB4.2)}$$

The vector makes an angle θ with the z-axis and an angle ϕ to the x-axis in the xy-plane. It follows that

$$v_x = v \sin\theta \cos\phi \quad v_y = v \sin\theta \sin\phi \quad v_z = v \cos\theta \qquad \text{(MB4.3a)}$$

and therefore that

$$\theta = \arccos(v_z/v) \qquad \phi = \arctan(v_y/v_x) \qquad \text{(MB4.3b)}$$

● A BRIEF ILLUSTRATION

The vector $v = 2i + 3j - k$ has magnitude

$$v = \{2^2 + 3^2 + (-1)^2\}^{1/2} = 14^{1/2} = 3.74$$

Its direction is given by

$$\theta = \arccos(-1/14^{1/2}) = 105.5° \quad \phi = \arctan(3/2) = 56.3°$$ ●

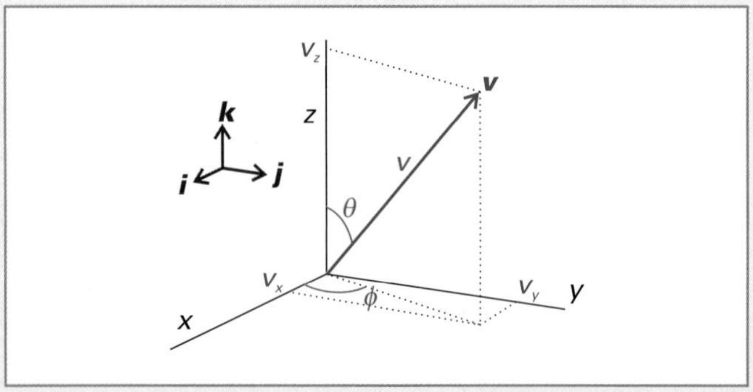

Fig. MB4.1 The vector v has components v_x, v_y, and v_z on the x-, y-, and z-axes, respectively. It has a magnitude v and makes an angle θ with the z-axis and an angle ϕ to the x-axis in the xy-plane.

MB4.2 Operations

Consider the two vectors

$$u = u_x i + u_y j + u_z k \qquad v = v_x i + v_y j + v_z k$$

The operations of addition, subtraction, and multiplication are as follows:

1. *Addition*:

$$v + u = (v_x + u_x)i + (v_y + u_y)j + (v_z + u_z)k \qquad \text{(MB4.4a)}$$

2. *Subtraction*:

$$v - u = (v_x - u_x)i + (v_y - u_y)j + (v_z - u_z)k \qquad \text{(MB4.4b)}$$

● A BRIEF ILLUSTRATION

Consider the vectors $u = i - 4j + k$ (of magnitude 4.24) and $v = -4i + 2j + 3k$ (of magnitude 5.39) Their sum is

$$u + v = (1-4)i + (-4+2)j + (1+3)k = -3i - 2j + 4k$$

The magnitude of the resultant vector is $29^{1/2} = 5.39$. The difference of the two vectors is

$$u - v = (1+4)i + (-4-2)j + (1-3)k = 5i - 6j - 2k$$

The magnitude of this resultant is 8.06. Note that in this case the difference is longer than either individual vector. ●

3. *Multiplication*:

(a) The **scalar product**, or *dot product*, of the two vectors u and v is

$$u \cdot v = u_x v_x + u_y v_y + u_z v_z \qquad \text{(MB4.4c)}$$

and is itself a scalar quantity. We can always choose a new coordinate system—we shall write it X, Y, Z—in which the Z-axis lies parallel to u, so $u = uK$, where K is the unit vector parallel to u. It then follows from eqn MB4.4c that $u \cdot v = uv_Z$. Then, with $v_Z = v\cos\theta$, where θ is the angle between u and v, we find

$$u \cdot v = uv\cos\theta \qquad \text{(MB4.4d)}$$

(b) The **vector product**, or *cross product*, of two vectors is

$$u \times v = \begin{vmatrix} i & j & k \\ u_x & u_y & u_z \\ v_x & v_y & v_z \end{vmatrix} = (u_y v_z - u_z v_y)i - (u_x v_z - u_z v_x)j + (u_x v_y - u_y v_x)k \qquad \text{(MB4.4e)}$$

(Determinants are discussed in *Mathematical background 5*.) Once again, choosing the coordinate system so that $u = uK$, leads to the simple expression:

$$u \times v = (uv\sin\theta)l \qquad \text{(MB4.4f)}$$

where θ is the angle between the two vectors and l is a unit vector perpendicular to both u and v, with a direction determined by

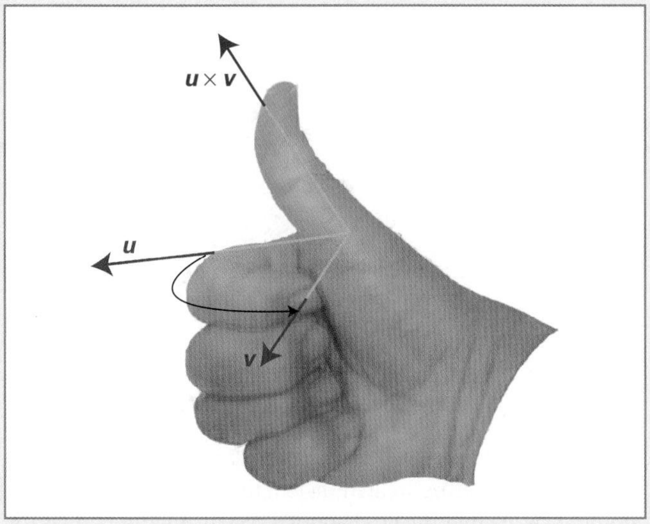

Fig. MB4.2 A depiction of the 'right-hand rule'. When the fingers of the right hand rotate $\boldsymbol{u}$ into $\boldsymbol{v}$, the thumb points in the direction of $\boldsymbol{u}\times\boldsymbol{v}$.

the 'right-hand rule' as in Fig. MB4.2. A special case is when each vector is a unit vector, for then

$$\boldsymbol{i}\times\boldsymbol{j}=\boldsymbol{k} \qquad \boldsymbol{j}\times\boldsymbol{k}=\boldsymbol{i} \qquad \boldsymbol{k}\times\boldsymbol{i}=\boldsymbol{j} \tag{MB4.5}$$

It is important to note that the order of vector multiplication is important and that $\boldsymbol{u}\times\boldsymbol{v}=-\boldsymbol{v}\times\boldsymbol{u}$.

A BRIEF ILLUSTRATION

The scalar and vector products of the two vectors in the previous *illustration*, $\boldsymbol{u}=\boldsymbol{i}-4\boldsymbol{j}+\boldsymbol{k}$ (of magnitude 4.24) and $\boldsymbol{v}=-4\boldsymbol{i}+2\boldsymbol{j}+3\boldsymbol{k}$ (of magnitude 5.39) are

$$\boldsymbol{u}\cdot\boldsymbol{v}=\{1\times(-4)\}+\{(-4)\times 2\}+\{1\times 3\}=-9$$

$$\begin{aligned}\boldsymbol{u}\times\boldsymbol{v}&=\begin{vmatrix}\boldsymbol{i}&\boldsymbol{j}&\boldsymbol{k}\\1&-4&1\\-4&2&3\end{vmatrix}\\&=\{(-4)(3)-(1)(2)\}\boldsymbol{i}-\{(1)(3)-(1)(-4)\}\boldsymbol{j}\\&\quad+\{(1)(2)-(-4)(-4)\}\boldsymbol{k}\\&=-14\boldsymbol{i}-7\boldsymbol{j}-14\boldsymbol{k}\end{aligned}$$

The vector product is a vector of magnitude 21.00 pointing in a direction perpendicular to the plane defined by the two individual vectors.

Self-test MB4.1 Determine the scalar and vector products of $\boldsymbol{v}=-\boldsymbol{i}+4\boldsymbol{j}+\boldsymbol{k}$ and $\boldsymbol{u}=2\boldsymbol{i}+3\boldsymbol{j}-\boldsymbol{k}$.

[$\boldsymbol{u}\cdot\boldsymbol{v}=9$, $\boldsymbol{u}\times\boldsymbol{v}=7\boldsymbol{i}-\boldsymbol{j}+11\boldsymbol{k}$]

MB4.3 The graphical representation of vector operations

Consider two vectors $\boldsymbol{v}$ and $\boldsymbol{u}$ making an angle θ (Fig. MB4.3). The first step in the addition of $\boldsymbol{u}$ to $\boldsymbol{v}$ consists of joining the tip

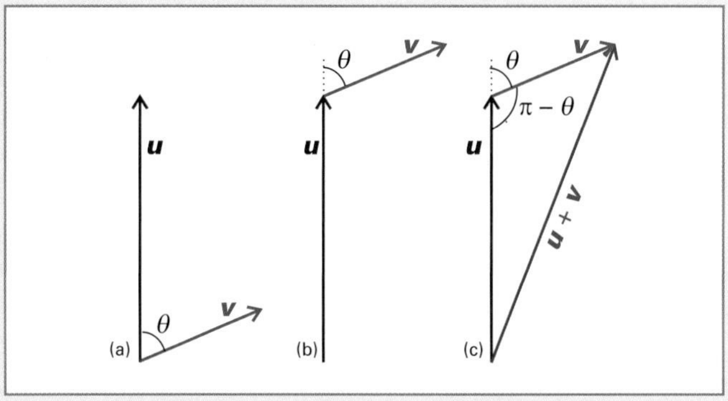

Fig. MB4.3 (a) The vectors $\boldsymbol{v}$ and $\boldsymbol{u}$ make an angle θ. (b) To add $\boldsymbol{u}$ to $\boldsymbol{v}$, we first join the head of $\boldsymbol{u}$ to the tail of $\boldsymbol{v}$, making sure that the angle θ between the vectors remains unchanged. (c) To finish the process, we draw the resultant vector by joining the tail of $\boldsymbol{u}$ to the head of $\boldsymbol{v}$.

(the 'head') of $\boldsymbol{u}$ to the starting point (the 'tail') of $\boldsymbol{v}$. In the second step, we draw a vector $\boldsymbol{v}_{res}$, the **resultant vector**, originating from the tail of $\boldsymbol{u}$ to the head of $\boldsymbol{v}$. Reversing the order of addition leads to the same result; that is, we obtain the same $\boldsymbol{v}_{res}$ whether we add $\boldsymbol{u}$ to $\boldsymbol{v}$ or $\boldsymbol{v}$ to $\boldsymbol{u}$. To calculate the magnitude of $\boldsymbol{v}_{res}$, we note that

$$\begin{aligned}v_{res}^2&=(\boldsymbol{u}+\boldsymbol{v})\cdot(\boldsymbol{u}+\boldsymbol{v})=\boldsymbol{u}\cdot\boldsymbol{u}+\boldsymbol{v}\cdot\boldsymbol{v}+2\boldsymbol{u}\cdot\boldsymbol{v}\\&=u^2+v^2+2uv\cos\theta'\end{aligned}$$

where θ' is the angle between $\boldsymbol{u}$ and $\boldsymbol{v}$. In terms of the angle $\theta=\pi-\theta'$ shown in the figure, and $\cos(\pi-\theta)=-\cos\theta$, we obtain the *law of cosines*:

$$v_{res}^2=u^2+v^2-2uv\cos\theta \tag{MB4.6}$$

for the relation between the lengths of the sides of a triangle.

Subtraction of $\boldsymbol{u}$ from $\boldsymbol{v}$ amounts to addition of $-\boldsymbol{u}$ to $\boldsymbol{v}$. It follows that in the first step of subtraction we draw $-\boldsymbol{u}$ by reversing the direction of $\boldsymbol{u}$ (Fig. MB4.4). Then, the second step consists of adding $-\boldsymbol{u}$ to $\boldsymbol{v}$ by using the strategy shown in the figure; we draw a resultant vector $\boldsymbol{v}_{res}$ by joining the tail of $-\boldsymbol{u}$ to the head of $\boldsymbol{v}$.

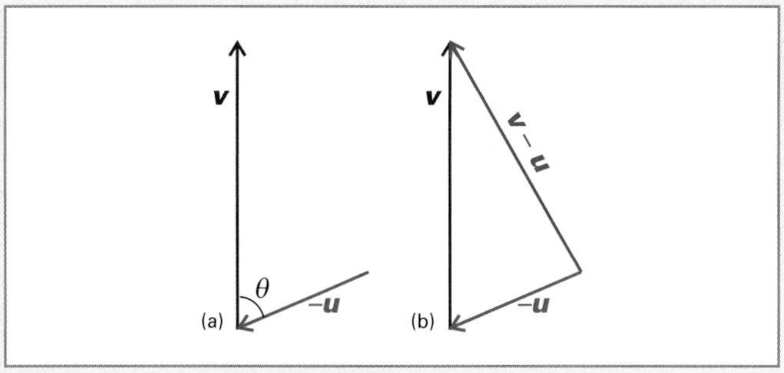

Fig. MB4.4 The graphical method for subtraction of the vector $\boldsymbol{u}$ from the vector $\boldsymbol{v}$ (as shown in Fig. MB4.3a) consists of two steps: (a) reversing the direction of $\boldsymbol{u}$ to form $-\boldsymbol{u}$, and (b) adding $-\boldsymbol{u}$ to $\boldsymbol{v}$.

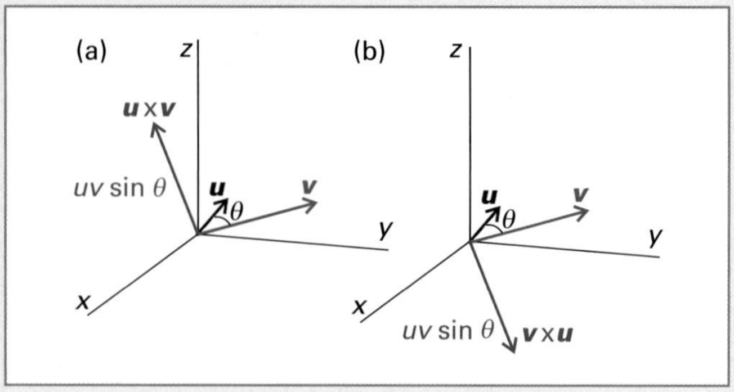

Fig. MB4.5 The direction of the cross products of two vectors $\boldsymbol{u}$ and $\boldsymbol{v}$ with an angle θ between them: (a) $\boldsymbol{u}\times\boldsymbol{v}$ and (b) $\boldsymbol{v}\times\boldsymbol{u}$. Note that the cross product, and the unit vector $\boldsymbol{l}$ of eqn MB4.4f, are perpendicular to both $\boldsymbol{u}$ and $\boldsymbol{v}$ but the direction depends on the order in which the product is taken. The magnitude of the cross product, in either case, is $uv\sin\theta$.

Vector multiplication is represented graphically by drawing a vector (using the right-hand rule) perpendicular to the plane defined by the vectors $\boldsymbol{u}$ and $\boldsymbol{v}$, as shown in Fig. MB4.5. Its length is equal to $uv\sin\theta$, where θ is the angle between $\boldsymbol{u}$ and $\boldsymbol{v}$.

MB4.4 Vector differentiation

The derivative $\mathrm{d}\boldsymbol{v}/\mathrm{d}t$, where the components v_x, v_y, and v_z are themselves functions of t, is

$$\frac{\mathrm{d}\boldsymbol{v}}{\mathrm{d}t}=\left(\frac{\mathrm{d}v_x}{\mathrm{d}t}\right)\boldsymbol{i}+\left(\frac{\mathrm{d}v_y}{\mathrm{d}t}\right)\boldsymbol{j}+\left(\frac{\mathrm{d}v_z}{\mathrm{d}t}\right)\boldsymbol{k} \qquad \text{(MB4.7)}$$

The derivatives of scalar and vector products are obtained using the rules of differentiating a product:

$$\frac{\mathrm{d}(\boldsymbol{u}\cdot\boldsymbol{v})}{\mathrm{d}t}=\boldsymbol{u}\cdot\frac{\mathrm{d}\boldsymbol{v}}{\mathrm{d}t}+\boldsymbol{v}\cdot\frac{\mathrm{d}\boldsymbol{u}}{\mathrm{d}t} \qquad \text{(MB4.8a)}$$

$$\frac{\mathrm{d}(\boldsymbol{u}\times\boldsymbol{v})}{\mathrm{d}t}=\boldsymbol{u}\times\frac{\mathrm{d}\boldsymbol{v}}{\mathrm{d}t}+\frac{\mathrm{d}\boldsymbol{u}}{\mathrm{d}t}\times\boldsymbol{v} \qquad \text{(MB4.8b)}$$

In the latter, note the importance of preserving the order of vectors.

The **gradient** of a function $f(x,y,z)$, denoted grad f or ∇f, is

$$\nabla f=\left(\frac{\partial f}{\partial x}\right)\boldsymbol{i}+\left(\frac{\partial f}{\partial y}\right)\boldsymbol{j}+\left(\frac{\partial f}{\partial z}\right)\boldsymbol{k} \qquad \text{(MB4.9)}$$

where partial derivatives were mentioned in *Mathematical background 1* and will be treated at length in *Mathematical background 8*. Note that the gradient of a scalar function is a vector. We can treat ∇ as a vector operator (in the sense that it operates on a function and results in a vector), and write

$$\nabla=\boldsymbol{i}\frac{\partial}{\partial x}+\boldsymbol{j}\frac{\partial}{\partial y}+\boldsymbol{k}\frac{\partial}{\partial z} \qquad \text{(MB4.10)}$$

The scalar product of ∇ and ∇f, using eqns MB4.9 and MB4.10, is

$$\nabla\cdot\nabla f=\left\{\boldsymbol{i}\frac{\partial}{\partial x}+\boldsymbol{j}\frac{\partial}{\partial y}+\boldsymbol{k}\frac{\partial}{\partial z}\right\}\cdot\left\{\left(\frac{\partial f}{\partial x}\right)\boldsymbol{i}+\left(\frac{\partial f}{\partial y}\right)\boldsymbol{j}+\left(\frac{\partial f}{\partial z}\right)\boldsymbol{k}\right\}$$

$$=\left(\frac{\partial^2 f}{\partial x^2}\right)+\left(\frac{\partial^2 f}{\partial y^2}\right)+\left(\frac{\partial^2 f}{\partial z^2}\right) \qquad \text{(MB4.11)}$$

Equation MB4.11 defines the **laplacian** ($\nabla^2=\nabla\cdot\nabla$) of a function.

The chemical bond

5

The concepts developed in Chapter 4, particularly that of orbitals, can be extended to a description of the electronic structures of molecules. There are two principal quantum mechanical theories of molecular electronic structure. In valence-bond theory the starting point is the concept of the shared electron pair. We see how to write the wavefunction for such a pair, and how it may be extended to account for the structures of a wide variety of molecules. The theory introduces the concepts of σ and π bonds, promotion, and hybridization that are used widely in chemistry. In molecular orbital theory (with which the bulk of the chapter is concerned), the concept of atomic orbital is extended to that of molecular orbital, which is a wavefunction that spreads over all the atoms in a molecule.

In this chapter we consider the origin of the strengths, numbers, and three-dimensional arrangement of chemical bonds between atoms. The quantum mechanical description of chemical bonding has become highly developed through the use of computers, and it is now possible to consider the structures of molecules of almost any complexity. We shall concentrate on the quantum mechanical description of the **covalent bond**, which was identified by G.N. Lewis (in 1916, before quantum mechanics was fully established) as an electron pair shared between two neighbouring atoms. We shall see, however, that the other principal type of bond, an **ionic bond**, in which the cohesion arises from the Coulombic attraction between ions of opposite charge, is also captured as a limiting case of a covalent bond between dissimilar atoms. In fact, although the Schrödinger equation might at first seem to obscure the fact, all chemical bonding can be traced to the interplay between the attraction of opposite charges, the repulsion of like charges, and the effect of changing kinetic energy as the electrons are confined to various regions when bonds form.

There are two major approaches to the calculation of molecular structure, **valence-bond theory** (VB theory) and **molecular orbital theory** (MO theory). Almost all modern computational work makes use of MO theory, and we concentrate on that theory in this chapter. Valence-bond theory, though, has left its imprint on the language of chemistry, and it is important to know the significance of terms that chemists use every day. Therefore, our discussion is organized as follows. First, we set out the concepts common to all levels of description. Then we present VB theory, which gives us a simple qualitative understanding of bond formation. Finally, we present the basic ideas of MO theory as a foundation for a full quantitative treatment in Chapter 6.

The Born–Oppenheimer approximation

All theories of molecular structure make the same simplification at the outset. Whereas the Schrödinger equation for a hydrogen atom can be solved exactly, an exact solution is not possible for any molecule because even the simplest consists of three particles (two nuclei and one electron). We therefore adopt the **Born–Oppenheimer approximation** in which it is supposed that the nuclei, being so much heavier than an electron, move relatively slowly and may be treated as stationary while the electrons move in their field. We can therefore think of the nuclei as being fixed at arbitrary locations, and then solve the Schrödinger equation for the wavefunction of the electrons alone.

The approximation is quite good for ground-state molecules, for calculations suggest that the nuclei in H_2 move through only about 1 pm while the electron speeds through 1000 pm, so the error of assuming that the nuclei are stationary is small. Exceptions to the validity of the approximation include certain excited states of polyatomic molecules and the ground states of cations; both types of species are important when considering photoelectron spectroscopy (Section 5.4) and mass spectrometry.

The Born–Oppenheimer approximation allows us to select an internuclear separation in a diatomic molecule and then to solve the Schrödinger equation for the electrons at that nuclear separation. Then we choose a different separation and repeat the calculation, and so on. In this way we can explore how the energy of the molecule varies with bond length and obtain a **molecular potential energy curve** (Fig. 5.1). More generally, in polyatomic molecules, where the energy also varies with angles and bond lengths, we obtain a **potential energy surface**. We need to consider the *potential* energy only because the kinetic energy of the stationary nuclei is zero. Once the potential energy curve of a diatomic molecule has been calculated or determined experimentally (by using the spectroscopic techniques described in Chapters 10 and 11), we can identify the **equilibrium bond length**, R_e, the internuclear separation at the minimum of the curve, and the **bond dissociation energy**, D_0, which is closely related to the depth, D_e, of the minimum below the energy of the infinitely widely separated and stationary atoms.

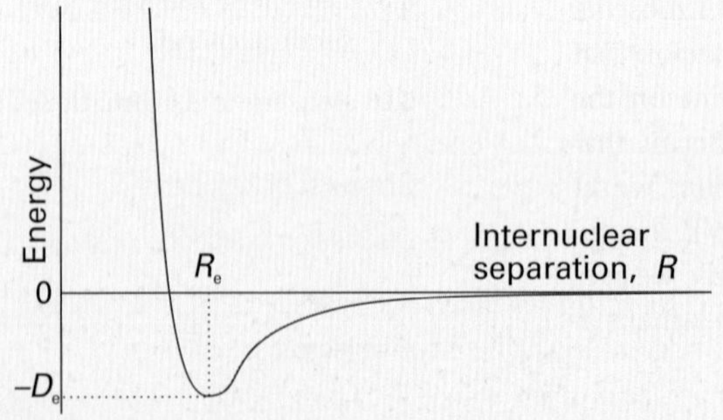

Fig. 5.1 A molecular potential energy curve. The equilibrium bond length corresponds to the energy minimum.

A note on good practice Be aware that the dissociation energy differs from the depth of the well by an energy equal to the zero-point vibrational energy of the bonded atoms: $D_0 = D_e - \frac{1}{2}\hbar\omega$, where ω is the vibrational frequency of the bond (Section 2.4).

Valence-bond theory

Valence-bond theory was the first quantum mechanical theory of bonding to be developed. The language it introduced, which includes concepts such as spin pairing, orbital overlap, σ and π bonds, and hybridization, is widely used throughout chemistry, especially in the description of the properties and reactions of organic compounds. Here we summarize essential topics of VB theory that are familiar from introductory chemistry and set the stage for the development of MO theory.

5.1 Homonuclear diatomic molecules

In VB theory, a bond is regarded as forming when an electron in an atomic orbital on one atom pairs its spin with that of an electron in an atomic orbital on another atom. To understand why this pairing leads to bonding, we have to examine the wavefunction for the two electrons that form the bond. We begin by considering the simplest possible two-electron chemical bond, the one in molecular hydrogen, H_2.

The spatial wavefunction for an electron on each of two widely separated H atoms is

$$\psi = \chi_{\mathrm{H1s_A}}(\boldsymbol{r}_1)\chi_{\mathrm{H1s_B}}(\boldsymbol{r}_2)$$

if electron 1 is on atom A and electron 2 is on atom B; in this chapter we use χ (chi) to denote atomic orbitals. For simplicity, we shall write this wavefunction as $\psi = A(1)B(2)$. When the atoms are close, it is not possible to know whether it is electron 1 that is on A or electron 2. An equally valid description is therefore $\psi = A(2)B(1)$, in which electron 2 is on A and electron 1 is on B. When two outcomes are equally probable, quantum mechanics instructs us to describe the true state of the system as a superposition of the wavefunctions for each possibility (Section 1.8), so a better description of the molecule than either wavefunction alone is the (unnormalized) linear combination

$$\psi = A(1)B(2) \pm A(2)B(1) \qquad (5.1)$$

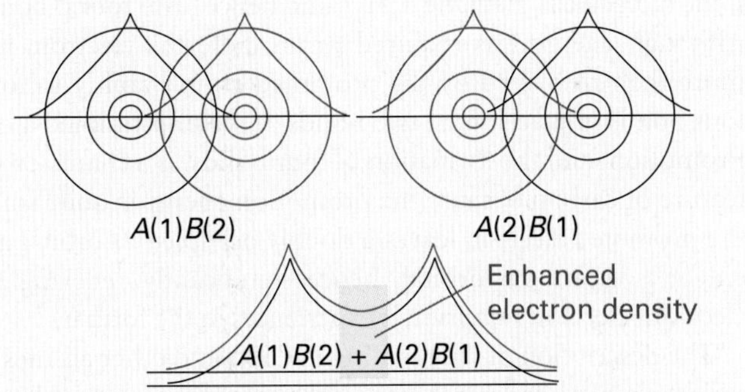

Fig. 5.2 It is very difficult to represent valence-bond wavefunctions because they refer to two electrons simultaneously. However, this figure is an attempt. The atomic orbital for electron 1 is represented by the black contours, and that of electron 2 is represented by the blue contours. The top two illustrations represent the contributions of $A(1)B(2)$ and $A(2)B(1)$ to the lower illustration, which represents the overall structure of the molecule.

It turns out that the combination with lower energy is the one with a + sign, so the valence-bond wavefunction of the H_2 molecule is

$$\psi = A(1)B(2) + A(2)B(1) \tag{5.2}$$

The formation of the bond in H_2 can be pictured as being due to the high probability that the two electrons will be found between the two nuclei and hence will bind them together. More formally, the wave pattern represented by the term $A(1)B(2)$ interferes constructively with the wave pattern represented by the contribution $A(2)B(1)$, and there is an enhancement in the value of the wavefunction in the internuclear region (Fig. 5.2).

The electron distribution described by the wavefunction in eqn 5.2 is called a **σ bond.** A σ bond has cylindrical symmetry around the internuclear axis, and is so called because, when viewed along the internuclear axis, it resembles a pair of electrons in an s orbital (and σ is the Greek equivalent of s).

A chemist's picture of a covalent bond is one in which the spins of two electrons pair as the atomic orbitals overlap. The origin of the role of spin is that the wavefunction given in eqn 5.2 can be formed only by a pair of electrons with opposed spins. Spin pairing is not an end in itself: as we show in the following *Justification*, it is a means of achieving a wavefunction (and the probability distribution it implies) that corresponds to a low energy.

Justification 5.1 *Electron pairing in VB theory*

The Pauli principle requires the wavefunction of two electrons to change sign when the labels of the electrons are interchanged (see *Justification 4.7*). The total VB wavefunction for two electrons is

$$\psi(1,2) = \{A(1)B(2) + A(2)B(1)\}\sigma(1,2)$$

where σ represents the spin component of the wavefunction. When the labels 1 and 2 are interchanged, this wavefunction becomes

$$\begin{aligned}\psi(2,1) &= \{A(2)B(1) + A(1)B(2)\}\sigma(2,1)\\ &= \{A(1)B(2) + A(2)B(1)\}\sigma(2,1)\end{aligned}$$

The Pauli principle requires that $\psi(2,1) = -\psi(1,2)$, which is satisfied only if $\sigma(2,1) = -\sigma(1,2)$. The combination of two spins that has this property is

$$\sigma_{-}(1,2) = (1/2^{1/2})\{\alpha(1)\beta(2) - \alpha(2)\beta(1)\}$$

which corresponds to paired electron spins (Section 4.4). Therefore, we conclude that the state of lower energy (and hence the formation of a chemical bond) is achieved if the electron spins are paired.

The VB description of H_2 can be applied to other homonuclear diatomic molecules, such as nitrogen, N_2. To construct the valence-bond description of N_2, we consider the valence electron configuration of each atom, which is $2s^2 2p_x^1 2p_y^1 2p_z^1$. It is conventional to take the z-axis to be the internuclear axis, so we can imagine each atom as having a $2p_z$ orbital pointing towards a $2p_z$ orbital on the other atom (Fig. 5.3), with the $2p_x$ and $2p_y$ orbitals perpendicular to the axis. A σ bond is then formed by spin pairing between the two electrons in the two $2p_z$ orbitals. Its spatial wavefunction is given by eqn 5.2, but now A and B stand for the two $2p_z$ orbitals.

The remaining 2p orbitals cannot merge to give σ bonds as they do not have cylindrical symmetry around the internuclear axis. Instead, they merge to form two π bonds. A **π bond** arises from the spin pairing of electrons in two p orbitals that approach side-by-side (Fig. 5.4). It is so called because, viewed along the internuclear axis, a π bond resembles a pair of electrons in a p orbital (and π is the Greek equivalent of p). There are two π bonds in N_2, one formed by spin pairing in two neighbouring $2p_x$ orbitals and the other by spin pairing in two neighbouring $2p_y$ orbitals. The overall bonding pattern in N_2 is therefore a σ bond plus two π bonds (Fig. 5.5), which is consistent with the Lewis structure :N≡N: for nitrogen.

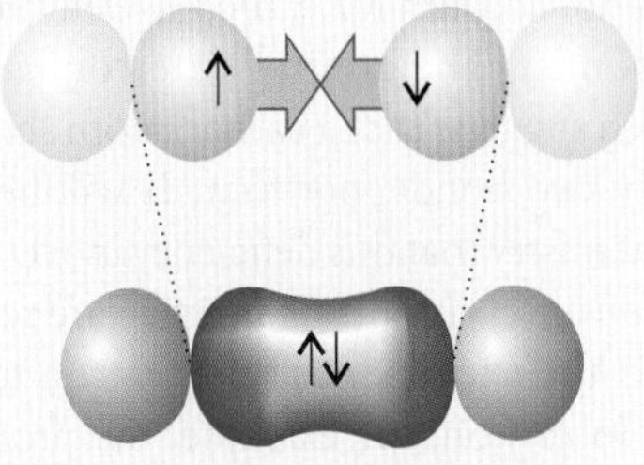

Fig. 5.3 The orbital overlap and spin pairing between electrons in two collinear p orbitals that results in the formation of a σ bond.

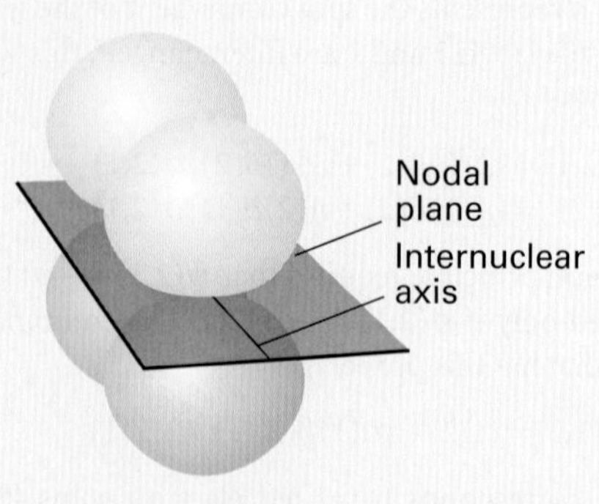

Fig. 5.4 A π bond results from orbital overlap and spin pairing between electrons in p orbitals with their axes perpendicular to the internuclear axis. The bond has two lobes of electron density separated by a nodal plane.

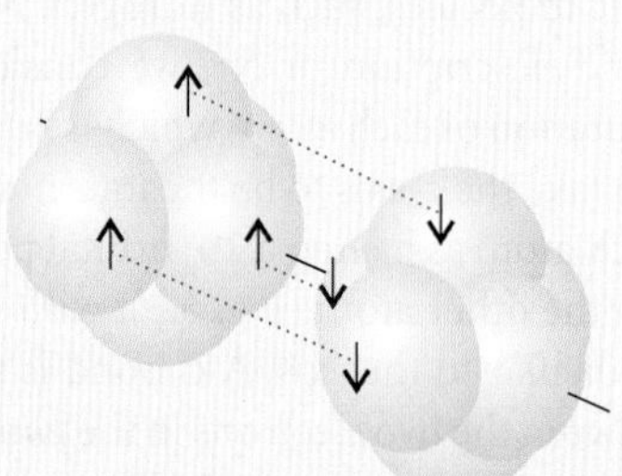

Fig. 5.5 The structure of bonds in a nitrogen molecule: there is one σ bond and two π bonds. As explained later, the overall electron density has cylindrical symmetry around the internuclear axis.

5.2 Polyatomic molecules

Each σ bond in a polyatomic molecule is formed by the spin pairing of electrons in atomic orbitals with cylindrical symmetry about the relevant internuclear axis. Likewise, π bonds are formed by pairing electrons that occupy atomic orbitals of the appropriate symmetry. However, it will be familiar from introductory chemistry that VB theory in this simple form does not lead to accurate descriptions of polyatomic molecules. For example, the theory does not account for carbon's tetravalence (its ability to form four bonds). The ground-state configuration of C is $2s^22p_x^12p_y^1$, which suggests that a carbon atom should be capable of forming only two bonds, not four. It will be familiar from introductory chemistry that this deficiency is overcome by allowing for *promotion* and *hybridization*, which are reviewed below.

Promotion is the excitation of an electron to an orbital of higher energy. In carbon, for example, the promotion of a 2s electron to a 2p orbital can be thought of as leading to the configuration $2s^12p_x^12p_y^12p_z^1$, with four unpaired electrons in separate orbitals. These electrons may pair with four electrons in orbitals provided by four other atoms (such as four H1s orbitals if the molecule is methane, CH_4), and hence form four σ bonds. Although energy was required to promote the electron, it is more than recovered by the promoted atom's ability to form four bonds in place of the two bonds of the unpromoted atom. Promotion, and the formation of four bonds, is a characteristic feature of carbon because the promotion energy is quite small: the promoted electron leaves a doubly occupied 2s orbital and enters a vacant 2p orbital, hence significantly relieving the electron–electron repulsion it experiences in the former.

The description of the bonding in CH_4 (and other alkanes) is still incomplete because it implies the presence of three σ bonds of one type (formed from H1s and C2p orbitals) and a fourth σ bond of a distinctly different character (formed from H1s and C2s). This problem is overcome by **hybridization**, which is predicated on the notion that the electron density distribution in the promoted atom is equivalent to the electron density in which each electron occupies a **hybrid orbital** formed by interference between the C2s and C2p orbitals. The origin of the hybridization can be appreciated by thinking of the four atomic orbitals centred on a nucleus as waves that interfere destructively and constructively in different regions, and give rise to four new shapes. However, we need to remember that promotion and hybridization are not 'real' processes in which an atom somehow becomes excited, forms hybrid orbitals, and then forms bonds: together, these processes help us account for the overall energy change that occurs when bonds form.

The specific linear combinations that give rise to four equivalent hybrid orbitals are

$$\begin{aligned} h_1 &= s + p_x + p_y + p_z \qquad & h_2 &= s - p_x - p_y + p_z \\ h_3 &= s - p_x + p_y - p_z \qquad & h_4 &= s + p_x - p_y - p_z \end{aligned} \tag{5.3}$$

As a result of the interference between the component orbitals, each hybrid orbital consists of a large lobe pointing in the direction of one corner of a regular tetrahedron (Fig. 5.6). The angle between the axes of the hybrid orbitals is the tetrahedral angle, 109.47°. Because each hybrid is built from one s orbital and three p orbitals, it is called an **sp^3 hybrid orbital.**

It is now easy to see how the valence-bond description of the CH_4 molecule leads to a tetrahedral molecule containing four equivalent C—H bonds. Each hybrid orbital of the promoted C atom contains a single unpaired electron; an H1s electron can pair with each one, giving rise to a σ bond pointing in a tetrahedral direction. For example, the (unnormalized) wavefunction for the bond formed by the hybrid orbital h_1 and the $1s_A$ orbital (with wavefunction that we shall denote A) is

$$\psi = h_1(1)A(2) + h_1(2)A(1)$$

Because each sp^3 hybrid orbital has the same composition, all four σ bonds are identical apart from their orientation in space (Fig. 5.7). A hybrid orbital has enhanced amplitude in the internuclear region, which arises from the constructive interference between the s orbital and the positive lobes of the p orbitals. As a result, the bond strength is greater than for a bond formed from

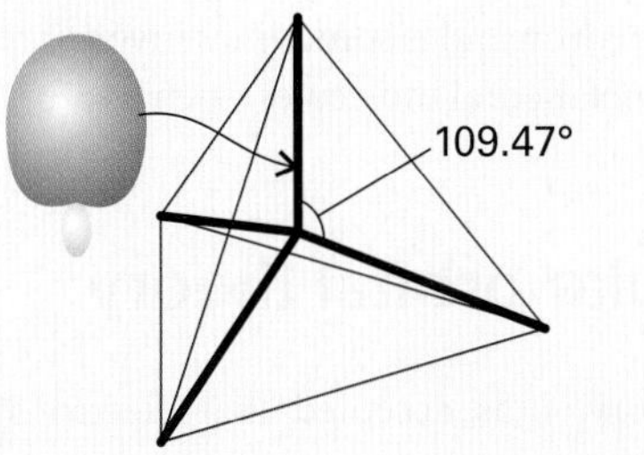

Fig. 5.6 An sp^3 hybrid orbital formed from the superposition of s and p orbitals on the same atom. There are four such hybrids: each one points towards the corner of a regular tetrahedron. The overall electron density remains spherically symmetrical.

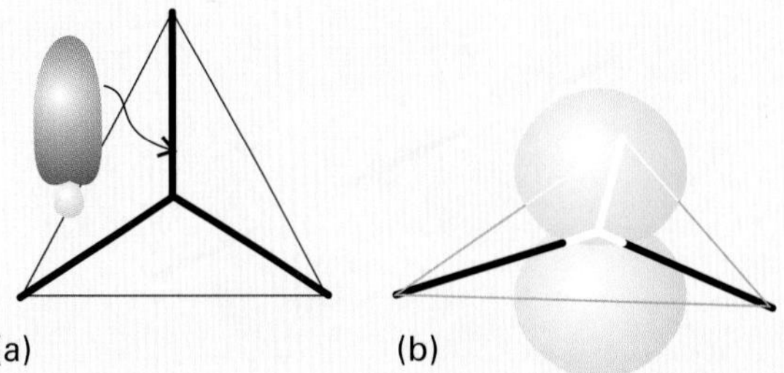

Fig. 5.8 (a) An s orbital and two p orbitals can be hybridized to form three equivalent orbitals that point towards the corners of an equilateral triangle. (b) The remaining, unhybridized p orbital is perpendicular to the plane.

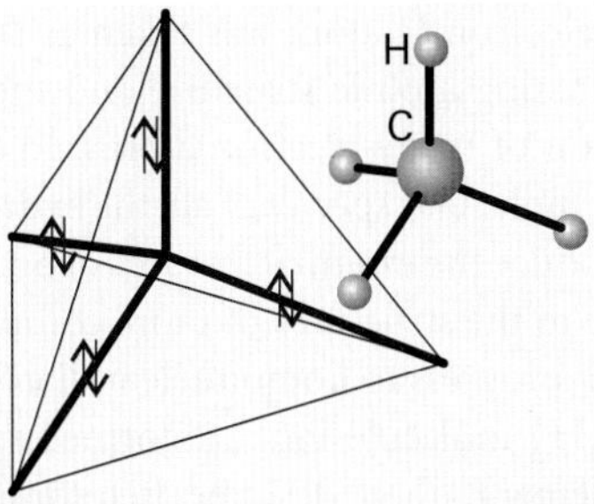

Fig. 5.7 Each sp^3 hybrid orbital forms a σ bond by overlap with an H1s orbital located at the corner of the tetrahedron. This model accounts for the equivalence of the four bonds in CH_4.

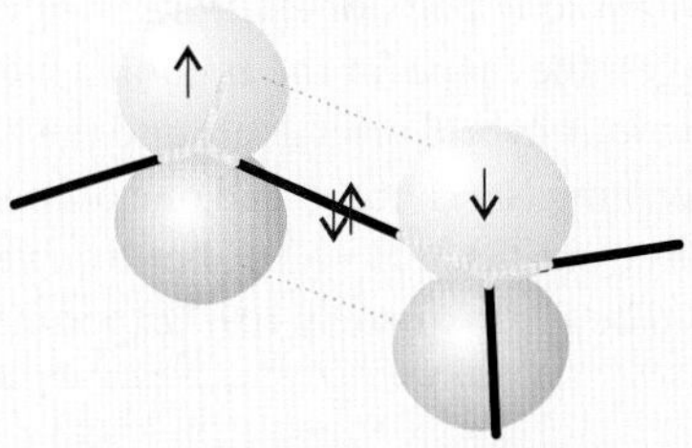

Fig. 5.9 A representation of the structure of a double bond in ethene; only the π bond is shown explicitly.

an s or p orbital alone. This increased bond strength is another factor that helps to repay the promotion energy.

Hybridization can also be used to describe the structure of an ethene molecule, $H_2C{=}CH_2$, and the torsional rigidity of double bonds. An ethene molecule is planar, with HCH and HCC bond angles close to 120°. To reproduce the σ bonding structure, we promote each C atom to a $2s^12p^3$ configuration. However, instead of using all four orbitals to form hybrids, we form **sp^2 hybrid orbitals:**

$$h_1 = s + 2^{1/2}p_y \qquad h_2 = s + (\tfrac{3}{2})^{1/2}p_x - (\tfrac{1}{2})^{1/2}p_y$$
$$h_3 = s - (\tfrac{3}{2})^{1/2}p_x - (\tfrac{1}{2})^{1/2}p_y \qquad (5.4)$$

that lie in a plane and point towards the corners of an equilateral triangle (Fig. 5.8). The third 2p orbital ($2p_z$) is not included in the hybridization; its axis is perpendicular to the plane in which the hybrids lie. As always in superpositions, the proportion of each orbital in the mixture is given by the *square* of the corresponding coefficient. Thus, in the first of these hybrids the ratio of s to p contributions is 1:2. Similarly, the total p contribution in each of h_2 and h_3 is $\frac{3}{2} + \frac{1}{2} = 2$, so the ratio for these orbitals is also 1:2. The different signs of the coefficients ensure that constructive interference takes place in different regions of space, so giving the patterns in the figure.

We can now describe the structure of $CH_2{=}CH_2$ as follows. The sp^2-hybridized C atoms each form three σ bonds by spin pairing with either the h_1 hybrid of the other C atom or with H1s orbitals. The σ framework therefore consists of C—H and C—C σ bonds at 120° to each other. When the two CH_2 groups lie in the same plane, the two electrons in the unhybridized p orbitals can pair and form a π bond (Fig. 5.9). The formation of this π bond locks the framework into the planar arrangement, for any rotation of one CH_2 group relative to the other leads to a weakening of the π bond (and consequently an increase in energy of the molecule).

A similar description applies to ethyne, HC≡CH, a linear molecule. Now the C atoms are **sp hybridized**, and the σ bonds are formed using hybrid atomic orbitals of the form

$$h_1 = s + p_z \qquad h_2 = s - p_z \qquad (5.5)$$

These two orbitals lie along the internuclear axis and point in opposite directions. The electrons in them pair either with an electron in the corresponding hybrid orbital on the other C atom or with an electron in one of the H1s orbitals. Electrons in the two remaining p orbitals on each atom, which are perpendicular to the molecular axis, pair to form two perpendicular π bonds (Fig. 5.10).

We can also generalize this discussion by noting that, in valence-bond theory, all single bonds are σ bonds, a double bond consists of one σ and one π bond, and a triple bond is composed of one σ and two π bonds.

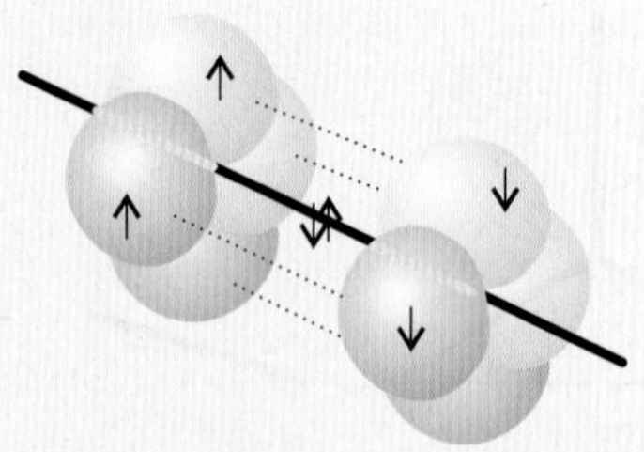

Fig. 5.10 A representation of the structure of a triple bond in ethyne; only the π bonds are shown explicitly. The overall electron density has cylindrical symmetry around the axis of the molecule.

Self-test 5.1 Hybrid orbitals do not always form bonds. They may also contain lone pairs of electrons. Use valence-bond theory to suggest the shape for the ammonia molecule, NH_3.

[There are four hybrid orbitals around the N atom, three of which forming N—H bonds and one occupied by a lone pair of electrons. Each H—N—H bond angle is predicted to be approximately 109° (experimental: 107°).]

Other hybridization schemes, particularly those involving d orbitals, are often invoked in elementary work to be consistent with other molecular geometries (Table 5.1). The hybridization of *N* atomic orbitals always results in the formation of *N* hybrid orbitals, which may either form bonds or may contain lone pairs of electrons. For example, sp^3d^2 hybridization results in six equivalent hybrid orbitals pointing towards the corners of a regular octahedron and is sometimes invoked to account for the structure of octahedral molecules, such as SF_6.

Table 5.1 Some hybridization schemes

Coordination number	Arrangement	Composition
2	Linear	sp, pd, sd
	Angular	sd
3	Trigonal planar	sp^2, p^2d
	Unsymmetrical planar	spd
	Trigonal pyramidal	pd^2
4	Tetrahedral	sp^3, sd^3
	Irregular tetrahedral	spd^2, p^3d, dp^3
	Square planar	p^2d^2, sp^2d
5	Trigonal bipyramidal	sp^3d, spd^3
	Tetragonal pyramidal	sp^2d^2, sd^4, pd^4, p^3d^2
	Pentagonal planar	p^2d^3
6	Octahedral	sp^3d^2
	Trigonal prismatic	spd^4, pd^5
	Trigonal antiprismatic	p^3d^3

Source: H. Eyring, J. Walter, and G.E. Kimball, *Quantum chemistry*, Wiley (1944).

Molecular orbital theory

In MO theory, it is accepted that electrons should not be regarded as belonging to particular bonds between 2 atoms (except of course in a diatomic molecule) but should be treated as spreading throughout the entire molecule. This theory has been more fully developed than VB theory and provides the language that is widely used in modern discussions of bonding. To introduce it, we follow the same strategy as in Chapter 4, where the one-electron H atom was taken as the fundamental species for discussing atomic structure and then developed it into a description of many-electron atoms. In this chapter we use the simplest molecular species of all, the hydrogen molecule-ion, H_2^+, to introduce the essential features of bonding, and then use it as a guide to the structures of more complex systems. To that end, we will progress to **homonuclear diatomic molecules**, which, like the H_2^+ molecule-ion, are formed from two atoms of the same element, then describe **heteronuclear diatomic molecules**, which are diatomic molecules formed from atoms of two different elements (such as CO and HCl), and end with a treatment of polyatomic molecules that forms the basis for modern computational models of molecular structure and chemical reactivity.

5.3 The hydrogen molecule-ion

The hamiltonian for the single electron in H_2^+ is

$$\hat{H} = -\frac{\hbar^2}{2m_e}\nabla_1^2 + V \qquad V = -\frac{e^2}{4\pi\varepsilon_0}\left(\frac{1}{r_{A1}} + \frac{1}{r_{B1}} - \frac{1}{R}\right) \tag{5.6}$$

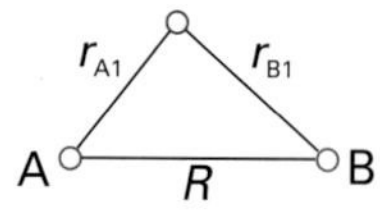

1

where r_{A1} and r_{B1} are the distances of the electron from the two nuclei (**1**) and R is the distance between the two nuclei. In the expression for V, the first two terms in parentheses are the attractive contribution from the interaction between the electron and the nuclei; the remaining term is the repulsive interaction between the nuclei.

The one-electron wavefunctions obtained by solving the Schrödinger equation $\hat{H}\psi = E\psi$ are called **molecular orbitals** (MO). A molecular orbital ψ gives, through the value of $|\psi|^2$, the probability distribution of the electron in the molecule. A molecular orbital is like an atomic orbital, but spreads throughout the molecule.

The Schrödinger equation can be solved analytically for H_2^+ (within the Born–Oppenheimer approximation), but the wavefunctions are very complicated functions; moreover, the solution cannot be extended to polyatomic systems. Therefore, we adopt a simpler procedure that, while more approximate, can be extended readily to other molecules.

(a) Linear combinations of atomic orbitals

If an electron can be found in an atomic orbital belonging to atom A and also in an atomic orbital belonging to atom B, then the overall wavefunction is a superposition of the two atomic orbitals:

$$\psi_{\pm} = N(A \pm B) \tag{5.7}$$

where, for H_2^+, A denotes χ_{H1s_A}, B denotes χ_{H1s_B}, and N is a normalization factor. The technical term for the superposition in eqn 5.7 is a **linear combination of atomic orbitals** (LCAO). An approximate molecular orbital formed from a linear combination of atomic orbitals is called an **LCAO-MO**. A molecular orbital that has cylindrical symmetry around the internuclear axis, such as the one we are discussing, is called a **σ orbital** because it resembles an s orbital when viewed along the axis and, more precisely, because it has zero orbital angular momentum around the internuclear axis.

Example 5.1 *Normalizing a molecular orbital*

Normalize the molecular orbital ψ_+ in eqn 5.7.

Method We need to find the factor N such that

$$\int \psi^* \psi \, d\tau = 1$$

To proceed, substitute the LCAO into this integral, and make use of the fact that the atomic orbitals are individually normalized.

Answer When we substitute the wavefunction, we find

$$\int \psi^* \psi \, d\tau = N^2 \left\{ \int A^2 \, d\tau + \int B^2 \, d\tau + 2\int AB \, d\tau \right\}$$

$$= N^2(1 + 1 + 2S)$$

where $S = \int AB \, d\tau$. For the integral to be equal to 1, we require

$$N = \frac{1}{\{2(1+S)\}^{1/2}}$$

In H_2^+, $S \approx 0.59$, so $N = 0.56$.

Self-test 5.2 Normalize the orbital ψ_- in eqn 5.7.

[$N = 1/\{2(1-S)\}^{1/2}$, so $N = 1.10$ in H_2^+]

Figure 5.11 shows the contours of constant amplitude for the two molecular orbitals in eqn 5.7, and Fig. 5.12 shows their boundary surfaces. Plots like these are readily obtained using commercially available software. The calculation is quite straightforward, because all we need do is feed in the mathematical forms of the two atomic orbitals and then let the program do the rest. In this case, we use

$$A = \frac{e^{-r_A/a_0}}{(\pi a_0^3)^{1/2}} \qquad B = \frac{e^{-r_B/a_0}}{(\pi a_0^3)^{1/2}} \tag{5.8}$$

and note that r_A and r_B are not independent (2), but related by the law of cosines (see the following *Comment*):

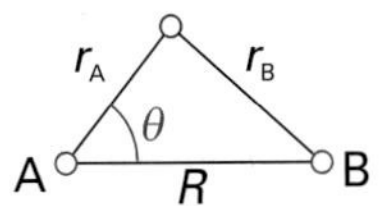

2

$$r_B = \{r_A^2 + R^2 - 2r_AR\cos\theta\}^{1/2} \tag{5.9}$$

To make this plot, we have taken $N^2 = 0.31$ (Example 5.1).

A brief comment The *law of cosines* states that, for a triangle with sides a, b, and c and angle C facing side c, then $c^2 = a^2 + b^2 - 2ab\cos C$.

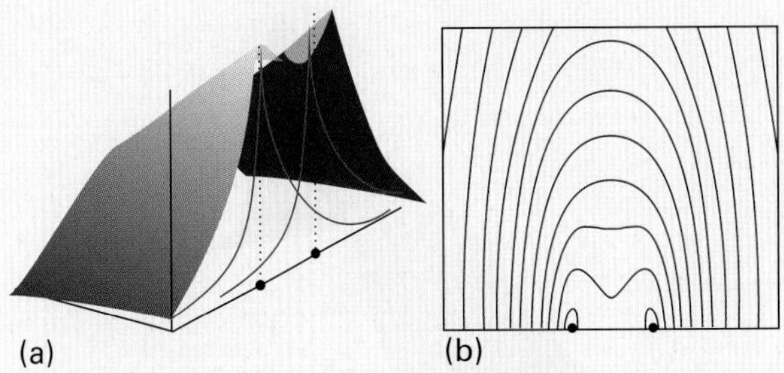

Fig. 5.11 (a) The amplitude of the bonding molecular orbital in a hydrogen molecule-ion in a plane containing the two nuclei and (b) a contour representation of the amplitude.

interActivity Plot the 1σ orbital for different values of the internuclear distance. Point to the features of the 1σ orbital that lead to bonding.

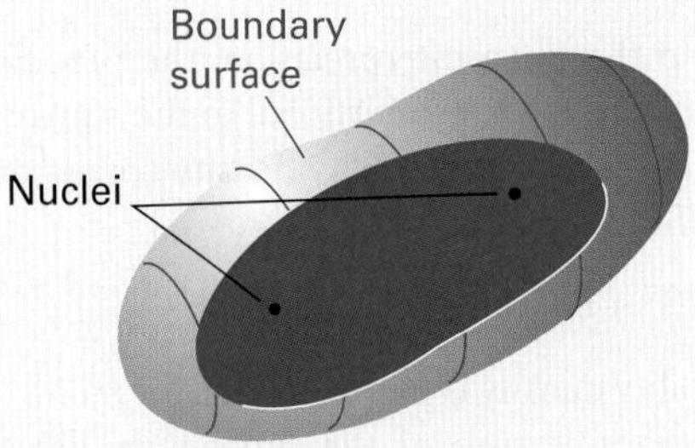

Fig. 5.12 A general indication of the shape of the boundary surface of a σ orbital.

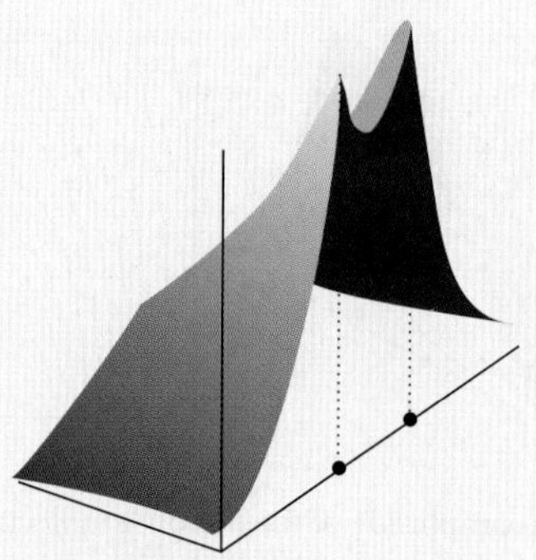

Fig. 5.13 The electron density calculated by forming the square of the wavefunction used to construct Fig. 5.11. Note the accumulation of electron density in the internuclear region.

(b) Bonding orbitals

According to the Born interpretation, the probability density of the electron in H_2^+ is proportional to the square modulus of its wavefunction. The probability density corresponding to the (real) wavefunction ψ_+ in eqn 5.7 is

$$\psi_+^2 = N^2(A^2 + B^2 + 2AB) \qquad (5.10)$$

This probability density is plotted in Fig. 5.13.

An important feature of the probability density becomes apparent when we examine the internuclear region, where both atomic orbitals have similar amplitudes. According to eqn 5.10, the total probability density is proportional to the sum of

- A^2, the probability density if the electron were confined to the atomic orbital A.
- B^2, the probability density if the electron were confined to the atomic orbital B.
- $2AB$, an extra contribution to the density.

This last contribution, the **overlap density**, is crucial, because it represents an enhancement of the probability of finding the electron in the internuclear region. The enhancement can be traced to the constructive interference of the two atomic orbitals: each has a positive amplitude in the internuclear region, so the total amplitude is greater there than if the electron were confined to a single atomic orbital.

We shall frequently make use of the result that *electrons accumulate in regions where atomic orbitals overlap and interfere constructively*. The conventional explanation is based on the notion that accumulation of electron density between the nuclei puts the electron in a position where it interacts strongly with both nuclei. Hence, the energy of the molecule is lower than that of the separate atoms, where each electron can interact strongly with only one nucleus. This conventional explanation, however, has been called into question because shifting an electron away from a nucleus into the internuclear region *raises* its potential energy. The modern (and still controversial) explanation does not emerge from the simple LCAO treatment given here. It seems that, at the same time as the electron shifts into the internuclear region, the atomic orbitals shrink. This orbital shrinkage improves the electron–nucleus attraction more than it is decreased by the migration to the internuclear region, so there is a net lowering of potential energy. The kinetic energy of the electron is also modified because the curvature of the wavefunction is changed, but the change in kinetic energy is dominated by the change in potential energy. Throughout the following discussion we ascribe the strength of chemical bonds to the accumulation of electron density in the internuclear region. We leave open the question whether in molecules more complicated than H_2^+ the true source of energy lowering is that accumulation itself or some indirect but related effect.

The σ orbital we have described is an example of a **bonding orbital**, an orbital that, if occupied, helps to bind two atoms together. Specifically, we label it 1σ as it is the σ orbital of lowest energy. An electron that occupies a σ orbital is called a **σ electron** and, if that is the only electron present in the molecule (as in the ground state of H_2^+), then we report the configuration of the molecule as $1\sigma^1$.

The energy $E_{1\sigma}$ of the 1σ orbital is (see Problem 5.23):

$$E_{1\sigma} = E_{\mathrm{H1s}} + \frac{j_0}{R} - \frac{J+K}{1+S} \qquad (5.11)$$

where j_0 is a convenient symbol for the commonly occurring term $e^2/4\pi\varepsilon_0$ and

$$S = \int AB\,\mathrm{d}\tau = \left\{1 + \frac{R}{a_0} + \frac{1}{3}\left(\frac{R}{a_0}\right)^2\right\}\mathrm{e}^{-R/a_0} \qquad (5.12a)$$

$$J = j_0\int \frac{A^2}{r_\mathrm{B}}\mathrm{d}\tau = \frac{j_0}{R}\left\{1 - \left(1 + \frac{R}{a_0}\right)\mathrm{e}^{-2R/a_0}\right\} \qquad (5.12b)$$

$$K = j_0\int \frac{AB}{r_\mathrm{B}}\mathrm{d}\tau = \frac{j_0}{a_0}\left(1 + \frac{R}{a_0}\right)\mathrm{e}^{-R/a_0} \qquad (5.12c)$$

We can interpret the preceding integrals as follows:

- All three integrals are positive and decline towards zero at large internuclear separations (S and K on account of the exponential term; J on account of the factor $1/R$).
- The integral J is a measure of the interaction between a nucleus and electron density centred on the other nucleus.
- The integral K is a measure of the interaction between a nucleus and the excess probability in the internuclear region arising from overlap.

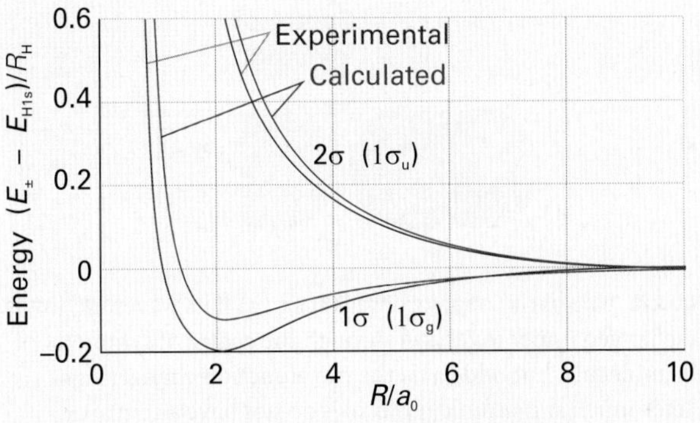

Fig. 5.14 The calculated and experimental molecular potential energy curves for a hydrogen molecule-ion showing the variation of the energy of the molecule as the bond length is changed.

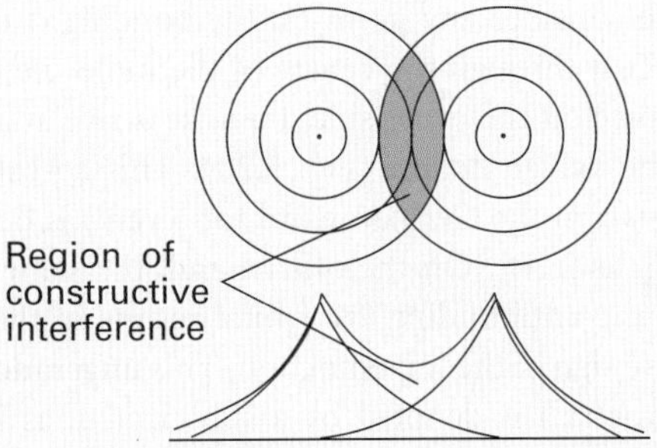

Fig. 5.15 A representation of the constructive interference that occurs when two H1s orbitals overlap and form a bonding σ orbital.

Figure 5.14 is a plot of $E_{1\sigma}$ against R relative to the energy of the separated atoms. The energy of the 1σ orbital decreases as the internuclear separation decreases from large values because electron density accumulates in the internuclear region as the constructive interference between the atomic orbitals increases (Fig. 5.15). However, at small separations there is too little space between the nuclei for significant accumulation of electron density there. In addition, the nucleus–nucleus repulsion (which is proportional to $1/R$) becomes large. As a result, the energy of the molecule rises at short distances, and there is a minimum in the potential energy curve. Calculations on H_2^+ give $R_e = 130$ pm and $D_e = 1.77$ eV (171 kJ mol^{-1}); the experimental values are 106 pm and 2.6 eV, so this simple LCAO-MO description of the molecule, while inaccurate, is not absurdly wrong.

(c) Antibonding orbitals

The linear combination ψ_- in eqn 5.7 corresponds to a higher energy than that of ψ_+. Because it is also a σ orbital we label it

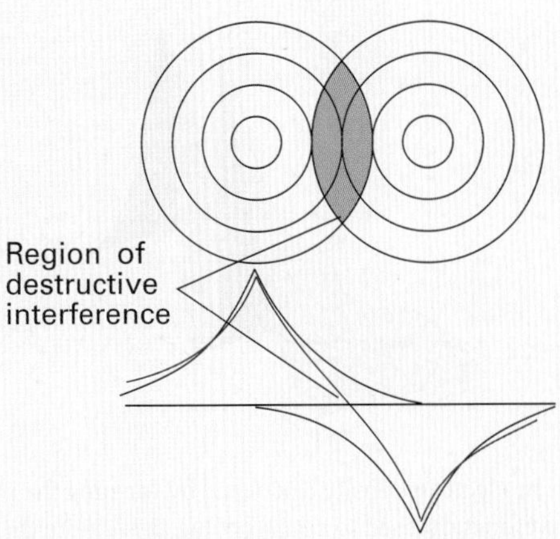

Fig. 5.16 A representation of the destructive interference that occurs when two H1s orbitals overlap and form an antibonding 2σ orbital.

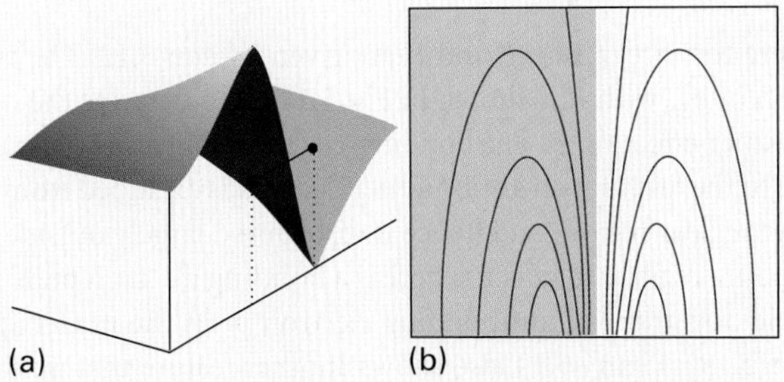

Fig. 5.17 (a) The amplitude of the antibonding molecular orbital in a hydrogen molecule-ion in a plane containing the two nuclei and (b) a contour representation of the amplitude. Note the internuclear node.

interActivity Plot the 2σ orbital for different values of the internuclear distance. Point to the features of the 2σ orbital that lead to antibonding.

2σ. This orbital has an internuclear nodal plane where A and B cancel exactly (Figs. 5.16 and 5.17). The probability density is

$$\psi_-^2 = N^2(A^2 + B^2 - 2AB) \tag{5.13}$$

There is a reduction in probability density between the nuclei due to the $-2AB$ term (Fig. 5.18); in physical terms, there is destructive interference where the two atomic orbitals overlap. The 2σ orbital is an example of an **antibonding orbital**, an orbital that, if occupied, contributes to a reduction in the cohesion between two atoms and helps to raise the energy of the molecule relative to the separated atoms.

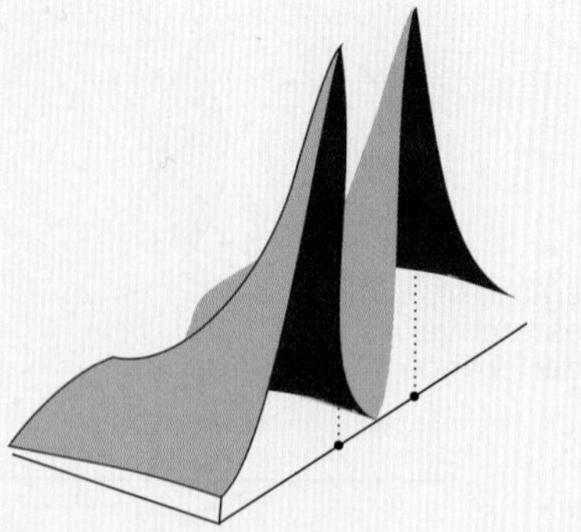

Fig. 5.18 The electron density calculated by forming the square of the wavefunction used to construct Fig. 5.17. Note the elimination of electron density from the internuclear region.

The energy $E_{2\sigma}$ of the 2σ antibonding orbital is given by (see Problem 5.23)

$$E_{2\sigma} = E_{\mathrm{H1s}} + \frac{j_0}{R} - \frac{J-K}{1-S} \qquad (5.14)$$

where the integrals S, J, and K are given by eqn 5.12. The variation of $E_{2\sigma}$ with R is shown in Fig. 5.14, where we see the destabilizing effect of an antibonding electron. The effect is partly due to the fact that an antibonding electron is excluded from the internuclear region, and hence is distributed largely outside the bonding region. In effect, whereas a bonding electron pulls two nuclei together, an antibonding electron pulls the nuclei apart (Fig. 5.19). Figure 5.14 also shows another feature that we draw on later: $|E_- - E_{\mathrm{H1s}}| > |E_+ - E_{\mathrm{H1s}}|$, which indicates that *the antibonding orbital is more antibonding than the bonding orbital is bonding*. This important conclusion stems in part from the presence of the nucleus–nucleus repulsion (the term j_0/R): this contribution raises the energy of both molecular orbitals. Antibonding orbitals are often labelled with an asterisk (*), so the 2σ orbital could also be denoted $2\sigma^*$ (and read '2 sigma star').

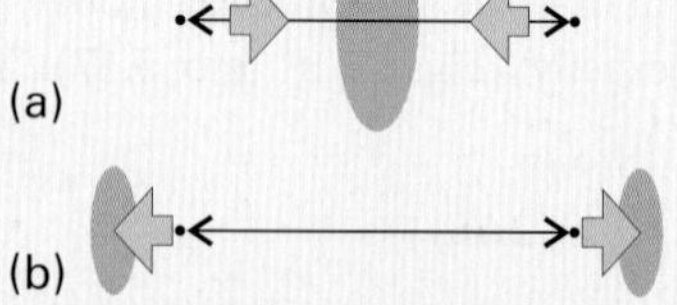

Fig. 5.19 A partial explanation of the origin of bonding and antibonding effects. (a) In a bonding orbital, the nuclei are attracted to the accumulation of electron density in the internuclear region. (b) In an antibonding orbital, the nuclei are attracted to an accumulation of electron density outside the internuclear region.

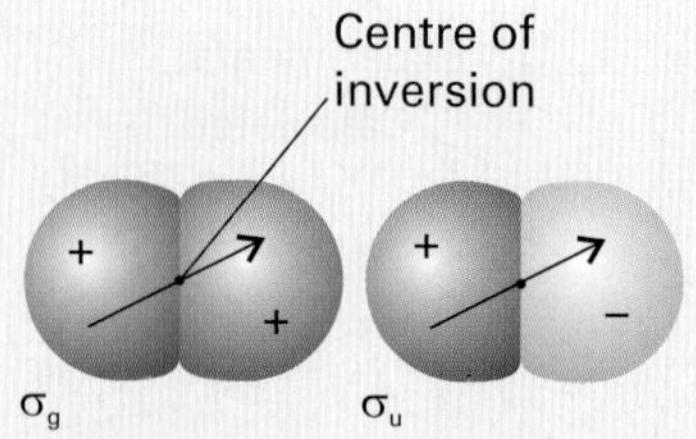

Fig. 5.20 The parity of an orbital is even (g) if its wavefunction is unchanged under inversion through the centre of symmetry of the molecule, but odd (u) if the wavefunction changes sign. Heteronuclear diatomic molecules do not have a centre of inversion, so for them the g, u classification is irrelevant.

For homonuclear diatomic molecules, it is helpful to describe a molecular orbital by identifying its **inversion symmetry**, the behaviour of the wavefunction when it is inverted through the centre (more formally, the centre of inversion) of the molecule. Thus, if we consider any point on the bonding σ orbital, and then project it through the centre of the molecule and out an equal distance on the other side, then we arrive at an identical value of the wavefunction (Fig. 5.20). This so-called **gerade symmetry** (from the German word for 'even') is denoted by a subscript g, as in σ_g. On the other hand, the same procedure applied to the antibonding 2σ orbital results in the same size but opposite sign of the wavefunction. This **ungerade symmetry** ('odd symmetry') is denoted by a subscript u, as in σ_u. This inversion symmetry classification is not applicable to diatomic molecules formed by atoms from two different elements (such as CO) because these molecules do not have a centre of inversion. When using the g,u notation, each set of orbitals of the same inversion symmetry are labelled separately, so whereas 1σ becomes $1\sigma_g$, its antibonding partner, which so far we have called 2σ, is the first orbital of a different symmetry, and is denoted $1\sigma_u$. The general rule is that *each set of orbitals of the same symmetry designation are labelled separately*.

A note on good practice When treating homonuclear diatomic molecules, we shall favour the more modern notation that focuses attention on the symmetry properties of the orbital. For all other molecules, on occasion we shall use asterisks to denote antibonding orbitals.

5.4 Homonuclear diatomic molecules

In Chapter 4 we used the hydrogenic atomic orbitals and the building-up principle to deduce the ground electronic configurations of many-electron atoms. We now do the same for many-electron diatomic molecules by using the H_2^+ molecular orbitals. The general procedure is to construct molecular orbitals by

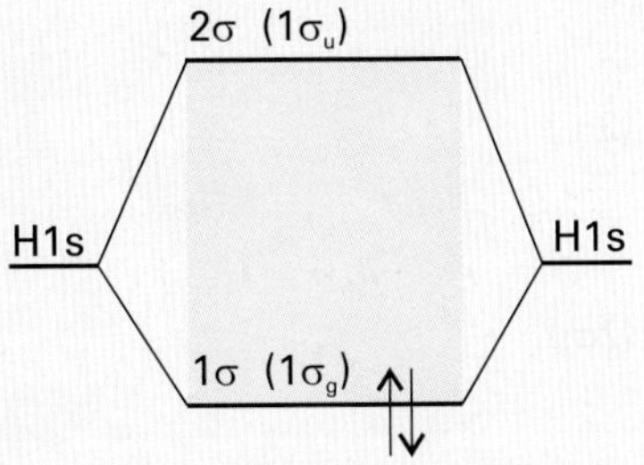

Fig. 5.21 A molecular orbital energy level diagram for orbitals constructed from the overlap of H1s orbitals; the separation of the levels corresponds to that found at the equilibrium bond length. The ground electronic configuration of H_2 is obtained by accommodating the two electrons in the lowest available orbital (the bonding orbital).

combining the available atomic orbitals. The electrons supplied by the atoms are then accommodated in the orbitals so as to achieve the lowest overall energy subject to the constraint of the Pauli exclusion principle, that no more than two electrons may occupy a single orbital (and then must be spin-paired). As in the case of atoms, if several degenerate molecular orbitals are available, we add the electrons singly to each individual orbital before doubly occupying any one orbital (because that minimizes electron–electron repulsions). We also take note of Hund's maximum multiplicity rule (Section 4.4) that, if electrons do occupy different degenerate orbitals, then a lower energy is obtained if they do so with parallel spins.

(a) σ orbitals

Consider H_2, the simplest many-electron diatomic molecule. Each H atom contributes a 1s orbital (as in H_2^+), so we can form the $1\sigma_g$ and $1\sigma_u$ orbitals from them, as we have seen already. At the experimental internuclear separation these orbitals will have the energies shown in Fig. 5.21, which is called a **molecular orbital energy level diagram.** Note that from two atomic orbitals we can build two molecular orbitals. In general, from N atomic orbitals we can build N molecular orbitals.

There are two electrons to accommodate, and both can enter $1\sigma_g$ by pairing their spins, as required by the Pauli principle (see the following *Justification*). The ground-state configuration is therefore $1\sigma_g^2$ and the atoms are joined by a bond consisting of an electron pair in a bonding σ orbital. This approach shows that an electron pair, which was the focus of Lewis's account of chemical bonding, represents the maximum number of electrons that can enter a bonding molecular orbital.

Justification 5.2 *Electron pairing in MO theory*

The spatial wavefunction for two electrons in a bonding molecular orbital ψ such as the bonding orbital in eqn 5.7, is $\psi_+(1)\psi_+(2)$. This two-electron wavefunction is obviously symmetric under interchange of the electron labels. To satisfy the Pauli principle, it must be multiplied by the antisymmetric spin state $\alpha(1)\beta(2) - \beta(1)\alpha(2)$ to give the overall antisymmetric state

$$\psi(1,2) = \psi_+(1)\psi_+(2)\{\alpha(1)\beta(2) - \beta(1)\alpha(2)\}$$

Because $\alpha(1)\beta(2) - \beta(1)\alpha(2)$ corresponds to paired electron spins, we see that two electrons can occupy the same molecular orbital (in this case, the bonding orbital) only if their spins are paired.

The same argument shows why He does not form diatomic molecules. Each He atom contributes a 1s orbital, so $1\sigma_g$ and $1\sigma_u$ molecular orbitals can be constructed. Although these orbitals differ in detail from those in H_2, the general shape is the same, and we can use the same qualitative energy level diagram in the discussion. There are four electrons to accommodate. Two can enter the $1\sigma_g$ orbital, but then it is full, and the next two must enter the $1\sigma_u$ orbital (Fig. 5.22). The ground electronic configuration of He_2 is therefore $1\sigma_g^2 1\sigma_u^2$. We see that there is one bond and one antibond. Because an antibond is slightly more antibonding than a bond is bonding, an He_2 molecule has a higher energy than the separated atoms, so it is unstable relative to the individual atoms. Diatomic helium 'molecules' have been prepared below 10 K: they consist of pairs of atoms held together by weak van der Waals forces of the type described in Chapter 8.

We shall now see how the concepts we have introduced apply to homonuclear diatomic molecules in general. In elementary treatments, only the orbitals of the valence shell are used to form molecular orbitals, so for molecules formed with atoms from Period 2 elements, only the 2s and 2p atomic orbitals are considered.

A general principle of molecular orbital theory is that *all orbitals of the appropriate symmetry contribute to a molecular orbital.* Thus, to build σ orbitals, we form linear combinations of all atomic orbitals that have cylindrical symmetry about the internuclear axis. These orbitals include the 2s orbitals on each

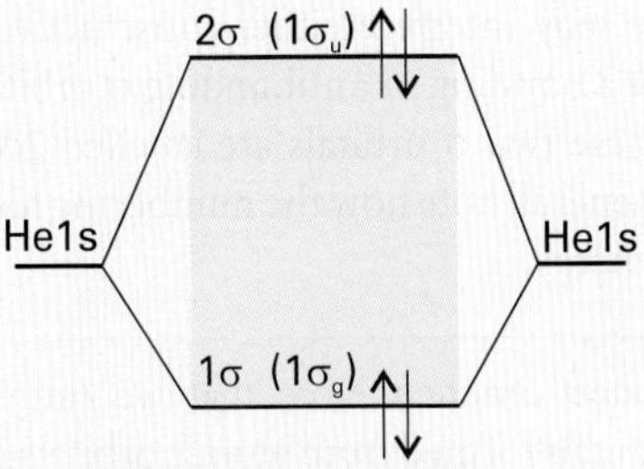

Fig. 5.22 The ground electronic configuration of the hypothetical four-electron molecule He_2 has two bonding electrons and two antibonding electrons. It has a higher energy than the separated atoms, and so is unstable.

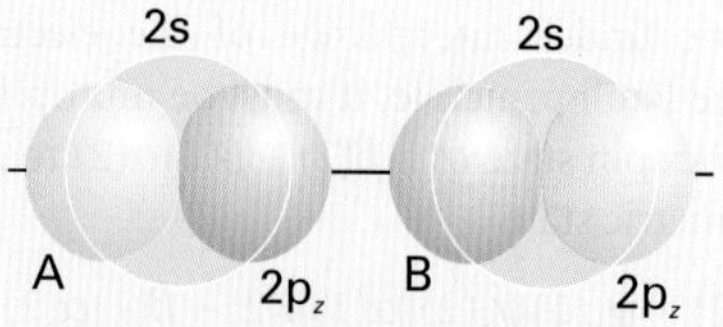

Fig. 5.23 According to molecular orbital theory, σ orbitals are built from all orbitals that have the appropriate symmetry. In homonuclear diatomic molecules of Period 2, that means that two 2s and two $2p_z$ orbitals should be used. From these four orbitals, four molecular orbitals can be built.

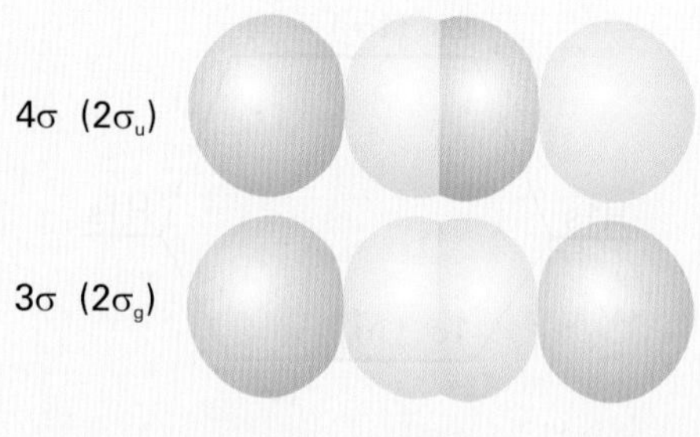

Fig. 5.24 A representation of the composition of bonding and antibonding σ orbitals built from the overlap of p orbitals. These illustrations are schematic.

atom and the $2p_z$ orbitals on the two atoms (Fig. 5.23). The general form of the σ orbitals that may be formed is therefore

$$\psi = c_{A2s}\chi_{A2s} + c_{B2s}\chi_{B2s} + c_{A2p_z}\chi_{A2p_z} + c_{B2p_z}\chi_{B2p_z} \qquad (5.15)$$

From these four atomic orbitals we can form four molecular orbitals of σ symmetry by an appropriate choice of the coefficients c.

The procedure for calculating the coefficients will be described in Section 5.5. At this stage we adopt a simpler route, and suppose that, because the 2s and $2p_z$ orbitals have distinctly different energies, they may be treated separately. That is, the four σ orbitals fall approximately into two sets, one consisting of two molecular orbitals of the form

$$\psi = c_{A2s}\chi_{A2s} + c_{B2s}\chi_{B2s} \qquad (5.16a)$$

and another consisting of two orbitals of the form

$$\psi = c_{A2p_z}\chi_{A2p_z} + c_{B2p_z}\chi_{B2p_z} \qquad (5.16b)$$

Because atoms A and B are identical, the energies of their 2s orbitals are the same, so the coefficients are equal (apart from a possible difference in sign); the same is true of the $2p_z$ orbitals. Therefore, the two sets of orbitals have the form $\chi_{A2s} \pm \chi_{B2s}$ and $\chi_{A2p_z} \pm \chi_{B2p_z}$.

The 2s orbitals on the two atoms overlap to give a bonding and an antibonding σ orbital ($1\sigma_g$ and $1\sigma_u$, respectively) in exactly the same way as we have already seen for 1s orbitals. The two $2p_z$ orbitals directed along the internuclear axis overlap strongly. They may interfere either constructively or destructively, and give a bonding or antibonding σ orbital, respectively (Fig. 5.24). These two σ orbitals are labelled $2\sigma_g$ and $2\sigma_u$, respectively. In general, note how the numbering follows the order of increasing energy.

A note on good practice Note that we number only the molecular orbitals formed from atomic orbitals in the valence shell. In an alternative system of notation, $1\sigma_g$ and $1\sigma_u$ are used to designate the molecular orbitals formed from the core 1s orbitals of the atoms; the orbitals we are considering would then be labelled starting from 2.

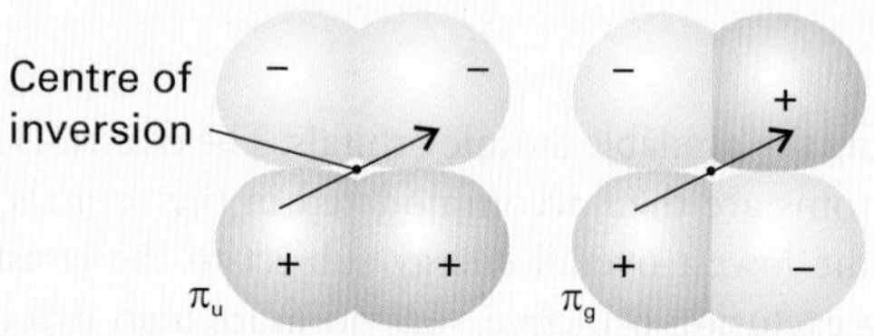

Fig. 5.25 A schematic representation of the structure of π bonding and antibonding molecular orbitals. The figure also shows that the bonding π orbital has odd parity, whereas the antibonding π orbital has even parity.

(b) π orbitals

Now consider the $2p_x$ and $2p_y$ orbitals of each atom. These orbitals are perpendicular to the internuclear axis and may overlap broadside-on. This overlap may be constructive or destructive, and results in a bonding or an antibonding **π orbital** (Fig. 5.25). The notation π is the analogue of p in atoms for, when viewed along the axis of the molecule, a π orbital looks like a p orbital, and has one unit of orbital angular momentum around the internuclear axis. The two $2p_x$ orbitals overlap to give a bonding and antibonding π_x orbital, and the two $2p_y$ orbitals overlap to give two π_y orbitals. The π_x and π_y bonding orbitals are degenerate; so too are their antibonding partners. We also see from Fig. 5.25 that a bonding π orbital has odd parity and is denoted π_u and an antibonding π orbital has even parity, denoted π_g.

(c) The overlap integral

The extent to which two atomic orbitals on different atoms overlap is measured by the **overlap integral**, S:

$$S = \int \chi_A^* \chi_B \, d\tau \qquad (5.17)$$

If the atomic orbital χ_A on A is small wherever the orbital χ_B on B is large, or vice versa, then the product of their amplitudes is everywhere small and the integral—the sum of these products—

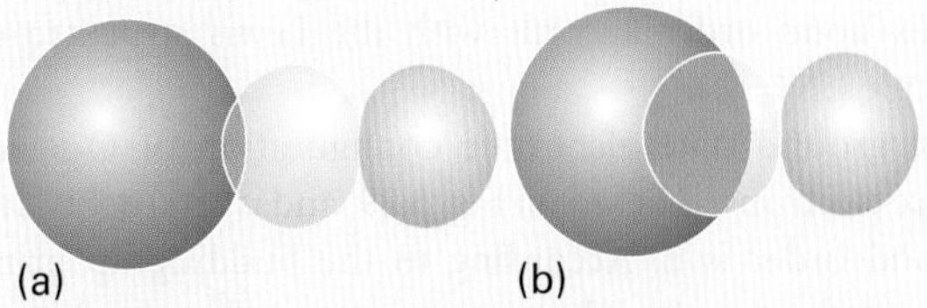

Fig. 5.26 (a) When two orbitals are on atoms that are far apart, the wavefunctions are small where they overlap, so S is small. (b) When the atoms are closer, both orbitals have significant amplitudes where they overlap, and S may approach 1. Note that S will decrease again as the two atoms approach more closely than shown here, because the region of negative amplitude of the p orbital starts to overlap the positive overlap of the s orbital. When the centres of the atoms coincide, $S = 0$.

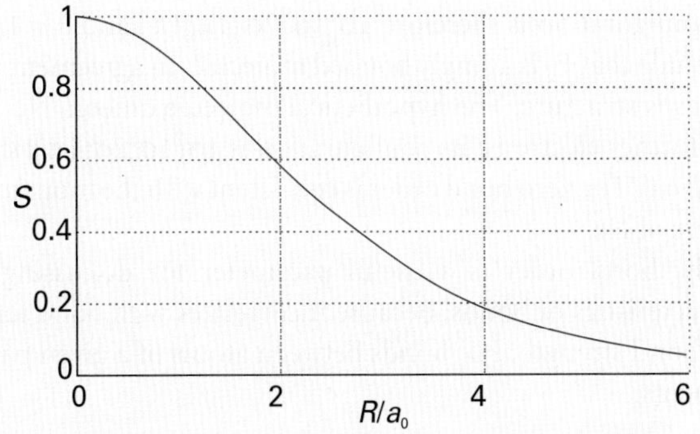

Fig. 5.27 The overlap integral, S, between two H1s orbitals as a function of their separation R.

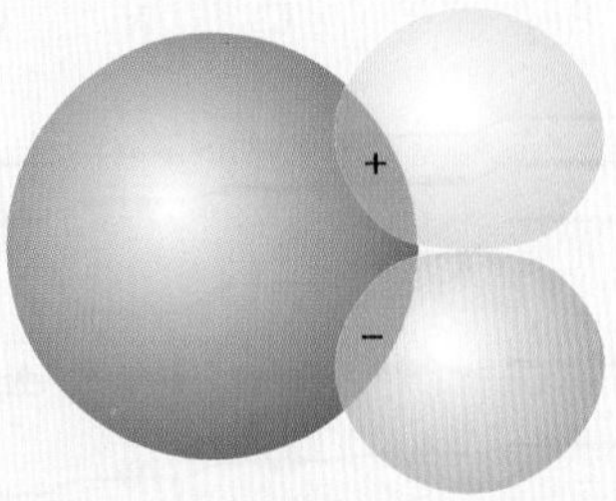

Fig. 5.28 A p orbital in the orientation shown here has zero net overlap ($S = 0$) with the s orbital at all internuclear separations.

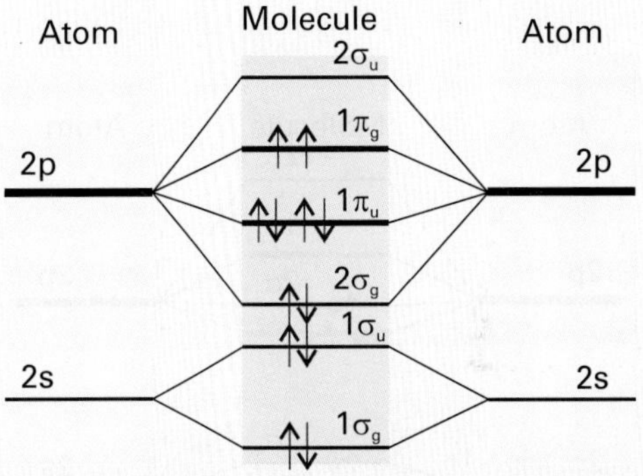

Fig. 5.29 The molecular orbital energy level diagram for homonuclear diatomic molecules. The lines in the middle are an indication of the energies of the molecular orbitals that can be formed by overlap of atomic orbitals. As remarked in the text, this diagram should be used for O_2 (the configuration shown) and F_2.

is small (Fig. 5.26). If χ_A and χ_B are simultaneously large in some region of space, then S may be large. If the two normalized atomic orbitals are identical (for instance, 1s orbitals on the same nucleus), then $S = 1$. In some cases, simple formulas can be given for overlap integrals and the variation of S with bond length plotted (Fig. 5.27). It follows that $S = 0.59$ for two H1s orbitals at the equilibrium bond length in H_2^+, which is an unusually large value. Typical values for orbitals with $n = 2$ are in the range 0.2 to 0.3.

Now consider the arrangement in which an s orbital is superimposed on a p_x orbital of a different atom (Fig. 5.28). The integral over the region where the product of orbitals is positive exactly cancels the integral over the region where the product of orbitals is negative, so overall $S = 0$ exactly. Therefore, there is no net overlap between the s and p orbitals in this arrangement.

(d) The electronic structures of homonuclear diatomic molecules

To construct the molecular orbital energy level diagram for Period 2 homonuclear diatomic molecules, we form eight molecular orbitals from the eight valence shell orbitals (four from each atom). In some cases, π orbitals are less strongly bonding than σ orbitals because their maximum overlap occurs off-axis. This relative weakness suggests that the molecular orbital energy level diagram ought to be as shown in Fig. 5.29. However, we must remember that we have assumed that 2s and $2p_z$ orbitals contribute to different sets of molecular orbitals, whereas in fact all four atomic orbitals contribute jointly to the four σ orbitals. Hence, there is no guarantee that this order of energies should prevail, and it is found experimentally (by spectroscopy) and by detailed calculation that the order varies along Period 2 (Fig. 5.30). The order shown in Fig. 5.31 is appropriate as far as N_2, and Fig. 5.29 applies for O_2 and F_2. The relative order is controlled by the separation of the 2s and 2p orbitals in the atoms, which increases across the group. The consequent switch in order occurs at about N_2.

With the orbitals established, we can deduce the ground configurations of the molecules by adding the appropriate number of electrons to the orbitals and following the building-up rules.

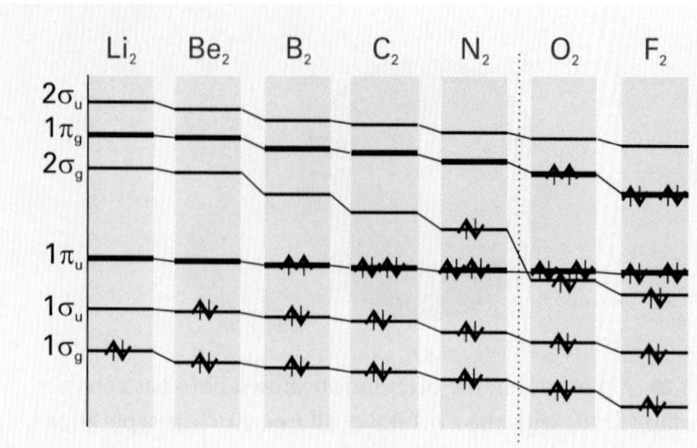

Fig. 5.30 The variation of the orbital energies of Period 2 homonuclear diatomics.

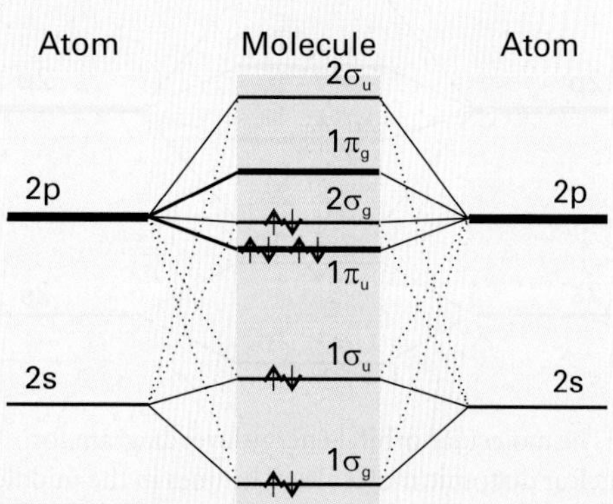

Fig. 5.31 An alternative molecular orbital energy level diagram for homonuclear diatomic molecules. As remarked in the text, this diagram should be used for diatomics up to and including N_2 (the configuration shown).

Anionic species (such as the peroxide ion, O_2^{2-}) need more electrons than the parent neutral molecules; cationic species (such as O_2^+) need fewer.

Consider N_2, which has 10 valence electrons. Two electrons pair, occupy, and fill the $1\sigma_g$ orbital; the next two occupy and fill the $1\sigma_u$ orbital. Six electrons remain. There are two 1π orbitals, so four electrons can be accommodated in them. The last two enter the $2\sigma_g$ orbital. Therefore, the ground-state configuration of N_2 is $1\sigma_g^2 1\sigma_u^2 1\pi_u^4 2\sigma_g^2$.

A measure of the net bonding in a diatomic molecule is its **bond order**, b:

$$b = \tfrac{1}{2}(n - n^*) \tag{5.18}$$

where n is the number of electrons in bonding orbitals and n^* is the number of electrons in antibonding orbitals. Thus each electron pair in a bonding orbital increases the bond order by 1 and each pair in an antibonding orbital decreases b by 1. For H_2, $b = 1$, corresponding to a single bond, H—H, between the two atoms. In He_2, $b = 0$, and there is no bond. In N_2, $b = \tfrac{1}{2}(8 - 2) = 3$. This bond order accords with the Lewis structure of the molecule (:N≡N:).

The ground-state electronic configuration of O_2, with 12 valence electrons, is based on Fig. 5.29, and is $1\sigma_g^2 1\sigma_u^2 2\sigma_g^2 1\pi_u^4 1\pi_g^2$. Its bond order is 2. According to the building-up principle, however, the two $1\pi_g$ electrons occupy different orbitals: one will enter $1\pi_{g,x}$ and the other will enter $1\pi_{g,y}$. Because the electrons are in different orbitals, they will have parallel spins. Therefore, we can predict that an O_2 molecule will have a net spin angular momentum $S = 1$ and, in the language introduced in Section 4.5, be in a triplet state. Because electron spin is the source of a magnetic moment, we can go on to predict that oxygen should be paramagnetic, a substance that tends to move into a magnetic field (Chapter 9). This prediction, which VB theory does not make, is confirmed by experiment.

An F_2 molecule has two more electrons than an O_2 molecule. Its configuration is therefore $1\sigma_g^2 1\sigma_u^2 2\sigma_g^2 1\pi_u^4 1\pi_g^4$ and $b = 1$. We conclude that F_2 is a singly bonded molecule, in agreement with its Lewis structure. The hypothetical molecule dineon, Ne_2, has two further electrons: its configuration is $1\sigma_g^2 1\sigma_u^2 2\sigma_g^2 1\pi_u^4 1\pi_g^2 2\sigma_u^2$ and $b = 0$. The zero bond order is consistent with the monatomic nature of Ne.

The bond order is a useful parameter for discussing the characteristics of bonds, because it correlates with bond length and bond strength. For bonds between atoms of a given pair of elements:

- The greater the bond order, the shorter the bond.
- The greater the bond order, the greater the bond strength.

Table 5.2 lists some typical bond lengths in diatomic and polyatomic molecules. The strength of a bond is measured by its bond dissociation energy, D_0, the energy required to separate the atoms to infinity. Table 5.3 lists some experimental values of dissociation energies.

Synoptic table 5.2 Bond lengths*

Bond	Order	R_e/pm
HH	1	74.14
NN	3	109.76
HCl	1	127.45
CH	1	*114*
CC	1	*154*
CC	2	*134*
CC	3	*120*

* More values will be found in the *Data section*. Numbers in italics are mean values for polyatomic molecules.

Synoptic table 5.3 Bond dissociation energies*

Bond	Order	$D_0/(\text{kJ mol}^{-1})$
HH	1	432.1
NN	3	941.7
HCl	1	427.7
CH	1	*435*
CC	1	*368*
CC	2	*720*
CC	3	*962*

* More values will be found in the *Data section*. Numbers in italics are mean values for polyatomic molecules.

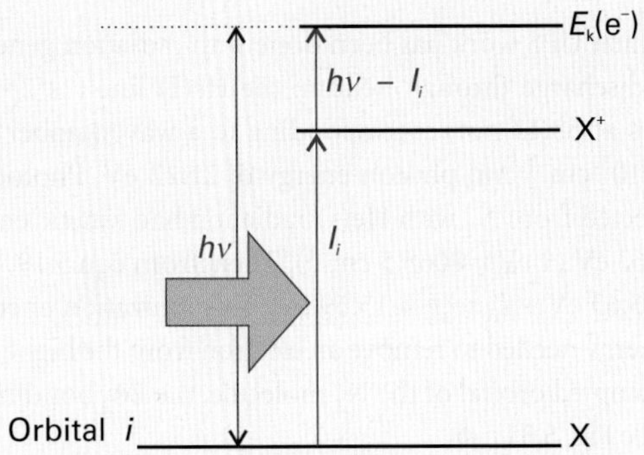

Fig. 5.32 An incoming photon carries an energy $h\nu$; an energy I_i is needed to remove an electron from an orbital i, and the difference appears as the kinetic energy of the electron.

Example 5.2 *Judging the relative bond strengths of molecules and ions*

Judge whether N_2^+ is likely to have a larger or smaller dissociation energy than N_2.

Method Because the molecule with the larger bond order is likely to have the larger dissociation energy, compare their electronic configurations and assess their bond orders.

Answer From Fig. 5.31, the electron configurations and bond orders are

$$N_2 \quad 1\sigma_g^2 1\sigma_u^2 1\pi_u^4 2\sigma_g^2 \quad b=3$$
$$N_2^+ \quad 1\sigma_g^2 1\sigma_u^2 1\pi_u^4 2\sigma_g^1 \quad b=2\tfrac{1}{2}$$

Because the cation has the smaller bond order, we expect it to have the smaller dissociation energy. The experimental dissociation energies are 942 kJ mol^{-1} for N_2 and 842 kJ mol^{-1} for N_2^+.

Self-test 5.3 Which can be expected to have the higher dissociation energy, F_2 or F_2^+? [F_2^+]

(e) Photoelectron spectroscopy

So far we have treated molecular orbitals as purely theoretical constructs, but is there experimental evidence for their existence? **Photoelectron spectroscopy** (PES) measures the ionization energies of molecules when electrons are ejected from different orbitals by absorption of a photon of the proper energy, and uses the information to infer the energies of molecular orbitals. The technique is also used to study solids, and in Chapter 21 we shall see the important information that it gives about species at or on surfaces.

Because energy is conserved when a photon ionizes a sample, the energy of the incident photon $h\nu$ must be equal to the sum of the ionization energy, I, of the sample and the kinetic energy of the **photoelectron**, the ejected electron (Fig. 5.32):

$$h\nu = \tfrac{1}{2}m_e v^2 + I \qquad (5.19a)$$

This equation (which is like the one used for the photoelectric effect, Section 1.2) can be refined by considering that photoelectrons may originate from one of a number of different orbitals, and each one has a different ionization energy. Hence, a series of different kinetic energies of the photoelectrons will be obtained, each one satisfying

$$h\nu = \tfrac{1}{2}m_e v^2 + I_i \qquad (5.19b)$$

where I_i is the ionization energy for ejection of an electron from an orbital i. Therefore, by measuring the kinetic energies of the photoelectrons, and knowing ν, these ionization energies can be determined. Photoelectron spectra are interpreted in terms of an approximation called **Koopmans' theorem**, which states that the ionization energy I_i is equal in magnitude to the orbital energy ε_i of the ejected electron (formally: $I_i = -\varepsilon_i$). That is, we can identify the ionization energy with the energy of the orbital from which it is ejected. Similarly, the energy of unfilled ('virtual orbitals') is related to the electron affinity. The theorem is only an approximation because it ignores the fact that the remaining electrons can adjust their distributions in response to changes in internuclear distances upon ionization.

The ionization energies of molecules are several electronvolts even for valence electrons, so it is essential to work in at least the ultraviolet region of the spectrum and with wavelengths of less than about 200 nm. Its use gives rise to the technique of **ultraviolet photoelectron spectroscopy** (UPS). When core electrons are being studied, photons of even higher energy are needed to expel them: X-rays are used, and the technique is denoted **X-ray photoelectron spectroscopy** (XPS).

● A BRIEF ILLUSTRATION

Much UPS work has been done with radiation generated by a discharge through helium: the He(I) line ($1s^12p^1 \rightarrow 1s^2$) lies at 58.43 nm, corresponding to a wavenumber of 1.711×10^5 cm^{-1} and photon energy of 21.22 eV. Photoelectrons ejected from N_2 with He(I) radiation had kinetic energies of 5.63 eV (1 eV = 8065.5 cm^{-1}). Then, from eqn 5.19, 21.22 eV = 5.63 eV + I_i, so I_i = 15.59 eV. This ionization energy is the energy needed to remove an electron from the highest energy occupied orbital of the N_2 molecule, the $2\sigma_g$ bonding orbital (see Fig. 5.31). ●

Self-test 5.4 Under the same circumstances, photoelectrons are also detected at 4.53 eV. To what ionization energy does that correspond? Suggest an origin. [16.7 eV, $1\pi_u$]

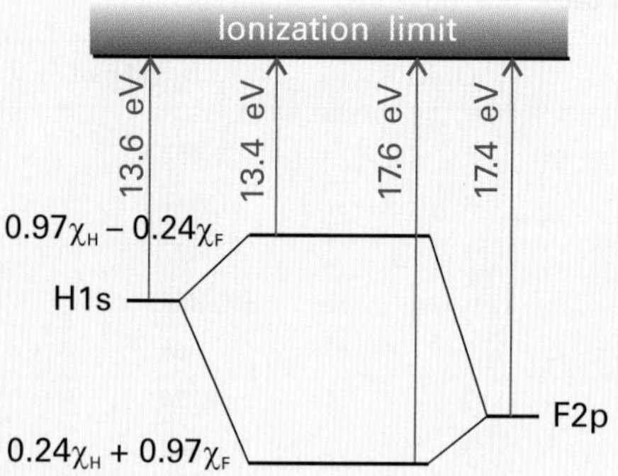

Fig. 5.33 The atomic orbital energy levels of H and F atoms and the molecular orbitals they form.

5.5 Heteronuclear diatomic molecules

The electron distribution in the covalent bond between the atoms in a heteronuclear diatomic molecule is not shared evenly because it is energetically favourable for the electron pair to be found closer to one atom than the other. This imbalance results in a **polar bond**, a covalent bond in which the electron pair is shared unequally by the two atoms. The bond in HF, for instance, is polar, with the electron pair closer to the F atom. The accumulation of the electron pair near the F atom results in that atom having a net negative charge, which is called a **partial negative charge** and denoted δ−. There is a matching **partial positive charge**, δ+, on the H atom.

(a) Polar bonds

A polar bond consists of two electrons in an orbital of the form

$$\psi = c_A A + c_B B \qquad (5.20)$$

with unequal coefficients. The proportion of the atomic orbital A in the bond is $|c_A|^2$ and that of B is $|c_B|^2$. A nonpolar bond has $|c_A|^2 = |c_B|^2$ and a pure ionic bond has one coefficient zero (so the species A^+B^- would have $c_A = 0$ and $c_B = 1$). The atomic orbital with the lower energy makes the larger contribution to the bonding molecular orbital. The opposite is true of the antibonding orbital, for which the dominant component comes from the atomic orbital with higher energy.

These points can be illustrated by considering HF, and judging the energies of the atomic orbitals from the ionization energies of the atoms. The general form of the molecular orbitals is

$$\psi = c_H \chi_H + c_F \chi_F \qquad (5.21)$$

where χ_H is an H1s orbital and χ_F is an F2p orbital. The H1s orbital lies 13.6 eV below the zero of energy (the separated proton and electron) and the F2p orbital lies at 17.4 eV (Fig. 5.33).

Hence, the bonding σ orbital in HF is mainly F2p and the antibonding σ orbital is mainly H1s orbital in character. The two electrons in the bonding orbital are most likely to be found in the F2p orbital, so there is a partial negative charge on the F atom and a partial positive charge on the H atom.

(b) Electronegativity

The charge distribution in bonds is commonly discussed in terms of the **electronegativity**, χ, of the elements involved (there should be little danger of confusing this use of χ with its use to denote an atomic orbital, which is another common convention). The electronegativity is a parameter introduced by Linus Pauling as a measure of the power of an atom to attract electrons to itself when it is part of a compound. Pauling used valence-bond arguments to suggest that an appropriate numerical scale of electronegativities could be defined in terms of bond dissociation energies, D, in electronvolts and proposed that the difference in electronegativities could be expressed as

$$|\chi_A - \chi_B| = \{D(\text{A—B}) - \tfrac{1}{2}[D(\text{A—A}) + D(\text{B—B})]\}^{1/2} \qquad (5.22)$$

Electronegativities based on this definition are called **Pauling electronegativities** (Table 5.4). The most electronegative elements are those close to fluorine; the least are those close to caesium. It is found that, the greater the difference in electronegativities, the greater the polar character of the bond. The difference for HF, for instance, is 1.78; a C—H bond, which is commonly regarded as almost nonpolar, has an electronegativity difference of 0.35.

The spectroscopist Robert Mulliken proposed an alternative definition of electronegativity. He argued that an element is likely to be highly electronegative if it has a high ionization energy (so it will not release electrons readily) and a high electron affinity (so it is energetically favourable to acquire electrons). The **Mulliken electronegativity scale** is therefore based on the definition

$$\chi_M = \tfrac{1}{2}(I + E_{ea}) \qquad (5.23)$$

Synoptic table 5.4* Pauling electronegativities

Element	χ_P
H	2.2
C	2.6
N	3.0
O	3.4
F	4.0
Cl	3.2
Cs	0.79

* More values will be found in the *Data section*.

where I is the ionization energy of the element and E_{ea} is its electron affinity (both in electronvolts, Section 4.4). The Mulliken and Pauling scales can be brought into close alignment. A reasonably reliable conversion between the two is $\chi_P = 1.35\chi_M^{1/2} - 1.37$.

(c) The variation principle

A more systematic way of discussing bond polarity and finding the coefficients in the linear combinations used to build molecular orbitals is provided by the **variation principle**:

> If an arbitrary wavefunction is used to calculate the energy, the value calculated is never less than the true energy.

This principle, which is established in the following *Justification*, is the basis of all modern molecular structure calculations (Chapter 6). The arbitrary wavefunction is called the **trial wavefunction**. The principle implies that, if we vary the coefficients in the trial wavefunction until the lowest energy is achieved (by evaluating the expectation value of the hamiltonian operator for each wavefunction), then those coefficients will be the best. We might get a lower energy if we use a more complicated wavefunction (for example, by taking a linear combination of several atomic orbitals on each atom), but we shall have the optimum (minimum energy) molecular orbital that can be built from the chosen **basis set**, the given set of atomic orbitals.

Justification 5.3 *The variation principle*

To justify the variation principle, consider a trial wavefunction written as a linear combination $\psi_{\text{trial}} = \sum_n c_n\psi_n$ of the true (but unknown), normalized, and orthogonal wavefunctions ψ_n that are solutions of the equations $\hat{H}\psi_n = E_n\psi_n$. The energy associated with this normalized trial wavefunction is then

$$\varepsilon = \int \psi^*_{\text{trial}}\hat{H}\psi_{\text{trial}}\,\mathrm{d}\tau$$

With E_0 the lowest energy for the system described by this basis set of ψ_n functions, the variation principle states that $\varepsilon \geq E_0$, or $\varepsilon - E_0 \geq 0$, for any ψ_{trial}. Now we must prove that this assertion is true.

The difference $\varepsilon - E_0$ may be written as the integral

$$\begin{aligned}\varepsilon - E_0 &= \overbrace{\int \psi^*_{\text{trial}}\hat{H}\psi_{\text{trial}}\,\mathrm{d}\tau}^{\varepsilon} - E_0\overbrace{\int \psi^*_{\text{trial}}\psi_{\text{trial}}\,\mathrm{d}\tau}^{=1}\\ &= \int \psi^*_{\text{trial}}\hat{H}\psi_{\text{trial}}\,\mathrm{d}\tau - \int \psi^*_{\text{trial}}E_0\psi_{\text{trial}}\,\mathrm{d}\tau\\ &= \int \psi^*_{\text{trial}}(\hat{H} - E_0)\psi_{\text{trial}}\,\mathrm{d}\tau\\ &= \int\left(\sum_n c^*_n\psi^*_n\right)(\hat{H} - E_0)\left(\sum_{n'} c_{n'}\psi_{n'}\right)\mathrm{d}\tau\\ &= \sum_{n,n'} c^*_n c_{n'}\int \psi^*_n(\hat{H} - E_0)\psi_{n'}\,\mathrm{d}\tau\end{aligned}$$

Because $\int\psi^*_n\hat{H}\psi_{n'}\mathrm{d}\tau = E_{n'}\int\psi^*_n\psi_{n'}\mathrm{d}\tau$ and $\int\psi^*_nE_0\psi_{n'}\mathrm{d}\tau = E_0\int\psi^*_n\psi_{n'}\mathrm{d}\tau$, we write

$$\int \psi^*_n(\hat{H} - E_0)\psi_{n'}\mathrm{d}\tau = (E_{n'} - E_0)\int \psi^*_n\psi_{n'}\mathrm{d}\tau$$

and

$$\varepsilon - E_0 = \sum_{n,n'} c^*_n c_{n'}(E_{n'} - E_0)\int \psi^*_n\psi_{n'}\mathrm{d}\tau$$

However, the functions ψ_n are orthogonal, so that $\int\psi^*_n\psi_{n'}\mathrm{d}\tau = 1$ if $n = n'$ and $\int\psi^*_n\psi_{n'}\mathrm{d}\tau = 0$ if $n \neq n'$. Therefore,

$$\varepsilon - E_0 = \sum_n c^*_n c_n(E_n - E_0)$$

We already know that $E_n - E_0 \geq 0$ and it is also true that $c^*_n c_n = |c_n|^2 > 0$. It follows that $\varepsilon - E_0 = \sum_n c^*_n c_n(E_n - E_0) \geq 0$, which is the result we set out to prove.

The method can be illustrated by the trial wavefunction in eqn 5.20. We show in the next *Justification* that the coefficients are given by the solutions of the two **secular equations**

$$(\alpha_A - E)c_A + (\beta - ES)c_B = 0 \tag{5.24a}$$

$$(\beta - ES)c_A + (\alpha_B - E)c_B = 0 \tag{5.24b}$$

The parameter α is called a **Coulomb integral**. It is a negative number that can be interpreted as the energy of the electron when it occupies A (for α_A) or B (for α_B). In a homonuclear diatomic molecule, $\alpha_A = \alpha_B$. The parameter β is called a **resonance integral** (for classical reasons). It vanishes when the orbitals do not overlap, and at equilibrium bond lengths it is normally negative.

Justification 5.4 *The variation principle applied to a heteronuclear diatomic molecule*

The trial wavefunction in eqn 5.20 is real but not normalized because at this stage the coefficients can take arbitrary values. Therefore, we can write $\psi^* = \psi$ but do not assume that $\int \psi^2 d\tau = 1$. The energy of the trial wavefunction is the expectation value of the energy operator (the hamiltonian operator, $\hat{H}$, Section 1.6):

$$E = \frac{\int \psi^* \hat{H} \psi \, d\tau}{\int \psi^* \psi \, d\tau} \qquad (5.25)$$

We must search for values of the coefficients in the trial function that minimize the value of E. This is a standard problem in calculus, and is solved by finding the coefficients for which

$$\frac{\partial E}{\partial c_A} = 0 \qquad \frac{\partial E}{\partial c_B} = 0$$

The first step is to express the two integrals in terms of the coefficients. The denominator is

$$\int \psi^2 \, d\tau = \int (c_A A + c_B B)^2 \, d\tau$$

$$= c_A^2 \int A^2 \, d\tau + c_B^2 \int B^2 \, d\tau + 2c_A c_B \int AB \, d\tau$$

$$= c_A^2 + c_B^2 + 2c_A c_B S$$

because the individual atomic orbitals are normalized and the third integral is the overlap integral S (eqn 5.17). The numerator is

$$\int \psi \hat{H} \psi \, d\tau = \int (c_A A + c_B B) \hat{H} (c_A A + c_B B) \, d\tau$$

$$= c_A^2 \int A\hat{H}A \, d\tau + c_B^2 \int B\hat{H}B \, d\tau + c_A c_B \int A\hat{H}B \, d\tau + c_A c_B \int B\hat{H}A \, d\tau$$

There are some complicated integrals in this expression, but we can combine them all into the parameters

$$\alpha_A = \int A\hat{H}A \, d\tau \qquad \alpha_B = \int B\hat{H}B \, d\tau \qquad (5.26)$$

$$\beta = \int A\hat{H}B \, d\tau = \int B\hat{H}A \, d\tau$$ (by the hermiticity of $\hat{H}$ and the reality of A and B)

Then

$$\int \psi \hat{H} \psi \, d\tau = c_A^2 \alpha_A + c_B^2 \alpha_B + 2c_A c_B \beta$$

The complete expression for E is

$$E = \frac{c_A^2 \alpha_A + c_B^2 \alpha_B + 2c_A c_B \beta}{c_A^2 + c_B^2 + 2c_A c_B S} \qquad (5.27)$$

Its minimum is found by differentiation with respect to the two coefficients and setting the results equal to 0. After some work, we obtain

$$\frac{\partial E}{\partial c_A} = \frac{2 \times (c_A \alpha_A - c_A E + c_B \beta - c_B S E)}{c_A^2 + c_B^2 + 2c_A c_B S} = 0$$

$$\frac{\partial E}{\partial c_B} = \frac{2 \times (c_B \alpha_B - c_B E + c_A \beta - c_A S E)}{c_A^2 + c_B^2 + 2c_A c_B S} = 0$$

For the derivatives to vanish, the numerators of these expressions must vanish. That is, we must find values of c_A and c_B that satisfy the conditions

$$c_A \alpha_A - c_A E + c_B \beta - c_B S E = (\alpha_A - E)c_A + (\beta - ES)c_B = 0$$

$$c_A \beta - c_A S E + c_B \alpha_B - c_B E = (\beta - ES)c_A + (\alpha_B - E)c_B = 0$$

which are the secular equations (eqn 5.24).

To solve the secular equations for the coefficients we need to know the energy E of the orbital. As for any set of simultaneous equations, the secular equations have a solution if the **secular determinant**, the determinant (see *Mathematical background 5*) of the coefficients, is zero; that is, if

$$\begin{vmatrix} \alpha_A - E & \beta - ES \\ \beta - ES & \alpha_B - E \end{vmatrix} = 0 \qquad (5.28)$$

This determinant expands to a quadratic equation in E (see the following *illustration*). Its two roots give the energies of the bonding and antibonding molecular orbitals formed from the atomic orbitals and, according to the variation principle, the lower root is the best energy achievable with the given basis set.

● A BRIEF ILLUSTRATION

To find the energies E of the bonding and antibonding orbitals of a homonuclear diatomic molecule, we need to know that a 2×2 determinant expands as follows:

$$\begin{vmatrix} a & b \\ c & d \end{vmatrix} = ad - bc$$

Setting $\alpha_A = \alpha_B = \alpha$ in eqn 5.28, we get

$$\begin{vmatrix} \alpha - E & \beta - ES \\ \beta - ES & \alpha - E \end{vmatrix} = (\alpha - E)^2 - (\beta - ES)^2 = 0$$

The solutions of this equation are

$$E_\pm = \frac{\alpha \pm \beta}{1 \pm S}$$ ●

The values of the coefficients in the linear combination are obtained by solving the secular equations using the two energies obtained from the secular determinant. The lower energy (E_+ in the *illustration*) gives the coefficients for the bonding molecular orbital; the upper energy (E_-) the coefficients for the antibonding molecular orbital. The secular equations give expressions for the ratio of the coefficients in each case, so we need a further equation in order to find their individual values. This equation is obtained by demanding that the best wavefunction should also be normalized. This condition means that, at this final stage, we must also ensure that

$$\int \psi^2 \, d\tau = c_A^2 + c_B^2 + 2c_A c_B S = 1 \qquad (5.29)$$

● **A BRIEF ILLUSTRATION**

To find the values of the coefficients c_A and c_B in the linear combination that corresponds to the energy E_+ from the previous *illustration*, we use eqn 5.27 (with $\alpha_A = \alpha_B = \alpha$) to write

$$E_+ = \frac{\alpha + \beta}{1 + S} = \frac{c_A^2\alpha + c_B^2\alpha + 2c_A c_B \beta}{c_A^2 + c_B^2 + 2c_A c_B S}$$

Now we use the normalization condition, eqn 5.29, to set $c_A^2 + c_B^2 + 2c_A c_B S = 1$, and so write

$$\frac{\alpha + \beta}{1 + S} = (c_A^2 + c_B^2)\alpha + 2c_A c_B \beta$$

This expression implies that

$$c_A^2 + c_B^2 = 2c_A c_B = \frac{1}{1+S} \quad \text{and} \quad |c_A| = \frac{1}{\{2(1+S)\}^{1/2}} \quad c_B = c_A$$

Proceeding in a similar way to find the coefficients in the linear combination that corresponds to the energy E_-, we write

$$E_- = \frac{\alpha - \beta}{1 - S} = (c_A^2 + c_B^2)\alpha + 2c_A c_B \beta$$

which implies that

$$c_A^2 + c_B^2 = -2c_A c_B = \frac{1}{1-S} \quad \text{and} \quad |c_A| = \frac{1}{\{2(1-S)\}^{1/2}}$$

$$c_B = -c_A$$ ●

(d) Two simple cases

The complete solutions of the secular equations are very cumbersome, even for 2 × 2 determinants, but there are two cases where the roots can be written down very simply.

We saw in the two *illustrations* we have worked through that, when the two atoms are the same, and we can write $\alpha_A = \alpha_B = \alpha$, the solutions are (choosing c_A to be positive)

$$E_+ = \frac{\alpha + \beta}{1 + S} \qquad c_A = \frac{1}{\{2(1+S)\}^{1/2}} \qquad c_B = c_A \qquad (5.30a)$$

$$E_- = \frac{\alpha - \beta}{1 - S} \qquad c_A = \frac{1}{\{2(1-S)\}^{1/2}} \qquad c_B = -c_A \qquad (5.30b)$$

In this case, the bonding orbital has the form

$$\psi_+ = \frac{A + B}{\{2(1+S)\}^{1/2}} \qquad (5.31a)$$

and its antibonding partner is

$$\psi_- = \frac{A - B}{\{2(1-S)\}^{1/2}} \qquad (5.31b)$$

in agreement with the discussion of homonuclear diatomics we have already given, but now with the normalization constant in place.

The second simple case is for a heteronuclear diatomic molecule but with $S = 0$ (a common approximation in elementary work). The secular determinant is then

$$\begin{vmatrix} \alpha_A - E & \beta \\ \beta & \alpha_B - E \end{vmatrix} = (\alpha_A - E)(\alpha_B - E) - \beta^2 = 0$$

The solutions can be expressed in terms of the parameter ζ (zeta), with

$$\zeta = \tfrac{1}{2}\arctan\frac{2|\beta|}{\alpha_B - \alpha_A} \qquad (5.32)$$

and are

$$E_- = \alpha_B - \beta\tan\zeta \qquad \psi_- = -A\sin\zeta + B\cos\zeta \qquad (5.33a)$$

$$E_+ = \alpha_A + \beta\tan\zeta \qquad \psi_+ = A\cos\zeta + B\sin\zeta \qquad (5.33b)$$

An important feature revealed by these solutions is that as the energy difference $|\alpha_B - \alpha_A|$ between the interacting atomic orbitals increases, the value of ζ decreases. We show in the following *Justification* that, when the energy difference is very large, in the sense that $|\alpha_B - \alpha_A| \gg \beta^2$, the energies of the resulting molecular orbitals differ only slightly from those of the atomic orbitals, which implies in turn that the bonding and antibonding effects are small. That is, *the strongest bonding and antibonding effects are obtained when the two contributing orbitals have closely similar energies.* The difference in energy between core and valence orbitals is the justification for neglecting the contribution of core orbitals to bonding. The core orbitals of one atom have a similar energy to the core orbitals of the other atom; but core–core interaction is largely negligible because the overlap between them (and hence the value of β) is so small.

Justification 5.5 *Bonding and antibonding effects in heteronuclear diatomic molecules*

When $x \ll 1$, we can write $\sin x \approx x$, $\cos x \approx 1$, $\tan x \approx x$, and $\arctan x = \tan^{-1}x \approx x$. It follows that when $|\alpha_B - \alpha_A| \gg 2|\beta|$ and $2|\beta|/|\alpha_B - \alpha_A| \ll 1$, we can write $\arctan 2|\beta|/|\alpha_B - \alpha_A| \approx 2|\beta|/|\alpha_B - \alpha_A|$ and, from eqn 5.32, $\zeta \approx |\beta|/(\alpha_B - \alpha_A)$. It follows that $\tan\zeta \approx |\beta|/(\alpha_B - \alpha_A)$. Noting that β is normally a negative number, so that $\beta/|\beta| = -1$, we can use eqn 5.33 to write

$$E_- = \alpha_B + \frac{\beta^2}{\alpha_B - \alpha_A} \qquad E_+ = \alpha_A - \frac{\beta^2}{\alpha_B - \alpha_A}$$

(In Problem 5.25 you are invited to derive these expressions in a different way.) It follows that, when the energy difference between the atomic orbitals is so large that $|\alpha_B - \alpha_A| \gg \beta^2$, the energies of the two molecular orbitals are $E_- \approx \alpha_B$ and $E_+ \approx \alpha_A$.

Now consider the behaviour of the wavefunctions in the limit of large $|\alpha_B - \alpha_A|$, when $\zeta \ll 1$. In this case, $\sin \zeta \approx \zeta$ and $\cos \zeta \approx 1$ and, from eqn 5.33, we write $\psi_- \approx B$ and $\psi_+ \approx A$. That is, the molecular orbitals are respectively almost pure B and almost pure A.

Example 5.3 *Calculating the molecular orbitals of HF*

Calculate the wavefunctions and energies of the σ orbitals in the HF molecule, taking $\beta = -1.0$ eV and the following ionization energies: H1s: 13.6 eV, F2s: 40.2 eV, F2p: 17.4 eV.

Method Because the F2p and H1s orbitals are much closer in energy than the F2s and H1s orbitals, to a first approximation neglect the contribution of the F2s orbital. To use eqn 5.32, we need to know the values of the Coulomb integrals α_H and α_F. Because these integrals represent the energies of the H1s and F2p electrons, respectively, by Koopmans' theorem they are approximately equal to (the negative of) the ionization energies of the atoms. Calculate ζ from eqn 5.32 (with A identified as F and B as H), and then write the wavefunctions by using eqn 5.33.

Answer Setting $\alpha_H = -13.6$ eV and $\alpha_F = -17.4$ eV gives $\tan 2\zeta = 0.53$; so $\zeta = 13.9°$. Then

$$E_- = -13.4\text{ eV} \qquad \psi_- = 0.97\chi_H - 0.24\chi_F$$
$$E_+ = -17.6\text{ eV} \qquad \psi_+ = 0.24\chi_H + 0.97\chi_F$$

Notice how the lower energy orbital (the one with energy −17.6 eV) has a composition that is more F2p orbital than H1s, and that the opposite is true of the higher energy, antibonding orbital.

Self-test 5.5 The ionization energy of Cl is 13.1 eV; find the form and energies of the σ orbitals in the HCl molecule using $\beta = -1.0$ eV.

$[E_- = -12.3\text{ eV}, \psi_- = -0.62\chi_H + 0.79\chi_{Cl};$
$E_+ = -14.4\text{ eV}, \psi_+ = 0.79\chi_H + 0.62\chi_{Cl}]$

IMPACT ON BIOCHEMISTRY

I5.1 The biochemical reactivity of O_2, N_2, and NO

We can now see how some of the concepts introduced in this chapter are applied to diatomic molecules that play a vital biochemical role. At sea level, air contains approximately 23.1 per cent O_2 and 75.5 per cent N_2 by mass. Molecular orbital theory predicts—correctly—that O_2 has unpaired electron spins and, consequently, is a reactive component of the Earth's atmosphere; its most important biological role is as an oxidizing agent. By contrast N_2, the major component of the air we breathe, is so stable (on account of the triple bond connecting the atoms) and unreactive that *nitrogen fixation*, the reduction of atmospheric N_2 to NH_3, is among the most thermodynamically demanding of biochemical reactions, in the sense that it requires a great deal of energy derived from metabolism. So taxing is the process that only certain bacteria and archaea are capable of carrying it out, making nitrogen available first to plants and other microorganisms in the form of ammonia. Only after incorporation into amino acids by plants does nitrogen adopt a chemical form that, when consumed, can be used by animals in the synthesis of proteins and other nitrogen-containing molecules.

The reactivity of O_2, while important for biological energy conversion, also poses serious physiological problems. During the course of the biological oxidation of glucose, a key set of reactions of metabolism, some electrons escape from the protein complexes that catalyse the process with the result that some O_2 is reduced to superoxide ion, O_2^-. The ground-state electronic configuration of O_2^- is $1\sigma_g^2 1\sigma_u^2 2\sigma_g^2 1\pi_u^4 1\pi_g^3$, so the ion is a radical with a bond order $b = \frac{3}{2}$. We predict that the superoxide ion is a reactive species that must be scavenged to prevent damage to cellular components. The enzyme superoxide dismutase protects cells by catalysing the disproportionation (or 'dismutation') of O_2^- into O_2 and H_2O_2:

$$2\,O_2^- + 2\,H^+ \rightarrow H_2O_2 + O_2$$

However, H_2O_2 (hydrogen peroxide), formed by the reaction above and by leakage of electrons out of the respiratory chain, is a powerful oxidizing agent and also harmful to cells. It is metabolized further by catalases and peroxidases. A catalase catalyses the reaction

$$2\,H_2O_2 \rightarrow 2\,H_2O + O_2$$

and a peroxidase reduces hydrogen peroxide to water by oxidizing an organic molecule. For example, the enzyme glutathione peroxidase catalyses the reaction

$$2\text{ glutathione}_{\text{red}} + H_2O_2 \rightarrow \text{glutathione}_{\text{ox}} + 2\,H_2O$$

There is growing evidence for the involvement of the damage caused by reactive oxygen species (ROS), such as O_2^-, H_2O_2, and ·OH (the hydroxyl radical), in the mechanism of ageing and in the development of cardiovascular disease, cancer, stroke, inflammatory disease, and other conditions. For this reason, much effort has been expended on studies of the biochemistry of *antioxidants*, substances that can either deactivate ROS directly (as glutathione does) or halt the progress of cellular damage through reactions with radicals formed by processes initiated by ROS. Important examples of antioxidants are vitamin C (ascorbic acid), vitamin E (α-tocopherol), and uric acid.

Nitric oxide (nitrogen monoxide, NO) is a small molecule that diffuses quickly between cells, carrying chemical messages that help initiate a variety of processes, such as regulation of blood pressure, inhibition of platelet aggregation, and defence against inflammation and attacks to the immune system. The

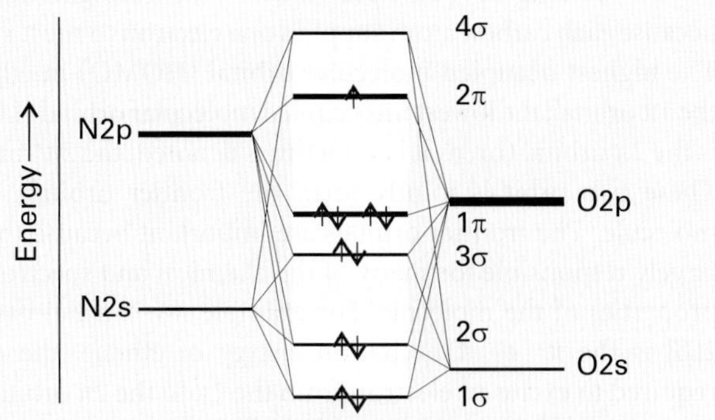

Fig. 5.34 The molecular orbital energy level diagram for NO.

molecule is synthesized from the amino acid arginine in a series of reactions catalysed by nitric oxide synthase and requiring O_2 and NADPH.

Figure 5.34 shows the bonding scheme in NO and illustrates a number of points we have made about heteronuclear diatomic molecules. The ground configuration is $1\sigma^2 2\sigma^2 3\sigma^2 1\pi^4 2\pi^1$. The 3σ and 1π orbitals are predominantly of O character as that is the more electronegative element. The highest energy occupied orbital is 2π, contains one electron, and has more N character than O character. It follows that NO is a radical with an unpaired electron that can be regarded as localized more on the N atom than on the O atom. The unoccupied orbital of lowest energy is 4σ, which is also localized predominantly on N.

Because NO is a radical, we expect it to be reactive. Its half-life is estimated at approximately 1–5 s, so it needs to be synthesized often in the cell. As we saw above, there is a biochemical price to be paid for the reactivity of biological radicals. Like O_2, NO participates in some reactions that are not beneficial to the cell. Indeed, the radicals O_2^- and NO combine to form the peroxynitrite ion:

$$NO\cdot + O_2^-\cdot \rightarrow ONOO^-$$

where we have shown the unpaired electrons explicitly. The peroxynitrite ion is a reactive oxygen species that damages proteins, DNA, and lipids, possibly leading to heart disease, amyotrophic lateral sclerosis (Lou Gehrig's disease), Alzheimer's disease, and multiple sclerosis. Note that the structure of the ion is consistent with the bonding scheme in Fig. 5.34: because the unpaired electron in NO is slightly more localized on the N atom, we expect that atom to form a bond with an O atom from the O_2^- ion.

Polyatomic molecules: the Hückel approximation

The molecular orbitals of polyatomic molecules are built in the same way as in diatomic molecules, the only difference being that we use more atomic orbitals to construct them. As for diatomic molecules, polyatomic molecular orbitals spread over the entire molecule. A molecular orbital has the general form

$$\psi = \sum_i c_i \chi_i \qquad (5.34)$$

where χ_i is an atomic orbital and the sum extends over all the valence orbitals of all the atoms in the molecule. The atomic orbitals used in this LCAO constitute the **basis set** of the calculation. To find the coefficients c_i, we set up the secular equations and the secular determinant, just as for diatomic molecules, solve the latter for the energies, and then use these energies in the secular equations to find the coefficients of the atomic orbitals for each molecular orbital.

The principal difference between diatomic and polyatomic molecules lies in the greater range of shapes that are possible: a diatomic molecule is necessarily linear, but a triatomic molecule, for instance, may be either linear or angular with a characteristic bond angle. The shape of a polyatomic molecule—the specification of its bond lengths and its bond angles—can be predicted by calculating the total energy of the molecule for a variety of nuclear positions, and then identifying the conformation that corresponds to the lowest energy.

Molecular orbital theory takes large molecules and extended aggregates of atoms, such as solid materials, into its stride. In this chapter we shall consider **conjugated hydrocarbons**, in which there is an alternation of single and double bonds along a chain of carbon atoms. Although the classification of an orbital as σ or π is strictly valid only in linear molecules, as will be familiar from introductory chemistry courses, it is also used to denote the local symmetry with respect to a given A—B bond axis. In Chapter 6, we shall develop more sophisticated techniques that are applicable to any molecule and form the basis for modern computational techniques in chemistry.

The π molecular orbital energy level diagrams of conjugated hydrocarbons can be constructed using a set of approximations suggested by Erich Hückel in 1931. In his approach, the π orbitals are treated separately from the σ orbitals, and the latter form a rigid framework that determines the general shape of the molecule. All the C atoms are treated identically, so all the Coulomb integrals α for the atomic orbitals that contribute to the π orbitals are set equal. For example, in ethene, we take the σ bonds as fixed, and concentrate on finding the energies of the single π bond and its companion antibond.

5.6 Ethene

We express the π orbitals as LCAOs of the C2p orbitals that lie perpendicular to the molecular plane. In ethene, for instance, we would write

$$\psi = c_A A + c_B B \qquad (5.35)$$

where the A is a C2p orbital on atom A, and so on. Next, the optimum coefficients and energies are found by the variation principle as explained in Section 5.5. That is, we have to solve the secular determinant, which in the case of ethene is eqn 5.28 with $\alpha_A = \alpha_B = \alpha$:

$$\begin{vmatrix} \alpha - E & \beta - ES \\ \beta - ES & \alpha - E \end{vmatrix} = 0 \qquad (5.36)$$

The roots of this determinant can be found very easily (they are the same as those in the first *illustration* from Section 5.5c). In a modern computation all the resonance integrals and overlap integrals would be included, but an indication of the molecular orbital energy level diagram can be obtained very readily if we make the following additional **Hückel approximations**:

- All overlap integrals are set equal to zero.
- All resonance integrals between non-neighbours are set equal to zero.
- All remaining resonance integrals are set equal (to β).

These approximations are obviously very severe, but they let us calculate at least a general picture of the molecular orbital energy levels with very little work. The assumptions result in the following structure of the secular determinant:

- All diagonal elements: $\alpha - E$.
- Off-diagonal elements between neighbouring atoms: β.
- All other elements: 0.

These approximations lead in the case of ethene to

$$\begin{vmatrix} \alpha - E & \beta \\ \beta & \alpha - E \end{vmatrix} = (\alpha - E)^2 - \beta^2 = 0 \qquad (5.37)$$

The roots of the equation are

$$E_{\pm} = \alpha \pm \beta \qquad (5.38)$$

The + sign corresponds to the bonding combination (β is negative) and the – sign corresponds to the antibonding combination (Fig. 5.35). We see the effect of neglecting overlap by comparing this result with eqn 5.30.

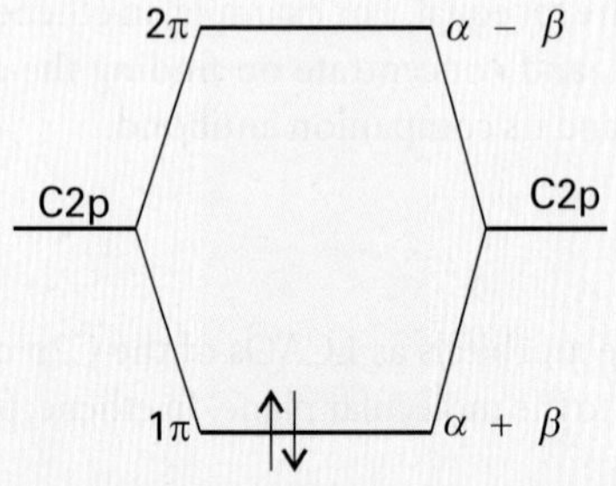

Fig. 5.35 The Hückel molecular orbital energy levels of ethene. Two electrons occupy the lower π orbital.

The building-up principle leads to the configuration $1\pi^2$, because each carbon atom supplies one electron to the π system. The **highest occupied molecular orbital** (HOMO) in ethene is the 1π orbital; the **lowest unoccupied molecular orbital** (LUMO) is the 2π orbital (or, as it is sometimes denoted, the 2π* orbital). These two orbitals jointly form the **frontier orbitals** of the molecule. The frontier orbitals are important because they are largely responsible for many of the chemical and spectroscopic properties of the molecule. For example, we can estimate that $2|\beta|$ is the $\pi^* \leftarrow \pi$ excitation energy of ethene, the energy required to excite an electron from the 1π to the 2π orbital. The constant β is often left as an adjustable parameter; an approximate value for π bonds formed from overlap of two C2p atomic orbitals is about −2.4 eV (−230 kJ mol^{-1}).

5.7 The matrix formulation of the Hückel method

In preparation for making Hückel theory more sophisticated and readily applicable to bigger molecules, we need to reformulate it in terms of matrices (see *Mathematical background* 5). We have seen that the secular equations that we have to solve for a two-atom system have the form

$$(H_{AA} - E_i S_{AA})c_{i,A} + (H_{AB} - E_i S_{AB})c_{i,B} = 0 \qquad (5.39a)$$

$$(H_{BA} - E_i S_{BA})c_{i,A} + (H_{BB} - E_i S_{BB})c_{i,B} = 0 \qquad (5.39b)$$

where $H_{ij} = \int \chi_i^* \hat{H} \chi_j \,d\tau$, $S_{ij} = \int \chi_i^* \chi_j \,d\tau$, and the eigenvalue E_i is that for a wavefunction of the form $\psi_i = c_{i,A}A + c_{i,B}B$. (These expressions generalize eqn 5.24, for which $S_{ii} = 1$.) There are two atomic orbitals, two eigenvalues, and two wavefunctions, so there are two pairs of secular equations, with the first corresponding to E_1 and ψ_1:

$$(H_{AA} - E_1 S_{AA})c_{1,A} + (H_{AB} - E_1 S_{AB})c_{1,B} = 0 \qquad (5.40a)$$

$$(H_{BA} - E_1 S_{BA})c_{1,A} + (H_{BB} - E_1 S_{BB})c_{1,B} = 0 \qquad (5.40b)$$

and another corresponding to E_2 and ψ_2:

$$(H_{AA} - E_2 S_{AA})c_{2,A} + (H_{AB} - E_2 S_{AB})c_{2,B} = 0 \qquad (5.40c)$$

$$(H_{BA} - E_2 S_{BA})c_{2,A} + (H_{BB} - E_2 S_{BB})c_{2,B} = 0 \qquad (5.40d)$$

The four expressions in eqn 5.40 can all be written as a single equation if we introduce the following four matrices

$$H = \begin{pmatrix} H_{AA} & H_{AB} \\ H_{BA} & H_{BB} \end{pmatrix} \quad S = \begin{pmatrix} S_{AA} & S_{AB} \\ S_{BA} & S_{BB} \end{pmatrix}$$
$$C = \begin{pmatrix} c_{1,A} & c_{2,A} \\ c_{1,B} & c_{2,B} \end{pmatrix} \quad E = \begin{pmatrix} E_1 & 0 \\ 0 & E_2 \end{pmatrix} \qquad (5.41)$$

where H is the **hamiltonian matrix** and S is the **overlap matrix**. Then the entire set of equations we have to solve can be expressed as

$$HC = SCE \qquad (5.42)$$

Self-test 5.6 Show by carrying out the necessary matrix multiplications that eqn 5.42 summarizes all four equations in eqn 5.40.

In the Hückel approximation, $H_{AA} = H_{BB} = \alpha$, $H_{AB} = H_{BA} = \beta$, and we neglect overlap, setting $S = 1$, the unit matrix (with 1 on the principal diagonal and 0 elsewhere). Then

$$HC = CE \qquad (5.43)$$

At this point, we multiply from the left by C^{-1}, the inverse of the matrix C, use $C^{-1}C = 1$, and find

$$C^{-1}HC = E \qquad (5.44)$$

In other words, to find the eigenvalues E_i, we have to find a transformation of H that makes it diagonal. This procedure is called **matrix diagonalization**. The diagonal elements then correspond to the eigenvalues E_i and the columns of the matrix C that brings about this diagonalization are the coefficients of the members of the basis set used in the calculation, and hence give us the composition of the molecular orbitals. If there are N orbitals in the basis set (there are only two in our example), then we have to diagonalize an $N \times N$ hamiltonian matrix H. Needless to say, all such heavy numerical work is done on a computer.

Example 5.4 *Finding the molecular orbitals by matrix diagonalization*

Set up and solve the matrix equations within the Hückel approximation for the π orbitals of butadiene (3).

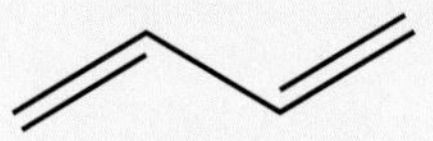

3 Butadiene

Method The matrices will be four-dimensional for this four-atom system with a basis set of four atomic orbitals, one on each atom. Ignore overlap, and construct the matrix H by using the Hückel parameters α and β. Find the matrix C that diagonalizes H: for this step, use mathematical software. Full details are given in *Mathematical background* 5.

Solution The hamiltonian matrix is

$$H = \begin{pmatrix} H_{11} & H_{12} & H_{13} & H_{14} \\ H_{21} & H_{22} & H_{23} & H_{24} \\ H_{31} & H_{32} & H_{33} & H_{34} \\ H_{41} & H_{42} & H_{43} & H_{44} \end{pmatrix} = \begin{pmatrix} \alpha & \beta & 0 & 0 \\ \beta & \alpha & \beta & 0 \\ 0 & \beta & \alpha & \beta \\ 0 & 0 & \beta & \alpha \end{pmatrix}$$

Mathematical software then diagonalizes this matrix to

$$E = \begin{pmatrix} \alpha + 1.62\beta & 0 & 0 & 0 \\ 0 & \alpha + 0.62\beta & 0 & 0 \\ 0 & 0 & \alpha - 0.62\beta & 0 \\ 0 & 0 & 0 & \alpha - 1.62\beta \end{pmatrix}$$

and the matrix that achieves the diagonalization is

$$C = \begin{pmatrix} 0.372 & 0.602 & 0.602 & -0.372 \\ 0.602 & 0.372 & -0.372 & 0.602 \\ 0.602 & -0.372 & -0.372 & -0.602 \\ 0.372 & -0.602 & 0.602 & 0.372 \end{pmatrix}$$

We can conclude that the energies and molecular orbitals are

$$E_1 = \alpha + 1.62\beta \quad \psi_1 = 0.372\chi_A + 0.602\chi_B + 0.602\chi_C + 0.372\chi_D$$
$$E_2 = \alpha + 0.62\beta \quad \psi_2 = 0.602\chi_A + 0.372\chi_B - 0.372\chi_C - 0.602\chi_D$$
$$E_3 = \alpha - 0.62\beta \quad \psi_3 = 0.602\chi_A - 0.372\chi_B - 0.372\chi_C + 0.602\chi_D$$
$$E_4 = \alpha - 1.62\beta \quad \psi_4 = -0.372\chi_A + 0.602\chi_B - 0.602\chi_C + 0.372\chi_D$$

where the C2p atomic orbitals are denoted by $\chi_A, \ldots, \chi_D$. Note that the orbitals are mutually orthogonal and, with overlap neglected, normalized.

Self-test 5.7 Repeat the exercise for the allyl radical, $\cdot CH_2$—CH=CH_2.

[$E = \alpha + 2^{1/2}\beta, \alpha, \alpha - 2^{1/2}\beta$; $\psi_1 = \frac{1}{2}\chi_A + (\frac{1}{2})^{1/2}\chi_B + \frac{1}{2}\chi_C$; $\psi_2 = (\frac{1}{2})^{1/2}\chi_A - (\frac{1}{2})^{1/2}\chi_C$; $\psi_3 = \frac{1}{2}\chi_A - (\frac{1}{2})^{1/2}\chi_B + \frac{1}{2}\chi_C$]

5.8 Butadiene and π-electron binding energy

As we saw in the preceding example, the energies of the four LCAO-MOs for butadiene are

$$E = \alpha \pm 1.62\beta, \qquad \alpha \pm 0.62\beta \qquad (5.45)$$

These orbitals and their energies are drawn in Fig. 5.36. Note that, the greater the number of internuclear nodes, the higher the energy of the orbital. There are four electrons to accommodate, so the ground-state configuration is $1\pi^2 2\pi^2$. The frontier orbitals of butadiene are the 2π orbital (the HOMO, which is largely bonding) and the 3π orbital (the LUMO, which is largely antibonding). 'Largely bonding' means that an orbital has both bonding and antibonding interactions between various neighbours, but the bonding effects dominate. 'Largely antibonding' indicates that the antibonding effects dominate.

An important point emerges when we calculate the total **π-electron binding energy**, E_π, the sum of the energies of each π electron, and compare it with what we find in ethene. In ethene the total energy is

$$E_\pi = 2(\alpha + \beta) = 2\alpha + 2\beta$$

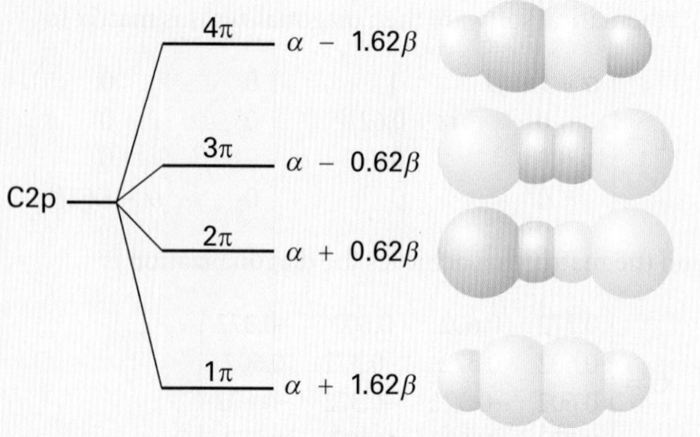

Fig. 5.36 The Hückel molecular orbital energy levels of butadiene and the top view of the corresponding π orbitals. The four p electrons (one supplied by each C) occupy the two lower π orbitals. Note that the orbitals are delocalized.

In butadiene it is

$$E_\pi = 2(\alpha + 1.62\beta) + 2(\alpha + 0.62\beta) = 4\alpha + 4.48\beta$$

Therefore, the energy of the butadiene molecule lies lower by 0.48β (about 110 kJ mol^{-1}) than the sum of two individual π bonds. This extra stabilization of a conjugated system is called the **delocalization energy**. A closely related quantity is the **π-bond formation energy**, E_{bf}, the energy released when a π bond is formed. Because the contribution of α is the same in the molecule as in the atoms, we can find the π-bond formation energy from the π-electron binding energy by writing

$$E_{bf} = E_\pi - N\alpha \tag{5.46}$$

where N is the number of carbon atoms in the molecule. The π-bond formation energy in butadiene, for instance, is 4.48β.

Example 5.5 *Estimating the delocalization energy*

Use the Hückel approximation to find the energies of the π orbitals of cyclobutadiene, and estimate the delocalization energy.

Method Set up the secular determinant using the same basis as for butadiene, but note that atoms A and D are also now neighbours. Then solve for the roots of the secular equation and assess the total π-bond energy. For the delocalization energy, subtract from the total π-bond energy the energy of two π bonds.

Answer The hamiltonian matrix is

$$H = \begin{pmatrix} \alpha & \beta & 0 & \beta \\ \beta & \alpha & \beta & 0 \\ 0 & \beta & \alpha & \beta \\ \beta & 0 & \beta & \alpha \end{pmatrix}$$

Diagonalization gives the energies of the orbitals as

$$E = \alpha + 2\beta, \quad \alpha, \quad \alpha, \quad \alpha - 2\beta$$

Four electrons must be accommodated. Two occupy the lowest orbital (of energy $\alpha + 2\beta$), and two occupy the doubly degenerate orbitals (of energy α). The total energy is therefore $4\alpha + 4\beta$. Two isolated π bonds would have an energy $4\alpha + 4\beta$; therefore, in this case, the delocalization energy is zero.

Self-test 5.8 Repeat the calculation for benzene.

[See next subsection]

5.9 Benzene and aromatic stability

The most notable example of delocalization conferring extra stability is benzene and the aromatic molecules based on its structure. Benzene is often expressed in a mixture of valence-bond and molecular orbital terms, with typically valence-bond language used for its σ framework and molecular orbital language used to describe its π electrons.

First, the valence-bond component. The six C atoms are regarded as sp^2 hybridized, with a single unhybridized perpendicular 2p orbital. One H atom is bonded by (Csp^2, H1s) overlap to each C atom, and the remaining hybrids overlap to give a regular hexagon of atoms (Fig. 5.37). The internal angle of a regular hexagon is 120°, so sp^2 hybridization is ideally suited for forming σ bonds. We see that benzene's hexagonal shape permits strain-free σ bonding.

Now consider the molecular orbital component of the description. The six C2p orbitals overlap to give six π orbitals that spread

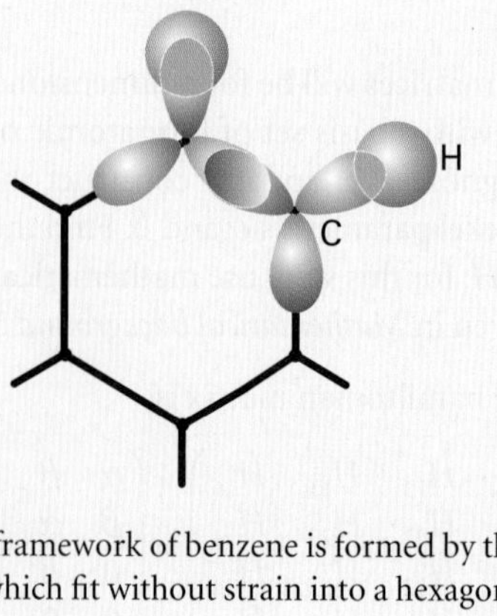

Fig. 5.37 The σ framework of benzene is formed by the overlap of Csp^2 hybrids, which fit without strain into a hexagonal arrangement.

all round the ring. Their energies are calculated within the Hückel approximation by diagonalizing the hamiltonian matrix

$$H = \begin{pmatrix} \alpha & \beta & 0 & 0 & 0 & \beta \\ \beta & \alpha & \beta & 0 & 0 & 0 \\ 0 & \beta & \alpha & \beta & 0 & 0 \\ 0 & 0 & \beta & \alpha & \beta & 0 \\ 0 & 0 & 0 & \beta & \alpha & \beta \\ \beta & 0 & 0 & 0 & \beta & \alpha \end{pmatrix}$$

The MO energies, the eigenvalues of this matrix, are simply

$$E = \alpha \pm 2\beta, \alpha \pm \beta, \alpha \pm \beta \qquad (5.47)$$

as shown in Fig. 5.38. The orbitals there have been given symmetry labels, which we explain in Chapter 7. Note that the lowest energy orbital is bonding between all neighbouring atoms, the highest energy orbital is antibonding between each pair of neighbours, and the intermediate orbitals are a mixture of bonding, nonbonding, and antibonding character between adjacent atoms.

We now apply the building-up principle to the π system. There are six electrons to accommodate (one from each C atom), so the three lowest orbitals (a_{2u} and the doubly degenerate pair e_{1g}) are fully occupied, giving the ground-state configuration $a_{2u}^2 e_{1g}^4$. A significant point is that the only molecular orbitals occupied are those with net bonding character.

The π-electron energy of benzene is

$$E_\pi = 2(\alpha + 2\beta) + 4(\alpha + \beta) = 6\alpha + 8\beta$$

If we ignored delocalization and thought of the molecule as having three isolated π bonds, it would be ascribed a π-electron energy of only $3(2\alpha + 2\beta) = 6\alpha + 6\beta$. The delocalization energy is therefore $2\beta \approx -460$ kJ mol^{-1}, which is considerably more than for butadiene. The π-bond formation energy in benzene is 8β.

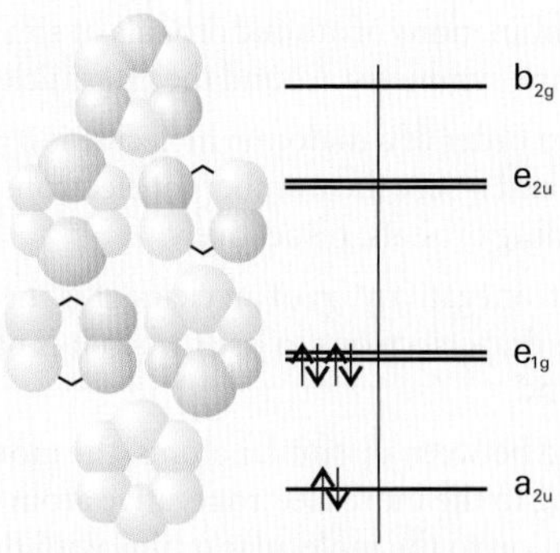

Fig. 5.38 The Hückel orbitals of benzene and the corresponding energy levels. The symmetry labels are explained in Chapter 7. The bonding and antibonding character of the delocalized orbitals reflects the numbers of nodes between the atoms. In the ground state, only the bonding orbitals are occupied.

This discussion suggests that aromatic stability can be traced to two main contributions. First, the shape of the regular hexagon is ideal for the formation of strong σ bonds: the σ framework is relaxed and without strain. Second, the π orbitals are such as to be able to accommodate all the electrons in bonding orbitals, and the delocalization energy is large.

The theoretical principles we have developed here are expanded and applied in subsequent chapters. In Chapter 6, we see how more sophisticated computational techniques based on the variation principle can lead to the prediction of molecular structure and reactivity. In Chapter 9, we apply the matrix formalism of the Hückel method to a discussion of the structure of solids.

Checklist of key ideas

☐ 1. In the Born–Oppenheimer approximation, nuclei are treated as stationary while electrons move around them.

☐ 2. In valence-bond theory (VB theory), a bond is regarded as forming when an electron in an atomic orbital on one atom pairs its spin with that of an electron in an atomic orbital on another atom.

☐ 3. A valence-bond wavefunction with cylindrical symmetry around the internuclear axis is a σ bond. A π bond arises from the merging of two p orbitals that approach side-by-side and the pairing of electrons that they contain.

☐ 4. Hybrid orbitals are mixtures of atomic orbitals on the same atom and are invoked in VB theory to explain molecular geometries.

☐ 5. In molecular orbital theory (MO theory), electrons are treated as spreading throughout the entire molecule.

☐ 6. A bonding orbital is a molecular orbital that, if occupied, contributes to the strength of a bond between two atoms. An antibonding orbital is a molecular orbital that, if occupied, decreases the strength of a bond between two atoms.

☐ 7. A σ molecular orbital has zero orbital angular momentum about the internuclear axis. A π molecular orbital has one unit of angular momentum around the internuclear axis; in a linear molecule, it has a nodal plane that includes the internuclear axis.

☐ 8. The electronic configurations of homonuclear diatomic molecules are shown in Figs. 5.29 and 5.31.

☐ 9. When constructing molecular orbitals, we need to consider only combinations of atomic orbitals of similar energies and of the same symmetry around the internuclear axis.

☐ 10. The bond order of a diatomic molecule is $b=\frac{1}{2}(n-n^*)$, where n and n^* are the numbers of electrons in bonding and antibonding orbitals, respectively.

☐ 11. The electronegativity, χ, of an element is the power of its atoms to draw electrons to itself when it is part of a compound.

☐ 12. In a bond between dissimilar atoms, the atomic orbital belonging to the more electronegative atom makes the larger contribution to the molecular orbital with the lowest energy. For the molecular orbital with the highest energy, the principal contribution comes from the atomic orbital belonging to the less electronegative atom.

☐ 13. The hamiltonian matrix, $\boldsymbol{H}$, is formed of all integrals $H_{ij}=\int\psi_i^*\hat{H}\psi_j\mathrm{d}\tau$. The overlap matrix, $\boldsymbol{S}$, is formed of all $S_{ij}=\int\psi_i^*\psi_j\mathrm{d}\tau$.

☐ 14. The variation principle states that, if an arbitrary wavefunction is used to calculate the energy, the value calculated is never less than the true energy.

☐ 15. In the Hückel method, all Coulomb integrals H_{ii} are set equal (to α), all overlap integrals S_{ij} for $i\neq j$ are set equal to zero, all resonance integrals H_{ij} between non-neighbours are set equal to zero, and all remaining resonance integrals are set equal (to β).

☐ 16. The π-electron binding energy is the sum of the energies of each π electron. The π-bond formation energy is the energy released when a π bond is formed. The delocalization energy is the extra stabilization of a conjugated system.

Discussion questions

5.1 State and compare the approximations on which the valence-bond and molecular orbital theories are based.

5.2 Discuss the role of the Born–Oppenheimer approximation in the calculation of a molecular potential energy curve or surface.

5.3 Write the Lewis structure of the peroxynitrite ion, ONOO—. Label each atom with its state of hybridization and specify the composition of each of the different types of bond.

5.4 Describe the various types of hybrid orbitals and how they are used to describe the bonding in alkanes, alkenes, and alkynes. How does hybridization explain that in allene, $CH_2{=}C{=}CH_2$, the two CH_2 groups lie in perpendicular planes?

5.5 Describe the Pauling and Mulliken electronegativity scales. Why should they be approximately in step?

5.6 Discuss the steps involved in the calculation of the energy of a system by using the variation principle. Are any assumptions involved?

5.7 Discuss the scope, consequences, and limitations of the approximations on which the Hückel method is based.

5.8 Distinguish between delocalization energy, π-electron binding energy, and π-bond formation energy. Explain how each concept is employed.

5.9 Use concepts of molecular orbital theory to describe the chemical and biochemical reactivity of O_2, N_2, and NO.

5.10 Draw diagrams to show the various orientations in which a p orbital and a d orbital on adjacent atoms may form bonding and antibonding molecular orbitals.

Exercises

5.1(a) Describe the structure of a P_2 molecule in valence-bond terms. Why is P_4 a more stable form of molecular phosphorus?

5.1(b) Describe the structures of SO_2 and SO_3 in terms of valence-bond theory.

5.2(a) Describe the bonding in 1,3-butadiene using hybrid orbitals.

5.2(b) Describe the bonding in 1,3-pentadiene using hybrid orbitals.

5.3(a) Show that the linear combinations $h_1=\mathrm{s}+\mathrm{p}_x+\mathrm{p}_y+\mathrm{p}_z$ and $h_2=\mathrm{s}-\mathrm{p}_x-\mathrm{p}_y+\mathrm{p}_z$ are mutually orthogonal.

5.3(b) Show that the linear combinations $h_1=(\sin\zeta)\mathrm{s}+(\cos\zeta)\mathrm{p}$ and $h_2=(\cos\zeta)\mathrm{s}-(\sin\zeta)\mathrm{p}$ are mutually orthogonal for all values of the angle ζ.

5.4(a) Normalize the sp^2 hybrid orbital $h=\mathrm{s}+2^{1/2}\mathrm{p}$ given that the s and p orbitals are each normalized to 1.

5.4(b) Normalize the linear combinations in Exercise 5.3b given that the s and p orbitals are each normalized to 1.

5.5(a) Give the ground-state electron configurations and bond orders of (a) Li_2, (b) Be_2, and (c) C_2.

5.5(b) Give the ground-state electron configurations of (a) F_2^-, (b) N_2, and (c) O_2^{2-}.

5.6(a) Give the ground-state electron configurations of (a) CO, (b) NO, and (c) CN^-.

5.6(b) Give the ground-state electron configurations of (a) XeF, (b) PN, and (c) O_2^-.

5.7(a) From the ground-state electron configurations of B_2 and C_2, predict which molecule should have the greater bond dissociation energy.

5.7(b) From the ground-state electron configurations of Li_2 and Be_2, predict which molecule should have the greater bond dissociation energy.

5.8(a) Which of the molecules N_2, NO, O_2, C_2, F_2, and CN would you expect to be stabilized by the addition of an electron to form AB^-?

5.8(b) Which of the molecules N_2, NO, O_2, C_2, F_2, and CN would you expect to be stabilized by the removal of an electron to form AB^+?

5.9(a) Sketch the molecular orbital energy level diagram for XeF and deduce its ground-state electron configuration. Is XeF likely to have a shorter bond length than XeF^+?

5.9(b) Sketch the molecular orbital energy level diagram for IF and deduce its ground-state electron configuration. Is IF likely to have a shorter bond length than IF^- or IF^+?

5.10(a) Use the electron configurations of NO^- and O_2^+ to predict which is likely to have the shorter bond length.

5.10(b) Arrange the species O_2^+, O_2, O_2^-, O_2^{2-} in order of increasing bond length.

5.11(a) For each of the species in Exercise 5.10b, specify which molecular orbital is the HOMO.

5.11(b) For each of the species in Exercise 5.10b, specify which molecular orbital is the LUMO.

5.12(a) Give the parities of the wavefunctions for the first four levels of a harmonic oscillator. How may the parity be expressed in terms of the quantum number v?

5.12(b) State the parities of the six π orbitals of benzene.

5.13(a) Normalize the molecular orbital $\psi = \psi_A + \lambda\psi_B$ in terms of the parameter λ and the overlap integral S.

5.13(b) A better description of the molecule in Exercise 5.13a might be obtained by including more orbitals on each atom in the linear combination. Normalize the molecular orbital $\psi = \psi_A + \lambda\psi_B + \lambda'\psi_B'$ in terms of the parameters λ and λ' and the appropriate overlap integrals S, where ψ_B and ψ_B' are mutually orthogonal orbitals on atom B.

5.14(a) Suppose that a molecular orbital has the (unnormalized) form $0.145A + 0.844B$. Find a linear combination of the orbitals A and B that is orthogonal to this combination and determine the normalization constants of both combinations using $S = 0.250$.

5.14(b) Suppose that a molecular orbital has the (unnormalized) form $0.727A + 0.144B$. Find a linear combination of the orbitals A and B that is orthogonal to this combination and determine the normalization constants of both combinations using $S = 0.117$.

5.15(a) What is the speed of a photoelectron ejected from an orbital of ionization energy 12.0 eV by a photon of radiation of wavelength 100 nm?

5.15(b) What is the speed of a photoelectron ejected from a molecule with radiation of energy 21 eV and known to come from an orbital of ionization energy 12 eV?

5.16(a) A reasonably reliable conversion between the Mulliken and Pauling electronegativity scales is given in Section 5.5b. Using Table 5.4, comment on how good the conversion formula is for Period 2 elements.

5.16(b) A reasonably reliable conversion between the Mulliken and Pauling electronegativity scales is given in Section 5.5b. Using Table 5.4, comment on how good the conversion formula is for Period 3 elements.

5.17(a) The languages of valence-bond theory and molecular orbital theory are commonly combined when discussing unsaturated organic compounds. Construct the molecular orbital energy level diagrams of ethene on the basis that the molecule is formed from the appropriately hybridized CH_2 or CH fragments.

5.17(b) The languages of valence-bond theory and molecular orbital theory are commonly combined when discussing unsaturated organic compounds. Construct the molecular orbital energy level diagrams of ethyne (acetylene) on the basis that the molecule is formed from the appropriately hybridized CH_2 or CH fragments.

5.18(a) Write down the secular determinants for (a) linear H_3, (b) cyclic H_3 within the Hückel approximation.

5.18(b) Write down the secular determinants for (a) butadiene, (b) cyclobutadiene within the Hückel approximation.

5.19(a) Predict the electron configurations of (a) the benzene anion, (b) the benzene cation. Estimate the π-electron binding energy in each case.

5.19(b) Predict the electron configurations of (a) the allyl radical, (b) the cyclobutadiene cation. Estimate the π-electron binding energy in each case.

5.20(a) Compute the delocalization energy and π-bond formation energy of (a) the benzene anion, (b) the benzene cation.

5.20(b) Compute the delocalization energy and π-bond formation energy of (a) the ally radical, (b) the cyclobutadiene cation.

5.21(a) Write down the secular determinants for (a) anthracene (**4**), (b) phenanthrene (**5**) within the Hückel approximation and using the C2p orbitals as the basis set.

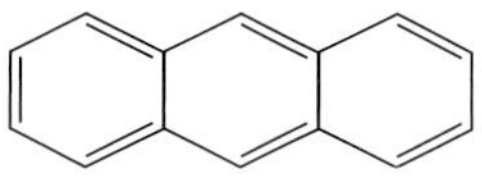

4 Anthracene

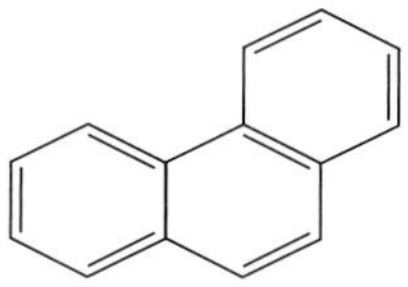

5 Phenanthrene

5.21(b) Write down the secular determinants for (a) azulene (**6**), (b) acenaphthalene (**7**) within the Hückel approximation and using the C2p orbitals as the basis set.

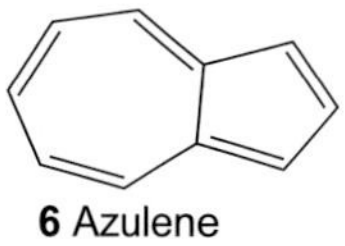

6 Azulene

7 Acenaphthalene

5.22(a) Use mathematical software to estimate the π-electron binding energy of (a) anthracene (**4**), (b) phenanthrene (**5**) within the Hückel approximation.

5.22(b) Use mathematical software to estimate the π-electron binding energy of (a) azulene (**6**), (b) acenaphthalene (**7**) within the Hückel approximation.

Problems*

Numerical problems

5.1 Calculate the molar energy of repulsion between two hydrogen nuclei at the separation in H_2 (74.1 pm). The result is the energy that must be overcome by the attraction from the electrons that form the bond. Does the gravitational attraction between them play any significant role? *Hint.* The gravitational potential energy of two masses is equal to $-Gm_1m_2/r$ where $G = 6.673 \times 10^{-11}$ Nm^2 kg^{-2} is the gravitational constant.

5.2 Show that if a wave cos *kx* centred on A (so that *x* is measured from A) interferes with a similar wave cos *k′x* centred on B (with *x* measured from B) a distance *R* away, then constructive interference occurs in the intermediate region when $k = k' = \pi/2R$ and destructive interference if $kR = \frac{1}{2}\pi$ and $k'R = \frac{3}{2}\pi$.

5.3 The overlap integral between two H1s orbitals on nuclei separated by a distance *R* is $S = \{1 + (R/a_0) + \frac{1}{3}(R/a_0)^2\}e^{-R/a_0}$. Plot this expression as a function of *R* and account for its variation.

5.4 Before doing the calculation below, sketch how the overlap between an s orbital and a 2p orbital directed towards it can be expected to depend on their separation. The overlap integral between an H1s orbital and an H2p orbital directed towards it on nuclei separated by a distance *R* is $S = (R/a_0)\{1 + (R/a_0) + \frac{1}{3}(R/a_0)^2\}e^{-R/a_0}$. Plot this function, and find the separation for which the overlap is a maximum.

5.5 Calculate the total amplitude of the normalized bonding and antibonding LCAO-MOs that may be formed from two H1s orbitals at a separation of 106 pm. Plot the two amplitudes for positions along the molecular axis both inside and outside the internuclear region.

5.6 Repeat the calculation in Problem 5.5 but plot the probability densities of the two orbitals. Then form the *difference density*, the difference between ψ^2 and $\frac{1}{2}(\psi_A^2 + \psi_B^2)$.

5.7‡ Use the $2p_x$ and $2p_z$ hydrogenic atomic orbitals to construct simple LCAO descriptions of 2pσ and 2pπ molecular orbitals. (a) Make a probability density plot, and both surface and contour plots of the *xz*-plane amplitudes of the $2p_z\sigma$ and $2p_z\sigma^*$ molecular orbitals. (b) Make surface and contour plots of the *xz*-plane amplitudes of the $2p_x\pi$ and $2p_x\pi^*$ molecular orbitals. Include plots for both an internuclear distance, *R*, of $10a_0$ and $3a_0$, where $a_0 = 52.9$ pm. Interpret the graphs, and explain why this graphical information is useful.

5.8 Imagine a small electron-sensitive probe of volume 1.00 pm^3 inserted into an H_2^+ molecule-ion in its ground state. Calculate the probability that it will register the presence of an electron at the following positions: (a) at nucleus A, (b) at nucleus B, (c) halfway between A and B, (c) at a point 20 pm along the bond from A and 10 pm perpendicularly. Do the same for the molecule-ion the instant after the electron has been excited into the antibonding LCAO-MO.

5.9 The energy of H_2^+ with internuclear separation *R* is given by the expression

$$E = E_H + \frac{e^2}{4\pi\varepsilon_0 R} - \frac{V_1 + V_2}{1 + S}$$

where E_H is the energy of an isolated H atom, V_1 is the attractive potential energy between the electron centred on one nucleus and the charge of the other nucleus; V_2 is the attraction between the overlap density and one of the nuclei; *S* is the overlap integral. The values are given below. Plot the molecular potential energy curve and find the bond dissociation energy (in electronvolts) and the equilibrium bond length.

R/a_0	0	1	2	3	4
V_1/E_h	1.000	0.729	0.472	0.330	0.250
V_2/E_h	1.000	0.736	0.406	0.199	0.092
S	1.000	0.858	0.587	0.349	0.189

where $E_h = 27.2$ eV, $a_0 = 52.9$ pm, and $E_H = -\frac{1}{2}E_h$.

5.10 The same data as in Problem 5.9 may be used to calculate the molecular potential energy curve for the antibonding orbital, which is given by

$$E = E_H + \frac{e^2}{4\pi\varepsilon_0 R} - \frac{V_1 - V_2}{1 - S}$$

Plot the curve. Examine whether occupation of the bonding orbital with one electron (as calculated in the preceding problem) has a greater or lesser bonding effect than occupation of the antibonding orbital with one electron. Is that true at all internuclear separations?

5.11‡ The LCAO-MO approach described in the text can be used to introduce numerical methods needed in quantum chemistry. In this problem we evaluate the overlap, Coulomb, and resonance integrals numerically and compare the results with the analytical equations (eqns 5.12). (a) Use the LCAO-MO wavefunction and the H_2^+ hamiltonian to derive equations 5.11 and 5.12 for the relevant integrals and use mathematical software or an electronic spreadsheet to evaluate the overlap, Coulomb, and resonance integrals numerically, and the total energy for the $1s\sigma_g$ MO in the range $a_0 < R < 4a_0$. Compare the results obtained by numerical integration with results obtained analytically. (b) Use the results of the numerical integrations to draw a graph of the total energy, *E*(*R*), and determine the minimum of total energy, the equilibrium internuclear distance, and the dissociation energy (D_e).

5.12 In a particular photoelectron spectrum using 21.21 eV photons, electrons were ejected with kinetic energies of 11.01 eV, 8.23 eV, and 5.22 eV. Sketch the molecular orbital energy level diagram for the species, showing the ionization energies of the three identifiable orbitals.

5.13 Set up and solve the Hückel secular equations for the π electrons of CO_3^{2-}. Express the energies in terms of the Coulomb integrals α_O and α_C and the resonance integral β. Determine the delocalization energy of the ion.

5.14 In the 'free electron molecular orbital' (FEMO) theory, the electrons in a conjugated molecule are treated as independent particles in a box of length *L*. Sketch the form of the two occupied orbitals in butadiene predicted by this model and predict the minimum excitation energy of the molecule. The conjugated tetraene $CH_2{=}CHCH{=}CHCH{=}CHCH{=}CH_2$ can be treated as a box of length 8*R*, where $R \approx 140$ pm (as in this case, an extra half bond-length is often added at each end of the box). Calculate the minimum excitation energy of the molecule and sketch the HOMO and LUMO. Estimate the colour a sample of the compound is likely to appear in white light.

5.15 The FEMO theory (Problem 5.14) of conjugated molecules is rather crude and better results are obtained with simple Hückel theory. (a) For a linear conjugated polyene with each of *N* carbon atoms contributing an

* Problems denoted with the symbol ‡ were supplied by Charles Trapp, Carmen Giunta, and Marshall Cady.

electron in a 2p orbital, the energies E_k of the resulting π molecular orbitals are given by:

$$E_k = \alpha + 2\beta \cos\frac{k\pi}{N+1} \qquad k = 1, 2, 3, \ldots, N$$

Use this expression to determine a reasonable empirical estimate of the resonance integral β for the homologous series consisting of ethene, butadiene, hexatriene, and octatetraene given that $\pi\leftarrow\pi$ ultraviolet absorptions from the HOMO to the LUMO occur at 61 500, 46 080, 39 750, and 32 900 cm^{-1}, respectively. (b) Calculate the π-electron delocalization energy, $E_{deloc} = E_\pi - n(\alpha + \beta)$, of octatetraene, where E_π is the total π-electron binding energy and n is the total number of π electrons. (c) In the context of this Hückel model, the π molecular orbitals are written as linear combinations of the carbon 2p orbitals. The coefficient of the jth atomic orbital in the kth molecular orbital is given by:

$$c_{kj} = \left(\frac{2}{N+1}\right)^{1/2} \sin\frac{jk\pi}{N+1} \qquad j = 1, 2, 3, \ldots, N$$

Determine the values of the coefficients of each of the six 2p orbitals in each of the six π molecular orbitals of hexatriene. Match each set of coefficients (that is, each molecular orbital) with a value of the energy calculated with the expression given in part (a) of the molecular orbital. Comment on trends that relate the energy of a molecular orbital with its 'shape', which can be inferred from the magnitudes and signs of the coefficients in the linear combination that describes the molecular orbital.

5.16 For monocyclic conjugated polyenes (such as cyclobutadiene and benzene) with each of N carbon atoms contributing an electron in a 2p orbital, simple Hückel theory gives the following expression for the energies E_k of the resulting π molecular orbitals:

$$E_k = \alpha + 2\beta \cos\frac{2k\pi}{N} \qquad k = 0, \pm 1, \pm 2, \ldots, \pm N/2 \text{ (even } N)$$

$$k = 0, \pm 1, \pm 2, \ldots, \pm (N-1)/2 \text{ (odd } N)$$

(a) Calculate the energies of the π molecular orbitals of benzene and cyclooctatetraene. Comment on the presence or absence of degenerate energy levels. (b) Calculate and compare the delocalization energies of benzene (using the expression above) and hexatriene (see Problem 5.15). What do you conclude from your results? (c) Calculate and compare the delocalization energies of cyclooctatetraene and octatetraene. Are your conclusions for this pair of molecules the same as for the pair of molecules investigated in part (b)?

5.17 Set up the secular determinants for the molecules treated in Problem 5.15 and diagonalize them by using mathematical software. Use your results to show that the π molecular orbitals of linear polyenes obey the following rules:

- The π molecular orbital with lowest energy is delocalized over all carbon atoms in the chain.
- The number of nodal planes between C2p orbitals increases with the energy of the π molecular orbital.

5.18 Set up the secular determinants for cyclobutadiene, benzene, and cyclooctatetraene and diagonalize them by using mathematical software. Use your results to show that the π molecular orbitals of monocyclic polyenes with an even number of carbon atoms follow a pattern in which:

- The π molecular orbitals of lowest and highest energy are non-degenerate.
- The remaining π molecular orbitals exist as degenerate pairs.

5.19 Electronic excitation of a molecule may weaken or strengthen some bonds because bonding and antibonding characteristics differ between the HOMO and the LUMO. For example, a carbon–carbon bond in a linear polyene may have bonding character in the HOMO and antibonding character in the LUMO. Therefore, promotion of an electron from the HOMO to the LUMO weakens this carbon–carbon bond in the excited electronic state, relative to the ground electronic state. Consult Figs. 5.36 and 5.38 and discuss in detail any changes in bond order that accompany the $\pi\leftarrow\pi$ ultraviolet absorptions in butadiene and benzene.

Theoretical problems

5.20 An sp^2 hybrid orbital that lies in the xy-plane and makes an angle of 120° to the x-axis has the form

$$\psi = \frac{1}{3^{1/2}}\left(s - \frac{1}{2^{1/2}}p_x + \frac{3^{1/2}}{2^{1/2}}p_y\right)$$

Use hydrogenic atomic orbitals to write the explicit form of the hybrid orbital. Show that it has its maximum amplitude in the direction specified.

5.21 In Problems 5.9 and 5.10 you were invited to use a numerical procedure to explore the relative bonding and antibonding characteristics of molecular orbitals. In this problem, proceed as far as you can analytically by setting up an expression for $\frac{1}{2}(E(\text{antibond}) + E(\text{bond}))$ and explore its consequences. At some stage, you will have to proceed numerically. Show that the antibonding orbital is more antibonding than the bonding orbital is bonding at most internuclear separations. At what separation is that no longer true?

5.22 Show, if overlap is ignored, (a) that any molecular orbital expressed as a linear combination of two atomic orbitals may be written in the form $\psi = \psi_A \cos\theta + \psi_B \sin\theta$, where θ is a parameter that varies between 0 and π, and (b) that, if ψ_A and ψ_B are orthogonal and normalized to 1, then ψ is also normalized to 1. (c) To what values of θ do the bonding and antibonding orbitals in a homonuclear diatomic molecule correspond?

5.23 Derive eqns 5.11 and 5.14 by working with the normalized LCAO-MOs for the H_2^+ molecule-ion (Section 5.3a). Proceed by evaluating the expectation value of the hamiltonian for the ion. Make use of the fact that A and B each individually satisfy the Schrödinger equation for an isolated H atom.

5.24 Suppose that the function $\psi = Ae^{-ar^2}$, with A being the normalization constant and a being an adjustable parameter, is used as a trial wavefunction for the 1s orbital of the hydrogen atom. Show that

$$E = \frac{3a\hbar^2}{2\mu} - 2e^2\left(\frac{2a}{\pi}\right)^{1/2}$$

where e is the fundamental charge, and μ is the reduced mass for the H atom. What is the minimum energy associated with this trial wavefunction?

5.25 We saw in Section 5.5 that, to find the energies of the bonding and antibonding orbitals of a heteronuclear diatomic molecule, we need to solve the secular determinant

$$\begin{vmatrix} \alpha_A - E & \beta \\ \beta & \alpha_B - E \end{vmatrix} = 0$$

where $\alpha_A \neq \alpha_B$ and we have taken $S = 0$. Equations 5.33a and 5.33b give the general solution to this problem. Here, we shall develop the result for the case $(\alpha_B - \alpha_A)^2 \gg \beta^2$. (a) Begin by showing that

$$E_{\pm} = \frac{\alpha_A + \alpha_B}{2} \pm \frac{\alpha_A - \alpha_B}{2}\left[1 + \frac{4\beta^2}{(\alpha_A - \alpha_B)^2}\right]^{1/2}$$

where E_+ and E_- are the energies of the bonding and antibonding molecular orbitals, respectively. (b) Now use the expansion

$$(1+x)^{1/2} = 1 + \frac{x}{2} - \frac{x^3}{8} + \cdots$$

to show that

$$E_- = \alpha_B + \frac{\beta^2}{\alpha_B - \alpha_A} \qquad E_+ = \alpha_A - \frac{\beta^2}{\alpha_B - \alpha_A}$$

which is the limiting result used in *Justification 5.5*.

5.26 Explore the consequences of not setting $S = 0$ in the preceding problem.

5.27‡ Prove that for an open chain of N conjugated carbons the characteristic polynomial of the secular determinant (the polynomial obtained by expanding the determinant), $P_N(x)$, where $x = (\alpha - E)/\beta$, obeys the recurrence relation $P_N = xP_{N-1} - P_{N-2}$, with $P_1 = x$ and $P_0 = 1$.

5.28 You will recall from your previous study of chemistry that the standard potential is a measure of the thermodynamic tendency of an atom, ion, or molecule to accept an electron (see also Chapter 17). Studies indicate that there is a correlation between the LUMO energy and the standard potential of aromatic hydrocarbons. Do you expect the standard potential to increase or decrease as the LUMO energy decreases? Explain your answer.

Applications: to astrophysics and biology

5.29‡ In Exercise 5.18a you were invited to set up the Hückel secular determinant for linear and cyclic H_3. The same secular determinant applies to the molecular ions H_3^+ and D_3^+. The molecular ion H_3^+ was discovered as long ago as 1912 by J.J. Thomson but the equilateral triangular structure was confirmed by M.J. Gaillard *et al.* much more recently (*Phys. Rev.* **A17**, 1797 (1978)). The molecular ion H_3^+ is the simplest polyatomic species with a confirmed existence and plays an important role in chemical reactions occurring in interstellar clouds that may lead to the formation of water, carbon monoxide, and ethyl alcohol. The H_3^+ ion has also been found in the atmospheres of Jupiter, Saturn, and Uranus. (a) Solve the Hückel secular equations for the energies of the cyclic H_3 system in terms of the parameters α and β, draw an energy level diagram for the orbitals, and determine the binding energies of H_3^+, H_3, and H_3^-. (b) Accurate quantum mechanical calculations by G.D. Carney and R.N. Porter (*J. Chem. Phys.* **65**, 3547 (1976)) give the dissociation energy for the process $H_3^+ \rightarrow H + H + H^+$ as 849 kJ mol^{-1}. From this information and data in Table 5.3, calculate the enthalpy of the reaction $H^+(g) + H_2(g) \rightarrow H_3^+(g)$. (c) From your equations and the information given, calculate a value for the resonance integral β in H_3^+. Then go on to calculate the binding energies of the other H_3 species in (a).

5.30‡ There is some indication that other hydrogen ring compounds and ions in addition to H_3 and D_3 species may play a role in interstellar chemistry. According to J. S. Wright and G.A. DiLabio (*J. Phys. Chem.* **96**, 10793 (1992)), H_5^-, H_6, and H_7^+ are particularly stable whereas H_4 and H_5^+ are not. Confirm these statements by Hückel calculations.

5.31 Here we develop a molecular orbital theory treatment of the peptide group (**8**), which links amino acids in proteins, and establish the features that stabilize its planar conformation.

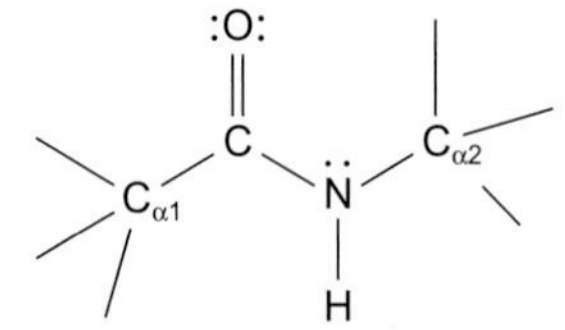

8 The peptide group

(a) It will be familiar from introductory chemistry that valence-bond theory explains the planar conformation by invoking delocalization of the π bond over the oxygen, carbon, and nitrogen atoms by resonance:

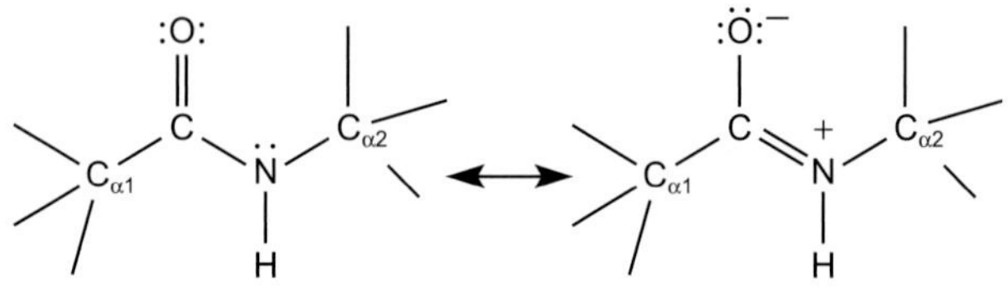

9 **10**

It follows that we can model the peptide group using molecular orbital theory by making LCAO-MOs from 2p orbitals perpendicular to the plane defined by the O, C, and N atoms. The three combinations have the form:

$$\psi_1 = a\psi_O + b\psi_C + c\psi_N \qquad \psi_2 = d\psi_O - e\psi_N \qquad \psi_3 = f\psi_O - g\psi_C + h\psi_N$$

where the coefficients a to h are all positive. Sketch the orbitals ψ_1, ψ_2, and ψ_3 and characterize them as bonding, non-bonding, or antibonding molecular orbitals. In a non-bonding molecular orbital, a pair of electrons resides in an orbital confined largely to one atom and not appreciably involved in bond formation. (b) Show that this treatment is consistent only with a planar conformation of the peptide link. (c) Draw a diagram showing the relative energies of these molecular orbitals and determine the occupancy of the orbitals. *Hint.* Convince yourself that there are four electrons to be distributed among the molecular orbitals. (d) Now consider a non-planar conformation of the peptide link, in which the O2p and C2p orbitals are perpendicular to the plane defined by the O, C, and N atoms, but the N2p orbital lies on that plane. The LCAO-MOs are given by

$$\psi_4 = a\psi_O + b\psi_C \qquad \psi_5 = e\psi_N \qquad \psi_6 = f\psi_O - g\psi_C$$

Just as before, sketch these molecular orbitals and characterize them as bonding, non-bonding, or antibonding. Also, draw an energy level diagram and determine the occupancy of the orbitals. (e) Why is this arrangement of atomic orbitals consistent with a non-planar conformation for the peptide link? (f) Does the bonding MO associated with the planar conformation have the same energy as the bonding MO associated with the non-planar conformation? If not, which bonding MO is lower in energy? Repeat the analysis for the non-bonding and antibonding molecular orbitals. (g) Use your results from parts (a)–(f) to construct arguments that support the planar model for the peptide link.

5.32 Carbon is an essential building block of complex biological structures. It can form covalent bonds with many other elements, such as hydrogen, nitrogen, oxygen, sulfur, and, more importantly, other carbon atoms. As a consequence, such networks as long carbon–carbon chains (as in lipids) and chains of peptide links can form readily. Furthermore, carbon atoms can form chains and rings containing single, double, or triple C—C bonds. Such a variety of bonding options leads to the intricate molecular architectures of proteins, nucleic acids, and cell membranes. But the balance of bond strengths is critical to biology: bonds need to be sufficiently strong to maintain the structure of the cell and yet need to be susceptible to dissociation and rearrangement during biochemical reactions. Discuss how the following properties of carbon explain the bonding features that make it an ideal biological building block: (i) among the elements in Period 2, carbon has intermediate values of the ionization energy, electron affinity, and electronegativity; (ii) carbon has four valence electrons; and (iii) a C—C bond is stronger than a C—N or C—O bond.

MATHEMATICAL BACKGROUND 5

Matrices

A **matrix** is an array of numbers that is a generalization of an ordinary number. We shall consider only square matrices, which have the numbers arranged in the same number of rows and columns. By using matrices, we can manipulate large numbers of ordinary numbers simultaneously. A **determinant** is a particular combination of the numbers that appear in a matrix and is used to manipulate the matrix.

Matrices may be combined together by addition or multiplication according to generalizations of the rules for ordinary numbers. Although we describe below the key algebraic procedures involving matrices, it is important to note that most numerical matrix manipulations are now carried out with mathematical software. You are encouraged to use such software if it is available to you.

MB5.1 Definitions

Consider a square matrix M of n^2 numbers arranged in n columns and n rows. These n^2 numbers are the **elements** of the matrix, and may be specified by stating the row, r, and column, c, at which they occur. Each element is therefore denoted M_{rc}. A **diagonal matrix** is a matrix in which the only nonzero elements lie on the major diagonal (the diagonal from M_{11} to M_{nn}). Thus, the matrix

$$D=\begin{pmatrix}1 & 0 & 0\\ 0 & 2 & 0\\ 0 & 0 & 1\end{pmatrix}$$

is a 3×3 diagonal square matrix. The condition may be written

$$M_{rc}=m_r\delta_{rc} \qquad \text{(MB5.1)}$$

where δ_{rc} is the **Kronecker delta** (*Further information 1.1*), which is equal to 1 for $r=c$ and to 0 for $r\neq c$. In the above example, $m_1=1$, $m_2=2$, and $m_3=1$. The **unit matrix**, **1** (and occasionally **I**), is a special case of a diagonal matrix in which all nonzero elements are 1.

The **transpose** of a matrix M is denoted M^{T} and is defined by

$$M^{\mathrm{T}}_{mn}=M_{nm} \qquad \text{(MB5.2)}$$

That is, the element in row n, column m of the original matrix becomes the element in row m, column n of the transpose (in effect, the elements are reflected across the diagonal). The **determinant**, $|M|$, of the matrix M is a real number arising from a specific procedure for taking sums and differences of products of matrix elements. For example, a 2×2 determinant is evaluated as

$$\begin{vmatrix}a & b\\ c & d\end{vmatrix}=ad-bc \qquad \text{(MB5.3a)}$$

and a 3×3 determinant is evaluated by expanding it as a sum of 2×2 determinants:

$$\begin{vmatrix}a & b & c\\ d & e & f\\ g & h & i\end{vmatrix}=a\begin{vmatrix}e & f\\ h & i\end{vmatrix}-b\begin{vmatrix}d & f\\ g & i\end{vmatrix}+c\begin{vmatrix}d & e\\ g & h\end{vmatrix} \qquad \text{(MB5.3b)}$$
$$=a(ei-fh)-b(di-fg)+c(dh-eg)$$

Note the sign change in alternate columns (b occurs with a negative sign in the expansion). An important property of a determinant is that, if any two rows or any two columns are interchanged, then the determinant changes sign.

● A BRIEF ILLUSTRATION

The matrix

$$M=\begin{pmatrix}1 & 2\\ 3 & 4\end{pmatrix}$$

is a 2×2 matrix with the elements $M_{11}=1$, $M_{12}=2$, $M_{21}=3$, and $M_{22}=4$. Its transpose is

$$M^{\mathrm{T}}=\begin{pmatrix}1 & 3\\ 2 & 4\end{pmatrix}$$

and its determinant is

$$|M|=\begin{vmatrix}1 & 2\\ 3 & 4\end{vmatrix}=1\times 4-2\times 3=-2$$ ●

MB5.2 Matrix addition and multiplication

Two matrices M and N may be added to give the sum $S=M+N$, according to the rule

$$S_{rc}=M_{rc}+N_{rc} \qquad \text{(MB5.4)}$$

That is, corresponding elements are added.

Two matrices may also be multiplied to give the product $P=MN$ according to the rule

$$P_{rc}=\sum_n M_{rn}N_{nc} \qquad \text{(MB5.5)}$$

These procedures are illustrated in Fig. MB5.1. It should be noticed that in general $MN\neq NM$, and matrix multiplication is in general non-commutative (that is, depends on the order of multiplication).

● A BRIEF ILLUSTRATION

Consider the matrices

$$M=\begin{pmatrix}1 & 2\\ 3 & 4\end{pmatrix} \quad \text{and} \quad N=\begin{pmatrix}5 & 6\\ 7 & 8\end{pmatrix}$$

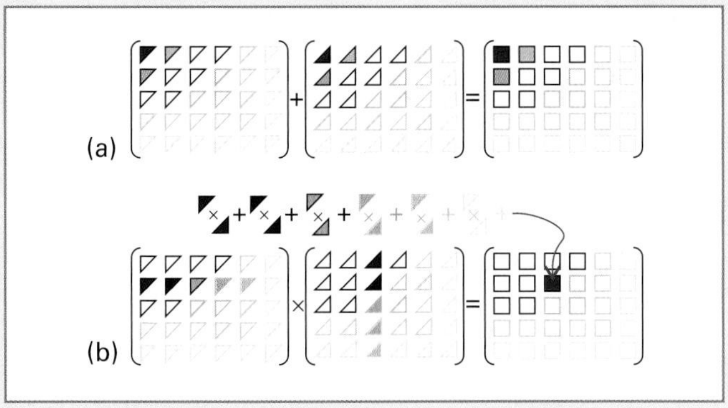

Fig. MB5.1 (a) The addition of two matrices: adding the matrix elements represented by triangles of the same colour in the matrices to the left of the equal sign gives the matrix element represented by the corresponding square in the matrix to the right of the equal sign. (b) The multiplication of matrices: to evaluate the matrix element represented by the red square, multiply the matrix elements represented by the triangles of the same colour in the intersecting row and column and add these products together.

Their sum is

$$S = \begin{pmatrix} 1 & 2 \\ 3 & 4 \end{pmatrix} + \begin{pmatrix} 5 & 6 \\ 7 & 8 \end{pmatrix} = \begin{pmatrix} 6 & 8 \\ 10 & 12 \end{pmatrix}$$

and their product is

$$P = \begin{pmatrix} 1 & 2 \\ 3 & 4 \end{pmatrix}\begin{pmatrix} 5 & 6 \\ 7 & 8 \end{pmatrix}$$

$$= \begin{pmatrix} 1\times 5 + 2\times 7 & 1\times 6 + 2\times 8 \\ 3\times 5 + 4\times 7 & 3\times 6 + 4\times 8 \end{pmatrix} = \begin{pmatrix} 19 & 22 \\ 43 & 50 \end{pmatrix} \bullet$$

The **inverse** of a matrix M is denoted M^{-1}, and is defined so that

$$MM^{-1} = M^{-1}M = 1 \qquad \text{(MB5.6)}$$

The inverse of a matrix can be constructed by using mathematical software, but in simple cases the following procedure can be carried through without much effort:

1. Form the determinant of the matrix, $|M|$.
2. Form the transpose of the matrix, M^{T}.
3. Form the matrix of co-factors, M', of M^{T}.
4. Construct the inverse as $M^{-1} = M'/|M|$

In step 3, the element M'_{rc} of the matrix M' is the **co-factor** of the element M^{T}_{rc} of the transposed matrix M^{T}. To form this co-factor, form the determinant from M^{T} with the row r and column c struck out and multiply the determinant by $(-1)^{r+c}$. Then construct M' by repeating this operation for all the elements of M^{T}. At this point you will certainly come to appreciate the usefulness of mathematical software, which performs all these operations automatically!

● A BRIEF ILLUSTRATION

Consider the matrix M from the first *illustration* in this section. We know from that *illustration* that:

1. The determinant of M is $|M| = -2$.
2. The transpose of M is $M^{\mathrm{T}} = \begin{pmatrix} 1 & 3 \\ 2 & 4 \end{pmatrix}$.
3. The matrix of co-factors of M^{T} is $M' = \begin{pmatrix} 4 & -2 \\ -3 & 1 \end{pmatrix}$.

For example, to form the element M'_{11}, we strike out the 1 3 in the first row of M^{T} and the 1 2 in the first column, and then form the determinant of what is left, just the number 4, which, upon multiplication by $(-1)^{1+1}$, becomes the 11 element of M'.

4. It follows that the inverse of M is

$$M^{-1} = \frac{M'}{|M|} = \frac{1}{-2}\begin{pmatrix} 4 & -2 \\ -3 & 1 \end{pmatrix} = \begin{pmatrix} -2 & 1 \\ \frac{3}{2} & -\frac{1}{2} \end{pmatrix} \bullet$$

MB5.3 Eigenvalue equations

An **eigenvalue equation** is an equation of the form

$$Mx = \lambda x \qquad \text{(MB5.7a)}$$

where M is a square matrix with n rows and n columns, λ is a constant, the **eigenvalue**, and x is the **eigenvector**, an $n \times 1$ (column) matrix that satisfies the conditions of the eigenvalue equation and has the form:

$$x = \begin{pmatrix} x_1 \\ x_2 \\ \vdots \\ x_n \end{pmatrix}$$

In general, there are n eigenvalues $\lambda^{(i)}$, $i = 1, 2, \ldots n$, and n corresponding eigenvectors $x^{(i)}$. We write eqn MB5.7a as (noting that $1x = x$)

$$(M - \lambda 1)x = 0 \qquad \text{(MB5.7b)}$$

Equation MB5.7b has a solution only if the determinant $|M - \lambda 1|$ of the coefficients of the matrix $M - \lambda 1$ is zero. It follows that the n eigenvalues may be found from the solution of the **secular equation**:

$$|M - \lambda 1| = 0 \qquad \text{(MB5.8)}$$

A brief comment If the inverse of the matrix $M - \lambda 1$ exists, then, from eqn MB5.7b, $(M - \lambda 1)^{-1}(M - \lambda 1)x = x = 0$, a trivial solution. For a nontrivial solution, $(M - \lambda 1)^{-1}$ must not exist, which is the case if eqn MB5.8 holds.

● **A BRIEF ILLUSTRATION**

Once again we use the matrix M in the first *illustration*, and write eqn MB5.7 as

$$\begin{pmatrix}1 & 2\\ 3 & 4\end{pmatrix}\begin{pmatrix}x_1\\ x_2\end{pmatrix}=\lambda\begin{pmatrix}x_1\\ x_2\end{pmatrix}$$

rearranged into $\begin{pmatrix}1-\lambda & 2\\ 3 & 4-\lambda\end{pmatrix}\begin{pmatrix}x_1\\ x_2\end{pmatrix}=0$

From the rules of matrix multiplication, the latter form expands into

$$\begin{pmatrix}(1-\lambda)x_1+2x_2\\ 3x_1+(4-\lambda)x_2\end{pmatrix}=0$$

which is simply a statement of the two simultaneous equations

$$(1-\lambda)x_1+2x_2=0 \text{ and } 3x_1+(4-\lambda)x_2=0$$

The condition for these two equations to have solutions is

$$|M-\lambda 1|=\begin{vmatrix}1-\lambda & 2\\ 3 & 4-\lambda\end{vmatrix}=(1-\lambda)(4-\lambda)-6=0$$

This condition corresponds to the quadratic equation

$$\lambda^2-5\lambda-2=0$$

with solutions $\lambda=+5.372$ and $\lambda=-0.372$, the two eigenvalues of the original equation. ●

The n eigenvalues found by solving the secular equations are used to find the corresponding eigenvectors. To do so, we begin by considering an $n\times n$ matrix X which will be formed from the eigenvectors corresponding to all the eigenvalues. Thus, if the eigenvalues are $\lambda_1, \lambda_2, \ldots$, and the corresponding eigenvectors are

$$\boldsymbol{x}^{(1)}=\begin{pmatrix}x_1^{(1)}\\ x_2^{(1)}\\ \vdots\\ x_n^{(1)}\end{pmatrix}\quad \boldsymbol{x}^{(2)}=\begin{pmatrix}x_1^{(2)}\\ x_2^{(2)}\\ \vdots\\ x_n^{(2)}\end{pmatrix}\quad \text{etc.} \qquad \text{(MB5.9a)}$$

the matrix X is

$$X=(\boldsymbol{x}^{(1)},\boldsymbol{x}^{(2)},\ldots \boldsymbol{x}^{(n)})=\begin{pmatrix}x_1^{(1)} & x_1^{(2)} & \cdots & x_1^{(n)}\\ x_2^{(1)} & x_2^{(2)} & \cdots & x_2^{(n)}\\ \vdots & \vdots & & \vdots\\ x_n^{(1)} & x_n^{(2)} & \cdots & x_n^{(n)}\end{pmatrix} \qquad \text{(MB5.9b)}$$

Similarly, we form an $n\times n$ matrix $\boldsymbol{\Lambda}$ with the eigenvalues λ along the diagonal and zeroes elsewhere:

$$\boldsymbol{\Lambda}=\begin{pmatrix}\lambda_1 & 0 & \cdots & 0\\ 0 & \lambda_2 & \cdots & 0\\ \vdots & \vdots & & \vdots\\ 0 & 0 & \cdots & \lambda_n\end{pmatrix} \qquad \text{(MB5.10)}$$

Now all the eigenvalue equations $M\boldsymbol{x}^{(i)}=\lambda_i\boldsymbol{x}^{(i)}$ may be confined into the single matrix equation

$$MX=X\Lambda \qquad \text{(MB5.11)}$$

● **A BRIEF ILLUSTRATION**

In the preceding *illustration* we established that if $M=\begin{pmatrix}1 & 2\\ 3 & 4\end{pmatrix}$ then $\lambda_1=+5.372$ and $\lambda_2=-0.372$, with eigenvectors $\boldsymbol{x}^{(1)}=\begin{pmatrix}x_1^{(1)}\\ x_2^{(1)}\end{pmatrix}$ and $\boldsymbol{x}^{(2)}=\begin{pmatrix}x_1^{(2)}\\ x_2^{(2)}\end{pmatrix}$, respectively. We form

$$X=\begin{pmatrix}x_1^{(1)} & x_1^{(2)}\\ x_2^{(1)} & x_2^{(2)}\end{pmatrix}\qquad \boldsymbol{\Lambda}=\begin{pmatrix}5.372 & 0\\ 0 & -0.372\end{pmatrix}$$

The expression $MX=X\Lambda$ becomes

$$\begin{pmatrix}1 & 2\\ 3 & 4\end{pmatrix}\begin{pmatrix}x_1^{(1)} & x_1^{(2)}\\ x_2^{(1)} & x_2^{(2)}\end{pmatrix}=\begin{pmatrix}x_1^{(1)} & x_1^{(2)}\\ x_2^{(1)} & x_2^{(2)}\end{pmatrix}\begin{pmatrix}5.372 & 0\\ 0 & -0.372\end{pmatrix}$$

which expands to

$$\begin{pmatrix}x_1^{(1)}+2x_2^{(1)} & x_1^{(2)}+2x_2^{(2)}\\ 3x_1^{(1)}+4x_2^{(1)} & 3x_1^{(2)}+4x_2^{(2)}\end{pmatrix}=\begin{pmatrix}5.372x_1^{(1)} & -0.372x_1^{(2)}\\ 5.372x_2^{(1)} & -0.372x_2^{(2)}\end{pmatrix}$$

This is a compact way of writing the four equations

$$x_1^{(1)}+2x_2^{(1)}=5.372x_1^{(1)} \qquad x_1^{(2)}+2x_2^{(2)}=-0.372x_1^{(2)}$$

$$3x_1^{(1)}+4x_2^{(1)}=5.372x_2^{(1)} \qquad 3x_1^{(2)}+4x_2^{(2)}=-0.372x_2^{(2)}$$

corresponding to the two original simultaneous equations and their two roots. ●

Finally, we form X^{-1} from X and multiply eqn MB5.11 by it from the left:

$$X^{-1}MX=X^{-1}X\Lambda=\Lambda \qquad \text{(MB5.12)}$$

A structure of the form $X^{-1}MX$ is called a **similarity transformation.** In this case the similarity transformation $X^{-1}MX$ makes M diagonal (because $\boldsymbol{\Lambda}$ is diagonal). It follows that, if the matrix X that causes $X^{-1}MX$ to be diagonal is known, then the problem is solved: the diagonal matrix so produced has the eigenvalues as its only nonzero elements, and the matrix X used to bring about the transformation has the corresponding eigenvectors as its columns. As will be appreciated once again, the solutions of eigenvalue equations are best found by using mathematical software.

● **A BRIEF ILLUSTRATION**

To apply the similarity transformation, eqn MB5.12, to the matrix $\begin{pmatrix}1 & 2\\ 3 & 4\end{pmatrix}$ from the preceding *illustration* it is best to use mathematical software to find the form of X. The result is

$$X = \begin{pmatrix} 0.416 & 0.825 \\ 0.909 & -0.566 \end{pmatrix}$$

This result can be verified by carrying out the multiplication

$$X^{-1}MX = \begin{pmatrix} 0.574 & 0.837 \\ 0.922 & -0.422 \end{pmatrix}\begin{pmatrix} 1 & 2 \\ 3 & 4 \end{pmatrix}\begin{pmatrix} 0.416 & 0.825 \\ 0.909 & -0.566 \end{pmatrix}$$

$$= \begin{pmatrix} 5.372 & 0 \\ 0 & -0.372 \end{pmatrix}$$

The result is indeed the diagonal matrix $\boldsymbol{\Lambda}$ calculated in the preceding *illustration*. It follows that the eigenvectors $\boldsymbol{x}^{(1)}$ and $\boldsymbol{x}^{(2)}$ are

$$\boldsymbol{x}^{(1)} = \begin{pmatrix} 0.416 \\ 0.909 \end{pmatrix} \quad \text{and} \quad \boldsymbol{x}^{(2)} = \begin{pmatrix} 0.825 \\ -0.566 \end{pmatrix} \bullet$$

6 Computational chemistry

In this chapter we extend the description of the electronic structure of molecules presented in Chapter 5 by introducing methods that harness the power of computers to calculate electronic wavefunctions and energies. These calculations are among the most useful tools used by chemists for the prediction of molecular structure and reactivity. The computational methods we discuss handle the electron–electron repulsion term in the Schrödinger equation in different ways. One such method, the Hartree–Fock method, treats electron–electron interactions in an average and approximate way. This approach typically requires the numerical evaluation of a large number of integrals. Semiempirical methods set these integrals to zero or to values determined experimentally. In contrast, *ab initio* methods attempt to evaluate the integrals numerically, leading to a more precise treatment of electron–electron interactions. Configuration interaction and Møller–Plesset perturbation theory are used to account for electron correlation, the tendency of electrons to avoid one another. Another computational approach, density functional theory, focuses on electron probability densities rather than on wavefunctions. The chapter concludes by comparing results from different electronic structure methods with experimental data and by describing some of the wide range of chemical and physical properties of molecules that can be computed.

The field of **computational chemistry**, the use of computers to predict molecular structure and reactivity, has grown in the past few decades due to the tremendous advances in computer hardware and to the development of efficient software packages. The latter are now applied routinely to compute molecular properties in a wide variety of chemical applications, including pharmaceuticals and drug design, atmospheric and environmental chemistry, nanotechnology, and materials science. Many software packages have sophisticated graphical interfaces that permit the visualization of results. The maturation of the field of computational chemistry was recognized by the awarding of the 1998 Nobel Prize in Chemistry to J.A. Pople and W. Kohn for their contributions to the development of computational techniques for the elucidation of molecular structure and reactivity.

The central challenge

The goal of electronic structure calculations in computational chemistry is the solution of the electronic Schrödinger equation, $\hat{H}\Psi = E\Psi$, where E is the electronic energy and Ψ is the many-electron wavefunction, a function of the coordinates of all the electrons and the nuclei. To make progress, we invoke at the outset the Born–Oppenheimer approximation and the separation of electronic and nuclear motion (Chapter 5). The electronic hamiltonian $\hat{H}$ is

$$\hat{H} = -\frac{\hbar^2}{2m_e}\sum_i^{N_e}\nabla_i^2 - \sum_i^{N_e}\sum_I^{N_n}\frac{Z_I e^2}{4\pi\varepsilon_0 r_{Ii}} + \frac{1}{2}\sum_{i\neq j}^{N_e}\frac{e^2}{4\pi\varepsilon_0 r_{ij}} \quad (6.1a)$$

where

- the first term is the kinetic energy of the N_e electrons;
- the second term is the potential energy of attraction between each electron and each of the N_n nuclei, with electron i at a distance r_{Ii} from nucleus I of charge $Z_I e$;
- the final term is the potential energy of repulsion between two electrons separated by r_{ij}.

The factor of $\frac{1}{2}$ in the final sum ensures that each repulsion is counted only once. The combination $e^2/4\pi\varepsilon_0$ occurs throughout computational chemistry, and we denote it j_0. Then the hamiltonian becomes

$$\hat{H} = -\frac{\hbar^2}{2m_e}\sum_i^{N_e}\nabla_i^2 - j_0\sum_i^{N_e}\sum_I^{N_n}\frac{Z_I}{r_{Ii}} + \frac{1}{2}j_0\sum_{i\neq j}^{N_e}\frac{1}{r_{ij}} \quad (6.1b)$$

We shall use the following labels:

Species	Label	Number used
Electrons	i and $j = 1, 2, \ldots$	N_e
Nuclei	$I = A, B, \ldots$	N_n
Molecular orbitals, ψ	$m = a, b, \ldots$	
Atomic orbitals used to construct the molecular orbitals (the 'basis'), χ	$o = 1, 2, \ldots$	N_b

Another general point is that the theme we develop in the sequence of *illustrations* in this chapter is aimed at showing explicitly how to use the equations that we have presented, and thereby give them a sense of reality. To do so, we shall take the simplest possible many-electron molecule, dihydrogen (H_2). Some of the techniques we introduce do not need to be applied to this simple molecule, but they serve to illustrate them in a simple manner and introduce problems that successive sections show how to solve. One consequence of choosing to develop a story in relation to H_2, we have to confess, is that not all the *illustrations* are actually as brief as we would wish; but we decided that it was more important to show the details of each little calculation than to adhere strictly to our normal use of the term 'brief'.

● A BRIEF ILLUSTRATION

The notation we use for the description of H_2 is shown in Fig. 6.1. For this two-electron ($N_e = 2$), two-nucleus ($N_n = 2$) molecule the hamiltonian is

$$\hat{H} = -\frac{\hbar^2}{2m_e}(\nabla_1^2 + \nabla_2^2) - j_0\left(\frac{1}{r_{A1}} + \frac{1}{r_{B1}} + \frac{1}{r_{A2}} + \frac{1}{r_{B2}}\right) + \frac{j_0}{r_{12}}$$

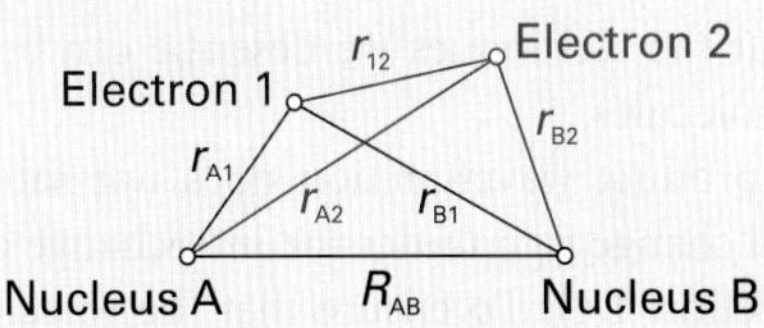

Fig. 6.1 The notation used for the description of molecular hydrogen, introduced in the *brief illustration* preceding Section 6.1 and used throughout the text.

To keep the notation simple, we introduce the one-electron operator

$$h_i = -\frac{\hbar^2}{2m_e}\nabla_i^2 - j_0\left(\frac{1}{r_{Ai}} + \frac{1}{r_{Bi}}\right) \quad (6.2)$$

which should be recognized as the hamiltonian for electron i in an H_2^+ molecule-ion. Then

$$\hat{H} = h_1 + h_2 + \frac{j_0}{r_{12}} \quad (6.3)$$

We see that the hamiltonian for H_2 is essentially that of each electron in an H_2^+-like molecule-ion but with the addition of the electron–electron repulsion term. ●

It is hopeless to expect to find analytical solutions with a hamiltonian of the complexity of that shown in eqn 6.1, even for H_2, and the whole thrust of computational chemistry is to formulate and implement numerical procedures that give ever more reliable results.

6.1 The Hartree–Fock formalism

The electronic wavefunction of a many-electron molecule is a function of the positions of all the electrons, $\Psi(\boldsymbol{r}_1,\boldsymbol{r}_2,\ldots)$. To formulate one very widely used approximation, we build on the material in Chapter 5, where we saw that in the MO description of H_2 we supposed that each electron occupies an orbital and that the overall wavefunction can be written $\psi(\boldsymbol{r}_1)\psi(\boldsymbol{r}_2)\ldots$. Note that this **orbital approximation** is quite severe and loses many of the details of the dependence of the wavefunction on the relative locations of the electrons. We do the same here, with two small changes of notation. To simplify the appearance of the expressions we write $\psi(\boldsymbol{r}_1)\psi(\boldsymbol{r}_2)\ldots$ as $\psi(1)\psi(2)\ldots$. Next, we suppose that electron 1 occupies a molecular orbital ψ_a with spin α, electron 2 occupies the same orbital with spin β, and so on, and hence write the many-electron wavefunction Ψ as the product $\Psi = \psi_a^\alpha(1)\psi_a^\beta(2)\ldots$. The combination of a molecular orbital and a spin function, such as $\psi_a^\alpha(1)$, is the spinorbital introduced in Section 4.4; for example, the spinorbital ψ_a^α should be interpreted as the product of the spatial wavefunction ψ_a and the spin state α, so $\psi_a^\alpha(1) = \psi_a(1)\alpha(1)$, and likewise for the other spinorbitals. We shall consider only closed-shell molecules but the techniques we describe can be extended to open-shell molecules.

A simple product wavefunction does not satisfy the Pauli principle and change sign under the interchange of any pair of electrons (Section 4.4). To ensure that the wavefunction does satisfy the principle, we modify it to a sum of all possible permutations, using plus and minus signs appropriately:

$$\Psi = \psi_a^{\alpha}(1)\psi_a^{\beta}(2)\cdots\psi_z^{\beta}(N_e) - \psi_a^{\alpha}(2)\psi_a^{\beta}(1)\cdots\psi_z^{\beta}(N_e) + \cdots \tag{6.4}$$

There are $N_e!$ terms in this sum, and the entire sum can be represented by the Slater determinant (Section 4.4):

$$\Psi = \frac{1}{\sqrt{N_e!}}\begin{vmatrix} \psi_a^{\alpha}(1) & \psi_a^{\beta}(1) & \cdots & \cdots & \psi_z^{\beta}(1) \\ \psi_a^{\alpha}(2) & \psi_a^{\beta}(2) & \cdots & \cdots & \psi_z^{\beta}(2) \\ \vdots & \vdots & & & \vdots \\ \vdots & \vdots & & & \vdots \\ \psi_a^{\alpha}(N_e) & \psi_a^{\beta}(N_e) & \cdots & \cdots & \psi_z^{\beta}(N_e) \end{vmatrix} \tag{6.5a}$$

The factor $1/\sqrt{N_e!}$ ensures that the wavefunction is normalized if the component molecular orbitals ψ_m are normalized. To save the tedium of writing out large determinants, the wavefunction is normally written by using only its principal diagonal:

$$\Psi = (1/N_e!)^{1/2}|\psi_a^{\alpha}(1)\psi_a^{\beta}(2)\cdots\psi_z^{\beta}(N_e)| \tag{6.5b}$$

● A BRIEF ILLUSTRATION

The Slater determinant for H_2 ($N_e = 2$) is

$$\Psi = \frac{1}{\sqrt{2}}\begin{vmatrix} \psi_a^{\alpha}(1) & \psi_a^{\beta}(1) \\ \psi_a^{\alpha}(2) & \psi_a^{\beta}(2) \end{vmatrix}$$

$$= \frac{1}{\sqrt{2}}\{\psi_a^{\alpha}(1)\psi_a^{\beta}(2) - \psi_a^{\beta}(1)\psi_a^{\alpha}(2)\}$$

$$= \frac{1}{\sqrt{2}}\psi_a(1)\psi_a(2)\{\alpha(1)\beta(2) - \beta(1)\alpha(2)\}$$

where both electrons occupy the molecular wavefunction ψ_a. We should recognize the spin factor as that corresponding to a singlet state (eqn 4.31b, $\sigma_- = (1/\sqrt{2})\{\alpha\beta - \beta\alpha\}$), so Ψ corresponds to two spin-paired electrons in ψ_a. ●

According to the variation principle (Section 5.5), the best form of Ψ is the one that corresponds to the lowest achievable energy as the ψ are varied, that is, we need the wavefunctions ψ that will minimize the expectation value $\int\Psi^*\hat{H}\Psi\,d\tau$. Because the electrons interact with one another, a variation in the form of ψ_a, for instance, will affect what will be the best form of all the other ψs, so finding the best form of the ψs is a far from trivial problem. However, D.R. Hartree and V. Fock showed that the optimum ψs each satisfy an at first sight very simple set of equations:

$$f_1\psi_a(1) = \varepsilon_a\psi_a(1) \tag{6.6}$$

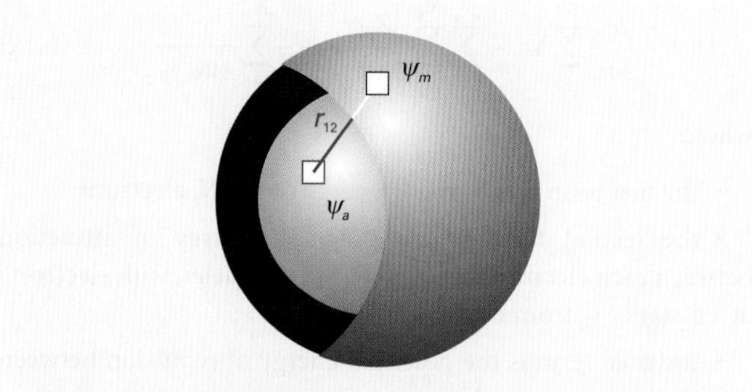

Fig. 6.2 A schematic interpretation of the physical interpretation of the Coulomb repulsion term, eqn 6.7a. An electron in orbital ψ_a experiences repulsion from an electron in orbital ψ_m where it has probability density $|\psi_m|^2$.

where f_1 is called the **Fock operator**. This is the equation to solve to find ψ_a; there are analogous equations for all the other occupied orbitals. This Schrödinger-like equation has the form we should expect (but its formal derivation is quite involved). Thus, f_1 has the following structure:

f_1 = core hamiltonian for electron 1 (h_1)
+ average Coulomb repulsion from electrons 2, 3, . . . ($V_{Coulomb}$)
+ average correction due to spin correlation ($V_{Exchange}$)
$= h_1 + V_{Coulomb} + V_{Exchange}$

By the **core hamiltonian** we mean the one-electron hamiltonian h_1 defined by eqn 6.2 and representing the energy of electron 1 in the field of the nuclei. The Coulomb repulsion from all the other electrons contributes a term that acts as follows (Fig. 6.2):

$$J_m(1)\psi_a(1) = j_0\int\psi_a(1)\frac{1}{r_{12}}\psi_m{}^*(2)\psi_m(2)\,d\tau_2 \tag{6.7a}$$

This integral represents the repulsion experienced by electron 1 in orbital ψ_a from electron 2 in orbital ψ_m, where it is distributed with probability density $\psi_m{}^*\psi_m$. There are two electrons in each orbital, so we can expect a total contribution of the form

$$V_{Coulomb}\psi_a(1) = 2\sum_m J_m(1)\psi_a(1)$$

where the sum is over all the occupied orbitals, including orbital a. You should be alert to the fact that counting 2 for the orbital with $m = a$ is incorrect, because electron 1 interacts only with the second electron in the orbital, not with itself. This error will be corrected in a moment. The spin correlation term takes into account the fact that electrons of the same spin tend to avoid each other (Section 4.4), which reduces the net Coulomb interaction between them. This contribution has the following form:

$$K_m(1)\psi_a(1) = j_0\int\psi_m(1)\frac{1}{r_{12}}\psi_m{}^*(2)\psi_a(2)\,d\tau_2 \tag{6.7b}$$

For a given electron 1 there is only one electron of the same spin in all the occupied orbitals, so we can expect a total contribution of the form

$$V_{\text{Exchange}}\psi_a(1) = -\sum_m K_m(1)\psi_a(1)$$

The negative sign reminds us that spin-correlation keeps electrons apart, and so reduces their classical, Coulombic repulsion. By collecting terms, we arrive at a specific expression for the effect of the Fock operator:

$$f_1\psi_a(1) = h_1\psi_a(1) + \sum_m \{2J_m(1) - K_m(1)\}\psi_a(1) \qquad (6.8)$$

with the sum extending over all the occupied orbitals. Note that $K_a(1)\psi_a(1) = J_a(1)\psi_a(1)$, so the term in the sum with $m = a$ loses one of its $2J_a$, which is the correction that avoids the electron repelling itself, which we referred to above.

Equation 6.8 reveals a second principal approximation of the Hartree–Fock formalism (the first being its dependence on the orbital approximation). Instead of electron 1 (or any other electron) responding to the instantaneous positions of the other electrons in the molecule through terms of the form $1/r_{1j}$, it responds to an *averaged* location of the other electrons through integrals of the kind that appear in eqn 6.7. When we look for reasons why the formalism gives poor results, this approximation is a principal reason; it is addressed in Section 6.6.

Although eqn 6.6 is the equation we have to solve to find ψ_a, eqn 6.7 reveals that it is necessary to know all the other occupied wavefunctions in order to set up the operators J and K and hence to find ψ_a. To make progress with this difficulty, we can guess the initial form of all the one-electron wavefunctions, use them in the definition of the Coulomb and exchange operators, and solve the Hartree–Fock equations. That process is then continued using the newly found wavefunctions until each cycle of calculation leaves the energies ε_m and wavefunctions ψ_m unchanged to within a chosen criterion. This is the origin of the term **self-consistent field** (SCF) for this type of procedure in general and of **Hartree–Fock self-consistent field** (HF-SCF) for the approach based on the orbital approximation.

● A BRIEF ILLUSTRATION

We continue with the H_2 example. According to eqn 6.6, the Hartree–Fock equation for ψ_a is $f_1\psi_a(1) = \varepsilon_a\psi_a(1)$ with

$$f_1\psi_a(1) = h_1\psi_a(1) + 2J_a(1)\psi_a(1) - K_a(1)\psi_a(1)$$

because there is only one term in the sum (there is only one occupied orbital). In this expression

$$J_a(1)\psi_a(1) = K_a(1)\psi_a(1) = j_0\int \psi_a(1)\frac{1}{r_{12}}\psi_a{}^*(2)\psi_a(2)\,\mathrm{d}\tau_2$$

The equation to solve is therefore

$$-\frac{\hbar^2}{2m_e}\nabla_1^2\psi_a(1) - j_0\left(\frac{1}{r_{A1}} + \frac{1}{r_{B1}}\right)\psi_a(1)$$

$$+j_0\int \psi_a(1)\frac{1}{r_{12}}\psi_a{}^*(2)\psi_a(2)\,\mathrm{d}\tau_2 = \varepsilon_a\psi_a(1)$$

This equation for ψ_a must be solved self-consistently (and numerically) because the integral that governs the form of ψ_a requires us to know ψ_a already. In the following examples we shall illustrate some of the procedures that have been adopted. ●

6.2 The Roothaan equations

The difficulty with the HF-SCF procedure lies in the numerical solution of the Hartree–Fock equations, an onerous task even for powerful computers. As a result, a modification of the technique was needed before the procedure could be of use to chemists. We saw in Chapter 5 how molecular orbitals are constructed as linear combinations of atomic orbitals. This simple approach was adopted in 1951 by C.C.J. Roothaan and G.G. Hall independently, who found a way to convert the Hartree–Fock equations for the molecular orbitals into equations for the coefficients that appear in the LCAO used to simulate the molecular orbital. Thus, they wrote (as we did in eqn 5.34)

$$\psi_m = \sum_{o=1}^{N_b} c_{om}\chi_o \qquad (6.9)$$

where c_{om} are unknown coefficients and the χ_o are the atomic orbitals (which we take to be real). Note that this approximation is in addition to those underlying the Hartree–Fock equations because the basis is finite and so cannot reproduce the molecular orbital exactly. The size of the basis set (N_b) is not necessarily the same as the number of atomic nuclei in the molecule (N_n), because we might use several atomic orbitals on each nucleus (such as the four 2s and 2p orbitals of a carbon atom). From N_b basis functions, we obtain N_b linearly independent molecular orbitals ψ.

We show in *Justification 6.1* that the use of a linear combination like in eqn 6.9 leads to a set of simultaneous equations for the coefficients called the **Roothaan equations**. These equations are best summarized in matrix form by writing

$$\boldsymbol{Fc} = \boldsymbol{Sc\varepsilon} \qquad (6.10)$$

where F is the $N_b \times N_b$ matrix with elements

$$F_{oo'} = \int \chi_o(1)f_1\chi_{o'}(1)\,\mathrm{d}\tau_1 \qquad (6.11a)$$

S is the $N_b \times N_b$ matrix of overlap integrals:

$$S_{oo'} = \int \chi_o(1)\chi_{o'}(1)\mathrm{d}\tau_1 \qquad (6.11b)$$

and c is an $N_b \times N_b$ matrix of all the coefficients we have to find:

$$\boldsymbol{c} = \begin{pmatrix} c_{1a} & c_{1b} & \cdots & c_{1N_b} \\ c_{2a} & c_{2b} & \cdots & c_{2N_b} \\ \vdots & \vdots & \vdots & \vdots \\ c_{N_b a} & c_{N_b b} & \cdots & c_{N_b N_b} \end{pmatrix} \qquad (6.11c)$$

The first column is the set of coefficients for ψ_a, the second column for ψ_b, and so on. Finally, $\boldsymbol{\varepsilon}$ is a diagonal matrix of orbital energies ε_a, ε_b, . . . :

$$\boldsymbol{\varepsilon} = \begin{pmatrix} \varepsilon_a & 0 & \cdots & 0 \\ 0 & \varepsilon_b & \cdots & 0 \\ \vdots & \vdots & \vdots & \vdots \\ 0 & 0 & \cdots & \varepsilon_{N_b} \end{pmatrix} \qquad (6.11d)$$

Justification 6.1 *The Roothaan equations*

To construct the Roothaan equations we substitute the linear combination of eqn 6.9 into eqn 6.6, which gives

$$f_1 \sum_{o=1}^{N_b} c_{om}\chi_o(1) = \varepsilon_m \sum_{o=1}^{N_b} c_{om}\chi_o(1)$$

Now multiply from the left by $\chi_{o'}(1)$ and integrate over the coordinates of electron 1:

$$\sum_{o=1}^{N_b} c_{om} \overbrace{\int \chi_{o'}(1) f_1 \chi_o(1)\mathrm{d}\tau_1}^{F_{o'o}} = \varepsilon_m \sum_{o=1}^{N_b} c_{om} \overbrace{\int \chi_{o'}(1)\chi_o(1)\mathrm{d}\tau_1}^{S_{o'o}}$$

That is,

$$\sum_{o=1}^{N_b} F_{o'o}c_{om} = \varepsilon_m \sum_{o=1}^{N_b} S_{o'o}c_{om}$$

This expression has the form of the matrix equation in eqn 6.10.

● **A BRIEF ILLUSTRATION**

In this *illustration* we show how to set up the Roothaan equations for H_2. To do so, we adopt a basis set of real, normalized functions χ_A and χ_B, centred on nuclei A and B, respectively. We can think of these functions as H1s orbitals on each nucleus, but they could be more general than that, and in a later *illustration* we shall make a computationally more friendly choice. The two possible linear combinations corresponding to eqn 6.9 are

$$\psi_a = c_{Aa}\chi_A + c_{Ba}\chi_B \qquad \psi_b = c_{Ab}\chi_A + c_{Bb}\chi_B$$

so the matrix c is

$$\boldsymbol{c} = \begin{pmatrix} c_{Aa} & c_{Ab} \\ c_{Ba} & c_{Bb} \end{pmatrix}$$

and the overlap matrix S is

$$\boldsymbol{S} = \begin{pmatrix} S_{AA} & S_{AB} \\ S_{BA} & S_{BB} \end{pmatrix} = \begin{pmatrix} 1 & S \\ S & 1 \end{pmatrix} \quad \text{with} \quad S = \int \chi_A(1)\chi_B(1)\mathrm{d}\tau_1$$

The Fock matrix is

$$\boldsymbol{F} = \begin{pmatrix} F_{AA} & F_{AB} \\ F_{BA} & F_{BB} \end{pmatrix} \quad \text{with} \quad F_{oo'} = \int \chi_o(1) f_1 \chi_{o'}(1)\mathrm{d}\tau_1$$

We shall explore the explicit form of the elements of F in a later *illustration*; for now, we just regard them as variable quantities. The Roothaan equations are therefore

$$\begin{pmatrix} F_{AA} & F_{AB} \\ F_{BA} & F_{BB} \end{pmatrix}\begin{pmatrix} c_{Aa} & c_{Ab} \\ c_{Ba} & c_{Bb} \end{pmatrix} = \begin{pmatrix} 1 & S \\ S & 1 \end{pmatrix}\begin{pmatrix} c_{Aa} & c_{Ab} \\ c_{Ba} & c_{Bb} \end{pmatrix}\begin{pmatrix} \varepsilon_a & 0 \\ 0 & \varepsilon_b \end{pmatrix}$$ ●

● **ANOTHER BRIEF ILLUSTRATION**

In this continuation of the preceding *illustration*, we establish the simultaneous equations corresponding to the Roothaan equations we have just established. After multiplying out the matrices constructed in the preceding *illustration*, we obtain

$$\begin{pmatrix} F_{AA}c_{Aa} + F_{AB}c_{Ba} & F_{AA}c_{Ab} + F_{AB}c_{Bb} \\ F_{BA}c_{Aa} + F_{BB}c_{Ba} & F_{BA}c_{Ab} + F_{BB}c_{Bb} \end{pmatrix}$$

$$= \begin{pmatrix} \varepsilon_a c_{Aa} + S\varepsilon_a c_{Ba} & \varepsilon_b c_{Ab} + S\varepsilon_b c_{Bb} \\ \varepsilon_a c_{Ba} + S\varepsilon_a c_{Aa} & \varepsilon_b c_{Bb} + S\varepsilon_b c_{Ab} \end{pmatrix}$$

On equating matching elements, we obtain the following four simultaneous equations:

$$F_{AA}c_{Aa} + F_{AB}c_{Ba} = \varepsilon_a c_{Aa} + S\varepsilon_a c_{Ba}$$
$$F_{BA}c_{Aa} + F_{BB}c_{Ba} = \varepsilon_a c_{Ba} + S\varepsilon_a c_{Aa}$$
$$F_{AA}c_{Ab} + F_{AB}c_{Bb} = \varepsilon_b c_{Ab} + S\varepsilon_b c_{Bb}$$
$$F_{BA}c_{Ab} + F_{BB}c_{Bb} = \varepsilon_b c_{Bb} + S\varepsilon_b c_{Ab}$$

Thus, to find the coefficients for the molecular orbital ψ_a, we need to solve the first and second equations, which we can write as

$$(F_{AA} - \varepsilon_a)c_{Aa} + (F_{AB} - S\varepsilon_a)c_{Ba} = 0$$
$$(F_{BA} - S\varepsilon_a)c_{Aa} + (F_{BB} - \varepsilon_a)c_{Ba} = 0$$

There is a similar pair of equations (the third and fourth) for the coefficients in ψ_b. ●

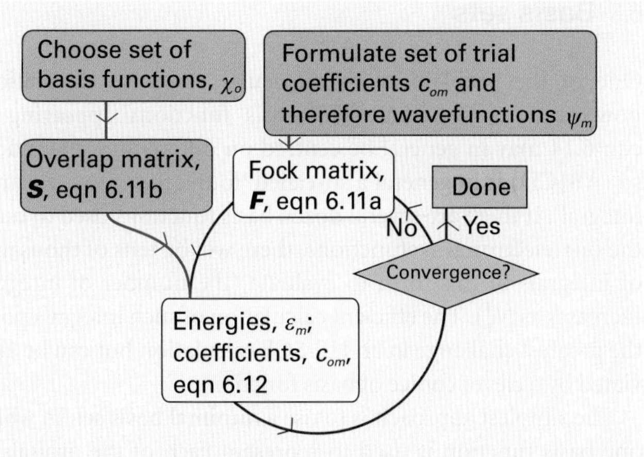

Fig. 6.3 The iteration procedure for a Hartree–Fock self-consistent field calculation.

If we write the Roothaan equations as $(\boldsymbol{F}-\boldsymbol{S\varepsilon})\boldsymbol{c}=0$ we see that they are simply a collection of N_b simultaneous equations for the coefficients. This point was demonstrated explicitly in the preceding *illustration*. Therefore, they have a solution only if

$$|\boldsymbol{F}-\boldsymbol{\varepsilon S}|=0 \qquad (6.12)$$

In principle, we can find the orbital energies that occur in $\boldsymbol{\varepsilon}$ by looking for the roots of this secular equation and then using those energies to find the coefficients that make up the matrix $\boldsymbol{c}$ by solving the Roothaan equations. There is a catch, though: the elements of $\boldsymbol{F}$ depend on the coefficients (through the presence of J and K in the expression for f_1). Therefore, we have to proceed iteratively: we guess an initial set of values for $\boldsymbol{c}$, solve the secular equation for the orbital energies, use them to solve the Roothaan equations for $\boldsymbol{c}$, and compare the resulting values with the ones we started with. In general they will be different, so we use those new values in another cycle of calculation, and continue until convergence has been achieved (Fig. 6.3).

● A BRIEF ILLUSTRATION

The two simultaneous equations for the coefficients in ψ_a obtained in the previous *illustration* have a solution if

$$\begin{vmatrix} F_{AA}-\varepsilon & F_{AB}-S\varepsilon \\ F_{BA}-S\varepsilon & F_{BB}-\varepsilon \end{vmatrix}=0$$

The determinant expands to give the following equation:

$$(F_{AA}-\varepsilon)(F_{BB}-\varepsilon)-(F_{AB}-S\varepsilon)(F_{BA}-S\varepsilon)=0$$

On collecting terms, we arrive at

$$(1-S^2)\varepsilon^2-(F_{AA}+F_{BB}-SF_{AB}-SF_{BA})\varepsilon + (F_{AA}F_{BB}-F_{AB}F_{BA})=0$$

This is a quadratic equation for the orbital energies ε, and may be solved by using the quadratic formula. Thus, if we summarize the equation as $a\varepsilon^2+b\varepsilon+c=0$, then

$$\varepsilon=\frac{-b\pm(b^2-4ac)^{1/2}}{2a}$$

With these energies established and taking the lower of the two energies to be ε_a since ψ_a is occupied in ground-state H_2, we can construct the coefficients by using the relation

$$c_{Aa}=-\frac{F_{AB}-S\varepsilon_a}{F_{AA}-\varepsilon_a}c_{Ba}$$

in conjunction with the normalization condition $c_{Aa}^2+c_{Ba}^2+2c_{Aa}c_{Ba}S=1$. (For this homonuclear diatomic molecule, there is, of course, a much simpler method of arriving at $c_{Aa}=c_{Ba}$.) ●

The principal outstanding problem is the form of the elements of the Fock matrix $\boldsymbol{F}$ and its dependence on the LCAO coefficients. The explicit form of $F_{oo'}$ is

$$\begin{aligned} F_{oo'}=&\int \chi_o(1)h_1\chi_{o'}(1)\mathrm{d}\tau_1 \\ &+2j_0\sum_m\int \chi_o(1)\chi_{o'}(1)\frac{1}{r_{12}}\psi_m(2)\psi_m(2)\mathrm{d}\tau_1\mathrm{d}\tau_2 \\ &-j_0\sum_m\int \chi_o(1)\psi_m(1)\frac{1}{r_{12}}\psi_m(2)\chi_{o'}(2)\mathrm{d}\tau_1\mathrm{d}\tau_2 \end{aligned} \qquad (6.13)$$

where the sums are over the occupied molecular orbitals. The dependence of $\boldsymbol{F}$ on the coefficients can now be seen to arise from the presence of the ψ_m in the two integrals, for these molecular orbitals depend on the coefficients in their LCAOs.

● A BRIEF ILLUSTRATION

At this point we are ready to tackle the matrix elements that occur in the treatment of H_2, using the LCAOs set up in a previous *illustration*. As we saw there, we need the four matrix elements F_{AA}, F_{AB}, F_{BA}, and F_{BB}. We show here how to evaluate F_{AA}. Only one molecular orbital is occupied (ψ_a), so eqn 6.13 becomes

$$\begin{aligned} F_{AA}=&\int \chi_A(1)h_1\chi_A(1)\mathrm{d}\tau_1 \\ &+2j_0\int \chi_A(1)\chi_A(1)\frac{1}{r_{12}}\psi_a(2)\psi_a(2)\mathrm{d}\tau_1\mathrm{d}\tau_2 \\ &-j_0\int \chi_A(1)\psi_a(1)\frac{1}{r_{12}}\psi_a(2)\chi_A(2)\mathrm{d}\tau_1\mathrm{d}\tau_2 \end{aligned}$$

With $\psi_a = c_{Aa}\chi_A + c_{Ba}\chi_B$, the second integral on the right is

$$j_0 \int \chi_A(1)\chi_A(1)\frac{1}{r_{12}}\psi_a(2)\psi_a(2)d\tau_1 d\tau_2$$

$$= j_0 \int \chi_A(1)\chi_A(1)\frac{1}{r_{12}}\{c_{Aa}\chi_A(2) + c_{Ba}\chi_B(2)\}$$

$$\times \{c_{Aa}\chi_A(2) + c_{Ba}\chi_B(2)\}d\tau_1 d\tau_2$$

$$= c_{Aa}c_{Aa}j_0 \int \chi_A(1)\chi_A(1)\frac{1}{r_{12}}\chi_A(2)\chi_A(2)d\tau_1 d\tau_2$$

$$+ c_{Aa}c_{Ba}j_0 \int \chi_A(1)\chi_A(1)\frac{1}{r_{12}}\chi_A(2)\chi_B(2)d\tau_1 d\tau_2 + \cdots$$

From now on we shall use the notation

$$(AB|CD) = j_0 \int \chi_A(1)\chi_B(1)\frac{1}{r_{12}}\chi_C(2)\chi_D(2)d\tau_1 d\tau_2 \quad (6.14)$$

Integrals like this are fixed throughout the calculation because they depend only on the choice of basis, so they can be tabulated once and for all and then used whenever required. Our task later in this chapter will be to see how they are evaluated. For the time being, we can treat them as constants. In this notation, the integral we are evaluating becomes

$$j_0 \int \chi_A(1)\chi_A(1)\frac{1}{r_{12}}\psi_a(2)\psi_a(2)d\tau_1 d\tau_2$$

$$= c_{Aa}c_{Aa}(AA|AA) + 2c_{Aa}c_{Ba}(AA|BA) + c_{Ba}c_{Ba}(AA|BB)$$

(We have used $(AA|BA) = (AA|AB)$.) There is a similar term for the third integral, and overall

$$F_{AA} = E_A + c_{Aa}^2(AA|AA) + 2c_{Aa}c_{Ba}(AA|BA) + c_{Ba}^2\{2(AA|BB) - (AB|BA)\}$$

where

$$E_A = \int \chi_A(1)h_1\chi_A(1)d\tau_1 \quad (6.15)$$

is the energy of an electron in orbital χ_A based on nucleus A, taking into account its interaction with both nuclei. Similar expressions may be derived for the other three matrix elements of F. The crucial point, though, is that we now see how F depends on the coefficients that we are trying to find. ●

Self-test 6.1 Construct the element F_{AB} using the same basis.

$$\left[F_{AB} = \int \chi_A(1)h_1\chi_B(1)d\tau_1 + c_{Aa}^2(BA|AA) + c_{Aa}c_{Ba}\{3(BA|AB) - (AA|BB)\} + c_{Ba}^2(BA|BB)\right]$$

6.3 Basis sets

One of the problems with molecular structure calculations now becomes apparent. The basis functions appearing in eqn 6.14 may in general be centred on different atomic nuclei so (AB|CD) is in general a so-called 'four-centre, two-electron integral'. If there are several dozen basis functions used to build the one-electron wavefunctions, there will be tens of thousands of integrals of this form to evaluate (the number of integrals increases as N_b^4). The efficient calculation of such integrals poses the greatest challenge in an HF-SCF calculation but can be alleviated by a clever choice of basis functions.

The simplest approach is to use a **minimal basis set**, in which one basis function is used to represent each of the orbitals in an elementary valence theory treatment of the molecule, that is, we include in the basis set one function each for H and He (to simulate a 1s orbital), five functions each for Li to Ne (for the 1s, 2s, and three 2p orbitals), nine functions each for Na to Ar, and so on. For example, a minimal basis set for CH_4 consists of nine functions: four basis functions to represent the four H1s orbitals, and one basis function each for the 1s, 2s, $2p_x$, $2p_y$, and $2p_z$ orbitals of carbon. Unfortunately, minimal basis set calculations frequently yield results that are far from agreement with experiment.

Significant improvements in the agreement between electronic structure calculations and experiment can often be achieved by increasing the number of basis set functions. In a **double-zeta** (DZ) basis set, each basis function in the minimal basis is replaced by two functions; in a **triple-zeta** (TZ) basis set, by three functions. For example, a double-zeta basis for H_2O consists of fourteen functions: a total of four basis functions to represent the two H1s orbitals, and two basis set functions each for the 1s, 2s, $2p_x$, $2p_y$, and $2p_z$ orbitals of oxygen. In a **split-valence** (SV) basis set, each inner-shell (core) atomic orbital is represented by one basis set function and each valence atomic orbital by two basis set functions; an SV calculation for H_2O, for instance, uses thirteen basis set functions. Further improvements to the accuracy of electronic structure calculations can often be achieved by including **polarization functions** in the basis; these functions represent atomic orbitals with higher values of the orbital angular momentum quantum number l than considered in an elementary valence theory treatment. For example, polarization functions in a calculation for CH_4 include basis functions representing d orbitals on carbon or p orbitals on hydrogen. Polarization functions often lead to improved results because atomic orbitals are distorted (or polarized) by adjacent atoms when bonds form in molecules.

One of the earliest choices for basis set functions was that of **Slater-type orbitals** (STO) centred on each of the atomic nuclei in the molecule and of the form

$$\chi = Nr^a e^{-br} Y_{lm_l}(\theta,\phi) \quad (6.16)$$

N is a normalization constant, a and b are (non-negative) parameters, Y_{lm_l} is a spherical harmonic (Table 3.2), and (r,θ,ϕ) are the spherical polar coordinates describing the location of the electron relative to the atomic nucleus. Several such basis functions are typically centred on each atom, with each basis function characterized by a unique set of values of a, b, l, and m_l. The values of a and b generally vary with the element and there are several rules for assigning reasonable values. For molecules containing hydrogen, there is an STO centred on each proton with $a = 0$ and $b = 1/a_0$, which simulates the correct behaviour of the 1s orbital at the nucleus (see eqn 4.14). However, using the STO basis set in HF-SCF calculations on molecules with three or more atoms requires the evaluation of so many two-electron integrals (AB|CD) that the procedure becomes computationally impractical.

The introduction of **Gaussian-type orbitals** (GTO) by S.F. Boys largely overcame the problem. Cartesian Gaussian functions centred on atomic nuclei have the form

$$\chi = Nx^iy^jz^k\mathrm{e}^{-\alpha r^2} \tag{6.17}$$

where (x,y,z) are the Cartesian coordinates of the electron at a distance r from the nucleus, (i,j,k) are a set of non-negative integers, and α is a positive constant. An **s-type** Gaussian has $i = j = k = 0$; a **p-type** Gaussian has $i + j + k = 1$; a **d-type** Gaussian has $i + j + k = 2$ and so on. Figure 6.4 shows contour plots for various Gaussian-type orbitals. The advantage of GTOs is that the product of two Gaussian functions on different centres is equivalent to a single Gaussian function located at a point between the two centres (Fig. 6.5). Therefore, two-electron integrals on three and four different atomic centres can be reduced to integrals over two different centres, which are much easier to evaluate numerically.

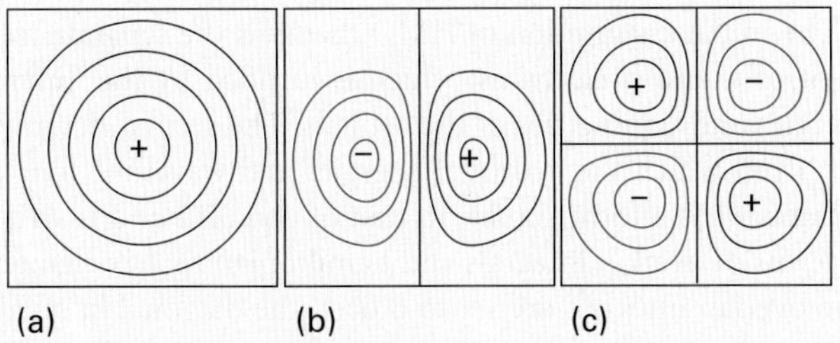

Fig. 6.4 Contour plots for Gaussian-type orbitals. (a) s-type Gaussian, e^{-r^2}; (b) p-type Gaussian $x\mathrm{e}^{-r^2}$; (c) d-type Gaussian, $xy\mathrm{e}^{-r^2}$.

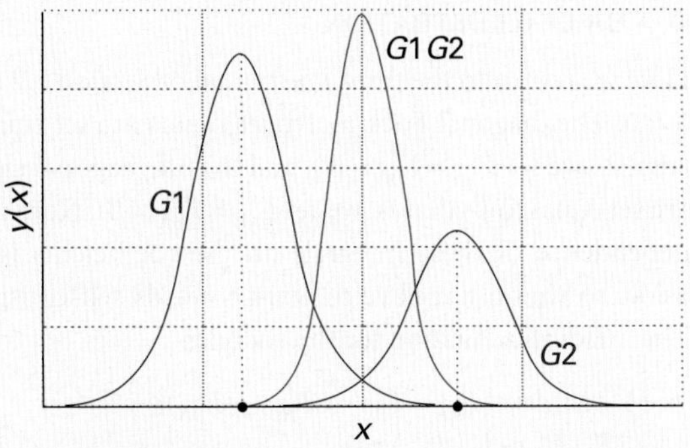

Fig. 6.5 The product of two Gaussian functions on different centres is itself a Gaussian function located at a point between the two contributing Gaussians. The scale of the product has been increased relative to that of its two components.

● **A BRIEF ILLUSTRATION**

There are no four-centre integrals in H_2, but we can illustrate the principle by considering one of the two-centre integrals that appear in the Fock matrix and, to be definite, we consider

$$(\mathrm{AB}|\mathrm{AB}) = j_0\int \chi_\mathrm{A}(1)\chi_\mathrm{B}(1)\frac{1}{r_{12}}\chi_\mathrm{A}(2)\chi_\mathrm{B}(2)\mathrm{d}\tau_1\mathrm{d}\tau_2$$

We choose an s-type Gaussian basis and write

$$\chi_\mathrm{A}(1) = N\mathrm{e}^{-\alpha|\boldsymbol{r}_1-\boldsymbol{R}_\mathrm{A}|^2} \qquad \chi_\mathrm{B}(1) = N\mathrm{e}^{-\alpha|\boldsymbol{r}_1-\boldsymbol{R}_\mathrm{B}|^2}$$

where $\boldsymbol{r}_1$ is the coordinate of electron 1 and $\boldsymbol{R}_I$ is the coordinate of nucleus I. The product of two such Gaussians, one centred on A and one centred on B, for electron 1, is

$$\chi_\mathrm{A}(1)\chi_\mathrm{B}(1) = N^2\mathrm{e}^{-\alpha|\boldsymbol{r}_1-\boldsymbol{R}_\mathrm{A}|^2}\mathrm{e}^{-\alpha|\boldsymbol{r}_1-\boldsymbol{R}_\mathrm{B}|^2} = N^2\mathrm{e}^{-\alpha\{|\boldsymbol{r}_1-\boldsymbol{R}_\mathrm{A}|^2+|\boldsymbol{r}_1-\boldsymbol{R}_\mathrm{B}|^2\}}$$

By using the relation

$$|\boldsymbol{r}-\boldsymbol{R}|^2 = (\boldsymbol{r}-\boldsymbol{R})\cdot(\boldsymbol{r}-\boldsymbol{R}) = |\boldsymbol{r}|^2 + |\boldsymbol{R}|^2 - 2\boldsymbol{r}\cdot\boldsymbol{R}$$

we can confirm that

$$|\boldsymbol{r}_1-\boldsymbol{R}_\mathrm{A}|^2 + |\boldsymbol{r}_1-\boldsymbol{R}_\mathrm{B}|^2 = \tfrac{1}{2}R^2 + 2|\boldsymbol{r}_1-\boldsymbol{R}_0|^2$$

where $\boldsymbol{R}_0 = \frac{1}{2}(\boldsymbol{R}_\mathrm{A} + \boldsymbol{R}_\mathrm{B})$ is the midpoint of the molecule and $R = |\boldsymbol{R}_\mathrm{A} - \boldsymbol{R}_\mathrm{B}|$ is the bond length. Hence

$$\chi_\mathrm{A}(1)\chi_\mathrm{B}(1) = N^2\mathrm{e}^{-\frac{1}{2}\alpha R^2}\mathrm{e}^{-2\alpha|\boldsymbol{r}_1-\boldsymbol{R}_0|^2}$$

The product $\chi_\mathrm{A}(2)\chi_\mathrm{B}(2)$ is the same, except for the index on $\boldsymbol{r}$. Therefore, the two-centre, two-electron integral (AB|AB) reduces to

$$(\mathrm{AB}|\mathrm{AB}) = N^4 j_0\mathrm{e}^{-\alpha R^2}\int \mathrm{e}^{-2\alpha|\boldsymbol{r}_1-\boldsymbol{R}_0|^2}\frac{1}{r_{12}}\mathrm{e}^{-2\alpha|\boldsymbol{r}_2-\boldsymbol{R}_0|^2}\mathrm{d}\tau_1\mathrm{d}\tau_2$$

This is a single-centre two-electron integral, with both exponential functions spherically symmetrical Gaussians centred on the midpoint of the bond, and much faster to evaluate than the original two-centre integral. ●

Some of the basis sets that employ Gaussian functions and are commonly used in electronic structure calculations are given in Table 6.1. An STO-*N*G basis is a minimal basis set in which each basis function is itself a linear combination of N Gaussians; the STO in the name of the basis reflects the fact that each linear

Table 6.1 Basis set designations and example basis sets for H_2O

General basis	Example basis	Basis functions
STO-*N*G	STO-3G	For each O 1s, 2s, $2p_x$, $2p_y$, $2p_z$ and H 1s orbital: One function, a linear combination of 3 Gaussians
m-*np*G	6-31G	For O 1s orbital: One linear combination of 6 Gaussians
		For each O 2s, $2p_x$, $2p_y$, $2p_z$ and H 1s orbital: 2 functions: – One Gaussian function – One linear combination of 3 Gaussians
m-*np*G*	6-31G*	6-31G plus d-type polarization functions on O
m-*np*G**	6-31G**	6-31G* plus p-type polarization functions on each H
m-*npq*G	6-311G	6-31G plus an additional Gaussian for each O 2s, $2p_x$, $2p_y$, $2p_z$ and H 1s orbital
m-*npq*+G	6-311+G	6-311G plus diffuse s- and p-type Gaussians on O
m-*npq*++G	6-311++G	6-311+G plus diffuse Gaussians on each H
m-*npq*+G*	6-311+G*	6-311+G plus d-type polarization functions on O
m-*npq*+G**	6-311+G**	6-311+G* plus p-type polarization functions on each H

combination is chosen by a least-squares fit to a Slater-type function. An *m*-*np*G basis is a split-valence basis set in which each core atomic orbital is represented by one function (a linear combination of *m* Gaussians) and each valence orbital is represented by two basis functions, one a linear combination of *n* Gaussians and the other of *p* Gaussian functions. The addition of d-type polarization functions for non-hydrogen atoms to the *m*-*np*G basis yields an *m*-*np*G* basis; further addition of p-type polarization functions for hydrogen atoms results in an *m*-*np*G** basis set. In an *m*-*npq*G basis, each valence atomic orbital is represented by three basis functions, linear combinations of *n*, *p*, and *q* Gaussians, respectively. Addition of diffuse (small α-valued, eqn 6.17) s- and p-type Gaussians on non-hydrogen atoms results in an *m*-*npq*+G basis set; additional diffuse functions to hydrogen, *m*-*npq*++G. A considerable amount of work has gone into the development of efficient basis sets and this is still an active area of research.

We have arrived at the point where we can see that the Hartree–Fock approach, coupled with the use of basis set functions, requires the evaluation of a large number of integrals. There are two approaches commonly taken at this point. In **semiempirical methods**, the integrals encountered are either set to zero or estimated from experimental data. In ***ab initio* methods**, an attempt is made to evaluate the integrals numerically, using as input only the values of fundamental constants and atomic numbers of the atoms present in the molecule.

The first approach: semiempirical methods

In semiempirical methods, many of the integrals that occur in a calculation are estimated by appealing to spectroscopic data or physical properties such as ionization energies, or by using a series of rules to set certain integrals equal to zero. These methods are applied routinely to molecules containing large numbers of atoms because of their computational speed but there is often a sacrifice in the accuracy of the results.

6.4 The Hückel method revisited

Semiempirical methods were first developed for conjugated π systems, the most famous semiempirical procedure being Hückel molecular orbital theory (HMO theory, Section 5.6).

The initial assumption of HMO theory is the separate treatment of π and σ electrons, which is justified by the different energies and symmetries of the orbitals. The secular determinant, from which the π-orbital energies and wavefunctions are obtained, has a form similar to that of eqn 6.12 and is written in terms of overlap integrals and hamiltonian matrix elements. The overlap integrals are set to 0 or 1, the diagonal hamiltonian matrix elements are set to a parameter α, and off-diagonal elements either to 0 or the parameter β. The HMO approach is useful for qualitative, rather than quantitative, discussions of conjugated π systems because it treats repulsions between electrons very poorly.

● A BRIEF ILLUSTRATION

Here we return to the third *illustration* of Section 6.2 and set $S = 0$. The diagonal Fock matrix elements are set equal to α (that is, we set $F_{AA} = F_{BB} = \alpha$), and the off-diagonal elements are set equal to β (that is, we set $F_{AB} = F_{BA} = \beta$). Note that the dependence of these integrals on the coefficients is swept aside, so we do not have to work towards self-consistency. The quadratic equation for the energies

$$(1 - S^2)\varepsilon^2 - (F_{AA} + F_{BB} - SF_{AB} - SF_{BA})\varepsilon + (F_{AA}F_{BB} - F_{AB}F_{BA}) = 0$$

becomes simply

$$\varepsilon^2 - 2\alpha\varepsilon + \alpha^2 - \beta^2 = 0$$

and the roots are $\varepsilon = \alpha \pm \beta$, exactly as we found in Section 5.6. ●

6.5 Differential overlap

In the second most primitive and severe approach, called **complete neglect of differential overlap** (CNDO), all two-electron integrals of the form (AB|CD) are set to zero unless χ_A and χ_B are the same, and likewise for χ_C and χ_D. That is, only integrals of the form (AA|CC) survive and they are often taken to be parameters with values adjusted until the calculated energies are in agreement with experiment. The origin of the term 'differential overlap' is that what we normally take to be a measure of 'overlap' is the integral $\int \chi_A \chi_B d\tau$. The differential of an integral of a function is the function itself, so in this sense the 'differential' overlap is the product $\chi_A \chi_B$. The implication is that we then simply compare orbitals: if they are the same, the integral is retained; if different, it is discarded.

● A BRIEF ILLUSTRATION

The expression for F_{AA} derived in the final *illustration* in Section 6.2 is

$$F_{AA} = E_A + c_{Aa}^2(AA|AA) + 2c_{Aa}c_{Ba}(AA|BA) + c_{Ba}^2\{2(AA|BB) - (AB|BA)\}$$

The last integral has the form

$$(AB|BA) = j_0 \int \chi_A(1)\chi_B(1)\frac{1}{r_{12}}\chi_B(2)\chi_A(2)d\tau_1 d\tau_2$$

The 'differential overlap' term $\chi_A(1)\chi_B(1)$ is set equal to zero, so in the CNDO approximation the integral is set equal to zero. The same is true of the integral (AA|BA). It follows that we write

$$F_{AA} \approx E_A + c_{Aa}^2(AA|AA) + 2c_{Ba}^2(AA|BB)$$

and identify the surviving two two-electron integrals as empirical parameters. ●

Self-test 6.2 Apply the CNDO approximation to F_{AB} for the same system.

$$[F_{AB} = \int \chi_A(1)h_1\chi_B(1)d\tau_1 - c_{Aa}c_{Bb}(AA|BB)]$$

More recent semiempirical methods make less draconian decisions about which integrals are to be ignored, but they are all descendants of the early CNDO technique. Whereas CNDO sets integrals of the form (AB|AB) to zero for all different χ_A and χ_B, **intermediate neglect of differential overlap** (INDO) does not neglect the (AB|AB) for which different basis functions χ_A and χ_B are centred on the same nucleus. Because these integrals are important for explaining energy differences between terms corresponding to the same electronic configuration, INDO is much preferred over CNDO for spectroscopic investigations. A still less severe approximation is **neglect of diatomic differential overlap** (NDDO) in which (AB|CD) is neglected only when χ_A and χ_B are centred on different nuclei or when χ_C and χ_D are centred on different nuclei.

There are other semiempirical methods, with names such as **modified intermediate neglect of differential overlap** (MINDO), **modified neglect of differential overlap** (MNDO), Austin model 1 (AM1), PM3, and **pairwise distance directed Gaussian** (PDDG). In each case, the values of integrals are either set to zero or set to parameters with values that have been determined by attempting to optimize agreement with experiment, such as measured values of enthalpies of formation, dipole moments, and ionization energies. MINDO is useful for the study of hydrocarbons; it tends to give more accurate computed results than MNDO but it gives poor results for systems with hydrogen bonds. AM1, PM3, and PDDG are improved versions of MNDO.

The second approach: *ab initio* methods

In *ab initio* methods, the two-electron integrals are evaluated numerically. However, even for small molecules, Hartree–Fock calculations with large basis sets and efficient and accurate calculation of two-electron integrals can give very poor results because they are rooted in the orbital approximation and the average effect of the other electrons on the electron of interest. Thus, the true wavefunction for H_2 is a function of the form $\Psi(\boldsymbol{r}_1,\boldsymbol{r}_2)$, with a complicated behaviour as $\boldsymbol{r}_1$ and $\boldsymbol{r}_2$ vary and perhaps approach one another. This complexity is lost when we write the wavefunction as a simple product of two functions, $\psi(\boldsymbol{r}_1)\psi(\boldsymbol{r}_2)$ and treat each electron as moving in the average field of the other electrons. That is, the approximations of the Hartree–Fock method imply that no attempt is made to take into account **electron correlation**, the tendency of electrons to stay apart in order to minimize their mutual repulsion. Most modern work in electronic structure, such as the approaches discussed in the following two sections as well as more sophisticated approaches that are beyond the scope of this text, tries to take electron correlation into account.

6.6 Configuration interaction

When we work through the formalism described so far using a basis set of N_b orbitals, we generate N_b molecular orbitals. However, if there are N_e electrons to accommodate, in the ground state only $\frac{1}{2}N_e$ of these N_b orbitals are occupied, leaving $N_b - \frac{1}{2}N_e$ so-called **virtual orbitals** unoccupied. The ground state is

$$\Psi_0 = (1/N_e!)^{1/2}|\psi_a^\alpha(1)\psi_a^\beta(2)\psi_b^\alpha(3)\psi_b^\beta(4)\cdots\psi_u^\beta(N_e)|$$

where ψ_u is the HOMO (Section 5.6). We can envisage transferring an electron from an occupied orbital to a virtual orbital ψ_v, and forming the corresponding **singly excited determinant**, such as

$$\Psi_1 = (1/N_e!)^{1/2}|\psi_a^{\alpha}(1)\psi_a^{\beta}(2)\psi_b^{\alpha}(3)\psi_v^{\beta}(4)\cdots\psi_u^{\beta}(N_e)|$$

Here a β electron, 'electron 4', has been promoted from ψ_b into ψ_v, but there are many other possible choices. We can also envisage doubly excited determinants, and so on. Each of the Slater determinants constructed in this way is called a **configuration state function** (CSF).

Now we come to the point of introducing these CSFs. In 1959 P.-O. Löwdin proved that the *exact* wavefunction (within the Born–Oppenheimer approximation) can be expressed as a linear combination of CSFs found from the exact solution of the Hartree–Fock equations:

$$\Psi = C_0\Psi_0(\equiv) + C_1\Psi_1(\equiv) + C_2\Psi_2(\equiv) + \cdots \qquad (6.18)$$

The inclusion of CSFs to improve the wavefunction in this way is called **configuration interaction** (CI). Configuration interaction can, at least in principle, yield the exact ground-state wavefunction and energy and thus accounts for the electron correlation neglected in Hartree–Fock methods. However, the wavefunction and energy are exact only if an infinite number of CSFs are used in the expansion in eqn 6.18; in practice, we are resigned to using a finite number of CSFs.

● A BRIEF ILLUSTRATION

We can begin to appreciate why CI improves the wavefunction of a molecule by considering H_2 again. We saw in the first *illustration* in Section 6.1 that, after expanding the Slater determinant, the ground state is

$$\Psi_0 = \psi_a(1)\psi_a(2)\sigma_-(1,2)$$

where $\sigma_-(1,2)$ is the singlet spin state wavefunction. We also know that if we use a minimal basis set and ignore overlap, we can write $\psi_a = (1/\sqrt{2})\{\chi_A + \chi_B\}$. Therefore

$$\begin{aligned}\Psi_0 &= \tfrac{1}{2}\{\chi_A(1) + \chi_B(1)\}\{\chi_A(2) + \chi_B(2)\}\sigma_-(1,2)\\ &= \tfrac{1}{2}\{\chi_A(1)\chi_A(2) + \chi_A(1)\chi_B(2) + \chi_B(1)\chi_A(2)\\ &\quad + \chi_B(1)\chi_B(2)\}\sigma_-(1,2)\end{aligned}$$

We can see a deficiency in this wavefunction: there are equal probabilities of finding both electrons on A (the first term) or on B (the fourth term) as there are for finding one electron on A and the other on B (the second and third terms). That is, electron correlation has not been taken into account and we can expect the calculated energy to be too high.

From two basis functions we can construct two molecular orbitals: we denote the second one $\psi_b = (1/\sqrt{2})\{\chi_A - \chi_B\}$. We need not consider the singly excited determinant constructed by moving one electron from ψ_a to ψ_b because it will be of ungerade symmetry and therefore not contribute to the gerade ground state of dihydrogen. A doubly excited determinant based on ψ_b would be

$$\begin{aligned}\Psi_2 &= \psi_b(1)\psi_b(2)\sigma_-(1,2)\\ &= \tfrac{1}{2}\{\chi_A(1) - \chi_B(1)\}\{\chi_A(2) - \chi_B(2)\}\sigma_-(1,2)\\ &= \tfrac{1}{2}\{\chi_A(1)\chi_A(2) - \chi_A(1)\chi_B(2) - \chi_B(1)\chi_A(2)\\ &\quad + \chi_B(1)\chi_B(2)\}\sigma_-(1,2)\end{aligned}$$

If we were simply to subtract one CSF from the other, the outer terms would cancel and we would be left with

$$\Psi_0 - \Psi_2 = \{\chi_A(1)\chi_B(2) + \chi_B(1)\chi_A(2)\}\sigma_-(1,2)$$

According to this wavefunction, the two electrons will never be found on the same atom: we have overcompensated for electron configuration. The obvious middle-ground is to form the linear combination $\Psi = C_0\Psi_0 + C_2\Psi_2$ and look for the values of the coefficients that minimize the energy. ●

The *illustration* shows that even a limited amount of CI can introduce some electron correlation; full CI—using orbitals built from a finite basis and allowing for all possible excitations—will take electron correlation into account more fully. The optimum procedure, using orbitals that form an infinite basis and allowing all excitations, is computationally impractical.

The optimum expansion coefficients in eqn 6.18 are found by using the variation principle; as in *Justification 6.1* for the Hartree–Fock method, application of the variation principle for CI results in a set of simultaneous equations for the expansion coefficients.

● A BRIEF ILLUSTRATION

If we take the linear combination $\Psi = C_0\Psi_0 + C_2\Psi_2$, the usual procedure for the variation method (Section 5.5) leads to the secular equation $|H - ES| = 0$, from which we can find the improved energy. Specifically:

$$H = \begin{pmatrix} H_{00} & H_{02} \\ H_{20} & H_{22}\end{pmatrix} \qquad H_{MN} = \int \Psi_M \hat{H} \Psi_N \, d\tau_1 d\tau_2$$

$$S = \begin{pmatrix} S_{00} & S_{02} \\ S_{20} & S_{22}\end{pmatrix} \qquad S_{MN} = \int \Psi_M \Psi_N \, d\tau_1 d\tau_2$$

and the secular equation we must solve to find E is (note that $S_{02} = S_{20}$ and that $H_{02} = H_{20}$ due to hermiticity)

$$\begin{vmatrix} H_{00} - ES_{00} & H_{02} - ES_{02} \\ H_{20} - ES_{20} & H_{22} - ES_{22}\end{vmatrix} = (H_{00} - ES_{00})(H_{22} - ES_{22}) - (H_{02} - ES_{02})^2 = 0$$

which is easily rearranged into a quadratic equation for E. As usual, the problem boils down to an evaluation of various integrals that appear in the matrix elements.

The molecular orbitals ψ_a and ψ_b are orthogonal, so S is diagonal and, provided ψ_a and ψ_b are normalized, $S_{00} = S_{22} = 1$. To evaluate the hamiltonian matrix elements, we first write the hamiltonian as in eqn 6.3 ($\hat{H} = h_1 + h_2 + j_0/r_{12}$), where h_1 and h_2 are the core hamiltonians for electrons 1 and 2, respectively, and so

$$H_{00} = \int \Psi_0 \left(h_1 + h_2 + \frac{j_0}{r_{12}} \right) \Psi_0 \, d\tau_1 d\tau_2$$

The first term in this integral (noting that the spin states are normalized) is:

$$\int \Psi_0 h_1 \Psi_0 \, d\tau_1 d\tau_2 = \int \psi_a(1)\psi_a(2) h_1 \psi_a(1)\psi_a(2) \, d\tau_1 d\tau_2$$

$$= \int \psi_a(1) h_1 \psi_a(1) \, d\tau_1$$

Similarly,

$$\int \Psi_0 h_2 \Psi_0 \, d\tau_1 d\tau_2 = \int \psi_a(2) h_2 \psi_a(2) \, d\tau_2$$

For the electron–electron repulsion term, using the notation of eqn 6.14,

$$\int \Psi_0 \frac{j_0}{r_{12}} \Psi_0 \, d\tau_1 d\tau_2 = j_0 \int \psi_a(1)\psi_a(1) \frac{1}{r_{12}} \psi_a(2)\psi_a(2) \, d\tau_1 d\tau_2$$
$$= c_{Aa}^4 (AA|AA) + c_{Aa}^3 c_{Ba} (AA|AB) + \cdots + c_{Ba}^4 (BB|BB)$$

Expressions of a similar kind can be developed for the other three elements of H, so the optimum energy can be found by substituting the calculated values of the coefficients and the integrals into the expression for the roots of the quadratic equation for E. The coefficients in the CI expression for Ψ can then be found in the normal way by using the lowest value of E and solving the secular equations. ●

6.7 Many-body perturbation theory

The application of perturbation theory to a molecular system of interacting electrons and nuclei is called **many-body perturbation theory**. Recall from discussions of perturbation theory in Chapter 2 (see eqn 2.31) that the hamiltonian is expressed as a sum of a simple, 'model' hamiltonian, $\hat{H}^{(0)}$, and a perturbation $\hat{H}^{(1)}$. Because we wish to find the correlation energy, a natural choice for the model hamiltonian are the Fock operators of the HF-SCF method and for the perturbation we take the difference between the Fock operators and the true many-electron hamiltonian (eqn 6.1). That is,

$$\hat{H} = \hat{H}^{(0)} + \hat{H}^{(1)} \qquad \text{with } \hat{H}^{(0)} = \sum_{i=1}^{N_e} f_i \tag{6.19}$$

Because the core hamiltonian in the Fock operator in eqn 6.8 cancels the one-electron terms in the full hamiltonian, the perturbation is the difference between the instantaneous interaction between the electrons (the third term in eqn 6.1) and the average interaction (as represented by the operators J and K in the Fock operator). Thus, for electron 1

$$\hat{H}^{(1)}(1) = \sum_i \frac{j_0}{r_{1i}} - \sum_m \{2J_m(1) - K_m(1)\} \tag{6.20}$$

where the first sum (the true interaction) is over all the electrons other than electron 1 itself and the second sum (the average interaction) is over all the occupied orbitals. This choice was first made by C. Møller and M.S. Plesset in 1934 and the method is called **Møller–Plesset perturbation theory** (MPPT). Applications of MPPT to molecular systems were not undertaken until the 1970s and the rise of sufficient computing power.

As usual in perturbation theory, the true wavefunction is written as a sum of the eigenfunction of the model hamiltonian and higher-order correction terms. The **correlation energy**, the difference between the true energy and the HF energy, is given by energy corrections that are second order and higher. If we suppose that the true wavefunction of the system is given by a sum of CSFs like that in eqn 6.18, then (see eqn 2.35)

$$E_0^{(2)} = \sum_{M \neq 0} \frac{|\int \Psi_M \hat{H}^{(1)} \Psi_0 \, d\tau|^2}{E_0^{(0)} - E_M^{(0)}} \tag{6.21}$$

According to **Brillouin's theorem**, only doubly excited Slater determinants have nonzero $\hat{H}^{(1)}$ matrix elements and hence only they make a contribution to $E_0^{(2)}$. The identification of the second-order energy correction with the correlation energy is the basis of the MPPT method denoted MP2. The extension of MPPT to include third- and fourth-order energy corrections are denoted MP3 and MP4, respectively.

● A BRIEF ILLUSTRATION

According to Brillouin's theorem, and for our simple model of H_2 built from two basis orbitals, we write

$$\Psi = C_0\Psi_0 + C_2\Psi_2 \qquad \text{with } \Psi_0 = \psi_a(1)\psi_a(2)\sigma\_(1,2)$$
$$\Psi_2 = \psi_b(1)\psi_b(2)\sigma\_(1,2)$$

The only matrix element we need for the sum in eqn 6.21 is

$$\int \Psi_2 \hat{H}^{(1)} \Psi_0 \, d\tau_1 d\tau_2 = j_0 \int \psi_b(1)\psi_b(2) \frac{1}{r_{12}} \psi_a(1)\psi_a(2) \, d\tau_1 d\tau_2$$

All the integrals over terms based on J and K are zero because these are one-electron operators and so either $\psi_a(1)$ or $\psi_a(2)$ is left unchanged and its orthogonality to ψ_b ensures that the integral vanishes. We now expand each molecular orbital in terms of the basis functions χ_A and χ_B, and obtain

$$\int \Psi_2 \hat{H}^{(1)} \Psi_0 \, d\tau_1 d\tau_2 = c_{Ab}^2 c_{Aa}^2 (AA|AA)$$
$$+ c_{Ab} c_{Bb} c_{Aa}^2 (BA|AA) + \cdots + c_{Bb}^2 c_{Ab}^2 (BB|BB)$$

If we ignore overlap the coefficients are all equal to $\pm 1/\sqrt{2}$, and if we use symmetries like (AA|AB) = (AA|BA) and (AA|AB) = (BB|BA), this expression simplifies to

$$\int \Psi_2 \hat{H}^{(1)} \Psi_0 \mathrm{d}\tau_1 \mathrm{d}\tau_2 = \tfrac{1}{2}\{(\mathrm{AA}|\mathrm{AA}) - (\mathrm{AA}|\mathrm{BB})\}$$

It follows that the second-order estimate of the correlation energy is

$$E_0^{(2)} = \frac{\frac{1}{4}\{(\mathrm{AA}|\mathrm{AA}) - (\mathrm{AA}|\mathrm{BB})\}^2}{E_0^{(0)} - E_2^{(0)}}$$

$$= \frac{\{(\mathrm{AA}|\mathrm{AA}) - (\mathrm{AA}|\mathrm{BB})\}^2}{8(\varepsilon_a - \varepsilon_b)}$$

The term (AA|AA) – (AA|BB) is the difference in repulsion energy between both electrons being confined to one atom and each being on a different atom. ●

The third approach: density functional theory

A technique that has gained considerable ground in recent years to become one of the most widely used procedures for the calculation of molecular structure is **density functional theory** (DFT). Its advantages include less demanding computational effort, less computer time, and—in some cases, particularly for d-metal complexes—better agreement with experimental values than is obtained from Hartree–Fock based methods.

6.8 The Kohn–Sham equations

The central focus of DFT is not the wavefunction but the electron probability density, ρ (Section 1.5). The 'functional' part of the name comes from the fact that the energy of the molecule is a function of the electron density and the electron density is itself a function of the positions of the electrons, $\rho(\boldsymbol{r})$. In mathematics a function of a function is called a *functional*, and in this specific case we write the energy as the functional $E[\rho]$. We have encountered a functional before but did not use this terminology: the expectation value of the hamiltonian is the energy expressed as a functional of the wavefunction, for a single value of the energy, $E[\psi]$, is associated with each function ψ. An important point to note is that because $E[\psi]$ is an integral of $\psi H \psi$ over all space, it has contributions from the whole range of values of ψ.

Simply from the structure of the hamiltonian in eqn 6.1 we can suspect that the energy of a molecule can be expressed as contributions from the kinetic energy, the electron–nucleus interaction, and the electron–electron interaction. The first two contributions depend on the electron density distribution. The electron–electron interaction is likely to depend on the same quantity, but we have to be prepared for there to be a modification of the classical electron–electron interaction due to electron exchange (the contribution which in Hartree–Fock theory is expressed by K). That the exchange contribution can be expressed in terms of the electron density is not at all obvious, but in 1964 P. Hohenberg and W. Kohn were able to prove that the exact ground-state energy of an N_e-electron molecule is uniquely determined by the electron probability density. They showed that it is possible to write

$$E[\rho] = E_{\text{Classical}}[\rho] + E_{\text{XC}}[\rho] \tag{6.22}$$

where $E_{\text{Classical}}[\rho]$ is the sum of the contributions of kinetic energy, electron–nucleus interactions, and the classical electron–electron potential energy, and $E_{\text{XC}}[\rho]$ is the **exchange–correlation energy**. This term takes into account all the non-classical electron–electron effects due to spin and applies small corrections to the kinetic energy part of $E_{\text{Classical}}$ that arise from electron–electron interactions. The Hohenberg–Kohn theorem guarantees the existence of $E_{\text{XC}}[\rho]$ but—like so many existence theorems in mathematics—gives no clue about how it should be calculated.

The first step in the implementation of this approach is to calculate the electron density. The relevant equations were deduced by Kohn and L.J. Sham in 1965, who showed that ρ can be expressed as a contribution from each electron present in the molecule, and written

$$\rho(\boldsymbol{r}) = \sum_{i=1}^{N_e} |\psi_i(\boldsymbol{r})|^2 \tag{6.23}$$

ψ_i is called a **Kohn–Sham orbital** and is a solution of the **Kohn–Sham equation**, which closely resembles the form of the Schrödinger equation (on which it is based). For a two-electron system,

$$h_1\psi_i(1) + j_0 \int \frac{\rho(2)}{r_{12}} \mathrm{d}\tau_2 \psi_i(1) + V_{\text{XC}}(1)\psi_i(1) = \varepsilon_i \psi_i(1) \tag{6.24}$$

The first term is the usual core term, the second term is the classical interaction between electron 1 and electron 2, and the third term takes exchange effects into account and is called the **exchange–correlation potential**. The ε_i are the Kohn–Sham orbital energies.

6.9 The exchange–correlation energy

The exchange–correlation potential plays a central role in DFT and can be calculated once we know the exchange–correlation energy $E_{\text{XC}}[\rho]$ by forming the following 'functional derivative':

$$V_{\text{XC}}(\boldsymbol{r}) = \frac{\delta E_{\text{XC}}[\rho]}{\delta \rho} \tag{6.25}$$

A functional derivative is defined like an ordinary derivative, but we have to remember that $E_{XC}[\rho]$ is a quantity that gets its value from the entire range of values of $\rho(\boldsymbol{r})$, not just from a single point. Thus, when $\boldsymbol{r}$ undergoes a small change d$\boldsymbol{r}$, the density changes by $\delta\rho$ to $\rho(\boldsymbol{r}+\mathrm{d}\boldsymbol{r})$ at each point and $E_{XC}[\rho]$ undergoes a change that is the sum (integral) of all such changes:

$$\delta E_{XC}[\rho] = \int \frac{\delta E_{XC}[\rho]}{\delta\rho}\delta\rho\,\mathrm{d}\boldsymbol{r} = \int V_{XC}(\boldsymbol{r})\delta\rho\,\mathrm{d}\boldsymbol{r}$$

Note that V_{XC} is an ordinary function of $\boldsymbol{r}$, not a functional: it is the local contribution to the integral that defines the global dependence of $E_{XC}[\rho]$ on $\delta\rho$ throughout the range of integration.

● A BRIEF ILLUSTRATION

The greatest challenge in density functional theory is to find an accurate expression for the exchange–correlation energy. One widely used but approximate form for $E_{XC}[\rho]$ is based on the model of a uniform electron gas, a hypothetical electrically neutral system in which electrons move in a space of continuous and uniform distribution of positive charge. For a uniform electron gas, the exchange–correlation energy can be written as the sum of an exchange contribution and a correlation contribution. The latter is a complicated functional that is beyond the scope of this chapter; we ignore it here. Then the exchange–correlation energy is

$$E_{XC}[\rho] = \int A\rho^{4/3}\mathrm{d}\boldsymbol{r} \qquad \text{with } A = -(9/8)(3/\pi)^{1/2}j_0$$

When the density changes from $\rho(\boldsymbol{r})$ to $\rho(\boldsymbol{r}) + \delta\rho(\boldsymbol{r})$ at each point (Fig. 6.6), the functional changes from $E_{XC}[\rho]$ to $E_{XC}[\rho+\delta\rho]$:

$$E_{XC}[\rho+\delta\rho] = \int A(\rho+\delta\rho)^{4/3}\mathrm{d}\boldsymbol{r}$$

The integrand can be expanded in a Taylor series (*Mathematical background 1*) and, discarding terms of order $\delta\rho^2$ and higher, we obtain:

$$E_{XC}[\rho+\delta\rho] = \int (A\rho^{4/3} + \tfrac{4}{3}A\rho^{1/3}\delta\rho)\mathrm{d}\boldsymbol{r}$$
$$= E_{XC}[\rho] + \int \tfrac{4}{3}A\rho^{1/3}\delta\rho\,\mathrm{d}\boldsymbol{r}$$

Therefore, the differential δE_{XC} of the functional (the difference $E_{XC}[\rho+\delta\rho] - E_{XC}[\rho]$ that depends linearly on $\delta\rho$) is

$$\delta E_{XC}[\rho] = \int \tfrac{4}{3}A\rho^{1/3}\delta\rho\,\mathrm{d}\boldsymbol{r}$$

and therefore

$$V_{XC}(\boldsymbol{r}) = \tfrac{4}{3}A\rho(\boldsymbol{r})^{1/3} = -\tfrac{3}{2}(3/\pi)^{1/3}j_0\rho(\boldsymbol{r})^{1/3} \qquad (6.26)$$ ●

Self-test 6.3 Find the exchange–correlation potential if the exchange–correlation energy is given by $E_{XC}[\rho] = \int B\rho^2\mathrm{d}\boldsymbol{r}$.
$[V_{XC}(\boldsymbol{r}) = 2B\rho(\boldsymbol{r})]$

The Kohn–Sham equations must be solved iteratively and self-consistently (Fig. 6.7). First, we guess the electron density; it is common to use a superposition of atomic electron probability densities. Second, the exchange–correlation potential is calculated by assuming an approximate form of the dependence of the exchange–correlation energy on the electron density and evaluating the functional derivative. Next, the Kohn–Sham equations are solved to obtain an initial set of Kohn–Sham

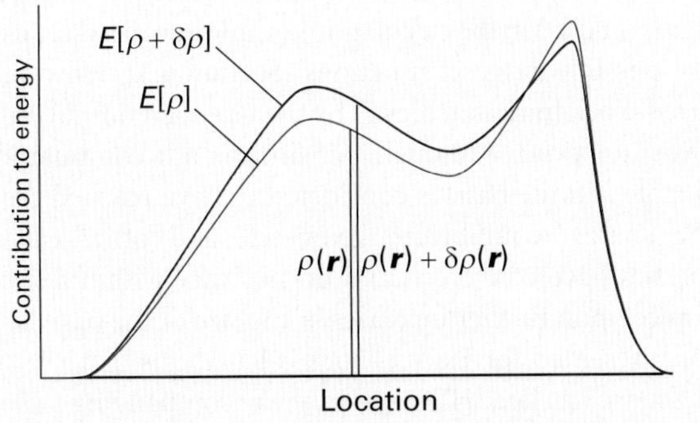

Fig. 6.6 The change in the exchange–correlation energy functional from $E_{XC}[\rho]$ to $E_{XC}[\rho+\delta\rho]$ (the area under each curve) as the density changes from ρ to $\rho+\delta\rho$ at each point $\boldsymbol{r}$.

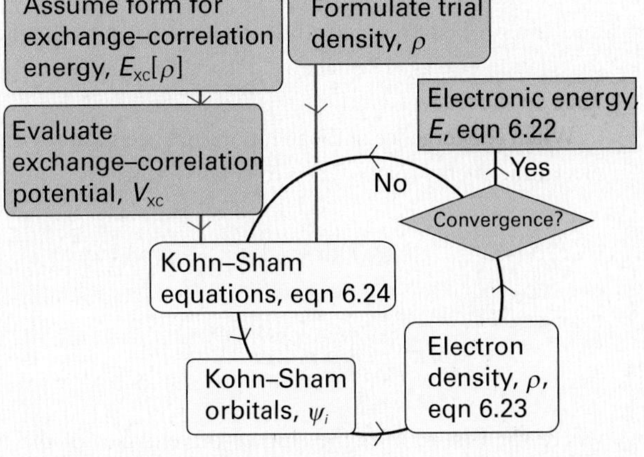

Fig. 6.7 The iteration procedure for solving the Kohn–Sham equations in density functional theory.

orbitals. This set of orbitals is used to obtain a better approximation to the electron probability density (from eqn 6.23) and the process is repeated until the density remains constant to within some specified tolerance. The electronic energy is then computed by using eqn 6.22.

As is the case for the Hartree–Fock one-electron wavefunctions, the Kohn–Sham orbitals can be expanded using a set of basis functions; solving eqn 6.24 then amounts to finding the coefficients in the expansion. Various basis functions, including Slater-type and Gaussian-type orbitals, can be used. Whereas Hartree–Fock methods have computational times that scale as N_b^4, DFT methods scale as N_b^3. Therefore, DFT methods are computationally more efficient, though not necessarily more accurate, than HF methods.

● A BRIEF ILLUSTRATION

In applying DFT to molecular hydrogen, we begin by assuming that the electron density is a sum of atomic electron densities arising from the presence of electrons in the atomic orbitals χ_A and χ_B (which may be STOs or GTOs) and write $\rho(\boldsymbol{r}) = |\chi_A|^2 + |\chi_B|^2$ for each electron. For the exchange–correlation energy E_{XC} we use the form appropriate to a uniform electron gas and the corresponding exchange–correlation potential derived in the previous *illustration* (eqn 6.26).

The Kohn–Sham orbital for the molecule is a solution of

$$\left\{h_1 + j_0\int\frac{\rho(r_2)}{r_{12}}\mathrm{d}r_2 - \tfrac{3}{2}(3/\pi)^{1/3}j_0\rho(r_1)^{1/3}\right\}\psi_1(r_1) = \varepsilon_1\psi_1(r_1)$$

We insert the $\rho(\boldsymbol{r}_1)$ and $\rho(\boldsymbol{r}_2)$ we have assumed and solve this equation numerically for ψ_1. Once we have that orbital, we replace our original guess at the electron density by $\rho(\boldsymbol{r}) = |\psi_1(\boldsymbol{r})|^2$. This density is then substituted back into the Kohn–Sham equation to obtain an improved function $\psi_1(\boldsymbol{r})$ and the process repeated until the density and exchange–correlation energy are unchanged to within a specified tolerance on successive iterations.

When convergence of the iterations has been achieved, the electronic energy (eqn 6.22) is calculated from

$$E[\rho] = 2\int\psi_1(r)h_1\psi_1(r)\mathrm{d}r + j_0\int\rho(r_1)\frac{1}{r_{12}}\rho(r_2)\mathrm{d}r_1\mathrm{d}r_2$$
$$- \tfrac{9}{8}(3/\pi)^{1/3}j_0\int\rho(r)^{4/3}\mathrm{d}r$$

where the first term is the sum of the energies of the two electrons in the field of the two nuclei, the second term is the electron–electron repulsion, and the final term includes the correction due to nonclassical electron–electron effects. ●

Current achievements

Electronic structure calculations provide valuable information about a wide range of important physical and chemical properties. One of the most important is the equilibrium molecular geometry, the arrangement of atoms that results in the lowest energy for the molecule. The calculation of equilibrium bond lengths and bond angles supplements experimental data obtained from structural studies such as X-ray crystallography (Section 9.3), electron diffraction (Section 9.4), and microwave spectroscopy (Section 10.3). Furthermore, analyses of the molecular potential energy curve can yield vibrational frequencies for comparison with results from infrared spectroscopy (Section 10.6) as well as molecular dipole moments.

6.10 Comparison of calculations and experiments

The choice of an electronic structure method to solve a chemical problem is not usually an easy task. Both the chemical accuracy associated with the method and the cost of the calculation (in terms of computational speed and memory) must be taken into account. An *ab initio* method such as full CI or MP2, each of which is capable of yielding accurate results on a molecule with a small number of atoms and electrons, is often computationally impractical for many-electron molecules. In contrast, a semiempirical or DFT calculation might make an electronic structure calculation on the large molecule feasible but with an accompanying sacrifice in reliability. Indeed, no single methodology has been found to be applicable to all molecules. However, the promise that computational chemistry has to enhance our ability to predict chemical and physical properties of a wide range of molecules is sufficient to drive further development of electronic structure methods.

First consider molecular hydrogen, the subject of most of the *illustrations* in this chapter. To compare results from different electronic structure methods, we need to say a few words about the basis set used in the calculations. A minimal basis set uses the fewest possible basis set functions (Section 6.3). However, the **Hartree–Fock limit** is achieved by the use of an infinite number of basis functions. Although this limit is not computationally attainable, a finite basis is considered to have reached the limit if the energy, equilibrium geometry, and other calculated properties have converged and do not vary within a specified tolerance upon further increases in the size of the basis set. The results presented for Hartee–Fock calculations that use such a basis set are labelled 'HF limit' in the accompanying tables. (In practice, the 'HF-limit' in the tables corresponds to a 6-311+G** basis; see Table 6.1.) So that we can compare different electronic structure methods more directly, we report literature results where the same or a similar basis set was used in the different

Table 6.2 Comparison of methods for small H-containing molecules

	Expt	HF limit	MNDO	PM3	CI*	MP2	DFT
R(H—H)/pm	74.2	73.6	66.3	69.9	73.9	73.8	76.7
R(O—H)/pm in H_2O	95.8	94.3	94.3	95.1	95.2	96.0	97.1
Bond angle/° in H_2O	104.5	106.4	106.8	107.7	104.9	103.5	105.1
Dipole moment, $\mu(H_2O)$/D†	1.85	2.2	1.8	1.7	1.9	2.2	2.2

* For dihydrogen, full CI. For water, CI with inclusion of singly and doubly excited determinants.
† 1 D (debye) = 3.336×10^{-30} C m.

types of calculations. The density functional theory calculations to which we refer all used the exchange–correlation potential for a uniform electron gas, including the correlation component neglected in the first *illustration* of Section 6.9.

Table 6.2 compares the equilibrium bond length for dihydrogen determined from various electronic structure methods; the equilibrium geometry corresponds to the minimum in the calculated molecular potential energy. Not surprisingly, the CI and MPPT *ab initio* methods are the most accurate. However, for this simple molecule the Hartree–Fock result is also within chemical accuracy (about 1 pm); the semiempirical methods do not fare as well by comparison but, as also shown in Table 6.2, MNDO and PM3 are more accurate for calculations on water than on dihydrogen. The CI and MP2 methods also achieve chemical accuracy (within 1°) for the bond angle in water. As for the dipole moment of water, the semiempirical methods are found to be slightly more accurate than the Hartree–Fock and density-functional calculations.

Table 6.3 shows some results from semiempirical, MPPT, and DFT calculations of carbon–carbon bond lengths in a variety of small organic molecules as well as the C=C stretching wavenumbers in the alkenes. As we shall see in Section 10.7, vibrational wavenumbers depend on the 'force constants' for displacements from the equilibrium geometry, and they in turn depend on the second derivatives of the potential energy with respect to the displacement. The methods generally do a good job of predicting bond lengths of the single and double bonds and, even though the semiempirical methods do not perform as well in calculating vibrational wavenumbers, the results from Table 6.3 do give us a reasonable level of confidence in the predictive abilities of DFT and semiempirical calculations.

Table 6.3 Comparison of methods for small organic molecules

	Expt	PM3	MP2	DFT
R(C—C)/pm				
propane	152.6	151.2	152.9	151.2
cyclobutane	154.8	154.2	155.0	153.7
R(C=C)/pm				
propene	131.8	132.8	134.1	133.0
cyclobutene	133.2	134.9	135.2	134.1
$\tilde{\nu}$(C=C stretch)/cm^{-1}				
propene	1656	1862	1698	1680
cyclobutene	1570	1772	1598	1600

Confidence in DFT and semiempirical methods becomes particularly important when the cost of computations makes *ab initio* methods impractical; such is the case for typical inorganic and organometallic compounds. Hartree–Fock methods generally perform poorly for d-metal complexes and *ab initio* methods can be prohibitively costly. However, DFT and semiempirical methods (such as PM3, which includes parameters for most d metals) have vastly improved the performance of applications of electronic structure theory to inorganic chemistry.

6.11 Applications to larger molecules

In the area of thermodynamics, computational chemistry is becoming the technique of choice for estimating the enthalpies of formation (Section 14.8) of molecules with complex three-dimensional structures. It also opens the way to exploring the effect of solvation on enthalpies of formation by calculating the enthalpy of formation in the gas phase and then including several solvent molecules around the solute molecule. The numerical results should currently be treated as only estimates with the primary purpose of predicting whether interactions with the solvent increase or decrease the enthalpy of formation. As an example, consider the amino acid glycine, which can exist in a neutral (NH_2CH_2COOH) or zwitterionic ($^+NH_3CH_2CO_2^-$) form. It has been found computationally that, whereas in the gas phase the neutral form has a lower enthalpy of formation than the zwitterion, in water the opposite is true because of strong interactions between the polar solvent and the charges in the zwitterion. Therefore, we might suspect that the zwitterionic form is the predominant one in polar media, as is confirmed by protonation/deprotonation calculations of the type carried out in introductory chemistry courses.

Computational chemistry can be used to predict trends in electrochemical properties, such as reduction potentials (Section 17.6). Several experimental and computational studies of aromatic hydrocarbons indicate that decreasing the energy of the lowest unoccupied molecular orbital (LUMO) enhances the ability of the molecule to accept an electron into the LUMO, with an attendant increase in the value of the molecule's reduction potential. The effect is also observed in quinones and flavins, which are co-factors involved in biological electron transfer reactions. For example, stepwise substitution of the hydrogen atoms in *p*-benzoquinone by methyl groups ($-CH_3$) results in a systematic increase in the energy of the LUMO and a decrease in the reduction potential for formation of the semiquinone radical (**1**):

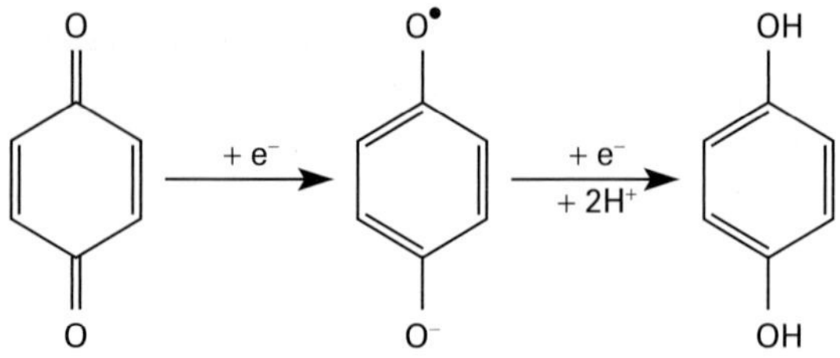

1 Semiquinone radical

The reduction potentials of naturally occurring quinones are also modified by the presence of different substituents, a strategy that imparts specific functions to specific quinones. For example, the substituents in coenzyme Q are largely responsible for positioning its reduction potential so that the molecule can function as an electron shuttle between specific proteins in the respiratory chain (*Impact I17.3*).

The electronic structure calculations described in this chapter provide insight into spectroscopic properties by correlating the absorption wavelengths and the energy gap between the LUMO and the HOMO in a series of molecules. For example, consider the linear polyenes shown in Table 6.4, all of which absorb in the UV region. The table shows that, as expected, the wavelength of the lowest-energy electronic transition decreases as the HOMO–LUMO energy difference increases. The smallest HOMO–LUMO gap and greatest transition wavelength is found for octatetraene, the longest polyene in the group. The wavelength of the transition increases with increasing number of conjugated double bonds in linear polyenes and extrapolation of the trend suggests that a sufficiently long linear polyene should absorb light in the visible region. This is indeed the case for β-carotene (**2**), which absorbs light with $\lambda \approx 450$ nm. The ability of β-carotene to absorb visible light is part of the strategy employed by plants to harvest solar energy for use in photosynthesis (*Impact I19.1*).

Table 6.4 Electronic structure calculations and spectroscopic data

Polyene	ΔE/eV*	λ/nm
══ (C_2H_4)	18.1	163
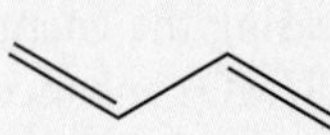	14.5	217
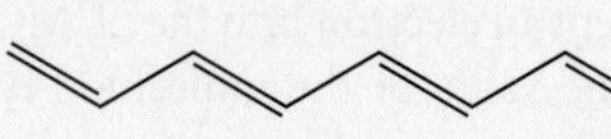	12.7	252
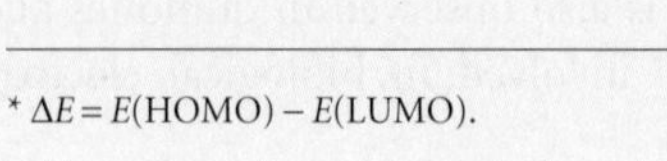	11.8	304

* $\Delta E = E(\text{HOMO}) - E(\text{LUMO})$.

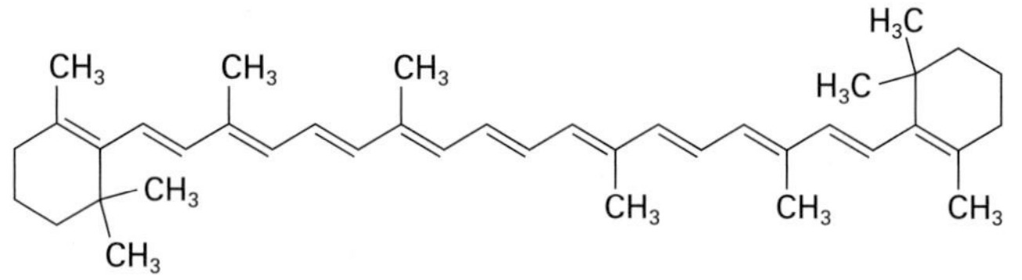

2 β-carotene

IMPACT ON NANOSCIENCE
I6.1 The structures of nanoparticles

Semiconductor oxides, such as TiO_2 and ZnO, are a major area of current research because they can act as *photocatalysts*, substances that accelerate chemical reactions upon absorption of light. Reactions that can be enhanced by photocatalysts include the splitting of water into H_2 and O_2, and the decomposition of pollutants. Among the most popular photocatalytic materials is TiO_2 due to its low cost and catalytic efficiency. The method of preparation of the bulk oxide has a strong influence on its catalytic properties and experiments that attempt to control the form of its crystal lattice have been undertaken widely. Similarly, there is widespread interest in controlling the structure and photocatalytic properties of TiO_2 on the nanometre scale. Computational studies on small clusters of TiO_2 particles can provide insight into effects of size on photochemical properties of nanometre-sized materials, the nature of oxide–substrate interactions, and the growth of larger aggregates.

The most stable form of bulk TiO_2 at atmospheric pressure and room temperature is rutile (Fig. 6.8), in which each titanium atom is surrounded by six oxygen atoms and each O is

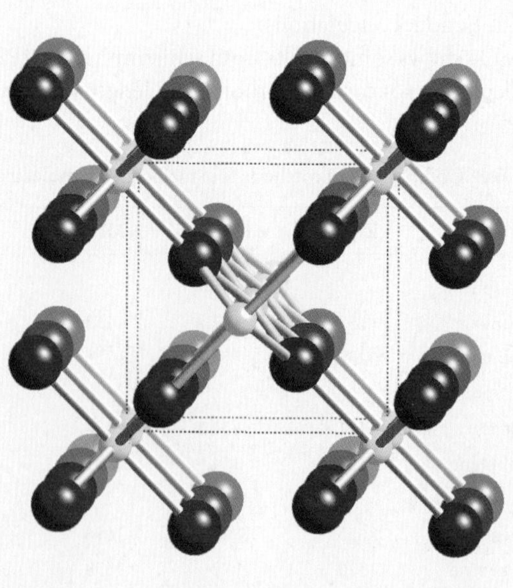

Fig. 6.8 The rutile structure of TiO_2 (blue spheres: Ti; red spheres: O).

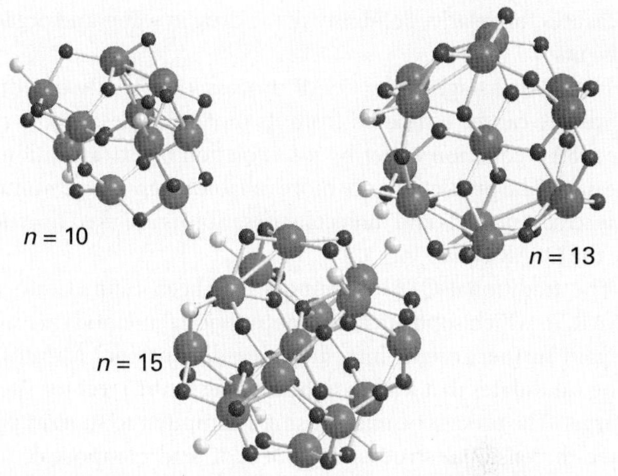

Fig. 6.9 Stable geometries for Ti_nO_{2n} clusters, with $n = 10$, 13, and 15, determined from density functional theory calculations. [From S. Hamad *et al. J. Phys. Chem. B*, 2005, **109**, 15741.]

surrounded by three Ti atoms. Each octahedron composed of the six O atoms around the Ti centre shares two edges with other octahedrons. Some experimental studies on TiO_2 nanoparticles suggest that the nanostructure is anatase, an elongated form of rutile in which the octahedrons share four edges. Other structural distortions appear to be possible as the particle size decreases.

A recent computational study on small Ti_nO_{2n} clusters with $n = 1$–15 has identified the most stable structures for nanoparticles with sizes less than 1 nm. To accomplish the challenging computational task of finding the most probable cluster structures, density functional theory was used to evaluate the energy as a function of geometry and specialized minimization algorithms were used to find equilibrium structures. The calculations revealed compact equilibrium structures with coordination numbers of the Ti atoms increasing with particle size. These structures were found not to be related to anatase. For $Ti_{11}O_{22}$ up to $Ti_{15}O_{30}$, the largest nanoparticle studied, the structures with lower energies consisted of a central octahedron surrounded by square base pyramids, trigonal bipyramids, and tetrahedra (Fig. 6.9). The DFT calculations revealed that structures with a small number of square base pyramids are particularly stable. The stable structures found for the various cluster sizes can be used in further computational work to study the effects of nanostructure on the photochemical properties of TiO_2.

IMPACT ON MEDICINE
I6.2 Molecular recognition and drug design

A drug is a small molecule or protein that binds to a specific receptor site of a target molecule, such as a larger protein or nucleic acid, and inhibits the progress of disease. To devise efficient therapies, we need to know how to characterize and optimize both the three-dimensional structure of the drug and the molecular interactions between the drug and its target.

The binding of a ligand, or *guest*, to a biopolymer, or *host*, is also governed by molecular interactions. Examples of biological *host–guest complexes* include enzyme–substrate complexes, antigen–antibody complexes, and drug–receptor complexes. In all these cases, a site on the guest contains functional groups that can interact with complementary functional groups of the host. Many specific intermolecular contacts must in general be made in a biological host–guest complex and, as a result, a guest binds only to hosts that are chemically similar. The strict rules governing molecular recognition of a guest by a host control every biological process, from metabolism to immunological response, and provide important clues for the design of effective drugs for the treatment of disease.

A full assessment of molecular recognition between a drug and its target requires knowledge of the full spectrum of interactions discussed in Chapter 8. But we can already anticipate some of the factors that optimize the formation of host–guest complexes. For example, a hydrogen bond donor group of the guest must be positioned near a hydrogen bond acceptor group of the host for tight binding to occur. We also expect that an electron-poor region in a host should interact strongly with an electron-rich region of a guest. Computational studies of the types described in this chapter can identify regions of a molecule that have high or low electron densities. Furthermore, graphical representation of numerical results allows for direct visualization of molecular properties, such as the distribution of electron density, thereby enhancing our ability to predict the nature of intermolecular contacts between host and guest.

Consider a protein host with the amino acid serine in a site that binds guests. Electronic structure methods on the serine molecule can provide electronic wavefunctions and electron probability densities at any point in the molecule. From the electron probability densities and the charges of the atomic nuclei, one can compute the electric potential (*Fundamentals F.6*) at any point in the molecule (except at the nuclei themselves). The resulting electric potential can be displayed as an **electrostatic potential surface** (an 'elpot surface') in which net positive potential is shown in one colour and net negative potential is shown in another, with intermediate gradations of colour. Such an elpot surface for serine ($NH_2CH(CH_2OH)COOH$) is shown in Fig. 6.10 where net positive potential is shown in blue and net negative potential in red. The electron-rich regions of the amino acid are susceptible to attack by an electropositive species and the electron-poor regions to attack by an electronegative species.

There are two main strategies for the discovery of a drug. In *structure-based design*, new drugs are developed on the basis of the known structure of the receptor site of a known target. However, in many cases a number of so-called *lead compounds* are known to have some biological activity but little information is available about the target. To design a molecule with improved

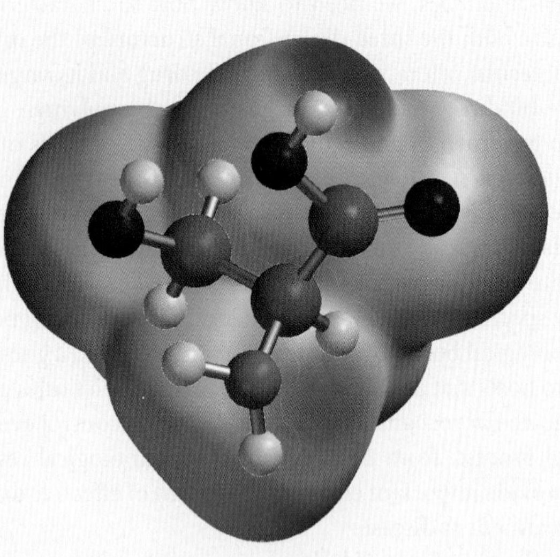

Fig. 6.10 An electrostatic potential surface for the amino acid serine. Positive charge is shown in blue and negative charge in red, with intermediate gradations of colour. The red regions of the molecule are electron-rich and the blue regions are electron-poor.

pharmacological efficacy, **quantitative structure–activity relationships** (QSAR) are often established by correlating data on activity of lead compounds with molecular properties, also called *molecular descriptors*, which can be determined either experimentally or computationally.

In broad terms, the first stage of the QSAR method consists of compiling molecular descriptors for a very large number of lead compounds. Descriptors such as molar mass, molecular dimensions and volume, and relative solubility in water and nonpolar solvents are available from routine experimental procedures. Quantum mechanical descriptors determined by calculations of the type described in this chapter include bond orders and HOMO and LUMO energies.

In the second stage of the process, biological activity is expressed as a function of the molecular descriptors. An example of a QSAR equation is:

$$\text{Activity} = c_0 + c_1 d_1 + c_2 d_1^2 + c_3 d_2 + c_4 d_2^2 + \cdots \qquad (6.27)$$

where d_i is the value of the descriptor and c_i is a coefficient calculated by fitting the data by regression analysis. The quadratic terms account for the fact that biological activity can have a maximum or minimum value at a specific descriptor value. For example, a molecule might not cross a biological membrane and become available for binding to targets in the interior of the cell if it is too hydrophilic, in which case it will not partition into the hydrophobic layer of the cell membrane (see *Impact I16.1* for details of membrane structure), or too hydrophobic, for then it may bind too tightly to the membrane. It follows that the activity will peak at some intermediate value of a parameter that measures the relative solubility of the drug in water and organic solvents.

In the final stage of the QSAR process, the activity of a drug candidate can be estimated from its molecular descriptors and the QSAR equation either by interpolation or extrapolation of the data. The predictions are more reliable when a large number of lead compounds and molecular descriptors are used to generate the QSAR equation.

The traditional QSAR technique has been refined into **3D QSAR**, in which sophisticated computational methods are used to gain further insight into the three-dimensional features of drug candidates that lead to tight binding to the receptor site of a target. The process begins by using a computer to superimpose three-dimensional structural models of lead compounds and looking for common features, such as similarities in shape, location of functional groups, and electrostatic potential plots. The key assumption of the method is that common structural features are indicative of molecular properties that enhance binding of the drug to the receptor. The collection of superimposed molecules is then placed inside a three-dimensional grid of points. An atomic probe, typically an sp^3-hybridized carbon atom, visits each grid point and two energies of interaction are calculated: E_{steric}, the steric energy reflecting interactions between the probe and electrons in uncharged regions of the drug, and E_{elec}, the electrostatic energy arising from interactions between the probe and a region of the molecule carrying a partial charge. The measured equilibrium constant for binding of the

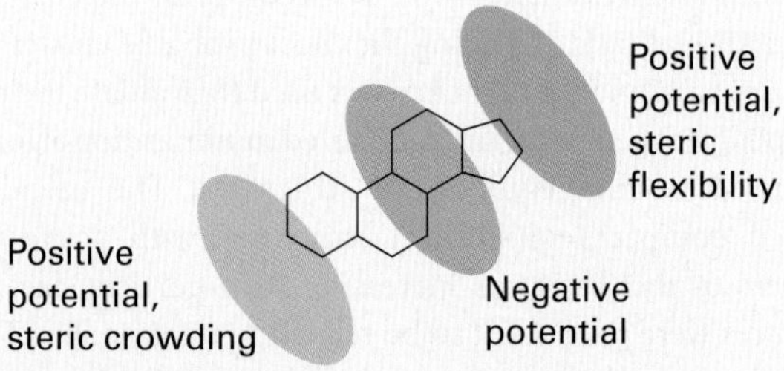

Fig. 6.11 A 3D QSAR analysis of the binding of steroids, molecules with the carbon skeleton shown, to human corticosteroid-binding globulin (CBG). The ellipses indicate areas in the protein's binding site with positive or negative electrostatic potentials and with little or much steric crowding. It follows from the calculations that addition of large substituents near the left-hand side of the molecule (as it is drawn on the page) leads to poor affinity of the drug to the binding site. Also, substituents that lead to the accumulation of negative electrostatic potential at either end of the drug are likely to show enhanced affinity for the binding site. [Adapted from P. Krogsgaard-Larsen, T. Liljefors, U. Madsen (ed.), *Textbook of drug design and discovery*, Taylor & Francis, London (2002).]

drug to the target, K_{bind}, is then assumed to be related to the interaction energies at each point $\boldsymbol{r}$ by the 3D QSAR equation

$$\log K_{\text{bind}} = c_0 + \sum_{r} \{c_s(\boldsymbol{r})E_{\text{steric}}(\boldsymbol{r}) + c_e(\boldsymbol{r})E_{\text{elec}}(\boldsymbol{r})\} \quad (6.28)$$

where the $c(\boldsymbol{r})$ are coefficients calculated by regression analysis, with the coefficients c_s and c_e reflecting the relative importance of steric and electrostatic interactions, respectively, at the grid point $\boldsymbol{r}$. Visualization of the regression analysis is facilitated by colouring each grid point according to the magnitude of the coefficients. Figure 6.11 shows results of a 3D QSAR analysis of the binding of steroids, molecules with the carbon skeleton shown, to human corticosteroid-binding globulin (CBG). Indeed, we see that the technique lives up to the promise of opening a window into the chemical nature of the binding site even when its structure is not known.

The QSAR and 3D QSAR methods, though powerful, have limited power: the predictions are only as good as the data used in the correlations are both reliable and abundant. However, the techniques have been used successfully to identify compounds that deserve further synthetic elaboration, such as addition or removal of functional groups, and testing.

Checklist of key ideas

- ☐ 1. A spinorbital is the product of a molecular orbital and a spin function.
- ☐ 2. The Hartree–Fock (HF) method uses a single Slater determinant, built from molecular orbitals that satisfy the HF equations, to represent the ground-state electronic wavefunction.
- ☐ 3. The Hartree–Fock equations involve the Fock operator, which consists of the core hamiltonian and terms representing the average Coulomb repulsion (J) and average correction due to spin correlation (K). The equations must be solved self-consistently.
- ☐ 4. The Hartree–Fock method neglects electron correlation, the tendency of electrons to avoid one another to minimize repulsion.
- ☐ 5. The Roothaan equations are a set of simultaneous equations, written in matrix form, that result from using a basis set of functions to expand the molecular orbitals.
- ☐ 6. In a minimal basis set, one basis set function represents each of the valence orbitals of the molecule.
- ☐ 7. Slater-type orbitals (STO) and Gaussian-type orbitals (GTO) centred on each of the atomic nuclei are commonly used as basis set functions; the product of two Gaussians on different centres is a single Gaussian function located between the centres.
- ☐ 8. In semiempirical methods, the two-electron integrals are set to zero or to empirical parameters; *ab initio* methods attempt to evaluate the integrals numerically.
- ☐ 9. The Hückel method is a simple semiempirical method for conjugated π systems.
- ☐ 10. In the complete neglect of differential overlap (CNDO) approximation, two-electron integrals are set to zero unless the two basis set functions for electron 1 are the same and the two basis functions for electron 2 are the same.
- ☐ 11. Other semiempirical methods include INDO (intermediate neglect of differential overlap), NDDO (neglect of diatomic differential overlap), MINDO (modified intermediate neglect of differential overlap), MNDO (modified neglect of differential overlap), AM1, and PM3.
- ☐ 12. Virtual orbitals are molecular orbitals that are unoccupied in the HF ground-state electronic wavefunction.
- ☐ 13. A singly excited determinant is formed by transferring an electron from an occupied orbital to a virtual orbital, a doubly excited determinant by transferring two electrons, and so on. Each of these Slater determinants (including the HF wavefunction) is a configuration state function (CSF).
- ☐ 14. Configuration interaction (CI) expresses the exact electronic wavefunction as a linear combination of configuration state functions.
- ☐ 15. Configuration interaction and Møller–Plesset perturbation theory are two popular *ab initio* methods that accommodate electron correlation.
- ☐ 16. Full CI uses molecular orbitals built from a finite basis set and allows for all possible excited determinants.
- ☐ 17. Many-body perturbation theory is the application of perturbation theory to a molecular system of interacting electrons and nuclei.
- ☐ 18. Møller–Plesset perturbation theory (MPPT) uses the sum of the Fock operators from the HF method as the simple, model hamiltonian $\hat{H}^{(0)}$.
- ☐ 19. According to Brillouin's theorem, only doubly excited determinants contribute to the second-order energy correction.
- ☐ 20. In density functional theory (DFT), the electronic energy is written as a functional of the electron probability density.
- ☐ 21. The exchange–correlation energy takes into account nonclassical electron–electron effects.
- ☐ 22. The electron density is computed from the Kohn–Sham orbitals, the solutions to the Kohn–Sham (KS) equations. The latter equations are solved self-consistently.
- ☐ 23. The exchange–correlation potential is the functional derivative of the exchange–correlation energy.

☐24. One commonly used but approximate form for the exchange–correlation energy is based on the model of an electron gas, a hypothetical system in which electrons move in a uniform distribution of positive charge.

☐25. Both the chemical accuracy and the computational cost of a particular method should be considered when deciding which electronic structure method to use in a given application.

☐26. The Hartree–Fock limit refers to an infinite basis set or, in practical terms, a finite basis set for which the energy and equilibrium geometry of the molecule do not vary as the size of the basis set is increased.

Discussion questions

6.1 Describe the physical significance of each of the terms that appears in the Fock operator.

6.2 Explain why the Hartree–Fock formalism does not account for electron correlation but the methods of configuration interaction and many-body perturbation theory do.

6.3 Outline the computational steps used in the Hartree–Fock self-consistent field approach to electronic structure calculations.

6.4 Explain how the Roothaan equations arise in the Hartree–Fock method. What additional approximations do they represent?

6.5 Discuss the role of basis set functions in electronic structure calculations. What are some commonly used basis sets? Why are polarization functions often included?

6.6 Explain why the use of Gaussian-type orbitals is generally preferred over the use of Slater-type orbitals in basis sets.

6.7 Distinguish between semiempirical, *ab initio*, and density functional theory methods of electronic structure determination.

6.8 Discuss how virtual orbitals are useful in CI and MPPT electronic structure calculations.

6.9 Is DFT a semiempirical method? Justify your answer.

Exercises

6.1(a) Write down the electronic hamiltonian for the helium atom.

6.1(b) Write down the electronic hamiltonian for the lithium atom.

6.2(a) Write the expression for the potential energy contribution to the electronic hamiltonian for LiH.

6.2(b) Write the expression for the potential energy contribution to the electronic hamiltonian for BeH_2.

6.3(a) Write down the electronic hamiltonian for HeH^+.

6.3(b) Write down the electronic hamiltonian for LiH^{2+}.

6.4(a) Write down the Slater determinant for the ground state of HeH^+.

6.4(b) Write down the Slater determinant for the ground state of LiH^{2+}.

6.5(a) Write down the Hartree–Fock equation for HeH^+.

6.5(b) Write down the Hartree–Fock equation for LiH^{2+}.

6.6(a) Set up the Roothaan equations for HeH^+ and establish the simultaneous equations corresponding to the Roothaan equations. Adopt a basis set of two real normalized functions, one centred on H and one on He; denote the molecular orbitals ψ_a and ψ_b.

6.6(b) Set up the Roothaan equations for LiH^{2+} and establish the simultaneous equations corresponding to the Roothaan equations. Adopt a basis set of two real normalized functions, one centred on H and one on Li; denote the molecular orbitals ψ_a and ψ_b.

6.7(a) Construct the elements F_{AA} and F_{AB} for the species HeH^+ and express them in terms of the notation in eqn 6.14.

6.7(b) Construct the elements F_{AA} and F_{AB} for the species LiH^{2+} and express them in terms of the notation in eqn 6.14.

6.8(a) Using the integral notation in eqn 6.14, identify all of the four-centre two-electron integrals that are equal to (AA|AB).

6.8(b) Using the integral notation in eqn 6.14, identify all of the four-centre two-electron integrals that are equal to (BB|BA).

6.9(a) How many basis functions are needed in an electronic structure calculation on CH_3Cl using a (a) minimal basis set, (b) split-valence basis set, (c) double-zeta basis set?

6.9(a) How many basis functions are needed in an electronic structure calculation on CH_2Cl_2 using a (a) minimal basis set, (b) split-valence basis set, (c) double-zeta basis set?

6.10(a) What is the general mathematical form of a p-type Gaussian?

6.10(b) What is the general mathematical form of a d-type Gaussian?

6.11(a) A one-dimensional Gaussian (in x) has the form $e^{-\alpha x^2}$ or $x^n e^{-\alpha x^2}$; one-dimensional Gaussians in y and z have similar forms. Show that the s-type Gaussian (eqn 6.17) can be written as a product of three one-dimensional Gaussians.

6.11(b) A one-dimensional Gaussian (in x) has the form $e^{-\alpha x^2}$ or $x^n e^{-\alpha x^2}$; one-dimensional Gaussians in y and z have similar forms. Show that a p-type Gaussian (eqn 6.17) can be written as a product of three one-dimensional Gaussians.

6.12(a) Show that the product of s-type Gaussians on He and H in HeH^+ is a Gaussian at an intermediate position. Note that the Gaussians have different exponents.

6.12(b) Show that the product of s-type Gaussians on Li and H in LiH^{2+} is a Gaussian at an intermediate position. Note that the Gaussians have different exponents.

6.13(a) How many basis functions are needed in an electronic structure calculation on CH_3Cl using a (a) 6-31G* basis set, (b) 6-311G** basis set, (c) 6-311++G basis set?

6.13(b) How many basis functions are needed in an electronic structure calculation on CH_2Cl_2 using a (a) 6-31G* basis set, (b) 6-311G** basis set, (c) 6-311++G basis set?

6.14(a) Identify the quadratic equation for the coefficient of the basis function centred on H in HeH^+ starting from the Fock matrix and making the Hückel approximations.

6.14(b) Identify the quadratic equation for the coefficient of the basis function centred on H in LiH^{2+} starting from the Fock matrix and making the Hückel approximations.

6.15(a) Identify the two-electron integrals that are set to zero in the semiempirical method known as (a) CNDO, (b) INDO.

6.15(b) Identify the two-electron integrals that are set to zero in the semiempirical method known as NDDO.

6.16(a) In a Hartree–Fock calculation on the silicon atom using 20 basis set functions, how many of the molecular orbitals generated would be unoccupied and could be used as virtual orbitals in a configuration interaction calculation?

6.16(b) In a Hartree–Fock calculation on the sulfur atom using 20 basis set functions, how many of the molecular orbitals generated would be unoccupied and could be used as virtual orbitals in a configuration interaction calculation?

6.17(a) Give an example of a singly excited determinant in a CI calculation of H_2.

6.17(b) Give an example of a doubly excited determinant in a CI calculation of H_2.

6.18(a) Using eqn 6.18, write down the expression for the ground-state wavefunction in a CI calculation on HeH^+ involving the ground-state determinant and a singly excited determinant.

6.18(b) Using eqn 6.18, write down the expression for the ground-state wavefunction in a CI calculation on LiH^{2+} involving the ground-state determinant and a doubly excited determinant.

6.19(a) The second-order energy correction (eqn 6.21) in MPPT arises from the doubly excited determinant (the $M = 2$ term). Derive an expression for the integral that appears in the numerator of eqn 6.21 in terms of the integrals (AB|CD) for HeH^+.

6.19(b) The second-order energy correction (eqn 6.21) in MPPT arises from the doubly excited determinant (the $M = 2$ term). Derive an expression for the integral that appears in the numerator of eqn 6.21 in terms of the integrals (AB|CD) for LiH^{2+}.

6.20(a) Which of the following are functionals: (a) $d(x^3)/dx$, (b) $d(x^3)/dx$ evaluated at $x = 1$, (c) $\int x^3 dx$, (d) $\int_1^3 x^3 dx$?

6.20(b) Which of the following are functionals: (a) $d(3x^2)/dx$, (b) $d(3x^3)/dx$ evaluated at $x = 4$, (c) $\int 3x^2 dx$, (d) $\int_1^3 3x^2 dx$?

6.21(a) Using eqn 6.23, write the expression for the electron density in terms of the Kohn–Sham orbitals in a DFT calculation on LiH.

6.21(b) Using eqn 6.23, write the expression for the electron density in terms of the Kohn–Sham orbitals in a DFT calculation on BeH_2.

6.22(a) Write the two Kohn–Sham equations for the Kohn–Sham orbitals in a DFT calculation on HeH^+. Use the exchange–correlation potential of eqn 6.26.

6.22(b) Write the two Kohn–Sham equations for the Kohn–Sham orbitals in a DFT calculation on LiH^{2+}. Use the exchange–correlation potential of eqn 6.26.

6.23(a) Which of the following basis sets should give a result closer to the Hartree–Fock limit in an electronic structure calculation on ethanol, C_2H_5OH: (a) double-zeta, (b) split-valence, (c) triple zeta?

6.23(b) Which of the following basis sets should give a result closer to the Hartree–Fock limit in an electronic structure calculation on methanol CH_3OH: (a) 4-31G, (b) 6-311+G**, (c) 6-31G*?

Problems*

Many of the following problems call on the use of commercially available software. Use versions that are available with this text or the software recommended by your instructor.

Numerical problems

6.1 Using appropriate electronic structure software, perform Hartree–Fock self-consistent field calculations for the ground electronic states of H_2 and F_2 using (a) 6-31G* and (b) 6-311+G** basis sets. Determine ground-state energies and equilibrium geometries. Compare computed equilibrium bond lengths to experimental values.

6.2 Using approprite electronic structure software and a basis set of your choice or on the advice of your instructor, perform calculations for: (a) the ground electronic state of H_2; (b) the ground electronic state of F_2; (c) the first electronic state of H_2; (d) the first electronic state of F_2. Determine energies and equilibrium geometries and compare to experimental values where possible.

6.3 Use the AM1 and PM3 semiempirical methods to compute the equilibrium bond lengths and standard enthalpies of formation of (a) ethanol, C_2H_5OH, (b) 1,4-dichlorobenzene, $C_6H_4Cl_2$. Compare with experimental values and suggest reasons for any discrepancies.

6.4 Molecular orbital calculations based on semiempirical (Section 6.5), *ab initio*, and DFT methods describe the spectroscopic properties of conjugated molecules better than simple Hückel theory (Section 6.4). (a) Using the computational method of your choice (semiempirical, *ab initio*, or density functional methods), calculate the energy separation between the HOMO and LUMO of ethene, butadiene, hexatriene, and octatetraene. (b) Plot the HOMO–LUMO energy separations against the experimental frequencies for $\pi^* \leftarrow \pi$ ultraviolet absorptions for these molecules (61 500, 46 080, 39 750, and 32 900 cm^{-1}, respectively). Use mathematical software to find the polynomial equation that best fits the data. (c) Use your polynomial fit from part (b) to estimate the wavenumber and wavelength of the $\pi^* \leftarrow \pi$ ultraviolet absorption of decapentaene from the calculated HOMO–LUMO energy separation. (d) Discuss why the calibration procedure of part (b) is necessary.

* Problems denoted with the symbol ‡ were supplied by Charles Trapp, Carmen Giunta, and Marshall Cady.

6.5 Molecular electronic structure methods may be used to estimate the standard enthalpy of formation of molecules in the gas phase. (a) Using a semiempirical method of your choice, calculate the standard enthalpy of formation of ethene, butadiene, hexatriene, and octatetraene in the gas phase. (b) Consult a database of thermochemical data, and, for each molecule in part (a), calculate the difference between the calculated and experimental values of the standard enthalpy of formation. (c) A good thermochemical database will also report the uncertainty in the experimental value of the standard enthalpy of formation. Compare experimental uncertainties with the relative errors calculated in part (b) and discuss the reliability of your chosen semiempirical method for the estimation of thermochemical properties of linear polyenes.

6.6‡ Luo *et al.* (*J. Chem. Phys.* **98**, 3564 (1993)) reported experimental observation of He_2, a species that had escaped detection for a long time. The observation required temperatures in the neighbourhood of 1 mK. Perform configuration interaction and MPPT electronic structure calculations and compute the equilibrium bond length R_e of the dimer as well as the energy of the dimer at R_e relative to the separated He + He atomic limit. (High level, accurate computational studies suggest that the well depth for He_2 is about 0.0151 zJ at a distance R_e of about 297 pm.)

6.7 An important quantity in nuclear magnetic resonance spectroscopy (Chapter 12), which should be familiar from ^{13}C-NMR spectra of organic molecules, is the chemical shift; this experimentally determined quantity is influenced by the details of the electronic structure near the ^{13}C nucleus of interest. Consider the following series of molecules: benzene, methylbenzene, trifluoromethylbenzene, benzonitrile, and nitrobenzene in which the substituents *para* to the C atom of interest are H, CH_3, CF_3, CN, and NO_2, respectively. (a) Using the computational method of your choice, calculate the net charge at the C atom *para* to these substituents in the series of organic molecules given above. (b) It is found empirically that the ^{13}C chemical shift of the *para* C atom increases in the order: methylbenzene, benzene, trifluoromethylbenzene, benzonitrile, nitrobenzene. Is there a correlation between the behaviour of the ^{13}C chemical shift and the computed net charge on the ^{13}C atom? (This problem is revisited in Problem 12.17.)

Theoretical problems

6.8 Show that the Slater determinant in eqn 6.5a is normalized assuming that the spinorbitals from which it is constructed are orthogonal and normalized.

6.9 In a configuration interaction calculation on the ground ^{2}S state of Li, which of the following Slater determinants can contribute to the ground-state wavefunction?

(a) $|\psi_{1s}^{\alpha}\psi_{1s}^{\beta}\psi_{2s}^{\alpha}|$ (b) $|\psi_{1s}^{\alpha}\psi_{1s}^{\beta}\psi_{2s}^{\beta}|$ (c) $|\psi_{1s}^{\alpha}\psi_{1s}^{\beta}\psi_{2p}^{\alpha}|$

(d) $|\psi_{1s}^{\alpha}\psi_{2p}^{\alpha}\psi_{2p}^{\beta}|$ (e) $|\psi_{1s}^{\alpha}\psi_{3d}^{\alpha}\psi_{3d}^{\beta}|$ (f) $|\psi_{1s}^{\alpha}\psi_{2s}^{\alpha}\psi_{3s}^{\alpha}|$

6.10 In a configuration interaction calculation on the excited $^3\Sigma_u^+$ electronic state of H_2, which of the following Slater determinants can contribute to the excited-state wavefunction?

(a) $|1\sigma_g^{\alpha}1\sigma_u^{\alpha}|$ (b) $|1\sigma_g^{\alpha}1\pi_u^{\alpha}|$ (c) $|1\sigma_u^{\alpha}1\pi_g^{\beta}|$

(d) $|1\sigma_g^{\beta}2\sigma_u^{\beta}|$ (e) $|1\pi_u^{\alpha}1\pi_g^{\alpha}|$ (f) $|1\pi_u^{\beta}2\pi_u^{\beta}|$

6.11 Use MPPT to obtain an expression for the ground-state wavefunction corrected to first order in the perturbation.

6.12 It is often necessary during the course of an electronic structure calculation to take derivatives of the basis functions with respect to nuclear coordinates. Show that the derivative of an s-type Gaussian with respect to *x* yields a p-type Gaussian and that the derivative of a p-type Gaussian ($i = 1, j = k = 0$ in eqn 6.17) yields a sum of s- and d-type Gaussians.

6.13 (a) In a continuation of Exercise 6.6a for HeH^+, proceed to determine the energies of the two molecular orbitals as well as the relation between the two coefficients for ψ_a and the relation between the two coefficients for ψ_b. (b) Repeat for LiH^{2+} (in a continuation of Exercise 6.6b).

6.14 (a) Continuing the Hartree–Fock calculation on HeH^+ in Problem 6.13(a), give the expressions for all four of the elements of the Fock matrix in terms of four-centre, two-electron integrals; the latter are defined in eqn 6.14. (b) Repeat for LiH^{2+} (in a continuation of Problem 6.13(b)).

6.15 (a) In a continuation of Problem 6.14(a) for HeH^+, use Hückel molecular orbital theory to express the energies of the molecular orbitals in terms of α and β. (b) Repeat for LiH^{2+} (in a continuation of Problem 6.14(b)).

6.16 (a) Using the expressions for the four elements of the Fock matrix for HeH^+ determined in Problem 6.14(a), show how these expressions simplify if the CNDO semiempirical method is used. (b) Repeat for LiH^{2+}, beginning with the expressions determined in Problem 6.14(b).

6.17 Consider a four-centre integral in an electronic structure calculation on NH_3 involving s-type Gaussian functions centred on each atomic nucleus. Show that the four-centre, two-electron integral reduces to an integral over two different centres.

6.18 (a) Show why configuration interaction gives an improved ground-state wavefunction for HeH^+ compared to the Hartree–Fock ground-state wavefunction. Use a minimal basis set and ignore overlap. Follow along the lines of the argument presented in the first *illustration* in Section 6.6 but recognize the complication introduced by the fact that HeH^+ does not have inversion symmetry. (b) Repeat for LiH^{2+}.

6.19 In the second *illustration* of Section 6.6, the secular equation for a CI calculation on molecular hydrogen using the ground-state Slater determinant and the doubly excited determinant was presented as well as the expression for one of the hamiltonian matrix elements. Develop similar expressions for the remaining hamiltonian matrix elements.

6.20 Show that in MPPT first-order energy corrections do not contribute to the correlation energy.

6.21 Prove Brillouin's theorem, which states that the hamiltonian matrix elements between the ground-state Hartree–Fock Slater determinant and singly excited determinants are zero.

6.22 Derive an expression for the second-order estimate of the correlation energy for H_2 if, in a CI calculation using a minimal basis set, the overlap between the two basis set functions is not ignored but set equal to a constant *S*.

6.23 Find the DFT exchange–correlation potential if the exchange–correlation energy is given by $\int C\rho^{5/3}d\mathbf{r}$.

Applications: to biology

6.24 Molecular orbital calculations may be used to predict trends in the standard potentials of conjugated molecules, such as the quinones and flavins, that are involved in biological electron transfer reactions (Section 20.8). It is commonly assumed that decreasing the energy of the LUMO enhances the ability of a molecule to accept an electron into the LUMO, with an accompanying increase in the value of the molecule's standard potential. Furthermore, a number of studies indicate that there is a linear correlation between the LUMO energy and the reduction potential of aromatic hydrocarbons. (a) The standard potentials at pH 7 for the one-electron reduction of methyl-substituted 1,4-benzoquinones (**3**) to their respective semiquinone radical anions are:

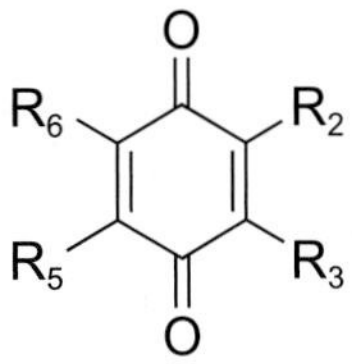

3

R_2	R_3	R_5	R_6	$E^⊖/V$
H	H	H	H	0.078
CH_3	H	H	H	0.023
CH_3	H	CH_3	H	−0.067
CH_3	CH_3	CH_3	H	−0.165
CH_3	CH_3	CH_3	CH_3	−0.260

Using the computational method of your choice (semiempirical, *ab initio*, or density functional theory methods), calculate E_{LUMO}, the energy of the LUMO of each substituted 1,4-benzoquinone, and plot E_{LUMO} against $E^⊖$. Do your calculations support a linear relation between E_{LUMO} and $E^⊖$? (b) The 1,4-benzoquinone for which $R_2 = R_3 = CH_3$ and $R_5 = R_6 = OCH_3$ is a suitable model of ubiquinone, a component of the respiratory electron transport chain (*Impact I17.3*). Determine E_{LUMO} of this quinone and then use your results from part (a) to estimate its standard potential. (c) The 1,4-benzoquinone for which $R_2 = R_3 = R_5 = CH_3$ and $R_6 = H$ is a suitable model of plastoquinone, a component of the photosynthetic electron transport chain (*Impact I19.1*). Determine E_{LUMO} of this quinone and then use your results from part (a) to estimate its standard potential. Is plastoquinone expected to be a better or worse oxidizing agent than ubiquinone? (d) Based on your predictions and on basic concepts of biological electron transport (Section 20.8), suggest a reason why ubiquinone is used in respiration and plastoquinone is used in photosynthesis.

6.25 This problem gives a simple example of a quantitative structure–activity relation (QSAR), *Impact I6.2*, showing how to predict the affinity of non-polar groups for hydrophobic sites in the interior of proteins. (a) Consider a family of hydrocarbons R—H. The hydrophobicity constants, π, for $R = CH_3$, CH_2CH_3, $(CH_2)_2CH_3$, $(CH_2)_3CH_3$, and $(CH_2)_4CH_3$ are, respectively, 0.5, 1.0, 1.5, 2.0, and 2.5. Use these data to predict the π value for $(CH_2)_6CH_3$. (b) The equilibrium constants K_I for the dissociation of inhibitors (**4**) from the enzyme chymotrypsin were measured for different substituents R:

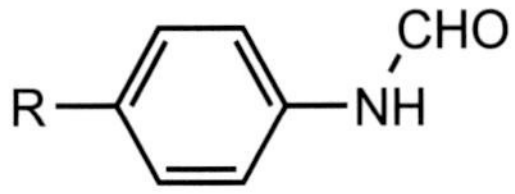

4

R	CH_3CO	CN	NO_2	CH_3	Cl
π	−0.20	−0.025	0.33	0.5	0.9
$\log K_I$	−1.73	−1.90	−2.43	−2.55	−3.40

Plot $\log K_I$ against π. Does the plot suggest a linear relationship? If so, what are the slope and intercept to the $\log K_I$ axis of the line that best fits the data? (c) Predict the value of K_I for the case R = H.

7 Molecular symmetry

In this chapter we sharpen the concept of 'shape' into a precise definition of 'symmetry', and show that symmetry may be discussed systematically. We see how to classify any molecule according to its symmetry and how to use this classification to discuss molecular properties. After describing the symmetry properties of molecules themselves, we turn to a consideration of the effect of symmetry transformations on orbitals and see that their transformation properties can be used to set up a labelling scheme. These symmetry labels are used to identify integrals that necessarily vanish. One important integral is the overlap integral between two orbitals. By knowing which atomic orbitals may have nonzero overlap, we can decide which ones can contribute to molecular orbitals. We also see how to select linear combinations of atomic orbitals that match the symmetry of the nuclear framework. Finally, by considering the symmetry properties of integrals, we see that it is possible to derive the selection rules that govern spectroscopic transitions.

The systematic discussion of symmetry is called **group theory**. Much of group theory is a summary of common sense about the symmetries of objects. However, because group theory is systematic, its rules can be applied in a straightforward, mechanical way. In most cases the theory gives a simple, direct method for arriving at useful conclusions with the minimum of calculation, and this is the aspect we stress here. In some cases, though, they lead to unexpected results.

A **group** in mathematics is a collection of transformations that satisfy four criteria. Thus, if we write the transformations as R, R', . . . , (which we can think of as reflections, rotations, and so on), then they form a group if:

1. One of the transformations is the identity (that is: 'do nothing').

2. For every transformation R, the inverse transformation R^{-1} is included in the collection so that the combination RR^{-1} (the transformation R^{-1} followed by R) is equivalent to the identity.

3. The combination RR' (the transformation R' followed by R) is equivalent to a single member of the collection of transformations.

4. The combination $R(R'R'')$, the transformation $(R'R'')$ followed by R, is equivalent to $(RR')R''$, the transformation R'' followed by (RR').

These criteria will be made more concrete in the following section. We are aware that not everyone will use this chapter, and have written the following chapters so that group theory, though often included (for it is so powerful), can easily be stepped around.

The symmetry elements of objects

Some objects are 'more symmetrical' than others. A sphere is more symmetrical than a cube because it looks the same after it has been rotated through any angle about any diameter. A cube looks the same only if it is rotated through certain angles about specific axes, such as 90°, 180°, or 270° about an axis passing through the centres of any of its opposite faces (Fig. 7.1), or by 120° or 240° about an axis passing through any of its opposite corners. Similarly, an NH_3 molecule is 'more symmetrical' than an H_2O molecule because NH_3 looks the same after rotations of 120° or 240° about the axis shown in Fig. 7.2, whereas H_2O looks the same only after a rotation of 180°.

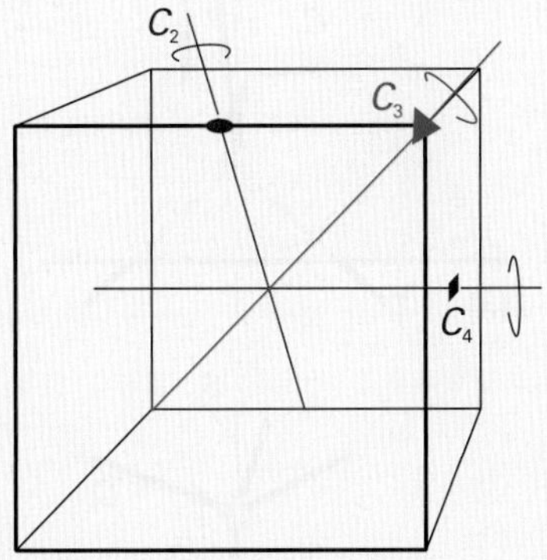

Fig. 7.1 Some of the symmetry elements of a cube. The twofold, threefold, and fourfold axes are labelled with the conventional symbols.

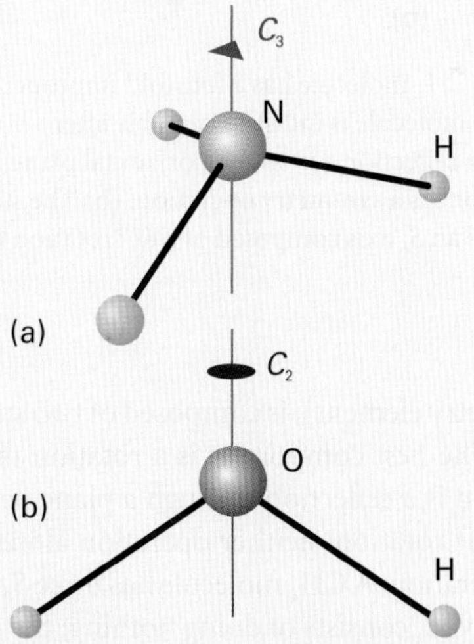

Fig. 7.2 (a) An NH_3 molecule has a threefold (C_3) axis and (b) an H_2O molecule has a twofold (C_2) axis. Both have other symmetry elements too.

An action that leaves an object looking the same after it has been carried out is called a **symmetry operation**. Typical symmetry operations include rotations, reflections, and inversions. There is a corresponding **symmetry element** for each symmetry operation, which is the point, line, or plane with respect to which the symmetry operation is performed. For instance, a rotation (a symmetry operation) is carried out around an axis (the corresponding symmetry element). We shall see that we can classify molecules by identifying all their symmetry elements, and grouping together molecules that possess the same set of symmetry elements. This procedure, for example, puts the trigonal pyramidal species NH_3 and SO_3^{2-} into one group and the angular species H_2O and SO_2 into another group.

7.1 Operations and symmetry elements

The classification of objects according to symmetry elements corresponding to operations that leave at least one common point unchanged gives rise to the **point groups**. There are five kinds of symmetry operation (and five kinds of symmetry element) of this kind. When we consider crystals (Chapter 9), we shall meet symmetries arising from translation through space. These more extensive groups are called **space groups.**

An ***n*-fold rotation** (the operation) about an ***n*-fold axis of symmetry**, C_n (the corresponding element), is a rotation through 360°/n. An H_2O molecule has one twofold axis, C_2. An NH_3 molecule has one threefold axis, C_3, with which is associated two symmetry operations, one being 120° rotation in a clockwise sense and the other 120° rotation in a counter-clockwise sense. There is only one twofold rotation associated with a C_2 axis because clockwise and counter-clockwise 180° rotations are identical. A pentagon has a C_5 axis, with two rotations (one clockwise, the other counter-clockwise) through 72° associated with it. It also has an axis denoted C_5^2, corresponding to two successive C_5 rotations; there are two such operations, one through 144° in a clockwise sense and the other through 144° in a counter-clockwise sense. A cube has three C_4 axes, four C_3 axes, and six C_2 axes. However, even this high symmetry is exceeded by a sphere, which possesses an infinite number of symmetry axes (along any diameter) of all possible integral values of n. If a molecule possesses several rotation axes, then the one (or more) with the greatest value of n is called the **principal axis.** The principal axis of a benzene molecule is the sixfold axis perpendicular to the hexagonal ring (**1**).

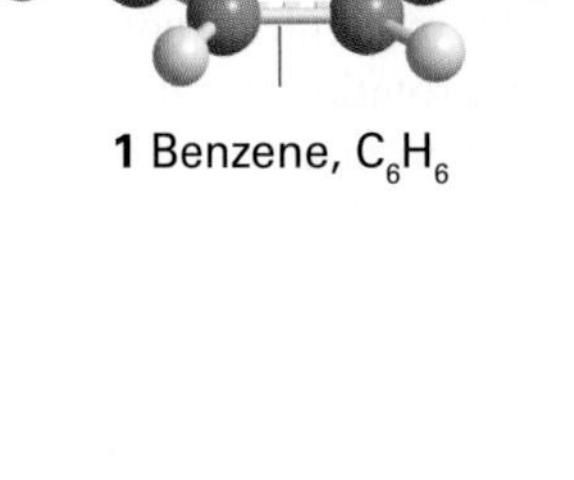

1 Benzene, C_6H_6

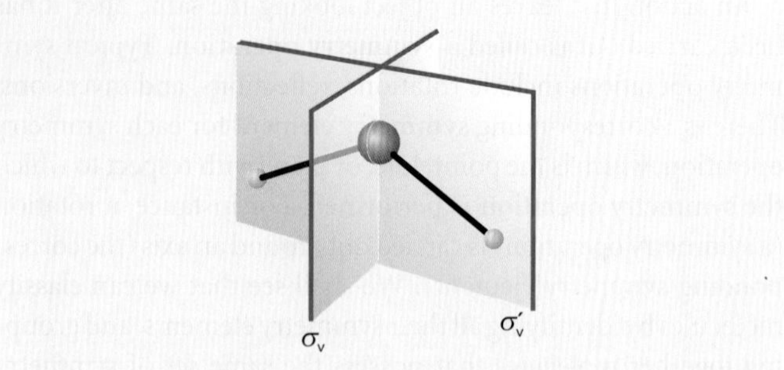

Fig. 7.3 An H_2O molecule has two mirror planes. They are both vertical (that is, contain the principal axis), so are denoted σ_v and σ_v'.

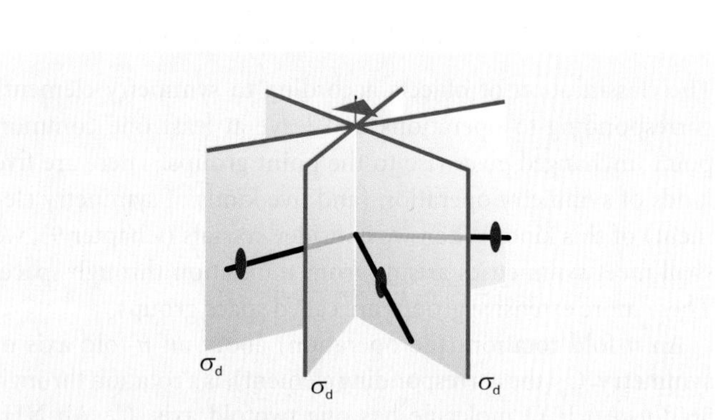

Fig. 7.4 Dihedral mirror planes (σ_d) bisect the C_2 axes perpendicular to the principal axis.

A **reflection** (the operation) in a **mirror plane**, σ (the element), may contain the principal axis of a molecule or be perpendicular to it. If the plane contains the principal axis, it is called 'vertical' and denoted σ_v. An H_2O molecule has two vertical planes of symmetry (Fig. 7.3) and an NH_3 molecule has three. A vertical mirror plane that bisects the angle between two C_2 axes is called a 'dihedral plane' and is denoted σ_d (Fig. 7.4). When the plane of symmetry is perpendicular to the principal axis it is called 'horizontal' and denoted σ_h. A C_6H_6 molecule has a C_6 principal axis and a horizontal mirror plane (as well as several other symmetry elements).

In an **inversion** (the operation) through a **centre of symmetry**, *i* (the element), we imagine taking each point in a molecule, moving it to the centre of the molecule, and then moving it out the same distance on the other side; that is, the point (x, y, z) is taken into the point $(-x, -y, -z)$. Neither an H_2O molecule nor an NH_3 molecule has a centre of inversion, but a sphere and a cube do have one. A C_6H_6 molecule does have a centre of inversion and so does a regular octahedron (Fig. 7.5); a regular tetrahedron and a CH_4 molecule do not.

An ***n*-fold improper rotation** (the operation) about an ***n*-fold axis of improper rotation** or an ***n*-fold improper rotation axis**,

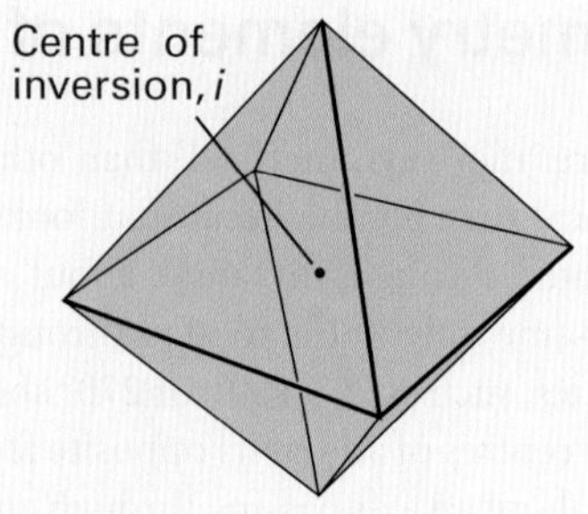

Fig. 7.5 A regular octahedron has a centre of inversion (*i*).

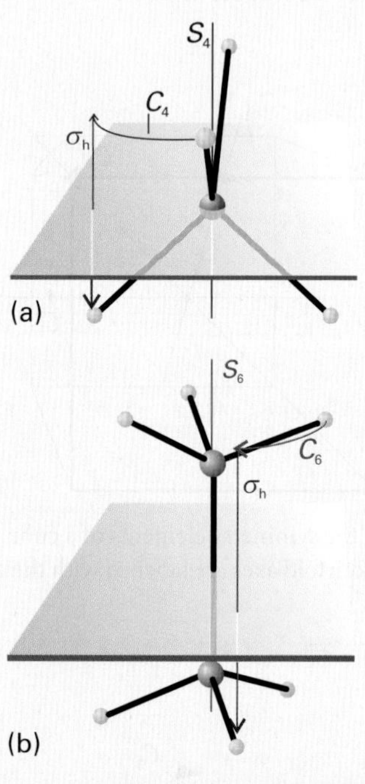

Fig. 7.6 (a) A CH_4 molecule has a fourfold improper rotation axis (S_4): the molecule is indistinguishable after a 90° rotation followed by a reflection across the horizontal plane, but neither operation alone is a symmetry operation. (b) The staggered form of ethane has an S_6 axis composed of a 60° rotation followed by a reflection.

S_n (the symmetry element), is composed of two successive transformations. The first component is a rotation through 360°/*n*, and the second is a reflection through a plane perpendicular to the axis of that rotation; neither operation alone needs to be a symmetry operation. A CH_4 molecule has three S_4 axes (Fig. 7.6).

The **identity**, *E*, consists of doing nothing; the corresponding symmetry element is the entire object. Because every molecule is indistinguishable from itself if nothing is done to it, every object possesses at least the identity element. One reason for including the identity is that some molecules have only this symmetry element (2); another reason is to ensure that the symmetry elements fulfil the criteria for forming a group.

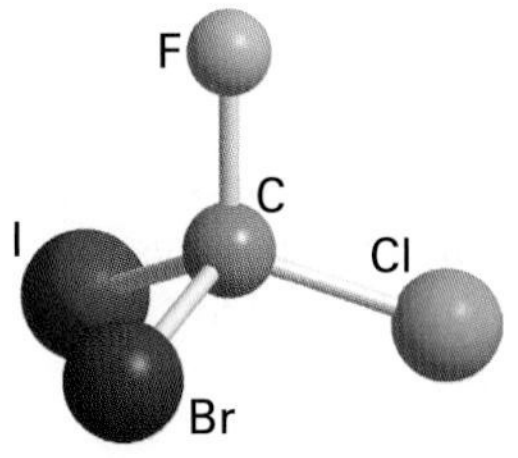

2 CBrClFI

7.2 The symmetry classification of molecules

To classify molecules according to their symmetries, we list their symmetry elements and collect together molecules with the same list of elements. This procedure puts CH_4 and CCl_4, which both possess the same symmetry elements as a regular tetrahedron, into the same group, and H_2O into another group.

The name of the group to which a molecule belongs is determined by the symmetry elements it possesses. There are two systems of notation (Table 7.1). The **Schoenflies system** (in which a name looks like C_{4v}) is more common for the discussion of individual molecules, and the **Hermann–Mauguin system**, or **International system** (in which a name looks like 4*mm*), is used almost exclusively in the discussion of crystal symmetry. The identification of a molecule's point group according to the Schoenflies system is simplified by referring to the flow diagram in Fig. 7.7 and the shapes shown in Fig. 7.8.

Table 7.1 The notation for point groups

C_i	$\bar{1}$								
C_s	*m*								
C_1	1	C_2	2	C_3	3	C_4	4	C_6	6
		C_{2v}	2*mm*	C_{3v}	3*m*	C_{4v}	4*mm*	C_{6v}	6*mm*
		C_{2h}	2*m*	C_{3h}	$\bar{6}$	C_{4h}	4/*m*	C_{6h}	6/*m*
		D_2	222	D_3	32	D_4	422	D_6	622
		D_{2h}	*mmm*	D_{3h}	$\bar{6}2m$	D_{4h}	4/*mmm*	D_{6h}	6/*mmm*
		D_{2d}	$\bar{4}2m$	D_{3d}	$\bar{3}m$	S_4	$\bar{4}/m$	S_6	$\bar{3}$
T	23	T_d	$\bar{4}3m$	T_h	*m*3				
O	432	O_h	*m*3*m*						

In the International system (or Hermann–Mauguin system) for point groups, a number *n* denotes the presence of an *n*-fold axis and *m* denotes a mirror plane. A slash (/) indicates that the mirror plane is perpendicular to the symmetry axis. It is important to distinguish symmetry elements of the same type but of different classes, as in 4/*mmm*, in which there are three classes of mirror plane. A bar over a number indicates that the element is combined with an inversion. The only groups listed here are the so-called 'crystallographic point groups'.

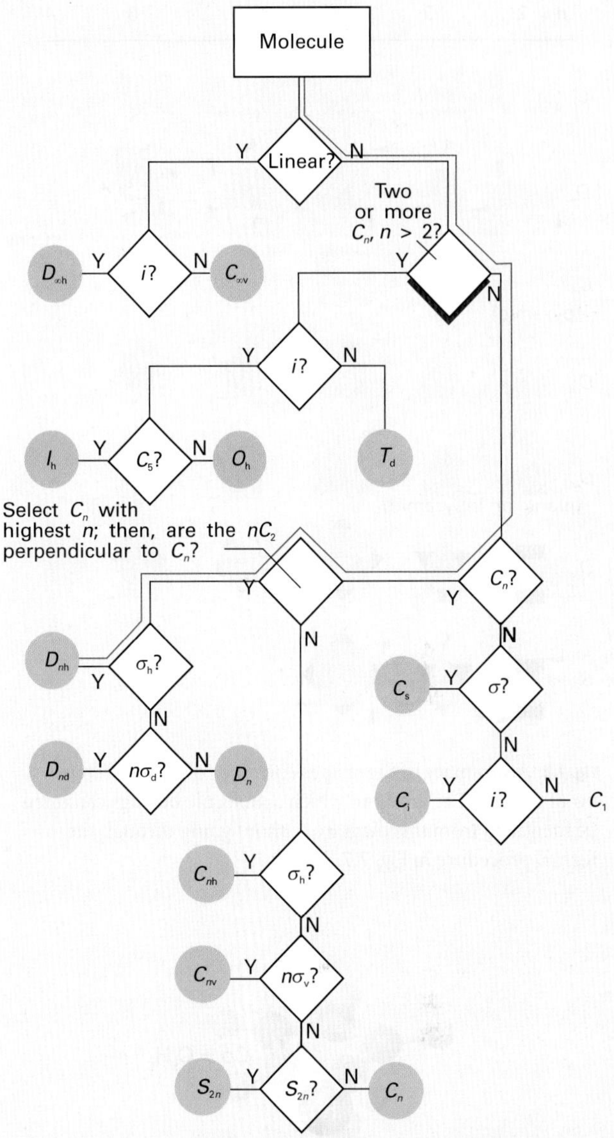

Fig. 7.7 A flow diagram for determining the point group of a molecule. Start at the top and answer the question posed in each diamond (Y = yes, N = no).

● A BRIEF ILLUSTRATION

To identify the point group to which a ruthenocene molecule (3) belongs we use the flow diagram in Fig. 7.7. The path to trace is shown by a blue line; it ends at D_{nh}. Because the molecule has a fivefold axis, it belongs to the group D_{5h}. If the rings were staggered, as they are in an excited state of ferrocene that lies 4 kJ mol^{-1} above the ground state (4), the horizontal reflection plane would be absent, but dihedral planes would be present. ●

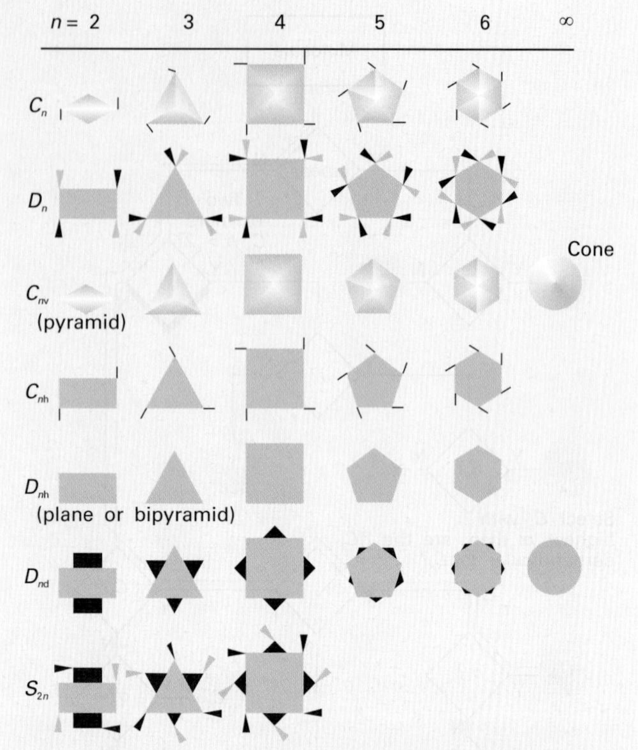

Fig. 7.8 A summary of the shapes corresponding to different point groups. The group to which a molecule belongs can often be identified from this diagram without going through the formal procedure in Fig. 7.7.

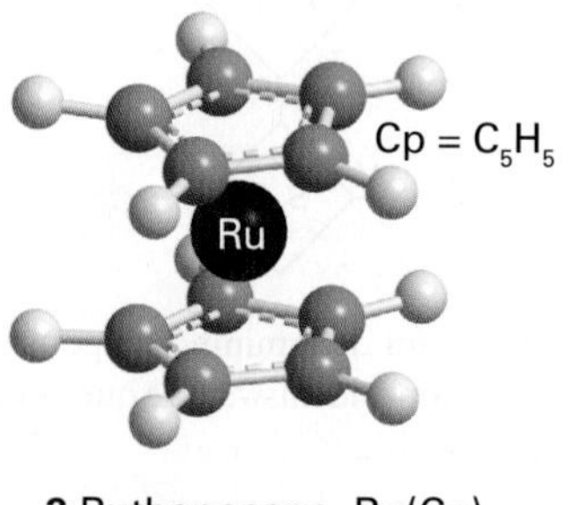

3 Ruthenocene, $Ru(Cp)_2$

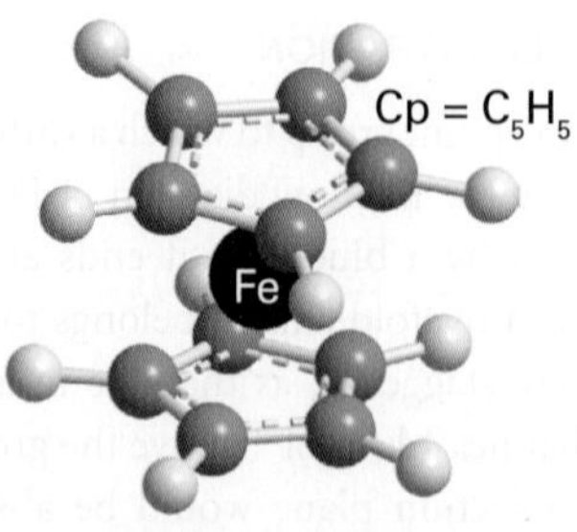

4 Ferrocene, $Fe(Cp)_2$

Self-test 7.1 Classify the pentagonal antiprismatic excited state of ferrocene (**4**). [D_{5d}]

(a) The groups C_1, C_i, and C_s

A molecule belongs to the group C_1 if it has no element other than the identity, as in (**2**). It belongs to C_i if it has the identity and the inversion alone (**5**), and to C_s if it has the identity and a mirror plane alone (**6**).

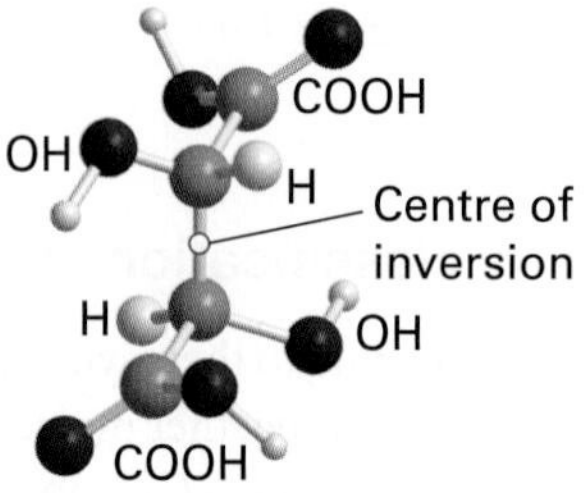

5 Meso-tartaric acid, HOOCCH(OH)CH(OH)COOH

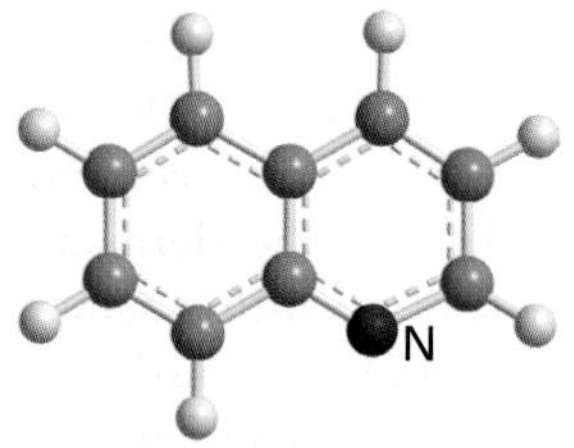

6 Quinoline, C_9H_7N

(b) The groups C_n, C_{nv}, and C_{nh}

A molecule belongs to the group C_n if it possesses an n-fold axis. Note that the symbol C_n is now playing a triple role: as the label of a symmetry element, a symmetry operation, and a group. For example, an H_2O_2 molecule has the elements E and C_2 (**7**), so it belongs to the group C_2.

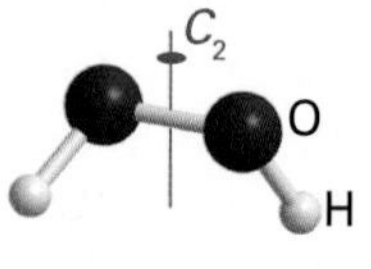

7 Hydrogen peroxide, H_2O_2

If, in addition to the identity and a C_n axis, a molecule has n vertical mirror planes σ_v, then it belongs to the group C_{nv}. An H_2O molecule, for example, has the symmetry elements E, C_2, and $2\sigma_v$, so it belongs to the group C_{2v}. An NH_3 molecule has the elements E, C_3, and $3\sigma_v$, so it belongs to the group C_{3v}. A heteronuclear diatomic molecule such as HCl belongs to the group $C_{\infty v}$ because rotations around the axis by any angle and

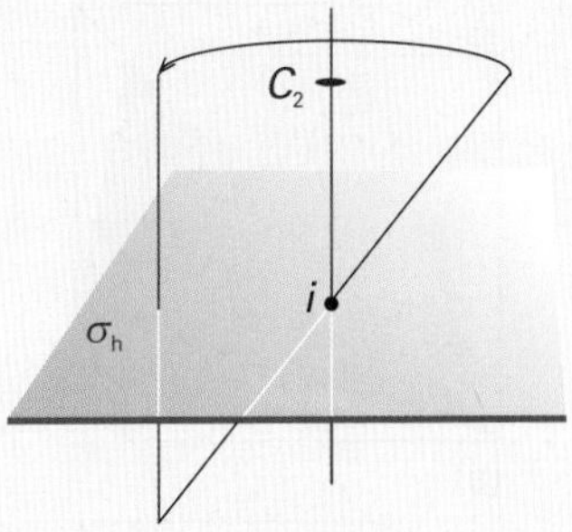

Fig. 7.9 The presence of a twofold axis and a horizontal mirror plane jointly imply the presence of a centre of inversion in the molecule.

reflections in all the infinite number of planes that contain the axis are symmetry operations. Other members of the group $C_{\infty v}$ include the linear OCS molecule and a cone.

Objects that, in addition to the identity and an n-fold principal axis, also have a horizontal mirror plane σ_h belong to the groups C_{nh}. An example is *trans*-CHCl=CHCl (**8**), which has the elements E, C_2, and σ_h, so belongs to the group C_{2h}; the molecule $B(OH)_3$ in the conformation shown in (**9**) belongs to the group C_{3h}. The presence of certain symmetry elements may be implied by the presence of others: thus, in C_{2h} the operations C_2 and σ_h jointly imply the presence of a centre of inversion (Fig. 7.9).

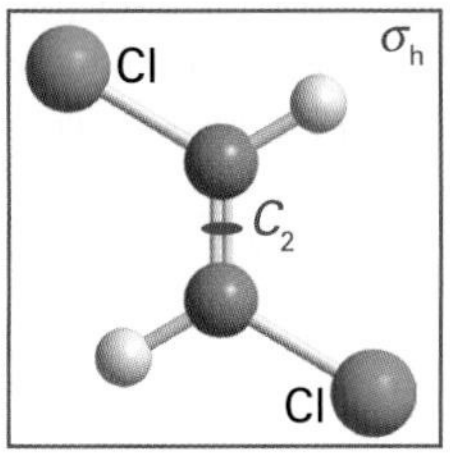

8 *trans*-CHCl=CHCl

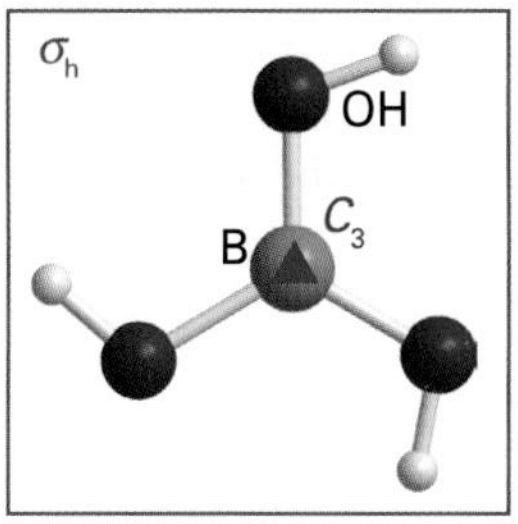

9 $B(OH)_3$

(c) The groups D_n, D_{nh}, and D_{nd}

We see from Fig. 7.7 that a molecule that has an n-fold principal axis and n twofold axes perpendicular to C_n belongs to the group D_n. A molecule belongs to D_{nh} if it also possesses a horizontal mirror plane. The planar trigonal BF_3 molecule has the elements E, C_3, $3C_2$, and σ_h (with one C_2 axis along each B—F bond), so belongs to D_{3h} (**10**). The C_6H_6 molecule has the elements E, C_6, $3C_2$, $3C_2'$, and σ_h together with some others that these elements imply, so it belongs to D_{6h}. Three of the C_2 axes bisect C—C bonds and the other three pass through vertices of the hexagon formed by the carbon framework of the molecule and the prime on $3C_2'$ indicates that the three C_2 axes are different from the other three C_2 axes. All homonuclear diatomic molecules, such as N_2, belong to the group $D_{\infty h}$ because all rotations around the axis are symmetry operations, as are end-to-end rotation and end-to-end reflection; $D_{\infty h}$ is also the group of the linear OCO and HCCH molecules and of a uniform cylinder. Other examples of D_{nh} molecules are shown in (**11**), (**12**), and (**13**).

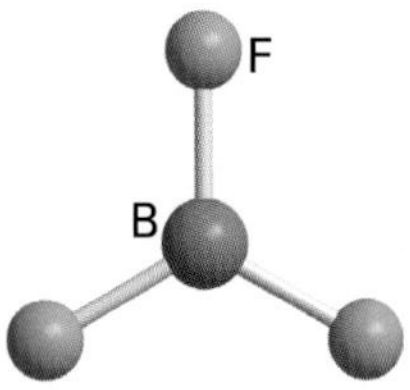

10 Boron trifluoride, BF_3

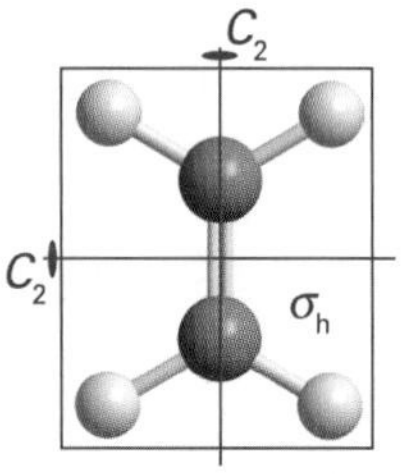

11 Ethene, $CH_2{=}CH_2$ (D_{2h})

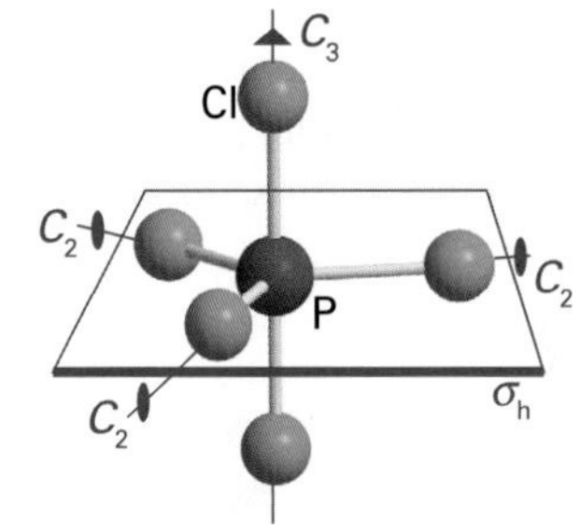

12 Phosphorus pentachloride, PCl_5 (D_{3h})

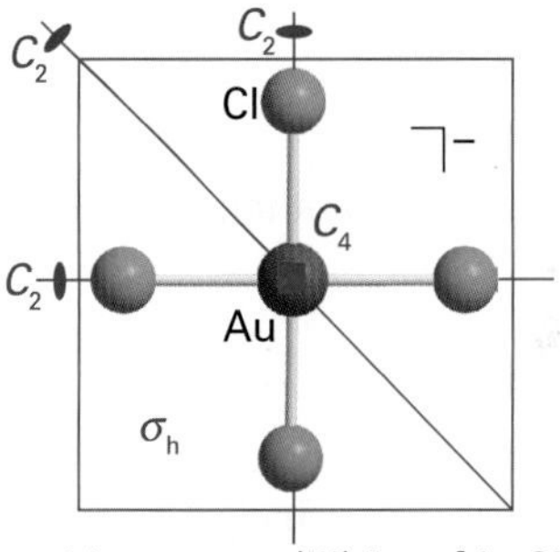

13 Tetrachloroaurate(III) ion, $[AuCl_4]^-$, (D_{4h})

A molecule belongs to the group D_{nd} if in addition to the elements of D_n it possesses n dihedral mirror planes σ_d. The twisted, 90° allene (**14**) belongs to D_{2d}, and the staggered conformation of ethane (**15**) belongs to D_{3d}.

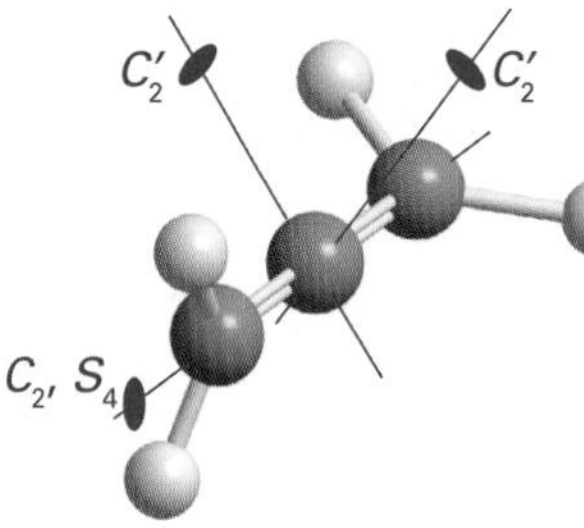

14 Allene, C_3H_4 (D_{2d})

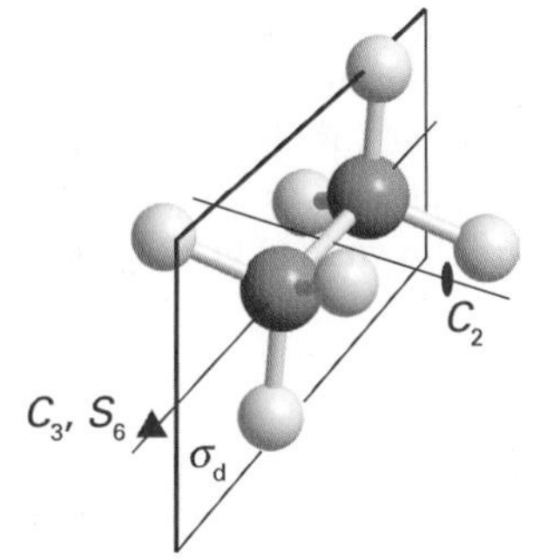

15 Ethane, C_2H_6 (D_{3d})

(d) The groups S_n

Molecules that have not been classified into one of the groups mentioned so far, but which possess one S_n axis, belong to the group S_n. An example is tetraphenylmethane, which belongs to the point group S_4 (**16**). Molecules belonging to S_n with $n > 4$ are rare. Note that the group S_2 is the same as C_i, so such a molecule will already have been classified as C_i.

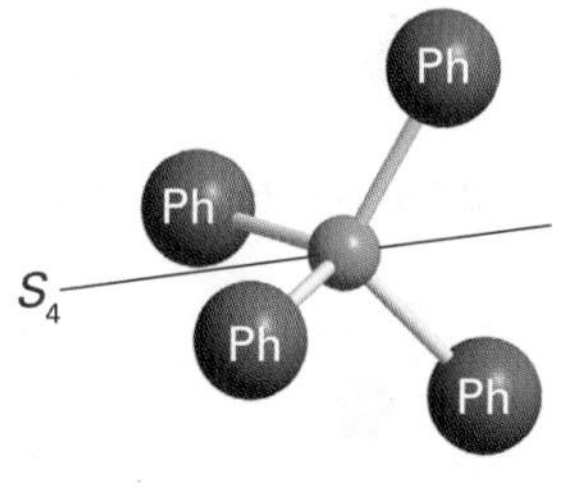

16 Tetraphenylmethane, $C(C_6H_5)_4$ (S_4)

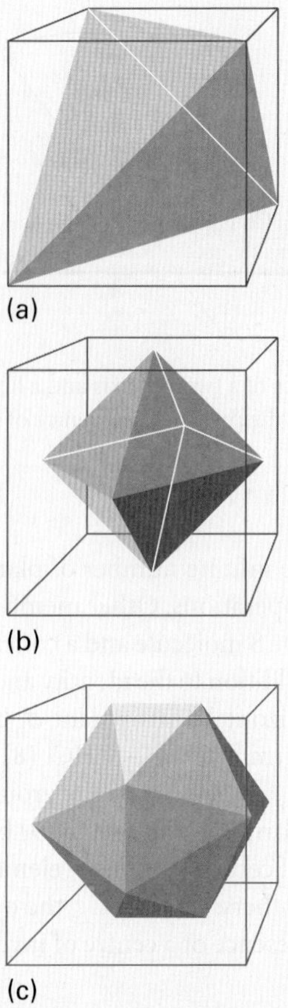

Fig. 7.10 (a) Tetrahedral, (b) octahedral, and (c) icosahedral molecules are drawn in a way that shows their relation to a cube: they belong to the cubic groups T_d, O_h, and I_h, respectively.

(e) The cubic groups

A number of very important molecules (for example, CH_4 and SF_6) possess more than one principal axis. Most belong to the **cubic groups**, and in particular to the **tetrahedral groups** T, T_d, and T_h (Fig. 7.10a) or to the **octahedral groups** O and O_h (Fig. 7.10b). A few icosahedral (20-faced) molecules belonging to the **icosahedral group**, I (Fig. 7.10c), are also known: they include some of the boranes and buckminsterfullerene, C_{60} (**17**).

17 Buckminsterfullerene, C_{60} (I)

The groups T_d and O_h are the groups of the regular tetrahedron (for instance, CH_4) and the regular octahedron (for instance, SF_6), respectively. If the object possesses the rotational symmetry of the tetrahedron or the octahedron, but none of their planes of reflection, then it belongs to the simpler groups T or O (Fig. 7.11). The group T_h is based on T but also contains a centre of inversion (Fig. 7.12).

(f) The full rotation group

The **full rotation group**, R_3 (the 3 refers to rotation in three dimensions), consists of an infinite number of rotation axes with all possible values of n. A sphere and an atom belong to R_3, but no molecule does. Exploring the consequences of R_3 is a very important way of applying symmetry arguments to atoms, and is an alternative approach to the theory of orbital angular momentum.

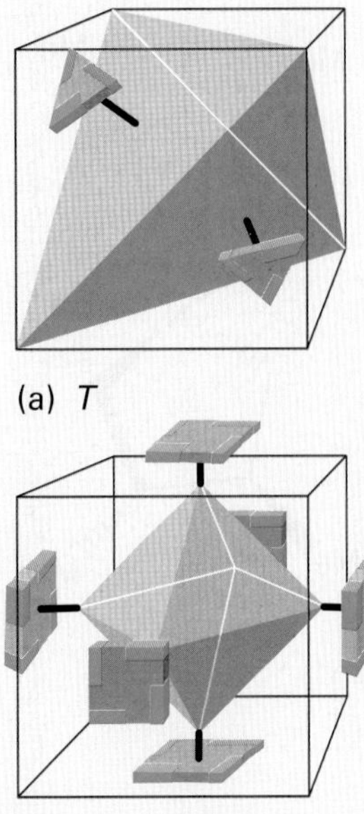

Fig. 7.11 Shapes corresponding to the point groups (a) T and (b) O. The presence of the decorated slabs reduces the symmetry of the object from T_d and O_h, respectively.

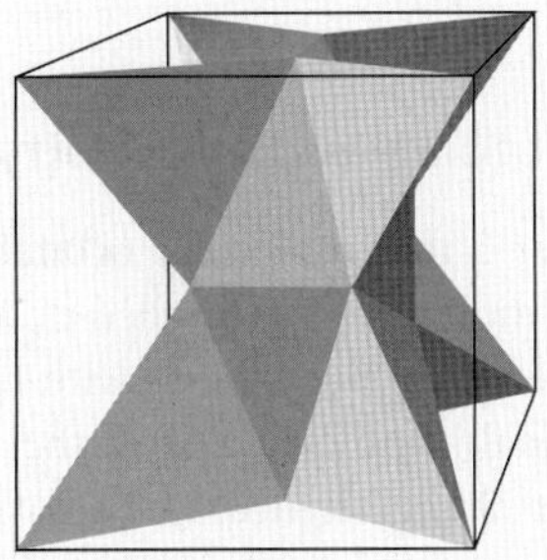

Fig. 7.12 The shape of an object belonging to the group T_h.

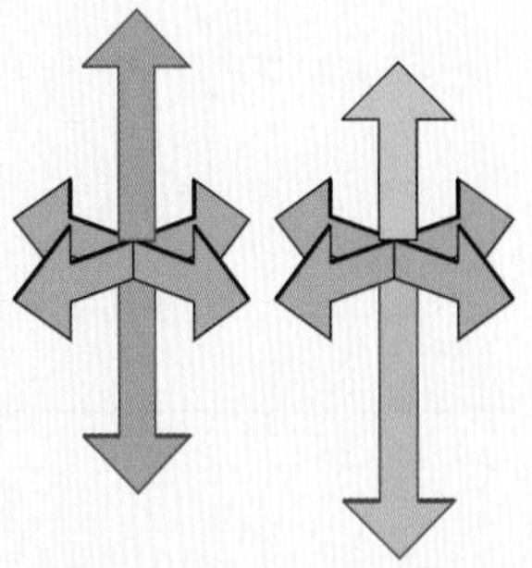

Fig. 7.13 (a) A molecule with a C_n axis cannot have a dipole perpendicular to the axis, but (b) it may have one parallel to the axis. The arrows represent local contributions to the overall electric dipole, such as may arise from bonds between pairs of neighbouring atoms with different electronegativities.

7.3 Some immediate consequences of symmetry

Some statements about the properties of a molecule can be made as soon as its point group has been identified.

(a) Polarity

A **polar molecule** is one with a permanent electric dipole moment (HCl, O_3, and NH_3 are examples). If the molecule belongs to the group C_n with $n > 1$, it cannot possess a charge distribution with a dipole moment perpendicular to the symmetry axis because the symmetry of the molecule implies that any dipole that exists in one direction perpendicular to the axis is cancelled by an opposing dipole (Fig. 7.13a). For example, the perpendicular component of the dipole associated with one O—H bond in H_2O is cancelled by an equal but opposite component of the dipole of the second O—H bond, so any dipole that the molecule has must be parallel to the twofold symmetry axis. However, as the group makes no reference to operations relating the two ends of the molecule, a charge distribution may exist that results in a dipole along the axis (Fig. 7.13b), and H_2O has a dipole moment parallel to its twofold symmetry axis. The same remarks apply generally to the group C_{nv}, so molecules belonging to any of the C_{nv} groups may be polar. In all the other groups, such as C_{3h}, D, etc., there are symmetry operations that take one end of the molecule into the other. Therefore, as well as having no dipole perpendicular to the axis, such molecules can have none along the axis, for otherwise these additional operations would not be symmetry operations. We can conclude that *only molecules belonging to the groups C_n, C_{nv}, and C_s may have a permanent electric dipole moment.*

For C_n and C_{nv}, that dipole moment must lie along the symmetry axis. Thus ozone, O_3, which is angular and belongs to the group C_{2v}, may be polar (and is), but carbon dioxide, CO_2, which is linear and belongs to the group $D_{\infty h}$, is not.

(b) Chirality

A **chiral molecule** (from the Greek word for 'hand') is a molecule that cannot be superimposed on its mirror image. An **achiral molecule** is a molecule that can be superimposed on its mirror image. Chiral molecules are **optically active** in the sense that they rotate the plane of polarized light (a property discussed in more detail in Section 11.4). A chiral molecule and its mirror-image partner constitute an **enantiomeric pair** of isomers and rotate the plane of polarization in equal but opposite directions.

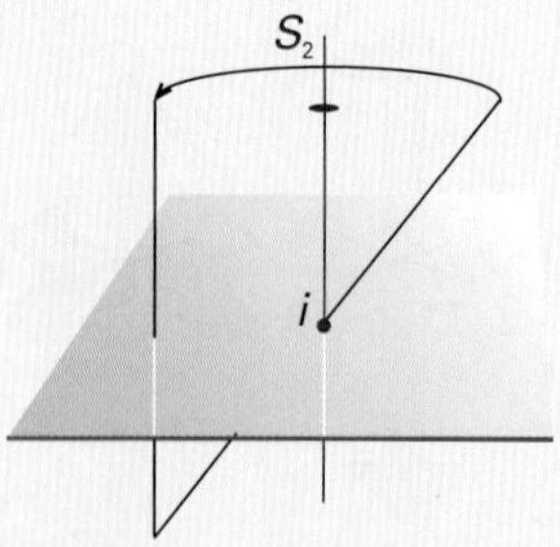

Fig. 7.14 Some symmetry elements are implied by the other symmetry elements in a group. Any molecule containing an inversion also possesses at least an S_2 element because i and S_2 are equivalent.

A molecule may be chiral, and therefore optically active, only if it does not possess an axis of improper rotation, S_n. We need to be aware that an S_n improper rotation axis may be present under a different name, and be implied by other symmetry elements that are present. For example, molecules belonging to the groups C_{nh} possess an S_n axis implicitly because they possess both C_n and σ_h, which are the two components of an improper rotation axis. Any molecule containing a centre of inversion, i, also possesses an S_2 axis, because i is equivalent to C_2 in conjunction with σ_h, and that combination of elements is S_2 (Fig. 7.14). It follows that all molecules with centres of inversion are achiral and hence optically inactive. Similarly, because $S_1 = \sigma$, it follows that any molecule with a mirror plane is achiral.

A molecule may be chiral if it does not have a centre of inversion or a mirror plane, which is the case with the amino acid alanine (**18**), but not with glycine (**19**). However, a molecule may be achiral even though it does not have a centre of inversion. For example, the S_4 species (**20**) is achiral and optically inactive: though it lacks i (that is, S_2) it does have an S_4 axis.

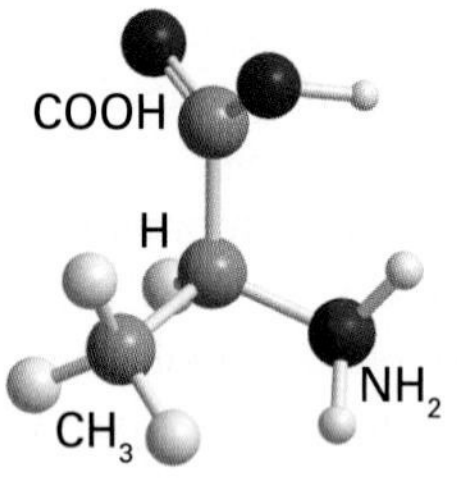

18 L-Alanine, $NH_2CH(CH_3)COOH$

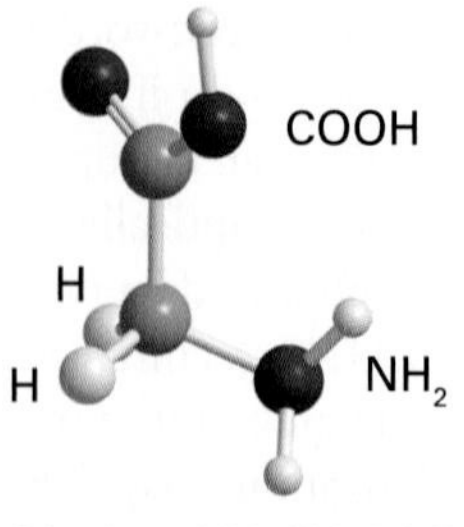

19 Glycine, NH_2CH_2COOH

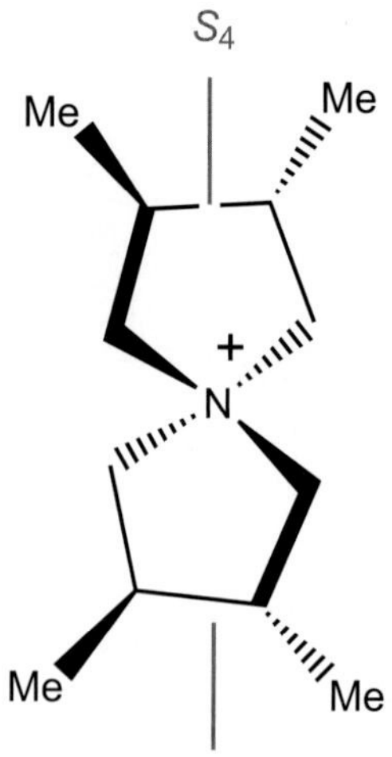

20

Applications

We shall now turn our attention away from the symmetries of molecules themselves and direct it towards the symmetry characteristics of orbitals that belong to the various atoms in a molecule. This material will enable us to discuss the formulation and labelling of molecular orbitals and selection rules in spectroscopy.

7.4 Character tables and symmetry labels

We saw in Chapter 5 that molecular orbitals of diatomic and linear polyatomic molecules are labelled σ, π, etc. These labels refer to the symmetries of the orbitals with respect to rotations around the principal symmetry axis of the molecule. Thus, a σ orbital does not change sign under a rotation through any angle, a π orbital changes sign when rotated by 180°, and so on (Fig. 7.15). The symmetry classifications σ and π can also be assigned to individual atomic orbitals in a linear molecule. For

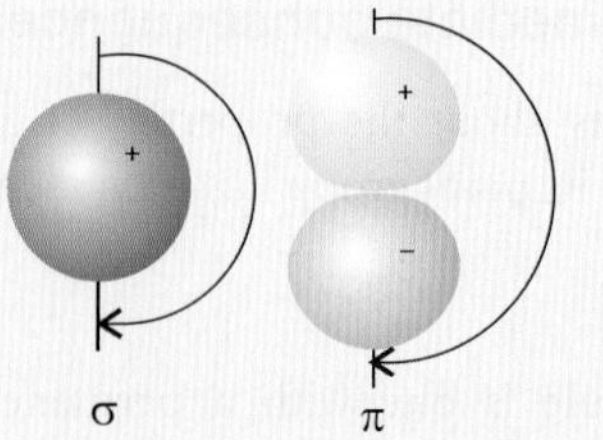

Fig. 7.15 A rotation through 180° about the internuclear axis (perpendicular to the page) leaves the sign of a σ orbital unchanged but the sign of a π orbital is changed. In the language introduced in this chapter, the characters of the C_2 rotation are +1 and −1 for the σ and π orbitals, respectively.

example, we can speak of an individual p_z orbital as having σ symmetry if the *z*-axis lies along the bond, because p_z is cylindrically symmetrical about the bond. This labelling of orbitals according to their behaviour under rotations can be generalized and extended to non-linear polyatomic molecules, where there may be reflections and inversions to take into account as well as rotations.

(a) Representations and characters

Labels analogous to σ and π are used to denote the symmetries of orbitals in polyatomic molecules. These labels look like a, a_1, e, e_g, and we first encountered them in Fig. 5.38 in connection with the molecular orbitals of benzene. As we shall see, they indicate the behaviour of the orbitals under the symmetry operations of the relevant point group of the molecule.

A label is assigned to an orbital by referring to the **character table** of the group, a table that characterizes the different symmetry types possible in the point group. Thus, to assign the labels σ and π, we use the following table:

	C_2
σ	+1
π	−1

This table is a fragment of the full character table for a linear molecule. The entry +1 shows that the orbital remains the same and the entry −1 shows that the orbital changes sign under the operation C_2 at the head of the column (as illustrated in Fig. 7.15). So, to assign the label σ or π to a particular orbital, we compare the orbital's behaviour with the information in the character table. Exactly the same procedure is used to assign labels to orbitals of molecules that belong to more elaborate point groups.

The entries in a complete character table are derived by using the formal techniques of group theory and are called **characters**, χ (chi). These numbers characterize the essential features of each symmetry type in a way that we can illustrate by considering the C_{2v} molecule SO_2 and the valence p_x orbitals on each atom, which we shall denote p_S, p_A, and p_B (Fig. 7.16). To understand their origin and significance, we need to introduce a hugely important new topic.

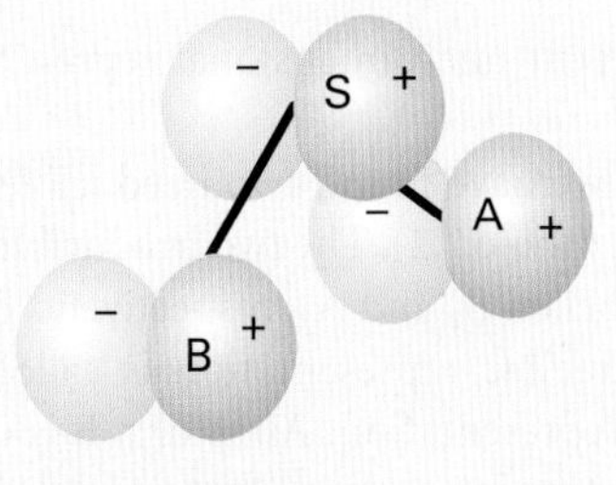

Fig. 7.16 The three p_x orbitals that are used to illustrate the construction of a matrix representation in a C_{2v} molecule (SO_2).

Under σ_v, the change $(p_S, p_B, p_A) \leftarrow (p_S, p_A, p_B)$ takes place. We can express this transformation by using matrix multiplication (*Mathematical background 5*):

$$(p_S, p_B, p_A) = (p_S, p_A, p_B)\begin{pmatrix}1 & 0 & 0\\ 0 & 0 & 1\\ 0 & 1 & 0\end{pmatrix} = (p_S, p_A, p_B)D(\sigma_v) \quad (7.1a)$$

The matrix $D(\sigma_v)$ is called a **representative** of the operation σ_v. Representatives take different forms according to the **basis**, the set of orbitals, that has been adopted.

We can use the same technique to find matrices that reproduce the other symmetry operations. For instance, C_2 has the effect $(-p_S, -p_B, -p_A) \leftarrow (p_S, p_A, p_B)$, and its representative is

$$D(C_2) = \begin{pmatrix}-1 & 0 & 0\\ 0 & 0 & -1\\ 0 & -1 & 0\end{pmatrix} \quad (7.1b)$$

The effect of σ'_v is $(-p_S, -p_A, -p_B) \leftarrow (p_S, p_A, p_B)$, and its representative is

$$D(\sigma'_v) = \begin{pmatrix}-1 & 0 & 0\\ 0 & -1 & 0\\ 0 & 0 & -1\end{pmatrix} \quad (7.1c)$$

The identity operation has no effect on the basis, so its representative is the 3 × 3 unit matrix:

$$D(E) = \begin{pmatrix}1 & 0 & 0\\ 0 & 1 & 0\\ 0 & 0 & 1\end{pmatrix} \quad (7.1d)$$

The set of matrices that represents *all* the operations of the group is called a **matrix representation**, Γ (upper-case gamma), of the group for the basis that has been chosen. We denote this three-dimensional representation by $\Gamma^{(3)}$. The matrices of a representation multiply together in the same way as the operations they represent. Thus, if for any two operations R and R' we know that $RR' = R''$, then $D(R)D(R') = D(R'')$ for a given basis.

● A BRIEF ILLUSTRATION

In the group C_{2v}, a twofold rotation followed by a reflection in a mirror plane is equivalent to a reflection in the second mirror plane: specifically, $\sigma'_v C_2 = \sigma_v$. When we use the representatives specified above, we find

$$D(\sigma'_v)D(C_2) = \begin{pmatrix}-1 & 0 & 0\\ 0 & -1 & 0\\ 0 & 0 & -1\end{pmatrix}\begin{pmatrix}-1 & 0 & 0\\ 0 & 0 & -1\\ 0 & -1 & 0\end{pmatrix}$$

$$= \begin{pmatrix}1 & 0 & 0\\ 0 & 0 & 1\\ 0 & 1 & 0\end{pmatrix} = D(\sigma_v)$$

This multiplication reproduces the group multiplication. The same is true of all pairs of representative multiplications, so the four matrices form a representation of the group. ●

The discovery of a matrix representation of the group means that we have found a link between symbolic manipulations of operations and algebraic manipulations of numbers.

The character of an operation in a particular matrix representation is the sum of the diagonal elements of the representative of that operation. Thus, in the basis we are illustrating, the characters of the representatives are

$D(E)$	$D(C_2)$	$D(\sigma_v)$	$D(\sigma_v')$
3	−1	1	−3

The character of an operation depends on the basis.

Inspection of the representatives shows that they are all of **block-diagonal form:**

$$D = \begin{pmatrix} \blacksquare & 0 & 0 \\ 0 & \blacksquare & \blacksquare \\ 0 & \blacksquare & \blacksquare \end{pmatrix} \tag{7.2}$$

The block-diagonal form of the representatives shows us that the symmetry operations of C_{2v} never mix p_S with the other two functions. Consequently, the basis can be cut into two parts, one consisting of p_S alone and the other of (p_A, p_B). It is readily verified that the p_S orbital itself is a basis for the one-dimensional representation

$$D(E) = 1 \qquad D(C_2) = -1 \qquad D(\sigma_v) = 1 \qquad D(\sigma_v') = -1$$

which we shall call $\Gamma^{(1)}$. The remaining two basis functions are a basis for the two-dimensional representation $\Gamma^{(2)}$:

$$D(E) = \begin{pmatrix} 1 & 0 \\ 0 & 1 \end{pmatrix} \qquad D(C_2) = \begin{pmatrix} 0 & -1 \\ -1 & 0 \end{pmatrix}$$

$$D(\sigma_v) = \begin{pmatrix} 0 & 1 \\ 1 & 0 \end{pmatrix} \qquad D(\sigma_v') = \begin{pmatrix} -1 & 0 \\ 0 & -1 \end{pmatrix}$$

These matrices are the same as those of the original three-dimensional representation, except for the loss of the first row and column. We say that the original three-dimensional representation has been **reduced** to the 'direct sum' of a one-dimensional representation 'spanned' by p_S, and a two-dimensional representation spanned by (p_A, p_B). This reduction is consistent with the common sense view that the central orbital plays a role different from the other two. We denote the reduction symbolically by writing

$$\Gamma^{(3)} = \Gamma^{(1)} + \Gamma^{(2)} \tag{7.3}$$

The one-dimensional representation $\Gamma^{(1)}$ cannot be reduced any further, and is called an **irreducible representation** of the group (an 'irrep'). We can demonstrate that the two-dimensional representation $\Gamma^{(2)}$ is reducible (for this basis in this group) by switching attention to the linear combinations $p_1 = p_A + p_B$ and $p_2 = p_A - p_B$. These combinations are sketched in Fig. 7.17. The representatives in the new basis can be constructed from the old by noting, for example, that under σ_v, $(p_B, p_A) \leftarrow (p_A, p_B)$. In this way we find the following representation in the new basis:

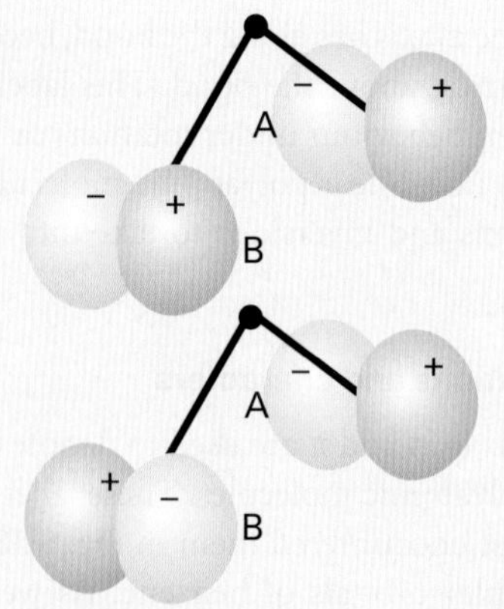

Fig. 7.17 Two symmetry-adapted linear combinations of the basis orbitals shown in Fig. 7.16. The two combinations each span a one-dimensional irreducible representation, and their symmetry species are different.

$$D(E) = \begin{pmatrix} 1 & 0 \\ 0 & 1 \end{pmatrix} \qquad D(C_2) = \begin{pmatrix} -1 & 0 \\ 0 & 1 \end{pmatrix}$$

$$D(\sigma_v) = \begin{pmatrix} 1 & 0 \\ 0 & -1 \end{pmatrix} \qquad D(\sigma_v') = \begin{pmatrix} -1 & 0 \\ 0 & -1 \end{pmatrix}$$

The new representatives are all in block-diagonal form, and the two combinations are not mixed with each other by any operation of the group. We have therefore achieved the reduction of $\Gamma^{(2)}$ to the sum of two one-dimensional representations. Thus, p_1 spans

$$D(E) = 1 \qquad D(C_2) = -1 \qquad D(\sigma_v) = 1 \qquad D(\sigma_v') = -1$$

which is the same one-dimensional representation as that spanned by p_S, and p_2 spans

$$D(E) = 1 \qquad D(C_2) = 1 \qquad D(\sigma_v) = -1 \qquad D(\sigma_v') = -1$$

which is a different one-dimensional representation; we shall denote it $\Gamma^{(1)\prime}$.

At this point we have found two irreducible representations of the group C_{2v} (Table 7.2). The two irreducible representations are normally labelled B_1 and A_2, respectively. The label used to specify an irreducible representation is called the **symmetry species** of that representation. An A or a B is used to denote a one-dimensional representation; A is used if the character under the principal rotation is +1, and B is used if the character is −1.

Table 7.2 The C_{2v} character table*

C_{2v}, 2*mm*	E	C_2	σ_v	σ'_v	$h=4$	
A_1	1	1	1	1	z	z^2, y^2, x^2
A_2	1	1	−1	−1		xy
B_1	1	−1	1	−1	x	zx
B_2	1	−1	−1	1	y	yz

* More character tables are given in the *Resource section*.

Subscripts are used to distinguish the irreducible representations if there is more than one of the same type: A_1 is reserved for the representation with character 1 for all operations. When higher dimensional irreducible representations are permitted, E denotes a two-dimensional irreducible representation and T a three-dimensional irreducible representation; all the irreducible representations of C_{2v} are one-dimensional.

There are in fact only two more species of irreducible representations of this group, for a surprising theorem of group theory states that

$$\text{Number of symmetry species} = \text{number of classes} \qquad (7.4)$$

Symmetry operations fall into the same **class** if they are of the same type (for example, rotations) and can be transformed into one another by a symmetry operation of the group. In C_{2v}, for instance, there are four classes (four columns in the character table), so there are only four species of irreducible representation. The character table in Table 7.2 therefore shows the characters of all the irreducible representations of this group. The two threefold rotations in C_{3v} belong to the same class because one can be converted into the other by a reflection (Fig. 7.18); the three reflections all belong to the same class because each can be rotated into another by a threefold rotation. The formal

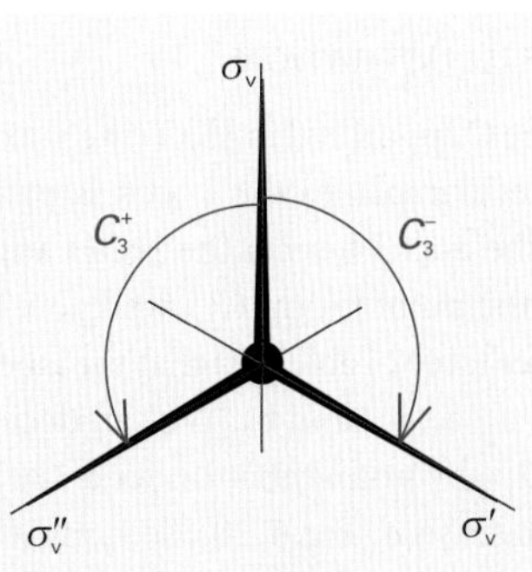

Fig. 7.18 Symmetry operations in the same class are related to one another by the symmetry operations of the group. Thus, the three mirror frames shown here are related by threefold rotations, and the two rotations shown here are related by reflection in σ_v.

definition of a class is that two operations R and R' belong to the same class if there is a member S of the group such that

$$R' = S^{-1}RS \qquad (7.5)$$

where S^{-1} is the inverse of S.

● A BRIEF ILLUSTRATION

To show that C_3^+ and C_3^- belong to the same class in C_{3v} (which intuitively we know to be the case), take $S = \sigma_v$. The reciprocal of a reflection is the reflection itself, so $\sigma_v^{-1} = \sigma_v$. It follows that

$$\sigma_v^{-1}C_3^+\sigma_v = \sigma_v C_3^+ \sigma_v = \sigma_v \sigma'_v = C_3^-$$

Therefore, C_3^+ and C_3^- are related by an equation of the form of eqn 7.5 and hence belong to the same class. ●

Self-test 7.2 Show that the two reflections of the group C_{2v} fall into different classes.

(b) The structure of character tables

In general, the columns in a character table are labelled with the symmetry operations of the group. For instance, for the group C_{3v} the columns are headed E, C_3, and σ_v (Table 7.3). The numbers multiplying each operation are the numbers of members of each class. The total number of operations in a group is called the **order**, h, of the group. The order of the group C_{3v}, for instance, is 6. The rows under the labels for the operations summarize the symmetry properties of the orbitals. They are labelled with the symmetry species (the analogues of the labels σ and π).

(c) Character tables and orbital degeneracy

The character of the identity operation E tells us the degeneracy of the orbitals. Thus, in a C_{3v} molecule, any orbital with a symmetry label A_1 or A_2 is non-degenerate. Any doubly degenerate pair of orbitals in C_{3v} must be labelled E because, in this group, only E symmetry species have characters greater than 1. (Take care to distinguish the identity element E (italic, a column heading) from the symmetry label E (roman, a row label).)

Table 7.3 The C_{3v} character table*

C_{3v}, 3*m*	E	$2C_3$	$3\sigma_v$	$h=6$	
A_1	1	1	1	z	z^2, x^2+y^2
A_2	1	1	−1		
E	2	−1	0	(x, y)	$(xy, x^2-y^2), (yz, zx)$

* More character tables are given in the *Resource section*.

Because there are no characters greater than 2 in the column headed E in C_{3v}, we know that there can be no triply degenerate orbitals in a C_{3v} molecule. This last point is a powerful result of group theory, for it means that, with a glance at the character table of a molecule, we can state the maximum possible degeneracy of its orbitals.

Example 7.1 *Using a character table to judge degeneracy*

Can a trigonal planar molecule such as BF_3 have triply degenerate orbitals? What is the minimum number of atoms from which a molecule can be built that does display triple degeneracy?

Method First, identify the point group, and then refer to the corresponding character table in the *Resource section*. The maximum number in the column headed by the identity E is the maximum orbital degeneracy possible in a molecule of that point group. For the second part, consider the shapes that can be built from two, three, etc. atoms, and decide which number can be used to form a molecule that can have orbitals of symmetry species T.

Answer Trigonal planar molecules belong to the point group D_{3h}. Reference to the character table for this group shows that the maximum degeneracy is 2, as no character exceeds 2 in the column headed E. Therefore, the orbitals cannot be triply degenerate. A tetrahedral molecule (symmetry group T) has an irreducible representation with a T symmetry species. The minimum number of atoms needed to build such a molecule is four (as in P_4, for instance).

Self-test 7.3 A buckminsterfullerene molecule, C_{60} (17), belongs to the icosahedral point group. What is the maximum possible degree of degeneracy of its orbitals? [5]

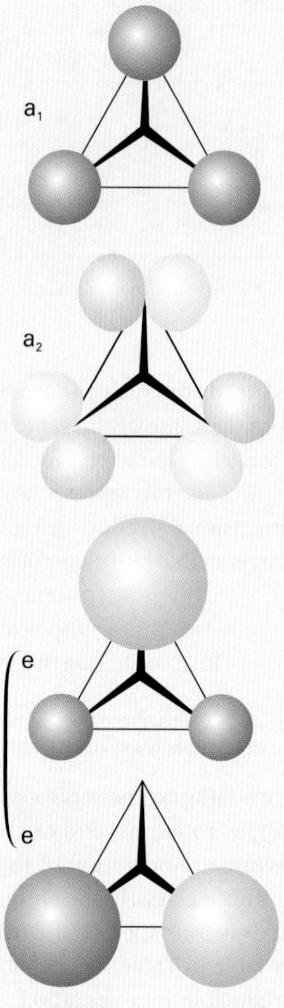

Fig. 7.19 Typical linear combinations of orbitals in a C_{3v} molecule.

(d) Characters and operations

The characters in the rows labelled A and B and in the columns headed by symmetry operations other than the identity E indicate the behaviour of an orbital under the corresponding operations: a +1 indicates that an orbital is unchanged, and a −1 indicates that it changes sign. It follows that we can identify the symmetry label of the orbital by comparing the changes that occur to an orbital under each operation, and then comparing the resulting +1 or −1 with the entries in a row of the character table for the point group concerned. By convention, irreducible representations are labelled with upper-case roman letters (such as A_1 and E) and the orbitals to which they apply are labelled with the lower-case equivalents (so an orbital of symmetry species A_1 is called an a_1 orbital). Examples of each type of orbital for molecules belonging to the C_{3v} point group are shown in Fig. 7.19.

● A BRIEF ILLUSTRATION

Consider the $O2p_x$ orbital in H_2O (the x-axis is perpendicular to the molecular plane; the y-axis is parallel to the H—H direction; the z-axis bisects the HOH angle). Because H_2O belongs to the point group C_{2v}, we know by referring to the C_{2v} character table (Table 7.2) that the labels available for the orbitals are a_1, a_2, b_1, and b_2. We can decide the appropriate label for $O2p_x$ by noting that under a 180° rotation (C_2) the orbital changes sign (Fig. 7.20), so it must be either B_1 or B_2, as only these two symmetry types have character −1 under C_2. The $O2p_x$ orbital also changes sign under the reflection σ_v', which identifies it as B_1. As we shall see, any molecular orbital built from this atomic orbital will also be a b_1 orbital. Similarly, $O2p_y$ changes sign under C_2 but not under σ_v'; therefore, it can contribute to b_2 orbitals. ●

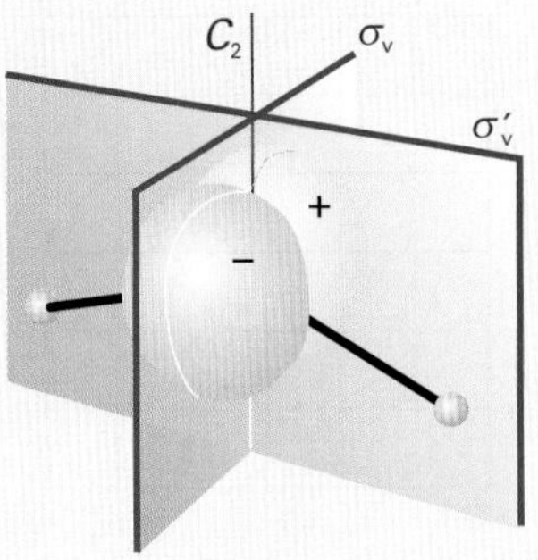

Fig. 7.20 A p_x orbital on the central atom of a C_{2v} molecule and the symmetry elements of the group.

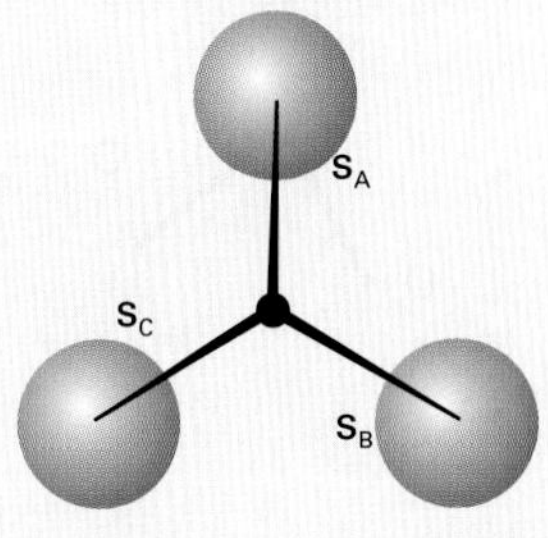

Fig. 7.22 The three H1s orbitals used to construct linear combinations in a C_{3v} molecule such as NH_3.

For the rows labelled E or T (which refer to the behaviour of sets of doubly and triply degenerate orbitals, respectively), the characters in a row of the table are the sums of the characters summarizing the behaviour of the individual orbitals in the basis. Thus, if one member of a doubly degenerate pair remains unchanged under a symmetry operation but the other changes sign (Fig. 7.21), then the entry is reported as $\chi = 1 - 1 = 0$. Care must be exercised with these characters because the transformations of orbitals can be quite complicated; nevertheless, the sums of the individual characters are integers.

The behaviour of s, p, and d orbitals on a central atom under the symmetry operations of the molecule is so important that the symmetry species of these orbitals are generally indicated in a character table. To make these allocations, we look at the symmetry species of x, y, and z, which appear on the right-hand side of the character table. Thus, the position of z in Table 7.3 shows that p_z (which is proportional to $zf(r)$), has symmetry species A_1 in C_{3v}, whereas p_x and p_y (which are proportional to $xf(r)$ and $yf(r)$, respectively) are jointly of E symmetry. In technical terms, we say that p_x and p_y jointly **span** an irreducible representation of symmetry species E. An s orbital on the central atom always spans the fully symmetrical irreducible representation (typically labelled A_1 but sometimes A'_1) of a group as it is unchanged under all symmetry operations.

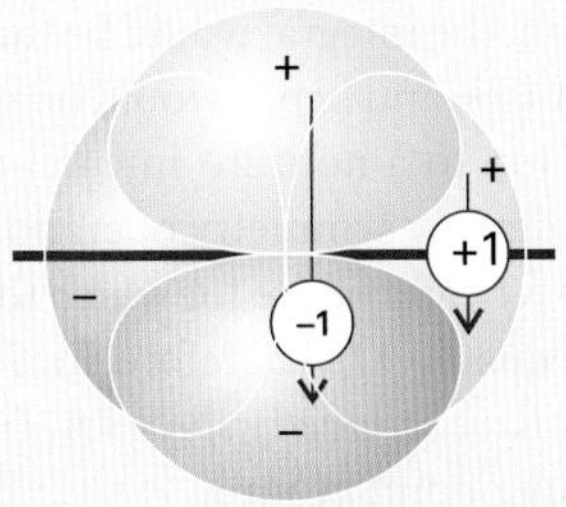

Fig. 7.21 The two orbitals shown here have different properties under reflection through the mirror plane: one changes sign (character –1), the other does not (character +1).

The five d orbitals of a shell are represented by xy for d_{xy}, etc. and are also listed on the right of the character table. We can see at a glance that, in C_{3v}, d_{xy} and $d_{x^2-y^2}$ on a central atom jointly belong to E and hence form a doubly degenerate pair.

(e) The classification of linear combinations of orbitals

So far, we have dealt with the symmetry classification of individual orbitals. The same technique may be applied to linear combinations of orbitals on atoms that are related by symmetry transformations of the molecule, such as the combination $\psi_1 = \psi_A + \psi_B + \psi_C$ of the three H1s orbitals in the C_{3v} molecule NH_3 (Fig. 7.22). This combination remains unchanged under a C_3 rotation and under any of the three vertical reflections of the group, so its characters are

$$\chi(E) = 1 \qquad \chi(C_3) = 1 \qquad \chi(\sigma_v) = 1$$

Comparison with the C_{3v} character table shows that ψ_1 is of symmetry species A_1, and therefore that it contributes to a_1 molecular orbitals in NH_3.

Example 7.2 *Identifying the symmetry species of orbitals*

Identify the symmetry species of the orbital $\psi = \psi_A - \psi_B$ in a C_{2v} NO_2 molecule, where ψ_A is an O2p_x orbital on one O atom and ψ_B that on the other O atom.

Method The negative sign in ψ indicates that the sign of ψ_B is opposite to that of ψ_A. We need to consider how the combination changes under each operation of the group, and then write the character as +1, –1, or 0 as specified above. Then we compare the resulting characters with each row in the character table for the point group, and hence identify the symmetry species.

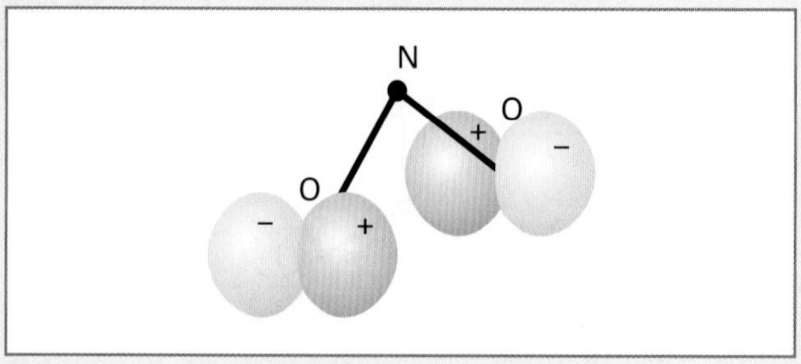

Fig. 7.23 One linear combination of $O2p_x$ orbitals in the C_{2v} NO_2 molecule.

Answer The combination is shown in Fig. 7.23. Under C_2, ψ changes into itself, implying a character of +1. Under the reflection σ_v, both orbitals change sign, so $\psi \to -\psi$, implying a character of −1. Under σ_v', $\psi \to -\psi$, so the character for this operation is also −1. The characters are therefore

$$\chi(E) = 1 \qquad \chi(C_2) = 1 \qquad \chi(\sigma_v) = -1 \qquad \chi(\sigma_v') = -1$$

These values match the characters of the A_2 symmetry species, so ψ can contribute to an a_2 orbital.

Self-test 7.4 Consider $PtCl_4^-$, in which the Cl ligands form a square planar array of point group D_{4h} (21). Identify the symmetry type of the combination $\psi_A - \psi_B + \psi_C - \psi_D$ where ψ is a chlorine s orbital. [B_{2g}]

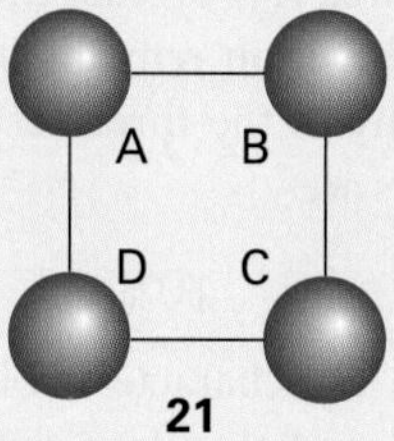

21

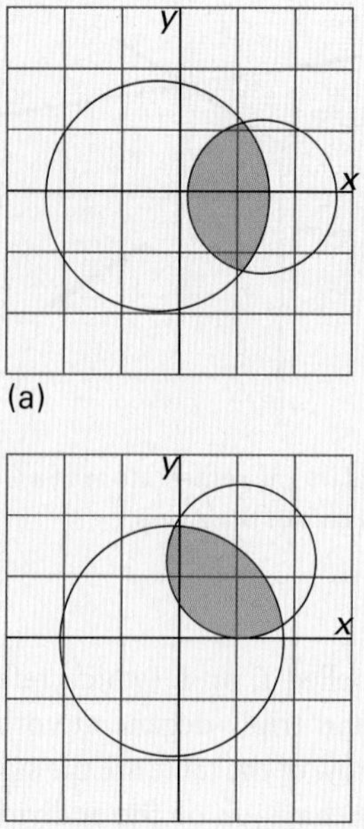

Fig. 7.24 The value of an integral I (for example, an area) is independent of the coordinate system used to evaluate it. That is, I is a basis of a representation of symmetry species A_1 (or its equivalent).

7.5 Vanishing integrals and orbital overlap

Suppose we had to evaluate the integral

$$I = \int f_1 f_2 \, d\tau \tag{7.6}$$

where f_1 and f_2 are functions. For example, f_1 might be an atomic orbital A on one atom and f_2 an atomic orbital B on another atom, in which case I would be their overlap integral. If we knew that the integral is zero, we could say at once that a molecular orbital does not result from (A, B) overlap in that molecule. We shall now see that character tables provide a quick way of judging whether an integral is necessarily zero.

(a) The criteria for vanishing integrals

The key point in dealing with the integral I is that the value of any integral, and of an overlap integral in particular, is independent of the orientation of the molecule (Fig. 7.24). In group theory we express this point by saying that I *is invariant under any symmetry operation of the molecule*, and that each operation brings about the trivial transformation $I \to I$. Because the volume element $d\tau$ is invariant under any symmetry operation, it follows that the integral is nonzero only if the integrand itself, the product $f_1 f_2$, is unchanged by any symmetry operation of the molecular point group. If the integrand changed sign under a symmetry operation, the integral would be the sum of equal and opposite contributions, and hence would be zero. It follows that the only contribution to a nonzero integral comes from functions for which under any symmetry operation of the molecular point group $f_1 f_2 \to f_1 f_2$, and hence for which the characters of the operations are all equal to +1. Therefore, for I not to be zero, *the integrand $f_1 f_2$ must have symmetry species A_1* (or its equivalent in the specific molecular point group).

We use the following procedure to deduce the symmetry species spanned by the product $f_1 f_2$ and hence to see whether it does indeed span A_1.

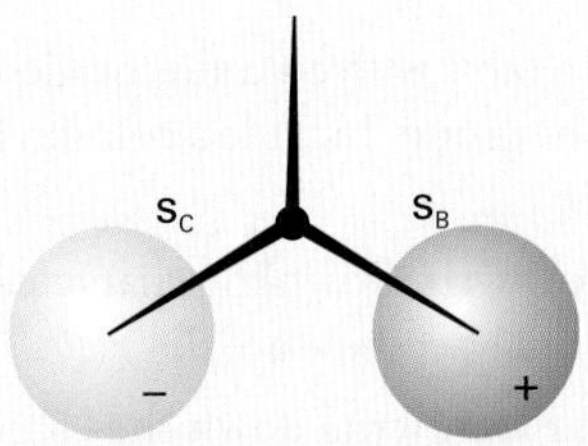

Fig. 7.25 A symmetry-adapted linear combination that belongs to the symmetry species E in a C_{3v} molecule such as NH_3. This combination can form a molecular orbital by overlapping with the p_x orbital on the central atom (the orbital with its axis parallel to the width of the page; see Fig. 7.28c).

1. Decide on the symmetry species of the individual functions f_1 and f_2 by reference to the character table, and write their characters in two rows in the same order as in the table.

2. Multiply the numbers in each column, writing the results in the same order.

3. Inspect the row so produced, and see if it can be expressed as a sum of characters from each column of the group. The integral must be zero if this sum does not contain A_1.

● A BRIEF ILLUSTRATION

If f_1 is the s_N orbital in NH_3 and f_2 is the linear combination $s_3 = s_B - s_C$ (Fig. 7.25) then, because s_N spans A_1 and s_3 is a member of the basis spanning E, we write

f_1:	1	1	1
f_2:	2	−1	0
f_1f_2:	2	−1	0

The characters 2, −1, 0 are those of E alone, so the integrand does not span A_1. It follows that the integral must be zero. Inspection of the form of the functions (see Fig. 7.25) shows why this is so: s_3 has a node running through s_N. ●

Self-test 7.5 Show that s_1 (the combination $s_A + s_B + s_C$ shown in Fig. 7.22) and s_N may have nonzero overlap.

A short cut that works when f_1 and f_2 are bases for irreducible representations of a group is to note their symmetry species: if they are different, then the integral of their product must vanish; if they are the same, then the integral may be nonzero.

It is important to note that group theory is specific about when an integral must be zero, but integrals that it allows to be nonzero may be zero for reasons unrelated to symmetry. For example, the N—H distance in ammonia may be so great that the (s_1,s_N) overlap integral is zero simply because the orbitals are so far apart.

Example 7.3 *Deciding if an integral must be zero (1)*

May the integral of the function $f = xy$ be nonzero when evaluated over a region the shape of an equilateral triangle centred on the origin (Fig. 7.26)?

Method First, note that an integral over a single function f is included in the previous discussion if we take $f_1 = f$ and $f_2 = 1$ in eqn 7.6. Therefore, we need to judge whether f alone belongs to the symmetry species A_1 (or its equivalent) in the point group of the system. To decide that, we identify the point group and then examine the character table to see whether f belongs to A_1 (or its equivalent).

Answer An equilateral triangle has the point-group symmetry D_{3h}. If we refer to the character table of the group, we see that xy is a member of a basis that spans the irreducible representation E′. Therefore, its integral must be zero, because the integrand has no component that spans A_1'.

Self-test 7.6 Can the function $x^2 + y^2$ have a nonzero integral when integrated over a regular pentagon centred on the origin? [Yes, Fig. 7.27]

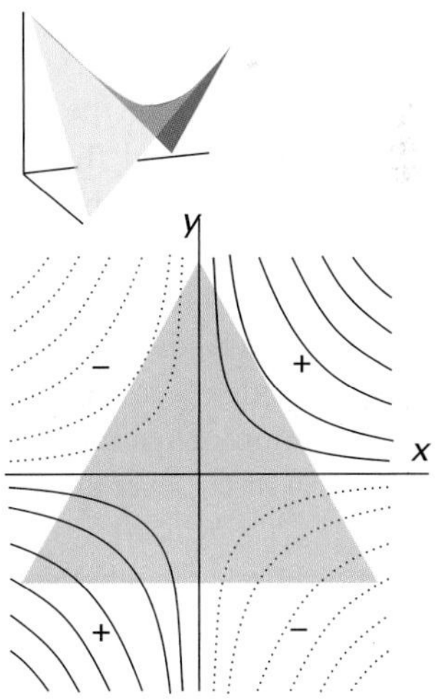

Fig. 7.26 The integral of the function $f = xy$ over the tinted region is zero. In this case, the result is obvious by inspection, but group theory can be used to establish similar results in less obvious cases. The insert shows the shape of the function in three dimensions.

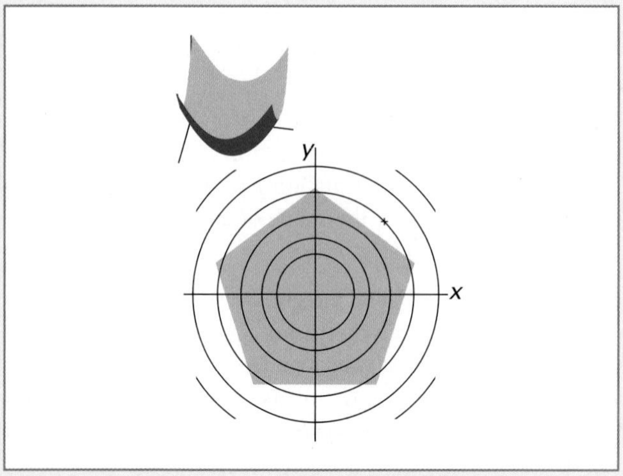

Fig. 7.27 The integration of a function over a pentagonal region. The insert shows the shape of the function in three dimensions.

In many cases, the product of functions f_1 and f_2 spans a sum of irreducible representations. For instance, in C_{2v} we may find the characters 2, 0, 0, −2 when we multiply the characters of f_1 and f_2 together. In this case, we note that these characters are the sum of the characters for A_2 and B_1:

	E	C_{2v}	σ_v	σ'_v
A_2	1	1	−1	−1
B_1	1	−1	1	−1
$A_2 + B_1$	2	0	0	−2

To summarize this result we write the symbolic expression $A_2 \times B_1 = A_2 + B_1$, which is called the **decomposition of a direct product**. This expression is symbolic. The × and + signs in this expression are not ordinary multiplication and addition signs: formally, they denote technical procedures with matrices called a 'direct product' and a 'direct sum'. Because the sum on the right does not include a component that is a basis for an irreducible representation of symmetry species A_1, we can conclude that the integral of $f_1 f_2$ over all space is zero in a C_{2v} molecule.

Whereas the decomposition of the characters 2, 0, 0, −2 can be done by inspection in this simple case, in other cases and more complex groups the decomposition is often far from obvious. For example, if we found the characters 8, −2, −6, 4, it might not be obvious that the sum contains A_1. Group theory, however, provides a systematic way of using the characters of the representation spanned by a product to find the symmetry species of the irreducible representations. The formal recipe is

$$n(\Gamma) = \frac{1}{h}\sum_R \chi^{(\Gamma)}(R)\chi(R) \qquad (7.7)$$

We implement this expression as follows:

1. Write down a table with columns headed by the symmetry operations, R, of the group. Include a column for every element, not just the classes.
2. In the first row write down the characters of the representation we want to analyse; these are the $\chi(R)$.
3. In the second row, write down the characters of the irreducible representation Γ we are interested in; these are the $\chi^{(\Gamma)}(R)$.
4. Multiply the two rows together, add the products together, and divide by the order of the group, h.

The resulting number, $n(\Gamma)$ is the number of times Γ occurs in the decomposition.

● A BRIEF ILLUSTRATION

To find whether A_1 does indeed occur in the product with characters 8, −2, −6, 4 in C_{2v}, we draw up the following table:

	E	C_{2v}	σ_v	σ'_v	$h = 4$ (the order of the group)
$f_1 f_2$	8	−2	−6	4	(the characters of the product)
A_1	1	1	1	1	(the symmetry species we are interested in)
	8	−2	−6	4	(the product of the two sets of characters)

The sum of the numbers in the last line is 4; when that number is divided by the order of the group, we get 1, so A_1 occurs once in the decomposition. When the procedure is repeated for all four symmetry species, we find that $f_1 f_2$ spans $A_1 + 2A_2 + 5B_2$. ●

Self-test 7.7 Does A_2 occur among the symmetry species of the irreducible representations spanned by a product with characters 7, −3, −1, 5 in the group C_{2v}? [No]

(b) Orbitals with nonzero overlap

The rules just given let us decide which atomic orbitals may have nonzero overlap in a molecule. We have seen that s_N may have nonzero overlap with s_1 (the combination $s_A + s_B + s_C$), so bonding and antibonding molecular orbitals can form from (s_N, s_1) overlap (Fig. 7.28). The general rule is that *only orbitals of the same symmetry species may have nonzero overlap*, so only orbitals of the same symmetry species form bonding and antibonding combinations. It should be recalled from Chapter 5 that the selection of atomic orbitals that had mutual nonzero overlap is the central and initial step in the construction of molecular orbitals by the LCAO procedure. We are therefore at the point of contact between group theory and the material introduced in

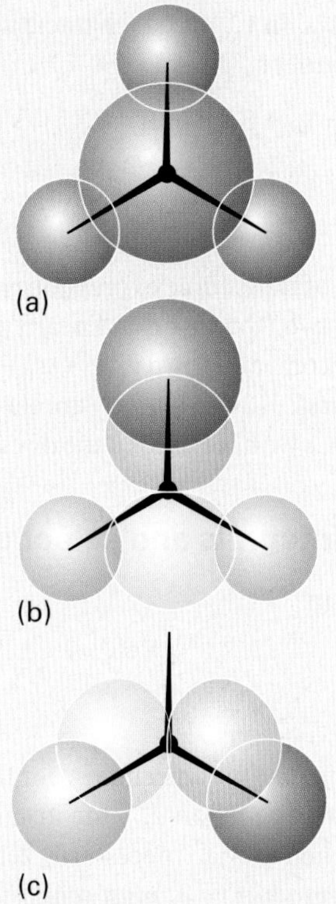

Fig. 7.28 Orbitals of the same symmetry species may have nonvanishing overlap. This diagram illustrates the three bonding orbitals that may be constructed from (N2s, H1s) and (N2p, H1s) overlap in a C_{3v} molecule. (a) a_1; (b) and (c) the two components of the doubly degenerate e orbitals. (There are also three antibonding orbitals of the same species.)

that chapter. The molecular orbitals formed from a particular set of atomic orbitals with nonzero overlap are labelled with the lower-case letter corresponding to the symmetry species. Thus, the (s_N, s_1)-overlap orbitals are called a_1 orbitals (or a_1^*, if we wish to emphasize that they are antibonding).

The linear combinations $s_2 = 2s_A - s_B - s_C$ and $s_3 = s_B - s_C$ have symmetry species E. Does the N atom have orbitals that have nonzero overlap with them (and give rise to e molecular orbitals)? Intuition (as supported by Figs. 7.28b and c) suggests that $N2p_x$ and $N2p_y$ should be suitable. We can confirm this conclusion by noting that the character table shows that, in C_{3v}, the functions x and y jointly belong to the symmetry species E. Therefore, $N2p_x$ and $N2p_y$ also belong to E, so may have nonzero overlap with s_2 and s_3. This conclusion can be verified by multiplying the characters and finding that the product of characters can be expressed as the decomposition $E \times E = A_1 + A_2 + E$. The two e orbitals that result are shown in Fig. 7.28 (there are also two antibonding e orbitals).

We can see the power of the method by exploring whether any d orbitals on the central atom can take part in bonding. As explained earlier, reference to the C_{3v} character table shows that d_{z^2} has A_1 symmetry and that the pairs $(d_{x^2-y^2}, d_{xy})$ and (d_{yz}, d_{zx}) each transform as E. It follows that molecular orbitals may be formed by (s_1, d_{z^2}) overlap and by overlap of the s_2, s_3 combinations with the E d orbitals. Whether or not the d orbitals are in fact important is a question group theory cannot answer because the extent of their involvement depends on energy considerations, not symmetry.

Example 7.4 *Determining which orbitals can contribute to bonding*

The four H1s orbitals of methane span $A_1 + T_2$. With which of the C atom orbitals can they overlap? What bonding pattern would be possible if the C atom had d orbitals available?

Method Refer to the T_d character table (in the *Resource section*) and look for s, p, and d orbitals spanning A_1 or T_2.

Answer An s orbital spans A_1, so it may have nonzero overlap with the A_1 combination of H1s orbitals. The C2p orbitals span T_2, so they may have nonzero overlap with the T_2 combination. The d_{xy}, d_{yz}, and d_{zx} orbitals span T_2, so they may overlap the same combination. Neither of the other two d orbitals span A_1 (they span E), so they remain nonbonding orbitals. It follows that in methane there are (C2s, H1s)-overlap a_1 orbitals and (C2p, H1s)-overlap t_2 orbitals. The C3d orbitals might contribute to the latter. The lowest energy configuration is probably $a_1^2 t_2^6$, with all bonding orbitals occupied.

Self-test 7.8 Consider the octahedral SF_6 molecule, with the bonding arising from overlap of S orbitals and a 2p orbital on each F directed towards the central S atom. The latter span $A_{1g} + E_g + T_{1u}$. What S orbitals have nonzero overlap? Suggest what the ground-state configuration is likely to be.

[$3s(A_{1g})$, $3p(T_{1u})$, $3d(E_g)$; $a_{1g}^2 t_{1u}^6 e_g^4$]

(c) Symmetry-adapted linear combinations

So far, we have only asserted the forms of the linear combinations (such as s_1, etc.) that have a particular symmetry. Group theory also provides machinery that takes an arbitrary **basis**, or set of atomic orbitals (s_A, etc.), as input and generates combinations of the specified symmetry. Because these combinations are adapted to the symmetry of the molecule, they are called **symmetry-adapted linear combinations** (SALC). Symmetry-adapted linear combinations are the building blocks of LCAO molecular orbitals, for they include combinations such as those used to construct molecular orbitals in benzene. The construction of SALCs is the first step in any molecular orbital treatment of molecules.

The technique for building SALCs is derived by using the full power of group theory and involves the use of a **projection**

operator, $P^{(\Gamma)}$, an operator that takes one of the basis orbitals and generates from it—projects from it—a SALC of the symmetry species Γ:

$$P^{(\Gamma)} = \frac{1}{h}\sum_R \chi^{(\Gamma)}(R)R \qquad (7.8)$$

To implement this rule, do the following:

1. Write each basis orbital at the head of a column and in successive rows show the effect of each operation R on each orbital. Treat each operation individually.

2. Multiply each member of the column by the character, $\chi^{(\Gamma)}(R)$, of the corresponding operation.

3. Add together all the orbitals in each column with the factors as determined in (2).

4. Divide the sum by the order of the group, h.

● A BRIEF ILLUSTRATION

From the (s_N, s_A, s_B, s_C) basis in NH_3 we form the following table:

	s_N	s_A	s_B	s_C
E	s_N	s_A	s_B	s_C
C_3^+	s_N	s_B	s_C	s_A
C_3^-	s_N	s_C	s_A	s_B
σ_v	s_N	s_A	s_C	s_B
σ_v'	s_N	s_B	s_A	s_C
σ_v''	s_N	s_C	s_B	s_A

To generate the A_1 combination, we take the characters for A_1 (1, 1, 1, 1, 1, 1); then rules 2 and 3 lead to $\psi \propto s_N + s_N + \cdots = 6s_N$. The order of the group (the number of elements) is 6, so the combination of A_1 symmetry that can be generated from s_N is s_N itself. Applying the same technique to the column under s_A gives

$$\psi = \tfrac{1}{6}(s_A + s_B + s_C + s_A + s_B + s_C) = \tfrac{1}{3}(s_A + s_B + s_C)$$

The same combination is built from the other two columns, so they give no further information. The combination we have just formed is the s_1 combination we used before (apart from the numerical factor). ●

We now form the overall molecular orbital by forming a linear combination of all the SALCs of the specified symmetry species. In this case, therefore, the a_1 molecular orbital is

$$\psi = c_N s_N + c_1 s_1$$

This is as far as group theory can take us. The coefficients are found by solving the Schrödinger equation; they do not come directly from the symmetry of the system.

We run into a problem when we try to generate an SALC of symmetry species E, because, for representations of dimension 2 or more, the rules generate sums of SALCs. This problem can be illustrated as follows. In C_{3v}, the E characters are 2, −1, −1, 0, 0, 0, so the column under s_N gives

$$\psi = \tfrac{1}{6}(2s_N - s_N - s_N + 0 + 0 + 0) = 0$$

The other columns give

$$\tfrac{1}{6}(2s_A - s_B - s_C) \qquad \tfrac{1}{6}(2s_B - s_A - s_C) \qquad \tfrac{1}{6}(2s_C - s_B - s_A)$$

However, any one of these three expressions can be expressed as a sum of the other two (they are not 'linearly independent'). The difference of the second and third gives $\tfrac{1}{2}(s_B - s_C)$, and this combination and the first, $\tfrac{1}{6}(2s_A - s_B - s_C)$ are the two (now linearly independent) SALCs we have used in the discussion of e orbitals.

7.6 Vanishing integrals and selection rules

Integrals of the form

$$I = \int f_1 f_2 f_3 \,\mathrm{d}\tau \qquad (7.9)$$

are also common in quantum mechanics for they include matrix elements of operators (Section 2.7, for instance), and it is important to know when they are necessarily zero. For the integral to be nonzero, *the product $f_1f_2f_3$ must span A_1 (or its equivalent) or contain a component that spans A_1.* To test whether this is so the characters of all three functions are multiplied together in the same way as in the rules set out above.

Example 7.5 *Deciding if an integral must be zero (2)*

Does the integral $\int(3d_{z^2})x(3d_{xy})\mathrm{d}\tau$ vanish in a C_{2v} molecule?

Method We must refer to the C_{2v} character table (Table 7.2) and the characters of the irreducible representations spanned by $3z^2 - r^2$ (the form of the d_{z^2} orbital), x, and xy; then we can use the procedure set out above (with one more row of multiplication).

Answer We draw up the following table:

	E	C_2	σ_v	σ_v'	
$f_3 = d_{xy}$	1	1	−1	−1	A_2
$f_2 = x$	1	−1	1	−1	B_1
$f_1 = d_{z^2}$	1	1	1	1	A_1
$f_1f_2f_3$	1	−1	−1	1	

The characters are those of B_2. Therefore, the integral is necessarily zero.

Self-test 7.9 Does the integral $\int(2p_x)(2p_y)(2p_z)\mathrm{d}\tau$ necessarily vanish in an octahedral environment? [Yes]

We saw in Chapter 4, and will see in more detail in Chapters 10 and 11, that the intensity of a spectral line arising from a molecular transition between some initial state with wavefunction ψ_i and a final state with wavefunction ψ_f depends on the (electric) transition dipole moment, μ_{fi}. The z-component of this vector is defined through

$$\mu_{z,fi} = -e\int \psi_f^* z \psi_i \, d\tau \qquad [7.10]$$

where $-e$ is the charge of the electron. The transition moment has the form of the integral in eqn 7.9 so, once we know the symmetry species of the states, we can use group theory to formulate the selection rules for the transitions.

The following chapters will show many more examples of the systematic use of symmetry, including the analysis of molecular vibrations. We shall see that the techniques of group theory greatly simplify the analysis of molecular structure and spectra.

Example 7.6 *Deducing a selection rule*

Is $p_x \rightarrow p_y$ an allowed transition in a tetrahedral environment?

Method We must decide whether the product $p_y q p_x$, with $q = x, y$, or z, spans A_1 by using the T_d character table.

Answer The procedure works out as follows:

	E	$8C_3$	$3C_2$	$6\sigma_d$	$6S_4$	
$f_3(p_y)$	3	0	−1	−1	1	T_2
$f_2(q)$	3	0	−1	1	−1	T_2
$f_1(p_x)$	3	0	−1	−1	1	T_2
$f_1f_2f_3$	27	0	−1	−1	1	

We can use the decomposition procedure described in Section 7.5a to deduce that A_1 occurs (once) in this set of characters, so $p_x \rightarrow p_y$ is allowed.

A more detailed analysis (using the matrix representatives rather than the characters) shows that only $q = z$ gives an integral that may be nonzero, so the transition is z-polarized. That is, the electromagnetic radiation involved in the transition has a component of its electric vector in the z-direction.

Self-test 7.10 What are the allowed transitions, and their polarizations, of an electron in a b_1 orbital in a C_{4v} molecule?

[$b_1 \rightarrow b_1(z)$; $b_1 \rightarrow e(x,y)$]

Checklist of key ideas

- ☐ 1. A symmetry operation is an action that leaves an object looking the same after it has been carried out.
- ☐ 2. A symmetry element is a point, line, or plane with respect to which a symmetry operation is performed.
- ☐ 3. A point group is a group of symmetry operations that leaves at least one common point unchanged. A space group is a group of symmetry operations that includes translation through space.
- ☐ 4. The notation for point groups commonly used for molecules and solids is summarized in Table 7.1.
- ☐ 5. To be polar, a molecule must belong to C_n, C_{nv}, or C_s (and have no higher symmetry).
- ☐ 6. A molecule may be chiral only if it does not possess an axis of improper rotation, S_n.
- ☐ 7. A representative $D(X)$ is a matrix that brings about the transformation of the basis under the operation X. The basis is the set of functions on which the representative acts.
- ☐ 8. A character, χ, is the sum of the diagonal elements of a matrix representative.
- ☐ 9. A character table characterizes the different symmetry types possible in the point group.
- ☐ 10. In a reduced representation all the matrices have block-diagonal form. An irreducible representation cannot be reduced further.
- ☐ 11. Symmetry species are the labels for the irreducible representations of a group.
- ☐ 12. Decomposition of the direct product is the reduction of a product of symmetry species to a sum of symmetry species, $\Gamma \times \Gamma' = \Gamma^{(1)} + \Gamma^{(2)} + \cdots$
- ☐ 13. For an integral $\int f_1 f_2 \, d\tau$ to be nonzero, the integrand $f_1 f_2$ must have the symmetry species A_1 (or its equivalent in the specific molecular point group).
- ☐ 14. A symmetry-adapted linear combination (SALC) is a combination of atomic orbitals adapted to the symmetry of the molecule and used as the building blocks for LCAO molecular orbitals.
- ☐ 15. Allowed and forbidden spectroscopic transitions can be identified by considering the symmetry criteria for the non-vanishing of the transition moment between the initial and final states.

Discussion questions

7.1 Explain how a molecule is assigned to a point group.

7.2 List the symmetry operations and the corresponding symmetry elements of the point groups.

7.3 Explain what is meant by a 'group'.

7.4 State and explain the symmetry criteria that allow a molecule to be polar.

7.5 State the symmetry criteria that allow a molecule to be optically active.

7.6 Explain what is meant by (a) a representative and (b) a representation in the context of group theory.

7.7 Explain the construction and content of a character table.

7.8 Explain what is meant by the reduction of a representation to a direct sum of representations.

7.9 Discuss the significance of the letters and subscripts used to denote the symmetry species of a representation.

7.10 Identify and list four applications of character tables.

Exercises

7.1(a) The CH_3Cl molecule belongs to the point group C_{3v}. List the symmetry elements of the group and locate them in a drawing of the molecule.

7.1(b) The CCl_4 molecule belongs to the point group T_d. List the symmetry elements of the group and locate them in a drawing of the molecule.

7.2(a) Identify the group to which the naphthalene molecule belongs and locate the symmetry elements in a drawing of the molecule.

7.2(b) Identify the group to which the anthracene molecule belongs and locate the symmetry elements in a drawing of the molecule.

7.3(a) Identify the point groups to which the following objects belong: (a) a sphere, (b) an isosceles triangle, (c) an equilateral triangle, (d) an unsharpened cylindrical pencil.

7.3(b) Identify the point groups to which the following objects belong: (a) a sharpened cylindrical pencil, (b) a three-bladed propellor, (c) a four-legged table, (d) yourself (approximately).

7.4(a) List the symmetry elements of the following molecules and name the point groups to which they belong: (a) NO_2, (b) N_2O, (c) $CHCl_3$, (d) $CH_2{=}CH_2$.

7.4(b) List the symmetry elements of the following molecules and name the point groups to which they belong: (a) furan (22), (b) γ-pyran (23), (c) 1,2,5-trichlorobenzene.

22 Furan

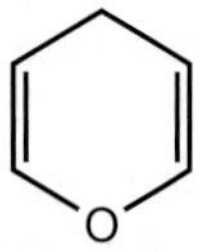

23 γ-Pyran

7.5(a) Assign (a) *cis*-dichloroethene and (b) *trans*-dichloroethene to point groups.

7.5(b) Assign the following molecules to point groups: (a) HF, (b) IF_7 (pentagonal bipyramid), (c) XeO_2F_2 (see-saw), (d) $Fe_2(CO)_9$ (24), (e) cubane, C_8H_8, (f) tetrafluorocubane, $C_8H_4F_4$ (25).

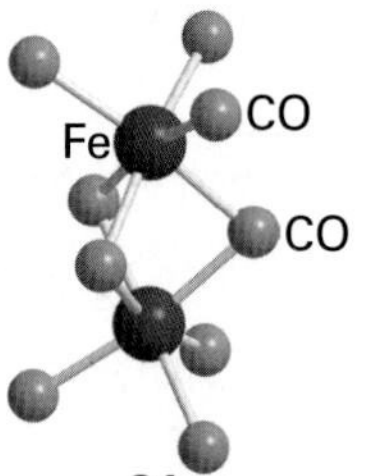

24

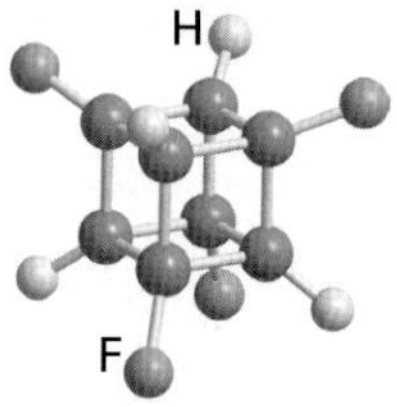

25

7.6(a) Which of the following molecules may be polar: (a) pyridine, (b) nitroethane, (c) gas-phase $HgBr_2$, (d) $B_3N_3H_6$?

7.6(b) Which of the following molecules may be polar: (a) CH_3Cl, (b) $HW_2(CO)_{10}$, (c) $SnCl_4$?

7.7(a) Identify the point groups to which all isomers of dichloronaphthalene belong.

7.7(b) Identify the point groups to which all isomers of dichloroanthracene belong.

7.8(a) Use as a basis the valence p_z orbitals on each atom in BF_3 to find the representative of the operation σ_h. Take z as perpendicular to the molecular plane.

7.8(b) Use as a basis the valence p_z orbitals on each atom in BF_3 to find the representative of the operation C_3. Take z as perpendicular to the molecular plane.

7.9(a) Use the matrix representatives of the operations σ_h and C_3 in a basis of valence p_z orbitals on each atom in BF_3 to find the operation and its representative resulting from $\sigma_h C_3$. Take z as perpendicular to the molecular plane.

7.9(b) Use the matrix representatives of the operations σ_h and C_3 in a basis of valence p_z orbitals on each atom in BF_3 to find the operation and its representative resulting from $C_3\sigma_h$. Take z as perpendicular to the molecular plane.

7.10(a) Show that all three C_2 operations in the group D_{3h} belong to the same class.

7.10(b) Show that all three σ_v operations in the group D_{3h} belong to the same class.

7.11(a) Use symmetry properties to determine whether or not the integral $\int p_x z p_z d\tau$ is necessarily zero in a molecule with symmetry C_{2v}.

7.11(b) Use symmetry properties to determine whether or not the integral $\int p_x z p_z d\tau$ is necessarily zero in a molecule with symmetry D_{3h}.

7.12(a) Is the transition $A_1 \rightarrow A_2$ forbidden for electric dipole transitions in a C_{3v} molecule?

7.12(b) Is the transition $A_{1g} \rightarrow E_{2u}$ forbidden for electric dipole transitions in a D_{6h} molecule?

7.13(a) Show that the function xy has symmetry species B_2 in the group C_{4v}.

7.13(b) Show that the function xyz has symmetry species A_1 in the group D_2.

7.14(a) Can molecules belonging to the point groups D_{2h} or C_{3h} be chiral? Explain your answer.

7.14(b) Can molecules belonging to the point groups T_h or T_d be chiral? Explain your answer.

7.15(a) What is the maximum degeneracy of a particle confined to the interior of an octahedral hole in a crystal?

7.15(b) What is the maximum degeneracy of a particle confined to the interior of an icosahedral nanoparticle?

7.16(a) What is the maximum possible degree of degeneracy of the orbitals in benzene?

7.16(b) What is the maximum possible degree of degeneracy of the orbitals in 1,4-dichlorobenzene?

7.17(a) Consider the C_{2v} molecule NO_2. The combination $p_x(A) - p_x(B)$ of the two O atoms (with x perpendicular to the plane) spans A_2. Is there any orbital of the central N atom that can have a nonzero overlap with that combination of O orbitals? What would be the case in SO_2, where 3d orbitals might be available?

7.17(b) Consider the C_{3v} ion NO_3^-. Is there any orbital of the central N atom that can have a nonzero overlap with the combination $2p_z(A) - p_z(B) - p_z(C)$ of the three O atoms (with z perpendicular to the plane). What would be the case in SO_3, where 3d orbitals might be available?

7.18(a) The ground state of NO_2 is A_1 in the group C_{2v}. To what excited states may it be excited by electric dipole transitions, and what polarization of light is it necessary to use?

7.18(b) The ClO_2 molecule (which belongs to the group C_{2v}) was trapped in a solid. Its ground state is known to be B_1. Light polarized parallel to the y-axis (parallel to the OO separation) excited the molecule to an upper state. What is the symmetry species of that state?

7.19(a) A set of basis functions is found to span a reducible representation of the group C_{4v} with characters 4, 1, 1, 3, 1 (in the order of operations in the character table in the *Resource section*). What irreducible representations does it span?

7.19(b) A set of basis functions is found to span a reducible representation of the group D_2 with characters 6, −2, 0, 0 (in the order of operations in the character table in the *Resource section*). What irreducible representations does it span?

7.20(a) What states of (a) benzene, (b) naphthalene may be reached by electric dipole transitions from their (totally symmetrical) ground states?

7.20(b) What states of (a) anthracene, (b) coronene (**26**) may be reached by electric dipole transitions from their (totally symmetrical) ground states?

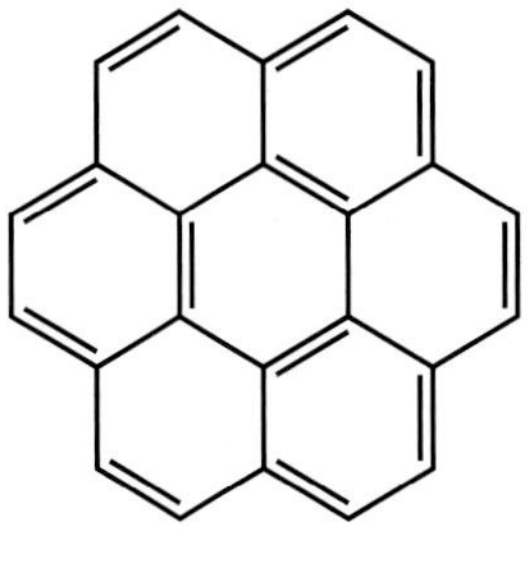

26 Coronene

7.21(a) Write $f_1 = \sin\theta$ and $f_2 = \cos\theta$, and show by symmetry arguments using the group C_s that the integral of their product over a symmetrical range around $\theta = 0$ is zero.

7.21(b) Write $f_1 = x$ and $f_2 = 3x^2 - 1$, and show by symmetry arguments using the group C_s that the integral of their product over a symmetrical range around $x = 0$ is zero.

Problems*

7.1 List the symmetry elements of the following molecules and name the point groups to which they belong: (a) staggered CH_3CH_3, (b) chair and boat cyclohexane, (c) B_2H_6, (d) $[Co(en)_3]^{3+}$, where en is ethylenediamine (1,2-diaminoethane; ignore its detailed structure), (e) crown-shaped S_8. Which of these molecules can be (i) polar, (ii) chiral?

7.2‡ In the square-planar complex anion [*trans*-$Ag(CF_3)_2(CN)_2]^-$, the Ag—CN groups are collinear. (a) Assume free rotation of the CF_3 groups (that is, disregarding the AgCF and AgCH angles) and name the point group of this complex ion. (b) Now suppose the CF_3 groups cannot rotate freely (because the ion was in a solid, for example). Structure (27) shows a plane that bisects the NC—Ag—CN axis and is perpendicular to it. Name the point group of the complex if each CF_3 group has a CF bond in that plane (so the CF_3 groups do not point to either CN group preferentially) and the CF_3 groups are (i) staggered, (ii) eclipsed.

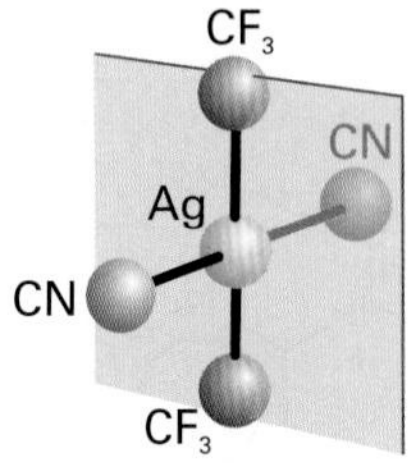

27

* Problems denoted with the symbol ‡ were supplied by Charles Trapp and Carmen Giunta.

7.3 The group C_{2h} consists of the elements E, C_2, σ_h, i. Construct the group multiplication table and find an example of a molecule that belongs to the group.

7.4 The group D_{2h} has a C_2 axis perpendicular to the principal axis and a horizontal mirror plane. Show that the group must therefore have a centre of inversion.

7.5 Consider the H_2O molecule, which belongs to the group C_{2v}. Take as a basis the two H1s orbitals and the four valence orbitals of the O atom and set up the 6×6 matrices that represent the group in this basis. Confirm by explicit matrix multiplication that the group multiplications (a) $C_2\sigma_v = \sigma_v'$ and (b) $\sigma_v\sigma_v' = C_2$. Confirm, by calculating the traces of the matrices: (a) that symmetry elements in the same class have the same character, (b) that the representation is reducible, and (c) that the basis spans $3A_1 + B_1 + 2B_2$.

7.6 Confirm that the z-component of orbital angular momentum is a basis for an irreducible representation of A_2 symmetry in C_{3v}.

7.7 Find the representatives of the operations of the group T_d in a basis of four H1s orbitals, one at each apex of a regular tetrahedron (as in CH_4).

7.8 Confirm that the representatives constructed in Problem 7.7 reproduce the group multiplications $C_3^+C_3^- = E$, $S_4C_3 = S_4'$, and $S_4C_3 = \sigma_d$.

7.9 The (one-dimensional) matrices $D(C_3) = 1$ and $D(C_2) = 1$, and $D(C_3) = 1$ and $D(C_2) = -1$ both represent the group multiplication $C_3C_2 = C_6$ in the group C_{6v} with $D(C_6) = +1$ and -1, respectively. Use the character table to confirm these remarks. What are the representatives of σ_v and σ_d in each case?

7.10 Construct the multiplication table of the Pauli spin matrices, $\boldsymbol{\sigma}$, and the 2×2 unit matrix:

$$\sigma_x = \begin{pmatrix} 0 & 1 \\ 1 & 0 \end{pmatrix} \quad \sigma_y = \begin{pmatrix} 0 & -i \\ i & 0 \end{pmatrix} \quad \sigma_z = \begin{pmatrix} 1 & 0 \\ 0 & -1 \end{pmatrix} \quad \sigma_0 = \begin{pmatrix} 1 & 0 \\ 0 & 1 \end{pmatrix}$$

Do the four matrices form a group under multiplication?

7.11 What irreducible representations do the four H1s orbitals of CH_4 span? Are there s and p orbitals of the central C atom that may form molecular orbitals with them? Could d orbitals, even if they were present on the C atom, play a role in orbital formation in CH_4?

7.12 Suppose that a methane molecule became distorted to (a) C_{3v} symmetry by the lengthening of one bond, (b) C_{2v} symmetry, by a kind of scissors action in which one bond angle opened and another closed slightly. Would more d orbitals become available for bonding?

7.13‡ B.A. Bovenzi and G.A. Pearse, Jr. (*J. Chem. Soc. Dalton Trans.* 2763 (1997)) synthesized coordination compounds of the tridentate ligand pyridine-2,6-diamidoxime ($C_7H_9N_5O_2$, **28**). Reaction with $NiSO_4$ produced a complex in which two of the essentially planar ligands are bonded at right angles to a single Ni atom. Name the point group and the symmetry operations of the resulting $[Ni(C_7H_9N_5O_2)_2]^{2+}$ complex cation.

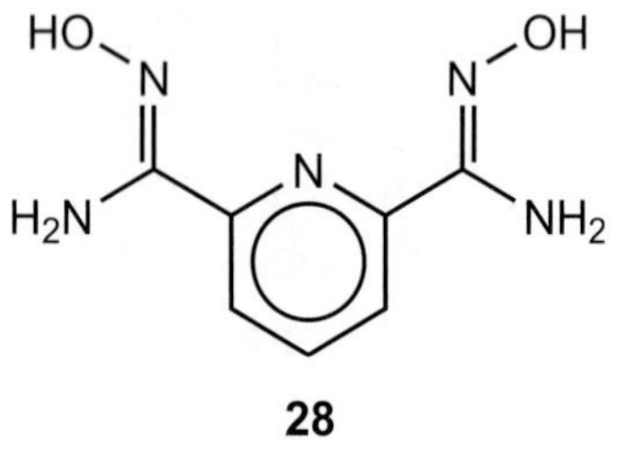

28

7.14‡ A computational study by C.J. Marsden (*Chem. Phys. Letts.* **245**, 475 (1995)) of AM_x compounds, where A is in Group 14 of the periodic table and M is an alkali metal, shows several deviations from the most symmetric structures for each formula. For example, most of the AM_4 structures were not tetrahedral but had two distinct values for MAM bond angles. They could be derived from a tetrahedron by a distortion shown in (**29**). (a) What is the point group of the distorted tetrahedron? (b) What is the symmetry species of the distortion considered as a vibration in the new, less symmetric group? Some AM_6 structures are not octahedral, but could be derived from an octahedron by translating a C—M—C axis as in (**30**). (c) What is the point group of the distorted octahedron? (d) What is the symmetry species of the distortion considered as a vibration in the new, less symmetric group?

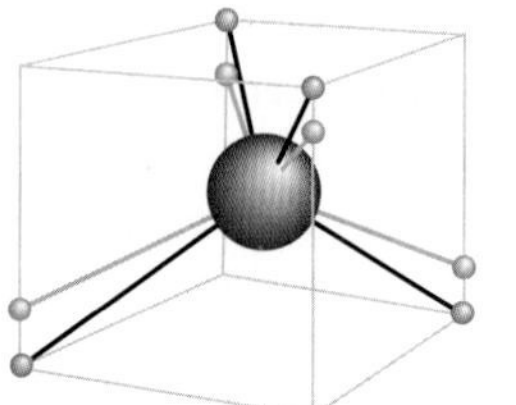

29

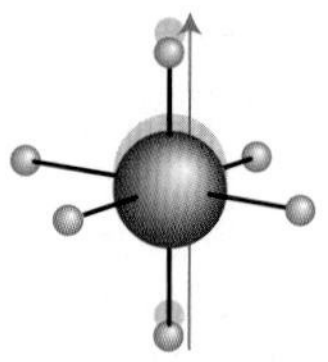

30

7.15 The algebraic forms of the f orbitals are a radial function multiplied by one of the factors (a) $z(5z^2 - 3r^2)$, (b) $y(5y^2 - 3r^2)$, (c) $x(5x^2 - 3r^2)$, (d) $z(x^2 - y^2)$, (e) $y(x^2 - z^2)$, (f) $x(z^2 - y^2)$, (g) xyz. Identify the irreducible representations spanned by these orbitals in (a) C_{2v}, (b) C_{3v}, (c) T_d, (d) O_h. Consider a lanthanoid ion at the centre of (a) a tetrahedral complex, (b) an octahedral complex. What sets of orbitals do the seven f orbitals split into?

7.16 Does the product $3x^2 - 1$ necessarily vanish when integrated over (a) a cube, (b) a tetrahedron, (c) a hexagonal prism, each centred on the origin?

7.17 The NO_2 molecule belongs to the group C_{2v}, with the C_2 axis bisecting the ONO angle. Taking as a basis the N2s, N2p, and O2p orbitals, identify the irreducible representations they span, and construct the symmetry-adapted linear combinations.

7.18 Construct the symmetry-adapted linear combinations of C2p$_z$ orbitals for benzene, and use them to calculate the Hückel secular determinant. This procedure leads to equations that are much easier to solve than using the original orbitals, and show that the Hückel orbitals are those specified in Section 5.9.

7.19 The phenanthrene molecule (**31**) belongs to the group C_{2v} with the C_2 axis in the plane of the molecule. (a) Classify the irreducible representations spanned by the carbon 2p$_z$ orbitals and find their symmetry-adapted linear combinations. (b) Use your results from part (a) to calculate the Hückel secular determinant. (c) What states of phenanthrene may be reached by electric dipole transitions from its (totally symmetrical) ground state?

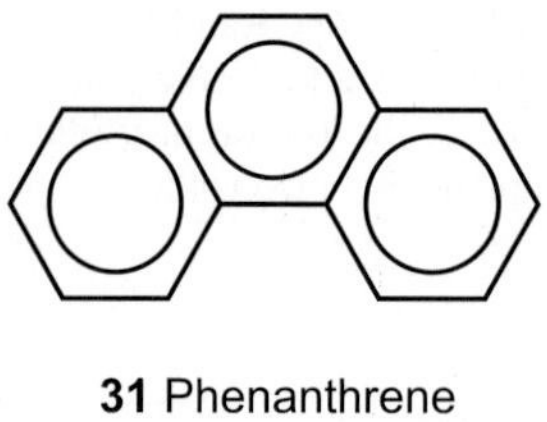

31 Phenanthrene

7.20‡ In a spectroscopic study of C_{60}, Negri *et al.* (*J. Phys. Chem.* **100**, 10849 (1996)) assigned peaks in the fluorescence spectrum. The molecule has icosahedral symmetry (I_h). The ground electronic state is A_{1g}, and the lowest-lying excited states are T_{1g} and G_g. (a) Are photon-induced transitions allowed from the ground state to either of these excited states? Explain your answer. (b) What if the molecule is distorted slightly so as to remove its centre of inversion?

7.21 In the square-planar XeF_4 molecule, consider the symmetry-adapted linear combination $p_1 = p_A - p_B + p_C - p_D$ where p_A, p_B, p_C, and p_D are $2p_z$ atomic orbitals on the fluorine atoms (clockwise labelling of the F atoms). Using the reduced point group D_4 rather than the full symmetry point group of the molecule, determine which of the various s, p, and d atomic orbitals on the central Xe atom can form molecular orbitals with p_1.

Applications: to astrophysics and biology

7.21‡ The H_3^+ molecular ion, which plays an important role in chemical reactions occurring in interstellar clouds, is known to be equilateral triangular. (a) Identify the symmetry elements and determine the point group of this molecule. (b) Take as a basis for a representation of this molecule the three H1s orbitals and set up the matrices that group in this basis. (c) Obtain the group multiplication table by explicit multiplication of the matrices. (d) Determine if the representation is reducible and, if so, give the irreducible representations obtained.

7.22 Some linear polyenes, of which β-carotene is an example, are important biological co-factors that participate in processes as diverse as the absorption of solar energy in photosynthesis (*Impact I19.2*) and protection against harmful biological oxidations. Use as a model of β-carotene a linear polyene containing 22 conjugated C atoms. (a) To what point group does this model of β-carotene belong? (b) Classify the irreducible representations spanned by the carbon $2p_z$ orbitals and find their symmetry-adapted linear combinations. (c) Use your results from part (b) to calculate the Hückel secular determinant. (d) What states of this model of β-carotene may be reached by electric dipole transitions from its (totally symmetrical) ground state?

7.23 The chlorophylls that participate in photosynthesis (*Impact I19.2*) and the haem groups of cytochromes are derived from the porphine dianion group (**32**), which belongs to the D_{4h} point group. The ground electronic state is A_{1g} and the lowest-lying excited state is E_u. Is a photon-induced transition allowed from the ground state to the excited state? Explain your answer.

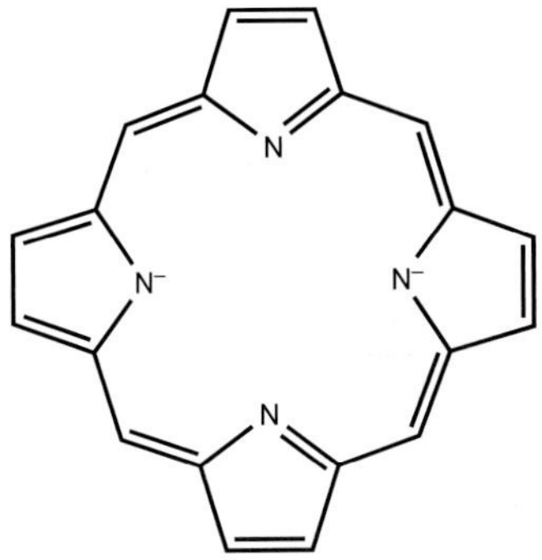

32

8 Molecular assemblies

Atoms, small molecules, and macromolecules can form large assemblies that are held together by interactions between their constituents. We begin this chapter with an examination of these molecular interactions, interpreting them in terms of electric properties of molecules, such as electric dipole moments and polarizabilities. Then we describe interactions in gases and liquids.

Molecular interactions are responsible for the unique properties of substances, especially in condensed phases. We shall see that small and sometimes fleeting imbalances of charge distributions in molecules allow them to interact with one another and with externally applied fields. One result of this interaction is the cohesion of molecules to form the bulk phases of matter. The interaction between ions is treated in Chapter 9 (for solids) and Chapter 16 (for solutions).

Interactions between molecules

A **van der Waals interaction** is an attractive interaction between closed-shell molecules that depends on the distance r between the molecules as $1/r^6$. In addition, there are interactions between ions and the partial charges of polar molecules and repulsive interactions that prevent the complete collapse of matter to nuclear densities. The repulsive interactions arise from Coulombic repulsions and, indirectly, from the Pauli principle and the exclusion of electrons from regions of space where the orbitals of neighbouring species overlap.

8.1 Interactions between partial charges

Atoms in molecules in general have partial charges. If these charges were separated by a vacuum, they would attract or repel each other in accord with Coulomb's law (see *Fundamentals F.6*), and we would write

$$V = \frac{Q_1Q_2}{4\pi\varepsilon_0 r} \tag{8.1a}$$

where Q_1 and Q_2 are the partial charges and r is their separation. However, we should take into account the possibility that other parts of the molecule, or other molecules, lie between the charges, and decrease the strength of the interaction. We therefore write

$$V = \frac{Q_1Q_2}{4\pi\varepsilon r} \tag{8.1b}$$

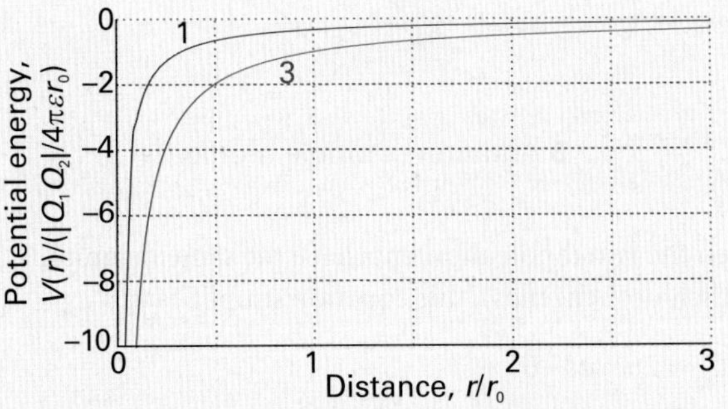

Fig. 8.1 The Coulomb potential for two opposite charges and its dependence on their separation. The two curves correspond to different relative permittivities ($\varepsilon_r = 1$ for a vacuum, 3 for a fluid).

where ε is the permittivity of the medium lying between the charges. As explained in *Fundamentals F.6*, the permittivity is usually expressed as a multiple of the vacuum permittivity by writing $\varepsilon = \varepsilon_r\varepsilon_0$, where ε_r is the relative permittivity. The effect of the medium can be very large: for water $\varepsilon_r = 78$, so the potential energy of two charges separated by bulk water is reduced by nearly two orders of magnitude compared to the value it would have if the charges were separated by a vacuum (Fig. 8.1).

● A BRIEF ILLUSTRATION

The energy of interaction between a partial charge of −0.36 (that is, $Q_1 = -0.36e$) on the N atom of an amide group (1) and the partial charge of −0.38 ($Q_2 = -0.38e$) on the carbonyl O atom at a distance of 3.0 nm on the assumption that the medium between them is a vacuum is

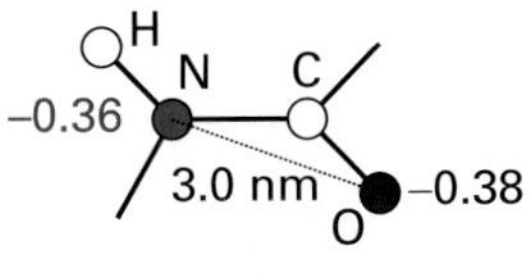

1

$$V = \frac{(-0.36e)\times(-0.38e)}{4\pi\varepsilon_0 \times (3.0\ \text{nm})}$$

$$= \frac{0.36\times 0.38\times(1.602\times 10^{-19}\ \text{C})^2}{4\pi\times(8.854\times 10^{-12}\ \text{J}^{-1}\ \text{C}^2\ \text{m}^{-1})\times(3.0\times 10^{-9}\ \text{m})}$$

$$= 1.1\times 10^{-20}\ \text{J}$$

This energy (after multiplication by Avogadro's constant) corresponds to 6.3 kJ mol^{-1}. However, if the medium has a 'typical' relative permittivity of 3.5, then the interaction energy is reduced to 1.8 kJ mol^{-1}. For bulk water as the medium, with the H_2O molecules able to rotate in response to a field, the energy of interaction would be reduced by a factor of 78, to only 0.081 kJ mol^{-1}. ●

8.2 Electric dipole moments

An **electric dipole** consists of two electric charges +Q and −Q separated by a distance *R*. This arrangement of charges is represented by a vector $\boldsymbol{\mu}$ (2). The magnitude of $\boldsymbol{\mu}$ is $\mu = QR$ and, although the SI unit of dipole moment is coulomb metre (C m), it is still commonly reported in the non-SI unit debye, D, named after Peter Debye, a pioneer in the study of dipole moments of molecules, where

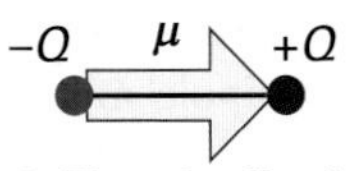

2 Electric dipole

$$1\ \text{D} = 3.335\ 64\times 10^{-30}\ \text{C m}$$

The dipole moment of a pair of charges +*e* and −*e* separated by 100 pm is 1.6×10^{-29} C m, corresponding to 4.8 D. Dipole moments of small molecules are typically about 1 D.

A **polar molecule** is a molecule with a permanent electric dipole moment. The permanent dipole moment stems from the partial charges on the atoms in the molecule that arise from differences in electronegativity or other features of bonding (Section 5.5). Nonpolar molecules acquire an induced dipole moment in an electric field on account of the distortion the field causes in their electronic distributions and nuclear positions; however, this induced moment is only temporary and disappears as soon as the perturbing field is removed. Polar molecules also have their existing dipole moments temporarily modified by an applied field.

All heteronuclear diatomic molecules are polar, and typical values of μ include 1.08 D for HCl and 0.42 D for HI (Table 8.1). Molecular symmetry is of the greatest importance in deciding whether a polyatomic molecule is polar or not. Indeed, molecular symmetry is more important than the question of whether or not the atoms in the molecule belong to the same element. Homonuclear polyatomic molecules may be polar if they have low symmetry and the atoms are in inequivalent positions. For instance, the angular molecule ozone, O_3 (3), is homonuclear; however, it is polar because the central O atom is different from the outer

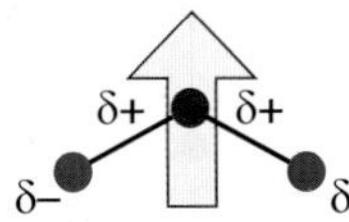

3 Ozone, O_3

Synoptic table 8.1* Dipole moments (μ) and polarizability volumes (α')

	μ/D	$\alpha'/(10^{-30}\ m^3)$
CCl_4	0	10.3
H_2	0	0.819
H_2O	1.85	1.48
HCl	1.08	2.63
HI	0.42	5.45

* More values are given in the *Data section*.

two (it is bonded to two atoms; they are bonded only to one); moreover, the dipole moments associated with each bond make an angle to each other and do not cancel. Heteronuclear polyatomic molecules may be nonpolar if they have high symmetry, because individual bond dipoles may then cancel. The heteronuclear linear triatomic molecule CO_2, for example, is nonpolar because, although there are partial charges on all three atoms, the dipole moment associated with the OC bond points in the opposite direction to the dipole moment associated with the CO bond, and the two cancel (**4**).

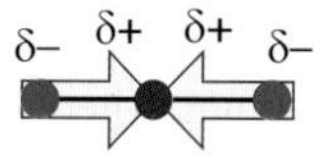

4 Carbon dioxide, CO_2

To a first approximation, it is possible to resolve the dipole moment of a polyatomic molecule into contributions from various groups of atoms in the molecule and the directions in which these individual contributions lie (Fig. 8.2). Thus, 1,4-dichlorobenzene is nonpolar by symmetry on account of the cancellation of two equal but opposing C—Cl moments (exactly as in carbon dioxide). 1,2-Dichlorobenzene, however, has a dipole moment which is approximately the resultant of two chlorobenzene dipole moments arranged at 60° to each other. This technique of 'vector addition' (see *Mathematical background 4*) can be applied with fair success to other series of related molecules, and the resultant μ_{res} of two dipole moments μ_1 and μ_2 that make an angle θ to each other (**5**) is approximately

$$\mu_{res} \approx (\mu_1^2 + \mu_2^2 + 2\mu_1\mu_2 \cos\theta)^{1/2} \qquad (8.2a)$$

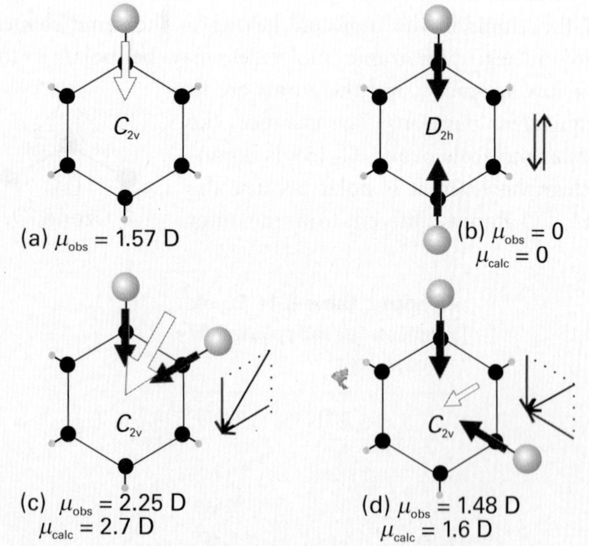

Fig. 8.2 The resultant dipole moments (yellow) of the dichlorobenzene isomers (b to d) can be obtained approximately by vectorial addition of two chlorobenzene dipole moments (1.57 D).

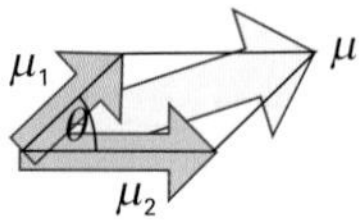

5 Addition of dipole moments

When the two dipole moments have the same magnitude (as in the dichlorobenzenes), this equation simplifies to

$$\mu_{res} \approx 2\mu_1 \cos \tfrac{1}{2}\theta \qquad (8.2b)$$

Self-test 8.1 Estimate the ratio of the electric dipole moments of *ortho* (1,2-) and *meta* (1,3-) disubstituted benzenes.

[$\mu(ortho)/\mu(meta) = 1.7$]

A better approach to the calculation of dipole moments is to take into account the locations and magnitudes of the partial charges on all the atoms. These partial charges are included in the output of many molecular structure software packages (Chapter 6). To calculate the *x*-component, for instance, we need to know the partial charge on each atom and the atom's *x*-coordinate relative to a point in the molecule and form the sum

$$\mu_x = \sum_J Q_J x_J \qquad (8.3a)$$

Here Q_J is the partial charge of atom J, x_J is the *x*-coordinate of atom J, and the sum is over all the atoms in the molecule. Analogous expressions are used for the *y*- and *z*-components. For an electrically neutral molecule, the origin of the coordinates is arbitrary, so it is best chosen to simplify the measurements. In common with all vectors, the magnitude of μ is related to the three components μ_x, μ_y, and μ_z by

$$\mu = (\mu_x^2 + \mu_y^2 + \mu_z^2)^{1/2} \qquad (8.3b)$$

Example 8.1 *Calculating a molecular dipole moment*

Estimate the electric dipole moment of the amide group shown in (**6**) by using the partial charges (as multiples of *e*) and the locations (*x*, *y*, *z*) of the atoms shown in pm.

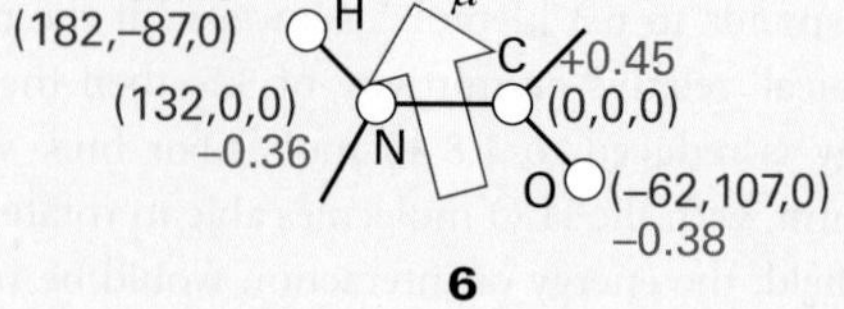

6

Table 8.2 Partial charges in polypeptides

Atom	Partial charge/e
C(=O)	+0.45
C(—CO)	+0.06
H(—C)	+0.02
H(—N)	+0.18
H(—O)	+0.42
N	−0.36
O	−0.38

Method We use eqn 8.3a to calculate each of the components of the dipole moment and then eqn 8.3b to assemble the three components into the magnitude of the dipole moment. Note that the partial charges are multiples of the fundamental charge, $e = 1.602 \times 10^{-19}$ C.

Answer The expression for μ_x is

$$\begin{aligned}\mu_x &= (-0.36e) \times (132\text{ pm}) + (0.45e) \times (0\text{ pm}) \\ &\quad + (0.18e) \times (182\text{ pm}) + (-0.38e) \times (-62.0\text{ pm}) \\ &= 8.8e\text{ pm} \\ &= 8.8 \times (1.602 \times 10^{-19}\text{ C}) \times (10^{-12}\text{ m}) \\ &= 1.4 \times 10^{-30}\text{ C m}\end{aligned}$$

corresponding to $\mu_x = +0.42$ D. The expression for μ_y is:

$$\begin{aligned}\mu_y &= (-0.36e) \times (0\text{ pm}) + (0.45e) \times (0\text{ pm}) \\ &\quad + (0.18e) \times (-87\text{ pm}) + (-0.38e) \times (107\text{ pm}) \\ &= -56e\text{ pm} = -9.0 \times 10^{-30}\text{ C m}\end{aligned}$$

It follows that $\mu_y = -2.7$ D. Therefore, because $\mu_z = 0$,

$$\mu = \{(0.42\text{ D})^2 + (-2.7\text{ D})^2\}^{1/2} = 2.7\text{ D}$$

We can find the orientation of the dipole moment by arranging an arrow of length 2.7 units of length to have x, y, and z components of 0.42, −2.7, and 0 units; the orientation is superimposed on (**6**).

Self-test 8.2 Calculate the electric dipole moment of formaldehyde, using the information in (**7**). [2.3 D]

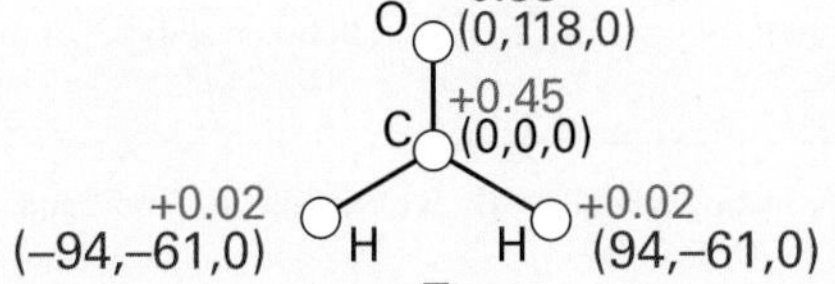

7

8.3 Interactions between dipoles

Most of the discussion in this section is based on the Coulombic potential energy of interaction between two charges (eqn 8.1a). We can easily adapt this expression to find the potential energy of a point charge and a dipole and to extend it to the interaction between two dipoles.

(a) The potential energy of interaction

We show in the following *Justification* that the potential energy of interaction between a point dipole $\mu_1 = Q_1 l$ and the point charge Q_2 in the arrangement shown in (**8**) is

$$V = -\frac{\mu_1 Q_2}{4\pi\varepsilon_0 r^2} \tag{8.4}$$

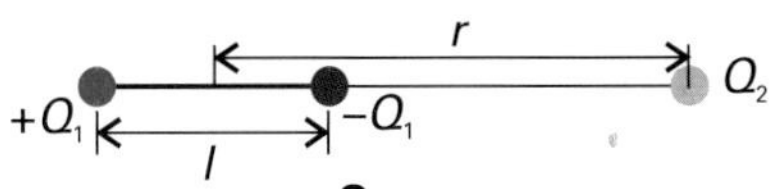

8

With μ in coulomb metres, Q_2 in coulombs, and r in metres, V is obtained in joules. A **point dipole** is a dipole in which the separation between the charges is much smaller than the distance at which the dipole is being observed: $l \ll r$. The potential energy rises towards zero (the value at infinite separation of the charge and the dipole) more rapidly (as $1/r^2$) than that between two point charges (which varies as $1/r$) because, from the viewpoint of the point charge, the partial charges of the dipole seem to merge and cancel as the distance r increases (Fig. 8.3).

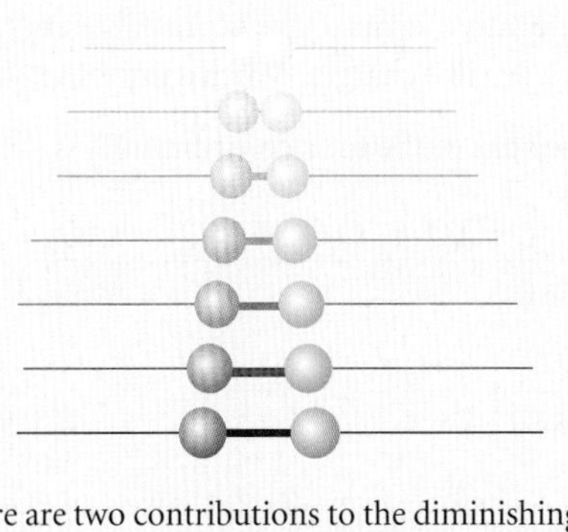

Fig. 8.3 There are two contributions to the diminishing field of an electric dipole with distance (here seen from the side). The potential of the charges decreases (shown here by a fading intensity) and the two charges appear to merge, so their combined effect approaches zero more rapidly than by the distance effect alone.

Justification 8.1 *The interaction between a point charge and a point dipole*

The sum of the potential energies of repulsion between like charges and attraction between opposite charges in the orientation shown in (**8**) is

$$V = \frac{1}{4\pi\varepsilon_0}\left(-\frac{Q_1Q_2}{r-\frac{1}{2}l} + \frac{Q_1Q_2}{r+\frac{1}{2}l}\right) = \frac{Q_1Q_2}{4\pi\varepsilon_0 r}\left(-\frac{1}{1-x} + \frac{1}{1+x}\right)$$

where $x = l/2r$. Because $l \ll r$ for a point dipole, this expression can be simplified by expanding the terms in x and retaining only the leading term:

$$V = \frac{Q_1Q_2}{4\pi\varepsilon_0 r}\{-(1+x+\cdots)+(1-x+\cdots)\}$$

$$\approx -\frac{2xQ_1Q_2}{4\pi\varepsilon_0 r} = -\frac{Q_1Q_2 l}{4\pi\varepsilon_0 r^2}$$

(Expansions are treated in *Mathematical background 1*.) With $\mu_1 = Q_1 l$, this expression becomes eqn 8.4. This expression should be multiplied by $\cos\theta$ when the point charge lies at an angle θ to the axis of the dipole.

Example 8.2 *Calculating the interaction energy of two dipoles*

Calculate the potential energy of interaction of two dipoles in the arrangement shown in (**9**) when their separation is r.

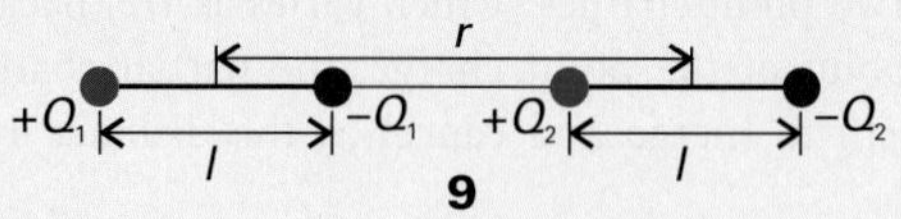

9

Method We proceed in exactly the same way as in *Justification 8.1*, but now the total interaction energy is the sum of four pairwise terms, two attractions between opposite charges, which contribute negative terms to the potential energy, and two repulsions between like charges, which contribute positive terms.

Answer The sum of the four contributions is

$$V = \frac{1}{4\pi\varepsilon_0}\left(-\frac{Q_1Q_2}{r+l} + \frac{Q_1Q_2}{r} + \frac{Q_1Q_2}{r} - \frac{Q_1Q_2}{r-l}\right)$$

$$= -\frac{Q_1Q_2}{4\pi\varepsilon_0 r}\left(\frac{1}{1+x} - 2 + \frac{1}{1-x}\right)$$

with $x = l/r$. As before, provided $l \ll r$ we can expand the two terms in x and retain only the first surviving term, which is equal to $2x^2$. This step results in the expression

$$V = -\frac{2x^2Q_1Q_2}{4\pi\varepsilon_0 r}$$

Therefore, because $\mu_1 = Q_1 l$ and $\mu_2 = Q_2 l$, the potential energy of interaction in the alignment shown in (**9**) is

$$V = -\frac{\mu_1\mu_2}{2\pi\varepsilon_0 r^3}$$

This interaction energy approaches zero more rapidly (as $1/r^3$) than for the previous case: now both interacting entities appear neutral to each other at large separations. See *Further information 8.1* for the general expression.

Self-test 8.3 Derive an expression for the potential energy when the dipoles are in the arrangement shown in (**10**).
$[V = \mu_1\mu_2/4\pi\varepsilon_0 r^3]$

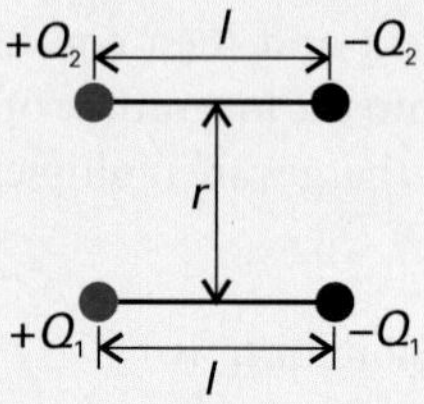

10

Table 8.3 summarizes the various expressions for the interaction of charges and dipoles. It is quite easy to extend the formulas given there to obtain expressions for the energy of interaction of higher **multipoles**, or arrays of point charges (Fig. 8.4). Specifically, an ***n*-pole** is an array of point charges with an n-pole moment but no lower moment. Thus, a **monopole** ($n = 1$) is a point charge, and the monopole moment is what we normally call the overall charge. A dipole ($n = 2$), as we have seen, is an array of charges

Table 8.3 Multipole interaction potential energies

Interaction type	Distance dependence of potential energy	Typical energy/ (kJ mol^{-1})	Comment
Ion–ion	$1/r$	250	Only between ions*
Ion–dipole	$1/r^2$	15	
Dipole–dipole	$1/r^3$	2	Between stationary polar molecules
	$1/r^6$	0.6	Between rotating polar molecules
London (dispersion)	$1/r^6$	2	Between all types of molecules

The energy of a hydrogen bond A—H···B is typically 20 kJ mol^{-1} and occurs on contact for A, B = O, N, or F.
* Electrolyte solutions are treated in Chapter 16, ionic solids in Chapter 19.

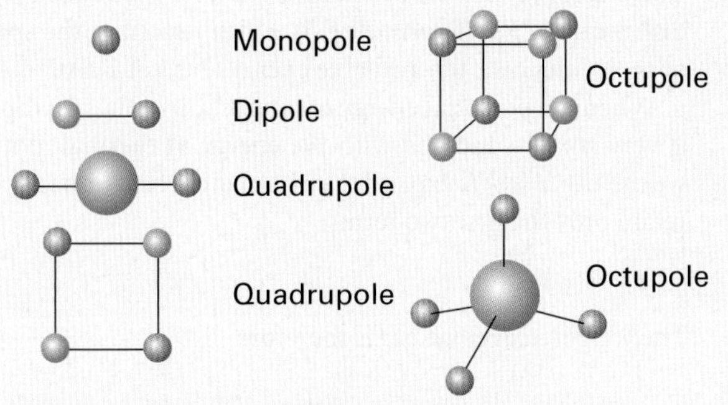

Fig. 8.4 Typical charge arrays corresponding to electric multipoles. The field arising from an arbitrary finite charge distribution can be expressed as the superposition of the fields arising from a superposition of multipoles.

that has no monopole moment (no net charge). A **quadrupole** ($n = 3$) consists of an array of point charges that has neither net charge nor dipole moment (as for CO_2 molecules, **11**). An **octupole** ($n = 4$) consists of an array of point charges that sum to zero and which has neither a dipole moment nor a quadrupole moment (as for CH_4 molecules, **12**). The feature to remember is that the interaction energy falls off more rapidly the higher the order of the multipole. For the interaction of an n-pole with an m-pole, the potential energy varies with distance as

$$V \propto \frac{1}{r^{n+m-1}} \tag{8.5}$$

11 Carbon dioxide, CO_2

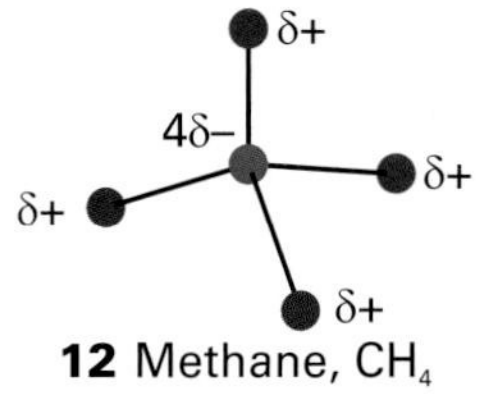

12 Methane, CH_4

The reason for the even steeper decrease with distance is the same as before: the array of charges appears to blend together into neutrality more rapidly with distance the higher the number of individual charges that contribute to the multipole. Note that a given molecule may have a charge distribution that corresponds to a superposition of several different multipoles.

(b) The electric field

The same kind of argument as that used to derive expressions for the potential energy can be used to establish the distance dependence of the strength of the electric field generated by a dipole. We shall need this expression when we calculate the dipole moment induced in one molecule by another.

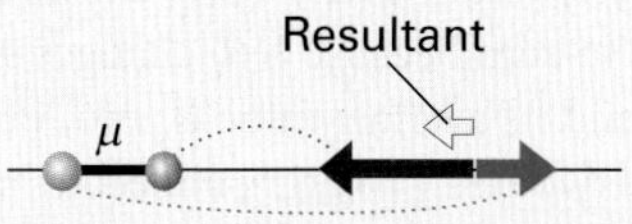

Fig. 8.5 The electric field of a dipole is the sum of the opposing fields from the positive and negative charges, each of which is proportional to $1/r^2$. The difference, the net field, is proportional to $1/r^3$.

The starting point for the calculation is the strength of the electric field generated by a point electric charge (see *Fundamentals F.6*):

$$\mathcal{E} = \frac{Q}{4\pi\varepsilon_0 r^2} \tag{8.6}$$

The electric field is actually a vector, and we cannot simply add and subtract magnitudes without taking into account the directions of the fields. In the cases we consider, this will not be a complication because the two charges of the dipoles will be collinear and give rise to fields in the same direction. For the point-dipole arrangement shown in Fig. 8.5, the same procedure that was used to derive the potential energy gives

$$\mathcal{E} = \frac{\mu}{2\pi\varepsilon_0 r^3} \tag{8.7}$$

The electric field of a multipole (in this case a dipole) decreases more rapidly with distance (as $1/r^3$ for a dipole) than a monopole (a point charge).

(c) Dipole–dipole interactions

The potential energy of interaction between two polar molecules is a complicated function of their relative orientation. When the two dipoles are parallel (as in **13**), the potential energy is simply (see *Further information 8.1*)

$$V = \frac{\mu_1\mu_2 f(\theta)}{4\pi\varepsilon_0 r^3} \qquad f(\theta) = 1 - 3\cos^2\theta \tag{8.8}$$

This expression applies to polar molecules in a fixed, parallel, orientation in a solid.

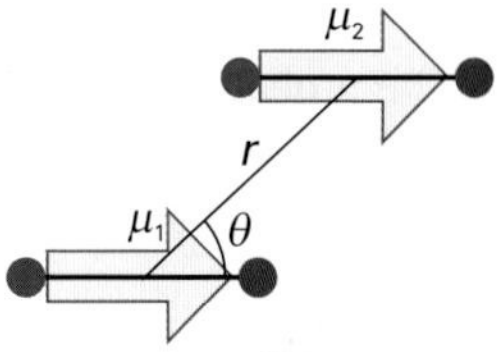

13

In a fluid of freely rotating molecules, the interaction between dipoles averages to zero because $f(\theta)$ changes sign as the orientation changes, and its average value is zero. Physically, the like partial charges of two freely rotating molecules are close together as much as the two opposite charges, and the repulsion of the former is cancelled by the attraction of the latter.

A brief comment The average (or mean value) of a function $f(x)$ over the range from $x = a$ to $x = b$ is

$$\langle f \rangle = \frac{1}{b-a}\int_a^b f(x)\mathrm{d}x$$

The volume element in spherical polar coordinates (see Table 3.1) is proportional to $\sin\theta\,\mathrm{d}\theta$, and θ ranges from 0 to π. Therefore the average value of $(1 - 3\cos^2\theta)$ is $(1/\pi)$

$$\int_0^\pi (1 - 3\cos^2\theta)\sin\theta\,\mathrm{d}\theta = 0.$$

The interaction energy of two *freely* rotating dipoles is zero. However, because their mutual potential energy depends on their relative orientation, the molecules do not in fact rotate completely freely, even in a gas. In fact, the lower energy orientations are marginally favoured, so there is a nonzero average interaction between polar molecules. We show in the following *Justification* that the average potential energy of two rotating molecules that are separated by a distance r is

$$\langle V \rangle = -\frac{C}{r^6} \qquad C = \frac{2\mu_1^2\mu_2^2}{3(4\pi\varepsilon_0)^2kT} \tag{8.9}$$

This expression describes the **Keesom interaction**, and is the first of the contributions to the van der Waals interaction.

Justification 8.2 *The Keesom interaction*

The detailed calculation of the Keesom interaction energy is quite complicated, but the form of the final answer can be constructed quite simply. First, we note that the average interaction energy of two polar molecules rotating at a fixed separation r is given by

$$\langle V \rangle = \frac{\mu_1\mu_2\langle f \rangle}{4\pi\varepsilon_0 r^3}$$

where $\langle f \rangle$ now includes a weighting factor in the averaging that is equal to the probability that a particular orientation will be adopted. This probability is given by $p \propto \mathrm{e}^{-E/kT}$, with E interpreted as the potential energy of interaction of the two dipoles in that orientation. That is,

$$p \propto \mathrm{e}^{-V/kT} \qquad V = \frac{\mu_1\mu_2 f}{4\pi\varepsilon_0 r^3}$$

This expression for the probability is a form of the Boltzmann distribution (*Fundamentals F.5*) which describes the spread of molecules over the available energy levels at $T > 0$.

When the potential energy of interaction of the two dipoles is very small compared with the energy of thermal motion, we can use $V \ll kT$, expand the exponential function in p, and retain only the first two terms:

$$p \propto 1 - V/kT + \cdots$$

The weighted average of f is therefore

$$\langle f \rangle = p\langle f \rangle_0 = \langle f \rangle_0 - \frac{\mu_1\mu_2}{4\pi\varepsilon_0 kTr^3}\langle f^2 \rangle_0 + \cdots$$

where $\langle \cdots \rangle_0$ denotes an unweighted spherical average. The spherical average of f is zero, so the first term vanishes. However, the average value of f^2 is nonzero because f^2 is positive at all orientations, so we can write

$$\langle V \rangle = -\frac{\mu_1^2\mu_2^2\langle f^2 \rangle_0}{(4\pi\varepsilon_0)^2kTr^6}$$

The average value $\langle f^2 \rangle_0$ turns out to be $\frac{2}{3}$ when the calculation is carried through in detail. The final result is that quoted in eqn 8.9.

The important features of eqn 8.9 are its negative sign (the average interaction is attractive), the dependence of the average interaction energy on the inverse sixth power of the separation (which identifies it as a van der Waals interaction), and its inverse dependence on the temperature. The last feature reflects the way that the greater thermal motion overcomes the mutual orientating effects of the dipoles at higher temperatures. The inverse sixth power arises from the inverse third power of the interaction potential energy that is weighted by the energy in the Boltzmann term, which is also proportional to the inverse third power of the separation.

At 25°C the average interaction energy for pairs of molecules with $\mu = 1$ D is about -0.06 kJ mol^{-1} when the separation is 0.5 nm. This energy should be compared with the average molar kinetic energy of $\frac{3}{2}RT = 3.7$ kJ mol^{-1} at the same temperature, given by the equipartition theorem (see *Fundamentals F.5*). The interaction energy is also much smaller than the energies involved in the making and breaking of chemical bonds.

8.4 Induced dipole moments

An applied electric field can distort a molecule as well as align its permanent electric dipole moment. The **induced dipole moment**, μ^*, is generally proportional to the field strength, $\mathcal{E}$, and we write

$$\mu^* = \alpha\mathcal{E} \tag{8.10}$$

(See Section 11.7 for exceptions to eqn 8.10.) The constant of proportionality α is the **polarizability** of the molecule. The greater the polarizability, the larger is the induced dipole moment for a given applied field. In a formal treatment, we should use vector quantities and allow for the possibility that the induced dipole moment might not lie parallel to the applied field, but for simplicity we discuss polarizabilities in terms of (scalar) magnitudes.

(a) Polarizability volumes

Polarizability has the units (coulomb metre)^2 per joule ($C^2\ m^2\ J^{-1}$). That collection of units is awkward, so α is often expressed as a **polarizability volume**, α', by using the relation

$$\alpha' = \frac{\alpha}{4\pi\varepsilon_0} \qquad [8.11]$$

where ε_0 is the vacuum permittivity. Because the units of $4\pi\varepsilon_0$ are coulomb-squared per joule per metre ($C^2\ J^{-1}\ m^{-1}$), it follows that α' has the dimensions of volume (hence its name). Polarizability volumes are similar in magnitude to actual molecular volumes (of the order of $10^{-30}\ m^3$, $10^{-3}\ nm^3$, 1 Å^3). When using older compilations of data, it is useful to note that polarizability volumes have the same numerical values as the 'polarizabilities' reported using c.g.s. electrical units, so the tabulated values previously called 'polarizabilities' can be used directly.

Some experimental polarizability volumes of molecules are given in Table 8.1. As shown in the following *Justification*, polarizability volumes correlate with the HOMO–LUMO separations in atoms and molecules. The electron distribution can be distorted readily if the LUMO lies close to the HOMO in energy, so the polarizability is then large. If the LUMO lies high above the HOMO, an applied field cannot perturb the electron distribution significantly, and the polarizability is low. Molecules with small HOMO–LUMO gaps are typically large, with numerous electrons.

Justification 8.3 *Polarizabilities and molecular structures*

When an electric field is increased by $d\mathcal{E}$, the energy of a molecule changes by $-\mu d\mathcal{E}$ and, if the molecule is polarizable, we interpret μ as the induced moment μ^* (eqn 8.10). Therefore, the change in energy when the field is increased from 0 to $\mathcal{E}$ is

$$\Delta E = -\int_0^{\mathcal{E}} \mu^* d\mathcal{E} = -\int_0^{\mathcal{E}} \alpha\mathcal{E} d\mathcal{E} = -\tfrac{1}{2}\alpha\mathcal{E}^2$$

The contribution to the hamiltonian when a dipole moment is exposed to an electric field $\mathcal{E}$ in the z-direction is

$$\hat{H}^{(1)} = -\hat{\mu}_z\mathcal{E}$$

Comparison of these two expressions suggests that we should use second-order perturbation theory to calculate the energy of the system in the presence of the field, because then we shall obtain an expression proportional to $\mathcal{E}^2$. According to eqn 2.35, the second-order contribution to the energy is

$$E^{(2)} = \sum_{n\neq 0} \frac{|\int \psi_n{}^* \hat{H}^{(1)} \psi_0 d\tau|^2}{E_0^{(0)} - E_n^{(0)}} = \mathcal{E}^2 \sum_{n\neq 0} \frac{|\int \psi_n{}^* \hat{\mu}_z \psi_0 d\tau|^2}{E_0^{(0)} - E_n^{(0)}}$$

$$= \mathcal{E}^2 \sum_{n\neq 0} \frac{|\mu_{z,0n}|^2}{E_0^{(0)} - E_n^{(0)}}$$

where $\mu_{z,0n}$ is the *transition* electric dipole moment in the z-direction (eqn 4.24) and where ψ_i and $E_i^{(0)}$ are the wavefunctions and energies, respectively, in the absence of the electric field. By comparing the two expressions for the energy, we conclude that the polarizablity of the molecule in the z-direction is

$$\alpha = 2\sum_{n\neq 0} \frac{|\mu_{z,0n}|^2}{E_n^{(0)} - E_0^{(0)}} \qquad (8.12)$$

The content of eqn 8.12 can be appreciated by approximating the excitation energies by a mean value ΔE (an indication of the HOMO–LUMO separation), and supposing that the most important transition dipole moment is approximately equal to the charge of an electron multiplied by the radius, R, of the molecule. Then

$$\alpha \approx \frac{2e^2R^2}{\Delta E}$$

This expression shows that α increases with the size of the molecule and with the ease with which it can be excited (the smaller the value of ΔE).

If the excitation energy is approximated by the energy needed to remove an electron to infinity from a distance R from a single positive charge, we can write $\Delta E \approx e^2/4\pi\varepsilon_0 R$. When this expression is substituted into the equation above, both sides are divided by $4\pi\varepsilon_0$, and the factor of 2 ignored in this approximation, we obtain $\alpha' \approx R^3$, which is of the same order of magnitude as the molecular volume.

(b) Dipole–induced-dipole interactions

A polar molecule with dipole moment μ_1 can induce a dipole μ_2^* in a neighbouring polarizable molecule (Fig. 8.6). The induced dipole interacts with the permanent dipole of the first molecule, and the two are attracted together. The average interaction energy when the separation of the molecules is r is

$$V = -\frac{C}{r^6} \qquad C = \frac{\mu_1^2\alpha_2'}{4\pi\varepsilon_0} \qquad (8.13)$$

where α_2' is the polarizability volume of molecule 2 and μ_1 is the permanent dipole moment of molecule 1. Note that the C in this expression is different from the C in eqn 8.9 and other expressions below: we are using the same symbol in C/r^6 to emphasize the similarity of form of each expression.

The dipole–induced-dipole interaction energy is independent of the temperature because thermal motion has no effect on the

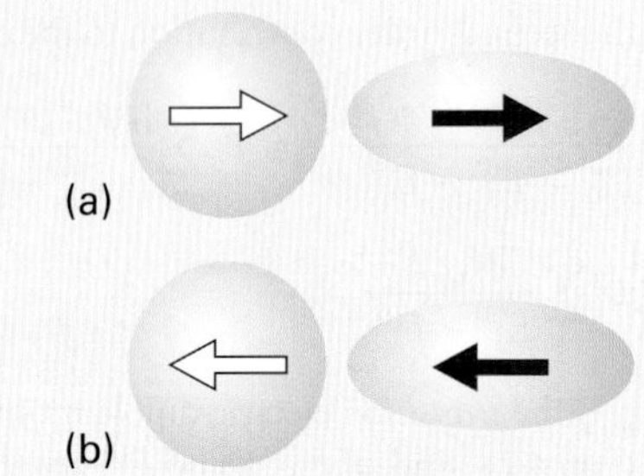

Fig. 8.6 (a) A polar molecule (full arrow) can induce a dipole (outline arrow) in a nonpolar molecule, and (b) the orientation of the latter follows that of the former, so the interaction does not average to zero.

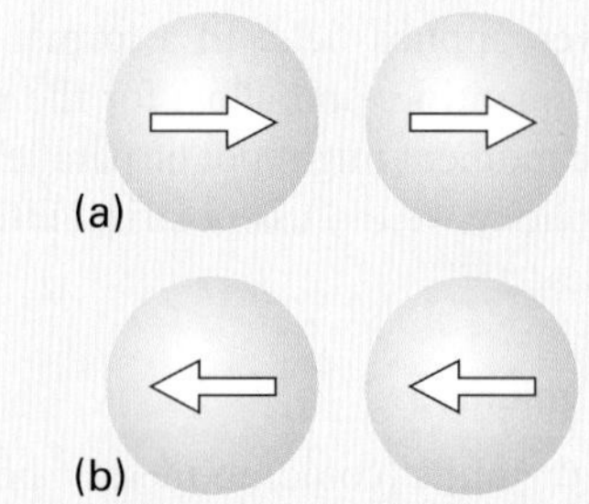

Fig. 8.7 (a) In the dispersion interaction, an instantaneous dipole on one molecule induces a dipole on another molecule, and the two dipoles then interact to lower the energy. (b) The two instantaneous dipoles are correlated and, although they occur in different orientations at different instants, the interaction does not average to zero.

averaging process. Moreover, like the dipole–dipole interaction, the potential energy depends on $1/r^6$: this distance dependence stems from the $1/r^3$ dependence of the field (and hence the magnitude of the induced dipole) and the $1/r^3$ dependence of the potential energy of interaction between the permanent and induced dipoles. For a molecule with $\mu = 1$ D (such as HCl) near a molecule of polarizability volume $\alpha' = 10 \times 10^{-30}$ m^3 (such as benzene, Table 8.1), the average interaction energy is about -0.8 kJ mol^{-1} when the separation is 0.3 nm.

(c) Induced-dipole–induced-dipole interactions

Nonpolar molecules (including closed-shell atoms, such as Ar) attract one another even though neither has a permanent dipole moment. The abundant evidence for the existence of interactions between them is the formation of condensed phases of nonpolar substances, such as the condensation of hydrogen or argon to a liquid at low temperatures and the fact that benzene is a liquid at normal temperatures.

The interaction between nonpolar molecules arises from the transient dipoles that all molecules possess as a result of fluctuations in the instantaneous positions of electrons. To appreciate the origin of the interaction, suppose that the electrons in one molecule flicker into an arrangement that gives the molecule an instantaneous dipole moment μ_1^*. This dipole generates an electric field that polarizes the other molecule, and induces in that molecule an instantaneous dipole moment μ_2^*. The two dipoles attract each other and the potential energy of the pair is lowered. Although the first molecule will go on to change the size and direction of its instantaneous dipole, the electron distribution of the second molecule will follow, that is, the two dipoles are correlated in direction (Fig. 8.7). Because of this correlation, the attraction between the two instantaneous dipoles does not average to zero, and gives rise to an induced-dipole–induced-dipole interaction. This interaction is called either the **dispersion interaction** or the **London interaction** (for Fritz London, who first described it).

Polar molecules also interact by a dispersion interaction: such molecules also possess instantaneous dipoles, the only difference being that the time average of each fluctuating dipole does not vanish, but corresponds to the permanent dipole. Such molecules therefore interact both through their permanent dipoles and through the correlated, instantaneous fluctuations in these dipoles.

The strength of the dispersion interaction depends on the polarizability of the first molecule because the instantaneous dipole moment μ_1^* depends on the looseness of the control that the nuclear charge exercises over the outer electrons. The strength of the interaction also depends on the polarizability of the second molecule, for that polarizability determines how readily a dipole can be induced by another molecule. The actual calculation of the dispersion interaction is quite involved, but a reasonable approximation to the interaction energy is given by the **London formula:**

$$V = -\frac{C}{r^6} \qquad C = \tfrac{3}{2}\alpha_1'\alpha_2'\frac{I_1 I_2}{I_1 + I_2} \tag{8.14}$$

where I_1 and I_2 are the ionization energies of the two molecules (Table 4.3). This interaction energy is also proportional to the inverse sixth power of the separation of the molecules, which identifies it as a third contribution to the van der Waals interaction. The dispersion interaction generally dominates all the interactions between molecules other than hydrogen bonds.

● A BRIEF ILLUSTRATION

For two CH_4 molecules, we can substitute $\alpha' = 2.6 \times 10^{-30}$ m^3 and $I \approx 700$ kJ mol^{-1} to obtain $V = -5$ kJ mol^{-1} for $r = 0.3$ nm. A very rough check on this figure is the enthalpy of vaporization of methane, which is 8.2 kJ mol^{-1}. However, this comparison is insecure, partly because the enthalpy of vaporization is a many-body quantity and partly because the long-distance assumption breaks down. ●

8.5 Hydrogen bonding

The interactions described so far are universal in the sense that they are possessed by all molecules independent of their specific identity. However, there is a type of interaction possessed by molecules that have a particular constitution. A **hydrogen bond** is an attractive interaction between two species that arises from a link of the form A—H···B, where A and B are highly electronegative elements and B possesses a lone pair of electrons. Hydrogen bonding is conventionally regarded as being limited to N, O, and F but, if B is an anionic species (such as Cl^-), it may also participate in hydrogen bonding. There is no strict cutoff for an ability to participate in hydrogen bonding, but N, O, and F participate most effectively.

The formation of a hydrogen bond can be regarded either as the approach between a partial positive charge of H and a partial negative charge of B or as a particular example of delocalized molecular orbital formation in which A, H, and B each supply one atomic orbital from which three molecular orbitals are constructed (Fig. 8.8). Experimental evidence and theoretical arguments have been presented in favour of both views and the matter has not yet been resolved. The electrostatic interaction model can be understood readily in terms of the discussion in Section 8.3. Here we develop the molecular orbital model.

If the A—H bond is regarded as formed from the overlap of an orbital on A, ψ_A, and a hydrogen 1s orbital, ψ_H, and the lone pair on B occupies an orbital on B, ψ_B, then when the two molecules are close together, we can build three molecular orbitals from the three basis orbitals:

$$\psi = c_A\psi_A + c_H\psi_H + c_B\psi_B$$

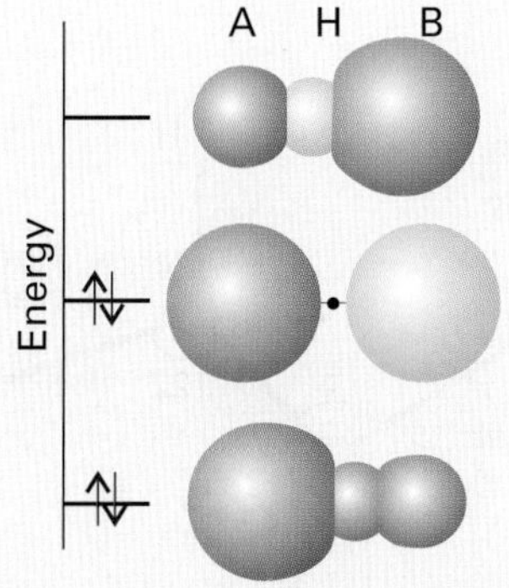

Fig. 8.8 The molecular orbital interpretation of the formation of an A—H···B hydrogen bond. From the three A, H, and B orbitals, three molecular orbitals can be formed (their relative contributions are represented by the sizes of the spheres). Only the two lower energy orbitals are occupied, and there may therefore be a net lowering of energy compared with the separate AH and B species.

One of the molecular orbitals is bonding, one almost non-bonding, and the third antibonding. These three orbitals need to accommodate four electrons (two from the original A—H bond and two from the lone pair of B), so two enter the bonding orbital and two enter the non-bonding orbital. Because the antibonding orbital remains empty, the net effect—depending on the precise location of the almost non-bonding orbital—may be a lowering of energy.

In practice, the strength of the bond is found to be about 20 kJ mol^{-1}. Because the bonding depends on orbital overlap, it is virtually a contact-like interaction that is turned on when the H of AH touches B and is zero as soon as the contact is broken. If hydrogen bonding is present, it dominates the other intermolecular interactions. The properties of liquid and solid water, for example, are dominated by the hydrogen bonding between H_2O molecules. The structure of DNA and hence the transmission of genetic information is crucially dependent on the strength and number of hydrogen bonds between base pairs. The structural evidence for hydrogen bonding comes from noting that the internuclear distance between formally non-bonded atoms is less than their van der Waals contact distance, which suggests that a dominating attractive interaction is present. For example, the O—O distance in O—H···O is expected to be 280 pm on the basis of van der Waals radii, but is found to be 270 pm in typical compounds. Moreover, the H···O distance is expected to be 260 pm but is found to be only 170 pm.

Hydrogen bonds may be either symmetric or unsymmetric. In a symmetric hydrogen bond, the H atom lies midway between the two other atoms. This arrangement is rare, but occurs in $F—H···F^-$, where both bond lengths are 120 pm. More common is the unsymmetrical arrangement, where the A—H bond is shorter than the H···B bond. Electrostatic arguments, treating A—H···B as an array of point charges (partial negative charges on A and B, partial positive on H), suggest that the lowest energy is achieved when the bond is linear, because then the two partial negative charges are furthest apart (see Problem 8.11). The experimental evidence from structural studies supports a linear or near-linear arrangement.

IMPACT ON BIOCHEMISTRY
I8.1 Proteins and nucleic acids

Polymers are macromolecules synthesized by stringing together and (in some cases) cross-linking smaller units known as **monomers**. There are macromolecules everywhere, inside us and outside us. Some are natural: they include cellulose, proteins, and deoxyribonucleic acid (DNA). Others are synthetic: they include polymers such as nylon and polystyrene. Here we describe the important role that hydrogen bonding plays in determining the shapes adopted by proteins and nucleic acids.

The concept of the 'structure' of a macromolecule takes on different meanings at the different levels at which we think about the arrangement of the chain or network of monomers. The term **configuration** refers to the structural features that can be changed only by breaking chemical bonds and forming new ones. Thus, the chains —A—B—C— and —A—C—B— have different configurations. The term **conformation** refers to the spatial arrangement of the different parts of a chain, and one conformation can be changed into another by rotating one part of a chain around a bond.

The **primary structure** of a macromolecule is the sequence of small molecular residues making up the polymer. For example, a protein is a **polypeptide** composed of linked α-amino acids, $NH_2CHRCOOH$, where R is one of about 20 groups. The monomers are strung together by the **peptide link**, —CONH—.

The **secondary structure** of a macromolecule is the (often local) spatial arrangement of a chain. The secondary structure of an isolated molecule of polyethylene is a random coil, whereas that of a protein is a highly organized arrangement determined largely by hydrogen bonds, and taking the form of random coils, helices (Fig. 8.9a), or sheets in various segments of the molecule.

The **tertiary structure** is the overall three-dimensional structure of a macromolecule. For instance, the hypothetical protein shown in Fig. 8.9b has helical regions connected by short random-coil sections. The helices interact to form a compact tertiary structure.

The **quaternary structure** of a macromolecule is the manner in which large molecules are formed by the aggregation of others. Figure 8.10 shows how four molecular subunits, each with a specific tertiary structure, aggregate together. Quaternary structure can be very important in biology. For example, the oxygen-transport protein haemoglobin consists of four subunits that work together to take up and release O_2.

For a protein to function correctly, it needs to have a well defined conformation. For example, an enzyme has its greatest catalytic efficiency only when it is in a specific conformation. The amino acid sequence of a protein contains the necessary information to create the active conformation of the protein from a newly synthesized random coil. However, the prediction of the conformation from the primary structure, the so-called *protein folding problem*, is extraordinarily difficult and is still the focus of much research.

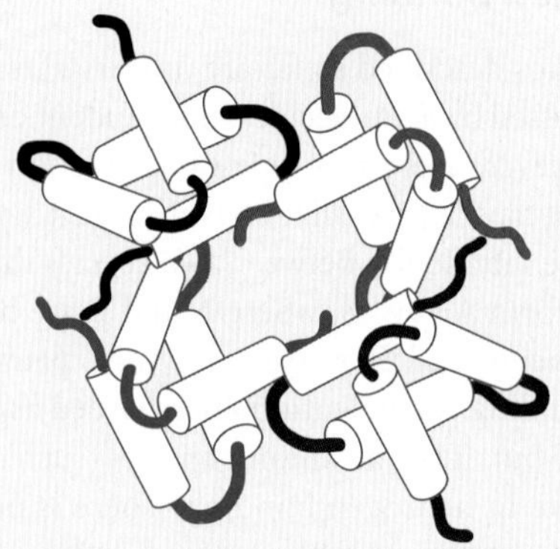

Fig. 8.10 Several subunits with specific tertiary structures pack together, providing an example of quaternary structure.

The origin of the secondary structures of proteins is found in the rules formulated by Linus Pauling and Robert Corey in 1951. The essential feature is the stabilization of structures by hydrogen bonds involving the peptide link. The latter can act both as a donor of the H atom (the NH part of the link) and as an acceptor (the CO part). The **Corey–Pauling rules** are as follows (Fig. 8.11):

1. The four atoms of the peptide link lie in a relatively rigid plane, which arises from delocalization of π electrons over the O, C, and N atoms and the maintenance of maximum overlap of their p orbitals (see Problem 8.30).

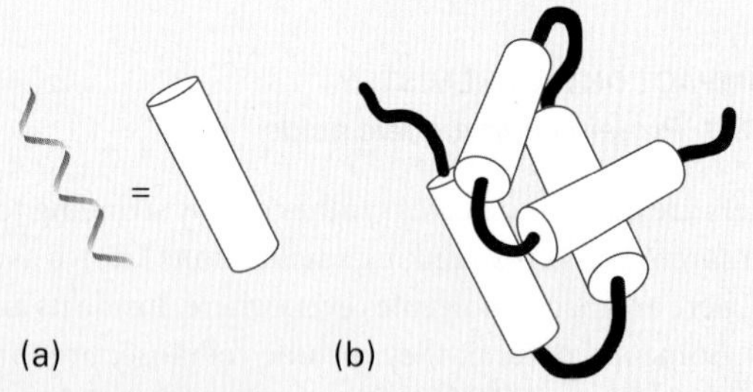

Fig. 8.9 (a) A polymer adopts a highly organized helical conformation, an example of a secondary structure. The helix is represented as a cylinder. (b) Several helical segments connected by short random coils pack together, providing an example of tertiary structure.

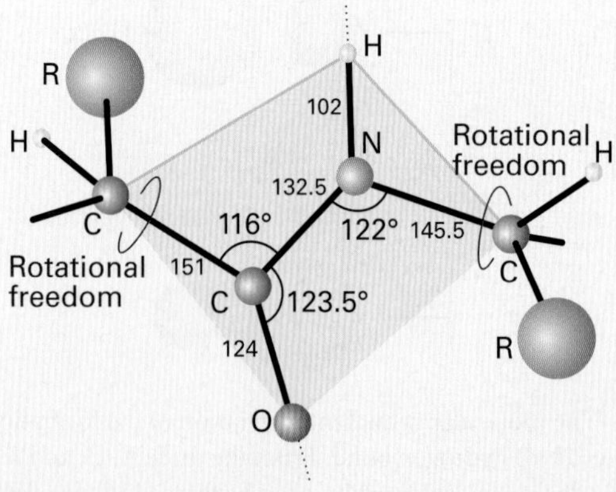

Fig. 8.11 The dimensions that characterize the peptide link. The C—NH—CO—C atoms define a plane (the C—N bond has partial double-bond character), but there is rotational freedom around the C—CO and N—C bonds.

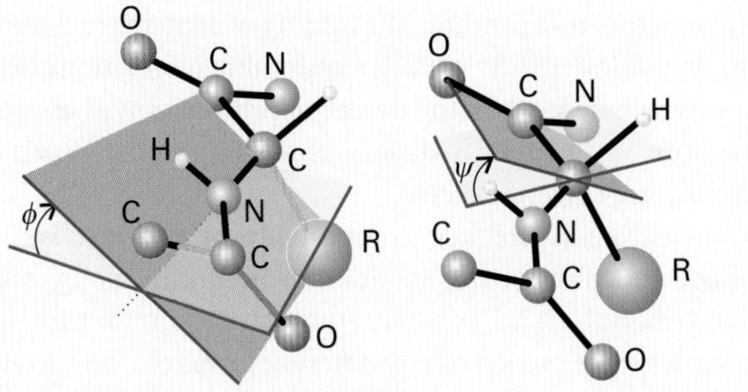

Fig. 8.12 The definition of the torsional angles ψ and ϕ between two peptide units. In α-L-polypeptide the chain is in its all-trans form, with $\psi = \phi = 180°$.

2. The N, H, and O atoms of a hydrogen bond lie in a straight line (with displacements of H tolerated up to not more than 30° from the N—O vector).

3. All NH and CO groups are engaged in hydrogen bonding.

The rules are satisfied by two structures. One, in which hydrogen bonding between peptide links leads to a helical structure, is a *helix*, which can be arranged as either a right- or a left-handed screw. The other, in which hydrogen bonding between peptide links leads to a planar structure, is a *sheet*; this form is the secondary structure of the protein fibroin, the constituent of silk.

There is a barrier to internal rotation of one bond relative to another (just like the barrier to internal rotation in ethane). Because the planar peptide link is relatively rigid, the geometry of a polypeptide chain can be specified by the two angles that two neighbouring planar peptide links make to each other. Figure 8.12 shows the two angles ϕ and ψ commonly used to specify this relative orientation. The sign convention is that a positive angle means that the front atom must be rotated clockwise to bring it into an eclipsed position relative to the rear atom. For an all-*trans* form of the chain, all ϕ and ψ are 180°. A helix is obtained when all the ϕ are equal and when all the ψ are equal. For a right-handed helix, all $\phi = -57°$ and all $\psi = -47°$. For a left-handed helix, both angles are positive. The torsional contribution to the total potential energy of the molecule is

$$V_{\text{torsion}} = A(1 + \cos 3\phi) + B(1 + \cos 3\psi) \qquad (8.15)$$

in which A and B are constants of the order of 1 kJ mol^{-1}. Because only two angles are needed to specify the conformation of a helix, and they range from −180° to +180°, the torsional potential energy of the entire molecule can be represented on a **Ramachandran plot**, a contour diagram in which one axis represents ϕ and the other represents ψ.

A right-handed **α-helix** is illustrated in Fig. 8.13. Each turn of the helix contains 3.6 amino acid residues, so the period of the helix corresponds to 5 turns (18 residues). The pitch of a single turn (the distance between points separated by 360°) is 544 pm.

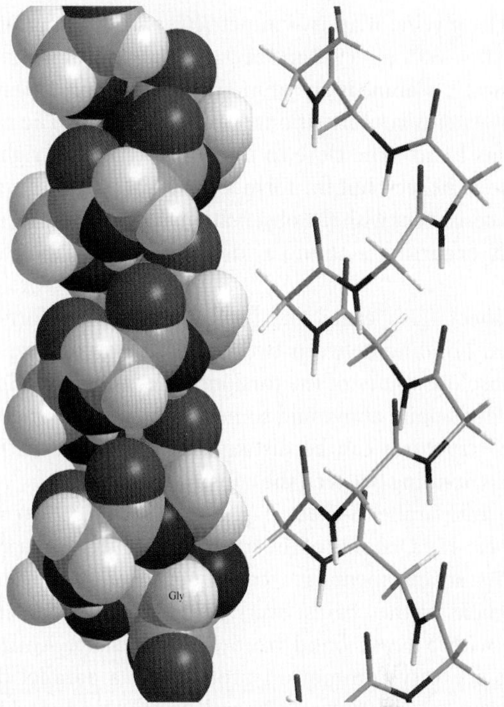

Fig. 8.13 The polypeptide α helix, with poly-L-glycine as an example. There are 3.6 residues per turn, and a translation along the helix of 150 pm per residue, giving a pitch of 540 pm. The diameter (ignoring side chains) is about 600 pm.

The N—H···O bonds lie parallel to the axis and link every fourth group (so residue i is linked to residues $i - 4$ and $i + 4$). All the R groups point away from the major axis of the helix.

Figure 8.14 shows the Ramachandran plots for the helical form of polypeptide chains formed from the nonchiral amino

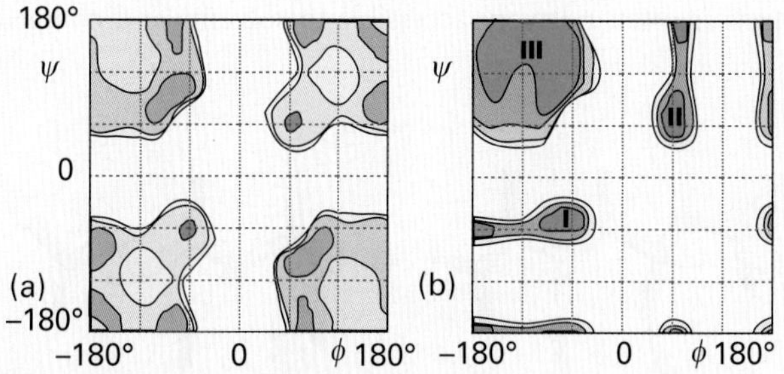

Fig. 8.14 Contour plots of potential energy against the torsional angles ψ and ϕ, also known as Ramachandran plots, for (a) a glycyl residue of a polypeptide chain and (b) an alanyl residue. The darker the shading is, the lower the potential energy. The glycyl diagram is symmetrical, but regions **I** and **II** in the alanyl diagram correspond to right- and left-handed helices, are unsymmetrical, and the minimum in region I lies lower than that in region **II**. (After D.A. Brant and P.J. Flory, *J. Mol. Biol.* 23, 47 (1967).)

acid glycine (R = H) and the chiral amino acid L-alanine (R = CH_3). The glycine map is symmetrical, with minima of equal depth at $\phi = -80°$, $\psi = +90°$ and at $\phi = +80°$, $\psi = -90°$. In contrast, the map for L-alanine is unsymmetrical, and there are three distinct low-energy conformations (marked I, II, III). The minima of regions I and II lie close to the angles typical of right- and left-handed helices, but the former has a lower minimum. This result is consistent with the observation that polypeptides of the naturally occurring L-amino acids tend to form right-handed helices.

A **β-sheet** (also called the **β-pleated sheet**) is formed by hydrogen bonding between two extended polypeptide chains (large absolute values of the torsion angles ϕ and ψ). Some of the R groups point above and some point below the sheet. Two types of structures can be distinguished from the pattern of hydrogen bonding between the constituent chains.

In an **antiparallel β-sheet** (Fig. 8.15a), $\phi = -139°$, $\psi = 113°$, and the N—H—O atoms of the hydrogen bonds form a straight line. This arrangement is a consequence of the antiparallel arrangement of the chains: every N—H bond on one chain is aligned with a C—O bond from another chain. Antiparallel β-sheets are very common in proteins. In a **parallel β-sheet** (Fig. 8.15b), $\phi = -119°$, $\psi = 113°$, and the N—H—O atoms of the hydrogen bonds are not perfectly aligned. This arrangement is a result of the parallel arrangement of the chains: each N—H bond on one chain is aligned with a N—H bond of another chain and, as a result, each C—O bond of one chain is aligned with a C—O bond of another chain. These structures are not common in proteins.

Although we do not know all the rules that govern protein folding, a few general conclusions may be drawn from X-ray diffraction studies of water-soluble natural proteins and synthetic polypeptides. In an aqueous environment, the chains fold in such a way as to place nonpolar R groups in the interior (which is often not very accessible to solvent) and charged R groups on the surface (in direct contact with the polar solvent). Other factors that promote the folding of proteins include covalent disulfide (—S—S—) links, Coulombic interactions between ions (which depend on the degree of protonation of groups and therefore on the pH), hydrogen bonding, van der Waals interactions, and solvent effects.

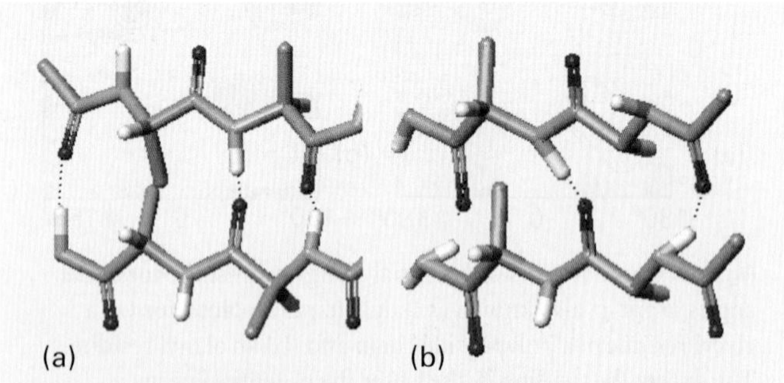

Fig. 8.15 The two types of β-sheets: (a) antiparallel ($\phi = -139°$, $\psi = 113°$), in which the N—H—O atoms of the hydrogen bonds form a straight line; (b) parallel ($\phi = -119°$, $\psi = 113°$ in which the N—H—O atoms of the hydrogen bonds are not perfectly aligned.

Nucleic acids are key components of the mechanism of storage and transfer of genetic information in biological cells. Deoxyribonucleic acid contains the instructions for protein synthesis, which is carried out by different forms of ribonucleic acid (RNA). In this section, we discuss the main structural features of DNA and RNA.

Both DNA and RNA are **polynucleotides** (**14**), in which base–sugar–phosphate units are linked by phosphodiester bonds. In RNA the sugar is β-D-ribose and in DNA it is β-D-2-deoxyribose (as shown in **15**). The most common bases are adenine (A, **16**), cytosine (C, **17**), guanine (G, **18**), thymine (T, found in DNA only, **19**), and uracil (U, found in RNA only, **20**). At physiological pH, each phosphate group of the chain carries a negative charge and the bases are deprotonated and neutral. This charge distribution leads to two important properties. One is that the polynucleotide chain is a **polyelectrolyte**, a macromolecule with many different charged sites, with a large and negative overall surface charge. The second is that the bases can interact by hydrogen bonding, as shown for A—T (**21**) and C—G base pairs (**22**). The secondary and tertiary structures of DNA and RNA arise primarily from the pattern of this hydrogen bonding between bases of one or more chains.

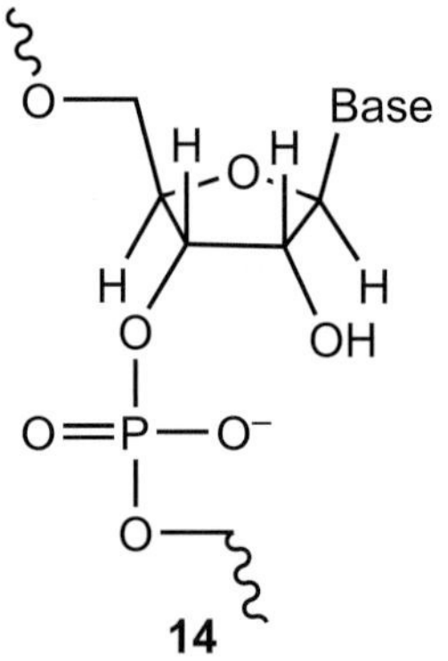

14

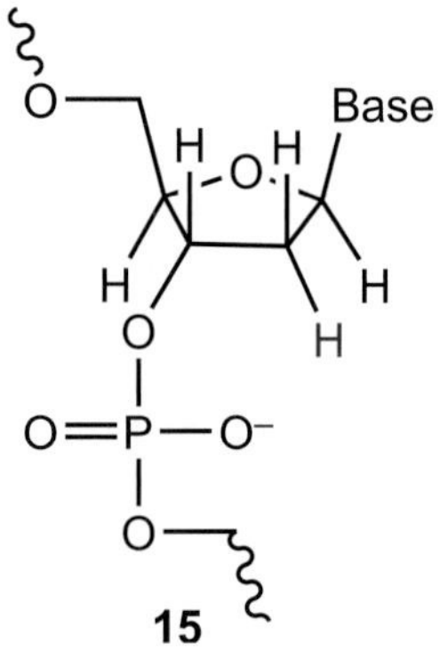

15

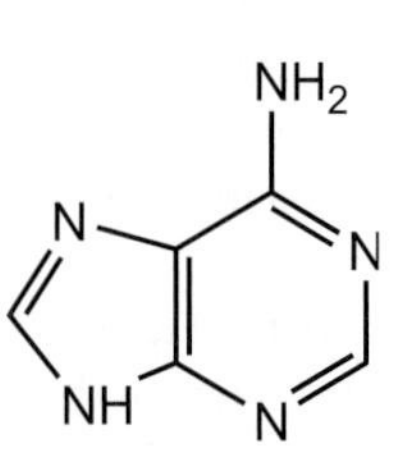

16 Adenine, A

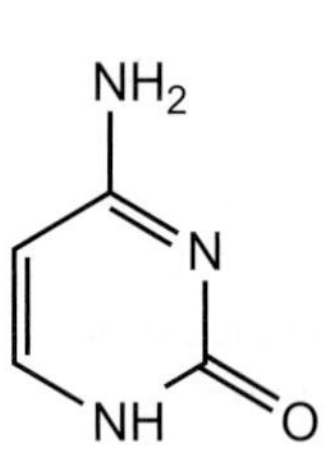

17 Cytosine, C

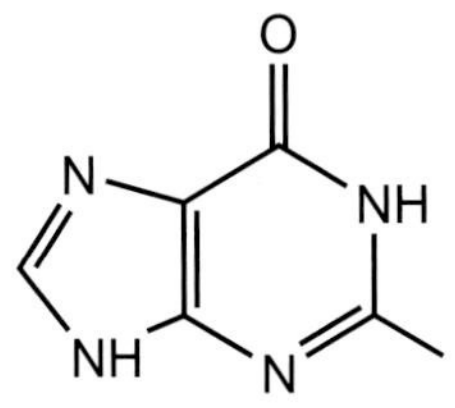

18 Guanine, G

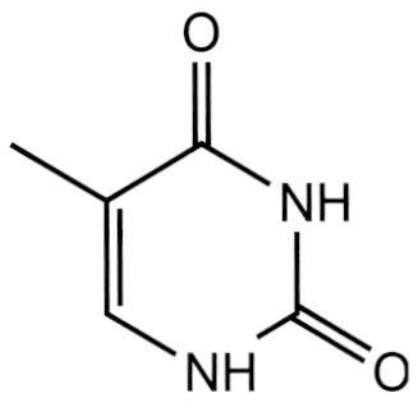

19 Thymine, T

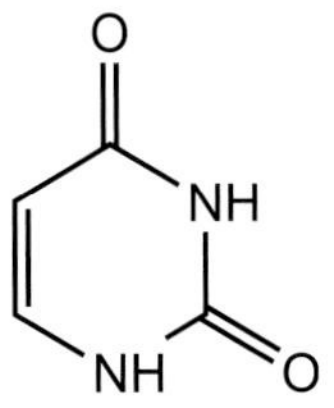

20 Uracil, U

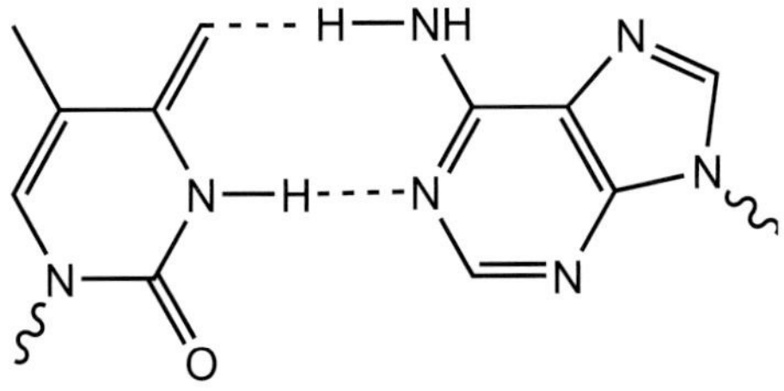

21 A–T base pair

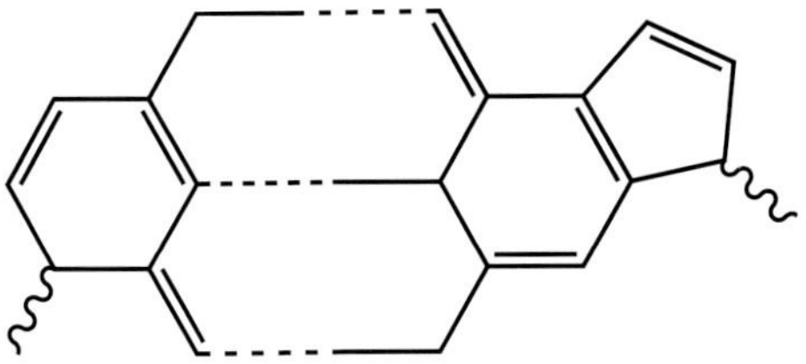

22 C–G base pair

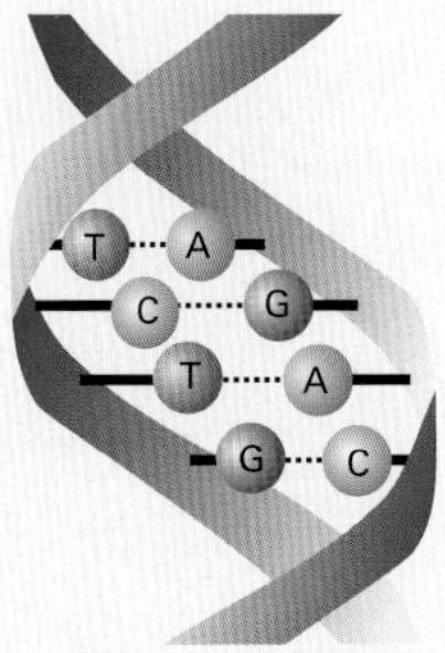

Fig. 8.16 DNA double helix, in which two polynucleotide chains are linked together by hydrogen bonds between adenine (A) and thymine (T) and between cytosine (C) and guanine (G).

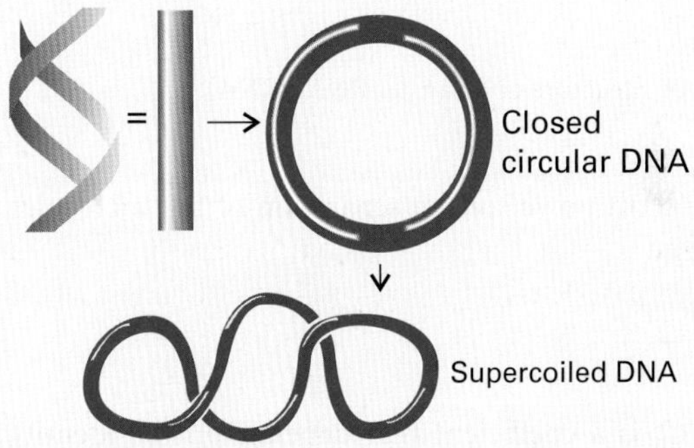

Fig. 8.17 A long section of DNA may form closed circular DNA (ccDNA) by covalent linkage of the two ends of the chain. Twisting of ccDNA leads to the formation of supercoiled DNA.

In DNA, two polynucleotide chains wind around each other to form a double helix (Fig. 8.16). The chains are held together by links involving A—T and C—G base pairs that lie parallel to each other and perpendicular to the major axis of the helix. The structure is stabilized further by **base-stacking**, in which dispersion interactions bring together the planar π systems of bases. In B-DNA, the most common form of DNA found in biological cells, the helix is right-handed with a diameter of 2.0 nm and a pitch of 3.4 nm. Long stretches of DNA can fold further into a variety of tertiary structures. Two examples are shown in Fig. 8.17. Supercoiled DNA is found in the chromosome and can be visualized as the twisting of closed circular DNA (ccDNA), much like the twisting of a rubber band.

The extra —OH group in β-D-ribose imparts enough steric strain to a polynucleotide chain so that stable double helices cannot form in RNA. Therefore, RNA exists primarily as single chains that can fold into complex structures by formation of A—U and G—C base pairs. One example of this is the structure of transfer RNA (tRNA), shown schematically in Fig. 8.18, in which base-paired regions are connected by loops and coils. Transfer RNAs help assemble polypeptide chains during protein synthesis in the cell.

8.6 The total interaction

We shall consider molecules that are unable to participate in hydrogen bond formation. The total attractive interaction energy between rotating molecules is then the sum of the three van der Waals contributions discussed above. (Only the dispersion interaction contributes if both molecules are nonpolar.) In a fluid phase, all three contributions to the potential energy vary as the

Fig. 8.18 Structure of a transfer RNA (tRNA).

inverse sixth power of the separation of the molecules, so we may write

$$V = -\frac{C_6}{r^6} \qquad (8.16)$$

where C_6 is a coefficient that depends on the identity of the molecules.

Although attractive interactions between molecules are often expressed as in eqn 8.16, we must remember that this equation has only limited validity. First, we have taken into account only dipolar interactions of various kinds, for they have the longest range and are dominant if the average separation of the molecules is large. However, in a complete treatment we should also consider quadrupolar and higher-order multipole interactions, particularly if the molecules do not have permanent electric dipole moments. Secondly, the expressions have been derived by assuming that the molecules can rotate reasonably freely. That is not the case in most solids, and in rigid media the dipole–dipole interaction is proportional to $1/r^3$ because the Boltzmann averaging procedure is irrelevant when the molecules are trapped into a fixed orientation.

A different kind of limitation is that eqn 8.16 relates to the interactions of pairs of molecules. There is no reason to suppose that the energy of interaction of three (or more) molecules is the sum of the pairwise interaction energies alone. The total dispersion energy of three closed-shell atoms, for instance, is given approximately by the **Axilrod–Teller formula**:

$$V = -\frac{C_6}{r_{AB}^6} - \frac{C_6}{r_{BC}^6} - \frac{C_6}{r_{CA}^6} + \frac{C'}{(r_{AB}r_{BC}r_{CA})^3} \qquad (8.17a)$$

where

$$C' = a(3\cos\theta_A \cos\theta_B \cos\theta_C + 1) \qquad (8.17b)$$

The parameter a is approximately equal to $\frac{3}{4}\alpha' C_6$; the angles θ are the internal angles of the triangle formed by the three atoms (**23**). The term in C' (which represents the non-additivity of the pairwise interactions) is negative for a linear arrangement of atoms (so that arrangement is stabilized) and positive for an equilateral triangular cluster. It is found that the three-body term contributes about 10 per cent of the total interaction energy in liquid argon.

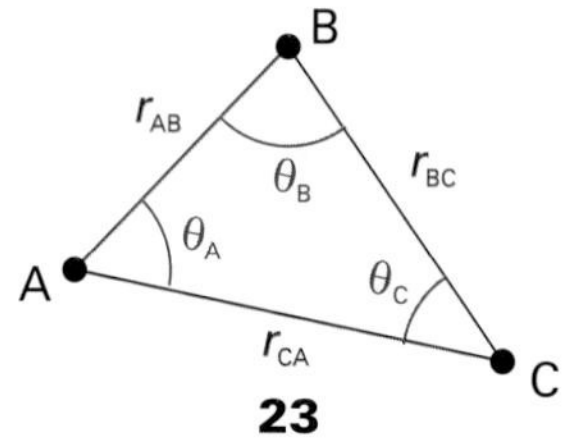

23

When molecules are squeezed together, the nuclear and electronic repulsions and the rising electronic kinetic energy begin to dominate the attractive forces. The repulsions increase steeply with decreasing separation in a way that can be deduced only by very extensive, complicated molecular structure calculations of the kind described in Chapter 6 (Fig. 8.19).

In many cases, however, progress can be made by using a greatly simplified representation of the potential energy, where the details are ignored and the general features expressed by a few adjustable parameters. One such approximation is the **hard-sphere potential**, in which it is assumed that the potential energy rises abruptly to infinity as soon as the particles come within a separation d:

$$V = \infty \text{ for } r \le d \qquad V = 0 \text{ for } r > d \qquad (8.18a)$$

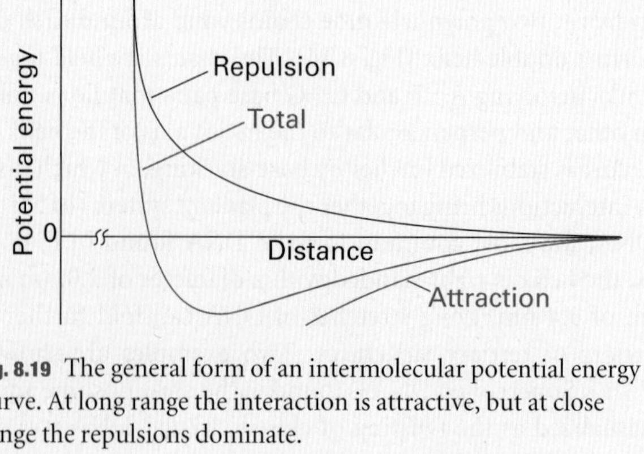

Fig. 8.19 The general form of an intermolecular potential energy curve. At long range the interaction is attractive, but at close range the repulsions dominate.

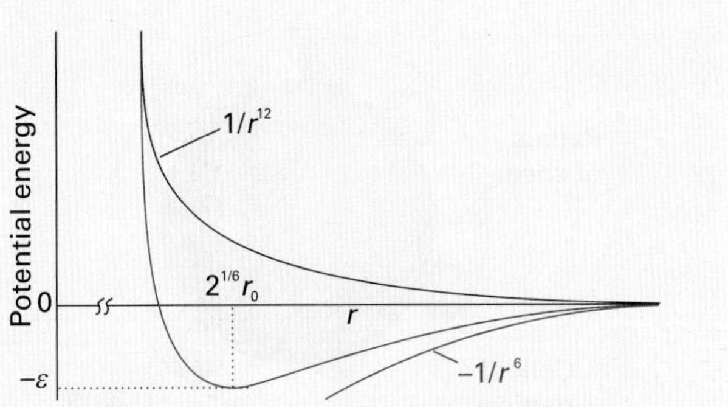

Fig. 8.20 The Lennard-Jones potential, and the relation of the parameters to the features of the curve. The light lines are the two contributions.

This very simple potential is surprisingly useful for assessing a number of properties. Another widely used approximation is the **Mie potential**:

$$V = \frac{C_n}{r^n} - \frac{C_m}{r^m} \tag{8.18b}$$

with $n > m$. The first term represents repulsions and the second term attractions. The **Lennard-Jones potential** is a special case of the Mie potential with $n = 12$ and $m = 6$ (Fig. 8.20); it is often written in the form

$$V = 4\varepsilon\left\{\left(\frac{r_0}{r}\right)^{12} - \left(\frac{r_0}{r}\right)^{6}\right\} \tag{8.18c}$$

The two parameters are ε, the depth of the well (not to be confused with the symbol of the permittivity of a medium used in Section 8.1), and r_0, the separation at which $V = 0$ (Table 8.4). The well minimum occurs at $r_e = 2^{1/6}r_0$ (see Problem 8.8). Although the Lennard-Jones potential has been used in many calculations, there is plenty of evidence to show that $1/r^{12}$ is a very poor representation of the repulsive potential, and that an exponential form, e^{-r/r_0}, is greatly superior. An exponential function is more faithful to the exponential decay of atomic wavefunctions at large distances, and hence to the overlap that is responsible for repulsion. The potential with an exponential repulsive term and a $1/r^6$ attractive term is known as an **exp-6 potential**.

Synoptic table 8.4* Lennard-Jones (12,6) parameters

	(ε/k)/K	r_0/pm
Ar	111.84	362.3
CCl_4	378.86	624.1
N_2	91.85	391.9
Xe	213.96	426.0

* More values are given in the *Data section*.

With the advent of **atomic force microscopy** (AFM), in which the force between a molecular sized probe and a surface is monitored (see *Impact I2.1*), it has become possible to measure directly the forces acting between molecules. The force, F, is the negative slope of potential, so for a Lennard-Jones potential between individual molecules we write

$$F = -\frac{dV}{dr} = \frac{24\varepsilon}{r_0}\left\{2\left(\frac{r_0}{r}\right)^{13} - \left(\frac{r_0}{r}\right)^{7}\right\} \tag{8.19}$$

The net attractive force is greatest (from $dF/dr = 0$) at $r = (\frac{26}{7})^{1/6}r_0$, or $1.244r_0$, and at that distance is equal to $-144(\frac{7}{26})^{7/6}\varepsilon/13r_0$, or $-2.396\varepsilon/r_0$. For typical parameters, the magnitude of this force is about 10 pN.

IMPACT ON NANOSCIENCE

I8.2 Colloidal nanoparticles

Much of the material discussed in this chapter applies to aggregates of particles that form by **self-assembly**, the spontaneous formation of complex structures of molecules held together by molecular interactions. A **colloid**, or **disperse phase**, is a dispersion of small particles of one material in another. In this context, 'small' means something less than about 500 nm in diameter (about the wavelength of visible light). In general, colloidal particles are aggregates of numerous atoms or molecules, but are too small to be seen with an ordinary optical microscope.

The name given to the colloid depends on the two phases involved. A **sol** is a dispersion of a solid in a liquid (such as clusters of gold atoms in water) or of a solid in a solid (such as ruby glass, which is a gold-in-glass sol, and achieves its colour by light scattering). An **aerosol** is a dispersion of a liquid in a gas (like fog and many sprays) or a solid in a gas (such as smoke): the particles are often large enough to be seen with a microscope. An **emulsion** is a dispersion of a liquid in a liquid (such as milk).

A further classification of colloids is as **lyophilic**, or solvent-attracting, and **lyophobic**, solvent-repelling. If the solvent is water, the terms **hydrophilic** and **hydrophobic**, respectively, are used instead. Lyophobic colloids include the metal sols. Lyophilic colloids generally have some chemical similarity to the solvent, such as —OH groups on the surface able to form hydrogen bonds. A **gel** is a semirigid mass of a lyophilic sol in which all the dispersion medium has penetrated into the sol particles.

The preparation of aerosols can be as simple as sneezing (which produces an imperfect aerosol). Laboratory and commercial methods make use of several techniques. Material (for example, quartz) may be ground in the presence of the dispersion medium. Passing a heavy electric current through a cell may lead to the sputtering (crumbling) of an electrode into colloidal particles. Arcing between electrodes immersed in the support medium also produces a colloid. Chemical precipitation sometimes results in a colloid. A precipitate (for example, silver iodide) already formed may be dispersed by the addition of a peptizing agent (for example, potassium iodide). Clays may be peptized by alkalis, the OH^- ion being the active agent.

Emulsions are normally prepared by shaking the two components together vigorously, although some kind of emulsifying agent usually has to be added to stabilize the product. This emulsifying agent may be a soap (the salt of a long-chain carboxylic acid) or other **surfactant** (surface active) species, or a lyophilic sol that forms a protective film around the dispersed phase. In milk, which is an emulsion of fats in water, the emulsifying agent is casein, a protein containing phosphate groups. It is clear from the formation of cream on the surface of milk that casein is not completely successful in stabilizing milk: the dispersed fats coalesce into oily droplets that float to the surface. This coagulation may be prevented by ensuring that the emulsion is dispersed very finely initially: intense agitation with ultrasonics brings this dispersion about, the product being 'homogenized' milk.

One way to form an aerosol is to tear apart a spray of liquid with a jet of gas. The dispersal is aided if a charge is applied to the liquid, for then electrostatic repulsions help to blast it apart into droplets. This procedure may also be used to produce emulsions, for the charged liquid phase may be directed into another liquid.

Colloidal particles attract one another over large distances by the dispersion interaction, so there is a long-range force tending to collapse them down into a single blob. Several factors oppose the long-range dispersion attraction. There may be a protective film at the surface of the colloid particles that stabilizes the interface and cannot be penetrated when two particles touch. For example, the surface atoms of a platinum sol in water react chemically and are turned into $—(OH)_3H_3$, and this layer encases the particle like a shell. A fat can be emulsified by a soap because the long hydrocarbon tails of the soap molecules penetrate the oil droplet but the $—CO_2^-$ head groups (or other hydrophilic groups in detergents) surround the surface, form hydrogen bonds with water, and give rise to a shell of negative charge that repels a possible approach from another similarly charged particle.

Apart from the physical stabilization of disperse systems, certain factors slow down their collapse. Chief among them is the existence of an electric charge on the surfaces of the colloidal particles. On account of this charge, ions of opposite charge tend to cluster nearby.

Two regions of charge must be distinguished. First, there is a fairly immobile layer of ions that stick tightly to the surface of the colloidal particle, and which may include water molecules (if that is the support medium). The radius of the sphere that captures this rigid layer is called the **radius of shear**, and is the major factor determining the mobility of the particles (Fig. 8.21).

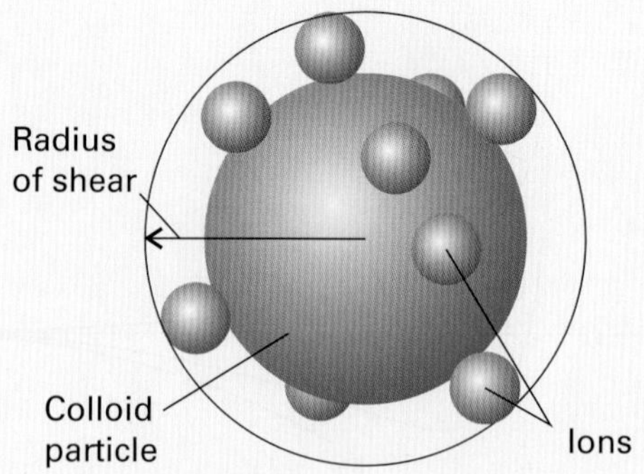

Fig. 8.21 The definition of the radius of shear for a colloidal particle. The spheres are ions attached to the surface of the particle.

The electric potential at the radius of shear relative to its value in the distant, bulk medium is called the **electrokinetic potential**, ζ (zeta). The charged unit attracts an oppositely charged ionic atmosphere. The inner shell of charge and the outer atmosphere jointly constitute the **electric double layer**.

At high concentrations of ions of high charge number, the atmosphere is dense and the potential falls to its bulk value within a short distance. In this case there is little electrostatic repulsion to hinder the close approach of two colloid particles. As a result, **flocculation**, the aggregation of the colloidal particles, occurs as a consequence of the van der Waals forces. Flocculation is often reversible, and should be distinguished from **coagulation**, which is the irreversible collapse of the colloid into a bulk phase. When river water containing colloidal clay flows into the sea, the brine induces coagulation and is a major cause of silting in estuaries.

Metal oxide and sulfide sols have charges that depend on the pH; sulfur and the noble metals tend to be negatively charged. Naturally occurring macromolecules also acquire a charge when dispersed in water, and an important feature of proteins and other natural macromolecules is that their overall charge depends on the pH of the medium. For instance, in acid environments protons attach to basic groups and the net charge of the macromolecule is positive; in basic media the net charge is negative as a result of proton loss. At the **isoelectric point**, the pH is such that there is no net charge on the macromolecule.

The primary role of the electric double layer is to slow down the collapse into a bulk phase. Colliding colloidal particles break through the double layer and coalesce only if the collision is sufficiently energetic to disrupt the layers of ions and solvating molecules, or if thermal motion has stirred away the surface accumulation of charge. This kind of disruption of the double layer may occur at high temperatures, which is one reason why sols precipitate when they are heated. The protective role of the double layer is the reason why it is important not to remove all the ions (other than those needed to ensure overall electrical

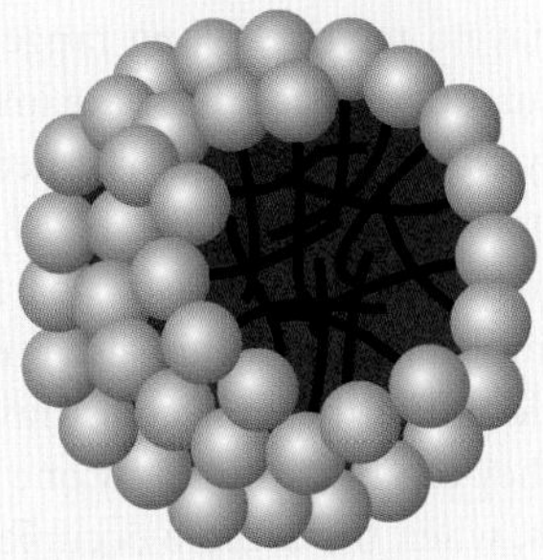

Fig. 8.22 A schematic version of a spherical micelle. The hydrophilic groups are represented by spheres and the hydrophobic hydrocarbon chains are represented by the stalks; these stalks are mobile.

neutrality) when a colloid is being purified, and why proteins coagulate most readily at their isoelectric point.

Surfactant molecules or ions can cluster together as **micelles**, which are colloid-sized clusters of molecules, for their hydrophobic tails tend to congregate, and their hydrophilic heads provide protection (Fig. 8.22).

Micelles form only above the **critical micelle concentration** (CMC) and above the **Krafft temperature**. The CMC is detected by noting a pronounced change in physical properties of the solution, particularly the molar conductivity (Fig. 8.23). There is no abrupt change in properties at the CMC; rather, there is a transition region corresponding to a range of concentrations around the CMC where physical properties vary smoothly but non-linearly with the concentration. The hydrocarbon interior of a micelle is like a droplet of oil. Nuclear magnetic resonance shows that the hydrocarbon tails are mobile, but slightly more restricted than in the bulk. Micelles are important in industry and biology on account of their solubilizing function: matter can be transported by water after it has been dissolved in their hydrocarbon interiors. For this reason, micellar systems are used as detergents, for organic synthesis, froth flotation, and petroleum recovery.

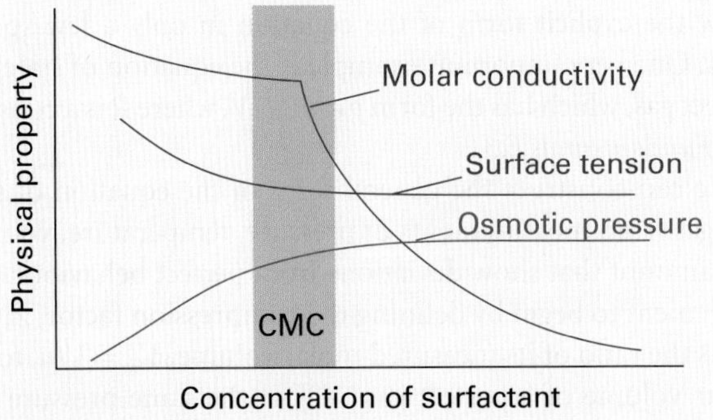

Fig. 8.23 The typical variation of some physical properties of an aqueous solution of sodium dodecylsulfate close to the critical micelle concentration (CMC).

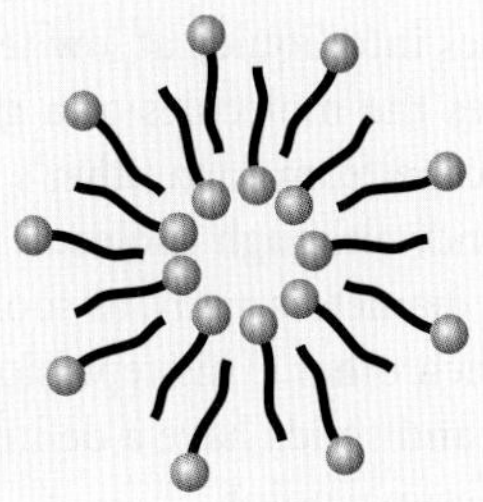

Fig. 8.24 The cross-sectional structure of a spherical liposome.

Non-ionic surfactant molecules may cluster together in clumps of 1000 or more, but ionic species tend to be disrupted by the electrostatic repulsions between head groups and are normally limited to groups of less than about 100. The micelle population is often polydisperse, and the shapes of the individual micelles vary with concentration. Spherical micelles do occur, but micelles are more commonly flattened spheres close to the CMC.

Under certain experimental conditions, a **liposome** may form, with an inward pointing inner surface of molecules surrounded by an outward pointing outer layer (Fig. 8.24). Liposomes may be used to carry nonpolar drug molecules in blood. In concentrated solutions micelles formed from surfactant molecules may take the form of long cylinders and stack together in reasonably close-packed (hexagonal) arrays. These orderly arrangements of micelles are called **lyotropic mesomorphs** and, more colloquially, 'liquid crystalline phases'.

Some micelles at concentrations well above the CMC form extended parallel sheets, called **lamellar micelles**, two molecules thick. The individual molecules lie perpendicular to the sheets, with hydrophilic groups on the outside in aqueous solution and on the inside in nonpolar media. Such lamellar micelles show a close resemblance to biological membranes, and are often a useful model on which to base investigations of biological structures. We discuss biological membranes more fully in *Impact I16.1*.

Gases and liquids

The form of matter with the least order is a gas. In a perfect gas there are no intermolecular interactions and the distribution of molecules is completely random. In a real gas there are weak attractions and repulsions that have minimal effect on the relative locations of the molecules but that cause deviations from the perfect gas law for the dependence of pressure on the volume, temperature, and amount (see *Fundamentals F.3*).

The attractions between molecules are responsible for the condensation of gases into liquids at low temperatures. At low enough temperatures the molecules of a gas have insufficient kinetic energy to escape from each other's attraction and they stick together. Second, although molecules attract each other when they are a few diameters apart, as soon as they come into contact they repel each other. This repulsion is responsible for the fact that liquids and solids have a definite bulk and do not collapse to an infinitesimal point.

In the following sections we build on the concepts just outlined. First, we describe molecular interactions in gases and see how the perfect gas equation of state is modified to take these interactions into account. Then, we explore liquids and see that it is possible to speak of a 'structure' for this state of matter.

8.7 Molecular interactions in gases

Real gases do not obey the perfect gas law exactly. Deviations from the law are particularly important at high pressures and low temperatures, especially when a gas is on the point of condensing to liquid.

Real gases show deviations from the perfect gas law because molecules interact with one another. Repulsive forces between molecules assist expansion and attractive forces assist compression. Repulsive forces are significant only when molecules are almost in contact: they are short-range interactions, even on a scale measured in molecular diameters (Fig. 8.25). Because they are short-range interactions, repulsions can be expected to be important only when the average separation of the molecules is small. This is the case at high pressure, when many molecules occupy a small volume. On the other hand, attractive intermolecular forces have a relatively long range and are effective over several molecular diameters. They are important when the molecules are fairly close together but not necessarily touching (at the intermediate separations in Fig. 8.25). Attractive forces are ineffective when the molecules are far apart (well to the right in Fig. 8.25). Intermolecular forces are also important when the temperature is so low that the molecules travel with such low mean speeds that they can be captured by one another.

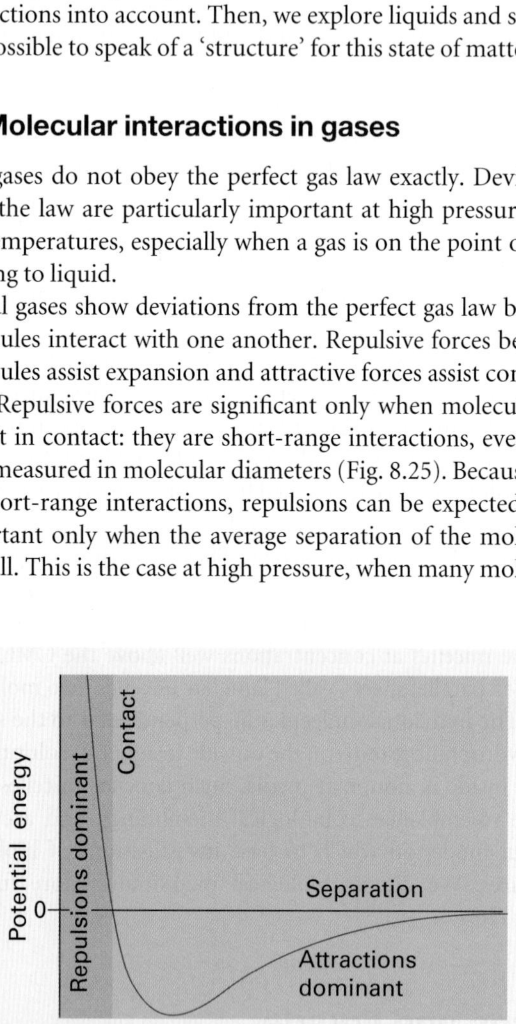

Fig. 8.25 The variation of the potential energy of two molecules on their separation. High positive potential energy (at very small separations) indicates that the interactions between them are strongly repulsive at these distances. At intermediate separations, where the potential energy is negative, the attractive interactions dominate. At large separations (on the right) the potential energy is zero and there is no interaction between the molecules.

At low pressures, when the sample occupies a large volume, the molecules are so far apart for most of the time that the intermolecular forces play no significant role, and the gas behaves virtually perfectly. At moderate pressures, when the average separation of the molecules is only a few molecular diameters, the attractive forces dominate the repulsive forces. In this case, the gas can be expected to be more compressible than a perfect gas because the forces help to draw the molecules together. At high pressures, when the average separation of the molecules is small, the repulsive forces dominate and the gas can be expected to be less compressible because now the forces help to drive the molecules apart.

(a) The virial equation of state

The physical state of a sample of a substance is defined by its physical properties. The state of a pure gas, for example, is specified by giving its volume, V, amount of substance, n, pressure, p, and temperature, T. However, it has been established experimentally that it is sufficient to specify only three of these variables, for then the fourth variable is fixed. That is, it is an experimental fact that each substance is described by an **equation of state**, an equation that interrelates these four variables.

The general form of an equation of state is

$$p = f(T, V, n)$$

This equation tells us that, if we know the values of n, T, and V for a particular substance, then the pressure has a fixed value. Each substance is described by its own equation of state, but we know the explicit form of the equation in only a few special cases. One very important example is the equation of state of a perfect gas, which has the form $p = nRT/V$, where R is a constant (see *Fundamentals F.3*).

We can construct the general form of the equation of state of a gas from measurements of pressure, temperature, volume, and amount that show deviations from perfect behaviour. It is convenient to begin by defining the **compression factor**, Z, of a gas as the ratio of its measured molar volume, $V_m = V/n$, to the molar volume of a perfect gas, V_m°, at the same pressure and temperature:

$$Z = \frac{V_m}{V_m^\circ} \qquad [8.20]$$

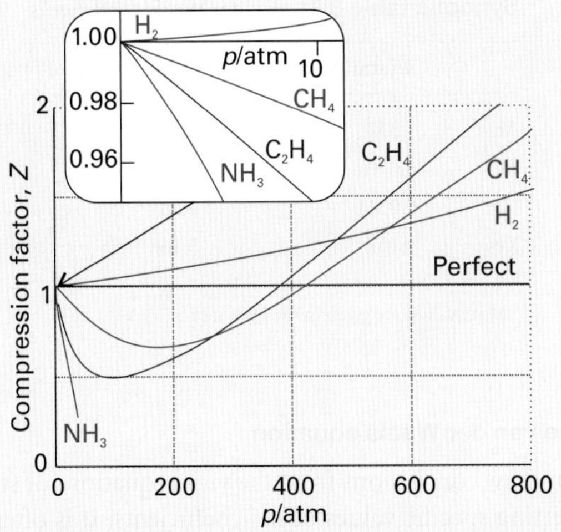

Fig. 8.26 The variation of the compression factor, Z, with pressure for several gases at 0°C. A perfect gas has $Z=1$ at all pressures. Notice that, although the curves approach 1 as $p \to 0$, they do so with different slopes.

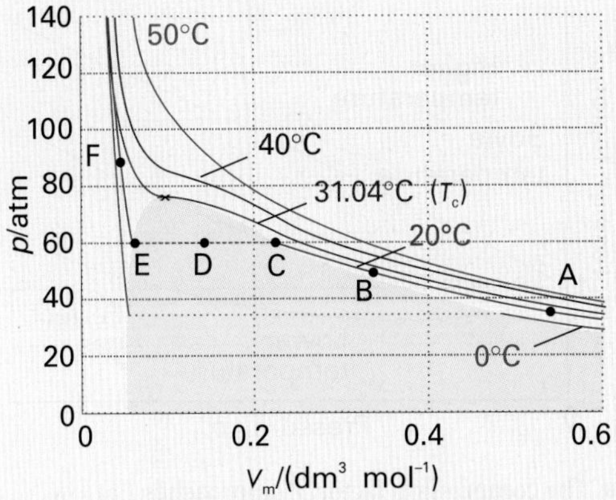

Fig. 8.27 Experimental isotherms of carbon dioxide at several temperatures. The 'critical isotherm', the isotherm at the critical temperature, is at 31.04°C. The critical point is marked with a star. Above the critical temperature, a gas cannot be liquefied by the application of pressure alone.

Because the molar volume of a perfect gas is equal to RT/p, an equivalent expression is $Z = pV_m/RT$, which we can write as

$$pV_m = RTZ \tag{8.21}$$

For a perfect gas $Z = 1$ under all conditions, so it follows that deviation of Z from 1 is a measure of departure from perfect behaviour.

Some experimental values of Z are plotted in Fig. 8.26. At very low pressures, all the gases shown have $Z \approx 1$ and behave nearly perfectly. At high pressures, all the gases have $Z > 1$, signifying that they have a larger molar volume than a perfect gas. Repulsive forces are now dominant. At intermediate pressures, most gases have $Z < 1$, indicating that the attractive forces are reducing the molar volume relative to that of a perfect gas.

Figure 8.27 shows the experimental *isotherms*, plots of data (in this case pressure and volume data) obtained at constant temperature, for carbon dioxide. At large molar volumes and high temperatures the real-gas isotherms do not differ greatly from perfect-gas isotherms. The small differences suggest that the perfect gas law is in fact the first term in an expression of the form

$$pV_m = RT(1 + B'p + C'p^2 + \cdots) \tag{8.22a}$$

This expression is an example of a common procedure in physical chemistry, in which a simple law that is known to be a good first approximation (in this case $pV_m = RT$) is treated as the first term in a series in powers of a variable (in this case p). A more convenient expansion for many applications is

$$pV_m = RT\left(1 + \frac{B}{V_m} + \frac{C}{V_m^2} + \cdots\right) \tag{8.22b}$$

These two expressions are two versions of the **virial equation of state** (the name *virial* comes from the Latin word for force). By comparing the expression with eqn 8.21 we see that the term in parentheses can be identified with the compression factor, Z.

The coefficients $B, C, \ldots$ (sometimes denoted $B_2, B_3, \ldots$) depend on the temperature and are the second, third, . . . **virial coefficients** (Table 8.5); the first virial coefficient is 1. The third virial coefficient, C, is usually less important than the second coefficient, B, in the sense that at typical molar volumes $C/V_m^2 \ll B/V_m$.

We can use the virial equation to demonstrate the important point that, although the equation of state of a real gas may coincide with the perfect gas law as $p \to 0$, not all its properties necessarily coincide with those of a perfect gas in that limit. Consider, for example, the value of dZ/dp, the slope of the graph of compression factor against pressure. For a perfect gas $dZ/dp = 0$ (because $Z = 1$ at all pressures), but for a real gas from eqns 8.21 and 8.22a we obtain

$$\frac{dZ}{dp} = B' + 2pC' + \cdots \to B' \text{ as } p \to 0 \tag{8.23a}$$

Synoptic table 8.5* Second virial coefficients, $B/(\text{cm}^3\ \text{mol}^{-1})$

	Temperature	
	273 K	600 K
Ar	−21.7	11.9
CO_2	−142	−12.4
N_2	−10.5	21.7
Xe	−153.7	−19.6

* More values are given in the *Data section*.

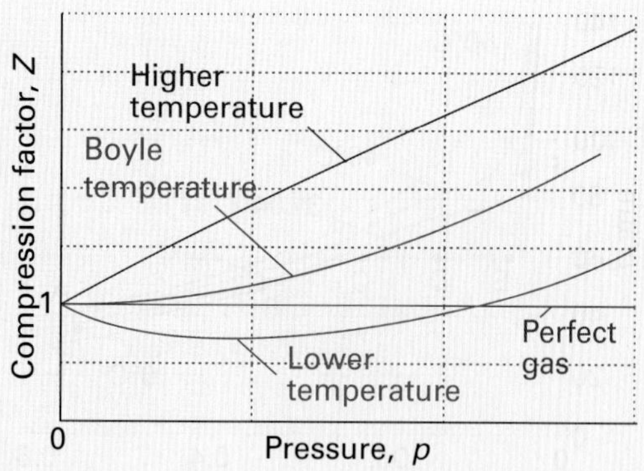

Fig. 8.28 The compression factor, Z, approaches 1 at low pressures, but does so with different slopes. For a perfect gas, the slope is zero, but real gases may have either positive or negative slopes, and the slope may vary with temperature. At the Boyle temperature, the slope is zero and the gas behaves perfectly over a wider range of conditions than at other temperatures.

However, B' is not necessarily zero, so the slope of Z with respect to p does not necessarily approach 0 (the perfect gas value), as we can see in Fig. 8.26. By a similar argument,

$$\frac{dZ}{d(1/V_m)} \to B \text{ as } V_m \to \infty, \text{ corresponding to } p \to 0 \quad (8.23b)$$

Because the virial coefficients depend on the temperature, there may be a temperature at which $Z \to 1$ with zero slope at low pressure or high molar volume (Fig. 8.28). At this temperature, which is called the **Boyle temperature**, T_B, the properties of the real gas do coincide with those of a perfect gas as $p \to 0$. According to eqn 8.23b, Z has zero slope as $p \to 0$ if $B = 0$, so we can conclude that $B = 0$ at the Boyle temperature. It then follows from eqn 8.21 that $pV_m \approx RT_B$ over a more extended range of pressures than at other temperatures because the first term after 1 (that is, B/V_m) in the virial equation is zero and C/V_m^2 and higher terms are negligibly small. For helium $T_B = 22.64$ K; for air $T_B = 346.8$ K; more values are given in Table 8.6.

Synoptic table 8.6* Boyle temperatures of gases

	T_B/K
Ar	411.5
CO_2	714.8
He	22.64
O_2	405.9

* More values are given in the *Data section*.

Synoptic table 8.7* van der Waals coefficients

	a/(atm dm^6 mol^{-2})	b/(10^{-2} dm^3 mol^{-1})
Ar	1.337	3.20
CO_2	3.610	4.29
He	0.0341	2.38
Xe	4.137	5.16

* More values are given in the *Data section*.

(b) The van der Waals equation

We can draw conclusions from the virial equations of state only by inserting specific values of the coefficients. It is often useful to have a broader, if less precise, view of all gases. Therefore, we introduce the approximate equation of state suggested by J.D. van der Waals in 1873. This equation is an excellent example of an expression that can be obtained by thinking scientifically about a mathematically complicated but physically simple problem, that is, it is a good example of 'model building'.

The **van der Waals equation** is

$$p = \frac{nRT}{V - nb} - a\left(\frac{n}{V}\right)^2 \quad (8.24a)$$

and a derivation is given in the following *Justification*. The equation is often written in terms of the molar volume $V_m = V/n$ as

$$p = \frac{RT}{V_m - b} - \frac{a}{V_m^2} \quad (8.24b)$$

The (positive) constants a and b are called the **van der Waals coefficients**. They are characteristic of each gas but independent of the temperature (Table 8.7).

Justification 8.4 *The van der Waals equation of state*

The repulsive interactions between molecules are taken into account by supposing that they cause the molecules to behave as small but impenetrable spheres. The nonzero volume of the molecules implies that instead of moving in a volume V they are restricted to a smaller volume $V - nb$, where nb is approximately the total volume taken up by the molecules themselves. This argument suggests that the perfect gas law $p = nRT/V$ should be replaced by

$$p = \frac{nRT}{V - nb}$$

when repulsions are significant. The closest distance of two hard-sphere molecules of radius r, and volume $V_{\text{molecule}} = \frac{4}{3}\pi r^3$, is $2r$, so the volume excluded is $\frac{4}{3}\pi(2r)^3$, or $8V_{\text{molecule}}$. The volume excluded per molecule is one-half this volume, or $4V_{\text{molecule}}$, so $b \approx 4V_{\text{molecule}}N_A$.

The pressure depends on both the frequency of collisions with the walls and the force of each collision. Both the frequency of the collisions and their force are reduced by the attractive forces, which act with a strength proportional to the molar concentration, n/V, of molecules in the sample. Therefore, because both the frequency and the force of the collisions are reduced by the attractive forces, the pressure is reduced in proportion to the square of this concentration. If the reduction of pressure is written as $-a(n/V)^2$, where a is a positive constant characteristic of each gas, the combined effect of the repulsive and attractive forces is the van der Waals equation of state as expressed in eqn 8.24.

In this *Justification* we have built the van der Waals equation using vague arguments about the volumes of molecules and the effects of forces. The equation can be derived in other ways, but the present method has the advantage that it shows how to derive the form of an equation out of general ideas. The derivation also has the advantage of keeping imprecise the significance of the coefficients a and b: they are much better regarded as empirical parameters than as precisely defined molecular properties (but see Section 14.10 for a precise thermodynamic interpretation of a).

Example 8.3 *Using the van der Waals equation to estimate a molar volume*

Estimate the molar volume of CO_2 at 500 K and 100 atm by treating it as a van der Waals gas.

Method To express eqn 8.24b as an equation for the molar volume, we multiply both sides by $(V_m - b)V_m^2$, to obtain

$$(V_m - b)V_m^2 p = RTV_m^2 - (V_m - b)a$$

and, after division by p, collect powers of V_m to obtain

$$V_m^3 - \left(b + \frac{RT}{p}\right)V_m^2 + \left(\frac{a}{p}\right)V_m - \frac{ab}{p} = 0$$

Although closed expressions for the roots of a cubic equation can be given, they are very complicated. Unless analytical solutions are essential, it is usually more expedient to solve such equations with commercial software.

Answer According to Table 8.7, $a = 3.610\ \mathrm{dm^6\ atm\ mol^{-2}}$ and $b = 4.29 \times 10^{-2}\ \mathrm{dm^3\ mol^{-1}}$. Under the stated conditions, $RT/p = 0.410\ \mathrm{dm^3\ mol^{-1}}$. The coefficients in the equation for V_m are therefore

$$b + RT/p = 0.453\ \mathrm{dm^3\ mol^{-1}}$$
$$a/p = 3.61 \times 10^{-2}\ (\mathrm{dm^3\ mol^{-1}})^2$$
$$ab/p = 1.55 \times 10^{-3}\ (\mathrm{dm^3\ mol^{-1}})^3$$

Therefore, on writing $x = V_m/(\mathrm{dm^3\ mol^{-1}})$, the equation to solve is

$$x^3 - 0.453x^2 + (3.61 \times 10^{-2})x - (1.55 \times 10^{-3}) = 0$$

The acceptable root is $x = 0.366$, which implies that $V_m = 0.366\ \mathrm{dm^3\ mol^{-1}}$. For a perfect gas under these conditions, the molar volume is $0.410\ \mathrm{dm^3\ mol^{-1}}$.

Self-test 8.4 Calculate the molar volume of argon at 100°C and 100 atm on the assumption that it is a van der Waals gas. [$0.298\ \mathrm{dm^3\ mol^{-1}}$]

It is too optimistic to expect a single, simple expression to be the true equation of state of all substances, and accurate work on gases must resort to the virial equation, use tabulated values of the coefficients at various temperatures, and analyse the systems numerically. The advantage of the van der Waals equation, however, is that it is analytical (that is, expressed symbolically) and allows us to draw some general conclusions about real gases. When the equation fails we must use one of the other equations of state that have been proposed (some are listed in Table 8.8), invent a new one, or go back to the virial equation.

That having been said, we can begin to judge the reliability of the equation by comparing the isotherms it predicts with the experimental isotherms in Fig. 8.27. Some calculated isotherms are shown in Figs. 8.29 and 8.30. Apart from the oscillations they do resemble experimental isotherms quite well. The oscillations, the **van der Waals' loops**, are unrealistic because they suggest that under some conditions an increase of pressure results in an increase of volume. Therefore they are replaced by horizontal lines drawn so the loops define equal areas above and below the lines: this procedure is called the **Maxwell construction** (24). The van der Waals coefficients, such as those in Table 8.7, are found by fitting the calculated curves to the experimental curves.

Table 8.8 Selected equations of state

	Equation	Critical constants p_c	V_c	T_c
Perfect gas	$p = \frac{RT}{V_m}$			
van der Waals	$p = \frac{RT}{V_m - b} - \frac{a}{V_m^2}$	$\frac{a}{27b^2}$	$3b$	$\frac{8a}{27bR}$
Berthelot	$p = \frac{RT}{V_m - b} - \frac{a}{TV_m^2}$	$\frac{1}{12}\left(\frac{2aR}{3b^3}\right)^{1/2}$	$3b$	$\frac{2}{3}\left(\frac{2a}{3bR}\right)^{1/2}$
Dieterici	$p = \frac{RTe^{-a/RTV_m}}{V_m - b}$	$\frac{a}{4e^2b^2}$	$2b$	$\frac{a}{4bR}$
Virial	$p = \frac{RT}{V_m}\left\{1 + \frac{B(T)}{V_m} + \frac{C(T)}{V_m^2} + \cdots\right\}$			

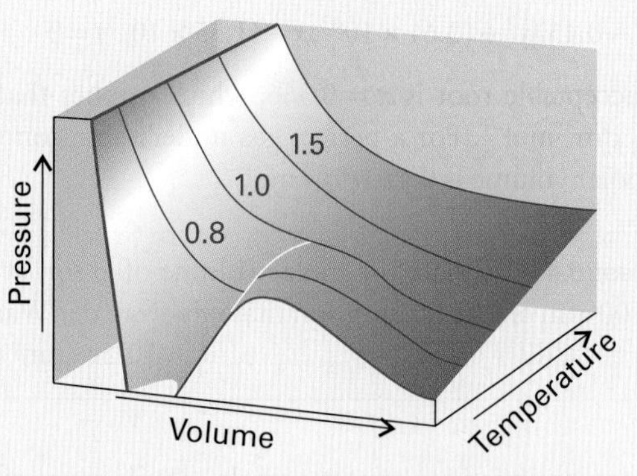

Fig. 8.29 The surface of possible states allowed by the van der Waals equation. The curves are labelled with the reduced temperature, $T_r = T/T_c$.

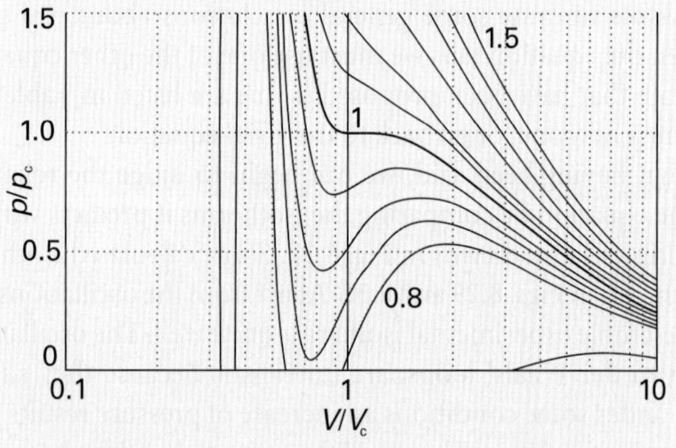

Fig. 8.30 Isotherms calculated by using the van der Waals equation of state. The axes are labelled with the 'reduced pressure', p/p_c, and 'reduced volume', V/V_c, where $p_c = a/27b^2$ and $V_c = 3b$. The individual isotherms are labelled with the 'reduced temperature', T/T_c, where $T_c = 8a/27Rb$. The van der Waals' loops are normally replaced by horizontal straight lines.

interActivity Calculate the molar volume of chlorine gas on the basis of the van der Waals equation of state at 250 K and 150 kPa and calculate the percentage difference from the value predicted by the perfect gas equation.

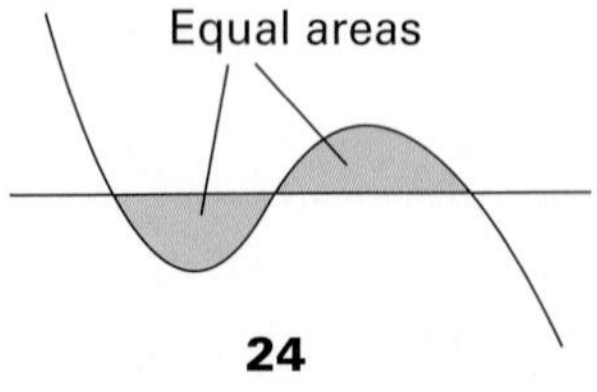

24

The principal features of the van der Waals equation can be summarized as follows.

1. Perfect gas isotherms are obtained at high temperatures and large molar volumes.

When the temperature is high, RT may be so large that the first term in eqn 8.24b greatly exceeds the second. Furthermore, if the molar volume is large in the sense $V_m \gg b$, then the denominator $V_m - b \approx V_m$. Under these conditions, the equation reduces to $p = RT/V_m$, the perfect gas equation.

2. Liquids and gases coexist when cohesive and dispersing effects are in balance.

The van der Waals loops occur when both terms in eqn 8.24b have similar magnitudes. The first term arises from the kinetic energy of the molecules and their repulsive interactions; the second represents the effect of the attractive interactions.

(c) Experimental studies of molecular interactions in gases

Molecular interactions in the gas phase can be studied in **molecular beams**, which consist of a collimated, narrow stream of molecules travelling though an evacuated vessel. The beam is directed towards other molecules, and the scattering that occurs on impact is related to the intermolecular interactions.

The primary experimental information from a molecular beam experiment is the fraction of the molecules in the incident beam that is scattered into a particular direction; we limit discussion here to non-reactive scattering. The fraction is normally expressed in terms of dI, the rate at which molecules are scattered into a cone that represents the area covered by the 'eye' of the detector (Fig. 8.31). This rate is reported as the **differential scattering cross-section**, σ, the constant of proportionality between the value of dI and the intensity, I, of the incident beam, the number density of target molecules, $\mathcal{N}$, and the infinitesimal path length dx through the sample:

$$dI = \sigma I \mathcal{N} dx \tag{8.25}$$

The value of σ (which has the dimensions of area) depends on the **impact parameter**, b, the initial perpendicular separation of the paths of the colliding molecules (Fig. 8.32), and the details of the intermolecular potential. The role of the impact parameter is most easily seen by considering the impact of two hard spheres (Fig. 8.33). If $b = 0$, the lighter projectile is on a trajectory that leads to a head-on collision, so the only scattering intensity is detected when the detector is at $\theta = \pi$. When the impact parameter

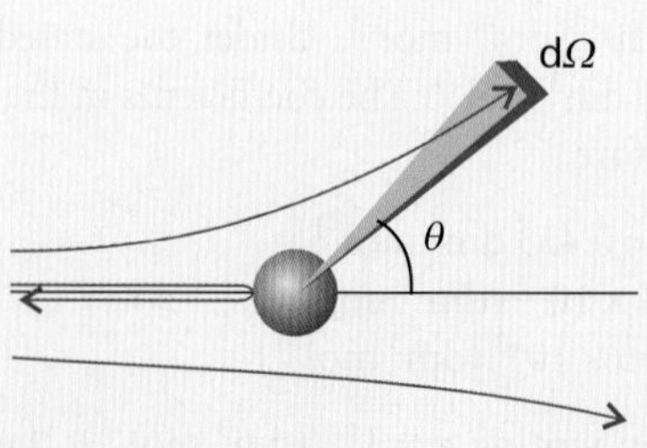

Fig. 8.31 The definition of the solid angle, dΩ, for scattering.

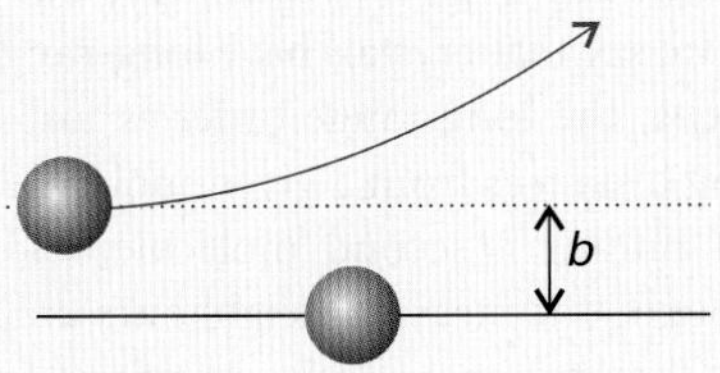

Fig. 8.32 The definition of the impact parameter, b, as the perpendicular separation of the initial paths of the particles.

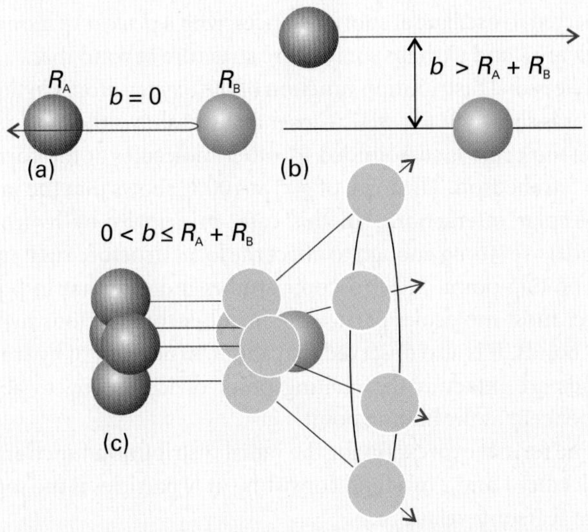

Fig. 8.33 Three typical cases for the collisions of two hard spheres: (a) $b = 0$, giving backward scattering; (b) $b > R_A + R_B$, giving forward scattering; (c) $0 < b \leq R_A + R_B$, leading to scattering into one direction on a ring of possibilities. (The target molecule is taken to be so heavy that it remains virtually stationary.)

is so great that the spheres do not make contact ($b > R_A + R_B$), there is no scattering and the scattering cross-section is zero at all angles except $\theta = 0$. Glancing blows, with $0 < b \leq R_A + R_B$, lead to scattering intensity in cones around the forward direction.

The scattering pattern of real molecules, which are not hard spheres, depends on the details of the intermolecular potential, including the anisotropy that is present when the molecules are non-spherical. The scattering also depends on the relative speed of approach of the two particles: a very fast particle might pass through the interaction region without much deflection, whereas a slower one on the same path might be temporarily captured and undergo considerable deflection (Fig. 8.34). The variation of the scattering cross-section with the relative speed of approach should therefore give information about the strength and range of the intermolecular potential.

A further point is that the outcome of collisions is determined by quantum, not classical, mechanics. The wave nature of the particles can be taken into account, at least to some extent, by drawing all classical trajectories that take the projectile particle from source to detector, and then considering the effects of interference between them.

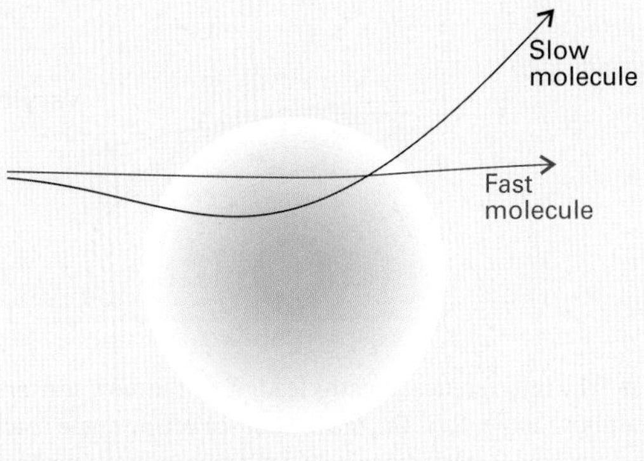

Fig. 8.34 The extent of scattering may depend on the relative speed of approach as well as the impact parameter. The dark central zone represents the repulsive core; the fuzzy outer zone represents the long-range attractive potential.

Two quantum mechanical effects are of great importance. A particle with a certain impact parameter might approach the attractive region of the potential in such a way that the particle is deflected towards the repulsive core (Fig. 8.35), which then repels it out through the attractive region to continue its flight in the forward direction. Some molecules, however, also travel in the forward direction because they have impact parameters so large that they are undeflected. The wavefunctions of the particles that take the two types of path interfere, and the intensity in the forward direction is modified. The effect is called **quantum oscillation.** The same phenomenon accounts for the optical 'glory effect', in which a bright halo can sometimes be seen surrounding an illuminated object. (The coloured rings around the shadow of an aircraft cast on clouds by the Sun, and often seen in flight, are an example of an optical glory.)

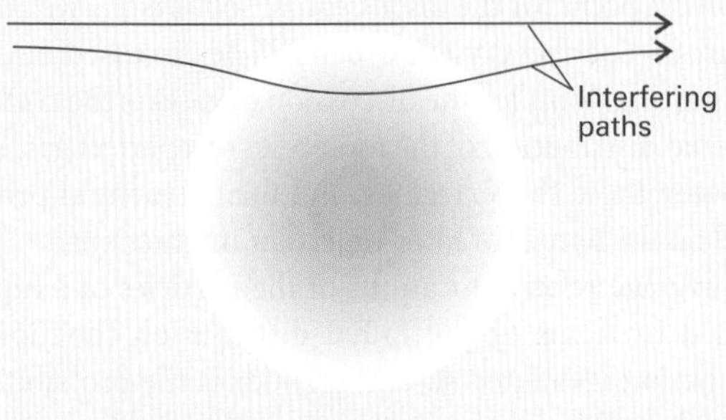

Fig. 8.35 Two paths leading to the same destination will interfere quantum mechanically; in this case they give rise to quantum oscillations in the forward direction.

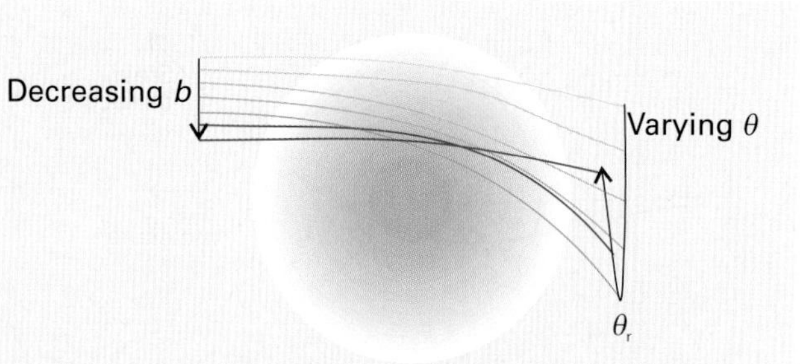

Fig. 8.36 The interference of paths leading to rainbow scattering. The rainbow angle, θ_r, is the maximum scattering angle reached as b is decreased. Interference between the numerous paths at that angle modifies the scattering intensity markedly.

The second quantum effect we need consider is the observation of a strongly enhanced scattering in a nonforward direction. This effect is called **rainbow scattering** because the same mechanism accounts for the appearance of an optical rainbow. The origin of the phenomenon is illustrated in Fig. 8.36. As the impact parameter decreases, there comes a stage at which the scattering angle passes through a maximum and the interference between the paths results in a strongly scattered beam. The **rainbow angle**, θ_r, is the angle for which $d\theta/db = 0$ and the scattering is strong.

Another phenomenon that can occur in certain beams is the capturing of one species by another. The vibrational temperature in supersonic beams is so low that **van der Waals molecules** may be formed, which are complexes of the form AB in which A and B are held together by van der Waals forces or hydrogen bonds. Large numbers of such molecules have been studied spectroscopically, including ArHCl, $(HCl)_2$, $ArCO_2$, and $(H_2O)_2$. More recently, van der Waals clusters of water molecules have been pursued as far as $(H_2O)_6$. The study of their spectroscopic properties gives detailed information about the intermolecular potentials involved.

8.8 Molecular interactions in liquids

The starting point for the discussion of solids is the well ordered structure of a perfect crystal, which will be discussed in Chapter 9. The starting point for the discussion of gases is the completely disordered distribution of the molecules of a perfect gas. Liquids lie between these two extremes, and their structural properties depend on the nature of intermolecular interactions.

The average relative locations of the particles of a liquid are expressed in terms of the **radial distribution function**, $g(r)$. This function is defined so that $g(r)r^2dr$ is the probability that a molecule will be found in the range dr at a distance r from another molecule. In a perfect crystal, $g(r)$ is a periodic array of sharp spikes, representing the certainty (in the absence of defects and thermal motion) that molecules (or ions) lie at definite locations. This regularity continues out to the edges of the crystal, so we say that crystals have **long-range order**. When the crystal melts, the long-range order is lost and, wherever we look at long distances from a given molecule, there is equal probability of finding a second molecule. Close to the first molecule, though, the nearest neighbours might still adopt approximately their original relative positions and, even if they are displaced by newcomers, the new particles might adopt their vacated positions. It is still possible to detect a sphere of nearest neighbours at a distance r_1, and perhaps beyond them a sphere of next-nearest neighbours at r_2. The existence of this **short-range order** means that the radial distribution function can be expected to oscillate at short distances, with a peak at r_1, a smaller peak at r_2, and perhaps some more structure beyond that.

The radial distribution function of the oxygen atoms in liquid water is shown in Fig. 8.37. Closer analysis shows that any given H_2O molecule is surrounded by other molecules at the corners of a tetrahedron. The form of $g(r)$ at 100°C shows that the intermolecular interactions (in this case, principally by hydrogen bonds) are strong enough to affect the local structure right up to the boiling point. Spectroscopic studies indicate that in liquid water most molecules participate in either three or four hydrogen bonds. It is also observed that about 90 per cent of hydrogen bonds are intact at the melting point of ice, falling to about 20 per cent at the boiling point.

The formal expression for the radial distribution function for molecules 1 and 2 in a fluid consisting of N particles is the somewhat fearsome relation

$$g(r_{12}) = \frac{\int\int\cdots\int e^{-\beta V_N} d\tau_3 d\tau_4 \cdots d\tau_N}{N^2 \int\int\cdots\int e^{-\beta V_N} d\tau_1 d\tau_2 \cdots d\tau_N} \qquad (8.26)$$

where $\beta = 1/kT$ and V_N is the N-particle potential energy, which depends on intermolecular interactions. There are several ways

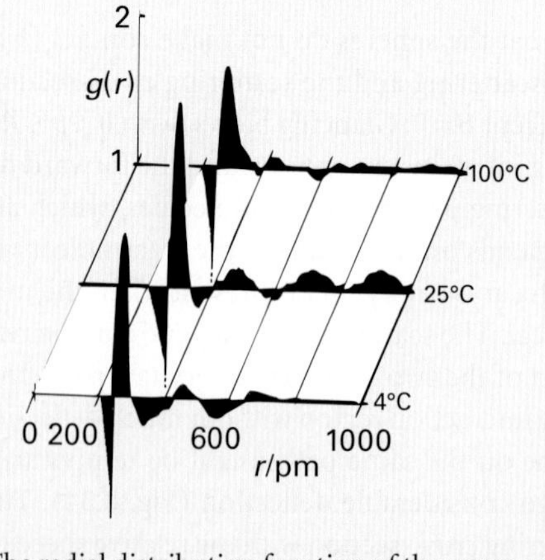

Fig. 8.37 The radial distribution function of the oxygen atoms in liquid water at three temperatures. Note the expansion as the temperature is raised. (A.H. Narten *et al., Discuss. Faraday. Soc.* 43, 97 (1967).)

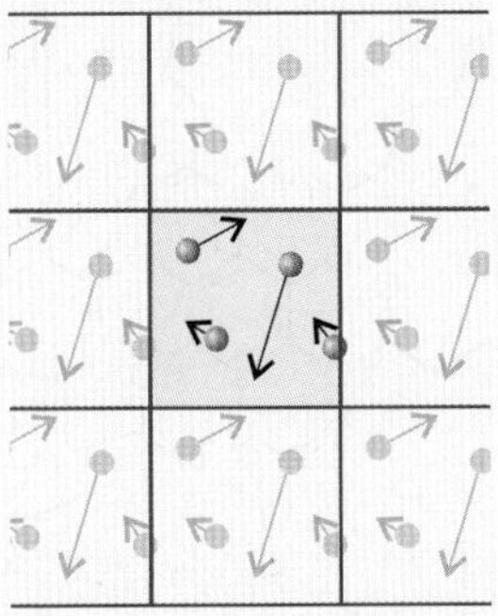

Fig. 8.38 In a two-dimensional simulation of a liquid that uses periodic boundary conditions, when one particle leaves the cell its mirror image enters through the opposite face.

of building the intermolecular potential into the calculation of $g(r)$. Numerical methods take a box of about 10^3 particles (the number increases as computers grow more powerful), and the rest of the liquid is simulated by surrounding the box with replications of the original box (Fig. 8.38). Then, whenever a particle leaves the box through one of its faces, its image arrives through the opposite face. When calculating the interactions of a molecule in a box, it interacts with all the molecules in the box and all the periodic replications of those molecules and itself in the other boxes.

In the **Monte Carlo method**, the particles in the box are moved through small but otherwise random distances, and the change in total potential energy of the N particles in the box, ΔV_N, is calculated using one of the intermolecular potentials discussed in this chapter. Whether or not this new configuration is accepted is then judged from the following rules:

1. If the potential energy is not greater than before the change, then the configuration is accepted.

If the potential energy is greater than before the change, then it is necessary to check if the new configuration is reasonable and can exist in equilibrium with configurations of lower potential energy at a given temperature. To make progress, we use the result that, at equilibrium, the ratio of populations of two states with energy separation ΔV_N is given by the Boltzmann distribution, here written as $e^{-\Delta V_N/kT}$. Because we are testing the viability of a configuration with a higher potential energy than the previous configuration in the calculation, $\Delta V_N > 0$ and the exponential factor varies between 0 and 1. In the Monte Carlo method, the second rule, therefore, is:

2. The exponential factor is compared with a random number between 0 and 1; if the factor is larger than the random number, then the configuration is accepted; if the factor is not larger, the configuration is rejected.

The configurations generated with Monte Carlo calculations can be used to construct $g(r)$ simply by counting the number of pairs of particles with a separation r and averaging the result over the whole collection of configurations.

In the **molecular dynamics** approach, the history of an initial arrangement is followed by calculating the trajectories of all the particles under the influence of the intermolecular potentials. To appreciate what is involved, we consider the motion of a particle in one dimension. We show in the following *Justification* that after a time interval Δt the position of a particle changes from x_{i-1} to a new value x_i given by

$$x_i = x_{i-1} + v_{i-1}\Delta t \tag{8.27a}$$

where v_{i-1} is the velocity of the atom when it was at x_{i-1}, its location at the start of the interval. The velocity at x_i is related to v_{i-1}, the velocity at the start of the interval, by

$$v_i = v_{i-1} - m^{-1}\left.\frac{dV_N(x)}{dx}\right|_{x_{i-1}}\Delta t \tag{8.27b}$$

where the derivative of the potential energy $V_N(x)$ is evaluated at x_{i-1}. The time interval Δt is approximately 1 fs (10^{-15} s), which is shorter than the average time between collisions. The calculation of x_i and v_i is then repeated for tens of thousands of such steps. The time-consuming part of the calculation is the evaluation of the net force on the molecule arising from all the other molecules present in the system.

Justification 8.5 *Particle trajectories according to molecular dynamics*

Consider a particle of mass m moving along the x direction with an initial velocity v_1 given by

$$v_1 = \frac{\Delta x}{\Delta t}$$

If the initial and new positions of the atom are x_1 and x_2, then $\Delta x = x_2 - x_1$ and

$$x_2 = x_1 + v_1\Delta t$$

The particle moves under the influence of a force arising from interactions with other atoms in the molecule. From Newton's second law of motion, we write the force F_1 at x_1 as

$$F_1 = ma_1$$

where the acceleration a_1 at x_1 is given by $a_1 = \Delta v/\Delta t$. If the initial and new velocities are v_1 and v_2, then $\Delta v = v_2 - v_1$ and

$$v_2 = v_1 + a_1\Delta t = v_1 + \frac{F_1}{m}\Delta t$$

Because $F = -dV/dx$, the force acting on the atom is related to the potential energy of interaction with other nearby atoms, the potential energy $V_N(x)$, by

$$F_1 = -\left.\frac{dV_N(x)}{dx}\right|_{x_1}$$

where the derivative is evaluated at x_1. It follows that

$$v_2 = v_1 - m^{-1}\left.\frac{dV_N(x)}{dx}\right|_{x_1} \Delta t$$

This expression generalizes to eqn 8.27b for the calculation of a velocity v_i from a previous velocity v_{i-1}.

Self-test 8.5 Consider a particle of mass m connected to a stationary wall with a spring of force constant k. Write an expression for the velocity of this particle once it is set into motion in the x direction from an equilibrium position x_0.

$[v_i = v_{i-1} + (k/m)(x_{i-1} - x_0)\Delta t]$

A molecular dynamics calculation gives a series of snapshots of the liquid, and $g(r)$ can be calculated as before. The temperature of the system is inferred by computing the mean kinetic energy of the particles and using the equipartition result that

$$\langle \tfrac{1}{2}mv_q^2\rangle = \tfrac{1}{2}kT \qquad (8.28)$$

for each coordinate q.

IMPACT ON MATERIALS SCIENCE
I8.3 Liquid crystals

A *mesophase* is a phase intermediate between solid and liquid. A mesophase may arise when molecules are markedly non-spherical, such as being long and thin (25) or disc-like (26). When the solid melts, some aspects of the long-range order characteristic of the solid may be retained, and the new phase may be a *liquid crystal*, a substance having liquid-like imperfect long-range order in at least one direction in space but positional or orientational order in at least one other direction. *Calamitic liquid crystals* (from the Greek word for reed) are made from long and thin molecules whereas *discotic liquid crystals* are made from disc-like molecules. A *thermotropic* liquid crystal displays a transition to the liquid crystalline phase as the temperature is changed. A *lyotropic* liquid crystal is a solution that undergoes a transition to the liquid crystalline phase as the composition is changed.

One type of retained long-range order gives rise to a *smectic phase* (from the Greek word for soapy) in which the molecules align themselves in layers (Fig. 8.39). Other materials, and some smectic liquid crystals at higher temperatures, lack the layered structure but retain a parallel alignment; this mesophase is called a *nematic phase* (from the Greek for thread, which refers to the observed defect structure of the phase). In the *cholesteric phase* (from the Greek for bile solid) the molecules lie in sheets at angles that change slightly between each sheet. That is, they form helical structures with a pitch that depends on the temperature. As a result, cholesteric liquid crystals diffract light and have colours that depend on the temperature. Disc-like molecules such as (26) can form nematic and *columnar* mesophases. In the latter, the aromatic rings stack one on top of the other and are separated by very small distances (less than 0.5 nm).

The optical properties of nematic liquid crystals are anisotropic, meaning that they depend on the relative orientation of the molecular assemblies with respect to the polarization of the incident beam of light. Nematic liquid crystals also respond in special ways to electric fields. Together, these unique optical and

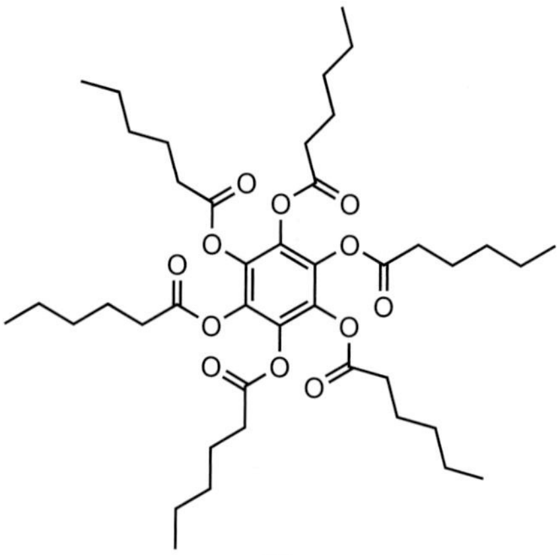

26

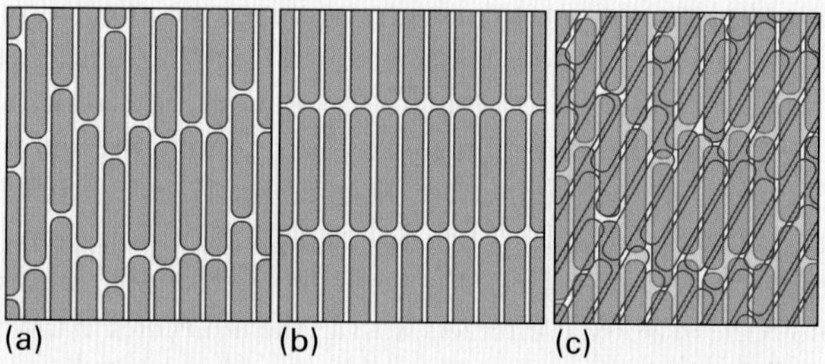

Fig. 8.39 The arrangement of molecules in (a) the nematic phase, (b) the smectic phase, and (c) the cholesteric phase of liquid crystals. In the cholesteric phase, the stacking of layers continues to give a helical arrangement of molecules.

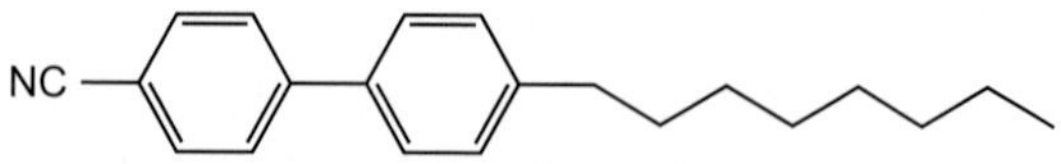

25

electrical properties form the basis of operation of liquid crystal displays (LCDs). In a 'twisted nematic' LCD, the liquid crystal is held between two flat plates about 10 μm apart. The inner surface of each plate is coated with a transparent conducting material, such as indium–tin oxide. The plates also have a surface that causes the liquid crystal to adopt a particular orientation at its interface and are typically set at 90° to each other but 270° in a 'supertwist' arrangement. The entire assembly is set between two polarizers, optical filters that allow light of one specific plane of polarization to pass. The incident light passes through the outer polarizer; then its plane of polarization is rotated as it passes through the twisted nematic and, depending on the setting of the second polarizer, the light will pass through. When a potential difference is applied across the cell, the helical arrangement is lost and the plane of the light is no longer rotated and will be blocked by the second polarizer.

Checklist of key ideas

- ☐ 1. A van der Waals interaction between closed-shell molecules is inversely proportional to the sixth power of their separation.
- ☐ 2. The permittivity is the quantity ε in the Coulomb potential energy, $V = Q_1Q_2/4\pi\varepsilon r$.
- ☐ 3. A polar molecule is a molecule with a permanent electric dipole moment; the magnitude of a dipole moment is the product of the partial charge and the separation.
- ☐ 4. The potential energy of the dipole–dipole interaction between two fixed (non-rotating) molecules is proportional to $\mu_1\mu_2/r^3$ and that between molecules that are free to rotate is proportional to $\mu_1^2\mu_2^2/kTr^6$.
- ☐ 5. The polarizability is a measure of the ability of an electric field to induce a dipole moment in a molecule ($\mu = \alpha\mathcal{E}$).
- ☐ 6. The dipole–induced-dipole interaction between two molecules is proportional to $\mu_1^2\alpha_2/r^6$, where α is the polarizability.
- ☐ 7. The potential energy of the dispersion (or London) interaction is proportional to $\alpha_1\alpha_2/r^6$.
- ☐ 8. A hydrogen bond is an interaction of the form A—H···B, where A and B are N, O, or F.
- ☐ 9. The Lennard-Jones (12,6) potential, $V = 4\varepsilon\{(r_0/r)^{12} - (r_0/r)^6\}$, is a model of the total intermolecular potential energy.
- ☐ 10. In real gases, molecular interactions affect the equation of state; the true equation of state is expressed in terms of virial coefficients $B, C, \ldots$: $pV_m = RT(1 + B/V_m + C/V_m^2 + \cdots)$.
- ☐ 11. The van der Waals equation of state is an approximation to the true equation of state in which attractions are represented by a parameter a and repulsions are represented by a parameter b: $p = nRT/(V - nb) - a(n/V)^2$.
- ☐ 12. A molecular beam is a collimated, narrow stream of molecules travelling though an evacuated vessel. Molecular beam techniques are used to investigate molecular interactions in gases.
- ☐ 13. The radial distribution function, $g(r)$, is defined so that $g(r)r^2\mathrm{d}r$ is the probability that a molecule will be found in the range $\mathrm{d}r$ at a distance r from another molecule in a liquid.

Further information

Further information 8.1 *The dipole–dipole interaction*

An important problem in physical chemistry is the calculation of the potential energy of interaction between two point dipoles with moments $\boldsymbol{\mu}_1$ and $\boldsymbol{\mu}_2$, separated by a vector $\boldsymbol{r}$. From classical electromagnetic theory, the potential energy of $\boldsymbol{\mu}_2$ in the electric field $\boldsymbol{\mathcal{E}}_1$ generated by $\boldsymbol{\mu}_1$ is given by the dot (scalar) product

$$V = -\boldsymbol{\mathcal{E}}_1 \cdot \boldsymbol{\mu}_2 \qquad (8.29)$$

To calculate $\boldsymbol{\mathcal{E}}_1$, we consider a distribution of point charges Q_i located at x_i, y_i, and z_i from the origin, that is, at the location $\boldsymbol{r}_i$. The Coulomb potential ϕ due to this distribution at a point $\boldsymbol{r}$ with coordinates x, y, and z is:

$$\phi = \sum_i \frac{Q_i}{4\pi\varepsilon_0} \frac{1}{\{(x - x_i)^2 + (y - y_i)^2 + (z - z_i)^2\}^{1/2}} \qquad (8.30)$$

If we suppose that all the charges are close to the origin (in the sense that $r_i \ll r$), we can use a Taylor expansion (*Mathematical background 1*) to write

$$\phi(\boldsymbol{r}) = \sum_i \frac{Q_i}{4\pi\varepsilon_0} \times \left\{ \frac{1}{r} + \left(\frac{\{(x - x_i)^2 + (y - y_i)^2 + (z - z_i)^2\}^{1/2}}{x_i} \right)_{x_i=0} x_i + \cdots \right\} \qquad (8.31)$$

$$= \sum_i \frac{Q_i}{4\pi\varepsilon_0} \left\{ \frac{1}{r} + \frac{x x_i}{r^3} + \cdots \right\}$$

where the unwritten terms include those arising from derivatives with respect to y_i and z_i and higher derivatives. If the charge distribution is electrically neutral, the first term disappears because $\sum_i Q_i = 0$. Next we note that $\sum_i Q_i x_i = \mu_x$, and likewise for the *y*- and *z*-components. That is,

$$\phi = \frac{1}{4\pi\varepsilon_0 r^3}(\mu_x x + \mu_y y + \mu_z z) = \frac{1}{4\pi\varepsilon_0 r^3}\boldsymbol{\mu}_1 \cdot \boldsymbol{r} \qquad (8.32)$$

The electric field strength is $\mathcal{E} = \nabla\phi$, so

$$\mathcal{E}_1 = \frac{1}{4\pi\varepsilon_0}\nabla\frac{\boldsymbol{\mu}_1 \cdot \boldsymbol{r}}{r^3} = -\frac{\boldsymbol{\mu}_1}{4\pi\varepsilon_0 r^3} - \frac{\boldsymbol{\mu}_1 \cdot \boldsymbol{r}}{4\pi\varepsilon_0}\nabla\frac{1}{r^3} \qquad (8.33)$$

It follows from eqns 8.29 and 8.33 that

$$V = \frac{\boldsymbol{\mu}_1 \cdot \boldsymbol{\mu}_2}{4\pi\varepsilon_0 r^3} - 3\frac{(\boldsymbol{\mu}_1 \cdot \boldsymbol{r})(\boldsymbol{\mu}_2 \cdot \boldsymbol{r})}{4\pi\varepsilon_0 r^5} \qquad (8.34)$$

For the arrangement shown in (**13**), in which $\boldsymbol{\mu}_1 \cdot \boldsymbol{r} = \mu_1 r \cos\theta$ and $\boldsymbol{\mu}_2 \cdot \boldsymbol{r} = \mu_2 r \cos\theta$, eqn 8.34 becomes:

$$V = \frac{\mu_1\mu_2 f(\theta)}{4\pi\varepsilon_0 r^3} \qquad f(\theta) = 1 - 3\cos^2\theta \qquad (8.35)$$

which is eqn 8.8.

Further information 8.2 *The basic principles of molecular beams*

The basic arrangement for a molecular beam experiment is shown in Fig. 8.40. If the pressure of vapour in the source is increased so that the mean free path of the molecules in the emerging beam is much shorter than the diameter of the pinhole, many collisions take place even outside the source. The net effect of these collisions, which give rise to **hydrodynamic flow**, is to transfer momentum into the direction of the beam. The molecules in the beam then travel with very similar speeds, so further downstream few collisions take place between them. This condition is called **molecular flow**. Because the spread in speeds is so small, the molecules are effectively in a state of very low translational temperature (Fig. 8.41). The translational temperature may reach as low as 1 K. Such jets are called **supersonic** because the average speed of the molecules in the jet is much greater than the speed of sound for the molecules that are not part of the jet.

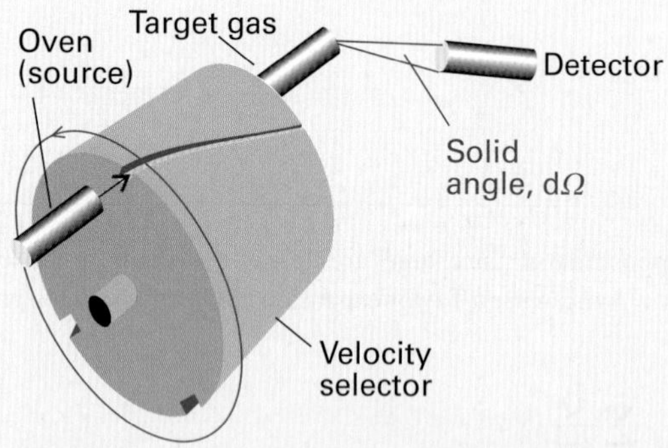

Fig. 8.40 The basic arrangement of a molecular beam apparatus. The atoms or molecules emerge from a heated source, and pass through the velocity selector, a rotating slotted cylinder. Molecules emanating from the source travel in a beam towards the rotating channels. Only if the speed of a molecule is such as to carry it along the channel that rotates into its path will it collide with the target gas. The scattering occurs from the target gas (which might take the form of another beam), and the flux of particles entering the detector set at some angle is recorded.

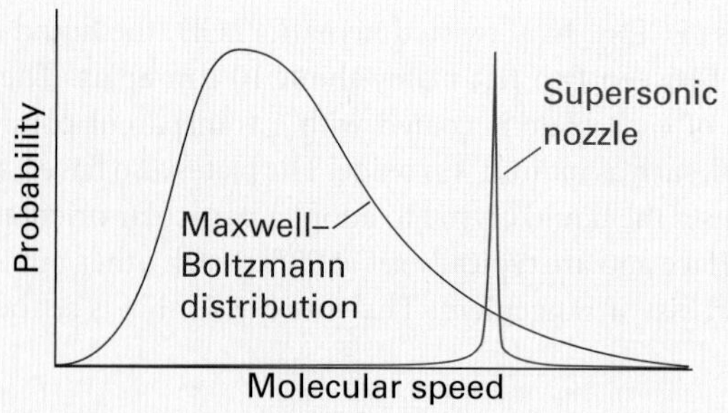

Fig. 8.41 The shift in the mean speed and the width of the distribution brought about by use of a supersonic nozzle.

A supersonic jet can be converted into a more parallel **supersonic beam** if it is 'skimmed' in the region of hydrodynamic flow and the excess gas pumped away. A skimmer consists of a conical nozzle shaped to avoid any supersonic shock waves spreading back into the gas and so increasing the translational temperature (Fig. 8.42). A jet or beam may also be formed by using helium or neon as the principal gas, and injecting molecules of interest into it in the hydrodynamic region of flow.

The low translational temperature of the molecules is reflected in the low rotational and vibrational temperatures of the molecules. In this context, a rotational or vibrational temperature means the temperature that should be used in the Boltzmann distribution to reproduce the observed populations of the states. However, as rotational modes equilibrate more slowly, and vibrational modes equilibrate even more slowly, the rotational and vibrational populations of the species correspond to somewhat higher temperatures, of the order of 10 K for rotation and 100 K for vibrations.

The target gas may be either a bulk sample or another molecular beam. The latter **crossed beam technique** gives a lot of information because the states of both the target and projectile molecules may be controlled. The intensity of the incident beam is measured by the

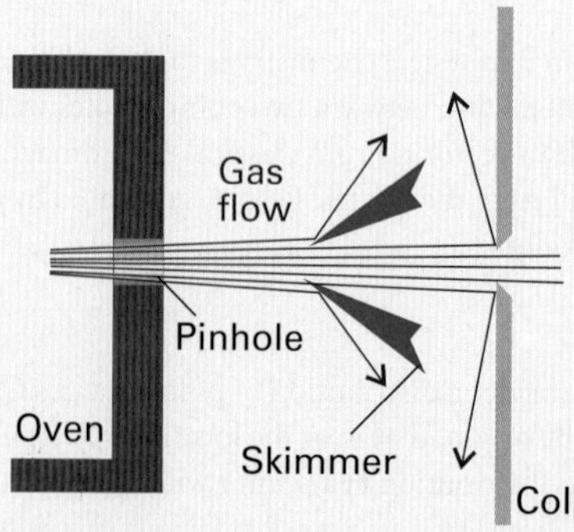

Fig. 8.42 A supersonic nozzle skims off some of the molecules of the beam and leads to a beam with well defined velocity.

incident beam flux, I, which is the number of particles passing through a given area in a given interval divided by the area and the duration of the interval.

The detectors may consist of a chamber fitted with a sensitive pressure gauge, a bolometer (a detector that responds to the incident energy by making use of the temperature-dependence of resistance) or an ionization detector, in which the incoming molecule is first ionized and then detected electronically. The state of the scattered molecules may also be determined spectroscopically, and is of interest when the collisions change their vibrational or rotational states.

Discussion questions

8.1 Explain how the permanent dipole moment and the polarizability of a molecule arise.

8.2 Identify the terms in and limit the generality of the following expressions: (a) $V=-Q_2\mu_1/4\pi\varepsilon_0 r^2$, (b) $V=-Q_2\mu_1\cos\theta/4\pi\varepsilon_0 r^2$, and (c) $V=\mu_2\mu_1(1-3\cos^2\theta)/4\pi\varepsilon_0 r^3$.

8.3 Draw examples of the arrangement of electrical charges that correspond to a monopole, dipole, quadrupole, and octupole and suggest a reason for the different distance dependencies of their electric fields.

8.4 Account for the theoretical conclusion that many attractive interactions between molecules vary with their separation as $1/r^6$.

8.5 Describe the formation of a hydrogen bond in terms of (a) electrostatic interactions and (b) molecular orbitals. How would you identify the better model?

8.6 Account for the formation of colloidal particles in terms of the balance between attractive and repulsive interactions between constituent atoms or molecules.

8.7 Explain why the critical micelle concentration of sodium dodecyl sulfate in aqueous solution decreases when sodium chloride is added.

8.8 Explain how the compression factor of a real gas varies with pressure and temperature and describe how it reveals information about intermolecular interactions in real gases.

8.9 Describe and criticize the formulation of the van der Waals equation.

8.10 Describe how molecular beams are used to investigate intermolecular potentials.

8.11 Compare and contrast the general features of the radial distribution function for a perfect crystal and a liquid (such as water).

8.12 Distinguish between the smectic, nematic, cholesteric, and columnar phases of a liquid crystal.

8.13 Some polymers have unusual properties. For example, Kevlar (27) is strong enough to be the material of choice for bulletproof vests and is stable at temperatures up to 600 K. What molecular interactions contribute to the formation and thermal stability of this polymer?

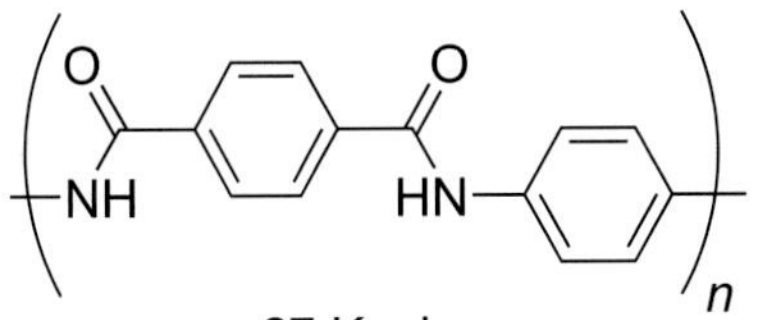

27 Kevlar

Exercises

8.1(a) Which of the following molecules may be polar: ClF_3, O_3, H_2O_2?

8.1(b) Which of the following molecules may be polar: SO_3, XeF_4, SF_4?

8.2(a) Calculate the resultant of two dipole moments of magnitude 1.5 D and 0.80 D that make an angle of 109.5° to each other.

8.2(b) Calculate the resultant of two dipole moments of magnitude 2.5 D and 0.50 D that make an angle of 120° to each other.

8.3(a) Calculate the magnitude and direction of the dipole moment of the following arrangement of charges in the xy-plane: $3e$ at (0, 0), $-e$ at (0.32 nm, 0), and $-2e$ at an angle of 20° from the x-axis and a distance of 0.23 nm from the origin.

8.3(b) Calculate the magnitude and direction of the dipole moment of the following arrangement of charges in the xy-plane: $4e$ at (0, 0), $-2e$ at (162 pm, 0), and $-2e$ at an angle of 300° from the x-axis and a distance of 143 pm from the origin.

8.4(a) Calculate the molar energy required to reverse the direction of an H_2O molecule located 100 pm from a Li^+ ion. Take the dipole moment of water as 1.85 D.

8.4(b) Calculate the molar energy required to reverse the direction of an HCl molecule located 300 pm from a Mg^{2+} ion. Take the dipole moment of HCl as 1.08 D.

8.5(a) Calculate the potential energy of the interaction between two linear quadrupoles when they are collinear and their centres are separated by a distance r.

8.5(b) Calculate the potential energy of the interaction between two linear quadrupoles when they are parallel and separated by a distance r.

8.6(a) The polarizability volume of H_2O is $1.48\times10^{-30}\ m^3$; calculate the dipole moment of the molecule (in addition to the permanent dipole moment) induced by an applied electric field of strength $1.0\ kV\ cm^{-1}$.

8.6(b) The polarizability volume of NH_3 is $2.22\times10^{-30}\ m^3$; calculate the dipole moment of the molecule (in addition to the permanent dipole moment) induced by an applied electric field of strength $15.0\ kV\ m^{-1}$.

8.7(a) Estimate the energy of the dispersion interaction (use the London formula) for two He atoms separated by 1.0 nm. Relevant data can be found in the *Data section*.

8.7(b) Estimate the energy of the dispersion interaction (use the London formula) for two Ar atoms separated by 1.0 nm. Relevant data can be found in the *Data section*.

8.8(a) How much energy (in kJ mol^{-1}) is required to break the hydrogen bond in a vacuum ($\varepsilon_r = 1$)? Use the electrostatic model of the hydrogen bond.

8.8(b) How much energy (in kJ mol^{-1}) is required to break the hydrogen bond in water ($\varepsilon_r \approx 80.0$)? Use the electrostatic model of the hydrogen bond.

8.9(a) What pressure would 3.15 g of nitrogen gas in a vessel of volume 2.05 dm^3 exert at 273 K if it obeyed the virial equation of state (8.22b)? What would be the pressure if it were a perfect gas?

8.9(b) What pressure would 4.56 g of carbon dioxide gas in a vessel of volume 2.25 dm^3 exert at 273 K if it obeyed the virial equation of state (8.22b)? What would be the pressure if it were a perfect gas?

8.10(a) What pressure would 4.56 g of carbon dioxide gas in a vessel of volume 2.25 dm^3 exert at its Boyle temperature?

8.10(b) What pressure would 3.01 g of oxygen gas in a vessel of volume 2.20 dm^3 exert at its Boyle temperature?

8.11(a) (a) What pressure would 131 g of xenon gas in a vessel of volume 1.0 dm^3 exert at 25°C if it behaved as a perfect gas? (b) What pressure would it exert if it behaved as a van der Waals gas?

8.11(b) (a) What pressure would 25 g of argon gas in a vessel of volume 1.5 dm^3 exert at 30°C if it behaved as a perfect gas? (b) What pressure would it exert if it behaved as a van der Waals gas?

8.12(a) Express the van der Waals parameters $a = 0.751$ atm dm^6 mol^{-2} and $b = 0.0226$ dm^3 mol^{-1} in SI base units.

8.12(b) Express the van der Waals parameters $a = 1.32$ atm dm^6 mol^{-2} and $b = 0.0436$ dm^3 mol^{-1} in SI base units.

8.13(a) A gas at 250 K and 12 atm has a molar volume 8.0 per cent smaller than that calculated from the perfect gas law. Calculate (a) the compression factor under these conditions and (b) the molar volume of the gas. Which are dominating in the sample, the attractive or the repulsive forces?

8.13(b) A gas at 350 K and 15 atm has a molar volume 15 per cent larger than that calculated from the perfect gas law. Calculate (a) the compression factor under these conditions and (b) the molar volume of the gas. Which are dominating in the sample, the attractive or the repulsive forces?

8.14(a) In an industrial process, nitrogen is heated to 500 K at a constant volume of 1.000 m^3. The gas enters the container at 300 K and 100 atm. The mass of the gas is 92.4 kg. Use the van der Waals equation to determine the approximate pressure of the gas at its working temperature of 500 K. For nitrogen, $a = 1.39$ dm^6 atm mol^{-2}, $b = 0.0391$ dm^3 mol^{-1}.

8.14(b) Cylinders of compressed gas are typically filled to a pressure of 200 bar. For oxygen, what would be the molar volume at this pressure and 25°C based on (a) the perfect gas equation, (b) the van der Waals equation? For oxygen, $a = 1.360$ dm^6 atm mol^{-2}, $b = 3.183 \times 10^{-2}$ dm^3 mol^{-1}.

8.15(a) Use the van der Waals parameters for chlorine to estimate its Boyle temperature.

8.15(b) Use the van der Waals parameters for hydrogen sulfide to estimate its Boyle temperature.

8.16(a) Use the van der Waals parameters for chlorine to estimate the radius of a Cl_2 molecule regarded as a sphere.

8.16(b) Use the van der Waals parameters for hydrogen sulfide to estimate the radius of a H_2S molecule regarded as a sphere.

8.17(a) A certain gas obeys the van der Waals equation with $a = 0.50$ m^6 Pa mol^{-2}. Its volume is found to be 5.00×10^{-4} m^3 mol^{-1} at 273 K and 3.0 MPa. From this information calculate the van der Waals constant b.

8.17(b) A certain gas obeys the van der Waals equation with $a = 0.76$ m^6 Pa mol^{-2}. Its volume is found to be 4.00×10^{-4} m^3 mol^{-1} at 288 K and 4.0 MPa. From this information calculate the van der Waals constant b.

8.18(a) What is the compression factor for the gas described in Exercise 8.17a at the prevailing temperature and pressure?

8.18(b) What is the compression factor for the gas described in Exercise 8.17b at the prevailing temperature and pressure?

Problems*

Numerical problems

8.1 Suppose an H_2O molecule ($\mu = 1.85$ D) approaches an anion. What is the favourable orientation of the molecule? Calculate the electric field (in volts per metre) experienced by the anion when the water dipole is (a) 1.0 nm, (b) 0.3 nm, (c) 30 nm from the ion.

8.2 The electric dipole moment of toluene (methylbenzene) is 0.4 D. Estimate the dipole moments of the three xylenes (dimethylbenzene). Which answer can you be sure about?

8.3 Plot the magnitude of the electric dipole moment of hydrogen peroxide as the H—O—O—H (azimuthal) angle ϕ changes from 0 to 2π. Use the dimensions shown in (28).

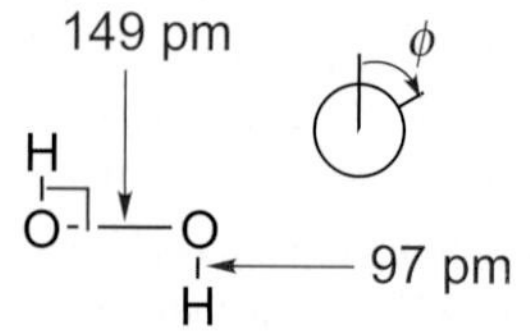

28

8.4 Acetic acid vapour contains a proportion of planar, hydrogen-bonded dimers (29). The apparent dipole moment of molecules in pure gaseous acetic acid increases with increasing temperature. Suggest an interpretation of this observation.

* Problems denoted with the symbol ‡ were supplied by Charles Trapp and Carmen Giunta.

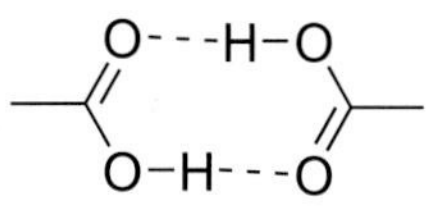

29

8.5 An H_2O molecule is aligned by an external electric field of strength 1.0 kV m^{-1} and an Ar atom ($\alpha' = 1.66 \times 10^{-30}$ m^3) is brought up slowly from one side. At what separation is it energetically favourable for the H_2O molecule to flip over and point towards the approaching Ar atom?

8.6 The relative permittivity of a substance is large if its molecules are polar or highly polarizable. The quantitative relation between the relative permittivity, the polarizability, and the permanent dipole moment of the molecule is expressed by the *Debye equation*

$$\frac{\varepsilon_r - 1}{\varepsilon_r + 2} = \frac{\rho N_A}{3M\varepsilon_0}\left(\alpha + \frac{\mu^2}{3kT}\right)$$

where ρ is the mass density of the sample, and M is the molar mass of the molecules. The relative permittivity of camphor (**30**) was measured at a series of temperatures with the results given below. Determine the dipole moment and the polarizability volume of the molecule. *Hint.* Plot the data in such a way that a fit to a straight line yields the permanent dipole moment and polarizability from the slope and y-intercept, respectively

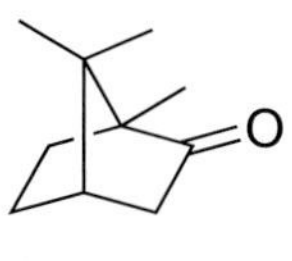

30 Camphor

θ/°C	0	20	40	60	80	100	120	140	160	200
ρ/(g cm^{-3})	0.99	0.99	0.99	0.99	0.99	0.99	0.97	0.96	0.95	0.91
ε_r	12.5	11.4	10.8	10.0	9.50	8.90	8.10	7.60	7.11	6.21

8.7 The magnitude of the electric field at a distance r from a point charge Q is equal to $Q/4\pi\varepsilon_0 r^2$. How close to a water molecule (of polarizability volume 1.48×10^{-30} m^3) must a proton approach before the dipole moment it induces is equal to the permanent dipole moment of the molecule (1.85 D)?

8.8 Show that the minimum in the Lennard-Jones potential well occurs at the separation $r_e = 2^{1/6} r_0$.

8.9‡ Nelson *et al.* (*Science* **238**, 1670 (1987)) examined several weakly bound gas-phase complexes of ammonia in search of examples in which the H atoms in NH_3 formed hydrogen bonds, but found none. For example, they found that the complex of NH_3 and CO_2 has the carbon atom nearest the nitrogen (299 pm away): the CO_2 molecule is at right angles to the C–N 'bond', and the H atoms of NH_3 are pointing away from the CO_2. The permanent dipole moment of this complex is reported as 1.77 D. If the N and C atoms are the centres of the negative and positive charge distributions, respectively, what is the magnitude of those partial charges (as multiples of e)?

8.10 Given that $F = -dV/dr$, calculate the distance dependence of the force acting between two non-bonded groups of atoms in a polymer chain that have a London dispersion interaction with each other.

8.11 Consider the arrangement shown in **31** for a system consisting of an O—H group and an O atom, and then use the electrostatic model of the hydrogen bond to calculate the dependence of the molar potential energy of interaction on the angle θ.

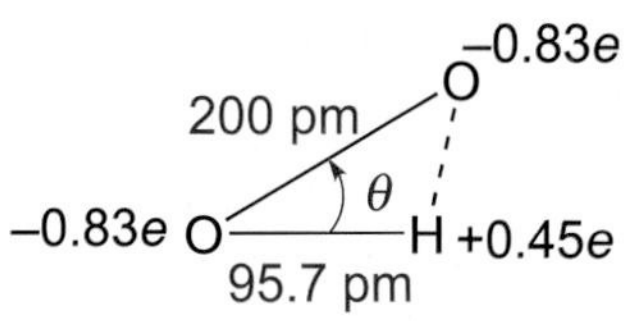

31

8.12 The theory of the stability of lyophobic dispersions was developed by B. Derjaguin and L. Landau and independently by E. Verwey and J. T. G. Overbeek, and is known as the *DLVO theory*. It assumes that there is a balance between the repulsive interaction between the charges of the electric double layers on neighbouring particles and the attractive interactions arising from van der Waals interactions between the molecules in the particles. The potential energy arising from the repulsion of double layers on particles of radius a has the form

$$V_{\text{repulsion}} = +\frac{Aa^2\zeta^2}{R}e^{-s/r_D}$$

where A is a constant, ζ is the electrokinetic potential, R is the separation of centres, s is the separation of the surfaces of the two particles, and r_D is the thickness of the double layer. This expression is valid for small particles with a thick double layer ($a \ll r_D$). When the double layer is thin ($r_D \ll a$), the expression is replaced by

$$V_{\text{repulsion}} = \tfrac{1}{2}Aa\zeta^2\ln(1 + e^{-s/r_D})$$

The potential energy arising from the attractive interaction has the form

$$V_{\text{attraction}} = -\frac{B}{s}$$

where B is another constant. (a) Plot $V_{\text{total}} = V_{\text{repulsion}} + V_{\text{attraction}}$ against s for the case $a \ll r_D$, with $A = 5.0 \times 10^{-7}$ J m^{-1} V^{-2}, $a = 1.0 \times 10^{-9}$ m, $r_D = 1.0 \times 10^{-7}$ m, $\zeta = 0.050$ V, and $B = 1.0 \times 10^{-28}$ J m. Coagulation is signalled by a sharp dip in this plot. Identify the portion of the plot that may be ascribed to the onset of coagulation. (b) Generate a series of plots of V_{total} against s for cases where $r_D \ll a$. Flocculation is signalled by secondary minima in these plots. Find the ratio r_D/a below which both coagulation and flocculation occur.

8.13 Suppose that 10.0 mol C_2H_6(g) is confined to 4.860 dm^3 at 27°C. Use the perfect gas and van der Waals equations of state to calculate the compression factor based on these calculations. For ethane, $a = 5.489$ dm^6 atm mol^{-2}, $b = 0.06380$ dm^3 mol^{-1}.

8.14 At 300 K and 20 atm, the compression factor of a gas is 0.86. Calculate (a) the volume occupied by 8.2 mmol of the gas under these conditions and (b) an approximate value of the second virial coefficient B at 300 K.

8.15 At 273 K measurements on argon gave $B = -21.7$ cm^3 mol^{-1} and $C = 1200$ cm^6 mol^{-2}, where B and C are the second and third virial coefficients in the expansion of Z in powers of $1/V_m$. Assuming that the perfect gas law holds sufficiently well for the estimation of the second and third terms of the expansion, calculate the compression factor of argon at 100 atm and 273 K. From your result, estimate the molar volume of argon under these conditions.

8.16‡ The second virial coefficient of methane can be approximated by the empirical equation $B'(T) = a + be^{-c/T^2}$, where $a = -0.1993$ bar^{-1}, $b = 0.2002$ bar^{-1}, and $c = 1131$ K^2 with 300 K < T < 600 K. What is the Boyle temperature of methane?

8.17‡ A substance as elementary and well known as argon still receives research attention. A review of thermodynamic properties of argon has been published (R.B. Stewart and R.T. Jacobsen, *J. Phys. Chem. Ref. Data* **18**, 639 (1989)) which included the following 300 K isotherm.

p/MPa	0.4000	0.5000	0.6000	0.8000	1.000
V_m/(dm^3 mol^{-1})	6.2208	4.9736	4.1423	3.1031	2.4795
p/MPa	1.500	2.000	2.500	3.000	4.000
V_m/(dm^3 mol^{-1})	1.6483	1.2328	0.98357	0.81746	0.60998

(a) Compute the second virial coefficient, B, at this temperature. (b) Use non-linear curve-fitting software to compute the third virial coefficient, C, at this temperature.

Theoretical problems

8.18 The potential energy of a CH_3 group in ethane as it is rotated around the C—C bond can be written $V = \frac{1}{2}V_0(1 + \cos 3\phi)$, where ϕ is the azimuthal angle (**32**) and $V_0 = 11.6$ kJ mol^{-1}. (a) What is the change in potential energy between the *trans* and fully eclipsed conformations? (b) Show that for small variations in angle, the torsional (twisting) motion around the C—C bond can be expected to be that of a harmonic oscillator. (c) Estimate the vibrational frequency of this torsional oscillation.

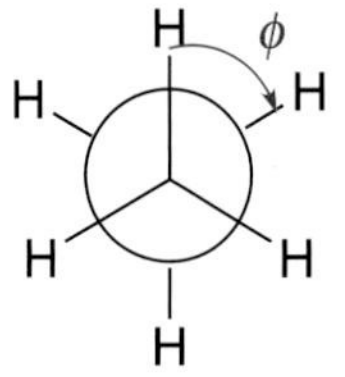

32

8.19 Use eqn 8.12 to calculate the polarizability of a one-dimensional harmonic oscillator in its ground state when the field is applied (a) perpendicular to, (b) parallel to the oscillator. You will need the following matrix elements:

$$\mu_{z,01} = \mu_{z,10} = e\left(\frac{\hbar}{2m\omega}\right)^{1/2}$$

All other matrix elements are zero. Suggest a reason why the polarizability is independent of the mass of the oscillator.

8.20 Use eqn 8.12 to compute the polarizibility of a hydrogen atom. For simplicity, confine the sum in eqn 8.12 to the np_z orbitals and use the following matrix element between the np_z and 1s orbitals:

$$\mu_{z,0n} = \mu_{z,n0} = -ea_0\left(\frac{2^8 n^7(n-1)^{2n-5}}{3(n+1)^{2n+5}}\right)^{1/2}$$

8.21 Show that the mean interaction energy of N atoms of diameter d interacting with a potential energy of the form C_6/R^6 is given by $U = -2N^2C_6/3Vd^3$, where V is the volume in which the molecules are confined and all effects of clustering are ignored. Hence, find a connection between the van der Waals parameter a and C_6, from $n^2a/V^2 = (\partial U/\partial V)_T$.

8.22 Suppose you distrusted the Lennard-Jones (12,6) potential for assessing a particular polypeptide conformation, and replaced the repulsive term by an exponential function of the form $e^{-r/\sigma}$. (a) Sketch the form of the potential energy and locate the distance at which it is a minimum. (b) Identify the distance at which the exponential-6 potential is a minimum.

8.23 Show that the van der Waals equation leads to values of $Z < 1$ and $Z > 1$, and identify the conditions for which these values are obtained.

8.24 Express the van der Waals equation of state as a virial expansion in powers of $1/V_m$ and obtain expressions for B and C in terms of the parameters a and b. The expansion you will need is $(1-x)^{-1} = 1 + x + x^2 + \cdots$. Measurements on argon gave $B = -21.7$ cm^3 mol^{-1} and $C = 1200$ cm^6 mol^{-2} for the virial coefficients at 273 K. What are the values of a and b in the corresponding van der Waals equation of state?

8.25 The second virial coefficient B' can be obtained from measurements of the density ρ of a gas at a series of pressures. Show that the graph of p/ρ against p should be a straight line with slope proportional to B'.

8.26 Derive an expression for the compression factor of a gas that obeys the equation of state $p(V - nb) = nRT$, where b and R are constants. If the pressure and temperature are such that $V_m = 10b$, what is the numerical value of the compression factor?

8.27 Consider the collision between a hard-sphere molecule of radius R_1 and mass m, and an infinitely massive impenetrable sphere of radius R_2. Plot the scattering angle θ as a function of the impact parameter b. Carry out the calculation using simple geometrical considerations.

8.28 The dependence of the scattering characteristics of atoms on the energy of the collision can be modelled as follows. We suppose that the two colliding atoms behave as impenetrable spheres, as in Problem 8.27, but that the effective radius of the heavy atoms depends on the speed v of the light atom. Suppose its effective radius depends on v as $R_2 e^{-v/v^*}$, where v^* is a constant. Take $R_1 = \frac{1}{2}R_2$ for simplicity and an impact parameter $b = \frac{1}{2}R_2$, and plot the scattering angle as a function of (a) speed, (b) kinetic energy of approach.

8.29 The *cohesive energy density*, $\mathcal{U}$, is defined as U/V, where U is the mean potential energy of attraction within the sample and V its volume. Show that $\mathcal{U} = \frac{1}{2}\mathcal{N}\int V(R)\,d\tau$, where $\mathcal{N}$ is the number density of the molecules and $V(R)$ is their attractive potential energy and where the integration ranges from d to infinity and over all angles. Go on to show that the cohesive energy density of a uniform distribution of molecules that interact by a van der Waals attraction of the form $-C_6/R^6$ is equal to $(2\pi/3)(N_A^2/d^3M^2)\rho^2C_6$, where ρ is the mass density of the solid sample and M is the molar mass of the molecules.

Applications: to biochemistry

8.30 Here we develop a molecular orbital theory treatment of the peptide link. (a) Use VB theory to explain why the peptide link is planar. (b) Taking a hint from VB theory, we can suspect that delocalization of the π bond between the oxygen, carbon, and nitrogen atoms can be modelled by making LCAO-MOs from 2p orbitals perpendicular to the plane defined by the atoms. The three combinations have the form:

$$\psi_1 = a\psi_O + b\psi_C + c\psi_N \qquad \psi_2 = d\psi_O - e\psi_N \qquad \psi_3 = f\psi_O - g\psi_C + h\psi_N$$

where the coefficients a through h are all positive. Sketch the orbitals ψ_1, ψ_2, and ψ_3 and characterize them as bonding, non-bonding, or antibonding molecular orbitals. (i) Show that this treatment is consistent only with a planar conformation of the peptide link. (ii) Draw a diagram showing the relative energies of these molecular orbitals and determine the occupancy of the orbitals. *Hint.* Convince yourself that there are four electrons to be distributed among the molecular orbitals. (iii) Now consider a non-planar conformation of the peptide link, in which the O2p and C2p orbitals are perpendicular to the plane defined by the O, C, and N atoms, but the N2p orbital lies on that plane. The LCAO-MOs are given by

$$\psi_4 = a\psi_O + b\psi_C \qquad \psi_5 = e\psi_N \qquad \psi_6 = f\psi_O - g\psi_C$$

Just as before, sketch these molecular orbitals and characterize them as bonding, non-bonding, or antibonding. Also, draw an energy level diagram and determine the occupancy of the orbitals. (iv) Why is this arrangement of atomic orbitals consistent with a non-planar conformation for the peptide link? (v) Does the bonding MO associated with the planar conformation have the same energy as the bonding MO associated with the non-planar conformation? If not, which bonding MO is lower in energy? Repeat the analysis for the non-bonding and antibonding molecular orbitals. (vi) Use your results from parts (i)–(v) to construct arguments that support the planar model for the peptide link.

8.31 Phenylalanine (Phe, **33**) is a naturally occurring amino acid. What is the energy of interaction between its phenyl group and the electric dipole moment of a neighbouring peptide group? Take the distance between the groups as 4.0 nm and treat the phenyl group as a benzene molecule. The dipole moment of the peptide group is $\mu = 2.7$ D and the polarizability volume of benzene is $\alpha' = 1.04 \times 10^{-29}$ m^3.

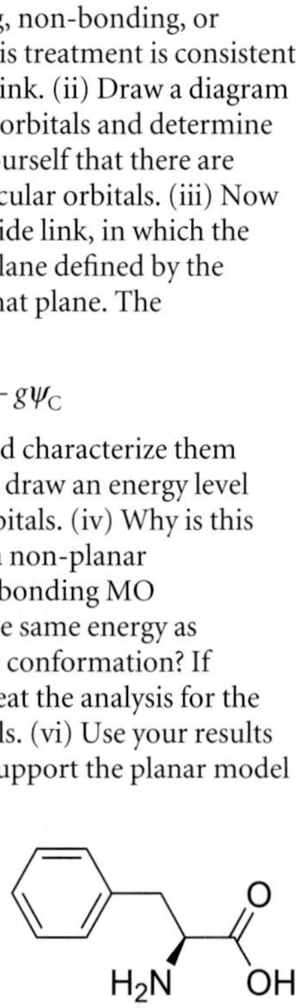

33 Phenylalanine

8.32 Now consider the London interaction between the phenyl groups of two Phe residues (see Problem 8.31). (a) Estimate the potential energy of interaction between two such rings (treated as benzene molecules) separated by 4.0 nm. For the ionization energy, use $I = 5.0$ eV. (b) Given that force is the negative slope of the potential, calculate the distance dependence of the force acting between two non-bonded groups of atoms, such as the phenyl groups of Phe, in a polypeptide chain that can have a London dispersion interaction with each other. What is the separation at which the force between the phenyl groups (treated as benzene molecules) of two Phe residues is zero? (*Hint.* Calculate the slope by considering the potential energy at r and $r + \delta r$, with $\delta r \ll r$, and evaluating $\{V(r+\delta r) - V(r)\}/\delta r$. At the end of the calculation, let δr become vanishingly small).

8.33 Molecular orbital calculations may be used to predict structures of intermolecular complexes. Hydrogen bonds between purine and pyrimidine bases are responsible for the double helix structure of DNA. Consider methyladenine (**34**, with $R = CH_3$) and methylthymine (**35**, with $R = CH_3$) as models of two bases that can form hydrogen bonds in DNA. (a) Using molecular modelling software and the computational method of your choice, calculate the atomic charges of all atoms in methyladenine and methylthymine. (b) Based on your tabulation of atomic charges, identify the atoms in methyladenine and methylthymine that are likely to participate in hydrogen bonds. (c) Draw all possible adenine–thymine pairs that can be linked by hydrogen bonds, keeping in mind that linear arrangements of the A—H···B fragments are preferred in DNA. For this step, you may want to use your molecular modelling software to align the molecules properly. (d) Which of the pairs that you drew in part (c) occur naturally in DNA molecules? (e) Repeat parts (a)–(d) for cytosine and guanine, which also form base pairs in DNA.

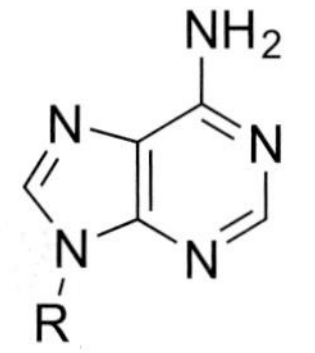

34

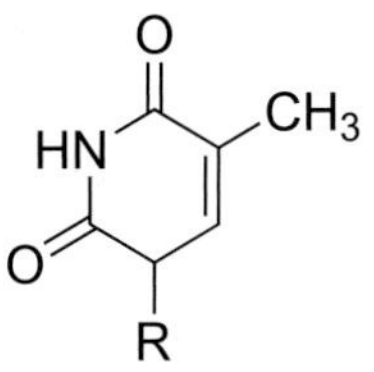

35

8.34 Molecular orbital calculations may be used to predict the dipole moments of molecules. (a) Using molecular modelling software and the computational method of your choice, calculate the dipole moment of the peptide link, modelled as a *trans-N*-methylacetamide (**36**). Plot the energy of interaction between these dipoles against the angle θ for $r = 3.0$ nm. (b) Compare the maximum value of the dipole–dipole interaction energy from part (a) to 20 kJ mol^{-1}, a typical value for the energy of a hydrogen-bonding interaction in biological systems.

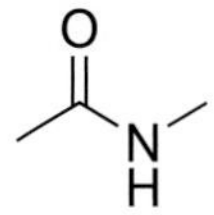

36 *trans-N*-methylacetamide

8.35 In this problem you will use molecular modelling software to gain some appreciation for the complexity of the calculations that lead to plots such as those in Fig. 8.14. A model for the protein is the dipeptide (**37**) in which the terminal methyl groups replace the rest of the polypeptide chain. (a) Draw three initial conformers of the dipeptide with R = H: one with $\phi = +75°$, $\psi = -65°$, a second with $\phi = \psi = +180°$, and a third with $\phi = +65°$, $\psi = +35°$. Use molecular modelling software to optimize the geometry of each conformer and measure the total potential energy and the final ϕ and ψ angles in each case. Did all of the initial conformers converge to the same final conformation? If not, what do these final conformers represent? Rationalize any observed differences in total potential energy of the final conformers. (b) Use the approach in part (a) to investigate the case $R = CH_3$, with the same three initial conformers as starting points for the calculations. Rationalize any similarities and differences between the final conformers of the dipeptides with R = H and $R = CH_3$. *Hint.* Although any molecular mechanics routine will work satisfactorily, one based on the AMBER force field is strongly recommended, as it is optimized for calculations on biopolymers.

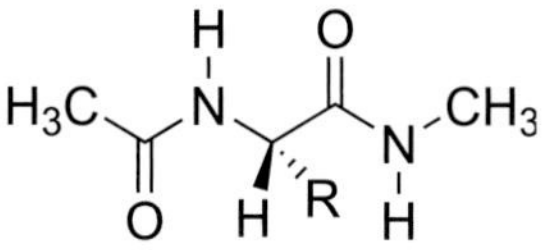

37

9 Solids

First, we see how to describe the regular arrangement of atoms in crystals and the symmetry of their arrangement. Then we consider the basic principles of X-ray diffraction and see how the diffraction pattern can be interpreted in terms of the distribution of electron density in a unit cell. X-ray diffraction leads to information about the structures of metallic, ionic, and molecular solids, and we review some typical results and their rationalization in terms of atomic and ionic radii. With structures established, we move on to the properties of solids, and see how their mechanical, electrical, optical, and magnetic properties stem from the properties of their constituent atoms and molecules.

The solid state includes most of the materials that make modern technology possible. It includes the many varieties of steel used in architecture and engineering, the semiconductors and metallic conductors that are used in information technology and power distribution, the ceramics that increasingly are replacing metals, and the synthetic and natural polymers that are used in the textile industry and in the fabrication of many of the common objects of the modern world. The properties of solids stem, of course, from the arrangement and properties of the constituent atoms, and one of the challenges of this chapter is to see how a wide range of bulk properties, including rigidity, electrical conductivity, and optical and magnetic properties, stems from the properties of atoms. One crucial aspect of this link is the pattern in which the atoms (and molecules) are stacked together, and we start this chapter with an examination of how the structures of solids are described and determined.

Crystal lattices

Early in the history of modern science it was suggested that the regular external form of crystals implied an internal regularity of their constituents. In this section we see how to describe the arrangement of atoms inside crystals.

9.1 Lattices and unit cells

A crystal is built up from regularly repeating 'structural motifs', which may be atoms, molecules, or groups of atoms, molecules, or ions. A **space lattice** is the pattern formed by points representing the locations of these motifs (Fig. 9.1). The space lattice is, in effect, an abstract scaffolding for the crystal structure. More formally, a space lattice is a three-dimensional, infinite array of points, each of which is surrounded in an identical way by its neighbours, and which defines the basic structure of the crystal.

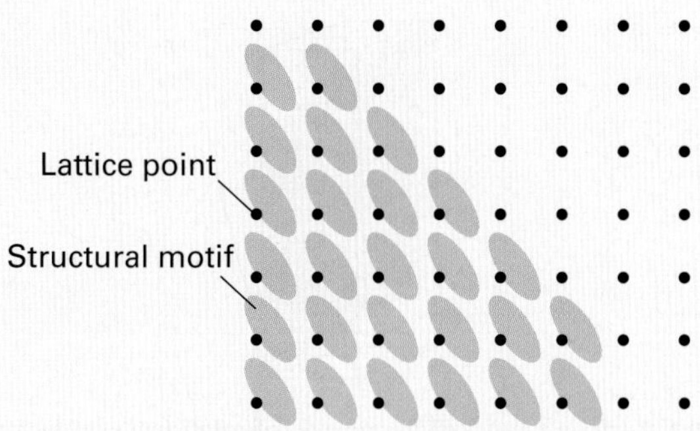

Fig. 9.1 Each lattice point specifies the location of a structural motif (for example, a molecule or a group of molecules). The crystal lattice is the array of lattice points; the crystal structure is the collection of structural motifs arranged according to the lattice.

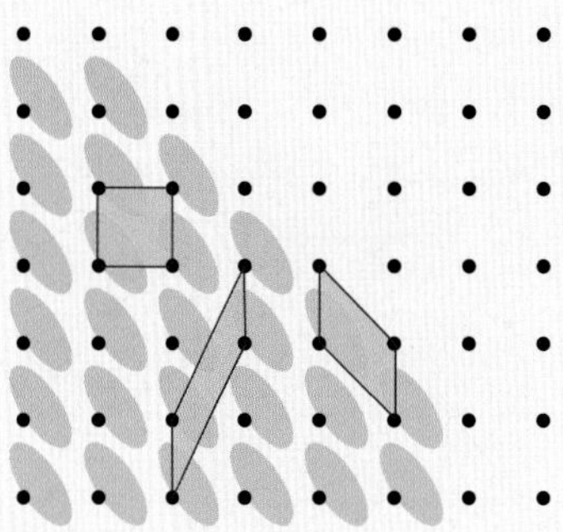

Fig. 9.3 A unit cell can be chosen in a variety of ways, as shown here. It is conventional to choose the cell that represents the full symmetry of the lattice. In this rectangular lattice, the rectangular unit cell would normally be adopted.

In some cases there may be a structural motif centred on each lattice point, but that is not necessary. The crystal structure itself is obtained by associating with each lattice point an identical structural motif.

The **unit cell** is an imaginary parallelepiped (parallel-sided figure) that contains one unit of the translationally repeating pattern (Fig. 9.2). A unit cell can be thought of as the fundamental region from which the entire crystal may be constructed by purely translational displacements (like bricks in a wall). A unit cell is commonly formed by joining neighbouring lattice points by straight lines (Fig. 9.3). Such unit cells are called **primitive**. It is sometimes more convenient to draw larger **non-primitive unit cells** that also have lattice points at their centres or on pairs of opposite faces. An infinite number of different unit cells can describe the same lattice, but the one with sides that have the shortest lengths and that are most nearly perpendicular to one another is normally chosen. The lengths of the sides of a unit cell are denoted a, b, and c, and the angles between them are denoted α, β, and γ (Fig. 9.4).

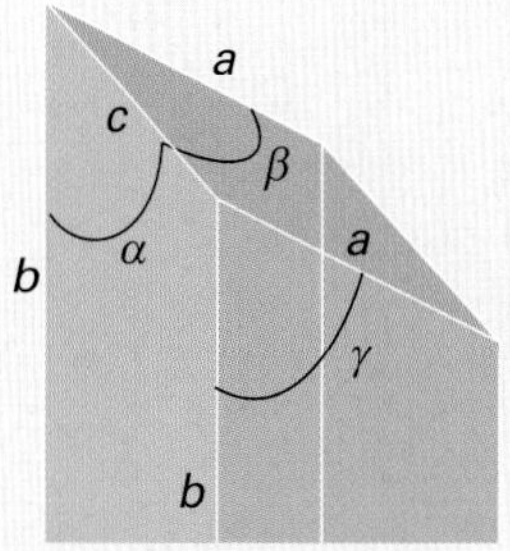

Fig. 9.4 The notation for the sides and angles of a unit cell. Note that the angle α lies in the plane (b,c) and perpendicular to the axis a.

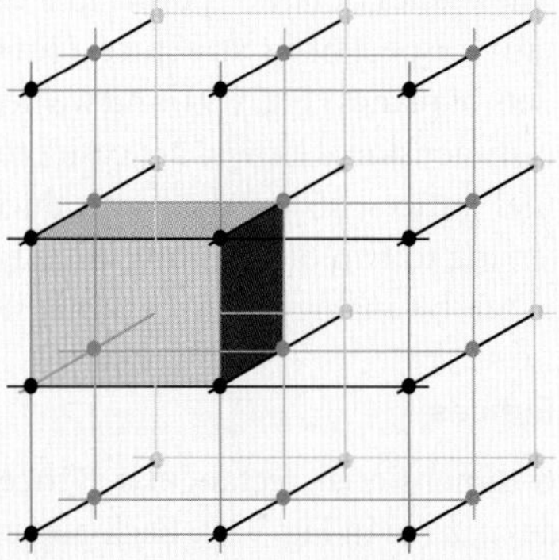

Fig. 9.2 A unit cell is a parallel-sided (but not necessarily rectangular) figure from which the entire crystal structure can be constructed by using only translations (not reflections, rotations, or inversions).

A brief comment A *symmetry operation* is an action (such as a rotation, reflection, or inversion) that leaves an object looking the same after it has been carried out. There is a corresponding *symmetry element* for each symmetry operation, which is the point, line, or plane with respect to which the symmetry operation is performed. For instance, an *n-fold rotation* (the symmetry operation) about an *n-fold axis of symmetry* (the corresponding symmetry element) is a rotation through $360°/n$. See Chapter 7 for a more detailed discussion of symmetry.

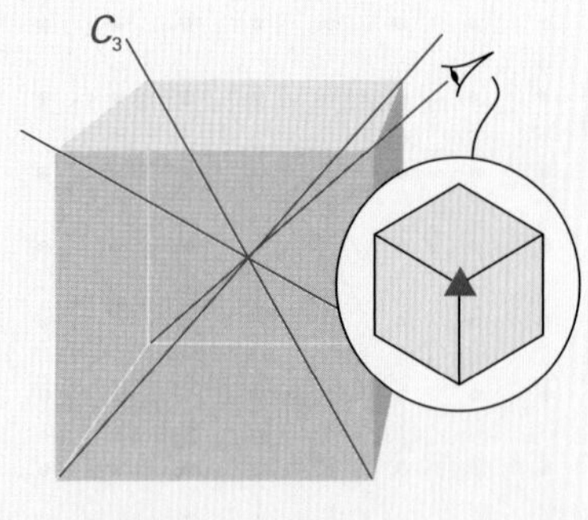

Fig. 9.5 A unit cell belonging to the cubic system has four threefold axes, denoted C_3, arranged tetrahedrally. The insert shows the threefold symmetry.

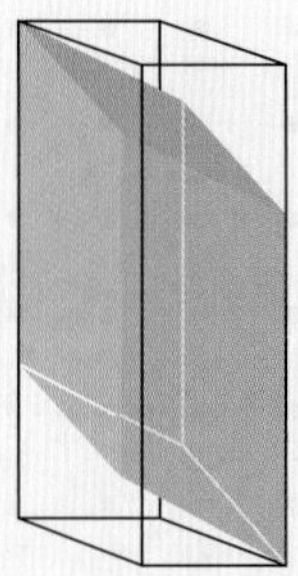

Fig. 9.7 A triclinic unit cell has no axes of rotational symmetry.

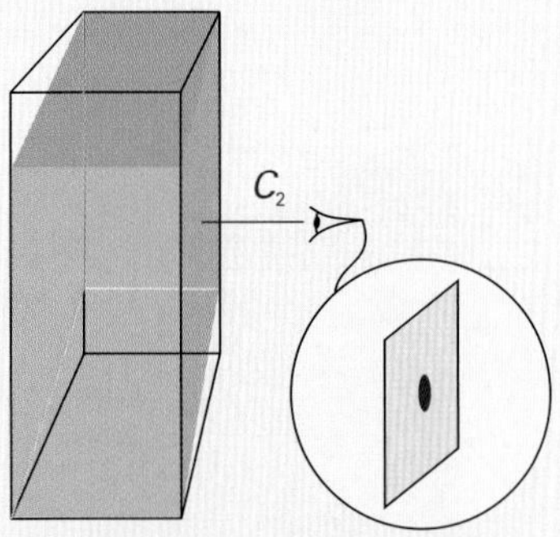

Fig. 9.6 A unit cell belonging to the monoclinic system has a twofold axis (denoted C_2 and shown in more detail in the insert).

Table 9.1 The seven crystal systems

System	Essential symmetries
Triclinic	None
Monoclinic	One C_2 axis
Orthorhombic	Three perpendicular C_2 axes
Rhombohedral	One C_3 axis
Tetragonal	One C_4 axis
Hexagonal	One C_6 axis
Cubic	Four C_3 axes in a tetrahedral arrangement

Unit cells are classified into seven **crystal systems** by noting the rotational symmetry elements they possess. A *cubic unit cell*, for example, has four threefold axes in a tetrahedral array (Fig. 9.5). A *monoclinic unit cell* has one twofold axis; the unique axis is by convention the *b* axis (Fig. 9.6). A *triclinic unit cell* has no rotational symmetry, and typically all three sides and angles are different (Fig. 9.7). Table 9.1 lists the **essential symmetries**, the elements that must be present for the unit cell to belong to a particular crystal system.

There are only 14 distinct space lattices in three dimensions. These **Bravais lattices** are illustrated in Fig. 9.8. It is conventional to portray these lattices by primitive unit cells in some cases and by non-primitive unit cells in others. A **primitive unit cell** (with lattice points only at the corners) is denoted P. A **body-centred unit cell** (I) also has a lattice point at its centre. A **face-centred unit cell** (F) has lattice points at its corners and also at the centres of its six faces. A **side-centred unit cell** (A, B, or C) has lattice points at its corners and at the centres of two opposite faces. For simple structures, it is often convenient to choose an atom belonging to the structural motif, or the centre of a molecule, as the location of a lattice point or the vertex of a unit cell, but that is not a necessary requirement.

9.2 The identification of lattice planes

The spacing of the planes of lattice points in a crystal is an important quantitative aspect of its structure. However, there are many different sets of planes (Fig. 9.9), and we need to be able to label them. Two-dimensional lattices are easier to visualize than three-dimensional lattices, so we shall introduce the concepts involved by referring to two dimensions initially, and then extend the conclusions by analogy to three dimensions.

(a) The Miller indices

Consider a two-dimensional rectangular lattice formed from a unit cell of sides a, b (as in Fig. 9.9). Each plane in the illustration (except the plane passing through the origin) can be distinguished by the distances at which it intersects the a and b axes. One way to label a plane would therefore be to quote the smallest intersection distances. For example, we could denote a representative plane of each type in the illustration as $(1a,1b)$,

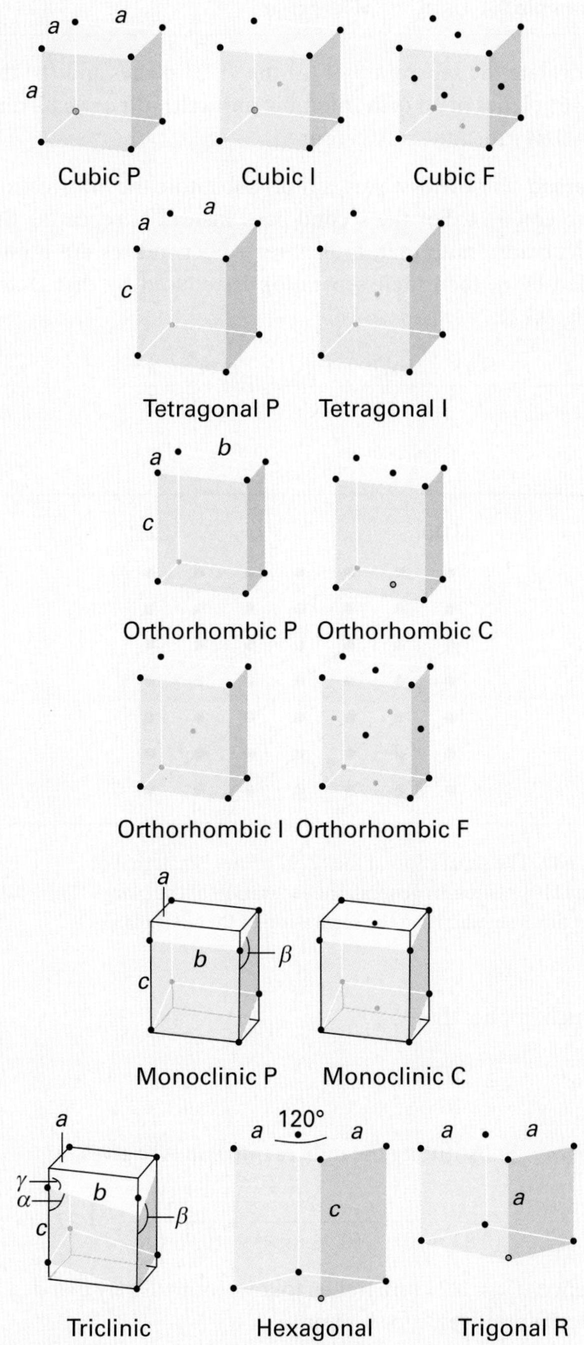

Fig. 9.8 The fourteen Bravais lattices. The points are lattice points, and are not necessarily occupied by atoms. P denotes a primitive unit cell (R is used for a trigonal lattice), I a body-centred unit cell, F a face-centred unit cell, and C (or A or B) a cell with lattice points on two opposite faces.

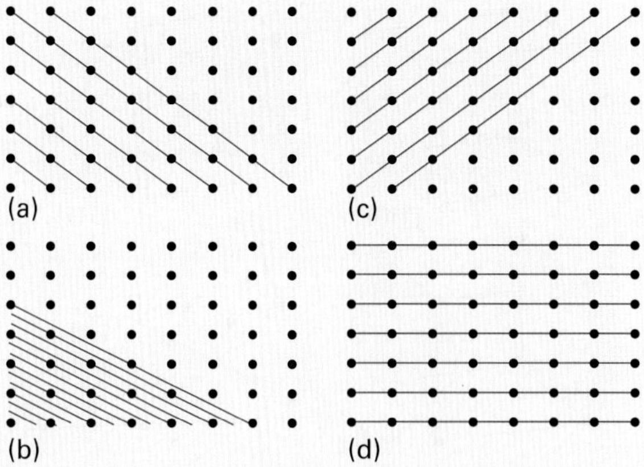

Fig. 9.9 Some of the planes that can be drawn through the points of a rectangular space lattice and their corresponding Miller indices (*hkl*): (a) (110), (b) (230), (c) ($\bar{1}$10), and (d) (010).

$(\frac{1}{2}a, \frac{1}{3}b)$, $(-1a, 1b)$, and $(\infty a, 1b)$. However, if we agree to quote distances along the axes as multiples of the lengths of the unit cell, then we can label the planes more simply as (1, 1), $(\frac{1}{2}, \frac{1}{3})$, (−1, 1), and (∞, 1). If the lattice in Fig. 9.9 is the top view of a three-dimensional orthorhombic lattice in which the unit cell has a length *c* in the *z*-direction, all four sets of planes intersect the *z*-axis at infinity. Therefore, the full labels are (1, 1, ∞), $(\frac{1}{2}, \frac{1}{3}, \infty)$, (−1, 1, ∞), and (∞, 1, ∞).

The presence of fractions and infinity in the labels is inconvenient. They can be eliminated by taking the reciprocals of the labels. As we shall see, taking reciprocals turns out to have further advantages. The **Miller indices**, (*hkl*), are the reciprocals of intersection distances (with fractions cleared by multiplying through by an appropriate factor, if taking the reciprocal results in a fraction). For example, the (1, 1, ∞) planes in Fig. 9.9a are the (110) planes in the Miller notation. Similarly, the $(\frac{1}{2}, \frac{1}{3}, \infty)$ planes are denoted (230). Negative indices are written with a bar over the number, and Fig. 9.9c shows the ($\bar{1}$10) planes. The Miller indices for the four types of plane in Fig. 9.9 are therefore (110), (230), ($\bar{1}$10), and (010). Figure 9.10 shows a three-dimensional representation of a selection of planes, including one in a lattice with non-orthogonal axes.

The notation (*hkl*) denotes an *individual* plane. To specify a *set* of parallel planes we use the notation {*hkl*}. Thus, we speak of the (110) plane in a lattice, and the set of all {110} planes that lie parallel to the (110) plane. A helpful feature to remember is that, the smaller the absolute value of *h* in {*hkl*}, the more nearly parallel the set of planes is to the *a* axis (the {*h*00} planes are an exception). The same is true of *k* and the *b* axis and *l* and the *c* axis. When $h = 0$, the planes intersect the *a* axis at infinity, so the {0*kl*} planes are parallel to the *a* axis. Similarly, the {*h*0*l*} planes are parallel to *b* and the {*hk*0} planes are parallel to *c*.

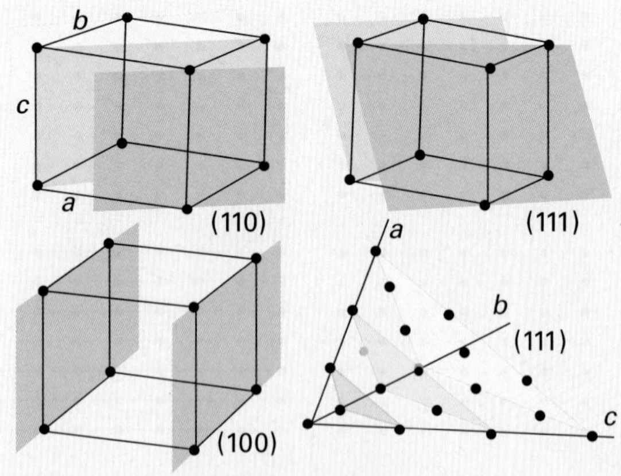

Fig. 9.10 Some representative planes in three dimensions and their Miller indices. Note that a 0 indicates that a plane is parallel to the corresponding axis, and that the indexing may also be used for unit cells with non-orthogonal axes.

(b) The separation of planes

The Miller indices are very useful for expressing the separation of planes. The separation of the $\{hk0\}$ planes in the square lattice shown in Fig. 9.11 is given by

$$\frac{1}{d_{hk0}^2}=\frac{h^2+k^2}{a^2} \quad \text{or} \quad d_{hk0}=\frac{a}{(h^2+k^2)^{1/2}} \tag{9.1a}$$

By extension to three dimensions, the separation of the $\{hkl\}$ planes of a cubic lattice is given by

$$\frac{1}{d_{hkl}^2}=\frac{h^2+k^2+l^2}{a^2} \quad \text{or} \quad d_{hkl}=\frac{a}{(h^2+k^2+l^2)^{1/2}} \tag{9.1b}$$

The corresponding expression for a general orthorhombic lattice is the generalization of this expression:

$$\frac{1}{d_{hkl}^2}=\frac{h^2}{a^2}+\frac{k^2}{b^2}+\frac{l^2}{c^2} \tag{9.1c}$$

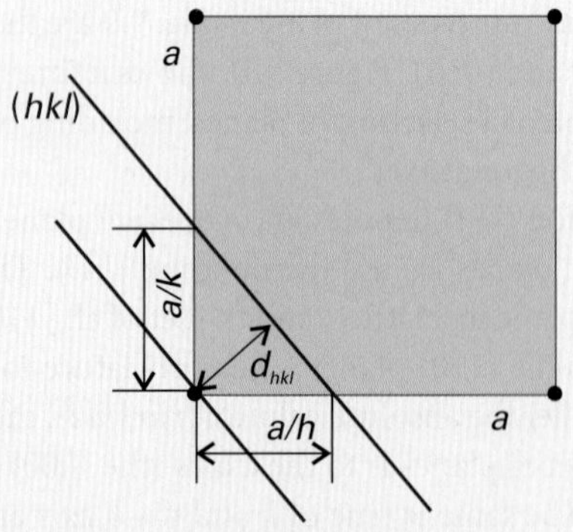

Fig. 9.11 The dimensions of a unit cell and their relation to the plane passing through the lattice points.

Example 9.1 *Using the Miller indices*

Calculate the separation of (a) the $\{123\}$ planes and (b) the $\{246\}$ planes of an orthorhombic unit cell with $a = 0.82$ nm, $b = 0.94$ nm, and $c = 0.75$ nm.

Method For the first part, simply substitute the information into eqn 9.1c. For the second part, instead of repeating the calculation, note that, if all three Miller indices are multiplied by n, then their separation is reduced by that factor (Fig. 9.12):

$$\frac{1}{d_{nh,nk,nl}^2}=\frac{(nh)^2}{a^2}+\frac{(nk)^2}{b^2}+\frac{(nl)^2}{c^2}=n^2\left(\frac{h^2}{a^2}+\frac{k^2}{b^2}+\frac{l^2}{c^2}\right)=\frac{n^2}{d_{hkl}^2}$$

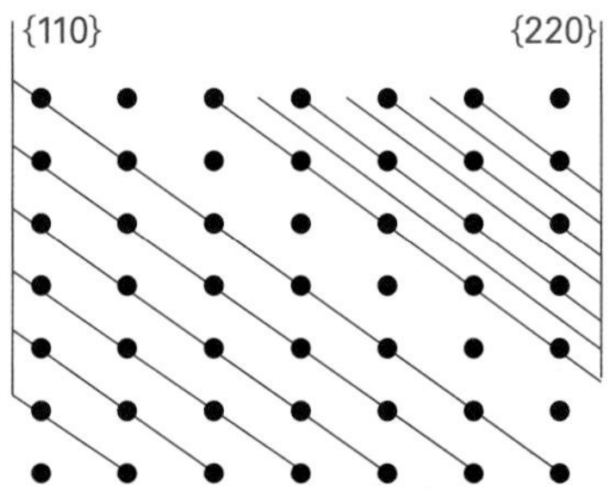

Fig. 9.12 The separation of the $\{220\}$ planes is half that of the $\{110\}$ planes. In general, the separation of the planes $\{nh,nk,nl\}$ is n times smaller than the separation of the $\{hkl\}$ planes.

which implies that

$$d_{nh,nk,nl}=\frac{d_{hkl}}{n}$$

Answer Substituting the indices into eqn 9.1c gives

$$\frac{1}{d_{123}^2}=\frac{1^2}{(0.82\text{ nm})^2}+\frac{2^2}{(0.94\text{ nm})^2}+\frac{3^2}{(0.75\text{ nm})^2}=22\text{ nm}^{-2}$$

Hence, $d_{123} = 0.21$ nm. It then follows immediately that d_{246} is one-half this value, or 0.11 nm.

A note on good practice It is always sensible to look for analytical relations between quantities rather than to evaluate expressions numerically each time, for that emphasizes the relations between quantities (and avoids unnecessary work).

Self-test 9.1 Calculate the separation of (a) the $\{133\}$ planes and (b) the $\{399\}$ planes in the same lattice.

[0.19 nm, 0.063 nm]

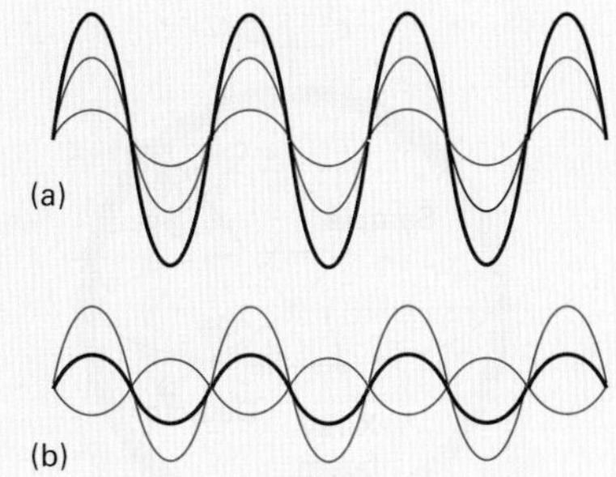

Fig. 9.13 When two waves are in the same region of space they interfere. Depending on their relative phase, they may interfere (a) constructively, to give an enhanced amplitude, or (b) destructively, to give a smaller amplitude. The component waves are shown in green and blue and the resultant in purple.

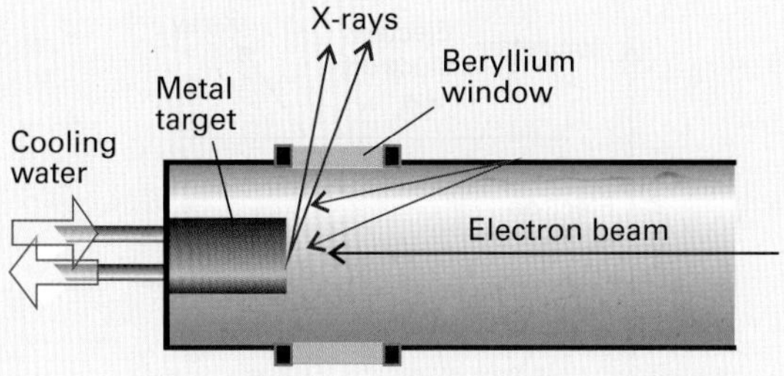

Fig. 9.14 X-rays are generated by directing an electron beam on to a cooled metal target. Beryllium is transparent to X-rays (on account of the small number of electrons in each atom) and is used for the windows.

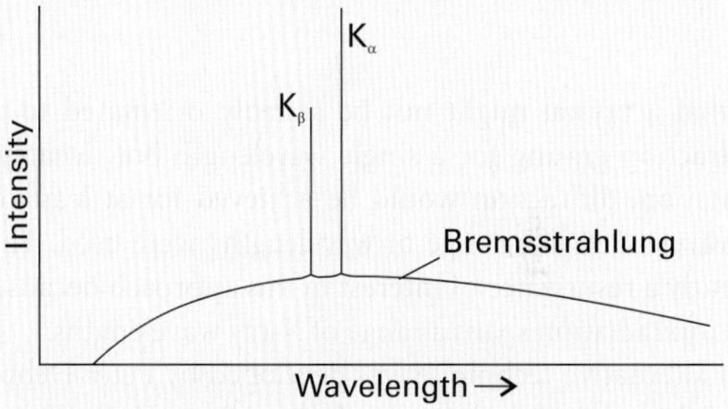

Fig. 9.15 The X-ray emission from a metal consists of a broad, featureless Bremsstrahlung background, with sharp transitions superimposed on it. The label K indicates that the radiation comes from a transition in which an electron falls into a vacancy in the K shell of the atom.

9.3 The investigation of structure

A characteristic property of waves is that they interfere with one another, giving a greater displacement where peaks or troughs coincide and a smaller displacement where peaks coincide with troughs (Fig. 9.13). According to classical electromagnetic theory, the intensity of electromagnetic radiation is proportional to the square of the amplitude of the waves. Therefore, the regions of constructive or destructive interference show up as regions of enhanced or diminished intensities. The phenomenon of **diffraction** is the interference caused by an object in the path of waves, and the pattern of varying intensity that results is called the **diffraction pattern**. Diffraction occurs when the dimensions of the diffracting object are comparable to the wavelength of the radiation.

(a) X-ray diffraction

Wilhelm Röntgen discovered X-rays in 1895. Seventeen years later, Max von Laue suggested that they might be diffracted when passed through a crystal, for by then he had realized that their wavelengths are comparable to the separation of lattice planes. This suggestion was confirmed almost immediately by Walter Friedrich and Paul Knipping and has grown since then into a technique of extraordinary power. The bulk of this section will deal with the determination of structures using X-ray diffraction. The mathematical procedures necessary for the determination of structure from X-ray diffraction data are enormously complex, but such is the degree of integration of computers into the experimental apparatus that the technique is almost fully automated, even for large molecules and complex solids. The analysis is aided by molecular modelling techniques, which can guide the investigation towards a plausible structure.

X-rays are electromagnetic radiation with wavelengths of the order of 10^{-10} m. They are typically generated by bombarding a metal with high-energy electrons (Fig. 9.14). The electrons decelerate as they plunge into the metal and generate radiation with a continuous range of wavelengths called **Bremsstrahlung** (*Bremse* is German for deceleration, *Strahlung* for ray). Superimposed on the continuum are a few high-intensity, sharp peaks (Fig. 9.15). These peaks arise from collisions of the incoming electrons with the electrons in the inner shells of the atoms. A collision expels an electron from an inner shell, and an electron of higher energy drops into the vacancy, emitting the excess energy as an X-ray photon (Fig. 9.16). If the electron falls into a K shell (a shell with $n = 1$), the X-rays are classified as **K-radiation**, and similarly for transitions into the L ($n = 2$) and M ($n = 3$) shells. Strong, distinct lines are labelled K_α, K_β, and so on. Increasingly, X-ray diffraction makes use of the radiation available from synchrotron sources, for its high intensity greatly enhances the sensitivity of the technique.

von Laue's original method consisted of passing a broad-band beam of X-rays into a single crystal, and recording the diffraction pattern photographically. The idea behind the approach

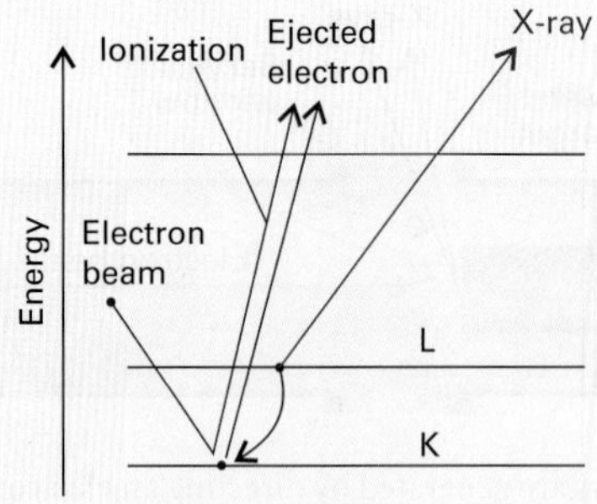

Fig. 9.16 The processes that contribute to the generation of X-rays. An incoming electron collides with an electron (in the K shell), and ejects it. Another electron (from the L shell in this illustration) falls into the vacancy and emits its excess energy as an X-ray photon.

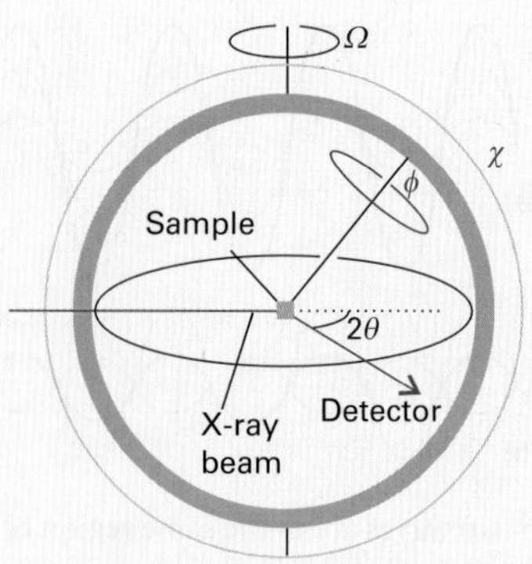

Fig. 9.18 A four-circle diffractometer. The settings of the orientations (ϕ, χ, θ, and Ω) of the components are controlled by computer; each (*hkl*) reflection is monitored in turn, and their intensities are recorded.

was that a crystal might not be suitably orientated to act as a diffraction grating for a single wavelength but, whatever its orientation, diffraction would be achieved for at least one of the wavelengths if a range of wavelengths were used. There is currently a resurgence of interest in this approach because synchrotron radiation spans a range of X-ray wavelengths.

An alternative technique was developed by Peter Debye and Paul Scherrer and independently by Albert Hull. They used monochromatic radiation and a powdered sample. When the sample is a powder, at least some of the crystallites will be orientated so as to give rise to diffraction. In modern powder diffractometers the intensities of the reflections are monitored electronically as the detector is rotated around the sample in a plane containing the incident ray (Fig. 9.17). Powder diffraction is used to identify a sample of a solid substance by comparison of the positions of the diffraction lines and their intensities with a

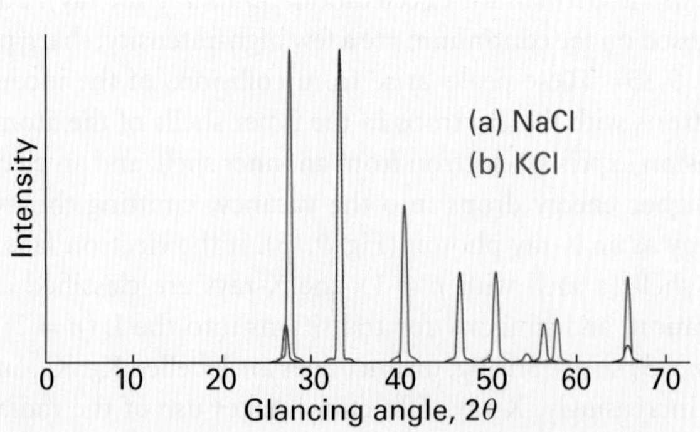

Fig. 9.17 X-ray powder photographs of (a) NaCl, (b) KCl and the indexed reflections. The smaller number of lines in (b) is a consequence of the similarity of the K^+ and Cl^- scattering factors, as discussed later in the chapter.

data bank. Powder diffraction is also used to determine phase diagrams, for different solid phases result in different diffraction patterns, and to determine the relative amounts of each phase present in a mixture. The technique is also used for the initial determination of the dimensions and symmetries of unit cells.

The method developed by the Braggs (William and his son Lawrence, who later jointly won the Nobel Prize) is the foundation of almost all modern work in X-ray crystallography. They used a single crystal and a monochromatic beam of X-rays, and rotated the crystal until a reflection was detected. There are many different sets of planes in a crystal, so there are many angles at which a reflection occurs. The raw data consist of the angles at which reflections are observed and their intensities.

Single-crystal diffraction patterns are measured by using a **four-circle diffractometer** (Fig. 9.18). An integrated computer determines the unit cell dimensions and the angular settings of the diffractometer's four circles that are needed to observe any particular intensity peak in the diffraction pattern. At each setting, the diffraction intensity is measured, and background intensities are assessed by making measurements at slightly different settings. Computing techniques are now available that lead not only to automatic indexing but also to the automated determination of the shape, symmetry, and size of the unit cell. Moreover, several techniques are now available for sampling large amounts of data, including area detectors and image plates, which sample whole regions of diffraction patterns simultaneously.

(b) Bragg's law

An early approach to the analysis of diffraction patterns produced by crystals was to regard a lattice plane as a semi-transparent mirror and to model a crystal as a stack of reflecting lattice planes of separation d (Fig. 9.19). The model makes it easy

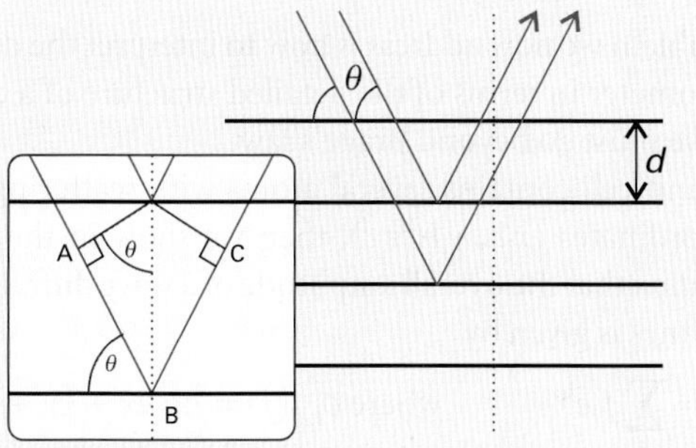

Fig. 9.19 The conventional derivation of Bragg's law treats each lattice plane as a plane reflecting the incident radiation. The path lengths differ by AB + BC, which depends on the glancing angle, θ. Constructive interference (a 'reflection') occurs when AB + BC is equal to an integer number of wavelengths.

to calculate the angle the crystal must make to the incoming beam of X-rays for constructive interference to occur. It has also given rise to the name **reflection** to denote an intense beam arising from constructive interference.

Consider the reflection of two parallel rays of the same wavelength by two adjacent planes of a lattice, as shown in Fig. 9.19. One ray strikes a point on the upper plane but the other ray must travel an additional distance AB before striking the plane immediately below. Similarly, the reflected rays will differ in path length by a distance BC. The net path length difference of the two rays is then

$$AB + BC = 2d\sin\theta$$

where θ is the **glancing angle**. For many glancing angles the path-length difference is not an integer number of wavelengths, and the waves interfere largely destructively. However, when the path-length difference is an integer number of wavelengths ($AB + BC = n\lambda$), the reflected waves are in phase and interfere constructively. It follows that a reflection should be observed when the glancing angle satisfies **Bragg's law**:

$$n\lambda = 2d\sin\theta \qquad (9.2a)$$

Reflections with $n = 2, 3, \ldots$ are called second order, third order, and so on; they correspond to path-length differences of 2, 3, . . . wavelengths. In modern work it is normal to absorb the n into d, to write Bragg's law as

$$\lambda = 2d\sin\theta \qquad (9.2b)$$

and to regard the nth-order reflection as arising from the $\{nh,nk,nl\}$ planes (see Example 9.1).

The primary use of Bragg's law is in the determination of the spacing between the layers in the lattice for, once the angle θ corresponding to a reflection has been determined, d may readily be calculated.

● **A BRIEF ILLUSTRATION**

A first-order reflection from the {111} planes of a cubic crystal was observed at a glancing angle of 11.2° when $Cu(K_\alpha)$ X-rays of wavelength 154 pm were used. According to eqn 9.2, the {111} planes responsible for the diffraction have separation $d_{111} = \lambda/2\sin\theta$. The separation of the {111} planes of a cubic lattice of side a is given by eqn 9.1 as $d_{111} = a/3^{1/2}$. Therefore,

$$a = \frac{3^{1/2}\lambda}{2\sin\theta} = \frac{3^{1/2}\times(154\text{ pm})}{2\sin 11.2°} = 687\text{ pm}$$ ●

Self-test 9.2 Calculate the angle at which the same crystal will give a reflection from the {123} planes. [24.8°]

Some types of unit cell give characteristic and easily recognizable patterns of lines. For example, in a cubic lattice of unit cell dimension a the spacing is given by eqn 9.1b, so the angles at which the $\{hkl\}$ planes give first-order reflections are given by

$$\sin\theta = (h^2 + k^2 + l^2)^{1/2}\frac{\lambda}{2a}$$

The reflections are then predicted by substituting the values of h, k, and l:

$\{hkl\}$	{100}	{110}	{111}	{200}	{210}
$h^2+k^2+l^2$	1	2	3	4	5
$\{hkl\}$	{211}	{220}	{300}	{221}	{310} . . .
$h^2+k^2+l^2$	6	8	9	9	10 . . .

Notice that 7 (and 15, . . .) is missing because the sum of the squares of three integers cannot equal 7 (or 15, . . .). Therefore the pattern has absences that are characteristic of the cubic P lattice.

Self-test 9.3 Normally, experimental procedures measure 2θ rather than θ itself. A diffraction examination of the element polonium gave lines at the following values of 2θ (in degrees) when 71.0 pm Mo X-rays were used: 12.1, 17.1, 21.0, 24.3, 27.2, 29.9, 34.7, 36.9, 38.9, 40.9, 42.8. Identify the unit cell and determine its dimensions. [cubic P; a = 337 pm.]

(c) Scattering factors

To prepare the way to discussing modern methods of structural analysis we need to note that the scattering of X-rays is caused by the oscillations an incoming electromagnetic wave generates in the electrons of atoms, and heavy atoms give rise to stronger scattering than light atoms. This dependence on the number of electrons is expressed in terms of the **scattering factor**, f, of the element. If the scattering factor is large, then the atoms scatter

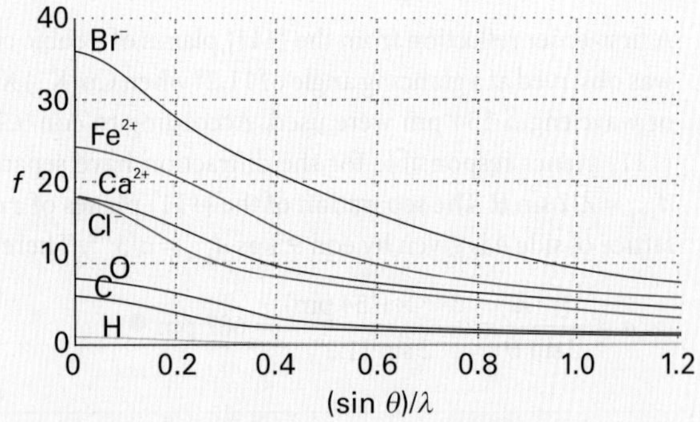

Fig. 9.20 The variation of the scattering factor of atoms and ions with atomic number and angle. The scattering factor in the forward direction (at $\theta = 0$, and hence at $(\sin\theta)/\lambda = 0$) is equal to the number of electrons present in the species.

X-rays strongly. The scattering factor of an atom is related to the electron density distribution in the atom, $\rho(r)$, by

$$f = 4\pi\int_0^\infty \rho(r)\frac{\sin kr}{kr}r^2\mathrm{d}r \qquad k = \frac{4\pi}{\lambda}\sin\theta \tag{9.3}$$

The value of f is greatest in the forward direction and smaller for directions away from the forward direction (Fig. 9.20). The detailed analysis of the intensities of reflections must take this dependence on direction into account (in single crystal studies as well as for powders). We show in the following *Justification* that, in the forward direction (for $\theta = 0$), f is equal to the total number of electrons in the atom.

Justification 9.1 *The forward scattering factor*

As $\theta \to 0$, so $k \to 0$. Because $\sin x = x - \frac{1}{6}x^3 + \cdots$,

$$\lim_{x\to 0}\frac{\sin x}{x} = \lim_{x\to 0}\frac{x - \frac{1}{6}x^3 + \cdots}{x} = \lim_{x\to 0}(1 - \tfrac{1}{6}x^2 + \cdots) = 1$$

The factor $(\sin kr)/kr$ is therefore equal to 1 for forward scattering. It follows that in the forward direction

$$f = 4\pi\int_0^\infty \rho(r)r^2\mathrm{d}r$$

The integral over the electron density ρ (the number of electrons in an infinitesimal region divided by the volume of the region) multiplied by the volume element $4\pi r^2\mathrm{d}r$ is the total number of electrons, N_e, in the atom. Hence, in the forward direction, $f = N_e$. For example, the scattering factors of Na^+, K^+, and Cl^- are 10, 18, and 18, respectively. The scattering factor is smaller in nonforward directions because $(\sin kr)/kr < 1$ for $\theta > 0$.

(d) The electron density

The problem we now address is how to interpret the data from a diffractometer in terms of the detailed structure of a crystal. To do so, we must go beyond Bragg's law.

If a unit cell contains several atoms with scattering factors f_j and coordinates (x_ja, y_jb, z_jc), then we show in the following *Justification* that the overall amplitude of a wave diffracted by the $\{hkl\}$ planes is given by

$$F_{hkl} = \sum_j f_j e^{i\phi_{hkl}(j)} \qquad \text{where } \phi_{hkl}(j) = 2\pi(hx_j + ky_j + lz_j) \tag{9.4}$$

The sum is over all the atoms in the unit cell. The quantity F_{hkl} is called the **structure factor**.

Justification 9.2 *The structure factor*

We begin by showing that, if in the unit cell there is an A atom at the origin and a B atom at the coordinates (xa, yb, zc), where x, y, and z lie in the range 0 to 1, then the phase difference, ϕ_{hkl}, between the (hkl) reflections of the A and B atoms is equal to $2\pi(hx + ky + lz)$.

Consider the crystal shown schematically in Fig. 9.21. The reflection corresponds to two waves from adjacent A planes; for the wavelength and angle of incidence shown, there is constructive interference and hence a strong (100) reflection when the phase difference of the waves is 2π. If there is a B atom at a fraction x of the distance between the two A planes, then it gives rise to a wave with a phase difference $2\pi x$ relative to an A reflection. To see this conclusion note that, if $x = 0$, there is no phase difference; if $x = \frac{1}{2}$ the phase difference is π; if $x = 1$, the B atom lies where the lower A atom is and the phase difference is 2π. Now consider a (200) reflection. There is now a $2 \times 2\pi$ difference between the waves from the two A

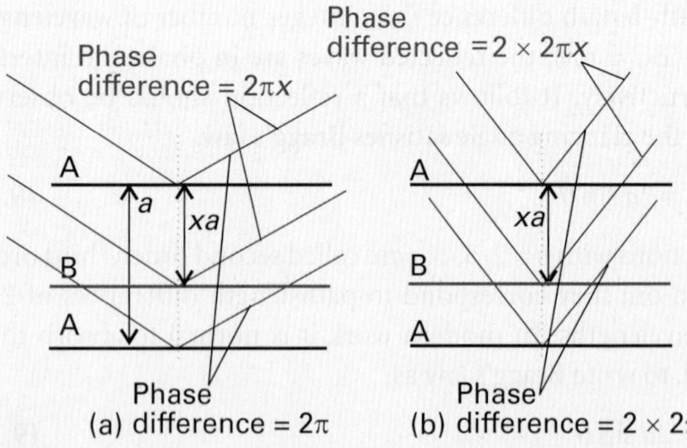

Fig. 9.21 Diffraction from a crystal containing two kinds of atoms. (a) For a (100) reflection from the A planes, there is a phase difference of 2π between waves reflected by neighbouring planes. (b) For a (200) reflection, the phase difference is 4π. The reflection from a B plane at a fractional distance xa from an A plane has a phase that is x times these phase differences.

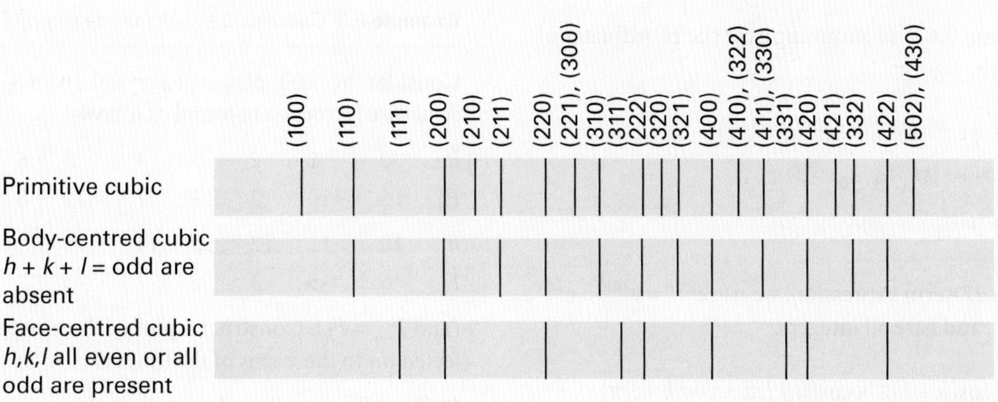

Fig. 9.22 The powder diffraction patterns and the systematic absences of three versions of a cubic cell. Comparison of the observed pattern with patterns like these enables the unit cell to be identified. The locations of the lines give the cell dimensions.

layers and, if B were to lie at $x = 0.5$, it would give rise to a wave that differed in phase by 2π from the wave from the upper A layer. Thus, for a general fractional position x, the phase difference for a (200) reflection is $2 \times 2\pi x$. For a general (h00) reflection, the phase difference is therefore $h \times 2\pi x$. For three dimensions, this result generalizes to eqn 9.4.

The A and B reflections interfere destructively when the phase difference is π, and the total intensity is zero if the atoms have the same scattering power. For example, if the unit cells are cubic I with a B atom at $x = y = z = \frac{1}{2}$, then the A,B phase difference is $(h + k + l)\pi$. Therefore, all reflections for odd values of $h + k + l$ vanish because the waves are displaced in phase by π. Hence the diffraction pattern for a cubic I lattice can be constructed from that for the cubic P lattice (a cubic lattice without points at the centre of its unit cells) by striking out all reflections with odd values of $h + k + l$. Recognition of these **systematic absences** in a powder spectrum immediately indicates a cubic I lattice (Fig. 9.22).

If the amplitude of the waves scattered from A is f_A at the detector, that of the waves scattered from B is $f_B e^{i\phi_{hkl}}$, with ϕ_{hkl} the phase difference given in eqn 9.4. The total amplitude at the detector is therefore

$$F_{hkl} = f_A + f_B e^{i\phi_{hkl}}$$

Because the intensity is proportional to the square modulus of the amplitude of the wave, the intensity, I_{hkl}, at the detector is

$$I_{hkl} \propto F^*_{hkl} F_{hkl} = (f_A + f_B e^{-i\phi_{hkl}})(f_A + f_B e^{i\phi_{hkl}})$$

This expression expands to

$$I_{hkl} \propto f_A^2 + f_B^2 + f_A f_B (e^{i\phi_{hkl}} + e^{-i\phi_{hkl}}) = f_A^2 + f_B^2 + 2 f_A f_B \cos \phi_{hkl}$$

The cosine term either adds to or subtracts from $f_A^2 + f_B^2$ depending on the value of ϕ_{hkl}, which in turn depends on h, k, and l and x, y, and z. Hence, there is a variation in the intensities of the lines with different hkl.

Example 9.2 *Calculating a structure factor*

Calculate the structure factors for the unit cell in Fig. 9.23.

Method The structure factor is defined by eqn 9.4. To use this equation, consider the ions at the locations specified in Fig. 9.23. Write f^+ for the Na^+ scattering factor and f^- for the Cl^- scattering factor. Note that ions in the body of the cell contribute to the scattering with a strength f. However, ions on faces are shared between two cells (use $\frac{1}{2}f$), those on edges by four cells (use $\frac{1}{4}f$), and those at corners by eight cells (use $\frac{1}{8}f$). Two useful relations are

$$e^{i\pi} = -1 \qquad \cos \phi = \tfrac{1}{2}(e^{i\phi} + e^{-i\phi})$$

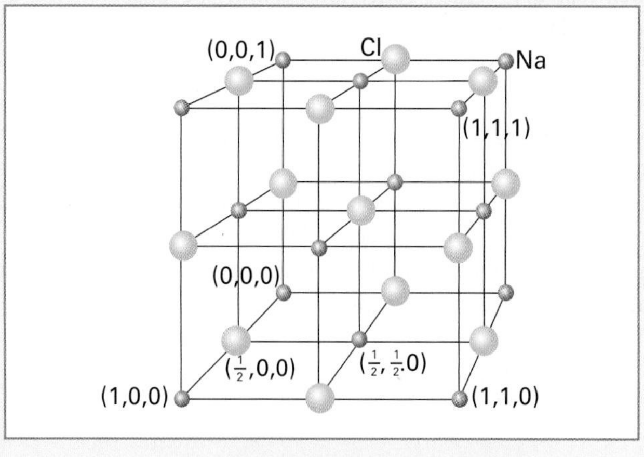

Fig. 9.23 The location of the atoms for the structure factor calculation in Example 9.2. The purple circles are Na^+; the blue circles are Cl^-.

Answer From eqn 9.4, and summing over the coordinates of all 27 atoms in the figure:

$$
\begin{aligned}
F_{hkl} = {} & f^+\left(\tfrac{1}{8} + \tfrac{1}{8}\mathrm{e}^{2\pi i l} + \cdots + \tfrac{1}{2}\mathrm{e}^{2\pi i(\frac{1}{2}h+\frac{1}{2}k+l)}\right) \\
& + f^-\left(\mathrm{e}^{2\pi i(\frac{1}{2}h+\frac{1}{2}k+\frac{1}{2}l)} + \tfrac{1}{4}\mathrm{e}^{2\pi i(\frac{1}{2}h)} + \cdots \right. \\
& \left. + \tfrac{1}{4}\mathrm{e}^{2\pi i(\frac{1}{2}h+l)}\right)
\end{aligned}
$$

To simplify this 27-term expression, we use $\mathrm{e}^{2\pi i h} = \mathrm{e}^{2\pi i k} = \mathrm{e}^{2\pi i l} = 1$ because h, k, and l are all integers:

$$
\begin{aligned}
F_{hkl} = {} & f^+\{1 + \cos(h+k)\pi + \cos(h+l)\pi + \cos(k+l)\pi\} \\
& + f^-\{(-1)^{h+k+l} + \cos k\pi + \cos l\pi + \cos h\pi\}
\end{aligned}
$$

Then, because $\cos h\pi = (-1)^h$,

$$
\begin{aligned}
F_{hkl} = {} & f^+\{1 + (-1)^{h+k} + (-1)^{h+l} + (-1)^{l+k}\} \\
& + f^-\{(-1)^{h+k+l} + (-1)^h + (-1)^k + (-1)^l\}
\end{aligned}
$$

Now note that:

if h, k, and l are all even, $F_{hkl} = f^+\{1+1+1+1\} + f^-\{1+1+1+1\} = 4(f^+ + f^-)$

if h, k, and l are all odd, $F_{hkl} = 4(f^+ - f^-)$

if one index is odd and two are even, or vice versa, $F_{hkl} = 0$

The hkl all-odd reflections are less intense than the hkl all-even. For $f^+ = f^-$, which is the case for identical atoms in a cubic P arrangement, the hkl all-odd have zero intensity, corresponding to the 'systematic absences' of cubic P unit cells.

Self-test 9.4 Which reflections cannot be observed for a cubic I lattice? [for $h+k+l$ odd, $F_{hkl} = 0$]

The intensity of the (hkl) reflection is proportional to $|F_{hkl}|^2$, so in principle we can determine the structure factors experimentally by taking the square root of the corresponding intensities (but see below). Then, once we know all the structure factors F_{hkl}, we can calculate the electron density distribution, $\rho(\boldsymbol{r})$, in the unit cell by using the expression

$$\rho(\boldsymbol{r}) = \frac{1}{V}\sum_{hkl} F_{hkl}\mathrm{e}^{-2\pi i(hx+ky+lz)} \tag{9.5}$$

where V is the volume of the unit cell. Equation 9.5 is called a **Fourier synthesis** of the electron density. Fourier transforms occur throughout chemistry in a variety of guises, and are described in more detail in *Mathematical background 6*.

Example 9.3 *Calculating an electron density by Fourier synthesis*

Consider the $\{h00\}$ planes of a crystal. In an X-ray analysis the structure factors were found as follows:

h:	0	1	2	3	4	5	6	7	8	9
F_h	16	−10	2	−1	7	−10	8	−3	2	−3
h:	10	11	12	13	14	15				
F_h	6	−5	3	−2	2	−3				

(and $F_{-h} = F_h$). Construct a plot of the electron density projected on to the x-axis of the unit cell.

Method Because $F_{-h} = F_h$, it follows from eqn 9.5 that

$$
\begin{aligned}
V\rho(x) &= \sum_{h=-\infty}^{\infty} F_h\mathrm{e}^{-2\pi i h x} = F_0 + \sum_{h=1}^{\infty}(F_h\mathrm{e}^{-2\pi i h x} + F_{-h}\mathrm{e}^{2\pi i h x}) \\
&= F_0 + \sum_{h=1}^{\infty} F_h(\mathrm{e}^{-2\pi i h x} + \mathrm{e}^{2\pi i h x}) = F_0 + 2\sum_{h=1}^{\infty} F_h \cos 2\pi h x
\end{aligned}
$$

and we evaluate the sum (truncated at $h = 15$) for points $0 \le x \le 1$ by using mathematical software.

Answer The results are plotted in Fig. 9.24 (blue line). The positions of three atoms can be discerned very readily. The more terms there are included, the more accurate the density plot. Terms corresponding to high values of h (short wavelength cosine terms in the sum) account for the finer details of the electron density; low values of h account for the broad features.

Self-test 9.5 Use mathematical software to experiment with different structure factors (including changing signs as well as amplitudes). For example, use the same values of F_h as above, but with positive signs for all values of h. [Fig. 9.24 (purple line)]

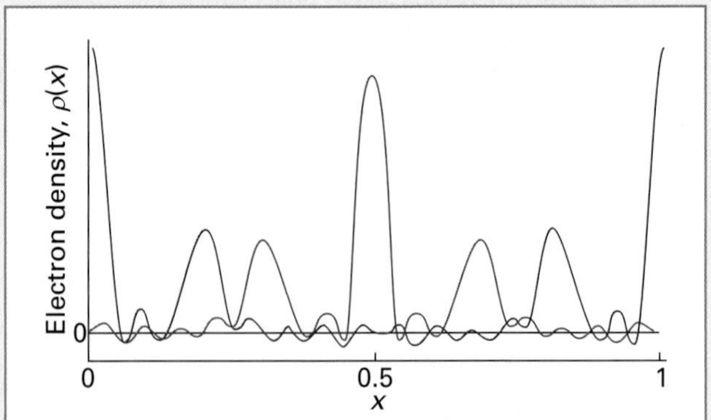

Fig. 9.24 The plot of the electron density calculated in Example 9.3 (blue) and Self-test 9.5 (purple).

interActivity If you do not have access to mathematical software, perform the calculations suggested in Self-test 9.5 by using the interactive applets found in the text's web site.

(e) The phase problem

A problem with the procedure outlined above is that the observed intensity I_{hkl} is proportional to the square modulus $|F_{hkl}|^2$, so we do not know whether to use $+|F_{hkl}|$ or $-|F_{hkl}|$ in the sum in eqn 9.5. In fact, the difficulty is more severe for non-centrosymmetric unit cells, because if we write F_{hkl} as the complex number $|F_{hkl}|e^{i\alpha}$, where α is the phase of F_{hkl} and $|F_{hkl}|$ is its magnitude, then the intensity lets us determine $|F_{hkl}|$ but tells us nothing of its phase, which may lie anywhere from 0 to 2π. This ambiguity is called the **phase problem**; its consequences are illustrated by comparing the two plots in Fig. 9.24. Some way must be found to assign phases to the structure factors, for otherwise the sum for ρ cannot be evaluated and the method would be useless.

The phase problem can be overcome to some extent by a variety of methods. One procedure that is widely used for inorganic materials with a reasonably small number of atoms in a unit cell and for organic molecules with a small number of heavy atoms is the **Patterson synthesis**. Instead of the structure factors F_{hkl}, the values of $|F_{hkl}|^2$, which can be obtained without ambiguity from the intensities, are used in an expression that resembles eqn 9.5:

$$P(\boldsymbol{r}) = \frac{1}{V}\sum_{hkl}|F_{hkl}|^2 e^{-2\pi i(hx+ky+lz)} \quad (9.6)$$

The outcome of a Patterson synthesis is a map of the vector separations of the atoms (the distances and directions between atoms) in the unit cell. Thus, if atom A is at the coordinates (x_A, y_A, z_A) and atom B is at (x_B, y_B, z_B), then there will be a peak at $(x_A - x_B, y_A - y_B, z_A - z_B)$ in the Patterson map. There will also be a peak at the negative of these coordinates, because there is a vector from B to A as well as a vector from A to B. The height of the peak in the map is proportional to the product of the atomic numbers of the two atoms, Z_AZ_B. For example, if the unit cell has the structure shown in Fig. 9.25a, the Patterson synthesis would be the map shown in Fig. 9.25b, where the location of each spot relative to the origin gives the separation and relative orientation of each pair of atoms in the original structure.

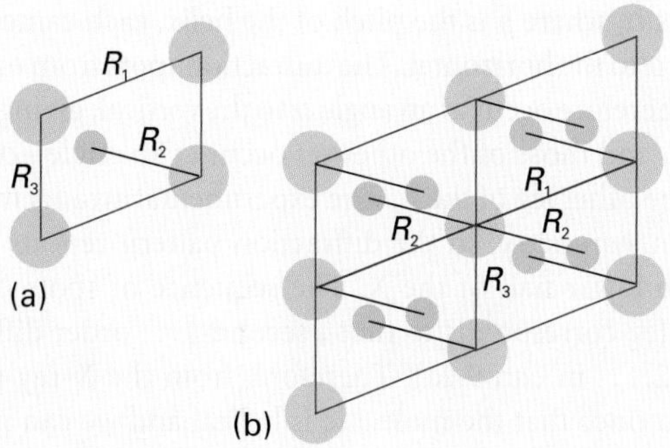

Fig. 9.25 The Patterson synthesis corresponding to the pattern in (a) is the pattern in (b). The distance and orientation of each spot from the origin gives the orientation and separation of one atom–atom separation in (a). Some of the typical distances and their contribution to (b) are shown as R_1, etc.

Heavy atoms dominate the scattering because their scattering factors are large, of the order of their atomic numbers, and their locations may be deduced quite readily. The sign of F_{hkl} can now be calculated from the locations of the heavy atoms in the unit cell, and to a high probability the phase calculated for them will be the same as the phase for the entire unit cell. To see why this is so, we have to note that a structure factor of a centrosymmetric cell has the form

$$F = (\pm)f_{\text{heavy}} + (\pm)f_{\text{light}} + (\pm)f_{\text{light}} + \cdots \quad (9.7)$$

where f_{heavy} is the scattering factor of the heavy atom and f_{light} the scattering factors of the light atoms. The f_{light} are all much smaller than f_{heavy}, and their phases are more or less random if the atoms are distributed throughout the unit cell. Therefore, the net effect of the f_{light} is to change F only slightly from f_{heavy}, and we can be reasonably confident that F will have the same sign as that calculated from the location of the heavy atom. This phase can then be combined with the observed $|F|$ (from the reflection intensity) to perform a Fourier synthesis of the full electron density in the unit cell, and hence to locate the light atoms as well as the heavy atoms.

Modern structural analyses make extensive use of **direct methods**. Direct methods are based on the possibility of treating the atoms in a unit cell as being virtually randomly distributed (from the radiation's point of view), and then to use statistical techniques to compute the probabilities that the phases have a particular value. It is possible to deduce relations between some structure factors and sums (and sums of squares) of others, which have the effect of constraining the phases to particular values (with high probability, so long as the structure factors are large). For example, the **Sayre probability relation** has the form

$$\text{sign of } F_{h+h',k+k',l+l'} \text{ is probably equal to (sign of } F_{hkl}) \times (\text{sign of } F_{h'k'l'}) \quad (9.8)$$

For example, if F_{122} and F_{232} are both large and negative, then it is highly likely that F_{354}, provided it is large, will be positive.

(f) Structure refinement

In the final stages of the determination of a crystal structure, the parameters describing the structure (atom positions, for instance) are adjusted systematically to give the best fit between the observed intensities and those calculated from the model of the structure deduced from the diffraction pattern. This process is called **structure refinement**. Not only does the procedure give accurate positions for all the atoms in the unit cell, but it also gives an estimate of the errors in those positions and in the bond lengths and angles derived from them. The procedure also provides information on the vibrational amplitudes of the atoms.

IMPACT ON BIOCHEMISTRY

I9.1 X-ray crystallography of biological macromolecules

X-ray crystallography is the deployment of X-ray diffraction techniques for the determination of the location of all the atoms in molecules as complicated as biopolymers. Bragg's law helps us understand the features of one of the most seminal X-ray images of all time, the characteristic X-shaped pattern obtained by Rosalind Franklin and Maurice Wilkins from strands of DNA and used by James Watson and Francis Crick in their construction of the double-helix model of DNA (Fig. 9.26). To interpret this image by using the Bragg law we have to be aware that it was obtained by using a fibre consisting of many DNA molecules oriented with their axes parallel to the axis of the fibre, with X-rays incident from a perpendicular direction. All the molecules in the fibre are parallel (or nearly so), but are randomly distributed in the perpendicular directions; as a result, the diffraction pattern exhibits the periodic structure parallel to the fibre axis superimposed on a general background of scattering from the distribution of molecules in the perpendicular directions.

There are two principal features in Fig. 9.26: the strong 'meridional' scattering upward and downward by the fibre and the X-shaped distribution at smaller scattering angles. Because scattering through large angles occurs for closely spaced features (from $\lambda = 2d \sin \theta$, if d is small, then θ must be large to preserve the equality), we can infer that the meridional scattering arises from closely spaced components and that the inner X-shaped pattern arises from features with a longer periodicity. Because the meridional pattern occurs at a distance of about 10 times that of the innermost spots of the X-pattern, the large-scale structure is about 10 times bigger than the small-scale structure. From the geometry of the instrument, the wavelength of the radiation, and Bragg's law, we can infer that the periodicity of the small-scale feature is 340 pm whereas that of the large-scale feature is 3400 pm (that is, 3.4 nm).

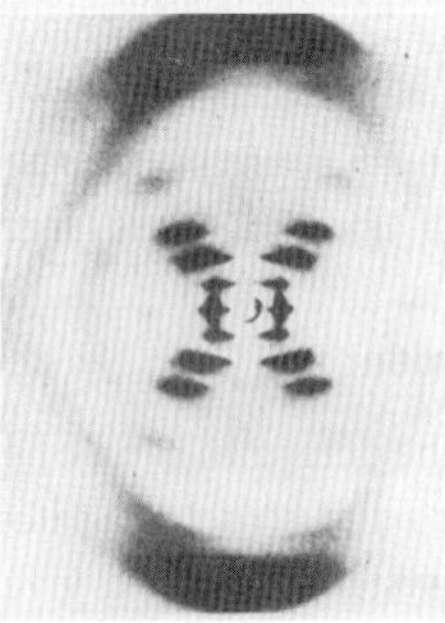

Fig. 9.26 The X-ray diffraction pattern obtained from a fibre of B-DNA. The black dots are the reflections, the points of maximum constructive interference, that are used to determine the structure of the molecule. (Adapted from an illustration that appears in J.P. Glusker and K.N. Trueblood, *Crystal structure analysis: a primer*. Oxford University Press (1972).)

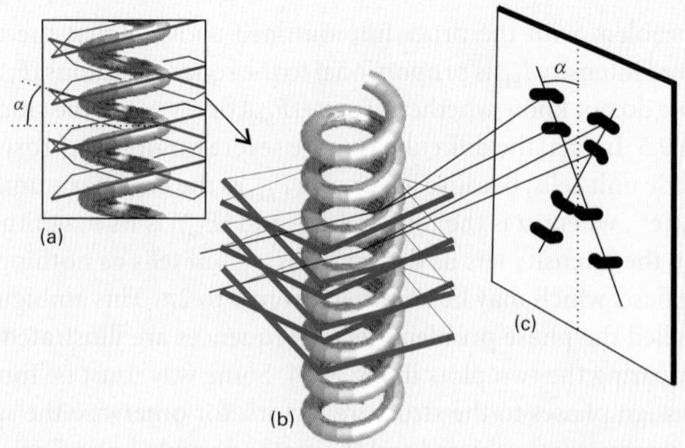

Fig. 9.27 The origin of the X pattern characteristic of diffraction by a helix. (a) A helix can be thought of as consisting of an array of planes at an angle α together with an array of planes at an angle $-\alpha$. (b) The diffraction spots from one set of planes appear at an angle α to the vertical, giving one leg of the X, and those of the other set appear at an angle $-\alpha$, giving rise to the other leg of the X. The lower half of the X appears because the helix has up-down symmetry in this arrangement. (c) The sequence of spots outward along a leg of the X corresponds to first-, second-, . . . order diffraction ($n = 1, 2, \ldots$).

To see that the cross is characteristic of a helix, look at Fig. 9.27. Each turn of the helix defines two planes, one orientated at an angle α to the horizontal and the other at $-\alpha$. As a result, to a first approximation, a helix can be thought of as consisting of an array of planes at an angle α together with an array of planes at an angle $-\alpha$ with a separation within each set determined by the pitch of the helix. Thus, a DNA molecule is like two arrays of planes, each set corresponding to those treated in the derivation of Bragg's law, with a perpendicular separation $d = p \cos \alpha$, where p is the pitch of the helix, each canted at the angles $\pm\alpha$ to the horizontal. The diffraction spots from one set of planes therefore occur at an angle α to the vertical, giving one leg of the X, and those of the other set occur at an angle $-\alpha$, giving rise to the other leg of the X. The experimental arrangement has up-down symmetry, so the diffraction pattern repeats to produce the lower half of the X. The sequence of spots outward along a leg corresponds to first-, second-, . . . order diffraction ($n = 1, 2, \ldots$ in eqn 9.2a). Therefore, from the X-ray pattern, we see at once that the molecule is helical and we can measure the angle α directly, and find $\alpha = 40°$. Finally, with the angle α and the pitch p determined, we can determine the radius r of the helix from $\tan \alpha = p/r$, from which it follows that $r = (3.4\ \text{nm})/\tan 40° = 4.1\ \text{nm}$.

To derive the relation between the helix and the cross-like pattern we have ignored the detailed structure of the helix,

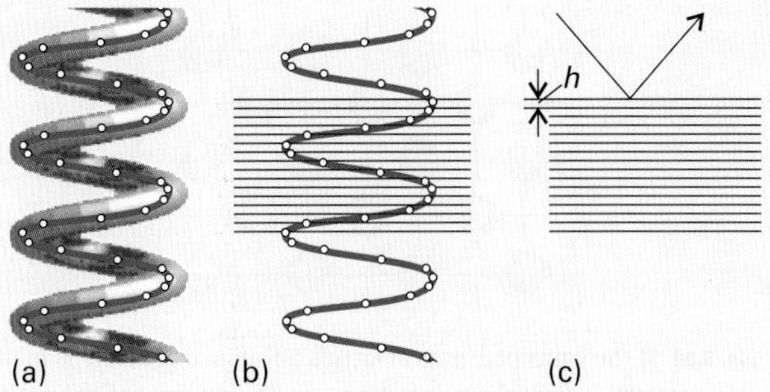

Fig. 9.28 The effect of the internal structure of the helix on the X-ray diffraction pattern. (a) The residues of the macromolecule are represented by points. (b) Parallel planes passing through the residues are perpendicular to the axis of the molecule. (c) The planes give rise to strong diffraction with an angle that allows us to determine the layer spacing h from $\lambda = 2h \sin \theta$.

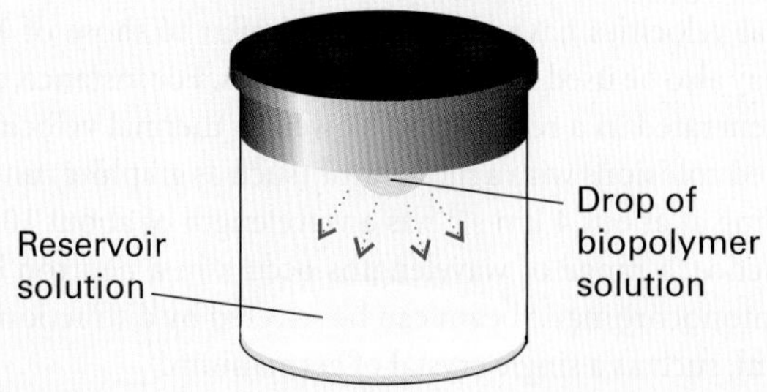

Fig. 9.29 In a common implementation of the vapour diffusion method of biopolymer crystallization, a single drop of biopolymer solution hangs above a reservoir solution that is very concentrated in a non-volatile solute. Solvent evaporates from the more dilute drop until the vapour pressure of water in the closed container reaches a constant equilibrium value. In the course of evaporation (denoted by the downward arrows), the biopolymer solution becomes more concentrated and, at some point, crystals may form.

the fact that it is a periodic array of nucleotide bases, not a smooth wire. In Fig. 9.28 we represent the bases by points, and see that there is an additional periodicity of separation h, forming planes that are perpendicular to the axis of the molecule (and the fibre). These planes give rise to the strong meridional diffraction with an angle that allows us to determine the layer spacing from Bragg's law in the form $\lambda = 2h \sin \theta$ as $h = 340$ pm.

The success of modern biochemistry in explaining such processes as DNA replication, protein biosynthesis, and enzyme catalysis is a direct result of developments in preparatory, instrumental, and computational procedures that have led to the determination of large numbers of structures of biological macromolecules by techniques based on X-ray diffraction. Most work is now done not on fibres but on crystals, in which the large molecules lie in orderly ranks. A technique that works well for charged proteins consists of adding large amounts of a salt, such as $(NH_4)_2SO_4$, to a buffer solution containing the biopolymer. The increase in the ionic strength of the solution decreases the solubility of the protein to such an extent that the protein precipitates, sometimes as crystals that are amenable to analysis by X-ray diffraction. A common strategy for inducing crystallization involves the gradual removal of solvent from a biopolymer solution by *vapour diffusion*. In one implementation of the method, a single drop of biopolymer solution hangs above an aqueous solution (the reservoir), as shown in Fig. 9.29. If the reservoir solution is more concentrated in a non-volatile solute (for example, a salt) than is the biopolymer solution, then solvent will evaporate slowly from the drop. At the same time, the concentration of biopolymer in the drop increases gradually until crystals begin to form.

Special techniques are used to crystallize hydrophobic proteins, such as those spanning the bilayer of a cell membrane. In such cases, surfactant molecules, which like phospholipids contain polar head groups and hydrophobic tails, are used to encase the protein molecules and make them soluble in aqueous buffer solutions. Vapour diffusion may then be used to induce crystallization.

After suitable crystals are obtained, X-ray diffraction data are collected and analysed as described in the previous sections. The three-dimensional structures of a very large number of biological polymers have been determined in this way. However, the techniques discussed so far give only static pictures and are not useful in studies of dynamics and reactivity. This limitation stems from the fact that the Bragg rotation method requires stable crystals that do not change structure during the lengthy data acquisition times required. However, special time-resolved X-ray diffraction techniques have become available in recent years and it is now possible to make exquisitely detailed measurements of atomic motions during chemical and biochemical reactions.

Time-resolved X-ray diffraction techniques make use of synchrotron sources, which can emit intense polychromatic pulses of X-ray radiation with pulse widths varying from 100 ps to 200 ps (1 ps = 10^{-12} s). Instead of the Bragg method, the Laue method is used because many reflections can be collected simultaneously, rotation of the sample is not required, and data acquisition times are short. However, good diffraction data cannot be obtained from a single X-ray pulse and reflections from several pulses must be averaged together. In practice, this averaging dictates the time resolution of the experiment, which is commonly tens of microseconds or less.

9.4 Neutron and electron diffraction

According to the de Broglie relation (eqn 1.3, $\lambda = h/p$), particles have wavelengths and may therefore undergo diffraction.

Neutrons generated in a nuclear reactor and then slowed to thermal velocities have wavelengths similar to those of X-rays and may also be used for diffraction studies. For instance, a neutron generated in a reactor and slowed to thermal velocities by repeated collisions with a moderator (such as graphite) until it is travelling at about 4 km s^{-1} has a wavelength of about 100 pm. In practice, a range of wavelengths occurs in a neutron beam, but a monochromatic beam can be selected by diffraction from a crystal, such as a single crystal of germanium.

Example 9.4 *Calculating the typical wavelength of thermal neutrons*

Calculate the typical wavelength of neutrons that have reached thermal equilibrium with their surroundings at 373 K.

Method We need to relate the wavelength to the temperature. There are two linking steps. First, the de Broglie relation expresses the wavelength in terms of the linear momentum. Then the linear momentum can be expressed in terms of the kinetic energy, the mean value of which is given in terms of the temperature by the equipartition theorem (Fundamentals F.5).

Answer From the equipartition principle, we know that the mean translational kinetic energy of a neutron at a temperature T travelling in the x-direction is $E_k = \frac{1}{2}kT$. The kinetic energy is also equal to $p^2/2m$, where p is the momentum of the neutron and m is its mass. Hence, $p = (mkT)^{1/2}$. It follows from the de Broglie relation $\lambda = h/p$ that the neutron's wavelength is

$$\lambda = \frac{h}{(mkT)^{1/2}}$$

Therefore, at 373 K,

$$\lambda = \frac{6.626\times10^{-34}\ \text{J s}}{\{(1.675\times10^{-27}\ \text{kg})\times(1.381\times10^{-23}\ \text{J K}^{-1})\times(373\ \text{K})\}^{1/2}}$$

$$= \frac{6.626\times10^{-34}}{(1.675\times1.381\times373\times10^{-50})^{1/2}}\ \frac{\text{J s}}{(\text{kg}^2\ \text{m}^2\ \text{s}^{-2})^{1/2}}$$

$$= 2.26\times10^{-10}\ \text{m} = 226\ \text{pm}$$

where we have used 1 J = 1 kg m^2 s^{-2}.

Self-test 9.6 Calculate the temperature needed for the average wavelength of the neutrons to be 100 pm. [1.90×10^3 K]

Neutron diffraction differs from X-ray diffraction in two main respects. First, the scattering of neutrons is a nuclear phenomenon. Neutrons pass through the extranuclear electrons of atoms and interact with the nuclei through the 'strong force' that is responsible for binding nucleons together. As a result, the intensity with which neutrons are scattered is independent of the number of electrons and neighbouring elements in the periodic table may scatter neutrons with markedly different intensities. Neutron diffraction can be used to distinguish atoms of elements such as Ni and Co that are present in the same compound and to study order–disorder phase transitions in FeCo. A second difference is that neutrons possess a magnetic moment due to their spin. This magnetic moment can couple to the magnetic fields of atoms or ions in a crystal (if the ions have unpaired electrons) and modify the diffraction pattern. One consequence is that neutron diffraction is well suited to the investigation of magnetically ordered lattices in which neighbouring atoms may be of the same element but have different orientations of their electronic spin (Fig. 9.30).

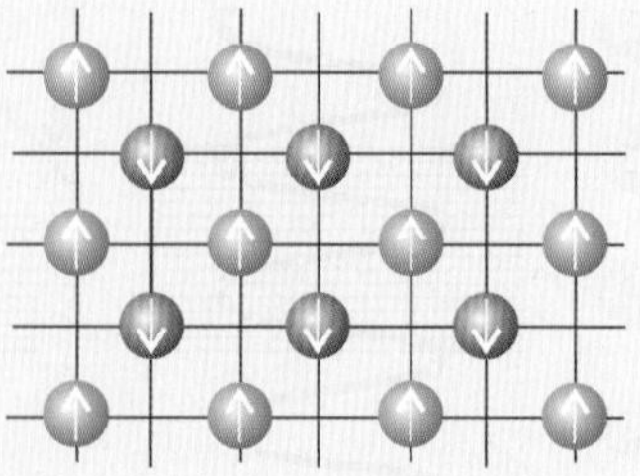

Fig. 9.30 If the spins of atoms at lattice points are orderly, as in this material, where the spins of one set of atoms are aligned antiparallel to those of the other set, neutron diffraction detects two interpenetrating simple cubic lattices on account of the magnetic interaction of the neutron with the atoms, but X-ray diffraction would see only a single bcc lattice (see Section 9.5b).

Electrons accelerated through a potential difference of 40 kV have wavelengths of about 6 pm, and so are also suitable for diffraction studies. However, their main application is to the study of surfaces, and we postpone their discussion until Chapter 21.

Crystal structure

The bonding within a solid may be of various kinds. Simplest of all (in principle) are elemental metals, where electrons are delocalized over arrays of identical cations and bind them together into a rigid but ductile and malleable whole.

9.5 Metallic solids

Most metallic elements crystallize in one of three simple forms, two of which can be explained in terms of hard spheres packing together in the closest possible arrangement.

(a) Close packing

Figure 9.31 shows a **close-packed** layer of identical spheres, one with maximum utilization of space. A close-packed three-

Fig. 9.31 The first layer of close-packed spheres used to build a three-dimensional close-packed structure.

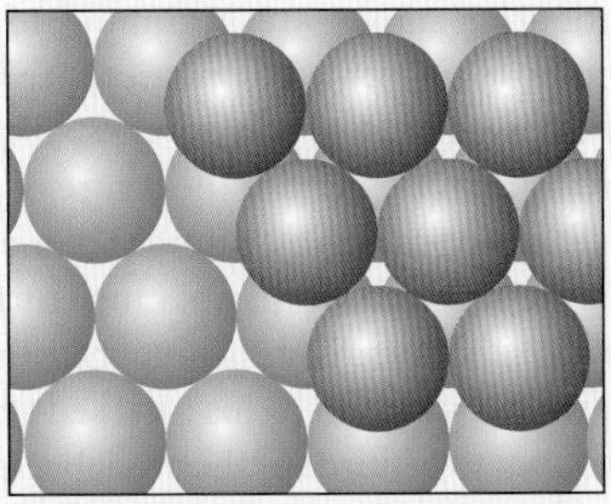

Fig. 9.32 The second layer of close-packed spheres occupies the dips of the first layer. The two layers are the AB component of the close-packed structure.

dimensional structure is obtained by stacking such close-packed layers on top of one another. However, this stacking can be done in different ways, which result in close-packed **polytypes**, or structures that are identical in two dimensions (the close-packed layers) but differ in the third dimension.

In all polytypes, the spheres of the second close-packed layer lie in the depressions of the first layer (Fig. 9.32). The third layer may be added in either of two ways. In one, the spheres are placed so that they reproduce the first layer (Fig. 9.33a), to give an ABA pattern of layers. Alternatively, the spheres may be placed over the gaps in the first layer (Fig. 9.33b), so giving an ABC pattern. Two polytypes are formed if the two stacking patterns are repeated in the vertical direction. If the ABA pattern is repeated, to give the sequence of layers ABABAB . . . , the spheres are **hexagonally close-packed** (hcp). Alternatively, if the ABC pattern is repeated, to give the sequence ABCABC . . . , the spheres are **cubic close-packed** (ccp). We can see the origins of these names by referring to Fig. 9.34. The ccp structure gives rise to a face-centred unit cell, so may also be denoted cubic F (or fcc, for face-centred cubic). It is also possible to have random sequences of layers; however, the hcp and ccp polytypes are the most important. Table 9.2 lists some elements possessing these structures.

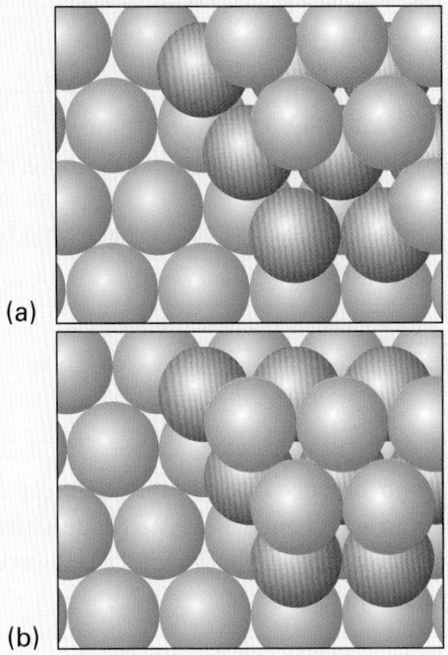

Fig. 9.33 (a) The third layer of close-packed spheres might occupy the dips lying directly above the spheres in the first layer, resulting in an ABA structure, which corresponds to hexagonal close packing. (b) Alternatively, the third layer might lie in the dips that are not above the spheres in the first layer, resulting in an ABC structure, which corresponds to cubic close packing.

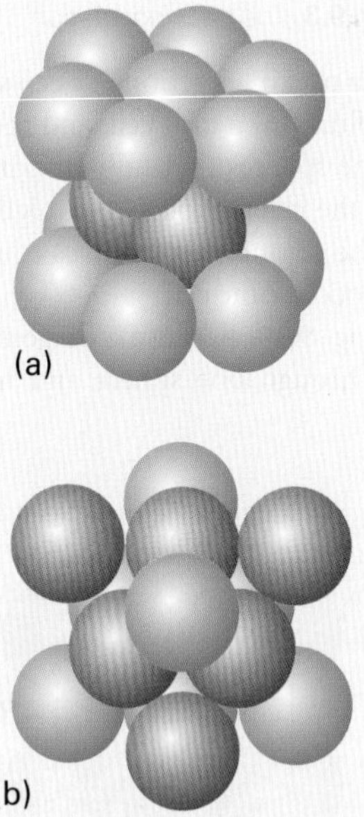

Fig. 9.34 A fragment of the structure shown in Fig. 9.33 revealing the (a) hexagonal (b) cubic symmetry. The tints on the spheres are the same as for the layers in Fig. 9.33.

Table 9.2 The crystal structures of some elements

Structure	Element
hcp*	Be, Cd, Co, He, Mg, Sc, Ti, Zn
fcc* (ccp, cubic F)	Ag, Al, Ar, Au, Ca, Cu, Kr, Ne, Ni, Pd, Pb, Pt, Rh, Rn, Sr, Xe
bcc (cubic I)	Ba, Cs, Cr, Fe, K, Li, Mn, Mo, Rb, Na, Ta, W, V
cubic P	Po

* Close-packed structures.

A note on good practice Strictly speaking, ccp refers to a close-packed arrangement whereas fcc refers to the lattice type of the common representation of ccp. However, this distinction is rarely made.

The compactness of close-packed structures is indicated by their **coordination number**, the number of atoms immediately surrounding any selected atom, which is 12 in all cases. Another measure of their compactness is the **packing fraction**, the fraction of space occupied by the spheres, which is 0.740 (see the following *Justification*). That is, in a close-packed solid of identical hard spheres, only 26.0 per cent of the volume is empty space. The fact that many metals are close-packed accounts for their high densities.

Justification 9.3 *The packing fraction*

To calculate a packing fraction of a ccp structure, we first calculate the volume of a unit cell, and then calculate the total volume of the spheres that fully or partially occupy it. The first part of the calculation is a straightforward exercise in geometry. The second part involves counting the fraction of spheres that occupy the cell.

Refer to Fig. 9.35. Because a diagonal of any face passes completely through one sphere and halfway through two other spheres, its length is $4R$. The length of a side is therefore $8^{1/2}R$ and the volume of the unit cell is $8^{3/2}R^3$. Because each cell contains the equivalent of $6 \times \frac{1}{2} + 8 \times \frac{1}{8} = 4$ spheres, and the volume of each sphere is $\frac{4}{3}\pi R^3$, the total occupied volume is $\frac{16}{3}\pi R^3$. The fraction of space occupied is therefore $\frac{16}{3}\pi R^3/8^{3/2}R^3 = \frac{16}{3}\pi/8^{3/2}$, or 0.740. Because an hcp structure has the same coordination number, its packing fraction is the same. The packing fractions of structures that are not close-packed are calculated similarly (see Exercise 9.20).

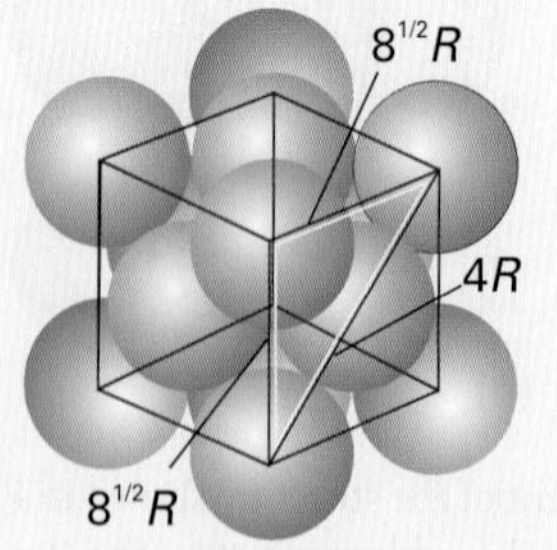

Fig. 9.35 The calculation of the packing fraction of a ccp unit cell.

(b) Less closely packed structures

As shown in Table 9.2, a number of common metals adopt structures that are less than close-packed. The departure from close packing suggests that factors such as specific covalent bonding between neighbouring atoms are beginning to influence the structure and impose a specific geometrical arrangement. One such arrangement results in a cubic I (bcc, for body-centred cubic) structure, with one sphere at the centre of a cube formed by eight others. The coordination number of a bcc structure is only 8, but there are six more atoms not much further away than the eight nearest neighbours. The packing fraction of 0.68 is not much smaller than the value for a close-packed structure (0.74), and shows that about two-thirds of the available space is actually occupied.

9.6 Ionic solids

Two questions arise when we consider ionic solids: the relative locations adopted by the ions and the energetics of the resulting structure.

(a) Structure

When crystals of compounds of monatomic ions (such as NaCl and MgO) are modelled by stacks of hard spheres it is essential to allow for the different ionic radii (typically with the cations smaller than the anions) and different charges. The coordination number of an ion is the number of nearest neighbours of opposite charge; the structure itself is characterized as having (N_+, N_-) **coordination**, where N_+ is the coordination number of the cation and N_- that of the anion.

Even if, by chance, the ions have the same size, the problems of ensuring that the unit cells are electrically neutral makes it impossible to achieve 12-coordinate close-packed ionic structures. As a result, ionic solids are generally less dense than metals. The best packing that can be achieved is the (8,8)-coordinate **caesium-chloride structure** in which each cation is surrounded by eight anions and each anion is surrounded by eight cations (Fig. 9.36). In this structure, an ion of one charge occupies the centre of a cubic unit cell with eight counter ions at its corners. The structure is adopted by CsCl itself and also by CaS, CsCN (with some distortion), and CuZn.

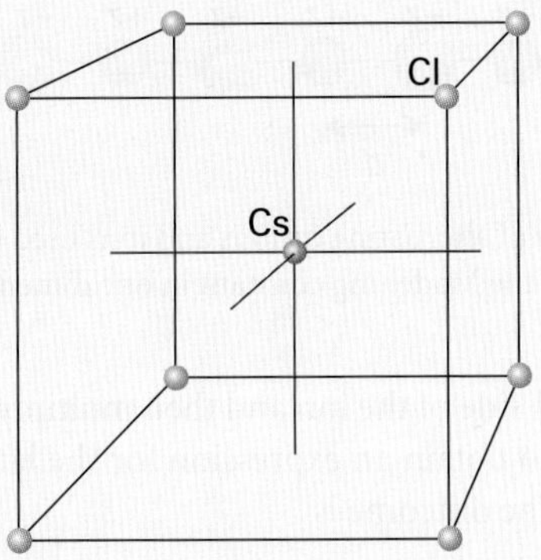

Fig. 9.36 The caesium-chloride structure consists of two interpenetrating simple cubic arrays of ions, one of cations and the other of anions, so that each cube of ions of one kind has a counter ion at its centre.

When the radii of the ions differ more than in CsCl, even eight-coordinate packing cannot be achieved. One common structure adopted is the (6,6)-coordinate **rock-salt structure** typified by NaCl (Fig. 9.37). In this structure, each cation is surrounded by six anions and each anion is surrounded by six cations. The rock-salt structure can be pictured as consisting of two interpenetrating slightly expanded cubic F (fcc) arrays, one composed of cations and the other of anions. This structure is adopted by NaCl itself and also by several other MX compounds, including KBr, AgCl, MgO, and ScN.

The switch from the caesium-chloride structure to the rock-salt structure is related to the value of the **radius ratio**, γ:

$$\gamma = \frac{r_{\text{smaller}}}{r_{\text{larger}}} \qquad [9.9]$$

The two radii are those of the larger and smaller ions in the crystal. The **radius-ratio rule**, which is derived by considering the geometrical problem of packing the maximum number of hard spheres of one radius around a hard sphere of a different radius, can be summarized as follows:

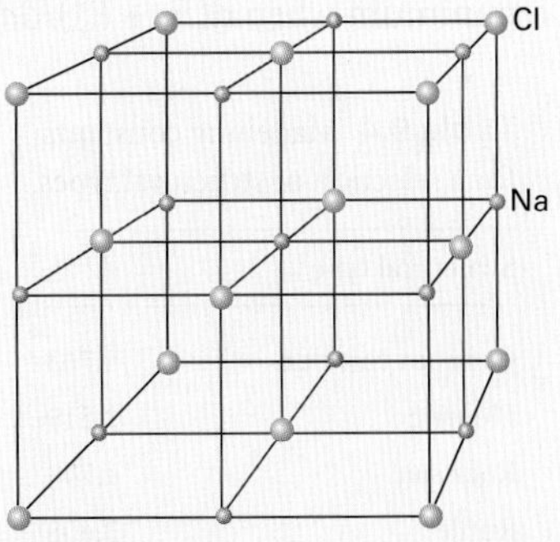

Fig. 9.37 The rock-salt (NaCl) structure consists of two mutually interpenetrating slightly expanded face-centred cubic arrays of ions. The entire assembly shown here is the unit cell.

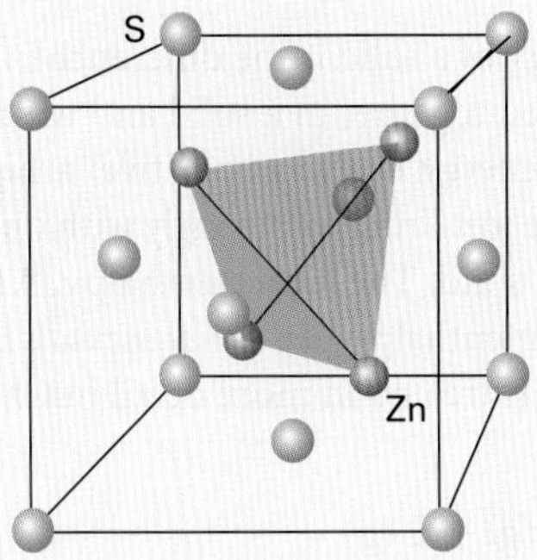

Fig. 9.38 The structure of the sphalerite form of ZnS showing the location of the Zn atoms in the tetrahedral holes formed by the array of S atoms. (Not shown is an S atom at the centre of the cube inside the tetrahedron of Zn atoms.)

Radius ratio	Structural type
$\gamma < 2^{1/2} - 1 = 0.414$	Sphalerite (or zinc blende, Fig. 9.38)
$2^{1/2} - 1 = 0.414 < \gamma < 0.732$	Rock-salt
$\gamma > 3^{1/2} - 1 = 0.732$	Caesium-chloride

The deviation of a structure from that expected on the basis of this rule is often taken to be an indication of a shift from ionic towards covalent bonding; a major source of unreliability, though, is the arbitrariness of ionic radii and their variation with coordination number.

Ionic radii are derived from the distance between centres of adjacent ions in a crystal. However, we need to apportion the total distance between the two ions by defining the radius of one ion and then inferring the radius of the other ion. One scale that is widely used is based on the value 140 pm for the radius of the O^{2-} ion (Table 9.3). Other scales are also available (such as one based on F^- for discussing halides), and it is essential not to mix values from different scales. Because ionic radii are so arbitrary, predictions based on them must be viewed cautiously.

Synoptic table 9.3* Ionic radii, r/pm

Na^+	102(6†), 116(8)
K^+	138(6), 151(8)
F^-	128(2), 131(4)
Cl^-	181 (close packing)

* More values are given in the *Data section*.
† Coordination number.

(b) Energetics

The **lattice energy** of a solid is the difference in potential energy of the ions packed together in a solid and widely separated as a gas. The lattice energy is always positive; a high lattice energy indicates that the ions interact strongly with one another to give a tightly bonded solid. The **lattice enthalpy**, ΔH_L, is the change in standard molar enthalpy (see *Fundamentals F.4* for a review of enthalpy, which is treated in more detail in Chapter 14) for the process

$$MX(s) \rightarrow M^+(g) + X^-(g)$$

and its equivalent for other charge types and stoichiometries. The lattice enthalpy is equal to the lattice energy at $T = 0$; at normal temperatures they differ by only a few kilojoules per mole, and the difference is normally neglected.

Each ion in a solid experiences electrostatic attractions from all the other oppositely charged ions and repulsions from all the other like-charged ions. The total Coulombic potential energy is the sum of all the electrostatic contributions. Each cation is surrounded by anions, and there is a large negative contribution from the attraction of the opposite charges. Beyond those nearest neighbours, there are cations that contribute a positive term to the total potential energy of the central cation. There is also a negative contribution from the anions beyond those cations, a positive contribution from the cations beyond them, and so on to the edge of the solid. These repulsions and attractions become progressively weaker as the distance from the central ion increases, but the net outcome of all these contributions is a lowering of energy.

First, consider a simple one-dimensional model of a solid consisting of a long line of uniformly spaced alternating cations and anions, with d the distance between their centres, the sum of the ionic radii (Fig. 9.39). If the charge numbers of the ions have the same absolute value (+1 and −1, or +2 and −2, for instance), then $z_1 = +z$, $z_2 = -z$, and $z_1z_2 = -z^2$. The potential energy of the central ion is calculated by summing all the terms, with negative terms representing attractions to oppositely charged ions and positive terms representing repulsions from like-charged ions. For the interaction with ions extending in a line to the right of the central ion, the lattice energy is

$$E_P = \frac{1}{4\pi\varepsilon_0} \times \left(-\frac{z^2e^2}{d} + \frac{z^2e^2}{2d} - \frac{z^2e^2}{3d} + \frac{z^2e^2}{4d} - \cdots \right)$$

$$= -\frac{z^2e^2}{4\pi\varepsilon_0 d}\left(1 - \frac{1}{2} + \frac{1}{3} - \frac{1}{4} + \cdots\right)$$

$$= -\frac{z^2e^2}{4\pi\varepsilon_0 d} \times \ln 2$$

We have used the relation $1 - \frac{1}{2} + \frac{1}{3} - \frac{1}{4} + \cdots = \ln 2$. Finally, we multiply E_P by 2 to obtain the total energy arising from interactions on each side of the ion and then multiply by Avogadro's constant, N_A, to obtain an expression for the lattice energy per mole of ions. The outcome is

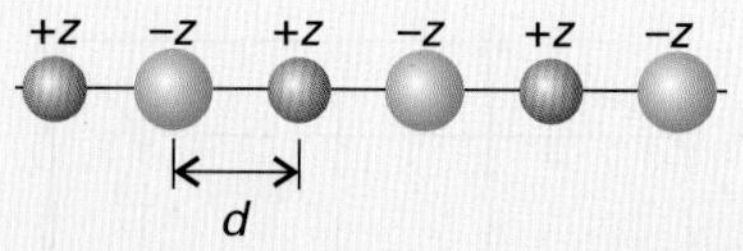

Fig. 9.39 A line of alternating cations and ions used in the calculation of the Madelung constant in one dimension.

$$E_P = -2\ln 2 \times \frac{z^2 N_A e^2}{4\pi\varepsilon_0 d}$$

with $d = r_{cation} + r_{anion}$. This energy is negative, corresponding to a net attraction. The calculation we have just performed can be extended to three-dimensional arrays of ions with different charges:

$$E_P = -A \times \frac{|z_1 z_2| N_A e^2}{4\pi\varepsilon_0 d} \qquad (9.10)$$

The factor A is a positive numerical constant called the **Madelung constant**; its value depends on how the ions are arranged about one another. For ions arranged in the same way as in sodium chloride, $A = 1.748$. Table 9.4 lists Madelung constants for other common structures.

There are also repulsions arising from the overlap of the atomic orbitals of the ions and the role of the Pauli principle. These repulsions are taken into account by supposing that, because wavefunctions decay exponentially with distance at large distances from the nucleus, and repulsive interactions depend on the overlap of orbitals, the repulsive contribution to the potential energy has the form

$$E_P^* = N_A C' e^{-d/d^*} \qquad (9.11)$$

with C' and d^* constants; the latter is commonly taken to be 34.5 pm. The total potential energy is the sum of E_P and E_P^*, and passes through a minimum when $d(E_P + E_P^*)/dd = 0$ (Fig. 9.40).

Table 9.4 Madelung constants for a selection of structural types

Structural type	A
Caesium chloride	1.763
Fluorite	2.519
Rock salt	1.748
Rutile	2.408
Sphalerite	1.638
Wurtzite	1.641

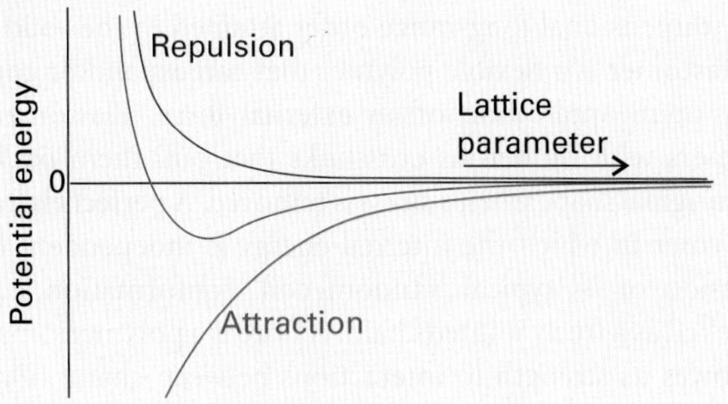

Fig. 9.40 The contributions to the total potential energy of an ionic crystal.

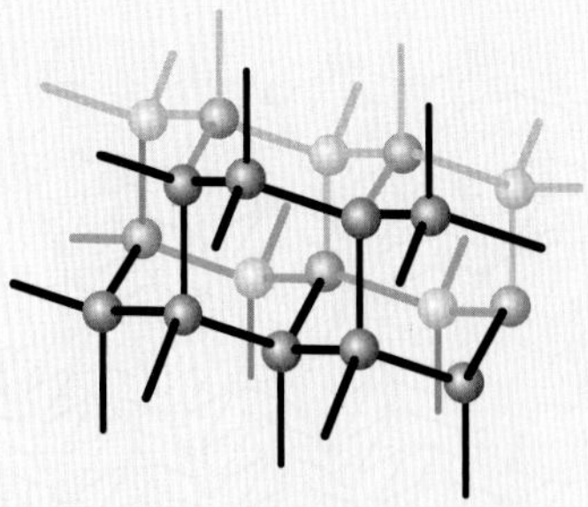

Fig. 9.41 A fragment of the structure of diamond. Each C atom is tetrahedrally bonded to four neighbours. This framework-like structure results in a rigid crystal.

A short calculation leads to the following expression for the minimum total potential energy (see Problem 9.24):

$$E_{\mathrm{P,min}} = -\frac{N_A |z_A z_B| e^2}{4\pi\varepsilon_0 d}\left(1-\frac{d^*}{d}\right)A \qquad (9.12)$$

This expression is called the **Born–Mayer equation**. Provided we ignore zero-point contributions to the energy, we can identify the negative of this potential energy with the lattice energy. We see that large lattice energies are expected when the ions are highly charged (so $|z_A z_B|$ is large) and small (so d is small).

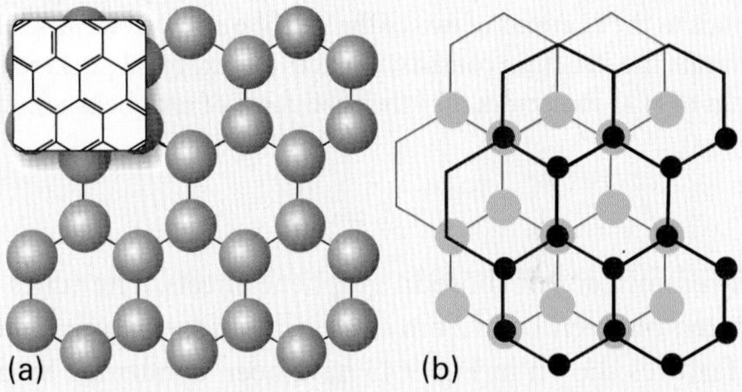

Fig. 9.42 Graphite consists of flat planes of hexagons of carbon atoms lying above one another. (a) The arrangement of carbon atoms in a sheet; (b) the relative arrangement of neighbouring sheets. When impurities are present, the planes can slide over one another easily.

9.7 Molecular solids and covalent networks

X-ray diffraction studies of solids reveal a huge amount of information, including interatomic distances, bond angles, stereochemistry, and vibrational parameters. In this section we can do no more than hint at the diversity of types of solids found when molecules pack together or atoms link together in extended networks.

In **covalent network solids**, covalent bonds in a definite spatial orientation link the atoms in a network extending through the crystal. The demands of directional bonding, which have only a small effect on the structures of many metals, now override the geometrical problem of packing spheres together, and elaborate and extensive structures may be formed. Examples include silicon, red phosphorus, boron nitride, and—very importantly—diamond, graphite, and carbon nanotubes, which we discuss in detail.

Diamond and graphite are two allotropes of carbon. In diamond each sp^3-hybridized carbon is bonded tetrahedrally to its four neighbours (Fig. 9.41). The network of strong C—C bonds is repeated throughout the crystal and, as a result, diamond is very hard (in fact, the hardest known substance). In graphite, σ bonds between sp^2-hybridized carbon atoms form hexagonal rings that, when repeated throughout a plane, give rise to 'graphene' sheets (Fig. 9.42). Because the sheets can slide against each other when impurities are present, graphite is used widely as a lubricant.

Carbon nanotubes are narrow cylinders of carbon atoms that have high tensile strength and electrical conductivity. They may be synthesized by condensing a carbon plasma either in the presence or absence of a catalyst. The simplest structural motif is called a *single-walled nanotube* (SWNT) and is shown in Fig. 9.43. In an SWNT, sp^2-hybridized carbon atoms form hexagonal

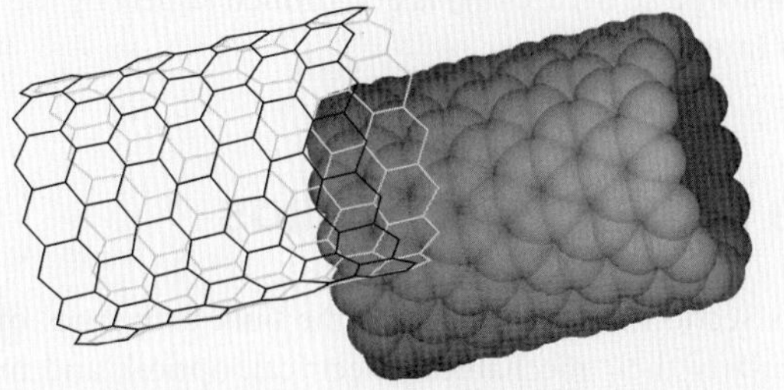

Fig. 9.43 In a single-walled nanotube (SWNT), sp^2-hybridized carbon atoms form hexagonal rings that grow as tubes with diameters between 1 and 2 nm and lengths of several micrometres.

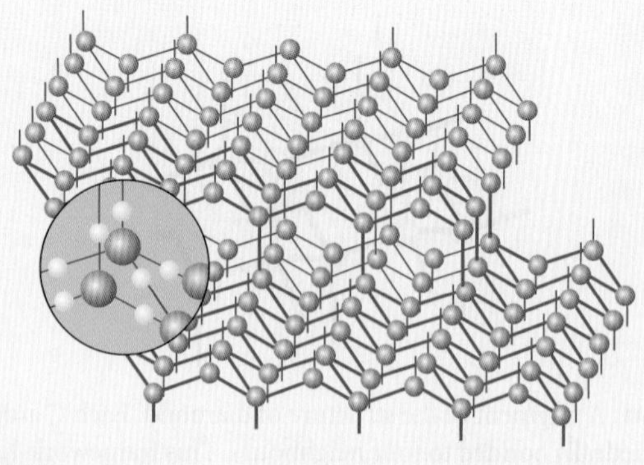

Fig. 9.44 A fragment of the crystal structure of ice (ice-I). Each O atom is at the centre of a tetrahedron of four O atoms at a distance of 276 pm. The central O atom is attached by two short O—H bonds to two H atoms and by two long hydrogen bonds to the H atoms of two of the neighbouring molecules. Overall, the structure consists of planes of hexagonal puckered rings of H_2O molecules (like the chair form of cyclohexane).

rings reminiscent of those in graphene sheets. The tubes have diameters between 1 and 2 nm and lengths of several micrometres. The features shown in Fig. 9.43 have been confirmed by direct visualization with scanning tunnelling microscopy (*Impact I2.1*). A *multi-walled nanotube* (MWNT) consists of several concentric SWNTs and its diameter varies between 2 and 25 nm.

Molecular solids, which are the subject of the overwhelming majority of modern structural determinations, are held together by van der Waals interactions (Chapter 8). The observed crystal structure is nature's solution to the problem of condensing objects of various shapes into an aggregate of minimum energy (actually, for $T > 0$, of minimum Gibbs energy). The prediction of the structure is difficult, but software specifically designed to explore interaction energies can now make reasonably reliable predictions. The problem is made more complicated by the role of hydrogen bonds, which in some cases dominate the crystal structure, as in ice (Fig. 9.44), but in others (for example, in phenol) distort a structure that is determined largely by the van der Waals interactions.

The properties of solids

In this section we consider how the bulk properties of solids, particularly their mechanical, electrical, optical, and magnetic properties, stem from the properties of their constituent atoms. The rational fabrication of modern materials depends crucially on an understanding of this link.

Synthetic polymers, which have greatly extended the types of material that are available, are classified broadly as *elastomers*, *fibres*, and *plastics*, depending on their **crystallinity**, the degree of three-dimensional long-range order attained in the solid state. An **elastomer** is a flexible polymer that can expand or contract easily upon application of an external force. Elastomers are polymers with numerous crosslinks that pull them back into their original shape when a stress is removed. A **perfect elastomer**, a polymer in which the internal energy is independent of the extension of its typical 'random coil' conformation, can be modelled as a freely jointed chain. A **fibre** is a polymeric material that owes its strength to interactions between chains. One example is nylon-66 (**1**). Under certain conditions, nylon-66 can be prepared in a state of high crystallinity, in which hydrogen bonding between the amide links of neighbouring chains results in an ordered array. A **plastic** is a polymer that can attain only a limited degree of crystallinity and as a result is neither as strong as a fibre nor as flexible as an elastomer. Certain materials, such as nylon-66, can be prepared either as a fibre or as a plastic. A sample of plastic nylon-66 may be visualized as consisting of crystalline hydrogen-bonded regions of varying size interspersed amongst amorphous, random coil regions. A single type of polymer may exhibit more than one characteristic for, to display fibrous character, the polymers need to be aligned; if the chains are not aligned, then the substance may be plastic. That is the case with nylon, poly(vinyl chloride), and the siloxanes.

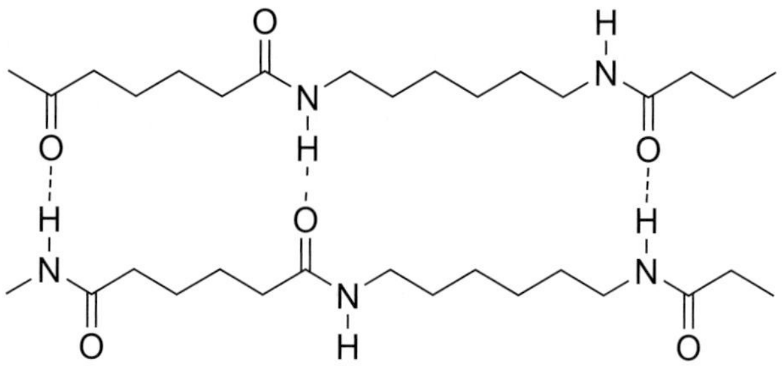

1 Nylon-66

9.8 Mechanical properties

The fundamental concepts for the discussion of the mechanical properties of solids are stress and strain. The **stress** on an object is the applied force divided by the area to which it is applied. The **strain** is the resulting distortion of the sample. The general field of the relations between stress and strain is called **rheology**.

Stress may be applied in a number of different ways. Thus, **uniaxial stress** is a simple compression or extension in one direction (Fig. 9.45); **hydrostatic stress** is a stress applied simultaneously in all directions, as in a body immersed in a fluid. A **pure shear** is a stress that tends to push opposite faces of the sample in opposite directions. A sample subjected to a small stress typically undergoes **elastic deformation** in the sense that it recovers its original shape when the stress is removed. For low stresses, the strain is linearly proportional to the stress. The response

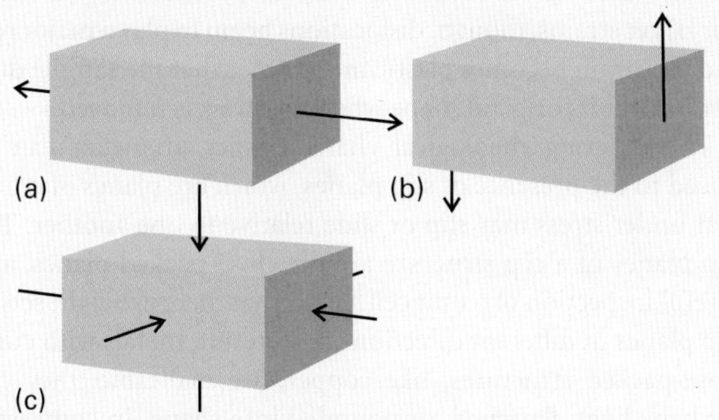

Fig. 9.45 Types of stress applied to a body. (a) Uniaxial stress; (b) shear stress; (c) hydrostatic pressure.

becomes non-linear at high stresses but may remain elastic. Above a certain threshold, the strain becomes **plastic** in the sense that recovery does not occur when the stress is removed. Plastic deformation occurs when bond breaking takes place and, in pure metals, typically takes place through the agency of dislocations. Brittle solids, such as ionic solids, exhibit sudden fracture as the stress focused by cracks causes them to spread catastrophically.

The response of a solid to an applied stress is commonly summarized by a number of coefficients of proportionality known as 'moduli':

$$\text{Young's modulus: } E = \frac{\text{normal stress}}{\text{normal strain}} \qquad [9.13a]$$

$$\text{Bulk modulus: } K = \frac{\text{pressure}}{\text{fractional change in volume}} \qquad [9.13b]$$

$$\text{Shear modulus: } G = \frac{\text{shear stress}}{\text{shear strain}} \qquad [9.13c]$$

where 'normal stress' refers to stretching and compression of the material, as shown in Fig 9.46a and 'shear stress' refers to the stress depicted in Fig 9.46b. The bulk modulus is the inverse of the **isothermal compressibility,** κ,

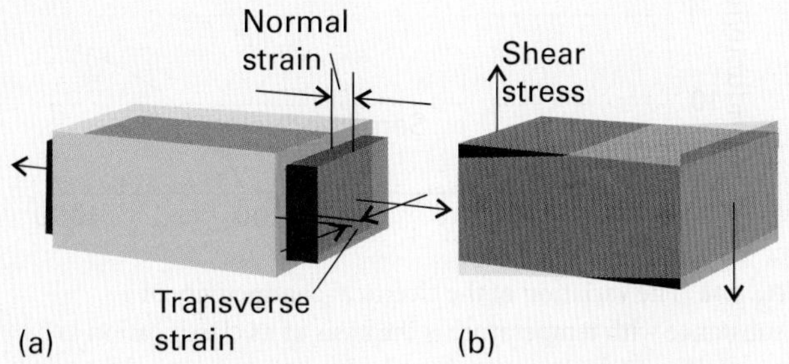

Fig. 9.46 (a) Normal stress and the resulting strain. (b) Shear stress. Poisson's ratio indicates the extent to which a body changes shape when subjected to a uniaxial stress.

$$K = \frac{1}{\kappa} \qquad \kappa = -\frac{1}{V}\left(\frac{\partial V}{\partial p}\right)_T \qquad [9.14]$$

A third ratio indicates how the sample changes its shape:

$$\text{Poisson's ratio: } \nu_P = \frac{\text{transverse strain}}{\text{normal strain}} \qquad [9.15]$$

The moduli are interrelated:

$$G = \frac{E}{2(1+\nu_P)} \qquad K = \frac{E}{3(1-2\nu_P)} \qquad (9.16)$$

If neighbouring molecules interact by a Lennard-Jones potential (Section 8.6), then we show in the following *Justification* that the bulk modulus and the compressibility of the solid are related to the Lennard-Jones parameter ε (the depth of the potential well) by

$$K = \frac{8N_A\varepsilon}{V_m} \qquad (9.17)$$

We see that the bulk modulus is large (the solid stiff) if the potential well represented by the Lennard-Jones potential is deep and the solid is dense (its molar volume small).

Justification 9.4 *The bulk modulus*

At $T = 0$, the thermodynamic relation between the internal energy U and the entropy S and volume V, $dU = TdS - pdV$ (see Section 15.7), becomes $dU = -pdV$, so $p = -(\partial U/\partial V)_T$ and

$$K = \frac{1}{\kappa} = -\frac{V}{(\partial V/\partial p)_T} = -V\left(\frac{\partial p}{\partial V}\right)_T = V\left(\frac{\partial^2 U}{\partial V^2}\right)_T$$

The volume of a sample is proportional to r^3, where r is the interatomic separation, so we write $V = ar^3$ with a a constant. Hence $dV/dr = 3ar^2$. Since, at constant temperature,

$$\left(\frac{\partial U}{\partial V}\right) = \left(\frac{\partial U}{\partial r}\right)\left(\frac{\partial r}{\partial V}\right)$$

we find

$$\left(\frac{\partial^2 U}{\partial V^2}\right) = \frac{\partial}{\partial V}\left\{\left(\frac{\partial U}{\partial r}\right)\left(\frac{\partial r}{\partial V}\right)\right\} = \left(\frac{\partial^2 U}{\partial r^2}\right)\left(\frac{\partial r}{\partial V}\right)^2 + \left(\frac{\partial U}{\partial r}\right)\left(\frac{\partial^2 r}{\partial V^2}\right)$$

and since, at the equilibrium volume of the sample (that is, at the equilibrium interatomic separation), $(\partial U/\partial r) = 0$,

$$\left(\frac{\partial^2 U}{\partial V^2}\right) = \left(\frac{\partial^2 U}{\partial r^2}\right)\left(\frac{\partial r}{\partial V}\right)^2 = \left(\frac{\partial^2 U}{\partial r^2}\right) \times \left(\frac{1}{3ar^2}\right)^2$$

Therefore, the bulk modulus at equilibrium volume is

$$K = \frac{V}{9a^2r^4}\left(\frac{\partial^2 U}{\partial r^2}\right)_T$$

The internal energy at $T = 0$ can be expressed in terms of the Lennard-Jones potential as

$$U = 4N\varepsilon\left\{\left(\frac{r_0}{r}\right)^{12} - \left(\frac{r_0}{r}\right)^{6}\right\}$$

with $r = 2^{1/6}r_0$ at the equilibrium interatomic separation (when $dU/dr = 0$). Therefore, at this separation

$$\left(\frac{\partial^2 U}{\partial r^2}\right)_T = 4N\varepsilon\left\{156\frac{r_0^{12}}{r^{14}} - 42\frac{r_0^{6}}{r^{8}}\right\} \overset{\text{at } r=2^{1/6}r_0}{=} \frac{4N\varepsilon}{r_0^2}\left\{\frac{156}{2^{7/3}} - \frac{42}{2^{4/3}}\right\}$$

It follows that the bulk modulus at the equilibrium separation ($r = 2^{1/6}r_0$) is

$$K = \frac{V}{9a^2(2^{1/6}r_0)^4} \times \frac{4N\varepsilon}{r_0^2}\left\{\frac{156}{2^{7/3}} - \frac{42}{2^{4/3}}\right\} = \frac{4VN\varepsilon}{a^2r_0^6}$$

Finally, we recognize that $ar_0^3 = ar^3/2^{1/2} = V/2^{1/2}$, and obtain

$$K = \frac{4VN\varepsilon}{(V/2^{1/2})^2} = \frac{8\varepsilon}{V_{mol}}$$

where $V_{mol} = V/N$ is the molecular volume. Multiplication of the numerator and denominator by Avogadro's constant and noting that $N_A V_{mol} = V_m$, the molar volume, then gives eqn 9.17.

The typical behaviour of a solid under stress is illustrated in Fig. 9.47. For small strains, the stress–strain relation is a Hooke's law of force, with the strain directly proportional to the stress. For larger strains, though, dislocations begin to play a major role and the strain becomes plastic in the sense that the sample does not recover its original shape when the stress is removed.

The differing rheological characteristics of metals can be traced to the presence of **slip planes**, which are planes of atoms that under stress may slip or slide relative to one another. The slip planes of a ccp structure are the close-packed planes, and careful inspection of a unit cell shows that there are eight sets of slip planes in different directions. As a result, metals with cubic close-packed structures, like copper, are malleable: they can easily be bent, flattened, or pounded into shape. In contrast, a hexagonal close-packed structure has only one set of slip planes; therefore metals with hexagonal close packing, like zinc or cadmium, tend to be brittle.

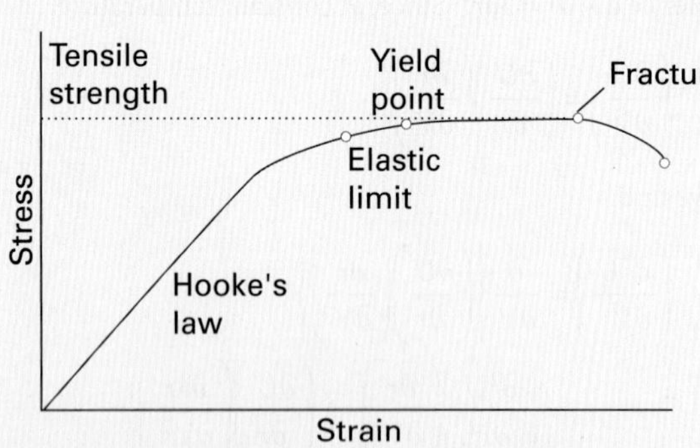

Fig. 9.47 At small strains, a body obeys Hooke's law (stress proportional to strain) and is elastic (recovers its shape when the stress is removed). At high strains, the body is no longer elastic and may yield and become plastic. At even higher strains, the solid fails (at its limiting tensile strength) and finally fractures.

9.9 Electrical properties

We shall confine attention to electronic conductivity, but note that some ionic solids display ionic conductivity. Two types of solid are distinguished by the temperature dependence of their electrical conductivity (Fig. 9.48):

A **metallic conductor** is a substance with a conductivity that decreases as the temperature is raised.

A **semiconductor** is a substance with a conductivity that increases as the temperature is raised.

A semiconductor generally has a lower conductivity than that typical of metals, but the magnitude of the conductivity is not the criterion of the distinction. It is conventional to classify semiconductors with very low electrical conductivities, such as most synthetic polymers, as **insulators**. We shall use this term, but it should be appreciated that it is one of convenience rather

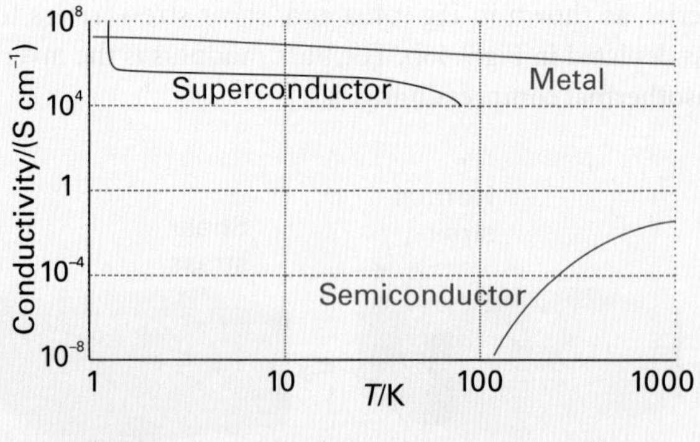

Fig. 9.48 The variation of the electrical conductivity of a substance with temperature is the basis of its classification as a metallic conductor, a semiconductor, or a superconductor. Conductivity is expressed in siemens per metre (S m^{-1} or, as here, S cm^{-1}), where 1 S = 1 Ω^{-1} (the resistance is expressed in ohms, Ω).

than one of fundamental significance. A **superconductor** is a solid that conducts electricity without resistance.

(a) The formation of bands

The central aspect of solids that determines their electrical properties is the distribution of their electrons. There are two models of this distribution. In one, the **nearly free-electron approximation**, the valence electrons are assumed to be trapped in a box with a periodic potential, with low energy corresponding to the locations of cations. In the **tight-binding approximation**, the valence electrons are assumed to occupy molecular orbitals delocalized throughout the solid. The latter model is more in accord with the discussion in the foregoing chapters, and we confine our attention to it.

We shall consider a one-dimensional solid, which consists of a single, infinitely long line of atoms. At first sight, this model may seem too restrictive and unrealistic. However, not only does it give us the concepts we need to understand conductivity in three-dimensional, macroscopic samples of metals and semiconductors, it is also the starting point for the description of long and thin structures, such as the carbon nanotubes discussed earlier in the chapter.

Suppose that each atom has one s orbital available for forming molecular orbitals. We can construct the LCAO-MOs of the solid by adding N atoms in succession to a line, and then infer the electronic structure by using the building-up principle. One atom contributes one s orbital at a certain energy (Fig. 9.49). When a second atom is brought up it overlaps the first and forms bonding and antibonding orbitals. The third atom overlaps its nearest neighbour (and only slightly the next-nearest) and, from these three atomic orbitals, three molecular orbitals are formed: one is fully bonding, one fully antibonding, and the intermediate orbital is non-bonding between neighbours. The fourth atom leads to the formation of a fourth molecular orbital. At this stage, we can begin to see that the general effect of bringing up successive atoms is to spread the range of energies covered by the molecular orbitals, and also to fill in the range of energies with more and more orbitals (one more for each atom). When N atoms have been added to the line, there are N molecular orbitals covering a band of energies of finite width, and the Hückel secular determinant (Section 5.7) is

$$\begin{vmatrix} \alpha-E & \beta & 0 & 0 & 0 & \cdots & 0 \\ \beta & \alpha-E & \beta & 0 & 0 & \cdots & 0 \\ 0 & \beta & \alpha-E & \beta & 0 & \cdots & 0 \\ 0 & 0 & \beta & \alpha-E & \beta & \cdots & 0 \\ 0 & 0 & 0 & \beta & \alpha-E & \cdots & 0 \\ \vdots & \vdots & \vdots & \vdots & \vdots & \cdots & \vdots \\ 0 & 0 & 0 & 0 & 0 & \cdots & \alpha-E \end{vmatrix} = 0$$

where β is now the (s,s) resonance integral. The theory of determinants applied to such a symmetrical example as this (technically a 'tridiagonal determinant') leads to the following expression for the roots:

$$E_k = \alpha + 2\beta\cos\frac{k\pi}{N+1} \qquad k = 1, 2, \ldots, N \qquad (9.18)$$

When N is infinitely large, the difference between neighbouring energy levels (the energies corresponding to k and $k+1$) is infinitely small, but, as we show in the following *Justification*, the band still has finite width overall:

$$E_N - E_1 \rightarrow 4\beta \text{ as } N \rightarrow \infty \qquad (9.19)$$

We can think of this band as consisting of N different molecular orbitals, the lowest-energy orbital ($k = 1$) being fully bonding, and the highest-energy orbital ($k = N$) being fully antibonding between adjacent atoms (Fig. 9.50). Similar bands form in three-dimensional solids.

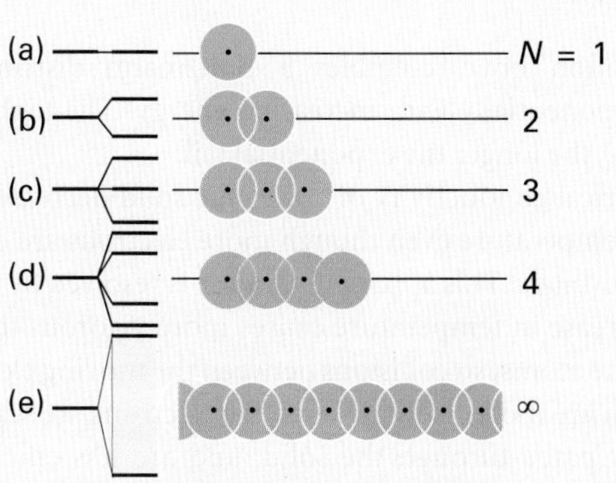

Fig. 9.49 The formation of a band of N molecular orbitals by successive addition of N atoms to a line. Note that the band remains of finite width as N becomes infinite and, although it looks continuous, it consists of N different orbitals.

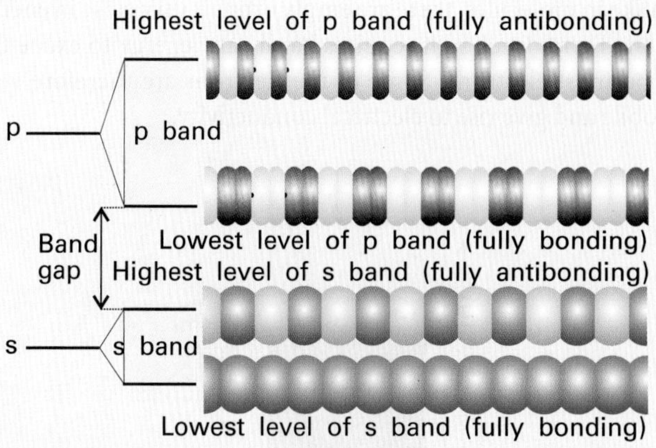

Fig. 9.50 The overlap of s orbitals gives rise to an s band and the overlap of p orbitals gives rise to a p band. In this case, the s and p orbitals of the atoms are so widely spaced that there is a band gap. In many cases the separation is less and the bands overlap.

Justification 9.5 *The width of a band*

The energy of the level with $k = 1$ is

$$E_1 = \alpha + 2\beta \cos \frac{\pi}{N+1}$$

As N becomes infinite, the cosine term becomes $\cos 0 = 1$. Therefore, in this limit $E_1 = \alpha + 2\beta$. When k has its maximum value of N,

$$E_N = \alpha + 2\beta \cos \frac{N\pi}{N+1}$$

As N approaches infinity, we can ignore the 1 in the denominator, and the cosine term becomes $\cos \pi = -1$. Therefore, in this limit $E_N = \alpha - 2\beta$. The difference between the upper and lower energies of the band is therefore 4β.

The band formed from overlap of s orbitals is called the **s band.** If the atoms have p orbitals available, the same procedure leads to a **p band** (as shown in the upper half of Fig. 9.50). If the atomic p orbitals lie higher in energy than the s orbitals, then the p band lies higher than the s band, and there may be a **band gap**, a range of energies to which no orbital corresponds. However, the s and p bands may also be contiguous or even overlap (as is the case for the 3s and 3p bands in magnesium).

(b) The occupation of orbitals

Now consider the electronic structure of a solid formed from atoms each able to contribute one electron (for example, the alkali metals). There are N atomic orbitals and therefore N molecular orbitals packed into an apparently continuous band. There are N electrons to accommodate.

At $T = 0$, only the lowest $\frac{1}{2}N$ molecular orbitals are occupied (Fig. 9.51), and the HOMO is called the **Fermi level**. However, unlike in molecules, there are empty orbitals very close in energy to the Fermi level, so it requires hardly any energy to excite the uppermost electrons. Some of the electrons are therefore very mobile and give rise to electrical conductivity.

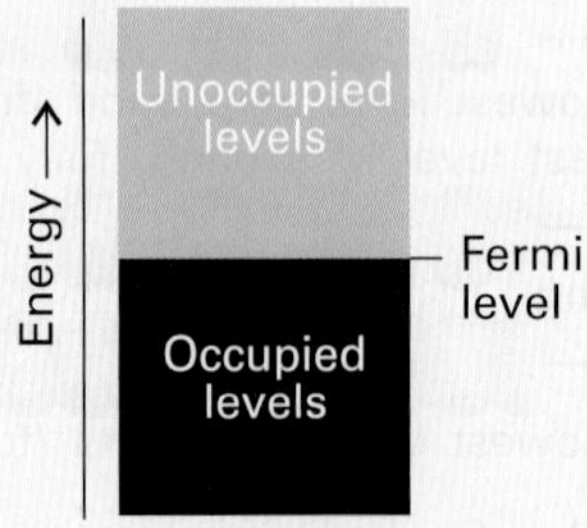

Fig. 9.51 When N electrons occupy a band of N orbitals, it is only half full and the electrons near the Fermi level (the top of the occupied levels) are mobile.

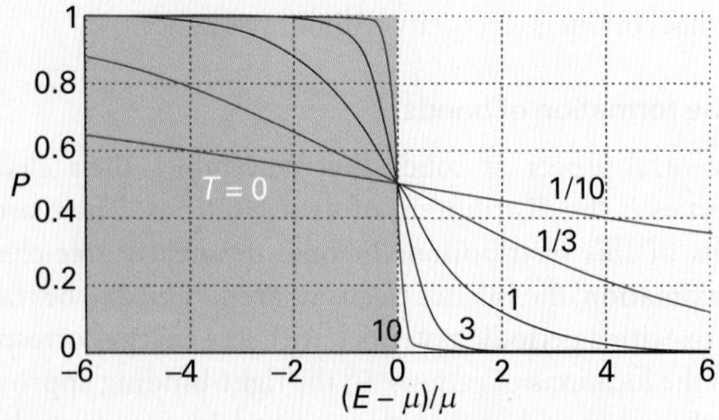

Fig. 9.52 The Fermi–Dirac distribution, which gives the population of the levels at a temperature T. The high-energy tail decays exponentially towards zero. The curves are labelled with the value of μ/kT. The tinted green region shows the occupation of levels at $T = 0$.

interActivity Express the population P as a function of the variables $(E - \mu)/\mu$ and μ/kT and then display the set of curves shown in Fig. 9.52 as a single surface.

At temperatures above absolute zero, electrons can be excited by the thermal motion of the atoms. The population, P, of the orbitals is given by the **Fermi–Dirac distribution**, a version of the Boltzmann distribution that takes into account the effect of the Pauli principle:

$$P = \frac{1}{e^{(E-\mu)/kT} + 1} \tag{9.20}$$

The quantity μ is the **chemical potential**, which in this context is the energy of the level for which $P = \frac{1}{2}$ (note that the chemical potential decreases as the temperature increases, see Problem 9.32).

The shape of the Fermi–Dirac distribution is shown in Fig. 9.52. For energies well above μ, the 1 in the denominator can be neglected, and then

$$P \approx e^{-(E-\mu)/kT} \tag{9.21}$$

The population now resembles a Boltzmann distribution, decaying exponentially with increasing energy. The higher the temperature, the longer the exponential tail.

The electrical conductivity of a metallic solid decreases with increasing temperature even though more electrons are excited into empty orbitals. This apparent paradox is resolved by noting that the increase in temperature causes more vigorous thermal motion of the atoms, so collisions between the moving electrons and an atom are more likely. That is, the electrons are scattered out of their paths through the solid, and are less efficient at transporting charge.

(c) Insulators and semiconductors

When each atom provides two electrons, the $2N$ electrons fill the N orbitals of the s band. The Fermi level now lies at the top of the

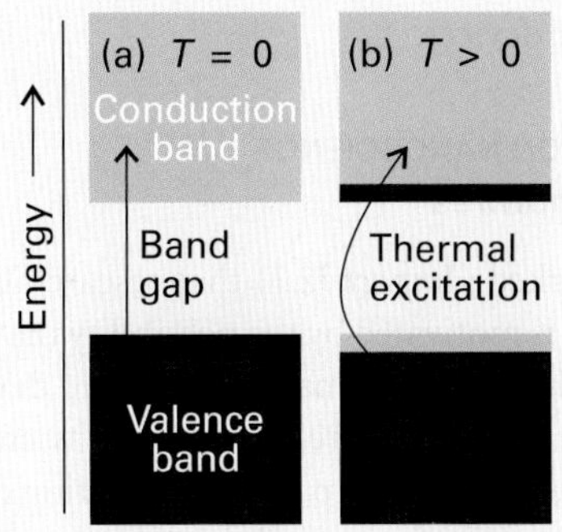

Fig. 9.53 (a) When $2N$ electrons are present, the band is full and the material is an insulator at $T=0$. (b) At temperatures above $T=0$, electrons populate the levels of the upper *conduction band* and the solid is a semiconductor.

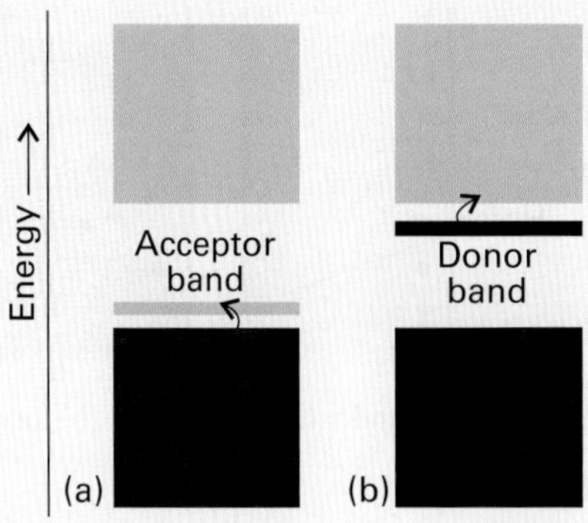

Fig. 9.54 (a) A dopant with fewer electrons than its host can form a narrow band that accepts electrons from the valence band. The holes in the band are mobile and the substance is a p-type semiconductor. (b) A dopant with more electrons than its host forms a narrow band that can supply electrons to the conduction band. The electrons it supplies are mobile and the substance is an n-type semiconductor.

band (at $T=0$), and there is a gap before the next band begins (Fig. 9.53). As the temperature is increased, the tail of the Fermi–Dirac distribution extends across the gap, and electrons leave the lower band, which is called the **valence band**, and populate the empty orbitals of the upper band, which is called the **conduction band**. As a consequence of electron promotion, positively charged 'holes' are left in the valence band. The holes and promoted electrons are now mobile, and the solid is an electrical conductor. In fact, it is a semiconductor, because the electrical conductivity depends on the number of electrons that are promoted across the gap, and that number increases as the temperature is raised. If the gap is large, though, very few electrons will be promoted at ordinary temperatures and the conductivity will remain close to zero, resulting in an insulator. Thus, the conventional distinction between an insulator and a semiconductor is related to the size of the band gap and is not an absolute distinction like that between a metal (incomplete bands at $T=0$) and a semiconductor (full bands at $T=0$).

Figure 9.53 depicts conduction in an **intrinsic semiconductor**, in which semiconduction is a property of the band structure of the pure material. Examples of intrinsic semiconductors include silicon and germanium. A **compound semiconductor** is an intrinsic semiconductor that is a combination of different elements, such as GaN, CdS, and many d-metal oxides. An **extrinsic semiconductor** is one in which charge carriers are present as a result of the replacement of some atoms (to the extent of about 1 in 10^9) by **dopant** atoms, the atoms of another element. If the dopants can trap electrons, they withdraw electrons from the filled band, leaving holes which allow the remaining electrons to move (Fig. 9.54a). This procedure gives rise to **p-type semiconductivity**, the p indicating that the holes are positive relative to the electrons in the band. An example is silicon doped with indium. We can picture the semiconduction as arising from the transfer of an electron from a Si atom to a neighbouring In atom. The electrons at the top of the silicon valence band are now mobile, and carry current through the solid. Alternatively, a dopant might carry excess electrons (for example, phosphorus atoms introduced into germanium), and these additional electrons occupy otherwise empty bands, giving **n-type semiconductivity**, where n denotes the negative charge of the carriers (Fig. 9.54b).

Now we consider the properties of a **p–n junction**, the interface of a p-type and n-type semiconductor. Consider the application of a 'reverse bias' to the junction, in the sense that a negative electrode is attached to the p-type semiconductor and a positive electrode is attached to the n-type semiconductor (Fig. 9.55a). Under these conditions, the positively charged holes in the p-type semiconductor are attracted to the negative electrode and the negatively charged electrons in the n-type semiconductor are attracted to the positive electrode. As a consequence, charge does not flow across the junction. Now consider the application of a 'forward bias' to the junction, in the sense that the positive electrode is attached to the p-type semiconductor and the negative electrode is attached to the n-type semiconductor (Fig. 9.55b). Now charge flows across the junction, with electrons in the n-type semiconductor moving toward the positive electrode and holes moving in the opposite direction. It follows that a p–n junction affords a great deal of control over the magnitude and direction of current through a material. This control is essential for the operation of transistors and diodes, which are key components of modern electronic devices.

As electrons and holes move across a p–n junction under forward bias, they recombine and release energy. However, as long as the forward bias continues to be applied, the flow of charge from the electrodes to the semiconductors will replenish them with electrons and holes, so the junction will sustain a current.

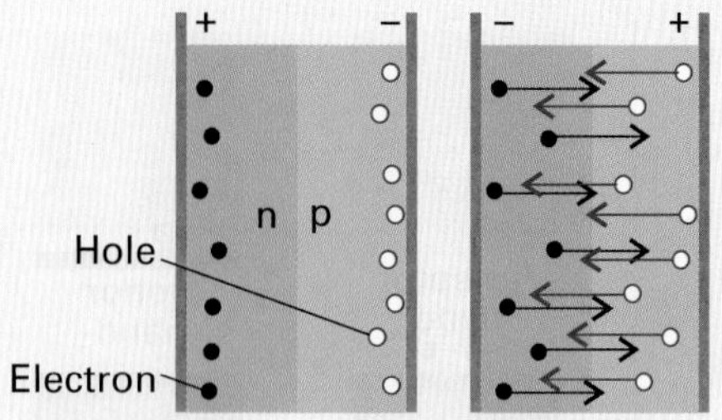

Fig. 9.55 A p–n junction under (a) reverse bias, (b) forward bias.

In some solids, the energy of electron–hole recombination is released as heat and the device becomes warm. This is the case for silicon semiconductors, and is one reason why computers need efficient cooling systems.

IMPACT ON TECHNOLOGY
I9.2 Conducting polymers

A variety of newly developed macromolecular materials have electrical conductivities that rival those of silicon-based semiconductors. Here we shall examine **conducting polymers**, in which extensively conjugated double bonds facilitate electron conduction along the polymer chain.

One example of a conducting polymer is polyacetylene (Fig. 9.56). Whereas the delocalized π bonds do suggest that electrons can move up and down the chain, the electrical conductivity of polyacetylene increases significantly when it is partially oxidized by I_2 and other strong oxidants. Oxidation leads to the formation of a **polaron**, a partially localized cation radical that travels through the chain, as shown in the illustration. Further oxidation of the polymer forms either **bipolarons**, a di-cation that moves as a unit through the chain, or two separate cation radicals that move independently.

Conducting polymers are currently used in a number of devices, such as electrodes in batteries, electrolytic capacitors, and sensors. Recent studies of photon emission by conducting polymers may lead to new technologies for light-emitting diodes and flat-panel displays. Conducting polymers also show promise as molecular wires that can be incorporated into nanometre-sized electronic devices.

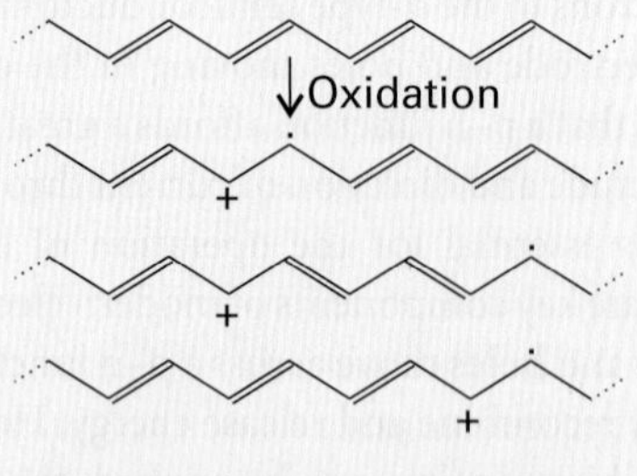

Fig. 9.56 The mechanism of migration of a partially localized cation radical, or polaron, in polyacetylene.

IMPACT ON NANOSCIENCE
I9.3 Nanowires

We have remarked (*Impact I2.1*) that research on nanometre-sized materials is motivated by the possibility that they will form the basis for cheaper and smaller electronic devices. The synthesis of *nanowires*, nanometre-sized atomic assemblies that conduct electricity, is a major step in the fabrication of nanodevices. An important type of nanowire is based on carbon nanotubes, which, like graphite, can conduct electrons through delocalized π molecular orbitals that form from unhydridized 2p orbitals on carbon. Recent studies have shown a correlation between structure and conductivity in single-walled nanotubes (SWNTs) that does not occur in graphite. The SWNT in Fig. 9.43 is a semiconductor. If the hexagons are rotated by 90° about their sixfold axis, the resulting SWNT is a metallic conductor.

Carbon nanotubes are promising building blocks not only because they have useful electrical properties but also because they have unusual mechanical properties. For example, an SWNT has a Young's modulus that is approximately five times larger and a tensile strength that is approximately 375 times larger than that of steel.

Silicon nanowires can be made by focusing a pulsed laser beam on to a solid target composed of silicon and iron. The beam ejects Fe and Si atoms from the surface of the target, forming a vapour that can condense into liquid $FeSi_n$ nanoclusters at sufficiently low temperatures. Solid silicon and liquid $FeSi_n$ coexist at temperatures higher than 1473 K. Hence, it is possible to precipitate solid silicon from the mixture if the experimental conditions are controlled to maintain the $FeSi_n$ nanoclusters in a liquid state that is supersaturated with silicon. The silicon precipitate consists of nanowires with diameters of about 10 nm and lengths greater than 1 μm.

Nanowires are also fabricated by *molecular beam epitaxy* (MBE), in which gaseous atoms or molecules are sprayed on to a crystalline surface in an ultra-high vacuum chamber. The result is the formation of highly ordered structures. Through careful control of the chamber temperature and of the spraying process, it is possible to deposit thin films on to a surface or to create nanometre-sized assemblies with specific shapes. Germanium nanowires can be made by this method (see Fig. 2.15).

9.10 Optical properties

From the discussion in earlier chapters, we are already familiar with the factors that determine the energy and intensity of light absorbed by atoms and molecules in the gas phase and in solution. Now we consider the effects on the electronic absorption spectrum of bringing atoms or molecules together into a solid.

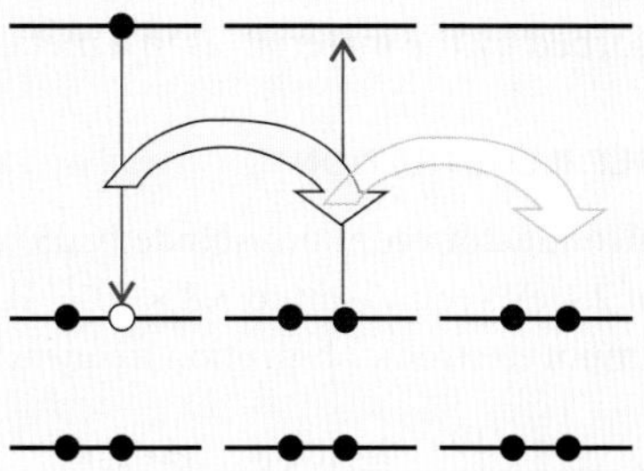

Fig. 9.57 The electron–hole pair shown on the left can migrate through a solid lattice as the excitation hops from molecule to molecule. The mobile excitation is called an exciton.

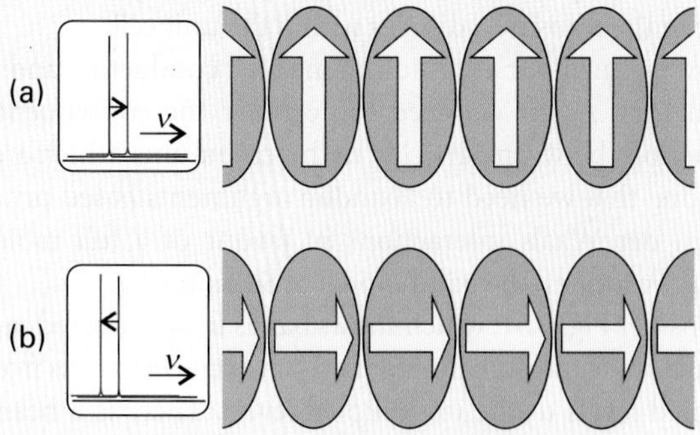

Fig. 9.58 (a) The alignment of transition dipoles (the yellow arrows) is energetically unfavourable, and the exciton absorption is shifted to higher energy (higher frequency). (b) The alignment is energetically favourable for a transition in this orientation, and the exciton band occurs at a lower frequency than in the isolated molecules.

Consider an electronic excitation of a molecule (or an ion) in a crystal. If the excitation corresponds to the removal of an electron from one orbital of a molecule and its elevation to an orbital of higher energy, then the excited state of the molecule can be envisaged as the coexistence of an electron and a hole. This electron–hole pair, a particle-like **exciton**, migrates from molecule to molecule in the crystal (Fig. 9.57). Exciton formation causes spectral lines to shift, split, and change intensity.

The electron and the hole jump together from molecule to molecule as they migrate. A migrating excitation of this kind is called a **Frenkel exciton**. The electron and hole can also be on different molecules, but in each other's vicinity. A migrating excitation of this kind, which is now spread over several molecules (more usually ions), is a **Wannier exciton**.

Frenkel excitons are more common in molecular solids. Their migration implies that there is an interaction between the species that constitute the crystal, for otherwise the excitation on one unit could not move to another. This interaction affects the energy levels of the system. The strength of the interaction governs the rate at which an exciton moves through the crystal: a strong interaction results in fast migration, and a vanishingly small interaction leaves the exciton localized on its original molecule. The specific mechanism of interaction that leads to exciton migration is the interaction between the transition dipole moments of the excitation. Thus, an electric dipole transition in a molecule is accompanied by a shift of charge, and the transition dipole exerts a force on an adjacent molecule. The latter responds by shifting its charge. This process continues and the excitation migrates through the crystal.

The energy shift arising from the interaction between transition dipoles can be understood in terms of their electrostatic interaction. An all-parallel arrangement of the dipoles (Fig. 9.58a) is energetically unfavourable, so the absorption occurs at a higher frequency than in the isolated molecule, corresponding to a *blue shift* in the absorption. Conversely, a head-to-tail alignment of transition dipoles (Fig. 9.58b) is energetically favourable, and the transition occurs at a lower frequency than in the isolated molecules, corresponding to a *red shift*.

If there are N molecules per unit cell, there are N **exciton bands** in the spectrum (if all of them are allowed). The splitting between the bands is the **Davydov splitting.** To understand the origin of the splitting, consider the case $N = 2$ with the molecules arranged as in Fig. 9.59. Let the transition dipoles be along the length of the molecules. The radiation stimulates the collective excitation of the transition dipoles that are in-phase between neighbouring unit cells. *Within* each unit cell the transition dipoles may be arrayed in the two different ways shown in the figure. Since the two orientations correspond to different interaction energies, with interaction being repulsive in one and attractive in the other, the two transitions appear in the spectrum at two bands of different frequencies. The magnitude of

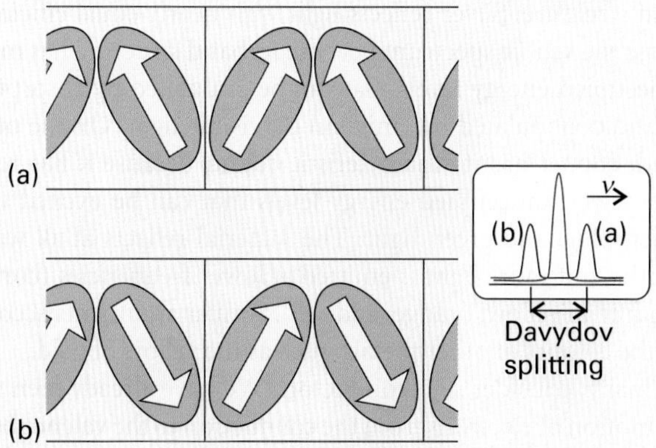

Fig. 9.59 When the transition moments within a unit cell may lie in different relative directions, as depicted in (a) and (b), the energies of the transitions are shifted and give rise to the two bands labelled (a) and (b) in the spectrum. The separation of the bands is the Davydov splitting.

the Davydov splitting is determined by the energy of interaction between the transition dipoles within the unit cell.

Now we turn our attention to metallic conductors and semiconductors. Again we need to consider the consequences of interactions between particles, in this case atoms, which are now so strong that we need to abandon arguments based primarily on van der Waals interactions in favour of a full molecular orbital treatment, the band model of Section 9.9.

Consider Fig. 9.51, which shows bands in an idealized metallic conductor. The absorption of light can excite electrons from the occupied levels to the unoccupied levels. There is a near continuum of unoccupied energy levels above the Fermi level, so we expect to observe absorption over a wide range of frequencies. In metals, the bands are sufficiently wide that radiation from the radiofrequency to the middle of the ultraviolet region of the electromagnetic spectrum is absorbed (metals are transparent to very high-frequency radiation, such as X-rays and γ-rays). Because this range of absorbed frequencies includes the entire visible spectrum, we expect that all metals should appear black. However, we know that metals are lustrous (that is, they reflect light) and some are coloured (that is, they absorb light of only certain wavelengths), so we need to extend our model.

To explain the lustrous appearance of a smooth metal surface, we need to realize that the absorbed energy can be re-emitted very efficiently as light, with only a small fraction of the energy being released to the surroundings as heat. Because the atoms near the surface of the material absorb most of the radiation, emission also occurs primarily from the surface. In essence, if the sample is excited with visible light, then visible light will be reflected from the surface, accounting for the lustre of the material.

The perceived colour of a metal depends on the frequency range of reflected light which, in turn, depends on the frequency range of light that can be absorbed and, by extension, on the band structure. Silver reflects light with nearly equal efficiency across the visible spectrum because its band structure has many unoccupied energy levels that can be populated by absorption of, and depopulated by emission of, visible light. On the other hand, copper has its characteristic colour because it has relatively fewer unoccupied energy levels that can be excited with violet, blue, and green light. The material reflects at all wavelengths, but more light is emitted at lower frequencies (corresponding to yellow, orange, and red). Similar arguments account for the colours of other metals, such as the yellow of gold.

Finally, consider semiconductors. We have already seen that promotion of electrons from the conduction to the valence band of a semiconductor can be the result of thermal excitation, if the band gap E_g is comparable to the energy that can be supplied by heating. In some materials, the band gap is very large and electron promotion can occur only by excitation with electromagnetic radiation. However, we see from Fig. 9.53 that there is a frequency $\nu_{min} = E_g/h$ below which light absorption cannot occur. Above this frequency threshold, a wide range of frequencies can be absorbed by the material, as in a metal.

● A BRIEF ILLUSTRATION

The semiconductor cadmium sulfide (CdS) has a band gap energy of 2.4 eV (equivalent to 3.8×10^{-19} J). It follows that the minimum electronic absorption frequency is

$$\nu_{min} = \frac{3.8 \times 10^{-19}\ \text{J}}{6.626 \times 10^{-34}\ \text{J}} = 5.8 \times 10^{14}\ \text{s}^{-1}$$

This frequency, of 5.8×10^{14} Hz, corresponds to a wavelength of 517 nm (green light). Lower frequencies, corresponding to yellow, orange, and red, are not absorbed and consequently CdS appears yellow-orange. ●

Self-test 9.7 Predict the colours of the following materials, given their band-gap energies (in parentheses): GaAs (1.43 eV), HgS (2.1 eV), and ZnS (3.6 eV).

[Black, red, and colourless]

9.11 Magnetic properties

The magnetic properties of metallic solids and semiconductors depend strongly on the band structures of the material. Here we confine our attention largely to magnetic properties that stem from collections of individual molecules or ions such as d-metal complexes. Much of the discussion applies to liquid and gas phase samples as well as to solids.

(a) Magnetic susceptibility

The magnetic and electric properties of molecules and solids are analogous. For instance, some molecules possess permanent magnetic dipole moments, and an applied magnetic field can induce a magnetic moment, with the result that the entire solid sample becomes magnetized. The analogue of the electric polarization, P, is the **magnetization**, $\mathcal{M}$, the average molecular magnetic dipole moment multiplied by the number density of molecules in the sample. The magnetization induced by a field of strength $\mathcal{H}$ is proportional to $\mathcal{H}$, and we write

$$\mathcal{M} = \chi\mathcal{H} \qquad [9.22]$$

where χ is the dimensionless **volume magnetic susceptibility**. A closely related quantity is the **molar magnetic susceptibility**, χ_m:

$$\chi_m = \chi V_m \qquad [9.23]$$

where V_m is the molar volume of the substance (we shall soon see why it is sensible to introduce this quantity). The **magnetic induction**, $\mathcal{B}$, is related to the applied field strength and the magnetization by

$$\mathcal{B} = \mu_0(\mathcal{H} + \mathcal{M}) = \mu_0(1 + \chi)\mathcal{H} \qquad [9.24]$$

where μ_0 is the **vacuum permeability**, a fundamental constant with the defined value $\mu_0 = 4\pi \times 10^{-7}$ J C^{-2} m^{-1} s^2. The magnetic induction, which is also called the *magnetic flux density*, can be thought of as the density of magnetic lines of force permeating the medium. This density is increased if $\mathcal{M}$ adds to $\mathcal{H}$ (when $\chi > 0$), but the density is decreased if $\mathcal{M}$ opposes $\mathcal{H}$ (when $\chi < 0$). Materials for which $\chi > 0$ are called **paramagnetic.** Those for which $\chi < 0$ are called **diamagnetic.** As we shall see, a paramagnetic material consists of ions or molecules with unpaired electrons, such as radicals and many d-metal complexes; a diamagnetic substance (a far more common property) is one with no unpaired electrons.

Just as polar molecules in fluid phases contribute a term proportional to $\mu^2/3kT$ to the electric polarization of a medium (the Debye equation, Problem 8.6), so molecules with a permanent magnetic dipole moment of magnitude m contribute to the magnetization an amount proportional to $m^2/3kT$. However, unlike for polar molecules, this contribution to the magnetization is obtained even for paramagnetic species trapped in solids, because the direction of the spin of the electrons is typically not coupled to the orientation of the molecular framework and so contributes even when the nuclei are stationary. An applied field can also induce a magnetic moment by stirring up currents in the electron distribution like those responsible for the chemical shift in NMR (Section 12.5). The constant of proportionality between the induced moment and the applied field is called the **magnetizability,** ξ (xi), and the magnetic analogue of the Debye equation is

$$\chi = \mathcal{N}\mu_0\left(\xi + \frac{m^2}{3kT}\right) \qquad (9.25a)$$

We can now see why it is convenient to introduce χ_m, because the product of the number density $\mathcal{N}$ and the molar volume is Avogadro's constant, N_A:

$$\mathcal{N}V_m = \frac{NV_m}{V} = \frac{nN_AV_m}{nV_m} = N_A$$

Hence

$$\chi_m = N_A\mu_0\left(\xi + \frac{m^2}{3kT}\right) \qquad (9.25b)$$

and the density dependence of the susceptibility (which occurs in eqn 9.25a via $\mathcal{N} = N_A\rho/M$) has been eliminated. The expression for χ_m is in agreement with the empirical **Curie law:**

$$\chi_m = A + \frac{C}{T} \qquad (9.26)$$

with $A = N_A\mu_0\xi$ and $C = N_A\mu_0 m^2/3k$. As indicated above and in contrast to electric moments, this expression applies to solids as well as fluid phases.

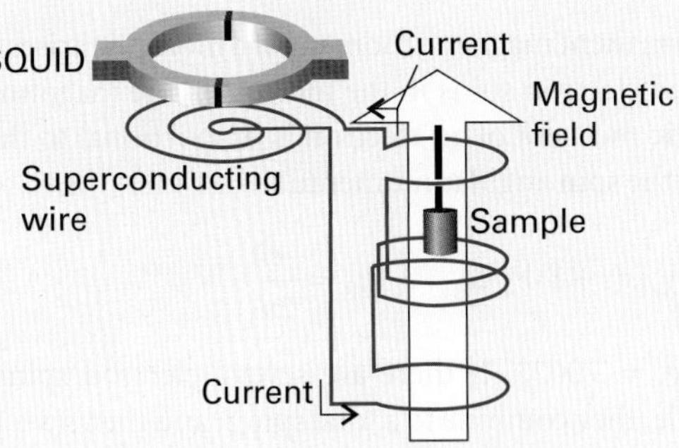

Fig. 9.60 The arrangement used to measure magnetic susceptibility with a SQUID. The sample is moved upwards in small increments and the potential difference across the SQUID is measured.

The magnetic susceptibility is traditionally measured with a **Gouy balance.** This instrument consists of a sensitive balance from which the sample hangs in the form of a narrow cylinder and lies between the poles of a magnet. If the sample is paramagnetic, it is drawn into the field, and its apparent weight is greater than when the field is off. A diamagnetic sample tends to be expelled from the field and appears to weigh less when the field is turned on. The balance is normally calibrated against a sample of known susceptibility. The modern version of the determination makes use of a **superconducting quantum interference device** (SQUID, Fig. 9.60). A SQUID takes advantage of the quantization of magnetic flux and the property of current loops in superconductors that, as part of the circuit, include a weakly conducting link through which electrons must tunnel. The current that flows in the loop in a magnetic field depends on the value of the magnetic flux, and a SQUID can be exploited as a very sensitive magnetometer.

Table 9.5 lists some experimental values. A typical paramagnetic volume susceptibility is about 10^{-3}, and a typical diamagnetic volume susceptibility is about $(-)10^{-5}$. The permanent magnetic moment can be extracted from susceptibility measurements by plotting χ against $1/T$.

Synoptic table 9.5* Magnetic susceptibilities at 298 K

	$\chi/10^{-6}$	$\chi_m/(10^{-10}$ m^3 mol$^{-1})$
H_2O(l)	−9.02	−1.63
NaCl(s)	−16	−3.8
Cu(s)	−9.7	−0.69
$CuSO_4 \cdot 5H_2O$(s)	+167	+183

* More values are given in the *Data section*.

(b) The permanent magnetic moment

The permanent magnetic moment of a molecule arises from any unpaired electron spins in the molecule. The magnitude of the magnetic moment of an electron is proportional to the magnitude of the spin angular momentum, $\{s(s+1)\}^{1/2}\hbar$.

$$m = g_e\{s(s+1)\}^{1/2}\mu_B \qquad \mu_B = \frac{e\hbar}{2m_e} \qquad (9.27)$$

where $g_e = 2.0023$. If there are several electron spins in each molecule, they combine to a total spin S, and then $s(s+1)$ should be replaced by $S(S+1)$. It follows that the spin contribution to the molar magnetic susceptibility is

$$\chi_m = \frac{N_A g_e^2 \mu_0 \mu_B^2 S(S+1)}{3kT} \qquad (9.28)$$

This expression shows that the susceptibility is positive, so the spin magnetic moments contribute to the paramagnetic susceptibilities of materials. The contribution decreases with increasing temperature because the thermal motion randomizes the spin orientations. In practice, a contribution to the paramagnetism also arises from the orbital angular momenta of electrons: we have discussed the spin-only contribution.

● A BRIEF ILLUSTRATION

Consider a complex salt with three unpaired electrons per complex cation at 298 K, of mass density 3.24 g cm^{-3}, and molar mass 200 g mol^{-1}. First note that

$$\frac{N_A g_e^2 \mu_0 \mu_B^2}{3k} = 6.3001 \times 10^{-6}\ \text{m}^3\ \text{K}^{-1}\ \text{mol}^{-1}$$

Consequently,

$$\chi_m = 6.3001 \times 10^{-6} \times \frac{S(S+1)}{T/\text{K}}\ \text{m}^3\ \text{mol}^{-1}$$

Substitution of the data with $S = \frac{3}{2}$ gives $\chi_m = 7.9 \times 10^{-8}$ m^3 mol^{-1}. Note that the density is not needed at this stage. To obtain the volume magnetic susceptibility, the molar susceptibility is divided by the molar volume $V_m = M/\rho$, where ρ is the mass density. In this *illustration*, $V_m = 61.7$ cm^3 mol^{-1}, so $\chi = 1.3 \times 10^{-3}$. ●

At low temperatures, some paramagnetic solids make a phase transition to a state in which large domains of spins align with parallel orientations due to the exchange interactions between them. This cooperative alignment gives rise to a very strong magnetization and is called **ferromagnetism** (Fig. 9.61). In other cases, exchange interactions lead to alternating spin orientations: the spins are locked into a low-magnetization arrangement to give an **antiferromagnetic phase**. The ferromagnetic phase has a nonzero magnetization in the absence of an applied field, but the antiferromagnetic phase has a zero magnetization because the spin magnetic moments cancel. The ferromagnetic transition occurs at the **Curie temperature**, and the antiferromagnetic transition occurs at the **Néel temperature**. Which type of cooperative behaviour occurs depends on the details of the band structure of the solid.

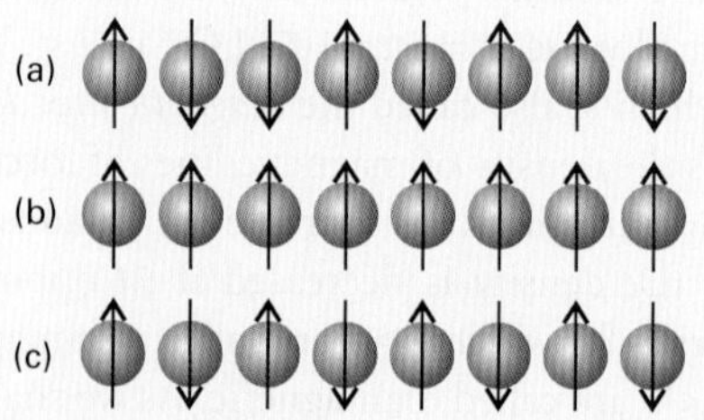

Fig 9.61 (a) In a paramagnetic material, the electron spins are aligned at random in the absence of an applied magnetic field. (b) In a ferromagnetic material, the electron spins are locked into a parallel alignment over large domains. (c) In an antiferromagnetic material, the electron spins are locked into an antiparallel arrangement. The latter two arrangements survive even in the absence of an applied field.

(c) Induced magnetic moments

An applied magnetic field induces the circulation of electronic currents. These currents give rise to a magnetic field that usually opposes the applied field, so the substance is diamagnetic. In a few cases the induced field augments the applied field, and the substance is then paramagnetic.

The great majority of molecules with no unpaired electron spins are diamagnetic. In these cases, the induced electron currents occur within the orbitals of the molecule that are occupied in its ground state. In the few cases in which molecules are paramagnetic despite having no unpaired electrons, the induced electron currents flow in the opposite direction because they can make use of unoccupied orbitals that lie close to the HOMO in energy. This orbital paramagnetism can be distinguished from spin paramagnetism by the fact that it is temperature independent: this is why it is called **temperature-independent paramagnetism** (TIP).

We can summarize these remarks as follows. All molecules have a diamagnetic component to their susceptibility, but it is dominated by spin paramagnetism if the molecules have unpaired electrons. In a few cases (where there are low-lying excited states) TIP is strong enough to make the molecules paramagnetic even though their electrons are paired.

9.12 Superconductors

The resistance to flow of electrical current of a normal metallic conductor decreases smoothly with temperature but never vanishes. However, certain solids known as **superconductors** conduct electricity without resistance below a critical temperature, T_c. Following the discovery in 1911 that mercury is a superconductor below 4.2 K, the normal boiling point of liquid helium, physicists and chemists made slow but steady progress in the discovery of superconductors with higher values of T_c. Metals, such as tungsten, mercury, and lead, tend to have T_c values below about 10 K. Intermetallic compounds, such as Nb_3X (X = Sn, Al, or Ge), and alloys, such as Nb/Ti and Nb/Zr, have intermediate T_c values ranging between 10 K and 23 K. In 1986, **high-temperature superconductors** (HTSC) were discovered. Several *ceramics*, inorganic powders that have been fused and hardened by heating to a high temperature, containing oxocuprate motifs, Cu_mO_n, are now known with T_c values well above 77 K, the boiling point of the inexpensive refrigerant liquid nitrogen. For example, $HgBa_2Ca_2Cu_2O_8$ has $T_c = 153$ K.

Superconductors have unique magnetic properties. Some superconductors, classed as *type I*, show abrupt loss of superconductivity when an applied magnetic field exceeds a critical value $\mathcal{H}_c$ characteristic of the material. It is observed that the value of $\mathcal{H}_c$ depends on temperature and T_c as

$$\mathcal{H}_c(T) = \mathcal{H}_c(0)\left(1 - \frac{T^2}{T_c^2}\right) \tag{9.29}$$

where $\mathcal{H}_c(0)$ is the value of $\mathcal{H}_c$ as $T \rightarrow 0$. Type I superconductors are also completely diamagnetic below $\mathcal{H}_c$, meaning that the magnetic field does not penetrate into the material. This complete exclusion of a magnetic field from a material is known as the *Meissner effect*, which can be visualized by the levitation of a superconductor above a magnet. *Type II* superconductors, which include the HTSCs, show a gradual loss of superconductivity and diamagnetism with increasing magnetic field.

There is a degree of periodicity in the elements that exhibit superconductivity. The metals iron, cobalt, nickel, copper, silver, and gold do not display superconductivity, nor do the alkali metals. It is observed that, for simple metals, ferromagnetism and superconductivity never coexist, but in some of the oxocuprate superconductors ferromagnetism and superconductivity can coexist. One of the most widely studied oxocuprate superconductors $YBa_2Cu_3O_7$ (informally known as '123' on account of the proportions of the metal atoms in the compound) has the structure shown in Fig. 9.62. The square-pyramidal CuO_5 units arranged as two-dimensional layers and the square-planar CuO_4 units arranged in sheets are common structural features of oxocuprate HTSCs.

The mechanism of superconduction is well understood for low-temperature materials but there is as yet no settled explanation of high-temperature superconductivity. The central concept of low-temperature superconduction is the existence of a *Cooper pair*, a pair of electrons that exists on account of the indirect electron–electron interactions fostered by the nuclei of the atoms in the lattice. Thus, if one electron is in a particular region of a solid, the nuclei there move toward it to give a distorted local structure (Fig. 9.63). Because that local distortion is rich in positive charge, it is favourable for a second electron to join the first. Hence, there is a virtual attraction between the two electrons, and they move together as a pair. The local distortion can be easily disrupted by thermal motion of the ions in the solid, so the virtual attraction occurs only at very low temperatures. A Cooper pair undergoes less scattering than an individual electron as it travels through the solid because the distortion caused by one electron can attract back the other electron should it be scattered out of its path in a collision. Because the Cooper pair is stable against scattering, it can carry charge freely through the solid, and hence give rise to superconduction.

The Cooper pairs responsible for low-temperature superconductivity are likely to be important in HTSCs, but the mechanism for pairing is hotly debated. There is evidence implicating

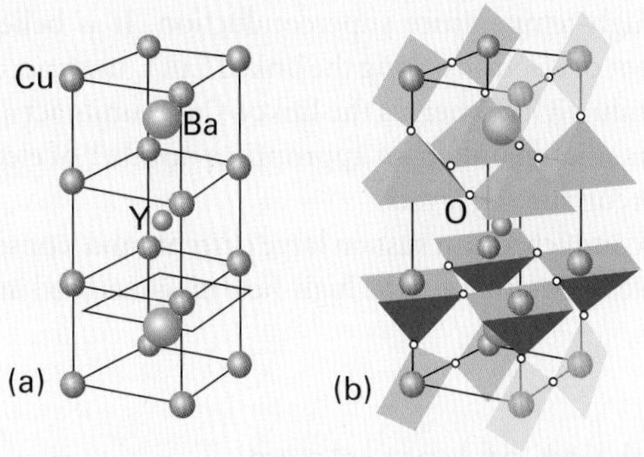

Fig. 9.62 Structure of the $YBa_2Cu_3O_7$ superconductor. (a) Metal atom positions. (b) The polyhedra show the positions of oxygen atoms and indicate that the metal ions are in square-planar and square-pyramidal coordination environments.

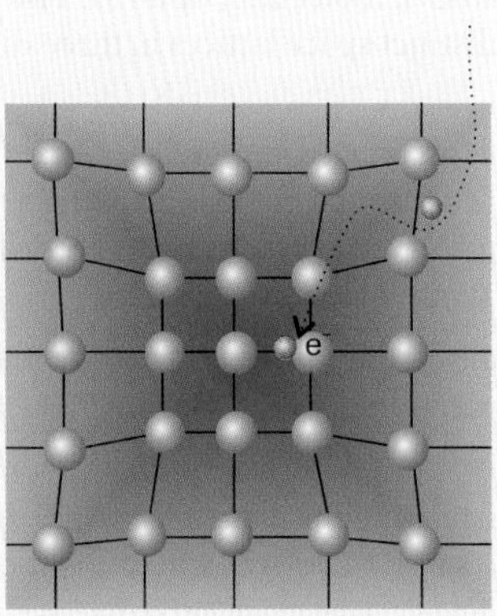

Fig. 9.63 The formation of a Cooper pair. One electron distorts the crystal lattice and the second electron has a lower energy if it goes to that region. These electron–lattice interactions effectively bind the two electrons into a pair.

the arrangement of CuO_5 layers and CuO_4 sheets in the mechanism of high-temperature superconduction. It is believed that movement of electrons along the linked CuO_4 units accounts for superconductivity, whereas the linked CuO_5 units act as 'charge reservoirs' that maintain an appropriate number of electrons in the superconducting layers.

Superconductors can sustain large currents and, consequently, are excellent materials for the high-field magnets used in modern NMR spectroscopy (Chapter 12). However, the potential uses of superconducting materials are not limited to the field of chemical instrumentation. For example, HTSCs with T_c values near ambient temperature would be very efficient components of an electrical power transmission system, in which energy loss due to electrical resistance would be minimized. The appropriate technology is not yet available, but research in this area of materials science is active.

Checklist of key ideas

- ☐ 1. Solids are classified as metallic, ionic, covalent, and molecular.
- ☐ 2. A space lattice is the pattern formed by points representing the locations of structural motifs (atoms, molecules, or groups of atoms, molecules, or ions). The Bravais lattices are the 14 distinct space lattices in three dimensions (Fig. 9.8).
- ☐ 3. A unit cell is an imaginary parellelepiped that contains one unit of a translationally repeating pattern. Unit cells are classified into seven crystal systems according to their rotational symmetries.
- ☐ 4. Crystal planes are specified by a set of Miller indices (*hkl*) and the separation of neighbouring planes in a rectangular lattice is given by $1/d_{hkl}^2 = h^2/a^2 + k^2/b^2 + l^2/c^2$.
- ☐ 5. Bragg's law relating the glancing angle θ to the separation of lattice planes is $\lambda = 2d \sin \theta$, where λ is the wavelength of the radiation.
- ☐ 6. The scattering factor is a measure of the ability of an atom to diffract radiation (eqn 9.3).
- ☐ 7. The structure factor is the overall amplitude of a wave diffracted by the {*hkl*} planes (eqn 9.4). Fourier synthesis is the construction of the electron density distribution from structure factors (eqn 9.5).
- ☐ 8. A Patterson synthesis is a map of interatomic vectors obtained by Fourier analysis of diffraction intensities (eqn 9.6).
- ☐ 9. Structure refinement is the adjustment of structural parameters to give the best fit between the observed intensities and those calculated from the model of the structure deduced from the diffraction pattern.
- ☐ 10. Many elemental metals have close-packed structures with coordination number 12; close-packed structures may be either cubic (ccp) or hexagonal (hcp).
- ☐ 11. Representative ionic structures include the caesium-chloride, rock-salt, and zinc-blende structures.
- ☐ 12. The radius-ratio rule may be used cautiously to predict which of these three structures is likely (eqn 9.9 and Section 9.6a).
- ☐ 13. The lattice enthalpy is the change in enthalpy (per mole of formula units) accompanying the complete separation of the components of the solid. The electrostatic contribution to the lattice enthalpy is expressed by the Born–Mayer equation (eqn 9.12).
- ☐ 14. A covalent network solid is a solid in which covalent bonds in a definite spatial orientation link the atoms in a network extending through the crystal. A molecular solid is a solid consisting of discrete molecules held together by van der Waals interactions.
- ☐ 15. The mechanical properties of a solid are discussed in terms of the relationship between stress, the applied force divided by the area to which it is applied, and strain, the distortion of a sample resulting from an applied stress.
- ☐ 16. The response of a solid to an applied stress is summarized by the Young's modulus (eqn 9.13a), the bulk modulus (eqn 9.13b), the shear modulus (eqn 9.13c), and Poisson's ratio (eqn 9.15).
- ☐ 17. Electronic conductors are classified as metallic conductors or semiconductors according to the temperature dependence of their conductivities. An insulator is a semiconductor with a very low electrical conductivity.
- ☐ 18. According to the band theory, electrons occupy molecular orbitals formed from the overlap of atomic orbitals: full bands are called valence bands and empty bands are called conduction bands. The occupation of the orbitals in a solid is given by the Fermi–Dirac distribution (eqn 9.20).
- ☐ 19. Semiconductors are classified as p-type or n-type according to whether conduction is due to holes in the valence band or electrons in the conduction band.
- ☐ 20. The spectroscopic properties of molecular solids can be understood in terms of the formation and migration of excitons, electron–hole pairs, from molecule to molecule.
- ☐ 21. The spectroscopic properties of metallic conductors and semiconductors can be understood in terms of the light-induced promotion of electrons from valence bands to conduction bands.
- ☐ 22. A bulk sample exposed to a magnetic field of strength $\mathcal{H}$ acquires a magnetization, $\mathcal{M} = \chi\mathcal{H}$, where χ is the dimensionless volume magnetic susceptibility. When

$\chi < 0$, the material is diamagnetic and moves out of a magnetic field. When $\chi > 0$, the material is paramagnetic and moves into a magnetic field.

☐23. The temperature dependence of χ_m is given by the Curie law $\chi_m = A + C/T$, where $A = N_A\mu_0\xi$, $C = N_A\mu_0 m^2/3k$, and ξ is the magnetizability, a measure of the extent to which a magnetic dipole moment may be induced in a molecule.

☐24. Ferromagnetism is the cooperative alignment of electron spins in a material and gives rise to strong magnetization. Antiferromagnetism results from alternating spin orientations in a material and leads to weak magnetization.

☐25. Temperature-independent paramagnetism arises from induced electron currents within the orbitals of a molecule that are occupied in its ground state.

☐26. Superconductors conduct electricity without resistance below a critical temperature T_c. Type I superconductors show abrupt loss of superconductivity when an applied magnetic field exceeds a critical value $\mathcal{H}_c$ characteristic of the material. They are also completely diamagnetic below $\mathcal{H}_c$. Type II superconductors show a gradual loss of superconductivity and diamagnetism with increasing magnetic field.

Discussion questions

9.1 Describe the relationship between the space lattice and unit cell.

9.2 Explain how planes of lattice points are labelled.

9.3 Describe the procedure for identifying the type and size of a cubic unit cell.

9.4 What is meant by a systematic absence? How do they arise?

9.5 Explain the general features of the X-ray diffraction pattern of a helical molecule. How would the pattern change as the pitch of the helix is increased?

9.6 Discuss what is meant by 'scattering factor'. How is it related to the number of electrons in the atoms scattering X-rays?

9.7 Describe the consequences of the phase problem in determining structure factors and how the problem is overcome.

9.8 To what extent is the hard-sphere model of metallic solids inaccurate?

9.9 Describe the caesium-chloride and rock-salt structures in terms of the occupation of holes in expanded close-packed lattices.

9.10 Describe the distinguishing characteristics of elastomers, fibres, and plastics.

9.11 Explain the origin of Davydov splitting in the exciton bands of a crystal.

9.12 Describe the characteristics of the Fermi–Dirac distribution.

Exercises

9.1(a) Equivalent lattice points within the unit cell of a Bravais lattice have identical surroundings. What points within a body-centred cubic unit cell are equivalent to the point $(0, \frac{1}{2}, 0)$?

9.1(b) Equivalent lattice points within the unit cell of a Bravais lattice have identical surroundings. What points within a face-centred cubic unit cell are equivalent to the point $(\frac{1}{2}, 0, \frac{1}{2})$?

9.2(a) Show that the volume of a monoclinic unit cell is $V = abc \sin\beta$.

9.2(b) Derive an expression for the volume of a hexagonal unit cell.

9.3(a) Find the Miller indices of the planes that intersect the crystallographic axes at the distances $(3a, 2b, c)$ and $(2a, \infty b, \infty c)$.

9.3(b) Find the Miller indices of the planes that intersect the crystallographic axes at the distances $(-a, 2b, -c)$ and $(a, 4b, -4c)$.

9.4(a) Calculate the separations of the planes {112}, {110}, and {224} in a crystal in which the cubic unit cell has side 562 pm.

9.4(b) Calculate the separations of the planes {123}, {222}, and {246} in a crystal in which the cubic unit cell has side 712 pm.

9.5(a) What are the values of the glancing angle (θ) of the first three diffraction lines of bcc iron (atomic radius 126 pm) when the X-ray wavelength is 72 pm?

9.5(b) What are the values of the glancing angle (θ) of the first three diffraction lines of fcc gold {atomic radius 144 pm) when the X-ray wavelength is 129 pm?

9.6(a) Copper K_α radiation consists of two components of wavelengths 154.433 pm and 154.051 pm. Calculate the difference in glancing angles (θ) of the diffraction lines arising from the two components in a powder diffraction pattern from planes of separation 77.8 pm.

9.6(b) A synchrotron source produces X-radiation at a range of wavelengths. Consider two components of wavelengths 93.222 and 95.123 pm. Calculate the separation of the glancing angles (θ) arising from the two components in a powder diffraction pattern from planes of separation 82.3 pm.

9.7(a) What is the value of the scattering factor in the forward direction for Br^-?

9.7(b) What is the value of the scattering factor in the forward direction for Mg^{2+}?

9.8(a) The orthorhombic unit cell of $NiSO_4$ has the dimensions $a = 634$ pm, $b = 784$ pm, and $c = 516$ pm, and the density of the solid is estimated as 3.9 g cm^{-3}. Determine the number of formula units per unit cell and calculate a more precise value of the density.

9.8(b) An orthorhombic unit cell of a compound of molar mass 135.01 g mol^{-1} has the dimensions $a = 589$ pm, $b = 822$ pm, and $c = 798$ pm. The density of the solid is estimated as 2.9 g cm^{-3}. Determine the number of formula units per unit cell and calculate a more precise value of the density.

9.9(a) The unit cells of $SbCl_3$ are orthorhombic with dimensions $a = 812$ pm, $b = 947$ pm, and $c = 637$ pm. Calculate the spacing, d, of the (321) planes.

9.9(b) An orthorhombic unit cell has dimensions $a = 769$ pm, $b = 891$ pm, and $c = 690$ pm. Calculate the spacing, d, of the (312) planes.

9.10(a) Potassium nitrate crystals have orthorhombic unit cells of dimensions $a = 542$ pm, $b = 917$ pm, and $c = 645$ pm. Calculate the glancing angles for the (100), (010), and (111) reflections using Cu K_α radiation (154 pm).

9.10(b) Calcium carbonate crystals in the form of aragonite have orthorhombic unit cells of dimensions $a = 574.1$ pm, $b = 796.8$ pm, and $c = 495.9$ pm. Calculate the glancing angles for the (100), (010), and (111) reflections using radiation of wavelength 83.42 pm (from aluminium).

9.11(a) Copper(I) chloride forms cubic crystals with four formula units per unit cell. The only reflections present in a powder photograph are those with either all even indices or all odd indices. What is the (Bravais) lattice type of the unit cell?

9.11(b) A powder diffraction photograph from tungsten shows lines that index as (110), (200), (211), (220), (310), (222), (321), (400), . . . Identify the (Bravais) lattice type of the unit cell.

9.12(a) The coordinates, in units of a, of the atoms in a primitive cubic unit cell are (0, 0, 0), (0, 1, 0), (0, 0, 1), (0, 1, 1), (1, 0, 0), (1, 1, 0), (1, 0, 1), and (1, 1, 1). Calculate the structure factors F_{hkl} when all the atoms are identical.

9.12(b) The coordinates, in units of a, of the atoms in a body-centred cubic unit cell are (0, 0, 0), (0, 1, 0), (0, 0, 1), (0, 1, 1), (1, 0, 0), (1, 1, 0), (1, 0, 1), (1, 1, 1), and $(\frac{1}{2}, \frac{1}{2}, \frac{1}{2})$. Calculate the structure factors F_{hkl} when all the atoms are identical.

9.13(a) Calculate the structure factors for a face-centred cubic structure (C) in which the scattering factors of the ions on the two faces are twice that of the ions at the corners of the cube.

9.13(b) Calculate the structure factors for a body-centred cubic structure in which the scattering factor of the central ion is twice that of the ions at the corners of the cube.

9.14(a) In an X-ray investigation, the following structure factors were determined (with $F_{-h00} = F_{h00}$):

h	0	1	2	3	4	5	6	7	8	9
F_{h00}	10	−10	8	−8	6	−6	4	−4	2	−2

Construct the electron density along the corresponding direction.

9.14(b) In an X-ray investigation, the following structure factors were determined (with $F_{-h00} = F_{h00}$):

h	0	1	2	3	4	5	6	7	8	9
F_{h00}	10	10	4	4	6	6	8	8	10	10

Construct the electron density along the corresponding direction.

9.15(a) Construct the Patterson synthesis from the information in Exercise 9.14a.

9.15(b) Construct the Patterson synthesis from the information in Exercise 9.14b.

9.16(a) In a Patterson synthesis, the spots correspond to the lengths and directions of the vectors joining the atoms in a unit cell. Sketch the pattern that would be obtained for a planar, triangular isolated BF_3 molecule.

9.16(b) In a Patterson synthesis, the spots correspond to the lengths and directions of the vectors joining the atoms in a unit cell. Sketch the pattern that would be obtained from the C atoms in an isolated benzene molecule.

9.17(a) What velocity should neutrons have if they are to have wavelength 65 pm?

9.17(b) What velocity should electrons have if they are to have wavelength 105 pm?

9.18(a) Calculate the wavelength of neutrons that have reached thermal equilibrium by collision with a moderator at 350 K.

9.18(b) Calculate the wavelength of electrons that have reached thermal equilibrium by collision with a moderator at 380 K.

9.19(a) Calculate the packing fraction for close-packed cylinders. (For a generalization of this Exercise, see Problem 9.21.)

9.19(b) Calculate the packing fraction for equilateral triangular rods stacked as shown in (2).

2

9.20(a) Calculate the packing fractions of (a) a primitive cubic unit cell, (b) a bcc unit cell, (c) an fcc unit cell composed of identical hard spheres.

9.20(b) Calculate the atomic packing factor for a side-centred (C) cubic unit cell.

9.21(a) From the data in Table 9.3 determine the radius of the smallest cation that can have (a) sixfold and (b) eightfold coordination with the Cl^- ion.

9.21(b) From the data in Table 9.3 determine the radius of the smallest cation that can have (a) sixfold and (b) eightfold coordination with the Rb^+ ion.

9.22(a) Does titanium expand or contract as it transforms from hcp to body-centred cubic? The atomic radius of titanium is 145.8 pm in hcp but 142.5 pm in bcc.

9.22(b) Does iron expand or contract as it transforms from hcp to bcc? The atomic radius of iron is 126 in hcp but 122 pm in bcc.

9.23(a) Calculate the lattice enthalpy of CaO from the following data:

	$\Delta H/(\text{kJ mol}^{-1})$
Sublimation of Ca(s)	+178
Ionization of Ca(g) to Ca^{2+}(g)	+1735
Dissociation of O_2(g)	+249
Electron attachment to O(g)	−141
Electron attachment to O^-(g)	+844
Formation of CaO(s) from Ca(s) and O_2(g)	−635

9.23(b) Calculate the lattice enthalpy of $MgBr_2$ from the following data:

	$\Delta H/(\text{kJ mol}^{-1})$
Sublimation of Mg(s)	+148
Ionization of Mg(g) to Mg^{2+}(g)	+2187
Vaporization of Br_2(l)	+31
Dissociation of Br_2(g)	+193
Electron attachment to Br(g)	−331
Formation of $MgBr_2$(s) from Mg(s) and Br_2(l)	−524

9.24(a) Young's modulus for polyethylene at room temperature is 1.2 GPa. What strain will be produced when a mass of 1.5 kg is suspended from a polyethylene thread of diameter 2.0 mm?

9.24(b) Young's modulus for iron at room temperature is 215 GPa. What strain will be produced when a mass of 15.0 kg is suspended from an iron wire of diameter 0.15 mm?

9.25(a) Poisson's ratio for polyethylene is 0.45. What change in volume takes place when a cube of polyethylene of volume 1.0 cm^3 is subjected to a uniaxial stress that produces a strain of 2.0 per cent?

9.25(b) Poisson's ratio for lead is 0.41. What change in volume takes place when a cube of lead of volume 100 cm^3 is subjected to a uniaxial stress that produces a strain of 1.5 per cent?

9.26(a) Is arsenic-doped germanium a p-type or n-type semiconductor?

9.26(b) Is gallium-doped germanium a p-type or n-type semiconductor?

9.27(a) The promotion of an electron from the valence band into the conduction band in pure TIO_2 by light absorption requires a wavelength of less than 350 nm. Calculate the energy gap in electronvolts between the valence and conduction bands.

9.27(b) The band gap in silicon is 1.12 eV. Calculate the maximum wavelength of electromagnetic radiation that results in promotion of electrons from the valence to the conduction band.

9.28(a) The magnetic moment of $CrCl_3$ is $3.81\mu_B$. How many unpaired electrons does the Cr possess?

9.28(b) The magnetic moment of Mn^{2+} in its complexes is typically $5.3\mu_B$. How many unpaired electrons does the ion possess?

9.29(a) Calculate the molar susceptibility of benzene given that its volume susceptibility is -8.8×10^{-6} and its density 0.879 g cm^{-3} at 25°C.

9.29(b) Calculate the molar susceptibility of cyclohexane given that its volume susceptibility is -7.9×10^{-7} and its density 811 kg m^{-3} at 25°C.

9.30(a) Data on a single crystal of MnF_2 give $\chi_m = 0.1463$ cm^3 mol^{-1} at 294.53 K. Determine the effective number of unpaired electrons in this compound and compare your result with the theoretical value.

9.30(b) Data on a single crystal of $NiSO_4{\cdot}7H_2O$ give $\chi_m = 5.03\times10^{-8}$ m^3 mol^{-1} at 298 K. Determine the effective number of unpaired electrons in this compound and compare your result with the theoretical value.

9.31(a) Estimate the spin-only molar susceptibility of $CuSO_4{\cdot}5H_2O$ at 25°C.

9.31(b) Estimate the spin-only molar susceptibility of $MnSO_4{\cdot}4H_2O$ at 298 K.

9.32(a) Lead has $T_c = 7.19$ K and $\mathcal{H}_c = 63.9$ kA m^{-1}. At what temperature does lead become superconducting in a magnetic field of 20 kA m^{-1}?

9.32(b) Tin has $T_c = 3.72$ K and $\mathcal{H}_c = 25$ kA m^{-1}. At what temperature does lead become superconducting in a magnetic field of 15 kA m^{-1}?

Problems*

Numerical problems

9.1 In the early days of X-ray crystallography there was an urgent need to know the wavelengths of X-rays. One technique was to measure the diffraction angle from a mechanically ruled grating. Another method was to estimate the separation of lattice planes from the measured density of a crystal. The density of NaCl is 2.17 g cm^{-3} and the (100) reflection using Pd K_α radiation occurred at 6.0°. Calculate the wavelength of the X-rays.

9.2 The element polonium crystallizes in a cubic system. Bragg reflections, with X-rays of wavelength 154 pm, occur at $\sin\theta = 0.225$, 0.316, and 0.388 from the (100), (110), and (111) sets of planes. The separation between the sixth and seventh lines observed in the powder diffraction pattern is larger than that between the fifth and sixth lines. Is the unit cell simple, body-centred, or face-centred? Calculate the unit cell dimension.

9.3 Elemental silver reflects X-rays of wavelength 154.18 pm at angles of 19.076°, 22.171°, and 32.256°. However, there are no other reflections at angles of less than 33°. Assuming a cubic unit cell, determine its type and dimension. Calculate the mass density of silver.

9.4 In their book *X-rays and crystal structures* (which begins 'It is now two years since Dr. Laue conceived the idea . . .') the Braggs give a number of simple examples of X-ray analysis. For instance, they report that the reflection from (100) planes in KCl occurs at 5° 23′, but for NaCl it occurs at 6° 0′ for X-rays of the same wavelength. If the side of the NaCl unit cell is 564 pm, what is the side of the KCl unit cell? The densities of KCl and NaCl are 1.99 g cm^{-3} and 2.17 g cm^{-3} respectively. Do these values support the X-ray analysis?

9.5 Calculate the thermal expansion coefficient, $\alpha = (\partial V/\partial T)_p/V$, of diamond given that the (111) reflection shifts from 22.0403° to 21.9664° on heating a crystal from 100 K to 300 K and 154.0562 pm X-rays are used.

9.6 The carbon–carbon bond length in diamond is 154.45 pm. If diamond were considered to be a close-packed structure of hard spheres with radii equal to half the bond length, what would be its expected density? The diamond lattice is face-centred cubic and its actual density is 3.516 g cm^{-3}. Can you explain the discrepancy?

9.7 The volume of a monoclinic unit cell is $abc\sin\beta$ (see Exercise 9.2a). Naphthalene has a monoclinic unit cell with two molecules per cell and sides in the ratio 1.377:1:1.436. The angle β is 122° 49′ and the density of the solid is 1.152 g cm^{-3}. Calculate the dimensions of the cell.

* Problems denoted with the symbol ‡ were supplied by Charles Trapp and Carmen Giunta.

9.8 Fully crystalline polyethylene has its chains aligned in an orthorhombic unit cell of dimensions 740 pm × 493 pm × 253 pm. There are two repeating CH_2CH_2 units per unit cell. Calculate the theoretical mass density of fully crystalline polyethylene. The actual density ranges from 0.92 to 0.95 g cm^{-3}.

9.9‡ B.A. Bovenzi and G.A. Pearse, Jr. (*J. Chem. Soc. Dalton Trans.* 2793 (1997)) synthesized coordination compounds of the tridentate ligand pyridine-2,6-diamidoxime (**3**, $C_7H_7N_3O_4$). The compound that they isolated from the reaction of the ligand with $CuSO_4$(aq) did not contain a $[Cu(C_7H_7N_3O_4)_2]^{2+}$ complex cation as expected. Instead, X-ray diffraction analysis revealed a linear polymer of formula $[Cu(Cu(C_7H_7N_3O_4)(SO_4)\cdot 2H_2O]_n$, which features bridging sulfate groups. The unit cell was primitive monoclinic with $a = 1.0427$ nm, $b = 0.8876$ nm, $c = 1.3777$ nm, and $\beta = 93.254°$. The mass density of the crystals is 2.024 g cm^{-3}. How many monomer units are there in the unit cell?

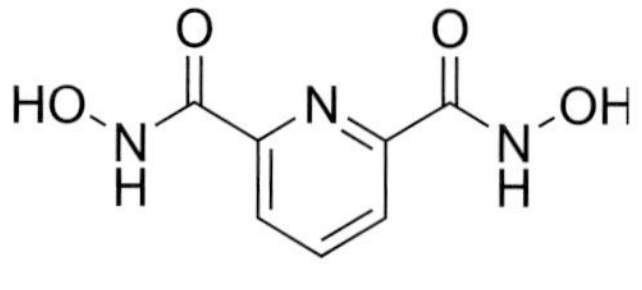

3 Pyridine-2,6-diamidoxime

9.10‡ D. Sellmann *et al.*, *Inorg. Chem.* **36**, 1397 (1997)) describe the synthesis and reactivity of the ruthenium nitrido compound $[N(C_4H_9)_4][Ru(N)(S_2C_6H_4)_2]$. The ruthenium complex anion has the two 1,2-benzenedithiolate ligands (**4**) at the base of a rectangular pyramid and the nitrido ligand at the apex. Compute the mass density of the compound given that it crystallizes into an orthorhombic unit cell with $a = 3.6881$ nm, $b = 0.9402$ nm, and $c = 1.7652$ nm and eight formula units per cell. Replacing the ruthenium with osmium results in a compound with the same crystal structure and a unit cell with a volume less than 1 per cent larger. Estimate the mass density of the osmium analogue.

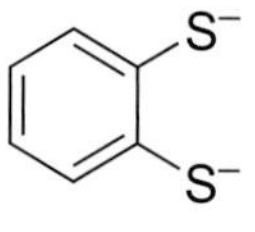

4 Benzenedithiolate

9.11 Use the Born–Mayer equation for the lattice enthalpy and a Born–Haber cycle to show that formation of CaCl is an exothermic process (the sublimation enthalpy of Ca(s) is 176 kJ mol^{-1}). Show that an explanation for the nonexistence of CaCl can be found in the reaction enthalpy for the reaction 2 CaCl(s) → Ca(s) + $CaCl_2$(s).

9.12 In an intrinsic semiconductor, the band gap is so small that the Fermi–Dirac distribution results in some electrons populating the conduction band. It follows from the exponential form of the Fermi–Dirac distribution that the conductance G, the inverse of the resistance (with units of siemens, 1 S = 1 Ω^{-1}), of an intrinsic semiconductor should have an Arrhenius-like temperature dependence, shown in practice to have the form $G = G_0 e^{-E_g/2kT}$, where E_g is the band gap. The conductance of a sample of germanium varied with temperature as indicated below. Estimate the value of E_g.

T/K	312	354	420
G/S	0.0847	0.429	2.86

9.13‡ J.J. Dannenberg *et al.* (*J. Phys. Chem.* **100**, 9631 (1996)) carried out theoretical studies of organic molecules consisting of chains of unsaturated four-membered rings. The calculations suggest that such compounds have large numbers of unpaired spins, and that they should therefore have unusual magnetic properties. For example, the lowest-energy state of the compound shown as (**5**) is computed to have $S = 3$, but the energies of $S = 2$ and $S = 4$ structures are each predicted to be 50 kJ mol^{-1} higher in energy. Compute the molar magnetic susceptibility of these three low-lying levels at 298 K. Estimate the molar susceptibility at 298 K if each level is present in proportion to its Boltzmann factor (effectively assuming that the degeneracy is the same for all three of these levels).

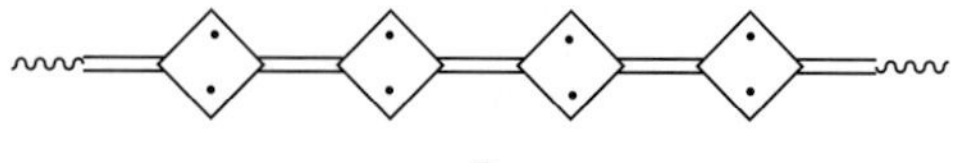

5

9.14‡ P.G. Radaelli *et al.* (*Science* **265**, 380 (1994)) reported the synthesis and structure of a material that becomes superconducting at temperatures below 45 K. The compound is based on a layered compound $Hg_2Ba_2YCu_2O_{8-\delta}$, which has a tetragonal unit cell with $a = 0.386\ 06$ nm and $c = 2.8915$ nm; each unit cell contains two formula units. The compound is made superconducting by partially replacing Y by Ca, accompanied by a change in unit cell volume by less than 1 per cent. Estimate the Ca content x in superconducting $Hg_2Ba_2Y_{1-x}Ca_xCu_2O_{7.55}$ given that the mass density of the compound is 7.651 g cm^{-3}.

Theoretical problems

9.15 Show that the separation of the (hkl) planes in an orthorhombic crystal with sides a, b, and c is given by eqn 9.1.

9.16 Show that the volume of a triclinic unit cell of sides a, b, and c and angles α, β, and γ is

$$V = abc(1 - \cos^2\alpha - \cos^2\beta - \cos^2\gamma + 2\cos\alpha\cos\beta\cos\gamma)^{1/2}$$

Use this expression to derive expressions for monoclinic and orthorhombic unit cells. For the derivation, it may be helpful to use the result from vector analysis that $V = \boldsymbol{a}\cdot\boldsymbol{b}\times\boldsymbol{c}$ and to calculate V^2 initially. The compound Rb_3TlF_6 has a *tetragonal* unit cell with dimensions $a = 651$ pm and $c = 934$ pm. Calculate the volume of the unit cell.

9.17 Use mathematical software to draw a graph of the scattering factor f against $(\sin\theta)/\lambda$ for an atom of atomic number Z for which $\rho(r) = 3Z/4\pi R^3$ for $0 \le r \le R$ and $\rho(r) = 0$ for $r > R$, with R a parameter that represents the radius of the atom. Explore how f varies with Z and R.

9.18 Calculate the scattering factor for a hydrogenic atom of atomic number Z in which the single electron occupies (a) the 1s orbital, (b) the 2s orbital. Radial wavefunctions are given in Table 4.1. Plot f as a function of $(\sin\theta)/\lambda$. *Hint.* Interpret $4\pi\rho(r)r^2$ as the radial distribution function $P(r)$ of eqn 4.15.

9.19 Explore how the scattering factor of Problem 9.18 changes when the actual 1s wavefunction of a hydrogenic atom is replaced by a Gaussian function.

9.20 Calculate the atomic packing factor for diamond.

9.21 Rods of elliptical cross-section with semi-major and -minor axes a and b are close-packed as shown in (**6**). What is the packing fraction? Draw a graph of the packing fraction against the eccentricity ε of the ellipse. For an ellipse with semi-major axis a and semi-minor axis b, $\varepsilon = (1 - b^2/a^2)^{1/2}$.

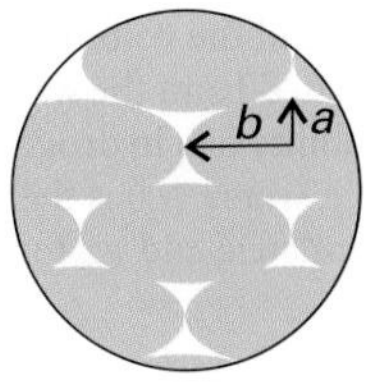

6

9.22 The coordinates of the four I atoms in the unit cell of KIO_4 are $(0, 0, 0)$, $(0, \frac{1}{2}, \frac{1}{2})$, $(\frac{1}{2}, \frac{1}{2}, \frac{1}{2})$, $(\frac{1}{2}, 0, \frac{3}{4})$. By calculating the phase of the I reflection in the structure factor, show that the I atoms contribute no net intensity to the (114) reflection.

9.23 The coordinates, in units of a, of the A atoms, with scattering factor f_A, in a cubic lattice are (0, 0, 0), (0, 1, 0), (0, 0, 1), (0, 1, 1), (1, 0, 0), (1, 1, 0), (1, 0, 1), and (1, 1, 1). There is also a B atom, with scattering factor f_B, at $(\frac{1}{2}, \frac{1}{2}, \frac{1}{2})$. Calculate the structure factors F_{hkl} and predict the form of the powder diffraction pattern when (a) $f_A = f, f_B = 0$, (b) $f_B = \frac{1}{2}f_A$, and (c) $f_A = f_B = f$.

9.24 Derive the Born–Mayer equation (eqn 9.12) by calculating the energy at which $\mathrm{d}(E_P + E_P^*)/\mathrm{d}d = 0$, with E_P and E_P^* given by eqns 9.10 and 9.11, respectively.

9.25 Suppose that ions are arranged in a (somewhat artificial) two-dimensional lattice like the fragment shown in (7). Calculate the Madelung constant for this array.

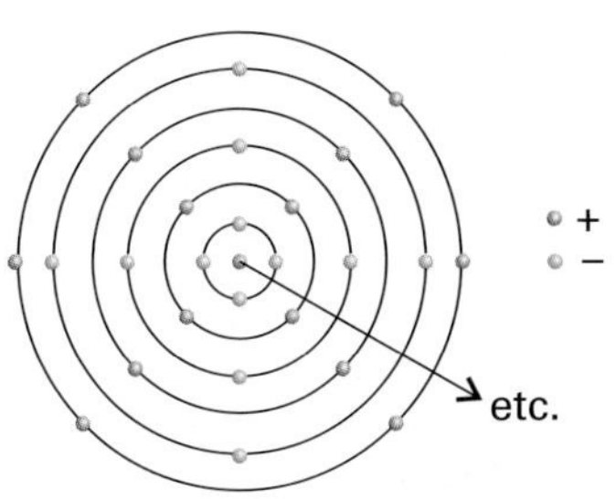

7

9.26 For an isotropic substance, the moduli and Poisson's ratio may be expressed in terms of two parameters λ and μ called the *Lamé constants*:

$$E = \frac{\mu(3\lambda + 2\mu)}{\lambda + \mu} \qquad K = \frac{3\lambda + 2\mu}{3} \qquad G = \mu \qquad v_P = \frac{\lambda}{2(\lambda + \mu)}$$

Use the Lamé constants to confirm the relations between G, K, and E given in eqn 9.16.

9.27 *Justification 9.4* showed how to relate the bulk modulus of a solid to the parameters that appear in a Lennard-Jones potential. What relation can be deduced for (a) a general Mie potential (eqn 8.18b) and (b) an exp-6 potential? *Hint.* For an exp-6 potential, $r_{min} = 1.63r_0$.

9.28 When energy levels in a band form a continuum, the density of states $\rho(E)$, the number of levels in an energy range divided by the width of the range, may be written as $\rho(E) = \mathrm{d}k/\mathrm{d}E$, where $\mathrm{d}k$ is the change in the quantum number k and $\mathrm{d}E$ is the energy change. (a) Use eqn 9.18 to show that

$$\rho(E) = -\frac{(N+1)/2\beta\pi}{\left\{1 - \left(\frac{E-\alpha}{2\beta}\right)^2\right\}^{1/2}}$$

where k, N, α, and β have the meanings described in Section 9.9. (b) Use this expression to show that $\rho(E)$ becomes infinite as E approaches $\alpha \pm 2\beta$. That is, show that the density of states increases towards the edges of the bands in a one-dimensional metallic conductor.

9.29 The treatment in Problem 9.28 applies only to one-dimensional solids. In three dimensions, the variation of density of states is more like that shown in (**8**). Account for the fact that in a three-dimensional solid the greatest density of states is near the centre of the band and the lowest density is at the edges.

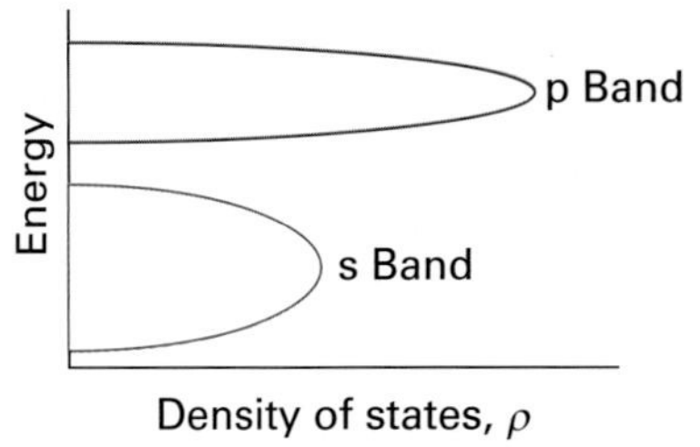

8

9.30 The energy levels of N atoms in the tight-binding Hückel approximation (Section 9.9) are the roots of a tridiagonal determinant (eqn 9.18):

$$E_k = \alpha + 2\beta\cos\frac{k\pi}{N+1} \qquad k = 1, 2, \ldots, N$$

If the atoms are arranged in a ring, the solutions are the roots of a 'cyclic' determinant:

$$E_k = \alpha + 2\beta\cos\frac{2k\pi}{N} \qquad k = 0, \pm 1, \pm 2, \ldots, \pm\tfrac{1}{2}N$$

(for N even). Discuss the consequences, if any, of joining the ends of an initially straight length of material.

9.31 In this and the following problem we explore some of the properties of the Fermi–Dirac distribution, eqn 9.20. First, we need to be aware that the distribution gives the probability of occupation of a given energy state, and to calculate the number of electrons in an energy range E to $E + \mathrm{d}E$ we need to multiply $P(E)$ by the number of states in that range, $\rho(E)\mathrm{d}E$, where $\rho(E)$ is the density of states at the energy E. Therefore, the total number of electrons, N_e, in the sample is

$$N_e = \int_0^\infty \rho(E)P(E)\mathrm{d}E$$

For a three-dimensional solid of volume V, it turns out that $\rho(E) = CE^{1/2}$, with $C = 4\pi V(2m_e/h^2)^{3/2}$. Show that at $T = 0$,

$$P(E) = 1 \text{ for } E < \mu \qquad P(E) = 0 \text{ for } E > \mu$$

and deduce that $\mu(0) = (3\mathcal{N}/8\pi)^{2/3}(h^2/2m_e)$, where $\mathcal{N} = N_e/V$, the number density of electrons in the solid. Evaluate $\mu(0)$ for sodium (where each atom contributes one electron).

9.32 By inspection of eqn 9.20 and the expression for N_e in Problem 9.31 (and without attempting to evaluate the integral explicitly), show that, in order for N_e to remain constant as the temperature is raised, the chemical potential must decrease in value from $\mu(0)$.

9.33 Show that if a substance responds non-linearly to two sources of radiation, one of frequency ω_1 and the other of frequency ω_2, then it may give rise to radiation of the sum and difference of the two frequencies. This non-linear optical phenomenon is known as *frequency mixing* and is used to expand the wavelength range of lasers in laboratory applications, such as spectroscopy and photochemistry.

9.34 In the following sequence of problems we investigate quantitatively the spectra of molecular solids. We begin by considering a dimer, with each monomer having a single transition with transition dipole moment $\mu_{\rm mon}$ and wavenumber $\tilde{\nu}_{\rm mon}$. We assume that the ground-state wavefunctions are not perturbed as a result of dimerization. We then write the dimer excited state wavefunctions Ψ_i as linear combinations of the excited state wavefunctions ψ_1 and ψ_2 of the monomer: $\Psi_i = c_j\psi_1 + c_k\psi_2$. Now we write the hamiltonian matrix with diagonal elements set to the energy between the excited and ground state of the monomer (which, expressed as a wavenumber, is simply $\tilde{\nu}_{\rm mon}$), and off-diagonal elements corresponding to the energy of interaction between the transition dipoles. Using the arrangement in (9), write this interaction energy (as a wavenumber) as:

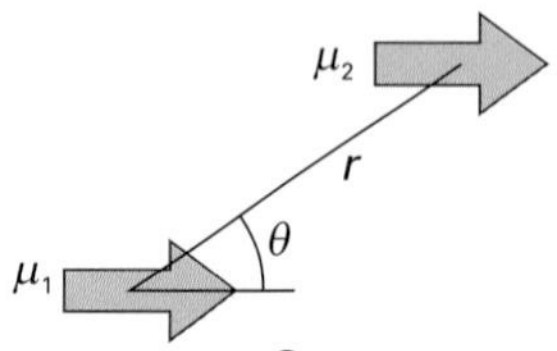

9

$$\beta = \frac{\mu_{\rm mon}^2}{4\pi\varepsilon_0 hcr^3}(1 - 3\cos^2\theta)$$

It follows that the hamiltonian matrix is

$$H = \begin{pmatrix} \tilde{\nu}_{\rm mon} & \beta \\ \beta & \tilde{\nu}_{\rm mon} \end{pmatrix}$$

The eigenvalues of the matrix are the dimer transition wavenumbers $\tilde{\nu}_1$ and $\tilde{\nu}_2$. The eigenvectors are the wavefunctions for the excited states of the dimer and have the form $\begin{pmatrix} c_j \\ c_k \end{pmatrix}$. The intensity of absorption of incident radiation is proportional to the square of the transition dipole moment (Section 4.3). The monomer transition dipole moment is $\mu_{\rm mon} = \int\psi_1^*\hat{\mu}\psi_0 {\rm d}\tau = \int\psi_2^*\hat{\mu}\psi_0 {\rm d}\tau$, where ψ_0 is the wavefunction of the monomer ground state. Assume that the dimer ground state may also be described by ψ_0 and show that the transition dipole moment μ_i of each dimer transition is given by $\mu_i = \mu_{\rm mon}(c_j + c_k)$.

9.35 (a) Consider a dimer of monomers with $\mu_{\rm mon} = 4.00$ D, $\tilde{\nu}_{\rm mon} = 25\,000$ cm^{-1}, and $r = 0.5$ nm. How do the transition wavenumbers $\tilde{\nu}_1$ and $\tilde{\nu}_2$ vary with the angle θ? The relative intensities of the dimer transitions may be estimated by calculating the ratio μ_2^2/μ_1^2. How does this ratio vary with the angle θ? (b) Now expand the treatment given above to a chain of N monomers ($N = 5$, 10, 15, and 20), with $\mu_{\rm mon} = 4.00$ D, $\tilde{\nu}_{\rm mon} = 25\,000$ cm^{-1}, and $r = 0.5$ nm. For simplicity, assume that $\theta = 0$ and that only nearest neighbours interact with interaction energy V. For example, the hamiltonian matrix for the case $N = 4$ is

$$H = \begin{pmatrix} \tilde{\nu}_{\rm mon} & V & 0 & 0 \\ V & \tilde{\nu}_{\rm mon} & V & 0 \\ 0 & V & \tilde{\nu}_{\rm mon} & V \\ 0 & 0 & V & \tilde{\nu}_{\rm mon} \end{pmatrix}$$

How does the wavenumber of the lowest energy transition vary with size of the chain? How does the transition dipole moment of the lowest energy transition vary with the size of the chain?

9.36 The magnetizability, ξ, and the volume and molar magnetic susceptibilities can be calculated from the wavefunctions of molecules. For instance, the magnetizability of a hydrogenic atom is given by the expression $\xi = -(e^2/6m_e)\langle r^2\rangle$, where $\langle r^2\rangle$ is the (expectation) mean value of r^2 in the atom. Calculate ξ and $\chi_{\rm m}$ for the ground state of a hydrogenic atom.

9.37 Nitrogen dioxide, a paramagnetic compound, is in equilibrium with its dimer, dinitrogen tetroxide, a diamagnetic compound. Derive an expression in terms of the equilibrium constant, K, for the dimerization to show how the molar susceptibility varies with the pressure of the sample. Suggest how the susceptibility might be expected to vary as the temperature is changed at constant pressure.

9.38 An NO molecule has thermally accessible electronically excited states. It also has an unpaired electron, and so may be expected to be paramagnetic. However, its ground state is not paramagnetic because the magnetic moment of the orbital motion of the unpaired electron almost exactly cancels the spin magnetic moment. The first excited state (at 121 cm^{-1}) is paramagnetic because the orbital magnetic moment adds to, rather than cancels, the spin magnetic moment. The upper state has a magnetic moment of $2\mu_{\rm B}$. Because the upper state is thermally accessible, the paramagnetic susceptibility of NO shows a pronounced temperature dependence even near room temperature. Calculate the molar paramagnetic susceptibility of NO and plot it as a function of temperature.

Applications to: biochemistry and nanoscience

9.39 Although the crystallization of large biological molecules may not be as readily accomplished as that of small molecules, their crystal lattices are no different. Tobacco seed globulin forms face-centred cubic crystals with unit cell dimension of 12.3 nm and a density of 1.287 g cm^{-3}. Determine its molar mass.

9.40 What features in an X-ray diffraction pattern suggest a helical conformation for a biological macromolecule? Use Fig. 9.26 to deduce as much quantitative information as you can about the shape and size of a DNA molecule.

9.41 A transistor is a semiconducting device that is commonly used either as a switch or an amplifier of electrical signals. Prepare a brief report on the design of a nanometre-sized transistor that uses a carbon nanotube as a component. A useful starting point is the work summarized by Tans *et al.* (*Nature* **393**, 49 (1998)).

9.42 The tip of a scanning tunnelling microscope can be used to move atoms on a surface. The movement of atoms and ions depends on their ability to leave one position and stick to another, and therefore on the energy changes that occur. As an illustration, consider a two-dimensional square lattice of univalent positive and negative ions separated by 200 pm, and consider a cation on top of this array. Calculate, by direct summation, its Coulombic interaction when it is in an empty lattice point directly above an anion.

MATHEMATICAL BACKGROUND 6

Fourier series and Fourier transforms

Some of the most versatile mathematical functions are the trigonometric functions sine and cosine. As a result, it is often very helpful to express a general function as a linear combination of these functions and then to carry out manipulations on the resulting series. Because sines and cosines have the form of waves, the linear combinations often have a straightforward physical interpretation. Throughout this discussion, the function $f(x)$ is real.

MB6.1 Fourier series

A *Fourier series* is a linear combination of sines and cosines that replicates a periodic function:

$$f(x) = \tfrac{1}{2}a_0 + \sum_{n=1}^{\infty}\left\{a_n \cos\frac{n\pi x}{L} + b_n \sin\frac{n\pi x}{L}\right\} \qquad \text{(MB6.1)}$$

A periodic function is one that repeats periodically, such that $f(x + 2L) = f(x)$, where $2L$ is the repeat length. Although it is perhaps not surprising that sines and cosines can be used to replicate continuous functions, it turns out that—with certain limitations—they can also be used to replicate discontinuous functions too. The coefficients in eqn MB6.1 are found by making use of the orthogonality of the sine and cosine functions

$$\int_{-L}^{L} \sin\frac{m\pi x}{L}\cos\frac{n\pi x}{L}\,dx = 0 \qquad \text{(MB6.2a)}$$

and the integrals

$$\int_{-L}^{L} \sin\frac{m\pi x}{L}\sin\frac{n\pi x}{L}\,dx = \int_{-L}^{L} \cos\frac{m\pi x}{L}\cos\frac{n\pi x}{L}\,dx = L\delta_{mn} \qquad \text{(MB6.2b)}$$

where $\delta_{mn} = 1$ if $m = n$ and 0 if $m \neq n$. Thus, multiplication of both sides of eqn MB6.1 by $\cos(k\pi x/L)$ and integration from $-L$ to L gives an expression for the coefficient a_k, and multiplication by $\sin(k\pi x/L)$ and integration likewise gives an expression for b_k:

$$a_k = \frac{1}{L}\int_{-L}^{L} f(x)\cos\frac{k\pi x}{L}\,dx \qquad k = 0, 1, 2, \ldots$$

$$b_k = \frac{1}{L}\int_{-L}^{L} f(x)\sin\frac{k\pi x}{L}\,dx \qquad k = 1, 2, \ldots \qquad \text{(MB6.3)}$$

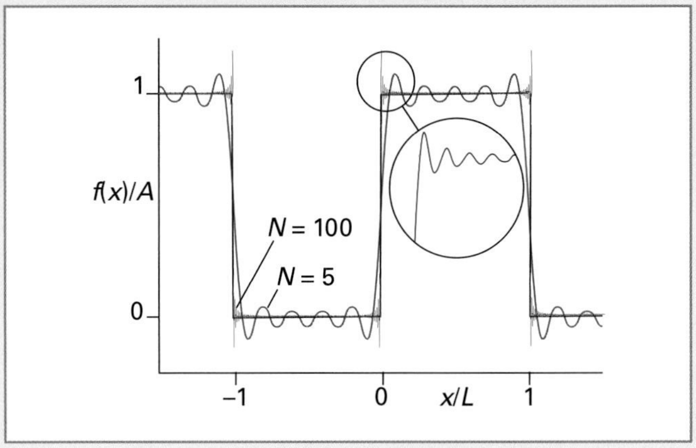

Fig. MB6.1 A square wave and two successive approximations by Fourier series ($N = 5$ and $N = 100$). The inset shows a magnification of the $N = 100$ approximation.

● A BRIEF ILLUSTRATION

Figure MB6.1 shows a graph of a square wave of amplitude A that is periodic between $-L$ and L. The mathematical form of the wave is

$$f(x) = \begin{cases} -A & -L \le x < 0 \\ +A & 0 \le x < L \end{cases}$$

The coefficients a are all zero because $f(x)$ is antisymmetric ($f(-x) = -f(x)$) whereas all the cosine functions are symmetric ($\cos(-x) = \cos(x)$) and so cosine waves make no contribution to the sum. The coefficients b are obtained from

$$b_k = \frac{1}{L}\int_{-L}^{L} f(x)\sin\frac{k\pi x}{L}\,dx = \frac{1}{L}\int_{-L}^{0} (-A)\sin\frac{k\pi x}{L}\,dx + \frac{1}{L}\int_{0}^{L} A\sin\frac{k\pi x}{L}\,dx = \frac{2A}{k\pi}\{1 - (-1)^k\}$$

The final expression has been formulated to acknowledge that the two integrals cancel when k is even but add together when k is odd. Therefore,

$$f(x) = \frac{2A}{\pi}\sum_{k=1}^{N}\frac{1-(-1)^k}{k}\sin\frac{k\pi x}{L} = \frac{4A}{\pi}\sum_{n=1}^{N}\frac{1}{2n-1}\sin\frac{(2n-1)\pi x}{L}$$

with $N \to \infty$. The sum over n is the same as the sum over k; in the latter, terms with k even are all zero. This function is plotted in Fig. MB6.1 for two values of N to show how the series becomes more faithful to the original function as N increases. ●

Self-test MB6.1 Repeat the analysis for a saw-tooth wave, $f(x) = Ax$ in the range $-L \le x < L$ and $f(x + 2L) = f(x)$ elsewhere. Use graphing software to depict the result.

$$[f(x) = (2AL/\pi)\sum_{n=1}^{\infty}\{(-1)^{n+1}/n\}\sin(n\pi x/L)\text{, Fig. MB6.2}]$$

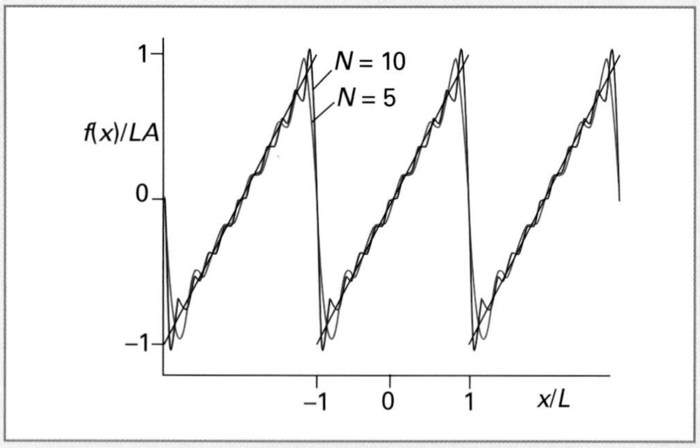

Fig. MB6.2 A saw-tooth function and its representation as a Fourier series with two successive approximations ($N=5$ in blue and $N=10$ in red).

MB6.2 Finite approximations and Parseval's theorem

Close inspection of Fig. MB6.1 (a region has been magnified) shows that the Fourier series has artefacts where the original wave is discontinuous; these artefacts persist even as $N \to \infty$. They disappear in regions where the function $f(x)$ is piecewise continuous (in regions where $f(x)$ does not jump from one value to another, as in the horizontal parts of the square wave as distinct from the end-points of each horizontal section). For these well behaved regions a useful result is obtained by considering how closely a finite sum (like the successive approximations depicted in Fig. MB6.1 to the square wave) approaches the original function. Thus, in place of eqn MB6.1, we write

$$f_N(x) = \tfrac{1}{2}a_0 + \sum_{n=1}^{N}\left\{a_n \cos\frac{n\pi x}{L} + b_n \sin\frac{n\pi x}{L}\right\} \qquad \text{(MB6.4)}$$

and examine the mean square error in this approximation to $f(x)$ for a finite value of N:

$$\int_{-L}^{L} \{f(x) - f_N(x)\}^2 \mathrm{d}x$$

$$= \int_{-L}^{L} f(x)^2 \mathrm{d}x - 2\int_{-L}^{L} f(x) f_N(x)\mathrm{d}x + \int_{-L}^{L} f_N(x)^2 \mathrm{d}x$$

Because the integral on the left is non-negative (that is, greater than or equal to zero), the sum of the three integrals on the right is also non-negative. However, we may use the orthogonality relations in eqn MB6.2 to write

$$\int_{-L}^{L} f(x) f_N(x)\mathrm{d}x$$

$$= \int_{-L}^{L} \left\{\tfrac{1}{2}a_0 + \sum_{m=1}^{\infty}\left(a_m \cos\frac{m\pi x}{L} + b_m \sin\frac{m\pi x}{L}\right)\right\}$$

$$\times \left\{\tfrac{1}{2}a_0 + \sum_{n=1}^{N}\left(a_n \cos\frac{n\pi x}{L} + b_n \sin\frac{n\pi x}{L}\right)\right\}\mathrm{d}x$$

$$= \tfrac{1}{2}La_0^2 + L\sum_{n=1}^{N}(a_n^2 + b_n^2)$$

The final sum terminates at N because the integrals over the cosine and sine terms in the infinite sum that have no matching terms in the finite sum are all zero. The integral over $f_N(x)^2$ therefore has the same value, and so we can conclude that

$$\frac{1}{L}\int_{-L}^{L} f(x)^2 \mathrm{d}x \geq \tfrac{1}{2}a_0^2 + \sum_{n=1}^{N}(a_n^2 + b_n^2) \qquad \text{(MB6.5a)}$$

If now we allow N to become infinite, the inequality is replaced by the equality and we arrive at *Parseval's theorem*:

$$\frac{1}{L}\int_{-L}^{L} f(x)^2 \mathrm{d}x = \tfrac{1}{2}a_0^2 + \sum_{n=1}^{\infty}(a_n^2 + b_n^2) \qquad \text{(MB6.5b)}$$

This theorem is useful in reverse: because it relates a sum to an integral, it is sometimes possible to evaluate a sum by evaluating the corresponding integral. Note that it is essential that the range of integration does not include any discontinuities of the function $f(x)$.

● **A BRIEF ILLUSTRATION**

In Self-test MB6.1 we derived the Fourier series for the function $f(x) = x$, which is piecewise continuous in the range $-L \leq x < L$. The integral of the square of this function in this range is

$$\int_{-L}^{L} x^2 \mathrm{d}x = \tfrac{2}{3}L^3$$

The coefficients in the Fourier series for x are $a_n = 0$ and $b_n = 2(-1)^{n+1}L/n\pi$. Therefore, from Parseval's theorem,

$$\tfrac{2}{3}L^2 = \sum_{n=1}^{\infty}\frac{4L^2}{n^2\pi^2}$$

from which we can conclude that

$$\sum_{n=1}^{\infty}\frac{1}{n^2} = \tfrac{1}{6}\pi^2$$ ●

Self-test MB6.2 Evaluate the sum of inverse fourth powers of the integers. (*Hint.* Use $f(x) = x^2$.) [$\sum_{n=1}^{\infty} 1/n^4 = \pi^4/90$]

MB6.3 Fourier transforms

The Fourier series in eqn MB6.1 can be expressed in a more succinct manner if we allow the coefficients to be complex numbers and make use of *de Moivre's relation*

$$e^{in\pi x/L} = \cos\frac{n\pi x}{L} + i\sin\frac{n\pi x}{L} \quad \text{(MB6.6)}$$

for then we may write

$$f(x) = \sum_{n=-\infty}^{\infty} c_n e^{in\pi x/L} \qquad c_n = \frac{1}{2L}\int_{-L}^{L} f(x)e^{-in\pi x/L}dx \quad \text{(MB6.7)}$$

This complex formalism is well suited to the extension of this discussion to functions with periods that become infinite. If a period is infinite, we are effectively dealing with a non-periodic function, such as e^{-x}.

We write $\delta k = \pi/L$ and consider the limit as $L \to \infty$ and therefore $\delta k \to 0$: that is, eqn MB6.7 becomes

$$f(x) = \lim_{L\to\infty}\sum_{n=-\infty}^{\infty}\left\{\frac{1}{2L}\int_{-L}^{L} f(x')e^{-in\pi x'/L}dx'\right\}e^{in\pi x/L}$$

$$= \lim_{\delta k\to 0}\sum_{n=-\infty}^{\infty}\left\{\frac{\delta k}{2\pi}\int_{-\pi/\delta k}^{\pi/\delta k} f(x')e^{-in\delta k x'}dx'\right\}e^{in\delta k x} \quad \text{(MB6.8)}$$

$$= \frac{1}{2\pi}\lim_{\delta k\to 0}\sum_{n=-\infty}^{\infty}\left\{\int_{-\infty}^{\infty} f(x')e^{-in\delta k(x'-x)}dx'\right\}\delta k$$

In the last line we have anticipated that the limits of the integral will become infinite. At this point we should recognize that a formal definition of an integral is the sum of the value of a function at a series of infinitely spaced points multiplied by the separation of each point (Fig. MB6.3):

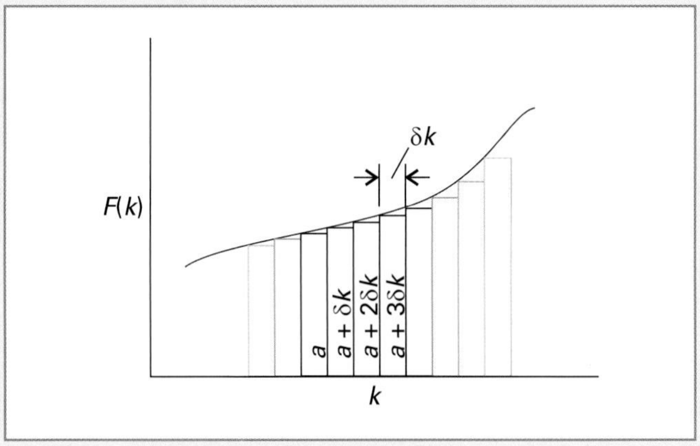

Fig. MB6.3 The formal definition of an integral as the sum of the value of a function at a series of infinitely spaced points multiplied by the separation of each point.

$$\int_a^b F(k)dk = \lim_{\delta k\to 0}\sum_{n=-\infty}^{\infty} F(n\delta k)\delta k \quad \text{(MB6.9)}$$

Exactly this form appears on the right-hand side of eqn MB6.8, so we can write that equation as

$$f(x) = \frac{1}{2\pi}\int_{-\infty}^{\infty}\tilde{f}(k)e^{ikx}dk \qquad \text{where } \tilde{f}(k) = \int_{-\infty}^{\infty} f(x')e^{-ikx'}dx' \quad \text{(MB6.10)}$$

We call the function $\tilde{f}(k)$ the *Fourier transform* of $f(x)$; the original function $f(x)$ is the *inverse Fourier transform* of $\tilde{f}(k)$.

● **A BRIEF ILLUSTRATION**

The Fourier transform of the symmetrical exponential function $f(x) = e^{-a|x|}$ is

$$\tilde{f}(k) = \int_{-\infty}^{\infty} e^{-a|x|-ikx}dx$$

$$= \int_{-\infty}^{0} e^{ax-ikx}dx + \int_{0}^{\infty} e^{-ax-ikx}dx$$

$$= \frac{1}{a-ik} + \frac{1}{a+ik} = \frac{2a}{a^2+k^2}$$

The original function and its Fourier transform are drawn in Fig. MB6.4. ●

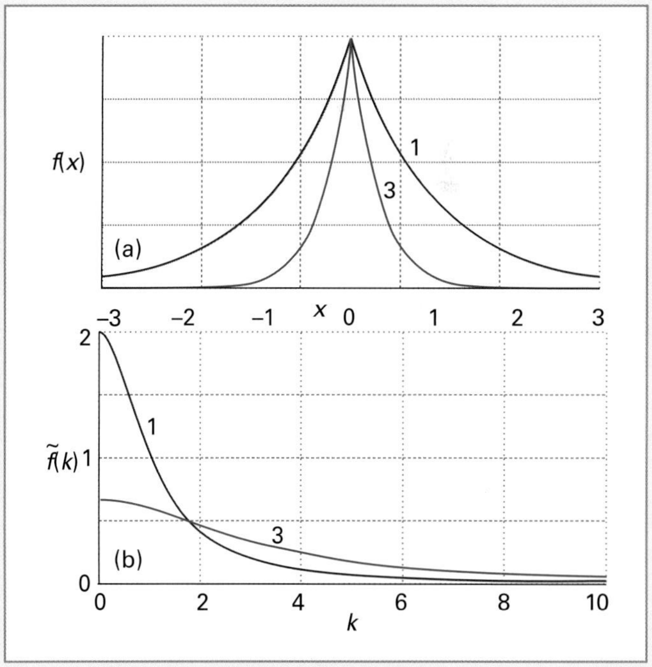

Fig. MB6.4 (a) The symmetrical exponential function $f(x) = e^{-a|x|}$ and (b) its Fourier transform for two values of the decay constant a. Note how the function with the more rapid decay has a Fourier transform richer in short-wavelength (high k) components.

Self-test MB6.3 Evaluate the Fourier transform of the Gaussian function $e^{-a^2x^2}$. $[\tilde{f}(k) = (\pi/a^2)^{1/2}e^{-k^2/4a^2}]$

The physical interpretation of eqn MB6.10 is that $f(x)$ is expressed as a superposition of harmonic (sine and cosine) functions of wavelength $\lambda = 2\pi/k$, and that the weight of each constituent function is given by the Fourier transform at the corresponding value of k. This interpretation is consistent with the calculation in the *illustration*. As we see from Fig. MB6.4, when the exponential function falls away rapidly, the Fourier transform is extended to high values of k, corresponding to a significant contribution from short-wavelength waves. When the exponential function decays only slowly, the most significant contributions to the superposition come from long-wavelength components, which is reflected in the Fourier transform, with its predominance of small-k contributions in this case. In general, a slowly varying function has a Fourier transform with significant contributions from small-k components.

MB6.4 The convolution theorem

A final point concerning the properties of Fourier transforms is the *convolution theorem*, which states that, if a function is the 'convolution' of two other functions, that is, if

$$F(x) = \int_{-\infty}^{\infty} f_1(x')f_2(x - x')\mathrm{d}x' \qquad \text{(MB6.11a)}$$

then the Fourier transform of $F(x)$ is the product of the Fourier transforms of its component functions:

$$\tilde{F}(k) = \tilde{f}_1(k)\tilde{f}_2(k) \qquad \text{(MB6.11b)}$$

● **A BRIEF ILLUSTRATION**

If $F(x)$ is the convolution of two Gaussian functions,

$$F(x) = \int_{-\infty}^{\infty} e^{-a^2x'^2}e^{-b^2(x-x')^2}\mathrm{d}x'$$

then from Self-test MB6.3 we can immediately write its transform as

$$\tilde{F}(k) = \left(\frac{\pi}{a^2}\right)^{1/2} e^{-k^2/4a^2}\left(\frac{\pi}{b^2}\right)^{1/2} e^{-k^2/4b^2} = \frac{\pi}{ab}e^{-(k^2/4)(1/a^2+1/b^2)}$$ ●

PART 3
Molecular spectroscopy

We now begin our study of molecular spectroscopy, the analysis of the electromagnetic radiation emitted, absorbed, or scattered by molecules. The starting point for the discussion in the next three chapters is the observation summarized in Parts 1 and 2 that photons of radiation ranging from the infrared to the ultraviolet bring information to us about molecules as a result of their electronic, vibrational, and rotational transitions. In Chapters 10 and 11 we describe techniques used to study these transitions and see how electronic transitions prepare atoms and molecules for such important light-induced processes as those associated with laser action. In Chapter 12 we see that the combined effect of an external magnetic field and molecular excitation with photons in the radiofrequency or microwave ranges leads to important spectroscopic techniques, collectively known as magnetic resonance spectroscopy, that are now common in chemistry, biochemistry, and medicine. In short, molecular spectra are complicated but contain a great deal of information, including bond lengths, bond angles, and bond strengths, that can be used to analyse systems ranging in size from diatomic molecules to living organisms. Along the way, we also see how spectra complement information on molecular structure obtained from the diffraction techniques discussed in Chapter 9.

Another outcome of spectroscopy is the determination of the energy levels available to electrons, atoms, and molecules. These energy levels will turn out to be crucial to understanding and predicting the thermodynamic properties of bulk matter that we treat in later chapters of the text.

Rotational and vibrational spectra

10

The general strategy we adopt in the chapter is to set up expressions for the energy levels of molecules and then apply selection rules and considerations of populations to infer the form of spectra. Rotational energy levels are considered first, and we see how to derive expressions for their values and how to interpret rotational spectra in terms of molecular dimensions. Not all molecules can occupy all rotational states: we see the experimental evidence for this restriction and its explanation in terms of nuclear spin and the Pauli principle. Next, we consider the vibrational energy levels of diatomic molecules, and see that we can use the properties of harmonic oscillators developed in Chapter 2. Then we consider polyatomic molecules and find that their vibrations may be discussed as though they consisted of a set of independent harmonic oscillators, so the same approach as employed for diatomic molecules may be used. We also see that the symmetry properties of the vibrations of polyatomic molecules are helpful for deciding which modes of vibration can be studied spectroscopically.

The origin of spectral lines in molecular spectroscopy is the absorption, emission, or scattering of a photon when the energy of a molecule changes. The difference from atomic spectroscopy is that the energy of a molecule can change not only as a result of electronic transitions but also because it can undergo changes of rotational and vibrational state. Molecular spectra are therefore more complex than atomic spectra. However, they also contain information relating to more properties, and their analysis leads to values of bond strengths, lengths, and angles. They also provide a way of determining a variety of molecular properties, particularly molecular dimensions, shapes, and dipole moments.

Pure rotational spectra, in which only the rotational state of a molecule changes, can be observed in the gas phase. Although vibrational spectra from condensed-phase samples do not display a structure due to accompanying rotational transitions, the spectra of gaseous samples do show such structure. Electronic spectra, which are described in Chapter 11, show features arising from simultaneous vibrational and, in the gas phase, rotational transitions (Fig. 10.1). The simplest way of dealing with these complexities is to tackle each type of transition in turn, and then to see how simultaneous changes affect the appearance of spectra.

Pure rotational spectra

The general strategy we adopt for discussing molecular spectra and the information they contain is to find expressions for the energy levels of molecules and then to

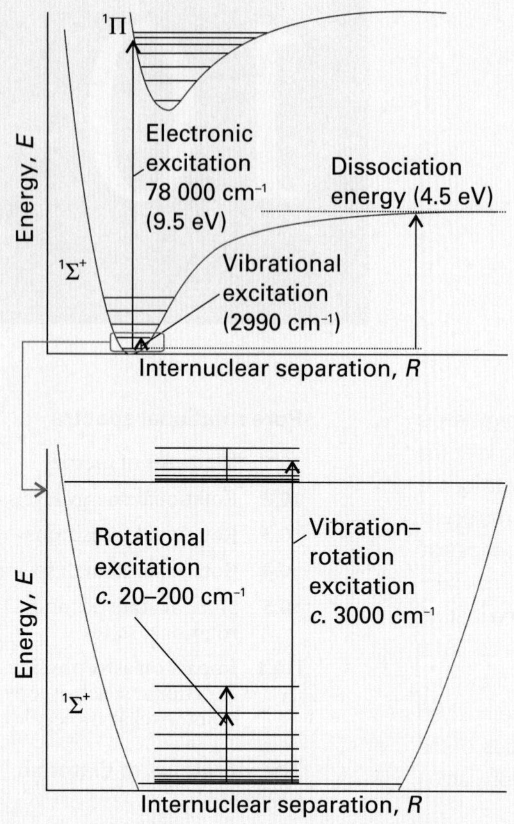

Fig. 10.1 A molecular spectrum displays transitions due to electronic excitation, vibrational excitation, and rotational excitation. This figure gives an indication of the magnitudes of each type of transition for HCl. The features illustrated here are developed in this and the following chapter.

calculate the transition frequencies by applying the selection rules. We then predict the appearance of the spectrum by taking into account the transition moments and the populations of the states. In this section we illustrate the strategy by considering the rotational states of molecules.

10.1 Moments of inertia

The key molecular parameter we shall need is the **moment of inertia**, I, of the molecule (Section 3.3). The moment of inertia of a molecule is defined as the mass of each atom multiplied by the square of its perpendicular distance from the rotational axis through the centre of mass of the molecule (Fig. 10.2):

$$I = \sum_i m_i x_i^2 \qquad [10.1]$$

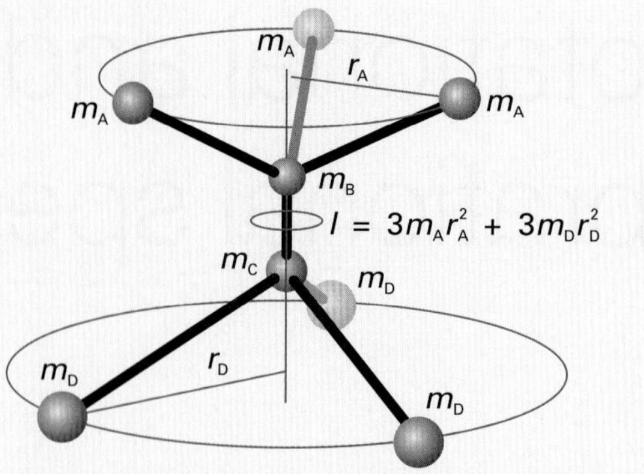

Fig. 10.2 The definition of moment of inertia. In this molecule there are three identical atoms attached to the B atom and three different but mutually identical atoms attached to the C atom. In this example, the centre of mass lies on an axis passing through the B and C atoms, and the perpendicular distances are measured from this axis.

where x_i is the perpendicular distance of the atom i from the selected axis of rotation. The moment of inertia depends on the masses of the atoms present and the molecular geometry, so we can suspect (and later shall see explicitly) that rotational spectroscopy will give information about bond lengths and bond angles.

In general, the rotational properties of any molecule can be expressed in terms of the **principal moments of inertia**, the moments of inertia about three perpendicular axes set in the molecule (Fig. 10.3). The convention is to label these moments of inertia I_a, I_b, and I_c, with the axes chosen so that $I_c \geq I_b \geq I_a$. For linear molecules, the moment of inertia around the internuclear axis is effectively zero. The explicit expressions for the principal moments of inertia of some symmetrical molecules are given in Table 10.1.

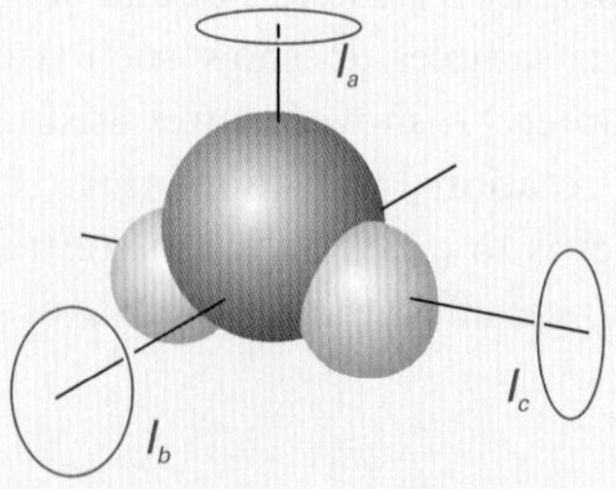

Fig. 10.3 An asymmetric rotor has three different moments of inertia; all three rotation axes coincide at the centre of mass of the molecule.

Table 10.1 Moments of inertia*

1. *Diatomic molecules*

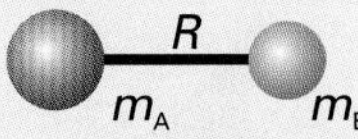

$$I = \mu R^2 \qquad \mu = \frac{m_A m_B}{m}$$

2. *Triatomic linear rotors*

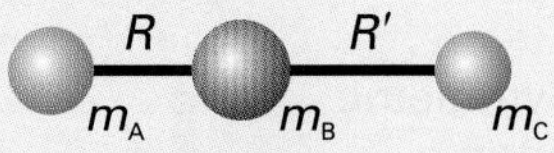

$$I = m_A R^2 + m_C R'^2 - \frac{(m_A R - m_C R')^2}{m}$$

$$I = 2m_A R^2$$

3. *Symmetric rotors*

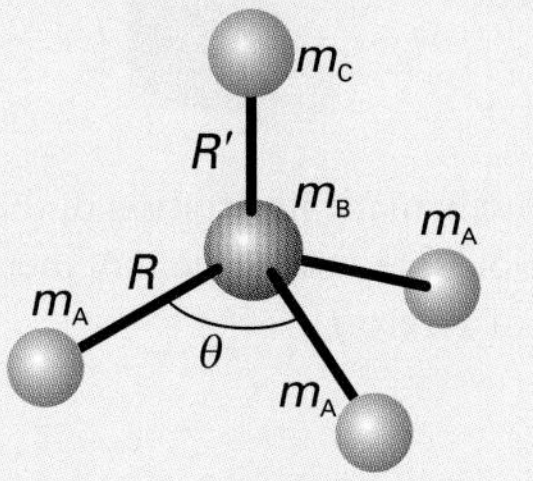

$$I_{\parallel} = 2m_A(1-\cos\theta)R^2$$

$$I_{\perp} = m_A(1-\cos\theta)R^2 + \frac{m_A}{m}(m_B+m_C)(1+2\cos\theta)R^2 + \frac{m_C}{m}\{(3m_A+m_B)R' + 6m_A R[\tfrac{1}{3}(1+2\cos\theta)]^{1/2}\}R'$$

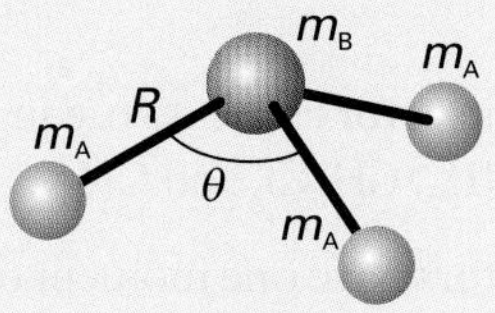

$$I_{\parallel} = 2m_A(1-\cos\theta)R^2$$

$$I_{\perp} = m_A(1-\cos\theta)R^2 + \frac{m_A m_B}{m}(1+2\cos\theta)R^2$$

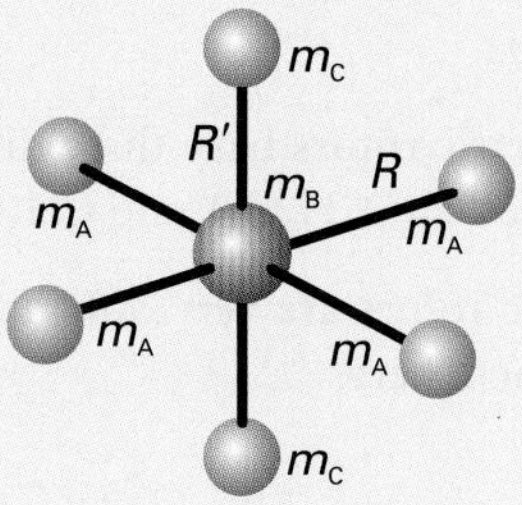

$$I_{\parallel} = 4m_A R^2$$

$$I_{\perp} = 2m_A R^2 + 2m_C R'^2$$

4. *Spherical rotors*

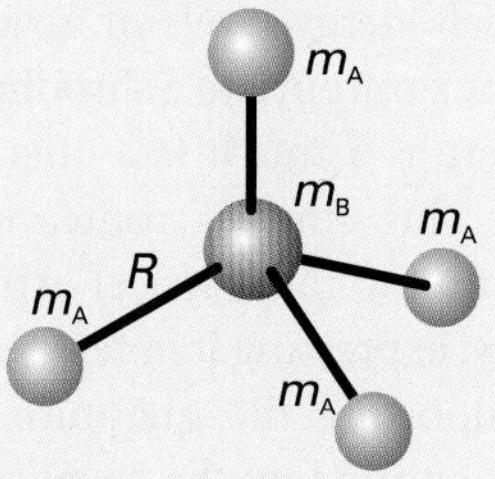

$$I = \tfrac{8}{3} m_A R^2$$

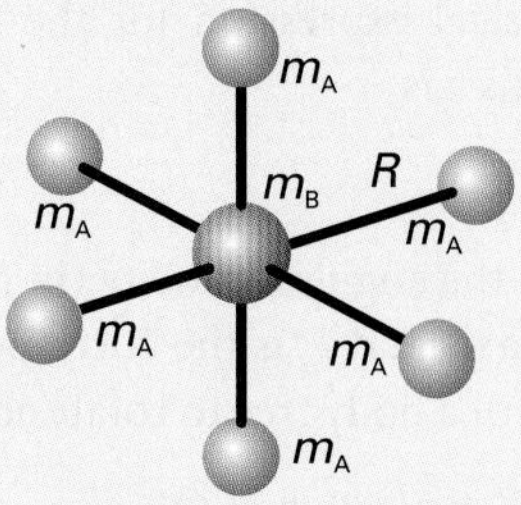

$$I = 4m_A R^2$$

* In each case, m is the total mass of the molecule.

Example 10.1 *Calculating the moment of inertia of a molecule*

Calculate the moment of inertia of an H_2O molecule around the axis defined by the bisector of the HOH angle (1). The HOH bond angle is 104.5° and the bond length is 95.7 pm.

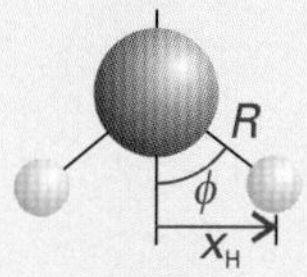

1

Method According to eqn 10.1, the moment of inertia is the sum of the masses multiplied by the squares of their distances from the axis of rotation. The latter can be expressed by using trigonometry and the bond angle and bond length.

Answer From eqn 10.1,

$$I = m_H x_H^2 + m_O x_O^2 + m_H x_H^2 = m_H x_H^2 + 0 + m_H x_H^2 = 2m_H x_H^2$$

If the bond angle of the molecule is denoted 2ϕ and the bond length is R, trigonometry gives $x_H = R \sin \phi$. It follows that

$$I = 2m_H R^2 \sin^2\phi$$

Substitution of the data gives

$$I = 2 \times (1.67 \times 10^{-27}\ \text{kg}) \times (9.57 \times 10^{-11}\ \text{m})^2 \times \sin^2 52.3° = 1.91 \times 10^{-47}\ \text{kg m}^2$$

Note that the mass of the O atom makes no contribution to the moment of inertia for this mode of rotation as the atom is immobile while the H atoms circulate around it.

A note on good practice The mass to use in the calculation of the moment of inertia is the actual atomic mass (in kilograms, noting the precise nuclide present), not the element's molar mass, which is an average over a typical isotopic composition.

Self-test 10.1 Calculate the moment of inertia of a $CH^{35}Cl_3$ molecule around a rotational axis that contains the C—H bond. The C—Cl bond length is 177 pm and the HCCl angle is 107°; $m(^{35}Cl) = 34.97m_u$. [4.99×10^{-45} kg m^2]

We shall suppose initially that molecules are **rigid rotors**, bodies that do not distort under the stress of rotation. Rigid rotors can be classified into four types (Fig. 10.4):

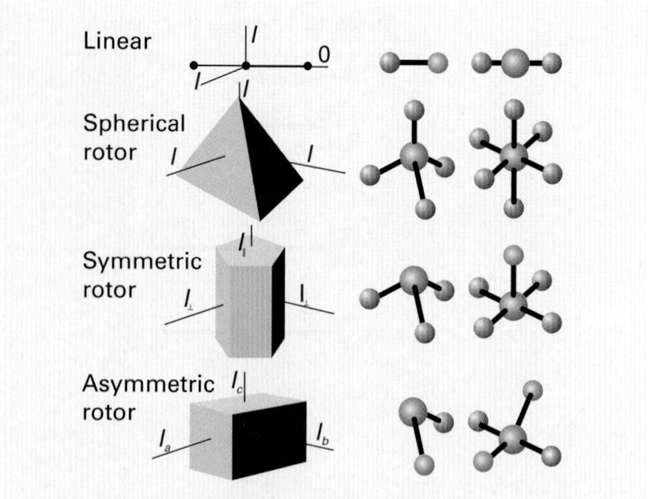

Fig. 10.4 A schematic illustration of the classification of rigid rotors. (Note that the moment of inertia around the axis of a linear rotor is zero.)

Spherical rotors have three equal moments of inertia (examples: CH_4, SiH_4, and SF_6).

Symmetric rotors have two equal moments of inertia (examples: NH_3, CH_3Cl, and CH_3CN).

Linear rotors have one moment of inertia (the one about the molecular axis) equal to zero (examples: CO_2, HCl, OCS, and HC≡CH).

Asymmetric rotors have three different moments of inertia (examples: H_2O, H_2CO, and CH_3OH).

Asymmetric rotors are very difficult to treat and we shall not deal with them here.

10.2 Rotational energy levels

The rotational energy levels of a rigid rotor may be obtained by solving the appropriate Schrödinger equation. Fortunately, however, there is a much less onerous short cut to the exact expressions. This simpler route, which is described below, depends on noting the classical expression for the energy of a rotating body, expressing it in terms of the angular momentum, and then importing the quantum mechanical properties of angular momentum into the equations.

The classical expression for the energy of a body rotating about an axis a is

$$E_a = \tfrac{1}{2}I_a\omega_a^2 \qquad (10.2)$$

where ω_a is the angular velocity (in radians per second, rad s^{-1}) about that axis and I_a is the corresponding moment of inertia. The energy of a body free to rotate about three axes is

$$E = \tfrac{1}{2}I_a\omega_a^2 + \tfrac{1}{2}I_b\omega_b^2 + \tfrac{1}{2}I_c\omega_c^2$$

Because the classical angular momentum about the axis a is $J_a = I_a\omega_a$, with similar expressions for the other axes, it follows that

$$E = \frac{J_a^2}{2I_a} + \frac{J_b^2}{2I_b} + \frac{J_c^2}{2I_c} \qquad (10.3)$$

This is the key equation. We described the quantum mechanical properties of angular momentum in Section 3.4, and can now make use of them in conjunction with this equation to obtain the rotational energy levels.

(a) Spherical rotors

When all three moments of inertia are equal to some value I, as in CH_4 and SF_6, the classical expression for the energy is

$$E = \frac{J_a^2 + J_b^2 + J_c^2}{2I} = \frac{\mathcal{J}^2}{2I}$$

where $\mathcal{J}^2 = J_a^2 + J_b^2 + J_c^2$ is the square of the magnitude of the angular momentum. We can immediately find the corresponding quantum expression by making the replacement

$$\mathcal{J}^2 \rightarrow J(J+1)\hbar^2 \qquad J = 0, 1, 2, \ldots$$

Therefore, the energy of a spherical rotor is confined to the values

$$E_J = J(J+1)\frac{\hbar^2}{2I} \qquad J = 0, 1, 2, \ldots \qquad (10.4)$$

The resulting ladder of energy levels is illustrated in Fig. 10.5. Note that the rotor may have zero energy because $J = 0$ is allowed: the cyclic boundary conditions on the wavefunction permit a solution that has a constant value at all angles (specifically $Y_{0,0}(\theta,\phi) = 1/2\pi^{1/2}$) and hence corresponds to zero rotational kinetic energy. The energy is normally expressed in terms of the **rotational constant**, $\tilde{B}$, of the molecule, where

$$hc\tilde{B} = \frac{\hbar^2}{2I} \qquad \text{so} \qquad \tilde{B} = \frac{\hbar}{4\pi cI} \qquad [10.5]$$

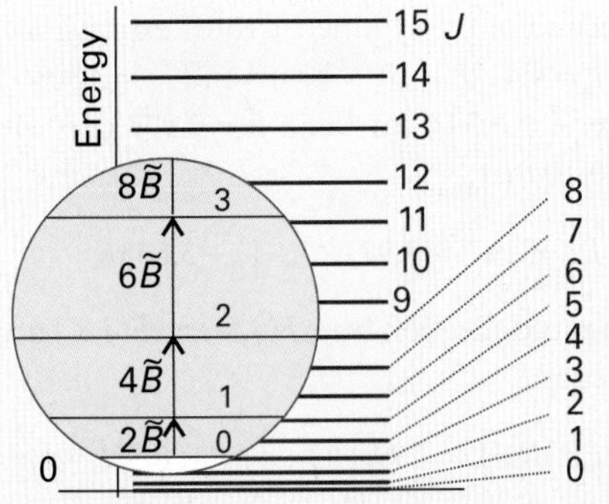

Fig. 10.5 The rotational energy levels of a linear or spherical rotor. Note that the energy separation between neighbouring levels increases as J increases.

The expression for the energy is then

$$E_J = hc\tilde{B}J(J+1) \qquad J = 0, 1, 2, \ldots \qquad (10.6)$$

The rotational constant as defined by eqn 10.5 is a wavenumber (as indicated by the tilde). The energy of a rotational state is normally reported as the **rotational term**, $\tilde{F}(J)$, a wavenumber, by division by hc:

$$\tilde{F}(J) = E_J/hc = \tilde{B}J(J+1) \qquad (10.7)$$

The definition of $\tilde{B}$ as a wavenumber is convenient for the discussion of vibration–rotation spectra. However, for pure rotational spectroscopy it is more common to define it as a frequency and to remove the tilde. Then $B = \hbar/4\pi I$, the energy is $E = hBJ(J+1)$, and the rotational term is $BJ(J+1)$.

The separation of adjacent levels is

$$\begin{aligned}\tilde{F}(J+1) - \tilde{F}(J) &= \tilde{B}(J+1)(J+2) - \tilde{B}J(J+1)\\ &= 2\tilde{B}(J+1)\end{aligned} \qquad (10.8)$$

Because the rotational constant decreases as I increases, we see that large, heavy molecules have closely spaced rotational energy levels. We can estimate the magnitude of the separation by considering CCl_4: from the bond lengths and masses of the atoms we find $I = 4.85 \times 10^{-45}$ kg m^2, and hence $\tilde{B} = 0.0577$ cm^{-1}.

The angular momentum of the molecule has a component on an external, laboratory-fixed axis. This component is quantized, and its permitted values are $M_J\hbar$, with $M_J = 0, \pm1, \ldots, \pm J$, giving $2J+1$ values in all (Fig. 10.6). The quantum number M_J does not appear in the expression for the energy because the orientation of the rotating molecule in space does not affect its energy, but it is necessary for a complete specification of the state of the rotor. An additional feature is that the angular momentum has

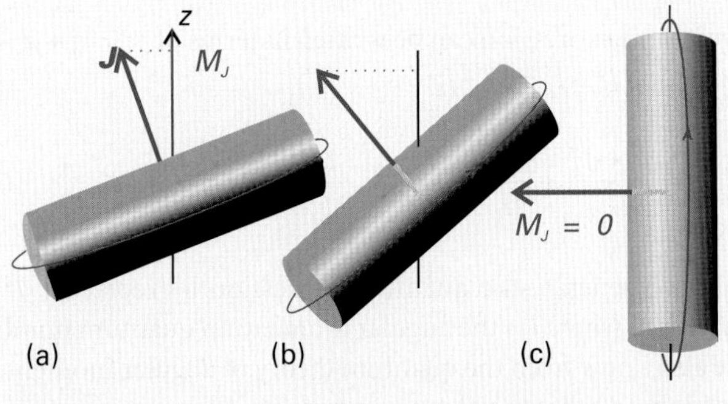

Fig. 10.6 The significance of the quantum number M_J. (a) When M_J is close to its maximum value, J, most of the molecular rotation is around the laboratory z-axis. (b) An intermediate value of M_J. (c) When $M_J = 0$ the molecule has no angular momentum about the z-axis. All three diagrams correspond to a state with $K = 0$; there are corresponding diagrams for different values of K, in which the angular momentum makes a different angle to the molecule's principal axis.

a component of angular momentum on an arbitrary axis set in the molecule (such as any of the C—H bonds of CH_4). This component is also quantized, and has the values $K\hbar$ with $K = J, J-1, \ldots, -J$. The energy is independent of which of those values it takes. Therefore, as well as having a $(2J+1)$-fold degeneracy arising from its orientation in space, the rotor also has a $(2J+1)$-fold degeneracy arising from its orientation with respect to an arbitrary axis in the molecule. The overall degeneracy of a spherical rotor with quantum number J is therefore $(2J+1)^2$. This degeneracy increases very rapidly: when $J = 10$, for instance, there are 441 states of the same energy.

(b) Symmetric rotors

In symmetric rotors, two moments of inertia are equal but different from the third (as in CH_3Cl, NH_3, and C_6H_6); the unique axis of the molecule is its **principal axis** (or *figure axis*). We shall write the unique moment of inertia (that about the principal axis) as $I_\parallel$ and the other two as $I_\perp$. If $I_\parallel > I_\perp$, the rotor is classified as **oblate** (2, like a pancake, and C_6H_6); if $I_\parallel < I_\perp$ it is classified as **prolate** (3, like a cigar, and CH_3Cl). The classical expression for the energy, eqn 10.3, becomes

$$E = \frac{J_b^2 + J_c^2}{2I_\perp} + \frac{J_a^2}{2I_\parallel}$$

2 Oblate

3 Prolate

Again, this expression can be written in terms of $\mathcal{J}^2 = J_a^2 + J_b^2 + J_c^2$ by writing $J_b^2 + J_c^2 = \mathcal{J}^2 - J_a^2$:

$$E = \frac{\mathcal{J}^2 - J_a^2}{2I_\perp} + \frac{J_a^2}{2I_\parallel} = \frac{\mathcal{J}^2}{2I_\perp} + \left(\frac{1}{2I_\parallel} - \frac{1}{2I_\perp}\right)J_a^2 \qquad (10.9)$$

Now we generate the quantum expression by replacing $\mathcal{J}^2$ by $J(J+1)\hbar^2$, where J is the angular momentum quantum number. We also know from the quantum theory of angular momentum (Section 3.4) that the component of angular momentum about any axis is restricted to the values $K\hbar$, with $K = 0, \pm1, \ldots, \pm J$. (Recall from above that K is the quantum number used to signify a component on a molecular axis; M_J is reserved for a component on an externally defined axis.) Therefore, we also replace J_a^2 by $K^2\hbar^2$. It follows that the rotational terms are

$$\tilde{F}(J,K) = \tilde{B}J(J+1) + (\tilde{A} - \tilde{B})K^2 \qquad J = 0, 1, 2, \ldots \quad K = 0, \pm1, \ldots, \pm J \qquad (10.10)$$

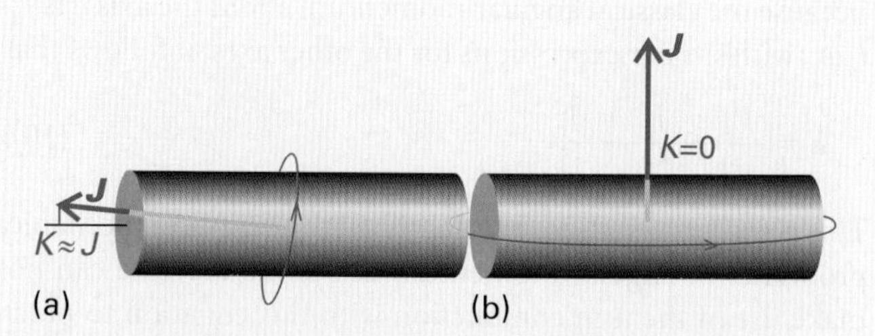

Fig. 10.7 The significance of the quantum number K. (a) When $|K|$ is close to its maximum value, J, most of the molecular rotation is around the figure axis. (b) When $K = 0$ the molecule has no angular momentum about its principal axis: it is undergoing end-over-end rotation.

with

$$\tilde{A} = \frac{\hbar}{4\pi c I_\parallel} \qquad \tilde{B} = \frac{\hbar}{4\pi c I_\perp} \qquad [10.11]$$

Equation 10.10 matches what we should expect for the dependence of the energy levels on the two distinct moments of inertia of the molecule. When $K = 0$, there is no component of angular momentum about the principal axis, and the energy levels depend only on $I_\perp$ (Fig. 10.7). When $K = \pm J$, almost all the angular momentum arises from rotation around the principal axis, and the energy levels are determined largely by $I_\parallel$. The sign of K does not affect the energy because opposite values of K correspond to opposite senses of rotation, and the energy does not depend on the sense of rotation.

A BRIEF ILLUSTRATION

A $^{14}NH_3$ molecule is a symmetric rotor with bond length 101.2 pm and HNH bond angle 106.7°. Substitution of $m_A = 1.0078m_u$, $m_B = 14.0031m_u$, $R = 101.2$ pm, and $\theta = 106.7°$ into the second of the symmetric rotor expressions in Table 10.1 gives $I_\parallel = 4.4128 \times 10^{-47}$ kg m^2 and $I_\perp = 2.8059 \times 10^{-47}$ kg m^2. Hence, $\tilde{A} = 6.344$ cm^{-1} and $\tilde{B} = 9.977$ cm^{-1}. It follows from eqn 10.10 that

$$\tilde{F}(J,K)/\text{cm}^{-1} = 9.977J(J+1) - 3.633K^2$$

Upon multiplication by c, $F(J,K) = c\tilde{F}(J,K)$ acquires units of frequency:

$$F(J,K)/\text{GHz} = 299.1J(J+1) - 108.9K^2$$

For $J = 1$, the energy needed for the molecule to rotate mainly about its figure axis ($K = \pm J$) is equivalent to 16.32 cm^{-1} (489.3 GHz), but end-over-end rotation ($K = 0$) corresponds to 19.95 cm^{-1} (598.1 GHz). ●

Self-test 10.2 A $CH_3{}^{35}Cl$ molecule has a C—Cl bond length of 178 pm, a C—H bond length of 111 pm, and an HCH angle of 110.5°. Calculate its rotational energy terms.

[$\tilde{F}(J,K)/\text{cm}^{-1} = 0.472J(J+1) + 4.56K^2$; also $F(J,K)/\text{GHz} = 14.1J(J+1) + 137K^2$]

The energy of a symmetric rotor depends on J and K, and each level except those with $K = 0$ is doubly degenerate: the states with K and $-K$ have the same energy because the direction of rotation about the internal figure axis does not affect the energy. The angular momentum continues to have any of $2J + 1$ components $M_J\hbar$ on an external axis, representing the different orientations that the rotation of the molecule may have in space, but as its energy is independent of that orientation all $2J + 1$ values of M_J correspond to the same energy. It follows that a symmetric rotor level is $2(2J + 1)$-fold degenerate for $K \neq 0$ and $(2J + 1)$-fold degenerate for $K = 0$.

(c) Linear rotors

For a linear rotor (such as CO_2, HCl, and C_2H_2), in which the nuclei are regarded as mass points, the rotation occurs only about an axis perpendicular to the line of atoms and there is zero angular momentum around the line. Therefore, the component of angular momentum around the figure axis of a linear rotor is identically zero, and $K \equiv 0$ in eqn 10.10. The rotational terms of a linear molecule are therefore

$$\tilde{F}(J) = \tilde{B}J(J+1) \qquad J = 0, 1, 2, \ldots \tag{10.12}$$

This expression is the same as eqn 10.7 but we have arrived at it in a significantly different way: here $K \equiv 0$ but for a spherical rotor $\tilde{A} = \tilde{B}$. Although a linear rotor has K fixed at 0, the angular momentum may still have $2J + 1$ components on the laboratory axis, so the degeneracy of a level with quantum number J is $2J + 1$.

(d) Centrifugal distortion

We have treated molecules as rigid rotors. However, the atoms of rotating molecules are subject to centrifugal forces that tend to distort the molecular geometry and change the moments of inertia (Fig. 10.8). The effect of centrifugal distortion on a diatomic molecule is to stretch the bond and hence to increase the moment of inertia. As a result, centrifugal distortion reduces the rotational constant and consequently the energy levels are slightly closer than the rigid-rotor expressions predict. The effect is usually taken into account by subtracting a term from the energy and writing

$$\tilde{F}(J) = \tilde{B}J(J+1) - \tilde{D}_J J^2(J+1)^2 \tag{10.13}$$

The parameter $\tilde{D}_J$ (a wavenumber) is the **centrifugal distortion constant**. It is large when the bond is easily stretched. In most

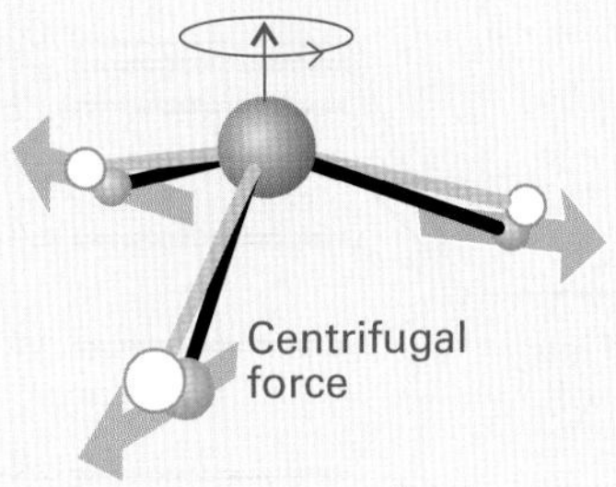

Fig. 10.8 The effect of rotation on a molecule. The centrifugal force arising from rotation distorts the molecule, opening out bond angles and stretching bonds slightly. The effect is to increase the moment of inertia of the molecule and hence to decrease its rotational constant.

cases, $\tilde{D} < 10^{-4}\tilde{B}$ (for HCl, for instance, $\tilde{D} = 0.0004$ cm^{-1}) and most molecules can be treated as rigid rotors.

The centrifugal distortion constant of a diatomic molecule is related to the vibrational wavenumber of the bond, $\tilde{\nu}$ (which, as we shall see later, is a measure of its stiffness), through the approximate relation (see Problem 10.26)

$$\tilde{D}_J = \frac{4\tilde{B}^3}{\tilde{\nu}^2} \tag{10.14}$$

Hence the observation of the convergence of the rotational levels as J increases can be interpreted in terms of the rigidity of the bond. A similar parameter, of the form $-D_K K^4$, is used to represent the effect of centrifugal distortion arising from the axial rotation of symmetric rotors.

10.3 Rotational transitions

Typical values of $\tilde{B}$ for small molecules are in the region of 0.1 to 10 cm^{-1} (for example, 0.356 cm^{-1} for NF_3 and 10.59 cm^{-1} for HCl), so rotational transitions lie in the microwave region of the spectrum and the observation of pure rotational transitions in the gas phase is a branch of **microwave spectroscopy**.

(a) Techniques

Microwave radiation is generated in a *klystron* or a *Gunn diode*. A klystron makes use of electrons that are accelerated back and forth across a gap in a resonant cavity. A Gunn diode is a device that makes use of electron transitions between the partly filled bands of a semiconductor in the presence of an electric field. The microwaves are detected by a *crystal diode* consisting of a tungsten tip in contact with a semiconductor. Modulation of the transmitted intensity—which results in signals that are easier to detect and process—can be achieved by varying the energy levels with an oscillating electric field. In this **Stark modulation**, an electric field of about 10^5 V m^{-1} and a frequency of between 10 and 100 kHz is applied to the sample.

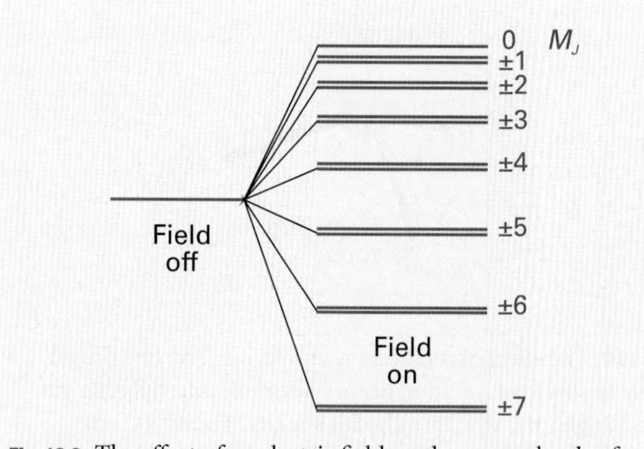

Fig. 10.9 The effect of an electric field on the energy levels of a polar linear rotor. All levels are doubly degenerate except that with $M_J = 0$.

The application of an electric field has a further use. The **Stark effect** is the partial removal of the degeneracy associated with the quantum number M_J (the orientation of the rotation in space) when an electric field is applied to a polar molecule (Fig. 10.9). For a linear rotor in an electric field $\mathcal{E}$, the energy of the state $|J,M_J\rangle$ is given by

$$E(J,M_J) = hc\tilde{B}J(J+1) + a(J,M_J)\mu^2\mathcal{E}^2 \qquad (10.15a)$$

where (after a lengthy calculation involving perturbation theory, which we do not reproduce here)

$$a(J,M_J) = \frac{J(J+1) - 3M_J^2}{2hc\tilde{B}J(J+1)(2J-1)(2J+3)} \qquad (10.15b)$$

with $M_J = 0, \pm1, \ldots, \pm J$. One application of the Stark effect is therefore to note the number of lines into which a transition is split and then to infer from that number the value of J. Another application is to determine the permanent electric dipole moment, μ. The technique is limited to molecules that are sufficiently volatile to be studied by rotational spectroscopy. However, as spectra can be recorded for samples at pressures of only about 1 Pa and special techniques (such as using an intense laser beam or an electrical discharge) can be used to vaporize even some quite non-volatile substances, a wide variety of samples may be studied. Sodium chloride, for example, can be studied as diatomic NaCl molecules at high temperatures.

On the assumption that all instrumental factors have been optimized, the highest resolution is obtained when the sample is gaseous and at such low pressure that collisions between the molecules are infrequent. We saw in Section 4.3 that collisions between atoms reduce the lifetimes of excited states and therefore broaden their emission and absorption lines: the same is true of collisions between molecules. That low pressures must be used has the implication that path lengths through the sample must be long in order for the absorption intensities to be significant.

A microwave spectrum displays the *net* rate of absorption between rotational states, the difference between the absorption that the incident radiation stimulates between rotational states, such as $J+1 \leftarrow J$, and the emission that the same radiation stimulates from excited states that are already occupied, in this case $J+1 \rightarrow J$. For a given intensity and frequency of incident radiation, a single stimulated absorption $J+1 \leftarrow J$ and a single stimulated emission $J+1 \rightarrow J$ have the same rate but as states of low energy are more highly populated than those of higher energy (at thermal equilibrium), the rate of upward transitions exceeds that of downward transitions and there is a net absorption of energy. Excited states may also make spontaneous transitions to lower states even in the absence of stimulating radiation, but the rate of these transitions is negligible for rotations. The quantitative relation between the rates of stimulated absorption and emission and spontaneous emission and their frequency dependence was established by Einstein; his conclusions are summarized in *Further information 10.1*.

(b) Selection rules

A **gross selection rule** specifies the general features a molecule must have if it is to have a spectrum of a given kind. A detailed study of the transition moment leads to the **specific selection rules** that express the allowed transitions in terms of the changes in quantum numbers. We have already encountered examples of specific selection rules when discussing atomic spectra (Section 4.3), such as the rule $\Delta l = \pm1$ for the orbital angular momentum quantum number.

The gross selection rule for rotational transitions is that a molecule must have a permanent electric dipole moment. That is, *for a molecule to give a pure rotational spectrum, it must be polar.* The classical basis of this rule is that a polar molecule appears to possess a fluctuating dipole when rotating, but a nonpolar molecule does not (Fig. 10.10). That fluctuating dipole drives the electromagnetic field into oscillation and energy is emitted into the surroundings; conversely, for absorption, the oscillating electromagnetic field drives the dipole moment into rotation. Homonuclear diatomic molecules and symmetrical linear molecules such as CO_2 are nonpolar and hence rotationally inactive.

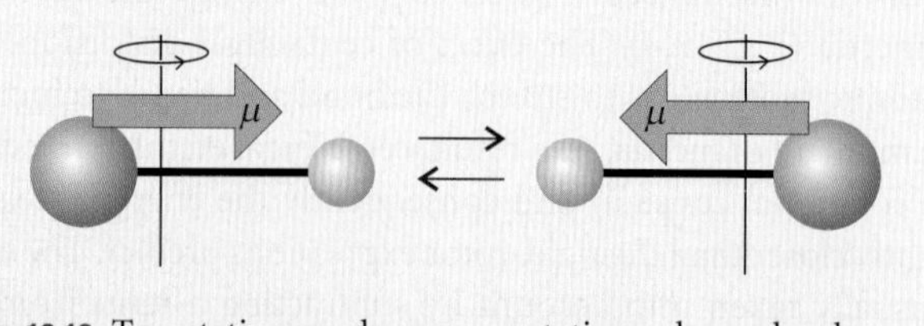

Fig. 10.10 To a stationary observer, a rotating polar molecule looks like an oscillating dipole that can stir the electromagnetic field into oscillation (and vice versa for absorption). This picture is the classical origin of the gross selection rule for rotational transitions.

Spherical rotors cannot have electric dipole moments unless they become distorted by rotation, so they are also inactive except in special cases. An example of a spherical rotor that does become sufficiently distorted for it to acquire a dipole moment is SiH_4, which has a dipole moment of about 8.3 mD by virtue of its rotation when $J \approx 10$ (for comparison, HCl has a permanent dipole moment of 1.1 D; molecular dipole moments and their units are discussed in Section 8.2). The pure rotational spectrum of SiH_4 has been detected by using long path lengths (10 m) through high-pressure (4 atm) samples.

● A BRIEF ILLUSTRATION

Of the molecules N_2, CO_2, OCS, H_2O, $CH_2{=}CH_2$, C_6H_6, only OCS and H_2O are polar, so only these two molecules have microwave spectra. ●

Self-test 10.3 Which of the molecules H_2, NO, N_2O, CH_4 can have a pure rotational spectrum? [NO, N_2O]

The specific rotational selection rules are found by evaluating the transition dipole moment between rotational states. We show in *Further information 10.2* that, for a linear molecule, the transition moment vanishes unless the following conditions are fulfilled:

$$\Delta J = \pm 1 \qquad \Delta M_J = 0, \pm 1$$

The transition $\Delta J = +1$ corresponds to absorption and the transition $\Delta J = -1$ corresponds to emission. The allowed change in J in each case arises from the conservation of angular momentum when a photon, a spin-1 particle, is emitted or absorbed (Fig. 10.11). For symmetric rotors, an additional selection rule states that $\Delta K = 0$. To understand this rule, consider the symmetric rotor NH_3, where the electric dipole moment lies parallel to the figure axis. Such a molecule cannot be accelerated into different states of rotation around the figure axis by the absorption of radiation, so $\Delta K = 0$.

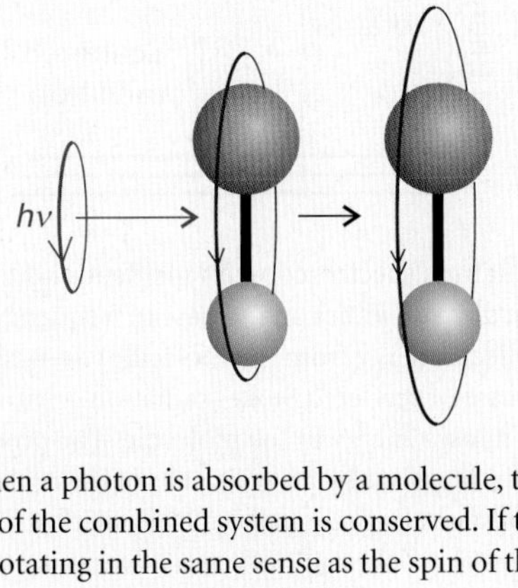

Fig. 10.11 When a photon is absorbed by a molecule, the angular momentum of the combined system is conserved. If the molecule is rotating in the same sense as the spin of the incoming photon, then J increases by 1.

When the transition moment is evaluated for all possible relative orientations of the molecule to the line of flight of the photon, it is found that the net $J+1 \leftrightarrow J$ transition intensity is proportional to

$$|\mu_{J+1,J}|^2 = \left(\frac{J+1}{2J+1}\right)\mu_0^2 \rightarrow \tfrac{1}{2}\mu_0^2 \qquad \text{for } J \gg 1 \qquad (10.16)$$

where μ_0 is the permanent electric dipole moment of the molecule. The intensity is proportional to the square of μ_0, so strongly polar molecules give rise to much more intense rotational lines than less polar molecules: they interact with the incident radiation more strongly.

(c) The appearance of rotational spectra

When these selection rules are applied to the expressions for the energy levels of a rigid symmetric or linear rotor, it follows that the wavenumbers of the allowed $J+1 \leftarrow J$ absorptions are

$$\begin{aligned}\tilde{\nu}(J+1 \leftarrow J) &= \tilde{F}(J+1) - \tilde{F}(J) \\ &= \tilde{B}(J+1)(J+2) - \tilde{B}J(J+1) \\ &= 2\tilde{B}(J+1)\end{aligned} \qquad (10.17a)$$

with $J = 0, 1, 2, \ldots$ When centrifugal distortion is taken into account, the corresponding expression is

$$\tilde{\nu}(J+1 \leftarrow J) = 2\tilde{B}(J+1) - 4\tilde{D}_J(J+1)^3 \qquad (10.17b)$$

However, because the second term is typically very small compared with the first, the appearance of the spectrum closely resembles that predicted from eqn 10.17a.

● A BRIEF ILLUSTRATION

The $^{14}NH_3$ molecule is a polar symmetric rotor, so the selection rules $\Delta J = \pm 1$ and $\Delta K = 0$ apply. For absorption, $\Delta J = +1$ and we can use eqn 10.17a. For $^{14}NH_3$ we have already calculated $\tilde{B} = 9.977\ cm^{-1}$. We can therefore draw up the following table for the $J+1 \leftarrow J$ transitions.

J	0	1	2	3	...
$\tilde{\nu}/cm^{-1}$	19.95	39.91	59.86	79.82	...
ν/GHz	598.1	1197	1795	2393	...

The line spacing is $19.95\ cm^{-1}$ (598.1 GHz). ●

Self-test 10.4 Repeat the problem for $C^{35}ClH_3$ (see Self-test 10.2 for details).
[Lines of separation $0.944\ cm^{-1}$ (28.3 GHz)]

A note on good practice For the discussion of spectroscopic transitions, the upper state is written first. So $X \rightarrow Y$ is an emission and $X \leftarrow Y$ is an absorption, where X and Y specify the states in some way (such as by giving the value of the rotational quantum number J, as we have done above).

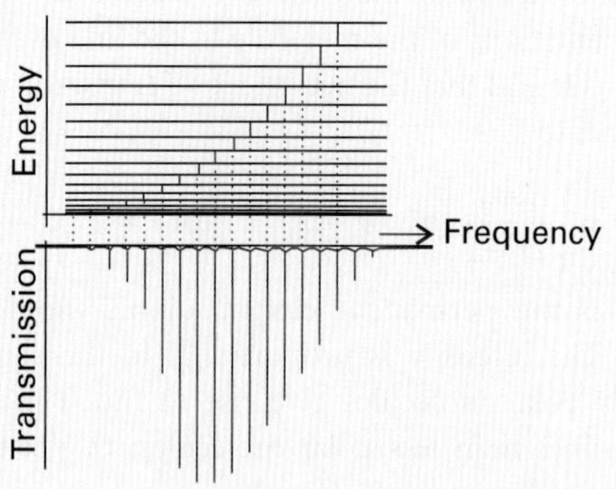

Fig. 10.12 The rotational energy levels of a linear rotor, the transitions allowed by the selection rule $\Delta J = \pm 1$, and a typical pure rotational absorption spectrum (displayed here in terms of the radiation transmitted through the sample). The intensities reflect the populations of the initial level in each case and the strengths of the transition dipole moments.

The form of the spectrum predicted by eqn 10.17a is shown in Fig. 10.12. The most significant feature is that it consists of a series of lines with wavenumbers $2\tilde{B}$, $4\tilde{B}$, $6\tilde{B}$, . . . and of separation $2\tilde{B}$. The measurement of the line spacing gives $\tilde{B}$, and hence the moment of inertia perpendicular to the principal axis of the molecule. Because the masses of the atoms are known, it is a simple matter to deduce the bond length of a diatomic molecule. However, in the case of a polyatomic molecule such as OCS or NH_3, the analysis gives only a single quantity, I, and it is not possible to infer both bond lengths (in OCS) or the bond length and bond angle (in NH_3). This difficulty can be overcome by using isotopomers (isotopically substituted molecules), such as ABC and A′BC; then, by assuming that $R(\text{A—B}) = R(\text{A}'\text{—B})$, both A—B and B—C bond lengths can be extracted from the two measured moments of inertia. A famous example of this procedure is the study of OCS; the actual calculation is worked through in Problem 10.7. The assumption that bond lengths are unchanged by isotopic substitution is only an approximation, but it is a good approximation in most cases.

The intensities of spectral lines increase with increasing J and pass through a maximum before tailing off as J becomes large. The most important reason for the maximum in intensity is the existence of a maximum in the population of rotational levels. The Boltzmann distribution (*Fundamentals F.5*) implies that the population of each state decays exponentially with increasing J, but the degeneracy of the levels increases. Specifically, the population of a rotational energy level J is given by the Boltzmann expression

$$N_J \propto N g_J e^{-E_J/kT} \qquad (10.18)$$

where N is the total number of molecules and g_J is the degeneracy of the level J. The value of J corresponding to a maximum of this expression is found by treating J as a continuous variable, differentiating with respect to J, and then setting the result equal to zero. The result is (see Problem 10.29)

$$J_{max} \approx \left(\frac{kT}{2hc\tilde{B}}\right)^{1/2} - \tfrac{1}{2} \qquad (10.19)$$

For a typical molecule (for example, OCS, with $\tilde{B} = 0.2\ \text{cm}^{-1}$) at room temperature, $kT \approx 1000hc\tilde{B}$, so $J_{max} \approx 30$. However, it must be recalled that the intensity of each transition also depends on the value of J (eqn 10.16) and on the population difference between the two states involved in the transition. Hence the value of J corresponding to the most intense line is not quite the same as the value of J for the most highly populated level.

10.4 Rotational Raman spectra

In **Raman spectroscopy** molecular energy levels are explored by examining the frequencies present in inelastically scattered radiation. 'Inelastic' in this context means with the acquisition or loss of energy. About 1 in 10^7 of the incident photons collide with the molecules, give up some of their energy, and emerge with a lower energy. These scattered photons constitute the lower-frequency **Stokes radiation** from the sample (Fig. 10.13). Other incident photons may collect energy from the molecules (if they are already excited), and emerge as higher-frequency **anti-Stokes radiation.** The component of radiation scattered without change of frequency is called **Rayleigh radiation.**

(a) Techniques

In a typical Raman spectroscopy experiment, a monochromatic incident laser beam is passed through the sample and the radiation

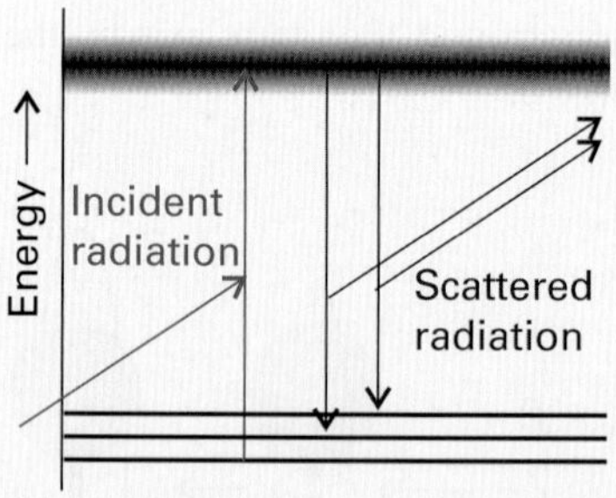

Fig. 10.13 In Raman spectroscopy, an incident photon is scattered from a molecule with either an increase in frequency (if the radiation collects energy from the molecule) or—as shown here for the case of scattered Stokes radiation—with a lower frequency if it loses energy to the molecule. The process can be regarded as taking place by an excitation of the molecule to a wide range of states (represented by the shaded band) and the subsequent return of the molecule to a lower state; the net energy change is then carried away by the photon.

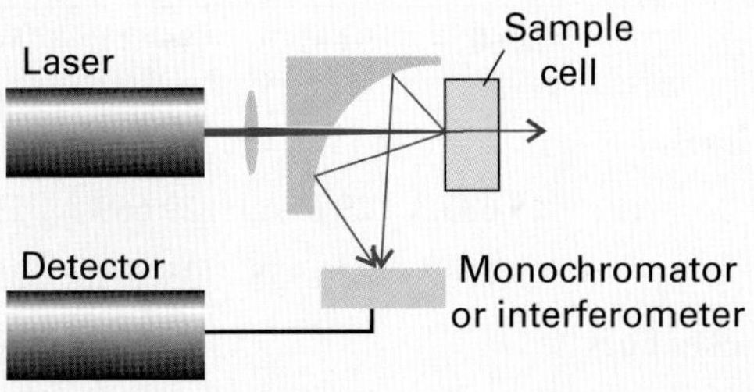

Fig. 10.14 A common arrangement adopted in Raman spectroscopy. A laser beam first passes through a lens and then through a small hole in a mirror with a curved reflecting surface. The focused beam strikes the sample and scattered light is both deflected and focused by the mirror. The spectrum is analysed by a monochromator or an interferometer.

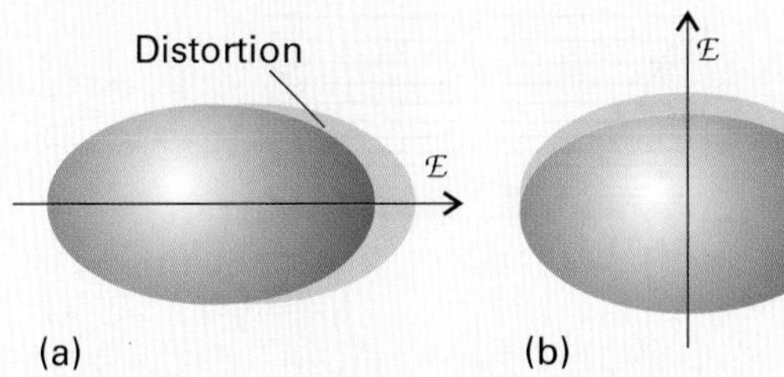

Fig. 10.15 An electric field applied to a molecule results in its distortion, and the distorted molecule acquires a contribution to its dipole moment (even if it is nonpolar initially). The polarizability may be different when the field is applied (a) parallel or (b) perpendicular to the molecular axis (or, in general, in different directions relative to the molecule); if that is so, then the molecule has an anisotropic polarizability.

scattered from the sample is monitored (Fig. 10.14). Lasers are used as the source of the incident radiation because an intense beam increases the intensity of scattered radiation. The monochromaticity of laser radiation is also a great advantage, for it makes possible the observation of frequencies of scattered light that differs only slightly from that of the incident radiation. Such high resolution is particularly useful for observing the rotational structure of Raman lines. The monochromaticity of laser radiation also allows observations to be made very close to absorption frequencies.

The use of a non-divergent incident beam also implies that the detector can be designed to collect only the radiation that has been scattered by the sample and can be screened much more effectively against stray light, which can obscure the Raman signal. The availability of non-divergent beams also makes possible a qualitatively different kind of spectroscopy in which Raman transitions are observed very close to the direction of propagation of the incident beam. This configuration is employed in the technique called **stimulated Raman spectroscopy**. In this form of spectroscopy, the Stokes and anti-Stokes radiation in the forward direction are powerful enough to undergo more scattering and hence give up or acquire more quanta of energy from the molecules in the sample. This multiple scattering results in lines of frequency $\nu_i \pm 2\nu_M$, $\nu_i \pm 3\nu_M$, . . . , where ν_i is the frequency of the incident radiation and ν_M the frequency of a molecular excitation.

(b) Selection rules

The gross selection rule for rotational Raman transitions is *that the molecule must be anisotropically polarizable*. We begin by explaining what this means; a formal derivation of the rule is given in *Further information 10.2*.

The distortion of a molecule in an electric field is determined by its polarizability, α (Section 8.4). More precisely, if the strength of the field is $\mathcal{E}$, then the molecule acquires an induced dipole moment of magnitude $\mu = \alpha\mathcal{E}$ in addition to any permanent dipole moment it may already have. An atom is *isotropically* polarizable. That is, the same distortion is induced whatever the direction of the applied field. The polarizability of a spherical rotor is also isotropic. However, nonspherical rotors have polarizabilities that do depend on the direction of the field relative to the molecule, so these molecules are *anisotropically* polarizable (Fig. 10.15). The electron distribution in H_2, for example, is more distorted when the field is applied parallel to the bond than when it is applied perpendicular to it, and we write $\alpha_{\parallel} > \alpha_{\perp}$.

All linear molecules and diatomic molecules (whether homonuclear or heteronuclear) have anisotropic polarizabilities, and so are rotationally Raman active. This activity is one reason for the importance of rotational Raman spectroscopy, for the technique can be used to study many of the molecules that are inaccessible to microwave spectroscopy. Spherical rotors such as CH_4 and SF_6, however, are rotationally Raman inactive as well as microwave inactive. This inactivity does not mean that such molecules are never found in rotationally excited states. Molecular collisions do not have to obey such restrictive selection rules, and hence collisions between molecules can result in the population of any rotational state.

We show in *Further information 10.2* that the specific rotational Raman selection rules are

Linear rotors: $\Delta J = 0, \pm 2$

Symmetric rotors: $\Delta J = 0, \pm 1, \pm 2$; $\quad \Delta K = 0$

The $\Delta J = 0$ transitions do not lead to a shift of the scattered photon's frequency in pure rotational Raman spectroscopy, and contribute to the unshifted Rayleigh radiation.

To predict the form of the Raman spectrum of a linear rotor we apply the selection rule $\Delta J = \pm 2$ to the rotational energy levels (Fig. 10.16). When the molecule makes a transition with $\Delta J = +2$, the scattered radiation leaves the molecule in a higher rotational state, so the wavenumber of the incident radiation, initially $\tilde{\nu}_i$, is

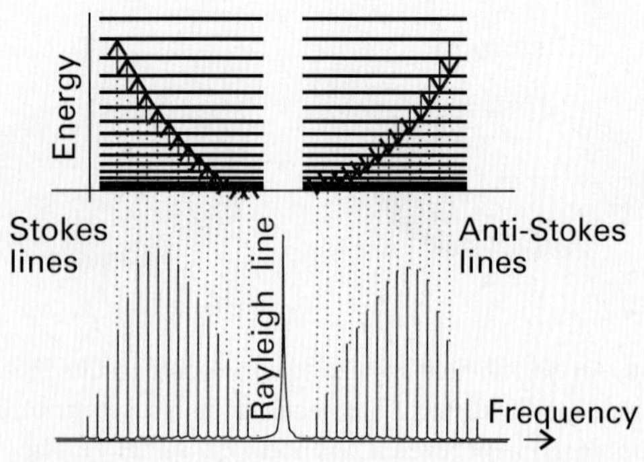

Fig. 10.16 The rotational energy levels of a linear rotor and the transitions allowed by the $\Delta J = \pm 2$ Raman selection rules. The form of a typical rotational Raman spectrum is also shown. The Rayleigh line is much stronger than depicted in the figure; it is shown as a weaker line to improve visualization of the Raman lines.

decreased. These transitions account for the Stokes lines in the spectrum:

$$\tilde{\nu}(J+2 \leftarrow J) = \tilde{\nu}_i - \{\tilde{F}(J+2) - \tilde{F}(J)\} = \tilde{\nu}_i - 2\tilde{B}(2J+3) \tag{10.20a}$$

The Stokes lines appear at lower frequency than the incident radiation and at displacements $6\tilde{B}$, $10\tilde{B}$, $14\tilde{B}$, ... from $\tilde{\nu}_i$ for $J = 0, 1, 2, \ldots$. When the molecule makes a transition with $\Delta J = -2$, the scattered photon emerges with increased energy. These transitions account for the anti-Stokes lines of the spectrum:

$$\tilde{\nu}(J-2 \leftarrow J) = \tilde{\nu}_i + \{\tilde{F}(J) - \tilde{F}(J-2)\} = \tilde{\nu}_i + 2\tilde{B}(2J-1) \tag{10.20b}$$

The anti-Stokes lines occur at displacements of $6\tilde{B}$, $10\tilde{B}$, $14\tilde{B}$, ... (for $J = 2, 3, 4, \ldots$; $J = 2$ is the lowest state that can contribute under the selection rule $\Delta J = -2$) at higher frequency than the incident radiation. The separation of adjacent lines in both the Stokes and the anti-Stokes regions is $4\tilde{B}$, so from its measurement $I_\perp$ can be determined and then used to find the bond lengths exactly as in the case of microwave spectroscopy.

Example 10.2 *Predicting the form of a Raman spectrum*

Predict the form of the rotational Raman spectrum of $^{14}N_2$, for which $\tilde{B} = 1.99$ cm^{-1}, when it is exposed to monochromatic 336.732 nm laser radiation.

Method The molecule is rotationally Raman active because end-over-end rotation modulates its polarizability as viewed by a stationary observer. The Stokes and anti-Stokes lines are given by eqn 10.20.

Answer Because λ_i = 336.732 nm corresponds to $\tilde{\nu}_i$ = 29 697.2 cm^{-1}, eqns 10.20a and 10.20b give the following line positions:

J	2←0	3←1	4←2	5←3
Stokes lines				
$\tilde{\nu}$/cm^{-1}	29 685.3	29 677.3	29 669.3	29 661.4
λ/nm	336.867	336.958	337.048	337.139
Anti-Stokes lines			2→0	3→1
$\tilde{\nu}$/cm^{-1}			29 709.1	29 717.1
λ/nm			336.597	336.507

There will be a strong central line at 336.732 nm accompanied on either side by lines of increasing and then decreasing intensity (as a result of transition moment and population effects). The spread of the entire spectrum is very small, so the incident light must be highly monochromatic.

Self-test 10.5 Repeat the calculation for the rotational Raman spectrum of NH_3 ($\tilde{B} = 9.977$ cm^{-1}).

[Stokes lines at 29 637.3, 29 597.4, 29 557.5, 29 517.6 cm^{-1}, anti-Stokes lines at 29 757.1, 29 797.0 cm^{-1}]

10.5 Nuclear statistics and rotational states

If eqn 10.20 is used in conjunction with the rotational Raman spectrum of CO_2, the rotational constant that is calculated is inconsistent with other measurements of C—O bond lengths. The results are consistent if it is supposed that the molecule can exist only in states with even values of J, so the Stokes lines are 2 ← 0, 4 ← 2, ... and not 2 ← 0, 3 ← 1, ...

The explanation of the missing lines is the Pauli principle and the fact that O nuclei are spin-0 bosons (bosons, recall from Section 4.4, are particles with integer spin quantum numbers): just as the Pauli principle excludes certain electronic states, so too does it exclude certain molecular rotational states. The form of the Pauli principle given in Section 4.4 states that, when two identical bosons are exchanged, the overall wavefunction must remain unchanged in every respect, including sign. In particular, when a CO_2 molecule rotates through 180°, two identical O nuclei are interchanged, so the overall wavefunction of the molecule must remain unchanged. However, inspection of the form of the rotational wavefunctions (which have the same *angular* form as the s, p, etc. orbitals of atoms, because in each case they correspond to angular momentum states) shows that they change sign by $(-1)^J$ under such a rotation (Fig. 10.17). Therefore, only even values of J are permissible for CO_2, and hence the Raman spectrum shows only alternate lines.

The selective existence of rotational states that stems from the Pauli principle is a manifestation of **nuclear statistics**. Nuclear statistics must be taken into account whenever a rotation

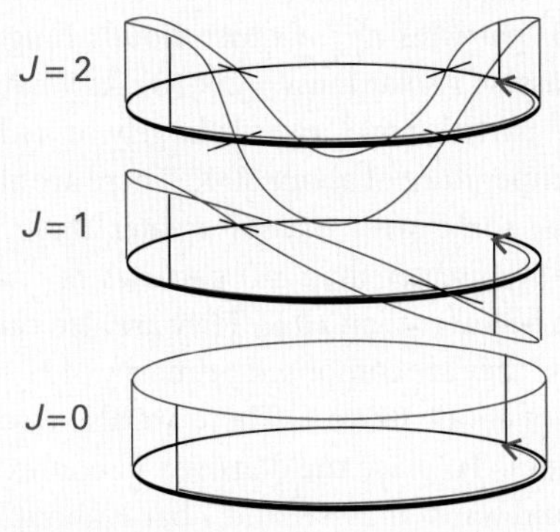

Fig. 10.17 The symmetries of rotational wavefunctions (shown here, for simplicity as a two-dimensional rotor) under a rotation through 180°. Wavefunctions with J even do not change sign; those with J odd do change sign.

interchanges equivalent nuclei. However, the consequences are not always as simple as for CO_2 because there are complicating features when the nuclei have nonzero spin: there may be several different relative nuclear spin orientations consistent with even values of J and a different number of spin orientations consistent with odd values of J. For molecular hydrogen and fluorine, for instance, with their two identical spin-$\frac{1}{2}$ nuclei, we show in the following *Justification* that there are three times as many ways of achieving a state with odd J than with even J, and there is a corresponding 3:1 alternation in intensity in their rotational Raman spectra (Fig. 10.18). In general, for a homonuclear diatomic molecule with nuclei of spin I, the numbers of ways of achieving states of odd and even J are in the ratio

$$\frac{\text{Number of ways of achieving odd } J}{\text{Number of ways of achieving even } J} = \begin{cases} (I+1)/I & \text{for half-integral spin nuclei} \\ I/(I+1) & \text{for integral spin nuclei} \end{cases} \qquad (10.21)$$

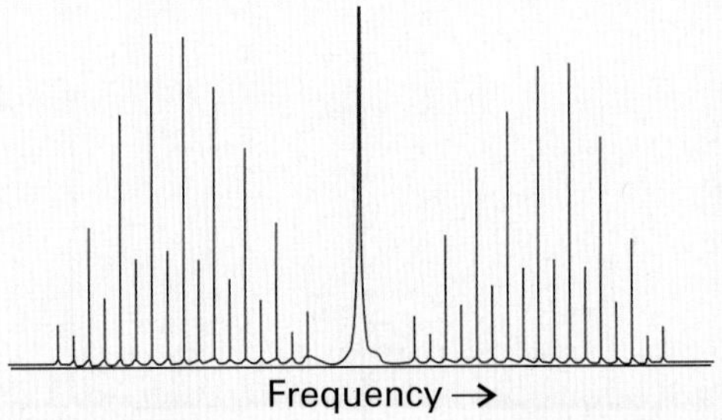

Fig. 10.18 The rotational Raman spectrum of a diatomic molecule with two identical spin-$\frac{1}{2}$ nuclei shows an alternation in intensity as a result of nuclear statistics. The Rayleigh line is much stronger than depicted in the figure; it is shown as a weaker line to improve visualization of the Raman lines.

For hydrogen, $I = \frac{1}{2}$, and the ratio is 3:1. For N_2, with $I = 1$, the ratio is 1:2.

Justification 10.1 *The effect of nuclear statistics on rotational spectra*

Hydrogen nuclei are fermions (particles with half-integer spin quantum number; in their case $I = \frac{1}{2}$), so the Pauli principle requires the overall wavefunction to change sign under particle interchange. However, the rotation of an H_2 molecule through 180° has a more complicated effect than merely relabelling the nuclei, because it interchanges their spin states too if the nuclear spins are paired (↑ ↓) but not if they are parallel (↑ ↑).

For the overall wavefunction of the molecule to change sign when the spins are parallel, the rotational wavefunction must change sign. Hence, only odd values of J are allowed. In contrast, if the nuclear spins are paired, their wavefunction is $\alpha(A)\beta(B) - \alpha(B)\beta(A)$, which changes sign when α and β are exchanged in order to bring about a simple A↔B interchange overall (Fig. 10.19). Therefore, for the overall wavefunction to change sign in this case requires the rotational wavefunction *not* to change sign. Hence, only even values of J are allowed if the nuclear spins are paired.

As there are three nuclear spin states with parallel spins (just like the triplet state of two parallel electrons, as in Fig. 4.25), but only one state with paired spins (the analogue of the singlet state of two electrons, see Fig. 4.19), it follows that the populations of the odd J and even J states should be in the ratio of 3:1, and hence the intensities of transitions originating in these levels will be in the same ratio.

Different relative nuclear spin orientations change into one another only very slowly, so an H_2 molecule with parallel nuclear spins remains distinct from one with paired nuclear spins for long periods. The two forms of hydrogen can be separated by

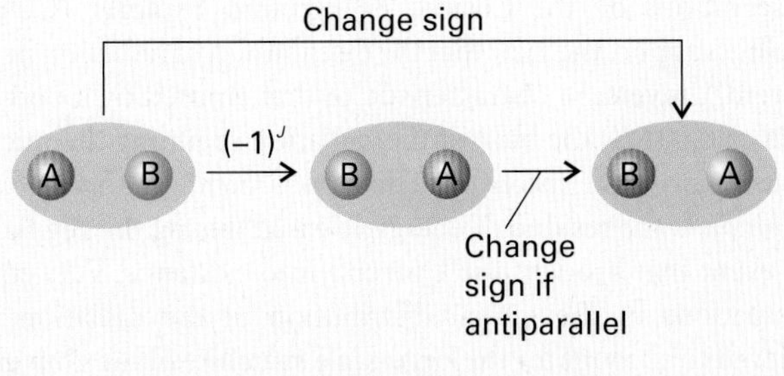

Fig. 10.19 The interchange of two identical fermion nuclei results in the change in sign of the overall wavefunction. The relabelling can be thought of as occurring in two steps: the first is a rotation of the molecule; the second is the interchange of unlike spins (represented by the different colours of the nuclei). The wavefunction changes sign in the second step if the nuclei have antiparallel spins.

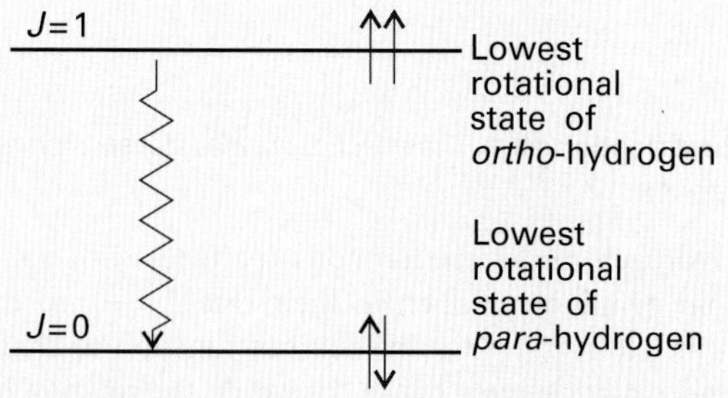

Fig. 10.20 When hydrogen is cooled, the molecules with parallel nuclear spins accumulate in their lowest available rotational state, the one with $J = 1$. They can enter the lowest rotational state ($J = 0$) only if the spins change their relative orientation and become antiparallel. This is a slow process under normal circumstances, so energy is slowly released.

physical techniques, and stored. The form with parallel nuclear spins is called ***ortho*-hydrogen** and the form with paired nuclear spins is called ***para*-hydrogen**. Because *ortho*-hydrogen cannot exist in a state with $J = 0$, it continues to rotate at very low temperatures and has an effective rotational zero-point energy (Fig. 10.20). This energy is of some concern to manufacturers of liquid hydrogen, for the slow conversion of *ortho*-hydrogen into *para*-hydrogen (which can exist with $J = 0$) as nuclear spins slowly realign releases rotational energy, which vaporizes the liquid. Techniques are used to accelerate the conversion of *ortho*-hydrogen to *para*-hydrogen to avoid this problem. One such technique is to pass hydrogen over a metal surface: the molecules adsorb on the surface as atoms, which then recombine in the lower energy *para*-hydrogen form.

IMPACT ON ASTROPHYSICS

I10.1 Rotational spectroscopy of interstellar molecules

Observations by the Cosmic Background Explorer (COBE) satellite support the view that the distribution of radiation in the current Universe is characteristic of that emitted by a body at 2.726 ± 0.001 K, the bulk of the radiation spanning the microwave region of the spectrum. This *cosmic microwave background radiation* is the residue of energy released during the Big Bang, the event that brought the Universe into existence. Very small fluctuations in the spatial distribution of the radiation are believed to account for the large-scale structure of the Universe.

The interstellar space in our galaxy is a little warmer than the cosmic background and consists largely of dust grains and gas clouds. The dust grains are carbon-based compounds and silicates of aluminium, magnesium, and iron, in which are embedded trace amounts of methane, water, and ammonia. Interstellar clouds are significant because it is from them that new stars, and consequently new planets, are formed. The hottest clouds are plasmas with temperatures of up to 10^6 K and densities of only about 3×10^3 particles m^{-3}. Colder clouds range from 0.1 to 1000 solar masses (1 solar mass = 2×10^{30} kg), have a density of about 5×10^5 particles m^{-3}, consist largely of hydrogen atoms, and have a temperature of about 80 K. There are also colder and denser clouds, some with masses greater than 500 000 solar masses, densities greater than 10^9 particles m^{-3}, and temperatures that can be lower than 10 K. They are also called *molecular clouds*, because they are composed primarily of H_2 and CO gas in a proportion of about 10^5 to 1. There are also trace amounts of larger molecules. To place the densities in context, the particle density of liquid water at 298 K and 1 bar is about 3×10^{28} m^{-3}.

It follows from the Boltzmann distribution and the low temperature of a molecular cloud that the vast majority of a cloud's molecules are in their vibrational and electronic ground states. However, rotational excited states are populated at 10–100 K and decay by spontaneous emission. As a result, the spectrum of the cloud in the radiofrequency and microwave regions consists of sharp lines corresponding to rotational transitions (Fig. 10.21). The emitted light is collected by Earth-bound or space-borne radiotelescopes, telescopes with antennas and detectors optimized for the collection and analysis of radiation in the microwave–radiowave range of the spectrum. Earth-bound radiotelescopes are often located at the tops of high mountains, as atmospheric water vapour can reabsorb microwave radiation from space and hence interfere with the measurement.

Over 100 interstellar molecules have been identified by their rotational spectra, often by comparing radiotelescope data with spectra obtained in the laboratory or calculated by computational methods. The experiments have revealed the presence of trace amounts (with abundances of less than 10^{-8} relative to hydrogen) of neutral molecules, ions, and radicals. Examples of neutral molecules include hydrides, oxides (including water), sulfides, halogenated compounds, nitriles, hydrocarbons, aldehydes, alcohols, ethers, ketones, and amides. The largest molecule detected so far by rotational spectroscopy is the nitrile $HC_{11}N$.

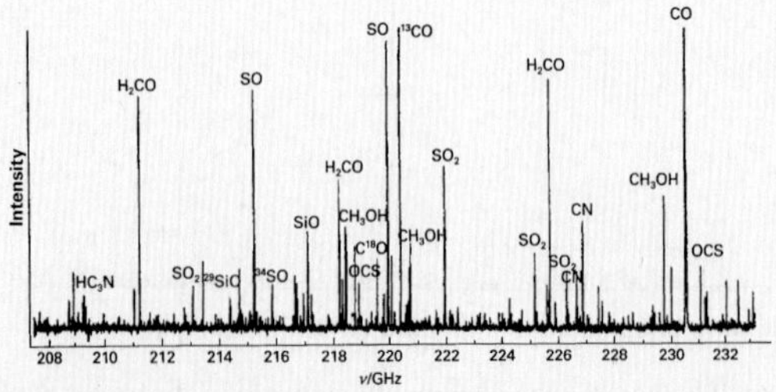

Fig. 10.21 Rotational spectrum of the Orion nebula, showing spectral fingerprints of diatomic and polyatomic molecules present in the interstellar cloud. Adapted from G.A. Blake *et al.* *Astrophys. J.* 315, 621 (1987).

The vibrations of diatomic molecules

The vibrations of molecules are responsible for the absorption in the infrared regions of the electromagnetic spectrum, so **infrared spectroscopy** is the detection of vibrational transitions. After a discussion of the experimental techniques used in this region, we adopt the same strategy as for microwave spectra: we find expressions for the energy levels, establish the selection rules, and then discuss the appearance of the spectra. We shall also see how the simultaneous excitation of rotation modifies the appearance of a vibrational spectrum.

10.6 Techniques

For far infrared radiation with $35\ \text{cm}^{-1} < \tilde{\nu} < 200\ \text{cm}^{-1}$, a typical source is a mercury arc inside a quartz envelope, most of the radiation being generated by the hot quartz. Either a *Nernst filament* or a *globar* is used as a source of mid-infrared radiation with $200\ \text{cm}^{-1} < \tilde{\nu} < 4000\ \text{cm}^{-1}$. The Nernst filament consists of a ceramic filament of lanthanoid oxides that is heated to temperatures ranging from 1200 to 2000 K. The globar consists of a rod of silicon carbide, which is heated electrically to about 1500 K.

The dispersing element, which selects particular wavelengths of radiation from the source, is typically a diffraction grating consisting of a glass or ceramic plate into which fine grooves have been cut and covered with a reflective aluminium coating. The grating causes interference between waves reflected from its surface, and constructive interference occurs when

$$n\lambda = d(\sin\theta - \sin\phi) \qquad (10.22)$$

where $n = 1, 2, \ldots$ is the *diffraction order*, λ is the wavelength of the diffracted radiation, d is the distance between grooves, θ is the angle of incidence of the beam, and ϕ is the angle of emergence of the beam (Fig. 10.22). In a **monochromator**, a narrow exit slit allows only a narrow range of wavelengths to reach the detector (Fig. 10.23). Turning the grating around an axis perpendicular to the incident and diffracted beams allows different wavelengths to be analysed; in this way, the absorption spectrum is built up one narrow wavelength range at a time. Typically, the grating is swept through an angle that investigates only the first order of diffraction ($n = 1$).

In a **Fourier transform spectrometer**, the diffraction grating is replaced by a *Michelson interferometer*, which works by splitting the beam from the sample into two and introducing a varying path length, p, into one of them (Fig. 10.24). When the two components recombine, there is a phase difference between them, and they interfere either constructively or destructively depending on the difference in path lengths. The detected signal oscillates as the two components alternately come into and out

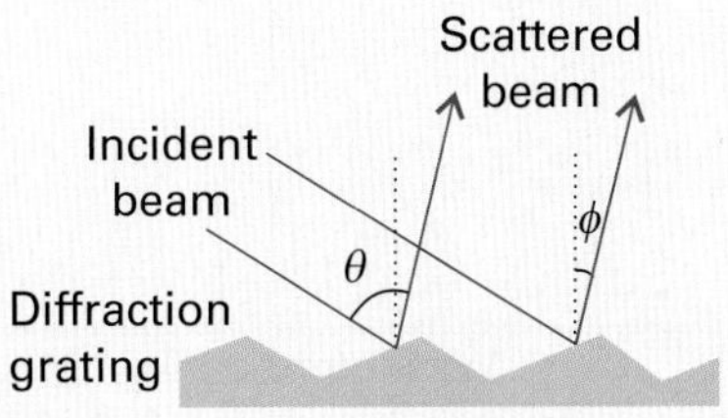

Fig. 10.22 One common dispersing element is a diffraction grating, which separates wavelengths spatially as a result of the scattering of light by fine grooves cut into a coated piece of glass. When a polychromatic light beam strikes the surface at an angle θ, several light beams of different wavelengths emerge at different angles ϕ.

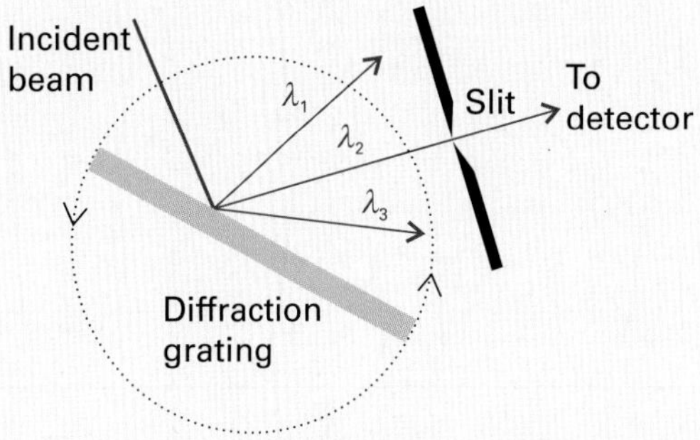

Fig. 10.23 A polychromatic beam is dispersed by a diffraction grating into three component wavelengths λ_1, λ_2, and λ_3. In the configuration shown, only radiation with λ_2 passes through a narrow slit and reaches the detector. Rotating the diffraction grating as shown by the double arrows allows λ_1 or λ_3 to reach the detector.

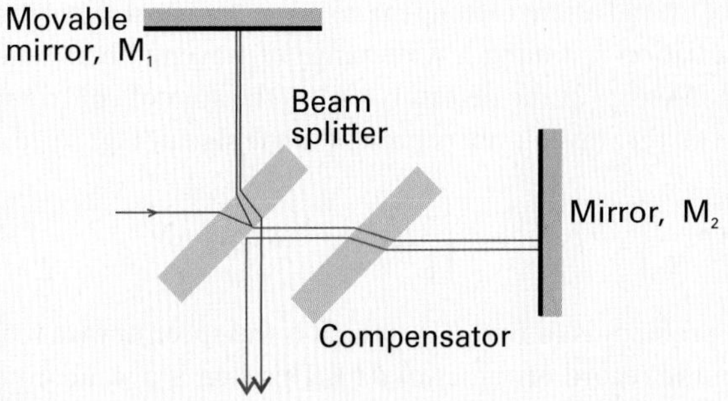

Fig. 10.24 A Michelson interferometer. The beam-splitting element divides the incident beam into two beams with a path difference that depends on the location of the mirror M_1. The compensator ensures that both beams pass through the same thickness of material.

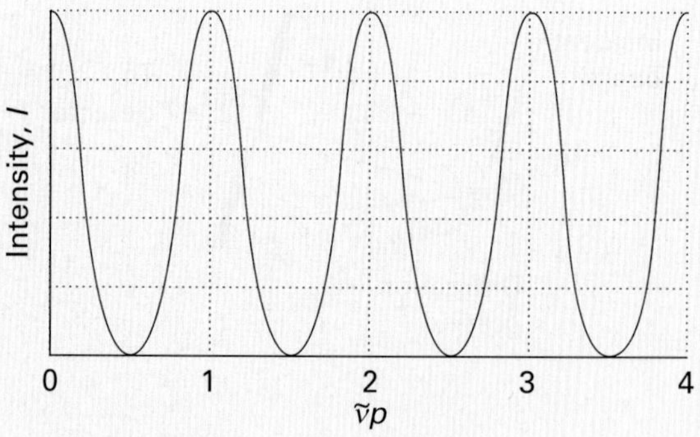

Fig. 10.25 An interferogram produced as the path length p is changed in the interferometer shown in Fig. 10.24. Only a single frequency component is present in the signal, so the graph is a plot of the function $I(p) = I_0(1 + \cos 2\pi\tilde{\nu}p)$, where I_0 is the intensity of the radiation.

interActivity Referring to Fig. 10.24, the mirror M_1 moves in finite distance increments, so the path difference p is also incremented in finite steps. Explore the effect of increasing the step size on the shape of the interferogram for a monochromatic beam of wavenumber $\tilde{\nu}$ and intensity I_0. That is, draw plots of $I(p)/I_0$ against $\tilde{\nu}p$, each with a different number of data points spanning the same total distance path taken by the movable mirror M_1.

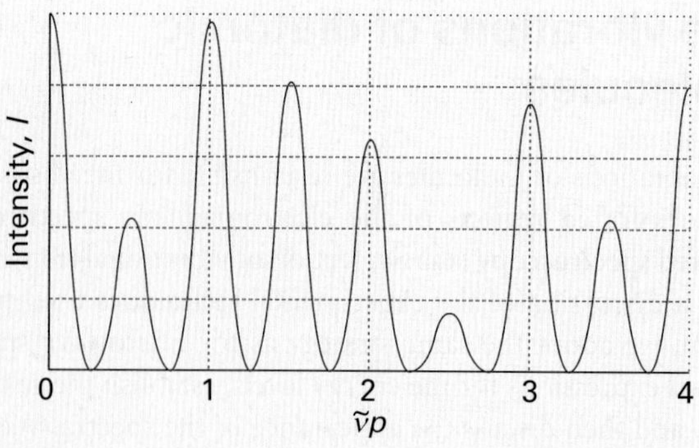

Fig. 10.26 An interferogram obtained when several (in this case, three) frequencies are present in the radiation.

interActivity For a signal consisting of only a few monochromatic beams, the integral in eqn 10.24 can be replaced by a sum over the finite number of wavenumbers. Use this information to draw your own version of Fig. 10.26. Then, go on to explore the effect of varying the wavenumbers and intensities of the three components of the radiation on the shape of the interferogram.

of phase as the path difference is changed (Fig. 10.25). If the radiation has wavenumber $\tilde{\nu}$, the intensity of the detected signal due to radiation in the range of wavenumbers $\tilde{\nu}$ to $\tilde{\nu} + \mathrm{d}\tilde{\nu}$, which we denote $I(p,\tilde{\nu})\mathrm{d}\tilde{\nu}$, varies with both path length p and wavenumber $\tilde{\nu}$ as

$$I(p,\tilde{\nu})\mathrm{d}\tilde{\nu} = I(\tilde{\nu})(1 + \cos 2\pi\tilde{\nu}p)\mathrm{d}\tilde{\nu} \qquad (10.23)$$

where $I(\tilde{\nu})$ is the variation of intensity solely with wavenumber. Hence, the interferometer converts the presence of a particular wavenumber component in the signal into a variation in intensity of the radiation reaching the detector. An actual signal consists of radiation spanning a wide range of wavenumbers, and the total intensity at the detector, $I(p)$, is the sum of contributions from all the wavenumbers present in the signal (Fig. 10.26):

$$I(p) = \int_0^\infty I(p,\tilde{\nu})\mathrm{d}\tilde{\nu} = \int_0^\infty I(\tilde{\nu})(1 + \cos 2\pi\tilde{\nu}p)\mathrm{d}\tilde{\nu} \qquad (10.24)$$

The problem is to find $I(\tilde{\nu})$, which is the spectrum we require, from the record of values of $I(p)$. This step is a standard technique of mathematics (see *Mathematical background* 6), and is the 'Fourier transformation' step from which this form of spectroscopy takes its name. Specifically:

$$I(\tilde{\nu}) = 4\int_0^\infty \{I(p) - \tfrac{1}{2}I(0)\}\cos 2\pi\tilde{\nu}p\,\mathrm{d}p \qquad (10.25)$$

where $I(0)$ is given by eqn 10.23 with $p = 0$. This integration is carried out numerically in a computer connected to the spectrometer, and the output, $I(\tilde{\nu})$, is the transmission spectrum of the sample (Fig. 10.27).

A major advantage of the Fourier transform procedure is that all the radiation emitted by the source is monitored continuously. This is in contrast to a spectrometer in which a monochromator discards most of the generated radiation. As a result, Fourier transform spectrometers have a higher sensitivity than conventional spectrometers. The resolution of Fourier transform

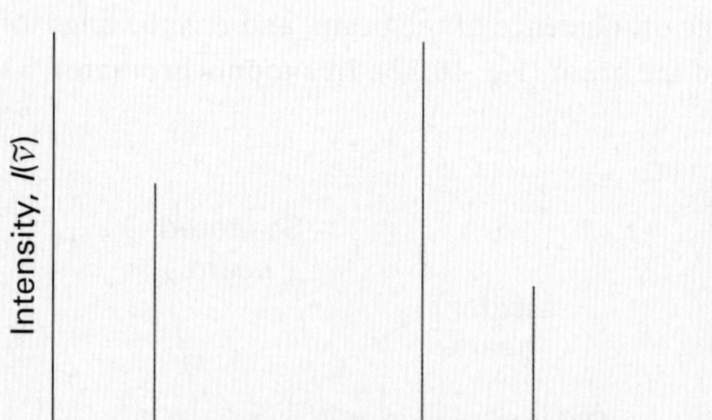

Fig. 10.27 The three frequency components and their intensities that account for the appearance of the interferogram in Fig. 10.26. This spectrum is the Fourier transform of the interferogram, and is a depiction of the contributing frequencies.

interActivity Calculate the Fourier transforms of the functions you generated in the previous interActivity.

spectrometers is determined by the maximum path length difference, p_{max}, of the interferometer:

$$\Delta\tilde{\nu} = \frac{1}{2p_{max}} \qquad (10.26)$$

To achieve a resolution of 0.1 cm^{-1} requires a maximum path length difference of 5 cm.

Detectors may consist of a single radiation sensing element or of several small elements arranged in one- or two-dimensional arrays. The most common detectors found in commercial infrared spectrometers are sensitive in the mid-infrared region. In a *photovoltaic device* the potential difference across a semiconductor changes upon its exposure to infrared radiation. In a *pyroelectric device* the capacitance is sensitive to temperature and hence the presence of infrared radiation.

10.7 Molecular vibrations

We base our discussion on Fig. 10.28, which shows a typical potential energy curve (as in Fig. 5.1) of a diatomic molecule. In regions close to R_e (at the minimum of the curve) the potential energy can be approximated by a parabola, so we can write

$$V = \tfrac{1}{2}kx^2 \qquad x = R - R_e \qquad (10.27)$$

where k is the **force constant** of the bond. The steeper the walls of the potential (the stiffer the bond), the greater the force constant.

To see the connection between the shape of the molecular potential energy curve and the value of k, note that we can expand the potential energy around its minimum by using a Taylor series (*Mathematical background 1*):

$$V(x) = V(0) + \left(\frac{dV}{dx}\right)_0 x + \frac{1}{2}\left(\frac{d^2V}{dx^2}\right)_0 x^2 + \cdots \qquad (10.28a)$$

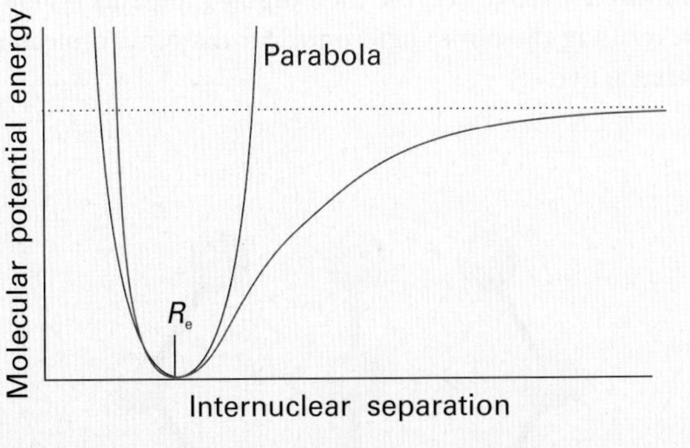

Fig. 10.28 A molecular potential energy curve can be approximated by a parabola near the bottom of the well. The parabolic potential leads to harmonic oscillations. At high excitation energies the parabolic approximation is poor (the true potential is less confining) and it is totally wrong near the dissociation limit.

The term $V(0)$ can be set arbitrarily to zero. The first derivative of V is 0 at the minimum. Therefore, the first surviving term is proportional to the square of the displacement. For small displacements we can ignore all the higher terms, and so write

$$V(x) \approx \frac{1}{2}\left(\frac{d^2V}{dx^2}\right)_0 x^2 \qquad (10.28b)$$

Therefore, the first approximation to a molecular potential energy curve is a parabolic potential, and by comparison with eqn 10.27 we can identify the force constant as

$$k = \left(\frac{d^2V}{dx^2}\right)_0 \qquad [10.29]$$

We see that, if the potential energy curve is sharply curved close to its minimum, then k will be large. Conversely, if the potential energy curve is wide and shallow, then k will be small (Fig. 10.29). Molecular force constants are readily calculated by using one of the computational techniques described in Chapter 6 either directly or by computing molecular energies at a series of bond lengths close to equilibrium and fitting a parabola to the values (Fig. 10.30).

The Schrödinger equation for the relative motion of two atoms of masses m_1 and m_2 with a parabolic potential energy is

$$-\frac{\hbar^2}{2m_{eff}}\frac{d^2\psi}{dx^2} + \tfrac{1}{2}kx^2\psi = E\psi \qquad (10.30)$$

where m_{eff} is the **effective mass:**

$$m_{eff} = \frac{m_1 m_2}{m_1 + m_2} \qquad [10.31]$$

These equations are derived in the same way as in *Further information 4.1*, but here the separation of variables procedure is used to separate the relative motion of the atoms from the motion of the molecule as a whole.

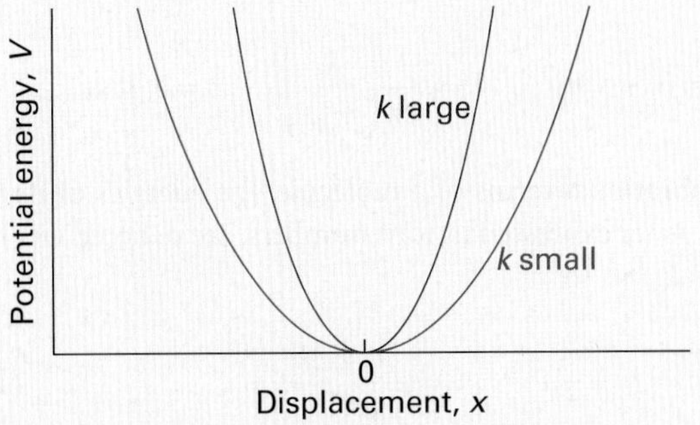

Fig. 10.29 The force constant is a measure of the curvature of the potential energy close to the equilibrium extension of the bond. A strongly confining well (one with steep sides, a stiff bond) corresponds to high values of k.

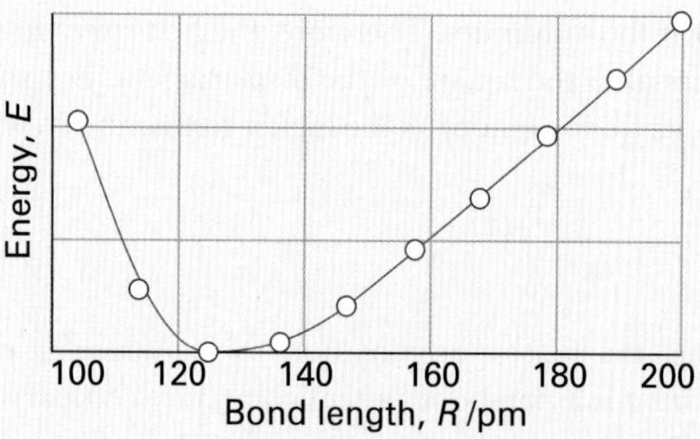

Fig. 10.30 A pointwise calculation of the energy of an HCl molecule in the Hartree–Fock approximation gives the shape of the molecular potential energy curve close to the equilibrium bond length (126 pm). The force constant of the bond is obtained by evaluating the second derivative of the curve at the equilibrium bond length.

interActivity Use molecular software to construct similar curves for the other hydrogen halides and identify the order of the stiffness of the H—Hal bonds.

A note on good practice Distinguish 'effective mass' from 'reduced mass'. The former is a measure of the amount of matter that moves during a vibration; the latter is the quantity that appears in the separation of internal and overall motion. For a diatomic molecule the effective mass is given by the same mathematical expression as for the reduced mass (eqn 10.31), but that is not true in general and each vibrational mode of a polyatomic molecule has a different effective mass. However, many people do not make this distinction, and you will often find the effective mass referred to as the reduced mass.

The Schrödinger equation in eqn 10.30 is the same as eqn 2.20 for a particle of mass m undergoing harmonic motion. Therefore, we can use the results of Section 2.4 to write down the permitted vibrational energy levels:

$$E_v = (v+\tfrac{1}{2})h\nu \qquad \nu = \frac{1}{2\pi}\left(\frac{k}{m_{\text{eff}}}\right)^{1/2} \qquad v = 0, 1, 2, \ldots \quad (10.32)$$

The **vibrational terms** of a molecule, the energies of its vibrational states expressed in wavenumbers, are denoted $\tilde{G}(v)$, with $\tilde{G}(v) = E_v/hc$, so

$$\tilde{G}(v) = (v+\tfrac{1}{2})\tilde{\nu} \qquad \tilde{\nu} = \frac{1}{2\pi c}\left(\frac{k}{m_{\text{eff}}}\right)^{1/2} \quad (10.33)$$

The vibrational wavefunctions are the same as those discussed in Section 2.5.

It is important to note that the vibrational terms depend on the effective mass of the molecule, not directly on its total mass. This dependence is physically reasonable for, if atom 1 were as heavy as a brick wall, then we would find $m_{\text{eff}} \approx m_2$, the mass of the lighter atom. The vibration would then be that of a light atom relative to that of a stationary wall (this is approximately the case in HI, for example, where the I atom barely moves and $m_{\text{eff}} \approx m_{\text{H}}$). For a homonuclear diatomic molecule $m_1 = m_2$, and the effective mass is half the total mass: $m_{\text{eff}} = \tfrac{1}{2}m$.

● A BRIEF ILLUSTRATION

An HCl molecule has a force constant of 516 N m^{-1}, a reasonably typical value for a single bond. The effective mass of $^{1}H^{35}Cl$ is 1.63×10^{-27} kg (note that this mass is very close to the mass of the hydrogen atom, 1.67×10^{-27} kg, so the Cl atom is acting like a brick wall). These values imply $\omega = 5.63 \times 10^{14}\ \text{s}^{-1}$, $\nu = 89.5$ THz (1 THz $= 10^{12}$ Hz), $\tilde{\nu} = 2987\ \text{cm}^{-1}$, $\lambda = 3.35\ \mu\text{m}$. These characteristics correspond to electromagnetic radiation in the infrared region. ●

10.8 Selection rules

The gross selection rule for a change in vibrational state brought about by absorption or emission of radiation is that *the electric dipole moment of the molecule must change when the atoms are displaced relative to one another*. Such vibrations are said to be **infrared active**. The classical basis of this rule is that the molecule can drive the electromagnetic field into oscillation if its dipole changes as it vibrates, and vice versa (Fig. 10.31); its formal basis is given in *Further information 10.2*. Note that the molecule need not have a permanent dipole: the rule requires only a change in dipole moment, possibly from zero. Some vibrations do not affect the molecule's dipole moment (e.g. the stretching motion of a homonuclear diatomic molecule), so they neither absorb nor generate radiation: such vibrations are said to be **infrared inactive**. Homonuclear diatomic molecules are infrared inactive because their dipole moments remain zero however long the bond; heteronuclear diatomic molecules are infrared active.

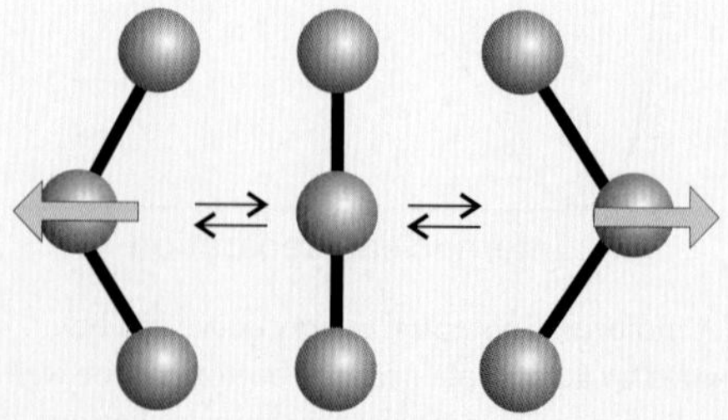

Fig. 10.31 The oscillation of a molecule, even if it is nonpolar, may result in an oscillating dipole that can interact with the electromagnetic field.

● A BRIEF ILLUSTRATION

Of the molecules N_2, CO_2, OCS, H_2O, $CH_2{=}CH_2$, and C_6H_6, all except N_2 possess at least one vibrational mode that results in a change of dipole moment, so all except N_2 can show a vibrational absorption spectrum. Not all the modes of complex molecules are vibrationally active. For example, the symmetric stretch of CO_2, in which the O—C—O bonds stretch and contract symmetrically, is inactive because it leaves the dipole moment unchanged (at zero). ●

Self-test 10.6 Which of the molecules H_2, NO, N_2O, and CH_4 have infrared active vibrations? [NO, N_2O, CH_4]

The specific selection rule, which is obtained from an analysis of the expression for the transition moment and the properties of integrals over harmonic oscillator wavefunctions (as shown in *Further information 10.2*), is

$$\Delta v = \pm 1$$

Transitions for which $\Delta v = +1$ correspond to absorption and those with $\Delta v = -1$ correspond to emission.

It follows from eqn 10.33 and the specific selection rules that the wavenumbers of allowed vibrational transitions, which are denoted $\Delta\tilde{G}(v)$ for the transition $v+1 \leftarrow v$, are

$$\Delta\tilde{G}(v) = \tilde{G}(v+1) - \tilde{G}(v) = \tilde{\nu} \qquad (10.34)$$

As we have seen, $\tilde{\nu}$ lies in the infrared region of the electromagnetic spectrum, so vibrational transitions absorb and generate infrared radiation.

At room temperature $kT/hc \approx 200\ \mathrm{cm^{-1}}$, and most vibrational wavenumbers are significantly greater than $200\ \mathrm{cm^{-1}}$. It follows from the Boltzmann distribution that almost all the molecules will be in their vibrational ground states initially. Hence, the dominant spectral transition will be the **fundamental transition**, $1 \leftarrow 0$. As a result, the spectrum is expected to consist of a single absorption line. If the molecules are formed in a vibrationally excited state, such as when vibrationally excited HF molecules are formed in the reaction $H_2 + F_2 \rightarrow 2\,HF^*$, the transitions $5 \rightarrow 4$, $4 \rightarrow 3, \ldots$ may also appear (in emission). In the harmonic approximation, all these lines lie at the same frequency, and the spectrum is also a single line. However, as we shall now show, the breakdown of the harmonic approximation causes the transitions to lie at slightly different frequencies, so several lines are observed.

10.9 Anharmonicity

The vibrational terms in eqn 10.33 are only approximate because they are based on a parabolic approximation to the actual potential energy curve. A parabola cannot be correct at all extensions because it does not allow a bond to dissociate, nor can it be correct at short extensions where the molecular potential energy curve rises very steeply as the nuclei come close together. At high vibrational excitations the swing of the atoms (more precisely, the spread of the vibrational wavefunction) allows the molecule to explore regions of the potential energy curve where the parabolic approximation is poor and additional terms in the Taylor expansion of V (eqn 10.28a) must be retained. The motion then becomes **anharmonic**, in the sense that the restoring force is no longer proportional to the displacement. Because the actual curve is less confining than a parabola, we can anticipate that the energy levels become less widely spaced at high excitations.

One approach to the calculation of the energy levels in the presence of anharmonicity is to use a function that resembles the true potential energy more closely. The **Morse potential energy** is

$$V = hc\tilde{D}_e\{1 - e^{-a(R-R_e)}\}^2 \qquad a = \left(\frac{m_{\mathrm{eff}}\pi\nu^2}{\hbar c\tilde{D}_e}\right)^{1/2} \qquad (10.35)$$

where $\tilde{D}_e$ is the depth of the potential minimum (Fig. 10.32). Near the well minimum the variation of V with displacement resembles a parabola (as can be checked by expanding the exponential as far as the first term) but, unlike a parabola, eqn 10.35 allows for dissociation at large displacements. The Schrödinger equation can be solved for the Morse potential and the permitted terms are

$$\tilde{G}(v) = (v+\tfrac{1}{2})\tilde{\nu} - (v+\tfrac{1}{2})^2 x_e\tilde{\nu}, \quad x_e = \frac{a^2\hbar}{4\pi m_{\mathrm{eff}}\nu} = \frac{\tilde{\nu}}{4\tilde{D}_e} \qquad (10.36)$$

The dimensionless parameter x_e is called the **anharmonicity constant**. The number of vibrational levels of a Morse oscillator is finite, and $v = 0, 1, 2, \ldots, v_{\max}$, as shown in Fig. 10.33 (see also Problem 10.31). The second term in the expression for $\tilde{G}$ subtracts from the first with increasing effect as v increases, and hence gives rise to the convergence of the levels at high quantum

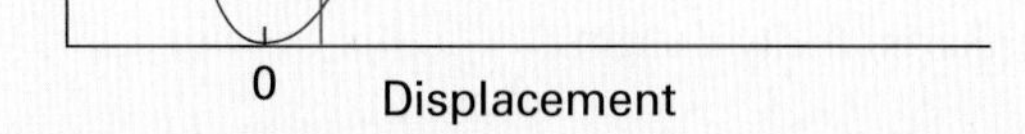

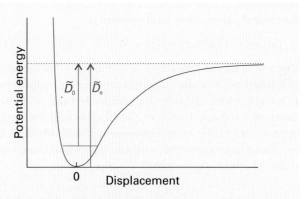

Fig. 10.32 The dissociation energy of a molecule, $\tilde{D}_0$, differs from the depth of the potential well, $\tilde{D}_e$, on account of the zero-point energy of the vibrations of the bond.

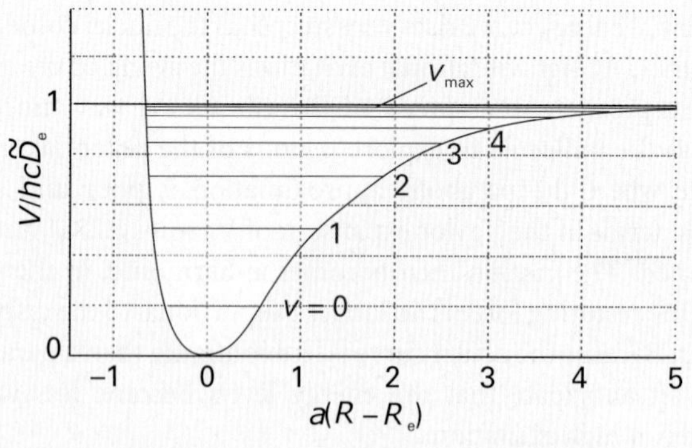

Fig. 10.33 The Morse potential energy curve reproduces the general shape of a molecular potential energy curve. The corresponding Schrödinger equation can be solved, and the values of the energies obtained. The number of bound levels is finite.

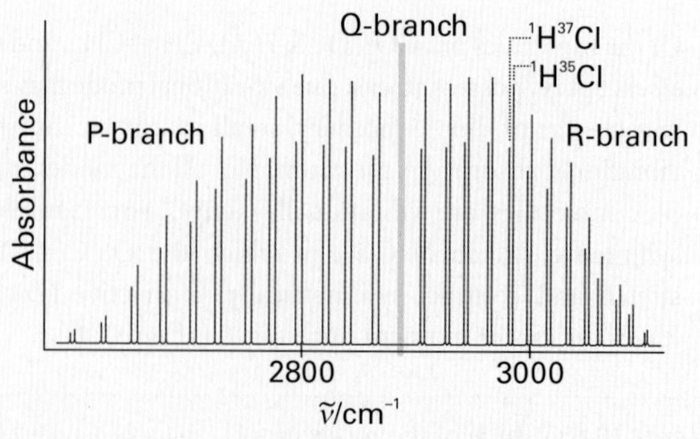

Fig. 10.34 A high-resolution vibration–rotation spectrum of HCl. The lines appear in pairs because $H^{35}Cl$ and $H^{37}Cl$ both contribute (the ^{35}Cl isotope is more abundant than the ^{37}Cl isotope). There is no Q branch, because $\Delta J = 0$ is forbidden for this molecule.

numbers as a result of the energy levels getting closer together as v increases.

Although the Morse oscillator is quite useful theoretically, in practice the more general expression

$$\tilde{G}(v) = (v+\tfrac{1}{2})\tilde{\nu} - (v+\tfrac{1}{2})^2 x_e\tilde{\nu} + (v+\tfrac{1}{2})^3 y_e\tilde{\nu} + \cdots \qquad (10.37)$$

where x_e, y_e, . . . are empirical dimensionless constants characteristic of the molecule, is used to fit the experimental data and to find the dissociation energy of the molecule. When anharmonicities are present, the wavenumbers of transitions with $\Delta v = +1$ are

$$\Delta\tilde{G}(v) = \tilde{\nu} - 2(v+1)x_e\tilde{\nu} + \cdots \qquad (10.38)$$

Equation 10.38 shows that, when $x_e > 0$, the transitions move to lower wavenumbers as v increases.

Anharmonicity also accounts for the appearance of additional weak absorption lines corresponding to the transitions $2 \leftarrow 0, 3 \leftarrow 0, \ldots$, even though these first, second, . . . **overtones** are forbidden by the selection rule $\Delta v = \pm 1$. The first overtone, for example, gives rise to an absorption at

$$\tilde{G}(v+2) - \tilde{G}(v) = 2\tilde{\nu} - 2(2v+3)x_e\tilde{\nu} + \cdots \qquad (10.39)$$

The reason for the appearance of overtones is that the selection rule is derived from the properties of harmonic oscillator wavefunctions, which are only approximately valid when anharmonicity is present. Therefore, the selection rule is also only an approximation. For an anharmonic oscillator, all values of Δv are allowed, but transitions with $\Delta v > 1$ are allowed only weakly if the anharmonicity is slight.

10.10 Vibration–rotation spectra

Each line of the high resolution vibrational spectrum of a gas-phase heteronuclear diatomic molecule is found to consist of a large number of closely spaced components (Fig. 10.34). Hence, molecular spectra are often called **band spectra**. The separation between the components is less than 10 cm^{-1}, which suggests that the structure is due to rotational transitions accompanying the vibrational transition. A rotational change should be expected because classically we can think of the transition as leading to a sudden increase or decrease in the instantaneous bond length. Just as ice-skaters rotate more rapidly when they bring their arms in, and more slowly when they throw them out, so the molecular rotation is either accelerated or retarded by a vibrational transition.

(a) Spectral branches

A detailed analysis of the quantum mechanics of simultaneous vibrational and rotational changes shows that the rotational quantum number J changes by ± 1 during the vibrational transition of a diatomic molecule. If the molecule also possesses angular momentum about its axis, as in the case of the electronic orbital angular momentum of the paramagnetic molecule NO in its ground-state configuration . . . π^1, then the selection rules also allow $\Delta J = 0$.

The appearance of the vibration–rotation spectrum of a diatomic molecule can be discussed in terms of the combined vibration–rotation terms, $\tilde{S}$:

$$\tilde{S}(v,J) = \tilde{G}(v) + \tilde{F}(J) \qquad (10.40)$$

If we ignore anharmonicity and centrifugal distortion,

$$\tilde{S}(v,J) = (v+\tfrac{1}{2})\tilde{\nu} + \tilde{B}J(J+1) \qquad (10.41)$$

In a more detailed treatment, $\tilde{B}$ is allowed to depend on the vibrational state because as v increases the bond length increases slightly and the moment of inertia changes. We shall continue with the simple expression initially.

When the vibrational transition $v+1 \leftarrow v$ occurs, J changes by ± 1 and in some cases by 0 (when $\Delta J = 0$ is allowed). The

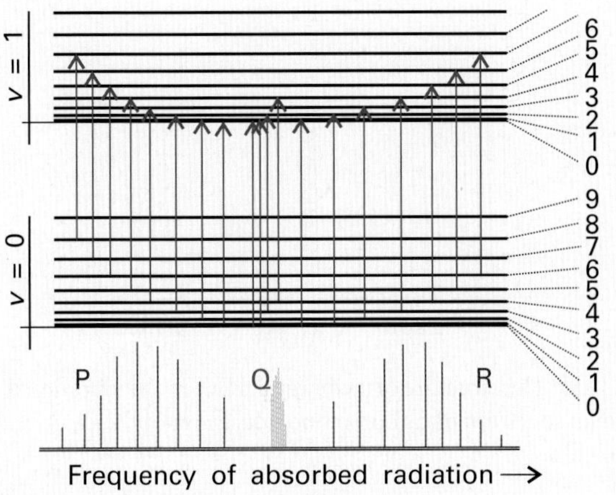

Fig. 10.35 The formation of P, Q, and R branches in a vibration–rotation spectrum. The intensities reflect the populations of the initial rotational levels.

absorptions then fall into three groups called **branches** of the spectrum. The **P branch** consists of all transitions with $\Delta J = -1$:

$$\begin{aligned}\tilde{\nu}_P(J) &= \tilde{S}(v+1, J-1) - \tilde{S}(v,J) \\ &= \{(v+\tfrac{3}{2})\tilde{\nu} + \tilde{B}(J-1)J\} - \{(v+\tfrac{1}{2})\tilde{\nu} + \tilde{B}J(J+1)\} \\ &= \tilde{\nu} - 2\tilde{B}J \end{aligned} \quad (10.42a)$$

This branch consists of lines at $\tilde{\nu} - 2\tilde{B}$, $\tilde{\nu} - 4\tilde{B}$, . . . with an intensity distribution reflecting both the populations of the rotational levels and the magnitude of the $J - 1 \leftarrow J$ transition moment (Fig. 10.35). Even though the rotational quantum number decreases in the transition, the transition still corresponds to absorption because the change in vibrational energy is so large. The **Q branch** consists of all lines with $\Delta J = 0$, and its wavenumbers are all

$$\tilde{\nu}_Q(J) = \tilde{S}(v+1, J) - \tilde{S}(v,J) = \tilde{\nu} \quad (10.42b)$$

for all values of J. This branch, when it is allowed (as in NO), appears at the vibrational transition wavenumber. In Fig. 10.35 there is a gap at the expected location of the Q branch because it is forbidden in HCl. The **R branch** consists of lines with $\Delta J = +1$:

$$\tilde{\nu}_R(J) = \tilde{S}(v+1, J+1) - \tilde{S}(v,J) = \tilde{\nu} + 2\tilde{B}(J+1) \quad (10.42c)$$

This branch consists of lines displaced from $\tilde{\nu}$ to high wavenumber by $2\tilde{B}$, $4\tilde{B}$, ...

The separation between the lines in the P and R branches of a vibrational transition gives the value of $\tilde{B}$. Therefore, the bond length can be deduced without needing to take a pure rotational microwave spectrum. However, the latter is more precise.

(b) Combination differences

The rotational constant of the vibrationally excited state, $\tilde{B}_1$ (in general, $\tilde{B}_v$), is in fact slightly smaller than that of the ground vibrational state, because the anharmonicity of the vibration results in a slightly extended bond in the upper state. As a result, the Q branch (if it exists) consists of a series of closely spaced lines. The lines of the R branch converge slightly as J increases; and those of the P branch diverge:

$$\begin{aligned}\tilde{\nu}_Q(J) &= \tilde{\nu} + (\tilde{B}_1 - \tilde{B}_0)J(J+1) \\ \tilde{\nu}_R(J) &= \tilde{\nu} + (\tilde{B}_1 + \tilde{B}_0)(J+1) + (\tilde{B}_1 - \tilde{B}_0)(J+1)^2 \end{aligned} \quad (10.43)$$

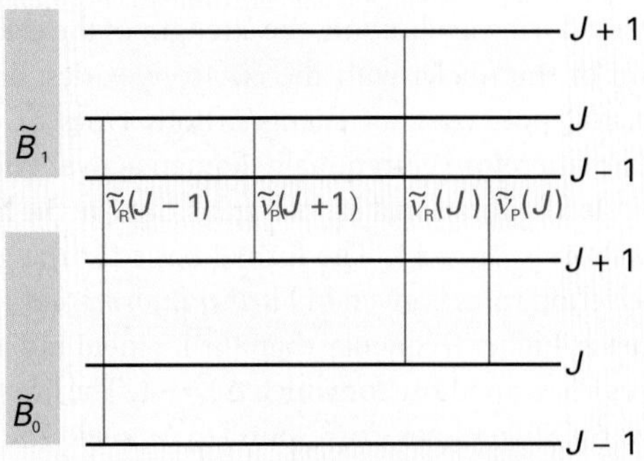

Fig. 10.36 The method of combination differences makes use of the fact that some transitions share a common level.

To determine the two rotational constants individually, we use the method of **combination differences.** This procedure is used widely in spectroscopy to extract information about a particular state and involves setting up expressions for the difference in the wavenumbers of transitions to a common state. As can be seen from Fig. 10.36, the transitions $\tilde{\nu}_R(J-1)$ and $\tilde{\nu}_P(J+1)$ have a common upper state, and from eqn 10.43 it follows that

$$\begin{aligned}\tilde{\nu}_R(J-1) - \tilde{\nu}_P(J+1) &= \{\tilde{\nu} + (\tilde{B}_1 + \tilde{B}_0)J + (\tilde{B}_1 - \tilde{B}_0)J^2\} - \\ &\quad \{\tilde{\nu} - (\tilde{B}_1 + \tilde{B}_0)(J+1) + (\tilde{B}_1 - \tilde{B}_0)(J+1)^2\} \\ &= 4\tilde{B}_0(J+\tfrac{1}{2}) \end{aligned} \quad (10.44a)$$

Therefore, a plot of the combination difference against $J + \frac{1}{2}$ should be a straight line of slope $4\tilde{B}_0$, so the rotational constant of the molecule in the state $v = 0$ can be determined. (Any deviation from a straight line is a consequence of centrifugal distortion, so that effect can be investigated too.) Similarly, $\tilde{\nu}_R(J)$ and $\tilde{\nu}_P(J)$ have a common lower state, and hence their combination difference gives information about the upper state:

$$\tilde{\nu}_R(J) - \tilde{\nu}_P(J) = 4\tilde{B}_1(J+\tfrac{1}{2}) \quad (10.44b)$$

The two rotational constants of $^1H^{35}Cl$ found in this way are $\tilde{B}_0 = 10.440\ cm^{-1}$ and $\tilde{B}_1 = 10.136\ cm^{-1}$.

10.11 Vibrational Raman spectra of diatomic molecules

The gross selection rule for vibrational Raman transitions is that *the polarizability should change as the molecule vibrates.* As homonuclear and heteronuclear diatomic molecules expand and contract during a vibration, the strength of the electrostatic interaction of the nuclei with the electrons varies, and hence the molecular polarizability changes. Both types of diatomic molecule are therefore vibrationally Raman active. The specific selection rule for vibrational Raman transitions in the harmonic approximation is $\Delta v = \pm 1$. The formal basis for the gross and specific selection rules is given in *Further information 10.2*.

The lines at higher frequency than the incident radiation, the anti-Stokes lines, are those for which $\Delta v = -1$. The lines at lower frequency, the Stokes lines, correspond to $\Delta v = +1$. The intensities of the anti-Stokes and Stokes lines are governed largely by the Boltzmann populations of the vibrational states involved in the transition. It follows that anti-Stokes lines are usually weak because very few molecules are in an excited vibrational state initially.

In gas-phase spectra, the Stokes and anti-Stokes lines have a branch structure arising from the simultaneous rotational transitions that accompany the vibrational excitation (Fig. 10.37). The selection rules are $\Delta J = 0, \pm 2$ (as in pure rotational Raman spectroscopy), and give rise to the **O branch** ($\Delta J = -2$), the **Q branch** ($\Delta J = 0$), and the **S branch** ($\Delta J = +2$):

$$\begin{aligned}\tilde{\nu}_O(J) &= \tilde{\nu}_i - \tilde{\nu} - 2\tilde{B} + 4\tilde{B}J \\ \tilde{\nu}_Q(J) &= \tilde{\nu}_i - \tilde{\nu} \\ \tilde{\nu}_S(J) &= \tilde{\nu}_i - \tilde{\nu} - 6\tilde{B} - 4\tilde{B}J\end{aligned} \qquad (10.45)$$

where $\tilde{\nu}_i$ is the wavenumber of the incident radiation. Note that, unlike in infrared spectroscopy, a Q branch is obtained for all linear molecules. The spectrum of CO, for instance, is shown in Fig. 10.38: the structure of the Q branch arises from the differences in rotational constants of the upper and lower vibrational states.

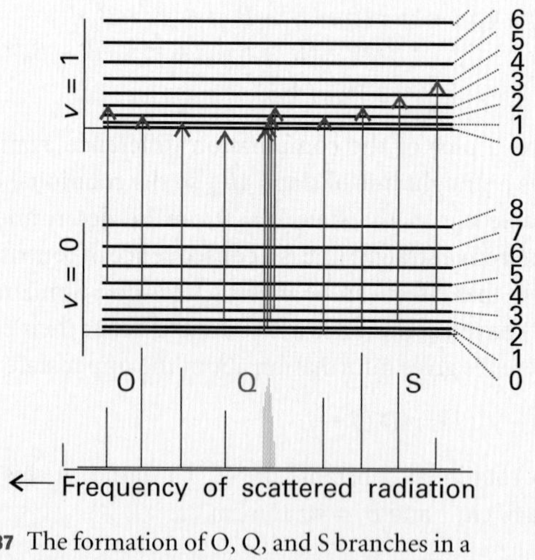

Fig. 10.37 The formation of O, Q, and S branches in a vibration–rotation Raman spectrum of a linear rotor. Note that the frequency scale runs in the opposite direction to that in Fig. 10.35, because the higher energy transitions (on the right) extract more energy from the incident beam and leave it at lower frequency.

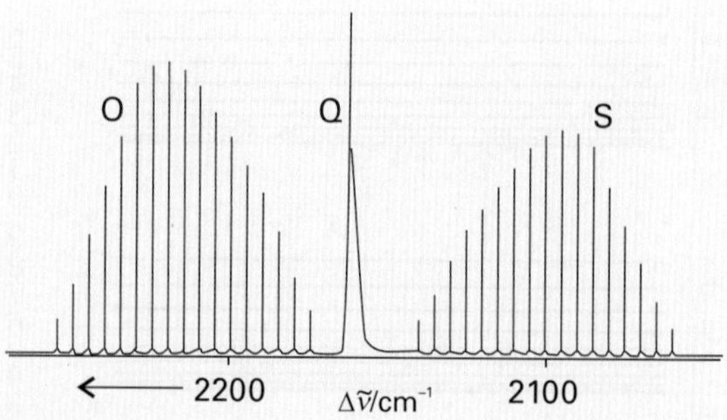

Fig. 10.38 The structure of a vibrational line in the vibrational Raman spectrum of carbon monoxide, showing the O, Q, and S branches.

Synoptic table 10.2* Properties of diatomic molecules

	$\tilde{\nu}$/cm^{-1}	R_e/pm	$\tilde{B}$/cm^{-1}	k/(N m^{-1})	D_0/(kJ mol^{-1})
$^1H_2$	4400	74	60.86	575	432
$^1H^{35}Cl$	2991	127	10.59	516	428
$^1H^{127}I$	2308	161	6.51	314	295
$^{35}Cl_2$	560	199	0.244	323	239

* More values are given in the *Data section*.

The information available from vibrational Raman spectra adds to that from infrared spectroscopy because homonuclear diatomics can also be studied. The spectra can be interpreted in terms of the force constants, dissociation energies, and bond lengths, and some of the information obtained is included in Table 10.2.

The vibrations of polyatomic molecules

There is only one mode of vibration for a diatomic molecule, the bond stretch. In polyatomic molecules there are several modes of vibration because all the bond lengths and angles may change and the vibrational spectra are very complex. Nonetheless, we shall see that infrared and Raman spectroscopy can be used to obtain information about the structure of systems as large as animal and plant tissues. Interstellar space can also be investigated with vibrational spectroscopy by using a combination of telescopes and infrared detectors. The experiments are conducted primarily in space-borne telescopes because the Earth's atmosphere absorbs a great deal of infrared radiation (see *Impact I10.2*). In most cases, absorption by an interstellar species is detected against the background of infrared radiation emitted by a nearby star. The data can detect the presence of gaseous and solid water, CO, and CO_2 in molecular clouds. In certain cases, infrared emission can be detected, but these events are rare because interstellar space is too cold and does not provide enough energy to promote a significant number of molecules to vibrational excited states. However, infrared emissions can be observed if molecules are occasionally excited by high-energy photons emitted by hot stars in the vicinity of the cloud. For example, the polycyclic aromatic hydrocarbons hexabenzocoronene ($C_{42}H_{18}$, **4**) and circumcoronene ($C_{54}H_{18}$, **5**) have been identified from characteristic infrared emissions.

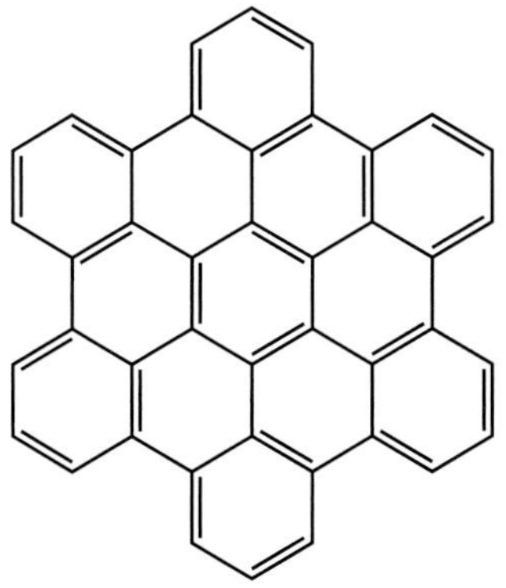

4 Hexabenzocoronene

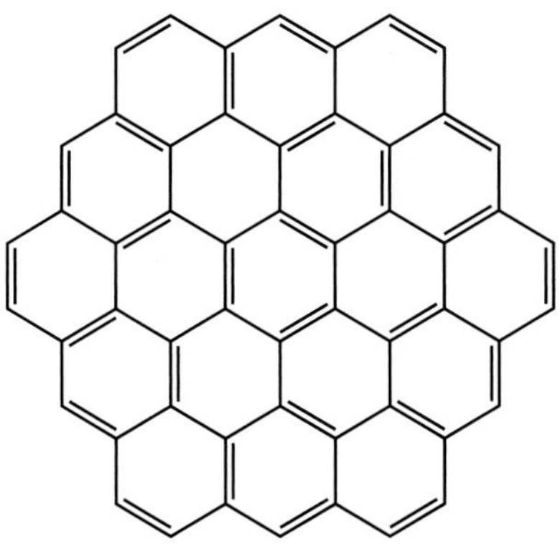

5 Circumcoronene

10.12 Normal modes

As shown in the *Justification* below, for a non-linear molecule that consists of N atoms, there are $3N - 6$ modes of vibration. If the molecule is linear, there are $3N - 5$ vibrational modes.

A BRIEF ILLUSTRATION

Water, H_2O, is a non-linear triatomic molecule, so $N = 3$. The total number of degrees of freedom is $3N = 9$. It follows that $N_{vib} = 9 - 6 = 3$, so it has three modes of vibration (and three modes of rotation); CO_2 is a linear triatomic molecule, so $N_{vib} = 9 - 5 = 4$, so it has four modes of vibration (and only two modes of rotation). Even a middle-sized molecule such as naphthalene ($C_{10}H_8$, $N = 18$) has $3 \times 18 - 6 = 48$ distinct modes of vibration. ●

Justification 10.2 *The number of vibrational modes*

The total number of coordinates needed to specify the locations of N atoms is $3N$. Each atom may change its location by varying one of its three coordinates (x, y, and z), so the total number of displacements available—the number of 'degrees of freedom'—is $3N$. These displacements can be grouped together in a physically sensible way. For example, three coordinates are needed to specify the location of the centre of mass of the molecule, so three of the $3N$ displacements correspond to the translational motion of the molecule as a whole. The remaining $3N - 3$ are non-translational 'internal' modes of the molecule.

Two angles are needed to specify the orientation of a linear molecule in space: in effect, we need to give only the latitude and longitude of the direction in which the molecular axis is pointing (Fig. 10.39a). However, three angles are needed for a non-linear molecule because we also need to specify the orientation of the molecule around the direction defined by the latitude and longitude (Fig. 10.39b). Therefore, two (linear) or three (non-linear) of the $3N - 3$ internal displacements are

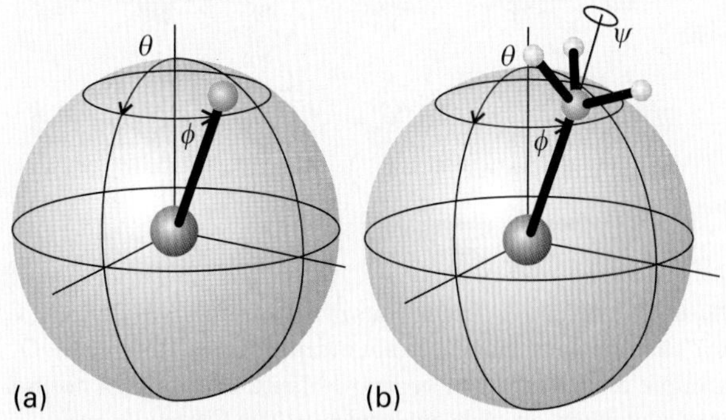

Fig. 10.39 (a) The orientation of a linear molecule requires the specification of two angles. (b) The orientation of a non-linear molecule requires the specification of three angles.

rotational. This leaves $3N-5$ (linear) or $3N-6$ (non-linear) displacements of the atoms relative to one another: these are the vibrational modes. It follows that the number of modes of vibration N_{vib} is $3N-5$ for linear molecules and $3N-6$ for non-linear molecules.

One choice for two of the four modes of CO_2 might be the ones in Fig. 10.40a. This illustration shows the stretching of one bond (the mode ν_L) and the stretching of the other (ν_R). The description has a disadvantage: when one CO bond vibration is excited, the motion of the C atom sets the other CO bond in motion, so energy flows backwards and forwards between ν_L and ν_R and the modes are not independent.

The description of the vibrational motion is much simpler if linear combinations of ν_L and ν_R are taken. For example, one combination is ν_1 in Fig. 10.40b: this mode is the **symmetric stretch**. In this mode, the C atom is buffeted simultaneously from each side and the motion continues indefinitely. Another mode is ν_3, the **antisymmetric stretch**, in which the two O atoms always move in the same direction and opposite to that of the C atom. Both modes are independent in the sense that, if one is excited, then it does not excite the other. They are two of the 'normal modes' of the molecule, its independent, collective vibrational displacements. The two other normal modes are the bending modes ν_2 (there are two independent bending modes, in perpendicular planes, but they have the same frequency and are given the same label). In general, a **normal mode** is an independent, synchronous motion of atoms or groups of atoms that may be excited without leading to the excitation of any other normal mode and without involving translation or rotation of the molecule as a whole.

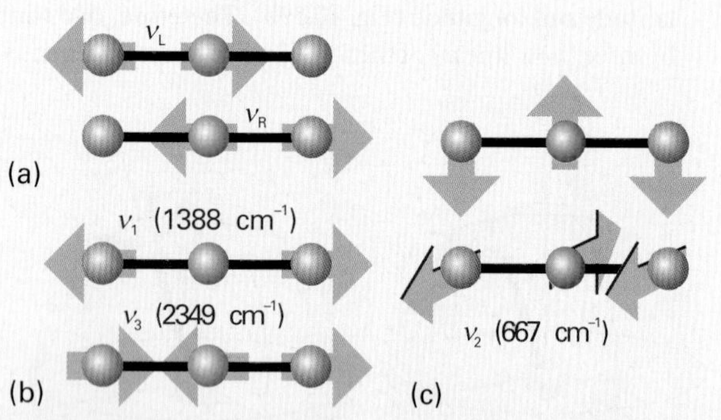

Fig. 10.40 Alternative descriptions of the vibrations of CO_2. (a) The stretching modes are not independent, and if one C—O group is excited the other begins to vibrate. They are not normal modes of vibration of the molecule. (b) The symmetric and antisymmetric stretches are independent, and one can be excited without affecting the other: they are normal modes. (c) The two perpendicular bending motions are also normal modes.

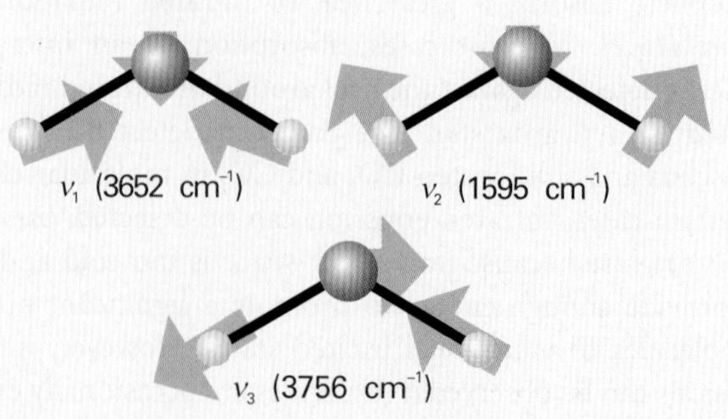

Fig. 10.41 The three normal modes of H_2O. The mode ν_2 is predominantly bending, and occurs at lower wavenumber than the other two.

The four normal modes of CO_2, and the N_{vib} normal modes of polyatomics in general, are the key to the description of molecular vibrations. Each normal mode, q, behaves like an independent harmonic oscillator (if anharmonicities are neglected), so each has a series of terms

$$\tilde{G}_q(v) = (v + \tfrac{1}{2})\tilde{\nu}_q \qquad \tilde{\nu}_q = \frac{1}{2\pi c}\left(\frac{k_q}{m_q}\right)^{1/2} \tag{10.46}$$

where $\tilde{\nu}_q$ is the wavenumber of mode q and depends on the force constant k_q for the mode and on the effective mass m_q of the mode. The effective mass of the mode is a measure of the mass that is swung about by the vibration and in general is a complicated function of the masses of the atoms. For example, in the symmetric stretch of CO_2, the C atom is stationary, and the effective mass depends on the masses of only the O atoms. In the antisymmetric stretch and in the bends, all three atoms move, so all contribute to the effective mass. The three normal modes of H_2O are shown in Fig. 10.41: note that the predominantly bending mode (ν_2) has a lower frequency than the others, which are predominantly stretching modes. It is generally the case that the frequencies of bending motions are lower than those of stretching modes because the potential energy changes more slowly with angle than with the stretching of a bond, so the force constant is lower. One point that must be appreciated is that only in special cases (such as the CO_2 molecule) are the normal modes purely stretches or purely bends. In general, a normal mode is a composite motion of simultaneous stretching and bending of bonds. Another point in this connection is that heavy atoms generally move less than light atoms in normal modes.

10.13 Infrared absorption spectra of polyatomic molecules

The gross selection rule for infrared activity is that *the motion corresponding to a normal mode should be accompanied by a change of dipole moment*. Deciding whether this is so can sometimes be done by inspection. For example, the symmetric stretch of CO_2 leaves the dipole moment unchanged (at zero, see Fig. 10.40b), so this mode is infrared inactive. The antisymmetric stretch, however, changes the dipole moment because the molecule becomes unsymmetrical as it vibrates, so this mode is infrared active. Because the dipole moment change is parallel to the principal axis, the transitions arising from this mode are classified as **parallel bands** in the spectrum. Both bending modes are infrared active: they are accompanied by a changing dipole perpendicular to the principal axis, so transitions involving them lead to a **perpendicular band** in the spectrum. The latter bands eliminate the linearity of the molecule, and as a result a Q branch is observed; the parallel band of a linear molecule does not have a Q branch.

The active modes are subject to the specific selection rule $\Delta v_q = \pm 1$ in the harmonic approximation, so the wavenumber of the fundamental transition (the 'first harmonic') of each active mode is $\tilde{\nu}_q$. From the analysis of the spectrum, a picture may be constructed of the stiffness of various parts of the molecule: that is, we can establish its **force field**, the set of force constants corresponding to all the displacements of the atoms. The force field may also be estimated by using the semiempirical, *ab initio*, and DFT computational techniques described in Chapter 6. Superimposed on the simple force-field scheme are the complications arising from anharmonicities and the effects of molecular rotation. Very often the sample is a liquid or a solid, and the molecules are unable to rotate freely. In a liquid, for example, a molecule may be able to rotate through only a few degrees before it is struck by another, so it changes its rotational state frequently.

The lifetimes of rotational states in liquids are very short, so in most cases the rotational energies are ill-defined. Collisions occur at a rate of about 10^{13} s^{-1} and, even allowing for only a 10 per cent success rate in knocking the molecule into another rotational state, a lifetime broadening of more than 1 cm^{-1} can easily result. The rotational structure of the vibrational spectrum is blurred by this effect, so the infrared spectra of molecules in condensed phases usually consist of broad lines spanning the entire range of the resolved gas-phase spectrum, and showing no branch structure.

One very important application of infrared spectroscopy to condensed phase samples, and for which the blurring of the rotational structure by random collisions is a welcome simplification, is to chemical analysis. The vibrational spectra of different groups in a molecule give rise to absorptions at characteristic frequencies because a normal mode of even a very large molecule is often dominated by the motion of a small group of atoms. The intensities of the vibrational bands that can be identified with the motions of small groups are also transferable between molecules. Consequently, the molecules in a sample can often be identified by examining its infrared spectrum and referring to a table of characteristic frequencies and intensities (Table 10.3).

Synoptic table 10.3* Typical vibrational wavenumbers

Vibration type	$\tilde{\nu}/cm^{-1}$
C—H stretch	2850–2960
C—H bend	1340–1465
C—C stretch, bend	700–1250
C=C stretch	1620–1680

* More values are given in the *Data section*.

IMPACT ON ENVIRONMENTAL SCIENCE
I10.2 Climate change[1]

Solar energy strikes the top of the Earth's atmosphere at a rate of 343 W m^{-2}. About 30 per cent of this energy is reflected back into space by the Earth or the atmosphere. The Earth–atmosphere system absorbs the remaining energy and re-emits it into space as black-body radiation, with most of the intensity being carried by infrared radiation in the range 200–2500 cm^{-1} (4–50 μm). The Earth's average temperature is maintained by an energy balance between solar radiation absorbed by the Earth and black-body radiation emitted by the Earth.

The trapping of infrared radiation by certain gases in the atmosphere is known as the *greenhouse effect*, so called because it warms the Earth as if the planet were enclosed in a huge greenhouse. The result is that the natural greenhouse effect raises the average surface temperature well above the freezing point of water and creates an environment in which life is possible. The major constituents to the Earth's atmosphere, O_2 and N_2, do not contribute to the greenhouse effect because homonuclear diatomic molecules cannot absorb infrared radiation. However, the minor atmospheric gases water vapour and CO_2 do absorb infrared radiation and hence are responsible for the greenhouse effect (Fig. 10.42). Water vapour absorbs strongly in the ranges 1300–1900 cm^{-1} (5.3–7.7 μm; the bending mode, ν_2) and 3550–3900 cm^{-1} (2.6–2.8 μm; the mainly stretching modes, ν_1 and ν_3), whereas CO_2 shows strong absorption in the ranges 500–725 cm^{-1} (14–20 μm; the bending mode, ν_2) and 2250–2400 cm^{-1} (4.2–4.4 μm; the antisymmetric stretch, ν_3).

Increases in the levels of greenhouse gases, which also include methane, dinitrogen oxide, ozone, and certain chlorofluorocarbons, as a result of human activity have the potential to enhance the natural greenhouse effect, leading to significant warming of the planet and other consequences for the environment. This problem is referred to as *climate change*, which we now explore in some detail.

[1] This section is based on a similar contribution initially prepared by Loretta Jones and appearing in *Chemical principles*, Peter Atkins and Loretta Jones, W.H. Freeman and Co., New York (2005).

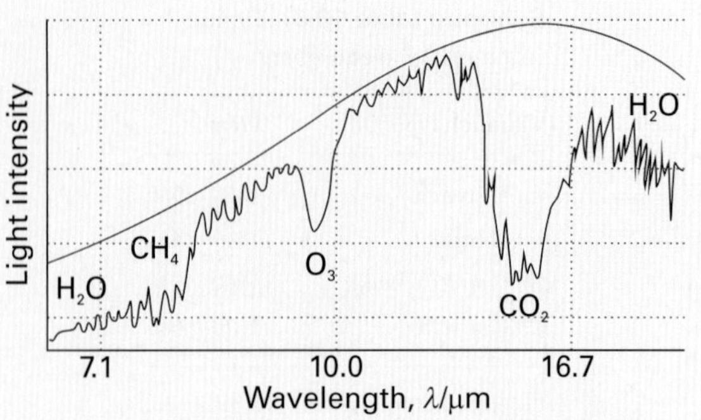

Fig. 10.42 The intensity of infrared radiation that would be lost from Earth in the absence of greenhouse gases is shown by the blue line. The brown line is the intensity of the radiation actually emitted. The maximum wavelength of radiation absorbed by each greenhouse gas is indicated.

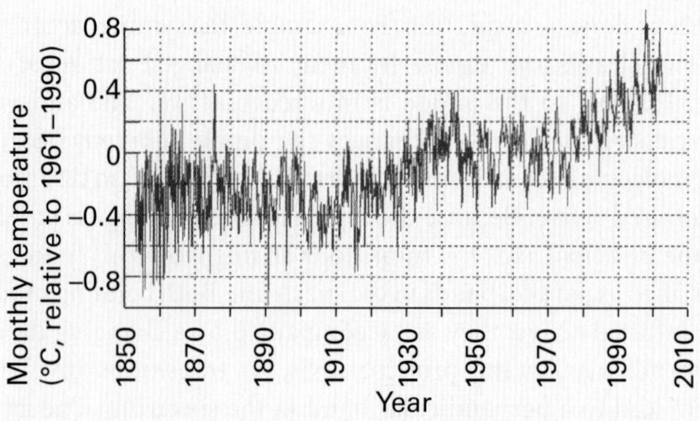

Fig. 10.43 The average change in surface temperature of the Earth from 1855 to 2002.

The concentration of water vapour in the atmosphere has remained steady over time on account of its equilibration with the oceans, but concentrations of some other greenhouse gases are rising. From about the year 1000 until about 1750, the CO_2 concentration remained fairly stable, but, since then, it has increased by 28 per cent. The concentration of methane, CH_4, has more than doubled during this time and is now at its highest level for 160 000 years (160 ka; a is the SI unit denoting 1 year). Studies of air pockets in ice cores taken from Antarctica show that increases in the concentration of both atmospheric CO_2 and CH_4 over the past 160 ka correlate well with increases in the global surface temperature.

Human activities are primarily responsible for the rising concentrations of atmospheric CO_2 and CH_4. Most of the atmospheric CO_2 comes from the burning of hydrocarbon fuels, which began on a large scale with the Industrial Revolution in the middle of the nineteenth century. The additional methane comes mainly from the petroleum industry and from agriculture.

The temperature of the surface of the Earth has increased by about 0.5 K since the late nineteenth century (Fig. 10.43). If we continue to rely on hydrocarbon fuels and current trends in population growth and energy are not reversed then, by the middle of the twenty-first century, the concentration of CO_2 in the atmosphere will be about twice its value prior to the Industrial Revolution. The Intergovernmental Panel on Climate Change (IPCC) has estimated that by the year 2100 the Earth will undergo an increase in temperature of 3 K. Furthermore, the rate of temperature change is likely to be greater than at any time in the last 10 ka. To place a temperature rise of 3 K in perspective, it is useful to consider that the average temperature of the Earth during the last ice age was only 6 K colder than at present. Just as cooling the planet (for example, during an ice age) can lead to detrimental effects on ecosystems, so too can a dramatic warming of the globe. One example of a significant change in the environment caused by a temperature increase of 3 K is a rise in sea level by about 0.5 m, which is sufficient to alter weather patterns and submerge currently coastal ecosystems.

Computer projections for the next 200 years predict further increases in atmospheric CO_2 levels and suggest that, to maintain CO_2 at its current concentration, we would have to reduce hydrocarbon fuel consumption immediately by about 50 per cent. Clearly, in order to reverse global warming trends, we need to develop alternatives to fossil fuels, such as hydrogen (which can be used in fuel cells, *Impact I17.2*) and solar energy technologies.

10.14 **Vibrational Raman spectra of polyatomic molecules**

The normal modes of vibration of molecules are Raman active if they are accompanied by a changing polarizability. Specifically, a mode is Raman active if $(\partial\alpha/\partial Q)_0$ is nonzero, where Q is the displacement corresponding to the normal mode. It is sometimes quite difficult to judge by inspection when this is so. The symmetric stretch of CO_2, for example, alternately expands and contracts the molecule: this motion changes the polarizability of the molecule, so the mode is Raman active. The polarizability of CO_2 does change as the molecule bends but, as the variation is symmetrical with respect to the angle changing from 180°, $(\partial\alpha/\partial Q)_0 = 0$ and the mode is inactive.

A more exact treatment of infrared and Raman activity of normal modes leads to the **exclusion rule**:

> If the molecule has a centre of symmetry (that is, is centrosymmetric), then no modes can be both infrared and Raman active.

(A mode may be inactive in both.) Because it is often possible to judge intuitively if a mode changes the molecular dipole moment, we can use this rule to identify modes that are not Raman active. The rule applies to CO_2 but to neither H_2O nor

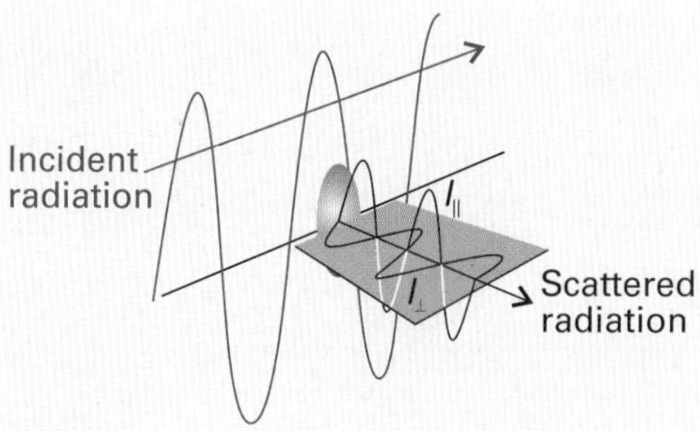

Fig. 10.44 The definition of the planes used for the specification of the depolarization ratio, ρ, in Raman scattering.

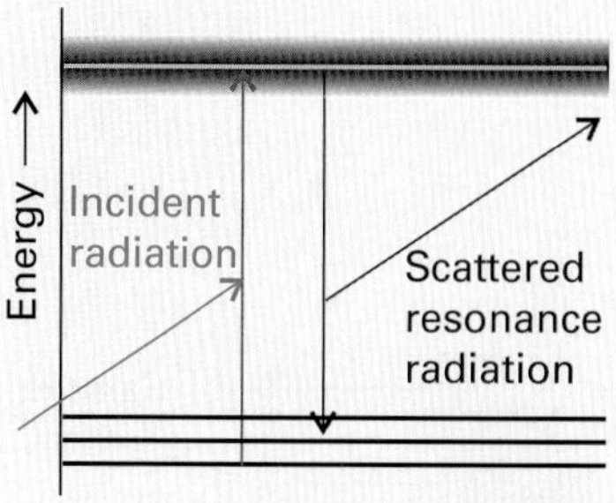

Fig. 10.45 In the *resonance Raman effect*, the incident radiation has a frequency corresponding to an actual electronic excitation of the molecule. A photon is emitted when the excited state returns to a state close to the ground state.

CH_4 because they have no centre of symmetry. In general, it is necessary to use group theory to predict whether a mode is infrared or Raman active (Section 10.15).

(a) Depolarization

The assignment of Raman lines to particular vibrational modes is aided by using plane-polarized incident radiation and noting the state of polarization of the scattered radiation. The **depolarization ratio**, ρ, of a line is the ratio of the intensities, I, of the scattered radiation with polarizations perpendicular and parallel to the plane of polarization of the incident radiation:

$$\rho = \frac{I_{\perp}}{I_{\parallel}} \qquad [10.47]$$

To measure ρ, the intensity of a Raman line is measured with a polarizing filter (a 'half-wave plate') first parallel and then perpendicular to the polarization of the incident beam. If the emergent radiation is not polarized, then both intensities are the same and ρ is close to 1; if the radiation retains its initial polarization, then $I_{\perp} = 0$, so $\rho = 0$ (Fig. 10.44). A line is classified as **depolarized** if it has ρ close to or greater than 0.75 and as **polarized** if $\rho < 0.75$. Only totally symmetrical vibrations give rise to polarized lines in which the incident polarization is largely preserved. Vibrations that are not totally symmetrical give rise to depolarized lines because the incident radiation can give rise to radiation in the perpendicular direction too.

(b) Resonance Raman spectra

A modification of the basic Raman effect involves using incident radiation that nearly coincides with the frequency of an electronic transition of the sample (Fig. 10.45). The technique is then called **resonance Raman spectroscopy**. It is characterized by a much greater intensity in the scattered radiation. Furthermore, because it is often the case that only a few vibrational modes contribute to the more intense scattering, the spectrum is greatly simplified.

Resonance Raman spectroscopy is used to study biological molecules that absorb strongly in the ultraviolet and visible regions of the spectrum. Examples include the pigments β-carotene and chlorophyll, which capture solar energy during plant photosynthesis (see *Impact I19.2*). The resonance Raman spectra of Fig. 10.46 show vibrational transitions from only the few pigment molecules that are bound to very large proteins dissolved in an aqueous buffer solution. This selectivity arises from the fact that water (the solvent), amino acid residues, and the peptide group do not have electronic transitions at the laser wavelengths used in the experiment, so their *conventional* Raman spectra are weak compared to the enhanced spectra of the pigments. Comparison of the spectra in Figs. 10.46a and 10.46b also shows that, with proper choice of excitation wavelength, it is possible to examine individual classes of pigments bound

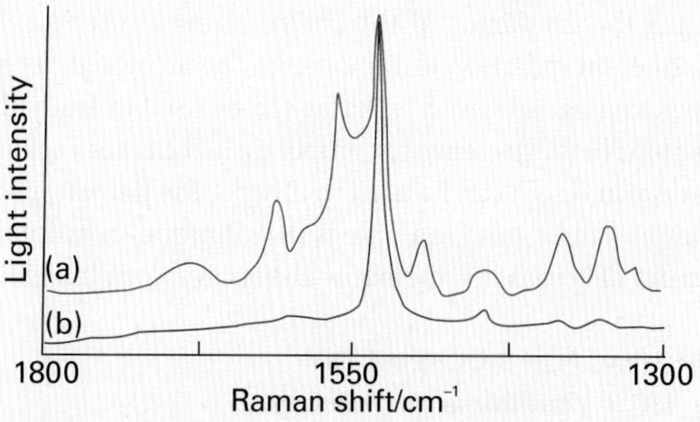

Fig. 10.46 The resonance Raman spectra of a protein complex that is responsible for some of the initial electron transfer events in plant photosynthesis. (a) Laser excitation of the sample at 407 nm shows Raman bands due to both chlorophyll *a* and β-carotene bound to the protein because both pigments absorb light at this wavelength. (b) Laser excitation at 488 nm shows Raman bands from β-carotene only because chlorophyll *a* does not absorb light very strongly at this wavelength. (Adapted from D.F. Ghanotakis *et al. Biochim. Biophys. Acta* **974**, 44 (1989).)

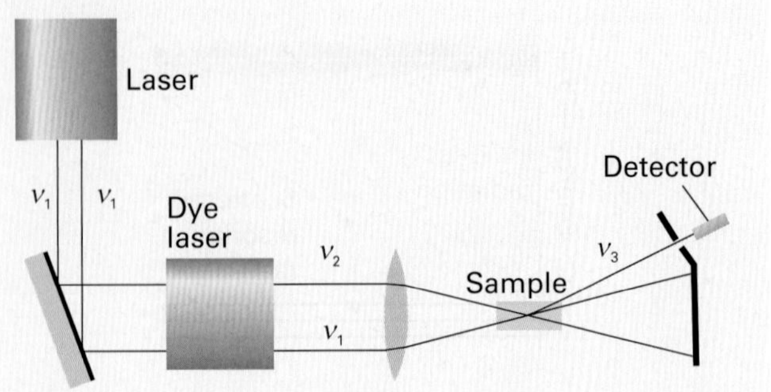

Fig. 10.47 The experimental arrangement for the CARS experiment.

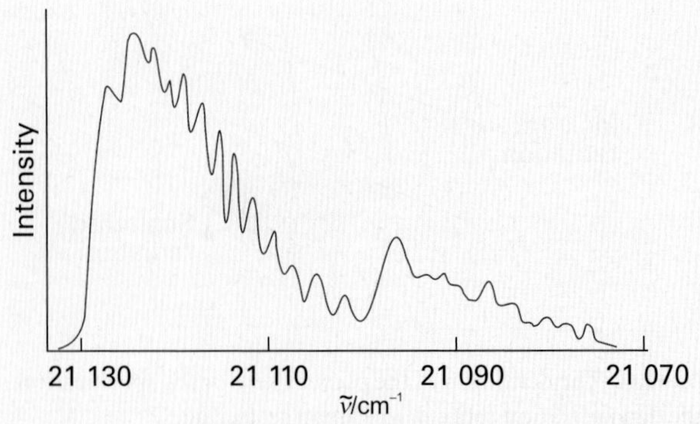

Fig. 10.48 CARS spectrum of a methane–air flame at 2104 K. The peaks correspond to the Q branch of the vibration–rotation spectrum of N_2 gas. (Adapted from J.F. Verdieck *et al. J. Chem. Ed.* 59, 495 (1982)).

to the same protein: excitation at 488 nm, where β-carotene absorbs strongly, shows vibrational bands from β-carotene only, whereas excitation at 407 nm, where chlorophyll *a* and β-carotene absorb, reveals features from both types of pigments.

(c) Coherent anti-Stokes Raman spectroscopy

The intensity of Raman transitions may be enhanced by **coherent anti-Stokes Raman spectroscopy** (CARS, Fig. 10.47). The technique relies on the fact that, if two laser beams of frequencies ν_1 and ν_2 pass through a sample, then they may mix together and give rise to coherent radiation of several different frequencies, one of which is $\nu' = 2\nu_1 - \nu_2$. Suppose that ν_2 is varied until it matches any Stokes line from the sample, such as the one with frequency $\nu_1 - \Delta\nu$; then the coherent emission will have frequency

$$\nu' = 2\nu_1 - (\nu_1 - \Delta\nu) = \nu_1 + \Delta\nu \qquad (10.48)$$

which is the frequency of the corresponding anti-Stokes line. This coherent radiation forms a narrow beam of high intensity.

An advantage of CARS is that it can be used to study Raman transitions in the presence of competing incoherent background radiation, and so it can be used to observe the Raman spectra of species in flames. One example is the vibration–rotation CARS spectrum of N_2 gas in a methane–air flame shown in Fig. 10.48.

IMPACT ON BIOCHEMISTRY
I10.3 Vibrational microscopy

Optical microscopes can now be combined with infrared and Raman spectrometers and the vibrational spectra of specimens as small as single biological cells obtained. The techniques of *vibrational microscopy* provide details of cellular events that cannot be observed with traditional light or electron microscopy.

The principles behind the operation of infrared and Raman microscopes are simple: radiation illuminates a small area of the sample, and the transmitted, reflected, or scattered light is first collected by a microscope and then analysed by a spectrometer. The sample is then moved by very small increments along a plane perpendicular to the direction of illumination and the process is repeated until vibrational spectra for all sections of the sample are obtained. The size of a sample that can be studied by vibrational microscopy depends on a number of factors, such as the area of illumination and the intensity and wavelength of the incident radiation. Up to a point, the smaller the area that is illuminated, the smaller the area from which a spectrum can be obtained. High intensity is required to increase the rate of arrival of photons at the detector from small illuminated areas. For this reason, lasers and synchrotron radiation are the preferred radiation sources.

In a conventional light microscope, an image is constructed from a pattern of diffracted light waves that emanate from the illuminated object. As a result, some information about the specimen is lost by destructive interference of scattered light waves. Ultimately, this *diffraction limit* prevents the study of samples that are much smaller than the wavelength of light used as a probe. In practice, two objects will appear as distinct images under a microscope if the distance between their centres is greater than the *Airy radius*, $r_{\text{Airy}} = 0.61\lambda/a$, where λ is the wavelength of the incident beam of radiation and a is the numerical aperture of the objective lens, the lens that collects light scattered by the object. The numerical aperture of the objective lens is defined as $a = n_r \sin\alpha$, where n_r is the refractive index of the lens material and the angle α is the half-angle of the widest cone of scattered light that can collected by the lens (so the lens collects light beams sweeping a cone with angle 2α). Use of the best equipment makes it possible to probe areas as small as 10 μm^2 by vibrational microscopy.

Figure 10.49 shows the infrared spectra of a single mouse cell, living and dying. Both spectra have features at 1545 cm^{-1} and 1650 cm^{-1} that are due to the peptide carbonyl groups of

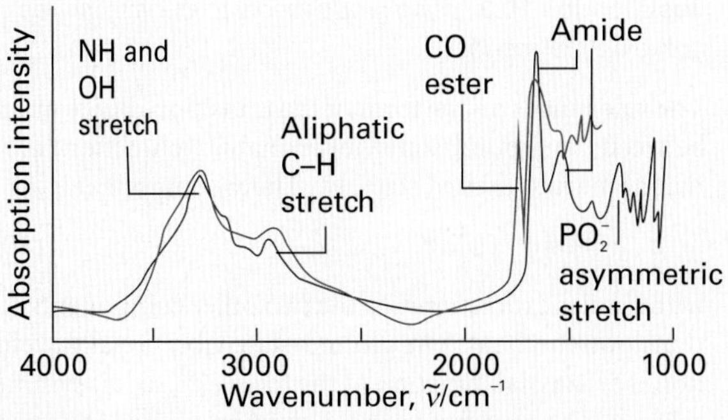

Fig. 10.49 Infrared absorption spectra of a single mouse cell: (purple) living cell, (blue) dying cell. (Adapted from N. Jamin *et al. Proc. Natl. Acad. Sci.USA* 95, 4837 (1998).)

proteins and a feature at 1240 cm^{-1} that is due to the phosphodiester (PO_2^-) groups of lipids. The dying cell shows an additional absorption at 1730 cm^{-1}, which is due to the ester carbonyl group from an unidentified compound. From a plot of the intensities of individual absorption features as a function of position in the cell, it has been possible to map the distribution of proteins and lipids during cell division and cell death.

Vibrational microscopy has also been used in biomedical and pharmaceutical laboratories. Examples include the determination of the size and distribution of a drug in a tablet, the observation of conformational changes in proteins of cancerous cells upon administration of anti-tumour drugs, and the measurement of differences between diseased and normal tissue, such as diseased arteries and the white matter from brains of multiple sclerosis patients.

10.15 Symmetry aspects of molecular vibrations

One of the most powerful ways of dealing with normal modes, especially of complex molecules, is to classify them according to their symmetries. Each normal mode must belong to one of the symmetry species of the molecular point group, as discussed in Chapter 7.

Example 10.3 *Identifying the symmetry species of a normal mode*

Establish the symmetry species of the normal mode vibrations of CH_4, which belongs to the group T_d.

Method The first step in the procedure is to identify the symmetry species of the irreducible representations spanned by all the $3N$ displacement coordinates of the atoms, using the characters of the molecular point group. Find these characters by counting 1 if the displacement is unchanged under a symmetry operation, −1 if it changes sign, and 0 if it is changed into some other displacement. Next, subtract the symmetry species of the translations. Translational displacements span the same symmetry species as x, y, and z, so they can be obtained from the right-most column of the character table. Finally, subtract the symmetry species of the rotations, which are also given in the character table (and denoted there by R_x, R_y, or R_z).

Answer There are $3 \times 5 = 15$ degrees of freedom, of which $(3 \times 5) - 6 = 9$ are vibrations. Refer to Fig. 10.50. Under E, no displacement coordinates are changed, so the character is 15. Under C_3, no displacements are left unchanged, so the character is 0. Under the C_2 indicated, the z-displacement of the central atom is left unchanged, whereas its x- and y-components both change sign. Therefore $\chi(C_2) = 1 - 1 - 1 + 0 + 0 + \cdots = -1$. Under the S_4 indicated, the z-displacement of the central atom is reversed, so $\chi(S_4) = -1$. Under σ_d, the x- and z-displacements of C, H_3, and H_4 are left unchanged and their y-displacements are reversed; hence $\chi(\sigma_d) = (1 + 1 - 1) + (1 + 1 - 1) + (1 + 1 - 1) = 3$. The characters are therefore 15, 0, −1, −1, 3. By decomposing the direct product (Section 7.5), we find that this representation spans $A_1 + E + T_1 + 3T_2$. The translations span T_2; the rotations span T_1. Hence, the nine vibrations span $A_1 + E + 2T_2$. The A_1 mode is non-degenerate, the E mode is doubly degenerate, and each T_2 mode is triply degenerate, so accounting for $1 + 2 + 6 = 9$ modes in all. The modes themselves are shown in Fig. 10.51. We shall see that symmetry analysis gives a quick way of deciding which modes are active.

Self-test 10.7 Establish the symmetry species of the normal modes of H_2O. [$2A_1 + B_2$]

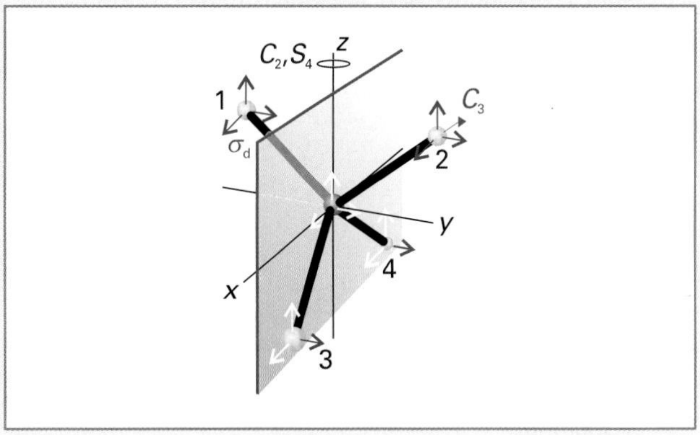

Fig. 10.50 The atomic displacements of CH_4 and the symmetry elements used to calculate the characters.

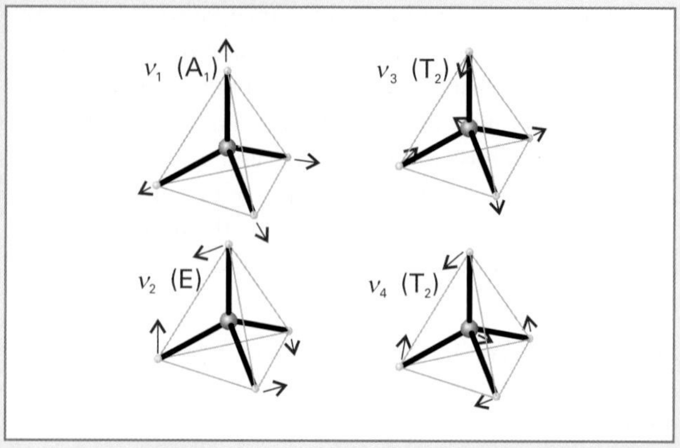

Fig. 10.51 Typical normal modes of vibration of a tetrahedral molecule. There are in fact two modes of symmetry species E and three modes of each T_2 symmetry species.

(a) Infrared activity of normal modes

It is best to use group theory to judge the activities of more complex modes of vibration. This is easily done by checking the character table of the molecular point group for the symmetry species of the irreducible representations spanned by x, y, and z, for their species are also the symmetry species of the components of the electric dipole moment. Then apply the following rule, which is developed in the following *Justification*:

> If the symmetry species of a normal mode is the same as any of the symmetry species of x, y, or z, then the mode is infrared active.

● A BRIEF ILLUSTRATION

We found in Example 10.3 that the symmetry species of the normal modes of CH_4 are $A_1 + E + 2T_2$. The displacements x, y, and z span T_2 in the group T_d. Therefore, only the T_2 modes are infrared active. The distortions accompanying these modes lead to a changing dipole moment. The A_1 mode, which is inactive, is the symmetrical 'breathing' mode of the molecule. ●

Self-test 10.8 Which of the normal modes of H_2O are infrared active? [All three]

Justification 10.3 *Using group theory to identify infrared active normal modes*

The rule hinges on the form of the transition dipole moment between the ground-state vibrational wavefunction, ψ_0, and that of the first excited state, ψ_1. The x-component is

$$\mu_{x,10} = -e\int \psi_1^* x \psi_0 \,d\tau \qquad (10.49)$$

with similar expressions for the two other components of the transition moment. The ground-state vibrational wavefunction is a Gaussian function of the form e^{-x^2}, so it is symmetrical in x. The wavefunction for the first excited state gives a non-vanishing integral only if it is proportional to x, for then the integrand is proportional to x^2 rather than to xy or xz. Consequently, the excited-state wavefunction must have the same symmetry as the displacement x.

(b) Raman activity of normal modes

Group theory provides an explicit recipe for judging the Raman activity of a normal mode. In this case, the symmetry species of the quadratic forms (x^2, xy, etc.) listed in the character table are noted (they transform in the same way as the polarizability), and then we use the following rule:

> If the symmetry species of a normal mode is the same as the symmetry species of a quadratic form, then the mode is Raman active.

● A BRIEF ILLUSTRATION

We established in Example 10.3 that the symmetry species of the normal modes of CH_4 are $A_1 + E + 2T_2$. Because the quadratic forms span $A_1 + E + T_2$ in the group T_d, all the normal modes are Raman active. Only the T_2 modes are both infrared and Raman active. This leaves the A_1 and E modes to be assigned in the Raman spectrum. The A_1 can be identified by noting that it is fully depolarized. Hence, all three modes can be identified. ●

Self-test 10.9 Which of the vibrational modes of H_2O are Raman active? [All three]

Checklist of key ideas

- ☐ 1. A rigid rotor is a body that does not distort under the stress of rotation.
- ☐ 2. The principal axis (figure axis) is the unique axis of a symmetric top. In an oblate top, $I_{\parallel} > I_{\perp}$. In a prolate top, $I_{\parallel} < I_{\perp}$.
- ☐ 3. The centrifugal distortion constant, D_J, takes into account centrifugal distortion.
- ☐ 4. For a molecule to give a pure rotational spectrum, it must be polar. The specific rotational selection rules are $\Delta J = \pm 1$, $\Delta M_J = 0, \pm 1, \Delta K = 0$.

- ☐ 5. For a molecule to give a rotational Raman spectra it must be anisotropically polarizable. The specific selection rules are: (i) linear rotors, $\Delta J = 0, \pm 2$; (ii) symmetric rotors, $\Delta J = 0, \pm 1, \pm 2$; $\Delta K = 0$.
- ☐ 6. The appearance of rotational spectra is affected by nuclear statistics, the selective occupation of rotational states that stems from the Pauli principle.
- ☐ 7. For a molecule to be infrared active its electric dipole moment must change when the atoms are displaced relative to one another. The specific selection rule for a harmonic oscillator is: $\Delta v = \pm 1$.
- ☐ 8. Morse potential energy is a model for the discussion of anharmonic motion, oscillatory motion in which the restoring force is not proportional to the displacement.
- ☐ 9. The P branch of a vibrational transition consists of vibration–rotation infrared transitions with $\Delta J = -1$; the Q branch has transitions with $\Delta J = 0$; the R branch has transitions with $\Delta J = +1$.
- ☐ 10. For a molecule to be vibrationally Raman active, its polarizability must change as it vibrates. The specific selection rule for a harmonic oscillator is $\Delta v = \pm 1$.
- ☐ 11. A normal mode is an independent, synchronous motion of atoms or groups of atoms that may be excited without leading (in the harmonic approximation) to the excitation of any other normal mode. The number of normal modes is $3N - 6$ (for non-linear molecules) or $3N - 5$ (linear molecules).
- ☐ 12. The exclusion rule states that if the molecule has a centre of symmetry, then no modes can be both infrared and Raman active.
- ☐ 13. The depolarization ratio, ρ, is the ratio of the intensities, I, of the scattered light with polarizations perpendicular and parallel to the plane of polarization of the incident radiation, $\rho = I_\perp / I_\parallel$.
- ☐ 14. Resonance Raman spectroscopy is a Raman technique in which the frequency of the incident radiation nearly coincides with the frequency of an electronic transition of the sample.
- ☐ 15. A normal mode is infrared active if its symmetry species is the same as any of the symmetry species of x, y, or z. A normal mode is Raman active if its symmetry species is the same as the symmetry species of a quadratic form.

Further information

Further information 10.1 *The Einstein coefficients*

Stimulated absorption is the transition from a low energy state to one of higher energy that is driven by the electromagnetic field oscillating at the transition frequency. The rate of transition, w, is proportional to $\mathcal{E}^2$, and therefore to the intensity of the incident radiation (because the intensity is proportional to $\mathcal{E}^2$). Therefore, the more intense the incident radiation, the stronger the absorption by the sample (Fig. 10.52). Einstein wrote this transition rate as

$$w = B\rho \qquad (10.50)$$

The constant B (which should not be confused with the rotational constant) is the **Einstein coefficient of stimulated absorption** and $\rho\, d\nu$ is the energy density of radiation in the frequency range ν to $\nu + d\nu$, where

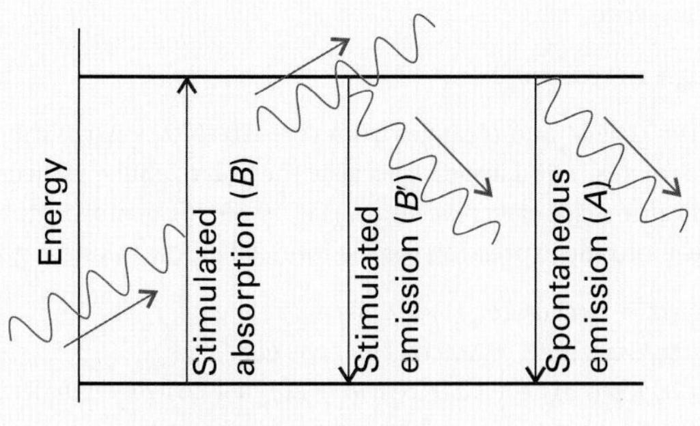

Fig. 10.52 The transitions treated by Einstein in his theory of stimulated and spontaneous processes.

ν is the frequency of the transition. When the molecule is exposed to black-body radiation from a source of temperature T, ρ is given by the *Planck distribution*:

$$\rho = \frac{8\pi h\nu^3/c^3}{e^{h\nu/kT} - 1} \qquad (10.51)$$

For the time being, we can treat B as an empirical parameter that characterizes the transition: if B is large, then a given intensity of incident radiation will induce transitions strongly and the sample will be strongly absorbing. The **total rate of absorption**, W, the number of molecules excited during an interval divided by the duration of the interval, is the transition rate of a single molecule multiplied by the number of molecules N in the lower state: $W = Nw$.

Einstein considered that the radiation was also able to induce the molecule in the upper state to undergo a transition to the lower state, and hence to generate a photon of frequency ν. Thus, he wrote the rate of this stimulated emission as

$$w' = B'\rho \qquad (10.52)$$

where B' is the **Einstein coefficient of stimulated emission.** Note that only radiation of the same frequency as the transition can stimulate an excited state to fall to a lower state. However, he realized that stimulated emission was not the only means by which the excited state could generate radiation and return to the lower state, and suggested that an excited state could undergo **spontaneous emission** at a rate that was independent of the intensity of the radiation (of any frequency) that is already present. Einstein therefore wrote the total rate of transition from the upper to the lower state as

$$w' = A + B'\rho \qquad (10.53)$$

The constant A is the **Einstein coefficient of spontaneous emission**. The overall rate of emission is

$$W' = N'(A + B'\rho) \quad (10.54)$$

where N' is the population of the upper state.

At thermal equilibrium, N and N' do not change over time. This condition is reached when the total rates of emission and absorption are equal, so

$$NB\rho = N'(A + B'\rho)$$

This expression rearranges into

$$\rho = \frac{N'A}{NB - N'B'} = \frac{A/B}{N/N' - B'/B} = \frac{A/B}{e^{h\nu/kT} - B'/B}$$

We have used the Boltzmann expression (*Fundamentals F.5*) for the ratio of populations of states of energies E and E' in the last step:

$$\frac{N'}{N} = e^{-h\nu/kT} \qquad h\nu = E' - E$$

This result has the same form as the Planck distribution, which describes the radiation density at thermal equilibrium. Indeed, when we compare the two expressions for ρ, we can conclude that $B' = B$ and that A is related to B by

$$A = \left(\frac{8\pi h\nu^3}{c^3}\right)B \quad (10.55)$$

Spontaneous emission can be largely ignored at the relatively low frequencies of rotational and vibrational transitions, and the intensities of these transitions can be discussed in terms of stimulated emission and absorption.

Further information 10.2 *Selection rules for rotational and vibrational spectroscopy*

The Born–Oppenheimer approximation allows us to write the internal wavefunction of a molecule as the product of an electronic part, with the label ε, a vibrational part, with quantum number v, and a rotational part, which for a diatomic molecule can be represented by the spherical harmonics $Y_{J,M_J}(\theta,\phi)$ (Section 3.4). To simplify the form of the integrals that will soon follow, we are using the Dirac bracket notation introduced in *Further information 1.1*. The transition dipole moment for a spectroscopic transition is

$$\boldsymbol{\mu}_{fi} = \langle \varepsilon_f v_f J_f M_{J,f} | \hat{\boldsymbol{\mu}} | \varepsilon_i v_i J_i M_{J,i} \rangle \quad (10.56)$$

and our task is to explore conditions for which this integral vanishes or has a nonzero value.

(a) Microwave spectra

During a pure rotational transition the molecule does not change electronic or vibrational states, so that $\langle \varepsilon_f v_f | = \langle \varepsilon_i v_i | = \langle \varepsilon v |$ and we identify $\boldsymbol{\mu}_{\varepsilon v} = \langle \varepsilon v | \hat{\boldsymbol{\mu}} | \varepsilon v \rangle$ with the *permanent* electric dipole moment of the molecule in the state εv. Equation 10.56 becomes

$$\boldsymbol{\mu}_{fi} = \langle J_f M_{J,f} | \hat{\boldsymbol{\mu}}_{\varepsilon v} | J_i M_{J,i} \rangle \quad (10.57)$$

The electric dipole moment has components $\mu_{\varepsilon v,x}$, $\mu_{\varepsilon v,y}$, and $\mu_{\varepsilon v,z}$, which, in spherical polar coordinates, are written in terms of μ_0, the magnitude of the permanent electric dipole moment vector of the molecule, and the angles θ and ϕ as

$$\mu_{\varepsilon v,x} = \mu_0 \sin\theta\cos\phi \qquad \mu_{\varepsilon v,y} = \mu_0 \sin\theta\sin\phi \qquad \mu_{\varepsilon v,z} = \mu_0\cos\theta \quad (10.58)$$

Here, we have taken the z-axis to be coincident with the figure axis. The transition dipole moment has three components, given by:

$$\begin{aligned} \mu_{fi,x} &= \mu_0 \langle J_f M_{J,f} | \sin\theta\cos\phi | J_i M_{J,i} \rangle \\ \mu_{fi,y} &= \mu_0 \langle J_f M_{J,f} | \sin\theta\sin\phi | J_i M_{J,i} \rangle \\ \mu_{fi,z} &= \mu_0 \langle J_f M_{J,f} | \cos\theta | J_i M_{J,i} \rangle \end{aligned} \quad (10.59)$$

We see immediately that the molecule must have a permanent dipole moment in order to have a microwave spectrum. This is the gross selection rule for microwave spectroscopy.

For the specific selection rules we need to examine the conditions for which the integrals do not vanish, and we must consider each component. For the z-component, we simplify the integral by using $\cos\theta \propto Y_{1,0}$ (Table 3.2). It follows that

$$\mu_{fi,z} \propto \langle J_f M_{J,f} | Y_{1,0} | J_i M_{J,i} \rangle \quad (10.60)$$

As pointed out in *Justification 4.4*, an important 'triple integral' involving the spherical harmonics is

$$\int_0^{\pi}\int_0^{2\pi} Y_{l'',m_l''}(\theta,\phi)^* Y_{l',m_l'}(\theta,\phi) Y_{l,m_l}(\theta,\phi)\sin\theta \, d\theta \, d\phi = 0 \quad (10.61)$$

unless $m_l'' = m_l' + m_l$ and lines of length l'', l', and l can form a triangle. Therefore, the transition moment vanishes unless $J_f - J_i = \pm 1$ and $M_{J,f} - M_{J,i} = 0$. These are two of the selection rules stated in Section 10.3.

For the x- and y-components, we use $\cos\phi = \frac{1}{2}(e^{i\phi} + e^{-i\phi})$ and $\sin\phi = -\frac{1}{2}i(e^{i\phi} - e^{-i\phi})$ to write $\sin\theta\cos\phi \propto Y_{1,1} + Y_{1,-1}$ and $\sin\theta\sin\phi \propto Y_{1,1} - Y_{1,-1}$. It follows that

$$\mu_{fi,x} \propto \langle J_f M_{J,f} | Y_{1,1} + Y_{1,-1} | J_i M_{J,i} \rangle \qquad \mu_{fi,y} \propto \langle J_f M_{J,f} | Y_{1,1} - Y_{1,-1} | J_i M_{J,i} \rangle \quad (10.62)$$

According to the properties of the spherical harmonics, these integrals vanish unless $J_f - J_i = \pm 1$ and $M_{J,f} - M_{J,i} = \pm 1$. This completes the selection rules of Section 10.3.

(b) Rotational Raman spectra

We shall develop the origin of the gross and specific selection rules for rotational Raman spectroscopy by using a diatomic molecule as an example. The incident electric field of a wave of electromagnetic radiation of frequency ω_i induces a molecular dipole moment that is given by

$$\mu_{ind} = \alpha\mathcal{E}(t) = \alpha\mathcal{E}\cos\omega_i t \quad (10.63)$$

If the molecule is rotating at a circular frequency ω_R, to an external observer its polarizability is also time dependent (if it is anisotropic), and we can write

$$\alpha = \alpha_0 + \Delta\alpha\cos 2\omega_R t \quad (10.64)$$

where $\Delta\alpha = \alpha_{\parallel} - \alpha_{\perp}$ and α ranges from $\alpha_0 + \Delta\alpha$ to $\alpha_0 - \Delta\alpha$ as the molecule rotates. The 2 appears because the polarizability returns to its initial value twice each revolution (Fig 10.53). Substituting this expression into the expression for the induced dipole moment gives

$$\begin{aligned} \mu_{ind} &= (\alpha_0 + \Delta\alpha\cos 2\omega_R t) \times (\mathcal{E}\cos\omega_i t) \\ &= \alpha_0\mathcal{E}\cos\omega_i t + \mathcal{E}\Delta\alpha\cos 2\omega_R t\cos\omega_i t \\ &= \alpha_0\mathcal{E}\cos\omega_i t + \tfrac{1}{2}\mathcal{E}\Delta\alpha\{\cos(\omega_i + 2\omega_R)t + \cos(\omega_i - 2\omega_R)t\} \end{aligned} \quad (10.65)$$

This calculation shows that the induced dipole has a component oscillating at the incident frequency (which generates Rayleigh radiation), and that it also has two components at $\omega_i \pm 2\omega_R$, which

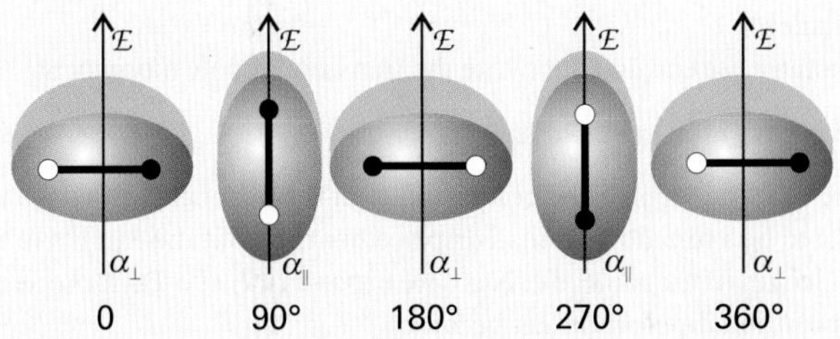

Fig. 10.53 The distortion induced in a molecule by an applied electric field returns to its initial value after a rotation of only 180° (that is, twice a revolution). This is the origin of the $\Delta J = \pm 2$ selection rule in rotational Raman spectroscopy.

give rise to the shifted Raman lines. These lines appear only if $\Delta\alpha \neq 0$; hence the polarizability must be anisotropic for there to be Raman lines. This is the gross selection rule for rotational Raman spectroscopy. We also see that the distortion induced in the molecule by the incident electric field returns to its initial value after a rotation of 180° (that is, twice a revolution). This is the origin of the specific selection rule $\Delta J = \pm 2$.

We now use a quantum mechanical formalism to understand the selection rules. First, we write the x-, y-, and z-components of the induced dipole moment as

$$\mu_{\text{ind},x} = \mu_x \sin\theta\cos\phi \quad \mu_{\text{ind},y} = \mu_y \sin\theta\sin\phi \quad \mu_{\text{ind},z} = \mu_z\cos\theta \quad (10.66a)$$

where μ_x, μ_y, and μ_z are the components of the electric dipole moment of the molecule and the z-axis is coincident with the molecular figure axis. The incident electric field also has components along the x-, y-, and z-axes:

$$\mathcal{E}_x = \mathcal{E}\sin\theta\cos\phi \qquad \mathcal{E}_y = \mathcal{E}\sin\theta\sin\phi \qquad \mathcal{E}_z = \mathcal{E}\cos\theta \qquad (10.66b)$$

From eqn 10.65 and the preceding equations, it follows that

$$\begin{aligned}\mu_{\text{ind}} &= \alpha_\perp \mathcal{E}_x \sin\theta\cos\phi + \alpha_\perp \mathcal{E}_y \sin\theta\sin\phi + \alpha_\parallel \mathcal{E}_z\cos\theta \\ &= \alpha_\perp \mathcal{E}\sin^2\theta + \alpha_\parallel \mathcal{E}\cos^2\theta \end{aligned} \qquad (10.67)$$

By using the spherical harmonic $Y_{2,0}(\theta,\phi) = (5/16\pi)^{1/2}(3\cos^2\theta - 1)$ and the relation $\sin^2\theta = 1 - \cos^2\theta$, it follows that:

$$\mu_{\text{ind}} = \left\{\tfrac{1}{3}\alpha_\parallel + \tfrac{2}{3}\alpha_\perp + \tfrac{4}{3}\left(\frac{\pi}{5}\right)^{1/2}\Delta\alpha Y_{2,0}(\theta,\phi)\right\}\mathcal{E} \qquad (10.68)$$

For a transition between two rotational states, we calculate the integral $\langle J_f M_{J,f}|\mu_{\text{ind}}|J_i M_{J,i}\rangle$, which has two components:

$$(\tfrac{1}{3}\alpha_\parallel + \tfrac{2}{3}\alpha_\perp)\langle J_f M_{J,f}|J_i M_{J,i}\rangle \qquad \text{and} \qquad \mathcal{E}\Delta\alpha\langle J_f M_{J,f}|Y_{2,0}|J_i M_{J,i}\rangle$$

According to the integral in eqn 10.61, the first integral vanishes unless $J_f - J_i = 0$ and the second integral vanishes unless $J_f - J_i = \pm 2$ and $\Delta\alpha \neq 0$. These are the gross and specific selection rules for linear rotors.

(c) Infrared spectra

The gross selection rule for infrared spectroscopy is based on an analysis of the transition dipole moment $\langle v_f|\hat{\mu}|v_i\rangle$, which arises from eqn 10.56 when the molecule does not change electronic or rotational states. For simplicity, we shall consider a one-dimensional oscillator (like a diatomic molecule). The electric dipole moment operator depends on the location of all the electrons and all the nuclei in the molecule, so it varies as the internuclear separation changes (Fig 10.54). If we think of the dipole moment as arising from two partial charges $\pm\delta q$ separated by a distance $R = R_e + x$, we can write its variation with displacement from the equilibrium separation, x, as

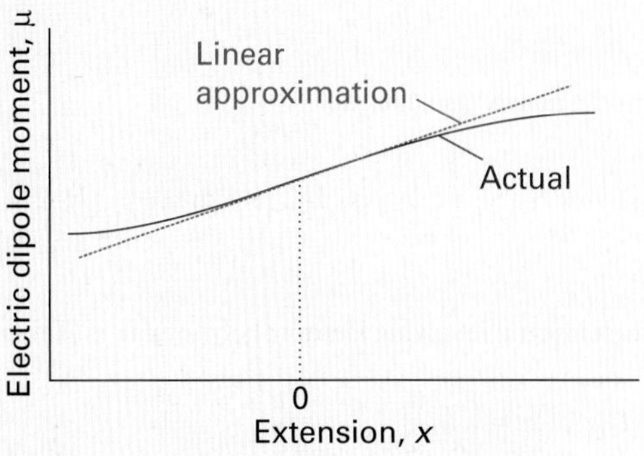

Fig. 10.54 The electric dipole moment of a heteronuclear diatomic molecule varies as shown by the purple curve. For small displacements the change in dipole moment is proportional to the displacement.

$$\hat{\mu} = R\delta q = R_e\delta q + x\delta q = \mu_0 + x\delta q \qquad (10.69)$$

where μ_0 is the electric dipole moment operator when the nuclei have their equilibrium separation. It then follows that, with $f \neq i$,

$$\langle v_f|\hat{\mu}|v_i\rangle = \mu_0\langle v_f|v_i\rangle + \delta q\langle v_f|x|v_i\rangle \qquad (10.70a)$$

The term proportional to μ_0 is zero because the states with different values of v are orthogonal. It follows that the transition dipole moment is

$$\langle v_f|\hat{\mu}|v_i\rangle = \langle v_f|x|v_i\rangle\delta q \qquad (10.70b)$$

Because (from eqn 10.69)

$$q = \frac{d\mu}{dx}$$

we can write the transition dipole moment more generally as

$$\langle v_f|\hat{\mu}|v_i\rangle = \langle v_f|x|v_i\rangle\left(\frac{d\mu}{dx}\right) \qquad (10.71)$$

and we see that the right-hand side is zero unless the dipole moment varies with displacement (so that $d\mu/dx$ is nonzero). This is the gross selection rule for infrared spectroscopy.

The specific selection rule is determined by considering the value of $\langle v_f|x|v_i\rangle$. We need to write out the wavefunctions in terms of the Hermite polynomials given in Section 2.5 and then to use their properties (Example 2.6 should be reviewed, for it gives further details of the calculation). We note that $x = \alpha y$ with $\alpha = (\hbar^2/m_{\text{eff}}k)^{1/4}$ (eqn 2.24; note that in this context α is not the polarizability). Then we write

$$\langle v_f|x|v_i\rangle = N_{v_f}N_{v_i}\int_{-\infty}^{\infty} H_{v_f}xH_{v_i}e^{-y^2}dx = \alpha^2 N_{v_f}N_{v_i}\int_{-\infty}^{\infty} H_{v_f}yH_{v_i}e^{-y^2}dy \qquad (10.72)$$

To evaluate the integral we use the recursion relation (Table 2.1)

$$yH_v = vH_{v-1} + \tfrac{1}{2}H_{v+1}$$

which turns the matrix element into

$$\langle v_f|x|v_i\rangle = \alpha^2 N_{v_f} N_{v_i}\left\{v_i\int_{-\infty}^{\infty} H_{v_f}H_{v_i-1}e^{-y^2}dy + \tfrac{1}{2}\int_{-\infty}^{\infty} H_{v_f}H_{v_i+1}e^{-y^2}dy\right\} \qquad (10.73)$$

An important integral involving Hermite polynomials is (Table 2.1)

$$\int_{-\infty}^{\infty} H_{v'}H_v e^{-y^2}dy = \begin{cases} 0 & \text{if } v' \neq v \\ \pi^{1/2}2^v v! & \text{if } v' = v \end{cases}$$

It follows that the first integral is zero unless $v_f = v_i - 1$ and the second is zero unless $v_f = v_i + 1$. It follows that the transition dipole moment is zero unless $\Delta v = \pm 1$.

(d) Vibrational Raman spectra

The gross selection rule for vibrational Raman spectroscopy is based on an analysis of the transition dipole moment $\langle \varepsilon v_f|\hat{\mu}|\varepsilon v_i\rangle$, which is written from eqn 10.56 by using the Born–Oppenheimer approximation and neglecting the effect of rotation and electron spin. For simplicity, we consider a one-dimensional harmonic oscillator (like a diatomic molecule).

First, we use eqn 10.63 to write the transition dipole moment as

$$\mu_{fi} = \langle \varepsilon v_f|\hat{\mu}|\varepsilon v_i\rangle = \langle \varepsilon v_f|\alpha|\varepsilon v_i\rangle\mathcal{E} = \langle v_f|\alpha(x)|v_i\rangle\mathcal{E} \qquad (10.74)$$

where $\alpha(x) = \langle \varepsilon|\alpha|\varepsilon\rangle$ is the polarizability of the molecule, which we expect to be a function of small displacements x from the equilibrium bond length of the molecule. Next, we expand $\alpha(x)$ as a Taylor series, so the transition dipole moment becomes

$$\mu_{fi} = \left\langle v_f\left|\alpha(0) + \left(\frac{d\alpha}{dx}\right)_0 x + \cdots\right|v_i\right\rangle\mathcal{E}$$

$$= \langle v_f|v_i\rangle\alpha(0)\mathcal{E} + \left(\frac{d\alpha}{dx}\right)_0 \langle v_f|x|v_i\rangle\mathcal{E} + \cdots \qquad (10.75)$$

The term containing $\langle v_f|v_i\rangle$ vanishes for $f \neq i$ because the harmonic oscillator wavefunctions are orthogonal. Therefore, the vibration is Raman active if $(d\alpha/dx)_0 \neq 0$ and $\langle v_f|x|v_i\rangle \neq 0$. Therefore, the polarizability of the molecule must change during the vibration; this is the gross selection rule of Raman spectroscopy. Also, we already know that $\langle v_f|x|v_i\rangle \neq 0$ if $v_f - v_i = \pm 1$; this is the specific selection rule of Raman spectroscopy.

Discussion questions

10.1 Account for the rotational degeneracy of the various types of rigid rotor. Would their lack of rigidity affect your conclusions?

10.2 Discuss the differences between an oblate and a prolate symmetric rotor and give several examples of each.

10.3 Discuss the origin of the Stark effect. What are some applications of the Stark effect?

10.4 Discuss the physical origins of the gross selection rules for microwave and infrared spectroscopy.

10.5 Discuss the physical origins of the gross selection rules for rotational and vibrational Raman spectroscopy.

10.6 Account for the existence of rotational zero-point energy in molecular hydrogen.

10.7 Discuss the role of nuclear statistics in the occupation of energy levels in $^1H^{12}C{\equiv}^{12}C^1H$, $^1H^{13}C{\equiv}^{13}C^1H$, and $^2H^{12}C{\equiv}^{12}C^2H$. For nuclear spin data, see Table 12.2.

10.8 Discuss the strengths and limitations of the parabolic and Morse functions as descriptors of the potential energy curve of a diatomic molecule.

10.9 Discuss the effect of vibrational excitation on the rotational constant of a diatomic molecule.

10.10 How is the method of combination differences used in rotation–vibration spectroscopy to determine rotational constants?

10.11 Suppose that you wish to characterize the normal modes of benzene in the gas phase. Why is it important to obtain both infrared absorption and Raman spectra of your sample?

Exercises

The masses of nuclides are listed in the tables at the start of the *Data section.*

10.1(a) Calculate the moment of inertia around the C_2 axis (the bisector of the OOO angle) and the corresponding rotational constant of a $^{16}O_3$ molecule (bond angle 117°; OO bond length 128 pm).

10.1(b) Calculate the moment of inertia around the C_3 axis (the threefold symmetry axis) and the corresponding rotational constant of a $^{31}P^1H_3$ molecule (bond angle 93.5°; PH bond length 142 pm).

10.2(a) Plot the expressions for the two moments of inertia of a symmetric top version of an AB_4 molecule (Table 10.1) with equal bond lengths but with the angle θ increasing from 90° to the tetrahedral angle.

10.2(b) Plot the expressions for the two moments of inertia of a symmetric top version of an AB_4 molecule (Table 10.1) with θ equal to the tetrahedral angle but with one A—B bond varying. *Hint.* Write $\rho = R'_{AB}/R_{AB}$, and allow ρ to vary from 2 to 1.

10.3(a) Classify the following rotors: (a) O_3, (b) CH_3CH_3, (c) XeO_4, (d) $FeCp_2$ (Cp denotes the cyclopentadienyl group, C_5H_5).

10.3(b) Classify the following rotors: (a) $CH_2{=}CH_2$, (b) SO_3, (c) ClF_3, (d) N_2O.

10.4(a) Calculate the frequency and wavenumber of the $J = 3 \leftarrow 2$ transition in the pure rotational spectrum of $^{14}N^{16}O$. The equilibrium bond length is 115 pm. Does the frequency increase or decrease if centrifugal distortion is considered?

10.4(b) Calculate the frequency and wavenumber of the $J = 2 \leftarrow 1$ transition in the pure rotational spectrum of $^{12}C^{16}O$. The equilibrium bond length is 112.81 pm. Does the frequency increase or decrease if centrifugal distortion is considered?

10.5(a) The wavenumber of the $J = 3 \leftarrow 2$ rotational transition of $^1H^{35}Cl$ considered as a rigid rotor is 63.56 cm^{-1}; what is the H—Cl bond length?

10.5(b) The wavenumber of the $J = 1 \leftarrow 0$ rotational transition of $^1H^{81}Br$ considered as a rigid rotor is 16.93 cm^{-1}; what is the H—Br bond length?

10.6(a) The spacing of lines in the microwave spectrum of $^{27}Al^1H$ is 12.604 cm^{-1}; calculate the moment of inertia and bond length of the molecule.

10.6(b) The spacing of lines in the microwave spectrum of $^{35}Cl^{19}F$ is 1.033 cm^{-1}; calculate the moment of inertia and bond length of the molecule.

10.7(a) Determine the HC and CN bond lengths in HCN from the rotational constants $B(^1H^{12}C^{14}N) = 44.316$ GHz and $B(^2H^{12}C^{14}N) = 36.208$ GHz.

10.7(b) Determine the CO and CS bond lengths in OCS from the rotational constants $B(^{16}O^{12}C^{32}S) = 6081.5$ MHz, $B(^{16}O^{12}C^{34}S) = 5932.8$ MHz.

10.8(a) The wavenumber of the incident radiation in a Raman spectrometer is 20 487 cm^{-1}. What is the wavenumber of the scattered Stokes radiation for the $J = 2 \leftarrow 0$ transition of $^{14}N_2$?

10.8(b) The wavenumber of the incident radiation in a Raman spectrometer is 20 623 cm^{-1}. What is the wavenumber of the scattered Stokes radiation for the $J = 4 \leftarrow 2$ transition of $^{16}O_2$?

10.9(a) The rotational Raman spectrum of $^{35}Cl_2$ shows a series of Stokes lines separated by 0.9752 cm^{-1} and a similar series of anti-Stokes lines. Calculate the bond length of the molecule.

10.9(b) The rotational Raman spectrum of $^{19}F_2$ shows a series of Stokes lines separated by 3.5312 cm^{-1} and a similar series of anti-Stokes lines. Calculate the bond length of the molecule.

10.10(a) Which of the following molecules may show a pure rotational microwave absorption spectrum: (a) H_2, (b) HCl, (c) CH_4, (d) CH_3Cl, (e) CH_2Cl_2?

10.10(b) Which of the following molecules may show a pure rotational microwave absorption spectrum: (a) H_2O, (b) H_2O_2, (c) NH_3, (d) N_2O?

10.11(a) Which of the following molecules may show a pure rotational Raman spectrum: (a) H_2, (b) HCl, (c) CH_4, (d) CH_3Cl?

10.11(b) Which of the following molecules may show a pure rotational Raman spectrum: (a) CH_2Cl_2, (b) CH_3CH_3, (c) SF_6, (d) N_2O?

10.12(a) What is the ratio of weights of populations due to the effects of nuclear statistics for $^{35}Cl_2$?

10.12(b) What is the ratio of weights of populations due to the effects of nuclear statistics for $^{12}C^{32}S_2$? What effect would be observed when ^{12}C is replaced by ^{13}C? For nuclear spin data, see Table 12.2.

10.13(a) An object of mass 100 g suspended from the end of a rubber band has a vibrational frequency of 2.0 Hz. Calculate the force constant of the rubber band.

10.13(b) An object of mass 1.0 g suspended from the end of a spring has a vibrational frequency of 10.0 Hz. Calculate the force constant of the spring.

10.14(a) Calculate the percentage difference in the fundamental vibrational wavenumbers of $^{23}Na^{35}Cl$ and $^{23}Na^{37}Cl$ on the assumption that their force constants are the same.

10.14(b) Calculate the percentage difference in the fundamental vibrational wavenumbers of $^1H^{35}Cl$ and $^2H^{37}Cl$ on the assumption that their force constants are the same.

10.15(a) The wavenumber of the fundamental vibrational transition of $^{35}Cl_2$ is 564.9 cm^{-1}. Calculate the force constant of the bond.

10.15(b) The wavenumber of the fundamental vibrational transition of $^{79}Br^{81}Br$ is 323.2 cm^{-1}. Calculate the force constant of the bond.

10.16(a) The hydrogen halides have the following fundamental vibrational wavenumbers: 4141.3 cm^{-1} ($^1H^{19}F$); 2988.9 cm^{-1} ($^1H^{35}Cl$); 2649.7 cm^{-1} ($^1H^{81}Br$); 2309.5 cm^{-1} ($^1H^{127}I$). Calculate the force constants of the hydrogen–halogen bonds.

10.16(b) From the data in Exercise 10.16a, predict the fundamental vibrational wavenumbers of the deuterium halides.

10.17(a) For $^{16}O_2$, $\Delta\tilde{G}$ values for the transitions $v = 1 \leftarrow 0$, $2 \leftarrow 0$, and $3 \leftarrow 0$ are, respectively, 1556.22, 3088.28, and 4596.21 cm^{-1}. Calculate $\tilde{\nu}$ and x_e. Assume y_e to be zero.

10.17(b) For $^{14}N_2$, $\Delta\tilde{G}$ values for the transitions $v = 1 \leftarrow 0$, $2 \leftarrow 0$, and $3 \leftarrow 0$ are, respectively, 2345.15, 4661.40, and 6983.73 cm^{-1}. Calculate $\tilde{\nu}$ and x_e. Assume y_e to be zero.

10.18(a) Infrared absorption by $^1H^{81}Br$ gives rise to an R branch from $v = 0$. What is the wavenumber of the line originating from the rotational state with $J = 2$? Use the information in Table 10.2.

10.18(b) Infrared absorption by $^1H^{127}I$ gives rise to an R branch from $v = 0$. What is the wavenumber of the line originating from the rotational state with $J = 2$? Use the information in Table 10.2.

10.19(a) Which of the following molecules may show infrared absorption spectra: (a) H_2, (b) HCl, (c) CO_2, (d) H_2O?

10.19(b) Which of the following molecules may show infrared absorption spectra: (a) CH_3CH_3, (b) CH_4, (c) CH_3Cl, (d) N_2?

10.20(a) How many normal modes of vibration are there for the following molecules: (a) H_2O, (b) H_2O_2, (c) C_2H_4?

10.20(b) How many normal modes of vibration are there for the following molecules: (a) C_6H_6, (b) $C_6H_5CH_3$, (c) HC≡C—C≡C—H?

10.21(a) How many vibrational modes are there for the molecule NC—(C≡C—C≡C—)$_{10}$CN detected in an interstellar cloud?

10.21(b) How many vibrational modes are there for the molecule NC—(C≡C—C≡C—)$_8$CN detected in an interstellar cloud?

10.22(a) Write an expression for the vibrational term for the ground vibrational state of H_2O in terms of the wavenumbers of the normal modes. Neglect anharmonicities as in eqn 10.46.

10.22(b) Write an expression for the vibrational term for the ground vibrational state of SO_2 in terms of the wavenumbers of the normal modes. Neglect anharmonicities as in eqn 10.46.

10.23(a) Which of the three vibrations of an AB_2 molecule are infrared or Raman active when it is (a) angular, (b) linear?

10.23(b) Which of the vibrations of an AB_3 molecule are infrared or Raman active when it is (a) trigonal planar, (b) trigonal pyramidal?

10.24(a) Consider the vibrational mode that corresponds to the uniform expansion of the benzene ring. Is it (a) Raman, (b) infrared active?

10.24(b) Consider the vibrational mode that corresponds to the boat-like bending of a benzene ring. Is it (a) Raman, (b) infrared active?

10.25(a) The molecule CH_2Cl_2 belongs to the point group C_{2v}. The displacements of the atoms span $5A_1 + 2A_2 + 4B_1 + 4B_2$. What are the symmetries of the normal modes of vibration?

10.25(b) A carbon disulfide molecule belongs to the point group $D_{\infty h}$. The nine displacements of the three atoms span $A_{1g} + A_{1u} + A_{2g} + 2E_{1u} + E_{1g}$. What are the symmetries of the normal modes of vibration?

10.26(a) Which of the normal modes of CH_2Cl_2 (Exercise 10.25a) are infrared active? Which are Raman active?

10.26(b) Which of the normal modes of carbon disulfide (Exercise 10.25b) are infrared active? Which are Raman active?

10.27(a) Calculate the ratio of the Einstein coefficients of spontaneous and stimulated emission, *A* and *B*, for transitions with the following characteristics: (a) 70.8 pm X-rays, (b) 500 nm visible light, (c) 3000 cm^{-1} infrared radiation.

10.27(b) Calculate the ratio of the Einstein coefficients of spontaneous and stimulated emission, *A* and *B*, for transitions with the following characteristics: (a) 500 MHz radiofrequency radiation, (e) 3.0 cm microwave radiation.

Problems*

Numerical problems

10.1 The rotational constant of NH_3 is 298 GHz. Compute the separation of the pure rotational spectrum lines in GHz, cm^{-1}, and mm, and show that the value of *B* is consistent with an N—H bond length of 101.4 pm and a bond angle of 106.78°.

10.2 The rotational constant for CO is 1.9314 cm^{-1} and 1.6116 cm^{-1} in the ground and first excited vibrational states, respectively. By how much does the internuclear distance change as a result of this transition?

10.3 Pure rotational Raman spectra of gaseous C_6H_6 and C_6D_6 yield the following rotational constants: $\tilde{B}(C_6H_6) = 0.18960\ cm^{-1}$, $\tilde{B}(C_6D_6) = 0.15681\ cm^{-1}$. The moments of inertia of the molecules about any axis perpendicular to the C_6 axis were calculated from these data as $I(C_6H_6) = 1.4759 \times 10^{-45}\ kg\ m^2$, $I(C_6D_6) = 1.7845 \times 10^{-45}\ kg\ m^2$. Calculate the CC, CH, and CD bond lengths.

10.4 Rotational absorption lines from $^1H^{35}Cl$ gas were found at the following wavenumbers (R.L. Hausler and R.A. Oetjen, *J. Chem. Phys.* **21**, 1340 (1953)): 83.32, 104.13, 124.73, 145.37, 165.89, 186.23, 206.60, 226.86 cm^{-1}. Calculate the moment of inertia and the bond length of the molecule. Predict the positions of the corresponding lines in $^2H^{35}Cl$.

10.5 Is the bond length in HCl the same as that in DCl? The wavenumbers of the $J = 1 \leftarrow 0$ rotational transitions for $H^{35}Cl$ and $^2H^{35}Cl$ are 20.8784 and 10.7840 cm^{-1}, respectively. Accurate atomic masses are $1.007\ 825m_u$ and $2.0140m_u$ for ^1H and ^2H, respectively. The mass of ^{35}Cl is $34.96885m_u$. Based on this information alone, can you conclude that the bond lengths are the same or different in the two molecules?

10.6 Thermodynamic considerations suggest that the copper monohalides CuX should exist mainly as polymers in the gas phase, and indeed it proved difficult to obtain the monomers in sufficient abundance to detect spectroscopically. This problem was overcome by flowing the halogen gas over copper heated to 1100 K (Manson *et al.* *J. Chem. Phys.* **63**, 2724 (1975)). For CuBr the $J = 13$–14, 14–15, and 15–16 transitions occurred at 84 421.34, 90 449.25, and 96 476.72 MHz, respectively. Calculate the rotational constant and bond length of CuBr.

10.7 The microwave spectrum of $^{16}O^{12}CS$ gave absorption lines (in GHz) as follows:

J	1	2	3	4
^{32}S	24.325 92	36.488 82	48.651 64	60.814 08
^{34}S	23.732 33		47.46240	

Using the expressions for moments of inertia in Table 10.1 and assuming that the bond lengths are unchanged by substitution, calculate the CO and CS bond lengths in OCS.

10.8 The average spacing between the rotational lines of the P and R branches of $^{12}C_2{}^1H_2$ and $^{12}C_2{}^2H_2$ is 2.352 cm^{-1} and 1.696 cm^{-1}, respectively. Estimate the CC and CH bond lengths.

10.9 Absorptions in the $v = 1 \leftarrow 0$ vibration–rotation spectrum of $^1H^{35}Cl$ were observed at the following wavenumbers (in cm^{-1}):

2998.05	2981.05	2963.35	2944.99	2925.92
2906.25	2865.14	2843.63	2821.59	2799.00

Assign the rotational quantum numbers and use the method of combination differences to determine the rotational constants of the two vibrational levels.

10.10 Equation 10.17b may be rearranged into

$$\tilde{\nu}(J+1 \leftarrow J)\,/\,2(J+1) = \tilde{B} - 2\tilde{D}_J(J+1)^2$$

which is the equation of a straight line when the left-hand side is plotted against $(J+1)^2$. The following wavenumbers of transitions (in cm^{-1}) were observed for $^{12}C^{16}O$:

J:	0	1	2	3	4
	3.845 033	7.689 919	11.534 510	15.378 662	19.222 223

Determine $\tilde{B}$, $\tilde{D}_J$, and the equilibrium bond length of CO.

10.11‡ In a study of the rotational spectrum of the linear FeCO radical, Tanaka *et al.* (*J. Chem. Phys.* **106**, 6820 (1997)) report the following $J+1 \leftarrow J$ transitions:

J	24	25	26	27	28	29
$\tilde{\nu}/m^{-1}$	214 777.7	223 379.0	231 981.2	240 584.4	249 188.5	257 793.5

Evaluate the rotational constant of the molecule. Also, estimate the value of *J* for the most highly populated rotational energy level at 298 K and at 100 K.

* Problems denoted with the symbol ‡ were supplied by Charles Trapp, Carmen Giunta, and Marshall Cady.

10.12 The vibrational levels of NaI lie at the wavenumbers 142.81, 427.31, 710.31, and 991.81 cm^{-1}. Show that they fit the expression $(v+\frac{1}{2})\tilde{\nu}-(v+\frac{1}{2})^2x_e\tilde{\nu}$, and deduce the force constant, zero-point energy, and dissociation energy of the molecule.

10.13 The rotational constant for a diatomic molecule in the vibrational state v typically fits the expression $\tilde{B}_v = \tilde{B}_e - a(v+\frac{1}{2})$. For the interhalogen molecule IF it is found that $\tilde{B}_e = 0.279\ 71\ cm^{-1}$ and $a = 0.187\ m^{-1}$ (note the change of units). Calculate $\tilde{B}_0$ and $\tilde{B}_1$ and use these values to calculate the wavenumbers of the $J' \rightarrow 3$ transitions of the P and R branches. You will need the following additional information: $\tilde{\nu} = 610.258\ cm^{-1}$ and $x_e\tilde{\nu} = 3.141\ cm^{-1}$. Estimate the dissociation energy of the IF molecule.

10.14 Predict the shape of the nitronium ion, NO_2^+, from its Lewis structure and the VSEPR model. It has one Raman active vibrational mode at 1400 cm^{-1}, two strong IR active modes at 2360 and 540 cm^{-1}, and one weak IR mode at 3735 cm^{-1}. Are these data consistent with the predicted shape of the molecule? Assign the vibrational wavenumbers to the modes from which they arise.

10.15 Provided higher order terms are neglected, eqn 10.38 for the vibrational wavenumbers of an anharmonic oscillator, $\Delta\tilde{G}(v) = \tilde{\nu} - 2(v+1)x_e\tilde{\nu}$, is the equation of a straight line when the left-hand side is plotted against $v+1$. Use the following data on CO to determine the values of $\tilde{\nu}$ and $x_e\tilde{\nu}$ for CO:

v	0	1	2	3	4
$\Delta\tilde{G}(v)$ / cm^{-1}	2143.1	2116.1	2088.9	2061.3	2033.5

10.16 At low resolution, the strongest absorption band in the infrared absorption spectrum of $^{12}C^{16}O$ is centred at 2150 cm^{-1}. Upon closer examination at higher resolution, this band is observed to be split into two sets of closely spaced peaks, one on each side of the centre of the spectrum at 2143.26 cm^{-1}. The separation between the peaks immediately to the right and left of the centre is 7.655 cm^{-1}. Make the harmonic oscillator and rigid rotor approximations and calculate from these data: (a) the vibrational wavenumber of a CO molecule, (b) its molar zero-point vibrational energy, (c) the force constant of the CO bond, (d) the rotational constant $\tilde{B}$, and (e) the bond length of CO.

10.17 The HCl molecule is quite well described by the Morse potential with $D_e = 5.33$ eV, $\tilde{\nu} = 2989.7\ cm^{-1}$, and $x_e\tilde{\nu} = 52.05\ cm^{-1}$. Assuming that the potential is unchanged on deuteration, predict the dissociation energies (D_0) of (a) HCl, (b) DCl.

10.18 The Morse potential (eqn 10.35) is very useful as a simple representation of the actual molecular potential energy. When RbH was studied, it was found that $\tilde{\nu} = 936.8\ cm^{-1}$ and $x\tilde{\nu} = 14.15\ cm^{-1}$. Plot the potential energy curve from 50 pm to 800 pm around $R_e = 236.7$ pm. Then go on to explore how the rotation of a molecule may weaken its bond by allowing for the kinetic energy of rotation of a molecule and plotting $V^* = V + hc\tilde{B}J(J+1)$ with $\tilde{B} = \hbar / 4\pi c\mu R^2$. Plot these curves on the same diagram for $J = 40$, 80, and 100, and observe how the dissociation energy is affected by the rotation. (Taking $\tilde{B} = 3.020\ cm^{-1}$ at the equilibrium bond length will greatly simplify the calculation.)

10.19‡ Luo *et al.* (*J. Chem. Phys.* **98**, 3564 (1993)) reported experimental observation of the He_2 complex, a species that had escaped detection for a long time. The fact that the observation required temperatures in the neighbourhood of 1 mK is consistent with computational studies that suggest that $hc\tilde{D}_e$ for He_2 is about 1.51×10^{-23} J, $hc\tilde{D}_0 \approx 2 \times 10^{-26}$ J, and R_e about 297 pm. (See Problem 6.6.) (a) Estimate the fundamental vibrational wavenumber, force constant, moment of inertia, and rotational constant based on the harmonic oscillator and rigid-rotor approximations. (b) Such a weakly bound complex is hardly likely to be rigid. Estimate the vibrational wavenumber and anharmonicity constant based on the Morse potential.

10.20 Use appropriate electronic structure software to perform MP2 and DFT calculations on H_2O and CO_2 using (a) 6-31G and (b) 6-31G basis sets in each case. (a) Compute ground-state energies, equilibrium geometries, and vibrational frequencies for each molecule. (b) Compute the dipole moment of H_2O; the experimental value is 1.854 D. (c) Compare computed values to experiment and suggest reasons for any discrepancies.

10.21 As mentioned in Section 10.13, the semiempirical, *ab initio*, and DFT methods discussed in Chapter 6 can be used to estimate the force field of a molecule. The molecule's vibrational spectrum can be simulated, and it is then possible to determine the correspondence between a vibrational frequency and the atomic displacements that give rise to a normal mode. (a) Using molecular modelling software and the computational method of your choice (semiempirical, *ab initio*, or DFT methods), calculate the fundamental vibrational wavenumbers and visualize the vibrational normal modes of SO_2 in the gas phase. (b) The experimental values of the fundamental vibrational wavenumbers of SO_2 in the gas phase are 525 cm^{-1}, 1151 cm^{-1}, and 1336 cm^{-1}. Compare the calculated and experimental values. Even if agreement is poor, is it possible to establish a correlation between an experimental value of the vibrational wavenumber and a specific vibrational normal mode?

10.22 Consider the molecule CH_3Cl. (a) To what point group does the molecule belong? (b) How many normal modes of vibration does the molecule have? (c) What are the symmetries of the normal modes of vibration for this molecule? (d) Which of the vibrational modes of this molecule are infrared active? (e) Which of the vibrational modes of this molecule are Raman active?

10.23 Suppose that three conformations are proposed for the non-linear molecule H_2O_2 (**6**, **7**, and **8**). The infrared absorption spectrum of gaseous H_2O_2 has bands at 870, 1370, 2869, and 3417 cm^{-1}. The Raman spectrum of the same sample has bands at 877, 1408, 1435, and 3407 cm^{-1}. All bands correspond to fundamental vibrational wavenumbers and you may assume that: (i) the 870 and 877 cm^{-1} bands arise from the same normal mode, and (ii) the 3417 and 3407 cm^{-1} bands arise from the same normal mode. (a) If H_2O_2 were linear, how many normal modes of vibration would it have? (b) Give the symmetry point group of each of the three proposed conformations of non-linear H_2O_2. (c) Determine which of the proposed conformations is inconsistent with the spectroscopic data. Explain your reasoning.

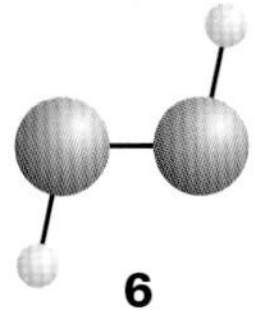
6

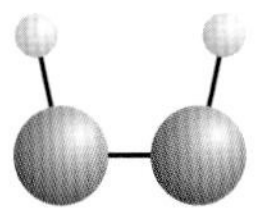
7

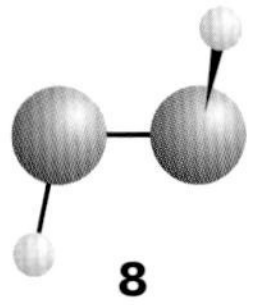
8

Theoretical problems

10.24 Show that the moment of inertia of a diatomic molecule composed of atoms of masses m_A and m_B and bond length R is equal to $m_{eff}R^2$, where $m_{eff} = m_A m_B/(m_A + m_B)$.

10.25 Confirm the expression given in Table 10.1 for the moment of inertia of a linear ABC molecule. *Hint*. Begin by locating the centre of mass.

10.26 Derive eqn 10.14 for the centrifugal distortion constant D_J of a diatomic molecule of effective mass m_{eff}. Treat the bond as an elastic spring with force constant k and equilibrium length r_e that is subjected to a centrifugal distortion to a new length r_c. Begin the derivation by letting the particles experience a restoring force of magnitude $k(r_c - r_e)$ that is countered perfectly by a centrifugal force $m_{eff}\omega^2 r_c$, where ω is the angular velocity of the rotating molecule. Then introduce quantum mechanical effects by writing the angular momentum as $\{J(J+1)\}^{1/2}\hbar$. Finally, write an expression for the energy of the rotating molecule, compare it with eqn 10.13, and infer an expression for $\tilde{D}_J$.

10.27 The rotational terms of a symmetric top, allowing for centrifugal distortion, are commonly written

$$\tilde{F}(J,K) = \tilde{B}J(J+1) + (\tilde{A} - \tilde{B})K^2 - \tilde{D}_J J^2(J+1)^2 - \tilde{D}_{JK}J(J+1)K^2 - \tilde{D}_K K^4$$

Derive an expression for the wavenumbers of the allowed rotational transitions. The following transition frequencies (in gigahertz, GHz) were observed for CH_3F:

51.0718 102.1426 102.1408 153.2103 153.2076

Determine the values of as many constants in the expression for the rotational terms as these values permit.

10.28 In the group theoretical language developed in Chapter 7, a spherical rotor is a molecule that belongs to a cubic or icosahedral point group, a symmetric rotor is a molecule with at least a threefold axis of symmetry, and an asymmetric rotor is a molecule without a threefold (or higher) axis. Linear molecules are linear rotors. Classify each of the following molecules as a spherical, symmetric, linear, or asymmetric rotor and justify your answers with group theoretical arguments: (a) CH_4, (b) CH_3CN, (c) CO_2, (d) CH_3OH, (e) benzene, (f) pyridine.

10.29 Derive an expression for the value of J corresponding to the most highly populated rotational energy level of a diatomic rotor at a temperature T remembering that the degeneracy of each level is $2J+1$. Evaluate the expression for ICl (for which $\tilde{B} = 0.1142$ cm^{-1}) at 25°C. Repeat the problem for the most highly populated level of a spherical rotor, taking note of the fact that each level is $(2J+1)^2$-fold degenerate. Evaluate the expression for CH_4 (for which $\tilde{B} = 5.24$ cm^{-1}) at 25°C.

10.30 The moments of inertia of the linear mercury(II) halides are very large, so the O and S branches of their vibrational Raman spectra show little rotational structure. Nevertheless, the peaks of both branches can be identified and have been used to measure the rotational constants of the molecules (R.J.H. Clark and D.M. Rippon, *J. Chem. Soc. Faraday Trans. II*, **69**, 1496 (1973)). Show, from a knowledge of the value of J corresponding to the intensity maximum, that the separation of the peaks of the O and S branches is given by the Placzek–Teller relation $\delta\tilde{\nu} = (32BkT/hc)^{1/2}$. The following widths were obtained at the temperatures stated:

	$HgCl_2$	$HgBr_2$	HgI_2
θ/°C	282	292	292
$\delta\tilde{\nu}$/cm^{-1}	23.8	15.2	11.4

Calculate the bond lengths in the three molecules.

10.31 Confirm that a Morse oscillator has a finite number of bound states, the states with $V < hc\tilde{D}_e$. Determine the value of v_{max} for the highest bound state.

10.32 Suppose that the out-of-plane distortion of a planar molecule AX_n is described by a potential energy $V(h) = V_0(1 - e^{-bh^4})$, where h is the distance by which the central atom is displaced. Sketch this potential energy as a function of h (allow h to be both negative and positive). What could be said about (a) the force constant, (b) the vibrations? Sketch the form of the ground-state wavefunction.

10.33 The analysis of combination differences summarized in Section 10.10 considered the R and P branches. Extend the analysis to the O and S branches of a Raman spectrum.

Applications: to biology, environmental science, and astrophysics

10.34 The protein haemerythrin is responsible for binding and carrying O_2 in some invertebrates. Each protein molecule has two Fe^{2+} ions that are in very close proximity and work together to bind one molecule of O_2. The Fe_2O_2 group of oxygenated haemerythrin is coloured and has an electronic absorption band at 500 nm. The resonance Raman spectrum of oxygenated haemerythrin obtained with laser excitation at 500 nm has a band at 844 cm^{-1} that has been attributed to the O—O stretching mode of bound $^{16}O_2$. (a) Why is resonance Raman spectroscopy and not infrared spectroscopy the method of choice for the study of the binding of O_2 to haemerythrin? (b) Proof that the 844 cm^{-1} band arises from a bound O_2 species may be obtained by conducting experiments on samples of haemerythrin that have been mixed with $^{18}O_2$, instead of $^{16}O_2$. Predict the fundamental vibrational wavenumber of the ^{18}O—^{18}O stretching mode in a sample of haemerythrin that has been treated with $^{18}O_2$. (c) The fundamental vibrational wavenumbers for the O—O stretching modes of O_2, O_2^- (superoxide anion), and O_2^{2-} (peroxide anion) are 1555, 1107, and 878 cm^{-1}, respectively. Explain this trend in terms of the electronic structures of O_2, O_2^-, and O_2^{2-}. *Hint.* Review Section 5.4. What are the bond orders of O_2, O_2^-, and O_2^{2-}? (d) Based on the data given above, which of the following species best describes the Fe_2O_2 group of haemerythrin: $Fe_2^{2+}O_2$, $Fe^{2+}Fe^{3+}O_2^-$, or $Fe_2^{3+}O_2^{2-}$? Explain your reasoning. (e) The resonance Raman spectrum of haemerythrin mixed with $^{16}O^{18}O$ has two bands that can be attributed to the O—O stretching mode of bound oxygen. Discuss how this observation may be used to exclude one or more of the four proposed schemes (**9–12**) for binding of O_2 to the Fe_2 site of haemerythrin.

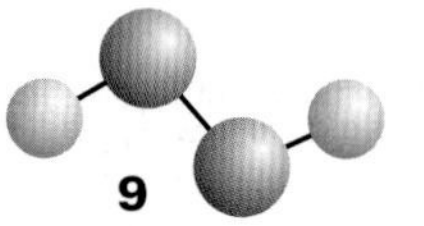

9

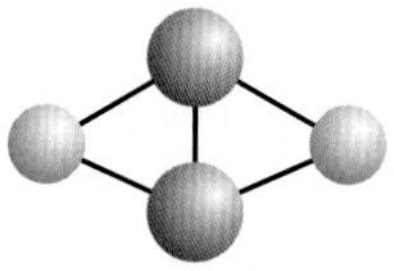

10

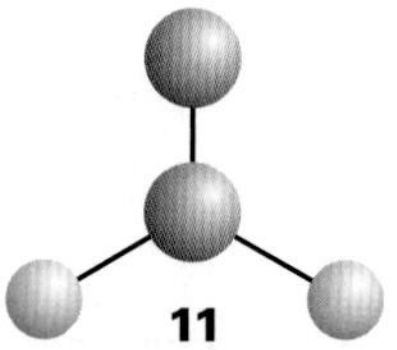

11

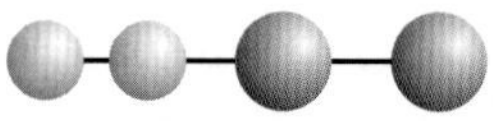

12

10.35 In *confocal Raman microscopy*, light must pass through several holes of very small diameter before reaching the detector. In this way light that is out of focus does not interfere with an image that is in focus. Consult monographs, journal articles, and reliable internet resources, such as those listed in the web site for this text, and write a brief report (similar in length and depth of coverage to one of the many *Impact* sections in this text) on the advantages and disadvantages of confocal Raman microscopy over conventional Raman microscopy in the study of biological systems. *Hint.* A good place to start is: P. Colarusso *et al.*, in *Encyclopedia of spectroscopy and spectrometry* (ed. J.C. Lindon, G.E. Tranter, and J.L. Holmes), **3**, 1945 Academic Press, San Diego (2000).

10.36‡ A mixture of carbon dioxide (2.1 per cent) and helium at 1.00 bar and 298 K in a gas cell of length 10 cm has an infrared absorption band centred at 2349 cm^{-1} with absorbances, $A(\tilde{\nu})$, described by:

$$A(\tilde{\nu}) = \frac{a_1}{1 + a_2(\tilde{\nu} - a_3)^2} + \frac{a_4}{1 + a_5(\tilde{\nu} - a_6)^2}$$

where the coefficients are $a_1 = 0.932$, $a_2 = 0.005050\ \text{cm}^2$, $a_3 = 2333\ \text{cm}^{-1}$, $a_4 = 1.504$, $a_5 = 0.01521\ \text{cm}^2$, $a_6 = 2362\ \text{cm}^{-1}$. (a) Draw graphs of $A(\tilde{\nu})$ and $\varepsilon(\tilde{\nu})$. What is the origin of both the band and the band width? What are the allowed and forbidden transitions of this band? (b) Calculate the transition wavenumbers and absorbances of the band with a simple harmonic oscillator–rigid rotor model and compare the result with the experimental spectra. The CO bond length is 116.2 pm. (c) Within what height, h, is basically all the infrared emission from the Earth in this band absorbed by atmospheric carbon dioxide? The mole fraction of CO_2 in the atmosphere is 3.3×10^{-4} and $T/\text{K} = 288 - 0.0065(h/\text{m})$ below 10 km. Draw a surface plot of the atmospheric transmittance of the band as a function of both height and wavenumber.

10.37 A. Dalgarno, in Chemistry in the interstellar medium, *Frontiers of Astrophysics*, E.H. Avrett (ed.), Harvard University Press, Cambridge (1976), notes that, although both CH and CN spectra show up strongly in the interstellar medium in the constellation Ophiuchus, the CN spectrum has become the standard for the determination of the temperature of the cosmic microwave background radiation. Demonstrate through a calculation why CH would not be as useful for this purpose as CN. The rotational constant $\tilde{B}_0$ for CH is $14.190\ \text{cm}^{-1}$.

10.38‡ There is a gaseous interstellar cloud in the constellation Ophiuchus that is illuminated from behind by the star ζ-Ophiuci. Analysis of the electronic–vibrational–rotational absorption lines shows the presence of CN molecules in the interstellar medium. A strong absorption line in the ultraviolet region at $\lambda = 387.5$ nm was observed corresponding to the transition $J = 0$–1. Unexpectedly, a second strong absorption line with 25 per cent of the intensity of the first was found at a slightly longer wavelength ($\Delta\lambda = 0.061$ nm) corresponding to the transition $J = 1$–1 (here allowed). Calculate the temperature of the CN molecules. Gerhard Herzberg, who was later to receive the Nobel Prize for his contributions to spectroscopy, calculated the temperature as 2.3 K. Although puzzled by this result, he did not realize its full significance. If he had, his prize might have been for the discovery of the cosmic microwave background radiation.

10.39‡ The H_3^+ ion has recently been found in the interstellar medium and in the atmospheres of Jupiter, Saturn, and Uranus. The rotational energy levels of H_3^+, an oblate symmetric rotor, are given by eqn 10.10, with $\tilde{C}$ replacing $\tilde{A}$, when centrifugal distortion and other complications are ignored. Experimental values for vibrational–rotational constants are $\tilde{\nu}(E') = 2521.6\ \text{cm}^{-1}$, $\tilde{B} = 43.55\ \text{cm}^{-1}$, and $\tilde{C} = 20.71\ \text{cm}^{-1}$. (a) Show that for a non-linear planar molecule (such as H_3^+) that $I_C = 2I_B$. The rather large discrepancy with the experimental values is due to factors ignored in eqn 10.10. (b) Calculate an approximate value of the H—H bond length in H_3^+. (c) The value of R_e obtained from the best quantum mechanical calculations by J.B. Anderson (*J. Chem. Phys.* **96**, 3702 (1991)) is 87.32 pm. Use this result to calculate the values of the rotational constants $\tilde{B}$ and $\tilde{C}$. (d) Assuming that the geometry and force constants are the same in D_3^+ and H_3^+, calculate the spectroscopic constants of D_3^+. The molecular ion D_3^+ was first produced by Shy *et al.* (*Phys. Rev. Lett.* **45**, 535 (1980)) who observed the $\nu_2(E')$ band in the infrared.

10.40 The space immediately surrounding stars, the *circumstellar space*, is significantly warmer because stars are very intense black-body emitters with temperatures of several thousand kelvin. Discuss how such factors as cloud temperature, particle density, and particle velocity may affect the rotational spectrum of CO in an interstellar cloud. What new features in the spectrum of CO can be observed in gas ejected from and still near a star with temperatures of about 1000 K, relative to gas in a cloud with temperatures of about 10 K? Explain how these features may be used to distinguish between circumstellar and interstellar material on the basis of the rotational spectrum of CO.

11 Electronic spectroscopy

Simple analytical expressions for the electronic energy levels of molecules cannot be given, so this chapter concentrates on the qualitative features of electronic transitions. A common theme throughout the chapter is that electronic transitions occur within a stationary nuclear framework. We pay particular attention to spontaneous radiative decay processes, which include fluorescence and phosphorescence. A specially important example of stimulated radiative decay is that responsible for the action of lasers, and we see how this stimulated emission may be achieved and employed.

The energies needed to change the electron distributions of molecules are of the order of several electronvolts (1 eV is equivalent to about 8000 cm^{-1} or 100 kJ mol^{-1}). Consequently, the photons emitted or absorbed when such changes occur lie in the visible and ultraviolet regions of the electromagnetic spectrum (Table 11.1).

We begin our study of electronic spectroscopy with a brief survey of common experimental techniques. Then, with help from quantum theory, we write the selection rules for molecular electronic spectroscopy. After describing absorption and emission processes, we turn our attention to lasers, which have revolutionized physical chemistry in recent years. As we have already seen in Chapter 10, lasers have brought unprecedented precision to molecular spectroscopy. They have also made it possible to study chemical reactions on a femtosecond timescale. We shall see the principles of their action in this chapter and will encounter their applications throughout the rest of the book.

Experimental techniques

Molecular electronic spectroscopy is commonly conducted by monitoring the absorption of electromagnetic radiation by molecules in their electronic ground states, for no other electronic states are occupied at normal temperatures. Emission is observed in certain cases, such as in chemiluminescence, when the products of a reaction are created in electronically excited states, and in the monitoring of fluorescence and phosphorescence (Section 11.5).

11.1 Spectrometers

Figure 11.1 shows the general layouts of an absorption spectrometer operating in the ultraviolet and visible ranges. Radiation from an appropriate source is directed toward a sample. In most spectrometers, the radiation transmitted by the sample is collected

Synoptic table 11.1* Colour, frequency, and energy of light

Colour	λ/nm	$\nu/(10^{14}\ \mathrm{Hz})$	$E/(\mathrm{kJ\ mol^{-1}})$
Infrared	>1000	<3.0	<120
Red	700	4.3	170
Yellow	580	5.2	210
Blue	470	6.4	250
Ultraviolet	<300	>10	>400

* More values are given in the *Data section*.

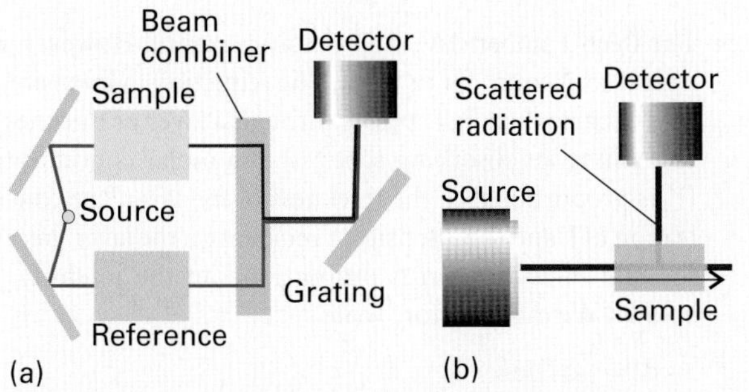

Fig. 11.1 Two examples of spectrometers: (a) the layout of an absorption spectrometer, used primarily for studies in the ultraviolet and visible ranges, in which the exciting beams of radiation pass alternately through a sample and a reference cell, and the detector is synchronized with them so that the relative absorption can be determined, and (b) a simple emission spectrometer, where light emitted or scattered by the sample is detected at right angles to the direction of propagation of an incident beam of radiation.

by mirrors or lenses and strikes a dispersing element that separates it into different wavelengths. The intensity of the radiation at each wavelength is then analysed by a suitable detector.

The source may be a *quartz–tungsten–halogen lamp* consisting of a tungsten filament that, when heated to about 3000 K, emits radiation in the range 320 nm < λ < 2500 nm. A *gas discharge lamp* is a common source of ultraviolet and visible radiation. In a *xenon discharge lamp*, an electrical discharge excites xenon atoms to excited states, which then emit ultraviolet radiation. At pressures exceeding 1 kPa, the output consists of sharp lines superimposed on a broad, intense background due to emission from a mixture of ions formed by the electrical discharge. These high-pressure xenon lamps have emission profiles similar to that of a black body heated to 6000 K. In a *deuterium lamp*, excited D_2 molecules emit intense radiation in the range 160–400 nm.

After passing through a monochromator, the intensity of light at each wavelength can be measured by a detector suitable for work in the ultraviolet and visible ranges. A common device is a *photomultiplier tube* (PMT), in which the photoelectric effect (Section 1.2) is used to generate an electrical signal proportional to the intensity of radiation that strikes the detector. A less sensitive alternative to the PMT is the *photodiode*, a solid-state device that conducts electricity when struck by photons because light-induced electron transfer reactions in the detector material create mobile charge carriers (negatively charged electrons and positively charged 'holes'). In an *avalanche photodiode*, the photo-generated electrons are accelerated through a very large electrical potential difference. The high-energy electrons then collide with other atoms in the solid and ionize them, thus creating an avalanche of secondary charge carriers and increasing the sensitivity of the device to incident photons.

A *charge-coupled device* (CCD) is a two-dimensional array of several million small photodiode detectors. With a CCD, a wide range of wavelengths that emerge from the sample are detected simultaneously, thus eliminating the need to measure light intensity one narrow wavelength range at a time. CCD detectors are the imaging devices in digital cameras, but are also used widely in spectroscopy to monitor absorption, emission, and Raman scattering.

11.2 The Beer–Lambert law

The ratio of the transmitted intensity, I, to the incident intensity, I_0, at a given frequency is called the **transmittance**, T, of the sample at that frequency:

$$T = \frac{I}{I_0} \qquad [11.1]$$

It is found empirically that the transmitted intensity varies with the length, l, of the sample and the molar concentration, [J], of the absorbing species J in accord with the **Beer–Lambert law** (see the following *Justification*):

$$I = I_0 10^{-\varepsilon[\mathrm{J}]l} \qquad (11.2)$$

The quantity ε is called the **molar absorption coefficient** (formerly, and still widely, the 'extinction coefficient'). The molar absorption coefficient depends on the frequency of the incident radiation and is greatest where the absorption is most intense. Its dimensions are 1/(concentration × length), and it is normally convenient to express it in cubic decimetres per mole per centimetre ($\mathrm{dm^3\ mol^{-1}\ cm^{-1}}$). Alternative units are square centimetres per mole ($\mathrm{cm^2\ mol^{-1}}$). This change of units demonstrates that ε may be regarded as a molar cross-section for absorption, and the greater the cross-sectional area of the molecule for absorption, the greater its ability to block the passage of the incident radiation.

To simplify eqn 11.2, we introduce the **absorbance**, A, of the sample at a given wavenumber as

$$A = \log\frac{I_0}{I} \quad \text{or} \quad A = -\log T \qquad [11.3]$$

Then the Beer–Lambert law becomes

$$A = \varepsilon[\mathrm{J}]l \qquad (11.4)$$

The product $\varepsilon[\mathrm{J}]l$ was known formerly as the *optical density* of the sample. Equation 11.4 suggests that, to achieve sufficient absorption, path lengths through gaseous samples must be very long, of the order of metres, because concentrations are low. Long path lengths are achieved by multiple passage of the beam between parallel mirrors at each end of the sample cavity. Conversely, path lengths through liquid samples can be significantly shorter, of the order of millimetres or centimetres.

Example 11.1 *The molar absorption coefficient of tryptophan*

Radiation of wavelength 280 nm passed through 1.0 mm of a solution that contained an aqueous solution of the amino acid tryptophan at a concentration of 0.50 mmol dm^{-3}. The light intensity is reduced to 54 per cent of its initial value (so $T = 0.54$). Calculate the absorbance and the molar absorption coefficient of tryptophan at 280 nm. What would be the transmittance through a cell of thickness 2.0 mm?

Method From eqns 11.3 and 11.4 we write

$$A = -\log T = \varepsilon[\mathrm{J}]l$$

so it follows that

$$\varepsilon = -\frac{\log T}{[\mathrm{J}]l}$$

For the transmittance through the thicker cell, we use $T = 10^{-A}$ and the value of ε calculated here.

Answer The molar absorption coefficient is

$$\varepsilon = -\frac{\log 0.54}{(5.0\times10^{-4}\ \mathrm{mol\ dm^{-3}})\times(1.0\ \mathrm{mm})}$$

$$= 5.4\times10^{2}\ \mathrm{dm^{3}\ mol^{-1}\ mm^{-1}}$$

These units are convenient for the rest of the calculation, so we do not combine the units dm^3 and mm^{-1}. The absorbance is

$$A = -\log 0.54 = 0.27$$

The absorbance of a sample of length 2.0 mm is

$$A = (5.4\times10^{2}\ \mathrm{dm^{3}\ mol^{-1}\ mm^{-1}})\times(5.0\times10^{-4}\ \mathrm{mol\ dm^{-3}})\times(2.0\ \mathrm{mm}) = 0.54$$

It follows that the transmittance is now

$$T = 10^{-A} = 10^{-0.54} = 0.29$$

That is, the emergent light is reduced to 29 per cent of its incident intensity.

Self-test 11.1 The transmittance of an aqueous solution that contained the amino acid tyrosine at a molar concentration of 0.10 mmol dm^{-3} was measured as 0.14 at 240 nm in a cell of length 5.0 mm. Calculate the molar absorption coefficient of tyrosine at that wavelength, and the absorbance of the solution. What would be the transmittance through a cell of length 1.0 mm?

[1.7×10^{3} dm^3 mol^{-1} cm^{-1}, $A = 0.85$, $T = 0.72$]

Justification 11.1 *The Beer–Lambert law*

The Beer–Lambert law is an empirical result. However, it is simple to account for its form. The reduction in intensity, dI, that occurs when light passes through a layer of thickness dl containing an absorbing species J at a molar concentration [J] is proportional to the thickness of the layer, the concentration of J, and the intensity, I, incident on the layer (because the rate of absorption is proportional to the intensity, see below). We can therefore write

$$\mathrm{d}I = -\kappa[\mathrm{J}]I\mathrm{d}l$$

where κ (kappa) is the proportionality coefficient, or equivalently

$$\frac{\mathrm{d}I}{I} = -\kappa[\mathrm{J}]\mathrm{d}l$$

This expression applies to each successive layer into which the sample can be regarded as being divided. Therefore, to obtain the intensity that emerges from a sample of thickness l when the intensity incident on one face of the sample is I_0, we sum all the successive changes:

$$\int_{I_0}^{I}\frac{\mathrm{d}I}{I} = -\kappa\int_{0}^{l}[\mathrm{J}]\mathrm{d}l$$

If the concentration is uniform, [J] is independent of location, and the expression integrates to

$$\ln\frac{I}{I_0} = -\kappa[\mathrm{J}]l$$

This expression gives the Beer–Lambert law when the logarithm is converted to base 10 by using $\ln x = (\ln 10)\log x$ and replacing κ by $\varepsilon \ln 10$.

The maximum value of the molar absorption coefficient, $\varepsilon_{\max}$, is an indication of the intensity of a transition. However, as absorption bands generally spread over a range of wavenumbers, quoting the absorption coefficient at a single wavenumber

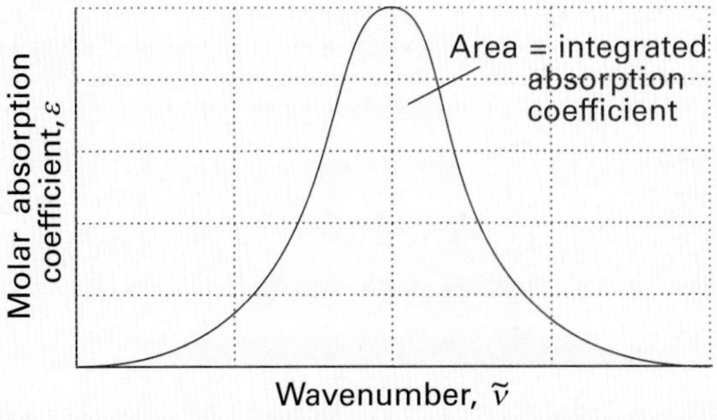

Fig. 11.2 The integrated absorption coefficient of a transition is the area under a plot of the molar absorption coefficient against the wavenumber of the incident radiation.

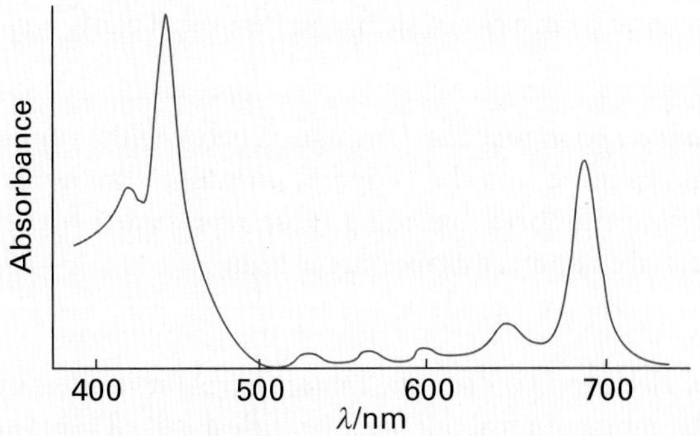

Fig. 11.3 The absorption spectrum of chlorophyll in the visible region. Note that it absorbs in the red and blue regions, and that green light is not absorbed.

might not give a true indication of the intensity of a transition. The **integrated absorption coefficient**, $\mathcal{A}$, is the sum of the absorption coefficients over the entire band (Fig. 11.2), and corresponds to the area under the plot of the molar absorption coefficient against wavenumber:

$$\mathcal{A} = \int_{\text{band}} \varepsilon(\tilde{\nu})\,d\tilde{\nu} \qquad [11.5]$$

For lines of similar widths, the integrated absorption coefficients are proportional to the heights of the lines.

The characteristics of electronic transitions

In the ground state of a molecule the nuclei are at equilibrium in the sense that they experience no net force from the electrons and other nuclei in the molecule. Immediately after an electronic transition they are subjected to different forces and the molecule may respond by starting to vibrate. The resulting vibrational structure of electronic transitions can be resolved for gaseous samples, but in a liquid or solid the lines usually merge together and result in a broad, almost featureless band (Fig. 11.3). Superimposed on the vibrational transitions that accompany the electronic transition of a molecule in the gas phase is an additional branch structure that arises from rotational transitions. The electronic spectra of gaseous samples are therefore very complicated but rich in information.

11.3 The electronic spectra of diatomic molecules

We examine some general features of electronic transitions by using diatomic molecules as examples. We begin by assigning term symbols to ground and excited electronic states. Then we use the symmetry designations to formulate selection rules. Finally, we examine the origin of vibrational structure in electronic spectra.

(a) Term symbols

The term symbols of linear molecules (the analogues of the symbols ^{2}P, etc. for atoms) are constructed in a similar way to those for atoms, but now we must pay attention to the component of total orbital angular momentum about the internuclear axis, $\Lambda\hbar$. The value of $|\Lambda|$ is denoted by the symbols Σ, Π, Δ, . . . for $|\Lambda| = 0, 1, 2 \ldots$, respectively. These labels are the analogues of S, P, D, . . . for atoms. The value of Λ is the sum of the values of λ, the quantum number for the component $\lambda\hbar$ of orbital angular momentum of an individual electron around the internuclear axis. A single electron in a σ orbital has $\lambda = 0$: the orbital is cylindrically symmetrical and has no angular nodes when viewed along the internuclear axis. Therefore, if that is the only electron present, $\Lambda = 0$. The term symbol for ground-state H_2^+ is therefore Σ.

As in atoms, we use a left superscript with the value of $2S+1$ to denote the multiplicity of the term. The component of total spin angular momentum about the internuclear axis is denoted Σ, where $\Sigma = S, S-1, S-2, \ldots, -S$. For H_2^+, because there is only one electron, $S = s = \frac{1}{2}$ ($\Sigma = \pm\frac{1}{2}$) and the term symbol is $^2\Sigma$, a doublet term. The overall parity of the term is added as a right subscript. For H_2^+, the parity of the only occupied orbital is g (Section 5.3), so the term itself is also g, and in full dress is $^2\Sigma_g$. If there are several electrons, the overall parity is calculated by using

$$\mathrm{g}\times\mathrm{g}=\mathrm{g} \qquad \mathrm{u}\times\mathrm{u}=\mathrm{g} \qquad \mathrm{u}\times\mathrm{g}=\mathrm{u} \qquad (11.6)$$

These rules are generated by interpreting g as +1 and u as −1. The term symbol for the ground state of any closed-shell homonuclear diatomic molecule is $^1\Sigma_g$ because the spin is zero (a singlet term in which all electrons paired), there is no orbital angular momentum from a closed shell, and the overall parity is g.

A note on good practice Distinguish between the (upright) term symbol Σ and the (sloping) quantum number Σ. All quantum numbers, including those represented by Greek letters, are sloping. All labels are upright.

A π electron in a diatomic molecule has one unit of orbital angular momentum about the internuclear axis ($\lambda = \pm 1$) and, if it is the only electron outside a closed shell, gives rise to a Π term. If there are two π electrons (as in the ground state of O_2, with configuration $1\pi_u^4 1\pi_g^2$) then the term symbol may be either Σ (if the electrons are travelling in opposite directions, which is the case if they occupy different π orbitals, one with $\lambda = +1$ and the other with $\lambda = -1$) or Δ (if they are travelling in the same direction, which is the case if they occupy the same π orbital, both $\lambda = +1$, for instance). For O_2, the two π electrons occupy different orbitals with parallel spins (a triplet term), so the ground term is $^3\Sigma$. The overall parity of the molecule is

$$(\text{closed shell}) \times \text{g} \times \text{g} = \text{g}$$

The term symbol is therefore $^3\Sigma_g$.

For Σ terms, a ± superscript denotes the behaviour of the molecular wavefunction under reflection in a plane containing the nuclei (Fig. 11.4). If, for convenience, we think of O_2 as having one electron in $1\pi_{g,x}$, which changes sign under reflection in the *yz*-plane (with *z* as the internuclear axis), and the other electron in $1\pi_{g,y}$, which does not change sign under reflection in the same plane, then the overall reflection symmetry is

$$(\text{closed shell}) \times (+) \times (-) = (-)$$

and the full term symbol of the ground electronic state of O_2 is $^3\Sigma_g^-$. Table 11.2 and Fig. 11.5 summarize the configurations, term symbols, and energies of the ground and some excited states of O_2.

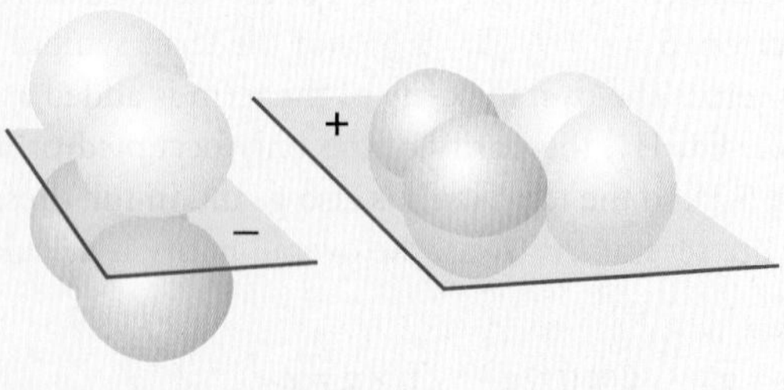

Fig. 11.4 The + or – on a term symbol refers to the overall symmetry of a configuration under reflection in a plane containing the two nuclei.

Table 11.2 Properties of O_2 in its lower electronic states*

Configuration†	Term	Relative energy/cm^{-1}	$\tilde{\nu}$/cm^{-1}	R_e/pm
$\pi_u^2\pi_u^2\pi_g^1\pi_g^1$	$^3\Sigma_g^-$	0	1580	120.74
$\pi_u^2\pi_u^2\pi_g^2\pi_g^0$	$^1\Delta_g$	7882.39	1509	121.55
$\pi_u^2\pi_u^2\pi_g^1\pi_g^1$	$^1\Sigma_g^+$	13 120.9	1433	122.68
$\pi_u^2\pi_u^1\pi_g^2\pi_g^1$	$^3\Sigma_u^+$	35 713	819	142
$\pi_u^2\pi_u^1\pi_g^2\pi_g^1$	$^3\Sigma_u^-$	49 363	700	160

* Adapted from G. Herzberg, *Spectra of diatomic molecules*, van Nostrand, New York (1950) and D.C. Harris and M.D. Bertolucci, *Symmetry and spectroscopy: an introduction to vibrational and electronic spectroscopy*, Dover, New York (1989).

† The configuration $\pi_u^2\pi_u^1\pi_g^2\pi_g^1$ should also give rise to a $^3\Delta_u$ term, but electronic transitions to or from this state have not been observed.

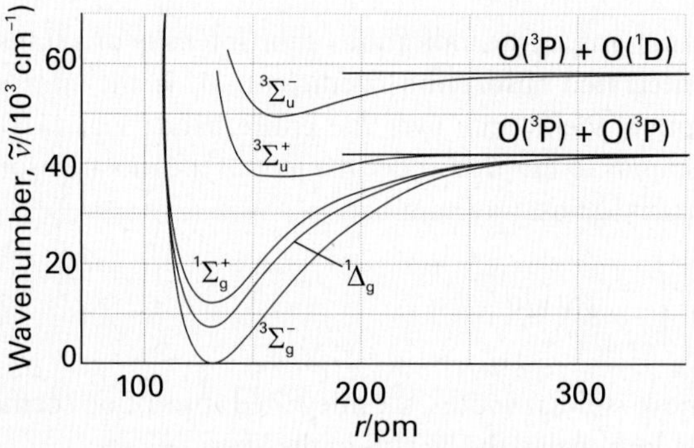

Fig. 11.5 The electronic states of dioxygen.

● **A BRIEF ILLUSTRATION**

The term symbol for the excited state of O_2 formed by placing two electrons in a $1\pi_{g,x}$ (or in a $1\pi_{g,y}$) orbital is $^1\Delta_g$ because $|\Lambda| = 2$ (two electrons in the same π orbital), the spin is zero (all electrons are paired), and the overall parity is (closed shell) × g × g = g. ●

(b) Selection rules

A number of selection rules govern which transitions will be observed in the electronic spectrum of a molecule. The selection rules concerned with changes in angular momentum are

$$\Delta\Lambda = 0, \pm 1 \qquad \Delta S = 0 \qquad \Delta\Sigma = 0 \qquad \Delta\Omega = 0, \pm 1$$

where $\Omega = \Lambda + \Sigma$ is the quantum number for the component of total angular momentum (orbital and spin) around the internuclear axis (Fig. 11.6). As in atoms (Section 4.3), the origins of these rules are conservation of angular momentum during a transition and the fact that a photon has a spin of 1.

There are two selection rules concerned with changes in symmetry. First, for Σ terms, only $\Sigma^+ \leftrightarrow \Sigma^+$ and $\Sigma^- \leftrightarrow \Sigma^-$ transitions

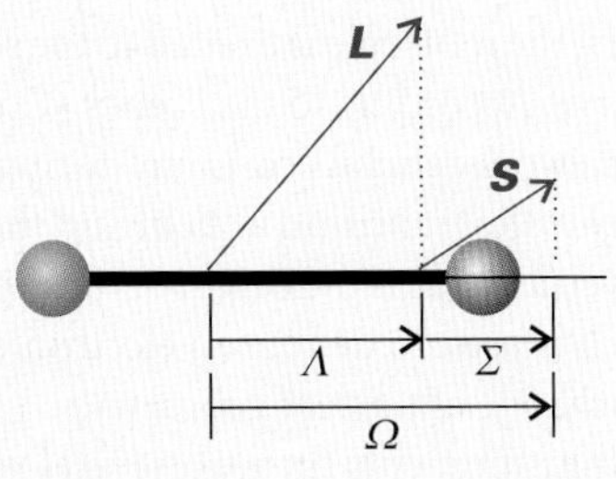

Fig. 11.6 The coupling of spin and orbital angular momenta in a linear molecule: only the components along the internuclear axis are conserved.

are allowed. Second, the **Laporte selection rule** for centrosymmetric molecules (those with a centre of inversion) and atoms states that:

> The only allowed transitions are transitions that are accompanied by a change of parity.

That is, u → g and g → u transitions are allowed, but g → g and u → u transitions are forbidden.

Justification 11.2 *The Laporte selection rule*

The last two selection rules result from the fact that the electric-dipole transition moment

$$\mu_{fi} = \int \psi_f^* \mu \psi_i d\tau$$

vanishes unless the integrand is invariant under all symmetry operations of the molecule. The three components of the dipole moment operator transform like x, y, and z, and are all u. Therefore, for a g → g transition, the overall parity of the transition dipole moment is g × u × g = u, so it must be zero. Likewise, for a u → u transition, the overall parity is u × u × u = u, so the transition dipole moment must also vanish. Hence, transitions without a change of parity are forbidden. The z-component of the dipole moment operator, the only component of μ responsible for $\Sigma \leftrightarrow \Sigma$ transitions, has (+) symmetry. Therefore, for a (+) ↔ (−) transition, the overall symmetry of the transition dipole moment is (+) × (+) × (−) = (−), so it must be zero. Therefore, for Σ terms, $\Sigma^+ \leftrightarrow \Sigma^-$ transitions are not allowed.

A forbidden g → g transition can become allowed if the centre of symmetry is eliminated by an asymmetrical vibration, such as the one shown in Fig. 11.7. When the centre of symmetry is lost, g → g and u → u transitions are no longer parity-forbidden and become weakly allowed. A transition that derives its intensity from an asymmetrical vibration of a molecule is called a **vibronic transition.**

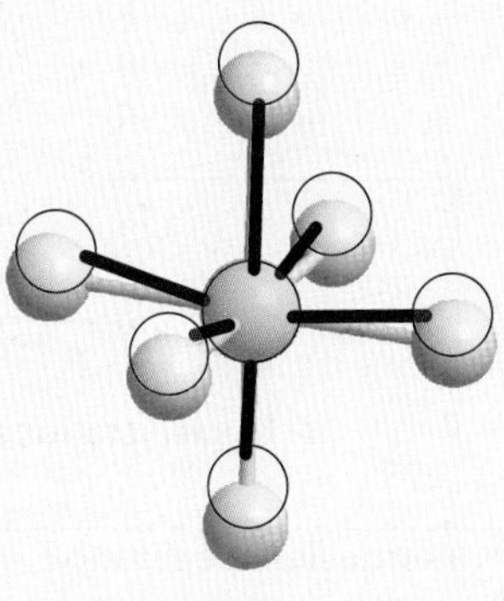

Fig. 11.7 A d–d transition is parity-forbidden because it corresponds to a g–g transition. However, a vibration of the molecule can destroy the inversion symmetry of the molecule and the g,u classification no longer applies. The removal of the centre of symmetry gives rise to a vibronically allowed transition.

Self-test 11.2 Which of the following electronic transitions are allowed in O_2: ${}^3\Sigma_g^- \leftrightarrow {}^1\Delta_g$, ${}^3\Sigma_g^- \leftrightarrow {}^1\Sigma_g^+$, ${}^3\Sigma_g^- \leftrightarrow {}^3\Delta_u$, ${}^3\Sigma_g^- \leftrightarrow {}^3\Sigma_u^+$, ${}^3\Sigma_g^- \leftrightarrow {}^3\Sigma_u^-$? [${}^3\Sigma_g^- \leftrightarrow {}^3\Sigma_u^-$]

(c) Vibrational structure

To account for the vibrational structure in electronic spectra of molecules (Fig. 11.8), we apply the **Franck–Condon principle:**

> Because the nuclei are so much more massive than the electrons, an electronic transition takes place very much faster than the nuclei can respond.

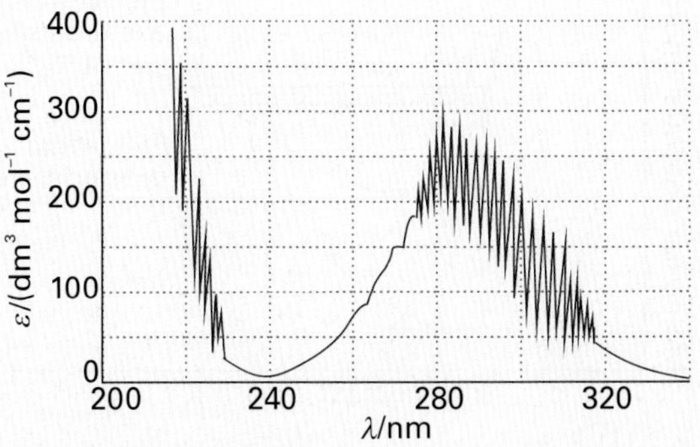

Fig. 11.8 The electronic spectra of some molecules show significant vibrational structure. Shown here is the ultraviolet spectrum of gaseous SO_2 at 298 K. As explained in the text, the sharp lines in this spectrum are due to transitions from a lower electronic state to different vibrational levels of a higher electronic state.

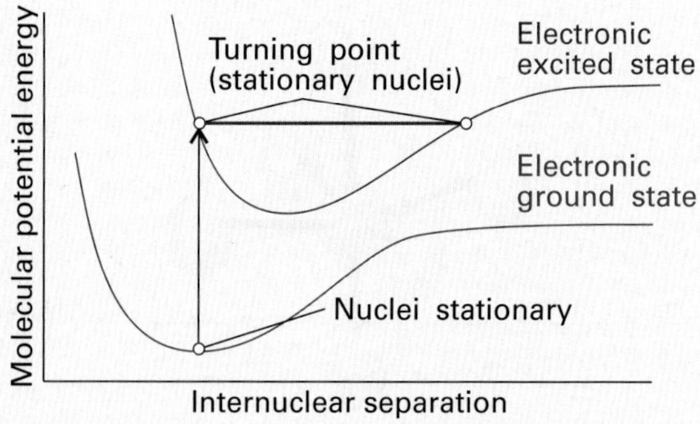

Fig. 11.9 According to the Franck–Condon principle, the most intense vibronic transition is from the ground vibrational state to the vibrational state lying vertically above it. Transitions to other vibrational levels also occur, but with lower intensity.

As a result of the transition, electron density is rapidly built up in new regions of the molecule and removed from others. The initially stationary nuclei suddenly experience a new force field, to which they respond by beginning to vibrate, and (in classical terms) swing backwards and forwards from their original separation (which was maintained during the rapid electronic excitation). The stationary equilibrium separation of the nuclei in the initial electronic state therefore becomes a stationary turning point in the final electronic state (Fig. 11.9).

The quantum mechanical version of the Franck–Condon principle refines this picture. Before the absorption, the molecule is in the lowest vibrational state of its lowest electronic state (Fig. 11.10); the most probable location of the nuclei is at their equilibrium separation, R_e. The electronic transition is most likely to take place when the nuclei have this separation. When the transition occurs, the molecule is excited to the state represented by the upper curve. According to the Franck–Condon principle, the nuclear framework remains constant during this excitation, so we may imagine the transition as being up the vertical line in Fig. 11.10. The vertical line is the origin of the expression **vertical transition**, which is used to denote an electronic transition that occurs without change of nuclear geometry.

The vertical transition cuts through several vibrational levels of the upper electronic state. The level marked * is the one in which the nuclei are most probably at the same initial separation R_e (because the vibrational wavefunction has maximum amplitude there), so this vibrational state is the most probable state for the termination of the transition. However, it is not the only accessible vibrational state because several nearby states have an appreciable probability of the nuclei being at the separation R_e. Therefore, transitions occur to all the vibrational states in this region, but most intensely to the state with a vibrational wavefunction that peaks most strongly near R_e.

The vibrational structure of the spectrum depends on the relative horizontal position of the two potential energy curves, and a long **vibrational progression**, a lot of vibrational structure, is stimulated if the upper potential energy curve is appreciably displaced horizontally from the lower. The upper curve is usually displaced to greater equilibrium bond lengths because electronically excited states usually have more antibonding character than electronic ground states.

The separation of the vibrational lines of an electronic absorption spectrum depends on the vibrational energies of the *upper* electronic state. Hence, electronic absorption spectra may be used to assess the force fields and dissociation energies of electronically excited molecules.

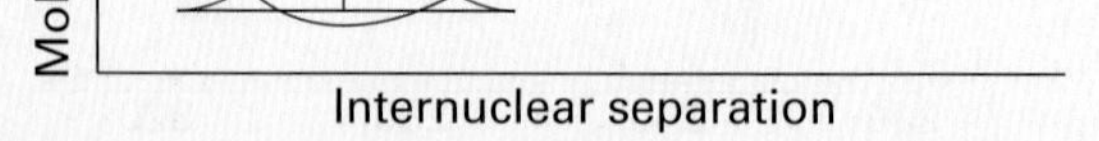

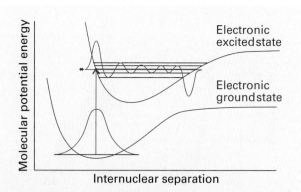

Fig. 11.10 In the quantum mechanical version of the Franck–Condon principle, the molecule undergoes a transition to the upper vibrational state that most closely resembles the vibrational wavefunction of the vibrational ground state of the lower electronic state. The two wavefunctions shown here have the greatest overlap integral of all the vibrational states of the upper electronic state and hence are most closely similar.

(d) Franck–Condon factors

The quantitative form of the Franck–Condon principle is derived from the expression for the transition dipole moment, $\mu_{fi} = \langle f|\mu|i\rangle$, written in the notation introduced in *Further information 1.1*. The dipole moment operator is a sum over all nuclei and electrons in the molecule:

$$\mu = -e\sum_i r_i + e\sum_I Z_I R_I \tag{11.7}$$

where the vectors are the distances from the centre of charge of the molecule. The intensity of the transition is proportional to the square modulus, $|\mu_{fi}|^2$, of the magnitude of the transition dipole moment (eqn 4.24), and we show in the following *Justification* that this intensity is proportional to the square modulus of the overlap integral, $S(v_f,v_i)$, between the vibrational states of the initial and final electronic states. This overlap integral is a measure of the match between the vibrational wavefunctions in the upper and lower electronic states: $S = 1$ for a perfect match and $S = 0$ when there is no similarity.

Justification 11.3 *The Franck–Condon approximation*

The transition dipole moment is written as $\mu_{fi} = \langle f|\mu|i\rangle$. The overall state of the molecule consists of an electronic part, $|\varepsilon\rangle$, and a vibrational (and real) part, $|\upsilon\rangle$. Therefore, within the Born–Oppenheimer approximation in which the electronic part does not explicitly depend on nuclear positions, the transition dipole moment factorizes as follows:

$$\mu_{fi} = \left\langle \varepsilon_f \upsilon_f \left| \left\{ -e\sum_i r_i + e\sum_I Z_I R_I \right\} \right| \varepsilon_i \upsilon_i \right\rangle$$

$$= -e\sum_i \langle \varepsilon_f|r_i|\varepsilon_i\rangle\langle \upsilon_f|\upsilon_i\rangle + e\sum_I Z_I\langle \varepsilon_f|\varepsilon_i\rangle\langle \upsilon_f|R_I|\upsilon_i\rangle$$

The second term on the right of the second row is zero, because $\langle \varepsilon_f|\varepsilon_i\rangle = 0$ for two different electronic states (they are orthogonal). Therefore,

$$\mu_{fi} = -e\sum_i \langle \varepsilon_f|r_i|\varepsilon_i\rangle\langle \upsilon_f|\upsilon_i\rangle = \mu_{\varepsilon_f,\varepsilon_i} S(\upsilon_f,\upsilon_i) \qquad (11.8a)$$

where

$$\mu_{\varepsilon_f,\varepsilon_i} = -e\sum_i \langle \varepsilon_f|r_i|\varepsilon_i\rangle \quad S(\upsilon_f,\upsilon_i) = \langle \upsilon_f|\upsilon_i\rangle = \int \psi_{\upsilon_f}\psi_{\upsilon_i}\,d\tau \qquad (11.8b)$$

The matrix element μ_{fi} is the electric-dipole transition moment arising from the redistribution of electrons (and a measure of the 'kick' this redistribution gives to the electromagnetic field, and vice versa for absorption). The factor $S(\upsilon_f,\upsilon_i)$, is the overlap integral between the vibrational state $|\upsilon_i\rangle$ in the initial electronic state of the molecule, and the vibrational state $|\upsilon_f\rangle$ in the final electronic state of the molecule.

Because the transition intensity is proportional to the square of the magnitude of the transition dipole moment, the intensity of an absorption is proportional to $|S(\upsilon_f,\upsilon_i)|^2$, which is known as the **Franck–Condon factor** for the transition. It follows that, the greater the overlap of the vibrational state wavefunction in the upper electronic state with the vibrational wavefunction in the lower electronic state, the greater the absorption intensity of that particular simultaneous electronic and vibrational transition. This conclusion is the basis of the illustration in Fig. 11.10, where we see that the vibrational wavefunction of the ground state has the greatest overlap with the vibrational states that have peaks at similar bond lengths in the upper electronic state.

Example 11.2 *Calculating a Franck–Condon factor*

Consider the transition from one electronic state to another, their bond lengths being R_e and R_e' and their force constants equal. Calculate the Franck–Condon factor for the 0–0 transition and show that the transition is most intense when the bond lengths are equal.

Method We need to calculate $S(0,0)$, the overlap integral of the two ground-state vibrational wavefunctions, and then take its square. The difference between harmonic and anharmonic vibrational wavefunctions is negligible for $\upsilon = 0$, so harmonic oscillator wavefunctions can be used (Table 2.1).

Answer We use the (real) wavefunctions

$$\psi_0 = \left(\frac{1}{\alpha\pi^{1/2}}\right)^{1/2} e^{-x^2/2\alpha^2} \qquad \psi_0' = \left(\frac{1}{\alpha\pi^{1/2}}\right)^{1/2} e^{-x'^2/2\alpha^2}$$

where $x = R - R_e$ and $x' = R - R_e'$, with $\alpha = (\hbar^2/mk)^{1/4}$ (Section 2.5a). The overlap integral is

$$S(0,0) = \int_{-\infty}^{\infty} \psi_0'\psi_0\,dR = \frac{1}{\alpha\pi^{1/2}}\int_{-\infty}^{\infty} e^{-(x^2+x'^2)/2\alpha^2}\,dx$$

We now write $\alpha z = R - \frac{1}{2}(R_e + R_e')$, and manipulate this expression into

$$S(0,0) = \frac{1}{\pi^{1/2}} e^{-(R_e-R_e')^2/4\alpha^2}\int_{-\infty}^{\infty} e^{-z^2}\,dz$$

The value of the integral is $\pi^{1/2}$. Therefore, the overlap integral is

$$S(0,0) = e^{-(R_e-R_e')^2/4\alpha^2}$$

and the Franck–Condon factor is

$$S(0,0)^2 = e^{-(R_e-R_e')^2/2\alpha^2}$$

This factor is equal to 1 when $R_e' = R_e$ and decreases as the equilibrium bond lengths diverge from each other (Fig. 11.11).

For Br_2, $R_e = 228$ pm and there is an upper state with $R_e' = 266$ pm. Taking the vibrational wavenumber as 250 cm^{-1} gives $S(0, 0)^2 = 5.1 \times 10^{-10}$, so the intensity of the 0–0 transition is only 5.1×10^{-10} what it would have been if the potential curves had been directly above each other.

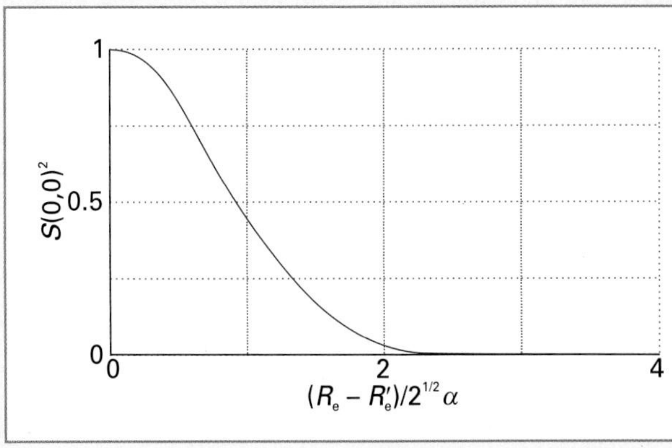

Fig. 11.11 The Franck–Condon factor for the arrangement discussed in Example 11.2.

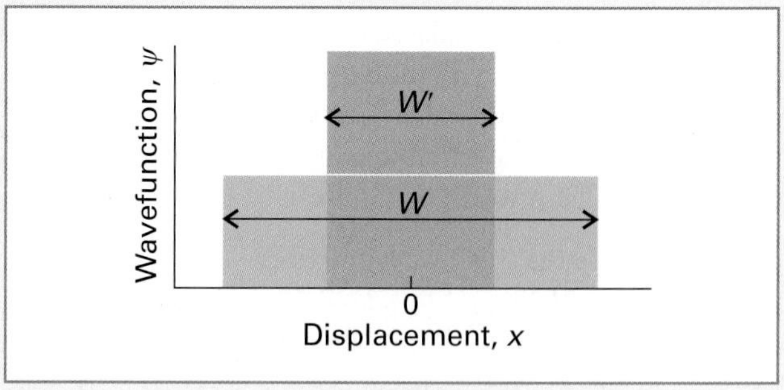

Fig. 11.12 The model wavefunctions used in *Self-test 11.3*.

Self-test 11.3 Suppose the vibrational wavefunctions can be approximated by rectangular functions of width W and W', centred on the equilibrium bond lengths (Fig. 11.12). Find the corresponding Franck–Condon factors when the centres are coincident and $W' < W$. [$S^2 = W'/W$]

(e) Rotational structure

Just as in vibrational spectroscopy, where a vibrational transition is accompanied by rotational excitation, so rotational transitions accompany the vibrational excitation that accompanies electronic excitation. We therefore see P, Q, and R branches for each vibrational transition, and the electronic transition has a very rich structure. However, the principal difference is that electronic excitation can result in much larger changes in bond length than vibrational excitation causes alone, and the rotational branches have a more complex structure than in vibration–rotation spectra.

We suppose that the rotational constants of the electronic ground and excited states are $\tilde{B}$ and $\tilde{B}'$, respectively. The rotational energy levels of the initial and final states are

$$E(J) = hc\tilde{B}J(J+1) \qquad E(J') = hc\tilde{B}'J'(J'+1)$$

and the rotational transitions occur at the following positions relative to the vibrational transition of wavenumber $\tilde{\nu}$ that they accompany:

$$\text{P branch } (\Delta J=-1): \tilde{\nu}_P(J) = \tilde{\nu} - (\tilde{B}'+\tilde{B})J + (\tilde{B}'-\tilde{B})J^2 \quad (11.9a)$$

$$\text{Q branch } (\Delta J=0): \tilde{\nu}_Q(J) = \tilde{\nu} + (\tilde{B}'-\tilde{B})J(J+1) \quad (11.9b)$$

$$\text{R branch } (\Delta J=+1): \tilde{\nu}_R(J) = \tilde{\nu} + (\tilde{B}'+\tilde{B})(J+1) + (\tilde{B}'-\tilde{B})(J+1)^2 \quad (11.9c)$$

First, suppose that the bond length in the electronically excited state is greater than that in the ground state; then $\tilde{B}' < \tilde{B}$ and $\tilde{B}' - \tilde{B}$ is negative. In this case the lines of the R branch converge with increasing J and, when J is such that $|\tilde{B}' - \tilde{B}|(J+1) > \tilde{B}' + \tilde{B}$, the lines start to appear at successively decreasing wavenumbers. That is, the R branch has a **band head** (Fig. 11.13a). When the bond is shorter in the excited state than in the ground state, $\tilde{B}' > \tilde{B}$ and $\tilde{B}' - \tilde{B}$ is positive. In this case, the lines of the P branch begin to converge and go through a head when J is such that $|\tilde{B}' - \tilde{B}|J > \tilde{B}' + \tilde{B}$ (Fig. 11.13b).

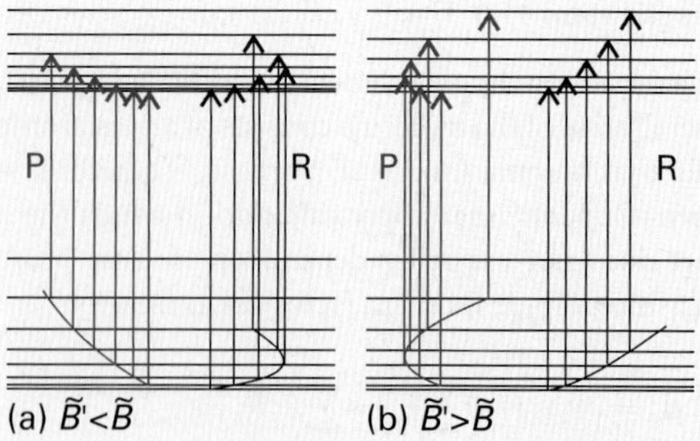

Fig. 11.13 When the rotational constants of a diatomic molecule differ significantly in the initial and final states of an electronic transition, the P and R branches show a head. (a) The formation of a head in the R branch when $\tilde{B}' < \tilde{B}$; (b) the formation of a head in the P branch when $\tilde{B}' > \tilde{B}$.

11.4 The electronic spectra of polyatomic molecules

The absorption of a photon can often be traced to the excitation of specific types of electrons or to electrons that belong to a small group of atoms in a polyatomic molecule. For example, when a carbonyl group (>C=O) is present, an absorption at about 290 nm is normally observed, although its precise location depends on the nature of the rest of the molecule. Groups with characteristic optical absorptions are called **chromophores** (from the Greek for 'colour bringer'), and their presence often accounts for the colours of substances (Table 11.3).

(a) d–d transitions

In a free atom, all five d orbitals of a given shell are degenerate. In a d-metal complex, where the immediate environment of the atom is no longer spherical, the d orbitals are not all degenerate, and electrons can absorb energy by making transitions between

Synoptic table 11.3* Absorption characteristics of some groups and molecules

Group	$\tilde{\nu}$/cm^{-1}	λ_{max}/nm	ε/(dm^3 mol^{-1} cm^{-1})
C=C (π*←π)	61 000	163	15 000
	57 300	174	5500
C=O (π*←n)	35 000–37 000	270–290	10–20
H_2O (π*←n)	60 000	167	7000

* More values are given in the *Data section*.

them. We show in the following *Justification* that in an octahedral complex, such as $[Ti(OH_2)_6]^{3+}$ (1), the five d orbitals of the central atom are split into two sets (2), a triply degenerate set labelled t_{2g} and a doubly degenerate set labelled e_g. The three t_{2g} orbitals lie below the two e_g orbitals; the difference in energy is denoted Δ_O and called the **ligand-field splitting parameter** (the O denoting octahedral symmetry).

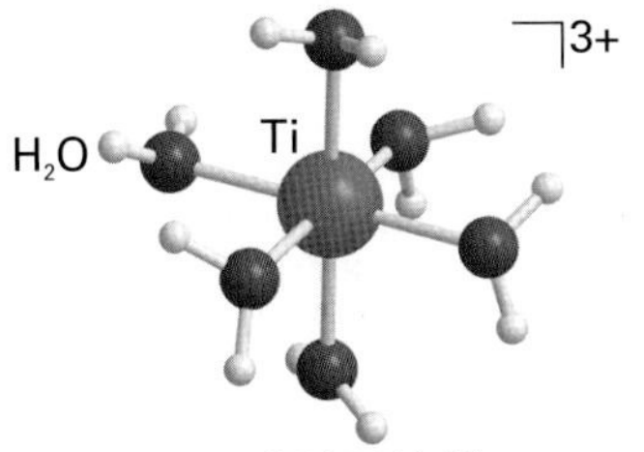

1 $[Ti(OH_2)_6]^{3+}$

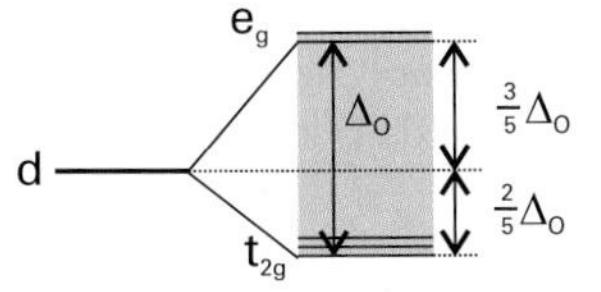

2

Justification 11.4 *The splitting of d-orbitals in an octahedral d-metal complex*

In an octahedral d-metal complex, six identical ions or molecules, the *ligands*, are at the vertices of a regular octahedron, with the metal ion at its centre. The ligands can be regarded as point negative charges that are repelled by the d electrons of the central ion. Figure 11.14 shows the consequence of this arrangement: the five d orbitals fall into two groups, with $d_{x^2-y^2}$ and d_{z^2} pointing directly towards the ligand positions, and d_{xy}, d_{yz}, and d_{zx} pointing between them. An electron occupying an orbital of the former group has a less favourable potential energy than when it occupies any of the three orbitals of the other group, and so the d orbitals split into the two sets shown in (2) with an energy difference Δ_O: a triply degenerate set comprising the d_{xy}, d_{yz}, and d_{zx} orbitals and labelled t_{2g}, and a doubly degenerate set comprising the $d_{x^2-y^2}$ and d_{z^2} orbitals and labelled e_g.

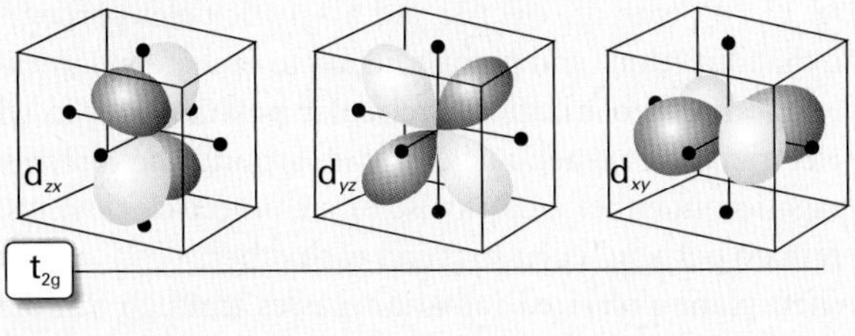

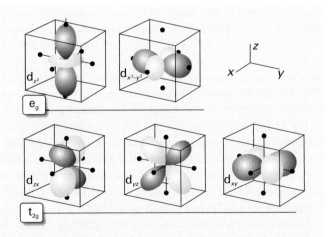

Fig. 11.14 The classification of d orbitals in an octahedral environment.

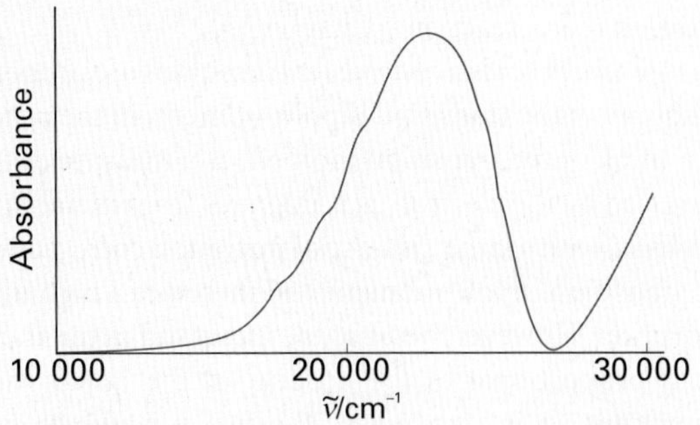

Fig. 11.15 The electronic absorption spectrum of $[Ti(OH_2)_6]^{3+}$ in aqueous solution.

The d orbitals also divide into two sets in a tetrahedral complex, but in this case the e orbitals lie below the t_2 orbitals and their separation is written Δ_T. Neither Δ_O nor Δ_T is large, so transitions between the two sets of orbitals typically occur in the visible region of the spectrum. The transitions are responsible for many of the colours that are so characteristic of d-metal complexes. As an example, the spectrum of $[Ti(OH_2)_6]^{3+}$ near 20 000 cm^{-1} (500 nm) is shown in Fig. 11.15, and can be ascribed to the promotion of its single d electron from a t_{2g} orbital to an e_g orbital. The wavenumber of the absorption maximum suggests that $\Delta_O \approx 20\,000$ cm^{-1} for this complex, which corresponds to about 2.5 eV.

According to the Laporte rule (Section 11.3), d–d transitions are parity-forbidden in octahedral complexes because they are $g \rightarrow g$ transitions (more specifically e_g–t_{2g} transitions). However, d–d transitions become weakly allowed as vibronic transitions as a result of coupling to asymmetrical vibrations such as that shown in Fig. 11.7.

(b) Charge-transfer transitions

A complex may absorb radiation as a result of the transfer of an electron from the ligands into the d orbitals of the central atom, or vice versa. In such **charge-transfer transitions** the electron moves through a considerable distance, which means that the transition dipole moment may be large and the absorption is correspondingly intense. This mode of chromophore activity is shown by the permanganate ion, MnO_4^-, and accounts for its intense violet colour (which arises from strong absorption within the range 420–700 nm). In this oxoanion, the electron migrates from an orbital that is largely confined to the O atom ligands to an orbital that is largely confined to the Mn atom. It is therefore an example of a **ligand-to-metal charge-transfer transition** (LMCT). The reverse migration, a **metal-to-ligand charge-transfer transition** (MLCT), can also occur. An example is the transfer of a d electron into the antibonding π orbitals of an aromatic ligand. The resulting excited state may have a very long lifetime if the electron is extensively delocalized over several aromatic rings, and such species can participate in photochemically induced redox reactions (Chapter 19).

The intensities of charge-transfer transitions are proportional to the square of the transition dipole moment, in the usual way. We can think of the transition moment as a measure of the distance moved by the electron as it migrates from metal to ligand or vice versa, with a large distance of migration corresponding to a large transition dipole moment and therefore a high intensity of absorption. However, because the integrand in the transition dipole is proportional to the product of the initial and final wavefunctions, it is zero unless the two wavefunctions have nonzero values in the same region of space. Therefore, although large distances of migration favour high intensities, the diminished overlap of the initial and final wavefunctions for large separations of metal and ligands favours low intensities (see Problem 11.22). We encounter similar considerations when we examine electron transfer reactions (*Impact I20.1*), which can be regarded as a special type of charge-transfer transition.

(c) $\pi^*\leftarrow\pi$ and $\pi^*\leftarrow$n transitions

Absorption by a C=C double bond results in the excitation of a π electron into an antibonding π^* orbital (Fig. 11.16). The chromophore activity is therefore due to a $\pi^*\leftarrow\pi$ transition (which is normally read 'π to π-star transition'). Its energy is about 7 eV for an unconjugated double bond, which corresponds to an absorption at 180 nm (in the ultraviolet). When the double bond is part of a conjugated chain, the energies of the molecular orbitals lie closer together and the $\pi^*\leftarrow\pi$ transition moves to longer wavelength; it may even lie in the visible region if the conjugated system is long enough. An important example of a $\pi^*\leftarrow\pi$ transition is provided by the photochemical mechanism of vision (*Impact I11.1*).

The transition responsible for absorption in carbonyl compounds can be traced to the lone pairs of electrons on the O atom. The Lewis concept of a 'lone pair' of electrons is represented in molecular orbital theory by a pair of electrons in an orbital confined largely to one atom and not appreciably involved in bond formation. One of these electrons may be excited

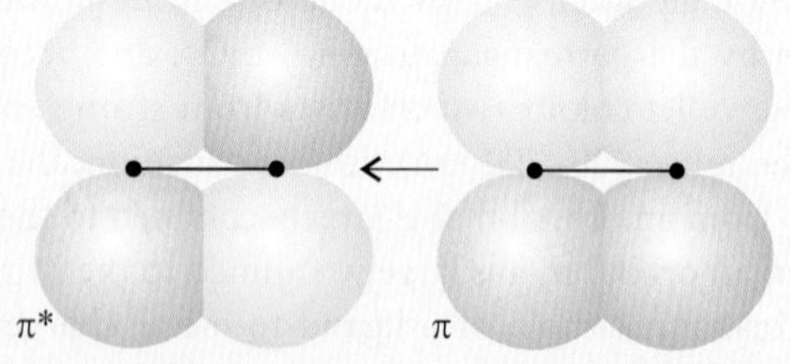

Fig. 11.16 A C=C double bond acts as a chromophore. One of its important transitions is the $\pi^*\leftarrow\pi$ transition illustrated here, in which an electron is promoted from a π orbital to the corresponding antibonding orbital.

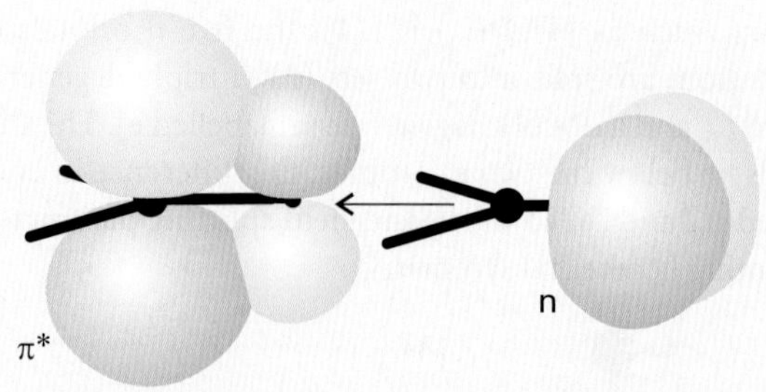

Fig. 11.17 A carbonyl group (C=O) acts as a chromophore primarily on account of the excitation of a nonbonding O lone-pair electron to an antibonding CO π^* orbital.

into an empty π^* orbital of the carbonyl group (Fig. 11.17), which gives rise to a $\pi^*\leftarrow$n transition (an 'n to π-star transition'). Typical absorption energies are about 4 eV (290 nm). Because $\pi^*\leftarrow$n transitions in carbonyls are symmetry forbidden, the absorptions are weak.

(d) Multiphoton processes

The large number of photons in an incident beam generated by a laser gives rise to a qualitatively different branch of spectroscopy, for the photon density is so high that more than one photon may be absorbed by a single molecule and give rise to **multiphoton processes**. One application of multiphoton processes is that states inaccessible by conventional one-photon spectroscopy become observable because the overall transition occurs with no change of parity. For example, in one-photon spectroscopy, only g $\leftrightarrow$ u transitions are observable; in two-photon spectroscopy, however, the overall outcome of absorbing two photons is a g $\rightarrow$ g or a u $\rightarrow$ u transition.

(e) Circular dichroism spectroscopy

Electronic spectra can reveal additional details of molecular structure when experiments are conducted with **polarized light**, electromagnetic radiation with electric and magnetic fields that oscillate only in certain directions. Light is **plane polarized** when the electric and magnetic fields each oscillate in a single plane (Fig. 11.18). The plane of polarization may be oriented in any direction around the direction of propagation (the *x*-direction in Fig. 11.18), with the electric and magnetic fields perpendicular to that direction (and perpendicular to each other). An alternative mode of polarization is **circular polarization**, in which the electric and magnetic fields rotate around the direction of propagation in either a clockwise or a counter-clockwise sense but remain perpendicular to it and each other.

When plane-polarized radiation passes through samples of certain kinds of matter, the plane of polarization is rotated around the direction of propagation. This rotation is the familiar phenomenon of optical activity, observed when the molecules in the

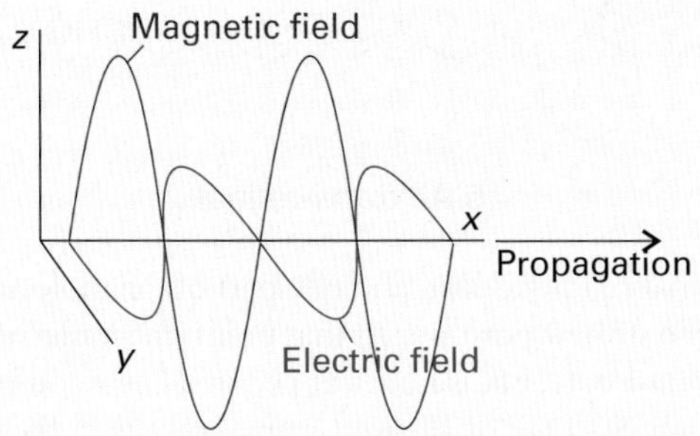

Fig. 11.18 Electromagnetic radiation consists of a wave of electric and magnetic fields perpendicular to the direction of propagation (in this case the *x*-direction), and mutually perpendicular to each other. This illustration shows a plane-polarized wave, with the electric and magnetic fields oscillating in the *xy*- and *xz*-planes, respectively.

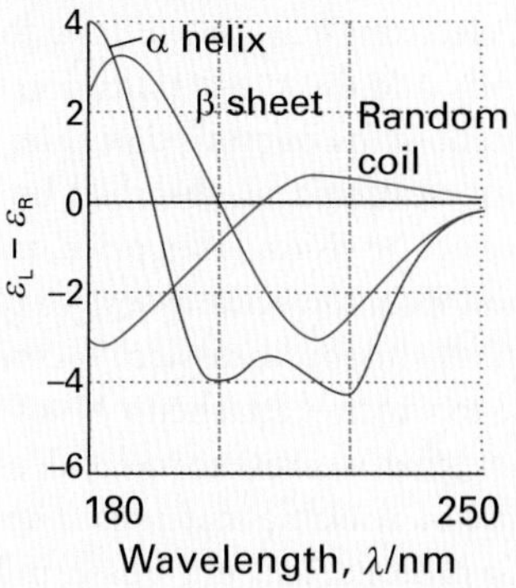

Fig. 11.20 Representative CD spectra of polypeptides. Random coils, α-helices, and β-sheets have different CD features in the spectral region where the peptide link absorbs.

sample are chiral (Section 7.3). Chiral molecules have a second characteristic: they absorb left and right circularly polarized light to different extents. In a circularly polarized ray of light, the electric field describes a helical path as the wave travels through space (Fig. 11.19), and the rotation may be either clockwise or counter-clockwise. The differential absorption of left- and right-circularly polarized light is called **circular dichroism**. In terms of the absorbances for the two components, A_L and A_R, the circular dichroism of a sample of molar concentration [J] is reported as

$$\Delta\varepsilon = \varepsilon_L - \varepsilon_R = \frac{A_L - A_R}{[J]l} \qquad (11.10)$$

where l is the path length of the sample.

Circular dichroism (CD) spectroscopy provides a great deal of information about the secondary structure of a biological polymer, with the spectrum of the polymer chain arising from the chirality of individual monomer units and, in addition, a contribution from the three-dimensional structure of the polymer itself. Consider a helical polypeptide. Not only are the individual monomer units chiral, but so is the helix. Therefore, we expect the α-helix to have a unique CD spectrum. Because β-sheets and random coils also have distinguishable spectral features (Fig. 11.20), circular dichroism is a very important technique for the study of protein conformation.

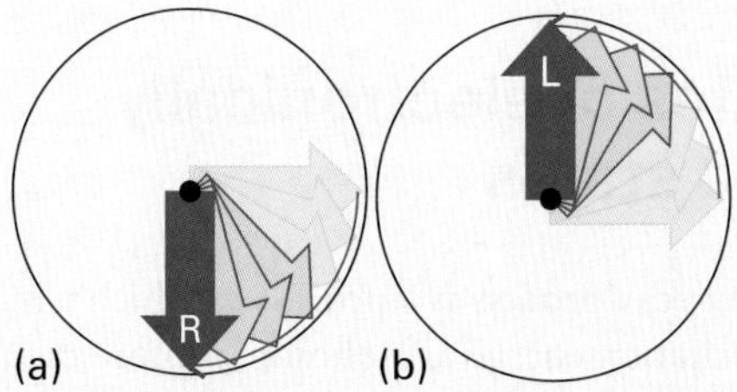

Fig. 11.19 In circularly polarized light, the electric field at different points along the direction of propagation rotates. The arrays of arrows in these illustrations show the view of the electric field when looking toward the oncoming ray: (a) right-circularly polarized, (b) left-circularly polarized light.

IMPACT ON BIOCHEMISTRY

I11.1 Vision

The eye is an exquisite photochemical organ that acts as a transducer, converting radiant energy into electrical signals that travel along neurons. Here we concentrate on the events taking place in the human eye, but similar processes occur in all animals. Indeed, a single type of protein, rhodopsin, is the primary receptor for light throughout the animal kingdom, which indicates that vision emerged very early in evolutionary history, no doubt because of its enormous value for survival.

Photons enter the eye through the cornea, pass through the ocular fluid that fills the eye, and fall on the retina. The ocular fluid is principally water, and passage of light through this medium is largely responsible for the *chromatic aberration* of the eye, the blurring of the image as a result of different frequencies being brought to slightly different focuses. The chromatic aberration is reduced to some extent by the tinted region called the *macular pigment* that covers part of the retina. The pigments in this region are the carotene-like xanthophylls (3), which absorb some of the blue light and hence help to sharpen the image. They also protect the photoreceptor molecules from too great a flux of

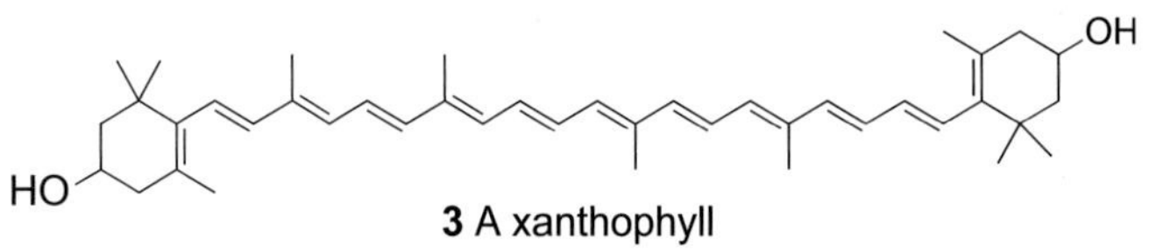

3 A xanthophyll

potentially dangerous high energy photons. The xanthophylls have delocalized electrons that spread along the chain of conjugated double bonds, and the $\pi^* \leftarrow \pi$ transition lies in the visible.

About 57 per cent of the photons that enter the eye reach the retina; the rest are scattered or absorbed by the ocular fluid. Here the primary act of vision takes place, in which the chromophore of a rhodopsin molecule absorbs a photon in another $\pi^* \leftarrow \pi$ transition. A rhodopsin molecule consists of an opsin protein molecule to which is attached a 11-*cis*-retinal molecule. The latter resembles half a carotene molecule, showing Nature's economy in its use of available materials. The attachment is by the formation of a protonated Schiff's base, utilizing the —CHO group of the chromophore and the terminal NH_2 group of the sidechain of a lysine residue from opsin (**4**). The free 11-*cis*-retinal molecule absorbs in the ultraviolet, but attachment to the opsin protein molecule shifts the absorption into the visible region. The rhodopsin molecules are situated in the membranes of special cells (the 'rods' and the 'cones') that cover the retina. The opsin molecule is anchored into the cell membrane by two hydrophobic groups and largely surrounds the chromophore (Fig. 11.21).

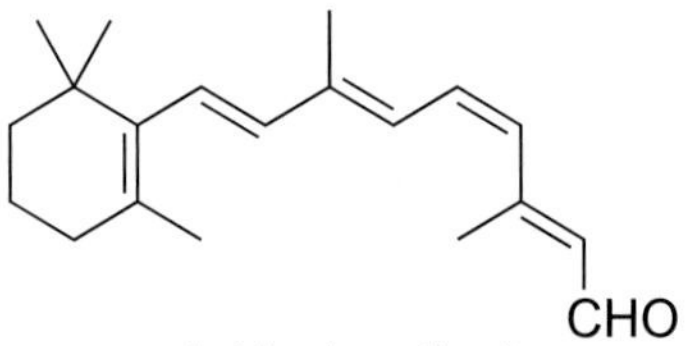

4 11-*cis*-retinal

Immediately after the absorption of a photon, the 11-*cis*-retinal molecule undergoes photoisomerization into all-*trans*-retinal (**5**). Photoisomerization takes about 200 fs and about 67 pigment molecules isomerize for every 100 photons that are absorbed. The process occurs because the $\pi^* \leftarrow \pi$ excitation of an electron loosens one of the π bonds (the one indicated by the arrow in **5**),

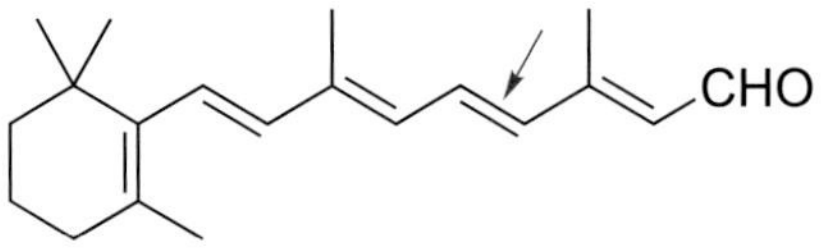

5 All-*trans*-retinal

its torsional rigidity is lost, and one part of the molecule swings round into its new position. At that point, the molecule returns to its ground state, but is now trapped in its new conformation. The straightened tail of all-*trans*-retinal results in the molecule taking up more space than 11-*cis*-retinal did, so the molecule presses against the coils of the opsin molecule that surrounds it. In about 0.25–0.50 ms from the initial absorption event, the rhodopsin molecule is activated both by the isomerization of retinal and deprotonation of its Schiff's base tether to opsin, forming an intermediate known as *metarhodopsin II*.

In a sequence of biochemical events known as the *biochemical cascade*, metarhodopsin II activates the protein transducin, which in turn activates a phosphodiesterase enzyme that hydrolyses cyclic guanine monophosphate (cGMP) to GMP. The reduction in the concentration of cGMP causes ion channels, proteins that mediate the movement of ions across biological membranes, to close and the result is a sizeable change in the transmembrane potential (see *Impact I17.1* for a discussion of transmembrane potentials). The pulse of electric potential travels through the optical nerve and into the optical cortex, where it is interpreted as a signal and incorporated into the web of events we call 'vision'.

The resting state of the rhodopsin molecule is restored by a series of nonradiative chemical events powered by ATP. The process involves the escape of all-*trans*-retinal as all-*trans*-retinol (in which —CHO has been reduced to $-CH_2OH$) from the opsin molecule by a process catalysed by the enzyme rhodopsin kinase and the attachment of another protein molecule, arrestin. The free all-*trans*-retinol molecule now undergoes enzyme-catalysed isomerization into 11-*cis*-retinol followed by dehydrogenation to form 11-*cis*-retinal, which is then delivered back into an opsin molecule. At this point, the cycle of excitation, photoisomerization, and regeneration is ready to begin again.

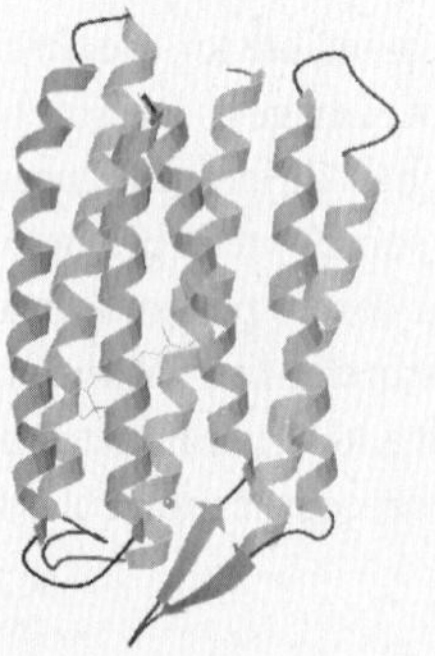

Fig. 11.21 The structure of the rhodopsin molecule, consisting of an opsin protein to which is attached a 11-*cis*-retinal molecule embedded in the space surrounded by the helical regions.

The fates of electronically excited states

A **radiative decay process** is a process in which a molecule discards its excitation energy as a photon. A more common fate is **nonradiative decay**, in which the excess energy is transferred into the vibration, rotation, and translation of the molecule and those surrounding it. This thermal degradation converts the excitation energy completely into thermal motion of the environment (that is, to 'heat'). An excited molecule may also take part in a chemical reaction, as we discuss in Chapter 19.

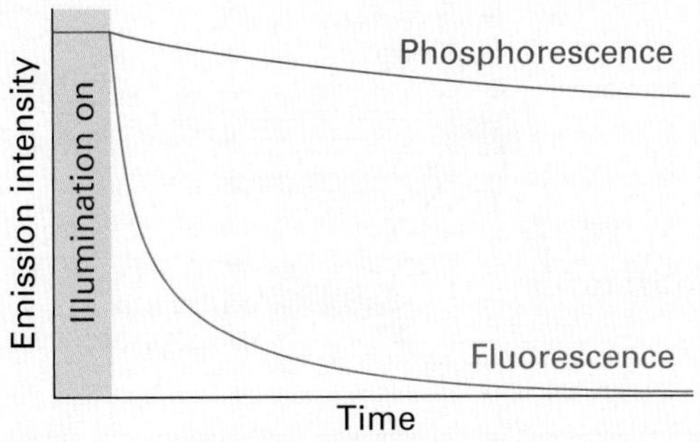

Fig. 11.22 The empirical (observation-based) distinction between fluorescence and phosphorescence is that the former is extinguished very quickly after the exciting source is removed, whereas the latter continues with relatively slowly diminishing intensity.

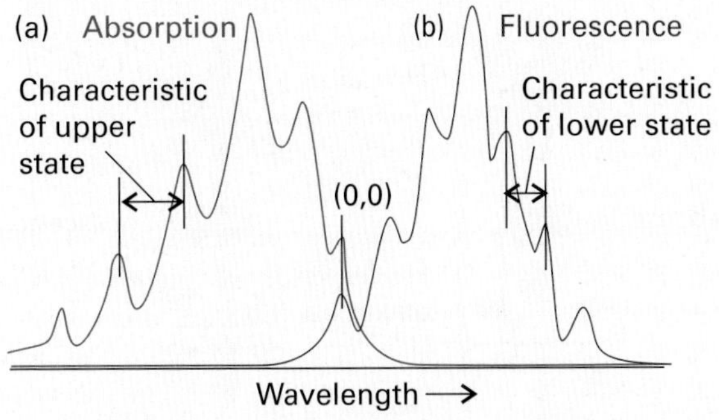

Fig. 11.24 An absorption spectrum (a) shows a vibrational structure characteristic of the upper state. A fluorescence spectrum (b) shows a structure characteristic of the lower state; it is also displaced to lower frequencies (but the 0–0 transitions are coincident) and resembles a mirror image of the absorption.

11.5 Fluorescence and phosphorescence

In **fluorescence**, spontaneous emission of radiation occurs within a few nanoseconds after the exciting radiation is extinguished (Fig. 11.22). In **phosphorescence**, the spontaneous emission may persist for long periods (even hours, but characteristically seconds or fractions of seconds). The difference suggests that fluorescence is a fast conversion of absorbed radiation into re-emitted energy, and that phosphorescence involves the storage of energy in a reservoir from which it slowly leaks.

(a) Fluorescence

Figure 11.23 shows the sequence of steps involved in fluorescence. The initial absorption takes the molecule to an excited electronic state, and if the absorption spectrum were monitored

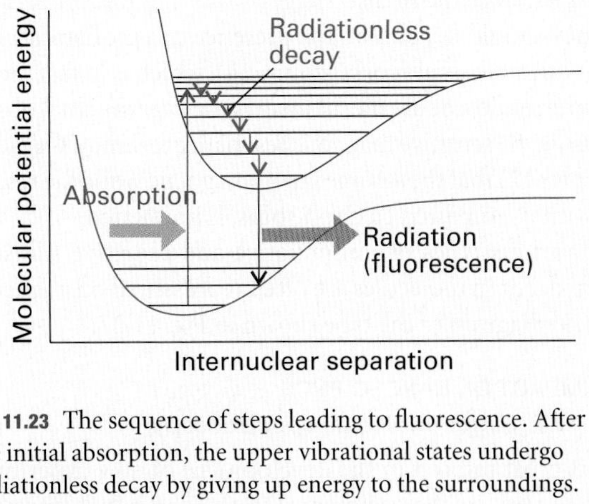

Fig. 11.23 The sequence of steps leading to fluorescence. After the initial absorption, the upper vibrational states undergo radiationless decay by giving up energy to the surroundings. A radiative transition then occurs from the vibrational ground state of the upper electronic state.

it would look like the one shown in Fig. 11.24a. The excited molecule can redistribute its energy and also is subjected to collisions with the surrounding molecules. As it gives up energy nonradiatively the molecule steps down the ladder of vibrational levels to the lowest vibrational level of the electronically excited molecular state. The surrounding molecules, however, might now be unable to accept the larger energy difference needed to lower the molecule to the ground electronic state. It might therefore survive long enough to undergo spontaneous emission, and emit the remaining excess energy as radiation. The downward electronic transition is vertical (in accord with the Franck–Condon principle) and the fluorescence spectrum has a vibrational structure characteristic of the *lower* electronic state (Fig. 11.24b).

Provided they can be seen, the 0–0 absorption and fluorescence transitions can be expected to be coincident. The absorption spectrum arises from 1–0, 2–0, . . . transitions that occur at progressively higher wavenumber and with intensities governed by the Franck–Condon principle. The fluorescence spectrum arises from 0–0, 0–1, . . . *downward* transitions that hence occur with decreasing wavenumbers. The 0–0 absorption and fluorescence peaks are not always exactly coincident, however, because the solvent may interact differently with the solute in the ground and excited states (for instance, the hydrogen bonding pattern might differ). Because the solvent molecules do not have time to rearrange during the transition, the absorption occurs in an environment characteristic of the solvated ground state; however, the fluorescence occurs in an environment characteristic of the solvated excited state (Fig. 11.25).

Fluorescence occurs at lower frequencies (longer wavelengths) than the incident radiation because the emissive transition occurs after some vibrational energy has been discarded into the surroundings. The vivid oranges and greens of fluorescent dyes are

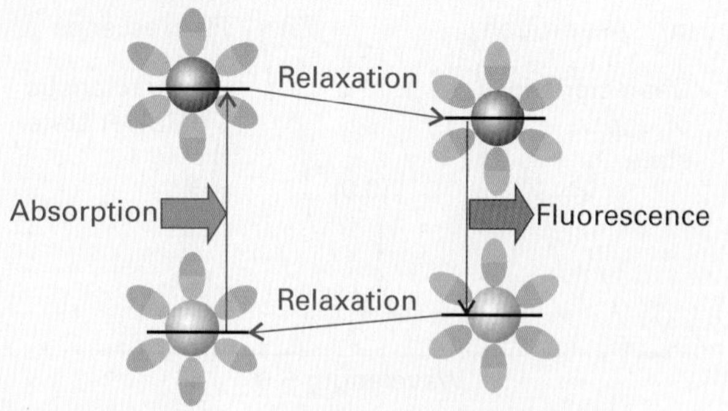

Fig. 11.25 The solvent can shift the fluorescence spectrum relative to the absorption spectrum. On the left we see that the absorption occurs with the solvent (the ellipses) in the arrangement characteristic of the ground electronic state of the molecule (the sphere). However, before fluorescence occurs, the solvent molecules relax into a new arrangement, and that arrangement is preserved during the subsequent radiative transition.

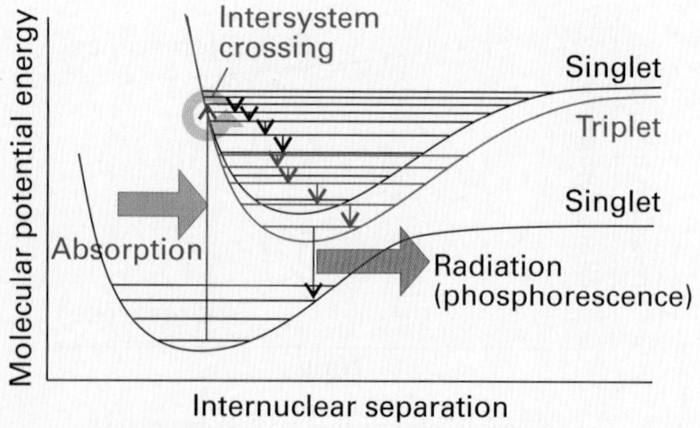

Fig. 11.26 The sequence of steps leading to phosphorescence. The important step is the intersystem crossing, the switch from a singlet state to a triplet state brought about by spin–orbit coupling. The triplet state acts as a slowly radiating reservoir because the return to the ground state is spin-forbidden.

an everyday manifestation of this effect: they absorb in the ultraviolet and blue, and fluoresce in the visible. The mechanism also suggests that the intensity of the fluorescence ought to depend on the ability of the solvent molecules to accept the electronic and vibrational quanta. It is indeed found that a solvent composed of molecules with widely spaced vibrational levels (such as water) can in some cases accept the large quantum of electronic energy and so extinguish, or 'quench', the fluorescence. We examine the mechanisms of fluorescence quenching in Section 19.9.

(b) Phosphorescence

Figure 11.26 shows the sequence of events leading to phosphorescence for a molecule with a singlet ground state. The first steps are the same as in fluorescence, but the presence of a triplet excited state plays a decisive role. The singlet and triplet excited states share a common geometry at the point where their potential energy curves intersect. Hence, if there is a mechanism for unpairing two electron spins (and achieving the conversion of $\uparrow\downarrow$ to $\uparrow\uparrow$), the molecule may undergo **intersystem crossing**, a nonradiative transition between states of different multiplicity, and become a triplet state. We saw in the discussion of atomic spectra (Section 4.5) that singlet–triplet transitions may occur in the presence of spin–orbit coupling, and the same is true in molecules. We can expect intersystem crossing to be important when a molecule contains a moderately heavy atom (such as S), because then the spin–orbit coupling is large.

If an excited molecule crosses into a triplet state, it continues to deposit energy into the surroundings. However, it is now stepping down the triplet's vibrational ladder, and at the lowest energy level it is trapped because the triplet state is at a lower energy than the corresponding singlet (recall Hund's rule, Section 4.4). The solvent cannot absorb the final, large quantum of electronic excitation energy, and the molecule cannot radiate its energy because return to the ground state is spin-forbidden. The radiative transition, however, is not totally forbidden because the spin–orbit coupling that was responsible for the intersystem crossing also breaks the selection rule. The molecules are therefore able to emit weakly, and the emission may continue long after the original excited state was formed.

The mechanism accounts for the observation that the excitation energy seems to get trapped in a slowly leaking reservoir. It also suggests (as is confirmed experimentally) that phosphorescence should be most intense from solid samples: energy transfer is then less efficient and intersystem crossing has time to occur as the singlet excited state steps slowly past the intersection point. The mechanism also suggests that the phosphorescence efficiency should depend on the presence of a moderately heavy atom (with strong spin–orbit coupling), which is in fact the case. The confirmation of the mechanism is the experimental observation (using the sensitive magnetic resonance techniques described in Chapter 12) that the sample is paramagnetic while the reservoir state, with its unpaired electron spins, is populated.

The various types of nonradiative and radiative transitions that can occur in molecules are often represented on a schematic **Jablonski diagram** of the type shown in Fig. 11.27.

IMPACT ON NANOSCIENCE

I11.2 Single-molecule spectroscopy

There is great interest in the development of new experimental probes of very small specimens. On the one hand, our understanding of biochemical processes, such as enzymatic catalysis, protein folding, and the insertion of DNA into a cell's nucleus,

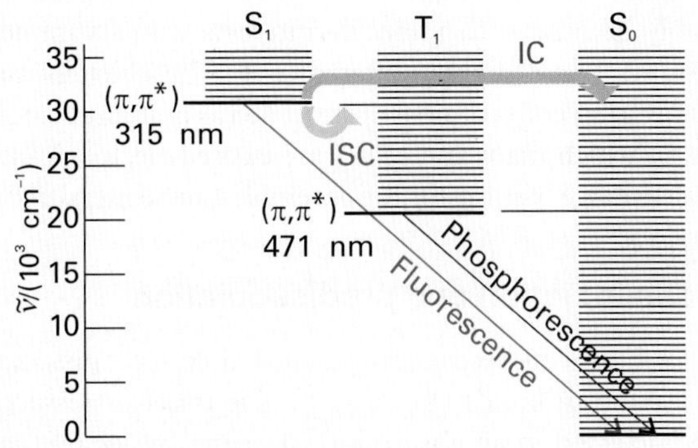

Fig. 11.27 A Jablonski diagram (here, for naphthalene) is a simplified portrayal of the relative positions of the electronic energy levels of a molecule. Vibrational levels of states of a given electronic state lie above each other, but the relative horizontal locations of the columns bear no relation to the nuclear separations in the states. The ground vibrational states of each electronic state are correctly located vertically but the other vibrational states are shown only schematically. (IC: internal conversion, Section 11.6; ISC: intersystem crossing.)

will be enhanced if it is possible to visualize individual biopolymers at work in cells. On the other hand, techniques that can probe the structure, dynamics, and reactivity of single molecules are needed to advance research on nanometre-sized materials. We saw in *Impact I10.3* that it is possible to obtain the vibrational spectrum of samples with areas of more than 10 μm^2. **Fluorescence microscopy**, in which the distribution of fluorescence intensity within an illuminated area is detected with a microscope, has also been used for many years to image small specimens, such as biological cells, but the diffraction limit prevents the visualization of samples that are smaller than the wavelength of light used as a probe. Most molecules—including biological polymers—have dimensions that are much smaller than visible wavelengths, so special techniques had to be developed to increase the resolution of light microscopy techniques so that even single molecules can be observed.

Apart from a small number of co-factors, such as the chlorophylls and flavins, the majority of the building blocks of proteins and nucleic acids do not fluoresce strongly. Four notable exceptions are the amino acids tryptophan ($\lambda_{abs} \approx 280$ nm and $\lambda_{fluor} \approx 348$ nm in water), tyrosine ($\lambda_{abs} \approx 274$ nm and $\lambda_{fluor} \approx 303$ nm in water), and phenylalanine ($\lambda_{abs} \approx 257$ nm and $\lambda_{fluor} \approx 282$ nm in water), and the oxidized form of the sequence serine–tyrosine–glycine (**6**) found in the green fluorescent protein (GFP) of certain jellyfish. The wild type of GFP from *Aequora victoria* absorbs strongly at 395 nm and emits maximally at 509 nm. The visualization of biological cells with fluorescence microscopy is achieved by detecting light emitted by a large number of fluorescent molecules attached to proteins, nucleic acids, and membranes. A common fluorescent label is GFP. With proper

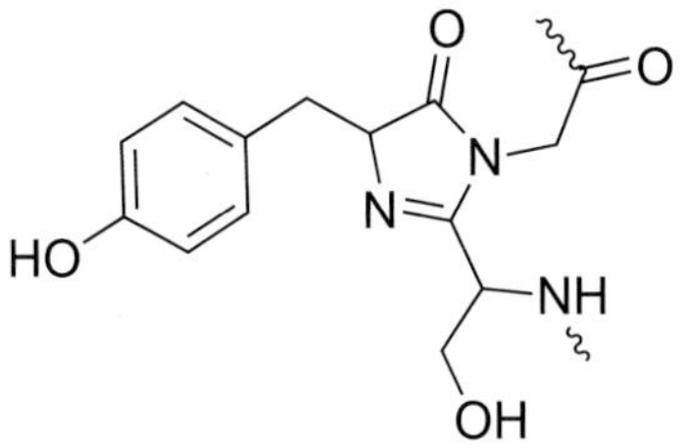

6 The chromophore of GFP

filtering to remove light due to Rayleigh scattering of the incident beam, it is possible to collect light from the sample that contains only fluorescence from the label. However, great care is required to eliminate fluorescent impurities from the sample.

The bulk of the work done in the field of single-molecule spectroscopy is based on fluorescence microscopy with laser excitation. The laser is the radiation source of choice because it provides the high excitance required to increase the rate of arrival of photons on to the detector from small illuminated areas. Two techniques are commonly used to circumvent the diffraction limit. First, the concentration of the sample is kept so low that, on average, only one fluorescent molecule is in the illuminated area. Second, special strategies are used to illuminate very small volumes. In **near-field scanning optical microscopy** (NSOM), a very thin metal-coated optical fibre is used to deliver light to a small area. It is possible to construct fibres with tip diameters in the range of 50 to 100 nm, which are indeed smaller than visible wavelengths. The fibre tip is placed very close to the sample, in a region known as the *near field*, where, according to classical physics, photons do not diffract. Figure 11.28 shows the image of a 4.5 μm × 4.5 μm sample of oxazine 720 dye molecules embedded in a polymer film and obtained with NSOM by measuring the fluorescence intensity as the tip is scanned over the film surface. Each peak corresponds to a single dye molecule.

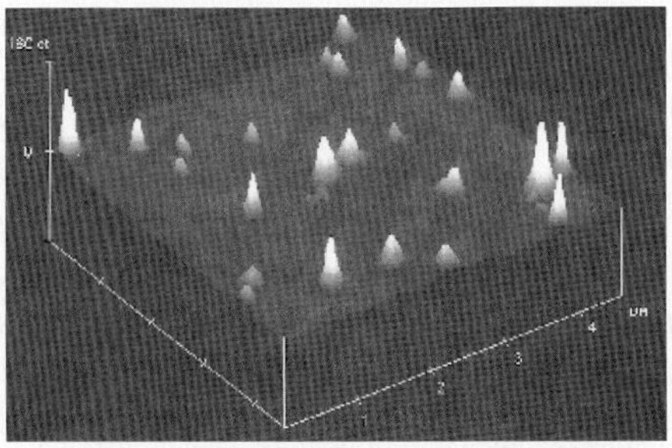

Fig. 11.28 Image of a 4.5 μm × 4.5 μm sample of oxazine-720 dye molecules embedded in a polymer film and obtained with NSOM. Each peak corresponds to a single dye molecule. (Reproduced with permission from X.S. Xie, *Acc. Chem. Res.* 1996, **29**, 598.)

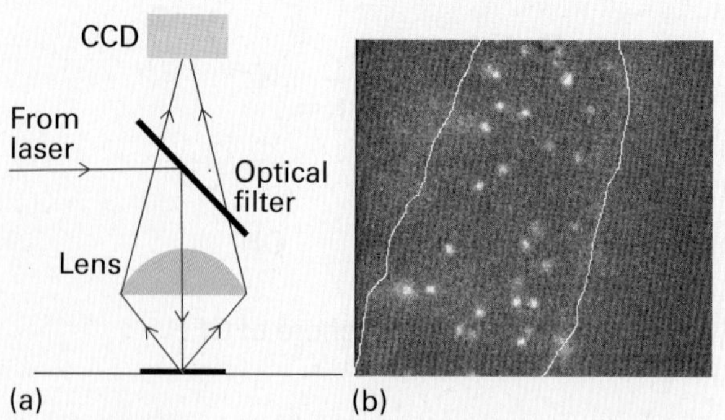

Fig. 11.29 (a) Layout of an epifluorescence microscope. Laser radiation is diverted to a sample by a special optical filter that reflects radiation with a specified wavelength (in this case the laser excitation wavelength) but transmits radiation with other wavelengths (in this case, wavelengths at which the fluorescent label emits). A CCD detector analyses the spatial distribution of the fluorescence signal from the illuminated area. (b) Observation of fluorescence from single MHC proteins that have been labelled with a fluorescent marker and are bound to the surface of a cell (the area shown has dimensions of 12 μm × 12 μm). (Image provided by Professor W.E. Moerner, Stanford University, USA.)

In **far-field confocal microscopy**, laser light focused by an objective lens is used to illuminate about 1 μm^3 of a very dilute sample placed beyond the near field. This illumination scheme is limited by diffraction and, as a result, data from far-field microscopy have less structural detail than data from NSOM. However, far-field microscopes are very easy to construct and the technique can be used to probe single molecules as long as there is one molecule, on average, in the illuminated area.

In the **wide-field epifluorescence method**, a two-dimensional array detector (Section 11.1) detects fluorescence excited by a laser and scattered back from the sample (Fig. 11.29a). If the fluorescing molecules are well separated in the specimen, then it is possible to obtain a map of the distribution of fluorescent molecules in the illuminated area. For example, Fig. 11.29b shows how epifluorescence microscopy can be used to observe single molecules of the major histocompatibility (MHC) protein on the surface of a cell.

Though still a relatively new technique, single-molecule spectroscopy has already been used to address important problems in chemistry and biology. Nearly all of the techniques discussed in this text measure the average value of a property in a large ensemble of molecules. Single-molecule methods allow a chemist to study the nature of distributions of physical and chemical properties in an ensemble of molecules. For example, it is possible to measure the fluorescence lifetime of a molecule by moving the laser focus to a location on the sample that contains a molecule and then measuring the fluorescence intensity after excitation with a pulsed laser. Such studies have shown that not every molecule in a sample has the same fluorescence lifetime, probably because each molecule interacts with its immediate environment in a slightly different way. These details are not apparent from conventional measurements of fluorescence lifetimes, in which many molecules are excited electronically and only an average lifetime for the ensemble can be measured.

11.6 Dissociation and predissociation

Another fate for an electronically excited molecule is **dissociation**, the breaking of bonds (Fig. 11.30). The onset of dissociation can be detected in an absorption spectrum by seeing that the vibrational structure of a band terminates at a certain energy. Absorption occurs in a continuous band above this **dissociation limit** because the final state is an unquantized translational motion of the fragments. Locating the dissociation limit is a valuable way of determining the bond dissociation energy.

In some cases, the vibrational structure disappears but resumes at higher photon energies. This **predissociation** can be interpreted in terms of the molecular potential energy curves shown in Fig. 11.31. When a molecule is excited to a vibrational level, its electrons may undergo a redistribution that results in it undergoing an **internal conversion**, a radiationless conversion to another state of the same multiplicity. An internal conversion occurs most readily at the point of intersection of the two molecular potential energy curves, because there the nuclear geometries of the two states are the same. The state into which the molecule converts may be dissociative, so the states near the intersection have a finite lifetime, and hence their energies are imprecisely defined (see Section 4.3). As a result, the absorption spectrum is blurred in the vicinity of the intersection. When the incoming photon brings enough energy to excite the molecule to a vibrational level high above the intersection, the internal conversion does not occur (the nuclei are unlikely to have the

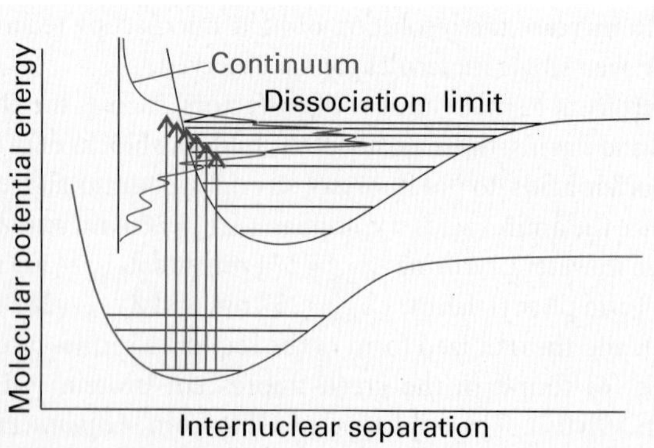

Fig. 11.30 When absorption occurs to unbound states of the upper electronic state, the molecule dissociates and the absorption is a continuum. Below the dissociation limit the electronic spectrum shows a normal vibrational structure.

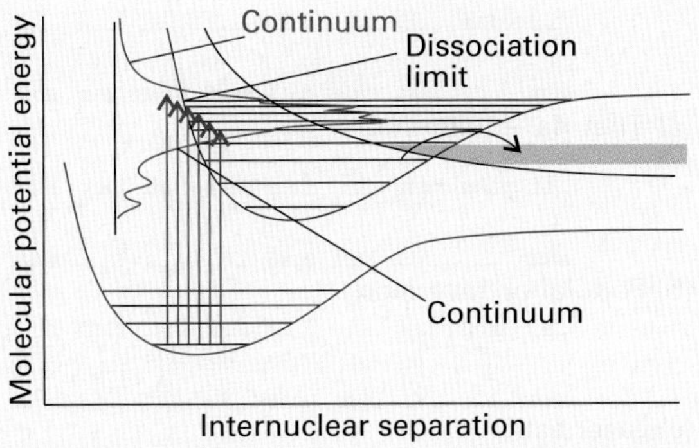

Fig. 11.31 When a dissociative state crosses a bound state, as in the upper part of the figure, molecules excited to levels near the crossing may dissociate. This process is called predissociation, and is detected in the spectrum as a loss of vibrational structure that resumes at higher frequencies.

same geometry). Consequently, the levels resume their well-defined, vibrational character with correspondingly well-defined energies, and the line structure resumes on the high-frequency side of the blurred region.

11.7 General principles of laser action

Lasers have transformed chemistry as much as they have transformed the everyday world. They lie very much on the frontier of physics and chemistry, for their operation depends on details of optics and, in some cases, of solid-state processes. In this section, we discuss the mechanisms of laser action, and then explore their applications in chemistry.

The word laser is an acronym formed from light amplification by stimulated emission of radiation. In stimulated emission, an excited state is stimulated to emit a photon by radiation of the same frequency; the more photons that are present, the greater the probability of the emission. The essential feature of laser action is positive-feedback: the more photons present of the appropriate frequency, the more photons of that frequency that will be stimulated to form.

(a) Population inversion

One requirement of laser action is the existence of a **metastable excited state**, an excited state with a long enough lifetime for it to participate in stimulated emission. Another requirement is the existence of a greater population in the metastable state than in the lower state where the transition terminates, for then there will be a net emission of radiation. Because at thermal equilibrium the opposite is true, it is necessary to achieve a **population inversion** in which there are more molecules in the upper state than in the lower.

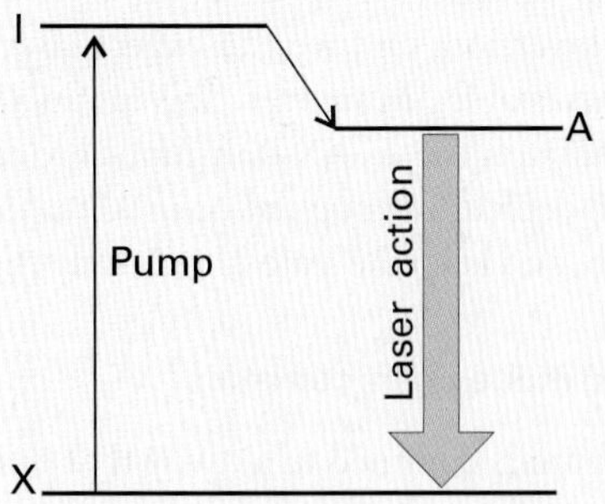

Fig. 11.32 The transitions involved in one kind of three-level laser. The pumping pulse populates the intermediate state I, which in turn populates the laser state A. The laser transition is the stimulated emission A←X.

One way of achieving population inversion is illustrated in Fig. 11.32. The molecule is excited to an intermediate state I, which then gives up some of its energy nonradiatively and changes into a lower state A; the laser transition is the return of A to the ground state X. Because three energy levels are involved overall, this arrangement leads to a **three-level laser**. In practice, I consists of many states, all of which can convert to the upper of the two laser states A. The I←X transition is stimulated with an intense flash of light in the process called **pumping**. The pumping is often achieved with an electric discharge through xenon or with the light of another laser. The conversion of I to A should be rapid, and the laser transitions from A to X should be relatively slow.

The disadvantage of this three-level arrangement is that it is difficult to achieve population inversion, because so many ground-state molecules must be converted to the excited state by the pumping action. The arrangement adopted in a **four-level laser** simplifies this task by having the laser transition terminate in a state A′ other than the ground state (Fig. 11.33). Because A′

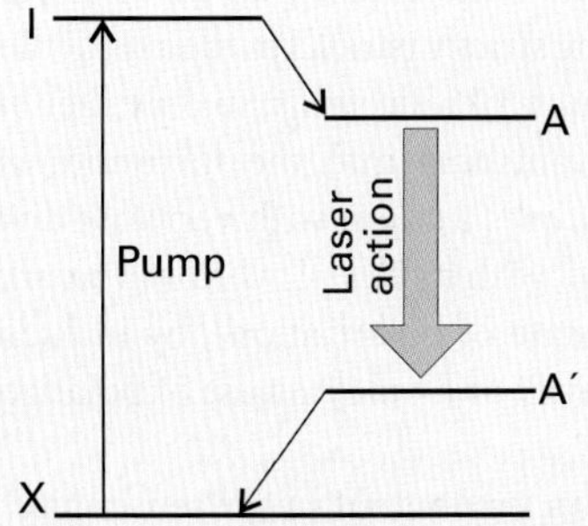

Fig. 11.33 The transitions involved in a four-level laser. Because the laser transition terminates in an excited state (A′), the population inversion between A and A′ is much easier to achieve.

is unpopulated initially, any population in A corresponds to a population inversion, and we can expect laser action if A is sufficiently metastable. Moreover, this population inversion can be maintained if the A′→X transitions are rapid, for these transitions will deplete any population in A′ that stems from the laser transition, and keep the state A′ relatively empty.

(b) Cavity and mode characteristics

The laser medium is confined to a cavity that ensures that only certain photons of a particular frequency, direction of travel, and state of polarization are generated abundantly. The cavity is essentially a region between two mirrors, which reflect the light back and forth. This arrangement can be regarded as a version of the particle in a box, with the particle now being a photon. As in the treatment of a particle in a box (Section 2.2), the only wavelengths that can be sustained satisfy

$$n \times \tfrac{1}{2}\lambda = L \qquad (11.11)$$

where n is an integer and L is the length of the cavity. That is, only an integral number of half-wavelengths fit into the cavity; all other waves undergo destructive interference with themselves. In addition, not all wavelengths that can be sustained by the cavity are amplified by the laser medium (many fall outside the range of frequencies of the laser transitions), so only a few contribute to the laser radiation. These wavelengths are the **resonant modes** of the laser.

Photons with the correct wavelength for the resonant modes of the cavity and the correct frequency to stimulate the laser transition are highly amplified. One photon might be generated spontaneously, and travel through the medium. It stimulates the emission of another photon, which in turn stimulates more (Fig. 11.34). The cascade of energy builds up rapidly, and soon the cavity is an intense reservoir of radiation at all the resonant modes it can sustain. Some of this radiation can escape if one of the mirrors is partially transmitting.

The resonant modes of the cavity have various natural characteristics and to some extent may be selected. Only photons that are travelling strictly parallel to the axis of the cavity undergo more than a couple of reflections, so only they are amplified, all others simply vanishing into the surroundings. Hence, laser light generally forms a beam with very low divergence. It may also be polarized, with its electric vector in a particular plane (or in some other state of polarization), by including a polarizing filter into the cavity or by making use of polarized transitions in a solid medium.

Laser radiation is **coherent** in the sense that the electromagnetic waves are all in step. In **spatial coherence** the waves are in step across the cross-section of the beam emerging from the cavity. In **temporal coherence** the waves remain in step along the beam. The latter is normally expressed in terms of a **coherence length**, l_C, the distance over which the waves remain coherent,

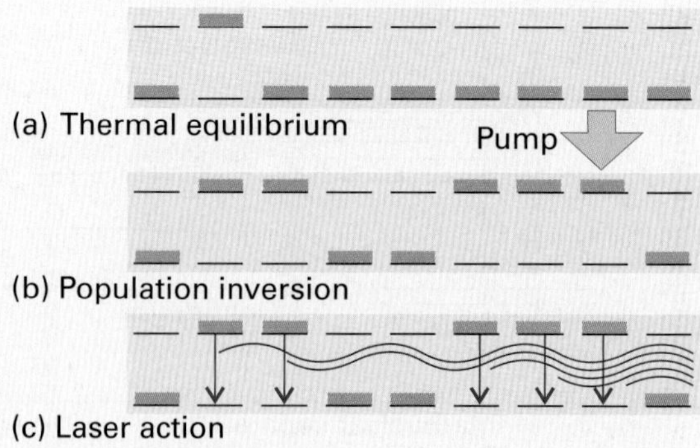

Fig. 11.34 A schematic illustration of the steps leading to laser action. (a) The Boltzmann population of states with more atoms in the ground state. (b) When the initial state absorbs, the populations are inverted (the atoms are pumped to the excited state). (c) A cascade of radiation then occurs, as one emitted photon stimulates another atom to emit, and so on. The radiation is coherent (phases in step).

and is related to the range of wavelengths, $\Delta\lambda$, present in the beam:

$$l_C = \frac{\lambda^2}{2\Delta\lambda} \qquad (11.12)$$

If the beam were perfectly monochromatic, with strictly one wavelength present, $\Delta\lambda$ would be zero and the waves would remain in step for an infinite distance. When many wavelengths are present, the waves get out of step in a short distance and the coherence length is small. A typical light bulb gives out light with a coherence length of only about 400 nm; a He–Ne laser with $\Delta\lambda \approx 2$ pm has a coherence length of about 10 cm.

(c) Q-switching

A laser can generate radiation for as long as the population inversion is maintained. A laser can operate continuously when heat is easily dissipated, for then the population of the upper level can be replenished by pumping. When overheating is a problem, the laser can be operated only in pulses, perhaps of microsecond or millisecond duration, so that the medium has a chance to cool or the lower state discard its population. However, it is sometimes desirable to have pulses of radiation rather than a continuous output, with a lot of power concentrated into a brief pulse. One way of achieving pulses is by **Q-switching**, the modification of the resonance characteristics of the laser cavity. The name comes from the 'Q-factor' used as a measure of the quality of a resonance cavity in microwave engineering.

Example 11.3 *Relating the power and energy of a laser*

A laser rated at 0.10 J can generate radiation in 3.0 ns pulses at a pulse repetition rate of 10 Hz. Assuming that the pulses are rectangular, calculate the peak power output and the average power output of this laser.

Method The power output is the energy released in an interval divided by the duration of the interval, and is expressed in watts ($1\ \mathrm{W} = 1\ \mathrm{J\ s^{-1}}$). To calculate the peak power output, P_{peak}, we divide the energy released during the pulse by the duration of the pulse. The average power output, $P_{average}$, is the total energy released by a large number of pulses divided by the duration of the time interval over which the total energy was measured. So, the average power is simply the energy released by one pulse multiplied by the pulse repetition rate.

Answer From the data,

$$P_{peak} = \frac{0.10\ \mathrm{J}}{3.0\times10^{-9}\ \mathrm{s}} = 3.3\times10^{7}\ \mathrm{J\ s^{-1}}$$

That is, the peak power output is 33 MW. The pulse repetition rate is 10 Hz, so ten pulses are emitted by the laser for every second of operation. It follows that the average power output is

$$P_{average} = 0.10\ \mathrm{J}\times10\ \mathrm{s^{-1}} = 1.0\ \mathrm{J\ s^{-1}} = 1.0\ \mathrm{W}$$

The peak power is much higher than the average power because this laser emits light for only 30 ns during each second of operation.

Self-test 11.4 Calculate the peak power and average power output of a laser with a pulse energy of 2.0 mJ, a pulse duration of 30 ps, and a pulse repetition rate of 38 MHz.

[$P_{peak} = 67$ MW, $P_{average} = 76$ kW]

The aim of Q-switching is to achieve a healthy population inversion in the absence of the resonant cavity, then to plunge the population-inverted medium into a cavity, and hence to obtain a sudden pulse of radiation. The switching may be achieved by impairing the resonance characteristics of the cavity in some way while the pumping pulse is active, and then suddenly to improve them (Fig. 11.35). One technique is to use a **Pockels cell**, which is an electro-optical device based on the ability of some crystals, such as those of potassium dihydrogenphosphate (KH_2PO_4), to convert plane-polarized light to circularly polarized light when an electrical potential difference is applied. If a Pockels cell is made part of a laser cavity, then its action and the change in polarization that occurs when light is reflected from a mirror convert light polarized in one plane into reflected light polarized in the perpendicular plane (Fig. 11.36). As a result, the reflected light does not stimulate more emission. However, if the cell is suddenly turned off, the polarization effect is extinguished and all the energy stored in the cavity can emerge as an intense pulse of stimulated emission. An alternative technique is to use a **saturable absorber**, typically a solution of a dye that loses its ability to absorb when many of its molecules have been excited by intense radiation. The dye then suddenly becomes transparent and the cavity becomes resonant. In practice, Q-switching can give pulses of about 5 ns duration.

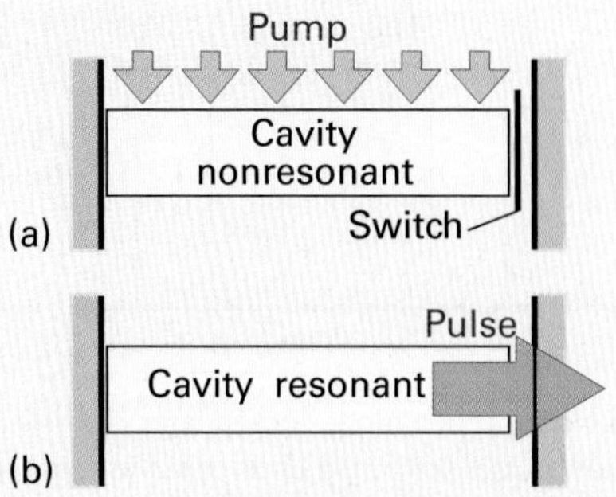

Fig. 11.35 The principle of Q-switching. (a) The excited state is populated while the cavity is nonresonant. (b) Then the resonance characteristics are suddenly restored, and the stimulated emission emerges in a giant pulse.

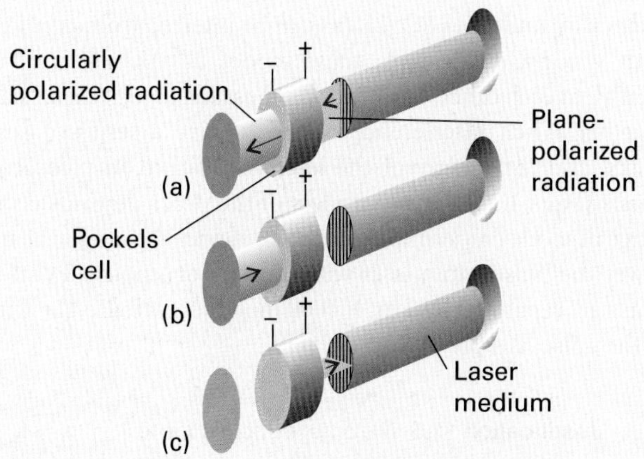

Fig. 11.36 The principle of a Pockels cell. When light passes through a cell that is 'on', its plane of polarization is rotated and so the laser cavity is nonresonant (its Q-factor is reduced). In this sequence, (a) the plane-polarized ray becomes circularly polarized, (b) is reflected, and (c) emerges from the Pockels cell with perpendicular plane polarization. When the cell is turned off, no change of polarization occurs, and the cavity becomes resonant.

(d) Mode locking

The technique of **mode locking** can produce pulses of picosecond duration and less. A laser radiates at a number of different

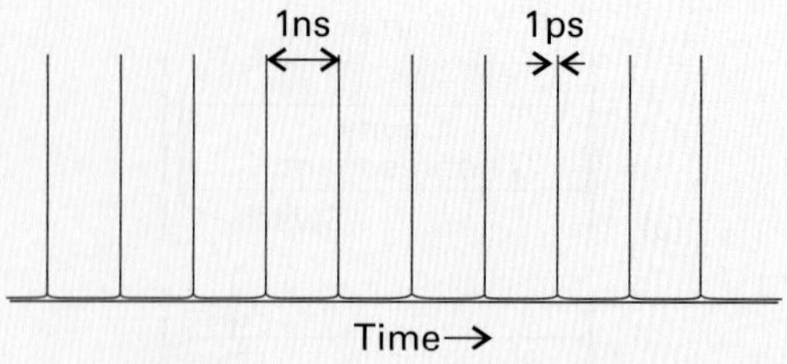

Fig. 11.37 The output of a mode-locked laser consists of a stream of very narrow pulses separated by an interval equal to the time it takes for light to make a round trip inside the cavity.

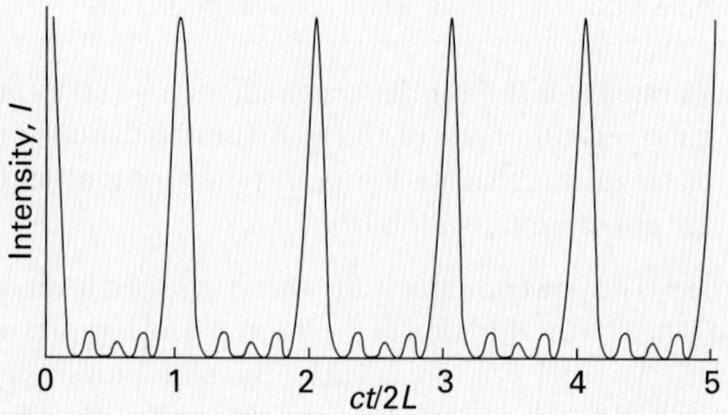

Fig. 11.38 The function derived in *Justification 11.5* showing in more detail the structure of the pulses generated by a mode-locked laser.

frequencies, depending on the precise details of the resonance characteristics of the cavity and in particular on the number of half-wavelengths of radiation that can be trapped between the mirrors (the cavity modes). The resonant modes differ in frequency by multiples of $c/2L$ (as can be inferred from eqn 11.11 with $\nu = c/\lambda$). Normally, these modes have random phases relative to each other. However, it is possible to lock their phases together. Then interference occurs to give a series of sharp peaks, and the energy of the laser is obtained in picosecond bursts (Fig. 11.37). The sharpness of the peaks depends on the range of modes superimposed, and the wider the range, the narrower the pulses. In a laser with a cavity of length 30 cm, the peaks are separated by 2 ns. If 1000 modes contribute, the width of the pulses is 4 ps.

Justification 11.5 *The origin of mode locking*

The general expression for a (complex) wave of amplitude $\mathcal{E}_0$ and frequency ω is $\mathcal{E}_0 e^{i\omega t}$. Therefore, each wave that can be supported by a cavity of length L has the form

$$\mathcal{E}_n(t) = \mathcal{E}_0 e^{2\pi i(\nu + nc/2L)t}$$

where ν is the lowest frequency. A wave formed by superimposing N modes with $n = 0, 1, \ldots, N-1$ has the form

$$\mathcal{E}(t) = \Sigma_n \mathcal{E}_n(t) = \mathcal{E}_0 e^{2\pi i\nu t} \sum_{n=0}^{N-1} e^{i\pi nct/L}$$

The sum is a geometrical progression:

$$\sum_{n=0}^{N-1} e^{i\pi nct/L} = 1 + e^{i\pi ct/L} + e^{2i\pi ct/L} + \cdots$$

$$= \frac{\sin(N\pi ct/2L)}{\sin(\pi ct/2L)} \times e^{(N-1)i\pi ct/2L}$$

The intensity, I, of the radiation is proportional to the square modulus of the total amplitude, so

$$I \propto \mathcal{E}^*\mathcal{E} = \mathcal{E}_0^2 \frac{\sin^2(N\pi ct/2L)}{\sin^2(\pi ct/2L)}$$

This function is shown in Fig. 11.38. We see that it is a series of peaks with maxima separated by $t = 2L/c$, the round-trip transit time of the light in the cavity, and that the peaks become sharper as N is increased.

Mode locking is achieved by varying the Q-factor of the cavity periodically at the frequency $c/2L$. The modulation can be pictured as the opening of a shutter in synchrony with the round-trip travel time of the photons in the cavity, so only photons making the journey in that time are amplified. The modulation can be achieved by linking a prism in the cavity to a transducer driven by a radiofrequency source at a frequency $c/2L$. The transducer sets up standing-wave vibrations in the prism and modulates the loss it introduces into the cavity. We also see below that the unique optical properties of some materials can be exploited to bring about mode locking.

(e) Non-linear optical phenomena

Non-linear optical phenomena arise from changes in the optical properties of a material in the presence of an intense electric field from electromagnetic radiation. Here we explore two phenomena that not only can be studied conveniently with intense laser beams but are commonly used in the laboratory to modify the output of lasers for specific experiments.

In **frequency doubling**, or **second harmonic generation**, an intense laser beam is converted to radiation with twice (and in general a multiple) of its initial frequency as it passes though a suitable material. It follows that frequency doubling and tripling of a Nd-YAG laser, which emits radiation at 1064 nm, produce green light at 532 nm and ultraviolet radiation at 355 nm, respectively.

We can account for frequency doubling by examining how a substance responds non-linearly to incident radiation of frequency $\omega = 2\pi\nu$. Radiation of a particular frequency arises from oscillations of an electric dipole at that frequency and the incident electric field $\mathcal{E}$ induces an electric dipole of magnitude μ in the substance. At low light intensity, most materials respond linearly, in the sense that $\mu = \alpha\mathcal{E}$, where α is the polarizability (see Section 8.4). To allow for non-linear response by some materials at high light intensity, we can write

$$\mu = \alpha\mathcal{E} + \tfrac{1}{2}\beta\mathcal{E}^2 + \cdots \quad (11.13)$$

where the coefficient β is the **hyperpolarizability** of the material. The non-linear term $\beta\mathcal{E}^2$ can be expanded as follows if we suppose that the incident electric field is $\mathcal{E}_0 \cos \omega t$:

$$\beta\mathcal{E}^2 = \beta\mathcal{E}_0^2 \cos^2\omega t = \tfrac{1}{2}\beta\mathcal{E}_0^2(1 + \cos 2\omega t) \quad (11.14)$$

Hence, the non-linear term contributes an induced electric dipole that oscillates at the frequency 2ω and which can act as a source of radiation of that frequency. Common materials that can be used for frequency doubling in laser systems include crystals of potassium dihydrogenphosphate (KH_2PO_4), lithium niobate ($LiNbO_3$), and β-barium borate (β-BaB_2O_4).

Another important non-linear optical phenomenon is the **optical Kerr effect**, which arises from a change in refractive index of a well chosen medium, the **Kerr medium**, when it is exposed to intense laser pulses. Because a beam of light changes direction when it passes from a region of one refractive index to a region with a different refractive index, changes in refractive index result in the self-focusing of an intense laser pulse as it travels through the Kerr medium (Fig. 11.39).

The optical Kerr effect is used as a mechanism of mode-locking lasers. A Kerr medium is included in the cavity and next to it is a small aperture. The procedure makes use of the fact that the **gain**, the growth in intensity, of a frequency component of the radiation in the cavity is very sensitive to amplification and, once a particular frequency begins to grow, it can quickly dominate. When the power inside the cavity is low, a portion of the photons will be blocked by the aperture, creating a significant loss. A spontaneous fluctuation in intensity—a bunching of photons—may begin to turn on the optical Kerr effect and the changes in the refractive index of the Kerr medium will result in a **Kerr lens**, which is the self-focusing of the laser beam. The bunch of photons can pass through and travel to the far end of the cavity, amplifying as it goes. The Kerr lens immediately disappears (if the medium is well chosen), but is re-created when the intense pulse returns from the mirror at the far end. In this way, that particular bunch of photons may grow to considerable intensity because it alone is stimulating emission in the cavity. Sapphire is an example of a Kerr medium that facilitates the mode locking of titanium sapphire lasers, resulting in very short laser pulses of duration in the femtosecond range.

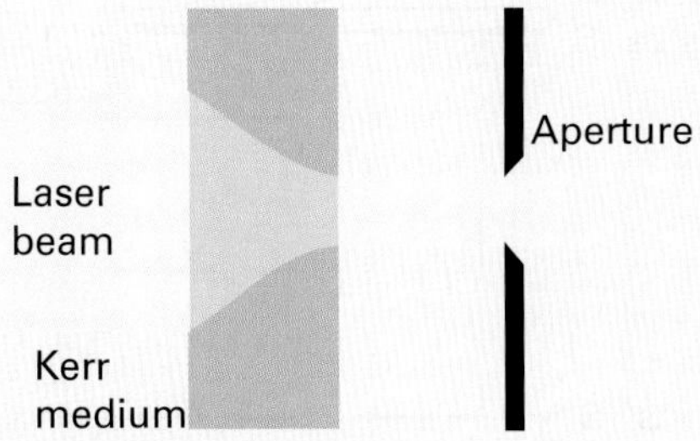

Fig. 11.39 An illustration of the Kerr effect. An intense laser beam is focused inside a Kerr medium and passes through a small aperture in the laser cavity. This effect may be used to mode-lock a laser, as explained in the text.

In addition to being useful laboratory tools, non-linear optical materials are also finding many applications in the telecommunications industry, which is becoming ever more reliant on optical signals transmitted through optical fibres to carry voice and data. Judicious use of non-linear phenomena leads to more ways in which the properties of optical signals, and hence the information they carry, can be manipulated.

(f) Time-resolved spectroscopy

The ability of lasers to produce pulses of very short duration is particularly useful in chemistry when we want to monitor processes in time. *Q*-switched lasers produce nanosecond pulses, which are generally fast enough to study reactions with rates controlled by the speed with which reactants can move through a fluid medium. However, when we want to study the rates at which energy is converted from one mode to another within a molecule, we need femtosecond and picosecond pulses. These timescales are available from mode-locked lasers.

In **time-resolved spectroscopy**, laser pulses are used to obtain the absorption, emission, or Raman spectrum of reactants, intermediates, products, and even transition states of reactions. It is also possible to study energy transfer, molecular rotations, vibrations, and conversion from one mode of motion to another. We shall see some of the information obtained from time-resolved spectroscopy in Chapters 19 and 20. Here, we describe some of the experimental techniques that employ pulsed lasers.

The arrangement shown in Fig. 11.40 is often used to study ultrafast chemical reactions that can be initiated by light, such as the initial events of vision (*Impact I11.1*). A strong and short laser pulse, the *pump*, promotes a molecule A to an excited electronic state A* that can either emit a photon (as fluorescence or phosphorescence) or react with another species B to yield a product C:

$$A + h\nu \rightarrow A^* \qquad \text{(absorption)}$$

$$A^* \rightarrow A \qquad \text{(emission)}$$

$$A^* + B \rightarrow [AB] \rightarrow C \qquad \text{(reaction)}$$

Here [AB] denotes either an intermediate or an activated complex. The rates of appearance and disappearance of the various species are determined by observing time-dependent changes in the absorption spectrum of the sample during the course of the reaction. This monitoring is done by passing a weak pulse of

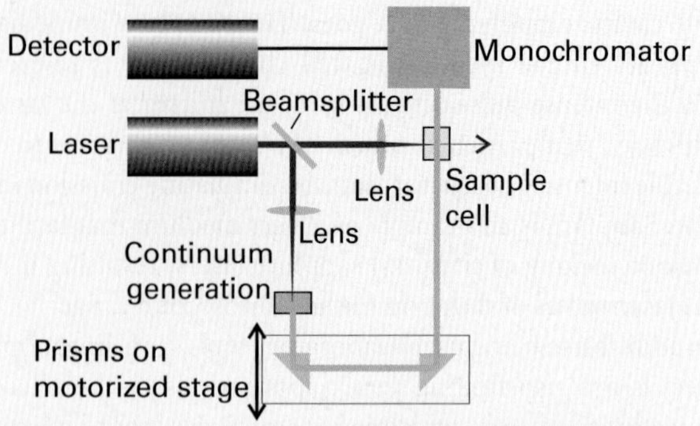

Fig. 11.40 A configuration used for time-resolved absorption spectroscopy, in which the same pulsed laser is used to generate a monochromatic pump pulse and, after continuum generation in a suitable liquid, a 'white' light probe pulse. The time delay between the pump and probe pulses may be varied by moving the motorized stage in the direction shown by the double arrow.

white light, the *probe*, through the sample at different times after the laser pulse. Pulsed 'white' light can be generated directly from the laser pulse by the phenomenon of **continuum generation**, in which focusing an ultrafast laser pulse on a vessel containing water, carbon tetrachloride, or sapphire results in an outgoing beam with a wide distribution of frequencies. A time delay between the strong laser pulse and the 'white' light pulse can be introduced by allowing one of the beams to travel a longer distance before reaching the sample. For example, a difference in travel distance of $\Delta d = 3$ mm corresponds to a time delay $\Delta t = \Delta d/c \approx 10$ ps between two beams, where c is the speed of light. The relative distances travelled by the two beams in Fig. 11.40 are controlled by directing the 'white' light beam to a motorized stage carrying a pair of mirrors.

Variations of the arrangement in Fig. 11.40 allow for the observation of fluorescence decay kinetics of A* and time-resolved Raman spectra during the course of the reaction. The fluorescence lifetime of A* can be determined by exciting A as before and measuring the decay of the fluorescence intensity after the pulse with a fast photodetector system. In this case, continuum generation is not necessary. Time-resolved resonance Raman spectra of A, A*, B, [AB], or C can be obtained by initiating the reaction with a strong laser pulse of a certain wavelength and then, some time later, irradiating the sample with another laser pulse that can excite the resonance Raman spectrum of the desired species. Also in this case continuum generation is not necessary. Instead, the Raman excitation beam may be generated in a dye laser (Section 11.8).

11.8 Examples of practical lasers

Figure 11.41 summarizes the requirements for an efficient laser. In practice, the requirements can be satisfied by using a variety

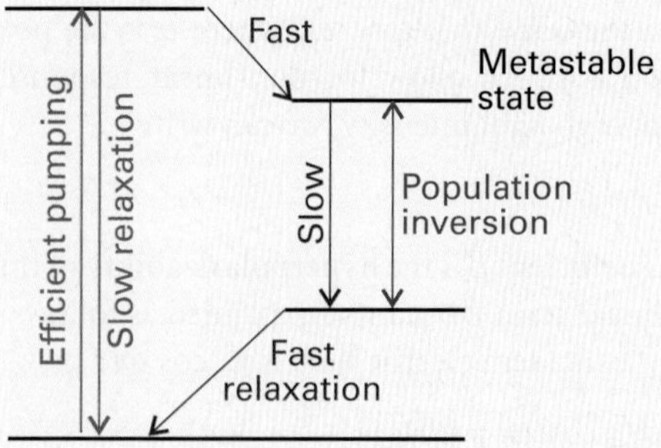

Fig. 11.41 A summary of the features needed for efficient laser action.

of different systems, and this section reviews some that are commonly available. We also include some lasers that operate by using other than electronic transitions.

(a) Gas lasers

Because gas lasers can be cooled by a rapid flow of the gas through the cavity, they can be used to generate high powers. The pumping is normally achieved using a gas that is different from the gas responsible for the laser emission itself.

In the **helium–neon laser** the active medium is a mixture of helium and neon in a mole ratio of about 5:1 (Fig. 11.42). The initial step is the excitation of an He atom to the metastable $1s^12s^1$ configuration by using an electric discharge (the collisions of electrons and ions cause transitions that are not restricted by electric-dipole selection rules). The excitation energy of this transition happens to match an excitation energy of neon, and during an He–Ne collision efficient transfer of energy may occur, leading to the production of highly excited, metastable Ne atoms with unpopulated intermediate states. Laser action generating 633 nm radiation (among about 100 much weaker other lines) then occurs.

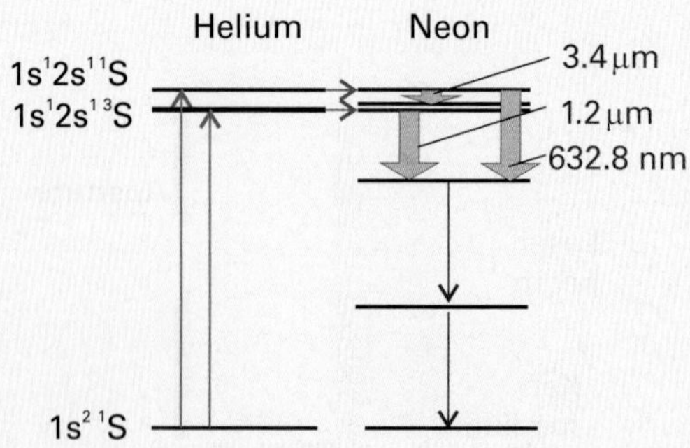

Fig. 11.42 The transitions involved in a helium–neon laser. The pumping (of the neon) depends on a coincidental matching of the helium and neon energy separations, so excited He atoms can transfer their excess energy to Ne atoms during a collision.

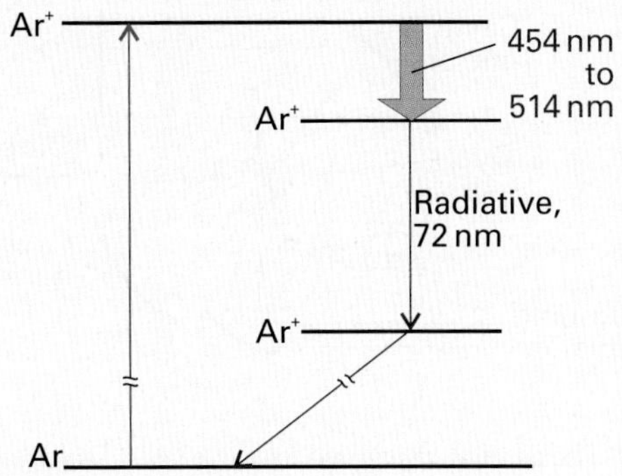

Fig. 11.43 The transitions involved in an argon-ion laser.

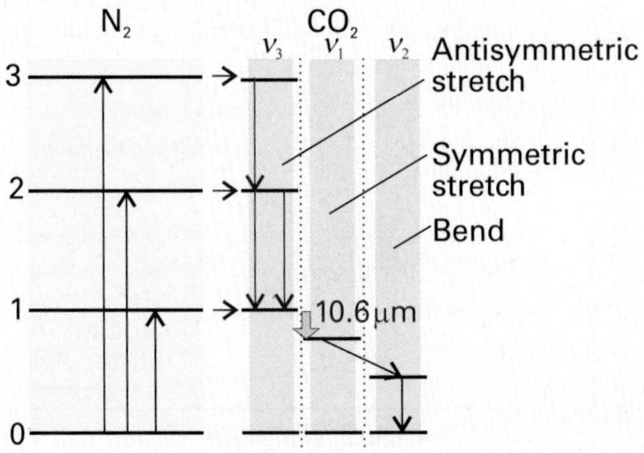

Fig. 11.44 The transitions involved in a carbon dioxide laser. The pumping also depends on the coincidental matching of energy separations; in this case the vibrationally excited N_2 molecules have excess energies that correspond to a vibrational excitation of the antisymmetric stretch of CO_2. The laser transition is from $v_3 = 1$ to $v_1 = 1$.

The **argon-ion laser** (Fig. 11.43), one of a number of 'ion lasers', consists of argon at about 1 Torr, through which is passed an electric discharge. The discharge results in the formation of Ar^+ and Ar^{2+} ions in excited states, which undergo a laser transition to a lower state. These ions then revert to their ground states by emitting hard ultraviolet radiation (at 72 nm), and are then neutralized by a series of electrodes in the laser cavity. One of the design problems is to find materials that can withstand this damaging residual radiation. There are many lines in the laser transition because the excited ions may make transitions to many lower states, but two strong emissions from Ar^+ are at 488 nm (blue) and 514 nm (green); other transitions occur elsewhere in the visible region, in the infrared, and in the ultraviolet. The **krypton-ion laser** works similarly. It is less efficient, but gives a wider range of wavelengths, the most intense being at 647 nm (red), but it can also generate yellow, green, and violet lines. Both lasers are widely used in laser light shows (for this application argon and krypton are often used simultaneously in the same cavity) as well as laboratory sources of high-power radiation.

The **carbon dioxide laser** works on a slightly different principle (Fig. 11.44), for its radiation (between 9.2 μm and 10.8 μm, with the strongest emission at 10.6 μm, in the infrared) arises from vibrational transitions. Most of the working gas is nitrogen, which becomes vibrationally excited by electronic and ionic collisions in an electric discharge. The vibrational levels happen to coincide with the ladder of antisymmetric stretch (ν_3, see Fig. 10.44) energy levels of CO_2, which pick up the energy during a collision. Laser action then occurs from the lowest excited level of ν_3 to the lowest excited level of the symmetric stretch (ν_1), which has remained unpopulated during the collisions. This transition is allowed by anharmonicities in the molecular potential energy. Some helium is included in the gas to help remove energy from this state and maintain the population inversion.

In a **nitrogen laser**, the efficiency of the stimulated transition (at 337 nm, in the ultraviolet, the transition $C^3\Pi_u \rightarrow B^3\Pi_g$) is so great that a single passage of a pulse of radiation is enough to generate laser radiation and mirrors are unnecessary: such lasers are said to be **superradiant**.

(b) Chemical and exciplex lasers

Chemical reactions may also be used to generate molecules with nonequilibrium, inverted populations. For example, the photolysis of Cl_2 leads to the formation of Cl atoms, which attack H_2 molecules in the mixture and produce HCl and H. The latter then attacks Cl_2 to produce vibrationally excited ('hot') HCl molecules. Because the newly formed HCl molecules have nonequilibrium vibrational populations, laser action can result as they return to lower states. Such processes are remarkable examples of the direct conversion of chemical energy into coherent electromagnetic radiation.

The population inversion needed for laser action is achieved in a more underhand way in **exciplex lasers**, for in these (as we shall see) the lower state does not effectively exist. This odd situation is achieved by forming an **exciplex**, a combination of two atoms that survives only in an excited state and that dissociates as soon as the excitation energy has been discarded. (The term 'excimer laser' is also widely encountered and used loosely when 'exciplex laser' is more appropriate. An exciplex has the form AB* whereas an excimer, an excited dimer, is AA*.) An exciplex can be formed in a mixture of xenon, chlorine, and neon (which acts as a buffer gas). An electric discharge through the mixture produces excited Cl atoms, which attach to the Xe atoms to give the exciplex XeCl*. The exciplex survives for about 10 ns, which is time for it to participate in laser action at 308 nm (in the ultraviolet). As soon as XeCl* has discarded a photon, the atoms separate because the molecular potential energy curve of the ground state is dissociative, and the ground state of the exciplex

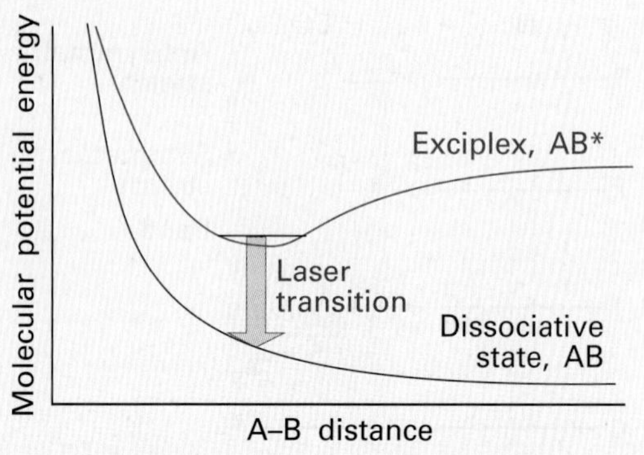

Fig. 11.45 The molecular potential energy curves for an exciplex. The species can survive only as an excited state, because on discarding its energy it enters the lower, dissociative state. Because only the upper state can exist, there is never any population in the lower state.

cannot become populated (Fig. 11.45). The KrF^* exciplex laser is another example: it produces radiation at 249 nm.

(c) Dye lasers

Gas lasers and most solid state lasers operate at discrete frequencies and, although the frequency required may be selected by suitable optics, the laser cannot be tuned continuously. The tuning problem is overcome by using a titanium sapphire laser (see below) or a **dye laser**, which has broad spectral characteristics because the solvent broadens the vibrational structure of the transitions into bands. Hence, it is possible to scan the wavelength continuously (by rotating the diffraction grating in the cavity) and achieve laser action at any chosen wavelength within the fluorescence spectrum of the dye. A commonly used dye is rhodamine 6G in methanol (Fig. 11.46). As the gain is very high, only a short length of the optical path need be through the dye. The excited states of the active medium, the dye, are

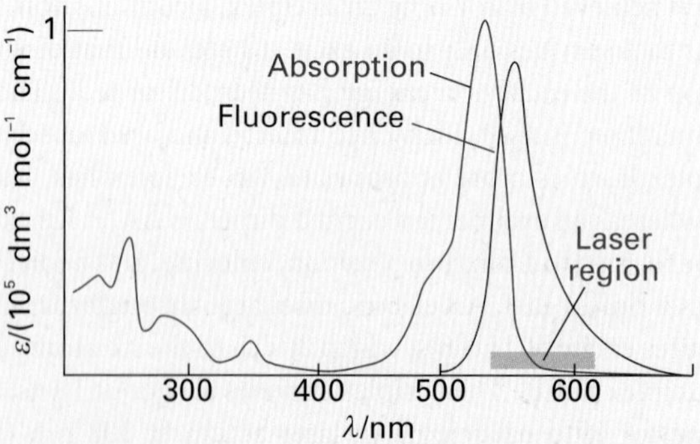

Fig. 11.46 The optical absorption spectrum of the dye rhodamine 6G and the region used for laser action.

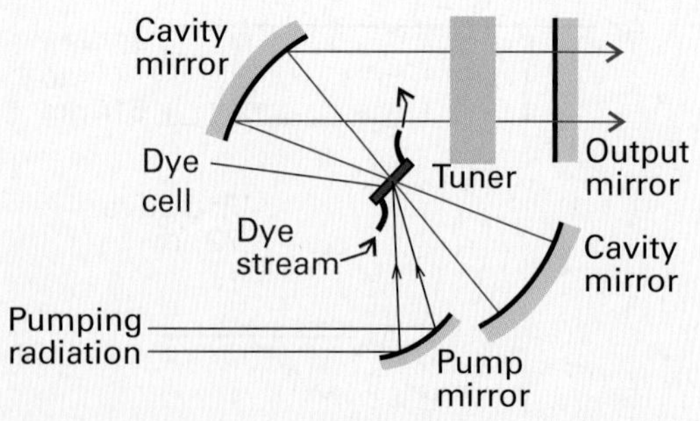

Fig. 11.47 The configuration used for a dye laser. The dye is flowed through the cell inside the laser cavity. The flow helps to keep it cool and prevents degradation.

sustained by another laser or a flash lamp, and the dye solution is flowed through the laser cavity to avoid thermal degradation (Fig. 11.47).

(d) Light emission by solid-state lasers and light-emitting diodes

In Section 9.10 we explored light emission in solids. Here we focus our attention on ionic crystals and semiconductors used in the design of lasers and light-emitting diodes.

The **neodymium laser** is an example of a four-level laser, in which the laser transition terminates in a state other than the ground state of the laser material (Fig. 11.48). In one form it consists of Nd^{3+} ions at low concentration in yttrium aluminium garnet (YAG, specifically $Y_3Al_5O_{12}$), and is then known as a **Nd-YAG laser**. The population inversion results from pumping a majority of the Nd^{3+} ions into an excited state by using an intense flash from another source, followed by a radiationless transition to another excited state. The pumping flash need not be monochromatic because the upper level actually consists of several states spanning a band of frequencies. A neodymium laser operates at a number of wavelengths in the infrared, the band at 1064 nm being most common. The transition at 1064 nm is very

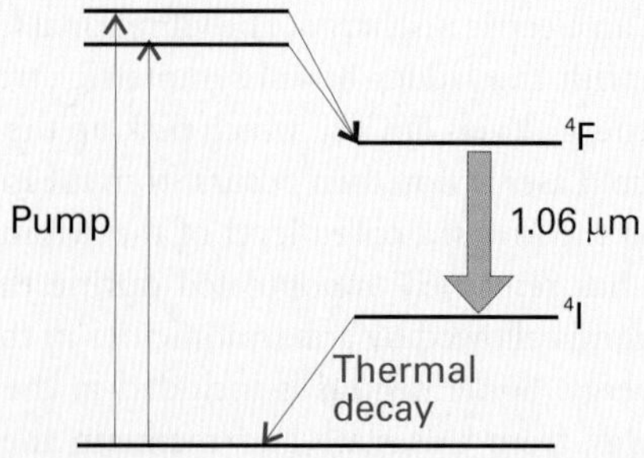

Fig. 11.48 The transitions involved in a neodymium laser. The laser action takes place between the ^4F and ^4I excited states.

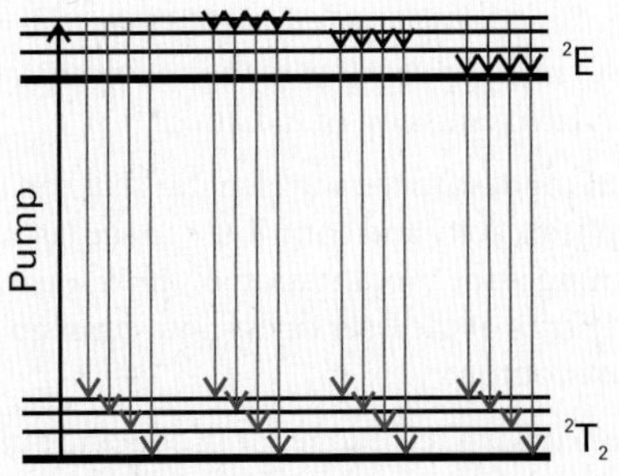

Fig. 11.49 The transitions involved in a titanium sapphire laser. The laser medium consists of sapphire (Al_2O_3) doped with Ti^{3+} ions. Monochromatic light from a pump laser induces a $^2E \leftarrow {}^2T_2$ transition in a Ti^{3+} ion that resides in a site with octahedral symmetry. After radiationless vibrational excitation in the ^2E state, laser emission occurs from a very large number of closely spaced vibronic states of the medium. As a result, the titanium sapphire laser emits radiation over a broad spectrum that spans from about 700 nm to about 1000 nm.

efficient and the laser is capable of substantial power output, either in continuous or pulsed (by *Q*-switching or mode locking as discussed in Section 11.7) modes of operation.

The **titanium sapphire laser** consists of Ti^{3+} ions at low concentration in a crystal of sapphire (Al_2O_3). The electronic absorption spectrum of Ti^{3+} ion in sapphire is very similar to that shown in Fig. 11.15, with a broad absorption band centred at around 500 nm that arises from vibronically allowed d–d transitions of the Ti^{3+} ion in an octahedral environment provided by oxygen atoms of the host lattice. As a result, the emission spectrum of Ti^{3+} in sapphire is also broad and laser action occurs over a wide range of wavelengths (Fig. 11.49). Therefore, the titanium sapphire laser is an example of a **vibronic laser**, in which the laser transitions originate from vibronic transitions in the laser medium. The titanium sapphire laser is usually pumped by another laser, such as a Nd-YAG laser or an argon-ion laser, and can be operated in either a continuous or pulsed fashion. Mode-locked titanium sapphire lasers produce energetic (20 mJ to 1 J) and very short (20–100 fs, 1 fs = 10^{-15} s) pulses. When considered together with broad wavelength tunability (700–1000 nm), these features of the titanium sapphire laser justify its wide use in modern spectroscopy and photochemistry.

The unique electrical properties of p–n junctions between semiconductors (Section 9.9) can be put to good use in optical devices. In some materials, most notably gallium arsenide, GaAs, energy from electron–hole recombination is released not as heat but is carried away by photons as electrons move across the junction under forward bias. Practical **light-emitting diodes** of this kind are widely used in electronic displays. The wavelength of emitted light depends on the band gap of the semiconductor. Gallium arsenide itself emits infrared light, but the band gap is widened by incorporating phosphorus, and a material of composition approximately $GaAs_{0.6}P_{0.4}$ emits light in the red region of the spectrum.

A light-emitting diode is not a laser, because no resonance cavity and stimulated emission are involved. In **diode lasers**, light emission due to electron–hole recombination is employed as the basis of laser action. The population inversion can be sustained by sweeping away the electrons that fall into the holes of the p-type semiconductor, and a resonant cavity can be formed by using the high refractive index of the semiconducting material and cleaving single crystals so that the light is trapped by the abrupt variation of refractive index. One widely used material is $Ga_{1-x}Al_xAs$, which produces infrared laser radiation and is widely used in compact-disc (CD) players.

High-power diode lasers are also used to pump other lasers. One example is the pumping of Nd:YAG lasers by $Ga_{0.91}Al_{0.09}As$/$Ga_{0.7}Al_{0.3}As$ diode lasers. The Nd:YAG laser is often used to pump yet another laser, such as a Ti:sapphire laser. As a result, it is now possible to construct a laser system for steady-state or time-resolved spectroscopy entirely out of solid-state components.

Checklist of key ideas

☐ 1. Emission spectroscopy is based on the detection of a transition from a state of high energy to a state of lower energy; absorption spectroscopy is based on the detection of the net absorption of nearly monochromatic incident radiation as the radiation is swept over a range of frequencies.

☐ 2. The Beer–Lambert law is $I(\tilde{\nu}) = I_0(\tilde{\nu})10^{-\varepsilon(\tilde{\nu})[J]l}$, where $I(\tilde{\nu})$ is the transmitted intensity, $I_0(\tilde{\nu})$ is the incident intensity, and $\varepsilon(\tilde{\nu})$ is the molar absorption coefficient.

☐ 3. The transmittance, $T = I/I_0$, and the absorbance, A, of a sample at a given wavenumber are related by $A = -\log T$.

☐ 4. The integrated absorption coefficient, $\mathcal{A}$, is the sum of the absorption coefficients over the entire band, $\mathcal{A} = \int_{\text{band}} \varepsilon(\tilde{\nu})d\tilde{\nu}$.

☐ 5. The selection rules for electronic transitions that are concerned with changes in angular momentum are: $\Delta\Lambda = 0, \pm1, \Delta S = 0, \Delta\Sigma = 0, \Delta\Omega = 0, \pm1$.

☐ 6. The Laporte selection rule (for centrosymmetric molecules) states that the only allowed transitions are transitions that are accompanied by a change of parity.

☐ 7. The Franck–Condon principle states that, because the nuclei are so much more massive than the electrons, an electronic transition takes place very much faster than the nuclei can respond.

☐ 8. The intensity of an electronic transition is proportional to the Franck–Condon factor, the quantity $|S(v_f,v_i)|^2$, with $S(v_f,v_i) = \langle v_f | v_i \rangle$.

☐ 9. Examples of electronic transitions include d–d transitions in d-metal complexes, charge-transfer transitions (a transition in which an electron moves from metal to ligand or from ligand to metal in a complex), $\pi^* \leftarrow \pi$, and $\pi^* \leftarrow n$ transitions.

☐ 10. A Jablonski diagram is a schematic diagram of the various types of nonradiative and radiative transitions that can occur in molecules.

☐ 11. Fluorescence is the spontaneous emission of radiation arising from a transition between states of the same multiplicity.

☐ 12. Phosphorescence is the spontaneous emission of radiation arising from a transition between states of different multiplicity.

☐ 13. Intersystem crossing is a nonradiative transition between states of different multiplicity.

☐ 14. Internal conversion is a nonradiative transition between states of the same multiplicity.

☐ 15. Laser action depends on the achievement of population inversion, an arrangement in which there are more molecules in an upper state than in a lower state, and the stimulated emission of radiation.

☐ 16. Resonant modes are the wavelengths that can be sustained by an optical cavity and contribute to the laser action. *Q*-switching is the modification of the resonance characteristics of the laser cavity and, consequently, of the laser output.

☐ 17. Mode locking is a technique for producing pulses of picosecond duration and less by matching the phases of many resonant cavity modes.

☐ 18. Non-linear optical phenomena arise from changes in the optical properties of a material in the presence of an intense field from electromagnetic radiation. Examples include second harmonic generation and the optical Kerr effect.

☐ 19. Practical lasers include gas, dye, chemical, exciplex, and solid-state lasers.

☐ 20. Applications of lasers in chemistry include laser light scattering, multiphoton spectroscopy, Raman spectroscopy, precision-specified transitions, time-resolved spectroscopy, and single-molecule spectroscopy.

Discussion questions

11.1 Explain the origin of the term symbol ${}^3\Sigma_g^-$ for the ground state of dioxygen.

11.2 Explain the basis of the Franck–Condon principle and how it leads to the formation of a vibrational progression.

11.3 How do the band heads in P and R branches arise? Could the Q branch show a head?

11.4 Explain how colour can arise from molecules.

11.5 Suppose that you are a colour chemist and had been asked to intensify the colour of a dye without changing the type of compound, and that the dye in question was a polyene. (a) Would you choose to lengthen or to shorten the chain? (b) Would the modification to the length shift the apparent colour of the dye towards the red or the blue?

11.6 Describe the mechanism of fluorescence. In what respects is a fluorescence spectrum not the exact mirror image of the corresponding absorption spectrum?

11.7 What is the evidence for the correctness of the mechanism of fluorescence?

11.8 Describe the principles of (a) continuous-wave, and (b) pulsed laser action.

11.9 What features of laser radiation are applied in chemistry? Discuss two applications of lasers in chemistry.

Exercises

11.1(a) The molar absorption coefficient of a substance dissolved in hexane is known to be 723 $dm^3\ mol^{-1}\ cm^{-1}$ at 260 nm. Calculate the percentage reduction in intensity when light of that wavelength passes through 2.50 mm of a solution of concentration 4.25 $mmol\ dm^{-3}$.

11.1(b) The molar absorption coefficient of a substance dissolved in hexane is known to be 227 $dm^3\ mol^{-1}\ cm^{-1}$ at 290 nm. Calculate the percentage reduction in intensity when light of that wavelength passes through 2.00 mm of a solution of concentration 2.52 $mmol\ dm^{-3}$.

11.2(a) A solution of an unknown component of a biological sample when placed in an absorption cell of path length 1.00 cm transmits 18.1 per cent of light of 320 nm incident upon it. If the concentration of the component is 0.139 $mmol\ dm^{-3}$, what is the molar absorption coefficient?

11.2(b) When light of wavelength 400 nm passes through 2.5 mm of a solution of an absorbing substance at a concentration 0.717 mmol dm^{-3}, the transmission is 61.5 per cent. Calculate the molar absorption coefficient of the solute at this wavelength and express the answer in $cm^2\ mol^{-1}$.

11.3(a) The molar absorption coefficient of a solute at 540 nm is 386 $dm^3\ mol^{-1}\ cm^{-1}$. When light of that wavelength passes through a 5.00 mm cell containing a solution of the solute, 38.5 per cent of the light was absorbed. What is the concentration of the solution?

11.3(b) The molar absorption coefficient of a solute at 440 nm is 423 $dm^3\ mol^{-1}\ cm^{-1}$. When light of that wavelength passes through a 6.50 mm cell containing a solution of the solute, 48.3 per cent of the light was absorbed. What is the concentration of the solution?

11.4(a) The absorption associated with a particular transition begins at 220 nm, peaks sharply at 270 nm, and ends at 300 nm. The maximum value of the molar absorption coefficient is $2.21 \times 10^4\ dm^3\ mol^{-1}\ cm^{-1}$. Estimate the integrated absorption coefficient of the transition assuming a triangular lineshape (see eqn 11.5).

11.4(b) The absorption associated with a certain transition begins at 156 nm and ends at 275 nm. The maximum value of the molar absorption coefficient is $3.35 \times 10^4\ dm^3\ mol^{-1}\ cm^{-1}$. Estimate the integrated absorption coefficient of the transition assuming an inverted parabolic lineshape (Fig. 11.50; use eqn 11.5).

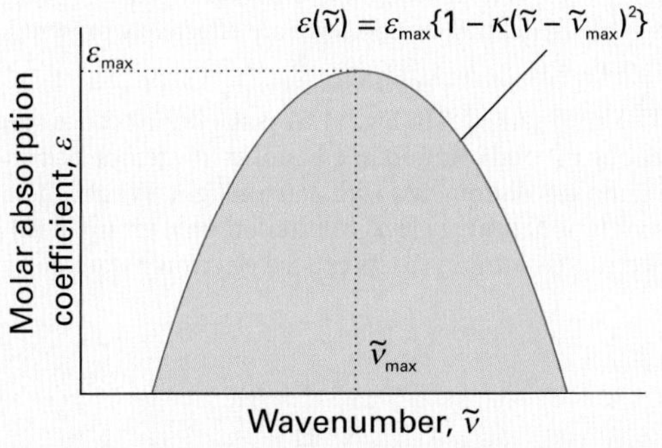

Fig. 11.50 A model parabolic absorption lineshape.

11.5(a) The following data were obtained for the absorption by Br_2 in carbon tetrachloride using a 2.0 mm cell. Calculate the molar absorption coefficient of bromine at the wavelength employed:

$[Br_2]/(mol\ dm^{-3})$	0.0010	0.0050	0.0100	0.0500
T/(per cent)	81.4	35.6	12.7	3.0×10^{-3}

11.5(b) The following data were obtained for the absorption by a dye dissolved in methylbenzene using a 2.50 mm cell. Calculate the molar absorption coefficient of the dye at the wavelength employed:

$[dye]/(mol\ dm^{-3})$	0.0010	0.0050	0.0100	0.0500
T/(per cent)	68	18	3.7	1.03×10^{-5}

11.6(a) A 2.0-mm cell was filled with a solution of benzene in a non-absorbing solvent. The concentration of the benzene was 0.010 $mol\ dm^{-3}$ and the wavelength of the radiation was 256 nm (where there is a maximum in the absorption). Calculate the molar absorption coefficient of benzene at this wavelength given that the transmission was 48 per cent. What will the transmittance be in a 4.0-mm cell at the same wavelength?

11.6(b) A 5.00-mm cell was filled with a solution of a dye. The concentration of the dye was 18.5 $mmol\ dm^{-3}$. Calculate the molar absorption coefficient of the dye at this wavelength given that the transmission was 29 per cent. What will the transmittance be in a 2.50-mm cell at the same wavelength?

11.7(a) A swimmer enters a gloomier world (in one sense) on diving to greater depths. Given that the mean molar absorption coefficient of sea water in the visible region is $6.2 \times 10^{-5}\ dm^3\ mol^{-1}\ cm^{-1}$, calculate the depth at which a diver will experience (a) half the surface intensity of light, (b) one-tenth the surface intensity.

11.7(b) Given that the maximum molar absorption coefficient of a molecule containing a carbonyl group is 30 $dm^3\ mol^{-1}\ cm^{-1}$ near 280 nm, calculate the thickness of a sample that will result in (a) half the initial intensity of radiation, (b) one-tenth the initial intensity.

11.8(a) The term symbol for one of the excited states of H_2 is $^3\Pi_u$. Use the building-up principle to find the excited-state configuration to which this term symbol corresponds.

11.8(b) The term symbol for the ground state of N_2^+ is $^2\Pi_g$. Use the building-up principle to find the excited-state configuration to which this term symbol corresponds.

11.9(a) One of the excited states of the C_2 molecule has the valence electron configuration $1\sigma_g^2 1\sigma_u^2 1\pi_u^3 1\pi_g^1$. Give the multiplicity and parity of the term.

11.9(b) One of the excited states of the C_2 molecule has the valence electron configuration $1\sigma_g^2 1\sigma_u^2 1\pi_u^2 1\pi_g^2$. Give the multiplicity and parity of the term.

11.10(a) Which of the following transitions are electric-dipole allowed: (a) $^2\Pi \leftrightarrow {}^2\Pi$, (b) $^1\Sigma \leftrightarrow {}^1\Sigma$, (c) $\Sigma \leftrightarrow \Delta$, (d) $\Sigma^+ \leftrightarrow \Sigma^-$, (e) $\Sigma^+ \leftrightarrow \Sigma^+$?

11.10(b) Which of the following transitions are electric-dipole allowed: (a) $^1\Sigma_g^+ \leftrightarrow {}^1\Sigma_u^+$, (b) $^3\Sigma_g^+ \leftrightarrow {}^3\Sigma_u^+$, (c) $t_{2g} \leftrightarrow e_g$, (d) $\pi^* \leftrightarrow n$?

11.11(a) The ground-state wavefunction of a certain molecule is described by the vibrational wavefunction $\psi_0 = N_0 e^{-ax^2}$. Calculate the Franck–Condon factor for a transition to a vibrational state described by the wavefunction $\psi_v = N_v e^{-b(x-x_0)^2}$, with $b = a/2$.

11.11(b) The ground-state wavefunction of a certain molecule is described by the vibrational wavefunction $\psi_0 = N_0 e^{-ax^2}$. Calculate the Franck–Condon factor for a transition to a vibrational state described by the wavefunction $\psi_v = N_v x e^{-b(x-x_0)^2}$, with $b = a/2$.

11.12(a) Suppose that the ground vibrational state of a molecule is modelled by using the particle-in-a-box wavefunction $\psi_0 = (2/L)^{1/2} \sin(\pi x/L)$ for $0 \le x \le L$ and 0 elsewhere. Calculate the Franck–Condon factor for a transition to a vibrational state described by the wavefunction $\psi_v = (2/L)^{1/2} \sin\{\pi(x - L/4)/L\}$ for $L/4 \le x \le 5L/4$ and 0 elsewhere.

11.12(b) Suppose that the ground vibrational state of a molecule is modelled by using the particle-in-a-box wavefunction $\psi_0 = (2/L)^{1/2} \sin(\pi x/L)$ for $0 \le x \le L$ and 0 elsewhere. Calculate the Franck–Condon factor for a transition to a vibrational state described by the wavefunction $\psi_v = (2/L)^{1/2} \sin\{\pi(x - L/4)/L\}$ for $L/2 \le x \le 3L/2$ and 0 elsewhere.

11.13(a) Use eqn 11.9a to infer the value of J corresponding to the location of the band head of the P branch of a transition.

11.13(b) Use eqn 11.9c to infer the value of J corresponding to the location of the band head of the R branch of a transition.

11.14(a) The following parameters describe the electronic ground state and an excited electronic state of SnO: $\tilde{B} = 0.3540\ cm^{-1}$, $\tilde{B}' = 0.3101\ cm^{-1}$. Which branch of the transition between them shows a head? At what value of J will it occur?

11.14(b) The following parameters describe the electronic ground state and an excited electronic state of BeH: $\tilde{B} = 10.308\ \text{cm}^{-1}$, $\tilde{B}' = 10.470\ \text{cm}^{-1}$. Which branch of the transition between them shows a head? At what value of J will it occur?

11.15(a) The R branch of the ${}^1\Pi_u \leftarrow {}^1\Sigma_g^+$ transition of H_2 shows a band head at the very low value of $J = 1$. The rotational constant of the ground state is $60.80\ \text{cm}^{-1}$. What is the rotational constant of the upper state? Has the bond length increased or decreased in the transition?

11.15(b) The P branch of the ${}^2\Pi \leftarrow {}^2\Sigma^+$ transition of CdH shows a band head at $J = 25$. The rotational constant of the ground state is $5.437\ \text{cm}^{-1}$. What is the rotational constant of the upper state? Has the bond length increased or decreased in the transition?

11.16(a) The complex ion $[Fe(H_2O)_6]^{3+}$ has an electronic absorption spectrum with a maximum at 700 nm. Estimate a value of Δ_O for the complex.

11.16(b) The complex ion $[Fe(CN)_6]^{3-}$ has an electronic absorption spectrum with a maximum at 305 nm. Estimate a value of Δ_O for the complex.

11.17(a) Suppose that we can model a charge-transfer transition in a one-dimensional system as a process in which a rectangular wavefunction that is nonzero in the range $0 \le x \le a$ makes a transition to another rectangular wavefunction that is nonzero in the range $\frac{1}{2}a \le x \le b$. Evaluate the transition moment $\int \psi_f x \psi_i \mathrm{d}x$.

11.17(b) Suppose that we can model a charge-transfer transition in a one-dimensional system as a process in which an electron described by a rectangular wavefunction that is nonzero in the range $0 \le x \le a$ makes a transition to another rectangular wavefunction that is nonzero in the range $ca \le x \le a$ where $0 \le c < 1$. Evaluate the transition moment $\int \psi_f x \psi_i \mathrm{d}x$ and explore its dependence on c.

11.18(a) Suppose that we can model a charge-transfer transition in a one-dimensional system as a process in which a Gaussian wavefunction centred on $x = 0$ and width a makes a transition to another Gaussian wavefunction of the same width centred on $x = \frac{1}{2}a$. Evaluate the transition moment $\int \psi_f x \psi_i \mathrm{d}x$.

11.18(b) Suppose that we can model a charge-transfer transition in a one-dimensional system as a process in which an electron described by a Gaussian wavefunction centred on $x = 0$ and width a makes a transition to another Gaussian wavefunction of width $a/2$ and centred on $x = 0$. Evaluate the transition moment $\int \psi_f x \psi_i \mathrm{d}x$.

11.19(a) The compound $CH_3CH{=}CHCHO$ has a strong absorption in the ultraviolet at $46\,950\ \text{cm}^{-1}$ and a weak absorption at $30\,000\ \text{cm}^{-1}$. Justify these features in terms of the structure of the molecule.

11.19(b) The two compounds 2,3-dimethyl-2-butene (**7**) and 2,5-dimethyl-2,4-hexadiene (**8**) are to be distinguished by their ultraviolet absorption spectra. The maximum absorption in one compound occurs at 192 nm and in the other at 243 nm. Match the maxima to the compounds and justify the assignment.

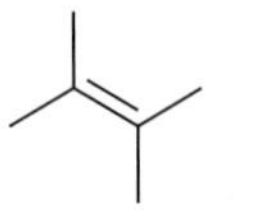

7 2,3-Dimethyl-2-butene

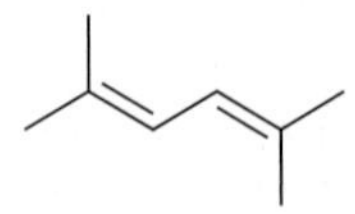

8 2,5-Dimethyl-2,4-hexadiene

11.20(a) Propanone (acetone, $(CH_3)_2CO$), has a strong absorption at 189 nm and a weaker absorption at 280 nm. Justify these features and assign the ultraviolet absorption transitions.

11.20(b) 3-Buten-2-one (**9**) has a strong absorption at 213 nm and a weaker absorption at 320 nm. Justify these features and assign the ultraviolet absorption transitions.

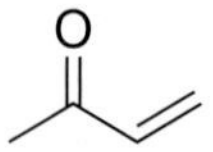

9 3-Butene-2-one

11.21(a) Suppose a molecule responds to the square of the electric field intensity that is incident upon it, and that the field is proportional to $\cos \omega t$. Show that the effect is equivalent to an incident field that is oscillating at 2ω.

11.21(b) Suppose a molecule responds to the third power of the electric field intensity that is incident upon it, and that the field is proportional to $\cos \omega t$. What frequency components are effectively present in the incident radiation?

11.22(a) The line marked A in Fig. 11.51 is the fluorescence spectrum of benzophenone in solid solution in ethanol at low temperatures observed when the sample is illuminated with 360 nm light. What can be said about the vibrational energy levels of the carbonyl group in (a) its ground electronic state and (b) its excited electronic state?

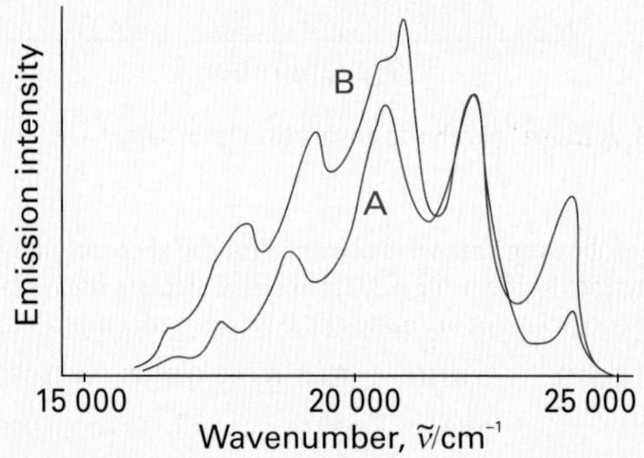

Fig. 11.51 The fluorescence and phosphorescence spectra of two solutions.

11.22(b) When naphthalene is illuminated with 360 nm light it does not absorb, but the line marked B in Fig. 11.51 is the phosphorescence spectrum of a solid solution of a mixture of naphthalene and benzophenone in ethanol. Now a component of fluorescence from naphthalene can be detected. Account for this observation.

11.23(a) The oxygen molecule absorbs ultraviolet radiation in a transition from its ${}^3\Sigma_g^-$ ground electronic state to an excited state that is energetically close to a dissociative ${}^5\Pi_u$ state. The absorption band has a relatively large experimental linewidth. Account for this observation.

11.23(b) The hydrogen molecule absorbs ultraviolet radiation in a transition from its ${}^1\Sigma_g^+$ ground electronic state to an excited state that is energetically close to a dissociative ${}^1\Sigma_u^+$ state. The absorption band has a relatively large experimental linewidth. Account for this observation.

11.24(a) Consider a laser cavity of length 30 cm. What are the allowed wavelengths and frequencies of the resonant modes?

11.24(b) Consider a laser cavity of length 1.0 m. What are the allowed wavelengths and frequencies of the resonant modes?

11.25(a) A pulsed laser rated at 0.10 mJ can generate radiation with peak power output of 5.0 MW and average power output of 7.0 kW. What are the pulse duration and repetition rate?

11.25(b) A pulsed laser rated at 20.0 μJ can generate radiation with peak power output of 100 kW and average power output of 0.40 mW. What are the pulse duration and repetition rate?

11.26(a) Use mathematical software or an electronic spreadsheet to simulate the output of a mode-locked laser (that is, plots such as that shown in Fig. 11.38) for $L = 30$ cm and $N = 100$ and 1000.

11.26(b) Use mathematical software or an electronic spreadsheet to simulate the output of a mode-locked laser (that is, plots such as that shown in Fig. 11.38) for $L = 1.0$ cm and $N = 50$ and 500.

11.27(a) How might you use a Q-switched Nd:YAG laser in the study of a very fast chemical reaction that can be initiated by absorption of light with $\lambda = 266$ nm?

11.27(b) How might you use a mode-locked titanium sapphire laser in the study of a very fast chemical reaction that can be initiated by absorption of light with $\lambda = 400$ nm?

Problems*

Numerical problems

11.1 A *Dubosq colorimeter* consists of a cell of fixed path length and a cell of variable path length. By adjusting the length of the latter until the transmission through the two cells is the same, the concentration of the second solution can be inferred from that of the former. Suppose that a plant dye of concentration 25 μg dm^{-3} is added to the fixed cell, the length of which is 1.55 cm. Then a solution of the same dye, but of unknown concentration, is added to the second cell. It is found that the same transmittance is obtained when the length of the second cell is adjusted to 1.18 cm. What is the concentration of the second solution?

11.2 In many cases it is possible to assume that an absorption band has a Gaussian lineshape (one proportional to e^{-x^2}) centred on the band maximum. Assume such a lineshape, and show that $\mathcal{A} = \int \varepsilon(\tilde{\nu})\mathrm{d}\tilde{\nu} \approx 1.0645\varepsilon_{max}\Delta\tilde{\nu}_{1/2}$, where $\Delta\tilde{\nu}_{1/2}$ is the width at half-height. The absorption spectrum of azoethane ($CH_3CH_2N_2$) between 24 000 cm^{-1} and 34 000 cm^{-1} is shown in Fig. 11.52. First, estimate $\mathcal{A}$ for the band by assuming that it is Gaussian. Then use mathematical software to fit a polynomial to the absorption band (or a Gaussian), and integrate the result analytically.

11.3‡ Dojahn *et al.* (*J. Phys. Chem.* **100**, 9649 (1996)) characterized the potential energy curves of the ground and electronic states of homonuclear diatomic halogen anions. These anions have a ${}^2\Sigma_u^+$ ground state and ${}^2\Pi_g$, ${}^2\Pi_u$, and ${}^2\Sigma_g^+$ excited states. To which of the excited states are electric-dipole transitions allowed? Explain your conclusion.

11.4 The vibrational wavenumber of the oxygen molecule in its electronic ground state is 1580 cm^{-1}, whereas that in the excited state (B ${}^3\Sigma_u^-$), to which there is an allowed electronic transition, is 700 cm^{-1}. Given that the separation in energy between the minima in their respective potential energy curves of these two electronic states is 6.175 eV, what is the wavenumber of the lowest energy transition in the band of transitions originating from the $v = 0$ vibrational state of the electronic ground state to this excited state? Ignore any rotational structure or anharmonicity.

11.5 We are now ready to understand more deeply the features of photoelectron spectra (Section 5.4). Figure 11.53 shows the photoelectron spectrum of HBr. Disregarding for now the fine structure, the HBr lines fall into two main groups. The least tightly bound electrons (with the lowest ionization energies and hence highest kinetic energies when ejected) are those in the lone pairs of the Br atom. The next ionization energy lies at 15.2 eV, and corresponds to the removal of an electron from the HBr σ bond. (a) The spectrum shows that ejection of a σ electron is accompanied by a considerable amount of vibrational excitation. Use the Franck–Condon principle to account for this

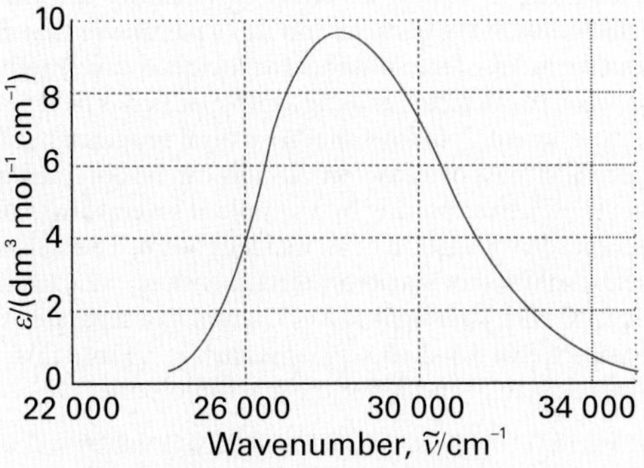

Fig. 11.52 The absorption spectrum of azoethane.

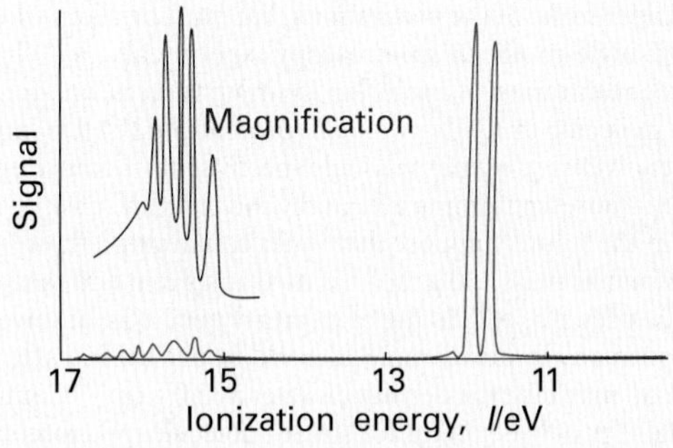

Fig. 11.53 The photoelectron spectrum of HBr.

* Problems denoted with the symbol ‡ were supplied by Charles Trapp and Carmen Giunta.

observation. (b) Go on to explain why the lack of much vibrational structure in the other band is consistent with the nonbonding role of the $Br4p_x$ and $Br4p_y$ lone-pair electrons.

11.6 The highest kinetic energy electrons in the photoelectron spectrum of H_2O using 21.22 eV radiation are at about 9 eV and show a large vibrational spacing of 0.41 eV. The symmetric stretching mode of the neutral H_2O molecule lies at 3652 cm^{-1}. (a) What conclusions can be drawn from the nature of the orbital from which the electron is ejected? (b) In the same spectrum of H_2O, the band near 14.0 eV shows a long vibrational series with spacing 0.125 eV. The bending mode of H_2O lies at 1596 cm^{-1}. What conclusions can you draw about the characteristics of the orbital occupied by the photoelectron?

11.7 A lot of information about the energy levels and wavefunctions of small inorganic molecules can be obtained from their ultraviolet spectra. An example of a spectrum with considerable vibrational structure, that of gaseous SO_2 at 25°C, is shown in Fig. 11.8. Estimate the integrated absorption coefficient for the transition. What electronic states are accessible from the A_1 ground state of this C_{2v} molecule by electric-dipole transitions?

11.8 Aromatic hydrocarbons and I_2 form complexes from which charge-transfer electronic transitions are observed. The hydrocarbon acts an electron donor and I_2 as an electron acceptor. The energies $h\nu_{max}$ of the charge transfer transitions for a number of hydrocarbon–I_2 complexes are given below:

Hydrocarbon	Benzene	Biphenyl	Naphthalene	Phenanthrene	Pyrene	Anthracene
$h\nu_{max}$/eV	4.184	3.654	3.452	3.288	2.989	2.890

Investigate the hypothesis that there is a correlation between the energy of the HOMO of the hydrocarbon (from which the electron comes in the charge-transfer transition) and $h\nu_{max}$. Use one of the molecular electronic structure methods discussed in Chapter 6 to determine the energy of the HOMO of each hydrocarbon in the data set.†

11.9 The fluorescence spectrum of anthracene vapour shows a series of peaks of increasing intensity with individual maxima at 440 nm, 410 nm, 390 nm, and 370 nm followed by a sharp cutoff at shorter wavelengths. The absorption spectrum rises sharply from zero to a maximum at 360 nm with a trail of peaks of lessening intensity at 345 nm, 330 nm, and 305 nm. Account for these observations.

11.10 A certain molecule fluoresces at a wavelength of 400 nm with a half-life of 1.0 ns. It phosphoresces at 500 nm. If the ratio of the transition probabilities for stimulated emission for the $S^* \rightarrow S$ to the $T \rightarrow S$ transitions is 1.0×10^5, what is the half-life of the phosphorescent state?

11.11 Consider some of the precautions that must be taken when conducting single-molecule spectroscopy experiments. (a) What is the molar concentration of a solution in which there is, on average, one solute molecule in 1.0 μm^3 (1.0 fL) of solution? (b) It is important to use pure solvents in single-molecule spectroscopy because optical signals from fluorescent impurities in the solvent may mask optical signals from the solute. Suppose that water containing a fluorescent impurity of molar mass 100 g mol^{-1} is used as solvent and that analysis indicates the presence of 0.10 mg of impurity per 1.0 kg of solvent. On average, how many impurity molecules will be present in 1.0 μm^3 of solution? You may take the density of water as 1.0 g cm^{-3}. Comment on the suitability of this solvent for single-molecule spectroscopy experiments.

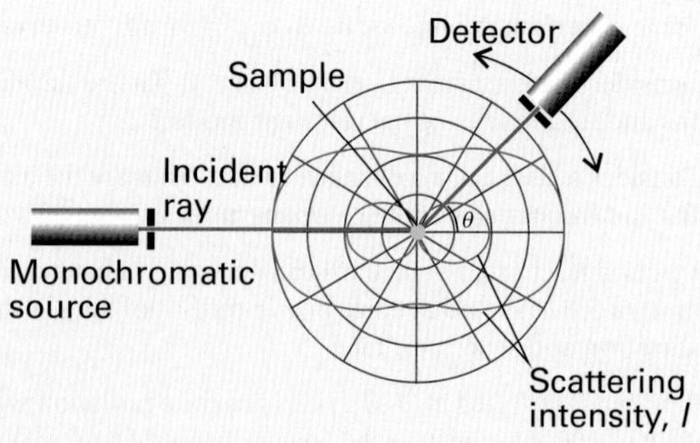

Fig. 11.54 A typical experimental arrangement of a laser light scattering measurement.

11.12 Light-induced degradation of molecules, also called *photobleaching*, is a serious problem in single-molecule spectroscopy. A molecule of a fluorescent dye commonly used to label biopolymers can withstand about 10^6 excitations by photons before light-induced reactions destroy its π system and the molecule no longer fluoresces. For how long will a single dye molecule fluoresce while being excited by 1.0 mW of 488 nm radiation from a continuous-wave argon ion laser? You may assume that the dye has an absorption spectrum that peaks at 488 nm and that every photon delivered by the laser is absorbed by the molecule.

11.13 *Laser light scattering* is a technique that uses the fact that the intensity of light scattered—by Rayleigh scattering—by a particle is proportional to the molar mass of the particle and to λ^{-4}, so shorter wavelength radiation is scattered more intensely than longer wavelengths. Consider the experimental arrangement shown in Fig. 11.54 for the measurement of light scattering from solutions of macromolecules. Typically, the sample is irradiated with monochromatic light from a laser. The intensity of scattered light is then measured as a function of the angle θ that the line of propagation of the laser beam makes with a line from the sample to the detector. For dilute solutions of a spherical macromolecule with a diameter much smaller than the wavelength of incident radiation, the intensity, I_θ, of light scattered by a sample of mass concentration c_M (units: kg m^{-3}) is given by

$$\frac{I_0}{I_\theta} = \frac{1}{Kc_M M} + \left(\frac{16\pi^2 R^2}{5\lambda^2}\right)\left(\frac{I_0}{I_\theta}\sin^2 \tfrac{1}{2}\theta\right)$$

where I_0 is the intensity of the incident laser radiation, M is the molar mass, R is the radius of the particle, and K is a parameter that depends on the refractive index of the solution, the incident wavelength, and the distance between the detector and the sample, which is held constant during the experiment. It follows that structural properties, such as size and the molar mass of a macromolecule, can be obtained from measurements of light scattering by a sample at several angles θ relative to the direction of propagation of an incident beam. The following data for an aqueous solution of a macromolecule with $c_M = 2.0$ kg m^{-3} were obtained at 20°C with laser light at $\lambda = 532$ nm. In a separate experiment, it was determined that $K = 2.40 \times 10^{-2}$ mol m^3 kg^{-2}. From this information, calculate R and M for the macromolecule.

θ/°	15.0	45.0	70.0	85.0	90.0
$10^2 \times I_0/I_\theta$	4.20	4.37	4.63	4.83	4.90

† The web site contains links to molecular modelling freeware and to other sites where you may perform molecular orbital calculations directly from your web browser.

11.14 Matrix-assisted laser desorption/ionization (MALDI) is a type of mass spectrometry, a technique in which the sample is first ionized in the gas phase and then the mass-to-charge number ratios (m/z) of all ions are measured. *MALDI-TOF mass spectrometry*, so called because the MALDI technique is coupled to a time-of-flight (TOF) ion detector, is used widely in the determination of the molar masses of macromolecules. In a MALDI-TOF mass spectrometer, the macromolecule is first embedded in a solid matrix that often consists of an organic acid such as 2,5-dihydroxybenzoic acid, nicotinic acid, or α-cyanocarboxylic acid. This sample is then irradiated with a laser pulse. The pulse of electromagnetic energy ejects matrix ions, cations, and neutral macromolecules, thus creating a dense gas plume above the sample surface. The macromolecule is ionized by collisions and complexation with H^+ cations, resulting in molecular ions of varying charges. The spectrum of a mixture of polymers consists of multiple peaks arising from molecules with different molar masses. A MALDI-TOF mass spectrum consists of two intense features at $m/z = 9912$ and $4554\ \mathrm{g\ mol^{-1}}$. Does the sample contain one or two distinct biopolymers? Explain your answer.

Theoretical problems

11.15 The Beer–Lambert law is derived on the basis that the concentration of absorbing species is uniform. Suppose, instead, that the concentration falls exponentially as $[J] = [J]_0 e^{-x/\lambda}$. Derive an expression for the variation of I with sample length; suppose that $l \gg \lambda$.

11.16 It is common to make measurements of absorbance at two wavelengths and use them to find the individual concentrations of two components A and B in a mixture. Show that the molar concentrations of A and B are

$$[A] = \frac{\varepsilon_{B2}A_1 - \varepsilon_{B1}A_2}{(\varepsilon_{A1}\varepsilon_{B2} - \varepsilon_{A2}\varepsilon_{B1})l} \qquad [B] = \frac{\varepsilon_{A1}A_2 - \varepsilon_{A2}A_1}{(\varepsilon_{A1}\varepsilon_{B2} - \varepsilon_{A2}\varepsilon_{B1})l}$$

where A_1 and A_2 are absorbances of the mixture at wavelengths λ_1 and λ_2, and the molar extinction coefficients of A (and B) at these wavelengths are ε_{A1} and ε_{A2} (and ε_{B1} and ε_{B2}).

11.17 When pyridine is added to a solution of iodine in carbon tetrachloride the 520 nm band of absorption shifts toward 450 nm. However, the absorbance of the solution at 490 nm remains constant: this feature is called an *isosbestic point*. Show that an isosbestic point should occur when two absorbing species are in equilibrium.

11.18 Spin angular momentum is conserved when a molecule dissociates into atoms. What atom multiplicities are permitted when (a) an O_2 molecule, (b) an N_2 molecule dissociates into atoms?

11.19 Assume that the electronic states of the π electrons of a conjugated molecule can be approximated by the wavefunctions of a particle in a one-dimensional box, and that the dipole moment can be related to the displacement along this length by $\mu = -ex$. Show that the transition probability for the transition $n = 1 \rightarrow n = 2$ is nonzero, whereas that for $n = 1 \rightarrow n = 3$ is zero. *Hint.* The following relations will be useful:

$$\sin x \sin y = \tfrac{1}{2}\cos(x-y) - \tfrac{1}{2}\cos(x+y)$$

$$\int x \cos ax \, dx = \frac{1}{a^2}\cos ax + \frac{x}{a}\sin ax$$

11.20 1,3,5-hexatriene (a kind of 'linear' benzene) was converted into benzene itself. On the basis of a free-electron molecular orbital model (in which hexatriene is treated as a linear box and benzene as a ring), would you expect the lowest energy absorption to rise or fall in energy?

11.21 Use a group theoretical argument to decide which of the following transitions are electric-dipole allowed: (a) the $\pi^* \leftarrow \pi$ transition in ethene, (b) the $\pi^* \leftarrow n$ transition in a carbonyl group in a C_{2v} environment.

11.22 Estimate the transition dipole moment of a charge-transfer transition modelled as the migration of an electron from a H1s orbital on one atom to another H1s orbital on an atom a distance R away. Approximate the transition moment by $-eRS$ where S is the overlap integral of the two orbitals. Sketch the transition moment as a function of R using the curve for S given in Fig. 5.27. Why does the intensity of a charge-transfer transition fall to zero as R approaches 0 and infinity?

11.23 Show that the intensity of fluorescence emission from a sample of J is proportional to [J] and l. To do so, consider a sample of J that is illuminated with a beam of intensity $I_0(\tilde{\nu})$ at the wavenumber $\tilde{\nu}$. Before fluorescence can occur, a fraction of $I_0(\tilde{\nu})$ must be absorbed and an intensity $I(\tilde{\nu})$ will be transmitted. However, not all of the absorbed intensity is emitted and the intensity of fluorescence depends on the fluorescence quantum yield, ϕ_f, the efficiency of photon emission. The fluorescence quantum yield ranges from 0 to 1 and is proportional to the ratio of the integral of the fluorescence spectrum over the integrated absorption coefficient. Because of a Stokes shift of magnitude $\Delta\tilde{\nu}_{\mathrm{Stokes}}$, fluorescence occurs at a wavenumber $\tilde{\nu}_f$, with $\tilde{\nu}_f + \Delta\tilde{\nu}_{\mathrm{Stokes}} = \tilde{\nu}$. It follows that the fluorescence intensity at $\tilde{\nu}_f$, $I_f(\tilde{\nu}_f)$, is proportional to ϕ_f and to the intensity of exciting radiation that is absorbed by J, $I_{\mathrm{abs}}(\tilde{\nu}) = I_0(\tilde{\nu}) - I(\tilde{\nu})$. (a) Use the Beer–Lambert law (eqn 11.2) to express $I_{\mathrm{abs}}(\tilde{\nu})$ in terms of $I_0(\tilde{\nu})$, [J], l, and $\varepsilon(\tilde{\nu})$, the molar absorption coefficient of J at $\tilde{\nu}$. (b) Use your result from part (a) to show that $I_f(\tilde{\nu}_f) \propto I_0(\tilde{\nu})\varepsilon(\tilde{\nu})\phi_f[J]l$.

Applications: to biochemistry, environmental science, and astrophysics

11.24 Figure 11.55 shows the UV-visible absorption spectra of a selection of amino acids. Suggest reasons for their different appearances in terms of the structures of the molecules.

11.25 The flux of visible photons reaching Earth from the North Star is about $4 \times 10^3\ \mathrm{mm^{-2}\ s^{-1}}$. Of these photons, 30 per cent are absorbed or scattered by the atmosphere and 25 per cent of the surviving photons are scattered by the surface of the cornea of the eye. A further 9 per cent are absorbed inside the cornea. The area of the pupil at night is about $40\ \mathrm{mm^2}$ and the response time of the eye is about 0.1 s. Of the photons passing through the pupil, about 43 per cent are absorbed in the ocular medium. How many photons from the North Star are focused on to the retina in 0.1 s? For a continuation of this story, see R.W. Rodieck, *The first steps in seeing*, Sinauer, Sunderland (1998).

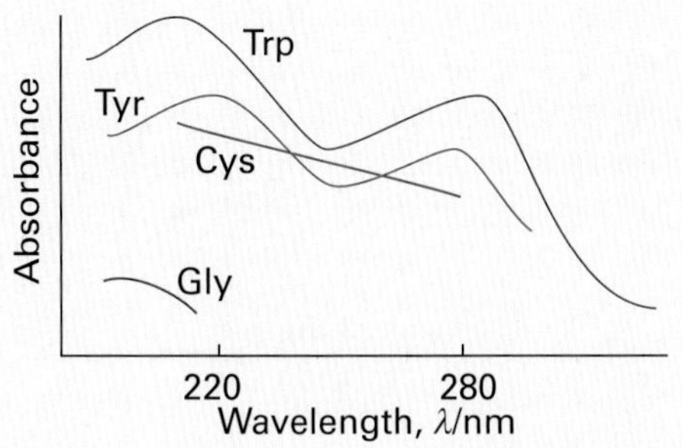

Fig.11.55 Electronic absorption spectra of selected amino acids.

11.26 Use molecule (**10**) as a model of the *trans* conformation of the chromophore found in rhodopsin. In this model, the methyl group bound to the nitrogen atom of the protonated Schiff's base replaces the protein. (a) Using molecular modelling software and the computational method of your instructor's choice, calculate the energy separation between the HOMO and LUMO of (**10**). (b) Repeat the calculation for the 11-*cis* form of (**10**). (c) Based on your results from parts (a) and (b), do you expect the experimental frequency for the $\pi^* \leftarrow \pi$ visible absorption of the *trans* form of (**10**) to be higher or lower than that for the 11-*cis* form of (**10**)?

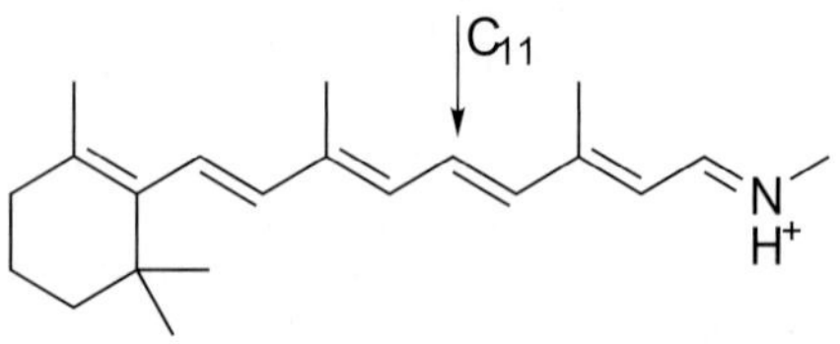

10

11.27‡ Ozone absorbs ultraviolet radiation in a part of the electromagnetic spectrum energetic enough to disrupt DNA in biological organisms and that is absorbed by no other abundant atmospheric constituent. This spectral range, denoted UV-B, spans the wavelengths of about 290 nm to 320 nm. The molar extinction coefficient of ozone over this range is given in the table below (W.B. DeMore *et al.*, *Chemical kinetics and photochemical data for use in stratospheric modeling: Evaluation Number 11*, JPL Publication 94–26 (1994)).

λ/nm	292.0	296.3	300.8	305.4	310.1	315.0	320.0
ε/(dm^3 mol^{-1} cm^{-1})	1512	865	477	257	135.9	69.5	34.5

Compute the integrated absorption coefficient of ozone over the wavelength range 290–320 nm. (*Hint.* $\varepsilon(\tilde{\nu})$ can be fitted to an exponential function quite well.)

11.28‡ G.C.G. Wachewsky *et al.* (*J. Phys. Chem.* **100**, 11559 (1996)) examined the UV absorption spectrum of CH_3I, a species of interest in connection with stratospheric ozone chemistry. They found the integrated absorption coefficient to be dependent on temperature and pressure to an extent inconsistent with internal structural changes in isolated CH_3I molecules; they explained the changes as due to dimerization of a substantial fraction of the CH_3I, a process that would naturally be pressure and temperature dependent. (a) Compute the integrated absorption coefficient over a triangular lineshape in the range 31 250 to 34 483 cm^{-1} and a maximal molar absorption coefficient of 150 dm^3 mol^{-1} cm^{-1} at 31 250 cm^{-1}. (b) Suppose 1 per cent of the CH_3I units in a sample at 2.4 Torr and 373 K exists as dimers. Compute the absorbance expected at 31 250 cm^{-1} in a sample cell of length 12.0 cm. (c) Suppose 18 per cent of the CH_3I units in a sample at 100 Torr and 373 K exists as dimers. Compute the absorbance expected at 31 250 cm^{-1} in a sample cell of length 12.0 cm; compute the molar absorption coefficient that would be inferred from this absorbance if dimerization was not considered.

11.29‡ One of the principal methods for obtaining the electronic spectra of unstable radicals is to study the spectra of comets, which are almost entirely due to radicals. Many radical spectra have been found in comets, including that due to CN. These radicals are produced in comets by the absorption of far ultraviolet solar radiation by their parent compounds. Subsequently, their fluorescence is excited by sunlight of longer wavelength. The spectra of comet Hale–Bopp (C/1995 O1) have been the subject of many recent studies. One such study is that of the fluorescence spectrum of CN in the comet at large heliocentric distances by R.M. Wagner and D.G. Schleicher (*Science* **275**, 1918 (1997)), in which the authors determine the spatial distribution and rate of production of CN in the coma. The (0–0) vibrational band is centred on 387.6 nm and the weaker (1–1) band with relative intensity 0.1 is centred on 386.4 nm. The band heads for (0–0) and (0–1) are known to be 388.3 and 421.6 nm, respectively. From these data, calculate the energy of the excited S_1 state relative to the ground S_0 state, the vibrational wavenumbers and the difference in the vibrational wavenumbers of the two states, and the relative populations of the $v = 0$ and $v = 1$ vibrational levels of the S_1 state. Also estimate the effective temperature of the molecule in the excited S_1 state. Only eight rotational levels of the S_1 state are thought to be populated. Is that observation consistent with the effective temperature of the S_1 state?

12

Magnetic resonance

The chapter begins with an account of conventional nuclear magnetic resonance that shows how the resonance frequency of a magnetic nucleus is affected by its own electronic environment and the presence of other magnetic nuclei in its vicinity. Then we turn to the modern versions of NMR, which are based on the use of pulses of electromagnetic radiation and the processing of the resulting signal by Fourier transform techniques. The experimental techniques for electron paramagnetic resonance resemble those used in the early days of NMR. The information obtained is very useful for the determination of the properties of species with unpaired electrons.

When two pendulums share a slightly flexible support and one is set in motion, the other is forced into oscillation by the motion of the common axle. As a result, energy flows between the two pendulums. The energy transfer occurs most efficiently when the frequencies of the two pendulums are identical. The condition of strong effective coupling when the frequencies of two oscillators are identical is called **resonance**. Resonance is the basis of a number of everyday phenomena, including the response of radios to the weak oscillations of the electromagnetic field generated by a distant transmitter. Resonance also provides a way of explaining spectroscopic transitions of any type, for they depend on matching the oscillations of an electromagnetic field to a set of energy levels and observing the strong absorption that occurs at resonance. Historically, spectroscopic techniques that measure transitions between nuclear or electron spin states and that, for historical reasons, have carried the term 'resonance' in their names share a common feature: in the original formulation of the technique, the energy levels themselves are brought into resonance with a fixed-frequency electromagnetic field rather than vice versa.

The effect of magnetic fields on electrons and nuclei

The Stern–Gerlach experiment (Section 3.5) provided evidence for electron spin. It turns out that many nuclei also possess spin angular momentum. Orbital and spin angular momenta give rise to magnetic moments, and to say that electrons and nuclei have magnetic moments means that, to some extent, they behave like small magnets. First, we establish how the energies of electrons and nuclei depend on the strength of an external field. Then we see how to use this dependence to study the structure and dynamics of complex molecules.

12.1 The energies of electrons in magnetic fields

Classically, the energy of a magnetic moment $\boldsymbol{\mu}$ in a magnetic field $\mathcal{B}$ is equal to the scalar product

$$E=-\boldsymbol{\mu}\cdot\mathcal{B} \tag{12.1}$$

A brief comment More formally, $\mathcal{B}$ is the magnetic induction and is measured in tesla, T; 1 T = 1 kg s^{-2} A^{-1}. The (non-SI) unit gauss, G, is also occasionally still used: 1 T = 10^4 G.

Quantum mechanically, we write the hamiltonian as

$$\hat{H}=-\hat{\boldsymbol{\mu}}\cdot\mathcal{B} \tag{12.2}$$

To write an expression for $\hat{\boldsymbol{\mu}}$, we recall from Section 4.5 that the magnetic moment is proportional to the angular momentum. For an electron possessing orbital angular momentum we write

$$\hat{\boldsymbol{\mu}}=\gamma_e\hat{\boldsymbol{l}} \quad \text{and} \quad \hat{H}=-\gamma_e\mathcal{B}\cdot\hat{\boldsymbol{l}} \tag{12.3}$$

where $\hat{\boldsymbol{l}}$ is the orbital angular momentum operator and

$$\gamma_e=-\frac{e}{2m_e} \qquad [12.4]$$

γ_e is called the **magnetogyric ratio** of the electron. The negative sign (arising from the sign of the electron's charge) shows that the orbital moment is antiparallel to its orbital angular momentum (as was depicted in Fig 4.26).

For a magnetic field $\mathcal{B}_0$ along the z-direction, eqn 12.3 becomes

$$\hat{\mu}_z=\gamma_e\hat{l}_z \quad \text{and} \quad \hat{H}=-\gamma_e\mathcal{B}_0\hat{l}_z=-\hat{\mu}_z\mathcal{B}_0 \tag{12.5a}$$

Because the eigenvalues of the operator $\hat{l}_z$ are $m_l\hbar$, the z-component of the orbital magnetic moment and the energy of interaction are, respectively,

$$\mu_z=\gamma_e m_l\hbar \quad \text{and} \quad E=-\gamma_e m_l\hbar\mathcal{B}_0=m_l\mu_B\mathcal{B}_0 \tag{12.5b}$$

where the **Bohr magneton, μ_B,** is

$$\mu_B=-\gamma_e\hbar=\frac{e\hbar}{2m_e}=9.724\times10^{-24}\,\text{J}\,\text{T}^{-1} \qquad [12.6]$$

The Bohr magneton is often regarded as the fundamental quantum of magnetic moment.

The spin magnetic moment of an electron, which has a spin quantum number $s=\frac{1}{2}$ (Section 3.5), is also proportional to its spin angular momentum. However, instead of eqn 12.3, the spin magnetic moment and hamiltonian operators are, respectively,

$$\hat{\boldsymbol{\mu}}=g_e\gamma_e\hat{\boldsymbol{s}} \quad \text{and} \quad \hat{H}=-g_e\gamma_e\mathcal{B}\cdot\hat{\boldsymbol{s}} \tag{12.7}$$

where $\hat{\boldsymbol{s}}$ is the spin angular momentum operator and the extra factor g_e is called the **g-value** of the electron: $g_e = 2.002\ 319\ldots$. The g-value arises from relativistic effects and from interactions of the electron with the electromagnetic fluctuations of the vacuum that surrounds the electron. For a magnetic field $\mathcal{B}_0$ in the z-direction,

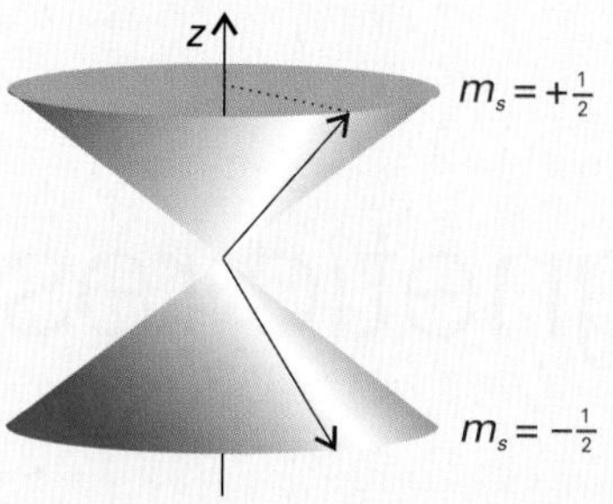

Fig. 12.1 The interactions between the m_s states of an electron and an external magnetic field may be visualized as the precession of the vectors representing the angular momentum.

$$\hat{\mu}_z=g_e\gamma_e\hat{s}_z \quad \text{and} \quad \hat{H}=-g_e\gamma_e\mathcal{B}_0\hat{s}_z \tag{12.8a}$$

Because the eigenvalues of the operator $\hat{s}_z$ are $m_s\hbar$ with $m_s=+\frac{1}{2}$ (α) and $m_s=-\frac{1}{2}$ (β), it follows that the energies of an electron spin in a magnetic field are

$$\mu_z=g_e\gamma_e m_s\hbar \quad \text{and} \quad E_{m_s}=-g_e\gamma_e m_s\hbar\mathcal{B}_0=g_e\mu_B m_s\mathcal{B}_0 \tag{12.8b}$$

with $m_s=\pm\frac{1}{2}$.

In the absence of a magnetic field, the states with different values of m_l and m_s are degenerate. When a field is present, the degeneracy is removed: the state with $m_s=+\frac{1}{2}$ moves up in energy by $\frac{1}{2}g_e\mu_B\mathcal{B}_0$ and the state with $m_s=-\frac{1}{2}$ moves down by $\frac{1}{2}g_e\mu_B\mathcal{B}_0$. The different energies arising from an interaction with an external field are sometimes represented on the vector model by picturing the vectors as **precessing**, or sweeping round their cones (Fig. 12.1), with the rate of precession equal to the **Larmor frequency**, ν_L:

$$\nu_L=\frac{\gamma_e\mathcal{B}_0}{2\pi} \tag{12.9}$$

Equation 12.9 shows that the Larmor frequency increases with the strength of the magnetic field. For a field of 1 T, the Larmor frequency is 30 GHz.

12.2 The energies of nuclei in magnetic fields

The spin quantum number, I, of a nucleus is a fixed characteristic property of a nucleus and is either an integer or a half-integer (Table 12.1). A nucleus with spin quantum number I has the following properties:

1. An angular momentum of magnitude $\{I(I+1)\}^{1/2}\hbar$.

2. A component of angular momentum $m_I\hbar$ on a specified axis ('the z-axis'), where $m_I=I, I-1, \ldots, -I$.

Table 12.1 Nuclear constitution and the nuclear spin quantum number*

Number of protons	Number of neutrons	I
even	even	0
odd	odd	integer (1, 2, 3, . . .)
even	odd	half-integer ($\frac{1}{2}, \frac{3}{2}, \frac{5}{2}, \ldots$)
odd	even	half-integer ($\frac{1}{2}, \frac{3}{2}, \frac{5}{2}, \ldots$)

* The spin of a nucleus may be different if it is in an excited state; throughout this chapter we deal only with the ground state of nuclei.

3. If $I > 0$, a magnetic moment with a constant magnitude and an orientation that is determined by the value of m_I.

According to the second property, the spin, and hence the magnetic moment, of the nucleus may lie in $2I + 1$ different orientations relative to an axis. A proton has $I = \frac{1}{2}$ and its spin may adopt either of two orientations; a ^{14}N nucleus has $I = 1$ and its spin may adopt any of three orientations; both ^{12}C and ^{16}O have $I = 0$ and hence zero magnetic moment.

The energy of interaction between a nucleus with a magnetic moment $\boldsymbol{\mu}$ and an external magnetic field $\boldsymbol{\mathcal{B}}$ may be calculated by using operators analogous to those of eqn 12.3:

$$\hat{\mu} = \gamma \hat{I} \qquad \text{and} \qquad \hat{H} = -\gamma \boldsymbol{\mathcal{B}} \cdot \hat{I} \tag{12.10a}$$

where γ is the magnetogyric ratio of the specified nucleus, an empirically determined characteristic arising from the internal structure of the nucleus (Table 12.2). The corresponding energies are

$$E_{m_I} = -\mu_z \mathcal{B}_0 = -\gamma \hbar \mathcal{B}_0 m_I \tag{12.10b}$$

As for electrons, the nuclear spin may be pictured as precessing around the direction of the applied field at a rate proportional to the applied field. For protons, a field of 1 T corresponds to a Larmor frequency (eqn 12.9, with γ_e replaced by γ) of about 40 MHz.

The magnetic moment of a nucleus is sometimes expressed in terms of the **nuclear *g*-factor**, g_I, a characteristic of the nucleus, and the **nuclear magneton**, μ_N, a quantity independent of the nucleus, by using

$$\gamma \hbar = g_I \mu_N \qquad \mu_N = \frac{e\hbar}{2m_p} = 5.051 \times 10^{-27}\,\text{J T}^{-1} \tag{12.11}$$

where m_p is the mass of the proton. Nuclear g-factors vary between −6 and +6 (see Table 12.2): positive values of g_I and γ denote a magnetic moment that is parallel to the spin; negative values indicate that the magnetic moment and spin are antiparallel. For the remainder of this chapter we shall assume that γ is positive, as is the case for the majority of nuclei. In such cases, the states with negative values of m_I lie above states with positive values of m_I. The nuclear magneton is about 2000 times smaller than the Bohr magneton, so nuclear magnetic moments—and consequently the energies of interaction with magnetic fields—are about 2000 times weaker than the electron spin magnetic moment.

12.3 Magnetic resonance spectroscopy

In its original form, the magnetic resonance experiment is the resonant absorption of radiation by nuclei or unpaired electrons in a magnetic field. From eqn 12.8, the separation between the $m_s = -\frac{1}{2}$ and $m_s = +\frac{1}{2}$ levels of an electron spin in a magnetic field $\mathcal{B}_0$ is

$$\Delta E = E_\alpha - E_\beta = g_e \mu_B \mathcal{B}_0 \tag{12.12a}$$

If the sample is exposed to radiation of frequency ν, the energy separations come into resonance with the radiation when the frequency satisfies the **resonance condition** (Fig. 12.2):

$$h\nu = g_e \mu_B \mathcal{B}_0 \tag{12.12b}$$

At resonance there is strong coupling between the electron spins and the radiation, and strong absorption occurs as the spins make the transition $\beta \rightarrow \alpha$. **Electron paramagnetic resonance** (EPR), or **electron spin resonance** (ESR), is the study of molecules

Synoptic table 12.2* Nuclear spin properties

Nuclide	Natural abundance/%	Spin I	g-factor, g_I	Magnetogyric ratio, $\gamma/(10^7\,\text{T}^{-1}\,\text{s}^{-1})$	NMR frequency at 1 T, ν/MHz
^{1}n		$\frac{1}{2}$	−3.826	−18.32	29.164
^{1}H	99.98	$\frac{1}{2}$	5.586	26.75	42.576
^{2}H	0.02	1	0.857	4.11	6.536
^{13}C	1.11	$\frac{1}{2}$	1.405	6.73	10.708
^{14}N	99.64	1	0.404	1.93	3.078

* More values are given in the *Data section*.

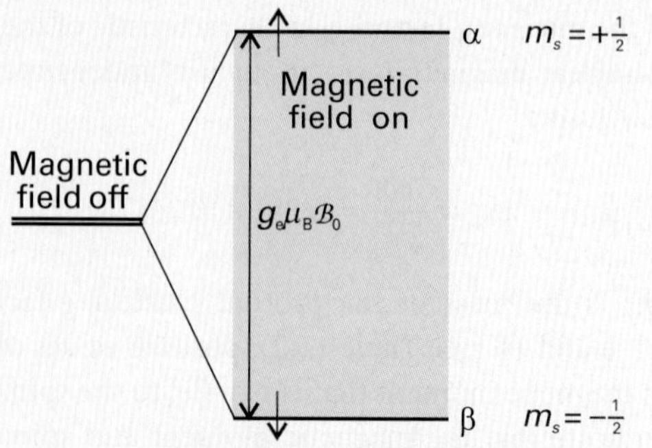

Fig. 12.2 Electron spin levels in a magnetic field. Note that the β state is lower in energy than the α state (because the magnetogyric ratio of an electron is negative). Resonance is achieved when the frequency of the incident radiation matches the frequency corresponding to the energy separation.

and ions containing unpaired electrons by observing the magnetic fields at which they come into resonance with monochromatic radiation. Magnetic fields of about 0.3 T (the value used in most commercial EPR spectrometers) correspond to resonance with an electromagnetic field of frequency 10 GHz (10^{10} Hz) and wavelength 3 cm. Because 3 cm radiation falls in the microwave region of the electromagnetic spectrum, EPR is a microwave technique.

The energy separation between the $m_I=+\frac{1}{2}$ (↑ or α) and the $m_I=-\frac{1}{2}$ (↓ or β) states of **spin-$\frac{1}{2}$ nuclei**, which are nuclei with $I=\frac{1}{2}$, is

$$\Delta E = E_\beta - E_\alpha = \tfrac{1}{2}\gamma\hbar\mathcal{B}_0 - (-\tfrac{1}{2}\gamma\hbar\mathcal{B}_0) = \gamma\hbar\mathcal{B}_0 \qquad (12.13a)$$

and resonant absorption occurs when the resonance condition (Fig. 12.3)

$$h\nu = \gamma\hbar\mathcal{B}_0 \qquad (12.13b)$$

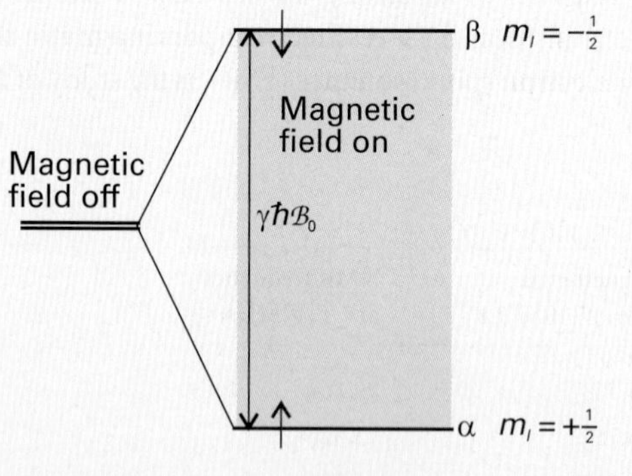

Fig. 12.3 The nuclear spin energy levels of a spin-$\frac{1}{2}$ nucleus with positive magnetogyric ratio (for example, ^{1}H or ^{13}C) in a magnetic field. Resonance occurs when the energy separation of the levels matches the energy of the photons in the electromagnetic field.

is fulfilled. Because $\gamma\hbar\mathcal{B}_0/h$ is the Larmor frequency of the nucleus, this resonance occurs when the frequency of the electromagnetic field matches the Larmor frequency ($\nu = \nu_L$). In its simplest form, **nuclear magnetic resonance** (NMR) is the study of the properties of molecules containing magnetic nuclei by applying a magnetic field and observing the frequency of the resonant electromagnetic field. Larmor frequencies of nuclei at the fields normally employed (about 12 T) typically lie in the radiofrequency region of the electromagnetic spectrum (close to 500 MHz), so NMR is a radiofrequency technique.

For much of this chapter we consider spin-$\frac{1}{2}$ nuclei, but NMR is applicable to nuclei with any non-zero spin. As well as protons, which are the most common nuclei studied by NMR, spin-$\frac{1}{2}$ nuclei include ^{13}C, ^{19}F, and ^{31}P.

Nuclear magnetic resonance

Although the NMR technique is simple in concept, NMR spectra can be highly complex. However, they have proved invaluable in chemistry, for they reveal so much structural information. A magnetic nucleus is a very sensitive, non-invasive probe of the surrounding electronic structure.

12.4 The NMR spectrometer

An NMR spectrometer consists of the appropriate sources of radiofrequency electromagnetic radiation and a magnet that can produce a uniform, intense field. Most instruments feature a **superconducting magnet** capable of producing fields of the order of 10 T and more (Fig. 12.4). The sample is placed in the cylindrically wound magnet. In some cases the sample is rotated rapidly to average out magnetic inhomogeneities. However, sample spinning can lead to irreproducible results, and is often avoided. Although a superconducting magnet operates at the

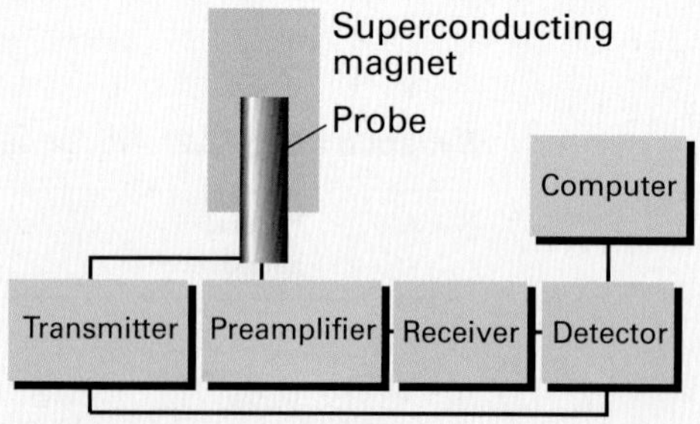

Fig. 12.4 The layout of a typical NMR spectrometer. The link from the transmitter to the detector indicates that the high frequency of the transmitter is subtracted from the high frequency received signal to give a low frequency signal for processing.

temperature of liquid helium (4 K), the sample itself is normally at room temperature.

The intensity of an NMR transition depends on a number of factors. We show in the following *Justification* that

$$\text{Intensity} \propto (N_\alpha - N_\beta)\mathcal{B}_0 \qquad (12.14a)$$

where N_α and N_β are, respectively, the numbers of α and β spins, and

$$N_\alpha - N_\beta \approx \frac{N\gamma\hbar\mathcal{B}_0}{2kT} \qquad (12.14b)$$

with N the total number of spins ($N = N_\alpha + N_\beta$). It follows that decreasing the temperature increases the intensity by increasing the population difference. By combining these two equations we see that the intensity is proportional to $\mathcal{B}_0^2$, so NMR transitions can be enhanced significantly by increasing the strength of the applied magnetic field. We shall also see (Section 12.6) that the use of high magnetic fields simplifies the appearance of spectra and so allows them to be interpreted more readily. We also conclude that absorptions of nuclei with large magnetogyric ratios (^{1}H, for instance) are more intense than those with small magnetogyric ratios (^{13}C, for instance)

Justification 12.1 *Intensities in NMR spectra*

From the general considerations of transition intensities in *Further information 10.1*, we know that the rate of absorption of electromagnetic radiation is proportional to the population of the lower energy state (N_α in the case of a proton NMR transition) and the rate of stimulated emission is proportional to the population of the upper state (N_β). At the low frequencies typical of magnetic resonance, we can neglect spontaneous emission as it is very slow. Therefore, the net rate of absorption is proportional to the difference in populations, and we can write

$$\text{Net rate of absorption} \propto N_\alpha - N_\beta$$

The intensity of absorption, the rate at which energy is absorbed, is proportional to the product of the rate of absorption (the rate at which photons are absorbed) and the energy of each photon, and the latter is proportional to the frequency ν of the incident radiation (through $E = h\nu$). At resonance, this frequency is proportional to the applied magnetic field (through $\nu = \nu_L = \gamma\mathcal{B}_0/2\pi$), so we can write

$$\text{Intensity of absorption} \propto (N_\alpha - N_\beta)\mathcal{B}_0$$

as in eqn 12.14a. To write an expression for the population difference, we use the Boltzmann distribution (*Fundamentals F.5* and Chapter 13) to write the ratio of populations as

$$\frac{N_\beta}{N_\alpha} = e^{-\Delta E/kT} \approx 1 - \frac{\Delta E}{kT} = 1 - \frac{\gamma\hbar\mathcal{B}_0}{kT}$$

A brief comment The expansion of an exponential function used here is $e^{-x} = 1 - x + \frac{1}{2}x^2 - \cdots$. If $x \ll 1$, then $e^{-x} \approx 1 - x$.

where $\Delta E = E_\beta - E_\alpha$. The expansion of the exponential term is appropriate for $\Delta E \ll kT$, a condition usually met for nuclear spins. It follows after rearrangement that

$$\frac{N_\alpha - N_\beta}{N_\alpha + N_\beta} = \frac{N_\alpha(1 - N_\beta/N_\alpha)}{N_\alpha(1 + N_\beta/N_\alpha)} = \frac{1 - N_\beta/N_\alpha}{1 + N_\beta/N_\alpha}$$
$$\approx \frac{1 - (1 - \gamma\hbar\mathcal{B}_0/kT)}{1 + (1 - \gamma\hbar\mathcal{B}_0/kT)} \approx \frac{\gamma\hbar\mathcal{B}_0/kT}{2}$$

Then, with $N_\alpha + N_\beta = N$, the total number of spins, we obtain eqn 12.14b.

12.5 The chemical shift

Nuclear magnetic moments interact with the *local* magnetic field. The local field may differ from the applied field because the latter induces electronic orbital angular momentum (that is, the circulation of electronic currents) which gives rise to a small additional magnetic field $\delta\mathcal{B}$ at the nuclei. This additional field is proportional to the applied field, and it is conventional to write

$$\delta\mathcal{B} = -\sigma\mathcal{B}_0 \qquad [12.15]$$

where the dimensionless quantity σ is called the **shielding constant** of the nucleus (σ is usually positive but may be negative). The ability of the applied field to induce an electronic current in the molecule, and hence affect the strength of the resulting local magnetic field experienced by the nucleus, depends on the details of the electronic structure near the magnetic nucleus of interest, so nuclei in different chemical groups have different shielding constants. The calculation of reliable values of the shielding constant is very difficult, but trends in it are quite well understood and we concentrate on them.

(a) The δ scale of chemical shifts

Because the total local field is

$$\mathcal{B}_{loc} = \mathcal{B}_0 + \delta\mathcal{B} = (1 - \sigma)\mathcal{B}_0 \qquad (12.16)$$

the nuclear Larmor frequency is

$$\nu_L = \frac{\gamma\mathcal{B}_{loc}}{2\pi} = (1 - \sigma)\frac{\gamma\mathcal{B}_0}{2\pi} \qquad (12.17)$$

This frequency is different for nuclei in different environments. Hence, different nuclei, even of the same element, come into resonance at different frequencies.

It is conventional to express the resonance frequencies in terms of an empirical quantity called the **chemical shift**, which is related to the difference between the resonance frequency, ν, of the nucleus in question and that of a reference standard, ν°:

$$\delta = \frac{\nu - \nu^\circ}{\nu^\circ} \times 10^6 \qquad [12.18]$$

The standard for protons is the proton resonance in tetramethylsilane ($Si(CH_3)_4$, commonly referred to as TMS), which bristles with protons and dissolves without reaction in many liquids. Other references are used for other nuclei. For ^{13}C, the reference frequency is the ^{13}C resonance in TMS; for ^{31}P it is the ^{31}P resonance in 85 per cent $H_3PO_4(aq)$. The advantage of the δ-scale is that shifts reported on it are independent of the applied field (because both numerator and denominator are proportional to the applied field).

● A BRIEF ILLUSTRATION

From eqn 12.18,

$$\nu - \nu^\circ = \nu^\circ \delta \times 10^{-6}$$

A nucleus with $\delta = 1.00$ in a spectrometer operating at 500 MHz will have a shift relative to the reference equal to

$$\nu - \nu^\circ = (500\ \text{MHz}) \times 1.00 \times 10^{-6} = 500\ \text{Hz}$$

In a spectrometer operating at 100 MHz, the shift relative to the reference would be only 100 Hz. ●

A note on good practice In much of the literature, chemical shifts are reported in 'parts per million', ppm, in recognition of the factor of 10^6 in the definition. This practice is unnecessary and can be confusing.

The relation between δ and σ is obtained by substituting eqn 12.16 into eqn 12.18:

$$\delta = \frac{(1-\sigma)\mathcal{B}_0 - (1-\sigma^\circ)\mathcal{B}_0}{(1-\sigma^\circ)\mathcal{B}_0} \times 10^6$$

$$= \frac{\sigma^\circ - \sigma}{1-\sigma^\circ} \times 10^6 \approx (\sigma^\circ - \sigma) \times 10^6 \qquad (12.19)$$

where σ° is the shielding constant for the reference standard. We see that as the shielding, σ, gets smaller, δ *increases*. Therefore, we speak of nuclei with large chemical shift as being strongly **deshielded**. Some typical chemical shifts are given in Fig. 12.5. As can be seen from the figure, the nuclei of different elements have very different ranges of chemical shifts. The ranges exhibit the variety of electronic environments of the nuclei in molecules: the more massive the element, the greater the number of electrons around the nucleus and hence the greater the range of shieldings.

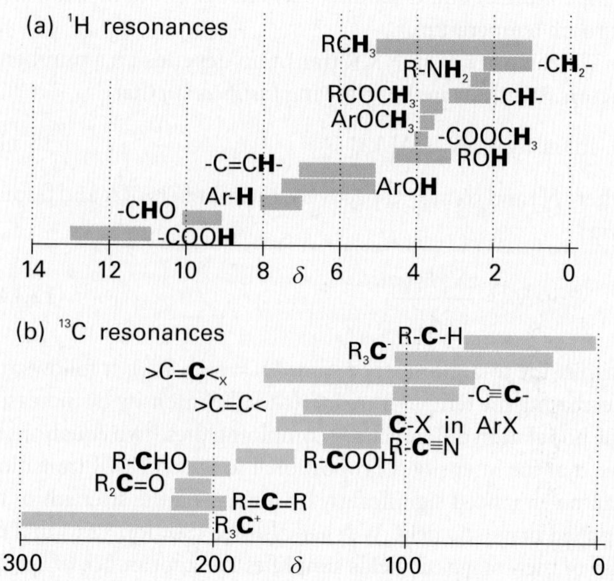

Fig. 12.5 The range of typical chemical shifts for (a) ^1H resonances and (b) ^{13}C resonances.

By convention, NMR spectra are plotted with δ increasing from right to left. Consequently, in a given applied magnetic field the Larmor frequency also increases from right to left. In the original continuous wave (CW) spectrometers, in which the radiofrequency was held constant and the magnetic field varied (a 'field sweep experiment'), the spectrum was displayed with the applied magnetic field increasing from left to right: a nucleus with a small chemical shift experiences a relatively low local magnetic field, so it needs a higher applied magnetic field to bring it into resonance with the radiofrequency field. Consequently, the right-hand (low chemical shift) end of the spectrum became known as the 'high field end' of the spectrum.

(b) Resonance of different groups of nuclei

The existence of a chemical shift explains the general features of the spectrum of ethanol shown in Fig.12.6. The CH_3 protons form one group of nuclei with $\delta \approx 1$. The two CH_2 protons are in a different part of the molecule, experience a different local magnetic field, and resonate at $\delta \approx 3.6$. Finally, the OH proton is in another environment, and has a chemical shift of $\delta \approx 4$. The increasing value of δ (that is, the decrease in shielding) is consistent with the electron-withdrawing power of the O atom: it reduces the electron density of the OH proton most, and that proton is strongly deshielded. It reduces the electron density of the distant methyl protons least, and those nuclei are least deshielded.

The relative intensities of the signals (the areas under the absorption lines) can be used to help distinguish which group

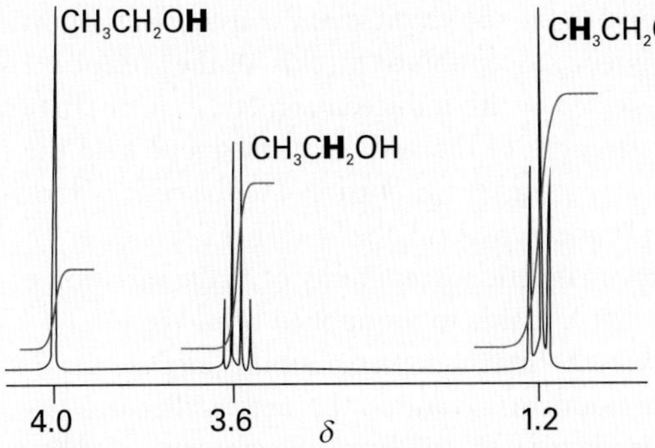

Fig. 12.6 The ^{1}H-NMR spectrum of ethanol. The bold letters denote the protons giving rise to the resonance peak, and the step-like curve is the integrated signal.

of lines corresponds to which chemical group. The determination of the area under an absorption line is referred to as the **integration** of the signal (just as any area under a curve may be determined by mathematical integration). Data analysis software performs this integration and the values are typically displayed as the height of step-like curves drawn on the spectrum, as shown in Fig. 12.6. In ethanol the group intensities are in the ratio 3:2:1 because there are three CH_3 protons, two CH_2 protons, and one OH proton in each molecule. Counting the number of magnetic nuclei as well as noting their chemical shifts helps to identify a compound present in a sample.

(c) The origin of shielding constants

The calculation of shielding constants is difficult, even for small molecules, for it requires detailed information about the distribution of electron density in the ground and excited states and the excitation energies of the molecule. Nevertheless, considerable success has been achieved with the calculation for diatomic molecules and small molecules such as H_2O and CH_4 and even large molecules, such as proteins, are within the scope of some types of calculation. Nevertheless, it is easier to understand the different contributions to chemical shifts by studying the large body of empirical information now available for large molecules.

The empirical approach supposes that the observed shielding constant is the sum of three contributions:

$$\sigma = \sigma(\text{local}) + \sigma(\text{neighbour}) + \sigma(\text{solvent}) \qquad (12.20)$$

The **local contribution**, σ(local), is essentially the contribution of the electrons of the atom that contains the nucleus in question. The **neighbouring group contribution**, σ(neighbour), is the contribution from the groups of atoms that form the rest of the molecule. The **solvent contribution**, σ(solvent), is the contribution from the solvent molecules.

(d) The local contribution

It is convenient to regard the local contribution to the shielding constant as the sum of a **diamagnetic contribution**, σ_d, and a **paramagnetic contribution**, σ_p:

$$\sigma(\text{local}) = \sigma_d + \sigma_p \qquad (12.21)$$

A diamagnetic contribution to σ(local) opposes the applied magnetic field and shields the nucleus in question. A paramagnetic contribution to σ(local) reinforces the applied magnetic field and deshields the nucleus in question. Therefore, $\sigma_d > 0$ and $\sigma_p < 0$. The total local contribution is positive if the diamagnetic contribution dominates, and is negative if the paramagnetic contribution dominates.

The diamagnetic contribution arises from the ability of the applied field to generate a circulation of charge in the ground-state electron distribution of the atom. The circulation generates a magnetic field that opposes the applied field and hence shields the nucleus. The magnitude of σ_d depends on the electron density close to the nucleus and can be calculated from the **Lamb formula**:

$$\sigma_d = \frac{e^2\mu_0}{12\pi m_e}\left\langle \frac{1}{r} \right\rangle \qquad (12.22)$$

where μ_0 is the vacuum permeability (a fundamental constant, see inside the front cover) and r is the electron–nucleus distance.

● A BRIEF ILLUSTRATION

To calculate σ_d for the proton in a free H atom, we need to calculate the expectation value of $1/r$ for a hydrogen 1s orbital. Wavefunctions are given in Table 4.1, and a useful integral is given in Example 4.4. Because $d\tau = r^2 dr \sin\theta d\theta d\phi$, we can write

$$\left\langle \frac{1}{r} \right\rangle = \int \frac{\psi^*\psi}{r} d\tau$$
$$= \frac{1}{\pi a_0^3}\int_0^{2\pi} d\phi \int_0^{\pi} \sin\theta d\theta \int_0^{\infty} re^{-2r/a_0} dr$$
$$= \frac{4}{a_0^3}\int_0^{\infty} re^{-2r/a_0} dr = \frac{1}{a_0}$$

Therefore,

$$\sigma_d = \frac{e^2\mu_0}{12\pi m_e a_0}$$

With the values of the fundamental constants inside the front cover, this expression evaluates to 1.78×10^{-5}. ●

The diamagnetic contribution is the only contribution in closed-shell free atoms. It is also the only contribution to the

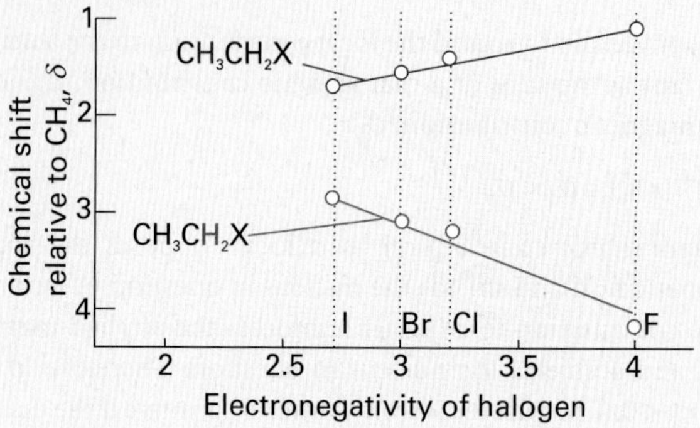

Fig. 12.7 The variation of chemical shielding with electronegativity. The shifts for the methylene protons agree with the trend expected with increasing electronegativity. However, to emphasize that chemical shifts are subtle phenomena, notice that the trend for the methyl protons is opposite to that expected. For these protons another contribution (the magnetic anisotropy of C—H and C—X bonds) is dominant.

local shielding for electron distributions that have spherical or cylindrical symmetry. Thus, it is the only contribution to the local shielding from inner cores of atoms, for cores remain spherical even though the atom may be a component of a molecule and its valence electron distribution highly distorted. The diamagnetic contribution is broadly proportional to the electron density of the atom containing the nucleus of interest. It follows that the shielding is decreased if the electron density on the atom is reduced by the influence of an electronegative atom nearby. That reduction in shielding translates into an increase in deshielding, and hence to an increase in the chemical shift δ as the electronegativity of a neighbouring atom increases (Fig. 12.7).

The local paramagnetic contribution, σ_p, arises from the ability of the applied field to force electrons to circulate through the molecule by making use of orbitals that are unoccupied in the ground state. It is zero in free atoms and around the axes of linear molecules (such as ethyne, HC≡CH) where the electrons can circulate freely and a field applied along the internuclear axis is unable to force them into other orbitals. We can expect large paramagnetic contributions from small atoms in molecules with low-lying excited states. In fact, the paramagnetic contribution is the dominant local contribution for atoms other than hydrogen.

(e) Neighbouring group contributions

The neighbouring group contribution arises from the currents induced in nearby groups of atoms. Consider the influence of the neighbouring group X on the proton H in a molecule such as H—X. The applied field generates currents in the electron distribution of X and gives rise to an induced magnetic moment proportional to the applied field; the constant of proportionality is the magnetic susceptibility, χ (chi), of the group X. The proton H is affected by this induced magnetic moment in two ways. First, the strength of the additional magnetic field the proton experiences is inversely proportional to the cube of the distance r between H and X. Second, the field at H depends on the anisotropy of the magnetic susceptibility of X, the variation of χ with the angle that X makes to the applied field. We assume that the magnetic susceptibility of X has two components, $\chi_{\parallel}$ and $\chi_{\perp}$ that are parallel and perpendicular to the axis of symmetry of X, respectively. The axis of symmetry of X makes an angle θ to the vector connecting X to H (**1**, where X is represented by the ellipse and H is represented by the circle).

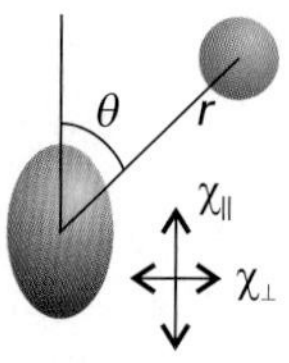

1

To examine the effect of anisotropy of the magnetic susceptibility of X on the shielding constant, consider the case $\theta = 0$ for a molecule H—X that is free to tumble (**2** and **3**). Some of the time the H—X axis will be perpendicular to the applied field and then only $\chi_{\perp}$ will contribute to the induced magnetic moment that shields X from the applied field. The result is deshielding of the proton H, or σ(neighbour) < 0 (**2**). When the applied field is parallel to the H—X axis, only $\chi_{\parallel}$ contributes to the induced magnetic moment at X. The result is shielding of the proton H (**3**). We conclude that, as the molecule tumbles and the H—X axis takes all possible angles with respect to the applied field, the effects of anisotropic magnetic susceptibility do not average to zero because $\chi_{\parallel} \neq \chi_{\perp}$.

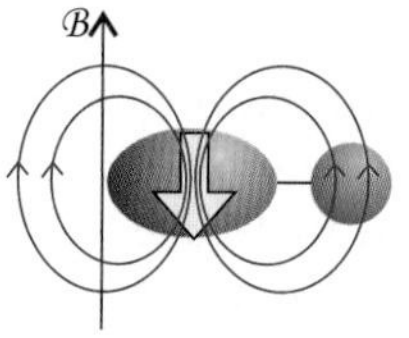

2

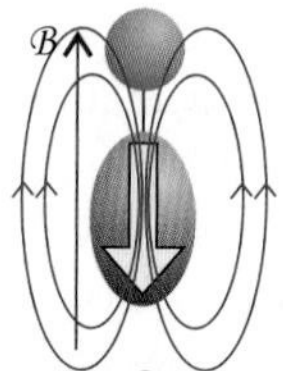

3

Self-test 12.1 For a tumbling H—X molecule, show that when $\theta = 90°$: (a) contributions from the $\chi_{\perp}$ component lead to shielding of H, or σ(neighbour) > 0, and (b) contributions from the $\chi_{\parallel}$ component lead to deshielding of H, or σ(neighbour) < 0. Comparison between the $\theta = 0$ and $\theta = 90°$ cases shows that the patterns of shielding and deshielding by neighbouring groups depend not only on differences between $\chi_{\parallel}$ and $\chi_{\perp}$, but also the angle θ.

[Draw diagrams similar to 2 and 3 where the $\chi_{\perp}$ component is parallel to the H—X axis and then analyse the problem as above.]

To a good approximation, the shielding constant σ(neighbour) depends on the distance r, the difference $\chi_{\parallel} - \chi_{\perp}$, as

$$\sigma(\text{neighbour}) \propto (\chi_{\parallel} - \chi_{\perp})\left(\frac{1 - 3\cos^2\theta}{r^3}\right) \qquad (12.23)$$

where $\chi_{\parallel}$ and $\chi_{\perp}$ are both negative for a diamagnetic group X. Equation 12.23 shows that the neighbouring group contribution may be positive or negative according to the relative magnitudes of the two magnetic susceptibilities and the relative orientation of the nucleus with respect to X. The latter effect is easy to anticipate: if $54.7° < \theta < 125.3°$, then $1 - 3\cos^2\theta$ is positive, but it is negative otherwise (Fig. 12.8).

A special case of a neighbouring group effect is found in aromatic compounds. The strong anisotropy of the magnetic susceptibility of the benzene ring is ascribed to the ability of the field to induce a **ring current**, a circulation of electrons around the ring, when it is applied perpendicular to the molecular plane. Protons in the plane are deshielded (Fig. 12.9), but any that happen to lie above or below the plane (as members of substituents of the ring) are shielded.

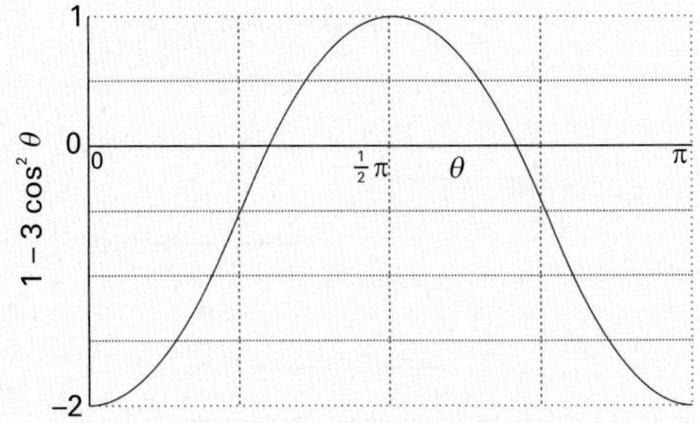

Fig. 12.8 The variation of the function $1 - 3\cos^2\theta$ with the angle θ.

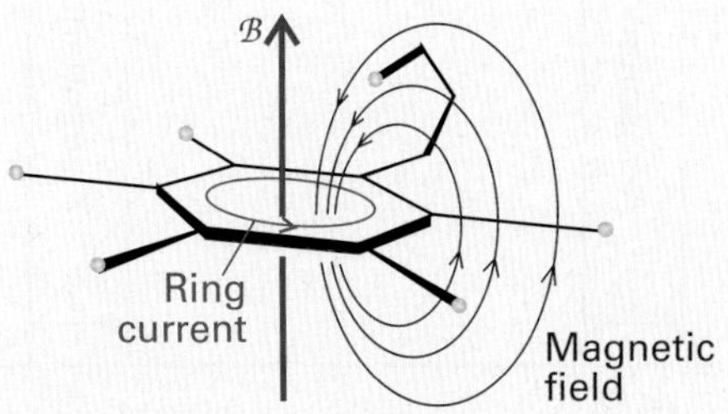

Fig. 12.9 The shielding and deshielding effects of the ring current induced in the benzene ring by the applied field. Protons attached to the ring are deshielded but a proton attached to a substituent that projects above the ring is shielded.

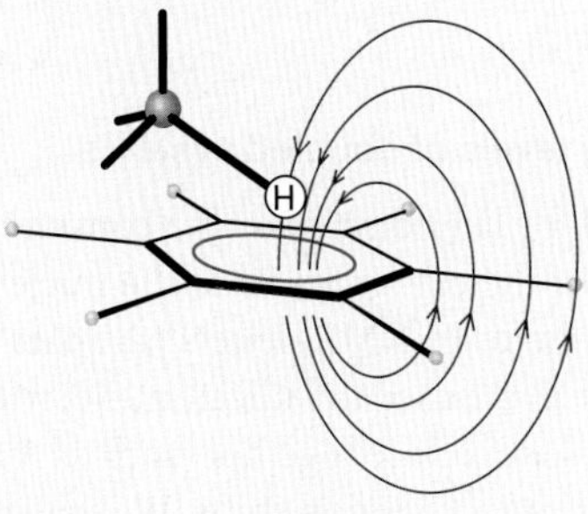

Fig. 12.10 An aromatic solvent (benzene here) can give rise to local currents that shield or deshield a proton in a solute molecule. In this relative orientation of the solvent and solute, the proton on the solute molecule is shielded.

(f) The solvent contribution

A solvent can influence the local magnetic field experienced by a nucleus in a variety of ways. Some of these effects arise from specific interactions between the solute and the solvent (such as hydrogen-bond formation and other forms of Lewis acid–base complex formation). The anisotropy of the magnetic susceptibility of the solvent molecules, especially if they are aromatic, can also be the source of a local magnetic field. Moreover, if there are steric interactions that result in a loose but specific interaction between a solute molecule and a solvent molecule, then protons in the solute molecule may experience shielding or deshielding effects according to their location relative to the solvent molecule (Fig. 12.10). We shall see that the NMR spectra of species that contain protons with widely different chemical shifts are easier to interpret than those in which the shifts are similar, so the appropriate choice of solvent may help to simplify the appearance and interpretation of a spectrum.

12.6 The fine structure

The splitting of resonances into individual lines in Fig. 12.6 is called the **fine structure** of the spectrum. It arises because each magnetic nucleus may contribute to the local field experienced by the other nuclei and so modify their resonance frequencies. The strength of the interaction is expressed in terms of the **scalar coupling constant**, J, and reported in hertz (Hz). The scalar coupling constant is so called because the energy of interaction it describes is proportional to the scalar product of the two interacting spins: $E \propto \boldsymbol{I}_1 \cdot \boldsymbol{I}_2$. The constant of proportionality in this expression is $hJ/\hbar^2$, because each angular momentum is proportional to $\hbar$.

Spin coupling constants are independent of the strength of the applied field because they do not depend on the latter for their ability to generate local fields. If the resonance line of a particular nucleus is split by a certain amount by a second nucleus, then the resonance line of the second nucleus is split by the first to the same extent.

(a) The energy levels of coupled systems

It will be useful for later discussions to consider an NMR spectrum in terms of the energy levels of the nuclei and the transitions between them. In NMR, letters far apart in the alphabet (typically A and X) are used to indicate nuclei with very different chemical shifts; letters close together (such as A and B) are used for nuclei with similar chemical shifts. We shall consider first an AX system, a molecule that contains two spin-$\frac{1}{2}$ nuclei A and X with very different chemical shifts in the sense that the difference in chemical shift corresponds to a frequency that is large compared to their spin–spin coupling. We consider AB systems in Section 12.6(f).

The energy level diagram for a single spin-$\frac{1}{2}$ nucleus and its single transition were shown in Fig. 12.3, and nothing more needs to be said. For a spin-$\frac{1}{2}$ AX system there are four spin states:

$$\alpha_A\alpha_X \qquad \alpha_A\beta_X \qquad \beta_A\alpha_X \qquad \beta_A\beta_X$$

The energy depends on the orientation of the spins in the external magnetic field, and if spin–spin coupling is neglected

$$\begin{aligned} E &= -\gamma\hbar(1-\sigma_A)\mathcal{B}m_A - \gamma\hbar(1-\sigma_X)\mathcal{B}m_X \\ &= -h\nu_A m_A - h\nu_X m_X \end{aligned} \tag{12.24}$$

where ν_A and ν_X are the Larmor frequencies of A and X and m_A and m_X are their quantum numbers. Recall that $m_s = +\frac{1}{2}$ and $-\frac{1}{2}$ for α and β spins, respectively. This expression gives the four lines on the left of Fig. 12.11. The spin–spin coupling depends on the relative orientation of the two nuclear spins, so it is proportional to the product $m_A m_X$. Therefore, the energy including spin–spin coupling is

$$E = -h\nu_A m_A - h\nu_X m_X + hJm_A m_X \tag{12.25}$$

If $J > 0$, a lower energy is obtained when $m_A m_X < 0$, which is the case if one spin is α and the other is β. A higher energy is obtained if both spins are α or both spins are β. The opposite is true if $J < 0$. The resulting energy level diagram (for $J > 0$) is shown on the right of Fig. 12.11. We see that the αα and ββ states are both raised by $\frac{1}{4}hJ$ and that the αβ and βα states are both lowered by $\frac{1}{4}hJ$.

When a transition of nucleus A occurs, nucleus X remains unchanged. Therefore, the A resonance is a transition for which $\Delta m_A = +1$ and $\Delta m_X = 0$. There are two such transitions, one in which $\beta_A \leftarrow \alpha_A$ occurs when the X nucleus is α_X, and the other in which $\beta_A \leftarrow \alpha_A$ occurs when the X nucleus is β_X. They are shown in Fig. 12.11 and in a slightly different form in Fig. 12.12. The energies of the transitions are

$$\Delta E = h\nu_A \pm \tfrac{1}{2}hJ \tag{12.26a}$$

Therefore, the A resonance consists of a doublet of separation J centred on the chemical shift of A (Fig. 12.13). Similar remarks

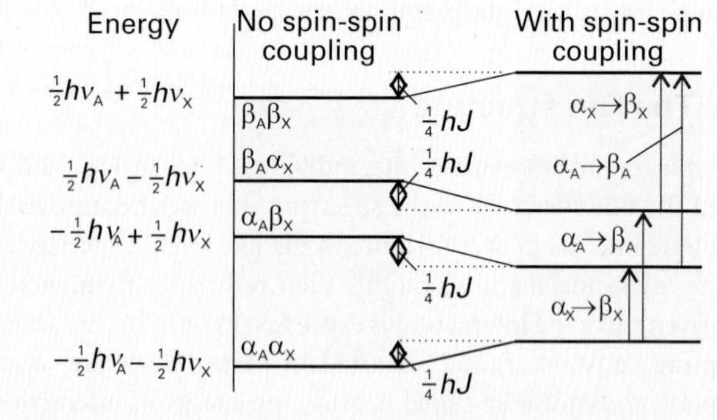

Fig. 12.11 The energy levels of an AX system. The four levels on the left are those of the two spins in the absence of spin–spin coupling. The four levels on the right show how a positive spin–spin coupling constant affects the energies. The transitions shown are for β ← α of A or X, the other nucleus (X or A, respectively) remaining unchanged. We have exaggerated the effect for clarity in practice, the splitting caused by spin–spin coupling is much smaller than that caused by the applied field.

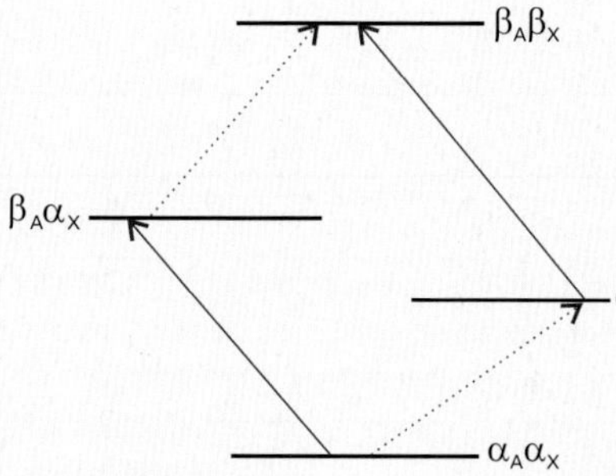

Fig. 12.12 An alternative depiction of the energy levels and transitions shown in Fig. 12.11. Once again, we have exaggerated the effect of spin–spin coupling.

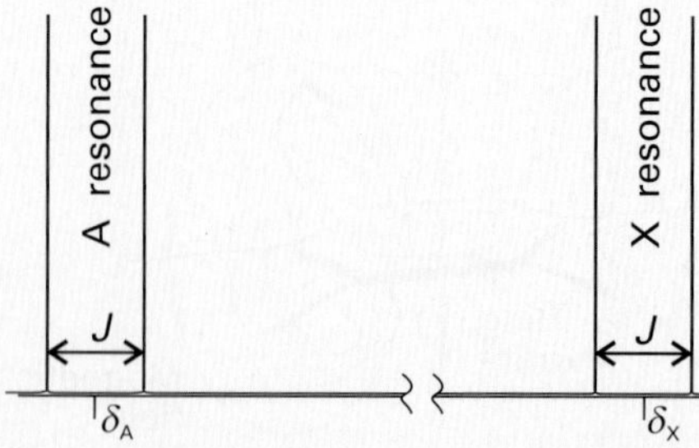

Fig. 12.13 The effect of spin–spin coupling on an AX spectrum. Each resonance is split into two lines separated by J. The pairs of resonances are centred on the chemical shifts of the protons in the absence of spin–spin coupling.

apply to the X resonance, which consists of two transitions according to whether the A nucleus is α or β (as shown in Fig. 12.12). The transition energies are

$$\Delta E = h\nu_X \pm \tfrac{1}{2}hJ \qquad (12.26b)$$

It follows that the X resonance also consists of two lines of separation J, but they are centred on the chemical shift of X (as shown in Fig. 12.13).

(b) Patterns of coupling

We have seen that, in an AX system, spin–spin coupling will result in four lines in the NMR spectrum. Instead of a single line from A, we get a doublet of lines separated by J and centred on the chemical shift characteristic of A. The same splitting occurs in the X resonance: instead of a single line, the resonance is a doublet with splitting J (the same value as for the splitting of A) centred on the chemical shift characteristic of X. These features are summarized in Fig. 12.13.

A subtle point is that the X resonance in an AX_n species (such as an AX_2 or AX_3 species) is also a doublet with splitting J. As we shall explain below, *a group of equivalent nuclei resonates like a single nucleus*. The only difference for the X resonance of an AX_n species is that the intensity is n times as great as that of an AX species (Fig. 12.14). The A resonance in an AX_n species, though, is quite different from the A resonance in an AX species. For example, consider an AX_2 species with two equivalent X nuclei. The resonance of A is split into a doublet of separation J by one X, and each line of that doublet is split again by the same amount by the second X (Fig. 12.15). This splitting results in three lines in the intensity ratio 1:2:1 (because the central frequency can be obtained in two ways). The A resonance of an A_nX_2 species would also be a 1:2:1 triplet of splitting J, the only difference being that the intensity of the A resonance would be n times as great as that of AX_2.

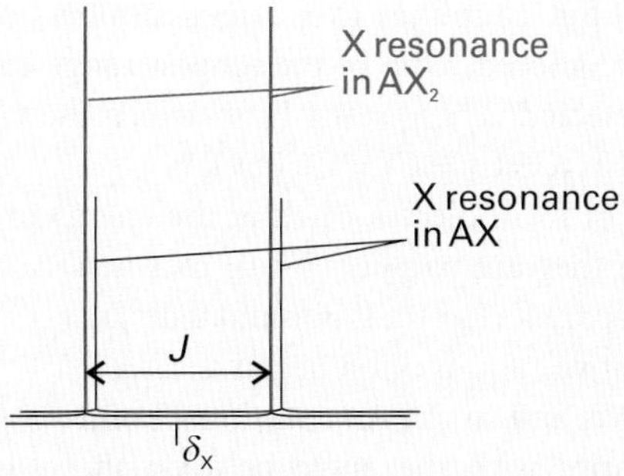

Fig. 12.14 The X resonance of an AX_2 species is also a doublet, because the two equivalent X nuclei behave like a single nucleus; however, the overall absorption is twice as intense as that of an AX species.

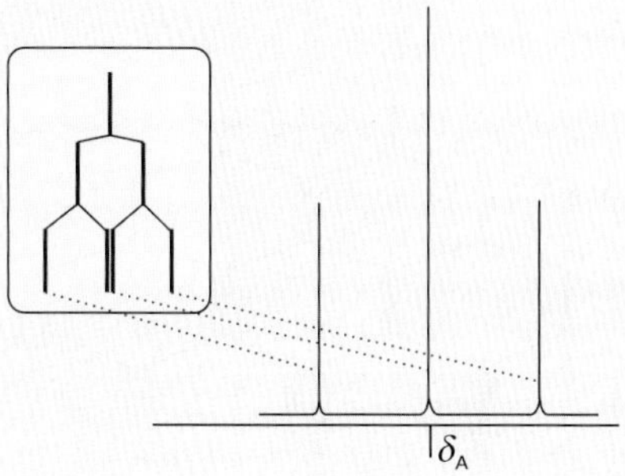

Fig. 12.15 The origin of the 1:2:1 triplet in the A resonance of an AX_2 species. The resonance of A is split into two by coupling with one X nucleus (as shown in the inset), and then each of those two lines is split into two by coupling to the second X nucleus. Because each X nucleus causes the same splitting, the two central transitions are coincident and give rise to an absorption line of double the intensity of the outer lines.

Three equivalent X nuclei (an AX_3 species) split the resonance of A into four lines of intensity ratio 1:3:3:1 and separation J (Fig. 12.16). The X resonance, though, is still a doublet of separation J. In general, n equivalent spin-$\tfrac{1}{2}$ nuclei split the resonance of a nearby spin or group of equivalent spins into $n + 1$ lines with an intensity distribution given by 'Pascal's triangle' in which each entry is the sum of the two entries immediately above (4). The easiest way of constructing the pattern

1
1 1
1 2 1
1 3 3 1
1 4 6 4 1
1 5 10 10 5 1

4

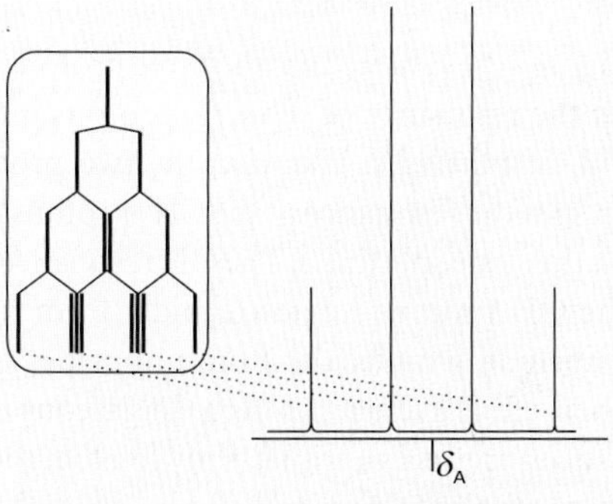

Fig. 12.16 The origin of the 1:3:3:1 quartet in the A resonance of an AX_3 species. The third X nucleus splits each of the lines shown in Fig. 12.15 for an AX_2 species into a doublet, and the intensity distribution reflects the number of transitions that have the same energy.

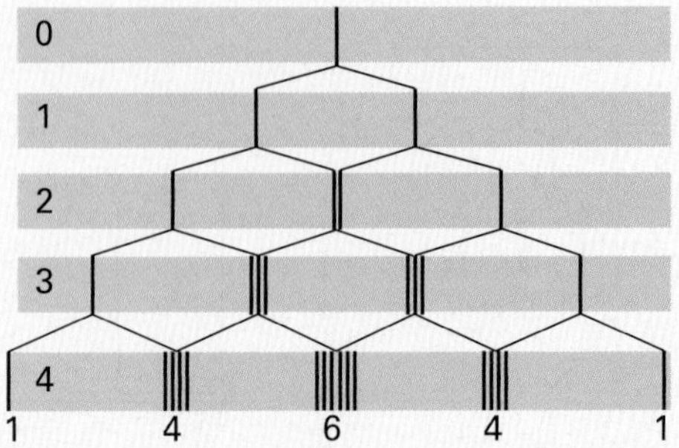

Fig. 12.17 The intensity distribution of the A resonance of an AX_n resonance can be constructed by considering the splitting caused by 1,2, . . . n protons, as in Figs. 12.15 and 12.16. The resulting intensity distribution has a binomial distribution and is given by the integers in the corresponding row of Pascal's triangle. Note that, although the lines have been drawn side-by-side for clarity, the members of each group are coincident. Four protons, in AX_4, split the A resonance into a 1:4:6:4:1 quintet.

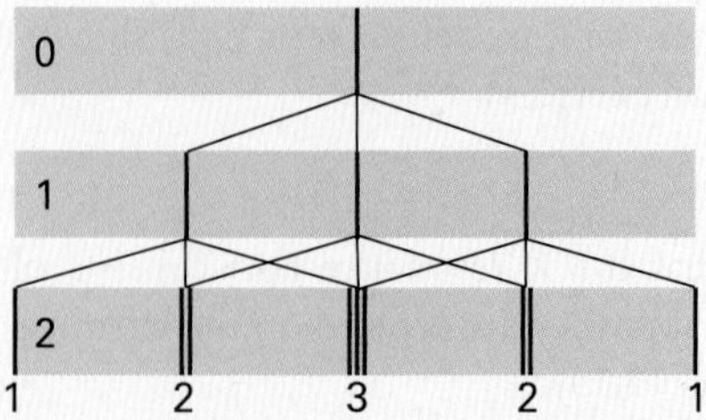

Fig. 12.18 The intensity distribution arising from spin–spin interaction with nuclei with $I = 1$ can be constructed similarly, but each successive nucleus splits the lines into three equal intensity components. Two equivalent spin-1 nuclei give rise to a 1:2:3:2:1 quintet.

of fine structure is to draw a diagram in which successive rows show the splitting of a subsequent proton. The procedure is illustrated in Fig. 12.17 and was used in Figs. 12.15 and 12.16. It is easily extended to molecules containing nuclei with $I > \frac{1}{2}$ (Fig. 12.18).

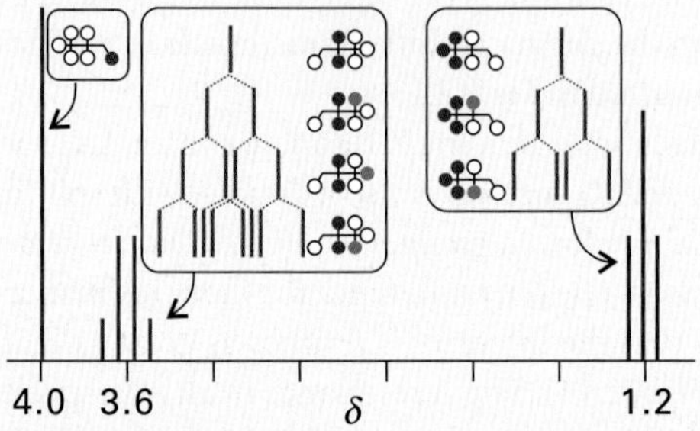

Fig. 12.19 A diagrammatic representation of the ^{1}H-NMR spectrum of ethanol.

Example 12.1 *Accounting for the fine structure in a spectrum*

Account for the fine structure in the NMR spectrum of the C—H protons of ethanol.

Method Consider how each group of equivalent protons (for instance, three methyl protons) splits the resonances of the other groups of protons. There is no splitting within groups of equivalent protons. Each splitting pattern can be decided by referring to Pascal's triangle.

Answer Figure 12.19 shows that the three protons of the CH_3 group split the resonance of the CH_2 protons into a 1:3:3:1 quartet with a splitting J. Likewise, the two protons of the CH_2 group split the resonance of the CH_3 protons into a 1:2:1 triplet with the same splitting J. The OH resonance is not split because the OH protons migrate rapidly from molecule to molecule (including molecules from impurities in the sample) and their effect averages to zero. In gaseous ethanol the OH resonance appears as a triplet, showing that the CH_2 protons interact with the OH proton.

Self-test 12.2 What fine structure can be expected for the protons in $^{14}NH_4^+$? The spin quantum number of nitrogen-14 is 1. [1:1:1 triplet from N]

(c) The magnitudes of coupling constants

The scalar coupling constant of two nuclei separated by N bonds is denoted ^NJ, with subscripts for the types of nuclei involved. Thus, $^1J_{CH}$ is the coupling constant for a proton joined directly to a ^{13}C atom, and $^2J_{CH}$ is the coupling constant when the same two nuclei are separated by two bonds (as in ^{13}C—C—H). A typical value of $^1J_{CH}$ is in the range 120 to 250 Hz; $^2J_{CH}$ is between −10 and +20 Hz. Both ^3J and ^4J can give detectable effects in a spectrum, but couplings over larger numbers of bonds can generally be ignored. One of the longest range couplings that has been detected is $^9J_{HH} = 0.4$ Hz between the CH_3 and CH_2 protons in $CH_3C{\equiv}CC{\equiv}CC{\equiv}CCH_2OH$.

The sign of J_{XY} indicates whether the energy of two spins is lower when they are parallel ($J < 0$) or when they are antiparallel ($J > 0$). It is found that $^1J_{CH}$ is often positive, $^2J_{HH}$ is often negative, $^3J_{HH}$ is often positive, and so on. An additional point is that J varies with the angle between the bonds (Fig. 12.20). Thus, a $^3J_{HH}$ coupling constant is often found to depend on the dihedral angle ϕ (**5**) according to the **Karplus equation**:

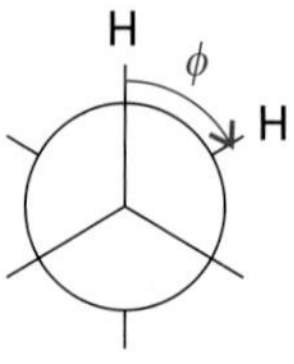

5

$$J = A + B\cos\phi + C\cos 2\phi \qquad (12.27)$$

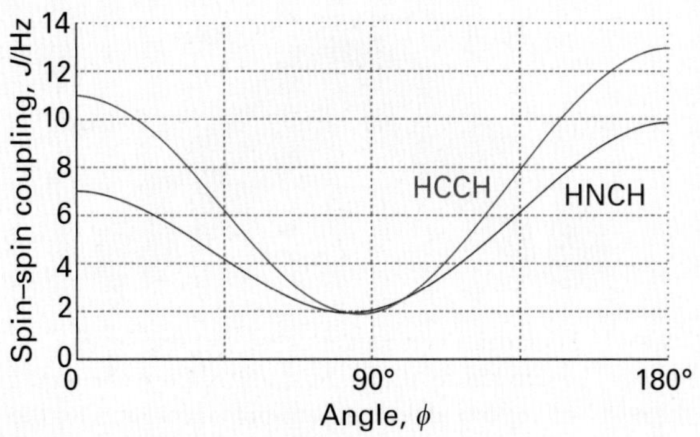

Fig. 12.20 The variation of the spin–spin coupling constant with angle predicted by the Karplus equation for an HCCH group and an HNCH group.

interActivity Draw a family of curves showing the variation of $^3J_{HH}$ with ϕ for which $A = +7.0$ Hz, $B = -1.0$ Hz, and C varies slightly from a typical value of +5.0 Hz. What is the effect of changing the value of the parameter C on the shape of the curve? In a similar fashion, explore the effect of the values of A and B on the shape of the curve.

with A, B, and C empirical constants with values close to +7 Hz, −1 Hz, and +5 Hz, respectively, for an HCCH fragment. It follows that the measurement of $^3J_{HH}$ in a series of related compounds can be used to determine their conformations. The coupling constant $^1J_{CH}$ also depends on the hybridization of the C atom, as the following values indicate:

	sp	sp^2	sp^3
$^1J_{CH}$/Hz:	250	160	125

(d) The origin of spin–spin coupling

Spin–spin coupling is a very subtle phenomenon, and it is better to treat J as an empirical parameter than to use calculated values. However, we can get some insight into its origins, if not its precise magnitude—or always reliably its sign—by considering the magnetic interactions within molecules.

A nucleus with spin projection m_I gives rise to a magnetic field with z-component $\mathcal{B}_{nuc}$ at a distance R, with, to a good approximation,

$$\mathcal{B}_{nuc} = -\frac{\gamma\hbar\mu_0}{4\pi R^3}(1-3\cos^2\theta)m_I \tag{12.28}$$

The angle θ is defined in (6). The magnitude of this field is about 0.1 mT when $R = 0.3$ nm, corresponding to a splitting of resonance signal of about 10^4 Hz, and is of the order of magnitude of the splitting observed in solid samples.

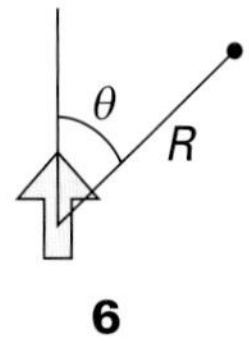

6

In a liquid, the angle θ sweeps over all values as the molecule tumbles, and $1 - 3\cos^2\theta$ averages to zero. Hence the direct dipolar interaction between spins cannot account for the fine structure of the spectra of rapidly tumbling molecules. The direct interaction does make an important contribution to the spectra of solid samples and is a very useful indirect source of structure information through its involvement in spin relaxation (Section 12.9).

A brief comment The average (or mean value) of a function $f(x)$ over the range $x = a$ to $x = b$ is $\int_a^b f(x)\mathrm{d}x/(b-a)$. The volume element in polar coordinates is proportional to $\sin\theta\,\mathrm{d}\theta$, and θ ranges from 0 to π. Therefore the average value of $(1 - 3\cos^2\theta)$ is $\int_0^\pi (1-3\cos^2\theta)\sin\theta\,\mathrm{d}\theta/\pi = 0$.

Spin–spin coupling in molecules in solution can be explained in terms of the **polarization mechanism**, in which the interaction is transmitted through the bonds. The simplest case to consider is that of $^1J_{XY}$ where X and Y are spin-$\frac{1}{2}$ nuclei joined by an electron-pair bond (Fig. 12.21). The coupling mechanism depends on the fact that in some atoms it is favourable for the nucleus and a nearby electron spin to be parallel (both α or both β), but in others it is favourable for them to be antiparallel (one α and the other β). The electron–nucleus coupling is magnetic in origin, and may be either a dipolar interaction between the magnetic moments of the electron and nuclear spins or a **Fermi contact interaction**. A pictorial description of the Fermi contact interaction is as follows. First, we regard the magnetic moment of the nucleus as arising from the circulation of a current in a tiny loop with a radius similar to that of the nucleus (Fig. 12.22). Far from the nucleus the field generated by this loop is indistinguishable from the field generated by a point magnetic dipole. Close to the loop, however, the field differs from that of a point dipole. The magnetic interaction between this non-dipolar field and the electron's magnetic moment is the contact interaction. The lines of force depicted in Fig. 12.22 correspond to those for a proton with α spin. The lower energy state of an electron spin in such a field is the β state. In conclusion, the contact interaction

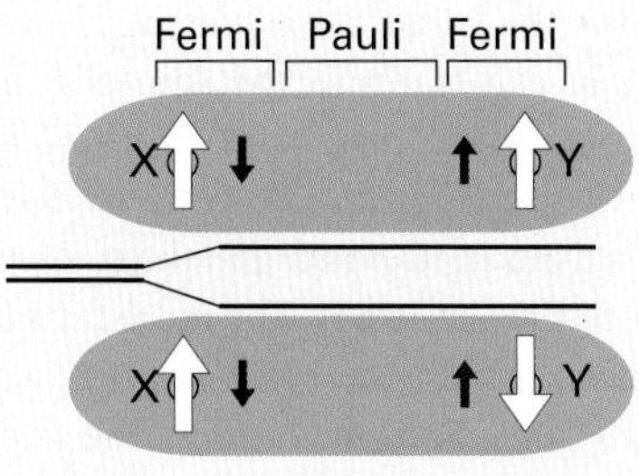

Fig. 12.21 The polarization mechanism for spin–spin coupling ($^1J_{HH}$). The two arrangements have slightly different energies. In this case, J is positive, corresponding to a lower energy when the nuclear spins are antiparallel.

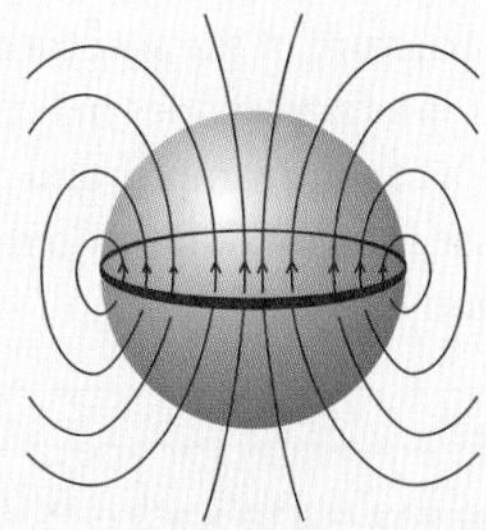

Fig. 12.22 The origin of the Fermi contact interaction. From far away, the magnetic field pattern arising from a ring of current (representing the rotating charge of the nucleus, the green sphere) is that of a point dipole. However, if an electron can sample the field close to the region indicated by the sphere, the field distribution differs significantly from that of a point dipole. For example, if the electron can penetrate the sphere, then the spherical average of the field it experiences is not zero.

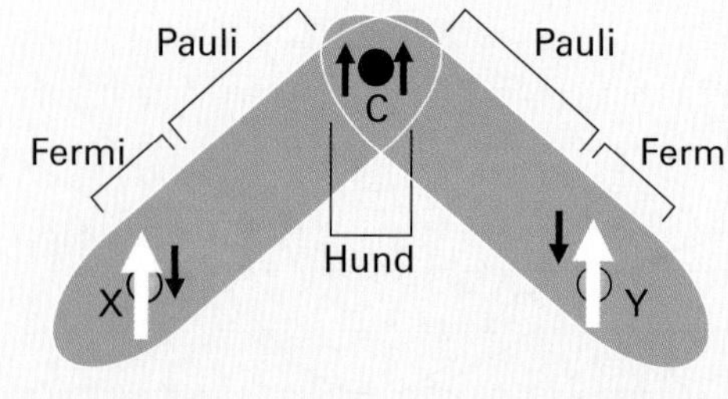

Fig. 12.23 The polarization mechanism for $^2J_{HH}$ spin–spin coupling. The spin information is transmitted from one bond to the next by a version of the mechanism that accounts for the lower energy of electrons with parallel spins in different atomic orbitals (Hund's rule of maximum multiplicity). In this case, $J < 0$, corresponding to a lower energy when the nuclear spins are parallel.

depends on the very close approach of an electron to the nucleus and hence can occur only if the electron occupies an s orbital (which is the reason why $^1J_{CH}$ depends on the hybridization ratio). We shall suppose that it is energetically favourable for an electron spin and a nuclear spin to be antiparallel (as is the case for a proton and an electron in a hydrogen atom).

If the X nucleus is α, a β electron of the bonding pair will tend to be found nearby (because that is energetically favourable for it). The second electron in the bond, which must have α spin if the other is β, will be found mainly at the far end of the bond (because electrons tend to stay apart to reduce their mutual repulsion). Because it is energetically favourable for the spin of Y to be antiparallel to an electron spin, a Y nucleus with β spin has a lower energy, and hence a lower Larmor frequency, than a Y nucleus with α spin. The opposite is true when X is β, for now the α spin of Y has the lower energy. In other words, the antiparallel arrangement of nuclear spins lies lower in energy than the parallel arrangement as a result of their magnetic coupling with the bond electrons. That is, $^1J_{CH}$ is positive.

To account for the value of $^2J_{XY}$, as in H—C—H, we need a mechanism that can transmit the spin alignments through the central C atom (which may be ^{12}C, with no nuclear spin of its own). In this case (Fig. 12.23), an X nucleus with α spin polarizes the electrons in its bond, and the α electron is likely to be found closer to the C nucleus. The more favourable arrangement of two electrons on the same atom is with their spins parallel (Hund's rule, Section 4.4), so the more favourable arrangement is for the α electron of the neighbouring bond to be close to the C nucleus. Consequently, the β electron of that bond is more likely to be found close to the Y nucleus, and therefore that nucleus will have a lower energy if it is α. Hence, according to this mechanism, the lower Larmor frequency of Y will be obtained if its spin is parallel to that of X. That is, $^2J_{HH}$ is negative.

The coupling of nuclear spin to electron spin by the Fermi contact interaction is most important for proton spins, but it is not necessarily the most important mechanism for other nuclei. These nuclei may also interact by a dipolar mechanism with the electron magnetic moments and with their orbital motion, and there is no simple way of specifying whether J will be positive or negative.

(e) Equivalent nuclei

A group of nuclei are **chemically equivalent** if they are related by a symmetry operation of the molecule and have the same chemical shifts. Chemically equivalent nuclei are nuclei that would be regarded as 'equivalent' according to ordinary chemical criteria. Nuclei are **magnetically equivalent** if, as well as being chemically equivalent, they also have identical spin–spin interactions with any other magnetic nuclei in the molecule.

The difference between chemical and magnetic equivalence is illustrated by CH_2F_2 and $H_2C{=}CF_2$. In each of these molecules the protons are chemically equivalent: they are related by symmetry and undergo the same chemical reactions. However, although the protons in CH_2F_2 are magnetically equivalent, those in $CH_2{=}CF_2$ are not. One proton in the latter has a *cis* spin-coupling interaction with a given F nucleus, whereas the other proton has a *trans* interaction with that F nucleus. In constrast, in CH_2F_2 both protons are connected to a given F nucleus by identical bonds, so there is no distinction between them. Strictly speaking, the CH_3 protons in ethanol (and other compounds) are magnetically inequivalent on account of their different interactions with the CH_2 protons in the next group. However, they are in practice made magnetically equivalent by the rapid rotation of the CH_3 group, which averages out any differences. Magnetically inequivalent species can give very complicated spectra (for

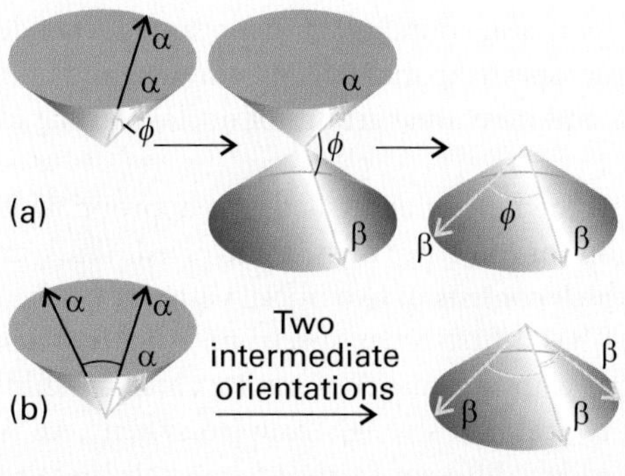

Fig. 12.24 (a) A group of two equivalent nuclei realigns as a group, without change of angle between the spins, when a resonant absorption occurs. Hence it behaves like a single nucleus and the spin–spin coupling between the individual spins of the group is undetectable. (b) Three equivalent nuclei also realign as a group without change of their relative orientations.

instance, the proton and ^{19}F spectra of $H_2C{=}CF_2$ each consist of 12 lines), and we shall not consider them further.

An important feature of chemically equivalent magnetic nuclei is that, although they do couple together, the coupling has no effect on the appearance of the spectrum. The reason for the invisibility of the coupling is set out in the following *Justification*, but qualitatively it is that all allowed nuclear spin transitions are *collective* reorientations of groups of equivalent nuclear spins that do not change the relative orientations of the spins within the group (Fig. 12.24). Then, because the relative orientations of nuclear spins are not changed in any transition, the magnitude of the coupling between them is undetectable. Hence, an isolated CH_3 group gives a single, unsplit line because all the allowed transitions of the group of three protons occur without change of their relative orientations.

Justification 12.2 *The energy levels of an A_2 system*

Consider an A_2 system of two chemically equivalent spin-$\frac{1}{2}$ nuclei. First, consider the energy levels in the absence of spin–spin coupling. There are four spin states, which (just as for two electrons) can be classified according to their total spin I (the analogue of S for two electrons) and their total projection M_I on the z-axis. The states are analogous to those we developed for two electrons in singlet and triplet states:

Spins parallel, $I=1$:	$M_I=+1$	$\alpha\alpha$
	$M_I=0$	$(1/2^{1/2})\{\alpha\beta+\beta\alpha\}$
	$M_I=-1$	$\beta\beta$
Spins paired, $I=0$:	$M_I=0$	$(1/2^{1/2})\{\alpha\beta-\beta\alpha\}$

The effect of a magnetic field on these four states is shown in Fig. 12.25: the energies of the two states with $M_I=0$ are unchanged by the field because they are composed of equal proportions of α and β spins.

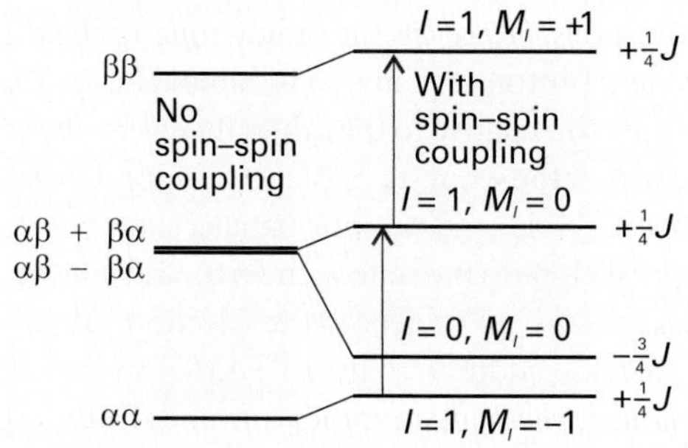

Fig. 12.25 The energy levels of an A_2 system in the absence of spin–spin coupling are shown on the left. When spin–spin coupling is taken into account, the energy levels on the right are obtained. Note that the three states with total nuclear spin $I=1$ correspond to parallel spins and give rise to the same increase in energy (J is positive); the one state with $I=0$ (antiparallel nuclear spins) has a lower energy in the presence of spin–spin coupling. The only allowed transitions are those that preserve the angle between the spins, and so take place between the three states with $I=1$. They occur at the same resonance frequency as they would have in the absence of spin–spin coupling.

A brief comment As in Section 4.4, the states specified here have a definite resultant, and hence a well defined value of I. The + sign in $\alpha\beta+\beta\alpha$ signifies an in-phase alignment of spins and $I=1$; the − sign in $\alpha\beta-\beta\alpha$ signifies an alignment out of phase by π, and hence $I=0$.

As remarked in Section 12.6a, the spin–spin coupling energy is proportional to the scalar product of the vectors representing the spins, $E=(hJ/\hbar^2)\boldsymbol{I}_1\cdot\boldsymbol{I}_2$. The scalar product can be expressed in terms of the total nuclear spin by noting that

$$I^2=(\boldsymbol{I}_1+\boldsymbol{I}_2)\cdot(\boldsymbol{I}_1+\boldsymbol{I}_2)=I_1^2+I_2^2+2\boldsymbol{I}_1\cdot\boldsymbol{I}_2$$

rearranging this expression to

$$\boldsymbol{I}_1\cdot\boldsymbol{I}_2=\tfrac{1}{2}\{I^2-I_1^2-I_2^2\}$$

and replacing the magnitudes by their quantum mechanical values:

$$\boldsymbol{I}_1\cdot\boldsymbol{I}_2=\tfrac{1}{2}\{I(I+1)-I_1(I_1+1)-I_2(I_2+1)\}\hbar^2$$

Then, because $I_1=I_2=\frac{1}{2}$, it follows that

$$E=\tfrac{1}{2}hJ\{I(I+1)-\tfrac{3}{2}\}$$

For parallel spins, $I=1$ and $E=+\frac{1}{4}hJ$; for antiparallel spins $I=0$ and $E=-\frac{3}{4}hJ$, as in Fig. 12.25. We see that three of the states move in energy in one direction and the fourth (the one with antiparallel spins) moves three times as much in the opposite direction. The resulting energy levels are shown on the right in Fig. 12.25.

The NMR spectrum of the A_2 species arises from transitions between the levels. However, the radiofrequency field affects the two equivalent protons equally, so it cannot change the orientation of one proton relative to the other; therefore, the transitions take place within the set of states that correspond to parallel spin (those labelled $I = 1$), and no spin-parallel state can change to a spin-antiparallel state (the state with $I = 0$). Put another way, the allowed transitions are subject to the selection rule $\Delta I = 0$. This selection rule is in addition to the rule $\Delta M_I = \pm 1$ that arises from the conservation of angular momentum and the unit spin of the photon. The allowed transitions are shown in Fig. 12.25: we see that there are only two transitions, and that they occur at the same resonance frequency that the nuclei would have in the absence of spin–spin coupling. Hence, the spin–spin coupling interaction does not affect the appearance of the spectrum.

(f) Strongly coupled nuclei

NMR spectra are usually much more complex than the foregoing simple analysis suggests. We have described the extreme case in which the differences in chemical shifts are much greater than the spin–spin coupling constants. In such cases it is simple to identify groups of magnetically equivalent nuclei and to think of the groups of nuclear spins as reorientating relative to each other. The spectra that result are called **first-order spectra**.

Transitions cannot be allocated to definite groups when the differences in their chemical shifts are comparable to their spin–spin coupling interactions. The complicated spectra that are then obtained are called **strongly coupled spectra** (or 'second-order spectra') and are much more difficult to analyse (Fig. 12.26).

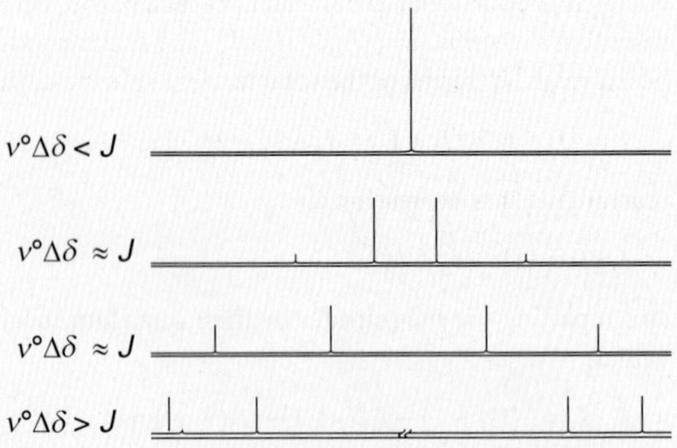

Fig. 12.26 The NMR spectra of an A_2 system (top) and an AX system (bottom) are simple 'first-order' spectra. At intermediate relative values of the chemical shift difference and the spin–spin coupling, complex 'strongly coupled' spectra are obtained. Note how the inner two lines of the bottom spectrum move together, grow in intensity, and form the single central line of the top spectrum. The two outer lines diminish in intensity and are absent in the top spectrum.

Because the difference in resonance frequencies increases with field, but spin–spin coupling constants are independent of it, a second-order spectrum may become simpler (and first-order) at high fields and individual groups of nuclei become identifiable again.

A clue to the type of analysis that is appropriate is given by the notation for the types of spins involved. Thus, an AX spin system (which consists of two nuclei with a large chemical shift difference) has a first-order spectrum. An AB system, on the other hand (with two nuclei of similar chemical shifts), gives a spectrum typical of a strongly coupled system. An AX system may have widely different Larmor frequencies because A and X are nuclei of different elements (such as ^{13}C and ^{1}H), in which case they form a **heteronuclear spin system**. AX may also denote a **homonuclear spin system** in which the nuclei are of the same element but in markedly different environments.

12.7 Conformational conversion and exchange processes

The appearance of an NMR spectrum is changed if magnetic nuclei can jump rapidly between different environments. Consider a molecule, such as *N,N*-dimethylformamide, that can jump between conformations; in its case, the methyl shifts depend on whether they are *cis* or *trans* to the carbonyl group (Fig. 12.27). When the jumping rate is low, the spectrum shows two sets of lines, one each from molecules in each conformation. When the interconversion is fast, the spectrum shows a single line at the mean of the two chemical shifts. At intermediate inversion rates, the line is very broad. This maximum broadening occurs when the lifetime, τ, of a conformation gives rise to a linewidth that is comparable to the difference of resonance frequencies, $\delta\nu$, and both broadened lines blend together into a very broad line. Coalescence of the two lines occurs when

$$\tau = \frac{\sqrt{2}}{\pi\delta\nu} \qquad (12.29)$$

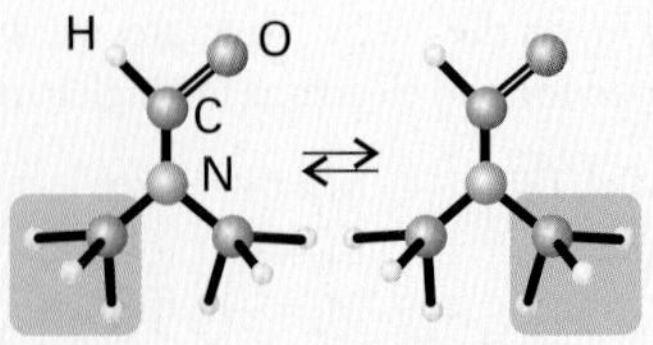

Fig. 12.27 When a molecule changes from one conformation to another, the positions of its protons are interchanged and jump between magnetically distinct environments.

● A BRIEF ILLUSTRATION

The NO group in *N,N*-dimethylnitrosamine, $(CH_3)_2N{-}NO$, rotates about the N—N bond and, as a result, the magnetic environments of the two CH_3 groups are interchanged. The two CH_3 resonances are separated by 390 Hz in a 600 MHz spectrometer. With $\delta\nu = 390$ Hz,

$$\tau = \frac{\sqrt{2}}{\pi \times (390\ \mathrm{s}^{-1})} = 1.2\ \mathrm{ms}$$

It follows that the signal will collapse to a single line when the interconversion rate exceeds about 830 s^{-1}. You should recall from introductory chemistry that the rate constant k_r of a chemical process depends on temperature according to the Arrhenius equation, $k_r = Ae^{-E_a/RT}$, with E_a as the activation energy, or energy barrier (see also Chapter 20). It follows that the dependence of the rate of exchange on the temperature can be used to determine the energy barrier to interconversion. ●

Self-test 12.3 What would you deduce from the observation of a single line from the same molecule in a 300 MHz spectrometer? [Conformation lifetime less than 2.3 ms]

A similar explanation accounts for the loss of fine structure in solvents able to exchange protons with the sample. For example, hydroxyl protons are able to exchange with water protons. When this **chemical exchange** occurs, a molecule ROH with an α-spin proton (we write this ROH_α) rapidly converts to ROH_β and then perhaps to ROH_α again because the protons provided by the solvent molecules in successive exchanges have random spin orientations. Therefore, instead of seeing a spectrum composed of contributions from both ROH_α and ROH_β molecules (that is, a spectrum showing a doublet structure due to the OH proton), we see a spectrum that shows no splitting caused by coupling of the OH proton (as in Fig. 12.6). The effect is observed when the lifetime of a molecule due to this chemical exchange is so short that the lifetime broadening is greater than the doublet splitting. Because this splitting is often very small (a few hertz), a proton must remain attached to the same molecule for longer than about 0.1 s for the splitting to be observable. In water, the exchange rate is much faster than that, so alcohols show no splitting from the OH protons. In dry dimethylsulfoxide (DMSO), the exchange rate may be slow enough for the splitting to be detected.

Pulse techniques in NMR

The common method of detecting the energy separation between nuclear spin states is more sophisticated than simply looking for the frequency at which resonance occurs. One of the best analogies that has been suggested to illustrate the preferred way of observing an NMR spectrum is that of detecting the spectrum of vibrations of a bell. We could stimulate the bell with a gentle vibration at a gradually increasing frequency, and note the frequencies at which it resonated with the stimulation. A lot of time would be spent getting zero response when the stimulating frequency was between the bell's vibrational modes. However, if we were simply to hit the bell with a hammer, we would immediately obtain a clang composed of all the frequencies that the bell can produce. The equivalent in NMR is to monitor the radiation nuclear spins emit as they return to equilibrium after the appropriate stimulation. The resulting **Fourier-transform NMR** gives greatly increased sensitivity, opening up much of the periodic table to the technique. Moreover, multiple-pulse FT-NMR gives chemists unparalleled control over the information content and display of spectra. We need to understand how the equivalent of the hammer blow is delivered and how the signal is monitored and interpreted. These features are generally expressed in terms of the vector model of angular momentum introduced in Section 3.4.

12.8 The magnetization vector

Consider a sample composed of many identical spin-$\frac{1}{2}$ nuclei. As we saw in Section 3.4, an angular momentum can be represented by a vector of length $\{I(I+1)\}^{1/2}$ units with a component of length m_I units along the *z*-axis. As the uncertainty principle does not allow us to specify the *x*- and *y*-components of the angular momentum, all we know is that the vector lies somewhere on a cone around the *z*-axis. For $I = \frac{1}{2}$, the length of the vector is $\frac{1}{2}\sqrt{3}$ and it makes an angle of 55° to the *z*-axis (Fig. 12.28).

In the absence of a magnetic field, the sample consists of equal numbers of α and β nuclear spins with their vectors lying at random angles on the cones. These angles are unpredictable, and at this stage we picture the spin vectors as stationary. The **magnetization**, *M*, of the sample, its net nuclear magnetic moment, is zero (Fig. 12.29a).

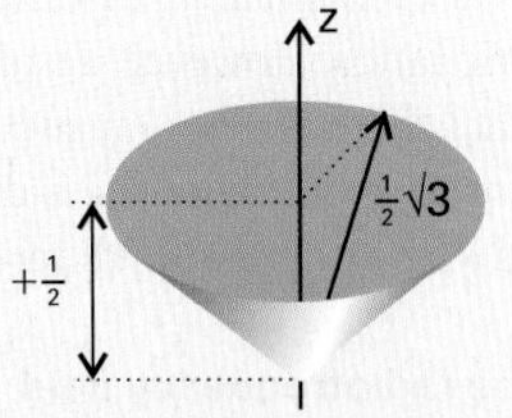

Fig. 12.28 The vector model of angular momentum for a single spin-$\frac{1}{2}$ nucleus. The angle around the *z*-axis is indeterminate.

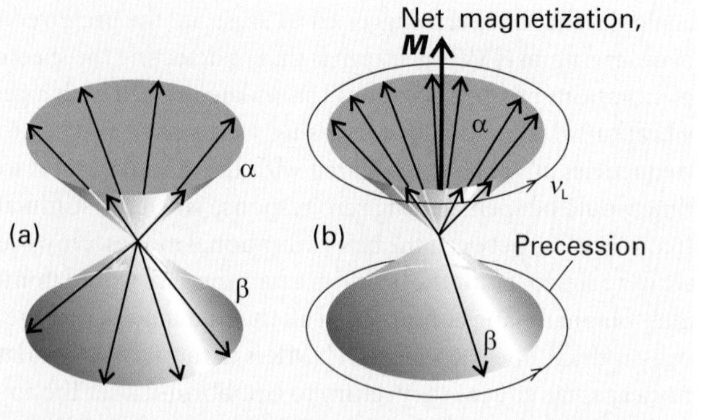

Fig. 12.29 The magnetization of a sample of spin-$\frac{1}{2}$ nuclei is the resultant of all their magnetic moments. (a) In the absence of an externally applied field, there are equal numbers of α and β spins at random angles around the *z*-axis (the field direction) and the magnetization is zero. (b) In the presence of a field, the spins precess around their cones (that is, there is an energy difference between the α and β states) and there are slightly more α spins than β spins. As a result, there is a net magnetization along the *z*-axis.

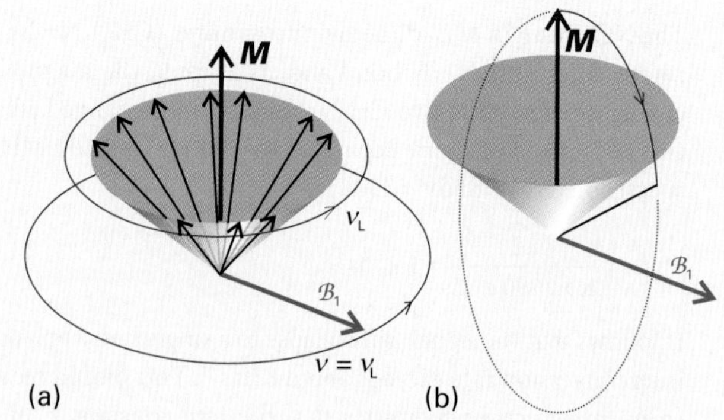

Fig. 12.30 (a) In a resonance experiment, a circularly polarized radiofrequency magnetic field $\mathcal{B}_1$ is applied in the *xy*-plane (the magnetization vector lies along the *z*-axis). (b) If we step into a frame rotating at the radiofrequency, $\mathcal{B}_1$ appears to be stationary, as does the magnetization *M* if the Larmor frequency is equal to the radiofrequency. When the two frequencies coincide, the magnetization vector of the sample rotates around the direction of the $\mathcal{B}_1$ field.

(a) The effect of the static field

Two changes occur in the magnetization when a magnetic field is present. First, the energies of the two orientations change, the α spins moving to low energy and the β spins to high energy (provided $\gamma > 0$). At 10 T, the Larmor frequency for protons is 427 MHz, and in the vector model the individual vectors are pictured as precessing at this rate. This motion is a pictorial representation of the difference in energy of the spin states, not an actual representation of reality. As the field is increased, the Larmor frequency increases and the precession becomes faster. Therefore, states of high energy are represented by precession at higher Larmor frequencies. Secondly, the populations of the two spin states (the numbers of α and β spins) at thermal equilibrium change, and there will be more α spins than β spins. Because $h\nu_L/kT \approx 7 \times 10^{-5}$ for protons at 300 K and 10 T, it follows from the Boltzmann distribution that $N_\beta/N_\alpha = e^{-h\nu_L/kT}$ is only slightly less than 1. That is, there is only a tiny imbalance of populations, and it is even smaller for other nuclei with their smaller magnetogyric ratios. However, despite its smallness, the imbalance means that there is a net magnetization that we can represent by a vector *M* pointing in the *z*-direction and with a length proportional to the population difference (Fig. 12.29b).

(b) The effect of the radiofrequency field

We now consider the effect of a radiofrequency field circularly polarized in the *xy*-plane, so that the magnetic component of the electromagnetic field (the only component we need to consider) is rotating around the *z*-direction, the direction of the applied field $\mathcal{B}_0$, in the same sense as the Larmor precession. The strength of the rotating magnetic field is $\mathcal{B}_1$. Suppose we choose the frequency of this field to be equal to the Larmor frequency of the spins, $\nu_L = (\gamma/2\pi)\mathcal{B}_0$; this choice is equivalent to selecting the resonance condition in the conventional experiment. The nuclei now experience a steady $\mathcal{B}_1$ field because the rotating magnetic field is in step with the precessing spins (Fig. 12.30a). Just as the spins precess about the strong static field $\mathcal{B}_0$ at a frequency $\gamma\mathcal{B}_0/2\pi$, so under the influence of the field $\mathcal{B}_1$ they precess about $\mathcal{B}_1$ at a frequency $\gamma\mathcal{B}_1/2\pi$.

To interpret the effects of radiofrequency pulses on the magnetization, it is often useful to look at the spin system from a different perspective. If we were to imagine stepping on to a platform, a so-called **rotating frame**, that rotates around the direction of the applied field at the radiofrequency, then the nuclear magnetization appears stationary if the radiofrequency is the same as the Larmor frequency (Fig. 12.30b). If the $\mathcal{B}_1$ field is applied in a pulse of duration $\pi/2\gamma\mathcal{B}_1$, the magnetization tips through 90° in the rotating frame and we say that we have applied a **90° pulse**, or a '$\pi/2$ pulse' (Fig. 12.31a). The duration of the pulse depends on the strength of the $\mathcal{B}_1$ field, but is typically of the order of microseconds.

Now imagine stepping out of the rotating frame. To an external observer (the role played by a radiofrequency coil) in this stationary frame, the magnetization vector is now rotating at the Larmor frequency in the *xy*-plane (Fig. 12.31b). The rotating magnetization induces in the coil a signal that oscillates at the Larmor frequency and that can be amplified and processed. In practice, the processing takes place after subtraction of a constant high frequency component (the radiofrequency

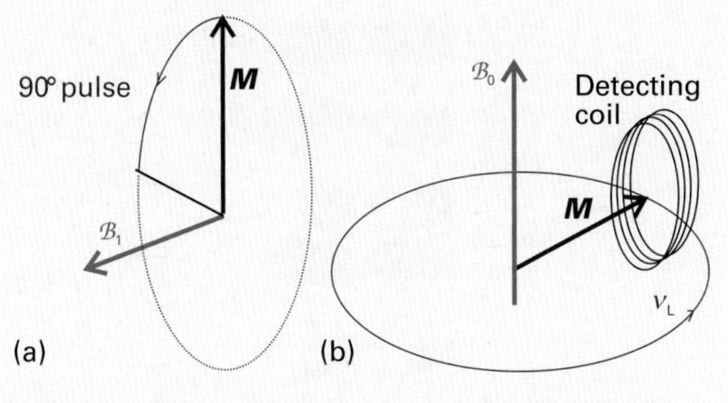

Fig. 12.31 (a) If the radiofrequency field is applied for a certain time, the magnetization vector is rotated into the xy-plane. (b) To an external stationary observer (the coil), the magnetization vector is rotating at the Larmor frequency, and can induce a signal in the coil.

used for $\mathcal{B}_1$), so that all the signal manipulation takes place at frequencies of a few kilohertz.

As time passes, the individual spins move out of step (partly because they are precessing at slightly different rates, as we shall explain later), so the magnetization vector shrinks exponentially with a time constant T_2 and induces an ever weaker signal in the detector coil. The form of the signal that we can expect is therefore the oscillating-decaying **free-induction decay** (FID) shown in Fig. 12.32. The y-component of the magnetization varies as

$$M_y(t) = M_0 \cos(2\pi\nu_L t)\, e^{-t/T_2} \qquad (12.30)$$

We have considered the effect of a pulse applied at exactly the Larmor frequency. However, virtually the same effect is obtained off resonance, provided that the pulse is short and built from a range of frequencies close to ν_L. A short pulse will certainly contain the resonance frequency of the nuclei in the sample and will effect the rotation of the magnetization into the xy-plane. Note that we do not need to know the Larmor frequency beforehand: the short pulse is the analogue of the hammer blow on the bell, exciting a range of frequencies. The detected signal shows that a particular resonance frequency is present.

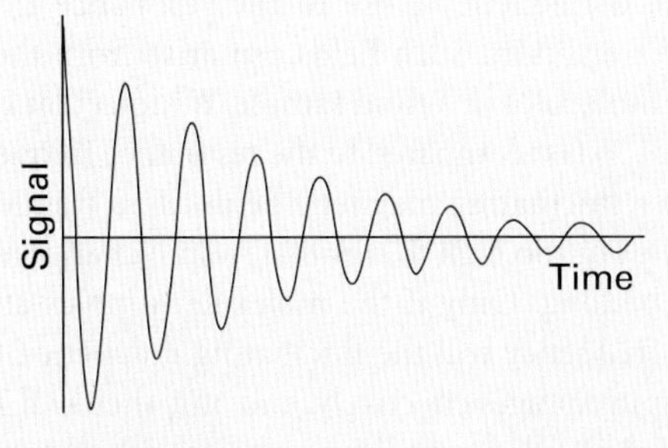

Fig. 12.32 A simple free-induction decay of a sample of spins with a single resonance frequency.

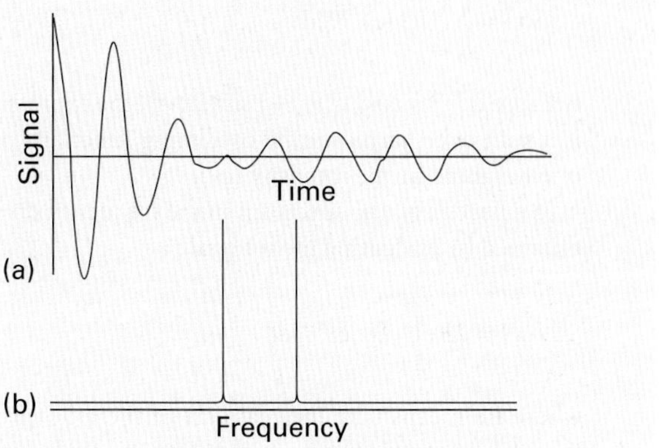

Fig. 12.33 (a) A free induction decay signal of a sample of AX species and (b) its analysis into its frequency components.

interActivity The *Living graphs* section of the text's web site has an applet that allows you to calculate and display the FID curve from an AX system. Explore the effect on the shape of the FID curve of changing the chemical shifts (and therefore the Larmor frequencies) of the A and X nuclei.

(c) Time- and frequency-domain signals

We can think of the magnetization vector of a homonuclear AX spin system with $J = 0$ as consisting of two parts, one formed by the A spins and the other by the X spins. When the 90° pulse is applied, both magnetization vectors are rotated into the xy-plane. However, because the A and X nuclei precess at different frequencies, they induce two signals in the detector coils, and the overall FID curve may resemble that in Fig. 12.33a. The composite FID curve is the analogue of the struck bell emitting a rich tone composed of all the frequencies at which it can vibrate.

The problem we must address is how to recover the resonance frequencies present in a free-induction decay. We know that the FID curve is a sum of oscillating functions, so the problem is to analyse it into its component frequencies by carrying out a Fourier transformation, as shown in *Justification 12.3*. When the signal in Fig. 12.33a is transformed in this way, we get the frequency-domain spectrum shown in Fig. 12.33b. One line represents the Larmor frequency of the A nuclei and the other that of the X nuclei.

Justification 12.3 *Fourier transformation of the FID curve*

The analysis of the FID curve is achieved by the standard mathematical technique of Fourier transformation, which we explored in *Mathematical background 6*. We start by noting that the signal $S(t)$ in the time domain, the total FID curve, is the sum (more precisely, the integral) over all the contributing frequencies

$$S(t) = \int_{-\infty}^{\infty} I(\nu)e^{-2\pi i\nu t}d\nu \qquad (12.31)$$

Because $e^{2\pi i\nu t} = \cos(2\pi\nu t) + i\sin(2\pi\nu t)$, the expression above is a sum over harmonically oscillating functions, with each one weighted by the intensity $I(\nu)$.

We need $I(\nu)$, the spectrum in the frequency domain; it is obtained by evaluating the integral

$$I(\nu) = 2\,\mathrm{Re}\int_{0}^{\infty} S(t)e^{2\pi i\nu t}dt \qquad (12.32)$$

where Re means take the real part of the following expression. This integral is very much like an overlap integral: it gives a nonzero value if $S(t)$ contains a component that matches the oscillating function $e^{2\pi i\nu t}$. The integration is carried out at a series of frequencies ν on a computer that is built into the spectrometer.

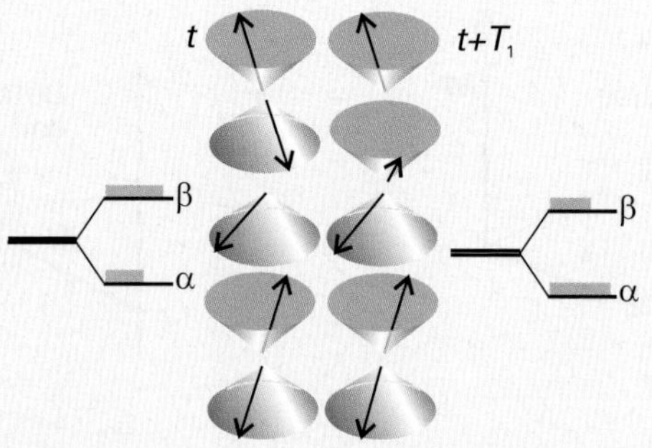

Fig. 12.35 In longitudinal relaxation the spins relax back towards their thermal equilibrium populations. On the left we see the precessional cones representing spin-$\frac{1}{2}$ angular momenta, and they do not have their thermal equilibrium populations (there are more β-spins than α-spins). On the right, which represents the sample a long time after a time T_1 has elapsed, the populations are those characteristic of a Boltzmann distribution. In actuality, T_1 is the time constant for relaxation to the arrangement on the right and $T_1 \ln 2$ is the half-life of the arrangement on the left.

The FID curve in Fig. 12.34 is obtained from a sample of ethanol. The frequency-domain spectrum obtained from it by Fourier transformation is the one that we have already discussed (Fig. 12.6). We can now see why the FID curve in Fig. 12.34 is so complex: it arises from the precession of a magnetization vector that is composed of eight components, each with a characteristic frequency.

12.9 Spin relaxation

There are two reasons why the component of the magnetization vector in the xy-plane shrinks. Both reflect the fact that the nuclear spins are not in thermal equilibrium with their surroundings (for then M lies parallel to z). The return to equilibrium is the process called **spin relaxation**.

(a) Longitudinal and transverse relaxation

At thermal equilibrium the spins have a Boltzmann distribution, with more α spins than β spins; however, a magnetization vector in the xy-plane immediately after a 90° pulse has equal numbers of α and β spins.

Now consider the effect of a **180° pulse**, which may be visualized in the rotating frame as a flip of the net magnetization vector from one direction along the z-axis to the opposite direction. That is, the 180° pulse leads to population inversion of the spin system, which now has more β spins than α spins. After the pulse, the populations revert to their thermal equilibrium values exponentially. As they do so, the z-component of magnetization reverts to its equilibrium value M_0 with a time constant called the **longitudinal relaxation time**, T_1 (Fig. 12.35):

$$M_z(t) - M_0 \propto e^{-t/T_1} \qquad (12.33)$$

Because this relaxation process involves giving up energy to the surroundings (the 'lattice') as β spins revert to α spins, the time constant T_1 is also called the **spin–lattice relaxation time**. Spin–lattice relaxation is caused by local magnetic fields that fluctuate at a frequency close to the resonance frequency of the α → β transition. Such fields can arise from the tumbling motion of molecules in a fluid sample. If molecular tumbling is too slow or too fast compared to the resonance frequency, it will give rise to a fluctuating magnetic field with a frequency that is either too low or too high to stimulate a spin change from β to α, so T_1 will be long. Only if the molecule tumbles at about the resonance frequency will the fluctuating magnetic field be able to induce spin changes effectively, and only then will T_1 be short. The rate of molecular tumbling increases with temperature and with reducing viscosity of the solvent, so we can expect a dependence like that shown in Fig. 12.36.

Fig. 12.34 A free-induction decay signal of a sample of ethanol. Its Fourier transform is the frequency-domain spectrum shown in Fig. 12.6. The total length of the image corresponds to about 1 s.

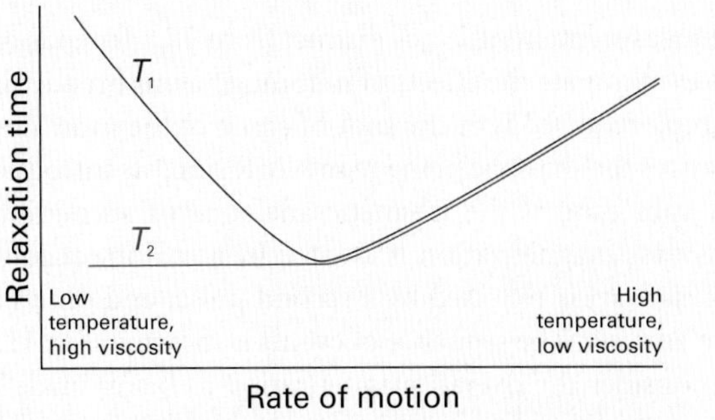

Fig. 12.36 The variation of the two relaxation times with the rate at which the molecules move (either by tumbling or migrating through the solution). The horizontal axis can be interpreted as representing temperature or viscosity. Note that at rapid rates of motion, the two relaxation times coincide.

A second aspect of spin relaxation is the fanning-out of the spins in the *xy*-plane if they precess at different rates (Fig. 12.37). The magnetization vector is large when all the spins are bunched together immediately after a 90° pulse. However, this orderly bunching of spins is not at equilibrium and, even if there were no spin–lattice relaxation, we would expect the individual spins to spread out until they were uniformly distributed with all possible angles around the *z*-axis. At that stage, the component of the magnetization vector in the plane would be zero. The randomization of the spin directions occurs exponentially with a time constant called the **transverse relaxation time,** T_2:

$$M_y(t) \propto e^{-t/T_2} \tag{12.34}$$

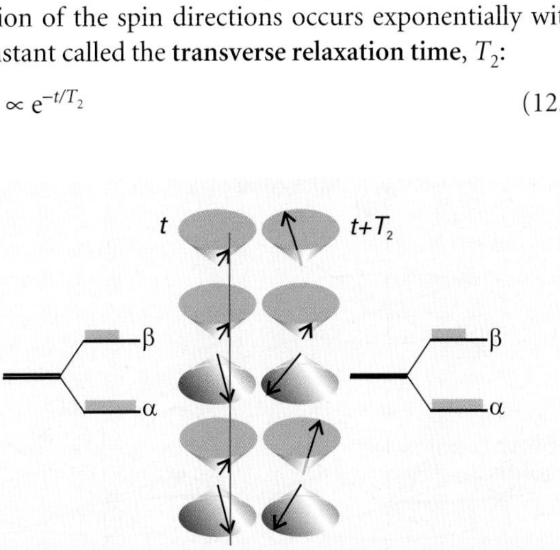

Fig. 12.37 The transverse relaxation time, T_2, is the time constant for the phases of the spins to become randomized (another condition for equilibrium) and to change from the orderly arrangement shown on the left to the disorderly arrangement on the right (long after a time T_2 has elapsed). Note that the populations of the states remain the same; only the relative phase of the spins relaxes. In actuality, T_2 is the time constant for relaxation to the arrangement on the right and $T_2 \ln 2$ is the half-life of the arrangement on the left.

Because the relaxation involves the relative orientation of the spins, T_2 is also known as the **spin–spin relaxation time.** Any relaxation process that changes the balance between α and β spins will also contribute to this randomization, so the time constant T_2 is almost always less than or equal to T_1.

Local magnetic fields also affect spin–spin relaxation. When the fluctuations are slow, each molecule lingers in its local magnetic environment and the spin orientations randomize quickly around the applied field direction. If the molecules move rapidly from one magnetic environment to another, the effects of differences in local magnetic field average to zero: individual spins do not precess at very different rates, they can remain bunched for longer, and spin–spin relaxation does not take place as quickly. In other words, slow molecular motion corresponds to short T_2 and fast motion corresponds to long T_2 (as shown in Fig. 12.36). Calculations show that, when the motion is fast, $T_2 \approx T_1$.

If the *y*-component of magnetization decays with a time constant T_2, the spectral line is broadened (Fig. 12.38), and its width at half-height becomes

$$\Delta\nu_{1/2} = \frac{1}{\pi T_2} \tag{12.35}$$

Typical values of T_2 in proton NMR are of the order of seconds, so linewidths of around 0.1 Hz can be anticipated, in broad agreement with observation.

So far, we have assumed that the equipment, and in particular the magnet, is perfect, and that the differences in Larmor frequencies arise solely from interactions within the sample. In practice, the magnet is not perfect, and the field is different at different locations in the sample. The inhomogeneity broadens

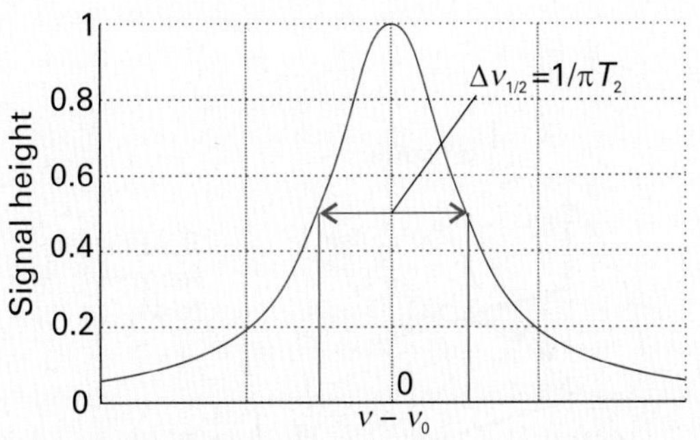

Fig. 12.38 A Lorentzian absorption line. The width at half-height is inversely proportional to the parameter T_2 and the longer the transverse relaxation time, the narrower the line.

interActivity The *Living graphs* section of the text's web site has an applet that allows you to calculate and display Lorenztian absorption lines. Explore the effect of the parameter T_2 on the width and the maximal intensity of a Lorentzian line. Rationalize your observations.

the resonance, and in most cases this **inhomogeneous broadening** dominates the broadening we have discussed so far. It is common to express the extent of inhomogeneous broadening in terms of an **effective transverse relaxation time**, T_2^*, by using a relation like eqn 12.35, but writing

$$T_2^* = \frac{1}{\Delta\nu_{1/2}} \qquad [12.36]$$

where $\Delta\nu_{1/2}$ is the observed width at half-height of a line with a Lorenztian shape of the form $I \propto 1/(1+x^2)$. For instance, a width of 10 Hz corresponds to $T_2^* = 32$ ms.

(b) The measurement of T_1

The longitudinal relaxation time T_1 can be measured by the **inversion recovery technique**. The first step is to apply a 180° pulse to the sample. A 180° pulse is achieved by applying the $\mathcal{B}_1$ field for twice as long as for a 90° pulse, so the magnetization vector precesses through 180° and points in the $-z$ direction (Fig. 12.39). No signal can be seen at this stage because there is no component of magnetization in the xy-plane (where the coil can detect it). The β spins begin to relax back into α spins, and the magnetization vector first shrinks exponentially, falling through zero to its thermal equilibrium value, M_z. After an interval τ, a 90° pulse is applied that rotates the magnetization into the xy-plane, where it generates an FID signal. The frequency-domain spectrum is then obtained by Fourier transformation.

The intensity of the spectrum obtained in this way depends on the length of the magnetization vector that is rotated into the xy-plane. The length of that vector changes exponentially as the interval between the two pulses is increased, so the intensity of the spectrum also changes exponentially with increasing τ. We can therefore measure T_1 by fitting an exponential curve to the series of spectra obtained with different values of τ.

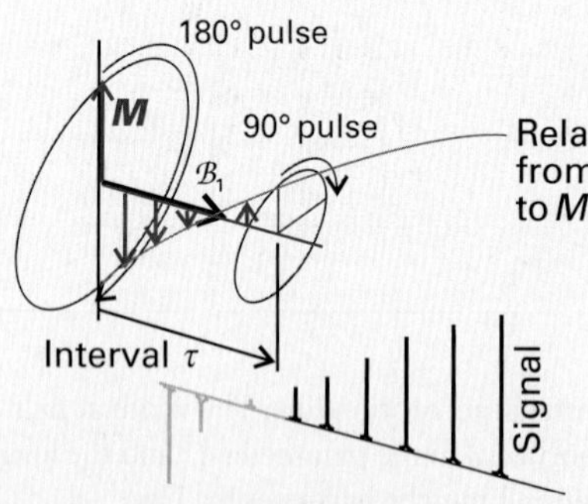

Fig. 12.39 The result of applying a 180° pulse to the magnetization in the rotating frame and the effect of a subsequent 90° pulse. The amplitude of the frequency-domain spectrum varies with the interval between the two pulses because spin–lattice relaxation has time to occur.

(c) Spin echoes

The measurement of T_2 (as distinct from T_2^*) depends on being able to eliminate the effects of inhomogeneous broadening. The cunning required is at the root of some of the most important advances that have been made in NMR since its introduction.

A **spin echo** is the magnetic analogue of an audible echo: transverse magnetization is created by a radiofrequency pulse, decays away, is reflected by a second pulse, and grows back to form an echo. The sequence of events is shown in Fig. 12.40. We can consider the overall magnetization as being made up of a number of different magnetizations, each of which arises from a **spin packet** of nuclei with very similar precession frequencies. The spread in these frequencies arises because the applied field $\mathcal{B}_0$ is inhomogeneous, so different parts of the sample experience different fields. The precession frequencies also differ if there is more than one chemical shift present. As will be seen, the importance of a spin echo is that it can suppress the effects of both field inhomogeneities and chemical shifts.

First, a 90° pulse is applied to the sample. We follow events by using the rotating frame, in which $\mathcal{B}_1$ is stationary along the x-axis and causes the magnetization to be into the xy-plane. The spin packets now begin to fan out because they have different Larmor frequencies, with some above the radiofrequency and some below. The detected signal depends on the resultant of the spin-packet magnetization vectors, and decays with a time-constant T_2^* because of the combined effects of field inhomogeneity and spin–spin relaxation.

After an evolution period τ, a 180° pulse is applied to the sample; this time, about the y-axis of the rotating frame (the axis

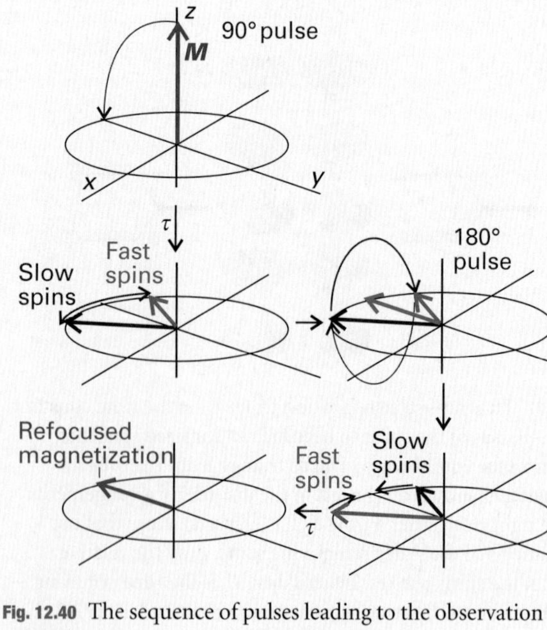

Fig. 12.40 The sequence of pulses leading to the observation of a spin echo.

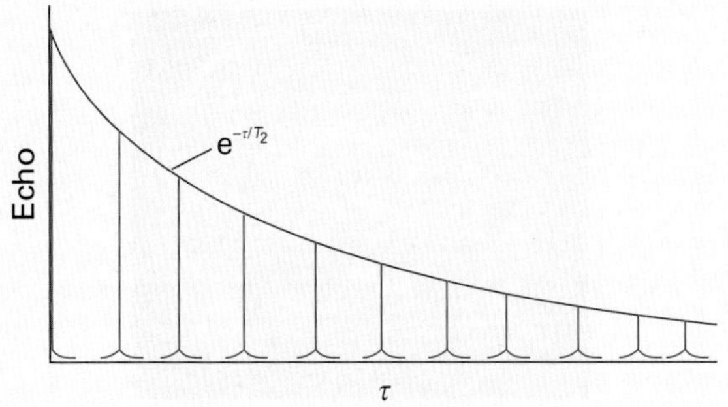

Fig. 12.41 The exponential decay of spin echoes can be used to determine the transverse relaxation time.

of the pulse is changed from x to y by a 90° phase shift of the radiofrequency radiation). The pulse rotates the magnetization vectors of the faster spin packets into the positions previously occupied by the slower spin packets, and vice versa. Thus, as the vectors continue to precess, the fast vectors are now behind the slow; the fan begins to close up again, and the resultant signal begins to grow back into an echo. At time 2τ, all the vectors will once more be aligned along the y-axis, and the fanning out caused by the field inhomogeneity is said to have been **refocused**: the spin echo has reached its maximum. Because the effects of field inhomogeneities have been suppressed by the refocusing, the echo signal will have been attenuated by the factor $e^{-2\tau/T_2}$ caused by spin–spin relaxation alone. After the time 2τ, the magnetization will continue to precess, fanning out once again, giving a resultant that decays with time constant T_2^*.

The important feature of the technique is that the size of the echo is independent of any local fields that remain constant during the two τ intervals. If a spin packet is 'fast' because it happens to be composed of spins in a region of the sample that experiences higher than average fields, then it remains fast throughout both intervals, and what it gains on the first interval it loses on the second interval. Hence, the size of the echo is independent of inhomogeneities in the magnetic field, for these remain constant. The true transverse relaxation arises from fields that vary on a molecular distance scale, and there is no guarantee that an individual 'fast' spin will remain 'fast' in the refocusing phase: the spins within the packets therefore spread with a time constant T_2. Hence, the effects of the true relaxation are not refocused, and the size of the echo decays with the time constant T_2 (Fig. 12.41).

IMPACT ON MEDICINE

I12.1 Magnetic resonance imaging

One of the most striking applications of nuclear magnetic resonance is in medicine. *Magnetic resonance imaging* (MRI) is a portrayal of the concentrations of protons in a solid object. The technique relies on the application of specific pulse sequences to an object in an inhomogeneous magnetic field.

If an object containing hydrogen nuclei (a tube of water or a human body) is placed in an NMR spectrometer and exposed to a *homogeneous* magnetic field, then a single resonance signal will be detected. Now consider a flask of water in a magnetic field that varies linearly in the z-direction according to $\mathcal{B}_0 + G_z z$, where $G_z = (\partial \mathcal{B}_0/\partial z)_0$ is the field gradient along the z-direction (Fig. 12.42). Then the water protons will be resonant at the frequencies

$$\nu_L(z) = \frac{\gamma}{2\pi}(\mathcal{B}_0 + G_z z) \qquad (12.37)$$

(Similar equations may be written for gradients along the x- and y-directions.) Application of a 90° radiofrequency pulse with $\nu = \nu_L(z)$ will result in a signal with an intensity that is proportional to the numbers of protons at the position z. This is an example of *slice selection*, the application of a selective 90° pulse that excites nuclei in a specific region, or slice, of the sample. It follows that the intensity of the NMR signal will be a projection of the numbers of protons on a line parallel to the field gradient. The image of a three-dimensional object such as a flask of water can be obtained if the slice selection technique is applied at different orientations (as shown in Fig. 12.42). In *projection reconstruction*, the projections can be analysed on a computer to reconstruct the three-dimensional distribution of protons in the object.

In practice, the NMR signal is not obtained by direct analysis of the FID curve after application of a single 90° pulse. Instead, spin echoes are often detected with several variations of the 90°–τ–180° pulse sequence (Section 12.9c). In *phase encoding*, field gradients are applied during the evolution period and the detection period of a spin-echo pulse sequence. The first step consists of a 90° pulse that results in slice selection along the

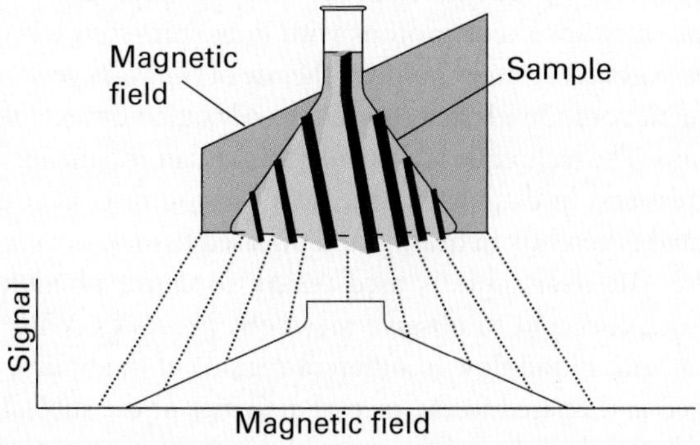

Fig. 12.42 In a magnetic field that varies linearly over a sample, all the protons within a given slice (that is, at a given field value) come into resonance and give a signal of the corresponding intensity. The resulting intensity pattern is a map of the numbers in all the slices, and portrays the shape of the sample. Changing the orientation of the field shows the shape along the corresponding direction, and computer manipulation can be used to build up the three-dimensional shape of the sample.

z-direction. The second step consists of application of a *phase gradient*, a field gradient along the y-direction, during the evolution period. At each position along the gradient, a spin packet will precess at a different Larmor frequency due to chemical shift effects and the field inhomogeneity, so each packet will dephase to a different extent by the end of the evolution period. We can control the extent of dephasing by changing the duration of the evolution period, so Fourier transformation on τ gives information about the location of a proton along the y-direction. (For technical reasons, it is more common to vary the magnitude of the phase gradient.) For each value of τ, the next steps are application of the 180° pulse and then of a *read gradient*, a field gradient along the x-direction, during detection of the echo. Protons at different positions along x experience different fields and will resonate at different frequencies. Therefore Fourier transformation of the FID gives different signals for protons at different positions along x.

A common problem with the techniques described above is image contrast, which must be optimized in order to show spatial variations in water content in the sample. One strategy for solving this problem takes advantage of the fact that the relaxation times of water protons are shorter for water in biological tissues than for the pure liquid. Furthermore, relaxation times from water protons are also different in healthy and diseased tissues. A *T_1-weighted image* is obtained by repeating the spin echo sequence before spin–lattice relaxation can return the spins in the sample to equilibrium. Under these conditions, differences in signal intensities are directly related to differences in T_1. A *T_2-weighted image* is obtained by using an evolution period τ that is relatively long. Each point on the image is an echo signal that behaves in the manner shown in Fig. 12.41, so signal intensities are strongly dependent on variations in T_2. However, allowing so much of the decay to occur leads to weak signals even for those protons with long spin–spin relaxation times. Another strategy involves the use of *contrast agents*, paramagnetic compounds that shorten the relaxation times of nearby protons. The technique is particularly useful in enhancing image contrast and in diagnosing disease if the contrast agent is distributed differently in healthy and diseased tissues.

The MRI technique is used widely to detect physiological abnormalities and to observe metabolic processes. With *functional MRI*, blood flow in different regions of the brain can be studied and related to the mental activities of the subject. The technique is based on differences in the magnetic properties of deoxygenated and oxygenated haemoglobin, the iron-containing protein that transports O_2 in red blood cells. The more paramagnetic deoxygenated haemoglobin affects the proton resonances of tissue differently from the oxygenated protein. Because there is greater blood flow in active regions of the brain than in inactive regions, changes in the intensities of proton resonances due to changes in levels of oxygenated haemoglobin can be related to brain activity.

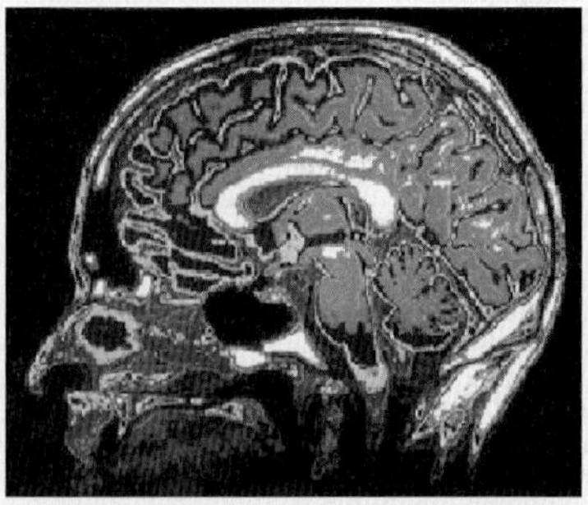

Fig. 12.43 The great advantage of MRI is that it can display soft tissue, such as in this cross-section through a patient's head. (Courtesy of the University of Manitoba.)

The special advantage of MRI is that it can image *soft* tissues (Fig. 12.43), whereas X-rays are largely used for imaging hard, bony structures and abnormally dense regions, such as tumours. In fact, the invisibility of hard structures in MRI is an advantage, as it allows the imaging of structures encased by bone, such as the brain and the spinal cord. X-rays are known to be dangerous on account of the ionization they cause; the high magnetic fields used in MRI may also be dangerous but, apart from anecdotes about the extraction of loose fillings from teeth, there is no convincing evidence of their harmfulness and the technique is considered safe.

12.10 Spin decoupling

Carbon-13 is a **dilute-spin species** in the sense that it is unlikely that more than one ^{13}C nucleus will be found in any given small molecule (provided the sample has not been enriched with that isotope; the natural abundance of ^{13}C is only 1.1 per cent). Even in large molecules, although more than one ^{13}C nucleus may be present, it is unlikely that they will be close enough to give an observable splitting. Hence, it is not normally necessary to take into account ^{13}C–^{13}C spin–spin coupling within a molecule.

Protons are **abundant-spin species** in the sense that a molecule is likely to contain many of them. If we were observing a ^{13}C-NMR spectrum, we would obtain a very complex spectrum on account of the coupling of the one ^{13}C nucleus with many of the protons that are present. To avoid this difficulty, ^{13}C-NMR spectra are normally observed using the technique of **proton decoupling.** Thus, if the CH_3 protons of ethanol are irradiated with a second, strong, resonant radiofrequency pulse, they undergo rapid spin reorientations and the ^{13}C nucleus senses an average orientation. As a result, its resonance is a single line and not a 1:3:3:1 quartet. Proton decoupling has the additional advantage of enhancing sensitivity, because the intensity is concentrated into a single transition frequency instead of being spread over several transition frequencies (see Section 12.11). If care is taken to ensure that the other parameters on which the strength of the signal depends are kept constant, the intensities of proton-decoupled spectra are proportional to the number of ^{13}C nuclei present. The technique is widely used to characterize synthetic polymers.

12.11 The nuclear Overhauser effect

We have seen already that one advantage of protons in NMR is their high magnetogyric ratio, which results in relatively large Boltzmann population differences and hence greater resonance intensities than for most other nuclei. In the steady-state **nuclear Overhauser effect** (NOE), spin relaxation processes involving internuclear dipole–dipole interactions are used to transfer this population advantage to another nucleus (such as ^{13}C or another proton), so that the latter's resonances are modified. In a dipole–dipole interaction between two nuclei, one nucleus influences the behaviour of another nucleus in much the same way that the orientation of a bar magnet is influenced by the presence of another bar magnet nearby.

To understand the effect, we consider the populations of the four levels of a homonuclear (for instance, proton) AX system; these were shown in Fig. 12.12. At thermal equilibrium, the population of the $\alpha_A\alpha_X$ level is the greatest, and that of the $\beta_A\beta_X$ level is the least; the other two levels have the same energy and an intermediate population. The thermal equilibrium absorption intensities reflect these populations as shown in Fig. 12.44. Now consider the combined effect of spin relaxation and keeping the X spins saturated. When we saturate the X transition, the populations of the X levels are equalized ($N_{\alpha X} = N_{\beta X}$) and all transitions involving $\alpha_X \leftrightarrow \beta_X$ spin flips are no longer observed. At this stage there is no change in the populations of the A levels. If that were all that was to happen, all we would see would be the loss of the X resonance and no effect on the A resonance.

Now consider the effect of spin relaxation. Relaxation can occur in a variety of ways if there is a dipolar interaction between the A and X spins. One possibility is for the magnetic field acting between the two spins to cause them both to flop from β to α, so the $\alpha_A\alpha_X$ and $\beta_A\beta_X$ states regain their thermal equilibrium populations. However, the populations of the $\alpha_A\beta_X$ and $\beta_A\alpha_X$ levels remain unchanged at the values characteristic of saturation. As we see from Fig. 12.45, the population difference between the states joined by transitions of A is now greater than at equilibrium, so the resonance absorption is enhanced. Another possibility is for the dipolar interaction between the two spins to cause α to flip to β and β to flop to α. This transition equilibrates the populations of $\alpha_A\beta_X$ and $\beta_A\alpha_X$ but leaves the $\alpha_A\alpha_X$ and $\beta_A\beta_X$ populations unchanged. Now we see from the figure that the population differences in the states involved in the A transitions are decreased, so the resonance absorption is diminished.

Which effect wins? Does the NOE enhance the A absorption or does it diminish it? As in the discussion of relaxation times in Section 12.9, the efficiency of the intensity-enhancing $\beta_A\beta_X \leftrightarrow \alpha_A\alpha_X$ relaxation is high if the dipole field oscillates at a frequency close to the transition frequency, which in this case is about 2ν;

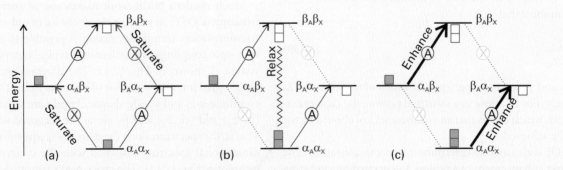

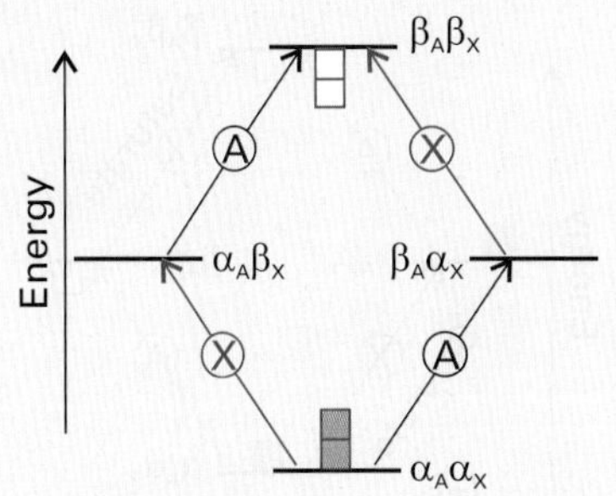

Fig. 12.44 The energy levels of an AX system and an indication of their relative populations. Each grey square above the line represents an excess population and each white square below the line represents a population deficit. The transitions of A and X are marked.

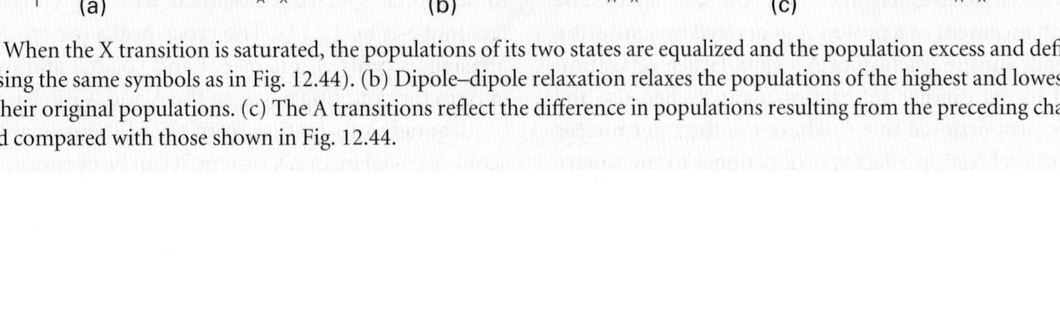

Fig. 12.45 (a) When the X transition is saturated, the populations of its two states are equalized and the population excess and deficit become as shown (using the same symbols as in Fig. 12.44). (b) Dipole–dipole relaxation relaxes the populations of the highest and lowest states, and they regain their original populations. (c) The A transitions reflect the difference in populations resulting from the preceding changes, and are enhanced compared with those shown in Fig. 12.44.

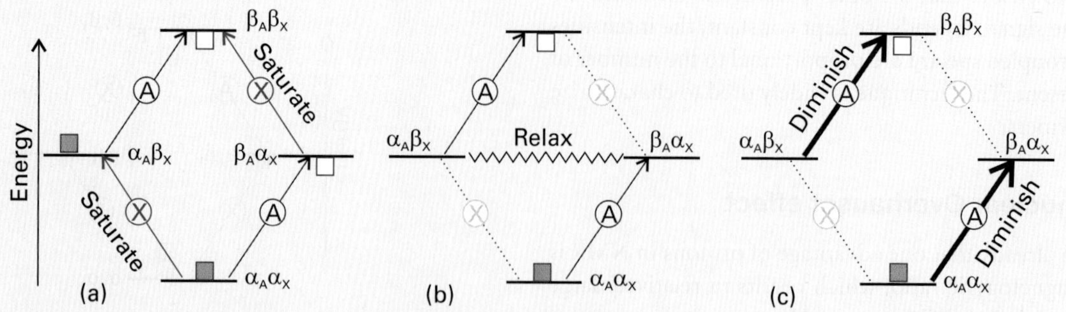

Fig. 12.46 (a) When the X transition is saturated, just as in Fig. 12.45 the populations of its two states are equalized and the population excess and deficit become as shown. (b) Dipole–dipole relaxation relaxes the populations of the two intermediate states, and they regain their original populations. (c) The A transitions reflect the difference in populations resulting from the preceding changes, and are diminished compared with those shown in Fig. 12.44.

likewise, the efficiency of the intensity-diminishing $\alpha_A\beta_X \leftrightarrow \beta_A\alpha_X$ relaxation is high if the dipole field is stationary (as there is no frequency difference between the initial and final states). A large molecule rotates so slowly that there is very little motion at 2ν, so we expect an intensity decrease (Fig. 12.46). A small molecule rotating rapidly can be expected to have substantial motion at 2ν, and a consequent enhancement of the signal. In practice, the enhancement lies somewhere between the two extremes and is reported in terms of the parameter η (eta), where

$$\eta = \frac{I_A - I_A^\circ}{I_A^\circ} \qquad [12.38]$$

Here I_A° and I_A are the intensities of the NMR signals due to nucleus A before and after application of the long ($> T_1$) radiofrequency pulse that saturates transitions due to the X nucleus. When A and X are nuclei of the same species, such as protons, η lies between -1 (diminution) and $+\frac{1}{2}$ (enhancement). However, η also depends on the values of the magnetogyric ratios of A and X. In the case of maximal enhancement it is possible to show that

$$\eta = \frac{\gamma_X}{2\gamma_A} \qquad (12.39)$$

where γ_A and γ_X are the magnetogyric ratios of nuclei A and X, respectively. For ^{13}C close to a saturated proton, the ratio evaluates to 1.99, which shows that an enhancement of about a factor of 2 can be achieved.

The NOE is also used to determine interproton distances. The Overhauser enhancement of a proton A generated by saturating a spin X depends on the fraction of A's spin–lattice relaxation that is caused by its dipolar interaction with X. Because the dipolar field is proportional to r^{-3}, where r is the internuclear distance, and the relaxation effect is proportional to the square of the field, and therefore to r^{-6}, the NOE may be used to determine the geometries of molecules in solution. The determination of the structure of a small protein in solution involves the use of several hundred NOE measurements, effectively casting a net over the protons present. The enormous importance of this procedure is that we can determine the conformation of biological macromolecules in an aqueous environment and do not need to try to make the single crystals that are essential for an X-ray diffraction investigation (Chapter 9).

12.12 Two-dimensional NMR

An NMR spectrum contains a great deal of information and, if many protons are present, is very complex. Even a first-order spectrum is complex, for the fine structure of different groups of lines can overlap. The complexity would be reduced if we could use two axes to display the data, with resonances belonging to different groups lying at different locations on the second axis. This separation is essentially what is achieved in **two-dimensional NMR**.

Much modern NMR work makes use of **correlation spectroscopy** (COSY) in which a clever choice of pulses and Fourier transformation techniques makes it possible to determine all spin–spin couplings in a molecule. A typical outcome for an AX system is shown in Fig. 12.47. The diagram shows peaks of equal signal intensity on a plot of intensity against the frequency coordinates ν_1 and ν_2. The **diagonal peaks** are signals centred on (δ_A,δ_A) and (δ_X,δ_X) and lie along the diagonal where $\nu_1 = \nu_2$. That is, the spectrum along the diagonal is equivalent to the one-dimensional spectrum obtained with the conventional NMR technique (Fig. 12.13). The **cross-peaks** (or *off-diagonal peaks*) are signals centred on (δ_A,δ_X) and (δ_X,δ_A) and owe their existence to the coupling between the A and X nuclei.

Although information from two-dimensional NMR spectroscopy is trivial in an AX system, it can be of enormous help in the

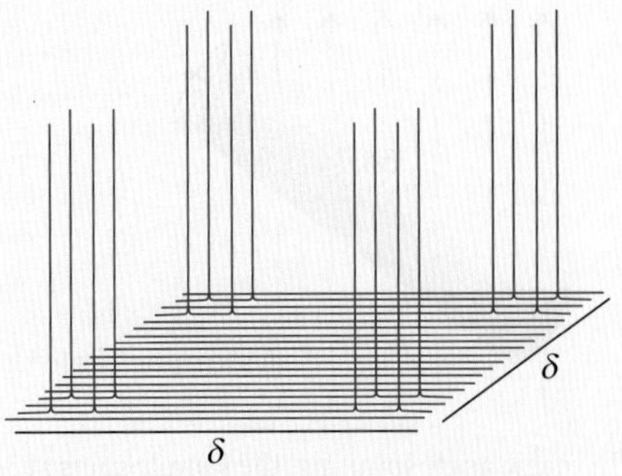

Fig. 12.47 The COSY spectrum of an AX spin system.

interpretation of more complex spectra, leading to a map of the couplings between spins and to the determination of the bonding network in complex molecules. Indeed, the spectrum of a synthetic or biological polymer that would be impossible to interpret in one-dimensional NMR can often be interpreted reasonably rapidly by two-dimensional NMR.

● A BRIEF ILLUSTRATION

Figure 12.48 is a portion of the COSY spectrum of the amino acid isoleucine (7), showing the resonances associated with the protons bound to the carbon atoms. We begin the assignment process by considering which protons should be interacting by spin–spin coupling. From the known molecular structure, we conclude that: (i) the C_a—H proton is coupled only to the C_b—H proton, (ii) the C_b—H protons are coupled to the C_a—H, C_c—H, and C_d—H protons, and (iii) the inequivalent C_d—H protons are coupled to the C_b—H and C_e—H protons. We now note that:

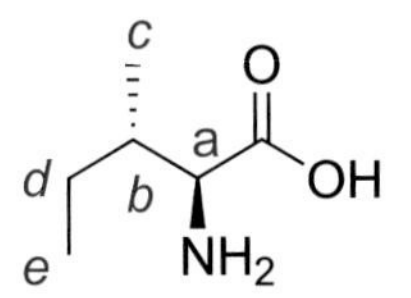

7 Isoleucine

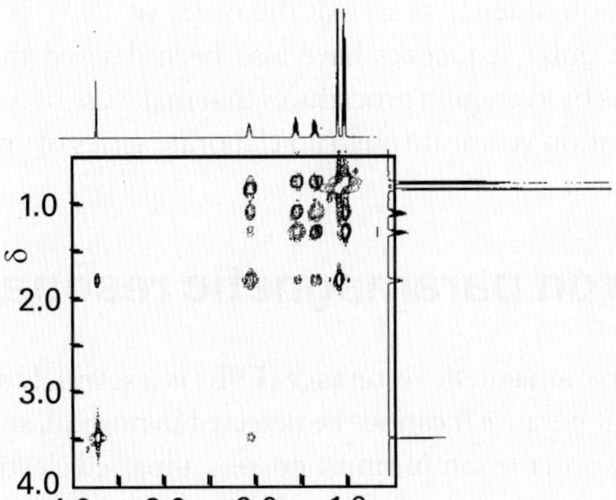

Fig. 12.48 Proton COSY spectrum of isoleucine. (The illustration and corresponding spectrum are adapted from K.E. van Holde *et al. Principles of physical biochemistry*, Prentice Hall, Upper Saddle River (1998).)

• The resonance with $\delta = 3.6$ shares a cross-peak with only one other resonance at $\delta = 1.9$, which in turn shares cross-peaks with resonances at $\delta = 1.4$, 1.2, and 0.9. This identification is consistent with the resonances at $\delta = 3.6$ and 1.9 corresponding to the C_a—H and C_b—H protons, respectively.

• The proton with resonance at $\delta = 0.8$ is not coupled to the C_b—H protons, so we assign the resonance at $\delta = 0.8$ to the C_e—H protons.

• The resonances at $\delta = 1.4$ and 1.2 do not share cross-peaks with the resonance at $\delta = 0.9$.

In the light of the expected couplings, we assign the resonance at $\delta = 0.9$ to the C_c—H protons and the resonances at $\delta = 1.4$ and 1.2 to the inequivalent C_d—H protons. ●

We have seen that the nuclear Overhauser effect can provide information about internuclear distances through analysis of enhancement patterns in the NMR spectrum before and after saturation of selected resonances. In **nuclear Overhauser effect spectroscopy** (NOESY) a map of all possible NOE interactions is obtained by again using a proper choice of radiofrequency pulses and Fourier transformation techniques. Like a COSY spectrum, a NOESY spectrum consists of a series of diagonal peaks that correspond to the one-dimensional NMR spectrum of the sample. The off-diagonal peaks indicate which nuclei are close enough to each other to give rise to a nuclear Overhauser effect. NOESY data reveal internuclear distances up to about 0.5 nm.

12.13 Solid-state NMR

The principal difficulty with the application of NMR to solids is the low resolution characteristic of solid samples. Nevertheless, there are good reasons for seeking to overcome these difficulties. They include the possibility that a compound of interest is unstable in solution or that it is insoluble, so conventional solution NMR cannot be employed. Moreover, many species are intrinsically interesting as solids, and it is important to determine their structures and dynamics. Synthetic polymers are particularly interesting in this regard, and information can be obtained about the arrangement of molecules, their conformations, and the motion of different parts of the chain. This kind of information is crucial to an interpretation of the bulk properties of the polymer in terms of its molecular characteristics. Similarly, inorganic substances, such as the zeolites that are used as molecular sieves and shape-selective catalysts, can be studied using solid-state NMR, and structural problems can be resolved that cannot be tackled by X-ray diffraction.

Problems of resolution and linewidth are not the only features that plague NMR studies of solids, but the rewards are so great that considerable efforts have been made to overcome them and have achieved notable success. Because molecular rotation has almost ceased (except in special cases, including 'plastic crystals' in which the molecules continue to tumble), spin–lattice relaxation times are very long but spin–spin relaxation times are very short. Hence, in a pulse experiment, there is a need for lengthy delays—of several seconds—between successive pulses so that the spin system has time to revert to equilibrium. Even gathering the murky information may therefore be a lengthy process. Moreover, because lines are so broad, very high powers of radiofrequency radiation may be required to achieve saturation. Whereas solution pulse NMR uses transmitters of a few tens of watts, solid-state NMR may require transmitters rated at several hundreds of watts.

(a) The origins of linewidths in solids

There are two principal contributions to the linewidths of solids. One is the direct magnetic dipolar interaction between nuclear spins. As we saw in the discussion of spin–spin coupling, a nuclear magnetic moment will give rise to a local magnetic field, which points in different directions at different locations around the nucleus. If we are interested only in the component parallel to the direction of the applied magnetic field (because only this component has a significant effect), then we can use a classical expression to write the magnitude of the local magnetic field as

$$\mathcal{B}_{\text{loc}} = -\frac{\gamma\hbar\mu_0 m_I}{4\pi R^3}(1 - 3\cos^2\theta) \qquad (12.40)$$

Unlike in solution, this field is not motionally averaged to zero. Many nuclei may contribute to the total local field experienced by a nucleus of interest, and different nuclei in a sample may experience a wide range of fields. Typical dipole fields are of the order of 10^{-3} T, which corresponds to splittings and linewidths of the order of 10^4 Hz.

A second source of linewidth is the anisotropy of the chemical shift. We have seen that chemical shifts arise from the ability of the applied field to generate electron currents in molecules. In general, this ability depends on the orientation of the molecule relative to the applied field. In solution, when the molecule is tumbling rapidly, only the average value of the chemical shift is relevant. However, the anisotropy is not averaged to zero for stationary molecules in a solid, and molecules in different orientations have resonances at different frequencies. The chemical shift anisotropy also varies with the angle between the applied field and the principal axis of the molecule as $1 - 3\cos^2\theta$.

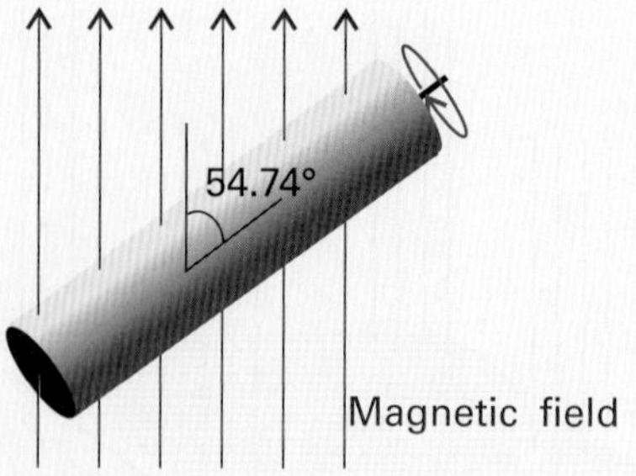

Fig. 12.49 In magic angle spinning, the sample spins at 54.74° (that is, arccos $(1/3)^{1/2}$) to the applied magnetic field. Rapid motion at this angle averages dipole–dipole interactions and chemical shift anisotropies to zero.

(b) The reduction of linewidths

Fortunately, there are techniques available for reducing the linewidths of solid samples. One technique, **magic-angle spinning** (MAS), takes note of the $1 - 3\cos^2\theta$ dependence of both the dipole–dipole interaction and the chemical shift anisotropy. The 'magic angle' is the angle at which $1 - 3\cos^2\theta = 0$, and corresponds to 54.74°. In the technique, the sample is spun at high speed at the magic angle to the applied field (Fig. 12.49). All the dipolar interactions and the anisotropies average to the value they would have at the magic angle, but at that angle they are zero. The difficulty with MAS is that the spinning frequency must not be less than the width of the spectrum, which is of the order of kilohertz. However, gas-driven sample spinners that can be rotated at up to 25 kHz are now routinely available, and a considerable body of work has been done.

Pulsed techniques similar to those described in the previous section may also be used to reduce linewidths. The dipolar field of protons, for instance, may be reduced by a decoupling procedure. However, because the range of coupling strengths is so large, radiofrequency power of the order of 1 kW is required. Elaborate pulse sequences have also been devised that reduce linewidths by averaging procedures that make use of twisting the magnetization vector through an elaborate series of angles.

Electron paramagnetic resonance

Electron paramagnetic resonance (EPR) is less widely applicable than NMR because it cannot be detected in normal, spin-paired molecules and the sample must possess unpaired electron spins. It is used to study radicals formed during chemical reactions or by radiation, radicals that act as probes of biological structure, many d-metal complexes, and molecules in triplet states (such as those involved in phosphorescence, Section 11.5). The sample may be a gas, a liquid, or a solid, but the free rotation of molecules in the gas phase gives rise to complications.

12.14 The EPR spectrometer

Both Fourier-transform (FT) and continuous wave (CW) EPR spectrometers are available. The FT-EPR instrument is based on the concepts developed in Section 12.8, except that pulses of microwaves are used to excite electron spins in the sample. The layout of the more common CW-EPR spectrometer is shown in Fig. 12.50. It consists of a microwave source (a klystron or a Gunn oscillator), a cavity in which the sample is inserted in a glass or quartz container, a microwave detector, and an electromagnet with a field that can be varied in the region of 0.3 T. The EPR spectrum is obtained by monitoring the microwave absorption as the field is changed, and a typical spectrum (of the benzene radical anion, $C_6H_6^-$) is shown in Fig. 12.51. The peculiar appearance of the spectrum, which is in fact the first-derivative of the absorption, arises from the detection technique, which is sensitive to the slope of the absorption curve (Fig. 12.52).

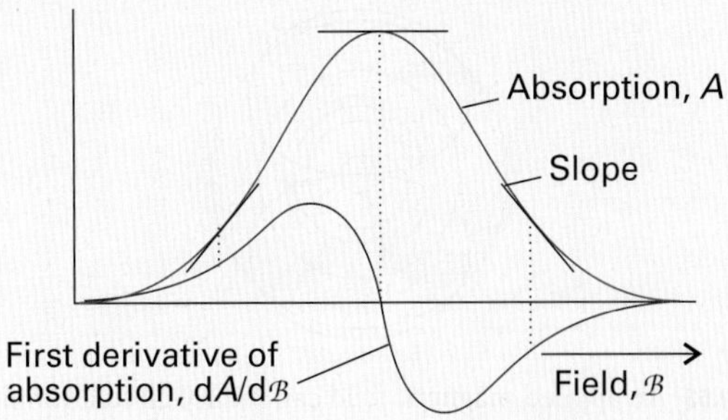

Fig. 12.52 When phase-sensitive detection is used, the signal is the first derivative of the absorption intensity. Note that the peak of the absorption corresponds to the point where the derivative passes through zero.

12.15 The *g*-value

Equation 12.12 gives the resonance frequency for a transition between the $m_s = -\frac{1}{2}$ and the $m_s = +\frac{1}{2}$ levels of a 'free' electron in terms of the *g*-value $g_e \approx 2.0023$. The magnetic moment of an unpaired electron in a radical also interacts with an external field, but the *g*-value is different from that for a free electron because of local magnetic fields induced by the molecular framework of the radical. Consequently, the resonance condition is normally written as

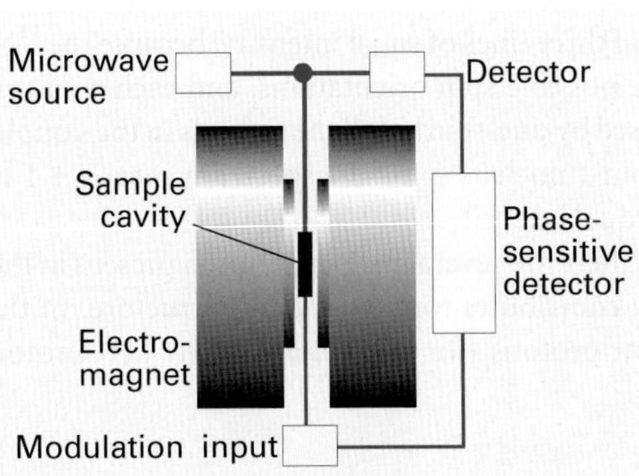

Fig. 12.50 The layout of a continuous-wave EPR spectrometer. A typical magnetic field is 0.3 T, which requires 9 GHz (3 cm) microwaves for resonance.

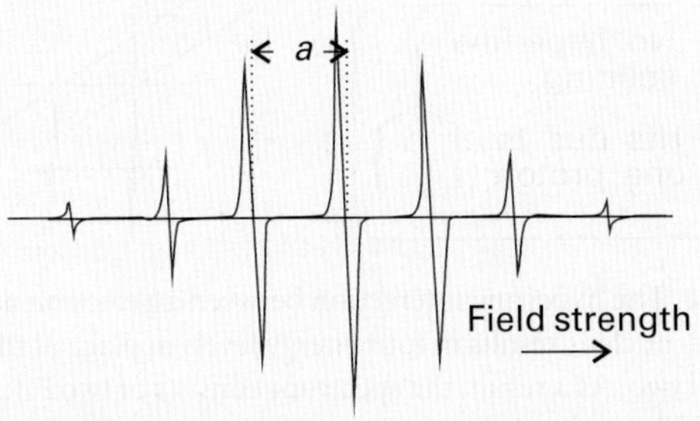

Fig. 12.51 The EPR spectrum of the benzene radical anion, $C_6H_6^-$, in fluid solution. *a* is the hyperfine splitting of the spectrum (Section 12.16); the centre of the spectrum is determined by the *g*-value of the radical.

$$h\nu = g\mu_B\mathcal{B}_0 \tag{12.41}$$

where *g* is the **g-value** of the radical.

● A BRIEF ILLUSTRATION

The centre of the EPR spectrum of the methyl radical occurred at 329.40 mT in a spectrometer operating at 9.2330 GHz (radiation belonging to the X band of the microwave region). Its *g*-value is therefore

$$g = \frac{h\nu}{\mu_B\mathcal{B}} = \frac{(6.626\,08 \times 10^{-34}\,\text{J s}) \times (9.2330 \times 10^{9}\,\text{s}^{-1})}{(9.2740 \times 10^{-24}\,\text{J T}^{-1}) \times (0.32940\,\text{T})} = 2.0027$$

●

Self-test 12.4 At what magnetic field would the methyl radical come into resonance in a spectrometer operating at 34.000 GHz (radiation belonging to the Q band of the microwave region)? [1.213 T]

The *g*-value in a molecular environment (a radical or a d-metal complex) is related to the ease with which the applied field can stir up currents through the molecular framework and the strength of the magnetic field the currents generate. Therefore, the *g*-value gives some information about electronic structure and plays a similar role in EPR to that played by shielding constants in NMR.

Electrons can migrate through the molecular framework by making use of excited states (Fig. 12.53). This additional path

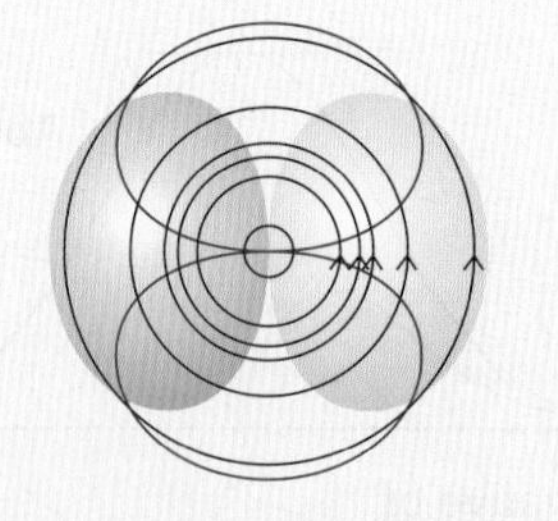

Fig. 12.53 An applied magnetic field can induce circulation of electrons that makes use of excited state orbitals (shown with a blue outline).

for circulation of electrons gives rise to a local magnetic field that adds to the applied field. Therefore, we expect the ease of stirring up currents to be inversely proportional to the separation of energy levels, ΔE, in the molecule. As we saw in Section 4.5, the strength of the field generated by electronic currents in atoms (and analogously in molecules) is related to the extent of coupling between spin and orbital angular momenta. That is, the local field strength is proportional to the molecular spin–orbit coupling constant, ξ.

We can conclude from the discussion above that the g-value of a radical or d-metal complex differs from g_e, the 'free-electron' g-value, by an amount that is proportional to $\xi/\Delta E$. This proportionality is widely observed. Many organic radicals have g-values close to 2.0027 and inorganic radicals have g-values typically in the range 1.9 to 2.1. The g-values of paramagnetic d-metal complexes often differ considerably from g_e, varying from 0 to 6, because in them ΔE is small (on account of the splitting of d orbitals brought about by interactions with ligands, as we saw in Section 11.4).

Just as in the case of the chemical shift in NMR spectroscopy, the g-value is anisotropic, that is, its magnitude depends on the orientation of the radical with respect to the applied field. In solution, when the molecule is tumbling rapidly, only the average value of the g-value is observed. Therefore, anisotropy of the g-value is observed only for radicals trapped in solids.

12.16 Hyperfine structure

The most important feature of EPR spectra is their **hyperfine structure**, the splitting of individual resonance lines into components. In general in spectroscopy, the term 'hyperfine structure' means the structure of a spectrum that can be traced to interactions of the electrons with nuclei other than as a result of the latter's point electric charge. The source of the hyperfine structure in EPR is the magnetic interaction between the electron spin and the magnetic dipole moments of the nuclei present in the radical.

(a) The effects of nuclear spin

Consider the effect on the EPR spectrum of a single H nucleus located somewhere in a radical. The proton spin is a source of magnetic field, and depending on the orientation of the nuclear spin, the field it generates adds to or subtracts from the applied field. The total local field is therefore

$$\mathcal{B}_{\text{loc}} = \mathcal{B} + am_I \qquad m_I = \pm\tfrac{1}{2} \tag{12.42}$$

where a is the **hyperfine coupling constant.** Half the radicals in a sample have $m_I = +\frac{1}{2}$, so half resonate when the applied field satisfies the condition

$$h\nu = g\mu_B(\mathcal{B} + \tfrac{1}{2}a), \text{ or} \qquad \mathcal{B} = \frac{h\nu}{g\mu_B} - \tfrac{1}{2}a \tag{12.43a}$$

The other half (which have $m_I = -\frac{1}{2}$) resonate when

$$h\nu = g\mu_B(\mathcal{B} - \tfrac{1}{2}a), \text{ or} \qquad \mathcal{B} = \frac{h\nu}{g\mu_B} + \tfrac{1}{2}a \tag{12.43b}$$

Therefore, instead of a single line, the spectrum shows two lines of half the original intensity separated by a and centred on the field determined by g (Fig. 12.54).

If the radical contains an ^{14}N atom ($I = 1$), its EPR spectrum consists of three lines of equal intensity, because the ^{14}N nucleus has three possible spin orientations, and each spin orientation is possessed by one-third of all the radicals in the sample. In general, a spin-I nucleus splits the spectrum into $2I + 1$ hyperfine lines of equal intensity.

When there are several magnetic nuclei present in the radical, each one contributes to the hyperfine structure. In the case of equivalent protons (for example, the two CH_2 protons in the

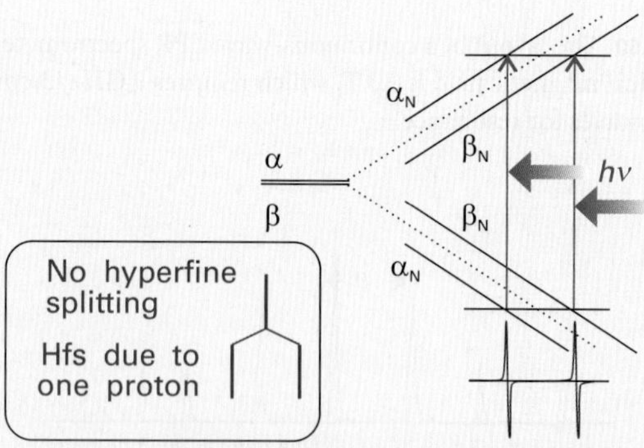

Fig. 12.54 The hyperfine interaction between an electron and a spin-$\frac{1}{2}$ nucleus results in four energy levels in place of the original two. As a result, the spectrum consists of two lines (of equal intensity) instead of one. The intensity distribution can be summarized by a simple stick diagram. The diagonal lines show the energies of the states as the applied field is increased, and resonance occurs when the separation of states matches the fixed energy of the microwave photon.

radical CH_3CH_2) some of the hyperfine lines are coincident. It is not hard to show that, if the radical contains N equivalent protons, then there are $N+1$ hyperfine lines with a binomial intensity distribution (the intensity distribution given by Pascal's triangle). The spectrum of the benzene radical anion in Fig. 12.51, which has seven lines with intensity ratio 1:6:15:20:15:6:1, is consistent with a radical containing six equivalent protons. More generally, if the radical contains N equivalent nuclei with spin quantum number I, then there are $2NI+1$ hyperfine lines with an intensity distribution based on a modified version of Pascal's triangle as shown in the following *Example*.

Example 12.2 *Predicting the hyperfine structure of an EPR spectrum*

A radical contains one ^{14}N nucleus ($I=1$) with hyperfine constant 1.61 mT and two equivalent protons ($I=\frac{1}{2}$) with hyperfine constant 0.35 mT. Predict the form of the EPR spectrum.

Method We should consider the hyperfine structure that arises from each type of nucleus or group of equivalent nuclei in succession. So, split a line with one nucleus; then each of those lines is split by a second nucleus (or group of nuclei), and so on. It is best to start with the nucleus with the largest hyperfine splitting; however, any choice could be made, and the order in which nuclei are considered does not affect the conclusion.

Answer The ^{14}N nucleus gives three hyperfine lines of equal intensity separated by 1.61 mT. Each line is split into doublets of spacing 0.35 mT by the first proton, and each line of these doublets is split into doublets with the same 0.35 mT splitting (Fig. 12.55). The central lines of each split doublet coincide, so the proton splitting gives 1:2:1 triplets of internal splitting 0.35 mT. Therefore, the spectrum consists of three equivalent 1:2:1 triplets.

Self-test 12.5 Predict the form of the EPR spectrum of a radical containing three equivalent ^{14}N nuclei. [Fig. 12.56]

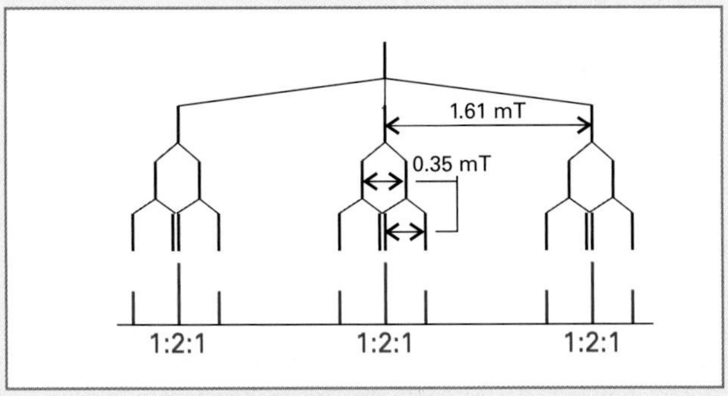

Fig. 12.55 The analysis of the hyperfine structure of radicals containing one ^{14}N nucleus ($I=1$) and two equivalent protons.

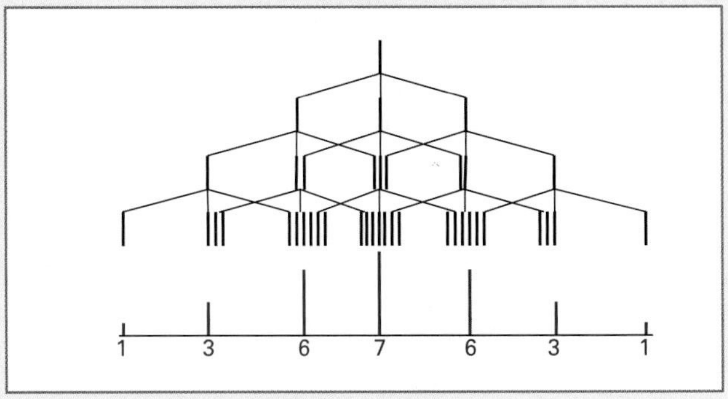

Fig. 12.56 The analysis of the hyperfine structure of radicals containing three equivalent ^{14}N nuclei.

The hyperfine structure of an EPR spectrum is a kind of fingerprint that helps to identify the radicals present in a sample. Moreover, because the magnitude of the splitting depends on the distribution of the unpaired electron near the magnetic nuclei present, the spectrum can be used to map the molecular orbital occupied by the unpaired electron. For example, because the hyperfine splitting in $C_6H_6^-$ is 0.375 mT, and one proton is close to a C atom with one-sixth the unpaired electron spin density (because the electron is spread uniformly around the ring), the hyperfine splitting caused by a proton in the electron spin entirely confined to a single adjacent C atom should be $6\times 0.375\ \text{mT} = 2.25\ \text{mT}$. If in another aromatic radical we find a hyperfine splitting constant a, then the **spin density**, ρ, the probability that an unpaired electron is on the atom, can be calculated from the **McConnell equation**:

$$a = Q\rho \qquad (12.44)$$

with $Q = 2.25$ mT. In this equation, ρ is the spin density on a C atom and a is the hyperfine splitting observed for the H atom to which it is attached.

● A BRIEF ILLUSTRATION

The hyperfine structure of the EPR spectrum of the radical anion (naphthalene)$^-$ can be interpreted as arising from two groups of four equivalent protons. Those at the 1, 4, 5, and 8 positions in the ring have $a = 0.490$ mT and those in the 2, 3, 6, and 7 positions have $a = 0.183$ mT. The densities obtained by using the McConnell equation are 0.22 and 0.08, respectively (**8**). ●

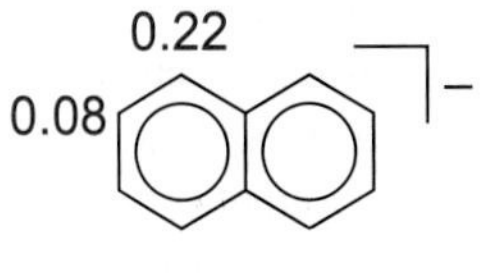

8

Self-test 12.6 The spin density in (anthracene)$^-$ is shown in (9). Predict the form of its EPR spectrum.

[A 1:2:1 triplet of splitting 0.43 mT split into a 1:4:6:4:1 quintet of splitting 0.22 mT, split into a 1:4:6:4:1 quintet of splitting 0.11 mT, $3 \times 5 \times 5 = 75$ lines in all]

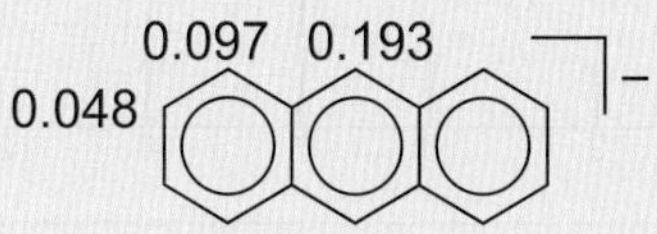

9

Synoptic table 12.3* Hyperfine coupling constants for atoms, a/mT

Nuclide	Isotropic coupling	Anisotropic coupling
^{1}H	50.8 (1s)	
^{2}H	7.8 (1s)	
^{14}N	55.2 (2s)	4.8 (2p)
^{19}F	1720 (2s)	108.4 (2p)

* More values are given in the *Data section*.

(b) The origin of the hyperfine interaction

The hyperfine interaction is an interaction between the magnetic moments of the unpaired electron and the nuclei. There are two contributions to the interaction.

An electron in a p orbital does not approach the nucleus very closely, so it experiences a field that appears to arise from a point magnetic dipole. The resulting interaction is called the **dipole–dipole interaction**. The contribution of a magnetic nucleus to the local field experienced by the unpaired electron is given by an expression like that in eqn 12.40. A characteristic of this type of interaction is that it is anisotropic. Furthermore, just as in the case of NMR, the dipole–dipole interaction averages to zero when the radical is free to tumble. Therefore, hyperfine structure due to the dipole–dipole interaction is observed only for radicals trapped in solids.

An s electron is spherically distributed around a nucleus and so has zero average dipole–dipole interaction with the nucleus even in a solid sample. However, because an s electron has a nonzero probability of being at the nucleus, it is incorrect to treat the interaction as one between two point dipoles. An s electron has a Fermi contact interaction with the nucleus, which as we saw in Section 12.6d is a magnetic interaction that occurs when the point dipole approximation fails. The contact interaction is isotropic (that is, independent of the radical's orientation), and consequently is shown even by rapidly tumbling molecules in fluids (provided the spin density has some s character).

The dipole–dipole interactions of p electrons and the Fermi contact interaction of s electrons can be quite large. For example, a 2p electron in a nitrogen atom experiences an average field of about 3.4 mT from the ^{14}N nucleus. A 1s electron in a hydrogen atom experiences a field of about 50 mT as a result of its Fermi contact interaction with the central proton. More values are listed in Table 12.3. The magnitudes of the contact interactions in radicals can be interpreted in terms of the s orbital character of the molecular orbital occupied by the unpaired electron, and the dipole–dipole interaction can be interpreted in terms of the p character. The analysis of hyperfine structure therefore gives information about the composition of the orbital, and especially the hybridization of the atomic orbitals (see Problem 12.13).

We still have the source of the hyperfine structure of the $C_6H_6^-$ anion and other aromatic radical anions to explain. The sample is fluid, and as the radicals are tumbling the hyperfine structure cannot be due to the dipole–dipole interaction. Moreover, the protons lie in the nodal plane of the π orbital occupied by the unpaired electron, so the structure cannot be due to a Fermi contact interaction. The explanation lies in a **polarization mechanism** similar to the one responsible for spin–spin coupling in NMR. There is a magnetic interaction between a proton (the H nucleus) and one of the electrons in the C—H bond, which results in that electron tending to be found with a greater probability near the proton (Fig. 12.57). The electron with opposite spin is therefore more likely to be close to the C atom at the other end of the bond. The unpaired electron on the C atom has a lower energy if it is parallel to that electron (Hund's rule favours parallel electrons on atoms), so the unpaired electron can detect the spin of the proton indirectly. Calculation using this model leads to a hyperfine interaction in agreement with the observed value of 2.25 mT.

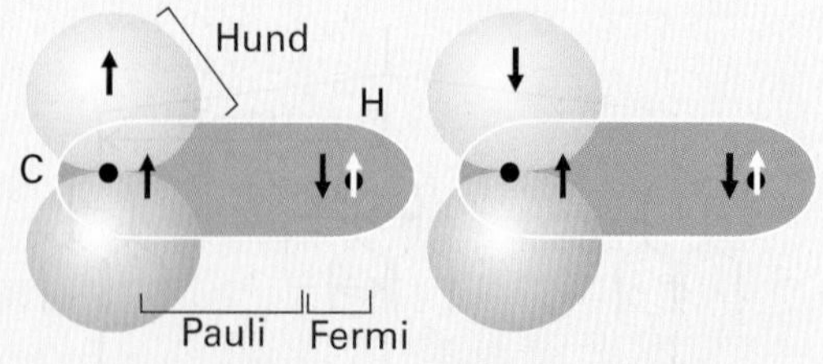

Fig. 12.57 The polarization mechanism for the hyperfine interaction in π-electron radicals. The arrangement in (a) is lower in energy than that in (b), so there is an effective coupling between the unpaired electron and the proton.

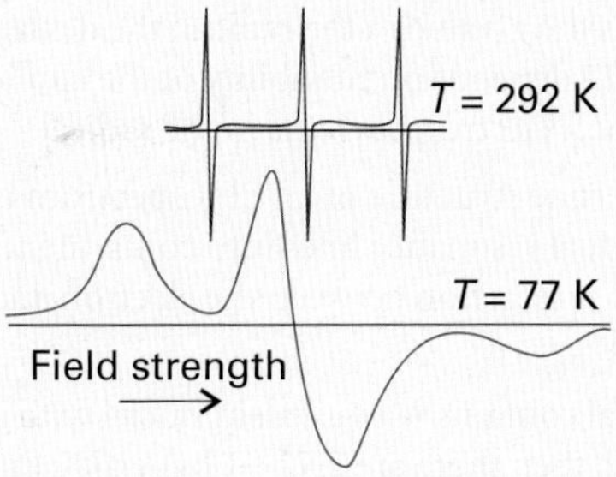

Fig. 12.58 ESR spectra of the di-*tert*-butyl nitroxide radical at 292 K (top) and 77 K (bottom). (Adapted from J.R. Bolton, in *Biological applications of electron spin resonance*, H.M. Swartz, J.R. Bolton, and D.C. Borg (ed.), Wiley, New York (1972).)

IMPACT ON BIOCHEMISTRY

I12.2 Spin probes

We saw in Sections 12.15 and 12.16 that anisotropy of the *g*-value and of the nuclear hyperfine interactions can be observed when a radical is immobilized in a solid. Figure 12.58 shows the variation of the lineshape of the EPR spectrum of the di-*tert*-butyl nitroxide radical (**10**) with temperature. At 292 K, the radical tumbles freely and isotropic hyperfine coupling to the ^{14}N nucleus gives rise to three sharp peaks. At 77 K, motion of the radical is restricted. Both isotropic and anisotropic hyperfine couplings determine the appearance of the spectrum, which now consists of three broad peaks.

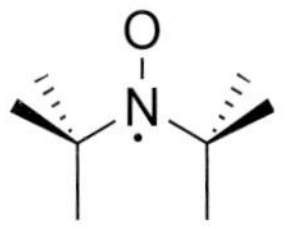
10

A *spin probe* (or *spin label*) is a radical that interacts with a biopolymer and with an EPR spectrum that reports on the dynamical properties of the biopolymer. The ideal spin probe is one with a spectrum that broadens significantly as its motion is restricted to a relatively small extent. Nitroxide spin probes have been used to show that the hydrophobic interiors of biological membranes, once thought to be rigid, are in fact very fluid and individual lipid molecules move laterally through the sheet-like structure of the membrane.

Just as chemical exchange can broaden proton NMR spectra (Section 12.7), electron exchange between two radicals can broaden EPR spectra. Therefore, the distance between two spin probe molecules may be measured from the linewidths of their EPR spectra. The effect can be used in a number of biochemical studies. For example, the kinetics of association of two polypeptides labelled with the synthetic amino acid 2,2,6,6,-tetramethylpiperidine-1-oxyl-4-amino-4-carboxylic acid (**11**) may be studied by measuring the change in linewidth of the label with time. Alternatively, the thermodynamics of association may be studied by examining the temperature dependence of the linewidth.

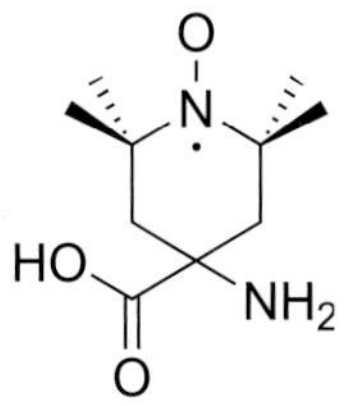
11

Checklist of key ideas

☐ 1. The energy of an electron in a magnetic field $\mathcal{B}_0$ is $E_{m_s} = -g_e\gamma_e\hbar\mathcal{B}_0 m_s$, where γ_e is the magnetogyric ratio of the electron. The energy of a nucleus in a magnetic field $\mathcal{B}_0$ is $E_{m_I} = -\gamma\hbar\mathcal{B}_0 m_I$, where γ is the nuclear magnetogyric ratio.

☐ 2. The resonance condition for an electron in a magnetic field is $h\nu = g_e\mu_B\mathcal{B}_0$. The resonance condition for a nucleus in a magnetic field is $h\nu = \gamma\hbar\mathcal{B}_0$.

☐ 3. The intensity of an NMR or EPR transition increases with the difference in population of α and β states and the strength of the applied magnetic field (as $\mathcal{B}_0^2$).

☐ 4. The chemical shift of a nucleus is the difference between its resonance frequency and that of a reference standard; chemical shifts are reported on the δ scale, in which $\delta = (\nu - \nu^\circ) \times 10^6/\nu^\circ$.

☐ 5. The fine structure of an NMR spectrum is the splitting of the groups of resonances into individual lines; the strength of the interaction is expressed in terms of the spin–spin coupling constant, *J*.

☐ 6. *N* equivalent spin-$\frac{1}{2}$ nuclei split the resonance of a nearby spin or group of equivalent spins into $N + 1$ lines with an intensity distribution given by Pascal's triangle.

☐ 7. Spin–spin coupling in molecules in solution can be explained in terms of the polarization mechanism, in which the interaction is transmitted through the bonds.

☐ 8. The Fermi contact interaction is a magnetic interaction that depends on the very close approach of an electron to the nucleus and can occur only if the electron occupies an s orbital.

☐ 9. Coalescence of the two lines occurs in conformational interchange or chemical exchange when the lifetime, τ, of the states is related to their resonance frequency difference, $\delta\nu$, by $\tau = 2^{1/2}/\pi\delta\nu$.

☐ 10. In Fourier-transform NMR, the spectrum is obtained by mathematical analysis of the free-induction decay of magnetization, the response of nuclear spins in a sample to the application of one or more pulses of radiofrequency radiation.

☐ 11. Spin relaxation is the nonradiative return of a spin system to an equilibrium distribution of populations in which the transverse spin orientations are random; the system returns exponentially to the equilibrium population distribution with a time constant called the spin–lattice relaxation time, T_1.

☐12. The spin–spin relaxation time, T_2, is the time constant for the exponential return of the system to random transverse spin orientations.

☐13. In proton decoupling of ^{13}C-NMR spectra, protons are made to undergo rapid spin reorientations and the ^{13}C nucleus senses an average orientation. As a result, its resonance is a single line and not a group of lines.

☐14. The nuclear Overhauser effect (NOE) is the modification of one resonance by the saturation of another.

☐15. In two-dimensional NMR, spectra are displayed in two axes, with resonances belonging to different groups lying at different locations on the second axis.

☐16. Magic-angle spinning (MAS) is a technique in which the NMR linewidths in a solid sample are reduced by spinning at an angle of 54.74° to the applied magnetic field.

☐17. The EPR resonance condition is written $h\nu = g\mu_B\mathcal{B}$, where g is the g-value of the radical; the deviation of g from $g_e = 2.0023$ depends on the ability of the applied field to induce local electron currents in the radical.

☐18. The hyperfine structure of an EPR spectrum is its splitting of individual resonance lines into components by the magnetic interaction between the electron and nuclei with spin.

☐19. If a radical contains N equivalent nuclei with spin quantum number I, then there are $2NI + 1$ hyperfine lines with an intensity distribution given by a modified version of Pascal's triangle.

☐20. The hyperfine structure due to a hydrogen attached to an aromatic ring is converted to spin density, ρ, on the neighbouring carbon atom by using the McConnell equation: $a = Q\rho$ with $Q = 2.25$ mT.

Discussion questions

12.1 To determine the structures of macromolecules by NMR spectroscopy, chemists use spectrometers that operate at the highest available fields and frequencies. Justify this choice.

12.2 Discuss in detail the origins of the local, neighbouring group, and solvent contributions to the shielding constant.

12.3 Discuss in detail the effects of a 90° pulse and of a 180° pulse on a system of spin-$\frac{1}{2}$ nuclei in a static magnetic field.

12.4 Suggest a reason why the relaxation times of ^{13}C nuclei are typically much longer than those of ^{1}H nuclei.

12.5 Suggest a reason why the spin–lattice relaxation time of a small molecule (like benzene) in a mobile, deuterated hydrocarbon solvent increases whereas that of a large molecule (like a polymer) decreases.

12.6 Discuss the origin of the nuclear Overhauser effect and how it can be used to measure distances between protons in a biopolymer.

12.7 Discuss the origins of diagonal and cross-peaks in the COSY spectrum of an AX system.

12.8 Discuss how the Fermi contact interaction and the polarization mechanism contribute to spin–spin couplings in NMR and hyperfine interactions in EPR.

12.9 Explain how the EPR spectrum of an organic radical can be used to identify and map the molecular orbital occupied by the unpaired electron.

Exercises

12.1(a) Given that g is a dimensionless number, what are the units of γ expressed in tesla and hertz?

12.1(b) Given that g is a dimensionless number, what are the units of γ expressed in SI base units?

12.2(a) For a proton, what are the magnitudes of the spin angular momentum and its allowed components along the z-axis? What are the possible orientations of the angular momentum in terms of the angle it makes with the z-axis?

12.2(b) For a ^{14}N nucleus, what are the magnitudes of the spin angular momentum and its allowed components along the z-axis? What are the possible orientations of the angular momentum in terms of the angle it makes with the z-axis?

12.3(a) What is the resonance frequency of a proton in a magnetic field of 13.5 T?

12.3(b) What is the resonance frequency of a ^{19}F nucleus in a magnetic field of 17.1 T?

12.4(a) ^{33}S has a nuclear spin of $\frac{3}{2}$ and a nuclear g-factor of 0.4289. Calculate the energies of the nuclear spin states in a magnetic field of 6.800 T.

12.4(b) ^{14}N has a nuclear spin of 1 and a nuclear g-factor of 0.404. Calculate the energies of the nuclear spin states in a magnetic field of 10.50 T.

12.5(a) Calculate the frequency separation of the nuclear spin levels of a ^{13}C nucleus in a magnetic field of 15.4 T given that the magnetogyric ratio is 6.73×10^{-7} T^{-1} s^{-1}.

12.5(b) Calculate the frequency separation of the nuclear spin levels of a ^{14}N nucleus in a magnetic field of 14.4 T given that the magnetogyric ratio is 1.93×10^{-7} T^{-1} s^{-1}.

12.6(a) In which of the following systems is the energy level separation larger: (a) a proton in a 600 MHz NMR spectrometer, (b) a deuteron in the same spectrometer?

12.6(b) In which of the following systems is the energy level separation larger: (a) a ^{14}N nucleus in (for protons) a 600 MHz NMR spectrometer, (b) an electron in a radical in a field of 0.300 T?

12.7(a) Calculate the magnetic field needed to satisfy the resonance condition for unshielded ^{11}B nuclei in a 90.0 MHz radiofrequency field.

12.7(b) Calculate the magnetic field needed to satisfy the resonance condition for unshielded protons in a 400.0 MHz radiofrequency field.

12.8(a) Use the information in Table 12.2 to predict the magnetic fields at which (a) ^1H, (b) ^2H, (c) ^{13}C come into resonance at (i) 500 MHz, (ii) 800 MHz.

12.8(b) Use the information in Table 12.2 to predict the magnetic fields at which (a) ^{14}N, (b) ^{19}F, and (c) ^{31}P come into resonance at (i) 400 MHz, (ii) 750 MHz.

12.9(a) Calculate the relative population differences ($\delta N/N$, where δN denotes a small difference $N_\alpha - N_\beta$) for protons in fields of (a) 0.30 T, (b) 1.5 T, and (c) 10 T at 25°C.

12.9(b) Calculate the relative population differences ($\delta N/N$, where δN denotes a small difference $N_\alpha - N_\beta$) for ^{13}C nuclei in fields of (a) 0.50 T, (b) 2.5 T, and (c) 15.5 T at 25°C.

12.10(a) The first generally available NMR spectrometers operated at a frequency of 60 MHz; today it is not uncommon to use a spectrometer that operates at 800 MHz. What are the relative population differences of ^{13}C spin states in these two spectrometers at 25°C?

12.10(b) What are the relative population differences of ^{19}F spin states in spectrometers operating at 60 MHz and 450 MHz at 25°C?

12.11(a) What are the relative values of the chemical shifts observed for nuclei in the spectrometers mentioned in Exercise 12.10a in terms of (a) δ values, (b) frequencies?

12.11(b) What are the relative values of the chemical shifts observed for nuclei in the spectrometers mentioned in Exercise 12.10b in terms of (a) δ values, (b) frequencies?

12.12(a) The chemical shift of the CH_3 protons in acetaldehyde (ethanal) is $\delta = 2.20$ and that of the CHO proton is 9.80. What is the difference in local magnetic field between the two regions of the molecule when the applied field is (a) 1.5 T, (b) 15 T?

12.12(b) The chemical shift of the CH_3 protons in diethyl ether is $\delta = 1.16$ and that of the CH_2 protons is 3.36. What is the difference in local magnetic field between the two regions of the molecule when the applied field is (a) 1.9 T, (b) 16.5 T?

12.13(a) Sketch the appearance of the ^1H-NMR spectrum of acetaldehyde (ethanal) using $J = 2.90$ Hz and the data in Exercise 12.12a in a spectrometer operating at (a) 250 MHz, (b) 800 MHz.

12.13(b) Sketch the appearance of the ^1H-NMR spectrum of diethyl ether using $J = 6.97$ Hz and the data in Exercise 12.12b in a spectrometer operating at (a) 400 MHz, (b) 650 MHz.

12.14(a) A proton jumps between two sites with $\delta = 2.7$ and $\delta = 4.8$. At what rate of interconversion will the two signals collapse to a single line in a spectrometer operating at 550 MHz?

12.14(b) A proton jumps between two sites with $\delta = 4.2$ and $\delta = 5.5$. At what rate of interconversion will the two signals collapse to a single line in a spectrometer operating at 350 MHz?

12.15(a) Sketch the form of the ^{19}F-NMR spectra of a natural sample of $^{10}BF_4^-$ and $^{11}BF_4^-$.

12.15(b) Sketch the form of the ^{31}P-NMR spectra of a sample of $^{31}PF_6^-$.

12.16(a) From the data in Table 12.2, predict the frequency needed for ^{19}F-NMR in an NMR spectrometer designed to observe proton resonance at 800 MHz. Sketch the proton and ^{19}F resonances in the NMR spectrum of FH_2^+.

12.16(b) From the data in Table 12.2, predict the frequency needed for ^{31}P-NMR in an NMR spectrometer designed to observe proton resonance at 500 MHz. Sketch the proton and ^{31}P resonances in the NMR spectrum of PH_4^+.

12.17(a) Sketch the form of an $A_3M_2X_4$ spectrum, where A, M, and X are protons with distinctly different chemical shifts and $J_{AM} > J_{AX} > J_{MX}$.

12.17(b) Sketch the form of an $A_2M_2X_5$ spectrum, where A, M, and X are protons with distinctly different chemical shifts and $J_{AM} > J_{AX} > J_{MX}$.

12.18(a) Which of the following molecules have sets of nuclei that are chemically but not magnetically equivalent: (a) CH_3CH_3, (b) $CH_2{=}CH_2$?

12.18(b) Which of the following molecules have sets of nuclei that are chemically but not magnetically equivalent: (a) $CH_2{=}C{=}CF_2$, (b) *cis*- and *trans*-$[Mo(CO)_4(PH_3)_2]$?

12.19(a) The duration of a 90° or 180° pulse depends on the strength of the $\mathcal{B}_1$ field. If a 90° pulse requires 10 μs, what is the strength of the $\mathcal{B}_1$ field? How long would the corresponding 180° pulse require?

12.19(b) The duration of a 90° or 180° pulse depends on the strength of the $\mathcal{B}_1$ field. If a 180° pulse requires 12.5 μs, what is the strength of the $\mathcal{B}_1$ field? How long would the corresponding 90° pulse require?

12.20(a) What magnetic field would be required in order to use an EPR X-band spectrometer (9 GHz) to observe ^1H-NMR and a 300 MHz spectrometer to observe EPR?

12.20(b) Some commercial EPR spectrometers use 8 mm microwave radiation (the Q band). What magnetic field is needed to satisfy the resonance condition?

12.21(a) The centre of the EPR spectrum of atomic hydrogen lies at 329.12 mT in a spectrometer operating at 9.2231 GHz. What is the *g*-value of the electron in the atom?

12.21(b) The centre of the EPR spectrum of atomic deuterium lies at 330.02 mT in a spectrometer operating at 9.2482 GHz. What is the *g*-value of the electron in the atom?

12.22(a) A radical containing two equivalent protons shows a three-line spectrum with an intensity distribution 1:2:1. The lines occur at 330.2 mT, 332.5 mT, and 334.8 mT. What is the hyperfine coupling constant for each proton? What is the *g*-value of the radical given that the spectrometer is operating at 9.319 GHz?

12.22(b) A radical containing three equivalent protons shows a four-line spectrum with an intensity distribution 1:3:3:1. The lines occur at 331.4 mT, 333.6 mT, 335.8 mT, and 338.0 mT. What is the hyperfine coupling constant for each proton? What is the *g*-value of the radical given that the spectrometer is operating at 9.332 GHz?

12.23(a) A radical containing two inequivalent protons with hyperfine constants 2.0 mT and 2.6 mT gives a spectrum centred on 332.5 mT. At what fields do the hyperfine lines occur and what are their relative intensities?

12.23(b) A radical containing three inequivalent protons with hyperfine constants 2.11 mT, 2.87 mT, and 2.89 mT gives a spectrum centred on 332.8 mT. At what fields do the hyperfine lines occur and what are their relative intensities?

12.24(a) Predict the intensity distribution in the hyperfine lines of the EPR spectra of (a) $\cdot CH_3$, (b) $\cdot CD_3$.

12.24(b) Predict the intensity distribution in the hyperfine lines of the EPR spectra of (a) $\cdot CH_2H_3$, (b) $\cdot CD_2CD_3$.

12.25(a) The benzene radical anion has $g = 2.0025$. At what field should you search for resonance in a spectrometer operating at (a) 9.313 GHz, (b) 33.80 GHz?

12.25(b) The naphthalene radical anion has $g = 2.0024$. At what field should you search for resonance in a spectrometer operating at (a) 9.501 GHz, (b) 34.77 GHz?

12.26(a) The EPR spectrum of a radical with a single magnetic nucleus is split into four lines of equal intensity. What is the nuclear spin of the nucleus?

12.26(b) The EPR spectrum of a radical with two equivalent nuclei of a particular kind is split into five lines of intensity ratio 1:2:3:2:1. What is the spin of the nuclei?

12.27(a) Sketch the form of the hyperfine structures of radicals XH_2 and XD_2, where the nucleus X has $I = \frac{5}{2}$.

12.27(b) Sketch the form of the hyperfine structures of radicals XH_3 and XD_3, where the nucleus X has $I = \frac{3}{2}$.

Problems*

Numerical problems

12.1‡ The relative sensitivity of NMR lines for equal numbers of different nuclei at constant temperature for a given frequency is $R_\nu \propto (I+1)\mu^3$ whereas for a given field is $R_{\mathcal{B}} \propto \{(I+1)/I^2\}\mu^3$. (a) From the data in Table 12.2, calculate these sensitivities for the deuteron, ^{13}C, ^{14}N, ^{19}F, and ^{31}P relative to the proton. (b) Derive the equation for $R_{\mathcal{B}}$ from the equation for R_ν.

12.2 Two groups of protons have $\delta = 4.0$ and $\delta = 5.2$ and are interconverted by a conformational change of a fluxional molecule. In a 60 MHz spectrometer the spectrum collapsed into a single line at 280 K but at 300 MHz the collapse did not occur until the temperature had been raised to 300 K. What is the activation energy of the interconversion?

12.3‡ Suppose that the FID in Fig. 12.32 was recorded in a 400 MHz spectrometer, and that the interval between maxima in the oscillations in the FID is 0.12 s. What are the Larmor frequency of the nuclei and the spin–spin relaxation time?

12.4 To gain some appreciation for the numerical work done by computers interfaced to NMR spectrometers, perform the following calculations. (a) The total FID $F(t)$ of a signal containing many frequencies, each corresponding to a different nucleus, is given by

$$F(t) = \sum_j S_{0j}\cos(2\pi\nu_{Lj}t)e^{-t/T_{2j}}$$

where, for each nucleus j, S_{0j} is the maximum intensity of the signal, ν_{Lj} is the Larmor frequency, and T_{2j} is the spin–spin relaxation time. Plot the FID for the case

$S_{01} = 1.0$ $\quad \nu_{L1} = 50$ MHz $\quad T_{21} = 0.50$ μs

$S_{02} = 3.0$ $\quad \nu_{L2} = 10$ MHz $\quad T_{22} = 1.0$ μs

(b) Explore how the shape of the FID curve changes with changes in the Larmor frequency and the spin–spin relaxation time. (c) Use mathematical software to calculate and plot the Fourier transforms of the FID curves you calculated in parts (a) and (b). How do spectral linewidths vary with the value of T_2? *Hint.* This operation can be performed with the 'fast Fourier transform' routine available in most mathematical software packages. Please consult the package's user manual for details.

12.5 (a) In many instances it is possible to approximate the NMR lineshape by using a *Lorenztian function* of the form

$$I_{\text{Lorenztian}}(\omega) = \frac{S_0T_2}{1 + T_2^2(\omega - \omega_0)^2}$$

where $I(\omega)$ is the intensity as a function of the angular frequency $\omega = 2\pi\nu$, ω_0 is the resonance frequency, S_0 is a constant, and T_2 is the spin–spin relaxation time. Confirm that for this lineshape the full-width at half-height is $2/T_2$. (b) Under certain circumstances, NMR lines are Gaussian functions of the frequency, given by

$$I_{\text{Gaussian}}(\omega) = S_0T_2e^{-T_2^2(\omega-\omega_0)^2}$$

Confirm that for the Gaussian lineshape the full-width at half-height is equal to $2(\ln 2)^{1/2}/T_2$. (c) Compare and contrast the shapes of Lorenztian and Gaussian lines by plotting two lines with the same values of S_0, T_2, and ω_0.

12.6‡ Various versions of the Karplus equation (eqn 12.27) have been used to correlate data on vicinal proton coupling constants in systems of the type $R_1R_2CHCHR_3R_4$. The original version, (M. Karplus, *J. Am. Chem. Soc.* **85**, 2870 (1963)), is $^3J_{HH} = A\cos^2\phi_{HH} + B$. When $R_3 = R_4 = H$, $^3J_{HH} = 7.3$ Hz; when $R_3 = CH_3$ and $R_4 = H$, $^3J_{HH} = 8.0$ Hz; when $R_3 = R_4 = CH_3$, $^3J_{HH} = 11.2$ Hz. Assume that only staggered conformations are important and determine which version of the Karplus equation fits the data better.

12.7‡ It might be unexpected that the Karplus equation, which was first derived for $^3J_{HH}$ coupling constants, should also apply to vicinal coupling between the nuclei of metals such as tin. T.N. Mitchell and B. Kowall (*Magn. Reson. Chem.* **33**, 325 (1995)) have studied the relation between $^3J_{HH}$ and $^3J_{SnSn}$ in compounds of the type $Me_3SnCH_2CHRSnMe_3$ and find that $^3J_{SnSn} = 78.86\,^3J_{HH} + 27.84$ Hz. (a) Does this result support a Karplus type equation for tin? Explain your reasoning. (b) Obtain the Karplus equation for $^3J_{SnSn}$ and plot it as a function of the dihedral angle. (c) Draw the preferred conformation.

12.8 Figure 12.59 shows the proton COSY spectrum of 1-nitropropane. Account for the appearance of off-diagonal peaks in the spectrum.

12.9 It is possible to produce very high magnetic fields over small volumes by special techniques. What would be the resonance frequency of an electron spin in an organic radical in a field of 1.0 kT? How does this frequency compare to typical molecular rotational, vibrational, and electronic energy-level separations?

* Problems denoted with the symbol ‡ were supplied by Charles Trapp and Carmen Giunta.

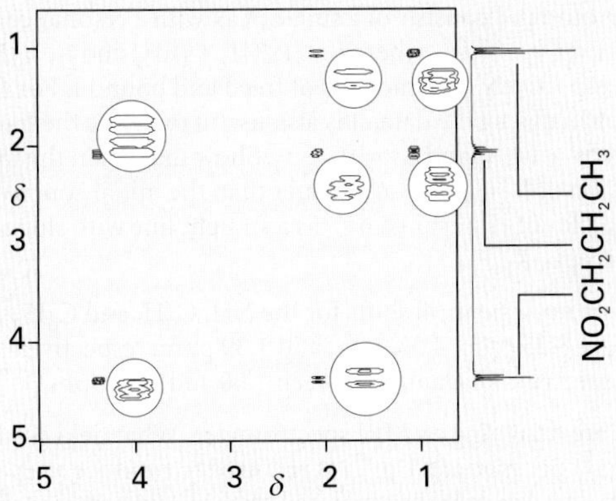

Fig. 12.59 The COSY spectrum of 1-nitropropane ($NO_2CH_2CH_2CH_3$). The circles show enhanced views of the spectral features. (Spectrum provided by Prof. G. Morris.)

12.10 The angular NO_2 molecule has a single unpaired electron and can be trapped in a solid matrix or prepared inside a nitrite crystal by radiation damage of NO_2^- ions. When the applied field is parallel to the OO direction the centre of the spectrum lies at 333.64 mT in a spectrometer operating at 9.302 GHz. When the field lies along the bisector of the ONO angle, the resonance lies at 331.94 mT. What are the *g*-values in the two orientations?

12.11 The hyperfine coupling constant in $\cdot CH_3$ is 2.3 mT. Use the information in Table 12.3 to predict the splitting between the hyperfine lines of the spectrum of $\cdot CD_3$. What are the overall widths of the hyperfine spectra in each case?

12.12 The *p*-dinitrobenzene radical anion can be prepared by reduction of *p*-dinitrobenzene. The radical anion has two equivalent N nuclei ($I = 1$) and four equivalent protons. Predict the form of the EPR spectrum using $a(N) = 0.148$ mT and $a(H) = 0.112$ mT.

12.13 When an electron occupies a 2s orbital on an N atom it has a hyperfine interaction of 55.2 mT with the nucleus. The spectrum of NO_2 shows an isotropic hyperfine interaction of 5.7 mT. For what proportion of its time is the unpaired electron of NO_2 occupying a 2s orbital? The hyperfine coupling constant for an electron in a 2p orbital of an N atom is 3.4 mT. In NO_2 the anisotropic part of the hyperfine coupling is 1.3 mT. What proportion of its time does the unpaired electron spend in the 2p orbital of the N atom in NO_2? What is the total probability that the electron will be found on (a) the N atoms, (b) the O atoms? What is the hybridization ratio of the N atom? Does the hybridization support the view that NO_2 is angular?

12.14 The hyperfine coupling constants observed in the radical anions (**12**), (**13**), and (**14**) are shown (in millitesla, mT). Use the value for the benzene radical anion to map the probability of finding the unpaired electron in the π orbital on each C atom.

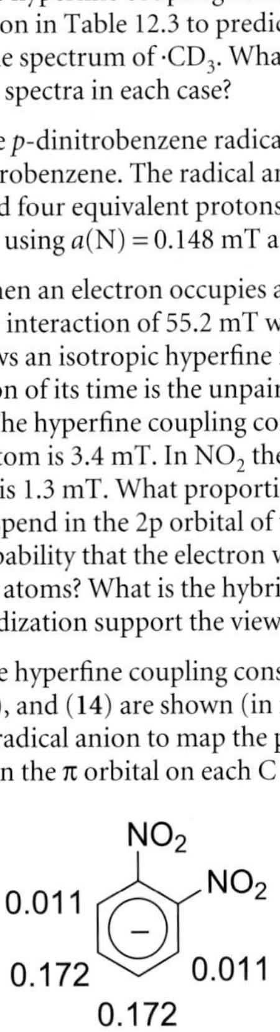

12

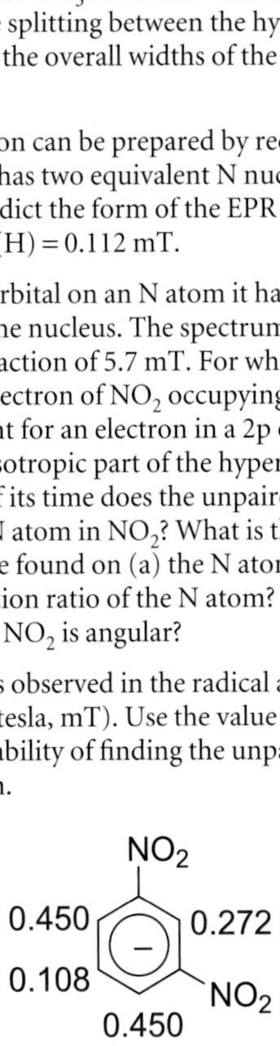

13

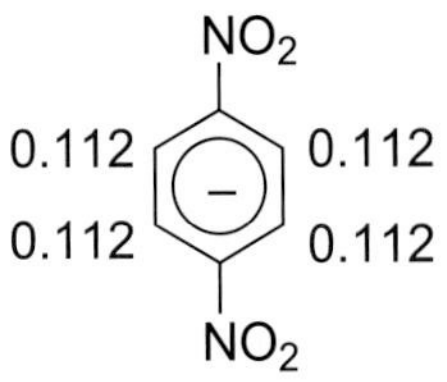

14

12.15 Sketch the EPR spectra of the di-*tert*-butyl nitroxide radical (**15**) at 292 K in the limits of very low concentration (at which electron exchange is negligible), moderate concentration (at which electron exchange effects begin to be observed), and high concentration (at which electron exchange effects predominate).

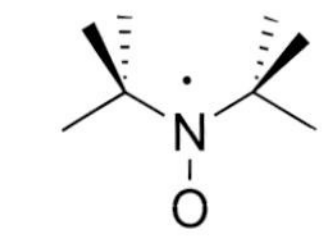

15 di-*tert*-Butyl nitroxide

Theoretical problems

12.16 Calculate σ_d for a hydrogenic atom with atomic number *Z*.

12.17 In Problem 6.7 you were asked to use molecular electronic structure methods to investigate the hypothesis that the magnitude of the ^{13}C chemical shift correlates with the net charge on a ^{13}C atom by calculating the net charge at the C atom *para* to the substituents in the following series of molecules: methylbenzene, benzene, trifluoromethylbenzene, benzonitrile, nitrobenzene. The ^{13}C chemical shifts of the *para* C atoms in each of the molecules that you examined are given below:

Substituent	CH_3	H	CF_3	CN	NO_2
δ	128.4	128.5	128.9	129.1	129.4

(a) Is there a linear correlation between net charge and ^{13}C chemical shift of the *para* C atom in this series of molecules? (b) If you did find a correlation in part (a), use the concepts developed in Chapter 12 to explain the physical origins of the correlation.

12.18 The *z*-component of the magnetic field at a distance *R* from a magnetic moment parallel to the *z*-axis is given by eqn 12.28. In a solid, a proton at a distance *R* from another can experience such a field and the measurement of the splitting it causes in the spectrum can be used to calculate *R*. In gypsum, for instance, the splitting in the H_2O resonance can be interpreted in terms of a magnetic field of 0.715 mT generated by one proton and experienced by the other. What is the separation of the protons in the H_2O molecule?

12.19 In a liquid, the dipolar magnetic field averages to zero: show this result by evaluating the average of the field given in eqn 12.28. *Hint.* The surface area element is $\sin\theta\, d\theta d\phi$ in polar coordinates.

12.20 In a liquid crystal (*Impact I8.3*) a molecule might not rotate freely in all directions and the dipolar interaction might not average to zero. Suppose a molecule is trapped so that, although the vector separating two protons may rotate freely around the *z*-axis, the colatitude may vary only between 0 and θ'. Use mathematical software to average the dipolar field over this restricted range of orientation and confirm that the average vanishes when θ' is equal to π (corresponding to free rotation over a sphere). What is the average value of the local dipolar field for the H_2O molecule in Problem 12.18 if it is dissolved in a liquid crystal that enables it to rotate up to $\theta' = 30°$?

12.21 The shape of a spectral line, $I(\omega)$, is related to the free induction decay signal $G(t)$ by

$$I(\omega) = a\,\mathrm{Re}\int_0^\infty G(t)\mathrm{e}^{\mathrm{i}\omega t}\mathrm{d}t$$

where a is a constant and 'Re' means take the real part of what follows. Calculate the lineshape corresponding to an oscillating, decaying function $G(t) = \cos \omega t\, \mathrm{e}^{-t/\tau}$.

12.22 In the language of Problem 12.21, show that, if $G(t) = (a \cos \omega_1 t + b \cos \omega_2 t)\mathrm{e}^{-t/\tau}$, then the spectrum consists of two lines with intensities proportional to a and b and located at $\omega = \omega_1$ and ω_2, respectively.

12.23 Show that the coupling constant as expressed by the Karplus equation passes through a minimum when $\cos \phi = B/4C$.

12.24 EPR spectra are commonly discussed in terms of the parameters that occur in the *spin-hamiltonian*, a hamiltonian operator that incorporates various effects involving spatial operators (like the orbital angular momentum) into operators that depend on the spin alone. Show that if you use $\hat{H} = -g_e\gamma_e\mathcal{B}_0\hat{s}_z - \gamma_e\mathcal{B}_0\hat{l}_z$ as the true hamiltonian, then, from second-order perturbation theory, the eigenvalues of the spin are the same as those of the spin-hamiltonian $\hat{H}_{\text{spin}} = -g\gamma_e\mathcal{B}_0\hat{s}_z$ (note the g in place of g_e) and find an expression for g.

Applications: to biochemistry and medicine

12.25 Interpret the following features of the NMR spectra of hen lysozyme: (a) saturation of a proton resonance assigned to the side chain of methionine-105 changes the intensities of proton resonances assigned to the side chains of tryptophan-28 and tyrosine-23; (b) saturation of proton resonances assigned to tryptophan-28 does not affect the spectrum of tyrosine-23.

12.26 NMR spectroscopy may be used to determine the equilibrium constant for dissociation of a complex between a small molecule, such as an enzyme inhibitor I, and a protein, such as an enzyme E:

$$\mathrm{EI} \rightleftharpoons \mathrm{E} + \mathrm{I} \qquad K_\mathrm{I} = [\mathrm{E}][\mathrm{I}]/[\mathrm{EI}]$$

In the limit of slow chemical exchange, the NMR spectrum of a proton in I would consist of two resonances: one at ν_I for free I and another at ν_EI for bound I. When chemical exchange is fast, the NMR spectrum of the same proton in I consists of a single peak with a resonance frequency ν given by $\nu = f_\mathrm{I}\nu_\mathrm{I} + f_\mathrm{EI}\nu_\mathrm{EI}$, where $f_\mathrm{I} = [\mathrm{I}]/([\mathrm{I}] + [\mathrm{EI}])$ and $f_\mathrm{EI} = [\mathrm{EI}]/([\mathrm{I}] + [\mathrm{EI}])$ are, respectively, the fractions of free I and bound I. For the purposes of analysing the data, it is also useful to define the frequency differences $\delta\nu = \nu - \nu_\mathrm{I}$ and $\Delta\nu = \nu_\mathrm{EI} - \nu_\mathrm{I}$. Show that when the initial concentration of I, $[\mathrm{I}]_0$, is much greater than the initial concentration of E, $[\mathrm{E}]_0$, a plot of $[\mathrm{I}]_0$ against $\delta\nu^{-1}$ is a straight line with slope $[\mathrm{E}]_0\Delta\nu$ and y-intercept $-K_\mathrm{I}$.

12.27 The proton chemical shifts for the NH, $C_\alpha H$, and $C_\beta H$ groups of alanine are 8.25 ppm, 4.35 ppm, and 1.39 ppm, respectively. Sketch the COSY spectrum of alanine between 1.00 and 8.50 ppm.

12.28 You are designing an MRI spectrometer. What field gradient (in microtesla per metre, $\mu\mathrm{T\,m^{-1}}$) is required to produce a separation of 100 Hz between two protons separated by the long diameter of a human kidney (taken as 8 cm) given that they are in environments with $\delta = 3.4$? The radiofrequency field of the spectrometer is at 400 MHz and the applied field is 9.4 T.

12.29 Suppose a uniform disc-shaped organ is in a linear field gradient, and that the MRI signal is proportional to the number of protons in a slice of width δx at each horizontal distance x from the centre of the disc. Sketch the shape of the absorption intensity for the MRI image of the disc before any computer manipulation has been carried out.

12.30 The computational techniques described in Chapter 6 may be used to predict the spin density distribution in a radical. Recent EPR studies have shown that the amino acid tyrosine participates in a number of biological electron transfer reactions, including the processes of water oxidation to O_2 in plant photosystem II and of O_2 reduction to water in cytochrome *c* oxidase (*Impact I17.3*). During the course of these electron transfer reactions, a tyrosine radical forms, with spin density delocalized over the side chain of the amino acid. (a) The phenoxy radical shown in (**16**) is a suitable model of the tyrosine radical. Using molecular modelling software and the computational method of your choice (semiempirical or *ab initio* methods), calculate the spin densities at the O atom and at all of the C atoms in (**16**). (b) Predict the form of the EPR spectrum of (**16**).

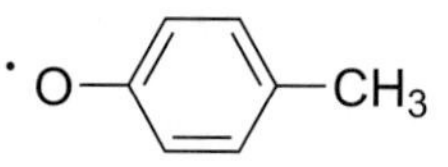

16 Phenoxy radical

PART 4
Molecular thermodynamics

One great river of physical chemistry consists of the concepts and applications of quantum theory that we have been considering in the preceding chapters. Another great river consists of the concepts and applications of thermodynamics, the theory of the transformation of energy into its different forms. Thermodynamics was originally formulated by considering the bulk properties of matter and finding relations—some of them unexpected—between various bulk properties. Thus, it was discovered, as we shall see in this part of the book, that observations on the heat output or requirements of chemical reactions could be used to predict their spontaneous direction and be used to calculate their equilibrium constants. However, although thermodynamics can be presented solely in terms of bulk properties, it is immeasurably enriched by drawing on what we have learned about the energy levels of atoms and molecules. That is the approach we adopt here, where we see how the quantum mechanical properties of matter presented so far, and in particular the energy levels available to molecules, underlie its thermodynamic properties.

The key concept in this account is the Boltzmann distribution, which we derive and describe in Chapter 13. This very simple, universal concept applies to all kinds of matter and all kinds of energy levels, and is the foundation both of structure (as we see in this part) and change (as we see in the following part). With the Boltzmann distribution established, we turn to the First Law of thermodynamics (essentially the conservation of energy) and then to the Second Law (essentially the increase in entropy), and include a short discussion of the Third Law, which is the concept that makes it possible to use calorimetric measurements to calculate equilibrium constants. In all three cases, we show how these phenomenological laws (laws based on observation) are illuminated by considering the distribution of molecules over their available energy levels.

With the laws established, we apply them to two principal problems. The first is physical equilibrium, where matter retains its identity but undergoes changes of state. The second, the culmination of this part, is chemical equilibrium and electrochemistry, where matter changes from one form to another.

The Boltzmann distribution

13

The material in this chapter provides the link between the microscopic properties of matter and its bulk properties. Two key ideas are introduced. The first is the Boltzmann distribution, which is used to predict the populations of states in systems at thermal equilibrium. In this chapter we see its derivation in terms of the distribution of particles over available states. The derivation leads naturally to the introduction of the partition function. We see how to interpret the partition function and how to calculate it in a number of simple cases and use it to calculate the mean energy of each mode of motion of a molecule. In the final part of the chapter, we generalize the discussion to include systems that are composed of assemblies of interacting particles. Very similar equations are developed to those in the first part of the chapter, but they are much more widely applicable.

The crucial step in going from the quantum mechanics of individual molecules to the properties of bulk samples, the province of **statistical thermodynamics**, is to recognize that the latter deals with the *average* behaviour of large numbers of molecules. For example, the pressure of a gas depends on the average force exerted by its molecules and there is no need to specify which molecules happen to be striking the wall at any instant. Nor is it necessary to consider the fluctuations in the pressure as different numbers of molecules collide with the wall at different moments. The fluctuations in pressure are very small compared with the steady pressure: it is highly improbable that there will be a sudden lull in the number of collisions, or a sudden surge. Fluctuations in other bulk properties also occur, but for large numbers of particles they are very much smaller than the mean values.

It will be helpful to keep in mind the analogies between the material of this chapter and that in the preceding chapters. In earlier chapters we have seen that the wavefunction contains all the *dynamical* information about an atom or molecule. In this chapter we introduce its analogue, the 'partition function', which contains all the *thermodynamic* information about a bulk sample of matter. In preceding chapters we have seen how observables are related to expectation values and calculated from wavefunctions; here we shall see how the thermodynamic properties introduced in the following chapters are related to average values and calculated from partition functions.

A brief comment This chapter introduces a lot of symbols, often with subtly different meanings. For your convenience, they are collected together in the table at the end of the chapter.

The distribution of molecular states

We consider a system composed of N molecules. Although the total energy is constant at E, it is not possible to be definite about how that energy is shared between the molecules. Collisions result in the ceaseless redistribution of energy not only between the molecules but also between the quantum states that each molecule occupies. The closest we can come to a description of the distribution of energy is to report the **population** of a state, the average number of molecules that occupy it, and to say that on average there are N_i molecules in a state of energy ε_i. The populations of the states remain almost constant, but the precise identities of the molecules in each state may change at every collision.

The problem we address in this section is the calculation of the populations of states for any type of molecule in any mode of motion at any temperature. The only restriction is that the molecules should be independent, in the sense that the total energy of the system is a sum of their individual energies. We are discounting (at this stage) the possibility that in a real system a contribution to the total energy may arise from interactions between molecules. We also adopt the **principle of equal *a priori* probabilities**, the assumption that all possibilities for the distribution of energy are equally probable. '*A priori*' means in this context loosely 'as far as one knows'. We have no reason to presume otherwise than that, for a collection of molecules at thermal equilibrium, vibrational states of a certain energy, for instance, are as likely to be populated as rotational states of the same energy.

One very important conclusion that will emerge from the following analysis is that the overwhelmingly most probable populations of the available states depend on a single parameter, the 'temperature'. That is, the work we do here provides a molecular justification for the concept of temperature and some insight into this crucially important quantity.

13.1 Configurations and weights

Any individual molecule may exist in states with energies ε_0, ε_1, ... For reasons that will become clear, we shall always take the lowest available state as the zero of energy (that is, we set $\varepsilon_0 = 0$), and measure all other energies relative to that state. To obtain the actual energy of the system we may have to add a constant to the energy calculated on this basis. For example, if we are considering the vibrational contribution to the energy, then we must add the total zero-point energy of any oscillators in the system.

(a) Instantaneous configurations

At any instant there will be N_0 molecules in the state with energy ε_0, N_1 with ε_1, and so on, with $N_0 + N_1 + \cdots = N$, the total number of molecules in the system. The specification of the set of populations $N_0, N_1, \ldots$ in the form $\{N_0, N_1, \ldots\}$ is a statement of the instantaneous **configuration** of the system. The instantaneous configuration fluctuates with time because the populations change. We can picture a large number of different instantaneous configurations. One, for example, might be $\{N, 0, 0, \ldots\}$, corresponding to every molecule being in the ground state. Another might be $\{N-2, 2, 0, 0, \ldots\}$, in which two molecules are in the first excited state. The latter configuration is intrinsically more likely to be found than the former because it can be achieved in more ways: $\{N, 0, 0, \ldots\}$ can be achieved in only one way, but $\{N-2, 2, 0, \ldots\}$ can be achieved in $\frac{1}{2}N(N-1)$ different ways (Fig. 13.1; see *Justification 13.1*). At this stage in the argument, we are ignoring the requirement that the total energy of the system should be constant (the second configuration has a higher energy than the first). The constraint of total energy is imposed later in this section.

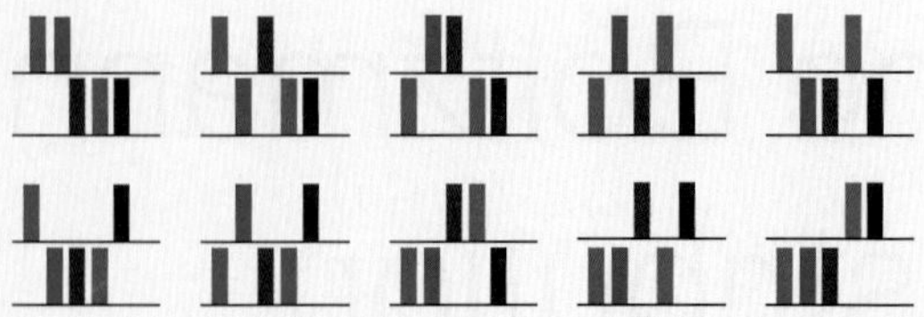

Fig. 13.1 Whereas a configuration $\{5, 0, 0, \ldots\}$ can be achieved in only one way, a configuration $\{3, 2, 0, \ldots\}$ can be achieved in the ten different ways shown here, where the tinted blocks represent different molecules.

If, as a result of collisions, the system were to fluctuate between the configurations $\{N, 0, 0, \ldots\}$ and $\{N-2, 2, 0, \ldots\}$, it would almost always be found in the second, more likely configuration (especially if N were large). In other words, a system free to switch between the two configurations would show properties characteristic almost exclusively of the second configuration. A general configuration $\{N_0, N_1, \ldots\}$ can be achieved in $\mathcal{W}$ different ways, where $\mathcal{W}$ is called the **weight** of the configuration. The weight of the configuration $\{N_0, N_1, \ldots\}$ is given by the expression

$$\mathcal{W} = \frac{N!}{N_0!N_1!N_2!\cdots} \tag{13.1}$$

with $x! = x(x-1)\ldots 1$ and by definition $0! = 1$. Equation 13.1 is a generalization of the formula $\mathcal{W} = \frac{1}{2}N(N-1)$, and reduces to it for the configuration $\{N-2, 2, 0, \ldots\}$.

● A BRIEF ILLUSTRATION

To calculate the number of ways of distributing 20 identical objects with the arrangement 1, 0, 3, 5, 10, 1, we note that the configuration is $\{1, 0, 3, 5, 10, 1\}$ with $N = 20$; therefore the weight is

$$\mathcal{W} = \frac{20!}{1!0!3!5!10!1!} = 9.31 \times 10^8$$ ●

Self-test 13.1 Calculate the weight of the configuration in which 20 objects are distributed in the arrangement 0, 1, 5, 0, 8, 0, 3, 2, 0, 1. [4.19×10^{10}]

Justification 13.1 *The weight of a configuration*

First, consider the weight of the configuration $\{N-2, 2, 0, 0, \ldots\}$. One candidate for promotion to an upper state can be selected in N ways. There are $N-1$ candidates for the second choice, so the total number of choices is $N(N-1)$. However, we should not distinguish the choice (Jack, Jill) from the choice (Jill, Jack) because they lead to the same configurations. Therefore, only half the choices lead to distinguishable configurations, and the total number of distinguishable choices is $\frac{1}{2}N(N-1)$.

Now we generalize this remark. Consider the number of ways of distributing N balls into bins. The first ball can be selected in N different ways, the next ball in $N-1$ different ways for the balls remaining, and so on. Therefore, there are $N(N-1)\ldots 1 = N!$ ways of selecting the balls for distribution over the bins. However, if there are N_0 balls in the bin labelled ε_0, there would be $N_0!$ different ways in which the same balls could have been chosen (Fig. 13.2). Similarly, there are $N_1!$ ways in which the N_1 balls in the bin labelled ε_1 can be chosen, and so on. Therefore, the total number of distinguishable ways of distributing the balls so that there are N_0 in bin ε_0, N_1 in bin ε_1, etc. regardless of the order in which the balls were chosen is $N!/N_0!N_1!\ldots$, which is the content of eqn 13.1.

It will turn out to be more convenient to deal with the natural logarithm of the weight, ln $\mathcal{W}$, rather than with the weight itself. We shall therefore need the expression

$$\begin{aligned}\ln \mathcal{W} &= \ln \frac{N!}{N_0!N_1!N_2!\cdots} = \ln N! - \ln(N_0!N_1!N_2!\cdots) \\ &= \ln N! - (\ln N_0! + \ln N_1! + \ln N_2! + \cdots) \\ &= \ln N! - \sum_i \ln N_i!\end{aligned}$$

where in the first line we have used $\ln(x/y) = \ln x - \ln y$ and in the second $\ln xy = \ln x + \ln y$. One reason for introducing ln $\mathcal{W}$ is that it is easier to make approximations. In particular, we can simplify the factorials by using *Stirling's approximation* in the form

$$\ln x! \approx x \ln x - x \qquad (13.2)$$

Then the approximate expression for the weight is

$$\begin{aligned}\ln \mathcal{W} &= (N \ln N - N) - \sum_i (N_i \ln N_i - N_i) \\ &= N \ln N - \sum_i N_i \ln N_i\end{aligned} \qquad (13.3)$$

Fig. 13.2 The 18 molecules shown here can be distributed into four receptacles (distinguished by the three vertical lines) in 18! different ways. However, 3! of the selections that put three molecules in the first receptacle are equivalent, 6! that put six molecules into the second receptacle are equivalent, and so on. Hence the number of distinguishable arrangements is 18!/3!6!5!4!.

The final form of eqn 13.3 is derived by noting that the sum of N_i is equal to N, so the second and fourth terms on the right in the first line cancel.

A brief comment A more accurate form of Stirling's approximation is

$$x! \approx (2\pi)^{1/2} x^{x+\frac{1}{2}} e^{-x}$$

and is in error by less than 1 per cent when x is greater than about 10. We deal with far larger values of x, and the simplified version in eqn 13.2 is adequate.

(b) The most probable distribution

We have seen that the configuration $\{N-2, 2, 0, \ldots\}$ dominates $\{N, 0, 0, \ldots\}$, and it should be easy to believe that there may be other configurations that have a much greater weight than both. We shall see, in fact, that there is a configuration with so great a weight that it overwhelms all the rest in importance to such an extent that the system will almost always be found in it. The properties of the system will therefore be characteristic of that particular dominating configuration. This dominating configuration can be found by looking for the values of N_i that lead to a maximum value of $\mathcal{W}$. Because $\mathcal{W}$ is a function of all the N_i, we can do this search by varying the N_i and looking for the values that correspond to $d\mathcal{W} = 0$ (just as in the search for the maximum of any function), or equivalently a maximum value of ln $\mathcal{W}$. However, there are two difficulties with this procedure.

The first difficulty is that the only permitted configurations are those corresponding to the specified, constant, total energy of the system. This requirement rules out many configurations; $\{N, 0, 0, \ldots\}$ and $\{N-2, 2, 0, \ldots\}$, for instance, have different energies (unless ε_0 and ε_1 are degenerate), so both cannot occur in the same isolated system. It follows that, in looking for the configuration with the greatest weight, we must ensure that the configuration also satisfies the condition

$$\text{Constant total energy: } \sum_i N_i \varepsilon_i = E \tag{13.4}$$

where E is the total energy of the system.

The second constraint is that, because the total number of molecules present is also fixed (at N), we cannot arbitrarily vary all the populations simultaneously. Thus, increasing the population of one state by 1 demands that the population of another state must be reduced by 1. Therefore, the search for the maximum value of $\mathcal{W}$ is also subject to the condition

$$\text{Constant total number of molecules: } \sum_i N_i = N \tag{13.5}$$

We show in *Further information 13.1* that the populations in the configuration of greatest weight, subject to the two constraints in eqns 13.4 and 13.5, depend on the energy of the state according to the **Boltzmann distribution:**

$$\frac{N_i}{N} = \frac{e^{-\beta\varepsilon_i}}{\sum_i e^{-\beta\varepsilon_i}} \tag{13.6a}$$

Equation 13.6a is the justification of the remark that a single parameter, here denoted β, determines the most probable populations of the states of the system. We shall confirm in Section 15.9 (but see the comment below) and anticipate throughout this chapter that

$$\beta = \frac{1}{kT} \tag{13.6b}$$

where T is the thermodynamic temperature and k is Boltzmann's constant. In other words, *the temperature is the unique parameter that governs the most probable populations of states of a system at thermal equilibrium.*

If we are interested only in the relative populations of states, the sum in the denominator of the Boltzmann distribution need not be evaluated, because it cancels when the ratio is taken:

$$\frac{N_i}{N_j} = \frac{e^{-\beta\varepsilon_i}}{e^{-\beta\varepsilon_j}} = e^{-\beta(\varepsilon_i - \varepsilon_j)} \tag{13.7}$$

This simple expression is enormously important for understanding a wide range of chemical phenomena and is the form in which the Boltzmann distribution is commonly employed (we saw one application of it in the discussion of the intensities of spectral transitions in Section 10.5). It tells us that the relative population of two states falls off exponentially with their difference in energy.

A brief comment Two chapters might seem a long time to take on trust the fact that $\beta = 1/kT$. That $\beta \propto 1/T$ is plausible is demonstrated by noting from eqn 13.7 that for a given energy separation the ratio of populations N_1/N_0 decreases as β increases, which is what is expected as the temperature decreases. At $T = 0$ ($\beta = \infty$) all the population is in the ground state and the ratio is zero.

Example 13.1 *Calculating the relative populations of rotational states*

Calculate the relative populations of the $J = 1$ and $J = 0$ rotational states of HCl at 25°C.

Method There is a trap for the unwary here: although the ground state is non-degenerate, the level with $J = 1$ is triply degenerate ($M_J = 0, \pm 1$) and, because eqn 13.7 gives the relative populations of *states*, to get the relative populations of *levels*, as distinct from states, we need to multiply the ratio by 3 when three states are equally occupied. The energy of a state with quantum number J is $hc\tilde{B}J(J+1)$. Use $\tilde{B} = 10.591\ \text{cm}^{-1}$. A useful relation is $kT/hc = 207.224\ \text{cm}^{-1}$ at 298.15 K.

Answer The energy separation of states with $J = 1$ and $J = 0$ is

$$\varepsilon_1 - \varepsilon_0 = 2hc\tilde{B}$$

The ratio of the population of a state with $J = 1$ and any one of its three states M_J to the population of the single state with $J = 0$ is therefore

$$\frac{N_{1,M_J}}{N_0} = e^{-2hc\tilde{B}\beta}$$

and the relative populations of the *levels* is

$$\frac{N_1}{N_0} = 3e^{-2hc\tilde{B}\beta}$$

because there are three states with $J = 1$. Insertion of $hc\tilde{B}\beta = 0.05111$ then gives

$$\frac{N_1}{N_0} = 3e^{-2\times 0.05111} = 2.708$$

We see that, because the $J = 1$ level is degenerate, it has a higher population than the level with $J = 0$, despite being of higher energy.

Self-test 13.2 What is the ratio of the populations of the levels with $J = 2$ and $J = 1$ at the same temperature? [1.359]

A note on good practice As the example illustrates, it is very important to take note of whether you are asked for the relative populations of individual states or of a (possibly degenerate) energy level.

13.2 The molecular partition function

The Boltzmann distribution contains a much richer variety of information than the relative populations of states. We can start to extract this information by writing the distribution as

$$p_i = \frac{e^{-\beta\varepsilon_i}}{q} \tag{13.8}$$

where p_i is the fraction of molecules in the state i, $p_i = N_i/N$, and q is the **molecular partition function**:

$$q = \sum_{\text{states } i} e^{-\beta\varepsilon_i} \qquad [13.9a]$$

As we have already indicated, it may happen that several states have the same energy, and so give the same contribution to this sum. If, for example, g_i states have the same energy ε_i (so the level is g_i-fold degenerate), we could write

$$q = \sum_{\text{levels } i} g_i e^{-\beta\varepsilon_i} \tag{13.9b}$$

where the sum is now over energy levels (sets of states with the same energy), not individual states.

Example 13.2 *Writing a partition function*

Write an expression for the partition function of a linear molecule (such as HCl) treated as a rigid rotor.

Method To use eqn 13.9 we need to know (a) the energies of the levels, (b) the degeneracies, the number of states that belong to each level. Whenever calculating a partition function, we express the energies of the levels relative to 0 for the state of lowest energy. The energy levels of a rigid linear rotor were derived in Section 10.2.

Answer The energy levels of a linear rotor are $hc\tilde{B}J(J+1)$, with $J = 0, 1, 2, \ldots$. The state of lowest energy has zero energy, so no adjustment need be made to the energies given by this expression. Each level consists of $2J+1$ degenerate states. Therefore,

$$q = \sum_{J=0}^{\infty} \overbrace{(2J+1)}^{g_J} e^{-\beta \overbrace{hc\tilde{B}J(J+1)}^{\varepsilon_J}}$$

The sum can be evaluated numerically by supplying the value of $\tilde{B}$ (from spectroscopy or calculation) and the temperature: see Example 13.5 below for an explicit illustration. For reasons explained in Section 13.3, this expression applies only to unsymmetrical linear rotors (for instance, HCl, not CO_2).

Self-test 13.3 Write the partition function for a two-level system, the lower state (at energy 0) being non-degenerate, and the upper state (at an energy ε) doubly degenerate.

[$q = 1 + 2e^{-\beta\varepsilon}$]

Some insight into the significance of a partition function can be obtained by considering how q depends on the temperature. When T is close to zero, the parameter $\beta = 1/kT$ is close to infinity. Then every term except one in the sum defining q is zero because each one has the form e^{-x} with $x \to \infty$. The exception is the term with $\varepsilon_0 \equiv 0$ (or the g_0 terms at zero energy if the ground state is g_0-fold degenerate), because then $\varepsilon_0/kT \equiv 0$ whatever the temperature, including zero. As there is only one surviving term when $T = 0$, and its value is g_0, it follows that

$$\lim_{T\to 0} q = g_0$$

That is, at $T = 0$, the partition function is equal to the degeneracy of the ground state.

Now consider the case when T is so high that for each term in the sum $\varepsilon_j/kT \approx 0$. Because $e^{-x} = 1$ when $x = 0$, each term in the sum now contributes 1. It follows that the sum is equal to the number of molecular states, which in general is infinite:

$$\lim_{T\to\infty} q = \infty$$

In some idealized cases, the molecule may have only a finite number of states; then the upper limit of q is equal to the number of states. For example, if we were considering only the spin energy levels of a radical in a magnetic field, then there would be only two states ($m_s = \pm\frac{1}{2}$). The partition function for such a system can therefore be expected to rise towards 2 as T is increased towards infinity.

We see that *the molecular partition function gives an indication of the number of states that are thermally accessible to a molecule at the temperature of the system*. At $T = 0$, only the ground level is accessible and $q = g_0$. At very high temperatures, virtually all states are accessible, and q is correspondingly large.

Example 13.3 *Evaluating the partition function for a harmonic oscillator*

Evaluate the partition function for a harmonic oscillator.

Method As we saw in Section 2.4, a harmonic oscillator has an infinite number of equally spaced non-degenerate energy levels (Fig. 13.3). Therefore, we expect the partition function to increase from 1 at $T = 0$ and approach infinity as T goes to ∞. Take the separation of energy levels to be $\varepsilon = h\nu$, so the individual levels lie at $0, \varepsilon, 2\varepsilon, \ldots$ relative to the ground state. To evaluate eqn 13.9 explicitly, note that

$$1 + x + x^2 + \cdots = \frac{1}{1-x}$$

A brief comment The sum of the infinite series $S = 1 + x + x^2 + \cdots$ is obtained by multiplying both sides by x, which gives $xS = x + x^2 + x^3 + \cdots = S - 1$ and hence $S = 1/(1-x)$.

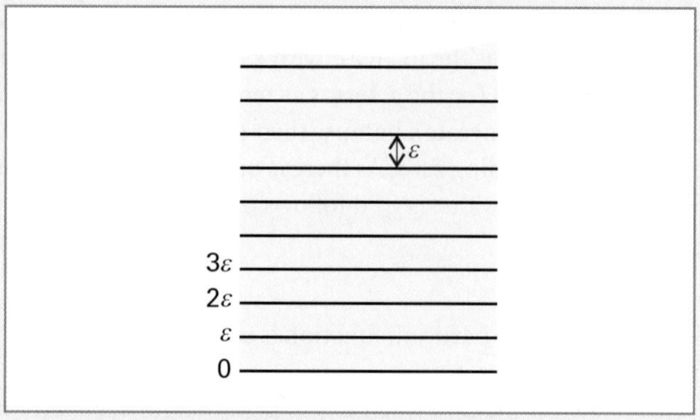

Fig. 13.3 The equally spaced infinite array of energy levels used in the calculation of the partition function. A harmonic oscillator has the same spectrum of levels.

Answer The partition function is

$$q = 1 + e^{-\beta\varepsilon} + e^{-2\beta\varepsilon} + \cdots$$
$$= 1 + e^{-\beta\varepsilon} + (e^{-\beta\varepsilon})^2 + \cdots$$
$$= \frac{1}{1 - e^{-\beta\varepsilon}}$$

This expression is plotted in Fig. 13.4: notice that, as anticipated, q rises from 1 to infinity as the temperature is raised.

Self-test 13.4 Find and plot an expression for the partition function of a system with one state at zero energy and another state at the energy ε. [$q = 1 + e^{-\beta\varepsilon}$, Fig. 13.5]

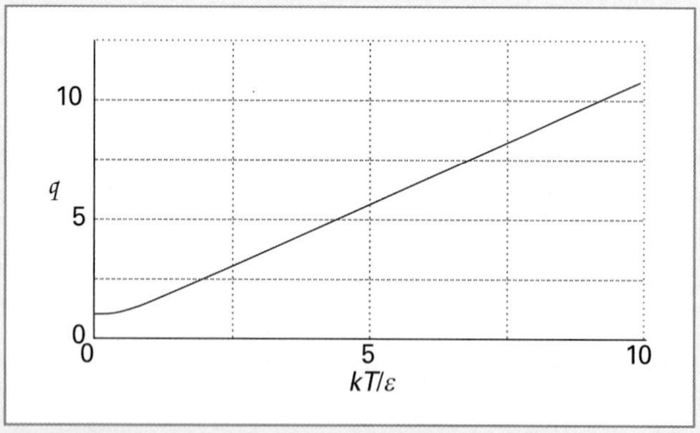

Fig. 13.4 The partition function for the system shown in Fig.13.3 (a harmonic oscillator) as a function of temperature.

interActivity Plot the partition function of a harmonic oscillator against temperature for several values of the energy separation ε. How does q vary with temperature when T is high, in the sense that $kT \gg \varepsilon$ (or $\beta\varepsilon \ll 1$)?

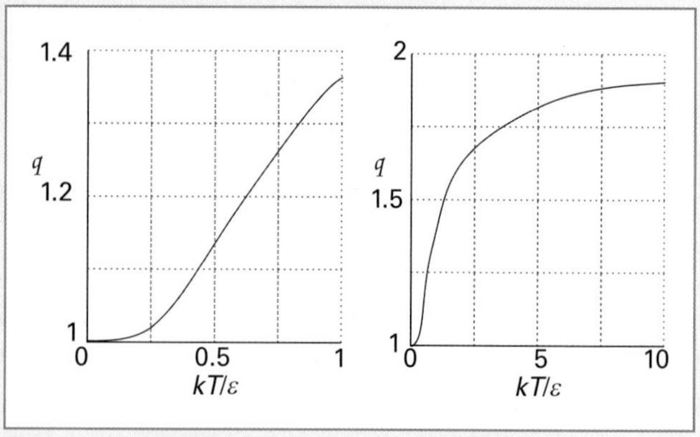

Fig. 13.5 The partition function for a two-level system as a function of temperature. The two graphs differ in the scale of the temperature axis to show the approach to 1 as $T \to 0$ and the slow approach to 2 as $T \to \infty$.

interActivity Consider a three-level system with levels 0, ε, and 2ε. Plot the partition function against kT/ε.

It follows from eqn 13.9 and the expression for q derived in Example 13.3 for the uniform ladder of states of spacing ε characteristic of a harmonic oscillator that

$$q = \frac{1}{1 - e^{-\beta\varepsilon}} \tag{13.10}$$

We can conclude that the fraction of molecules in the state with energy ε_i is

$$p_i = \frac{e^{-\beta\varepsilon_i}}{q} = (1 - e^{-\beta\varepsilon})e^{-\beta\varepsilon_i} \tag{13.11}$$

Figure 13.6 shows how p_i varies with temperature. At very low temperatures, where q is close to 1, only the lowest state is significantly populated. As the temperature is raised, the population breaks out of the lowest state, and the upper states become progressively more highly populated. At the same time, the partition function rises from 1 and its value gives an indication of the range of states populated. The name 'partition function' reflects the sense in which q measures how the total number of molecules is distributed—partitioned—over the available states.

The corresponding expressions for a two-level system derived in Self-test 13.4 are

$$p_0 = \frac{1}{1 + e^{-\beta\varepsilon}} \qquad p_1 = \frac{e^{-\beta\varepsilon}}{1 + e^{-\beta\varepsilon}} \tag{13.12}$$

These functions are plotted in Fig. 13.7. Notice how the populations tend towards equality ($p_0 = \frac{1}{2}$, $p_1 = \frac{1}{2}$) as $T \to \infty$. A common error is to suppose that when $T = \infty$ all the molecules in the

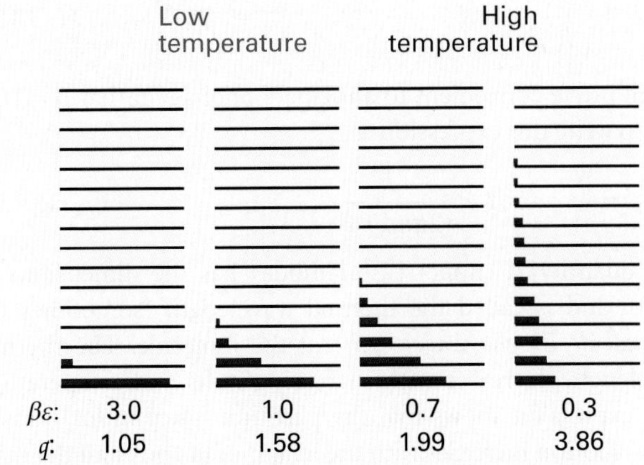

Fig. 13.6 The populations of the energy levels of the system shown in Fig. 13.3 at different temperatures, and the corresponding values of the partition function calculated in Example 13.3. Note that $\beta = 1/kT$.

interActivity To visualize the content of Fig. 13.6 in a different way, plot the functions p_0, p_1, p_2, and p_3 against kT/ε.

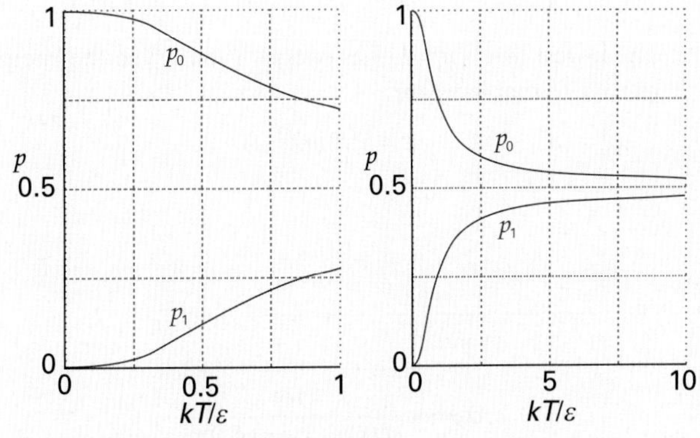

Fig. 13.7 The fraction of populations of the two states of a two-level system as a function of temperature (eqn 13.12). Note that as the temperature approaches infinity, the populations of the two states become equal (and the fractions both approach 0.5).

interActivity Consider a three-level system with levels 0, ε, and 2ε. Plot the functions p_0, p_1, and p_2 against kT/ε.

system will be found in the upper energy state; however, we see from eqn 13.12 that, as $T \to \infty$, the populations of states become equal. The same conclusion is true of multilevel systems too: as $T \to \infty$, all states become equally populated.

Example 13.4 *Using the partition function to calculate a population*

Calculate the proportion of I_2 molecules in their ground, first excited, and second excited vibrational states at 25°C. The vibrational wavenumber is 214.6 cm^{-1}.

Method Vibrational energy levels have a constant separation (in the harmonic approximation, Section 2.4), so the partition function is given by eqn 13.10 and the populations by eqn 13.11. To use the latter equation, we identify the index i with the quantum number v, and calculate p_v for $v = 0$, 1, and 2. At 298.15 K, $kT/hc = 207.224\ cm^{-1}$.

Answer First, we note that

$$\beta\varepsilon = \frac{hc\tilde{\nu}}{kT} = \frac{214.6\ cm^{-1}}{207.224\ cm^{-1}} = 1.036$$

Then it follows from eqn 13.11 that the populations are

$$p_v = (1 - e^{-\beta\varepsilon})e^{-v\beta\varepsilon} = 0.645e^{-1.036v}$$

Therefore, $p_0 = 0.645$, $p_1 = 0.229$, $p_2 = 0.081$. The I—I bond is not stiff and the atoms are heavy: as a result, the vibrational energy separations are small and at room temperature several vibrational levels are significantly populated, as may be seen in the electronic absorption spectrum, where transitions are observed that originate from several vibrational levels. The value of the partition function, $q = 1.55$, reflects this small but significant spread of populations.

Self-test 13.5 At what temperature would the $v = 1$ level of I_2 have (a) half the population of the ground state, (b) the same population as the ground state? [(a) 445 K, (b) infinite]

13.3 Contributions to the molecular partition function

The energy of a molecule is the sum of contributions from its different modes of motion:

$$\varepsilon_i = \varepsilon_i^{T} + \varepsilon_i^{R} + \varepsilon_i^{V} + \varepsilon_i^{E} \qquad (13.13)$$

where T denotes translation, R rotation, V vibration, and E the electronic contribution. The electronic contribution is not actually a 'mode of motion', but it is convenient to include it here. The separation of terms in eqn 13.13 is only approximate (except for translation) because the modes are not completely independent, but in most cases it is satisfactory. The separation of the electronic and vibrational motions is justified provided only the ground electronic state is occupied (for otherwise the vibrational characteristics depend on the electronic state) and, for the electronic ground state, that the Born–Oppenheimer approximation is valid (Chapter 5). The separation of the vibrational and rotational modes is justified to the extent that the rotational constant is independent of the vibrational state.

Given that the energy is a sum of independent contributions, the partition function factorizes into a product of contributions:

$$q = \sum_i e^{-\beta\varepsilon_i} = \sum_{i\,\text{(all states)}} e^{-\beta\varepsilon_i^{T} - \beta\varepsilon_i^{R} - \beta\varepsilon_i^{V} - \beta\varepsilon_i^{E}} \tag{13.14}$$

$$= \sum_{i\,\text{(translational)}} \sum_{i\,\text{(rotational)}} \sum_{i\,\text{(vibrational)}} \sum_{i\,\text{(electronic)}} e^{-\beta\varepsilon_i^{T} - \beta\varepsilon_i^{R} - \beta\varepsilon_i^{V} - \beta\varepsilon_i^{E}}$$

$$= \left(\sum_{i\,\text{(translational)}} e^{-\beta\varepsilon_i^{T}}\right)\left(\sum_{i\,\text{(rotational)}} e^{-\beta\varepsilon_i^{R}}\right)\left(\sum_{i\,\text{(vibrational)}} e^{-\beta\varepsilon_i^{V}}\right)\left(\sum_{i\,\text{(electronic)}} e^{-\beta\varepsilon_i^{E}}\right)$$

$$= q^{T}q^{R}q^{V}q^{E}$$

This factorization means that we can investigate each contribution separately. In general, exact analytical expressions for partition functions cannot be obtained. However, closed approximate expressions can often be found and prove to be very important for understanding chemical phenomena; they are derived in the following sections and collected in Table 13.1.

(a) The translational contribution

The translational partition function for a particle of mass m free to move in a one-dimensional container of length X can be evaluated by making use of the fact that the separation of energy levels is very small and that large numbers of states are accessible at normal temperatures. As shown in the *Justification* below, in this case

$$q_X^{T} = \left(\frac{2\pi m}{h^2\beta}\right)^{1/2} X \tag{13.15a}$$

It will prove convenient to anticipate once again that $\beta = 1/kT$ and to write this expression as

$$q_X^{T} = \frac{X}{\Lambda} \qquad \Lambda = \frac{h}{(2\pi mkT)^{1/2}} \tag{13.15b}$$

The quantity Λ (upper-case lambda) has the dimensions of length and is called the **thermal wavelength** (sometimes the *thermal de Broglie wavelength*) of the molecule. The thermal wavelength decreases with increasing mass and temperature. This expression shows that the partition function for translational motion increases with the length of the box and the mass of the particle, for in each case the separation of the energy levels becomes smaller and more levels become thermally accessible. For a given mass and length of the box, the partition function also increases with increasing temperature (decreasing β), because more states become accessible.

Justification 13.2 *The partition function for a particle in a one-dimensional box*

The energy levels of a molecule of mass m in a container of length X are given by eqn 2.5 with $L = X$:

$$E_n = \frac{n^2h^2}{8mX^2} \qquad n = 1, 2, \ldots$$

The lowest level ($n = 1$) has energy $h^2/8mX^2$, so the energies relative to that level are

$$\varepsilon_n = (n^2 - 1)\varepsilon \qquad \varepsilon = h^2/8mX^2$$

Table 13.1 Contributions to the molecular partition function

Mode	Expression	Value
Translation	$q^{T} = \dfrac{V}{\Lambda^3} \qquad \Lambda = \dfrac{h}{(2\pi mkT)^{1/2}}$	$\Lambda/\text{pm} = \dfrac{1749}{(T/\text{K})^{1/2}(M/\text{g mol}^{-1})^{1/2}}$
Rotation		
Linear molecules	$q^{R} = \dfrac{kT}{\sigma hc\tilde{B}} = \dfrac{T}{\sigma\theta_R} \qquad \theta_R = \dfrac{hc\tilde{B}}{k}$	$q^{R} = \dfrac{0.6950}{\sigma} \times \dfrac{T/\text{K}}{(\tilde{B}/\text{cm}^{-1})}$
Non-linear molecules	$q^{R} = \dfrac{1}{\sigma}\left(\dfrac{kT}{hc}\right)^{3/2}\left(\dfrac{\pi}{\tilde{A}\tilde{B}\tilde{C}}\right)^{1/2}$	$q^{R} = \dfrac{1.027}{\sigma} \times \dfrac{(T/\text{K})^{3/2}}{(\tilde{A}\tilde{B}\tilde{C}/\text{cm}^{-3})^{1/2}}$
Vibration	$q^{V} = \dfrac{1}{1 - e^{-hc\tilde{\nu}/kT}} = \dfrac{1}{1 - e^{-\theta_V/T}} \qquad \theta_V = \dfrac{hc\tilde{\nu}}{k} = \dfrac{h\nu}{k}$	
	For $T \gg \theta_V$, $q^{V} = \dfrac{kT}{hc\tilde{\nu}} = \dfrac{T}{\theta_V}$	$q^{V} = 0.695 \times \dfrac{T/\text{K}}{\tilde{\nu}/\text{cm}^{-1}}$
Electronic	$q^{E} = g_0$ [+ higher terms] where g_0 is the degeneracy of the electronic ground state	

Note that $\beta = 1/kT$.

The sum to evaluate is therefore

$$q_X^{\mathrm{T}} = \sum_{n=1}^{\infty} \mathrm{e}^{-(n^2-1)\beta\varepsilon}$$

The translational energy levels are very close together in a container the size of a typical laboratory vessel; therefore, the sum can be approximated by an integral:

$$q_X^{\mathrm{T}} = \int_1^{\infty} \mathrm{e}^{-(n^2-1)\beta\varepsilon}\mathrm{d}n \approx \int_0^{\infty} \mathrm{e}^{-n^2\beta\varepsilon}\mathrm{d}n$$

The extension of the lower limit to $n = 0$ and the replacement of $n^2 - 1$ by n^2 introduces negligible error but turns the integral into standard form. We make the substitution $x^2 = n^2\beta\varepsilon$, implying $\mathrm{d}n = \mathrm{d}x/(\beta\varepsilon)^{1/2}$, and therefore that

$$q_X^{\mathrm{T}} = \left(\frac{1}{\beta\varepsilon}\right)^{1/2} \overbrace{\int_0^{\infty} \mathrm{e}^{-x^2}\mathrm{d}x}^{\pi^{1/2}/2} = \left(\frac{1}{\beta\varepsilon}\right)^{1/2}\left(\frac{\pi^{1/2}}{2}\right) = \left(\frac{2\pi m}{h^2\beta}\right)^{1/2} X$$

The total energy of a molecule free to move in three dimensions is the sum of its translational energies in all three directions:

$$\varepsilon_{n_1n_2n_3} = \varepsilon_{n_1}^{(X)} + \varepsilon_{n_2}^{(Y)} + \varepsilon_{n_3}^{(Z)} \qquad (13.16)$$

where n_1, n_2, and n_3 are the quantum numbers for motion in the x-, y-, and z-directions, respectively. Therefore, because $\mathrm{e}^{a+b+c} = \mathrm{e}^a\mathrm{e}^b\mathrm{e}^c$, the partition function factorizes as follows:

$$q^{\mathrm{T}} = \sum_{\text{all } n} \mathrm{e}^{-\beta\varepsilon_{n_1}^{(X)}-\beta\varepsilon_{n_2}^{(Y)}-\beta\varepsilon_{n_3}^{(Z)}} = \sum_{\text{all } n} \mathrm{e}^{-\beta\varepsilon_{n_1}^{(X)}}\mathrm{e}^{-\beta\varepsilon_{n_2}^{(Y)}}\mathrm{e}^{-\beta\varepsilon_{n_3}^{(Z)}}$$

$$= \left(\sum_{n_1} \mathrm{e}^{-\beta\varepsilon_{n_1}^{(X)}}\right)\left(\sum_{n_2} \mathrm{e}^{-\beta\varepsilon_{n_2}^{(Y)}}\right)\left(\sum_{n_3} \mathrm{e}^{-\beta\varepsilon_{n_3}^{(Z)}}\right) \qquad (13.17)$$

$$= q_X^{\mathrm{T}} q_Y^{\mathrm{T}} q_Z^{\mathrm{T}}$$

Equation 13.15 gives the partition function for translational motion in the x-direction. The only change for the other two directions is to replace the length X by the length Y or Z. Hence the partition function for motion in three dimensions is

$$q^{\mathrm{T}} = \left(\frac{2\pi m}{h^2\beta}\right)^{3/2} XYZ \qquad (13.18a)$$

The product of lengths XYZ is the volume, V, of the container, so we can write

$$q^{\mathrm{T}} = \frac{V}{\Lambda^3} \qquad (13.18b)$$

with Λ as defined in eqn 13.15b. As in the one-dimensional case, the partition function increases with the mass of the particle (as $m^{3/2}$) and the volume of the container (as V); for a given mass and volume, the partition function increases with temperature (as $T^{3/2}$). As in one dimension, $q^{\mathrm{T}} \to \infty$ as $T \to \infty$ because an infinite number of states becomes accessible as the temperature is raised. Even at room temperature $q^{\mathrm{T}} \approx 2 \times 10^{28}$ for an O_2 molecule in a vessel of volume 100 cm^3.

● A BRIEF ILLUSTRATION

To calculate the translational partition function of an H_2 molecule confined to a 100 cm^3 vessel at 25°C we use $m = 2.016m_u$; then

$$\Lambda = \frac{6.626 \times 10^{-34}\ \mathrm{J\,s}}{\{2\pi \times (2.016 \times 1.6605 \times 10^{-27}\ \mathrm{kg}) \times (1.381 \times 10^{-23}\ \mathrm{J\,K^{-1}}) \times (298\ \mathrm{K})\}^{1/2}}$$

$$= 7.12 \times 10^{-11}\ \mathrm{m}$$

where we have used 1 J = 1 kg m^2 s^{-2}. Therefore,

$$q = \frac{1.00 \times 10^{-4}\ \mathrm{m}^3}{(7.12 \times 10^{-11}\ \mathrm{m})^3} = 2.77 \times 10^{26}$$

About 10^{26} quantum states are thermally accessible, even at room temperature and for this light molecule. Many states are occupied if the thermal wavelength (which in this case is 71.2 pm) is small compared with the linear dimensions of the container. ●

Self-test 13.6 Calculate the translational partition function for a D_2 molecule under the same conditions.

[$q = 7.8 \times 10^{26}$, $2^{3/2}$ times larger]

The validity of the approximations that led to eqn 13.18 can be expressed in terms of the average separation, d, of the particles in the container. Because q is the total number of accessible states, the average number of states per molecule is q/N. For this quantity to be large, we require $V/N\Lambda^3 \gg 1$. However, V/N is the volume occupied by a single particle, and therefore the average separation of the particles is $d = (V/N)^{1/3}$. The condition for there being many states available per molecule is therefore $d^3/\Lambda^3 \gg 1$, and therefore $d \gg \Lambda$. That is, for eqn 13.18 to be valid, *the average separation of the particles must be much greater than their thermal wavelength*. For H_2 molecules at 1 bar and 298 K, the average separation is 3 nm, which is significantly larger than their thermal wavelength (71.2 pm).

The validity of eqn 13.18 can be expressed in a different way by noting that the approximations that led to it are valid if many states are occupied, which requires V/Λ^3 to be large. That will be so if Λ is small compared with the linear dimensions of the container. For H_2 at 25°C, $\Lambda = 71$ pm, which is far smaller than any conventional container is likely to be (but comparable to pores in zeolites or cavities in clathrates). For O_2, a heavier molecule, $\Lambda = 18$ pm.

(b) The rotational contribution

The partition function of a nonsymmetrical (AB) linear rotor is

$$q^{R} = \sum_{J}(2J+1)e^{-\beta hc\tilde{B}J(J+1)} \qquad (13.19)$$

The direct method of calculating q^R is to substitute the experimental values of the rotational energy levels into this expression and to sum the series numerically.

Example 13.5 *Evaluating the rotational partition function explicitly*

Evaluate the rotational partition function of $^{1}H^{35}Cl$ at 25°C, given that $\tilde{B} = 10.591\ cm^{-1}$.

Method We use eqn 13.19 and evaluate it term by term. Once again, we use $kT/hc = 207.224\ cm^{-1}$ at 298.15 K. The sum is readily evaluated by using mathematical software.

Answer To show how successive terms contribute, we draw up the following table by using $hc\tilde{B}/kT = 0.051\ 11$ (Fig. 13.8):

J	0	1	2	3	4	··· 10
$(2J+1)e^{-0.05111J(J+1)}$	1	2.71	3.68	3.79	3.24	··· 0.08

The sum required by eqn 13.19 (the sum of the numbers in the second row of the table) is 19.9, hence $q^R = 19.9$ at this temperature. Taking J up to 50 gives $q^R = 19.902$. Notice that about ten J-levels are significantly populated but the number of populated *states* is larger on account of the $(2J+1)$-fold degeneracy of each level. We shall shortly encounter the approximation that $q^R \approx kT/hc\tilde{B}$, which in the present case gives $q^R = 19.6$, in good agreement with the exact value and with much less work.

Self-test 13.7 Evaluate the rotational partition function for HCl at 0°C. [18.26]

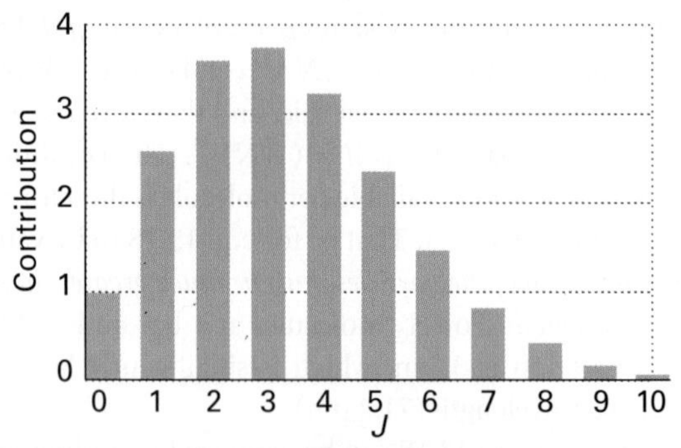

Fig. 13.8 The contributions to the rotational partition function of an HCl molecule at 25°C. The vertical axis is the value of $(2J+1)e^{-\beta hc\tilde{B}J(J+1)}$. Successive terms (which are proportional to the populations of the levels) pass through a maximum because the population of individual states decreases exponentially, but the degeneracy of the levels increases with J.

Synoptic table 13.2* Rotational and vibrational temperatures

Molecule	Mode	θ_V/K	θ_R/K
H_2		6330	88
HCl		4300	15.2
I_2		39	0.053
CO_2	ν_1	1997	0.561
	ν_2	3380	
	ν_3	960	

* For more values, see Table 10.2 in the *Data section* and use $hc/k = 1.439$ K cm.

At room temperature $kT/hc \approx 200\ cm^{-1}$. The rotational constants of many molecules are close to 1 cm^{-1} (Tables 10.2 and 13.2) and often smaller (though the very light H_2 molecule, for which $\tilde{B} = 60.9\ cm^{-1}$, is one exception). It follows that many rotational levels are populated at normal temperatures. When this is the case, the partition function may be approximated by

$$\text{Linear rotors:} \quad q^{R} = \frac{kT}{hc\tilde{B}} \qquad (13.20a)$$

$$\text{Non-linear rotors:} \quad q^{R} = \left(\frac{kT}{hc}\right)^{3/2}\left(\frac{\pi}{\tilde{A}\tilde{B}\tilde{C}}\right)^{1/2} \qquad (13.20b)$$

where $\tilde{A}$, $\tilde{B}$, and $\tilde{C}$ are the rotational constants of the molecule expressed as wavenumbers. However, before using these expressions, read on (to eqns 13.21 and 13.22).

Justification 13.3 *The rotational contribution to the molecular partition function*

When many rotational states are occupied and kT is much larger than the separation between neighbouring states, the sum in the partition function can be approximated by an integral, much as we did for translational motion:

$$q^{R} = \int_{0}^{\infty}(2J+1)e^{-\beta hc\tilde{B}J(J+1)}dJ$$

This integral can be evaluated without much effort by making the substitution $x = \beta hc\tilde{B}J(J+1)$, so that $dx/dJ = \beta hc\tilde{B}(2J+1)$ and therefore $(2J+1)dJ = dx/\beta hc\tilde{B}$. Then

$$q^{R} = \frac{1}{\beta hc\tilde{B}}\int_{0}^{\infty}e^{-x}dx = \frac{1}{\beta hc\tilde{B}}$$

which (because $\beta = 1/kT$) is eqn 13.20. The calculation for a non-linear molecule is along the same lines, but slightly trickier. It, and the origin of the factor $\pi^{1/2}$ in eqn 13.20b, is described in *Further information 13.2*.

A useful way of expressing the temperature above which the rotational approximation is valid is to introduce the **characteristic rotational temperature**, $\theta_R = hc\tilde{B}/k$. Then 'high temperature' means $T \gg \theta_R$ and under these conditions the rotational partition function of a linear molecule is simply T/θ_R. Some typical values of θ_R are shown in Table 13.2. The value for H_2 is abnormally high and we must be careful with the approximation for this molecule.

The general conclusion at this stage is that molecules with large moments of inertia (and hence small rotational constants and low characteristic rotational temperatures) have large rotational partition functions. The large value of q^R reflects the closeness in energy (compared with kT) of the rotational levels in large, heavy molecules, and the large number of rotational states that are accessible at normal temperatures.

We must take care, however, not to include too many rotational states in the sum. For a homonuclear diatomic molecule or a symmetrical linear molecule (such as CO_2 or HC≡CH), a rotation through 180° results in an indistinguishable state of the molecule. Hence, the number of thermally accessible states is only half the number that can be occupied by a heteronuclear diatomic molecule, where rotation through 180° does result in a distinguishable state. Therefore, for a symmetrical linear molecule,

$$q^R = \frac{kT}{2hc\tilde{B}} = \frac{T}{2\theta_R} \qquad (13.21a)$$

The equations for symmetrical and nonsymmetrical molecules can be combined into a single expression by introducing the **symmetry number**, σ, which is the number of indistinguishable orientations of the molecule. Then

$$q^R = \frac{kT}{\sigma hc\tilde{B}} = \frac{T}{\sigma\theta_R} \qquad (13.21b)$$

For a heteronuclear diatomic molecule $\sigma = 1$; for a homonuclear diatomic molecule or a symmetrical linear molecule, $\sigma = 2$.

Justification 13.4 *The origin of the symmetry number*

The quantum mechanical origin of the symmetry number is the Pauli principle, which forbids the occupation of certain states. We saw in Section 10.5, for example, that H_2 may occupy rotational states with even J only if its nuclear spins are paired (*para*-hydrogen), and odd J states only if its nuclear spins are parallel (*ortho*-hydrogen). There are three states of *ortho*-H_2 to each value of J (because there are three parallel spin states of the two nuclei).

To set up the rotational partition function we note that 'ordinary' molecular hydrogen is a mixture of one part *para*-H_2 (with only its even-J rotational states occupied) and three parts *ortho*-H_2 (with only its odd-J rotational states occupied). Therefore, the average partition function per molecule is

$$q^R = \tfrac{1}{4}\sum_{\text{even } J}(2J+1)e^{-\beta hc\tilde{B}J(J+1)} + \tfrac{3}{4}\sum_{\text{odd } J}(2J+1)e^{-\beta hc\tilde{B}J(J+1)}$$

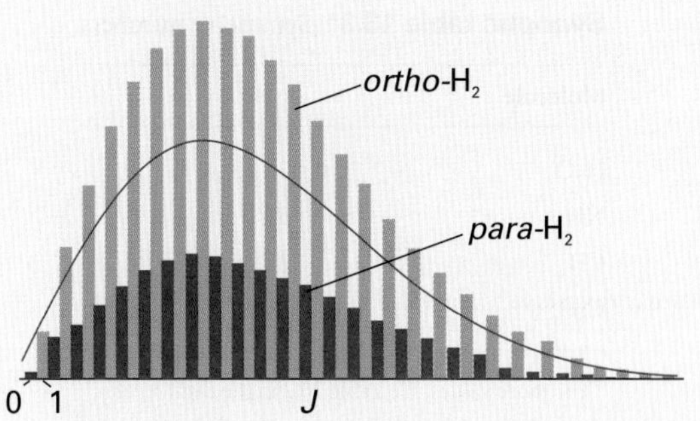

Fig. 13.9 The values of the individual terms $(2J+1)e^{-\beta hc\tilde{B}J(J+1)}$ contributing to the mean partition function of a 3:1 mixture of *ortho*- and *para*-H_2. The partition function is the sum of all these terms. At high temperatures, the sum is approximately equal to the sum of the terms over all values of J, each with a weight of $\frac{1}{2}$. This is the sum of the contributions indicated by the curve.

The odd-J states are more heavily weighted than the even-J states (Fig. 13.9). From the figure we see that we would obtain approximately the same answer for the partition function (the sum of all the populations) if each J term contributed half its normal value to the sum. That is, the last equation can be approximated as

$$q^R = \tfrac{1}{2}\sum_J (2J+1)e^{-\beta hc\tilde{B}J(J+1)}$$

and this approximation is very good when many terms contribute (at high temperatures, $T \gg 88$ K).

The same type of argument may be used for linear symmetrical molecules in which identical bosons are interchanged by rotation (such as CO_2). As pointed out in Section 10.5, if the nuclear spin of the bosons is 0, then only even-J states are admissible. Because only half the rotational states are occupied, the rotational partition function is only half the value of the sum obtained by allowing all values of J to contribute (Fig. 13.10).

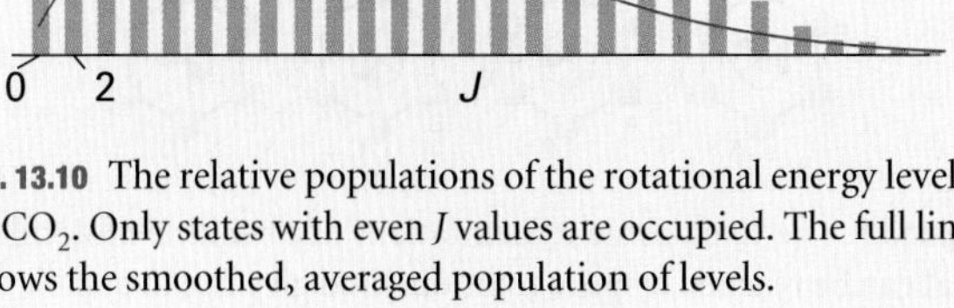

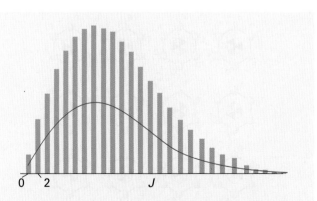

Fig. 13.10 The relative populations of the rotational energy levels of CO_2. Only states with even J values are occupied. The full line shows the smoothed, averaged population of levels.

Synoptic table 13.3* Symmetry numbers

Molecule	σ
H_2O	2
NH_3	3
CH_4	12
C_6H_6	12

* For more values, see Table 10.2 in the *Data section*.

The same care must be exercised for other types of symmetrical molecule, and for a non-linear molecule we write

$$q^R = \frac{1}{\sigma}\left(\frac{kT}{hc}\right)^{3/2}\left(\frac{\pi}{\tilde{A}\tilde{B}\tilde{C}}\right)^{1/2} \qquad (13.22)$$

Some typical values of the symmetry numbers are given in Table 13.3. The value $\sigma(H_2O) = 2$ reflects the fact that a 180° rotation about the bisector of the H—O—H angle interchanges two indistinguishable atoms. In NH_3, there are three indistinguishable orientations around the axis shown in (**1**). For CH_4, any of three 120° rotations about any of its four C—H bonds leaves the molecule in an indistinguishable state (**2**), so the symmetry number is 3 × 4 = 12. For benzene, any of six orientations around the axis perpendicular to the plane of the molecule leaves it apparently unchanged (Fig. 13.11), as does a rotation of 180° around any of six axes in the plane of the molecule (three of which pass along each C—H bond and the remaining three pass through each C—C bond in the plane of the molecule). For the way that group theory is used to identify the value of the symmetry number, see Problem 13.18.

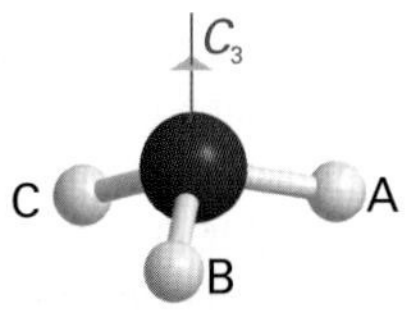

1

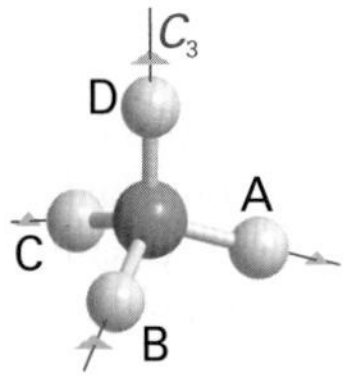

2

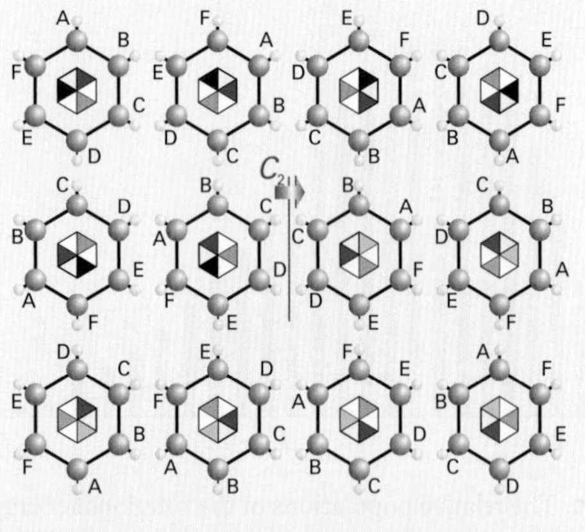

Fig. 13.11 The 12 equivalent orientations of a benzene molecule that can be reached by pure rotations, and give rise to a symmetry number of 12.

(c) The vibrational contribution

The vibrational partition function of a molecule is calculated by substituting the measured vibrational energy levels into the exponentials appearing in the definition of q^V, and summing them numerically. In a polyatomic molecule each normal mode (Section 10.12) has its own partition function (provided the anharmonicities are so small that the modes are independent). The overall vibrational partition function is the product of the individual partition functions, and we can write $q^V = q^V(1)q^V(2)\cdots$, where $q^V(K)$ is the partition function for the Kth normal mode and is calculated by direct summation of the observed spectroscopic levels.

If the vibrational excitation is not too great, the harmonic approximation may be made, and the vibrational energy levels written as

$$E_v = (v + \tfrac{1}{2})hc\tilde{\nu} \qquad v = 0, 1, 2, \ldots \qquad (13.23)$$

If, as usual, we measure energies from the zero-point level, then the permitted values are $\varepsilon_v = vhc\tilde{\nu}$ and the partition function is

$$q^V = \sum_v e^{-\beta vhc\tilde{\nu}} = \sum_v (e^{-\beta hc\tilde{\nu}})^v$$

(because $e^{ax} = (e^x)^a$). We evaluated an essentially identical sum in Example 13.3 and in this case can write:

$$q^V = \frac{1}{1 - e^{-\beta hc\tilde{\nu}}} \qquad (13.24)$$

This function is plotted in Fig. 13.12. In a polyatomic molecule, each normal mode gives rise to a partition function of this form.

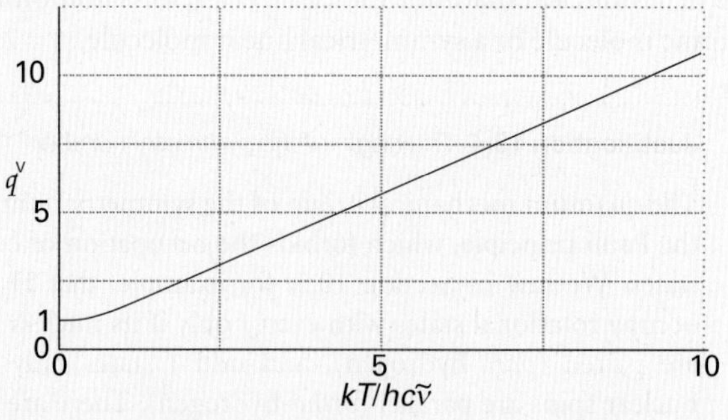

Fig. 13.12 The vibrational partition function of a molecule in the harmonic approximation. Note that the partition function is linearly proportional to the temperature when the temperature is high ($T \gg \theta_V$).

interActivity Plot the temperature dependence of the vibrational contribution to the molecular partition function for several values of the vibrational wavenumber. Estimate from your plots the temperature above which the harmonic oscillator is in the 'high temperature' limit.

Example 13.6 *Calculating a vibrational partition function*

The wavenumbers of the three normal modes of H_2O are 3656.7 cm^{-1}, 1594.8 cm^{-1}, and 3755.8 cm^{-1}. Evaluate the vibrational partition function at 1500 K.

Method Use eqn 13.24 for each mode, and then form the product of the three contributions. At 1500 K, kT/hc = 1042.6 cm^{-1}.

Answer We draw up the following table displaying the contributions of each mode:

Mode:	1	2	3
$\tilde{\nu}$/cm^{-1}	3656.7	1594.8	3755.8
$hc\tilde{\nu}/kT$	3.507	1.530	3.602
q^V	1.031	1.276	1.028

The overall vibrational partition function is therefore

$$q^V = 1.031 \times 1.276 \times 1.028 = 1.352$$

The three normal modes of H_2O are at such high wavenumbers that even at 1500 K most of the molecules are in their vibrational ground state. However, there may be so many normal modes in a large molecule that their excitation may be significant even though each mode is not appreciably excited. For example, a non-linear molecule containing 10 atoms has $3N - 6 = 24$ normal modes (Section 10.12). If we assume a value of about 1.1 for the vibrational partition function of one normal mode, the overall vibrational partition function is about $q^V \approx (1.1)^{24} = 9.8$, which indicates significant vibrational excitation relative to a smaller molecule, such as H_2O.

Self-test 13.8 Repeat the calculation for CO_2, where the vibrational wavenumbers are 1388 cm^{-1}, 667.4 cm^{-1}, and 2349 cm^{-1}, the second being the doubly degenerate bending mode. [6.79]

In many molecules the vibrational wavenumbers are so great that $\beta hc\tilde{\nu} > 1$. For example, the lowest vibrational wavenumber of CH_4 is 1306 cm^{-1}, so $\beta hc\tilde{\nu} = 6.3$ at room temperature. Most C—H stretches normally lie in the range 2850 to 2960 cm^{-1}, so for them $\beta hc\tilde{\nu} \approx 14$. In these cases, $e^{-\beta hc\tilde{\nu}}$ in the denominator of q^V is very close to zero (for example, $e^{-6.3} = 0.002$), and the vibrational partition function for a single mode is very close to 1 ($q^V = 1.002$ when $\beta hc\tilde{\nu} = 6.3$), implying that only the zero-point level is significantly occupied.

Now consider the case of bonds with such low vibrational frequencies that $\beta hc\tilde{\nu} \ll 1$. When this condition is satisfied, the partition function may be approximated by expanding the exponential ($e^x = 1 + x + \cdots$):

$$q^V = \frac{1}{1-(1-\beta hc\tilde{\nu}+\cdots)}$$

That is, for weak bonds at high temperatures,

$$q^V = \frac{1}{\beta hc\tilde{\nu}} = \frac{kT}{hc\tilde{\nu}} \qquad (13.25)$$

The temperatures for which eqn 13.25 is valid can be expressed in terms of the **characteristic vibrational temperature,** $\theta_V = hc\tilde{\nu}/k$ (Table 13.2). The value for H_2 is abnormally high because the atoms are so light and the vibrational frequency is correspondingly high. In terms of the vibrational temperature, 'high temperature' means $T \gg \theta_V$ and, when this condition is satisfied, $q^V = T/\theta_V$ (the analogue of the rotational expression).

(d) The electronic contribution

Electronic energy separations from the ground state are usually very large, so for most cases $q^E = 1$. An important exception arises in the case of atoms and molecules having electronically degenerate ground states, in which case $q^E = g^E$, where g^E is the degeneracy of the electronic ground state. Alkali metal atoms, for example, have doubly degenerate ground states (corresponding to the two orientations of their electron spin), so $q^E = 2$.

Some atoms and molecules have low-lying electronically excited states. (At high enough temperatures, all atoms and molecules have thermally accessible excited states.) An example is NO, which has a configuration of the form . . . π^1 (see *Impact I5.1*). The orbital angular momentum may take two orientations with respect to the molecular axis (corresponding to circulation clockwise or counter-clockwise around the axis), and the spin angular momentum may also take two orientations, giving four states in all (Fig. 13.13). The energy of the two states in which the orbital and spin momenta are parallel (giving the $^2\Pi_{3/2}$ term) is

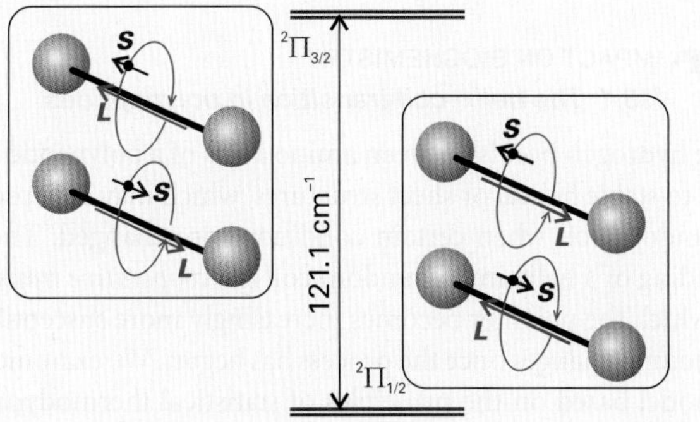

Fig. 13.13 The doubly degenerate ground electronic level of NO (with the spin and orbital angular momenta around the axis in opposite directions) and the doubly degenerate first excited level (with the spin and orbital momenta parallel). The upper level is thermally accessible at room temperature.

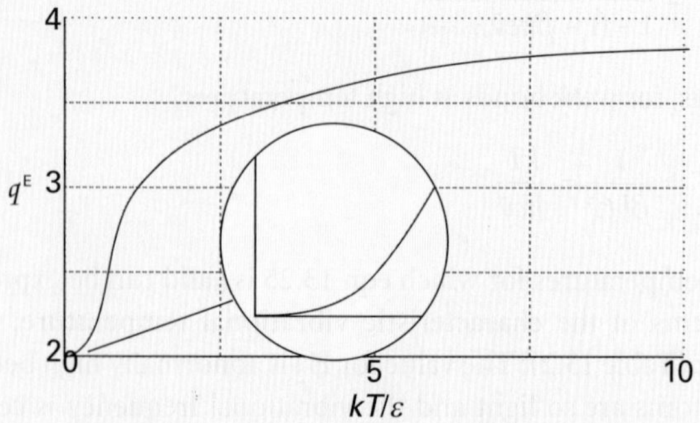

Fig. 13.14 The variation with temperature of the electronic partition function of an NO molecule. Note that the curve resembles that for a two-level system (Fig. 13.5), but rises from 2 (the degeneracy of the lower level) and approaches 4 (the total number of states) at high temperatures.

interActivity Plot the temperature dependence of the electronic partition function for several values of the energy separation ε between two doubly degenerate levels. From your plots, estimate the temperature at which the population of the excited level begins to increase sharply.

slightly greater than that of the two other states in which they are antiparallel (giving the $^2\Pi_{1/2}$ term). The separation, which arises from spin–orbit coupling, is only 121 cm^{-1}. Hence, at normal temperatures, all four states are thermally accessible. If we denote the energies of the two levels as $E_{1/2} = 0$ and $E_{3/2} = \varepsilon$, the partition function is

$$q^E = \sum_{\text{energy levels } j} g_j e^{-\beta\varepsilon_j} = 2 + 2e^{-\beta\varepsilon} \qquad (13.26)$$

Figure 13.14 shows the variation of this function with temperature. At $T=0$, $q^E = 2$, because only the doubly degenerate ground state is accessible. At high temperatures, q^E approaches 4 because all four states are accessible. At 25°C, $q^E = 3.1$.

IMPACT ON BIOCHEMISTRY

I13.1 The helix–coil transition in polypeptides

The hydrogen bonds between amino acids of a polypeptide give rise to stable helical or sheet structures, which may collapse into a random coil when certain conditions are changed. The unwinding of a helix into a random coil is a *cooperative transition*, in which the polymer becomes increasingly more susceptible to structural changes once the process has begun. We examine here a model based on the principles of statistical thermodynamics that accounts for the cooperativity of the helix–coil transition in polypeptides.

To calculate the fraction of polypeptide molecules present as helix or coil we need to set up the partition function for the various states of the molecule. To illustrate the approach, consider a short polypeptide with four amino acid residues, each labelled *h* if it contributes to a helical region and *c* if it contributes to a random coil region. We suppose that conformations *hhhh* and *cccc* contribute terms q_0 and q_4, respectively, to the partition function q. Then we assume that each of the four conformations with one *c* amino acid (such as *hchh*) contributes q_1. Similarly, each of the six states with two *c* amino acids contributes a term q_2, and each of the four states with three *c* amino acids contributes a term q_3. The partition function is then

$$q = q_0 + 4q_1 + 6q_2 + 4q_3 + q_4 = q_0\left(1 + \frac{4q_1}{q_0} + \frac{6q_2}{q_0} + \frac{4q_3}{q_0} + \frac{q_4}{q_0}\right)$$

We shall now suppose that each partition function differs from q_0 only by the energy of each conformation relative to *hhhh*, and write

$$\frac{q_i}{q_0} = e^{-(\varepsilon_i - \varepsilon_0)/kT}$$

Next, we suppose that the conformational transformations are non-cooperative, in the sense that the energy associated with changing one *h* amino acid into one *c* amino acid has the same value regardless of how many *h* or *c* amino acid residues are in the reactant or product state and regardless of where in the chain the conversion occurs. That is, we suppose that the difference in energy between c^ih^{4-i} and $c^{i+1}h^{3-i}$ has the same value γ for all i. This assumption implies that $\varepsilon_i - \varepsilon_0 = i\gamma$ and therefore that

$$q = q_0(1 + 4s + 6s^2 + 4s^3 + s^4) \qquad s = e^{-\gamma/kT} = e^{-\Gamma/RT} \qquad (13.27)$$

where $\Gamma = N_A\gamma$ and s is called the *stability parameter*. The term in parentheses has the form of the binomial expansion of $(1+s)^4$.

$$\frac{q}{q_0} = \sum_{i=0}^{4} C(4,i)s^i \qquad \text{with} \qquad C(4,i) = \frac{4!}{(4-i)!i!} \qquad (13.28)$$

which we interpret as the number of ways in which a state with i *c* amino acids can be formed. The extension of this treatment to take into account a longer chain of residues is now straightforward: we simply replace the upper limit of 4 in the sum by n:

$$\frac{q}{q_0} = \sum_{i=0}^{n} C(n,i)s^i \qquad (13.29)$$

A cooperative transformation is more difficult to accommodate, and depends on building a model of how neighbours facilitate each other's conformational change. In the simple *zipper model*, conversion from *h* to *c* is allowed only if a residue adjacent to the one undergoing the conversion is already a *c* residue. Thus, the zipper model allows a transition of the type . . . *hhhch* . . . → . . . *hhhcc* . . . , but not a transition of the type . . . *hhhch* . . . → . . . *hchch*. . . . The only exception to this rule is, of course, the very first conversion from *h* to *c* in a fully helical chain. Cooperativity is included in the zipper model by assuming that

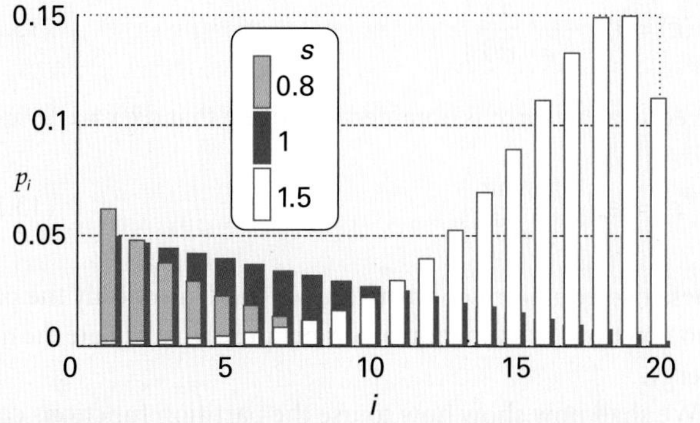

Fig. 13.15 The distribution of p_i, the fraction of molecules that has a number i of c amino acids for $s = 0.8$ ($\langle i\rangle = 1.1$), 1.0 ($\langle i\rangle = 3.8$), and 1.5 ($\langle i\rangle = 15.9$), with $\sigma = 5.0\times10^{-3}$.

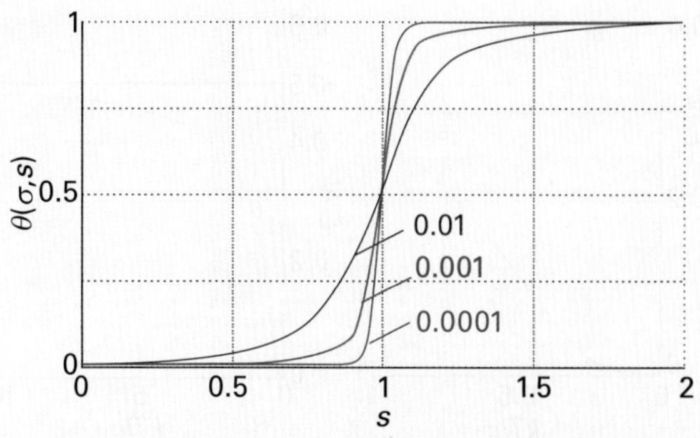

Fig. 13.16 Plots of the degree of conversion θ against s for several values of σ. The curves show the sigmoidal shape characteristic of cooperative behaviour.

the first conversion from h to c, called the *nucleation step*, is less favourable than the remaining conversions and replacing s for that step by σs, where $\sigma \ll 1$. Each subsequent step is called a *propagation step* and has a stability parameter s. In this case it is found that

$$\frac{q}{q_0} = 1 + \frac{\sigma s[s^{n+1} - (n+1)s^n + 1]}{(s-1)^2} \tag{13.30}$$

The fraction $p_i = q_i/q$ of molecules that has a number i of c amino acids is $p_i = [(n-i+1)\sigma s^i]/(q/q_0)$ for $i > 0$ and the mean value of i is then $\langle i\rangle = \sum_i ip_i$. Figure 13.15 shows the distribution of p_i for various values of s with $\sigma = 5.0\times10^{-3}$. We see that most of the polypeptide chains remain largely helical when $s < 1$ and that most of the chains exist largely as random coils when $s > 1$. When $s = 1$, there is a more widespread distribution of length of random coil segments.

A more sophisticated model for the helix–coil transition must allow for helical segments to form in different regions of a long polypeptide chain, with the nascent helices being separated by shrinking coil segments. Calculations based on this more complete *Zimm–Bragg model* give

$$\theta = \tfrac{1}{2}\left(1 + \frac{(s-1)+2\sigma}{[(s-1)^2+4s\sigma]^{1/2}}\right) \tag{13.31}$$

Figure 13.16 shows plots of θ against s for several values of σ. The curves show the sigmoidal shape characteristic of cooperative behaviour. There is a sudden surge of transition to a random coil as s passes through 1 and, the smaller the parameter σ, the greater the sharpness and hence the greater the cooperativity of the transition. That is, the harder it is to get coil formation started, the sharper the transition from helix to coil.

13.4 The mean energy

The importance of the molecular partition function is that it contains all the information needed to calculate the bulk properties of a system of independent particles. In this respect, as we remarked in the introduction, q plays a role for bulk matter very similar to that played by the wavefunction in quantum mechanics for individual molecules: q is a kind of thermal wavefunction.

We shall begin to unfold the importance of q by showing how to derive an expression for the mean energy of a molecule in a system of independent particles. The mean energy of a molecule, $\langle\varepsilon\rangle$, relative to its energy in its ground state, is the total energy of the system divided by the total number of molecules:

$$\langle\varepsilon\rangle = \frac{E}{N} = \frac{1}{N}\sum_i N_i\varepsilon_i \tag{13.32}$$

Because the most probable configuration is so strongly dominating, we can use the Boltzmann distribution for the ratio N_i/N and write

$$\langle\varepsilon\rangle = \frac{1}{q}\sum_i \varepsilon_i e^{-\beta\varepsilon_i} \tag{13.33}$$

To manipulate this expression into a form involving only q we note that

$$\varepsilon_i e^{-\beta\varepsilon_i} = -\frac{d}{d\beta}e^{-\beta\varepsilon_i}$$

It follows that

$$\langle\varepsilon\rangle = -\frac{1}{q}\sum_i \frac{d}{d\beta}e^{-\beta\varepsilon_i} = -\frac{1}{q}\frac{d}{d\beta}\sum_i e^{-\beta\varepsilon_i} = -\frac{1}{q}\frac{dq}{d\beta} \tag{13.34}$$

● A BRIEF ILLUSTRATION

From the two-level partition function $q = 1 + e^{-\beta\varepsilon}$, we can deduce that the mean energy at a temperature T is

$$\langle\varepsilon\rangle = -\left(\frac{1}{1+e^{-\beta\varepsilon}}\right)\frac{d}{d\beta}(1+e^{-\beta\varepsilon}) = \frac{\varepsilon e^{-\beta\varepsilon}}{1+e^{-\beta\varepsilon}} = \frac{\varepsilon}{1+e^{\beta\varepsilon}}$$

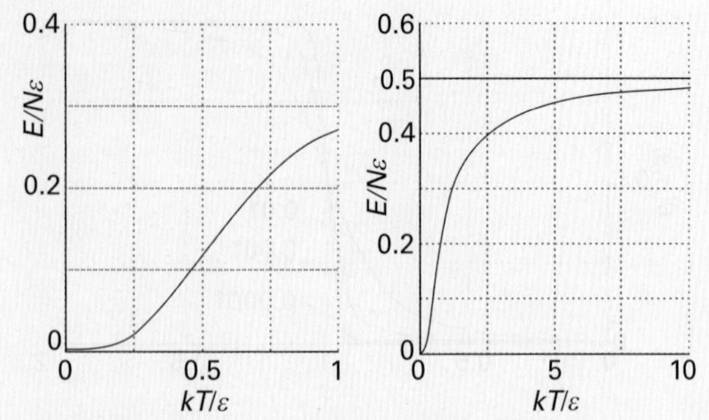

Fig. 13.17 The total energy of a two-level system (expressed as a multiple of $N\varepsilon$) as a function of temperature, on two temperature scales. The graph on the left shows the slow rise away from zero energy at low temperatures; the slope of the graph at $T = 0$ is 0 (that is, the heat capacity is zero at $T = 0$). The graph on the right shows the slow rise to 0.5 as $T \to \infty$ as both states become equally populated (see Fig. 13.7).

interActivity Draw graphs similar to those in Fig. 13.17 for a three-level system with levels 0, ε, and 2ε.

This function is plotted in Fig. 13.17. Notice how the mean energy is zero at $T = 0$, when only the lower state (at the zero of energy) is occupied, and rises to $\frac{1}{2}\varepsilon$ as $T \to \infty$, when the two levels become equally populated. ●

A brief comment To reinforce the analogy between statistical thermodynamics and quantum mechanics, note the resemblance of eqn 13.34 written as

$$\langle\varepsilon\rangle q = -\frac{dq}{d\beta}$$

to the time-dependent Schrödinger equation written as

$$\hat{H}\psi = -\frac{\partial\psi}{\partial i\hbar t}$$

There are several points in relation to eqn 13.34 that need to be made. Because $\varepsilon_0 = 0$, (remember that we measure all energies from the lowest available level), $\langle\varepsilon\rangle$ should be interpreted as the value of the mean energy relative to its ground-state (zero-point) energy. If the lowest energy of the molecule is in fact ε_{gs} rather than 0, then the true mean energy is $\varepsilon_{gs} + \langle\varepsilon\rangle$. For instance, for an harmonic oscillator, we would set ε_{gs} equal to the zero-point energy, $\frac{1}{2}hc\tilde{\nu}$. Secondly, because the partition function may depend on variables other than the temperature (for example, the volume), the derivative with respect to β in eqn 13.34 is actually a *partial* derivative with these other variables held constant. The complete expression relating the molecular partition function to the mean energy of a molecule is therefore

$$\langle\varepsilon\rangle = \varepsilon_{gs} - \frac{1}{q}\left(\frac{\partial q}{\partial\beta}\right)_V \qquad (13.35a)$$

An equivalent form is obtained by noting that $dx/x = d\ln x$:

$$\langle\varepsilon\rangle = \varepsilon_{gs} - \left(\frac{\partial \ln q}{\partial\beta}\right)_V \qquad (13.35b)$$

These two equations confirm that we need know only the partition function (as a function of temperature) to calculate the mean energy.

We shall now show how to use the partition functions calculated in Section 13.3 to calculate the mean energy of each mode of motion of independent, non-interacting molecules.

(a) The translational contribution

For a one-dimensional system of length X, for which $q^{\mathrm{T}} = X/\Lambda$, with $\Lambda = h(\beta/2\pi m)^{1/2}$, we note that Λ is a constant multiplied by $\beta^{1/2}$, and obtain

$$\langle\varepsilon^{\mathrm{T}}\rangle = -\frac{1}{q^{\mathrm{T}}}\left(\frac{\partial q^{\mathrm{T}}}{\partial\beta}\right)_V = -\frac{\Lambda}{X}\left(\frac{\partial}{\partial\beta}\frac{X}{\Lambda}\right)_V$$

$$= -\frac{\text{constant}\times\beta^{1/2}}{X}\times X\times\frac{d}{d\beta}\left(\frac{1}{\text{constant}\times\beta^{1/2}}\right)$$

$$= -\beta^{1/2}\frac{d}{d\beta}\left(\frac{1}{\beta^{1/2}}\right) = \frac{1}{2\beta} = \tfrac{1}{2}kT \qquad (13.36a)$$

For a molecule free to move in three dimensions, the analogous calculation leads to

$$\langle\varepsilon^{\mathrm{T}}\rangle = \tfrac{3}{2}kT \qquad (13.36b)$$

The **equipartition theorem** of classical mechanics is consistent with this result and provides a useful short cut. First, we need to know that a 'quadratic contribution' to the energy means a contribution that can be expressed as the square of a variable, such as the position or the velocity. For example, the kinetic energy of an atom of mass m as it moves through three-dimensional space is

$$E_{\mathrm{k}} = \tfrac{1}{2}mv_x^2 + \tfrac{1}{2}mv_y^2 + \tfrac{1}{2}mv_z^2$$

and there are three quadratic contributions to its energy. The equipartition theorem then states that for a collection of particles at thermal equilibrium at a temperature T, *the average value of each quadratic contribution to the energy is the same and equal to* $\frac{1}{2}kT$, *where k is Boltzmann's constant.* (For the molar energy, we multiply by Avogadro's constant and use $N_{\mathrm{A}}k = R$.) The equipartition theorem is a conclusion from classical mechanics and is applicable only when the effects of quantization can be ignored. In practice, it can be used for molecular translation and rotation but not vibration.

A note on good practice You will commonly see the equipartition theorem expressed in terms of the 'degrees of freedom' rather than quadratic contributions. That can be misleading and is best avoided, for a single vibrational degree of freedom has two quadratic contributions (the kinetic energy and the potential energy).

According to the equipartition theorem, the average energy of each term in the expression above is $\frac{1}{2}kT$. Therefore, the mean energy of the atoms is $\frac{3}{2}kT$ and their molar energy is $\frac{3}{2}RT$ (because $N_A k = R$). At 25°C, $\frac{3}{2}RT = 3.7$ kJ mol^{-1}, so translational motion contributes about 4 kJ mol^{-1} to the molar internal energy of a gaseous sample of atoms or molecules (the remaining contribution arises from the internal structure of the atoms and molecules).

(b) The rotational contribution

The mean rotational energy of a linear molecule is obtained from the partition function given in eqn 13.19. When the temperature is low ($T < \theta_R$), the series must be summed term by term, which for a heteronuclear diatomic molecule or other nonsymmetrical linear molecule gives

$$q^R = 1 + 3e^{-2\beta hc\tilde{B}} + 5e^{-6\beta hc\tilde{B}} + \cdots$$

Hence, because

$$\frac{dq^R}{d\beta} = -6hc\tilde{B}e^{-2\beta hc\tilde{B}} - 30hc\tilde{B}e^{-6\beta hc\tilde{B}} - \cdots$$

$$= -hc\tilde{B}(6e^{-2\beta hc\tilde{B}} + 30e^{-6\beta hc\tilde{B}} + \cdots)$$

we find

$$\langle \varepsilon^R \rangle = \frac{hc\tilde{B}(6e^{-2\beta hc\tilde{B}} + 30e^{-6\beta hc\tilde{B}} + \cdots)}{1 + 3e^{-2\beta hc\tilde{B}} + 5e^{-6\beta hc\tilde{B}} + \cdots} \quad (13.37a)$$

This function is plotted in Fig. 13.18. At high temperatures ($T \gg \theta_R$), q^R is given by eqn 13.21a and

$$\langle \varepsilon^R \rangle = -\frac{1}{q^R}\frac{dq^R}{d\beta} = -\sigma hc\beta\tilde{B}\frac{d}{d\beta}\frac{1}{\sigma hc\beta\tilde{B}}$$

$$= -\sigma hc\beta\tilde{B}\left(-\frac{1}{\sigma hc\beta^2\tilde{B}}\right) = \frac{1}{\beta} = kT \quad (13.37b)$$

(q^R is independent of V, so the partial derivatives have been replaced by complete derivatives.) The high-temperature result, which is valid when many rotational states are occupied, is also in agreement with the equipartition theorem, for the classical expression for the energy of a linear rotor is $E_k = \frac{1}{2}I_\perp\omega_a^2 + \frac{1}{2}I_\perp\omega_b^2$ and therefore has two quadratic contributions. (There is no rotation around the line of atoms.) It follows from the equipartition theorem that the mean rotational energy is $2 \times \frac{1}{2}kT = kT$.

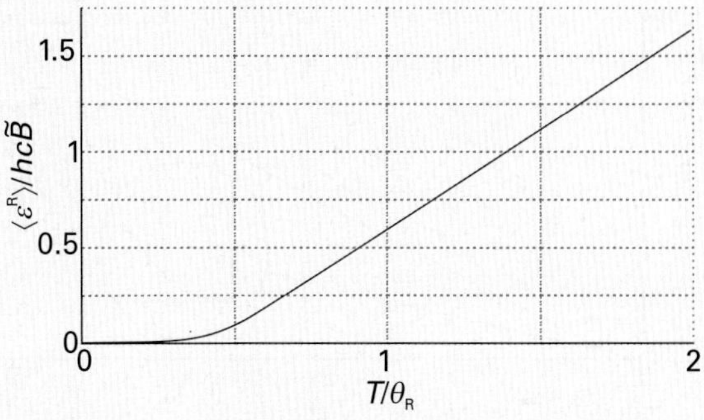

Fig. 13.18 The mean rotational energy of a nonsymmetrical linear rotor as a function of temperature. At high temperatures ($T \gg \theta_R$), the energy is linearly proportional to the temperature, in accord with the equipartition theorem.

interActivity Plot the temperature dependence of the mean rotational energy for several values of the rotational constant (for reasonable values of the rotational constant, see the *Data resource section*). From your plots, estimate the temperature at which the mean rotational energy begins to increase sharply.

(c) The vibrational contribution

The vibrational partition function in the harmonic approximation is given in eqn 13.24. Because q^V is independent of the volume, it follows that

$$\frac{dq^V}{d\beta} = \frac{d}{d\beta}\left(\frac{1}{1 - e^{-\beta hc\tilde{\nu}}}\right) = -\frac{hc\tilde{\nu}e^{-\beta hc\tilde{\nu}}}{(1 - e^{-\beta hc\tilde{\nu}})^2} \quad (13.38)$$

and hence from

$$\langle \varepsilon^V \rangle = -\frac{1}{q^V}\frac{dq^V}{d\beta} = -(1 - e^{-\beta hc\tilde{\nu}})\left\{-\frac{hc\tilde{\nu}e^{-\beta hc\tilde{\nu}}}{(1 - e^{-\beta hc\tilde{\nu}})^2}\right\}$$

$$= \frac{hc\tilde{\nu}e^{-\beta hc\tilde{\nu}}}{1 - e^{-\beta hc\tilde{\nu}}}$$

that

$$\langle \varepsilon^V \rangle = \frac{hc\tilde{\nu}}{e^{\beta hc\tilde{\nu}} - 1} \quad (13.39)$$

The zero-point energy, $\frac{1}{2}hc\tilde{\nu}$, can be added to the right-hand side if the mean energy is to be measured from 0 rather than the lowest attainable level (the zero-point level). The variation of the mean energy with temperature is illustrated in Fig. 13.19. At high temperatures, when $T \gg \theta_V$, or $\beta hc\tilde{\nu} \ll 1$, the exponential function can be expanded ($e^x = 1 + x + \cdots$) and all but the leading terms discarded. This approximation leads to

$$\langle \varepsilon^V \rangle = \frac{hc\tilde{\nu}}{(1 + \beta hc\tilde{\nu} + \cdots) - 1} \approx \frac{1}{\beta} = kT \quad (13.40)$$

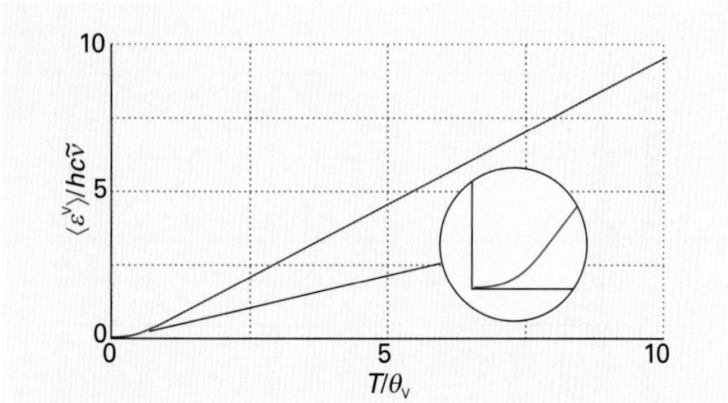

Fig. 13.19 The mean vibrational energy of a molecule in the harmonic approximation as a function of temperature. At high temperatures ($T \gg \theta_V$), the energy is linearly proportional to the temperature, in accord with the equipartition theorem.

interActivity Plot the temperature dependence of the mean vibrational energy for several values of the vibrational wavenumber (for reasonable values of the vibrational wavenumber, see the *Data section*). From your plots, estimate the temperature at which the mean vibrational energy begins to increase sharply.

This result is in agreement with the value predicted by the classical equipartition theorem, because the energy of a one-dimensional oscillator is $E = \frac{1}{2}mv_x^2 + \frac{1}{2}kx^2$ and the mean energy of each quadratic term is $\frac{1}{2}kT$.

(d) The overall contribution

Atoms without low-lying excited states have only translational degrees of freedom, so their mean energy is given by eqn 13.36b. When a gas consists of polyatomic molecules, we need to take into account the effect of rotation and vibration. A linear molecule, such as N_2 and CO_2, can rotate around two axes perpendicular to the line of the atoms, so it has two rotational modes of motion, each contributing a term $\frac{1}{2}kT$ to the mean energy. Therefore, the mean rotational energy is kT. By adding the translational and rotational contributions, we obtain

$$\langle \varepsilon \rangle = \varepsilon_{gs} + \tfrac{5}{2}kT \quad \text{(linear molecule, translation and rotation only)} \qquad (13.41a)$$

where ε_{gs} is the vibrational ground-state energy, typically the only state significantly populated at normal temperatures. A non-linear molecule, such as CH_4 or water, can rotate around three axes and, again, each mode of motion contributes a term $\frac{1}{2}kT$ to the mean energy. Therefore, the mean rotational energy is $\frac{3}{2}kT$. That is,

$$\langle \varepsilon \rangle = \varepsilon_{gs} + 3kT \quad \text{(non-linear molecule, translation and rotation only)} \qquad (13.41b)$$

The mean energy now increases twice as rapidly with temperature compared with the monatomic gas.

The energy of interacting molecules in condensed phases also has a contribution from the potential energy of their interaction. However, no simple expressions can be written down in general. Nevertheless, the crucial molecular point is that, as the temperature of a system is raised, the mean energy increases as the various modes of motion become more highly excited.

The canonical partition function

In this section we see how to generalize our conclusions to include systems composed of interacting molecules. We shall also see how to obtain the molecular partition function from the more general form of the partition function developed here. This section just sets up the general approach: we shall show how it is applied when we deal with real gases (Section 14.10).

13.5 The canonical ensemble

The crucial new concept we need when treating systems of interacting particles is the 'ensemble'. Like so many scientific terms, the term has basically its normal meaning of 'collection', but it has been sharpened and refined into a precise significance.

(a) The concept of ensemble

To set up an ensemble, we take a closed system of specified volume, composition, and temperature, and think of it as replicated $\tilde{N}$ times (Fig. 13.20). All the identical closed systems are regarded as being in thermal contact with one another, so

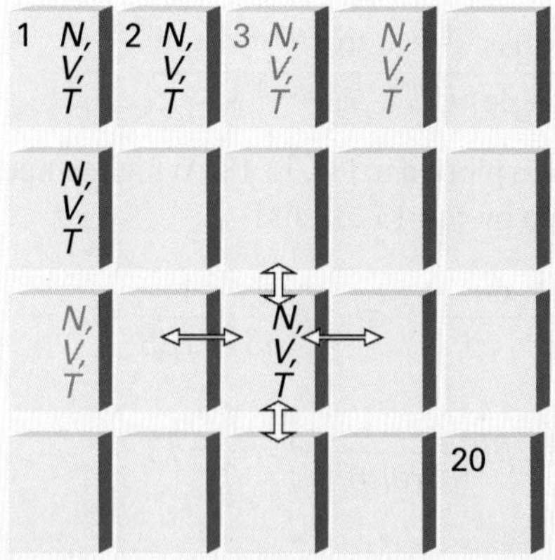

Fig. 13.20 A representation of the canonical ensemble, in this case for $\tilde{N} = 20$. The individual replications of the actual system all have the same composition and volume. They are all in mutual thermal contact, and so all have the same temperature. Energy may be transferred between them as heat, and so they do not all have the same energy. The total energy $\tilde{E}$ of all 20 replications is a constant because the ensemble is isolated overall.

they can exchange energy. The total energy of all the systems is $\tilde{E}$ and, because they are in thermal equilibrium with one another, they all have the same temperature, T. The volume of each member of the ensemble is the same, so the energy levels available to the molecules are the same in each system, and each member contains the same number of molecules, so there is a fixed number of molecules to distribute within each system. This imaginary collection of replications of the actual system with a common temperature is called the **canonical ensemble**.

The word 'canon' means 'according to a rule'. There are two other important ensembles. In the **microcanonical ensemble** the condition of constant temperature is replaced by the requirement that all the systems should have exactly the same energy: each system is individually isolated. In the **grand canonical ensemble** the volume and temperature of each system is the same, but they are open, which means that matter can be imagined as able to pass between the systems; the composition of each one may fluctuate, but now the property known as the chemical potential (which is described in Section 16.3) is the same in each system:

Microcanonical ensemble: N, V, E common
Canonical ensemble: N, V, T common
Grand canonical ensemble: μ, V, T common

The microcanonical ensemble was silently the basis of the discussion earlier in this chapter; we shall not consider the grand canonical ensemble explicitly.

The important point about an ensemble is that it is a collection of *imaginary* replications of the system, so we are free to let the number of members be as large as we like; when appropriate, we can let $\tilde{N}$ become infinite. The number of members of the ensemble in a state with energy E_i is denoted $\tilde{N}_i$, and we can speak of the configuration of the ensemble (by analogy with the configuration of the system used in Section 13.1) and its weight, $\tilde{\mathcal{W}}$. Note that $\tilde{N}$ is unrelated to N, the number of molecules in the actual system; $\tilde{N}$ is the number of imaginary replications of that system.

(b) Dominating configurations

Just as in Section 13.1, some of the configurations of the canonical ensemble will be very much more probable than others. For instance, it is very unlikely that the whole of the total energy, $\tilde{E}$, will accumulate in one system. By analogy with the earlier discussion, we can anticipate that there will be a dominating configuration, and that we can evaluate the thermodynamic properties by taking the average over the ensemble using that single, most probable, configuration. In the **thermodynamic limit** of $\tilde{N} \to \infty$, this dominating configuration is overwhelmingly the most probable, and it dominates the properties of the system virtually completely.

The quantitative discussion follows the argument in Section 13.1 with the modification that N and N_i are replaced by $\tilde{N}$ and $\tilde{N}_i$. The weight of a configuration $\{\tilde{N}_0, \tilde{N}_1, \ldots\}$ is

$$\tilde{\mathcal{W}} = \frac{\tilde{N}!}{\tilde{N}_0!\tilde{N}_1!\cdots} \qquad (13.42)$$

The configuration of greatest weight, subject to the constraints that the total energy of the ensemble is constant at $\tilde{E}$ and that the total number of members is fixed at $\tilde{N}$, is given by the **canonical distribution**:

$$\frac{\tilde{N}_i}{\tilde{N}} = \frac{e^{-\beta E_i}}{Q} \qquad Q = \sum_i e^{-\beta E_i} \qquad (13.43)$$

where the sum is over all members of the ensemble, each one having an energy E_i. The quantity Q, which is a function of the temperature, is called the **canonical partition function**.

(c) Fluctuations from the most probable distribution

The canonical distribution in eqn 13.43 is only apparently an exponentially decreasing function of the energy of the system. We must appreciate that the equation gives the probability of occurrence of members in a single state i of the entire system of energy E_i. There may in fact be numerous states with almost identical energies. For example, in a gas the identities of the molecules moving slowly or quickly can change without necessarily affecting the total energy. The density of states, the number of states in an energy range divided by the width of the range (Fig. 13.21), is a very sharply increasing function of energy. It follows that the probability of a member of an ensemble having a specified energy (as distinct from being in a specified state) is given by eqn 13.43, a sharply decreasing function, multiplied by a sharply increasing function (Fig. 13.22). Therefore, the overall distribution is a sharply peaked function. We conclude that most members of the ensemble have an energy very close to the mean value.

13.6 The mean energy of a system

Just as the molecular partition function can be used to calculate the mean value of a molecular property, so the canonical partition function can be used to calculate the mean energy of an entire system composed of molecules (that might or might not be interacting with one another). Thus, Q is more general than q because it does not assume that the molecules are

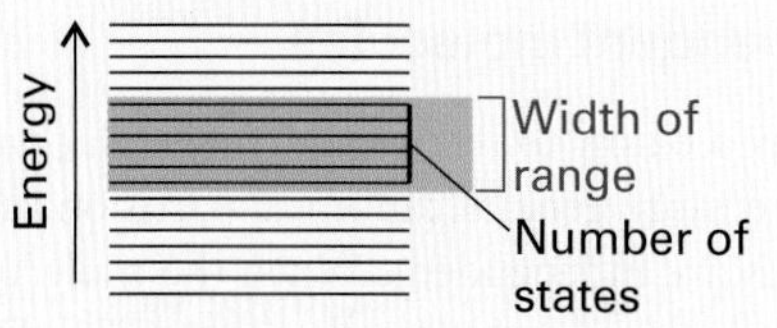

Fig. 13.21 The energy density of states is the number of states in an energy range divided by the width of the range.

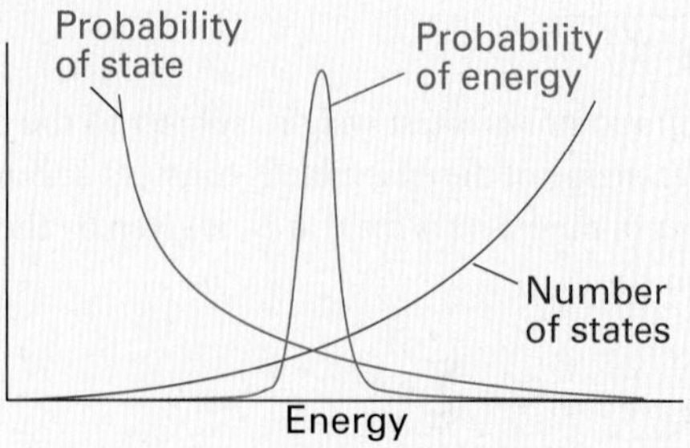

Fig. 13.22 To construct the form of the distribution of members of the canonical ensemble in terms of their energies, we multiply the probability that any one is in a state of given energy, eqn 13.43, by the number of states corresponding to that energy (a steeply rising function). The product is a sharply peaked function at the mean energy, which shows that almost all the members of the ensemble have that energy.

independent. We can therefore use Q to discuss the properties of condensed phases and real gases where molecular interactions are important.

If the total energy of the ensemble is $\tilde{E}$, and there are $\tilde{N}$ members, the average energy of a member is $\langle E\rangle = \tilde{E}/\tilde{N}$. Because the fraction, $\tilde{p}_i$, of members of the ensemble in a state i with energy E_i is given by the analogue of eqn 13.8 as

$$\tilde{p}_i = \frac{e^{-\beta E_i}}{Q} \tag{13.44}$$

it follows that

$$\langle E\rangle = \frac{\tilde{E}}{\tilde{N}} = \frac{1}{Q}\sum_i E_i e^{-\beta E_i} \tag{13.45}$$

By the same argument that led to eqn 13.34,

$$\langle E\rangle = -\frac{1}{Q}\left(\frac{\partial Q}{\partial \beta}\right)_V = -\left(\frac{\partial \ln Q}{\partial \beta}\right)_V \tag{13.46}$$

As in the case of the mean molecular energy, we must add to this expression the ground-state energy of the entire system if that is not zero.

13.7 Independent molecules

We shall now see how to recover the molecular partition function from the more general canonical partition function when the molecules are independent. When the molecules are independent and distinguishable (in the sense to be described), the relation between Q and q is

$$Q = q^N \tag{13.47}$$

Justification 13.5 *The relation between Q and q*

The total energy of a collection of N independent molecules is the sum of the energies of the molecules. Therefore, we can write the total energy of a state i of the system as

$$E_i = \varepsilon_i(1) + \varepsilon_i(2) + \cdots + \varepsilon_i(N)$$

In this expression, $\varepsilon_i(1)$ is the energy of molecule 1 when the system is in the state i, $\varepsilon_i(2)$ the energy of molecule 2 when the system is in the same state i, and so on. The canonical partition function is then

$$Q = \sum_i e^{-\beta\varepsilon_i(1)-\beta\varepsilon_i(2)-\cdots-\beta\varepsilon_i(N)}$$

The sum over the states of the system can be reproduced by letting each molecule enter all its own individual states (although we meet an important proviso shortly). Therefore, instead of summing over the states i of the system, we can sum over all the individual states j of molecule 1, all the states j of molecule 2, and so on. This rewriting of the original expression leads to

$$Q = \left(\sum_j e^{-\beta\varepsilon_j}\right)\left(\sum_j e^{-\beta\varepsilon_j}\right)\cdots\left(\sum_j e^{-\beta\varepsilon_j}\right) = \left(\sum_j e^{-\beta\varepsilon_j}\right)^N = q^N$$

If all the molecules are identical and free to move through space, we cannot distinguish them and the relation $Q = q^N$ is not valid. Suppose that molecule 1 is in some state a, molecule 2 is in b, and molecule 3 is in c, then one member of the ensemble has an energy $E = \varepsilon_a + \varepsilon_b + \varepsilon_c$. This member, however, is indistinguishable from one formed by putting molecule 1 in state b, molecule 2 in state c, and molecule 3 in state a, or some other permutation. There are six such permutations in all, and $N!$ in general. In the case of indistinguishable molecules, it follows that we have counted too many states in going from the sum over system states to the sum over molecular states, so writing $Q = q^N$ overestimates the value of Q. The detailed argument is quite involved, but at all except very low temperatures it turns out that the correction factor is $1/N!$. Therefore:

- For distinguishable independent molecules: $Q = q^N$ (13.48a)
- For indistinguishable independent molecules: $Q = q^N/N!$ (13.48b)

For molecules to be indistinguishable, they must be of the same kind: an Ar atom is never indistinguishable from a Ne atom. Their identity, however, is not the only criterion. Each identical molecule in a crystal lattice, for instance, can be 'named' with a set of coordinates. Identical molecules in a lattice can therefore be treated as distinguishable because their sites are distinguishable, and we use eqn 13.48a. On the other hand, identical molecules in a gas are free to move to different locations, and there is no way of keeping track of the identity of a given molecule; we therefore use eqn 13.48b.

Checklist of key ideas

☐ 1. The instantaneous configuration of a system of N molecules is the specification of the set of populations $N_0, N_1, \ldots$ of the energy levels $\varepsilon_0, \varepsilon_1, \ldots$. The weight $\mathcal{W}$ of a configuration is given by $\mathcal{W} = N!/N_0!N_1!\ldots$.

☐ 2. The Boltzmann distribution gives the numbers of molecules in each state of a system at any temperature: $N_i = Ne^{-\beta\varepsilon_i}/q$, $\beta = 1/kT$.

☐ 3. The molecular partition function is defined as $q = \sum_i e^{-\beta\varepsilon_i}$ and is an indication of the number of thermally accessible states at the temperature of interest.

☐ 4. If the energy of a molecule is given by the sum of translational, rotational, vibrational, and electronic contributions, then the molecular partition function is a product of contributions from the different modes.

☐ 5. The translational partition function of a molecule freely moving in a volume V is given by $q^{\mathrm{T}} = V/\Lambda^3$ where Λ is the thermal wavelength defined in eqn 13.15b.

☐ 6. The rotational partition function for a nonsymmetrical linear rotor is given by eqn 13.19. When $T \gg \theta_{\mathrm{R}}$, where θ_{R} is the characteristic rotational temperature $hc\tilde{B}/k$, the rotational partition function for a linear rotor is $q^{\mathrm{R}} = T/\sigma\theta_{\mathrm{R}}$, where σ is the symmetry number.

☐ 7. The vibrational partition function for a harmonic oscillator is given by eqn 13.24.

☐ 8. Since electronic energy separations from the ground state are usually very big, in most cases q^{E} is the degeneracy of the electronic ground state.

☐ 9. The mean energy of a molecule can be expressed in terms of the molecular partition function as given in eqn 13.35b.

☐ 10. The mean translational energy of a molecule free to move in three dimensions is $\frac{3}{2}kT$.

☐ 11. The canonical ensemble is an imaginary collection of replications of the actual system with a common temperature.

☐ 12. The canonical distribution is given by $\tilde{N}_i/\tilde{N} = e^{-\beta E_i}/Q$ where $Q = \sum_i e^{-\beta E_i}$ is the canonical partition function.

☐ 13. The mean energy of the system is $E_{\mathrm{gs}} - (\partial \ln Q/\partial\beta)_V$ where E_{gs} is the ground-state energy of the entire system.

☐ 14. For distinguishable independent molecules we write $Q = q^N$. For indistinguishable independent molecules we write $Q = q^N/N!$.

Table 13.4 List of symbols

Symbol	Meaning
Microcanonical ensemble (N, V, E common); think: molecule	
N_i	Number of molecules in the state of energy ε_i
ε_i	Relative energy of the state i ($\varepsilon_0 = 0$)
N	Total number of molecules in a system
E	Total energy of a system
q	Molecular partition function
$\langle\varepsilon\rangle$	Mean energy of molecules, relative to $\varepsilon_0 = 0$, $\langle\varepsilon\rangle = E/N$
$\mathcal{W}$	Weight of a configuration of the system
Canonical ensemble (N, V, T common); think: entire system	
$\tilde{N}_i$	Number of members of the ensemble in a state with energy E_i
E_i	Energy of the state of a system
$\tilde{N}$	Total number of replications of the system
$\tilde{E}$	Total energy of the ensemble
Q	Canonical partition function
$\langle E\rangle$	Mean energy of a system, $\langle E\rangle = \tilde{E}/\tilde{N}$, relative to $E_0 = 0$
$\mathcal{W}$	Weight of a configuration of the ensemble

Further information

Further information 13.1 *The derivation of the Boltzmann distribution*

We remarked in Section 13.1 that $\ln \mathcal{W}$ is easier to handle than $\mathcal{W}$. Therefore, to find the form of the Boltzmann distribution, we look for the condition for $\ln \mathcal{W}$ being a maximum rather than dealing directly with $\mathcal{W}$. Because $\ln \mathcal{W}$ depends on all the N_i, when a configuration changes and the N_i change to $N_i + dN_i$, the function $\ln \mathcal{W}$ changes to $\ln \mathcal{W} + d\ln \mathcal{W}$, where

$$d\ln\mathcal{W} = \sum_i \left(\frac{\partial \ln\mathcal{W}}{\partial N_i}\right) dN_i$$

All this expression states is that a change in $\ln \mathcal{W}$ is the sum of contributions arising from changes in each value of N_i. At a maximum, $d\ln \mathcal{W} = 0$. However, when the N_i change, they do so subject to the two constraints

$$\sum_i \varepsilon_i dN_i = 0 \qquad \sum_i dN_i = 0 \qquad (13.49)$$

The first constraint recognizes that the total energy must not change, and the second recognizes that the total number of molecules must not change. These two constraints prevent us from solving $d\ln \mathcal{W} = 0$ simply by setting all $(\partial \ln \mathcal{W}/\partial N_i) = 0$ because the dN_i are not all independent.

The way to take constraints into account was devised by the French mathematician Lagrange, and is called the **method of undetermined multipliers.**[1] All we need here is the rule that a constraint should be multiplied by a constant and then added to the main variation equation. The variables are then treated as though they were all independent, and the constants are evaluated at the end of the calculation.

We employ the technique as follows. The two constraints in eqn 13.49 are multiplied by the constants $-\beta$ and α, respectively (the minus sign in $-\beta$ has been included for future convenience), and then added to the expression for d ln $\mathcal{W}$:

$$\mathrm{d}\ln\mathcal{W} = \sum_i\left(\frac{\partial\ln\mathcal{W}}{\partial N_i}\right)\mathrm{d}N_i + \alpha\sum_i \mathrm{d}N_i - \beta\sum_i \varepsilon_i \mathrm{d}N_i$$

$$= \sum_i\left\{\left(\frac{\partial\ln\mathcal{W}}{\partial N_i}\right) + \alpha - \beta\varepsilon_i\right\}\mathrm{d}N_i$$

All the $\mathrm{d}N_i$ are now treated as independent. Hence the only way of satisfying d ln $\mathcal{W}$ = 0 is to require that, for each i,

$$\frac{\partial\ln\mathcal{W}}{\partial N_i} + \alpha - \beta\varepsilon_i = 0 \qquad (13.50)$$

when the N_i have their most probable values.

Equation 13.3 for $\mathcal{W}$ is

$$\ln\mathcal{W} = N\ln N - \sum_i N_i \ln N_i$$

There is a small housekeeping step to take before differentiating ln $\mathcal{W}$ with respect to N_i: this equation is identical to

$$\ln\mathcal{W} = N\ln N - \sum_j N_j \ln N_j$$

because all we have done is to change the 'name' of the states from i to j. This step makes sure that we do not confuse the i in the differentiation variable (N_i) with the i in the summation. Now differentiation of this expression gives

$$\frac{\partial\ln\mathcal{W}}{\partial N_i} = \frac{\partial(N\ln N)}{\partial N_i} - \sum_j\frac{\partial(N_j\ln N_j)}{\partial N_i} \qquad (13.51)$$

The derivative of the first term is obtained as follows:

$$\frac{\partial(N\ln N)}{\partial N_i} = \left(\frac{\partial N}{\partial N_i}\right)\ln N + N\left(\frac{\partial\ln N}{\partial N_i}\right)$$

$$= \ln N + \frac{\partial N}{\partial N_i} = \ln N + 1 \qquad (13.52)$$

The ln N in the first term on the right in the second line arises because $N = N_1 + N_2 + \cdots$ and so the derivative of N with respect to any of the N_i is 1: that is, $\partial N/\partial N_i = 1$. The second term on the right in the second line arises because $\partial(\ln N)/\partial N_i = (1/N)\partial N/\partial N_i$. The final 1 is then obtained in the same way as in the preceding remark, by using $\partial N/\partial N_i = 1$.

For the derivative of the second term we first note that

$$\frac{\partial\ln N_j}{\partial N_i} = \frac{1}{N_j}\left(\frac{\partial N_j}{\partial N_i}\right) \qquad (13.53)$$

Morever, if $i \neq j$, N_j is independent of N_i, so $\partial N_j/\partial N_i = 0$. However, if $i = j$,

$$\frac{\partial N_j}{\partial N_i} = \frac{\partial N_j}{\partial N_j} = 1 \qquad (13.54)$$

Therefore,

$$\frac{\partial N_j}{\partial N_i} = \delta_{ij} \qquad (13.55)$$

with δ_{ij} the Kronecker delta ($\delta_{ij} = 1$ if $i = j$, $\delta_{ij} = 0$ otherwise). Then

$$\sum_j\frac{\partial(N_j\ln N_j)}{\partial N_i} = \sum_j\left\{\left(\frac{\partial N_j}{\partial N_i}\right)\ln N_j + N_j\left(\frac{\partial\ln N_j}{\partial N_i}\right)\right\}$$

$$= \sum_j\left\{\left(\frac{\partial N_j}{\partial N_i}\right)\ln N_j + \left(\frac{\partial N_j}{\partial N_i}\right)\right\}$$

$$= \sum_j\left(\frac{\partial N_j}{\partial N_i}\right)(\ln N_j + 1)$$

$$= \sum_j \delta_{ij}(\ln N_j + 1) = \ln N_i + 1$$

and therefore

$$\frac{\partial\ln\mathcal{W}}{\partial N_i} = -(\ln N_i + 1) + (\ln N + 1) = -\ln\frac{N_i}{N} \qquad (13.56)$$

It follows from eqn 13.50 that

$$-\ln\frac{N_i}{N} + \alpha - \beta\varepsilon_i = 0$$

and therefore that

$$\frac{N_i}{N} = \mathrm{e}^{\alpha-\beta\varepsilon_i} \qquad (13.57)$$

At this stage we note that

$$N = \sum_i N_i = \sum_i N\mathrm{e}^{\alpha-\beta\varepsilon_i} = N\mathrm{e}^{\alpha}\sum_i \mathrm{e}^{-\beta\varepsilon_i}$$

Because the N cancels on each side of this equality, it follows that

$$\mathrm{e}^{\alpha} = \frac{1}{\sum_j \mathrm{e}^{-\beta\varepsilon_j}} \qquad (13.58)$$

and

$$\frac{N_i}{N} = \mathrm{e}^{\alpha-\beta\varepsilon_i} = \mathrm{e}^{\alpha}\mathrm{e}^{-\beta\varepsilon_i} = \frac{1}{\sum_j \mathrm{e}^{-\beta\varepsilon_j}}\mathrm{e}^{-\beta\varepsilon_i}$$

which is eqn 13.6a.

[1] For a detailed account, see our *Physical chemistry* (2007).

Further information 13.2 *The partition functions of polyatomic rotors*

The energies of a symmetric rotor are

$$E_{J,K,M_J} = hc\tilde{B}J(J+1) + hc(\tilde{A}-\tilde{B})K^2$$

with $J = 0, 1, 2, \ldots, K = J, J-1, \ldots, -J$, and $M_J = J, J-1, \ldots, -J$. Instead of considering these ranges, we can cover the same values by allowing K to range from $-\infty$ to ∞, with J confined to $|K|, |K|+1, \ldots, \infty$ for each value of K (Fig. 13.23). Because the energy is independent of M_J, and there are $2J+1$ values of M_J for each value of J, each value of J is $(2J+1)$-fold degenerate. It follows that the partition function

$$q = \sum_{J=0}^{\infty}\sum_{K=-J}^{J}\sum_{M_J=-J}^{J} e^{-E_{J,K,M_J}/kT}$$

can be written equivalently as

$$\begin{aligned} q &= \sum_{K=-\infty}^{\infty}\sum_{J=|K|}^{\infty}(2J+1)e^{-E_{J,K,M_J}/kT} \\ &= \sum_{K=-\infty}^{\infty}\sum_{J=|K|}^{\infty}(2J+1)e^{-hc\{\tilde{B}J(J+1)+(\tilde{A}-\tilde{B})K^2\}/kT} \qquad (13.59) \\ &= \sum_{K=-\infty}^{\infty} e^{-\{hc(\tilde{A}-\tilde{B})/kT\}K^2}\sum_{J=|K|}^{\infty}(2J+1)e^{-hc\tilde{B}J(J+1)/kT} \end{aligned}$$

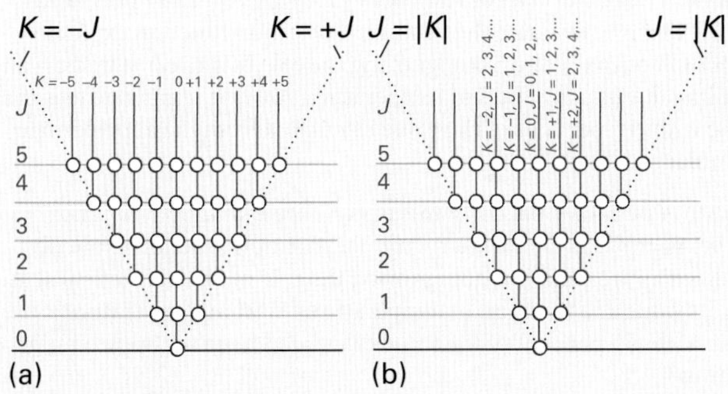

Fig. 13.23 (a) The sum over $J = 0, 1, 2, \ldots$ and $K = J, J-1, \ldots, -J$ (depicted by the circles) can be covered (b) by allowing K to range from $-\infty$ to ∞, with J confined to $|K|, |K|+1, \ldots, \infty$ for each value of K.

Now we assume that the temperature is so high that numerous states are occupied and that the sums may be approximated by integrals. Then

$$q = \int_{-\infty}^{\infty} e^{-\{hc(\tilde{A}-\tilde{B})/kT\}K^2}\int_{|K|}^{\infty}(2J+1)e^{-hc\tilde{B}J(J+1)/kT}\,dJdK \qquad (13.60)$$

As before, the integral over J can be recognized as the integral of the derivative of a function, which is the function itself, so

$$\begin{aligned} \int_{|K|}^{\infty}(2J+1)e^{-hc\tilde{B}J(J+1)/kT}dJ &= \int_{|K|}^{\infty}\left(-\frac{kT}{hc\tilde{B}}\right)\frac{d}{dJ}e^{-hc\tilde{B}J(J+1)/kT}dJ \\ &= \left(-\frac{kT}{hc\tilde{B}}\right)e^{-hc\tilde{B}J(J+1)/kT}\Big|_{|K|}^{\infty} \\ &= \left(\frac{kT}{hc\tilde{B}}\right)e^{-hc\tilde{B}|K|(|K|+1)/kT} \qquad (13.61) \\ &\approx \left(\frac{kT}{hc\tilde{B}}\right)e^{-hc\tilde{B}K^2/kT} \end{aligned}$$

In the last line we have supposed that $|K| \gg 1$ for most contributions. Now we can write

$$\begin{aligned} q &= \frac{kT}{hc\tilde{B}}\int_{-\infty}^{\infty} e^{-\{hc(\tilde{A}-\tilde{B})/kT\}K^2}e^{-hc\tilde{B}K^2/kT}dK \\ &= \frac{kT}{hc\tilde{B}}\int_{-\infty}^{\infty} e^{-\{hc\tilde{A}/kT\}K^2}dK \qquad (13.62) \\ &= \left(\frac{kT}{hc\tilde{B}}\right)\left(\frac{kT}{hc\tilde{A}}\right)^{1/2}\overbrace{\int_{-\infty}^{\infty}e^{-x^2}dx}^{\pi^{1/2}} \\ &= \left(\frac{kT}{hc}\right)^{3/2}\left(\frac{\pi}{\tilde{A}\tilde{B}^2}\right)^{1/2} \end{aligned}$$

For an asymmetric rotor, one of the $\tilde{B}$s is replaced by $\tilde{C}$, to give eqn 13.20b.

Discussion questions

13.1 Describe the physical significance of the partition function.

13.2 Describe how the internal energy of a system composed of two levels varies with temperature.

13.3 Discuss the relationship between 'population', 'configuration', and 'weight'. What is the significance of the most probable configuration?

13.4 What are the significance and importance of the principle of equal *a priori* probabilities?

13.5 What is temperature?

13.6 What is the difference between a 'state' and an 'energy level'? Why is it important to make this distinction?

13.7 Discuss the conditions under which energies predicted from the equipartition theorem coincide with energies computed by using partition functions.

13.8 Explain what is meant by an *ensemble* and why it is useful in statistical thermodynamics.

13.9 Under what circumstances may identical particles be regarded as distinguishable?

13.10 Discuss the role of the Boltzmann distribution in chemistry.

Exercises

13.1(a) Calculate the weight of the configuration in which 16 objects are distributed in the arrangement 0, 1, 2, 3, 8, 0, 0, 0, 0, 2.

13.1(b) Calculate the weight of the configuration in which 21 objects are distributed in the arrangement 6, 0, 5, 0, 4, 0, 3, 0, 2, 0, 0, 1.

13.2(a) Evaluate 8! by using (a) the exact formula, (b) Stirling's approximation, (c) the improved version of Stirling's approximation.

13.2(b) Evaluate 10! by using (a) the exact formula, (b) Stirling's approximation, (c) the improved version of Stirling's approximation.

13.3(a) What are the relative populations of the states of a two-level system when the temperature is infinite?

13.3(b) What are the relative populations of the states of a two-level system as the temperature approaches zero?

13.4(a) What is the temperature of a two-level system of energy separation equivalent to 400 cm^{-1} when the population of the upper state is one-third that of the lower state?

13.4(b) What is the temperature of a two-level system of energy separation equivalent to 300 cm^{-1} when the population of the upper state is one-half that of the lower state?

13.5(a) Calculate the relative populations of a linear rotor in the levels with $J = 0$ and $J = 5$, given that $\tilde{B} = 2.71$ cm^{-1} and a temperature of 298 K.

13.5(b) Calculate the relative populations of a spherical rotor in the levels with $J = 0$ and $J = 5$, given that $\tilde{B} = 2.71$ cm^{-1} and a temperature of 298 K.

13.6(a) A certain molecule has a non-degenerate excited state lying at 540 cm^{-1} above the non-degenerate ground state. At what temperature will 10 per cent of the molecules be in the upper state?

13.6(b) A certain molecule has a doubly degenerate excited state lying at 360 cm^{-1} above the non-degenerate ground state. At what temperature will 15 per cent of the molecules be in the upper level?

13.7(a) Calculate (a) the thermal wavelength, (b) the translational partition function at (i) 300 K and (ii) 3000 K of a molecule of molar mass 150 g mol^{-1} in a container of volume 1.00 cm^3.

13.7(b) Calculate (a) the thermal wavelength, (b) the translational partition function of a Ne atom in a cubic box of side 1.00 cm at (i) 300 K and (ii) 3000 K.

13.8(a) Calculate the ratio of the translational partition functions of H_2 and He at the same temperature and volume.

13.8(b) Calculate the ratio of the translational partition functions of Ar and Ne at the same temperature and volume.

13.9(a) The bond length of O_2 is 120.75 pm. Use the high-temperature approximation to calculate the rotational partition function of the molecule at 300 K.

13.9(b) The bond length of N_2 is 109.75 pm. Use the high-temperature approximation to calculate the rotational partition function of the molecule at 300 K.

13.10(a) The NOF molecule is an asymmetric rotor with rotational constants 3.1752 cm^{-1}, 0.3951 cm^{-1}, and 0.3505 cm^{-1}. Calculate the rotational partition function of the molecule at (a) 25°C, (b) 100°C.

13.10(b) The H_2O molecule is an asymmetric rotor with rotational constants 27.877 cm^{-1}, 14.512 cm^{-1}, and 9.285 cm^{-1}. Calculate the rotational partition function of the molecule at (a) 25°C, (b) 100°C.

13.11(a) The rotational constant of CO is 1.931 cm^{-1}. Evaluate the rotational partition function explicitly (without approximation) and plot its value as a function of temperature. At what temperature is the value within 5 per cent of the value calculated from the approximate formula?

13.11(b) The rotational constant of HI is 6.511 cm^{-1}. Evaluate the rotational partition function explicitly (without approximation) and plot its value as a function of temperature. At what temperature is the value within 5 per cent of the value calculated from the approximate formula?

13.12(a) The rotational constant of CH_4 is 5.241 cm^{-1}. Evaluate the rotational partition function explicitly (without approximation but ignoring the role of nuclear statistics) and plot its value as a function of temperature. At what temperature is the value within 5 per cent of the value calculated from the approximate formula?

13.12(b) The rotational constant of CCl_4 is 0.0572 cm^{-1}. Evaluate the rotational partition function explicitly (without approximation but ignoring the role of nuclear statistics) and plot its value as a function of temperature. At what temperature is the value within 5 per cent of the value calculated from the approximate formula?

13.13(a) The rotational constants of CH_3Cl are $\tilde{A} = 5.097$ cm^{-1} and $\tilde{B} = 0.443$ cm^{-1}. Evaluate the rotational partition function explicitly (without approximation but ignoring the role of nuclear statistics) and plot its value as a function of temperature. At what temperature is the value within 5 per cent of the value calculated from the approximate formula?

13.13(b) The rotational constants of NH_3 are $\tilde{A} = 6.196$ cm^{-1} and $\tilde{B} = 9.444$ cm^{-1}. Evaluate the rotational partition function explicitly (without approximation but ignoring the role of nuclear statistics) and plot its value as a function of temperature. At what temperature is the value within 5 per cent of the value calculated from the approximate formula?

13.14(a) Give the symmetry number for each of the following molecules: (a) CO, (b) O_2, (c) H_2S, (d) SiH_4, and (e) $CHCl_3$.

13.14(b) Give the symmetry number for each of the following molecules: (a) CO_2, (b) O_3, (c) SO_3, (d) SF_6, and (e) Al_2Cl_6.

13.15(a) Estimate the rotational partition function of ethene at 25°C given that $\tilde{A} = 4.828$ cm^{-1}, $\tilde{B} = 1.0012$ cm^{-1}, and $\tilde{C} = 0.8282$ cm^{-1}. Take the symmetry number into account.

13.15 (b) Evaluate the rotational partition function of pyridine, C_5H_5N, at room temperature given that $\tilde{A} = 0.2014$ cm^{-1}, $\tilde{B} = 0.1936$ cm^{-1}, $\tilde{C} = 0.0987$ cm^{-1}. Take the symmetry number into account.

13.16(a) The vibrational wavenumber of Br_2 is 323.2 cm^{-1}. Evaluate the vibrational partition function (without approximation) and plot its value as a function of temperature. At what temperature is the value within 5 per cent of the value calculated from the approximate formula?

13.16(b) The vibrational wavenumber of I_2 is 214.5 cm^{-1}. Evaluate the vibrational partition function (without approximation) and plot its value as a function of temperature. At what temperature is the value within 5 per cent of the value calculated from the approximate formula?

13.17(a) Calculate the vibrational partition function of CS_2 at 500 K given the wavenumbers 658 cm^{-1} (symmetric stretch), 397 cm^{-1} (bend; two modes), 1535 cm^{-1} (asymmetric stretch).

13.17(b) Calculate the vibrational partition function of HCN at 900 K given the wavenumbers 3311 cm^{-1} (symmetric stretch), 712 cm^{-1} (bend; two modes), 2097 cm^{-1} (asymmetric stretch).

13.18(a) Calculate the vibrational partition function of CCl_4 at 500 K given the wavenumbers 459 cm^{-1} (symmetric stretch, A), 217 cm^{-1} (deformation, E), 776 cm^{-1} (deformation, T), 314 cm^{-1} (deformation, T).

13.18(b) Calculate the vibrational partition function of CI_4 at 500 K given the wavenumbers 178 cm^{-1} (symmetric stretch, A), 90 cm^{-1} (deformation, E), 555 cm^{-1} (deformation, T), 125 cm^{-1} (deformation, T).

13.19(a) A certain atom has a fourfold degenerate ground level, a non-degenerate electronically excited level at 2500 cm^{-1}, and a twofold degenerate level at 3500 cm^{-1}. Calculate the partition function of these electronic states at 1900 K. What is the relative population of each level at 1900 K?

13.19(b) A certain atom has a triply degenerate ground level, a non-degenerate electronically excited level at 850 cm^{-1}, and a fivefold degenerate level at 1100 cm^{-1}. Calculate the partition function of these electronic states at 2000 K. What is the relative population of each level at 2000 K?

13.20(a) Compute the mean energy at 298 K of a two-level system of energy separation equivalent to 500 cm^{-1}.

13.20(b) Compute the mean energy at 400 K of a two-level system of energy separation equivalent to 600 cm^{-1}.

13.21(a) Evaluate, by explicit summation, the mean rotational energy of CO and plot its value as a function of temperature. Use the data in Exercise 13.11a. At what temperature is the equipartition value within 5 per cent of the accurate value?

13.21(b) Evaluate, by explicit summation, the mean rotational energy of HI and plot its value as a function of temperature. Use the data in Exercise 13.11b. At what temperature is the equipartition value within 5 per cent of the accurate value?

13.22(a) Evaluate, by explicit summation, the mean rotational energy of CH_4 and plot its value as a function of temperature. Use the data in Exercise 13.12a. At what temperature is the equipartition value within 5 per cent of the accurate value?

13.22(b) Evaluate, by explicit summation, the mean rotational energy of CCl_4 and plot its value as a function of temperature. Use the data in Exercise 13.12b. At what temperature is the equipartition value within 5 per cent of the accurate value?

13.23(a) Evaluate, by explicit summation, the mean rotational energy of CH_3Cl and plot its value as a function of temperature. Use the data in Exercise 13.13a. At what temperature is the equipartition value within 5 per cent of the accurate value?

13.23(b) Evaluate, by explicit summation, the mean rotational energy of NH_3 and plot its value as a function of temperature. Use the data in Exercise 13.13b. At what temperature is the equipartition value within 5 per cent of the accurate value?

13.24(a) Evaluate, by explicit summation, the mean vibrational energy of Br_2 and plot its value as a function of temperature. Use the data in Exercise 13.16a. At what temperature is the equipartition value within 5 per cent of the accurate value?

13.24(b) Evaluate, by explicit summation, the mean vibrational energy of I_2 and plot its value as a function of temperature. Use the data in Exercise 13.16b. At what temperature is the equipartition value within 5 per cent of the accurate value?

13.25(a) Evaluate, by explicit summation, the mean vibrational energy of CS_2 and plot its value as a function of temperature. Use the data in Exercise 13.17a. At what temperature is the equipartition value within 5 per cent of the accurate value?

13.25(b) Evaluate, by explicit summation, the mean vibrational energy of HCN and plot its value as a function of temperature. Use the data in Exercise 13.17b. At what temperature is the equipartition value within 5 per cent of the accurate value?

13.26(a) Evaluate, by explicit summation, the mean vibrational energy of CCl_4 and plot its value as a function of temperature. Use the data in Exercise 13.18a. At what temperature is the equipartition value within 5 per cent of the accurate value?

13.26(b) Evaluate, by explicit summation, the mean vibrational energy of CI_4 and plot its value as a function of temperature. Use the data in Exercise 13.18b. At what temperature is the equipartition value within 5 per cent of the accurate value?

13.27(a) Calculate the mean contribution to the electronic energy at 1900 K for a sample composed of the atoms specified in Exercise 13.19a.

13.27(b) Calculate the mean contribution to the electronic energy at 2000 K for a sample composed of the atoms specified in Exercise 13.19b.

13.28(a) An electron spin can adopt either of two orientations in a magnetic field, and its energies are $\pm\mu_B\mathcal{B}$, where μ_B is the Bohr magneton. Deduce an expression for the partition function and mean energy of the electron and sketch the variation of the functions with $\mathcal{B}$. Calculate the relative populations of the spin states at (a) 4.0 K, (b) 298 K when $\mathcal{B} = 1.0$ T.

13.28(b) A nitrogen nuclear spin can adopt any of three orientations in a magnetic field, and its energies are 0, $\pm\gamma\hbar\mathcal{B}$, where γ_N is the magnetogyric ratio of the nucleus. Deduce an expression for the partition function and mean energy of the nucleus and sketch the variation of the functions with $\mathcal{B}$. Calculate the relative populations of the spin states at (a) 1.0 K, (b) 298 K when $\mathcal{B} = 20.0$ T.

13.29(a) Identify the systems for which it is essential to include a factor of $1/N!$ on going from Q to q: (a) a sample of helium gas, (b) a sample of carbon monoxide gas, (c) a solid sample of carbon monoxide, (d) water vapour.

13.29(b) Identify the systems for which it is essential to include a factor of $1/N!$ on going from Q to q: (a) a sample of carbon dioxide gas, (b) a sample of graphite, (c) a sample of diamond, (d) ice.

Problems*

Numerical problems

13.1 A sample consisting of five molecules has a total energy 5ε. Each molecule is able to occupy states of energy $j\varepsilon$, with $j = 0, 1, 2, \ldots$ (a) Calculate the weight of the configuration in which the energy is distributed evenly over the available molecules. (b) Draw up a table with columns headed by the energy of the states and write beneath them all configurations that are consistent with the total energy. Calculate the weights of each configuration and identify the most probable configurations.

13.2 A sample of nine molecules is numerically tractable but on the verge of being thermodynamically significant. Draw up a table of configurations for $N = 9$, total energy 9ε in a system with energy levels $j\varepsilon$ (as in Problem 13.1). Before evaluating the weights of the configurations, guess (by looking for the most 'exponential' distribution of populations) which of the configurations will turn out to be the most probable. Go on to calculate the weights and identify the most probable configuration.

13.3 Use mathematical software to evaluate $\mathcal{W}$ for $N = 20$ for a series of distributions over a uniform ladder of energy levels, ensuring that the total energy is constant. Identify the configuration of greatest weight and compare it to the distribution predicted by the Boltzmann expression. Explore what happens as the value of the total energy is changed.

13.4 This problem is also best done using mathematical software. Equation 13.10 is the partition function for a harmonic oscillator. Consider a Morse oscillator (Section 10.9) in which the energy levels are given by eqn 10.36.

$$E_v = (v + \tfrac{1}{2})hc\tilde{\nu} - (v + \tfrac{1}{2})^2 hcx_e\tilde{\nu}$$

Evaluate the partition function for this oscillator, remembering (1) to measure energies from the lowest level and (2) to note that there is only a finite number of levels. Plot the partition function against temperature for a variety of values of x_e and—on the same graph—compare your results with those for a harmonic oscillator.

13.5 Explore the conditions under which the 'integral' approximation for the translational partition function is not valid by considering the translational partition function of an H atom in a one-dimensional box of side comparable to that of a typical nanoparticle, 100 nm. Estimate the temperature at which, according to the integral approximation, $q = 10$ and evaluate the exact partition function at that temperature.

13.6 An electron trapped in an infinitely deep spherical well of radius R, such as may be encountered in the investigation of nanoparticles, has energies given by the expression $E_{nl} = \hbar^2 X_{nl}^2 / 2m_e R^2$, with X_{nl} the value obtained by searching for the zeroes of the spherical Bessel functions. The first six values (with a degeneracy of the corresponding energy level equal to $2l + 1$) are as follows:

n	l	X_{nl}
1	0	3.142
1	1	4.493
1	2	5.763
2	0	6.283
1	3	6.988
2	1	7.725

Evaluate the partition function and mean energy of an electron as a function of temperature. Choose the temperature range and radius to be so low that only these six energy levels need be considered. *Hint.* Remember to measure energies from the lowest level.

13.7 A certain atom has a doubly degenerate ground level pair and an upper level of four degenerate states at 450 cm^{-1} above the ground level. In an atomic beam study of the atoms it was observed that 30 per cent of the atoms were in the upper level, and the translational temperature of the beam was 300 K. Are the electronic states of the atoms in thermal equilibrium with the translational states?

13.8 (a) Calculate the electronic partition function of a tellurium atom at (i) 298 K, (ii) 5000 K by direct summation using the following data:

Term	Degeneracy	Wavenumber/cm^{-1}
Ground	5	0
1	1	4707
2	3	4751
3	5	10 559

(b) What proportion of the Te atoms are in the ground term and in the term labelled 2 at the two temperatures?

13.9 The four lowest electronic levels of a Ti atom are: $^3F_2$, $^3F_3$, $^3F_4$, and $^5F_1$, at 0, 170, 387, and 6557 cm^{-1}, respectively. There are many other electronic states at higher energies. The boiling point of titanium is 3287°C. What are the relative populations of these levels at the boiling point? *Hint.* The degeneracies of the levels are $2J + 1$.

13.10 The NO molecule has a doubly degenerate excited electronic level 121.1 cm^{-1} above the doubly degenerate electronic ground term. Calculate and plot the electronic partition function of NO from $T = 0$ to 1000 K. Evaluate (a) the term populations and (b) the mean electronic energy at 300 K.

13.11‡ J. Sugar and A. Musgrove (*J. Phys. Chem. Ref. Data* 22, 1213 (1993)) have published tables of energy levels for germanium atoms and cations from Ge^+ to Ge^{+31}. The lowest-lying energy levels in neutral Ge are as follows:

	$^3P_0$	$^3P_1$	$^3P_2$	$^1D_2$	$^1S_0$
(E/hc)/cm^{-1}	0	557.1	1410.0	7125.3	16367.3

Calculate the electronic partition function at 298 K and 1000 K by direct summation. *Hint.* The degeneracy of a level J is $2J + 1$.

13.12 The pure rotational microwave spectrum of HCl has absorption lines at the following wavenumbers (in cm^{-1}): 21.19, 42.37, 63.56, 84.75, 105.93, 127.12, 148.31, 169.49, 190.68, 211.87, 233.06, 254.24, 275.43, 296.62, 317.80, 338.99, 360.18, 381.36, 402.55, 423.74, 444.92, 466.11, 487.30, 508.48. Calculate the rotational partition function at 25°C by direct summation.

13.13 Calculate, by explicit summation, the vibrational partition function and the vibrational contribution to the energy of I_2 molecules at (a) 100 K, (b) 298 K given that its vibrational energy levels lie at the following wavenumbers above the zero-point energy level: 0, 213.30, 425.39, 636.27, 845.93 cm^{-1}. What proportion of I_2 molecules are in the ground and first two excited levels at the two temperatures?

* Problems denoted with the symbol ‡ were supplied by Charles Trapp and Carmen Giunta.

Theoretical problems

13.14 Explore the consequences of using the full version of Stirling's approximation, $x! \approx (2\pi)^{1/2}x^{x+1/2}e^{-x}$, in the development of the expression for the configuration of greatest weight. Does the more accurate approximation have a significant effect on the form of the Boltzmann distribution?

13.15 The most probable configuration is characterized by a parameter we know as the 'temperature'. The temperatures of the system specified in Problems 13.1 and 13.2 must be such as to give a mean value of ε for the energy of each molecule and a total energy $N\varepsilon$ for the system. (a) Show that the temperature can be obtained by plotting p_j against j, where p_j is the (most probable) fraction of molecules in the state with energy $j\varepsilon$. Apply the procedure to the system in Problem 13.2. What is the temperature of the system when ε corresponds to 50 cm^{-1}? (b) Choose configurations other than the most probable, and show that the same procedure gives a worse straight line, indicating that a temperature is not well-defined for them.

13.16 Consider a system with energy levels $\varepsilon_j = j\varepsilon$ and N molecules. (a) Show that if the mean energy per molecule is $a\varepsilon$, then the temperature is given by

$$\beta = \frac{1}{\varepsilon}\ln\left(1+\frac{1}{a}\right)$$

Evaluate the temperature for a system in which the mean energy is ε, taking ε equivalent to 50 cm^{-1}. (b) Calculate the molecular partition function q for the system when its mean energy is $a\varepsilon$.

13.17 Deduce an expression for the root mean square energy, $\langle\varepsilon^2\rangle^{1/2}$, in terms of the partition function and hence an expression for the root mean square deviation from the mean, $\Delta\varepsilon = (\langle\varepsilon^2\rangle - \langle\varepsilon\rangle^2)^{1/2}$. Evaluate the resulting expression for a harmonic oscillator.

13.18 A formal way of arriving at the value of the symmetry number is to note that σ is the order (the number of elements) of the *rotational subgroup* of the molecule, the point group of the molecule with all but the identity and the rotations removed. Identify the symmetry number of (a) H_2O, (b) NH_3, (c) CH_4, (d) benzene. In each case, specify the elements of the subgroup.

13.19‡ For gases, the canonical partition function, Q, is related to the molecular partition function q by $Q = q^N/N!$. Use the expression for q and general thermodynamic relations to derive the perfect gas law $pV = nRT$.

Applications: to atmospheric science, astrophysics, and biochemistry

13.20‡ The variation of the atmospheric pressure p with altitude h is predicted by the *barometric formula* to be $p = p_0 e^{-h/H}$ where p_0 is the pressure at sea level and $H = RT/Mg$ with M the average molar mass of air and T the average temperature. Obtain the barometric formula from the Boltzmann distribution. Recall that the potential energy of a particle at height h above the surface of the Earth is mgh. Convert the barometric formula from pressure to number density, $\mathcal{N}$. Compare the relative number densities, $\mathcal{N}(h)/\mathcal{N}(0)$, for O_2 and H_2O at $h = 8.0$ km, a typical cruising altitude for commercial aircraft.

13.21‡ Planets lose their atmospheres over time unless they are replenished. A complete analysis of the overall process is very complicated and depends upon the radius of the planet, temperature, atmospheric composition, and other factors. Prove that the atmosphere of planets cannot be in an equilibrium state by demonstrating that the Boltzmann distribution leads to a uniform finite number density as $r \to \infty$. *Hint.* Recall that in a gravitational field the potential energy is $V(r) = -GMm/r$, where G is the gravitational constant, M is the mass of the planet, and m the mass of the particle.

13.22‡ Consider the electronic partition function of a perfect atomic hydrogen gas at a density of 1.99×10^{-4} kg m^{-3} and 5780 K. These are the mean conditions within the Sun's photosphere, the surface layer of the Sun that is about 190 km thick. (a) Show that this partition function, which involves a sum over an infinite number of quantum states that are solutions to the Schrödinger equation for an isolated atomic hydrogen atom, is infinite. (b) Develop a theoretical argument for truncating the sum and estimate the maximum number of quantum states that contribute to the sum. (c) Calculate the equilibrium probability that an atomic hydrogen electron is in each quantum state. Are there any general implications concerning electronic states that will be observed for other atoms and molecules? Is it wise to apply these calculations in the study of the Sun's photosphere?

13.23 Consider a protein P with four distinct sites, with each site capable of binding one ligand L. Show that the possible varieties ('configurations') of the species PL_i (with PL_0 denoting P) are given by the binomial coefficients $C(4,i)$.

13.24 Complete some of the derivations in the discussion of the helix–coil transition in polypeptides (*Impact I13.1*). (a) Show that, within the tenets of the zipper model,

$$q = 1 + \sum_{i=1}^{n}(n-i+1)\sigma s^i$$

and that $(n - i + 1)$ is the number of ways in which an allowed state with a number i of c amino acids can be formed. (b) Go on to show that

$$\frac{q}{q_0} = 1 + \sigma(n+1)\sum_{i=1}^{n}s^i - \sigma\sum_{i=1}^{n}is^i$$

(c) Use the relations

$$\sum_{i=1}^{n}x^i = \frac{x^{n+1}-x}{x-1} \qquad \sum_{i=1}^{n}ix^i = \frac{x}{(x-1)^2}[nx^{n+1}-(n+1)x^n+1]$$

to show that

$$\frac{q}{q_0} = 1 + \frac{\sigma s[s^{n+1}-(n+1)s^n+1]}{(s-1)^2}$$

which is eqn 13.30. (d) Using the zipper model, show that $\theta = (1/n)\mathrm{d}(\ln q)/\mathrm{d}(\ln s)$. *Hint.* As a first step, show that $\sum_i i(n-i+1)\sigma s^i = s(\mathrm{d}q/\mathrm{d}s)$.

13.25 Here you will use the zipper model discussed in *Impact I13.1* to explore the helix–coil transition in polypeptides. (a) Investigate the effect of the parameter s on the distribution of random coil segments in a polypeptide with $n = 20$ by plotting p_i, the fraction of molecules with a number i of amino acids in a coil region, against i for $s = 0.8$, 1.0, and 1.5, with $\sigma = 5.0 \times 10^{-2}$. Compare your results with those shown in Fig. 13.15 and discuss the significance of any effects you discover. (b) The average value of i given by $\langle i\rangle = \sum_i ip_i$. Use the results of the zipper model to calculate $\langle i\rangle$ for all the combinations of s and σ used in Fig. 13.16 and part (a).

MATHEMATICAL BACKGROUND 7

Probability theory

Probability theory deals with quantities and events that are distributed randomly and shows how to calculate average values of various kinds. We shall consider variables that take discrete values (as in a one-dimensional random walk with a fixed step length) and continuous values (as in the diffusion of a particle through a fluid).

MB7.1 Discrete distributions

We denote a variable X and the discrete values that it may take x_i, $i = 1, 2, \ldots, N$. If the probability that x_i occurs is p_i, then the **mean value** (or *expectation value*) of X is

$$\langle X\rangle = \sum_{i=1}^{N} x_i p_i \qquad \text{(MB7.1a)}$$

The mean values of higher powers of X may be computed similarly:

$$\langle X^n\rangle = \sum_{i=1}^{N} x_i^n p_i \qquad \text{(MB7.1b)}$$

Although the mean is a useful measure, it is important to know the width in the scatter of outcomes around the mean. There are two related measures: one is the **variance**, $V(X)$, and the other is the **standard deviation**, $\sigma(X)$, the square-root of the variance:

$$V(X) = \langle X^2\rangle - \langle X\rangle^2 \qquad \text{(MB7.2a)}$$

$$\sigma(X) = \{V(X)\}^{1/2} = \{\langle X^2\rangle - \langle X\rangle^2\}^{1/2} \qquad \text{(MB7.2b)}$$

In certain cases, the probabilities can be expressed in a simple way, depending on the nature of the events being considered.

(a) The binomial distribution

In a **Bernoulli trial**, the outcome of an observation is one of a mutually exclusive pair (such as heads or tails in a coin toss) and successive trials are independent (so that getting 'heads' on one toss does not influence the following toss). Suppose the probability of outcome 1 is p and that of the alternative outcome 2 is q, with $p + q = 1$. For a fair coin, $p = q = \frac{1}{2}$. Then one series of $N = 12$ trials might be

thhtthhthttht Probability of occurrence $= p^5q^7$, and in general p^nq^{N-n}

However, if the order in which heads come up is unimportant, there are 12!/5!7! ways of achieving 5 heads in 12 tosses, and in general $N!/n!(N-n)!$ ways of achieving n 'heads' in N trials. The probability of getting exactly n 'heads' is therefore the product of p^nq^{N-n} and the number of ways of distributing n 'heads' over N trials:

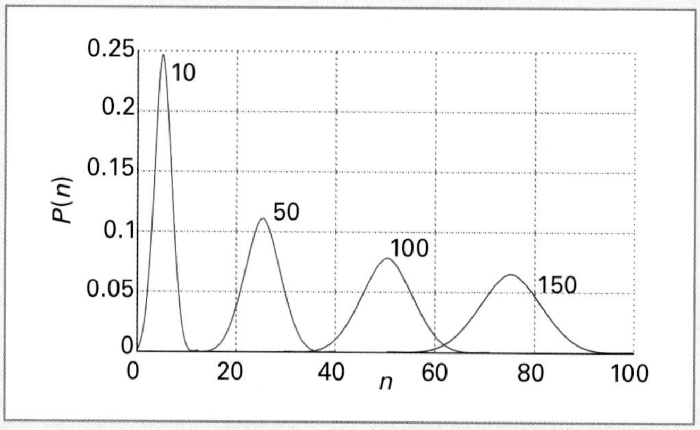

Fig. MB7.1 The binomial distribution for different values of N and $p = q = \frac{1}{2}$.

$$P(n) = \binom{N}{n} p^nq^{N-n} \quad \text{where} \quad \binom{N}{n} = \frac{N!}{n!(N-n)!} \qquad \text{(MB7.3)}$$

The symbol $\binom{N}{n}$ is the **binomial coefficient**, as it occurs in the binomial expansion (*Mathematical background 1*):

$$(x+y)^N = \sum_{n=0}^{N} \binom{N}{n} x^n y^{N-n} \qquad \text{(MB7.4)}$$

As a result, the expression $P(n)$ is called the **binomial distribution** (Fig. MB7.1).

● **A BRIEF ILLUSTRATION**

We can use the binomial distribution to determine the mean number of times that heads will be obtained in a series of N trials (this is the average value of n, denoted $\langle n\rangle$):

$$\langle n\rangle = \sum_{n=0}^{N} nP(n) = \sum_{n=0}^{N} n\binom{N}{n} p^nq^{N-n} = \sum_{n=0}^{N} n\binom{N}{n} p^n(1-p)^{N-n}$$

A useful procedure for evaluating sums in which n is a factor is to:

1. Introduce a dummy variable a.
2. Express the sum as a derivative with respect to a, using $na^n = a\mathrm{d}a^n/\mathrm{d}a$ to eliminate the factor n; if a power of n occurs, apply $\mathrm{d}/\mathrm{d}a$ the appropriate number of times.
3. Express the resulting sum, perhaps by using eqn MB7.4.
4. Evaluate the derivative.
5. Finally, set $a = 1$.

Application of this procedure to the expression for $\langle n\rangle$ above gives:

$$\langle n\rangle \overset{\text{Step 1}}{=} \sum_{n=0}^{N} n\binom{N}{n}a^n p^n(1-p)^{N-n}$$

$$\overset{\text{Step 2}}{=} a\frac{\mathrm{d}}{\mathrm{d}a}\sum_{n=0}^{N}\binom{N}{n}a^n p^n(1-p)^{N-n}$$

$$\overset{\text{Step 3}}{=} a\frac{\mathrm{d}}{\mathrm{d}a}\{ap+(1-p)\}^N$$

$$\overset{\text{Step 4}}{=} Nap\{ap+(1-p)\}^{N-1} \xrightarrow{\text{Step 5: set } a=1} Np$$

For example the mean number of times heads is obtained for a fair coin ($p=\tfrac{1}{2}$) in 10 trials is 5. ●

Self-test MB7.1 Evaluate the standard deviation $\sigma(n)$. $[\sigma(n)=\{Np(1-p)\}^{1/2}]$

Note that, although the width of the distribution increases (as $N^{1/2}$), its value relative to the mean decreases (as $N^{1/2}/N = 1/N^{1/2}$). For tosses of a fair coin, when $p=\tfrac{1}{2}$, $\langle n\rangle = \tfrac{1}{2}N$ (half the tosses turn up heads), and $\sigma(n)=\tfrac{1}{2}N^{1/2}$.

(b) The Poisson distribution

In another important type of trial, an event either takes place or does not, such as an excited molecule dissociating into fragments or (more mundanely) a bus arriving. At first sight it appears that we cannot assign a meaning to the number of times an event does not occur (how many times did the bus not arrive in an interval?). However, we can still assign a probability that an event occurs by imagining an interval of time Δt that is divided into N regions, each of duration $\Delta t/N$ (or, similarly, regions of space $\Delta x/N$) that are so small that the probability that two or more events occurs in it is negligible and we have a 'heads' for the event occurring in that brief interval or tiny region and a 'tails' if it does not.

If the events occur at random, the probability of an event occurring within this tiny interval is proportional to the length of the interval, and we can write $p=\Delta t/N\tau$, where $1/\tau$ is a constant of proportionality (we give it a physical meaning later). It then follows that in a set of Bernoulli trials (where 'heads' now corresponds to 'did occur' and 'tails' to 'did not occur') the total probability that n events occur in the interval $\Delta t = N(\Delta t/N)$ is just the probability of getting n 'heads' in a total interval that spans N of the tiny intervals:

$$P(n)=\binom{N}{n}p^n(1-p)^{N-n}=\binom{N}{n}\left(\frac{\Delta t}{N\tau}\right)^n\left(1-\frac{\Delta t}{N\tau}\right)^{N-n}$$

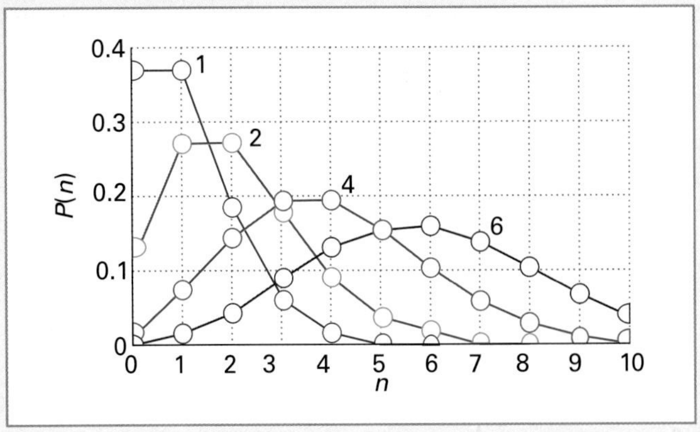

Fig. MB7.2 The Poisson distribution for different values of $\Delta t/\tau$.

If then we suppose that N is very large and make use of the relation

$$\lim_{x\to 0}(1+x)^{1/x}=\mathrm{e} \qquad \text{(MB7.5)}$$

then in a slightly involved but straightforward calculation (which we do not reproduce here) we arrive at the **Poisson distribution**:

$$P(n)=\frac{(\Delta t/\tau)^n}{n!}\mathrm{e}^{-\Delta t/\tau} \qquad \text{(MB7.6)}$$

for the probability that n events will occur in the interval Δt (Fig. MB7.2). As shown in the following *brief illustration*, the mean number of events in the interval Δt is just $\Delta t/\tau$, so τ can be interpreted as the average time between events.

● A BRIEF ILLUSTRATION

To calculate the average number of events that occur in an interval Δt, we need to evaluate

$$\langle n\rangle=\sum_{n=0}^{\infty}nP(n)=\sum_{n=0}^{\infty}\frac{n(\Delta t/\tau)^n}{n!}\mathrm{e}^{-\Delta t/\tau}=\mathrm{e}^{-\Delta t/\tau}\sum_{n=1}^{\infty}\frac{n(\Delta t/\tau)^n}{(n-1)!}$$

$$=(\Delta t/\tau)\mathrm{e}^{-\Delta t/\tau}\sum_{n=1}^{\infty}\frac{(\Delta t/\tau)^{n-1}}{(n-1)!}=(\Delta t/\tau)\mathrm{e}^{-\Delta t/\tau}\sum_{n=0}^{\infty}\frac{(\Delta t/\tau)^n}{n!}$$

$$=(\Delta t/\tau)\mathrm{e}^{-\Delta t/\tau}\mathrm{e}^{\Delta t/\tau}=\Delta t/\tau$$

where we have used the Taylor series expansion (eqn MB1.7b) of e^x. ●

Self-test MB7.2 Deduce an expression for the standard deviation of n. $[\sigma(n)=(\Delta t/\tau)^{1/2}]$

(c) The Gaussian distribution

Suppose that a variable can take positive and negative integer values centred on zero and that to reach a certain value n the system jumps to the left or right by taking steps of length λ at random for a total of N steps. The number of ways of taking

N_R steps to the right and N_L to the left (with $N = N_R + N_L$) in any one such trial of N steps is

$$W = \frac{N!}{N_R!N_L!} \qquad \text{(MB7.7)}$$

Then, because there are 2^N possible choices of direction in the course of N steps, the probability of being n steps from the origin with $n = N_R - N_L$ in a trial of N steps, is

$$P(n) = \frac{W}{2^N} = \frac{N!}{N_R!N_L!2^N} = \frac{N!}{\left(\frac{N+n}{2}\right)!\left(\frac{N-n}{2}\right)!2^N} \qquad \text{(MB7.8)}$$

A point of some subtlety and which we draw on later is that, if N is even, then n must be even (you cannot end up an odd number of steps from the origin if you take an even number of steps; think about $N = 4$); similarly, if N is odd, then n must be odd too (you cannot end up an even number of steps from the origin if you take an odd number of steps; think about $N = 5$). Because it then follows that $N + n$ and $N - n$ are both even numbers, the factorials we have to evaluate in $P(n)$ are of whole numbers.

To develop this expression in the case of large numbers of steps, we take logarithms and use Stirling's approximation

$$\ln x! \approx \ln(2\pi)^{1/2} + (x + \tfrac{1}{2}) \ln x - x \qquad \text{(MB7.9)}$$

This approximation leads, after a fair amount of algebra, to

$$\ln P(n) = \ln\left(\frac{2}{(2\pi N)^{1/2}}\right) - \tfrac{1}{2}(N - n + 1)\ln\left(1 - \frac{n}{N}\right) - \tfrac{1}{2}(N + n + 1)\ln\left(1 + \frac{n}{N}\right) \qquad \text{(MB7.10)}$$

Now, we allow N to become very large and use $\ln(1+x) \approx x - \tfrac{1}{2}x^2$, and obtain

$$\ln P(n) = \ln\left(\frac{2}{(2N\pi)^{1/2}}\right) - \frac{n^2}{2N}$$

and thence obtain the **Gaussian distribution**

$$P(n) = \frac{2}{(2N\pi)^{1/2}} e^{-n^2/2N} \qquad \text{(MB7.11)}$$

with, remember, n even if N is even and n odd if N is odd. This bell-shaped curve is illustrated in Fig. MB7.3 for N even and N odd. The manipulation of the Gaussian distribution is to use it to discuss a continuous function, as we demonstrate in the following section.

MB7.2 Continuous distributions

A **continuous distribution** is a distribution in which the variable can take on a continuum of values. One of the most important examples can be developed from the Gaussian distribution of the preceding section as follows.

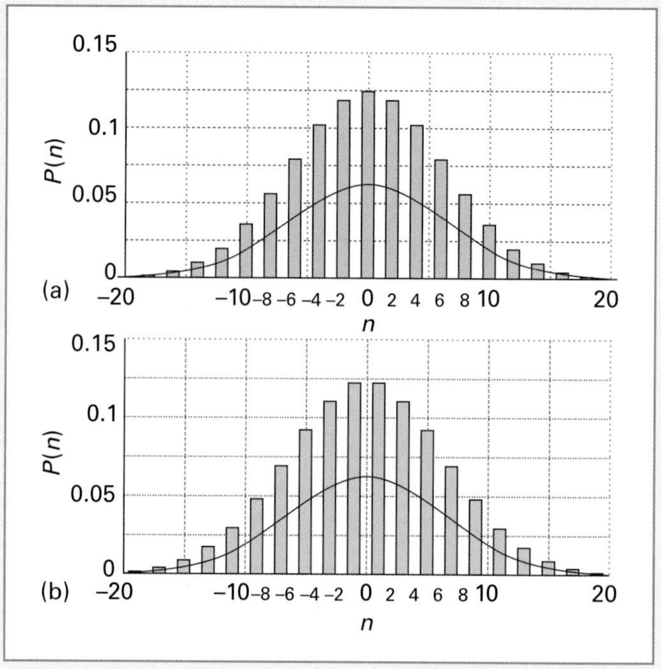

Fig. MB7.3 The Gaussian distribution for (a) $N = 40$ and (b) $N = 41$. In each case the outer wings of the distribution have been trimmed at $n = \pm 20$. The bell-shaped curve is the average of the distribution, and corresponds to the Gaussian distribution function for a continuous variable.

When N is very large and the x_i values are so closely spaced that the observable X can be regarded as varying continuously, it is useful to express the probability that X can have a value between x and $x + dx$ as

$$dP(X) = \rho(x)dx \qquad \text{(MB7.12)}$$

where the function $\rho(x)$ is the **probability density**, a measure of the distribution of the probability values over x, and dx is an infinitesimally small interval of x values. It follows that the probability that X has a value between $x = a$ and $x = b$ is the integral of the expression above evaluated between a and b:

$$P(X; a \le X \le b) = \int_a^b \rho(x)dx \qquad \text{(MB7.13)}$$

The mean value of X continuously varying between $-\infty$ and ∞ is given by

$$\langle X \rangle = \int_{-\infty}^{+\infty} x\rho(x)dx \qquad \text{(MB7.14)}$$

with analogous expressions for the variance and standard deviation. This expression is similar to that written for the case of discrete values of X, with $\rho(x)dx$ as the probability term and integration over the closely spaced x values replacing summation over widely spaced x_i. The mean value of a function $g(X)$ can be calculated with a formula similar to that for $\langle X \rangle$:

$$\langle g(X) \rangle = \int_{-\infty}^{+\infty} g(x)\rho(x)\mathrm{d}x \qquad \text{(MB7.15)}$$

To derive the Gaussian version of the probability density we use the random walk model in Section MB7.1 and write $x = n\lambda$, allowing λ to be very small and n to be very large and effectively continuous. If the width of dx spreads over a sufficiently wide range of points then, instead of dealing with a distribution like that in Fig. MB7.3a or like that in Fig. MB7.3b, we can deal with the average of the two, as shown by the curves superimposed on the distribution. That is, for a continuous distribution we use

$$P(n) = \frac{1}{(2N\pi)^{1/2}}\mathrm{e}^{-n^2/2N} \qquad \text{(MB7.16)}$$

The total probability of being in the range $\mathrm{d}x = \lambda \mathrm{d}n$ at $x = n\lambda$ is therefore

$$\rho(x)\mathrm{d}x = \frac{1}{(2N\pi)^{1/2}}\mathrm{e}^{-n^2/2N}\mathrm{d}n = \frac{1}{(2N\pi)^{1/2}}\mathrm{e}^{-x^2/2N\lambda^2}\frac{\mathrm{d}x}{\lambda}$$

It follows that

$$\rho(x) = \frac{1}{(2\pi N\lambda^2)^{1/2}}\mathrm{e}^{-x^2/2N\lambda^2} \qquad \text{(MB7.17a)}$$

This expression is commonly expressed as the **Gaussian distribution function** (or *normal distribution function*)

$$\rho(x) = \left(\frac{2}{(2\pi\sigma^2)}\right)^{1/2} \mathrm{e}^{-x^2/2\sigma^2} \qquad \text{(MB7.17b)}$$

where $\sigma = N^{1/2}\lambda$ turns out to be the standard deviation of the distribution (see below).

● **A BRIEF ILLUSTRATION**

If the bell-shaped curve of the Gaussian distribution function is centred on $\langle x \rangle$, the distribution becomes

$$\rho(x) = \left(\frac{2}{2\pi\sigma^2}\right)^{1/2} \mathrm{e}^{-(x-\langle x \rangle)^2/2\sigma^2}$$

To evaluate the mean square value of x we write

$$\langle x^2 \rangle = \int_{-\infty}^{\infty} x^2\rho(x)\mathrm{d}x = \left(\frac{1}{2\pi\sigma^2}\right)^{1/2}\int_{-\infty}^{\infty} x^2\mathrm{e}^{-(x-\langle x \rangle)^2/2\sigma^2}\mathrm{d}x$$

We make the substitution $z = (x - \langle x \rangle)/(2\sigma^2)^{1/2}$, so $\mathrm{d}x = (2\sigma^2)^{1/2}\mathrm{d}z$ and $x^2 = 2\sigma^2 z^2 + 2(2\sigma^2)^{1/2}z\langle x \rangle + \langle x \rangle^2$, and obtain

$$\langle x^2 \rangle = \left(\frac{1}{2\pi\sigma^2}\right)^{1/2}(2\sigma^2)^{1/2}\left\{\int_{-\infty}^{\infty} 2\sigma^2 z^2 \mathrm{e}^{-z^2}\mathrm{d}z + 2(2\sigma^2)^{1/2}\langle x \rangle \int_{-\infty}^{\infty} z\mathrm{e}^{-z^2}\mathrm{d}z + \langle x \rangle^2 \int_{-\infty}^{\infty} \mathrm{e}^{-z^2}\mathrm{d}z\right\}$$

$$= \frac{1}{\pi^{1/2}}\{\pi^{1/2}\sigma^2 + \langle x \rangle^2 \pi^{1/2}\} = \sigma^2 + \langle x \rangle^2$$ ●

Self-test MB7.3 Determine the standard deviation of the normal distribution. [Equal to the width parameter σ]

14 The First Law of thermodynamics

This chapter introduces some of the basic concepts of thermodynamics. It concentrates on the conservation of energy—the experimental observation that energy can be neither created nor destroyed—and shows how the principle of the conservation of energy can be used to assess the energy changes that accompany physical and chemical processes. Much of this chapter examines the means by which a system can exchange energy with its surroundings in terms of the work it may do or the heat that it may produce, and shows how these concepts can be understood at a molecular level. Another target concept of the chapter is enthalpy, which is a very useful bookkeeping property for keeping track of the heat output or requirements of physical processes and chemical reactions taking place at constant pressure. We also begin to unfold some of the power of thermodynamics and statistical thermodynamics by showing how to establish relations between different properties of a system. We shall see that one very useful aspect of thermodynamics is that a property can be measured indirectly by measuring others and then combining their values.

The discussion of the Boltzmann distribution in Chapter 13 allows us to discuss the mean energy of a collection of molecules. That energy can be used to bring about a variety of processes, such as heating the surroundings or causing an electric current to flow through a circuit. In other words, the energy stored by a collection of molecules may be transformed into a variety of forms. **Thermodynamics** is the study of these transformations of energy. The historical development of thermodynamics was in terms of observations on the properties of bulk samples. It can still be explained in that way, but we shall see that our understanding of its concepts is greatly enriched by drawing on molecular concepts, and in particular the Boltzmann distribution.

The internal energy

For the purposes of thermodynamics, the universe is divided into two parts, the system and its surroundings. The **system** is the part of the world in which we have a special interest. It may be a reaction vessel, an engine, an electrochemical cell, a biological cell, and so on. The **surroundings** comprise the region outside the system and are where we make our measurements. The type of system depends on the characteristics of the boundary that divides it from the surroundings (Fig. 14.1). If matter can be transferred through the boundary between the system and its surroundings the system is classified as **open**. If matter cannot pass through the boundary the system is classified as **closed**. Both open and closed systems can exchange energy with their surroundings. For example, a closed system can expand and thereby raise a weight in

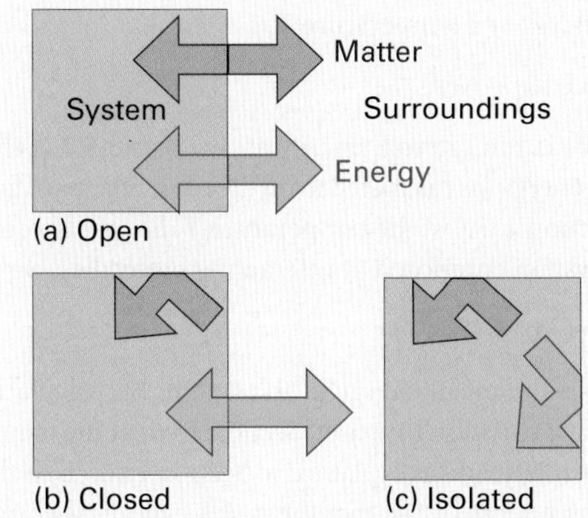

Fig. 14.1 (a) An open system can exchange matter and energy with its surroundings. (b) A closed system can exchange energy with its surroundings, but it cannot exchange matter. (c) An isolated system can exchange neither energy nor matter with its surroundings.

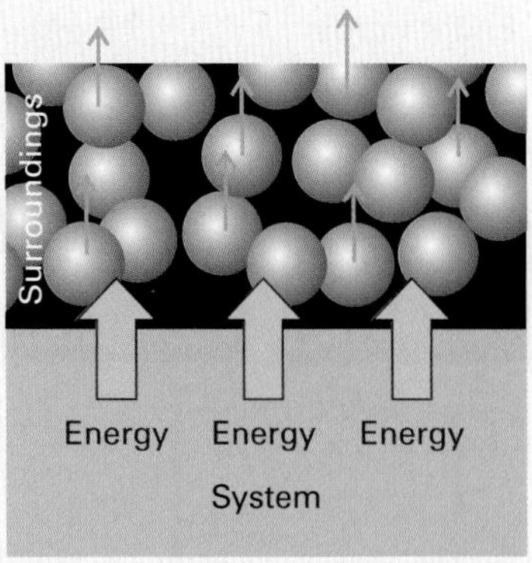

Fig. 14.2 When a system does work, it stimulates orderly motion in the surroundings. For instance, the atoms shown here may be part of a weight that is being raised. The ordered motion of the atoms in a falling weight does work on the system.

the surroundings; it may also transfer energy to them if they are at a lower temperature. An **isolated system** can exchange neither energy nor matter with its surroundings.

14.1 Work, heat, and energy

A fundamental concept in thermodynamics is work: **work** is done when motion takes place against an opposing force. Doing work is equivalent to raising a weight somewhere in the surroundings. An example of doing work is the expansion of a gas that pushes out a piston and raises a weight. A chemical reaction that drives an electric current through a resistance also does work, because the same current could be driven through a motor and used to raise a weight.

In molecular terms, *work is the transfer of energy that makes use of organized motion* (Fig. 14.2). When a weight is raised or lowered, its atoms move in an organized way (up or down). The atoms in a spring move in an orderly way when it is wound; the electrons in an electric current move in an orderly direction when it flows. When a system does work it causes atoms or electrons in its surroundings to move in an organized way. Likewise, when work is done *on* a system, molecules in the surroundings are used to transfer energy to it in an organized way, as the atoms in a weight are lowered or a current of electrons is passed.

The **energy** of a system is its capacity to do work. When work is done on an otherwise isolated system (for instance, by compressing a gas or winding a spring), the capacity of the system to do work is increased; in other words, the energy of the system is increased. When the system does work (when the piston moves out or the spring unwinds), the energy of the system is reduced and it can do less work than before.

Experiments have shown that the energy of a system may be changed by means other than work itself. When the energy of a system changes as a result of a temperature difference between the system and its surroundings we say that energy has been transferred as **heat**. When a heater is immersed in a beaker of water (the system), the capacity of the system to do work increases because hot water can be used to do more work than the same amount of cold water.

An **exothermic process** is one that releases energy as heat. All combustion reactions are exothermic. An **endothermic process** is one in which energy is acquired as heat. An example of an endothermic process is the vaporization of water. To avoid a lot of awkward language, we say that in an exothermic process 'heat is released' and in an endothermic process 'heat is absorbed'. However, it must never be forgotten that heat is a process (the transfer of energy as a result of a temperature difference), not a thing. When an endothermic process takes place in a **diathermic** (thermally conducting) container, heat flows into the system from the surroundings. When an exothermic process takes place in a diathermic container, heat flows into the surroundings. When an endothermic process takes place in an **adiabatic** (thermally insulating) container, it results in a lowering of temperature of the system; an exothermic process results in a rise of temperature. These features are summarized in Fig. 14.3.

In molecular terms, heating is the transfer of energy that makes use of *random molecular motion*. The random motion of molecules is called **thermal motion**. The thermal motion of the molecules in the hot surroundings stimulates the molecules in the cooler system to move more vigorously and, as a result, the energy of the system is increased. When a system heats its surroundings, molecules of the system stimulate the thermal motion of the molecules in the surroundings (Fig. 14.4).

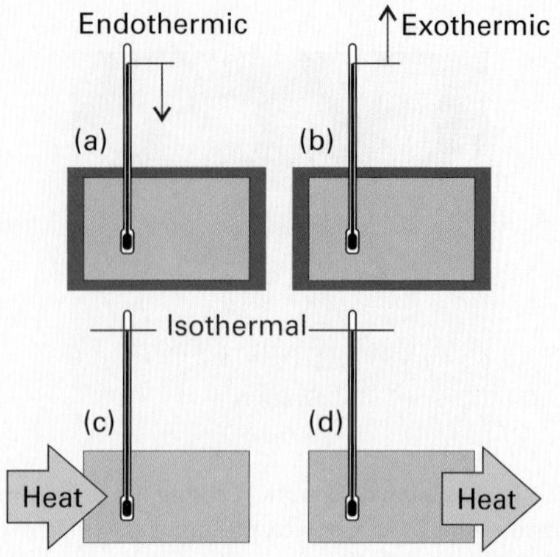

Fig. 14.3 (a) When an endothermic process occurs in an adiabatic system, the temperature falls; (b) if the process is exothermic, then the temperature rises. (c) When an endothermic process occurs in a diathermic container, energy enters as heat from the surroundings, and the system remains at the same temperature. (d) If the process is exothermic, then energy leaves as heat, and the process is isothermal.

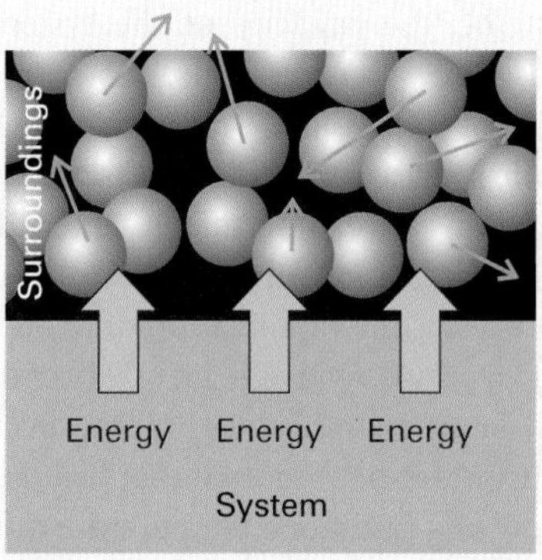

Fig. 14.4 When energy is transferred to the surroundings as heat, the transfer stimulates random motion of the atoms in the surroundings. Transfer of energy from the surroundings to the system makes use of random motion (thermal motion) in the surroundings.

14.2 The First Law

In thermodynamics, the total energy of a system is called its **internal energy**, U. The internal energy is the total kinetic and potential energy of the molecules in the system. It is 'internal' in the sense that it does not include the kinetic energy arising from the motion of the system as a whole, such as its kinetic energy as it accompanies the Earth on its orbit round the Sun.

For a system composed of N independent molecules, the internal energy at a temperature T is

$$U(T) = U(0) + N\langle\varepsilon\rangle \qquad (14.1a)$$

where $U(0)$ is the internal energy at $T = 0$ and $\langle\varepsilon\rangle$ is the mean molecular energy (as calculated from the molecular partition function in Section 13.4) at the temperature T. For the more general case of a system composed of interacting molecules, we write

$$U(T) = U(0) + \langle E\rangle \qquad (14.1b)$$

where $\langle E\rangle$ is the mean energy of the system (as calculated from the canonical partition function, Section 13.6) at the temperature T. More formally, and as explained in Section 13.6, $\langle E\rangle$ is the mean value of the energy of the members of a canonical ensemble in the thermodynamic limit of $\tilde{N} \to \infty$.

● A BRIEF ILLUSTRATION

The molar internal energy is obtained by setting $N = nN_A$, where N_A is Avogadro's constant, and dividing the total internal energy by the amount of molecules, n, in the sample:

$$U_m(T) = U_m(0) + N_A\langle\varepsilon\rangle$$

We saw in Section 13.4, that the mean energy of a molecule due to its translational motion is $\frac{3}{2}kT$; therefore

$$U_m(T) = U_m(0) + \tfrac{3}{2}N_AkT = U_m(0) + \tfrac{3}{2}RT$$

At 25°C, $RT = 2.48$ kJ mol^{-1}, so the translational motion contributes 3.72 kJ mol^{-1} to the molar internal energy of gases. ●

Self-test 14.1 Calculate the molar internal energy of carbon dioxide at 25°C, taking into account its translational and rotational degrees of freedom. $[U_m(T) = U_m(0) + \tfrac{5}{2}RT]$

We denote by ΔU the change in internal energy when a system changes from an initial state i with internal energy U_i to a final state f of internal energy U_f:

$$\Delta U = U_f - U_i \qquad [14.2]$$

The internal energy of a closed system may be changed either by doing work on the system or by heating it. Whereas we may know how the energy transfer has occurred (because we can see if a weight has been raised or lowered in the surroundings, indicating transfer of energy by doing work, or if ice has melted in the surroundings, indicating transfer of energy as heat), the system is blind to the mode employed. *Heat and work are equivalent ways of changing a system's internal energy.* A system is like a bank: it accepts deposits in either currency, but stores its reserves as internal energy. If we write w for the work done on a system, q for the energy transferred as heat to a system, and ΔU for the resulting change in internal energy, then

$$\Delta U = q + w \qquad (14.3)$$

This equation employs the 'acquisitive convention', in which $w > 0$ or $q > 0$ if energy is transferred to the system as work or heat and $w < 0$ or $q < 0$ if energy is lost from the system as work or heat. In other words, we view the flow of energy as work or heat from the system's perspective.

A brief comment Be careful to distinguish q, the symbol for energy transferred as heat, from $\mathcal{q}$, the symbol for the partition function. Later you will also need to distinguish p, for pressure, from $\mathcal{p}$, for population. We use script symbols for statistical properties.

It is an experimental fact that we cannot use a system to do work, leave it isolated for a while, and then return to it to find its internal energy restored to its original value and ready to provide the same amount of work again. Despite the great amount of effort that has been spent trying to build a 'perpetual motion machine', a device that would be an exception to this rule by producing work without using fuel, no one has ever succeeded in building one. In other words, eqn 14.3 is a *complete* statement of how changes in internal energy may be achieved in a closed system: the only way to change the internal energy of a closed system is to transfer energy into it as heat or as work. If the system is isolated, then even that ability is eliminated, and the internal energy cannot change at all. This conclusion is known as the **First Law of thermodynamics**, which states:

The internal energy of an isolated system is constant.

Equation 14.3 is the mathematical statement of this law, for it implies the equivalence of heat and work as modes of transfer of energy and the fact that the internal energy is constant in an isolated system (for which $q = 0$ and $w = 0$). The First Law is closely related to the conservation of energy (*Fundamentals F.6*) but goes beyond it: the concept of heat does not apply to the single particles treated in classical mechanics.

● A BRIEF ILLUSTRATION

If an electric motor produced 15 kJ of energy each second as mechanical work and lost 2 kJ as heat to the surroundings, then the change in the internal energy of the motor each second is

$$\Delta U = -2\ \text{kJ} - 15\ \text{kJ} = -17\ \text{kJ}$$

Suppose that, when a spring was wound, 100 J of work was done on it but 15 J escaped to the surroundings as heat. The change in internal energy of the spring is

$$\Delta U = +100\ \text{kJ} - 15\ \text{kJ} = +85\ \text{kJ}$$ ●

According to the First Law, if an isolated system has a certain internal energy at one instant and is inspected again later, then it will be found to have exactly the same internal energy. Therefore, if a second system consisting of exactly the same amount of substance in exactly the same state as the first (and therefore indistinguishable from the first system) is inspected, it too would have the same internal energy as the first system. We summarize this conclusion by saying that the internal energy is a **state function**, a property that depends only on the current state of the system and is independent of how that state was prepared. The pressure, volume, temperature, and density of a system are also state functions.

14.3 Expansion work

The way can now be opened to powerful methods of calculation by switching attention to infinitesimal changes of state (such as an infinitesimal change in temperature) and infinitesimal changes in the internal energy $\mathrm{d}U$. Then, if the work done on a system is $\mathrm{d}w$ and the energy supplied to it as heat is $\mathrm{d}q$, in place of eqn 14.3 we write

$$\mathrm{d}U = \mathrm{d}q + \mathrm{d}w \tag{14.4}$$

To use this expression we must be able to relate $\mathrm{d}q$ and $\mathrm{d}w$ to events taking place in the surroundings.

We begin by discussing the important case of **expansion work**, the work arising from a change in volume. This type of work includes the work done by a gas as it expands and drives back the atmosphere. Many chemical reactions result in the generation or consumption of gases (for instance, the thermal decomposition of calcium carbonate or the combustion of octane), and the thermodynamic characteristics of a reaction depend on the work it can do. The term 'expansion work' also includes work associated with negative changes of volume, that is, compression.

(a) The general expression for work

The calculation of expansion work (*Fundamentals F.6*) starts from the definition used in physics, which states that the work required to move an object a distance $\mathrm{d}z$ against an opposing force of magnitude F is

$$\mathrm{d}w = -F\mathrm{d}z \qquad [14.5]$$

The negative sign tells us that when the system moves an object against an opposing force, the internal energy of the system doing the work will decrease. Now consider the arrangement shown in Fig. 14.5, in which one wall of a system is a massless, frictionless, rigid, perfectly fitting piston of area A. If the external pressure is p_{ex}, the magnitude of the force acting on the outer face of the piston is $F = p_{ex}A$. When the system expands through a distance $\mathrm{d}z$ against an external pressure p_{ex}, it follows that the work done is $\mathrm{d}w = -p_{ex}A\mathrm{d}z$. But $A\mathrm{d}z$ is the change in volume, $\mathrm{d}V$, in the course of the expansion. Therefore, the work done when the system expands by $\mathrm{d}V$ against a pressure p_{ex} is

$$\mathrm{d}w = -p_{ex}\mathrm{d}V \tag{14.6}$$

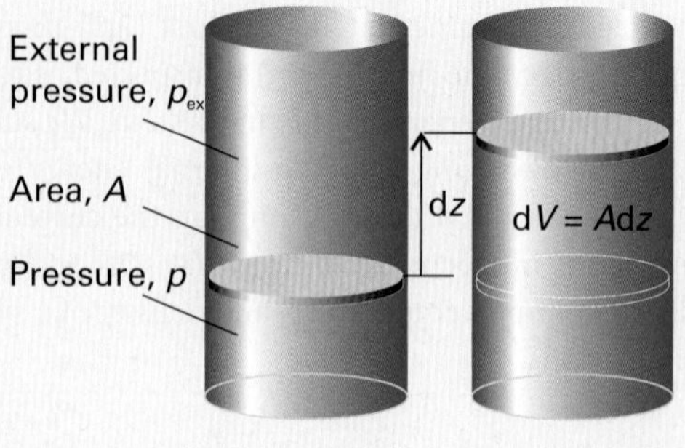

Fig. 14.5 When a piston of area A moves out through a distance dz, it sweeps out a volume $dV = Adz$. The external pressure p_{ex} is equivalent to a weight pressing on the piston, and the force opposing expansion is $F = p_{ex}A$.

Table 14.1 Varieties of work*

Type of work	dw	Comment	Units†
Expansion	$-p_{ex}dV$	p_{ex} is the external pressure	Pa
		dV is the change in volume	m^3
Surface expansion	$\gamma d\sigma$	γ is the surface tension	$N\,m^{-1}$
		$d\sigma$ is the change in area	m^2
Extension	fdl	f is the tension	N
		dl is the change in length	m
Electrical	ϕdQ	ϕ is the electric potential	V
		dQ is the change in charge	C

* In general, the work done on a system can be expressed in the form $dw = -Fdz$, where F is a 'generalized force' and dz is a 'generalized displacement'.
† For work in joules. Note that $1\,N\,m = 1\,J$ and $1\,V\,C = 1\,J$.

To obtain the total work done when the volume changes from V_i to V_f we integrate this expression between the initial and final volumes:

$$w = -\int_{V_i}^{V_f} p_{ex}\,dV \qquad (14.7)$$

The force acting on the piston, $p_{ex}A$, is equivalent to a weight that is raised as the system expands.

If the system is compressed instead, then the same weight is lowered in the surroundings and eqn 14.7 can still be used, but now $V_f < V_i$. It is important to note that it is still the *external* pressure that determines the magnitude of the work. This somewhat perplexing conclusion seems to be inconsistent with the fact that the gas *inside* the container is opposing the compression. However, when a gas is compressed, the ability of the *surroundings* to do work is diminished by an amount determined by the weight that is lowered, and it is this energy that is transferred into the system.

Other types of work (for example, electrical work), which we shall call **non-expansion work**, have analogous expressions, with each one the product of an intensive factor (the pressure, for instance) and an extensive factor (the change in volume). Some are collected in Table 14.1. For the present we continue with the work associated with changing the volume, the expansion work, and see what we can extract from eqns 14.6 and 14.7.

(b) Expansion against constant pressure

Now suppose that the external pressure is constant throughout the expansion. For example, the piston may be pressed on by the atmosphere, which exerts the same pressure throughout the expansion. A chemical example of this condition is the expansion of a gas formed in a chemical reaction. We can evaluate eqn 14.7 by taking the constant p_{ex} outside the integral:

$$w = -p_{ex}\int_{V_i}^{V_f} dV = -p_{ex}(V_f - V_i)$$

Therefore, if we write the change in volume as $\Delta V = V_f - V_i$,

$$w = -p_{ex}\Delta V \qquad (14.8)$$

This result is illustrated graphically in Fig. 14.6, which makes use of the fact that an integral can be interpreted as an area. The magnitude of w, denoted $|w|$, is equal to the area beneath the horizontal line at $p = p_{ex}$ lying between the initial and final volumes.

When the external pressure is zero (as for expansion into a vacuum), the system undergoes **free expansion.** When $p_{ex} = 0$, eqn 14.8 implies that the expansion work is zero, that is,

$$\textit{Free expansion: } w = 0 \qquad (14.9)$$

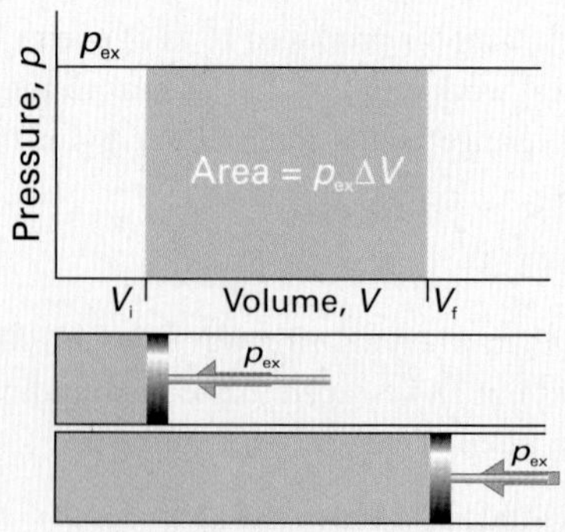

Fig. 14.6 The work done by a gas when it expands against a constant external pressure, p_{ex}, is equal to the shaded area in this example of an indicator diagram.

(c) Reversible expansion

A **reversible change** in thermodynamics is a change that can be reversed by an infinitesimal modification of a variable. The key word 'infinitesimal' sharpens the everyday meaning of the word 'reversible' as something that can change direction. We say that a system is in **equilibrium** with its surroundings if an infinitesimal change in the conditions in opposite directions results in opposite changes in its state. One example of reversibility is the **thermal equilibrium** of two systems with the same temperature. The transfer of energy as heat between the two is reversible because, if the temperature of either system is lowered infinitesimally, then energy flows into the system with the lower temperature. If the temperature of either system at thermal equilibrium is raised infinitesimally, then energy flows out of the hotter (higher temperature) system.

Suppose a gas is confined by a piston and that the external pressure, p_{ex}, is set equal to the pressure, p, of the confined gas. Such a system is in **mechanical equilibrium** with its surroundings because an infinitesimal change in the external pressure in either direction causes changes in volume in opposite directions. If the external pressure is reduced infinitesimally, then the gas expands slightly. If the external pressure is increased infinitesimally, then the gas contracts slightly. In either case the change is reversible in the thermodynamic sense. If, on the other hand, the external pressure differs measurably from the internal pressure, then changing p_{ex} infinitesimally will not decrease it below the pressure of the gas, so will not change the direction of the process. Such a system is not in mechanical equilibrium with its surroundings and the expansion is thermodynamically irreversible.

To achieve reversible expansion we set p_{ex} equal to p at each stage of the expansion. In practice, this equalization could be achieved by gradually removing weights from the piston so that the downward force due to the weights always matched the changing upward force due to the pressure of the gas. When we set $p_{ex} = p$, eqn 14.6 becomes

$$\mathrm{d}w = -p_{ex}\mathrm{d}V = -p\mathrm{d}V \qquad (14.10)_{rev}$$

(Equations valid only for reversible processes are labelled with a subscript rev.) Although the pressure inside the system appears in this expression for the work, it does so only because p_{ex} has been set equal to p to ensure reversibility. The total work of reversible expansion is therefore

$$w = -\int_{V_i}^{V_f} p\mathrm{d}V \qquad (14.11)_{rev}$$

We can evaluate the integral once we know how the pressure of the confined gas depends on its volume. Equation 14.11 is the link with the material covered in Chapter 8 for, if we know the equation of state of the gas, then we can express p in terms of V and evaluate the integral.

(d) Isothermal reversible expansion

Consider the **isothermal** (constant temperature) reversible expansion of a perfect gas. The expansion is made isothermal by keeping the system in thermal contact with its surroundings (which may be a constant-temperature bath). Because the equation of state is $pV = nRT$, we know that at each stage $p = nRT/V$, with V the volume at that stage of the expansion. The temperature T is constant in an isothermal expansion, so (together with n and R) it may be taken outside the integral. It follows that the work of reversible isothermal expansion of a perfect gas from V_i to V_f at a temperature T is

$$w = -nRT\int_{V_i}^{V_f} \frac{\mathrm{d}V}{V} = -nRT\ln\frac{V_f}{V_i} \qquad (14.12)^{\circ}_{rev}$$

where we have used the standard integral

$$\int_a^b \frac{1}{x}\mathrm{d}x = (\ln x + \text{constant})\Big|_a^b = \ln\frac{b}{a}.$$

Expressions that are valid only for perfect gases will be labelled, as here, with a ° sign.

When the final volume is greater than the initial volume, as in an expansion, the logarithm in eqn 14.12 is positive and hence $w < 0$. In this case, the system has done work on the surroundings and the internal energy of the system has decreased as a result. We shall see later that there is a compensating influx of energy as heat, so overall the internal energy is constant for the isothermal expansion of a perfect gas. The equations also show that more work is done for a given change of volume when the temperature is increased. The greater pressure of the confined gas then needs a higher opposing pressure to ensure reversibility. We cannot obtain more work than for the reversible process because increasing the external pressure even infinitesimally at any stage results in compression. We may infer from this discussion that, because some pushing power is wasted when $p > p_{ex}$, the maximum work available from a system operating between specified initial and final states and passing along a specified path is obtained when the change takes place reversibly. As in the case of constant external pressure, the work done is equal to the area under the isotherm, in this case representing the balanced internal and external pressures (Fig. 14.7).

The logarithmic term in eqn 14.12 can be explained in molecular terms by noting that, in general, an infinitesimal change in the total energy ($N\langle\varepsilon\rangle = \sum_i \varepsilon_i N_i$) of a collection of independent molecules arises from a change in the energy levels ε_i they occupy and a change in the populations N_i of those levels:

$$N\mathrm{d}\langle\varepsilon\rangle = \sum_i \varepsilon_i \mathrm{d}N_i + \sum_i N_i \mathrm{d}\varepsilon_i \qquad (14.13)$$

We cannot in general identify one of the terms on the right with the work done by the system, except in the case of a *reversible*

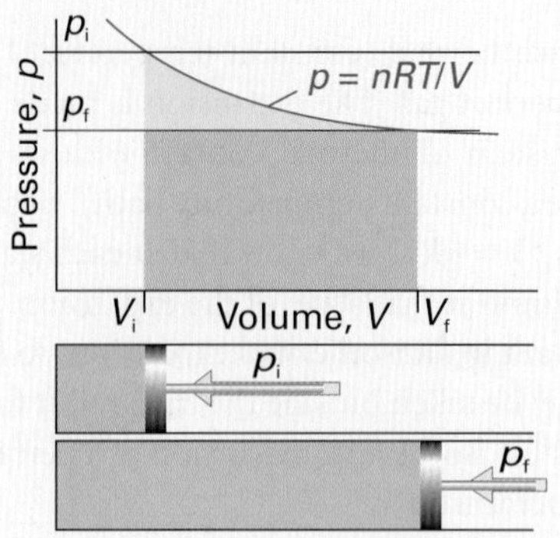

Fig. 14.7 The work done by a perfect gas when it expands reversibly and isothermally is equal to the area under the isotherm $p = nRT/V$. The work done during the irreversible expansion against the same final pressure is equal to the rectangular area shown slightly darker. Note that the reversible work is greater than the irreversible work.

interActivity Calculate the work of isothermal reversible expansion of 1.0 mol $CO_2(g)$ at 298 K from 1.0 m^3 to 3.0 m^3 on the basis that it obeys the van der Waals equation of state.

change, when the external force exerted against the system as it expands is matched to the force exerted by the molecules inside the system. In that case we can use the properties of the system to calculate the external opposing force (just as we replaced p_{ex} by p). When the volume of the system remains the same, the only change in internal energy arises from heating and from the consequent redistribution of populations of the unchanging energy levels. Therefore, the first term in eqn 14.13 can be identified with dq. When the volume is allowed to change reversibly, work is done and the translational energy levels change. Therefore, the work done by the system in the course of a reversible, isothermal expansion can be identified with the second term on the right:

$$\mathrm{d}w = \sum_i N_i \mathrm{d}\varepsilon_i \qquad (14.14)_{\text{rev,isothermal}}$$

We show in the following *Justification* that, when this expression is applied to a collection of independent molecules in a box, we obtain the logarithmic term in eqn 14.12.

Justification 14.1 *The molecular expression for the work of isothermal, reversible expansion*

We know from Section 2.2 that the energy levels of a particle in a box relative to its ground state are

$$\varepsilon_i = \frac{\gamma_i}{L^2} \qquad \gamma_i = (i^2 - 1)\frac{h^2}{8m} \qquad i = 1, 2, \ldots \qquad (14.15)$$

The change in the energy of a level with quantum number i when the length of a one-dimensional box changes by dL is

$$\mathrm{d}\varepsilon_i = \left(\frac{\mathrm{d}\varepsilon_i}{\mathrm{d}L}\right)\mathrm{d}L = -\frac{2\gamma_i}{L^3}\mathrm{d}L = -2\varepsilon_i\frac{\mathrm{d}L}{L} \qquad (14.16)$$

Equation 14.14 is therefore

$$\mathrm{d}w = \sum_i N_i \mathrm{d}\varepsilon_i = -\frac{2\mathrm{d}L}{L}\sum_i N_i\varepsilon_i = -\frac{2\mathrm{d}L}{L}N\langle\varepsilon\rangle$$

Provided the temperature is constant (so that $\langle\varepsilon\rangle = \frac{1}{2}kT$ is constant), we can integrate this expression between the initial and final lengths of the box:

$$w = \int_i^f \sum_i N_i \mathrm{d}\varepsilon_i = -2N\langle\varepsilon\rangle\int_{L_i}^{L_f}\frac{\mathrm{d}L}{L} = -2N\langle\varepsilon\rangle\ln\frac{L_f}{L_i}$$

However, for this one-dimensional system, $\langle\varepsilon\rangle = \frac{1}{2}kT$, so

$$w = -NkT\ln\frac{L_f}{L_i} = -nRT\ln\frac{L_f}{L_i}$$

We have used $N = nN_A$ and $N_Ak = R$. In three dimensions, the ratio of lengths is replaced by the ratio of volumes, as in the classical calculation, and we recover eqn 14.12.

14.4 Heat transactions

In general, the change in internal energy of a closed system is

$$\mathrm{d}U = \mathrm{d}q + \mathrm{d}w = \mathrm{d}q + \mathrm{d}w_{\text{exp}} + \mathrm{d}w_{\text{e}} \qquad (14.17)$$

where, as usual, dq is the energy transferred as heat and dw the energy transferred as work: dw_e is non-expansion work, that is, work in addition (e for 'extra') to the expansion work, dw_{exp}. For instance, dw_e might be the electrical work of driving a current through a circuit. A system kept at constant volume can do no expansion work, so dw_{exp} = 0. If the system is also incapable of doing any other kind of work (if it is not, for instance, an electrochemical cell connected to an electric motor), then dw_e = 0 too. Under these circumstances:

$$\mathrm{d}U = \mathrm{d}q \quad \text{(at constant volume, no additional work)} \qquad (14.18a)$$

We express this relation by writing dU = dq_V, where the subscript implies a change at constant volume. For a measurable change,

$$\Delta U = q_V \quad \text{(at constant volume, no additional work)} \qquad (14.18b)$$

It follows that by measuring the energy supplied to a constant-volume system as heat ($q > 0$) or obtained from it as heat ($q < 0$) when it undergoes a change of state, we are in fact measuring the change in its internal energy.

In molecular terms, the influx of energy as heat does not change the energy levels of a system, but does modify their populations. That is, eqn 14.13 becomes

$$N\mathrm{d}\langle\varepsilon\rangle = \sum_i \varepsilon_i \mathrm{d}N_i \quad \text{(at constant volume, no additional work)} \qquad (14.19a)$$

and for a measurable change

$$\Delta U = \sum_i \varepsilon_i \Delta N_i \quad \text{(at constant volume, no additional work)} \qquad (14.19b)$$

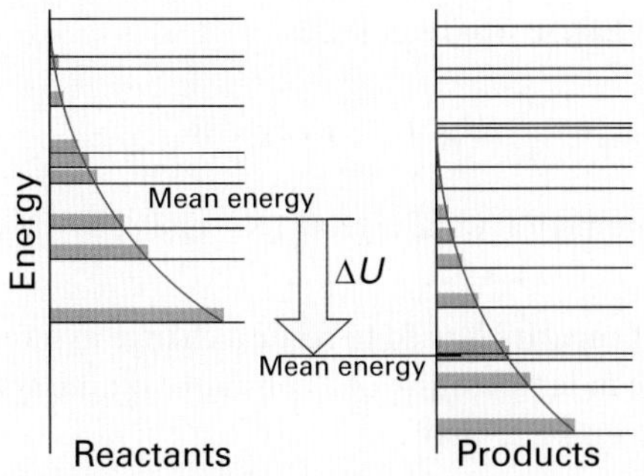

Fig. 14.8 The change in internal energy when a chemical reaction takes place is equal to the change in mean energy between products and reactants taking into account the Boltzmann distribution of populations. The blue lines are the distributions, and correspond to the same temperature.

The change in populations is due to a change in temperature, which redistributes the molecules over the fixed energy levels.

(a) Calorimetry

Equation 14.18 is more general than at first it might look, for it applies to a chemical reaction as well as to a system composed of a single species. To interpret it in terms of the reaction A → B, we imagine the energy levels of the molecules A and B as forming a single ladder of levels (Fig. 14.8). It should be recalled from Chapter 13 that the principle of equal *a priori* probabilities, on which the Boltzmann distribution is based, ignores the specific types of energy levels, treating all kinds equally: that blindness applies to the energy levels of different species too. At the start of the reaction, only the levels belonging to A are occupied; at the end of a complete reaction, only the levels belonging to B are occupied (later we see that equilibrium corresponds to a Boltzmann distribution over both sets of levels.) The redistribution of populations corresponds to the ΔN_i in eqn 14.19b:

$$\Delta U = \sum_i \varepsilon_i \Delta N_i = \sum_{i,\text{products}} \varepsilon_i N_i - \sum_{i,\text{reactants}} \varepsilon_i N_i \qquad (14.20)$$

Calorimetry is the study of heat transfer during physical and chemical processes. A **calorimeter** is a device for measuring energy transferred as heat. The most common device for measuring ΔU is an **adiabatic bomb calorimeter** (Fig. 14.9). The process we wish to study—which may be a chemical reaction—is initiated inside a constant-volume container, the 'bomb'. The bomb is immersed in a stirred water bath, and the whole device is the calorimeter. The calorimeter is also immersed in an outer water bath. The water in the calorimeter and of the outer bath are both monitored and adjusted to the same temperature. This arrangement ensures that there is no net loss of heat from the calorimeter to the surroundings (the bath) and hence that the calorimeter is adiabatic.

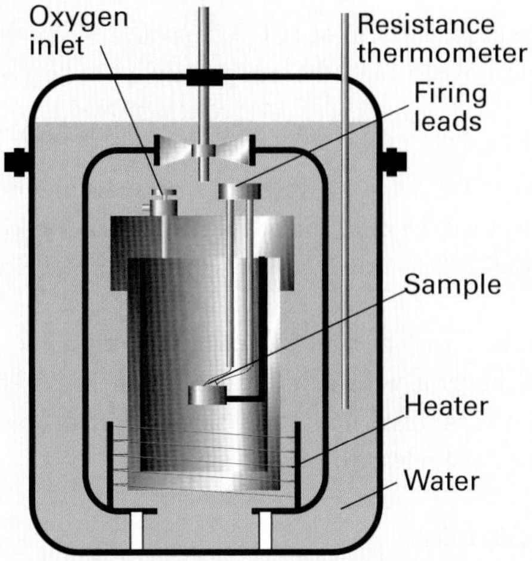

Fig. 14.9 A constant-volume bomb calorimeter. The 'bomb' is the central vessel, which is strong enough to withstand high pressures. The calorimeter (for which the heat capacity must be known) is the entire assembly shown here. To ensure adiabaticity, the calorimeter is immersed in a water bath with a temperature continuously readjusted to that of the calorimeter at each stage of the combustion.

The change in temperature, ΔT, of the calorimeter is proportional to the heat that the reaction releases or absorbs. Therefore, by measuring ΔT we can determine q_V and hence find ΔU. The conversion of ΔT to q_V is best achieved by calibrating the calorimeter using a process of known heat output, observing the temperature rise it produces, and determining the **calorimeter constant**, the constant C in the relation

$$q = C\Delta T \qquad (14.21)$$

The calorimeter constant may also be measured electrically by passing a constant current, I, from a source of known potential difference, $\mathcal{V}$, through a heater for a known period of time, t, for then

$$q = I\mathcal{V}t \qquad (14.22)$$

A brief comment Electrical charge is measured in *coulombs*, C. The motion of charge gives rise to an electric current, I, measured in coulombs per second, or *amperes*, A, where $1\ \text{A} = 1\ \text{C s}^{-1}$. If a constant current I flows through a potential difference $\mathcal{V}$ (measured in volts, V), the total energy supplied in an interval t is

$$\text{Energy supplied} = I\mathcal{V}t$$

Because $1\ \text{A V s} = 1\ (\text{C s}^{-1})\ \text{V s} = 1\ \text{C V} = 1\ \text{J}$, the energy is obtained in joules with the current in amperes, the potential difference in volts, and the time in seconds.

● **A BRIEF ILLUSTRATION**

If we pass a current of 10.0 A from a 12 V supply for 300 s, then from eqn 14.22 the energy supplied as heat is

$$q = (10.0\ \text{A}) \times (12\ \text{V}) \times (300\ \text{s}) = 3.6 \times 10^4\ \text{A V s} = 36\ \text{kJ}$$

because 1 A V s = 1 J. If the observed rise in temperature is 5.5 K, then the calorimeter constant is $C = (36\ \text{kJ})/(5.5\ \text{K}) = 6.5\ \text{kJ K}^{-1}$. ●

Alternatively, C may be determined by burning a known mass of substance (benzoic acid is often used) that has a known heat output. With C known, it is simple to interpret an observed temperature rise as a release of heat.

(b) Heat capacity

The internal energy of a substance increases when its temperature is raised and the Boltzmann distribution populates higher energy levels. The increase depends on the conditions under which the heating takes place and for the present we suppose that the sample is confined to a constant volume. For example, the sample may be a gas in a container of fixed volume. If the internal energy is plotted against temperature, then a curve like that in Fig. 14.10 may be obtained. The slope of the tangent to the curve at any temperature is called the **heat capacity** of the system at that temperature. The **heat capacity at constant volume** is denoted C_V and is defined formally as

$$C_V = \left(\frac{\partial U}{\partial T}\right)_V \qquad [14.23]$$

● **A BRIEF ILLUSTRATION**

The heat capacity of a monatomic perfect gas can be calculated by inserting the expression for the internal energy, $U_m = U_m(0) + \frac{3}{2}RT$ (see the *brief illustration* following eqn 14.1), so from eqn 14.23

$$C_{V,m} = \frac{\partial}{\partial T}(U_m(0) + \tfrac{3}{2}RT) = \tfrac{3}{2}R$$

The numerical value is 12.47 J K^{-1} mol^{-1}. ●

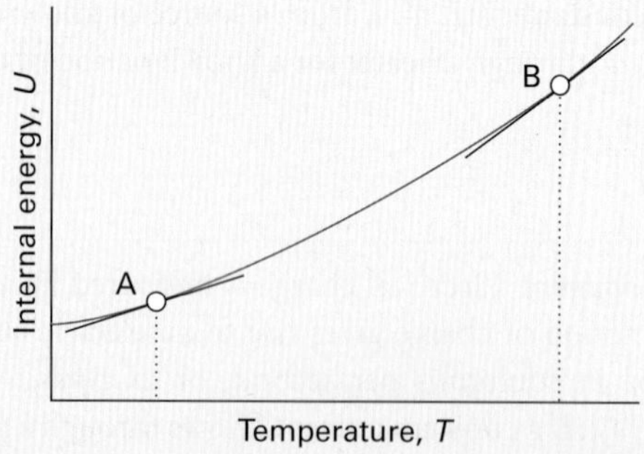

Fig. 14.10 The internal energy of a system increases as the temperature is raised; this graph shows its variation as the system is heated at constant volume. The slope of the tangent to the curve at any temperature is the heat capacity at constant volume at that temperature. Note that, for the system illustrated, the heat capacity is greater at B than at A.

The heat capacity is used to relate a change in internal energy to a change in temperature of a constant-volume system. It follows from eqn 14.23 that

$$dU = C_V dT \qquad \text{(at constant volume)} \qquad (14.24a)$$

That is, at constant volume, an infinitesimal change in temperature brings about an infinitesimal change in internal energy, and the constant of proportionality is C_V. If the heat capacity is independent of temperature over the range of temperatures of interest, a measurable change of temperature, ΔT, brings about a measurable increase in internal energy, ΔU, where

$$\Delta U = C_V \Delta T \qquad \text{(at constant volume)} \qquad (14.24b)$$

Because a change in internal energy can be identified with the heat supplied at constant volume (eqn 14.18), the last equation can also be written

$$q_V = C_V \Delta T \qquad (14.24c)$$

This relation provides a simple way of measuring the heat capacity of a sample: a measured quantity of energy is transferred as heat to the sample (electrically, for example), and the resulting increase in temperature is monitored. The ratio of the energy transferred as heat to the temperature rise it causes ($q_V/\Delta T$) is the constant-volume heat capacity of the sample.

The molecular basis of heat capacities provides considerable insight into this important property. We proceed in two steps: first we show in the *Justification* below that eqn 14.19 for the relation between the change in internal energy and the change in level populations can be used to deduce that the heat capacity is proportional to the spread of occupied energy levels in the sense that

$$C_V = \frac{N\Delta\varepsilon^2}{kT^2} \qquad \Delta\varepsilon^2 = \langle\varepsilon^2\rangle - \langle\varepsilon\rangle^2 \qquad (14.25)$$

Then we show how the heat capacity may be calculated from a knowledge of the structure of molecules.

Justification 14.2 *The heat capacity and the width of a population distribution*

We start with eqn 13.6a ($N_i = Ne^{-\beta\varepsilon_i}/q$) and calculate dN_i/dT, or—what is equivalent and easier—$dN_i/d\beta$:

$$\frac{dN_i}{d\beta} = N\frac{d}{d\beta}\frac{e^{-\beta\varepsilon_i}}{q} = -\frac{N\varepsilon_i e^{-\beta\varepsilon_i}}{q} - \frac{dq}{d\beta}\frac{Ne^{-\beta\varepsilon_i}}{q^2}$$

where q denotes the partition function. Because $e^{-\beta\varepsilon_i}/q = N_i/N$ and $-(1/q)(dq/d\beta) = \langle\varepsilon\rangle$,

$$\frac{dN_i}{d\beta} = -N_i\varepsilon_i + \frac{N\langle\varepsilon\rangle e^{-\beta\varepsilon_i}}{q}$$

$$= -N_i\varepsilon_i + N_i\langle\varepsilon\rangle$$

Therefore, with use of eqn 13.32,

$$\frac{dU}{d\beta} = N\frac{d\langle\varepsilon\rangle}{d\beta} = \sum_i \varepsilon_i \frac{dN_i}{d\beta}$$

$$= -\sum_i N_i\varepsilon_i^2 + \langle\varepsilon\rangle\sum_i N_i\varepsilon_i = -N\langle\varepsilon^2\rangle + N\langle\varepsilon\rangle^2$$

$$= -N\{\langle\varepsilon^2\rangle - \langle\varepsilon\rangle^2\}$$

That is, provided the volume is constant and the system is closed (so that there is no change in its composition), the change in its internal energy with temperature is

$$\frac{dU}{dT} = \frac{dU}{d\beta}\frac{d\beta}{dT} = -N\{\langle\varepsilon^2\rangle - \langle\varepsilon\rangle^2\}\frac{d\beta}{dT}$$

$$= \frac{N\{\langle\varepsilon^2\rangle - \langle\varepsilon\rangle^2\}}{kT^2}$$

It follows that we can identify the constant-volume heat capacity with the 'variance' of the energy, the mean-square deviation of the energy from its mean value, as expressed in eqn 14.25.

Equation 14.25 shows that the heat capacity rises as the spread in occupied energy levels increases. When all the molecules occupy the ground state, there is no difference between the mean of the square of the energies and the square of the mean energy, so the heat capacity is then zero. As the system is heated, this equality no longer holds and, provided the numerator in eqn 14.25 increases more rapidly than the denominator, the heat capacity rises. In certain cases, the numerator increases as T^2, in which case C_V is independent of temperature (provided $T > 0$).

Example 14.1 *Calculating the mean square deviation of energies of a perfect gas*

Calculate the value of $\langle\varepsilon^2\rangle$ for a one-dimensional perfect monatomic gas and then show that its heat capacity is independent of temperature.

Method For a one-dimensional system, we know that $\langle\varepsilon\rangle = \frac{1}{2}kT$, and can evaluate the mean square energy by using the same approximations used to derive the translational partition function in Section 13.3.

Answer To evaluate the mean square energy, we write

$$\langle\varepsilon^2\rangle = \frac{1}{N}\sum_i N_i\varepsilon_i^2 = \frac{1}{q}\sum_i \varepsilon_i^2 e^{-\beta\varepsilon_i} \approx \frac{1}{q}\int_0^\infty \left(\frac{i^2h^2}{8mL^2}\right)^2 e^{-\beta i^2h^2/8mL^2}\,di$$

As in *Justification 13.2*, we have replaced $i^2 - 1$ by i^2 in the expression for the energy, because most occupied levels have $i \gg 1$. We now make the substitution

$$i = \left(\frac{8mL^2}{\beta h^2}\right)^{1/2} x$$

and use the standard integral

$$\int_0^\infty x^4 e^{-x^2}\,dx = \tfrac{3}{8}\pi^{1/2}$$

and the translational partition function

$$q = \left(\frac{2\pi m}{h^2\beta}\right)^{1/2} L$$

to obtain

$$\langle\varepsilon^2\rangle = \frac{3}{4\beta^2} = \tfrac{3}{4}k^2T^2$$

The mean energy is $\frac{1}{2}kT$, so

$$C_V = \frac{N\{\frac{3}{4}k^2T^2 - \frac{1}{4}k^2T^2\}}{kT^2} = \tfrac{1}{2}Nk = \tfrac{1}{2}nR$$

where once again we have used $N = nN_A$ and $N_Ak = R$.

Self-test 14.2 Use eqn 14.25 to show that the heat capacity of a two-level system is zero at infinite temperature.

[$C_V \propto 1/T^2$ as $T \to \infty$]

Although eqn 14.25 gives some insight into the origin of heat capacities, it is not the easiest route to their calculation. To calculate the heat capacity at constant volume, all we need do is to evaluate the internal energy as a function of temperature and then form the derivative. In most cases it is easier to evaluate the derivative with respect to β, and therefore to evaluate

$$C_V = \left(\frac{\partial U}{\partial T}\right)_V = \frac{d\beta}{dT}\left(\frac{\partial U}{\partial \beta}\right)_V \tag{14.26}$$

$$= -\frac{N}{kT^2}\left(\frac{\partial\langle\varepsilon\rangle}{\partial\beta}\right)_V = \frac{N}{kT^2}\left(\frac{\partial^2 \ln q}{\partial\beta^2}\right)_V$$

We have used $d\beta/dT = -1/kT^2$, $dU = Nd\langle\varepsilon\rangle$, and $\langle\varepsilon\rangle = -(\partial \ln q/\partial\beta)_V$. In Problem 14.23 you are invited to confirm that this expression can be converted into eqn 14.25. Once we know the partition function, all we need to do is to evaluate its second derivative with respect to β.

Example 14.2 *The vibrational contribution to the heat capacity*

Derive an expression for the heat capacity of a harmonic oscillator. At the same time derive an expression for the spread of populations ($\Delta\varepsilon^2$) and plot both functions against temperature.

Method The partition function is given in eqn 13.24 and the expression derived for the mean energy is given in eqn 13.39. It is simplest to differentiate the latter expression with respect to β, and express the resulting expression as a function of T. Compare that expression with eqn 14.25 to identify $\Delta\varepsilon^2$.

Answer The mean energy of a harmonic oscillator (relative to its ground state, eqn 13.39) is

$$\langle\varepsilon^{V}\rangle = \frac{hc\tilde{\nu}}{e^{\beta hc\tilde{\nu}} - 1}$$

Therefore,

$$\frac{d\langle\varepsilon^{V}\rangle}{d\beta} = -\frac{(hc\tilde{\nu})^2 e^{\beta hc\tilde{\nu}}}{(e^{\beta hc\tilde{\nu}} - 1)^2}$$

and consequently, by using eqn 14.26 in the form $C_V = -Nk(1/kT)^2(\partial\langle\varepsilon\rangle/\partial\beta)_V$,

$$C_V = Nk\left(\frac{hc\tilde{\nu}}{kT}\right)^2 \frac{e^{\beta hc\tilde{\nu}}}{(e^{\beta hc\tilde{\nu}} - 1)^2} = Nk\left(\frac{hc\tilde{\nu}}{kT}\right)^2 \frac{e^{hc\tilde{\nu}/kT}}{(e^{hc\tilde{\nu}/kT} - 1)^2}$$

This function is plotted in Fig. 14.11. Comparison with eqn 14.25 lets us infer that

$$\Delta\varepsilon^2 = \frac{(hc\tilde{\nu})^2 e^{hc\tilde{\nu}/kT}}{(e^{hc\tilde{\nu}/kT} - 1)^2}$$

This function is also plotted in Fig. 14.12. Note that, at high temperatures ($T \gg hc\tilde{\nu}/k$), the spread is

$$\Delta\varepsilon^2 = \frac{(hc\tilde{\nu})^2[1 + hc\tilde{\nu}/kT + \cdots]}{([1 + hc\tilde{\nu}/kT + \cdots] - 1)^2} \approx \frac{(hc\tilde{\nu})^2}{(hc\tilde{\nu}/kT)^2} = (kT)^2$$

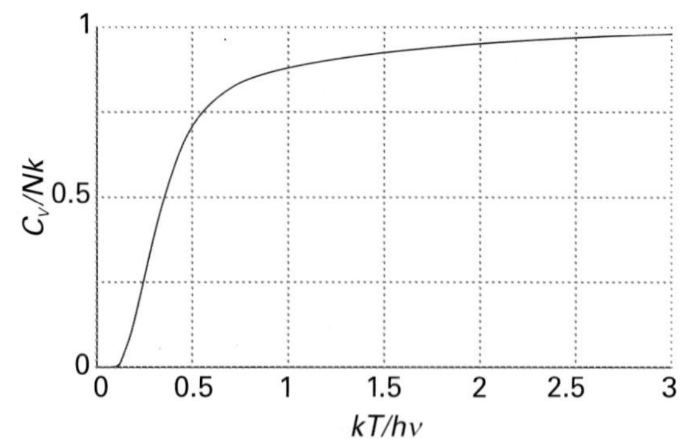

Fig. 14.11 The heat capacity of a harmonic oscillator as a function of temperature.

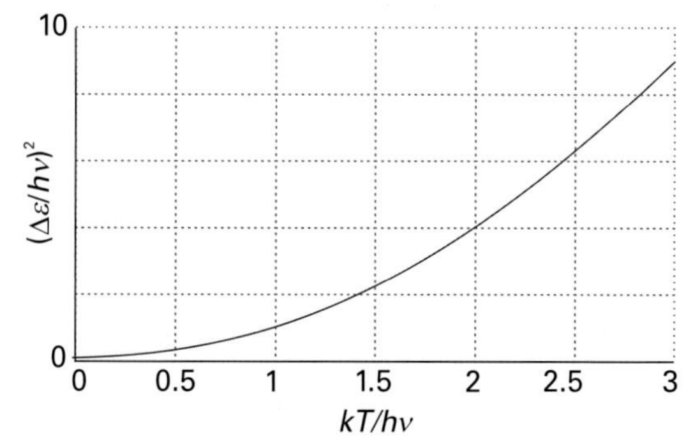

Fig. 14.12 The mean square fluctuations in the energy of a collection of harmonic oscillators as a function of temperature.

and the spread increases quadratically. That dependence is cancelled by the T^2 in the denominator of eqn 14.25, and the heat capacity approaches a constant value (of Nk) at high temperatures.

Self-test 14.3 Repeat the analysis for a two-level system.

[$C_V = Nkf \to 0, \Delta\varepsilon^2 = (kT)^2 f \to \frac{1}{4}\varepsilon^2$; $f = (\varepsilon/kT)^2 e^{\varepsilon/kT}/(1 + e^{\varepsilon/kT})^2$, Fig. 14.13]

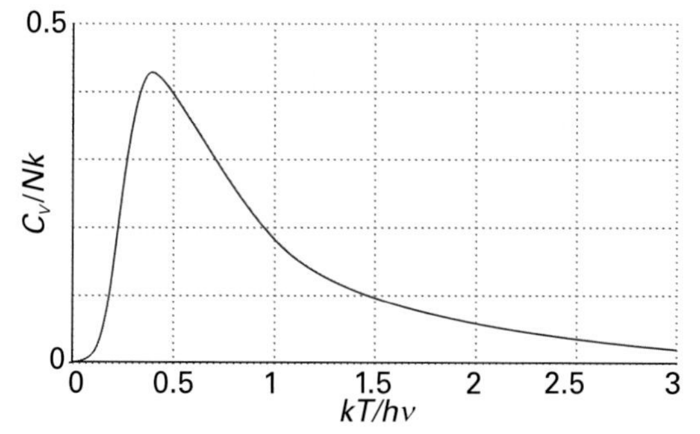

Fig. 14.13 The heat capacity of a collection of two-level systems as a function of temperature.

When the temperature is high enough for the rotations of the molecules to be highly excited (when $T \gg \theta_R$, where the rotational temperature θ_R is defined in Section 13.3 as $\theta_R = hc\tilde{B}/k$), we can use the equipartition value kT for the mean rotational energy (for a linear rotor) and $N_A k = R$ to obtain $C_{V,m} = R$. For non-linear molecules, the mean rotational energy rises to $\frac{3}{2}kT$, so the molar rotational heat capacity rises to $\frac{3}{2}R$ when $T \gg \theta_R$. Only the lowest rotational state is occupied when the temperature is very low, and then rotation does not contribute to the heat capacity. We can calculate the rotational heat capacity at

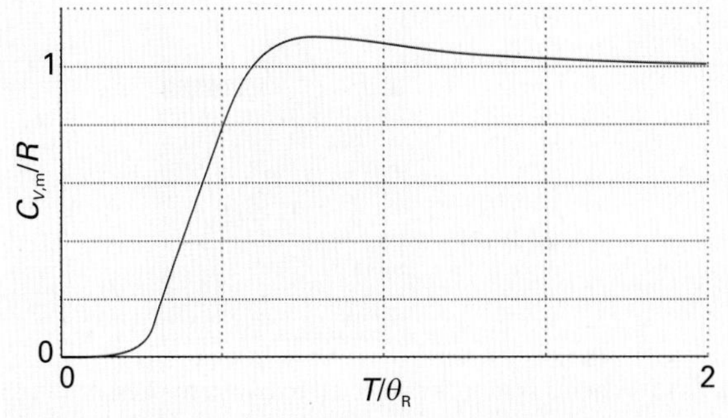

Fig. 14.14 The temperature dependence of the rotational contribution to the heat capacity of a linear molecule.

interActivity The *Living graphs* section of the text's web site has applets for the calculation of the temperature dependence of the rotational contribution to the heat capacity. Explore the effect of the rotational constant on the plot of $C^R_{V,m}$ against T.

intermediate temperatures by differentiating the equation for the mean rotational energy (eqn 13.37a). The resulting (untidy) expression, which is plotted in Fig. 14.14, shows that the contribution rises from zero (when $T = 0$) to the equipartition value (when $T \gg \theta_R$). Because the translational contribution is always present, we can expect the molar heat capacity of a gas of diatomic molecules ($C^T_{V,m} + C^R_{V,m}$) to rise from $\frac{3}{2}R$ to $\frac{5}{2}R$ as the temperature is increased above θ_R. Problem 14.32 explores how the overall shape of the curve can be traced to the sum of thermal excitations between all the available rotational energy levels (Fig. 14.15).

Molecular vibrations contribute to the heat capacity, but only when the temperature is high enough for them to be significantly excited. The equipartition mean energy is kT for each mode, so the maximum contribution to the molar heat capacity is R. However, it is very unusual for the vibrations to be so highly excited that equipartition is valid, and it is more appropriate to use the full expression for the vibrational heat capacity, which was derived in Example 14.2. The curve in Fig. 14.16 shows how the vibrational heat capacity depends on temperature. Note that, even when the temperature is only slightly above the characteristic vibrational temperature $\theta_V = hc\tilde{\nu}/k$ of the normal mode, the heat capacity is close to its equipartition value.

The total heat capacity of a molecular substance is the sum of each contribution (Fig. 14.17). When equipartition is valid

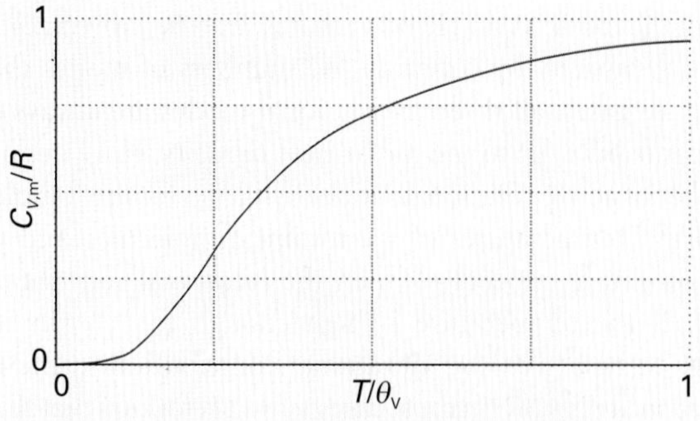

Fig. 14.16 The temperature dependence of the vibrational heat capacity of a molecule in the harmonic approximation calculated by using the full expression derived in Example 14.2. Note that the heat capacity is within 10 per cent of its classical value for temperatures greater than θ_V.

interActivity The *Living graphs* section of the text's web site has applets for the calculation of the temperature dependence of the vibrational contribution to the heat capacity. Explore the effect of the vibrational wavenumber on the plot of $C^V_{V,m}$ against T.

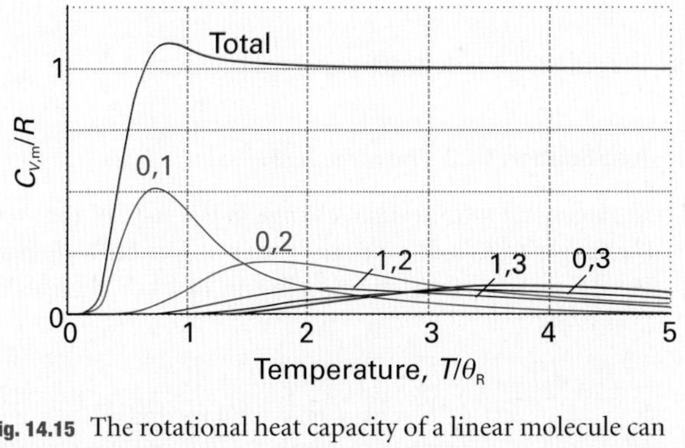

Fig. 14.15 The rotational heat capacity of a linear molecule can be regarded as the sum of contributions from a collection of two-level systems, in which the rise in temperature stimulates transitions between J levels, some of which are shown here. The calculation on which this illustration is based is sketched in Problem 14.32.

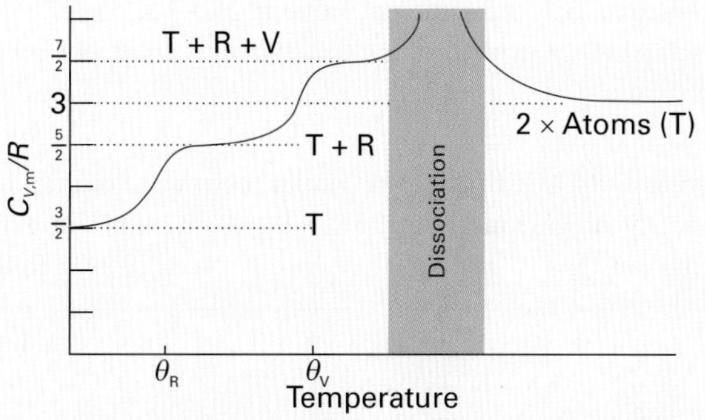

Fig. 14.17 The general features of the temperature dependence of the heat capacity of diatomic molecules are as shown here. Each mode becomes active when its characteristic temperature is exceeded. The heat capacity becomes very large when the molecule dissociates because the energy is used to cause dissociation and not to raise the temperature. Then it falls back to the translation-only value of the atoms.

(when the temperature is well above the characteristic temperature of the mode M = T, R, or V; that is, $T \gg \theta_M$) we can estimate the heat capacity by counting the numbers of modes that are active. In gases, all three translational modes are always active and contribute $\frac{3}{2}R$ to the molar heat capacity. If we denote the number of active rotational modes by ν_R^* (so for most molecules at normal temperatures $\nu_R^* = 2$ for linear molecules, and $\nu_R^* = 3$ for non-linear molecules), then the rotational contribution is $\frac{1}{2}\nu_R^* R$. If the temperature is high enough for ν_V^* vibrational modes to be active, the vibrational contribution to the molar heat capacity is $\nu_V^* R$. In most cases $\nu_V^* \approx 0$. It follows that the total molar heat capacity is

$$C_{V,m} = \tfrac{1}{2}(3 + \nu_R^* + 2\nu_V^*)R \qquad (14.27)$$

Example 14.3 *Estimating the molar heat capacity of a gas*

Estimate the molar constant-volume heat capacity of water vapour at 100°C. The vibrational wavenumbers of H_2O are 3656.7 cm^{-1}, 1594.8 cm^{-1}, and 3755.8 cm^{-1} and the rotational constants are 27.9, 14.5, and 9.3 cm^{-1}.

Method We need to assess whether the rotational and vibrational modes are active by computing their characteristic temperatures from the data (to do so, use $hc/k = 1.439$ cm K).

Answer The characteristic temperatures (in round numbers) of the vibrations are 5300 K, 2300 K, and 5400 K; the vibrations are therefore not significantly excited at 373 K. The three rotational modes have characteristic temperatures 40 K, 21 K, and 13 K, so they are fully excited, like the three translational modes. The translational contribution is $\frac{3}{2}R = 12.5$ J K^{-1} mol^{-1}. Fully excited rotations ($\nu_R^* = 3$) contribute a further $\frac{3}{2}R =$ 12.5 J K^{-1} mol^{-1}. Therefore, a value close to 25 J K^{-1} mol^{-1} is predicted. The experimental value is 26.1 J K^{-1} mol^{-1}. The discrepancy is probably due to deviations from perfect gas behaviour.

Self-test 14.4 Estimate the molar constant-volume heat capacity of gaseous I_2 at 25°C ($\tilde{B} = 0.037$ cm^{-1} and $\tilde{\nu} =$ 214.5 cm^{-1}). [29 J K^{-1} mol^{-1}]

14.5 Enthalpy

When the system is free to change its volume, some of the energy supplied as heat to the system is returned to the surroundings as expansion work (Fig. 14.18), so dU is less than dq. However, we shall now show that in this case the energy supplied as heat at constant pressure is equal to the change in another thermodynamic property of the system, the enthalpy.

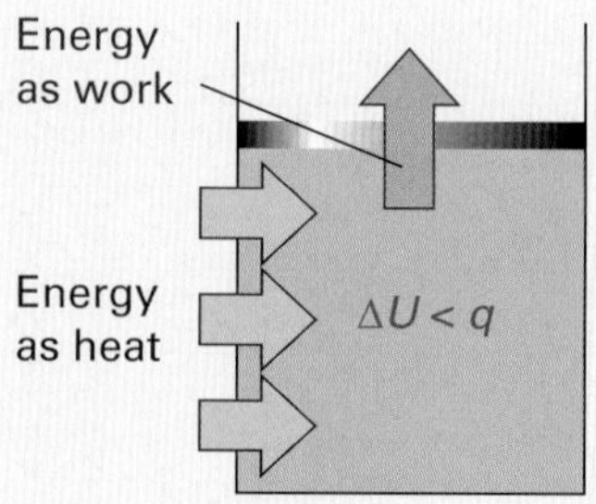

Fig. 14.18 When a system is subjected to constant pressure and is free to change its volume, some of the energy supplied as heat may escape back into the surroundings as work. In such a case, the change in internal energy is smaller than the energy supplied as heat.

(a) The definition of enthalpy

The **enthalpy**, H, is defined as

$$H = U + pV \qquad [14.28]$$

where p is the pressure of the system and V is its volume. Because U, p, and V are all state functions, the enthalpy is a state function too. As is true of any state function, the change in enthalpy, ΔH, between any pair of initial and final states is independent of the path between them.

We show in the following *Justification* that eqn 14.28 implies that *the change in enthalpy is equal to the energy supplied as heat at constant pressure* (provided the system does no additional work):

$$\mathrm{d}H = \mathrm{d}q \quad \text{(at constant pressure, no additional work)} \qquad (14.29a)$$

For a measurable change,

$$\Delta H = q_p \qquad (14.29b)$$

where the subscript p denotes constant pressure.

Justification 14.3 *The relation $\Delta H = q_p$*

For a general infinitesimal change in the state of the system, U changes to $U + \mathrm{d}U$, p changes to $p + \mathrm{d}p$, and V changes to $V + \mathrm{d}V$ so, from the definition in eqn 14.28, H changes from $U + pV$ to

$$H + \mathrm{d}H = (U + \mathrm{d}U) + (p + \mathrm{d}p)(V + \mathrm{d}V)$$
$$= U + \mathrm{d}U + pV + p\mathrm{d}V + V\mathrm{d}p + \mathrm{d}p\mathrm{d}V$$

The last term is the product of two infinitesimally small quantities and can therefore be neglected. As a result, after recognizing $U + pV = H$ on the right, we find that H changes to

$$H + \mathrm{d}H = H + \mathrm{d}U + p\mathrm{d}V + V\mathrm{d}p$$

and hence that

$$dH = dU + pdV + Vdp$$

If we now substitute $dU = dq + dw$ into this expression, we get

$$dH = dq + dw + pdV + Vdp$$

If the system is in mechanical equilibrium with its surroundings at a pressure p and does only expansion work, we can write $dw = -pdV$ and obtain

$$dH = dq + Vdp$$

Now we impose the condition that the heating occurs at constant pressure by writing $dp = 0$. Then

$$dH = dq \quad \text{(at constant pressure, no additional work)}$$

as in eqn 14.29a.

The result expressed in eqn 14.29 states that, when a system is subjected to a constant pressure, and only expansion work can occur, the change in enthalpy is equal to the energy supplied as heat. For example, if we supply 36 kJ of energy through an electric heater immersed in an open beaker of water, then the enthalpy of the water increases by 36 kJ and we write $\Delta H = +36$ kJ.

(b) The measurement of an enthalpy change

A calorimeter for studying processes at constant pressure is called an **isobaric calorimeter**. A simple example is a thermally insulated vessel open to the atmosphere: the heat released in the reaction is monitored by measuring the change in temperature of the contents. For a combustion reaction an **adiabatic flame calorimeter** may be used to measure ΔT when a given amount of substance burns in a supply of oxygen (Fig. 14.19). Another route to ΔH is to measure the internal energy change by using a bomb calorimeter, and then to convert ΔU to ΔH. Because solids and liquids have small molar volumes, for them pV_m is so small that the molar enthalpy and molar internal energy are almost identical ($H_m = U_m + pV_m \approx U_m$). Consequently, if a process involves only solids or liquids, the values of ΔH and ΔU are almost identical. Physically, such processes are accompanied by a very small change in volume, the system does negligible work on the surroundings when the process occurs, so the energy supplied as heat stays entirely within the system. The most sophisticated way, however, is to use a **differential scanning calorimeter** (DSC), as explained in *Impact I14.1* at the end of this section. Changes in enthalpy and internal energy may also be measured by noncalorimetric methods (Chapter 17).

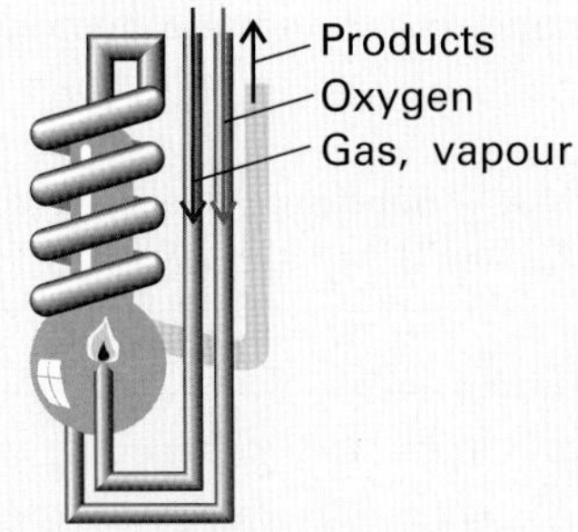

Fig. 14.19 A constant-pressure flame calorimeter consists of this component immersed in a stirred water bath. Combustion occurs as a known amount of reactant is passed through to fuel the flame, and the rise of temperature is monitored.

Equation 14.28 is a definition and applies to all substances. In the special case of a perfect gas we can write $pV = nRT$ in the definition of H and then with $U = U(0) + N\langle\varepsilon\rangle$ obtain

$$H = U + pV = U + nRT \qquad (14.30)^\circ$$
$$= U(0) + N\{\langle\varepsilon\rangle + kT\}$$

The first line of this relation implies that the change of enthalpy in a reaction that produces or consumes gas is

$$\Delta H = \Delta U + \Delta n_g RT \qquad (14.31)^\circ$$

where Δn_g is the change in the amount of gas molecules in the reaction.

● A BRIEF ILLUSTRATION

In the reaction $2\,H_2(g) + O_2(g) \rightarrow 2\,H_2O(l)$, 3 mol of gas-phase molecules is replaced by 2 mol of liquid-phase molecules, so $\Delta n_g = -3$ mol. Therefore, at 298 K, when $RT = 2.5\ \text{kJ mol}^{-1}$, the enthalpy and internal energy changes taking place in the system are related by

$$\Delta H - \Delta U = (-3\ \text{mol}) \times RT \approx -7.5\ \text{kJ}$$ ●

Example 14.4 *Calculating a change in enthalpy*

Water is heated to boiling under a pressure of 1.0 atm. When an electric current of 0.50 A from a 12 V supply is passed for 300 s through a resistance in thermal contact with it, it is found that 0.798 g of water is vaporized. Calculate the molar internal energy and enthalpy changes at the boiling point (373.15 K).

Method Because the vaporization occurs at constant pressure, the enthalpy change is equal to the heat supplied by the heater. Therefore, the strategy is to calculate the energy supplied as heat (from $q = I\mathcal{V}t$), express that as an enthalpy change, and then convert the result to a molar enthalpy change by division by the amount of H_2O molecules vaporized. To convert from enthalpy change to internal energy change, we assume that the vapour is a perfect gas and use eqn 14.31.

Answer The enthalpy change is

$$\Delta H = q_p = (0.50\ \text{A}) \times (12\ \text{V}) \times (300\ \text{s}) = +(0.50 \times 12 \times 300)\ \text{J}$$

Here we have used 1 A V s = 1 J. Because 0.798 g of water is (0.798 g)/(18.02 g mol^{-1}) = (0.798/18.02) mol H_2O, the enthalpy of vaporization per mole of H_2O is

$$\Delta H_m = +\frac{0.50 \times 12 \times 300 \text{ J}}{(0.798/18.02) \text{ mol}} = +41 \text{ kJ mol}^{-1}$$

In the process $H_2O(l) \rightarrow H_2O(g)$ the change in the amount of gas molecules is $\Delta n_g = +1$ mol, so

$$\Delta U_m = \Delta H_m - RT = +38 \text{ kJ mol}^{-1}$$

Notice that the internal energy change is smaller than the enthalpy change because energy has been used to drive back the surrounding atmosphere to make room for the vapour.

Self-test 14.5 The molar enthalpy of vaporization of benzene at its boiling point (353.25 K) is 30.8 kJ mol^{-1}. What is the molar internal energy change? For how long would the same 12 V source need to supply a 0.50 A current in order to vaporize a sample of mass 10.0 g? [+27.9 kJ mol^{-1}, 660 s]

(c) The variation of enthalpy with temperature

The enthalpy of a substance increases as its temperature is raised. The relation between the increase in enthalpy and the increase in temperature depends on the conditions (for example, constant pressure or constant volume). The most important condition is constant pressure, and the slope of the tangent to a plot of enthalpy against temperature at constant pressure is called the **heat capacity at constant pressure**, C_p, at a given temperature (Fig. 14.20). More formally:

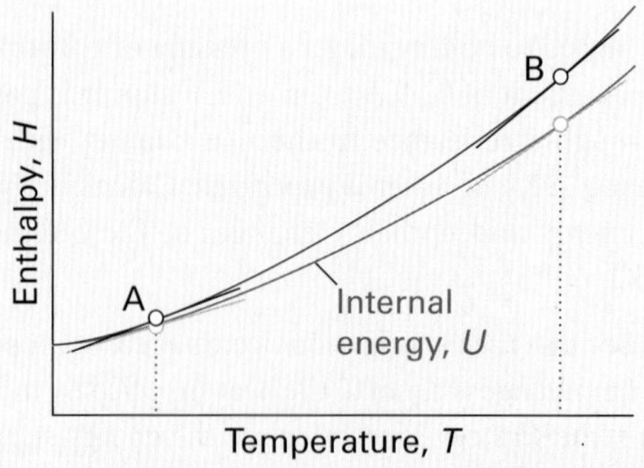

Fig. 14.20 The slope of the tangent to a curve of the enthalpy of a system subjected to a constant pressure plotted against temperature is the constant-pressure heat capacity. The slope may change with temperature, in which case the heat capacity varies with temperature. Thus, the heat capacities at A and B are different. For gases, at a given temperature the slope of enthalpy versus temperature is steeper than that of internal energy versus temperature, and $C_{p,m}$ is larger than $C_{V,m}$.

$$C_p = \left(\frac{\partial H}{\partial T}\right)_p \qquad [14.32]$$

The heat capacity at constant pressure is the analogue of the heat capacity at constant volume, and is an extensive property. As in the case of C_V, if the system can change its composition it is necessary to distinguish between equilibrium and fixed-composition values. All applications in this chapter refer to pure substances, so this complication can be ignored. The **molar heat capacity at constant pressure**, $C_{p,m}$, is the heat capacity per mole of material; it is an intensive property.

The heat capacity at constant pressure is used to relate the change in enthalpy to a change in temperature. For infinitesimal changes of temperature,

$$dH = C_p dT \quad \text{(at constant pressure)} \qquad (14.33a)$$

If the heat capacity is constant over the range of temperatures of interest, then for a measurable increase in temperature

$$\Delta H = C_p \Delta T \quad \text{(at constant pressure)} \qquad (14.33b)$$

Because an increase in enthalpy can be equated to the energy supplied as heat at constant pressure, the practical form of the latter equation is

$$q_p = C_p \Delta T \qquad (14.33c)$$

This expression shows us how to measure the heat capacity of a sample: a measured quantity of energy is supplied as heat under conditions of constant pressure (as in a sample exposed to the atmosphere and free to expand), and the temperature rise is monitored.

The variation of heat capacity with temperature can sometimes be ignored if the temperature range is small; this approximation is highly accurate for a monatomic perfect gas (for instance, one of the noble gases at low pressure). However, when it is necessary to take the variation into account, a convenient approximate empirical expression is

$$C_{p,m} = a + bT + \frac{c}{T^2} \qquad (14.34)$$

The empirical parameters a, b, and c are independent of temperature (Table 14.2).

Synoptic table 14.2* Temperature variation of molar heat capacities, $C_{p,m}/(\text{J K}^{-1}\text{ mol}^{-1}) = a + bT + c/T^2$

	a	$b/(10^{-3}\text{ K}^{-1})$	$c/(10^5\text{ K}^2)$
C(s, graphite)	16.86	4.77	−8.54
CO_2(g)	44.22	8.79	−8.62
H_2O(l)	75.29	0	0
N_2(g)	28.58	3.77	−0.50

* More values are given in the *Data section*.

Example 14.5 *Evaluating an increase in enthalpy with temperature*

What is the change in molar enthalpy of N_2 when it is heated from 25°C to 100°C? Use the heat capacity information in Table 14.2.

Method The heat capacity of N_2 changes with temperature, so we cannot use eqn 14.33b (which assumes that the heat capacity of the substance is constant). Therefore, we must use eqn 14.33a, substitute eqn 14.34 for the temperature dependence of the heat capacity, and integrate the resulting expression from 25°C to 100°C.

Answer For convenience, we denote the two temperatures T_1 (298 K) and T_2 (373 K). The integral we require is

$$\int_{H_m(T_1)}^{H_m(T_2)} dH_m = \int_{T_1}^{T_2} \left(a + bT + \frac{c}{T^2} \right) dT$$

Now we use the integral

$$\int x^n \, dx = \frac{x^{n+1}}{n+1} + \text{constant}$$

to obtain

$$H_m(T_2) - H_m(T_1) = a(T_2 - T_1) + \tfrac{1}{2}b(T_2^2 - T_1^2) - c\left(\frac{1}{T_2} - \frac{1}{T_1}\right)$$

Substitution of the numerical data results in

$$H_m(373\text{ K}) = H_m(298\text{ K}) + 2.20 \text{ kJ mol}^{-1}$$

If we had assumed a constant heat capacity of 29.14 J K^{-1} mol^{-1} (the value given by eqn 14.34 at 25°C), we would have found that the two enthalpies differed by 2.19 kJ mol^{-1}.

Self-test 14.6 At very low temperatures the heat capacity of a solid is proportional to T^3, and we can write $C_p = aT^3$. What is the change in enthalpy of such a substance when it is heated from 0 to a temperature T (with T close to 0)?

$[\Delta H = \tfrac{1}{4}aT^4]$

Most systems expand when heated at constant pressure. Such systems do work on the surroundings and therefore some of the energy supplied to them as heat escapes back to the surroundings as work. As a result, the temperature of the system rises less than when the heating occurs at constant volume. A smaller increase in temperature implies a larger heat capacity, so we conclude that in most cases the heat capacity at constant pressure of a system is larger than its heat capacity at constant volume. We show later (Section 14.10) that there is a simple relation between the two heat capacities of a perfect gas:

$$C_p - C_V = nR \qquad (14.35)°$$

It follows that the molar heat capacity of a perfect gas is about 8 J K^{-1} mol^{-1} larger at constant pressure than at constant volume. Because the heat capacity at constant volume of a monatomic gas is about 12 J K^{-1} mol^{-1}, the difference is highly significant and must be taken into account. However, the molecular origin of C_p is essentially the same as that of C_V, the additional term being due to the energy lost by expansion. For substances other than perfect gases, the forces between atoms play a role in determining the magnitude of the work of expansion and the expression for the difference between C_p and C_V is more complicated.[1]

IMPACT ON BIOCHEMISTRY

I14.1 Differential scanning calorimetry

A *differential scanning calorimeter* (DSC) measures the energy transferred as heat to or from a sample at constant pressure during a physical or chemical change. The term 'differential' refers to the fact that the behaviour of the sample is compared to that of a reference material which does not undergo a physical or chemical change during the analysis. The term 'scanning' refers to the fact that the temperatures of the sample and reference material are increased, or scanned, during the analysis.

A DSC consists of two small compartments that are heated electrically at a constant rate. The temperature, T, at time t during a linear scan is $T = T_0 + \alpha t$, where T_0 is the initial temperature and α is the temperature scan rate (in kelvin per second, K s^{-1}). A computer controls the electrical power output in order to maintain the same temperature in the sample and reference compartments throughout the analysis (Fig. 14.21).

The temperature of the sample changes significantly relative to that of the reference material if a chemical or physical process involving the transfer of energy as heat occurs in the sample

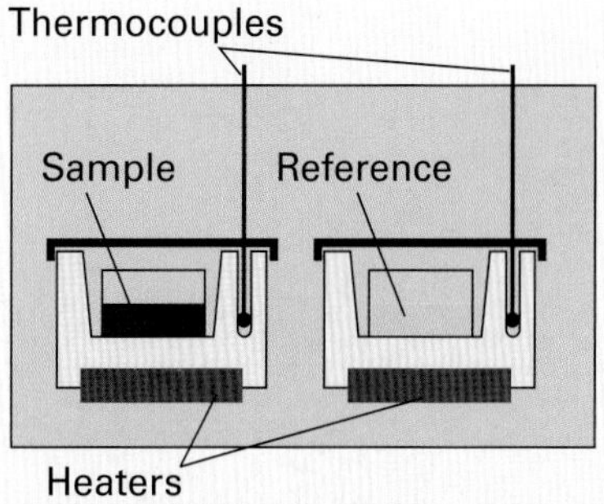

Fig. 14.21 A differential scanning calorimeter. The sample and a reference material are heated in separate but identical metal heat sinks. The output is the difference in power needed to maintain the heat sinks at equal temperatures as the temperature rises.

[1] For details, see our *Physical chemistry* (2007).

during the scan. To maintain the same temperature in both compartments, excess energy is transferred as heat to or from the sample during the process. For example, an endothermic process lowers the temperature of the sample relative to that of the reference and, as a result, the sample must be heated more strongly than the reference in order to maintain equal temperatures.

If no physical or chemical change occurs in the sample at temperature T, we use eqn 14.33c to write the heat transferred to the sample as $q_p = C_p\Delta T$, where $\Delta T = T - T_0$ and we have assumed that C_p is independent of temperature. The chemical or physical process requires the transfer of $q_p + q_{p,\text{ex}}$, where $q_{p,\text{ex}}$ is excess energy transferred as heat, to attain the same change in temperature of the sample. We interpret $q_{p,\text{ex}}$ in terms of an apparent change in the heat capacity at constant pressure of the sample, C_p, during the temperature scan. Then we write the heat capacity of the sample as $C_p + C_{p,\text{ex}}$, and

$$q_p + q_{p,\text{ex}} = (C_p + C_{p,\text{ex}})\Delta T$$

It follows that

$$C_{p,\text{ex}} = \frac{q_{p,\text{ex}}}{T} = \frac{q_{p,\text{ex}}}{\alpha t} = \frac{P_{\text{ex}}}{\alpha}$$

where $P_{\text{ex}} = q_{p,\text{ex}}/t$ is the excess electrical power necessary to equalize the temperature of the sample and reference compartments.

A DSC trace, a *thermogram*, consists of a plot of P_{ex} or $C_{p,\text{ex}}$ against T (Fig. 14.22). Broad peaks in the thermogram indicate processes requiring transfer of energy as heat. From eqn 14.33a, the enthalpy change associated with the process is

$$\Delta H = \int_{T_1}^{T_2} C_{p,\text{ex}}\,\mathrm{d}T$$

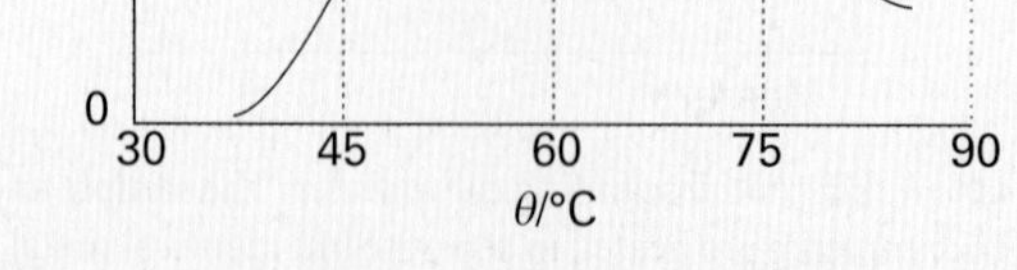

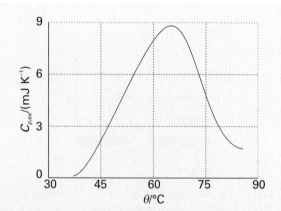

Fig. 14.22 A thermogram for the protein ubiquitin at pH = 2.45. The protein retains its native structure up to about 45°C and then undergoes an endothermic conformational change. (Adapted from B. Chowdhry and S. LeHarne, *J. Chem. Educ.* 74, 236 (1997).)

where T_1 and T_2 are, respectively, the temperatures at which the process begins and ends. This relation shows that the enthalpy change is the area under the curve of $C_{p,\text{ex}}$ versus T. With a DSC, enthalpy changes may be determined in samples of masses as low as 0.5 mg, which is a significant advantage over bomb or flame calorimeters, which require several grams of material.

Differential scanning calorimetry is used in the chemical industry to characterize polymers and in the biochemistry laboratory to assess the stability of proteins, nucleic acids, and membranes. Large molecules, such as synthetic or biological polymers, attain complex three-dimensional structures due to intra- and intermolecular interactions, such as hydrogen bonding and hydrophobic interactions (Section 8.5). Disruption of these interactions is an endothermic process that can be studied with a DSC. For example, the thermogram shown in Fig. 14.22 indicates that the protein ubiquitin retains its native structure up to about 45°C. At higher temperatures, the protein undergoes an endothermic conformational change that results in the loss of its three-dimensional structure. The same principles also apply to the study of structural integrity and stability of synthetic polymers, such as plastics.

14.6 Adiabatic changes

Work is done when a perfect gas expands adiabatically (without a transfer of energy as heat) but, because no heat enters the system, the internal energy falls and therefore the temperature of the working gas also falls. In molecular terms, the kinetic energy of the molecules falls as work is done, so their average speed decreases, and hence the temperature falls.

The change in internal energy of a perfect gas when the temperature is changed from T_i to T_f and the volume is changed from V_i to V_f can be expressed as the sum of two steps (Fig. 14.23). In the first step, only the volume changes and the temperature is held constant at its initial value. However, because the internal energy of a perfect gas is independent of the volume the molecules occupy, the overall change in internal energy arises solely from the second step, the change in temperature at constant volume. Provided the heat capacity is independent of temperature, this change is

$$\Delta U = C_V(T_f - T_i) = C_V\Delta T$$

Because the expansion is adiabatic, we know that $q = 0$; because $\Delta U = q + w$, it then follows that $\Delta U = w_{\text{ad}}$. The subscript 'ad' denotes an adiabatic process. Therefore, by equating the two values we have derived for ΔU, we obtain

$$w_{\text{ad}} = C_V\Delta T \qquad (14.36)$$

That is, the work done during an adiabatic expansion of a perfect gas is proportional to the temperature difference between the initial and final states. That is exactly what we expect on molecular grounds, because the mean kinetic energy is proportional

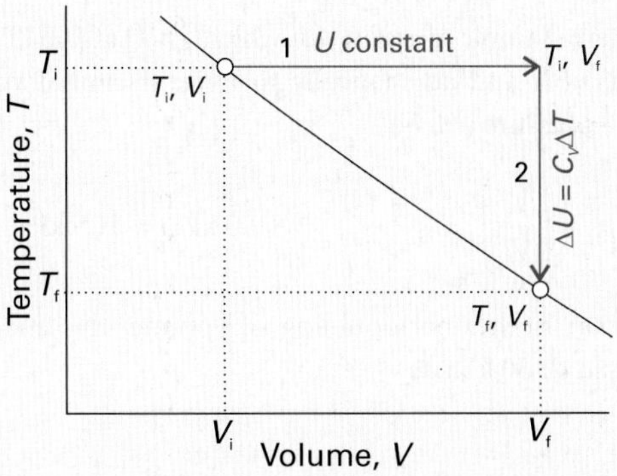

Fig. 14.23 To achieve a change of state from one temperature and volume to another temperature and volume, we may consider the overall change as composed of two steps. In the first step, the system expands at constant temperature; there is no change in internal energy if the system consists of a perfect gas. In the second step, the temperature of the system is reduced at constant volume. The overall change in internal energy is the sum of the changes for the two steps.

to T, so a change in internal energy arising from temperature alone is also expected to be proportional to ΔT. In *Further information 14.1* we show that the initial and final temperatures of a perfect gas that undergoes reversible adiabatic expansion (reversible expansion in a thermally insulated container) can be calculated from

$$T_f = T_i\left(\frac{V_i}{V_f}\right)^{1/c} \qquad (14.37a)^{\circ}_{\text{rev}}$$

where $c = C_{V,m}/R$, or equivalently

$$V_iT_i^c = V_fT_f^c \qquad (14.37b)^{\circ}_{\text{rev}}$$

This result is often summarized in the form VT^c = constant.

● A BRIEF ILLUSTRATION

Consider the adiabatic, reversible expansion of 0.020 mol Ar, initially at 25°C, from 0.50 dm^3 to 1.00 dm^3. The molar heat capacity of argon at constant volume is 12.48 J K^{-1} mol^{-1}, so $c = 1.501$. Therefore, from eqn 14.37a,

$$T_f = (298\text{ K}) \times \left(\frac{0.50\text{ dm}^3}{1.00\text{ dm}^3}\right)^{1/1.501} = 188\text{ K}$$

It follows that $\Delta T = -110$ K, and therefore, from eqn 14.36, that

$$w = \{(0.020\text{ mol}) \times (12.48\text{ J K}^{-1}\text{ mol}^{-1})\} \times (-110\text{ K}) = -27\text{ J}$$

Note that temperature change is independent of the amount of gas but the work is not. ●

Self-test 14.7 Calculate the final temperature, the work done, and the change of internal energy when ammonia is used in a reversible adiabatic expansion from 0.50 dm^3 to 2.00 dm^3, the other initial conditions being the same.

[194 K, −56 J, −56 J]

We can get some insight into the origin of the temperature dependence expressed by eqn 14.37 and understand adiabatic changes at a molecular level, by considering the reversible, adiabatic expansion of particles in a one-dimensional box. In quantum mechanics, an **adiabatic process** is one that occurs so slowly that the system follows a single changing state of a system and, in contrast to an impulsive change, does not jump into a linear combination of other states. We can explore the consequence of this model in statistical mechanics, by supposing that the molecules that occupy a given level of the box all remain in that level as the box expands reversibly and adiabatically (Fig. 14.24). That is, we suppose that all the N_i remain constant even though the energy levels are changing. For the populations to remain the same even though the energy levels are getting closer together, the temperature must fall, so our task is to see how that changing temperature must vary with the length of the box. In other words, we must look for a solution of

$$\frac{dN_i}{dL} = \frac{d}{dL}\frac{Ne^{-\beta\varepsilon_i}}{q} = 0 \qquad (14.38)$$

with β and ε_i, and therefore q too, functions of L. Because $\beta\varepsilon_i = \gamma_i/kL^2T$ (in the notation used in *Justification 14.1*), the population N_i is independent of the length if at all stages of the expansion

$$LT^{1/2} = \text{constant} \qquad (14.39)$$

However, for a one-dimensional system, the molar heat capacity is $\frac{1}{2}R$, so this solution is the one-dimensional version of eqn 14.37b.

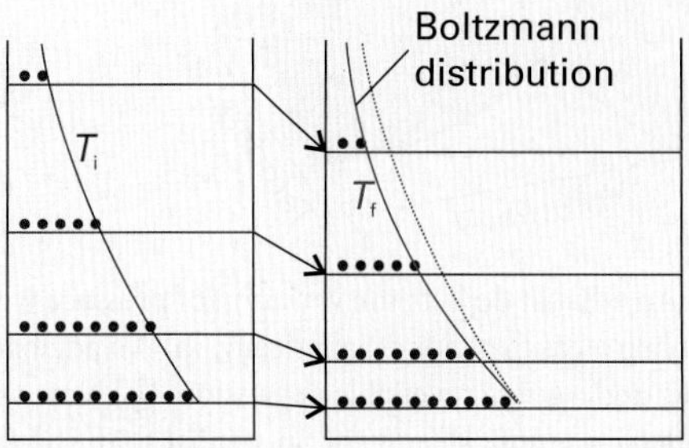

Fig. 14.24 In a reversible adiabatic expansion, the populations of the quantum states remain constant, which corresponds to a lowering of the temperature if the Boltzmann distribution is to continue to match the same distribution but now over the changed energy levels.

The result we have obtained shows what to imagine is happening at a molecular level during a reversible adiabatic expansion of a perfect gas. As shown in Fig. 14.24, the populations of each level remain constant as the levels fall in energy. However, for that to be the case the temperature must fall. The precise dependence of the temperature that guarantees this constancy is exactly the condition expressed in eqn 14.37.

We can now develop eqn 14.37. We show in *Further information 14.1* that the pressure of a perfect gas that undergoes reversible adiabatic expansion from a volume V_i to a volume V_f is related to its initial pressure by

$$p_f V_f^{\gamma} = p_i V_i^{\gamma} \qquad (14.40)^{\circ}_{\text{rev}}$$

where $\gamma = C_{p,m}/C_{V,m}$. This result is summarized in the form pV^{γ} = constant. For a monatomic perfect gas, $C_{V,m} = \frac{3}{2}R$, and from eqn 14.35 $C_{p,m} = \frac{5}{2}R$; so $\gamma = \frac{5}{3}$. For a gas of non-linear polyatomic molecules (which can rotate as well as translate), $C_{V,m} = 3R$, so $\gamma = \frac{4}{3}$.

The curves of pressure versus volume for adiabatic change are known as **adiabats**, and one for a reversible path is illustrated in Fig. 14.25. Because $\gamma > 1$, an adiabat falls more steeply ($p \propto 1/V^{\gamma}$) than the corresponding isotherm ($p \propto 1/V$). The physical reason for the difference is that, in an isothermal expansion, energy flows into the system as heat and maintains the temperature; as a result, the pressure does not fall as much as in an adiabatic expansion.

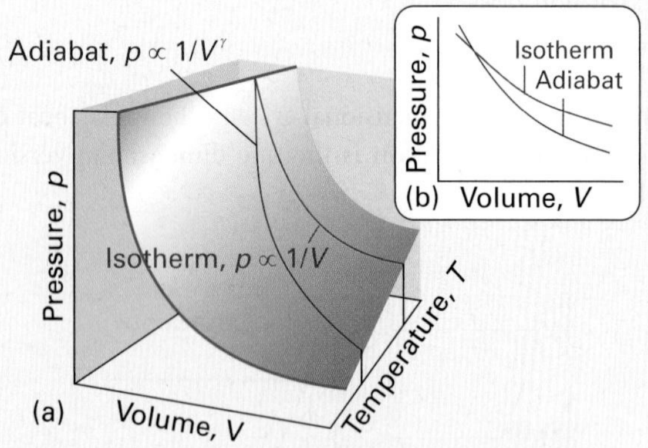

Fig. 14.25 An adiabat depicts the variation of pressure with volume when a gas expands adiabatically. (a) An adiabat for a perfect gas undergoing reversible expansion. (b) Note that the pressure declines more steeply for an adiabat than it does for an isotherm because the temperature decreases in the former.

interActivity Explore how the parameter γ affects the dependence of the pressure on the relative volume. Does the pressure–volume dependence become stronger or weaker with increasing volume?

● A BRIEF ILLUSTRATION

When a sample of argon (for which $\gamma = \frac{5}{3}$) at 100 kPa expands reversibly and adiabatically to twice its initial volume the final pressure will be

$$p_f = \left(\frac{V_i}{V_f}\right)^{\gamma} p_i = \left(\frac{1}{2}\right)^{5/3} \times (100\ \text{kPa}) = 31.5\ \text{kPa}$$

For an isothermal doubling of volume, the final pressure would be 50 kPa. ●

Thermochemistry

The study of the energy transferred as heat during the course of chemical reactions is called **thermochemistry**. Thermochemistry is a branch of thermodynamics because a reaction vessel and its contents form a system, and chemical reactions result in the exchange of energy between the system and the surroundings. Thus we can use calorimetry to measure the energy supplied or discarded as heat by a reaction, and can identify q with a change in internal energy (if the reaction occurs at constant volume) or a change in enthalpy (if the reaction occurs at constant pressure). Conversely, if we know the ΔU or ΔH for a reaction, we can predict the energy (transferred as heat) the reaction can produce.

We have already remarked that a process that releases heat is classified as exothermic and one that absorbs heat is classified as endothermic. Because the release of energy into the surroundings signifies a decrease in the enthalpy of a system (at constant pressure), we can now see that an exothermic process at constant pressure is one for which $\Delta H < 0$. Conversely, because the absorption of energy as heat from the surroundings results in an increase in enthalpy, an endothermic process at constant pressure has $\Delta H > 0$.

14.7 Standard enthalpy changes

Changes in enthalpy are normally reported for processes taking place under a set of standard conditions. In most of our discussions we shall consider the **standard enthalpy change**, $\Delta H^{\ominus}$, the change in enthalpy for a process in which the initial and final substances are in their standard states:

The **standard state** of a substance is its pure form at 1 bar.

(The standard state of substances in solution is described in Section 16.6.) Standard enthalpy changes may be reported for any temperature. However, the conventional temperature for reporting thermodynamic data is 298.15 K (corresponding to 25.00°C). Unless otherwise mentioned, all thermodynamic data in this text are for this conventional temperature.

Synoptic table 14.3* Standard enthalpies of fusion and vaporization at the transition temperature, $\Delta_{trs}H^{\ominus}/(\text{kJ mol}^{-1})$

	T_f/K	Fusion	T_b/K	Vaporization
Ar	83.8	1.188	87.29	6.506
C_6H_6	278.61	10.59	353.2	30.8
H_2O	273.15	6.008	373.15	40.656
				44.016 at 298 K
He	3.5	0.021	4.22	0.084

* More values are given in the *Data section*.

(a) Enthalpies of physical change

The standard enthalpy change that accompanies a change of physical state is called the **standard enthalpy of transition** and is denoted $\Delta_{trs}H^{\ominus}$ (Table 14.3). The **standard enthalpy of vaporization,** $\Delta_{vap}H^{\ominus}$, is one example. Another is the **standard enthalpy of fusion,** $\Delta_{fus}H^{\ominus}$, the standard enthalpy change accompanying the conversion of a solid to a liquid, as in

$$H_2O(s) \rightarrow H_2O(l) \qquad \Delta_{fus}H^{\ominus}(273\text{ K}) = +6.01\text{ kJ mol}^{-1}$$

As in this case, it is sometimes convenient to know the standard enthalpy change at the transition temperature as well as at the conventional temperature. The different types of enthalpy changes encountered in thermochemistry are summarized in Table 14.4.

Table 14.4 Enthalpies of transition

Transition	Process	Symbol*
Transition	Phase α → phase β	$\Delta_{trs}H$
Fusion	s → l	$\Delta_{fus}H$
Vaporization	l → g	$\Delta_{vap}H$
Sublimation	s → g	$\Delta_{sub}H$
Mixing	Pure → mixture	$\Delta_{mix}H$
Solution	Solute → solution	$\Delta_{sol}H$
Hydration	$X^{\pm}(g) \rightarrow X^{\pm}(aq)$	$\Delta_{hyd}H$
Atomization	Species(s, l, g) → atoms(g)	$\Delta_{at}H$
Ionization	$X(g) \rightarrow X^+(g) + e^-(g)$	$\Delta_{ion}H$
Electron gain	$X(g) + e^-(g) \rightarrow X^-(g)$	$\Delta_{eg}H$
Reaction	Reactants → products	$\Delta_r H$
Combustion	Compounds(s, l, g) + O_2(g) → CO_2(g), H_2O(l, g)	$\Delta_c H$
Formation	Elements → compound	$\Delta_f H$
Activation	Reactants → activated complex	$\Delta^{\ddagger}H$

* IUPAC recommendations. In common usage, the transition subscript is often attached to ΔH, as in ΔH_{trs}.

We shall meet them again in various locations throughout the text.

A note on good practice The attachment of the name of the transition to the symbol Δ, as in $\Delta_{trs}H$, is the modern convention. However, the older convention, ΔH_{trs}, is still widely used. The new convention is more logical because the subscript identifies the type of change, not the physical observable related to the change.

Because enthalpy is a state function, a change in enthalpy is independent of the path between the specified initial and final states of the system. There are two immediate consequences of this path-independence. One is that the enthalpy of an overall transition, such as sublimation (the direct conversion from solid to vapour) may be expressed as the sum of the enthalpies of fusion and vaporization (at the same temperature, 1).

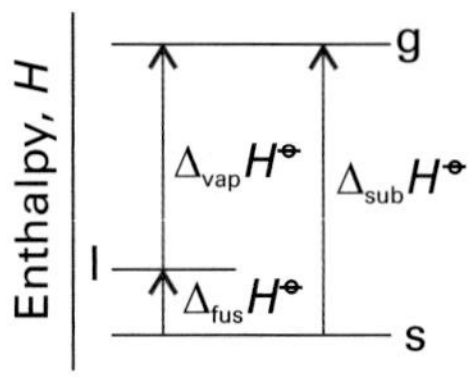

1

$$\Delta_{sub}H^{\ominus} = \Delta_{fus}H^{\ominus} + \Delta_{vap}H^{\ominus} \qquad (14.41)$$

An immediate conclusion is that, because almost all enthalpies of fusion are positive (helium is the exception), apart from helium the enthalpy of sublimation of a substance is always greater than its enthalpy of vaporization (at a given temperature). The second consequence is that the standard enthalpy changes of a forward process and its reverse differ only in sign (2):

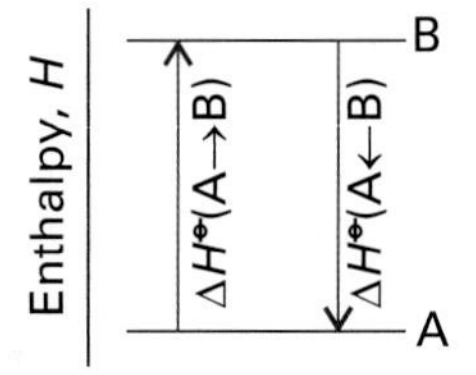

2

$$\Delta H^{\ominus}(A \rightarrow B) = -\Delta H^{\ominus}(B \rightarrow A) \qquad (14.42)$$

For instance, because the enthalpy of vaporization of water is +44 kJ mol^{-1} at 298 K, its enthalpy of condensation at that temperature is −44 kJ mol^{-1}.

(b) Enthalpies of chemical change

The **standard enthalpy of reaction,** $\Delta_r H^{\ominus}$, is the difference in standard molar enthalpies of the products and reactants weighted by their stoichiometric coefficients in the chemical equation. Thus, for the reaction

$$2\,A + B \rightarrow 3\,C + D$$

the standard reaction enthalpy is

$$\Delta_r H^{\ominus} = \{3H_m^{\ominus}(C) + H_m^{\ominus}(D)\} - \{2H_m^{\ominus}(A) + H_m^{\ominus}(B)\}$$

and in general

$$\Delta_r H^{\ominus} = \sum_{\text{Products}} \nu H_m^{\ominus} - \sum_{\text{Reactants}} \nu H_m^{\ominus} \qquad (14.43a)$$

corresponding to the process

> Reactants in their standard states → products in their standard states

Except in the case of ionic reactions in solution, the enthalpy changes accompanying mixing of the pure reactants and separation of the products into their pure form (as implied by 'standard state') are insignificant in comparison with the contribution from the reaction itself.

> ***A note on good practice*** The units of the standard reaction enthalpy $\Delta_r H^\ominus$ are kilojoules per mole. The 'per mole' refers to 'per mole of reaction events' of the reaction as written, such as per 2 mol A or per mol B that are consumed, or per 3 mol C or per mol D that are formed.

Equation 14.43a may be written in a more compact manner by adopting the convention that the stoichiometric coefficients of products are positive and those of reactants are negative. For instance, in the reaction 2 A + B → 3 C + D, the coefficients are $\nu_A = -2$, $\nu_B = -1$, $\nu_C = 3$, and $\nu_D = 1$. Then eqn 14.43a becomes

$$\Delta_r H^\ominus = \sum_J \nu_J H_m^\ominus(J) \qquad (14.43b)$$

It will be familiar from introductory chemistry that, just as standard enthalpies of transition may be combined, standard enthalpies of reactions can also be combined to obtain the enthalpy of another reaction. This application of the First Law is called **Hess's law** (3):

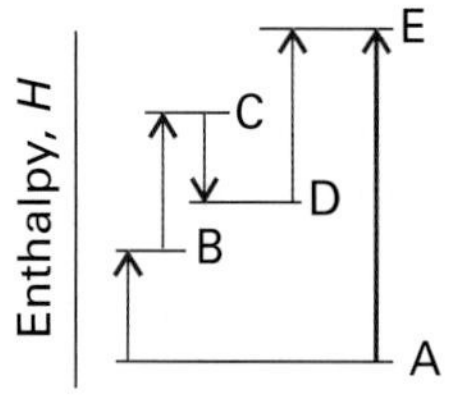

3

> The standard enthalpy of an overall reaction is the sum of the standard enthalpies of the individual reactions into which a reaction may be divided.

The individual steps need not be realizable in practice: they may be hypothetical reactions, the only requirement being that their chemical equations should balance and that all numerical values should refer to the same temperature. Review exercises will be found at the end of the chapter.

14.8 Standard enthalpies of formation

The **standard enthalpy of formation**, $\Delta_f H^\ominus$, of a substance is the standard reaction enthalpy for the formation of the compound from its elements in their reference states. The **reference state** of an element is its most stable state at the specified temperature and 1 bar. For example, at 298 K the reference state of nitrogen is a gas of N_2 molecules, that of mercury is liquid mercury, that of carbon is graphite, and that of tin is the white (metallic) form. There is one exception to this general prescription: the reference state of phosphorus is taken to be white phosphorus despite this allotrope not being the most stable form but simply the most reproducible form of the element. Standard enthalpies of formation are expressed as enthalpies per mole of molecules or (for ionic substances) formula units of the compound. The standard enthalpy of formation of liquid benzene at 298 K, for example, refers to the reaction

$$6\,C(s, graphite) + 3\,H_2(g) \rightarrow C_6H_6(l)$$

and is +49.0 kJ mol^{-1}. The standard enthalpies of formation of elements in their reference states are zero at all temperatures because they are the enthalpies of such 'null' reactions as $N_2(g) \rightarrow N_2(g)$. Some enthalpies of formation are listed in Tables 14.5 and 14.6.

The standard enthalpy of formation of ions in solution poses a special problem because it is impossible to prepare a solution of cations alone or of anions alone. This problem is solved by defining one ion, conventionally the hydrogen ion, to have zero standard enthalpy of formation at all temperatures:

$$\Delta_f H^\ominus(H^+, aq) = 0 \qquad [14.44]$$

Thus, if the enthalpy of formation of HBr(aq) is found to be −122 kJ mol^{-1}, then the whole of that value is ascribed to the

Synoptic table 14.5* Standard enthalpies of formation and combustion of organic compounds at 298 K

	$\Delta_f H^\ominus$/(kJ mol^{-1})	$\Delta_c H^\ominus$/(kJ mol^{-1})
Benzene, $C_6H_6(l)$	+49.0	−3268
Ethane, $C_2H_6(g)$	−84.7	−1560
Glucose, $C_6H_{12}O_6(s)$	−1274	−2808
Methane, $CH_4(g)$	−74.8	−890
Methanol, $CH_3OH(l)$	−238.7	−726

* More values are given in the *Data section.*

Synoptic table 14.6* Standard enthalpies of formation of inorganic compounds at 298 K

	$\Delta_f H^\ominus$/(kJ mol^{-1})
$H_2O(l)$	−285.83
$H_2O(g)$	−241.82
$NH_3(g)$	−46.11
$N_2H_4(l)$	+50.63
$NO_2(g)$	33.18
$N_2O_4(g)$	+9.16
NaCl(s)	−411.15
KCl(s)	−436.75

* More values are given in the *Data section.*

formation of $Br^-(aq)$, and we write $\Delta_f H^{\ominus}(Br^-, aq) = -122$ kJ mol^{-1}. That value may then be combined with, for instance, the enthalpy of formation of AgBr(aq) to determine the value of $\Delta_f H^{\ominus}(Ag^+, aq)$, and so on. In essence, this definition adjusts the actual values of the enthalpies of formation of ions by a fixed amount, which is chosen so that the standard value for one of them, $H^+(aq)$, has the value zero. Tabulated values are based on a set of experimental values and adjusted to give the best fit to the entire set rather than just isolated pairs of measurements.

(a) The reaction enthalpy in terms of enthalpies of formation

Conceptually, we can regard a reaction as proceeding by decomposing the reactants into their elements and then forming those elements into the products. The value of $\Delta_r H^{\ominus}$ for the overall reaction is the sum of these 'unforming' and forming enthalpies. Because 'unforming' is the reverse of forming, the enthalpy of an unforming step is the negative of the enthalpy of formation (**4**). Hence, in the enthalpies of formation of substances, we have enough information to calculate the enthalpy of any reaction by using

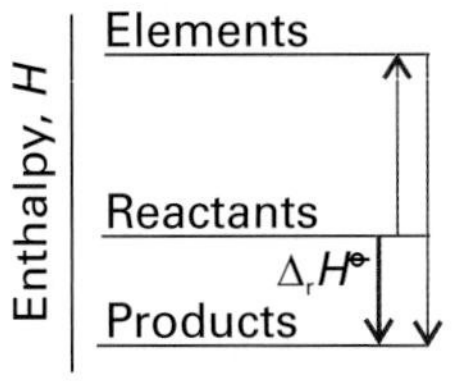

4

$$\Delta_r H^{\ominus} = \sum_J \nu_J \Delta_f H^{\ominus}(J) \qquad (14.45)$$

● A BRIEF ILLUSTRATION

The standard reaction enthalpy of $2\ HN_3(l) + 2\ NO(g) \rightarrow H_2O_2(l) + 4\ N_2(g)$ is calculated as follows:

$$\begin{aligned}\Delta_r H^{\ominus} &= \Delta_f H^{\ominus}(H_2O_2,l) + 4\Delta_f H^{\ominus}(N_2,g) - 2\Delta_f H^{\ominus}(HN_3,l) \\ &\quad - 2\Delta_f H^{\ominus}(NO,g) \\ &= \{-187.78 + 4(0) - 2(264.0) - 2(90.25)\}\ \text{kJ mol}^{-1} \\ &= -896.3\ \text{kJ mol}^{-1}\end{aligned}$$

Therefore, 896.3 kJ of energy is transferred to the surroundings as heat at constant pressure per mole of reaction events; that is per 2 mol HN_3 or per 2 mol NO that are consumed or per mol H_2O_2 or per 4 mol N_2 that are formed. ●

(b) Enthalpies of formation and molecular modelling

We have seen how to construct standard reaction enthalpies by combining standard enthalpies of formation. The question that now arises is whether we can construct standard enthalpies of formation from a knowledge of the chemical constitution of the species. The short answer is that there is no thermodynamically exact way of expressing enthalpies of formation in terms of contributions from individual atoms and bonds. In the past, approximate procedures based on **mean bond enthalpies**, ΔH(A—B), the average molar enthalpy change associated with the breaking of a specific A—B bond,

$$\text{A—B(g)} \rightarrow \text{A(g)} + \text{B(g)}$$

have been used. However, this procedure is notoriously unreliable, in part because the ΔH(A—B) are average values for a series of related compounds. Nor does the approach distinguish between geometrical isomers, where the same atoms and bonds may be present but experimentally the enthalpies of formation might be significantly different.

Computer-aided molecular modelling has largely displaced this more primitive approach. Commercial software packages use the principles developed in Chapter 6 to calculate the standard enthalpy of formation of a molecule drawn on the computer screen. As pointed out there, the parameters used in a variety of semiempirical approaches are optimized for the computation of enthalpies of formation. The techniques can be applied to different conformations of the same molecule. In the case of methylcyclohexane, for instance, the calculated conformational energy difference ranges from 5.9 to 7.9 kJ mol^{-1}, with the equatorial conformer having the lower standard enthalpy of formation. These estimates compare favourably with the experimental value of 7.5 kJ mol^{-1}. However, good agreement between calculated and experimental values is relatively rare. Computational methods almost always predict correctly which conformer is more stable but do not always predict the correct magnitude of the conformational energy difference.

14.9 The temperature dependence of reaction enthalpies

The standard enthalpies of many important reactions have been measured at different temperatures. However, in the absence of this information, standard reaction enthalpies at different temperatures may be calculated from heat capacities and the reaction enthalpy at some other temperature (Fig. 14.26). In many

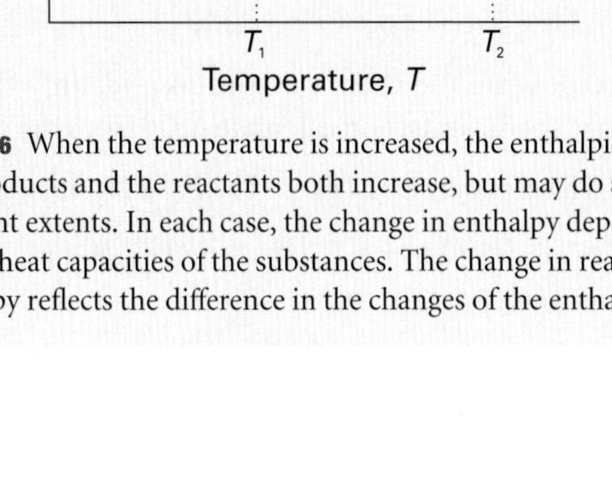

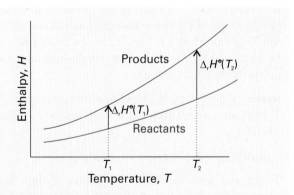

Fig. 14.26 When the temperature is increased, the enthalpies of the products and the reactants both increase, but may do so to different extents. In each case, the change in enthalpy depends on the heat capacities of the substances. The change in reaction enthalpy reflects the difference in the changes of the enthalpies.

cases heat capacity data are more accurate than reaction enthalpies so, providing the information is available, the procedure we are about to describe is more accurate than a direct measurement of a reaction enthalpy at an elevated temperature.

It follows from eqn 14.33a that, when a substance is heated from T_1 to T_2, its enthalpy changes from $H(T_1)$ to

$$H(T_2) = H(T_1) + \int_{T_1}^{T_2} C_p \, dT \qquad (14.46)$$

(We have assumed that no phase transition takes place in the temperature range of interest.) Because this equation applies to each substance in the reaction, the standard reaction enthalpy changes from $\Delta_r H^\circ(T_1)$ to

$$\Delta_r H^\circ(T_2) = \Delta_r H^\circ(T_1) + \int_{T_1}^{T_2} \Delta_r C_p^\circ \, dT \qquad (14.47)$$

where $\Delta_r C_p^\circ$ is the difference of the molar heat capacities of products and reactants under standard conditions weighted by the stoichiometric coefficients that appear in the chemical equation:

$$\Delta_r C_p^\circ = \sum_J \nu_J C_{p,m}^\circ(J) \qquad [14.48]$$

Equation 14.47 is known as **Kirchhoff's law**. It is normally a good approximation to assume that $\Delta_r C_p$ is independent of the temperature, at least over reasonably limited ranges, as illustrated in the following example. Although the individual heat capacities may vary, their difference varies less significantly. In some cases the temperature dependence of heat capacities is taken into account by using eqn 14.34.

Example 14.6 *Using Kirchhoff's law*

The standard enthalpy of formation of gaseous H_2O at 298 K is -241.82 kJ mol^{-1}. Estimate its value at 100°C given the following values of the molar heat capacities at constant pressure: $H_2O(g)$: 33.58 J K^{-1} mol^{-1}; $H_2(g)$: 28.82 J K^{-1} mol^{-1}; $O_2(g)$: 29.36 J K^{-1} mol^{-1}. Assume that the heat capacities are independent of temperature.

Method When $\Delta_r C_p^\circ$ is independent of temperature in the range T_1 to T_2, the integral in eqn 14.47 evaluates to $(T_2 - T_1)\Delta_r C_p^\circ$. Therefore,

$$\Delta_r H^\circ(T_2) = \Delta_r H^\circ(T_1) + (T_2 - T_1)\Delta_r C_p^\circ$$

To proceed, write the chemical equation, identify the stoichiometric coefficients, and calculate $\Delta_r C_p^\circ$ from the data.

Answer The reaction is $H_2(g) + \frac{1}{2}O_2(g) \rightarrow H_2O(g)$, so

$$\begin{aligned}\Delta_r C_p^\circ &= C_{p,m}^\circ(H_2O,g) - C_{p,m}^\circ(H_2,g) - \tfrac{1}{2}C_{p,m}^\circ(O_2,g)\\ &= -9.92\ \text{J K}^{-1}\,\text{mol}^{-1}\end{aligned}$$

(Notice: joules not kilojoules.) It then follows that

$$\begin{aligned}\Delta_f H^\circ(373\ \text{K}) &= -241.82\ \text{kJ mol}^{-1} + (75\ \text{K}) \times (-9.94\ \text{J K}^{-1}\,\text{mol}^{-1})\\ &= -242.6\ \text{kJ mol}^{-1}\end{aligned}$$

Self-test 14.8 Estimate the standard enthalpy of formation of cyclohexane at 400 K from the data in Table 14.6.

[-163 kJ mol^{-1}]

Properties of the internal energy and the enthalpy

Thermodynamic considerations extend far beyond the relatively elementary applications in thermochemistry. In particular, the mathematical properties of state functions can be used to draw far-reaching conclusions about the relations between physical properties and establish connections that might be completely unexpected. The practical importance of these results is that we can combine measurements of different properties to obtain the value of a property we require.

The crucial point we build on in the following sections is that the fact that a property X is a state function (as is the case for U and H) implies that dX is an exact differential, in the sense described in *Mathematical background 8*.

14.10 Changes in internal energy

The internal energy U is a state function, so dU is an exact differential. We begin to unfold the consequences by exploring a closed system of constant composition (the only type of system considered in the rest of this chapter). The internal energy U can be regarded as a function of V, T, and p but, because there is an equation of state, stating the values of two of the variables fixes the value of the third (for instance, $p = nRT/V$ for a perfect gas). Therefore, it is possible to write U in terms of just two independent variables: V and T, p and T, or p and V. Expressing U as a function of volume and temperature fits the purpose of our discussion.

(a) Changes in internal energy at constant temperature

When V changes by dV and T changes by dT, the internal energy changes by

$$dU = \left(\frac{\partial U}{\partial V}\right)_T dV + \left(\frac{\partial U}{\partial T}\right)_V dT \qquad (14.49)$$

We have already met $(\partial U/\partial T)_V$ in eqn 14.23 where we saw that it is the constant-volume heat capacity, C_V. The other coefficient, $(\partial U/\partial V)_T$, plays a major role in thermodynamics because it is a

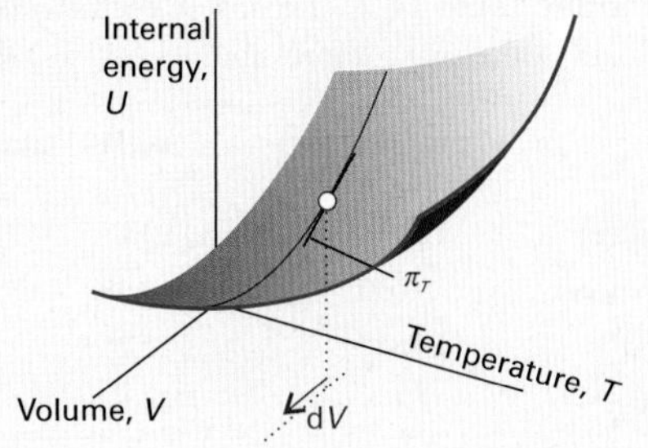

Fig. 14.27 The internal pressure, π_T, is the slope of U with respect to V with the temperature T held constant.

measure of the variation of the internal energy of a substance as its volume is changed at constant temperature (Fig. 14.27). We shall denote it π_T and, because it has the same dimensions as pressure, call it the **internal pressure**:

$$\pi_T = \left(\frac{\partial U}{\partial V}\right)_T \qquad [14.50]$$

In terms of the notation C_V and π_T, eqn 14.49 becomes

$$dU = \pi_T dV + C_V dT \qquad (14.51)$$

When there are no interactions between the molecules, the internal energy is independent of their separation and hence independent of the volume of the sample. Therefore, for a perfect gas we can write $\pi_T = 0$. The statement $\pi_T = 0$ (that is, the internal energy is independent of the volume occupied by the sample) can be taken to be the definition of a perfect gas, for later we shall see that it implies the equation of state $pV = nRT$. If the internal energy increases ($dU > 0$) as the volume of the sample expands isothermally ($dV > 0$), which is the case when there are attractive forces between the particles, then a plot of internal energy against volume slopes upwards and $\pi_T > 0$ (Fig. 14.28).

To make contact with the statistical thermodynamic discussion in Chapter 13 we need to use the canonical partition function Q to calculate the internal pressure because only that formulation allows us to include the effects of intermolecular interactions. From the expression for U in terms of Q ($\langle E\rangle = -(\partial \ln Q/\partial\beta)_V$ and $U = U(0) + \langle E\rangle$), we can write

$$\pi_T = -\left(\frac{\partial}{\partial V}\left(\frac{\partial \ln Q}{\partial \beta}\right)_V\right)_T \qquad (14.52)$$

To develop this expression, we need to find a way to build an intermolecular potential energy into the expression for Q. The total kinetic energy of a gas is the sum of the kinetic energies of the individual molecules. Therefore, even in a real gas the canonical partition function factorizes into a part arising from the kinetic energy, which is the same as for the perfect gas

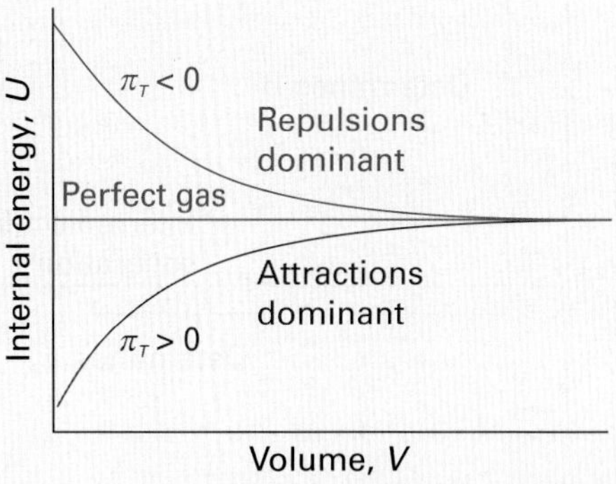

Fig. 14.28 For a perfect gas, the internal energy is independent of the volume (at constant temperature). If attractions are dominant in a real gas, the internal energy increases with volume because the molecules become farther apart on average. If repulsions are dominant, the internal energy decreases as the gas expands.

($Q = V^N/\Lambda^{3N}N!$, where Λ is the thermal wavelength, eqn 13.18), and a factor called the **configuration integral**, Z, which depends on the intermolecular potentials. We therefore write

$$Q = \frac{Z}{\Lambda^{3N}} \qquad (14.53)$$

with Z replacing $V^N/N!$. It then follows that

$$\begin{aligned}\pi_T &= -\left(\frac{\partial}{\partial V}\left(\frac{\partial \ln(Z/\Lambda^{3N})}{\partial \beta}\right)_V\right)_T \\ &= -\left(\frac{\partial}{\partial V}\left(\frac{\partial \ln Z}{\partial \beta}\right)_V\right)_T - \left(\frac{\partial}{\partial V}\left(\frac{\partial \ln(1/\Lambda^{3N})}{\partial \beta}\right)_V\right)_T \\ &= -\left(\frac{\partial}{\partial V}\left(\frac{\partial \ln Z}{\partial \beta}\right)_V\right)_T = -\left(\frac{\partial}{\partial V}\left(\frac{1}{Z}\frac{\partial Z}{\partial \beta}\right)_V\right)_T\end{aligned} \qquad (14.54)$$

In the second line, although the derivative of Λ with respect to temperature and hence β is nonzero, Λ is independent of volume, so the derivative with respect to volume is zero.

For a real gas of atoms (for which the intermolecular interactions are isotropic), Z is related to the total potential energy E_P of interaction of all the particles by

$$Z = \frac{1}{N!}\int e^{-\beta E_P} d\tau_1 d\tau_2 \cdots d\tau_N \qquad (14.55)$$

where $d\tau_i$ is the volume element for atom i. The physical origin of this term is that the probability of occurrence of each arrangement of molecules possible in the sample is given by a Boltzmann distribution in which the exponent is given by the potential energy corresponding to that arrangement.

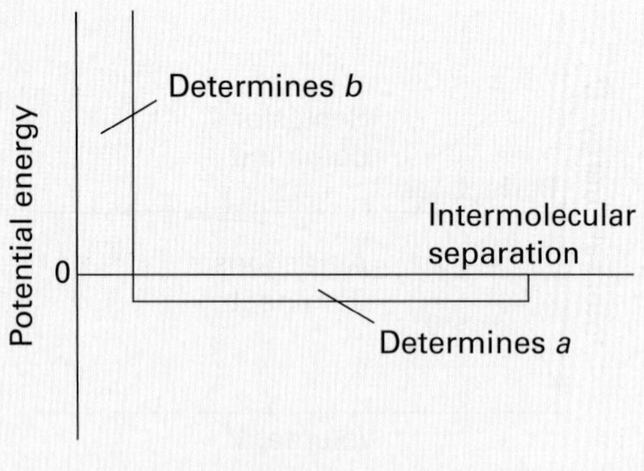

Fig. 14.29 The van der Waals equation of state can be derived on the basis that the intermolecular potential energy has a hard core surrounded by a long-range, shallow attractive well.

Equation 14.55 is very difficult to manipulate in practice, even for quite simple intermolecular potentials, except for a perfect gas for which $E_P = 0$ and hence

$$Z = \frac{1}{N!}\int \mathrm{d}\tau_1 \mathrm{d}\tau_2 \cdots \mathrm{d}\tau_N = \frac{V^N}{N!} \qquad (14.56)°$$

Because V is held constant in the evaluation of π_T, it follows from eqn 14.54 that for a perfect gas $\pi_T = 0$.

If, however, the potential has the form of a central hard sphere surrounded by a shallow attractive well (Fig. 14.29), then detailed calculation, which is too involved to reproduce here, leads to

$$\pi_T = \frac{n^2 a}{V^2} \qquad (14.57)$$

where a is a constant that is proportional to the area under the attractive part of the potential. In Section 15.8 we shall see that exactly the same expression is implied by the van der Waals equation of state. At this point we can conclude that, if there are attractive interactions between molecules in a gas, then its internal energy increases as it expands isothermally (because $\pi_T > 0$, and the slope of U with respect to V is positive). The energy rises because, at greater average separations, the molecules spend less time in regions where they interact favourably.

(b) Changes in internal energy at constant pressure

Partial derivatives have many useful properties and some that we shall draw on frequently are reviewed in *Mathematics background 8*. Skilful use of them can often turn some unfamiliar quantity into a quantity that can be recognized, interpreted, and measured.

As an example, suppose we want to find out how the internal energy varies with temperature when the pressure of the system is kept constant. If we divide both sides of eqn 14.51 by $\mathrm{d}T$ and impose the condition of constant pressure on the resulting differentials, so $\mathrm{d}U/\mathrm{d}T$ on the left becomes $(\partial U/\partial T)_p$, we obtain

Synoptic table 14.7* Expansion coefficients (α) and isothermal compressibilities (κ_T) at 298 K

	$\alpha/(10^{-4}\ \mathrm{K}^{-1})$	$\kappa_T/(10^{-6}\ \mathrm{atm}^{-1})$
Benzene	12.4	92.1
Diamond	0.030	0.187
Lead	0.861	2.21
Water	2.1	49.6

* More values are given in the *Data section*.

$$\left(\frac{\partial U}{\partial T}\right)_p = \pi_T\left(\frac{\partial V}{\partial T}\right)_p + C_V$$

It is usually sensible in thermodynamics to inspect the output of a manipulation like this to see if it contains any recognizable physical quantity. The partial derivative on the right in this expression is the slope of the plot of volume against temperature (at constant pressure). This property is normally tabulated as the **expansion coefficient**, α, of a substance, which is defined as

$$\alpha = \frac{1}{V}\left(\frac{\partial V}{\partial T}\right)_p \qquad [14.58]$$

and physically is the fractional change in volume that accompanies a rise in temperature. A large value of α means that the volume of the sample responds strongly to changes in temperature. Table 14.7 lists some experimental values of α and of the **isothermal compressibility**, κ_T (kappa), which is defined as

$$\kappa_T = -\frac{1}{V}\left(\frac{\partial V}{\partial p}\right)_T \qquad [14.59]$$

The isothermal compressibility is a measure of the fractional change in volume when the pressure is increased by a small amount; the negative sign in the definition ensures that the compressibility is a positive quantity, because an increase of pressure, implying a positive $\mathrm{d}p$, brings about a reduction of volume, a negative $\mathrm{d}V$.

Example 14.7 *Calculating the expansion coefficient of a gas*

Derive an expression for the expansion coefficient of a perfect gas.

Method The expansion coefficient is defined in eqn 14.58. To use this expression, substitute the expression for V in terms of T obtained from the equation of state for the gas. As implied by the subscript in eqn 14.58, the pressure, p, is treated as a constant.

Answer Because $pV = nRT$, we can write

$$\alpha = \frac{1}{V}\left(\frac{\partial(nRT/p)}{\partial T}\right)_p = \frac{1}{V}\times\frac{nR}{p}\frac{dT}{dT} = \frac{nR}{pV} = \frac{1}{T}$$

The higher the temperature, the less responsive is the volume of a perfect gas to a change in temperature.

Self-test 14.9 Derive an expression for the isothermal compressibility of a perfect gas. [$\kappa_T = 1/p$]

When we introduce the definition of α into the equation for $(\partial U/\partial T)_p$, we obtain

$$\left(\frac{\partial U}{\partial T}\right)_p = \alpha\pi_T V + C_V \quad (14.60)$$

This equation is entirely general (provided the system is closed and its composition is constant). It expresses the dependence of the internal energy on the temperature at constant pressure in terms of C_V, which can be measured in one experiment, in terms of α, which can be measured in another, and in terms of the quantity π_T. For a perfect gas, $\pi_T = 0$, so then

$$\left(\frac{\partial U}{\partial T}\right)_p = C_V \quad (14.61)°$$

That is, although the constant-volume heat capacity of a perfect gas is defined as the slope of a plot of internal energy against temperature at constant volume, for a perfect gas C_V is also the slope at constant pressure.

Equation 14.61 provides an easy way to derive the relation between C_p and C_V for a perfect gas expressed in eqn 14.35. Thus, we can use it to express both heat capacities in terms of derivatives at constant pressure:

$$C_p - C_V = \left(\frac{\partial H}{\partial T}\right)_p - \left(\frac{\partial U}{\partial T}\right)_p \quad (14.62)°$$

Then we introduce $H = U + pV = U + nRT$ into the first term, which results in

$$C_p - C_V = \left(\frac{\partial U}{\partial T}\right)_p + nR - \left(\frac{\partial U}{\partial T}\right)_p = nR \quad (14.63)°$$

which is eqn 14.35.

14.11 The Joule–Thomson effect

We can carry out a similar set of operations on the enthalpy, $H = U + pV$. The quantities U, p, and V are all state functions; therefore H is also a state function and dH is an exact differential. It turns out that H is a useful thermodynamic function when the pressure is under our control: we saw a sign of that in the relation $\Delta H = q_p$ (eqn 14.29b). We shall therefore regard H as a function of p and T, and adapt the argument in Section 14.10 to find an expression for the variation of H with temperature at constant volume. As set out in the following *Justification*, we find that, for a closed system of constant composition,

$$dH = -\mu C_p dp + C_p dT \quad (14.64)$$

where the **Joule–Thomson coefficient**, μ (mu), is defined as

$$\mu = \left(\frac{\partial T}{\partial p}\right)_H \quad [14.65]$$

This relation will prove useful for relating the heat capacities at constant pressure and volume and for a discussion of the liquefaction of gases.

Justification 14.4 *The variation of enthalpy with pressure and temperature*

By the same argument that led to eqn 14.49 but with H regarded as a function of p and T we can write

$$dH = \left(\frac{\partial H}{\partial p}\right)_T dp + \left(\frac{\partial H}{\partial T}\right)_p dT \quad (14.66)$$

The second partial derivative is C_p; our task here is to express $(\partial H/\partial p)_T$ in terms of recognizable quantities. The chain relation (see *Mathematics background 8*) lets us write

$$\left(\frac{\partial H}{\partial p}\right)_T = -\frac{1}{(\partial p/\partial T)_H(\partial T/\partial H)_p}$$

and both partial derivatives can be brought into the numerator by using the reciprocal identity twice:

$$\left(\frac{\partial H}{\partial p}\right)_T = -\frac{(\partial T/\partial p)_H}{(\partial T/\partial H)_p} = -\left(\frac{\partial T}{\partial p}\right)_H\left(\frac{\partial H}{\partial T}\right)_p = -\mu C_p \quad (14.67)$$

We have used the definitions of the constant-pressure heat capacity, C_p, and the Joule–Thomson coefficient, μ (eqn 14.65). Equation 14.64 now follows directly.

The analysis of the Joule–Thomson coefficient is central to the technological problems associated with the liquefaction of gases. We need to be able to interpret it physically and to measure it. As shown in the *Justification* below, the cunning required to impose the constraint of constant enthalpy, so that the process is **isenthalpic**, was supplied by Joule and William Thomson (later Lord Kelvin). They let a gas expand through a porous barrier from one constant pressure to another, and monitored the difference of temperature that arose from the expansion (Fig. 14.30). The whole apparatus was insulated so that the

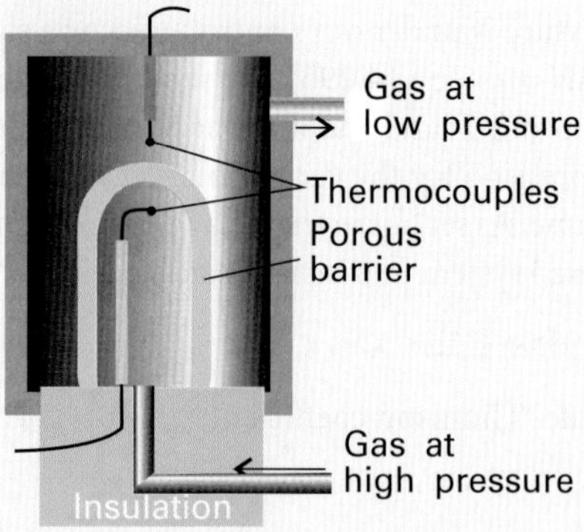

Fig. 14.30 The apparatus used for measuring the Joule–Thomson effect. The gas expands through the porous barrier, which acts as a throttle, and the whole apparatus is thermally insulated. As explained in the text, this arrangement corresponds to an isenthalpic expansion (expansion at constant enthalpy). Whether the expansion results in a heating or a cooling of the gas depends on the conditions.

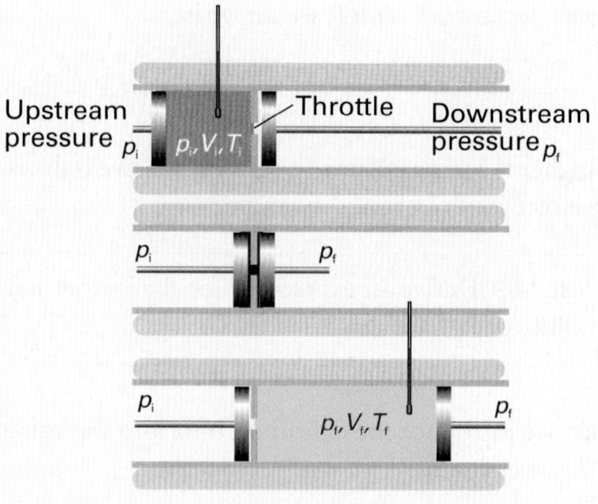

Fig. 14.31 The thermodynamic basis of Joule–Thomson expansion. The pistons represent the upstream and downstream gases, which maintain constant pressures either side of the throttle. The transition from the top diagram to the bottom diagram, which represents the passage of a given amount of gas through the throttle, occurs without change of enthalpy.

process was adiabatic. They observed a lower temperature on the low pressure side, the difference in temperature being proportional to the pressure difference they maintained. This cooling by isenthalpic expansion is now called the **Joule–Thomson effect.**

Justification 14.5 *The Joule–Thomson effect*

Here we show that the experimental arrangement results in expansion at constant enthalpy. Because all changes to the gas occur adiabatically,

$$q = 0, \text{ which implies } \Delta U = w$$

Consider the work done as the gas passes through the barrier. We focus on the passage of a fixed amount of gas from the high pressure side, where the pressure is p_i, the temperature T_i, and the gas occupies a volume V_i (Fig. 14.31). The gas emerges on the low pressure side, where the same amount of gas has a pressure p_f, a temperature T_f, and occupies a volume V_f. The gas on the left is compressed isothermally by the upstream gas acting as a piston. The relevant pressure is p_i and the volume changes from V_i to 0; therefore, the work done on the gas is

$$w_1 = -p_i(0 - V_i) = p_iV_i$$

The gas expands isothermally on the right of the barrier (but possibly at a different constant temperature) against the pressure p_f provided by the downstream gas acting as a piston to be driven out. The volume changes from 0 to V_f, so the work done on the gas in this stage is

$$w_2 = -p_f(V_f - 0) = -p_fV_f$$

The total work done on the gas is the sum of these two quantities, or

$$w = w_1 + w_2 = p_iV_i - p_fV_f$$

It follows that the change of internal energy of the gas as it moves adiabatically from one side of the barrier to the other is

$$U_f - U_i = w = p_iV_i - p_fV_f$$

Reorganization of this expression gives

$$U_f + p_fV_f = U_i + p_iV_i, \text{ or } H_f = H_i$$

Therefore, the expansion occurs without change of enthalpy.

The property measured in the experiment is the ratio of the temperature change to the change of pressure, $\Delta T/\Delta p$. Adding the constraint of constant enthalpy and taking the limit of small Δp implies that the thermodynamic quantity measured is $(\partial T/\partial p)_H$, which is the Joule–Thomson coefficient, μ. In other words, the physical interpretation of μ is that it is the ratio of the change in temperature to the change in pressure when a gas expands under conditions that ensure there is no change in enthalpy.

The modern method of measuring μ is indirect, and involves measuring the **isothermal Joule–Thomson coefficient**, the quantity

$$\mu_T = \left(\frac{\partial H}{\partial p}\right)_T \qquad [14.68]$$

which is the slope of a plot of enthalpy against pressure at constant temperature (Fig. 14.32). Comparing eqns 14.67 and 14.68, we see that the two coefficients are related by:

$$\mu_T = -C_p\mu \qquad (14.69)$$

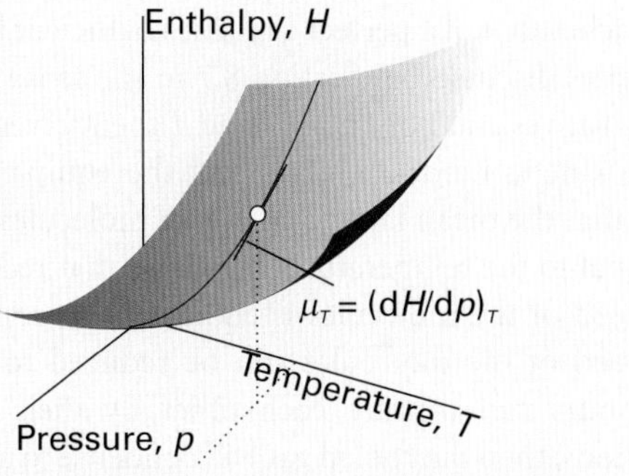

Fig. 14.32 The isothermal Joule–Thomson coefficient is the slope of the enthalpy with respect to changing pressure, the temperature being held constant.

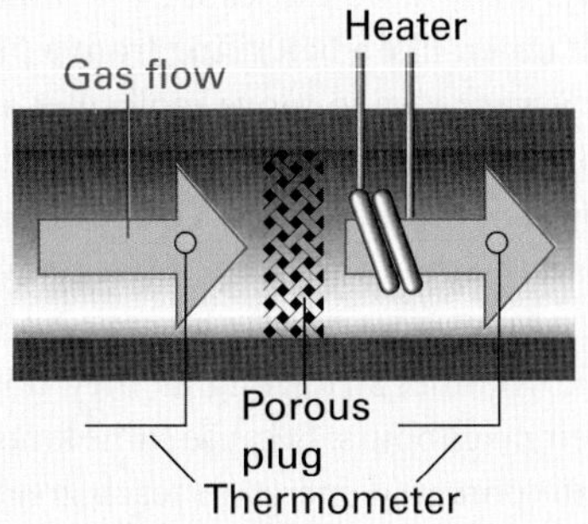

Fig. 14.33 A schematic diagram of the apparatus used for measuring the isothermal Joule–Thomson coefficient. The electrical heating required to offset the cooling arising from expansion is interpreted as ΔH and used to calculate $(\partial H/\partial p)_T$, which is then converted to μ as explained in the text.

To measure μ_T, the gas is pumped continuously at a steady pressure through a heat exchanger (which brings it to the required temperature) and then through a porous plug inside a thermally insulated container. The steep pressure drop is measured, and the cooling effect is exactly offset by an electric heater placed immediately after the plug (Fig. 14.33). The energy provided by the heater is monitored. Because the energy transferred as heat can be identified with the value of ΔH for the gas (because $\Delta H = q_p$), and the pressure change Δp is known, we can find μ_T from the limiting value of $\Delta H/\Delta p$ as $\Delta p \rightarrow 0$, and then convert it to μ. Table 14.8 lists some values obtained in this way.

Real gases have nonzero Joule–Thomson coefficients. Depending on the identity of the gas, the pressure, the relative magnitudes of the attractive and repulsive intermolecular forces, and the temperature, the sign of the coefficient may be either positive or negative (Fig. 14.34). A positive sign implies that dT is negative when dp is negative, in which case the gas cools on expansion. Gases that show a heating effect ($\mu < 0$) at one temperature show a cooling effect ($\mu > 0$) when the temperature passes through an **inversion temperature**, T_I (Table 14.8, Fig. 14.35). As indicated

Synoptic table 14.8* Inversion temperatures, normal freezing and boiling points, and Joule–Thomson coefficients at 1 atm and 298 K

	T_I/K	T_f/K	T_b/K	μ/(K atm^{-1})
Ar	723	83.8	87.3	
CO_2	1500		194.7s	+1.11 at 300 K
He	40		4.2	−0.062
N_2	621	63.3	77.4	+0.27

* More values are given in the *Data section*.
s: sublimes

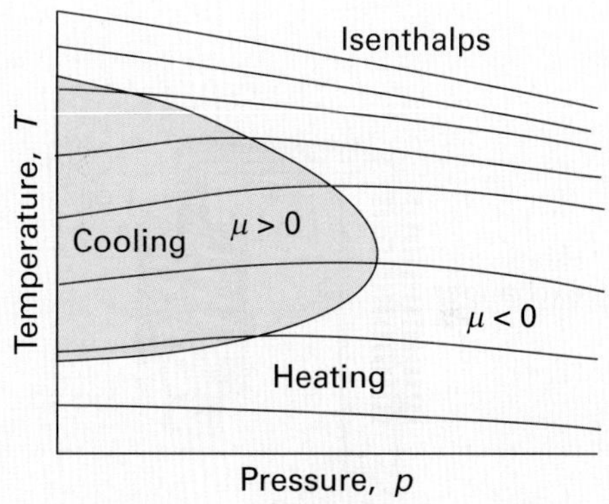

Fig. 14.34 The sign of the Joule–Thomson coefficient, μ, depends on the conditions. Inside the boundary, the shaded area, it is positive and outside it is negative. The temperature corresponding to the boundary at a given pressure is the 'inversion temperature' of the gas at that pressure. For a given pressure, the temperature must be below a certain value if cooling is required, but if it becomes too low, the boundary is crossed again and heating occurs. Reduction of pressure under adiabatic conditions moves the system along one of the isenthalps, or curves of constant enthalpy. The inversion temperature curve runs through the points of the isenthalps where their slope changes from negative to positive.

in Fig. 14.35, a gas typically has two inversion temperatures at a given pressure, one at high temperature and the other at low.

The 'Linde refrigerator' makes use of Joule–Thompson expansion to liquefy gases (Fig. 14.36). The gas at high pressure is allowed to expand through a throttle, it cools, and is circulated past the incoming gas. That gas is cooled, and its subsequent expansion cools it still further. There comes a stage when the circulating gas becomes so cold that it condenses to a liquid.

For a perfect gas, $\mu = 0$; hence, the temperature of a perfect gas is unchanged by Joule–Thomson expansion. This characteristic points clearly to the involvement of intermolecular forces in determining the size of the effect. However, the Joule–Thomson coefficient of a real gas does not necessarily approach zero as the

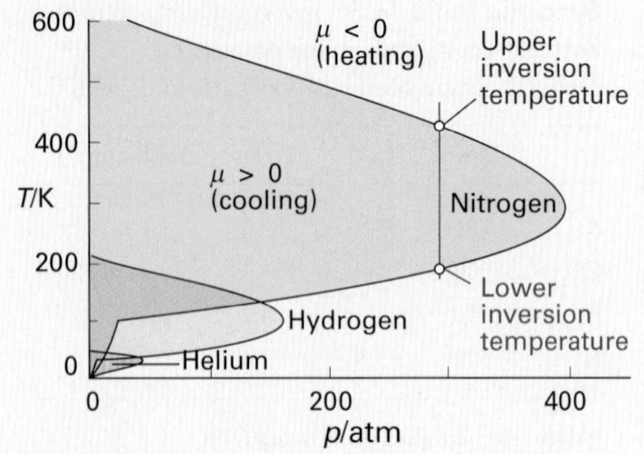

Fig. 14.35 The inversion temperatures for three real gases, nitrogen, hydrogen, and helium.

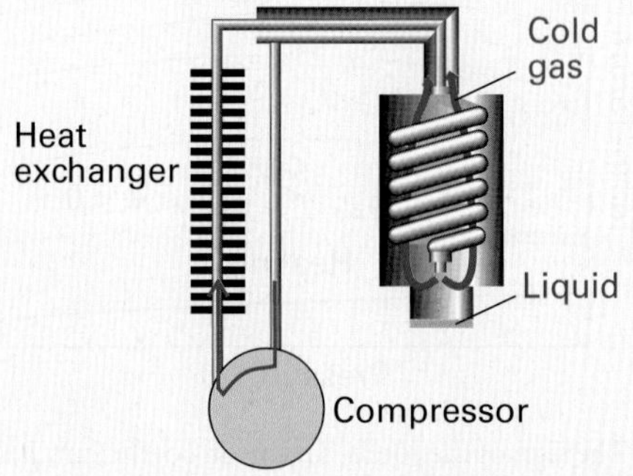

Fig. 14.36 The principle of the Linde refrigerator is shown in this diagram. The gas is recirculated, and so long as it is beneath its inversion temperature it cools on expansion through the throttle. The cooled gas cools the high-pressure gas, which cools still further as it expands. Eventually liquefied gas drips from the throttle.

pressure is reduced even though the equation of state of the gas approaches that of a perfect gas. The coefficient behaves like the properties discussed in Section 8.7 in the sense that it depends on derivatives and not on p, V, and T themselves.

The kinetic model of gases and the equipartition theorem imply that the mean kinetic energy of molecules in a gas is proportional to the temperature. It follows that reducing the average speed of the molecules is equivalent to cooling the gas. If the speed of the molecules can be reduced to the point that neighbours can capture each other by their intermolecular attractions, then the cooled gas will condense to a liquid. To slow the gas molecules, we make use of an effect similar to that seen when a ball is thrown into the air: as it rises it slows in response to the gravitational attraction of the Earth and its kinetic energy is converted into potential energy. Molecules in a real gas attract each other (the attraction is not gravitational, but the effect is the same). It follows that, if we can cause the molecules to move apart from each other, like a ball rising from a planet, then they should slow. It is very easy to move molecules apart from each other: we simply allow the gas to expand, which increases the average separation of the molecules. To cool a gas, therefore, we allow it to expand without allowing any energy to enter from outside as heat. As the gas expands, the molecules move apart to fill the available volume, struggling as they do so against the attraction of their neighbours. Because some kinetic energy must be converted into potential energy to reach greater separations, the molecules travel more slowly as their separation increases. This sequence of molecular events explains the Joule–Thomson effect: the cooling of a real gas by adiabatic expansion. The cooling effect, which corresponds to $\mu > 0$, is observed under conditions when attractive interactions are dominant, because the molecules have to climb apart against the attractive force in order for them to travel more slowly. For molecules under conditions when repulsions are dominant, the Joule–Thomson effect results in the gas becoming warmer, or $\mu < 0$.

Checklist of key ideas

- ☐ 1. Thermodynamics is the study of the transformations of energy.
- ☐ 2. The system is the part of the world in which we have a special interest. The surroundings is the region outside the system where we make our measurements.
- ☐ 3. An open system has a boundary through which matter can be transferred. A closed system has a boundary through which matter cannot be transferred. An isolated system has a boundary through which neither matter nor energy can be transferred.
- ☐ 4. Energy is the capacity to do work. The internal energy is the total energy of a system.
- ☐ 5. Work is the transfer of energy by motion against an opposing force, $dw = -Fdz$. Heat is the transfer of energy as a result of a temperature difference between the system and the surroundings. In molecular terms, heating makes use of thermal motion, that is, disorderly molecular motion.
- ☐ 6. An exothermic process releases energy as heat; an endothermic process absorbs energy as heat.
- ☐ 7. A state function is a property that depends only on the current state of the system and is independent of how that state has been prepared.
- ☐ 8. The First Law of thermodynamics states that the internal energy of an isolated system is constant, $\Delta U = q + w$.

☐ 9. Expansion work is the work of expansion (or compression) of a system, $dw = -p_{ex}dV$. The work of free expansion is $w = 0$. The work of expansion against a constant external pressure is $w = -p_{ex}\Delta V$. The work of isothermal reversible expansion of a perfect gas is $w = -nRT\ln(V_f/V_i)$.

☐ 10. A reversible change is a change that can be reversed by an infinitesimal modification of a variable. A system is in equilibrium with its surroundings if an infinitesimal change in conditions in an opposite direction results in an opposite change in state.

☐ 11. Maximum work is achieved in a reversible change.

☐ 12. In a reversible isothermal expansion, the work done is due to a change in the energy levels occupied by the molecules of the system.

☐ 13. Calorimetry is the study of heat transfers during physical and chemical processes.

☐ 14. At constant volume, with no additional work, $\Delta U = q_V = \sum_i \varepsilon_i \Delta n_i$.

☐ 15. The heat capacity at constant volume is defined as $C_V = (\partial U/\partial T)_V$. The heat capacity at constant pressure, $C_p = (\partial H/\partial T)_p$. For a perfect gas, the heat capacities are related by $C_p - C_V = nR$.

☐ 16. The total molar heat capacity at constant volume is $C_{V,m} = \frac{1}{2}(3 + \nu_R^* + 2\nu_V^*)R$ where ν_R^* and ν_V^* are the numbers of active rotational and vibrational modes, respectively.

☐ 17. The enthalpy is defined as $H = U + pV$. The enthalpy change is the energy transferred as heat at constant pressure, $\Delta H = q_p$.

☐ 18. Thermochemistry is the study of the energy transferred as heat during chemical reactions.

☐ 19. The standard enthalpy change is the change in enthalpy for a process in which the initial and final substances are in their standard states. The standard state is the pure substance at 1 bar.

☐ 20. Hess's law states that the standard enthalpy of an overall reaction is the sum of the standard enthalpies of the individual reactions into which a reaction may be divided.

☐ 21. The standard enthalpy of formation ($\Delta_f H^\ominus$) is the standard reaction enthalpy for the formation of the compound from its elements in their reference states. The reference state is the most stable state of an element at the specified temperature and 1 bar.

☐ 22. The standard reaction enthalpy may be estimated by combining enthalpies of formation, $\Delta_r H^\ominus = \sum_{\text{Products}} \nu\Delta_f H^\ominus - \sum_{\text{Reactants}} \nu\Delta_f H^\ominus$.

☐ 23. The temperature dependence of the reaction enthalpy is given by Kirchhoff's law, $\Delta_r H^\ominus(T_2) = \Delta_r H^\ominus(T_1) + \int_{T_1}^{T_2} \Delta_r C_p^\ominus dT$.

☐ 24. The internal pressure is defined as $\pi_T = (\partial U/\partial V)_T$. For a perfect gas, $\pi_T = 0$. For a real gas, π_T can be expressed in terms of the volume- and temperature-dependencies of the configuration integral Z.

☐ 25. For a perfect gas, the expansion coefficient $\alpha = 1/T$ and the isothermal compressibility $\kappa_T = 1/p$.

☐ 26. The Joule–Thomson effect is the cooling of a gas by isenthalpic expansion.

☐ 27. The Joule–Thomson coefficient is defined as $\mu = (\partial T/\partial p)_H$. The isothermal Joule–Thomson coefficient is defined as $\mu_T = (\partial H/\partial p)_T = -C_p\mu$. For a perfect gas, $\mu = 0$.

☐ 28. The inversion temperature is the temperature at which the Joule–Thomson coefficient changes sign.

Further information

Further information 14.1 *Adiabatic processes*

Consider a stage in a reversible adiabatic expansion when the pressure inside and out is p. The work done when the gas expands by dV is $dw = -pdV$; however, for a perfect gas, $dU = C_V dT$ since $\pi_T = 0$ in eqn 14.51 for a perfect gas. Therefore, because for an adiabatic change ($dq = 0$) $dU = dw + dq = dw$, we can equate these two expressions for dU and write

$$C_V dT = -pdV$$

We are dealing with a perfect gas, so we can replace p by nRT/V and obtain

$$\frac{C_V dT}{T} = -\frac{nRdV}{V}$$

To integrate this expression we note that T is equal to T_i when V is equal to V_i, and is equal to T_f when V is equal to V_f at the end of the expansion. Therefore,

$$C_V \int_{T_i}^{T_f} \frac{dT}{T} = -nR \int_{V_i}^{V_f} \frac{dV}{V}$$

(We are taking C_V to be independent of temperature.) Then, because $\int dx/x = \ln x + \text{constant}$, we obtain

$$C_V \ln\frac{T_f}{T_i} = -nR\ln\frac{V_f}{V_i}$$

Because $\ln(x/y) = -\ln(y/x)$, this expression rearranges to

$$\frac{C_V}{nR}\ln\frac{T_f}{T_i} = \ln\frac{V_i}{V_f}$$

With $c = C_V/nR$ we obtain (because $\ln x^a = a\ln x$)

$$\ln\left(\frac{T_f}{T_i}\right)^c = \ln\left(\frac{V_i}{V_f}\right)$$

which implies that $(T_f/T_i)^c = (V_i/V_f)$ and, upon rearrangement, eqn 14.37.

The initial and final states of a perfect gas satisfy the perfect gas law regardless of how the change of state takes place, so we can use $pV = nRT$ to write

$$\frac{p_i V_i}{p_f V_f} = \frac{T_i}{T_f}$$

However, we have just shown that

$$\frac{T_i}{T_f} = \left(\frac{V_f}{V_i}\right)^{1/c} = \left(\frac{V_f}{V_i}\right)^{\gamma-1}$$

where we use the definition of the heat capacity ratio $\gamma = C_{p,m}/C_{V,m}$ and the fact that, for a perfect gas, $C_{p,m} - C_{V,m} = R$ (the molar version of eqn 14.35). Then we combine the two expressions, to obtain

$$\frac{p_i}{p_f} = \frac{V_f}{V_i} \times \left(\frac{V_f}{V_i}\right)^{\gamma-1} = \left(\frac{V_f}{V_i}\right)^{\gamma}$$

which rearranges to $p_i V_i^{\gamma} = p_f V_f^{\gamma}$, which is eqn 14.40.

Discussion questions

14.1 Describe and distinguish the various uses of the words 'system' and 'state' in physical chemistry.

14.2 Describe the distinction between heat and work in thermodynamic and molecular terms, the latter in terms of populations and energy levels.

14.3 Give examples of state functions and discuss why they play a critical role in thermodynamics.

14.4 Explain the difference between the change in internal energy and the change in enthalpy accompanying a chemical or physical process.

14.5 Describe two calorimetric methods for the determination of enthalpy changes that accompany chemical processes.

14.6 Distinguish between 'standard state' and 'reference state', and indicate their applications.

14.7 Suggest (with explanation) how the internal energy of a van der Waals gas should vary with volume at constant temperature.

14.8 Use concepts of statistical thermodynamics to describe the molecular features that determine the magnitude of the constant-volume molar heat capacity of a molecular substance.

14.9 Use concepts of statistical thermodynamics to describe the molecular features that lead to the equations of state of perfect and real gases.

14.10 Explain why a perfect gas does not have an inversion temperature.

Exercises

Assume all gases are perfect unless stated otherwise. Unless otherwise stated, thermochemical data are for 298.15 K.

14.1(a) Use the equipartition theorem to estimate the molar internal energy of (a) I_2, (b) CH_4, (c) C_6H_6 in the gas phase at 25°C.

14.1(b) Use the equipartition theorem to estimate the molar internal energy of (a) O_3, (b) C_2H_6, (c) SO_2 in the gas phase at 25°C.

14.2(a) Which of (a) pressure, (b) temperature, (c) work, (d) enthalpy are state functions?

14.2(b) Which of (a) volume, (b) heat, (c) internal energy, (d) density are state functions?

14.3(a) Calculate the work needed for a 60 kg person to climb through 6.0 m (a) on the surface of the Earth and (b) the Moon ($g = 1.60$ m s^{-2}).

14.3(b) Calculate the work needed for a bird of mass 150 g to fly to a height of 75 m from the surface of the Earth.

14.4(a) A chemical reaction takes place in a container of cross-sectional area 50 cm^2. As a result of the reaction, a piston is pushed out through 15 cm against an external pressure of 1.0 atm. Calculate the work done by the system.

14.4(b) A chemical reaction takes place in a container of cross-sectional area 75.0 cm^2. As a result of the reaction, a piston is pushed out through 25.0 cm against an external pressure of 150 kPa. Calculate the work done by the system.

14.5(a) A sample consisting of 1.00 mol Ar is expanded isothermally at 20°C from 10.0 dm^3 to 30.0 dm^3 (a) reversibly, (b) against a constant external pressure equal to the final pressure of the gas, and (c) freely (against zero external pressure). For the three processes calculate q, w, and ΔU.

14.5(b) A sample consisting of 2.00 mol He is expanded isothermally at 0°C from 5.0 dm^3 to 20.0 dm^3 (a) reversibly, (b) against a constant external pressure equal to the final pressure of the gas, and (c) freely (against zero external pressure). For the three processes calculate q, w, and ΔU.

14.6(a) When 229 J of energy is supplied as heat to 3.0 mol Ar(g), the temperature of the sample increases by 2.55 K. Calculate the molar heat capacities at constant volume and constant pressure of the gas.

14.6(b) When 178 J of energy is supplied as heat to 1.9 mol of gas molecules, the temperature of the sample increases by 1.78 K. Calculate the molar heat capacities at constant volume and constant pressure of the gas.

14.7(a) A sample consisting of 1.00 mol of perfect gas atoms, for which $C_{V,m} = \frac{3}{2}R$, initially at $p_1 = 1.00$ atm and $T_1 = 300$ K, is heated reversibly to 400 K at constant volume. Calculate the final pressure, ΔU, q, and w.

14.7(b) A sample consisting of 2.00 mol of perfect gas molecules, for which $C_{V,m} = \frac{5}{2}R$, initially at $p_1 = 111$ kPa and $T_1 = 277$ K, is heated reversibly to 356 K at constant volume. Calculate the final pressure, ΔU, q, and w.

14.8(a) The constant-pressure heat capacity of a sample of a perfect gas was found to vary with temperature according to the expression $C_p/(\mathrm{J\,K^{-1}}) = 20.17 + 0.3665(T/\mathrm{K})$. Calculate q, w, and ΔU when the temperature is raised from 25°C to 100°C for 1.00 mol gas molecules (a) at constant pressure, (b) at constant volume.

14.8(b) The constant-pressure heat capacity of a sample of a perfect gas was found to vary with temperature according to the expression $C_p/(\mathrm{J\,K^{-1}}) = 20.17 + 0.4001(T/\mathrm{K})$. Calculate q, w, and ΔU when the temperature is raised from 0°C to 200°C for 1.00 mol gas molecules (a) at constant pressure, (b) at constant volume.

14.9(a) When 3.0 mol O_2 is heated at a constant pressure of 3.25 atm, its temperature increases from 260 K to 285 K. Given that the molar heat capacity of O_2 at constant pressure is 29.4 J K^{-1} mol^{-1}, calculate q, ΔH, and ΔU.

14.9(b) When 2.0 mol CO_2 is heated at a constant pressure of 1.25 atm, its temperature increases from 250 K to 277 K. Given that the molar heat capacity of CO_2 at constant pressure is 37.11 J K^{-1} mol^{-1}, calculate q, ΔH, and ΔU.

14.10(a) The ground level of Cl is $^2P_{3/2}$ and a $^2P_{1/2}$ level lies 881 cm^{-1} above it. Calculate the electronic contribution to the molar constant-volume heat capacity of Cl atoms at (a) 500 K and (b) 900 K.

14.10(b) The first electronically excited state of O_2 is $^1\Delta_g$ and lies 7918.1 cm^{-1} above the ground state, which is $^3\Sigma_g^-$. Calculate the electronic contribution to the molar constant-volume heat capacity of O_2 at 400 K.

14.11(a) Use the equipartition principle to estimate the values of $\gamma = C_p/C_V$ for gaseous ammonia and methane. Do this calculation with and without the vibrational contribution to the energy. Which is closer to the expected experimental value at 25°C?

14.11(b) Use the equipartition principle to estimate the value of $\gamma = C_p/C_V$ for carbon dioxide. Do this calculation with and without the vibrational contribution to the energy. Which is closer to the expected experimental value at 25°C?

14.12(a) What is the root mean square deviation of the molecular energy of argon atoms at 298 K?

14.12(b) What is the root mean square deviation of the molecular energy of carbon dioxide molecules at 298 K?

14.13(a) Calculate the final temperature of a sample of argon of mass 12.0 g that is expanded reversibly and adiabatically from 1.0 dm^3 at 273.15 K to 3.0 dm^3.

14.13(b) Calculate the final temperature of a sample of carbon dioxide of mass 16.0 g that is expanded reversibly and adiabatically from 500 cm^3 at 298.15 K to 2.00 dm^3.

14.14(a) A sample consisting of 1.0 mol of perfect gas molecules with $C_V = 20.8$ J K^{-1} is initially at 4.25 atm and 300 K. It undergoes reversible adiabatic expansion until its pressure reaches 2.50 atm. Calculate the final volume and temperature and the work done.

14.14(b) A sample consisting of 2.5 mol of perfect gas molecules with $C_{p,\mathrm{m}} = 20.8$ J K^{-1} mol^{-1} is initially at 240 kPa and 325 K. It undergoes reversible adiabatic expansion until its pressure reaches 150 kPa. Calculate the final volume and temperature and the work done.

14.15(a) A sample of carbon dioxide of mass 2.45 g at 27.0°C is allowed to expand reversibly and adiabatically from 500 cm^3 to 3.00 dm^3. What is the work done by the gas?

14.15(b) A sample of nitrogen of mass 3.12 g at 23.0°C is allowed to expand reversibly and adiabatically from 400 cm^3 to 2.00 dm^3. What is the work done by the gas?

14.16(a) Calculate the final pressure of a sample of carbon dioxide that expands reversibly and adiabatically from 67.4 kPa and 0.50 dm^3 to a final volume of 2.00 dm^3. Take $\gamma = 1.4$.

14.16(b) Calculate the final pressure of a sample of water vapour that expands reversibly and adiabatically from 97.3 Torr and 400 cm^3 to a final volume of 5.0 dm^3. Take $\gamma = 1.3$.

14.17(a) For tetrachloromethane, $\Delta_{\mathrm{vap}}H^\ominus = 30.0$ kJ mol^{-1}. Calculate q, w, ΔH, and ΔU when 0.75 mol CCl_4(l) is vaporized at 250 K and 750 Torr.

14.17(b) For ethanol, $\Delta_{\mathrm{vap}}H^\ominus = 43.5$ kJ mol^{-1}. Calculate q, w, ΔH, and ΔU when 1.75 mol C_2H_5OH(l) is vaporized at 260 K and 765 Torr.

14.18(a) The standard enthalpy of formation of ethylbenzene is −12.5 kJ mol^{-1}. Calculate its standard enthalpy of combustion.

14.18(b) The standard enthalpy of formation of phenol is −165.0 kJ mol^{-1}. Calculate its standard enthalpy of combustion.

14.19(a) The standard enthalpy of combustion of cyclopropane is −2091 kJ mol^{-1} at 25°C. From this information and enthalpy of formation data for CO_2(g) and H_2O(g), calculate the enthalpy of formation of cyclopropane. The enthalpy of formation of propene is +20.42 kJ mol^{-1}. Calculate the enthalpy of isomerization of cyclopropane to propene.

14.19(b) From the following data, determine $\Delta_f H^\ominus$ for diborane, B_2H_6(g), at 298 K:

$$(1)\ B_2H_6(g) + 3\,O_2(g) \rightarrow B_2O_3(s) + 3\,H_2O(g) \quad \Delta_r H^\ominus = -1941\ \mathrm{kJ\,mol^{-1}}$$

$$(2)\ 2\,B(s) + \tfrac{3}{2}O_2(g) \rightarrow B_2O_3(s) \quad \Delta_r H^\ominus = -2368\ \mathrm{kJ\,mol^{-1}}$$

$$(3)\ H_2(g) + \tfrac{1}{2}O_2(g) \rightarrow H_2O(g) \quad \Delta_r H^\ominus = -241.8\ \mathrm{kJ\,mol^{-1}}$$

14.20(a) Given that the standard enthalpy of formation of HCl(aq) is −167 kJ mol^{-1}, what is the value of $\Delta_f H^\ominus(Cl^-, aq)$?

14.20(b) Given that the standard enthalpy of formation of HI(aq) is −55 kJ mol^{-1}, what is the value of $\Delta_f H^\ominus(I^-, aq)$?

14.21(a) When 120 mg of naphthalene, $C_{10}H_8$(s), was burned in a bomb calorimeter the temperature rose by 3.05 K. Calculate the calorimeter constant. By how much will the temperature rise when 150 mg of phenol, C_6H_5OH(s), is burned in the calorimeter under the same conditions?

14.21(b) When 2.25 mg of anthracene, $C_{14}H_{10}$(s), was burned in a bomb calorimeter the temperature rose by 1.75 K. Calculate the calorimeter constant. By how much will the temperature rise when 125 mg of phenol, C_6H_5OH(s), is burned in the calorimeter under the same conditions? ($\Delta_c H^\ominus(C_{14}H_{10}, s) = -7061$ kJ mol^{-1}.)

14.22(a) Given the reactions (1) and (2) below, determine (a) $\Delta_r H^\ominus$ and $\Delta_r U^\ominus$ for reaction (3), (b) $\Delta_f H^\ominus$ for both HCl(g) and H_2O(g) all at 298 K.

$$(1)\ H_2(g) + Cl_2(g) \rightarrow 2\,HCl(g) \quad \Delta_r H^\ominus = -184.62\ \mathrm{kJ\,mol^{-1}}$$

$$(2)\ 2\,H_2(g) + O_2(g) \rightarrow 2\,H_2O(g) \quad \Delta_r H^\ominus = -483.64\ \mathrm{kJ\,mol^{-1}}$$

$$(3)\ 4\,HCl(g) + O_2(g) \rightarrow 2\,Cl_2(g) + 2\,H_2O(g)$$

14.22(b) Given the reactions (1) and (2) below, determine (a) $\Delta_r H^\ominus$ and $\Delta_r U^\ominus$ for reaction (3), (b) $\Delta_f H^\ominus$ for both HI(g) and H_2O(g) all at 298 K.

$$(1)\ H_2(g) + I_2(s) \rightarrow 2\,HI(g) \quad \Delta_r H^\ominus = +52.96\ \mathrm{kJ\,mol^{-1}}$$

$$(2)\ 2\,H_2(g) + O_2(g) \rightarrow 2\,H_2O(g) \quad \Delta_r H^\ominus = -483.64\ \mathrm{kJ\,mol^{-1}}$$

$$(3)\ 4\,HI(g) + O_2(g) \rightarrow 2\,I_2(s) + 2\,H_2O(g)$$

14.23(a) For the reaction $C_2H_5OH(l) + 3\,O_2(g) \rightarrow 2\,CO_2(g) + 3\,H_2O(g)$, $\Delta_r U^\ominus = -1373$ kJ mol^{-1} at 298 K. Calculate $\Delta_r H^\ominus$.

14.23(b) For the reaction $2\,C_6H_5COOH(s) + 15\,O_2(g) \rightarrow 14\,CO_2(g) + 6\,H_2O(g)$, $\Delta_r U^\ominus = -772.7$ kJ mol^{-1} at 298 K. Calculate $\Delta_r H^\ominus$.

14.24(a) From the data in Tables 14.5 and 14.6, calculate $\Delta_r H^\ominus$ and $\Delta_r U^\ominus$ at (a) 298 K, (b) 478 K for the reaction C(graphite) + H_2O(g) → CO(g) + H_2(g). Assume all heat capacities to be constant over the temperature range of interest.

14.24(b) Calculate $\Delta_r H^\ominus$ and $\Delta_r U^\ominus$ at 298 K and $\Delta_r H^\ominus$ at 427 K for the hydrogenation of ethyne (acetylene) to ethene (ethylene) from the enthalpy of combustion and heat capacity data in Tables 14.5 and 14.6. Assume the heat capacities to be constant over the temperature range involved.

14.25(a) Estimate $\Delta_r H^\ominus$(500 K) for the combustion of methane, $CH_4(g) + 2\,O_2(g) \rightarrow CO_2(g) + 2\,H_2O(g)$ by using the data on the temperature dependence of heat capacities in Table 14.2.

14.25(b) Estimate $\Delta_r H^\ominus$(478 K) for the combustion of naphthalene, $C_{10}H_8(g) + 12\,O_2(g) \rightarrow 10\,CO_2(g) + 4\,H_2O(g)$ by using the data on the temperature dependence of heat capacities in Table 14.2.

14.26(a) Set up a thermodynamic cycle for determining the enthalpy of hydration of Mg^{2+} ions using the following data: enthalpy of sublimation of Mg(s), +167.2 kJ mol^{-1}; first and second ionization enthalpies of Mg(g), 7.646 eV and 15.035 eV; dissociation enthalpy of Cl_2(g), +241.6 kJ mol^{-1}; electron gain enthalpy of Cl(g), −3.78 eV; enthalpy of solution of $MgCl_2$(s), −150.5 kJ mol^{-1}; enthalpy of hydration of Cl^-(g), −383.7 kJ mol^{-1}.

14.26(b) Set up a thermodynamic cycle for determining the enthalpy of hydration of Ca^{2+} ions using the following data: enthalpy of sublimation of Ca(s), +178.2 kJ mol^{-1}; first and second ionization enthalpies of Ca(g), 589.7 kJ mol^{-1} and 1145 kJ mol^{-1}; enthalpy of vaporization of bromine, +30.91 kJ mol^{-1}; dissociation enthalpy of Br_2(g), +192.9 kJ mol^{-1}; electron gain enthalpy of Br(g), −331.0 kJ mol^{-1}; enthalpy of solution of $CaBr_2$(s), −103.1 kJ mol^{-1}; enthalpy of hydration of Br^-(g), +97.5 kJ mol^{-1}.

14.27(a) When a certain freon used in refrigeration was expanded adiabatically from an initial pressure of 32 atm and 0°C to a final pressure of 1.00 atm, the temperature fell by 22 K. Calculate the Joule–Thomson coefficient, μ, at 0°C, assuming it remains constant over this temperature range.

14.27(b) When a vapour at 22 atm and 5°C was allowed to expand adiabatically to a final pressure of 1.00 atm, the temperature fell by 10 K. Calculate the Joule–Thomson coefficient, μ, at 5°C, assuming it remains constant over this temperature range.

14.28(a) Estimate the internal pressure, π_T, of water vapour at 1.00 bar and 298 , treating it as a van der Waals gas. *Hint.* Simplify the approach by estimating the molar volume by treating the gas as perfect.

14.28(b) Estimate the internal pressure, π_T, of sulfur dioxide at 1.00 bar and 298 K, treating it as a van der Waals gas. *Hint.* Simplify the approach by estimating the molar volume by treating the gas as perfect.

14.29(a) For a van der Waals gas, $\pi_T = a/V_m^2$. Calculate ΔU_m for the isothermal expansion of nitrogen gas from an initial volume of 1.00 dm^3 to 20.00 dm^3 at 298 K. What are the values of q and w?

14.29(b) Repeat Exercise 14.29a for argon, from an initial volume of 1.00 dm^3 to 30.00 dm^3 at 298 K.

14.30(a) The volume of a certain liquid varies with temperature as

$$V = V'\{0.75 + 3.9 \times 10^{-4}(T/\mathrm{K}) + 1.48 \times 10^{-6}(T/\mathrm{K})^2\}$$

where V' is its volume at 300 K. Calculate its expansion coefficient, α, at 320 K.

14.30(b) The volume of a certain liquid varies with temperature as

$$V = V'\{0.77 + 3.7 \times 10^{-4}(T/\mathrm{K}) + 1.52 \times 10^{-6}(T/\mathrm{K})^2\}$$

where V' is its volume at 298 K. Calculate its expansion coefficient, α, at 310 K.

14.31(a) The isothermal compressibility of water at 293 K is 4.96×10^{-5} atm^{-1}. Calculate the pressure that must be applied in order to increase its density by 0.10 per cent.

14.31(b) The isothermal compressibility of lead at 293 K is 2.21×10^{-6} atm^{-1}. Calculate the pressure that must be applied in order to increase its density by 0.10 per cent.

14.32(a) Given that $\mu = 0.25$ K atm^{-1} for nitrogen, calculate the value of its isothermal Joule–Thomson coefficient. Calculate the energy that must be supplied as heat to maintain constant temperature when 10.0 mol N_2 flows through a throttle in an isothermal Joule–Thomson experiment and the pressure drop is 85 atm.

14.32(b) Given that $\mu = 1.11$ K atm^{-1} for carbon dioxide, calculate the value of its isothermal Joule–Thomson coefficient. Calculate the energy that must be supplied as heat to maintain constant temperature when 10.0 mol CO_2 flows through a throttle in an isothermal Joule–Thomson experiment and the pressure drop is 75 atm.

Problems*

Numerical problems

14.1‡ In 2006, the Intergovernmental Panel on Climate Change (IPCC) considered a global average temperature rise of 1.0–3.5°C likely by the year 2100, with 2.0°C its best estimate. Predict the average rise in sea level due to thermal expansion of sea water based on temperature rises of 1.0°C, 2.0°C, and 3.5°C given that the volume of the Earth's oceans is 1.37×10^9 km^3 and their surface area is 361×10^6 km^2, and state the approximations that go into the estimates.

14.2 The following data show how the standard molar constant-pressure heat capacity of sulfur dioxide varies with temperature. By how much does the standard molar enthalpy of SO_2(g) increase when the temperature is raised from 298.15 K to 1500 K?

T/K	300	500	700	900	1100	1300	1500
$C^\ominus_{p,m}$/(J K^{-1} mol^{-1})	39.909	46.490	50.829	53.407	54.993	56.033	56.759

14.3 The following data show how the standard molar constant-pressure heat capacity of ammonia depends on the temperature. Use mathematical software to fit an expression of the form of eqn 14.34 to the data and determine the values of a, b, and c. Explore whether it would be better to express the data as $C_{p,m} = \alpha + \beta T + \gamma T^2$, and determine the values of these coefficients.

* Problems denoted with the symbol ‡ were supplied by Charles Trapp, Carmen Giunta, and Marshall Cady.

T/K	300	400	500	600	700	800	900	1000
$C^{\ominus}_{p,m}/(\mathrm{J\,K^{-1}\,mol^{-1}})$	35.678	38.674	41.994	45.229	48.269	51.112	53.769	56.244

14.4 The constant-volume heat capacity of a gas can be measured by observing the decrease in temperature when it expands adiabatically and reversibly. The value of $\gamma = C_p/C_V$ can be inferred if the decrease in pressure is also measured and the constant-pressure heat capacity deduced by combining the two values. A fluorocarbon gas was allowed to expand reversibly and adiabatically to twice its volume; as a result, the temperature fell from 298.15 K to 248.44 K and its pressure fell from 202.94 kPa to 81.840 kPa. Evaluate C_p.

14.5 The NO molecule has a doubly degenerate electronic ground state and a doubly degenerate excited state at 121.1 cm^{-1}. Calculate the electronic contribution to the molar heat capacity of the molecule at (a) 100 K, (b) 298 K, and (c) 600 K.

14.6 Explore whether a magnetic field can influence the heat capacity of a paramagnetic molecule by calculating the electronic contribution to the heat capacity of an NO_2 molecule in a magnetic field. Estimate the total constant-volume heat capacity by using equipartition, and calculate the percentage change in heat capacity brought about by a 10.0 T magnetic field at (a) 100 K, (b) 298 K.

14.7 The energy levels of a CH_3 group attached to a larger fragment are given by the expression for a particle on a ring, provided the group is rotating freely. What is the high-temperature contribution to the heat capacity of such a freely rotating group at 25°C? The moment of inertia of CH_3 about its threefold rotation axis (the axis that passes through the C atom and the centre of the equilateral triangle formed by the H atoms) is 5.341×10^{-47} kg m^2.

14.8 Calculate the temperature dependence of the heat capacity of p-H_2 (in which only rotational states with even values of J are populated) at low temperatures on the basis that its rotational levels $J = 0$ and $J = 2$ constitute a system that resembles a two-level system except for the degeneracy of the upper level. Use $\tilde{B} = 60.864$ cm^{-1} and sketch the heat capacity curve. The experimental heat capacity of p-H_2 does in fact show a peak at low temperatures.

14.9‡ In a spectroscopic study of buckminsterfullerene C_{60}, F. Negri *et al.* (*J. Phys. Chem.* **100**, 10849 (1996)) reviewed the wavenumbers of all the vibrational modes of the molecule. The wavenumber for the single A_u mode is 976 cm^{-1}; wavenumbers for the four threefold degenerate T_{1u} modes are 525, 578, 1180, and 1430 cm^{-1}; wavenumbers for the five threefold degenerate T_{2u} modes are 354, 715, 1037, 1190, and 1540 cm^{-1}; wavenumbers for the six fourfold degenerate G_u modes are 345, 757, 776, 963, 1315, and 1410 cm^{-1}; and wavenumbers for the seven fivefold degenerate H_u modes are 403, 525, 667, 738, 1215, 1342, and 1566 cm^{-1}. How many modes have a vibrational temperature θ_V below 1000 K? Estimate the molar constant-volume heat capacity of C_{60} at 1000 K, counting as active all modes with θ_V below this temperature.

14.10 A sample consisting of 2.0 mol CO_2 occupies a fixed volume of 15.0 dm^3 at 300 K. When it is supplied with 2.35 kJ of energy as heat its temperature increases to 341 K. Assume that CO_2 is described by the van der Waals equation of state, and calculate w, ΔU, and ΔH.

14.11 Calculate the work done during the isothermal reversible expansion of a van der Waals gas. Account physically for the way in which the coefficients a and b appear in the final expression. Plot on the same graph the indicator diagrams (graphs of pressure against volume) for the isothermal reversible expansion of (a) a perfect gas, (b) a van der Waals gas in which $a = 0$ and $b = 5.11 \times 10^{-2}$ dm^3 mol^{-1}, and (c) $a = 4.2$ dm^6 atm mol^{-2} and $b = 0$. The values selected exaggerate the imperfections but give rise to significant effects on the indicator diagrams. Take $V_i = 1.0$ dm^3, $n = 1.0$ mol, and $T = 298$ K.

14.12 A sample of the sugar D-ribose ($C_5H_{10}O_5$) of mass 0.727 g was placed in a calorimeter and then ignited in the presence of excess oxygen. The temperature rose by 0.910 K. In a separate experiment in the same calorimeter, the combustion of 0.825 g of benzoic acid, for which the internal energy of combustion is −3251 kJ mol^{-1}, gave a temperature rise of 1.940 K. Calculate the enthalpy of formation of D-ribose.

14.13 The standard enthalpy of formation of bis(benzene)chromium was measured in a calorimeter. It was found for the reaction $Cr(C_6H_6)_2(s) \rightarrow Cr(s) + 2\,C_6H_6(g)$ that $\Delta_r U^{\ominus}(583\text{ K}) = +8.0$ kJ mol^{-1}. Find the corresponding reaction enthalpy and estimate the standard enthalpy of formation of the compound at 583 K. The constant-pressure molar heat capacity of benzene is 136.1 J K^{-1} mol^{-1} in its liquid range and 81.67 J K^{-1} mol^{-1} as a gas.

14.14‡ From the enthalpy of combustion data in Table 14.5 for the alkanes methane through octane, test the extent to which the relation $\Delta_c H^{\ominus} = k\{(M/(\mathrm{g\,mol^{-1}})\}^n$ holds and find the numerical values for k and n. Predict $\Delta_c H^{\ominus}$ for decane and compare to the known value.

14.15 As described in Chapter 6, the thermochemical properties of hydrocarbons are commonly investigated by using molecular modelling methods. (a) Use software to predict $\Delta_c H^{\ominus}$ values for the alkanes methane through pentane. To calculate $\Delta_c H^{\ominus}$ values, estimate the standard enthalpy of formation of $C_nH_{2n+2}(g)$ by performing semiempirical calculations (for example, AM1 or PM3 methods) and use experimental standard enthalpy of formation values for $CO_2(g)$ and $H_2O(l)$. (b) Compare your estimated values with the experimental values of $\Delta_c H^{\ominus}$ (Table 14.5) and comment on the reliability of the molecular modelling method. (c) Test the extent to which the relation $\Delta_c H^{\ominus} = k\{(M/(\mathrm{g\,mol^{-1}})\}^n$ holds (Problem 14.14) and determine the numerical values of k and n.

14.16‡ Kolesov *et al.* reported the standard enthalpy of combustion and of formation of crystalline C_{60} based on calorimetric measurements (V.P. Kolesov *et al.*, *J. Chem. Thermodynamics* **28**, 1121 (1996)). In one of their runs, they found the standard specific internal energy of combustion to be −36.0334 kJ g^{-1} at 298.15 K. Compute $\Delta_c H^{\ominus}$ and $\Delta_f H^{\ominus}$ of C_{60}.

14.17‡ A thermodynamic study of $DyCl_3$ (E.H.P. Cordfunke *et al.*, *J. Chem. Thermodynamics* **28**, 1387 (1996)) determined its standard enthalpy of formation from the following information

(1) $DyCl_3(s) \rightarrow DyCl_3(\text{aq, in 4.0 M HCl})$ $\quad \Delta_r H^{\ominus} = -180.06$ kJ mol^{-1}

(2) $Dy(s) + 3\,HCl(\text{aq, 4.0 M}) \rightarrow$
$DyCl_3(\text{aq, in 4.0 M HCl(aq)}) + \frac{3}{2}H_2(g)$ $\quad \Delta_r H^{\ominus} = -699.43$ kJ mol^{-1}

(3) $\frac{1}{2}H_2(g) + \frac{1}{2}Cl_2(g) \rightarrow HCl(\text{aq, 4.0 M})$ $\quad \Delta_r H^{\ominus} = -158.31$ kJ mol^{-1}

Determine $\Delta_f H^{\ominus}(DyCl_3, s)$ from these data.

14.18‡ Silylene (SiH_2) is a key intermediate in the thermal decomposition of silicon hydrides such as silane (SiH_4) and disilane (Si_2H_6). H.K. Moffat *et al.* (*J. Phys. Chem.* **95**, 145 (1991)) report $\Delta_f H^{\ominus}(SiH_2) = +274$ kJ mol^{-1}. If $\Delta_f H^{\ominus}(SiH_4) = +34.3$ kJ mol^{-1} and $\Delta_f H^{\ominus}(Si_2H_6) = +80.3$ kJ mol^{-1}, compute the standard enthalpies of the following reactions:

(a) $SiH_4\,(g) \rightarrow SiH_2\,(g) + H_2\,(g)$

(b) $Si_2H_6\,(g) \rightarrow SiH_2\,(g) + SiH_4\,(g)$

14.19‡ Treat carbon monoxide as a perfect gas and apply equilibrium statistical thermodynamics to the study of its properties, as specified below, in the temperature range 100–1000 K at 1 bar. $\tilde{\nu} = 2169.8$ cm^{-1}, $\tilde{B} = 1.931$ cm^{-1}, and $hcD_0 = 11.09$ eV; neglect anharmonicity and centrifugal distortion. (a) Examine the probability distribution of molecules over available rotational and vibrational states. (b) Explore

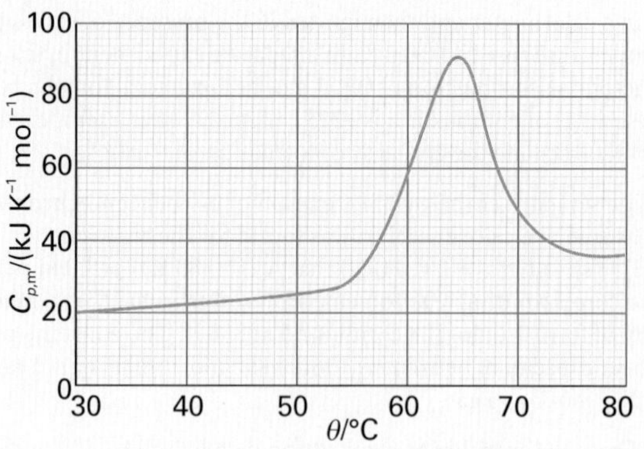

Fig. 14.37 Experimental DSC scan of hen white lysozyme.

numerically the differences, if any, between the rotational molecular partition function as calculated with the discrete energy distribution and that calculated with the classical, continuous energy distribution. (c) Calculate the individual contributions to $U_m(T) - U_m(100\text{ K})$ and $C_{V,m}(T)$ made by the translational, rotational, and vibrational degrees of freedom.

14.20 As remarked in Problem 14.3, it is sometimes appropriate to express the temperature dependence of the heat capacity by the empirical expression $C_{p,m} = \alpha + \beta T + \gamma T^2$. Use this expression to estimate the standard enthalpy of combustion of methane at 350 K. Use the following data:

	α/(J K^{-1} mol^{-1})	β/(mJ K^{-2} mol^{-1})	γ/(μJ K^{-3} mol^{-1})
CH_4(g)	14.16	75.5	−17.99
CO_2(g)	26.86	6.97	−0.82
O_2(g)	25.72	12.98	−3.862
H_2O(g)	30.36	9.61	1.184

14.21 Figure 14.37 shows the experimental DSC scan of hen white lysozyme (G. Privalov *et al.*, *Anal. Biochem.* **79**, 232 (1995)) converted to joules (from calories). Determine the enthalpy of unfolding of this protein by integration of the curve and the change in heat capacity accompanying the transition.

Theoretical problems

14.22 In the realm of nanotechnology, even translational quantization may have significant consequences. Suppose an electron is trapped in a tiny one-dimensional well, where only about 10 states are thermally accessible. Derive an expression for (a) the heat capacity, (b) the root-mean square spread in energies ($\Delta\varepsilon$), at such low temperatures (without making the 'continuum' approximation), and plot the heat capacity as a function of temperature. Can you identify a 'characteristic temperature' for the system?

14.23 Show that eqn. 14.26 can be converted into eqn 14.25.

14.24 The energies of the first six levels of a particle in a spherical cavity are specified in Problem 13.6. Suppose that these levels are the only ones that are thermally accessible, and derive an expression for (a) the heat capacity, (b) the root-mean square spread in energies ($\Delta\varepsilon$), and plot the former as a function of temperature.

14.25 Derive an expression for the rotational contribution to the heat capacity of a linear rotor without making the high-temperature approximation, and plot $C_{V,m}$ against T/θ_R, where the 'rotational temperature' is $\theta_R = hc\tilde{B}/k$. Ignore the role of nuclear statistics.

14.26 Are there thermal consequences of nuclear statistics that even the Victorians might have noticed? Explore the consequences, by direct summation of energy levels, of nuclear statistics for the molar heat capacities of *ortho*- and *para*-hydrogen (see Section 10.5).

14.27 In one of the earliest applications of quantum theory, Einstein sought to account for the decrease in heat capacity with decreasing temperature that had been observed. He supposed that each of the atoms in a monatomic solid could vibrate in three dimensions with a frequency ν. Deduce the Einstein formula and plot $C_{V,m}$ against T/θ_E, where the 'Einstein temperature' is $\theta_E = h\nu/k$.

14.28 Debye improved on Einstein's model by considering the *collective* modes of the atoms in the solid. Why would that lead to a higher heat capacity at all temperatures? He took the Einstein formula (Problem 14.27), multiplied it by a factor that represents the number of vibrational modes in the range ν to $\nu + d\nu$, and then integrated the resulting expression from $\nu = 0$ up to a maximum value ν_{max}. The result is

$$C_{V,m} = 9R\left(\frac{T}{\theta_D}\right)^3 \int_0^{\theta_D/T} \frac{x^4 e^x}{(e^x - 1)^2}dx$$

where the 'Debye temperature' is $\theta_D = h\nu_{max}/k$. Use mathematical software to plot $C_{V,m}$ against T/θ_D. Show that, when $T \ll \theta_D$, the heat capacity follows the 'Debye T^3 law' (see Self-test 14.6). You will need the standard integral

$$\int_0^{\infty} \frac{x^4 e^x}{(e^x - 1)^2}dx = \frac{\pi^4}{15}$$

14.29‡ For H_2 at very low temperatures, only translational motion contributes to the heat capacity. At temperatures above $\theta_R = hc\tilde{B}/k$, the rotational contribution to the heat capacity becomes significant. At still higher temperatures, above $\theta_V = h\nu/k$, the vibrations contribute. But at this latter temperature, dissociation of the molecule into the atoms must be considered. (a) Explain the origin of the expressions for θ_R and θ_V, and calculate their values for hydrogen. (b) Obtain an expression for the molar constant-pressure heat capacity of hydrogen at all temperatures taking into account the dissociation of hydrogen. (c) Make a plot of the molar constant-pressure heat capacity as a function of temperature in the high temperature region where dissociation of the molecule is significant.

14.30 Although expressions like $\varepsilon = -d\ln q/d\beta$ are useful for formal manipulations in statistical thermodynamics, and for expressing thermodynamic functions in neat formulas, they are sometimes more trouble than they are worth in practical applications. When presented with a table of energy levels, it is often much more convenient to evaluate the following sums directly:

$$q = \sum_j e^{-\beta\varepsilon_j} \qquad \dot{q} = \sum_j \beta\varepsilon_j e^{-\beta\varepsilon_j} \qquad \ddot{q} = \sum_j (\beta\varepsilon_j)^2 e^{-\beta\varepsilon_j}$$

where $\dot{q}$ and $\ddot{q}$ represent the first and second derivatives of q with respect to β. (a) Derive expressions for the internal energy and heat capacity in terms of these three functions. (b) Apply the technique to the calculation of the electronic contribution to the constant-volume molar heat capacity of magnesium vapour at 5000 K using the following data:

Term	^1S	$^3P_0$	$^3P_1$	$^3P_2$	$^1P_1$	^3S
Degeneracy	1	1	3	5	3	3
$\tilde{\nu}$/cm^{-1}	0	21850	21870	21911	35051	41197

14.31 Calculate the values of q, $\dot{q}$, and $\ddot{q}$ (Problem 14.30) for the rotational states of (a) HCl, $\tilde{B} = 10.593\ \text{cm}^{-1}$ and (b) CCl_4, $\tilde{B} = 5.797\ \text{m}^{-1}$ (be alert to the units!).

14.32 Show how the heat capacity of a linear rotor is related to the following sum:

$$\zeta(\beta) = \frac{1}{q^2}\sum_{J,J'}[\{\varepsilon(J) - \varepsilon(J')\}^2 g(J)g(J')e^{-\beta\{\varepsilon(J)+\varepsilon(J')\}}$$

by

$$C = \tfrac{1}{2}k\beta^2\zeta(\beta)$$

where the $\varepsilon(J)$ are the rotational energy levels and $g(J)$ their degeneracies. Then go on to show graphically that the total contribution to the heat capacity of a linear rotor can be regarded as a sum of contributions due to transitions $0\rightarrow1$, $0\rightarrow2$, $1\rightarrow2$, $1\rightarrow3$, etc. In this way, construct Fig. 14.15 for the rotational heat capacities of a linear molecule.

14.33 The 'bump' in the low temperature variation of the heat capacity with temperature seen in Fig. 14.15 is more pronounced in the case of *para*-hydrogen, where only even values of J are allowed. Adapt the expression derived in Problem 14.32 to the cases of *ortho*- and *para*-hydrogen, and construct graphs of their heat capacities.

14.34 Set up a calculation like that in Problem 14.32 to analyse the vibrational contribution to the heat capacity in terms of excitations between levels and illustrate your results graphically in terms of a diagram like that in Fig. 14.15.

14.35 Equation 14.38 in the form

$$\frac{d}{dL}\frac{e^{-\beta\varepsilon_i}}{q(\beta)} = 0$$

is a differential equation for β (and therefore T) as a function of L, but we solved it there by inspection. Solve the equation formally and confirm that eqn 14.39 is a solution.

14.36 The heat capacity ratio of a gas determines the speed of sound in it through the formula $c_s = (\gamma RT/M)^{1/2}$, where $\gamma = C_p/C_V$ and M is the molar mass of the gas. Deduce an expression for the speed of sound in a perfect gas of (a) diatomic, (b) linear triatomic, (c) non-linear triatomic molecules at high temperatures (with translation and rotation active). Estimate the speed of sound in air at 25°C.

14.37 (a) Express $(\partial C_V/\partial V)_T$ as a second derivative of U and find its relation to $(\partial U/\partial V)_T$ and $(\partial C_p/\partial p)_T$ as a second derivative of H and find its relation to $(\partial H/\partial p)_T$. (b) From these relations show that $(\partial C_V/\partial V)_T = 0$ and $(\partial C_p/\partial p)_T = 0$ for a perfect gas.

14.38 (a) Derive the relation $C_V = -(\partial U/\partial V)_T(\partial V/\partial T)_U$ from the expression for the total differential of $U(T,V)$ and (b) starting from the expression for the total differential of $H(T,p)$, express $(\partial H/\partial p)_T$ in terms of C_p and the Joule–Thomson coefficient, μ.

14.39 Starting from the expression $C_p - C_V = T(\partial p/\partial T)_V(\partial V/\partial T)_p$, use the appropriate relations between partial derivatives to show that

$$C_p - C_V = \frac{T(\partial V/\partial T)_p^2}{(\partial V/\partial p)_T}$$

Evaluate $C_p - C_V$ for a perfect gas.

14.40 (a) By direct differentiation of $H = U + pV$, obtain a relation between $(\partial H/\partial U)_p$ and $(\partial U/\partial V)_p$. (b) Confirm that $(\partial H/\partial U)_p = 1 + p(\partial V/\partial U)_p$ by expressing $(\partial H/\partial U)_p$ as the ratio of two derivatives with respect to volume and then using the definition of enthalpy.

14.41 (a) Write expressions for dV and dp given that V is a function of p and T and p is a function of V and T. (b) Deduce expressions for $d\ln V$ and $d\ln p$ in terms of the expansion coefficient and the isothermal compressibility.

14.42‡ A gas obeying the equation of state $p(V - nb) = nRT$ is subjected to a Joule–Thomson expansion. Will the temperature increase, decrease, or remain the same?

14.43 Use the fact that $(\partial U/\partial V)_T = a/V_m^2$ for a van der Waals gas to show that $\mu C_{p,m} \approx (2a/RT) - b$ by using the definition of μ and appropriate relations between partial derivatives. *Hint.* Use the approximation $pV_m \approx RT$ when it is justifiable to do so.

14.44 Rearrange the van der Waals equation of state to give an expression for T as a function of p and V (with n constant). Calculate $(\partial T/\partial p)_V$ and confirm that $(\partial T/\partial p)_V = 1/(\partial p/\partial T)_V$. Go on to confirm Euler's chain relation (*Mathematical background 8*).

14.45 Calculate the isothermal compressibility and the expansion coefficient of a van der Waals gas. Show, using Euler's chain relation (*Mathematical background 8*), that $\kappa_T R = \alpha(V_m - b)$.

14.46 The speed of sound, c_s, in a gas of molar mass M is related to the ratio of heat capacities γ by $c_s = (\gamma RT/M)^{1/2}$. Show that $c_s = (\gamma p/\rho)^{1/2}$, where ρ is the mass density of the gas. Calculate the speed of sound in argon at 25°C.

14.47‡ A gas obeys the equation of state $V_m = RT/p + aT^2$ and its constant-pressure heat capacity is given by $C_{p,m} = A + BT + Cp$, where a, A, B, and C are constants independent of T and p. Obtain expressions for (a) the Joule–Thomson coefficient and (b) its constant-volume heat capacity.

14.48 The statistical properties of a two-level system enable us to give formal significance to negative thermodynamic temperatures. From the Boltzmann distribution for such a system, show that the temperature may be defined as

$$T = \frac{\varepsilon/k}{\ln(N_-/N_+)}$$

where ε is the energy separation and N_+ and N_- are the populations of the upper and lower states, respectively. Find the corresponding expression for $\beta = 1/kT$. It follows that, if the system can be contrived to have $N_- < N_+$, then $T < 0$. Go on to plot graphs of the partition function, internal energy, and heat capacity of the system in the range $-\infty < kT/\varepsilon < \infty$. Observe that there are discontinuities in the graphs. These discontinuities are eliminated by plotting the properties against β in the range $-\infty < \varepsilon\beta < \infty$: do so. For further investigations of negative temperatures, see Problem 15.27.

Applications: to biology and environmental science

14.49 It is possible to see with the aid of a powerful microscope that a long piece of double-stranded DNA is flexible, with the distance between the ends of the chain adopting a wide range of values. This flexibility is important because it allows DNA to adopt very compact conformations as it is packaged in a chromosome. It is convenient to visualize a long piece of DNA as a *freely jointed chain*, a chain of N small, rigid units of length l that are free to make any angle with respect to each other. The length l, the *persistence length*, is approximately 45 nm, corresponding to approximately 130 base pairs. You will now explore the work associated with extending a DNA molecule. (a) Suppose that a DNA molecule resists being extended from an equilibrium, more compact conformation with a restoring force $F = -k_F x$, where x is the difference in the end-to-end distance of the chain from an equilibrium value and k_F is the force constant. Use this model to write an expression for the work that must be done to extend a DNA molecule by a distance x. Draw a graph of your conclusion. (b) A better model of a DNA molecule is the *one-dimensional freely jointed chain*, in which a rigid unit of length l can only make an angle of 0° or 180° with an adjacent unit. In this case, the restoring force of a chain extended by $x = nl$ is given by

$$F = \frac{kT}{2l}\ln\left(\frac{1+\nu}{1-\nu}\right) \qquad \nu = n/N$$

where k is Boltzmann's constant. (i) What is the magnitude of the force that must be applied to extend a DNA molecule with $N = 200$ by 90 nm? (ii) Plot the restoring force against ν, noting that ν can be either positive or negative. How is the variation of the restoring force with end-to-end distance different from that predicted by Hooke's law? (iii) Keep in mind that the difference in end-to-end distance from an equilibrium value is $x = nl$ and, consequently, $dx = l dn = Nl d\nu$, and write an expression for the work of extending a DNA molecule. (iv) Calculate the work of extending a DNA molecule from $\nu = 0$ to $\nu = 1.0$. *Hint.* You must integrate the expression for w. The task can be accomplished easily with mathematical software. (c) Show that for small extensions of the chain, when $\nu \ll 1$, the restoring force is given by

$$F \approx \frac{\nu kT}{l} = \frac{nkT}{Nl}$$

(d) Is the variation of the restoring force with extension of the chain given in part (c) different from that predicted by Hooke's law? Explain your answer.

14.50 An average human produces about 10 MJ of heat each day through metabolic activity. If a human body were an isolated system of mass 65 kg with the heat capacity of water, what temperature rise would the body experience? Human bodies are actually open systems, and the main mechanism of heat loss is through the evaporation of water. What mass of water should be evaporated each day to maintain constant temperature?

14.51 Glucose and fructose are simple sugars with the molecular formula $C_6H_{12}O_6$. Sucrose, or table sugar, is a complex sugar with molecular formula $C_{12}H_{22}O_{11}$ that consists of a glucose unit covalently bound to a fructose unit (a water molecule is eliminated as a result of the reaction between glucose and fructose to form sucrose). (a) Calculate the energy released as heat when a typical table sugar cube of mass 1.5 g is burned in air. (b) To what height could you climb on the energy a table sugar cube provides assuming 25 per cent of the energy is available for work? (c) The mass of a typical glucose tablet is 2.5 g. Calculate the energy released as heat when a glucose tablet is burned in air. (d) To what height could you climb on the energy a cube provides assuming 25 per cent of the energy is available for work?

14.52 In biological cells that have a plentiful supply of oxygen, glucose is oxidized completely to CO_2 and H_2O by a process called *aerobic oxidation*. Muscle cells may be deprived of O_2 during vigorous exercise and, in that case, one molecule of glucose is converted to two molecules of lactic acid ($CH_3CH(OH)COOH$) by a process called *anaerobic glycolysis*. (a) When 0.3212 g of glucose was burned in a bomb calorimeter of calorimeter constant 641 J K^{-1}, the temperature rose by 7.793 K. Calculate (i) the standard molar enthalpy of combustion, (ii) the standard internal energy of combustion, and (iii) the standard enthalpy of formation of glucose. (b) What is the biological advantage (in kilojoules per mole of energy released as heat) of complete aerobic oxidation compared with anaerobic glycolysis to lactic acid?

14.53‡ Concerns over the harmful effects of chlorofluorocarbons on stratospheric ozone have motivated a search for new refrigerants. One such alternative is 2,2-dichloro-1,1,1-trifluoroethane (refrigerant 123). Younglove and McLinden published a compendium of thermophysical properties of this substance (B.A. Younglove and M. McLinden, *J. Phys. Chem. Ref. Data* **23**, 7 (1994)), from which properties such as the Joule–Thomson coefficient μ can be computed. (a) Compute μ at 1.00 bar and 50°C given that $(\partial H/\partial p)_T = -3.29 \times 10^3$ J MPa^{-1} mol^{-1} and $C_{p,m} = 110.0$ J K^{-1} mol^{-1}. (b) Compute the temperature change that would accompany adiabatic expansion of 2.0 mol of this refrigerant from 1.5 bar to 0.5 bar at 50°C.

14.54‡ Another alternative refrigerant (see preceding problem) is 1,1,1,2-tetrafluoroethane (refrigerant HFC-134a). A compendium of thermophysical properties of this substance has been published (R. Tillner-Roth and H.D. Baehr, *J. Phys. Chem. Ref. Data* **23**, 657 (1994)) from which properties such as the Joule–Thomson coefficient μ can be computed. (a) Compute μ at 0.100 MPa and 300 K from the following data (all referring to 300 K):

p/MPa	0.080	0.100	0.12
Specific enthalpy/(kJ kg^{-1})	426.48	426.12	425.76

(The specific constant-pressure heat capacity is 0.7649 kJ K^{-1} kg^{-1}.) (b) Compute μ at 1.00 MPa and 350 K from the following data (all referring to 350 K):

p/MPa	0.80	1.00	1.2
Specific enthalpy/(kJ kg^{-1})	461.93	459.12	456.15

(The specific constant-pressure heat capacity is 1.0392 kJ K^{-1} kg^{-1}.)

MATHEMATICAL BACKGROUND 8

Multivariate calculus

A property of a system typically depends on a number of variables, such as the pressure depending on the amount, volume, and temperature according to an equation of state, $p = f(n,T,V)$. To understand how these properties vary with the conditions we need to understand how to manipulate their derivatives. This is the field of **multivariate calculus**, the calculus of several variables.

MB8.1 Partial derivatives

A **partial derivative** of a function of more than one variable, such as $f(x,y)$, is the slope of the function with respect to one of the variables, all the other variables being held constant (Fig. MB8.1). Although a partial derivative shows how a function changes when one variable changes, it may be used to determine how the function changes when more than one variable changes by an infinitesimal amount. Thus, if f is a function of x and y then, when x and y change by dx and dy, respectively, f changes by

$$df = \left(\frac{\partial f}{\partial x}\right)_y dx + \left(\frac{\partial f}{\partial y}\right)_x dy \qquad \text{(MB8.1)}$$

where the symbol ∂ is used (instead of d) to denote a partial derivative and the subscript on the parentheses indicates which variable is being held constant. The quantity df is also called the **differential** of f. Successive partial derivatives may be taken in any order:

$$\left(\frac{\partial}{\partial y}\left(\frac{\partial f}{\partial x}\right)_y\right)_x = \left(\frac{\partial}{\partial x}\left(\frac{\partial f}{\partial y}\right)_x\right)_y \qquad \text{(MB8.2)}$$

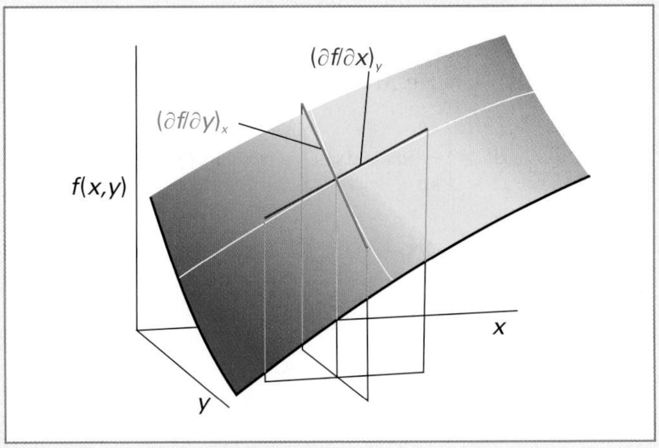

Fig. MB8.1 A function of two variables, $f(x,y)$, as depicted by the coloured surface and the two partial derivatives, $(\partial f/\partial x)_y$ and $(\partial f/\partial y)_x$, the slope of the function parallel to the x- and y-axes, respectively. The function plotted here is $f(x,y) = ax^3y + by^2$ with $a = 1$ and $b = -2$.

● A BRIEF ILLUSTRATION

Suppose that $f(x,y) = ax^3y + by^2$ (the function plotted in Fig. MB8.1), then

$$\left(\frac{\partial f}{\partial x}\right)_y = 3ax^2y \qquad \left(\frac{\partial f}{\partial y}\right)_x = ax^3 + 2by$$

Then, when x and y undergo infinitesimal changes, f changes by

$$df = 3ax^2y\,dx + (ax^3 + 2by)dy$$

To verify that the order of taking the second partial derivative is irrelevant, we form

$$\left(\frac{\partial}{\partial y}\left(\frac{\partial f}{\partial x}\right)_y\right)_x = \left(\frac{\partial(3ax^2y)}{\partial y}\right)_x = 3ax^2$$

$$\left(\frac{\partial}{\partial x}\left(\frac{\partial f}{\partial y}\right)_x\right)_y = \left(\frac{\partial(ax^3 + 2by)}{\partial x}\right)_y = 3ax^2 \; ●$$

Self test MB8.1 Evaluate df for $f(x,y) = 2x^2 \sin 3y$ and verify that the order of taking the second partial derivative is irrelevant. [$df = 4x \sin 3y\,dx + 6x^2 \cos 3y\,dy$]

In the following, z is a variable on which x and y depend (for example, x, y, and z might correspond to p, V, and T).

Relation 1. When x is changed at constant z:

$$\left(\frac{\partial f}{\partial x}\right)_z = \left(\frac{\partial f}{\partial x}\right)_y + \left(\frac{\partial f}{\partial y}\right)_x\left(\frac{\partial y}{\partial x}\right)_z \qquad \text{(MB8.3a)}$$

Relation 2

$$\left(\frac{\partial y}{\partial x}\right)_z = \frac{1}{(\partial x/\partial y)_z} \qquad \text{(MB8.3b)}$$

Relation 3

$$\left(\frac{\partial x}{\partial y}\right)_z = -\left(\frac{\partial x}{\partial z}\right)_y\left(\frac{\partial z}{\partial y}\right)_x \qquad \text{(MB8.3c)}$$

By combining this relation and Relation 2 we obtain the **Euler chain relation:**

$$\left(\frac{\partial y}{\partial x}\right)_z\left(\frac{\partial x}{\partial z}\right)_y\left(\frac{\partial z}{\partial y}\right)_x = -1 \qquad \text{(MB8.4)}$$

MB8.2 Exact differentials

The relation in eqn MB8.2 is the basis of a test for an **exact differential**, that is, the test of whether

$$df = g(x,y)dx + h(x,y)dy \qquad \text{(MB8.5)}$$

has the form in eqn MB8.1. If it has that form, then g can be identified with $(\partial f/\partial x)_y$ and h can be identified with $(\partial f/\partial y)_x$, and therefore eqn MB8.2 becomes

$$\left(\frac{\partial g}{\partial y}\right)_x = \left(\frac{\partial h}{\partial x}\right)_y \qquad \text{(MB8.6)}$$

● A BRIEF ILLUSTRATION

Suppose, instead of the form $df = 3ax^2y dx + (ax^3 + 2by)dy$ in the previous *brief illustration* we were presented with the expression

$$df = \overbrace{3ax^2y}^{g(x,y)}dx + \overbrace{(ax^2 + 2by)}^{h(x,y)}dy$$

with ax^2 in place of ax^3. To test whether this is an exact differential, we form

$$\left(\frac{\partial g}{\partial y}\right)_x = \left(\frac{\partial(3ax^2y)}{\partial y}\right)_x = 3ax^2$$

$$\left(\frac{\partial h}{\partial x}\right)_y = \left(\frac{\partial(ax^2 + 2by)}{\partial x}\right)_y = 2ax$$

These two expressions are not equal, so this form of df is not an exact differential and there is not a corresponding integrated function of the form $f(x,y)$. ●

Self-test MB8.2 Determine whether the expression $df = (2y - x^3)dx + xdy$ is an exact differential. [No]

If df is exact, then we can do two things: (1) from a knowledge of the functions g and h we can reconstruct the function f; (2) be confident that the integral of df between specified limits is independent of the path between those limits. The first conclusion is best demonstrated with a specific example.

● A BRIEF ILLUSTRATION

We consider the differential $df = 3ax^2ydx + (ax^3 + 2by)dy$, which we know to be exact. Because $(\partial f/\partial x)_y = 3ax^2y$, we can integrate with respect to x with y held constant, to obtain

$$f = \int df = \int 3ax^2\, y dx = 3ay \int x^2\, dx = ax^3y + k$$

where the 'constant' of integration k may depend on y (which has been treated as a constant in the integration), but not on x. To find $k(y)$, we note that $(\partial f/\partial y)_x = ax^3 + 2by$, and therefore

$$\left(\frac{\partial f}{\partial y}\right)_x = \left(\frac{\partial(ax^3y + k)}{\partial y}\right)_x = ax^3 + \frac{dk}{dy} = ax^3 + 2by$$

Therefore

$$\frac{dk}{dy} = 2by$$

from which it follows that $k = by^2 + \text{constant}$. We have found, therefore, that

$$f(x,y) = ax^3y + by^2 + \text{constant}$$

which, apart from the constant, is the original function in the first *brief illustration*. The value of the constant is pinned down by stating the boundary conditions; thus, if it is known that $f(0,0) = 0$, then the constant is zero. ●

Self-test MB8.3 Confirm that $df = 3x^2 \cos y\, dx - x^3 \sin y\, dy$ is exact and find the function $f(x,y)$. [$f = x^3 \cos y$]

To demonstrate that the integral of df is independent of the path is now straightforward. Because df is a differential, its integral between the limits a and b is

$$\int_a^b df = f(b) - f(a)$$

The value of the integral depends only on the values at the end-points and is independent of the path between them. If df is not an exact differential, the function f does not exist, and this argument no longer holds. In such cases, the integral of df does depend on the path.

● A BRIEF ILLUSTRATION

Consider the inexact differential (the expression with ax^2 in place of ax^3):

$$df = 3ax^2ydx + (ax^2 + 2by)dy$$

Suppose we integrate df from (0,0) to (2,2) along the two paths shown in Fig. MB8.2. Along Path 1,

$$\int_{\text{Path 1}} df = \int_{0,0}^{2,0} 3ax^2ydx + \int_{2,0}^{2,2} (ax^2 + 2by)dy$$

$$= 0 + 4a\int_0^2 dy + 2b\int_0^2 ydy = 8a + 4b$$

whereas along Path 2,

$$\int_{\text{Path 2}} df = \int_{0,2}^{2,2} 3ax^2ydx + \int_{0,0}^{0,2} (ax^2 + 2by)dy$$

$$= 6a\int_0^2 x^2\, dx + 0 + 2b\int_0^2 ydy = 16a + 4b$$

The two integrals are not the same. ●

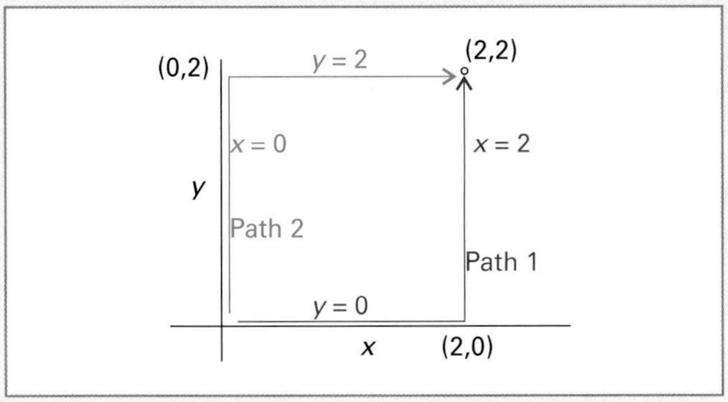

Fig. MB8.2 The two integration paths referred to in the *brief illustration*.

Self-test MB8.4 Confirm that the two paths do give the same value for the exact differential in the first *brief illustration*.

[Both paths: $16a+4b$]

An inexact differential may sometimes be converted into an exact differential by multiplication by a factor known as an *integrating factor*. A physical example is the integrating factor $1/T$ that converts the inexact differential dq_{rev} into the exact differential dS in thermodynamics (Chapter 15).

● A BRIEF ILLUSTRATION

We have seen that the differential $df=3ax^2ydx+ax^2dy$ is inexact. Suppose we multiply df by x^my^n and write $x^my^ndf=df'$; then we obtain

$$df' = \overbrace{3ax^{m+2}y^{n+1}}^{g(x,y)}dx + \overbrace{ax^{m+2}y^n}^{h(x,y)}dy$$

We evaluate the following two partial derivatives:

$$\left(\frac{\partial g}{\partial y}\right)_x = \left(\frac{\partial(3ax^{m+2}y^{n+1})}{\partial y}\right)_x = 3a(n+1)x^{m+2}y^n$$

$$\left(\frac{\partial h}{\partial x}\right)_y = \left(\frac{\partial(ax^{m+2}y^n)}{\partial x}\right)_y = a(m+2)x^{m+1}y^n$$

For the new differential to be exact, these two partial derivatives must be equal, so we write

$$3a(n+1)x^{m+2}y^n = a(m+2)x^{m+1}y^n$$

which simplifies to

$$3(n+1)x = m+2$$

The only solution that is independent of x is $n=-1$ and $m=-2$. It follows that

$$df' = 3adx + (a/y)dy$$

is an exact differential. By the procedure already illustrated, its integrated form is $f'(x,y)=3ax+a\ln y+\text{constant}$. ●

Self-test MB8.5 Find an integrating factor of the form x^my^n for the inexact differential $df=(2y-x^3)dx+xdy$ and the integrated form of f'. [$df'=xdf$, $f'=yx^2-\frac{1}{5}x^5+\text{constant}$]

15 The Second Law of thermodynamics

The purpose of this chapter is to explain the origin of the spontaneity of physical and chemical change. To do so, we introduce the property known as the entropy and show how it is related to the distribution of molecules over the available states and can be calculated from the partition function. Then we show that the entropy may be defined in terms of the heat transferred to a system, and hence may be calculated from calorimetric information. These two approaches to the definition and calculation of the entropy are brought together by the Third Law of thermodynamics, which is also introduced in this chapter. The chapter also introduces and shows how to calculate from thermodynamic and spectroscopic data a major subsidiary thermodynamic property, the Gibbs energy. This property lets us express the spontaneity of a process in terms of the properties of the system alone. The Gibbs energy also enables us to predict the maximum non-expansion work that a process can do. The chapter concludes with an analysis of how the Gibbs energy changes with temperature and pressure, two dependencies that play a central role in the following chapters.

Some things happen naturally; some things don't. A gas expands to fill the available volume, a hot body cools to the temperature of its surroundings, and a chemical reaction runs in one direction rather than another. Some aspect of the world determines the **spontaneous** direction of change, the direction of change that does not require work to be done to bring it about. A gas can be confined to a smaller volume, an object can be cooled by using a refrigerator, and some reactions can be driven in reverse (as in the electrolysis of water). However, none of these processes is spontaneous; each one must be brought about by doing work. An important point, though, is that throughout this text 'spontaneous' must be interpreted as a natural *tendency* that may or may not be realized in practice. Thermodynamics is silent on the rate at which a spontaneous change in fact occurs, and some spontaneous processes (such as the conversion of diamond to graphite) may be so slow that the tendency is never realized in practice, whereas others (such as the expansion of a gas into a vacuum) are almost instantaneous.

The recognition of two classes of process, spontaneous and nonspontaneous, is summarized by the **Second Law of thermodynamics.** This law may be expressed in a variety of equivalent ways. One statement was formulated by Kelvin:

> No process is possible in which the *sole* result is the absorption of heat from a reservoir and its complete conversion into work.

For example, it has proved impossible to construct an engine like that shown in Fig. 15.1, in which heat is drawn from a hot reservoir and completely converted into work. All real heat engines have both a hot source and a cold sink; some energy is always discarded into the cold sink as heat and not converted into work. The Kelvin

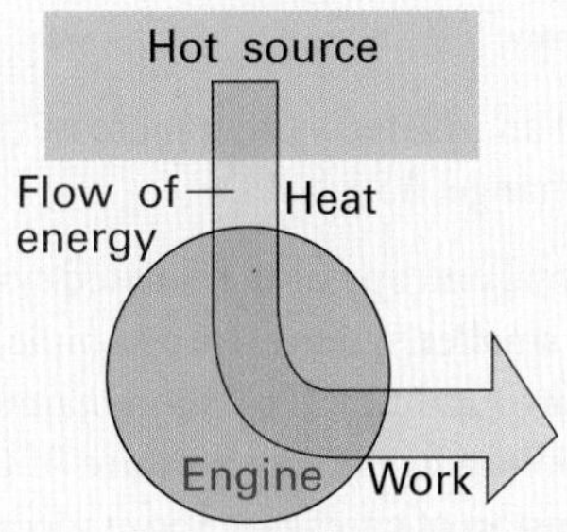

Fig. 15.1 The Kelvin statement of the Second Law denies the possibility of the process illustrated here, in which heat is changed completely into work, there being no other change. The process is not in conflict with the First Law because energy is conserved.

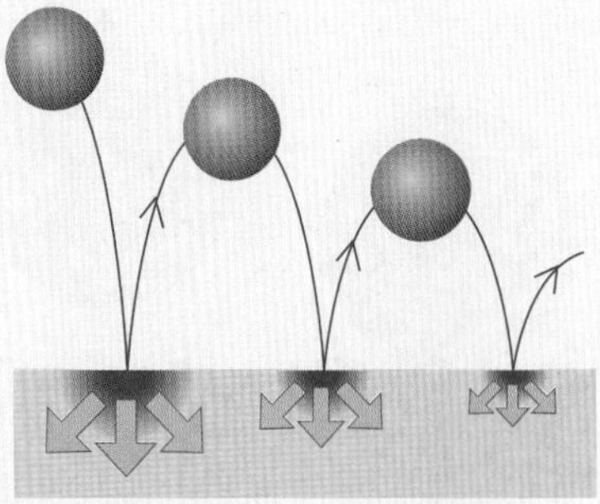

Fig. 15.2 The direction of spontaneous change for a ball bouncing on a floor. On each bounce some of its energy is degraded into the thermal motion of the atoms of the floor, and that energy disperses. The reverse has never been observed to take place on a macroscopic scale.

statement is a generalization of another everyday observation, that a ball at rest on a surface has never been observed to leap spontaneously upwards. An upward leap of the ball would be equivalent to the conversion of heat from the surface into work.

The direction of spontaneous change

What determines the direction of spontaneous change? It is not the total energy of the isolated system. The First Law of thermodynamics states that energy is conserved in any process, and we cannot disregard that law now and say that everything tends towards a state of lower energy: the total energy of an isolated system is constant.

Is it perhaps the energy of the *system* that tends towards a minimum? Two arguments show that this cannot be so. First, a perfect gas expands spontaneously into a vacuum, yet its internal energy remains constant as it does so. Secondly, if the energy of a system does happen to decrease during a spontaneous change, the energy of its surroundings must increase by the same amount (by the First Law). The increase in energy of the surroundings is just as spontaneous a process as the decrease in energy of the system.

When a change occurs, the total energy of an isolated system remains constant but it is parcelled out in different ways. Can it be, therefore, that the direction of change is related to the *distribution* of energy? We shall see that this idea is the key, and that spontaneous changes are always accompanied by the random dispersal of energy.

15.1 The dispersal of energy

We can begin to understand the role of the distribution of energy by thinking about a ball (the system) bouncing on a floor (the surroundings). The ball does not rise as high after each bounce because there are inelastic losses in the materials of the ball and floor. The kinetic energy of the ball's overall motion is spread out into the energy of thermal motion of its particles and those of the floor that it hits. The direction of spontaneous change is towards a state in which the ball is at rest with all its energy dispersed as the disorderly thermal motion of molecules in the air and spread over the atoms of the virtually infinite floor (Fig. 15.2).

A ball resting on a warm floor has never been observed to start bouncing. For bouncing to begin, something rather special would need to happen. In the first place, some of the thermal motion of the atoms in the floor would have to accumulate in a single small object, the ball. This accumulation requires a spontaneous localization of energy from the myriad vibrations of the atoms of the floor into the much smaller number of atoms that constitute the ball (Fig. 15.3). Furthermore, whereas the thermal motion is random, for the ball to move upwards its atoms must all move in the same direction. The localization of random, disorderly motion as concerted, ordered motion is so unlikely that we can dismiss it as virtually impossible except on the very small scale characteristic of 'Brownian motion', the jittering motion of small particles suspended in water.

We appear to have found the signpost of spontaneous change: *we look for the direction of change that leads to the random dispersal of the total energy of the isolated system.* This principle accounts for the direction of change of the bouncing ball, because its energy is spread out as thermal motion of the atoms of the floor. The reverse process is not spontaneous because it is highly improbable that energy will become not only localized but also localized as uniform motion of the ball's atoms. A gas does not contract spontaneously because to do so the random motion of its molecules, which distributes their kinetic energy throughout the container, would have to take them all into the same region of the container, thereby localizing the energy. The opposite

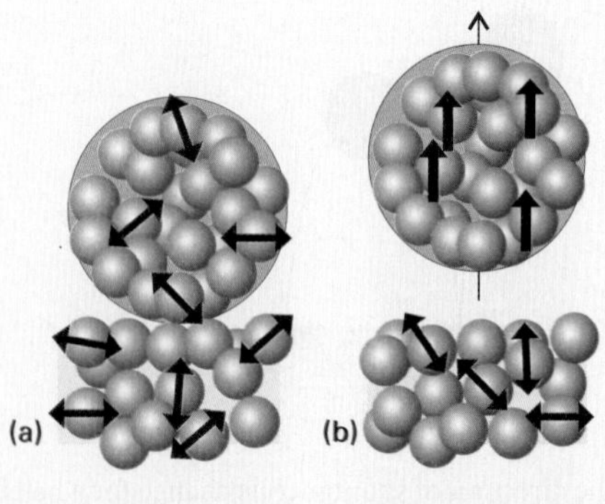

Fig. 15.3 The molecular interpretation of the irreversibility expressed by the Second Law. (a) A ball resting on a warm surface; the atoms are undergoing thermal motion (vibration, in this instance), as indicated by the arrows. (b) For the ball to fly upwards, some of the random vibrational motion would have to change into coordinated, directed motion. Such a conversion is highly improbable.

change, spontaneous expansion, is a natural consequence of energy becoming more widely dispersed as the gas molecules occupy a larger volume. An object does not spontaneously become warmer than its surroundings because it is highly improbable that the jostling of randomly vibrating atoms in the surroundings will lead to the localization of thermal motion in the object. The opposite change, the spreading of the object's energy into the surroundings as thermal motion, is natural.

It may seem very puzzling that the spreading out of energy and matter, the collapse into disorder, can lead to the formation of such ordered structures as crystals or proteins. Nevertheless, in due course, we shall see that the tendency of energy and matter to disperse in disorder accounts for change in all its forms.

15.2 Entropy

The First Law of thermodynamics led to the introduction of the internal energy, U. The internal energy is a state function that lets us assess whether a change is permissible: only those changes may occur for which the internal energy of an isolated system remains constant. The law that is used to identify the signpost of spontaneous change, the Second Law of thermodynamics, may also be expressed in terms of another state function, the **entropy**, S. We shall see that the entropy (which we shall define shortly, but is a measure of the extent to which energy is distributed in a disorderly manner) lets us assess whether one state is accessible from another by a spontaneous change. The First Law uses the internal energy to identify *permissible changes*; the Second Law uses the entropy to identify the *spontaneous changes* among those permissible changes.

The Second Law of thermodynamics can be expressed in terms of the entropy:

> The entropy of an isolated system increases in the course of a spontaneous change: $\Delta S_{tot} > 0$

where S_{tot} is the total entropy of the isolated system, which itself may consist of a smaller system (for example, a beaker of hot water) and its surroundings. Thermodynamically irreversible processes (like cooling to the temperature of the surroundings and the free expansion of gases) are spontaneous processes, and hence must be accompanied by an increase in total entropy.

(a) The statistical definition of entropy

If it is true, as we claimed in Chapter 13, that a partition function contains all the thermodynamic information about a system, then it must be possible to use it to calculate the entropy as well as the internal energy. Because entropy is related to the dispersal of energy and the partition function is a measure of the number of thermally accessible states, we can be confident that the two are indeed related.

Ludwig Boltzmann looked for a definition of entropy that was a measure of the dispersal of energy and was a state function, extensive, and increased in the course of a spontaneous change. He suggested that an appropriate definition is what we now call the **Boltzmann formula** for the entropy:

$$S = k \ln \mathcal{W} \qquad [15.1]$$

where k is Boltzmann's constant and $\mathcal{W}$ is the weight of the most probable configuration of the system. The quantity S as defined in eqn 15.1 is clearly a state function, as the weight of the most probable configuration is independent of how the system was prepared. That S is extensive can be seen by considering a system as being composed of two parts with entropies $S_1 = k \ln \mathcal{W}_1$ and $S_2 = k \ln \mathcal{W}_2$, respectively. The total weight of the entire system (the total number of ways of achieving a configuration) is the product of the weights of the two component parts, $\mathcal{W} = \mathcal{W}_1\mathcal{W}_2$, and so the total entropy of the system is

$$S = k \ln \mathcal{W}_1\mathcal{W}_2 = k \ln \mathcal{W}_1 + k \ln \mathcal{W}_2 = S_1 + S_2$$

Because the total entropy is the sum of the entropies of its component parts, the entropy is extensive.

That S is the signpost of spontaneous change is plausible because an isolated system in an arbitrary initial configuration tends to collapse into the distribution with the greatest weight, so the entropy as defined in eqn 15.1 increases in a spontaneous change. Informally, we can imagine the system as exploring all the distributions available to it, with certain distributions achieved far more often than others. To an external observer, the system migrates into a configuration corresponding to the overwhelmingly dominant distribution.

To express this conclusion more formally, we first suppose that the two parts of an entire system are not in equilibrium with each other. Each part of the system is in internal equilibrium, and so $\mathcal{W}_1$ and $\mathcal{W}_2$ have their respective maximum values, the combined weight is $\mathcal{W}_i = \mathcal{W}_1\mathcal{W}_2$, and the initial entropy is

$$S_i = k \ln \mathcal{W}_i = k \ln \mathcal{W}_1\mathcal{W}_2 = k \ln \mathcal{W}_1 + k \ln \mathcal{W}_2 = S_1 + S_2$$

When the two parts are allowed to interact, the weight becomes $\mathcal{W}_f$, which—if any change occurs at all—will become larger than $\mathcal{W}_i$ as the total system explores all the available configurations and settles into a configuration of greatest weight. The entropy becomes

$$S_f = k \ln \mathcal{W}_f > k \ln \mathcal{W}_i = S_i$$

That is, because $\ln \mathcal{W}_f$ is greater than $\ln \mathcal{W}_i$, the entropy of the final state is greater than that of the initial state, as we sought to demonstrate.

A note on good practice Equation 15.1 shows that the units of entropy are the same as those of Boltzmann's constant (joules per kelvin, J K^{-1}). The molar entropy, $S_m = S/n$, therefore has the units joules per kelvin per mole (J K^{-1} mol^{-1}), the same as the gas constant. The standard molar entropy, $S^{\ominus}_m$, is the molar entropy under standard conditions (pure, 1 bar).

For eqn 15.1 to be a useful route to the calculation of the entropy, we need to express it in terms of the partition function. To do so, we substitute the expression for $\ln \mathcal{W}$ given in Section 13.1 into eqn 15.1 and, as shown in the following *Justification*, for distinguishable particles obtain

$$S = \frac{U - U(0)}{T} + Nk\ln q \qquad (15.2a)$$

and for indistinguishable particles

$$S = \frac{U - U(0)}{T} + Nk\ln \frac{qe}{N} \qquad (15.2b)$$

(The presence of the exponential e in the logarithm is explained in the *Justification*.) Apart from changes of detail, we see that the entropy increases as the number of thermally accessible states (as measured by q) increases, just as we should expect.

Justification 15.1 *The statistical entropy*

For a system composed of N distinguishable molecules, eqn 13.3 is

$$\ln \mathcal{W} = N\ln N - \sum_i N_i \ln N_i$$

Then, because $N = \sum_i N_i$, we can write eqn 15.1 ($S = k \ln \mathcal{W}$) as

$$S = k\sum_i (N_i \ln N - N_i \ln N_i) = -k\sum_i N_i \ln \frac{N_i}{N}$$

The value of N_i/N for the most probable distribution is given by the Boltzmann distribution:

$$\ln \frac{N_i}{N} = \ln\left(\frac{e^{-\beta\varepsilon_i}}{q}\right) = -\beta\varepsilon_i - \ln q$$

Therefore,

$$S = -k\left\{-\beta\sum_i N_i\varepsilon_i - \sum_i N_i \ln q\right\} = Nk\beta\langle\varepsilon\rangle + Nk\ln q$$

Finally, because $N\langle\varepsilon\rangle = U - U(0)$ and $\beta = 1/kT$, we obtain eqn 15.2a.

To treat a system composed of N *in*distinguishable molecules, we need to reduce the weight $\mathcal{W}$ by a factor of $1/N!$, because the $N!$ permutations of the molecules among the states result in the same state of the system. Then, because $\ln(\mathcal{W}/N!) = \ln \mathcal{W} - \ln N!$, the equation in the first line of this *Justification* becomes

$$\begin{aligned}\ln \mathcal{W} &= N\ln N - \sum_i N_i \ln N_i - \ln N! \\ &= N\ln N - \sum_i N_i \ln N_i - N\ln N + N \\ &= -\sum_i N_i \ln N_i + N\end{aligned}$$

where we have used Stirling's approximation in the second line to write $\ln N! = N \ln N - N$. Then the same calculation as above leads to eqn 15.2b if we note that Nk can be written $Nk \ln e$ and $Nk \ln q + Nk \ln e = Nk \ln qe$.

Equation 15.2 is in terms of the molecular partition function and is too restrictive to accommodate interactions between molecules. As in Chapter 13, to accommodate interacting particles we have to use the canonical partition function Q and the weight $\mathcal{W}$ of the most probable configuration of the canonical ensemble, $\tilde{W}$. However, because $\tilde{W} = W^{\tilde{N}}$ (each member of the ensemble is independent of the others, so we can multiply together their weights to get the overall weight of the ensemble) we can use $W = \tilde{W}^{1/\tilde{N}}$ in the Boltzmann formula and obtain

$$S = k\ln \tilde{W}^{1/\tilde{N}} = \frac{k}{\tilde{N}}\ln \tilde{W} \qquad (15.3)$$

The number of members of the ensemble, $\tilde{N}$, goes to infinity (to achieve the thermodynamic limit). The entropy can be expressed in terms of Q by the same argument as in *Justification 15.1*, and we obtain

$$S = \frac{U - U(0)}{T} + k\ln Q \qquad (15.4)$$

This expression reduces to eqn 15.2 when we write $Q = q^N$ for distinguishable non-interacting particles and $Q = q^N/N!$ for indistinguishable non-interacting particles.

The expressions we have derived for the entropy accord with what we should expect for entropy if it is a measure of the spread of the populations of molecules over the available states. For instance, we show in the following *Justification* that the **Sackur–Tetrode equation** for the molar entropy of a monatomic gas is

$$S_m = R\ln\left(\frac{V_m e^{5/2}}{N_A \Lambda^3}\right) \qquad (15.5a)$$

where Λ is the thermal wavelength introduced in eqn 13.15b ($\Lambda = h/(2\pi mkT)^{1/2}$). To calculate the standard molar entropy, we note that $V_m = RT/p$, and set $p = p^{\ominus}$:

$$S_m^{\ominus} = R\ln\left(\frac{RTe^{5/2}}{p^{\ominus}N_A\Lambda^3}\right) = R\ln\left(\frac{kTe^{5/2}}{p^{\ominus}\Lambda^3}\right) \qquad (15.5b)$$

We have used $R/N_A = k$. These expressions are based on the high-temperature approximation of the partition functions, which assumes that many levels are occupied; therefore, they do not apply when T is equal to or very close to zero.

● A BRIEF ILLUSTRATION

To calculate the standard molar entropy of gaseous argon at 25°C, we use eqn 15.5b with $\Lambda = h/(2\pi mkT)^{1/2}$. The mass of an Ar atom is $m = 39.95m_u$. At 25°C, its thermal wavelength is 16.0 pm (by the same kind of calculation as the *brief illustration* in Section 13.3) and $kT = 4.12 \times 10^{-21}$ J. Therefore,

$$S_m^{\ominus} = R\ln\left\{\frac{e^{5/2}\times(4.12\times10^{-21}\text{ J})}{(10^5\text{ N m}^{-2})\times(1.60\times10^{-11}\text{ m})^3}\right\}$$
$$= 18.6R = 155\text{ J K}^{-1}\text{ mol}^{-1}$$

We can anticipate, on the basis of the number of accessible states for a lighter molecule, that the standard molar entropy of Ne is likely to be smaller than for Ar; its actual value is 17.60*R* at 298 K. ●

Self-test 15.1 Calculate the translational contribution to the standard molar entropy of H_2 at 25°C. [14.2*R*]

The implications of these equations are as follows:

• Because the molecular mass appears in the numerator (because it appears in the denominator of Λ), the molar entropy of a perfect gas of heavy molecules is greater than that of a perfect gas of light molecules under the same conditions.

We can understand this feature in terms of the energy levels of a particle in a box being closer together for heavy particles than for light particles, so more states are thermally accessible.

• Because the molar volume appears in the numerator, the molar entropy increases with the molar volume of the gas.

The reason is similar: large containers have more closely spaced energy levels than small containers, so once again more states are thermally accessible.

• Because the temperature appears in the numerator (because, like m, it appears in the denominator of Λ), the molar entropy increases with increasing temperature.

The reason for this behaviour is that more energy levels become accessible as the temperature is raised.

Justification 15.2 *The Sackur–Tetrode equation*

We start with eqn 15.2b for a collection of indistinguishable particles and write $N = nN_A$, where N_A is Avogadro's constant. The only mode of motion for a gas of atoms is translation, and we saw in the *brief illustration* in Section 14.2 that $U - U(0) = \frac{3}{2}nRT$. The partition function is $q = V/\Lambda^3$ (eqn 13.18), where Λ is the thermal wavelength. Therefore,

$$S = \tfrac{3}{2}nR + nR\ln\left(\frac{Ve}{nN_A\Lambda^3}\right) = nR\left(\ln e^{3/2} + \ln\frac{V_m e}{N_A\Lambda^3}\right)$$
$$= nR\ln\left(\frac{V_m e^{5/2}}{N_A\Lambda^3}\right)$$

where $V_m = V/n$ is the molar volume of the gas and we have used $\frac{3}{2} = \ln e^{3/2}$. Division of both sides by n then results in eqn 15.5a.

The Sackur–Tetrode equation implies that, when a monatomic perfect gas expands isothermally from V_i to V_f, its molar entropy changes by

$$\Delta S_m = R\ln(aV_f) - R\ln(aV_i) = R\ln\frac{V_f}{V_i} \qquad (15.6)°$$

where aV is the collection of quantities inside the logarithm of eqn 15.5a. This expression, which is plotted in Fig. 15.4, shows that the molar entropy of a perfect gas increases logarithmically with volume (because more energy levels become accessible as they become less widely spaced), and that the increase is independent of the identity of the gas (because the factor a has cancelled).

We can get more insight into the properties of the entropy by using eqn 15.2 to calculate its value for types of motion other than translation. For instance, the vibrational contribution to the molar entropy, S_m^V, is obtained by combining the expression for the molecular partition function (eqn 13.24, $q = 1/(1 - e^{-\beta\varepsilon})$, with $\varepsilon = hc\tilde{\nu}$) with the expression for the mean energy (eqn 13.39, $\langle\varepsilon\rangle = \varepsilon/(e^{\beta\varepsilon} - 1)$), to obtain

$$S_m^V = R\left\{\frac{\beta\varepsilon}{e^{\beta\varepsilon}-1} - \ln(1 - e^{-\beta\varepsilon})\right\} \qquad (15.7)$$

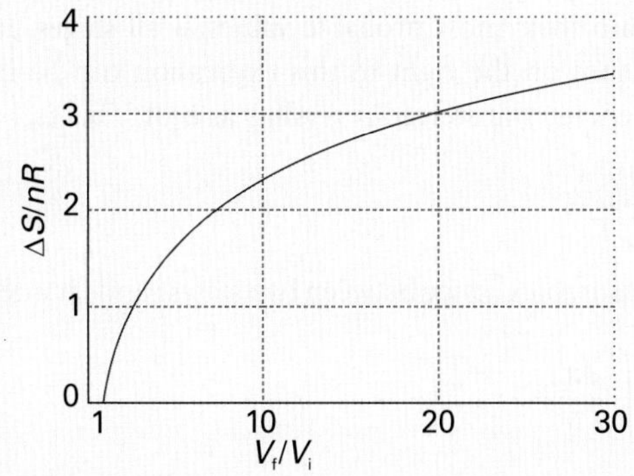

Fig. 15.4 The logarithmic increase in entropy of a perfect gas as it expands isothermally.

interActivity Evaluate the change in entropy that accompanies the expansion of 1.00 mol CO_2(g) from 0.001 m^3 to 0.010 m^3 at 298 K, treated as a van der Waals gas.

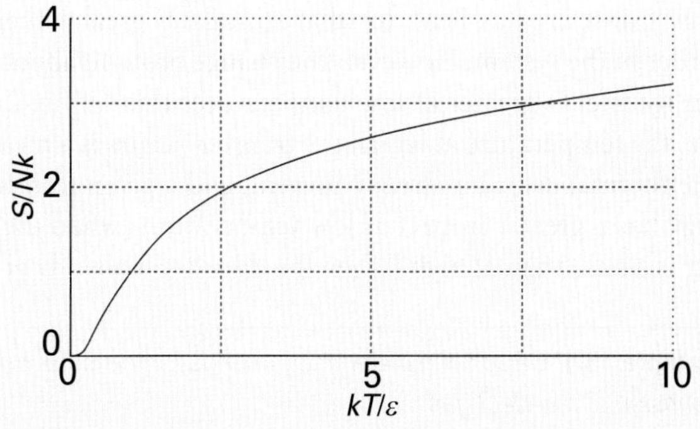

Fig. 15.5 The temperature variation of the entropy of a collection of harmonic oscillators (expressed here as a multiple of Nk). The entropy approaches zero as $T \to 0$, and increases without limit as $T \to \infty$.

interActivity Plot the function dS/dT, the temperature coefficient of the entropy, against kT/ε. Is there a temperature at which this coefficient passes through a maximum? If you find a maximum, explain its physical origins.

This function is plotted in Fig. 15.5. As usual, it is helpful to interpret it, with the graph in mind:

- Both terms multiplying R become zero as β becomes infinite, so the entropy is zero at $T = 0$.
- The molar entropy rises as the temperature is increased as more vibrational states become accessible.
- The molar entropy is higher at a given temperature for molecules with heavy atoms or low force constant than one with light atoms or high force constant.

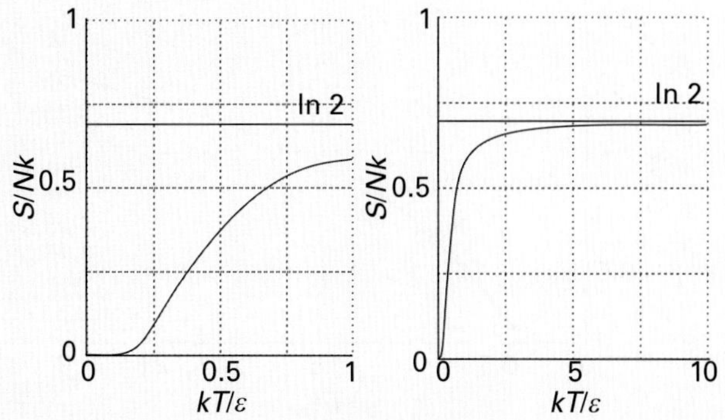

Fig. 15.6 The temperature variation of the entropy of a two-level system (expressed as a multiple of Nk). As $T \to \infty$, the two states become equally populated and S approaches $Nk \ln 2$.

interActivity Draw graphs similar to those in Fig. 15.6 for a three-level system with levels 0, ε, and 2ε.

The vibrational energy levels are closer together in the former case than in the latter, so more are thermally accessible. For I_2 at 25°C, for instance, $\beta\varepsilon = 1.036$ (Self-test 14.4), so $S_m^V = 8.38\ \mathrm{J\ K^{-1}\ mol^{-1}}$.

Self-test 15.2 Evaluate the molar entropy of N two-level systems and plot the resulting expression. What is the entropy when the two states are equally thermally accessible?
[$S/Nk = \beta\varepsilon/(1 + e^{\beta\varepsilon}) + \ln(1 + e^{-\beta\varepsilon})$; see Fig. 15.6; $S = Nk \ln 2$]

The rotational contribution to the molar entropy, S_m^R, can also be calculated once we know the molecular partition function. For a linear molecule, the high-temperature limit of q is $kT/\sigma hc\tilde{B}$ (eqn 13.21b) and the equipartition theorem gives the rotational contribution to the molar internal energy as RT; therefore, from eqn 15.2a, the contribution at high temperatures is

$$S_m^R = R + R\ln\frac{kT}{\sigma hc\tilde{B}} = R\left\{1 + \ln\frac{kT}{\sigma hc\tilde{B}}\right\} \qquad (15.8)$$

This function is plotted in Fig. 15.7. We see that:

- The rotational contribution to the entropy increases with temperature because more rotational states become accessible.
- The rotational contribution is large when $\tilde{B}$ is small, because then the rotational energy levels are close together.

Thus, large, heavy molecules have a large rotational contribution to their entropy. The rotational contribution for Cl_2 at 25°C, for instance, is $58.6\ \mathrm{J\ K^{-1}\ mol^{-1}}$ whereas that for H_2 is only $12.7\ \mathrm{J\ K^{-1}\ mol^{-1}}$. We can regard Cl_2 as a more rotationally disordered gas than H_2, in the sense that at a given temperature Cl_2 occupies a greater number of rotational states than H_2 does.

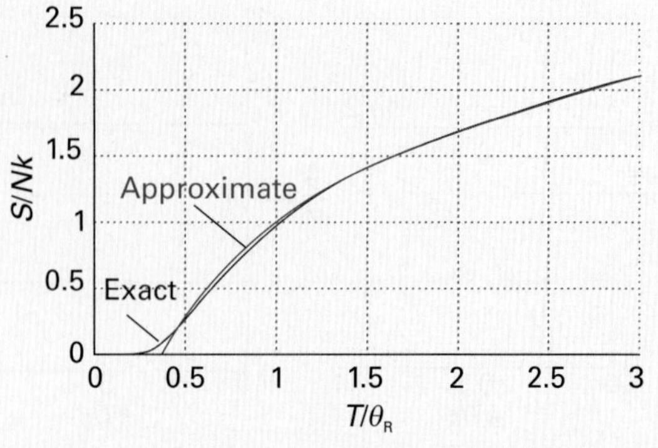

Fig. 15.7 The variation of the rotational contribution to the entropy of a linear molecule ($\sigma = 1$) using the high-temperature approximation and the exact expression (the latter evaluated up to $J = 20$).

Equation 15.8 is valid at high temperatures ($T \gg \theta_R$); to track the rotational contribution down to low temperatures it would be necessary to use the full form of the rotational partition function (Section 13.3; see Problem 15.1); the resulting curve has the form shown in Fig. 15.7. We see, in fact, that the approximate curve matches the exact curve very well for T/θ_R greater than about 1.

(b) The thermodynamic definition of entropy

We shall now use the Boltzmann formula for the entropy to motivate a thermodynamic definition. First, we note that, according to eqn 15.1, a change in entropy arises from a change in populations of the available states:

$$\mathrm{d}S = k\,\mathrm{d}\ln \mathcal{W} = k\sum_i \left(\frac{\partial \ln \mathcal{W}}{\partial N_i}\right)\mathrm{d}N_i$$

We established in eqn 13.50 that for the most probable distribution

$$\frac{\partial \ln \mathcal{W}}{\partial N_i} = \beta\varepsilon_i - \alpha$$

and therefore it follows that

$$\mathrm{d}S = k\beta\sum_i \varepsilon_i \mathrm{d}N_i - k\alpha\sum_i \mathrm{d}N_i$$

The second sum, over changes in populations, is zero for a system with a fixed number of particles, and so

$$\mathrm{d}S = k\beta\sum_i \varepsilon_i \mathrm{d}N_i = \frac{1}{T}\sum_i \varepsilon_i \mathrm{d}N_i$$

However, for a system that is heated, there is no change in the energy levels themselves and the change in energy of the system is due only to the change in populations of those unchanging energy levels. Moreover, if the heating is reversible, the populations retain their most probable values at all stages. It follows that the sum on the right in this expression can be identified with the energy transferred reversibly as heat. That is,

$$\mathrm{d}S = \frac{\mathrm{d}q_{\text{rev}}}{T} \qquad [15.9]$$

For a measurable change between two states i and f it integrates to

$$\Delta S = \int_{\mathrm{i}}^{\mathrm{f}} \frac{\mathrm{d}q_{\text{rev}}}{T} \qquad (15.10)$$

That is, to calculate the difference in entropy between any two states of a system from thermodynamic data as distinct from the spectroscopic data used to calculate changes in the statistical entropy, we find a *reversible* path between them, and integrate the energy supplied as heat at each stage of the path divided by the temperature at which the heating occurs.

In classical thermodynamics, eqn 15.9 is taken as the thermodynamic *definition* of entropy and has a straightforward physical interpretation. The change in entropy is proportional to the energy transferred as heat, for that stimulates greater thermal disorder in the system. However, the change brought about by a given transfer is greater at low temperatures than at high. In a sense, the temperature at which the transfer occurs is a measure of the thermal disorder already present, and a given transfer of energy has a greater impact at low temperature (when the disorder is small) than at high (when the disorder is already great).

Example 15.1 *Calculating the entropy change for the isothermal expansion of a perfect gas*

Calculate the entropy change of a sample of perfect gas when it expands isothermally from a volume V_i to a volume V_f.

Method Equation 15.10 instructs us to find the energy supplied as heat for a reversible path between the stated initial and final states regardless of the actual manner in which the process in fact takes place. A simplification is that the expansion is isothermal, so the temperature is a constant and may be taken outside the integral. The energy absorbed as heat during a reversible isothermal expansion of a perfect gas can be calculated from $\Delta U = q + w$ and $\Delta U = 0$, which implies that $q = -w$ in general and therefore that $q_{\text{rev}} = -w_{\text{rev}}$ for a reversible change. The work of reversible isothermal expansion was calculated in Section 14.3.

Answer Because the temperature is constant, eqn 15.10 becomes

$$\Delta S = \frac{1}{T}\int_{\mathrm{i}}^{\mathrm{f}} \mathrm{d}q_{\text{rev}} = \frac{q_{\text{rev}}}{T}$$

From eqn 14.12 and $q_{rev} = -w_{rev}$, we know that

$$q_{rev} = -w_{rev} = nRT\ln\frac{V_f}{V_i}$$

It follows that

$$\Delta S = nR\ln\frac{V_f}{V_i}$$

exactly as obtained by statistical arguments (eqn 15.6). As an illustration of this formula, when the volume occupied by 1.00 mol of any perfect gas molecules is doubled at any constant temperature, $V_f/V_i = 2$ and

$$\Delta S = (1.00\ \text{mol}) \times (8.3145\ \text{J K}^{-1}\ \text{mol}^{-1}) \times \ln 2 = +5.76\ \text{J K}^{-1}$$

Self-test 15.3 Calculate the change in entropy when the pressure of a perfect gas is changed isothermally from p_i to p_f. [$\Delta S = nR\ln(p_i/p_f)$]

We can use the definition in eqn 15.9 to formulate an expression for the change in entropy of the surroundings, ΔS_{sur}, which would be very difficult to do statistically. Consider an infinitesimal transfer of heat dq_{sur} to the surroundings. The surroundings consist of a reservoir of constant pressure and temperature, so the energy supplied to them as heat can be identified with the change in their enthalpy, dH_{sur}. The enthalpy is a state function and dH_{sur} is an exact differential. As we have seen, these properties imply that dH_{sur} is independent of how the change is brought about and in particular is independent of whether the process is reversible or irreversible. The same remarks therefore apply to dq_{sur}, to which dH_{sur} is equal. Therefore, we can adapt the definition in eqn 15.9 to write

$$dS_{sur} = \frac{dq_{sur,rev}}{T_{sur}} = \frac{dq_{sur}}{T_{sur}} \qquad (15.11a)$$

Furthermore, because the temperature of the surroundings is constant however much heat enters them (they have infinite heat capacity), for a measurable change

$$\Delta S_{sur} = \frac{q_{sur}}{T_{sur}} \qquad (15.11b)$$

That is, regardless of how the change is brought about in the system, reversibly or irreversibly, we can calculate the change of entropy of the surroundings by dividing the heat transferred by the temperature at which the transfer takes place.

Equation 15.11 makes it very simple to calculate the changes in entropy of the surroundings that accompany any process. For instance, for any adiabatic change, $q_{sur} = 0$, so

For an adiabatic change: $\Delta S_{sur} = 0$ (15.12)

This expression is true however the change takes place, reversibly or irreversibly.

● A BRIEF ILLUSTRATION

To calculate the entropy change in the surroundings when 1.00 mol $H_2O(l)$ is formed from its elements under standard conditions at 298 K, we use $\Delta H^{\ominus} = -286$ kJ from Table 14.6. The energy released as heat is supplied to the surroundings, so $q_{sur} = +286$ kJ. Therefore,

$$\Delta S_{sur} = \frac{2.86 \times 10^5\ \text{J}}{298\ \text{K}} = +960\ \text{J K}^{-1}$$

This strongly exothermic reaction results in an increase in the entropy of the surroundings as energy is released into them as heat. ●

Self-test 15.4 Calculate the entropy change in the surroundings when 1.00 mol $N_2O_4(g)$ is formed from 2.00 mol $NO_2(g)$ under standard conditions at 298 K. [$-192\ \text{J K}^{-1}$]

(c) The entropy as a state function

Entropy is a state function. That is evident from the statistical definition, but is it true of the thermodynamic definition? To prove that it is, we need to show that the integral of dS is independent of path. To do so, it is sufficient to prove that the integral of eqn 15.10 around an arbitrary cycle is zero, for that guarantees that the entropy is the same at the initial and final states of the system regardless of the path taken between them (Fig. 15.8). That is, we need to show that

$$\oint \frac{dq_{rev}}{T} = 0 \qquad (15.13)$$

where the symbol $\oint$ denotes integration around a closed path. There are three steps in the argument:

- First, to show that eqn 15.13 is true for a special cycle (a 'Carnot cycle') involving a perfect gas.

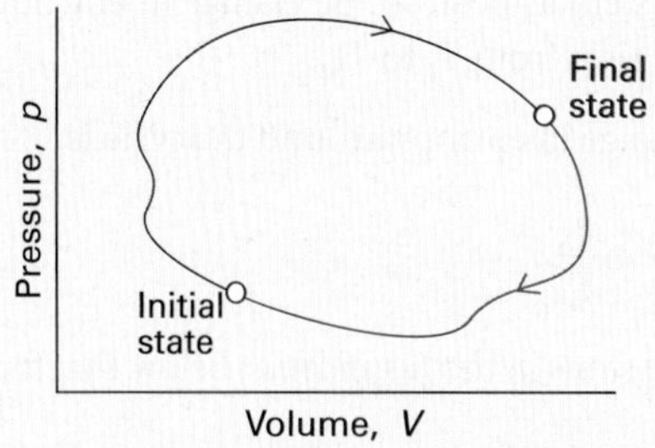

Fig. 15.8 In a thermodynamic cycle, the overall change in a state function (from the initial state to the final state and then back to the initial state again) is zero.

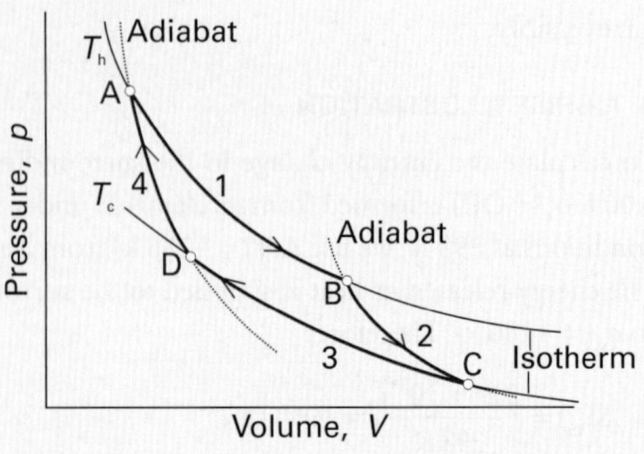

Fig. 15.9 The basic structure of a Carnot cycle. In step 1, there is isothermal reversible expansion at the temperature T_h. Step 2 is a reversible adiabatic expansion in which the temperature falls from T_h to T_c. In Step 3 there is an isothermal reversible compression at T_c, and that isothermal step is followed by an adiabatic reversible compression, which restores the system to its initial state.

- Then to show that the result is true whatever the working substance.
- Finally, to show that the result is true for any cycle.

A **Carnot cycle**, which is named after the French engineer Sadi Carnot, consists of four reversible stages (Fig. 15.9):

1. Reversible isothermal expansion from A to B at T_h; the entropy change is q_h/T_h, where q_h is the heat supplied from the hot source.

2. Reversible adiabatic expansion from B to C. No energy leaves the system as heat, so the change in entropy is zero. In the course of this expansion, the temperature falls from T_h to T_c, the temperature of the cold sink.

3. Reversible isothermal compression from C to D at T_c. Energy is released as heat to the cold sink; the corresponding change in entropy is q_c/T_c; in this expression q_c is negative.

4. Reversible adiabatic compression from D to A. No energy enters the system as heat, so the change in entropy is zero. The temperature rises from T_c to T_h.

The total change in entropy around the cycle is

$$\oint dS = \frac{q_h}{T_h} + \frac{q_c}{T_c}$$

However, we show in the *Justification* below that for a perfect gas

$$\frac{q_h}{q_c} = -\frac{T_h}{T_c} \qquad (15.14)_{rev}$$

Substitution of this relation into the preceding equation gives zero on the right, which is what we wanted to prove.

Justification 15.3 *Heating accompanying reversible adiabatic expansion*

This *Justification* is based on the fact that the two temperatures in eqn 15.14 lie on the same adiabat in Fig. 15.9. As explained in Example 15.1, for a perfect gas:

$$q_h = nRT_h \ln\frac{V_B}{V_A} \qquad q_c = nRT_c \ln\frac{V_D}{V_C}$$

From the relations between temperature and volume for reversible adiabatic processes (eqn 14.37, summarized as $VT^c = \text{constant}$ with $c = C_{V,m}/R$):

$$V_A T_h^c = V_D T_c^c \qquad V_C T_c^c = V_B T_h^c$$

Multiplication of the first of these expressions by the second gives

$$V_A V_C T_h^c T_c^c = V_D V_B T_h^c T_c^c$$

which simplifies to

$$\frac{V_A}{V_B} = \frac{V_D}{V_C}$$

Consequently,

$$q_c = nRT_c \ln\frac{V_D}{V_C} = nRT_c \ln\frac{V_A}{V_B} = -nRT_c \ln\frac{V_B}{V_A}$$

and therefore

$$\frac{q_h}{q_c} = \frac{nRT_h \ln(V_B/V_A)}{-nRT_c \ln(V_B/V_A)} = -\frac{T_h}{T_c}$$

as in eqn 15.14.

In the second step we need to show that eqn 15.14 applies to any material, not just a perfect gas (which is why, in anticipation, we have not labelled it with a °). We begin this step of the argument by introducing the **efficiency**, η (eta), of a heat engine:

$$\eta = \frac{\text{work performed}}{\text{heat absorbed}} = \frac{|w|}{q_h} \qquad [15.15]$$

The definition implies that, the greater the work output for a given supply of heat from the hot reservoir, the greater is the efficiency of the engine. We can express the definition in terms of the heat transactions alone, because (as shown in Fig. 15.10) the energy supplied as work by the engine is the difference between the energy supplied as heat by the hot reservoir and returned to the cold reservoir:

$$\eta = \frac{|q_h| - |q_c|}{|q_h|} = 1 - \frac{|q_c|}{|q_h|} \qquad (15.16)$$

We have used absolute values because keeping track of signs can be tricky. It then follows from eqn 15.14 in the form $|q_c|/|q_h| = T_c/T_h$ that

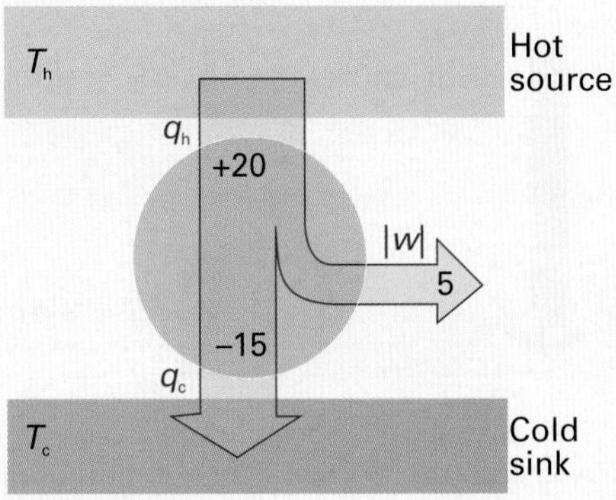

Fig. 15.10 Suppose an energy q_h (for example, 20 kJ) is supplied to the engine and q_c is lost from the engine (for example, $q_c = -15$ kJ) and discarded into the cold reservoir. The work done by the engine is equal to $q_h + q_c$ (for example, 20 kJ + (−15 kJ) = 5 kJ). The efficiency is the work done divided by the energy supplied as heat from the hot source.

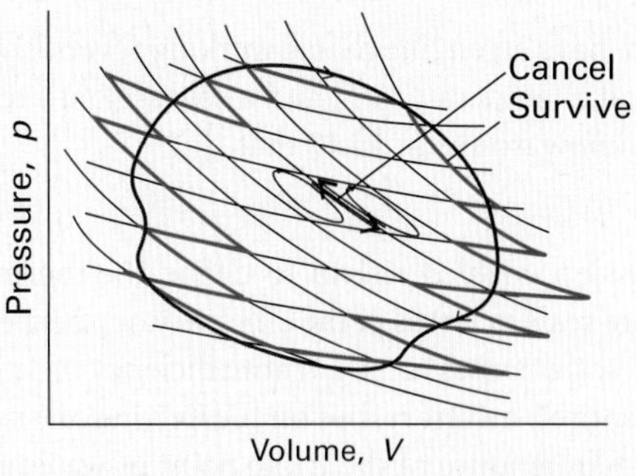

Fig. 15.12 A general cycle can be divided into small Carnot cycles. The match is exact in the limit of infinitesimally small cycles. Paths cancel in the interior of the collection, and only the perimeter, an increasingly good approximation to the true cycle as the number of cycles increases, survives. Because the entropy change around every individual cycle is zero, the integral of the entropy around the perimeter is zero too.

$$\eta_{\text{rev}} = 1 - \frac{T_c}{T_h} \qquad (15.17)_{\text{rev}}$$

Now we are ready to generalize this conclusion. The Second Law of thermodynamics implies that *all reversible engines have the same efficiency regardless of their construction.* To see the truth of this statement, suppose two reversible engines are coupled together and run between the same two reservoirs (Fig. 15.11). The working substances and details of construction of the two engines are entirely arbitrary. Initially, suppose that engine A is more efficient than engine B, and that we choose a setting of the controls that causes engine B to take energy as heat from the cold reservoir and to release a certain quantity of energy as heat into the hot reservoir. However, because engine A is more efficient than engine B, not all the work that A produces is needed for this process, and the difference can be used to do work. The net result is that the cold reservoir is unchanged, work has been done, and the hot reservoir has lost a certain amount of energy. This outcome is contrary to the Kelvin statement of the Second Law, because some heat has been converted directly into work, and so the initial assumption that engines A and B can have different efficiencies must be false. It follows that the relation between the heat transfers and the temperatures must also be independent of the working material, and therefore that eqn 15.17 is true for any substance involved in a Carnot cycle.

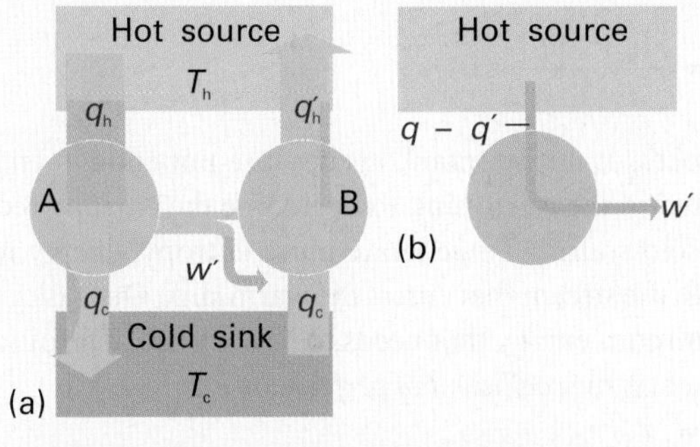

Fig. 15.11 (a) The demonstration of the equivalence of the efficiencies of all reversible engines working between the same thermal reservoirs is based on the flow of energy represented in this diagram. (b) The net effect of the processes is the conversion of heat into work without there being a need for a cold sink: this is contrary to the Kelvin statement of the Second Law.

For the final step in the argument, we note that any reversible cycle can be approximated as a collection of Carnot cycles and the cyclic integral around an arbitrary path is the sum of the integrals around each of the Carnot cycles (Fig. 15.12). This approximation becomes exact as the individual cycles are allowed to become infinitesimal. The entropy change around each individual cycle is zero, so the sum of entropy changes for all the cycles is zero. However, in the sum, the entropy change along any individual path is cancelled by the entropy change along the path it shares with the neighbouring cycle. Therefore, all the entropy changes cancel except for those along the perimeter of the overall cycle. That is,

$$\sum_{\text{all}} \frac{q_{\text{rev}}}{T} = \sum_{\text{perimeter}} \frac{q_{\text{rev}}}{T} = 0$$

In the limit of infinitesimal cycles, the non-cancelling edges of the Carnot cycles match the overall cycle exactly, and the sum becomes an integral. Equation 15.13 then follows immediately. This result implies that dS is an exact differential and therefore that S is a state function.

(d) The thermodynamic temperature

Suppose we have an engine that is working reversibly between a hot source at a temperature T_h and a cold sink at a temperature T; then we know from eqn 15.17 that

$$T = (1 - \eta)T_h \qquad (15.18)$$

This expression enabled Kelvin to define the **thermodynamic temperature scale** in terms of the efficiency of a heat engine. The zero of the scale occurs for a Carnot efficiency of 1. The size of the unit is entirely arbitrary, but on the Kelvin scale is defined by setting the temperature of the triple point of water as 273.16 K exactly (Section 16.1). Then, if the heat engine has a hot source at the triple point of water, the temperature of the cold sink (the object we want to measure) is found by measuring the efficiency of the engine.

(e) The Clausius inequality

We now show that the thermodynamic definition of entropy, like the statistical definition, is consistent with the Second Law. To begin, we recall that more energy flows as work under reversible conditions than under irreversible conditions. That is $-dw_{rev} \geq -dw$, or $dw - dw_{rev} \geq 0$. Because the internal energy is a state function, its change is the same for irreversible and reversible paths between the same two states, so we can also write:

$$dU = dq + dw = dq_{rev} + dw_{rev}$$

It follows that $dq_{rev} - dq = dw - dw_{rev} \geq 0$, or $dq_{rev} \geq dq$, and therefore that $dq_{rev}/T \geq dq/T$. Now we use the thermodynamic definition of the entropy (eqn 15.9; $dS = dq_{rev}/T$) to write

$$dS \geq \frac{dq}{T} \qquad (15.19)$$

This expression is the **Clausius inequality**. It will prove to be of great importance for the discussion of the spontaneity of chemical reactions, as we shall see in Section 15.5.

● A BRIEF ILLUSTRATION

Consider the transfer of energy as heat from one system—the hot source—at a temperature T_h to another system—the cold sink—at a temperature T_c (Fig. 15.13). When $|dq|$ leaves the hot source (so $dq_h < 0$), the Clausius inequality implies that $dS \geq dq_h/T_h$. When $|dq|$ enters the cold sink the Clausius inequality implies that $dS \geq dq_c/T_c$ (with $dq_c > 0$). Overall, therefore,

$$dS \geq \frac{dq_h}{T_h} + \frac{dq_c}{T_c}$$

However, $dq_h = -dq_c$, so

$$dS \geq -\frac{dq_c}{T_h} + \frac{dq_c}{T_c} = dq_c\left(\frac{1}{T_c} - \frac{1}{T_h}\right)$$

which is positive (because $dq_c > 0$ and $T_h \geq T_c$). Hence, cooling (the transfer of heat from hot to cold) is spontaneous, as we know from experience. ●

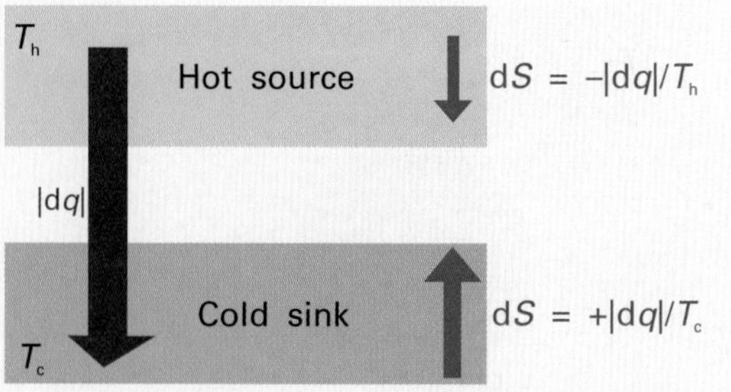

Fig. 15.13 When energy leaves a hot reservoir as heat, the entropy of the reservoir decreases. When the same quantity of energy enters a cooler reservoir, the entropy increases by a larger amount. Hence, overall there is an increase in entropy and the process is spontaneous. Relative changes in entropy are indicated by the sizes of the arrows.

We now suppose that the system is isolated from its surroundings, so that $dq = 0$. The Clausius inequality implies that

$$dS \geq 0 \qquad (15.20)$$

and we conclude that *in an isolated system the entropy cannot decrease when a spontaneous change occurs*. This statement captures the content of the Second Law and is the thermodynamic version of the statistical argument at the beginning of Section 15.1.

IMPACT ON TECHNOLOGY
I15.1 Refrigeration

Our discussion so far is the basis of the thermodynamic assessment of the power needed to cool objects in refrigerators. First, we consider the work required to cool an object, and refer to Fig. 15.14.

When heat $|q_c|$ is removed from a cool source at a temperature T_c and then deposited in a warmer sink at a temperature T_h as in a typical refrigerator, the change in entropy is

$$\Delta S = -\frac{|q_c|}{T_c} + \frac{|q_c|}{T_h} < 0$$

The process is not spontaneous because not enough entropy is generated in the warm sink to overcome the entropy reduction of the cold source. To generate more entropy, energy must be added to the stream that enters the warm sink. Our task is to find the minimum energy that needs to be supplied. The outcome is expressed as the *coefficient of performance*, c:

$$c = \frac{\text{energy transferred as heat}}{\text{energy transferred as work}} = \frac{|q_c|}{|w|} \qquad [15.21]$$

The less the work that is required to achieve a given transfer, the greater the coefficient of performance and the more efficient is the refrigerator.

Because $|q_c|$ is removed from the cold source, and the work $|w|$ is added to the energy stream, the energy deposited as heat in

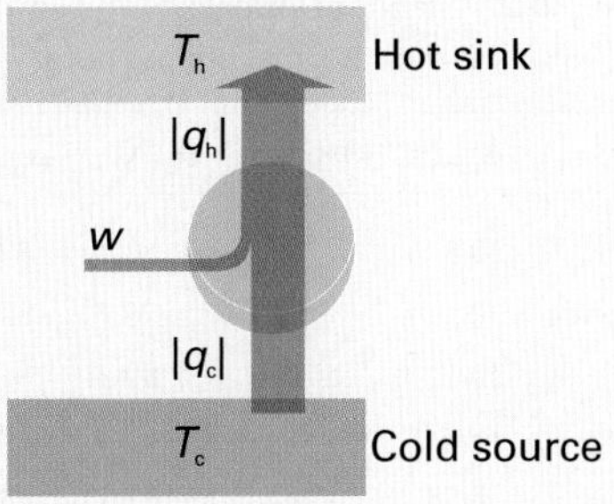

Fig. 15.14 If energy is added (as work) to the energy stream, the increase in entropy of the hot reservoir can be changed to match the loss in the cold source.

the hot sink is $|q_h| = |q_c| + |w|$, so $|w| = |q_h| - |q_c|$. The next step is easier to take if we consider $1/c$ rather than c itself:

$$\frac{1}{c} = \frac{|w|}{|q_c|} = \frac{|q_h| - |q_c|}{|q_c|} = \frac{|q_h|}{|q_c|} - 1$$

We can now use eqn 15.14 in the form $|q_h|/|q_c| = T_h/T_c$ to express this result in terms of the temperatures alone, which is possible if the transfer is performed reversibly. This substitution leads to

$$\frac{1}{c} = \frac{T_h}{T_c} - 1 = \frac{T_h - T_c}{T_c}$$

and therefore to

$$c = \frac{T_c}{T_h - T_c} \tag{15.22}$$

for the thermodynamically optimum coefficient of performance. For a refrigerator withdrawing heat from ice-cold water ($T_c = 273$ K) in a typical environment ($T_h = 293$ K), $c = 14$, so to remove 10 kJ (enough to freeze 30 g of water) requires transfer of at least 0.71 kJ as work. Practical refrigerators, of course, have a lower coefficient of performance.

The work to *maintain* a low temperature is also relevant to the design of refrigerators. No thermal insulation is perfect, so there is always a flow of energy as heat into the sample at a rate proportional to the temperature difference. If the rate at which energy leaks in is written $A(T_h - T_c)$, where A is a constant that depends on the size of the sample and the details of the insulation, then the minimum power, P, required to maintain the original temperature difference by pumping out that energy by heating the surroundings is

$$P = \frac{1}{c} \times A(T_h - T_c) = A \times \frac{(T_h - T_c)^2}{T_c} \tag{15.23}$$

We see that the power increases as the square of the temperature difference we are trying to maintain. For this reason, air-conditioners are much more expensive to run on hot days than on mild days.

15.3 Entropy changes accompanying specific processes

We now see how to calculate the entropy changes that accompany a variety of basic processes. In each case we develop the argument using classical thermodynamics, but relate each result to the statistical definition. You should always keep in mind that the entropy of a system increases as more states become accessible, either through a rise in temperature (which extends the tail of the Boltzmann distribution) or because there is a change in the energy levels that brings more of them into reach.

(a) Expansion

We have already established (Example 15.1 and eqn 15.6) that the change in entropy of a perfect gas that expands isothermally from V_i to V_f is

$$\Delta S = nR\ln\frac{V_f}{V_i} \tag*{(15.24)°}$$

This increase is due to the decreasing separation of energy levels as the container expands, which makes more states accessible. Because S is a state function, the value of ΔS *of the system* is independent of the path between the initial and final states, so this expression applies whether the change of state occurs reversibly or irreversibly. The logarithmic dependence of entropy on volume was illustrated in Fig. 15.4.

The *total* change in entropy, the sum of the changes in the system and the surroundings, however, does depend on how the expansion takes place. For any process $dq_{sur} = -dq$, and for a reversible change we use the expression in Example 15.1; consequently, from eqn 15.11b

$$\Delta S_{sur} = \frac{q_{sur}}{T} = -\frac{q_{rev}}{T} = -nR\ln\frac{V_f}{V_i} \tag*{(15.25)$^{\circ}_{rev}$}$$

This change is the negative of the change in the system, so we can conclude that $\Delta S_{tot} = 0$, which is what we should expect for a reversible process. If the isothermal expansion occurs freely ($w = 0$) and irreversibly, then $q = 0$ (because $\Delta U = 0$). Consequently, $\Delta S_{sur} = 0$, and the total entropy change is given by eqn 15.24 itself:

$$\Delta S_{tot} = nR\ln\frac{V_f}{V_i} \tag*{(15.26)°}$$

In this case, $\Delta S_{tot} > 0$, as we expect for an irreversible process.

(b) Phase transition

The entropy of a substance changes when a substance freezes or boils as a result of changes in the orderliness with which the molecules pack together and the extent to which the energy is localized or dispersed. For example, when a substance vaporizes, a compact condensed phase changes into a widely dispersed gas and we can expect the entropy of the substance to increase considerably. The entropy of a solid also increases when it melts to a liquid and when that liquid turns into a gas. Although we can *understand* the changes of entropy in these statistical, molecular terms, it is far easier to use classical thermodynamics to calculate numerical values, and we shall do that here.

Consider a system and its surroundings at the **normal transition temperature**, T_{trs}, the temperature at which two phases are in equilibrium at 1 atm. This temperature is 0°C (273 K) for ice in equilibrium with liquid water at 1 atm, and 100°C (373 K) for water in equilibrium with its vapour at 1 atm. At the transition temperature, any transfer of energy as heat between the system and its surroundings is reversible because the two phases in the system are in equilibrium. Because at constant pressure $q = \Delta_{trs}H$, the change in molar entropy of the system is

$$\Delta_{trs}S = \frac{\Delta_{trs}H}{T_{trs}} \quad (15.27)$$

Recall from Section 14.7 that $\Delta_{trs}H$ is an enthalpy change per mole of substance; so $\Delta_{trs}S$ is also a molar quantity with units joules per kelvin per mole (J K^{-1} mol^{-1}). If the phase transition is exothermic ($\Delta_{trs}H < 0$, as in freezing or condensing), then the entropy change is negative. This decrease in entropy is consistent with localization of matter and energy that accompanies the formation of a solid from a liquid or a liquid from a gas. If the transition is endothermic ($\Delta_{trs}H > 0$, as in melting and vaporization), then the entropy change is positive, which is consistent with dispersal of energy and matter in the system.

Table 15.1 lists some experimental entropies of transition. Table 15.2 lists in more detail the standard entropies of vaporization of several liquids at their normal boiling points. An interesting feature of the data is that a wide range of liquids gives approximately the same standard entropy of vaporization (about 85 J K^{-1} mol^{-1}): this empirical observation is called **Trouton's rule**. The explanation of Trouton's rule is that a similar change in volume occurs (with an accompanying change in the number of accessible microstates) when any liquid evaporates and becomes a gas at 1 bar. Hence, all liquids can be expected to have similar standard entropies of vaporization.

Liquids that show significant deviations from Trouton's rule do so on account of strong molecular interactions that result in a partial ordering of their molecules. As a result, there is a greater change in disorder when the liquid turns into a vapour than for a fully disordered liquid. An example is water, where the large entropy of vaporization reflects the presence of structure arising from hydrogen bonding in the liquid. Hydrogen bonds tend to organize the molecules in the liquid so that they are less random than, for example, the molecules in liquid hydrogen sulfide (in which there is no hydrogen bonding).

Methane has an unusually low entropy of vaporization. A part of the reason is that the entropy of the gas itself is slightly low (186 J K^{-1} mol^{-1} at 298 K); the entropy of N_2 under the same conditions is 192 J K^{-1} mol^{-1}. As we have seen (Section 15.2), small, light molecules have a low rotational contribution to their entropy because the rotational levels are far apart.

Synoptic table 15.1* Standard entropies (and temperatures) of phase transitions, $\Delta_{trs}S^{\ominus}/(\text{J K}^{-1}\text{ mol}^{-1})$

	Fusion (at T_f)	Vaporization (at T_b)
Argon, Ar	14.17 (at 83.8 K)	74.53 (at 87.3 K)
Benzene, C_6H_6	38.00 (at 279 K)	87.19 (at 353 K)
Water, H_2O	22.00 (at 273.15 K)	109.1 (at 373.15 K)
Helium, He	4.8 (at 1.8 K and 30 bar)	19.9 (at 4.22K)

* More values are given in the *Data section*.

Synoptic table 15.2* Standard enthalpies and entropies of vaporization of liquids at their normal boiling point

	$\Delta_{vap}H^{\ominus}/(\text{kJ mol}^{-1})$	$\theta_b/°\text{C}$	$\Delta_{vap}S^{\ominus}/(\text{J K}^{-1}\text{ mol}^{-1})$
Benzene	30.8	80.1	87.2
Carbon tetrachloride	30	76.7	85.8
Cyclohexane	30.1	80.7	85.1
Hydrogen sulfide	18.7	−60.4	87.9
Methane	8.18	−161.5	73.2
Water	40.7	100.0	109.1

* More values are given in the *Data section*.

● A BRIEF ILLUSTRATION

There is no hydrogen bonding in liquid bromine and Br_2 is a heavy molecule that is unlikely to display unusual behaviour in the gas phase, so it is safe to use Trouton's rule. To predict the standard molar enthalpy of vaporization of bromine given that it boils at 59.2°C, we use the rule in the form

$$\Delta_{vap}H^{\ominus} = T_b \times (85\text{ J K}^{-1}\text{ mol}^{-1})$$

Substitution of the data then gives

$$\Delta_{vap}H^{\ominus} = (332.4\text{ K}) \times (85\text{ J K}^{-1}\text{ mol}^{-1}) = +2.8 \times 10^4\text{ J mol}^{-1}$$
$$= +28\text{ kJ mol}^{-1}$$

The experimental value is +29.45 kJ mol^{-1}. ●

Self-test 15.5 Predict the enthalpy of vaporization of ethane from its normal boiling point, −88.6°C [16 kJ mol^{-1}]

(c) Heating

We have already seen how the translational, vibrational, and rotational contributions to the statistical entropy increase with temperature as more states become accessible. However, we can use the thermodynamic expression (eqn 15.10) in the form

$$S(T_f) = S(T_i) + \int_{T_i}^{T_f} \frac{dq_{rev}}{T} \tag{15.28}$$

to calculate the entropy of a system at a temperature T_f from a knowledge of its entropy at a temperature T_i and the heat supplied to change its temperature from one value to the other. We shall be particularly interested in the entropy change when the system is subjected to constant pressure (such as from the atmosphere) during the heating. Then, from the definition of constant-pressure heat capacity (eqns 14.32 and 14.33), $dq_{rev} = C_p dT$ provided the system is doing no non-expansion work. Consequently, at constant pressure:

$$S(T_f) = S(T_i) + \int_{T_i}^{T_f} \frac{C_p\, dT}{T} \tag{15.29}$$

The same expression applies at constant volume, but with C_p replaced by C_V. When C_p is independent of temperature in the temperature range of interest, it can be taken outside the integral and we obtain

$$S(T_f) = S(T_i) + C_p \int_{T_i}^{T_f} \frac{dT}{T} = S(T_i) + C_p \ln\frac{T_f}{T_i} \tag{15.30}$$

with a similar expression for heating at constant volume. The logarithmic dependence of entropy on temperature is illustrated in Fig. 15.15 and stems from the logarithmic dependence of the entropy on the partition function (eqn 15.2) and hence on the number of accessible levels.

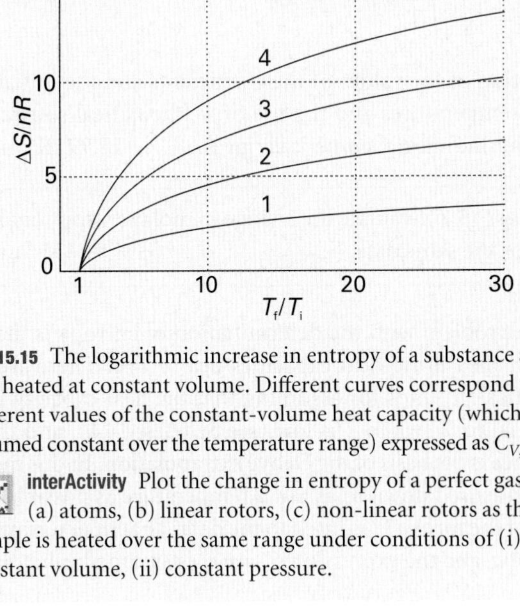

Fig. 15.15 The logarithmic increase in entropy of a substance as it is heated at constant volume. Different curves correspond to different values of the constant-volume heat capacity (which is assumed constant over the temperature range) expressed as $C_{V,m}/R$.

interActivity Plot the change in entropy of a perfect gas of (a) atoms, (b) linear rotors, (c) non-linear rotors as the sample is heated over the same range under conditions of (i) constant volume, (ii) constant pressure.

Example 15.2 *Calculating the entropy change*

Calculate the entropy change when argon at 25°C and 1.00 bar in a container of volume 0.500 dm^3 is allowed to expand to 1.000 dm^3 and is simultaneously heated to 100°C.

Method Because S is a state function, we are free to choose the most convenient path from the initial state. One such path is reversible isothermal expansion to the final volume, followed by reversible heating at constant volume to the final temperature. The entropy change in the first step is given by eqn 15.24 and that of the second step, provided C_V is independent of temperature, by eqn 15.30 (with C_V in place of C_p). In each case we need to know n, the amount of gas molecules, and can calculate it from the perfect gas equation and the data for the initial state from $n = p_iV_i/RT_i$. The heat capacity at constant volume is given by the equipartition theorem as $\frac{3}{2}R$. (The equipartition theorem is reliable for monatomic gases: for others and in general use experimental data like that in Tables 14.5 and 14.6. If necessary, convert $C_{p,m}$ to $C_{V,m}$ by using the relation $C_{p,m} - C_{V,m} = R$.)

Answer Because $n = p_iV_i/RT_i$, from eqn 15.24

$$\Delta S(\text{Step 1}) = \left(\frac{p_iV_i}{RT_i}\right) \times R \ln\frac{V_f}{V_i} = \frac{p_iV_i}{T_i}\ln\frac{V_f}{V_i}$$

The entropy change in the second step, from 298 K to 373 K at constant volume, from eqn 15.30 with C_V replacing C_p, is

$$\Delta S(\text{Step 2}) = \left(\frac{p_iV_i}{RT_i}\right) \times \tfrac{3}{2} R \ln\frac{T_f}{T_i} = \frac{p_iV_i}{T_i}\ln\left(\frac{T_f}{T_i}\right)^{3/2}$$

The overall entropy change, the sum of these two changes, is

$$\Delta S = \frac{p_iV_i}{T_i}\ln\frac{V_f}{V_i} + \frac{p_iV_i}{T_i}\ln\left(\frac{T_f}{T_i}\right)^{3/2} = \frac{p_iV_i}{T_i}\ln\left\{\frac{V_f}{V_i}\left(\frac{T_f}{T_i}\right)^{3/2}\right\}$$

At this point we substitute the data and obtain (by using 1 Pa m^3 = 1 J)

$$\Delta S = \frac{(1.00\times10^5\ \text{Pa})\times(0.500\times10^{-3}\ \text{m}^3)}{298\ \text{K}}\ln\left\{\frac{1.000}{0.500}\left(\frac{373}{298}\right)^{3/2}\right\}$$

$$= +0.173\ \text{J K}^{-1}$$

A note on good practice It is sensible to proceed as generally as possible before inserting numerical data so that, if required, the formula can be used for other data; the procedure also minimizes rounding errors.

Self-test 15.6 Calculate the entropy change when the same initial sample is compressed to 50.0 cm^3 and cooled to −25°C. [−0.433 J K^{-1}]

(d) The calorimetric measurement of entropy

We have seen how to calculate the entropy of a substance from spectroscopic data. Now we see how it may be measured calorimetrically. The entropy of a system at a temperature T is related to its entropy at $T = 0$ by measuring its heat capacity C_p at different temperatures and evaluating the integral in eqn 15.29, taking care to add the entropy of transition ($\Delta_{trs}H/T_{trs}$) for each phase transition between $T = 0$ and the temperature of interest. For example, if a substance melts at T_f and boils at T_b, then its entropy above its boiling temperature is given by

$$S(T) = S(0) + \int_0^{T_f} \frac{C_p(s)\,dT}{T} + \frac{\Delta_{fus}H}{T_f} + \int_{T_f}^{T_b} \frac{C_p(l)\,dT}{T} + \frac{\Delta_{vap}H}{T_b} + \int_{T_b}^{T} \frac{C_p(g)\,dT}{T} \qquad (15.31)$$

where we have indicated the phases for the heat capacities. All the properties required, except $S(0)$, can be measured calorimetrically, and the integrals can be evaluated either graphically or, as is now more usual, by fitting a polynomial to the data and integrating the polynomial analytically or numerically. The former procedure is illustrated in Fig. 15.16: the area under the curve of C_p/T against T is the integral required. Because $dT/T = d\ln T$, an alternative procedure is to evaluate the area under a plot of C_p against $\ln T$.

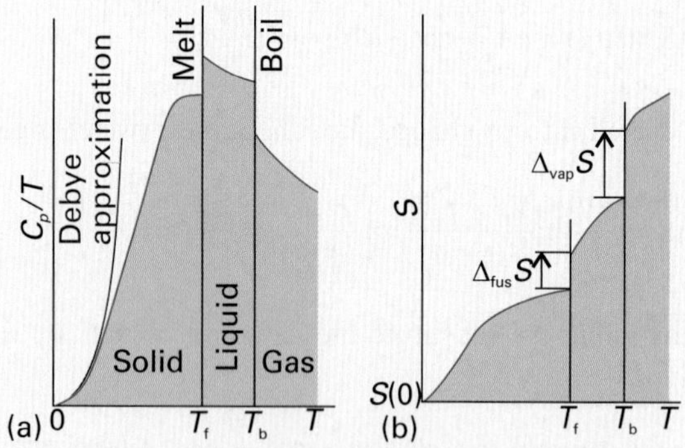

Fig. 15.16 The calculation of the entropy from heat capacity data. (a) The variation of C_p/T with the temperature for a sample. (b) The entropy, which is equal to the area beneath the curve in (a) up to the corresponding temperature, plus the entropy of each phase transition passed.

interActivity Allow for the temperature dependence of the heat capacity by writing $C = a + bT + c/T^2$, and plot the change in entropy for different values of the three coefficients (including negative values of c).

Example 15.3 *Evaluating a change in entropy*

Over a small range in temperature the molar constant-pressure heat capacity of carbon dioxide gas varies with temperature as $C_{p,m} = a + bT + c/T^2$ with the parameters $a = 44.22$ J K^{-1} mol^{-1}, $b = 8.79$ mJ K^{-2} mol^{-1}, and $c = 862$ kJ K mol^{-1}. What is the change in molar entropy when carbon dioxide is heated from 0°C to 100°C?

Method The general expression for the change in entropy with temperature is eqn 15.29, so we begin by substituting the polynomial expression for $C_{p,m}$ into that equation and simplifying the resulting expression as far as possible. Then there are two routes forward. One is to evaluate the resulting integrals analytically. The other is to feed the integral into mathematical software, and let it do the hard work. We demonstrate the latter approach here. For the numerical evaluation, take $T_i = 273$ K (0°C) and $T_f = 373$ K (100°C).

Solution We substitute $C_{p,m} = a + bT + c/T^2$ into eqn 15.29 and obtain

$$\Delta S_m = \int_{T_i}^{T_f} \frac{C_{p,m}\,dT}{T} = \int_{T_i}^{T_f} \left(a + bT + \frac{c}{T^2}\right)\frac{dT}{T} = \int_{T_i}^{T_f} \left(\frac{a}{T} + b + \frac{c}{T^3}\right)dT$$

Insertion of the values of the coefficients and the initial and final temperatures and the use of mathematical software to evaluate the integral numerically gives $\Delta S_m = +11.99$ J K^{-1} mol^{-1}.

Self-test 15.7 Evaluate the change in molar entropy analytically for the same data. [+11.99 J K^{-1} mol^{-1}]

One problem with the determination of entropy is the difficulty of measuring heat capacities near $T = 0$. There are good theoretical grounds for assuming that the heat capacity is proportional to T^3 when T is low (see Self-test 14.6), and this dependence is the basis of the **Debye extrapolation**. In this method, C_p is measured down to as low a temperature as possible, and a curve of the form aT^3 is fitted to the data. That fit determines the value of a, and the expression $C_p = aT^3$ is assumed valid down to $T = 0$.

Example 15.4 *Calculating the entropy at low temperatures*

The molar constant-pressure heat capacity of a certain solid at 4.2 K is 0.43 J K^{-1} mol^{-1}. What is its molar entropy at that temperature?

Method Because the temperature is so low, we can assume that the heat capacity varies with temperature as aT^3, in which case we can use eqn 15.29 to calculate the entropy at a temperature T in terms of the entropy at $T = 0$ and the constant a. When the integration is carried out, it turns out that the result can be expressed in terms of the heat capacity at the temperature T, so the data can be used directly to calculate the entropy.

Answer The integration required is

$$S(T) = S(0) + \int_0^T \frac{aT^3\,\mathrm{d}T}{T} = S(0) + a\int_0^T T^2\,\mathrm{d}T = S(0) + \tfrac{1}{3}aT^3$$

However, because aT^3 is the heat capacity at the temperature T,

$$S(T) = S(0) + \tfrac{1}{3}C_p(T)$$

from which it follows that

$$S_\mathrm{m}(4.2\ \mathrm{K}) = S_\mathrm{m}(0) + 0.14\ \mathrm{J\ K^{-1}\ mol^{-1}}$$

Self-test 15.8 For metals, there is also a contribution to the heat capacity from the electrons, which is linearly proportional to T when the temperature is low. Find its contribution to the entropy at low temperatures.

[$S(T) = S(0) + C_p(T)$]

15.4 The Third Law of thermodynamics

At $T = 0$, all thermal motion has been quenched, and in a perfect crystal all the atoms or ions are in a regular, uniform array. The localization of matter and the absence of thermal motion suggest that such materials also have zero entropy. This conclusion is consistent with the statistical definition of entropy, because $S = 0$ if there is only one way of arranging the molecules and only one state is accessible (the ground state) and $\mathcal{W} = 1$. However, because the argument that we used to relate the statistical definition to the thermodynamic definition was in terms of *changes* of entropy, there remains the possibility that the definition in eqn 15.9 differs from the statistical entropy by a constant that might be different for each substance. That it does not is the domain of the Third Law and an experimental observation known as the 'Nernst heat theorem'.

(a) The Nernst heat theorem

The experimental observation that turns out to be consistent with the view that the entropy of a regular array of molecules is zero at $T = 0$ is summarized by the **Nernst heat theorem**:

> The entropy change accompanying any physical or chemical transformation approaches zero as the temperature approaches zero: $\Delta S \to 0$ as $T \to 0$ provided all the substances involved are perfectly ordered.

● A BRIEF ILLUSTRATION

Consider the entropy of the transition between orthorhombic sulfur, S(α), and monoclinic sulfur, S(β), which can be calculated from the transition enthalpy (−402 J mol^{-1}) at the transition temperature (369 K):

$$\Delta_\mathrm{trs}S = S_\mathrm{m}(\alpha) - S_\mathrm{m}(\beta) = \frac{(-402\ \mathrm{J\ mol^{-1}})}{369\ \mathrm{K}} = -1.09\ \mathrm{J\ K^{-1}\ mol^{-1}}$$

The two individual entropies can also be determined by measuring the heat capacities from $T = 0$ up to $T = 369$ K. It is found that $S_\mathrm{m}(\alpha) = S_\mathrm{m}(\alpha,0) + 37$ J K^{-1} mol^{-1} and $S_\mathrm{m}(\beta) = S_\mathrm{m}(\beta,0) + 38$ J K^{-1} mol^{-1}. These two values imply that at the transition temperature

$$\Delta_\mathrm{trs}S = S_\mathrm{m}(\alpha,0) - S_\mathrm{m}(\beta,0) - 1\ \mathrm{J\ K^{-1}\ mol^{-1}}$$

On comparing this value with the one above, we conclude that $S_\mathrm{m}(\alpha,0) - S_\mathrm{m}(\beta,0) \approx 0$, in accord with the theorem. ●

It follows from the Nernst theorem that, if we arbitrarily ascribe the value zero to the entropies of elements in their perfect crystalline form at $T = 0$, then all perfect crystalline compounds also have zero entropy at $T = 0$ (because the change in entropy that accompanies the formation of the compounds, like the entropy of all transformations at that temperature, is zero). This conclusion is summarized by the **Third Law of thermodynamics**:

> The entropy of all perfect crystalline substances is zero at $T = 0$.

As far as thermodynamics is concerned, choosing this common value as zero is then a matter of convenience. The molecular interpretation of entropy, however, justifies the value $S = 0$ at $T = 0$.

In most cases, $\mathcal{W} = 1$ at $T = 0$ because there is only one way of achieving the lowest total energy: put all the molecules into the same, lowest state. Therefore, $S = 0$ at $T = 0$, in accord with the Third Law of thermodynamics. In certain cases, though, $\mathcal{W}$ may differ from 1 at $T = 0$. This is the case if there is no energy advantage in adopting a particular orientation even at absolute zero. For instance, for a diatomic molecule AB there may be almost no energy difference between the arrangements . . . AB AB AB . . . and . . . BA AB BA . . . , so $\mathcal{W} > 1$ even at $T = 0$. If $S > 0$ at

$T = 0$ we say that the substance has a **residual entropy**. Ice has a residual entropy of 3.4 J K^{-1} mol^{-1}. It stems from the arrangement of the hydrogen bonds between neighbouring water molecules: a given O atom has two short O—H bonds and two long O···H bonds to its neighbours, but there is a degree of randomness as to which two bonds are short and which two are long.

(b) Third Law entropies

Entropies reported on the basis that $S(0) = 0$ are called **Third Law entropies** (and often just 'entropies'). When the substance is in its standard state at the temperature T, the **standard (Third Law) entropy** is denoted $S^{\ominus}(T)$. A list of molar values at 298 K is given in Table 15.3.

The **standard reaction entropy**, $\Delta_r S^{\ominus}$, is defined, like the standard reaction enthalpy, as the difference between the molar entropies of the pure, separated products and the pure, separated reactants, all substances being in their standard states at the specified temperature:

$$\Delta_r S^{\ominus} = \sum_J \nu_J S_m^{\ominus}(J) \qquad (15.32)$$

where we have used the notation introduced in Section 14.7. Standard reaction entropies are likely to be positive if there is a net formation of gas in a reaction, and are likely to be negative if there is a net consumption of gas.

Synoptic table 15.3* Standard Third Law entropies at 298 K

	$S_m^{\ominus}$/(J K^{-1} mol^{-1})
Solids:	
Graphite, C(s)	5.7
Diamond, C(s)	2.4
Sucrose, $C_{12}H_{22}O_{11}$(s)	360.2
Iodine, I_2(s)	116.1
Liquids:	
Benzene, C_6H_6(l)	173.3
Water, H_2O(l)	69.9
Mercury, Hg(l)	76.0
Gases:	
Methane, CH_4(g)	186.3
Carbon dioxide, CO_2(g)	213.7
Hydrogen, H_2(g)	130.7
Helium, He	126.2
Ammonia, NH_3(g)	192.4

* More values are given in the *Data section*.

● A BRIEF ILLUSTRATION

To calculate the standard reaction entropy of $H_2(g) + \frac{1}{2}O_2(g) \rightarrow H_2O(l)$ at 25°C, we use the data in Table 14.6 of the *Data section* to write

$$\begin{aligned}\Delta_r S^{\ominus} &= S_m^{\ominus}(H_2O, l) - S_m^{\ominus}(H_2, g) - \tfrac{1}{2}S_m^{\ominus}(O_2, g)\\ &= 69.9\ \text{J K}^{-1}\ \text{mol}^{-1} - 130.7\ \text{J K}^{-1}\ \text{mol}^{-1}\\ &\quad - \tfrac{1}{2}(205.1)\ \text{J K}^{-1}\ \text{mol}^{-1}\\ &= -163.4\ \text{J K}^{-1}\ \text{mol}^{-1}\end{aligned}$$

The negative value is consistent with the conversion of two gases to a compact liquid. ●

A note on good practice Do not make the mistake of setting the standard molar entropies of elements equal to zero: they have nonzero values (provided $T > 0$), as we have already discussed.

Self-test 15.9 Calculate the standard reaction entropy for the combustion of methane to carbon dioxide and liquid water at 25°C. [−243 J K^{-1} mol^{-1}]

Just as in the discussion of enthalpies in Section 14.8, where we acknowledged that solutions of cations cannot be prepared in the absence of anions, the standard molar entropies of ions in solution are reported on a scale in which the standard entropy of the H^+ ions in water is taken as zero at all temperatures:

$$S^{\ominus}(H^+, aq) = 0 \qquad [15.33]$$

The values based on this choice are included in Table 14.6 in the *Data section*. Because the entropies of ions in water are values relative to the hydrogen ion in water, they may be either positive or negative. A positive entropy means that an ion has a higher molar entropy than H^+ in water and a negative entropy means that the ion has a lower molar entropy than H^+ in water. For instance, the standard molar entropy of Cl^-(aq) is +57 J K^{-1} mol^{-1} and that of Mg^{2+}(aq) is −138 J K^{-1} mol^{-1}. In terms of the language of 'partial molar' quantities to be introduced in Chapter 16, the entropies of ions in solution are actually *partial molar entropies*, for their values include the consequences of their presence on the organization of the solvent molecules around them. Thus, the entropies vary as expected on the basis that they are related to the degree to which the ions order the water molecules around them in the solution. Small, highly charged ions induce local structure in the surrounding water, and the disorder of the solution is decreased more than in the case of large, singly charged ions. The absolute, Third Law standard molar entropy of the proton in water can be estimated by proposing a model of the structure it induces, and there is some agreement on the value −21 J K^{-1} mol^{-1}. The negative value indicates that the proton induces order in the solvent.

Concentrating on the system

Entropy is the basic concept for discussing the direction of natural change, but to use it we have to analyse changes in both the system and its surroundings. We have seen that it is always very simple to calculate the entropy change in the surroundings, and we shall now see that it is possible to devise a simple method for taking that contribution into account automatically. This approach focuses our attention on the system and simplifies discussions. Moreover, it is the foundation of all the applications of chemical thermodynamics that follow.

15.5 The Helmholtz and Gibbs energies

Consider a system in thermal equilibrium with its surroundings at a temperature T. When a change in the system occurs and there is a transfer of energy as heat between the system and the surroundings, the Clausius inequality (eqn 15.19, $\mathrm{d}S \geq \mathrm{d}q/T$) reads

$$\mathrm{d}S - \frac{\mathrm{d}q}{T} \geq 0 \tag{15.34}$$

We can develop this inequality in two ways according to the conditions (of constant volume or constant pressure) under which the process occurs.

(a) Criteria for spontaneity

First, consider heating at constant volume. Then, in the absence of non-expansion work, we can write $\mathrm{d}q_V = \mathrm{d}U$; consequently

$$\mathrm{d}S - \frac{\mathrm{d}U}{T} \geq 0 \tag{15.35a}$$

The importance of the inequality in this form is that it expresses the criterion for spontaneous change solely in terms of the state functions of the system. The inequality is easily rearranged to

$$T\mathrm{d}S \geq \mathrm{d}U \qquad \text{(constant } V\text{, no additional work)} \tag{15.35b}$$

(Recall that 'additional work' is work other than expansion work.) At either constant internal energy ($\mathrm{d}U = 0$) or constant entropy ($\mathrm{d}S = 0$), this expression becomes, respectively,

$$\mathrm{d}S_{U,V} \geq 0 \qquad \mathrm{d}U_{S,V} \leq 0 \tag{15.35c}$$

where the subscripts indicate the constant conditions.

Equation 15.35(c) expresses the criteria for spontaneous change in terms of properties relating to the system. The first inequality states that, in a system at constant volume and constant internal energy (such as an isolated system), the entropy increases in a spontaneous change. That statement is essentially the content of the Second Law and corresponds to the tendency of an isolated system to collapse into its most probable distribution and never into a less probable one. The second inequality is less obvious, for it says that, if the entropy and volume of the system are constant, then the internal energy must decrease in a spontaneous change. Do not interpret this criterion as a tendency of the system to sink to lower energy. It is a disguised statement about entropy, and should be interpreted as implying that, if the entropy of the system is unchanged, then there must be an increase in entropy of the surroundings, which can be achieved only if the energy of the system decreases as energy flows out as heat.

When energy is transferred as heat at constant pressure, and there is no work other than expansion work, we can write $\mathrm{d}q_p = \mathrm{d}H$ and obtain

$$T\mathrm{d}S \geq \mathrm{d}H \qquad \text{(constant } p\text{, no additional work)} \tag{15.36a}$$

At either constant enthalpy or constant entropy this inequality becomes, respectively,

$$\mathrm{d}S_{H,p} \geq 0 \qquad \mathrm{d}H_{S,p} \leq 0 \tag{15.36b}$$

The interpretations of these inequalities are similar to those of eqn 15.35(c). The entropy of the system at constant pressure must increase if its enthalpy remains constant (for there can then be no change in entropy of the surroundings). Alternatively, the enthalpy must decrease if the entropy of the system is constant, for then it is essential to have an increase in entropy of the surroundings.

Because eqns 15.35b and 15.36a have the forms $\mathrm{d}U - T\mathrm{d}S \leq 0$ and $\mathrm{d}H - T\mathrm{d}S \leq 0$, respectively, they can be expressed more simply by introducing two more thermodynamic quantities. One is the **Helmholtz energy**, A, which is defined as

$$A = U - TS \qquad [15.37]$$

This relation implies that $A(0) = U(0)$, so substitution for U and S by using eqn 15.4 ($S = \{U - U(0)\}/T + k \ln Q$) leads to the very simple expression

$$A - A(0) = -kT \ln Q \tag{15.38a}$$

For a system composed of independent, indistinguishable molecules (as in a perfect gas), we may replace Q by $q^N/N!$ and obtain

$$A - A(0) = -kT \ln \frac{q^N}{N!} = -NkT \ln q + kT \ln N!$$

Then, by using Stirling's approximation and expressing NkT as $NkT \ln \mathrm{e}$,

$$\begin{aligned} A - A(0) &= -NkT \ln q + kT(N \ln N - N) \\ &= -NkT \ln q + NkT \ln N - NkT \ln \mathrm{e} \\ &= -NkT \ln \frac{\mathrm{e}q}{N} \end{aligned} \tag{15.38b}$$

We see that $A - A(0)$ is essentially proportional to the logarithm of the molecular partition function, and therefore that it is (negatively) large when many energy levels are thermally accessible.

The other function we introduce is the **Gibbs energy**, G:

$$G = H - TS \qquad [15.39]$$

Because $H = U + pV$ and $A = U - TS$, we can write $G = A + pV$ and from eqn 15.38a the statistical thermodynamic expression for the Gibbs energy is

$$G - G(0) = -kT \ln Q + pV$$

For a gas of independent particles, pV can be replaced by nRT and Q by $q^N/N!$ It follows that

$$\begin{aligned} G - G(0) &= -NkT \ln q + kT \ln N! + nRT \\ &= -nRT \ln q + kT(N \ln N - N) + nRT \\ &= -nRT \ln \frac{q}{N} \end{aligned} \qquad (15.40)$$

where we have used Stirling's approximation to write $\ln N!$ as $N \ln N - N$ and noted that $-NkT$ cancels nRT because $Nk = nN_A k = nR$. All the symbols in these definitions refer to the system.

As we shall see, it is easier to give precise molecular interpretations of the Helmholtz energy than of the Gibbs energy, just as it is easier to give precise molecular interpretations of the internal energy than the enthalpy. However, it will also turn out that the Gibbs energy is more important than the Helmholtz energy in chemistry, just as enthalpy is more important than the internal energy. Fortunately, because (like H and U) G differs from A by the addition of the term pV, we may use—with caution in some cases—the molecular interpretation of A to interpret most of the properties of G.

When the state of the system changes at constant temperature, the two properties change as follows:

$$\text{(a) } dA = dU - TdS \qquad \text{(b) } dG = dH - TdS \qquad (15.41)$$

When we introduce eqns 15.35b and 15.36a, respectively, we obtain the criteria of spontaneous change as

$$\text{(a) } dA_{T,V} \le 0 \qquad \text{(b) } dG_{T,p} \le 0 \qquad (15.42)$$

These inequalities are the most important conclusions from thermodynamics for chemistry. They are developed in subsequent sections and chapters.

(b) Some remarks on the Helmholtz energy

According to eqn 15.42, a change in a system at constant temperature and volume is spontaneous if $dA_{T,V} < 0$. That is, a change under these conditions is spontaneous if it corresponds to a decrease in the Helmholtz energy. Such systems move spontaneously towards states of lower A if a path is available. The criterion of equilibrium, when neither the forward nor reverse process has a tendency to occur, is

$$dA_{T,V} = 0 \qquad (15.43)$$

The expressions $dA = dU - TdS$ and $dA < 0$ are sometimes interpreted as follows. A negative value of dA is favoured by a negative value of dU and a positive value of TdS. This observation suggests that the tendency of a system to move to lower A is due to its tendency to move towards states of lower internal energy and higher entropy. However, this interpretation is false (even though it is a good rule of thumb for remembering the expression for dA) because the tendency to lower A is solely a tendency towards states of greater overall entropy. Systems change spontaneously if in doing so the total entropy of the system and its surroundings increases, not because they tend to lower internal energy. The form of dA may give the impression that systems favour lower energy, but that is misleading: dS is the entropy change of the system, $-dU/T$ is the entropy change of the surroundings (when the volume of the system is constant), and their total tends to a maximum.

(c) Maximum work

It turns out that A carries a greater significance than being simply a signpost of spontaneous change: *the change in the Helmholtz function is equal to the maximum work accompanying a process*:

$$dw_{max} = dA \qquad (15.44)$$

As a result, A is sometimes called the 'maximum work function', or the 'work function'. (*Arbeit* is the German word for work; hence the symbol A.)

Justification 15.4 *Maximum work*

To demonstrate that maximum work can be expressed in terms of the changes in Helmholtz energy, we combine the Clausius inequality $dS \ge dq/T$ in the form $TdS \ge dq$ with the First Law, $dU = dq + dw$, and obtain

$$dU \le TdS + dw$$

(dU is smaller than the term of the right because we are replacing dq by TdS, which in general is larger.) This expression rearranges to $dU - TdS \le dw$ and therefore to

$$dw \ge dU - TdS$$

Now recall that a large negative w means that a lot of energy has been transferred *from* the system as work—the system has done a lot of work. It follows that the most negative value of dw, and therefore the maximum energy that can be obtained from the system as work, must correspond to the equals sign in this expression because a higher (less negative) value of dw implies that less work has been done. Therefore,

$$dw_{max} = dU - TdS$$

This maximum work is done only when the path is traversed reversibly (because then the equality applies). Because at constant temperature $dA = dU - TdS$, we conclude that $dw_{max} = dA$.

The relation between A and maximum work can be understood by noting that, in terms of the molecular partition function and eqn 15.38b ($A = A(0) - NkT\ln(eq/N)$),

$$\Delta A = -NkT\ln\frac{eq_f}{N} + NkT\ln\frac{eq_i}{N} = -NkT\ln\frac{q_f}{q_i} \qquad (15.45)$$

However, because q for translational motion is proportional to V (recall that $q = V/\Lambda^3$) we immediately find that

$$w_{max} = -NkT\ln\frac{V_f}{V_i} \qquad (15.46)$$

which is the same expression as we found for the reversible, isothermal expansion of a perfect gas. We can now see that reversible expansion produces maximum work because it corresponds to the progressive change of the distribution of molecules through a sequence of equilibrium states, all corresponding to the same temperature.

When a macroscopic isothermal change takes place in the system, eqn 15.44 becomes

$$w_{max} = \Delta A \qquad (15.47)$$

with

$$\Delta A = \Delta U - T\Delta S \qquad (15.48)$$

This expression shows that in some cases, depending on the sign of $T\Delta S$, not all the change in internal energy may be available for doing work. If the change occurs with a decrease in entropy (of the system), in which case $T\Delta S < 0$, then the right-hand side of this equation is not as negative as ΔU itself, and consequently the maximum work is less than ΔU. For the change to be spontaneous, some of the energy must escape as heat in order to generate enough entropy in the surroundings to overcome the reduction in entropy in the system (Fig. 15.17). In this case, Nature is demanding a tax on the internal energy as it is converted into work. This is the origin of the alternative name 'Helmholtz free energy' for A, because ΔA is that part of the change in internal energy that we are free to use to do work.

Further insight into the relation between the work that a system can do and the Helmholtz energy is to recall that work is energy transferred to the surroundings as the uniform motion of atoms. We can interpret the expression $A = U - TS$ as showing that A is the total internal energy of the system, U, less a contribution that is stored as energy of thermal motion (the quantity TS). Because energy stored in random thermal motion cannot be used to achieve uniform motion in the surroundings, only the part of U that is not stored in that way, the quantity $U - TS$, is available for conversion into work.

If the change occurs with an increase of entropy of the system (in which case $T\Delta S > 0$), the right-hand side of the equation is more negative than ΔU. In this case, the maximum work that can be obtained from the system is greater than ΔU. The explanation of this apparent paradox is that the system is not isolated and energy may flow in as heat as work is done. Because the entropy of the system increases, we can afford a reduction of the entropy of the surroundings yet still have, overall, a spontaneous process. Therefore, some energy (no more than the value of $T\Delta S$) may leave the surroundings as heat and contribute to the work the change is generating (Fig. 15.18). Nature is now providing a tax refund.

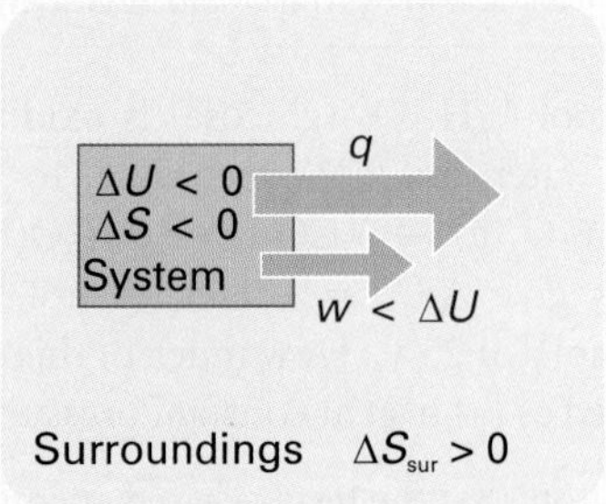

Fig. 15.17 In a system not isolated from its surroundings, the work done may be different from the change in internal energy. Moreover, the process is spontaneous if overall the entropy of the global, isolated system increases. In the process depicted here, the entropy of the system decreases, so that of the surroundings must increase in order for the process to be spontaneous, which means that energy must pass from the system to the surroundings as heat. Therefore, less work than ΔU can be obtained.

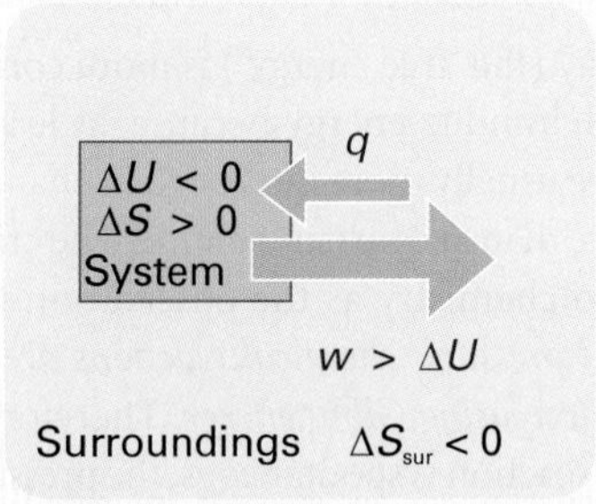

Fig. 15.18 In this process, the entropy of the system increases; hence we can afford to lose some entropy of the surroundings. That is, some of their energy may be lost as heat to the system. This energy can be returned to them as work. Hence the work done can exceed ΔU.

Example 15.5 *Calculating the maximum available work*

When 1.000 mol $C_6H_{12}O_6$ (glucose) is oxidized to carbon dioxide and water at 25°C according to the equation $C_6H_{12}O_6(s) + 6\,O_2(g) \rightarrow 6\,CO_2(g) + 6\,H_2O(l)$ calorimetric measurements give $\Delta_r U^\ominus = -2808$ kJ mol^{-1} and $\Delta_r S^\ominus = +259.1$ J K^{-1} mol^{-1} at 25°C. How much of this energy change can be extracted as (a) heat at constant pressure, (b) work?

Method We know that the heat released at constant pressure is equal to the value of ΔH, so we need to relate $\Delta_r H^\ominus$ to $\Delta_r U^\ominus$, which is given. To do so, we suppose that all the gases involved are perfect, and use eqn 14.31 ($\Delta H = \Delta U + \Delta n_g RT$). For the maximum work available from the process we use eqn 15.47.

Answer (a) Because $\Delta n_g = 0$, we know that $\Delta_r H^\ominus = \Delta_r U^\ominus = -2808$ kJ mol^{-1}. Therefore, at constant pressure, the energy available as heat is 2808 kJ mol^{-1}. (b) Because $T = 298$ K, the value of $\Delta_r A^\ominus$ is

$$\Delta_r A^\ominus = \Delta_r U^\ominus - T\Delta_r S^\ominus = -2885 \text{ kJ mol}^{-1}$$

Therefore, the combustion of 1.000 mol $C_6H_{12}O_6$ can be used to produce up to 2885 kJ of work. The maximum work available is greater than the change in internal energy on account of the positive entropy of reaction (which is partly due to the generation of a large number of small molecules from one big one). The system can therefore draw in energy from the surroundings (so reducing their entropy) and make it available for doing work.

Self-test 15.10 Repeat the calculation for the combustion of 1.000 mol $CH_4(g)$ under the same conditions, using data from Tables 14.5 and 14.6. [$|q_p| = 890$ kJ, $|w_{max}| = 813$ kJ]

(d) Some remarks on the Gibbs energy

The Gibbs energy (the 'free energy') is more common in chemistry than the Helmholtz energy because, at least in laboratory chemistry, we are usually more interested in changes occurring at constant pressure than at constant volume. The criterion $dG_{T,p} < 0$ carries over into chemistry as the observation that *at constant temperature and pressure, chemical reactions are spontaneous in the direction of decreasing Gibbs energy*. Therefore, if we want to know whether a reaction is spontaneous, the pressure and temperature being constant, we assess the change in the Gibbs energy. If G decreases as the reaction proceeds, then the reaction has a spontaneous tendency to convert the reactants into products. If G increases, then the reverse reaction is spontaneous.

The existence of spontaneous endothermic reactions provides an illustration of the role of G. In such reactions, H increases, the system rises spontaneously to states of higher enthalpy, and $dH > 0$. Because the reaction is spontaneous we know that $dG < 0$ despite $dH > 0$; it follows that the entropy of the system increases so much that TdS outweighs dH in $dG = dH - TdS$. Endothermic reactions are therefore driven by the increase of entropy of the system, and this entropy change overcomes the reduction of entropy brought about in the surroundings by the inflow of heat into the system ($dS_{sur} = -dH/T$ at constant pressure).

(e) Maximum non-expansion work

The analogue of the maximum work interpretation of ΔA, and the origin of the name 'free energy', can be found for ΔG. In the following *Justification* we show that, at constant temperature and pressure, the maximum additional (non-expansion) work, $w_{add,max}$, is given by the change in Gibbs energy:

$$dw_{add,max} = dG \qquad (15.49a)$$

The corresponding expression for a measurable change is

$$w_{add,max} = \Delta G \qquad (15.49b)$$

This expression is particularly useful for assessing the electrical work that may be produced by fuel cells and electrochemical cells, and we shall see many applications of it.

Justification 15.5 *Maximum non-expansion work*

Because $H = U + pV$, for a general change in conditions, the change in enthalpy is

$$dH = dq + dw + d(pV)$$

The corresponding change in Gibbs energy ($G = H - TS$) is

$$dG = dH - TdS - SdT = dq + dw + d(pV) - TdS - SdT$$

When the change is isothermal we can set $dT = 0$; then

$$dG = dq + dw + d(pV) - TdS$$

When the change is reversible, $dw = dw_{rev}$ and $dq = dq_{rev} = TdS$, so for a reversible, isothermal process

$$dG = TdS + dw_{rev} + d(pV) - TdS = dw_{rev} + d(pV)$$

The work consists of expansion work, which for a reversible change is given by $-pdV$, and possibly some other kind of work (for instance, the electrical work of pushing electrons through a circuit or of raising a column of liquid); this additional work we denote dw_{add}. Therefore, with $d(pV) = pdV + Vdp$,

$$dG = (-pdV + dw_{add,rev}) + pdV + Vdp = dw_{add,rev} + Vdp$$

If the change occurs at constant pressure (as well as constant temperature), we can set $dp = 0$ and obtain $dG = dw_{add,rev}$. Therefore, at constant temperature and pressure, $dw_{add,rev} = dG$. However, because the process is reversible, the work done must now have its maximum value, so eqn 15.49 follows.

Example 15.6 *Calculating the maximum non-expansion work of a reaction*

How much energy is available for sustaining muscular and nervous activity from the combustion of 1.00 mol of glucose molecules under standard conditions at 37°C (blood temperature)? The standard entropy of reaction is +295.1 J K^{-1} mol^{-1}.

Method The non-expansion work available from the reaction is equal to the change in standard Gibbs energy for the reaction ($\Delta_r G^\ominus$, a quantity defined more fully below). To calculate this quantity, it is legitimate to ignore the temperature-dependence of the reaction enthalpy, to obtain $\Delta_r H^\ominus$ from Tables 14.5 and 14.6, and to substitute the data into $\Delta_r G^\ominus = \Delta_r H^\ominus - T\Delta_r S^\ominus$.

Answer Because the standard reaction enthalpy is −2808 kJ mol^{-1}, it follows that the standard reaction Gibbs energy is

$$\Delta_r G^\ominus = -2808\ \text{kJ mol}^{-1} - (310\ \text{K}) \times (259.1\ \text{J K}^{-1}\ \text{mol}^{-1})$$
$$= -2888\ \text{kJ mol}^{-1}$$

Therefore, $w_{\text{add,max}} = -2888$ kJ for the combustion of 1 mol glucose molecules, and the reaction can be used to do up to 2888 kJ of non-expansion work. To place this result in perspective, consider that a person of mass 70 kg needs to do 2.1 kJ of work to climb vertically through 3.0 m; therefore, at least 0.13 g of glucose is needed to complete the task (and in practice significantly more).

Self-test 15.11 How much non-expansion work can be obtained from the combustion of 1.00 mol CH_4(g) under standard conditions at 298 K? Use $\Delta_r S^\ominus = -243$ J K^{-1} mol^{-1}.
[818 kJ]

15.6 Standard molar Gibbs energies

Standard entropies and enthalpies of reaction can be combined to obtain the **standard Gibbs energy of reaction** (or 'standard reaction Gibbs energy'), $\Delta_r G^\ominus$:

$$\Delta_r G^\ominus = \Delta_r H^\ominus - T\Delta_r S^\ominus \qquad [15.50]$$

The standard Gibbs energy of reaction is the difference in standard molar Gibbs energies of the products and reactants in their standard states at the temperature specified for the reaction as written. As in the case of standard reaction enthalpies, it is convenient to define the **standard Gibbs energies of formation**, $\Delta_f G^\ominus$, the standard reaction Gibbs energy for the formation of a compound from its elements in their reference states (Section 14.8). Standard Gibbs energies of formation of the elements in their reference states are zero, because their formation is a 'null' reaction. A selection of values for compounds is given in Table 15.4. From the values there, it is a simple matter to obtain the standard Gibbs energy of reaction by taking the appropriate combination:

Synoptic table 15.4* Standard Gibbs energies of formation (at 298 K)

	$\Delta_f G^\ominus$/(kJ mol^{-1})
Diamond, C(s)	+2.9
Benzene, C_6H_6(l)	+124.3
Methane, CH_4(g)	−50.7
Carbon dioxide, CO_2(g)	−394.4
Water, H_2O(l)	−237.1
Ammonia, NH_3(g)	−16.5
Sodium chloride, NaCl(s)	−384.1

* More values are given in the *Data section*.

$$\Delta_r G^\ominus = \sum_J \nu_J \Delta_f G^\ominus(J) \qquad (15.51)$$

● A BRIEF ILLUSTRATION

To calculate the standard Gibbs energy of the reaction CO(g) $+\frac{1}{2}O_2(g) \rightarrow CO_2(g)$ at 25°C, we write

$$\Delta_r G^\ominus = \Delta_f G^\ominus(CO_2, g) - \Delta_f G^\ominus(CO, g) - \tfrac{1}{2}\Delta_f G^\ominus(O_2, g)$$
$$= -394.4\ \text{kJ mol}^{-1} - (-137.2)\ \text{kJ mol}^{-1} - \tfrac{1}{2}(0)$$
$$= -257.2\ \text{kJ mol}^{-1}$$ ●

Self-test 15.12 Calculate the standard reaction Gibbs energy for the combustion of CH_4(g) at 298 K.
[−818 kJ mol^{-1}]

Just as we did in Section 14.8, where we acknowledged that solutions of cations cannot be prepared without their accompanying anions, we define one ion, conventionally the hydrogen ion, to have zero standard Gibbs energy of formation at all temperatures:

$$\Delta_f G^\ominus(H^+, aq) = 0 \qquad [15.52]$$

In essence, this definition adjusts the actual values of the Gibbs energies of formation of ions by a fixed amount, which is chosen so that the standard value for one of them, H^+(aq), has the value zero. Then for the reaction

$$\tfrac{1}{2}H_2(g) + \tfrac{1}{2}Cl_2(g) \rightarrow H^+(aq) + Cl^-(aq)$$

for which $\Delta_r G^\ominus = -131.23$ kJ mol^{-1} we can write

$$\Delta_r G^\ominus = \Delta_f G^\ominus(\mathrm{H^+, aq}) + \Delta_f G^\ominus(\mathrm{Cl^-, aq}) = \Delta_f G^\ominus(\mathrm{Cl^-, aq})$$

and hence identify $\Delta_f G^\ominus(\mathrm{Cl^-, aq})$ as -131.23 kJ mol^{-1}. All the Gibbs energies of formation of ions tabulated in the *Data section* were calculated in the same way.

● A BRIEF ILLUSTRATION

With the value of $\Delta_f G^\ominus(\mathrm{Cl^-, aq})$ established, we can find the value of $\Delta_f G^\ominus(\mathrm{Ag^+, aq})$ from

$$\mathrm{Ag(s)} + \tfrac{1}{2}\mathrm{Cl_2(g)} \rightarrow \mathrm{Ag^+(aq)} + \mathrm{Cl^-(aq)}$$

for which $\Delta_r G^\ominus = -54.12$ kJ mol^{-1}. It follows that $\Delta_f G^\ominus(\mathrm{Ag^+, aq}) = +77.11$ kJ mol^{-1}. ●

The factors responsible for the magnitude of the Gibbs energy of formation of an ion in solution can be identified by analysing it in terms of a thermodynamic cycle. As an illustration, we consider the standard Gibbs energies of formation of $\mathrm{Cl^-}$ in water, which is -131 kJ mol^{-1}. We do so by treating the formation reaction

$$\tfrac{1}{2}\mathrm{H_2(g)} + \tfrac{1}{2}\mathrm{X_2(g)} \rightarrow \mathrm{H^+(aq)} + \mathrm{X^-(aq)}$$

as the outcome of the sequence of steps shown in Fig. 15.19 (with values taken from the *Data section*). The sum of the Gibbs energies for all the steps around a closed cycle is zero, so

$$\Delta_f G^\ominus(\mathrm{Cl^-, aq}) = 1287\ \mathrm{kJ\ mol^{-1}} + \Delta_{solv} G^\ominus(\mathrm{H^+}) + \Delta_{solv} G^\ominus(\mathrm{Cl^-})$$

A brief comment The standard Gibbs energies of formation of the gas-phase ions are unknown. We have therefore used ionization energies and electron affinities and have assumed that any differences from the Gibbs energies arising from conversion to enthalpy and the inclusion of entropies to obtain Gibbs energies in the formation of $\mathrm{H^+}$ are cancelled by the corresponding terms in the electron gain of X. The conclusions from the cycles are therefore only approximate.

An important point to note is that the value of $\Delta_f G^\ominus$ of an ion X is not determined by the properties of X alone but includes contributions from the dissociation, ionization, and hydration of hydrogen.

Gibbs energies of solvation of individual ions may be estimated from an equation derived by Max Born, who identified $\Delta_{solv} G^\ominus$ with the electrical work of transferring an ion from a vacuum into the solvent treated as a continuous dielectric of relative permittivity ε_r. The resulting **Born equation**, which is derived in *Further information 15.1*, is

$$\Delta_{solv} G^\ominus = -\frac{z_i^2 e^2 N_A}{8\pi\varepsilon_0 r_i}\left(1 - \frac{1}{\varepsilon_r}\right) \qquad (15.53a)$$

where z_i is the charge number of the ion and r_i its radius (N_A is Avogadro's constant). Note that $\Delta_{solv} G^\ominus < 0$, and that $\Delta_{solv} G^\ominus$ is strongly negative for small, highly charged ions in media of high relative permittivity. For water at 25°C,

$$\Delta_{solv} G^\ominus = -\frac{z_i^2}{(r_i/\mathrm{pm})} \times (6.86 \times 10^4\ \mathrm{kJ\ mol^{-1}}) \qquad (15.53b)$$

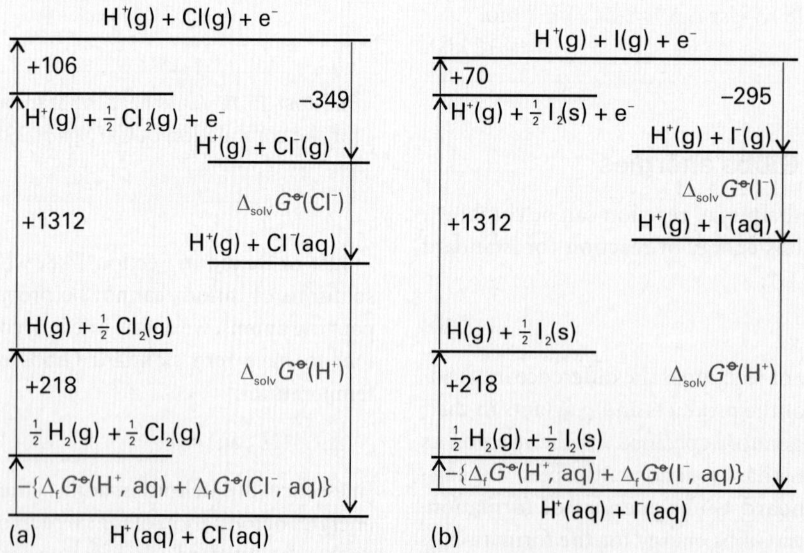

Fig. 15.19 The thermodynamic cycles for the discussion of the Gibbs energies of solvation (hydration) and formation of (a) chloride ions, (b) iodide ions in aqueous solution. The sum of the changes in Gibbs energies around the cycle sum to zero because G is a state function. All values are in kilojoules per mole.

● A BRIEF ILLUSTRATION

To see how closely the Born equation reproduces the experimental data, we calculate the difference in the values of $\Delta_f G^{\ominus}$ for Cl^- and I^- in water, for which $\varepsilon_r = 78.54$ at 25°C, given their radii as 181 pm and 220 pm (Table 9.3), respectively; the difference is

$$\Delta_{solv}G^{\ominus}(Cl^-) - \Delta_{solv}G^{\ominus}(I^-) = -\left(\frac{1}{181} - \frac{1}{220}\right) \times (6.86 \times 10^4 \text{ kJ mol}^{-1}) = -67 \text{ kJ mol}^{-1}$$

This estimated difference is in good agreement with the experimental difference, which is -61 kJ mol^{-1}. ●

Self-test 15.13 Estimate the value of $\Delta_{solv}G^{\ominus}(Cl^-, aq) - \Delta_{solv}G^{\ominus}(Br^-, aq)$ from experimental data and from the Born equation.
[-26 kJ mol^{-1} experimental; -29 kJ mol^{-1} calculated]

Calorimetry (for ΔH directly, and for S via heat capacities) is only one of the ways of determining Gibbs energies. They may also be obtained from equilibrium constants and electrochemical measurements (Chapter 17), and for gases they may be calculated using data from spectroscopic observations to evaluate the molecular partition function and then using eqn 15.40. Indeed, many of the Gibbs energies of formation are calculated in this way.

Combining the First and Second Laws

The First and Second Laws of thermodynamics are both relevant to the behaviour of matter, and we can bring the whole force of thermodynamics to bear on a problem by setting up a formulation that combines them.

15.7 The fundamental equation

We have seen that the First Law of thermodynamics may be written $dU = dq + dw$. For a reversible change in a closed system of constant composition, and in the absence of any additional (non-expansion) work, we may set $dw_{rev} = -pdV$ and (from the definition of entropy) $dq_{rev} = TdS$, where p is the pressure of the system and T its temperature. Therefore, for a reversible change in a closed system,

$$dU = TdS - pdV \tag{15.54}$$

However, because dU is an exact differential, its value is independent of path. Therefore, the same value of dU is obtained whether the change is brought about irreversibly or reversibly. Consequently, *eqn 15.54 applies to any change—reversible or irreversible—of a closed system that does no additional (non-expansion) work*. We shall call this combination of the First and Second Laws the **fundamental equation of thermodynamics**.

The fact that the fundamental equation applies to both reversible and irreversible changes may be puzzling at first sight. The reason is that only in the case of a reversible change may TdS be identified with dq and $-pdV$ with dw. When the change is irreversible, $TdS > dq$ (the Clausius inequality) and $-pdV > dw$. The sum of dw and dq remains equal to the sum of TdS and $-pdV$, provided the composition is constant.

15.8 Properties of the internal energy

Equation 15.54 shows that the internal energy of a closed system changes in a simple way when either S or V is changed ($dU \propto dS$ and $dU \propto dV$). These simple proportionalities suggest that U should be regarded as a function of S and V. We could regard U as a function of other variables, such as S and p or T and V, because they are all interrelated, but the simplicity of the fundamental equation suggests that $U(S,V)$ is the best choice.

The *mathematical* consequence of U being a function of S and V is that we can express an infinitesimal change dU in terms of changes dS and dV by

$$dU = \left(\frac{\partial U}{\partial S}\right)_V dS + \left(\frac{\partial U}{\partial V}\right)_S dV \tag{15.55}$$

The two partial derivatives are the slopes of the plots of U against S and V, respectively. When this *mathematical* expression is compared to the *thermodynamic* relation, eqn 15.54, we see that, for systems of constant composition,

$$\left(\frac{\partial U}{\partial S}\right)_V = T \qquad \left(\frac{\partial U}{\partial V}\right)_S = -p \tag{15.56}$$

The first of these two equations is a purely thermodynamic definition of temperature as the ratio of the changes in the internal energy (a First Law concept) and entropy (a Second Law concept) of a constant-volume, closed, constant-composition system. We are beginning to generate relations between the properties of a system and to discover the power of thermodynamics for establishing unexpected relations.

Because the fundamental equation, eqn 15.54, is an expression for an exact differential, the functions multiplying dS and dV (namely T and $-p$) must pass the test for exact differentials set out in *Mathematical background 8* (that $df = gdx + hdy$ is exact if $(\partial g/\partial y)_x = (\partial h/\partial x)_y$). That is, it must be the case that

$$\left(\frac{\partial T}{\partial V}\right)_S = -\left(\frac{\partial p}{\partial S}\right)_V \tag{15.57}$$

We have generated a relation between quantities that, at first sight, would not seem to be related and which is certainly very difficult to justify on a molecular basis.

Table 15.5 The Maxwell relations

From U:	$\left(\frac{\partial T}{\partial V}\right)_S = -\left(\frac{\partial p}{\partial S}\right)_V$
From H:	$\left(\frac{\partial T}{\partial p}\right)_S = \left(\frac{\partial V}{\partial S}\right)_p$
From A:	$\left(\frac{\partial p}{\partial T}\right)_V = \left(\frac{\partial S}{\partial V}\right)_T$
From G:	$\left(\frac{\partial V}{\partial T}\right)_p = -\left(\frac{\partial S}{\partial p}\right)_T$

Equation 15.57 is an example of a **Maxwell relation**. However, apart from being unexpected, it does not look particularly interesting. Nevertheless, it does suggest that there may be other similar relations that are more useful. Indeed, we can use the fact that H, G, and A are all state functions to derive three more Maxwell relations. The argument to obtain them runs in the same way in each case: because H, G, and A are state functions, the expressions for dH, dG, and dA satisfy relations that yield expressions like eqn 15.57. All four relations are listed in Table 15.5. As an example of their use, we show in the following *Justification* that the internal pressure $\pi_T = (\partial U/\partial V)_T$ introduced in Section 14.10 may be converted into

$$\pi_T = T\left(\frac{\partial p}{\partial T}\right)_V - p \qquad (15.58)$$

This relation is called a **thermodynamic equation of state** because it is an expression for pressure in terms of a variety of thermodynamic properties of the system.

Justification 15.6 *The thermodynamic equation of state*

We obtain an expression for the coefficient π_T by dividing both sides of eqn 15.55 by dV, imposing the constraint of constant temperature, which gives

$$\left(\frac{\partial U}{\partial V}\right)_T = \left(\frac{\partial U}{\partial S}\right)_V\left(\frac{\partial S}{\partial V}\right)_T + \left(\frac{\partial U}{\partial V}\right)_S$$

Next, we introduce the two relations in eqn 15.56 and the definition of π_T to obtain

$$\pi_T = T\left(\frac{\partial S}{\partial V}\right)_T - p$$

The third Maxwell relation in Table 15.5 turns $(\partial S/\partial V)_T$ into $(\partial p/\partial T)_V$, which completes the proof of eqn 15.58.

Example 15.7 *Deriving a thermodynamic relation*

Show thermodynamically that $\pi_T = 0$ for a perfect gas, and derive its value for a van der Waals gas.

Method Proving a result 'thermodynamically' means basing it entirely on general thermodynamic relations and equations of state, without drawing on molecular arguments (such as the existence of intermolecular forces). We know that for a perfect gas, $p = nRT/V$, so this relation should be used in eqn 15.58. Similarly, the van der Waals equation, eqn 8.24, should be used instead for the second part of the question.

Answer For a perfect gas we write

$$\left(\frac{\partial p}{\partial T}\right)_V = \left(\frac{\partial(nRT/V)}{\partial T}\right)_V = \frac{nR}{V}$$

Then, eqn 15.58 becomes

$$\pi_T = \frac{nRT}{V} - p = 0$$

The equation of state of a van der Waals gas is

$$p = \frac{nRT}{V - nb} - a\frac{n^2}{V^2}$$

Because a and b are independent of temperature,

$$\left(\frac{\partial p}{\partial T}\right)_V = \frac{nR}{V - nb}$$

Therefore, from eqn 15.58,

$$\pi_T = \frac{nRT}{V - nb} - \frac{nRT}{V - nb} + a\frac{n^2}{V^2} = a\frac{n^2}{V^2}$$

exactly as we obtained in Section 14.10 on the basis of molecular arguments and the formulation of a model intermolecular potential with an attractive region proportional to a.

Self-test 15.14 Calculate π_T for a gas that obeys the virial equation of state (eqn 8.22).
[$\pi_T = RT^2(\partial B/\partial T)_V/V_m^2 + \cdots$]

15.9 Properties of the Gibbs energy

The same arguments that we have used for U can be used for the Gibbs energy $G = H - TS$. They lead to expressions showing how G varies with pressure and temperature that are important for discussing phase transitions and chemical reactions.

(a) General considerations

When the system undergoes a change of state, G may change because H, T, and S all change. As in *Justification 15.5*, we write for infinitesimal changes in each property

$$dG = dH - d(TS) = dH - TdS - SdT$$

Because $H = U + pV$, we know that

$$dH = dU + d(pV) = dU + pdV + Vdp$$

and therefore

$$dG = dU + pdV + Vdp - TdS - SdT$$

For a closed system doing no non-expansion work, we can replace dU by the fundamental equation $dU = TdS - pdV$ and obtain

$$dG = TdS - pdV + pdV + Vdp - TdS - SdT$$

Four terms now cancel on the right, and we conclude that for a closed system in the absence of non-expansion work and at constant composition

$$dG = Vdp - SdT \qquad (15.59)$$

This expression, which shows that a change in G is proportional to a change in p or T, suggests that G may be best regarded as a function of p and T. It confirms that G is an important quantity in chemistry because the pressure and temperature are usually the variables under our control. In other words, G carries around the combined consequences of the First and Second Laws in a way that makes it particularly suitable for chemical applications.

The same argument that led to eqn 15.56, when applied to the exact differential $dG = Vdp - SdT$, now gives

$$\left(\frac{\partial G}{\partial T}\right)_p = -S \qquad \left(\frac{\partial G}{\partial p}\right)_T = V \qquad (15.60)$$

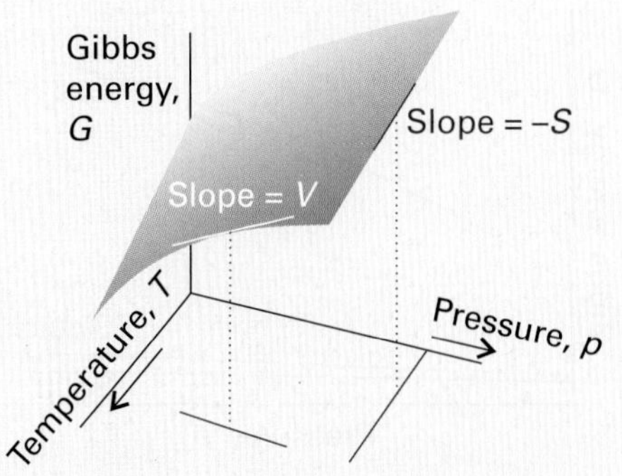

Fig. 15.20 The variation of the Gibbs energy of a system with temperature and pressure. The slope of the former is equal to the negative of the entropy of the system and that of the latter is equal to the volume.

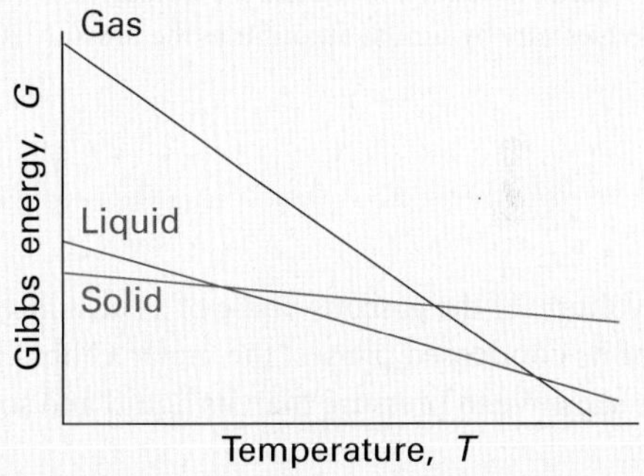

Fig. 15.21 The variation of the Gibbs energy with the temperature is determined by the entropy. Because the entropy of the gaseous phase of a substance is greater than that of the liquid phase, and the entropy of the solid phase is smallest, the Gibbs energy changes most steeply for the gas phase, followed by the liquid phase, and then the solid phase of the substance.

These relations show how the Gibbs energy varies with temperature and pressure (Fig. 15.20). The first implies that:

- Because $S > 0$ for all substances, G always *decreases* when the temperature is raised (at constant pressure and composition).

Insofar as G is the negative logarithm of the molecular partition function (eqn 15.40), we can understand this behaviour on the basis that q increases with temperature as more states become thermally accessible, and therefore $-\ln q$ becomes more negative.

- Because $(\partial G/\partial T)_p$ becomes more negative as S increases, G decreases most sharply when the entropy of the system is large.

A large entropy implies that many states are occupied, and therefore that many are thermally accessible. Such a system is more sensitive to changes in temperature than one in which only a small number of states are accessible. In macroscopic terms, the Gibbs energy of the gaseous phase of a substance, which has a high molar entropy, is more sensitive to temperature than its liquid and solid phases (Fig. 15.21). Similarly, the second relation implies that:

- Because $V > 0$ for all substances, G always *increases* when the pressure of the system is increased (at constant temperature and composition).

We can understand this behaviour on the basis that q decreases with decreasing volume as states move apart and become less thermally accessible, and therefore $-\ln q$ becomes less negative.

- Because $(\partial G/\partial p)_T$ increases with V, G is more sensitive to pressure when the volume of the system is large.

Once again, a large volume implies closely spaced translational energy levels, and therefore a responsiveness to change. Because

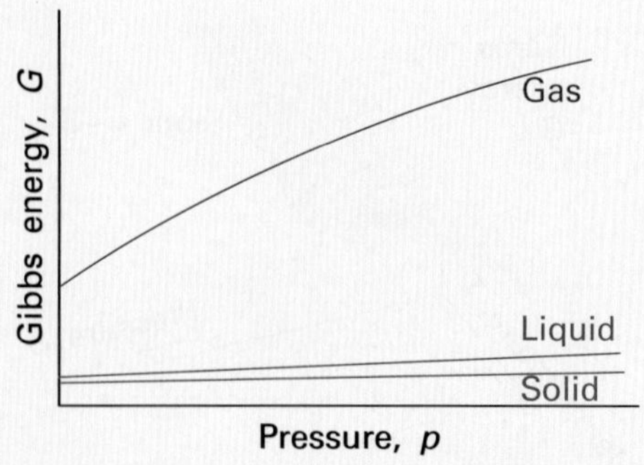

Fig. 15.22 The variation of the Gibbs energy with the pressure is determined by the volume of the sample. Because the volume of the gaseous phase of a substance is greater than that of the same amount of liquid phase, and the entropy of the solid phase is smallest (for most substances), the Gibbs energy changes most steeply for the gas phase, followed by the liquid phase, and then the solid phase of the substance. Because the volumes of the solid and liquid phases of a substance are similar, their molar Gibbs energies vary by similar amounts as the pressure is changed.

the molar volume of the gaseous phase of a substance is greater than that of its condensed phases, the molar Gibbs energy of a gas is more sensitive to pressure than its liquid and solid phases (Fig. 15.22).

(b) The pressure from the partition function

By exactly the same kind of argument that led to eqn 15.56, we can compare the two equations for the variation of A with temperature and volume. From the definition of A,

$$dA = -p dV - S dT$$

and from the general mathematical form

$$dA = \left(\frac{\partial A}{\partial V}\right)_T dV + \left(\frac{\partial A}{\partial T}\right)_V dT$$

it follows that

$$\left(\frac{\partial A}{\partial T}\right)_V = -S \qquad \left(\frac{\partial A}{\partial V}\right)_T = -p$$

The second of these relations gives us a valuable route to the calculation of the pressure from the partition function, for we already know how to calculate A. Thus, when eqn 15.38a is substituted into this expression, we obtain the simple relation

$$p = kT\left(\frac{\partial \ln Q}{\partial V}\right)_T \tag{15.61}$$

● A BRIEF ILLUSTRATION

To show that eqn 15.61 reduces to the perfect gas equation of state in the case of a gas of independent particles, we write $Q = q^N/N!$, to obtain

$$p = kT\left(\frac{\partial}{\partial V}(\ln q^N - \ln N!)\right)_T = NkT\left(\frac{\partial \ln q}{\partial V}\right)_T$$

because only q (and not N) depends on V. Moreover, because $\ln q = \ln q^T q^R q^V = \ln q^T + \ln q^R q^V$, and both q^R and q^V are independent of volume, we need consider only the translational partition function. Thus we use $q = V/\Lambda^3$ to obtain

$$\left(\frac{\partial \ln q}{\partial V}\right)_T = \frac{1}{q}\left(\frac{\partial q}{\partial V}\right)_T = \frac{\Lambda^3}{V}\left(\frac{\partial (V/\Lambda^3)}{\partial V}\right)_T = \frac{1}{V}$$

When this relation is substituted into the preceding one and we use $N = nN_A$ and $N_A k = R$, we find $p = NkT/V = nRT/V$, the equation of state of a perfect gas. ●

Self-test 15.15 Obtain the corresponding expression for a real gas in terms of the configuration integral Z introduced in eqn 14.53. [$p = kT(\partial \ln Z/\partial V)_T$]

At long last we are at the point where we can confirm that $\beta = 1/kT$, which we have asked you to accept since the Lagrangian multiplier $-\beta$ was first introduced in Chapter 13. Thus, if we had not made the identification $\beta = 1/kT$ earlier, we would have found

$$p = \frac{1}{\beta}\left(\frac{\partial \ln Q}{\partial V}\right)_T$$

in place of eqn 15.61. The work in the preceding *brief illustration* would have led to

$$p = \frac{N}{\beta V}$$

which requires us to identify β as $1/kT$ in order to recover the perfect gas law.

(c) The variation of the Gibbs energy with temperature

Although eqn 15.60 expresses the variation of G in terms of the entropy, we can express it in terms of the enthalpy by using the definition of G to write $S = (H - G)/T$. Then

$$\left(\frac{\partial G}{\partial T}\right)_p = \frac{G - H}{T} \tag{15.62a}$$

In the following *Justification* we show that an alternative form of this equation is the **Gibbs–Helmholtz equation:**

$$\left(\frac{\partial}{\partial T}\frac{G}{T}\right)_p = -\frac{H}{T^2} \tag{15.62b}$$

This expression shows that, if we know the enthalpy of the system, then we know how G/T varies with temperature.

Justification 15.7 *The Gibbs–Helmholtz equation*

First, we reorganize eqn 15.62a into

$$\left(\frac{\partial G}{\partial T}\right)_p - \frac{G}{T} = -\frac{H}{T}$$

Then we combine the two terms on the left by noting that

$$\left(\frac{\partial}{\partial T}\frac{G}{T}\right)_p = \frac{1}{T}\left(\frac{\partial G}{\partial T}\right)_p + G\frac{\mathrm{d}}{\mathrm{d}T}\frac{1}{T} = \frac{1}{T}\left(\frac{\partial G}{\partial T}\right)_p - \frac{G}{T^2}$$

$$= \frac{1}{T}\left\{\left(\frac{\partial G}{\partial T}\right)_p - \frac{G}{T}\right\}$$

When we substitute eqn 15.60, that $(\partial G/\partial T)_p = -S$, into this expression, we obtain

$$\left(\frac{\partial}{\partial T}\frac{G}{T}\right)_p = \frac{1}{T}\left\{-S - \frac{G}{T}\right\} = \frac{1}{T}\left\{-S - \frac{H - TS}{T}\right\} = -\frac{H}{T^2}$$

which is eqn 15.62b.

The Gibbs–Helmholtz equation is most useful when it is applied to changes, including changes of physical state and chemical reactions at constant pressure. Then, because $\Delta G = G_f - G_i$ for the change of Gibbs energy between the final and initial states and because the equation applies to both G_f and G_i, we can write

$$\left(\frac{\partial}{\partial T}\frac{\Delta G}{T}\right)_p = -\frac{\Delta H}{T^2} \tag{15.63}$$

This equation shows that, if we know the change in enthalpy of a system that is undergoing some kind of transformation (such as vaporization or reaction), then we know how the corresponding change in Gibbs energy varies with temperature. As we shall see, this is a crucial piece of information in chemistry.

(d) The variation of the Gibbs energy with pressure

To find the Gibbs energy at one pressure in terms of its value at another pressure, the temperature being constant, we set $\mathrm{d}T = 0$ in eqn 15.59, which gives $\mathrm{d}G = V\mathrm{d}p$, and integrate:

$$G(p_f) = G(p_i) + \int_{p_i}^{p_f} V\,\mathrm{d}p \tag{15.64a}$$

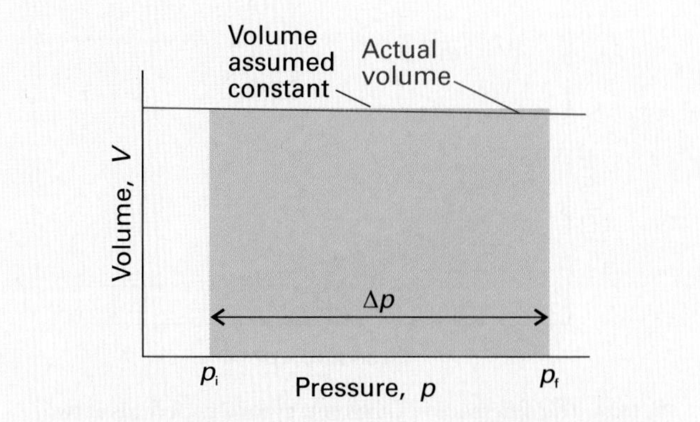

Fig. 15.23 The difference in Gibbs energy of a solid or liquid at two pressures is equal to the rectangular area shown. We have assumed that the variation of volume with pressure is negligible.

For molar quantities,

$$G_m(p_f) = G_m(p_i) + \int_{p_i}^{p_f} V_m\,\mathrm{d}p \tag{15.64b}$$

This expression is applicable to any phase of matter, but to evaluate it we need to know how the molar volume, V_m, depends on the pressure.

The molar volume of a condensed phase changes only slightly as the pressure changes (Fig. 15.23), so we can treat V_m as a constant and take it outside the integral:

$$G_m(p_f) = G_m(p_i) + V_m\int_{p_i}^{p_f} \mathrm{d}p = G_m(p_i) + (p_f - p_i)V_m \tag{15.65}$$

Under normal laboratory conditions $(p_f - p_i)V_m$ is very small and may be neglected. Hence, we may usually suppose that the Gibbs energies of solids and liquids are independent of pressure.

Self-test 15.16 Calculate the change in G_m for ice at −10°C, with density 917 kg m^{-3}, when the pressure is increased from 1.0 bar to 2.0 bar. [+2.0 J mol^{-1}]

The molar volumes of gases are large, so the Gibbs energy of a gas depends strongly on the pressure: the closeness of its translational energy levels makes the partition function highly responsive to external influences. Furthermore, because the volume also varies markedly with the pressure, we cannot treat it as a constant in the integral in eqn 15.64b (Fig. 15.24). For a perfect gas we substitute $V_m = RT/p$ into the integral, treat RT as a constant, and find

$$G_m(p_f) = G_m(p_i) + RT\int_{p_i}^{p_f} \frac{\mathrm{d}p}{p} = G_m(p_i) + RT\ln\frac{p_f}{p_i} \tag{15.66°}$$

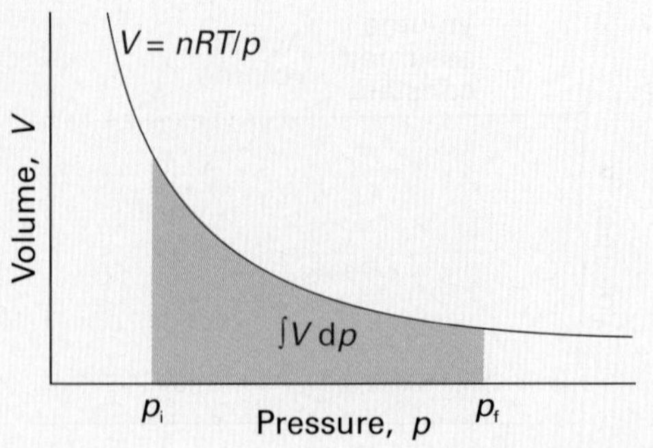

Fig. 15.24 The difference in Gibbs energy for a perfect gas at two pressures is equal to the area shown below the perfect-gas isotherm.

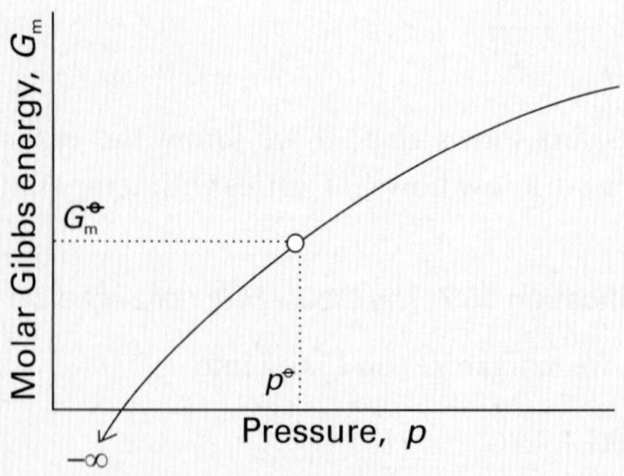

Fig. 15.25 The molar Gibbs energy of a perfect gas is proportional to ln p, and the standard state is reached at $p^\ominus$. Note that as $p \rightarrow 0$, the molar Gibbs energy becomes negatively infinite.

interActivity Show how the first derivative of G, $(\partial G/\partial p)_T$, varies with pressure, and plot the resulting expression over a pressure range. What is the physical significance of $(\partial G/\partial p)_T$?

This expression shows that, when the pressure is increased tenfold at room temperature, the molar Gibbs energy increases by $RT\ln 10 \approx 6\ \text{kJ mol}^{-1}$. It also follows from this equation that if we set $p_i = p^\ominus$ (the standard pressure of 1 bar), then the molar Gibbs energy of a perfect gas at a pressure p (set $p_f = p$) is related to its standard value by

$$G_m(p) = G_m^\ominus + RT\ln\frac{p}{p^\ominus} \qquad (15.67)°$$

The logarithmic dependence of the molar Gibbs energy on the pressure predicted by eqn 15.67 is illustrated in Fig. 15.25.

Self-test 15.17 Calculate the change in the molar Gibbs energy of water vapour (treated as a perfect gas) when the pressure is increased isothermally from 1.0 bar to 2.0 bar at 298 K. Note that, whereas the change in molar Gibbs energy for a condensed phase (Self-test 15.16) is a few joules per mole, this change is of the order of kilojoules per mole.

[$+1.7\ \text{kJ mol}^{-1}$]

Checklist of key ideas

☐ 1. The Second Law in terms of entropy: The entropy of an isolated system increases in the course of a spontaneous change: $\Delta S_{tot} > 0$.

☐ 2. The thermodynamic definition of entropy is $dS = dq_{rev}/T$. The statistical definition of entropy is given by the Boltzmann formula, $S = k \ln \mathcal{W}$.

☐ 3. The Clausius inequality is $dS \geq dq/T$.

☐ 4. The entropy of transition at the transition temperature, $\Delta_{trs}S = \Delta_{trs}H/T_{trs}$.

☐ 5. Trouton's rule states that many normal liquids have approximately the same standard entropy of vaporization (about $85\ \text{J K}^{-1}\ \text{mol}^{-1}$) at their normal boiling point.

☐ 6. The variation of entropy with temperature is given by eqn 15.29.

☐ 7. Third Law of thermodynamics: The entropy of all perfect crystalline substances is zero at $T = 0$.

☐ 8. The Helmholtz energy is $A = U - TS$; the Gibbs energy is $G = H - TS$.

☐ 9. The criteria of spontaneity are: (a) $dS_{U,V} > 0$ and $dU_{S,V} < 0$, or (b) $dA_{T,V} < 0$ and $dG_{T,p} < 0$.

☐ 10. The criterion of equilibrium at constant temperature and volume, $dA_{T,V} = 0$. The criterion of equilibrium at constant temperature and pressure, $dG_{T,p} = 0$.

☐ 11. The standard Gibbs energy of reaction is $\Delta_r G^\ominus = \Delta_r H^\ominus - T\Delta_r S^\ominus$.

☐ 12. The maximum work and the Helmholtz energy are related by $w_{max} = \Delta A$. The maximum additional (non-expansion) work and the Gibbs energy are related by $w_{add,max} = \Delta G$.

☐ 13. The standard Gibbs energy of formation ($\Delta_f G^\ominus$) is the standard reaction Gibbs energy for the formation of a compound from its elements in their reference states.

☐ 14. The fundamental equation of thermodynamics is $dU = TdS - pdV$.

☐ 15. The Gibbs energy is best described as a function of pressure and temperature, $dG = Vdp - SdT$, so the variation of Gibbs energy with pressure and that with temperature are, respectively, $(\partial G/\partial p)_T = V$ and $(\partial G/\partial T)_p = -S$.

☐ 16. The temperature dependence of the Gibbs energy is given by the Gibbs–Helmholtz equation, eqn 15.62.

☐ 17. For a perfect gas, the variation of the Gibbs energy with pressure is given by eqn 15.66.

Further information

Further information 15.1 *The Born equation*

The electrical concepts required in this derivation are reviewed in *Fundamentals F.6*. The strategy of the calculation is to identify the Gibbs energy of solvation with the work of transferring an ion from a vacuum into the solvent. That work is calculated by taking the difference of the work of charging an ion when it is in the solution and the work of charging the same ion when it is in a vacuum.

The Coulomb interaction between two charges Q_1 and Q_2 separated by a distance r is described by the **Coulombic potential energy**:

$$V = \frac{Q_1Q_2}{4\pi\varepsilon r}$$

where ε is the medium's permittivity. The permittivity of a vacuum is $\varepsilon_0 = 8.854 \times 10^{-12}\ \mathrm{J^{-1}\ C^2\ m^{-1}}$. The relative permittivity (formerly called the 'dielectric constant') of a substance is defined as $\varepsilon_r = \varepsilon/\varepsilon_0$. Ions do not interact as strongly in a solvent of high relative permittivity (such as water, with $\varepsilon_r = 80$ at 293 K) as they do in a solvent of lower relative permittivity (such as ethanol, with $\varepsilon_r = 25$ at 293 K). The potential energy of a charge Q_1 in the presence of a charge Q_2 can be expressed in terms of the **Coulomb potential**, ϕ:

$$V = Q_1\phi \qquad \phi = \frac{Q_2}{4\pi\varepsilon r}$$

We model an ion as a sphere of radius r_i immersed in a medium of permittivity ε. It turns out that, if the charge of the sphere is Q, the electric potential, ϕ, at its surface is the same as the potential due to a point charge at its centre, so we can use the last expression and write

$$\phi = \frac{Q}{4\pi\varepsilon r_i}$$

The work of bringing up a charge dQ to the sphere is ϕdQ. Therefore, the total work of charging the sphere from 0 to z_ie is

$$w = \int_0^{z_ie} \phi\,\mathrm{d}Q = \frac{1}{4\pi\varepsilon r_i}\int_0^{z_ie} Q\,\mathrm{d}Q = \frac{z_i^2e^2}{8\pi\varepsilon r_i}$$

This electrical work of charging, when multiplied by Avogadro's constant, is the molar Gibbs energy for charging the ions.

The work of charging an ion in a vacuum is obtained by setting $\varepsilon = \varepsilon_0$, the vacuum permittivity. The corresponding value for charging the ion in a medium is obtained by setting $\varepsilon = \varepsilon_r\varepsilon_0$, where ε_r is the relative permittivity of the medium. It follows that the change in molar Gibbs energy that accompanies the transfer of ions from a vacuum to a solvent is the difference of these two quantities:

$$\begin{aligned}\Delta_{\mathrm{solv}}G^{\ominus} &= \frac{z_i^2e^2N_A}{8\pi\varepsilon r_i} - \frac{z_i^2e^2N_A}{8\pi\varepsilon_0 r_i} \\ &= \frac{z_i^2e^2N_A}{8\pi\varepsilon_r\varepsilon_0 r_i} - \frac{z_i^2e^2N_A}{8\pi\varepsilon_0 r_i} \\ &= -\frac{z_i^2e^2N_A}{8\pi\varepsilon_0 r_i}\left(1 - \frac{1}{\varepsilon_r}\right)\end{aligned}$$

which is eqn 15.53.

Discussion questions

15.1 The evolution of life requires the organization of a very large number of molecules into biological cells. Does the formation of living organisms violate the Second Law of thermodynamics? State your conclusion clearly and present detailed arguments to support it.

15.2 The following expressions have been used to establish criteria for spontaneous change: $\mathrm{d}A_{T,V} < 0$ and $\mathrm{d}G_{T,p} < 0$. Discuss the origin, significance, and applicability of each criterion.

15.3 Discuss the significance of the terms 'dispersal' and 'disorder' in the context of the Second Law.

15.4 Discuss the relationship between the thermodynamic and statistical definitions of entropy.

15.5 Justify the differences between the partition-function expression for the entropy for distinguishable particles and the expression for indistinguishable particles.

15.6 Discuss the relationships between the various formulations of the Second Law of thermodynamics.

15.7 Account for the temperature and volume dependence of the entropy of a perfect gas in terms of the Boltzmann distribution.

15.8 Account for deviations from Trouton's rule for liquids such as water and ethanol. Is their entropy of vaporization larger or smaller than 85 J K^{-1} mol^{-1}? Why?

15.9 Discuss why the standard entropies of ions in solution may be positive, negative, or zero.

15.10 Suggest a physical interpretation of the dependence of the Gibbs energy on the temperature.

15.11 Under what circumstances, and why, can the spontaneity of a process be discussed in terms of the properties of the system alone?

Exercises

Assume that all gases are perfect and that data refer to 298.15 K unless otherwise stated.

15.1(a) During a hypothetical process, the entropy of a system increases by 125 $J K^{-1}$ while the entropy of the surroundings decreases by 125 $J K^{-1}$. Is the process spontaneous?

15.1(b) During a hypothetical process, the entropy of a system increases by 105 $J K^{-1}$ while the entropy of the surroundings decreases by 95 $J K^{-1}$. Is the process spontaneous?

15.2(a) Which of F_2(g) and I_2(g) is likely to have the higher standard molar entropy at 298 K?

15.2(b) Which of H_2O(g) and CO_2(g) is likely to have the higher standard molar entropy at 298 K?

15.3(a) Calculate the standard molar entropy at 298 K of (a) gaseous helium, (b) gaseous xenon.

15.3(b) Calculate the translational contribution to the standard molar entropy at 298 K of (a) H_2O(g), (b) CO_2(g).

15.4(a) At what temperature is the standard molar entropy of helium equal to that of xenon at 298 K?

15.4(b) At what temperature is the translational contribution to the standard molar entropy of CO_2(g) equal to that of H_2O(g) at 298 K?

15.5(a) Calculate the rotational partition function of H_2O at 298 K from its rotational constants 27.878 cm^{-1}, 14.509 cm^{-1}, and 9.287 cm^{-1} and use your result to calculate the rotational contribution to the molar entropy of gaseous water at 25°C.

15.5(b) Calculate the rotational partition function of SO_2 at 298 K from its rotational constants 2.02736 cm^{-1}, 0.34417 cm^{-1}, and 0.293535 cm^{-1} and use your result to calculate the rotational contribution to the molar entropy of sulfur dioxide at 25°C.

15.6(a) The ground state of the Co^{2+} ion in $CoSO_4 \cdot 7H_2O$ may be regarded as $^4T_{9/2}$. The entropy of the solid at temperatures below 1 K is derived almost entirely from the electron spin. Estimate the molar entropy of the solid at these temperatures.

15.6(b) Estimate the contribution of the spin to the molar entropy of a solid sample of a d-metal complex with $S = \frac{5}{2}$.

15.7(a) Calculate the change in entropy when 15 g of carbon dioxide gas is allowed to expand from 1.0 dm^3 to 3.0 dm^3 at 300 K.

15.7(b) Calculate the change in entropy when 4.00 g of nitrogen is allowed to expand from 500 cm^3 to 750 cm^3 at 300 K.

15.8(a) Predict the standard molar entropy of methanoic acid (formic acid, HCOOH) at (a) 298 K, (b) 500 K. The normal modes occur at wavenumbers 3570, 2943, 1770, 1387, 1229, 1105, 625, 1033, 638 cm^{-1}.

15.8(b) Predict the standard molar entropy of ethyne at (a) 298 K, (b) 500 K. The normal modes (and their degeneracies in parentheses) occur at wavenumbers 612(2), 729(2), 1974, 3287, and 3374 cm^{-1}.

15.9(a) Calculate the rotational contribution to the molar entropy of (a) H_2, (b) Cl_2 at 298 K. Use $\tilde{B}(H_2) = 60.864\ cm^{-1}$ and $\tilde{B}(Cl_2) = 0.2441\ cm^{-1}$.

15.9(b) Calculate the rotational contribution to the molar entropy of (a) CO_2, (b) CS_2 at 298 K. Use $\tilde{B}(CO_2) = 0.3902\ cm^{-1}$ and $\tilde{B}(CS_2) = 0.1091\ cm^{-1}$.

15.10(a) A certain ideal heat engine uses water at the triple point as the hot source and an organic liquid as the cold sink. It withdraws 10.00 kJ of heat from the hot source and generates 3.00 kJ of work. What is the temperature of the organic liquid?

15.10(b) A certain ideal heat engine uses water at the triple point as the hot source and an organic liquid as the cold sink. It withdraws 2.71 kJ of heat from the hot source and generates 0.71 kJ of work. What is the temperature of the organic liquid?

15.11(a) Calculate the change in entropy when 100 kJ of energy is transferred reversibly and isothermally as heat to a large block of copper at (a) 0°C, (b) 50°C.

15.11(b) Calculate the change in entropy when 250 kJ of energy is transferred reversibly and isothermally as heat to a large block of lead at (a) 20°C, (b) 100°C.

15.12(a) Predict the enthalpy of vaporization of benzene from its normal boiling point, 80.1°C.

15.12(b) Predict the enthalpy of vaporization of cyclohexane from its normal boiling point, 80.7°C.

15.13(a) Calculate the molar entropy of a constant-volume sample of neon at 500 K given that it is 146.22 $J K^{-1} mol^{-1}$ at 298 K.

15.13(b) Calculate the molar entropy of a constant-volume sample of argon at 250 K given that it is 154.84 $J K^{-1} mol^{-1}$ at 298 K.

15.14(a) Calculate ΔS (for the system) when the state of 3.00 mol of perfect gas atoms, for which $C_{p,m} = \frac{5}{2}R$, is changed from 25°C and 1.00 atm to 125°C and 5.00 atm. How do you rationalize the sign of ΔS?

15.14(b) Calculate ΔS (for the system) when the state of 2.00 mol diatomic perfect gas molecules, for which $C_{p,m} = \frac{7}{2}R$, is changed from 25°C and 1.50 atm to 135°C and 7.00 atm. How do you rationalize the sign of ΔS?

15.15(a) A sample consisting of 3.00 mol of diatomic perfect gas molecules at 200 K is compressed reversibly and adiabatically until its temperature reaches 250 K. Given that $C_{V,m} = 27.5\ J K^{-1} mol^{-1}$, calculate ΔS.

15.15(b) A sample consisting of 2.00 mol of diatomic perfect gas molecules at 250 K is compressed reversibly and adiabatically until its temperature reaches 300 K. Given that $C_{V,m} = 27.5\ J K^{-1} mol^{-1}$, calculate ΔS.

15.16(a) Calculate ΔS_{tot} when two copper blocks, each of mass 1.00 kg, one at 50°C and the other at 0°C, are placed in contact in an isolated container. The specific heat capacity of copper is 0.385 $J K^{-1} g^{-1}$ and may be assumed constant over the temperature range involved.

15.16(b) Calculate ΔS_{tot} when two iron blocks, each of mass 10.0 kg, one at 100°C and the other at 25°C, are placed in contact in an isolated container. The specific heat capacity of iron is 0.449 $J K^{-1} g^{-1}$ and may be assumed constant over the temperature range involved.

15.17(a) Calculate the change in the entropies of the system and the surroundings, and the total change in entropy, when a sample of nitrogen gas of mass 14 g at 298 K and 1.00 bar doubles its volume in (a) an isothermal reversible expansion, (b) an isothermal irreversible expansion against $p_{ex} = 0$, and (c) an adiabatic reversible expansion.

15.17(b) Calculate the change in the entropies of the system and the surroundings, and the total change in entropy, when the volume of a sample of argon gas of mass 21 g at 298 K and 1.50 bar increases from 1.20 dm^3 to 4.60 dm^3 in (a) an isothermal reversible expansion, (b) an isothermal irreversible expansion against $p_{ex} = 0$, and (c) an adiabatic reversible expansion.

15.18(a) The enthalpy of vaporization of chloroform ($CHCl_3$) is 29.4 kJ mol^{-1} at its normal boiling point of 334.88 K. Calculate (a) the entropy of vaporization of chloroform at this temperature and (b) the entropy change of the surroundings.

15.18(b) The enthalpy of vaporization of methanol is 35.27 kJ mol^{-1} at its normal boiling point of 64.1°C. Calculate (a) the entropy of vaporization of methanol at this temperature and (b) the entropy change of the surroundings.

15.19(a) Calculate the change in entropy of the system when 10.0 g of ice at −10.0°C is converted into water vapour at 115.0°C and at a constant pressure of 1 bar. The constant-pressure molar heat capacity of $H_2O(s)$ and $H_2O(l)$ is 75.291 J K^{-1} mol^{-1} and that of $H_2O(g)$ is 33.58 J K^{-1} mol^{-1}. Enthalpies of phase transitions are given in Table 14.3.

15.19(b) Calculate the change in entropy of the system when 15.0 g of ice at −12.0°C is converted to water vapour at 105.0°C at a constant pressure of 1 bar. For data, see the preceding exercise.

15.20(a) Calculate the residual molar entropy of a solid in which the molecules can adopt (a) three, (b) five, (c) six orientations of equal energy at $T = 0$.

15.20(b) Suppose that the hexagonal molecule $C_6H_nF_{6-n}$ has a residual entropy on account of the similarity of the H and F atoms. Calculate the residual entropy for each value of n.

15.21(a) Calculate the standard reaction entropy at 298 K of

(a) $2\,CH_3CHO(g) + O_2(g) \rightarrow 2\,CH_3COOH(l)$

(b) $2\,AgCl(s) + Br_2(l) \rightarrow 2\,AgBr(s) + Cl_2(g)$

(c) $Hg(l) + Cl_2(g) \rightarrow HgCl_2(s)$

15.21(b) Calculate the standard reaction entropy at 298 K of

(a) $Zn(s) + Cu^{2+}(aq) \rightarrow Zn^{2+}(aq) + Cu(s)$

(b) $C_{12}H_{22}O_{11}(s) + 12\,O_2(g) \rightarrow 12\,CO_2(g) + 11\,H_2O(l)$

15.22(a) Combine the reaction entropies calculated in Exercise 15.21a with the reaction enthalpies, and calculate the standard reaction Gibbs energies at 298 K.

15.22(b) Combine the reaction entropies calculated in Exercise 15.21b with the reaction enthalpies, and calculate the standard reaction Gibbs energies at 298 K.

15.23(a) Use standard Gibbs energies of formation to calculate the standard reaction Gibbs energies at 298 K of the reactions in Exercise 15.21a.

15.23(b) Use standard Gibbs energies of formation to calculate the standard reaction Gibbs energies at 298 K of the reactions in Exercise 15.21b.

15.24(a) Calculate the standard Gibbs energy of the reaction $4\,HI(g) + O_2(g) \rightarrow 2\,I_2(s) + 2\,H_2O(l)$ at 298 K from the standard entropies and enthalpies of formation given in the *Data section*.

15.24(b) Calculate the standard Gibbs energy of the reaction $CO(g) + CH_3CH_2OH(l) \rightarrow CH_3CH_2COOH(l)$ at 298 K from the standard entropies and enthalpies of formation given in the *Data section*.

15.25(a) The standard enthalpy of combustion of ethyl acetate ($CH_3COOC_2H_5$) is −2231 kJ mol^{-1} at 298 K and its standard molar entropy is 259.4 J K^{-1} mol^{-1}. Calculate the standard Gibbs energy of formation of the compound at 298 K.

15.25(b) The standard enthalpy of combustion of the amino acid glycine (NH_2CH_2COOH) is −969 kJ mol^{-1} at 298 K and its standard molar entropy is 103.5 J K^{-1} mol^{-1}. Calculate the standard Gibbs energy of formation of glycine at 298 K.

15.26(a) Calculate the maximum non-expansion work per mole that may be obtained from a fuel cell in which the chemical reaction is the combustion of methane at 298 K.

15.26(b) Calculate the maximum non-expansion work per mole that may be obtained from a fuel cell in which the chemical reaction is the combustion of propane at 298 K.

15.27(a) Suppose that 2.5 mmol $N_2(g)$ occupies 42 cm^3 at 300 K and expands isothermally to 600 cm^3. Calculate ΔG for the process.

15.27(b) Suppose that 6.0 mmol Ar(g) occupies 52 cm^3 at 298 K and expands isothermally to 122 cm^3. Calculate ΔG for the process.

15.28(a) The change in the Gibbs energy of a certain constant-pressure process was found to fit the expression $\Delta G/J = -85.40 + 36.5(T/K)$. Calculate the value of ΔS for the process.

15.28(b) The change in the Gibbs energy of a certain constant-pressure process was found to fit the expression $\Delta G/J = -73.1 + 42.8(T/K)$. Calculate the value of ΔS for the process.

15.29(a) Calculate the change in Gibbs energy of 35 g of ethanol (mass density 0.789 g cm^{-3}) when the pressure is increased isothermally from 1 atm to 3000 atm.

15.29(b) Calculate the change in Gibbs energy of 25 g of methanol (mass density 0.791 g cm^{-3}) when the pressure is increased isothermally from 100 kPa to 100 MPa.

15.30(a) Estimate the change in the Gibbs energy of 1.0 dm^3 of octane when the pressure acting on it is increased from 1.0 atm to 100 atm. The mass density of octane is 0.703 g cm^{-3}.

15.30(b) Estimate the change in the Gibbs energy of 100 cm^3 of water when the pressure acting on it is increased from 100 kPa to 500 kPa. The mass density of water is 0.997 g cm^{-3}.

15.31(a) Calculate the change in the molar Gibbs energy of hydrogen gas when its pressure is increased isothermally from 1.0 atm to 100.0 atm at 298 K.

15.31(b) Calculate the change in the molar Gibbs energy of oxygen when its pressure is increased isothermally from 50.0 kPa to 100.0 kPa at 500 K.

15.32(a) A CO_2 molecule is linear, and its vibrational wavenumbers are 1388.2 cm^{-1}, 667.4 cm^{-1}, and 2349.2 cm^{-1}, the last being doubly degenerate and the others non-degenerate. The rotational constant of the molecule is 0.3902 cm^{-1}. Calculate the rotational and vibrational contributions to the molar Gibbs energy at 298 K.

15.32(b) An O_3 molecule is angular, and its vibrational wavenumbers are 1110 cm^{-1}, 705 cm^{-1}, and 1042 cm^{-1}. The rotational constants of the molecule are 3.553 cm^{-1}, 0.4452 cm^{-1}, and 0.3948 cm^{-1}. Calculate the rotational and vibrational contributions to the molar Gibbs energy at 298 K.

15.33(a) The ground level of Cl is $^2P_{3/2}$ and a $^2P_{1/2}$ level lies 881 cm^{-1} above it. Calculate the electronic contribution to the molar Gibbs energy of Cl atoms at (a) 500 K and (b) 900 K.

15.33(b) The first electronically excited state of O_2 is $^1\Delta_g$ and lies 7918.1 cm^{-1} above the ground state, which is $^3\Sigma_g^-$. Calculate the electronic contribution to the molar Gibbs energy of O_2 at 500 K.

Problems*

Assume that all gases are perfect and that data refer to 298 K unless otherwise stated.

Numerical problems

15.1 Use the accurate expression for the rotational partition function calculated in Problem 13.12 for HCl(g) to calculate the rotational contribution to the molar entropy over a range of temperature and plot the contribution as a function of temperature.

15.2 Calculate the difference in molar entropy (a) between liquid water and ice at −5°C, (b) between liquid water and its vapour at 95°C and 1.00 atm. The differences in heat capacities on melting and on vaporization are 37.3 J K^{-1} mol^{-1} and −41.9 J K^{-1} mol^{-1}, respectively. Distinguish between the entropy changes of the sample, the surroundings, and the total system, and discuss the spontaneity of the transitions at the two temperatures.

15.3 The molar heat capacity of chloroform (trichloromethane, $CHCl_3$) in the range 240 K to 330 K is given by $C_{p,m}/(\text{J K}^{-1}\text{ mol}^{-1}) = 91.47 + 7.5\times10^{-2}\,(T/\text{K})$. In a particular experiment, 1.00 mol $CHCl_3$ is heated from 273 K to 300 K. Calculate the change in molar entropy of the sample.

15.4 A block of copper of mass 2.00 kg ($C_{p,m} = 24.44$ J K^{-1} mol^{-1}) and temperature 0°C is introduced into an insulated container in which there is 1.00 mol H_2O(g) at 100°C and 1.00 atm. (a) Assuming all the steam is condensed to water, what will be the final temperature of the system, the heat transferred from water to copper, and the entropy change of the water, copper, and the total system? (b) In fact, some water vapour is present at equilibrium. From the vapour pressure of water at the temperature calculated in (a), and assuming that the heat capacities of both gaseous and liquid water are constant and given by their values at that temperature, obtain an improved value of the final temperature, the heat transferred, and the various entropies. (*Hint.* You will need to make plausible approximations.)

15.5 Consider a perfect gas contained in a cylinder and separated by a frictionless adiabatic piston into two sections A and B. All changes in B are isothermal, that is, a thermostat surrounds B to keep its temperature constant. There is 2.00 mol of the gas in each section. Initially $T_A = T_B = 300$ K, $V_A = V_B = 2.00$ dm^3. Energy is supplied as heat to section A and the piston moves to the right reversibly until the final volume of section B is 1.00 L. Calculate (a) ΔS_A and ΔS_B, (b) ΔA_A and ΔA_B, (c) ΔG_A and ΔG_B, (d) ΔS of the total system and its surroundings. If numerical values cannot be obtained, indicate whether the values should be positive, negative, or zero or are indeterminate from the information given. (Assume $C_{V,m} = 20$ J K^{-1} mol^{-1}.)

15.6 A Carnot cycle uses 1.00 mol of a monatomic perfect gas as the working substance from an initial state of 10.0 atm and 600 K. It expands isothermally to a pressure of 1.00 atm (step 1), and then adiabatically to a temperature of 300 K (step 2). This expansion is followed by an isothermal compression (step 3), and then an adiabatic compression (step 4) back to the initial state. Determine the values of q, w, ΔU, ΔH, ΔS, ΔS_{tot}, and ΔG for each stage of the cycle and for the cycle as a whole. Express your answer as a table of values.

15.7 A sample consisting of 1.00 mol of perfect gas molecules at 27°C is expanded isothermally from an initial pressure of 3.00 atm to a final pressure of 1.00 atm in two ways: (a) reversibly, and (b) against a constant external pressure of 1.00 atm. Determine the values of q, w, ΔU, ΔH, ΔS, ΔS_{sur}, ΔS_{tot} for each path.

15.8 The standard molar entropy of NH_3(g) is 192.45 J K^{-1} mol^{-1} at 298 K, and its heat capacity is given by eqn 14.34 with the coefficients given in Table 14.2. Calculate the standard molar entropy at (a) 100°C and (b) 500°C.

15.9 A block of copper of mass 500 g and initially at 293 K is in thermal contact with an electric heater of resistance 1.00 kΩ and negligible mass. A current of 1.00 A is passed for 15.0 s. Calculate the change in entropy of the copper, taking $C_{p,m} = 24.4$ J K^{-1} mol^{-1}. The experiment is then repeated with the copper immersed in a stream of water that maintains its temperature at 293 K. Calculate the change in entropy of the copper and the water in this case.

15.10 Find an expression for the change in entropy when two blocks of the same substance and of equal mass, one at the temperature T_h and the other at T_c, are brought into thermal contact and allowed to reach equilibrium. Evaluate the change for two blocks of copper, each of mass 500 g, with $C_{p,m} = 24.4$ J K^{-1} mol^{-1}, taking $T_h = 500$ K and $T_c = 250$ K.

15.11 A gaseous sample consisting of 1.00 mol molecules is described by the equation of state $pV_m = RT(1 + Bp)$. Initially at 373 K, it undergoes Joule–Thomson expansion from 100 atm to 1.00 atm. Given that $C_{p,m} = \frac{5}{2}R$, $\mu = 0.21$ K atm^{-1}, $B = -0.525(\text{K}/T)$ atm^{-1} and that these are constant over the temperature range involved, calculate ΔT and ΔS for the gas.

15.12 The expressions that apply to the treatment of refrigerators also describe the behaviour of heat pumps, where warmth is obtained from the back of a refrigerator while its front is being used to cool the outside world. Heat pumps are popular home heating devices because they are very efficient. Compare heating of a room at 295 K by each of two methods: (a) direct conversion of 1.00 kJ of electrical energy in an electrical heater, and (b) use of 1.00 kJ of electrical energy to run a reversible heat pump with the outside at 260 K. Discuss the origin of the difference in the energy delivered to the interior of the house by the two methods.

15.13 The molar heat capacity of lead varies with temperature as follows:

T/K	10	15	20	25	30	50
$C_{p,m}$/(J K^{-1} mol^{-1})	2.8	7.0	10.8	14.1	16.5	21.4
T/K	70	100	150	200	250	298
$C_{p,m}$/(J K^{-1} mol^{-1})	23.3	24.5	25.3	25.8	26.2	26.6

Calculate the standard Third Law entropy of lead at (a) 0°C and (b) 25°C.

15.14 From standard enthalpies of formation, standard entropies, and standard heat capacities available from tables in the *Data section*, calculate: (a) the standard enthalpies and entropies at 298 K and 398 K for the reaction $CO_2(g) + H_2(g) \rightarrow CO(g) + H_2O(g)$. Assume that the heat capacities are constant over the temperature range involved.

* Problems denoted with the symbol ‡ were supplied by Charles Trapp, Carmen Giunta, and Marshall Cady.

15.15 The molar heat capacity of anhydrous potassium hexacyanoferrate(II) varies with temperature as follows:

T/K	$C_{p,m}$/(J K^{-1} mol^{-1})	T/K	$C_{p,m}$/(J K^{-1} mol^{-1})
10	2.09	100	179.6
20	14.43	110	192.8
30	36.44	150	237.6
40	62.55	160	247.3
50	87.03	170	256.5
60	111.0	180	265.1
70	131.4	190	273.0
80	149.4	200	280.3
90	165.3		

Calculate the molar enthalpy relative to its value at $T = 0$ and the Third Law entropy at each of these temperatures.

15.16 The compound 1,3,5-trichloro-2,4,6-trifluorobenzene is an intermediate in the conversion of hexachlorobenzene to hexafluorobenzene, and its thermodynamic properties have been examined by measuring its heat capacity over a wide temperature range (R.L. Andon and J.F. Martin, *J. Chem. Soc. Faraday Trans. I*, 871 (1973)). Some of the data are as follows:

T/K	14.14	16.33	20.03	31.15	44.08	64.81
$C_{p,m}$/(J K^{-1} mol^{-1})	9.492	12.70	18.18	32.54	46.86	66.36
T/K	100.90	140.86	183.59	225.10	262.99	298.06
$C_{p,m}$/(J K^{-1} mol^{-1})	95.05	121.3	144.4	163.7	180.2	196.4

Calculate the molar enthalpy relative to its value at $T = 0$ and the Third Law molar entropy of the compound at these temperatures.

15.17‡ Given that $S_m^\ominus = 29.79$ J K^{-1} mol^{-1} for bismuth at 100 K and the following tabulated heat capacity data (D.G. Archer, *J. Chem. Eng. Data* **40**, 1015 (1995)), compute the standard molar entropy of bismuth at 200 K.

T/K	100	120	140	150	160	180	200
$C_{p,m}$/(J K^{-1} mol^{-1})	23.00	23.74	24.25	24.44	24.61	24.89	25.11

Compare the value to the value that would be obtained by taking the heat capacity to be constant at 24.44 J K^{-1} mol^{-1} over this range.

15.18 Calculate $\Delta_r G^\ominus(375\text{ K})$ for the reaction $2\,CO(g) + O_2(g) \rightarrow 2\,CO_2(g)$ from the value of $\Delta_r G^\ominus(298\text{ K})$, $\Delta_r H^\ominus(298\text{ K})$, and the Gibbs–Helmholtz equation.

15.19 Estimate the standard reaction Gibbs energy of $N_2(g) + 3\,H_2(g) \rightarrow 2\,NH_3(g)$ at (a) 500 K, (b) 1000 K from its value at 298 K.

15.20 Calculate the standard molar entropy of $N_2(g)$ at 298 K from its rotational constant $\tilde{B} = 1.9987$ cm^{-1} and its vibrational wavenumber $\tilde{\nu} = 2358$ cm^{-1}. The thermochemical value is 192.1 J K^{-1} mol^{-1}. What does this suggest about the solid at $T = 0$?

15.21‡ J.G. Dojahn *et al.* (*J. Phys. Chem.* **100**, 9649 (1996)) characterized the potential energy curves of the ground and electronic states of homonuclear diatomic halogen anions. The ground state of F_2^- is $^2\Sigma_u^+$ with a fundamental vibrational wavenumber of 450.0 cm^{-1} and equilibrium internuclear distance of 190.0 pm. The first two excited states are at 1.609 and 1.702 eV above the ground state. Compute the standard molar entropy of F_2^- at 298 K.

15.22‡ Treat carbon monoxide as a perfect gas and apply equilibrium statistical thermodynamics to the study of its properties, as specified below, in the temperature range 100–1000 K at 1 bar. $\tilde{\nu} = 2169.8$ cm^{-1}, $\tilde{B} = 1.931$ cm^{-1}, and $D_0 = 11.09$ eV; neglect anharmonicity and centrifugal distortion. (a) Examine the probability distribution of molecules over available rotational and vibrational states. (b) Explore numerically the differences, if any, between the rotational molecular partition function as calculated with the discrete energy distribution with that calculated with the classical, continuous energy distribution. (c) Calculate the individual contributions to $U_m(T) - U_m(100\text{ K})$, $C_{V,m}(T)$, and $S_m(T) - S_m(100\text{ K})$ made by the translational, rotational, and vibrational degrees of freedom.

15.23 At 298 K the standard enthalpy of combustion of sucrose is −5797 kJ mol^{-1} and the standard Gibbs energy of the reaction is −6333 kJ mol^{-1}. Estimate the additional non-expansion work that may be obtained by raising the temperature to blood temperature, 37°C.

15.24‡ R. Viswanathan *et al.* (*J. Phys. Chem.* **100**, 10784 (1996)) studied the thermodynamic properties of several boron–silicon gas-phase species experimentally and theoretically. These species can occur in the high-temperature chemical vapour deposition (CVD) of silicon-based semiconductors. Among the computations they reported was computation of the Gibbs energy of BSi(g) at several temperatures based on a $^4\Sigma^-$ ground state with equilibrium internuclear distance of 190.5 pm and fundamental vibrational wavenumber of 772 cm^{-1} and a $^2P_0$ first excited level 8000 cm^{-1} above the ground level. Compute the standard molar Gibbs energy $G_m^\ominus(2000\text{ K}) - G_m^\ominus(0)$.

15.25 Suppose that an internal combustion engine runs on octane, for which the enthalpy of combustion is −5512 kJ mol^{-1} and take the mass of 1 gallon of fuel as 3 kg. What is the maximum height, neglecting all forms of friction, to which a car of mass 1000 kg can be driven on 1.00 gallon of fuel given that the engine cylinder temperature is 2000°C and the exit temperature is 800°C?

Theoretical problems

15.26 The energy levels of a Morse oscillator are given in eqn 10.36. Set up the expression for the molar entropy of a collection of Morse oscillators and plot it as a function of temperature for a series of anharmonicities. Take into account only the finite number of bound states. On the same graph plot the entropy of an harmonic oscillator and investigate how the two diverge.

15.27 Explore how the entropy of a collection of two-level systems behaves when the temperature is formally allowed to become negative (recall Problem 14.48). You should also construct a graph in which the temperature is replaced by the variable $\beta = 1/kT$. Account for the appearance of the graphs physically.

15.28 Deduce the result $(\partial U/\partial S)_V = T$ and then use the calculation on which Problem 15.27 is based to draw a graph of U against S (or vice versa) to identify the temperature. *Hint.* Use mathematical software to construct the graph.

15.29 According to Newton's law of cooling, the rate of change of temperature is proportional to the temperature difference between the system and its surroundings. Given that $S(T) - S(T_i) = C \ln(T/T_i)$, where T_i is the initial temperature and C the heat capacity, deduce an expression for the rate of change of entropy of the system as it cools.

15.30 Represent the Carnot cycle on a temperature–entropy diagram and show that the area enclosed by the cycle is equal to the work done.

15.31 The cycle involved in the operation of an internal combustion engine is called the *Otto cycle*. Air can be considered to be the working substance and can be assumed to be a perfect gas. The cycle consists of the following steps: (1) Reversible adiabatic compression from A to B, (2) reversible constant-volume pressure increase from B to C due to the combustion of a small amount of fuel, (3) reversible adiabatic expansion from C to D, and (4) reversible and constant-volume pressure decrease back to state A. Determine the change in entropy (of the system and of the surroundings) for each step of the cycle and determine an expression for the efficiency of the cycle, assuming that the heat is supplied in step 2. Evaluate the efficiency for a compression ratio of 10:1. Assume that in state A, $V = 4.00$ L, $p = 1.00$ atm, and $T = 300$ K, that $V_A = 10V_B$, $p_C/p_B = 5$, and that $C_{p,m} = \frac{7}{2}R$.

15.32 Prove that two reversible adiabatic paths can never cross. Assume that the energy of the system under consideration is a function of temperature only. (*Hint.* Suppose that two such paths can intersect, and complete a cycle with the two paths plus one isothermal path. Consider the changes accompanying each stage of the cycle and show that they conflict with the Kelvin statement of the Second Law.)

15.33 Derive an expression for the molar entropy of a monatomic solid on the basis of the Einstein and Debye models and plot the molar entropy against the temperature (use T/θ in each case, with θ the Einstein or Debye temperature). Use the following expressions for the temperature dependence of the heat capacities:

$$\text{Einstein: } C_{V,m}(T) = 3Rf_E(T) \qquad f_E(T) = \left(\frac{\theta_E}{T}\right)^2\left(\frac{e^{\theta_E/2T}}{e^{\theta_E/T}-1}\right)^2$$

$$\text{Debye: } C_{V,m}(T) = 3Rf_D(T) \qquad f_D(T) = 3\left(\frac{T}{\theta_D}\right)^2\int_0^{\theta_D/T}\frac{x^4e^x}{(e^x-1)^2}dx$$

Use mathematical software to evaluate the appropriate expressions.

15.34 Two empirical equations of state of a real gas are as follows:

$$\text{van der Waals: } p = \frac{RT}{V_m - b} - \frac{a}{V_m^2}$$

$$\text{Dieterici: } p = \frac{RTe^{-a/RTV_m}}{V_m - b}$$

Evaluate $(\partial S/\partial V)_T$ for each gas. For an isothermal expansion, for which kind of gas (also consider a perfect gas) will ΔS be greatest? Explain your conclusion.

15.35 Two of the four Maxwell relations were derived in the text, but two were not. Complete their derivation by showing that $(\partial S/\partial V)_T = (\partial p/\partial T)_V$ and $(\partial T/\partial p)_S = (\partial V/\partial S)_p$.

15.36 (a) Use the Maxwell relations to express the derivatives $(\partial S/\partial V)_T$, $(\partial V/\partial S)_p$, $(\partial p/\partial S)_V$, and $(\partial V/\partial S)_p$ in terms of the heat capacities, the expansion coefficient $\alpha = (1/V)(\partial V/\partial T)_p$, and the isothermal compressibility, $\kappa_T = -(1/V)(\partial V/\partial p)_T$. (b) The Joule coefficient, μ_J, is defined as $\mu_J = (\partial T/\partial V)_U$. Show that $\mu_J C_V = p - \alpha T/\kappa_T$.

15.37 Suppose that S is regarded as a function of p and T. Show that $TdS = C_p dT - \alpha TV dp$. Hence, show that the energy transferred as heat when the pressure on an incompressible liquid or solid is increased by Δp is equal to $-\alpha TV\Delta p$, where $\alpha = (1/V)(\partial V/\partial T)_p$. Evaluate q when the pressure acting on 100 cm^3 of mercury at 0°C is increased by 1.0 kbar. ($\alpha = 1.82 \times 10^{-4}$ K^{-1}.)

15.38 Derive the Sackur–Tetrode equation for a monatomic gas confined to a two-dimensional surface, and hence derive an expression for the standard molar entropy of condensation to form a mobile surface film.

15.39 In Problem 14.30 you were invited to consider the expressions

$$q = \sum_j e^{-\beta\varepsilon_j} \qquad \dot{q} = \sum_j \beta\varepsilon_j e^{-\beta\varepsilon_j} \qquad \ddot{q} = \sum_j (\beta\varepsilon_j)^2 e^{-\beta\varepsilon_j}$$

in the context of the First Law. To see that these expressions are also relevant to the Second Law, derive an expression for the entropy in terms of these three functions. (b) Apply the technique to the calculation of the electronic contribution to the standard molar entropy of magnesium vapour at 5000 K using the following data:

Term	^{1}S	$^3P_0$	$^3P_1$	$^3P_2$	$^1P_1$	^{3}S
Degeneracy	1	1	3	5	3	3
$\tilde{\nu}/cm^{-1}$	0	21850	21870	21911	35051	41197

15.40 To calculate the work required to lower the temperature of an object, we need to consider how the coefficient of performance c changes with the temperature of the object. (a) Find an expression for the work of cooling an object from T_i to T_f when the refrigerator is in a room at a temperature T_h. *Hint.* Write $dw = dq/c(T)$, relate dq to dT through the heat capacity C_p, and integrate the resulting expression. Assume that the heat capacity is independent of temperature in the range of interest. (b) Use the result in part (a) to calculate the work needed to freeze 250 g of water in a refrigerator at 293 K. How long will it take when the refrigerator operates at 100 W?

15.41 Calculate the molar internal energy, molar entropy, and molar Helmholtz energy of a collection of harmonic oscillators and plot your expressions as a function of T/θ_V, where $\theta_V = h\tilde{\nu}/k$.

15.42 Equation 15.58 expresses the internal pressure π_T in terms of the pressure and its derivative with respect to temperature. Express π_T in terms of the molecular partition function.

15.43 Identify as many arguments as you can that confirm the relation $\beta = 1/kT$. Present arguments that show that β is a more appropriate parameter for expressing the temperature than T itself. What is the status of k as a fundamental constant?

15.44 Explore the consequences of replacing the equation of state of a perfect gas by the van der Waals equation of state for the pressure dependence of the molar Gibbs energy (eqn 15.64). Proceed in three steps. First, consider the case when $a = 0$ and only repulsions are significant. Then consider the case when $b = 0$ and only attractions are significant. For the latter, you should consider making the approximation that the attractions are weak. Finally, explore the full expression by using mathematical software. In each case plot your results graphically and account physically for the deviations from the perfect gas expression.

Applications: to biology and environmental science

15.45 An average human DNA molecule has 5×10^8 binucleotides (rungs on the DNA ladder) of four different kinds. If each rung were a random choice of one of these four possibilities, what would be the residual entropy associated with this typical DNA molecule?

15.46 The protein lysozyme unfolds at a transition temperature of 75.5°C and the standard enthalpy of transition is 509 kJ mol^{-1}. Calculate the entropy of unfolding of lysozyme at 25.0°C, given that the difference in the constant-pressure heat capacities upon unfolding is 6.28 kJ K^{-1} mol^{-1} and can be assumed to be independent of temperature. *Hint.* Imagine that the transition at 25.0°C occurs in three steps: (i) heating of the folded protein from 25.0°C to the transition temperature, (ii) unfolding at the transition temperature, and (iii) cooling of the unfolded protein to 25.0°C. Because the entropy is a state function, the entropy change at 25.0°C is equal to the sum of the entropy changes of the steps.

15.47 In biological cells, the energy released by the oxidation of foods is stored in adenosine triphosphate (ATP or ATP^{4-}). The essence of ATP's action is its ability to lose its terminal phosphate group by hydrolysis and to form adenosine diphosphate (ADP or ADP^{3-}):

$$ATP^{4-}(aq) + H_2O(l) \rightarrow ADP^{3-}(aq) + HPO_4^{2-}(aq) + H_3O^+(aq)$$

At pH = 7.0 and 37°C (310 K, blood temperature) the enthalpy and Gibbs energy of hydrolysis are $\Delta_r H = -20$ kJ mol^{-1} and $\Delta_r G = -31$ kJ mol^{-1}, respectively. Under these conditions, the hydrolysis of 1 mol ATP^{4-}(aq) results in the extraction of up to 31 kJ of energy that can be used to do non-expansion work, such as the synthesis of proteins from amino acids, muscular contraction, and the activation of neuronal circuits in our brains. (a) Calculate and account for the sign of the entropy of hydrolysis of ATP at pH = 7.0 and 310 K. (b) Suppose that the radius of a typical biological cell is 10 μm and that inside it 10^6 ATP molecules are hydrolysed each second. What is the power density of the cell in watts per cubic metre (1 W = 1 J s^{-1})? A computer battery delivers about 15 W and has a volume of 100 cm^3. Which has the greater power density, the cell or the battery? (c) The formation of glutamine from glutamate and ammonium ions requires 14.2 kJ mol^{-1} of energy input. It is driven by the hydrolysis of ATP to ADP mediated by the enzyme glutamine synthetase. How many moles of ATP must be hydrolysed to form 1 mol glutamine?

15.48‡ The molecule Cl_2O_2, which is believed to participate in the seasonal depletion of ozone over Antarctica, has been studied by several means. Birk *et al.* (*J. Chem. Phys.* **91**, 6588 (1989)) report its rotational constants as $A = 13109.4$, $B = 2409.8$, and $C = 2139.7$ MHz. They also report that its rotational spectrum indicates a molecule with a symmetry number of 2.19. J. Jacobs *et al.* (*J. Amer. Chem. Soc.* 1106 (1994)) report its vibrational wavenumbers as 753, 542, 310, 127, 646, and 419 cm^{-1}. Compute $G_m^\ominus(200\text{ K}) - G_m^\ominus(0)$ of Cl_2O_2.

15.49‡ Nitric acid hydrates have received much attention as possible catalysts for heterogeneous reactions that bring about the Antarctic ozone hole. Worsnop *et al.* (*Science* **259**, 71 (1993)) investigated the thermodynamic stability of these hydrates under conditions typical of the polar winter stratosphere. They report thermodynamic data for the sublimation of mono-, di-, and trihydrates to nitric acid and water vapours, $HNO_3{\cdot}nH_2O\text{ (s)} \rightarrow HNO_3\text{ (g)} + nH_2O\text{ (g)}$, for $n = 1$, 2, and 3. Given $\Delta_r G^\ominus$ and $\Delta_r H^\ominus$ for these reactions at 220 K, use the Gibbs–Helmholtz equation to compute $\Delta_r G^\ominus$ at 190 K.

n	1	2	3
$\Delta_r G^\ominus/(\text{kJ mol}^{-1})$	46.2	69.4	93.2
$\Delta_r H^\ominus/(\text{kJ mol}^{-1})$	127	188	237

16 Physical equilibria

The discussion of the phase transitions of pure substances is among the simplest applications of thermodynamics to chemistry. We shall see that a phase diagram is a map of the pressures and temperatures at which each phase of a substance is the most stable. First, we describe the interpretation of empirically determined phase diagrams for a selection of one- and two-component systems. Then we turn to a consideration of the factors that determine the positions and shapes of the boundaries between the regions on a phase diagram. This chapter also introduces the chemical potential, a property that is at the centre of discussions of phase transitions and chemical reactions. The underlying principle to keep in mind is that at equilibrium the chemical potential of a species is the same in every phase. We see, by making use of the experimental observations known as Raoult's and Henry's laws, how to express the chemical potential of a substance in terms of the composition of a mixture and calculate the effect of a solute on certain thermodynamic properties of a solution. Finally, we see how to express the chemical potential of a substance in a real mixture in terms of the activity, and effective concentration, and see how to express the activity in terms of various models of molecular and ionic interaction.

Vaporization, melting, and the conversion of graphite to diamond are all examples of changes of phase without change of chemical composition. In this chapter we describe such processes thermodynamically, using as the guiding principle the tendency of systems at constant temperature and pressure to minimize their Gibbs energy. We also describe a systematic way of discussing the physical changes mixtures undergo when they are heated or cooled and when their compositions are changed and see how to use phase diagrams. To set the scene for this discussion, we first review how empirical information about phase equilibria is summarized graphically. Then we examine how these phase diagrams may be interpreted and in some cases modelled in terms of thermodynamic quantities.

Phase diagrams

A **phase** of a substance is a form of matter that is uniform throughout in chemical composition and physical state. Thus, we speak of solid, liquid, and gas phases of a substance, and of its various solid phases, such as the white and black allotropes of phosphorus. A **phase transition**, the spontaneous conversion of one phase into another phase, occurs at a characteristic temperature for a given pressure. Thus, at 1 atm, ice is the stable phase of water below 0°C, but above 0°C liquid water is more stable. This difference indicates that below 0°C the Gibbs energy decreases as liquid water changes into ice and that above 0°C the Gibbs energy decreases as ice changes into liquid water. An empirically constructed **phase diagram** of a substance shows

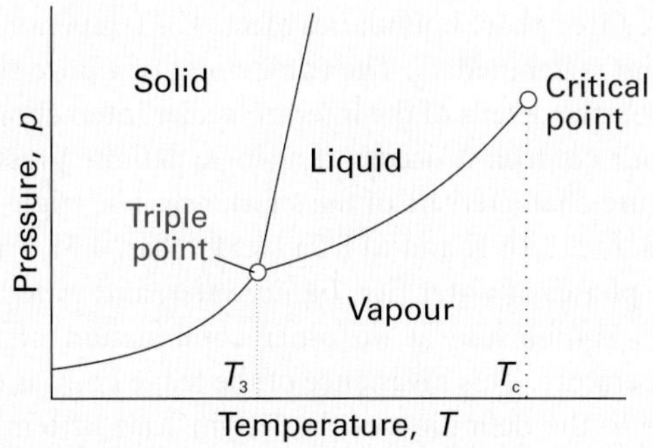

Fig. 16.1 The general regions of pressure and temperature where solid, liquid, or gas is stable (that is, has minimum molar Gibbs energy) are shown on this phase diagram. For example, the solid phase is the most stable phase at low temperatures and high pressures. In the following paragraphs we locate the precise boundaries between the regions.

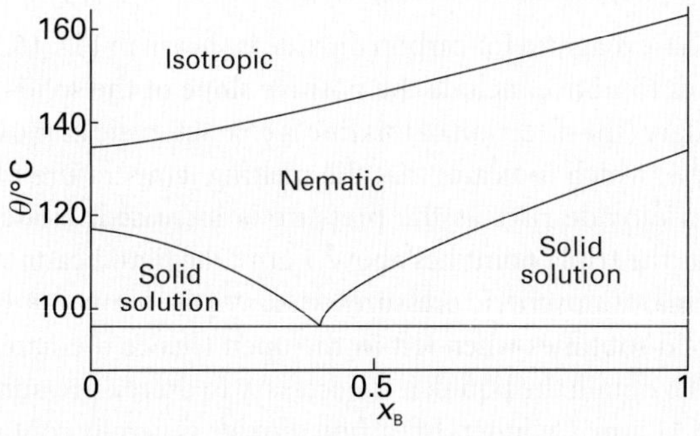

Fig. 16.2 A typical phase diagram for a mixture, in this case of two liquid crystalline materials showing the conditions under which various phases are the most stable. The two components are 4,4′-dimethoxyazobenzene (A) and 4,4′-diethoxyazobenzene (B).

the regions of pressure and temperature at which its various phases are thermodynamically stable (Fig. 16.1). The lines separating the regions, which are called **phase boundaries**, show the values of p and T at which two phases coexist in equilibrium. Similar diagrams can be constructed for mixtures, where the regions summarize, for instance, the compositions and temperatures at which the various phases are stable and the boundaries summarize the conditions under which these phases are in equilibrium with one another (Fig. 16.2).

All the properties we describe in this chapter can be traced to the intermolecular interactions that bind the molecules together. Which particular phase is the most stable at a given pressure and at $T = 0$, where entropy plays no role, corresponds to the lowest energy that can be achieved as the molecules pack together. At higher temperatures entropy does play a role, and the substance adopts the molecular arrangement that corresponds to lowest Gibbs energy at the prevailing pressure. Although it is difficult to extend the computational techniques of Chapter 6 to solids and liquids, that can be done, and phase transitions can in some cases be predicted. However, that is beyond the scope of this text: here for the most part we shall simply refer to the qualitative features of intermolecular forces.

There are two further preliminary points. One is that one of the most celebrated results in chemical thermodynamics, the phase rule, can be used as a basis for discussing the implications of phase diagrams, but it is not essential. It is described in *Further information 16.1*. The second is that it is sometimes useful to classify phase transitions into different types. The Ehrenfest classification is described in *Further information 16.2*.

16.1 One-component systems

We begin by considering a liquid sample of a pure substance in a closed vessel. The pressure of a vapour in equilibrium with the liquid is called the **vapour pressure** of the substance (Fig. 16.3). Therefore, the liquid–vapour phase boundary in a phase diagram shows how the vapour pressure of the liquid varies with temperature. Similarly, the solid–vapour phase boundary shows the temperature variation of the **sublimation vapour pressure**, the vapour pressure of the solid phase. The vapour pressure of a substance increases with temperature because at higher temperatures more molecules have sufficient energy to escape from their neighbours.

(a) Phase characteristics

When a liquid is heated in an *open* vessel, the liquid vaporizes from its surface as molecules acquire enough kinetic energy to escape from their neighbours. At the temperature at which its vapour pressure would be equal to the external pressure, vapour can form throughout the bulk of the liquid and can expand freely into the surroundings. The condition of free vaporization throughout the liquid is called **boiling**. The temperature at which the vapour pressure of a liquid is equal to the external pressure is called the **boiling temperature** at that pressure. For the special case of an external pressure of 1 atm, the boiling temperature is called the **normal boiling point**, T_b. With the replacement of 1 atm by 1 bar as standard pressure, there is some

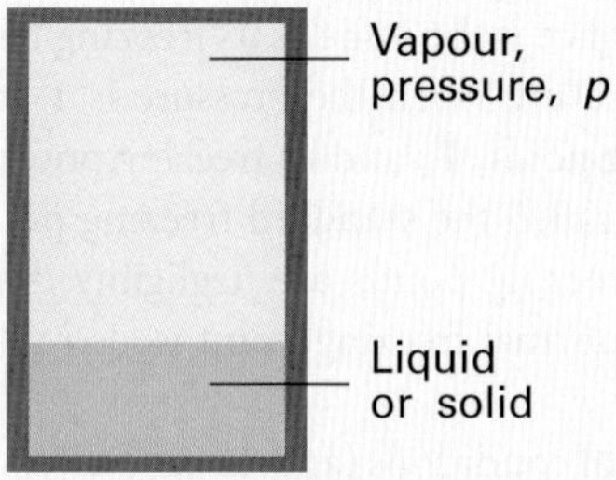

Fig. 16.3 The vapour pressure of a liquid or solid is the pressure exerted by the vapour in equilibrium with the condensed phase.

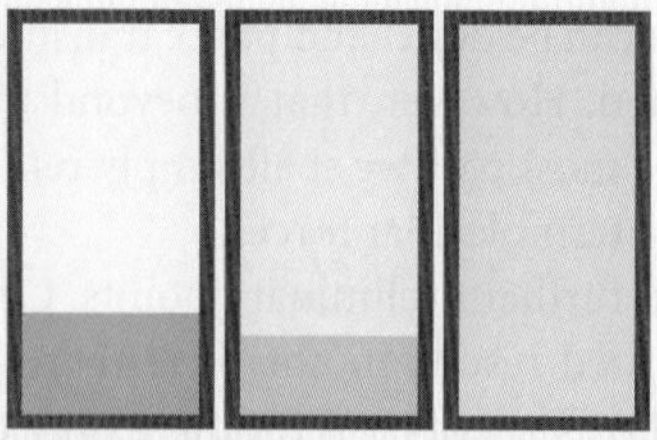

Fig. 16.4 (a) A liquid in equilibrium with its vapour. (b) When a liquid is heated in a sealed container, the density of the vapour phase increases and that of the liquid decreases slightly. There comes a stage, (c), at which the two densities are equal and the interface between the fluids disappears. This disappearance occurs at the critical temperature. The container needs to be strong: the critical temperature of water is 374°C and the vapour pressure is then 218 atm.

advantage in using the **standard boiling point** instead: this is the temperature at which the vapour pressure reaches 1 bar. Because 1 bar is slightly less than 1 atm (1.00 bar = 0.987 atm), the standard boiling point of a liquid is slightly lower than its normal boiling point. The normal boiling point of water is 100.0°C; its standard boiling point is 99.6°C.

Boiling does not occur when a liquid is heated in a partly filled, rigid, closed vessel. Instead, the vapour pressure, and hence the density of the vapour, rise as the temperature is raised (Fig. 16.4). At the same time, the density of the liquid decreases slightly as a result of its expansion. There comes a stage when the density of the vapour is equal to that of the remaining liquid and the surface between the two phases disappears. The temperature at which the surface disappears is the **critical temperature**, T_c, of the substance. The vapour pressure at the critical temperature is called the **critical pressure**, p_c. At and above the critical temperature, a single uniform phase called a **supercritical fluid** fills the container and an interface no longer exists. That is, above the critical temperature, the liquid phase of the substance does not exist.

The temperature at which, under a specified pressure, the liquid and solid phases of a substance coexist in equilibrium is called the **melting temperature.** Because a substance melts at exactly the same temperature as it freezes, the melting temperature of a substance is the same as its **freezing temperature.** The freezing temperature when the pressure is 1 atm is called the **normal freezing point,** T_f, and its freezing point when the pressure is 1 bar is called the **standard freezing point.** The normal and standard freezing points are negligibly different for most purposes. The normal freezing point is also called the **normal melting point.**

There is a set of conditions (and, in general, several sets) under which three different phases of a single substance (such as solid, liquid, and vapour) all simultaneously coexist in equilibrium. These conditions are represented by the **triple point**, a point at which the three phase boundaries meet. The temperature at the triple point is denoted T_3. The triple point of a pure substance depends on the details of the intermolecular interactions and is outside our control: it occurs at a single definite pressure and temperature characteristic of the substance. The triple point of water lies at 273.16 K and 611 Pa (6.11 mbar, 4.58 Torr), and the three phases of water (ice, liquid water, and water vapour) coexist in equilibrium at no other combination of pressure and temperature. This invariance of the triple point is the basis of its use in the definition of the Kelvin scale of temperature (Section 15.2).

As we can see from Fig. 16.1, the triple point marks the lowest pressure at which a liquid phase of a substance can exist. If (as is common) the slope of the solid–liquid phase boundary is as shown in the diagram, then the triple point also marks the lowest temperature at which the liquid can exist; the critical temperature is the upper limit.

(b) Three typical phase diagrams

The phase diagram for carbon dioxide is shown in Fig. 16.5. The features to notice include the positive slope of the solid–liquid boundary (the direction of this line is characteristic of most substances), which indicates that the melting temperature of solid carbon dioxide rises as the pressure is increased. Notice also that, as the triple point lies above 1 atm, the liquid cannot exist at normal atmospheric pressures whatever the temperature, and the solid sublimes when left in the open (hence the name 'dry ice'). To obtain the liquid, it is necessary to exert a pressure of at least 5.11 atm. Cylinders of carbon dioxide generally contain the liquid or compressed gas; at 25°C that implies a vapour pressure of 67 atm if both gas and liquid are present in equilibrium. When the gas squirts through the throttle it cools by the Joule–

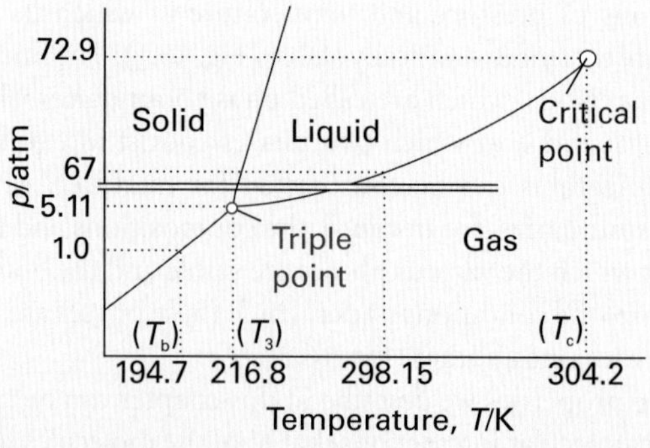

Fig. 16.5 The experimental phase diagram for carbon dioxide. Note that, as the triple point lies at pressures well above atmospheric, liquid carbon dioxide does not exist under normal conditions (a pressure of at least 5.1 atm must be applied).

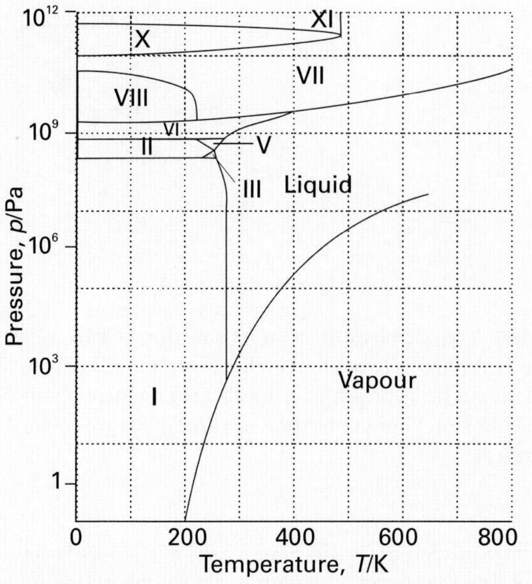

Fig. 16.6 The experimental phase diagram for water showing the different solid phases. Metastable phases are not included.

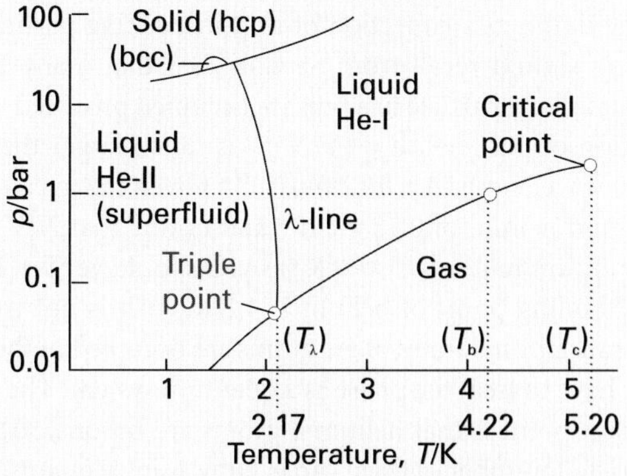

Fig. 16.7 The phase diagram for helium (^4He). The λ-line marks the conditions under which the two liquid phases are in equilibrium. Helium-II is the superfluid phase. Note that a pressure of over 20 bar must be exerted before solid helium can be obtained. The labels hcp and bcc denote different solid phases in which the atoms pack together differently: hcp denotes hexagonal closed packing and bcc denotes body-centred cubic (see Section 9.5 for a description of these structures).

Thomson effect (Section 14.11) so, when it emerges into a region where the pressure is only 1 atm, it condenses into a finely divided snow-like solid.

Figure 16.6 is the phase diagram for water. The liquid–vapour boundary in the phase diagram summarizes how the vapour pressure of liquid water varies with temperature. It also summarizes how the boiling temperature varies with pressure: we simply read off the temperature at which the vapour pressure is equal to the prevailing atmospheric pressure. The solid–liquid boundary shows how the melting temperature varies with the pressure. Its very steep slope indicates that enormous pressures are needed to bring about significant changes. Notice that the line has a negative slope up to 2 kbar, which means that the melting temperature falls as the pressure is raised. The reason for this almost unique behaviour can be traced to the decrease in volume that occurs on melting, and hence it being more favourable for the solid to transform into the liquid as the pressure is raised. The decrease in volume is a result of the very open molecular structure of ice: the water molecules are held apart, as well as together, by the hydrogen bonds between them but the structure partially collapses on melting and the liquid is denser than the solid.

Figure 16.6 shows that water has one liquid phase but many different **polymorphs**, or different solid phases, other than ordinary ice ('ice I'). This polymorphic richness is due in large measure to the adaptability of the directional characteristics of hydrogen-bonding interactions, which allow the oxygen atoms to adopt slightly different arrangements throughout the solid in response to changes in pressure and temperature. Some of the phases melt at high temperatures. Ice VII, for instance, melts at 100°C but exists only above 25 kbar. Note that many more triple points occur in the diagram other than the one where vapour, liquid, and ice I coexist. Each one occurs at a definite pressure and temperature that cannot be changed. The polymorphs of ice differ in the arrangement of the water molecules: under the influence of very high pressures, hydrogen bonds buckle and the H_2O molecules adopt different arrangements. They may be responsible for the advance of glaciers, for ice at the bottom of glaciers experiences very high pressures where it rests on jagged rocks.

Figure 16.7 shows the phase diagram of helium. Helium behaves unusually at low temperatures. For instance, the solid and gas phases of helium are never in equilibrium however low the temperature: the atoms are so light that they vibrate with a large-amplitude motion even at very low temperatures and the solid simply shakes itself apart. Solid helium can be obtained, but only by holding the atoms together by applying pressure. Pure helium-4 has two liquid phases. The phase marked He-I in the diagram behaves like a normal liquid; the other phase, He-II, is a **superfluid**; it is so called because it flows without viscosity. Provided we discount the liquid crystalline substances discussed in *Impact I8.3*, helium is the only known substance with a liquid–liquid boundary, shown as the **λ-line** (lambda line) in Fig. 16.7. The phase diagram of helium-3 differs from the phase diagram of helium-4, but it also possesses a superfluid phase. Helium-3 is unusual in that the entropy of the liquid is lower than that of the solid, and melting is exothermic.

The existence of superfluid helium is a quantum phenomenon that manifests itself on a macroscopic scale. Because the interatomic forces in helium are so weak, as a first approximation we can treat the liquid as a collection of non-interacting particles in a box. You should recall from Section 13.3 that many translational states of a particle in a box are occupied provided that the separation of particles, $d = (V/N)^{1/3}$ is much larger than their thermal wavelength, $\Lambda = h/(2\pi mkT)^{1/2}$. Given the mass density, ρ, of liquid helium of 0.15 g cm^{-3} and noting that $N/V = \rho/m$, this condition requires $T \gg 6$ K (as you should verify). But the normal boiling point of helium is 4.2 K, so it is not true that many translational states of He atoms are occupied in the liquid and we have to treat the phase as a quantum system. The second important point is that helium-4 atoms are bosons, so that an unrestricted number of them can occupy a single quantum state (Section 4.4). The current view is that helium-II consists of two components. In this **two-fluid model**, below 2.17 K the liquid consists of a normal liquid component and a superfluid component, with the proportions changing as the temperature is lowered and becoming entirely superfluid at $T = 0$. Although it is tempting to think of the superfluid phase as consisting of all the atoms in the lowest energy state (corresponding to $n = 1$ for a particle in a box) and zero linear momentum, that is not quite right, for neutron scattering experiments have shown that only about 10 per cent of the atoms have zero linear momentum at $T = 0$, despite the phase then being entirely superfluid. The ground state is in fact much more complicated, with correlated pairs of atoms with zero overall linear momentum. Helium-3 forms a superfluid phase despite being a spin-$\frac{1}{2}$ fermion. In its case, pairs of atoms act jointly (like pairs of electrons in superconductivity, Section 9.12), and each pair behaves like a single spin-0 boson.

16.2 Two-component systems

If two components are present in a mixture, there are three variables to consider: the pressure, the temperature, and the composition. Hence, one form of the phase diagram is a map of pressures and compositions at which each phase is stable. Alternatively, the pressure could be held constant and the phase diagram depicted in terms of temperature and composition.

(a) Liquid–vapour systems

Figure 16.8 is a typical **pressure–composition diagram** at a fixed temperature. All the points above the diagonal line in the graph correspond to a system under such high pressure that it contains only a liquid phase (the applied pressure is higher than the vapour pressure). All points below the lower curve correspond to a system under such low pressure that it contains only a vapour phase (the applied pressure is lower than the vapour pressure). Points that lie between the two lines correspond to a system in which there are two phases present, one a liquid and the other a vapour. To see this interpretation, consider the effect of lowering the pressure on a liquid mixture of overall composition a in Fig. 16.8. The changes to the system do not affect the

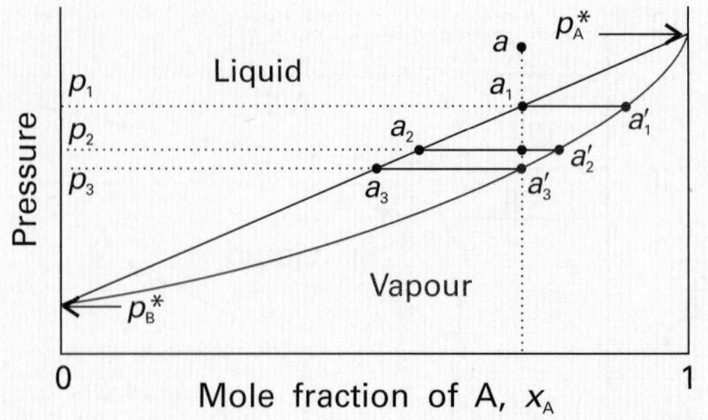

Fig. 16.8 A typical phase diagram for a mixture of two volatile liquids. A point between the two lines corresponds to both liquid and vapour being present; outside that region there is only one phase present. Vapour pressures of pure liquids are denoted p_A^* and p_B^*.

overall composition, so the state of the system moves down the vertical line that passes through a. This vertical line is called an **isopleth** (from the Greek words for 'equal abundance'). Until the point a_1 is reached (when the pressure has been reduced to p_1), the sample consists of a single liquid phase. At a_1 the liquid can exist in equilibrium with its vapour of composition a_1'. A line joining two points representing phases in equilibrium is called a **tie line**. The composition of the liquid is the same as initially (a_1 lies on the isopleth through a), so we have to conclude that at this pressure there is virtually no vapour present; however, the tiny amount of vapour that is present has the composition a_1'.

To use a phase diagram to find the relative amounts of two phases α and β that are in equilibrium when the system is in a two-phase region, we measure the distances l_α and l_β along the horizontal tie line, and then use the **lever rule** (Fig. 16.9):

$$n_\alpha l_\alpha = n_\beta l_\beta \qquad (16.1)$$

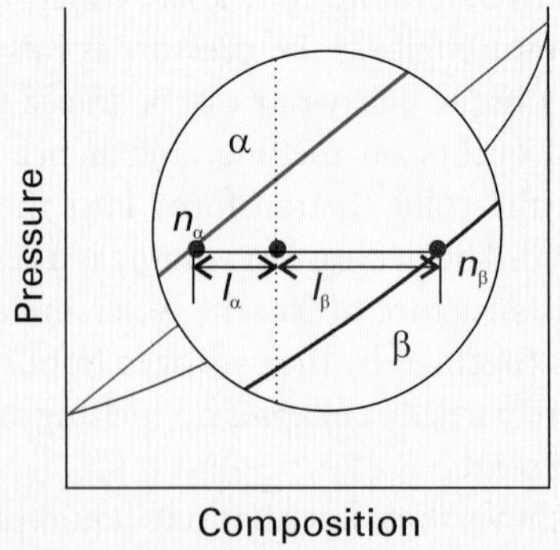

Fig. 16.9 The lever rule. The distances l_α and l_β are used to find the proportions of the amounts of phases α (such as vapour) and β (for example, liquid) present at equilibrium. The lever rule is so called because a similar rule relates the masses at two ends of a lever to their distances from a pivot ($m_\alpha l_\alpha = m_\beta l_\beta$ for balance).

Here n_α is the amount of phase α and n_β the amount of phase β. In the case illustrated in Fig. 16.9, because $l_\beta \approx 2l_\alpha$, the amount of phase α is about twice the amount of phase β.

● A BRIEF ILLUSTRATION

At p_1 in Fig. 16.8, the ratio l_{vap}/l_{liq} is almost infinite for this tie line, so n_{liq}/n_{vap} is also almost infinite, and there is only a trace of vapour present. When the pressure is reduced to p_2, the value of l_{vap}/l_{liq} is about 0.3, so $n_{liq}/n_{vap} \approx 0.3$ and the amount of liquid is about 0.3 times the amount of vapour. When the pressure has been reduced to p_3, the sample is almost completely gaseous and because $l_{vap}/l_{liq} \approx 0$ we conclude that there is only a trace of liquid present. ●

Justification 16.1 *The lever rule*

To prove the lever rule we write $n = n_\alpha + n_\beta$ and the overall amount of A as nx_A, where x_A is the mole fraction of A in the mixture. The amount of A is also the sum of its amounts in the two phases:

$$nx_A = n_\alpha x_{A,\alpha} + n_\beta x_{A,\beta}$$

where $x_{A,\alpha}$ is the mole fraction of A in phase α and $x_{A,\beta}$ is its mole fraction in phase β. Since also

$$nx_A = (n_\alpha + n_\beta)x_A = n_\alpha x_A + n_\beta x_A$$

by equating these two expressions it follows that

$$n_\alpha(x_{A,\alpha} - x_A) = n_\beta(x_A - x_{A,\beta})$$

which corresponds to eqn 16.1.

A **temperature–composition diagram** is a phase diagram in which the boundaries show the composition of the phases that are in equilibrium at various temperatures (and at a fixed pressure, typically 1 atm). An example is shown in Fig. 16.10. Note that the liquid phase now lies in the lower part of the diagram. The region between the lines in Fig. 16.10 is a two-phase region; the regions outside the phase lines correspond to a single phase.

Consider what happens when a liquid of composition a_1 is heated. It boils when the temperature reaches T_2. Then the liquid has composition a_2 (the same as a_1) and the vapour (which is present only as a trace) has composition a_2'. The vapour is richer in the more volatile component A (the component with the lower boiling point). From the location of a_2', we can state the vapour's composition at the boiling point, and from the location of the tie line joining a_2 and a_2' we can read off the boiling temperature (T_2) of the original liquid mixture.

In a **simple distillation**, the vapour is withdrawn and condensed. This technique is used to separate a volatile liquid from a non-volatile solute or solid. In **fractional distillation**, the boiling and condensation cycle is repeated successively. This technique is used to separate volatile liquids. We can follow the changes

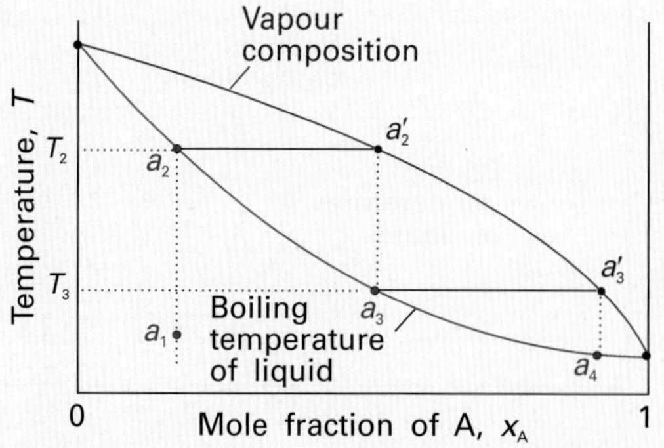

Fig. 16.10 The temperature–composition diagram corresponding to an ideal mixture with the component A more volatile than component B. Successive boilings and condensations of a liquid originally of composition a_1 lead to a condensate that is pure A. The separation technique is called fractional distillation.

that occur by seeing what happens when the first condensate of composition a_3 is reheated. The phase diagram shows that this mixture boils at T_3 and yields a vapour of composition a_3', which is even richer in the more volatile component. That vapour is drawn off, and the first drop condenses to a liquid of composition a_4. The cycle can then be repeated until in due course almost pure A is obtained.

Although many liquids have temperature–composition phase diagrams resembling the ideal version in Fig. 16.10, in a number of important cases there are marked deviations. A maximum in the phase diagram (Fig. 16.11) may occur when the favourable interactions between A and B molecules reduce the vapour pressure of the mixture below the ideal value: in effect, the A–B interactions stabilize the liquid. Examples of this behaviour include trichloromethane/propanone and nitric acid/water mixtures. Phase diagrams showing a minimum (Fig. 16.12) indicate that the mixture is destabilized relative to the ideal solution, the

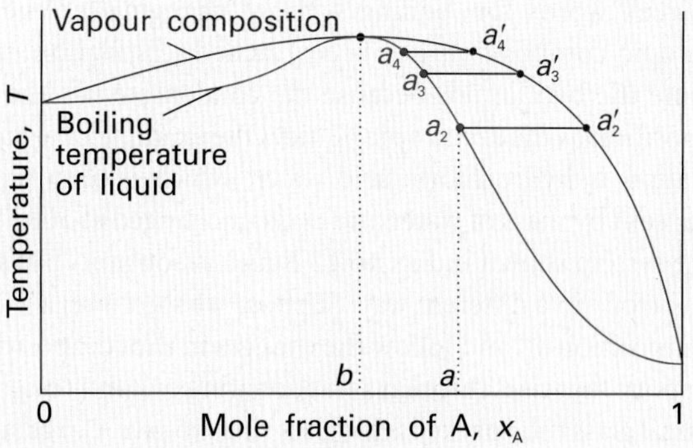

Fig. 16.11 A high-boiling azeotrope. When the liquid of composition *a* is distilled, the composition of the remaining liquid changes towards *b* but no further.

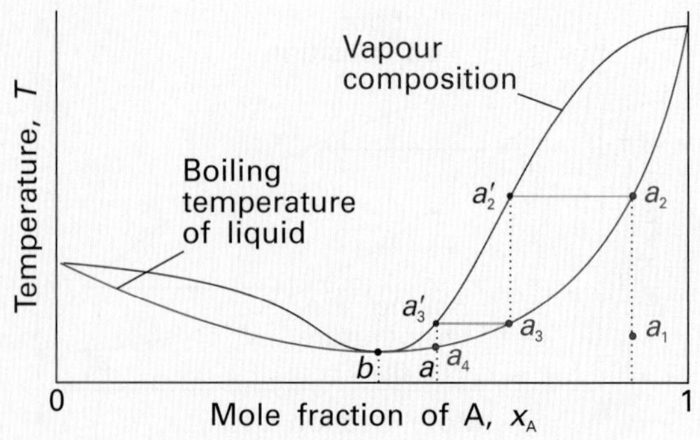

Fig. 16.12 A low-boiling azeotrope. When the mixture at a is fractionally distilled, the vapour in equilibrium in the fractionating column moves towards b and then remains unchanged.

A–B interactions then being unfavourable. Examples include dioxane/water and ethanol/water mixtures.

Deviations from ideality are not always so strong as to lead to a maximum or minimum in the phase diagram, but when they do there are important consequences for distillation. Consider a liquid of composition a on the right of the maximum in Fig. 16.11. The vapour (at a_2') of the boiling mixture (at a_2) is richer in A. If that vapour is removed (and condensed elsewhere), then the remaining liquid will move to a composition that is richer in B, such as that represented by a_3, and the vapour in equilibrium with this mixture will have composition a_3'. As that vapour is removed, the composition of the boiling liquid shifts to a point such as a_4, and the composition of the vapour shifts to a_4'. Hence, as evaporation proceeds, the composition of the remaining liquid shifts towards B as A is drawn off. The boiling point of the liquid rises, and the vapour becomes richer in B. When so much A has been evaporated that the liquid has reached the composition b, the vapour has the same composition as the liquid. Evaporation then occurs without change of composition. The mixture is said to form an **azeotrope** (from the Greek words for 'boiling without changing'). When the azeotropic composition has been reached, distillation cannot separate the two liquids because the condensate has the same composition as the azeotropic liquid. One example of azeotrope formation is hydrochloric acid/water, which is azeotropic at 80 per cent by mass of water and boils unchanged at 108.6°C.

The system shown in Fig. 16.12 is also azeotropic, but shows its azeotropy in a different way. Suppose we start with a mixture of composition a_1, and follow the changes in the composition of the vapour that rises through a fractionating column (essentially a vertical glass tube packed with glass rings to give a large surface area). The mixture boils at a_2 to give a vapour of composition a_2'. This vapour condenses in the column to a liquid of the same composition (now marked a_3). That liquid reaches equilibrium with its vapour at a_3', which condenses higher up the tube to give a liquid of the same composition, which we now call a_4. The fractionation therefore shifts the vapour towards the azeotropic composition at b, but not beyond, and the azeotropic vapour emerges from the top of the column. An example is ethanol/water, which boils unchanged when the water content is 4 per cent by mass and the temperature is 78°C.

(b) Liquid–liquid systems

Now we consider temperature–composition diagrams for systems that consist of pairs of **partially miscible** liquids, which are liquids that do not mix in all proportions at all temperatures. An example is hexane and nitrobenzene.

Suppose a small amount of a liquid B is added to a sample of another liquid A at a temperature T'. It dissolves completely, and the binary system remains a single phase. As more B is added, a stage comes at which no more dissolves. The sample now consists of two phases in equilibrium with each other, the most abundant one consisting of A saturated with B, the minor one a trace of B saturated with A. In the temperature–composition diagram drawn in Fig. 16.13, the composition of the former is represented by the point a' and that of the latter by the point a''. The relative abundances of the two phases are given by the lever rule. When more B is added, A dissolves in it slightly. However, the amount of one phase increases at the expense of the other. A stage is reached when so much B is present that it can dissolve all the A, and the system reverts to a single phase. The addition of more B now simply dilutes the solution, and from then on it remains a single phase.

The composition of the two phases at equilibrium varies with the temperature. For hexane and nitrobenzene, raising the

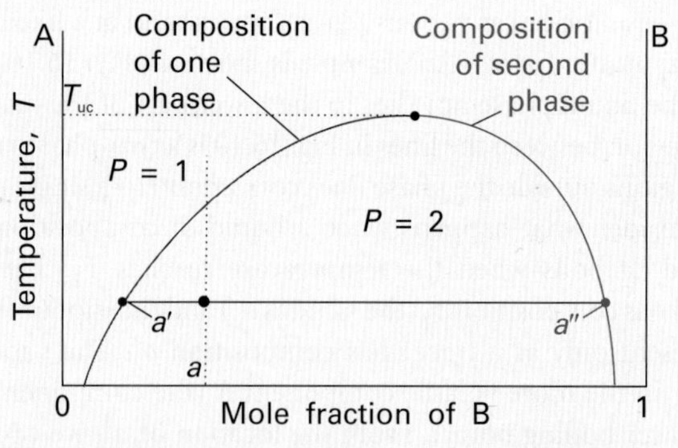

Fig. 16.13 The temperature–composition diagram for a system composed of two partially miscible liquids. The region below the curve corresponds to the compositions and temperatures at which the liquids are partially miscible. The upper critical temperature, T_{uc}, is the temperature above which the two liquids are miscible in all proportions. In this and subsequent phase diagrams P is the number of phases. This phase diagram is for A = hexane and B = nitrobenzene, for which T_{uc} = 294 K.

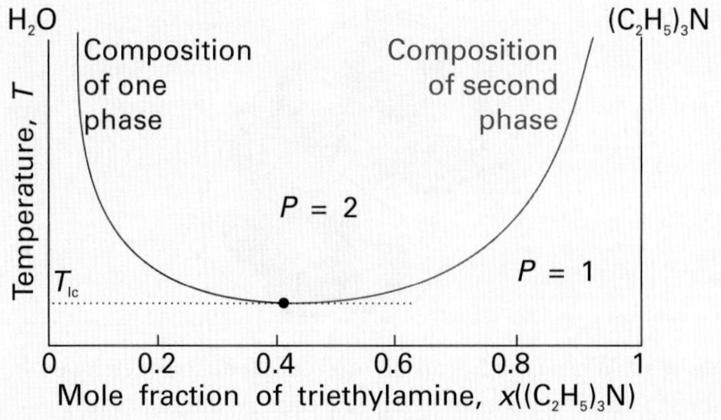

Fig. 16.14 The temperature–composition diagram for water and triethylamine. This system shows a lower critical temperature at 292 K. The labels indicate the interpretation of the boundaries.

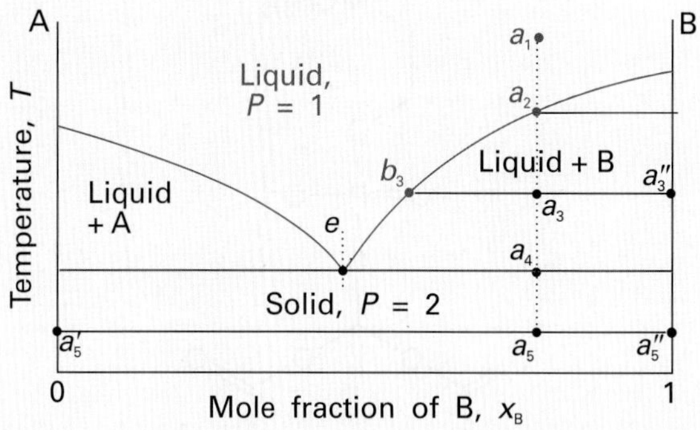

Fig. 16.15 The temperature–composition phase diagram for two almost immiscible solids and their completely miscible liquids. The isopleth through e corresponds to the eutectic composition, the mixture with lowest melting point.

temperature increases their miscibility. The two-phase system therefore becomes less extensive, because each phase in equilibrium is richer in its minor component: the A-rich phase is richer in B and the B-rich phase is richer in A. The phase diagram is constructed by repeating the observations at different temperatures and drawing the envelope of the two-phase region.

The **upper critical solution temperature** (or upper *consolute temperature*), T_{uc}, is the highest temperature at which phase separation occurs. Above the upper critical temperature the two components are fully miscible. This temperature exists because the greater thermal motion overcomes any potential energy advantage in molecules of one type being close together. One example is the nitrobenzene/hexane system shown in Fig. 16.13. Some systems show a **lower critical solution temperature** (or *lower consolute temperature*), T_{lc}, below which they mix in all proportions and above which they form two phases. An example is water and triethylamine (Fig. 16.14). In this case, at low temperatures the two components are more miscible because they form a weak complex; at higher temperatures the complexes break up and the two components are less miscible.

(c) Liquid–solid systems

Finally, we consider a simple example of a two-component mixture that forms both liquid and solid phases. The interpretation of the phase diagram follows the same principles as before. Thus, consider the two-component liquid of composition a_1 in Fig. 16.15. The changes that occur may be expressed as follows.

1. $a_1 \rightarrow a_2$. The system enters the two-phase region labelled 'Liquid + B'. Pure solid B begins to come out of solution and the remaining liquid becomes richer in A.

2. $a_2 \rightarrow a_3$. More of the solid forms, and the relative amounts of the solid and liquid (which are in equilibrium) are given by the lever rule. At this stage there are roughly equal amounts of each. The liquid phase is richer in A than before (its composition is given by b_3) because some B has been deposited.

3. $a_3 \rightarrow a_4$. At the end of this step, there is less liquid than at a_3, and its composition is given by e. This liquid now freezes to give a two-phase system of pure B and pure A.

The isopleth at e in Fig. 16.15 corresponds to the **eutectic** composition, the mixture with the lowest melting point (the name comes from the Greek words for 'easily melted'). A liquid with the eutectic composition freezes at a single temperature, without previously depositing solid A or B. A solid with the eutectic composition melts, without change of composition, at the lowest temperature of any mixture. Solutions of composition to the right of e deposit B as they cool, and solutions to the left deposit A: only the eutectic mixture (apart from pure A or pure B) solidifies at a single definite temperature without gradually unloading one or other of the components from the liquid.

IMPACT ON BIOCHEMISTRY
I16.1 Biological membranes

At this stage of your study of physical chemistry we are ready to use several concepts explored earlier in the text to understand rather complex processes. In Chapter 8 we encountered molecular aggregates, such as colloids and micelles, with special structural and physical properties. We also encountered mesophases, which are intermediate between liquids and solids. Now that we understand the molecular basis of phase transitions between liquids and solids, we can expand the discussion to include the phases in which molecular aggregates exist. We shall focus on biological membranes, which are special forms of micelles.

As we saw in *Impact I8.2*, lamellar micelles, extended parallel sheets two molecules thick, are convenient models of membranes of biological cells, but actual membranes are highly sophisticated structures. The basic structural element of a membrane is a phospholipid, such as phosphatidyl choline (1), which contains long hydrocarbon chains (typically in the range C_{14}–C_{24}) and a variety of polar groups, such as $-CH_2CH_2N(CH_3)_3^+$.

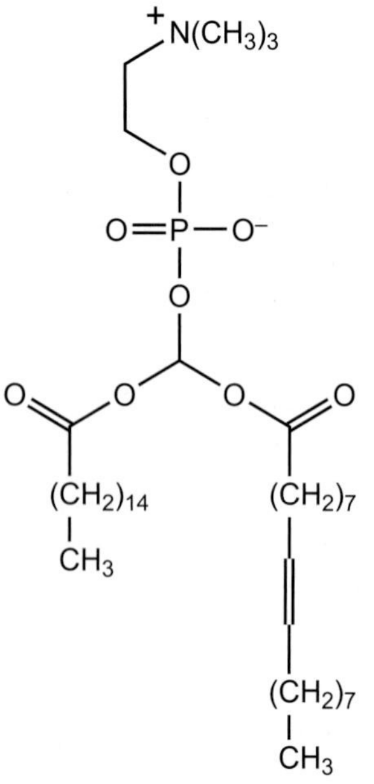

1 Phosphatidyl choline

The hydrophobic chains stack together to form an extensive bilayer about 5 nm across. The lipid molecules form layers instead of micelles because the hydrocarbon chains are too bulky to allow packing into nearly spherical clusters.

Interspersed among the phospholipids of biological membranes are sterols, such as cholesterol (**2**), which is largely hydrophobic but does contain a hydrophilic —OH group. Sterols, which are present in different proportions in different types of cells, prevent the hydrophobic chains of lipids from 'freezing' into a gel and, by disrupting the packing of the chains, spread the melting point of the membrane over a range of temperatures.

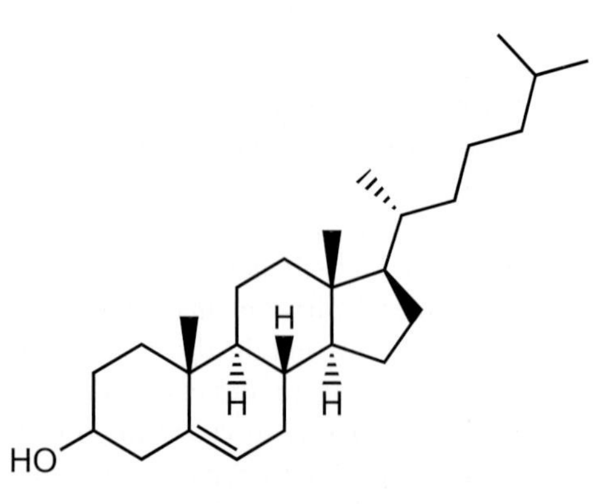

2 Cholesterol

Peripheral proteins are proteins attached to the bilayer. **Integral proteins** are proteins immersed in the mobile but viscous bilayer. These proteins may span the depth of the bilayer and consist of tightly packed α-helices or, in some cases, β-sheets containing hydrophobic residues that sit comfortably within the hydrocarbon region of the bilayer.

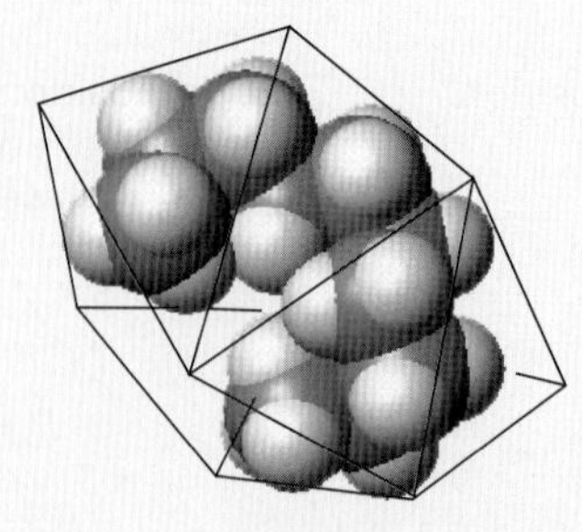

Fig. 16.16 When a hydrocarbon molecule is surrounded by water, the H_2O molecules form a clathrate cage. As a result of this acquisition of structure, the entropy of the water decreases, so the dispersal of the hydrocarbon into the water is entropy-opposed; its coalescence is entropy-favoured.

Before we can discuss the phases of biological membranes and the transitions between them, we need to understand why hydrophobic chains come together to form a micelle or membrane. To make progress, we explore the behaviour of nonpolar solutes in polar solvents. Nonpolar molecules do dissolve slightly in polar solvents, but strong interactions between solute and solvent are not possible and as a result it is found that each individual solute molecule is surrounded by a solvent cage (Fig. 16.16). Experiments indicate that the transfer of a nonpolar hydrocarbon solute from a nonpolar solvent to water, a polar solvent, has $\Delta_{\text{transfer}}G > 0$, as expected on the basis of the increase in polarity of the solvent, but exothermic ($\Delta_{\text{transfer}}H < 0$). Therefore, it is a large decrease in the entropy of the system ($\Delta_{\text{transfer}}S < 0$) that accounts for the positive Gibbs energy of transfer. For example, the process $CH_4(\text{in } CCl_4) \rightarrow CH_4(\text{aq})$ has $\Delta_{\text{transfer}}G = +12\ \text{kJ mol}^{-1}$, $\Delta_{\text{transfer}}H = -10\ \text{kJ mol}^{-1}$, and $\Delta_{\text{transfer}}S = -75\ \text{J K}^{-1}\ \text{mol}^{-1}$ at 298 K. Hydrophobic substances are characterized by a positive Gibbs energy of transfer from a nonpolar to a polar solvent.

At the molecular level, formation of a solvent cage around a hydrophobic molecule involves the formation of new hydrogen bonds among solvent molecules. This is an exothermic process and accounts for the negative values of $\Delta_{\text{transfer}}H$. On the other hand, the increase in order associated with formation of a very large number of small solvent cages decreases the entropy of the system and accounts for the negative values of $\Delta_{\text{transfer}}S$. However, when many solute molecules cluster together, fewer (though larger) cages are required and more solvent molecules are free to move. The net effect of formation of large clusters of hydrophobic molecules is then a decrease in the organization of the solvent and therefore a net *increase* in entropy of the system. This increase in entropy of the solvent is large enough to render spontaneous the association of hydrophobic molecules in a polar solvent.

The increase in entropy that results from fewer structural demands on the solvent placed by the clustering of nonpolar molecules is the origin of the **hydrophobic interaction**, which tends to stabilize groupings of hydrophobic groups in micelles and biopolymers. The hydrophobic interaction is an example of an ordering process that is stabilized by a tendency toward greater disorder of the solvent.

The bilayer of a biological membrane is a highly mobile structure, as shown by EPR studies with spin-labelled phospholipids (*Impact I12.2*). Not only are the hydrocarbon chains ceaselessly twisting and turning in the region between the polar groups, but the phospholipid and cholesterol molecules migrate over the surface. It is better to think of the membrane as a viscous fluid rather than a permanent structure, with a viscosity about 100 times that of water. In common with diffusional behaviour in general (see Section 18.8), the average distance a phospholipid molecule diffuses is propotional to the square-root of the time; more precisely, for a molecule confined to a two-dimensional plane, the average distance travelled in a time t is equal to $(4Dt)^{1/2}$. Typically, a phospholipid molecule migrates through about 1 μm (the diameter of a cell) in about 1 min.

Integral proteins also move in the bilayer. In the **fluid mosaic model** shown in Fig. 16.17 the proteins are mobile, but their diffusion coefficients are much smaller than those of the lipids. In the **lipid raft model**, a number of lipid and cholesterol molecules form ordered structures, or 'rafts', that envelope proteins and help carry them to specific parts of the cell.

The mobility of the bilayer enables it to flow round a molecule close to the outer surface, to engulf it, and incorporate it into the cell by the process of *endocytosis*. Alternatively, material from the cell interior wrapped in cell membrane may coalesce with the cell membrane itself, which then withdraws and ejects the material in the process of *exocytosis*. The function of the proteins embedded in the bilayer, though, is to act as devices for transporting matter into and out of the cell in a more subtle manner. By providing hydrophilic channels through an otherwise alien hydrophobic environment, some proteins act as **ion channels** and **ion pumps** (*Impact I18.2*).

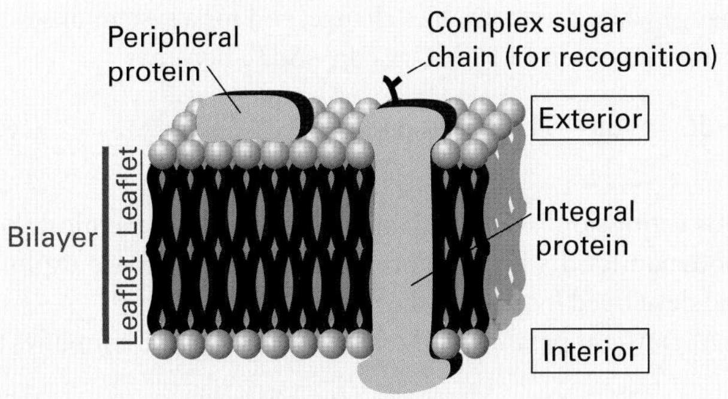

Fig. 16.17 In the fluid mosaic model of a biological cell membrane, integral proteins diffuse through the lipid bilayer. In the alternative lipid raft model, a number of lipid and cholesterol molecules envelop and transport the protein around the membrane.

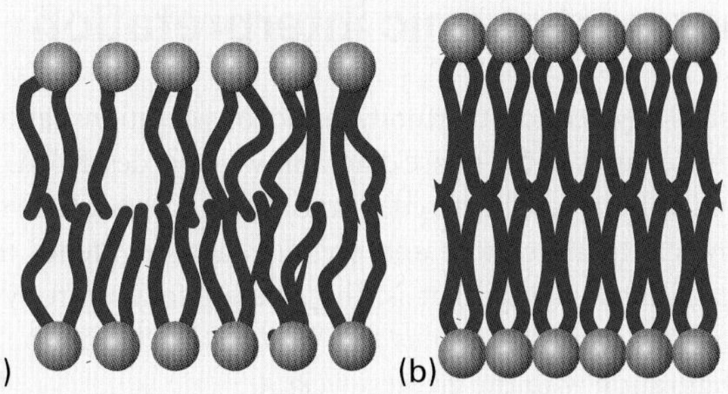

Fig. 16.18 A depiction of the variation with temperature of the flexibility of hydrocarbon chains in a lipid bilayer. (a) At physiological temperature, the bilayer exists as a liquid crystal, in which some order exists but the chains writhe. (b) At a specific temperature, the chains are largely frozen and the bilayer is said to exist as a gel.

All lipid bilayers undergo a transition from a state of high to low chain mobility at a temperature that depends on the structure of the lipid. To visualize the transition, we consider what happens to a membrane as we lower its temperature (Fig. 16.18). There is sufficient energy available at normal temperatures for limited bond rotation to occur and the flexible chains writhe. However, the membrane is still highly organized in the sense that the bilayer structure does not come apart and the system is best described as a liquid crystal (Fig. 16.18a). At lower temperatures, the amplitudes of the writhing motion decrease until a specific temperature is reached at which motion is largely frozen. The membrane is said to exist as a gel (Fig. 16.18b). Biological membranes exist as liquid crystals at physiological temperatures.

Phase transitions in membranes are often observed as 'melting' from gel to liquid crystal by differential scanning calorimetry (*Impact I14.1*). The data show relations between the structure of the lipid and the melting temperature. For example, the melting temperature increases with the length of the hydrophobic chain of the lipid. This correlation is reasonable, as we expect longer chains to be held together more strongly by hydrophobic interactions than shorter chains. It follows that stabilization of the gel phase in membranes of lipids with long chains results in relatively high melting temperatures. On the other hand, any structural elements that prevent alignment of the hydrophobic chains in the gel phase lead to low melting temperatures. Indeed, lipids containing unsaturated chains, those containing some C=C bonds, form membranes with lower melting temperatures than those formed from lipids with fully saturated chains, those consisting of C—C bonds only.

Thermodynamic interpretation

We shall now see how thermodynamic considerations can account for the features of the phase diagrams we have described. All our considerations will be based on the Gibbs energy of the system. At constant temperature and pressure, a system tends towards lowest Gibbs energy: that is, the greatest total entropy of the system and its surroundings and, in statistical terms, the overall configuration with the greatest weight.

To discuss the Gibbs energy of a system that in general consists of several components J, each of which contributes to the total Gibbs energy, we introduce the **chemical potential**, μ (mu):

$$\mu_J = \left(\frac{\partial G}{\partial n_J}\right)_{p,T,n'} \qquad [16.2]$$

The n' signifies that the abundances of all the other species in the mixture are held constant; the units of a chemical potential are joules (from G) per mole (from n). For a one-component system, $G = nG_m$, and the chemical potential is simply the molar Gibbs energy of the substance because

$$\mu = \left(\frac{\partial G}{\partial n}\right)_{p,T} = \left(\frac{\partial (nG_m)}{\partial n}\right)_{p,T} = G_m \qquad (16.3)$$

In general, the chemical potential of a substance in a mixture varies with the composition of the mixture because the environment of each type of molecule changes as the composition changes. When the composition of a binary (two-component) mixture of A and B molecules is nearly pure A, each A molecule is surrounded almost entirely by A molecules and μ_A has a value characteristic of this environment. When the mixture is almost pure B, each A molecule is surrounded almost entirely by B molecules and now μ_A has a different value, one characteristic of this environment. At intermediate compositions, the environment of A lies between these two extremes, and μ_A has the corresponding value.

The name 'chemical potential' is instructive and should be borne in mind. As we develop the concept, we shall see that μ is a measure of the potential—the capacity—that a substance has for producing change in a system. In this chapter, it reflects the potential of a substance to bring about physical change. In Chapter 17 we shall go on to see that μ is also the potential of a substance to bring about chemical change.

16.3 Properties of the chemical potential

We show in the following *Justification* that, with the chemical potential defined in eqn 16.2, the total Gibbs energy of a mixture is simply

$$G = \sum_J n_J \mu_J \qquad (16.4)$$

with the chemical potential of each component measured at the composition of the mixture. According to this equation, the chemical potential of a substance in a mixture is the contribution of that substance to the total Gibbs energy of the mixture.

Justification 16.2 *The total Gibbs energy of a mixture*

For simplicity, we shall consider a binary mixture of A and B. The definition in eqn 16.2 implies that, when the composition of the mixture at constant temperature and pressure is changed by the addition of dn_A of A and dn_B of B, then the total Gibbs energy of the mixture changes by

$$dG = \left(\frac{\partial G}{\partial n_A}\right)_{p,T,n_B} dn_A + \left(\frac{\partial G}{\partial n_B}\right)_{p,T,n_A} dn_B = \mu_A dn_A + \mu_B dn_B$$

Provided the relative proportions of A and B are held constant as they are added, the mixture has the same composition and the two chemical potentials are therefore constant. Therefore, to calculate the total Gibbs function, we integrate dG as n_A and n_B are raised simultaneously from 0 to their final values with the μ_J treated as constants:

$$G = \int_0^{n_A} \mu_A dn_A + \int_0^{n_B} \mu_B dn_B = \mu_A \int_0^{n_A} dn_A + \mu_B \int_0^{n_B} dn_B$$
$$= \mu_A n_A + \mu_B n_B$$

This expression generalizes to eqn 16.4. Although we have envisaged the two integrations as being linked (in order to preserve constant composition), because G is a state function the final result in eqn 16.4 is valid however the solution is in fact prepared.

(a) Changes in the Gibbs energy

Because chemical potentials depend on composition (and the pressure and temperature), the Gibbs energy of a mixture may change when these variables change, and for a system of several components eqn 15.59 ($dG = Vdp - SdT$) becomes

$$dG = Vdp - SdT + \sum_J \mu_J dn_J \qquad (16.5)$$

This expression is the **fundamental equation of chemical thermodynamics**. Its implications and consequences are explored and developed in this and the next chapter.

At constant pressure and temperature, eqn 16.5 simplifies to

$$dG = \sum_J \mu_J dn_J \qquad (16.6)$$

It follows that when the composition is changed infinitesimally we might expect (from eqn 16.4) G to change by

$$dG = \sum_J n_J d\mu_J + \sum_J \mu_J dn_J$$

However, we have seen that at constant pressure and temperature a change in Gibbs energy is in fact given by eqn 16.6. Because G is a state function, these two equations must be equal, which implies that, at constant temperature and pressure, changes in the chemical potentials of the components of a mixture at equilibrium must satisfy the **Gibbs–Duhem equation**:

$$\sum_J n_J d\mu_J = 0 \qquad (16.7)$$

The significance of this equation is that the chemical potential of one component of a mixture cannot change independently of the chemical potentials of the other components. In a binary mixture, if the chemical potential of one component increases, then the other must decrease, with the two changes related by $n_A d\mu_A + n_B d\mu_B = 0$, and therefore by

$$d\mu_B = -\frac{n_A}{n_B} d\mu_A \qquad (16.8)$$

Thus, if the composition of the mixture is such that $n_A = 2n_B$, and a small change in composition results in μ_A increasing by 1 J mol^{-1}, μ_B will decrease by 2 J mol^{-1}.

(b) The thermodynamic criterion of equilibrium

The importance of the chemical potential for the discussion of phase equilibria is that, *at equilibrium, the chemical potential of a substance is the same throughout a sample, regardless of how many phases are present.* When the liquid and solid phases of a single substance are in equilibrium, the chemical potential of the substance is the same in both phases and throughout each phase (Fig. 16.19). When two phases of a many-component mixture are in equilibrium the chemical potential of each component is the same in every phase.

To see the validity of this remark, consider a system in which the chemical potential of any one component is μ_1 at one location and μ_2 at another location. The locations may be in the same or in different phases. When an amount dn of the substance is transferred from one location to the other, the Gibbs energy of the system changes by $-\mu_1 dn$ when material is removed from location 1, and it changes by $+\mu_2 dn$ when that material is added to location 2. The overall change is therefore $dG = (\mu_2 - \mu_1)dn$. If the chemical potential at location 1 is higher than that at location 2, the transfer is accompanied by a decrease in G, and so has a spontaneous tendency to occur. The spontaneous direction is from high to low chemical potential and the most stable phase is the one of lowest chemical potential under the prevailing conditions. Only if $\mu_1 = \mu_2$ is there no change in G, and only then is the system at equilibrium.

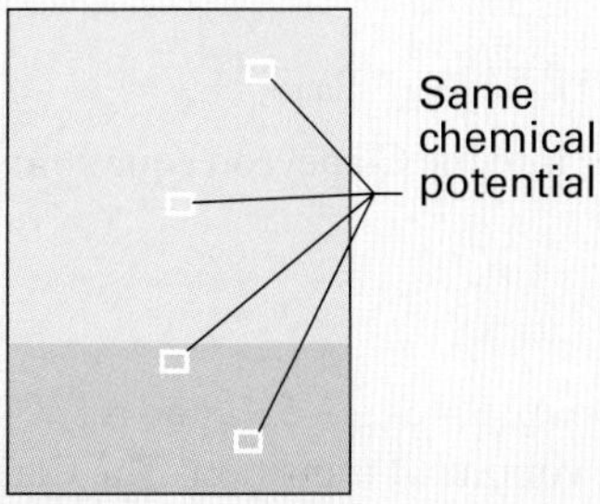

Fig. 16.19 When two or more phases are in equilibrium, the chemical potential of a substance (and, in a mixture, a component) is the same in each phase and is the same at all points in each phase.

(c) The response of the chemical potential to the conditions

The temperature dependence of the Gibbs energy is expressed in terms of the entropy of the system by eqn 15.60 ($(\partial G/\partial T)_p = -S$). Because, as we have seen, the chemical potential of a pure substance is just another name for its molar Gibbs energy, it follows that

$$\left(\frac{\partial \mu}{\partial T}\right)_p = -S_m \qquad (16.9)$$

This relation shows that, as the temperature is raised, the chemical potential of a pure substance decreases: $S_m > 0$ for all substances, so the slope of a plot of μ against T is negative. At first sight it might seem odd that the chemical potential, the capacity to bring about change, decreases as the temperature is raised. However, it should be recalled from Section 15.5 that an interpretation of the Gibbs energy is that it is the difference between the total energy and the energy stored chaotically: the latter increases with temperature, so the energy 'free' to do work decreases.

Equation 16.9 implies that the slope of a plot of μ against temperature is steeper for gases than for liquids, because $S_m(g) > S_m(l)$. The slope is also steeper for a liquid than the corresponding solid, because $S_m(l) > S_m(s)$ almost always. These features are illustrated in Fig. 16.20. The steep negative slope of $\mu(l)$ results in its falling below $\mu(s)$ when the temperature is high enough, and then the liquid becomes the stable phase: the solid melts. The chemical potential of the gas phase plunges steeply downwards as the temperature is raised (because the molar entropy of the vapour is so high), and there comes a temperature at which it lies lowest. Then the gas is the stable phase and vaporization is spontaneous.

As summarized in the phase diagrams shown earlier in the chapter, most substances melt at a higher temperature when subjected to pressure. It is as though the pressure is preventing the formation of the less dense liquid phase. Exceptions to this behaviour include water, for which the liquid is denser than the solid. Application of pressure to water encourages the formation of the liquid phase. That is, water freezes at a lower temperature when it is under pressure (see Fig. 16.6).

We can rationalize the response of melting temperatures to pressure as follows. The variation of the chemical potential of a

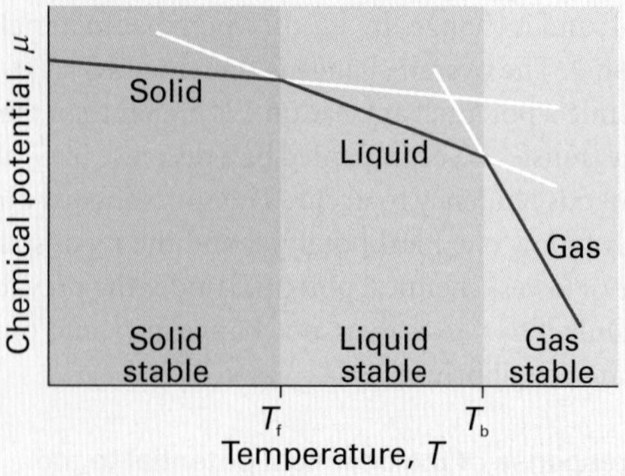

Fig. 16.20 The schematic temperature dependence of the chemical potential of the solid, liquid, and gas phases of a substance (in practice, the lines are curved). The phase with the lowest chemical potential at a specified temperature is the most stable one at that temperature. The transition temperatures, the melting and boiling temperatures (T_f and T_b, respectively), are the temperatures at which the chemical potentials of the two phases are equal.

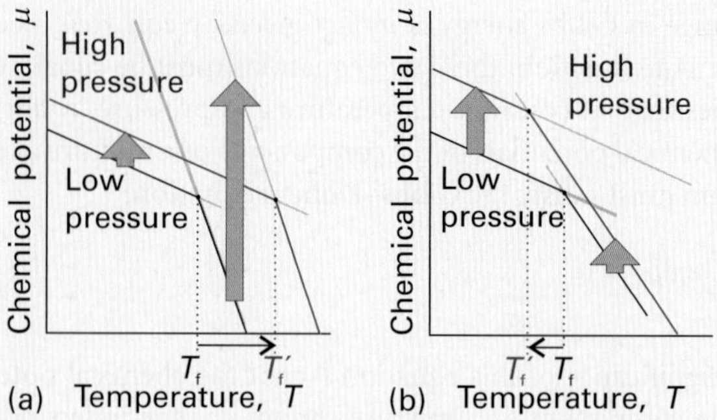

Fig. 16.21 The pressure dependence of the chemical potential of a substance depends on the molar volume of the phase. The lines show schematically the effect of increasing pressure on the chemical potential of the solid and liquid phases (in practice, the lines are curved), and the corresponding effects on the freezing temperatures. (a) In this case the molar volume of the solid is smaller than that of the liquid and $\mu(s)$ increases less than $\mu(l)$. As a result, the freezing temperature rises. (b) Here the molar volume is greater for the solid than the liquid (as for water), $\mu(s)$ increases more strongly than $\mu(l)$, and the freezing temperature is lowered.

pure substance with pressure is expressed (from the second of eqn 15.60, $(\partial G/\partial p)_T = V$) by

$$\left(\frac{\partial \mu}{\partial p}\right)_T = V_m \tag{16.10}$$

This equation shows that the slope of a plot of chemical potential against pressure is equal to the molar volume of the substance. An increase in pressure raises the chemical potential of any pure substance (because $V_m > 0$). In most cases, $V_m(l) > V_m(s)$ and the equation predicts that an increase in pressure increases the chemical potential of the liquid more than that of the solid. As shown in Fig. 16.21a, the effect of pressure in such a case is to raise the melting temperature slightly, as we saw for carbon dioxide in Fig. 16.5. For water, however, $V_m(l) < V_m(s)$, and an increase in pressure increases the chemical potential of the solid more than that of the liquid. In this case, the melting temperature is lowered slightly (Fig. 16.21b), as we saw for water in Fig. 16.6.

16.4 The structure of one-component phase diagrams

We can find the precise locations of the phase boundaries in a one-component phase diagram—the pressures and temperatures at which any two of its phases can coexist—by making use of the fact that, when two phases are in equilibrium, their chemical potentials must be equal. Therefore, where the phases α and β are in equilibrium,

$$\mu_\alpha(p,T) = \mu_\beta(p,T) \tag{16.11}$$

By solving this equation for p in terms of T, we get an equation for the phase boundary.

It turns out to be simplest to discuss the phase boundaries in terms of their slopes, dp/dT. Let p and T be changed infinitesimally, but in such a way that the two phases α and β remain in equilibrium. The chemical potentials of the phases are initially equal (the two phases are in equilibrium). They remain equal when the conditions are changed to another point on the phase boundary, where the two phases continue to be in equilibrium (Fig. 16.22). Therefore, the changes in the chemical potentials of the two phases must be equal and we can write $d\mu_\alpha = d\mu_\beta$. Because, $dG = Vdp - SdT$, we know that $d\mu = -S_m dT + V_m dp$ for each phase; it follows that

$$-S_{\alpha,m}dT + V_{\alpha,m}dp = -S_{\beta,m}dT + V_{\beta,m}dp$$

where $S_{\alpha,m}$ and $S_{\beta,m}$ are the molar entropies of the phases and $V_{\alpha,m}$ and $V_{\beta,m}$ are their molar volumes. Hence

$$(V_{\beta,m} - V_{\alpha,m})dp = (S_{\beta,m} - S_{\alpha,m})dT$$

which rearranges into the **Clapeyron equation**:

$$\frac{dp}{dT} = \frac{\Delta_{trs}S}{\Delta_{trs}V} \tag{16.12}$$

In this expression $\Delta_{trs}S = S_{\beta,m} - S_{\alpha,m}$ and $\Delta_{trs}V = V_{\beta,m} - V_{\alpha,m}$ are the entropy and volume of transition. The Clapeyron equation is an exact expression for the slope of the phase boundary and applies to any phase equilibrium of any pure substance. It implies that we can use thermodynamic data to predict the appearance of phase diagrams and to understand their form.

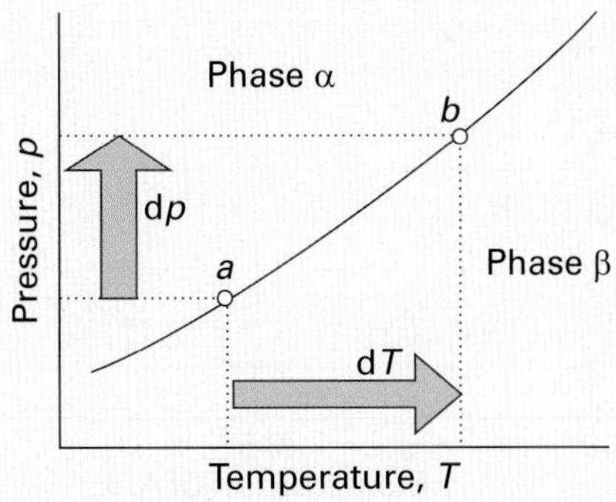

Fig. 16.22 When pressure is applied to a system in which two phases are in equilibrium (at *a*), the equilibrium is disturbed. It can be restored by changing the temperature, so moving the state of the system to *b*. It follows that there is a relation between d*p* and d*T* that ensures that the system remains in equilibrium as either variable is changed.

Melting (fusion) is accompanied by a molar enthalpy change $\Delta_{fus}H$ and occurs at a temperature T. The molar entropy of melting at T is therefore $\Delta_{fus}H/T$ (Section 15.3), and the Clapeyron equation becomes

$$\frac{dp}{dT} = \frac{\Delta_{fus}H}{T\Delta_{fus}V} \quad (16.13a)$$

where $\Delta_{fus}V$ is the change in molar volume that occurs on melting. The enthalpy of melting is positive (the only exception is helium-3) and the volume change is usually positive (water being an exception) and always small. Consequently, the slope dp/dT is steep and usually positive, as we saw in Fig. 16.5. The entropy of vaporization at a temperature T is equal to $\Delta_{vap}H/T$; the Clapeyron equation for the liquid–vapour boundary is therefore

$$\frac{dp}{dT} = \frac{\Delta_{vap}H}{T\Delta_{vap}V} \quad (16.13b)$$

The enthalpy of vaporization is positive; $\Delta_{vap}V$ is large and positive. Therefore, dp/dT is positive, but it is much smaller than for the solid–liquid boundary. It follows that dT/dp, the inverse of dp/dT, is large, and hence that the boiling temperature is more responsive to pressure than the freezing temperature, as shown in Figs. 16.5 and 16.6.

Example 16.1 *Estimating the effect of pressure on the boiling temperature*

Estimate the typical size of the effect of increasing pressure on the boiling point of a liquid.

Method To use eqn 16.13b we need to estimate the right-hand side. At the boiling point, the term $\Delta_{vap}H/T$ is Trouton's constant (Section 15.3). Because the molar volume of a gas is so much greater than the molar volume of a liquid, we can write

$$\Delta_{vap}V = V_m(g) - V_m(l) \approx V_m(g)$$

and take for $V_m(g)$ the molar volume of a perfect gas (at low pressures, at least).

Answer Trouton's constant has the value 85 J K^{-1} mol^{-1}. The molar volume of a perfect gas is about 25 dm^3 mol^{-1} at 1 atm and near but above room temperature. Therefore,

$$\frac{dp}{dT} \approx \frac{85\ \text{J K}^{-1}\ \text{mol}^{-1}}{2.5\times10^{-2}\ \text{m}^3\ \text{mol}^{-1}} = 3.4\times10^3\ \text{Pa K}^{-1}$$

We have used 1 J = 1 Pa m^3. This value corresponds to 0.034 atm K^{-1}, and hence to dT/dp = 29 K atm^{-1}. Therefore, a change of pressure of +0.1 atm can be expected to change a boiling temperature by about +3 K.

Self-test 16.1 Estimate dT/dp for water at its normal boiling point using the information in Table 15.2 and $V_m(g) = RT/p$.
[28 K atm^{-1}]

Because the molar volume of a gas is so much greater than the molar volume of a liquid, we can write $\Delta_{vap}V \approx V_m(g)$ (as in *Example 16.1*). Moreover, if the gas behaves perfectly, $V_m(g) = RT/p$. These two approximations turn the exact Clapeyron equation into

$$\frac{dp}{dT} = \frac{\Delta_{vap}H}{T(RT/p)} = \frac{p\Delta_{vap}H}{RT^2}$$

Because $dp/p = d\ln p$, this expression rearranges into the **Clausius–Clapeyron equation** for the variation of vapour pressure with temperature:

$$\frac{d\ln p}{dT} = \frac{\Delta_{vap}H}{RT^2} \quad (16.14)^\circ$$

Like the Clapeyron equation, the Clausius–Clapeyron equation is used to understand the location and shape of the liquid–vapour and solid–vapour phase boundaries shown in the one-component phase diagrams earlier in the chapter. The only difference between the solid–vapour boundary and liquid–vapour boundary is the replacement of the enthalpy of vaporization by the enthalpy of sublimation, $\Delta_{sub}H$. Because the enthalpy of sublimation is greater than the enthalpy of vaporization ($\Delta_{sub}H = \Delta_{fus}H + \Delta_{vap}H$ at a given temperature), the equation predicts a steeper slope for the sublimation curve than for the vaporization curve at similar temperatures, which is near where they meet at the triple point. This behaviour can be seen in Figs. 16.5 and 16.6.

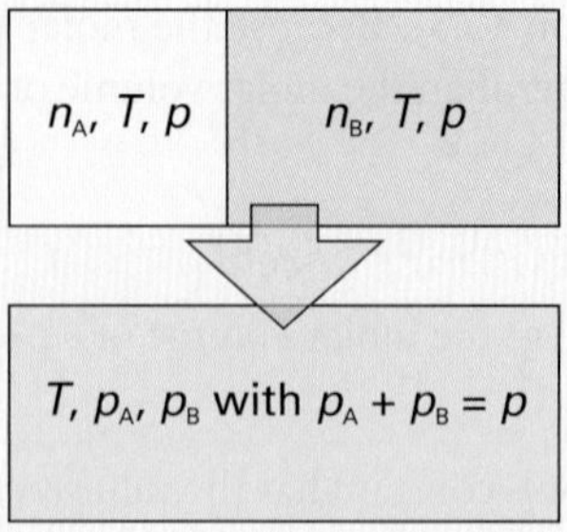

Fig. 16.23 The arrangement for calculating the thermodynamic functions of mixing of two perfect gases.

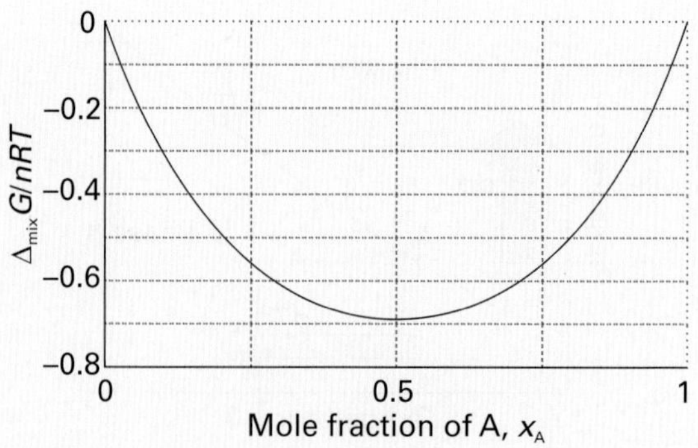

Fig. 16.24 The Gibbs energy of mixing of two perfect gases and (as discussed later) of two liquids that form an ideal solution. The Gibbs energy of mixing is negative for all compositions and temperatures, so perfect gases mix spontaneously in all proportions.

interActivity Draw graphs of $\Delta_{mix}G$ against x_A at different temperatures in the range 298 K to 500 K. For what value of x_A does $\Delta_{mix}G$ depend on temperature most strongly?

16.5 The structure of two-component phase diagrams

To start our investigation of the thermodynamics of mixtures, we consider the act of mixing itself, and begin with the simplest possible case, the mixing of two perfect gases.

(a) The mixing of perfect gases

Let the amounts of two perfect gases in the two containers be n_A and n_B; both are at a temperature T and a pressure p (Fig. 16.23). At this stage, the chemical potentials of the two gases have their 'pure' values. It then follows from eqn 15.67 ($G_m(p) = G_m^{\ominus} + RT\ln(p/p^{\ominus})$) with $\mu_J = G_m(J)$ that for each gas J

$$\mu_J = \mu_J^{\ominus} + RT\ln\frac{p}{p^{\ominus}} \qquad (16.15a)^{\circ}$$

where $\mu_J^{\ominus}$ is the **standard chemical potential**, the chemical potential of the pure gas J at 1 bar (that is, its standard molar Gibbs energy). It will be much simpler notationally if we agree to let p denote the pressure relative to $p^{\ominus}$, that is, to replace $p/p^{\ominus}$ by p, for then we can write

$$\mu_J = \mu_J^{\ominus} + RT\ln p \qquad \{16.15b\}^{\circ}$$

Equations for which this convention is used will be labelled {1}, {2}, . . . ; to use the equations, we have to remember to replace p by $p/p^{\ominus}$ again. In practice, that simply means using the numerical value of p in bars. The initial Gibbs energy of the total system is then given by eqn 16.4 as

$$G_i = n_A\mu_A + n_B\mu_B = n_A(\mu_A^{\ominus} + RT\ln p) + n_B(\mu_B^{\ominus} + RT\ln p)$$

After mixing, the partial pressures of the gases are p_A and p_B, with $p_A + p_B = p$. The total Gibbs energy changes to

$$G_f = n_A(\mu_A^{\ominus} + RT\ln p_A) + n_B(\mu_B^{\ominus} + RT\ln p_B)$$

The difference $G_f - G_i$, the **Gibbs energy of mixing**, $\Delta_{mix}G$, is therefore

$$\Delta_{mix}G = n_ART\ln\frac{p_A}{p} + n_BRT\ln\frac{p_B}{p} \qquad (16.16)^{\circ}$$

At this point we may replace n_J by x_Jn, where n is the total amount of A and B and $x_J = n_J/n$ is the mole fraction of J, and use the definition of partial pressure in terms of the mole fraction

$$p_J = x_Jp \qquad [16.17]$$

to write $p_J/p = x_J$ for each component, which gives

$$\Delta_{mix}G = nRT(x_A\ln x_A + x_B\ln x_B) \qquad (16.18)^{\circ}$$

Because mole fractions are never greater than 1, the logarithms in this equation are negative, and $\Delta_{mix}G < 0$ (Fig. 16.24). The conclusion that $\Delta_{mix}G$ is negative for all compositions confirms that perfect gases mix spontaneously in all proportions. However, the equation extends common sense by allowing us to discuss the process quantitatively.

Example 16.2 *Calculating a Gibbs energy of mixing*

A container is divided into two equal compartments (Fig. 16.25). One contains 3.0 mol H_2(g) at 25°C; the other contains 1.0 mol N_2(g) at 25°C. Calculate the Gibbs energy of mixing when the partition is removed. Assume perfect behaviour.

Method Equation 16.18 cannot be used directly because the two gases are initially at different pressures. We proceed by calculating the initial Gibbs energy from the chemical potentials. To do so, we need the pressure of each gas. Write the pressure of nitrogen as p; then the pressure of hydrogen as a multiple of p can be found from the gas laws. Next, calculate the Gibbs energy for the system when the partition is removed.

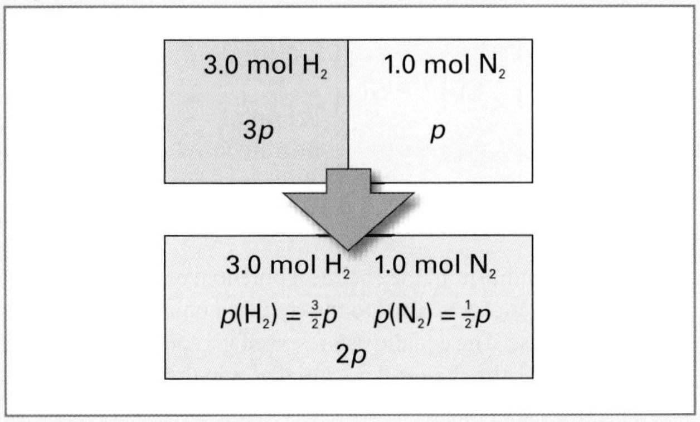

Fig. 16.25 The initial and final states considered in the calculation of the Gibbs energy of mixing of gases at different initial pressures.

The volume occupied by each gas doubles, so its initial partial pressure is halved.

Answer Given that the pressure of nitrogen is p, the pressure of hydrogen is $3p$; therefore, the initial Gibbs energy is

$$G_i = (3.0\text{ mol})\{\mu^{\ominus}(H_2) + RT\ln 3p\} + (1.0\text{ mol})\{\mu^{\ominus}(N_2) + RT\ln p\}$$

When the partition is removed and each gas occupies twice the original volume, the partial pressure of nitrogen falls to $\frac{1}{2}p$ and that of hydrogen falls to $\frac{3}{2}p$. Therefore, the Gibbs energy changes to

$$G_f = (3.0\text{ mol})\{\mu^{\ominus}(H_2) + RT\ln \tfrac{3}{2}p\} + (1.0\text{ mol})\{\mu^{\ominus}(N_2) + RT\ln \tfrac{1}{2}p\}$$

The Gibbs energy of mixing is the difference of these two quantities:

$$\Delta_{\text{mix}}G = (3.0\text{ mol})RT\ln\left(\frac{\frac{3}{2}p}{3p}\right) + (1.0\text{ mol})RT\ln\left(\frac{\frac{1}{2}p}{p}\right)$$
$$= -(3.0\text{ mol})RT\ln 2 - (1.0\text{ mol})\,RT\ln 2$$
$$= -(4.0\text{ mol})RT\ln 2 = -6.9\text{ kJ}$$

In this example, the value of $\Delta_{\text{mix}}G$ is the sum of two contributions: the mixing itself, and the changes in pressure of the two gases to their final total pressure, $2p$. When 3.0 mol H_2 mixes with 1.0 mol N_2 at the same pressure, with the volumes of the vessels adjusted accordingly, the change of Gibbs energy is −5.6 kJ.

Self-test 16.2 Suppose that 2.0 mol H_2 at 2.0 atm and 25°C and 4.0 mol N_2 at 3.0 atm and 25°C are mixed at constant volume. Calculate $\Delta_{\text{mix}}G$. What would be the value of $\Delta_{\text{mix}}G$ had the pressures been identical initially? [−9.7 kJ, −9.5 kJ]

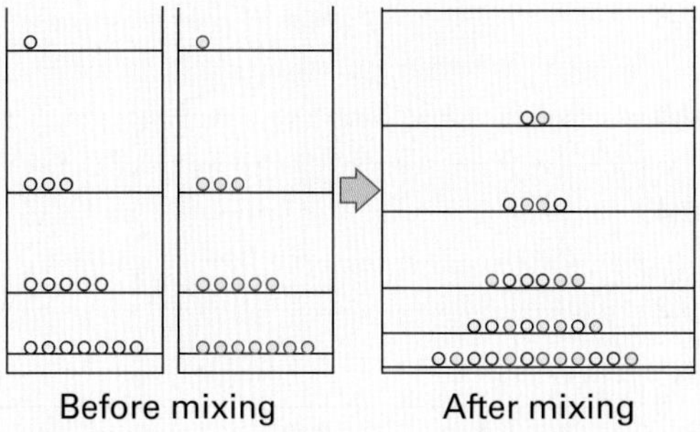

Fig. 16.26 The molecular interpretation of the entropy of mixing. Before mixing, each collection of gas molecules occupies the available energy levels with a Boltzmann distribution of populations. After mixing, more levels are accessible to the molecules as they occupy a greater volume (with the same temperature).

We can understand the spontaneity of mixing of perfect gases as the molecular tendency to populate the new energy levels that become available when the volume accessible to the molecules of each becomes larger (Fig. 16.26). In a sense, the mixing of perfect gases is simply their expansion into the available volume, since each one is blind to the presence of the other. This statistical picture can be expressed quantitatively by using the relation between the Gibbs energy and the molecular partition function in eqn 15.40 ($G = G(0) - nRT\ln(q/N)$). Thus, initially, when the partition function of gas J is $q_J = V_J/\Lambda_J^3$, where Λ_J is the thermal wavelength of J (eqn 13.15b, $\Lambda_J = h/(2\pi m_J kT)^{1/2}$), the total Gibbs energy is

$$G_i = G_A + G_B$$
$$= G_A(0) + G_B(0) - n_A RT\ln\frac{V_A}{N_A\Lambda_A^3} - n_B RT\ln\frac{V_B}{N_B\Lambda_B^3}$$
$$= G_A(0) + G_B(0) - n_A RT\ln\frac{kT}{p\Lambda_A^3} - n_B RT\ln\frac{kT}{p\Lambda_B^3} \qquad (16.19a)$$

In the second line we have used $V_J/N_J = kT/p$. In the final state of the system, each type of molecule has access to the total volume, $V_A + V_B$, and the partial pressures are p_J, so

$$G_f = G_A + G_B \qquad (16.19b)$$
$$= G_A(0) + G_B(0) - n_A RT\ln\frac{V_A + V_B}{N_A\Lambda_A^3} - n_B RT\ln\frac{V_A + V_B}{N_B\Lambda_B^3}$$
$$= G_A(0) + G_B(0) - n_A RT\ln\frac{kT}{p_A\Lambda_A^3} - n_B RT\ln\frac{kT}{p_B\Lambda_B^3}$$

We have used $(V_A + V_B)/N_J = kT/p_J$ in the second line. The difference of these two equations is eqn 16.16, and therefore eqn 16.18. This result confirms that spontaneous mixing is just

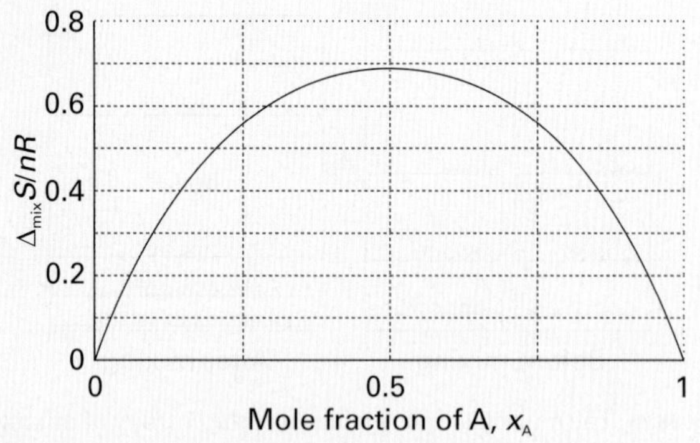

Fig. 16.27 The entropy of mixing of two perfect gases and (as discussed later) of two liquids that form an ideal solution. The entropy increases for all compositions and temperatures, so perfect gases mix spontaneously in all proportions. Because there is no transfer of heat to the surroundings when perfect gases mix, the entropy of the surroundings is unchanged. Hence, the graph also shows the total entropy change of the system plus the surroundings when perfect gases mix.

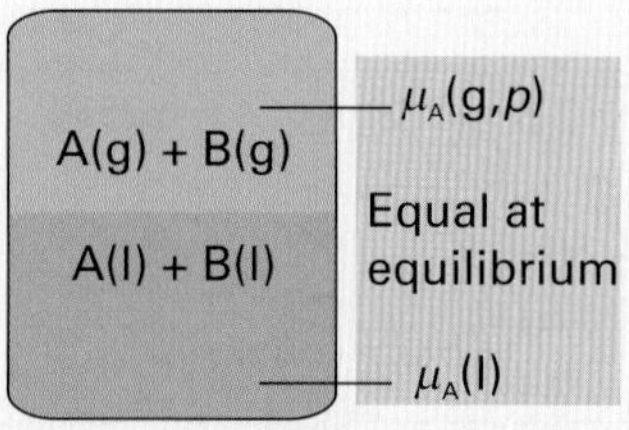

Fig. 16.28 At equilibrium, the chemical potential of the gaseous form of a substance A is equal to the chemical potential of its condensed phase. The equality is preserved if a solute is also present. Because the chemical potential of A in the vapour depends on its partial vapour pressure, it follows that the chemical potential of liquid A can be related to its partial vapour pressure.

the hunt of the molecules for the configuration of the greatest weight, their most probable distribution in the system.

Once we have an expression for the Gibbs energy of mixing we can find expressions for other mixing functions. Thus, because $(\partial G/\partial T)_{p,n} = -S$, it follows immediately from eqn 16.18 that, for a mixture of perfect gases initially at the same pressure, the entropy of mixing, $\Delta_{mix}S$, is

$$\Delta_{mix}S = -\left(\frac{\partial \Delta_{mix}G}{\partial T}\right)_{p,n_A,n_B} = -nR(x_A \ln x_A + x_B \ln x_B) \quad (16.20)^\circ$$

Because $\ln x < 0$, it follows that $\Delta_{mix}S > 0$ for all compositions (Fig. 16.27). For equal amounts of gas, for instance, we set $x_A = x_B = \frac{1}{2}$, and obtain $\Delta_{mix}S = nR \ln 2$, with n the total amount of gas molecules. This increase in entropy is what we expect when one gas disperses into the other and the disorder increases.

We can calculate **enthalpy of mixing**, $\Delta_{mix}H$, the enthalpy change accompanying mixing, of two perfect gases from $\Delta G = \Delta H - T\Delta S$. It follows from eqns 16.18 and 16.20 that

$$\Delta_{mix}H = 0 \quad (16.21)^\circ$$

The enthalpy of mixing is zero, as we should expect for a system in which there are no interactions between the molecules forming the gaseous mixture. It follows that the whole of the driving force for mixing comes from the increase in entropy of the system, because the entropy of the surroundings is unchanged.

(b) The mixing of liquids

To discuss the equilibrium properties of liquid mixtures we need to know how the Gibbs energy of a liquid varies with composition. To calculate its value, we use the fact that, at equilibrium, the chemical potential of a substance present as a vapour must be equal to its chemical potential in the liquid.

We shall denote quantities relating to pure substances by a superscript *, so the chemical potential of pure A is written μ_A^*, and as $\mu_A^*(l)$ when we need to emphasize that A is a liquid. Because the vapour pressure of the pure liquid is p_A^*, it follows from eqn 16.15 that the chemical potential of A in the vapour (treated as a perfect gas) is $\mu_A^\ominus + RT \ln p_A^*$ (with p_A to be interpreted as the relative pressure $p_A/p^\ominus$). These two chemical potentials are equal at equilibrium (Fig. 16.28), so we can write

$$\mu_A^* = \mu_A^\ominus + RT \ln p_A^* \quad \{16.22a\}$$

If another substance, a solute, is also present in the liquid, the chemical potential of A in the liquid is changed to μ_A, its vapour pressure is changed to p_A, and the chemical potential of the vapour becomes $\mu_A^\ominus + RT \ln p_A$. The vapour and solvent are still in equilibrium, so we can write

$$\mu_A = \mu_A^\ominus + RT \ln p_A \quad \{16.22b\}$$

Next, we combine these two equations to eliminate the standard chemical potential of the gas. To do so, we write eqn 16.22a as $\mu_A^\ominus = \mu_A^* - RT \ln p_A^*$ and substitute this expression into eqn 16.22b to obtain

$$\mu_A = \mu_A^* - RT \ln p_A^* + RT \ln p_A = \mu_A^* + RT \ln \frac{p_A}{p_A^*} \quad (16.23)$$

In the final step we draw on additional experimental information about the relation between the ratio of vapour pressures and the composition of the liquid. In a series of experiments on mixtures of closely related liquids (such as benzene and methylbenzene), the French chemist François Raoult found that the ratio of the partial vapour pressure of each component to its vapour pressure as a pure liquid, p_A/p_A^*, is approximately equal

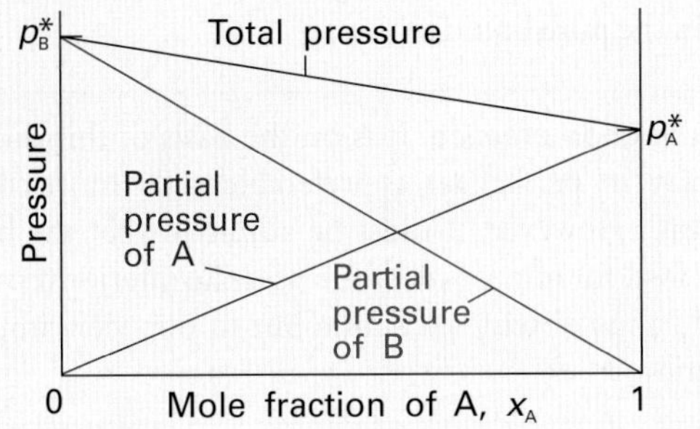

Fig. 16.29 The total vapour pressure and the two partial vapour pressures of an ideal binary mixture are proportional to the mole fractions of the components.

to the mole fraction of A in the liquid mixture. That is, he established what we now call **Raoult's law:**

$$p_A = x_A p_A^* \qquad (16.24)^\circ$$

This law is illustrated in Fig. 16.29. Some mixtures obey Raoult's law very well, especially when the components are structurally similar. Mixtures that obey the law throughout the composition range from pure A to pure B are called **ideal solutions**. When we write equations that are valid only for ideal solutions, we shall label them with a superscript °, as in eqn 16.24. Some solutions depart significantly from Raoult's law. Nevertheless, even in these cases the law is obeyed increasingly closely for the component in excess (the solvent) as it approaches purity. The law is therefore a good approximation for the properties of the solvent if the solution is dilute.

For an ideal solution, it follows from eqns 16.23 and 16.24 that

$$\mu_A = \mu_A^* + RT \ln x_A \qquad (16.25)^\circ$$

This important equation can be used as the *definition* of an ideal solution (so that it implies Raoult's law rather than stemming from it). It is in fact a better definition than eqn 16.24 because it does not assume that the vapour is a perfect gas.

In ideal solutions the solute, as well as the solvent, obeys Raoult's law. However, the English chemist William Henry found experimentally that for real solutions at low concentrations, although the vapour pressure of the solute is proportional to its mole fraction, the constant of proportionality is not the vapour pressure of the pure substance (Fig. 16.30). **Henry's law** is:

$$p_B = x_B K_B \qquad (16.26a)^\circ$$

In this expression x_B is the mole fraction of the solute and K_B is an empirical constant (with the dimensions of pressure) chosen so that the plot of the vapour pressure of B against its mole fraction is tangent to the experimental curve at $x_B = 0$. In practice, a polynomial is fitted to the vapour pressure data, and the slope of the curve at $x_B = 0$ is determined by differentiation or equi-

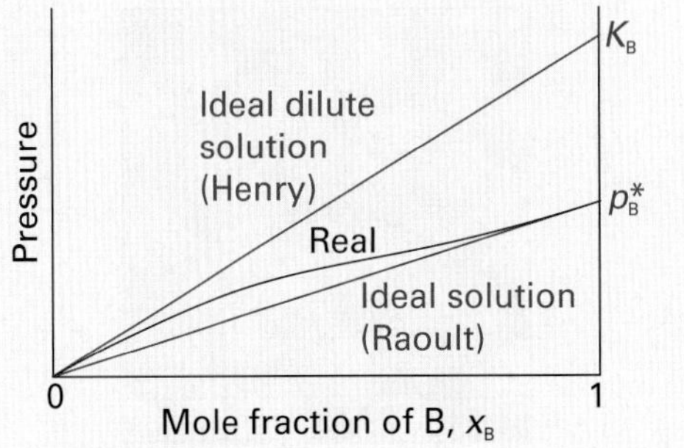

Fig. 16.30 When a component (the solvent) is nearly pure, it has a vapour pressure that is proportional to mole fraction with a slope p_B^* (Raoult's law). When it is the minor component (the solute) its vapour pressure is still proportional to the mole fraction, but the constant of proportionality is now K_B (Henry's law).

Synoptic table 16.1* Henry's law constants for gases in water at 298 K

	$K/(\text{kPa kg mol}^{-1})$
CO_2	3.01×10^3
H_2	1.28×10^5
N_2	1.56×10^5
O_2	7.92×10^4

* More values are given in the *Data section*.

valently by identifying the coefficient of the term that is proportional to x_B. In practical applications of Henry's law it is often more convenient to express it in terms of the molality, b, of the solute, in which case we write

$$p_B = b_B K_B \qquad (16.26b)^\circ$$

Table 16.1 gives a selection of values of the Henry's law constant for this convention.

Mixtures for which the solute obeys Henry's law and the solvent obeys Raoult's law are called **ideal-dilute solutions.** We shall also label equations with a superscript ° when they have been derived from Henry's law. The difference in behaviour of the solute and solvent at low concentrations (as expressed by Henry's and Raoult's laws, respectively) arises from the fact that in a dilute solution the solvent molecules are in an environment very much like the one they have in the pure liquid (Fig. 16.31). In contrast, the solute molecules are surrounded by solvent molecules, which is entirely different from their environment when pure. Thus, the solvent behaves like a slightly modified pure liquid, but the solute behaves entirely differently from its

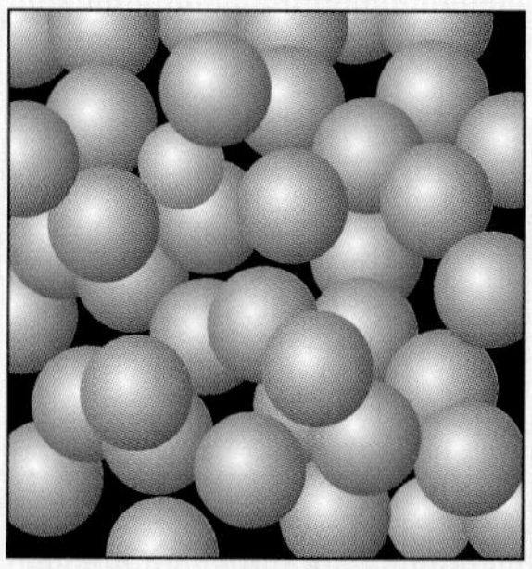

Fig. 16.31 In a dilute solution, the solvent molecules (the purple spheres) are in an environment that differs only slightly from that of the pure solvent. The solute particles (the blue spheres), however, are in an environment totally unlike that of the pure solute.

pure state unless the solvent and solute molecules happen to be very similar. In the latter case, the solute also obeys Raoult's law.

The Gibbs energy of mixing of two liquids to form an ideal solution is calculated in exactly the same way as for two gases. The total Gibbs energy before liquids are mixed is

$$G_i = n_A\mu_A^\star + n_B\mu_B^\star$$

When they are mixed, the individual chemical potentials are given by eqn 16.25 and the total Gibbs energy is

$$G_f = n_A\{\mu_A^\star + RT\ln x_A\} + n_B\{\mu_B^\star + RT\ln x_B\}$$

Consequently, the Gibbs energy of mixing is

$$\Delta_{mix}G = nRT\{x_A\ln x_A + x_B\ln x_B\} \qquad (16.27a)^\circ$$

where $n = n_A + n_B$. As for gases, it follows that the ideal entropy of mixing of two liquids is

$$\Delta_{mix}S = -nR\{x_A\ln x_A + x_B\ln x_B\} \qquad (16.27b)^\circ$$

and, because $\Delta_{mix}H = \Delta_{mix}G + T\Delta_{mix}S = 0$, the ideal enthalpy of mixing is zero. The ideal volume of mixing, the change in volume on mixing, is also zero because it follows from eqn 15.60 ($(\partial G/\partial p)_T = V$) that $\Delta_{mix}V = (\partial\Delta_{mix}G/\partial p)_T$, but $\Delta_{mix}G$ in eqn 16.27a is independent of pressure, so the derivative with respect to pressure is zero.

Equation 16.27 is the same as that for two perfect gases and all the conclusions drawn there are valid here: the driving force for mixing is the increasing entropy of the system as the molecules mingle and the enthalpy of mixing is zero. The variation of the Gibbs energy and entropy of mixing with composition is the same as that already depicted for gases in Figs. 16.24 and 16.27. It should be noted, however, that solution ideality means something different from gas perfection. In a perfect gas there are no forces acting between molecules. In ideal solutions there are interactions, but the average energy of A–B interactions in the mixture is the same as the average energy of A–A and B–B interactions in the pure liquids.

A note on good practice It is on the basis of this difference that the term 'perfect gas' is preferable to the more common 'ideal gas'. However, it must be admitted that the battle is almost lost, for almost everyone uses the inferior term 'ideal gas' and thus obscures a subtle but in our view important distinction.

(c) Colligative properties

The presence of a solute modifies the physical properties of the solvent, such as its vapour pressure, boiling point, and freezing point. It also introduces a new property, the 'osmotic pressure'. In dilute solutions these properties depend only on the number of solute particles present, not their identity. For this reason, they are called **colligative properties** (denoting 'depending on the collection').

All the colligative properties stem from the reduction of the chemical potential of the liquid solvent as a result of the presence of solute. For an ideal-dilute solution, the reduction is from $\mu_A^\star$ for the pure solvent to $\mu_A^\star + RT\ln x_A$ when a solute is present ($\ln x_A$ is negative because $x_A < 1$). There is no direct influence of the solute on the chemical potential of the solvent vapour and the solid solvent because the solute, which is assumed to be non-volatile and insoluble in the solid solvent, appears in neither the vapour nor the solid. As can be seen from Fig. 16.32, the reduction in chemical potential of the solvent implies that the liquid–vapour equilibrium occurs at a higher temperature (the boiling point is raised) and the solid–liquid equilibrium occurs at a lower temperature (the freezing point is lowered).

The molecular origin of the lowering of the chemical potential is the effect of the solute on the entropy of the solution. The

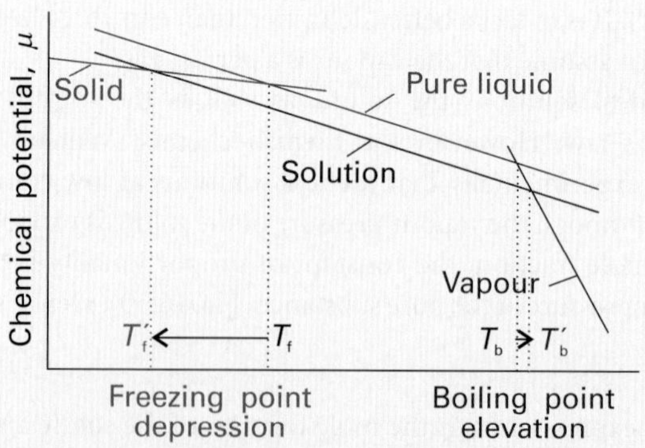

Fig. 16.32 The chemical potential of a solvent in the presence of a solute. The lowering of the liquid's chemical potential has a greater effect on the freezing point than on the boiling point because of the angles at which the lines intersect.

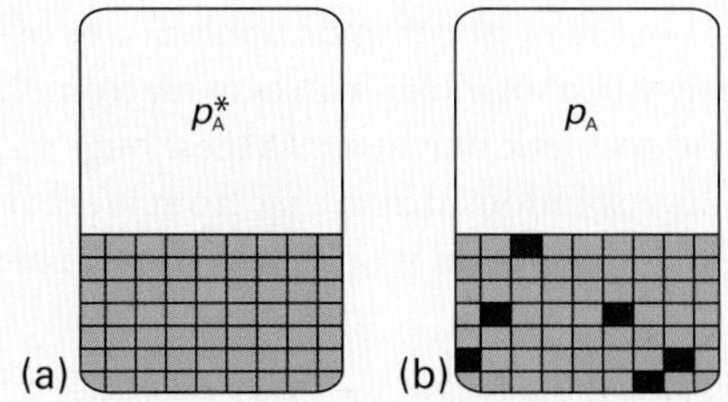

Fig. 16.33 The vapour pressure of a pure liquid represents a balance between the increase in disorder arising from vaporization and the decrease in disorder of the surroundings. (a) Here the structure of the liquid is represented highly schematically by the grid of squares. (b) When solute (the dark squares) is present, the disorder of the condensed phase is higher than that of the pure liquid, and there is a decreased tendency to acquire the disorder characteristic of the vapour.

pure liquid solvent has a characteristic entropy and its vapour pressure reflects the tendency of the system towards greater entropy, which can be achieved if the liquid vaporizes to form a gas. When a solute is present, there is an additional contribution to the entropy of the liquid, even in an ideal solution, because in a blind selection of molecules we cannot predict with certainty that we will draw a solvent or a solute molecule. Because the entropy of the liquid is already higher than that of the pure liquid, there is a weaker tendency to form the gas in the sense that less has to vaporize to maximize the total entropy (Fig. 16.33). As a result, the vapour pressure of the solvent is lowered and hence its boiling point is raised. Similarly, the enhanced molecular randomness of the solution opposes the tendency to freeze. Consequently, a lower temperature must be reached before equilibrium between solid and solution is achieved. Hence, the freezing point is lowered.

A brief comment It is commonly stated that the depression of freezing point is an important colligative property because, for instance, it accounts for the use of antifreeze in engines. Antifreeze, however, is used in much higher concentration than can justify its effect as a colligative property; its effect is to interfere with the solidification of water molecules. The use of salt on highways is also at concentrations far outside the range of this discussion. The effect of impurities on the melting points of organic compounds is more complex, as the impurity is dissolved in the solid, a situation excluded in the treatment of colligative properties. Boiling point elevation is too small to be of any practical significance. No one, except perhaps as a laboratory exercise, any longer uses the depression of freezing point or elevation of boiling point to determine molar mass.

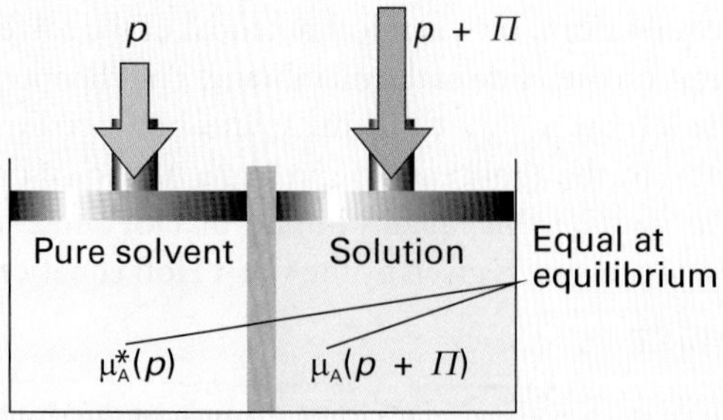

Fig. 16.34 The equilibrium involved in the calculation of osmotic pressure, Π, is between pure solvent A at a pressure p on one side of the semipermeable membrane and A as a component of the mixture on the other side of the membrane, where the pressure is $p + \Pi$.

The only colligative property of real importance is **osmosis** (from the Greek word for 'push'), the spontaneous passage of a pure solvent into a solution separated from it by a **semipermeable membrane**, a membrane permeable to the solvent but not to the solute (Fig. 16.34). The **osmotic pressure**, Π, is the pressure that must be applied to the solution to stop the influx of solvent. Important examples of osmosis include transport of fluids through cell membranes, dialysis, and **osmometry**, the determination of molar mass by the measurement of osmotic pressure. Osmometry is widely used to determine the molar masses of macromolecules.

In the simple arrangement shown in Fig. 16.35, the opposing pressure arises from the head of solution that the osmosis itself produces. Equilibrium is reached when the hydrostatic pressure of the column of solution matches the osmotic pressure. The complicating feature of this arrangement is that the entry of solvent into the solution results in its dilution, and so it is more difficult to treat than the arrangement in Fig. 16.34, in which there is no flow and the concentrations remain unchanged.

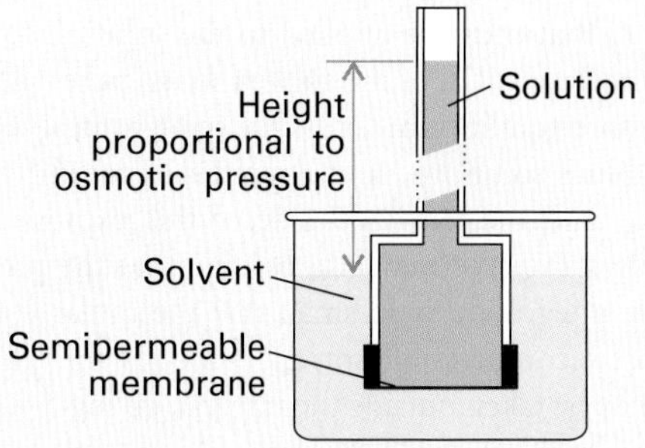

Fig. 16.35 In a simple version of the osmotic pressure experiment, A is at equilibrium on each side of the membrane when enough has passed into the solution to cause a hydrostatic pressure difference.

The thermodynamic treatment of osmosis depends on noting that, at equilibrium, the chemical potential of the solvent must be the same on each side of the membrane. The chemical potential of the solvent is lowered by the solute, but is restored to its 'pure' value by the application of pressure. As shown in the following *Justification*, this equality implies that for dilute solutions the osmotic pressure is given by the **van 't Hoff equation:**

$$\Pi = [\mathrm{B}]RT \qquad (16.28)°$$

where $[\mathrm{B}] = n_\mathrm{B}/V$ is the molar concentration of the solute.

Justification 16.3 *The van't Hoff equation*

On the pure solvent side the chemical potential of the solvent, which is at a pressure p, is $\mu_\mathrm{A}^\star(p)$. On the solution side, the chemical potential is lowered by the presence of the solute, which reduces the mole fraction of the solvent from 1 to x_A. However, the chemical potential of A is raised on account of the greater pressure, $p + \Pi$, that the solution experiences. At equilibrium the chemical potential of A is the same in both compartments, and we can write

$$\mu_\mathrm{A}^\star(p) = \mu_\mathrm{A}(x_\mathrm{A}, p + \Pi)$$

The presence of solute is taken into account in the normal way:

$$\mu_\mathrm{A}(x_\mathrm{A}, p + \Pi) = \mu_\mathrm{A}^\star(p + \Pi) + RT \ln x_\mathrm{A}$$

We saw in Section 15.9 (specifically eqn 15.64) how to take the effect of pressure into account. Thus, because $(\partial\mu_\mathrm{A}/\partial p)_T = V_{\mathrm{A,m}}$, for pure A

$$\mu_\mathrm{A}^\star(p + \Pi) = \mu_\mathrm{A}^\star(p) + \int_p^{p+\Pi} V_\mathrm{m}\,\mathrm{d}p$$

where V_m is the molar volume of the pure solvent A. When these three equations are combined we get

$$-RT \ln x_\mathrm{A} = \int_p^{p+\Pi} V_\mathrm{m}\,\mathrm{d}p$$

This expression enables us to calculate the additional pressure Π that must be applied to the solution to restore the chemical potential of the solvent to its 'pure' value and thus to restore equilibrium across the semipermeable membrane. For dilute solutions, $\ln x_\mathrm{A}$ may be replaced by $\ln(1 - x_\mathrm{B}) \approx -x_\mathrm{B}$, and the left-hand side of this expression becomes simply RTx_B. We may also assume that the pressure range in the integration is so small that the molar volume of the almost incompressible solvent is a constant. That being so, V_m may be taken outside the integral, giving

$$RTx_\mathrm{B} = \Pi V_\mathrm{m}$$

When the solution is dilute, $x_\mathrm{B} \approx n_\mathrm{B}/n_\mathrm{A}$. Moreover, because $n_\mathrm{A}V_\mathrm{m} = V$, the total volume of the solvent, the equation simplifies to eqn 16.28.

Because the effect of osmotic pressure is so readily measurable and large, one of the most common applications of osmometry is to the measurement of molar masses of macromolecules, such as proteins and synthetic polymers. As these huge molecules dissolve to produce solutions that are far from ideal, it is assumed that the van't Hoff equation is only the first term of a virial-like expansion:

$$\Pi = [\mathrm{J}]RT\{1 + B[\mathrm{J}] + \cdots\} \qquad (16.29)$$

where we have denoted the solute by J to avoid too many different Bs in this expression. The additional terms take the non-ideality into account; the empirical constant B is called the **osmotic virial coefficient**.

Example 16.3 *Using osmometry to determine the molar mass of a macromolecule*

The osmotic pressures of solutions of poly(vinyl chloride), PVC, in cyclohexanone at 298 K are given below. The pressures are expressed in terms of the heights of solution (of mass density $\rho = 0.980\ \mathrm{g\,cm^{-3}}$) in balance with the osmotic pressure. Determine the molar mass of the polymer.

$c/(\mathrm{g\,dm^{-3}})$	1.00	2.00	4.00	7.00	9.00
h/cm	0.28	0.71	2.01	5.10	8.00

Method The osmotic pressure is measured at a series of mass concentrations, c, and a plot of Π/c against c is used to determine the molar mass of the polymer. We use eqn 16.29 with $[\mathrm{J}] = c/M$ where c is the mass concentration of the polymer and M is its molar mass. The osmotic pressure is related to the hydrostatic pressure by $\Pi = \rho gh$ with $g = 9.81\ \mathrm{m\,s^{-2}}$. With these substitutions, eqn 16.29 becomes

$$\frac{h}{c} = \frac{RT}{\rho gM}\left(1 + \frac{Bc}{M} + \cdots\right) = \frac{RT}{\rho gM} + \left(\frac{RTB}{\rho gM^2}\right)c + \cdots$$

Therefore, to find M, plot h/c against c, and expect a straight line with intercept $RT/\rho gM$ at $c = 0$.

Answer The data give the following values for the quantities to plot:

$c/(\mathrm{g\,dm^{-3}})$	1.00	2.00	4.00	7.00	9.00
$(h/c)/(\mathrm{cm\,g^{-1}\,dm^3})$	0.28	0.36	0.503	0.729	0.889

The points are plotted in Fig. 16.36. The intercept from a least-squares analysis is at 0.21. Therefore,

$$M = \frac{RT}{\rho g} \times \frac{1}{0.21\ \mathrm{cm\,g^{-1}\,dm^3}}$$

$$= \frac{(8.3145\ \mathrm{J\,K^{-1}\,mol^{-1}}) \times (298\ \mathrm{K})}{(980\ \mathrm{kg\,m^{-3}}) \times (9.81\ \mathrm{m\,s^{-2}})} \times \frac{1}{2.1 \times 10^{-3}\ \mathrm{m^4\,kg^{-1}}}$$

$$= 1.2 \times 10^2\ \mathrm{kg\,mol^{-1}}$$

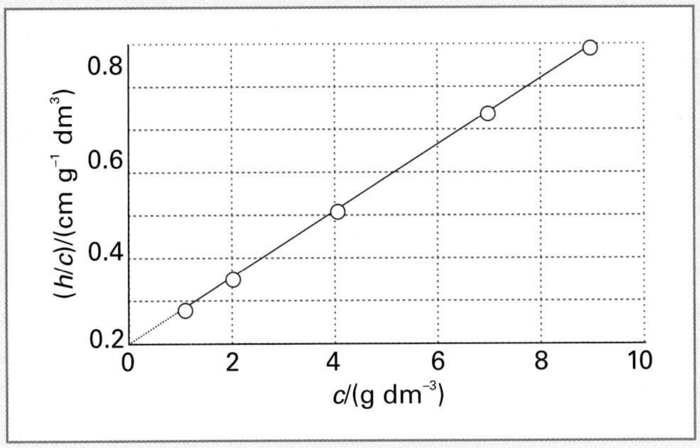

Fig. 16.36 The plot involved in the determination of molar mass by osmometry. The molar mass is calculated from the intercept at $c = 0$.

interActivity Determine the value of the osmotic virial coefficient B from these data.

where we have used $1\ \mathrm{kg\ m^2\ s^{-2}} = 1\ \mathrm{J}$. Molar masses of macromolecules are often reported in daltons (Da), with $1\ \mathrm{Da} = 1\ \mathrm{g\ mol^{-1}}$. The macromolecule in this example has a molar mass of about 120 kDa. Modern osmometers give readings of osmotic pressure in pascals, so the analysis of the data is more straightforward and eqn 16.29 can be used directly.

Self-test 16.3 The osmotic pressures of solutions of polystyrene in methylbenzene at 293 K are given below. Determine the molar mass of the polymer.

$c/(\mathrm{g\ dm^{-3}})$	1.00	2.00	4.00	7.00	9.00
Π/Pa	17.7	35.5	70.8	125.0	162.0

[140 kDa]

Care must be taken in the interpretation of molar masses determined by osmosis. Thus, although a pure protein is almost **monodisperse**, meaning that it has a single, definite molar mass (although there may be small variations, such as one amino acid replacing another, depending on the source of the sample), a synthetic polymer is **polydisperse**, in the sense that a sample is a mixture of molecules with various chain lengths and molar masses. The **number-average molar mass**, $\bar{M}_n$, is the value obtained by weighting each molar mass by the number of molecules of that mass present in the sample:

$$\bar{M}_n = \frac{1}{N}\sum_i N_i M_i \qquad [16.30a]$$

where N_i is the number of molecules with molar mass M_i and there are N molecules in all. The **weight-average molar mass**, $\bar{M}_w$ is the average calculated by weighting the molar masses of the molecules by the mass of each one present in the sample:

$$\bar{M}_w = \frac{1}{m}\sum_i m_i M_i \qquad [16.30b]$$

In this expression, m_i is the total mass of molecules of molar mass M_i and m is the total mass of the sample. Because $m_i = N_i M_i / N_A$, we can also express this average as

$$\bar{M}_w = \frac{\sum_i N_i M_i^2}{\sum_i N_i M_i} \qquad [16.30c]$$

This expression shows that the weight-average molar mass is proportional to the mean square molar mass. In practice, osmometry gives the number-average molar mass and light scattering experiments give the weight-average molar mass.

Example 16.4 *Calculating number- and weight-average molar masses*

Determine the number-average and the weight-average molar masses for a sample of poly(vinyl chloride) from the following data:

Molar mass interval/(kg mol⁻¹)	Average molar mass within interval/(kg mol⁻¹)	Mass of sample within interval/g
5–10	7.5	9.6
10–15	12.5	8.7
15–20	17.5	8.9
20–25	22.5	5.6
25–30	27.5	3.1
30–35	32.5	1.7

Method The relevant equations are eqns 16.30a and 16.30b. Calculate the two averages by weighting the molar mass within each interval by the number and mass, respectively, of the molecule in each interval. Obtain the numbers in each interval by dividing the mass of the sample in each interval by the average molar mass for that interval. Because number of molecules is proportional to amount of substance (the number of moles), the number-weighted average can be obtained directly from the amounts in each interval.

Answer The amounts in each interval are as follows:

Interval	5–10	10–15	15–20	20–25	25–30	30–35
Molar mass/(kg mol⁻¹)	7.5	12.5	17.5	22.5	27.5	32.5
Amount/(10⁻³ mol)	1.3	0.7	0.51	0.25	0.11	0.052
					Total:	2.92

The number-average molar mass is therefore

$$\bar{M}_n/(\text{kg mol}^{-1}) = \frac{1}{2.92}(1.3\times 7.5 + 0.70\times 12.5 + 0.51\times 17.5$$
$$+ 0.25\times 22.5 + 0.11\times 27.5 + 0.052\times 32.5)$$
$$= 13$$

where the factor 10^{-3} cancels. The weight-average molar mass is calculated directly from the data after noting that the total mass of the sample is 37.6 g:

$$\bar{M}_w/(\text{kg mol}^{-1}) = \frac{1}{37.6}(9.6\times 7.5 + 8.7\times 12.5 + 8.9\times 17.5$$
$$+ 5.6\times 22.5 + 3.1\times 27.5 + 1.7\times 32.5)$$
$$= 16$$

Note the significantly different values of the two averages. In this instance, $\bar{M}_w/\bar{M}_n = 1.2$.

Self-test 16.4 The *Z-average molar mass* is defined as

$$\bar{M}_Z = \frac{\sum_i N_i M_i^3}{\sum_i N_i M_i^2} \qquad [16.30d]$$

and can be interpreted in terms of the mean cubic molar mass. Evaluate the *Z*-average molar mass of the sample described in the Example. [19 kg mol^{-1}]

The ratio $\bar{M}_w/\bar{M}_n$ is called the **heterogeneity index** (or 'polydispersity index'). In the determination of protein molar masses we expect the various averages to be the same because the sample is monodisperse (unless there has been degradation). A synthetic polymer normally spans a range of molar masses and the different averages yield different values. Typical synthetic materials have $\bar{M}_w/\bar{M}_n \approx 4$. The term 'monodisperse' is conventionally applied to synthetic polymers in which this index is less than 1.1; commercial polyethylene samples might be much more heterogeneous, with a ratio close to 30. One consequence of a narrow molar mass distribution for synthetic polymers is often a higher degree of three-dimensional long-range order in the solid and therefore higher density and melting point. The spread of values is controlled by the choice of catalyst and reaction conditions. In practice, it is found that long-range order is determined more by structural factors (branching, for instance) than by molar mass.

IMPACT ON BIOCHEMISTRY

I16.2 Osmosis and the structure of biological cells

Osmosis helps biological cells maintain their structure. Cell membranes are semipermeable and allow water, small molecules, and hydrated ions to pass, while blocking the passage of biopolymers synthesized inside the cell. The difference in concentrations of solutes inside and outside the cell gives rise to an osmotic pressure, and water passes into the more concentrated solution in the interior of the cell, carrying small nutrient molecules. The influx of water also keeps the cell swollen, whereas dehydration causes the cell to shrink. These effects are important in everyday medical practice. To maintain the integrity of blood cells, solutions that are injected into the bloodstream for blood transfusions and intravenous feeding must be **isotonic** with the blood, meaning that they must have the same osmotic pressure as blood. If the injected solution is too dilute, or *hypotonic*, the flow of solvent into the cells, required to equalize the osmotic pressure, causes the cells to burst and die by a process called *haemolysis*. If the solution is too concentrated, or *hypertonic*, equalization of the osmotic pressure requires flow of solvent out of the cells, which shrink and die.

16.6 Real solutions

Real solutions are composed of particles for which A–A, A–B, and B–B interactions are all different. Not only may there be enthalpy and volume changes when liquids mix, but there may also be an additional contribution to the entropy arising from the way in which the molecules of one type might cluster together instead of mingling freely with the others. If the enthalpy change is large and positive or if the entropy change is adverse (because of a reorganization of the molecules that results in an orderly mixture), then the Gibbs energy might be positive for mixing. In that case, separation is spontaneous and the liquids may be immiscible. Alternatively, the liquids might be **partially miscible**, which means that they are miscible only over a certain range of compositions.

The thermodynamic properties of real solutions are expressed in terms of the **excess functions**, X^E, the difference between the observed thermodynamic function of mixing and the function for an ideal solution. The **excess entropy**, S^E, for example, is defined as

$$S^E = \Delta_{mix}S - \Delta_{mix}S^{ideal} \qquad [16.31]$$

where $\Delta_{mix}S^{ideal}$ is given by eqn 16.20. The excess enthalpy and volume are both equal to the observed enthalpy and volume of mixing, because the ideal values are zero in each case. Figure 16.37 shows two examples of the composition dependence of molar excess functions.

(a) A model system: the regular solution

Deviations of the excess energies from zero indicate the extent to which the solutions are non-ideal. In this connection a useful model system is the **regular solution**, a solution for which $H^E \neq 0$ but $S^E = 0$. We can think of a regular solution as one in which the two kinds of molecules are distributed randomly (as

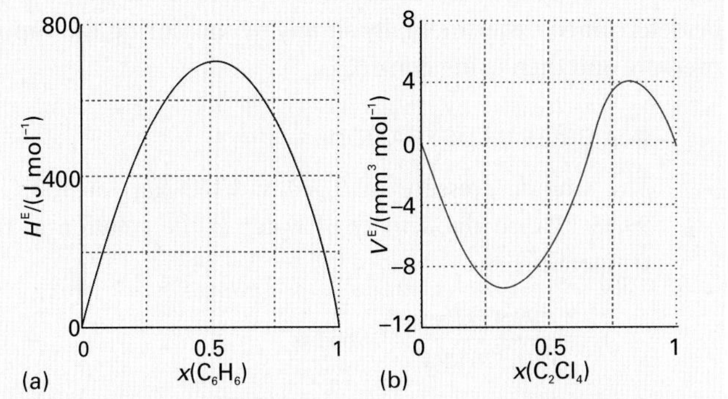

Fig. 16.37 Experimental excess functions at 25°C. (a) H^E for benzene/cyclohexane; this graph shows that the mixing is endothermic (because $\Delta_{mix}H = 0$ for an ideal solution). (b) The excess volume, V^E, for tetrachloroethene/cyclopentane; this graph shows that there is a contraction at low tetrachloroethene mole fractions, but an expansion at high mole fractions (because $\Delta_{mix}V = 0$ for an ideal mixture).

in an ideal solution) but have different energies of interactions with each other.

To express this model quantitatively, we suppose that the excess enthalpy depends on composition as

$$H^E = n\xi RTx_Ax_B \tag{16.32}$$

where ξ (xi) is a dimensionless parameter that is a measure of the energy of AB interactions relative to that of the AA and BB interactions; for an ideal solution, $\xi = 0$. The function given by eqn 16.32 is plotted in Fig. 16.38, and we see it resembles the experimental curve in Fig. 16.37. If $\xi < 0$, mixing is exothermic and the solute–solvent interactions are more favourable than

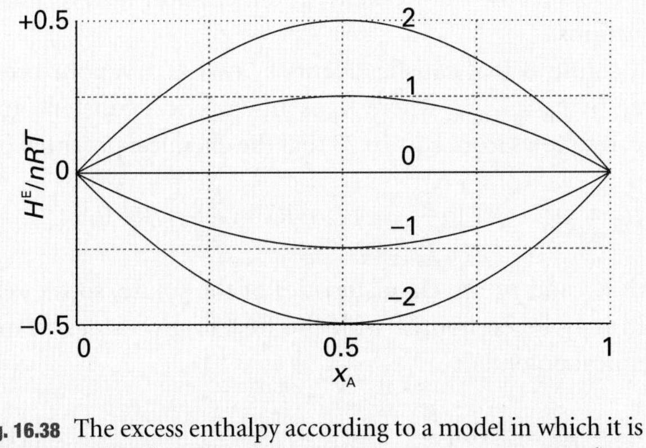

Fig. 16.38 The excess enthalpy according to a model in which it is proportional to ξx_Ax_B for different values of the parameter ξ.

interActivity Using the graph above, fix ξ and vary the temperature. For what value of x_A does the excess enthalpy depend on temperature most strongly?

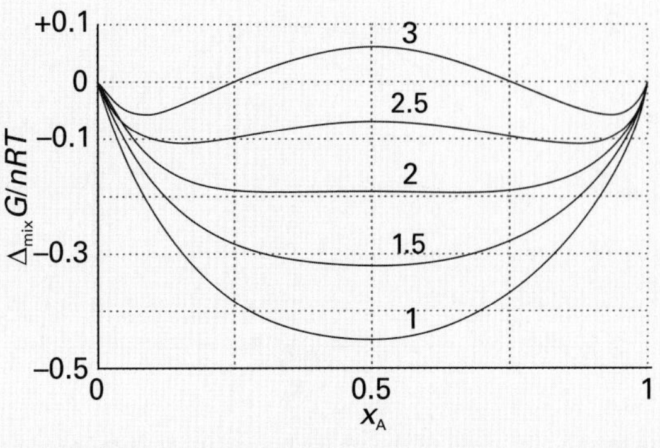

Fig. 16.39 The Gibbs energy of mixing for different values of the parameter ξ.

interActivity Using the graph above, fix ξ at 1.5 and vary the temperature. Is there a range of temperatures over which you observe phase separation?

the solvent–solvent and solute–solute interactions. If $\xi > 0$, then the mixing is endothermic. Because for a regular solution the entropy of mixing has its ideal value, the excess Gibbs energy is equal to the excess enthalpy, and the Gibbs energy of mixing is

$$\Delta_{mix}G = nRT\{x_A \ln x_A + x_B \ln x_B + \xi x_Ax_B\} \tag{16.33}$$

with $x_B = 1 - x_A$.

Figure 16.39 shows how $\Delta_{mix}G$ varies with composition for different values of ξ. The important feature is that for $\xi > 2$ the graph shows two minima separated by a maximum. The implication of this observation is that, provided $\xi > 2$, the system will separate spontaneously into two phases with compositions corresponding to the two minima, for that separation corresponds to a reduction in Gibbs energy. This behaviour is what is summarized by the experimentally determined two-component liquid mixture in Fig. 16.37.

We can take this analysis further and identify the upper critical solution temperature. The compositions corresponding to the two minima in Fig. 16.39 are obtained by looking for the conditions at which $\partial\Delta_{mix}G/\partial x_A = 0$, and a simple manipulation of eqn 16.33 shows that we have to solve

$$\ln\frac{x_A}{1 - x_A} + \xi(1 - 2x_A) = 0$$

A brief comment This expression is an example of a *transcendental equation*, an equation that does not have a solution that can be expressed in a closed form. The solutions can be found numerically by using mathematical software or by plotting the first term against the second and identifying the points of intersection as ξ is changed.

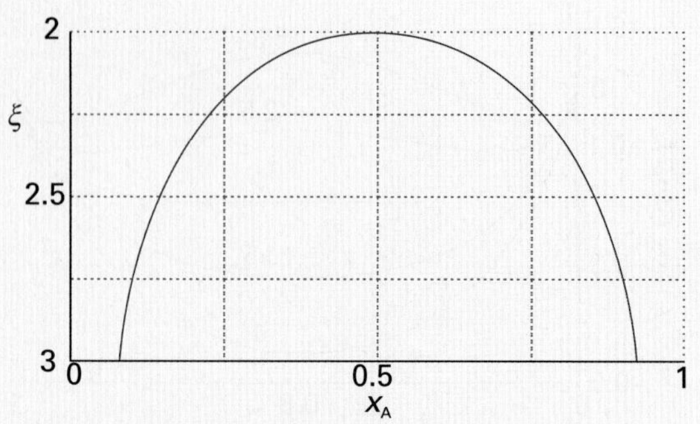

Fig. 16.40 The location of the phase boundary as computed on the basis of the ξ-parameter model.

The solutions are plotted in Fig. 16.40. We see that the two minima move together as ξ decreases and merge when $\xi = 2$. Because $H^E \propto \xi RT$, for constant excess enthalpy (corresponding to the effect of intermolecular forces being constant as the temperature is raised) a decrease in ξ can be interpreted as an increase in temperature, so the vertical axis in Fig. 16.40 can be interpreted as indicating the temperature, and so the topmost point of the curve corresponds to the upper critical solution temperature, as in Fig. 16.37. This model is developed further in Section 16.6d.

(b) The solvent activity

As we have seen, real solutions differ from ideal solutions as a result of differences in intermolecular interactions between their components and the manner in which the molecules aggregate. These differences are taken into account by replacing the concentrations in expressions for the chemical potential by effective concentrations known as 'activities'.

The general form of the chemical potential of a real or ideal solvent is given by a straightforward modification of eqn 16.23 (that $\mu_A = \mu_A^* + RT\ln(p_A/p_A^*)$), where p_A^* is the vapour pressure of pure A and p_A is the vapour pressure of A when it is a component of a solution. For an ideal solution, as we have seen, the solvent obeys Raoult's law at all concentrations and we can express this relation as eqn 16.25 (that is, as $\mu_A = \mu_A^* + RT\ln x_A$). The form of this relation can be preserved when the solution does not obey Raoult's law by writing

$$\mu_A = \mu_A^* + RT\ln a_A \qquad [16.34]$$

The quantity a_A is the **activity** of A, a kind of 'effective' mole fraction.

Because eqn 16.23 is true for both real and ideal solutions, we can conclude by comparing it with eqn 16.34 that

$$a_A = \frac{p_A}{p_A^*} \qquad (16.35)$$

Note—and this is a very important point—that it follows that the activity of a pure substance (when $p_A = p_A^*$) is 1. We see that there is nothing mysterious about the activity of a solvent: it can be determined experimentally simply by measuring the vapour pressure and then using eqn 16.35.

● A BRIEF ILLUSTRATION

The vapour pressure of 0.500 M KNO_3(aq) at 100°C is 99.95 kPa, so the activity of water in the solution at this temperature is

$$a_A = \frac{99.95\ \text{kPa}}{101.325\ \text{kPa}} = 0.9864 \; ●$$

Because all solvents obey Raoult's law (that $p_A/p_A^* = x_A$) increasingly closely as the concentration of solute approaches zero, the activity of the solvent approaches the mole fraction as $x_A \to 1$:

$$a_A \to x_A \text{ as } x_A \to 1 \qquad (16.36)$$

A convenient way of expressing this convergence is to introduce the **activity coefficient**, γ, by the definition

$$a_A = \gamma_A x_A \qquad \gamma_A \to 1 \text{ as } x_A \to 1 \qquad [16.37]$$

at all temperatures and pressures. The chemical potential of the solvent is then

$$\mu_A = \mu_A^* + RT\ln x_A + RT\ln \gamma_A \qquad (16.38)$$

and all the deviation from ideal behaviour is expressed by $RT\ln\gamma_A$. The standard state of the solvent, the pure liquid solvent at 1 bar, is established when $x_A = 1$.

(c) The solute activity

The problem with defining activity coefficients and standard states for solutes is that they approach ideal-dilute (Henry's law) behaviour as $x_B \to 0$, not as $x_B \to 1$ (corresponding to pure solute). We shall show how to set up the definitions for a solute that obeys Henry's law exactly, and then show how to allow for deviations.

A solute B that satisfies Henry's law has a vapour pressure given by $p_B = K_B x_B$, where K_B is an empirical constant. In this case, it follows from eqn 16.23 that the chemical potential of B is

$$\mu_B = \mu_B^* + RT\ln\frac{p_B}{p_B^*} = \mu_B^* + RT\ln\frac{K_B}{p_B^*} + RT\ln x_B$$

Both K_B and p_B^* are characteristics of the solute, so the second term may be combined with the first to give a new standard chemical potential:

$$\mu_B^{\ominus} = \mu_B^* + RT\ln\frac{K_B}{p_B^*} \qquad [16.39]$$

It then follows that the chemical potential of a solute in an ideal-dilute solution is related to its mole fraction by

$$\mu_B = \mu_B^{\ominus} + RT\ln x_B \qquad (16.40)^{\circ}$$

Table 16.2 Activities and standard states

Component	Basis	Standard state	Activity	Limits
Solid or liquid		Pure	$a = 1$	
Solvent	Raoult	Pure solvent	$a = p/p^*$, $a = \gamma x$	$\gamma \to 1$ as $x \to 1$ (pure solvent)
Solute	Henry	(1) A hypothetical state of the pure solute	$a = p/K$, $a = \gamma x$	$\gamma \to 1$ as $x \to 0$
		(2) A hypothetical state of the solute at molality $b^{\ominus} =$ 1 mol kg^{-1}	$a = \gamma b/b^{\ominus}$	$\gamma \to 1$ as $b \to 0$

In each case, $\mu = \mu^{\ominus} + RT \ln a$.

If the solution is ideal, $K_B = p_B^*$ and eqn 16.39 reduces to $\mu_B^{\ominus} = \mu_B^*$, as we should expect.

We now permit deviations from ideal-dilute, Henry's law behaviour. For the solute, we introduce a_B in place of x_B in eqn 16.40, and obtain

$$\mu_B = \mu_B^{\ominus} + RT \ln a_B \qquad [16.41]$$

The standard state remains unchanged in this last stage, and all the deviations from ideality are captured in the activity a_B. The value of the activity at any concentration can be obtained in the same way as for the solvent, but in place of eqn 16.35 we use

$$a_B = \frac{p_B}{K_B} \qquad (16.42)$$

As for the solvent, it is sensible to introduce an activity coefficient through

$$a_B = \gamma_B x_B \qquad [16.43]$$

Now all the deviations from ideality are captured in the activity coefficient γ_B. Because the solute obeys Henry's law as its concentration goes to zero, it follows that

$$a_B \to x_B \text{ and } \gamma_B \to 1 \text{ as } x_B \to 0 \qquad (16.44)$$

at all temperatures and pressures. Deviations of the solute from ideality disappear as zero concentration is approached.

The conventions for activities, activity coefficients, and standard states are summarized in Table 16.2.

Example 16.5 *Measuring activity*

Calculate the activity and activity coefficient of chloroform (trichloromethane) in acetone (propanone) at 25°C, treating it first as a solvent and then as a solute, given the following data for the partial pressures p_C of chloroform and p_A of acetone as a function of chloroform mole fraction x_C.

x_C	0	0.20	0.40	0.60	0.80	1
p_C/kPa	0	4.7	11	18.9	26.7	36.4
p_A/kPa	46.3	33.3	23.3	12.3	4.9	0

The Henry's law constant K_C is 22.0 kPa for chloroform.

Method For the activity of chloroform as a solvent (the Raoult's law activity), form $a_C = p_C/p_C^*$ and $\gamma_C = a_C/x_C$. For its activity as a solute (the Henry's law activity), form $a_C = p_C/K_C$ and $\gamma_C = a_C/x_C$.

Answer Because $p_C^* = 36.4$ kPa and $K_C = 22.0$ kPa, we can construct the following tables. For instance, at $x_C = 0.20$, in the Raoult's law case we find $a_C = (4.7 \text{ kPa})/(36.4 \text{ kPa}) = 0.13$ and $\gamma_C = 0.13/0.20 = 0.65$; likewise, in the Henry's law case, $a_C = (4.7 \text{ kPa})/(22.0 \text{ kPa}) = 0.21$ and $\gamma_C = 0.21/0.20 = 1.05$.

From Raoult's law (chloroform regarded as the solvent):

a_C	0	0.13	0.30	0.52	0.73	1.00
γ_C		0.65	0.75	0.87	0.91	1.00

From Henry's law (chloroform regarded as the solute):

a_C	0	0.21	0.50	0.86	1.21	1.65
γ_C	1	1.05	1.25	1.43	1.51	1.65

These values are plotted in Fig. 16.41. Notice that $\gamma_C \to 1$ as $x_C \to 1$ in the Raoult's law case, but that $\gamma_C \to 1$ as $x_C \to 0$ in the Henry's law case.

Self-test 16.5 Calculate the activities and activity coefficients ($K_A = 23.3$ kPa) for acetone according to the two conventions.

[At $x_A = 0.60$, for instance, $a_R = 0.50$; $\gamma_R = 0.83$; $a_H = 1.00$, $\gamma_H = 1.67$]

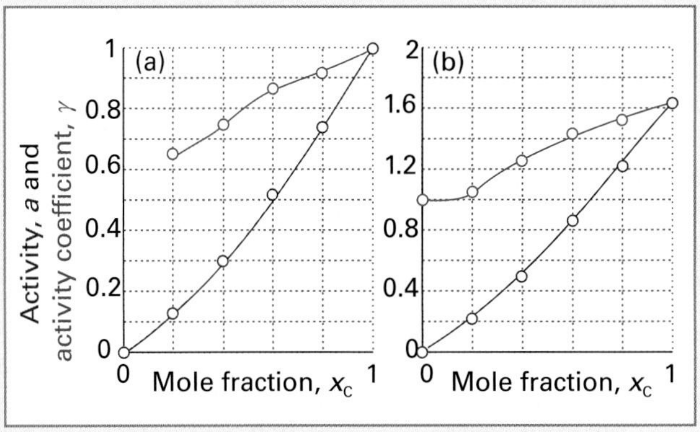

Fig. 16.41 The variation of activity and activity coefficient of chloroform (trichloromethane) and acetone (propanone) with composition according to (a) Raoult's law, (b) Henry's law.

(d) A model system: activities in a regular solution

The material on regular solutions presented in Section 16.6a gives further insight into the origin of deviations from Raoult's law and its relation to activity coefficients. We show in the following *Justification* that for a regular solution modelled by the parameter ξ the activity coefficients are given by the **Margules equations:**

$$\ln \gamma_A = \xi x_B^2 \qquad \ln \gamma_B = \xi x_A^2 \tag{16.45}$$

Justification 16.4 *The Margules equations*

The Gibbs energy of mixing to form a non-ideal solution is

$$\Delta_{mix}G = nRT\{x_A \ln a_A + x_B \ln a_B\}$$

This relation follows from the derivation of eqn 16.27a for ideal solutions with activities in place of mole fractions. When each activity is replaced by $\gamma_J x_J$, this expression becomes

$$\Delta_{mix}G = nRT\{x_A \ln x_A + x_B \ln x_B + x_A \ln \gamma_A + x_B \ln \gamma_B\}$$

Now we introduce the two expressions in eqn 16.45, and use $x_A + x_B = 1$, which gives

$$\begin{aligned}\Delta_{mix}G &= nRT\{x_A \ln x_A + x_B \ln x_B + \xi x_A x_B^2 + \xi x_B x_A^2\} \\ &= nRT\{x_A \ln x_A + x_B \ln x_B + \xi x_A x_B(x_A + x_B)\} \\ &= nRT\{x_A \ln x_A + x_B \ln x_B + \xi x_A x_B\}\end{aligned}$$

as required by eqn 16.33. Note, moreover, that the activity coefficients behave correctly for dilute solutions: $\gamma_A \to 1$ as $x_B \to 0$ and $\gamma_B \to 1$ as $x_A \to 0$.

At this point we can use the Margules equations to write the activity of A as

$$a_A = \gamma_A x_A = x_A e^{\xi x_B^2} = x_A e^{\xi(1-x_A)^2} \tag{16.46}$$

with a similar expression for a_B. The activity of A, though, is just the ratio of the vapour pressure of A in the solution to the vapour pressure of pure A (eqn 16.35, $a_A = p_A/p_A^*$), so we can write

$$p_A = \{x_A e^{\xi(1-x_A)^2}\}p_A^* \tag{16.47}$$

This function is plotted in Fig. 16.42. We see that $\xi = 0$, corresponding to an ideal solution, gives a straight line, in accord with Raoult's law (indeed, when $\xi = 0$, eqn 16.47 becomes $p_A = x_A p_A^*$, which is Raoult's law). Positive values of ξ (endothermic mixing, unfavourable solute–solvent interactions) give vapour pressures higher than ideal. Negative values of ξ (exothermic mixing, favourable solute–solvent interactions) give a lower vapour pressure. All the curves approach linearity and coincide with the Raoult's law line as $x_A \to 1$ and the exponential function in eqn 16.47 approaches 1. When $x_A \ll 1$, eqn 16.47 approaches

$$p_A = x_A e^{\xi} p_A^* \tag{16.48}$$

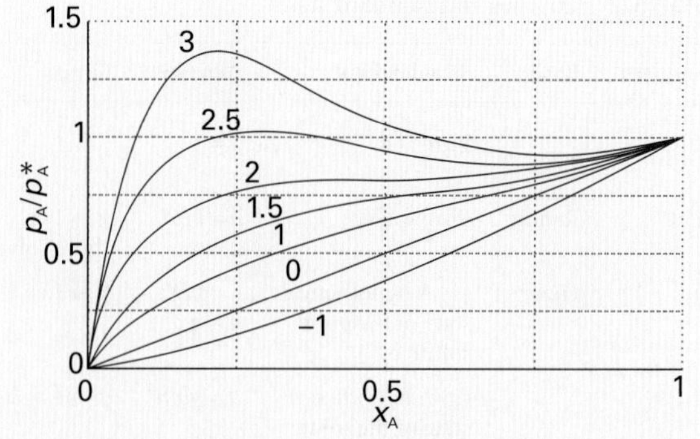

Fig. 16.42 The vapour pressure of a mixture based on a model in which the excess enthalpy is proportional to $\xi x_A x_B$. An ideal solution corresponds to $\xi = 0$ and gives a straight line, in accord with Raoult's law. Positive values of ξ give vapour pressures higher than ideal. Negative values of ξ give a lower vapour pressure.

interActivity Plot p_A/p_A^* against x_A with $\xi = 2.5$ by using eqn 16.24 and then eqn 16.47. Above what value of x_A do the values of p_A/p_A^* given by these equations differ by more than 10 per cent?

This expression has the form of Henry's law once we identify K with $e^{\xi}p_A^*$, which is different for each solute–solvent system.

(e) A model system: activities in an ionic solution

Solutions of ionic compounds are central to much of chemistry and we need to be able to discuss them thermodynamically. However, the Coulombic interactions between ions are so strong that the approximation of replacing activities by mole fractions or molalities is valid only in very dilute solutions (less than 10^{-3} mol kg^{-1} in total ion concentration) and in precise work activities themselves must be used. We need, therefore, to pay special attention to the activities of ions in solution, especially in preparation for the discussion of electrochemical phenomena.

If the chemical potential of a univalent cation M^+ is denoted μ_+ and that of a univalent anion X^- is denoted μ_-, the total molar Gibbs energy of the ions in the electrically neutral solution is the sum of these two quantities. The molar Gibbs energy of an *ideal* solution is

$$G_m^{ideal} = \mu_+^{ideal} + \mu_-^{ideal} \tag{16.49a°}$$

However, for a *real* solution of M^+ and X^- of the same molality,

$$\begin{aligned}G_m &= \mu_+ + \mu_- = \mu_+^{ideal} + \mu_-^{ideal} + RT\ln \gamma_+ + RT\ln \gamma_- \\ &= G_m^{ideal} + RT\ln \gamma_+\gamma_-\end{aligned} \tag{16.49b}$$

All the deviations from ideality are contained in the last term.

There is no experimental way of separating the product $\gamma_+\gamma_-$ into contributions from the cations and the anions. The best we can do experimentally is to assign responsibility for the nonideality equally to both kinds of ion. Therefore, for a 1,1-electrolyte, we introduce the **mean activity coefficient** as the geometric mean of the individual coefficients:

$$\gamma_\pm = (\gamma_+\gamma_-)^{1/2} \qquad [16.50]$$

and express the individual chemical potentials of the ions as

$$\mu_+ = \mu_+^{\text{ideal}} + RT\ln\gamma_\pm \qquad \mu_- = \mu_-^{\text{ideal}} + RT\ln\gamma_\pm \qquad (16.51)$$

The sum of these two chemical potentials is the same as before, eqn 16.49b, but now the non-ideality is shared equally.

A brief comment The geometric mean of x^p and y^q is $(x^py^q)^{1/(p+q)}$. For example, the geometric mean of x^2 and y^{-3} is $(x^2y^{-3})^{-1}$.

We can generalize this approach to the case of a compound M_pX_q that dissolves to give a solution of p cations and q anions from each formula unit. The molar Gibbs energy of the ions is the sum of their partial molar Gibbs energies:

$$G_m = p\mu_+ + q\mu_- = G_m^{\text{ideal}} + pRT\ln\gamma_+ + qRT\ln\gamma_- \qquad (16.52)$$

If we introduce the mean activity coefficient

$$\gamma_\pm = (\gamma_+^p\gamma_-^q)^{1/s} \qquad s = p+q \qquad [16.53]$$

and write the chemical potential of each ion as

$$\mu_i = \mu_i^{\text{ideal}} + RT\ln\gamma_\pm \qquad (16.54)$$

we get the same expression as in eqn 16.52 for G_m when we write

$$G_m = p\mu_+ + q\mu_- \qquad (16.55)$$

However, both types of ion now share equal responsibility for the non-ideality.

The long range and strength of the Coulombic interaction between ions means that it is likely to be primarily responsible for the departures from ideality in ionic solutions and to dominate all the other contributions to non-ideality. This domination is the basis of the **Debye–Hückel theory** of ionic solutions, which was devised by Peter Debye and Erich Hückel in 1923. We give here a qualitative account of the theory and its principal conclusions. The calculation itself, which is a profound example of how a seemingly intractable problem can be formulated and then resolved by drawing on physical insight, is described in *Further information 16.3.*

Oppositely charged ions attract one another. As a result, anions are more likely to be found near cations in solution, and vice versa (Fig. 16.43). Overall the solution is electrically neutral, but near any given ion there is an excess of counter ions (ions of opposite charge). Averaged over time, counter ions are more likely to be found near any given ion. This time-averaged, spherical haze around the central ion, in which counter ions outnumber ions of the same charge as the central ion, has a net charge equal in magnitude but opposite in sign to that on the central ion, and is called its **ionic atmosphere**. The energy, and therefore the chemical potential, of any given central ion is lowered as a result of its electrostatic interaction with its ionic atmosphere. This lowering of energy appears as the difference between the molar Gibbs energy G_m and the ideal value G_m^{ideal} of the solute, and hence can be identified with $RT\ln\gamma_\pm$. (That the Gibbs energy is involved rather than the internal energy is clarified in the formal derivation of the theory, where we see that we need to consider the electrical work of charging the ion, and we have seen that non-expansion work is equal to the change in Gibbs energy.) The stabilization of ions by their interaction with their ionic atmospheres is part of the explanation why chemists commonly use dilute solutions, in which the stabilization is less important, to achieve precipitation of ions from electrolyte solutions.

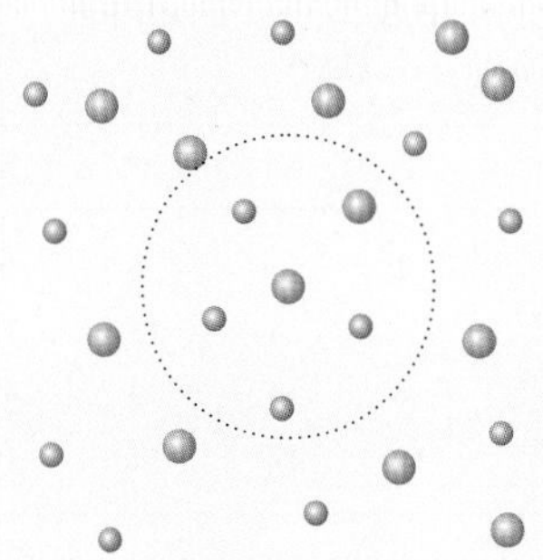

Fig. 16.43 The picture underlying the Debye–Hückel theory is of a tendency for anions to be found around cations, and of cations to be found around anions (one such local clustering region is shown by the circle). The ions are in ceaseless motion, and the diagram represents a snapshot of their motion. The solutions to which the theory applies are far less concentrated than shown here.

The model leads to the result that at very low concentrations the activity coefficient can be calculated from the **Debye–Hückel limiting law**

$$\log\gamma_\pm = -|z_+z_-|AI^{1/2} \qquad (16.56)$$

where $A = 0.509$ for an aqueous solution at 25°C and I is the dimensionless **ionic strength** of the solution:

$$I = \tfrac{1}{2}\sum_i z_i^2(b_i/b^\ominus) \qquad [16.57]$$

In this expression z_i is the charge number of an ion i (positive for cations and negative for anions) and b_i is its molality with $b^\ominus = 1\ \text{mol kg}^{-1}$. The ionic strength occurs widely wherever ionic solutions are discussed, as we shall see. The sum extends over all the ions present in the solution. For solutions consisting of two types of ion at molalities b_+ and b_-,

$$I = \tfrac{1}{2}(b_+z_+^2 + b_-z_-^2)/b^\ominus \qquad (16.58)$$

The ionic strength emphasizes the charges of the ions because the charge numbers occur as their squares. Table 16.3 summarizes the relation of ionic strength and molality in an easily usable form.

Table 16.3 Ionic strength and molality, $I = kb/b^{\ominus}$

k	X^-	X^{2-}	X^{3-}	X^{4-}
M^+	1	3	6	10
M^{2+}	3	4	15	12
M^{3+}	6	15	9	42
M^{4+}	10	12	42	16

For example, the ionic strength of an M_2X_3 solution of molality b, which is understood to give M^{3+} and X^{2-} ions in solution, is $15b/b^{\ominus}$.

Synoptic table 16.4* Mean activity coefficients in water at 298 K

$b/b^{\ominus}$	KCl	$CaCl_2$
0.001	0.966	0.888
0.01	0.902	0.732
0.1	0.770	0.524
1.0	0.607	0.725

* More values are given in the *Data section*.

● **A BRIEF ILLUSTRATION**

The mean activity coefficient of 5.0 mmol kg^{-1} KCl(aq) at 25°C is calculated by writing

$$I = \tfrac{1}{2}(b_+ + b_-)/b^{\ominus} = b/b^{\ominus}$$

where b is the molality of the solution (and $b_+ = b_- = b$). Then, from eqn 16.56,

$$\log \gamma_{\pm} = -0.509 \times (5.0 \times 10^{-3})^{1/2} = -0.036$$

Hence, $\gamma_{\pm} = 0.92$. The experimental value is 0.927. ●

Self-test 16.6 Calculate the ionic strength and the mean activity coefficient of 1.00 mmol kg^{-1} $CaCl_2$(aq) at 25°C.

[3.00×10^{-3}, 0.880]

The name 'limiting law' is applied to eqn 16.56 because ionic solutions of moderate molalities may have activity coefficients that differ from the values given by this expression, yet all solutions are expected to conform as $b \to 0$. Table 16.4 lists some experimental values of activity coefficients for salts of various valence types. Figure 16.44 shows some of these values plotted against $I^{1/2}$, and compares them with the theoretical straight lines calculated from the limiting law. The agreement at very low molalities (less than about 1 mmol kg^{-1}, depending on charge type) is impressive, and provides convincing evidence in support of the model. Nevertheless, the departures from the theoretical curves above these molalities are large, and show that the approximations are valid only at very low concentrations.

When the ionic strength of the solution is too high for the limiting law to be valid, the activity coefficient may be estimated from the **extended Debye–Hückel law**:

$$\log \gamma_{\pm} = -\frac{A|z_+z_-|I^{1/2}}{1 + BI^{1/2}} + CI \qquad (16.59)$$

where B and C are dimensionless constants. Although B can be interpreted as a measure of the closest approach of the ions, it (like C) is best regarded as an adjustable empirical parameter. A curve drawn in this way is shown in Fig. 16.45. It is clear that

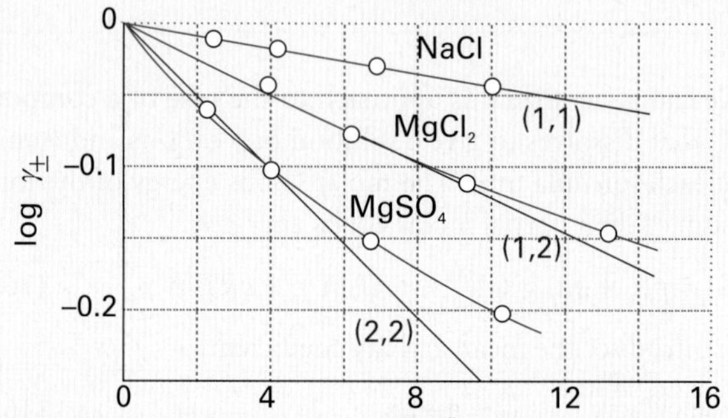

Fig. 16.44 An experimental test of the Debye–Hückel limiting law. Although there are marked deviations for moderate ionic strengths, the limiting slopes as $I \to 0$ are in good agreement with the theory, so the law can be used for extrapolating data to very low molalities.

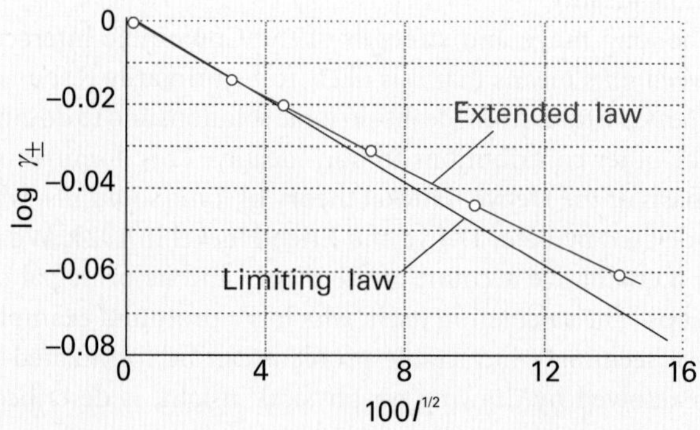

Fig. 16.45 The extended Debye–Hückel law gives agreement with experiment over a wider range of molalities (as shown here for a 1,1-electrolyte), but it fails at higher molalities.

interActivity Consider the plot of log $\gamma_{\pm}$ against $I^{1/2}$ with $B = 1.50$ and $C = 0$ as a representation of experimental data for a certain 1,1 electrolyte. Over what range of ionic strengths does the application of the limiting law lead to an error in the value of the activity coefficient of less than 10 per cent of the value predicted by the extended law?

eqn 16.59 accounts for some activity coefficients over a moderate range of dilute solutions (up to about 0.1 mol kg^{-1}); nevertheless it remains very poor near 1 mol kg^{-1}.

Current theories of activity coefficients for ionic solutes take an indirect route. They set up a theory for the dependence of the activity coefficient of the solvent on the concentration of the solute, and then use the Gibbs–Duhem equation (eqn 16.7) to estimate the activity coefficient of the solute. The results are reasonably reliable for solutions with molalities greater than about 0.1 mol kg^{-1} and are valuable for the discussion of mixed salt solutions, such as sea water.

Checklist of key ideas

☐ 1. A phase is a form of matter that is uniform throughout in chemical composition and physical state.

☐ 2. A transition temperature is the temperature at which the two phases are in equilibrium.

☐ 3. A phase diagram is a diagram showing the regions of pressure and temperature at which its various phases are thermodynamically stable.

☐ 4. The vapour pressure is the pressure of a vapour in equilibrium with the condensed phase.

☐ 5. The boiling temperature is the temperature at which the vapour pressure of a liquid is equal to the external pressure.

☐ 6. The critical temperature is the temperature at which a liquid surface disappears and above which a liquid does not exist whatever the pressure. The critical pressure is the vapour pressure at the critical temperature.

☐ 7. The melting temperature (or freezing temperature) is the temperature at which, under a specified pressure, the liquid and solid phases of a substance coexist in equilibrium.

☐ 8. The triple point is a point on a phase diagram at which the three phase boundaries meet and all three phases are in mutual equilibrium.

☐ 9. The lever rule allows for the calculation of the relative amounts of two phases in equilibrium: $n_\alpha l_\alpha = n_\beta l_\beta$.

☐ 10. An azeotrope is a mixture that boils without change of composition.

☐ 11. The upper critical solution temperature is the highest temperature at which phase separation occurs in a binary liquid mixture. The lower critical solution temperature is the temperature below which the components of a binary mixture mix in all proportions and above which they form two phases.

☐ 12. The chemical potential is defined as $\mu = (\partial G/\partial n)_{p,T,n'}$ and the total Gibbs energy of a mixture is $G = \sum_J n_J \mu_J$.

☐ 13. The chemical potential is uniform throughout a system at equilibrium.

☐ 14. The temperature dependence of the vapour pressure of a condensed phase is given by the Clausius–Clapeyron equation, eqn 16.14.

☐ 15. The fundamental equation of chemical thermodynamics, eqn 16.5, relates the change in Gibbs energy to changes in pressure, temperature, and composition.

☐ 16. The chemical potential of a perfect gas is $\mu = \mu^\ominus + RT\ln(p/p^\ominus)$, where $\mu^\ominus$ is the standard chemical potential, the chemical potential of the pure gas at 1 bar.

☐ 17. An ideal solution is a solution in which all components obey Raoult's law ($p_A = x_A p_A^*$) throughout the composition range.

☐ 18. An ideal-dilute solution is a solution for which the solute obeys Henry's law ($p_B = x_B K_B^*$) and the solvent obeys Raoult's law.

☐ 19. An excess function (X^E) is the difference between the observed thermodynamic function of mixing and the function for an ideal solution.

☐ 20. A regular solution is a solution for which $H^E \neq 0$ but $S^E = 0$.

☐ 21. A colligative property is a property that depends only on the number of solute particles present, not their identity.

☐ 22. Osmosis is the spontaneous passage of a pure solvent into a solution separated from it by a semipermeable membrane, a membrane permeable to the solvent but not to the solute.

☐ 23. The osmotic pressure is the pressure that must be applied to the solution to stop the influx of solvent.

☐ 24. The activity is defined as $a_A = p_A/p_A^*$.

☐ 25. The solvent activity is related to its chemical potential by $\mu_A = \mu_A^* + RT\ln a_A$. The activity may be written in terms of the activity coefficient $\gamma_A = a_A/x_A$.

☐ 26. The chemical potential of a solute in an ideal-dilute solution is given by $\mu_B = \mu_B^\ominus + RT\ln a_B$. The activity may be written in terms of the activity coefficient $\gamma_B = a_B/x_B$.

☐ 27. The mean activity coefficient is the geometric mean of the individual coefficients: $\gamma_\pm = (\gamma_+^p \gamma_-^q)^{1/(p+q)}$.

☐ 28. The Debye–Hückel theory of activity coefficients of electrolyte solutions is based on the assumption that Coulombic interactions between ions are dominant; a key idea of the theory is that of an ionic atmosphere.

Further information

Further information 16.1 *The phase rule*

The phase rule is a general relation between the variance, F, the number of components, C, and the number of phases at equilibrium, P, for a system of any composition:

$$F = C - P + 2 \tag{16.60}$$

A **constituent** is a chemical species (an ion or a molecule) that is present. Thus, a mixture of ethanol and water has two constituents. A solution of sodium chloride has three constituents: water, Na^+ ions, and Cl^- ions. The term constituent should be carefully distinguished from 'component', which has a more technical meaning. A **component** is a *chemically independent* constituent of a system. The number of components, C, in a system is the minimum number of independent species necessary to define the composition of all the phases present in the system. The **variance,** F, of a system is the number of intensive variables that can be changed independently without disturbing the number of phases in equilibrium. In a single-component, single-phase system ($C = 1, P = 1$), the pressure and temperature may be changed independently without changing the number of phases, so $F = 2$. We say that such a system is **bivariant**, or that it has two **degrees of freedom**. On the other hand, if two phases are in equilibrium (a liquid and its vapour, for instance) in a single-component system ($C = 1, P = 2$), the temperature (or the pressure) can be changed at will, but the change in temperature (or pressure) demands an accompanying change in pressure (or temperature) to preserve the number of phases in equilibrium. That is, the variance of the system has fallen to 1.

To see the origin of eqn 16.60, consider first the special case of a one-component system. For two phases in equilibrium, we can write $\mu_J(\alpha) = \mu_J(\beta)$. Each chemical potential is a function of the pressure and temperature, so

$$\mu_J(\alpha; p,T) = \mu_J(\beta; p,T)$$

This is an equation relating p and T, so only one of these variables is independent (just as the equation $x + y = 2$ is a relation for y in terms of x: $y = 2 - x$). That conclusion is consistent with $F = 1$. For three phases in mutual equilibrium,

$$\mu_J(\alpha; p,T) = \mu_J(\beta; p,T) = \mu_J(\gamma; p,T)$$

This relation is actually two equations for two unknowns ($\mu_J(\alpha; p,T) = \mu_J(\beta; p,T)$ and $\mu_J(\beta; p,T) = \mu_J(\gamma; p,T)$), and therefore has a solution only for a single value of p and T (just as the pair of equations $x + y = 2$ and $3x - y = 4$ has the single solution $x = \frac{3}{2}$ and $y = \frac{1}{2}$). That conclusion is consistent with $F = 0$. Four phases cannot be in mutual equilibrium in a one-component system because the three equalities

$$\mu_J(\alpha, p,T) = \mu_J(\beta; p,T) \quad \mu_J(\beta; p,T) = \mu_J(\gamma, p,T) \quad \mu_J(\gamma; p,T) = \mu_J(\delta; p,T)$$

are three equations for two unknowns (p and T) and are not consistent (just as $x + y = 2$, $3x - y = 4$, and $x + 4y = 6$ have no solution).

Now consider the general case. We begin by counting the total number of intensive variables. The pressure, p, and temperature, T, count as 2. We can specify the composition of a phase by giving the mole fractions of $C - 1$ components. We need specify only $C - 1$ and not all C mole fractions because $x_1 + x_2 + \cdots + x_C = 1$, and all mole fractions are known if all except one are specified. Because there are P phases, the total number of composition variables is $P(C - 1)$. At this stage, the total number of intensive variables is $P(C - 1) + 2$.

At equilibrium, the chemical potential of a component J must be the same in every phase:

$$\mu_J(\alpha) = \mu_J(\beta) = \cdots \text{ for } P \text{ phases}$$

That is, there are $P - 1$ equations of this kind to be satisfied for each component J. As there are C components, the total number of equations is $C(P - 1)$. Each equation reduces our freedom to vary one of the $P(C - 1) + 2$ intensive variables. It follows that the total number of degrees of freedom is

$$F = P(C - 1) + 2 - C(P - 1) = C - P + 2$$

which is eqn 16.60.

For a one-component system, such as pure water, $F = 3 - P$. When only one phase is present, $F = 2$ and both p and T can be varied independently without changing the number of phases. In other words, a single phase is represented by an *area* on a phase diagram. When two phases are in equilibrium $F = 1$, which implies that pressure is not freely variable if the temperature is set; indeed, at a given temperature, a liquid has a characteristic vapour pressure. It follows that the equilibrium of two phases is represented by a *line* in the phase diagram. Instead of selecting the temperature, we could select the pressure, but having done so the two phases would be in equilibrium at a single definite temperature. Therefore, freezing (or any other phase transition) occurs at a definite temperature at a given pressure.

When three phases are in equilibrium, $F = 0$ and the system is invariant. This special condition can be established only at a definite temperature and pressure that is characteristic of the substance and outside our control. The equilibrium of three phases is therefore represented by a *point*, the triple point, on a phase diagram. Four phases cannot be in equilibrium in a one-component system because F cannot be negative.

Further information 16.2 *The Ehrenfest classification*

Many familiar phase transitions, like fusion and vaporization, are accompanied by changes of enthalpy and volume. These changes have implications for the slopes of the chemical potentials of the phases at either side of the phase transition. Thus, at the transition from a phase α to another phase β,

$$\left(\frac{\partial \mu_\beta}{\partial p}\right)_T - \left(\frac{\partial \mu_\alpha}{\partial p}\right)_T = V_{\beta,m} - V_{\alpha,m} = \Delta_{trs}V \tag{16.61}$$

$$\left(\frac{\partial \mu_\beta}{\partial T}\right)_p - \left(\frac{\partial \mu_\alpha}{\partial T}\right)_p = -S_{\beta,m} + S_{\alpha,m} = -\Delta_{trs}S = -\frac{\Delta_{trs}H}{T_{trs}}$$

Because $\Delta_{trs}V$ and $\Delta_{trs}H$ are nonzero for melting and vaporization, it follows that for such transitions the slopes of the chemical potential plotted against either pressure or temperature are different on either side of the transition (Fig. 16.46a). In other words, the first derivatives of the chemical potentials with respect to pressure and temperature are discontinuous at the transition.

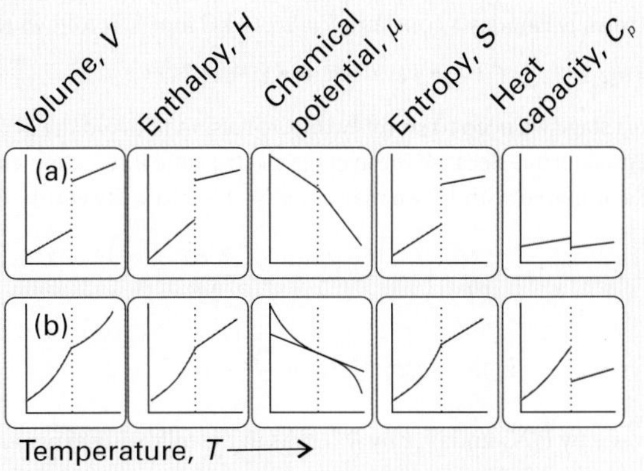

Fig. 16.46 The changes in thermodynamic properties accompanying (a) first-order and (b) second-order phase transitions.

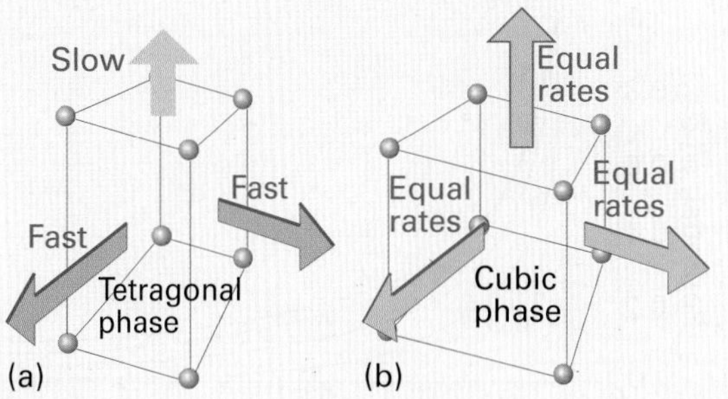

Fig. 16.47 One version of a second-order phase transition in which (a) a tetragonal phase expands more rapidly in two directions than a third, and hence becomes a cubic phase, which (b) expands uniformly in three directions as the temperature is raised. There is no rearrangement of atoms at the transition temperature, and hence no enthalpy of transition.

A transition for which the first derivative of the chemical potential with respect to temperature is discontinuous is classified as a **first-order phase transition**. The constant-pressure heat capacity, C_p, of a substance is the slope of a plot of the enthalpy with respect to temperature. At a first-order phase transition, H changes by a finite amount for an infinitesimal change of temperature. Therefore, at the transition the heat capacity is infinite. The physical reason is that heating drives the transition rather than raising the temperature. For example, boiling water stays at the same temperature even though heat is being supplied.

A **second-order phase transition** in the Ehrenfest sense is one in which the first derivative of μ with respect to temperature is continuous but its second derivative is discontinuous. A continuous slope of μ (a graph with the same slope on either side of the transition) implies that the volume and entropy (and hence the enthalpy) do not change at the transition (Fig. 16.46b). The heat capacity is discontinuous at the transition but does not become infinite there. An example of a second-order transition is the conducting–superconducting transition in metals at low temperatures. One type of second-order transition is associated with a change in symmetry of the crystal structure of a solid. Thus, suppose the arrangement of atoms in a solid is like that represented in Fig. 16.47a, with one dimension of the tetragonal unit cell longer than the other two, which are equal. Moreover, suppose the two shorter dimensions increase more than the long dimension when the temperature is raised. There may come a stage when the three dimensions become equal. At that point the crystal has cubic symmetry (Fig. 16.47b), and at higher temperatures it will expand equally in all three directions (because there is no longer any distinction between them). The tetragonal → cubic phase transition has occurred but, as it has not involved a discontinuity in the interaction energy between the atoms or the volume they occupy, the transition is not first order.

Further information 16.3 *The Debye–Hückel theory of ionic solutions*

Imagine a solution in which all the ions have their actual positions, but in which their Coulombic interactions have been turned off. The difference in molar Gibbs energy between the ideal and real solutions is equal to w_e, the electrical work of charging the system in this arrangement. For a salt M_pX_q, we write

$$w_e = \overbrace{(p\mu_+ + q\mu_-)}^{G_m} - \overbrace{(p\mu_+^{\text{ideal}} + q\mu_-^{\text{ideal}})}^{G_m^{\text{ideal}}}$$
$$= p(\mu_+ - \mu_+^{\text{ideal}}) + q(\mu_- - \mu_-^{\text{ideal}})$$

From eqn 16.54 we write

$$\mu_+ - \mu_+^{\text{ideal}} = \mu_- - \mu_-^{\text{ideal}} = RT \ln \gamma_\pm$$

So it follows that

$$\ln \gamma_\pm = \frac{w_e}{sRT} \qquad s = p + q \tag{16.62}$$

This equation tells us that to calculate a mean activity coefficient we must first find the final distribution of the ions and then the electrical work of charging them in that distribution.

The Coulomb potential at a distance r from an isolated ion of charge z_ie in a medium of permittivity ε is

$$\phi_i = \frac{Z_i}{r} \qquad Z_i = \frac{z_ie}{4\pi\varepsilon} \tag{16.63}$$

The ionic atmosphere causes the potential to decay with distance more sharply than this expression implies. Such shielding is a familiar problem in electrostatics, and its effect is taken into account by replacing the Coulomb potential by the **shielded Coulomb potential**, an expression of the form

$$\phi_i = \frac{Z_i}{r} e^{-r/r_D} \tag{16.64}$$

where r_D is called the **Debye length**. When r_D is large, the shielded potential is virtually the same as the unshielded potential. When r_D is small, the shielded potential is much smaller than the unshielded potential, even for short distances (Fig. 16.48).

To calculate r_D, we need to know how the **charge density**, ρ_i, of the ionic atmosphere, the charge in a small region divided by the volume of the region, varies with distance from the ion. This step draws on another

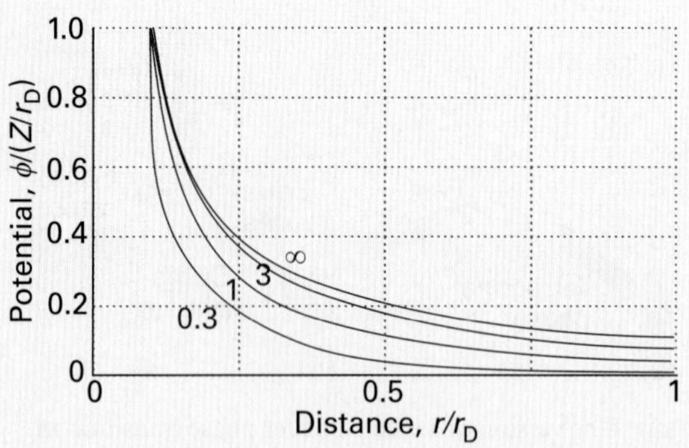

Fig. 16.48 The variation of the shielded Coulomb potential with distance for different values of the Debye length, r_D/a. The smaller the Debye length, the more sharply the potential decays to zero. In each case, a is an arbitrary unit of length.

interActivity Write an expression for the difference between the unshielded and shielded Coulomb potentials evaluated at r_D. Then plot this expression against r_D and provide a physical interpretation for the shape of the plot.

standard result of electrostatics, in which charge density and potential are related by **Poisson's equation**:

$$\nabla^2\phi = -\frac{\rho}{\varepsilon} \qquad (16.65)$$

where $\nabla^2 = (\partial^2/\partial x^2 + \partial^2/\partial y^2 + \partial^2/\partial z^2)$ is the Laplacian (*Mathematical background 4*). Because we are considering only a spherical ionic atmosphere, we can use a simplified form of this equation in which the charge density varies only with distance from the central ion:

$$\frac{1}{r^2}\frac{\mathrm{d}}{\mathrm{d}r}\left(r^2\frac{\mathrm{d}\phi_i}{\mathrm{d}r}\right) = -\frac{\rho_i}{\varepsilon}$$

Substitution of the expression for the shielded potential, eqn 16.64, results in

$$r_D^2 = -\frac{\varepsilon\phi_i}{\rho_i} \qquad (16.66)$$

To solve this equation we need to relate ρ_i and ϕ_i.

For the next step we draw on the fact that the energy of an ion depends on its closeness to the central ion, and then use the Boltzmann distribution to work out the probability that an ion will be found at each distance. The energy of an ion of charge z_je at a distance where it experiences the potential ϕ_i of the central ion i relative to its energy when it is far away in the bulk solution is its charge times the potential:

$$E = z_je\phi_i$$

Therefore, according to the Boltzmann distribution, the ratio of the molar concentration, c_j, of ions at a distance r and the molar concentration in the bulk, c_j°, where the energy is zero, is:

$$\frac{c_j}{c_j^\circ} = \mathrm{e}^{-E/kT}$$

The charge density, ρ_i, at a distance r from the ion i is the molar concentration of each type of ion multiplied by the charge per mole of ions, z_ieN_A. The quantity eN_A, the magnitude of the charge per mole of electrons, is Faraday's constant, $F = 96.48\ \mathrm{kC\ mol^{-1}}$. It follows that

$$\rho_i = c_+z_+F + c_-z_-F = c_+^\circ z_+F\mathrm{e}^{-z_+e\phi_i/kT} + c_-^\circ z_-F\mathrm{e}^{-z_-e\phi_i/kT} \qquad (16.67)$$

At this stage we need to simplify the expression to avoid the awkward exponential terms. Because the average electrostatic interaction energy is small compared with kT we may use $\mathrm{e}^{-x} \approx 1 - x$ to write eqn 16.67 as

$$\rho_i = c_+^\circ z_+F\left(1 - \frac{z_+e\phi_i}{kT} + \cdots\right) + c_-^\circ z_-F\left(1 - \frac{z_-e\phi_i}{kT} + \cdots\right)$$

$$= (c_+^\circ z_+ + c_-^\circ z_-)F - (c_+^\circ z_+^2 + c_-^\circ z_-^2)F\frac{e\phi_i}{kT} + \cdots$$

Replacing e by F/N_A and N_Ak by R results in the following expression:

$$\rho_i = (c_+^\circ z_+ + c_-^\circ z_-)F - (c_+^\circ z_+^2 + c_-^\circ z_-^2)\frac{F^2\phi_i}{RT} + \cdots \qquad (16.68)$$

The first term in the expansion is zero because it is the charge density in the bulk, uniform solution, and the solution is electrically neutral. The unwritten terms are assumed to be too small to be significant. The one remaining term can be expressed in terms of the ionic strength, eqn 16.57, by noting that in the dilute aqueous solutions we are considering there is little difference between molality and molar concentration, and $c \approx b\rho$, where ρ is the mass density of the solvent:

$$c_+^\circ z_+^2 + c_-^\circ z_-^2 \approx (b_+^\circ z_+^2 + b_-^\circ z_-^2)\rho = 2Ib^\ominus\rho$$

With these approximations, eqn 16.68 becomes

$$\rho_i = -\frac{2\rho F^2Ib^\ominus\phi_i}{RT}$$

We can now solve eqn 16.66 for r_D:

$$r_D = \left(\frac{\varepsilon RT}{2\rho F^2Ib^\ominus}\right)^{1/2} \qquad (16.69)$$

To calculate the activity coefficient we need to find the electrical work of charging the central ion when it is surrounded by its atmosphere. To do so, we need to know the potential at the ion due to its atmosphere, ϕ_{atmos}. This potential is the difference between the total potential, given by eqn 16.64, and the potential due to the central ion itself:

$$\phi_{\text{atmos}} = \phi - \phi_{\text{central ion}} = Z_i\left(\frac{\mathrm{e}^{-r/r_D}}{r} - \frac{1}{r}\right)$$

The potential at the central ion (at $r = 0$) is obtained by taking the limit of this expression as $r \to 0$ and is

$$\phi_{\text{atmos}}(0) = -\frac{Z_i}{r_D}$$

This expression shows us that the potential of the ionic atmosphere is equivalent to the potential arising from a single charge of equal magnitude but opposite sign to that of the central ion and located at a distance r_D from the ion. If the charge of the central ion were Q and not z_ie, then the potential due to its atmosphere would be

$$\phi_{\text{atmos}}(0) = -\frac{Q}{4\pi\varepsilon r_D}$$

The work of adding a charge $\mathrm{d}Q$ to a region where the electrical potential is $\phi_{\text{atmos}}(0)$ is

$$dw_e = \phi_{atmos}(0)dQ$$

Therefore, the total molar work of fully charging the ions is

$$w_e = N_A\int_0^{z_ie}\phi_{atmos}(0)dQ = -\frac{N_A}{4\pi\varepsilon r_D}\int_0^{z_ie}QdQ$$

$$= -\frac{N_A z_i^2 e^2}{8\pi\varepsilon r_D} = -\frac{z_i^2F^2}{8\pi\varepsilon N_A r_D}$$

where in the last step we have used $F = N_Ae$. It follows from eqn 16.62 that the mean activity coefficient of the ions is

$$\ln\gamma_\pm = \frac{pw_{e,+}+qw_{e,-}}{sRT} = -\frac{(pz_+^2+qz_-^2)F^2}{8\pi\varepsilon sN_ART r_D}$$

However, for neutrality $pz_+ + qz_- = 0$; therefore (for this step, multiply $pz_+ + qz_- = 0$ by z_+ and also, separately, by z_-; add the two expressions and rearrange the result by using $p + q = s$ and $z_+z_- = -|z_+z_-|$):

$$\ln\gamma_\pm = -\frac{|z_+z_-|F^2}{8\pi\varepsilon N_A RT r_D}$$

Replacing r_D with the expression in eqn 16.69 gives

$$\ln\gamma_\pm = -\frac{|z_+z_-|F^2}{8\pi\varepsilon N_ART}\left(\frac{2\rho F^2 Ib^\ominus}{\varepsilon RT}\right)^{1/2}$$

$$= -|z_+z_-|\left\{\frac{F^3}{4\pi N_A}\left(\frac{\rho b^\ominus}{2\varepsilon^3R^3T^3}\right)^{1/2}\right\}I^{1/2}$$

where we have grouped terms in such a way as to show that this expression is beginning to take the form of the limiting law in eqn 16.56. Indeed, conversion to common logarithms (by using $\ln x = \ln 10 \times \log x$) gives

$$\log\gamma_\pm = -|z_+z_-|\left\{\frac{F^3}{4\pi N_A\ln 10}\left(\frac{\rho b^\ominus}{2\varepsilon^3R^3T^3}\right)^{1/2}\right\}I^{1/2}$$

which is eqn 16.56 ($\log\gamma_\pm = -|z_+z_-|AI^{1/2}$) with

$$A = \frac{F^3}{4\pi N_A\ln 10}\left(\frac{\rho b^\ominus}{2\varepsilon^3R^3T^3}\right)^{1/2} \qquad (16.70)$$

Discussion questions

16.1 Discuss what will be observed as a sample of water is taken along a path in its phase diagram that encircles and is close to its critical point.

16.2 Consider the variation of the melting point with pressure observed for water and carbon dioxide. Provide a molecular explanation for differences in behaviour of these two substances.

16.3 Discuss the scope and applicability of the fundamental equation of chemical thermodynamics.

16.4 Compare the dependence of the chemical potential on temperature for a solid, liquid, and gas. How are these dependencies reflected in the phase diagram?

16.5 Interpret the forms of the Clapeyron and Clausius–Clapeyron equations, paying particular attention to the appearance in them of parameters such as the enthalpy of transition and the temperature.

16.6 Distinguish between a first-order phase transition, a second-order phase transition, and a λ-transition at both molecular and macroscopic levels.

16.7 Why does the chemical potential of a substance depend on the pressure even if the substance is incompressible?

16.8 Explain the origin of colligative properties in both thermodynamic and molecular terms.

16.9 Explain what is meant by a regular solution. Discuss how the magnitude and sign of the parameter ξ captures various features of real solutions.

16.10 Describe the general features of the Debye–Hückel theory of electrolyte solutions. Why is it only a limiting law?

16.11 Define the following terms: phase, constituent, component, and degree of freedom.

16.12 What factors determine the number of theoretical plates required to achieve a desired degree of separation in fractional distillation?

Exercises

16.1(a) Methylethyl ether (A) and diborane, B_2H_6 (B), form a compound that melts congruently at 133 K. The system exhibits two eutectics, one at 25 mol per cent B and 123 K and a second at 90 mol per cent B and 104 K. The melting points of pure A and B are 131 K and 110 K, respectively. Sketch the phase diagram for this system. Assume negligible solid–solid solubility.

16.1(b) Sketch the phase diagram of the system NH_3/N_2H_4 given that the two substances do not form a compound with each other, that NH_3 freezes at −78°C and N_2H_4 freezes at +2°C, and that a eutectic is formed when the mole fraction of N_2H_4 is 0.07 and that the eutectic melts at −80°C.

16.2(a) Figure 16.49 shows the phase diagram for water (A) and 2-methyl-1-propanol (B). Describe what will be observed when a mixture of composition $x_B = 0.8$ is heated, at each stage giving the number, composition, and relative amounts of the phases present.

16.2(b) Figure 16.50 is the phase diagram for silver and tin. Label the regions, and describe what will be observed when liquids of compositions *a* and *b* are cooled to 200 K.

16.3(a) The solid and liquid phases of a mixture of water (W) and ethanol (E) are in equilibrium. What are the relationships between the magnitudes of $\mu_W(s)$, $\mu_W(l)$, $\mu_E(s)$, and $\mu_E(l)$?

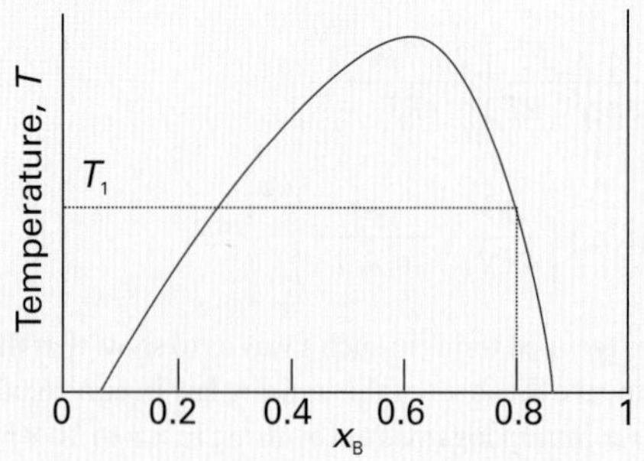

Fig. 16.49 A phase diagram for two partially miscible liquids.

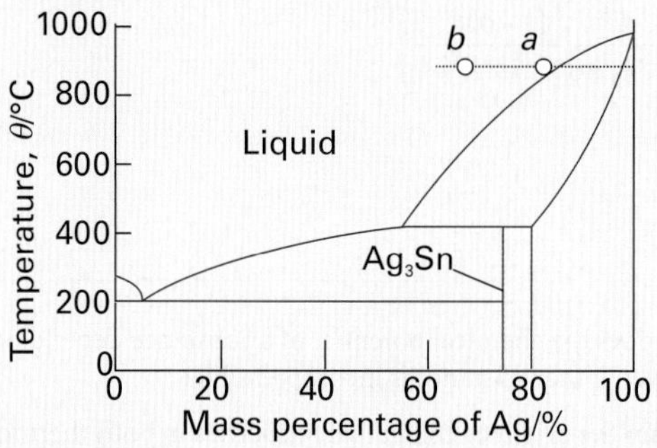

Fig. 16.50 The phase diagram of silver and tin.

16.3(b) The liquid and vapour phases of a mixture of benzene (B) and methylbenzene (M) are in equilibrium. What are the relationships between the magnitudes of $\mu_B(l)$, $\mu_B(g)$, $\mu_M(l)$, and $\mu_M(g)$?

16.4(a) A mixture of water and ethanol is prepared with a mole fraction of water of 0.60. If a small change in the mixture composition results in an increase in the chemical potential of water by 0.25 J mol^{-1}, by how much will the chemical potential of ethanol change?

16.4(b) A mixture of water and ethanol is prepared with a mole fraction of water of 0.40. If a small change in the mixture composition results in an increase in the chemical potential of ethanol by 0.35 J mol^{-1}, by how much will the chemical potential of water change?

16.5(a) By how much does the chemical potential of pure water change when the temperature of a sample is increased from 20°C to 25°C?

16.5(b) By how much does the chemical potential of pure octane change when the temperature of a sample is increased from 20°C to 25°C?

16.6(a) By how much does the chemical potential of pure water increase when the pressure on a sample is increased from 1.0 bar to 100 kbar? The mass density of water at 20°C is 0.997 g cm^{-3}.

16.6(b) By how much does the chemical potential of pure octane increase when the pressure on a sample is increased from 1.0 bar to 100 kbar? The mass density of octane at 20°C is 0.703 g cm^{-3}.

16.7(a) The vapour pressure of dichloromethane at 24.1°C is 53.3 kPa and its enthalpy of vaporization is 28.7 kJ mol^{-1}. Estimate the temperature at which its vapour pressure is 80.0 kPa.

16.7(b) The vapour pressure of a substance at 20.0°C is 48.4 kPa and its enthalpy of vaporization is 34.7 kJ mol^{-1}. Estimate the temperature at which its vapour pressure is 58.0 kPa.

16.8(a) The molar volume of a certain solid is 161.0 cm^3 mol^{-1} at 1.00 atm and 350.75 K, its melting temperature. The molar volume of the liquid at this temperature and pressure is 163.3 cm^3 mol^{-1}. At 100 atm the melting temperature changes to 351.26 K. Calculate the enthalpy and entropy of fusion of the solid.

16.8(b) The molar volume of a certain solid is 142.0 cm^3 mol^{-1} at 1.00 atm and 427.15 K, its melting temperature. The molar volume of the liquid at this temperature and pressure is 152.6 cm^3 mol^{-1}. At 1.2 MPa the melting temperature changes to 429.26 K. Calculate the enthalpy and entropy of fusion of the solid.

16.9(a) The vapour pressure of a liquid in the temperature range 200 K to 260 K was found to fit the expression $\ln(p/\text{Torr}) = 19.176 - 1501.8/(T/\text{K})$. Calculate the enthalpy of vaporization of the liquid.

16.9(b) The vapour pressure of a liquid in the temperature range 200 K to 260 K was found to fit the expression $\ln(p/\text{Torr}) = 17.461 - 2100.8/(T/\text{K})$. Calculate the enthalpy of vaporization of the liquid.

16.10(a) When benzene freezes at 5.5°C (the normal freezing point) its density changes from 0.879 g cm^{-3} to 0.891 g cm^{-3}. Its enthalpy of fusion is 10.59 kJ mol^{-1}. Estimate the freezing point of benzene at 10.0 kbar.

16.10(b) When a certain liquid freezes at −3.65°C (the normal freezing point) its density changes from 0.790 g cm^{-3} to 0.799 g cm^{-3}. Its enthalpy of fusion is 6.68 kJ mol^{-1}. Estimate the freezing point of the liquid at 100 MPa.

16.11(a) An open vessel containing (a) water (vapour pressure 3.2 kPa), (b) benzene (vapour pressure 13.1 kPa) stands in a laboratory measuring 4.0 m × 4.0 m × 3.0 m at 25°C. What mass of each substance will be found in the air if there is no ventilation?

16.11(b) An open vessel containing mercury (vapour pressure 0.23 Pa) stands in a laboratory measuring 6.0 m × 5.0 m × 3.5 m at 25°C. What mass of mercury will be found in the air if there is no ventilation?

16.12(a) The normal boiling point of carbon tetrachloride is 76.8°C. Estimate (a) its enthalpy of vaporization and (b) its vapour pressure at 25°C and 70°C.

16.12(b) The normal boiling point of hexane is 69.0°C. Estimate (a) its enthalpy of vaporization and (b) its vapour pressure at 25°C and 60°C.

16.13(a) Calculate the melting point of ice under a pressure of 50 bar. Assume that the density of ice under these conditions is approximately 0.92 g cm^{-3} and that of liquid water is 1.00 g cm^{-3}.

16.13(b) Calculate the melting point of ice under a pressure of 10 MPa. Assume that the density of ice under these conditions is approximately 0.915 g cm^{-3} and that of liquid water is 0.998 g cm^{-3}.

16.14(a) Consider a container of volume 10.0 dm^3 that is divided into two compartments of equal size. In the left compartment there is nitrogen at 1.5 bar and 25°C; in the right compartment there is hydrogen at the same temperature and pressure. Calculate the entropy and Gibbs energy of mixing when the partition is removed. Assume that the gases are perfect.

16.14(b) Consider a container of volume 350 cm^3 that is divided into two compartments of equal size. In the left compartment there is argon at 150 kPa and 0°C; in the right compartment there is neon at the same temperature and pressure. Calculate the entropy and Gibbs energy of mixing when the partition is removed. Assume that the gases are perfect.

16.15(a) Air is approximately 76 per cent N_2, 23 per cent O_2, and 1 per cent Ar by mass. Calculate the entropy, enthalpy, and Gibbs energy of mixing when it is prepared from the pure (and perfect) gases.

16.15(b) Calculate the Gibbs energy, entropy, and enthalpy of mixing when 25 g of hexane is mixed with 25 g of heptane at 298 K; treat the solution as ideal.

16.16(a) What proportions of hexane and heptane should be mixed (a) by mole fraction, (b) by mass in order to achieve the greatest entropy of mixing?

16.16(b) What proportions of benzene and ethylbenzene should be mixed (a) by mole fraction, (b) by mass in order to achieve the greatest entropy of mixing?

16.17(a) At 300 K, the partial vapour pressures of HCl (that is, the partial pressure of the HCl vapour) in liquid $GeCl_4$ are as follows:

x_{HCl}	0.005	0.012	0.019
p_{HCl}/kPa	32.0	76.9	121.8

Show that the solution obeys Henry's law in this range of mole fractions, and calculate Henry's law constant at 300 K.

16.17(b) At 310 K, the partial vapour pressures of a substance B dissolved in a liquid A are as follows:

x_B	0.010	0.015	0.020
p_B/kPa	82.0	122.0	166.1

Show that the solution obeys Henry's law in this range of mole fractions, and calculate Henry's law constant at 310 K.

16.18(a) The vapour pressure of pure liquid A at 300 K is 76.7 kPa and that of pure liquid B is 52.0 kPa. These two compounds form ideal liquid and gaseous mixtures. Consider the equilibrium composition of a mixture in which the mole fraction of A in the vapour is 0.350. Calculate the total pressure of the vapour and the composition of the liquid mixture.

16.18(b) The vapour pressure of pure liquid A at 293 K is 68.8 kPa and that of pure liquid B is 82.1 kPa. These two compounds form ideal liquid and gaseous mixtures. Consider the equilibrium composition of a mixture in which the mole fraction of A in the vapour is 0.612. Calculate the total pressure of the vapour and the composition of the liquid mixture.

16.19(a) Predict the partial vapour pressure of HCl above its solution in liquid germanium tetrachloride of molality 0.15 mol kg^{-1}. For data, see Exercise 16.17a.

16.19(b) Predict the partial vapour pressure of the component B above its solution in A in Exercise 16.17b when the molality of B is 0.15 mol kg^{-1}.

16.20(a) The vapour pressure of benzene is 53.3 kPa at 60.6°C, but it fell to 51.5 kPa when 19.0 g of an involatile organic compound was dissolved in 500 g of benzene. Calculate the molar mass of the compound.

16.20(b) The vapour pressure of 2-propanol is 50.00 kPa at 338.8°C, but it fell to 49.62 kPa when 8.69 g of an involatile organic compound was dissolved in 250 g of 2-propanol. Calculate the molar mass of the compound.

16.21(a) Use Henry's law and the data in Table 16.1 to calculate the solubility (as a molality) of CO_2 in water at 25°C when its partial pressure is (a) 0.10 atm, (b) 1.00 atm.

16.21(b) The mole fractions of N_2 and O_2 in air at sea level are approximately 0.78 and 0.21. Calculate the molalities of the solution formed in an open flask of water at 25°C.

16.22(a) The osmotic pressures of solutions of polystyrene in toluene were measured at 25°C and the pressure was expressed in terms of the height of the solvent of density 1.004 g cm^{-3}:

c/(g dm^{-3})	2.042	6.613	9.521	12.602
h/cm	0.592	1.910	2.750	3.600

Calculate the molar mass of the polymer.

16.22(b) The molar mass of an enzyme was determined by dissolving it in water, measuring the osmotic pressure at 20°C, and extrapolating the data to zero concentration. The following data were obtained:

c/(mg cm^{-3})	3.221	4.618	5.112	6.722
h/cm	5.746	8.238	9.119	11.990

Calculate the molar mass of the enzyme.

16.23(a) A sample consists of 30 per cent by mass of a dimer with $M = 30$ kg mol^{-1} and its monomer. What are the values of the number-average and weight-average molar masses? What is the value of the heterogeneity index?

16.23(b) A sample consists of 25 per cent by mass of a trimer with $M = 22$ kg mol^{-1} and its monomer. What are the values of the number-average and weight-average molar masses? What is the value of the heterogeneity index?

16.24(a) The maximum value of the excess enthalpy of mixing of a mixture of benzene and cyclohexane at 25°C is 700 J mol^{-1}. Identify the value of the parameter ξ in the expression that is used to model a regular solution. Can you expect phase separation?

16.24(b) The maximum value of the excess enthalpy of mixing of a mixture at 25°C is 1.4 kJ mol^{-1}. Identify the value of the parameter ξ in the expression that is used to model a regular solution. Can you expect phase separation?

16.25(a) Substances A and B are both volatile liquids with $p_A^* = 300$ Torr, $p_B^* = 250$ Torr, and $K_B = 200$ Torr (concentration expressed in mole fraction). When $x_A = 0.9$, $b_B = 2.22$ mol kg^{-1}, $p_A = 250$ Torr, and $p_B =$ 25 Torr. Calculate the activities and activity coefficients of A and B. Use the mole fraction, Raoult's law basis system for A and the Henry's law basis system (both mole fractions and molalities) for B.

16.25(b) Given that $p^*(H_2O) = 0.02308$ atm and $p(H_2O) = 0.02239$ atm in a solution in which 0.122 kg of a non-volatile solute ($M = 241$ g mol^{-1}) is dissolved in 0.920 kg water at 293 K, calculate the activity and activity coefficient of water in the solution.

16.26(a) By measuring the equilibrium between liquid and vapour phases of an acetone (A)–methanol (M) solution at 57.2°C at 1.00 atm, it was found that $x_A = 0.400$ when $y_A = 0.516$. Calculate the activities and activity coefficients of both components in this solution on the Raoult's law basis. The vapour pressures of the pure components at this temperature are: $p_A^* = 105$ kPa and $p_M^* = 73.5$ kPa.

16.26(b) By measuring the equilibrium between liquid and vapour phases of a solution at 30°C at 1.00 atm, it was found that $x_A = 0.220$ when $y_A = 0.314$. Calculate the activities and activity coefficients of both components in this solution on the Raoult's law basis. The vapour pressures of the pure components at this temperature are: $p_A^* = 73.0$ kPa and $p_B^* = 92.1$ kPa.

16.27(a) The maximum value of the excess enthalpy of mixing of two liquids that form a regular solution at 20°C is 800 J mol^{-1}. What are the activities of the components at that composition?

16.27(b) The maximum value of the excess enthalpy of mixing of two liquids that form a regular solution at 30°C is 1.4 kJ mol^{-1}. What are the activities of the components at that composition?

16.28(a) Calculate the ionic strength of a solution that is 0.15 mol kg^{-1} in KCl(aq) and 0.35 mol kg^{-1} in $CuSO_4$(aq).

16.28(b) Calculate the ionic strength of a solution that is 0.080 mol kg^{-1} in $K_3[Fe(CN)_6]$(aq), 0.030 mol kg^{-1} in KCl(aq), and 0.075 mol kg^{-1} in NaBr(aq).

16.29(a) Calculate the masses of (a) $Ca(NO_3)_2$ and, separately, (b) NaCl to add to a 0.250 mol kg^{-1} solution of KNO_3(aq) containing 800 g of solvent to raise its ionic strength to 0.450.

16.29(b) Calculate the masses of (a) KNO_3 and, separately, (b) $Ba(NO_3)_2$ to add to a 0.150 mol kg^{-1} solution of KNO_3(aq) containing 250 g of solvent to raise its ionic strength to 1.00.

16.30(a) Estimate the mean activity coefficient and activity of a solution that is 5.0 mmol kg^{-1} $CaCl_2$(aq) and 4.0 mmol kg^{-1} NaF(aq).

16.30(b) Estimate the mean activity coefficient and activity of a solution that is 2.5 mmol kg^{-1} NaCl(aq) and 5.5 mmol kg^{-1} $Ca(NO_3)_2$(aq).

16.31(a) The mean activity coefficients of HBr in three dilute aqueous solutions at 25°C are 0.930 (at 5.0 mmol kg^{-1}), 0.907 (at 10.0 mmol kg^{-1}), and 0.879 (at 20.0 mmol kg^{-1}). Estimate the value of B in the extended Debye–Hückel law if $C = 0$.

16.31(b) The mean activity coefficients of KCl in three dilute aqueous solutions at 25°C are 0.927 (at 5.0 mmol kg^{-1}), 0.902 (at 10.0 mmol kg^{-1}), and 0.816 (at 50.0 mmol kg^{-1}). Estimate the value of B in the extended Debye–Hückel law if $C = 0$.

Problems*

Numerical problems

16.1 The following temperature/composition data were obtained for a mixture of octane (O) and methylbenzene (M) at 1.00 atm, where x is the mole fraction in the liquid and y the mole fraction in the vapour at equilibrium.

θ/°C	110.9	112.0	114.0	115.8	117.3	119.0	121.1	123.0
x_M	0.908	0.795	0.615	0.527	0.408	0.300	0.203	0.097
y_M	0.923	0.836	0.698	0.624	0.527	0.410	0.297	0.164

The boiling points are 110.6°C and 125.6°C, for M and O, respectively. Plot the temperature/composition diagram for the mixture. What is the composition of the vapour in equilibrium with the liquid of composition (a) $x_M = 0.250$ and (b) $x_O = 0.250$?

16.2 Figure 16.51 shows the experimentally determined phase diagrams for the nearly ideal solution of hexane and heptane.
(a) Label the regions of the diagrams as to which phases are present.
(b) For an equimolar mixture of C_6H_{14} and C_7H_{16}, estimate the vapour pressure at 70°C when vaporization on reduction of the external pressure just begins. (c) What is the vapour pressure of the solution at 70°C when just one drop of liquid remains? (d) Estimate from the figures the mole fraction of hexane in the liquid and vapour phases for the conditions of part b. (e) What are the mole fractions for the conditions of part c?

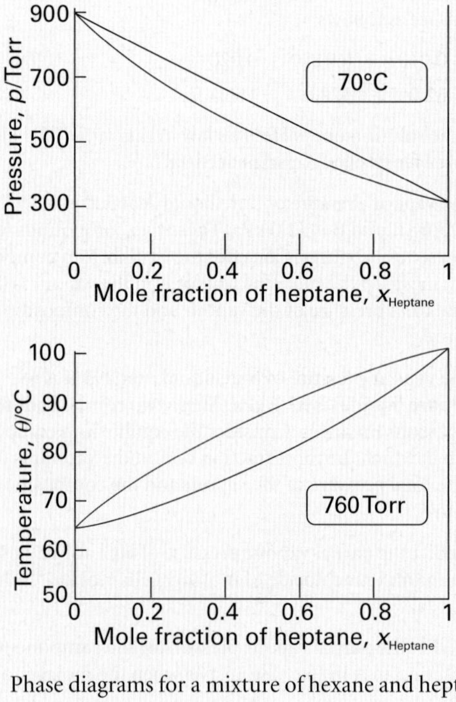

Fig. 16.51 Phase diagrams for a mixture of hexane and heptane.

16.3 Uranium tetrafluoride and zirconium tetrafluoride melt at 1035°C and 912°C, respectively. They form a continuous series of solid solutions with a minimum melting temperature of 765°C and composition $x(ZrF_4) = 0.77$. At 900°C, the liquid solution of composition $x(ZrF_4) = 0.28$ is in equilibrium with a solid solution of composition $x(ZrF_4) = 0.14$. At 850°C the two compositions are 0.87 and 0.90, respectively. Sketch the phase diagram for this system and state what is observed when a liquid of composition $x(ZrF_4) = 0.40$ is cooled slowly from 900°C to 500°C.

16.4 Hexane and perfluorohexane show partial miscibility below 22.7°C. The concentration at the upper critical temperature is $x = 0.355$, where x is the mole fraction of C_6F_{14}. At 22.0°C the two solutions in equilibrium have $x = 0.24$ and $x = 0.48$, respectively, and at 21.5°C the mole fractions are 0.22 and 0.51. Sketch the phase diagram. Describe the phase changes that occur when perfluorohexane is added to a fixed amount of hexane at (a) 23°C, (b) 22°C.

16.5 Methane (melting point 91 K) and tetrafluoromethane (melting point 89 K) do not form solid solutions with each other, and as liquids they are only partially miscible. The upper critical temperature of the

* Problems denoted by the symbol ‡ were supplied by Charles Trapp, Carmen Giunta, and Marshall Cady.

liquid mixture is 94 K at $x(CF_4) = 0.43$ and the eutectic temperature is 84 K at $x(CF_4) = 0.88$. At 86 K, the phase in equilibrium with the tetrafluoromethane-rich solution changes from solid methane to a methane-rich liquid. At that temperature, the two liquid solutions that are in mutual equilibrium have the compositions $x(CF_4) = 0.10$ and $x(CF_4) = 0.80$. Sketch the phase diagram.

16.6 The temperature dependence of the vapour pressure of solid sulfur dioxide can be approximately represented by the relation $\log(p/\text{Torr}) = 10.5916 - 1871.2/(T/\text{K})$ and that of liquid sulfur dioxide by $\log(p/\text{Torr}) = 8.3186 - 1425.7/(T/\text{K})$. Estimate the temperature and pressure of the triple point of sulfur dioxide.

16.7 The enthalpy of vaporization of a certain liquid is found to be 22.0 kJ mol^{-1} at 380 K, its normal boiling point. The molar volumes of the liquid and the vapour at the boiling point are 120 cm^3 mol^{-1} and 13.5 dm^3 mol^{-1}, respectively. (a) Estimate dp/dT from the Clapeyron equation and (b) the percentage error in its value if the Clausius–Clapeyron equation is used instead.

16.8 The enthalpy of fusion of mercury is 2.292 kJ mol^{-1}, and its normal freezing point is 234.3 K with a change in molar volume of +0.517 cm^{-3} mol^{-1} on melting. At what temperature will the bottom of a column of mercury (density 13.6 g cm^{-3}) of height 10.0 m be expected to freeze?

16.9 The vapour pressure, p, of a compound varies with temperature as follows:

T/K	250	270	280	290	300	310	330
p/kPa	0.036	0.514	1.662	4.918	13.43	34.1	182.9

What are (a) the normal boiling point and (b) the enthalpy of vaporization of the compound?

16.10 Construct the phase diagram for benzene near its triple point at 36 Torr and 5.50°C using the following data: $\Delta_{fus}H = 10.6$ kJ mol^{-1}, $\Delta_{vap}H = 30.8$ kJ mol^{-1}, $\rho(s) = 0.891$ g cm^{-3}, $\rho(l) = 0.879$ g cm^{-3}.

16.11‡ Sato *et al.* (*J. Polymer Sci., Polym. Phys.* **14**, 619 (1976)) have reported the data in the table below for the osmotic pressures of polychloroprene ($\rho = 1.25$ g cm^{-3}) in toluene ($\rho = 0.858$ g cm^{-3}) at 30°C. Determine the molar mass of polychloroprene and its second osmotic virial coefficient.

c/(mg cm^{-3})	1.33	2.10	4.52	7.18	9.87
Π/(N m^{-2})	30	51	132	246	390

16.12‡ In an investigation of thermophysical properties of methylbenzene (R.D. Goodwin, *J. Phys. Chem. Ref. Data* **18**, 1565 (1989)), Goodwin presented expressions for two coexistence curves (phase boundaries). The solid–liquid coexistence curve is given by

$$p/\text{bar} = p_3/\text{bar} + 1000 \times (5.60 + 11.727x)x$$

where $x = T/T_3 - 1$ and the triple point pressure and temperature are $p_3 = 0.4362$ μbar and $T_3 = 178.15$ K. The liquid–vapour curve is given by:

$$\ln(p/\text{bar}) = -10.418/y + 21.157 - 15.996y + 14.015y^2 - 5.0120y^3 + 4.7224(1-y)^{1.70}$$

where $y = T/T_c = T/(593.95\text{ K})$. (a) Plot the solid–liquid and liquid–vapour phase boundaries. (b) Estimate the standard melting point of methylbenzene. (c) Estimate the standard boiling point of methylbenzene. (d) Compute the standard enthalpy of vaporization of methylbenzene, given that the molar volumes of the liquid and vapour at the normal boiling point are 0.12 dm^3 mol^{-1} and 30.3 dm^3 mol^{-1}, respectively.

16.13‡ The following data have been obtained for the liquid–vapour equilibrium compositions of mixtures of nitrogen and oxygen at 100 kPa:

T/K	77.3	78	80	82	84	86	88	90.2
$x(O_2)$	0	10	34	54	70	82	92	100
$y(O_2)$	0	2	11	22	35	52	73	100
$p^*(O_2)$/Torr	154	171	225	294	377	479	601	760

Plot the data on a temperature–composition diagram and determine the extent to which it fits the predictions for an ideal solution by calculating the activity coefficients of O_2 at each composition.

16.14 The following table gives the mole fraction of methylbenzene (A) in liquid and gaseous mixtures with butanone at equilibrium at 303.15 K and the total pressure p. Take the vapour to be perfect and calculate the partial pressures of the two components. Plot them against their respective mole fractions in the liquid mixture and find the Henry's law constants for the two components.

x_A	0	0.0898	0.2476	0.3577	0.5194	0.6036
y_A	0	0.0410	0.1154	0.1762	0.2772	0.3393
p/kPa	36.066	34.121	30.900	28.626	25.239	23.402
x_A	0.7188	0.8019	0.9105	1		
y_A	0.4450	0.5435	0.7284	1		
p/kPa	20.6984	18.592	15.496	12.295		

16.15 The table below lists the vapour pressures of mixtures of iodoethane (I) and ethyl acetate (A) at 50°C. Find the activity coefficients of both components on (a) the Raoult's law basis, (b) the Henry's law basis with iodoethane as solute.

x_I	0	0.0579	0.1095	0.1918	0.2353	0.3718
p_I/kPa	0	3.73	7.03	11.7	14.05	20.72
p_A/kPa	37.38	35.48	33.64	30.85	29.44	25.05
x_I	0.5478	0.6349	0.8253	0.9093	1.0000	
p_I/kPa	28.44	31.88	39.58	43.00	47.12	
p_A/kPa	19.23	16.39	8.88	5.09	0	

16.16‡ Aminabhavi *et al.* examined mixtures of cyclohexane with various long-chain alkanes (*J. Chem. Eng. Data* **41**, 526 (1996)). Among their data are the following measurements of the density of a mixture of cyclohexane and pentadecane as a function of mole fraction of cyclohexane (x_c) at 298.15 K:

x_c	0.6965	0.7988	0.9004
ρ/(g cm^{-3})	0.7661	0.7674	0.7697

Fit a polynomial expression to the data (use mathematical software) and determine the excess volume of mixing and partial molar volumes, $V_J = (\partial V/\partial n_J)_{p,T}$, of the components. Plot your results.

16.17‡ Comelli and Francesconi examined mixtures of propionic acid with various other organic liquids at 313.15 K (*J. Chem. Eng. Data* **41**, 101 (1996)). They report the excess volume of mixing propionic acid with oxane as $V^E = x_1x_2\{a_0 + a_1(x_1 - x_2)\}$, where x_1 is the mole fraction of propionic acid, x_2 that of oxane, $a_0 = -2.4697$ cm^3 mol^{-1}, and $a_1 = 0.0608$ cm^3 mol^{-1}. The density of propionic acid at this temperature is 0.97174 g cm^{-3}; that of oxane is 0.86398 g cm^{-3}. (a) Derive an expression for the partial molar volume $V_J = (\partial V/\partial n_J)_{p,T}$ of each component at this temperature. (b) Compute the partial molar volume for each component in an equimolar mixture.

16.18‡ Francesconi *et al.* studied the liquid–vapour equilibria of trichloromethane and 1,2-epoxybutane at several temperatures (*J. Chem. Eng. Data* **41**, 310 (1996)). Among their data are the following measurements of the mole fractions of trichloromethane in the liquid phase (x_T) and the vapour phase (y_T) at 298.15 K as a function of pressure.

p/kPa	23.40	21.75	20.25	18.75	18.15	20.25	22.50	26.30
x	0	0.129	0.228	0.353	0.511	0.700	0.810	1
y	0	0.065	0.145	0.285	0.535	0.805	0.915	1

Compute the activity coefficients of both components on the basis of Raoult's law.

16.19 The mean activity coefficients for aqueous solutions of NaCl at 25°C are given below. Confirm that they support the Debye–Hückel limiting law and that an improved fit is obtained with the extended law.

$b/(\text{mmol kg}^{-1})$	1.0	2.0	5.0	10.0	20.0
$\gamma_\pm$	0.9649	0.9519	0.9275	0.9024	0.8712

Theoretical problems

16.20 Show that two phases are in thermal equilibrium only if their temperatures are the same.

16.21 Show that two phases are in mechanical equilibrium only if their pressures are equal.

16.22 The change in enthalpy is given by $dH = C_p dT + V dp$. The Clapeyron equation relates dp and dT at equilibrium, and so in combination the two equations can be used to find how the enthalpy changes along a phase boundary as the temperature changes and the two phases remain in equilibrium. Show that $d(\Delta H/T) = \Delta C_p \, d\ln T$.

16.23 Combine the barometric formula (Problem 13.20) for the dependence of the pressure on altitude with the Clausius–Clapeyron equation, and predict how the boiling temperature of a liquid depends on the altitude and the ambient temperature. Take the mean ambient temperature as 20°C and predict the boiling temperature of water at 3000 m.

16.24 Figure 16.20 is a schematic representation of how the chemical potentials of the solid, liquid, and gaseous phases of a substance vary with temperature. All have a negative slope, but it is unlikely that they are truly straight lines as indicated in the illustrations. Derive an expression for the curvatures (specifically, the second derivatives with respect to temperature) of these lines. Is there a restriction on the curvature of these lines? Which state of matter shows the greatest curvature?

16.25 Use the expressions relating the Gibbs energy and entropy of a perfect gas to the molecular partition function to show that (a) eqns 15.40 and 15.2 are consistent with eqn 16.9, (b) eqn 15.40 is consistent with eqn 16.10.

16.26 The excess Gibbs energy of a certain binary mixture is equal to $gRTx(1-x)$ where g is a constant and x is the mole fraction of a solute A. Find an expression for the chemical potential of A in the mixture and sketch its dependence on the composition.

16.27 The excess Gibbs energy of mixing of methylcyclohexane (MCH) and tetrahydrofuran (THF) at 303.15 K were found to fit the expression

$$G^E = RTx(1-x)\{0.4857 - 0.1077(2x-1) + 0.0191(2x-1)^2\}$$

where x is the mole fraction of the methylcyclohexane. Does this empirical expression support the view that the solution is regular? If it is not regular over the full range of compositions, is it regular over short regions? Can you adjust the regular model to accommodate the data?

16.28 Use the Gibbs–Duhem equation to show that the chemical potential of a component B in a binary mixture can be obtained if the chemical potential of the second component A is known for all compositions up to the one of interest. Do this by proving that

$$\mu_B = \mu_B^* - \int_{\mu_A^*}^{\mu_A} \frac{x_A}{1-x_A} d\mu_A$$

Go on to formulate a version of this expression for a regular solution in which the activity coefficients are represented by the Margules equations and the parameter ξ.

16.29 A similar expression to that derived in Problem 16.28 applies to the partial molar volume. Use the following data (which are for 298 K) to evaluate the integral graphically to find the partial molar volume of propanone in a propanone/trichloromethane mixture at $x = 0.500$.

$x(CHCl_3)$	0	0.194	0.385	0.559	0.788	0.889	1.000
$V_m/(\text{cm}^3 \text{ mol}^{-1})$	73.99	75.29	76.50	77.55	79.08	79.82	80.67

16.30 The 'osmotic coefficient', ϕ, is defined as $\phi = -(x_A/x_B)\ln a_A$. By writing $r = x_B/x_A$, and using the Gibbs–Duhem equation, show that it is possible to calculate the activity of B from the activities of A over a composition range by using the formula

$$\ln\left(\frac{a_B}{r}\right) = \phi - \phi(0) + \int_0^r \left(\frac{\phi - 1}{r}\right) dr$$

16.31 Show that the osmotic pressure of a real solution is given by $\Pi V = -RT\ln a_A$. Go on to show that, provided the concentration of the solution is low, this expression takes the form $\Pi V = \phi RT[B]$ and hence that the osmotic coefficient, ϕ, (which is defined in Problem 16.30) may be determined from osmometry.

16.32 Deduce an expression for the depression of freezing point of an ideal solution and show, subject to a series of approximations that you should specify, that the depression is proportional to the mole fraction of the solute, $\Delta T = K_f x_B$, with $K_f = RT_f^*/\Delta_{fus}H$, where T_f^* is the freezing temperature of the pure solvent and $\Delta_{fus}H$ its enthalpy of fusion. *Hint.* At the freezing temperature of the solvent, the chemical potential of the liquid solvent is equal to that of the solid solvent.

16.33 Repeat the preceding problem for the elevation of boiling point of a solvent in a solution.

16.34 Show that the freezing-point depression of a real solution in which the solvent of molar mass M has activity a_A obeys

$$\frac{d\ln a_A}{d(\Delta T)} = -\frac{M}{K_f}$$

and use the Gibbs–Duhem equation to show that

$$\frac{d\ln a_B}{d(\Delta T)} = -\frac{1}{b_B K_f}$$

where a_B is the solute activity and b_B is its molality. Use the Debye–Hückel limiting law to show that the osmotic coefficient (ϕ, Problem 16.30) is given by $\phi = 1 - \frac{1}{3}A'I$ with $A' = 2.303A$ and $I = b/b^\ominus$.

Applications: to biology and materials science

16.35 For the calculation of the solubility c of a gas in a solvent, it is often convenient to use the expression $c = Kp$, where K is the Henry's law constant. Breathing air at high pressures, such as in scuba diving, results in an increased concentration of dissolved nitrogen. The Henry's law constant for the solubility of nitrogen is 0.18 μg/(g H_2O atm). What mass of nitrogen is dissolved in 100 g of water saturated with air at 4.0 atm and 20°C? Compare your answer to that for 100 g of water saturated with air at 1.0 atm. (Air is 78.08 mole per cent N_2.) If nitrogen is four times as soluble in fatty tissues as in water, what is the increase in nitrogen concentration in fatty tissue in going from 1 atm to 4 atm?

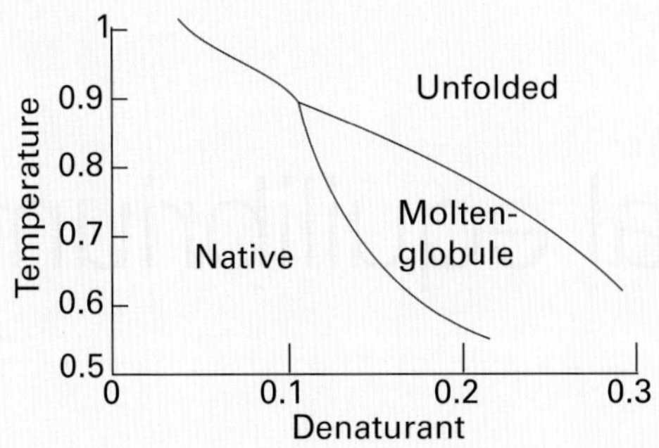

Fig. 16.52 A phase diagram for the conformations of a model protein.

16.36 The addition of a small amount of a salt, such as $(NH_4)_2SO_4$, to a solution containing a charged protein increases the solubility of the protein in water. This observation is called the *salting-in effect*. However, the addition of large amounts of salt can decrease the solubility of the protein to such an extent that the protein precipitates from solution. This observation is called the *salting-out effect* and is used widely by biochemists to isolate and purify proteins. Consider the equilibrium $PX_\nu(s) \rightleftharpoons P^{\nu+}(aq) + \nu X^-(aq)$, where $P^{\nu+}$ is a polycationic protein of charge $+\nu$ and X^- is its counter ion. Use Le Chatelier's principle and the physical principles behind the Debye–Hückel theory to provide a molecular interpretation for the salting-in and salting-out effects. *Hint.* Le Chatelier's principle should be familiar to you from introductory chemistry. For a review, see Section 17.4.

16.37 In a theoretical study of a protein, the temperature–composition diagram shown in Fig. 16.52 was obtained. It shows three structural regions: the native form, the unfolded form, and a 'molten globule' form, a partially unfolded but still compact form of the protein. (i) Is the molten globule form ever stable when the denaturant concentration is below 0.1? (ii) Describe what happens to the polymer as the native form is heated in the presence of denaturant at concentration 0.15.

16.38 Magnesium oxide and nickel oxide withstand high temperatures. However, they do melt when the temperature is high enough and the behaviour of mixtures of the two is of considerable interest to the ceramics industry. Draw the temperature–composition diagram for the system using the data below, where x is the mole fraction of MgO in the solid and y its mole fraction in the liquid.

θ/°C	1960	2200	2400	2600	2800
x	0	0.35	0.60	0.83	1.00
y	0	0.18	0.38	0.65	1.00

State (a) the melting point of a mixture with $x = 0.30$, (b) the composition and proportion of the phases present when a solid of composition $x = 0.30$ is heated to 2200°C, (c) the temperature at which a liquid of composition $y = 0.70$ will begin to solidify.

16.39‡ A substance as well known as methane still receives research attention because it is an important component of natural gas, a commonly used fossil fuel. Friend *et al.* have published a review of thermophysical properties of methane (*J. Phys. Chem. Ref. Data* 18, 583 (1989)), which included the following data describing the liquid–vapour phase boundary.

T/K	100	108	110	112	114	120
p/MPa	0.034	0.074	0.088	0.104	0.122	0.192
T/K	130	140	150	160	170	190
p/MPa	0.368	0.642	1.041	1.593	2.329	4.521

(a) Plot the liquid–vapour phase boundary. (b) Estimate the standard boiling point of methane. (c) Compute the standard enthalpy of vaporization of methane, given that the molar volumes of the liquid and vapour at the standard boiling point are 3.80×10^{-2} and 8.89 $dm^3\ mol^{-1}$, respectively.

16.40‡ Diamond is the hardest substance and the best conductor of heat yet characterized. For these reasons, it is used widely in industrial applications that require a strong abrasive. Unfortunately, it is difficult to synthesize diamond from the more readily available allotropes of carbon, such as graphite. To illustrate this point, calculate the pressure required for diamond to become thermodynamically more stable than graphite at 25°C. The following data apply to 25°C and 100 kPa. Assume the specific volume, V_s, and κ_T are constant with respect to pressure changes.

	Graphite	Diamond
$\Delta_f G^\ominus/(kJ\ mol^{-1})$	0	+2.8678
$V_s/(cm^3\ g^{-1})$	0.444	0.284
κ_T/kPa	3.04×10^{-8}	0.187×10^{-8}

16.41‡ Polymer scientists often report their data in rather strange units. For example, in the determination of molar masses of polymers in solution by osmometry, osmotic pressures are often reported in grams per square centimetre ($g\ cm^{-2}$) and concentrations in grams per cubic centimetre ($g\ cm^{-3}$). (a) With these choices of units, what would be the units of R in the van't Hoff equation? (b) The data in the table below on the concentration dependence of the osmotic pressure of polyisobutene in chlorobenzene at 25°C have been adapted from J. Leonard and H. Daoust (*J. Polymer Sci.* 57, 53 (1962)). From these data, determine the molar mass of polyisobutene by plotting Π/c against c. (c) Theta solvents are solvents for which the second osmotic coefficient is zero; for 'poor' solvents the plot is linear and for good solvents the plot is non-linear. From your plot, how would you classify chlorobenzene as a solvent for polyisobutene? Rationalize the result in terms of the molecular structure of the polymer and solvent. (d) Determine the second and third osmotic virial coefficients by fitting the curve to the virial form of the osmotic pressure equation. (e) Experimentally, it is often found that the virial expansion can be represented as

$$\Pi/c = RT/M\,(1 + B'c + gB'^2c'^2 + \cdots)$$

and in good solvents, the parameter g is often about 0.25. With terms beyond the second power ignored, obtain an equation for $(\Pi/c)^{1/2}$ and plot this quantity against c. Determine the second and third virial coefficients from the plot and compare to the values from the first plot. Does this plot confirm the assumed value of g?

$10^{-2}(\Pi/c)/(g\ cm^{-2}/g\ cm^{-3})$	2.6	2.9	3.6	4.3	6.0	12.0
$c/(g\ cm^{-3})$	0.0050	0.010	0.020	0.033	0.057	0.10
$10^{-2}(\Pi/c)/(g\ cm^{-2}/g\ cm^{-3})$	19.0	31.0	38.0	52	63	
$c/(g\ cm^{-3})$	0.145	0.195	0.245	0.27	0.29	

17 Chemical equilibrium

This chapter develops the concept of chemical potential and shows how it is used to account for the equilibrium composition of chemical reactions. The equilibrium composition corresponds to a minimum in the Gibbs energy plotted against the extent of reaction, and by locating this minimum we establish the relation between the equilibrium constant and the standard Gibbs energy of reaction. The thermodynamic formulation of equilibrium enables us to establish the quantitative effects of changes in the conditions. The principles of thermodynamics established in the preceding chapters can be applied to the description of the thermodynamic properties of reactions that take place in electrochemical cells, in which, as the reaction proceeds, it drives electrons through an external circuit. Thermodynamic arguments can be used to derive an expression for the electric potential of such cells and the potential can be related to their composition. There are two major topics developed in this connection. One is the definition and tabulation of standard potentials; the second is the use of these standard potentials to predict the equilibrium constants and other thermodynamic properties of chemical reactions.

Chemical reactions tend to move towards a dynamic equilibrium in which both reactants and products are present but have no further tendency to undergo net change. In some cases, the concentration of products in the equilibrium mixture is so much greater than that of the unchanged reactants that for all practical purposes the reaction is 'complete'. However, in many important cases the equilibrium mixture has significant concentrations of both reactants and products. In this chapter we see how to use thermodynamics to predict the equilibrium composition under any reaction conditions. Because many reactions of ions involve the transfer of electrons, they can be studied (and utilized) by allowing them to take place in an electrochemical cell. Measurements like those described in this chapter provide data that are very useful for discussing the characteristics of electrolyte solutions and of ionic equilibria in solution.

Spontaneous chemical reactions

We have seen that the direction of spontaneous change at constant temperature and pressure is towards lower values of the Gibbs energy, G. The idea is entirely general, and in this chapter we apply it to the discussion of chemical reactions.

17.1 The Gibbs energy minimum and the reaction Gibbs energy

We locate the equilibrium composition of a reaction mixture by calculating the Gibbs energy of the reaction mixture and identifying the composition that corresponds to minimum G.

Consider the equilibrium $A \rightleftharpoons B$. Even though this reaction looks trivial, there are many examples of it, such as the isomerization of pentane to 2-methylbutane and the conversion of L-alanine to D-alanine. Suppose an infinitesimal amount $d\xi$ of A turns into B; then the change in the amount of A present is $dn_A = -d\xi$ and the change in the amount of B present is $dn_B = +d\xi$. The quantity ξ (xi) is called the **extent of reaction**; it has the dimensions of amount of substance and is reported in moles. When the extent of reaction changes by a finite amount $\Delta\xi$, the amount of A present changes from $n_{A,0}$ to $n_{A,0} - \Delta\xi$ and the amount of B changes from $n_{B,0}$ to $n_{B,0} + \Delta\xi$. So, if initially 2.0 mol A is present and we wait until $\Delta\xi = +1.5$ mol, then the amount of A remaining will be 0.5 mol.

The **reaction Gibbs energy**, $\Delta_r G$, is defined as the slope of the graph of the Gibbs energy plotted against the extent of reaction:

$$\Delta_r G = \left(\frac{\partial G}{\partial \xi}\right)_{p,T} \qquad [17.1]$$

Although Δ normally signifies a *difference* in values, here Δ_r signifies a *derivative*, the slope of G with respect to ξ. However, to see that there is a close relationship with the normal usage, suppose the reaction advances by $d\xi$ at constant temperature and pressure. The corresponding change in Gibbs energy is

$$dG = \mu_A dn_A + \mu_B dn_B = -\mu_A d\xi + \mu_B d\xi = (\mu_B - \mu_A)d\xi$$

This equation can be reorganized into

$$\left(\frac{\partial G}{\partial \xi}\right)_{p,T} = \mu_B - \mu_A$$

That is,

$$\Delta_r G = \mu_B - \mu_A \qquad (17.2)$$

We see that $\Delta_r G$ can also be interpreted as the difference between the chemical potentials (the partial molar Gibbs energies) of the reactants and products *at the composition of the reaction mixture*.

Because chemical potential varies with composition, the slope of the plot of Gibbs energy against extent of reaction changes as the reaction proceeds. Moreover, because the reaction runs in the direction of decreasing G (that is, down the slope of G plotted against ξ), we see from eqn 17.2 that the reaction $A \rightarrow B$ is spontaneous when $\mu_A > \mu_B$, whereas the reverse reaction is spontaneous when $\mu_B > \mu_A$. The slope is zero, and the reaction is spontaneous in neither direction, when

$$\Delta_r G = 0 \qquad (17.3)$$

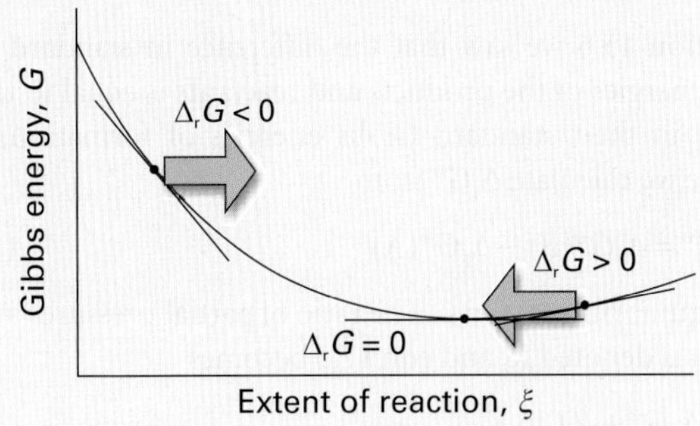

Fig. 17.1 As the reaction advances (represented by motion from left to right along the horizontal axis) the slope of the Gibbs energy changes. Equilibrium corresponds to zero slope, at the foot of the valley.

This condition occurs when $\mu_B = \mu_A$ (Fig. 17.1). It follows that, if we can find the composition of the reaction mixture that ensures $\mu_B = \mu_A$, then we can identify the composition of the reaction mixture at equilibrium.

We can express the spontaneity of a reaction at constant temperature and pressure in terms of the reaction Gibbs energy:

If $\Delta_r G < 0$, the forward reaction is spontaneous.
If $\Delta_r G > 0$, the reverse reaction is spontaneous.
If $\Delta_r G = 0$, the reaction is at equilibrium.

17.2 The thermodynamic description of equilibrium

With the background established, we are now ready to see how to apply thermodynamics to the description of chemical equilibrium.

(a) Perfect gas equilibria

When A and B are perfect gases we can use eqn 16.15 ($\mu = \mu^\ominus + RT \ln p$, with p interpreted as $p/p^\ominus$) to write

$$\Delta_r G = \mu_B - \mu_A = (\mu_B^\ominus + RT \ln p_B) - (\mu_A^\ominus + RT \ln p_A)$$

$$= \Delta_r G^\ominus + RT \ln \frac{p_B}{p_A} \qquad (17.4)^\circ$$

If we denote the ratio of partial pressures by Q, we obtain

$$\Delta_r G = \Delta_r G^\ominus + RT \ln Q \qquad Q = \frac{p_B}{p_A} \qquad (17.5)^\circ$$

The ratio Q is an example of a **reaction quotient**. It ranges from 0 when $p_B = 0$ (corresponding to pure A) to infinity when $p_A = 0$ (corresponding to pure B). The **standard reaction Gibbs energy**, $\Delta_r G^\ominus$, is defined (like the standard reaction enthalpy) as the difference in the standard molar Gibbs energies of the reactants and products. For our reaction

$$\Delta_r G^{\ominus} = G^{\ominus}_{B,m} - G^{\ominus}_{A,m} \qquad (17.6a)$$

In Section 15.6 we saw that the difference in standard molar Gibbs energies of the products and reactants is equal to the difference in their standard Gibbs energies of formation, so in practice we calculate $\Delta_r G^{\ominus}$ from

$$\Delta_r G^{\ominus} = \Delta_f G^{\ominus}(B) - \Delta_f G^{\ominus}(A) \qquad (17.6b)$$

At equilibrium $\Delta_r G = 0$. The ratio of partial pressures at equilibrium is denoted K, and eqn 17.5 becomes

$$0 = \Delta_r G^{\ominus} + RT \ln K$$

This expression rearranges to

$$RT \ln K = -\Delta_r G^{\ominus} \qquad K = \left(\frac{p_B}{p_A}\right)_{\text{equilibrium}} \qquad (17.7)^{\circ}$$

This relation is a special case of one of the most important equations in chemical thermodynamics: it is the link between tables of thermodynamic data, such as those in the *Data section* at the end of this volume, and the chemically important **equilibrium constant**, K.

We see from eqn 17.7 that, when $\Delta_r G^{\ominus} > 0$, the equilibrium constant $K < 1$. Therefore, at equilibrium the partial pressure of A exceeds that of B, which means that the reactant A is favoured in the equilibrium. When $\Delta_r G^{\ominus} < 0$, the equilibrium constant $K > 1$, so at equilibrium the partial pressure of B exceeds that of A. Now the product B is favoured in the equilibrium.

In molecular terms, the minimum in the Gibbs energy, which corresponds to $\Delta_r G = 0$, stems from the Gibbs energy of mixing of the two gases. Hence, an important contribution to the position of chemical equilibrium is the mixing of the products with the reactants as the products are formed.

To appreciate the role of mixing, consider a hypothetical reaction in which A molecules change into B molecules without mingling together. The Gibbs energy of the system changes from $G^{\ominus}(A)$ to $G^{\ominus}(B)$ in proportion to the amount of B that had been formed, and the slope of the plot of G against the extent of reaction is a constant and equal to $\Delta_r G^{\ominus}$ at all stages of the reaction (Fig. 17.2). There is no intermediate minimum in the graph. However, in fact, the newly produced B molecules do mix with the surviving A molecules. We have seen that the contribution of a mixing process to the change in Gibbs energy is given by eqn 16.18 ($\Delta_{mix} G = nRT(x_A \ln x_A + x_B \ln x_B)$). This expression makes a U-shaped contribution to the total change in Gibbs energy. As can be seen from Fig. 17.2, there is now an intermediate minimum in the Gibbs energy, and its position corresponds to the equilibrium composition of the reaction mixture.

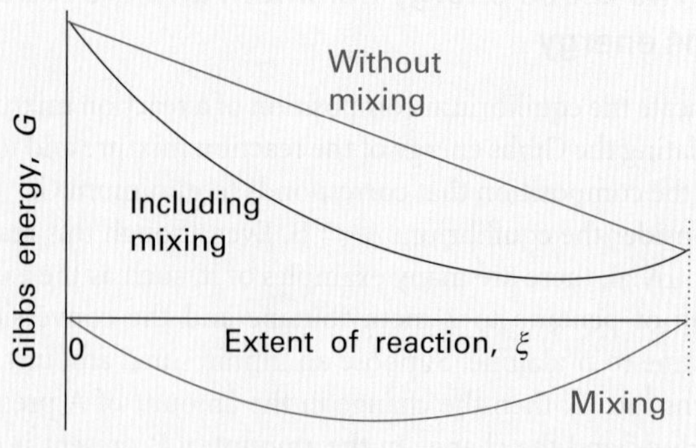

Fig. 17.2 If the mixing of reactants and products is ignored, then the Gibbs energy changes linearly from its initial value (pure reactants) to its final value (pure products) and the slope of the line is $\Delta_r G^{\ominus}$. However, as products are produced, there is a further contribution to the Gibbs energy arising from their mixing (lowest curve). The sum of the two contributions has a minimum. That minimum corresponds to the equilibrium composition of the system.

(b) The general case of a reaction

We can easily extend the argument that led to eqn 17.7 to a general reaction. First, we need to generalize the concept of extent of reaction.

In Section 14.7 we saw that a chemical reaction, such as $2\,A + B \rightarrow 3\,C + D$, can be expressed in the form

$$0 = \sum_J \nu_J J \qquad (17.8)$$

where J denotes the substances and the ν_J are the corresponding stoichiometric numbers in the chemical equation. In our example, these numbers have the values $\nu_A = -2$, $\nu_B = -1$, $\nu_C = +3$, and $\nu_D = +1$; as we saw in Section 14.7, a stoichiometric number is positive for products and negative for reactants. Then we define ξ so that, if it changes by $\Delta\xi$, then the change in the amount of any species J is $\nu_J \Delta\xi$.

A BRIEF ILLUSTRATION

In the notation of eqn 17.8, the stoichiometric numbers in the equation

$$N_2(g) + 3\,H_2(g) \rightarrow 2\,NH_3(g)$$

are $\nu_{N_2} = -1$, $\nu_{H_2} = -3$, and $\nu_{NH_3} = +2$. Therefore, if initially there is 10 mol N_2 present, then when the extent of reaction changes from $\xi = 0$ to $\xi = 1$ mol, implying that $\Delta\xi = +1$ mol, the amount of N_2 changes from 10 mol to 9 mol. All the N_2 has been consumed when $\xi = 10$ mol. When $\Delta\xi = +1$ mol, the amount of H_2 changes by $-3 \times (1\text{ mol}) = -3$ mol and the amount of NH_3 changes by $+2 \times (1\text{ mol}) = +2$ mol. ●

In the following *Justification*, we show that the Gibbs energy of reaction can always be written

$$\Delta_r G = \Delta_r G^{\ominus} + RT \ln Q \qquad (17.9)$$

with the standard reaction Gibbs energy calculated from

$$\Delta_r G^\ominus = \sum_{\text{Products}} \nu\Delta_f G^\ominus - \sum_{\text{Reactants}} \nu\Delta_f G^\ominus \tag{17.10a}$$

or, more formally,

$$\Delta_r G^\ominus = \sum_J \nu_J \Delta_f G^\ominus(J) \tag{17.10b}$$

The **reaction quotient**, Q, has the form

$$Q = \frac{\text{activities of products}}{\text{activities of reactants}} \tag{17.11a}$$

with each species raised to the power given by its stoichiometric coefficient. More formally, to write the general expression for Q we introduce the symbol $\prod$ to denote the product of what follows it (just as $\sum$ denotes the sum), and define Q as

$$Q = \prod_J a_J^{\nu_J} \qquad [17.11b]$$

Because reactants have negative stoichiometric numbers, they automatically appear as the denominator when the product is written out explicitly. Recall from Table 16.2 that, for pure solids and liquids, the activity is 1 under ordinary laboratory conditions, so such substances make no contribution to Q even though they may appear in the chemical equation.

● A BRIEF ILLUSTRATION

Consider the reaction 2 A + 3 B → C + 2 D, in which case $\nu_A = -2$, $\nu_B = -3$, $\nu_C = +1$, and $\nu_D = +2$. The reaction quotient is then

$$Q = a_A^{-2} a_B^{-3} a_C a_D^2 = \frac{a_C a_D^2}{a_A^2 a_B^3}$$ ●

Justification 17.1 *The dependence of the reaction Gibbs energy on the reaction quotient*

Consider the reaction with stoichiometric numbers ν_J. When the reaction advances by $d\xi$, the amounts of reactants and products change by $dn_J = \nu_J d\xi$. The resulting infinitesimal change in the Gibbs energy at constant temperature and pressure is

$$dG = \sum_J \mu_J dn_J = \sum_J \mu_J \nu_J d\xi = \left(\sum_J \nu_J \mu_J\right) d\xi$$

It follows that

$$\Delta_r G = \left(\frac{\partial G}{\partial \xi}\right)_{p,T} = \sum_J \nu_J \mu_J$$

To make further progress, we note that the chemical potential of a species J is related to its activity by eqn 16.41 ($\mu_J = \mu_J^\ominus + RT \ln a_J$). When this expression is substituted into the preceding equation we obtain

$$\Delta_r G = \overbrace{\sum_J \nu_J \mu_J^\ominus}^{\Delta_r G^\ominus} + RT \sum_J \nu_J \ln a_J$$

$$= \Delta_r G^\ominus + RT \sum_J \ln a_J^{\nu_J} = \Delta_r G^\ominus + RT \ln \overbrace{\prod_J a_J^{\nu_J}}^{Q}$$

$$= \Delta_r G^\ominus + RT \ln Q$$

with Q given by eqn 17.11b. We have used the relations

$$a \ln x = \ln x^a \qquad \sum_i \ln x_i = \ln\left(\prod_i x_i\right)$$

Now we conclude the argument based on eqn 17.9. At equilibrium, the slope of G is zero: $\Delta_r G = 0$. The activities then have their equilibrium values and we can write

$$K = \left(\prod_J a_J^{\nu_J}\right)_{\text{equilibrium}} \qquad [17.12]$$

This expression has the same form as Q, eqn 17.11, but is evaluated using *equilibrium* activities. From now on, we shall not write the 'equilibrium' subscript explicitly, and will rely on the context to make it clear that for K we use equilibrium values and for Q we use the values at the specified stage of the reaction.

An equilibrium constant K expressed in terms of activities is called a **thermodynamic equilibrium constant**. Note that, because activities are dimensionless numbers, the thermodynamic equilibrium constant is also dimensionless. In elementary applications, the activities that occur in eqn 17.12 are often replaced by the numerical values of molalities (that is, by replacing a_J by $b_J/b^\ominus$, where $b^\ominus = 1$ mol kg^{-1}), molar concentrations (that is, as $[J]/c^\ominus$, where $c^\ominus = 1$ mol dm^{-3}), or the numerical values of partial pressures (that is, by $p_J/p^\ominus$, where $p^\ominus = 1$ bar). In each case, the resulting expressions are only approximations. The approximation is particularly severe for electrolyte solutions, for in them activity coefficients differ from 1 even in very dilute solutions (Section 16.6e).

● A BRIEF ILLUSTRATION

The equilibrium constant for the heterogeneous equilibrium $CaCO_3(s) \rightleftharpoons CaO(s) + CO_2(g)$ is

$$K = a_{CaCO_3(s)}^{-1} a_{CaO(s)} a_{CO_2(g)} = \frac{\overbrace{a_{CaO(s)}}^{1} a_{CO_2(g)}}{\underbrace{a_{CaCO_3(s)}}_{1}} = a_{CO_2}$$

where the activities of CaO(s) and $CaCO_3$(s) are set to 1 because each substance appears in the reaction as a separate, pure phase. Provided the carbon dioxide can be treated as a perfect gas, we can go on to write

$$K \approx p_{CO_2}/p^{\ominus}$$

and conclude that in this case the equilibrium constant is the numerical value of the decomposition vapour pressure of calcium carbonate. ●

(c) Calculating equilibrium constants from thermodynamic data

If we set $\Delta_r G = 0$ in eqn 17.9 and replace Q by K, then we immediately obtain

$$RT \ln K = -\Delta_r G^{\ominus} \quad (17.13)$$

This is an exact and highly important thermodynamic relation, for it enables us to predict the equilibrium constant of any reaction from tables of thermodynamic data, and hence to predict the equilibrium composition of the reaction mixture.

Example 17.1 *Calculating an equilibrium constant 1*

Calculate the equilibrium constant for the ammonia synthesis reaction, $N_2(g) + 3\,H_2(g) \rightarrow 2\,NH_3(g)$, at 298 K and show how K is related to the partial pressures of the species at equilibrium when the overall pressure is low enough for the gases to be treated as perfect.

Method Calculate the standard reaction Gibbs energy from eqn 17.10 and convert it to the value of the equilibrium constant by using eqn 17.13. The expression for the equilibrium constant is obtained from eqn 17.12 and, because the gases are taken to be perfect, replace each activity by the ratio $p/p^{\ominus}$, where p is a partial pressure.

Answer The standard Gibbs energy of the reaction is

$$\begin{aligned}\Delta_r G^{\ominus} &= 2\Delta_f G^{\ominus}(NH_3, g) - \{\Delta_f G^{\ominus}(N_2, g) + 3\Delta_f G^{\ominus}(H_2, g)\} \\ &= 2\Delta_f G^{\ominus}(NH_3, g) = 2 \times (-16.5\ \text{kJ mol}^{-1})\end{aligned}$$

Then,

$$\ln K = -\frac{2 \times (-16.5 \times 10^3\ \text{J mol}^{-1})}{(8.3145\ \text{J K}^{-1}\ \text{mol}^{-1}) \times (298\ \text{K})} = \frac{2 \times 16.5 \times 10^3}{8.3145 \times 298}$$

Hence, $K = 6.1 \times 10^5$. This result is thermodynamically exact. The thermodynamic equilibrium constant for the reaction is

$$K = \frac{a_{NH_3}^2}{a_{N_2} a_{H_2}^3}$$

and this ratio has exactly the value we have just calculated. At low overall pressures, the activities can be replaced by the ratios $p/p^{\ominus}$, where p is a partial pressure, and an approximate form of the equilibrium constant is

$$K = \frac{(p_{NH_3}/p^{\ominus})^2}{(p_{N_2}/p^{\ominus})(p_{H_2}/p^{\ominus})^3} = \frac{p_{NH_3}^2 p^{\ominus 2}}{p_{N_2} p_{H_2}^3}$$

Self-test 17.1 Evaluate the equilibrium constant for $N_2O_4(g) \rightleftharpoons 2\,NO_2(g)$ at 298 K. [$K = 0.15$]

We can also express the thermodynamic equilibrium constant in terms of the mole fractions, x_J, or molalities, b_J, of the species. To do so, we need to know the activity coefficients, and then to use $a_J = \gamma_J x_J$ or $a_J = \gamma_J b_J/b^{\ominus}$ (recalling that the activity coefficients depend on the choice). For example, in the latter case, for an equilibrium of the form $A + B \rightleftharpoons C + D$, where all four species are solutes, we write

$$K = \frac{a_C a_D}{a_A a_B} = \frac{\gamma_C \gamma_D}{\gamma_A \gamma_B} \times \frac{b_C b_D}{b_A b_B} = K_\gamma K_b \quad (17.14)$$

The activity coefficients must be evaluated at the equilibrium composition of the mixture (for instance, by using one of the Debye–Hückel expressions, Section 16.6e), which may involve a complicated calculation, because the activity coefficients are known only if the equilibrium composition is already known. In elementary applications, and to begin the iterative calculation of the concentrations in a real example, the assumption is often made that the activity coefficients are all so close to unity that $K_\gamma = 1$. Then we obtain the result widely used in elementary chemistry that $K \approx K_b$, and equilibria are discussed in terms of molalities (or molar concentrations) themselves.

(d) Equilibria in biological systems

For biological systems it is appropriate to adopt the biological standard state, in which $a_{H^+} = 10^{-7}$ and $pH = -\log a_{H^+} = 7$. It follows from eqn 17.9 that the relation between the thermodynamic and biological standard Gibbs energies of reaction for a reaction of the form

$$A + \nu H^+(aq) \rightarrow P \quad (17.15a)$$

is

$$\Delta_r G^{\oplus} = \Delta_r G^{\ominus} + 7\nu RT \ln 10 \quad (17.15b)$$

Note that there is no difference between the two standard values if hydrogen ions are not involved in the reaction ($\nu = 0$).

● A BRIEF ILLUSTRATION

Consider the reaction

$$NADH(aq) + H^+(aq) \rightarrow NAD^+(aq) + H_2(g)$$

at 37°C, for which $\Delta_r G^{\ominus} = -21.8\ \text{kJ mol}^{-1}$; NADH is the reduced form of nicotinamide adenine dinucleotide and NAD^+

is its oxidized form; the molecules play an important role in the later stages of the respiratory process. It follows that because $\nu = 1$ and $7 \ln 10 = 16.1$,

$$\begin{aligned}\Delta_r G^{\oplus} &= -21.8 \text{ kJ mol}^{-1} \\ &\quad + 16.1 \times (8.3145 \times 10^{-3} \text{ kJ K}^{-1} \text{ mol}^{-1}) \times (310 \text{ K}) \\ &= +19.7 \text{ kJ mol}^{-1}\end{aligned}$$

Note that the biological standard value is opposite in sign (in this example) to the thermodynamic standard value: the much lower concentration of hydronium ions (by seven orders of magnitude) at pH = 7 in place of pH = 0, has resulted in the reverse reaction becoming spontaneous under these biologically standard conditions (with all other species at unit activity). ●

Self-test 17.2 For a particular reaction of the form $A \rightarrow B + 2 H^+$ in aqueous solution, it was found that $\Delta_r G^{\ominus} = +20$ kJ mol^{-1} at 28°C. Estimate the value of $\Delta_r G^{\oplus}$. [−61 kJ mol^{-1}]

(e) Exergonic and endergonic reactions

A reaction for which $\Delta_r G^{\ominus} < 0$ is called **exergonic** (from the Greek words for work producing). The name signifies that, because the process is spontaneous, it can be used to drive another process, such as another reaction, or used to do non-expansion work. A simple mechanical analogy is a pair of weights joined by a string (Fig. 17.3): the lighter of the pair of weights will be pulled up as the heavier weight falls down. Although the lighter weight has a natural tendency to move downward, its coupling to the heavier weight results in it being raised. In biological cells, the oxidation of carbohydrates acts as the heavy weight that drives other reactions forward and results in the formation of proteins from amino acids, muscle contraction, and brain activity. A reaction for which $\Delta_r G^{\ominus} > 0$ is called **endergonic** (signifying work consuming). The reaction can be made to occur only by doing work on it, such as electrolysing water to reverse its spontaneous formation reaction.

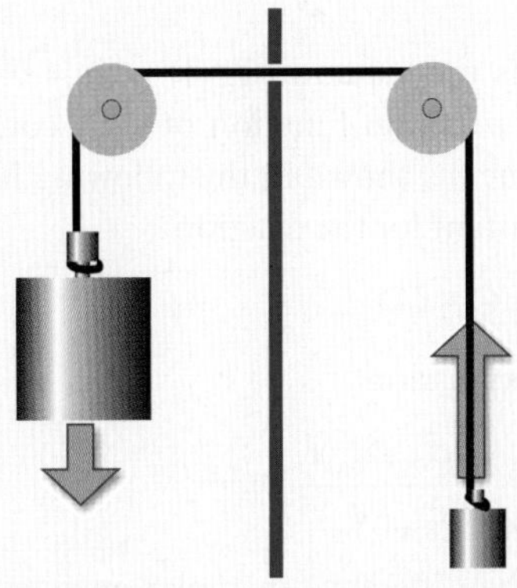

Fig. 17.3 If two weights are coupled as shown here, then the heavier weight will move the lighter weight in its nonspontaneous direction: overall, the process is still spontaneous. The weights are the analogues of two chemical reactions: a reaction with a large negative ΔG can force another reaction with a smaller ΔG to run in its nonspontaneous direction.

IMPACT ON BIOLOGY

I17.1 Energy conversion in biological cells

The whole of life's activities depends on the coupling of exergonic and endergonic reactions, for the oxidation of food drives other reactions forward. In biological cells, the energy released by the oxidation of foods is stored in adenosine triphosphate (ATP, **1**). The essence of the action of ATP is its ability to lose its terminal phosphate group by hydrolysis and to form adenosine diphosphate (ADP):

$$ATP(aq) + H_2O(l) \rightarrow ADP(aq) + P_i^-(aq) + H_3O^+(aq)$$

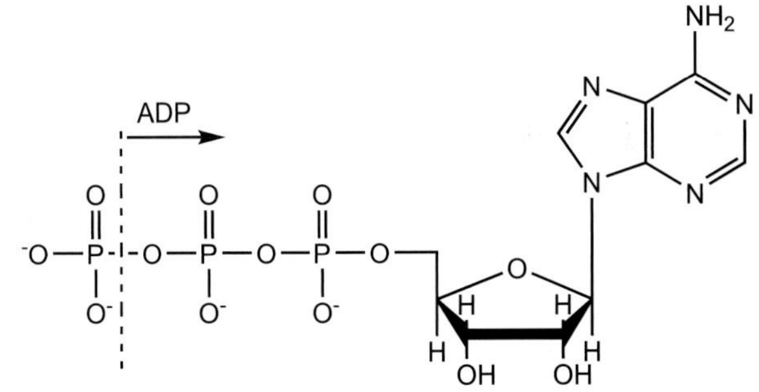

1 Adenosine triphosphate, ATP

where P_i^- denotes an inorganic phosphate group, such as $H_2PO_4^-$. The biological standard values for ATP hydrolysis at 37°C (310 K, blood temperature) are $\Delta_r G^{\oplus} = -31$ kJ mol^{-1}, $\Delta_r H^{\oplus} = -20$ kJ mol^{-1}, and $\Delta_r S^{\oplus} = +34$ J K^{-1} mol^{-1}. The hydrolysis is therefore exergonic ($\Delta_r G^{\oplus} < 0$) under these conditions and 31 kJ mol^{-1} is available for driving other reactions, such as the strongly endergonic biosynthesis of proteins from aminoacids. Moreover, because the reaction entropy is large, the reaction Gibbs energy is sensitive to temperature.

In view of its exergonicity the ADP—phosphate bond has been called a 'high-energy phosphate bond'. The name is intended to signify a high tendency to undergo reaction, and should not be confused with 'strong' bond. In fact, even in the biological sense it is not of very 'high energy'. The action of ATP depends on it being intermediate in activity. Thus ATP acts as a phosphate donor to a number of acceptors (for example, glucose), but is recharged by more powerful phosphate donors in a number of biochemical processes.

We now use the oxidation of glucose to CO_2 and H_2O by O_2 as an example of how the breakdown of foods is coupled to the formation of ATP in the cell. The process begins with *glycolysis*, a partial oxidation of glucose by nicotinamide adenine dinucleotide (NAD^+, **2**) to pyruvate ion, $CH_3COCO_2^-$, continues with the *citric acid cycle*, which oxidizes pyruvate to CO_2, and ends with the reactions of the *respiratory chain*, during which O_2 is reduced to H_2O (*Impact I17.3*).

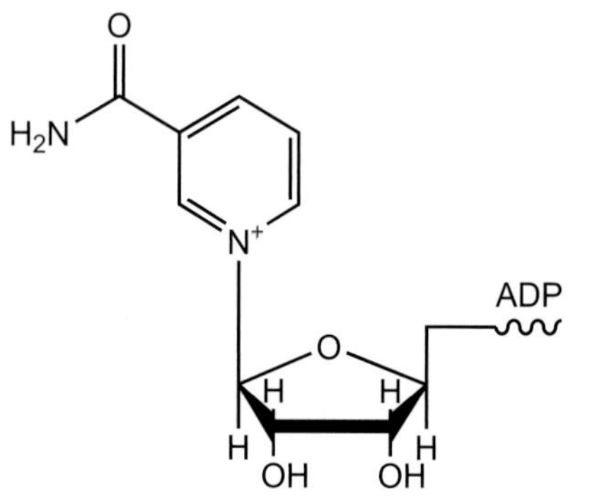

2 NAD+

Glycolysis occurs in the *cytosol*, the aqueous material encapsulated by the cell membrane, and consists of ten enzyme-catalysed reactions. At blood temperature, $\Delta_r G^{\ominus} = -147\ \mathrm{kJ\ mol^{-1}}$ for the oxidation of glucose by NAD^+ to pyruvate ions. The oxidation of one glucose molecule is coupled to the conversion of two ADP molecules to two ATP molecules, so the net reaction of glycolysis is:

$$\mathrm{C_6H_{12}O_6(aq) + 2\,NAD^+(aq) + 2\,ADP(aq) + 2\,P_i^-(aq) + 2\,H_2O(l)}$$
$$\mathrm{\rightarrow 2\,CH_3COCO_2^-(aq) + 2\,NADH(aq) + 2\,ATP(aq) + 2\,H_3O^+(aq)}$$

The standard reaction Gibbs energy is $(-147) - 2(-31)\ \mathrm{kJ\ mol^{-1}} = -85\ \mathrm{kJ\ mol^{-1}}$. The reaction is exergonic, and therefore spontaneous: the oxidation of glucose is used to 'recharge' the ATP.

In the presence of O_2, pyruvate is oxidized further during the citric acid cycle and oxidative phosphorylation, which occur in a special compartment of the cell called the *mitochondrion*. The citric acid cycle requires eight enzymes that couple the synthesis of ATP to the oxidation of pyruvate by NAD^+ and flavin adenine dinucleotide (FAD, **3**):

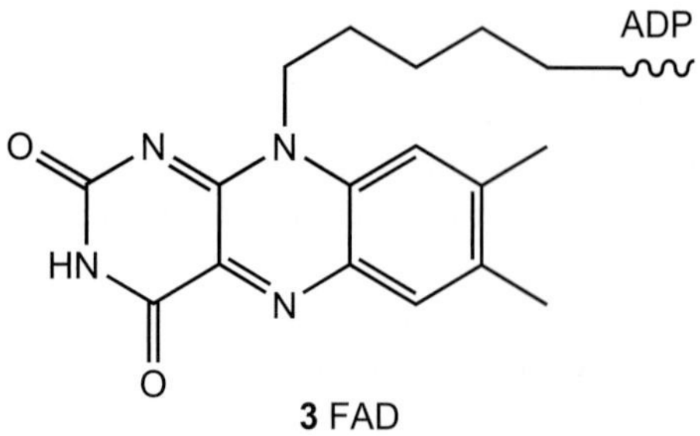

3 FAD

$$\mathrm{2\,CH_3COCO_2^-(aq) + 8\,NAD^+(aq) + 2\,FAD(aq) + 2\,ADP(aq)}$$
$$\mathrm{+ 2\,P_i^-(aq) + 8\,H_2O(l) \rightarrow 6\,CO_2(g) + 8\,NADH(aq)}$$
$$\mathrm{+ 4\,H_3O^+(aq) + 2\,FADH_2(aq) + 2\,ATP(aq)}$$

The NADH and $FADH_2$ go on to reduce O_2 during oxidative phosphorylation, which also produces ATP. The citric acid cycle and oxidative phosphorylation generate as many as 38 ATP molecules for each glucose molecule consumed. Each mole of ATP molecules extracts 31 kJ from the 2880 kJ supplied by 1 mol $C_6H_{12}O_6$ (180 g of glucose), so 1178 kJ is stored for later use.

17.3 The statistical description of equilibrium

The Gibbs energy of a gas of independent molecules is given by eqn 15.40 in terms of the molar partition function, $q_m = q/n$. The equilibrium constant K of a reaction is related to the standard Gibbs energy of reaction by eqn 17.13 ($\Delta_r G^{\ominus} = -RT \ln K$). It follows that we can combine these two equations to calculate the equilibrium constant. We shall consider gas-phase reactions in which the equilibrium constant is expressed in terms of the partial pressures of the reactants and products.

(a) The relation between *K* and the partition function

To find an expression for the standard reaction Gibbs energy we need expressions for the standard molar Gibbs energies, $G^{\ominus}/n$, of each species. For these expressions, we need the value of the molar partition function when $p = p^{\ominus}$ (where $p^{\ominus} = 1$ bar): we denote this **standard molar partition function** $q_m^{\ominus}$. Because only the translational component depends on the pressure, we can find $q_m^{\ominus}$ by evaluating the partition function with V replaced by $V_m^{\ominus}$, where $V_m^{\ominus} = RT/p^{\ominus}$. For a species J it follows that (eqn 15.40)

$$G_m^{\ominus}(\mathrm{J}) = G_m^{\ominus}(\mathrm{J},0) - RT \ln \frac{q_{\mathrm{J},m}^{\ominus}}{N_A} \qquad (17.16)^{\circ}$$

where $G_m^{\ominus}(\mathrm{J},0)$ is the value of $G_m^{\ominus}(\mathrm{J})$ at $T = 0$ and $q_{\mathrm{J,m}}^{\ominus}$ is the standard molar partition function of J. By combining expressions like this one (as shown in the following *Justification*), the equilibrium constant for the reaction

$$a\,\mathrm{A} + b\,\mathrm{B} \rightarrow c\,\mathrm{C} + d\,\mathrm{D}$$

is given by the expression

$$K = \frac{(q_{\mathrm{C,m}}^{\ominus}/N_A)^c (q_{\mathrm{D,m}}^{\ominus}/N_A)^d}{(q_{\mathrm{A,m}}^{\ominus}/N_A)^a (q_{\mathrm{B,m}}^{\ominus}/N_A)^b} e^{-\Delta_r E_0/RT} \qquad (17.17a)$$

where $\Delta_r E_0$ is the difference in molar energies of the ground states of the products and reactants (this term is defined more precisely in the *Justification*), and is calculated from the bond dissociation energies of the species (Fig. 17.4). In terms of the stoichiometric numbers, we would write

$$K = \left\{ \prod_{\mathrm{J}} \left(\frac{q_{\mathrm{J,m}}^{\ominus}}{N_A} \right)^{\nu_{\mathrm{J}}} \right\} e^{-\Delta_r E_0/RT} \qquad (17.17b)$$

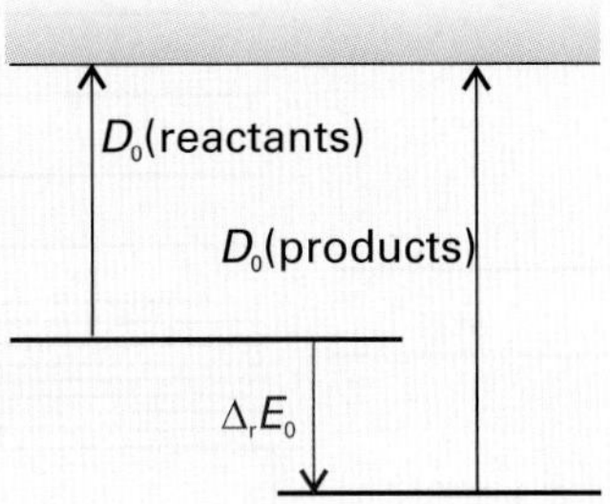

Fig. 17.4 The definition of $\Delta_r E_0$ for the calculation of equilibrium constants.

Justification 17.2 *The equilibrium constant in terms of the partition function 1*

The standard molar reaction Gibbs energy for the reaction is

$$\Delta_r G^\ominus = cG_m^\ominus(C) + dG_m^\ominus(D) - aG_m^\ominus(A) - bG_m^\ominus(B)$$
$$= cG_m^\ominus(C,0) + dG_m^\ominus(D,0) - aG_m^\ominus(A,0) - bG_m^\ominus(B,0)$$
$$- RT\left\{c\ln\frac{q_{C,m}^\ominus}{N_A} + d\ln\frac{q_{D,m}^\ominus}{N_A} - a\ln\frac{q_{A,m}^\ominus}{N_A} - b\ln\frac{q_{B,m}^\ominus}{N_A}\right\}$$

Because $G(0) = U(0)$, the first term on the right is

$$\Delta_r E_0 = cU_m^\ominus(C,0) + dU_m^\ominus(D,0) - aU_m^\ominus(A,0) - bU_m^\ominus(B,0) \quad (17.18a)$$

the reaction internal energy at $T = 0$ (a molar quantity). At $T = 0$ only the ground states of the species are accessible, so $\Delta_r E_0$ is the weighted difference between the molar ground-state energies:

$$\Delta_r E_0 = cE_m(C) + dE_m(D) - aE_m(A) - bE_m(B) \quad (17.18b)$$

Now we can write

$$\Delta_r G^\ominus = \Delta_r E_0 - RT\left\{\ln\left(\frac{q_{C,m}^\ominus}{N_A}\right)^c + \ln\left(\frac{q_{D,m}^\ominus}{N_A}\right)^d - \ln\left(\frac{q_{A,m}^\ominus}{N_A}\right)^a - \ln\left(\frac{q_{B,m}^\ominus}{N_A}\right)^b\right\}$$
$$= \Delta_r E_0 - RT\ln\frac{(q_{C,m}^\ominus/N_A)^c(q_{D,m}^\ominus/N_A)^d}{(q_{A,m}^\ominus/N_A)^a(q_{B,m}^\ominus/N_A)^b}$$
$$= -RT\left\{\frac{-\Delta_r E_0}{RT} + \ln\frac{(q_{C,m}^\ominus/N_A)^c(q_{D,m}^\ominus/N_A)^d}{(q_{A,m}^\ominus/N_A)^a(q_{B,m}^\ominus/N_A)^b}\right\}$$

At this stage we can pick out an expression for K by comparing this equation with $\Delta_r G^\ominus = -RT\ln K$, which gives

$$\ln K = -\frac{\Delta_r E_0}{RT} + \ln\frac{(q_{C,m}^\ominus/N_A)^c(q_{D,m}^\ominus/N_A)^d}{(q_{A,m}^\ominus/N_A)^a(q_{B,m}^\ominus/N_A)^b}$$

This expression is easily rearranged into eqn 17.17a by forming the exponential of both sides.

We shall illustrate the application of eqn 17.17 to an equilibrium in which a diatomic molecule X_2 dissociates into its atoms:

$$X_2(g) \rightleftharpoons 2\,X(g) \qquad K = \frac{p_X^2}{p_{X_2}p^\ominus}$$

According to eqn 17.17 (with $a = 1$, $b = 0$, $c = 2$, and $d = 0$):

$$K = \frac{(q_{X,m}^\ominus/N_A)^2}{q_{X_2,m}^\ominus/N_A}e^{-\Delta_r E_0/RT} = \frac{(q_{X,m}^\ominus)^2}{q_{X_2,m}^\ominus N_A}e^{-\Delta_r E_0/RT} \quad (17.19a)$$

with

$$\Delta_r E_0 = 2E_m(X) - E_m(X_2) = D_0(X{-}X) \quad (17.19b)$$

where $D_0(X{-}X)$ is the dissociation energy of the X—X bond. The standard molar partition functions of the atoms X are

$$q_{X,m}^\ominus = g_X\left(\frac{V_m^\ominus}{\Lambda_X^3}\right) = \frac{RTg_X}{p^\ominus\Lambda_X^3}$$

where g_X is the degeneracy of the electronic ground state of X and we have used $V_m^\ominus = RT/p^\ominus$. The diatomic molecule X_2 also has rotational and vibrational degrees of freedom, so its standard molar partition function is

$$q_{X_2,m}^\ominus = g_{X_2}\left(\frac{V_m^\ominus}{\Lambda_{X_2}^3}\right)q_{X_2}^R q_{X_2}^V = \frac{RTg_{X_2}q_{X_2}^R q_{X_2}^V}{p^\ominus\Lambda_{X_2}^3}$$

where g_{X_2} is the degeneracy of the electronic ground state of X_2. It follows from eqn 17.19 that the equilibrium constant is

$$K = \frac{kTg_X^2\Lambda_{X_2}^3}{p^\ominus g_{X_2}q_{X_2}^R q_{X_2}^V\Lambda_X^6}e^{-D_0/RT} \quad (17.20)$$

where we have used $R/N_A = k$. All the quantities in this expression can be calculated from spectroscopic data. Expressions for the Λs, the thermal wavelengths of the species, are collected in Table 13.1 and depend on the masses of the species and the temperature; the expressions for the rotational and vibrational partition functions are also available in Table 13.1 and depend on the rotational constant and vibrational wavenumber of the molecule.

Example 17.2 *Calculating an equilibrium constant 2*

Evaluate the equilibrium constant for the dissociation $Na_2(g) \rightleftharpoons 2\,Na(g)$ at 1000 K from the following data: $\tilde{B} = 0.1547$ cm^{-1}, $\tilde{\nu} = 159.2$ cm^{-1}, $D_0 = 70.4$ kJ mol^{-1}. The Na atoms have doublet ground terms.

Method The partition functions required are specified in eqn 17.20. They are evaluated by using the expressions in Table 13.1. For a homonuclear diatomic molecule, $\sigma = 2$. In the evaluation of $kT/p^\ominus$ use $p^\ominus = 10^5$ Pa and 1 Pa m^3 = 1 J.

Answer The partition functions and other quantities required are as follows:

$\Lambda(\mathrm{Na_2}) = 8.14$ pm $\quad \Lambda(\mathrm{Na}) = 11.5$ pm
$q^{\mathrm{R}}(\mathrm{Na_2}) = 2246 \quad q^{\mathrm{V}}(\mathrm{Na_2}) = 4.885$
$g(\mathrm{Na}) = 2 \quad g(\mathrm{Na_2}) = 1$

Then, from eqn 17.20,

$$K = \frac{(1.38\times10^{-23}\,\mathrm{J\,K^{-1}})\times(1000\,\mathrm{K})\times4\times(8.14\times10^{-12}\,\mathrm{m})^3}{(10^5\,\mathrm{Pa})\times2246\times4.885\times(1.15\times10^{-11}\,\mathrm{m})^6}\times e^{-8.47}$$

$$= 2.46$$

where we have used $1\ \mathrm{J} = 1\ \mathrm{kg\ m^2\ s^{-2}}$ and $1\ \mathrm{Pa} = 1\ \mathrm{kg\ m^{-1}\ s^{-1}}$.

Self-test 17.3 Evaluate K at 1500 K. [52]

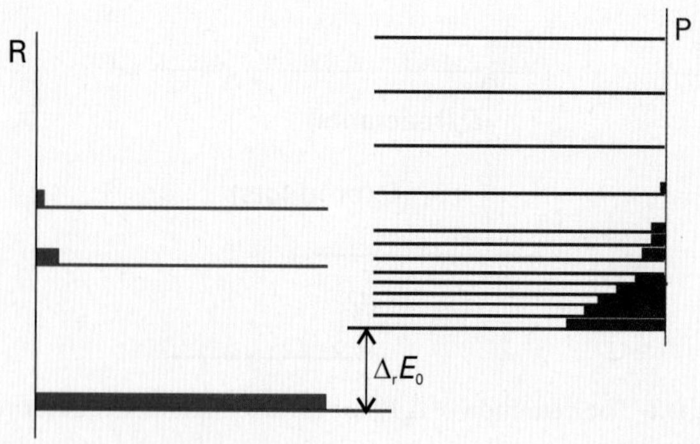

Fig. 17.6 It is important to take into account the densities of states of the molecules. Even though P might lie well above R in energy (that is, ΔE_0 is large and positive), P might have so many states that its total population dominates in the mixture. In classical thermodynamic terms, we have to take entropies into account as well as enthalpies when considering equilibria.

(b) Contributions to the equilibrium constant

We are now in a position to appreciate the physical basis of equilibrium constants. To see what is involved, consider a simple R ⇌ P gas-phase equilibrium (R for reactants, P for products).

Figure 17.5 shows two sets of energy levels; one set of states belongs to R, and the other belongs to P. The populations of the states are given by the Boltzmann distribution, and are independent of whether any given state happens to belong to R or to P. We can therefore imagine a single Boltzmann distribution spreading, without distinction, over the two sets of states. If the spacings of R and P are similar (as in Fig. 17.5), and P lies above R, the diagram indicates that R will dominate in the equilibrium mixture. However, if P has a high density of states (a large number of states in a given energy range, as in Fig. 17.6) then, even though its zero-point energy lies above that of R, the species P might still dominate at equilibrium.

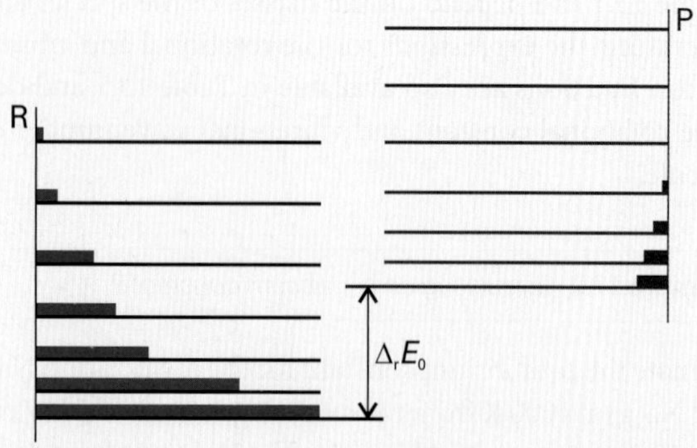

Fig. 17.5 The array of R(eactants) and P(roducts) energy levels. At equilibrium all are accessible (to differing extents, depending on the temperature), and the equilibrium composition of the system reflects the overall Boltzmann distribution of populations. As ΔE_0 increases, R becomes dominant.

It is quite easy to show (see the *Justification* below) that the ratio of numbers of R and P molecules at equilibrium is given by

$$\frac{N_{\mathrm{P}}}{N_{\mathrm{R}}} = \frac{q_{\mathrm{P}}}{q_{\mathrm{R}}} e^{-\Delta_r E_0/RT} \qquad (17.21a)$$

and therefore that the equilibrium constant for the reaction is

$$K = \frac{q_{\mathrm{P}}}{q_{\mathrm{R}}} e^{-\Delta_r E_0/RT} \qquad (17.21b)$$

just as would be obtained from eqn 17.17. For an R ⇌ P equilibrium, the V factors in the partition functions cancel, so the appearance of q in place of $q^{\ominus}$ has no effect. In the case of a more general reaction, the conversion from q to $q^{\ominus}$ comes about at the stage of converting the pressures that occur in K to numbers of molecules.

Justification 17.3 *The equilibrium constant in terms of the partition function 2*

The population in a state i of the composite (R,P) system is

$$n_i = \frac{Ne^{-\beta\varepsilon_i}}{q}$$

where N is the total number of molecules. The total number of R molecules is the sum of these populations taken over the states belonging to R; these states we label r with energies ε_r. The total number of P molecules is the sum over the states belonging to P; these states we label P with energies ε'_p (the prime is explained in a moment):

$$N_{\mathrm{R}} = \sum_r n_r = \frac{N}{q}\sum_r e^{-\beta\varepsilon_r} \qquad N_{\mathrm{P}} = \sum_p n_p = \frac{N}{q}\sum_p e^{-\beta\varepsilon'_p}$$

The sum over the states of R is its partition function, q_R, so

$$N_R = \frac{Nq_R}{q}$$

The sum over the states of P is also a partition function, but the energies are measured from the ground state of the combined system, which is the ground state of R. However, because $\varepsilon'_p = \varepsilon_p + \Delta\varepsilon_0$, where $\Delta\varepsilon_0$ is the separation of zero-point energies (as in Fig. 17.5),

$$N_P = \frac{N}{q}\sum_p e^{-\beta(\varepsilon_p+\Delta\varepsilon_0)} = \frac{N}{q}\left(\sum_p e^{-\beta\varepsilon_p}\right)e^{-\beta\Delta\varepsilon_0} = \frac{Nq_P}{q}e^{-\Delta_r E_0/RT}$$

The switch from $\Delta\varepsilon_0/k$ to $\Delta_r E_0/R$ in the last step is the conversion of molecular energies to molar energies: $E_0 = N_A\varepsilon_0$ and $R = N_A k$.

The equilibrium constant of the R $\rightleftharpoons$ P reaction is proportional to the ratio of the numbers of the two types of molecule. Therefore,

$$K = \frac{N_P}{N_R} = \frac{q_P}{q_R}e^{-\Delta_r E_0/RT}$$

as in eqn 17.21b.

The content of eqn 17.21 can be seen most clearly by exaggerating the molecular features that contribute to it. We shall suppose that R has only a single accessible level, which implies that $q_R = 1$. We also suppose that P has a large number of evenly, closely spaced levels (Fig. 17.7). The partition function of P is then $q_P = kT/\varepsilon$. In this model system, the equilibrium constant is

$$K = \frac{kT}{\varepsilon}e^{-\Delta_r E_0/RT} \tag{17.22}$$

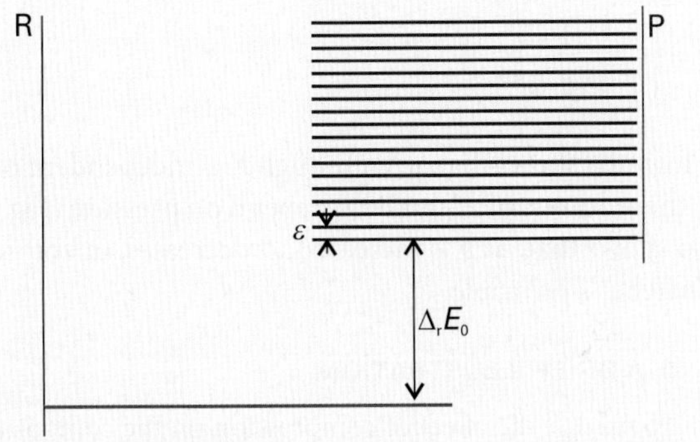

Fig. 17.7 The model used in the text for exploring the effects of energy separations and densities of states on equilibria. The products P can dominate provided ΔE_0 is not too large and P has an appreciable density of states.

When $\Delta_r E_0$ is very large, the exponential term dominates and $K \ll 1$, which implies that very little P is present at equilibrium. When $\Delta_r E_0$ is small but still positive, K can exceed 1 because the factor kT/ε may be large enough to overcome the small size of the exponential term. The size of K then reflects the predominance of P at equilibrium on account of its high density of states.

The model also shows why the Gibbs energy, G, and not just the enthalpy, determines the position of equilibrium. It shows that the density of states (and hence the entropy) of each species as well as their relative energies controls the distribution of populations and hence the value of the equilibrium constant. This competition is mirrored in eqn 17.13, as can be seen most clearly by using $\Delta_r G^\ominus = \Delta_r H^\ominus - T\Delta_r S^\ominus$ and writing it in the form

$$K = e^{-\Delta_r H^\ominus/RT}e^{\Delta_r S^\ominus/R} \tag{17.23}$$

Note that a positive reaction enthalpy results in a lowering of the equilibrium constant (that is, an endothermic reaction can be expected to have an equilibrium composition that favours the reactants). However, if there is positive reaction entropy, then the equilibrium composition may favour products, despite the endothermic character of the reaction.

The response of equilibria to the conditions

Equilibria respond to changes in pressure, temperature, and concentrations of reactants and products. The equilibrium constant for a reaction is not affected by the presence of a catalyst or an enzyme (a biological catalyst). As we shall see in detail in Chapter 21, catalysts increase the rate at which equilibrium is attained but do not affect its position. However, it is important to note that in industry reactions rarely reach equilibrium, partly on account of the rates at which reactants mix.

17.4 How equilibria respond to pressure

The equilibrium constant depends on the value of $\Delta_r G^\ominus$, which is defined at a single, standard pressure. The value of $\Delta_r G^\ominus$, and hence of K, is therefore independent of the pressure at which the equilibrium is actually established. Formally we may express this independence as

$$\left(\frac{\partial K}{\partial p}\right)_T = 0 \tag{17.24}$$

The conclusion that K is independent of pressure does not necessarily mean that the equilibrium composition is independent of the pressure, and its effect depends on how the pressure is applied. The pressure within a reaction vessel can be increased by injecting an inert gas into it. However, so long as the gases are perfect, this addition of gas leaves all the partial pressures of the reacting gases unchanged: the partial pressure of a perfect gas is the pressure it would exert if it were alone in the container, so the presence of another gas has no effect. It follows that pressurization by the addition of an inert gas has no effect on the equilibrium composition of the system (provided the gases are perfect). Alternatively, the pressure of the system may be increased by confining the gases to a smaller volume (that is, by compression). Now the individual partial pressures are changed but their ratio (as it appears in the equilibrium constant) remains the same. Consider, for instance, the perfect gas equilibrium A $\rightleftharpoons$ 2 B, for which the equilibrium constant is

$$K = \frac{p_B^2}{p_A p^\ominus}$$

The right-hand side of this expression remains constant only if an increase in p_A cancels an increase in the *square* of p_B. This relatively steep increase of p_A compared to p_B will occur if the equilibrium composition shifts in favour of A at the expense of B. Then the number of A molecules will increase as the volume of the container is decreased and its partial pressure will rise more rapidly than can be ascribed to a simple change in volume alone (Fig. 17.8).

The increase in the number of A molecules and the corresponding decrease in the number of B molecules in the equilibrium A $\rightleftharpoons$ 2 B is a special case of a principle proposed by the French chemist Henri Le Chatelier. **Le Chatelier's principle** states that:

> A system at equilibrium, when subjected to a disturbance, responds in a way that tends to minimize the effect of the disturbance.

The principle implies that, if a system at equilibrium is compressed, then the reaction will adjust so as to minimize the increase in pressure. This it can do by reducing the number of particles in the gas phase, which implies a shift A $\leftarrow$ 2 B.

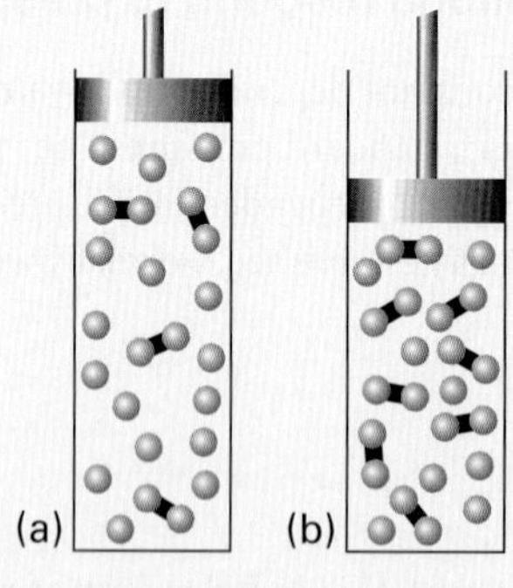

Fig. 17.8 When a reaction at equilibrium is compressed (from *a* to *b*), the reaction responds by reducing the number of molecules in the gas phase (in this case by producing the dimers represented by the linked spheres).

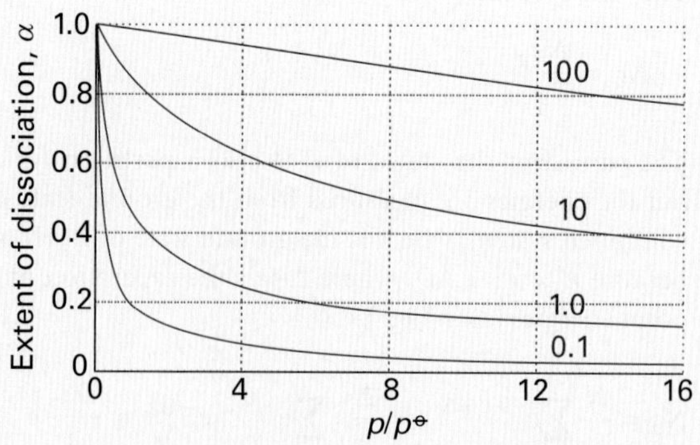

Fig. 17.9 The pressure dependence of the degree of dissociation, α, at equilibrium for an A(g) $\rightleftharpoons$ 2 B(g) reaction for different values of the equilibrium constant *K*. The value $\alpha = 0$ corresponds to pure A; $\alpha = 1$ corresponds to pure B.

interActivity Plot x_A and x_B against the pressure *p* for several values of the equilibrium constant *K*.

To treat the effect of compression quantitatively, we suppose that there is an amount *n* of A present initially (and no B). At equilibrium the amount of A is $(1 - \alpha)n$ and the amount of B is $2\alpha n$, where α is the extent of dissociation of A into 2B. It follows that the mole fractions present at equilibrium are

$$x_A = \frac{(1-\alpha)n}{(1-\alpha)n + 2\alpha n} = \frac{1-\alpha}{1+\alpha} \qquad x_B = \frac{2\alpha}{1+\alpha}$$

The equilibrium constant for the reaction is

$$K = \frac{p_B^2}{p_A p^\ominus} = \frac{x_B^2 p^2}{x_A p p^\ominus} = \frac{4\alpha^2(p/p^\ominus)}{1-\alpha^2}$$

which rearranges to

$$\alpha = \left(\frac{1}{1 + 4p/Kp^\ominus}\right)^{1/2} \qquad (17.25)$$

This formula shows that, even though *K* is independent of pressure, the amounts of A and B do depend on pressure (Fig. 17.9). It also shows that, as *p* is increased, α decreases, in accord with Le Chatelier's principle.

● A BRIEF ILLUSTRATION

To predict the effect of compression on the composition of the ammonia synthesis at equilibrium, $N_2(g) + 3\,H_2(g) \rightleftharpoons 2\,NH_3(g)$, we note that the number of gas molecules decreases (from 4 to 2). So, Le Chatelier's principle predicts that an increase in pressure will favour the product. The equilibrium constant, treating all gases as perfect, is

$$K = \frac{p_{NH_3}^2 p^{\ominus 2}}{p_{N_2} p_{H_2}^3} = \frac{x_{NH_3}^2 p^2 p^{\ominus 2}}{x_{N_2} x_{H_2}^3 p^4} = \frac{K_x p^{\ominus 2}}{p^2}$$

where K_x is the part of the equilibrium constant expression that contains the equilibrium mole fractions of reactants and products (note that, unlike K itself, K_x is not an equilibrium constant). Therefore, doubling the pressure must increase K_x by a factor of 4 to preserve the value of K. ●

Self-test 17.4 Predict the effect of a compression that results in a tenfold pressure increase on the equilibrium composition of the reaction $3\,N_2(g) + H_2(g) \rightleftharpoons 2\,N_3H(g)$.
[100-fold increase in K_x]

17.5 The response of equilibria to temperature

Le Chatelier's principle predicts that a system at equilibrium will tend to shift in the endothermic direction if the temperature is raised, for then energy is absorbed as heat and the rise in temperature is opposed. Conversely, an equilibrium can be expected to shift in the exothermic direction if the temperature is lowered, for then energy is released and the reduction in temperature is opposed. These conclusions can be summarized as follows:

Exothermic reactions ($\Delta_r H^\ominus < 0$): increased temperature favours the reactants.

Endothermic reactions ($\Delta_r H^\ominus > 0$): increased temperature favours the products.

We shall now justify these remarks and see how to express the changes quantitatively.

(a) The van 't Hoff equation

The **van 't Hoff equation**, which is derived in the *Justification* below, is an expression for the slope of a plot of the equilibrium constant (specifically, ln K) as a function of temperature. It may be expressed in either of two ways:

$$\text{(a)}\quad \frac{\mathrm{d}\ln K}{\mathrm{d}T} = \frac{\Delta_r H^\ominus}{RT^2} \qquad \text{(b)}\quad \frac{\mathrm{d}\ln K}{\mathrm{d}(1/T)} = -\frac{\Delta_r H^\ominus}{R} \qquad (17.26)$$

Justification 17.4 *The van't Hoff equation*

From eqn 17.13, we know that

$$\ln K = -\frac{\Delta_r G^\ominus}{RT}$$

Differentiation of ln K with respect to temperature then gives

$$\frac{\mathrm{d}\ln K}{\mathrm{d}T} = -\frac{1}{R}\frac{\mathrm{d}(\Delta_r G^\ominus/T)}{\mathrm{d}T}$$

The differentials are complete because K and $\Delta_r G^\ominus$ depend only on temperature, not on pressure. To develop this equation we use the Gibbs–Helmholtz equation (eqn 15.63) in the form

$$\frac{\mathrm{d}(\Delta_r G^\ominus/T)}{\mathrm{d}T} = -\frac{\Delta_r H^\ominus}{T^2}$$

where $\Delta_r H^\ominus$ is the standard reaction enthalpy at the temperature T. Combining the two equations gives the van't Hoff equation, eqn 17.26a. The second form of the equation is obtained by noting that

$$\frac{\mathrm{d}(1/T)}{\mathrm{d}T} = -\frac{1}{T^2}, \qquad \text{so } \mathrm{d}T = -T^2\mathrm{d}(1/T)$$

It follows that eqn 17.26a can be rewritten as

$$-\frac{\mathrm{d}\ln K}{T^2\mathrm{d}(1/T)} = \frac{\Delta_r H^\ominus}{RT^2}$$

which simplifies into eqn 17.26b.

Equation 17.26a shows that $\mathrm{d}\ln K/\mathrm{d}T < 0$ (and therefore that $\mathrm{d}K/\mathrm{d}T < 0$) for an exothermic reaction ($\Delta_r H^\ominus < 0$). A negative slope means that ln K, and therefore K itself, decreases as the temperature rises. Therefore, as asserted above, in the case of an exothermic reaction the equilibrium shifts away from products. The opposite occurs in the case of endothermic reactions.

Some insight into the thermodynamic basis of this behaviour comes from the expression $\Delta_r G^\ominus = \Delta_r H^\ominus - T\Delta_r S^\ominus$ written in the form $-\Delta_r G^\ominus/T = -\Delta_r H^\ominus/T + \Delta_r S^\ominus$. When the reaction is exothermic, $-\Delta_r H^\ominus/T$ corresponds to a positive change of entropy of the surroundings and favours the formation of products. When the temperature is raised, $-\Delta_r H^\ominus/T$ decreases, and the increasing entropy of the surroundings has a less important role. As a result, the equilibrium lies less to the right. When the reaction is endothermic, the principal factor is the increasing entropy of the reaction system. The importance of the unfavourable change of entropy of the surroundings is reduced if the temperature is raised (because then $\Delta_r H^\ominus/T$ is smaller), and the reaction is able to shift towards products.

From a molecular perspective, consider the typical arrangement of energy levels for an endothermic reaction as shown in Fig. 17.10a. When the temperature is increased, the Boltzmann distribution adjusts and the populations change as shown. The change corresponds to an increased population of the higher energy states at the expense of the population of the lower energy states. We see that the states that arise from the B molecules become more populated at the expense of the A molecules. Therefore, the total population of B states increases, and B becomes more abundant in the equilibrium mixture. Conversely, if the reaction is exothermic (Fig. 17.10b), then an increase in temperature increases the population of the A states (which start

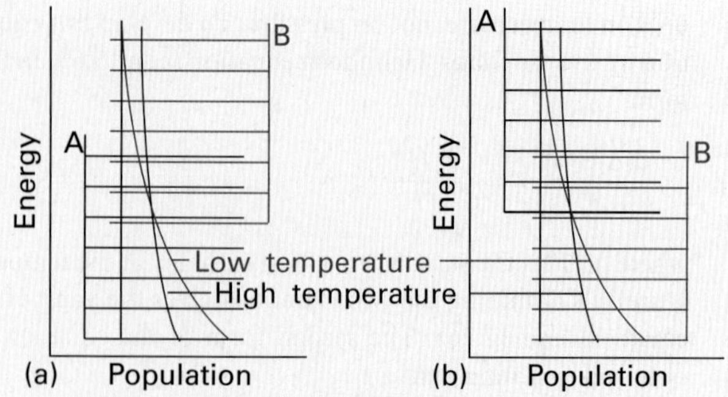

Fig. 17.10 The effect of temperature on a chemical equilibrium can be interpreted in terms of the change in the Boltzmann distribution with temperature and the effect of that change in the population of the species. (a) In an endothermic reaction, the population of B increases at the expense of A as the temperature is raised. (b) In an exothermic reaction, the opposite happens.

at higher energy) at the expense of the B states, so the reactants become more abundant.

Example 17.3 *Measuring a reaction enthalpy*

The data below show the temperature variation of the equilibrium constant of the reaction $Ag_2CO_3(s) \rightleftharpoons Ag_2O(s) + CO_2(g)$. Calculate the standard reaction enthalpy of the decomposition.

T/K	350	400	450	500
K	3.98×10^{-4}	1.41×10^{-2}	1.86×10^{-1}	1.48

Method It follows from eqn 17.26b that, provided the reaction enthalpy can be assumed to be independent of temperature, a plot of $-\ln K$ against $1/T$ should be a straight line of slope $\Delta_r H^{\ominus}/R$.

Answer We draw up the following table:

T/K	350	400	450	500
$(10^3\ \text{K})/T$	2.86	2.50	2.22	2.00
$-\ln K$	7.83	4.26	1.68	-0.39

These points are plotted in Fig. 17.11. The slope of the graph is $+9.6\times10^3$ K, so

$$\Delta_r H^{\ominus} = (+9.6\times10^3\ \text{K})\times R = +80\ \text{kJ mol}^{-1}$$

The temperature dependence of the equilibrium constant provides a non-calorimetric method of determining $\Delta_r H^{\ominus}$. A drawback is that the reaction enthalpy is actually temperature dependent, so the plot is not expected to be perfectly linear. However, the temperature dependence is weak in many cases, so the plot is reasonably straight. In practice, the method is not very accurate, but it is often the only method available.

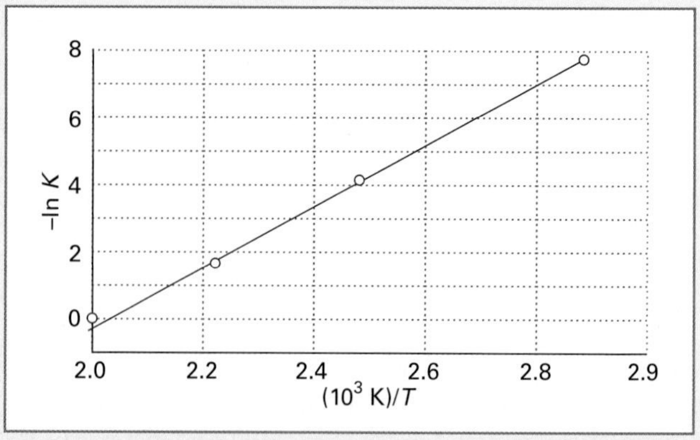

Fig. 17.11 When $-\ln K$ is plotted against $1/T$, a straight line is expected with slope equal to $\Delta_r H^{\ominus}/R$ if the standard reaction enthalpy does not vary appreciably with temperature. This is a non-calorimetric method for the measurement of reaction enthalpies.

interActivity The equilibrium constant of a reaction is found to fit the expression $\ln K = a + b/(T/\text{K}) + c/(T/\text{K})^3$ over a range of temperatures. (a) Write expressions for $\Delta_r H^{\ominus}$ and $\Delta_r S^{\ominus}$. (b) Plot $\ln K$ against T between 400 K and 600 K for $a = -2.0$, $b = -1.0\times10^3$, and $c = 2.0\times10^7$.

Self-test 17.5 The equilibrium constant of the reaction $2\,SO_2(g) + O_2(g) \rightleftharpoons 2\,SO_3(g)$ is 4.0×10^{24} at 300 K, 2.5×10^{10} at 500 K, and 3.0×10^4 at 700 K. Estimate the reaction enthalpy at 500 K. [-200 kJ mol^{-1}]

(b) The value of *K* at different temperatures

To find the value of the equilibrium constant at a temperature T_2 in terms of its value K_1 at another temperature T_1, we integrate eqn 17.26b between these two temperatures:

$$\ln K_2 - \ln K_1 = -\frac{1}{R}\int_{1/T_1}^{1/T_2} \Delta_r H^{\ominus}\,\mathrm{d}(1/T) \qquad (17.27)$$

If we suppose that $\Delta_r H^{\ominus}$ varies only slightly with temperature over the temperature range of interest, then we may take it outside the integral. It follows that

$$\ln K_2 - \ln K_1 = -\frac{\Delta_r H^{\ominus}}{R}\left(\frac{1}{T_2} - \frac{1}{T_1}\right) \qquad (17.28)$$

● A BRIEF ILLUSTRATION

To estimate the equilibrium constant for the synthesis of ammonia at 500 K from its value at 298 K (6.1×10^5 for the reaction $N_2(g) + 3\,H_2(g) \rightarrow 2\,NH_3(g)$) we use the standard reaction enthalpy, which can be obtained from Table 14.6

in the *Data section* by using $\Delta_r H^\ominus = 2\Delta_f H^\ominus(NH_3,g)$, and assume that its value is constant over the range of temperatures. Then, with $\Delta_r H^\ominus = -92.2\ kJ\ mol^{-1}$, from eqn 17.28 we find

$$\ln K_2 = \ln(6.1\times10^5) - \frac{(-92.2\times10^3\ J\ mol^{-1})}{8.3145\ J\ K^{-1}\ mol^{-1}}\left(\frac{1}{500\ K} - \frac{1}{298\ K}\right)$$
$$= -1.71$$

It follows that $K_2 = 0.18$, a lower value than at 298 K, as expected for this exothermic reaction. ●

Self-test 17.6 The equilibrium constant for $N_2O_4(g) \rightleftharpoons 2\ NO_2(g)$ was calculated in Self-test 17.1. Estimate its value at 100°C. [15]

Knowledge of the temperature dependence of the equilibrium constant for a reaction can be useful in the design of laboratory and industrial processes. For example, synthetic chemists can improve the yield of a reaction by changing the temperature of the reaction mixture. Also, reduction of a metal oxide with carbon or carbon monoxide results in the extraction of the metal when the process is carried out at a temperature for which $K \gg 1$.

Electrochemistry

We shall now see how the foregoing ideas, with certain changes of technical detail, can be used to describe the equilibrium properties of reactions taking place in electrochemical cells. The ability to make very precise measurements of currents and potential differences ('voltages') means that electrochemical methods can be used to determine thermodynamic properties of reactions that may be inaccessible by other methods.

An **electrochemical cell** consists of two **electrodes**, or metallic conductors, in contact with an **electrolyte**, an ionic conductor (which may be a solution, a liquid, or a solid). An electrode and its electrolyte comprise an **electrode compartment**. The two electrodes may share the same compartment. The various kinds of electrode are summarized in Table 17.1. Any 'inert metal' shown as part of the specification is present to act as a source or sink of electrons, but takes no other part in the reaction other than acting as a catalyst for it. If the electrolytes are different, the two compartments may be joined by a **salt bridge**, which is a tube containing a concentrated electrolyte solution (almost always potassium chloride in agar jelly) that completes the electrical circuit and enables the cell to function. A **galvanic cell** is an electrochemical cell that produces electricity as a result of the spontaneous reaction occurring inside it. An **electrolytic cell** is an electrochemical cell in which a nonspontaneous reaction is driven by an external source of current.

17.6 Half-reactions and electrodes

It will be familiar from introductory chemistry courses that **oxidation** is the removal of electrons from a species, a **reduction** is the addition of electrons to a species, and a **redox reaction** is a reaction in which there is a transfer of electrons from one species to another. The electron transfer may be accompanied by other events, such as atom or ion transfer, but the net effect is electron transfer and hence a change in oxidation number of an element. The **reducing agent** (or 'reductant') is the electron donor; the **oxidizing agent** (or 'oxidant') is the electron acceptor. It should also be familiar that any redox reaction may be expressed as the difference of two reduction **half-reactions**, which are conceptual reactions showing the gain of electrons. Even reactions that are not redox reactions may often be expressed as the difference of two reduction half-reactions. The reduced and oxidized species in a half-reaction form a **redox couple**. In general we write a couple as Ox/Red and the corresponding reduction half-reaction as

$$\text{Ox} + \nu e^- \rightarrow \text{Red} \qquad (17.29)$$

● A BRIEF ILLUSTRATION

The dissolution of silver chloride in water $AgCl(s) \rightarrow Ag^+(aq) + Cl^-(aq)$, which is not a redox reaction, can be written as the difference of the following two reduction half-reactions:

$$AgCl(s) + e^- \rightarrow Ag(s) + Cl^-(aq)$$
$$Ag^+(aq) + e^- \rightarrow Ag(s)$$

The redox couples are $AgCl/Ag,Cl^-$ and Ag^+/Ag, respectively. ●

Table 17.1 Varieties of electrode

Electrode type	Designation	Redox couple	Half-reaction
Metal/metal ion	$M(s)\vert M^+(aq)$	M^+/M	$M^+(aq) + e^- \rightarrow M(s)$
Gas	$Pt(s)\vert X_2(g)\vert X^+(aq)$	X^+/X_2	$X^+(aq) + e^- \rightarrow \frac{1}{2}X_2(g)$
	$Pt(s)\vert X_2(g)\vert X^-(aq)$	X_2/X^-	$\frac{1}{2}X_2(g) + e^- \rightarrow X^-(aq)$
Metal/insoluble salt	$M(s)\vert MX(s)\vert X^-(aq)$	$MX/M,X^-$	$MX(s) + e^- \rightarrow M(s) + X^-(aq)$
Redox	$Pt(s)\vert M^+(aq),M^{2+}(aq)$	M^{2+}/M^+	$M^{2+}(aq) + e^- \rightarrow M^+(aq)$

Self-test 17.7 Express the formation of H_2O from H_2 and O_2 in acidic solution (a redox reaction) as the difference of two reduction half-reactions.

$$[4\,H^+(aq) + 4\,e^- \rightarrow 2\,H_2(g),\; O_2(g) + 4\,H^+(aq) + 4\,e^- \rightarrow 2\,H_2O(l)]$$

We shall often find it useful to express the composition of an electrode compartment in terms of the reaction quotient, Q, for the half-reaction. This quotient is defined like the reaction quotient for the overall reaction, but the electrons are ignored.

● A BRIEF ILLUSTRATION

The reaction quotient for the reduction of O_2 to H_2O in acid solution, $O_2(g) + 4\,H^+(aq) + 4\,e^- \rightarrow 2\,H_2O(l)$, is

$$Q = \frac{a_{H_2O}^2}{a_{H^+}^4 a_{O_2}} \approx \frac{p^{\ominus}}{a_{H^+}^4 p_{O_2}}$$

The approximations used in the second step are that the activity of water is 1 (because the solution is dilute) and the oxygen behaves as a perfect gas, so $a_{O_2} \approx p_{O_2}/p^{\ominus}$. ●

Self-test 17.8 Write the half-reaction and the reaction quotient for a chlorine gas electrode.

$$[Cl_2(g) + 2\,e^- \rightarrow 2\,Cl^-(aq),\; Q = a_{Cl^-}^2/a_{Cl_2} \approx a_{Cl^-}^2 p^{\ominus}/p_{Cl_2}]$$

The reduction and oxidation processes responsible for the overall reaction in a cell are separated in space: oxidation takes place at one electrode and reduction takes place at the other. As the reaction proceeds, the electrons released in the oxidation $Red_1 \rightarrow Ox_1 + \nu e^-$ at one electrode travel through the external circuit and re-enter the cell through the other electrode. There they bring about reduction $Ox_2 + \nu e^- \rightarrow Red_2$. The electrode at which oxidation occurs is called the **anode**; the electrode at which reduction occurs is called the **cathode**. In a galvanic cell, the cathode has a higher potential than the anode: the species undergoing reduction, Ox_2, withdraws electrons from its electrode (the cathode, Fig. 17.12), so leaving a relative positive charge on it (corresponding to a high potential). At the anode, oxidation results in the transfer of electrons to the electrode, so giving it a relative negative charge (corresponding to a low potential).

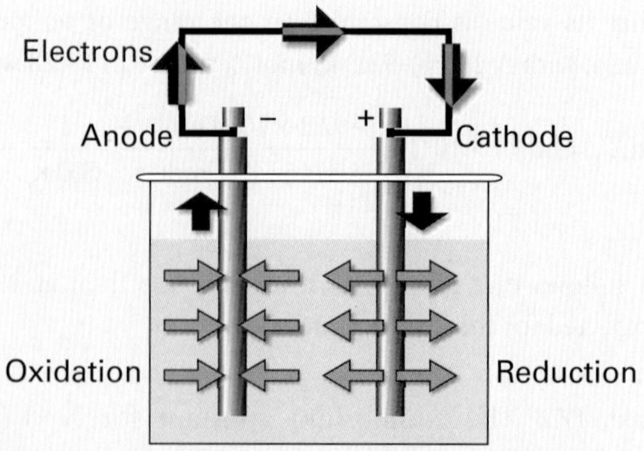

Fig. 17.12 When a spontaneous reaction takes place in a galvanic cell, electrons are deposited in one electrode (the site of oxidation, the anode) and collected from another (the site of reduction, the cathode), and so there is a net flow of current which can be used to do work. Note that the + sign of the cathode can be interpreted as indicating the electrode at which electrons enter the cell, and the – sign of the anode is where the electrons leave the cell.

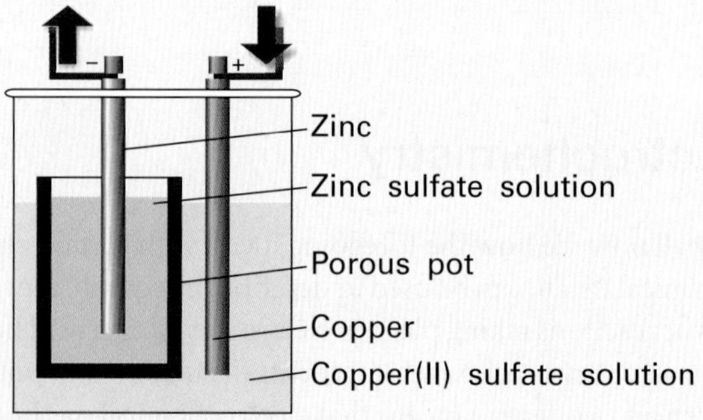

Fig. 17.13 One version of the Daniell cell. The copper electrode is the cathode and the zinc electrode is the anode. Electrons leave the cell from the zinc electrode and enter it again through the copper electrode.

17.7 Varieties of cells

The simplest type of cell has a single electrolyte common to both electrodes (as in Fig. 17.12). In some cases it is necessary to immerse the electrodes in different electrolytes, as in the 'Daniell cell' in which the redox couple at one electrode is Cu^{2+}/Cu and at the other is Zn^{2+}/Zn (Fig. 17.13). In an **electrolyte concentration cell**, the electrode compartments are identical except for the concentrations of the electrolytes. In an **electrode concentration cell**, the electrodes themselves have different concentrations, either because they are gas electrodes operating at different pressures or because they are amalgams (solutions in mercury) with different concentrations.

(a) Liquid junction potentials

In a cell with two different electrolyte solutions in contact, as in the Daniell cell, there is an additional source of potential difference across the interface of the two electrolytes. This potential is called the **liquid junction potential**, E_{lj}. Another example of a junction potential is that between different concentrations of hydrochloric acid. At the junction, the mobile H^+ ions diffuse into the more dilute solution. The bulkier Cl^- ions follow, but initially do so more slowly, which results in a potential difference

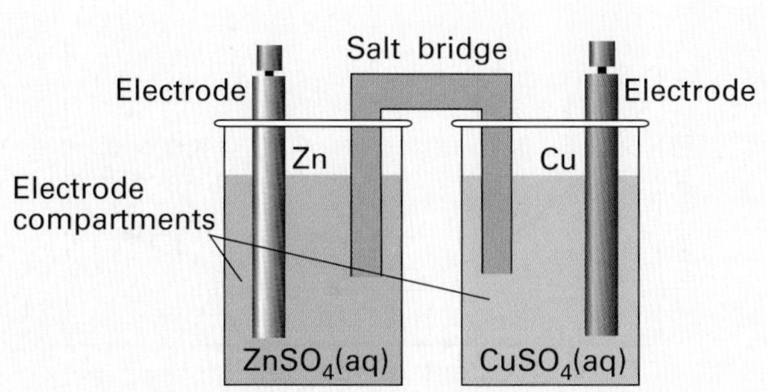

Fig. 17.14 The salt bridge, essentially an inverted U-tube full of concentrated salt solution in a jelly, has two opposing liquid junction potentials that almost cancel.

at the junction. The potential then settles down to a value such that, after that brief initial period, the ions diffuse at the same rates. Electrolyte concentration cells always have a liquid junction; electrode concentration cells do not.

The contribution of the liquid junction to the potential can be reduced (to about 1 to 2 mV) by joining the electrolyte compartments through a salt bridge (Fig. 17.14). The reason for the success of the salt bridge is that the liquid junction potentials at either end are largely independent of the concentrations of the two dilute solutions, and so nearly cancel.

(b) Notation

In the notation for cells, phase boundaries are denoted by a vertical bar. For example, the cell corresponding to the redox reaction $AgCl(s) + \frac{1}{2} H_2(g) \rightarrow Ag(s) + HCl(aq)$ and using a Pt electrode is denoted

$$Pt(s)|H_2(g)|HCl(aq)|AgCl(s)|Ag(s)$$

A liquid junction is denoted by ⁝, so the cell in Fig. 17.13, is denoted

$$Zn(s)|ZnSO_4(aq) \vdots CuSO_4(aq)|Cu(s)$$

A double vertical line, ||, denotes an interface for which it is assumed that the junction potential has been eliminated. Thus the cell in Fig. 17.14 is denoted

$$Zn(s)|ZnSO_4(aq)||CuSO_4(aq)|Cu(s)$$

An example of an electrolyte concentration cell in which the liquid junction potential is assumed to be eliminated is

$$Pt(s)|H_2(g)|HCl(aq, b_1)||HCl(aq, b_2)|H_2(g)|Pt(s)$$

17.8 The cell potential

The current produced by a galvanic cell arises from the spontaneous chemical reaction taking place inside it. The **cell reaction** is the reaction in the cell written on the assumption that the right-hand electrode is the cathode, and hence that the spontaneous reaction is one in which reduction is taking place in the right-hand compartment. Later we see how to predict if the right-hand electrode is in fact the cathode; if it is, then the cell reaction is spontaneous as written. If the left-hand electrode turns out to be the cathode, then the reverse of the corresponding cell reaction is spontaneous.

To write the cell reaction corresponding to a cell diagram, we first write the right-hand half-reaction as a reduction (because we have assumed that to be spontaneous). Then we subtract from it the left-hand reduction half-reaction (for, by implication, that electrode is the site of oxidation). Thus, in the cell $Zn(s)|ZnSO_4(aq)||CuSO_4(aq)|Cu(s)$ the two electrodes and their reduction half-reactions are

Right-hand electrode: $Cu^{2+}(aq) + 2\,e^- \rightarrow Cu(s)$
Left-hand electrode: $Zn^{2+}(aq) + 2\,e^- \rightarrow Zn(s)$

Hence, the overall cell reaction is the difference:

$$Cu^{2+}(aq) + Zn(s) \rightarrow Cu(s) + Zn^{2+}(aq)$$

(a) The Nernst equation

A cell in which the overall cell reaction has not reached chemical equilibrium can do electrical work as the reaction drives electrons through an external circuit. The work that a given transfer of electrons can accomplish depends on the potential difference between the two electrodes. When this potential difference is large, a given number of electrons travelling between the electrodes can do a large amount of electrical work. When the potential difference is small, the same number of electrons can do only a small amount of work. A cell in which the overall reaction is at equilibrium can do no work, and then its potential difference is zero.

According to the discussion in Section 15.5, we know that the maximum non-expansion work, which in the current context is electrical work, that a system (the cell) can do is given by eqn 15.49b ($w_{add,max} = \Delta G$), with ΔG identified (as we shall show) with the Gibbs energy of the cell reaction, $\Delta_r G$. It follows that, to draw thermodynamic conclusions from measurements of the work a cell can do, we must ensure that the cell is operating reversibly, for only then is it producing maximum work. Moreover, we saw in Section 17.1 that the reaction Gibbs energy is actually a property relating to a specified composition of the reaction mixture. Therefore, to make use of $\Delta_r G$ we must ensure that the cell is operating reversibly at a specific, constant composition. Both these conditions are achieved by measuring the potential difference generated by the cell when it is balanced by an exactly opposing source of potential so that the cell reaction occurs reversibly, the composition is constant, and no current flows: in effect, the cell reaction is poised for change, but not actually changing. The resulting potential difference is called the **cell potential**, E_{cell}.[1]

[1] The cell potential was formerly and is still widely called the *electromotive force* (emf). That name has fallen out of favour with the IUPAC because the cell potential is not a force.

As we show in the following *Justification*, the relation between the reaction Gibbs energy and the cell potential is

$$-\nu F E_{\text{cell}} = \Delta_r G \tag{17.30}$$

where F is Faraday's constant, $F = eN_A$, and ν is the stoichiometric coefficient of the electrons in the half-reactions into which the cell reaction can be divided. This equation is the key connection between electrical measurements on the one hand and thermodynamic properties on the other. It will be the basis of all that follows.

Justification 17.5 *The relation between the cell potential and the reaction Gibbs energy*

We consider the change in G when the cell reaction advances by an infinitesimal amount $d\xi$ at some composition. From eqn 17.1 we can write (at constant temperature and pressure)

$$dG = \Delta_r G d\xi$$

The maximum non-expansion (electrical) work that the reaction can do as it advances by $d\xi$ at constant temperature and pressure is therefore

$$dw_e = \Delta_r G d\xi$$

This work is infinitesimal, and the composition of the system is virtually constant when it occurs.

Suppose that the reaction advances by $d\xi$, then $\nu d\xi$ electrons must travel from the anode to the cathode. The total charge transported between the electrodes when this change occurs is $-\nu e N_A d\xi$ (because $\nu d\xi$ is the amount of electrons and the charge per mole of electrons is $-eN_A$). Hence, the total charge transported is $-\nu F d\xi$ because $eN_A = F$. The work done when an infinitesimal charge $-\nu F d\xi$ travels from the anode to the cathode is equal to the product of the charge and the potential difference E_{cell}:

$$dw_e = -\nu F E_{\text{cell}} d\xi$$

When we equate this relation to the one above ($dw_e = \Delta_r G d\xi$), the advancement $d\xi$ cancels, and we obtain eqn 17.30.

It follows from eqn 17.30 that, by knowing the reaction Gibbs energy at a specified composition, we can state the cell potential at that composition. Note that a negative reaction Gibbs energy, corresponding to a spontaneous cell reaction, corresponds to a positive cell potential. Another way of looking at the content of eqn 17.30 is that it shows that the driving power of a cell is proportional to the slope of the Gibbs energy with respect to the extent of reaction. It is plausible that a reaction that is far from equilibrium (when the slope is steep) has a strong tendency to drive electrons through an external circuit (Fig. 17.15). When the slope is close to zero (when the cell reaction is close to equilibrium), the cell potential is small.

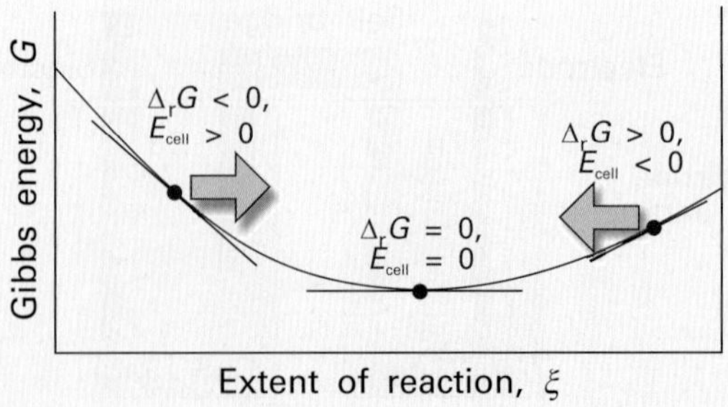

Fig. 17.15 A spontaneous reaction occurs in the direction of decreasing Gibbs energy. When expressed in terms of a cell potential, the spontaneous direction of change can be expressed in terms of the cell potential, E_{cell}. The reaction is spontaneous as written (from left to right in the figure) when $E_{\text{cell}} > 0$. The reverse reaction is spontaneous when $E_{\text{cell}} < 0$. When the cell reaction is at equilibrium, the cell potential is zero.

● A BRIEF ILLUSTRATION

Equation 17.30 provides an electrical method for measuring a reaction Gibbs energy at any composition of the reaction mixture: we simply measure the cell potential and convert it to $\Delta_r G$. Conversely, if we know the value of $\Delta_r G$ at a particular composition, then we can predict the cell potential. For example, if $\Delta_r G = -1 \times 10^2$ kJ mol^{-1} and $\nu = 1$, then

$$E_{\text{cell}} = -\frac{\Delta_r G}{\nu F} = -\frac{(-1\times 10^5 \text{ J mol}^{-1})}{1\times(9.6485\times 10^4 \text{ C mol}^{-1})} = 1 \text{ V}$$

where we have used 1 J = 1 C V. ●

We can go on to relate the cell potential to the activities of the participants in the cell reaction. We know that the reaction Gibbs energy is related to the composition of the reaction mixture by eqn 17.9 ($\Delta_r G = \Delta_r G^{\ominus} + RT \ln Q$); it follows, on division of both sides by $-\nu F$, that

$$E_{\text{cell}} = -\frac{\Delta_r G^{\ominus}}{\nu F} - \frac{RT}{\nu F}\ln Q$$

The first term on the right is written

$$E^{\ominus}_{\text{cell}} = -\frac{\Delta_r G^{\ominus}}{\nu F} \tag*{[17.31]}$$

and called the **standard cell potential**. That is, the standard cell potential is the standard reaction Gibbs energy expressed as a potential (in volts). It follows that

$$E_{\text{cell}} = E^{\ominus}_{\text{cell}} - \frac{RT}{\nu F}\ln Q \tag{17.32}$$

This equation for the cell potential in terms of the composition is called the **Nernst equation**; the dependence on composition

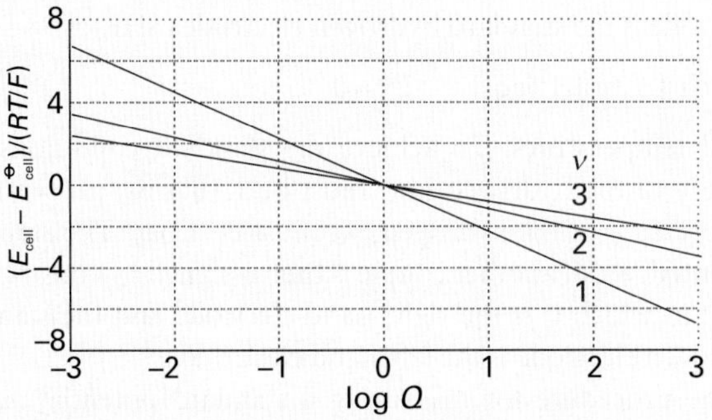

Fig. 17.16 The variation of cell potential with the value of the reaction quotient for the cell reaction for different values of ν (the number of electrons transferred). At 298 K, $RT/F = 25.69$ mV, so the vertical scale refers to multiples of this value.

interActivity Plot the variation of cell potential with the value of the reaction quotient for the cell reaction for different values of the temperature. Does the cell potential become more or less sensitive to composition as the temperature increases?

that it predicts is summarized in Fig. 17.16. One important application of the Nernst equation is to the determination of the pH of a solution and, with a suitable choice of electrodes, of the concentration of other ions.

We see from eqn 17.32 that the standard cell potential (which will shortly move to centre stage of the exposition) can be interpreted as the cell potential when all the reactants and products in the cell reaction are in their standard states, for then all activities are 1, so $Q = 1$ and $\ln Q = 0$. However, the fact that the standard cell potential is merely a disguised form of the standard reaction Gibbs energy (eqn 17.31) should always be kept in mind and underlies all its applications.

● A BRIEF ILLUSTRATION

Because RT/F = 25.7 mV at 25°C, a practical form of the Nernst equation is

$$E_{\text{cell}} = E^{\ominus}_{\text{cell}} - \frac{25.7\text{ mV}}{\nu}\ln Q$$

It then follows that, for a reaction in which $\nu = 1$, if Q is increased by a factor of 10, then the cell potential decreases by 59.2 mV. ●

(b) Cells at equilibrium

A special case of the Nernst equation has great importance in electrochemistry and provides a link to the earlier part of the chapter. Suppose the reaction has reached equilibrium; then $Q = K$, where K is the equilibrium constant of the cell reaction. However, a chemical reaction at equilibrium cannot do work, and hence it generates zero potential difference between the electrodes of a galvanic cell. Therefore, setting $E_{\text{cell}} = 0$ and $Q = K$ in the Nernst equation gives

$$\ln K = \frac{\nu F E^{\ominus}_{\text{cell}}}{RT} \qquad (17.33)$$

This very important equation (which could also have been obtained more directly by substituting eqn 17.31 into eqn 17.13) lets us predict equilibrium constants from measured standard cell potentials. However, before we use it extensively, we need to establish a further result (in Section 17.9, after the following *Impact* section).

● A BRIEF ILLUSTRATION

Because the standard potential of the Daniell cell is +1.10 V, the equilibrium constant for the cell reaction $Cu^{2+}(aq) + Zn(s) \rightarrow Cu(s) + Zn^{2+}(aq)$, for which $\nu = 2$, is $K = 1.6 \times 10^{37}$ at 298 K. We conclude that the displacement of copper by zinc goes virtually to completion. Note that a cell potential of about 1 V is easily measurable but corresponds to an equilibrium constant that would be impossible to measure by direct chemical analysis. ●

IMPACT ON ENGINEERING

I17.2 Fuel cells

A fuel cell operates like a conventional galvanic cell with the exception that the reactants are supplied from outside rather than forming an integral part of its construction. A fundamental and important example of a fuel cell is the *hydrogen/oxygen cell*, such as the ones used in the Apollo Moon missions. One of the electrolytes used is concentrated aqueous potassium hydroxide maintained at 200°C and 20–40 atm; the electrodes may be porous nickel in the form of sheets of compressed powder. The cathode reaction is the reduction

$$O_2(g) + 2\,H_2O(l) + 4\,e^- \rightarrow 4\,OH^-(aq) \qquad E^{\ominus} = +0.40\text{ V}$$

and the anode reaction is the oxidation

$$H_2(g) + 2\,OH^-(aq) \rightarrow 2\,H_2O(l) + 2\,e^-$$

For the corresponding reduction, $E^{\ominus} = -0.83$ V. Because the overall reaction

$$2\,H_2(g) + O_2(g) \rightarrow 2\,H_2O(l) \qquad E^{\ominus} = +1.23\text{ V}$$

is exothermic, it is less favourable thermodynamically at 200°C than at 25°C, so the cell potential is lower at the higher temperature. However, the increased pressure compensates for the increased temperature, and at 200°C and 40 atm $E \approx +1.2$ V.

A property that determines the efficiency of an electrode is the *current density*, the electric current flowing through a region of an electrode divided by the area of the region. One advantage of the hydrogen/oxygen system is the large *exchange current density*, the magnitude of the equal but opposite current densities when the electrode is at equilibrium, of the hydrogen reaction. Unfortunately, the oxygen reaction has an exchange current density of only about 0.1 nA cm^{-2}, which limits the current available from the cell. One way round the difficulty is to use a catalytic surface with a large surface area. One type of highly developed fuel cell has phosphoric acid as the electrolyte and operates with hydrogen and air at about 200°C; the hydrogen is obtained from a reforming reaction on natural gas

Anode: $2\,H_2(g) \rightarrow 4\,H^+(aq) + 4\,e^-$
Cathode: $O_2(g) + 4\,H^+(aq) + 4\,e^- \rightarrow 2\,H_2O(l)$

This fuel cell has shown promise for *combined heat and power systems* (CHP systems). In such systems, the waste heat is used to heat buildings or to do work. Efficiency in a CHP plant can reach 80 per cent. The power output of batteries of such cells has reached the order of 10 MW. Although hydrogen gas is an attractive fuel, it has disadvantages for mobile applications: it is difficult to store and dangerous to handle. One possibility for portable fuel cells is to store the hydrogen in carbon nanotubes. It has been shown that carbon nanofibres in herringbone patterns can store huge amounts of hydrogen and result in energy densities twice that of gasoline.

Cells with molten carbonate electrolytes at about 600°C can make use of natural gas directly. Until these materials have been developed, one attractive fuel is methanol, which is easy to handle and is rich in hydrogen atoms:

Anode: $CH_3OH(l) + 6\,OH^-(aq) \rightarrow 5\,H_2O(l) + CO_2(g) + 6\,e^-$
Cathode: $O_2(g) + 4\,e^- + 2\,H_2O(l) \rightarrow 4\,OH^-(aq)$

One disadvantage of methanol, however, is the phenomenon of 'electro-osmotic drag' in which protons moving through the polymer electrolyte membrane separating the anode and cathode carry water and methanol with them into the cathode compartment where the potential is sufficient to oxidize CH_3OH to CO_2, so reducing the efficiency of the cell. Solid ionic conducting oxide cells operate at about 1000°C and can use hydrocarbons directly as fuel.

17.9 Standard electrode potentials

A galvanic cell is a combination of two electrodes, and each one can be considered as making a characteristic contribution to the overall cell potential, which is then the difference of the two contributions:

$$\mathrm{Red_1,Ox_1||Red_2,Ox_2} \qquad E_{\rm cell} = E_2 - E_1 \tag{17.34}$$

Here E_2 is the potential of the right-hand electrode (of the cell as written) and E_1 is that of the left-hand electrode. Although it is not possible to measure the contribution of a single electrode, we can define the potential of one of the electrodes as zero and then assign values to others on that basis. The specially selected electrode is the **standard hydrogen electrode** (SHE):

$$\mathrm{Pt(s)|H_2(g)|H^+(aq)} \qquad E^{\ominus} = 0 \qquad [17.35]$$

at all temperatures. To achieve the standard conditions, the activity of the hydrogen ions must be 1 (that is, pH = 0) and the pressure of the hydrogen gas must be 1 bar. The **standard potential**, $E^{\ominus}$, of another couple is then assigned by constructing a cell in which it is the right-hand electrode and the standard hydrogen electrode is the left-hand electrode.

The procedure for measuring a standard potential can be illustrated by considering a specific case, the silver chloride electrode. The measurement is made on the 'Harned cell':

$\mathrm{Pt(s)|H_2(g)|HCl(aq)|AgCl(s)|Ag(s)}$
$\tfrac{1}{2}H_2(g) + AgCl(s) \rightarrow HCl(aq) + Ag(s)$

for which the Nernst equation is

$$E_{\rm cell} = E^{\ominus}_{\rm cell}(\mathrm{AgCl/Ag, Cl^-}) - \frac{RT}{F}\ln\frac{a_{\rm H^+}a_{\rm Cl^-}}{a_{\rm H_2}^{1/2}}$$

We shall set $a_{\rm H_2} = 1$ from now on, and for simplicity write the standard cell potential as $E^{\ominus}_{\rm cell}$; then

$$E_{\rm cell} = E^{\ominus}_{\rm cell} - \frac{RT}{F}\ln a_{\rm H^+}a_{\rm Cl^-}$$

The activities can be expressed in terms of the molality b of HCl(aq) through $a_{\rm H^+} = \gamma_\pm b/b^{\ominus}$ and $a_{\rm Cl^-} = \gamma_\pm b/b^{\ominus}$ (as we saw in Section 16.6e), so

$$E_{\rm cell} = E^{\ominus}_{\rm cell} - \frac{RT}{F}\ln b^2 - \frac{RT}{F}\ln\gamma_\pm^2$$

where for simplicity we have replaced $b/b^{\ominus}$ by b. This expression rearranges to

$$E_{\rm cell} + \frac{2RT}{F}\ln b = E^{\ominus}_{\rm cell} - \frac{2RT}{F}\ln\gamma_\pm \qquad \{17.36\}$$

From the Debye–Hückel limiting law for a 1,1-electrolyte (Section 16.6e; a 1,1-electrolyte is a solution of singly charged M^+ and X^- ions), we know that $\ln\gamma_\pm \propto -b^{1/2}$. The natural logarithm used here is proportional to the common logarithm that appears in eqn 16.56 (because $\ln x = \ln 10 \log x = 2.303 \log x$). Therefore, with the constant of proportionality in this relation written as $(F/2RT)C$, eqn 17.36 becomes

$$E_{\rm cell} + \frac{2RT}{F}\ln b = E^{\ominus}_{\rm cell} + Cb^{1/2} \qquad \{17.37\}$$

The expression on the left is evaluated at a range of molalities, plotted against $b^{1/2}$, and extrapolated to $b = 0$. The intercept at $b^{1/2} = 0$ is the value of $E^{\ominus}$ for the silver/silver-chloride electrode. In precise work, the $b^{1/2}$ term is brought to the left, and a higher order correction term from the extended Debye–Hückel law is used on the right.

● A BRIEF ILLUSTRATION

The potential of the cell Pt(s)|H_2(g, $p^⦵$)|HCl(aq, b)|AgCl(s)|Ag(s) at 25°C has the following values:

$b/(10^{-3}\,b^⦵)$	3.215	5.619	9.138	25.63
E_{cell}/V	0.52053	0.49257	0.46860	0.41824

To determine the standard cell potential we draw up the following table, using $2RT/F = 0.051\ 39$ V:

$b/(10^{-3}\,b^⦵)$	3.215	5.619	9.138	25.63
$\{b/(10^{-3}\,b^⦵)\}^{1/2}$	1.793	2.370	3.023	5.063
E_{cell}/V	0.52053	0.49257	0.46860	0.41824
$E_{cell}/V + 0.051\ 39 \ln b$	0.2256	0.2263	0.2273	0.2299

The data are plotted in Fig. 17.17; as can be seen, they extrapolate to $E^⦵_{cell} = 0.2232$ V, which we identify as the value of $E^⦵(Ag,AgCl,Cl^-)$. ●

Self-test 17.9 The data below are for the cell Pt(s)|H_2(g, $p^⦵$)|HBr(aq, b)|AgBr(s)|Ag(s) at 25°C. Determine the standard potential of the cell.

$b/(10^{-4}\,b^⦵)$	4.042	8.444	37.19
E_{cell}/V	0.47381	0.43636	0.36173

[0.071 V]

Table 17.2 lists the standard potentials of a number of electrodes at 298 K. An important feature of standard cell potentials and standard electrode potentials is that they are unchanged if the chemical equation for the cell reaction or a half-reaction is multiplied by a numerical factor. A numerical factor increases the value of the standard Gibbs energy for the reaction. However, it also increases the number of electrons transferred by the same factor, and by eqn 17.30 the value of $E^⦵$ remains unchanged. A practical consequence is that a cell potential is independent of the physical size of the cell. In other words, cell potential is an intensive property.

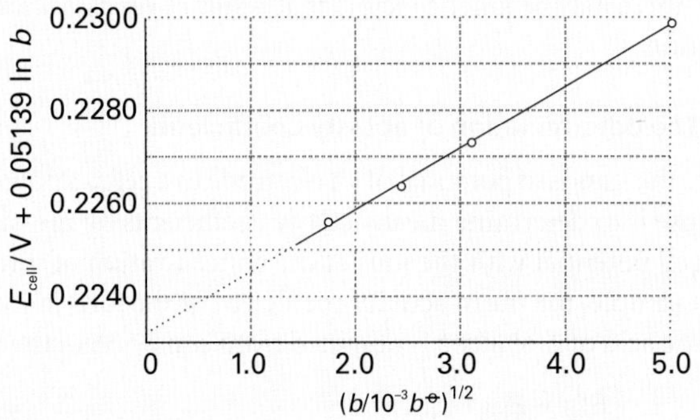

Fig. 17.17 The plot and the extrapolation used for the experimental measurement of a standard cell potential. The intercept at $b^{1/2} = 0$ is $E^⦵_{cell}$.

interActivity For the cell described by Fig. 17.17 and the corresponding *brief illustration*, plot a family of curves of E against $b/b^⦵$ for several values of the temperature T.

Synoptic table 17.2* Standard potentials at 298 K

Couple	$E^⦵$/V
$Ce^{4+}(aq) + e^- \rightarrow Ce^{3+}(aq)$	+1.61
$Cu^{2+}(aq) + 2\,e^- \rightarrow Cu(s)$	+0.34
$AgCl(s) + e^- \rightarrow Ag(s) + Cl^-(aq)$	+0.22
$H^+(aq) + e^- \rightarrow \frac{1}{2}H_2(g)$	0
$Zn^{2+}(aq) + 2\,e^- \rightarrow Zn(s)$	−0.76
$Na^+(aq) + e^- \rightarrow Na(s)$	−2.71

* More values are given in the *Data section*.

The standard potentials in Table 17.2 may be combined to give values for couples that are not listed there. However, to do so, we must take into account the fact that different couples may correspond to the transfer of different numbers of electrons. The procedure is illustrated in the following example.

Example 17.4 *Evaluating a standard potential from two others*

Given that the standard potentials of the Cu^{2+}/Cu and Cu^+/Cu couples are +0.340 V and +0.522 V, respectively, evaluate $E^⦵(Cu^{2+},Cu^+)$.

Method First, we note that reaction Gibbs energies may be added (as in a Hess's law analysis of reaction enthalpies). Therefore, we should convert the $E^⦵$ values to $\Delta G^⦵$ values by using eqn 17.30, add them appropriately, and then convert the overall $\Delta G^⦵$ to the required $E^⦵$ by using eqn 17.30 again. This roundabout procedure is necessary because, as we shall see, although the factor F cancels, the factor ν in general does not.

Answer The electrode reactions are as follows:

(a) $Cu^{2+}(aq) + 2\,e^- \rightarrow Cu(s)$
$E^⦵ = +0.340$ V, so $\Delta_r G^⦵ = -2(0.340\ V)F$

(b) $Cu^+(aq) + e^- \rightarrow Cu(s)$
$E^⦵ = +0.522$ V, so $\Delta_r G^⦵ = -(0.522\ V)F$

The required reaction is

(c) $Cu^{2+}(aq) + e^- \rightarrow Cu^+(aq)$ $\quad E^⦵ = -\Delta_r G^⦵/F$

Because (c) = (a) − (b), the standard Gibbs energy of reaction (c) is

$$\Delta_r G^⦵ = \Delta_r G^⦵(a) - \Delta_r G^⦵(b) = -(0.158\ V) \times F$$

Therefore, $E^{\ominus} = +0.158$ V. Note that the generalization of the calculation we just performed, since (a) = (b) + (c), is

$$\nu_a E^{\ominus}(a) = \nu_b E^{\ominus}(b) + \nu_c E^{\ominus}(c)$$

A note on good practice Whenever combining standard potentials to obtain the standard potential of a third couple, always work via the Gibbs energies because they are additive, whereas, in general, standard potentials are not.

Self-test 17.10 Calculate the standard potential of the Fe^{3+}/Fe couple from the values for the Fe^{3+}/Fe^{2+} and Fe^{2+}/Fe couples. [−0.037 V]

Table 17.3 The electrochemical series of the metals*

Least strongly reducing
Gold
Platinum
Silver
Mercury
Copper
(Hydrogen)
Lead
Tin
Nickel
Iron
Zinc
Chromium
Aluminium
Magnesium
Sodium
Calcium
Potassium
Most strongly reducing

* Additional metals can be inserted into the series by considering the data in Table 17.2.

17.10 Applications of standard potentials

Cell potentials are a convenient source of data on equilibrium constants and the Gibbs energies, enthalpies, and entropies of reactions. In practice the standard values of these quantities are the ones normally determined.

(a) The electrochemical series

We have seen that for two redox couples, Ox_1/Red_1 and Ox_2/Red_2, the cell potential is given by eqn 17.34; for their standard values

$$\mathrm{Red_1,Ox_1||Red_2,Ox_2} \qquad E^{\ominus}_{cell} = E^{\ominus}_2 - E^{\ominus}_1 \qquad (17.38a)$$

It follows that the cell reaction

$$\mathrm{Red_1 + Ox_2 \rightarrow Ox_1 + Red_2} \qquad (17.38b)$$

is spontaneous as written if $E^{\ominus}_{cell} > 0$, and therefore if $E^{\ominus}_2 > E^{\ominus}_1$. Because in the cell reaction Red_1 reduces Ox_2, we can conclude that

Red_1 has a thermodynamic tendency to reduce Ox_2 if $E^{\ominus}_1 < E^{\ominus}_2$

More briefly: low reduces high. The reactions of the electron transport chains of respiration are good applications of this principle (see *Impact I17.3*).

● A BRIEF ILLUSTRATION

Because $E^{\ominus}(Zn^{2+},Zn) = -0.76$ V $< E^{\ominus}(Cu^{2+},Cu) = +0.34$ V, zinc has a thermodynamic tendency to reduce Cu^{2+} ions in aqueous solution. ●

Table 17.3 shows a part of the **electrochemical series**, the metallic elements (and hydrogen) arranged in the order of their reducing power as measured by their standard potentials in aqueous solution. A metal low in the series (with a lower standard potential) can reduce the ions of metals with higher standard potentials. This conclusion is qualitative. The quantitative value of K is obtained by doing the calculations we have described previously. For example, to determine whether zinc can displace magnesium from aqueous solutions at 298 K, we note that zinc lies above magnesium in the electrochemical series, so zinc cannot reduce magnesium ions in aqueous solution. Zinc can reduce hydrogen ions, because hydrogen lies higher in the series. However, even for reactions that are thermodynamically favourable, there may be kinetic factors that result in very slow rates of reaction.

(b) The determination of activity coefficients

Once the standard potential of an electrode in a cell is known, we can use it to determine mean activity coefficients by measuring the cell potential with the ions at the concentration of interest. For example, the mean activity coefficient of the ions in hydrochloric acid of molality b is obtained from eqn 17.36 in the form

$$\ln \gamma_{\pm} = \frac{E^{\ominus}_{cell} - E_{cell}}{2RT/F} - \ln b \qquad \{17.39\}$$

once E_{cell} has been measured.

(c) The determination of equilibrium constants

The principal use for standard potentials is to calculate the standard potential of a cell formed from any two electrodes. To do so, we subtract the standard potential of the left-hand electrode from the standard potential of the right-hand electrode, $E^\ominus_{cell} = E^\ominus(\text{right}) - E^\ominus(\text{left})$. Because $\Delta G^\ominus = -\nu FE^\ominus_{cell}$, it then follows that if the result gives $E^\ominus_{cell} > 0$, then the corresponding cell reaction has $K > 1$.

● A BRIEF ILLUSTRATION

A disproportionation is a reaction in which a species is both oxidized and reduced. To study the disproportionation $2\,Cu^+(aq) \rightarrow Cu(s) + Cu^{2+}(aq)$ we combine the following electrodes:

Right-hand electrode: Cu(s)|Cu^+(aq)
$Cu^+(aq) + e^- \rightarrow Cu(aq) \quad E^\ominus = +0.52\text{ V}$

Left-hand electrode: Pt(s)|Cu^{2+}(aq),Cu^+(aq)
$Cu^{2+}(aq) + e^- \rightarrow Cu^+(s) \quad E^\ominus = +0.16\text{ V}$

where the standard potentials are measured at 298 K. The standard cell potential is therefore

$$E^\ominus_{cell} = +0.52\text{ V} - 0.16\text{ V} = +0.36\text{ V}$$

We can now calculate the equilibrium constant of the cell reaction. Because $\nu = 1$, from eqn 17.33

$$\ln K = \frac{0.36\text{ V}}{0.025693\text{ V}} = \frac{0.36}{0.025693}$$

Hence, $K = 1.2 \times 10^6$. ●

Self-test 17.11 Calculate the solubility constant (the equilibrium constant for the reaction $Hg_2Cl_2(s) \rightleftharpoons Hg_2^{2+}(aq) + 2\,Cl^-(aq)$) and the solubility of mercury(I) chloride at 298.15 K assuming activity coefficients of 1. *Hint.* The mercury(I) ion is the diatomic species Hg_2^{2+}.

[2.6×10^{-18}, 8.7×10^{-7} mol kg^{-1}]

(d) The determination of thermodynamic functions

The standard cell potential is related to the standard reaction Gibbs energy through eqn 17.30 ($\Delta_r G^\ominus = -\nu FE^\ominus_{cell}$). Therefore, by measuring $E^\ominus_{cell}$ we can obtain this important thermodynamic quantity. Its value can then be used to calculate the Gibbs energy of formation of ions by using the convention explained in Section 15.6.

The temperature coefficient of the standard cell potential, $dE^\ominus_{cell}/dT$, gives the standard entropy of the cell reaction. This conclusion follows from the thermodynamic relation $(\partial G/\partial T)_p = -S$ and eqn 17.30, which combine to give

$$\frac{dE^\ominus_{cell}}{dT} = \frac{\Delta_r S^\ominus}{\nu F} \qquad (17.40)$$

The derivative is complete because $E^\ominus_{cell}$, like $\Delta_r G^\ominus$, is independent of the pressure. Hence we have an electrochemical technique for obtaining standard reaction entropies and through them the entropies of ions in solution.

Finally, we can combine the results obtained so far and use them to obtain the standard reaction enthalpy:

$$\Delta_r H^\ominus = \Delta_r G^\ominus + T\Delta_r S^\ominus = -\nu F\left(E^\ominus_{cell} - T\frac{dE^\ominus_{cell}}{dT}\right) \qquad (17.41)$$

This expression provides a non-calorimetric method for measuring $\Delta_r H^\ominus$ and, through the convention $\Delta_f H^\ominus(H^+, aq) = 0$, the standard enthalpies of formation of ions in solution (Section 14.8). Thus, electrical measurements can be used to calculate all the thermodynamic properties with which this chapter began.

Example 17.5 *Using the temperature coefficient of the cell potential*

The standard potential of the cell Pt(s)|H_2(g)|HBr(aq)|AgBr(s)|Ag(s) was measured over a range of temperatures, and the data were fitted to the following polynomial:

$$E^\ominus_{cell}/\text{V} = 0.07131 - 4.99 \times 10^{-4}(T/\text{K} - 298) - 3.45 \times 10^{-6}(T/\text{K} - 298)^2$$

Evaluate the standard reaction Gibbs energy, enthalpy, and entropy at 298 K of the reaction $AgBr(s) + \frac{1}{2}H_2(g) \rightarrow Ag(s) + HBr(aq)$.

Method The standard Gibbs energy of reaction is obtained by using eqn 17.30 after evaluating $E^\ominus_{cell}$ at 298 K and by using 1 V C = 1 J. The standard entropy of reaction is obtained by using eqn 17.40, which involves differentiating the polynomial with respect to T and then setting $T = 298$ K. The reaction enthalpy is obtained by combining the values of the standard Gibbs energy and entropy.

Answer At $T = 298$ K, $E^\ominus_{cell} = +0.07131$ V, so

$$\Delta_r G^\ominus = -\nu FE^\ominus_{cell} = -(1) \times (9.6485 \times 10^4\text{ C mol}^{-1}) \times (+0.07131\text{ V}) = -6.880 \times 10^3\text{ V C mol}^{-1} = -6.880\text{ kJ mol}^{-1}$$

The temperature coefficient of the standard cell potential is

$$\frac{dE^\ominus_{cell}}{dT} = -4.99 \times 10^{-4}\text{ V K}^{-1} - 2(3.45 \times 10^{-6})(T/\text{K} - 298)\text{ V K}^{-1}$$

At $T = 298$ K this expression evaluates to

$$\frac{dE^\ominus_{cell}}{dT} = -4.99 \times 10^{-4}\text{ V K}^{-1}$$

So, from eqn 17.40, the reaction entropy is

$$\Delta_r S^\ominus = 1 \times (9.6485 \times 10^4\text{ C mol}^{-1}) \times (-4.99 \times 10^{-4}\text{ V K}^{-1}) = -48.1\text{ J K}^{-1}\text{ mol}^{-1}$$

It then follows that

$$\begin{aligned}\Delta_r H^{\ominus} &= \Delta_r G^{\ominus} + T\Delta_r S^{\ominus}\\ &= -6.880 \text{ kJ mol}^{-1} + (298 \text{ K}) \times (-0.0481 \text{ kJ K}^{-1} \text{ mol}^{-1})\\ &= -21.2 \text{ kJ mol}^{-1}\end{aligned}$$

One difficulty with this procedure lies in the accurate measurement of small temperature coefficients of the cell potential. Nevertheless, it is another example of the striking ability of thermodynamics to relate the apparently unrelated, in this case to relate electrical measurements to thermal properties.

Self-test 17.12 Predict the standard potential of the Harned cell at 303 K from tables of thermodynamic data.

[+0.219 V]

IMPACT ON BIOLOGY

I17.3 The respiratory chain

In the exergonic oxidation of glucose, 24 electrons are transferred from each $C_6H_{12}O_6$ molecule to six O_2 molecules. The electrons do not flow directly from glucose to O_2. During glycolysis and the citric acid cycle, glucose is oxidized to CO_2 by NAD^+ and FAD:

$$C_6H_{12}O_6(s) + 10\,NAD^+ + 2\,FAD + 4\,ADP + 4\,P_i^- + 4\,H_2O \rightarrow 6\,CO_2 + 10\,NADH + 2\,FADH_2 + 4\,ATP + 6\,H^+$$

In the *respiratory chain*, electrons from the powerful reducing agents NADH and $FADH_2$ pass through four membrane-bound protein complexes and two mobile electron carriers before reducing O_2 to H_2O.

The respiratory chain begins in complex I (NADH-Q oxidoreductase), where NADH is oxidized by coenzyme Q (Q, **4**) in a two-electron reaction:

$$H^+ + NADH + Q \xrightarrow{\text{complex I}} NAD^+ + QH_2$$
$$E^{\ominus}_{cell} = +0.42 \text{ V}, \Delta_r G^{\ominus} = -81 \text{ kJ mol}^{-1}$$

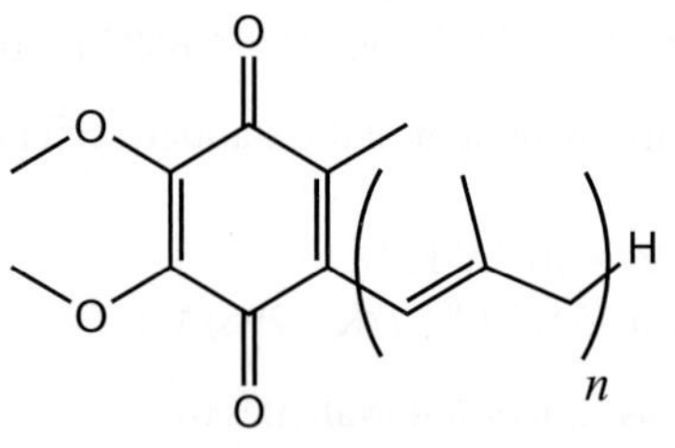

4 Coenzyme Q, Q

Additional Q molecules are reduced by $FADH_2$ in complex II (succinate-Q reductase):

$$FADH_2 + Q \xrightarrow{\text{complex II}} FAD + QH_2$$
$$E^{\ominus}_{cell} = +0.32 \text{ V}, \Delta_r G^{\ominus} = -62 \text{ kJ mol}^{-1}$$

Reduced Q migrates to complex III (Q-cytochrome *c* oxidoreductase), which catalyses the reduction of the protein cytochrome *c* (Cyt *c*). Cytochrome *c* contains the haem *c* group (**5**), the central iron ion of which can exist in oxidation states +3 and +2. The net reaction catalysed by complex III is

$$QH_2 + 2\,Fe^{3+}(\text{Cyt } c) \xrightarrow{\text{complex III}} Q + 2\,Fe^{2+}(\text{Cyt } c) + 2\,H^+$$
$$E^{\ominus}_{cell} = +0.15 \text{ V}, \Delta_r G^{\ominus} = -29 \text{ kJ mol}^{-1}$$

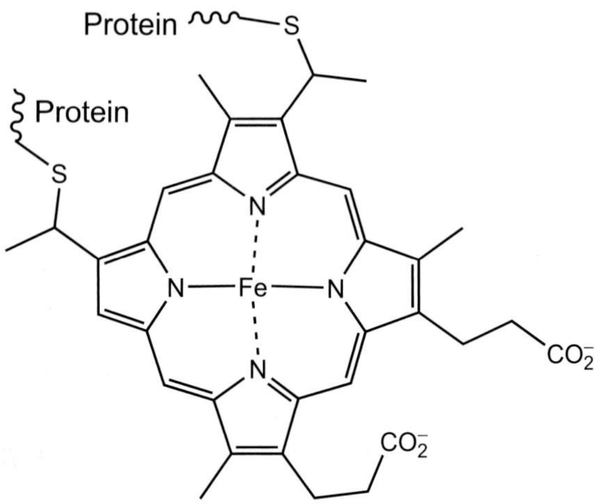

5 Haem (heme) *c*

Reduced cytochrome *c* carries electrons from complex III to complex IV (cytochrome *c* oxidase), where O_2 is reduced to H_2O:

$$2\,Fe^{2+}(\text{Cyt } c) + 2\,H^+ + \tfrac{1}{2}O_2 \xrightarrow{\text{complex IV}} 2\,Fe^{3+}(\text{Cyt } c) + H_2O$$
$$E^{\ominus}_{cell} = +0.56 \text{ V}, \Delta_r G^{\ominus} = -108 \text{ kJ mol}^{-1}$$

The reactions that occur in complexes I, III, and IV drive the synthesis of ATP during *oxidative phosphorylation*. To understand this process, we need to consider the structure of the mitochondrion, which is shown in Fig. 17.18. The protein complexes associated with the electron transport chain span the inner membrane and phosphorylation takes place in the intermembrane space. The Gibbs energy of the reactions in complexes I, III, and IV is first used to do the work of moving protons across the mitochondrial membrane. For example, the oxidation of NADH by Q in complex I is coupled to the transfer of four protons across the membrane. The coupling of electron transfer and proton pumping in complexes III and IV contribute further to a gradient of proton concentration across the membrane. Then the enzyme H^+-ATPase uses the energy stored in the proton gradient to phosphorylate ADP to ATP. Experiments show that 11 molecules of ATP are made for every three molecules of NADH and one molecule of $FADH_2$ that are oxidized by the respiratory chain. The ATP is then hydrolysed on demand to perform useful biochemical work throughout the cell.

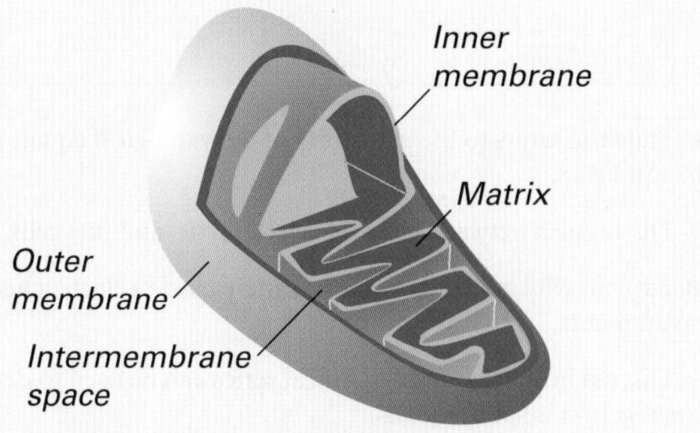

Fig. 17.18 The general structure of a mitochondrion.

The *chemiosmotic theory* proposed by Peter Mitchell in the late 1960s explains how H^+-ATPases synthesize ATP from ADP. The energy stored in a transmembrane proton gradient comes from two contributions. First, the difference in activity of H^+ ion results in a difference in molar Gibbs energy across the mitochrondrial membrane, which contributes a term $RT \ln(a_{H^+,in}/a_{H^+,out})$ to the stored energy. Second, there is a membrane potential difference $\Delta\phi = \phi_{in} - \phi_{out}$ that arises from differences in Coulombic interactions on each side of the membrane. The charge difference across a membrane per mole of H^+ ions is $N_A e$, or F, where $F = eN_A$. It follows from *Justification 17.5* that the molar Gibbs energy difference is then $F\Delta\phi$. Consequently, we write the total Gibbs energy stored by the combination of an activity gradient and a membrane potential gradient. After replacing activities by molar concentrations, we obtain

$$\Delta G_m = G_{m,in} - G_{m,out} = RT \ln \frac{[H^+]_{in}}{[H^+]_{out}} + F\Delta\phi$$

This equation also provides an estimate of the Gibbs energy available for phosphorylation of ADP. After using $\ln [H^+] = \ln 10 \times \log [H^+]$ and substituting $\Delta pH = pH_{in} - pH_{out} = -\log [H^+]_{in} + \log [H^+]_{out}$, it follows that

$$\Delta G_m = F\Delta\phi - (RT \ln 10)\Delta pH$$

In the mitochondrion, $\Delta pH \approx -1.4$ and $\Delta\phi \approx 0.14$ V, so $\Delta G_m \approx +21.5$ kJ mol^{-1}. Because 31 kJ mol^{-1} is needed for phosphorylation, we conclude that at least 2 mol H^+ (and probably more) must flow through the membrane for the phosphorylation of 1 mol ADP.

Checklist of key ideas

- ☐ 1. The extent of reaction (ξ) is defined such that, when the extent of reaction changes by $\Delta\xi$, the amount of A present changes from $n_{A,0}$ to $n_{A,0} - \Delta\xi$.
- ☐ 2. The reaction Gibbs energy is the slope of the graph of the Gibbs energy plotted against the extent of reaction; at equilibrium, $\Delta_r G = 0$.
- ☐ 3. An exergonic reaction is a reaction for which $\Delta_r G < 0$; such a reaction can be used to drive another process. An endergonic reaction is a reaction for which $\Delta_r G > 0$.
- ☐ 4. At an arbitrary stage of a reaction, $\Delta_r G = \Delta_r G^{\ominus} + RT \ln Q$.
- ☐ 5. The equilibrium constant (K) may be written in terms of $\Delta_r G^{\ominus}$ as $\Delta_r G^{\ominus} = -RT \ln K$.
- ☐ 6. The standard reaction Gibbs energy may be calculated from standard Gibbs energies of formation, eqn 17.10.
- ☐ 7. The thermodynamic equilibrium constant is related to the molecular partition functions by eqn 17.17.
- ☐ 8. Le Chatelier's principle states that a system at equilibrium, when subjected to a disturbance, responds in a way that tends to minimize the effect of the disturbance.
- ☐ 9. The temperature dependence of the equilibrium constant is given by the van't Hoff equation: eqn 17.26.
- ☐ 10. A galvanic cell is an electrochemical cell that produces electricity as a result of the spontaneous reaction occurring inside it. An electrolytic cell is an electrochemical cell in which a nonspontaneous reaction is driven by an external source of current.
- ☐ 11. Oxidation is the removal of electrons from a species; reduction is the addition of electrons to a species; a redox reaction is a reaction in which there is a transfer of electrons from one species to another.
- ☐ 12. An anode is the electrode at which oxidation occurs; a cathode is the electrode at which reduction occurs.
- ☐ 13. The cell potential is the potential difference of the cell when it is balanced by an exactly opposing source of potential so that the cell reaction occurs reversibly, the composition is constant, and no current flows.
- ☐ 14. The cell potential and the reaction Gibbs energy are related by $-\nu F E_{cell} = \Delta_r G$.
- ☐ 15. The standard cell potential is the standard reaction Gibbs energy expressed as a potential: $E^{\ominus}_{cell} = -\Delta_r G^{\ominus}/\nu F$.
- ☐ 16. The Nernst equation, eqn 17.32, expresses the cell potential in terms of the composition.
- ☐ 17. The equilibrium constant for a cell reaction is related to the standard cell potential by $\ln K = \nu F E^{\ominus}_{cell}/RT$.
- ☐ 18. The standard potential of a couple ($E^{\ominus}$) is the standard potential of a cell in which a couple forms the right-hand electrode and the standard hydrogen electrode is the left-hand electrode.
- ☐ 19. To calculate the standard cell potential, form the difference of electrode potentials: $E^{\ominus}_{cell} = E^{\ominus}(\text{right}) - E^{\ominus}(\text{left})$.
- ☐ 20. The temperature coefficient of the standard cell potential is given by $dE^{\ominus}/dT = \Delta_r S^{\ominus}/\nu F$.

Discussion questions

17.1 Explain how the mixing of reactants and products affects the position of chemical equilibrium.

17.2 Explain how a reaction that is not spontaneous may be driven forward by coupling to a spontaneous reaction.

17.3 Use concepts of statistical thermodynamics to describe the molecular features that determine the magnitudes of equilibrium constants and their variation with temperature.

17.4 Suggest how the thermodynamic equilibrium constant may respond differently to changes in pressure and temperature from the equilibrium constant expressed in terms of partial pressures.

17.5 Account for Le Chatelier's principle in terms of thermodynamic quantities. Can you think of a reason why the principle might fail?

17.6 State the limits to the generality of the van't Hoff equation, written as in eqn 17.28.

17.7 Distinguish between galvanic, electrolytic, and fuel cells.

17.8 Explain why salt bridges are routinely used in electrochemical cell measurements.

17.9 Discuss how the electrochemical series can be used to determine if a redox reaction is spontaneous.

17.10 Describe a method for the determination of the standard potential of a redox couple.

17.11 Describe at least one non-calorimetric experimental method for determining a standard reaction enthalpy.

Exercises

17.1(a) Write the expressions for the equilibrium constants of the following reactions in terms of (i) activities and (ii) where appropriate, the ratios $p/p^{\ominus}$ and the products $\gamma b/b^{\ominus}$:

(a) $CO(g) + Cl_2(g) \rightleftharpoons COCl(g) + Cl(g)$
(b) $2\,SO_2(g) + O_2(g) \rightleftharpoons 2\,SO_3(g)$
(c) $Fe(s) + PbSO_4(aq) \rightleftharpoons FeSO_4(aq) + Pb(s)$
(d) $Hg_2Cl_2(s) + H_2(g) \rightleftharpoons 2\,HCl(aq) + 2\,Hg(l)$
(e) $2\,CuCl(aq) \rightleftharpoons Cu(s) + CuCl_2(aq)$

17.1(b) Write the expressions for the equilibrium constants of the following reactions in terms of (i) activities and (ii) where appropriate, the ratios $p/p^{\ominus}$ and the products $\gamma b/b^{\ominus}$:

(a) $H_2(g) + Br_2(g) \rightleftharpoons 2\,HBr(g)$
(b) $2\,O_3(g) \rightleftharpoons 3\,O_2(g)$
(c) $2\,H_2(g) + O_2(g) \rightleftharpoons 2\,H_2O(l)$
(d) $H_2(g) + O_2(g) \rightleftharpoons H_2O_2(aq)$
(e) $H_2(g) + I_2(g) \rightleftharpoons 2\,HI(aq)$

17.2(a) Identify the stoichiometric numbers in the reaction $Hg_2Cl_2(s) + H_2(g) \rightarrow 2\,HCl(aq) + 2\,Hg(l)$.

17.2(b) Identify the stoichiometric numbers in the reaction $CH_4(g) + 2\,O_2(g) \rightarrow CO_2(g) + 2\,H_2O(l)$.

17.3(a) The standard reaction Gibbs energy of the isomerization of borneol ($C_{10}H_{17}OH$) to isoborneol in the gas phase at 503 K is +9.4 kJ mol^{-1}. Calculate the reaction Gibbs energy in a mixture consisting of 0.15 mol of borneol molecules and 0.30 mol of isoborneol molecules when the total pressure is 600 Torr. Under these conditions, is the isomerization of borneol spontaneous?

17.3(b) The standard reaction Gibbs energy of the isomerization of *cis*-2-butene to *trans*-2-butene in the gas phase at 298 K is −2.9 kJ mol^{-1}. Calculate the reaction Gibbs energy in a mixture consisting of 0.25 mol of *cis*-2-butene molecules and 0.95 mol of *trans*-2-butene molecules when the total pressure is 600 Torr. Under these conditions, is the isomerization of *cis*-2-butene spontaneous?

17.4(a) The equilibrium pressure of O_2 over solid silver and silver oxide, Ag_2O, at 298 K is 11.85 Pa. Calculate the standard Gibbs energy of formation of $Ag_2O(s)$ at 298 K.

17.4(b) The equilibrium pressure of H_2 over solid uranium and uranium hydride, UH_3, at 500 K is 139 Pa. Calculate the standard Gibbs energy of formation of $UH_3(s)$ at 500 K.

17.5(a) From information in the *Data section*, calculate the standard Gibbs energy and the equilibrium constant at (a) 298 K and (b) 400 K for the reaction $PbO(s) + CO(g) \rightleftharpoons Pb(s) + CO_2(g)$. Assume that the reaction enthalpy is independent of temperature.

17.5(b) From information in the *Data section*, calculate the standard Gibbs energy and the equilibrium constant at (a) 25°C and (b) 50°C for the reaction $CH_4(g) + 3\,Cl_2(g) \rightleftharpoons CHCl_3(l) + 3\,HCl(g)$. Assume that the reaction enthalpy is independent of temperature. $\Delta_f H^{\ominus}(CHCl_3) = -134.71$ kJ mol^{-1}, $\Delta_f G^{\ominus}(CHCl_3) = -73.66$ kJ mol^{-1}.

17.6(a) For $CaF_2(s) \rightleftharpoons Ca^{2+}(aq) + 2\,F^-(aq)$, $K = 3.9 \times 10^{-11}$ at 25°C and the standard Gibbs energy of formation of $CaF_2(s)$ is −1167 kJ mol^{-1}. Calculate the standard Gibbs energy of formation of $CaF_2(aq)$.

17.6(b) For $PbI_2(s) \rightleftharpoons Pb^{2+}(aq) + 2\,I^-(aq)$, $K = 1.4 \times 10^{-8}$ at 25°C and the standard Gibbs energy of formation of $PbI_2(s)$ is −173.64 kJ mol^{-1}. Calculate the standard Gibbs energy of formation of $PbI_2(aq)$.

17.7(a) In the gas-phase reaction $2\,A + B \rightleftharpoons 3\,C + 2\,D$, it was found that, when 1.00 mol A, 2.00 mol B, and 1.00 mol D were mixed and allowed to come to equilibrium at 25°C, the resulting mixture contained 0.90 mol C at a total pressure of 1.00 bar. Calculate (a) the mole fractions of each species at equilibrium, (b) K_x, (c) K, and (d) $\Delta_r G^{\ominus}$.

17.7(b) In the gas-phase reaction $A + B \rightleftharpoons C + 2\,D$, it was found that when 2.00 mol A, 1.00 mol B, and 3.00 mol D were mixed and allowed to come to equilibrium at 25°C, the resulting mixture contained 0.79 mol C at a total pressure of 1.00 bar. Calculate (a) the mole fractions of each species at equilibrium, (b) K_x, (c) K, and (d) $\Delta_r G^{\ominus}$.

17.8(a) The hydrolysis of ATP (*Impact I17.1*) is written as $ATP^{4-}(aq) + H_2O(l) \rightarrow ADP^{3-}(aq) + HPO_4^{2-}(aq) + H_3O^+(aq)$. For this reaction the standard reaction Gibbs energy is +10 kJ mol^{-1} at 298 K. What is the biological standard state value?

17.8(b) The overall reaction for the glycolysis reaction (*Impact I17.1*) is $C_6H_{12}O_6(aq) + 2\,NAD^+(aq) + 2\,ADP^{3-}(aq) + 2\,HPO_4^{2-}(aq) + 2\,H_2O(l) \rightarrow 2\,CH_3COCO_2^-(aq) + 2\,NADH(aq) + 2\,ATP^{4-}(aq) + 2\,H_3O^+(aq)$. For this reaction, the standard reaction Gibbs energy is -80.6 kJ mol^{-1} at 298 K. What is the biological standard state value?

17.9(a) Calculate the value of K for the reaction $I_2(g) \rightleftharpoons 2\,I(g)$ at 1000 K from the following data for I_2: $\tilde{\nu} = 214.36$ cm^{-1}, $\tilde{B} = 0.0373$ cm^{-1}, $D_e = 1.5422$ eV. The ground state of the I atoms is $^2P_{3/2}$, implying fourfold degeneracy.

17.9(b) Calculate the value of K at 298 K for the gas-phase isotopic exchange reaction $2\,^{79}Br^{81}Br \rightleftharpoons {}^{79}Br^{79}Br + {}^{81}Br^{81}Br$. The Br_2 molecule has a non-degenerate ground state, with no other electronic states nearby. Base the calculation on the wavenumber of the vibration of $^{79}Br^{81}Br$, which is 323.33 cm^{-1}.

17.10(a) Calculate the percentage change in K_x for the reaction $H_2CO(g) \rightleftharpoons CO(g) + H_2(g)$ when the total pressure is increased from 1.0 bar to 3.0 bar at constant temperature.

17.10(b) Calculate the percentage change in K_x for the reaction $CH_3OH(g) + NOCl(g) \rightleftharpoons HCl(g) + CH_3NO_2(g)$ when the total pressure is increased from 1.0 bar to 4.0 bar at constant temperature.

17.11(a) The standard reaction enthalpy of $Zn(s) + H_2O(g) \rightarrow ZnO(s) + H_2(g)$ is approximately constant at +224 kJ mol^{-1} from 920 K up to 1280 K. The standard reaction Gibbs energy is +33 kJ mol^{-1} at 1280 K. Estimate the temperature at which the equilibrium constant becomes greater than 1.

17.11(b) The standard enthalpy of a certain reaction is approximately constant at +125 kJ mol^{-1} from 800 K up to 1500 K. The standard reaction Gibbs energy is +22 kJ mol^{-1} at 1120 K. Estimate the temperature at which the equilibrium constant becomes greater than 1.

17.12(a) The equilibrium constant of the reaction $2\,C_3H_6(g) \rightleftharpoons C_2H_4(g) + C_4H_8(g)$ is found to fit the expression $\ln K = A + B/T + C/T^2$ between 300 K and 600 K, with $A = -1.04$, $B = -1088$ K, and $C = 1.51 \times 10^5$ K^2. Calculate the standard reaction enthalpy and standard reaction entropy at 450 K.

17.12(b) The equilibrium constant of a reaction is found to fit the expression $\ln K = A + B/T + C/T^3$ between 400 K and 600 K with $A = -2.01$, $B = -1170$ K, and $C = 2.2 \times 10^7$ K^3. Calculate the standard reaction enthalpy and standard reaction entropy at 500 K.

17.13(a) What is the standard enthalpy of a reaction for which the equilibrium constant is (a) doubled, (b) halved when the temperature is increased by 10 K at 298 K?

17.13(b) What is the standard enthalpy of a reaction for which the equilibrium constant is (a) doubled, (b) halved when the temperature is increased by 15 K at 310 K?

17.14(a) Estimate the temperature at which $CaCO_3$(calcite) decomposes.

17.14(b) Estimate the temperature at which $CuSO_4 \cdot 5H_2O$ undergoes dehydration.

17.15(a) Write the cell reaction and electrode half-reactions and calculate the standard cell potential of each of the following cells:

(a) $Zn(s)|ZnSO_4(aq)||AgNO_3(aq)|Ag(s)$
(b) $Cd(s)|CdCl_2(aq)||HNO_3(aq)|H_2(g)|Pt(s)$
(c) $Pt(s)|K_3[Fe(CN)_6](aq),K_4[Fe(CN)_6](aq)||CrCl_3(aq)|Cr(s)$
(d) $Pt(s)|Fe^{3+}(aq),Fe^{2+}(aq)||Sn^{4+}(aq),Sn^{2+}(aq)|Pt(s)$

17.15(b) Write the cell reaction and electrode half-reactions and calculate the standard cell potential of each of the following cells:

(a) $Pt(s)|K_3[Fe(CN)_6](aq),K_4[Fe(CN)_6](aq)||Mn^{2+}(aq),H^+(aq)|MnO_2(s)|Pt(s)$
(b) $Cu(s)|Cu^{2+}(aq)||Mn^{2+}(aq),H^+(aq)|MnO_2(s)|Pt(s)$
(c) $Pt(s)|Cl_2(g)|HCl(aq)||HBr(aq)|Br_2(l)|Pt(s)$
(d) $Fe(s)|Fe^{2+}(aq)||Mn^{2+}(aq),H^+(aq)|MnO_2(s)|Pt(s)$

17.16(a) Devise cells in which the following are the reactions and calculate the standard cell potential in each case:

(a) $Fe(s) + PbSO_4(aq) \rightarrow FeSO_4(aq) + Pb(s)$
(b) $Hg_2Cl_2(s) + H_2(g) \rightarrow 2\,HCl(aq) + 2\,Hg(l)$
(c) $2\,H_2(g) + O_2(g) \rightarrow 2\,H_2O(l)$

17.16(b) Devise cells in which the following are the reactions and calculate the standard cell potential in each case:

(a) $H_2(g) + O_2(g) \rightarrow H_2O_2(aq)$
(b) $H_2(g) + I_2(g) \rightarrow 2\,HI(aq)$
(c) $2\,CuCl(aq) \rightarrow Cu(s) + CuCl_2(aq)$

17.17(a) Consider the cell Ag|AgBr(s)|KBr(aq, 0.050 mol kg^{-1})|| $Cd(NO_3)_2$(aq, 0.010 mol kg^{-1})|Cd. (a) Write the cell reaction. (b) Write the Nernst equation for the cell. (c) Use the Debye–Hückel limiting law and the Nernst equation to estimate the cell potential at 25°C.

17.17(b) Consider the cell Pt|$H_2(g,p^{\ominus})$|HCl(aq, 0.010 mol kg^{-1})|AgCl(s)|Ag. (a) Write the cell reaction. (b) Write the Nernst equation for the cell. (c) Use the Debye–Hückel limiting law and the Nernst equation to estimate the cell potential at 25°C.

17.18(a) Calculate $\Delta_r G^{\ominus}$ for the cell reactions in Exercise 17.15a.

17.18(b) Calculate $\Delta_r G^{\ominus}$ for the cell reactions in Exercise 17.15b.

17.19(a) Calculate the equilibrium constants of the following reactions at 25°C from standard potential data:

(a) $Sn(s) + Sn^{4+}(aq) \rightleftharpoons 2\,Sn^{2+}(aq)$
(b) $Fe(s) + Hg(NO_3)_2(aq) \rightleftharpoons Hg(l) + Fe(NO_3)_2(aq)$

17.19(b) Calculate the equilibrium constants of the following reactions at 25°C from standard potential data:

(a) $Cd(s) + CuSO_4(aq) \rightleftharpoons Cu(s) + CdSO_4(aq)$
(b) $3\,Au^+(aq) \rightleftharpoons 2\,Au(s) + Au^{3+}(aq)$

17.20(a) Calculate the standard potential of the Ce^{4+}/Ce couple from the values for the Ce^{3+}/Ce and Ce^{4+}/Ce^{3+} couples.

17.20(b) Calculate the standard potential of the Au^{3+}/Au^+ couple from the values for the Au^{3+}/Au and Au^+/Au couples.

17.21(a) Can mercury produce zinc metal from aqueous zinc sulfate under standard conditions?

17.21(b) Can chlorine gas oxidize water to oxygen gas under standard conditions in basic solution?

17.22(a) The potential of the cell Ag|AgI(s)|AgI(aq)|Ag is +0.9509 V at 25°C. Calculate (a) the solubility of AgI and (b) its solubility.

17.22(b) The potential of the cell Bi|Bi_2S_3(s)|Bi_2S_3(aq)|Bi is -0.96 V at 25°C. Calculate (a) the solubility of Bi_2S_3 and (b) its solubility.

17.23(a) The standard potential of the cell $Pt(s)|H_2(g)|HCl(aq)|Hg_2Cl_2(s)|Hg(l)$ was found to be +0.2699 V at 293 K and +0.2669 V at 303 K. Evaluate the standard reaction Gibbs energy, enthalpy, and entropy at 298 K of the reaction $Hg_2Cl_2(s) + H_2(g) \rightarrow 2\,Hg(l) + 2\,HCl(aq)$.

17.23(b) The standard potential of the cell $Pt(s)|H_2(g)|HBr(aq)|AgBr(s)|Ag(s)$ was found to be +0.073 72 V at 293 K and +0.068 73 V at 303 K. Evaluate the standard reaction Gibbs energy, enthalpy, and entropy at 298 K of the reaction $AgBr(s) + \frac{1}{2}H_2(g) \rightarrow Ag(s) + HBr(aq)$.

Problems*

Numerical problems

17.1 The equilibrium constant for the reaction, $I_2(s) + Br_2(g) \rightleftharpoons 2\,IBr(g)$ is 0.164 at 25°C. (a) Calculate $\Delta_r G^\ominus$ for this reaction. (b) Bromine gas is introduced into a container with excess solid iodine. The pressure and temperature are held at 0.164 atm and 25°C. Find the partial pressure of IBr(g) at equilibrium. Assume that all the bromine is in the liquid form and that the vapour pressure of iodine is negligible. (c) In fact, solid iodine has a measurable vapour pressure at 25°C. In this case, how would the calculation have to be modified?

17.2 The standard Gibbs energy of formation of $NH_3(g)$ is -16.5 kJ mol^{-1} at 298 K. What is the reaction Gibbs energy when the partial pressures of the N_2, H_2, and NH_3 (treated as perfect gases) are 3.0 bar, 1.0 bar, and 4.0 bar, respectively? What is the spontaneous direction of the reaction in this case?

17.3 The degree of dissociation, α, is defined as the fraction of reactant that has decomposed; if the initial amount of reactant is n and the amount at equilibrium is n_{eq}, then $\alpha = (n - n_{eq})/n$. The standard Gibbs energy of reaction for the decomposition $H_2O(g) \rightarrow H_2(g) + \frac{1}{2}O_2(g)$ is $+118.08$ kJ mol^{-1} at 2300 K. What is the degree of dissociation of H_2O at 2300 K and 1.00 bar? *Hints.* The equilibrium constant is obtained from the standard Gibbs energy of reaction by using eqn 17.13, so the task is to relate the degree of dissociation, α, to K and then to find its numerical value. Proceed by expressing the equilibrium compositions in terms of α. For example, if an amount n H_2O is present initially, then an amount αn H_2O reacts to reach equilibrium and an amount$(1-\alpha)n$ $H_2O(g)$ is present at equilibrium. Then, solve for α in terms of K. Because the standard Gibbs energy of reaction is large and positive, we can anticipate that K will be small, and hence that $\alpha \ll 1$, which opens the way to making approximations to obtain its numerical value.

17.4 Calculate and plot as a function of temperature, in the range 300 K to 1000 K, the equilibrium constant for the reaction $CD_4(g) + HCl(g) \rightleftharpoons CHD_3(g) + DCl(g)$ using the following data (numbers in parentheses are degeneracies): $\tilde{\nu}(CHD_3)/cm^{-1} = 2993(1), 2142(1), 1003(3), 1291(2), 1036(2)$; $\tilde{\nu}(CD_4)/cm^{-1} = 2109(1), 1092(2), 2259(3), 996(3)$; $\tilde{\nu}(HCl)/cm^{-1} = 2991$; $\tilde{\nu}(DCl)/cm^{-1} = 2145$; $\tilde{B}(HCl)/cm^{-1} = 10.59$; $\tilde{B}(DCl)/cm^{-1} = 5.445$; $\tilde{B}(CHD_3)/cm^{-1} = 3.28$; $\tilde{A}(CHD_3)/cm^{-1} = 2.63$, $\tilde{B}(CD_4)/cm^{-1} = 2.63$.

17.5 The exchange of deuterium between acid and water is an important type of equilibrium, and we can examine it using spectroscopic data on the molecules. Calculate the equilibrium constant at (a) 298 K and (b) 800 K for the gas-phase exchange reaction $H_2O + DCl \rightleftharpoons HDO + HCl$ from the following data: $\tilde{\nu}(H_2O)/cm^{-1} = 3656.7, 1594.8, 3755.8$; $\tilde{\nu}(HDO)/cm^{-1} = 2726.7, 1402.2, 3707.5$; $\tilde{A}(H_2O)/cm^{-1} = 27.88$, $\tilde{B}(H_2O)/cm^{-1} = 14.51$, $\tilde{C}(H_2O)/cm^{-1} = 9.29$; $\tilde{A}(HDO)/cm^{-1} = 23.38$, $\tilde{B}(HDO)/cm^{-1} = 9.102$, $\tilde{C}(HDO)/cm^{-1} = 6.417$; $\tilde{B}(HCl)/cm^{-1} = 10.59$; $\tilde{B}(DCl)/cm^{-1} = 5.449$, $\tilde{\nu}(HCl)/cm^{-1} = 2991$; $\tilde{\nu}(DCl)/cm^{-1} = 2145$.

17.6 The dissociation vapour pressure of NH_4Cl at 427°C is 608 kPa but at 459°C it has risen to 1115 kPa. Calculate (a) the equilibrium constant, (b) the standard reaction Gibbs energy, (c) the standard enthalpy, (d) the standard entropy of dissociation, all at 427°C. Assume that the vapour behaves as a perfect gas and that $\Delta H^\ominus$ and $\Delta S^\ominus$ are independent of temperature in the range given.

17.7 Consider the dissociation of methane, $CH_4(g)$, into the elements $H_2(g)$ and C(s, graphite). (a) Given that $\Delta_f H^\ominus(CH_4, g) = -74.85$ kJ mol^{-1} and that, for the formation of methane from its elements, $\Delta_f S^\ominus(CH_4, g) = -80.67$ J K^{-1} mol^{-1} at 298 K, calculate the value of the equilibrium constant at 298 K. (b) Assuming that $\Delta_f H^\ominus$ is independent of temperature, calculate K at 50°C.

17.8 The equilibrium pressure of H_2 over U(s) and $UH_3(s)$ between 450 K and 715 K fits the expression $\ln(p/\text{Pa}) = A + B/T + C\ln(T/\text{K})$, with $A = 69.32$, $B = -1.464 \times 10^4$ K, and $C = -5.65$. Find an expression for the standard enthalpy of formation of $UH_3(s)$ and from it calculate $\Delta_r C_p^\ominus$.

17.9 The degree of dissociation, α_e (see Problem 17.3), of $CO_2(g)$ into CO(g) and $O_2(g)$ at high temperatures was found to vary with temperature as follows:

T/K	1395	1443	1498
$\alpha_e/10^{-4}$	1.44	2.50	4.71

Assuming $\Delta_r H^\ominus$ to be constant over this temperature range, calculate K, $\Delta_r G^\ominus$, $\Delta_r H^\ominus$, and $\Delta_r S^\ominus$. Make any justifiable approximations.

17.10 The standard reaction enthalpy for the decomposition of $CaCl_2 \cdot NH_3(s)$ into $CaCl_2(s)$ and $NH_3(g)$ is nearly constant at +78 kJ mol^{-1} between 350 K and 470 K. The equilibrium pressure of NH_3 in the presence of $CaCl_2 \cdot NH_3$ is 1.71 kPa at 400 K. Find an expression for the temperature dependence of $\Delta_r G^\ominus$ in the same range.

17.11 Calculate the equilibrium constant of the reaction $CO(g) + H_2(g) \rightleftharpoons H_2CO(g)$ given that for the production of liquid formaldehyde $\Delta_r G^\ominus = +28.95$ kJ mol^{-1} at 298 K and that the vapour pressure of formaldehyde is 1500 Torr at that temperature.

17.12 Acetic acid was evaporated in a container of volume 21.45 cm^3 at 437 K and at an external pressure of 101.9 kPa. The container was then sealed. The mass of acid present in the sealed container was 0.0519 g. The experiment was repeated with the same container but at 471 K, and it was found that 0.0380 g of acetic acid was present. Calculate the equilibrium constant for the dimerization of the acid in the vapour and the enthalpy of vaporization.

17.13 A sealed container was filled with 0.300 mol $H_2(g)$, 0.400 mol $I_2(g)$, and 0.200 mol HI(g) at 870 K and total pressure 1.00 bar. Calculate the amounts of the components in the mixture at equilibrium given that $K = 870$ for the reaction $H_2(g) + I_2(g) \rightleftharpoons 2\,HI(g)$.

17.14 The dissociation of I_2 can be monitored by measuring the total pressure, and three sets of results are as follows:

T/K	973	1073	1173
$100p$/atm	6.244	7.500	9.181
$10^4 n_I$	2.4709	2.4555	2.4366

where n_I is the amount of I atoms per mole of I_2 molecules in the mixture, which occupied 342.68 cm^3. Calculate the equilibrium constants of the dissociation and the standard enthalpy of dissociation at the mean temperature.

17.15‡ In a study of $Cl_2O(g)$ by photoelectron ionization (R.P. Thorn *et al.*, *J. Phys. Chem.* **100**, 141 78 (1996)) the authors report $\Delta_f H^\ominus(Cl_2O) = +77.2$ kJ mol^{-1}. They combined this measurement with literature data on the reaction $Cl_2O\,(g) + H_2O(g) \rightarrow 2\,HOCl(g)$, for which $K = 8.2 \times 10^{-2}$ and $\Delta_r S^\ominus = +16.38$ J K^{-1} mol^{-1}, and with readily available thermodynamic data on water vapour to report a value for $\Delta_f H^\ominus(HOCl)$. Calculate that value. All quantities refer to 298 K.

* Problems denoted with the symbol ‡ were supplied by Charles Trapp, Carmen Giunta, and Marshall Cady.

17.16‡ The 1980s saw reports of $\Delta_f H^{\ominus}(SiH_2)$ ranging from 243 to 289 kJ mol^{-1}. For example, the lower value was cited in the review article by R. Walsh (*Acc. Chem. Res.* **14**, 246 (1981)); Walsh later leant towards the upper end of the range (H.M. Frey *et al.*, *J. Chem. Soc., Chem. Commun.* 1189 (1986)). The higher value was reported in S.-K. Shin and J.L. Beauchamp, *J. Phys. Chem.* **90**, 1507 (1986). If the standard enthalpy of formation is uncertain by this amount, by what factor is the equilibrium constant for the formation of SiH_2 from its elements uncertain at (a) 298 K, (b) 700 K?

17.17‡ Suppose that an iron catalyst at a particular manufacturing plant produces ammonia in the most cost-effective manner at 450°C when the pressure is such that $\Delta_r G$ for the reaction $\frac{1}{2} N_2(g) + \frac{3}{2} H_2(g) \rightarrow NH_3(g)$ is equal to −500 J mol^{-1}. (a) What pressure is needed? (b) Now suppose that a new catalyst is developed that is most cost-effective at 400°C when the pressure gives the same value of $\Delta_r G$. What pressure is needed when the new catalyst is used? What are the advantages of the new catalyst? Assume that (i) all gases are perfect gases or that (ii) all gases are van der Waals gases. Isotherms of $\Delta_r G(T, p)$ in the pressure range 100 atm $\le p \le$ 400 atm are needed to derive the answer. (c) Do the isotherms you plotted confirm Le Chatelier's principle concerning the response of equilibrium to changes in temperature and pressure?

17.18 Given that $\Delta_r G^{\ominus} = -212.7$ kJ mol^{-1} for the reaction in the Daniell cell at 25°C, and $b(CuSO_4) = 1.0 \times 10^{-3}$ mol kg^{-1} and $b(ZnSO_4) = 3.0 \times 10^{-3}$ mol kg^{-1}, calculate (a) the ionic strengths of the solutions, (b) the mean activity coefficients in the compartments, (c) the reaction quotient, (d) the standard cell potential, and (e) the cell potential. (Take $\gamma_+ = \gamma_- = \gamma_\pm$ in the respective compartments.)

17.19 Although the hydrogen electrode may be simple conceptually, it is cumbersome to use and several substitutes have been devised. One of these alternatives is the quinhydrone electrode (quinhydrone, Q·QH_2, is a complex of quinone, $C_6H_4O_2$ = Q, and hydroquinone, $C_6H_4O_2H_2$ = QH_2). The electrode half-reaction is $Q(aq) + 2 H^+(aq) + 2 e^- \rightarrow QH_2(aq)$, $E^{\ominus} = +0.6994$ V. If the cell Hg|Hg_2Cl_2(s)|HCl(aq)|Q·QH_2|Au is prepared, and the measured cell potential is +0.190 V, what is the pH of the HCl solution? Assume that the Debye–Hückel limiting law is applicable.

17.20 Consider the cell, Zn(s)|$ZnCl_2$(0.0050 mol kg^{-1})|Hg_2Cl_2(s)|Hg(l), for which the cell reaction is $Hg_2Cl_2(s) + Zn(s) \rightarrow 2 Hg(l) + 2 Cl^-(aq) + Zn^{2+}(aq)$. Given that $E^{\ominus}(Zn^{2+},Zn) = -0.7628$ V, $E^{\ominus}(Hg_2Cl_2, Hg) = +0.2676$ V, and that the cell potential is +1.2272 V, (a) write the Nernst equation for the cell. Determine (b) the standard cell potential, (c) $\Delta_r G$, $\Delta_r G^{\ominus}$, and K for the cell reaction, (d) the mean activity and activity coefficient of $ZnCl_2$ from the measured cell potential, and (e) the mean activity coefficient of $ZnCl_2$ from the Debye–Hückel limiting law. (f) Given that $(\partial E/\partial T)_p = -4.52 \times 10^{-4}$ V K^{-1}, calculate $\Delta_r S$ and $\Delta_r H$.

17.21 The potential of the cell Pt|H_2(g, $p^{\ominus}$)|HCl(aq,b)|Hg_2Cl_2(s)|Hg(l) has been measured with high precision (G. J. Hills and D. J. G. Ives, *J. Chem. Soc.*, 311 (1951)) with the following results at 25°C:

b/(mmol kg^{-1})	1.6077	3.0769	5.0403	7.6938	10.9474
E/V	0.60080	0.56825	0.54366	0.52267	0.50532

Determine the standard potential of the cell and the mean activity coefficient of HCl at these molalities. (Make a least-squares fit of the data to the best straight line.)

17.22 Careful measurements of the potential of the cell Pt|H_2(g, $p^{\ominus}$)|NaOH(aq, 0.0100 mol kg^{-1}), NaCl(aq, 0.011 25 mol kg^{-1})|AgCl(s)|Ag have been reported (C.P. Bezboruah *et al.*, *J. Chem. Soc. Faraday Trans. I* **69**, 949 (1973)). Among the data is the following information:

θ/°C	20.0	25.0	30.0
E/V	1.04774	1.04864	1.04942

Calculate pK_w at these temperatures and the standard enthalpy and entropy of the autoprotolysis of water at 25.0°C.

17.23 Measurements of the potentials of cells of the type Ag|AgX(s)|MX(b_1)|M_xHg|MX(b_2)|AgX(s)|Ag, where M_xHg denotes an amalgam and the electrolyte is an alkali metal halide dissolved in ethylene glycol, have been reported (U. Sen, *J. Chem. Soc. Faraday Trans. I* **69**, 2006 (1973)) and some values for LiCl are given below. Estimate the activity coefficient at the concentration marked * and then use this value to calculate activity coefficients from the measured cell potential at the other concentrations. Base your answer on the following version of the extended Debye–Hückel law:

$$\log \gamma_\pm = -\frac{AI^{1/2}}{1 + BI^{1/2}} + CI$$

with $A = 1.461$, $B = 1.70$, $C = 0.20$, and $I = b/b^{\ominus}$. For b_2 = 0.091 41 mol kg^{-1}:

b_1/(mol kg^{-1})	0.0555	0.09141*	0.1652	0.2171	1.040	1.350
E/V	−0.0220	0.0000	0.0263	0.0379	0.1156	0.1336

17.24 The standard potential of the AgCl/Ag,Cl^- couple has been measured very carefully over a range of temperature (R.G. Bates and V.E. Bowers, *J. Res. Nat. Bur. Stand.* **53**, 283 (1954)) and the results were found to fit the expression

$$E^{\ominus}/V = 0.23659 - 4.8564 \times 10^{-4}(\theta/°C) - 3.4205 \times 10^{-6}(\theta/°C)^2 + 5.869 \times 10^{-9}(\theta/°C)^3$$

Calculate the standard Gibbs energy and enthalpy of formation of Cl^-(aq) and its entropy at 298 K.

17.25‡ The table below summarizes the potential observed for the cell Pd|H_2(g, 1 bar)|BH(aq, b), B(aq, b)|AgCl(s)|Ag. Each measurement is made at equimolar concentrations of 2-aminopyridinium chloride (BH) and 2-aminopyridine (B). The data are for 25°C and it is found that $E^{\ominus} = 0.222\,51$ V. Use the data to determine pK_a for the acid at 25°C and the mean activity coefficient ($\gamma_\pm$) of BH as a function of molality (b) and ionic strength (I). Use the extended Debye–Hückel equation for the mean activity coefficient in the form

$$-\log \gamma_\pm = \frac{AI^{1/2}}{1 + BI^{1/2}} - kb$$

where $A = 0.5091$ and B and k are parameters that depend upon the ions. Draw a graph of the mean activity coefficient with $b = 0.04$ mol kg^{-1} and $0 \le I \le 0.1$.

b/(mol kg^{-1})	0.01	0.02	0.03	0.04	0.05
$E^{\ominus}$(25°C)/V	0.74452	0.72853	0.71928	0.71314	0.70809
b/(mol kg^{-1})	0.06	0.07	0.08	0.09	0.10
$E^{\ominus}$(25°C)/V	0.70380	0.70059	0.69790	0.69571	0.69338

Hint. Use mathematical software or a spreadsheet.

Theoretical problems

17.26 Express the equilibrium constant of a gas-phase reaction A + 3 B $\rightleftharpoons$ 2 C in terms of the equilibrium value of the extent of reaction, ξ, given that initially A and B were present in stoichiometric proportions. Find an expression for ξ as a function of the total pressure, p, of the reaction mixture and sketch a graph of the expression obtained.

17.27 The equilibrium constant K calculated from thermodynamic data refers to activities. For gas-phase reactions, that means partial pressures (and explicitly, $p_J/p^{\ominus}$). However, in practical applications we might wish to discuss gas-phase reactions in terms of molar concentrations. The equilibrium constant is then denoted K_c and, for the equilibrium $a\,A(g) + b\,B(g) \rightleftharpoons c\,C(g) + d\,D(g)$, we write

$$K_c = \frac{[C]^c[D]^d}{[A]^a[B]^b}$$

with, as usual, the molar concentration [J] interpreted as $[J]/c^\ominus$ with $c^\ominus = 1\ \text{mol dm}^{-3}$. Show that

$$K = K_c \times \left(\frac{c^\ominus RT}{p^\ominus}\right)^{\Delta\nu_{\text{gas}}}$$

where $\Delta\nu_{\text{gas}} = c + d - (a + b)$.

17.28 Find an expression for the standard reaction Gibbs energy at a temperature T' in terms of its value at another temperature T and the coefficients a, b, and c in the expression for the molar heat capacity listed in Table 14.2. Evaluate the standard Gibbs energy of formation of $H_2O(l)$ at 372 K from its value at 298 K.

17.29 Show that, if the ionic strength of a solution of the sparingly soluble salt MX and the freely soluble salt NX is dominated by the concentration C of the latter, and if it is valid to use the Debye–Hückel limiting law, the solubility S' in the mixed solution is given by

$$S' = \frac{K_s e^{4.606AC^{1/2}}}{C}$$

when K_s is small (in a sense to be specified).

Applications: to biology, environmental science, and chemical engineering

17.30 The protein myoglobin (Mb) stores O_2 in muscle and the protein haemoglobin (Hb) transports O_2 in blood; haemoglobin is composed of four myoglobin-like molecules. Here we explore the chemical equilibria associated with binding of O_2 in these proteins. (a) First, consider the equilibrium between Mb and O_2:

$$\text{Mb(aq)} + O_2(g) \rightleftharpoons \text{MbO}_2(aq) \qquad K = \frac{[\text{MbO}_2]}{p[\text{Mb}]}$$

where p is the numerical value of the partial pressure (in torr) of O_2 gas. Show that the *fractional saturation*, s, the fraction of Mb molecules that are oxygenated, is

$$s = \frac{Kp}{1 + Kp}$$

and plot the dependence of s on p for $K = 0.05$. (b) Now consider the equilibria between Hb and O_2:

$$\text{Hb(aq)} + O_2(g) \rightleftharpoons \text{HbO}_2(aq) \qquad K_1 = \frac{[\text{HbO}_2]}{p[\text{Hb}]}$$

$$\text{HbO}_2(aq) + O_2(g) \rightleftharpoons \text{Hb(O}_2)_2(aq) \qquad K_2 = \frac{[\text{Hb(O}_2)_2]}{p[\text{HbO}_2]}$$

$$\text{Hb(O}_2)_2(aq) + O_2(g) \rightleftharpoons \text{Hb(O}_2)_3(aq) \qquad K_3 = \frac{[\text{Hb(O}_2)_3]}{p[\text{Hb(O}_2)_2]}$$

$$\text{Hb(O}_2)_3(aq) + O_2(g) \rightleftharpoons \text{Hb(O}_2)_4(aq) \qquad K_4 = \frac{[\text{Hb(O}_2)_4]}{p[\text{Hb(O}_2)_3]}$$

Show that

$$s = \frac{[\text{O2}]_{\text{bound}}}{4[\text{Hb}]_{\text{total}}} = \frac{(1 + 2K_2p + 3K_2K_3p^2 + 4K_2K_3K_4p^3)K_1p}{4(1 + K_1p + K_1K_2p^2 + K_1K_2K_3p^3 + K_1K_2K_3K_4p^4)}$$

and plot the dependence of s on p (in torr) for $K_1 = 0.01$, $K_2 = 0.02$, $K_3 = 0.04$, and $K_4 = 0.08$. *Hints.* To develop an expression for s, proceed as follows: (i) express $[\text{Hb(O}_2)_2]$ in terms of $[\text{HbO}_2]$ by using K_2, then express $[\text{HbO}_2]$ in terms of [Hb] by using K_1, and likewise for all the other concentrations of $\text{Hb(O}_2)_3$ and $\text{Hb(O}_2)_4$. (ii) Show that

$$[O_2]_{\text{bound}} = [\text{HbO}_2] + 2[\text{Hb(O}_2)_2] + 3[\text{Hb(O}_2)_3] + 4[\text{Hb(O}_2)_4]$$
$$= (1 + 2K_2p + 3K_2K_3p^2 + 4K_2K_3K_4p^3)K_1p[\text{Hb}]$$

$$[\text{Hb}]_{\text{total}} = (1 + K_1p + K_1K_2p^2 + K_1K_2K_3p^3 + K_1K_2K_3K_4p^4)[\text{Hb}]$$

(iii) Use the fact that each Hb molecule has four sites at which O_2 can attach. (c) The binding of O_2 to haemoglobin is an example of *cooperative binding*, in which the binding of a ligand (in this case O_2) to a biopolymer (in this case Hb) becomes more favourable thermodynamically (that is, the equilibrium constant increases) as the number of bound ligands increases up to the maximum number of binding sites. Which features of the plot from part (b) can be ascribed to cooperative binding of O_2 to Hb?

17.31 The curves you were asked to plot in Problem 17.30 may also be modelled mathematically by the equation

$$\log\frac{s}{1-s} = \nu \log p - \nu \log K$$

where s is the saturation, p is the partial pressure of O_2, K is a constant (not the binding constant for one ligand), and ν is the *Hill coefficient*, which varies from 1, for no cooperativity, to N for all-or-none binding of N ligands ($N = 4$ in Hb). The Hill coefficient for myoglobin is 1, and for haemoglobin it is 2.8. (a) Determine the constant K for both Mb and Hb from the graph of fractional saturation (at $s = 0.5$) and then calculate the fractional saturation of Mb and Hb for the following values of p/kPa: 1.0, 1.5, 2.5, 4.0, 8.0. (b) Calculate the value of s at the same p values assuming ν has the theoretical maximum value of 4.

17.32 Here we investigate the molecular basis for the observation that the hydrolysis of ATP is exergonic at pH = 7.0 and 310 K. (a) It is thought that the exergonicity of ATP hydrolysis is due in part to the fact that the standard entropies of hydrolysis of polyphosphates are positive. Why would an increase in entropy accompany the hydrolysis of a triphosphate group into a diphosphate and a phosphate group? (b) Under identical conditions, the Gibbs energies of hydrolysis of H_4ATP and MgATP^{2-}, a complex between the Mg^{2+} ion and ATP^{4-}, are less negative than the Gibbs energy of hydrolysis of ATP^{4-}. This observation has been used to support the hypothesis that electrostatic repulsion between adjacent phosphate groups is a factor that controls the exergonicity of ATP hydrolysis. Provide a rationale for the hypothesis and discuss how the experimental evidence supports it. Do these electrostatic effects contribute to the $\Delta_r H$ or $\Delta_r S$ terms that determine the exergonicity of the reaction? *Hint.* In the MgATP^{2-} complex, the Mg^{2+} ion and ATP^{4-} anion form two bonds: one that involves a negatively charged oxygen belonging to the terminal phosphate group of ATP^{4-} and another that involves a negatively charged oxygen belonging to the phosphate group adjacent to the terminal phosphate group of ATP^{4-}.

17.33 To get a sense of the effect of cellular conditions on the ability of ATP to drive biochemical processes, compare the standard Gibbs energy of hydrolysis of ATP to ADP with the reaction Gibbs energy in an environment at 37°C in which pH = 7.0 and the ATP, ADP, and P_i^- concentrations are all 1.0 μmol dm^{-3}.

17.34 Under biochemical standard conditions, aerobic respiration produces approximately 38 molecules of ATP per molecule of glucose that is completely oxidized. (a) What is the percentage efficiency of aerobic respiration under biochemical standard conditions? (b) The following conditions are more likely to be observed in a living cell: $p_{CO_2} = 5.3 \times 10^{-2}$ atm, $p_{O_2} = 0.132$ atm, [glucose] $= 5.6 \times 10^{-2}$ mol dm^{-3}, [ATP] = [ADP] = $[P_i] = 1.0 \times 10^{-4}$ mol dm^{-3}, pH = 7.4, $T = 310$ K. Assuming that activities can be replaced by the numerical values of molar concentrations, calculate the efficiency of aerobic respiration under these physiological conditions. (c) A typical diesel engine operates between $T_c = 873$ K and $T_h = 1923$ K with an efficiency that is approximately 75 per cent of the theoretical limit of $(1 - T_c/T_h)$ (see Section 15.2). Compare the efficiency of a typical diesel engine with that of aerobic respiration under typical physiological conditions (see part b). Why is biological energy conversion more or less efficient than energy conversion in a diesel engine?

17.35 In anaerobic bacteria, the source of carbon may be a molecule other than glucose and the final electron acceptor is some molecule other than O_2. Could a bacterium evolve to use the ethanol/nitrate pair instead of the glucose/O_2 pair as a source of metabolic energy?

17.36 If the mitochondrial electric potential between matrix and the intermembrane space were 70 mV, as is common for other membranes, how much ATP could be synthesized from the transport of 4 mol H^+, assuming the pH difference remains the same?

17.37 The standard potentials of proteins are not commonly measured by the methods described in this chapter because proteins often lose their native structure and function when they react on the surfaces of electrodes. In an alternative method, the oxidized protein is allowed to react with an appropriate electron donor in solution. The standard potential of the protein is then determined from the Nernst equation, the equilibrium concentrations of all species in solution, and the known standard potential of the electron donor. We shall illustrate this method with the protein cytochrome *c*. The one-electron reaction between cytochrome *c* (cyt) and 2,6-dichloroindophenol, D, can be followed spectrophotometrically because each of the four species in solution has a distinct colour, or absorption spectrum. We write the reaction as $\text{cyt}_{ox} + D_{red} \rightleftharpoons \text{cyt}_{red} + D_{ox}$, where the subscripts 'ox' and 'red' refer to oxidized and reduced states, respectively. (a) Consider $E^{\ominus}_{cyt}$ and $E^{\ominus}_{D}$ to be the standard potentials of cytochrome *c* and D, respectively. Show that, at equilibrium ('eq'), a plot of $\ln([D_{ox}]_{eq}/[D_{red}]_{eq})$ versus $\ln([\text{cyt}_{ox}]_{eq}/[\text{cyt}_{red}]_{eq})$ is linear with slope of one and *y*-intercept $F(E^{\ominus}_{cyt} - E^{\ominus}_{D})/RT$, where equilibrium activities are replaced by the numerical values of equilibrium molar concentrations. (b) The following data were obtained for the reaction between oxidized cytochrome *c* and reduced D in a pH = 6.5 buffer at 298 K. The ratios $[D_{ox}]_{eq}/[D_{red}]_{eq}$ and $[\text{cyt}_{ox}]_{eq}/[\text{cyt}_{red}]_{eq}$ were adjusted by titrating a solution containing oxidized cytochrome *c* and reduced D with a solution of sodium ascorbate, which is a strong reductant. From the data and the standard potential of D of 0.237 V, determine the standard potential of cytochrome *c* at pH = 6.5 and 298 K.

$[D_{ox}]_{eq}/[D_{red}]_{eq}$	0.00279	0.00843	0.0257	0.0497
$[\text{cyt}_{ox}]_{eq}/[\text{cyt}_{red}]_{eq}$	0.0106	0.0230	0.0894	0.197
$[D_{ox}]_{eq}/[D_{red}]_{eq}$	0.0748	0.238	0.534	
$[\text{cyt}_{ox}]_{eq}/[\text{cyt}_{red}]_{eq}$	0.335	0.809	1.39	

17.38‡ The dimerization of ClO in the Antarctic winter stratosphere is believed to play an important part in that region's severe seasonal depletion of ozone. The following equilibrium constants are based on measurements by P. A. Cox and G. D. Hayman (*Nature* **332**, 796 (1988)) on the reaction $2ClO\,(g) \rightarrow (ClO)_2\,(g)$.

T/K	233	248	258	268	273
K	4.13×10^8	5.00×10^7	1.45×10^7	5.37×10^6	3.20×10^6
T/K	280	288	295	303	
K	9.62×10^5	4.28×10^5	1.67×10^5	7.02×10^4	

(a) Derive the values of $\Delta_r H^{\ominus}$ and $\Delta_r S^{\ominus}$ for this reaction. (b) Compute the standard enthalpy of formation and the standard molar entropy of $(ClO)_2$ given $\Delta_f H^{\ominus}(ClO) = +101.8$ kJ mol^{-1} and $S^{\ominus}_m(ClO) = 226.6$ J K^{-1} mol^{-1}.

17.39 A fuel cell develops an electric potential from the chemical reaction between reagents supplied from an outside source. What is the potential of a cell fuelled by (a) hydrogen and oxygen, (b) the complete oxidation of benzene at 1.0 bar and 298 K?

17.40 A fuel cell is constructed in which both electrodes make use of the oxidation of methane. The left-hand electrode makes use of the complete oxidation of methane to carbon dioxide and water; the right-hand electrode makes use of the partial oxidation of methane to carbon monoxide and water. (a) Which electrode is the cathode? (b) What is the cell potential at 25°C when all gases are at 1 bar?

17.41‡ Nitric acid hydrates have received much attention as possible catalysts for heterogeneous reactions which bring about the Antarctic ozone hole. Worsnop *et al.* investigated the thermodynamic stability of these hydrates under conditions typical of the polar winter stratosphere (D.R. Worsnop *et al.*, *Science* **259**, 71 (1993)). Standard reaction Gibbs energies can be computed for the following reactions at 190 K from their data:

(i) $H_2O(g) \rightarrow H_2O(s)$ $\quad \Delta_r G^{\ominus} = -23.6$ kJ mol^{-1}

(ii) $H_2O(g) + HNO_3(g) \rightarrow HNO_3{\cdot}H_2O(s)$ $\quad \Delta_r G^{\ominus} = -57.2$ kJ mol^{-1}

(iii) $2\,H_2O(g) + HNO_3(g) \rightarrow HNO_3{\cdot}2H_2O(s)$ $\quad \Delta_r G^{\ominus} = -85.6$ kJ mol^{-1}

(iv) $3\,H_2O(g) + HNO_3(g) \rightarrow HNO_3{\cdot}3H_2O(s)$ $\quad \Delta_r G^{\ominus} = -112.8$ kJ mol^{-1}

Which solid is thermodynamically most stable at 190 K if $p_{H_2O} = 1.3 \times 10^{-7}$ bar and $p_{HNO_3} = 4.1 \times 10^{-10}$ bar? *Hint.* Try computing $\Delta_r G$ for each reaction under the prevailing conditions; if more than one solid forms spontaneously, examine $\Delta_r G$ for the conversion of one solid to another.

PART 5
Chemical dynamics

The final part of the book is concerned with another fertile area of study within physical chemistry, namely, the rates and mechanisms of chemical reactions. In Chapter 18 we consider the motion of molecules in gases and liquids and the motion of ions in solution. Our goal is to account for the rates at which molecules and energy migrate through gases by using the kinetic model and generalize the discussion to treat migration through various media by formulating the diffusion equation.

In Chapter 19 we explore rates of chemical reactions and experimental techniques for their measurement. A rate law provides both a summary of experimental observations on the rate of a reaction and insight into its mechanism, the sequence of elementary steps by which it occurs. We show how to derive rate laws for simple and complex reaction mechanisms, often by using a variety of common approximations.

In Chapter 20 we introduce several theories, some simple and others more sophisticated, that account for reaction rates and their temperature dependence. The simplest theory, collision theory, can be used only for simple gas-phase reactions and is based on the kinetic model of gases. A more sophisticated approach is provided by transition-state theory and makes use of concepts of statistical thermodynamics introduced in Part 4, building on the information about molecular energy levels that have either been determined spectroscopically or computed. Transition-state theory is useful for discussions of rates of reactions in solution, including electron transfer processes. The highest level of sophistication comes from the theoretical and experimental explorations of potential energy surfaces computed by using the techniques described in Part 2.

Of enormous importance in both industry and biology is the control of reaction rates by catalysis, which we discuss in Chapter 21. The themes developed there give us deep understanding of how reaction rates are optimized in biological and industrial processes.

18

Molecular motion

One of the simplest types of molecular motion to describe is the chaotic motion of molecules of a perfect gas. We see that a simple theory accounts for the pressure of a gas and the rates at which molecules and energy migrate through gases. Molecular mobility is particularly important in liquids. Another simple kind of motion is the net motion of ions in solution in the presence of an electric field. Molecular and ionic motion have common features and, by considering them from a more general viewpoint, we derive expressions that govern the migration of properties through matter. One of the most useful consequences of this general approach is the formulation of the diffusion equation, which is an equation that shows how matter and energy spread through media of various kinds. Finally, we build a simple model for all types of molecular motion, in which the molecules migrate in a series of small steps, and see that it accounts for many of the properties of migrating molecules in both gases and condensed phases.

The general approach we describe in this chapter provides techniques for discussing the motion of all kinds of particles in all kinds of fluids. We set the scene by considering a simple type of motion, that of molecules in a perfect gas, and go on to see that molecular motion in liquids shows a number of similarities. We shall concentrate on the **transport properties** of a substance, its ability to transfer matter, energy, or some other property from one place to another. Four examples of transport properties are:

Diffusion, the net transport of a species down a concentration gradient.

Thermal conduction, the net transport of heat (more formally, the energy of thermal motion) down a temperature gradient.

Electric conduction, the net transport of charged species down a potential gradient (as part of a complete electric circuit).

Viscosity, the net transfer of linear momentum down a velocity gradient.

It is convenient to include in the discussion **effusion**, the emergence of a gas from a container through a small hole.

Motion in gases

Here we present the kinetic model of a perfect gas as a starting point for the discussion of its transport properties. In the **kinetic model** of gases (which is sometimes called the *kinetic-molecular theory*, KMT) we assume that the only contribution to the energy of the gas is from the kinetic energies of the molecules. The kinetic model is one of the most remarkable—and arguably most beautiful—models in physical chemistry for, from a set of very slender assumptions, powerful quantitative conclusions can be reached.

18.1 The kinetic model of gases

The kinetic model is based on three assumptions:

1. The gas consists of molecules of mass m in ceaseless random motion obeying the laws of classical mechanics.
2. The size of the molecules is negligible, in the sense that their diameters are much smaller than the average distance travelled between collisions.
3. The molecules interact only through brief elastic collisions.

An **elastic collision** is a collision in which the total translational kinetic energy of the molecules is conserved.

(a) Pressure and molecular speeds

From the very economical assumptions of the kinetic model, we show in the following *Justification* that the pressure and volume of the gas are related by

$$pV = \tfrac{1}{3}nMc^2 \qquad (18.1)°$$

where $M = mN_A$, the molar mass of the molecules of mass m, and c is their **root mean square speed**, the square root of the mean of the squares of the speeds, v, of the molecules:

$$c = \langle v^2 \rangle^{1/2} \qquad [18.2]$$

Justification 18.1 *The pressure of a gas according to the kinetic model*

Consider the arrangement in Fig. 18.1. When a particle of mass m that is travelling with a component of velocity v_x parallel to the x-axis collides with the wall on the right and is reflected, its linear momentum changes from mv_x before the collision to $-mv_x$ after the collision (when it is travelling in the opposite direction). The x-component of momentum therefore changes by $2mv_x$ on each collision (the y- and z-components are unchanged). Many molecules collide with the wall in an interval Δt, and the total change of momentum is the product of the change in momentum of each molecule multiplied by the number of molecules that reach the wall during the interval.

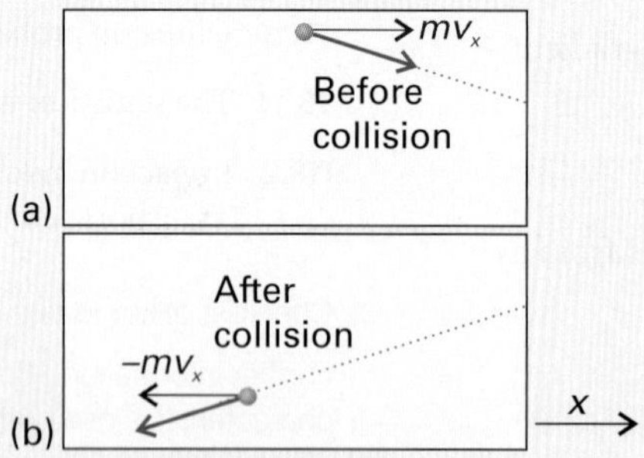

Fig. 18.1 The pressure of a gas arises from the impact of its molecules on the walls. In an elastic collision of a molecule with a wall perpendicular to the x-axis, the x-component of velocity is reversed but the y- and z-components are unchanged.

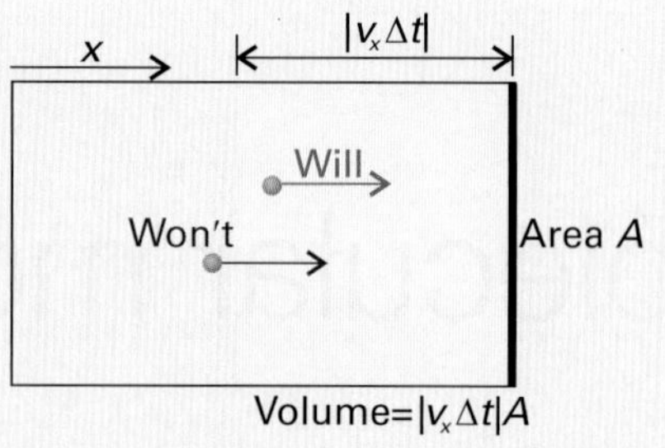

Fig. 18.2 A molecule will reach the wall on the right within an interval Δt if it is within a distance $v_x\Delta t$ of the wall and travelling to the right.

Because a molecule with velocity component v_x can travel a distance $v_x\Delta t$ along the x-axis in an interval Δt, all the molecules within a distance $v_x\Delta t$ of the wall will strike it if they are travelling towards it (Fig. 18.2). It follows that, if the wall has area A, then all the particles in a volume $A \times v_x\Delta t$ will reach the wall (if they are travelling towards it). The number density of particles is nN_A/V, where n is the total amount of molecules in the container of volume V and N_A is Avogadro's constant, so the number of molecules in the volume $Av_x\Delta t$ is $(nN_A/V) \times Av_x\Delta t$.

At any instant, half the particles are moving to the right and half are moving to the left. Therefore, the average number of collisions with the wall during the interval Δt is $\tfrac{1}{2}nN_AAv_x\Delta t/V$. The total momentum change in that interval is the product of this number and the change $2mv_x$:

$$\text{Momentum change} = \frac{nN_AAv_x\Delta t}{2V} \times 2mv_x$$

$$= \frac{nmAN_Av_x^2\Delta t}{V} = \frac{nMAv_x^2\Delta t}{V}$$

where $M = mN_A$.

Next, to find the force, we calculate the rate of change of momentum, which is this change of momentum divided by the interval Δt during which it occurs:

$$\text{Rate of change of momentum} = \frac{nMAv_x^2}{V}$$

This rate of change of momentum is equal to the force (by Newton's second law of motion). It follows that the pressure, the force divided by the area, is

$$\text{Pressure} = \frac{nMv_x^2}{V}$$

Not all the molecules travel with the same velocity, so the detected pressure, p, is the average (denoted $\langle \ldots \rangle$) of the quantity just calculated:

$$p = \frac{nM\langle v_x^2 \rangle}{V}$$

This expression already resembles the perfect gas equation of state.

To write an expression of the pressure in terms of the root mean square speed, c, we begin by writing the speed of a single molecule, v, as $v^2 = v_x^2 + v_y^2 + v_z^2$. Because the root mean square speed, c, is defined as $c = \langle v^2 \rangle^{1/2}$ (eqn 18.2), it follows that

$$c^2 = \langle v^2 \rangle = \langle v_x^2 \rangle + \langle v_y^2 \rangle + \langle v_z^2 \rangle$$

However, because the molecules are moving randomly, all three averages are the same. It follows that $c^2 = 3\langle v_x^2 \rangle$. Equation 18.1 follows immediately by substituting $\langle v_x^2 \rangle = \frac{1}{3}c^2$ into $p = nM\langle v_x^2 \rangle / V$.

Equation 18.1 is one of the key results of the kinetic model. We see that, if the root mean square speed of the molecules depends only on the temperature, then at constant temperature

$$pV = \text{constant}$$

which is the content of Boyle's law. Moreover, for eqn 18.1 to be the equation of state of a perfect gas, its right-hand side must be equal to nRT. It follows that the root mean square speed of the molecules in a gas at a temperature T must be

$$c = \left(\frac{3RT}{M}\right)^{1/2} \qquad (18.3)°$$

However, it is also possible to demonstrate that the right-hand side of eqn 18.1 is equal to nRT by appealing to the Boltzmann distribution. To do so, we proceed in two steps. First, in the following *Justification* we show that the fraction of molecules that have a speed in the range v to $v + dv$ is $f(v)dv$, where

$$f(v) = 4\pi\left(\frac{M}{2\pi RT}\right)^{3/2} v^2 e^{-Mv^2/2RT} \qquad (18.4)$$

The function $f(v)$ is called the **Maxwell–Boltzmann distribution of speeds.** Then we use this distribution to calculate the average value of v^2.

Justification 18.2 *The Maxwell distribution of speeds*

The Boltzmann distribution implies that the fraction of molecules with velocity components v_x, v_y, v_z is proportional to an exponential function of their kinetic energy, which is

$$E = \tfrac{1}{2}mv_x^2 + \tfrac{1}{2}mv_y^2 + \tfrac{1}{2}mv_z^2$$

Therefore, we can use the relation $a^{x+y+z+\cdots} = a^x a^y a^z \ldots$ to write

$$f = Ke^{-E/kT} = Ke^{-(\frac{1}{2}mv_x^2 + \frac{1}{2}mv_y^2 + \frac{1}{2}mv_z^2)/kT} = Ke^{-mv_x^2/2kT}e^{-mv_y^2/2kT}e^{-mv_z^2/2kT}$$

where K is a constant of proportionality (at constant temperature) and $f dv_x dv_y dv_z$ is the fraction of molecules in the velocity range v_x to $v_x + dv_x$, v_y to $v_y + dv_y$, and v_z to $v_z + dv_z$. We see that the fraction factorizes into three terms, one for each axis, and we can write $f = f(v_x)f(v_y)f(v_z)$ with

$$f(v_x) = K^{1/3}e^{-mv_x^2/2kT}$$

and likewise for the other two axes.

To determine the constant K, we note that a molecule must have a velocity component somewhere in the range $-\infty < v_x < \infty$, so

$$\int_{-\infty}^{\infty} f(v_x)dv_x = 1$$

Substitution of the expression for $f(v_x)$ then gives

$$1 = K^{1/3}\int_{-\infty}^{\infty} e^{-mv_x^2/2kT}dv_x = K^{1/3}\left(\frac{2\pi kT}{m}\right)^{1/2}$$

where we have used the standard integral

$$\int_{-\infty}^{\infty} e^{-ax^2}dx = \left(\frac{\pi}{a}\right)^{1/2}$$

Therefore, $K = (m/2\pi kT)^{3/2} = (M/2\pi RT)^{3/2}$, where M is the molar mass of the molecules ($m = M/N_A$ and $N_A k = R$). At this stage we know that

$$f(v_x) = \left(\frac{M}{2\pi RT}\right)^{1/2} e^{-Mv_x^2/2RT} \qquad (18.5)$$

The probability that a molecule has a velocity in the range v_x to $v_x + dv_x$, v_y to $v_y + dv_y$, v_z to $v_z + dv_z$ is

$$f(v_x)f(v_y)f(v_z)dv_x dv_y dv_z = \left(\frac{M}{2\pi RT}\right)^{3/2} e^{-Mv^2/2RT}dv_x dv_y dv_z$$

where $v^2 = v_x^2 + v_y^2 + v_z^2$. The probability $f(v)dv$ that the molecules have a speed in the range v to $v + dv$ regardless of direction is the sum of the probabilities that the velocity lies in any of the volume elements $dv_x dv_y dv_z$ forming a spherical shell of radius v and thickness dv (Fig. 18.3). The sum of the volume elements on the right-hand side of the last equation is the volume of this shell, $4\pi v^2 dv$. Therefore,

$$f(v) = 4\pi\left(\frac{M}{2\pi RT}\right)^{3/2} v^2 e^{-Mv^2/2RT}$$

which is eqn 18.4.

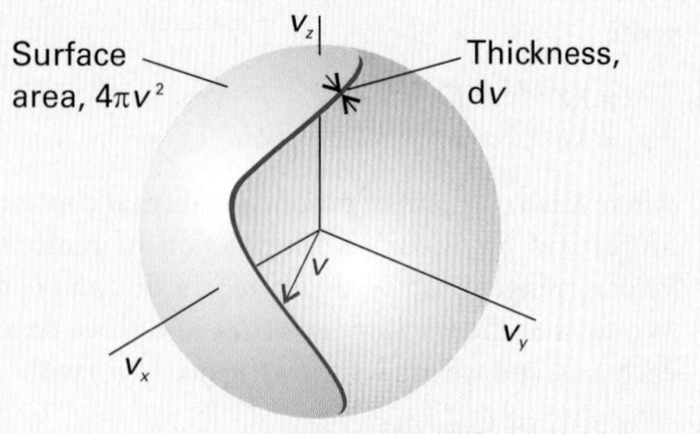

Fig. 18.3 To evaluate the probability that a molecule has a speed in the range v to $v + dv$, we evaluate the total probability that the molecule will have a speed that is anywhere on the surface of a sphere of radius $v = (v_x^2 + v_y^2 + v_z^2)^{1/2}$ by summing the probabilities that it is in a volume element $dv_x dv_y dv_z$ at a distance v from the origin.

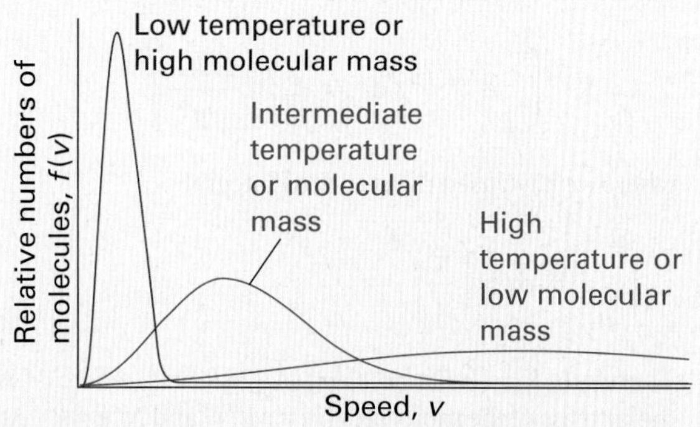

Fig. 18.4 The distribution of molecular speeds with temperature and molar mass. Note that the most probable speed (corresponding to the peak of the distribution) increases with temperature and with decreasing molar mass, and simultaneously the distribution becomes broader. We first saw this diagram as Fundamentals Fig. F.7.

interActivity (a) Plot different distributions by keeping the molar mass constant at 100 g mol^{-1} and varying the temperature of the sample between 200 K and 2000 K. (b) Use mathematical software or the *Living graph* applet from the text's web site to evaluate numerically the fraction of molecules with speeds in the range 100 m s^{-1} to 200 m s^{-1} at 300 K and 1000 K. (c) Based on your observations, provide a molecular interpretation of temperature.

The important features of the Maxwell–Boltzmann distribution are as follows (and are shown pictorially in Fig. 18.4):

1. Equation 18.4 includes a decaying exponential function, the term $e^{-Mv^2/2RT}$. Its presence implies that the fraction of molecules with very high speeds will be very small because e^{-x^2} becomes very small when x^2 is large.

2. The factor $M/2RT$ multiplying v^2 in the exponent is large when the molar mass, M, is large, so the exponential factor goes most rapidly towards zero when M is large. That is, heavy molecules are unlikely to be found with very high speeds.

3. The opposite is true when the temperature, T, is high: then the factor $M/2RT$ in the exponent is small, so the exponential factor falls towards zero relatively slowly as v increases. In other words, a greater fraction of the molecules can be expected to have high speeds at high temperatures than at low temperatures.

4. A factor v^2 (the term before the e) multiplies the exponential. This factor goes to zero as v goes to zero, so the fraction of molecules with very low speeds will also be very small.

5. The remaining factors (the term in parentheses in eqn 18.4 and the 4π) simply ensure that, when we sum the fractions over the entire range of speeds from zero to infinity, then we get 1.

Once we have the Maxwell–Boltzmann distribution, we can calculate the mean value of any power of the speed by evaluating the appropriate integral. For instance, to evaluate the fraction of molecules in a given narrow range of speeds, Δv, we evaluate $f(v)$ at the speed of interest, assume that $f(v)$ is constant over the narrow range, then multiply it by the width of the range of speeds of interest, that is, we form $f(v)\Delta v$. To use the distribution to calculate the fraction in a range of speeds that is too wide to be treated as infinitesimal, we evaluate the integral:

$$\text{Fraction in the range } v_1 \text{ to } v_2 = \int_{v_1}^{v_2} f(v)\mathrm{d}v \qquad (18.6)$$

This integral is the area under the graph of f as a function of v and, except in special cases, has to be evaluated numerically by using mathematical software (Fig. 18.5). To evaluate the average value of v^n we calculate

$$\langle v^n \rangle = \int_0^{\infty} v^n f(v)\mathrm{d}v \qquad (18.7)$$

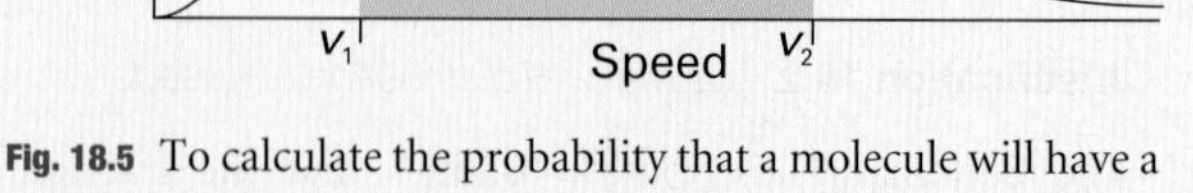

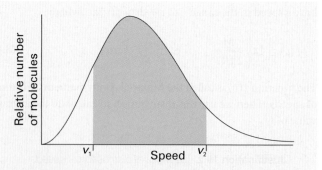

Fig. 18.5 To calculate the probability that a molecule will have a speed in the range v_1 to v_2, we integrate the distribution between those two limits; the integral is equal to the area of the curve between the limits, as shown shaded here.

In particular, straightforward integration with $n = 2$ results in eqn 18.3 for the mean square speed (c^2) of the molecules at a temperature T. We can conclude that *the root mean square speed of the molecules of a gas is proportional to the square root of the temperature and inversely proportional to the square root of the molar mass.* That is, the higher the temperature, the higher the root mean square speed of the molecules, and, at a given temperature, heavy molecules travel more slowly than light molecules. Sound waves are pressure waves, and for them to propagate the molecules of the gas must move to form regions of high and low pressure. Therefore, we should expect the root mean square speeds of molecules to be comparable to the speed of sound in air (340 m s^{-1}). The root mean square speed of N_2 molecules, for instance, is found from eqn 18.3 to be 515 m s^{-1} at 298 K.

Example 18.1 *Calculating the mean speed of molecules in a gas*

What is the mean speed, $\bar{c}$, of N_2 molecules in air at 25°C?

Method The mean speed, $\bar{c}$, is obtained by evaluating the integral

$$\bar{c} = \int_0^\infty v f(v)\mathrm{d}v$$

with f given in eqn 18.4.

Answer The integral required is

$$\bar{c} = 4\pi\left(\frac{M}{2\pi RT}\right)^{3/2}\int_0^\infty v^3 \mathrm{e}^{-Mv^2/2RT}\,\mathrm{d}v$$

$$= 4\pi\left(\frac{M}{2\pi RT}\right)^{3/2} \times \frac{1}{2}\left(\frac{2RT}{M}\right)^2 = \left(\frac{8RT}{\pi M}\right)^{1/2}$$

where we have used the standard result from tables of integrals (or software) that

$$\int_0^\infty x^3 \mathrm{e}^{-ax^2}\,\mathrm{d}x = \frac{1}{2a^2}$$

Substitution of the data then gives

$$\bar{c} = \left(\frac{8 \times (8.3145\ \mathrm{J\,K^{-1}\,mol^{-1}}) \times (298\ \mathrm{K})}{\pi \times (28.02 \times 10^{-3}\ \mathrm{kg\,mol^{-1}})}\right)^{1/2} = 475\ \mathrm{m\,s^{-1}}$$

We have used 1 J = 1 kg m^2 s^{-2}.

Self-test 18.1 Evaluate the root mean square speed of the molecules by integration. You will need the standard integral

$$\int_0^\infty x^4 \mathrm{e}^{-ax^2}\,\mathrm{d}x = \frac{3}{8a^2}\left(\frac{\pi}{a}\right)^{1/2}$$

[$c = (3RT/M)^{1/2}$, 515 m s^{-1} at 298 K]

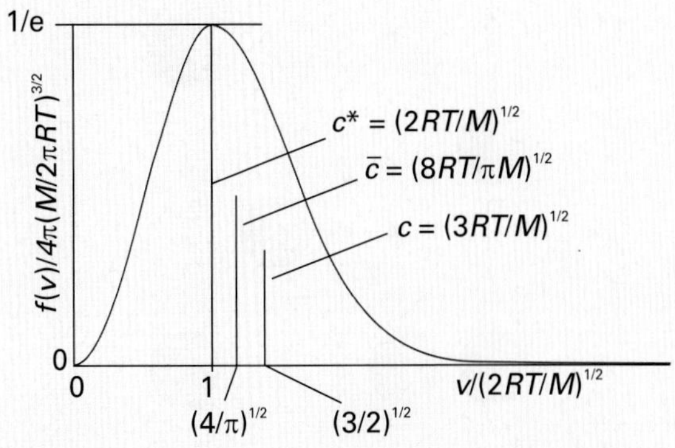

Fig. 18.6 A summary of the conclusions that can be deduced from the Maxwell distribution for molecules of molar mass M at a temperature T: c^* is the most probable speed, $\bar{c}$ is the mean speed, and c is the root mean square speed.

As shown in Example 18.1, we can use the Maxwell distribution to evaluate the **mean speed**, $\bar{c}$, of the molecules in a gas:

$$\bar{c} = \left(\frac{8RT}{\pi M}\right)^{1/2} \quad (18.8)$$

We can identify the **most probable speed**, c^*, from the location of the peak of the distribution:

$$c^* = \left(\frac{2RT}{M}\right)^{1/2} \quad (18.9)$$

The location of the peak of the distribution is found by differentiating f with respect to v and looking for the value of v at which the derivative is zero (other than at $v = 0$ and $v = \infty$); see Problem 8.21. Figure 18.6 summarizes these results.

The **mean relative speed**, $\bar{c}_{rel}$, the mean speed with which one molecule approaches another, can also be calculated from the distribution:

$$\bar{c}_{rel} = 2^{1/2}\bar{c} \quad (18.10)$$

This result is much harder to derive, but the diagram in Fig. 18.7 should help to show that it is plausible. The last result can also be generalized to the relative mean speed of two dissimilar molecules of masses m_A and m_B:

$$\bar{c}_{rel} = \left(\frac{8kT}{\pi\mu}\right)^{1/2} \qquad \mu = \frac{m_A m_B}{m_A + m_B} \quad (18.11)$$

Note that the molecular masses (not the molar masses) and Boltzmann's constant, $k = R/N_A$, appear in this expression; the quantity μ is called the **reduced mass** of the molecules. Equation 18.11 turns into eqn 18.10 when the molecules are identical (that is, $m_A = m_B = m$, so $\mu = \frac{1}{2}m$).

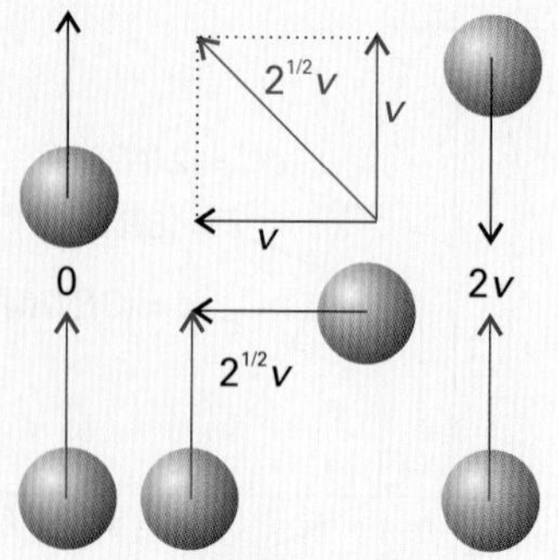

Fig. 18.7 A simplified version of the argument to show that the mean relative speed of molecules in a gas is related to the mean speed. When the molecules are moving in the same direction, the mean relative speed is zero; it is $2v$ when the molecules are approaching each other. A typical mean direction of approach is from the side, and the mean speed of approach is then $2^{1/2}v$. The last direction of approach is the most characteristic, so the mean speed of approach can be expected to be about $2^{1/2}v$. This value is confirmed by more detailed calculation.

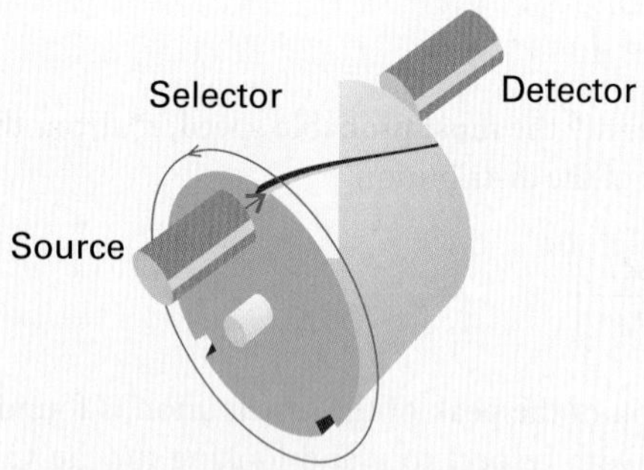

Fig. 18.8 A velocity selector. The molecules are produced in the source (which may be an oven with a small hole in one wall), and travel in a beam towards the rotating channels. Only if the speed of a molecule is such as to carry it along the channel that rotates into its path will it reach the detector. Thus, the number of slow molecules can be counted by rotating the discs slowly and the number of fast molecules counted by rotating the discs rapidly.

The Maxwell distribution has been verified experimentally. For example, molecular speeds can be measured directly with a velocity selector (Fig. 18.8). The spinning cylinder has channels that permit the passage of only those molecules moving through them at the appropriate speed, and the number of molecules can be determined by collecting them at a detector.

(b) The collision frequency

Although the kinetic-molecular theory assumes that the molecules are pointlike, we can count a 'hit' whenever the centres of two molecules come within a distance d of each other, where d, the **collision diameter**, is of the order of the actual diameters of the molecules (for impenetrable hard spheres d is the diameter). As we show in the following *Justification*, we can use the kinetic model to deduce that the **collision frequency**, z, the number of collisions made by one molecule divided by the time interval during which the collisions are counted, when there are N molecules in a volume V is

Synoptic table 18.1* Collision cross-sections

	σ/nm^2
C_6H_6	0.88
CO_2	0.52
He	0.21
N_2	0.43

* More values are given in the *Data section*.

$$z = \sigma\bar{c}_{\text{rel}}\mathcal{N} \qquad (18.12a)°$$

with $\mathcal{N} = N/V$ and $\bar{c}_{\text{rel}}$ given in eqn 18.10. The area $\sigma = \pi d^2$ is called the **collision cross-section** of the molecules. Some typical collision cross-sections are given in Table 18.1. In terms of the pressure,

$$z = \frac{\sigma\bar{c}_{\text{rel}}p}{kT} \qquad (18.12b)°$$

Justification 18.3 *Using the kinetic model to calculate the collision frequency*

We consider the positions of all the molecules except one to be frozen. Then we note what happens as one mobile molecule travels through the gas with a mean relative speed $\bar{c}_{\text{rel}}$ for a time Δt. In doing so it sweeps out a 'collision tube' of cross-sectional area $\sigma = \pi d^2$ and length $\bar{c}_{\text{rel}}\Delta t$, and therefore of volume $\sigma\bar{c}_{\text{rel}}\Delta t$ (Fig. 18.9). The number of stationary

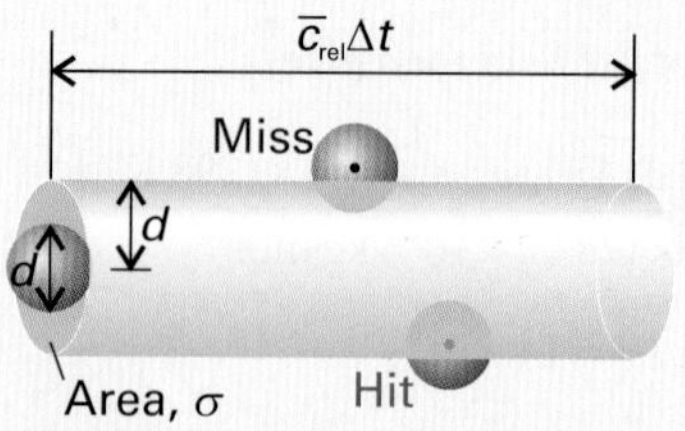

Fig. 18.9 In an interval Δt, a molecule of diameter d sweeps out a tube of radius d and length $\bar{c}_{\text{rel}}\Delta t$. As it does so it encounters other molecules with centres that lie within the tube, and each such encounter counts as one collision. In reality, the tube is not straight, but changes direction at each collision. Nevertheless, the volume swept out is the same, and this straightened version of the tube can be used as a basis of the calculation.

molecules with centres inside the collision tube is given by the volume of the tube multiplied by the number density $\mathcal{N} = N/V$, and is $\mathcal{N}\sigma\bar{c}_{rel}\Delta t$. The number of hits scored in the interval Δt is equal to this number, so the number of collisions divided by the time interval is $\mathcal{N}\sigma\bar{c}_{rel}$. The expression in terms of the pressure of the gas is obtained by using the perfect gas equation to write

$$\mathcal{N} = \frac{N}{V} = \frac{nN_A}{V} = \frac{pN_A}{RT} = \frac{p}{kT}$$

Equation 18.12a shows that, at constant volume, the collision frequency increases with increasing temperature. Equation 18.12b shows that, at constant temperature, the collision frequency is proportional to the pressure. Such a proportionality is plausible for, the greater the pressure, the greater the number density of molecules in the sample, and the rate at which they encounter one another is greater even though their average speed remains the same. For an N_2 molecule in a sample at 1 atm and 25°C, $z \approx 5 \times 10^9\ s^{-1}$, so a given molecule collides about 5×10^9 times each second. We are beginning to appreciate the timescale of events in gases.

(c) The mean free path

Once we have the collision frequency, we can calculate the **mean free path**, λ (lambda), the average distance a molecule travels between collisions. If a molecule collides with a frequency z, it spends a time $1/z$ in free flight between collisions, and therefore travels a distance $(1/z)\bar{c}$. It follows that the mean free path is

$$\lambda = \frac{\bar{c}}{z} \tag{18.13}$$

Substitution of the expression for z in eqn 18.12 gives

$$\lambda = \frac{kT}{2^{1/2}\sigma p} \tag{18.14}$$

Doubling the pressure reduces the mean free path by half. A typical mean free path in nitrogen gas at 1 atm is 70 nm, or about 10^3 molecular diameters. Although the temperature appears in eqn 18.14, in a sample of constant volume, the pressure is proportional to T, so T/p remains constant when the temperature is increased. Therefore, the mean free path is independent of the temperature in a sample of gas in a container of fixed volume. The distance between collisions is determined by the number of molecules present in the given volume, not by the speed at which they travel.

In summary, a typical gas (N_2 or O_2) at 1 atm and 25°C can be thought of as a collection of molecules travelling with a mean speed of about 500 m s^{-1}. Each molecule makes a collision within about 1 ns, and between collisions it travels about 10^3 molecular diameters. The kinetic model of gases is valid (and the gas behaves nearly perfectly) if the diameter of the molecules is much smaller than the mean free path ($d \ll \lambda$), for then the molecules spend most of their time far from one another.

18.2 Collisions with walls and surfaces

The key result for accounting for transport in the gas phase is the rate at which molecules strike an area (which may be an imaginary area embedded in the gas, or part of a real wall). The **collision flux**, Z_W, is the number of collisions with the area in a given time interval divided by the area and the duration of the interval. The collision frequency, the number of hits per second, is obtained by multiplication of the collision flux by the area of interest. We show in the following *Justification* that the collision flux is

$$Z_W = \frac{p}{(2\pi mkT)^{1/2}} \tag{18.15°}$$

A (VERY) BRIEF ILLUSTRATION

When $p = 100$ kPa (1.00 bar) and $T = 300$ K, $Z_W \approx 3 \times 10^{23}\ cm^{-2}\ s^{-1}$ for O_2.

Justification 18.4 *The collision flux*

Consider a wall of area A perpendicular to the x-axis (as in Fig. 18.2). If a molecule has $v_x > 0$ (that is, it is travelling in the direction of positive x), then it will strike the wall within an interval Δt if it lies within a distance $v_x\Delta t$ of the wall. Therefore, all molecules in the volume $Av_x\Delta t$, and with positive x-component of velocities, will strike the wall in the interval Δt. The total number of collisions in this interval is therefore the volume $Av_x\Delta t$ multiplied by the number density, $\mathcal{N}$, of molecules. However, to take account of the presence of a range of velocities in the sample, we must sum the result over all the positive values of v_x weighted by the probability distribution of velocities (eqn 18.5):

$$\text{Number of collisions} = \mathcal{N}A\Delta t\int_0^\infty v_x f(v_x)\mathrm{d}x$$

The collision flux is the number of collisions divided by A and Δt, so

$$Z_W = \mathcal{N}\int_0^\infty v_x f(v_x)\mathrm{d}v_x$$

Then, using the velocity distribution in eqn 18.5,

$$\int_0^\infty v_x f(v_x)\mathrm{d}v_x = \left(\frac{m}{2\pi kT}\right)^{1/2}\int_0^\infty v_x \mathrm{e}^{-mv_x^2/2kT}\mathrm{d}v_x = \left(\frac{kT}{2\pi m}\right)^{1/2}$$

where we have used the standard integral

$$\int_0^\infty x\mathrm{e}^{-ax^2}\mathrm{d}x = \frac{1}{2a}$$

Therefore,

$$Z_W = \mathcal{N}\left(\frac{kT}{2\pi m}\right)^{1/2} = \tfrac{1}{4}\bar{c}\mathcal{N} \qquad (18.16)°$$

where we have used eqn 18.8 in the form $\bar{c} = (8kT/\pi m)^{1/2}$, which implies that $\frac{1}{4}\bar{c} = (kT/2\pi m)^{1/2}$. Substitution of $\mathcal{N} = nN_A/V = p/kT$ gives eqn 18.15.

18.3 The rate of effusion

The essential empirical observations on effusion are summarized by **Graham's law of effusion**, which states that the rate of effusion is inversely proportional to the square root of the molar mass. The basis of this result is that, as remarked above, the mean speed of molecules is inversely proportional to $M^{1/2}$, so the rate at which they strike the area of the hole is also inversely proportional to $M^{1/2}$. However, by using the expression for the rate of collisions, we can obtain a more detailed expression for the rate of effusion and hence use effusion data more effectively.

When a gas at a pressure p and temperature T is separated from a vacuum by a small hole, the rate of escape of its molecules is equal to the rate at which they strike the area of the hole (which is the product of the area and collision flux). Therefore, for a hole of area A_0,

$$\text{Rate of effusion} = Z_W A_0 = \frac{pA_0}{(2\pi mkT)^{1/2}} = \frac{pA_0 N_A}{(2\pi MRT)^{1/2}} \qquad (18.17)°$$

where, in the last step, we have used $R = N_A k$ and $M = mN_A$. This rate is inversely proportional to $M^{1/2}$, in accord with Graham's law.

Equation 18.17 is the basis of the **Knudsen method** for the determination of the vapour pressures of liquids and solids, particularly of substances with very low vapour pressures. In this technique, a sample of the substance is enclosed in a cavity with a small hole and its mass is monitored as a function of time. The value of the vapour pressure, p, is then obtained by applying eqn 18.17.

Example 18.2 *Calculating the vapour pressure from a mass loss*

Caesium (m.p. 29°C, b.p. 686°C) was introduced into a container and heated to 500 K. When a hole of diameter 0.50 mm was opened in the container for 100 s, a mass loss of 385 mg was measured. Calculate the vapour pressure of liquid caesium at 500 K.

Method The pressure of vapour is constant inside the container despite the effusion of atoms because the hot liquid metal replenishes the vapour. The rate of effusion is therefore constant, and given by eqn 18.17. To express the rate in terms of mass, multiply the number of atoms that escape by the mass of each atom.

Answer The mass loss Δm in an interval Δt is related to the collision flux by $\Delta m = Z_W A_0 m \Delta t$, where A_0 is the area of the hole and m is the mass of one atom. It follows that

$$Z_W = \frac{\Delta m}{A_0 m \Delta t}$$

Because Z_W is related to the pressure by eqn 18.15, we can write

$$p = \left(\frac{2\pi RT}{M}\right)^{1/2} \frac{\Delta m}{A_0 \Delta t}$$

Because $M = 132.9\ \text{g mol}^{-1}$, substitution of the data gives $p = 8.7$ kPa (using $1\ \text{Pa} = 1\ \text{N m}^{-2} = 1\ \text{J m}^{-1}$), or 65 Torr.

Self-test 18.2 How long would it take 1.0 g of Cs atoms to effuse out of the oven under the same conditions? [260 s]

18.4 Transport properties of a perfect gas

Transport properties are commonly expressed in terms of a number of 'phenomenological' equations, equations that are empirical summaries of experimental observations. These phenomenological equations apply to all kinds of properties and media. In the following sections, we introduce the equations for the general case and then show how to calculate the parameters that appear in them.

(a) The phenomenological equations

The net rate of transport of a property is measured by its **flux**, J, the quantity of that property passing through a given area in a given time interval divided by the area and the duration of the interval. If matter is flowing (as in diffusion), we speak of a **matter flux** of so many molecules per square metre per second; if the property is the energy of thermal motion (as in thermal conduction), then we speak of the **energy flux** and express it in joules per square metre per second, and so on.

Experimental observations on transport properties show that the flux of a property is usually proportional to the first derivative of some other related property. For example, the flux of matter diffusing parallel to the z-axis of a container is found to be proportional to the first derivative of the concentration:

$$J(\text{matter}) \propto \frac{d\mathcal{N}}{dz} \qquad (18.18)$$

where $\mathcal{N}$ is the number density of particles with units number per metre cubed (m^{-3}). The SI units of J are number per square

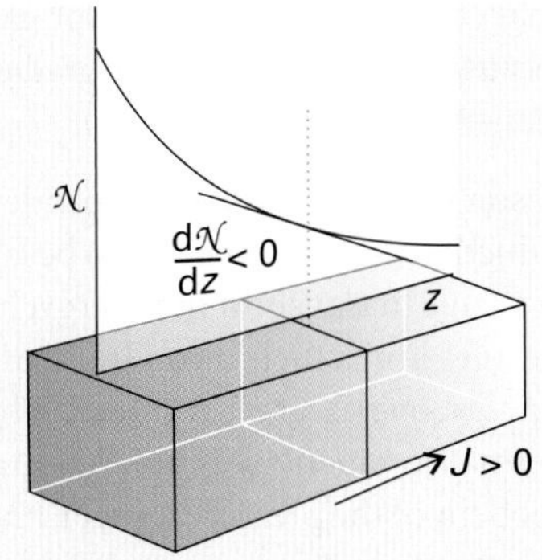

Fig. 18.10 The flux of particles down a concentration gradient. Fick's first law states that the flux of matter (the number of particles passing through an imaginary window in a given interval divided by the area of the window and the length of the interval) is proportional to the density gradient at that point.

metre per second ($m^{-2}\ s^{-1}$). The proportionality of the flux of matter to the concentration gradient is sometimes called **Fick's first law of diffusion**: the law implies that, if the concentration varies steeply with position, then diffusion will be fast. There is no net flux if the concentration is uniform ($d\mathcal{N}/dz = 0$). Similarly, the rate of thermal conduction (the flux of the energy associated with thermal motion) is found to be proportional to the temperature gradient:

$$J(\text{energy}) \propto \frac{dT}{dz} \tag{18.19}$$

The SI units of this flux are joules per square metre per second ($J\ m^{-2}\ s^{-1}$).

A positive value of J signifies a flux towards positive z; a negative value of J signifies a flux towards negative z. Because matter flows down a concentration gradient, from high concentration to low concentration, J is positive if $d\mathcal{N}/dz$ is negative (Fig. 18.10). Therefore, the coefficient of proportionality in eqn 18.18 must be negative, and we write it $-D$, with D a positive constant:

$$J(\text{matter}) = -D\frac{d\mathcal{N}}{dz} \tag{18.20}$$

The constant D is called the **diffusion coefficient**; its SI units are metre squared per second ($m^2\ s^{-1}$). Energy of thermal motion ('heat') migrates down a temperature gradient, and the same reasoning leads to

$$J(\text{energy}) = -\kappa\frac{dT}{dz} \tag{18.21}$$

where κ (kappa) is the **coefficient of thermal conductivity**. The SI units of κ are joules per kelvin per metre per second ($J\ K^{-1}\ m^{-1}\ s^{-1}$). Some experimental values are given in Table 18.2.

To see the connection between the flux of momentum and the viscosity, consider a fluid in a state of **Newtonian flow**, which

Synoptic table 18.2* Transport properties of gases at 1 atm

	$\kappa/(J\ K^{-1}\ m^{-1}\ s^{-1})$	$\eta/(\mu P)^\dagger$	
	273 K	273 K	293 K
Ar	0.0163	210	223
CO_2	0.0145	136	147
He	0.1442	187	196
N_2	0.0240	166	176

* More values are given in the *Data section*.
† $1\ \mu P = 10^{-7}\ kg\ m^{-1}\ s^{-1}$.

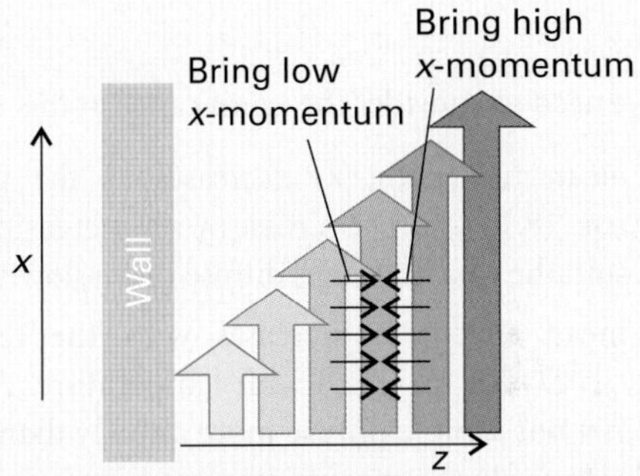

Fig. 18.11 The viscosity of a fluid arises from the transport of linear momentum. In this illustration the fluid is undergoing laminar flow, and particles bring their initial momentum when they enter a new layer. If they arrive with high x-component of momentum they accelerate the layer; if with low x-component of momentum they retard the layer.

can be imagined as occurring by a series of layers moving past one another (Fig. 18.11). The layer next to the wall of the vessel is stationary, and the velocity of successive layers varies linearly with distance, z, from the wall. Molecules ceaselessly move between the layers and bring with them the x-component of linear momentum they possessed in their original layer. A layer is retarded by molecules arriving from a more slowly moving layer because they have a low momentum in the x-direction. A layer is accelerated by molecules arriving from a more rapidly moving layer. We interpret the net retarding effect as the fluid's viscosity.

Because the retarding effect depends on the transfer of the x-component of linear momentum into the layer of interest, the viscosity depends on the flux of this x-component in the z-direction. The flux of the x-component of momentum is proportional to dv_x/dz because there is no net flux when all the layers move at the same velocity. We can therefore write

$$J(x\text{-component of momentum}) = -\eta\frac{dv_x}{dz} \tag{18.22}$$

The constant of proportionality, η (eta), is the **coefficient of viscosity** (or simply 'the viscosity'). Its SI units are kilogram per metre per second (kg m^{-1} s^{-1}, which is equivalent to Pa s). Viscosities are often reported in the non-SI unit poise (P), where 1 P = 10^{-1} kg m^{-1} s^{-1}. There are a variety of methods of determining viscosity, including monitoring the rate of flow of a fluid through a narrow tube. Some experimental values are given in Table 18.2.

(b) The transport parameters

As shown in *Further information 18.1* and summarized in Table 18.3, the kinetic model leads to expressions for the diffusional parameters of a perfect gas.

The diffusion coefficient is

$$D = \tfrac{1}{3}\lambda\bar{c} \qquad (18.23)^\circ$$

As usual, we need to consider the significance of this expression:

1. The mean free path, λ, decreases as the pressure is increased (eqn 18.14), so D decreases with increasing pressure and, as a result, the gas molecules diffuse more slowly.

2. The mean speed, $\bar{c}$, increases with the temperature (eqn 18.8), so D also increases with temperature. As a result, molecules in a hot sample diffuse more quickly than those in a cool sample (for a given concentration gradient).

3. Because the mean free path increases when the collision cross-section of the molecules decreases (eqn 18.14), the diffusion coefficient is greater for small molecules than for large molecules.

Similarly, according to the kinetic model of gases, the thermal conductivity of a perfect gas A having molar concentration [A] is given by the expression

$$\kappa = \tfrac{1}{3}\lambda\bar{c}C_{V,\mathrm{m}}[\mathrm{A}] \qquad (18.24)^\circ$$

where $C_{V,\mathrm{m}}$ is the molar heat capacity at constant volume. To interpret this expression, we note that:

1. Because λ is inversely proportional to the pressure, and hence inversely proportional to the molar concentration of the gas, the thermal conductivity is independent of the pressure.

2. The thermal conductivity is greater for gases with a high heat capacity because a given temperature gradient then corresponds to a greater energy gradient.

The physical reason for the pressure independence of κ is that the thermal conductivity can be expected to be large when many molecules are available to transport the energy, but the presence of so many molecules limits their mean free path and they cannot carry the energy over a great distance. These two effects balance. The thermal conductivity is indeed found experimentally to be independent of the pressure, except when the pressure is very low, when $\kappa \propto p$. At low pressures λ exceeds the dimensions of the apparatus, and the distance over which the energy is transported is determined by the size of the container and not by the other molecules present. The flux is still proportional to the number of carriers, but the length of the journey no longer depends on λ, so $\kappa \propto$ [A], which implies that $\kappa \propto p$.

Finally, the kinetic model leads to the following expression for the viscosity (see *Further information 18.1*):

$$\eta = \tfrac{1}{3}M\lambda\bar{c}[\mathrm{A}] \qquad (18.25)^\circ$$

where [A] is the molar concentration of the gas molecules and M is their molar mass. We can interpret this expression as follows:

1. Because $\lambda \propto 1/p$ (eqn 18.14) and $[\mathrm{A}] \propto p$, it follows that $\eta \propto \bar{c}$, independent of p. That is, the viscosity is independent of the pressure.

2. Because $\bar{c} \propto T^{1/2}$ (eqn 18.8), $\eta \propto T^{1/2}$. That is, the viscosity of a gas *increases* with temperature.

The physical reason for the pressure independence of the viscosity is the same as for the thermal conductivity: more molecules are available to transport the momentum, but they carry it less far on account of the decrease in mean free path. The increase of viscosity with temperature is explained when we remember that at high temperatures the molecules travel more quickly, so the flux of momentum is greater. By contrast, we shall see in Section 18.5, the viscosity of a liquid *decreases* with increase in temperature because intermolecular interactions must be overcome.

Table 18.3 Transport properties of perfect gases

Property	Transported quantity	Simple kinetic theory	Units
Diffusion	Matter	$D = \tfrac{1}{3}\lambda\bar{c}$	m^2 s^{-1}
Thermal conductivity	Energy	$\kappa = \tfrac{1}{3}\lambda\bar{c}C_{V,\mathrm{m}}[\mathrm{A}] = \dfrac{\bar{c}C_{V,\mathrm{m}}}{3\sqrt{2}\sigma N_A}$	J K^{-1} m^{-1} s^{-1}
Viscosity	Linear momentum	$\eta = \tfrac{1}{3}\lambda\bar{c}m\mathcal{N} = \dfrac{m\bar{c}}{3\sqrt{2}\sigma}$	kg m^{-1} s^{-1}

Motion in liquids

We outlined what is currently known about the structure of simple liquids in Section 8.8. Here we consider a particularly simple type of motion through a liquid, that of an ion, and see that the information that motion provides can be used to infer the behaviour of uncharged species too.

18.5 Experimental results

The motion of molecules in liquids can be studied experimentally by a variety of methods. Relaxation time measurements in NMR and EPR (Chapter 12) can be interpreted in terms of the mobilities of the molecules, and have been used to show that big molecules in viscous fluids typically rotate in a series of small (about 5°) steps, whereas small molecules in nonviscous fluids typically jump through about 1 radian (57°) in each step. Another important technique is **inelastic neutron scattering**, in which the energy neutrons collect or discard as they pass through a sample is interpreted in terms of the motion of its particles. The same technique is used to examine the internal dynamics of macromolecules.

More mundane than these experiments are viscosity measurements (Table 18.4). For a molecule to move in a liquid, it must acquire at least a minimum energy to escape from its neighbours. The probability that a molecule has at least an energy E_a is proportional to $e^{-E_a/RT}$, so the mobility of the molecules in the liquid should follow this type of temperature dependence. Because the coefficient of viscosity, η, is inversely proportional to the mobility of the particles, we should expect that

$$\eta \propto e^{E_a/RT} \qquad (18.26)$$

(Note the positive sign of the exponent.) This expression implies that the viscosity should decrease sharply with increasing temperature. Such a variation is found experimentally, at least over reasonably small temperature ranges (Fig. 18.12). The activation energy E_a typical of viscosity is comparable to the mean potential energy of intermolecular interactions.

Synoptic table 18.4* Viscosities of liquids at 298 K

	$\eta/(10^{-3}\ \mathrm{kg\ m^{-1}\ s^{-1}})$
Benzene	0.601
Mercury	1.55
Pentane	0.224
Water†	0.891

*More values are given in the *Data section*.
† The viscosity of water corresponds to 0.891 cP.

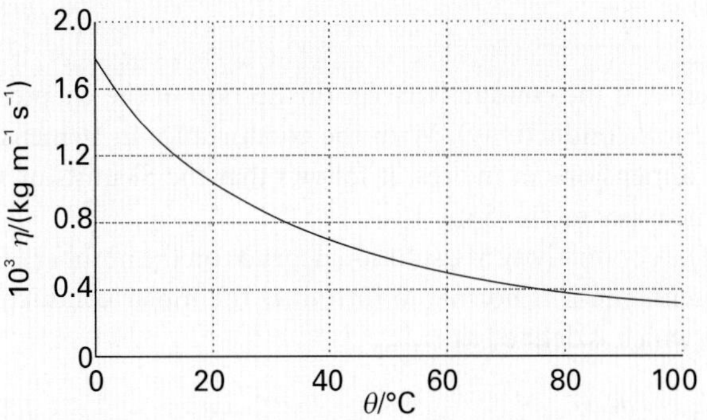

Fig. 18.12 The experimental temperature dependence of the viscosity of water. As the temperature is increased, more molecules are able to escape from the potential wells provided by their neighbours, and so the liquid becomes more fluid. A plot of ln η against $1/T$ is a straight line (over a small range) with positive slope.

One problem with the interpretation of viscosity measurements is that the change in density of the liquid as it is heated makes a pronounced contribution to the temperature variation of the viscosity. Thus, the temperature dependence of viscosity at constant volume, when the density is constant, is much less than that at constant pressure. The intermolecular interactions between the molecules of the liquid govern the magnitude of E_a, but the problem of calculating it is immensely difficult and still largely unsolved. At low temperatures, the viscosity of water decreases as the pressure is increased. This behaviour is consistent with the rupture of hydrogen bonds.

18.6 The conductivities of electrolyte solutions

Further insight into the nature of molecular motion can be obtained by studying the net transport of charged species through solution, for ions can be dragged through the solvent by the application of a potential difference between two electrodes immersed in the sample. By studying the transport of charge through electrolyte solutions it is possible to build up a picture of the events that occur in them and, in some cases, to extrapolate the conclusions to species that have zero charge, that is, to neutral molecules.

The fundamental measurement used to study the motion of ions is that of the electrical resistance, R, of the solution. The **conductance**, G, of a solution is the inverse of its resistance R: $G = 1/R$. As resistance is expressed in ohms, Ω, the conductance of a sample is expressed in Ω^{-1}. The reciprocal ohm used to be called the mho, but its SI designation is now the siemens, S, and $1\ \mathrm{S} = 1\ \Omega^{-1} = 1\ \mathrm{C\ V^{-1}\ s^{-1}}$. The conductance of a sample decreases with its length l and increases with its cross-sectional area A. We therefore write

$$G = \frac{\kappa A}{l} \quad (18.27)$$

where κ is the **conductivity** (in this section of the chapter, the *electrical* conductivity). With the conductance in siemens and the dimensions in metres, it follows that the SI units of κ are siemens per metre (S m^{-1}).

The conductivity of a solution depends on the number of ions present, and it is normal to introduce the **molar conductivity**, Λ_m, which is defined as

$$\Lambda_m = \frac{\kappa}{c} \quad [18.28]$$

where c is the molar concentration of the added electrolyte. The SI unit of molar conductivity is siemens metre-squared per mole (S m^2 mol^{-1}), and typical values are about 10 mS m^2 mol^{-1} (where 1 mS = 10^{-3} S).

The values of the molar conductivity as calculated by eqn 18.28 are found to vary with the concentration. One reason for this variation is that the number of ions in the solution might not be proportional to the concentration of the electrolyte. For instance, the concentration of ions in a solution of a weak electrolyte depends on the concentration of the solute in a complicated way, and doubling the concentration of the solute added does not double the number of ions. Secondly, because ions interact strongly with one another, the conductivity of a solution is not exactly proportional to the number of ions present.

In an extensive series of measurements during the nineteenth century, Friedrich Kohlrausch established the **Kohlrausch law**, that at low concentrations the molar conductivities of strong electrolytes vary linearly with the square root of the concentration:

$$\Lambda_m = \Lambda_m^\circ - \mathcal{K}c^{1/2} \quad (18.29)$$

He also established that Λ_m°, the **limiting molar conductivity**, the molar conductivity in the limit of zero concentration, is the sum of contributions from its individual ions. If the limiting molar conductivity of the cations is denoted λ_+ and that of the anions λ_-, then his **law of the independent migration of ions** states that

$$\Lambda_m^\circ = \nu_+\lambda_+ + \nu_-\lambda_- \quad (18.30)^\circ$$

where ν_+ and ν_- are the numbers of cations and anions per formula unit of electrolyte (for example, $\nu_+ = \nu_- = 1$ for HCl, NaCl, and $CuSO_4$, but $\nu_+ = 1$, $\nu_- = 2$ for $MgCl_2$).

18.7 The mobilities of ions

To interpret conductivity measurements we need to know why ions move at different rates, why they have different molar conductivities, and why the molar conductivities of strong electrolytes decrease with the square root of the molar concentration. The central idea in this section is that, although the motion of an ion remains largely random, the presence of an electric field biases its motion, and the ion undergoes net migration through the solution.

(a) The drift speed

When the potential difference between two planar electrodes a distance l apart is $\Delta\phi$, the ions in the solution between them experience a uniform electric field of magnitude

$$\mathcal{E} = \frac{\Delta\phi}{l} \quad (18.31)$$

In such a field, an ion of charge ze experiences a force of magnitude

$$\mathcal{F} = ze\mathcal{E} = \frac{ze\Delta\phi}{l} \quad (18.32)$$

where here and throughout the chapter we disregard the sign of the charge number and so avoid notational complications. A cation responds to the application of the field by accelerating towards the negative electrode and an anion responds by accelerating towards the positive electrode. However, this acceleration is short-lived. As the ion moves through the solvent it experiences a frictional retarding force, $\mathcal{F}_{fric}$, proportional to its speed. For a spherical particle of radius a travelling at a speed s, this force is given by **Stokes' law**, which was derived by considering the hydrodynamics of the passage of a sphere through a continuous fluid:

$$\mathcal{F}_{fric} = fs \qquad f = 6\pi\eta a \quad (18.33)$$

where η is the viscosity. In writing eqn 18.33, we assume that it applies on a molecular scale, and independent evidence from magnetic resonance suggests that it often gives at least the right order of magnitude.

The two forces act in opposite directions, and the ions quickly reach a terminal speed, the **drift speed**, when the accelerating force is balanced by the viscous drag. The net force is zero when

$$s = \frac{ze\mathcal{E}}{f} \quad (18.34)$$

It follows that the drift speed of an ion is proportional to the strength of the applied field. We write

$$s = u\mathcal{E} \quad [18.35]$$

where u is called the **mobility** of the ion (Table 18.5). Comparison of the last two equations shows that

$$u = \frac{ze}{f} = \frac{ze}{6\pi\eta a} \quad (18.36)$$

Synoptic table 18.5* Ionic mobilities in water at 298 K

	$u/(10^{-8}\ m^2\ s^{-1}\ V^{-1})$		$u/(10^{-8}\ m^2\ s^{-1}\ V^{-1})$
H^+	36.23	OH^-	20.64
Na^+	5.19	Cl^-	7.91
K^+	7.62	Br^-	8.09
Zn^{2+}	5.47	SO_4^{2-}	8.29

* More values are given in the *Data section*.

● A BRIEF ILLUSTRATION

For an order of magnitude estimate we can take $z = 1$ and a the radius of an ion such as Cs^+ (which might be typical of a smaller ion plus its hydration sphere), which is 170 pm. For the viscosity, we use $\eta = 1.0$ cP (1.0×10^{-3} kg m^{-1} s^{-1}, Table 18.4). Then $u \approx 5 \times 10^{-8}$ m^2 V^{-1} s^{-1}. This value means that, when there is a potential difference of 1 V across a solution of length 1 cm (so $\mathcal{E} = 100$ V m^{-1}), the drift speed is typically about 5 μm s^{-1}. That speed might seem slow, but not when expressed on a molecular scale, for it corresponds to an ion passing about 10^4 solvent molecules per second. ●

Because the drift speed governs the rate at which charged species are transported, we might expect the conductivity to decrease with increasing solution viscosity and ion size. Experiments confirm these predictions for bulky ions (such as R_4N^+ and RCO_2^-) but not for small ions. For example, the molar conductivities of the alkali metal ions in water increase from Li^+ to Cs^+ (Table 18.5) even though the ionic radii increase. The paradox is resolved when we realize that the radius a in the Stokes formula is the **hydrodynamic radius** (or 'Stokes radius') of the ion, its effective radius in the solution taking into account all the H_2O molecules it carries in its hydration shell. Small ions give rise to stronger electric fields than large ones (the electric field at the surface of a sphere of radius r is proportional to ze/r^2 and it follows that the smaller the radius the stronger the field), so small ions are more extensively solvated than big ions. Thus, an ion of small ionic radius may have a large hydrodynamic radius because it drags many solvent molecules through the solution as it migrates. The hydrating H_2O molecules are often very labile, however, and NMR and isotope studies have shown that the exchange between the coordination sphere of the ion and the bulk solvent is very rapid for ions of low charge but may be slow for ions of high charge (Fig. 18.13).

The proton, although it is very small, has a very high molar conductivity (Table 18.5)! Proton and ^{17}O-NMR show that the times characteristic of protons hopping from one molecule to the next are about 1.5 ps, which is comparable to the time that inelastic neutron scattering shows it takes a water molecule to reorientate through about 1 radian (1 to 2 ps). According to the **Grotthuss mechanism**, there is an effective motion of a proton that involves the rearrangement of bonds in a group of water molecules. However, the actual mechanism is still highly contentious. Attention now focuses on the $H_9O_4^+$ unit, in which the nearly trigonal planar H_3O^+ ion is linked to three strongly solvating H_2O molecules. This cluster of atoms is itself hydrated, but the hydrogen bonds in the secondary sphere are weaker than in the primary sphere. It is envisaged that the rate-determining step is the cleavage of one of the weaker hydrogen bonds of this secondary sphere (Fig. 18.14a). After this bond cleavage has taken place, and the released molecule has rotated through a few degrees (a process that takes about 1 ps), there is a rapid adjustment of bond lengths and angles in the remaining cluster, to form an $H_5O_2^+$ cation of structure $H_2O\cdots H^+\cdots OH_2$ (Fig. 18.14b).

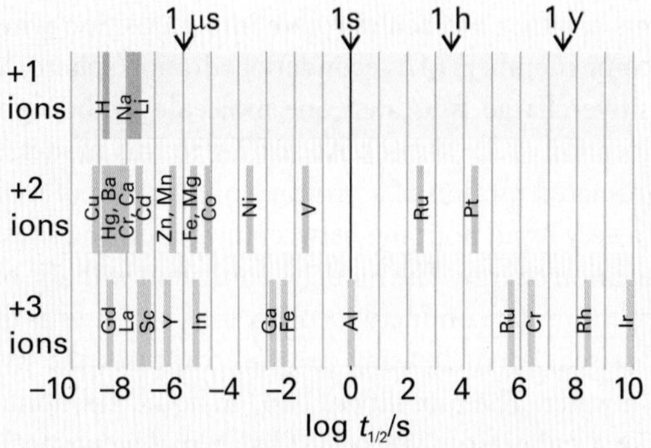

Fig. 18.13 The half-lives of water molecules in the hydration spheres of ions.

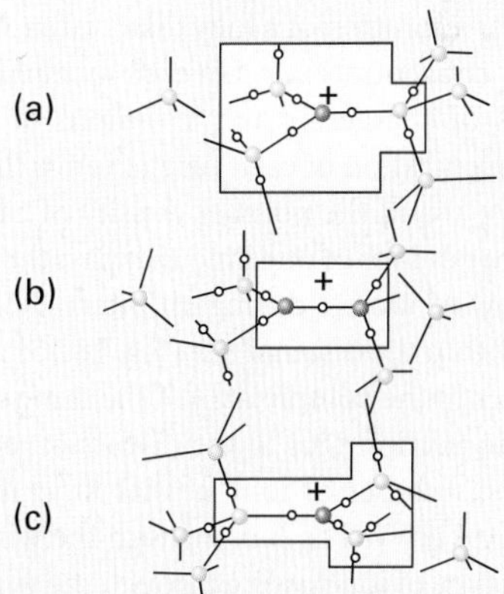

Fig. 18.14 The mechanism of conduction by water as proposed by N. Agmon (*Chem. Phys. Lett.* **244**, 456 (1995)). Proton transfer between neighbouring molecules occurs when one molecule rotates into such a position that an O—H···O hydrogen bond can flip into being an O···H—O hydrogen bond. See text for a description of the steps.

Shortly after this reorganization has occurred, a new $H_9O_4^+$ cluster forms as other molecules rotate into a position where they can become members of a secondary hydration sphere, but now the positive charge is located one molecule to the right of its initial location (Fig. 18.14c). According to this model, there is no coordinated motion of a proton along a chain of molecules, simply a very rapid hopping between neighbouring sites, with a low activation energy. The model is consistent with the observation that the molar conductivity of protons increases as the pressure is raised, for increasing pressure ruptures the hydrogen bonds in water. The mobility of NH_4^+ in liquid ammonia is also anomalous and presumably occurs by an analogous mechanism.

A brief comment The H_3O^+ ion is trigonal pyramidal in the gas phase but, as a result of its hydration, is nearly planar in water.

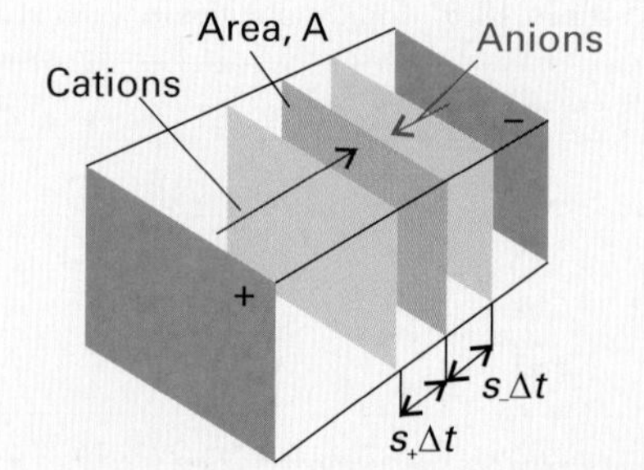

Fig. 18.15 In the calculation of the current, all the cations within a distance $s_+\Delta t$ (that is, those in the volume $s_+A\Delta t$) will pass through the area A. The anions in the corresponding volume the other side of the window will also contribute to the current similarly.

(b) Mobility and conductivity

Ionic mobilities provide a link between measurable and theoretical quantities. As a first step we establish in the following *Justification* the relation between an ion's mobility and its molar conductivity:

$$\lambda = zuF \qquad (18.37)°$$

where F is Faraday's constant ($F = N_A e$).

Justification 18.5 *The relation between ionic mobility and molar conductivity*

To keep the calculation simple, we ignore signs in the following, and concentrate on the magnitudes of quantities.

Consider a solution of a fully dissociated strong electrolyte at a molar concentration c. Let each formula unit give rise to ν_+ cations of charge z_+e and ν_- anions of charge z_-e. The molar concentration of each type of ion is therefore νc (with $\nu = \nu_+$ or ν_-), and the number density of each type is $\nu c N_A$. The number of ions of one kind that pass through an imaginary window of area A during an interval Δt is equal to the number within the distance $s\Delta t$ (Fig. 18.15), and therefore to the number in the volume $s\Delta tA$. (The same sort of argument was used in Section 18.1 in the discussion of the pressure of a gas.) The number of ions of that kind in this volume is equal to $s\Delta tA\nu cN_A$. The flux through the window (the number of this type of ion passing through the window divided by the area of the window and the duration of the interval) is therefore

$$J(\text{ions}) = \frac{s\Delta tA\nu cN_A}{A\Delta t} = s\nu cN_A$$

Each ion carries a charge ze, so the flux of charge is

$$J(\text{charge}) = zs\nu ceN_A = zs\nu cF$$

Because $s = u\mathcal{E}$, the flux is

$$J(\text{charge}) = zu\nu cF\mathcal{E}$$

The current, I, through the window due to the ions we are considering is the charge flux times the area:

$$I = JA = zu\nu cF\mathcal{E}A$$

Because the electric field is the potential gradient, $\Delta\phi/l$, we can write

$$I = \frac{zu\nu cFA\Delta\phi}{l}$$

Current and potential difference are related by Ohm's law, $\Delta\phi = IR$, so it follows that

$$I = \frac{\Delta\phi}{R} = G\Delta\phi = \frac{\kappa A\Delta\phi}{l}$$

where we have used eqn 18.27 in the form $\kappa = Gl/A$. Note that the proportionality of current to potential difference ($I \propto \Delta\phi$) is another example of a phenomenological flux equation like those introduced in Section 18.4. Comparison of the last two expressions gives $\kappa = zu\nu cF$. Division by the molar concentration of ions, νc, then results in eqn 18.37.

Equation 18.37 applies to the cations and to the anions. Therefore, for the solution itself in the limit of zero concentration (when there are no ionic interactions),

$$\Lambda_m^\circ = (z_+u_+\nu_+ + z_-u_-\nu_-)F \qquad (18.38a)°$$

For a symmetrical z:z electrolyte (for example, $CuSO_4$ with $z = 2$), this equation simplifies to

$$\Lambda_m^\circ = z(u_+ + u_-)F \qquad (18.38b)°$$

A BRIEF ILLUSTRATION

Earlier, we estimated the typical ionic mobility as 5×10^{-8} m^2 V^{-1} s^{-1}; so, with $z = 1$ for both the cation and anion, we can estimate that a typical limiting molar conductivity should be about 10 mS m^2 mol^{-1}, in accord with experiment. The experimental value for KCl, for instance, is 15 mS m^2 mol^{-1}.

IMPACT ON BIOCHEMISTRY
I18.1 Gel electrophoresis in genomics and proteomics

Advances in biotechnology are linked strongly to the development of physical techniques. The effort to characterize the entire genetic material, or **genome**, of organisms as simple as bacteria and as complex as *Homo sapiens* will lead to important new insights into the molecular mechanisms of disease, primarily through the discovery of previously unknown proteins encoded by the deoxyribonucleic acid (DNA) in genes. However, decoding genomic DNA will not always lead to accurate predictions of the amino acids present in biologically active proteins. Many proteins undergo chemical modification, such as cleavage into smaller proteins, after being synthesized in the cell. Moreover, it is known that one piece of DNA may encode more than one active protein. It follows that it is also important to describe the **proteome**, the full complement of functional proteins of an organism, by characterizing directly the proteins after they have been synthesized and processed in the cell.

The procedures of **genomics** and **proteomics**, the analysis of the genome and proteome, of complex organisms are time-consuming because of the very large number of molecules that must be characterized. For example, the human genome contains about 30 000 genes and the number of active proteins is likely to be much larger. Success in the characterization of the genome and proteome of any organism will depend on the deployment of very rapid techniques for the determination of the order in which molecular building blocks are linked covalently in DNA and proteins.

Many macromolecules, such as DNA, are charged and move in response to an electric field. This motion, with a drift speed given by eqn 18.34, is called **electrophoresis** and it depends on its net charge, size (and hence molar mass), and shape. The latter two factors are implied by the dependence of the drift speed on the frictional force. Consequently, experimental techniques based on electrophoresis are very useful to polymer chemists and biochemists in the characterization of macromolecules.

An important tool in genomics and proteomics is **gel electrophoresis**, in which biopolymers are separated on a slab of a porous gel, a semirigid dispersion of a solid in a liquid. Because the molecules must pass through the pores in the gel, the larger the macromolecule the less mobile it is in the electric field and, conversely, the smaller the macromolecule the more swiftly it moves through the pores. In this way, gel electrophoresis allows for the separation of components of a mixture according to their molar masses. Two common gel materials for the study of proteins and nucleic acids are agarose and cross-linked polyacrylamide. Agarose has large pores and is better suited for the study of large macromolecules, such as DNA and enzyme complexes. Polyacrylamide gels with varying pore sizes can be made by changing the concentration of acrylamide in the polymerization solution. In general, smaller pores form as the concentration of acrylamide is increased, making possible the separation of relatively small macromolecules by **polyacrylamide gel electrophoresis** (PAGE).

The separation of very large pieces of DNA, such as chromosomes, by conventional gel electrophoresis is not effective, making the analysis of genomic material rather difficult. Double-stranded DNA molecules are thin enough to pass through gel pores, but long and flexible DNA coils can become trapped in the pores and the result is impaired mobility along the direction of the applied electric field. This problem can be avoided with **pulsed-field electrophoresis**, in which a brief burst of the electric field is applied first along one direction and then along a perpendicular direction. In response to the switching back and forth between field directions, the DNA coils writhe about and eventually pass through the gel pores. In this way, the mobility of the macromolecule can be related to its molar mass.

We have seen that charge also determines the drift speed. For example, proteins of the same size but different net charge travel along the slab at different speeds. One way to avoid this problem and to achieve separation by molar mass is to denature the proteins in a controlled way. Sodium dodecyl sulfate is an anionic detergent that is very useful in this respect: it denatures proteins, whatever their initial shapes, into rods by forming a complex with them. Moreover, most protein molecules bind a constant number of ions, so the net charge per protein is well regulated. Under these conditions, different proteins in a mixture may be separated according to size only. The molar mass of each constituent protein is estimated by comparing its mobility in its rod-like complex form with a standard sample of known molar mass. However, molar masses obtained by this method, often referred to as **SDS-PAGE** when polyacrylamide gels are used, are not as accurate as those obtained by mass spectrometry.

Another technique that deals with the effect of charge on drift speed takes advantage of the fact that the overall charge of proteins and other biopolymers depends on the pH of the medium. For instance, in acidic environments protons attach to basic groups and the net charge is positive; in basic media the net charge is negative as a result of proton loss. At the **isoelectric point**, the pH is such that there is no net charge on the biopolymer. Consequently, the drift speed of a biopolymer depends on the pH of the medium, with $s = 0$ at the isoelectric point (Fig. 18.16). **Isoelectric focusing** is an electrophoresis method that exploits the dependence of drift speed on pH. In this technique, a mixture of proteins is dispersed in a medium with a pH gradient along the direction of an applied electric field. Each

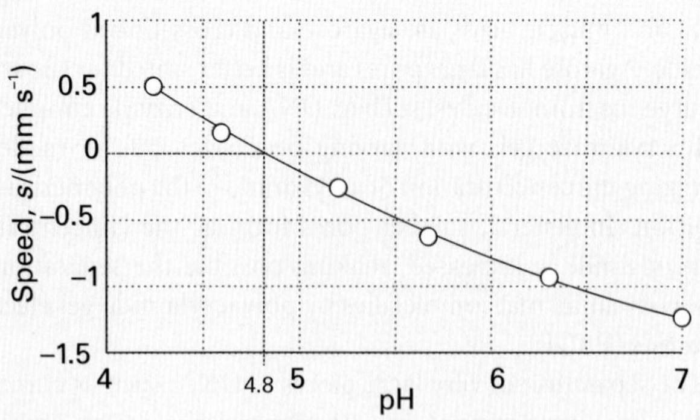

Fig. 18.16 The plot of drift speed of the protein bovine serum albumin in water against pH. The isoelectric point of the macromolecule corresponds to the pH at which the drift speed in the presence of an electric field is zero.

protein in the mixture will stop moving at a position in the gradient where the pH is equal to the isoelectric point. In this manner, the protein mixture can be separated into its components.

Diffusion

We are now in a position to extend the discussion of ionic motion to cover the migration of neutral molecules and of ions in the absence of an applied electric field. We shall do this by expressing ion motion in a more general way than hitherto, and will then discover that the same equations apply even when the charge on the particles is zero.

18.8 The thermodynamic view

We saw in Section 15.5 that, at constant temperature and pressure, the maximum non-expansion work that can be done per mole when a substance moves from a location where its chemical potential is μ to a location where its chemical potential is $\mu + d\mu$ is $dw = d\mu$. In a system in which the chemical potential depends on the position x,

$$dw = d\mu = \left(\frac{\partial \mu}{\partial x}\right)_{p,T} dx$$

We also saw in Chapter 14 (Table 14.1) that in general work can always be expressed in terms of an opposing force (which here we write $\mathcal{F}$), and that

$$dw = -\mathcal{F}\,dx$$

By comparing these two expressions, we see that the slope of the chemical potential can be interpreted as an effective force per mole of molecules. We write this **thermodynamic force** as

$$\mathcal{F} = -\left(\frac{\partial \mu}{\partial x}\right)_{p,T} \qquad [18.39]$$

There is not necessarily a real force pushing the particles down the slope of the chemical potential. As we shall see, the force may represent the spontaneous tendency of the molecules to disperse as a consequence of the Second Law and the hunt for maximum entropy.

In a solution in which the activity of the solute is a, the chemical potential is $\mu = \mu^{\ominus} + RT \ln a$. If the solution is not uniform the activity depends on the position and we can write

$$\mathcal{F} = -RT\left(\frac{\partial \ln a}{\partial x}\right)_{p,T} \qquad (18.40a)$$

If the solution is ideal, a may be replaced by the molar concentration c, and then

$$\mathcal{F} = -\frac{RT}{c}\left(\frac{\partial c}{\partial x}\right)_{p,T} \qquad (18.40b)^{\circ}$$

where we have also used the relation $d \ln y/dx = (1/y)(dy/dx)$.

In Section 18.4 we saw that Fick's first law of diffusion (that the particle flux is proportional to the concentration gradient) could be deduced from the kinetic model of gases. We shall now show that it can be deduced more generally and that it applies to the diffusion of species in condensed phases too. We suppose that the flux of diffusing particles is motion in response to a thermodynamic force arising from a concentration gradient. The particles reach a steady drift speed, s, when the thermodynamic force, $\mathcal{F}$, is matched by the viscous drag. This drift speed is proportional to the thermodynamic force, and we write $s \propto \mathcal{F}$. However, the particle flux, J, is proportional to the drift speed, and the thermodynamic force is proportional to the concentration gradient, dc/dx. The chain of proportionalities ($J \propto s$, $s \propto \mathcal{F}$, and $\mathcal{F} \propto dc/dx$) implies that $J \propto dc/dx$, which is the content of Fick's law.

(a) The Einstein relation

If we divide both sides of eqn 18.20 by Avogadro's constant, thereby converting numbers into amounts (numbers of moles), then Fick's law becomes

$$J = -D\frac{dc}{dx} \qquad (18.41)$$

In this expression, D is the diffusion coefficient and dc/dx is the slope of the molar concentration. The flux is related to the drift speed by

$$J = sc \qquad (18.42)$$

This relation follows from the argument that we have used several times before. Thus, all particles within a distance $s\Delta t$, and therefore in a volume $s\Delta tA$, can pass through a window of area A

in an interval Δt. Hence, the amount of substance that can pass through the window in that interval is $s\Delta tAc$. Therefore,

$$sc = -D\frac{dc}{dx}$$

If now we express dc/dx in terms of $\mathcal{F}$ by using eqn 18.40b, we find

$$s = -\frac{D}{c}\frac{dc}{dx} = \frac{D\mathcal{F}}{RT} \qquad (18.43)$$

Therefore, once we know the effective force and the diffusion coefficient, D, we can calculate the drift speed of the particles (and vice versa) whatever the origin of the force.

There is one case where we already know the drift speed and the effective force acting on a particle: an ion in solution has a drift speed $s = u\mathcal{E}$ when it experiences a force $N_A ez\mathcal{E}$ from an electric field of strength $\mathcal{E}$. Therefore, substituting these known values into eqn 18.43 and using $N_A e = F$ gives $u\mathcal{E} = DFz\mathcal{E}/RT$ and hence

$$u = \frac{zFD}{RT} \qquad (18.44)$$

This equation rearranges into the very important result known as the **Einstein relation** between the diffusion coefficient and the ionic mobility:

$$D = \frac{uRT}{zF} \qquad (18.45)°$$

On inserting the typical value $u = 5 \times 10^{-8}\ m^2\ s^{-1}\ V^{-1}$, we find $D \approx 1 \times 10^{-9}\ m^2\ s^{-1}$ at 25°C as a typical value of the diffusion coefficient of an ion in water.

(b) The Nernst–Einstein equation

The Einstein relation provides a link between the molar conductivity of an electrolyte and the diffusion coefficients of its ions. First, by using eqns 18.37 and 18.44 we write

$$\lambda = zuF = \frac{z^2DF^2}{RT} \qquad (18.46)°$$

for each type of ion. Then, from $\Lambda_m^\circ = \nu_+\lambda_+ + \nu_-\lambda_-$, the limiting molar conductivity is

$$\Lambda_m = (\nu_+z_+^2D_+ + \nu_-z_-^2D_-)\frac{F^2}{RT} \qquad (18.47)°$$

which is the **Nernst–Einstein equation**. An application of this equation is to the determination of ionic diffusion coefficients from conductivity measurements; another is to the prediction of conductivities using models of ionic diffusion (see below).

(c) The Stokes–Einstein equation

Equations 18.36 ($u = ez/f$) and 18.44 relate the mobility of an ion to the frictional force and to the diffusion coefficient, respectively. We can combine the two expressions into the **Stokes–Einstein equation**:

$$D = \frac{kT}{f} \qquad (18.48)$$

If the frictional force is described by Stokes's law, then we also obtain a relation between the diffusion coefficient and the viscosity of the medium:

$$D = \frac{kT}{6\pi\eta a} \qquad (18.49)$$

An important feature of eqn 18.48 (and of its special case, eqn 18.49) is that it makes no reference to the charge of the diffusing species. Therefore, the equation also applies in the limit of vanishingly small charge, that is, it also applies to neutral molecules. Consequently, we may use viscosity measurements to estimate the diffusion coefficients for electrically neutral molecules in solution (Table 18.6). It must not be forgotten, however, that both equations depend on the assumption that the viscous drag is proportional to the speed.

Synoptic table 18.6* Diffusion coefficients at 298 K

	$D/(10^{-9}\ m^2\ s^{-1})$
H^+ in water	9.31
I_2 in hexane	4.05
Na^+ in water	1.33
Sucrose in water	0.522

* More values are given in the *Data section*.

Example 18.3 *Interpreting the mobility of an ion*

Use the experimental value of the mobility to evaluate the diffusion coefficient, the limiting molar conductivity, and the hydrodynamic radius of a sulfate ion in aqueous solution at 298 K.

Method The starting point is the mobility of the ion, which is given in Table 18.5. The diffusion coefficient can then be determined from the Einstein relation, eqn 18.45. The ionic conductivity is related to the mobility by eqn 18.37. To estimate the hydrodynamic radius, a, of the ion, use the Stokes–Einstein relation to find f and the Stokes law to relate f to a.

Answer From Table 18.5, the mobility of SO_4^{2-} is $8.29 \times 10^{-8}\ m^2\ s^{-1}\ V^{-1}$. It follows from eqn 18.45 that

$$D = \frac{uRT}{zF} = 1.1 \times 10^{-9}\ m^2\ s^{-1}$$

From eqn 18.37 it follows that

$$\lambda = zuF = 16 \text{ mS m}^2 \text{ mol}^{-1}$$

Finally, from $f = 6\pi\eta a$ using 0.891 cP (or 8.91×10^{-4} kg m^{-1} s^{-1}) for the viscosity of water (Table 18.4):

$$a = \frac{kT}{6\pi\eta D} = 220 \text{ pm}$$

The bond length in SO_4^{2-} is 144 pm, so the radius calculated here is plausible and consistent with a small degree of solvation.

Self-test 18.3 Repeat the calculation for the NH_4^+ ion.
[1.96×10^{-9} m^2 s^{-1}, 7.4 mS m^2 mol^{-1}, 125 pm]

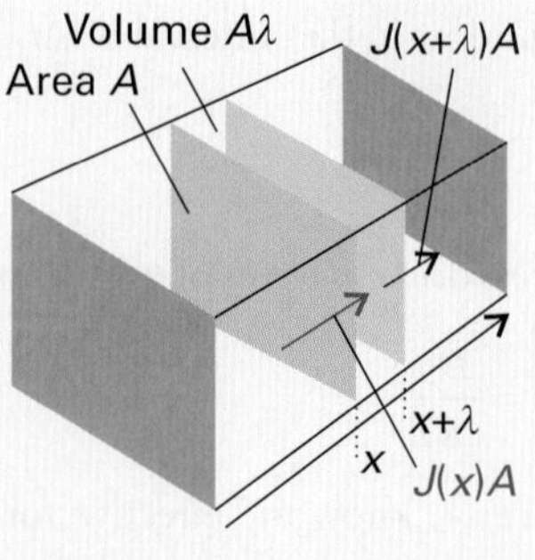

Fig. 18.17 The net flux in a region is the difference between the flux entering from the region of high concentration (on the left) and the flux leaving to the region of low concentration (on the right).

18.9 The diffusion equation

We now turn to the discussion of time-dependent diffusion processes, where we are interested in the spreading of inhomogeneities with time. One example is the temperature of a metal bar that has been heated at one end: if the source of heat is removed, then the bar gradually settles down into a state of uniform temperature. When the source of heat is maintained and the bar can radiate, it settles down into a steady state of non-uniform temperature. Another example (and one more relevant to chemistry) is the concentration distribution in a solvent to which a solute is added. We shall focus on the description of the diffusion of particles, but similar arguments apply to the diffusion of physical properties, such as temperature. Our aim is to obtain an equation for the rate of change of the concentration of particles in an inhomogeneous region.

The central equation of this section is the **diffusion equation**, also called 'Fick's second law of diffusion', which relates the rate of change of concentration at a point to the spatial variation of the concentration at that point:

$$\frac{\partial c}{\partial t} = D\frac{\partial^2 c}{\partial x^2} \qquad (18.50)$$

We show in the following *Justification* that the diffusion equation follows from Fick's first law of diffusion.

Justification 18.6 *The diffusion equation*

Consider a thin slab of cross-sectional area A that extends from x to $x + \lambda$ (Fig. 18.17). Let the concentration at x be c at the time t. The amount (in moles) of particles that enter the slab in the infinitesimal interval dt is $JAdt$, so the rate of increase in molar concentration inside the slab (which has volume $A\lambda$) on account of the flux from the left is

$$\frac{\partial c}{\partial t} = \frac{JAdt}{A\lambda dt} = \frac{J}{\lambda}$$

There is also an outflow through the right-hand window. The flux through that window is J', and the rate of change of concentration that results is

$$\frac{\partial c}{\partial t} = -\frac{J'Adt}{A\lambda dt} = -\frac{J'}{\lambda}$$

The net rate of change of concentration is therefore

$$\frac{\partial c}{\partial t} = \frac{J - J'}{\lambda}$$

Each flux is proportional to the concentration gradient at the window. So, by using Fick's first law, we can write

$$J - J' = -D\frac{\partial c}{\partial x} + D\frac{\partial c'}{\partial x}$$

$$= -D\frac{\partial c}{\partial x} + D\frac{\partial}{\partial x}\left\{c + \left(\frac{\partial c}{\partial x}\right)\lambda\right\} = D\lambda\frac{\partial^2 c}{\partial x^2}$$

When this relation is substituted into the expression for the rate of change of concentration in the slab, we get eqn 18.50.

The diffusion equation shows that the rate of change of concentration is proportional to the curvature (more precisely, to the second derivative) of the concentration with respect to distance. If the concentration changes sharply from point to point (if the distribution is highly wrinkled) then the concentration changes rapidly with time. Where the curvature is positive (a dip, Fig. 18.18), the change in concentration is positive; the dip tends to fill. Where the curvature is negative (a heap), the change in concentration is negative; the heap tends to spread. If the curvature is zero, then the concentration is constant in time. If the concentration decreases linearly with distance, then the concentration at any point is constant because the inflow of particles is exactly balanced by the outflow.

The diffusion equation can be regarded as a mathematical formulation of the intuitive notion that there is a natural tendency

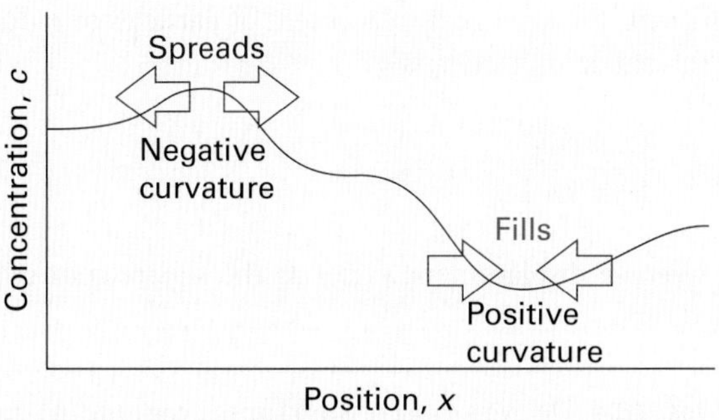

Fig. 18.18 Nature abhors a wrinkle. The diffusion equation tells us that peaks in a distribution (regions of negative curvature) spread and troughs (regions of positive curvature) fill in.

for the wrinkles in a distribution to disappear. More succinctly: Nature abhors a wrinkle.

(a) Diffusion with convection

The transport of particles arising from the motion of a streaming fluid is called **convection.** If for the moment we ignore diffusion, then the flux of particles through an area A in an interval Δt when the fluid is flowing at a velocity v can be calculated in the way we have used several times before (by counting the particles within a distance $v\Delta t$), and is

$$J = \frac{cAv\Delta t}{A\Delta t} = cv \tag{18.51}$$

This J is called the **convective flux**. The rate of change of concentration in a slab of thickness l and area A is, by the same argument as before and assuming that the velocity does not depend on the position,

$$\frac{\partial c}{\partial t} = \frac{J - J'}{l} = \left\{c - \left[c + \left(\frac{\partial c}{\partial x}\right)l\right]\right\}\frac{v}{l} = -v\frac{\partial c}{\partial x} \tag{18.52}$$

When both diffusion and convection occur, the total change of concentration in a region is the sum of the two effects, and the **generalized diffusion equation** is

$$\frac{\partial c}{\partial t} = D\frac{\partial^2 c}{\partial x^2} - v\frac{\partial c}{\partial x} \tag{18.53}$$

A further refinement, which is important in chemistry, is the possibility that the concentrations of particles may change as a result of reaction. When reactions are included in eqn 18.53 (Section 20.5), we get a powerful differential equation for discussing the properties of reacting, diffusing, convecting systems, which is the basis of reactor design in chemical industry and of the utilization of resources in living cells.

(b) Solutions of the diffusion equation

The diffusion equation is a second-order differential equation with respect to space and a first-order differential equation with respect to time. Therefore, we must specify two boundary conditions for the spatial dependence and a single initial condition for the time dependence.

As an illustration, consider a solvent in which the solute is initially coated on one surface of the container (for example, a layer of sugar on the bottom of a deep beaker of water). The single initial condition is that at $t = 0$ all N_0 particles are concentrated on the yz-plane (of area A) at $x = 0$. The two boundary conditions are derived from the requirements (1) that the concentration must everywhere be finite and (2) that the total amount (number of moles) of particles present is n_0 (with $n_0 = N_0/N_A$) at all times. These requirements imply that the flux of particles is zero at the top and bottom surfaces of the system. Under these conditions it is found that

$$c(x,t) = \frac{n_0}{A(\pi Dt)^{1/2}}\mathrm{e}^{-x^2/4Dt} \tag{18.54}$$

as may be verified by direct substitution (Problem 18.30). Figure 18.19 shows the shape of the concentration distribution at various times, and it is clear that the concentration spreads and tends to uniformity.

Another useful result is for a localized concentration of solute in a three-dimensional solvent (a sugar lump suspended in a large flask of water). The concentration of diffused solute is spherically symmetrical, and at a radius r is

$$c(r,t) = \frac{n_0}{8(\pi Dt)^{3/2}}\mathrm{e}^{-r^2/4Dt} \tag{18.55}$$

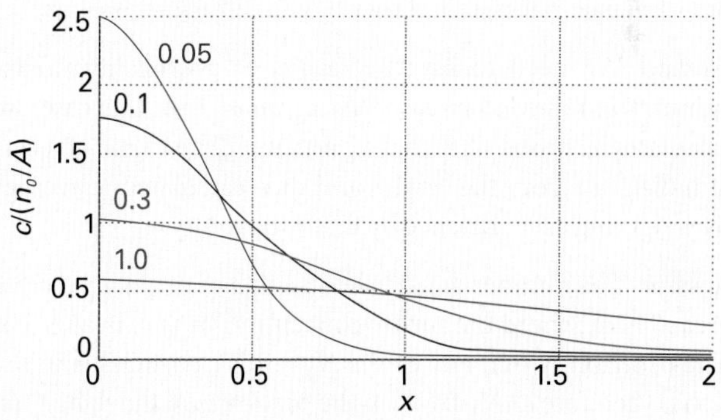

Fig. 18.19 The concentration profiles above a plane from which a solute is diffusing. The curves are plots of eqn 18.54 and are labelled with different values of Dt. The units of Dt and x are arbitrary, but are related so that Dt/x^2 is dimensionless. For example, if x is in metres, Dt would be in metres^2; so, for $D = 10^{-9}$ m^2 s^{-1}, $Dt = 0.1$ m^2 corresponds to $t = 10^8$ s.

interActivity Generate a family of curves similar to that shown in Fig. 18.19 but by using eqn 18.55, which describes diffusion in three dimensions.

Other chemically (and physically) interesting arrangements, such as transport of substances across biological membranes can be treated (*Impact I18.2*). In many cases the solutions are more cumbersome.

The solutions of the diffusion equation are useful for experimental determinations of diffusion coefficients. In the **capillary technique**, a capillary tube, open at one end and containing a solution, is immersed in a well-stirred larger quantity of solvent, and the change of concentration in the tube is monitored. The solute diffuses from the open end of the capillary at a rate that can be calculated by solving the diffusion equation with the appropriate boundary conditions, so D may be determined. In the **diaphragm technique**, the diffusion occurs through the capillary pores of a sintered glass diaphragm separating the well-stirred solution and solvent. The concentrations are monitored and then related to the solutions of the diffusion equation corresponding to this arrangement. Diffusion coefficients may also be measured by a number of techniques, including NMR spectroscopy.

18.10 Diffusion probabilities

The solutions of the diffusion equation can be used to predict the concentration of particles (or the value of some other physical quantity, such as the temperature in a nonuniform system) at any location. We can also use them to calculate the average displacement of the particles in a given time.

Example 18.4 *Calculating the average displacement*

Calculate the average displacement of particles in a time t in a one-dimensional system if they have a diffusion constant D.

Method We need to use the results of probability theory summarized in *Mathematical background 7*. In this case, we calculate the probability that a particle will be found at a certain distance from the origin, and then calculate the average by weighting each distance by that probability.

Answer The number of particles in a slab of thickness dx and area A at x, where the molar concentration is c, is $cAN_A\mathrm{d}x$. The probability that any of the $N_0 = n_0N_A$ particles is in the slab is therefore $cAN_A\mathrm{d}x/N_0$. If the particle is in the slab, it has travelled a distance x from the origin. Therefore, the average displacement of all the particles is the sum of each x weighted by the probability of its occurrence:

$$\langle x\rangle = \int_0^\infty \frac{xcAN_A}{N_0}\mathrm{d}x = \frac{1}{(\pi Dt)^{1/2}}\int_0^\infty xe^{-x^2/4Dt}\mathrm{d}x = 2\left(\frac{Dt}{\pi}\right)^{1/2}$$

where we have used the standard integral in *Justification 18.4*. If we use the Stokes–Einstein relation for the diffusion coefficient, the average displacement of particles of radius a in a solvent of viscosity η is

$$\langle x\rangle = \left(\frac{2kTt}{3\pi^2\eta a}\right)^{1/2}$$

The average displacement varies as the square root of the lapsed time.

Self-test 18.4 Derive an expression for the root mean square distance travelled by diffusing particles in a time t in a one-dimensional system. You will need the standard integral

$$\int_0^\infty x^2\mathrm{e}^{-ax^2}\mathrm{d}x = \frac{1}{4}\left(\frac{\pi}{a^3}\right)^{1/2}$$

$[\langle x^2\rangle^{1/2} = (2Dt)^{1/2}]$

As shown in Example 18.4, the average displacement of a diffusing particle in a time t in a one-dimensional system is

$$\langle x\rangle = 2\left(\frac{Dt}{\pi}\right)^{1/2} \qquad (18.56)$$

and the root mean square displacement in the same time is

$$\langle x^2\rangle^{1/2} = (2Dt)^{1/2} \qquad (18.57)$$

The latter is a valuable measure of the spread of particles when they can diffuse in both directions from the origin (for then $\langle x\rangle = 0$ at all times). The root mean square displacement of particles with a typical diffusion coefficient ($D = 5\times10^{-10}$ m^2 s^{-1}) is illustrated in Fig. 18.20, which shows how long it takes for diffusion to increase the net distance travelled on average to about 1 cm in an unstirred solution. The graph shows that diffusion is a very slow process (which is why solutions are stirred, to encourage mixing by convection). The diffusion of pheromones in still air is also very slow, and greatly accelerated by convection.

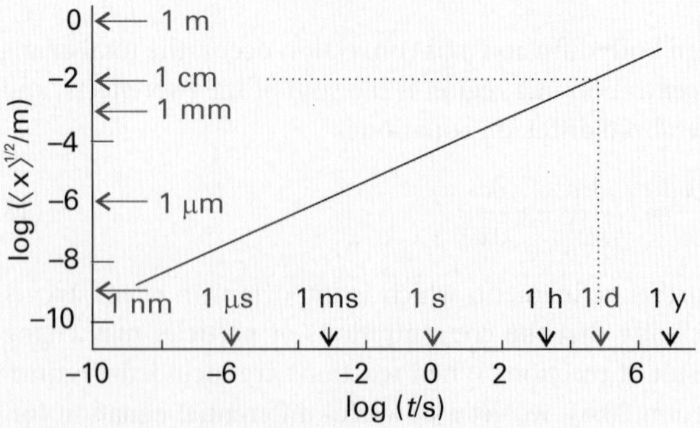

Fig. 18.20 The root mean square distance covered by particles with $D = 5\times10^{-10}$ m^2 s^{-1}. Note the great slowness of diffusion.

18.11 The statistical view

An intuitive picture of diffusion is of the particles moving in a series of small steps and gradually migrating from their original positions. We shall explore this idea using a model in which the particles can jump through a distance λ in a time τ. The total distance travelled by a particle in a time t is therefore $t\lambda/\tau$. However, the particle will not necessarily be found at that distance from the origin. The direction of each step may be different, and the net distance travelled must take the changing directions into account.

If we simplify the discussion by allowing the particles to travel only along a straight line (the x-axis), and for each step (to the left or the right) to be through the same distance λ, then we obtain the **one-dimensional random walk**. The same model can be used to discuss the random coil structures of denatured polymers (*Further information 18.2*).

We show in the following *Justification* that the probability of a particle being at a distance x from the origin after a time t is

$$P = \left(\frac{2\tau}{\pi t}\right)^{1/2} e^{-x^2\tau/2t\lambda^2} \tag{18.58}$$

Justification 18.7 *The one-dimensional random walk*

Consider a one-dimensional random walk in which each step is through a distance λ to the left or right. The net distance travelled after N steps is equal to the difference between the number of steps to the right (N_R) and to the left (N_L), and is $(N_R - N_L)\lambda$. We write $n = N_R - N_L$ and the total number of steps as $N = N_R + N_L$.

The number of ways of performing a walk with a given net distance of travel $n\lambda$ is the number of ways of making N_R steps to the right and N_L steps to the left, and is given by the binomial coefficient

$$W = \frac{N!}{N_L!N_R!} = \frac{N!}{\{\tfrac{1}{2}(N+n)\}!\{\tfrac{1}{2}(N-n)\}!}$$

The probability of the net distance walked being $n\lambda$ is

$$P = \frac{\text{number of paths with } N_R \text{ steps to the right}}{\text{total number of steps}}$$

$$= \frac{W}{2^N} = \frac{N!}{\{\tfrac{1}{2}(N+n)\}!\{\tfrac{1}{2}(N-n)\}!2^N}$$

The use of Stirling's approximation (Section 13.1) in the form

$$\ln x! \approx \ln(2\pi)^{1/2} + (x+\tfrac{1}{2})\ln x - x$$

gives (after quite a lot of algebra; see Problem 18.35)

$$\ln P = \ln\left(\frac{2}{\pi N}\right)^{1/2} - \tfrac{1}{2}(N+n+1)\ln\left(1+\frac{n}{N}\right) - \tfrac{1}{2}(N-n+1)\ln\left(1-\frac{n}{N}\right)$$

For small net distances ($n \ll N$) we can use the approximation $\ln(1\pm x) \approx \pm x - \tfrac{1}{2}x^2$, and so obtain

$$\ln P \approx \ln\left(\frac{2}{\pi N}\right)^{1/2} - \frac{n^2}{2N}$$

At this point, we note that the number of steps taken in a time t is $N = t/\tau$ and the net distance travelled from the origin is $x = n\lambda$. Substitution of these quantities into the expression for $\ln P$ gives

$$\ln P \approx \ln\left(\frac{2\tau}{\pi t}\right)^{1/2} - \frac{x^2\tau}{2t\lambda^2}$$

which, upon using $e^{\ln x} = x$ and $e^{x+y} = e^x e^y$, rearranges into eqn 18.58.

The differences of detail between eqns 18.54 and 18.58 arise from the fact that in the present calculation the particles can migrate in either direction from the origin. Moreover, they can be found only at discrete points separated by λ instead of being anywhere on a continuous line. The fact that the two expressions are so similar suggests that diffusion can indeed be interpreted as the outcome of a large number of steps in random directions.

We can now relate the coefficient D to the step length λ and the rate at which the jumps occur. Thus, by comparing the two exponents in eqn 18.54 and eqn 18.58 we can immediately write down the **Einstein–Smoluchowski equation**:

$$D = \frac{\lambda^2}{2\tau} \tag{18.59}$$

● A BRIEF ILLUSTRATION

Suppose that an SO_4^{2-} ion jumps through its own diameter each time it makes a move in an aqueous solution; then, because $D = 1.1 \times 10^{-9}\ \text{m}^2\ \text{s}^{-1}$ and $a = 220$ pm (as deduced from mobility measurements), it follows from $\lambda = 2a$ that $\tau = 90$ ps. Because τ is the time for one jump, the ion makes 1×10^{10} jumps per second. ●

The Einstein–Smoluchowski equation is the central connection between the microscopic details of particle motion and the macroscopic parameters relating to diffusion (for example, the diffusion coefficient and, through the Stokes–Einstein relation, the viscosity). It also brings us back full circle to the properties of the perfect gas. For if we interpret λ/τ as $\bar{c}$, the mean speed of the molecules, and interpret λ as a mean free path, then we can recognize in the Einstein–Smoluchowski equation the same expression as we obtained from the kinetic model of gases, eqn 18.23. That is, the diffusion of a perfect gas is a random walk with an average step size equal to the mean free path.

IMPACT ON BIOCHEMISTRY

I18.2 Transport across membranes

Controlled transport of molecules and ions across biological membranes is at the heart of a number of key cellular processes, such as the transmission of nerve impulses, the transfer of glucose into red blood cells, and the synthesis of ATP by oxidative phosphorylation (*Impact I17.3*). Here we examine in some detail the various ways in which ions cross the alien environment of the lipid bilayer.

Suppose that a membrane provides a barrier that slows down the transfer of molecules or ions into or out of the cell. We saw in *Impact I17.3* that the thermodynamic tendency to transport an ion through the membrane is partially determined by a concentration gradient (more precisely, an activity gradient) across the membrane, which results in a difference in molar Gibbs energy between the inside and the outside of the cell, and a transmembrane potential gradient, which is due to the different potential energy of the ions on each side of the bilayer. There is a tendency, called **passive transport**, for a species to move down concentration and membrane potential gradients. It is also possible to move a species against these gradients, but now the flow must be driven by an exergonic process, such as the hydrolysis of ATP. This process is called **active transport.**

Consider the passive transport of an uncharged species A across a lipid bilayer of thickness l. To simplify the problem, we will assume that the concentration of A is always maintained at $[A] = [A]_0$ on one surface of the membrane and at $[A] = 0$ on the other surface, perhaps by a perfect balance between the rate of the process that produces A on one side and the rate of another process that consumes A completely on the other side. This is one example of a steady-state assumption, which will be discussed in more detail in Section 19.6. Then $\partial[A]/\partial t = 0$ and eqn 18.50 simplifies to

$$D\frac{d^2[A]}{dx^2} = 0$$

where D is the diffusion coefficient and the steady-state assumption makes partial derivatives unnecessary. We use the boundary conditions $[A](0) = [A]_0$ and $[A](l) = 0$ to solve the differential equation above and the result, which may be verified by differentiation, is

$$[A](x) = [A]_0\left(1 - \frac{x}{l}\right)$$

which implies that the [A] decreases linearly inside the membrane. We now use Fick's first law to calculate the flux J of A through the membrane and the result is

$$J = D\frac{[A]_0}{l}$$

However, we need to modify this equation slightly to account for the fact that the concentration of A on the surface of a membrane is not always equal to the concentration of A measured in the bulk solution, which we assume to be aqueous. This difference arises from the significant difference in the solubility of A in an aqueous environment and in the solution–membrane interface. One way to deal with this problem is to define a *partition coefficient,* κ (kappa), as

$$\kappa = \frac{[A]_0}{[A]_s}$$

where $[A]_s$ is the concentration of A in the bulk aqueous solution. It follows that

$$J = \kappa D\frac{[A]_s}{l} \tag{18.60}$$

In spite of the assumptions that led to its final form, this equation describes adequately the passive transport of many nonelectrolytes through membranes of blood cells.

In many cases the flux is underestimated by the equation above and the implication is that the membrane is more permeable than expected. However, the permeability increases only for certain species and not others and this is evidence that transport can be mediated by carriers. One example is the transporter protein that carries glucose into cells.

A characteristic of a carrier C is that it binds to the transported species A and the dissociation of the AC complex is described by

$$AC \rightleftharpoons A + C \qquad K = \frac{[A][C]}{[AC]}$$

where we have used concentrations instead of activities. After writing $[C]_0 = [C] + [AC]$, where $[C]_0$ is the total concentration of carrier, it follows that

$$[AC] = \frac{[A][C]_0}{[A] + K}$$

We can now use eqn 18.60 to write an expression for the flux of the species AC through the membrane:

$$J = \frac{\kappa_{AC}D_{AC}[C]_0}{l}\frac{[A]}{[A]+K} = J_{max}\frac{[A]}{[A]+K}$$

where κ_{AC} and D_{AC} are the partition coefficient and diffusion coefficient of the species AC, respectively. We see from Fig. 18.21 that when $[A] \ll K$ the flux varies linearly with [A] and that the flux reaches a maximum value of $J_{max} = \kappa_{AC}D_{AC}[C]_0/l$ when $[A] \gg K$. This behaviour is characteristic of mediated transport.

The transport of ions into or out of a cell needs to be mediated (that is, facilitated by other species) because the hydrophobic environment of the membrane is inhospitable to ions. There are two mechanisms for ion transport: mediation by a carrier molecule and transport through a **channel former**, a protein

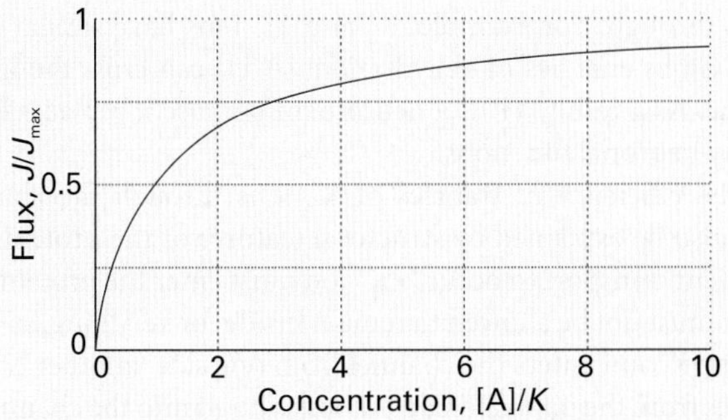

Fig. 18.21 The flux of the species AC through a membrane varies with the concentration of the species A. The behaviour shown in the figure and explained in the text is characteristic of mediated transport of A, with C as a carrier molecule.

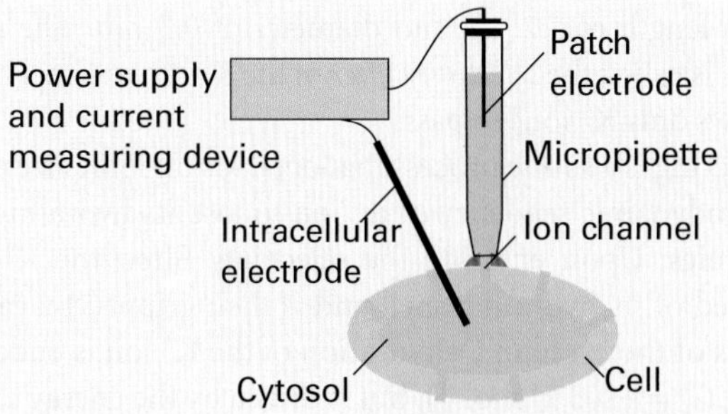

Fig. 18.22 A representation of the patch clamp technique for the measurement of ionic currents through membranes in intact cells. A section of membrane containing an ion channel is in tight contact with the tip of a micropipette containing an electrolyte solution and the patch electrode. An intracellular electronic conductor is inserted into the cytosol of the cell and the two conductors are connected to a power supply and current measuring device.

that creates a hydrophilic pore through which the ion can pass. An example of a channel former is the polypeptide gramicidin A, which increases the membrane permeability to cations such as H^+, K^+, and Na^+.

Ion channels are proteins that effect the movement of specific ions down a membrane potential gradient (see *Impact I17.3*). They are highly selective, so there is a channel protein for Ca^{2+}, another for Cl^-, and so on. The opening of the gate may be triggered by potential differences between the two sides of the membrane or by the binding of an *effector* molecule to a specific receptor site on the channel.

Ions such as H^+, Na^+, K^+, and Ca^{2+} are often transported actively across membranes by integral proteins called **ion pumps**. Ion pumps are molecular machines that work by adopting conformations that are permeable to one ion but not others depending on the state of phosphorylation of the protein. Because protein phosphorylation requires dephosphorylation of ATP, the conformational change that opens or closes the pump is endergonic and requires the use of energy stored during metabolism.

The structures of a number of channel proteins have been obtained by the now traditional X-ray diffraction techniques described in Chapter 9. Information about the flow of ions across channels and pumps is supplied by the **patch clamp technique**. One of many possible experimental arrangements is shown in Fig. 18.22. With mild suction, a 'patch' of membrane from a whole cell or a small section of a broken cell can be attached tightly to the tip of a micropipette filled with an electrolyte solution and containing an electronic conductor, the so-called *patch electrode*. A potential difference (the 'clamp') is applied between the patch electrode and an intracellular electronic conductor in contact with the cytosol of the cell. If the membrane is permeable to ions at the applied potential difference, a current flows through the completed circuit. Using narrow micropipette tips with diameters of less than 1 μm, ion currents of a few picoamperes (1 pA = 10^{-12} A) have been measured across sections of membranes containing only one ion channel protein.

A detailed picture of the mechanism of action of ion channels has emerged from analysis of patch clamp data and structural data. Here we focus on the K^+ ion channel protein, which, like all other mediators of ion transport, spans the membrane bilayer (Fig. 18.23). The pore through which ions move has a length of 3.4 nm and is divided into two regions: a wide region with a

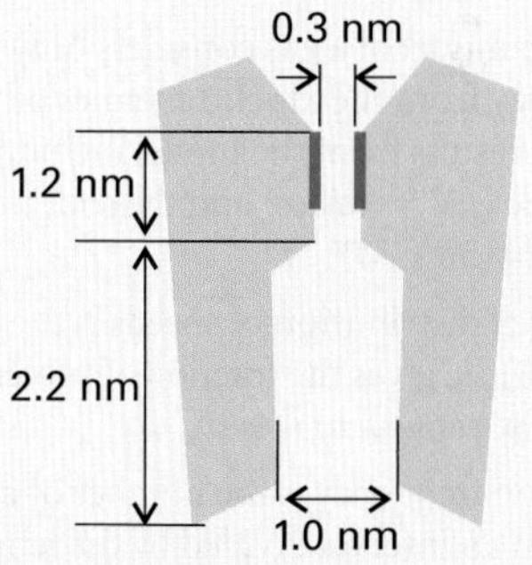

Fig. 18.23 A schematic representation of the cross-section of a membrane-spanning K^+ ion channel protein. The bulk of the protein is shown in light shades of beige. The pore through which ions move is divided into two regions: a wide region with a length of 2.2 nm and diameter of 1.0 nm, and a narrow region, the *selectivity filter*, with a length of 1.2 nm and diameter of 0.3 nm. The selectivity filter has a number of carbonyl groups (shown in dark green) that grip K^+ ions. As explained in the text, electrostatic repulsions between two bound K^+ ions 'encourage' ionic movement through the selectivity filter and across the membrane.

length of 2.2 nm and diameter of 1.0 nm, and a narrow region with a length of 1.2 nm and diameter of 0.3 nm. The narrow region is called the *selectivity filter* of the K^+ ion channel because it allows only K^+ ions to pass.

Filtering is a subtle process that depends on ionic size and the thermodynamic tendency of an ion to lose its hydrating water molecules. Upon entering the selectivity filter, the K^+ ion is stripped of its hydrating shell and is then gripped by carbonyl groups of the protein. Dehydration of the K^+ ion is endergonic ($\Delta_{\text{dehyd}}G^{\ominus} = +203\ \text{kJ mol}^{-1}$), but is driven by the energy of interaction between the ion and the protein. The Na^+ ion, though smaller than the K^+ ion, does not pass through the selectivity filter of the K^+ ion channel because interactions with the protein are not sufficient to compensate for the high Gibbs energy of dehydration of Na^+ ($\Delta_{\text{dehyd}}G^{\ominus} = +301\ \text{kJ mol}^{-1}$). More specifically, a dehydrated Na^+ ion is too small and cannot be held tightly by the protein carbonyl groups, which are positioned for ideal interactions with the larger K^+ ion. In its hydrated form, the Na^+ ion is too large (larger than a dehydrated K^+ ion), does not fit in the selectivity filter, and does not cross the membrane.

Though very selective, a K^+ ion channel can still let other ions pass through. For example, K^+ and Tl^+ ions have similar radii and Gibbs energies of dehydration, so Tl^+ can cross the membrane. As a result, Tl^+ is a neurotoxin because it replaces K^+ in many neuronal functions.

The efficiency of transfer of K^+ ions through the channel can also be explained by structural features of the protein. For efficient transport to occur, a K^+ ion must enter the protein, but then must not be allowed to remain inside for very long, so that, as one K^+ ion enters the channel from one side, another K^+ ion leaves from the opposite side. An ion is lured into the channel by water molecules about halfway through the length of the membrane. Consequently, the thermodynamic cost of moving an ion from an aqueous environment to the less hydrophilic interior of the protein is minimized. The ion is encouraged to leave the protein by electrostatic interactions in the selectivity filter, which can bind two K^+ ions simultaneously, usually with a bridging water molecule. Electrostatic repulsion prevents the ions from binding too tightly, minimizing the residence time of an ion in the selectivity filter, and maximizing the transport rate.

Checklist of key ideas

☐ 1. Diffusion is the migration of matter down a concentration gradient; thermal conduction is the migration of energy down a temperature gradient; electric conduction is the migration of electric charge along an electrical potential gradient; viscosity is the migration of linear momentum down a velocity gradient.

☐ 2. The kinetic model of a gas considers only the contribution to the energy from the kinetic energies of the molecules. Important results from the model include expressions for the pressure ($pV = \frac{1}{3}nMc^2$) and the root mean square speed ($c = \langle v^2 \rangle^{1/2} = (3RT/M)^{1/2}$).

☐ 3. The Maxwell distribution of speeds is the function which, through $f(v)\text{d}v$, gives the fraction of molecules that have speeds in the range v to $v + \text{d}v$.

☐ 4. The collision frequency is the number of collisions made by a molecule in an interval divided by the length of the interval: $z = \sigma \bar{c}_{\text{rel}} \mathcal{N}$, where the collision cross-section is $\sigma = \pi d^2$.

☐ 5. The mean free path is the average distance a molecule travels between collisions: $\lambda = \bar{c}/z$.

☐ 6. The collision flux, Z_W, is the number of collisions with an area in a given time interval divided by the area and the duration of the interval: $Z_W = p/(2\pi mkT)^{1/2}$.

☐ 7. Effusion is the emergence of a gas from a container through a small hole. Graham's law of effusion states that the rate of effusion is inversely proportional to the square root of the molar mass.

☐ 8. Flux J is the quantity of a property passing through a given area in a given time interval divided by the area and the duration of the interval.

☐ 9. Fick's first law of diffusion states that the flux of matter is proportional to the concentration gradient, $J(\text{matter}) = -D\text{d}\mathcal{N}/\text{d}z$, where D is the diffusion coefficient.

☐ 10. The ionic conductivity is the contribution of ions of one type to the molar conductivity: $\lambda = zuF$.

☐ 11. The diffusion equation is a relation between the rate of change of concentration at a point and the spatial variation of the concentration at that point: $\partial c/\partial t = D\partial^2 c/\partial x^2$.

☐ 12. In a one-dimensional random walk, the probability P that a molecule moves a distance x from the origin for a period t by taking small steps with size λ and time τ is: $P = (2\tau/\pi t)^{1/2} e^{-x^2\tau/2t\lambda^2}$.

Further information

Further information 18.1 *The transport characteristics of a perfect gas*

Here we derive expressions for the diffusion characteristics (specifically, the diffusion coefficient, the thermal conductivity, and the viscosity) of a perfect gas on the basis of the kinetic-molecular theory.

(a) The diffusion coefficient

Consider the arrangement depicted in Fig. 18.24. On average, the molecules passing through the area A at $z=0$ have travelled about one mean free path λ since their last collision. Therefore, the number density where they originated is $\mathcal{N}(z)$ evaluated at $z=-\lambda$. This number density is approximately

$$\mathcal{N}(-\lambda) = \mathcal{N}(0) - \lambda\left(\frac{d\mathcal{N}}{dz}\right)_0 \qquad (18.61)$$

where we have used a Taylor expansion of the form $f(x)=f(0)+(df/dx)_0x+\cdots$ truncated after the second term (see *Mathematical background 1*). The average number of impacts on the imaginary window of area A_0 during an interval Δt is $Z_WA_0\Delta t$, with $Z_W=\frac{1}{4}\mathcal{N}\bar{c}$ (eqn 18.16). Therefore, the flux from left to right, $J(L\rightarrow R)$, arising from the supply of molecules on the left, is

$$J(L\rightarrow R) = \frac{\frac{1}{4}A_0\mathcal{N}(-\lambda)\bar{c}\Delta t}{A_0\Delta t} = \tfrac{1}{4}\mathcal{N}(-\lambda)\bar{c} \qquad (18.62)$$

There is also a flux of molecules from right to left. On average, the molecules making the journey have originated from $z=+\lambda$ where the number density is $\mathcal{N}(\lambda)$. Therefore,

$$J(L\leftarrow R) = -\tfrac{1}{4}\mathcal{N}(\lambda)\bar{c} \qquad (18.63)$$

The average number density at $z=+\lambda$ is approximately

$$\mathcal{N}(\lambda) = \mathcal{N}(0) + \lambda\left(\frac{d\mathcal{N}}{dz}\right)_0 \qquad (18.64)$$

The net flux is

$$\begin{aligned} J_z &= J(L\rightarrow R) + J(L\leftarrow R) \\ &= \tfrac{1}{4}\bar{c}\left\{\left[\mathcal{N}(0) - \lambda\left(\frac{d\mathcal{N}}{dz}\right)_0\right] - \left[\mathcal{N}(0) + \lambda\left(\frac{d\mathcal{N}}{dz}\right)_0\right]\right\} \qquad (18.65) \\ &= -\tfrac{1}{2}\bar{c}\lambda\left(\frac{d\mathcal{N}}{dz}\right)_0 \end{aligned}$$

This equation shows that the flux is proportional to the first derivative of the concentration, in agreement with Fick's law.

At this stage it looks as though we can pick out a value of the diffusion coefficient by comparing eqns 18.65 and 18.41, so obtaining $D=\frac{1}{2}\lambda\bar{c}$. It must be remembered, however, that the calculation is quite crude, and is little more than an assessment of the order of magnitude of D. One aspect that has not been taken into account is illustrated in Fig. 18.25, which shows that, although a molecule may have begun its journey very close to the window, it could have a long flight before it gets there. Because the path is long, the molecule is likely to collide before reaching the window, so it ought to be added to the graveyard of other molecules that have collided. To take this effect into account involves a lot of work, but the end result is the appearance of a factor of $\frac{2}{3}$ representing the lower flux. The modification results in eqn 18.23.

(b) Thermal conductivity

According to the equipartition theorem (*Fundamentals F.5*), each molecule carries an average energy $\varepsilon = \nu kT$, where ν is a number of the order of 1. For monatomic particles, $\nu=\frac{3}{2}$. When one molecule passes through the imaginary window, it transports that energy on average. We suppose that the number density is uniform but that the temperature is not. On average, molecules arrive from the left after travelling a mean free path from their last collision in a hotter region, and therefore with a higher energy. Molecules also arrive from the right after travelling a

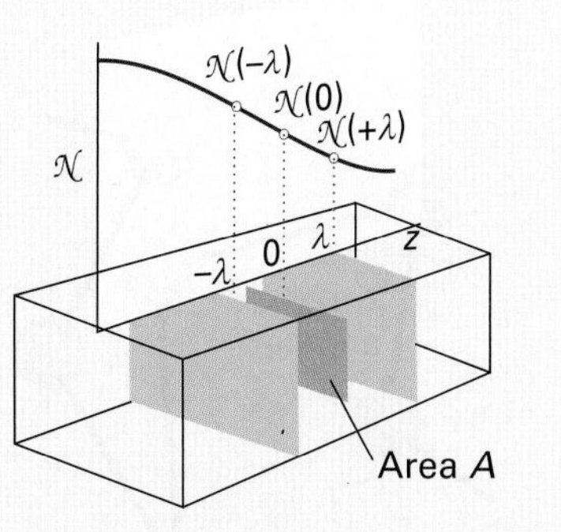

Fig. 18.24 The calculation of the rate of diffusion of a gas considers the net flux of molecules through a plane of area A as a result of arrivals from on average a distance λ away in each direction, where λ is the mean free path.

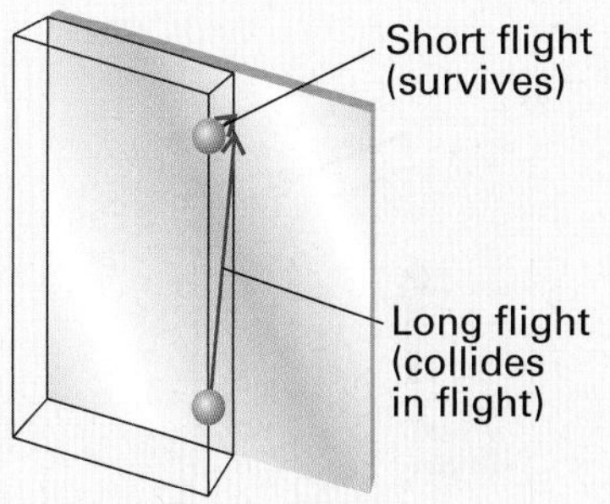

Fig. 18.25 One approximation ignored in the simple treatment is that some particles might make a long flight to the plane even though they are only a short perpendicular distance away, and therefore they have a higher chance of colliding during their journey.

mean free path from a cooler region. The two opposing energy fluxes are therefore

$$J(\mathrm{L} \rightarrow \mathrm{R}) = \tfrac{1}{4}\bar{c}\mathcal{N}\varepsilon(-\lambda) \qquad \varepsilon(-\lambda) = \nu k\left\{T - \lambda\left(\frac{\mathrm{d}T}{\mathrm{d}z}\right)_0\right\}$$

$$J(\mathrm{L} \leftarrow \mathrm{R}) = -\tfrac{1}{4}\bar{c}\mathcal{N}\varepsilon(\lambda) \qquad \varepsilon(\lambda) = \nu k\left\{T + \lambda\left(\frac{\mathrm{d}T}{\mathrm{d}z}\right)_0\right\} \qquad (18.66)$$

and the net flux is

$$J_z = J(\mathrm{L} \rightarrow \mathrm{R}) + J(\mathrm{L} \leftarrow \mathrm{R}) = -\tfrac{1}{2}\nu k\lambda\bar{c}\mathcal{N}\left(\frac{\mathrm{d}T}{\mathrm{d}z}\right)_0 \qquad (18.67)$$

As before, we multiply by $\frac{2}{3}$ to take long flight paths into account, and so arrive at

$$J_z = -\tfrac{1}{3}\nu k\lambda\bar{c}\mathcal{N}\left(\frac{\mathrm{d}T}{\mathrm{d}z}\right)_0 \qquad (18.68)$$

The energy flux is proportional to the temperature gradient, as we wanted to show. Comparison of this equation with eqn 18.21 shows that

$$\kappa = \tfrac{1}{3}\nu k\lambda\bar{c}\mathcal{N} \qquad (18.69)$$

Equation 18.24 then follows from $C_{V,\mathrm{m}} = \nu kN_\mathrm{A}$ for a perfect gas, where [A] is the molar concentration of A. For this step, we use $\mathcal{N} = N/V = nN_\mathrm{A}/V = N_\mathrm{A}[\mathrm{A}]$.

(c) Viscosity

Molecules travelling from the right in Fig. 18.26 (from a fast layer to a slower one) transport a momentum $mv_x(\lambda)$ to their new layer at $z = 0$; those travelling from the left transport $mv_x(-\lambda)$ to it. If it is assumed that the density is uniform, the collision flux is $\frac{1}{4}\mathcal{N}\bar{c}$. Those arriving from the right on average carry a momentum

$$mv_x(\lambda) = mv_x(0) + m\lambda\left(\frac{\mathrm{d}v_x}{\mathrm{d}z}\right)_0 \qquad (18.70\mathrm{a})$$

Those arriving from the left bring a momentum

$$mv_x(-\lambda) = mv_x(0) - m\lambda\left(\frac{\mathrm{d}v_x}{\mathrm{d}z}\right)_0 \qquad (18.70\mathrm{b})$$

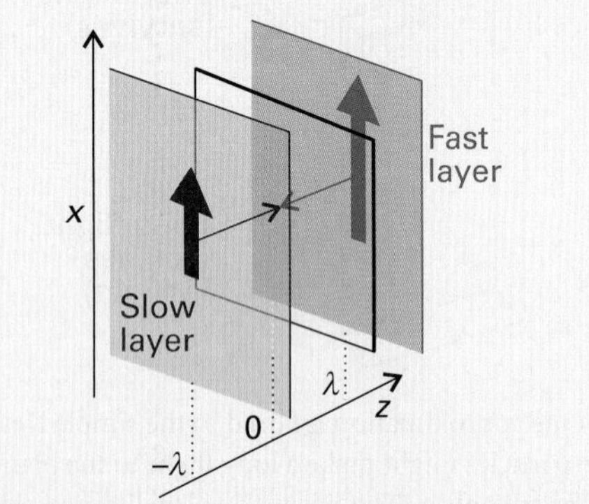

Fig. 18.26 The calculation of the viscosity of a gas examines the net x-component of momentum brought to a plane from faster and slower layers on average a mean free path away in each direction.

The net flux of x-momentum in the z-direction is therefore

$$J = \tfrac{1}{4}\mathcal{N}\bar{c}\left\{\left[mv_x(0) - m\lambda\left(\frac{\mathrm{d}v_x}{\mathrm{d}z}\right)_0\right] - \left[mv_x(0) + m\lambda\left(\frac{\mathrm{d}v_x}{\mathrm{d}z}\right)_0\right]\right\}$$
$$= -\tfrac{1}{2}\mathcal{N}m\lambda\bar{c}\left(\frac{\mathrm{d}v_x}{\mathrm{d}z}\right)_0 \qquad (18.71)$$

The flux is proportional to the velocity gradient, as we wished to show. Comparison of this expression with eqn 18.22, and multiplication by $\frac{2}{3}$ in the normal way, leads to

$$\eta = \tfrac{1}{3}\mathcal{N}m\lambda\bar{c} \qquad (18.72)$$

which can easily be converted into eqn 18.25 by using $\mathcal{N}m = nM$ and $[\mathrm{A}] = n/V$.

Further information 18.2 *Random coils*

The most likely conformation of a polymer chain of identical monomer units not capable of forming hydrogen bonds or any other type of specific bond is a **random coil**. Polyethylene is a simple example. The random coil model is a helpful starting point for estimating the orders of magnitude of the hydrodynamic properties of polymers and denatured proteins in solution. The mathematics of random coils is almost identical to that of the random walk, for a coil is like the track of a walk frozen in time.

The simplest model of a random coil is a **freely jointed chain**, in which any bond is free to make any angle with respect to the preceding one (Fig. 18.27). We assume that the residues occupy zero volume, so different parts of the chain can occupy the same region of space. The model is obviously an oversimplification because a bond is actually constrained to a cone of angles around a direction defined by its neighbour (Fig. 18.28). In a hypothetical one-dimensional freely jointed chain all the residues lie in a straight line, and the angle between neighbours is either 0° or 180°. The residues in a three-dimensional freely jointed chain are not restricted to lie in a line or a plane.

By exactly the same argument as in *Justification 18.7*, the probability, P, that the ends of a one-dimensional freely jointed chain composed of N units of length l are a distance nl apart is

$$P = \left(\frac{2}{\pi N}\right)^{1/2} \mathrm{e}^{-n^2/2N} \qquad (18.73)$$

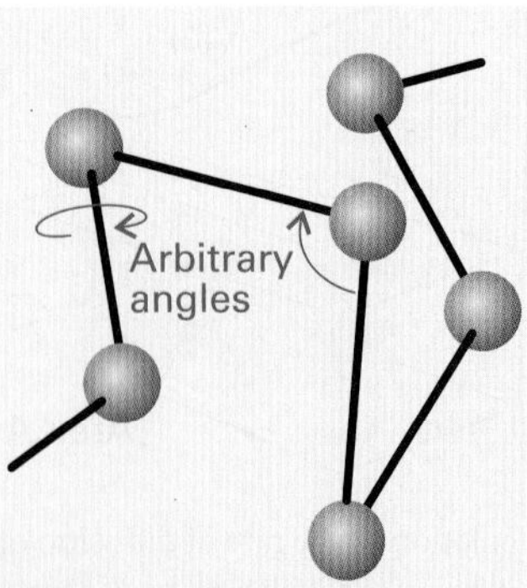

Fig. 18.27 A freely jointed chain is like a three-dimensional random walk, each step being in an arbitrary direction but of the same length.

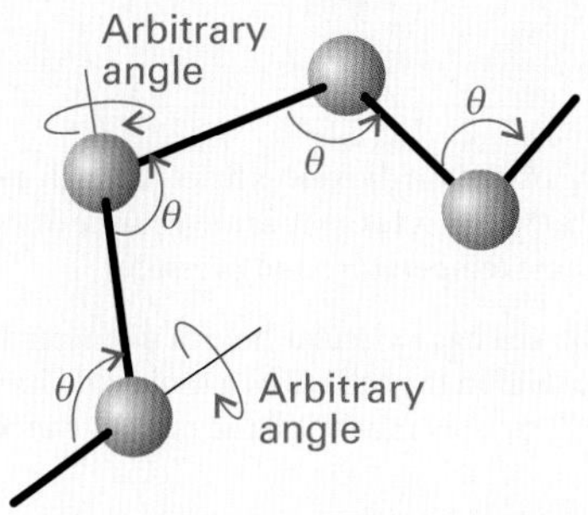

Fig. 18.28 A better description is obtained by fixing the bond angle (for example, at the tetrahedral angle) and allowing free rotation about a bond direction.

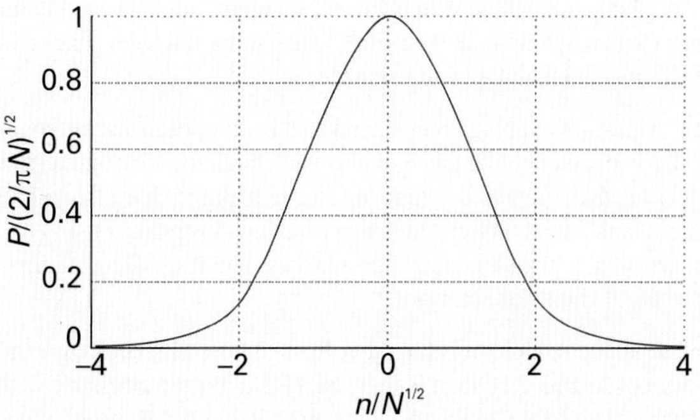

Fig. 18.29 The probability distribution for the separation of the ends of a one-dimensional random coil. The separation of the ends is nl, where l is the bond length.

This function is plotted in Fig. 18.29 and can be used to calculate the probability that the ends of a three-dimensional freely jointed chain lie in the range r to $r + \mathrm{d}r$. We write this probability as $f(r)\mathrm{d}r$, where

$$f(r) = 4\pi\left(\frac{a}{\pi^{1/2}}\right)^3 r^2 e^{-a^2r^2} \qquad a = \left(\frac{3}{2Nl^2}\right)^{1/2} \qquad (18.74)$$

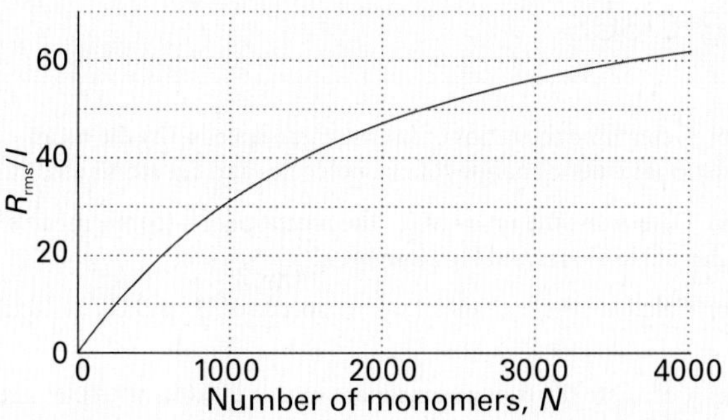

Fig. 18.30 The variation of the root mean square separation of the ends of a three-dimensional random coil, R_{rms}, with the number of monomers.

In some coils, the ends may be far apart whereas in others their separation is small. Here and elsewhere we are ignoring the fact that the chain cannot be longer than Nl. Although eqn 18.74 gives a nonzero probability for $R > Nl$, the values are so small that the errors in pretending that R can range up to infinity are negligible. An alternative interpretation of eqn 18.74 is to regard each coil in a sample as ceaselessly writhing from one conformation to another; then $f(r)\mathrm{d}r$ is the probability that at any instant the chain will be found with the separation of its ends between r and $r + \mathrm{d}r$.

There are several measures of the geometrical size of a random coil. The **contour length**, R_c, is the length of the macromolecule measured along its backbone from atom to atom (it corresponds to the total length of a random walk). For a polymer of N monomer units each of length l, the contour length is

$$R_c = Nl \qquad (18.75)$$

The **root mean square separation**, R_{rms}, is a measure of the average separation of the ends of a random coil (it corresponds to the root mean square distance reached from the origin of a radom walk): it is the square root of the mean value of R^2:

$$R_{rms} = N^{1/2}l \qquad (18.76)$$

We see that as the number of monomer units increases, the root mean square separation of its ends increases as $N^{1/2}$ (Fig. 18.30), and consequently its volume increases as $N^{3/2}$.

Discussion questions

18.1 Specify and analyse critically the assumptions that underlie the kinetic model of gases.

18.2 Provide molecular interpretations for the dependencies of the mean free path on the temperature, pressure, and size of gas molecules.

18.3 Provide a molecular interpretation for each of the following processes: diffusion, thermal conduction, electric conduction, and viscosity.

18.4 Discuss the difference between the hydrodynamic radius of an ion and its ionic radius and explain why a small ion can have a large hydrodynamic radius.

18.5 To what extent is the analogy between a random walk and a random coil valid?

18.6 Discuss the mechanism of proton conduction in water. How could the model be tested?

18.7 Describe the origin of the thermodynamic force. To what extent can it be regarded as an actual force?

18.8 Account physically for the form of the diffusion equation.

18.9 Identify and describe the processes responsible for the transport of ions across biological membranes.

Exercises

18.1(a) Determine the ratios of (a) the mean speeds, (b) the mean translational kinetic energies of H_2 molecules and Hg atoms at 20°C.

18.1(b) Determine the ratios of (a) the mean speeds, (b) the mean kinetic energies of He atoms and Hg atoms at 25°C.

18.2(a) Calculate the root mean square speeds of H_2 and O_2 molecules at 20°C.

18.2(b) Calculate the root mean square speeds of CO_2 molecules and He atoms at 20°C.

18.3(a) Use the Maxwell distribution of speeds to estimate the fraction of N_2 molecules at 400 K that have speeds in the range 200 to 210 m s^{-1}.

18.3(b) Use the Maxwell distribution of speeds to estimate the fraction of CO_2 molecules at 400 K that have speeds in the range 400 to 405 m s^{-1}.

18.4(a) Calculate the most probable speed, the mean speed, and the mean relative speed of CO_2 molecules in air at 20°C.

18.4(b) Calculate the most probable speed, the mean speed, and the mean relative speed of H_2 molecules in air at 20°C.

18.5(a) Assume that air consists of N_2 molecules with a collision diameter of 395 pm. Calculate (a) the mean speed of the molecules, (b) the mean free path, (c) the collision frequency in air at 1.0 atm and 25°C.

18.5(b) The best laboratory vacuum pump can generate a vacuum of about 1 nTorr. At 25°C and assuming that air consists of N_2 molecules with a collision diameter of 395 pm, calculate (a) the mean speed of the molecules, (b) the mean free path, (c) the collision frequency in the gas.

18.6(a) At what pressure does the mean free path of argon at 20°C become comparable to the diameter of a 100 cm^3 vessel that contains it? Take $\sigma = 0.36$ nm^2.

18.6(b) At what pressure does the mean free path of argon at 20°C become comparable to 10 times the diameters of the atoms themselves? Take $\sigma = 0.36$ nm^2.

18.7(a) At an altitude of 20 km the temperature is 217 K and the pressure 0.050 atm. What is the mean free path of N_2 molecules? ($\sigma = 0.43$ nm^2.)

18.7(b) At an altitude of 15 km the temperature is 217 K and the pressure 12.1 kPa. What is the mean free path of N_2 molecules? ($\sigma = 0.43$ nm^2.)

18.8(a) How many collisions does a single Ar atom make in 1.0 s when the temperature is 25°C and the pressure is (a) 10 atm, (b) 1.0 atm, (c) 1.0 μatm?

18.8(b) How many collisions per second does an N_2 molecule make at an altitude of 15 km? (See Exercise 18.7b for data.)

18.9(a) A solid surface with dimensions 5.0 mm × 4.0 mm is exposed to argon gas at 25 Pa and 300 K. How many collisions do the Ar atoms make with this surface in 100 s?

18.9(b) A solid surface with dimensions 2.0 cm × 10.0 cm is exposed to helium gas at 120 Pa and 1200 K. How many collisions do the He atoms make with this surface in 1.0 s?

18.10(a) If 125 cm^3 of hydrogen gas effuses through a small hole in 135 seconds, how long will it take the same volume of oxygen gas to effuse under the same temperature and pressure?

18.10(b) If 175 cm^3 of carbon dioxide effuses through a small hole in 255 seconds, how long will it take the same volume of sulfur dioxide to effuse under the same temperature and pressure?

18.11(a) An effusion cell has a circular hole of diameter 1.50 mm. If the molar mass of the solid in the cell is 300 g mol^{-1} and its vapour pressure is 0.735 Pa at 500 K, by how much will the mass of the solid decrease in a period of 1.00 h?

18.11(b) An effusion cell has a circular hole of diameter 1.00 mm. If the molar mass of the solid in the cell is 250 g mol^{-1} and its vapour pressure is 0.324 Pa at 425 K, by how much will the mass of the solid decrease in a week (use 1 week = 7 × 24 h)?

18.12(a) A manometer was connected to a bulb containing a gaseous sample under slight pressure. The gas was allowed to escape through a small pinhole, and the time for the manometer reading to drop from 74 cm to 20 cm was 152 s. When the experiment was repeated using nitrogen (for which $M = 28.02$ g mol^{-1}) the same fall took place in 45 s. Calculate the molar mass of the sample.

18.12(b) A manometer was connected to a bulb containing nitrogen under slight pressure. The gas was allowed to escape through a small pinhole, and the time for the manometer reading to drop from 75.1 cm to 32.5 cm was 22.5 s. When the experiment was repeated using a fluorocarbon gas, the same fall took place in 135.0 s. Calculate the molar mass of the fluorocarbon.

18.13(a) A space vehicle of internal volume 3.0 m^3 is struck by a meteor and a hole of radius 0.10 mm is formed. If the oxygen pressure within the vehicle is initially 80 kPa and its temperature 298 K, how long will the pressure take to fall to 70 kPa?

18.13(b) A container of internal volume 22.0 m^3 was punctured, and a hole of radius 0.050 mm was formed. If the nitrogen pressure within the vehicle is initially 122 kPa and its temperature 293 K, how long will the pressure take to fall to 105 kPa?

18.14(a) Calculate the thermal conductivity of argon ($C_{V,m} = 12.5$ J K^{-1} mol^{-1}, $\sigma = 0.36$ nm^2) at room temperature.

18.14(b) Calculate the thermal conductivity of nitrogen ($C_{V,m} = 20.8$ J K^{-1} mol^{-1}, $\sigma = 0.43$ nm^2) at room temperature.

18.15(a) Calculate the diffusion constant of argon at 20°C and (a) 1.00 Pa, (b) 100 kPa, (c) 10.0 MPa. If a pressure gradient of 1.0 bar m^{-1} is established in a pipe, what is the flow of gas due to diffusion?

18.15(b) Calculate the diffusion constant of nitrogen at 20°C and (a) 100.0 Pa, (b) 100 kPa, (c) 20.0 MPa. If a pressure gradient of 1.20 bar m^{-1} is established in a pipe, what is the flow of gas due to diffusion?

18.16(a) Calculate the flux of energy arising from a temperature gradient of 10.5 K m^{-1} in a sample of argon in which the mean temperature is 280 K.

18.16(b) Calculate the flux of energy arising from a temperature gradient of 8.5 K m^{-1} in a sample of hydrogen in which the mean temperature is 290 K.

18.17(a) Use the experimental value of the thermal conductivity of neon (Table 18.2) to estimate the collision cross-section of Ne atoms at 273 K.

18.17(b) Use the experimental value of the thermal conductivity of nitrogen (Table 18.2) to estimate the collision cross-section of N_2 molecules at 298 K.

18.18(a) In a double-glazed window, the panes of glass are separated by 1.0 cm. What is the rate of transfer of heat by conduction from the warm room (28°C) to the cold exterior (−15°C) through a window of area 1.0 m^2? What power of heater is required to make good the loss of heat?

18.18(b) Two sheets of copper of area 2.00 m^2 are separated by 5.00 cm. What is the rate of transfer of heat by conduction from the warm sheet (70°C) to the cold sheet (0°C)? What is the rate of loss of heat?

18.19(a) Use the experimental value of the coefficient of viscosity for neon (Table 18.2) to estimate the collision cross-section of Ne atoms at 273 K.

18.19(b) Use the experimental value of the coefficient of viscosity for nitrogen (Table 18.2) to estimate the collision cross-section of the molecules at 273 K.

18.20(a) Calculate the viscosity of air at (a) 273 K, (b) 298 K, (c) 1000 K. Take $\sigma \approx 0.40$ nm^2. (The experimental values are 173 μP at 273 K, 182 μP at 20°C, and 394 μP at 600°C.)

18.20(b) Calculate the viscosity of benzene vapour at (a) 273 K, (b) 298 K, (c) 1000 K. Take $\sigma \approx 0.88$ nm^2.

18.21(a) The viscosity of water at 20°C is 1.002 cP and 0.7975 cP at 30°C. What is the energy of activation for the transport process?

18.21(b) The viscosity of mercury at 20°C is 1.554 cP and 1.450 cP at 40°C. What is the energy of activation for the transport process?

18.22(a) The mobility of a chloride ion in aqueous solution at 25°C is 7.91×10^{-8} m^2 s^{-1} V^{-1}. Calculate the molar ionic conductivity.

18.22(b) The mobility of an acetate ion in aqueous solution at 25°C is 4.24×10^{-8} m^2 s^{-1} V^{-1}. Calculate the molar ionic conductivity.

18.23(a) The mobility of a Rb^+ ion in aqueous solution is 7.92×10^{-8} m^2 s^{-1} V^{-1} at 25°C. The potential difference between two electrodes placed in the solution is 25.0 V. If the electrodes are 7.00 mm apart, what is the drift speed of the Rb^+ ion?

18.23(b) The mobility of a Li^+ ion in aqueous solution is 4.01×10^{-8} m^2 s^{-1} V^{-1} at 25°C. The potential difference between two electrodes placed in the solution is 24.0 V. If the electrodes are 5.0 mm apart, what is the drift speed of the ion?

18.24(a) The limiting molar conductivities of NaI, $NaNO_3$, and $AgNO_3$ are 12.69 mS m^2 mol^{-1}, 12.16 mS m^2 mol^{-1}, and 13.34 mS m^2 mol^{-1}, respectively (all at 25°C). What is the limiting molar conductivity of AgI at this temperature?

18.24(b) The limiting molar conductivities of KF, KCH_3CO_2, and $Mg(CH_3CO_2)_2$ are 12.89 mS m^2 mol^{-1}, 11.44 mS m^2 mol^{-1}, and 18.78 mS m^2 mol^{-1}, respectively (all at 25°C). What is the limiting molar conductivity of MgF_2 at this temperature?

18.25(a) At 25°C the molar ionic conductivities of Li^+, Na^+, and K^+ are 3.87 mS m^2 mol^{-1}, 5.01 mS m^2 mol^{-1}, and 7.35 mS m^2 mol^{-1}, respectively. What are their mobilities?

18.25(b) At 25°C the molar ionic conductivities of F^-, Cl^-, and Br^- are 5.54 mS m^2 mol^{-1}, 7.635 mS m^2 mol^{-1}, and 7.81 mS m^2 mol^{-1}, respectively. What are their mobilities?

18.26(a) The mobility of a SO_4^{2-} ion in aqueous solution at 25°C is 8.19×10^{-8} m^2 s^{-1} V^{-1}. Calculate its diffusion coefficient in water at 25°C.

18.26(b) The mobility of an NH_4^+ ion in aqueous solution at 25°C is 7.63×10^{-8} m^2 s^{-1} V^{-1}. Calculate its diffusion coefficient in water at 25°C.

18.27(a) The diffusion coefficient of glucose in water at 25°C is 6.73×10^{-10} m^2 s^{-1}. Estimate the time required for a glucose molecule to undergo a root mean square displacement of 5.0 mm.

18.27(b) The diffusion coefficient of H_2O in water at 25°C is 2.26×10^{-9} m^2 s^{-1}. Estimate the time required for an H_2O molecule to undergo a root mean square displacement of 1.0 cm.

18.28(a) Estimate the effective radius of a sucrose molecule in water at 25°C given that its diffusion coefficient is 5.2×10^{-10} m^2 s^{-1} and that the viscosity of water is 1.00 cP.

18.28(b) Estimate the effective radius of a glycine molecule in water at 25°C given that its diffusion coefficient is 1.055×10^{-9} m^2 s^{-1} and that the viscosity of water is 1.00 cP.

18.29(a) The diffusion coefficient for Cl^- in water is 2.03×10^{-9} m^2 s^{-1}. How long does the ion take to jump through about one ionic diameter (approximately the fundamental jump length for translational motion)?

18.29(b) The diffusion coefficient for a glycine molecule in water is 1.064×10^{-9} m^2 s^{-1}. How long does a molecule take to jump through about one molecular diameter (approximately the fundamental jump length for translational motion)?

18.30(a) A layer of 20.0 g of sucrose is spread uniformly over a surface of area 5.0 cm^2 and covered in water to a depth of 20 cm. What will be the molar concentration of sucrose molecules at 10 cm above the original layer at (a) 10 s, (b) 24 h? Assume diffusion is the only transport process and take $D = 5.216 \times 10^{-9}$ m^2 s^{-1}.

18.30(b) A layer of 10.0 g of sucrose is spread uniformly over a surface of area 10.0 cm^2 and covered in hexane to a depth of 10 cm. What will be the molar concentration of sucrose molecules at 5.0 cm above the original layer at (a) 10 s, (b) 24 h? Assume diffusion is the only transport process and take $D = 4.05 \times 10^{-9}$ m^2 s^{-1}.

18.31(a) A certain polymer chain consists of 900 segments, each 1.05 nm long. If the chain were ideally flexible, what would be the root mean square separation of the ends of the chain?

18.31(b) A certain polymer chain consists of 800 segments, each 1.25 nm long. If the chain were ideally flexible, what would be the root mean square separation of the ends of the chain?

18.32(a) Calculate the contour length and the root mean square separation of the ends of a polyethylene molecule of molar mass 300 kg mol^{-1}.

18.32(b) Calculate the contour length and the root mean square separation of the ends of a polypropylene molecule of molar mass 250 kg mol^{-1}.

Problems*

Numerical problems

18.1 Instead of the arrangement in Fig. 18.8, the speed of molecules can also be measured with a rotating slotted-disc apparatus, which consists of five coaxial 5.0 cm diameter discs separated by 1.0 cm, the slots in their rims being displaced by 2.0° between neighbours. The relative intensities, I, of the detected beam of Kr atoms for two different temperatures and at a series of rotation rates were as follows:

ν/Hz	20	40	80	100	120
I (40 K)	0.846	0.513	0.069	0.015	0.002
I (100 K)	0.592	0.485	0.217	0.119	0.057

Find the distributions of molecular velocities, $f(v_x)$, at these temperatures, and check that they conform to the theoretical prediction for a one-dimensional system.

18.2‡ Fenghour *et al.* (*J. Phys. Chem. Ref. Data* **24**, 1649 (1995)) have compiled an extensive table of viscosity coefficients for ammonia in the liquid and vapour phases. Deduce the effective molecular diameter of NH_3 based on each of the following vapour-phase viscosity coefficients: (a) $\eta = 9.08 \times 10^{-6}$ kg m^{-1} s^{-1} at 270 K and 1.00 bar; (b) $\eta = 1.749 \times 10^{-5}$ kg m^{-1} s^{-1} at 490 K and 10.0 bar.

18.3 Calculate the ratio of the thermal conductivities of gaseous hydrogen at 300 K to gaseous hydrogen at 10 K. Be circumspect, and think about the modes of motion that are thermally active at the two temperatures.

18.4 A Knudsen cell was used to determine the vapour pressure of germanium at 1000°C. During an interval of 7200 s the mass loss through a hole of radius 0.50 mm amounted to 43 μg. What is the vapour pressure of germanium at 1000°C? Assume the gas to be monatomic.

18.5 The pressure of a Knudsen cell of volume V in which a vapour is confined (with no condensed phase to replenish the vapour phase) decays exponentially with a time constant $\tau = (2\pi M/RT)^{1/2}(V/A)$ (see Problem 18.28 for the derivation of a related expression). How long would it take the pressure of barium vapour in a cell with $V = 100$ cm^3 and $A = 0.10$ mm^2 at 1300°C to fall to 1/10 of its initial value?

18.6 The vapour pressure of zinc in the range 250°C to 419°C can be estimated from the expression $\log(p/\text{Torr}) = a - b/T$ with $a = 9.200$ and $b = 6947$ K. Calculate and plot the beam flux (in Zn atoms per second) emerging from a hole of radius 0.20 mm as the temperature of the oven containing solid zinc is raised from 250°C to 400°C.

18.7 The viscosity of benzene varies with temperature as shown in the following table. Use the data to infer the activation energy for viscosity (the parameter E_a in eqn 18.26).

θ/°C	10	20	30	40	50	60	70
η/cP	0.758	0.652	0.564	0.503	0.442	0.392	0.358

18.8 An empirical expression that reproduces the viscosity of water in the range 20–100°C is

$$\log(\eta/\eta_{20}) = \frac{1.3272(20 - \theta/°\text{C}) - 0.001053(20 - \theta/°\text{C})^2}{\theta/°\text{C} + 105}$$

Explore (by using mathematical software) the possibility of fitting an exponential curve to this expression and hence identifying an activation energy for the viscosity. This approach is taken further in Problem 18.29.

18.9 The conductivity of aqueous ammonium chloride at a series of concentrations is listed in the following table. Deduce the molar conductivity and determine the parameters that occur in the Kohlrausch law.

c/(mol dm^{-3})	1.334	1.432	1.529	1.672	1.725
κ/(mS cm^{-1})	131	139	147	156	164

18.10 Conductivities are often measured by comparing the resistance of a cell filled with the sample to its resistance when filled with some standard solution, such as aqueous potassium chloride. The conductivity of water is 76 mS m^{-1} at 25°C and the conductivity of 0.100 mol dm^{-3} KCl(aq) is 1.1639 S m^{-1}. A cell had a resistance of 33.21 Ω when filled with 0.100 mol dm^{-3} KCl(aq) and 300.0 Ω when filled with 0.100 mol dm^{-3} CH_3COOH. What is the molar conductivity of acetic acid at that concentration and temperature?

18.11 The resistances of a series of aqueous NaCl solutions, formed by successive dilution of a sample, were measured in a cell with cell constant (the constant C in the relation $\kappa = C/R$) equal to 0.2063 cm^{-1}. The following values were found:

c/(mol dm^{-3})	0.00050	0.0010	0.0050	0.010	0.020	0.050
R/Ω	3314	1669	342.1	174.1	89.08	37.14

Verify that the molar conductivity follows the Kohlrausch law and find the limiting molar conductivity. Determine the coefficient K. Use the value of K (which should depend only on the nature, not the identity, of the ions) and the information that $\lambda(Na^+) = 5.01$ mS m^2 mol^{-1} and $\lambda(I^-) = 7.68$ mS m^2 mol^{-1} to predict (a) the molar conductivity, (b) the conductivity, (c) the resistance it would show in the cell of 0.010 mol dm^{-3} NaI(aq) at 25°C.

18.12 What are the drift speeds of Li^+, Na^+, and K^+ in water when a potential difference of 100 V is applied across a 5.00-cm conductivity cell? How long would it take an ion to move from one electrode to the other? In conductivity measurements it is normal to use alternating current: what are the displacements of the ions in (a) centimetres, (b) solvent diameters, about 300 pm, during a half cycle of 2.0 kHz applied potential difference?

18.13‡ Bakale *et al.* (*J. Phys. Chem.* **100**, 12477 (1996)) measured the mobility of singly charged C_{60}^- ions in a variety of nonpolar solvents. In cyclohexane at 22°C, the mobility is 1.1×10^{-8} m^2 V^{-1} s^{-1}. Estimate the effective radius of the C_{60}^- ion. The viscosity of the solvent is 0.93×10^{-3} kg m^{-1} s^{-1}. Suggest a reason why there is a substantial difference between this number and the van der Waals radius of neutral C_{60}.

18.14 A dilute solution of potassium permanganate in water at 25°C was prepared. The solution was in a horizontal tube of length 10 cm, and at first there was a linear gradation of intensity of the purple solution from the left (where the concentration was 0.100 mol dm^{-3}) to the right (where the concentration was 0.050 mol dm^{-3}). What is the magnitude and sign of the thermodynamic force acting on the solute (a) close to the

* Problems denoted with the symbol ‡ were supplied by Charles Trapp, Carmen Giunta, and Marshall Cady.

left face of the container, (b) in the middle, (c) close to the right face. Give the force per mole and force per molecule in each case.

18.15 A dilute solution of potassium permanganate in water at 25°C was prepared. The solution was in a horizontal tube of length 10 cm, and at first there was a Gaussian distribution of concentration around the centre of the tube at $x = 0$, $c(x) = c_0e^{-ax^2}$, with $c_0 = 0.100$ mol dm^{-3} and $a = 0.10$ cm^{-2}. Determine the thermodynamic force acting on the solute as a function of location, x, and plot the result. Give the force per mole and force per molecule in each case. What do you expect to be the consequence of the thermodynamic force?

18.16 Instead of a Gaussian 'heap' of solute, as in Problem 18.15, suppose that there is a Gaussian dip, a distribution of the form $c(x) = c_0(1 - e^{-ax^2})$. Repeat the calculation in Problem 18.15 and its consequences.

18.17 Estimate the diffusion coefficients and the effective hydrodynamic radii of the alkali metal cations in water from their mobilities at 25°C. Estimate the approximate number of water molecules that are dragged along by the cations. Ionic radii are given in Table 9.3.

18.18 Nuclear magnetic resonance can be used to determine the mobility of molecules in liquids. A set of measurements on methane in carbon tetrachloride showed that its diffusion coefficient is 2.05×10^{-9} m^2 s^{-1} at 0°C and 2.89×10^{-9} m^2 s^{-1} at 25°C. Deduce what information you can about the mobility of methane in carbon tetrachloride.

18.19 A lump of sucrose of mass 10.0 g is suspended in the middle of a spherical flask of water of radius 10 cm at 25°C. What is the concentration of sucrose at the wall of the flask after (a) 1.0 h, (b) 1.0 week. Take $D = 5.22 \times 10^{-10}$ m^2 s^{-1}.

18.20 In a series of observations on the displacement of rubber latex spheres of radius 0.212 μm, the mean square displacements after selected time intervals were on average as follows:

t/s	30	60	90	120
$10^{12}\langle x^2\rangle$/m^2	88.2	113.5	128	144

These results were originally used to find the value of Avogadro's constant, but there are now better ways of determining N_A, so the data can be used to find another quantity. Find the effective viscosity of water at the temperature of this experiment (25°C).

Theoretical problems

18.21 Start from the Maxwell–Boltzmann distribution and derive an expression for the most probable speed of a gas of molecules at a temperature T. Go on to demonstrate the validity of the equipartition conclusion that the average translational kinetic energy of molecules free to move in three dimensions is $\frac{3}{2}kT$.

18.22 In Section 14.4 it was established that the heat capacity of a collection of molecules is proportional to the variance of their energy (the mean square deviation of the energy from its mean value). Use the Maxwell–Boltzmann distribution of speeds to calculate the translational contribution to the heat capacity of a gas by this approach.

18.23 Consider molecules that are confined to move in a plane (a two-dimensional gas). Calculate the distribution of speeds and determine the mean speed of the molecules at a temperature T.

18.24 A specially constructed velocity selector accepts a beam of molecules from an oven at a temperature T but blocks the passage of molecules with a speed greater than the mean. What is the mean speed of the emerging beam, relative to the initial value, treated as a one-dimensional problem?

18.25 What, according to the Maxwell–Boltzmann distribution, is the proportion of gas molecules having (a) more than, (b) less than the root mean square speed? (c) What are the proportions having speeds greater and smaller than the mean speed?

18.26 Calculate the fractions of molecules in a gas that have a speed in a range Δv at the speed nc^* relative to those in the same range at c^* itself? This calculation can be used to estimate the fraction of very energetic molecules (which is important for reactions). Evaluate the ratio for $n = 3$ and $n = 4$.

18.27 Derive an expression for $\langle v^n\rangle^{1/n}$ from the Maxwell–Boltzmann distribution of speeds. You will need the standard integrals

$$\int_0^\infty x^{2m+1}e^{-ax^2}dx = \frac{m!}{2a^{m+1}}$$

$$\int_0^\infty x^{2m}e^{-ax^2}dx = \frac{(2m-1)!!}{2^{m+1}a^m}\left(\frac{\pi}{a}\right)^{1/2}$$

where the 'double factorial' means $(2m - 1)!! = 1 \times 3 \times 5 \cdots \times (2m - 1)$.

18.28 Derive an expression that shows how the pressure of a gas inside an effusion oven (a heated chamber with a small hole in one wall) varies with time if the oven is not replenished as the gas escapes. Then show that $t_{1/2}$, the time required for the pressure to decrease to half its initial value, is independent of the initial pressure. *Hint.* Begin by setting up a differential equation relating dp/dt to $p = NkT/V$, and then integrating it.

18.29 In Section 20.1 we shall see that a general expression for the activation energy of a chemical reaction is $E_a = RT^2(d \ln k/dT)$. Confirm that the same expression may be used to extract the activation energy from eqn 18.26 for the viscosity and then apply the expression to deduce the temperature dependence of the activation energy when the viscosity of water is given by the empirical expression in Problem 18.8. Plot this activation energy as a function of temperature. Suggest an explanation of the temperature dependence of E_a.

18.30 Confirm that eqn 18.54 is a solution of the diffusion equation with the correct initial value.

18.31 Confirm that

$$c(x,t) = \frac{c_0}{(4\pi Dt)^{1/2}}e^{-(x-x_0-vt)^2/4Dt}$$

is a solution of the diffusion equation with convection (eqn 18.53) with all the solute concentrated at $x = x_0$ at $t = 0$ and plot the concentration profile at a series of times to show how the distribution spreads and its centroid drifts.

18.32 The thermodynamic force has a direction as well as a magnitude, and in a three-dimensional ideal system eqn 18.40 becomes $\mathcal{F} = -RT\nabla \ln c$. What is the thermodynamic force acting to bring about the diffusion summarized by eqn 18.55 (that of a solute initially suspended at the centre of a flask of solvent)? *Hint.* Use $\nabla = \boldsymbol{i}(\partial/\partial x) + \boldsymbol{j}(\partial/\partial y) + \boldsymbol{k}(\partial/\partial z)$.

18.33 The diffusion equation is valid when many elementary steps are taken in the time interval of interest; but the random walk calculation lets us discuss distributions for short times as well as for long. Use eqn 18.58 to calculate the probability of being six paces from the origin (that is, at $x = 6\lambda$) after (a) four, (b) six, (c) twelve steps.

18.34 Use mathematical software to calculate P in a one-dimensional random walk, and evaluate the probability of being at $x = 6\lambda$ for $N = 6$, 10, 14, . . . , 60. Compare the numerical value with the analytical value in the limit of a large number of steps. At what value of N is the discrepancy no more than 0.1 per cent?

18.35 Supply the intermediate mathematical steps in *Justification 18.7*.

18.36 Evaluate the radius of gyration of (a) a solid sphere of radius R, (b) a long, thin uniform rod of length l for rotation about an axis perpendicular to its long axis.

18.37 Derive eqn 18.76 for the root mean square separation of the ends of a random coil.

Applications to: astrophysics and biochemistry

18.38 Calculate the escape velocity (the minimum initial velocity that will take an object to infinity) from the surface of a planet of radius R. What is the value for (a) the Earth, $R = 6.37 \times 10^6$ m, $g = 9.81$ m s^{-2}, (b) Mars, $R = 3.38 \times 10^6$ m, $m_{\text{Mars}}/m_{\text{Earth}} = 0.108$. At what temperatures do H_2, He, and O_2 molecules have mean speeds equal to their escape speeds? What proportion of the molecules have enough speed to escape when the temperature is (a) 240 K, (b) 1500 K? Calculations of this kind are very important in considering the composition of planetary atmospheres.

18.39 The kinetic model of gases is valid when the size of the particles is negligible compared with their mean free path. It may seem absurd, therefore, to expect the kinetic theory and, as a consequence, the perfect gas law, to be applicable to the dense matter of stellar interiors. In the Sun, for instance, the density is 150 times that of liquid water at its centre and comparable to that of water about halfway to its surface. However, we have to realize that the state of matter is that of a *plasma*, in which the electrons have been stripped from the atoms of hydrogen and helium that make up the bulk of the matter of stars. As a result, the particles making up the plasma have diameters comparable to those of nuclei, or about 10 fm. Therefore, a mean free path of only 0.1 pm satisfies the criterion for the validity of the kinetic model and the perfect gas law. We can therefore use $pV = nRT$ as the equation of state for the stellar interior. (a) Calculate the pressure halfway to the centre of the Sun, assuming that the interior consists of ionized hydrogen atoms, the temperature is 3.6 MK, and the mass density is 1.20 g cm^{-3} (slightly higher than the density of water). (b) Combine the result from part (a) with the expression for the pressure from the kinetic model to show that the pressure of the plasma is related to its *kinetic energy density* $\rho_k = E_k/V$, the kinetic energy of the molecules in a region divided by the volume of the region, by $p = \frac{2}{3}\rho_k$. (c) What is the kinetic energy density halfway to the centre of the Sun? Compare your result with the (translational) kinetic energy density of the Earth's atmosphere on a warm day (25°C): 1.5×10^5 J m^{-3} (corresponding to 0.15 J cm^{-3}). (d) A star eventually depletes some of the hydrogen in its core, which contracts and results in higher temperatures. The increased temperature results in an increase in the rates of nuclear reaction, some of which result in the formation of heavier nuclei, such as carbon. The outer part of the star expands and cools to produce a red giant. Assume that halfway to the centre a red giant has a temperature of 3500 K, is composed primarily of fully ionized carbon atoms and electrons, and has a mass density of 1200 kg m^{-3}. What is the pressure at this point? (e) If the red giant in part (d) consisted of neutral carbon atoms, what would be the pressure at the same point under the same conditions?

18.40 Interstellar space is quite a different medium than the gaseous environments we commonly encounter on Earth. For instance, a typical density of the medium is about 1 atom cm^{-3} and that atom is typically H; the effective temperature due to stellar background radiation is about 10 000 K. Estimate the diffusion coefficient and thermal conductivity of H under these conditions. *Comment.* Energy is in fact transferred much more effectively by radiation.

18.41 The principal components of the atmosphere of the Earth are diatomic molecules, which can rotate as well as translate. Given that the translational kinetic energy density of the atmosphere is 0.15 J cm^{-3}, what is the total kinetic energy density, including rotation?

18.42 The diffusion coefficient of a particular kind of t-RNA molecule is $D = 1.0 \times 10^{-11}$ m^2 s^{-1} in the medium of a cell interior. How long does it take molecules produced in the cell nucleus to reach the walls of the cell at a distance 1.0 μm, corresponding to the radius of the cell?

18.43‡ In this problem, we examine a model for the transport of oxygen from air in the lungs to blood. First, show that, for the initial and boundary conditions $c(x,0) = c_o$, $(0 < x < \infty)$ and $c(0,t) = c_s$ $(0 \le t \le \infty)$ where c_o and c_s are constants, the concentration, $c(x,t)$, of a species is given by

$$c(x,t) = c_o + (c_s - c_o)\{1 - \text{erf}(\xi)\} \qquad \xi(x,t) = \frac{x}{(4Dt)^{1/2}}$$

where erf(ξ) is the error function

$$\text{erf}(\xi) = 1 - \frac{2}{\pi^{1/2}}\int_{\xi}^{\infty} e^{-x^2}\text{d}x$$

and the concentration $c(x,t)$ evolves by diffusion from the *yz*-plane of constant concentration, such as might occur if a condensed phase is absorbing a species from a gas phase. Now draw graphs of concentration profiles at several different times of your choice for the diffusion of oxygen into water at 298 K (when $D = 2.10 \times 10^{-9}$ m^2 s^{-1}) on a spatial scale comparable to passage of oxygen from lungs through alveoli into the blood. Use $c_o = 0$ and set c_s equal to the solubility of oxygen in water. *Hint.* Use mathematical software.

19

Chemical kinetics

This chapter, the first of a sequence that explores the rates of chemical reactions, begins with a discussion of the definition of reaction rate and outlines the techniques for its measurement. The results of such measurements show that reaction rates depend on the concentration of reactants (and products) in characteristic ways that can be expressed in terms of differential equations known as rate laws. The solutions of these equations are used to predict the concentrations of species at any time after the start of the reaction. The form of the rate law also provides insight into the series of elementary steps by which a reaction takes place. The key task in this connection is the construction of a rate law from a proposed mechanism and its comparison with experiment. Simple elementary steps have simple rate laws, and these rate laws can be combined together by invoking one or more approximations. These approximations include the concept of the rate-determining step of a reaction, the steady-state concentration of a reaction intermediate, and the existence of a pre-equilibrium. We go on to consider complex reaction mechanisms, focusing on polymerization reactions and photochemistry, in which reactions are initiated by light.

This chapter introduces the principles of **chemical kinetics**, the study of reaction rates, by showing how the rates of reactions may be measured and interpreted. The rate of a chemical reaction might depend on variables under our control, such as the pressure, the temperature, and the presence of a catalyst, and we may be able to optimize the rate by the appropriate choice of conditions. The study of reaction rates also leads to an understanding of the **mechanisms** of reactions, their analysis into a sequence of elementary steps.

Empirical chemical kinetics

The first steps in the kinetic analysis of reactions are to establish the stoichiometry of the reaction and identify any side reactions. The basic data of chemical kinetics are then the concentrations of the reactants and products at different times after a reaction has been initiated. The rates of most chemical reactions are sensitive to the temperature, so in conventional experiments the temperature of the reaction mixture must be held constant throughout the course of the reaction. This requirement puts severe demands on the design of an experiment. Gas-phase reactions, for instance, are often carried out in a vessel held in contact with a substantial block of metal. Liquid-phase reactions, including flow reactions, must be carried out in an efficient thermostat. Special efforts have to be made to study reactions at low temperatures, as in the study of the kinds of reactions that take place in interstellar clouds. Thus, supersonic expansion of the reaction gas can be used to attain temperatures as low as 10 K. For work in the liquid phase and the solid phase, very low temperatures are often reached by flowing cold liquid or cold gas around the reaction vessel. Alternatively, the entire reaction vessel is immersed in a thermally insulated container filled with a cryogenic liquid, such as liquid helium (for work at around 4 K) or liquid nitrogen (for work at around 77 K). Non-isothermal conditions are sometimes employed. For instance, the shelf-life of an expensive pharmaceutical may be explored by slowly raising the temperature of a single sample.

19.1 Experimental techniques

The method used to monitor concentrations depends on the species involved and the rapidity with which their concentrations change. Many reactions reach equilibrium over periods of minutes or hours, and several techniques may then be used to follow the changing concentrations.

(a) Monitoring the progress of a reaction

A reaction in which at least one component is a gas might result in an overall change in pressure in a system of constant volume, so its progress may be followed by recording the variation of pressure with time.

Example 19.1 *Monitoring the variation in pressure*

Predict how the total pressure varies during the isothermal gas-phase decomposition $2\,N_2O_5(g) \rightarrow 4\,NO_2(g) + O_2(g)$ in a constant-volume container.

Method The total pressure (at constant volume and temperature and assuming perfect gas behaviour) is proportional to the number of gas-phase molecules. Therefore, because 1 mol N_2O_5 gives rise to $\frac{5}{2}$ mol of gas molecules, we can expect the pressure to rise to $\frac{5}{2}$ times its initial value. To confirm this conclusion, express the progress of the reaction in terms of the fraction, α, of N_2O_5 molecules that have reacted.

Answer Let the initial pressure be p_0 and the initial amount of N_2O_5 molecules present be n. When a fraction α of the N_2O_5 molecules has decomposed, the amounts of the components in the reaction mixture are:

	N_2O_5	NO_2	O_2	Total
Amount:	$n(1-\alpha)$	$2\alpha n$	$\frac{1}{2}\alpha n$	$n(1+\frac{3}{2}\alpha)$

When $\alpha = 0$ the pressure is p_0, so at any stage the total pressure is

$$p = (1 + \tfrac{3}{2}\alpha)p_0$$

When the reaction is complete ($\alpha = 1$), the pressure will have risen to $\frac{5}{2}$ times its initial value.

Self-test 19.1 Repeat the calculation for $2\,NOBr(g) \rightarrow 2\,NO(g) + Br_2(g)$. $[p = (1 + \frac{1}{2}\alpha)p_0]$

Spectroscopy is widely applicable, and is especially useful when one substance in the reaction mixture has a strong characteristic absorption in a conveniently accessible region of the electromagnetic spectrum. For example, the progress of the reaction

$$H_2(g) + Br_2(g) \rightarrow 2\,HBr(g)$$

can be followed by measuring the absorption of visible light by bromine. A reaction that changes the number or type of ions present in a solution may be followed by monitoring the electrical conductivity of the solution. The replacement of neutral molecules by ionic products can result in dramatic changes in the conductivity, as in the reaction

$$(CH_3)_3CCl(aq) + H_2O(l) \rightarrow (CH_3)_3COH(aq) + H^+(aq) + Cl^-(aq)$$

If hydrogen ions are produced or consumed, the reaction may be followed by monitoring the pH of the solution.

Other methods of determining composition include emission spectroscopy, mass spectrometry, gas chromatography, nuclear magnetic resonance, and electron paramagnetic resonance (for reactions involving radicals or paramagnetic d-metal ions).

(b) Application of the techniques

In a **real-time analysis** the composition of the system is analysed while the reaction is in progress. Either a small sample is withdrawn or the bulk solution is monitored. In the **flow method** the reactants are mixed as they flow together in a chamber (Fig. 19.1). The reaction continues as the thoroughly mixed solutions flow through the outlet tube, and observation of the composition at different positions along the tube is equivalent to the observation of the reaction mixture at different times after mixing. The disadvantage of conventional flow techniques is that a large volume of reactant solution is necessary. This makes the study of fast reactions particularly difficult because to spread the reaction over a length of tube the flow must be rapid. This disadvantage is avoided by the **stopped-flow technique**, in which the reagents are mixed very quickly in a small chamber fitted with a syringe instead of an outlet tube (Fig. 19.2). The flow ceases when the plunger of the syringe reaches a stop, and the reaction continues in the mixed solutions. Observations, commonly using

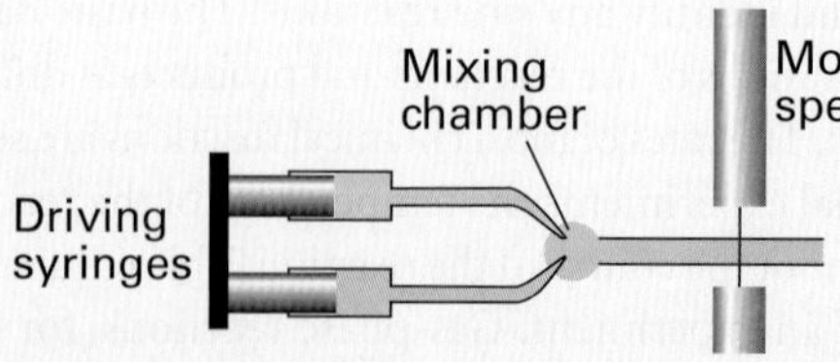

Fig. 19.1 The arrangement used in the flow technique for studying reaction rates. The reactants are injected into the mixing chamber at a steady rate. The location of the spectrometer corresponds to different times after initiation.

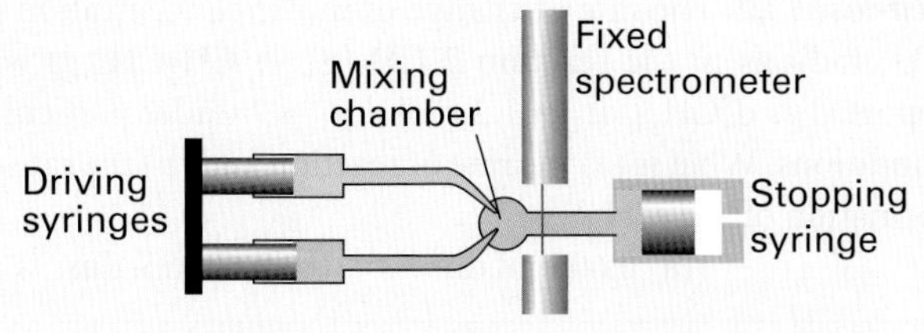

Fig. 19.2 In the stopped-flow technique the reagents are driven quickly into the mixing chamber by the driving syringes and then the time dependence of the concentrations is monitored.

spectroscopic techniques such as ultraviolet–visible absorption, circular dichroism, and fluorescence emission, are made on the sample as a function of time. The technique allows for the study of reactions that occur on the millisecond to second timescale. The suitability of the stopped-flow method to the study of small samples means that it is appropriate for many biochemical reactions, and it has been widely used to study the kinetics of protein folding and enzyme action (see *Impact I19.1*).

Very fast reactions can be studied by **flash photolysis**, in which the sample is exposed to a brief flash of light that initiates the reaction and then the contents of the reaction chamber are monitored. Most work is now done with lasers with photolysis pulse widths that range from femtoseconds to nanoseconds (Section 11.7). The apparatus used for flash photolysis studies is based on the experimental design for time-resolved spectroscopy (Section 11.7). Reactions occurring on a picosecond or femtosecond timescale may be monitored by using electronic absorption or emission, infrared absorption, or Raman scattering. The spectra are recorded at a series of times following laser excitation. The laser pulse can initiate the reaction by forming a reactive species, such as an excited electronic state of a molecule, a radical, or an ion. We discuss examples of excited state reactions in Section 19.9. An example of radical generation is the light-induced dissociation of $Cl_2(g)$ to yield Cl atoms that react with HBr to make HCl and Br according to the following sequence:

$$Cl_2 + h\nu \rightarrow Cl + Cl$$

$$Cl + HBr \rightarrow HCl^* + Br$$

$$HCl^* + M \rightarrow HCl + M$$

Here HCl* denotes a vibrationally excited 'hot' HCl molecule and M is a body (an unreactive molecule or the wall of the container) that removes the excess energy stored in HCl. A so-called 'third body' (M) is not always necessary for heteronuclear diatomic molecules because they can discard energy radiatively, but homonuclear diatomic molecules are vibrationally and rotationally inactive, and can discard energy only by collision.

In contrast to real-time analysis, **quenching methods** are based on stopping, or quenching, the reaction after it has been allowed to proceed for a certain time. In this way the composition is analysed at leisure and reaction intermediates may be trapped. These methods are suitable only for reactions that are slow enough for there to be little reaction during the time it takes to quench the mixture. In the **chemical quench flow method**, the reactants are mixed in much the same way as in the flow method but the reaction is quenched by another reagent, such as a solution of acid or base, after the mixture has travelled along a fixed length of the outlet tube. Different reaction times can be selected by varying the flow rate along the outlet tube. An advantage of the chemical quench flow method over the stopped-flow method is that spectroscopic fingerprints are not needed in order to measure the concentration of reactants and products. Once the reaction has been quenched, the solution may be examined by 'slow' techniques, such as gel electrophoresis, mass spectrometry, and chromatography. In the **freeze quench method**, the reaction is quenched by cooling the mixture within milliseconds and the concentrations of reactants, intermediates, and products are measured spectroscopically.

19.2 The rates of reactions

Reaction rates depend on the composition and the temperature of the reaction mixture. The next few sections look at these observations in more detail.

(a) The definition of rate

Consider a reaction of the form $A + 2\,B \rightarrow 3\,C + D$, in which at some instant the molar concentration of a participant J is [J] and the volume of the system is constant. The instantaneous **rate of consumption** of one of the reactants at a given time is $-d[R]/dt$, where R is A or B. This rate is a positive quantity (Fig. 19.3). The **rate of formation** of one of the products (C or D, which we denote P) is $d[P]/dt$ (note the difference in sign). This rate is also positive.

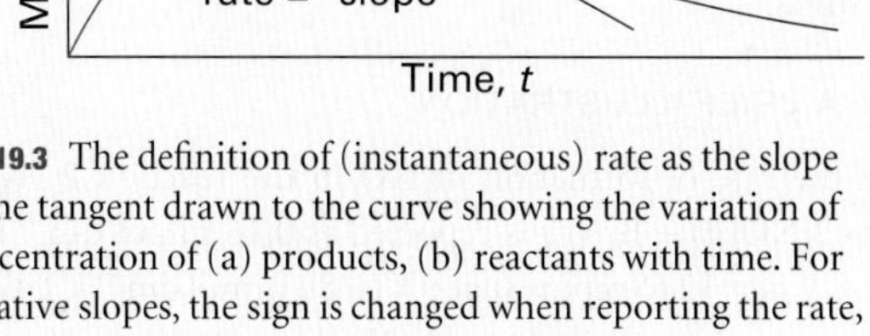

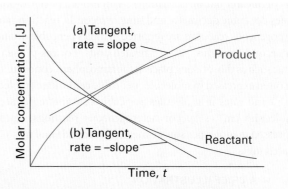

Fig. 19.3 The definition of (instantaneous) rate as the slope of the tangent drawn to the curve showing the variation of concentration of (a) products, (b) reactants with time. For negative slopes, the sign is changed when reporting the rate, so all reaction rates are positive.

It follows from the stoichiometry of the reaction A + 2 B → 3 C + D that

$$\frac{d[D]}{dt} = \frac{1}{3}\frac{d[C]}{dt} = -\frac{d[A]}{dt} = -\frac{1}{2}\frac{d[B]}{dt}$$

so there are several rates connected with the reaction. The undesirability of having different rates to describe the same reaction is avoided by using the extent of reaction, ξ (xi, the quantity introduced in Section 17.1):

$$\xi = \frac{n_J - n_{J,0}}{\nu_J} \tag{19.1}$$

where ν_J is the stoichiometric number of species J, and defining the unique **rate of reaction**, v, as the rate of change of the extent of reaction divided by the volume V:

$$v = \frac{1}{V}\frac{d\xi}{dt} \qquad [19.2]$$

It follows that

$$v = \frac{1}{\nu_J} \times \frac{1}{V}\frac{dn_J}{dt} \tag{19.3a}$$

(Remember that ν_J is negative for reactants and positive for products.) For a homogeneous reaction in a constant-volume system the volume V can be taken inside the differential and we use $[J] = n_J/V$ to write

$$v = \frac{1}{\nu_J}\frac{d[J]}{dt} \tag{19.3b}$$

For a heterogeneous reaction, we use the (constant) surface area, A, occupied by the species in place of V and use $\sigma_J = n_J/A$ to write

$$v = \frac{1}{\nu_J}\frac{d\sigma_J}{dt} \tag{19.3c}$$

In each case there is now a single rate for the entire reaction (for the chemical equation as written). With molar concentrations in moles per cubic decimetre and time in seconds, reaction rates of homogeneous reactions are reported in moles per cubic decimetre per second (mol dm^{-3} s^{-1}) or related units. For gas-phase reactions, such as those taking place in the atmosphere, concentrations are often expressed in molecules per cubic centimetre (molecules cm^{-3}) and rates in molecules per cubic centimetre per second (molecules cm^{-3} s^{-1}). For heterogeneous reactions, rates are expressed in moles per square metre per second (mol m^{-2} s^{-1}) or related units.

● **A BRIEF ILLUSTRATION**

If the rate of formation of NO in the reaction 2 NOBr(g) → 2 NO(g) + Br_2(g) is reported as 0.16 mmol dm^{-3} s^{-1}, we use $\nu_{NO} = +2$ to report that $v = 0.080$ mmol dm^{-3} s^{-1}. Because $\nu_{NOBr} = -2$ it follows that $d[NOBr]/dt = -0.16$ mmol dm^{-3} s^{-1}. The rate of consumption of NOBr is therefore 0.16 mmol dm^{-3} s^{-1}, or 9.6×10^{16} molecules cm^{-3} s^{-1}. ●

Self-test 19.2 The rate of change of molar concentration of CH_3 radicals in the reaction 2 CH_3(g) → CH_3CH_3(g) was reported as $d[CH_3]/dt = -1.2$ mol dm^{-3} s^{-1} under particular conditions. What is (a) the rate of reaction and (b) the rate of formation of CH_3CH_3?

[(a) 0.60 mol dm^{-3} s^{-1}, (b) 0.60 mol dm^{-3} s^{-1}]

(b) Rate laws and rate constants

The rate of reaction is often found to be proportional to the concentrations of the reactants raised to a power. For example, the rate of a reaction may be proportional to the molar concentrations of two reactants A and B, so we write

$$v = k_r[A][B] \tag{19.4}$$

with each concentration raised to the first power. The coefficient k_r is called the **rate constant** for the reaction. The rate constant is independent of the concentrations but depends on the temperature. An experimentally determined equation of this kind is called the **rate law** of the reaction. More formally, a rate law is an equation that expresses the rate of reaction as a function of the concentrations of all the species present in the overall chemical equation for the reaction at some time:

$$v = f([A],[B], \ldots) \qquad [19.5a]$$

For homogeneous gas-phase reactions, it is often more convenient to express the rate law in terms of partial pressures, which are related to molar concentrations by $p_J = RT[J]$. In this case, we write

$$v = f(p_A, p_B, \ldots) \qquad [19.5b]$$

The rate law of a reaction is determined experimentally, and cannot in general be inferred from the chemical equation for the reaction. The reaction of hydrogen and bromine, for example, has a very simple stoichiometry, H_2(g) + Br_2(g) → 2 HBr(g), but its rate law is complicated:

$$v = \frac{k_a[H_2][Br_2]^{3/2}}{[Br_2] + k_b[HBr]} \tag{19.6}$$

In certain cases the rate law does reflect the stoichiometry of the reaction; but that is either a coincidence or reflects a feature of the underlying reaction mechanism (see later).

A note on good (or, at least, our) practice We denote a general rate constant k_r to distinguish it from the Boltzmann constant k. In some texts k is used for the former and k_B for the latter. When expressing the rate constants in a more complicated rate law, such as that in eqn 19.6, we use k_a, k_b, and so on.

A practical application of a rate law is that, once we know the law and the value of the rate constant, we can predict the rate of reaction from the composition of the mixture. Moreover, as we shall see later, by knowing the rate law, we can go on to predict the composition of the reaction mixture at a later stage of the reaction. Moreover, a rate law is a guide to the mechanism of the reaction, for any proposed mechanism must be consistent with the observed rate law.

(c) Reaction order

Many reactions are found to have rate laws of the form

$$v = k_r[A]^a[B]^b \cdots \qquad (19.7)$$

The power to which the concentration of a species (a product or a reactant) is raised in a rate law of this kind is the **order** of the reaction with respect to that species. A reaction with the rate law in eqn 19.4 is first order in A and first order in B. The **overall order** of a reaction with a rate law like that in eqn 19.7 is the sum of the individual orders, $a + b + \cdots$. The rate law in eqn 19.4 is therefore second order overall.

A reaction need not have an integral order, and many gas-phase reactions do not. For example, a reaction having the rate law $v = k_r[A]^{1/2}[B]$ is half order in A, first order in B, and three-halves order overall. Some reactions obey a zero-order rate law, and therefore have a rate that is independent of the concentration of the reactant (so long as some is present). Thus, the catalytic decomposition of phosphine (PH_3) on hot tungsten at high pressures has the rate law $v = k_r$. The PH_3 decomposes at a constant rate until it has almost entirely disappeared. Zero-order reactions typically occur when there is a bottle-neck of some kind in the mechanism, as in heterogeneous reactions when the surface is saturated and the subsequent reaction slow and in a number of enzyme reactions when there is a large excess of substrate relative to the enzyme.

When a rate law is not of the form in eqn 19.7, the reaction does not have an overall order and may not even have definite orders with respect to each participant. Thus, although eqn 19.6 shows that the reaction of hydrogen and bromine is first order in H_2, the reaction has an indefinite order with respect to both Br_2 and HBr and has no overall order.

These remarks point to three problems. First, we must see how to identify the rate law and obtain the rate constant from the experimental data. We concentrate on this aspect in this chapter. Second, we must see how to construct reaction mechanisms that are consistent with the rate law. We shall introduce the techniques of doing so in this chapter and develop them further in Chapter 20. Third, we must account for the values of the rate constants and explain their temperature dependence. We shall see a little of what is involved in this chapter, but leave the details until Chapter 20.

(d) The determination of the rate law

The determination of a rate law is simplified by the **isolation method** in which the concentrations of all the reactants except one are in large excess. If B is in large excess, for example, then to a good approximation its concentration is constant throughout the reaction. Although the true rate law might be $v = k_r[A][B]$, we can approximate [B] by $[B]_0$, its initial value, and write

$$v = k_r'[A] \qquad k_r' = k_r[B]_0 \qquad (19.8)$$

which has the form of a first-order rate law. Because the true rate law has been forced into first-order form by assuming that the concentration of B is constant, eqn 19.8 is called a **pseudofirst-order rate law**. The dependence of the rate on the concentration of each of the reactants may be found by isolating them in turn (by having all the other substances present in large excess), and so constructing a picture of the overall rate law.

In the **method of initial rates**, which is often used in conjunction with the isolation method, the rate is measured at the beginning of the reaction for several different initial concentrations of reactants. We shall suppose that the rate law for a reaction with A isolated is $v = k_r'[A]^a$, then its initial rate, v_0, is given by the initial values of the concentration of A, and we write $v_0 = k_r'[A]_0^a$. Taking logarithms gives:

$$\log v_0 = \log k_r' + a \log[A]_0 \qquad (19.9)$$

For a series of initial concentrations, a plot of the logarithms of the initial rates against the logarithms of the initial concentrations of A should be a straight line with slope a.

Example 19.2 *Using the method of initial rates*

The recombination of iodine atoms in the gas phase in the presence of argon was investigated and the order of the reaction was determined by the method of initial rates. The initial rates of reaction of $2\,I(g) + Ar(g) \rightarrow I_2(g) + Ar(g)$ were as follows:

$[I]_0/(10^{-5}\ mol\ dm^{-3})$		1.0	2.0	4.0	6.0
$v_0/(mol\ dm^{-3}\ s^{-1})$	(a)	8.70×10^{-4}	3.48×10^{-3}	1.39×10^{-2}	3.13×10^{-2}
	(b)	4.35×10^{-3}	1.74×10^{-2}	6.96×10^{-2}	1.57×10^{-1}
	(c)	8.69×10^{-3}	3.47×10^{-2}	1.38×10^{-1}	3.13×10^{-1}

The Ar concentrations are (a) 1.0 mmol dm^{-3}, (b) 5.0 mmol dm^{-3}, and (c) 10.0 mmol dm^{-3}. Determine the orders of reaction with respect to the I and Ar atom concentrations and the rate constant.

Method Plot the logarithm of the initial rate, $\log v_0$, against $\log[I]_0$ for a given concentration of Ar, and, separately, against $\log [Ar]_0$ for a given concentration of I. The slopes

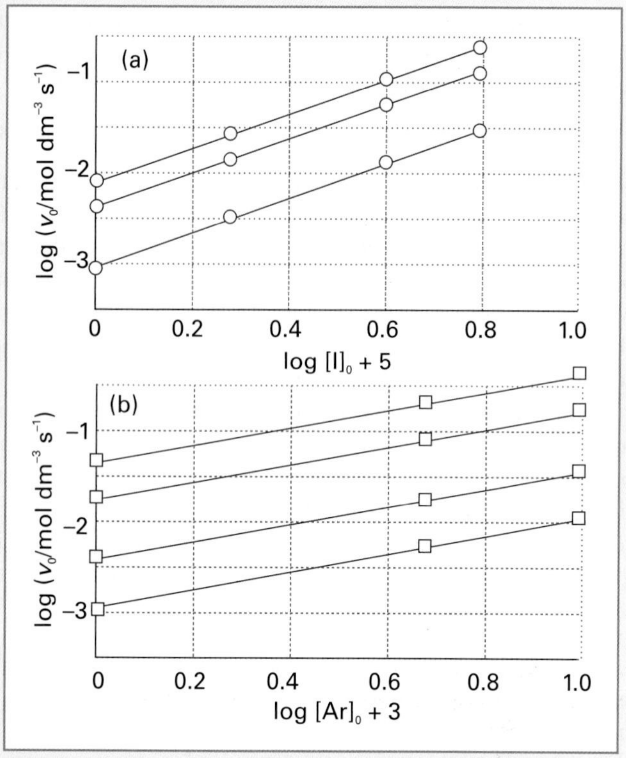

Fig. 19.4 The plot of log v_0 against (a) log$[\mathrm{I}]_0$ for a given $[\mathrm{Ar}]_0$, and (b) log$[\mathrm{Ar}]_0$ for a given $[\mathrm{I}]_0$.

of the two lines are the orders of reaction with respect to I and Ar, respectively. The intercepts with the vertical axis give log k_r' and, by using eqn 19.8, k_r.

Answer The plots are shown in Fig. 19.4. The slopes are 2 and 1, respectively, so the (initial) rate law is $v_0 = k_r[\mathrm{I}]_0^2[\mathrm{Ar}]_0$. This rate law signifies that the reaction is second order in [I], first order in [Ar], and third order overall. The intercept corresponds to $k_r = 9 \times 10^9$ mol^{-2} dm^6 s^{-1}.

A note on good practice The units of k_r come automatically from the calculation, and are always such as to convert the product of concentrations to a rate in concentration/time (for example, mol dm^{-3} s^{-1}).

Self-test 19.3 The initial rate of a reaction depended on concentration of a substance J as follows:

$[\mathrm{J}]_0$/(mmol dm^{-3})	5.0	8.2	17	30
v_0/(10^{-7} mol dm^{-3} s^{-1})	3.6	9.6	41	130

Determine the order of the reaction with respect to J and calculate the rate constant. [2, 1.4×10^{-2} dm^3 mol^{-1} s^{-1}]

The method of initial rates might not reveal the full rate law, for once the products have been generated they might participate in the reaction and affect its rate. For example, products participate in the synthesis of HBr, because eqn 19.6 shows that the full rate law depends on the concentration of HBr. To avoid this difficulty, the rate law should be fitted to the data throughout the reaction. The fitting may be done, in simple cases at least, by using a proposed rate law to predict the concentration of any component at any time, and comparing it with the data. A law should also be tested by observing whether the addition of products or, for gas-phase reactions, a change in the surface-to-volume ratio in the reaction chamber affects the rate.

19.3 Integrated rate laws

Because rate laws are differential equations, we must integrate them if we want to find the concentrations as a function of time. Even the most complex rate laws may be integrated numerically. However, in a number of simple cases analytical solutions, known as **integrated rate laws**, are easily obtained, and prove to be very useful. We examine a few of these simple cases here.

(a) First-order reactions

As shown in the following *Justification*, the integrated form of the first-order rate law

$$\frac{\mathrm{d}[\mathrm{A}]}{\mathrm{d}t} = -k_r[\mathrm{A}] \qquad (19.10\mathrm{a})$$

is

$$\ln\left(\frac{[\mathrm{A}]}{[\mathrm{A}]_0}\right) = -k_r t \qquad [\mathrm{A}] = [\mathrm{A}]_0 e^{-k_r t} \qquad (19.10\mathrm{b})$$

where $[\mathrm{A}]_0$ is the initial concentration of A (at $t = 0$).

Justification 19.1 *First-order integrated rate law*

First, we rearrange eqn 19.10a into

$$\frac{\mathrm{d}[\mathrm{A}]}{[\mathrm{A}]} = -k_r \mathrm{d}t$$

This expression can be integrated directly because k_r is a constant independent of t. Initially (at $t = 0$) the concentration of A is $[\mathrm{A}]_0$, and at a later time t it is [A], so we make these values the limits of the integrals and write

$$\int_{[\mathrm{A}]_0}^{[\mathrm{A}]} \frac{\mathrm{d}[\mathrm{A}]}{[\mathrm{A}]} = -k_r \int_0^t \mathrm{d}t$$

Because the integral of $1/x$ is $\ln x$, eqn 19.10b is obtained immediately.

Synoptic table 19.1* Kinetic data for first-order reactions

Reaction	Phase	θ/°C	k_r/s^{-1}	$t_{1/2}$
$2\,N_2O_5 \rightarrow 4\,NO_2 + O_2$	g	25	3.38×10^{-5}	5.70 h
	$Br_2(l)$	25	4.27×10^{-5}	4.51 h
$C_2H_6 \rightarrow 2\,CH_3$	g	700	5.36×10^{-4}	21.6 min

* More values are given in the *Data section*.

Equation 19.10b shows that, if $\ln([A]/[A]_0)$ is plotted against t, then a first-order reaction will give a straight line of slope $-k_r$. Some rate constants determined in this way are given in Table 19.1. The second expression in eqn 19.10b shows that in a first-order reaction the reactant concentration decreases exponentially with time with a rate determined by k_r (Fig. 19.5).

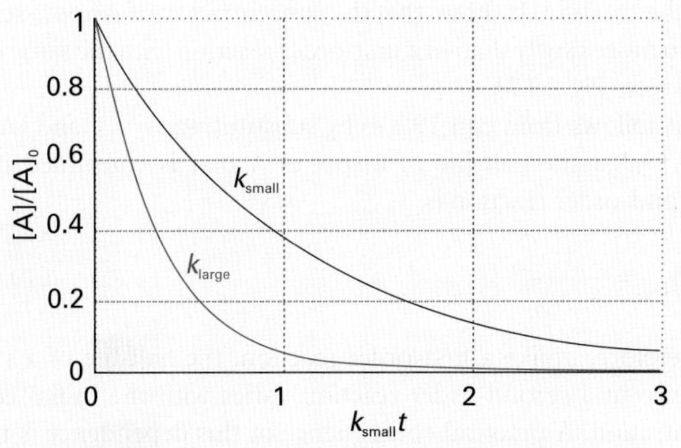

Fig. 19.5 The exponential decay of the reactant in a first-order reaction. The larger the rate constant, the more rapid the decay: here $k_{large} = 3k_{small}$.

interActivity For a first-order reaction of the form $A \rightarrow nB$ (with n possibly fractional), the concentration of the product varies with time as $[B] = n[B]_0(1 - e^{-k_r t})$. Plot the time dependence of [A] and [B] for the cases $n = 0.5$, 1, and 2.

Example 19.3 *Analysing a first-order reaction*

The variation in the partial pressure of azomethane with time was followed at 600 K, with the results given below. Confirm that the decomposition

$$CH_3N_2CH_3(g) \rightarrow CH_3CH_3(g) + N_2(g)$$

is first order in azomethane, and find the rate constant at 600 K.

t/s	0	1000	2000	3000	4000
p/Pa	10.9	7.63	5.32	3.71	2.59

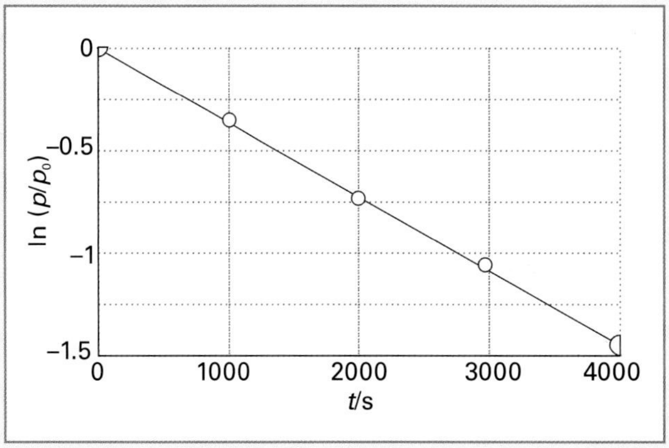

Fig. 19.6 The determination of the rate constant of a first-order reaction: a straight line is obtained when ln[A] (or, as here, $\ln p/p_0$) is plotted against t; the slope gives k_r.

Method As indicated in the text, to confirm that a reaction is first order, plot $\ln([A]/[A]_0)$ against time and expect a straight line. Because the partial pressure of a gas is proportional to its concentration, an equivalent procedure is to plot $\ln(p/p_0)$ against t. If a straight line is obtained, its slope can be identified with $-k_r$.

Answer We draw up the following table:

t/s	0	1000	2000	3000	4000
$\ln(p/p_0)$	1	−0.357	−0.717	−1.078	−1.437

Figure 19.6 shows the plot of $\ln(p/p_0)$ against t. The plot is straight, confirming a first-order reaction, and its slope is -3.6×10^{-4}. Therefore, $k_r = 3.6 \times 10^{-4}\ s^{-1}$.

A note on good practice Because the horizontal and vertical axes of graphs are labelled with pure numbers, the slope of a graph is always dimensionless. For a graph of the form $y = b + mx$ we can write $y = b + (m\ \text{units})(x/\text{units})$, where 'units' are the units of x, and identify the (dimensionless) slope with 'm units'. Then m = slope/units. In the present case, because the graph shown here is a plot of $\ln(p/p_0)$ against t/s (with 'units' = s) and k_r is the negative value of the slope of $\ln(p/p_0)$ against t itself, k_r = −slope/s.

Self-test 19.4 In a particular experiment, it was found that the concentration of N_2O_5 in liquid bromine varied with time as follows:

t/s	0	200	400	600	1000
$[N_2O_5]/(mol\ dm^{-3})$	0.110	0.073	0.048	0.032	0.014

Confirm that the reaction is first order in N_2O_5 and determine the rate constant. [$k_r = 2.1 \times 10^{-3}\ s^{-1}$]

(b) Half-lives and time constants

A useful indication of the rate of a first-order chemical reaction is the **half-life**, $t_{1/2}$, of a substance, the time taken for the concentration of a reactant to fall to half its initial value. The time for [A] to decrease from $[A]_0$ to $\frac{1}{2}[A]_0$ in a first-order reaction is given by eqn 19.10b as

$$k_r t_{1/2} = -\ln\left(\frac{\frac{1}{2}[A]_0}{[A]_0}\right) = -\ln\tfrac{1}{2} = \ln 2$$

Hence

$$t_{1/2} = \frac{\ln 2}{k_r} \qquad (19.11)$$

(ln 2 = 0.693.) The main point to note about this result is that, for a first-order reaction, the half-life of a reactant is independent of its initial concentration. Therefore, if the concentration of A at some *arbitrary* stage of the reaction is [A], then it will have fallen to $\frac{1}{2}$[A] after a further interval of $(\ln 2)/k_r$. Some half-lives are given in Table 19.1.

Another indication of the rate of a first-order reaction is the **time constant**, τ (tau), the time required for the concentration of a reactant to fall to 1/e of its initial value. From eqn 19.10b it follows that

$$k_r\tau = -\ln\left(\frac{[A]_0/e}{[A]_0}\right) = -\ln\frac{1}{e} = 1$$

That is, the time constant of a first-order reaction is the reciprocal of the rate constant:

$$\tau = \frac{1}{k_r} \qquad (19.12)$$

(c) Second-order reactions

We show in the *Justification* below that the integrated form of the second-order rate law

$$\frac{d[A]}{dt} = -k_r[A]^2 \qquad (19.13a)$$

is either of the following two forms:

$$\frac{1}{[A]} - \frac{1}{[A]_0} = k_r t \qquad (19.13b)$$

$$[A] = \frac{[A]_0}{1 + k_r t[A]_0} \qquad (19.13c)$$

where $[A]_0$ is the initial concentration of A (at $t = 0$).

Justification 19.2 *Second-order integrated rate law*

To integrate eqn 19.13a we rearrange it into

$$\frac{d[A]}{[A]^2} = -k_r dt$$

The concentration is $[A]_0$ at $t = 0$ and [A] at a general time t later. Therefore,

$$-\int_{[A]_0}^{[A]} \frac{d[A]}{[A]^2} = k_r \int_0^t dt$$

Because the integral of $1/x^2$ is $-1/x$, we obtain eqn 19.13b by substitution of the limits

$$\left.\frac{1}{[A]}\right|_{[A]_0}^{[A]} = \frac{1}{[A]} - \frac{1}{[A]_0} = k_r t$$

We can then rearrange this expression into eqn 19.13c.

Equation 19.13b shows that to test for a second-order reaction we should plot 1/[A] against t and expect a straight line. The slope of the graph is k_r. Some rate constants determined in this way are given in Table 19.2. The rearranged form, eqn 19.13c, lets us predict the concentration of A at any time after the start of the reaction. It shows that the concentration of A approaches zero more slowly than in a first-order reaction with the same initial rate (Fig. 19.7).

It follows from eqn 19.13b by substituting $t = t_{1/2}$ and $[A] = \frac{1}{2}[A]_0$ that the half-life of a species A that is consumed in a second-order reaction is

$$t_{1/2} = \frac{1}{k_r[A]_0} \qquad (19.14)$$

Therefore, unlike a first-order reaction, the half-life of a substance in a second-order reaction varies with the initial concentration. A practical consequence of this dependence is that species that decay by second-order reactions (which includes some environmentally harmful substances) may persist in low concentrations for long periods because their half-lives are long when their concentrations are low. In general, for an nth-order reaction (with n neither 0 nor 1) of the form A → products, the half-life is related to the rate constant and the initial concentration of A by (see Problem 19.19)

$$t_{1/2} = \frac{(2^{n-1} - 1)}{(n-1)k_r[A]_0^{n-1}} \qquad (19.15)$$

Synoptic table 19.2* Kinetic data for second-order reactions

Reaction	Phase	θ/°C	k_r/($dm^3\ mol^{-1}\ s^{-1}$)
$2\,NOBr \rightarrow 2\,NO + Br_2$	g	10	0.80
$2\,I \rightarrow I_2$	g	23	7×10^9
$CH_3Cl + CH_3O^-$	CH_3OH(l)	20	2.29×10^{-6}

* More values are given in the *Data section*.

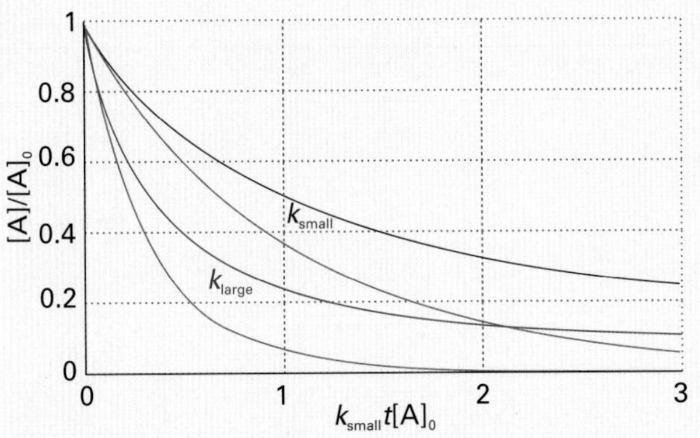

Fig. 19.7 The variation with time of the concentration of a reactant in a second-order reaction. The grey lines are the corresponding decays in a first-order reaction with the same initial rate. For this illustration, $k_{large} = 3k_{small}$.

interActivity For a second-order reaction of the form $A \rightarrow nB$ (with n possibly fractional), the concentration of the product varies with time as $[B] = nk_rt[A]_0^2/(1 + k_rt[A]_0)$. Plot the time dependence of [A] and [B] for the cases $n = 0.5$, 1, and 2.

Another type of second-order reaction is one that is first order in each of two reactants A and B:

$$\frac{d[A]}{dt} = -k_r[A][B] \tag{19.16}$$

Such a rate law cannot be integrated until we know how the concentration of B is related to that of A. For example, if the reaction is $A + B \rightarrow P$, where P denotes products, and the initial concentrations are $[A]_0$ and $[B]_0$, then it is shown in the *Justification* below that, at a time t after the start of the reaction, the concentrations satisfy the relation

$$\ln\left(\frac{[B]/[B]_0}{[A]/[A]_0}\right) = ([B]_0 - [A]_0)k_rt \tag{19.17}$$

Therefore, a plot of the expression on the left against t should be a straight line from which k_r can be obtained.

Justification 19.3 *Overall second-order rate law*

It follows from the reaction stoichiometry that, when the concentration of A has fallen to $[A]_0 - x$, the concentration of B will have fallen to $[B]_0 - x$ (because each A that disappears entails the disappearance of one B). It follows that

$$\frac{d[A]}{dt} = -k_r([A]_0 - x)([B]_0 - x)$$

Because $[A] = [A]_0 - x$, it follows that $d[A]/dt = -dx/dt$ and the rate law may be written as

$$\frac{dx}{dt} = k_r([A]_0 - x)([B]_0 - x)$$

The initial condition is that $x = 0$ when $t = 0$, so the integration required is

$$\int_0^x \frac{dx}{([A]_0 - x)([B]_0 - x)} = k_r\int_0^t dt$$

The integral on the right is simply k_rt. The integral on the left is evaluated by using the method of partial fractions in which we write

$$\frac{1}{(a-x)(b-x)} = \frac{1}{b-a}\left(\frac{1}{a-x} - \frac{1}{b-x}\right)$$

It follows that

$$\int \frac{dx}{(a-x)(b-x)} = \frac{1}{b-a}\left[\int \frac{dx}{a-x} - \int \frac{dx}{b-x}\right]$$

$$= \frac{1}{b-a}(\ln(b-x) - \ln(a-x)) + \text{constant}$$

$$= \frac{1}{b-a}\left(\ln\frac{1}{a-x} - \ln\frac{1}{b-x}\right) + \text{constant}$$

and therefore that

$$\int_0^x \frac{dx}{([A]_0 - x)([B]_0 - x)}$$

$$= \frac{1}{[B]_0 - [A]_0}\left\{\ln\left(\frac{[A]_0}{[A]_0 - x}\right) - \ln\left(\frac{[B]_0}{[B]_0 - x}\right)\right\}$$

This expression can be simplified and rearranged into eqn 19.17 by combining the two logarithms by using $\ln y - \ln z = \ln(y/z)$ and noting that $[A] = [A]_0 - x$ and $[B] = [B]_0 - x$. Similar calculations may be carried out to find the integrated rate laws for other orders, and some are listed in Table 19.3.

19.4 Reactions approaching equilibrium

Because all the laws considered so far disregard the possibility that the reverse reaction is important, none of them describes the overall rate when the reaction is close to equilibrium. At that stage the products may be so abundant that the reverse reaction must be taken into account. In practice, however, most kinetic studies are made on reactions that are far from equilibrium, and the reverse reactions are unimportant.

(a) First-order reactions close to equilibrium

We can explore the variation of the composition with time close to chemical equilibrium by considering the reaction in which A

Table 19.3 Integrated rate laws

Order	Reaction	Rate law*	$t_{1/2}$
0	$A \rightarrow P$	$v = k_r$ $k_r t = x$ for $0 \le x \le [A]_0$	$[A]_0/2k_r$
1	$A \rightarrow P$	$v = k_r[A]$ $k_r t = \ln \dfrac{[A]_0}{[A]_0 - x}$	$(\ln 2)/k_r$
2	$A \rightarrow P$	$v = k_r[A]^2$ $k_r t = \dfrac{x}{[A]_0([A]_0 - x)}$	$1/k_r[A]_0$
	$A + B \rightarrow P$	$v = k_r[A][B]$ $k_r t = \dfrac{1}{[B]_0 - [A]_0} \ln \dfrac{[A]_0([B]_0 - x)}{([A]_0 - x)[B]_0}$	
	$A + 2\,B \rightarrow P$	$v = k_r[A][B]$ $k_r t = \dfrac{1}{[B]_0 - 2[A]_0} \ln \dfrac{[A]_0([B]_0 - 2x)}{([A]_0 - x)[B]_0}$	
	$A \rightarrow P$ with autocatalysis	$v = k_r[A][P]$ $k_r t = \dfrac{1}{[A]_0 + [P]_0} \ln \dfrac{[A]_0([P]_0 + x)}{([A]_0 - x)[P]_0}$	
3	$A + 2\,B \rightarrow P$	$v = k_r[A][B]^2$ $k_r t = \dfrac{2x}{(2[A]_0 - [B]_0)([B]_0 - 2x)[B]_0} + \dfrac{1}{(2[A]_0 - [B]_0)^2} \ln \dfrac{[A_0]([B]_0 - 2x)}{([A]_0 - x)[B]_0}$	
$n \neq 1$	$A \rightarrow P$	$v = k_r[A]^n$ $k_r t = \dfrac{1}{n-1}\left\{\dfrac{1}{([A]_0 - x)^{n-1}} - \dfrac{1}{[A]_0^{n-1}}\right\}$	$\dfrac{2^{n-1} - 1}{(n-1)k_r[A]_0^{n-1}}$

* $x = [P]$ and $v = dx/dt$.

forms B and both forward and reverse reactions are first order (as in some isomerizations). The scheme we consider is

$$\begin{array}{ll} A \rightarrow B & v = k_r[A] \\ B \rightarrow A & v = k_r'[B] \end{array} \qquad (19.18)$$

The concentration of A is reduced by the forward reaction (at a rate $k_r[A]$) but it is increased by the reverse reaction (at a rate $k_r'[B]$). The net rate of change is therefore

$$\frac{d[A]}{dt} = -k_r[A] + k_r'[B]$$

If the initial concentration of A is $[A]_0$, and no B is present initially, then at all times $[A] + [B] = [A]_0$. Therefore,

$$\frac{d[A]}{dt} = -k_r[A] + k_r'([A]_0 - [A]) = -(k_r + k_r')[A] + k_r'[A]_0$$

The solution of this first-order differential equation (as may be checked by differentiation) is

$$[A] = \frac{k_r' + k_r e^{-(k_r + k_r')t}}{k_r' + k_r}[A]_0 \qquad (19.19)$$

Figure 19.8 shows the time dependence predicted by this equation.

As $t \rightarrow \infty$, the concentrations reach their equilibrium values, which are given by eqn 19.19 as:

$$[A]_{eq} = \frac{k_r'[A]_0}{k_r + k_r'} \qquad [B]_{eq} = [A]_0 - [A]_\infty = \frac{k_r[A]_0}{k_r + k_r'} \qquad (19.20)$$

It follows that the equilibrium constant of the reaction is

$$K = \frac{[B]_{eq}}{[A]_{eq}} = \frac{k_r}{k_r'} \qquad (19.21)$$

(This expression is approximate because thermodynamic equilibrium constants are expressed in terms of activities, not concentrations.) Exactly the same conclusion can be reached—more simply, in fact—by noting that, at equilibrium, the forward and reverse rates must be the same, so

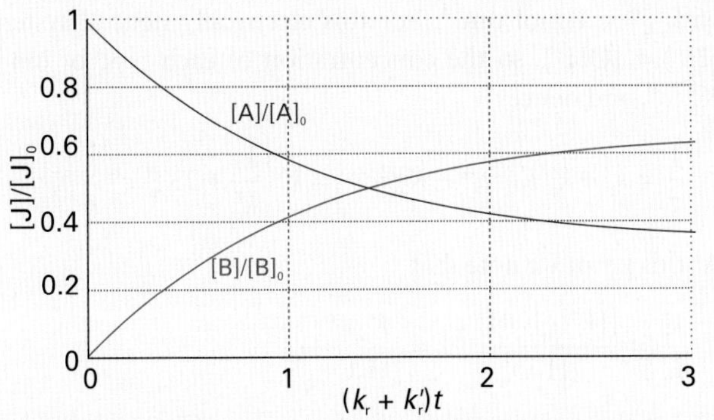

Fig. 19.8 The approach of concentrations to their equilibrium values as predicted by eqn 19.19 for a reaction $A \rightleftarrows B$ that is first order in each direction, and for which $k_r = 2k_r'$.

interActivity Set up the rate equations and plot the corresponding graphs for the approach to and equilibrium of the reaction scheme $A \rightleftarrows 2\,B$.

$$k_r[A]_{eq} = k_r'[B]_{eq}$$

This relation rearranges into eqn 19.21. The theoretical importance of eqn 19.21 is that it relates a thermodynamic quantity, the equilibrium constant, to quantities relating to rates. Its practical importance is that, if one of the rate constants can be measured, then the other may be obtained if the equilibrium constant is known.

For a more general reaction, the overall equilibrium constant can be expressed in terms of the rate constants for all the intermediate stages of the reaction mechanism:

$$K = \frac{k_a}{k_a'} \times \frac{k_b}{k_b'} \times \cdots \qquad (19.22)$$

where k_a, k_b, . . . are the rate constants for the individual steps and k_a', k_b', . . . are for the corresponding reverse steps.

(b) Relaxation methods

The term **relaxation** denotes the return of a system to equilibrium. It is used in chemical kinetics to indicate that an externally applied influence has shifted the equilibrium position of a reaction, normally suddenly, and that the reaction is adjusting to the equilibrium composition characteristic of the new conditions (Fig. 19.9). We shall consider the response of reaction rates to a **temperature jump**, a sudden change in temperature. We know from Section 17.5 that the equilibrium composition of a reaction depends on the temperature (provided $\Delta_r H^\ominus$ is nonzero), so a shift in temperature acts as a perturbation on the system. One way of achieving a temperature jump is to discharge a capacitor through a sample made conducting by the addition of ions, but laser or microwave discharges can also be used. Temperature jumps of between 5 and 10 K can be achieved in about 1 μs with electrical discharges. The high energy output of pulsed lasers

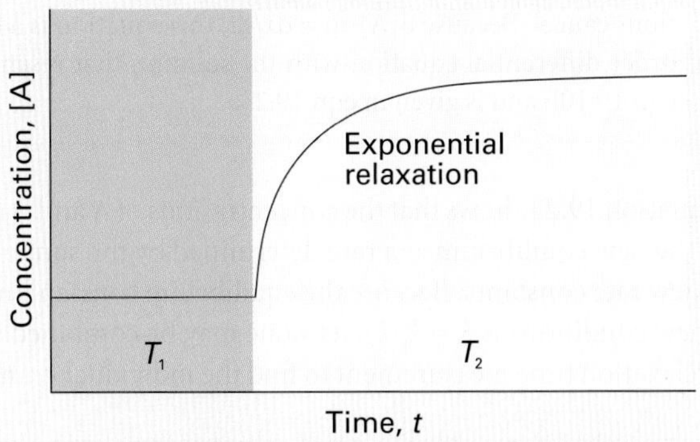

Fig. 19.9 The relaxation to the new equilibrium composition when a reaction initially at equilibrium at a temperature T_1 is subjected to a sudden change of temperature, which takes it to T_2.

(Section 11.7) is sufficient to generate temperature jumps of between 10 and 30 K within nanoseconds in aqueous samples. Some equilibria are also sensitive to pressure, and **pressure-jump techniques** may then also be used.

When a sudden temperature increase is applied to a simple $A \rightleftharpoons B$ equilibrium that is first order in each direction, we show in the following *Justification* that the composition relaxes exponentially to the new equilibrium composition:

$$x = x_0 e^{-t/\tau} \qquad \frac{1}{\tau} = k_r + k_r' \qquad (19.23)$$

where x_0 is the departure from equilibrium immediately after the temperature jump, x is the departure from equilibrium at the new temperature after a time t, and k_r and k_r' are the forward and reverse rate constants, respectively, at the new temperature.

Justification 19.4 *Relaxation to equilibrium*

When the temperature of a system at equilibrium is increased suddenly, the rate constants change from their earlier values to the new values k_r and k_r' characteristic of that temperature, but the concentrations of A and B remain for an instant at their old equilibrium values. As the system is no longer at equilibrium, it readjusts to the new equilibrium concentrations, which are now given by

$$k_r[A]_{eq} = k_r'[B]_{eq}$$

and it does so at a rate that depends on the new rate constants. We write the deviation of [A] from its new equilibrium value as x, so $[A] = x + [A]_{eq}$ and $[B] = [B]_{eq} - x$. The concentration of A then changes as follows:

$$\begin{aligned}\frac{d[A]}{dt} &= -k_r[A] + k_r'[B] \\ &= -k_r([A]_{eq} + x) + k_r'([B]_{eq} - x) \\ &= -(k_r + k_r')x\end{aligned}$$

because the two terms involving the equilibrium concentrations cancel. Because $d[A]/dt = dx/dt$, this equation is a first-order differential equation with the solution that resembles eqn 19.10b and is given in eqn 19.23.

Equation 19.23 shows that the concentrations of A and B relax into the new equilibrium at a rate determined by the sum of the two new rate constants. Because the equilibrium constant under the new conditions is $K \approx k_r/k_r'$, its value may be combined with the relaxation time measurement to find the individual k_r and k_r'.

Example 19.4 *Analysing a temperature-jump experiment*

The equilibrium constant for the autoprotolysis of water, $H_2O(l) \rightleftharpoons H^+(aq) + OH^-(aq)$, is $K_w = a(H^+)a(OH^-) = 1.008 \times 10^{-14}$ at 298 K. After a temperature jump, the reaction returns to equilibrium with a relaxation time of 37 μs at 298 K and pH ≈ 7. Given that the forward reaction is first order and the reverse is second order overall, calculate the rate constants for the forward and reverse reactions.

Method We need to derive an expression for the relaxation time, τ (the time constant for return to equilibrium), in terms of k_r (forward, first-order reaction) and k_r' (reverse, second-order reaction). Relate k_r and k_r' through the equilibrium constant, but be careful with units because K_w is dimensionless. Throughout the problem, be prepared to make approximations to simplify mathematical manipulations.

Answer The forward rate at the final temperature is $k_r[H_2O]$ and the reverse rate is $k_r'[H^+][OH^-]$. The net rate of deprotonation of H_2O is

$$\frac{d[H_2O]}{dt} = -k_r[H_2O] + k_r'[H^+][OH^-]$$

We write $[H_2O] = [H_2O]_{eq} + x$, $[H^+] = [H^+]_{eq} - x$, and $[OH^-] = [OH^-]_{eq} - x$, and obtain

$$\begin{aligned}\frac{dx}{dt} &= -\{k_r + k_r'([H^+]_{eq} + [OH^-]_{eq})\}x - k_r[H_2O]_{eq} \\ &\quad + k_r'[H^+]_{eq}[OH^-]_{eq} + k_r'x^2 \\ &\approx -\{k_r + k_r'([H^+]_{eq} + [OH^-]_{eq})\}x\end{aligned}$$

where we have neglected the term in x^2 and used the equilibrium condition $k_r[H_2O]_{eq} = k_r'[H^+]_{eq}[OH^-]_{eq}$ to eliminate the terms that are independent of x. It follows that

$$\frac{1}{\tau} = k_r + k_r'([H^+]_{eq} + [OH^-]_{eq})$$

At this point we note that

$$\begin{aligned}K_w &= a(H^+)a(OH^-) \approx ([H^+]_{eq}/c^\ominus)([OH^-]_{eq})/c^\ominus) \\ &= [H^+]_{eq}[OH^-]_{eq}/c^{\ominus 2}\end{aligned}$$

with $c^\ominus = 1$ mol dm^{-3}. For this electrically neutral system, $[H^+] = [OH^-]$, so the concentration of each type of ion is $K_w^{1/2}c^\ominus$, and hence

$$\frac{1}{\tau} = k_r + k_r'(K_w^{1/2}c^\ominus + K_w^{1/2}c^\ominus) = k_r'\left\{\frac{k_r}{k_r'} + 2K_w^{1/2}c^\ominus\right\}$$

At this point we note that

$$\frac{k_r}{k_r'} = \frac{[H^+]_{eq}[OH^-]_{eq}}{[H_2O]_{eq}} = \frac{K_w c^{\ominus 2}}{[H_2O]_{eq}}$$

The molar concentration of pure water is 55.6 mol dm^{-3}, so $[H_2O]_{eq}/c^\ominus = 55.6$. If we write $K = K_w/55.6 = 1.81 \times 10^{-16}$, we obtain

$$\frac{1}{\tau} = k_r'\{K + 2K_w^{1/2}\}c^\ominus$$

Hence,

$$\begin{aligned}k_r' &= \frac{1}{\tau(K + 2K_w^{1/2})c^\ominus} \\ &= \frac{1}{(3.7\times10^{-5}\ \mathrm{s})\times(2.0\times10^{-7})\times(1\ \mathrm{mol\ dm^{-3}})} \\ &= 1.4\times10^{11}\ \mathrm{dm^3\ mol^{-1}\ s^{-1}}\end{aligned}$$

It follows that

$$k_r = k_r'Kc^\ominus = 2.5\times10^{-5}\ \mathrm{s^{-1}}$$

The reaction is faster in ice, where $k_r' = 8.6\times10^{12}$ dm^3 mol^{-1} s^{-1}.

A note on good practice Notice how we keep track of units through the use of $c^\ominus$: K and K_w are dimensionless; k_r' is expressed in dm^3 mol^{-1} s^{-1} and k_r is expressed in s^{-1}.

Self-test 19.5 Derive an expression for the relaxation time of a concentration when the reaction $A + B \rightleftharpoons C + D$ is second order in both directions.

$[1/\tau = k_r([A] + [B])_{eq} + k_r'([C] + [D])_{eq}]$

Accounting for the rate laws

We now move on to the second stage of the analysis of kinetic data, their explanation in terms of a postulated reaction mechanism.

19.5 Elementary reactions

Most reactions occur in a sequence of steps called **elementary reactions**, each of which involves only a small number of molecules or ions. A typical elementary reaction is

$$H + Br_2 \rightarrow HBr + Br$$

Note that the phase of the species is not specified in the chemical equation for an elementary reaction, and the equation represents the specific process occurring to individual molecules. This equation, for instance, signifies that an H atom attacks a Br_2 molecule to produce an HBr molecule and a Br atom. The **molecularity** of an elementary reaction is the number of molecules coming together to react in an elementary reaction. In a **unimolecular reaction**, a single molecule shakes itself apart or its atoms into a new arrangement, as in the isomerization of cyclopropane (**1**) to propene (**2**). In a **bimolecular reaction**, a pair of molecules collide and exchange energy, atoms, or groups of atoms, or undergo some other kind of change. It is most important to distinguish molecularity from order: reaction order is an empirical quantity, and obtained from the experimental rate law; molecularity refers to an elementary reaction proposed as an individual step in a mechanism.

1 Cyclopropane

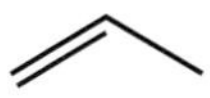
2 Propene

The rate law of a unimolecular elementary reaction is first order in the reactant:

$$A \rightarrow P \qquad \frac{d[A]}{dt} = -k_r[A] \qquad (19.24)$$

where P denotes products (several different species may be formed). A unimolecular reaction is first order because the number of A molecules that decay in a short interval is proportional to the number available to decay. (Ten times as many decay in the same interval when there are initially 1000 A molecules as when there are only 100 present.) Therefore, the rate of decomposition of A is proportional to its molar concentration.

An elementary bimolecular reaction has a second-order rate law:

$$A + B \rightarrow P \qquad \frac{d[A]}{dt} = -k_r[A][B] \qquad (19.25)$$

A bimolecular reaction is second order because its rate is proportional to the rate at which the reactant species meet, which in turn is proportional to their concentrations. Therefore, if we have evidence that a reaction is a single-step, bimolecular process, we can write down the rate law (and then go on to test it). Bimolecular elementary reactions are believed to account for many homogeneous reactions, such as the dimerizations of alkenes and dienes and reactions such as

$$CH_3I(alc) + CH_3CH_2O^-(alc) \rightarrow CH_3OCH_2CH_3(alc) + I^-(alc)$$

(where 'alc' signifies alcohol solution). There is evidence that the mechanism of this reaction is a single elementary step

$$CH_3I + CH_3CH_2O^- \rightarrow CH_3OCH_2CH_3 + I^-$$

This mechanism is consistent with the observed rate law $v = k_r[CH_3I][CH_3CH_2O^-]$.

We shall see below how to combine a series of simple steps together into a mechanism and how to arrive at the corresponding rate law. For the present we emphasize that *if the reaction is an elementary bimolecular process, then it has second-order kinetics but, if the kinetics are second order, then the reaction might be complex.* The postulated mechanism can be explored only by detailed detective work on the system, and by investigating whether side products or intermediates appear during the course of the reaction. Detailed analysis of this kind was one of the ways, for example, in which the reaction $H_2(g) + I_2(g) \rightarrow 2\,HI(g)$ was shown to proceed by a complex mechanism. For many years the reaction had been accepted on good, but insufficiently meticulous evidence, as a fine example of a simple bimolecular reaction, $H_2 + I_2 \rightarrow HI + HI$, in which atoms exchanged partners during a collision.

19.6 Consecutive elementary reactions

Some reactions proceed through the formation of an intermediate (I), as in the consecutive unimolecular reactions

$$A \xrightarrow{k_a} I \xrightarrow{k_b} P$$

An example is the decay of a radioactive family, such as

$$^{239}U \xrightarrow{23.5\ \text{min}} {}^{239}Np \xrightarrow{2.35\ \text{day}} {}^{239}Pu$$

(The times are half-lives.) We can discover the characteristics of this type of reaction by setting up the rate laws for the net rate of change of the concentration of each substance.

(a) The variation of concentrations with time

The rate of unimolecular decomposition of A is

$$\frac{d[A]}{dt} = -k_a[A] \qquad (19.26a)$$

and A is not replenished. The intermediate I is formed from A (at a rate $k_a[A]$) but decays to P (at a rate $k_b[I]$). The net rate of formation of I is therefore

$$\frac{d[I]}{dt} = k_a[A] - k_b[I] \qquad (19.26b)$$

The product P is formed by the unimolecular decay of I:

$$\frac{d[P]}{dt} = k_b[I] \qquad (19.26c)$$

We suppose that initially only A is present, and that its concentration is $[A]_0$.

The first of the rate laws, eqn 19.26a, is an ordinary first-order decay, so we can write

$$[A] = [A]_0 e^{-k_a t} \qquad (19.27a)$$

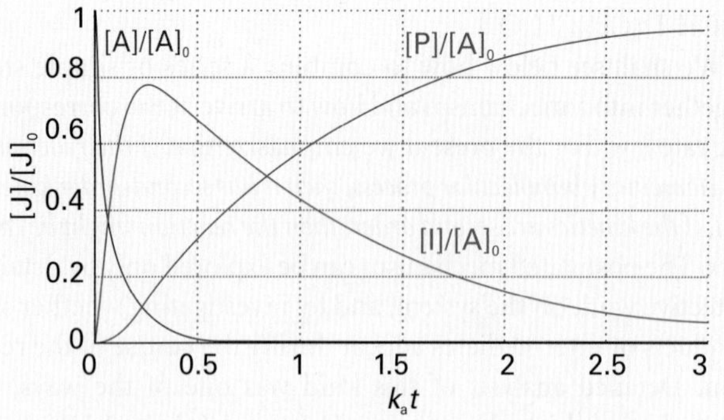

Fig. 19.10 The concentrations of A, I, and P in the consecutive reaction scheme $A \rightarrow I \rightarrow P$. The curves are plots of eqns 19.27a–c with $k_a = 10k_b$. If the intermediate I is in fact the desired product, it is important to be able to predict when its concentration is greatest; see Example 19.5.

interActivity Use mathematical software, an electronic spreadsheet, or the applets found in the *Living graphs* section of the text's web site to investigate the effects on [A], [I], [P], and t_{max} of decreasing the ratio k_a/k_b from 10 (as in Fig. 19.10) to 0.05. Compare your results with those shown in Fig. 19.12.

When this equation is substituted into eqn 19.26b, we obtain after rearrangement

$$\frac{d[I]}{dt} + k_b[I] = k_a[A]_0 e^{-k_a t}$$

This differential equation has a standard form (see *Mathematical background 2*) and, after setting $[I]_0 = 0$, the solution is

$$[I] = \frac{k_a}{k_b - k_a}(e^{-k_a t} - e^{-k_b t})[A]_0 \qquad (19.27b)$$

At all times $[A] + [I] + [P] = [A]_0$, so it follows that

$$[P] = \left\{1 + \frac{k_a e^{-k_b t} - k_b e^{-k_a t}}{k_b - k_a}\right\}[A]_0 \qquad (19.27c)$$

The concentration of the intermediate I rises to a maximum and then falls to zero (Fig. 19.10). The concentration of the product P rises from zero towards $[A]_0$.

Example 19.5 *Analysing consecutive reactions*

Suppose that in an industrial batch process a substance A produces the desired compound I, which goes on to decay to a worthless product C, each step of the reaction being first order. At what time will I be present in greatest concentration?

Method The time dependence of the concentration of I is given by eqn 19.27b. We can find the time at which [I] passes through a maximum, t_{max}, by calculating $d[I]/dt$ and setting the resulting rate equal to zero.

Answer It follows from eqn 19.27b that

$$\frac{d[I]}{dt} = -\frac{k_a[A]_0(k_a e^{-k_a t} - k_b e^{-k_b t})}{k_b - k_a}$$

This rate is equal to zero when

$$k_a e^{-k_a t} = k_b e^{-k_b t}$$

Therefore,

$$t_{max} = \frac{1}{k_a - k_b}\ln\frac{k_a}{k_b}$$

For a given value of k_a, as k_b increases both the time at which [I] is a maximum and the yield of I decrease.

Self-test 19.6 Calculate the maximum concentration of I and justify the last remark.
$[[I]_{max}/[A]_0 = (k_a/k_b)^c,\ c = k_b/(k_b - k_a)]$

(b) The steady-state approximation

One feature of the calculations so far has probably not gone unnoticed: there is a considerable increase in mathematical complexity as soon as the reaction mechanism has more than a couple of steps. A reaction scheme involving many steps is nearly always unsolvable analytically, and alternative methods of solution are necessary. One approach is to integrate the rate laws numerically. An alternative approach, which continues to be widely used because it leads to convenient expressions and more readily digestible results, is to make an approximation.

The **steady-state approximation**, which is also widely called the **quasi-steady-state approximation** (QSSA) to distinguish it from a true steady state, assumes that, after an initial **induction period**, an interval during which the concentrations of intermediates, I, rise from zero, and during the major part of the reaction, the rates of change of concentrations of all reaction intermediates are negligibly small (Fig. 19.11):

$$\frac{d[I]}{dt} \approx 0 \qquad (19.28)$$

This approximation greatly simplifies the discussion of reaction schemes. For example, when we apply the approximation to the consecutive first-order mechanism, we set $d[I]/dt = 0$ in eqn 19.26b, which then becomes

$$k_a[A] - k_b[I] \approx 0$$

Then

$$[I] \approx (k_a/k_b)[A] \qquad (19.29)$$

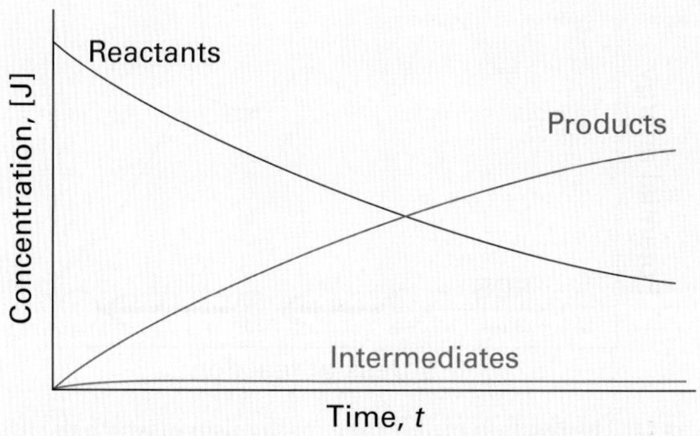

Fig. 19.11 The basis of the steady-state approximation. It is supposed that the concentrations of intermediates remain small and hardly change during most of the course of the reaction.

For this expression to be consistent with eqn 19.27b, we require $k_a/k_b \ll 1$ (so that, even though [A] does depend on the time, the dependence of [I] on the time is negligible). On substituting this value of [I] into eqn 19.26c, that equation becomes

$$\frac{d[P]}{dt} = k_b[I] \approx k_a[A] \qquad (19.30)$$

and we see that P is formed by a first-order decay of A, with a rate constant k_a, the rate constant of the slower step. We can write down the solution of this equation at once by substituting the solution for [A], eqn 19.27a, and integrating:

$$[P] = k_a[A]_0 \int_0^t e^{-k_a t} dt = (1 - e^{-k_a t})[A]_0 \qquad (19.31)$$

This is the same (approximate) result as before, eqn 19.27c, but much more quickly obtained. Figure 19.12 compares the approximate solutions found here with the exact solutions found

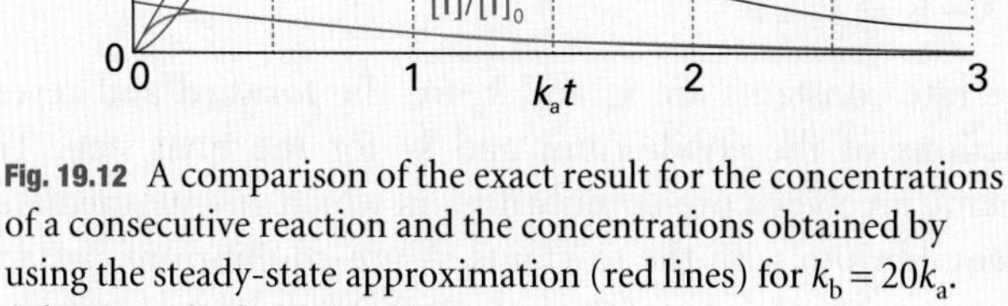

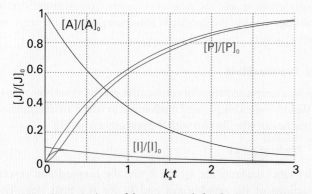

Fig. 19.12 A comparison of the exact result for the concentrations of a consecutive reaction and the concentrations obtained by using the steady-state approximation (red lines) for $k_b = 20k_a$. (The curve for [A] is unchanged.)

earlier: k_b does not have to be very much bigger than k_a for the approach to be reasonably accurate.

Example 19.6 *Using the steady-state approximation*

Devise the rate law for the decomposition of N_2O_5,

$$2\,N_2O_5(g) \rightarrow 4\,NO_2(g) + O_2(g)$$

on the basis of the following mechanism:

$N_2O_5 \rightarrow NO_2 + NO_3$	k_a
$NO_2 + NO_3 \rightarrow N_2O_5$	k_a'
$NO_2 + NO_3 \rightarrow NO_2 + O_2 + NO$	k_b
$NO + N_2O_5 \rightarrow NO_2 + NO_2 + NO_2$	k_c

A note on good practice Note that when writing the equation for an elementary reaction all the species are displayed individually, so we write $A \rightarrow B + B$, for instance, not $A \rightarrow 2\,B$.

Method First identify the intermediates (species that occur in the reaction steps but do not appear in the overall reaction) and write expressions for their net rates of formation. Then, all net rates of change of the concentrations of intermediates are set equal to zero and the resulting equations are solved algebraically.

Answer The intermediates are NO and NO_3; the net rates of change of their concentrations are

$$\frac{d[NO]}{dt} = k_b[NO_2][NO_3] - k_c[NO][N_2O_5] \approx 0$$

$$\frac{d[NO_3]}{dt} = k_a[N_2O_5] - k_a'[NO_2][NO_3] - k_b[NO_2][NO_3] \approx 0$$

The net rate of change of concentration of N_2O_5 is

$$\frac{d[N_2O_5]}{dt} = -k_a[N_2O_5] + k_a'[NO_2][NO_3] - k_c[NO][N_2O_5]$$

and replacing the concentrations of the intermediates by using the equations above gives

$$\frac{d[N_2O_5]}{dt} = -\frac{2k_a k_b[N_2O_5]}{k_a' + k_b}$$

Self-test 19.7 Derive the rate law for the decomposition of ozone in the reaction $2\,O_3(g) \rightarrow 3\,O_2(g)$ on the basis of the (incomplete) mechanism

$O_3 \rightarrow O_2 + O$	k_a
$O_2 + O \rightarrow O_3$	k_a'
$O + O_3 \rightarrow O_2 + O_2$	k_b

$[d[O_3]/dt = -2k_a k_b[O_3]^2/(k_a'[O_2] + k_b[O_3])]$

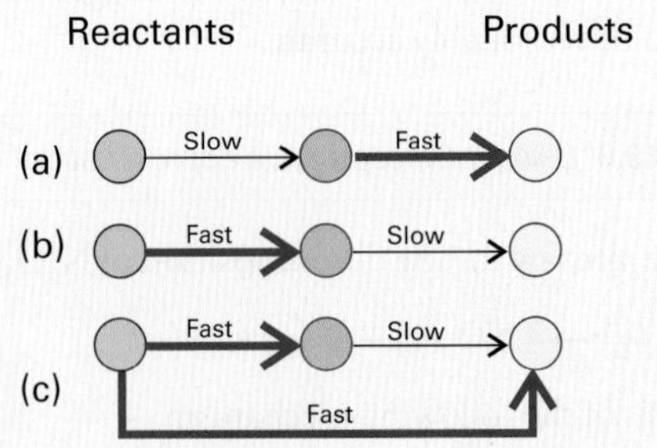

Fig. 19.13 In these diagrams of reaction schemes, heavy arrows represent fast steps and light arrows represent slow steps. (a) The first step is rate-determining; (b) the second step is rate-determining; (c) although one step is slow, it is not rate-determining because there is a fast route that circumvents it.

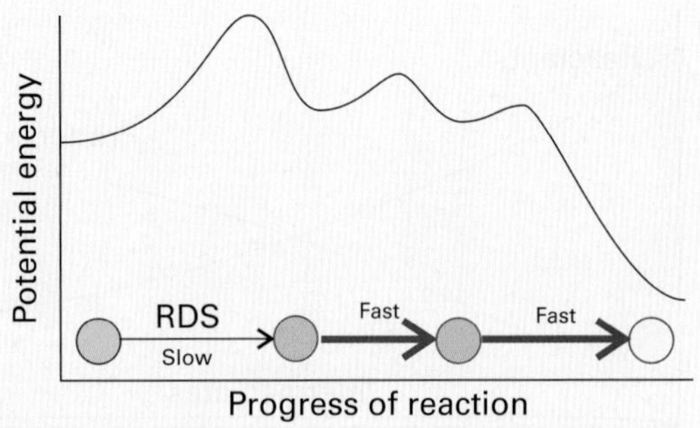

Fig. 19.14 The reaction profile for a mechanism in which the first step (RDS) is rate-determining.

(c) The rate-determining step

Equation 19.31 shows that, when $k_b \gg k_a$, then the formation of the final product P depends on only the *smaller* of the two rate constants. That is, the rate of formation of P depends on the rate at which I is formed, not on the rate at which I changes into P. For this reason, the step $A \rightarrow I$ is called the 'rate-determining step' of the reaction. Its existence has been likened to building a six-lane highway up to a single-lane bridge: the traffic flow is governed by the rate of crossing the bridge. Similar remarks apply to more complicated reaction mechanisms, and in general the **rate-determining step** is the slowest step in a mechanism and controls the overall rate of the reaction. The rate-determining step is not just the slowest step: it must be slow *and* be a crucial gateway for the formation of products. If a faster reaction can also lead to products, then the slowest step is irrelevant because the slow reaction can then be side-stepped (Fig. 19.13).

The rate law of a reaction that has a rate-determining step can often be written down almost by inspection. If the first step in a mechanism is rate-determining, then the rate of the overall reaction is equal to the rate of the first step because all subsequent steps are so fast that once the first intermediate is formed it results immediately in the formation of products. Figure 19.14 shows the reaction profile for a mechanism of this kind in which the slowest step is the one with the highest activation energy (the concept of activation energy should be familiar from introductory chemistry and is discussed in more detail in Section 20.1). Once over the initial barrier, the intermediates cascade into products. However, a rate-determining step may also stem from the low concentration of a crucial reactant and need not correspond to the step with highest activation barrier.

(d) Kinetic and thermodynamic control of reactions

In some cases reactants can give rise to a variety of products, as in nitrations of monosubstituted benzene, when various proportions of the *ortho-*, *meta-*, and *para-*substituted products are obtained, depending on the directing power of the original substituent. Suppose two products, P_1 and P_2, are produced by the following competing reactions:

$$A + B \rightarrow P_1 \quad \textit{Rate of formation of } P_1 = k_1[A][B]$$
$$A + B \rightarrow P_2 \quad \textit{Rate of formation of } P_2 = k_2[A][B]$$

The relative proportion in which the two products have been produced at a given stage of the reaction (before it has reached equilibrium) is given by the ratio of the two rates, and therefore to the two rate constants:

$$\frac{[P_2]}{[P_1]} = \frac{k_2}{k_1} \tag{19.32}$$

This ratio represents the **kinetic control** over the proportions of products, and is a common feature of the reactions encountered in organic chemistry where reactants are chosen that facilitate pathways favouring the formation of a desired product. If a reaction is allowed to reach equilibrium, then the proportion of products is determined by thermodynamic rather than kinetic considerations, and the ratio of concentrations is controlled by considerations of the standard Gibbs energies of all the reactants and products.

(e) Pre-equilibria

From a simple sequence of consecutive reactions we now turn to a slightly more complicated mechanism in which an intermediate I reaches an equilibrium with the reactants A and B:

$$A + B \rightleftharpoons I \rightarrow P$$

The rate constants are k_a and k_a' for the forward and reverse reactions of the equilibrium and k_b for the final step. This scheme involves a **pre-equilibrium**, in which an intermediate is in equilibrium with the reactants. A pre-equilibrium can arise when the rate of decay of the intermediate back into reactants is much faster than the rate at which it forms products; thus, the condition is possible when $k'_a \gg k_b$ but not when $k_b \gg k'_a$. Because we assume that A, B, and I are in equilibrium, we can write

$$K = \frac{[I]}{[A][B]} \qquad K = \frac{k_a}{k'_a} \tag{19.33}$$

In writing these equations, we are presuming that the rate of reaction of I to form P is too slow to affect the maintenance of the pre-equilibrium (see the example below). The rate of formation of P may now be written:

$$\frac{d[P]}{dt} = k_b[I] = k_b K[A][B] \tag{19.34}$$

This rate law has the form of a second-order rate law with a composite rate constant:

$$\frac{d[P]}{dt} = k_r[A][B] \qquad k_r = k_b K = \frac{k_a k_b}{k'_a} \tag{19.35}$$

Example 19.7 *Analysing a pre-equilibrium*

Repeat the pre-equilibrium calculation but without ignoring the fact that I is slowly leaking away as it forms P.

Method Begin by writing the net rates of change of the concentrations of the substances and then invoke the steady-state approximation for the intermediate I. Use the resulting expression to obtain the rate of change of the concentration of P.

Answer The net rates of change of P and I are

$$\frac{d[P]}{dt} = k_b[I]$$

$$\frac{d[I]}{dt} = k_a[A][B] - k'_a[I] - k_b[I] \approx 0$$

The second equation solves to

$$[I] \approx \frac{k_a[A][B]}{k'_a + k_b}$$

When we substitute this result into the expression for the rate of formation of P, we obtain

$$\frac{d[P]}{dt} \approx k_r[A][B] \qquad k_r = \frac{k_a k_b}{k'_a + k_b}$$

This expression reduces to that in eqn 19.35 when the rate constant for the decay of I into products is much smaller than that for its decay into reactants, $k_b \ll k'_a$.

Self-test 19.8 Show that the pre-equilibrium mechanism in which $2\,A \rightleftharpoons I$ (K) followed by $I + B \rightarrow P$ (k_b) results in an overall third-order reaction. $[d[P]/dt = k_b K[A]^2[B]]$

The kinetics of complex reactions

Many reactions take place by mechanisms that involve several elementary steps. Some take place at a useful rate only after absorption of light or if a catalyst is present. In the following sections we begin to see how to develop the ideas introduced above to deal with these special kinds of reactions. We leave the study of catalysis to Chapter 21 and focus here on the kinetic analysis of a special class of reactions in the gas phase, polymerization processes, and photochemical reactions.

19.7 The Lindemann–Hinshelwood mechanism of unimolecular reactions

A number of gas-phase reactions follow first-order kinetics, as in the isomerization of cyclopropane:

$$cyclo\text{-}C_3H_6 \rightarrow CH_3CH{=}CH_2 \qquad v = k_r[cyclo\text{-}C_3H_6] \tag{19.36}$$

The problem with the interpretation of first-order rate laws is that presumably a molecule acquires enough energy to react as a result of its collisions with other molecules. However, collisions are simple bimolecular events, so how can they result in a first-order rate law? First-order gas-phase reactions are widely called 'unimolecular reactions' because they also involve an elementary unimolecular step in which the reactant molecule changes into the product. This term must be used with caution, though, because the overall mechanism has bimolecular as well as unimolecular steps.

The first successful explanation of unimolecular reactions was provided by Frederick Lindemann in 1921 and then elaborated by Cyril Hinshelwood. In the **Lindemann–Hinshelwood mechanism** it is supposed that a reactant molecule A becomes energetically excited by collision with another A molecule (Fig. 19.15):

$$A + A \rightarrow A^* + A \qquad \frac{d[A^*]}{dt} = k_a[A]^2 \tag{19.37}$$

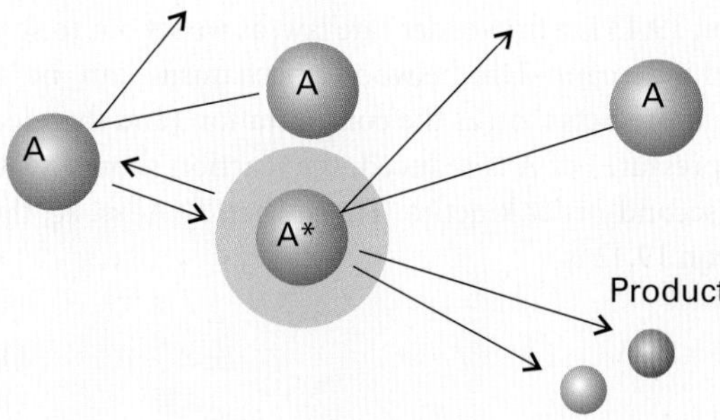

Fig. 19.15 A representation of the Lindemann–Hinshelwood mechanism of unimolecular reactions. The species A is excited by collision with A, and the excited A molecule (A*) may either be deactivated by a collision with A or go on to decay by a unimolecular process to form products.

The energized molecule (A^*) might lose its excess energy by collision with another molecule:

$$A + A^* \rightarrow A + A \qquad \frac{d[A^*]}{dt} = -k_a'[A][A^*] \tag{19.38}$$

Alternatively, the excited molecule might shake itself apart and form products P. That is, it might undergo the unimolecular decay

$$A^* \rightarrow P \qquad \frac{d[A^*]}{dt} = -k_b[A^*] \tag{19.39}$$

If the unimolecular step is slow enough to be the rate-determining step, the overall reaction will have first-order kinetics, as observed. This conclusion can be demonstrated explicitly by applying the steady-state approximation to the net rate of formation of A^*:

$$\frac{d[A^*]}{dt} = k_a[A]^2 - k_a'[A][A^*] - k_b[A^*] \approx 0 \tag{19.40}$$

This equation solves to

$$[A^*] = \frac{k_a[A]^2}{k_b + k_a'[A]} \tag{19.41}$$

so the rate law for the formation of P is

$$\frac{d[P]}{dt} = k_b[A^*] = \frac{k_a k_b[A]^2}{k_b + k_a'[A]} \tag{19.42}$$

At this stage the rate law is not first order. However, if the rate of deactivation by (A^*, A) collisions is much greater than the rate of unimolecular decay, in the sense that

$$k_a'[A^*][A] \gg k_b[A^*] \qquad \text{or} \qquad k_a'[A] \gg k_b$$

then we can neglect k_b in the denominator and obtain

$$\frac{d[P]}{dt} = k_r[A] \qquad k_r = \frac{k_a k_b}{k_a'} \tag{19.43}$$

Equation 19.43 is a first-order rate law, as we set out to show.

The Lindemann–Hinshelwood mechanism can be tested because it predicts that, as the concentration (and therefore the partial pressure) of A is reduced, the reaction should switch to overall second-order kinetics. Thus, when $k_a'[A] \ll k_b$, the rate law in eqn 19.42 is

$$\frac{d[P]}{dt} \approx k_a[A]^2 \tag{19.44}$$

The physical reason for the change of order is that at low pressures the rate-determining step is the bimolecular formation of A^*. If we write the full rate law in eqn 19.42 as

$$\frac{d[P]}{dt} = k_r[A] \qquad k_r = \frac{k_a k_b[A]}{k_b + k_a'[A]} \tag{19.45}$$

then the expression for the effective rate constant, k_r, can be rearranged to

$$\frac{1}{k_r} = \frac{k_a'}{k_a k_b} + \frac{1}{k_a[A]} \tag{19.46}$$

Hence, a test of the theory is to plot $1/k_r$ against $1/[A]$, and to expect a straight line. Deviations from this behaviour are common and in Chapter 20 we shall enhance the description of the mechanism to take into account experimental results over a range of pressures.

19.8 Polymerization kinetics

We explored the structure of synthetic polymers in Chapter 9. Now we give an overview of the mechanisms of the reactions that lead to the formation of this important class of macromolecules.

There are two major classes of polymerization process and the average molar mass of the product varies with time in distinctive ways. In **stepwise polymerization** any two monomers present in the reaction mixture can link together at any time and growth of the polymer is not confined to chains that are already forming (Fig. 19.16). As a result, monomers are removed early in the reaction and, as we shall see, the average molar mass of the product grows with time. In **chain polymerization** an activated monomer, M, attacks another monomer, links to it, then that unit attacks another monomer, and so on. The monomer is used up as it becomes linked to the growing chains (Fig. 19.17). High polymers are formed rapidly and only the yield, not the average molar mass, of the polymer is increased by allowing long reaction times.

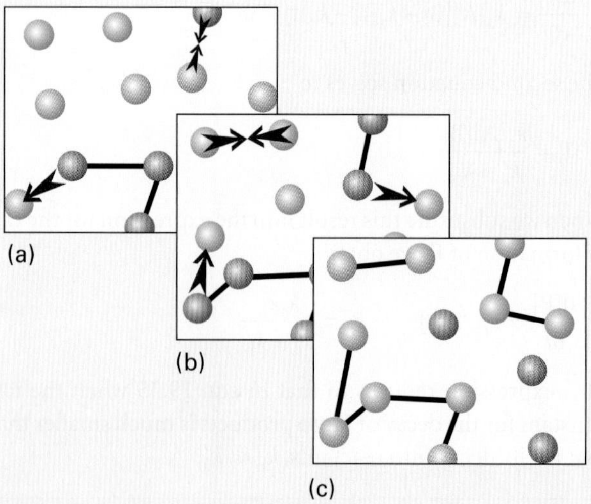

Fig. 19.16 In stepwise polymerization, growth can start at any pair of monomers, and so new chains begin to form throughout the reaction.

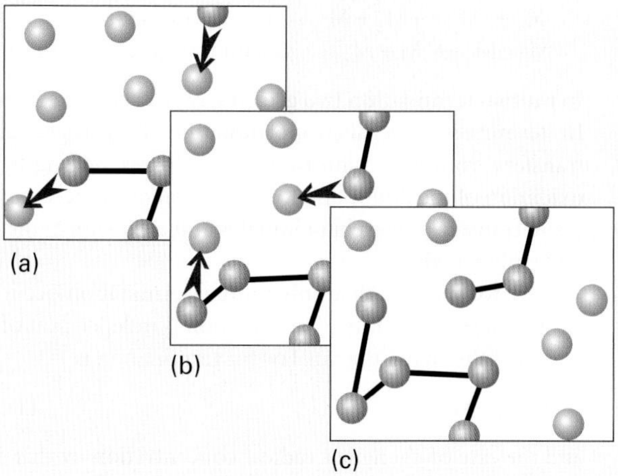

Fig. 19.17 The process of chain polymerization. Chains grow as each chain acquires additional monomers.

(a) Stepwise polymerization

Stepwise polymerization commonly proceeds by a condensation reaction, in which a small molecule (typically H_2O) is eliminated in each step. Stepwise polymerization is the mechanism of production of polyamides, as in the formation of nylon-66:

$$\begin{aligned}&H_2N(CH_2)_6NH_2 + HOOC(CH_2)_4COOH\\&\rightarrow H_2N(CH_2)_6NHCO(CH_2)_4COOH + H_2O\\&\rightarrow H{-}[NH(CH_2)_6NHCO(CH_2)_4CO]_n{-}OH\end{aligned}$$

Polyesters and polyurethanes are formed similarly (the latter without elimination). A polyester, for example, can be regarded as the outcome of the stepwise condensation of a hydroxyacid HO—M—COOH. We shall consider the formation of a polyester from such a monomer, and measure its progress in terms of the concentration of the —COOH groups in the sample (which we denote A), for these groups gradually disappear as the condensation proceeds. Because the condensation reaction can occur between molecules containing any number of monomer units, chains of many different lengths can grow in the reaction mixture.

In the absence of a catalyst, we can expect the condensation to be overall second order in the concentration of the —OH and —COOH (or A) groups, and write

$$\frac{d[A]}{dt} = -k_r[OH][A] \qquad (19.47a)$$

However, because there is one —OH group for each —COOH group, this equation is the same as

$$\frac{d[A]}{dt} = -k_r[A]^2 \qquad (19.47b)$$

If we assume that the rate constant for the condensation is independent of the chain length, then k_r remains constant throughout the reaction. The solution of this rate law is given by eqn 19.13c ($[A] = [A]_0/(1 + k_r t[A]_0)$). It follows that the fraction, p, of —COOH groups that have condensed at time t is

$$p = \frac{[A]_0 - [A]}{[A]_0} = \frac{k_r t[A]_0}{1 + k_r t[A]_0} \qquad (19.48)$$

Next, we calculate the **degree of polymerization**, which is defined as the average number of monomer residues per polymer molecule. This quantity is the ratio of the initial concentration of A, $[A]_0$, to the concentration of end groups, $[A]$, at the time of interest, because there is one —A group per polymer molecule. For example, if there were initially 1000 A groups and there are now only 10, each polymer must be 100 units long on average. Because we can express $[A]$ in terms of p (eqn 19.48), the average number of monomers per polymer molecule, $\langle N\rangle$, is

$$\langle N\rangle = \frac{[A]_0}{[A]} = \frac{1}{1-p} \qquad (19.49a)$$

This result is illustrated in Fig. 19.18. When we express p in terms of the rate constant k_r (eqn 19.48), we find

$$\langle N\rangle = 1 + k_r t[A]_0 \qquad (19.49b)$$

The average length grows linearly with time. Therefore, the longer a stepwise polymerization proceeds, the higher the average molar mass of the product.

(b) Chain polymerization

Many gas-phase reactions and liquid-phase polymerization reactions are **chain reactions**. In a chain reaction, a reaction intermediate produced in one step generates an intermediate

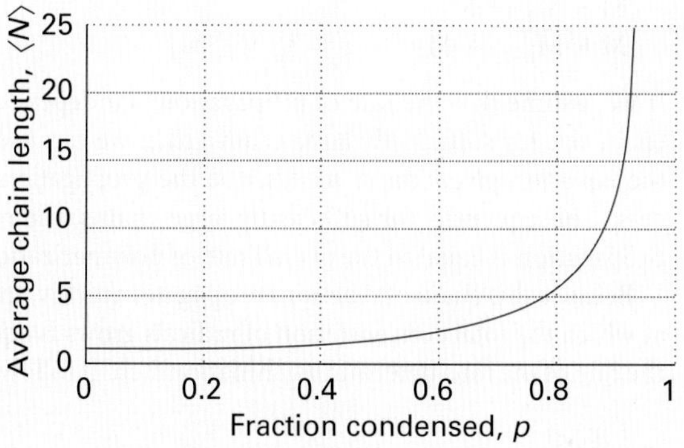

Fig. 19.18 The average chain length of a polymer as a function of the fraction of reacted monomers, p. Note that p must be very close to 1 for the chains to be long.

interActivity Plot the variation of p with time for a range of k_r values of your choice (take $[A]_0 = 1.0\ mol\ dm^{-3}$).

in a subsequent step, then that intermediate generates another intermediate, and so on. The intermediates in a chain reaction are called **chain carriers**. In a **radical chain reaction** the chain carriers are radicals (species with unpaired electrons).

Chain polymerization occurs by addition of monomers to a growing polymer, often by a radical chain process. It results in the rapid growth of an individual polymer chain for each activated monomer. Examples include the polymerizations of ethene, methyl methacrylate, and styrene, as in

$$—CH_2CHX\cdot + CH_2{=}CHX \rightarrow —CH_2CHXCH_2CHX\cdot$$

and subsequent reactions. The central feature of the kinetic analysis (which is summarized in the following *Justification*) is that the rate of polymerization is proportional to the square root of the initiator concentration:

$$v = k_r[I]^{1/2}[M] \qquad (19.50)$$

Justification 19.5 *The rate of chain polymerization*

There are three basic types of reaction step in a chain polymerization process:

(a) Initiation, in which the chain carriers are formed:

$$I \rightarrow R\cdot + R\cdot \qquad v_i = k_i[I]$$
$$M + R\cdot \rightarrow \cdot M_1 \qquad \text{(fast)}$$

where I is the initiator, R· the radical I forms, and $\cdot M_1$ is a monomer radical. We have shown a reaction in which a radical is produced, but in some polymerizations the initiation step leads to the formation of an ionic chain carrier. The rate-determining step is the formation of the radicals R· by homolysis of the initiator, so the rate of initiation is equal to the v_i given above.

(b) Propagation, in which a chain carrier attacks a monomer molecule and each attack leads to a new carrier:

$$M + \cdot M_1 \rightarrow \cdot M_2$$
$$M + \cdot M_2 \rightarrow \cdot M_3$$
$$M + \cdot M_{n-1} \rightarrow \cdot M_n \qquad v_p = k_p[M][\cdot M]$$

If we assume that the rate of propagation is independent of chain size for sufficiently large chains, then we can use only the equation given above to describe the propagation process. Consequently, for sufficiently large chains, the rate of propagation is equal to the overall rate of polymerization.

Because this chain of reactions propagates quickly, the rate at which the total concentration of radicals grows is equal to the rate of the rate-determining initiation step. It follows that

$$\left(\frac{d[\cdot M]}{dt}\right)_{\text{production}} = 2fk_i[I] \qquad (19.51)$$

where f is the fraction of radicals R· that successfully initiate a chain.

(c) Termination, in which radicals combine and end the chain:

$$\cdot M_n + \cdot M_m \rightarrow M_{n+m} \qquad \text{(mutual termination)}$$
$$\cdot M_n + \cdot M_m \rightarrow M_n + M_m \qquad \text{(disproportionation)}$$
$$M + \cdot M_n \rightarrow \cdot M + M_n \qquad \text{(chain transfer)}$$

In **mutual termination** two growing radical chains combine. In termination by **disproportionation** a hydrogen atom transfers from one chain to another, corresponding to the oxidation of the donor and the reduction of the acceptor. In **chain transfer**, a new chain initiates at the expense of the one currently growing.

Here we suppose that only mutual termination occurs. If we assume that the rate of termination is independent of the length of the chain, the rate law for termination is

$$v_t = k_t[\cdot M]^2 \qquad (19.52)$$

and the rate of change of radical concentration by this process is

$$\left(\frac{d[\cdot M]}{dt}\right)_{\text{depletion}} = -2k_t[\cdot M]^2 \qquad (19.53)$$

The steady-state approximation gives:

$$\frac{d[\cdot M]}{dt} = 2fk_i[I] - 2k_t[\cdot M]^2 = 0$$

The steady-state concentration of radical chains is therefore

$$[\cdot M] = \left(\frac{fk_i}{k_t}\right)^{1/2}[I]^{1/2} \qquad (19.54)$$

Because the rate of propagation of the chains is the negative of the rate at which the monomer is consumed, we can write $v_p = -d[M]/dt$ and

$$v_p = k_p[\cdot M][M] = k_p\left(\frac{fk_i}{k_t}\right)^{1/2}[I]^{1/2}[M] \qquad (19.55)$$

This rate is also the rate of polymerization, which has the form of eqn 19.50.

The **kinetic chain length**, ν, is the ratio of the number of monomer units consumed per activated centre produced in the initiation step:

$$\nu = \frac{\text{number of monomer units consumed}}{\text{number of activated centres produced}} \qquad [19.56a]$$

The kinetic chain length can be expressed in terms of the rate expressions in *Justification 19.5*. To do so, we recognize that monomers are consumed at the rate that chains propagate. Then,

$$\nu = \frac{\text{rate of propagation of chains}}{\text{rate of production of radicals}} \qquad (19.56b)$$

By making the steady-state approximation, we set the rate of production of radicals equal to the termination rate. Therefore,

Table 19.4 Examples of photochemical processes

Process	General form	Example
Ionization	$A^* \rightarrow A^+ + e^-$	$NO^* \xrightarrow{134\ nm} NO^+ + e^-$
Electron transfer	$A^* + B \rightarrow A^+ + B^-$ or $A^- + B^+$	$[Ru(bpy)_3^{2+}]^* + Fe^{3+} \xrightarrow{452\ nm} Ru(bpy)_3^{3+} + Fe^{2+}$
Dissociation	$A^* \rightarrow B + C$	$O_3^* \xrightarrow{1180\ nm} O_2 + O$
	$A^* + B{-}C \rightarrow A + B + C$	$Hg^*\ CH_4 \xrightarrow{254\ nm} Hg + CH_3 + H$
Addition	$2\,A^* \rightarrow B$	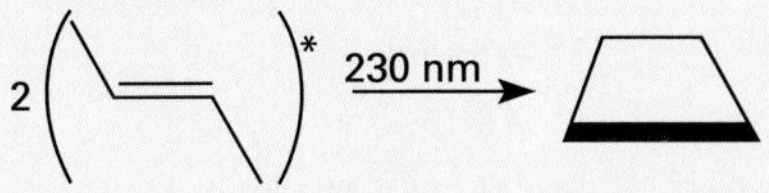
	$A^* + B \rightarrow AB$	
Abstraction	$A^* + B{-}C \rightarrow A{-}B + C$	$Hg^* + H_2 \xrightarrow{254\ nm} HgH + H$
Isomerization or rearrangement	$A^* \rightarrow A'$	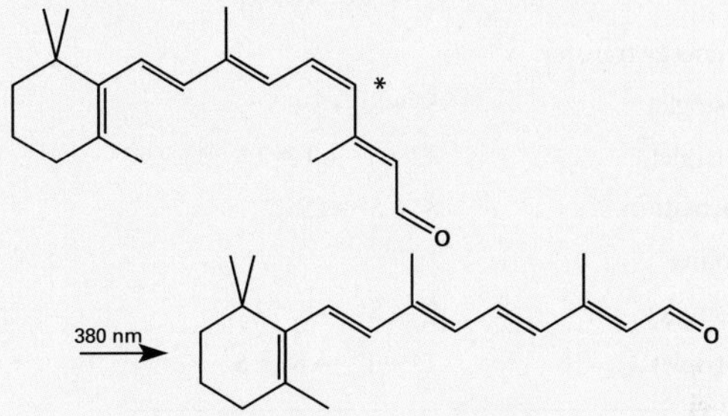

* Excited state.

we can write the expression for the kinetic chain length as

$$\nu = \frac{k_p[\cdot M][M]}{2k_t[\cdot M]^2} = \frac{k_p[M]}{2k_t[\cdot M]}$$

When we substitute the steady-state expression, eqn 19.54 for the radical concentration, we obtain

$$\nu = k_r[M][I]^{-1/2} \qquad k_r = \tfrac{1}{2}k_p(fk_ik_t)^{-1/2} \qquad (19.57)$$

Consider a polymer produced by a chain mechanism with mutual termination. In this case, the average number of monomers in a polymer molecule, $\langle N \rangle$, produced by the reaction is the sum of the numbers in the two combining polymer chains. The average number of units in each chain is ν. Therefore,

$$\langle N \rangle = 2\nu = 2k_r[M][I]^{-1/2} \qquad (19.58)$$

with k_r given in eqn 19.57. We see that, the slower the initiation of the chain (the smaller the initiator concentration and the smaller the initiation rate constant), the greater the kinetic chain length, and therefore the higher the average molar mass of the polymer.

19.9 Photochemistry

Many reactions can be initiated by the absorption of electromagnetic radiation. The most important of all are the photochemical processes that capture the radiant energy of the Sun. Some of these reactions lead to the heating of the atmosphere during the daytime by absorption of ultraviolet radiation. Others include the absorption of visible radiation during photosynthesis (*Impact I19.1*). Table 19.4 summarizes common photochemical reactions.

Photochemical processes are initiated by the absorption of radiation by at least one component of a reaction mixture. In a **primary process**, products are formed directly from the excited state of a reactant. Examples include fluorescence (Section 11.5) and the *cis–trans* photoisomerization of retinal (Table 19.4, see also *Impact I11.1*). Products of a **secondary process** originate from intermediates that are formed directly from the excited state of a reactant. Examples include photosynthesis (*Impact I19.1*).

Competing with the formation of photochemical products is a host of primary photophysical processes that can deactivate the excited state (Table 19.5). Electronic transitions caused by absorption of ultraviolet and visible radiation occur within 10^{-16}–10^{-15} s. We expect, then, that the upper limit for the rate constant of a first-order photochemical reaction is about $10^{16}\ s^{-1}$. Fluorescence is slower than absorption, with typical lifetimes of 10^{-12}–10^{-6} s. Therefore, the excited singlet state can initiate very fast photochemical reactions in the femtosecond (10^{-15} s) to picosecond (10^{-12} s) timescale. An example is the

Table 19.5 Common photophysical processes†

Primary absorption	$S + h\nu \rightarrow S^*$
Excited-state absorption	$S^* + h\nu \rightarrow S^{**}$
	$T^* + h\nu \rightarrow T^{**}$
Fluorescence	$S^* \rightarrow S + h\nu$
Stimulated emission	$S^* + h\nu \rightarrow S + 2h\nu$
Intersystem crossing (ISC)	$S^* \rightarrow T^*$
Phosphorescence	$T^* \rightarrow S + h\nu$
Internal conversion (IC)	$S^* \rightarrow S$
Collision-induced emission	$S^* + M \rightarrow S + M + h\nu$
Collisional deactivation	$S^* + M \rightarrow S + M$
	$T^* + M \rightarrow S + M$
Electronic energy transfer:	
Singlet–singlet	$S^* + S \rightarrow S + S^*$
Triplet–triplet	$T^* + T \rightarrow T + T^*$
Excimer formation	$S^* + S \rightarrow (SS)^*$
Energy pooling	
Singlet–singlet	$S^* + S^* \rightarrow S^{**} + S$
Triplet–triplet	$T^* + T^* \rightarrow S^* + S$

† S denotes a singlet state, T a triplet state, and M is a third-body.
* denotes an excited state.

initial event of vision (*Impact I11.1*). Typical intersystem crossing (ISC) and phosphorescence times for large organic molecules are 10^{-12}–10^{-4} s and 10^{-6}–10^{-1} s, respectively. As a consequence, excited triplet states are photochemically important. Indeed, because phosphorescence decay is several orders of magnitude slower than most typical reactions, species in excited triplet states can undergo a very large number of collisions with other reactants before deactivation.

(a) The primary quantum yield

We shall see that the rates of deactivation of the excited state by radiative, non-radiative, and chemical processes determine the yield of product in a photochemical reaction. The **primary quantum yield**, ϕ, is defined as the number of photophysical or photochemical events that lead to primary products divided by the number of photons absorbed by the molecule in the same interval:

$$\phi = \frac{\text{number of events}}{\text{number of photons absorbed}} \qquad [19.59a]$$

When we divide both the numerator and denominator of this expression by the time interval over which the events occurred, we see that the primary quantum yield is also the rate of radiation-induced primary events divided by the rate of photon absorption, I_{abs}:

$$\phi = \frac{\text{rate of process}}{\text{rate of photon absorption}} = \frac{v}{I_{abs}} \qquad [19.59b]$$

A molecule in an excited state must either decay to the ground state or form a photochemical product. Therefore, the total number of molecules deactivated by radiative processes, non-radiative processes, and photochemical reactions must be equal to the number of excited species produced by absorption of light. We conclude that the sum of primary quantum yields ϕ_i for *all* photophysical and photochemical events i must be equal to 1, regardless of the number of reactions involving the excited state. It follows that

$$\sum_i \phi_i = \sum_i \frac{v_i}{I_{abs}} = 1 \qquad (19.60)$$

It follows that for a decaying excited singlet state we write

$$\phi_f + \phi_{IC} + \phi_p = 1$$

where ϕ_f, ϕ_{IC}, and ϕ_p are the quantum yields of fluorescence, internal conversion, and phosphorescence, respectively (intersystem crossing from the singlet to the triplet state is taken into account with the measurement of ϕ_p). The quantum yield of photon emission by fluorescence and phosphorescence is $\phi_{emission} = \phi_f + \phi_p$, which is less than 1. If the excited singlet state also participates in a primary photochemical reaction with quantum yield ϕ_r, we write

$$\phi_f + \phi_{IC} + \phi_p + \phi_r = 1$$

We can now strengthen the link between reaction rates and primary quantum yield already established by eqns 19.59 and 19.60. By taking the constant I_{abs} out of the summation in eqn 19.60 and rearranging, we obtain $I_{abs} = \sum_i v_i$. Substituting this result into eqn 19.59b gives the general result

$$\phi_i = \frac{v_i}{\sum_i v_i} \qquad (19.61)$$

Therefore, the primary quantum yield may be determined directly from the experimental rates of *all* photophysical and photochemical processes that deactivate the excited state (Fig. 19.19).

(b) Mechanism of decay of excited singlet states

Consider the formation and decay of an excited singlet state in the absence of a chemical reaction:

Absorption:	$S + h\nu_i \rightarrow S^*$	$v_{abs} = I_{abs}$
Fluorescence:	$S^* \rightarrow S + h\nu_f$	$v_f = k_f[S^*]$
Internal conversion:	$S^* \rightarrow S$	$v_{IC} = k_{IC}[S^*]$
Intersystem crossing:	$S^* \rightarrow T^*$	$v_{ISC} = k_{ISC}[S^*]$

in which S is an absorbing species, S^* an excited singlet state, T^* an excited triplet state, and $h\nu_i$ and $h\nu_f$ are the energies of the incident and fluorescent photons, respectively. From the methods

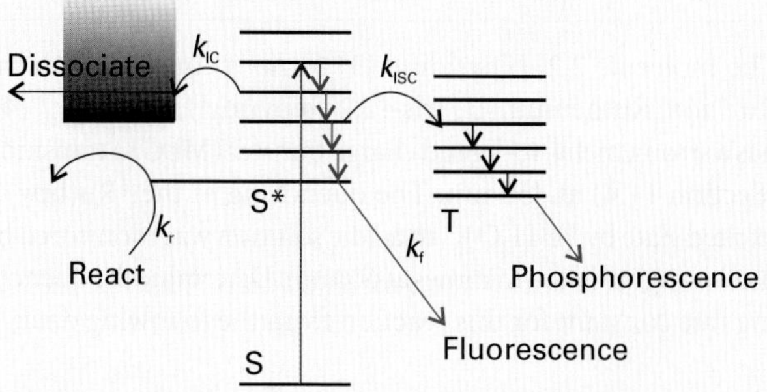

Fig. 19.19 A Jablonski diagram for deactivation of a singlet electronic state S*, with each process labelled by its corresponding rate constant: fluorescence, internal conversion, intersystem crossing, and photochemical reaction (k_f, k_{IC}, k_{ISC}, and k_r, respectively). Jablonski diagrams were introduced in Section 11.5.

developed in this chapter and the rates of the steps that form and destroy the excited singlet state S*, we write the rate of formation and decay of S* as:

$$\text{Rate of formation of } [S^*] = I_{abs}$$
$$\text{Rate of decay of } [S^*] = -k_f[S^*] - k_{ISC}[S^*] - k_{IC}[S^*] = -(k_f + k_{ISC} + k_{IC})[S^*]$$

It follows that the excited state decays by a first-order process, so, when the light is turned off, the concentration of S* varies with time *t* as:

$$[S^*]_t = [S^*]_0 e^{-t/\tau_0} \quad (19.62)$$

where the **observed lifetime**, τ_0, of the first excited singlet state is defined as

$$\tau_0 = \frac{1}{k_f + k_{ISC} + k_{IC}} \quad [19.63]$$

We show in the following *Justification* that the quantum yield of fluorescence is

$$\phi_f = \frac{k_f}{k_f + k_{ISC} + k_{IC}} \quad (19.64)$$

Justification 19.6 *The quantum yield of fluorescence*

Most fluorescence measurements are conducted by illuminating a relatively dilute sample with a continuous and intense beam of light. It follows that [S*] is small and constant, so we may invoke the steady-state approximation (Section 19.6) and write:

$$\frac{d[S^*]}{dt} = I_{abs} - k_f[S^*] - k_{ISC}[S^*] - k_{IC}[S^*] = I_{abs} - (k_f + k_{ISC} + k_{IC})[S^*] = 0$$

Consequently,

$$I_{abs} = (k_f + k_{ISC} + k_{IC})[S^*]$$

By using this expression and eqn 19.59b, the quantum yield of fluorescence is written as:

$$\phi_f = \frac{v_f}{I_{abs}} = \frac{k_f[S^*]}{(k_f + k_{ISC} + k_{IC})[S^*]}$$

which, by cancelling the [S*], simplifies to eqn 19.64.

The observed fluorescence lifetime can be measured with a pulsed laser technique (Section 11.7). First, the sample is excited with a short light pulse from a laser using a wavelength at which S absorbs strongly. Then, the exponential decay of the fluorescence intensity after the pulse is monitored. From eqn 19.63, it follows that

$$\tau_0 = \frac{1}{k_f + k_{ISC} + k_{IC}} = \left(\frac{k_f}{k_f + k_{ISC} + k_{IC}}\right) \times \frac{1}{k_f} = \frac{\phi_f}{k_f} \quad (19.65)$$

A BRIEF ILLUSTRATION

In water, the fluorescence quantum yield and observed fluorescence lifetime of tryptophan are $\phi_f = 0.20$ and $\tau_0 = 2.6$ ns, respectively. It follows from eqn 19.65 that the fluorescence rate constant k_f is

$$k_f = \frac{\phi_f}{\tau_0} = \frac{0.20}{2.6 \times 10^{-9}\ \text{s}} = 7.7 \times 10^7\ \text{s}^{-1}$$

(c) Quenching

The shortening of the lifetime of the excited state is called **quenching**. Quenching may be either a desired process, such as in energy or electron transfer, or an undesired side reaction that can decrease the quantum yield of a desired photochemical process. Quenching effects may be studied by monitoring the emission from the excited state that is involved in the photochemical reaction.

The addition of a quencher, Q, opens an additional channel for deactivation of S*:

$$\text{Quenching:} \quad S^* + Q \rightarrow S + Q \qquad v_Q = k_Q[Q][S^*]$$

The **Stern–Volmer equation**, which is derived in the *Justification* below, relates the fluorescence quantum yields $\phi_{f,0}$ and ϕ_f measured in the absence and presence, respectively, of a quencher Q at a molar concentration [Q]:

$$\frac{\phi_{f,0}}{\phi_f} = 1 + \tau_0 k_Q[Q] \quad (19.66)$$

This equation tells us that a plot of $\phi_{f,0}/\phi_f$ against [Q] should be a straight line with slope $\tau_0 k_Q$. Such a plot is called a **Stern–Volmer plot** (Fig. 19.20). The method may also be applied to the quenching of phosphorescence.

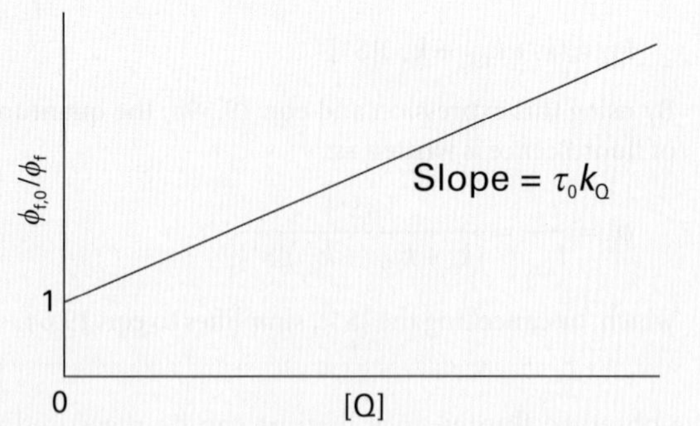

Fig. 19.20 The format of a Stern–Volmer plot and the interpretation of the slope in terms of the rate constant for quenching and the observed fluorescence lifetime in the absence of quenching.

Justification 19.7 *The Stern–Volmer equation*

With the addition of quenching, the steady-state approximation for [S*] now gives:

$$\frac{d[S^*]}{dt} = I_{abs} - (k_f + k_{ISC} + k_{IC} + k_Q[Q])[S^*] = 0$$

and the fluorescence quantum yield in the presence of the quencher is:

$$\phi_f = \frac{k_f}{k_f + k_{ISC} + k_{IC} + k_Q[Q]}$$

The quantum yield when [Q] = 0 is

$$\phi_{f,0} = \frac{k_f}{k_f + k_{ISC} + k_{IC}}$$

It follows that

$$\frac{\phi_{f,0}}{\phi_f} = \left(\frac{k_f}{k_f + k_{ISC} + k_{IC}}\right) \times \left(\frac{k_f + k_{ISC} + k_{IC} + k_Q[Q]}{k_f}\right)$$

$$= \frac{k_f + k_{ISC} + k_{IC} + k_Q[Q]}{k_f + k_{ISC} + k_{IC}}$$

$$= 1 + \frac{k_Q}{k_f + k_{ISC} + k_{IC}}[Q]$$

By using eqn 19.65, this expression simplifies to eqn 19.66.

Because the fluorescence intensity and lifetime are both proportional to the fluorescence quantum yield (specifically, from eqn 19.65, $\tau = \phi_f/k_f$), plots of $I_{f,0}/I_f$ and τ_0/τ (where the subscript 0 indicates a measurement in the absence of quencher) against [Q] should also be linear with the same slope and intercept as those shown for eqn 19.66.

Example 19.8 *Determining the quenching rate constant*

The molecule 2,2′-bipyridine (**3**) forms a complex with the Ru^{2+} ion. Ruthenium(II) tris-(2,2′-bipyridyl), $Ru(bpy)_3^{2+}$ (**4**), has a strong metal-to-ligand charge transfer (MLCT) transition (Section 11.4) at 450 nm. The quenching of the $^*Ru(bpy)_3^{2+}$ excited state by $Fe(H_2O)_6^{3+}$ in acidic solution was monitored by measuring emission lifetimes at 600 nm. Determine the quenching rate constant for this reaction from the following data:

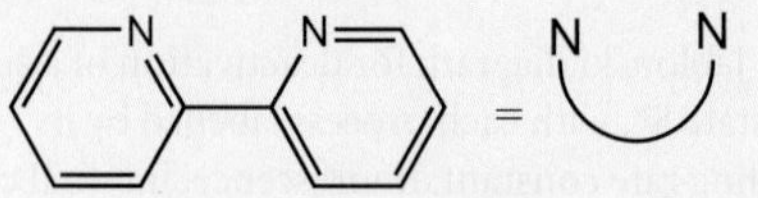

3 2,2′-Bipyridine (bpy)

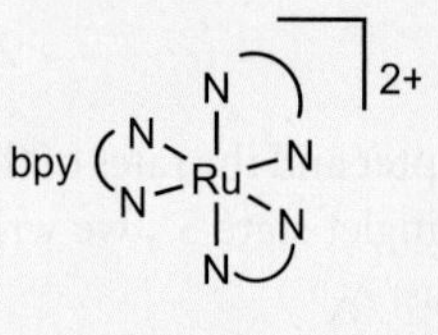

4 $[Ru(bpy)_3]^{2+}$

$[Fe(H_2O)_6^{3+}]/(10^{-4}\ mol\ dm^{-3})$	0	1.6	4.7	7	9.4
$\tau/(10^{-7}\ s)$	6	4.05	3.37	2.96	2.17

Method Re-write the Stern–Volmer equation (eqn 19.66) for use with lifetime data; then fit the data to a straight line.

Answer Upon substitution of τ_0/τ for $\phi_{f,0}/\phi_f$ in eqn 19.66 and after rearrangement, we obtain:

$$\frac{1}{\tau} = \frac{1}{\tau_0} + k_Q[Q]$$

Figure 19.21 shows a plot of $1/\tau$ versus $[Fe^{3+}]$ and the results of a fit to this equation. The slope of the line is 2.8×10^9, so $k_Q = 2.8 \times 10^9\ dm^3\ mol^{-1}\ s^{-1}$.

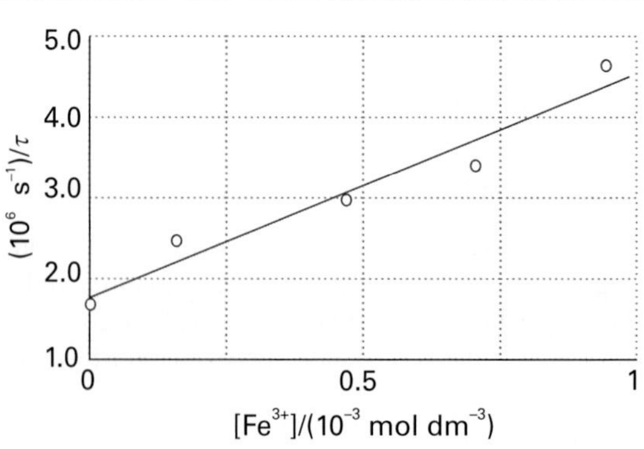

Fig. 19.21 The Stern–Volmer plot of the data for Example 19.8.

This example shows that measurements of emission lifetimes are preferred because they yield the value of k_Q directly. To determine the value of k_Q from intensity or quantum yield measurements, we need to make an independent measurement of τ_0.

Self-test 19.9 The quenching of tryptophan fluorescence by dissolved O_2 gas was monitored by measuring emission lifetimes at 348 nm in aqueous solutions. Determine the quenching rate constant for this process from the following data:

$[O_2]/(10^{-2}\ \text{mol dm}^{-3})$	0	2.3	5.5	8	10.8
$\tau/(10^{-9}\ \text{s})$	2.6	1.5	0.92	0.71	0.57

$[1.3 \times 10^{10}\ \text{dm}^3\ \text{mol}^{-1}\ \text{s}^{-1}]$

Three common mechanisms for bimolecular quenching of an excited singlet (or triplet) state are:

Collisional deactivation: $S^* + Q \rightarrow S + Q$

Electron transfer: $S^* + Q \rightarrow S^+ + Q^-$ or $S^- + Q^+$

Resonance energy transfer: $S^* + Q \rightarrow S + Q^*$

The quenching rate constant itself does not give much insight into the mechanism of quenching. For the system of Example 19.8, it is known that the quenching of the excited state of $Ru(bpy)_3^{2+}$ is a result of light-induced electron transfer to Fe^{3+}, but the quenching data do not allow us to prove the mechanism. However, there are some criteria that govern the relative efficiencies of collisional quenching, energy and electron transfer. Collisional quenching is particularly efficient when Q is a heavy species, such as iodide ion, which receives energy from S* and then decays primarily by internal conversion to the ground state. According to the **Marcus theory** of electron transfer, which was proposed by R.A. Marcus in 1965, the rates of electron transfer (from ground or excited states) depend on (see also Section 20.8):

1. The distance between the donor and acceptor, with electron transfer becoming more efficient as the distance between donor and acceptor decreases.

2. The reaction Gibbs energy, $\Delta_r G$, with electron transfer becoming more efficient as the reaction becomes more exergonic. For example, efficient photooxidation of S requires that the reduction potential of S* be lower than the reduction potential of Q.

3. The reorganization energy, the energy cost incurred by molecular rearrangements of donor, acceptor, and medium during electron transfer. The electron transfer rate is predicted to increase as this reorganization energy is matched closely by the reaction Gibbs energy.

Electron transfer can also be studied by time-resolved spectroscopy (Section 11.7). The oxidized and reduced products often have electronic absorption spectra distinct from those of their neutral parent compounds. Therefore, the rapid appearance of such known features in the absorption spectrum after excitation by a laser pulse may be taken as indication of quenching by electron transfer. In the following section we explore energy transfer in detail.

(d) Resonance energy transfer

We visualize the processes $S \rightarrow S^*$ and $S^* + Q \rightarrow S + Q^*$ as follows. The oscillating electric field of the incoming electromagnetic radiation induces an oscillating electric dipole moment in S. Energy is absorbed by S if the frequency of the incident radiation, ν, is such that $\nu = \Delta E_S/h$, where ΔE_S is the energy separation between the ground and excited electronic states of S and h is Planck's constant. This is the 'resonance condition' for absorption of radiation. The collapse of S* to S results in a transition dipole that can induce a transition in Q. That is, the oscillating dipole on S now can affect electrons bound to a nearby Q molecule by inducing an oscillating dipole moment in the latter. If the frequency of oscillation of the electric dipole moment in S is such that $\nu = \Delta E_Q/h$ then Q will absorb energy from S.

The efficiency, η_T, of resonance energy transfer is defined as

$$\eta_T = 1 - \frac{\phi_f}{\phi_{f,0}} \qquad [19.67]$$

According to the **Förster theory** of resonance energy transfer, which was proposed by T. Förster in 1959, energy transfer is efficient when:

1. The energy donor and acceptor are separated by a short distance (of the order of nanometres).

2. Photons emitted by the excited state of the donor can be absorbed directly by the acceptor.

We show in *Further information 19.1* that, for donor–acceptor systems that are held rigidly either by covalent bonds or by a protein 'scaffold', η_T increases with decreasing distance, R, according to

$$\eta_T = \frac{R_0^6}{R_0^6 + R^6} \qquad (19.68)$$

where R_0 is a parameter (with dimensions of distance) that is characteristic of each donor–acceptor pair. Equation 19.68 has been verified experimentally and values of R_0 are available for a number of donor–acceptor pairs (Table 19.6).

The emission and absorption spectra of molecules span a range of wavelengths, so the second requirement of the Förster theory is met when the emission spectrum of the donor molecule overlaps significantly with the absorption spectrum of the

Table 19.6 Values of R_0 for some donor–acceptor pairs*

Donor[†]	Acceptor	R_0/nm
Naphthalene	Dansyl	2.2
Dansyl	ODR	4.3
Pyrene	Coumarin	3.9
IEDANS	FITC	4.9
Tryptophan	IEDANS	2.2
Tryptophan	Haem (heme)	2.9

* Additional values may be found in J.R. Lacowicz, *Principles of fluorescence spectroscopy*, Kluwer Academic/Plenum, New York (1999).
[†] Abbreviations:
Dansyl: 5-dimethylamino-1-naphthalenesulfonic acid
FITC: fluorescein 5-isothiocyanate
IEDANS: 5-((((2-iodoacetyl)amino)ethyl)amino)naphthalene-1-sulfonic acid
ODR: octadecyl-rhodamine

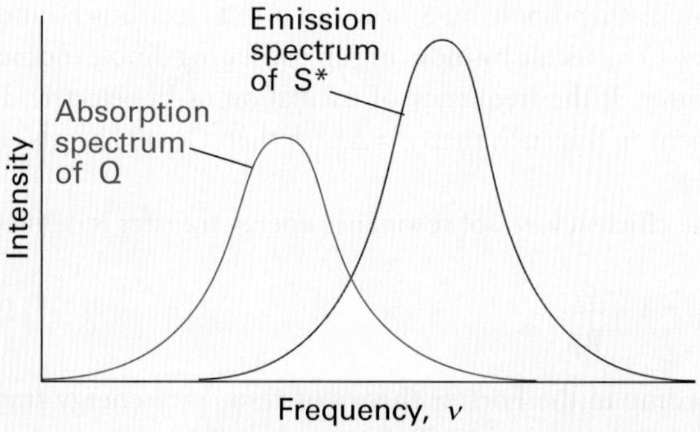

Fig. 19.22 According to the Förster theory, the rate of energy transfer from a molecule S* in an excited state to a quencher molecule Q is optimized at radiation frequencies in which the emission spectrum of S* overlaps with the absorption spectrum of Q, as shown in the shaded region.

acceptor. In the overlap region, photons emitted by the donor have the proper energy to be absorbed by the acceptor (Fig. 19.22).

In many cases, it is possible to prove that energy transfer is the dominant mechanism of quenching if the excited state of the acceptor fluoresces or phosphoresces at a characteristic wavelength. In a pulsed laser experiment, the rise in fluorescence intensity from Q* with a characteristic time that is the same as that for the decay of the fluorescence of S* is often taken as indication of energy transfer from S to Q.

Equation 19.68 forms the basis of **fluorescence resonance energy transfer** (FRET), in which the dependence of the energy transfer efficiency, η_T, on the distance, R, between energy donor and acceptor can be used to measure distances in biological systems. In a typical FRET experiment, a site on a biopolymer or membrane is labelled covalently with an energy donor and another site is labelled covalently with an energy acceptor. In certain cases, the donor or acceptor may be natural constituents of the system, such as amino acid groups, co-factors, or enzyme substrates. The distance between the labels is then calculated from the known value of R_0 and eqn 19.68. Several tests have shown that the FRET technique is useful for measuring distances ranging from 1 to 9 nm.

● **A BRIEF ILLUSTRATION**

When an amino acid on the surface of rhodopsin was labelled covalently with the energy donor 1,5-I AEDANS (**5**), the fluorescence quantum yield of the label decreased from 0.75 to 0.68 due to quenching by the visual pigment 11-*cis*-retinal (**6**). From eqn 19.67, we calculate $\eta_T = 1 - (0.68/0.75) = 0.093$ and from eqn 19.68 and the known value of $R_0 = 5.4$ nm for the 1,5-I AEDANS/11-*cis*-retinal pair we calculate $R = 7.9$ nm. Therefore, we take 7.9 nm to be the distance between the surface of the protein and 11-*cis*-retinal. ●

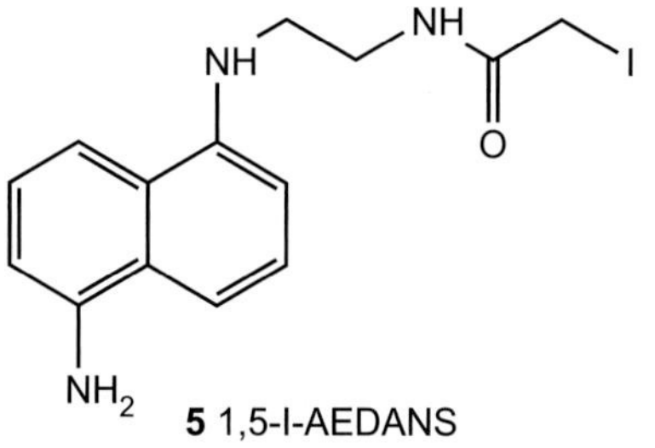

5 1,5-I-AEDANS

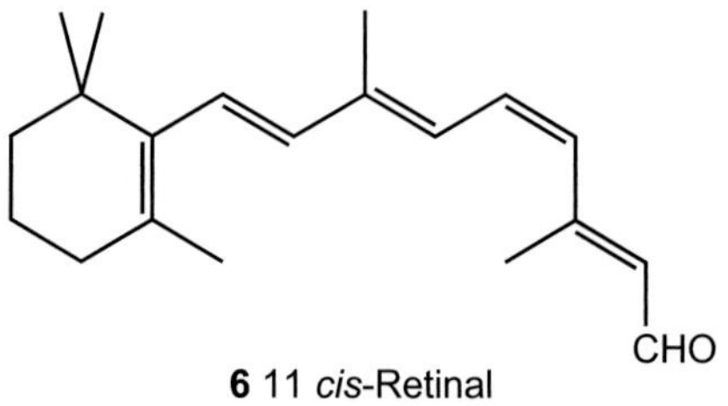

6 11 *cis*-Retinal

If donor and acceptor molecules diffuse in solution or in the gas phase, Förster theory predicts that the efficiency of quenching by energy transfer increases as the average distance travelled between collisions of donor and acceptor decreases. That is, the quenching efficiency increases with concentration of quencher, as predicted by the Stern–Volmer equation.

IMPACT ON BIOCHEMISTRY

I19.1 Harvesting of light during plant photosynthesis

A large proportion of solar radiation with wavelengths below 400 nm and above 1000 nm is absorbed by atmospheric gases such as ozone and O_2, which absorb ultraviolet radiation, and

CO_2 and H_2O, which absorb infrared radiation (*Impact I10.2*). As a result, plants, algae, and some species of bacteria evolved photosynthetic apparatus that captures visible and near-infrared radiation. Plants use radiation in the wavelength range of 400–700 nm to drive the endergonic reduction of CO_2 to glucose, with concomitant oxidation of water to O_2 ($\Delta_r G^{\oplus} = +2880$ kJ mol^{-1}; recall that the symbol $^{\oplus}$ means the biological standard state, Section 17.2), in essence the reverse of glycolysis and the citric acid cycle (*Impact I17.1*):

$$6\,CO_2(g) + 6\,H_2O(l) \underset{\text{glycolysis and the citric acid cycle}}{\overset{\text{photosynthesis}}{\rightleftarrows}} C_6H_{12}O_6(s) + 6\,O_2(g)$$

Electrons flow from reductant to oxidant via a series of electrochemical reactions that are coupled to the synthesis of ATP. The process takes place in the *chloroplast*, a special organelle of the plant cell, where chlorophylls *a* and *b* (**7**) and carotenoids (of which β-carotene (**8**) is an example) bind to integral proteins called *light harvesting complexes*, which absorb solar energy and transfer it to protein complexes known as *reaction centres*, where light-induced electron transfer reactions occur. The combination of a light harvesting complex and a reaction centre complex is called a *photosystem*. Plants have two photosystems that drive the reduction of $NADP^+$ (**9**) by water:

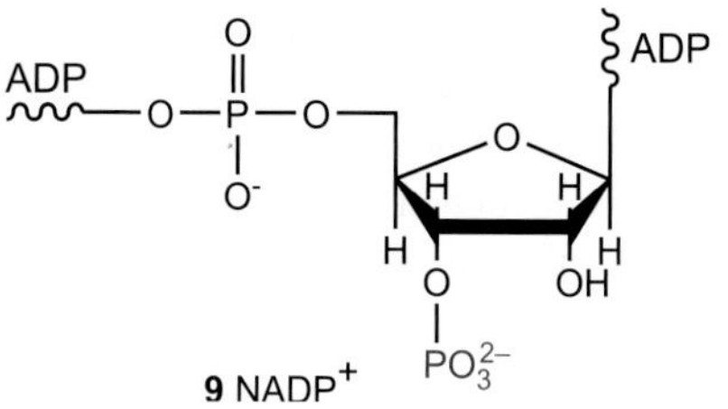

9 $NADP^+$

$$2\,H_2O + 2\,NADP^+ \xrightarrow{\text{light, photosystems I and II}} O_2 + 2\,NADPH + 2\,H^+$$

It is clear that energy from light is required to drive this reaction because, in the dark, $E^{\oplus} = -1.135$ V and $\Delta_r G^{\oplus} = +438.0$ kJ mol^{-1}.

Light harvesting complexes bind large numbers of pigments in order to provide a sufficiently large area for capture of radiation. In photosystems I and II, absorption of a photon raises a chlorophyll or carotenoid molecule to an excited singlet state and within 0.1–5 ps the energy hops to a nearby pigment by the Förster mechanism (Section 19.9). About 100–200 ps later, which corresponds to thousands of hops within the light harvesting complex, more than 90 per cent of the absorbed energy reaches the reaction centre. There, a chlorophyll *a* dimer becomes electronically excited and initiates ultrafast electron transfer reactions. For example, the transfer of an electron from the excited singlet state of P680, the chlorophyll dimer of the photosystem II reaction centre, to its immediate electron acceptor, a phaeophytin *a* molecule (a chlorophyll *a* molecule where the central Mg^{2+} ion is replaced by two protons, which are bound to two of the pyrrole nitrogen atoms in the ring), occurs within 3 ps. Once the excited state of P680 has been quenched efficiently by this first reaction, subsequent steps that lead to the oxidation of water occur more slowly, with reaction times varying from 200 ps to 1 ms. The electrochemical reactions within the photosystem I reaction centre also occur in this time interval.

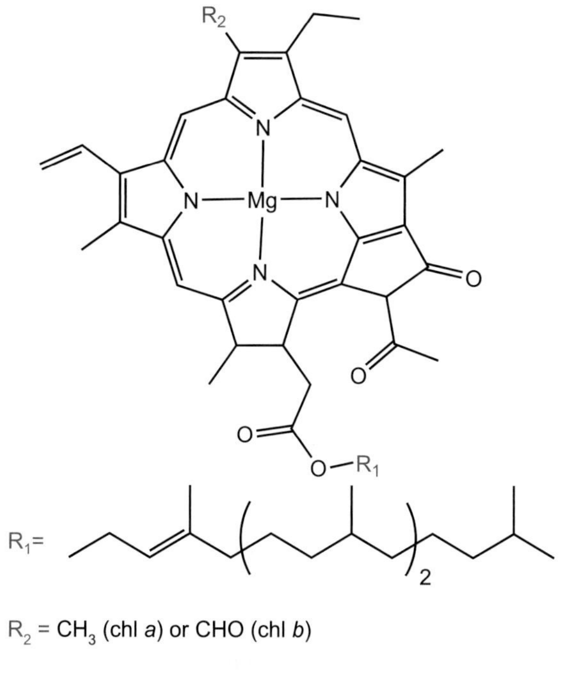

7 Chlorophyll *a* and *b*

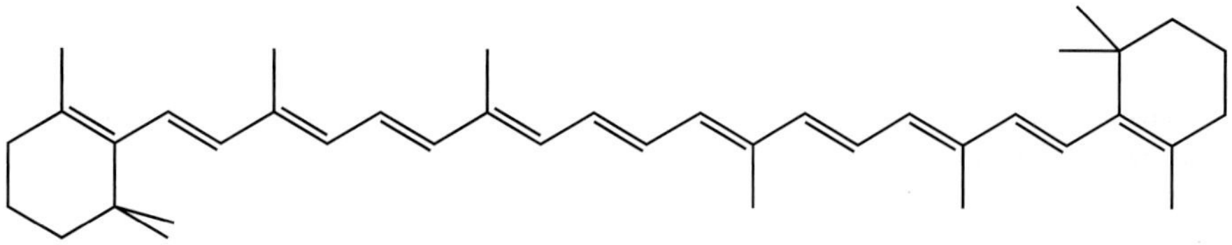

8 β-Carotene

We see that the initial energy and electron transfer events of photosynthesis are under tight kinetic control. Photosynthesis captures solar energy efficiently because the excited singlet state of chlorophyll is quenched rapidly by processes that occur with relaxation times that are much shorter than the fluorescence lifetime, which is typically about 1 ns in organic solvents at room temperature.

Working together, photosystem I and the enzyme ferredoxin: $NADP^+$ oxidoreductase catalyse the light-induced oxidation of $NADP^+$ to NADPH. The electrons required for this process come initially from P700, the chlorophyll dimer of the photosystem I reaction centre, in its excited state. The resulting $P700^+$ is then reduced by the mobile carrier plastocyanin (Pc), a protein in which the bound copper ion can exist in oxidation states +2 and +1. The net reaction is

$$\mathrm{NADP^+ + 2\,Cu^+(Pc) + H^+ \xrightarrow{\text{light, photosystem I}} NADPH + 2\,Cu^{2+}(Pc)}$$

Oxidized plastocyanin accepts electrons from reduced plastoquinone (PQ, **10**). The process is catalysed by the cytochrome b_6f complex, a membrane protein complex that resembles complex III of mitochondria (*Impact I17.3*):

$$\mathrm{PQH_2 + 2\,Cu^{2+}(Pc) \xrightarrow{\text{cyt } b_6f \text{ complex}} PQ + 2\,H^+ + 2\,Cu^+(Pc)}$$
$$E^{\oplus} = +0.370\ \mathrm{V}, \Delta_r G^{\oplus} = -71.4\ \mathrm{kJ\,mol^{-1}}$$

This reaction is sufficiently exergonic to drive the synthesis of ATP in the process known as *photophosphorylation*.

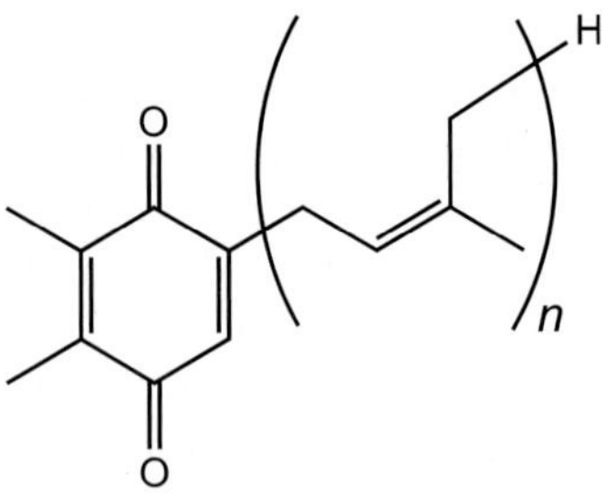

10 Plastoquinone

Plastoquinone is reduced by water in a process catalysed by light and photosystem II. The electrons required for the reduction of plastoquinone come initially from P680 in its excited state. The resulting $P680^+$ is then reduced ultimately by water. The net reaction is

$$\mathrm{H_2O + PQ \xrightarrow{\text{light, photosystem II}} \tfrac{1}{2}O_2 + PQH_2}$$

In this way, plant photosynthesis uses an abundant source of electrons (water) and of energy (the Sun) to drive the endergonic reduction of $NADP^+$, with concomitant synthesis of ATP (Fig. 19.23). Experiments show that, for each molecule of NADPH formed in the chloroplast of green plants, one molecule of ATP is synthesized.

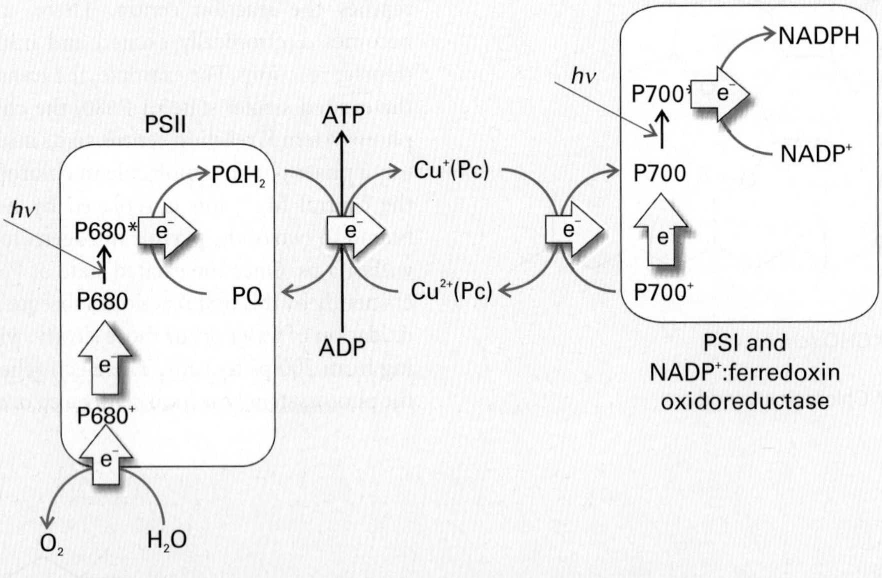

Fig. 19.23 In plant photosynthesis, light-induced electron transfer processes lead to the oxidation of water to O_2 and the reduction of $NADP^+$ to NADPH, with concomitant production of ATP. The energy stored in ATP and NADPH is used to reduce CO_2 to carbohydrate in a separate set of reactions. The scheme summarizes the general patterns of electron flow and does not show all the intermediate electron carriers in photosystems I and II, the cytochrome b_6f complex, and ferredoxin:$NADP^+$ oxidoreductase.

The ATP and NADPH molecules formed by the light-induced electron transfer reactions of plant photosynthesis participate directly in the reduction of CO_2 to glucose in the chloroplast:

$$6\,CO_2 + 12\,NADPH + 12\,ATP + 12\,H^+ \rightarrow C_6H_{12}O_6 + 12\,NADP^+ + 12\,ADP + 12\,P_i + 6\,H_2O$$

In summary, plant photosynthesis uses solar energy to transfer electrons from a poor reductant (water) to carbon dioxide. In the process, high energy molecules (carbohydrates, such as glucose) are synthesized in the cell. Animals feed on the carbohydrates derived from photosynthesis. During aerobic metabolism, the O_2 released by photosynthesis as a waste product is used to oxidize carbohydrates to CO_2, driving biological processes, such as biosynthesis, muscle contraction, cell division, and nerve conduction. Hence, the sustenance of life on Earth depends on a tightly regulated carbon–oxygen cycle that is driven by solar energy.

Checklist of key ideas

☐ 1. The rates of chemical reactions are measured by using techniques that monitor the concentrations of species present in the reaction mixture. Examples include real-time and quenching procedures, flow and stopped-flow techniques, and flash photolysis.

☐ 2. The instantaneous rate of a reaction is the slope of the tangent to the graph of concentration against time (expressed as a positive quantity).

☐ 3. A rate law is an expression for the reaction rate in terms of the concentrations of the species that occur in the overall chemical reaction.

☐ 4. For a rate law of the form $v = k_r[A]^a[B]^b \ldots$, the rate constant is k_r, the order with respect to A is a, and the overall order is $a + b + \cdots$.

☐ 5. An integrated rate law is an expression for the concentration of a reactant or product as a function of time (Table 19.3).

☐ 6. The half-life $t_{1/2}$ of a reaction is the time it takes for the concentration of a species to fall to half its initial value. The time constant τ is the time required for the concentration of a reactant to fall to 1/e of its initial value. For a first-order reaction, $t_{1/2} = (\ln 2)/k_r$ and $\tau = 1/k_r$.

☐ 7. The equilibrium constant for a reaction is equal to the ratio of the forward and reverse rate constants, $K = k_r/k_r'$.

☐ 8. In relaxation methods of kinetic analysis, the equilibrium position of a reaction is first shifted suddenly and then allowed to readjust to the equilibrium composition characteristic of the new conditions.

☐ 9. The mechanism of reaction is the sequence of elementary steps involved in a reaction.

☐ 10. The molecularity of an elementary reaction is the number of molecules coming together to react. An elementary unimolecular reaction has first-order kinetics; an elementary bimolecular reaction has second-order kinetics.

☐ 11. The rate-determining step is the slowest step in a reaction mechanism that controls the rate of the overall reaction.

☐ 12. In the steady-state approximation, it is assumed that the concentrations of all reaction intermediates remain constant and small throughout the reaction.

☐ 13. Provided a reaction has not reached equilibrium, the products of competing reactions are controlled kinetically, with $[P_2]/[P_1] = k_2/k_1$.

☐ 14. Pre-equilibrium is a state in which an intermediate is in equilibrium with the reactants and which arises when the rates of formation of the intermediate and its decay back into reactants are much faster than its rate of formation of products.

☐ 15. The Lindemann–Hinshelwood mechanism of 'unimolecular' reactions accounts for the first-order kinetics of gas-phase reactions.

☐ 16. In stepwise polymerization any two monomers in the reaction mixture can link together at any time and growth of the polymer is not confined to chains that are already forming. The longer a stepwise polymerization proceeds, the higher the average molar mass of the product.

☐ 17. In chain polymerization an activated monomer attacks another monomer and links to it. That unit attacks another monomer, and so on. The slower the initiation of the chain, the higher the average molar mass of the polymer.

☐ 18. The primary quantum yield of a photochemical reaction is the number of reactant molecules producing specified primary products for each photon absorbed; the overall quantum yield is the number of reactant molecules that react for each photon absorbed.

☐ 19. The observed fluorescence lifetime is related to the quantum yield, ϕ_f, and rate constant, k_f, of fluorescence by $\tau_0 = \phi_f/k_f$.

☐ 20. A Stern–Volmer plot is used to analyse the kinetics of fluorescence quenching in solution. It is based on the Stern–Volmer equation, $\phi_{f,0}/\phi_f = 1 + \tau_0 k_Q[Q]$.

☐ 21. Collisional deactivation, electron transfer, and resonance energy transfer are common fluorescence quenching processes. The rate constants of electron and resonance energy transfer decrease with increasing separation between donor and acceptor molecules.

Further information

Further information 19.1 *Förster theory of resonance energy transfer*

From the qualitative description given in Section 19.9, we conclude that resonance energy transfer arises from the interaction between two oscillating dipoles with moments μ_S and μ_Q. From Section 8.3, the energy of the dipole–dipole interaction, $V_{\text{dipole–dipole}}$, is

$$V_{\text{dipole–dipole}} \propto \frac{\mu_S \mu_Q}{R^3}$$

where R is the distance between the dipoles. We saw in *Further information 4.2* that the rate of a transition from a state i to a state f at a radiation frequency ν is proportional to the square modulus of the matrix element of the perturbation between the two states:

$$w_{f \leftarrow i} \propto |H^{(1)}_{fi}|^2$$

For energy transfer, the wavefunctions of the initial and final states may be denoted as $\psi^*_S \psi_Q$ and $\psi_S \psi^*_Q$, respectively, and $H^{(1)}$ may be written from $V_{\text{dipole–dipole}}$. It follows that the rate of energy transfer, w_T, at a fixed distance R is given by

$$w_T \propto \frac{1}{R^6} |\langle \psi_S \psi_{Q^*} | \mu_S \mu_Q | \psi_{S^*} \psi_Q \rangle|^2 = \frac{1}{R^6} |\langle \psi_S | \mu_S | \psi_{S^*} \rangle|^2 |\langle \psi_{Q^*} | \mu_Q | \psi_Q \rangle|^2$$

We have used the fact that the terms related to S are functions of coordinates that are independent of those for the functions related to Q. In the last expression, the integrals are squares of transition dipole moments at the radiation frequency ν, the first corresponding to emission of S* to S and the second to absorption of Q to Q*.

We interpret the expression for w_T as follows. The rate of energy transfer is proportional to R^{-6}, so it decreases sharply with increasing separation between the energy donor and acceptor. Furthermore, the energy transfer rate is optimized when both emission of radiation by S* and absorption of radiation by Q are efficient at the frequency ν. Because the absorption and emission spectra of large molecules in condensed phases are broad, it follows that the energy transfer rate is optimal at radiation frequencies in which the emission spectrum of the donor and the absorption spectrum of the acceptor overlap significantly.

In practice, it is more convenient to measure the efficiency of energy transfer and not the rate itself. In much the same way that we defined the quantum yield as a ratio of rates, we can also define the efficiency of energy transfer, η_T, as the ratio

$$\eta_T = \frac{w_T}{w_T + w_0} \qquad w_0 = (k_f + k_{IC} + k_{ISC})[S^*] \qquad (19.69)$$

where w_0 is the rate of deactivation of S* in the absence of the quencher. The efficiency may be expressed in terms of the experimental fluorescence quantum yields $\phi_{f,0}$ and ϕ_f of the donor in the absence and presence of the acceptor, respectively. To proceed, we use eqn 19.61 to write:

$$\phi_{f,0} = \frac{v_f}{w_0} \qquad \text{and} \qquad \phi_f = \frac{v_f}{w_0 + w_T}$$

where v_f is the rate of fluorescence. Substituting these results into eqn 19.69 gives, after a little algebra, eqn 19.67.

Alternatively, we can express w_0 in terms of the parameter R_0, the characteristic distance at which $w_T = w_0$ for a specified pair of S and Q (Table 19.6). By using $w_T \propto R^{-6}$ and $w_0 \propto R_0^{-6}$, we can rearrange the expression for η_T into eqn 19.68.

Discussion questions

19.1 Consult literature sources and list the observed ranges of timescales during which the following processes occur: radiative decay of excited electronic states, molecular rotational motion, molecular vibrational motion, proton transfer reactions, energy transfer between fluorescent molecules used in FRET analysis, electron transfer events between complex ions in solution, and collisions in liquids.

19.2 Describe the main features, including advantages and disadvantages, of the following experimental methods for determining the rate law of a reaction: the isolation method, the method of initial rates, and fitting data to integrated rate law expressions.

19.3 Distinguish between reaction order and molecularity.

19.4 Distinguish between zeroth-order, first-order, second-order, and pseudofirst-order reactions and illustrate how reaction orders may change under different circumstances.

19.5 Assess the validity of the following statement: the rate-determining step is the slowest step in a reaction mechanism.

19.6 Distinguish between a pre-equilibrium approximation and a steady-state approximation.

19.7 Distinguish between kinetic and thermodynamic control of a reaction. Suggest criteria for expecting one rather than the other.

19.8 Discuss the limitations of the generality of the expression $k_r = k_a k_b[A]/(k_b + k'_a[A])$ for the effective rate constant of a unimolecular reaction according to the Lindemann mechanism.

19.9 Bearing in mind distinctions between the mechanisms of stepwise and chain polymerization, describe ways in which it is possible to control the molar mass of a polymer by manipulating the kinetic parameters of polymerization.

19.10 Distinguish between the primary quantum yield and overall quantum yield of a chemical reaction. Describe an experimental procedure for the determination of the quantum yield.

19.11 Discuss experimental procedures that make it possible to differentiate between quenching by energy transfer, collisions, or electron transfer.

19.12 Discuss the factors that govern the rates of photo-induced electron transfer according to Marcus theory and that govern the rates of resonance energy transfer according to Förster theory. Can you find similarities between the two theories?

Exercises

19.1(a) Predict how the total pressure varies during the gas-phase reaction $2\,ICl(g) + H_2(g) \rightarrow I_2(g) + 2\,HCl(g)$ in a constant-volume container.

19.1(b) Predict how the total pressure varies during the gas-phase reaction $N_2(g) + 3\,H_2(g) \rightarrow 2\,NH_3(g)$ in a constant-volume container.

19.2(a) The rate of the reaction $A + 2\,B \rightarrow 3\,C + D$ was reported as $2.7\ mol\ dm^{-3}\ s^{-1}$. State the rates of formation and consumption of the participants.

19.2(b) The rate of the reaction $A + 3\,B \rightarrow C + 2\,D$ was reported as $2.7\ mol\ dm^{-3}\ s^{-1}$. State the rates of formation and consumption of the participants.

19.3(a) The rate of formation of C in the reaction $2\,A + B \rightarrow 2\,C + 3\,D$ is $2.7\ mol\ dm^{-3}\ s^{-1}$. State the reaction rate, and the rates of formation or consumption of A, B, and D.

19.3(b) The rate of consumption of B in the reaction $A + 3\,B \rightarrow C + 2\,D$ is $2.7\ mol\ dm^{-3}\ s^{-1}$. State the reaction rate, and the rates of formation or consumption of A, C, and D.

19.4(a) The rate law for the reaction in Exercise 19.2a was found to be $v = k_r[A][B]$. What are the units of k_r? Express the rate law in terms of the rates of formation and consumption of (a) A, (b) C.

19.4(b) The rate law for the reaction in Exercise 19.2b was found to be $v = k_r[A][B]^2$. What are the units of k_r? Express the rate law in terms of the rates of formation and consumption of (a) A, (b) C.

19.5(a) The rate law for the reaction in Exercise 19.3a was reported as $d[C]/dt = k_r[A][B][C]$. Express the rate law in terms of the reaction rate; what are the units for k_r?

19.5(b) The rate law for the reaction in Exercise 19.3b was reported as $d[C]/dt = k_r[A][B][C]^{-1}$. Express the rate law in terms of the reaction rate; what are the units for k_r?

19.6(a) If the rate laws are expressed with (a) concentrations in moles per decimetre cubed, (b) pressures in kilopascals, what are the units of the second-order and third-order rate constants?

19.6(b) If the rate laws are expressed with (a) concentrations in molecules per metre cubed, (b) pressures in pascals, what are the units of the second-order and third-order rate constants?

19.7(a) At 518°C, the rate of decomposition of a sample of gaseous acetaldehyde, initially at a pressure of 363 Torr, was $1.07\ Torr\ s^{-1}$ when 5.0 per cent had reacted and $0.76\ Torr\ s^{-1}$ when 20.0 per cent had reacted. Determine the order of the reaction.

19.7(b) At 400 K, the rate of decomposition of a gaseous compound initially at a pressure of 12.6 kPa, was $9.71\ Pa\ s^{-1}$ when 10.0 per cent had reacted and $7.67\ Pa\ s^{-1}$ when 20.0 per cent had reacted. Determine the order of the reaction.

19.8(a) At 518°C, the half-life for the decomposition of a sample of gaseous acetaldehyde (ethanal) initially at 363 Torr was 410 s. When the pressure was 169 Torr, the half-life was 880 s. Determine the order of the reaction.

19.8(b) At 400 K, the half-life for the decomposition of a sample of a gaseous compound initially at 55.5 kPa was 340 s. When the pressure was 28.9 kPa, the half-life was 178 s. Determine the order of the reaction.

19.9(a) The rate constant for the first-order decomposition of N_2O_5 in the reaction $2\,N_2O_5(g) \rightarrow 4\,NO_2(g) + O_2(g)$ is $k_r = 3.38 \times 10^{-5}\ s^{-1}$ at 25°C. What is the half-life of N_2O_5? What will be the pressure, initially 500 Torr, after (a) 50 s, (b) 20 min after initiation of the reaction?

19.9(b) The rate constant for the first-order decomposition of a compound A in the reaction $2\,A \rightarrow P$ is $k_r = 3.56 \times 10^{-7}\ s^{-1}$ at 25°C. What is the half-life of A? What will be the pressure, initially 33.0 kPa after (a) 50 s, (b) 20 min after initiation of the reaction?

19.10(a) A second-order reaction of the type $A + B \rightarrow P$ was carried out in a solution that was initially $0.075\ mol\ dm^{-3}$ in A and $0.050\ mol\ dm^{-3}$ in B. After 1.0 h the concentration of B had fallen to $0.020\ mol\ dm^{-3}$. (a) Calculate the rate constant. (b) What is the half-life of the reactants?

19.10(b) A second-order reaction of the type $A + 2\,B \rightarrow P$ was carried out in a solution that was initially $0.050\ mol\ dm^{-3}$ in A and $0.030\ mol\ dm^{-3}$ in B. After 1.0 h the concentration of B had fallen to $0.010\ mol\ dm^{-3}$. (a) Calculate the rate constant. (b) What is the half-life of the reactants?

19.11(a) The second-order rate constant for the reaction

$$CH_3COOC_2H_5(aq) + OH^-(aq) \rightarrow CH_3CO_2^-(aq) + CH_3CH_2OH(aq)$$

is $0.11\ dm^3\ mol^{-1}\ s^{-1}$. What is the concentration of ester ($CH_3COOC_2H_5$) after (a) 20 s, (b) 15 min when ethyl acetate is added to sodium hydroxide so that the initial concentrations are $[NaOH] = 0.060\ mol\ dm^{-3}$ and $[CH_3COOC_2H_5] = 0.110\ mol\ dm^{-3}$?

19.11(b) The second-order rate constant for the reaction $A + 2\,B \rightarrow C + D$ is $0.34\ dm^3\ mol^{-1}\ s^{-1}$. What is the concentration of C after (a) 20 s, (b) 15 min when the reactants are mixed with initial concentrations of $[A] = 0.027\ mol\ dm^{-3}$ and $[B] = 0.130\ mol\ dm^{-3}$?

19.12(a) A reaction $2\,A \rightarrow P$ has a second-order rate law with $k_r = 4.30 \times 10^{-4}\ dm^3\ mol^{-1}\ s^{-1}$. Calculate the time required for the concentration of A to change from $0.210\ mol\ dm^{-3}$ to $0.010\ mol\ dm^{-3}$.

19.12(b) A reaction $2\,A \rightarrow P$ has a third-order rate law with $k_r = 6.50 \times 10^{-4}\ dm^6\ mol^{-2}\ s^{-1}$. Calculate the time required for the concentration of A to change from $0.067\ mol\ dm^{-3}$ to $0.015\ mol\ dm^{-3}$.

19.13(a) The equilibrium $NH_3(aq) + H_2O(l) \rightleftharpoons NH_4^+(aq) + OH^-(aq)$ at 25°C is subjected to a temperature jump which slightly increased the concentration of $NH_4^+(aq)$ and $OH^-(aq)$. The measured relaxation time is 7.61 ns. The equilibrium constant for the system is 1.78×10^{-5} at 25°C, and the equilibrium concentration of $NH_3(aq)$ is $0.15\ mol\ dm^{-3}$. Calculate the rate constants for the forward and reversed steps.

19.13(b) The equilibrium $A \rightleftharpoons B + C$ at 25°C is subjected to a temperature jump which slightly increases the concentrations of B and C. The measured relaxation time is 3.0 μs. The equilibrium constant for the system is 2.0×10^{-16} at 25°C, and the equilibrium concentrations of B and C at 25°C are both $2.0 \times 10^{-4}\ mol\ dm^{-3}$. Calculate the rate constants for the forward and reverse steps.

19.14(a) The mechanism for the reaction of A_2 with B

$$A_2 \rightleftarrows A + A\ \text{(fast)}$$

$$A + B \rightarrow P\ \text{(slow)}$$

involves an intermediate A. Deduce the rate law for the reaction in two ways by (i) assuming a pre-equilibrium and (ii) making a steady-state approximation.

19.14(b) The reaction mechanism for renaturation of a double helix from its strands A and B:

$$A + B \rightleftharpoons \text{unstable helix (fast)}$$

$$\text{Unstable helix} \rightarrow \text{stable double helix (slow)}$$

involves an intermediate. Deduce the rate law for the reaction in two ways by (i) assuming a pre-equilibrium and (ii) making a steady-state approximation.

19.15(a) The effective rate constant for a gaseous reaction that has a Lindemann–Hinshelwood mechanism is $2.50 \times 10^{-4}\ s^{-1}$ at 1.30 kPa and $2.10 \times 10^{-5}\ s^{-1}$ at 12 Pa. Calculate the rate constant for the activation step in the mechanism.

19.15(b) The effective rate constant for a gaseous reaction that has a Lindemann–Hinshelwood mechanism is $1.7 \times 10^{-3}\ s^{-1}$ at 1.09 kPa and $2.2 \times 10^{-4}\ s^{-1}$ at 25 Pa. Calculate the rate constant for the activation step in the mechanism.

19.16(a) Calculate the fraction condensed and the degree of polymerization at $t = 5.00$ h of a polymer formed by a stepwise process with $k_r = 1.39\ dm^3\ mol^{-1}\ s^{-1}$ and an initial monomer concentration of $1.00 \times 10^{-2}\ mol\ dm^{-3}$.

19.16(b) Calculate the fraction condensed and the degree of polymerization at $t = 10.00$ h of a polymer formed by a stepwise process with $k_r = 2.80 \times 10^{-2}\ dm^3\ mol^{-1}\ s^{-1}$ and an initial monomer concentration of $5.00 \times 10^{-2}\ mol\ dm^{-3}$.

19.17(a) Consider a polymer formed by a chain process. By how much does the kinetic chain length change if the concentration of initiator increases by a factor of 3.6 and the concentration of monomer decreases by a factor of 4.2?

19.17(b) Consider a polymer formed by a chain process. By how much does the kinetic chain length change if the concentration of initiator decreases by a factor of 10.0 and the concentration of monomer increases by a factor of 5.0?

19.18(a) In a photochemical reaction $A \rightarrow 2\,B + C$, the quantum yield with 500 nm light is 2.1×10^2 mol einstein^{-1} (1 einstein = 1 mol photons). After exposure of 300 mmol of A to the light, 2.28 mmol of B is formed. How many photons were absorbed by A?

19.18(b) In a photochemical reaction $A \rightarrow B + C$, the quantum yield with 500 nm light is 1.2×10^2 mol einstein^{-1}. After exposure of 200 mmol A to the light, 1.77 mmol B is formed. How many photons were absorbed by A?

19.19(a) In an experiment to measure the quantum yield of a photochemical reaction, the absorbing substance was exposed to 490 nm light from a 100 W source for 45 min. The intensity of the transmitted light was 40 per cent of the intensity of the incident light. As a result of irradiation, 0.344 mol of the absorbing substance decomposed. Determine the quantum yield.

19.19(b) In an experiment to measure the quantum yield of a photochemical reaction, the absorbing substance was exposed to 320 nm radiation from a 87.5 W source for 28.0 min. The intensity of the transmitted light was 0.257 that of the incident light. As a result of irradiation, 0.324 mol of the absorbing substance decomposed. Determine the quantum yield.

19.20(a) Consider the quenching of an organic fluorescent species with $\tau_0 = 6.0$ ns by a d-metal ion with $k_Q = 3.0 \times 10^8\ dm^3\ mol^{-1}\ s^{-1}$. Predict the concentration of quencher required to decrease the fluorescence intensity of the organic species to 50 per cent of the unquenched value.

19.20(b) Consider the quenching of an organic fluorescent species with $\tau_0 = 3.5$ ns by a d-metal ion with $k_Q = 2.5 \times 10^9\ dm^3\ mol^{-1}\ s^{-1}$. Predict the concentration of quencher required to decrease the fluorescence intensity of the organic species to 75 per cent of the unquenched value.

19.21(a) An amino acid on the surface of a protein was labelled covalently with 1,5-I-AEDANS and another was labelled covalently with FITC. The fluorescence quantum yield of 1,5-I-AEDANS decreased by 10 per cent due to quenching by FITC. What is the distance between the amino acids? *Hint*. See Table 19.6.

19.21(b) An amino acid on the surface of an enzyme was labelled covalently with 1,5-I-AEDANS and it is known that the active site contains a tryptophan residue. The fluorescence quantum yield of tryptophan decreased by 15 per cent due to quenching by 1,5-IAEDANS. What is the distance between the active site and the surface of the enzyme? *Hint*. See Table 19.6.

Problems*

Numerical problems

19.1 The data below apply to the formation of urea from ammonium cyanate, $NH_4CNO \rightarrow NH_2CONH_2$. Initially 22.9 g of ammonium cyanate was dissolved in enough water to prepare 1.00 dm^3 of solution. Determine the order of the reaction, the rate constant, and the mass of ammonium cyanate left after 300 min.

t/min	0	20.0	50.0	65.0	150
m(urea)/g	0	7.0	12.1	13.8	17.7

19.2 The data below apply to the reaction, $(CH_3)_3CBr + H_2O \rightarrow (CH_3)_3COH + HBr$. Determine the order of the reaction, the rate constant, and the molar concentration of $(CH_3)_3CBr$ after 43.8 h.

t/h	0	3.15	6.20	10.00	18.30	30.80
$[(CH_3)_3CBr]/(10^{-2}\ mol\ dm^{-3})$	10.39	8.96	7.76	6.39	3.53	2.07

19.3 The thermal decomposition of an organic nitrile produced the following data:

$t/(10^3\ s)$	0	2.00	4.00	6.00	8.00	10.00	12.00
[nitrile]/(mol dm^{-3})	1.50	1.26	1.07	0.92	0.81	0.72	0.65

Determine the order of the reaction and the rate constant.

19.4 The following data have been obtained for the decomposition of $N_2O_5(g)$ at 67°C according to the reaction $2\,N_2O_5(g) \rightarrow 4\,NO_2(g) + O_2(g)$. Determine the order of the reaction, the rate constant, and the half-life. It is not necessary to obtain the result graphically; you may do a calculation using estimates of the rates of change of concentration.

t/min	0	1	2	3	4	5
$[N_2O_5]/(mol\ dm^{-3})$	1.000	0.705	0.497	0.349	0.246	0.173

* Problems denoted with the symbol ‡ were supplied by Charles Trapp, Carmen Giunta, and Marshall Cady.

19.5 The gas-phase decomposition of acetic acid at 1189 K proceeds by way of two parallel reactions:

(1) $CH_3COOH \rightarrow CH_4 + CO_2$ $\quad k_1 = 3.74\ s^{-1}$

(2) $CH_3COOH \rightarrow CH_2CO + H_2O$ $\quad k_2 = 4.65\ s^{-1}$

What is the maximum percentage yield of the ketene CH_2CO obtainable at this temperature?

19.6 Sucrose is readily hydrolysed to glucose and fructose in acidic solution. The hydrolysis is often monitored by measuring the angle of rotation of plane-polarized light passing through the solution. From the angle of rotation the concentration of sucrose can be determined. An experiment on the hydrolysis of sucrose in 0.50 M HCl(aq) produced the following data:

t/min	0	14	39	60	80	110	140	170	210
[sucrose]/(mol dm^{-3})	0.316	0.300	0.274	0.256	0.238	0.211	0.190	0.170	0.146

Determine the rate constant of the reaction and the half-life of a sucrose molecule.

19.7 The composition of a liquid phase reaction 2 A → B was followed by a spectrophotometric method with the following results:

t/min	0	10	20	30	40	∞
[B]/(mol dm^{-3})	0	0.089	0.153	0.200	0.230	0.312

Determine the order of the reaction and its rate constant.

19.8 The ClO radical decays rapidly by way of the reaction, $2\,ClO \rightarrow Cl_2 + O_2$. The following data have been obtained:

$t/(10^{-3}$ s)	0.12	0.62	0.96	1.60	3.20	4.00	5.75
[ClO]/(10^{-6} mol dm^{-3})	8.49	8.09	7.10	5.79	5.20	4.77	3.95

Determine the rate constant of the reaction and the half-life of a ClO radical.

19.9 Cyclopropane isomerizes into propene when heated to 500°C in the gas phase. The extent of conversion for various initial pressures has been followed by gas chromatography by allowing the reaction to proceed for a time with various initial pressures:

p_0/Torr	200	200	400	400	600	600
t/s	100	200	100	200	100	200
p/Torr	186	173	373	347	559	520

where p_0 is the initial pressure and p is the final pressure of cyclopropane. What are the order and rate constant for the reaction under these conditions?

19.10 The addition of hydrogen halides to alkenes has played a fundamental role in the investigation of organic reaction mechanisms. In one study (M.J. Haugh and D.R. Dalton, *J. Am. Chem. Soc.* **97**, 5674 (1975)), high pressures of hydrogen chloride (up to 25 atm) and propene (up to 5 atm) were examined over a range of temperatures and the amount of 2-chloropropane formed was determined by NMR. Show that, if the reaction A + B → P proceeds for a short time δt, the concentration of product follows $[P]/[A] = k_r[A]^{m-1}[B]^n\delta t$ if the reaction is mth-order in A and nth-order in B. In a series of runs the ratio of [chloropropane] to [propene] was independent of [propene] but the ratio of [chloropropane] to [HCl] for constant amounts of propene depended on [HCl]. For $\delta t \approx 100$ h (which is short on the timescale of the reaction) the latter ratio rose from zero to 0.05, 0.03, 0.01 for p(HCl) = 10 atm, 7.5 atm, 5.0 atm. What are the orders of the reaction with respect to each reactant?

19.11 Use mathematical software or an electronic spreadsheet to examine the time dependence of [I] in the reaction mechanism A → I → P (k_a, k_b). In all of the following calculations, use $[A]_0 = 1$ mol dm^{-3} and a time range of 0 to 5 s. (a) Plot [I] against t for $k_a = 10\ s^{-1}$ and $k_b = 1\ s^{-1}$. (b) Increase the ratio k_b/k_a steadily by decreasing the value of k_a and examine the plot of [I] against t at each turn. What approximation about d[I]/dt becomes increasingly valid?

19.12 Show that the following mechanism can account for the rate law of the reaction in Problem 19.10:

$HCl + HCl \rightleftharpoons (HCl)_2$ $\quad K_1$

$HCl + CH_3CH{=}CH_2 \rightleftharpoons$ complex $\quad K_2$

$(HCl)_2$ + complex $\rightarrow CH_3CHClCH_3 + 2\,HCl$ $\quad k_r$ (slow)

What further tests could you apply to verify this mechanism?

19.13 Consider the dimerization $2\,A \rightleftharpoons A_2$, with forward rate constant k_a and backward rate constant k_a'. (a) Derive the following expression for the relaxation time in terms of the total concentration of monomer, $[A]_{tot} = [A] + 2[A_2]$:

$$\frac{1}{\tau^2} = k_a'^2 + 8k_ak_a'[A]_{tot}$$

(b) Describe the computational procedures that lead to the determination of the rate constants k_a and k_a' from measurements of τ for different values of $[A]_{tot}$. (c) Use the data provided below and the procedure you outlined in part (b) to calculate the rate constants k_a and k_a', and the equilibrium constant K for formation of hydrogen-bonded dimers of 2-pyridone:

[P]/(mol dm^{-3})	0.500	0.352	0.251	0.151	0.101
τ/ns	2.3	2.7	3.3	4.0	5.3

19.14 In Problem 19.9 the isomerization of cyclopropane over a limited pressure range was examined. If the Lindemann mechanism of first-order reactions is to be tested we also need data at low pressures. These have been obtained (H.O. Pritchard *et al.*, *Proc. R. Soc.* **A217**, 563 (1953)):

p/Torr	84.1	11.0	2.89	0.569	0.120	0.067
$10^4\ k_r/s^{-1}$	2.98	2.23	1.54	0.857	0.392	0.303

Test the Lindemann theory with these data.

19.15 Dansyl chloride, which absorbs maximally at 330 nm and fluoresces maximally at 510 nm, can be used to label amino acids in fluorescence microscopy and FRET studies. Tabulated below is the variation of the fluorescence intensity of an aqueous solution of dansyl chloride with time after excitation by a short laser pulse (with I_0 the initial fluorescence intensity). The ratio of intensities is equal to the ratio of the rates of photon emission.

t/ns	5.0	10.0	15.0	20.0
I_f/I_0	0.45	0.21	0.11	0.05

(a) Calculate the observed fluorescence lifetime of dansyl chloride in water. (b) The fluorescence quantum yield of dansyl chloride in water is 0.70. What is the fluorescence rate constant?

19.16 When benzophenone is illuminated with ultraviolet light it is excited into a singlet state. This singlet changes rapidly into a triplet, which phosphoresces. Triethylamine acts as a quencher for the triplet. In an experiment in methanol as solvent, the phosphorescence intensity varied with amine concentration as shown below. A time-resolved laser spectroscopy experiment had also shown that the half-life of the fluorescence in the absence of quencher is 29 μs. What is the value of k_Q?

$[Q]/(\text{mol dm}^{-3})$	0.0010	0.0050	0.0100
$I_f/(\text{arbitrary units})$	0.41	0.25	0.16

19.17 An electronically excited state of Hg can be quenched by N_2 according to

$$Hg^* (g) + N_2 (g, v = 0) \rightarrow Hg (g) + N_2 (g, v = 1)$$

in which energy transfer from Hg^* excites N_2 vibrationally. Fluorescence lifetime measurements of samples of Hg with and without N_2 present are summarized below ($T = 300$ K):

$p_{N_2} = 0.0$ atm

Relative fluorescence intensity	1.000	0.606	0.360	0.22	0.135
$t/\mu s$	0.0	5.0	10.0	15.0	20.0

$p_{N_2} = 9.74 \times 10^{-4}$ atm

Relative fluorescence intensity	1.000	0.585	0.342	0.200	0.117
$t/\mu s$	0.0	3.0	6.0	9.0	12.0

You may assume that all gases are perfect. Determine the rate constant for the energy transfer process.

19.18 The Förster theory of resonance energy transfer and the basis for the FRET technique can be tested by performing fluorescence measurements on a series of compounds in which an energy donor and an energy acceptor are covalently linked by a rigid molecular linker of variable and known length. L. Stryer and R.P. Haugland (*Proc. Natl. Acad. Sci. USA* **58**, 719 (1967)) collected the following data on a family of compounds with the general composition dansyl-(L-prolyl)$_n$-naphthyl, in which the distance R between the naphthyl donor and the dansyl acceptor was varied from 1.2 nm to 4.6 nm by increasing the number of prolyl units in the linker:

R/nm	1.2	1.5	1.8	2.8	3.1	3.4	3.7	4.0	4.3	4.6
$1 - \eta_T$	0.99	0.94	0.87	0.82	0.74	0.65	0.40	0.28	0.24	0.16

Are the data described adequately by eqn 19.68? If so, what is the value of R_0 for the naphthyl–dansyl pair?

Theoretical problems

19.19 Show that $t_{1/2}$ is given by eqn 19.15 for a reaction that is nth order in A. Then deduce an expression for the time it takes for the concentration of a substance to fall to one-third the initial value in an nth-order reaction.

19.20 The equilibrium $A \rightleftharpoons B$ is first order in both directions. Derive an expression for the concentration of A as a function of time when the initial molar concentrations of A and B are $[A]_0$ and $[B]_0$. What is the final composition of the system?

19.21 Derive an integrated expression for a second-order rate law $v = k_r[A][B]$ for a reaction of stoichiometry $2\,A + 3\,B \rightarrow P$.

19.22 Derive the integrated form of a third-order rate law $v = k_r[A]^2[B]$ in which the stoichiometry is $2\,A + B \rightarrow P$ and the reactants are initially present in (a) their stoichiometric proportions, (b) with B present initially in twice the amount.

19.23 Set up the rate equations for the reaction mechanism:

$$A \overset{k_a, k'_a}{\rightleftharpoons} B \overset{k_b, k'_b}{\rightleftharpoons} C$$

Show that the mechanism is equivalent to

$$A \overset{k_r, k'_r}{\rightleftharpoons} C$$

under specified circumstances.

19.24 Show that the ratio $t_{1/2}/t_{3/4}$, where $t_{1/2}$ is the half-life and $t_{3/4}$ is the time for the concentration of A to decrease to $\frac{3}{4}$ of its initial value (implying that $t_{3/4} < t_{1/2}$), can be written as a function of n alone, and can therefore be used as a rapid assessment of the order of a reaction.

19.25 Derive an equation for the steady-state rate of the sequence of reactions $A \rightleftharpoons B \rightleftharpoons C \rightleftharpoons D$, with [A] maintained at a fixed value and the product D removed as soon as it is formed.

19.26 Consider the dimerization $2\,A \rightleftharpoons A_2$ with forward rate constant k_r and backward rate constant k'_r. Show that the relaxation time is:

$$\tau = \frac{1}{k'_r + 4k_r[A]_{eq}}$$

19.27 Express the root mean square deviation $\{\langle M^2\rangle - \langle M\rangle^2\}^{1/2}$ of the molar mass of a condensation polymer in terms of the fraction p, and deduce its time dependence.

19.28 Calculate the ratio of the mean cube molar mass to the mean square molar mass in terms of (a) the fraction p, (b) the chain length.

19.29 Calculate the average polymer length in a polymer produced by a chain mechanism in which termination occurs by a disproportionation reaction of the form $M\cdot + \cdot M \rightarrow M + {:}M$.

19.30 Derive an expression for the time dependence of the degree of polymerization for a stepwise polymerization in which the reaction is acid-catalysed by the —COOH acid functional group. The rate law is $d[A]/dt = -k_r[A]^2[OH]$.

19.31 Conventional equilibrium considerations do not apply when a reaction is being driven by light absorption. Thus the steady-state concentrations of products and reactants might differ significantly from equilibrium values. For instance, suppose the reaction $A \rightarrow B$ is driven by light absorption, and that its rate is I_a, but that the reverse reaction $B \rightarrow A$ is bimolecular and second order with a rate $k_r[B]^2$. What is the stationary state concentration of B? Why does this 'photostationary state' differ from the equilibrium state?

19.32 The photochemical chlorination of chloroform in the gas has been found to follow the rate law $d[CCl_4]/dt = k_r[Cl_2]^{1/2}I_a^{1/2}$. Devise a mechanism that leads to this rate law when the chlorine pressure is high.

Applications to: biochemistry and environmental science

19.33 Pharmacokinetics is the study of the rates of absorption and elimination of drugs by organisms. In most cases, elimination is slower than absorption and is a more important determinant of availability of a drug for binding to its target. A drug can be eliminated by many mechanisms, such as metabolism in the liver, intestine, or kidney followed by excretion of breakdown products through urine or faeces. As an example of pharmacokinetic analysis, consider the elimination of beta adrenergic blocking agents (beta blockers), drugs used in the treatment of hypertension. After intravenous administration of a beta blocker, the blood plasma of a patient was analysed for remaining drug and the data are shown below, where c is the drug concentration measured at a time t after the injection.

t/min	30	60	120	150	240	360	480
$c/(\text{ng cm}^{-3})$	699	622	413	292	152	60	24

(a) Is removal of the drug a first- or second-order process? (b) Calculate the rate constant and half-life of the process. *Comment.* An essential aspect of drug development is the optimization of the half-life of elimination, which needs to be long enough to allow the drug to find and act on its target organ but not so long that harmful side-effects become important.

19.34 Consider a mechanism for the helix–coil transition in polypeptides that begins in the middle of the chain:

$$hhhh\ldots \rightleftarrows hchh\ldots$$
$$hchh\ldots \rightleftarrows cccc\ldots$$

The first conversion from h to c, also called a nucleation step, is relatively slow, so neither step may be rate-determining. (a) Set up the rate equations for this mechanism. (b) Apply the steady-state approximation and show that, under these circumstances, the mechanism is equivalent to $hhhh\ldots \rightleftarrows cccc\ldots$.

19.35‡ The oxidation of HSO_3^- by O_2 in aqueous solution is a reaction of importance to the processes of acid rain formation and flue gas desulfurization. R.E. Connick *et al.* (*Inorg. Chem.* **34**, 4543 (1995)) report that the reaction $2\,HSO_3^- + O_2 \rightarrow 2\,SO_4^{2-} + 2\,H^+$ follows the rate law $v = k_r[HSO_3^-]^2[H^+]^2$. Given pH = 5.6 and an oxygen molar concentration of 2.4×10^{-4} mol dm^{-3} (both presumed constant), an initial HSO_3^- molar concentration of 5×10^{-5} mol dm^{-3}, and a rate constant of 3.6×10^{6} dm^9 mol^{-3} s^{-1}, what is the initial rate of reaction? How long would it take for HSO_3^- to reach half its initial concentration?

19.36 In light-harvesting complexes, the fluorescence of a chlorophyll molecule is quenched by nearby chlorophyll molecules. Given that for a pair of chlorophyll *a* molecules $R_0 = 5.6$ nm, by what distance should two chlorophyll *a* molecules be separated to shorten the fluorescence lifetime from 1 ns (a typical value for monomeric chlorophyll *a* in organic solvents) to 10 ps?

19.37‡ Ultraviolet radiation photolyses O_3 to O_2 and O. Determine the rate at which ozone is consumed by 305 nm radiation in a layer of the stratosphere of thickness 1 km. The quantum yield is 0.94 at 220 K, the concentration about 8×10^{-9} mol dm^{-3}, the molar absorption coefficient 260 dm^3 mol^{-1} cm^{-1}, and the flux of 305 nm radiation about 1×10^{14} photons cm^{-2} s^{-1}. (Data from W.B. DeMore *et al.*, *Chemical kinetics and photochemical data for use in stratospheric modeling: Evaluation Number 11*, JPL Publication 94–26 (1994).)

20 Molecular reaction dynamics

The rates of chemical reactions typically increase with temperature according to the Arrhenius equation. The simplest quantitative account of reaction rates and their temperature dependence is in terms of collision theory, which can be used only for the discussion of reactions between simple species in the gas phase. Reactions in solution, in general, are classified as diffusion controlled or activation controlled. In transition state theory, it is assumed that the reactant molecules form a complex that can be discussed in terms of the population of its energy levels. Transition state theory inspires a thermodynamic approach to reaction rates, in which the rate constant is expressed in terms of thermodynamic parameters. This approach is useful for parametrizing the rates of reactions in solution. For electron transfer reactions in particular, the rate depends on the distance between electron donor and acceptor, the standard Gibbs energy of reaction, and the energy associated with molecular rearrangements that accompany the transfer of charge. The highest level of sophistication is in terms of potential energy surfaces and the motion of molecules through these surfaces. As we shall see, such an approach gives an intimate picture of the events that occur when reactions occur and is open to experimental study.

Now we are at the heart of chemistry. Here we examine the details of what happens to molecules at the climax of reactions. Extensive changes of structure are taking place and energies the size of dissociation energies are being redistributed among bonds: old bonds are being ripped apart and new bonds are being formed.

The calculation of the rates of chemical processes from first principles is very difficult. Nevertheless, like so many intricate problems, the broad features can be established quite simply. Only when we enquire more deeply do the complications emerge. In this chapter we look at several approaches to the calculation of a rate constant for elementary bimolecular processes, ranging from electron transfer to chemical reactions involving bond breakage and formation. Although a great deal of information can be obtained from gas-phase reactions, many reactions of interest take place in solution, and we shall also see to what extent their rates can be predicted.

The temperature dependence of reaction rates

Chemical reactions usually speed up as the temperature is increased because the rate constants of most reactions increase as the temperature is raised. Throughout the chapter we shall see that the dependence of the rate constant on temperature is captured by various of the theoretical approaches to computing rate constants.

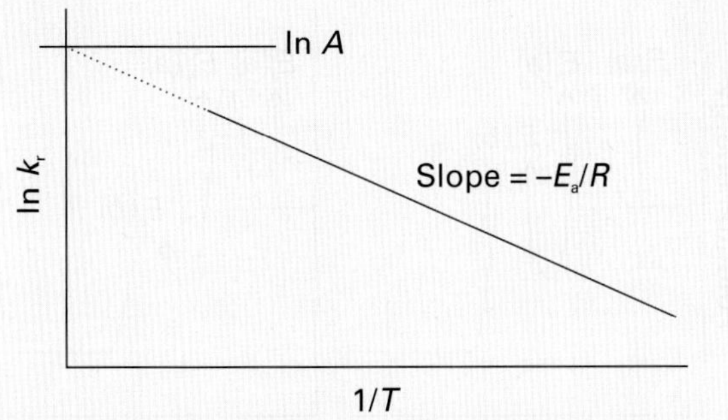

Fig. 20.1 A plot of $\ln k_r$ against $1/T$ is a straight line when the reaction follows the behaviour described by the Arrhenius equation (eqn 20.1). The slope gives $-E_a/R$ and the intercept at $1/T = 0$ gives $\ln A$.

20.1 The Arrhenius equation

It is found experimentally for many reactions that a plot of $\ln k_r$ against $1/T$ gives a straight line. This behaviour is normally expressed mathematically by introducing two parameters, one representing the intercept and the other the slope of the straight line, and writing the **Arrhenius equation**

$$\ln k_r = \ln A - \frac{E_a}{RT} \tag{20.1a}$$

This equation is more commonly written in the form

$$k_r = Ae^{-E_a/RT} \tag{20.1b}$$

The parameter A, which corresponds to the intercept of the line at $1/T = 0$ (at infinite temperature, Fig. 20.1), is called the **pre-exponential factor** or the 'frequency factor'. The parameter E_a, which is obtained from the slope of the line ($-E_a/R$), is called the **activation energy**. Collectively the two quantities are called the **Arrhenius parameters** (Table 20.1).

Synoptic table 20.1* Arrhenius parameters

(1) First-order reactions	$A/\mathrm{s^{-1}}$	$E_a/(\mathrm{kJ\ mol^{-1}})$
$CH_3NC \rightarrow CH_3CN$	3.98×10^{13}	160
$2\,N_2O_5 \rightarrow 4\,NO_2 + O_2$	4.94×10^{13}	103.4
(2) Second-order reactions	$A/(\mathrm{dm^3\ mol^{-1}\ s^{-1}})$	$E_a/(\mathrm{kJ\ mol^{-1}})$
$OH + H_2 \rightarrow H_2O + H$	8.0×10^{10}	42
$NaC_2H_5O + CH_3I$ in ethanol	2.42×10^{11}	81.6

* More values are given in the *Data section*.

Example 20.1 *Determining the Arrhenius parameters*

The rate of the second-order decomposition of acetaldehyde (ethanal, CH_3CHO) was measured over the temperature range 700–1000 K, and the rate constants are reported below. Find E_a and A.

T/K	700	730	760	790
$k_r/(\mathrm{dm^3\ mol^{-1}\ s^{-1}})$	0.011	0.035	0.105	0.343
T/K	810	840	910	1000
$k_r/(\mathrm{dm^3\ mol^{-1}\ s^{-1}})$	0.789	2.17	20.0	145

Method According to eqn 20.1a, the data can be analysed by plotting $\ln(k_r/\mathrm{dm^3\ mol^{-1}\ s^{-1}})$ against $1/(T/\mathrm{K})$, or more conveniently $(10^3\ \mathrm{K})/T$, and getting a straight line. We obtain the activation energy from the slope which equals $-E_a/R$; the intercept at $T = 0$ is $\ln(A/\mathrm{dm^3\ mol^{-1}\ s^{-1}})$.

Answer We draw up the following table:

$(10^3\ \mathrm{K})/T$	1.43	1.37	1.32	1.27
$\ln(k_r/\mathrm{dm^3\ mol^{-1}\ s^{-1}})$	−4.51	−3.35	−2.25	−1.07
$(10^3\ \mathrm{K})/T$	1.23	1.19	1.10	1.00
$\ln(k_r/\mathrm{dm^3\ mol^{-1}\ s^{-1}})$	−0.24	0.77	3.00	4.98

Now plot $\ln k_r$ against $1/T$ (Fig. 20.2). The least-squares fit results in a line with slope -22.7×10^3 K and intercept 27.7. Therefore,

$$E_a = -(-22.7 \times 10^3\ \mathrm{K}) \times (8.3145\ \mathrm{J\ K^{-1}\ mol^{-1}}) = 189\ \mathrm{kJ\ mol^{-1}}$$
$$A = e^{27.7}\ \mathrm{dm^3\ mol^{-1}\ s^{-1}} = 1.1 \times 10^{12}\ \mathrm{dm^3\ mol^{-1}\ s^{-1}}$$

A note on good practice Note that A has the same units as k_r. In practice, A is obtained from one of the mid-range data values rather than using a lengthy extrapolation.

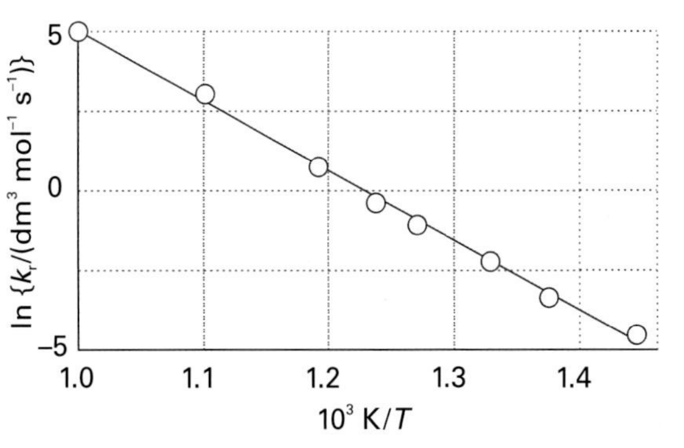

Fig. 20.2 The Arrhenius plot using the data in Example 20.1.

Self-test 20.1 Determine A and E_a from the following data:

T/K	300	350	400
$k_r/(\mathrm{dm^3\,mol^{-1}\,s^{-1}})$	7.9×10^6	3.0×10^7	7.9×10^7
T/K	450	500	
$k_r/(\mathrm{dm^3\,mol^{-1}\,s^{-1}})$	1.7×10^8	3.2×10^8	

[$8\times10^{10}\,\mathrm{dm^3\,mol^{-1}\,s^{-1}}$, 23 kJ mol^{-1}]

The fact that E_a is given by the slope of the plot of ln k_r against $1/T$ means that, the higher the activation energy, the stronger the temperature dependence of the rate constant (that is, the steeper the slope). *A high activation energy signifies that the rate constant depends strongly on temperature.* If a reaction has zero activation energy, its rate is independent of temperature. In some cases the activation energy is negative, which indicates that the rate decreases as the temperature is raised.

The temperature dependence of some reactions is non-Arrhenius, in the sense that a straight line is not obtained when ln k_r is plotted against $1/T$. However, it is still possible to define an activation energy at any temperature as

$$E_a = RT^2\left(\frac{d\ln k_r}{dT}\right) = -R\left(\frac{d\ln k_r}{d(1/T)}\right) \qquad [20.2]$$

This definition reduces to the earlier one (as the slope of a straight line) for a temperature-independent activation energy (see Problem 20.16). However, the definition in eqn 20.2 is more general than that in eqn 20.1, because it allows E_a to be obtained from the slope (at the temperature of interest) of a plot of ln k_r against $1/T$ even if the Arrhenius plot is not a straight line. Non-Arrhenius behaviour is sometimes a sign that quantum mechanical tunnelling is playing a significant role in the reaction.

20.2 The activation energy of a composite reaction

Although the rate of each step of a complex mechanism might increase with temperature and show Arrhenius behaviour, is that true of a composite reaction? To answer this question, we consider the high-pressure limit of the Lindemann–Hinshelwood mechanism as expressed in eqn 19.43. If each of the rate constants has an Arrhenius-like temperature dependence, we can use eqn 20.1b for each of them, and write

$$k_r = \frac{k_a k_b}{k_a'} = \frac{(A_a e^{-E_a(a)/RT})(A_b e^{-E_a(b)/RT})}{(A_a' e^{-E_a'(a)/RT})} = \frac{A_a A_b}{A_a'} e^{-\{E_a(a)+E_a(b)-E_a'(a)\}/RT} \qquad (20.3)$$

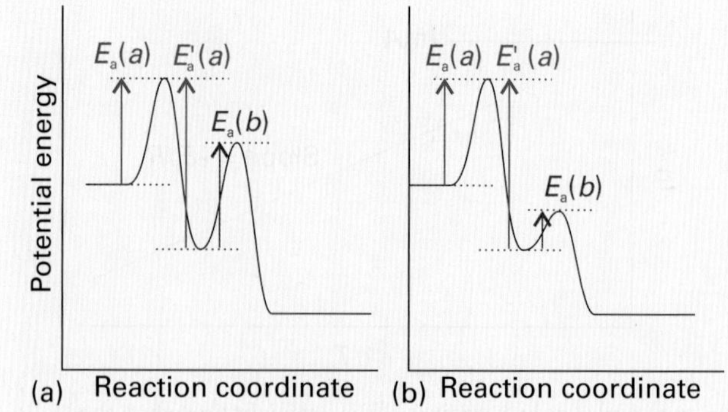

Fig. 20.3 For a reaction with a pre-equilibrium, there are three activation energies to take into account, two referring to the reversible steps of the pre-equilibrium and one for the final step. The relative magnitudes of the activation energies determine whether the overall activation energy is (a) positive or (b) negative.

That is, the composite rate constant k_r has an Arrhenius-like form with activation energy

$$E_a = E_a(a) + E_a(b) - E_a'(a) \qquad (20.4)$$

Provided $E_a(a) + E_a(b) > E_a'(a)$, the activation energy is positive and the rate increases with temperature. However, it is conceivable that $E_a(a) + E_a(b) < E_a'(a)$ (Fig. 20.3), in which case the activation energy is negative and the rate will *decrease* as the temperature is raised. There is nothing remarkable about this behaviour: all it means is that the reverse reaction (corresponding to the deactivation of A*) is so sensitive to temperature that its rate increases sharply as the temperature is raised, and depletes the steady-state concentration of A*.

The Lindemann–Hinshelwood mechanism is an unlikely candidate for the type of behaviour we have described because the deactivation of A* has only a small activation energy, but there are reactions with analogous mechanisms in which a negative activation energy is observed.

● A BRIEF ILLUSTRATION

The rate law for the oxidation of NO according to the mechanism

$$NO + NO \rightleftarrows (NO)_2 \qquad k_a, k_a'$$

$$(NO)_2 + O_2 \rightarrow NO_2 + NO_2 \qquad k_b$$

is

$$\frac{d[NO_2]}{dt} = \frac{2k_a k_b[NO]^2[O_2]}{k_a' + k_b[O_2]} \approx \frac{2k_a k_b}{k_a'}[NO]^2[O_2]$$

when $k_a' \gg k_b[O_2]$. Provided the activation energy of the step $(NO)_2 \rightarrow NO + NO$ is higher than the sum of the activation energies of the other two steps, the overall activation energy will be negative. The decrease in rate with increasing temperature can be understood in terms of a pre-equilibrium, in which $(NO)_2$ is formed in an exothermic reaction, so its concentration decreases as the temperature is raised and consequently NO_2 is formed more slowly. ●

Reactive encounters

In this section we consider two elementary approaches to the calculation of reaction rates, one relating to gas-phase reactions and the other to reactions in solution. Both approaches are based on the view that reactant molecules must meet, and that reaction takes place only if the molecules have a certain minimum energy. In the collision theory of bimolecular gas-phase reactions, products are formed only if the collision is sufficiently energetic; otherwise the colliding reactant molecules separate again. In solution, the reactant molecules may simply diffuse together and then acquire energy from their immediate surroundings while they are in contact.

20.3 Collision theory

In this section, we build on the material from Chapter 18 on the kinetic theory of gases and consider the bimolecular elementary reaction

$$\mathrm{A}+\mathrm{B}\rightarrow\mathrm{P} \qquad v=k_r[\mathrm{A}][\mathrm{B}] \tag{20.5}$$

where P denotes products. We aim to calculate the second-order rate constant k_r.

We can anticipate the general form of the expression for k_r by considering the physical requirements for reaction. We expect the rate v to be proportional to the rate of collisions, and therefore to the mean speed of the molecules, $\bar{c}\propto(T/M)^{1/2}$ where M is the molar mass of the molecules; we also expect the rate to be proportional to their collision cross-section, σ (Section 18.1), and to the number densities $\mathcal{N}_A$ and $\mathcal{N}_B$ of A and B:

$$v\propto\sigma(T/M)^{1/2}\mathcal{N}_A\mathcal{N}_B\propto\sigma(T/M)^{1/2}[\mathrm{A}][\mathrm{B}]$$

However, a collision will be successful only if the kinetic energy exceeds a minimum value which we denote E'. This requirement suggests that the rate should also be proportional to a Boltzmann factor of the form $e^{-E'/RT}$. Therefore,

$$v\propto\sigma(T/M)^{1/2}\,e^{-E'/RT}[\mathrm{A}][\mathrm{B}]$$

and we can anticipate, by writing the reaction rate in the form given in eqn 20.5, that

$$k_r\propto\sigma(T/M)^{1/2}\,e^{-E'/RT}$$

At this point, we begin to recognize the form of the Arrhenius equation, eqn 20.1b, and identify the minimum kinetic energy E' with the activation energy E_a of the reaction. This identification, however, should not be regarded as precise, since collision theory is only a rudimentary model of chemical reactivity.

Not every collision will lead to reaction even if the energy requirement is satisfied, because the reactants may need to collide in a certain relative orientation. This 'steric requirement' suggests that a further factor, P, should be introduced, and that

$$k_r\propto P\sigma(T/M)^{1/2}e^{-E_a/RT} \tag{20.6}$$

As we shall see in detail below, this expression has the form predicted by collision theory. It reflects three aspects of a successful collision:

$$k_r\propto \text{steric requirement}\times\text{encounter rate}\times\text{minimum energy requirement}$$

(a) Collision rates in gases

We have anticipated that the reaction rate, and hence k_r, depends on the frequency with which molecules collide. The **collision density**, Z_{AB}, is the number of (A,B) collisions in a region of the sample in an interval of time divided by the volume of the region and the duration of the interval. The frequency of collisions of a single molecule in a gas was calculated in Section 18.1. As shown in the following *Justification*, that result can be adapted to deduce that

$$Z_{AB}=\sigma\left(\frac{8kT}{\pi\mu}\right)^{1/2}N_A^2[\mathrm{A}][\mathrm{B}] \tag{20.7a}$$

where σ is the collision cross-section (Fig. 20.4)

$$\sigma=\pi d^2 \qquad d=\tfrac{1}{2}(d_A+d_B) \tag{20.7b}$$

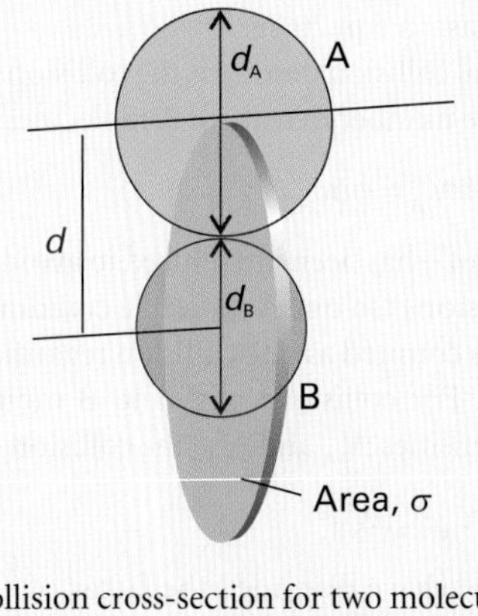

Fig. 20.4 The collision cross-section for two molecules can be regarded to be the area within which the projectile molecule (A) must enter around the target molecule (B) in order for a collision to occur. If the diameters of the two molecules are d_A and d_B, the radius of the target area is $d=\frac{1}{2}(d_A+d_B)$ and the cross-section is πd^2.

d_A and d_B are the diameters of A and B, respectively and μ is the reduced mass,

$$\mu = \frac{m_A m_B}{m_A + m_B} \tag{20.7c}$$

Similarly, the collision density for like molecules at a molar concentration [A] is

$$Z_{AA} = \sigma\left(\frac{4kT}{\pi m_A}\right)^{1/2} N_A^2[A]^2 \tag{20.8}$$

Collision densities may be very large. For example, in nitrogen at room temperature and atmospheric pressure, with $d = 280$ pm, $Z = 5 \times 10^{34}$ m^{-3} s^{-1}.

Justification 20.1 *The collision density*

It follows from eqn 18.12 that the collision frequency, z, for a single A molecule of mass m_A in a gas of other A molecules is

$$z = \sigma\bar{c}_{rel}\mathcal{N}_A \tag{20.9}$$

where $\mathcal{N}_A$ is the number density of A molecules and $\bar{c}_{rel}$ is their relative mean speed. As indicated in Section 18.1,

$$\bar{c}_{rel} = 2^{1/2}\bar{c} \qquad \bar{c} = \left(\frac{8kT}{\pi m}\right)^{1/2} \tag{20.10}$$

For future convenience, it is sensible to introduce $\mu = \frac{1}{2}m$ (for like molecules of mass m), and then to write

$$\bar{c}_{rel} = \left(\frac{8kT}{\pi\mu}\right)^{1/2} \tag{20.11}$$

This expression also applies to the mean relative speed of dissimilar molecules, provided that μ is interpreted as the reduced mass in eqn 20.7c.

The total collision density is the collision frequency multiplied by the number density of A molecules:

$$Z_{AA} = \tfrac{1}{2}z\mathcal{N}_A = \tfrac{1}{2}\sigma\bar{c}_{rel}\mathcal{N}_A^2 \tag{20.12a}$$

The factor of $\frac{1}{2}$ has been introduced to avoid double counting of the collisions (so one A molecule colliding with another A molecule is counted as one collision regardless of their actual identities). For collisions of A and B molecules present at number densities $\mathcal{N}_A$ and $\mathcal{N}_B$, the collision density is

$$Z_{AB} = \sigma\bar{c}_{rel}\mathcal{N}_A\mathcal{N}_B \tag{20.12b}$$

Note that we have discarded the factor of $\frac{1}{2}$ because now we are considering an A molecule colliding with any of the B molecules as a collision.

The number density of a species J is $\mathcal{N}_J = N_A[J]$, where [J] is their molar concentration and N_A is Avogadro's constant. Equations 20.7a and 20.8 then follow.

(b) The energy requirement

According to collision theory, the rate of change in the number density, $\mathcal{N}_A$, of A molecules is the product of the collision density and the probability that a collision occurs with sufficient energy. The latter condition can be incorporated by writing the collision cross-section σ as a function of the kinetic energy ε of approach of the two colliding species, and setting the cross-section, $\sigma(\varepsilon)$, equal to zero if the kinetic energy of approach is below a certain threshold value, ε_a. Later, we shall identify $N_A\varepsilon_a$ as E_a, the (molar) activation energy of the reaction. Then, for a collision between A and B with a specific relative speed of approach v_{rel} (not, at this stage, a mean value),

$$\frac{d\mathcal{N}_A}{dt} = -\sigma(\varepsilon)v_{rel}\mathcal{N}_A\mathcal{N}_B$$

or, in terms of molar concentrations,

$$\frac{d[A]}{dt} = -\sigma(\varepsilon)v_{rel}N_A[A][B] \tag{20.13}$$

The kinetic energy associated with the relative motion of the two particles takes the form $\varepsilon = \frac{1}{2}\mu v_{rel}^2$ when the centre-of-mass coordinates are separated from the internal coordinates of each particle. Therefore the relative speed is given by $v_{rel} = (2\varepsilon/\mu)^{1/2}$. At this point we recognize that a wide range of approach energies ε is present in a sample, so we should average the expression just derived over a Boltzmann distribution of energies $f(\varepsilon)$, and write

$$\frac{d[A]}{dt} = -\left\{\int_0^\infty \sigma(\varepsilon)v_{rel}f(\varepsilon)d\varepsilon\right\}N_A[A][B] \tag{20.14}$$

and hence recognize the rate constant as

$$k_r = N_A\int_0^\infty \sigma(\varepsilon)v_{rel}f(\varepsilon)d\varepsilon \tag{20.15}$$

Now suppose that the reactive collision cross-section is zero below ε_a. We show in the following *Justification* that, above ε_a, $\sigma(\varepsilon)$ varies as

$$\sigma(\varepsilon) = \left(1 - \frac{\varepsilon_a}{\varepsilon}\right)\sigma \tag{20.16}$$

with the energy-independent σ given by eqn 20.7b. This form of the energy dependence for $\sigma(\varepsilon)$ is broadly consistent with experimental determinations of the reaction between H and D_2 as determined by molecular beam measurements of the kind described later (Fig. 20.5).

Justification 20.2 *The collision cross-section*

Consider two colliding molecules A and B with relative speed v_{rel} and relative kinetic energy $\varepsilon = \frac{1}{2}\mu v_{rel}^2$ (Fig. 20.6). Intuitively we expect that a head-on collision between A and B will be

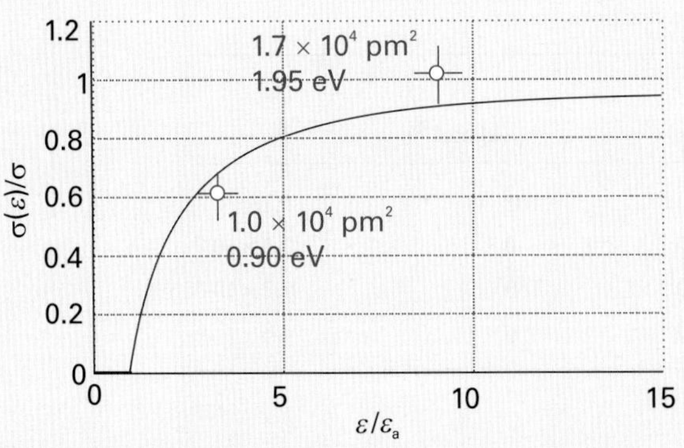

Fig. 20.5 The variation of the reactive cross-section with energy as expressed by eqn 20.16. The data points are from experiments on the reaction $H + D_2 \rightarrow HD + D$ (K. Tsukiyama *et al.*, *J. Chem. Phys.*, **84**, 1934 (1986)).

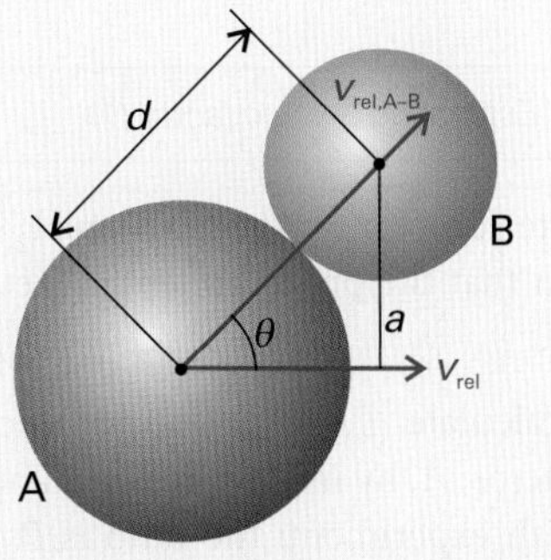

Fig. 20.6 The parameters used in the calculation of the dependence of the collision cross-section on the relative kinetic energy of two molecules A and B.

most effective in bringing about a chemical reaction. Therefore, $v_{rel,A-B}$, the magnitude of the relative velocity component parallel to an axis that contains the vector connecting the centres of A and B, must be large. From trigonometry and the definitions of the distances a and d, and the angle θ given in Fig. 20.6, it follows that

$$v_{rel,A-B} = v_{rel}\cos\theta = v_{rel}\left(\frac{d^2 - a^2}{d^2}\right)^{1/2}$$

We assume that only the kinetic energy associated with the head-on component of the collision, ε_{A-B}, can lead to a chemical reaction. After squaring both sides of this equation and multiplying by $\frac{1}{2}\mu$, it follows that

$$\varepsilon_{A-B} = \varepsilon\frac{d^2 - a^2}{d^2}$$

The existence of an energy threshold, ε_a, for the formation of products implies that there is a maximum value of a, a_{max}, above which reaction does not occur. Setting $a = a_{max}$ and $\varepsilon_{A-B} = \varepsilon_a$ gives

$$a_{max}^2 = \left(1 - \frac{\varepsilon_a}{\varepsilon}\right)d^2$$

Substitution of $\sigma(\varepsilon)$ for πa_{max}^2 and σ for πd^2 in the equation above gives eqn 20.16. Note that the equation can be used only when $\varepsilon > \varepsilon_a$.

With the energy dependence of the collision cross-section established, we can evaluate the integral in eqn 20.15. In the following *Justification* we show that

$$k_r \propto N_A\sigma\bar{c}_{rel}\,e^{-E_a/RT} \qquad (20.17)$$

Justification 20.3 *The rate constant*

The Maxwell–Boltzmann distribution of molecular speeds given in Section 18.1 may be expressed in terms of the kinetic energy, ε, by writing $\varepsilon = \frac{1}{2}\mu v^2$; then $dv = d\varepsilon/(2\mu\varepsilon)^{1/2}$ and eqn 18.4 becomes

$$f(v)dv = 4\pi\left(\frac{\mu}{2\pi kT}\right)^{3/2}\left(\frac{2\varepsilon}{\mu}\right)e^{-\varepsilon/kT}\frac{d\varepsilon}{(2\mu\varepsilon)^{1/2}}$$

$$= 2\pi\left(\frac{1}{\pi kT}\right)^{3/2}\varepsilon^{1/2}e^{-\varepsilon/kT}d\varepsilon = f(\varepsilon)d\varepsilon$$

The integral we need to evaluate is therefore

$$\int_0^\infty \sigma(\varepsilon)v_{rel}f(\varepsilon)d\varepsilon = 2\pi\left(\frac{1}{\pi kT}\right)^{3/2}\int_0^\infty \sigma(\varepsilon)\left(\frac{2\varepsilon}{\mu}\right)^{1/2}\varepsilon^{1/2}e^{-\varepsilon/kT}d\varepsilon$$

$$= \left(\frac{8}{\pi\mu kT}\right)^{1/2}\left(\frac{1}{kT}\right)\int_0^\infty \varepsilon\sigma(\varepsilon)e^{-\varepsilon/kT}d\varepsilon$$

To proceed, we introduce the approximation for $\sigma(\varepsilon)$ in eqn 20.16, and evaluate

$$\int_0^\infty \varepsilon\sigma(\varepsilon)e^{-\varepsilon/kT}d\varepsilon = \sigma\int_{\varepsilon_a}^\infty \varepsilon\left(1 - \frac{\varepsilon_a}{\varepsilon}\right)e^{-\varepsilon/kT}d\varepsilon = (kT)^2\sigma e^{-\varepsilon_a/kT}$$

We have made use of the fact that $\sigma = 0$ for $\varepsilon < \varepsilon_a$. It follows that

$$\int_0^\infty \sigma(\varepsilon)v_{rel}f(\varepsilon)d\varepsilon = \sigma\left(\frac{8kT}{\pi\mu}\right)^{1/2}e^{-\varepsilon_a/kT}$$

as in eqn 20.17 (with $\varepsilon_a/kT = E_a/RT$).

Equation 20.17 has the Arrhenius form $k_r = Ae^{-E_a/RT}$ provided the exponential temperature dependence dominates the weak square-root temperature dependence of the pre-exponential factor. It follows that we can identify (within the constraints of collision theory) the activation energy, E_a, with the minimum

Synoptic table 20.2* Arrhenius parameters for gas-phase reactions

	$A/(\mathrm{dm^3\ mol^{-1}\ s^{-1}})$			
	Experiment	Theory	$E_a/(\mathrm{kJ\ mol^{-1}})$	P
$2\,NOCl \rightarrow 2\,NO + Cl_2$	9.4×10^{9}	5.9×10^{10}	102	0.16
$2\,ClO \rightarrow Cl_2 + O_2$	6.3×10^{7}	2.5×10^{10}	0	2.5×10^{-3}
$H_2 + C_2H_4 \rightarrow C_2H_6$	1.24×10^{6}	7.4×10^{11}	180	1.7×10^{-6}
$K + Br_2 \rightarrow KBr + Br$	1.0×10^{12}	2.1×10^{11}	0	4.8

* More values are given in the *Data section.*

kinetic energy along the line of approach that is needed for reaction, and that the pre-exponential factor is a measure of the rate at which collisions occur in the gas.

(c) The steric requirement

The simplest procedure for calculating k_r is to use for σ the values obtained for non-reactive collisions (for example, typically those obtained from viscosity measurements) or from tables of molecular radii. Table 20.2 compares some values of the pre-exponential factor calculated in this way with values obtained from Arrhenius plots. One of the reactions shows fair agreement between theory and experiment, but for others there are major discrepancies. In some cases the experimental values are orders of magnitude smaller than those calculated, which suggests that the collision energy is not the only criterion for reaction and that some other feature, such as the relative orientation of the colliding species, is important. Moreover, one reaction in the table has a pre-exponential factor larger than theory, which seems to indicate that the reaction occurs more quickly than the particles collide!

We can accommodate the disagreement between experiment and theory by introducing a **steric factor**, P, and expressing the **reactive cross-section**, σ^*, as a multiple of the collision cross-section, $\sigma^* = P\sigma$ (Fig. 20.7). Then the rate constant becomes

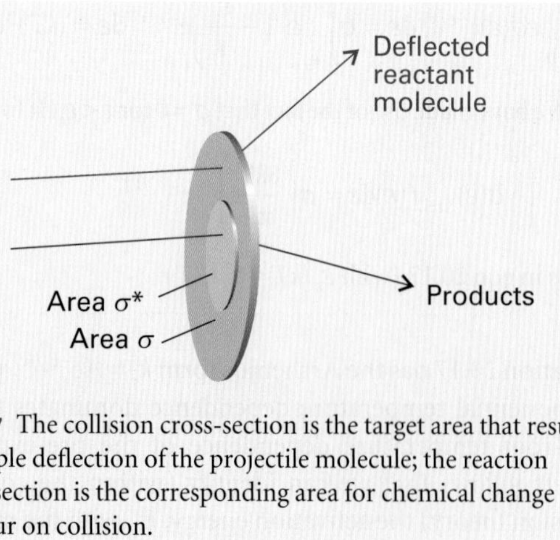

Fig. 20.7 The collision cross-section is the target area that results in simple deflection of the projectile molecule; the reaction cross-section is the corresponding area for chemical change to occur on collision.

$$k_r = P\sigma\left(\frac{8kT}{\pi\mu}\right)^{1/2} N_A e^{-E_a/RT} \qquad (20.18)$$

This expression has the form we anticipated in eqn 20.6. The steric factor is normally found to be several orders of magnitude smaller than 1.

Example 20.2 *Estimating a steric factor (1)*

Estimate the steric factor for the reaction $H_2 + C_2H_4 \rightarrow C_2H_6$ at 628 K given that the pre-exponential factor is 1.24×10^{6} $\mathrm{dm^3\ mol^{-1}\ s^{-1}}$.

Method To calculate P, we need to calculate the pre-exponential factor, A, by using eqn 20.18 and then compare the answer with experiment: the ratio is P. Table 18.1 lists collision cross-sections for non-reactive encounters. The best way to estimate the collision cross-section for dissimilar spherical species is to calculate the collision diameter for each one (from $\sigma = \pi d^2$), to calculate the mean of the two diameters, and then to calculate the cross-section for that mean diameter (eqn 20.7b). However, as neither species is spherical, a simpler but more approximate procedure is just to take the average of the two collision cross-sections.

Answer The reduced mass of the colliding pair with $m_1 = 2.016m_u$ for H_2 and $m_2 = 28.05m_u$ for C_2H_4, is

$$\mu = \frac{m_1 m_2}{m_1 + m_2} = 3.12 \times 10^{-27}\ \mathrm{kg}$$

Hence

$$\left(\frac{8kT}{\pi\mu}\right)^{1/2} = 2.66 \times 10^{3}\ \mathrm{m\ s^{-1}}$$

From Table 18.1, $\sigma(H_2) = 0.27\ \mathrm{nm^2}$ and $\sigma(C_2H_4) = 0.64\ \mathrm{nm^2}$, giving a mean collision cross-section of $\sigma = 0.46\ \mathrm{nm^2}$. Therefore,

$$A = \sigma\left(\frac{8kT}{\pi\mu}\right)^{1/2} N_A = 7.37 \times 10^{11}\ \mathrm{dm^3\ mol^{-1}\ s^{-1}}$$

Experimentally $A = 1.24 \times 10^6\ dm^3\ mol^{-1}\ s^{-1}$, so it follows that $P = 1.68 \times 10^{-6}$. The very small value of P is one reason why catalysts are needed to bring this reaction about at a reasonable rate. As a general guide, the more complex the molecules, the smaller the value of P.

Self-test 20.2 It is found for the reaction $NO + Cl_2 \rightarrow NOCl + Cl$ that $A = 4.0 \times 10^9\ dm^3\ mol^{-1}\ s^{-1}$ at 298 K. Use $\sigma(NO) = 0.42\ nm^2$ and $\sigma(Cl_2) = 0.93\ nm^2$ to estimate the P factor for the reaction. [0.018]

An example of a reaction for which it is possible to estimate the steric factor is $K + Br_2 \rightarrow KBr + Br$, with the experimental value $P = 4.8$. In this reaction, the distance of approach at which reaction occurs appears to be considerably larger than the distance needed for deflection of the path of the approaching molecules in a non-reactive collision. It has been proposed that the reaction proceeds by a **harpoon mechanism**. This brilliant name is based on a model of the reaction that pictures the K atom as approaching a Br_2 molecule and, when the two are close enough, an electron (the harpoon) flips across from K to Br_2. In place of two neutral particles there are now two ions, so there is a Coulombic attraction between them: this attraction is the line on the harpoon. Under its influence the ions move together (the line is wound in), the reaction takes place, and KBr + Br emerge. The harpoon extends the cross-section for the reactive encounter, and the reaction rate is greatly underestimated by taking for the collision cross-section the value for simple mechanical contact between K and Br_2.

Example 20.3 *Estimating a steric factor (2)*

Estimate the value of P for the harpoon mechanism by calculating the distance at which it becomes energetically favourable for the electron to leap from K to Br_2. Take the sum of the radii of the reactants (treating them as spherical) to be 400 pm.

Method We should begin by identifying all the contributions to the energy of interaction between the colliding species. There are three contributions to the energy of the process $K + Br_2 \rightarrow K^+ + Br_2^-$. The first is the ionization energy, I, of K. The second is the electron affinity, E_{ea}, of Br_2. The third is the Coulombic interaction energy between the ions when they have been formed: when their separation is R, this energy is $-e^2/4\pi\varepsilon_0 R$. The electron flips across when the sum of these three contributions changes from positive to negative (that is, when the sum is zero) and becomes energetically favourable.

Answer The net change in energy when the transfer occurs at a separation R is

$$E = I - E_{ea} - \frac{e^2}{4\pi\varepsilon_0 R}$$

The ionization energy I is larger than E_{ea}, so E becomes negative only when R has decreased to less than some critical value R^* given by

$$\frac{e^2}{4\pi\varepsilon_0 R^*} = I - E_{ea}$$

When the particles are at this separation, the harpoon shoots across from K to Br_2, so we can identify the reactive cross-section as $\sigma^* = \pi R^{*2}$. This value of σ^* implies that the steric factor is

$$P = \frac{\sigma^*}{\sigma} = \frac{R^{*2}}{d^2} = \left\{\frac{e^2}{4\pi\varepsilon_0 d(I - E_{ea})}\right\}^2$$

where $d = R(K) + R(Br_2)$, the sum of the radii of the spherical reactants. With $I = 420\ kJ\ mol^{-1}$ (corresponding to 7.0×10^{-19} J), $E_{ea} \approx 250\ kJ\ mol^{-1}$ (corresponding to 4.2×10^{-19} J), and $d = 400$ pm, we find $P = 4.2$, in good agreement with the experimental value (4.8).

Self-test 20.3 Estimate the value of P for the harpoon reaction between Na and Cl_2 for which $d \approx 350$ pm; take $E_{ea} \approx 230\ kJ\ mol^{-1}$. [2.2]

Example 20.3 illustrates two points about steric factors. First, the concept of a steric factor is not wholly useless because in some cases its numerical value can be estimated. Second, and more pessimistically, most reactions are much more complex than $K + Br_2$, and we cannot expect to obtain P so easily.

(d) The RRK model

The steric factor P can also be estimated for unimolecular gas-phase reactions, and its introduction brings the Lindemann–Hinshelwood mechanism (Section 19.7) into closer agreement with experimental results. For example, Fig. 20.8 shows a typical plot of experimental values of $1/k_r$ against $1/[A]$. The plot has a pronounced curvature, corresponding to a larger value of k_r (a smaller value of $1/k_r$) at high pressures (low $1/[A]$) than would be expected by a Lindemann–Hinshelwood extrapolation of the reasonably linear low pressure (high $1/[A]$) data.

The improved model was proposed in 1926 by O.K. Rice and H.C. Ramsperger and almost simultaneously by L.S. Kassel, and is now known as the **Rice–Ramsperger–Kassel model** (RRK model). The model has been elaborated, largely by R.A. Marcus, into the 'RRKM model'. Here we outline Kassel's original approach to the RRK model; the details are set out in *Further information 20.1*. The essential feature of the model is that, although a molecule might have enough energy to react, that

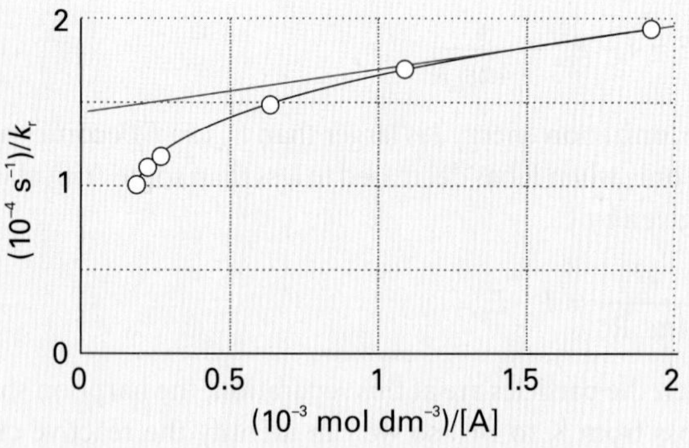

Fig. 20.8 The pressure dependence of the unimolecular isomerization of *trans*-CHD=CHD showing a pronounced departure from the straight line predicted by eqn 19.46 based on the Lindemann–Hinshelwood mechanism.

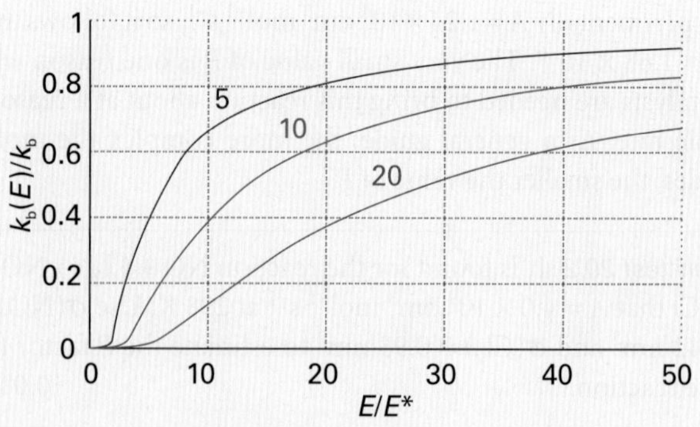

Fig. 20.9 The energy dependence of the rate constant given by eqn 20.19b for three values of *s*.

energy is distributed over all the modes of motion of the molecule, and reaction will occur only when enough of that energy has migrated into a particular location (such as a bond) in the molecule. We show in *Further information 20.1* that this distribution effect leads to a *P*-factor of the form

$$P = \left(1 - \frac{E^*}{E}\right)^{s-1} \tag{20.19a}$$

where *s* is the number of modes of motion over which the energy *E* may be dissipated and E^* is the energy required for the bond of interest to break. The resulting **Kassel form** of the unimolecular rate constant for the decay of A^* to products is

$$k_b(E) = \left(1 - \frac{E^*}{E}\right)^{s-1} k_b \qquad \text{for } E \geq E^* \tag{20.19b}$$

where k_b is the rate constant used in the original Lindemann theory.

The energy dependence of the rate constant given by eqn 20.19b is shown in Fig. 20.9 for various values of *s*. We see that the rate constant is smaller at a given excitation energy if *s* is large, as it takes longer for the excitation energy to migrate through all the oscillators of a large molecule and accumulate in the critical mode. As *E* becomes very large, however, the term in parentheses approaches 1, and $k_b(E)$ becomes independent of the energy and the number of oscillators in the molecule, as there is now enough energy to accumulate immediately in the critical mode regardless of the size of the molecule.

This approach to the calculation of *P* is limited to unimolecular reactions, so it is clear that we need a more powerful theory that lets us calculate rate constants for a wider variety of reactions. We go part of the way toward describing such a theory in Section 20.6.

20.4 Diffusion-controlled reactions

Encounters between reactants in solution occur in a very different manner from encounters in gases. Reactant molecules have to jostle their way through the solvent, so their encounter frequency is considerably less than in a gas. However, because a molecule also migrates only slowly away from a location, two reactant molecules that encounter each other stay near each other for much longer than in a gas. This lingering of one molecule near another on account of the hindering presence of solvent molecules is called the **cage effect**. Such an encounter pair may accumulate enough energy to react even though it does not have enough energy to do so when it first forms. The activation energy of a reaction is a much more complicated quantity in solution than in a gas because the encounter pair is surrounded by solvent and we need to consider the energy of the entire local assembly of reactant and solvent molecules.

(a) Classes of reaction

The complicated overall process can be divided into simpler parts by setting up a simple kinetic scheme. We suppose that the rate of formation of an encounter pair AB is first order in each of the reactants A and B:

$$A + B \rightarrow AB \qquad v = k_d[A][B]$$

As we shall see, k_d (where the d signifies diffusion) is determined by the diffusional characteristics of A and B. The encounter pair can break up without reaction or it can go on to form products P. If we suppose that both processes are pseudofirst-order reactions (with the solvent perhaps playing a role), then we can write

$$AB \rightarrow A + B \qquad v = k_d'[AB]$$

and

$$AB \rightarrow P \qquad v = k_a[AB]$$

The concentration of AB can now be found from the equation for the net rate of change of concentration of AB:

$$\frac{d[AB]}{dt} = k_d[A][B] - k_d'[AB] - k_a[AB] = 0$$

where we have applied the steady-state approximation. This expression solves to

$$[AB] = \frac{k_d[A][B]}{k_a + k_d'}$$

The rate of formation of products is therefore

$$\frac{d[P]}{dt} = k_a[AB] = k_r[A][B] \qquad k_r = \frac{k_a k_d}{k_a + k_d'} \tag{20.20}$$

Two limits can now be distinguished. If the rate of separation of the unreacted encounter pair is much slower than the rate at which it forms products, then $k_d' \ll k_a$ and the effective rate constant is

$$k_r \approx \frac{k_a k_d}{k_a} = k_d \tag{20.21a}$$

In this **diffusion-controlled limit**, the rate of reaction is governed by the rate at which the reactant molecules diffuse through the solvent. Because the combination of radicals involves very little activation energy, radical and atom recombination reactions are often diffusion-controlled. An **activation-controlled reaction** arises when a substantial activation energy is involved in the reaction AB → P. Then $k_a \ll k_d'$ and

$$k_r \approx \frac{k_a k_d}{k_d'} = k_a K \tag{20.21b}$$

where K is the equilibrium constant for A + B $\rightleftharpoons$ AB. In this limit, the reaction proceeds at the rate at which energy accumulates in the encounter pair from the surrounding solvent. Some experimental data are given in Table 20.3.

(b) Diffusion and reaction

The rate of a diffusion-controlled reaction is calculated by considering the rate at which the reactants diffuse together.

Synoptic table 20.3* Arrhenius parameters for reactions in solution

	Solvent	$A/(\text{dm}^3\ \text{mol}^{-1}\ \text{s}^{-1})$	$E_a/(\text{kJ mol}^{-1})$
$(CH_3)_3CCl$ solvolysis	Water	7.1×10^{16}	100
	Ethanol	3.0×10^{13}	112
	Chloroform	1.4×10^{4}	45
$CH_3CH_2Br + OH^-$	Ethanol	4.3×10^{11}	90

* More values are given in the *Data section*.

As shown in the following *Justification*, the rate constant for a reaction in which the two reactant molecules react if they come within a distance R^* of one another is

$$k_d = 4\pi R^* D N_A \tag{20.22}$$

where D is the sum of the diffusion coefficients of the two reactant species in the solution. It follows from this expression that an indication that a reaction is diffusion-controlled is that its rate constant is of the order of 10^9 dm^3 mol^{-1} s^{-1} or greater, as may be confirmed by taking values of R^* and D to be 100 nm and 10^{-9} m^2 s^{-1}, respectively.

Justification 20.4 *Solution of the radial diffusion equation*

The general form of the diffusion equation (Section 18.9) corresponding to motion in three dimensions is $D_B\nabla^2[B] = \partial[B]/\partial t$; therefore, the concentration of B when the system has reached a steady state ($\partial[B]/\partial t = 0$) satisfies $\nabla^2[B]_r = 0$, where the subscript r signifies a quantity that varies with the distance r. For a spherically symmetrical system, ∇^2 can be replaced by radial derivatives alone (see Table 1.1), so the equation satisfied by $[B]_r$ is

$$\frac{d^2[B]_r}{dr^2} + \frac{2}{r}\frac{d[B]_r}{dr} = 0$$

The general solution of this equation is

$$[B]_r = a + \frac{b}{r}$$

as may be verified by substitution. We need two boundary conditions to pin down the values of the two constants (a and b). One condition is that $[B]_r$ has its bulk value [B] as $r \to \infty$. The second condition is that the concentration of B is zero at $r = R^*$, the distance at which reaction occurs. It follows that $a = [B]$ and $b = -R^*[B]$, and hence that (for $r \geq R^*$)

$$[B]_r = \left(1 - \frac{R^*}{r}\right)[B] \tag{20.23}$$

Figure 20.10 illustrates the variation of concentration expressed by this equation.

The rate of reaction is the (molar) flux, J, of the reactant B towards A multiplied by the area of the spherical surface of radius R^*:

$$\text{Rate of reaction} = 4\pi R^{*2} J$$

From Fick's first law (eqn 18.41), the flux of B towards A is proportional to the concentration gradient, so at a radius R^*:

$$J = D_B\left(\frac{d[B]_r}{dr}\right)_{r=R^*} = \frac{D_B[B]}{R^*}$$

(A sign change has been introduced because we are interested in the flux towards decreasing values of r.) It follows that

$$\text{Rate of reaction} = 4\pi R^* D_B[B]$$

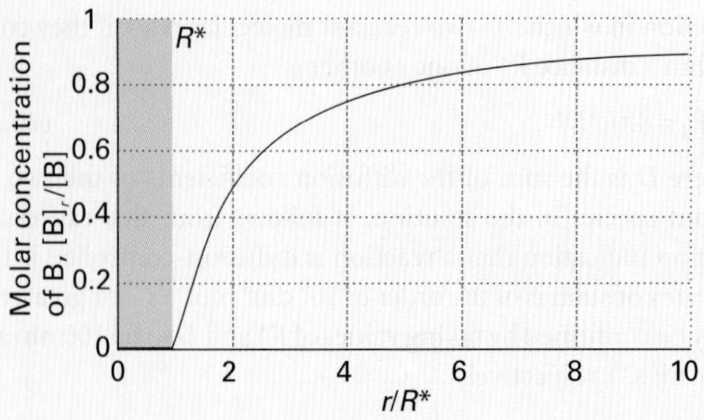

Fig. 20.10 The concentration profile for reaction in solution when a molecule B diffuses towards another reactant molecule and reacts if it reaches R^*.

The rate of the diffusion-controlled reaction is equal to the average flow of B molecules to all the A molecules in the sample. If the bulk concentration of A is [A], the number of A molecules in the sample of volume V is $N_A[A]V$; the global flow of all B to all A is therefore $4\pi R^* D_B N_A[A][B]V$. Because it is unrealistic to suppose that all A molecules are stationary, we replace D_B by the sum of the diffusion coefficients of the two species and write $D = D_A + D_B$. Then the rate of change of concentration of AB is

$$\frac{d[AB]}{dt} = 4\pi R^* D N_A[A][B]$$

Hence, the diffusion-controlled rate constant is as given in eqn 20.22.

We can take eqn 20.22 further by incorporating the Stokes–Einstein equation (eqn 18.48) relating the diffusion constant and the hydrodynamic radius R_A and R_B of each molecule in a medium of viscosity η:

$$D_A = \frac{kT}{6\pi\eta R_A} \qquad D_B = \frac{kT}{6\pi\eta R_B} \tag{20.24}$$

As these relations are approximate, little extra error is introduced if we write $R_A = R_B = \frac{1}{2}R^*$, which leads to

$$k_d = \frac{8RT}{3\eta} \tag{20.25}$$

(The R in this equation is the gas constant.) The radii have cancelled because, although the diffusion constants are smaller when the radii are large, the reactive collision radius is larger and the particles need travel a shorter distance to meet. In this approximation, the rate constant is independent of the identities of the reactants, and depends only on the temperature and the viscosity of the solvent.

● A BRIEF ILLUSTRATION

The rate constant for the recombination of I atoms in hexane at 298 K, when the viscosity of the solvent is 0.326 cP (with $1\ \text{P} = 10^{-1}\ \text{kg m}^{-1}\ \text{s}^{-1}$), is

$$k_d = \frac{8 \times (8.3145\ \text{J K}^{-1}\,\text{mol}^{-1}) \times (298\ \text{K})}{3 \times (3.26 \times 10^{-4}\ \text{kg m}^{-1}\,\text{s}^{-1})} = 2.0 \times 10^7\ \text{m}^3\,\text{mol}^{-1}\,\text{s}^{-1}$$

where we have used $1\ \text{J} = 1\ \text{kg m}^2\ \text{s}^{-2}$. Because $1\ \text{m}^3 = 10^3\ \text{dm}^3$, this result corresponds to $2.0 \times 10^{10}\ \text{dm}^3\ \text{mol}^{-1}\ \text{s}^{-1}$. The experimental value is $1.3 \times 10^{10}\ \text{dm}^3\ \text{mol}^{-1}\ \text{s}^{-1}$, so the agreement is very good considering the approximations involved. ●

20.5 The material balance equation

The diffusion of reactants plays an important role in many chemical processes, such as the diffusion of O_2 molecules into red blood corpuscles and the diffusion of a gas towards a catalyst. We can catch a glimpse of the kinds of calculations involved by considering the diffusion equation (Section 18.9) generalized to take into account the possibility that the diffusing, convecting molecules are also reacting.

(a) The formulation of the equation

Consider a small volume element in a chemical reactor (or a biological cell). The net rate at which J molecules enter the region by diffusion and convection is given by eqn 18.53:

$$\frac{\partial[J]}{\partial t} = D\frac{\partial^2[J]}{\partial x^2} - v\frac{\partial[J]}{\partial x} \tag{20.26}$$

where v is the velocity of flow of J. The net rate of change of molar concentration due to chemical reaction is

$$\frac{\partial[J]}{\partial t} = -k_r[J]$$

if we suppose that J disappears by a pseudofirst-order reaction. Therefore, the overall rate of change of the concentration of J is

$$\frac{\partial[J]}{\partial t} = \underbrace{D\frac{\partial^2[J]}{\partial x^2}}_{\text{Spread due to non-uniform concentration}} - \underbrace{v\frac{\partial[J]}{\partial x}}_{\text{Change due to convection}} - \underbrace{k_r[J]}_{\text{Loss due to reaction}} \tag{20.27}$$

Equation 20.27 is called the **material balance equation**. If the rate constant is large, then [J] will decline rapidly. However, if the diffusion constant is large, then the decline can be replenished as J diffuses rapidly into the region. The convection term, which may represent the effects of stirring, can sweep material either into or out of the region according to the signs of v and the concentration gradient $\partial[J]/\partial x$.

(b) Solutions of the equation

The material balance equation is a second-order partial differential equation and is far from easy to solve in general. Some idea of how it is solved can be obtained by considering the special case in which there is no convective motion (as in an unstirred reaction vessel):

$$\frac{\partial[\mathrm{J}]}{\partial t} = D\frac{\partial^2[\mathrm{J}]}{\partial x^2} - k_r[\mathrm{J}] \tag{20.28}$$

As may be verified by substitution (Problem 20.17), if the solution of this equation in the absence of reaction (that is, for $k_r = 0$) is [J], then the solution $[\mathrm{J}]^*$ in the presence of reaction ($k_r > 0$) is

$$[\mathrm{J}]^* = [\mathrm{J}]e^{-k_r t} \tag{20.29}$$

We have already met one solution of the diffusion equation in the absence of reaction: eqn 18.54 is the solution for a system in which initially a layer of n_0N_A molecules is spread over a plane of area A:

$$[\mathrm{J}] = \frac{n_0 e^{-x^2/4Dt}}{A(\pi Dt)^{1/2}} \tag{20.30}$$

When this expression is substituted into eqn 20.29 and the integral is evaluated numerically, we obtain the concentration of J as it diffuses away from its initial surface layer and undergoes reaction in the solution above (Fig. 20.11).

Even this relatively simple example has led to an equation that is difficult to solve, and only in some special cases can the full material balance equation be solved analytically. Most modern work on reactor design and cell kinetics uses numerical methods to solve the equation, and detailed solutions for realistic environments, such as vessels of different shapes (which influence the boundary conditions on the solutions) and with a variety of inhomogeneously distributed reactants, can be obtained reasonably easily.

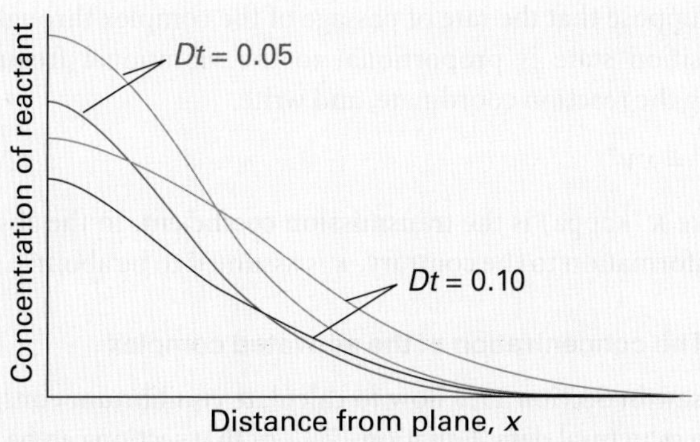

Fig. 20.11 The concentration profiles for a diffusing, reacting system (for example, a column of solution) in which one reactant is initially in a layer at $x = 0$. In the absence of reaction (grey lines) the concentration profiles are the same as in Fig. 18.19.

interActivity Use the interactive applet found in the *Living graphs* section of the text's web site to explore the effect of varying the value of the rate constant k_r on the spatial variation of [J] for a constant value of the diffusion constant D.

Transition state theory

During the course of a chemical reaction that begins with a collision between molecules of A and molecules of B, the potential energy of the system typically changes in a manner shown in Fig. 20.12. Although the illustration displays an exothermic reaction, a potential barrier is also common for endothermic reactions. As the reaction event proceeds, A and B come into contact, distort, and begin to exchange or discard atoms. The **reaction coordinate** is a representation of the atomic displacements, such as changes in interatomic distances and bond angles, that are directly involved in the formation of products from reactants. The potential energy rises to a maximum and the cluster of atoms that corresponds to the region close to the maximum is called the **activated complex**. After the maximum, the potential energy falls as the atoms rearrange in the cluster and reaches a value characteristic of the products. The climax of the reaction is at the peak of the potential energy, which can be identified with the activation energy E_a; however, as in collision theory, this identification should be regarded as approximate. Here two reactant molecules have come to such a degree of closeness and distortion that a small further distortion will send them in the direction of products. This crucial configuration is called the **transition state** of the reaction. Although some molecules entering the transition state might revert to reactants, if they pass through this configuration then it is inevitable that products will emerge from the encounter.

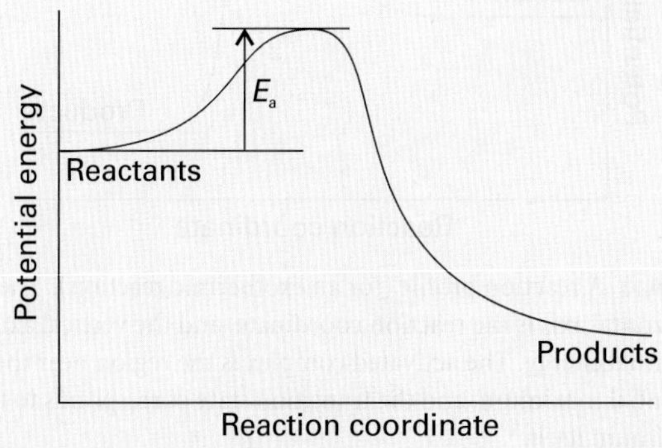

Fig. 20.12 A potential energy profile for an exothermic reaction. The height of the barrier between the reactants and products is the activation energy of the reaction.

A note on good practice The terms *activated complex* and *transition state* are often used as synonyms; however, it is best to preserve the distinction, with the former referring to the cluster of atoms and the latter to their critical configuration.

In **transition state theory** (which is also widely referred to as *activated complex theory*), the notion of the transition state is used in conjunction with concepts of statistical thermodynamics to provide a more detailed calculation of rate constants than that presented by collision theory (Section 20.3). Transition state theory has the advantage that a quantity corresponding to the steric factor appears automatically, and P does not need to be grafted on to an equation as an afterthought; it is an attempt to identify the principal features governing the size of a rate constant in terms of a model of the events that take place during the reaction. There are several approaches to the calculation of rate constants by transition state theory; here we present the simplest one.

20.6 The Eyring equation

Transition state theory pictures a reaction between A and B as proceeding through the formation of an activated complex, $C^\ddagger$, in a rapid pre-equilibrium (Fig. 20.13):[1]

$$A + B \rightleftharpoons C^\ddagger \qquad K^\ddagger = \frac{p_{C^\ddagger}p^\ominus}{p_A p_B} \tag{20.31}$$

where we have replaced the activity of each species by $p/p^\ominus$. When we express the partial pressures, p_J, in terms of the molar concentrations, [J], by using $p_J = RT[J]$, the concentration of activated complex is related to the (dimensionless) equilibrium constant by

$$[C^\ddagger] = \frac{RT}{p^\ominus}K^\ddagger[A][B] \tag{20.32}$$

The activated complex falls apart by unimolecular decay into products, P, with a rate constant $k^\ddagger$:

$$C^\ddagger \rightarrow P \qquad v = k^\ddagger[C^\ddagger] \tag{20.33}$$

It follows that

$$v = k_r[A][B] \qquad k_r = \frac{RT}{p^\ominus}k^\ddagger K^\ddagger \tag{20.34}$$

Our task is to calculate the unimolecular rate constant $k^\ddagger$ and the equilibrium constant $K^\ddagger$.

(a) The rate of decay of the activated complex

An activated complex can form products if it passes through the transition state. As the reactant molecules approach the activated complex region, some bonds are forming and shortening while others are lengthening and breaking; therefore, along the reaction coordinate, there is a vibration-like motion of the atoms in the activated complex. If this vibration-like motion occurs with a frequency $\nu^\ddagger$, then the frequency with which the cluster of atoms forming the complex approaches the transition state is also $\nu^\ddagger$. However, it is possible that not every oscillation along the reaction coordinate takes the complex through the transition state. For instance, the centrifugal effect of rotations might also be an important contribution to the breakup of the complex, and in some cases the complex might be rotating too slowly, or rotating rapidly but about the wrong axis. Therefore, we suppose that the rate of passage of the complex through the transition state is proportional to the vibrational frequency along the reaction coordinate, and write

$$k^\ddagger = \kappa\nu^\ddagger \tag{20.35}$$

where κ (kappa) is the **transmission coefficient.** In the absence of information to the contrary, κ is assumed to be about 1.

(b) The concentration of the activated complex

We saw in Section 17.3 how to calculate equilibrium constants from structural data. Equation 17.17 of that section can be used directly, which in this case gives

$$K^\ddagger = \frac{N_A q^\ominus_{C^\ddagger}}{q^\ominus_A q^\ominus_B}e^{-\Delta E_0/RT} \tag{20.36}$$

where

$$\Delta E_0 = E_0(C^\ddagger) - E_0(A) - E_0(B) \tag{20.37}$$

and the $q^\ominus_J$ are the standard molar partition functions, as defined in Section 17.3. Note that the units of N_A and the q_J are mol^{-1},

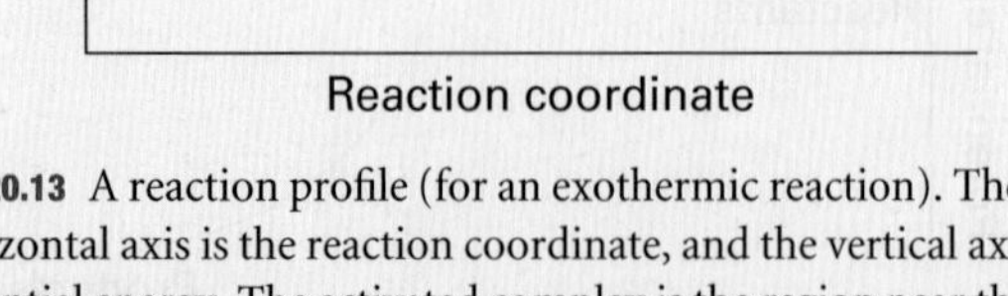

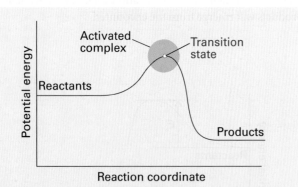

Fig. 20.13 A reaction profile (for an exothermic reaction). The horizontal axis is the reaction coordinate, and the vertical axis is potential energy. The activated complex is the region near the potential maximum, and the transition state corresponds to the maximum itself.

[1] This chapter inevitably puts heavy demands on the letter K; the various meanings are summarized in Table 20.4 at the end of the chapter.

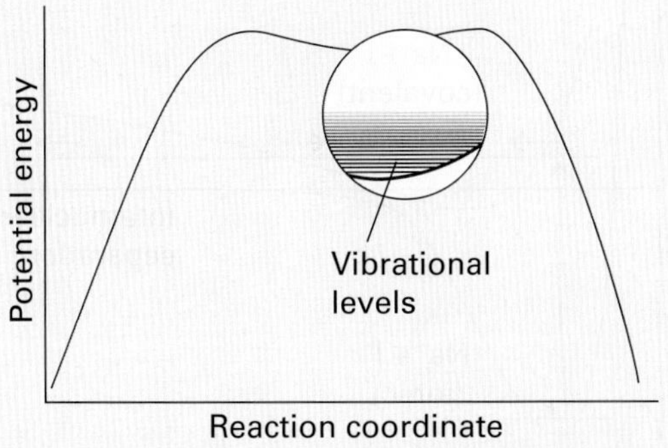

Fig. 20.14 In an elementary depiction of the activated complex close to the transition state, there is a broad, shallow dip in the potential energy surface along the reaction coordinate. The complex vibrates harmonically and almost classically in this well. However, this depiction is an oversimplification, for in many cases there is no dip at the top of the barrier, and the curvature of the potential energy, and therefore the force constant, is negative. Formally, the vibrational frequency is then imaginary. We ignore this problem here.

so $K^{\ddagger}$ is dimensionless (as is appropriate for an equilibrium constant).

In the final step of this part of the calculation, we focus attention on the partition function of the activated complex. We have already assumed that a vibration of the activated complex $C^{\ddagger}$ tips it through the transition state. The partition function for this vibration is (see eqn 13.10)

$$q = \frac{1}{1 - e^{-h\nu^{\ddagger}/kT}}$$

where $\nu^{\ddagger}$ is its frequency (the same frequency that determines $k^{\ddagger}$). This frequency is much lower than for an ordinary molecular vibration because the oscillation corresponds to the complex falling apart (Fig. 20.14), so the force constant is very low. Therefore, provided that $h\nu^{\ddagger}/kT \ll 1$, the exponential may be expanded and the partition function reduces to

$$q = \frac{1}{1 - \left(1 - \dfrac{h\nu^{\ddagger}}{kT} + \cdots\right)} \approx \frac{kT}{h\nu^{\ddagger}}$$

We can therefore write

$$q_{C^{\ddagger}} \approx \frac{kT}{h\nu^{\ddagger}} \bar{q}_{C^{\ddagger}} \tag{20.38}$$

where $\bar{q}$ denotes the partition function for all the other modes of the complex. The constant $K^{\ddagger}$ is therefore

$$K^{\ddagger} = \frac{kT}{h\nu^{\ddagger}} \bar{K}^{\ddagger} \qquad \bar{K}^{\ddagger} = \frac{N_A \bar{q}^{\ominus}_{C^{\ddagger}}}{q^{\ominus}_A q^{\ominus}_B} e^{-\Delta E_0/RT} \tag{20.39}$$

with $\bar{K}^{\ddagger}$ a kind of equilibrium constant, but with one vibrational mode of $C^{\ddagger}$ discarded.

(c) The rate constant

We can now combine all the parts of the calculation into

$$k_r = k^{\ddagger} \frac{RT}{p^{\ominus}} K^{\ddagger} = \kappa \nu^{\ddagger} \frac{kT}{h\nu^{\ddagger}} \frac{RT}{p^{\ominus}} \bar{K}^{\ddagger}$$

At this stage the unknown frequencies $\nu^{\ddagger}$ cancel, and after writing $\bar{K}^{\ddagger}_c = (RT/p^{\ominus})\bar{K}^{\ddagger}$, we obtain the **Eyring equation**:

$$k_r = \kappa \frac{kT}{h} \bar{K}^{\ddagger}_c \tag{20.40}$$

The factor $\bar{K}^{\ddagger}_c$ is given by eqn 20.39 and the definition $\bar{K}^{\ddagger}_c = (RT/p^{\ominus})\bar{K}^{\ddagger}$ in terms of the partition functions of A, B, and $C^{\ddagger}$, so in principle we now have an explicit expression for calculating the second-order rate constant for a bimolecular reaction in terms of the molecular parameters for the reactants and the activated complex and the quantity κ.

The partition functions for the reactants can normally be calculated quite readily by using either spectroscopic information about their energy levels or the approximate expressions set out in Table 13.1. The difficulty with the Eyring equation, however, lies in the calculation of the partition function of the activated complex: $C^{\ddagger}$ is difficult to investigate spectroscopically (but see Section 20.6e), and in general we need to make assumptions about its size, shape, and structure. We shall illustrate what is involved in one simple but significant case.

(d) The collision of structureless particles

Consider the case of two structureless particles A and B colliding to give an activated complex that resembles a diatomic molecule. Because the reactants J = A, B are structureless 'atoms', the only contributions to their partition functions are the translational terms:

$$q^{\ominus}_J = \frac{V^{\ominus}_m}{\Lambda^3_J} \qquad \Lambda_J = \frac{h}{(2\pi m_J kT)^{1/2}} \qquad V^{\ominus}_m = \frac{RT}{p^{\ominus}} \tag{20.41a}$$

The activated complex is a diatomic cluster of mass $m_C = m_A + m_B$ and moment of inertia I. It has one vibrational mode, but that mode corresponds to motion along the reaction coordinate and therefore does not appear in $\bar{q}_{C^{\ddagger}}$. It follows that the standard molar partition function of the activated complex, taking into account rotational and translational contributions, is

$$\bar{q}^{\ominus}_{C^{\ddagger}} = \left(\frac{2IkT}{\hbar^2}\right) \frac{V^{\ominus}_m}{\Lambda^3_{C^{\ddagger}}} \tag{20.41b}$$

The moment of inertia of a diatomic molecule of bond length r is μr^2, where $\mu = m_A m_B/(m_A + m_B)$ is the effective mass, so the expression for the rate constant is

$$k_r = \kappa \frac{kT}{h} \frac{RT}{p^{\ominus}} \left(\frac{N_A \Lambda_A^3 \Lambda_B^3}{\Lambda_{C^\ddagger}^3 V_m^{\ominus}} \right) \left(\frac{2IkT}{\hbar^2} \right) e^{-\Delta E_0/RT}$$

$$= \kappa \frac{kT}{h} N_A \left(\frac{\Lambda_A \Lambda_B}{\Lambda_{C^\ddagger}} \right)^3 \left(\frac{2IkT}{\hbar^2} \right) e^{-\Delta E_0/RT} \qquad (20.42)$$

$$= \kappa N_A \left(\frac{8kT}{\pi\mu} \right)^{1/2} \pi r^2 e^{-\Delta E_0/RT}$$

Finally, by identifying $\kappa\pi r^2$ as the reactive cross-section σ^*, we arrive at precisely the same expression as that obtained from simple collision theory (eqn 20.17).

(e) Observation and manipulation of the activated complex

The development of femtosecond pulsed lasers (Section 11.7) has made it possible to make observations on species that have such short lifetimes that in a number of respects they resemble an activated complex, which often survives for only a few picoseconds. In a typical experiment designed to detect an activated complex, a femtosecond laser pulse is used to excite a molecule to a dissociative state, and then the system is exposed to a second femtosecond pulse at an interval after the dissociating pulse. The frequency of the second pulse is set at an absorption of one of the free fragmentation products, so its absorption is a measure of the abundance of the dissociation product. For example, when ICN is dissociated by the first pulse, the emergence of CN from the photoactivated state can be monitored by watching the growth of the free CN absorption (or, more commonly, its laser-induced fluorescence). In this way it has been found that the CN signal remains zero until the fragments have separated by about 600 pm, which takes about 205 fs.

Some sense of the progress that has been made in the study of the intimate mechanism of chemical reactions can be obtained by considering the decay of the ion pair Na^+I^-. As shown in Fig. 20.15, excitation of the ionic species with a femtosecond laser pulse forms an excited state that corresponds to a covalently bonded NaI molecule. The system can be described with two potential energy surfaces, one largely 'ionic' and another 'covalent', which cross at an internuclear separation of 693 pm. A short laser pulse is composed of a wide range of frequencies, which excite many vibrational states of NaI simultaneously. Consequently, the electronically excited complex exists as a superposition of states, or a localized wavepacket, which oscillates between the 'covalent' and 'ionic' potential energy surfaces, as shown in Fig. 20.15. The complex can also dissociate, shown as movement of the wavepacket toward very long internuclear separation along the dissociative surface. However, not every outward-going swing leads to dissociation because there is a chance that the I atom can be harpooned again, in which case it fails to make good its escape. The dynamics of the system is probed by a second laser pulse with a frequency that corresponds to the absorption frequency of the free Na product or to the frequency at which Na absorbs when it is a part of the complex. The latter frequency depends on the Na···I distance, so an absorption (in practice, a laser-induced fluorescence) is obtained each time the wavepacket returns to that separation.

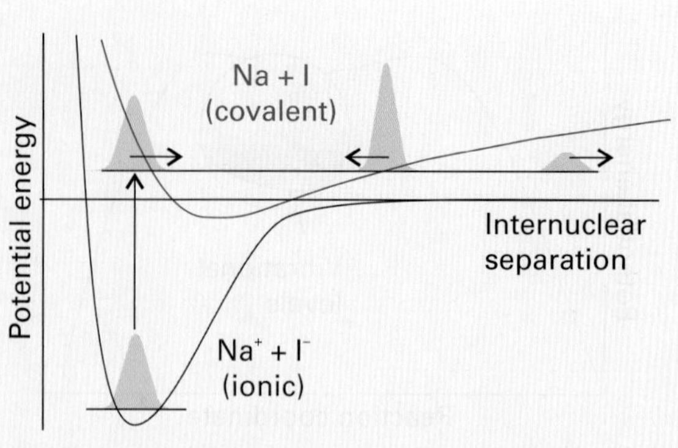

Fig. 20.15 Excitation of the ion pair Na^+I^- forms an excited state with covalent character. Also shown is movement between a 'covalent' surface (in green) and an 'ionic' surface (in purple) of the wavepacket formed by laser excitation. The non-crossing rule ensures that the two curves mix at their intersection; Na + I is the lower state at large distances.

A typical set of results is shown in Fig. 20.16. The bound Na absorption intensity shows up as a series of pulses that recur in about 1 ps, showing that the wavepacket oscillates with about that period. The decline in intensity shows the rate at which the complex can dissociate as the two atoms swing away from each other. The free Na absorption also grows in an oscillating manner, showing the periodicity of wavepacket oscillation, each swing of which gives it a chance to dissociate. The precise period of the oscillation in NaI is 1.25 ps, corresponding to a vibrational wavenumber of 27 cm^{-1} (recall that the activated complex theory

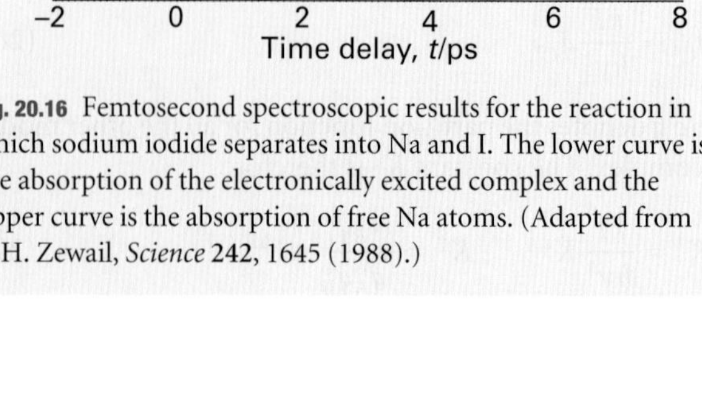

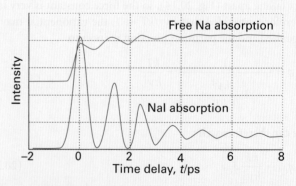

Fig. 20.16 Femtosecond spectroscopic results for the reaction in which sodium iodide separates into Na and I. The lower curve is the absorption of the electronically excited complex and the upper curve is the absorption of free Na atoms. (Adapted from A.H. Zewail, *Science* **242**, 1645 (1988).)

assumes that such a vibration has a very low frequency). The complex survives for about ten oscillations. In contrast, although the oscillation frequency of NaBr is similar, it barely survives one oscillation.

Femtosecond spectroscopy has also been used to examine analogues of the activated complex involved in bimolecular reactions. Thus, a molecular beam can be used to produce a van der Waals molecule (Section 8.7), such as IH···OCO. The HI bond can be dissociated by a femtosecond pulse, and the H atom is ejected towards the O atom of the neighbouring CO_2 molecule to form HOCO. Hence, the van der Waals molecule is a source of a species that resembles the activated complex of the reaction

$$H + CO_2 \rightarrow [HOCO]^{\ddagger} \rightarrow HO + CO$$

The probe pulse is tuned to the OH radical, which enables the evolution of $[HOCO]^{\ddagger}$ to be studied in real time. Femtosecond transition state spectroscopy has also been used to study more complex reactions, such as the Diels–Alder reaction, nucleophilic substitution reactions, and pericyclic addition and cleavage reactions. Biological processes that are open to study by femtosecond spectroscopy include the energy-converting processes of photosynthesis and the photostimulated processes of vision. In other experiments, the photoejection of carbon monoxide from myoglobin and the attachment of O_2 to the exposed site have been studied to obtain rate constants for the two processes.

The techniques used for the spectroscopic detection of transition states can also be used to control the outcome of a chemical reaction by direct manipulation of the transition state. Consider the reaction $I_2 + Xe \rightarrow XeI^* + I$, which occurs by a harpoon mechanism with a transition state denoted as $[Xe^+\cdots I^-\cdots I]$. The reaction can be initiated by exciting I_2 to an electronic state at least 52 460 cm^{-1} above the ground state and then followed by measuring the time dependence of the chemiluminescence of XeI^*. To exert control over the yield of the product, a pair of femtosecond pulses can be used to induce the reaction. The first pulse excites the I_2 molecule to a low energy and unreactive electronic state. We already know that excitation by a femtosecond pulse generates a wavepacket that can be treated as a particle travelling across the potential energy surface. In this case, the wavepacket does not have enough energy to react, but excitation by another laser pulse with the appropriate wavelength can provide the necessary additional energy. It follows that activated complexes with different geometries can be prepared by varying the time delay between the two pulses, as the partially localized wavepacket will be at different locations on the potential energy surface as it evolves after being formed by the first pulse. Because the reaction occurs by the harpoon mechanism, the product yield is expected to be optimal if the second pulse is applied when the wavepacket is at a point where the $Xe\cdots I_2$ distance is just right for electron transfer from Xe to I_2 to occur (see Example 20.3). This type of control of the I_2 + Xe reaction has been demonstrated.

20.7 Thermodynamic aspects

The statistical thermodynamic version of transition state theory rapidly runs into difficulties because only in some cases is anything known about the structure of the activated complex. However, the concepts that it introduces, principally that of an equilibrium between the reactants and the activated complex, have motivated a more general, empirical approach in which the activation process is expressed in terms of thermodynamic functions.

(a) Activation parameters

If we accept that $\bar{K}^{\ddagger}$ is an equilibrium constant (despite one mode of $C^{\ddagger}$ having been discarded), we can express it in terms of a **Gibbs energy of activation**, $\Delta^{\ddagger}G$, through the definition:

$$\Delta^{\ddagger}G = -RT\ln \bar{K}^{\ddagger} \qquad [20.43]$$

All the $\Delta^{\ddagger}X$ in this section are *standard* thermodynamic quantities, $\Delta^{\ddagger}X^{\ominus}$, but we shall omit the standard state sign to avoid overburdening the notation. Then the rate constant becomes

$$k_r = \kappa\frac{kT}{h}\frac{RT}{p^{\ominus}}e^{-\Delta^{\ddagger}G/RT} \qquad (20.44)$$

Because $G = H - TS$, the Gibbs energy of activation can be divided into an **entropy of activation**, $\Delta^{\ddagger}S$, and an **enthalpy of activation**, $\Delta^{\ddagger}H$, by writing

$$\Delta^{\ddagger}G = \Delta^{\ddagger}H - T\Delta^{\ddagger}S \qquad [20.45]$$

When eqn 20.45 is used in eqn 20.44 and κ is absorbed into the entropy term, we obtain

$$k_r = Be^{\Delta^{\ddagger}S/R}e^{-\Delta^{\ddagger}H/RT} \qquad B = \frac{kT}{h}\frac{RT}{p^{\ominus}} \qquad (20.46)$$

The formal definition (eqn 20.2) of activation energy, $E_a = RT^2(d\ln k_r/dT)$, then gives $E_a = \Delta^{\ddagger}H + 2RT$, so

$$k_r = e^2Be^{\Delta^{\ddagger}S/R}e^{-E_a/RT} \qquad (20.47a)$$

A brief comment For reactions of the type A + B → P in the gas phase, $E_a = \Delta^{\ddagger}H + 2RT$. For these reactions in solution, $E_a = \Delta^{\ddagger}H + RT$.

from which it follows that the Arrhenius factor A can be identified as

$$A = e^2Be^{\Delta^{\ddagger}S/R} \qquad (20.47b)$$

The entropy of activation is negative because throughout the system reactant species are combining to form reactive pairs. However, if there is a reduction in entropy below what would be expected for the simple encounter of A and B, then the Arrhenius factor A will be smaller than that expected on the

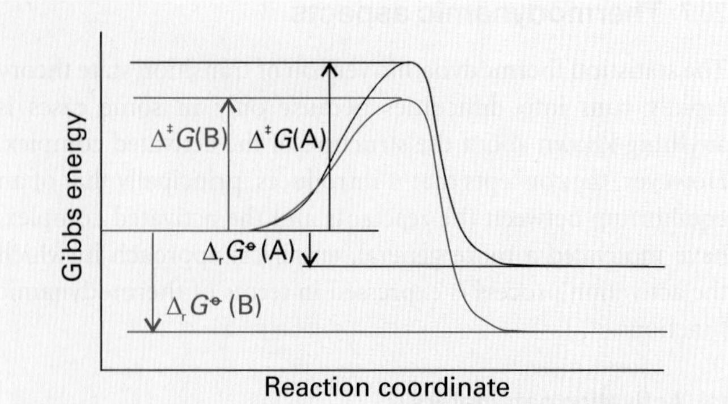

Fig. 20.17 For a related series of reactions, as the magnitude of the standard reaction Gibbs energy increases, so the activation barrier decreases. The approximate linear correlation between $\Delta^{\ddagger}G$ and $\Delta_r G^{\ominus}$ is the origin of linear free energy relations.

basis of simple collision theory. Indeed, we can identify that additional reduction in entropy, $\Delta^{\ddagger}S_{\text{steric}}$, as the origin of the steric factor of collision theory, and write

$$P = e^{\Delta^{\ddagger}S_{\text{steric}}/R} \tag{20.48}$$

Thus, the more complex the steric requirements of the encounter, the more negative the value of $\Delta^{\ddagger}S_{\text{steric}}$, and the smaller the value of P.

Gibbs energies, enthalpies, entropies, volumes, and heat capacities of activation are widely used to report experimental reaction rates, especially for organic reactions in solution. They are encountered when relationships between equilibrium constants and rates of reaction are explored using **correlation analysis**, in which ln K (which is equal to $-\Delta_r G^{\ominus}/RT$) is plotted against ln k_r (which is proportional to $-\Delta^{\ddagger}G/RT$). In many cases the correlation is linear, signifying that, as the reaction becomes thermodynamically more favourable, its rate constant increases (Fig. 20.17). This linear correlation is the origin of the alternative name **linear free energy relation** (LFER).

(b) Reactions between ions

The thermodynamic version of transition state theory simplifies the discussion of reactions in solution. The statistical thermodynamic theory is very complicated to apply because the solvent plays a role in the activated complex. In the thermodynamic approach we combine the rate law

$$\frac{d[P]}{dt} = k^{\ddagger}[C^{\ddagger}]$$

with the thermodynamic equilibrium constant (Section 17.2)

$$K = \frac{a_{C^{\ddagger}}}{a_A a_B} = K_\gamma \frac{[C^{\ddagger}]c^{\ominus}}{[A][B]} \qquad K_\gamma = \frac{\gamma_{C^{\ddagger}}}{\gamma_A \gamma_B}$$

Then

$$\frac{d[P]}{dt} = k_r[A][B] \qquad k_r = \frac{k^{\ddagger}K}{K_\gamma} \tag{20.49a}$$

If k_r° is the rate constant at zero ionic strength when the activity coefficients are 1 (that is, $k_r^{\circ} = k^{\ddagger}K$), we can write

$$k_r = \frac{k_r^{\circ}}{K_\gamma} \tag{20.49b}$$

At low concentrations the activity coefficients can be expressed in terms of the ionic strength, I, of the solution by using the Debye–Hückel limiting law (Section 16.6, particularly eqn 16.56). We assume here that both reactants A and B are 1,1-electrolytes so $z_+ = z_-$ for each; therefore

$$\log \gamma_A = -Az_A^2 I^{1/2} \qquad \log \gamma_B = -Az_B^2 I^{1/2} \tag{20.50a}$$

with $A = 0.509$ in aqueous solution at 298 K and z_A and z_B the charge numbers of A and B, respectively. Since the activated complex forms from reaction of one of the ions of A with one of the ions of B, the charge number of the activated complex is $z_A + z_B$ where z_J is positive for cations and negative for anions. Therefore

$$\log \gamma_{C^{\ddagger}} = -A(z_A + z_B)^2 I^{1/2} \tag{20.50b}$$

Inserting these relations into eqn 20.49b results in

$$\begin{aligned}\log k_r &= \log k_r^{\circ} - A\{z_A^2 + z_B^2 - (z_A + z_B)^2\}I^{1/2} \\ &= \log k_r^{\circ} + 2Az_A z_B I^{1/2}\end{aligned} \tag{20.51}$$

Equation 20.51 expresses the **kinetic salt effect**, the variation of the rate constant of a reaction between ions with the ionic strength of the solution (Fig. 20.18). If the reactant ions have the same sign (as in a reaction between cations or between anions), then increasing the ionic strength by the addition of inert ions increases the rate constant. The formation of a single, highly

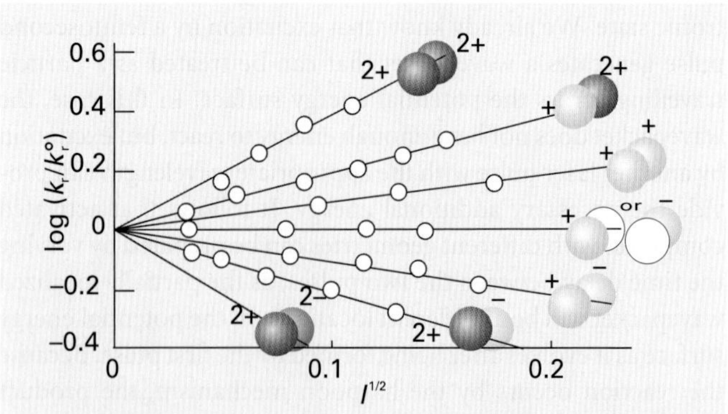

Fig. 20.18 Experimental tests of the kinetic salt effect for reactions in water at 298 K. The ion types are shown as spheres, and the slopes of the lines are those given by the Debye–Hückel limiting law and eqn 20.51.

charged ionic complex from two less highly charged ions is favoured by a high ionic strength because the new ion has a denser ionic atmosphere and interacts with that atmosphere more strongly. Conversely, ions of opposite charge react more slowly in solutions of high ionic strength. Now the charges cancel and the complex has a less favourable interaction with its atmosphere than the separated ions.

Example 20.4 *Analysing the kinetic salt effect*

The rate constant for the base (OH^-) hydrolysis of $[CoBr(NH_3)_5]^{2+}$ varies with ionic strength as tabulated below. What can be deduced about the charge of the activated complex in the rate-determining stage? We cannot assume without more evidence that it is a bimolecular process with an activated complex of charge +1.

I	0.0050	0.0100	0.0150	0.0200	0.0250	0.0300
k_r/k_r°	0.718	0.631	0.562	0.515	0.475	0.447

Method According to eqn 20.51, a plot of $\log(k_r/k_r^\circ)$ against $I^{1/2}$ will have a slope of $1.02z_Az_B$, from which we can infer the charges of the ions involved in the formation of the activated complex.

Answer Form the following table:

I	0.0050	0.0100	0.0150	0.0200	0.0250	0.0300
$I^{1/2}$	0.071	0.100	0.122	0.141	0.158	0.173
$\log(k_r/k_r^\circ)$	−0.14	−0.20	−0.25	−0.29	−0.32	−0.35

These points are plotted in Fig. 20.19. The slope of the (least squares) straight line is −2.04, indicating that $z_Az_B = -2$. Because $z_A = -1$ for the OH^- ion, if that ion is involved in the formation of the activated complex, then the charge number of the second ion is +2. This analysis suggests that the pentaamminebromocobalt(III) cation $[CoBr(NH_3)_5]^{2+}$ participates in the formation of the activated complex and that the charge of the activated complex is $-1 + 2 = +1$. Although we do not pursue the point here, you should be aware that the rate constant is also influenced by the relative permittivity of the medium.

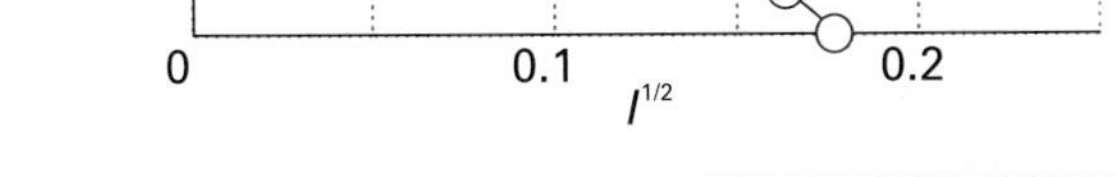

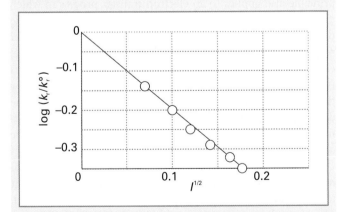

Fig. 20.19 The experimental ionic strength dependence of the rate constant of a hydrolysis reaction: the slope gives information about the charge types involved in the activated complex of the rate-determining step. See Example 20.4.

Self-test 20.4 An ion of charge number +1 is known to be involved in the activated complex of a reaction. Deduce the charge number of the other ion from the following data:

I	0.0050	0.010	0.015	0.020	0.025	0.030
k_r/k_r°	0.930	0.902	0.884	0.867	0.853	0.841

[−1]

20.8 Electron transfer in homogeneous systems

Here we apply the concepts of transition state theory and quantum theory to the study of a deceptively simple process, electron transfer between molecules in homogeneous systems. We describe a theoretical approach to the calculation of rate constants and discuss the theory in the light of experimental results on a variety of systems, including protein complexes. Electron transfer reactions between protein-bound co-factors or between proteins play an important role in a variety of biological processes, including oxidative phosphorylation, photosynthesis, nitrogen fixation, the reduction of atmospheric N_2 to NH_3 by certain microorganisms, and the mechanisms of action of oxidoreductases, which are enzymes that catalyse redox reactions. We shall see that relatively simple expressions may be used to predict the rates of electron transfer with reasonable accuracy.

(a) The rates of electron transfer processes

Consider electron transfer from a donor species D to an acceptor species A in solution. The overall reaction is

$$D + A \rightarrow D^+ + A^- \qquad v = k_r[D][A] \tag{20.52}$$

In the first step of the mechanism, D and A must diffuse through the solution and collide to form a complex DA, in which the donor and acceptor are separated by r, the distance between the outer surface of each species.

$$D + A \rightleftarrows DA \qquad (k_a, k_a') \tag{20.53a}$$

Next, electron transfer occurs within the DA complex to yield D^+A^-:

$$DA \rightleftarrows D^+A^- \qquad (k_{et}, k_{et}') \tag{20.53b}$$

The complex D^+A^- can also break apart and the ions diffuse through the solution:

$$D^+A^- \rightarrow D^+ + A^- \quad (k_d) \tag{20.53c}$$

We show in the following *Justification* that on the basis of this model

$$\frac{1}{k_r} = \frac{1}{k_a} + \frac{k'_a}{k_a k_{et}}\left(1 + \frac{k'_{et}}{k_d}\right) \tag{20.54}$$

Justification 20.5 *The rate constant for electron transfer in solution*

We begin by identifying the rate of the overall reaction (eqn 20.52) with the rate of formation of separated ions:

$$v = k_r[D][A] = k_d[D^+A^-]$$

There are two reaction intermediates, DA and D^+A^-, and we apply the steady-state approximation to both. From

$$\frac{d[D^+A^-]}{dt} = k_{et}[DA] - k'_{et}[D^+A^-] - k_d[D^+A^-] = 0$$

it follows that

$$[DA] = \frac{k'_{et} + k_d}{k_{et}}[D^+A^-]$$

and from

$$\frac{d[DA]}{dt} = k_a[D][A] - k'_a[DA] - k_{et}[DA] + k'_{et}[D^+A^-]$$

$$= k_a[D][A] - \left\{\frac{(k'_a + k_{et})(k'_{et} + k_d)}{k_{et}} - k'_{et}\right\}[D^+A^-] = 0$$

it follows that

$$[D^+A^-] = \frac{k_a k_{et}}{k'_a k'_{et} + k'_a k_d + k_d k_{et}}[D][A]$$

When we multiply this expression by k_d, we see that the resulting equation has the form of the rate of electron transfer, $v = k_r[D][A]$, with k_r given by

$$k_r = \frac{k_d k_a k_{et}}{k'_a k'_{et} + k'_a k_d + k_d k_{et}}$$

To obtain eqn 20.54, we divide the numerator and denominator on the right-hand side of this expression by $k_d k_{et}$ and solve for the reciprocal of k_r.

To gain insight into eqn 20.54 and the factors that determine the rate of electron transfer reactions in solution, we assume that the main decay route for D^+A^- is dissociation of the complex into separated ions, or $k_d \gg k'_{et}$. It follows that

$$\frac{1}{k_r} \approx \frac{1}{k_a}\left(1 + \frac{k'_a}{k_{et}}\right)$$

When $k_{et} \gg k'_a$, we see that $k_r \approx k_a$ and the rate of product formation is controlled by diffusion of D and A in solution, which fosters formation of the DA complex. When $k_{et} \ll k'_a$, we see that $k_r \approx (k_a/k'_a)k_{et}$ or, after using eqn 20.53a,

$$k_r \approx (k_a/k'_a)k_{et} = Kk_{et} \tag{20.55}$$

where K is the equilibrium constant for the diffusive encounter process in eqn 20.53a. We see that the process is controlled by k_{et} and therefore the activation energy of electron transfer in the DA complex. Using transition state theory (Section 20.7), we write

$$k_{et} = \kappa v^{\ddagger} e^{-\Delta^{\ddagger}G/RT} \tag{20.56}$$

where κ is the transmission coefficient, $v^{\ddagger}$ is the vibrational frequency with which the activated complex approaches the transition state, and $\Delta^{\ddagger}G$ is the Gibbs energy of activation. Our remaining task, therefore, is to find expressions for $\kappa v^{\ddagger}$ and $\Delta^{\ddagger}G$.

Our discussion concentrates on the following two key aspects of the theory of electron transfer processes, which was developed independently by R.A. Marcus, N.S. Hush, V.G. Levich, and R.R. Dogonadze:

1. Electrons are transferred by tunnelling through a potential energy barrier, the height of which is partly determined by the ionization energies of the DA and D^+A^- complexes. Electron tunnelling influences the magnitude of $\kappa v^{\ddagger}$.

2. The complex DA and the solvent molecules surrounding it undergo structural rearrangements prior to electron transfer. The energy associated with these rearrangements and the standard reaction Gibbs energy determine $\Delta^{\ddagger}G$.

(b) The role of electron tunnelling

We saw in Section 11.3 that, according to the Franck–Condon principle, electronic transitions are so fast that they can be regarded as taking place in a stationary nuclear framework. This principle also applies to an electron transfer process in which an electron migrates from one energy surface, representing the dependence of the energy of DA on its geometry, to another representing the energy of D^+A^-. We can represent the potential energy (and the Gibbs energy) surfaces of the two complexes (the reactant complex, DA, and the product complex, D^+A^-) by the parabolas characteristic of harmonic oscillators, with the displacement coordinate corresponding to the changing geometries (Fig. 20.20). This coordinate represents a collective mode of the donor, acceptor, and solvent.

According to the Franck–Condon principle, the nuclei do not have time to move when the system passes from the reactant to the product surface as a result of the transfer of an electron. Therefore, electron transfer can occur only after thermal fluctuations bring the geometry of DA to q^* in Fig 20.20, the value of the nuclear coordinate at which the two parabolas intersect.

The factor $\kappa v^{\ddagger}$ is a measure of the probability that the system will convert from reactants (DA) to products (D^+A^-) at q^* by

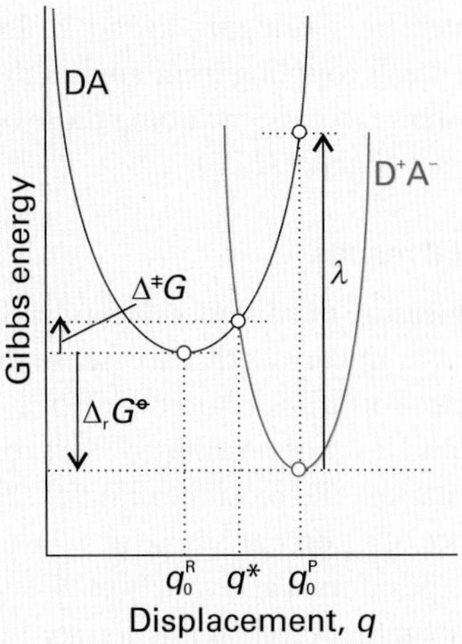

Fig. 20.20 The Gibbs energy surfaces of the complexes DA and D^+A^- involved in an electron transfer process are represented by parabolas characteristic of harmonic oscillators, with the displacement coordinate q corresponding to the changing geometries of the system. In the plot, q_0^R and q_0^P are the values of q at which the minima of the reactant and product parabolas occur, respectively. The parabolas intersect at $q = q^*$. The plots also portray the Gibbs energy of activation, $\Delta^‡G$, the standard reaction Gibbs energy, $\Delta_r G^⦵$, and the reorganization energy, λ (discussed in Section 20.8c).

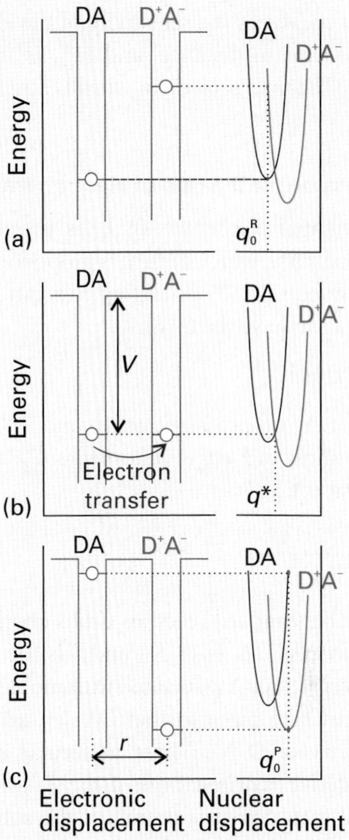

Fig. 20.21 Correspondence between the electronic energy levels (shown on the left) and the nuclear energy levels (shown on the right) for the DA and D^+A^- complexes involved in an electron transfer process. (a) At the nuclear configuration denoted by q_0^R, the electron to be transferred in DA is in an occupied electronic energy level (denoted by a blue circle) and the lowest unoccupied energy level of D^+A^- (denoted by an unfilled circle) is of too high an energy to be a good electron acceptor. (b) As the nuclei rearrange to a configuration represented by q^*, DA and D^+A^- become degenerate and electron transfer occurs by tunnelling through the barrier of height V and width r, the distance between the outer surfaces of the donor and acceptor. (c) The system relaxes to the equilibrium nuclear configuration of D^+A^- denoted by q_0^P, in which the lowest unoccupied electronic level of DA is higher in energy than the highest occupied electronic level of D^+A^-. (Adapted from R.A. Marcus and N. Sutin, *Biochim. Biophys. Acta* **811**, 265 (1985).)

electron transfer within the thermally excited DA complex. To understand the process, we must turn our attention to the effect that the rearrangement of nuclear coordinates has on electronic energy levels of DA and D^+A^- for a given distance r between D and A (Fig. 20.21). Initially, the electron to be transferred occupies the HOMO of D, and the overall energy of DA is lower than that of D^+A^- (Fig. 20.21a). As the nuclei rearrange to a configuration represented by q^* in Fig. 20.21b, the highest occupied electronic level of DA and the lowest unoccupied electronic level of D^+A^- become degenerate and electron transfer becomes energetically feasible. Over reasonably short distances r, the main mechanism of electron transfer is tunnelling through the potential energy barrier depicted in Fig. 20.21b. After an electron moves from the HOMO of D to the LUMO of A, the system relaxes to the configuration represented by q_0^P in Fig. 20.21c. As shown in the figure, now the energy of D^+A^- is lower than that of DA, reflecting the thermodynamic tendency for A to remain reduced and for D to remain oxidized.

The tunnelling event responsible for electron transfer is similar to that described in Section 2.3, except that in this case the electron tunnels from an electronic level of D, with wavefunction ψ_D, to an electronic level of A, with wavefunction ψ_A. We saw in Section 4.3 that the rate of an electronic transition from a level described by the wavefunction ψ_D to a level described by ψ_A is proportional to the square of the integral

$$H_{DA} = \int \psi_D \hat{h} \psi_A \, d\tau$$

where $\hat{h}$ is a hamiltonian that describes the coupling of the electronic wavefunctions. The probability of tunnelling through a potential barrier typically has an exponential dependence on distance, so we suspect that the distance dependence of H_{DA}^2 is

$$H_{DA}(r)^2 = H_{DA}^{o2} e^{-\beta r} \qquad (20.57)$$

where r is the edge-to-edge distance between D and A, β is a parameter that measures the sensitivity of the electronic coupling matrix element to distance, and H°_{DA} is the value of the electronic coupling matrix element when D and A are in contact ($r = 0$).

(c) The expression for the rate of electron transfer

The pre-exponential factor $\kappa\nu^{\ddagger}$ in k_{et} is proportional to the tunnelling probability, which in turn is proportional to $H_{DA}(r)^2$, as expressed by eqn 20.57. Therefore, we can expect the full expression for k_{et} to have the form

$$k_{et} = CH_{DA}(r)^2 e^{-\Delta^{\ddagger}G/RT} \tag{20.58}$$

with C a constant of proportionality and $H_{DA}(r)^2$ given by eqn 20.57. We show in *Further information 20.2* that the Gibbs energy of activation $\Delta^{\ddagger}G$ is

$$\Delta^{\ddagger}G = \frac{(\Delta_r G^{\ominus} + \lambda)^2}{4\lambda} \tag{20.59}$$

where $\Delta_r G^{\ominus}$ is the standard reaction Gibbs energy for the electron transfer process $DA \rightarrow D^+A^-$, and λ is the **reorganization energy**, the energy change associated with molecular rearrangements that must take place so that DA can take on the equilibrium geometry of D^+A^-. These molecular rearrangements include the relative reorientation of the D and A molecules in DA and the relative reorientation of the solvent molecules surrounding DA. Equation 20.59 shows that $\Delta^{\ddagger}G = 0$, with the implication that the reaction is not slowed down by an activation barrier, when $\Delta_r G^{\ominus} = -\lambda$, corresponding to the cancellation of the reorganization energy term by the standard reaction Gibbs energy.

The only missing piece of the expression for k_{et} is the value of the constant of proportionality C. Detailed calculation, which we do not repeat here, gives

$$C = \frac{1}{h}\left(\frac{\pi^3}{\lambda RT}\right)^{1/2} \tag{20.60}$$

Equation 20.58 has some limitations. For instance, it describes processes with weak electronic coupling between donor and acceptor. Weak coupling is observed when the electroactive species are sufficiently far apart that the wavefunctions ψ_A and ψ_D do not overlap extensively and the tunnelling is an exponential function of distance. An example of a weakly coupled system is the cytochrome *c*–cytochrome b_5 complex, in which the electroactive haem-bound iron ions shuttle between oxidation states +2 and +3 during electron transfer and are about 1.7 nm apart. Strong coupling is observed when the wavefunctions ψ_A and ψ_D overlap very extensively and, as well as other complications, the tunnelling probability is no longer a simple exponential function of distance. Examples of strongly coupled systems are mixed-valence, binuclear d-metal complexes with the general structure $L_mM^{n+}–B–M^{p+}L_{m'}$, in which the electroactive metal ions are separated by a bridging ligand B. In these systems, $r < 1.0$ nm. The weak coupling limit applies to a large number of electron transfer reactions, including those between proteins during metabolism (*Impact I17.3*).

(d) Experimental results

It is difficult to measure the distance dependence of k_{et} when the reactants are ions or molecules that are free to move in solution. In such cases, electron transfer occurs after a donor–acceptor complex forms and it is not possible to exert control over r, the edge-to-edge distance. The most meaningful experimental tests of the dependence of k_{et} on r are those in which the same donor and acceptor are positioned at a variety of distances, perhaps by covalent attachment to molecular linkers (see **1** for an example). Under these conditions, the term $e^{-\Delta^{\ddagger}G/RT}$ becomes a constant and, after taking the natural logarithm of eqn 20.58 and using eqn 20.57, we obtain

$$\ln k_{et} = -\beta r + \text{constant} \tag{20.61}$$

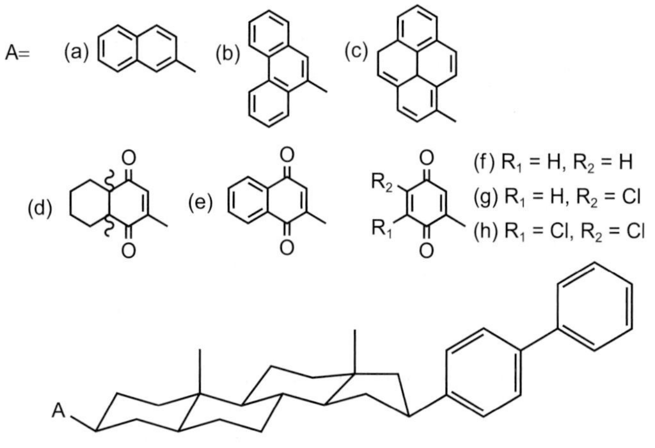

1 An electron donor–acceptor system

which implies that a plot of ln k_{et} against r should be a straight line with slope $-\beta$. The value of β depends on the medium through which the electron must travel from donor to acceptor. In a vacuum, $28\ \text{nm}^{-1} < \beta < 35\ \text{nm}^{-1}$, whereas $\beta \approx 9\ \text{nm}^{-1}$ when the intervening medium is a molecular link between donor and acceptor. Electron transfer between protein-bound co-factors can occur at distances of up to about 2.0 nm, a long distance on a molecular scale, corresponding to about 20 carbon atoms, with the protein providing an intervening medium between donor and acceptor.

The dependence of k_{et} on the standard reaction Gibbs energy has been investigated in systems where the edge-to-edge distance and the reorganization energy are constant for a series of reactions. Then, by using eqn 20.59 for $\Delta^{\ddagger}G$, eqn 20.58 becomes

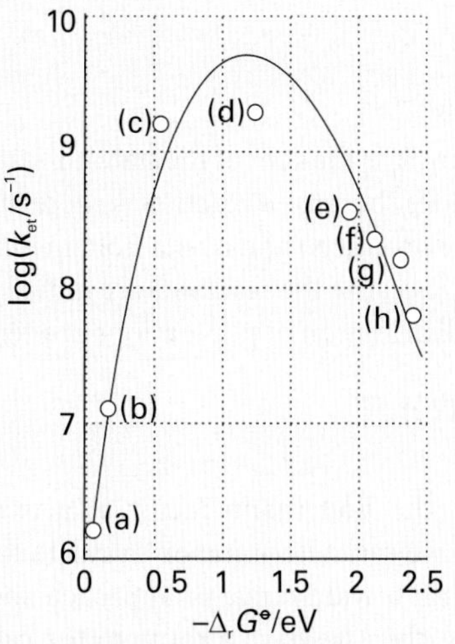

Fig. 20.22 Variation of $\log k_{et}$ with $-\Delta_r G^\ominus$ for a series of compounds with the structures given in (1). Kinetic measurements were conducted in 2-methyltetrahydrofuran and at 296 K. The distance between donor (the reduced biphenyl group) and the acceptor is constant for all compounds in the series because the molecular linker remains the same. Each acceptor has a characteristic standard reduction potential, so it follows that the standard Gibbs energy for the electron transfer process is different for each compound in the series. The line is a fit to a version of eqn 20.62 and the maximum of the parabola occurs at $-\Delta_r G^\ominus = \lambda = 1.2$ eV $= 1.2 \times 10^2$ kJ mol^{-1}. (Reproduced with permission from J.R. Miller *et al.*, *J. Am. Chem. Soc.* **106**, 3047 (1984).)

$$\ln k_{et} = -\frac{RT}{4\lambda}\left(\frac{\Delta_r G^\ominus}{RT}\right)^2 - \frac{1}{2}\left(\frac{\Delta_r G^\ominus}{RT}\right) + \text{constant} \qquad (20.62)$$

and a plot of $\ln k_{et}$ (or $\log k_{et}$) against $\Delta_r G^\ominus$ (or $-\Delta_r G^\ominus$) is predicted to be shaped like a downward parabola. Equation 20.62 implies that the rate constant increases as $\Delta_r G^\ominus$ decreases but only up to $-\Delta_r G^\ominus = \lambda$. Beyond that, the reaction enters the **inverted region**, in which the rate constant decreases as the reaction becomes more exergonic ($\Delta_r G^\ominus$ becomes more negative). The inverted region has been observed in a series of special compounds in which the electron donor and acceptor are linked covalently to a molecular spacer of known and fixed size (Fig. 20.22).

The dynamics of molecular collisions

We now come to the third and most detailed level of our examination of the factors that govern the rates of reactions.

20.9 Reactive collisions

Molecular beams allow us to study collisions between molecules in preselected energy states (for example, specific rotational and vibrational states), and can be used to determine the states of the products of a reactive collision. Information of this kind is essential if a full picture of the reaction is to be built, because the rate constant is an average over events in which reactants in different initial states evolve into products in their final states.

(a) Experimental probes of reactive collisions

Detailed experimental information about the intimate processes that occur during reactive encounters comes from molecular beams, especially crossed molecular beams (Fig. 20.23). The detector for the products of the collision of two beams can be moved to different angles, so the angular distribution of the products can be determined. Because the molecules in the incoming beams can be prepared with different energies (for example, with different translational energies by using rotating sectors and supersonic nozzles, with different vibrational energies by using selective excitation with lasers, and with different orientations by using electric fields), it is possible to study the dependence of the success of collisions on these variables and to study how they affect the properties of the outcoming product molecules.

One method for examining the energy distribution in the products is **infrared chemiluminescence**, in which vibrationally excited molecules emit infrared radiation as they return to their ground states. By studying the intensities of the infrared emission spectrum, the populations of the vibrational states of the products may be determined (Fig. 20.24). Another method makes use of **laser-induced fluorescence**. In this technique, a laser is used to excite a product molecule from a specific vibration-rotation level; the intensity of the fluorescence from the upper state is monitored and interpreted in terms of the population of

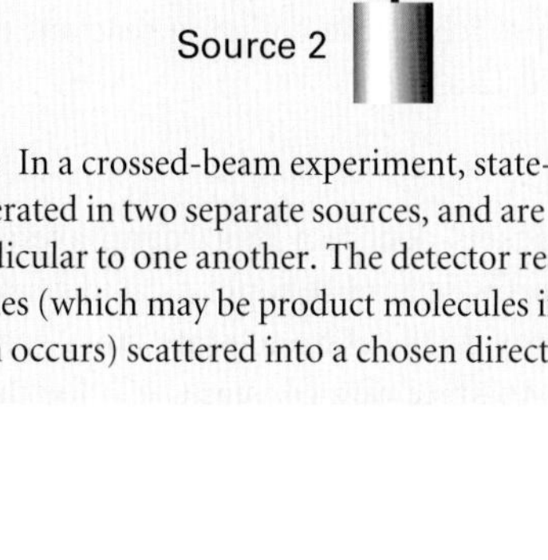

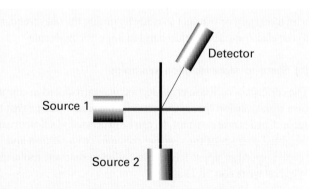

Fig. 20.23 In a crossed-beam experiment, state-selected molecules are generated in two separate sources, and are directed perpendicular to one another. The detector responds to molecules (which may be product molecules if a chemical reaction occurs) scattered into a chosen direction.

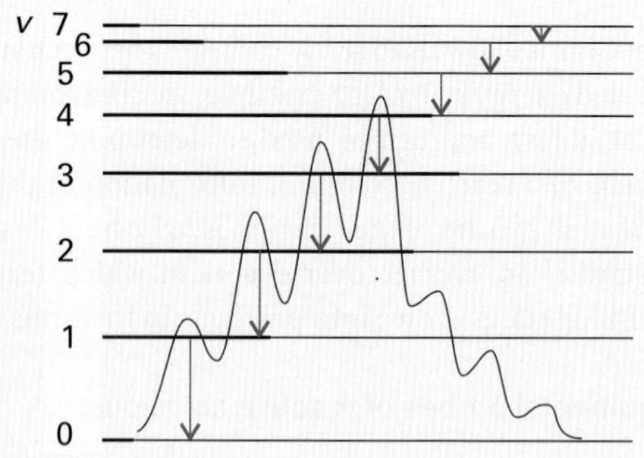

Fig. 20.24 Infrared chemiluminescence from CO produced in the reaction $O + CS \rightarrow CO + S$ arises from the non-equilibrium populations of the vibrational states of CO and the radiative relaxation to equilibrium.

the initial vibration-rotation state. When the molecules being studied do not fluoresce efficiently, coherent anti-Stokes Raman spectroscopy (CARS, Section 10.14) can be used to monitor the progress of reaction. **Multiphoton ionization** (MPI) techniques are also good alternatives for the study of weakly fluorescing molecules. In MPI, the absorption by a molecule of several photons from one or more pulsed lasers results in ionization if the total photon energy is greater than the ionization energy of the molecule. The angular distribution of products can be determined by **reaction product imaging**. In this technique, product ions are accelerated by an electric field towards a phosphorescent screen and the light emitted from specific spots where the ions struck the screen is imaged by a charge-coupled device (CCD). An important variant of MPI is **resonant multiphoton ionization** (REMPI), in which one or more photons promote a molecule to an electronically excited state and then additional photons are used to generate ions from the excited state. The power of REMPI lies in the fact that the experimenter can choose which reactant or product to study by tuning the laser frequency to the electronic absorption band of a specific molecule.

(b) State-to-state reaction dynamics

The concept of collision cross-section was introduced in connection with collision theory in Section 20.3, where we saw that the second-order rate constant, k_r, can be expressed as a Boltzmann-weighted average of the reactive collision cross-section and the relative speed of approach of the colliding reactant molecules. We shall write eqn 20.15 as

$$k_r = \langle \sigma v_{rel} \rangle N_A \qquad (20.63)$$

where the angle brackets denote a Boltzmann average. Molecular beam studies provide a more sophisticated version of this quantity, for they provide the **state-to-state cross-section**, $\sigma_{nn'}$, and hence the **state-to-state rate constant**, $k_{nn'}$ for the reactive transition from initial state n of the reactants to final state n' of the products:

$$k_{nn'} = \langle \sigma_{nn'} v_{rel} \rangle N_A \qquad (20.64)$$

The rate constant k_r is the sum of the state-to-state rate constants over all final states (because a reaction is successful whatever the final state of the products) and over a Boltzmann-weighted sum of initial states (because the reactants are initially present with a characteristic distribution of populations at a temperature T):

$$k_r = \sum_{n,n'} k_{nn'}(T) f_n(T) \qquad (20.65)$$

where $f_n(T)$ is the Boltzmann factor at a temperature T. It follows that, if we can determine or calculate the state-to-state cross-sections for a wide range of approach speeds and initial and final states, then we have a route to the calculation of the rate constant for the reaction.

20.10 Potential energy surfaces

One of the most important concepts for discussing beam results and calculating the state-to-state collision cross-section is the **potential energy surface** of a reaction, the potential energy as a function of the relative positions of all the atoms taking part in the reaction. Potential energy surfaces may be constructed from experimental data, with the techniques described in Section 20.9, and from results of quantum chemical calculations (Chapter 6). The theoretical method requires the systematic calculation of the energies of the system in a large number of geometrical arrangements. Special computational techniques, such as those described in Chapter 6, are used to take into account electron correlation, which arises from instantaneous interactions between electrons as they move closer to and farther from each other in a molecule or molecular cluster. Techniques that incorporate electron correlation accurately are very time-consuming and, consequently, only reactions between relatively simple particles, such as the reactions $H + H_2 \rightarrow H_2 + H$ and $H + H_2O \rightarrow OH + H_2$, are amenable to this type of theoretical treatment. An alternative is to use semiempirical methods, in which results of calculations and experimental parameters are used to construct the potential energy surface.

To illustrate the features of a potential energy surface we consider the collision between an H atom and an H_2 molecule. Detailed calculations show that the approach of an atom H_A along the H_B—H_C axis requires less energy for reaction than any other approach, so initially we confine our attention to a collinear approach. Two parameters are required to define the nuclear separations: one is the H_A—H_B separation R_{AB}, and the other is the H_B—H_C separation R_{BC}.

At the start of the encounter R_{AB} is infinite and R_{BC} is the H_2 equilibrium bond length. At the end of a successful reactive encounter R_{AB} is equal to the equilibrium bond length and R_{BC}

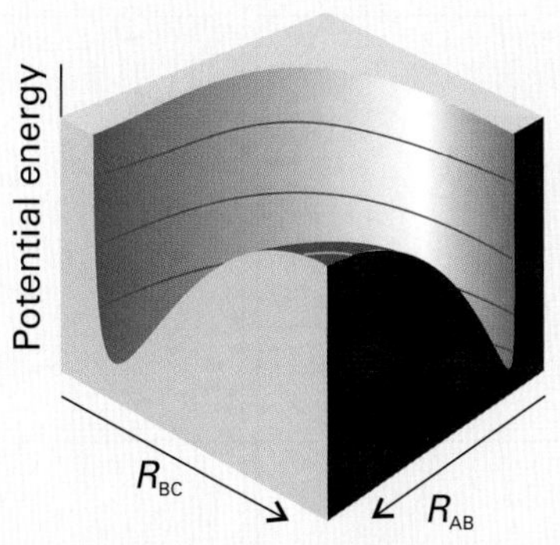

Fig. 20.25 The potential energy surface for the $H + H_2 \rightarrow H_2 + H$ reaction when the atoms are constrained to be collinear.

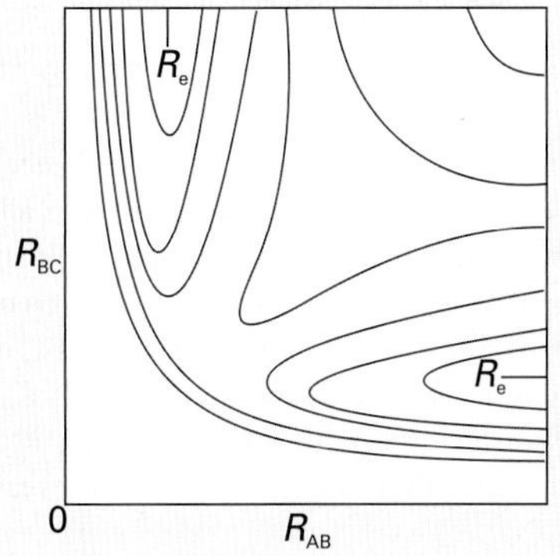

Fig. 20.26 The contour diagram (with contours of equal potential energy) corresponding to the surface in Fig. 20.25. R_e marks the equilibrium bond length of an H_2 molecule (strictly, it relates to the arrangement when the third atom is at infinity).

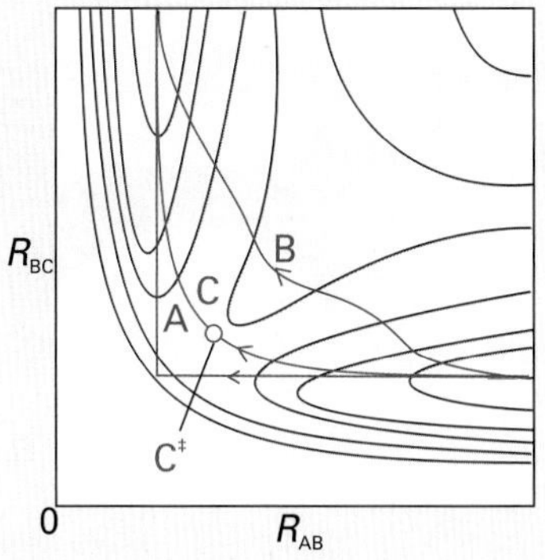

Fig. 20.27 Various trajectories through the potential energy surface shown in Fig. 20.26. Path A corresponds to a route in which R_{BC} is held constant as H_A approaches; path B corresponds to a route in which R_{BC} lengthens at an early stage during the approach of H_A; path C is the route along the floor of the potential valley.

is infinite. The total energy of the three-atom system depends on their relative separations, and can be found by doing an electronic structure calculation. The plot of the total energy of the system against R_{AB} and R_{BC} gives the potential energy surface of this collinear reaction (Fig. 20.25). This surface is normally depicted as a contour diagram (Fig. 20.26).

When R_{AB} is very large, the variation in potential energy represented by the surface as R_{BC} changes is that of an isolated H_2 molecule as its bond length is altered. A section through the surface at $R_{AB} = \infty$, for example, is the same as the H_2 bonding potential energy curve. At the edge of the diagram where R_{BC} is very large, a section through the surface is the molecular potential energy curve of an isolated H_AH_B molecule.

The actual path of the atoms in the course of the encounter depends on their total energy, the sum of their kinetic and potential energies. However, we can obtain an initial idea of the paths available to the system for paths that correspond to least potential energy. For example, consider the changes in potential energy as H_A approaches H_BH_C. If the H_B—H_C bond length is constant during the initial approach of H_A, then the potential energy of the H_3 cluster rises along the path marked A in Fig. 20.27. We see that the potential energy reaches a high value as H_A is pushed into the molecule and then decreases sharply as H_C breaks off and separates to a great distance. An alternative reaction path can be imagined (B) in which the H_B—H_C bond length increases while H_A is still far away. Both paths, although feasible if the molecules have sufficient initial kinetic energy, take the three atoms to regions of high potential energy in the course of the encounter.

The path of least potential energy is the one marked C, corresponding to R_{BC} lengthening as H_A approaches and begins to form a bond with H_B. The H_B—H_C bond relaxes at the demand of the incoming atom, and the potential energy climbs only as far as the saddle-shaped region of the surface, to the **saddle point** marked $C^{\ddagger}$. The encounter of least potential energy is one in which the atoms take route C up the floor of the valley, through the saddle point, and down the floor of the other valley as H_C recedes and the new H_A—H_B bond achieves its equilibrium length. This path is the reaction coordinate we met earlier in this chapter.

We can now make contact with the transition state theory of reaction rates. In terms of trajectories on potential surfaces with a total energy close to the saddle point energy, the transition state can be identified with a critical geometry such that every trajectory that goes through this geometry goes on to react (Fig. 20.28).

A brief comment Most trajectories on potential energy surfaces do not go directly over the saddle point and therefore, to result in a reaction, they require a total energy significantly higher than the saddle point energy. As a result, the experimentally determined activation energy is often much higher than the calculated saddle-point energy.

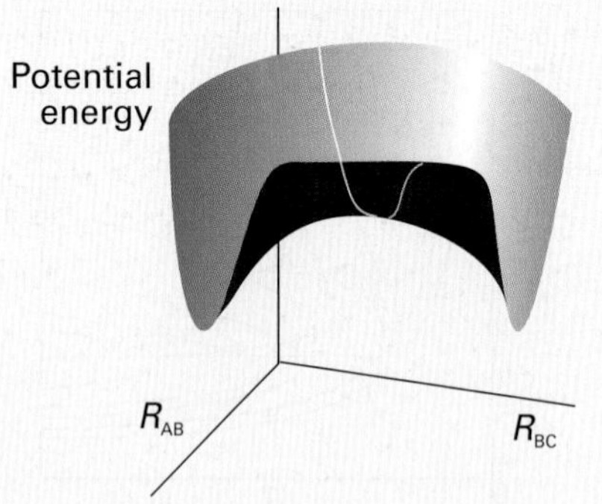

Fig. 20.28 The transition state is a set of configurations (here, marked by the line across the saddle point) through which successful reactive trajectories must pass.

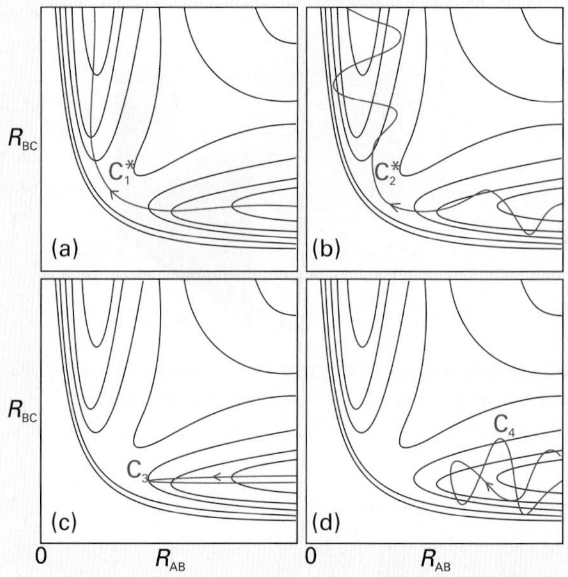

Fig. 20.29 Some successful (*) and unsuccessful encounters. (a) C_1^* corresponds to the path along the foot of the valley; (b) C_2^* corresponds to an approach of A to a vibrating BC molecule, and the formation of a vibrating AB molecule as C departs. (c) C_3 corresponds to A approaching a non-vibrating BC molecule, but with insufficient translational kinetic energy; (d) C_4 corresponds to A approaching a vibrating BC molecule, but still the energy, and the phase of the vibration, is insufficient for reaction.

20.11 Some results from experiments and calculations

To travel successfully from reactants to products, classically the incoming molecules must possess enough kinetic energy to be able to climb to the saddle point of the potential surface. We proceed with this classical argument, although quantum mechanical tunnelling can also play an important role in reactivity, particularly in hydrogen atom or electron transfer reactions. Therefore, the shape of the surface can be explored experimentally by changing the relative speed of approach (by selecting the beam velocity) and the degree of vibrational excitation and observing whether reaction occurs and whether the products emerge in a vibrationally excited state (Fig. 20.29). For example, one question that can be answered is whether it is better to smash the reactants together with a lot of translational kinetic energy or to ensure instead that they approach in highly excited vibrational states. Thus, is trajectory C_2^*, where the H_BH_C molecule is initially vibrationally excited, more efficient at leading to reaction than the trajectory C_1^*, in which the total energy is the same but reactants have a high translational kinetic energy?

(a) The direction of attack and separation

Figure 20.30 shows the results of a calculation of the potential energy as an H atom approaches an H_2 molecule from different angles, the H_2 bond being allowed to relax to the optimum length in each case. The potential barrier is least for collinear attack, as we assumed earlier. (But we must be aware that other lines of attack are feasible and contribute to the overall rate; see the *brief comment* above.) In contrast, Fig. 20.31 shows the potential energy changes that occur as a Cl atom approaches an HI molecule. The lowest barrier occurs for approaches within a cone of half-angle 30° surrounding the H atom. The relevance of this result to the calculation of the steric factor of collision

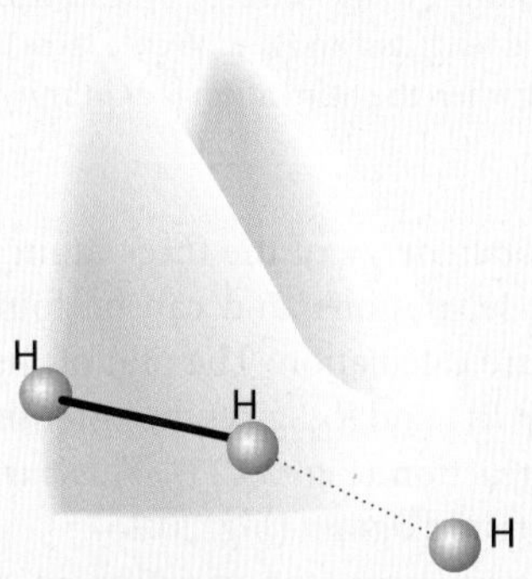

Fig. 20.30 An indication of how the anisotropy of the potential energy changes as H approaches H_2 with different angles of attack. The collinear attack has the lowest potential barrier to reaction. The surface indicates the potential energy profile along the reaction coordinate for each configuration.

theory should be noted: not every collision is successful, because not every one lies within the reactive cone.

If the collision is sticky, so that when the reactants collide they orbit around each other, the products can be expected to emerge in random directions because all memory of the approach

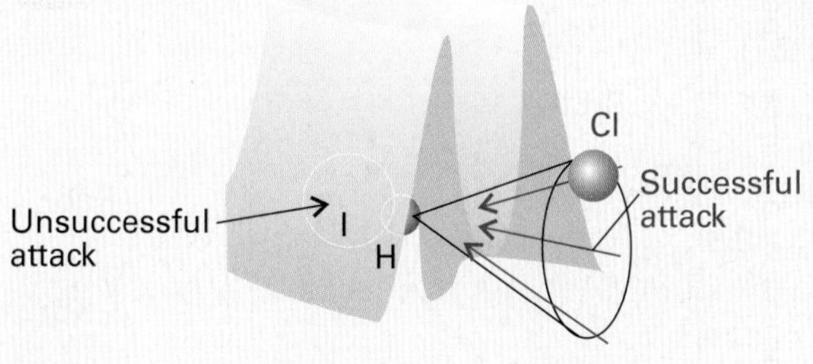

Fig. 20.31 The potential energy barrier for the approach of Cl to HI. In this case, successful encounters occur only when Cl approaches within a cone surrounding the H atom.

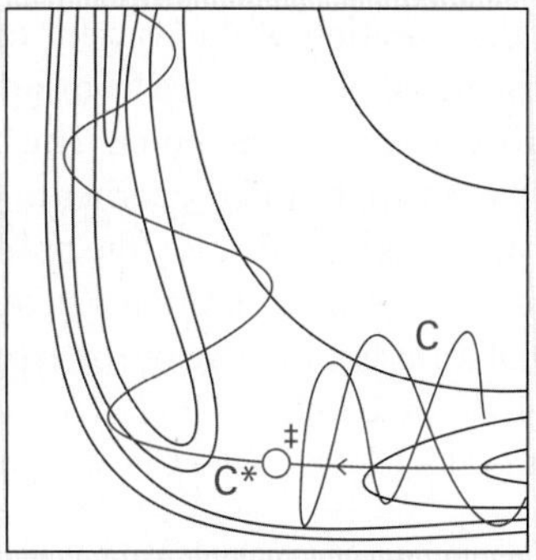

Fig. 20.32 An attractive potential energy surface. A successful encounter (C*) involves high translational kinetic energy and results in a vibrationally excited product.

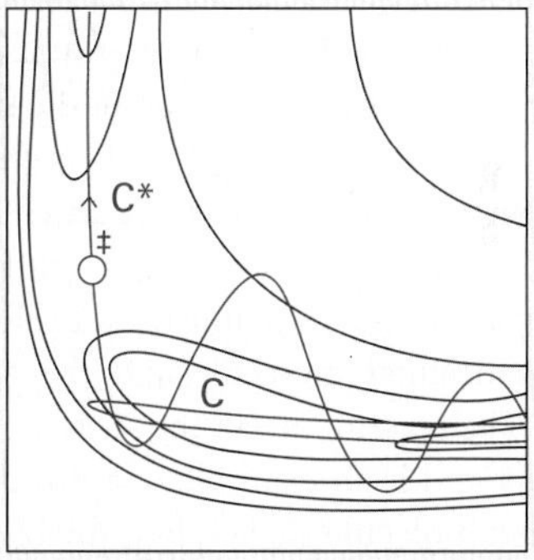

Fig. 20.33 A repulsive potential energy surface. A successful encounter (C*) involves initial vibrational excitation and the products have high translational kinetic energy. A reaction that is attractive in one direction is repulsive in the reverse direction.

direction has been lost. A rotation takes about 1 ps, so if the collision is over in less than that time the complex will not have had time to rotate and the products will be thrown off in a specific direction. In the collision of K and I_2, for example, most of the products are thrown off in the forward direction (forward and backward directions refer to directions in a centre-of-mass coordinate system with the origin at the centre of mass of the colliding reactants and collision occurring when molecules are at the origin). This product distribution is consistent with the harpoon mechanism (Section 20.3) because the transition takes place at long range. In contrast, the collision of K with CH_3I leads to reaction only if the molecules approach each other very closely. In this mechanism, K effectively bumps into a brick wall, and the KI product bounces out in the backward direction. The detection of this anisotropy in the angular distribution of products gives an indication of the distance and orientation of approach needed for reaction, as well as showing that the event is complete in less than 1 ps.

(b) Attractive and repulsive surfaces

Some reactions are very sensitive to whether the energy has been predigested into a vibrational mode or left as the relative translational kinetic energy of the colliding molecules. For example, if two HI molecules are hurled together with more than twice the activation energy of the reaction, then no reaction occurs if all the energy is translational. For $F + HCl \rightarrow Cl + HF$, for example, the reaction is about five times more efficient when the HCl is in its first vibrational excited state than when, although HCl has the same total energy, it is in its vibrational ground state.

The origin of these requirements can be found by examining the potential energy surface. Figure 20.32 shows an **attractive surface** in which the saddle point occurs early in the reaction coordinate. Figure 20.33 shows a **repulsive surface** in which the saddle point occurs late. A surface that is attractive in one direction is repulsive in the reverse direction.

Consider first the attractive surface. If the original molecule is vibrationally excited, then a collision with an incoming molecule takes the system along C. This path is bottled up in the region of the reactants, and does not take the system to the saddle point. If, however, the same amount of energy is present solely as translational kinetic energy, then the system moves along C* and travels smoothly over the saddle point into products. We can therefore conclude that reactions with attractive potential energy surfaces proceed more efficiently if the energy is in relative translational motion. Moreover, the potential surface shows that once past the saddle point the trajectory runs up the steep wall of the product valley, and then rolls from side to side as it falls to the foot of the valley as the products separate. In other words, the products emerge in a vibrationally excited state.

Now consider the repulsive surface. On trajectory C the collisional energy is largely in translation. As the reactants approach, the potential energy rises. Their path takes them up the opposing face of the valley, and they are reflected back into the reactant region. This path corresponds to an unsuccessful encounter, even though the energy is sufficient for reaction. On C* some of the energy is in the vibration of the reactant molecule and the motion causes the trajectory to weave from side to side up the valley as it approaches the saddle point. This motion may be sufficient to tip the system round the corner to the saddle point and then on to products. In this case, the product molecule is expected to be in an unexcited vibrational state. Reactions with repulsive potential surfaces can therefore be expected to proceed more efficiently if the excess energy is present as vibrations. This is the case with the $H + Cl_2 \rightarrow HCl + Cl$ reaction, for instance.

(c) Classical trajectories

A clear picture of the reaction event can be obtained by using classical mechanics to calculate the trajectories of the atoms taking place in a reaction from a set of initial conditions, such as velocities, relative orientations, and internal energies of the reacting particles. The initial values used for the internal energy reflect the quantization of electronic, vibrational, and rotational energies in molecules but the features of quantum mechanics are not used explicitly in the calculation of the trajectory.

Figure 20.34 shows the result of such a calculation of the positions of the three atoms in the reaction $H + H_2 \rightarrow H_2 + H$, the horizontal coordinate now being time and the vertical coordinate the separations. This illustration shows clearly the vibration of the original molecule and the approach of the attacking atom. The reaction itself, the switch of partners, takes place very rapidly and is an example of a **direct mode process.** The newly formed molecule shakes, but quickly settles down to steady, harmonic vibration as the expelled atom departs. In contrast, Fig. 20.35 shows an example of a **complex mode process,** in which the activated complex survives for an extended period. The reaction in the figure is the exchange reaction $KCl + NaBr \rightarrow KBr + NaCl$. The tetratomic activated complex survives for about 5 ps, during which time the atoms make about 15 oscillations before dissociating into products.

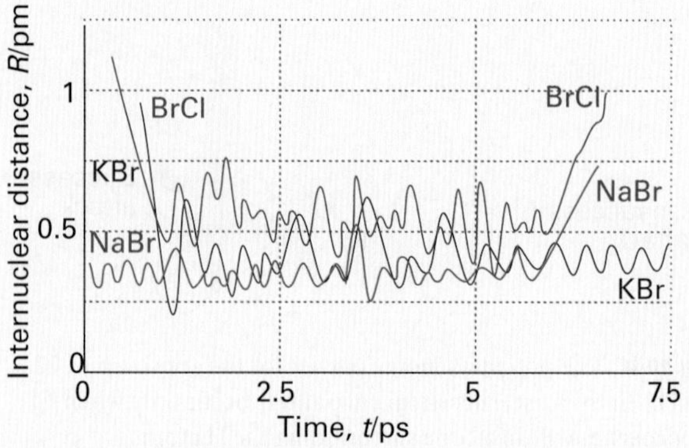

Fig. 20.35 An example of the trajectories calculated for a complex-mode reaction, $KCl + NaBr \rightarrow KBr + NaCl$, in which the collision cluster has a long lifetime. (P. Brumer and M. Karplus, *Faraday Disc. Chem. Soc.* 55, 80 (1973).)

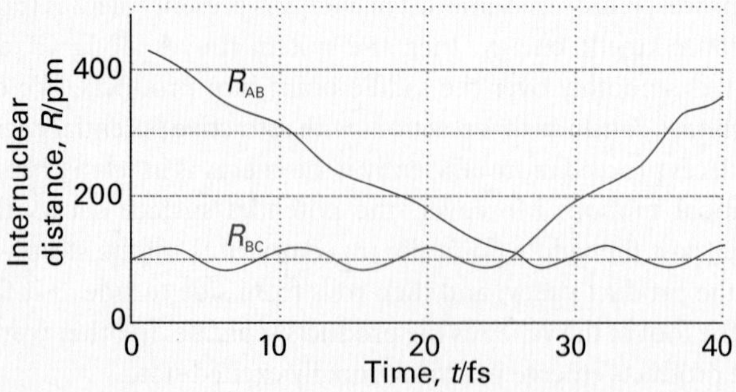

Fig. 20.34 The calculated trajectories for a reactive encounter between A and a vibrating BC molecule leading to the formation of a vibrating AB molecule. This direct-mode reaction is between H and H_2. (M. Karplus *et al., J. Chem. Phys.* **43**, 3258 (1965).)

(d) Quantum mechanical scattering theory

Classical trajectory calculations do not recognize the fact that the motion of atoms, electrons, and nuclei is governed by quantum mechanics. The concept of trajectory then fades and is replaced by the unfolding of a wavefunction that represents initially the reactants and finally products.

Complete quantum mechanical calculations of trajectories and rate constants are very onerous because it is necessary to take into account all the allowed electronic, vibrational, and rotational states populated by each atom and molecule in the system at a given temperature. It is common to define a 'channel' as a group of molecules in well-defined quantum mechanically allowed states. Then, at a given temperature, there are many channels that represent the reactants and many channels that represent possible products, with some transitions between channels being allowed but others not allowed. Furthermore, not every transition leads to a chemical reaction. For example, the process $H_2^* + OH \rightarrow H_2 + (OH)^*$, where the asterisk denotes an excited state, amounts to energy transfer between H_2 and OH, whereas the process $H_2^* + OH \rightarrow H_2O + H$ represents a chemical reaction. What complicates a quantum mechanical calculation of rate constants even in this simple four-atom system is that many reacting channels present at a given temperature can lead to the desired products $H_2O + H$, which themselves may be formed as many distinct channels. The **cumulative reaction probability,** $N(E)$, at a fixed total energy E is then written as

$$N(E) = \sum_{i,j} P_{ij}(E) \qquad (20.66)$$

where $P_{ij}(E)$ is the probability for a transition between a reactant channel i and a product channel j and the summation is over all possible transitions that lead to product. It is then possible to show that the rate constant is given by

$$k_r(T) = \frac{\int_0^\infty N(E)e^{-E/kT}dE}{hQ_R(T)} \qquad (20.67)$$

where $Q_R(T)$ is the partition function density (the partition function divided by the volume) of the reactants at the temperature T. The significance of eqn 20.67 is that it provides a direct connection between an experimental quantity, the rate constant, and a theoretical quantity, $N(E)$.

Table 20.4 Summary of uses of k

Symbol	Significance
k	Boltzmann's constant
k_r	Rate constant
k_r°	Rate constant at zero ionic strength
$k_a, k_b, \ldots$	Rate constants for individual steps
$k'_a, k'_b, \ldots$	Rate constants for individual reverse steps
$k^\ddagger$	Rate constant for unimolecular decay of activated complex
K	Equilibrium constant (dimensionless)
K_γ	Ratio of activity coefficients
$K^\ddagger$	Proportionality constant in transition state theory
κ	Transmission coefficient
$\bar{K}$	Equilibrium constant with one mode discarded
k_f	Force constant

Checklist of key ideas

☐ 1. The temperature dependence of the rate constant of a reaction typically follows the Arrhenius equation, $\ln k_r = \ln A - E_a/RT$.

☐ 2. The activation energy, the parameter E_a in the Arrhenius equation, is the minimum kinetic energy for reaction during a molecular encounter. The larger the activation energy, the more sensitive the rate constant is to the temperature.

☐ 3. In collision theory, it is supposed that the rate is proportional to the collision frequency, a steric factor, and the fraction of collisions that occur with at least the kinetic energy E_a along their lines of centres.

☐ 4. For unimolecular reactions, the steric factor can be computed by using the RRK model.

☐ 5. A reaction in solution may be diffusion-controlled if its rate is controlled by the rate at which reactant molecules encounter each other in solution. The rate of an activation-controlled reaction is controlled by the rate at which the encounter pair accumulates sufficient energy.

☐ 6. The material balance equation relates the overall rate of change of the concentration of a species to its rates of diffusion, convection, and reaction (eqn 20.27).

☐ 7. In transition state theory, it is supposed that an activated complex is in equilibrium with the reactants, and that the rate at which that complex forms products depends on the rate at which it passes through a transition state. The result is the Eyring equation, $k_r = \kappa(kT/h)\bar{K}_c^\ddagger$.

☐ 8. The rate constant may be parametrized in terms of the Gibbs energy, entropy, and enthalpy of activation.

☐ 9. The kinetic salt effect is the effect of an added inert salt on the rate of a reaction between ions.

☐ 10. The rate constant of electron transfer in a donor–acceptor complex depends on the distance between electron donor and acceptor, the standard reaction Gibbs energy, and the reorganization energy, λ.

☐ 11. Techniques for the study of reactive collisions include infrared chemiluminescence, laser-induced fluorescence, multiphoton ionization (MPI), reaction product imaging, and resonant multiphoton ionization (REMPI).

☐ 12. A potential energy surface maps the potential energy as a function of the relative positions of all the atoms taking part in a reaction. In an attractive surface, the saddle point (the highest point) occurs early on the reaction coordinate. In a repulsive surface, the saddle point occurs late on the reaction coordinate.

☐ 13. Femtosecond laser techniques can be used to probe directly the activated complex and to control the outcome of some chemical reactions.

Further information

Further information 20.1 *The RRK model of unimolecular reactions*

To set up the RRK model, we suppose that a molecule consists of s identical harmonic oscillators, each of which has frequency ν. In practice, of course, the vibrational modes of a molecule have different frequencies, but assuming that they are all the same is a good first approximation. Next, we suppose that the vibrations are excited to a total energy $E = nh\nu$ and then set out to calculate the number of ways N in which the energy can be distributed over the oscillators.

We can represent the n quanta as follows:

□□□□□□□□□□□□□□□□□□□□□□□□□□□□□□□□ . . .
□□□

These quanta must be put in s containers (the s oscillators), which can be represented by inserting $s-1$ walls, denoted by |. One such distribution is

□□|□□□□|□□||□□□|□□□□□□□□|□□□□|||□□□□□|
□□□□ . . . □|□□

The total number of arrangements of each quantum and wall (of which there are $n+s-1$ in all) is $(n+s-1)!$ where, as usual, $x! = x(x-1)\ldots 1$. However the $n!$ arrangements of the n quanta are indistinguishable, as are the $(s-1)!$ arrangements of the $s-1$ walls. Therefore, to find N we must divide $(n+s-1)!$ by these two factorials. It follows that

$$N = \frac{(n+s-1)!}{n!(s-1)!} \qquad (20.68)$$

The distribution of the energy throughout the molecule means that it is too sparsely spread over all the modes for any particular bond to be sufficiently highly excited to undergo dissociation. If we suppose that a bond will break if it is excited to at least an energy $E^\star = n^\star h\nu$, then the number of ways in which at least this energy can be localized in one bond is

$$N^\star = \frac{(n-n^\star+s-1)!}{(n-n^\star)!(s-1)!} \qquad (20.69)$$

To obtain this result, we isolate one critical oscillator as the one that undergoes dissociation if it has *at least* $n^\star$ of the quanta, leaving up to $n-n^\star$ quanta to be accommodated in the remaining $s-1$ oscillators (and therefore with $s-2$ walls in the partition in place of the $s-1$ walls we used above). We suppose that the critical oscillator consists of a single level plus an array of levels like the other oscillators, and that dissociation occurs however many quanta are in this latter array of levels, from 0 upwards. For example, in a system of five oscillators (other than the critical one) we might suppose that at least 6 quanta out of the 28 available must be present in the critical oscillator, then all the following partitions will result in dissociation:

□□□□□□|□□□□□|□□□□□□□□|□□□□||□□□□□
□□□□□□□|□□□□|□□□□□□□□|□□□□||□□□□□
□□□□□□□□|□□□|□□□□□□□□|□□□□||□□□□□
. . .

(The leftmost partition is the critical oscillator.) However, these partitions are equivalent to

□□□□□□ |□□□□□|□□□□□□□□|□□□□||□□□□□
□□□□□□ □|□□□□|□□□□□□□□|□□□□||□□□□□
□□□□□□ □□|□□□|□□□□□□□□|□□□□||□□□□□
. . .

and we see that we have the problem of permuting $28 - 6 = 22$ (in general, $n - n^\star$) quanta and 5 (in general, $s-1$) walls, and therefore a total of 27 (in general, $n - n^\star + s - 1$) objects. Therefore, the calculation is exactly like the one above for N, except that we have to find the number of distinguishable permutations of $n - n^\star$ quanta in s containers (and therefore $s-1$ walls). The number $N^\star$ is therefore obtained from eqn 20.68 by replacing n by $n - n^\star$.

From the preceding discussion we conclude that the probability that one specific oscillator will have undergone sufficient excitation to dissociate is the ratio $N^\star/N$, which is

$$P = \frac{N^\star}{N} = \frac{n!(n-n^\star+s-1)!}{(n-n^\star)!(n+s-1)!} \qquad (20.70)$$

Equation 20.70 is still awkward to use, even when written out in terms of its factors:

$$P = \frac{n(n-1)(n-2)\cdots 1}{(n-n^\star)(n-n^\star-1)\cdots 1} \times \frac{(n-n^\star+s-1)(n-n^\star+s-2)\cdots 1}{(n+s-1)(n+s-2)\cdots 1}$$

$$= \frac{(n-n^\star+s-1)(n-n^\star+s-2)\cdots(n-n^\star+1)}{(n+s-1)(n+s-2)\cdots(n+2)(n+1)}$$

However, because $s-1$ is small (in the sense $s-1 \ll n-n^\star$), we can approximate this expression by

$$P \approx \frac{(n-n^\star)(n-n^\star)\cdots(n-n^\star)_{s-1 \text{ factors}}}{(n)(n)\cdots(n)_{s-1 \text{ factors}}} = \left(\frac{n-n^\star}{n}\right)^{s-1}$$

Because the energy of the excited molecule is $E = nh\nu$ and the critical energy is $E^\star = n^\star h\nu$, this expression may be written

$$P = \left(1 - \frac{E^\star}{E}\right)^{s-1}$$

The dispersal of the energy of the collision reduces the rate constant below its simple 'Lindemann' form. To obtain the observed rate constant we should multiply the latter by the probability that the energy will in fact be localized in the bond of interest. Doing so gives eqn 20.19b, with the probability P identified with the P-factor of ordinary collision theory.

Further information 20.2 *The Gibbs energy of activation of electron transfer*

The simplest way to derive an expression for the Gibbs energy of activation of electron transfer processes is to construct a model in which the surfaces for DA (the 'reactant complex', denoted R) and for D^+A^- (the 'product complex', denoted P) are described by classical harmonic oscillators with identical reduced masses μ and angular frequencies ω, but displaced minima, as shown in Fig. 20.20. The molar Gibbs energies $G_{m,R}(q)$ and $G_{m,P}(q)$ of the reactant and product complexes, respectively, may be written as

$$G_{m,R}(q) = \tfrac{1}{2}N_A\mu\omega^2(q - q_0^R)^2 + G_{m,R}(q_0^R) \quad (20.71a)$$

$$G_{m,P}(q) = \tfrac{1}{2}N_A\mu\omega^2(q - q_0^P)^2 + G_{m,P}(q_0^P) \quad (20.71b)$$

where q_0^R and q_0^P are the values of q at which the minima of the reactant and product parabolas occur, respectively. The standard reaction Gibbs energy for the electron transfer process $DA \rightarrow D^+A^-$ is $\Delta_r G^\ominus = G_{m,P}(q_0^P) - G_{m,R}(q_0^R)$, the difference in standard molar Gibbs energy between the minima of the parabolas. In Fig. 20.20, $\Delta_r G^\ominus < 0$.

We also note that q^*, the value of q corresponding to the transition state of the complex, may be written in terms of the parameter α, the fractional change in q:

$$q^* = q_0^R + \alpha(q_0^P - q_0^R) \quad (20.72)$$

We see from Fig. 20.20 that $\Delta^\ddagger G = G_{m,R}(q^*) - G_{m,R}(q_0^R)$. It then follows from eqns 20.71a, 20.71b, and 20.72 that

$$\Delta^\ddagger G = \tfrac{1}{2}N_A\mu\omega^2(q^* - q_0^R)^2 = \tfrac{1}{2}N_A\mu\omega^2\{\alpha(q_0^P - q_0^R)\}^2 \quad (20.73)$$

We now define the reorganization energy, λ, as

$$\lambda = \tfrac{1}{2}N_A\mu\omega^2(q_0^P - q_0^R)^2 \quad (20.74)$$

which can be interpreted as $G_{m,R}(q_0^P) - G_{m,R}(q_0^R)$ and, consequently, as the (Gibbs) energy required to deform the equilibrium configuration of DA to the equilibrium configuration of D^+A^- (as shown in Fig. 20.20). It follows from eqns 20.73 and 20.74 that

$$\Delta^\ddagger G = \alpha^2\lambda \quad (20.75)$$

Because $G_{m,R}(q^*) = G_{m,P}(q^*)$, it follows from eqns 20.71b, 20.74, and 20.75 that

$$\alpha^2\lambda = \tfrac{1}{2}N_A\mu\omega^2\{(\alpha - 1)(q_0^P - q_0^R)\}^2 + \Delta_r G^\ominus = (\alpha - 1)\lambda + \Delta_r G^\ominus \quad (20.76)$$

which implies that

$$\alpha = \tfrac{1}{2}\left(\frac{\Delta_r G^\ominus}{\lambda} + 1\right) \quad (20.77)$$

By combining eqns 20.75 and 20.77, we obtain eqn 20.59. We can obtain an identical relation if we allow the harmonic oscillators to have different angular frequencies and hence different curvatures.

Discussion questions

20.1 Define the terms in and discuss the generality of the expression $\ln k_r = \ln A - E_a/RT$.

20.2 Is it possible for the activation energy of a reaction to be negative? Explain your conclusion and provide a molecular interpretation.

20.3 Discuss how the collision theory of gases builds upon the kinetic-molecular theory.

20.4 Describe the essential features of the harpoon mechanism.

20.5 Discuss the significance of the steric *P*-factor in the RRK model.

20.6 Distinguish between a diffusion-controlled reaction and an activation-controlled reaction. Do both have activation energies?

20.7 Describe in outline the formulation of the Eyring equation.

20.8 Explain the physical origin of the kinetic salt effect. What might be the effect of the relative permittivity of the medium?

20.9 Discuss how the following factors determine the rate of electron transfer in homogeneous systems: the distance between electron donor and acceptor, the standard Gibbs energy of the process, and the reorganization energy of redox active species and the surrounding medium.

20.10 Describe how the following techniques are used in the study of chemical dynamics: infrared chemiluminescence, laser-induced fluorescence, multiphoton ionization, resonant multiphoton ionization, reaction product imaging, and femtosecond spectroscopy.

20.11 Discuss the relationship between the saddle-point energy and the activation energy of a reaction.

20.12 A method for directing the outcome of a chemical reaction consists of using molecular beams to control the relative orientations of reactants during a collision. Consider the reaction $Rb + CH_3I \rightarrow RbI + CH_3$. How should CH_3I molecules and Rb atoms be oriented to maximize the production of RbI?

20.13 Consider a reaction with an attractive potential energy surface. Discuss how the initial distribution of reactant energy affects how efficiently the reaction proceeds. Repeat for a repulsive potential energy surface.

Exercises

20.1(a) The rate constant for the decomposition of a certain substance is 3.80×10^{-3} dm^3 mol^{-1} s^{-1} at 35°C and 2.67×10^{-2} dm^3 mol^{-1} s^{-1} at 50°C. Evaluate the Arrhenius parameters of the reaction.

20.1(b) The rate constant for the decomposition of a certain substance is 2.25×10^{-2} dm^3 mol^{-1} s^{-1} at 29°C and 4.01×10^{-2} dm^3 mol^{-1} s^{-1} at 37°C. Evaluate the Arrhenius parameters of the reaction.

20.2(a) The rate of a chemical reaction is found to triple when the temperature is raised from 24°C to 49°C. Determine the activation energy.

20.2(b) The rate of a chemical reaction is found to double when the temperature is raised from 25°C to 35°C. Determine the activation energy.

20.3(a) The mechanism of a composite reaction consists of a fast pre-equilibrium step with forward and reverse activation energies of 25 kJ mol^{-1} and 38 kJ mol^{-1}, respectively, followed by an elementary step of activation energy 10 kJ mol^{-1}. What is the activation energy of the composite reaction?

20.3(b) The mechanism of a composite reaction consists of a fast pre-equilibrium step with forward and reverse activation energies of 27 kJ mol^{-1} and 35 kJ mol^{-1}, respectively, followed by an elementary step of activation energy 15 kJ mol^{-1}. What is the activation energy of the composite reaction?

20.4(a) Calculate the collision frequency, z, and the collision density, Z, in ammonia, $R = 190$ pm, at 30°C and 120 kPa. What is the percentage increase when the temperature is raised by 10 K at constant volume?

20.4(b) Calculate the collision frequency, z, and the collision density, Z, in carbon monoxide, $R = 180$ pm at 30°C and 120 kPa. What is the percentage increase when the temperature is raised by 10 K at constant volume?

20.5(a) Collision theory depends on knowing the fraction of molecular collisions having at least the kinetic energy E_a along the line of flight. What is this fraction when (a) $E_a = 20$ kJ mol^{-1}, (b) $E_a = 100$ kJ mol^{-1} at (i) 350 K and (ii) 900 K?

20.5(b) Collision theory depends on knowing the fraction of molecular collisions having at least the kinetic energy E_a along the line of flight. What is this fraction when (a) $E_a = 15$ kJ mol^{-1}, (b) $E_a = 150$ kJ mol^{-1} at (i) 300 K and (ii) 800 K?

20.6(a) Calculate the percentage increase in the fractions in Exercise 20.5a when the temperature is raised by 10 K.

20.6(b) Calculate the percentage increase in the fractions in Exercise 20.5b when the temperature is raised by 10 K.

20.7(a) Use the collision theory of gas-phase reactions to calculate the theoretical value of the second-order rate constant for the reaction $H_2(g) + I_2(g) \rightarrow 2\,HI(g)$ at 650 K, assuming that it is elementary and bimolecular. The collision cross-section is 0.36 nm^2, the reduced mass is 3.32×10^{-27} kg, and the activation energy is 171 kJ mol^{-1}. (Assume a steric factor of 1.)

20.7(b) Use the collision theory of gas-phase reactions to calculate the theoretical value of the second-order rate constant for the reaction $D_2(g) + Br_2(g) \rightarrow 2\,DBr(g)$ at 450 K, assuming that it is elementary and bimolecular. Take the collision cross-section as 0.30 nm^2, the reduced mass as $3.930m_u$, and the activation energy as 200 kJ mol^{-1}. (Assume a steric factor of 1.)

20.8(a) A typical diffusion coefficient for small molecules in aqueous solution at 25°C is 6×10^{-9} m^2 s^{-1}. If the critical reaction distance is 0.5 nm, what value is expected for the second-order rate constant for a diffusion-controlled reaction?

20.8(b) Suppose that the typical diffusion coefficient for a reactant in aqueous solution at 25°C is 5.2×10^{-9} m^2 s^{-1}. If the critical reaction distance is 0.4 nm, what value is expected for the second-order rate constant for the diffusion-controlled reaction?

20.9(a) Calculate the magnitude of the diffusion-controlled rate constant at 298 K for a species in (a) water, (b) pentane. The viscosities are 1.00×10^{-3} kg m^{-1} s^{-1}, and 2.2×10^{-4} kg m^{-1} s^{-1}, respectively.

20.9(b) Calculate the magnitude of the diffusion-controlled rate constant at 298 K for a species in (a) decylbenzene, (b) concentrated sulfuric acid. The viscosities are 3.36 cP and 27 cP, respectively.

20.10(a) Calculate the magnitude of the diffusion-controlled rate constant at 320 K for the recombination of two atoms in water, for which $\eta = 0.89$ cP. Assuming the concentration of the reacting species is 1.5 mmol dm^{-3} initially, how long does it take for the concentration of the atoms to fall to half that value? Assume the reaction is elementary.

20.10(b) Calculate the magnitude of the diffusion-controlled rate constant at 320 K for the recombination of two atoms in benzene, for which $\eta = 0.601$ cP. Assuming the concentration of the reacting species is 2.0 mmol dm^{-3} initially, how long does it take for the concentration of the atoms to fall to half that value? Assume the reaction is elementary.

20.11(a) For the gaseous reaction $A + B \rightarrow P$, the reactive cross-section obtained from the experimental value of the pre-exponential factor is 9.2×10^{-22} m^2. The collision cross-sections of A and B estimated from the transport properties are 0.95 and 0.65 nm^2, respectively. Calculate the P-factor for the reaction.

20.11(b) For the gaseous reaction $A + B \rightarrow P$, the reactive cross-section obtained from the experimental value of the pre-exponential factor is 8.7×10^{-22} m^2. The collision cross-sections of A and B estimated from the transport properties are 0.88 and 0.40 nm^2, respectively. Calculate the P-factor for the reaction.

20.12(a) Two neutral species, A and B, with diameters 655 pm and 1820 pm, respectively, undergo the diffusion-controlled reaction $A + B \rightarrow P$ in a solvent of viscosity 2.93×10^{-3} kg m^{-1} s^{-1} at 40°C. Calculate the initial rate $d[P]/dt$ if the initial concentrations of A and B are 0.170 mol dm^{-3} and 0.350 mol dm^{-3}, respectively.

20.12(b) Two neutral species, A and B, with diameters 421 pm and 945 pm, respectively, undergo the diffusion-controlled reaction $A + B \rightarrow P$ in a solvent of viscosity 1.35 cP at 20°C. Calculate the initial rate $d[P]/dt$ if the initial concentrations of A and B are 0.155 mol dm^{-3} and 0.195 mol dm^{-3}, respectively.

20.13(a) The reaction of propylxanthate ion in acetic acid buffer solutions has the mechanism $A^- + H^+ \rightarrow P$. Near 30°C the rate constant is given by the empirical expression $k_r = (2.05 \times 10^{13})e^{-(8681\,K)/T}$ dm^3 mol^{-1} s^{-1}. Evaluate the energy and entropy of activation at 30°C.

20.13(b) The reaction $A^- + H^+ \rightarrow P$ has a rate constant given by the empirical expression $k_r = (6.92 \times 10^{12})e^{-(5925\,K)/T}$ dm^3 mol^{-1} s^{-1}. Evaluate the energy and entropy of activation at 25°C.

20.14(a) When the reaction in Exercise 20.13a occurs in a dioxane/water mixture that is 30 per cent dioxane by mass, the rate constant fits $k_r = (7.78 \times 10^{14})e^{-(9134\,K)/T}$ dm^3 mol^{-1} s^{-1} near 30°C. Calculate $\Delta^{\ddagger}G$ for the reaction at 30°C.

20.14(b) A rate constant is found to fit the expression $k_r = (4.98 \times 10^{13})e^{-(4972\,K)/T}$ dm^3 mol^{-1} s^{-1} near 25°C. Calculate $\Delta^{\ddagger}G$ for the reaction at 25°C.

20.15(a) The gas-phase association reaction between F_2 and IF_5 is first order in each of the reactants. The energy of activation for the reaction is 58.6 kJ mol^{-1}. At 65°C the rate constant is 7.84×10^{-3} kPa^{-1} s^{-1}. Calculate the entropy of activation at 65°C.

20.15(b) A gas-phase recombination reaction is first order in each of the reactants. The energy of activation for the reaction is 39.7 kJ mol^{-1}. At 65°C the rate constant is 0.35 m^3 s^{-1}. Calculate the entropy of activation at 65°C.

20.16(a) Calculate the entropy of activation for a collision between two structureless particles at 300 K, taking $M = 65$ g mol^{-1} and $\sigma^* = 0.35$ nm^2.

20.16(b) Calculate the entropy of activation for a collision between two structureless particles at 450 K, taking $M = 92$ g mol^{-1} and $\sigma^* = 0.45$ nm^2.

20.17(a) The pre-exponential factor for the gas-phase decomposition of a gas at low pressures is 4.6×10^{12} dm^3 mol^{-1} s^{-1} and its activation energy is 10.0 kJ mol^{-1}. What are (a) the entropy of activation, (b) the enthalpy of activation, (c) the Gibbs energy of activation at 298 K?

20.17(b) The pre-exponential factor for a gas-phase decomposition of a gas at low pressures is 2.3×10^{13} dm^3 mol^{-1} s^{-1} and its activation energy is 30.0 kJ mol^{-1}. What are (a) the entropy of activation, (b) the enthalpy of activation, (c) the Gibbs energy of activation at 298 K?

20.18(a) The rate constant of the reaction $H_2O_2(aq) + I^-(aq) + H^+(aq) \rightarrow H_2O(l) + HIO(aq)$ is sensitive to the ionic strength of the aqueous solution in which the reaction occurs. At 25°C, $k_r = 12.2\ dm^6\ mol^{-2}\ min^{-1}$ at an ionic strength of 0.0525. Use the Debye–Hückel limiting law to estimate the rate constant at zero ionic strength.

20.18(b) At 25°C, $k_r = 1.55\ dm^6\ mol^{-2}\ min^{-1}$ at an ionic strength of 0.0241 for a reaction in which the rate-determining step involves the encounter of two singly charged cations. Use the Debye–Hückel limiting law to estimate the rate constant at zero ionic strength.

20.19(a) For a pair of electron donor and acceptor at 298 K, $H_{DA}(r) = 0.04\ cm^{-1}$, $\Delta_r G^\ominus = -0.185$ eV, and $k_{et} = 37.5\ s^{-1}$. Estimate the value of the reorganization energy.

20.19(b) For a pair of electron donor and acceptor at 298 K, $k_{et} = 2.02 \times 10^5\ s^{-1}$ for $\Delta_r G^\ominus = -0.665$ eV. The standard reaction Gibbs energy changes to $\Delta_r G^\ominus = -0.975$ eV when a substituent is added to the electron acceptor and the rate constant for electron transfer changes to $k_{et} = 3.33 \times 10^6\ s^{-1}$. Assuming that the distance between donor and acceptor is the same in both experiments, estimate the values of $H_{DA}(r)$ and λ.

20.20(a) For a pair of electron donor and acceptor, $k_{et} = 2.02 \times 10^5\ s^{-1}$ when $r = 1.11$ nm and $k_{et} = 4.51 \times 10^4\ s^{-1}$ when $r = 1.23$ nm. Assuming that $\Delta_r G^\ominus$ and λ are the same in both experiments, estimate the value of β.

20.20(b) Refer to Exercise 20.20a. Estimate the value of k_{et} when $r = 1.59$ nm.

Problems*

Numerical problems

20.1 A first-order decomposition reaction is observed to have the following rate constants at the indicated temperatures. Estimate the activation energy.

$k_r/(10^{-3}\ s^{-1})$	2.46	45.1	576
θ/°C	0	20.0	40.0

20.2 The second-order rate constants for the reaction of oxygen atoms with aromatic hydrocarbons have been measured (R. Atkinson and J.N. Pitts, *J. Phys. Chem.* **79**, 295 (1975)). In the reaction with benzene the rate constants are $1.44 \times 10^7\ dm^3\ mol^{-1}\ s^{-1}$ at 300.3 K, $3.03 \times 10^7\ dm^3\ mol^{-1}\ s^{-1}$ at 341.2 K, and $6.9 \times 10^7\ dm^3\ mol^{-1}\ s^{-1}$ at 392.2 K. Find the pre-exponential factor and activation energy of the reaction.

20.3‡ P.W. Seakins, *et al.* (*J. Phys. Chem.* **96**, 9847 (1992)) measured the forward and reverse rate constants for the gas-phase reaction $C_2H_5(g) + HBr(g) \rightarrow C_2H_6(g) + Br(g)$ and used their findings to compute thermodynamic parameters for C_2H_5. The reaction is bimolecular in both directions with Arrhenius parameters $A = 1.0 \times 10^9\ dm^3\ mol^{-1}\ s^{-1}$, $E_a = -4.2\ kJ\ mol^{-1}$ for the forward reaction and $A = 1.4 \times 10^{11}\ dm^3\ mol^{-1}\ s^{-1}$, $E_a = 53.3\ kJ\ mol^{-1}$ for the reverse reaction. Compute $\Delta_f H^\ominus$, $S_m^\ominus$, and $\Delta_f G^\ominus$ of C_2H_5 at 298 K.

20.4 In the dimerization of methyl radicals at 25°C, the experimental pre-exponential factor is $2.4 \times 10^{10}\ dm^3\ mol^{-1}\ s^{-1}$. What are (a) the reactive cross-section, (b) the *P*-factor for the reaction if the C—H bond length is 154 pm?

20.5 Nitrogen dioxide reacts bimolecularly in the gas phase to give $2\ NO + O_2$. The temperature dependence of the second-order rate constant for the rate law $d[P]/dt = k_r[NO_2]^2$ is given below. What are the *P*-factor and the reactive cross-section for the reaction?

T/K	600	700	800	1000
$k_r/(cm^3\ mol^{-1}\ s^{-1})$	4.6×10^2	9.7×10^3	1.3×10^5	3.1×10^6

Take $\sigma = 0.60\ nm^2$.

20.6 The diameter of the methyl radical is about 308 pm. What is the maximum rate constant in the expression $d[C_2H_6]/dt = k_r[CH_3]^2$ for second-order recombination of radicals at room temperature? 10 per cent of a 1.0-dm³ sample of ethane at 298 K and 100 kPa is dissociated into methyl radicals. What is the minimum time for 90 per cent recombination?

20.7 The rates of thermolysis of a variety of *cis*- and *trans*-azoalkanes have been measured over a range of temperatures in order to settle a controversy concerning the mechanism of the reaction. In ethanol an unstable *cis*-azoalkane decomposed at a rate that was followed by observing the N_2 evolution, and this led to the rate constants listed below (P.S. Engel and D.J. Bishop, *J. Am. Chem. Soc.* **97**, 6754 (1975)). Calculate the enthalpy, entropy, energy, and Gibbs energy of activation at −20°C.

θ/°C	−24.82	−20.73	−17.02	−13.00	−8.95
$10^4 \times k_r/s^{-1}$	1.22	2.31	4.39	8.50	14.3

20.8 In an experimental study of a bimolecular reaction in aqueous solution, the second-order rate constant was measured at 25°C and at a variety of ionic strengths and the results are tabulated below. It is known that a singly charged ion is involved in the rate-determining step. What is the charge on the other ion involved?

$I/(mol\ kg^{-1})$	0.0025	0.0037	0.0045	0.0065	0.0085
$k_r/(dm^3\ mol^{-1}\ s^{-1})$	1.05	1.12	1.16	1.18	1.26

20.9 The rate constant of the reaction $I^-(aq) + H_2O_2(aq) \rightarrow H_2O(l) + IO^-(aq)$ varies slowly with ionic strength, even though the Debye–Hückel limiting law predicts no effect. Use the following data from 25°C to find the dependence of $\log k_r$ on the ionic strength:

$I/(mol\ kg^{-1})$	0.0207	0.0525	0.0925	0.1575
$k_r/(dm^3\ mol^{-1}\ min^{-1})$	0.663	0.670	0.679	0.694

Evaluate the limiting value of k_r at zero ionic strength. What does the result suggest for the dependence of $\log \gamma$ on ionic strength for a neutral molecule in an electrolyte solution?

20.10 The total cross-sections for reactions between alkali metal atoms and halogen molecules are given in the table below (R.D. Levine and R.B. Bernstein, *Molecular reaction dynamics*, Clarendon Press, Oxford, p. 72 (1974)). Assess the data in terms of the harpoon mechanism.

* Problems denoted with the symbol ‡ were supplied by Charles Trapp, Carmen Giunta and Marshall Cady.

σ^*/nm^2	Cl_2	Br_2	I_2
Na	1.24	1.16	0.97
K	1.54	1.51	1.27
Rb	1.90	1.97	1.67
Cs	1.96	2.04	1.95

Electron affinities are approximately 1.3 eV (Cl_2), 1.2 eV (Br_2), and 1.7 eV (I_2), and ionization energies are 5.1 eV (Na), 4.3 eV (K), 4.2 eV (Rb), and 3.9 eV (Cs).

20.11‡ M. Cyfert *et al.* (*Int. J. Chem. Kinet.* **28**, 103 (1996)) examined the oxidation of tris(1,10-phenanthroline)iron(II) by periodate in aqueous solution, a reaction that shows autocatalytic behaviour. To assess the kinetic salt effect, they measured rate constants at a variety of concentrations of Na_2SO_4 far in excess of reactant concentrations and reported the following data:

$b(Na_2SO_4)/(mol\ kg^{-1})$	0.2	0.15	0.1	0.05	0.025	0.0125	0.005
$k_r/(dm^{3/2}\ mol^{-1/2}\ s^{-1})$	0.462	0.430	0.390	0.321	0.283	0.252	0.224

What can be inferred about the charge of the activated complex of the rate-determining step?

20.12‡ J. Czarnowski and H.J. Schuhmacher (*Chem. Phys. Lett.* **17**, 235 (1972)) suggested the following mechanism for the thermal decomposition of F_2O in the reaction $2\,F_2O(g) \rightarrow 2\,F_2(g) + O_2(g)$:

(1) $F_2O + F_2O \rightarrow F + OF + F_2O$ $\quad k_a$

(2) $F + F_2O \rightarrow F_2 + OF$ $\quad k_b$

(3) $OF + OF \rightarrow O_2 + F + F$ $\quad k_c$

(4) $F + F + F_2O \rightarrow F_2 + F_2O$ $\quad k_d$

(a) Using the steady-state approximation, show that this mechanism is consistent with the experimental rate law $-d[F_2O]/dt = k_r[F_2O]^2 + k_r'[F_2O]^{3/2}$. (b) The experimentally determined Arrhenius parameters in the range 501–583 K are $A = 7.8 \times 10^{13}\ dm^3\ mol^{-1}\ s^{-1}$, $E_a/R = 1.935 \times 10^4$ K for k_r and $A = 2.3 \times 10^{10}\ dm^3\ mol^{-1}\ s^{-1}$, $E_a/R = 1.691 \times 10^4$ K for k_r'. At 540 K, $\Delta_f H^\ominus(F_2O) = +24.41\ kJ\ mol^{-1}$, $D(F{-}F) = 160.6\ kJ\ mol^{-1}$, and $D(O{-}O) = 498.2\ kJ\ mol^{-1}$. Estimate the bond dissociation energies of the first and second F—O bonds and the Arrhenius activation energy of reaction 2.

20.13‡ For the gas-phase reaction $A + A \rightarrow A_2$, the experimental rate constant, k_r, has been fitted to the Arrhenius equation with the pre-exponential factor $A = 4.07 \times 10^5\ dm^3\ mol^{-1}\ s^{-1}$ at 300 K and an activation energy of 65.43 kJ mol^{-1}. Calculate $\Delta^\ddagger S$, $\Delta^\ddagger H$, $\Delta^\ddagger U$, and $\Delta^\ddagger G$ for the reaction.

20.14‡ One of the most historically significant studies of chemical reaction rates was that by M. Bodenstein (*Z. physik. Chem.* **29**, 295 (1899)) of the gas-phase reaction $2\,HI(g) \rightarrow H_2(g) + I_2(g)$ and its reverse, with rate constants k_r and k_r', respectively. The measured rate constants as a function of temperature are:

T/K	647	666	683	700	716	781
$k_r/(22.4\ dm^3\ mol^{-1}\ min^{-1})$	0.230	0.588	1.37	3.10	6.70	105.9
$k_r'/(22.4\ dm^3\ mol^{-1}\ min^{-1})$	0.0140	0.0379	0.0659	0.172	0.375	3.58

Demonstrate that these data are consistent with the collision theory of bimolecular gas-phase reactions.

20.15 Consider the reaction $D + A \rightarrow D^+ + A^-$. The rate constant k_r may be determined experimentally or may be predicted by the *Marcus cross-relation*

$$k_r = (k_{DD}k_{AA}K)^{1/2}f$$

where k_{DD} and k_{AA} are the experimental rate constants for the electron self-exchange processes $^*D + D^+ \rightarrow {}^*D^+ + D$ and $^*A + A^+ \rightarrow {}^*A^+ + A$, respectively, and f is a function of $K = [D^+][A^-]/[D][A]$, k_{DD}, k_{AA}, and the collision frequencies (see Problem 20.27 for a derivation of the Marcus cross-relation). It is common to make the assumption that $f \approx 1$. Use the approximate form of the Marcus relation to estimate the rate constant for the reaction $Ru(bpy)_3^{3+} + Fe(H_2O)_6^{2+} \rightarrow Ru(bpy)_3^{2+} + Fe(H_2O)_6^{3+}$, where bpy stands for 4,4′-bipyridine. The following data are useful:

$Ru(bpy)_3^{3+} + e^- \rightarrow Ru(bpy)_3^{2+}$ $\quad E^\ominus = 1.26$ V

$Fe(H_2O)_6^{3+} + e^- \rightarrow Fe(H_2O)_6^{2+}$ $\quad E^\ominus = 0.77$ V

$^*Ru(bpy)_3^{3+} + Ru(bpy)_3^{2+} \rightarrow {}^*Ru(bpy)_3^{2+} + Ru(bpy)_3^{3+}$

$k_{Ru} = 4.0 \times 10^8\ dm^3\ mol^{-1}\ s^{-1}$

$^*Fe(H_2O)_6^{3+} + Fe(H_2O)_6^{2+} \rightarrow {}^*Fe(H_2O)_6^{2+} + Fe(H_2O)_6^{3+}$

$k_{Fe} = 4.2\ dm^3\ mol^{-1}\ s^{-1}$

Theoretical problems

20.16 Show that the definition of E_a given in eqn 20.2 reduces to eqn 20.1 for a temperature-independent activation energy.

20.17 Confirm that eqn 20.29 is a solution of eqn 20.28, where [J] is a solution of the same equation but with $k_r = 0$ and for the same initial conditions.

20.18 Evaluate $[J]^*$ numerically using mathematical software for integration in eqn 20.29, and explore the effect of increasing reaction rate constant on the spatial distribution of J.

20.19 Estimate the orders of magnitude of the partition functions involved in a rate expression. State the order of magnitude of q_m^T/N_A, q^R, q^V, q^E for typical molecules. Check that in the collision of two structureless molecules the order of magnitude of the pre-exponential factor is of the same order as that predicted by collision theory. Go on to estimate the *P*-factor for a reaction in which $A + B \rightarrow P$, and A and B are non-linear triatomic molecules.

20.20 Use the Debye–Hückel limiting law to show that changes in ionic strength can affect the rate of reaction catalysed by H^+ from the deprotonation of a weak acid. Consider the mechanism: $H^+(aq) + B(aq) \rightarrow P$, where H^+ comes from the deprotonation of the weak acid, HA. The weak acid has a fixed concentration. First show that log $[H^+]$, derived from the ionization of HA, depends on the activity coefficients of ions and thus depends on the ionic strength. Then find the relationship between log(rate) and log $[H^+]$ to show that the rate also depends on the ionic strength.

20.21 The Eyring equation can also be applied to physical processes. As an example, consider the rate of diffusion of an atom stuck to the surface of a solid. Suppose that in order to move from one site to another it has to reach the top of the barrier where it can vibrate classically in the vertical direction and in one horizontal direction, but vibration along the other horizontal direction takes it into the neighbouring site. Find an expression for the rate of diffusion, and evaluate it for W atoms on a tungsten surface ($E_a = 60\ kJ\ mol^{-1}$). Suppose that the vibration frequencies at the transition state are (a) the same as, (b) one-half the value for the adsorbed atom. What is the value of the diffusion coefficient D at 500 K? (Take the site separation as 316 pm and $\nu^\ddagger = 1 \times 10^{11}$ Hz.)

20.22 Suppose now that the adsorbed, migrating species treated in Problem 20.21 is a spherical molecule, and that it can rotate classically as well as vibrate at the top of the barrier, but that at the adsorption site itself it can only vibrate. What effect does this have on the diffusion constant? Take the molecule to be methane, for which $\tilde{B} = 5.24\ cm^{-1}$.

20.23 Show that the intensities of a molecular beam before and after passing through a chamber of length L containing inert scattering atoms are related by $I = I_0 e^{-\mathcal{N}\sigma L}$, where σ is the collision cross-section and $\mathcal{N}$ the number density of scattering atoms.

20.24 In a molecular beam experiment to measure collision cross-sections it was found that the intensity of a CsCl beam was reduced to 60 per cent of its intensity on passage through CH_2F_2 at 10 μTorr, but that when the target was Ar at the same pressure the intensity was reduced only by 10 per cent. What are the relative cross-sections of the two types of collision? Why is one much larger than the other?

20.25‡ Show that bimolecular reactions between non-linear molecules are much slower than between atoms even when the activation energies of both reactions are equal. Use transition state theory and make the following assumptions. (1) All vibrational partition functions are close to unity; (2) all rotational partition functions are approximately $1 \times 10^{1.5}$, which is a reasonable order of magnitude number; (3) the translational partition function for each species is 1×10^{26}.

20.26 This exercise gives some familiarity with the difficulties involved in predicting the structure of activated complexes. It also demonstrates the importance of femtosecond spectroscopy to our understanding of chemical dynamics because direct experimental observation of the activated complex removes much of the ambiguity of theoretical predictions. Consider the attack of H on D_2, which is one step in the $H_2 + D_2$ reaction. (a) Suppose that the H approaches D_2 from the side and forms a complex in the form of an isosceles triangle. Take the H–D distance as 30 per cent greater than in H_2 (74 pm) and the D–D distance as 20 per cent greater than in H_2. Let the critical coordinate be the antisymmetric stretching vibration in which one H–D bond stretches as the other shortens. Let all the vibrations be at about 1000 cm^{-1}. Estimate k_r for this reaction at 400 K using the experimental activation energy of about 35 kJ mol^{-1}. (b) Now change the model of the activated complex in part (a) and make it linear. Use the same estimated molecular bond lengths and vibrational frequencies to calculate k_r for this choice of model. (c) Clearly, there is much scope for modifying the parameters of the models of the activated complex. Use mathematical software or write and run a program that allows you to vary the structure of the complex and the parameters in a plausible way, and look for a model (or more than one model) that gives a value of k_r close to the experimental value, 4×10^5 dm^3 mol^{-1} s^{-1}.

20.27 Derive the approximate form of the Marcus cross-relation, $k_r = (k_{DD}k_{AA}K)^{1/2}$ (Problem 20.15), by following these steps. (a) Use eqn 20.59 to write expressions for $\Delta^{\ddagger}G$, $\Delta^{\ddagger}G_{DD}$, and $\Delta^{\ddagger}G_{AA}$, keeping in mind that $\Delta_r G^{\ominus} = 0$ for the electron self-exchange reactions. (b) Assume that the reorganization energy λ_{DA} for the reaction $D + A \rightarrow D^+ + A^-$ is the average of the reorganization energies λ_{DD} and λ_{AA} of the electron self-exchange reactions. Then show that, in the limit of small magnitude of $\Delta_r G^{\ominus}$, or $|\Delta_r G^{\ominus}| \ll \lambda_{DA}$,

$$\Delta^{\ddagger}G = \tfrac{1}{2}(\Delta^{\ddagger}G_{DD} + \Delta^{\ddagger}G_{AA} + \Delta_r G^{\ominus})$$

where $\Delta_r G^{\ominus}$ is the standard Gibbs energy for the reaction $D + A \rightarrow D^+ + A^-$. (c) Use an equation of the form of eqn 20.56 to write expressions for k_{DD} and k_{AA}. (d) Use eqn 20.56 and the result above to write an expression for k_r. (e) Complete the derivation by using the results from part (c), the relation $K = e^{-\Delta_r G^{\ominus}/RT}$, and assuming that all $\kappa\nu^{\ddagger}$ terms, which may be interpreted as collision frequencies, are identical.

Applications: to environmental science and biochemistry

20.28‡ R. Atkinson (*J. Phys. Chem. Ref. Data* **26**, 215 (1997)) has reviewed a large set of rate constants relevant to the atmospheric chemistry of volatile organic compounds. The recommended rate constant for the bimolecular association of O_2 with an alkyl radical R at 298 K is 4.7×10^9 dm^3 mol^{-1} s^{-1} for R = C_2H_5 and 8.4×10^9 dm^3 mol^{-1} s^{-1} for R = cyclohexyl. Assuming no energy barrier, compute the steric factor, *P*, for each reaction. (*Hint.* Obtain collision diameters from collision cross-sections of similar molecules in the *Data section.*)

20.29‡ The compound α-tocopherol, a form of vitamin E, is a powerful antioxidant that may help to maintain the integrity of biological membranes. R.H. Bisby and A.W. Parker (*J. Am. Chem. Soc.* **117**, 5664 (1995)) studied the reaction of photochemically excited duroquinone with the antioxidant in ethanol. Once the duroquinone was photochemically excited, a bimolecular reaction took place at a rate described as diffusion limited. (a) Estimate the rate constant for a diffusion-limited reaction in ethanol. (b) The reported rate constant was 2.77×10^9 dm^3 mol^{-1} s^{-1}; estimate the critical reaction distance if the sum of diffusion constants is 1×10^{-9} m^2 s^{-1}.

20.30 The study of conditions that optimize the association of proteins in solution guides the design of protocols for formation of large crystals that are amenable to analysis by the X-ray diffraction techniques discussed in Chapter 9. It is important to characterize protein dimerization because the process is considered to be the rate-determining step in the growth of crystals of many proteins. Consider the variation with ionic strength of the rate constant of dimerization in aqueous solution of a cationic protein P:

I/(mol kg^{-1})	0.0100	0.0150	0.0200	0.0250	0.0300	0.0350
k_r/k_r°	8.10	13.30	20.50	27.80	38.10	52.00

What can be deduced about the charge of P?

20.31 A useful strategy for the study of electron transfer in proteins consists of attaching an electroactive species to the protein's surface and then measuring k_{et} between the attached species and an electroactive protein co-factor. J.W. Winkler and H.B. Gray (*Chem. Rev.* **92**, 369 (1992)) summarize data for cytochrome *c* modified by replacement of the haem iron by a zinc ion, resulting in a zinc–porphyrin (ZnP) group in the interior of the protein, and by attachment of a ruthenium ion complex to a surface histidine aminoacid. The edge-to-edge distance between the electroactive species was thus fixed at 1.23 nm. A variety of ruthenium ion complexes with different standard potentials was used. For each ruthenium-modified protein, either the $Ru^{2+} \rightarrow ZnP^+$ or the $ZnP^* \rightarrow Ru^{3+}$, in which the electron donor is an electronically excited state of the zinc–porphyrin group formed by laser excitation, was monitored. This arrangement leads to different standard reaction Gibbs energies because the redox couples ZnP^+/ZnP and ZnP^+/ZnP^* have different standard potentials, with the electronically excited porphyrin being a more powerful reductant. Use the following data to estimate the reorganization energy for this system:

$-\Delta_r G^{\ominus}$/eV	0.665	0.705	0.745	0.975	1.015	1.055
$k_{et}/(10^6$ s$^{-1})$	0.657	1.52	1.12	8.99	5.76	10.1

20.32 The photosynthetic reaction centre of the purple photosynthetic bacterium *Rhodopseudomonas viridis* contains a number of bound co-factors that participate in electron transfer reactions. The following table shows data compiled by Moser *et al.* (*Nature* **355**, 796 (1992)) on the rate constants for electron transfer between different co-factors and their edge-to-edge distances:

Reaction	$BChl^- \rightarrow BPh$	$BPh^- \rightarrow BChl_2^+$	$BPh^- \rightarrow Q_A$	cyt $c_{559} \rightarrow BChl_2^+$
r/nm	0.48	0.95	0.96	1.23
k_{et}/s^{-1}	1.58×10^{12}	3.98×10^9	1.00×10^9	1.58×10^8
Reaction	$Q_A^- \rightarrow Q_B$	$Q_A^- \rightarrow BChl_2^+$		
r/nm	1.35	2.24		
k_{et}/s^{-1}	3.98×10^7	63.1		

(BChl, bacteriochlorophyll; $BChl_2$, bacteriochlorophyll dimer, functionally distinct from BChl; BPh, bacteriophaeophytin; Q_A and Q_B, quinone molecules bound to two distinct sites; cyt c_{559}, a cytochrome bound to the reaction centre complex). Are these data in agreement with the behaviour predicted by eqn 20.61? If so, evaluate the value of β.

20.33 The rate constant for electron transfer between a cytochrome *c* and the bacteriochlorophyll dimer of the reaction centre of the purple bacterium *Rhodobacter sphaeroides* (Problem 20.32) decreases with decreasing temperature in the range 300 K to 130 K. Below 130 K, the rate constant becomes independent of temperature. Account for these results.

20.34‡ Methane is a byproduct of a number of natural processes (such as digestion of cellulose in ruminant animals, anaerobic decomposition of organic waste matter) and industrial processes (such as food production and fossil fuel use). Reaction with the hydroxyl radical OH is the main path by which CH_4 is removed from the lower atmosphere. T. Gierczak *et al.* (*J. Phys. Chem.* **A 101**, 3125 (1997)) measured the rate constants for the elementary bimolecular gas-phase reaction of methane with the hydroxyl radical over a range of temperatures of importance to atmospheric chemistry. Deduce the Arrhenius parameters A and E_a from the following measurements:

T/K	295	223	218	213	206	200	195
$k_r/(10^6\ dm^3\ mol^{-1}\ s^{-1})$	3.55	0.494	0.452	0.379	0.295	0.241	0.217

20.35‡ As we saw in Problem 20.34, reaction with the hydroxyl radical OH is the main path by which CH_4, a byproduct of many natural and industrial processes, is removed from the lower atmosphere. T. Gierczak *et al.* (*J. Phys. Chem.* **A 101**, 3125 (1997)) measured the rate constants for the bimolecular gas-phase reaction $CH_4(g) + OH(g) \rightarrow CH_3(g) + H_2O(g)$ and found $A = 1.13 \times 10^9\ dm^3\ mol^{-1}\ s^{-1}$ and $E_a = 14.1\ kJ\ mol^{-1}$ for the Arrhenius parameters. (a) Estimate the rate of consumption of CH_4. Take the average OH concentration to be $3.0 \times 10^{-15}\ mol\ dm^{-3}$, that of CH_4 to be $4.0 \times 10^{-8}\ mol\ dm^{-3}$, and the temperature to be −10°C. (b) Estimate the global annual mass of CH_4 consumed by this reaction (which is slightly less than the amount introduced to the atmosphere) given an effective volume for the Earth's lower atmosphere of $4 \times 10^{21}\ dm^3$.

21

Catalysis

This chapter extends the material introduced in Chapters 19 and 20 by showing how to deal with catalysis. We begin with a description of homogeneous catalysis and apply the associated concepts to enzyme-catalysed reactions. We go on to consider heterogeneous catalysis by exploring the extent to which a solid surface is covered and the variation of the extent of coverage with pressure and temperature. Then we use this material to discuss how surfaces affect the rate and course of chemical change by acting as the site of catalysis.

A **catalyst** is a substance that accelerates a reaction but undergoes no net chemical change. The catalyst lowers the activation energy of the reaction by providing an alternative path that avoids the slow, rate-determining step of the uncatalysed reaction (Fig. 21.1). Catalysts can be very effective; for instance, the activation energy for the decomposition of hydrogen peroxide in solution is 76 kJ mol^{-1}, and the reaction is slow at room temperature. When a little iodide ion is added, the activation energy falls to 57 kJ mol^{-1} and the rate constant increases by a factor of 2000.

A **homogeneous catalyst** is a catalyst in the same phase as the reaction mixture. For example, the decomposition of hydrogen peroxide in aqueous solution is catalysed by bromide ion or catalase. **Enzymes**, which are biological catalysts, are very specific and can have a dramatic effect on the reactions they control. For example, the enzyme catalase reduces the activation energy for the decomposition of hydrogen peroxide to 8 kJ mol^{-1}, corresponding to an acceleration of the reaction by a factor of 10^{15} at 298 K.

A **heterogeneous catalyst** is a catalyst in a different phase from the reaction mixture. For example, the hydrogenation of ethene to ethane, a gas-phase reaction, is accelerated in the presence of a solid catalyst such as palladium, platinum, or nickel.

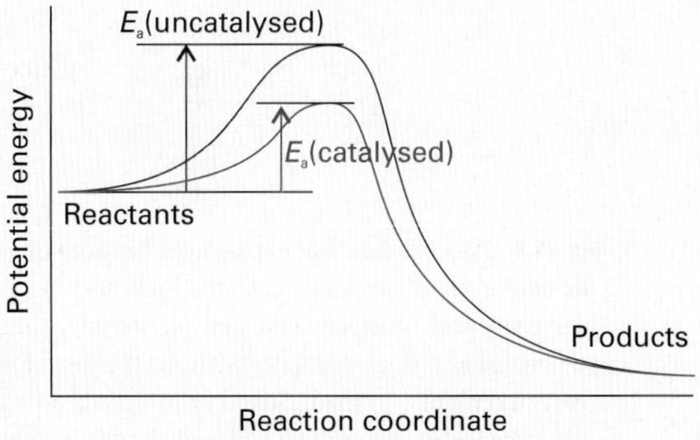

Fig. 21.1 A catalyst provides a different path with a lower activation energy. The result is an increase in the rate of formation of products.

The metal provides a surface upon which the reactants bind; this binding facilitates encounters between reactants and increases the rate of the reaction.

Homogeneous catalysis

We can obtain some idea of the mode of action of homogeneous catalysts by examining the kinetics of the bromide-catalysed decomposition of hydrogen peroxide:

$$2\,H_2O_2(aq) \rightarrow 2\,H_2O(l) + O_2(g)$$

The reaction is believed to proceed through the following pre-equilibrium:

$$H_3O^+ + H_2O_2 \rightleftharpoons H_3O_2^+ + H_2O \qquad K = \frac{[H_3O_2^+]}{[H_2O_2][H_3O^+]}$$

$$H_3O_2^+ + Br^- \rightarrow HOBr + H_2O \qquad v = k_a[H_3O_2^+][Br^-]$$

$$HOBr + H_2O_2 \rightarrow H_3O^+ + O_2 + Br^- \qquad \text{(fast)}$$

where we have set the activity of H_2O in the equilibrium constant equal to 1 and assumed that the thermodynamic properties of the other substances are ideal. The second step is rate-determining. Therefore, we can obtain the rate law of the overall reaction by setting the overall rate equal to the rate of the second step and using the equilibrium constant to express the concentration of $H_3O_2^+$ in terms of the reactants. The result is

$$\frac{d[O_2]}{dt} = k_r[H_2O_2][H_3O^+][Br^-]$$

with $k_r = k_aK$, in agreement with the observed dependence of the rate on the Br^- concentration and the pH of the solution. The observed activation energy is that of the effective rate constant k_aK.

21.1 Acid and base catalysis

The catalytic action of acids and bases should be well known from previous study of organic and inorganic chemistry. Here we summarize and generalize the main ideas.

In **acid catalysis** the crucial step is the transfer of a proton to the substrate:

$$X + HA \rightarrow HX^+ + A^- \qquad HX^+ \rightarrow \text{products}$$

Acid catalysis is the primary process in the solvolysis of esters and keto–enol tautomerism. In **base catalysis**, a proton is transferred from the substrate to a base:

$$XH + B \rightarrow X^- + BH^+ \qquad X^- \rightarrow \text{products}$$

Base catalysis is the primary step in the isomerization and halogenation of organic compounds, and of the Claisen and aldol condensation reactions.

21.2 Enzymes

Enzymes are homogeneous biological catalysts. These ubiquitous compounds are special proteins or nucleic acids that contain an **active site**, which is responsible for binding the **substrates**, the reactants, and processing them into products. As is true of any catalyst, the active site returns to its original state after the products are released. Many enzymes consist primarily of proteins, some featuring organic or inorganic co-factors in their active sites. However, certain RNA molecules can also be biological catalysts, forming *ribozymes*. A very important example of a ribozyme is the *ribosome*, a large assembly of proteins and catalytically active RNA molecules responsible for the synthesis of proteins in the cell.

The structure of the active site is specific to the reaction that it catalyses, with groups in the substrate interacting with groups in the active site through intermolecular interactions, such as hydrogen-bonding, electrostatic, or van der Waals interactions. Figure 21.2 shows two models that explain the binding of a substrate to the active site of an enzyme. In the **lock-and-key model**, the active site and substrate have complementary three-dimensional structures and dock perfectly without the need for major atomic rearrangements. Experimental evidence favours the **induced fit model**, in which binding of the substrate induces a conformational change in the active site. Only after the change does the substrate fit snugly in the active site.

Enzyme-catalysed reactions are prone to inhibition by molecules that interfere with the formation of product. Many drugs for the treatment of disease function by inhibiting enzymes. For example, an important strategy in the treatment of acquired immune deficiency syndrome (AIDS) involves the steady administration of a specially designed protease inhibitor. The drug

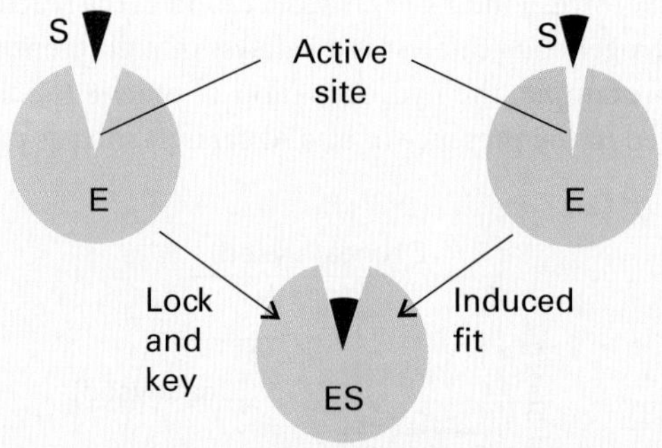

Fig. 21.2 Two models that explain the binding of a substrate to the active site of an enzyme. In the lock-and-key model, the active site and substrate have complementary three-dimensional structures and dock perfectly without the need for major atomic rearrangements. In the induced fit model, binding of the substrate induces a conformational change in the active site. The substrate fits well in the active site after the conformational change has taken place.

inhibits an enzyme that is key to the formation of the protein envelope surrounding the genetic material of the human immunodeficiency virus (HIV). Without a properly formed envelope, HIV cannot replicate in the host organism.

(a) The Michaelis–Menten mechanism of enzyme catalysis

Experimental studies of enzyme kinetics are typically conducted by monitoring the initial rate of product formation in a solution in which the enzyme is present at very low concentration. Indeed, enzymes are such efficient catalysts that significant accelerations may be observed even when their concentration is more than three orders of magnitude smaller than that of the substrate.

The principal features of many enzyme-catalysed reactions are as follows:

1. For a given initial concentration of substrate, $[\mathrm{S}]_0$, the initial rate of product formation is proportional to the total concentration of enzyme, $[\mathrm{E}]_0$.

2. For a given $[\mathrm{E}]_0$ and low values of $[\mathrm{S}]_0$, the rate of product formation is proportional to $[\mathrm{S}]_0$.

3. For a given $[\mathrm{E}]_0$ and high values of $[\mathrm{S}]_0$, the rate of product formation becomes independent of $[\mathrm{S}]_0$, reaching a maximum value known as the **maximum velocity**, $v_{\max}$.

The **Michaelis–Menten mechanism** accounts for these features (Fig. 21.3). According to this mechanism, an enzyme–substrate complex is formed in the first step and either the substrate is released unchanged or after modification to form products:

$$\begin{array}{ll} \mathrm{E+S \rightleftarrows ES} & k_a, k_a' \\ \mathrm{ES \rightarrow P+E} & k_b \end{array} \qquad (21.1)$$

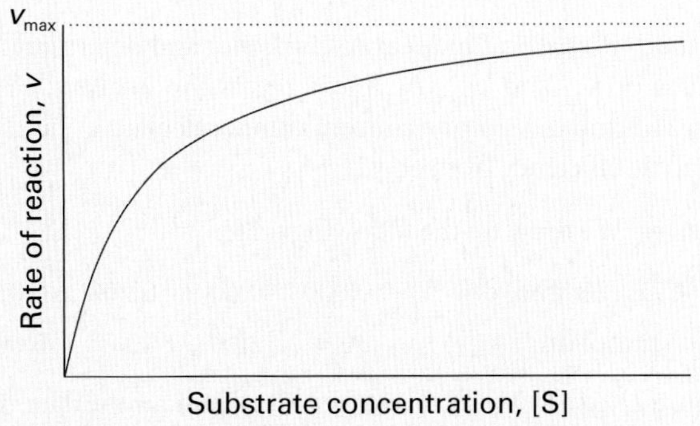

Fig. 21.3 The variation of the rate of an enzyme-catalysed reaction with substrate concentration. The approach to a maximum rate, $v_{\max}$, for large [S] is explained by the Michaelis–Menten mechanism.

interActivity Use the Michaelis–Menten equation to generate two families of curves showing the dependence of v on [S]: one in which K_M varies but $v_{\max}$ is constant, and another in which $v_{\max}$ varies but K_M is constant.

We show in the following *Justification* that this mechanism leads to the **Michaelis–Menten equation** for the rate of product formation

$$v = \frac{k_b[\mathrm{E}]_0}{1 + K_M/[\mathrm{S}]_0} \qquad (21.2)$$

where $K_M = (k_a' + k_b)/k_a$ is the **Michaelis constant**, characteristic of a given enzyme acting on a given substrate.

Justification 21.1 *The Michaelis–Menten equation*

The rate of product formation according to the Michaelis–Menten mechanism is

$$v = k_b[\mathrm{ES}]$$

We can obtain the concentration of the enzyme–substrate complex by invoking the steady-state approximation and writing

$$\frac{\mathrm{d[ES]}}{\mathrm{d}t} = k_a[\mathrm{E}][\mathrm{S}] - k_a'[\mathrm{ES}] - k_b[\mathrm{ES}] = 0$$

It follows that

$$[\mathrm{ES}] = \left(\frac{k_a}{k_a' + k_b}\right)[\mathrm{E}][\mathrm{S}]$$

where [E] and [S] are the concentrations of *free* enzyme and substrate, respectively. Now we define the Michaelis constant as

$$K_M = \frac{k_a' + k_b}{k_a} = \frac{[\mathrm{E}][\mathrm{S}]}{[\mathrm{ES}]}$$

and note that K_M has the same units as molar concentration. To express the rate law in terms of the concentrations of enzyme and substrate added, we note that $[\mathrm{E}]_0 = [\mathrm{E}] + [\mathrm{ES}]$. Moreover, because the substrate is typically in large excess relative to the enzyme, the free substrate concentration is approximately equal to the initial substrate concentration and we can write $[\mathrm{S}] \approx [\mathrm{S}]_0$. It then follows that:

$$[\mathrm{ES}] = \frac{[\mathrm{E}]_0}{1 + K_M/[\mathrm{S}]_0}$$

We obtain eqn 21.2 when we substitute this expression for [ES] into that for the rate of product formation ($v = k_b[\mathrm{ES}]$).

Equation 21.2 shows that, in accord with experimental observations (Fig. 21.3):

1. When $[\mathrm{S}]_0 \ll K_M$, the rate is proportional to $[\mathrm{S}]_0$:

$$v = \frac{k_b}{K_M}[\mathrm{S}]_0[\mathrm{E}]_0 \qquad (21.3a)$$

2. When $[\mathrm{S}]_0 \gg K_M$, the rate reaches its maximum value and is independent of $[\mathrm{S}]_0$:

$$v = v_{\max} = k_b[\mathrm{E}]_0 \qquad (21.3b)$$

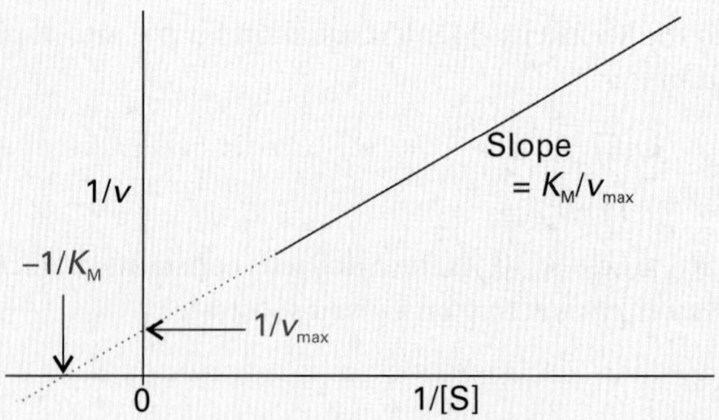

Fig. 21.4 A Lineweaver–Burk plot for the analysis of an enzyme-catalysed reaction that proceeds by a Michaelis–Menten mechanism and the significance of the intercepts and the slope.

Substitution of the definition of v_{max} into eqn 21.2 gives:

$$v = \frac{v_{max}}{1 + K_M/[S]_0} \tag{21.4a}$$

We can rearrange this expression into a form that is amenable to data analysis by linear regression:

$$\frac{1}{v} = \frac{1}{v_{max}} + \left(\frac{K_M}{v_{max}}\right)\frac{1}{[S]_0} \tag{21.4b}$$

A **Lineweaver–Burk plot** is a plot of $1/v$ against $1/[S]_0$, and according to eqn 21.4b it should yield a straight line with slope of K_M/v_{max}, a y-intercept at $1/v_{max}$, and an x-intercept at $-1/K_M$ (Fig. 21.4). The value of k_b is then calculated from the y-intercept and eqn 21.3b. However, the plot cannot give the individual rate constants k_a and k_a' that appear in the expression for K_M. The stopped-flow technique described in Section 19.1 can give the additional data needed, because we can find the rate of formation of the enzyme–substrate complex by monitoring the concentration after mixing the enzyme and substrate. This procedure gives a value for k_a, and k_a' is then found by combining this result with the values of k_b and K_M.

(b) The catalytic efficiency of enzymes

The **turnover frequency**, or **catalytic constant**, of an enzyme, k_{cat}, is the number of catalytic cycles (turnovers) performed by the active site in a given time interval divided by the duration of the interval. This quantity has units of a first-order rate constant and, in terms of the Michaelis–Menten mechanism, is numerically equivalent to k_b, the rate constant for release of product from the enzyme–substrate complex. It follows from the identification of k_{cat} with k_b and from eqn 21.3b that

$$k_{cat} = k_b = \frac{v_{max}}{[E]_0} \tag{21.5}$$

The **catalytic efficiency**, η (eta), of an enzyme is the ratio k_{cat}/K_M. The higher the value of η, the more efficient is the enzyme. We can think of the catalytic efficiency as the effective rate constant of the enzymatic reaction. From $K_M = (k_a' + k_b)/k_a$ and eqn 21.5, it follows that

$$\eta = \frac{k_{cat}}{K_M} = \frac{k_a k_b}{k_a' + k_b} \tag{21.6}$$

The efficiency reaches its maximum value of k_a when $k_b \gg k_a'$. Because k_a is the rate constant for the formation of a complex from two species that are diffusing freely in solution, the maximum efficiency is related to the maximum rate of diffusion of E and S in solution. This limit (Section 20.4) leads to rate constants of about 10^8–10^9 dm^3 mol^{-1} s^{-1} for molecules as large as enzymes at room temperature. The enzyme catalase has $\eta = 4.0 \times 10^8$ dm^3 mol^{-1} s^{-1} and is said to have attained 'catalytic perfection', in the sense that the rate of the reaction it catalyses is controlled only by diffusion: it acts as soon as a substrate makes contact.

Example 21.1 *Determining the catalytic efficiency of an enzyme*

The enzyme carbonic anhydrase catalyses the hydration of CO_2 in red blood cells to give bicarbonate (hydrogencarbonate) ion:

$$CO_2(g) + H_2O(l) \rightarrow HCO_3^-(aq) + H^+(aq)$$

The following data were obtained for the reaction at pH = 7.1, 273.5 K, and an enzyme concentration of 2.3 nmol dm^{-3}:

$[CO_2]_0$/(mmol dm^{-3})	1.25	2.5	5	20
rate/(mmol dm^{-3} s^{-1})	2.78×10^{-2}	5.00×10^{-2}	8.33×10^{-2}	1.67×10^{-1}

Determine the catalytic efficiency of carbonic anhydrase at 273.5 K.

Method Prepare a Lineweaver–Burk plot and determine the values of K_M and v_{max} by linear regression analysis. From eqn 21.5 and the enzyme concentration, calculate k_{cat} and the catalytic efficiency from eqn 21.6.

Answer We draw up the following table:

$1/([CO_2]_0/(\text{mmol dm}^{-3}))$	0.800	0.400	0.200	0.0500
$1/(v/(\text{mmol dm}^{-3}\text{ s}^{-1}))$	36.0	20.0	12.0	6.00

Figure 21.5 shows the Lineweaver–Burk plot for the data. The slope is 40.0 and the y-intercept is 4.00. Hence,

$$v_{max}/(\text{mmol dm}^{-3}\text{ s}^{-1}) = \frac{1}{\text{intercept}} = \frac{1}{4.00} = 0.250$$

and

$$K_M/(\text{mmol dm}^{-3}) = \text{slope} \times v_{max} = \frac{\text{slope}}{\text{intercept}} = \frac{40.0}{4.00} = 10.0$$

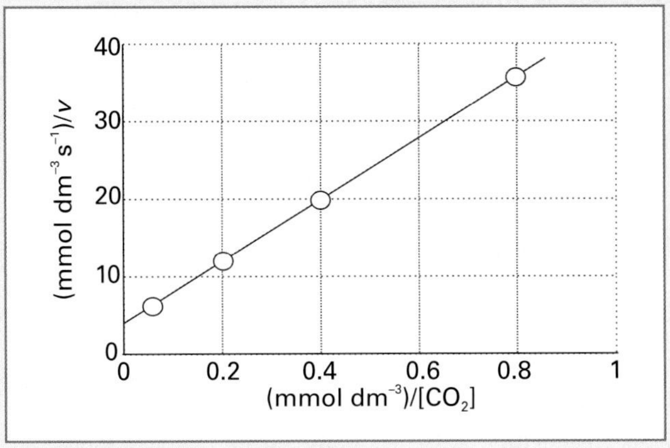

Fig. 21.5 The Lineweaver–Burk plot of the data for Example 21.1.

It follows that

$$k_{cat} = \frac{v_{max}}{[E]_0} = \frac{2.5\times10^{-4}\,\text{mol dm}^{-3}\,\text{s}^{-1}}{2.3\times10^{-9}\,\text{mol dm}^{-3}} = 1.1\times10^5\,\text{s}^{-1}$$

and

$$\eta = \frac{k_{cat}}{K_M} = \frac{1.1\times10^5\,\text{s}^{-1}}{1.0\times10^{-2}\,\text{mol dm}^{-3}} = 1.1\times10^7\,\text{dm}^3\,\text{mol}^{-1}\,\text{s}^{-1}$$

A note on good practice The slope and the intercept are unitless: we have remarked previously that all graphs should be plotted as pure numbers.

Self-test 21.1 The enzyme α-chymotrypsin is secreted in the pancreas of mammals and cleaves peptide bonds made between certain amino acids. Several solutions containing the small peptide *N*-glutaryl-L-phenylalanine-*p*-nitroanilide at different concentrations were prepared and the same small amount of α-chymotrypsin was added to each one. The following data were obtained on the initial rates of the formation of product:

$[S]_0/(\text{mmol dm}^{-3})$	0.334	0.450	0.667	1.00	1.33	1.67
$v/(\text{mmol dm}^{-3}\,\text{s}^{-1})$	0.152	0.201	0.269	0.417	0.505	0.667

Determine the maximum velocity and the Michaelis constant for the reaction.

$[v_{max} = 2.80\text{ mmol dm}^{-3}\,\text{s}^{-1}, K_M = 5.89\text{ mmol dm}^{-3}]$

(c) Mechanisms of enzyme inhibition

An inhibitor, I, decreases the rate of product formation from the substrate by binding to the enzyme, to the ES complex, or to the enzyme and ES complex simultaneously. The most general kinetic scheme for enzyme inhibition is then:

$$E + S \rightleftarrows ES \qquad k_a, k_a'$$

$$ES \rightarrow E + P \qquad k_b$$

$$EI \rightleftharpoons E + I \qquad K_I = \frac{[E][I]}{[EI]} \tag{21.7a}$$

$$ESI \rightleftharpoons ES + I \qquad K_I' = \frac{[ES][I]}{[ESI]} \tag{21.7b}$$

The lower the values of K_I and K_I' the more efficient are the inhibitors. The rate of product formation is always given by $v = k_b[ES]$, because only ES leads to product. As shown in the following *Justification*, the rate of reaction in the presence of an inhibitor is

$$v = \frac{v_{max}}{\alpha' + \alpha K_M/[S]_0} \tag{21.8a}$$

where $\alpha = 1 + [I]/K_I$ and $\alpha' = 1 + [I]/K_I'$. This equation is very similar to the Michaelis–Menten equation for the uninhibited enzyme (eqn 21.2) and is also amenable to analysis by a Lineweaver–Burk plot:

$$\frac{1}{v} = \frac{\alpha'}{v_{max}} + \left(\frac{\alpha K_M}{v_{max}}\right)\frac{1}{[S]_0} \tag{21.8b}$$

Justification 21.2 *Enzyme inhibition*

By mass balance, the total concentration of enzyme is:

$$[E]_0 = [E] + [EI] + [ES] + [ESI]$$

By using eqns 21.7a and 21.7b and the definitions

$$\alpha = 1 + \frac{[I]}{K_I} \quad \text{and} \quad \alpha' = 1 + \frac{[I]}{K_I'}$$

it follows that

$$[E]_0 = [E]\alpha + [ES]\alpha'$$

By using $K_M = [E][S]/[ES]$ and replacing [S] with $[S]_0$, we can write

$$[E]_0 = \frac{K_M[ES]}{[S]_0}\alpha + [ES]\alpha' = [ES]\left(\frac{\alpha K_M}{[S]_0} + \alpha'\right)$$

The expression for the rate of product formation is then:

$$v = k_b[ES] = \frac{k_b[E]_0}{\alpha K_M/[S]_0 + \alpha'}$$

which, after using eqn 21.5, gives eqn 21.8.

There are three major modes of inhibition that give rise to distinctly different kinetic behaviour. In **competitive inhibition**

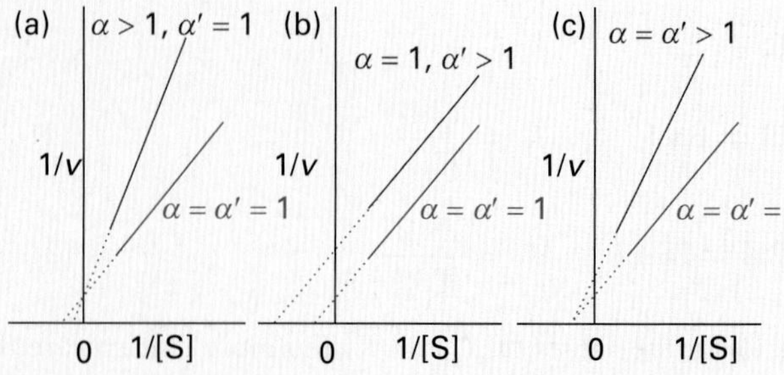

Fig. 21.6 Lineweaver–Burk plots characteristic of the three major modes of enzyme inhibition: (a) competitive inhibition, (b) uncompetitive inhibition, and (c) noncompetitive inhibition, showing the special case $\alpha = \alpha' > 1$.

interActivity Use eqn 21.8 to explore the effect of competitive, uncompetitive, and noncompetitive inhibition on the shapes of the plots of v against [S] for constant K_M and v_{max}.

the inhibitor binds only to the active site of the enzyme and thereby inhibits the attachment of the substrate. This condition corresponds to $\alpha > 1$ and $\alpha' = 1$ (because ESI does not form). The slope of the Lineweaver–Burk plot increases by a factor of α relative to the slope for data on the uninhibited enzyme ($\alpha = \alpha' = 1$). The y-intercept does not change as a result of competitive inhibition (Fig. 21.6a). In **uncompetitive inhibition** the inhibitor binds to a site of the enzyme that is removed from the active site, but only if the substrate is already present. The inhibition occurs because ESI reduces the concentration of ES, the active type of complex. In this case $\alpha = 1$ (because EI does not form) and $\alpha' > 1$. The y-intercept of the Lineweaver–Burk plot increases by a factor of α' relative to the y-intercept for data on the uninhibited enzyme but the slope does not change (Fig. 21.6b). In **non-competitive inhibition** (also called *mixed inhibition*) the inhibitor binds to a site other than the active site, and its presence reduces the ability of the substrate to bind to the active site. Inhibition occurs at both the E and ES sites. This condition corresponds to $\alpha > 1$ and $\alpha' > 1$. Both the slope and y-intercept of the Lineweaver–Burk plot increase upon addition of the inhibitor. Figure 21.6c shows the special case of $K_I = K_I'$ and $\alpha = \alpha'$, which results in intersection of the lines at the x-axis.

In all cases, the efficiency of the inhibitor may be obtained by determining K_M and v_{max} from a control experiment with uninhibited enzyme and then repeating the experiment with a known concentration of inhibitor. From the slope and y-intercept of the Lineweaver–Burk plot for the inhibited enzyme (eqn 21.8b), the mode of inhibition, the values of α or α', and the values of K_I or K_I' may be obtained.

Example 21.2 *Distinguishing between types of inhibition*

Five solutions of a substrate, S, were prepared with the concentrations given in the first column below and each one was divided into five equal volumes. The same concentration of enzyme was present in each one. An inhibitor, I, was then added in four different concentrations to the samples, and the initial rate of formation of product was determined with the results given below. Does the inhibitor act competitively or noncompetitively? Determine K_I and K_M.

	[I]/(mmol dm^{-3})					
$[S]_0$/(mmol dm^{-3})	0	0.20	0.40	0.60	0.80	
0.050	0.033	0.026	0.021	0.018	0.016	v/(μmol dm^{-3} s^{-1})
0.10	0.055	0.045	0.038	0.033	0.029	
0.20	0.083	0.071	0.062	0.055	0.050	
0.40	0.111	0.100	0.091	0.084	0.077	
0.60	0.126	0.116	0.108	0.101	0.094	

Method We draw a series of Lineweaver–Burk plots for different inhibitor concentrations. If the plots resemble those in Fig. 21.6a, then the inhibition is competitive. On the other hand, if the plots resemble those in Fig. 21.6c, then the inhibition is noncompetitive. To find K_I, we need to determine the slope at each value of [I], which is equal to $\alpha K_M/v_{max}$, or $K_M/v_{max} + K_M[I]/K_I v_{max}$, then plot this slope against [I]: the intercept at [I] = 0 is the value of K_M/v_{max} and the slope is $K_M/K_I v_{max}$.

Answer First, we draw up a table of $1/[S]_0$ and $1/v$ for each value of [I]:

	[I]/(mmol dm^{-3})					
$1/([S]_0$/(mmol dm^{-3}))	0	0.20	0.40	0.60	0.80	
20	30	38	48	56	62	$1/(v$/(μmol dm^{-3} s^{-1}))
10	18	22	26	30	34	
5.0	12	14	16	18	20	
2.5	9.01	10.0	11.0	11.9	13.0	
1.7	7.94	8.62	9.26	9.90	10.6	

The five plots (one for each [I]) are given in Fig. 21.7. We see that they pass through the same intercept on the vertical axis, so the inhibition is competitive. The mean of the (least squares) intercepts is 5.83, so v_{max} = 0.172 μmol dm^{-3} s^{-1} (note how it picks up the units for v in the data). The (least squares) slopes of the lines are as follows:

[I]/(mmol dm^{-3})	0	0.20	0.40	0.60	0.80
Slope	1.219	1.627	2.090	2.489	2.832

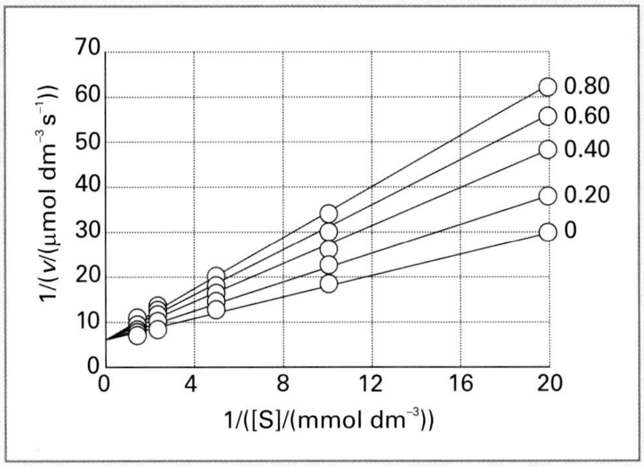

Fig. 21.7 Lineweaver–Burk plots for the data in Example 21.2. Each line corresponds to a different concentration of inhibitor.

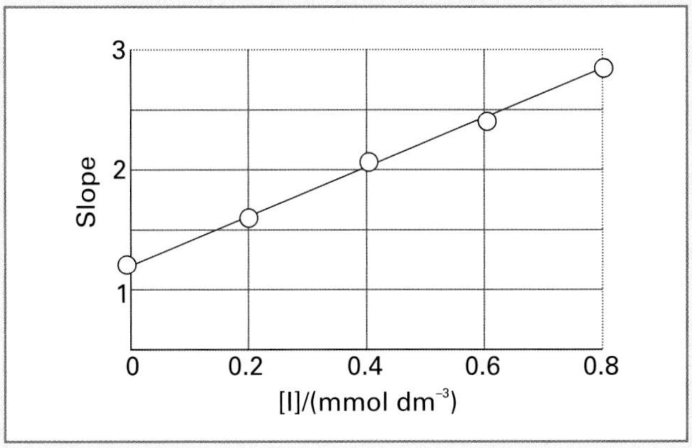

Fig. 21.8 Plot of the slopes of the plots in Fig. 21.7 against [I] based on the data in Example 21.2.

These values are plotted in Fig. 21.8. The intercept at [I] = 0 is 1.234, so $K_M = 0.212$ mmol dm^{-3}. The (least squares) slope of the line is 2.045, so

$$K_I/(\text{mmol dm}^{-3}) = \frac{K_M}{\text{slope} \times v_{max}} = \frac{0.212}{2.045 \times 0.172} = 0.603$$

Self-test 21.2 Repeat the question using the following data:

$[S]_0$/(mmol dm^{-3})	[I]/(mmol dm^{-3}) 0	0.20	0.40	0.60	0.80
0.050	0.020	0.015	0.012	0.0098	0.0084
0.10	0.035	0.026	0.021	0.017	0.015
0.20	0.056	0.042	0.033	0.028	0.024
0.40	0.080	0.059	0.047	0.039	0.034
0.60	0.093	0.069	0.055	0.046	0.039

(Values: v/(μmol dm^{-3} s^{-1}))

[Noncompetitive, $K_M = 0.30$ mmol dm^{-3}, $K_I = 0.57$ mmol dm^{-3}]

Heterogeneous catalysis

The remainder of this chapter is devoted to developing and applying concepts of structure and reactivity in heterogeneous catalysis. For simplicity, we consider only gas/solid systems. To understand the catalytic role of a solid surface we begin by describing its unique structural features. Then, because many reactions catalysed by surfaces involve reactants and products in the gas phase, we discuss **adsorption**, the attachment of particles to a solid surface, and **desorption**, the reverse process. Finally, we consider specific mechanisms of heterogeneous catalysis.

21.3 The growth and structure of surfaces

The substance that adsorbs to a surface is the **adsorbate** and the underlying material that we are concerned with in this section is the **adsorbent** or **substrate**.

(a) Surface growth

A simple picture of a perfect crystal surface is as a tray of oranges in a grocery store (Fig. 21.9). A gas molecule that collides with the surface can be imagined as a ping-pong ball bouncing erratically over the oranges. The molecule loses energy as it bounces under the influence of intermolecular forces, but it is likely to escape from the surface before it has lost enough kinetic energy to be trapped. The same is true, to some extent, of an ionic crystal in contact with a solution. There is little energy advantage for an ion in solution to discard some of its solvating molecules and stick at an exposed position on a flat surface.

The picture changes when the surface has defects, for then there are ridges of incomplete layers of atoms or ions. A typical type of surface defect is a **step** between two otherwise flat layers of atoms called **terraces** (Fig. 21.10). A step defect might itself have defects, including kinks. When an atom settles on a terrace it bounces across it under the influence of the intermolecular potential, and might come to a step or a corner formed by a kink. Instead of interacting with a single terrace atom, the molecule

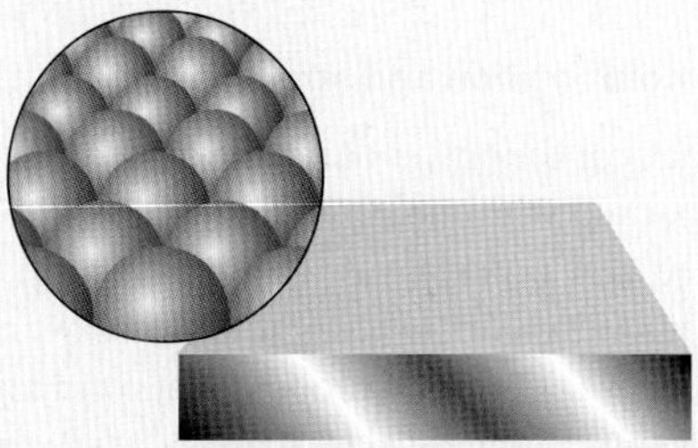

Fig. 21.9 A schematic diagram of the flat surface of a solid. This primitive model is largely supported by scanning tunnelling microscope images (see *Impact I2.1*).

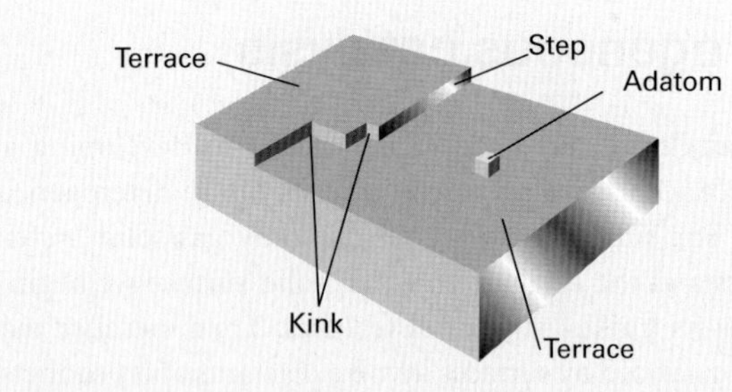

Fig. 21.10 Some of the kinds of defects that may occur on otherwise perfect terraces. Defects play an important role in surface growth and catalysis.

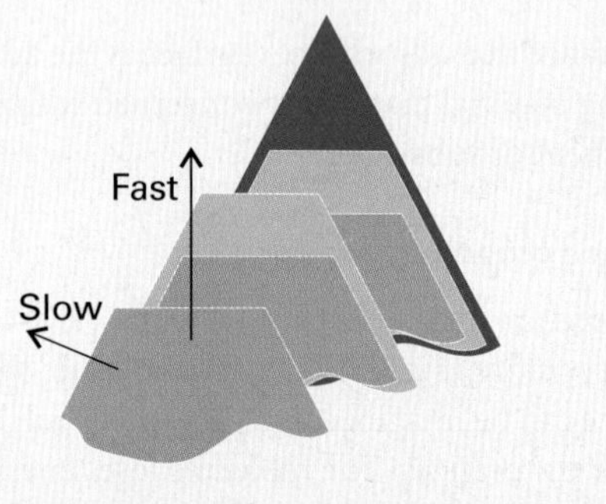

Fig. 21.11 The slower-growing faces of a crystal dominate its final external appearance. Three successive stages of the growth are shown.

now interacts with several, and the interaction may be strong enough to trap it. Likewise, when ions deposit from solution, the loss of the solvation interaction is offset by a strong Coulombic interaction between the arriving ions and several ions at the surface defect.

The rapidity of growth depends on the crystal plane concerned, and the slowest growing faces dominate the appearance of the crystal. This feature is explained in Fig. 21.11, where we see that, although the horizontal face grows forward most rapidly, it grows itself out of existence, and the more slowly growing faces survive.

(b) Surface composition and structure

Under normal conditions, a surface exposed to a gas is constantly bombarded with molecules and a freshly prepared surface is covered very quickly. Just how quickly can be estimated by using the kinetic theory of gases and the following expression for the **collision flux**, Z_W, the number of hits on a region of a surface during an interval divided by the area of the region and the duration of the interval:

$$Z_W = \frac{p}{(2\pi mkT)^{1/2}}$$

At first glance this expression seems to suggest that the collision flux decreases with increasing temperature, which seems unlikely because the molecules then move faster. However, it is always important to be circumspect when interpreting equations. In this case, we should note that the pressure increases with temperature ($p \propto T$ in a constant-volume container), so the overall temperature dependence is $Z_W \propto T^{1/2}$ and, as expected, the flux increases with temperature. A practical form of this equation is

$$Z_W = \frac{Z_0(p/\text{Pa})}{\{(T/\text{K})(M/(\text{g mol}^{-1}))\}^{1/2}} \quad \text{with } Z_0 = 2.63 \times 10^{24}\ \text{m}^{-2}\,\text{s}^{-1}$$

where M is the molar mass of the gas. For air ($M \approx 29$ g mol^{-1}) at 1 atm and 22°C the collision flux is 3×10^{27} m^{-2} s^{-1}. Because 1 m^2 of metal surface consists of about 10^{19} atoms, each atom is struck about 10^8 times each second. Even if only a few collisions leave a molecule adsorbed to the surface, the time for which a freshly prepared surface remains clean is very short.

The obvious way to retain cleanliness is to reduce the pressure. When it is reduced to 0.1 mPa (as in a simple vacuum system) the collision flux falls to about 10^{18} m^{-2} s^{-1}, corresponding to one hit per surface atom in each 0.1 s. Even that is too brief in most experiments, and in **ultra-high vacuum** (UHV) techniques pressures as low as 0.1 μPa (when $Z_W = 10^{15}$ m^{-2} s^{-1}) are reached on a routine basis and ones as low as 1 nPa (when $Z_W = 10^{13}$ m^{-2} s^{-1}) are reached with special care. These collision fluxes correspond to each surface atom being hit once every 10^5 to 10^6 s, or about once a day.

The chemical composition of a surface can be determined by a variety of ionization techniques. The same techniques can be used to detect any remaining contamination after cleaning and to detect layers of material adsorbed later in the experiment. One technique that may be used is **photoemission spectroscopy**, a derivative of the photoelectric effect, in which X-rays (for XPS) or hard (short wavelength) ultraviolet (for UPS) ionizing radiation is used to eject electrons from adsorbed species. The kinetic energies of the electrons ejected from their orbitals are measured and the pattern of energies is a fingerprint of the material present (Fig. 21.12). UPS, which examines electrons ejected from valence shells, is also used to establish the bonding characteristics and the details of valence shell electronic structures of substances on the surface. Its usefulness is its ability to reveal which orbitals of the adsorbate are involved in the bond to the substrate. For instance, the principal difference between the photoemission results on free benzene and benzene adsorbed on palladium is in the energies of the π electrons. This difference is interpreted as meaning that the C_6H_6 molecules lie parallel to the surface and are attached to it by their π orbitals. In contrast, pyridine (C_6H_5N) stands almost perpendicular to the surface, and is attached by a σ bond formed by the nitrogen lone pair.

A very important technique, which is widely used in the microelectronics industry, is **Auger electron spectroscopy** (AES). The

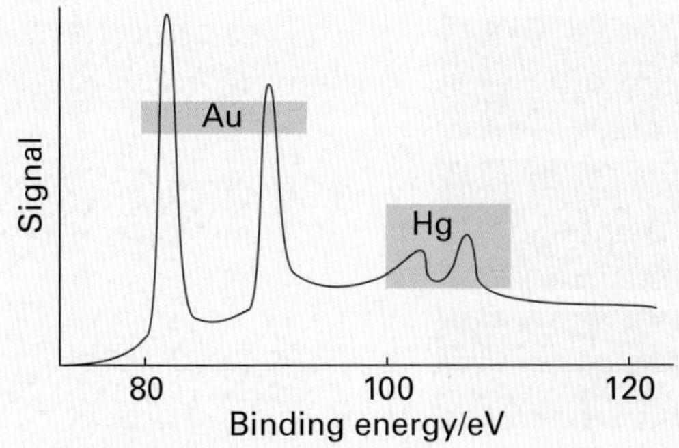

Fig. 21.12 The X-ray photoelectron emission spectrum of a sample of gold contaminated with a surface layer of mercury. (M.W. Roberts and C.S. McKee, *Chemistry of the metal–gas interface*, Oxford (1978).)

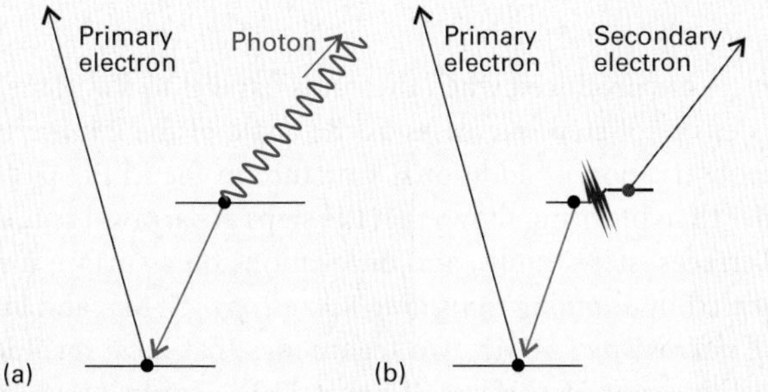

Fig. 21.13 When an electron is expelled from a solid (a) an electron of higher energy may fall into the vacated orbital and emit an X-ray photon to produce X-ray fluorescence. Alternatively (b) the electron falling into the orbital may give up its energy to another electron, which is ejected in the Auger effect.

Auger effect (pronounced oh-zhey) is the emission of a second electron after high energy radiation has expelled another electron. The first electron to depart leaves a hole in a low-lying orbital, and an upper electron falls into it. The energy released in this transition may result either in the generation of radiation, which is called **X-ray fluorescence** (Fig. 21.13a), or in the ejection of another electron (Fig. 21.13b). The latter is the secondary electron of the Auger effect. The energies of the secondary electrons are characteristic of the material present, so the Auger effect effectively takes a fingerprint of the sample. In practice, the Auger spectrum is normally obtained by irradiating the sample with an electron beam rather than electromagnetic radiation. In **scanning Auger electron microscopy** (SAM), the finely focused electron beam is scanned over the surface and a map of composition is compiled; the resolution can reach to below about 50 nm.

One of the most informative techniques for determining the arrangement of the atoms close to and adsorbed on the surface is **low energy electron diffraction** (LEED). This technique is like X-ray diffraction but using the wave character of electrons, and the sample is now the surface of a solid. The use of low-energy electrons (with energies in the range 10–200 eV, corresponding to wavelengths in the range 100–400 pm) ensures that the diffraction is caused only by atoms on and close to the surface. The experimental arrangement is shown in Fig. 21.14, and typical LEED patterns, obtained by photographing the fluorescent screen through the viewing port, are shown in Fig. 21.15.

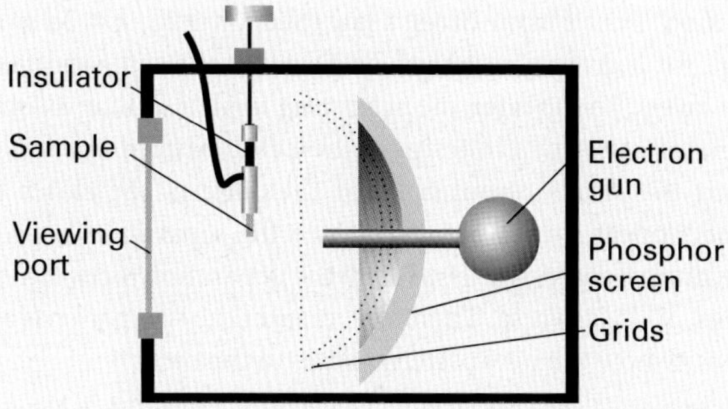

Fig. 21.14 A schematic diagram of the apparatus used for a LEED experiment. The electrons diffracted by the surface layers are detected by the fluorescence they cause on the phosphor screen.

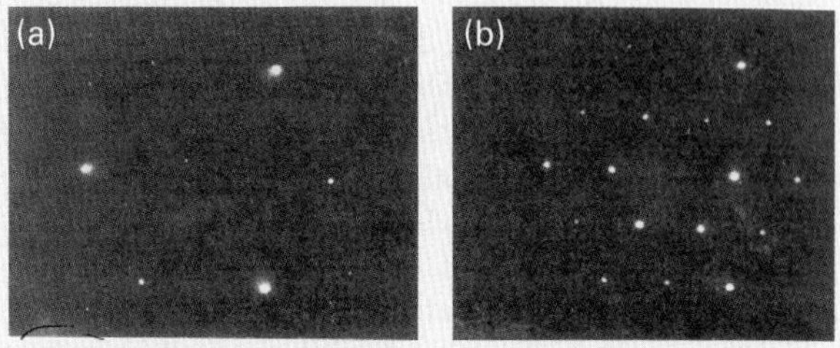

Fig. 21.15 LEED photographs of (a) a clean platinum surface and (b) after its exposure to propyne, $CH_3C{\equiv}CH$. (Photographs provided by Professor G.A. Somorjai.)

Example 21.3 *Interpreting a LEED pattern*

The LEED pattern from a clean unreconstructed (110) face of palladium is shown in (a) below. The reconstructed surface gives a LEED pattern shown as (b). What can be inferred about the structure of the reconstructed surface?

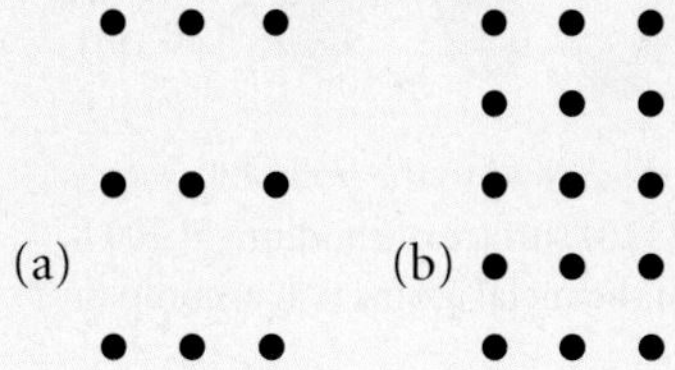

Method Recall from Bragg's law (Section 9.3), $\lambda = 2d \sin \theta$, that, for a given wavelength, the smaller the separation d of the layers, the greater the scattering angle (so that $2d \sin \theta$ remains constant). In terms of the LEED pattern, the farther apart the atoms responsible for the pattern, the closer the spots appear in the pattern. Twice the separation between the atoms corresponds to half the separation between the spots, and vice versa. Therefore, inspect the two patterns and identify how the new pattern relates to the old.

Solution The vertical separation between spots is unchanged, which indicates that the atoms remain in the same position in that dimension when reconstruction occurs. However, the horizontal spacing is halved, which suggests that the atoms are twice as far apart in that direction as they are in the unreconstructed surface.

Self-test 21.3 Sketch the LEED pattern for a surface that was reconstructed from that shown in (a) above by tripling the vertical separation.

LEED experiments show that the surface of a crystal rarely has exactly the same form as a slice through the bulk. As a general rule, it is found that metal surfaces are often simply truncations of the bulk lattice, but the distance between the top layer of atoms and the one below is contracted by around 5 per cent. Semiconductors generally have surfaces reconstructed to a depth of several layers. Reconstruction occurs in ionic solids. For example, in lithium fluoride the Li^+ and F^- ions close to the surface apparently lie on slightly different planes. An actual example of the detail that can now be obtained from refined LEED techniques is shown in Fig. 21.16 for $CH_3C—$ adsorbed on a (111) plane of rhodium.

The presence of terraces, steps, and kinks in a surface shows up in LEED patterns, and their surface density (the number of defects in a region divided by the area of the region) can be estimated. Three examples of how steps and kinks affect the pattern are shown in Fig. 21.17. The samples used were obtained by cleaving a crystal at different angles to a plane of atoms. Only

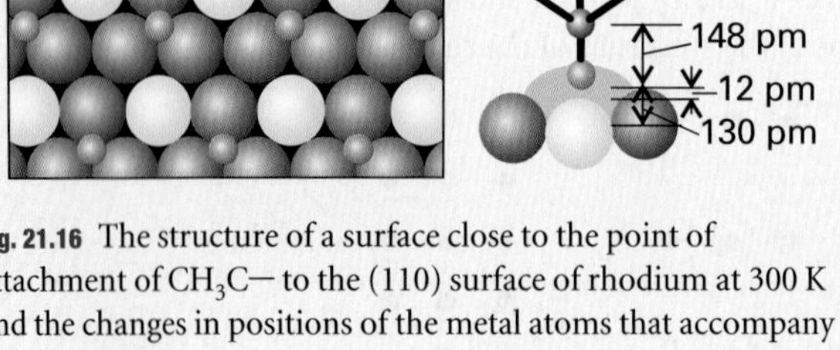

Fig. 21.16 The structure of a surface close to the point of attachment of $CH_3C—$ to the (110) surface of rhodium at 300 K and the changes in positions of the metal atoms that accompany chemisorption.

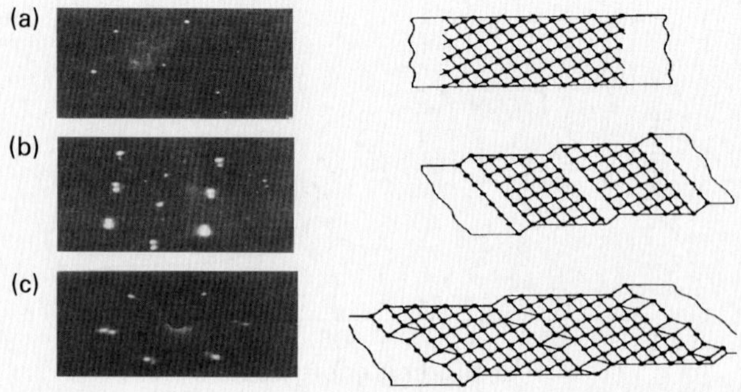

Fig. 21.17 LEED patterns may be used to assess the defect density of a surface. The photographs correspond to a platinum surface with (a) low defect density, (b) regular steps separated by about six atoms, and (c) regular steps with kinks. (Photographs provided by Professor G.A. Samorjai.)

terraces are produced when the cut is parallel to the plane, and the density of steps increases as the angle of the cut increases. The observation of additional structure in the LEED patterns, rather than blurring, shows that the steps are arrayed regularly.

Terraces, steps, kinks, and dislocations on a surface may be observed by *scanning tunnelling microscopy* (STM), and *atomic force microscopy* (AFM), two techniques that have revolutionized the study of surfaces (*Impact I2.1*). Figure 21.18 shows the dissociation of SiH_3 adsorbed on to a Si(001) surface into adsorbed SiH_2 units and H atoms.

21.4 The extent of adsorption

The extent of surface coverage is normally expressed as the **fractional coverage**, θ:

$$\theta = \frac{\text{number of adsorption sites occupied}}{\text{number of adsorption sites available}} \qquad [21.9]$$

The fractional coverage is often expressed in terms of the volume of adsorbate adsorbed by $\theta = V/V_\infty$, where V_∞ is the volume of adsorbate corresponding to complete monolayer coverage. The **rate of adsorption**, $d\theta/dt$, is the rate of change of surface coverage, and can be determined by observing the change of fractional coverage with time.

Among the principal techniques for measuring $d\theta/dt$ are flow methods, in which the sample itself acts as a pump because adsorption removes particles from the gas. One commonly used technique is therefore to monitor the rates of flow of gas into and out of the system: the difference is the rate of gas uptake by the sample. Integration of this rate then gives the fractional coverage at any stage. In **flash desorption** the sample is suddenly heated (electrically) and the resulting rise of pressure is interpreted in terms of the amount of adsorbate originally on the

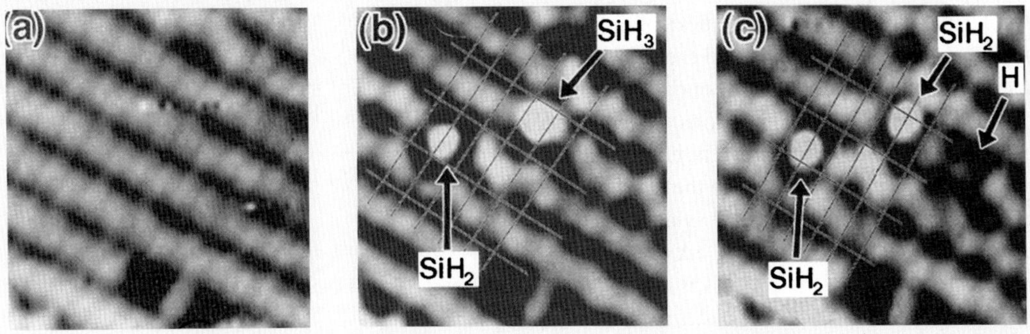

Fig. 21.18 Visualization by STM of the reaction $SiH_3 \rightarrow SiH_2 + H$ on a 4.7 nm × 4.7 nm area of a Si(001) surface. (a) The Si(001) surface before exposure to Si_2H_6(g). (b) Adsorbed Si_2H_6 dissociates into SiH_2(surface), on the left of the image, and SiH_3(surface), on the right. (c) After 8 min, SiH_3(surface) dissociates to SiH_2(surface) and H(surface). (Reproduced with permission from Y. Wang *et al. Surface Science* **64**, 311 (1994).)

sample. The interpretation may be confused by the desorption of a non-adsorbed compound (for example, WO_3 from oxygen on tungsten). **Gravimetry**, in which the sample is weighed on a microbalance during the experiment, can also be used. A common instrument for gravimetric measurements is the **quartz crystal microbalance** (QCM), in which the mass of a sample laid on the surface of a quartz crystal is related to changes in the latter's mechanical properties. The key principle behind the operation of a QCM is the ability of a quartz crystal to vibrate at a characteristic frequency when an oscillating electric field is applied. The vibrational frequency decreases when material is spread over the surface of the crystal and the change in frequency is proportional to the mass of material. Masses as small as a few nanograms (1 ng $= 10^{-9}$ g) can be measured reliably in this way.

(a) Physisorption and chemisorption

Molecules and atoms can attach to surfaces in two ways. In **physisorption** (an abbreviation of 'physical adsorption'), there is a van der Waals interaction (for example, a dispersion or a dipolar interaction) between the adsorbate and the substrate. Van der Waals interactions have a long range but are weak, and the energy released when a particle is physisorbed is of the same order of magnitude as the enthalpy of condensation. Such small energies can be absorbed as vibrations of the lattice and dissipated as thermal motion, and a molecule bouncing across the surface will gradually lose its energy and finally adsorb to it in the process called **accommodation**. The enthalpy of physisorption can be measured by monitoring the rise in temperature of a sample of known heat capacity, and typical values are in the region of -20 kJ mol^{-1} (Table 21.1). This small enthalpy change is insufficient to lead to bond breaking, so a physisorbed molecule retains its identity, although it might be distorted by the presence of the surface.

Synoptic table 21.1* Maximum observed enthalpies of physisorption

Adsorbate	$\Delta_{ad}H^{\ominus}/(\text{kJ mol}^{-1})$
CH_4	−21
H_2	−84
H_2O	−59
N_2	−21

* More values are given in the *Data section*.

In **chemisorption** (an abbreviation of 'chemical adsorption'), the molecules (or atoms) stick to the surface by forming a chemical (usually covalent) bond, and tend to find sites that maximize their coordination number with the substrate. The enthalpy of chemisorption is very much greater than that for physisorption, and typical values are in the region of -200 kJ mol^{-1} (Table 21.2). The distance between the surface and the closest adsorbate atom

Synoptic table 21.2* Enthalpies of chemisorption, $\Delta_{ad}H^{\ominus}/(\text{kJ mol}^{-1})$

Adsorbate	Adsorbent (substrate)		
	Cr	Fe	Ni
C_2H_4	−427	−285	−243
CO		−192	
H_2	−188	−134	
NH_3		−188	−155

* More values are given in the *Data section*.

is also typically shorter for chemisorption than for physisorption. A chemisorbed molecule may be torn apart at the demand of the unsatisfied valencies of the surface atoms, and the existence of molecular fragments on the surface as a result of chemisorption is one reason why solid surfaces catalyse reactions.

Except in special cases, chemisorption must be exothermic. A spontaneous process requires $\Delta G < 0$. Because the translational freedom of the adsorbate is reduced when it is adsorbed, ΔS is negative. Therefore, in order for $\Delta G = \Delta H - T\Delta S$ to be negative, ΔH must be negative (that is, the process is exothermic). Exceptions may occur if the adsorbate dissociates and has high translational mobility on the surface. For example, H_2 adsorbs endothermically on glass because there is a large increase of translational entropy accompanying the dissociation of the molecules into atoms that move quite freely over the surface. In this case, the entropy change in the process $H_2(g) \rightarrow 2\ H(glass)$ is sufficiently positive to overcome the small positive enthalpy change.

The enthalpy of adsorption depends on the extent of surface coverage, mainly because the adsorbate particles interact. If the particles repel each other (as for CO on palladium) the adsorption becomes less exothermic (the enthalpy of adsorption less negative) as coverage increases. Moreover, LEED studies show that such species settle on the surface in a disordered way until packing requirements demand order. If the adsorbate particles attract one another (as for O_2 on tungsten), then they tend to cluster together in islands, and growth occurs at the borders. These adsorbates also show order–disorder transitions when they are heated enough for thermal motion to overcome the particle–particle interactions, but not so much that they are desorbed.

(b) Adsorption isotherms

The free gas and the adsorbed gas are in dynamic equilibrium, and the fractional coverage of the surface depends on the pressure of the overlying gas. The variation of θ with pressure at a chosen temperature is called the **adsorption isotherm**.

The simplest physically plausible isotherm is based on three assumptions:

1. Adsorption cannot proceed beyond monolayer coverage.
2. All sites are equivalent and the surface is uniform (that is, the surface is perfectly flat on a microscopic scale).
3. The ability of a molecule to adsorb at a given site is independent of the occupation of neighbouring sites (that is, there are no interactions between adsorbed molecules).

The dynamic equilibrium is

$$A(g) + M(surface) \rightleftharpoons AM(surface)$$

with rate constants k_a for adsorption and k_d for desorption. The rate of change of surface coverage due to adsorption is proportional to the partial pressure p of A and the number of vacant sites $N(1-\theta)$, where N is the total number of sites:

$$\frac{d\theta}{dt} = k_a pN(1-\theta) \qquad (21.10a)$$

The rate of change of θ due to desorption is proportional to the number of adsorbed species, $N\theta$:

$$\frac{d\theta}{dt} = -k_d N\theta \qquad (21.10b)$$

At equilibrium there is no net change (that is, the sum of these two rates is zero), and solving for θ gives the **Langmuir isotherm**:

$$\theta = \frac{Kp}{1+Kp} \qquad K = \frac{k_a}{k_d} \qquad (21.11)$$

Example 21.4 *Using the Langmuir isotherm*

The data given below are for the adsorption of CO on charcoal at 273 K. Confirm that they fit the Langmuir isotherm, and find the constant K and the volume corresponding to complete coverage. In each case V has been corrected to 1.00 atm (101.325 kPa).

p/kPa	13.3	26.7	40.0	53.3	66.7	80.0	93.3
V/cm^3	10.2	18.6	25.5	31.5	36.9	41.6	46.1

Method From eqn 21.11,

$$Kp\theta + \theta = Kp$$

With $\theta = V/V_\infty$, where V_∞ is the volume corresponding to complete coverage, this expression can be rearranged into

$$\frac{p}{V} = \frac{p}{V_\infty} + \frac{1}{KV_\infty}$$

Hence, a plot of p/V against p should give a straight line of slope $1/V_\infty$ and intercept $1/KV_\infty$.

Answer The data for the plot are as follows:

p/kPa	13.3	26.7	40.0	53.3	66.7	80.0	93.3
$(p/\text{kPa})/(V/\text{cm}^3)$	1.30	1.44	1.57	1.69	1.81	1.92	2.02

The points are plotted in Fig. 21.19. The (least squares) slope is 0.00900, so $V_\infty = 111\ \text{cm}^3$. The intercept at $p = 0$ is 1.20, so

$$K = \frac{1}{(111\ \text{cm}^3)\times(1.20\ \text{kPa cm}^{-3})} = 7.51\times10^{-3}\ \text{kPa}^{-1}$$

Self-test 21.4 Repeat the calculation for the following data:

p/kPa	13.3	26.7	40.0	53.3	66.7	80.0	93.3
V/cm^3	10.3	19.3	27.3	34.1	40.0	45.5	48.0

[$128\ \text{cm}^3$, $6.69\times10^{-3}\ \text{kPa}^{-1}$]

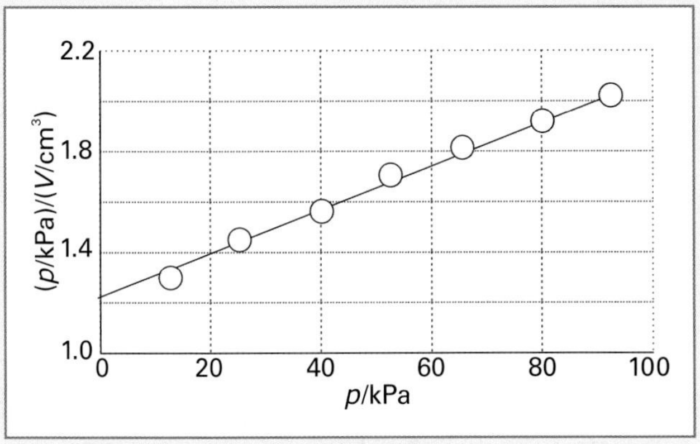

Fig. 21.19 The plot of the data in Example 21.4. As illustrated here, the Langmuir isotherm predicts that a straight line should be obtained when p/V is plotted against p.

For adsorption with dissociation, the rate of adsorption is proportional to the pressure and to the probability that both atoms will find sites, which is proportional to the square of the number of vacant sites,

$$\frac{d\theta}{dt} = k_a p\{N(1-\theta)\}^2 \tag{21.12a}$$

The rate of desorption is proportional to the frequency of encounters of atoms on the surface, and is therefore second order in the number of atoms present:

$$\frac{d\theta}{dt} = -k_d(N\theta)^2 \tag{21.12b}$$

The condition for no net change leads to the isotherm

$$\theta = \frac{(Kp)^{1/2}}{1+(Kp)^{1/2}} \tag{21.13}$$

The surface coverage now depends more weakly on pressure than for non-dissociative adsorption.

The shapes of the Langmuir isotherms with and without dissociation are shown in Figs. 21.20 and 21.21. The fractional coverage increases with increasing pressure, and approaches 1 only at very high pressure, when the gas is forced on to every available site of the surface. Different curves (and therefore different values of K) are obtained at different temperatures, and the temperature dependence of K can be used to determine the **isosteric enthalpy of adsorption**, $\Delta_{ad}H^{\ominus}$, the standard enthalpy of adsorption at a fixed surface coverage. To determine this quantity we recognize that K is essentially an equilibrium constant, and then use the van't Hoff equation (eqn 17.26) to write:

$$\left(\frac{\partial \ln K}{\partial T}\right)_\theta = \frac{\Delta_{ad}H^{\ominus}}{RT^2} \tag{21.14}$$

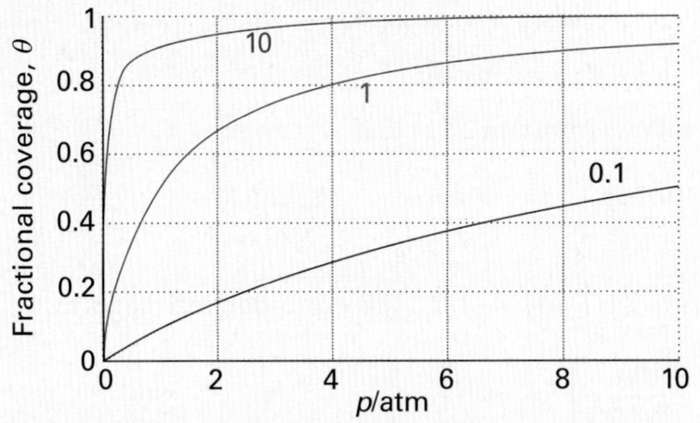

Fig. 21.20 The Langmuir isotherm for dissociative adsorption, $X_2(g) \rightarrow 2\,X(\text{surface})$, for different values of K.

interActivity Using eqn 21.13, generate a family of curves showing the dependence of $1/\theta$ on $1/p$ for several values of K.

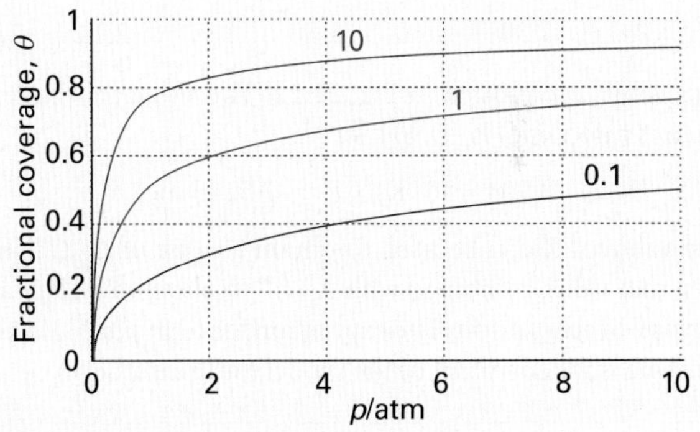

Fig. 21.21 The Langmuir isotherm for non-dissociative adsorption for different values of K.

interActivity Using eqn 21.11, generate a family of curves showing the dependence of $1/\theta$ on $1/p$ for several values of K. Taking these results together with those of the previous *interActivity*, discuss how plots of $1/\theta$ against $1/p$ can be used to distinguish between adsorption with and without dissociation.

Example 21.5 *Measuring the isosteric enthalpy of adsorption*

The data below show the pressures of CO needed for the volume of adsorption (corrected to 1.00 atm and 273 K) to be 10.0 cm³ using the same sample as in Example 21.4. Calculate the adsorption enthalpy at this surface coverage.

T/K	200	210	220	230	240	250
p/kPa	4.00	4.95	6.03	7.20	8.47	9.85

Method The Langmuir isotherm can be rearranged to

$$Kp = \frac{\theta}{1-\theta}$$

Therefore, when θ is constant,

$$\ln K + \ln p = \text{constant}$$

It follows from eqn 21.14 that

$$\left(\frac{\partial \ln p}{\partial T}\right)_\theta = -\left(\frac{\partial \ln K}{\partial T}\right)_\theta = -\frac{\Delta_{ad}H^{\ominus}}{RT^2}$$

With $d(1/T)/dT = -1/T^2$, this expression rearranges to

$$\left(\frac{\partial \ln p}{\partial(1/T)}\right)_\theta = \frac{\Delta_{ad}H^{\ominus}}{R}$$

Therefore, a plot of $\ln p$ against $1/T$ should be a straight line of slope $\Delta_{ad}H^{\ominus}/R$.

Answer We draw up the following table:

T/K	200	210	220	230	240	250
$10^3/(T/\text{K})$	5.00	4.76	4.55	4.35	4.17	4.00
$\ln(p/\text{kPa})$	1.39	1.60	1.80	1.97	2.14	2.29

The points are plotted in Fig. 21.22. The slope (of the least squares fitted line) is –0.904, so

$$\Delta_{ad}H^{\ominus} = -(0.904 \times 10^3\ \text{K}) \times R = -7.52\ \text{kJ mol}^{-1}$$

The value of K can be used to obtain a value of $\Delta_{ad}G^{\ominus}$, and then that value combined with $\Delta_{ad}H^{\ominus}$ to obtain the standard entropy of adsorption. The expression for $(\partial \ln p/\partial T)_\theta$ in this example is independent of the model for the isotherm.

Self-test 21.5 Repeat the calculation using the following data:

T/K	200	210	220	230	240	250
p/kPa	4.32	5.59	7.07	8.80	10.67	12.80

[–9.0 kJ mol^{-1}]

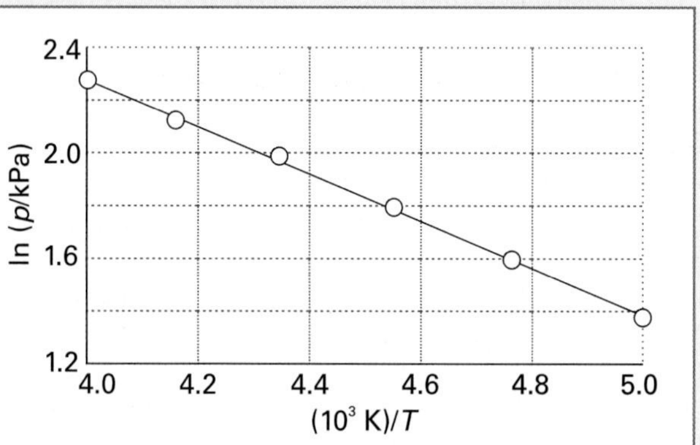

Fig. 21.22 The isosteric enthalpy of adsorption can be obtained from the slope of the plot of $\ln p$ against $1/T$, where p is the pressure needed to achieve the specified surface coverage. The data used are from Example 21.5.

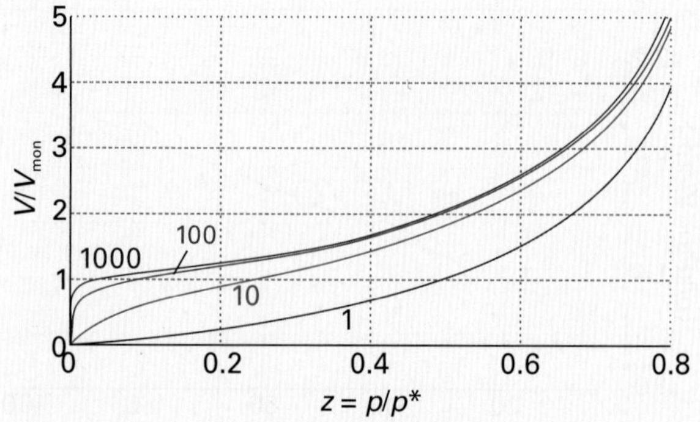

Fig. 21.23 Plots of the BET isotherm for different values of c. The value of V/V_{mon} rises indefinitely because the adsorbate may condense on the covered substrate surface.

interActivity Using eqn 21.15, generate a family of curves showing the dependence of $zV_{mon}/(1-z)V$ on z for different values of c.

If the initial adsorbed layer can act as a substrate for further (for example, physical) adsorption, then instead of the isotherm levelling off to some saturated value at high pressures, it can be expected to rise indefinitely. The most widely used isotherm dealing with multilayer adsorption was derived by Stephen Brunauer, Paul Emmett, and Edward Teller, and is called the **BET isotherm** (it is derived in *Further Information* 21.1):

$$\frac{V}{V_{mon}} = \frac{cz}{(1-z)\{1-(1-c)z\}} \quad \text{with } z = \frac{p}{p^*} \tag{21.15}$$

In this expression, p^* is the vapour pressure above a layer of adsorbate that is more than one molecule thick and which resembles a pure bulk liquid, V_{mon} is the volume corresponding to monolayer coverage, and c is a constant that is large when the enthalpy of desorption from a monolayer is large compared with the enthalpy of vaporization of the liquid adsorbate:

$$c = e^{(\Delta_{des}H^{\ominus} - \Delta_{vap}H^{\ominus})/RT} \tag{21.16}$$

Figure 21.23 illustrates the shapes of BET isotherms. They rise indefinitely as the pressure is increased because there is no limit to the amount of material that may condense when multilayer coverage may occur. A BET isotherm is not accurate at all pressures, but it is widely used in industry to determine the surface areas of solids.

Example 21.6 *Using the BET isotherm*

The data below relate to the adsorption of N_2 on rutile (TiO_2) at 75 K. Confirm that they fit a BET isotherm in the range of pressures reported, and determine V_{mon} and c.

p/kPa	0.160	1.87	6.11	11.67	17.02	21.92	27.29
V/mm^3	601	720	822	935	1046	1146	1254

At 75 K, $p^* = 76.0$ kPa. The volumes have been corrected to 1.00 atm and 273 K and refer to 1.00 g of substrate.

Method Equation 21.15 can be reorganized into

$$\frac{z}{(1-z)V} = \frac{1}{cV_{\text{mon}}} + \frac{(c-1)z}{cV_{\text{mon}}}$$

It follows that $(c-1)/cV_{\text{mon}}$ can be obtained from the slope of a plot of the expression on the left against z, and cV_{mon} can be found from the intercept at $z = 0$. The results can then be combined to give c and V_{mon}.

Answer We draw up the following table:

p/kPa	0.160	1.87	6.11	11.67	17.02	21.92	27.29
10^3z	2.11	24.6	80.4	154	224	288	359
$10^4z/(1-z)(V/\text{mm}^3)$	0.035	0.350	1.06	1.95	2.76	3.53	4.47

These points are plotted in Fig. 21.24. The least squares best line has an intercept at 0.0398, so

$$\frac{1}{cV_{\text{mon}}} = 3.98 \times 10^{-6}\,\text{mm}^{-3}$$

The slope of the line is 1.23×10^{-2}, so

$$\frac{c-1}{cV_{\text{mon}}} = (1.23 \times 10^{-2}) \times 10^3 \times 10^{-4}\,\text{mm}^{-3} = 1.23 \times 10^{-3}\,\text{mm}^{-3}$$

The solutions of these equations are $c = 310$ and $V_{\text{mon}} = 811\,\text{mm}^3$. At 1.00 atm and 273 K, 811 mm^3 corresponds to 3.6×10^{-5} mol, or 2.2×10^{19} atoms. Because each atom occupies an area of about 0.16 nm^2, the surface area of the sample is about 3.5 m^2.

Self-test 21.6 Repeat the calculation for the following data:

p/kPa	0.160	1.87	6.11	11.67	17.02	21.92	27.29
V/cm^3	235	559	649	719	790	860	950

[370, 615 cm^3]

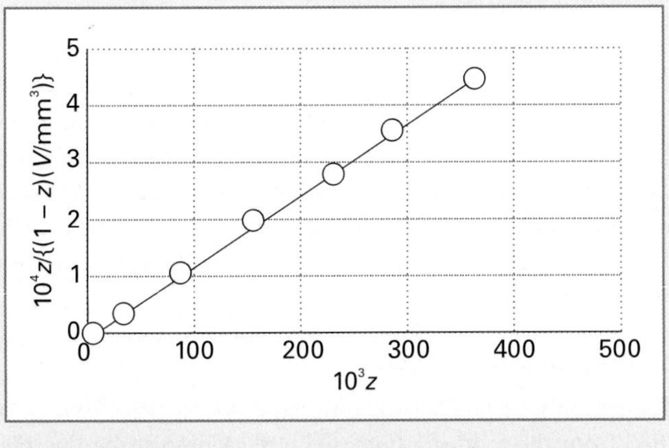

Fig. 21.24 The BET isotherm can be tested, and the parameters determined, by plotting $z/(1-z)V$ against $z = p/p^*$. The data are from Example 21.6.

When $c \gg 1$, the BET isotherm takes the simpler form

$$\frac{V}{V_{\text{mon}}} = \frac{1}{1-z} \qquad (21.17)$$

This expression is applicable to unreactive gases on polar surfaces, for which $c \approx 10^2$ because $\Delta_{\text{des}}H^\ominus$ is then significantly greater than $\Delta_{\text{vap}}H^\ominus$ (eqn 21.16). The BET isotherm fits experimental observations moderately well over restricted pressure ranges, but it errs by underestimating the extent of adsorption at low pressures and by overestimating it at high pressures.

An assumption of the Langmuir isotherm is the independence and equivalence of the adsorption sites. Deviations from the isotherm can often be traced to the failure of these assumptions. For example, the enthalpy of adsorption often becomes less negative as θ increases, which suggests that the energetically most favourable sites are occupied first. Various attempts have been made to take these variations into account. The **Temkin isotherm**,

$$\theta = c_1 \ln(c_2 p) \qquad (21.18)$$

where c_1 and c_2 are constants, corresponds to supposing that the adsorption enthalpy changes linearly with pressure. The **Freundlich isotherm**

$$\theta = c_1 p^{1/c_2} \qquad (21.19)$$

corresponds to a logarithmic change. This isotherm attempts to incorporate the role of substrate–substrate interactions on the surface.

Different isotherms agree with experiment more or less well over restricted ranges of pressure, but they remain largely empirical. Empirical, however, does not mean useless, for if the parameters of a reasonably reliable isotherm are known, reasonably reliable results can be obtained for the extent of surface coverage under various conditions. This kind of information is essential for any discussion of heterogeneous catalysis.

21.5 The rates of surface processes

The rates of surface processes may be studied by techniques described in Section 21.3. Another technique, **second harmonic generation** (SHG), is very important for the study of all types of surfaces, including thin films and liquid–gas interfaces. We saw in Section 11.7 that second harmonic generation is the conversion of an intense, pulsed laser beam to radiation with twice its initial frequency as it passes though a material. In addition to a number of crystals, surfaces are also suitable materials for SHG. Because pulsed lasers are the excitation sources, time-resolved measurements of the kinetics and dynamics of surface processes are possible over timescales as short as femtoseconds.

Figure 21.25 shows how the potential energy of a molecule varies with its distance from the substrate surface. As the molecule approaches the surface its energy falls as it becomes

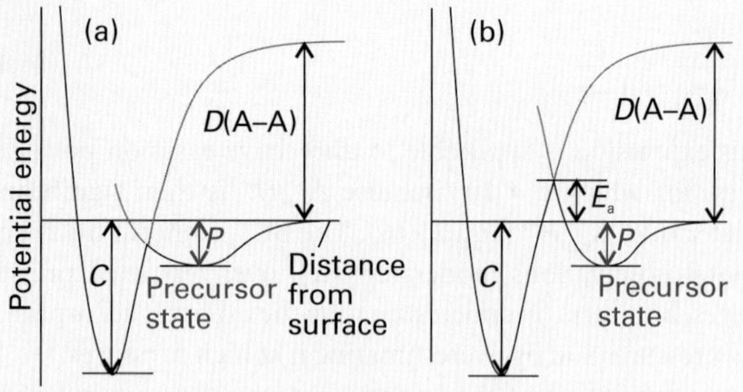

Fig. 21.25 The potential energy profiles for the dissociative chemisorption of an A_2 molecule. In each case, P is the enthalpy of (non-dissociative) physisorption and C that for chemisorption (at $T = 0$). The relative locations of the curves determine whether the chemisorption is (a) not activated or (b) activated.

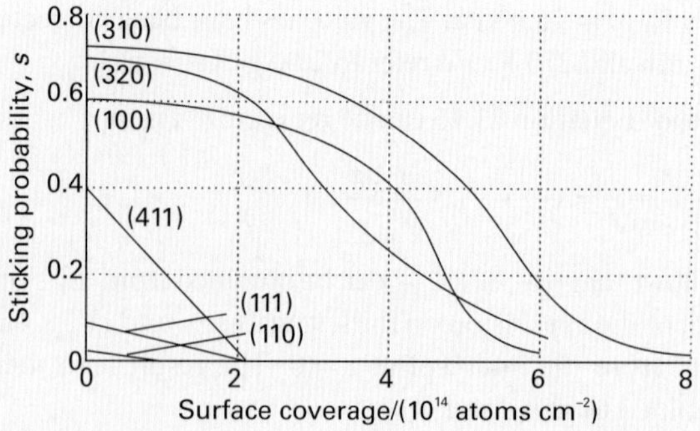

Fig. 21.26 The sticking probability of N_2 on various faces of a tungsten crystal and its dependence on surface coverage. Note the very low sticking probability for the (110) and (111) faces. (Data provided by Professor D.A. King.)

physisorbed into the **precursor state** for chemisorption. Dissociation into fragments often takes place as a molecule moves into its chemisorbed state, and after an initial increase of energy as the bonds stretch there is a sharp decrease as the adsorbate–substrate bonds reach their full strength. Even if the molecule does not fragment, there is likely to be an initial increase of potential energy as the molecule approaches the surface and the bonds adjust.

In most cases, therefore, we can expect there to be a potential energy barrier separating the precursor and chemisorbed states. This barrier, though, might be low, and might not rise above the energy of a distant, stationary particle (as in Fig. 21.25a). In this case, chemisorption is not an activated process and can be expected to be rapid. Many gas adsorptions on clean metals appear to be non-activated. In some cases the barrier rises above the zero axis (as in Fig. 21.25b); such chemisorptions are activated and slower than the non-activated kind. An example is H_2 on copper, which has an activation energy in the region of 20–40 kJ mol^{-1}.

One point that emerges from this discussion is that rates are not good criteria for distinguishing between physisorption and chemisorption. Chemisorption can be fast if the activation energy is small or zero, but it may be slow if the activation energy is large. Physisorption is usually fast, but it can appear to be slow if adsorption is taking place on a porous medium.

(a) The rate of adsorption

The rate at which a surface is covered by adsorbate depends on the ability of the substrate to dissipate the energy of the incoming particle as thermal motion as it crashes on to the surface. If the energy is not dissipated quickly, the particle migrates over the surface until a vibration expels it into the overlying gas or it reaches an edge. The proportion of collisions with the surface that successfully lead to adsorption is called the **sticking probability**, s:

$$s = \frac{\text{rate of adsorption of particles by the surface}}{\text{rate of collision of particles with the surface}} \qquad [21.20]$$

The denominator can be calculated from the kinetic model, and the numerator can be measured by observing the rate of change of pressure.

Values of s vary widely. For example, at room temperature CO has s in the range 0.1–1.0 for several d-metal surfaces, but for N_2 on rhenium $s < 10^{-2}$, indicating that more than a hundred collisions are needed before one molecule sticks successfully. Beam studies on specific crystal planes show a pronounced specificity: for N_2 on tungsten, s ranges from 0.74 on the (320) faces down to less than 0.01 on the (110) faces at room temperature. The sticking probability decreases as the surface coverage increases (Fig. 21.26). A simple assumption is that s is proportional to $1 - \theta$, the fraction uncovered, and it is common to write

$$s = (1 - \theta)s_0 \qquad (21.21)$$

where s_0 is the sticking probability on a perfectly clean surface. The results in Fig. 21.26 do not fit this expression because they show that s remains close to s_0 until the coverage has risen to about 3×10^{14} molecules cm^{-2}, and then falls steeply. The explanation is probably that the colliding molecule does not enter the chemisorbed state at once, but moves over the surface until it encounters an empty site.

(b) The rate of desorption

Desorption is always activated because the particles have to be lifted from the foot of a potential well. A physisorbed particle vibrates in its shallow potential well, and might shake itself off the surface after a short time. The temperature dependence of the first-order rate of departure can be expected to be Arrhenius-like, with an activation energy for desorption, E_d, comparable to the enthalpy of physisorption:

$$k_d = Ae^{-E_d/RT} \tag{21.22}$$

Therefore, the half-life for remaining on the surface has a temperature dependence

$$t_{1/2} = \frac{\ln 2}{k_d} = \tau_0 e^{E_d/RT} \qquad \tau_0 = \frac{\ln 2}{A} \tag{21.23}$$

(Note the positive sign in the exponent.) If we suppose that $1/\tau_0$ is approximately the same as the vibrational frequency of the weak particle–surface bond (about 10^{12} Hz) and $E_d \approx 25$ kJ mol^{-1}, then residence half-lives of around 10 ns are predicted at room temperature. Lifetimes close to 1 s are obtained only by lowering the temperature to about 100 K. For chemisorption, with E_d = 100 kJ mol^{-1} and guessing that $\tau_0 = 10^{-14}$ s (because the adsorbate–substrate bond is quite stiff), we expect a residence half-life of about 3×10^3 s (about an hour) at room temperature, decreasing to 1 s at about 350 K.

The desorption activation energy can be measured in several ways. However, we must be guarded in its interpretation because it often depends on the fractional coverage, and so may change as desorption proceeds. Moreover, the transfer of concepts such as 'reaction order' and 'rate constant' from bulk studies to surfaces is hazardous, and there are few examples of strictly first-order or second-order desorption kinetics (just as there are few integral-order reactions in the gas phase too).

If we disregard these complications, one way of measuring the desorption activation energy is to monitor the rate of increase in pressure when the sample is maintained at a series of temperatures, and to attempt to make an Arrhenius plot. A more sophisticated technique is **temperature programmed desorption** (TPD) or **thermal desorption spectroscopy** (TDS). The basic observation is a surge in desorption rate (as monitored by a mass spectrometer) when the temperature is raised linearly to the temperature at which desorption occurs rapidly, but once the desorption has occurred there is no more adsorbate to escape from the surface, so the desorption flux falls again as the temperature continues to rise. The TPD spectrum, the plot of desorption flux against temperature, therefore shows a peak, the location of which depends on the desorption activation energy. There are three maxima in the example shown in Fig. 21.27, indicating the presence of three sites with different activation energies.

In many cases only a single activation energy (and a single peak in the TPD spectrum) is observed. When several peaks are observed they might correspond to adsorption on different crystal planes or to multilayer adsorption. For instance, Cd atoms on tungsten show two activation energies, one of 18 kJ mol^{-1} and the other of 90 kJ mol^{-1}. The explanation is that the more tightly bound Cd atoms are attached directly to the substrate, and the less strongly bound are in a layer (or layers) above the primary overlayer. Another example of a system showing two desorption activation energies is CO on tungsten, the values being 120 kJ mol^{-1} and 300 kJ mol^{-1}. The explanation is believed to be the existence of two types of metal–adsorbate binding site, one involving a simple M—CO bond, the other adsorption with dissociation into individually adsorbed C and O atoms.

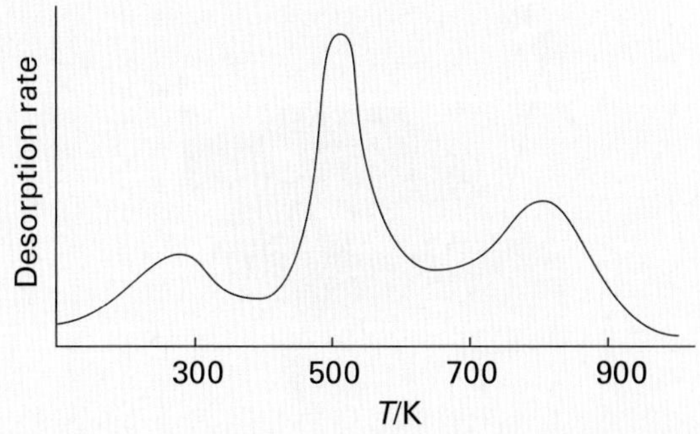

Fig. 21.27 The flash desorption spectrum of H_2 on the (100) face of tungsten. The three peaks indicate the presence of three sites with different adsorption enthalpies and therefore different desorption activation energies. (P.W. Tamm and L.D. Schmidt, *J. Chem. Phys.* **51**, 5352 (1969).)

(c) Mobility on surfaces

A further aspect of the strength of the interactions between adsorbate and substrate is the mobility of the adsorbate. Mobility is often a vital feature of a catalyst's activity, because a catalyst might be impotent if the reactant molecules adsorb so strongly that they cannot migrate. The activation energy for diffusion over a surface need not be the same as for desorption because the particles may be able to move through valleys between potential peaks without leaving the surface completely. In general, the activation energy for migration is about 10–20 per cent of the energy of the surface–adsorbate bond, but the actual value depends on the extent of coverage. The defect structure of the sample (which depends on the temperature) may also play a dominant role because the adsorbed molecules might find it easier to skip across a terrace than to roll along the foot of a step, and these molecules might become trapped in vacancies in an otherwise flat terrace. Diffusion may also be easier across one crystal face than another, and so the surface mobility depends on which lattice planes are exposed.

Diffusion characteristics of an adsorbate can be examined by using STM to follow the change in surface characteristics or by **field-ionization microscopy** (FIM), which portrays the electrical characteristics of a surface by using the ionization of noble gas atoms to probe the surface (Fig. 21.28). An individual atom is imaged, the temperature is raised, and then lowered after a definite interval. A new image is then recorded, and the new position of the atom measured (Fig. 21.29). A sequence of images

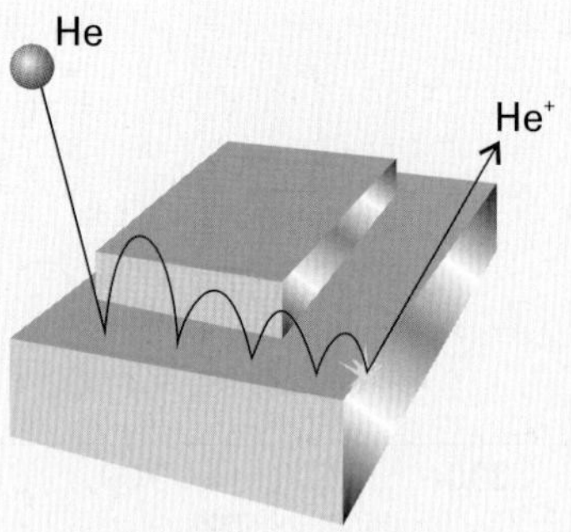

Fig. 21.28 The events leading to an FIM image of a surface. The He atom migrates across the surface until it is ionized at an exposed atom, when it is pulled off by the externally applied potential. (The bouncing motion is due to the intermolecular potential, not gravity!)

shows that the atom makes a random walk across the surface, and the diffusion coefficient, D, can be inferred from the mean distance, d, travelled in an interval τ by using the two-dimensional random walk expression $d = (D\tau)^{1/2}$. The value of D for different crystal planes at different temperatures can be determined directly in this way, and the activation energy for migration over each plane obtained from the Arrhenius-like expression

$$D = D_0 e^{-E_D/RT} \qquad (21.24)$$

where E_D is the activation energy for diffusion. Typical values for W atoms on tungsten have E_D in the range 57–87 kJ mol^{-1} and $D_0 \approx 3.8 \times 10^{-11}$ m^2 s^{-1}. For CO on tungsten, the activation energy falls from 144 kJ mol^{-1} at low surface coverage to 88 kJ mol^{-1} when the coverage is high.

IMPACT ON BIOCHEMISTRY
I21.1 Biosensor analysis

Biosensor analysis is a very sensitive and sophisticated optical technique that is now used routinely to measure the kinetics and thermodynamics of interactions between biopolymers. A biosensor detects changes in the optical properties of a surface in contact with a biopolymer.

The mobility of delocalized valence electrons accounts for the electrical conductivity of metals and these mobile electrons form a **plasma**, a dense gas of charged particles. Bombardment of the plasma by light or an electron beam can cause transient changes in the distribution of electrons, with some regions becoming slightly more dense than others. Coulomb repulsion in the regions of high density causes electrons to move away from each other, so lowering their density. The resulting oscillations in electron density, called **plasmons**, can be excited both in the bulk and on the surface of a metal. Plasmons in the bulk may be visualized as waves that propagate through the solid. A surface plasmon also propagates away from the surface, but the amplitude of the wave, also called an **evanescent wave**, decreases sharply with distance from the surface.

Biosensor analysis is based on the phenomenon of **surface plasmon resonance** (SPR), the absorption of energy from an incident beam of electromagnetic radiation by surface plasmons. Absorption, or 'resonance', can be observed with appropriate choice of the wavelength and angle of incidence of the excitation beam. It is common practice to use a monochromatic beam and to vary the angle of incidence θ (Fig. 21.30). The beam passes through a prism that strikes one side of a thin film of gold or silver. The angle corresponding to light absorption depends on the refractive index of the medium in direct contact with the opposing side of the metallic film. This variation of the resonance angle with the state of the surface arises from the ability of the evanescent wave to interact with material a short distance away from the surface.

As an illustration of biosensor analysis, we consider the association of two polymers, A and B. In a typical experiment, a stream of solution containing a known concentration of A flows above the surface to which B is chemisorbed. Figure 21.31 shows that the kinetics of binding of A to B may be followed by monitoring the time dependence of the SPR signal, denoted by R, which is typically the shift in resonance angle. The system is normally allowed to reach equilibrium, which is denoted by the plateau in Fig. 21.31. Then, a solution containing no A is flowed above the surface and the AB complex dissociates. Again, analysis of the decay of the SPR signal reveals the kinetics of dissociation of the AB complex.

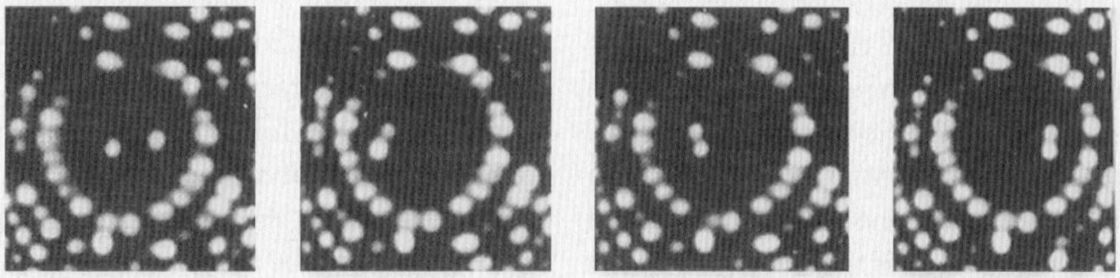

Fig. 21.29 FIM micrographs showing the migration of Re atoms on rhenium during 3 s intervals at 375 K. (Photographs provided by Professor G. Ehrlich.)

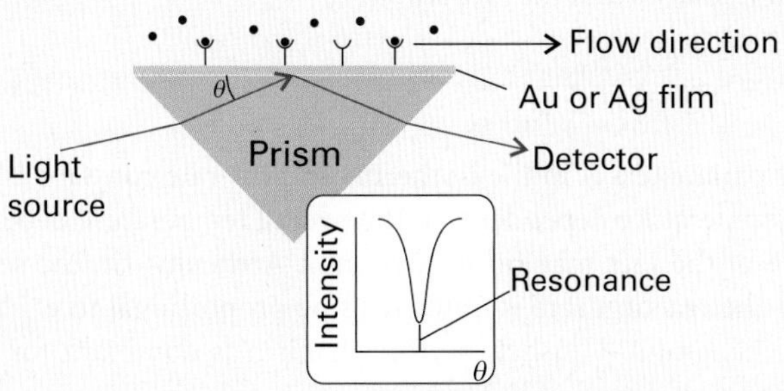

Fig. 21.30 The experimental arrangement for the observation of surface plasmon resonance, as explained in the text.

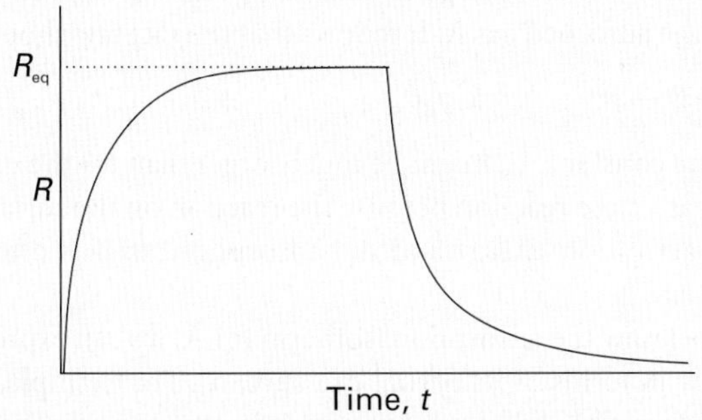

Fig. 21.31 The time dependence of a surface plasmon resonance signal, R, showing the effect of binding of a ligand to a biopolymer adsorbed on to a surface. Binding leads to an increase in R until an equilibrium value, R_{eq}, is obtained. Passing a solution containing no ligand over the surface leads to dissociation and decrease in R.

The equilibrium constant for formation of the AB complex can be measured directly from data of the type displayed in Fig. 21.31. Consider the equilibrium

$$A + B \rightleftharpoons AB \qquad K = \frac{k_{on}}{k_{off}}$$

where k_{on} and k_{off} are the rate constants for formation and dissociation of the AB complex, and K is the equilibrium constant for formation of the AB complex. It follows that

$$\frac{d[AB]}{dt} = k_{on}[A][B] - k_{off}[AB] \tag{21.25}$$

In a typical experiment, the flow rate of A is sufficiently high that $[A] = a_0$ is essentially constant. We can also write $[B] = b_0 - [AB]$ from mass-balance considerations, where b_0 is the total concentration of B. Finally, the SPR signal is often observed to be proportional to [AB]. The maximum value that R can have is $R_{max} \propto b_0$, which would be measured if all B molecules were ligated to A. We may then write

$$\frac{dR}{dt} = k_{on}a_0(R_{max} - R) - k_{off}R = k_{on}a_0R_{max} - (k_{on}a_0 + k_{off})R \tag{21.26}$$

At equilibrium $R = R_{eq}$ and $dR/dt = 0$. It follows (after some algebra) that

$$R_{eq} = \left(\frac{a_0K}{a_0K + 1}\right)R_{max} \tag{21.27}$$

Hence, the value of K can be obtained from measurements of R_{eq} for a series of a_0.

Biosensor analysis has been used in the study of thin films, metal/electrolyte surfaces, Langmuir–Blodgett films, and a number of biopolymer interactions, such as antibody–antigen and protein–DNA interactions. The most important advantage of the technique is its sensitivity; it is possible to measure the adsorption of nanograms of material on to a surface. For biological studies, the main disadvantage is the requirement for immobilization of at least one of the components of the system under study.

21.6 Mechanisms of heterogeneous catalysis

Many catalysts depend on **co-adsorption**, the adsorption of two or more species. One consequence of the presence of a second species may be the modification of the electronic structure at the surface of a metal. For instance, partial coverage of d-metal surfaces by alkali metals has a pronounced effect on the electron distribution and reduces the work function of the metal. Such modifiers can act as promoters (to enhance the action of catalysts) or as poisons (to inhibit catalytic action).

Figure 21.32 shows the potential energy curve for a reaction influenced by the action of a heterogeneous catalyst. Differences between Figs. 21.32 and 21.1 arise from the fact that heterogeneous catalysis normally depends on at least one reactant being

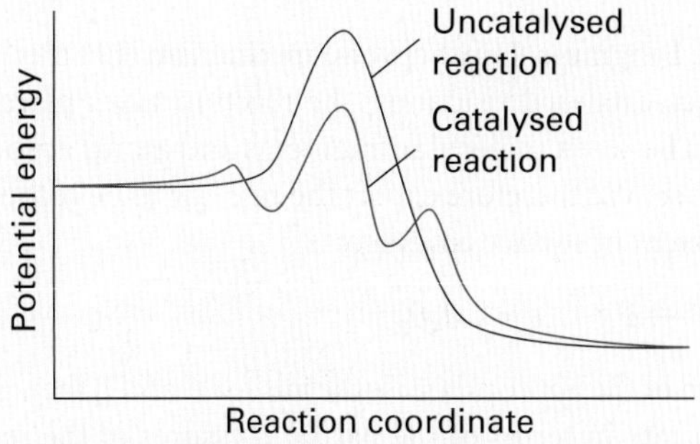

Fig. 21.32 The reaction profiles for catalysed and uncatalysed reactions. The catalysed reaction path includes activation energies for adsorption and desorption as well as an overall lower activation energy for the process.

adsorbed (usually chemisorbed) and modified to a form in which it readily undergoes reaction, and desorption of the products. Modification of the reactant often takes the form of a molecular fragmentation. In practice, the active phase is dispersed as very small particles of linear dimension less than 2 nm on a porous oxide support. **Shape-selective catalysts**, such as the zeolites (*Impact I21.2*), which have a pore size that can distinguish shapes and sizes at a molecular scale, have high internal specific surface areas, in the range of 100–500 $m^2\,g^{-1}$.

The decomposition of phosphine (PH_3) on tungsten is first order at low pressures and zeroth order at high pressures. To account for these observations, we write down a plausible rate law in terms of an adsorption isotherm and explore its form in the limits of high and low pressure. If the rate is supposed to be proportional to the surface coverage and we suppose that θ is given by the Langmuir isotherm, we would write

$$v = k_r\theta = \frac{k_r Kp}{1+Kp} \tag{21.28a}$$

where p is the pressure of phosphine. When the pressure is so low that $Kp \ll 1$, we can neglect Kp in the denominator and obtain

$$v = k_r Kp \tag{21.28b}$$

and the decomposition is first order. When $Kp \gg 1$, we can neglect the 1 in the denominator, whereupon the Kp terms cancel and we are left with

$$v = k_r \tag{21.28c}$$

and the decomposition is zeroth order.

Self-test 21.7 Suggest the form of the rate law for the deuteration of NH_3 in which D_2 adsorbs dissociatively but not extensively (that is, $Kp \ll 1$, with p the partial pressure of D_2), and NH_3 (with partial pressure p') adsorbs at different sites.
$[v = k_r(Kp)^{1/2}K'p'/(1+K'p')]$

In the **Langmuir–Hinshelwood mechanism** (LH mechanism) of surface-catalysed reactions, the reaction takes place by encounters between molecular fragments and atoms adsorbed on the surface. We therefore expect the rate law to be second order in the extent of surface coverage:

$$A + B \rightarrow P \qquad v = k_r\theta_A\theta_B \tag{21.29}$$

Insertion of the appropriate isotherms for A and B then gives the reaction rate in terms of the partial pressures of the reactants. For example, if A and B follow Langmuir isotherms, and adsorb without dissociation, so that

$$\theta_A = \frac{K_A p_A}{1+K_A p_A + K_B p_B} \qquad \theta_B = \frac{K_B p_B}{1+K_A p_A + K_B p_B} \tag{21.30}$$

then it follows that the rate law is

$$v = \frac{k_r K_A K_B p_A p_B}{(1+K_A p_A + K_B p_B)^2} \tag{21.31}$$

The parameters K in the isotherms and the rate constant k_r are all temperature dependent, so the overall temperature dependence of the rate may be strongly non-Arrhenius (in the sense that the reaction rate is unlikely to be proportional to $e^{-E_a/RT}$). The Langmuir–Hinshelwood mechanism is dominant for the catalytic oxidation of CO to CO_2.

In the **Eley–Rideal mechanism** (ER mechanism) of a surface-catalysed reaction, a gas-phase molecule collides with another molecule already adsorbed on the surface. The rate of formation of product is expected to be proportional to the partial pressure, p_B, of the non-adsorbed gas B and the extent of surface coverage, θ_A, of the adsorbed gas A. It follows that the rate law should be

$$A + B \rightarrow P \qquad v = k_r p_B\theta_A \tag{21.32}$$

The rate constant, k_r, might be much larger than for the uncatalysed gas-phase reaction because the reaction on the surface has a low activation energy and the adsorption itself is often not activated.

If we know the adsorption isotherm for A, we can express the rate law in terms of its partial pressure, p_A. For example, if the adsorption of A follows a Langmuir isotherm in the pressure range of interest, then the rate law would be

$$v = \frac{k_r K p_A p_B}{1+Kp_A} \tag{21.33}$$

If A were a diatomic molecule that adsorbed as atoms, we would substitute the isotherm given in eqn 21.13 instead.

According to eqn 21.33, when the partial pressure of A is high (in the sense $Kp_A \gg 1$) there is almost complete surface coverage, and the rate is equal to $k_r p_B$. Now the rate-determining step is the collision of B with the adsorbed fragments. When the pressure of A is low ($Kp_A \ll 1$), perhaps because of its reaction, the rate is equal to $k_r Kp_A p_B$; now the extent of surface coverage is important in the determination of the rate.

Almost all thermal surface-catalysed reactions are thought to take place by the LH mechanism, but a number of reactions with an ER mechanism have also been identified from molecular beam investigations. For example, the reaction between H(g) and D(ad) to form HD(g) is thought to be by an ER mechanism involving the direct collision and pick-up of the adsorbed D atom by the incident H atom. However, the two mechanisms should really be thought of as ideal limits, and all reactions lie somewhere between the two and show features of each one.

21.7 Catalytic activity at surfaces

It has become possible to investigate how the catalytic activity of a surface depends on its structure as well as its composition. For instance, the cleavage of C—H and H—H bonds appears to depend on the presence of steps and kinks, and a terrace often has only minimal catalytic activity. The reaction $H_2 + D_2 \rightarrow 2\ HD$ has been studied in detail. For this reaction, terrace sites are inactive but one molecule in ten reacts when it strikes a step. Although the step itself might be the important feature, it may be that the presence of the step merely exposes a more reactive crystal face (the step face itself). Likewise, the dehydrogenation of hexane to hexene depends strongly on the kink density, and it appears that kinks are needed to cleave C—C bonds. These observations suggest a reason why even small amounts of impurities may poison a catalyst: they are likely to attach to step and kink sites, and so impair the activity of the catalyst entirely. A constructive outcome is that the extent of dehydrogenation may be controlled relative to other types of reactions by seeking impurities that adsorb at kinks and act as specific poisons.

The activity of a catalyst depends on the strength of chemisorption as indicated by the 'volcano' curve in Fig. 21.33 (which is so called on account of its general shape). To be active, the catalyst should be extensively covered by adsorbate, which is the case if chemisorption is strong. On the other hand, if the strength of the substrate–adsorbate bond becomes too great, the activity declines either because the other reactant molecules cannot react with the adsorbate or because the adsorbate molecules are immobilized on the surface. This pattern of behaviour suggests that the activity of a catalyst should initially increase with strength of adsorption (as measured, for instance, by the enthalpy of adsorption) and then decline, and that the most active catalysts should be those lying near the summit of the volcano. Most active metals are those that lie close to the middle of the d-block.

Many metals are suitable for adsorbing gases, and the general order of adsorption strengths decreases along the series O_2, C_2H_2, C_2H_4, CO, H_2, CO_2, N_2. Some of these molecules adsorb dissociatively (for example, H_2). Elements from the d-block, such as iron, vanadium, and chromium, show a strong activity towards all these gases, but manganese and copper are unable to adsorb N_2 and CO_2. Metals towards the left of the periodic table (for example, magnesium and lithium) can adsorb (and, in fact, react with) only the most active gas (O_2). These trends are summarized in Table 21.3.

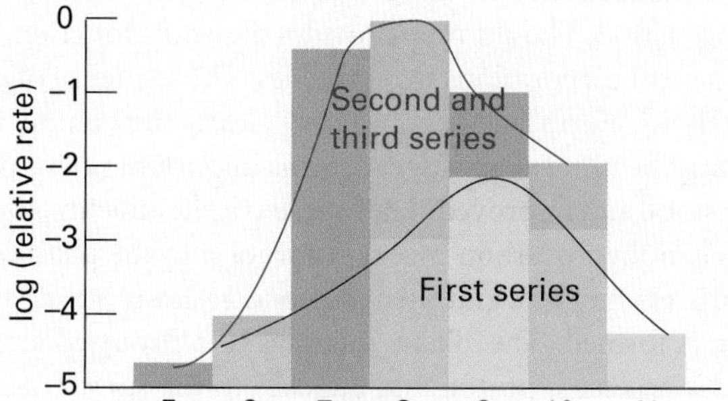

Fig. 21.33 A volcano curve of catalytic activity arises because, although the reactants must adsorb reasonably strongly, they must not adsorb so strongly that they are immobilized. The lower curve refers to the first series of d-block metals, the upper curve to the second and third series d-block metals. The group numbers relate to the periodic table inside the back cover.

Table 21.3 Chemisorption abilities*

	O_2	C_2H_2	C_2H_4	CO	H_2	CO_2	N_2
Ti, Cr, Mo, Fe	+	+	+	+	+	+	+
Ni, Co	+	+	+	+	+	+	−
Pd, Pt	+	+	+	+	+	−	−
Mn, Cu	+	+	+	+	±	−	−
Al, Au	+	+	+	−	−	−	−
Li, Na, K	+	+	−	−	−	−	−
Mg, Ag, Zn, Pb	+	−	−	−	−	−	−

* +, Strong chemisorption; ±, chemisorption; −, no chemisorption.

IMPACT ON TECHNOLOGY

I21.2 Catalysis in the chemical industry

Almost the whole of modern chemical industry depends on the development, selection, and application of catalysts (Table 21.4). All we can hope to do in this section is to give a brief indication of some of the problems involved. Other than the ones we consider, these problems include the danger of the catalyst being poisoned by byproducts or impurities, and economic considerations relating to cost and lifetime.

An example of catalytic action is found in the hydrogenation of alkenes. The alkene (**1**) adsorbs by forming two bonds with the surface (**2**), and on the same surface there may be adsorbed H atoms. When an encounter occurs, one of the alkene–surface bonds is broken (forming **3** or **4**) and later an encounter with a second H atom releases the fully hydrogenated hydrocarbon,

Table 21.4 Properties of catalysts

Catalyst	Function	Examples
Metals	Hydrogenation Dehydrogenation	Fe, Ni, Pt, Ag
Semiconducting oxides and sulfides	Oxidation Desulfurization	NiO, ZnO, MgO, Bi_2O_3/MoO_3, MoS_2
Insulating oxides	Dehydration	Al_2O_3, SiO_2, MgO
Acids	Polymerization Isomerization Cracking Alkylation	H_3PO_4, H_2SO_4, SiO_2/Al_2O_3, zeolites

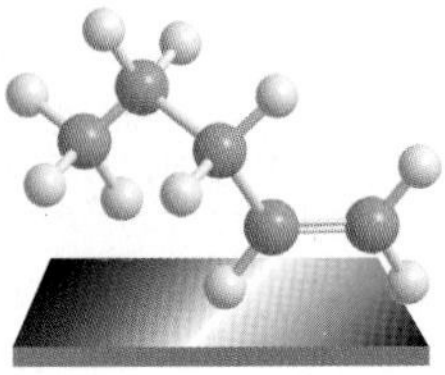

1

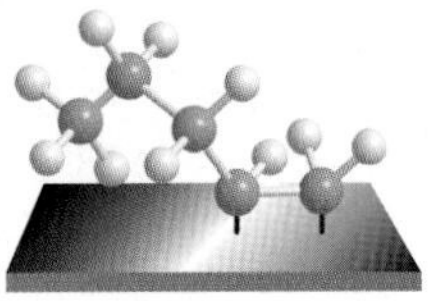

2

3

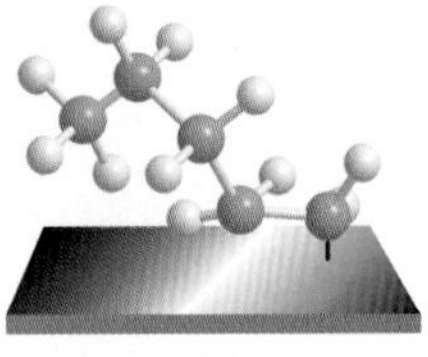

4

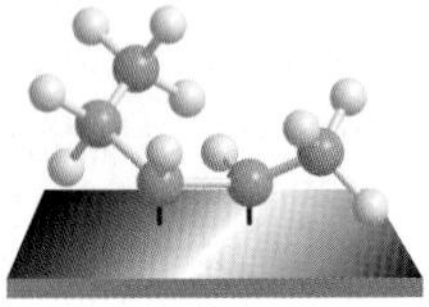

5

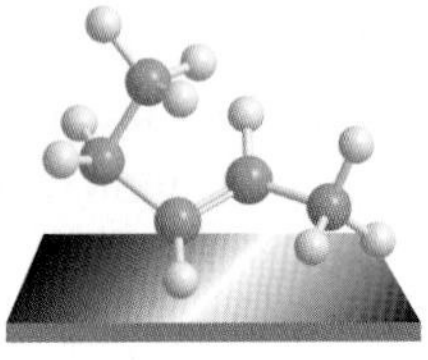

6

which is the thermodynamically more stable species. The evidence for a two-stage reaction is the appearance of different isomeric alkenes in the mixture. The formation of isomers comes about because, while the hydrocarbon chain is waving about over the surface of the metal, an atom in the chain might chemisorb again to form (5) and then desorb to (6), an isomer of the original molecule. The new alkene would not be formed if the two hydrogen atoms attached simultaneously.

A major industrial application of catalytic hydrogenation is to the formation of edible fats from vegetable and animal oils. Raw oils obtained from sources such as the soya bean have the structure $CH_2(OOCR)CH(OOCR')CH_2(OOCR'')$, where R, R′, and R″ are long-chain hydrocarbons with several double bonds. One disadvantage of the presence of many double bonds is that the oils are susceptible to atmospheric oxidation, and therefore are liable to become rancid. The geometrical configuration of the chains is responsible for the liquid nature of the oil, and in many applications a solid fat is at least much better and often necessary. Controlled partial hydrogenation of an oil with a catalyst, carefully selected so that hydrogenation is incomplete and so that the chains do not isomerize (finely divided nickel, in fact), is used on a wide scale to produce edible fats. The process, and the industry, is not made any easier by the seasonal variation of the number of double bonds in the oils.

Catalytic oxidation is also widely used in industry and in pollution control. Although in some cases it is desirable to achieve complete oxidation (as in the production of nitric acid from ammonia), in others partial oxidation is the aim. For example, the complete oxidation of propene to carbon dioxide and water is wasteful, but its partial oxidation to propenal (acrolein, $CH_2{=}CHCHO$) is the start of important industrial processes. Likewise, the controlled oxidations of ethene to ethanol, ethanal (acetaldehyde), and (in the presence of acetic acid and chlorine) to chloroethene (vinyl chloride, for the manufacture of PVC), are the initial stages of very important chemical industries.

Some of these oxidation reactions are catalysed by d-metal oxides of various kinds. The physical chemistry of oxide surfaces is very complex, as can be appreciated by considering what happens during the oxidation of propene to propenal on bismuth molybdate. The first stage is the adsorption of the propene molecule with loss of a hydrogen to form the propenyl (allyl) radical, $CH_2{=}CHCH_2$. An O atom in the surface can now transfer to this radical, leading to the formation of propenal and its desorption from the surface. The H atom also escapes with a surface O atom, and goes on to form H_2O, which leaves the surface. The surface is left with vacancies and metal ions in lower oxidation states. These vacancies are attacked by O_2 molecules in the overlying gas, which then chemisorb as O_2^- ions, so reforming the catalyst. This sequence of events, which is called the **Mars van Krevelen mechanism**, involves great upheavals of the surface, and some materials break up under the stress.

Many of the small organic molecules used in the preparation of all kinds of chemical products come from oil. These small building blocks of polymers, perfumes, and petrochemicals in general, are usually cut from the long-chain hydrocarbons drawn from the Earth as petroleum. The catalytically induced fragmentation of the long-chain hydrocarbons is called **cracking**, and is often brought about on silica–alumina catalysts. These catalysts act by forming unstable carbocations, which dissociate and rearrange to more highly branched isomers. These branched isomers burn more smoothly and efficiently in internal combustion engines, and are used to produce higher octane fuels.

Catalytic **reforming** uses a dual-function catalyst, such as a dispersion of platinum and acidic alumina. The platinum provides the metal function, and brings about dehydrogenation and hydrogenation. The alumina provides the acidic function, being able to form carbocations from alkenes. The sequence of events in catalytic reforming shows up very clearly the complications that must be unravelled if a reaction as important as this is to be understood and improved. The first step is the attachment of the long-chain hydrocarbon by chemisorption to the platinum. In this process first one and then a second H atom is lost, and an alkene is formed. The alkene migrates to a Brønsted acid site, where it accepts a proton and attaches to the surface as a carbocation. This carbocation can undergo several different reactions. It can break into two, isomerize into a more highly branched form, or undergo varieties of ring-closure. Then the adsorbed molecule loses a proton, escapes from the surface, and migrates (possibly through the gas) as an alkene to a metal part of the catalyst where it is hydrogenated. We end up with a rich selection of smaller molecules which can be withdrawn, fractionated, and then used as raw materials for other products.

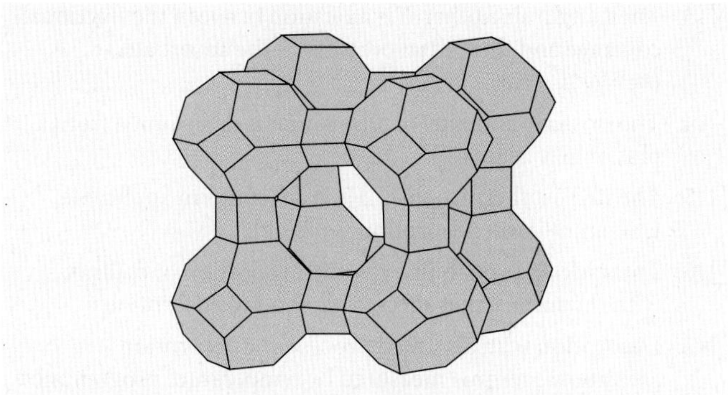

Fig. 21.34 A framework representation of the general layout of the Si, Al, and O atoms in a zeolite material. Each vertex corresponds to a Si or Al atom and each edge corresponds to the approximate location of an O atom. Note the large central pore, which can hold cations, water molecules, or other small molecules.

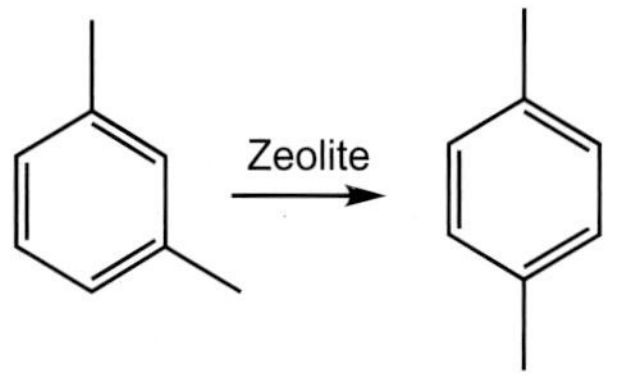

7 8

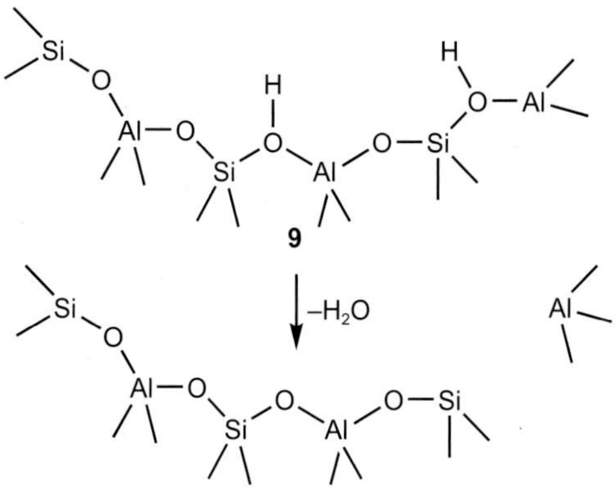

9

10

The concept of a solid surface has been extended with the availability of **microporous materials**, in which the surface effectively extends deep inside the solid. Zeolites are microporous aluminosilicates with the general formula $\{[M^{n+}]_{x/n}\cdot[H_2O]_m\}\{[AlO_2]_x[SiO_2]_y\}^{x-}$, where M^{n+} cations and H_2O molecules bind inside the cavities, or pores, of the Al—O—Si framework (Fig. 21.34). Small neutral molecules, such as CO_2, NH_3, and hydrocarbons (including aromatic compounds), can also adsorb to the internal surfaces and we shall see that this partially accounts for the utility of zeolites as catalysts.

Some zeolites for which $M = H^+$ are very strong acids and catalyse a variety of reactions that are of particular importance to the petrochemical industry. Examples include the dehydration of methanol to form hydrocarbons such as gasoline and other fuels:

$$x\,CH_3OH \xrightarrow{\text{zeolite}} (CH_2)_x + x\,H_2O$$

and the isomerization of *m*-xylene (**7**) to *p*-xylene (**8**). The catalytically important form of these acidic zeolites may be either a Brønsted acid (**9**) or a Lewis acid (**10**). Like enzymes, a zeolite catalyst with a specific composition and structure is very selective toward certain reactants and products because only molecules of certain sizes can enter and exit the pores in which catalysis occurs. It is also possible that zeolites derive their selectivity from the ability to bind and to stabilize only transition states that fit properly in the pores. The analysis of the mechanism of zeolyte catalysis is greatly facilitated by computer simulation of microporous systems, which shows how molecules fit in the pores, migrate through the connecting tunnels, and react at the appropriate active sites.

Checklist of key ideas

☐ 1. Catalysts are substances that accelerate reactions but undergo no net chemical change.

☐ 2. A homogeneous catalyst is a catalyst in the same phase as the reaction mixture. Enzymes are homogeneous, biological catalysts.

☐ 3. The Michaelis–Menten mechanism of enzyme kinetics accounts for the dependence of rate on the concentration of the substrate, $v = v_{max}[S]_0/([S]_0 + K_M)$.

☐ 4. A Lineweaver–Burk plot, based on $1/v = 1/v_{max} + (K_M/v_{max})(1/[S]_0)$, is used to determine the parameters that occur in the Michaelis–Menten mechanism.

☐ 5. In competitive inhibition of an enzyme, the inhibitor binds only to the active site of the enzyme and thereby inhibits the attachment of the substrate.

☐ 6. In uncompetitive inhibition the inhibitor binds to a site of the enzyme that is removed from the active site, but only if the substrate is already present.

☐ 7. In noncompetitive inhibition, the inhibitor binds to a site other than the active site, and its presence reduces the ability of the substrate to bind to the active site.

☐ 8. Adsorption is the attachment of molecules to a surface; the substance that adsorbs is the adsorbate and the underlying material is the adsorbent or substrate. The reverse of adsorption is desorption.

☐ 9. Techniques for studying surface composition and structure include scanning electron microscopy (SEM), scanning tunnelling microscopy (STM), photoemission spectroscopy, Auger electron spectroscopy (AES), and low energy electron diffraction (LEED).

☐ 10. The fractional coverage, θ, is the ratio of the number of occupied sites to the number of available sites.

☐ 11. Techniques for studying the rates of surface processes include flash desorption, biosensor analysis, second harmonic generation (SHG), gravimetry by using a quartz crystal microbalance (QCM), and molecular beam reactive scattering (MRS).

☐ 12. Physisorption is adsorption by a van der Waals interaction; chemisorption is adsorption by formation of a chemical (usually covalent) bond.

☐ 13. The Langmuir isotherm is a relation between the fractional coverage and the partial pressure of the adsorbate: $\theta = Kp/(1+Kp)$.

☐ 14. The isosteric enthalpy of adsorption is determined from a plot of ln K against $1/T$.

☐ 15. The BET isotherm, eqn 21.15, is an isotherm applicable when multilayer adsorption is possible.

☐ 16. The sticking probability, s, is the proportion of collisions with the surface that successfully lead to adsorption.

☐ 17. Desorption is an activated process; the desorption activation energy is measured by temperature programmed desorption (TPD) or thermal desorption spectroscopy (TDS).

☐ 18. In the Langmuir–Hinshelwood mechanism (LH mechanism) of surface-catalysed reactions, the reaction takes place by encounters between molecular fragments and atoms adsorbed on the surface.

☐ 19. In the Eley–Rideal mechanism (ER mechanism) of a surface-catalysed reaction, a gas-phase molecule collides with another molecule already adsorbed on the surface.

Further information

Further information 21.1 *The BET isotherm*

We suppose that at equilibrium a fraction θ_0 of the surface sites are unoccupied, a fraction θ_1 is covered by a monolayer, a fraction θ_2 is covered by a bilayer, and so on. The number of adsorbed molecules is therefore

$$N = N_{\text{sites}}(\theta_1 + 2\theta_2 + 3\theta_3 + \cdots)$$

where N_{sites} is the total number of sites. We now follow the derivation that led to the Langmuir isotherm (eqn 21.11) but allow for different rates of desorption from the substrate and the various layers:

First layer: Rate of adsorption $= Nk_{a,0}p\theta_0$ Rate of desorption $= Nk_{d,0}\theta_1$
At equilibrium $k_{a,0}p\theta_0 = k_{d,0}\theta_1$
Second layer: Rate of adsorption $= Nk_{a,1}p\theta_1$ Rate of desorption $= Nk_{d,1}\theta_2$
At equilibrium $k_{a,1}p\theta_1 = k_{d,1}\theta_2$
Third layer: Rate of adsorption $= Nk_{a,2}p\theta_2$ Rate of desorption $= Nk_{d,2}\theta_3$
At equilibrium $k_{a,2}p\theta_2 = k_{d,2}\theta_3$

and so on. We now suppose that, once a monolayer has been formed, all the rate constants involving adsorption and desorption from the physisorbed layers are the same, and write these equations as

$$k_{a,0}p\theta_0 = k_{d,0}\theta_1,\text{ so } \theta_1 = (k_{a,0}/k_{d,0})p\theta_0 = K_0p\theta_0$$

$$k_{a,1}p\theta_1 = k_{d,1}\theta_2,\text{ so } \theta_2 = (k_{a,1}/k_{d,1})p\theta_1 = (k_{a,0}/k_{d,0})(k_{a,1}/k_{d,1})p^2\theta_0 = K_0K_1p^2\theta_0$$

$$k_{a,1}p\theta_2 = k_{d,1}\theta_3,\text{ so } \theta_3 = (k_{a,1}/k_{d,1})p\theta_2 = (k_{a,0}/k_{d,0})(k_{a,1}/k_{d,1})^2p^3\theta_0 = K_0K_1^2p^3\theta_0$$

and so on, with $K_0 = k_{a,0}/k_{d,0}$ and $K_1 = k_{a,1}/k_{d,1}$ the equilibrium constants for adsorption to the substrate and an overlayer, respectively. Now, because $\theta_0 + \theta_1 + \theta_2 + \cdots = 1$, it follows that with

$$\begin{aligned}\theta_0 + K_0p\theta_0 + K_0K_1p^2\theta_0 + K_0K_1^2p^3\theta_0 + \cdots &= \theta_0 + K_0p\theta_0\{1 + K_1p + K_1^2p^2 + \cdots\}\\ &= \left\{1 + \frac{K_0p}{1-K_1p}\right\}\theta_0\\ &= \left\{\frac{1-K_1p+K_0p}{1-K_1p}\right\}\theta_0\end{aligned}$$

then, because this expression is equal to 1,

$$\theta_0 = \frac{1-K_1p}{1-(K_1-K_0)p}$$

In a similar way, we can write the number of adsorbed species as

$$\begin{aligned}N &= N_{\text{sites}}K_0p\theta_0 + 2N_{\text{sites}}K_0K_1p^2\theta_0 + \cdots\\ &= N_{\text{sites}}K_0p\theta_0(1 + 2K_1p + 3K_1^2p^2 + \cdots)\\ &= \frac{N_{\text{sites}}K_0p\theta_0}{(1-K_1p)^2}\end{aligned}$$

By combining the last two expressions, we obtain

$$N = \frac{N_{\text{sites}}K_0p}{(1-K_1p)^2} \times \frac{1-K_1p}{1-(K_1-K_0)p} = \frac{N_{\text{sites}}K_0p}{(1-K_1p)\{1-(K_1-K_0)p\}}$$

The ratio N/N_{sites} is equal to the ratio V/V_{mon}, where V is the total volume adsorbed and V_{mon} the volume adsorbed had there been complete monolayer coverage. The term K_1 is the equilibrium constant for the 'reaction' in which the 'reactant' is a molecule physisorbed on to adsorbed layers and the 'product' is the molecule in the vapour. This process is very much like the equilibrium $M(g) \rightleftharpoons M(l)$, for which $K = 1/p^*$, where p^* is the vapour pressure of the liquid. Therefore, with $K_1 = 1/p^*$, $z = p/p^*$, and $c = K_0/K_1$, the last equation becomes

$$\frac{V}{V_{mon}} = \frac{K_0 p}{(1 - p/p^*)\{1 - (1 - K_0/K_1)p/p^*\}} = \frac{cz}{(1 - z)\{1 - (1 - c)z\}}$$

as in eqn 21.15.

Discussion questions

21.1 Discuss the features, advantages, and limitations of the Michaelis–Menten mechanism of enzyme action.

21.2 Prepare a report on the application of the experimental strategies described in Chapter 19 to the study of enzyme-catalysed reactions. Devote some attention to the following topics: (a) the determination of reaction rates over a large timescale; (b) the determination of the rate constants and equilibrium constant of binding of substrate to an enzyme, and (c) the characterization of intermediates in a catalytic cycle. Your report should be similar in content and extent to one of the *Impact* sections found throughout this text.

21.3 A plot of the rate of an enzyme-catalysed reaction against temperature has a maximum, in an apparent deviation from the behaviour predicted by the Arrhenius relation (eqn 20.1). Suggest a molecular interpretation for this effect.

21.4 Distinguish between competitive, noncompetitive, and uncompetitive inhibition of enzymes. Discuss how these modes of inhibition may be detected experimentally.

21.5 Some enzymes are inhibited by high concentrations of their own products. (a) Sketch a plot of reaction rate against concentration of substrate for an enzyme that is prone to product inhibition.

21.6 (a) Distinguish between a step and a terrace. (b) Describe how steps and terraces can be formed by dislocations.

21.7 Drawing from knowledge you have acquired through the text, describe the advantages and limitations of each of the microscopy, diffraction, and scattering techniques designated by the acronyms AFM, FIM, LEED, SAM, SEM, and STM.

21.8 Distinguish between the following adsorption isotherms: Langmuir, BET, Temkin, and Freundlich and indicate when and why they are likely to be appropriate.

21.9 Describe the essential features of the Langmuir–Hinshelwood, Eley–Rideal, and Mars van Krevelen mechanisms for surface-catalysed reactions.

21.10 Account for the dependence of catalytic activity of a surface on the strength of chemisorption, as shown in Fig. 21.25.

Exercises

21.1(a) Consider the base-catalysed reaction

(1) $AH + B \rightleftharpoons BH^+ + A^-$ $\quad k_a, k_a'$, both fast

(2) $A^- + AH \rightarrow \text{product}$ $\quad k_b$, slow

Deduce the rate law.

21.1(b) Consider the acid-catalysed reaction

(1) $HA + H^+ \rightleftharpoons HAH^+$ $\quad k_a, k_a'$, both fast

(2) $HAH^+ + B \rightarrow BH^+ + AH$ $\quad k_b$, slow

Deduce the rate law.

21.2(a) The enzyme-catalysed conversion of a substrate at 25°C has a Michaelis constant of 0.046 mol dm^{-3}. The rate of the reaction is 1.04×10^{-3} mol dm^{-3} s^{-1} when the substrate concentration is 0.105 mol dm^{-3}. What is the maximum velocity of this reaction?

21.2(b) The enzyme-catalysed conversion of a substrate at 25°C has a Michaelis constant of 0.032 mol dm^{-3}. The rate of the reaction is 2.05×10^{-4} mol dm^{-3} s^{-1} when the substrate concentration is 0.875 mol dm^{-3}. What is the maximum velocity of this reaction?

21.3(a) The enzyme-catalysed conversion of a substrate at 25°C has a Michaelis constant of 0.015 mol dm^{-3} and a maximum velocity of 4.25×10^{-4} mol dm^{-3} s^{-1} when the enzyme concentration is 3.60×10^{-9} mol dm^{-3}. Calculate k_{cat} and η. Is the enzyme 'catalytically perfect'?

21.3(b) The enzyme-catalysed conversion of a substrate at 25°C has a Michaelis constant of 9.0×10^5 mol dm^{-3} and a maximum velocity of 2.24×10^{-5} mol dm^{-3} s^{-1} when the enzyme concentration is 1.60×10^{-9} mol dm^{-3}. Calculate k_{cat} and η. Is the enzyme 'catalytically perfect'?

21.4(a) Consider an enzyme-catalysed reaction that follows Michaelis–Menten kinetics with $K_M = 3.0 \times 10^{-3}$ mol dm^{-3}. What concentration of a competitive inhibitor characterized by $K_I = 2.0 \times 10^{-5}$ mol dm^{-3} will reduce the rate of formation of product by 50 per cent when the substrate concentration is held at 1.0×10^{-4} mol dm^{-3}?

21.4(b) Consider an enzyme-catalysed reaction that follows Michaelis–Menten kinetics with $K_M = 7.5 \times 10^{-4}$ mol dm^{-3}. What concentration of a competitive inhibitor characterized by $K_I = 5.6 \times 10^{-4}$ mol dm^{-3} will reduce the rate of formation of product by 75 per cent when the substrate concentration is held at 1.0×10^{-4} mol dm^{-3}?

21.5(a) Calculate the frequency of molecular collisions per square centimetre of surface in a vessel containing (a) hydrogen, (b) propane at 25°C when the pressure is (i) 100 Pa, (ii) 0.10 μTorr.

21.5(b) Calculate the frequency of molecular collisions per square centimetre of surface in a vessel containing (a) nitrogen, (b) methane at 25°C when the pressure is (i) 10.0 Pa, (ii) 0.150 μTorr.

21.6(a) What pressure of argon gas is required to produce a collision rate of 4.5×10^{20} s^{-1} at 425 K on a circular surface of diameter 1.5 mm?

21.6(b) What pressure of nitrogen gas is required to produce a collision rate of 5.00×10^{19} s^{-1} at 525 K on a circular surface of diameter 2.0 mm?

21.7(a) The LEED pattern from a clean unreconstructed (110) face of a metal is shown below. Sketch the LEED pattern for a surface that was reconstructed by doubling the vertical separation between the atoms.

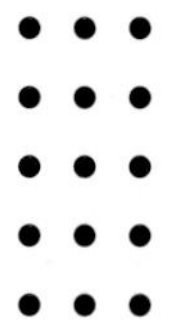

21.7(b) The LEED pattern from a clean unreconstructed (110) face of a metal is shown below. Sketch the LEED pattern for a surface that was reconstructed by tripling the horizontal separation between the atoms.

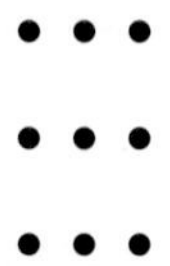

21.8(a) A monolayer of N_2 molecules is adsorbed on the surface of 1.00 g of an Fe/Al_2O_3 catalyst at 77 K, the boiling point of liquid nitrogen. Upon warming, the nitrogen occupies 3.86 cm^3 at 0°C and 760 Torr. What is the surface area of the catalyst?

21.8(b) A monolayer of CO molecules is adsorbed on the surface of 1.00 g of an Fe/Al_2O_3 catalyst at 77 K, the boiling point of liquid nitrogen. Upon warming, the carbon monoxide occupies 3.75 cm^3 at 0°C and 1.00 bar. What is the surface area of the catalyst?

21.9(a) The volume of oxygen gas at 0°C and 104 kPa adsorbed on the surface of 1.00 g of a sample of silica at 0°C was 0.286 cm^3 at 145.4 Torr and 1.443 cm^3 at 760 Torr. What is the value of V_{mon}?

21.9(b) The volume of gas at 20°C and 1.00 bar adsorbed on the surface of 1.50 g of a sample of silica at 0°C was 1.42 cm^3 at 56.4 kPa and 2.77 cm^3 at 108 kPa. What is the value of V_{mon}?

21.10(a) The enthalpy of adsorption of CO on a surface is found to be −120 kJ mol^{-1}. Estimate the mean lifetime of a CO molecule on the surface at 400 K.

21.10(b) The enthalpy of adsorption of ammonia on a nickel surface is found to be −155 kJ mol^{-1}. Estimate the mean lifetime of an NH_3 molecule on the surface at 500 K.

21.11(a) A certain solid sample adsorbs 0.44 mg of CO when the pressure of the gas is 26.0 kPa and the temperature is 300 K. The mass of gas adsorbed when the pressure is 3.0 kPa and the temperature is 300 K is 0.19 mg. The Langmuir isotherm is known to describe the adsorption. Find the fractional coverage of the surface at the two pressures.

21.11(b) A certain solid sample adsorbs 0.63 mg of CO when the pressure of the gas is 36.0 kPa and the temperature is 300 K. The mass of gas adsorbed when the pressure is 4.0 kPa and the temperature is 300 K is 0.21 mg. The Langmuir isotherm is known to describe the adsorption. Find the fractional coverage of the surface at the two pressures.

21.12(a) The adsorption of a gas is described by the Langmuir isotherm with $K = 0.75$ kPa^{-1} at 25°C. Calculate the pressure at which the fractional surface coverage is (a) 0.15, (b) 0.95.

21.12(b) The adsorption of a gas is described by the Langmuir isotherm with $K = 0.548$ kPa^{-1} at 25°C. Calculate the pressure at which the fractional surface coverage is (a) 0.20, (b) 0.75.

21.13(a) A solid in contact with a gas at 12 kPa and 25°C adsorbs 2.5 mg of the gas and obeys the Langmuir isotherm. The enthalpy change when 1.00 mmol of the adsorbed gas is desorbed is +10.2 J. What is the equilibrium pressure for the adsorption of 2.5 mg of gas at 40°C?

21.13(b) A solid in contact with a gas at 8.86 kPa and 25°C adsorbs 4.67 mg of the gas and obeys the Langmuir isotherm. The enthalpy change when 1.00 mmol of the adsorbed gas is desorbed is +12.2 J. What is the equilibrium pressure for the adsorption of the same mass of gas at 45°C?

21.14(a) Nitrogen gas adsorbed on charcoal to the extent of 0.921 cm^3 g^{-1} at 490 kPa and 190 K, but at 250 K the same amount of adsorption was achieved only when the pressure was increased to 3.2 MPa. What is the enthalpy of adsorption of nitrogen on charcoal?

21.14(b) Nitrogen gas adsorbed on a surface to the extent of 1.242 cm^3 g^{-1} at 350 kPa and 180 K, but at 240 K the same amount of adsorption was achieved only when the pressure was increased to 1.02 MPa. What is the enthalpy of adsorption of nitrogen on the surface?

21.15(a) In an experiment on the adsorption of oxygen on tungsten it was found that the same volume of oxygen was desorbed in 27 min at 1856 K and 2.0 min at 1978 K. What is the activation energy of desorption? How long would it take for the same amount to desorb at (a) 298 K, (b) 3000 K?

21.15(b) In an experiment on the adsorption of ethene on iron it was found that the same volume of the gas was desorbed in 1856 s at 873 K and 8.44 s at 1012 K. What is the activation energy of desorption? How long would it take for the same amount of ethene to desorb at (a) 298 K, (b) 1500 K?

21.16(a) The average time for which an oxygen atom remains adsorbed to a tungsten surface is 0.36 s at 2548 K and 3.49 s at 2362 K. What is the activation energy for chemisorption?

21.16(b) The average time for which a hydrogen atom remains adsorbed on a manganese surface is 35 per cent shorter at 1000 K than at 600 K. What is the activation energy for chemisorption?

21.17(a) For how long on average would an H atom remain on a surface at 400 K if its desorption activation energy is (a) 15 kJ mol^{-1}, (b) 150 kJ mol^{-1}? Take $\tau_0 = 0.10$ ps. For how long on average would the same atoms remain at 1000 K?

21.17(b) For how long on average would an atom remain on a surface at 298 K if its desorption activation energy is (a) 20 kJ mol^{-1}, (b) 200 kJ mol^{-1}? Take $\tau_0 = 0.12$ ps. For how long on average would the same atoms remain at 800 K?

21.18(a) Hydrogen iodide is very strongly adsorbed on gold but only slightly adsorbed on platinum. Assume the adsorption follows the Langmuir isotherm and predict the order of the HI decomposition reaction on each of the two metal surfaces.

21.18(b) Suppose it is known that ozone adsorbs on a particular surface in accord with a Langmuir isotherm. How could you use the pressure dependence of the fractional coverage to distinguish between adsorption (a) without dissociation, (b) with dissociation into $O + O_2$, (c) with dissociation into $O + O + O$?

Problems*

Numerical problems

21.1 The following results were obtained for the action of an ATPase on ATP at 20°C, when the concentration of the ATPase was 20 nmol dm^{-3}:

[ATP]/(μmol dm^{-3})	0.60	0.80	1.4	2.0	3.0
v/(μmol dm^{-3} s^{-1})	0.81	0.97	1.30	1.47	1.69

Determine the Michaelis constant, the maximum velocity of the reaction, the turnover number, and the catalytic efficiency of the enzyme.

21.2 There are different ways to represent and analyse data for enzyme-catalysed reactions. For example, in the *Eadie–Hofstee plot*, $v/[S]_0$ is plotted against v. Alternatively, in the *Hanes plot*, $v/[S]_0$ is plotted against $[S]_0$. (a) Using the simple Michaelis–Menten mechanism, derive relations between $v/[S]_0$ and v and between $v/[S]_0$ and $[S]_0$. (b) Discuss how the values of K_M and v_{max} are obtained from analysis of the Eadie–Hofstee and Hanes plots. (c) Determine the Michaelis constant and the maximum velocity of the reaction from Problem 21.1 by using Eadie–Hofstee and Hanes plots to analyse the data.

21.3 In general, the catalytic efficiency of an enzyme depends on the pH of the medium in which it operates. One way to account for this behaviour is to propose that the enzyme and the enzyme–substrate complex are active only in specific protonation states. This proposition can be summarized by the following mechanism:

$$\mathrm{EH} + \mathrm{S} \rightleftarrows \mathrm{ESH} \qquad k_a, k'_a$$

$$\mathrm{ESH} \rightarrow \mathrm{E} + \mathrm{P} \qquad k_b$$

$$\mathrm{EH} \rightleftharpoons \mathrm{E}^- + \mathrm{H}^+ \qquad K_{E,a} = \frac{[\mathrm{E}^-][\mathrm{H}^+]}{[\mathrm{EH}^+]}$$

$$\mathrm{EH}_2^+ \rightleftharpoons \mathrm{EH} + \mathrm{H}^+ \qquad K_{E,b} = \frac{[\mathrm{EH}][\mathrm{H}^+]}{[\mathrm{EH}_2^+]}$$

$$\mathrm{ESH} \rightleftharpoons \mathrm{ES}^- + \mathrm{H}^+ \qquad K_{ES,a} = \frac{[\mathrm{ES}^-][\mathrm{H}^+]}{[\mathrm{ESH}]}$$

$$\mathrm{ESH}_2 \rightleftharpoons \mathrm{ESH} + \mathrm{H}^+ \qquad K_{ES,b} = \frac{[\mathrm{ESH}][\mathrm{H}^+]}{[\mathrm{ESH}_2]}$$

in which only the EH and ESH forms are active. (a) For the mechanism above, show that

$$v = \frac{v'_{max}}{1 + K'_M[S]_0}$$

with

$$v'_{max} = \frac{v_{max}}{1 + \dfrac{[\mathrm{H}^+]}{K_{ES,b}} + \dfrac{K_{ES,a}}{[\mathrm{H}^+]}}$$

$$K'_M = K_M \frac{1 + \dfrac{[\mathrm{H}^+]}{K_{E,b}} + \dfrac{K_{E,a}}{[\mathrm{H}^+]}}{1 + \dfrac{[\mathrm{H}^+]}{K_{ES,b}} + \dfrac{K_{ES,a}}{[\mathrm{H}^+]}}$$

where v_{max} and K_M correspond to the form EH of the enzyme. (b) For pH values ranging from 0 to 14, plot v'_{max} against pH for a hypothetical reaction for which $v_{max} = 1.0 \times 10^{-6}$ mol dm^{-3} s^{-1}, $K_{ES,b} = 1.0 \times 10^{-6}$ mol dm^{-3}, and $K_{ES,a} = 1.0 \times 10^{-8}$. Is there a pH at which v_{max} reaches a maximum value? If so, determine the pH. (c) Redraw the plot in part (b) by using the same value of v_{max}, but $K_{ES,b} = 1.0 \times 10^{-4}$ mol dm^{-3} and $K_{ES,a} = 1.0 \times 10^{-10}$ mol dm^{-3}. Account for any differences between this plot and the plot from part (b).

21.4 The enzyme carboxypeptidase catalyses the hydrolysis of polypeptides and here we consider its inhibition. The following results were obtained when the rate of the enzymolysis of carbobenzoxy-glycyl-D-phenylalanine (CBGP) was monitored without inhibitor:

$[\mathrm{CBGP}]_0/(10^{-2}$ mol $dm^{-3})$	1.25	3.84	5.81	7.13
Relative reaction rate	0.398	0.669	0.859	1.000

(All rates in this problem were measured with the same concentration of enzyme and are relative to the rate measured when $[\mathrm{CBGP}]_0 = 0.0713$ mol dm^{-3} in the absence of inhibitor.) When 2.0×10^{-3} mol dm^{-3} phenylbutyrate ion was added to a solution containing the enzyme and substrate, the following results were obtained:

$[\mathrm{CBGP}]_0/(10^{-2}$ mol $dm^{-3})$	1.25	2.50	4.00	5.50
Relative reaction rate	0.172	0.301	0.344	0.548

In a separate experiment, the effect of 5.0×10^{-2} mol dm^{-3} benzoate ion was monitored and the results were:

$[\mathrm{CBGP}]_0/(10^{-2}$ mol $dm^{-3})$	1.75	2.50	5.00	10.00
Relative reaction rate	0.183	0.201	0.231	0.246

Determine the mode of inhibition of carboxypeptidase by the phenylbutyrate ion and benzoate ion.

21.5 The movement of atoms and ions on a surface depends on their ability to leave one position and stick to another, and therefore on the energy changes that occur. As an illustration, consider a two-dimensional square lattice of univalent positive and negative ions separated by 200 pm, and consider a cation on the upper terrace of this array. Calculate, by direct summation, its Coulombic interaction when it is in an empty lattice point directly above an anion. Now consider a high step in the same lattice, and let the cation move into the corner formed by the step and the terrace. Calculate the Coulombic energy for this position, and decide on the likely settling point for the cation.

21.6 In a study of the catalytic properties of a titanium surface it was necessary to maintain the surface free from contamination. Calculate the collision frequency per square centimetre of surface made by O_2 molecules at (a) 100 kPa, (b) 1.00 Pa and 300 K. Estimate the number of collisions made with a single surface atom in each second. The conclusions underline the importance of working at very low pressures (much lower than 1 Pa, in fact) in order to study the properties of uncontaminated surfaces. Take the nearest-neighbour distance as 291 pm.

21.7 Nickel is face-centred cubic with a unit cell of side 352 pm. What is the number of atoms per square centimetre exposed on a surface formed by (a) (100), (b) (110), (c) (111) planes? Calculate the frequency of

* Problems denoted with the symbol ‡ were supplied by Charles Trapp, Carmen Giunta, and Marshall Cady.

molecular collisions per surface atom in a vessel containing (a) hydrogen, (b) propane at 25°C when the pressure is (i) 100 Pa, (ii) 0.10 μTorr.

21.8 The data below are for the chemisorption of hydrogen on copper powder at 25°C. Confirm that they fit the Langmuir isotherm at low coverages. Then find the value of K for the adsorption equilibrium and the adsorption volume corresponding to complete coverage.

p/Pa	25	129	253	540	1000	1593
V/cm^3	0.042	0.163	0.221	0.321	0.411	0.471

21.9 The data for the adsorption of ammonia on barium fluoride are reported below. Confirm that they fit a BET isotherm and find values of c and V_{mon}.

(a) $\theta = 0$°C, $p^* = 429.6$ kPa:

p/kPa	14.0	37.6	65.6	79.2	82.7	100.7	106.4
V/cm^3	11.1	13.5	14.9	16.0	15.5	17.3	16.5

(b) $\theta = 18.6$°C, $p^* = 819.7$ kPa:

p/kPa	5.3	8.4	14.4	29.2	62.1	74.0	80.1	102.0
V/cm^3	9.2	9.8	10.3	11.3	12.9	13.1	13.4	14.1

21.10 The following data have been obtained for the adsorption of H_2 on the surface of 1.00 g of copper at 0°C. The volume of H_2 below is the volume that the gas would occupy at STP (0°C and 1 atm).

p/atm	0.050	0.100	0.150	0.200	0.250
V/cm^3	23.8	13.3	8.70	6.80	5.71

Determine the volume of H_2 necessary to form a monolayer and estimate the surface area of the copper sample. The density of liquid hydrogen is 0.708 g cm^{-3}.

21.11 The adsorption of solutes on solids from liquids often follows a Freundlich isotherm. Check the applicability of this isotherm to the following data for the adsorption of acetic acid on charcoal at 25°C and find the values of the parameters c_1 and c_2.

[acid]/(mol dm^{-3})	0.05	0.10	0.50	1.0	1.5
w_a/g	0.04	0.06	0.12	0.16	0.19

w_a is the mass adsorbed per gram of charcoal.

21.12 In some catalytic reactions the products may adsorb more strongly than the reacting gas. This is the case, for instance, in the catalytic decomposition of ammonia on platinum at 1000°C. As a first step in examining the kinetics of this type of process, show that the rate of ammonia decomposition should follow

$$\frac{dp_{NH_3}}{dt} = -k_c \frac{p_{NH_3}}{p_{H_2}}$$

in the limit of very strong adsorption of hydrogen. Start by showing that, when a gas J adsorbs very strongly and its pressure is p_J, the fraction of uncovered sites is approximately $1/Kp_J$. Solve the rate equation for the catalytic decomposition of NH_3 on platinum and show that a plot of $F(t) = (1/t)\ln(p/p_0)$ against $G(t) = (p - p_0)/t$, where p is the pressure of ammonia, should give a straight line from which k_c can be determined. Check the rate law on the basis of the data below, and find k_c for the reaction.

t/s	0	30	60	100	160	200	250
p/kPa	13.3	11.7	11.2	10.7	10.3	9.9	9.6

21.13‡ A. Akgerman and M. Zardkoohi (*J. Chem. Eng. Data* **41**, 185 (1996)) examined the adsorption of phenol from aqueous solution on to fly ash at 20°C. They fitted their observations to a Freundlich isotherm of the form $c_{ads} = Kc_{sol}^{1/n}$, where c_{ads} is the concentration of adsorbed phenol and c_{sol} is the concentration of aqueous phenol. Among the data reported are the following:

c_{sol}/(mg g^{-1})	8.26	15.65	25.43	31.74	40.00
c_{ads}/(mg g^{-1})	4.41	9.2	35.2	52.0	67.2

Determine the constants K and n. What further information would be necessary in order to express the data in terms of fractional coverage, θ?

21.14‡ C. Huang and W.P. Cheng (*J. Colloid Interface Sci.* **188**, 270 (1997)) examined the adsorption of the hexacyanoferrate(III) ion, $[Fe(CN)_6]^{3-}$, on γ-Al_2O_3 from aqueous solution. They modelled the adsorption with a modified Langmuir isotherm, obtaining the following values of K at pH = 6.5:

T/K	283	298	308	318
$10^{-11}K$	2.642	2.078	1.286	1.085

Determine the isosteric enthalpy of adsorption, $\Delta_{ads}H^\ominus$, at this pH. The researchers also reported $\Delta_{ads}S^\ominus = +146$ J mol^{-1} K^{-1} under these conditions. Determine $\Delta_{ads}G^\ominus$.

21.15‡ M.-G. Olivier and R. Jadot (*J. Chem. Eng. Data* **42**, 230 (1997)) studied the adsorption of butane on silica gel. They report the following amounts of adsorption (in moles per kilogram of silica gel) at 303 K:

p/kPa	31.00	38.22	53.03	76.38	101.97
n/(mol kg^{-1})	1.00	1.17	1.54	2.04	2.49
p/kPa	130.47	165.06	182.41	205.75	219.91
n/(mol kg^{-1})	2.90	3.22	3.30	3.35	3.36

Fit these data to a Langmuir isotherm, and determine the value of n that corresponds to complete coverage and the constant K.

21.16‡ The following data were obtained for the extent of adsorption, s, of acetone on charcoal from an aqueous solution of molar concentration, c, at 18°C.

c/(mmol dm^{-3})	15.0	23.0	42.0	84.0	165	390	800
s/(mmol acetone/g charcoal)	0.60	0.75	1.05	1.50	2.15	3.50	5.10

Which isotherm fits this data best: Langmuir, Freundlich, or Temkin?

Theoretical problems

21.17 Autocatalysis is the catalysis of a reaction by the products. For example, for a reaction A → P it may be found that the rate law is $v = k_r[A][P]$ and the reaction rate is proportional to the concentration of P. The reaction gets started because there are usually other reaction routes for the formation of some P initially, which then takes part in the autocatalytic reaction proper. (a) Integrate the rate equation for an autocatalytic reaction of the form A → P, with rate law $v = k_r[A][P]$, and show that

$$\frac{[P]}{[P]_0} = (b+1)\frac{e^{at}}{1+be^{at}}$$

where $a = ([A]_0 + [P]_0)k_r$ and $b = [P]_0/[A]_0$. *Hint.* Starting with the expression $v = -d[A]/dt = k_r[A][P]$, write $[A] = [A]_0 - x$, $[P] = [P]_0 + x$ and then write the expression for the rate of change of either species in terms of x. To integrate the resulting expression, use

$$\frac{1}{([A]_0 - x)([P]_0 + x)} = \frac{1}{[A]_0 + [P]_0}\left(\frac{1}{[A]_0 - x} + \frac{1}{[P]_0 + x}\right)$$

(b) Plot $[P]/[P]_0$ against at for several values of b. Discuss the effect of autocatalysis on the shape of a plot of $[P]/[P]_0$ against t by comparing your results with those for a first-order process, in which $[P]/[P]_0 = 1 - e^{-k_r t}$. (c) Show that for the autocatalytic process discussed in parts (a) and (b), the reaction rate reaches a maximum at $t_{max} = -(1/a)\ln b$. (d) An autocatalytic reaction $A \rightarrow P$ is observed to have the rate law $d[P]/dt = k_r[A]^2[P]$. Solve the rate law for initial concentrations $[A]_0$ and $[P]_0$. Calculate the time at which the rate reaches a maximum. (e) Another reaction with the stoichiometry $A \rightarrow P$ has the rate law $d[P]/dt = k_r[A][P]^2$; integrate the rate law for initial concentrations $[A]_0$ and $[P]_0$. Calculate the time at which the rate reaches a maximum.

21.18 Many biological and biochemical processes involve autocatalytic steps (Problem 21.17). In the SIR model of the spread and decline of infectious diseases the population is divided into three classes; the susceptibles, S, who can catch the disease, the infectives, I, who have the disease and can transmit it, and the removed class, R, who have either had the disease and recovered, are dead, are immune or isolated. The model mechanism for this process implies the following rate laws:

$$\frac{dS}{dt} = -rSI \qquad \frac{dI}{dt} = rSI - aI \qquad \frac{dR}{dt} = aI$$

What are the autocatalytic steps of this mechanism? Find the conditions on the ratio a/r that decide whether the disease will spread (an epidemic) or die out. Show that a constant population is built into this system, namely that $S + I + R = N$, meaning that the timescales of births, deaths by other causes, and migration are assumed large compared to that of the spread of the disease.

21.19 Michaelis and Menten derived their rate law by assuming a rapid pre-equilibrium of E, S, and ES. Derive the rate law in this manner, and identify the conditions under which it becomes the same as that based on the steady-state approximation (eqn 21.2).

21.20 For many enzymes, the mechanism of action involves the formation of two intermediates:

$$\begin{array}{ll} E + S \rightarrow ES & v = k_a[E][S] \\ ES \rightarrow E + S & v = k_a'[ES] \\ ES \rightarrow ES' & v = k_b[ES] \\ ES' \rightarrow E + P & v = k_c[ES'] \end{array}$$

Show that the rate of formation of product has the same form as that shown in eqn 21.4a

$$v = \frac{v_{max}}{1 + K_M/[S]_0}$$

but with v_{max} and K_M given by

$$v_{max} = \frac{k_b k_c [E]_0}{k_b + k_c} \qquad \text{and} \qquad K_M = \frac{k_c(k_a' + k_b)}{k_a(k_b + k_c)}$$

21.21 Some enzymes are inhibited by high concentrations of their own substrates. (a) Show that when substrate inhibition is important the reaction rate v is given by

$$v = \frac{v_{max}}{1 + K_M/[S]_0 + [S]_0/K_I}$$

where K_I is the equilibrium constant for dissociation of the inhibited enzyme–substrate complex. (b) What effect does substrate inhibition have on a plot of $1/v$ against $1/[S]_0$?

21.22 Although the attractive van der Waals interaction between individual molecules varies as R^{-6} the interaction of a molecule with a nearby solid (a homogeneous collection of molecules) varies as R^{-3}, where R is its vertical distance above the surface. Confirm this assertion. Calculate the interaction energy between an Ar atom and the surface of solid argon on the basis of a Lennard-Jones (6,12)-potential. Estimate the equilibrium distance of an atom above the surface.

21.23 Show that, for the association part of the surface plasmon resonance experiment in Fig. 21.31, $R(t) = R_{eq}(1 - e^{-k_r t})$ and write an expression for k_r. Then, derive an expression for $R(t)$ that applies to the dissociation part of the surface plasmon resonance experiment in Fig. 21.31.

Applications to: chemical engineering and environmental science

21.24 The designers of a new industrial plant wanted to use a catalyst code-named CR-1 in a step involving the fluorination of butadiene. As a first step in the investigation they determined the form of the adsorption isotherm. The volume of butadiene adsorbed per gram of CR-1 at 15°C varied with pressure as given below. Is the Langmuir isotherm suitable at this pressure?

p/kPa	13.3	26.7	40.0	53.3	66.7	80.0
V/cm^3	17.9	33.0	47.0	60.8	75.3	91.3

Investigate whether the BET isotherm gives a better description of the adsorption of butadiene on CR-1. At 15°C, p^*(butadiene) = 200 kPa. Find V_{mon} and c.

21.25‡ In a study relevant to automobile catalytic converters, C.E. Wartnaby *et al.* (*J. Phys. Chem.* **100**, 12483 (1996)) measured the enthalpy of adsorption of CO, NO, and O_2 on initially clean platinum (110) surfaces. They report $\Delta_{ads}H^{\ominus}$ for NO to be −160 kJ mol^{-1}. How much more strongly adsorbed is NO at 500°C than at 400°C?

21.26‡ The removal or recovery of volatile organic compounds (VOCs) from exhaust gas streams is an important process in environmental engineering. Activated carbon has long been used as an adsorbent in this process, but the presence of moisture in the stream reduces its effectiveness. M.-S. Chou and J.-H. Chiou (*J. Envir. Engrg.* ASCE **123**, 437(1997)) have studied the effect of moisture content on the adsorption capacities of granular activated carbon (GAC) for normal hexane and cyclohexane in air streams. From their data for dry streams containing cyclohexane, shown in the table below, they conclude that GAC obeys a Langmuir-type model in which $q_{VOC,RH=0} = abc_{VOC}/(1 + bc_{VOC})$, where $q = m_{VOC}/m_{GAC}$, RH denotes relative humidity, a is the maximum adsorption capacity, b is an affinity parameter, and p is the abundance in parts per million (ppm). The following table gives values of $q_{VOC,RH=0}$ for cyclohexane:

c/ppm	33.6°C	41.5°C	57.4°C	76.4°C	99°C
200	0.080	0.069	0.052	0.042	0.027
500	0.093	0.083	0.072	0.056	0.042
1000	0.101	0.088	0.076	0.063	0.045
2000	0.105	0.092	0.083	0.068	0.052
3000	0.112	0.102	0.087	0.072	0.058

(a) By linear regression of $1/q_{VOC,RH=0}$ against $1/c_{VOC}$, test the goodness of fit and determine values of a and b. (b) The parameters a and b can be related to $\Delta_{ads}H$, the enthalpy of adsorption, and Δ_bH, the difference in activation energy for adsorption and desorption of the VOC molecules, through Arrhenius-type equations of the form $a = k_a\exp(-\Delta_{ads}H/RT)$ and $b = k_b\exp(-\Delta_bH/RT)$. Test the goodness of fit of the data to these equations and obtain values for k_a, k_b, $\Delta_{ads}H$, and Δ_bH. (c) What interpretation might you give to k_a and k_b?

21.27‡ M.-S. Chou and J.-H. Chiou (*J. Envir. Engrg.*, ASCE **123**, 437(1997)) have studied the effect of moisture content on the adsorption capacities of granular activated carbon (GAC, Norit PK 1-3) for the volatile organic compounds (VOCs) normal hexane and cyclohexane in air streams. The following table shows the adsorption capacities ($q_{\text{water}} = m_{\text{water}}/m_{\text{GAC}}$) of GAC for pure water from moist air streams as a function of relative humidity (RH) in the absence of VOCs at 41.5°C.

RH	0.00	0.26	0.49	0.57	0.80	1.00
q_{water}	0.00	0.026	0.072	0.091	0.161	0.229

The authors conclude that the data at this and other temperatures obey a Freundlich-type isotherm, $q_{\text{water}} = k(\text{RH})^{1/n}$. (a) Test this hypothesis for their data at 41.5°C and determine the constants k and n. (b) Why might VOCs obey the Langmuir model, but water the Freundlich model? (c) When both water vapour and cyclohexane were present in the stream the values given in the table below were determined for the ratio $r_{\text{VOC}} = q_{\text{VOC}}/q_{\text{VOC,RH=0}}$ at 41.5°C.

RH	0.00	0.10	0.25	0.40	0.53	0.76	0.81
r_{VOC}	1.00	0.98	0.91	0.84	0.79	0.67	0.61

The authors propose that these data fit the equation $r_{\text{VOC}} = 1 - q_{\text{water}}$. Test their proposal and determine values for k and n and compare to those obtained in part (b) for pure water. Suggest reasons for any differences.

21.28‡ The release of petroleum products by leaky underground storage tanks is a serious threat to clean ground water. BTEX compounds (benzene, toluene, ethylbenzene, and xylenes) are of primary concern due to their ability to cause health problems at low concentrations. D.S. Kershaw *et al.* (*J. Geotech. Geoenvir. Engrg.* **123**, 324(1997)) have studied the ability of ground tyre rubber to sorb (adsorb and absorb) benzene and *o*-xylene. Though sorption involves more than surface interactions, sorption data is usually found to fit one of the adsorption isotherms. In this study, the authors have tested how well their data fit the linear ($q = Kc_{\text{eq}}$), Freundlich ($q = K_F c_{\text{eq}}^{1/n}$), and Langmuir ($q = K_L M c_{\text{eq}}/(1 + K_L c_{\text{eq}})$) type isotherms, where q is the mass of solvent sorbed per gram of ground rubber (in milligrams per gram), the Ks and M are empirical constants, and c_{eq} the equilibrium concentration of contaminant in solution (in milligrams per litre). (a) Determine the units of the empirical constants. (b) Determine which of the isotherms best fits the data in the table below for the sorption of benzene on ground rubber.

$c_{\text{eq}}/(\text{mg dm}^{-3})$	97.10	36.10	10.40	6.51	6.21	2.48
$q/(\text{mg g}^{-1})$	7.13	4.60	1.80	1.10	0.55	0.31

(c) Compare the sorption efficiency of ground rubber to that of granulated activated charcoal, which for benzene has been shown to obey the Freundlich isotherm in the form $q = 1.0c_{\text{eq}}^{1.6}$ with coefficient of determination $R^2 = 0.94$.

Resource section

Contents

The following is a directory of all tables in the text; those included in this *Resource section* are marked with an asterisk. The remainder will be found on the pages indicated.

The following tables reproduce and expand the data given in the short tables in the text, and follow their numbering. Standard states refer to a pressure of $p^\ominus = 1$ bar. The general references are as follows:

AIP: D.E. Gray (ed.), *American Institute of Physics handbook*. McGraw Hill, New York (1972).

E: J. Emsley, *The elements*. Oxford University Press (1991).

HCP: D.R. Lide (ed.), *Handbook of chemistry and physics*. CRC Press, Boca Raton (2000).

JL: A.M. James and M.P. Lord, *Macmillan's chemical and physical data*. Macmillan, London (1992).

KL: G.W.C. Kaye and T.H. Laby (ed.), *Tables of physical and chemical constants*. Longman, London (1973).

LR: G.N. Lewis and M. Randall, revised by K.S. Pitzer and L. Brewer, *Thermodynamics*. McGraw Hill, New York (1961).

NBS: *NBS tables of chemical thermodynamic properties*, published as *J. Phys. Chem. Reference Data*, **11**, Supplement 2 (1982).

RS: R.A. Robinson and R.H. Stokes, *Electrolyte solutions*, Butterworth, London (1959).

TDOC: J.B. Pedley, J.D. Naylor, and S.P. Kirby, *Thermochemical data of organic compounds*. Chapman & Hall, London (1986).

Part 1 Data section

Physical properties of selected materials

	$\rho/(\mathrm{g\,cm^{-3}})$ at 293 K†	T_f/K	T_b/K
Elements			
Aluminium(s)	2.698	933.5	2740
Argon(g)	1.381	83.8	87.3
Boron(s)	2.340	2573	3931
Bromine(l)	3.123	265.9	331.9
Carbon(s, gr)	2.260	3700s	
Carbon(s, d)	3.513		
Chlorine(g)	1.507	172.2	239.2
Copper(s)	8.960	1357	2840
Fluorine(g)	1.108	53.5	85.0
Gold(s)	19.320	1338	3080
Helium(g)	0.125		4.22
Hydrogen(g)	0.071	14.0	20.3
Iodine(s)	4.930	386.7	457.5
Iron(s)	7.874	1808	3023
Krypton(g)	2.413	116.6	120.8
Lead(s)	11.350	600.6	2013
Lithium(s)	0.534	453.7	1620
Magnesium(s)	1.738	922.0	1363
Mercury(l)	13.546	234.3	629.7
Neon(g)	1.207	24.5	27.1
Nitrogen(g)	0.880	63.3	77.4
Oxygen(g)	1.140	54.8	90.2
Phosphorus(s, wh)	1.820	317.3	553
Potassium(s)	0.862	336.8	1047
Silver(s)	10.500	1235	2485
Sodium(s)	0.971	371.0	1156
Sulfur(s, α)	2.070	386.0	717.8
Uranium(s)	18.950	1406	4018
Xenon(g)	2.939	161.3	166.1
Zinc(s)	7.133	692.7	1180

	$\rho/(\mathrm{g\,cm^{-3}})$ at 293 K†	T_f/K	T_b/K
Inorganic compounds			
$CaCO_3$(s, calcite)	2.71	1612	1171d
$CuSO_4\cdot 5H_2O$(s)	2.284	383($-H_2O$)	423($-5H_2O$)
HBr(g)	2.77	184.3	206.4
HCl(g)	1.187	159.0	191.1
HI(g)	2.85	222.4	237.8
H_2O(l)	0.997	273.2	373.2
D_2O(l)	1.104	277.0	374.6
NH_3(g)	0.817	195.4	238.8
KBr(s)	2.750	1003	1708
KCl(s)	1.984	1049	1773s
NaCl(s)	2.165	1074	1686
H_2SO_4(l)	1.841	283.5	611.2
Organic compounds			
Acetaldehyde, CH_3CHO(l)	0.788	152	293
Acetic acid, CH_3COOH(l)	1.049	289.8	391
Acetone, $(CH_3)_2CO$(l)	0.787	178	329
Aniline, $C_6H_5NH_2$(l)	1.026	267	457
Anthracene, $C_{14}H_{10}$(s)	1.243	490	615
Benzene, C_6H_6(l)	0.879	278.6	353.2
Carbon tetrachloride, CCl_4(l)	1.63	250	349.9
Chloroform, $CHCl_3$(l)	1.499	209.6	334
Ethanol, C_2H_5OH(l)	0.789	156	351.4
Formaldehyde, HCHO(g)		181	254.0
Glucose, $C_6H_{12}O_6$(s)	1.544	415	
Methane, CH_4(g)		90.6	111.6
Methanol, CH_3OH(l)	0.791	179.2	337.6
Naphthalene, $C_{10}H_8$(s)	1.145	353.4	491
Octane, C_8H_{18}(l)	0.703	216.4	398.8
Phenol, C_6H_5OH(s)	1.073	314.1	455.0
Sucrose, $C_{12}H_{22}O_{11}$(s)	1.588	457d	

d: decomposes; s: sublimes; Data: AIP, E, HCP, KL. † For gases, at their boiling points.

Masses and natural abundances of selected nuclides

Nuclide		m/m_u	Abundance/%
H	^{1}H	1.0078	99.985
	^{2}H	2.0140	0.015
He	^{3}He	3.0160	0.000 13
	^{4}He	4.0026	100
Li	^{6}Li	6.0151	7.42
	^{7}Li	7.0160	92.58
B	^{10}B	10.0129	19.78
	^{11}B	11.0093	80.22
C	^{12}C	12*	98.89
	^{13}C	13.0034	1.11
N	^{14}N	14.0031	99.63
	^{15}N	15.0001	0.37
O	^{16}O	15.9949	99.76
	^{17}O	16.9991	0.037
	^{18}O	17.9992	0.204
F	^{19}F	18.9984	100
P	^{31}P	30.9738	100
S	^{32}S	31.9721	95.0
	^{33}S	32.9715	0.76
	^{34}S	33.9679	4.22
Cl	^{35}Cl	34.9688	75.53
	^{37}Cl	36.9651	24.4
Br	^{79}Br	78.9183	50.54
	^{81}Br	80.9163	49.46
I	^{127}I	126.9045	100

* Exact value.

Table 4.2 Effective nuclear charge, $Z_{eff} = Z - \sigma$

	H							He
1s	1							1.6875
	Li	Be	B	C	N	O	F	Ne
1s	2.6906	3.6848	4.6795	5.6727	6.6651	7.6579	8.6501	9.6421
2s	1.2792	1.9120	2.5762	3.2166	3.8474	4.4916	5.1276	5.7584
2p			2.4214	3.1358	3.8340	4.4532	5.1000	5.7584
	Na	Mg	Al	Si	P	S	Cl	Ar
1s	10.6259	11.6089	12.5910	13.5745	14.5578	15.5409	16.5239	17.5075
2s	6.5714	7.3920	8.3736	9.0200	9.8250	10.6288	11.4304	12.2304
2p	6.8018	7.8258	8.9634	9.9450	10.9612	11.9770	12.9932	14.0082
3s	2.5074	3.3075	4.1172	4.9032	5.6418	6.3669	7.0683	7.7568
3p			4.0656	4.2852	4.8864	5.4819	6.1161	6.7641

Data: E. Clementi and D.L. Raimondi, *Atomic screening constants from SCF functions.* IBM Res. Note NJ-27 (1963). *J. Chem. Phys.* **38**, 2686 (1963).

Table 4.3 Ionization energies, $I/(\mathrm{kJ\ mol^{-1}})$

H							He
1312.0							2372.3
							5250.4
Li	Be	B	C	N	O	F	Ne
513.3	899.4	800.6	1086.2	1402.3	1313.9	1681	2080.6
7298.0	1757.1	2427	2352	2856.1	3388.2	3374	3952.2
Na	Mg	Al	Si	P	S	Cl	Ar
495.8	737.7	577.4	786.5	1011.7	999.6	1251.1	1520.4
4562.4	1450.7	1816.6	1577.1	1903.2	2251	2297	2665.2
		2744.6		2912			
K	Ca	Ga	Ge	As	Se	Br	Kr
418.8	589.7	578.8	762.1	947.0	940.9	1139.9	1350.7
3051.4	1145	1979	1537	1798	2044	2104	2350
		2963	2735				
Rb	Sr	In	Sn	Sb	Te	I	Xe
403.0	549.5	558.3	708.6	833.7	869.2	1008.4	1170.4
2632	1064.2	1820.6	1411.8	1794	1795	1845.9	2046
		2704	2943.0	2443			
Cs	Ba	Tl	Pb	Bi	Po	At	Rn
375.5	502.8	589.3	715.5	703.2	812	930	1037
2420	965.1	1971.0	1450.4	1610			
		2878	3081.5	2466			

Data: E.

Table 4.4 Electron affinities, $E_{ea}/(\mathrm{kJ\ mol^{-1}})$

H							He
72.8							−21
Li	Be	B	C	N	O	F	Ne
59.8	≤0	23	122.5	−7	141	322	−29
					−844		
Na	Mg	Al	Si	P	S	Cl	Ar
52.9	≤0	44	133.6	71.7	200.4	348.7	−35
					−532		
K	Ca	Ga	Ge	As	Se	Br	Kr
48.3	2.37	36	116	77	195.0	324.5	−39
Rb	Sr	In	Sn	Sb	Te	I	Xe
46.9	5.03	34	121	101	190.2	295.3	−41
Cs	Ba	Tl	Pb	Bi	Po	At	Rn
45.5	13.95	30	35.2	101	186	270	−41

Data: E.

Table 5.2 Bond lengths, R_e/pm

(a) Bond lengths in specific molecules

Br_2	228.3
Cl_2	198.75
CO	112.81
F_2	141.78
H_2^+	106
H_2	74.138
HBr	141.44
HCl	127.45
HF	91.680
HI	160.92
N_2	109.76
O_2	120.75

(b) Mean bond lengths from covalent radii*

H	37						
C	77(1)	N	74(1)	O	66(1)	F	64
	67(2)		65(2)		57(2)		
	60(3)						
Si	118	P	110	S	104(1)	Cl	99
					95(2)		
Ge	122	As	121	Se	104	Br	114
		Sb	141	Te	137	I	133

* Values are for single bonds except where indicated otherwise (values in parentheses). The length of an A—B covalent bond (of given order) is the sum of the corresponding covalent radii.

Table 5.3a Bond dissociation enthalpies, $\Delta H^{\ominus}$(A—B)/(kJ mol^{-1}) at 298 K*

Diatomic molecules

H—H	436	F—F	155	Cl—Cl	242	Br—Br	193	I—I	151
O=O	497	C=O	1076	N≡N	945				
H—O	428	H—F	565	H—Cl	431	H—Br	366	H—I	299

Polyatomic molecules

H—CH_3	435	H—NH_2	460	H—OH	492	H—C_6H_5	469
H_3C—CH_3	368	H_2C=CH_2	720	HC≡CH	962		
HO—CH_3	377	Cl—CH_3	352	Br—CH_3	293	I—CH_3	237
O=CO	531	HO—OH	213	O_2N—NO_2	54		

* To a good approximation bond dissociation enthalpies and dissociation energies are related by $\Delta H^{\ominus} = D_e + \frac{3}{2}RT$ with $D_e = D_0 + \frac{1}{2}\hbar\omega$. For precise values of D_0 for diatomic molecules, see Table 10.2.
Data: HCP, KL.

Table 5.3b Mean bond enthalpies, $\Delta H^{\ominus}(\text{A—B})/(\text{kJ mol}^{-1})$*

	H	C	N	O	F	Cl	Br	I	S	P	Si
H	436										
C	412	348(i)									
		612(ii)									
		838(iii)									
		518(a)									
N	388	305(i)	163(i)								
		613(ii)	409(ii)								
		890(iii)	946(iii)								
O	463	360(i)	157	146(i)							
		743(ii)		497(ii)							
F	565	484	270	185	155						
Cl	431	338	200	203	254	242					
Br	366	276				219	193				
I	299	238				210	178	151			
S	338	259			496	250	212		264		
P	322									201	
Si	318		374	466							226

* Mean bond enthalpies are such a crude measure of bond strength that they need not be distinguished from dissociation energies.
(i) Single bond, (ii) double bond, (iii) triple bond, (a) aromatic.
Data: HCP and L. Pauling, *The nature of the chemical bond.* Cornell University Press (1960).

Table 5.4 Pauling (*italics*) and Mulliken electronegativities

H							He
2.20							
3.06							
Li	Be	B	C	N	O	F	Ne
0.98	*1.57*	*2.04*	*2.55*	*3.04*	*3.44*	*3.98*	
1.28	1.99	1.83	2.67	3.08	3.22	4.43	4.60
Na	Mg	Al	Si	P	S	Cl	Ar
0.93	*1.31*	*1.61*	*1.90*	*2.19*	*2.58*	*3.16*	
1.21	1.63	1.37	2.03	2.39	2.65	3.54	3.36
K	Ca	Ga	Ge	As	Se	Br	Kr
0.82	*1.00*	*1.81*	*2.01*	*2.18*	*2.55*	*2.96*	*3.0*
1.03	1.30	1.34	1.95	2.26	2.51	3.24	2.98
Rb	Sr	In	Sn	Sb	Te	I	Xe
0.82	*0.95*	*1.78*	*1.96*	*2.05*	*2.10*	*2.66*	*2.6*
0.99	1.21	1.30	1.83	2.06	2.34	2.88	2.59
Cs	Ba	Tl	Pb	Bi			
0.79	*0.89*	*2.04*	*2.33*	*2.02*			

Data: Pauling values: A.L. Allred, *J. Inorg. Nucl. Chem.* **17**, 215 (1961); L.C. Allen and J.E. Huheey, *ibid.*, **42**, 1523 (1980). Mulliken values: L.C. Allen, *J. Am. Chem. Soc.* **111**, 9003 (1989). The Mulliken values have been scaled to the range of the Pauling values.

Table 8.1 Dipole moments (μ), polarizabilities (α), and polarizability volumes (α')

	$\mu/(10^{-30}$ C m)	μ/D	$\alpha'/(10^{-30}$ m^3)	$\alpha/(10^{-40}$ J^{-1} C^2 m^2)
Ar	0	0	1.66	1.85
C_2H_5OH	5.64	1.69		
$C_6H_5CH_3$	1.20	0.36		
C_6H_6	0	0	10.4	11.6
CCl_4	0	0	10.3	11.7
CH_2Cl_2	5.24	1.57	6.80	7.57
CH_3Cl	6.24	1.87	4.53	5.04
CH_3OH	5.70	1.71	3.23	3.59
CH_4	0	0	2.60	2.89
$CHCl_3$	3.37	1.01	8.50	9.46
CO	0.390	0.117	1.98	2.20
CO_2	0	0	2.63	2.93
H_2	0	0	0.819	0.911
H_2O	6.17	1.85	1.48	1.65
HBr	2.67	0.80	3.61	4.01
HCl	3.60	1.08	2.63	2.93
He	0	0	0.20	0.22
HF	6.37	1.91	0.51	0.57
HI	1.40	0.42	5.45	6.06
N_2	0	0	1.77	1.97
NH_3	4.90	1.47	2.22	2.47
1,2-$C_6H_4(CH_3)_2$	2.07	0.62		

Data: HCP and C.J.F. Böttcher and P. Bordewijk, *Theory of electric polarization*. Elsevier, Amsterdam (1978).

Table 8.4 Lennard-Jones (12,6)-potential parameters

	(ε/k)/K	r_0/pm
Ar	111.84	362.3
C_2H_2	209.11	463.5
C_2H_4	200.78	458.9
C_2H_6	216.12	478.2
C_6H_6	377.46	617.4
CCl_4	378.86	624.1
Cl_2	296.27	448.5
CO_2	201.71	444.4
F_2	104.29	357.1
Kr	154.87	389.5
N_2	91.85	391.9
O_2	113.27	365.4
Xe	213.96	426.0

Source: F. Cuadros, I. Cachadiña, and W. Ahamuda, *Molec. Engineering*, **6**, 319 (1996).

Table 8.5 Second virial coefficients, $B/(\mathrm{cm^3\ mol^{-1}})$

	100 K	273 K	373 K	600 K
Air	−167.3	−13.5	3.4	19.0
Ar	−187.0	−21.7	−4.2	11.9
CH_4		−53.6	−21.2	8.1
CO_2		−142	−72.2	−12.4
H_2	−2.0	13.7	15.6	
He	11.4	12.0	11.3	10.4
Kr		−62.9	−28.7	1.7
N_2	−160.0	−10.5	6.2	21.7
Ne	−6.0	10.4	12.3	13.8
O_2	−197.5	−22.0	−3.7	12.9
Xe		−153.7	−81.7	−19.6

Data: AIP, JL. The values relate to the expansion in eqn 8.22a of Section 8.7a; convert to eqn 8.22b using $B' = B/RT$.
For Ar at 273 K, $C = 1200\ \mathrm{cm^6\ mol^{-1}}$.

Table 8.6 Boyle temperatures of gases

	T_B/K
Ar	411.5
CH_4	510.0
CO_2	714.8
H_2	110.0
He	22.64
Kr	575.0
N_2	327.2
Ne	122.1
O_2	405.9
Xe	768.0

Data: AIP, KL.

Table 8.7 van der Waals coefficients

	$a/(\mathrm{atm\ dm^6\ mol^{-2}})$	$b/(10^{-2}\ \mathrm{dm^3\ mol^{-1}})$		$a/(\mathrm{atm\ dm^6\ mol^{-2}})$	$b/(10^{-2}\ \mathrm{dm^3\ mol^{-1}})$
Ar	1.337	3.20	H_2S	4.484	4.34
C_2H_4	4.552	5.82	He	0.0341	2.38
C_2H_6	5.507	6.51	Kr	5.125	1.06
C_6H_6	18.57	11.93	N_2	1.352	3.87
CH_4	2.273	4.31	Ne	0.205	1.67
Cl_2	6.260	5.42	NH_3	4.169	3.71
CO	1.453	3.95	O_2	1.364	3.19
CO_2	3.610	4.29	SO_2	6.775	5.68
H_2	0.2420	2.65	Xe	4.137	5.16
H_2O	5.464	3.05			

Data: HCP.

Table 9.3 Ionic radii, r/pm†

Li^{+}(4)	Be^{2+}(4)	B^{3+}(4)	N^{3-}	O^{2-}(6)	F^{-}(6)
59	27	12	171	140	133
Na^{+}(6)	Mg^{2+}(6)	Al^{3+}(6)	P^{3-}	S^{2-}(6)	Cl^{-}(6)
102	72	53	212	184	181
K^{+}(6)	Ca^{2+}(6)	Ga^{3+}(6)	As^{3-}(6)	Se^{2-}(6)	Br^{-}(6)
138	100	62	222	198	196
Rb^{+}(6)	Sr^{2+}(6)	In^{3+}(6)		Te^{2-}(6)	I^{-}(6)
149	116	79		221	220
Cs^{+}(6)	Ba^{2+}(6)	Tl^{3+}(6)			
167	136	88			

d-block elements (high-spin ions)

Sc^{3+}(6)	Ti^{4+}(6)	Cr^{3+}(6)	Mn^{3+}(6)	Fe^{2+}(6)	Co^{3+}(6)	Cu^{2+}(6)	Zn^{2+}(6)
73	60	61	65	63	61	73	75

† Numbers in parentheses are the coordination numbers of the ions. Values for ions without a coordination number stated are estimates.
Data: R.D. Shannon and C.T. Prewitt, *Acta Cryst.* **B25**, 925 (1969).

Table 9.5 Magnetic susceptibilities at 298 K

	$\chi/10^{-6}$	$\chi_m/(10^{-10}\,m^3\,mol^{-1})$
H_2O(l)	−9.02	−1.63
C_6H_6(l)	−8.8	−7.8
C_6H_{12}(l)	−10.2	−11.1
CCl_4(l)	−5.4	−5.2
NaCl(s)	−16	−3.8
Cu(s)	−9.7	−0.69
S(rhombic)	−12.6	−1.95
Hg(l)	−28.4	−4.21
Al(s)	+20.7	+2.07
Pt(s)	+267.3	+24.25
Na(s)	+8.48	+2.01
K(s)	+5.94	+2.61
$CuSO_4\cdot 5H_2O$(s)	+167	+183
$MnSO_4\cdot 4H_2O$(s)	+1859	+1835
$NiSO_4\cdot 7H_2O$(s)	+355	+503
$FeSO_4$(s)	+3743	+1558

Source: Principally HCP, with $\chi_m = \chi V_m = \chi\rho/M$.

Table 10.2 Properties of diatomic molecules

	$\tilde{\nu}$/cm^{-1}	θ/K	$\tilde{B}$/cm^{-1}	θ_R/K	R_e/pm	k/(N m^{-1})	D_0/(kJ mol^{-1})	σ
$^1H_2^+$	2321.8	3341	29.8	42.9	106	160	255.8	2
$^1H_2$	4400.39	6332	60.864	87.6	74.138	574.9	432.1	2
$^2H_2$	3118.46	4487	30.442	43.8	74.154	577.0	439.6	2
$^1H^{19}F$	4138.32	5955	20.956	30.2	91.680	965.7	564.4	1
$^1H^{35}Cl$	2990.95	4304	10.593	15.2	127.45	516.3	427.7	1
$^1H^{81}Br$	2648.98	3812	8.465	12.2	141.44	411.5	362.7	1
$^1H^{127}I$	2308.09	3321	6.511	9.37	160.92	313.8	294.9	1
$^{14}N_2$	2358.07	3393	1.9987	2.88	109.76	2293.8	941.7	2
$^{16}O_2$	1580.36	2274	1.4457	2.08	120.75	1176.8	493.5	2
$^{19}F_2$	891.8	1283	0.8828	1.27	141.78	445.1	154.4	2
$^{35}Cl_2$	559.71	805	0.2441	0.351	198.75	322.7	239.3	2
$^{12}C^{16}O$	2170.21	3122	1.9313	2.78	112.81	1903.17	1071.8	1
$^{79}Br^{81}Br$	323.2	465	0.0809	10.116	283.3	245.9	190.2	1

Data: AIP.

Table 10.3 Typical vibrational wavenumbers, $\tilde{\nu}$/cm^{-1}

C—H stretch	2850–2960
C—H bend	1340–1465
C—C stretch, bend	700–1250
C=C stretch	1620–1680
C≡C stretch	2100–2260
O—H stretch	3590–3650
H-bonds	3200–3570
C=O stretch	1640–1780
C≡N stretch	2215–2275
N—H stretch	3200–3500
C—F stretch	1000–1400
C—Cl stretch	600–800
C—Br stretch	500–600
C—I stretch	500
CO_3^{2-}	1410–1450
NO_3^-	1350–1420
NO_2^-	1230–1250
SO_4^{2-}	1080–1130
Silicates	900–1100

Data: L.J. Bellamy, *The infrared spectra of complex molecules* and *Advances in infrared group frequencies*. Chapman and Hall.

Table 11.1 Colour, frequency, and energy of light

Colour	λ/nm	ν/(10^{14} Hz)	$\tilde{\nu}$/(10^4 cm^{-1})	E/eV	E/(kJ mol^{-1})
Infrared	>1000	<3.00	<1.00	<1.24	<120
Red	700	4.28	1.43	1.77	171
Orange	620	4.84	1.61	2.00	193
Yellow	580	5.17	1.72	2.14	206
Green	530	5.66	1.89	2.34	226
Blue	470	6.38	2.13	2.64	254
Violet	420	7.14	2.38	2.95	285
Near ultraviolet	300	10.0	3.33	4.15	400
Far ultraviolet	<200	>15.0	>5.00	>6.20	>598

Data: J.G. Calvert and J.N. Pitts, *Photochemistry*. Wiley, New York (1966).

Table 11.3 Absorption characteristics of some groups and molecules

Group	$\tilde{\nu}_{max}/(10^4\ cm^{-1})$	λ_{max}/nm	$\varepsilon_{max}/(dm^3\ mol^{-1}\ cm^{-1})$
C=C ($\pi^* \leftarrow \pi$)	6.10	163	1.5×10^4
	5.73	174	5.5×10^3
C=O ($\pi^* \leftarrow n$)	3.7–3.5	270–290	10–20
—N=N—	2.9	350	15
	>3.9	<260	Strong
$—NO_2$	3.6	280	10
	4.8	210	1.0×10^4
$C_6H_5—$	3.9	255	200
	5.0	200	6.3×10^3
	5.5	180	1.0×10^5
$[Cu(OH_2)_6]^{2+}(aq)$	1.2	810	10
$[Cu(NH_3)_4]^{2+}(aq)$	1.7	600	50
H_2O ($\pi^* \leftarrow n$)	6.0	167	7.0×10^3

Table 12.2 Nuclear spin properties

Nuclide	Natural abundance %	Spin I	Magnetic moment μ/μ_N	g-value	$\gamma/(10^7\ T^{-1}\ s^{-1})$	NMR frequency at 1 T, ν/MHz
^1n*		$\frac{1}{2}$	−1.9130	−3.8260	−18.324	29.164
^1H	99.9844	$\frac{1}{2}$	2.792 85	5.5857	26.752	42.576
^2H	0.0156	1	0.857 44	0.857 44	4.1067	6.536
^3H*		$\frac{1}{2}$	2.978 96	−4.2553	−20.380	45.414
^{10}B	19.6	3	1.8006	0.6002	2.875	4.575
^{11}B	80.4	$\frac{3}{2}$	2.6886	1.7923	8.5841	13.663
^{13}C	1.108	$\frac{1}{2}$	0.7024	1.4046	6.7272	10.708
^{14}N	99.635	1	0.403 56	0.403 56	1.9328	3.078
^{17}O	0.037	$\frac{5}{2}$	−1.893 79	−0.7572	−3.627	5.774
^{19}F	100	$\frac{1}{2}$	2.628 87	5.2567	25.177	40.077
^{31}P	100	$\frac{1}{2}$	1.1316	2.2634	10.840	17.251
^{33}S	0.74	$\frac{3}{2}$	0.6438	0.4289	2.054	3.272
^{35}Cl	75.4	$\frac{3}{2}$	0.8219	0.5479	2.624	4.176
^{37}Cl	24.6	$\frac{3}{2}$	0.6841	0.4561	2.184	3.476

* Radioactive.

μ is the magnetic moment of the spin state with the largest value of m_I: $\mu = g_I\mu_N I$ and μ_N is the nuclear magneton (see inside front cover).

Data: KL and HCP.

Table 12.3 Hyperfine coupling constants for atoms, a/mT

Nuclide	Spin	Isotropic coupling	Anisotropic coupling
^{1}H	$\frac{1}{2}$	50.8(1s)	
^{2}H	1	7.8(1s)	
^{13}C	$\frac{1}{2}$	113.0(2s)	6.6(2p)
^{14}N	1	55.2(2s)	4.8(2p)
^{19}F	$\frac{1}{2}$	1720(2s)	108.4(2p)
^{31}P	$\frac{1}{2}$	364(3s)	20.6(3p)
^{35}Cl	$\frac{3}{2}$	168(3s)	10.0(3p)
^{37}Cl	$\frac{3}{2}$	140(3s)	8.4(3p)

Data: P.W. Atkins and M.C.R. Symons, *The structure of inorganic radicals*. Elsevier, Amsterdam (1967).

Table 13.2 Rotatimal and vibrational temperatures: see Table 10.2

Table 13.3 Symmetry numbers: see Table 10.2

Table 14.2 Temperature variation of molar heat capacities†

	a	$b/(10^{-3}\ K^{-1})$	$c/(10^{5}\ K^{2})$
Monatomic gases			
	20.78	0	0
Other gases			
Br_2	37.32	0.50	−1.26
Cl_2	37.03	0.67	−2.85
CO_2	44.22	8.79	−8.62
F_2	34.56	2.51	−3.51
H_2	27.28	3.26	0.50
I_2	37.40	0.59	−0.71
N_2	28.58	3.77	−0.50
NH_3	29.75	25.1	−1.55
O_2	29.96	4.18	−1.67
Liquids (from melting to boiling)			
$C_{10}H_8$, naphthalene	79.5	0.4075	0
I_2	80.33	0	0
H_2O	75.29	0	0
Solids			
Al	20.67	12.38	0
C (graphite)	16.86	4.77	−8.54
$C_{10}H_8$, naphthalene	−115.9	3.920×10^{3}	0
Cu	22.64	6.28	0
I_2	40.12	49.79	0
NaCl	45.94	16.32	0
Pb	22.13	11.72	0.96

† For $C_{p,m}/(J\ K^{-1}\ mol^{-1}) = a + bT + c/T^2$.
Source: LR.

Table 14.3 Standard enthalpies of fusion and vaporization at the transition temperature, $\Delta_{trs}H^\ominus/(kJ\ mol^{-1})$

	T_f/K	Fusion	T_b/K	Vaporization
Elements				
Ag	1234	11.30	2436	250.6
Ar	83.81	1.188	87.29	6.506
Br_2	265.9	10.57	332.4	29.45
Cl_2	172.1	6.41	239.1	20.41
F_2	53.6	0.26	85.0	3.16
H_2	13.96	0.117	20.38	0.916
He	3.5	0.021	4.22	0.084
Hg	234.3	2.292	629.7	59.30
I_2	386.8	15.52	458.4	41.80
N_2	63.15	0.719	77.35	5.586
Na	371.0	2.601	1156	98.01
O_2	54.36	0.444	90.18	6.820
Xe	161	2.30	165	12.6
K	336.4	2.35	1031	80.23
Inorganic compounds				
CO_2	217.0	8.33	194.6	25.23s
CS_2	161.2	4.39	319.4	26.74
H_2O	273.15	6.008	373.15	40.656
				44.016 at 298 K
H_2S	187.6	2.377	212.8	18.67
H_2SO_4	283.5	2.56		
NH_3	195.4	5.652	239.7	23.35
Organic compounds				
CH_4	90.68	0.941	111.7	8.18
CCl_4	250.3	2.47	349.9	30.00
C_2H_6	89.85	2.86	184.6	14.7
C_6H_6	278.61	10.59	353.2	30.8
C_6H_{14}	178	13.08	342.1	28.85
$C_{10}H_8$	354	18.80	490.9	51.51
CH_3OH	175.2	3.16	337.2	35.27
				37.99 at 298 K
C_2H_5OH	158.7	4.60	352	43.5

Data: AIP; s denotes sublimation.

Table 14.5 Thermodynamic data for organic compounds at 298 K

	$M/(g\ mol^{-1})$	$\Delta_fH^\ominus/(kJ\ mol^{-1})$	$\Delta_fG^\ominus/(kJ\ mol^{-1})$	$S_m^\ominus/(J\ K^{-1}\ mol^{-1})$†	$C_{p,m}^\ominus/(J\ K^{-1}\ mol^{-1})$	$\Delta_cH^\ominus/(kJ\ mol^{-1})$
C(s) (graphite)	12.011	0	0	5.740	8.527	−393.51
C(s) (diamond)	12.011	+1.895	+2.900	2.377	6.113	−395.40
CO_2(g)	44.040	−393.51	−394.36	213.74	37.11	
Hydrocarbons						
CH_4(g), methane	16.04	−74.81	−50.72	186.26	35.31	−890
CH_3(g), methyl	15.04	+145.69	+147.92	194.2	38.70	
C_2H_2(g), ethyne	26.04	+226.73	+209.20	200.94	43.93	−1300
C_2H_4(g), ethene	28.05	+52.26	+68.15	219.56	43.56	−1411
C_2H_6(g), ethane	30.07	−84.68	−32.82	229.60	52.63	−1560
C_3H_6(g), propene	42.08	+20.42	+62.78	267.05	63.89	−2058
C_3H_6(g), cyclopropane	42.08	+53.30	+104.45	237.55	55.94	−2091
C_3H_8(g), propane	44.10	−103.85	−23.49	269.91	73.5	−2220
C_4H_8(g), 1-butene	56.11	−0.13	+71.39	305.71	85.65	−2717
C_4H_8(g), *cis*-2-butene	56.11	−6.99	+65.95	300.94	78.91	−2710
C_4H_8(g), *trans*-2-butene	56.11	−11.17	+63.06	296.59	87.82	−2707
C_4H_{10}(g), butane	58.13	−126.15	−17.03	310.23	97.45	−2878
C_5H_{12}(g), pentane	72.15	−146.44	−8.20	348.40	120.2	−3537
C_5H_{12}(l)	72.15	−173.1				
C_6H_6(l), benzene	78.12	+49.0	+124.3	173.3	136.1	−3268

Table 14.5 (Continued)

	$M/(\mathrm{g\ mol^{-1}})$	$\Delta_f H^{\ominus}/(\mathrm{kJ\ mol^{-1}})$	$\Delta_f G^{\ominus}/(\mathrm{kJ\ mol^{-1}})$	$S_m^{\ominus}/(\mathrm{J\ K^{-1}\ mol^{-1}})$†	$C_{p,m}^{\ominus}/(\mathrm{J\ K^{-1}\ mol^{-1}})$	$\Delta_c H^{\ominus}/(\mathrm{kJ\ mol^{-1}})$
Hydrocarbons (Continued)						
$C_6H_6(g)$	78.12	+82.93	+129.72	269.31	81.67	−3302
$C_6H_{12}(l)$, cyclohexane	84.16	−156	+26.8	204.4	156.5	−3920
$C_6H_{14}(l)$, hexane	86.18	−198.7		204.3		−4163
$C_6H_5CH_3(g)$, methylbenzene (toluene)	92.14	+50.0	+122.0	320.7	103.6	−3953
$C_7H_{16}(l)$, heptane	100.21	−224.4	+1.0	328.6	224.3	
$C_8H_{18}(l)$, octane	114.23	−249.9	+6.4	361.1		−5471
$C_8H_{18}(l)$, iso-octane	114.23	−255.1				−5461
$C_{10}H_8(s)$, naphthalene	128.18	+78.53				−5157
Alcohols and phenols						
$CH_3OH(l)$, methanol	32.04	−238.66	−166.27	126.8	81.6	−726
$CH_3OH(g)$	32.04	−200.66	−161.96	239.81	43.89	−764
$C_2H_5OH(l)$, ethanol	46.07	−277.69	−174.78	160.7	111.46	−1368
$C_2H_5OH(g)$	46.07	−235.10	−168.49	282.70	65.44	−1409
$C_6H_5OH(s)$, phenol	94.12	−165.0	−50.9	146.0		−3054
Carboxylic acids, hydroxy acids, and esters						
$HCOOH(l)$, formic	46.03	−424.72	−361.35	128.95	99.04	−255
$CH_3COOH(l)$, acetic	60.05	−484.5	−389.9	159.8	124.3	−875
$CH_3COOH(aq)$	60.05	−485.76	−396.46	178.7		
$CH_3CO_2^-(aq)$	59.05	−486.01	−369.31	+86.6	−6.3	
$(COOH)_2(s)$, oxalic	90.04	−827.2			117	−254
$C_6H_5COOH(s)$, benzoic	122.13	−385.1	−245.3	167.6	146.8	−3227
$CH_3CH(OH)COOH(s)$, lactic	90.08	−694.0				−1344
$CH_3COOC_2H_5(l)$, ethyl acetate	88.11	−479.0	−332.7	259.4	170.1	−2231
Alkanals and alkanones						
$HCHO(g)$, methanal	30.03	−108.57	−102.53	218.77	35.40	−571
$CH_3CHO(l)$, ethanal	44.05	−192.30	−128.12	160.2		−1166
$CH_3CHO(g)$	44.05	−166.19	−128.86	250.3	57.3	−1192
$CH_3COCH_3(l)$, propanone	58.08	−248.1	−155.4	200.4	124.7	−1790
Sugars						
$C_6H_{12}O_6(s)$, α-D-glucose	180.16	−1274				−2808
$C_6H_{12}O_6(s)$, β-D-glucose	180.16	−1268	−910	212		
$C_6H_{12}O_6(s)$, β-D-fructose	180.16	−1266				−2810
$C_{12}H_{22}O_{11}(s)$, sucrose	342.30	−2222	−1543	360.2		−5645
Nitrogen compounds						
$CO(NH_2)_2(s)$, urea	60.06	−333.51	−197.33	104.60	93.14	−632
$CH_3NH_2(g)$, methylamine	31.06	−22.97	+32.16	243.41	53.1	−1085
$C_6H_5NH_2(l)$, aniline	93.13	+31.1				−3393
$CH_2(NH_2)COOH(s)$, glycine	75.07	−532.9	−373.4	103.5	99.2	−969

Data: NBS, TDOC. † Standard entropies of ions may be either positive or negative because the values are relative to the entropy of the hydrogen ion.

Table 14.6 Thermodynamic data for elements and inorganic compounds at 298 K

	$M/(\text{g mol}^{-1})$	$\Delta_f H^\ominus/(\text{kJ mol}^{-1})$	$\Delta_f G^\ominus/(\text{kJ mol}^{-1})$	$S_m^\ominus/(\text{J K}^{-1}\text{ mol}^{-1})$†	$C_{p,m}^\ominus/(\text{J K}^{-1}\text{ mol}^{-1})$
Aluminium (aluminum)					
Al(s)	26.98	0	0	28.33	24.35
Al(l)	26.98	+10.56	+7.20	39.55	24.21
Al(g)	26.98	+326.4	+285.7	164.54	21.38
Al^{3+}(g)	26.98	+5483.17			
Al^{3+}(aq)	26.98	−531	−485	−321.7	
Al_2O_3(s, α)	101.96	−1675.7	−1582.3	50.92	79.04
$AlCl_3$(s)	133.24	−704.2	−628.8	110.67	91.84
Argon					
Ar(g)	39.95	0	0	154.84	20.786
Antimony					
Sb(s)	121.75	0	0	45.69	25.23
SbH_3(g)	124.77	+145.11	+147.75	232.78	41.05
Arsenic					
As(s, α)	74.92	0	0	35.1	24.64
As(g)	74.92	+302.5	+261.0	174.21	20.79
As_4(g)	299.69	+143.9	+92.4	314	
AsH_3(g)	77.95	+66.44	+68.93	222.78	38.07
Barium					
Ba(s)	137.34	0	0	62.8	28.07
Ba(g)	137.34	+180	+146	170.24	20.79
Ba^{2+}(aq)	137.34	−537.64	−560.77	+9.6	
BaO(s)	153.34	−553.5	−525.1	70.43	47.78
$BaCl_2$(s)	208.25	−858.6	−810.4	123.68	75.14
Beryllium					
Be(s)	9.01	0	0	9.50	16.44
Be(g)	9.01	+324.3	+286.6	136.27	20.79
Bismuth					
Bi(s)	208.98	0	0	56.74	25.52
Bi(g)	208.98	+207.1	+168.2	187.00	20.79
Bromine					
Br_2(l)	159.82	0	0	152.23	75.689
Br_2(g)	159.82	+30.907	+3.110	245.46	36.02
Br(g)	79.91	+111.88	+82.396	175.02	20.786
Br^-(g)	79.91	−219.07			
Br^-(aq)	79.91	−121.55	−103.96	+82.4	−141.8
HBr(g)	90.92	−36.40	−53.45	198.70	29.142
Cadmium					
Cd(s, γ)	112.40	0	0	51.76	25.98
Cd(g)	112.40	+112.01	+77.41	167.75	20.79
Cd^{2+}(aq)	112.40	−75.90	−77.612	−73.2	

Table 14.6 (Continued)

	$M/(\mathrm{g\,mol^{-1}})$	$\Delta_f H^\ominus/(\mathrm{kJ\,mol^{-1}})$	$\Delta_f G^\ominus/(\mathrm{kJ\,mol^{-1}})$	$S_m^\ominus/(\mathrm{J\,K^{-1}\,mol^{-1}})$†	$C_{p,m}^\ominus/(\mathrm{J\,K^{-1}\,mol^{-1}})$
Cadmium (Continued)					
$CdO(s)$	128.40	−258.2	−228.4	54.8	43.43
$CdCO_3(s)$	172.41	−750.6	−669.4	92.5	
Caesium (cesium)					
$Cs(s)$	132.91	0	0	85.23	32.17
$Cs(g)$	132.91	+76.06	+49.12	175.60	20.79
$Cs^+(aq)$	132.91	−258.28	−292.02	+133.05	−10.5
Calcium					
$Ca(s)$	40.08	0	0	41.42	25.31
$Ca(g)$	40.08	+178.2	+144.3	154.88	20.786
$Ca^{2+}(aq)$	40.08	−542.83	−553.58	−53.1	
$CaO(s)$	56.08	−635.09	−604.03	39.75	42.80
$CaCO_3(s)$ (calcite)	100.09	−1206.9	−1128.8	92.9	81.88
$CaCO_3(s)$ (aragonite)	100.09	−1207.1	−1127.8	88.7	81.25
$CaF_2(s)$	78.08	−1219.6	−1167.3	68.87	67.03
$CaCl_2(s)$	110.99	−795.8	−748.1	104.6	72.59
$CaBr_2(s)$	199.90	−682.8	−663.6	130	
Carbon (for 'organic' compounds of carbon, see Table 14.5)					
$C(s)$ (graphite)	12.011	0	0	5.740	8.527
$C(s)$ (diamond)	12.011	+1.895	+2.900	2.377	6.113
$C(g)$	12.011	+716.68	+671.26	158.10	20.838
$C_2(g)$	24.022	+831.90	+775.89	199.42	43.21
$CO(g)$	28.011	−110.53	−137.17	197.67	29.14
$CO_2(g)$	44.010	−393.51	−394.36	213.74	37.11
$CO_2(aq)$	44.010	−413.80	−385.98	117.6	
$H_2CO_3(aq)$	62.03	−699.65	−623.08	187.4	
$HCO_3^-(aq)$	61.02	−691.99	−586.77	+91.2	
$CO_3^{2-}(aq)$	60.01	−677.14	−527.81	−56.9	
$CCl_4(l)$	153.82	−135.44	−65.21	216.40	131.75
$CS_2(l)$	76.14	+89.70	+65.27	151.34	75.7
$HCN(g)$	27.03	+135.1	+124.7	201.78	35.86
$HCN(l)$	27.03	+108.87	+124.97	112.84	70.63
$CN^-(aq)$	26.02	+150.6	+172.4	+94.1	
Chlorine					
$Cl_2(g)$	70.91	0	0	223.07	33.91
$Cl(g)$	35.45	+121.68	+105.68	165.20	21.840
$Cl^-(g)$	34.45	−233.13			
$Cl^-(aq)$	35.45	−167.16	−131.23	+56.5	−136.4
$HCl(g)$	36.46	−92.31	−95.30	186.91	29.12
$HCl(aq)$	36.46	−167.16	−131.23	56.5	−136.4
Chromium					
$Cr(s)$	52.00	0	0	23.77	23.35
$Cr(g)$	52.00	+396.6	+351.8	174.50	20.79

Table 14.6 (Continued)

	$M/(\mathrm{g\,mol^{-1}})$	$\Delta_f H^\ominus/(\mathrm{kJ\,mol^{-1}})$	$\Delta_f G^\ominus/(\mathrm{kJ\,mol^{-1}})$	$S_m^\ominus/(\mathrm{J\,K^{-1}\,mol^{-1}})$†	$C_{p,m}^\ominus/(\mathrm{J\,K^{-1}\,mol^{-1}})$
Chromium (Continued)					
CrO_4^{2-}(aq)	115.99	−881.15	−727.75	+50.21	
$Cr_2O_7^{2-}$(aq)	215.99	−1490.3	−1301.1	+261.9	
Copper					
Cu(s)	63.54	0	0	33.150	24.44
Cu(g)	63.54	+338.32	+298.58	166.38	20.79
Cu^+(aq)	63.54	+71.67	+49.98	+40.6	
Cu^{2+}(aq)	63.54	+64.77	+65.49	−99.6	
Cu_2O(s)	143.08	−168.6	−146.0	93.14	63.64
CuO(s)	79.54	−157.3	−129.7	42.63	42.30
$CuSO_4$(s)	159.60	−771.36	−661.8	109	100.0
$CuSO_4\cdot H_2O$(s)	177.62	−1085.8	−918.11	146.0	134
$CuSO_4\cdot 5H_2O$(s)	249.68	−2279.7	−1879.7	300.4	280
Deuterium					
D_2(g)	4.028	0	0	144.96	29.20
HD(g)	3.022	+0.318	−1.464	143.80	29.196
D_2O(g)	20.028	−249.20	−234.54	198.34	34.27
D_2O(l)	20.028	−294.60	−243.44	75.94	84.35
HDO(g)	19.022	−245.30	−233.11	199.51	33.81
HDO(l)	19.022	−289.89	−241.86	79.29	
Fluorine					
F_2(g)	38.00	0	0	202.78	31.30
F(g)	19.00	+78.99	+61.91	158.75	22.74
F^-(aq)	19.00	−332.63	−278.79	−13.8	−106.7
HF(g)	20.01	−271.1	−273.2	173.78	29.13
Gold					
Au(s)	196.97	0	0	47.40	25.42
Au(g)	196.97	+366.1	+326.3	180.50	20.79
Helium					
He(g)	4.003	0	0	126.15	20.786
Hydrogen (see also deuterium)					
H_2(g)	2.016	0	0	130.684	28.824
H(g)	1.008	+217.97	+203.25	114.71	20.784
H^+(aq)	1.008	0	0	0	0
H^+(g)	1.008	+1536.20			
H_2O(s)	18.015			37.99	
H_2O(l)	18.015	−285.83	−237.13	69.91	75.291
H_2O(g)	18.015	−241.82	−228.57	188.83	33.58
H_2O_2(l)	34.015	−187.78	−120.35	109.6	89.1
Iodine					
I_2(s)	253.81	0	0	116.135	54.44
I_2(g)	253.81	+62.44	+19.33	260.69	36.90

Table 14.6 (Continued)

	$M/(g\ mol^{-1})$	$\Delta_f H^{\ominus}/(kJ\ mol^{-1})$	$\Delta_f G^{\ominus}/(kJ\ mol^{-1})$	$S_m^{\ominus}/(J\ K^{-1}\ mol^{-1})$†	$C_{p,m}^{\ominus}/(J\ K^{-1}\ mol^{-1})$
Iodine (Continued)					
I(g)	126.90	+106.84	+70.25	180.79	20.786
I^-(aq)	126.90	−55.19	−51.57	+111.3	−142.3
HI(g)	127.91	+26.48	+1.70	206.59	29.158
Iron					
Fe(s)	55.85	0	0	27.28	25.10
Fe(g)	55.85	+416.3	+370.7	180.49	25.68
Fe^{2+}(aq)	55.85	−89.1	−78.90	−137.7	
Fe^{3+}(aq)	55.85	−48.5	−4.7	−315.9	
Fe_3O_4(s) (magnetite)	231.54	−1118.4	−1015.4	146.4	143.43
Fe_2O_3(s) (haematite)	159.69	−824.2	−742.2	87.40	103.85
FeS(s, α)	87.91	−100.0	−100.4	60.29	50.54
FeS_2(s)	119.98	−178.2	−166.9	52.93	62.17
Krypton					
Kr(g)	83.80	0	0	164.08	20.786
Lead					
Pb(s)	207.19	0	0	64.81	26.44
Pb(g)	207.19	+195.0	+161.9	175.37	20.79
Pb^{2+}(aq)	207.19	−1.7	−24.43	+10.5	
PbO(s, yellow)	223.19	−217.32	−187.89	68.70	45.77
PbO(s, red)	223.19	−218.99	−188.93	66.5	45.81
PbO_2(s)	239.19	−277.4	−217.33	68.6	64.64
Lithium					
Li(s)	6.94	0	0	29.12	24.77
Li(g)	6.94	+159.37	+126.66	138.77	20.79
Li^+(aq)	6.94	−278.49	−293.31	+13.4	68.6
Magnesium					
Mg(s)	24.31	0	0	32.68	24.89
Mg(g)	24.31	+147.70	+113.10	148.65	20.786
Mg^{2+}(aq)	24.31	−466.85	−454.8	−138.1	
MgO(s)	40.31	−601.70	−569.43	26.94	37.15
$MgCO_3$(s)	84.32	−1095.8	−1012.1	65.7	75.52
$MgCl_2$(s)	95.22	−641.32	−591.79	89.62	71.38
Mercury					
Hg(l)	200.59	0	0	76.02	27.983
Hg(g)	200.59	+61.32	+31.82	174.96	20.786
Hg^{2+}(aq)	200.59	+171.1	+164.40	−32.2	
Hg_2^{2+}(aq)	401.18	+172.4	+153.52	+84.5	
HgO(s)	216.59	−90.83	−58.54	70.29	44.06
Hg_2Cl_2(s)	472.09	−265.22	−210.75	192.5	102
$HgCl_2$(s)	271.50	−224.3	−178.6	146.0	
HgS(s, black)	232.65	−53.6	−47.7	88.3	

Table 14.6 (Continued)

	$M/(\mathrm{g\ mol^{-1}})$	$\Delta_f H^\ominus/(\mathrm{kJ\ mol^{-1}})$	$\Delta_f G^\ominus/(\mathrm{kJ\ mol^{-1}})$	$S_m^\ominus/(\mathrm{J\ K^{-1}\ mol^{-1}})$†	$C_{p,m}^\ominus/(\mathrm{J\ K^{-1}\ mol^{-1}})$
Neon					
Ne(g)	20.18	0	0	146.33	20.786
Nitrogen					
N_2(g)	28.013	0	0	191.61	29.125
N(g)	14.007	+472.70	+455.56	153.30	20.786
NO(g)	30.01	+90.25	+86.55	210.76	29.844
N_2O(g)	44.01	+82.05	+104.20	219.85	38.45
NO_2(g)	46.01	+33.18	+51.31	240.06	37.20
N_2O_4(g)	92.1	+9.16	+97.89	304.29	77.28
N_2O_5(s)	108.01	−43.1	+113.9	178.2	143.1
N_2O_5(g)	108.01	+11.3	+115.1	355.7	84.5
HNO_3(l)	63.01	−174.10	−80.71	155.60	109.87
HNO_3(aq)	63.01	−207.36	−111.25	146.4	−86.6
NO_3^-(aq)	62.01	−205.0	−108.74	+146.4	−86.6
NH_3(g)	17.03	−46.11	−16.45	192.45	35.06
NH_3(aq)	17.03	−80.29	−26.50	111.3	
NH_4^+(aq)	18.04	−132.51	−79.31	+113.4	79.9
NH_2OH(s)	33.03	−114.2			
HN_3(l)	43.03	+264.0	+327.3	140.6	43.68
HN_3(g)	43.03	+294.1	+328.1	238.97	98.87
N_2H_4(l)	32.05	+50.63	+149.43	121.21	139.3
NH_4NO_3(s)	80.04	−365.56	−183.87	151.08	84.1
NH_4Cl(s)	53.49	−314.43	−202.87	94.6	
Oxygen					
O_2(g)	31.999	0	0	205.138	29.355
O(g)	15.999	+249.17	+231.73	161.06	21.912
O_3(g)	47.998	+142.7	+163.2	238.93	39.20
OH^-(aq)	17.007	−229.99	−157.24	−10.75	−148.5
Phosphorus					
P(s, wh)	30.97	0	0	41.09	23.840
P(g)	30.97	+314.64	+278.25	163.19	20.786
P_2(g)	61.95	+144.3	+103.7	218.13	32.05
P_4(g)	123.90	+58.91	+24.44	279.98	67.15
PH_3(g)	34.00	+5.4	+13.4	210.23	37.11
PCl_3(g)	137.33	−287.0	−267.8	311.78	71.84
PCl_3(l)	137.33	−319.7	−272.3	217.1	
PCl_5(g)	208.24	−374.9	−305.0	364.6	112.8
PCl_5(s)	208.24	−443.5			
H_3PO_3(s)	82.00	−964.4			
H_3PO_3(aq)	82.00	−964.8			
H_3PO_4(s)	94.97	−1279.0	−1119.1	110.50	106.06
H_3PO_4(l)	94.97	−1266.9			
H_3PO_4(aq)	94.97	−1277.4	−1018.7	−222	

Table 14.6 (Continued)

	$M/(\mathrm{g\,mol^{-1}})$	$\Delta_f H^\ominus/(\mathrm{kJ\,mol^{-1}})$	$\Delta_f G^\ominus/(\mathrm{kJ\,mol^{-1}})$	$S_m^\ominus/(\mathrm{J\,K^{-1}\,mol^{-1}})$†	$C_{p,m}^\ominus/(\mathrm{J\,K^{-1}\,mol^{-1}})$
Phosphorus (Continued)					
PO_4^{3-}(aq)	94.97	−1277.4	−1018.7	−221.8	
P_4O_{10}(s)	283.89	−2984.0	−2697.0	228.86	211.71
P_4O_6(s)	219.89	−1640.1			
Potassium					
K(s)	39.10	0	0	64.18	29.58
K(g)	39.10	+89.24	+60.59	160.336	20.786
K^+(g)	39.10	+514.26			
K^+(aq)	39.10	−252.38	−283.27	+102.5	21.8
KOH(s)	56.11	−424.76	−379.08	78.9	64.9
KF(s)	58.10	−576.27	−537.75	66.57	49.04
KCl(s)	74.56	−436.75	−409.14	82.59	51.30
KBr(s)	119.01	−393.80	−380.66	95.90	52.30
KI(s)	166.01	−327.90	−324.89	106.32	52.93
Silicon					
Si(s)	28.09	0	0	18.83	20.00
Si(g)	28.09	+455.6	+411.3	167.97	22.25
SiO_2(s, α)	60.09	−910.94	−856.64	41.84	44.43
Silver					
Ag(s)	107.87	0	0	42.55	25.351
Ag(g)	107.87	+284.55	+245.65	173.00	20.79
Ag^+(aq)	107.87	+105.58	+77.11	+72.68	21.8
AgBr(s)	187.78	−100.37	−96.90	107.1	52.38
AgCl(s)	143.32	−127.07	−109.79	96.2	50.79
Ag_2O(s)	231.74	−31.05	−11.20	121.3	65.86
$AgNO_3$(s)	169.88	−129.39	−33.41	140.92	93.05
Sodium					
Na(s)	22.99	0	0	51.21	28.24
Na(g)	22.99	+107.32	+76.76	153.71	20.79
Na^+(aq)	22.99	−240.12	−261.91	+59.0	46.4
NaOH(s)	40.00	−425.61	−379.49	64.46	59.54
NaCl(s)	58.44	−411.15	−384.14	72.13	50.50
NaBr(s)	102.90	−361.06	−348.98	86.82	51.38
NaI(s)	149.89	−287.78	−286.06	98.53	52.09
Sulfur					
S(s, α) (rhombic)	32.06	0	0	31.80	22.64
S(s, β) (monoclinic)	32.06	+0.33	+0.1	32.6	23.6
S(g)	32.06	+278.81	+238.25	167.82	23.673
S_2(g)	64.13	+128.37	+79.30	228.18	32.47
S^{2-}(aq)	32.06	+33.1	+85.8	−14.6	
SO_2(g)	64.06	−296.83	−300.19	248.22	39.87
SO_3(g)	80.06	−395.72	−371.06	256.76	50.67

Table 14.6 (Continued)

	$M/(\text{g mol}^{-1})$	$\Delta_f H^\ominus/(\text{kJ mol}^{-1})$	$\Delta_f G^\ominus/(\text{kJ mol}^{-1})$	$S_m^\ominus/(\text{J K}^{-1}\text{ mol}^{-1})$†	$C_{p,m}^\ominus/(\text{J K}^{-1}\text{ mol}^{-1})$
Sulfur (Continued)					
H_2SO_4(l)	98.08	−813.99	−690.00	156.90	138.9
H_2SO_4(aq)	98.08	−909.27	−744.53	20.1	−293
SO_4^{2-}(aq)	96.06	−909.27	−744.53	+20.1	−293
HSO_4^-(aq)	97.07	−887.34	−755.91	+131.8	−84
H_2S(g)	34.08	−20.63	−33.56	205.79	34.23
H_2S(aq)	34.08	−39.7	−27.83	121	
HS^-(aq)	33.072	−17.6	+12.08	+62.08	
SF_6(g)	146.05	−1209	−1105.3	291.82	97.28
Tin					
Sn(s, β)	118.69	0	0	51.55	26.99
Sn(g)	118.69	+302.1	+267.3	168.49	20.26
Sn^{2+}(aq)	118.69	−8.8	−27.2	−17	
SnO(s)	134.69	−285.8	−256.9	56.5	44.31
SnO_2(s)	150.69	−580.7	−519.6	52.3	52.59
Xenon					
Xe(g)	131.30	0	0	169.68	20.786
Zinc					
Zn(s)	65.37	0	0	41.63	25.40
Zn(g)	65.37	+130.73	+95.14	160.98	20.79
Zn^{2+}(aq)	65.37	−153.89	−147.06	−112.1	46
ZnO(s)	81.37	−348.28	−318.30	43.64	40.25

Source: NBS. † Standard entropies of ions may be either positive or negative because the values are relative to the entropy of the hydrogen ion.

Table 14.7 Expansion coefficients, α, and isothermal compressibilities, κ_T

	$\alpha/(10^{-4}\text{ K}^{-1})$	$\kappa_T/(10^{-6}\text{ atm}^{-1})$
Liquids		
Benzene	12.4	92.1
Carbon tetrachloride	12.4	90.5
Ethanol	11.2	76.8
Mercury	1.82	38.7
Water	2.1	49.6
Solids		
Copper	0.501	0.735
Diamond	0.030	0.187
Iron	0.354	0.589
Lead	0.861	2.21

The values refer to 20°C.
Data: AIP(α), KL(κ_T).

Table 14.8 Inversion temperatures, normal freezing and boiling points, and Joule–Thomson coefficients at 1 atm and 298 K

	T_I/K	T_f/K	T_b/K	$\mu/(\text{K atm}^{-1})$
Air	603			0.189 at 50°C
Argon	723	83.8	87.3	
Carbon dioxide	1500	194.7s		1.11 at 300 K
Helium	40		4.22	−0.062
Hydrogen	202	14.0	20.3	−0.03
Krypton	1090	116.6	120.8	
Methane	968	90.6	111.6	
Neon	231	24.5	27.1	
Nitrogen	621	63.3	77.4	0.27
Oxygen	764	54.8	90.2	0.31

s: sublimes.
Data: AIP, JL, and M.W. Zemansky, *Heat and thermodynamics*. McGraw-Hill, New York (1957).

Table 15.1 Standard entropies (and temperatures) of phase transitions, $\Delta_{trs}S^\ominus/(\mathrm{J\,K^{-1}\,mol^{-1}})$

	Fusion (at T_f)	Vaporization (at T_b)
Ar	14.17 (at 83.8 K)	74.53 (at 87.3 K)
Br_2	39.76 (at 265.9 K)	88.61 (at 332.4 K)
C_6H_6	38.00 (at 278.6 K)	87.19 (at 353.2 K)
CH_3COOH	40.4 (at 289.8 K)	61.9 (at 391.4 K)
CH_3OH	18.03 (at 175.2 K)	104.6 (at 337.2 K)
Cl_2	37.22 (at 172.1 K)	85.38 (at 239.0 K)
H_2	8.38 (at 14.0 K)	44.96 (at 20.38 K)
H_2O	22.00 (at 273.2 K)	109.1 (at 373.2 K)
H_2S	12.67 (at 187.6 K)	87.75 (at 212.0 K)
He	4.8 (at 1.8 K and 30 bar)	19.9 (at 4.22 K)
N_2	11.39 (at 63.2 K)	75.22 (at 77.4 K)
NH_3	28.93 (at 195.4 K)	97.41 (at 239.73 K)
O_2	8.17 (at 54.4 K)	75.63 (at 90.2 K)

Data: AIP.

Table 15.2 The standard enthalpies and entropies of vaporization of liquids at their normal boiling point

	$\Delta_{vap}H^\ominus/(\mathrm{kJ\,mol^{-1}})$	$\theta_b/°C$	$\Delta_{vap}S^\ominus/(\mathrm{J\,K^{-1}\,mol^{-1}})$
Benzene	30.8	80.1	+87.2
Carbon disulfide	26.74	46.25	+83.7
Carbon tetrachloride	30.00	76.7	+85.8
Cyclohexane	30.1	80.7	+85.1
Decane	38.75	174	+86.7
Dimethyl ether	21.51	−23	+86
Ethanol	38.6	78.3	+110.0
Hydrogen sulfide	18.7	−60.4	+87.9
Mercury	59.3	356.6	+94.2
Methane	8.18	−161.5	+73.2
Methanol	35.21	65.0	+104.1
Water	40.7	100.0	+109.1

Data: JL.

Table 15.3 Standard Third-Law entropies at 298 K: see Tables 14.5 and 14.6

Table 15.4 Standard Gibbs energies of formation at 298 K: see Tables 14.5 and 14.6

Table 16.1 Henry's law constants for gases at 298 K, $K/(\text{kPa kg mol}^{-1})$

	Water	Benzene
CH_4	7.55×10^4	44.4×10^3
CO_2	3.01×10^3	8.90×10^2
H_2	1.28×10^5	2.79×10^4
N_2	1.56×10^5	1.87×10^4
O_2	7.92×10^4	

Data: converted from R.J. Silbey and R.A. Alberty, *Physical chemistry*. Wiley, New York (2001).

Table 16.4 Mean activity coefficients in water at 298 K

$b/b^{\ominus}$	HCl	KCl	$CaCl_2$	H_2SO_4	$LaCl_3$	$In_2(SO_4)_3$
0.001	0.966	0.966	0.888	0.830	0.790	
0.005	0.929	0.927	0.789	0.639	0.636	0.16
0.01	0.905	0.902	0.732	0.544	0.560	0.11
0.05	0.830	0.816	0.584	0.340	0.388	0.035
0.10	0.798	0.770	0.524	0.266	0.356	0.025
0.50	0.769	0.652	0.510	0.155	0.303	0.014
1.00	0.811	0.607	0.725	0.131	0.387	
2.00	1.011	0.577	1.554	0.125	0.954	

Data: RS, HCP, and S. Glasstone, *Introduction to electrochemistry*. Van Nostrand (1942).

Table 17.2 Standard potentials at 298 K. (a) In electrochemical order

Reduction half-reaction	$E^{\ominus}$/V	Reduction half-reaction	$E^{\ominus}$/V
Strongly oxidizing		$Sn^{4+} + 2e^- \rightarrow Sn^{2+}$	+0.15
$H_4XeO_6 + 2H^+ + 2e^- \rightarrow XeO_3 + 3H_2O$	+3.0	$NO_3^- + H_2O + 2e^- \rightarrow NO_2^- + 2OH^-$	+0.10
$F_2 + 2e^- \rightarrow 2F^-$	+2.87	$AgBr + e^- \rightarrow Ag + Br^-$	+0.0713
$O_3 + 2H^+ + 2e^- \rightarrow O_2 + H_2O$	+2.07	$Ti^{4+} + e^- \rightarrow Ti^{3+}$	0.00
$S_2O_8^{2-} + 2e^- \rightarrow 2SO_4^{2-}$	+2.05	$2H^+ + 2e^- \rightarrow H_2$	0, by definition
$Ag^{2+} + e^- \rightarrow Ag^+$	+1.98	$Fe^{3+} + 3e^- \rightarrow Fe$	−0.04
$Co^{3+} + e^- \rightarrow Co^{2+}$	+1.81	$O_2 + H_2O + 2e^- \rightarrow HO_2^- + OH^-$	−0.08
$H_2O_2 + 2H^+ + 2e^- \rightarrow 2H_2O$	+1.78	$Pb^{2+} + 2e^- \rightarrow Pb$	−0.13
$Au^+ + e^- \rightarrow Au$	+1.69	$In^+ + e^- \rightarrow In$	−0.14
$Pb^{4+} + 2e^- \rightarrow Pb^{2+}$	+1.67	$Sn^{2+} + 2e^- \rightarrow Sn$	−0.14
$2HClO + 2H^+ + 2e^- \rightarrow Cl_2 + 2H_2O$	+1.63	$AgI + e^- \rightarrow Ag + I^-$	−0.15
$Ce^{4+} + e^- \rightarrow Ce^{3+}$	+1.61	$Ni^{2+} + 2e^- \rightarrow Ni$	−0.23
$2HBrO + 2H^+ + 2e^- \rightarrow Br_2 + 2H_2O$	+1.60	$V^{3+} + e^- \rightarrow V^{2+}$	−0.26
$MnO_4^- + 8H^+ + 5e^- \rightarrow Mn^{2+} + 4H_2O$	+1.51	$Co^{2+} + 2e^- \rightarrow Co$	−0.28
$Mn^{3+} + e^- \rightarrow Mn^{2+}$	+1.51	$In^{3+} + 3e^- \rightarrow In$	−0.34
$Au^{3+} + 3e^- \rightarrow Au$	+1.40	$Tl^+ + e^- \rightarrow Tl$	−0.34
$Cl_2 + 2e^- \rightarrow 2Cl^-$	+1.36	$PbSO_4 + 2e^- \rightarrow Pb + SO_4^{2-}$	−0.36
$Cr_2O_7^{2-} + 14H^+ + 6e^- \rightarrow 2Cr^{3+} + 7H_2O$	+1.33	$Ti^{3+} + e^- \rightarrow Ti^{2+}$	−0.37
$O_3 + H_2O + 2e^- \rightarrow O_2 + 2OH^-$	+1.24	$Cd^{2+} + 2e^- \rightarrow Cd$	−0.40
$O_2 + 4H^+ + 4e^- \rightarrow 2H_2O$	+1.23	$In^{2+} + e^- \rightarrow In^+$	−0.40
$ClO_4^- + 2H^+ + 2e^- \rightarrow ClO_3^- + H_2O$	+1.23	$Cr^{3+} + e^- \rightarrow Cr^{2+}$	−0.41
$MnO_2 + 4H^+ + 2e^- \rightarrow Mn^{2+} + 2H_2O$	+1.23	$Fe^{2+} + 2e^- \rightarrow Fe$	−0.44
$Pt^{2+} + 2e^- \rightarrow Pt$	+1.20	$In^{3+} + 2e^- \rightarrow In^+$	−0.44
$Br_2 + 2e^- \rightarrow 2Br^-$	+1.09	$S + 2e^- \rightarrow S^{2-}$	−0.48
$Pu^{4+} + e^- \rightarrow Pu^{3+}$	+0.97	$In^{3+} + e^- \rightarrow In^{2+}$	−0.49
$NO_3^- + 4H^+ + 3e^- \rightarrow NO + 2H_2O$	+0.96	$O_2 + e^- \rightarrow O_2^-$	−0.56
$2Hg^{2+} + 2e^- \rightarrow Hg_2^{2+}$	+0.92	$U^{4+} + e^- \rightarrow U^{3+}$	−0.61
$ClO^- + H_2O + 2e^- \rightarrow Cl^- + 2OH^-$	+0.89	$Cr^{3+} + 3e^- \rightarrow Cr$	−0.74
$Hg^{2+} + 2e^- \rightarrow Hg$	+0.86	$Zn^{2+} + 2e^- \rightarrow Zn$	−0.76
$NO_3^- + 2H^+ + e^- \rightarrow NO_2 + H_2O$	+0.80	$Cd(OH)_2 + 2e^- \rightarrow Cd + 2OH^-$	−0.81
$Ag^+ + e^- \rightarrow Ag$	+0.80	$2H_2O + 2e^- \rightarrow H_2 + 2OH^-$	−0.83
$Hg_2^{2+} + 2e^- \rightarrow 2Hg$	+0.79	$Cr^{2+} + 2e^- \rightarrow Cr$	−0.91
$AgF + e^- \rightarrow Ag + F^-$	+0.78	$Mn^{2+} + 2e^- \rightarrow Mn$	−1.18
$Fe^{3+} + e^- \rightarrow Fe^{2+}$	+0.77	$V^{2+} + 2e^- \rightarrow V$	−1.19
$BrO^- + H_2O + 2e^- \rightarrow Br^- + 2OH^-$	+0.76	$Ti^{2+} + 2e^- \rightarrow Ti$	−1.63
$Hg_2SO_4 + 2e^- \rightarrow 2Hg + SO_4^{2-}$	+0.62	$Al^{3+} + 3e^- \rightarrow Al$	−1.66
$MnO_4^{2-} + 2H_2O + 2e^- \rightarrow MnO_2 + 4OH^-$	+0.60	$U^{3+} + 3e^- \rightarrow U$	−1.79
$MnO_4^- + e^- \rightarrow MnO_4^{2-}$	+0.56	$Be^{2+} + 2e^- \rightarrow Be$	−1.85
$I_2 + 2e^- \rightarrow 2I^-$	+0.54	$Sc^{3+} + 3e^- \rightarrow Sc$	−2.09
$I_3^- + 2e^- \rightarrow 3I^-$	+0.53	$Mg^{2+} + 2e^- \rightarrow Mg$	−2.36
$Cu^+ + e^- \rightarrow Cu$	+0.52	$Ce^{3+} + 3e^- \rightarrow Ce$	−2.48
$NiOOH + H_2O + e^- \rightarrow Ni(OH)_2 + OH^-$	+0.49	$La^{3+} + 3e^- \rightarrow La$	−2.52
$Ag_2CrO_4 + 2e^- \rightarrow 2Ag + CrO_4^{2-}$	+0.45	$Na^+ + e^- \rightarrow Na$	−2.71
$O_2 + 2H_2O + 4e^- \rightarrow 4OH^-$	+0.40	$Ca^{2+} + 2e^- \rightarrow Ca$	−2.87
$ClO_4^- + H_2O + 2e^- \rightarrow ClO_3^- + 2OH^-$	+0.36	$Sr^{2+} + 2e^- \rightarrow Sr$	−2.89
$[Fe(CN)_6]^{3-} + e^- \rightarrow [Fe(CN)_6]^{4-}$	+0.36	$Ba^{2+} + 2e^- \rightarrow Ba$	−2.91
$Cu^{2+} + 2e^- \rightarrow Cu$	+0.34	$Ra^{2+} + 2e^- \rightarrow Ra$	−2.92
$Hg_2Cl_2 + 2e^- \rightarrow 2Hg + 2Cl^-$	+0.27	$Cs^+ + e^- \rightarrow Cs$	−2.92
$AgCl + e^- \rightarrow Ag + Cl^-$	+0.22	$Rb^+ + e^- \rightarrow Rb$	−2.93
$Bi^{3+} + 3e^- \rightarrow Bi$	+0.20	$K^+ + e^- \rightarrow K$	−2.93
$Cu^{2+} + e^- \rightarrow Cu^+$	+0.16	$Li^+ + e^- \rightarrow Li$	−3.05

Table 17.2 Standard potentials at 298 K. (b) In alphabetical order

Reduction half-reaction	$E^\ominus$/V	Reduction half-reaction	$E^\ominus$/V
$Ag^+ + e^- \rightarrow Ag$	+0.80	$I_2 + 2e^- \rightarrow 2I^-$	+0.54
$Ag^{2+} + e^- \rightarrow Ag^+$	+1.98	$I_3^- + 2e^- \rightarrow 3I^-$	+0.53
$AgBr + e^- \rightarrow Ag + Br^-$	+0.0713	$In^+ + e^- \rightarrow In$	−0.14
$AgCl + e^- \rightarrow Ag + Cl^-$	+0.22	$In^{2+} + e^- \rightarrow In^+$	−0.40
$Ag_2CrO_4 + 2e^- \rightarrow 2Ag + CrO_4^{2-}$	+0.45	$In^{3+} + 2e^- \rightarrow In^+$	−0.44
$AgF + e^- \rightarrow Ag + F^-$	+0.78	$In^{3+} + 3e^- \rightarrow In$	−0.34
$AgI + e^- \rightarrow Ag + I^-$	−0.15	$In^{3+} + e^- \rightarrow In^{2+}$	−0.49
$Al^{3+} + 3e^- \rightarrow Al$	−1.66	$K^+ + e^- \rightarrow K$	−2.93
$Au^+ + e^- \rightarrow Au$	+1.69	$La^{3+} + 3e^- \rightarrow La$	−2.52
$Au^{3+} + 3e^- \rightarrow Au$	+1.40	$Li^+ + e^- \rightarrow Li$	−3.05
$Ba^{2+} + 2e^- \rightarrow Ba$	−2.91	$Mg^{2+} + 2e^- \rightarrow Mg$	−2.36
$Be^{2+} + 2e^- \rightarrow Be$	−1.85	$Mn^{2+} + 2e^- \rightarrow Mn$	−1.18
$Bi^{3+} + 3e^- \rightarrow Bi$	+0.20	$Mn^{3+} + e^- \rightarrow Mn^{2+}$	+1.51
$Br_2 + 2e^- \rightarrow 2Br^-$	+1.09	$MnO_2 + 4H^+ + 2e^- \rightarrow Mn^{2+} + 2H_2O$	+1.23
$BrO^- + H_2O + 2e^- \rightarrow Br^- + 2OH^-$	+0.76	$MnO_4^- + 8H^+ + 5e^- \rightarrow Mn^{2+} + 4H_2O$	+1.51
$Ca^{2+} + 2e^- \rightarrow Ca$	−2.87	$MnO_4^- + e^- \rightarrow MnO_4^{2-}$	+0.56
$Cd(OH)_2 + 2e^- \rightarrow Cd + 2OH^-$	−0.81	$MnO_4^{2-} + 2H_2O + 2e^- \rightarrow MnO_2 + 4OH^-$	+0.60
$Cd^{2+} + 2e^- \rightarrow Cd$	−0.40	$Na^+ + e^- \rightarrow Na$	−2.71
$Ce^{3+} + 3e^- \rightarrow Ce$	−2.48	$Ni^{2+} + 2e^- \rightarrow Ni$	−0.23
$Ce^{4+} + e^- \rightarrow Ce^{3+}$	+1.61	$NiOOH + H_2O + e^- \rightarrow Ni(OH)_2 + OH^-$	+0.49
$Cl_2 + 2e^- \rightarrow 2Cl^-$	+1.36	$NO_3^- + 2H^+ + e^- \rightarrow NO_2 + H_2O$	+0.80
$ClO^- + H_2O + 2e^- \rightarrow Cl^- + 2OH^-$	+0.89	$NO_3^- + 4H^+ + 3e^- \rightarrow NO + 2H_2O$	+0.96
$ClO_4^- + 2H^+ + 2e^- \rightarrow ClO_3^- + H_2O$	+1.23	$NO_3^- + H_2O + 2e^- \rightarrow NO_2^- + 2OH^-$	+0.10
$ClO_4^- + H_2O + 2e^- \rightarrow ClO_3^- + 2OH^-$	+0.36	$O_2 + 2H_2O + 4e^- \rightarrow 4OH^-$	+0.40
$Co^{2+} + 2e^- \rightarrow Co$	−0.28	$O_2 + 4H^+ + 4e^- \rightarrow 2H_2O$	+1.23
$Co^{3+} + e^- \rightarrow Co^{2+}$	+1.81	$O_2 + e^- \rightarrow O_2^-$	−0.56
$Cr^{2+} + 2e^- \rightarrow Cr$	−0.91	$O_2 + H_2O + 2e^- \rightarrow HO_2^- + OH^-$	−0.08
$Cr_2O_7^{2-} + 14H^+ + 6e^- \rightarrow 2Cr^{3+} + 7H_2O$	+1.33	$O_3 + 2H^+ + 2e^- \rightarrow O_2 + H_2O$	+2.07
$Cr^{3+} + 3e^- \rightarrow Cr$	−0.74	$O_3 + H_2O + 2e^- \rightarrow O_2 + 2OH^-$	+1.24
$Cr^{3+} + e^- \rightarrow Cr^{2+}$	−0.41	$Pb^{2+} + 2e^- \rightarrow Pb$	−0.13
$Cs^+ + e^- \rightarrow Cs$	−2.92	$Pb^{4+} + 2e^- \rightarrow Pb^{2+}$	+1.67
$Cu^+ + e^- \rightarrow Cu$	+0.52	$PbSO_4 + 2e^- \rightarrow Pb + SO_4^{2-}$	−0.36
$Cu^{2+} + 2e^- \rightarrow Cu$	+0.34	$Pt^{2+} + 2e^- \rightarrow Pt$	+1.20
$Cu^{2+} + e^- \rightarrow Cu^+$	+0.16	$Pu^{4+} + e^- \rightarrow Pu^{3+}$	+0.97
$F_2 + 2e^- \rightarrow 2F^-$	+2.87	$Ra^{2+} + 2e^- \rightarrow Ra$	−2.92
$Fe^{2+} + 2e^- \rightarrow Fe$	−0.44	$Rb^+ + e^- \rightarrow Rb$	−2.93
$Fe^{3+} + 3e^- \rightarrow Fe$	−0.04	$S + 2e^- \rightarrow S^{2-}$	−0.48
$Fe^{3+} + e^- \rightarrow Fe^{2+}$	+0.77	$S_2O_8^{2-} + 2e^- \rightarrow 2SO_4^{2-}$	+2.05
$[Fe(CN)_6]^{3-} + e^- \rightarrow [Fe(CN)_6]^{4-}$	+0.36	$Sc^{3+} + 3e^- \rightarrow Sc$	−2.09
$2H^+ + 2e^- \rightarrow H_2$	0, by definition	$Sn^{2+} + 2e^- \rightarrow Sn$	−0.14
$2H_2O + 2e^- \rightarrow H_2 + 2OH^-$	−0.83	$Sn^{4+} + 2e^- \rightarrow Sn^{2+}$	+0.15
$2HBrO + 2H^+ + 2e^- \rightarrow Br_2 + 2H_2O$	+1.60	$Sr^{2+} + 2e^- \rightarrow Sr$	−2.89
$2HClO + 2H^+ + 2e^- \rightarrow Cl_2 + 2H_2O$	+1.63	$Ti^{2+} + 2e^- \rightarrow Ti$	−1.63
$H_2O_2 + 2H^+ + 2e^- \rightarrow 2H_2O$	+1.78	$Ti^{3+} + e^- \rightarrow Ti^{2+}$	−0.37
$H_4XeO_6 + 2H^+ + 2e^- \rightarrow XeO_3 + 3H_2O$	+3.0	$Ti^{4+} + e^- \rightarrow Ti^{3+}$	0.00
$Hg_2^{2+} + 2e^- \rightarrow 2Hg$	+0.79	$Tl^+ + e^- \rightarrow Tl$	−0.34
$Hg_2Cl_2 + 2e^- \rightarrow 2Hg + 2Cl^-$	+0.27	$U^{3+} + 3e^- \rightarrow U$	−1.79
$Hg^{2+} + 2e^- \rightarrow Hg$	+0.86	$U^{4+} + e^- \rightarrow U^{3+}$	−0.61
$2Hg^{2+} + 2e^- \rightarrow Hg_2^{2+}$	+0.92	$V^{2+} + 2e^- \rightarrow V$	−1.19
$Hg_2SO_4 + 2e^- \rightarrow 2Hg + SO_4^{2-}$	+0.62	$V^{3+} + e^- \rightarrow V^{2+}$	−0.26
		$Zn^{2+} + 2e^- \rightarrow Zn$	−0.76

Table 18.1 Collision cross-sections, σ/nm^2

Ar	0.36
C_2H_4	0.64
C_6H_6	0.88
CH_4	0.46
Cl_2	0.93
CO_2	0.52
H_2	0.27
He	0.21
N_2	0.43
Ne	0.24
O_2	0.40
SO_2	0.58

Data: KL.

Table 18.2 Transport properties of gases at 1 atm

	$\kappa/(\text{J K}^{-1}\text{ m}^{-1}\text{ s}^{-1})$	$\eta/\mu\text{P}$	
	273 K	273 K	293 K
Air	0.0241	173	182
Ar	0.0163	210	223
C_2H_4	0.0164	97	103
CH_4	0.0302	103	110
Cl_2	0.079	123	132
CO_2	0.0145	136	147
H_2	0.1682	84	88
He	0.1442	187	196
Kr	0.0087	234	250
N_2	0.0240	166	176
Ne	0.0465	298	313
O_2	0.0245	195	204
Xe	0.0052	212	228

Data: KL.

Table 18.4 Viscosities of liquids at 298 K, $\eta/(10^{-3}\text{ kg m}^{-1}\text{ s}^{-1})$

Benzene	0.601
Carbon tetrachloride	0.880
Ethanol	1.06
Mercury	1.55
Methanol	0.553
Pentane	0.224
Sulfuric acid	27
Water†	0.891

† The viscosity of water over its entire liquid range is represented with less than 1 per cent error by the expression

$$\log(\eta_{20}/\eta) = A/B,$$
$$A = 1.370\,23(t-20) + 8.36\times10^{-4}(t-20)^2$$
$$B = 109 + t \qquad t = \theta/{}^\circ\text{C}$$

Convert kg m^{-1} s^{-1} to centipoise (cP) by multiplying by 10^3 (so $\eta \approx 1$ cP for water).
Data: AIP, KL.

Table 18.5 Ionic mobilities in water at 298 K, $u/(10^{-8}\text{ m}^2\text{ s}^{-1}\text{ V}^{-1})$

Cations		Anions	
Ag^+	6.24	Br^-	8.09
Ca^{2+}	6.17	$CH_3CO_2^-$	4.24
Cu^{2+}	5.56	Cl^-	7.91
H^+	36.23	CO_3^{2-}	7.46
K^+	7.62	F^-	5.70
Li^+	4.01	$[Fe(CN)_6]^{3-}$	10.5
Na^+	5.19	$[Fe(CN)_6]^{4-}$	11.4
NH_4^+	7.63	I^-	7.96
$[N(CH_3)_4]^+$	4.65	NO_3^-	7.40
Rb^+	7.92	OH^-	20.64
Zn^{2+}	5.47	SO_4^{2-}	8.29

Data: Principally Table 18.4 and $u = \lambda/zF$.

Table 18.6 Diffusion coefficients at 298 K, $D/(10^{-9}\ m^2\ s^{-1})$

Molecules in liquids				Ions in water			
I_2 in hexane	4.05	H_2 in CCl_4(l)	9.75	K^+	1.96	Br^-	2.08
in benzene	2.13	N_2 in CCl_4(l)	3.42	H^+	9.31	Cl^-	2.03
CCl_4 in heptane	3.17	O_2 in CCl_4(l)	3.82	Li^+	1.03	F^-	1.46
Glycine in water	1.055	Ar in CCl_4(l)	3.63	Na^+	1.33	I^-	2.05
Dextrose in water	0.673	CH_4 in CCl_4(l)	2.89			OH^-	5.03
Sucrose in water	0.5216	H_2O in water	2.26				
		CH_3OH in water	1.58				
		C_2H_5OH in water	1.24				

Data: AIP.

Table 19.1 Kinetic data for first-order reactions

	Phase	θ/°C	k_r/s^{-1}	$t_{1/2}$
$2\ N_2O_5 \rightarrow 4\ NO_2 + O_2$	g	25	3.38×10^{-5}	5.70 h
	HNO_3(l)	25	1.47×10^{-6}	131 h
	Br_2(l)	25	4.27×10^{-5}	4.51 h
$C_2H_6 \rightarrow 2\ CH_3$	g	700	5.36×10^{-4}	21.6 min
Cyclopropane $\rightarrow$ propene	g	500	6.71×10^{-4}	17.2 min
$CH_3N_2CH_3 \rightarrow C_2H_6 + N_2$	g	327	3.4×10^{-4}	34 min
Sucrose $\rightarrow$ glucose + fructose	aq(H^+)	25	6.0×10^{-5}	3.2 h

g: High pressure gas-phase limit.
Data: Principally K.J. Laidler, *Chemical kinetics*. Harper & Row, New York (1987); M.J. Pilling and P.W. Seakins, *Reaction kinetics*. Oxford University Press (1995); J. Nicholas, *Chemical kinetics*. Harper & Row, New York (1976). See also JL.

Table 19.2 Kinetic data for second-order reactions

	Phase	θ/°C	$k_r/(dm^3\ mol^{-1}\ s^{-1})$
$2\ NOBr \rightarrow 2\ NO + Br_2$	g	10	0.80
$2\ NO_2 \rightarrow 2\ NO + O_2$	g	300	0.54
$H_2 + I_2 \rightarrow 2\ HI$	g	400	2.42×10^{-2}
$D_2 + HCl \rightarrow DH + DCl$	g	600	0.141
$2\ I \rightarrow I_2$	g	23	7×10^9
	hexane	50	1.8×10^{10}
$CH_3Cl + CH_3O^-$	methanol	20	2.29×10^{-6}
$CH_3Br + CH_3O^-$	methanol	20	9.23×10^{-6}
$H^+ + OH^- \rightarrow H_2O$	water	25	1.35×10^{11}
	ice	−10	8.6×10^{12}

Data: Principally K.J. Laidler, *Chemical kinetics*. Harper & Row, New York (1987); M.J. Pilling and P.W. Seakins, *Reaction kinetics*. Oxford University Press (1995); J. Nicholas, *Chemical kinetics*. Harper & Row, New York (1976).

Table 20.1 Arrhenius parameters

First-order reactions	A/s^{-1}	$E_a/(kJ\ mol^{-1})$
Cyclopropane → propene	1.58×10^{15}	272
$CH_3NC \rightarrow CH_3CN$	3.98×10^{13}	160
cis-CHD=CHD → *trans*-CHD=CHD	3.16×10^{12}	256
Cyclobutane → $2\ C_2H_4$	3.98×10^{13}	261
$C_2H_5I \rightarrow C_2H_4 + HI$	2.51×10^{17}	209
$C_2H_6 \rightarrow 2\ CH_3$	2.51×10^{7}	384
$2\ N_2O_5 \rightarrow 4\ NO_2 + O_2$	4.94×10^{13}	103.4
$N_2O \rightarrow N_2 + O$	7.94×10^{11}	250
$C_2H_5 \rightarrow C_2H_4 + H$	1.0×10^{13}	167

Second-order, gas-phase	$A/(dm^3\ mol^{-1}\ s^{-1})$	$E_a/(kJ\ mol^{-1})$
$O + N_2 \rightarrow NO + N$	1×10^{11}	315
$OH + H_2 \rightarrow H_2O + H$	8×10^{10}	42
$Cl + H_2 \rightarrow HCl + H$	8×10^{10}	23
$2\ CH_3 \rightarrow C_2H_6$	2×10^{10}	ca. 0
$NO + Cl_2 \rightarrow NOCl + Cl$	4.0×10^{9}	85
$SO + O_2 \rightarrow SO_2 + O$	3×10^{8}	27
$CH_3 + C_2H_6 \rightarrow CH_4 + C_2H_5$	2×10^{8}	44
$C_6H_5 + H_2 \rightarrow C_6H_6 + H$	1×10^{8}	ca. 25

Second-order, solution	$A/(dm^3\ mol^{-1}\ s^{-1})$	$E_a/(kJ\ mol^{-1})$
$C_2H_5ONa + CH_3I$ in ethanol	2.42×10^{11}	81.6
$C_2H_5Br + OH^-$ in water	4.30×10^{11}	89.5
$C_2H_5I + C_2H_5O^-$ in ethanol	1.49×10^{11}	86.6
$C_2H_5Br + OH^-$ in ethanol	4.30×10^{11}	89.5
$CO_2 + OH^-$ in water	1.5×10^{10}	38
$CH_3I + S_2O_3^{2-}$ in water	2.19×10^{12}	78.7
Sucrose + H_2O in acidic water	1.50×10^{15}	107.9
$(CH_3)_3CCl$ solvolysis		
in water	7.1×10^{16}	100
in methanol	2.3×10^{13}	107
in ethanol	3.0×10^{13}	112
in acetic acid	4.3×10^{13}	111
in chloroform	1.4×10^{4}	45
$C_6H_5NH_2 + C_6H_5COCH_2Br$		
in benzene	91	34

Data: Principally J. Nicholas, *Chemical kinetics*. Harper & Row, New York (1976) and A.A. Frost and R.G. Pearson, *Kinetics and mechanism*. Wiley, New York (1961).

Table 20.2 Arrhenius parameters for gas-phase reactions

	$A/(dm^3\ mol^{-1}\ s^{-1})$			
	Experiment	Theory	$E_a/(kJ\ mol^{-1})$	P
$2\ NOCl \rightarrow 2\ NO + Cl_2$	9.4×10^9	5.9×10^{10}	102.0	0.16
$2\ NO_2 \rightarrow 2\ NO + O_2$	2.0×10^9	4.0×10^{10}	111.0	5.0×10^{-2}
$2\ ClO \rightarrow Cl_2 + O_2$	6.3×10^7	2.5×10^{10}	0.0	2.5×10^{-3}
$H_2 + C_2H_4 \rightarrow C_2H_6$	1.24×10^6	7.4×10^{11}	180	1.7×10^{-6}
$K + Br_2 \rightarrow KBr + Br$	1.0×10^{12}	2.1×10^{11}	0.0	4.8

Data: Principally M.J. Pilling and P.W. Seakins, *Reaction kinetics*. Oxford University Press (1995).

Table 20.3 Arrhenius parameters for reactions in solution. See Table 20.1

Table 21.1 Maximum observed enthalpies of physisorption, $\Delta_{ad}H^{\ominus}/(kJ\ mol^{-1})$

C_2H_2	−38	H_2	−84
C_2H_4	−34	H_2O	−59
CH_4	−21	N_2	−21
Cl_2	−36	NH_3	−38
CO	−25	O_2	−21
CO_2	−25		

Data: D.O. Haywood and B.M.W. Trapnell, *Chemisorption*. Butterworth (1964).

Table 21.2 Enthalpies of chemisorption, $\Delta_{ad}H^{\ominus}/(kJ\ mol^{-1})$

Adsorbate	Adsorbent (substrate)											
	Ti	Ta	Nb	W	Cr	Mo	Mn	Fe	Co	Ni	Rh	Pt
H_2		−188			−188	−167	−71	−134			−117	
N_2		−586						−293				
O_2						−720					−494	−293
CO	−640							−192	−176			
CO_2	−682	−703	−552	−456	−339	−372	−222	−225	−146	−184		
NH_3				−301				−188		−155		
C_2H_4		−577		−427	−427			−285		−243	−209	

Data: D.O. Haywood and B.M.W. Trapnell, *Chemisorption*. Butterworth (1964).

Part 2 Common operators in quantum mechanics

Observable	Operator	Representation*
Energy	$\hat{H}$ (hamiltonian)	$\hat{H} = -\frac{\hbar^2}{2m}\nabla^2 + \hat{V}$
Kinetic energy	$\hat{E}_k$	$\hat{E}_k = -\frac{\hbar^2}{2m}\nabla^2$
Potential energy	$\hat{V}$	$\hat{V} = V(\boldsymbol{r})\times$
Position (*x*-, *y*-, and *z*-components)	$\hat{x}, \hat{y}, \hat{z}$; in general $\hat{q}$	$\hat{q} = q\times$
Radial distance	$\hat{r}$	$\hat{r} = r\times$
Linear momentum (*x*-, *y*-, *z*-components)	$\hat{p}_x, \hat{p}_y, \hat{p}_z$; in general $\hat{p}_q$	$\hat{p}_q = \frac{\hbar}{\mathrm{i}}\frac{\partial}{\partial q}$
Square of linear momentum	$\hat{p}^2 = \hat{p}_x^2 + \hat{p}_y^2 + \hat{p}_z^2$	$\hat{p}^2 = -\hbar^2\nabla^2$
Electric dipole moment (*x*-, *y*-, *z*-components)	$\hat{\mu}_x, \hat{\mu}_y, \hat{\mu}_z$; in general $\hat{\mu}_q$	$\hat{\mu}_q = -eq\times$
Orbital angular momentum (*x*-, *y*-, *z*-components)	$\hat{l}_x, \hat{l}_y, \hat{l}_z$; in general $\hat{l}_q$	$\hat{l}_z = \frac{\hbar}{\mathrm{i}}\frac{\partial}{\partial\phi}$ and in general $\hat{\boldsymbol{l}} = \hat{\boldsymbol{r}} \times \hat{\boldsymbol{p}} = \frac{\hbar}{\mathrm{i}}\boldsymbol{r}\times\nabla$
Square of magnitude of the orbital angular momentum	$\hat{l}^2 = \hat{l}_x^2 + \hat{l}_y^2 + \hat{l}_z^2$	$\hat{l}^2 = -\hbar^2\Lambda^2$
Square of magnitude of the spin angular momentum and its *z*-component	$\hat{s}^2, \hat{s}_z$	–

* In the 'position representation', in which the position operators have a simple multiplicative form.

Part 3 Character tables

The groups C_1, C_s, C_i

C_1 (1)	E	$h=1$
A	1	

$C_s = C_h$ (m)	E	σ_h	$h=2$	
A′	1	1	x, y, R_z	x^2, y^2, z^2, xy
A″	1	−1	z, R_x, R_y	yz, xz

$C_i = S_2$ $(\bar{1})$	E	i	$h=2$	
A_g	1	1	R_x, R_y, R_z	$x^2, y^2, z^2, xy, xz, yz$
A_u	1	−1	x, y, z	

The groups C_{nv}

C_{2v}, $2mm$	E	C_2	σ_v	σ_v'	$h=4$	
A_1	1	1	1	1	z, z^2, x^2, y^2	
A_2	1	1	−1	−1	xy	R_z
B_1	1	−1	1	−1	x, xz	R_y
B_2	1	−1	−1	1	y, yz	R_x

C_{3v}, $3m$	E	$2C_3$	$3\sigma_v$	$h=6$	
A_1	1	1	1	z, z^2, x^2+y^2	
A_2	1	1	−1		R_z
E	2	−1	0	$(x, y), (xy, x^2-y^2)$ (xz, yz)	(R_x, R_y)

C_{4v}, $4mm$	E	C_2	$2C_4$	$2\sigma_v$	$2\sigma_d$	$h=8$	
A_1	1	1	1	1	1	z, z^2, x^2+y^2	
A_2	1	1	1	−1	−1		R_z
B_1	1	1	−1	1	−1	x^2-y^2	
B_2	1	1	−1	−1	1	xy	
E	2	−2	0	0	0	$(x, y), (xz, yz)$	(R_x, R_y)

C_{5v}	E	$2C_5$	$2C_5^2$	$5\sigma_v$	$h=10, \alpha=72°$	
A_1	1	1	1	1	z, z^2, x^2+y^2	
A_2	1	1	1	−1		R_z
E_1	2	$2\cos\alpha$	$2\cos 2\alpha$	0	$(x, y), (xz, yz)$	(R_x, R_y)
E_2	2	$2\cos 2\alpha$	$2\cos\alpha$	0	(xy, x^2-y^2)	

C_{6v}, $6mm$	E	C_2	$2C_3$	$2C_6$	$3\sigma_d$	$3\sigma_v$	$h=12$	
A_1	1	1	1	1	1	1	z, z^2, x^2+y^2	
A_2	1	1	1	1	−1	−1		R_z
B_1	1	−1	1	−1	−1	1		
B_2	1	−1	1	−1	1	−1		
E_1	2	−2	−1	1	0	0	$(x, y), (xz, yz)$	(R_x, R_y)
E_2	2	2	−1	−1	0	0	(xy, x^2-y^2)	

$C_{\infty v}$	E	$2C_\phi$†	$\infty\sigma_v$	$h=\infty$	
$A_1(\Sigma^+)$	1	1	1	z, z^2, x^2+y^2	
$A_2(\Sigma^-)$	1	1	−1		R_z
$E_1(\Pi)$	2	$2\cos\phi$	0	$(x, y), (xz, yz)$	(R_x, R_y)
$E_2(\Delta)$	2	$2\cos 2\phi$	0	(xy, x^2-y^2)	

† There is only one member of this class if $\phi=\pi$.

The groups D_n

D_2, 222	E	C_2^z	C_2^y	C_2^x	$h=4$	
A_1	1	1	1	1	x^2, y^2, z^2	
B_1	1	1	−1	−1	z, xy	R_z
B_2	1	−1	1	−1	y, xz	R_y
B_3	1	−1	−1	1	x, yz	R_x

D_3, 32	E	$2C_3$	$3C_2'$	$h=6$	
A_1	1	1	1	z^2, x^2+y^2	
A_2	1	1	−1	z	R_z
E	2	−1	0	$(x, y), (xz, yz), (xy, x^2-y^2)$	(R_x, R_y)

D_4, 422	E	C_2	$2C_4$	$2C_2'$	$2C_2''$	$h=8$	
A_1	1	1	1	1	1	z^2, x^2+y^2	
A_2	1	1	1	−1	−1	z	R_z
B_1	1	1	−1	1	−1	x^2-y^2	
B_2	1	1	−1	−1	1	xy	
E	2	−2	0	0	0	$(x, y), (xz, yz)$	(R_x, R_y)

The groups D_{nh}

D_{3h}, $\bar{6}2m$	E	σ_h	$2C_3$	$2S_3$	$3C_2'$	$3\sigma_v$	$h=12$	
A_1'	1	1	1	1	1	1	z^2, x^2+y^2	
A_2'	1	1	1	1	−1	−1		R_z
A_1''	1	−1	1	−1	1	−1		
A_2''	1	−1	1	−1	−1	1	z	
E′	2	2	−1	−1	0	0	$(x, y), (xy, x^2-y^2)$	
E″	2	−2	−1	1	0	0	(xz, yz)	(R_x, R_y)

D_{4h}, 4/*mmm*	E	$2C_4$	C_2	$2C_2'$	$2C_2''$	i	$2S_4$	σ_h	$2\sigma_v$	$2\sigma_d$	$h=16$	
A_{1g}	1	1	1	1	1	1	1	1	1	1	x^2+y^2, z^2	
A_{2g}	1	1	1	−1	−1	1	1	1	−1	−1		R_z
B_{1g}	1	−1	1	1	−1	1	−1	1	1	−1	x^2-y^2	
B_{2g}	1	−1	1	−1	1	1	−1	1	−1	1	xy	
E_g	2	0	−2	0	0	2	0	−2	0	0	(xz, yz)	(R_x, R_y)
A_{1u}	1	1	1	1	1	−1	−1	−1	−1	−1		
A_{2u}	1	1	1	−1	−1	−1	−1	−1	1	1	z	
B_{1u}	1	−1	1	1	−1	−1	1	−1	−1	1		
B_{2u}	1	−1	1	−1	1	−1	1	−1	1	−1		
E_u	2	0	−2	0	0	−2	0	2	0	0	(x, y)	

D_{5h}	E	$2C_5$	$2C_5^2$	$5C_2$	σ_h	$2S_5$	$2S_5^3$	$5\sigma_v$	$h=20$	$\alpha=72°$
A_1'	1	1	1	1	1	1	1	1	x^2+y^2, z^2	
A_2'	1	1	1	−1	1	1	1	−1		R_z
E_1'	2	$2\cos\alpha$	$2\cos 2\alpha$	0	2	$2\cos\alpha$	$2\cos 2\alpha$	0	(x, y)	
E_2'	2	$2\cos 2\alpha$	$2\cos\alpha$	0	2	$2\cos 2\alpha$	$2\cos\alpha$	0	(x^2-y^2, xy)	
A_1''	1	1	1	1	−1	−1	−1	−1		
A_2''	1	1	1	−1	−1	−1	−1	1	z	
E_1''	2	$2\cos\alpha$	$2\cos 2\alpha$	0	−2	$-2\cos\alpha$	$-2\cos 2\alpha$	0	(xz, yz)	(R_x, R_y)
E_2''	2	$2\cos 2\alpha$	$2\cos\alpha$	0	−2	$-2\cos 2\alpha$	$-2\cos\alpha$	0		

$D_{\infty h}$	E	$2C_\phi$	…	$\infty\sigma_v$	i	$2S_\infty$	…	$\infty C_2'$	$h=\infty$	
$A_{1g}(\Sigma_g^+)$	1	1	…	1	1	1	…	1	z^2, x^2+y^2	
$A_{1u}(\Sigma_u^+)$	1	1	…	1	−1	−1	…	−1	z	
$A_{2g}(\Sigma_g^-)$	1	1	…	−1	1	1	…	−1		R_z
$A_{2u}(\Sigma_u^-)$	1	1	…	−1	−1	−1	…	1		
$E_{1g}(\Pi_g)$	2	$2\cos\phi$	…	0	2	$-2\cos\phi$	…	0	(xz, yz)	(R_x, R_y)
$E_{1u}(\Pi_u)$	2	$2\cos\phi$	…	0	−2	$2\cos\phi$	…	0	(x, y)	
$E_{2g}(\Delta_g)$	2	$2\cos 2\phi$	…	0	2	$2\cos 2\phi$	…	0	(xy, x^2-y^2)	
$E_{2u}(\Delta_u)$	2	$2\cos 2\phi$	…	0	−2	$-2\cos 2\phi$	…	0		
⋮	⋮	⋮		⋮	⋮	⋮		⋮		

The cubic groups

T_d, $\bar{4}3m$	E	$8C_3$	$3C_2$	$6\sigma_d$	$6S_4$	$h=24$	
A_1	1	1	1	1	1	$x^2+y^2+z^2$	
A_2	1	1	1	−1	−1		
E	2	−1	2	0	0	$(3z^2-r^2, x^2-y^2)$	
T_1	3	0	−1	−1	1		(R_x, R_y, R_z)
T_2	3	0	−1	1	−1	$(x, y, z), (xy, xz, yz)$	

O_h $(m3m)$	E	$8C_3$	$6C_2$	$6C_4$	$3C_2\ (=C_4^2)$	i	$6S_4$	$8S_6$	$3\sigma_h$	$6\sigma_d$	$h=48$	
A_{1g}	1	1	1	1	1	1	1	1	1	1	$x^2+y^2+z^2$	
A_{2g}	1	1	−1	−1	1	1	−1	1	1	−1		
E_g	2	−1	0	0	2	2	0	−1	2	0	$(2z^2-x^2-y^2, x^2-y^2)$	
T_{1g}	3	0	−1	1	−1	3	1	0	−1	−1		(R_x, R_y, R_z)
T_{2g}	3	0	1	−1	−1	3	−1	0	−1	1	(xy, yz, zx)	
A_{1u}	1	1	1	1	1	−1	−1	−1	−1	−1		
A_{2u}	1	1	−1	−1	1	−1	1	−1	−1	1		
E_u	2	−1	0	0	2	−2	0	1	−2	0		
T_{1u}	3	0	−1	1	−1	−3	−1	0	1	1	(x, y, z)	
T_{2u}	3	0	1	−1	−1	−3	1	0	1	−1		

The icosahedral group

I	E	$12C_5$	$12C_5^2$	$20C_3$	$15C_2$	$h=60$	
A	1	1	1	1	1	$x^2+y^2+z^2$	
T_1	3	$\frac{1}{2}(1+\sqrt{5})$	$\frac{1}{2}(1-\sqrt{5})$	0	−1	(x, y, z)	(R_x, R_y, R_z)
T_2	3	$\frac{1}{2}(1-\sqrt{5})$	$\frac{1}{2}(1+\sqrt{5})$	0	−1		
G	4	−1	−1	1	0		
H	5	0	0	−1	1	$(2z^2-x^2-y^2, x^2-y^2, xy, yz, zx)$	

Further information: P.W. Atkins, M.S. Child, and C.S.G. Phillips, *Tables for group theory*. Oxford University Press (1970). In this source, which is available on the web (see p. *xi* for more details), other character tables such as D_2, D_4, D_{2d}, D_{3d}, and D_{5d} can be found.

Solutions to a) exercises

Fundamentals

F1.2a

	Example	Element	Ground-state Electronic Configuration
(a)	Group 2	Ca, calcium	$[Ar]4s^2$
(b)	Group 7	Mn, manganese	$[Ar]3d^54s^2$
(c)	Group 15	As, arsenic	$[Ar]3d^{10}4s^24p^3$

F2.1a A **single bond** is a shared pair of electrons between adjacent atoms within a molecule while a **multiple bond** involves the sharing of either two pairs of electrons (a double bond) or three pairs of electrons (a triple bond).

F2.6a (**a**) CO_2 is a linear, nonpolar molecule, (**b**) SO_2 is a bent, polar molecule, (**c**) N_2O is linear, polar molecule, (**d**) SF_4 is a seesaw molecule and it is a polar molecule

F2.7a In the order of increasing dipole moment: CO_2, N_2O, SF_4, SO_2

F3.3a (a) 0.543 mol, (b) 3.27×10^{23} molecules

F3.4a (a) 180 g, (b) 1.77 N

F3.5a 0.45 bar

F3.6a 0.44 atm

F3.7a (a) 1.47×10^{5} Pa, (b) 1.47 bar

F3.8a $T = 310.2$ K

F3.9a $\theta/°C = \frac{5}{9}\times(\theta_F/°F - 32)$ or $\theta_F/°F = \frac{9}{5}\times\theta/°C + 32$, $\theta_F = 173$ °F

F3.10a 105 kPa

F3.11a S_8

F3.12a 1.8 MPa

F3.13a 4.6×10^{5} Pa, 6.9×10^{5} Pa

F4.1a 27.2 K or 27.2 °C

F4.2a 128 J

F4.3a 2.4208 J K^{-1} g^{-1}

F4.4a 75.3 J K^{-1} mol^{-1}

F4.5a 2.48 kJ mol^{-1}

F4.6a 1.824 J mol^{-1}

F5.2a (a) 1.602×10^{-17}, (b) 2.092×10^{-2}

F5.4a 4.631×10^{-6}

F5.7a 1.07

F5.8a 1.25

F5.9a 0.47 kJ

F5.10a (a) 1.38 kJ, (b) 4.56 kJ

F5.11a 12.47 J mol^{-1} K^{-1}

F5.12a (a) 20.78 J mol^{-1} K^{-1}, (b) 24.94 J mol^{-1} K^{-1}

F6.1a (a) 9.81 m s^{-1}, (b) 29.4 m s^{-1}

F6.2a $s_{\text{terminal}} = \dfrac{ze\mathcal{E}}{6\pi\eta R}$

F6.4a $E = \frac{1}{2}kA^2$, $E = V_{\max} = \frac{1}{2}kA^2$

F6.6a $1.0\hbar$

F6.7a (a) 2.25×10^{-20} J, (b) 9.00×10^{-20} J

F6.8a 1.88×10^{8} m s^{-1}, 100 keV

F6.9a 1.15×10^{-18} J, 1.48×10^{-20} J

F6.10a −2.40 V

F6.11a 24.1 kJ, 28.8 °C

F7.1a 1.69×10^{4} cm^{-1}, 5.08×10^{14} Hz

F7.2a 2.26×10^{8} m s^{-1}

F7.3a 4.00 μm, 7.50×10^{13} Hz

F8.1a 1.45×10^{-6} m^3

F8.2a 11.2×10^{3} kg m^{-3}

F8.3a m^{-3}

F8.4a 207.1 cm^{-1}

F8.5a 0.08206 atm dm^3 K^{-1} mol^{-1}

F8.6a 101.325 J

F8.7a (a) kg m s^{-2}, (b) N

Chapter 1

E1.1a 3.37×10^{-19} J

E1.2a (a) 33 zJ, 20 kJ mol^{-1}, (b) 3.013×10^{-34} J, 0.20 nJ mol^{-1}

E1.3a (a) No kinetic energy and zero speed, (b) 0.452 a J, 996 km s^{-1}

E1.4a 2.23 μm

E1.5a 7.35×10^{5} m s^{-1}

E1.6a 6.96 keV, 6.96 keV

E1.8a (a) 3.0×10^{19} s^{-1}, (b) 7.4×10^{20} s^{-1}

E1.9a (a) 3.3×10^{-29} m, (b) 1.3×10^{-36} m, (c) 99.7 pm

E1.10a 7.27×10^{6} m s^{-1}, 150 V

E1.11a 1.71×10^{6} m s^{-1}

E1.15a $N = \left(\dfrac{1}{2\pi}\right)^{1/2}$

E1.16a $(1/2\pi)\,d\phi$

E1.17a $\dfrac{1}{2}$

E1.19a −4

E1.20a (a) ik

E1.23a $\dfrac{L}{2}$

E1.24a 0

E1.27a 52 pm

E1.28a (a) 1.1×10^{-28} m s^{-1}, (b) 1×10^{-27} m

E1.29a (a) 0, (b) 0, (c) $i\hbar$, (d) $2ix\hbar$, (e) $nix^{n-1}\hbar$

Chapter 2

E2.1a 3×10^{34} kg m s^{-1}, 5×10^{-38} J

E2.2a (a) Ae^{ikx}, 5.1×10^{9} m^{-1}, (b) 5.1×10^{11} m^{-1}

E2.3a (a) 1.8×10^{-19} J, 1.1 eV, 9.1×10^{4} cm^{-1}, 1.1×10^{2} kJ mol^{-1}, (b) 6.6×10^{-19} J, 4.1 eV, 3.3×10^{4} cm^{-1}, 4.0×10^{2} kJ mol^{-1}

E2.4a (a) 0.04, (b) 0

E2.5a 0, $\dfrac{h^2}{4L^2}$

E2.6a $\dfrac{\lambda^C}{8^{1/2}}$

E2.7a $x = \dfrac{L}{6}, \dfrac{L}{2}$ and $\dfrac{5L}{6}$

E2.8a 0.8

E2.9a 3.21×10^{-20} J

E2.10a $278\ \mathrm{N\ m^{-1}}$

E2.11a 2.63 μm

E2.12a 3.72 μm

E2.14a $\pm0.525\alpha$ or $\pm1.65\alpha$

E2.16a $565\ \mathrm{cm^{-1}}$

E2.17a 0.056

E2.18a $\langle E_\mathrm{k}\rangle = \dfrac{3\langle V\rangle}{2}$

E2.19a 0

E2.20a $\dfrac{\varepsilon}{2}$

E2.21a $\dfrac{\varepsilon}{2}$

E2.22a 0

E2.23a $\dfrac{-mg^2}{2\omega^2}$

Chapter 3

E3.1a (a) 3.6×10^{-19} J, 2.3 eV, $1.8\times10^{4}\ \mathrm{cm^{-1}}$, $2.2\times10^{2}\ \mathrm{kJ\ mol^{-1}}$, (b) 1.3×10^{-18} J, 8.3 eV, $6.7\times10^{4}\ \mathrm{cm^{-1}}$, $8.0\times10^{2}\ \mathrm{kJ\ mol^{-1}}$

E3.2a (a) $5.5\times10^{14}\ \mathrm{s^{-1}}$, 5.5×10^{-7} m, 550 nm, (b) $2.0\times10^{15}\ \mathrm{s^{-1}}$, 1.5×10^{-7} m, 150 nm

E3.3a (a) 5.5×10^{3} K, (b) $3.\bar{3}\times10^{3}$ K

E3.4a (a) 0.0016, (b) 0

E3.5a (1/4,1/6), (1/4,1/2), (1/4,5/6), (3/4,1/6), (3/4,1/2), (3/4,5/6)

E3.6a $\dfrac{x}{L}=0, \dfrac{1}{2}, 1 \quad \dfrac{y}{L}=0, \dfrac{1}{3}, \dfrac{2}{3}, 1$

E3.7a $0, \dfrac{h^2}{2L^2}$

E3.9a $n_1=1$, $n_2=4$

E3.10a 3

E3.11a 0.23 or 23 per cent

E3.12a 9.77×10^{10}, 1.27×10^{-31} J, 25.8 pm

E3.14a 1.49×10^{-34} J s, 0 or $\pm\hbar$

E3.15a (a) 0.955 radians or 54.7°, (b) π/4 radians or 45°, (c) π/2 radians or 90°

E3.16a 7

E3.17a 2.04×10^{-22} J, 9.75×10^{-4} m = 0.975 mm, microwave

E3.18a $8.89\times10^{10}\ \mathrm{m\ s^{-1}}$

Chapter 4

E4.1a 9.118×10^{-6} cm, 1.216×10^{-5} cm

E4.2a $3.292\times10^{5}\ \mathrm{cm^{-1}}$, 3.038×10^{-6} cm, $9.869\times10^{13}\ \mathrm{s^{-1}}$

E4.5a 14.0 eV

E4.6a 4

E4.7a (a) $g=1$, (b) $g=4$, (c) $g=16$

E4.8a $N=\dfrac{2}{a_0^{3/2}}$

E4.9a $4a_0$, $r=0$

E4.10a 1.058×10^{-10} m

E4.11a $2E_{1s}$, $-E_{1s}$, -4.3594 aJ, 2.1797 aJ

E4.12a $\langle r_{2s}\rangle = \dfrac{6a_0}{Z}$, $r^* = (3+\sqrt{5})\dfrac{a}{2}$

E4.13a $P_{2s}=4\pi r^2\,(1/4\pi)\times(1/243)\times(Z/a_0)^3\times(6-6+\rho+\rho^2)^2\,e^{-\rho}$, $0.74\,a_0/Z$, $4.19\,a_0/Z$, $13.08\,a_0/Z$

E4.14a xy plane, $\theta=\dfrac{\pi}{2}$, zy, $\theta=0$, xz, $\theta=0$

E4.15a (a) forbidden, (b) allowed, (c) allowed

E4.16a 0.999 999 944 × 680 nm

E4.17a (a) 27 ps, (b) 2.5 ps

E4.18a (a) $53\ \mathrm{cm^{-1}}$, (b) $0.53\ \mathrm{cm^{-1}}$

E4.21a
Sc: $[\mathrm{Ar}]4s^23d^1$
Ti: $[\mathrm{Ar}]4s^23d^2$
V: $[\mathrm{Ar}]4s^23d^3$
Cr: $[\mathrm{Ar}]4s^23d^4$ or $[\mathrm{Ar}]4s^13d^5$ (most probable)
Mn: $[\mathrm{Ar}]4s^23d^5$
Fe: $[\mathrm{Ar}]4s^23d^6$
Co: $[\mathrm{Ar}]4s^23d^7$
Ni: $[\mathrm{Ar}]4s^23d^8$
Cu: $[\mathrm{Ar}]4s^23d^9$ or $[\mathrm{Ar}]4s^13d^{10}$ (most probable)
Zn: $[\mathrm{Ar}]4s^23d^{10}$

E4.22a $[\mathrm{Kr}]4d^8$, $S=1, 0$, $m_s=-1, 0, 1$, $m_s=0$

E4.23a (a) 5/2, 3/2, (b) 7/2, 5/2

E4.24a $l=3$ or 2, $l=1$ or 0

E4.25a 3, 2, 1

E4.26a $L=1$, $S=1$, $J=2$

E4.27a $\dfrac{3}{2}, \dfrac{1}{2}$, and $\dfrac{1}{2}$, 4, 2, 2

E4.28a ${}^3D_3$, ${}^3D_2$, ${}^3D_1$, ${}^1D_2$, ${}^3D$ set of terms are the lower in energy

E4.29a (a) $J=1$, 3 states, (b) $J=5/2$, 3/2 respectively, 6 and 2 states respectively, (c) $J=1$, 3 states

E4.30a (a) ${}^2S_{1/2}$, (b) ${}^2D_{5/2}$, and ${}^2D_{3/2}$

E4.31a (a) allowed, (b) forbidden, (c) allowed

Chapter 5

E5.4a $N=3^{-1/2}$, $\psi=3^{-1/2}(\mathrm{s}+2^{1/2}\mathrm{p})$

E5.5a (a) $1\sigma_g^2$, $b=1$, (b) $1\sigma_g^21\sigma_u^2$, $b=0$, (c) $1\sigma_g^21\sigma_u^21\pi_u^4$, $b=2$

E5.6a (a) $1\sigma^22\sigma^21\pi^43\sigma^2$, $b=3$, (b) $1\sigma^22\sigma^23\sigma^21\pi^42\pi^1$, $b=2.5$, (c) $1\sigma^22\sigma^21\pi^43\sigma^2$, $b=3$

E5.7a C_2

E5.8a C_2 and CN

E5.10a O_2^+

E5.11a $2\pi_g$

E5.12a g, u, g, u, if v even, if odd u

E5.13a $N=\left(\dfrac{1}{1+2\lambda S+\lambda^2}\right)^{1/2}$

E5.14a $N=1.12$, $\psi_1=0.163\mathrm{A}+0.947\mathrm{B}$, $b=0.412$, $a=-1.02$, $\psi_2=-1.02\mathrm{A}+0.412\mathrm{B}$

E5.15a $4\times10^{5}\ \mathrm{m\ s^{-1}}$

E5.19a (a) $a_{2u}^2e_{1g}^4e_{2u}^1$, $7\alpha+7\beta$, (b) ${a_{2u}}^2{e_{1g}}^3$, $5\alpha+7\beta$

E5.20a 7β, 0, 7β, 2β

E5.22a (a) $14\alpha+19.31368\beta$, $14\alpha+19.44824\beta$, (b) $12\alpha+16.61894\beta$

Chapter 6

E6.5a $f_1\psi_a(1)=h_1\psi_a(1)+J_a(1)\psi_a(1)=\varepsilon_a\psi_a(1)$

E6.7a $F_{AA}=E_{He}+c_{Hea}^2(\mathrm{HeHe|HeHe})+2c_{Hea}c_{Ha}(\mathrm{HeHe|HeH})+c_{Ha}^2\{2(\mathrm{HeHe|HH})-(\mathrm{HeH|HeH})$

$F_{AB}=\int\chi_{He}(1)h_1\chi_H\mathrm{d}\tau_1+c_{Hea}^2(\mathrm{HeH|HeHe})+c_{Hea}c_{Ha}\{3(\mathrm{HeH|HeH})-(\mathrm{HeHe|HH})\}+c_{Ha}^2(\mathrm{HeH|HH})$

E6.9a (a) 17 basis functions, (b) 28 basis functions, (c) 34 basis functions

E6.13a (a) 40 basis functions, (b) 58 basis functions

E6.14a $$c_{\mathrm{H}a}=\left(\frac{1}{1+\left(\dfrac{\beta}{\alpha_{\mathrm{He}}-\varepsilon_a}\right)^2}\right)^{1/2},\ \varepsilon_a=\frac{\alpha_{\mathrm{He}}+\alpha_{\mathrm{H}}-(\{\alpha_{\mathrm{He}}-\alpha_{\mathrm{H}}\}^2+4\beta^2)^{1/2}}{2},$$

$$c_{\mathrm{H}a}=\left(\frac{1}{1+\left(\dfrac{\beta}{\alpha_{\mathrm{He}}-\varepsilon_a}\right)^2}\right)^{1/2}$$

E6.16a 13

E6.17a $$\Psi_1=(1/2)^{1/2}\begin{vmatrix}\psi_a^{\alpha}(1) & \psi_a^{\beta}(1)\\ \psi_b^{\alpha}(2) & \psi_b^{\beta}(2)\end{vmatrix}$$

E6.18a $$\Psi=C_0\Psi_0+C_1\Psi_1=C_0\begin{vmatrix}\psi_a^{\alpha}(1) & \psi_a^{\beta}(1)\\ \psi_a^{\alpha}(2) & \psi_a^{\beta}(2)\end{vmatrix}+C_1\begin{vmatrix}\psi_a^{\alpha}(1) & \psi_a^{\beta}(1)\\ \psi_b^{\alpha}(2) & \psi_b^{\beta}(2)\end{vmatrix}$$

E6.19a $\int\Psi_2\hat{H}^{(1)}\Psi_0\mathrm{d}\tau=c_{\mathrm{He}a}^2c_{\mathrm{He}b}^2(\mathrm{HeHe}|\mathrm{HeHe})+2(c_{\mathrm{He}a}^2c_{\mathrm{He}b}c_{\mathrm{H}b}+c_{\mathrm{He}a}c_{\mathrm{H}a}c_{\mathrm{He}b}^2)$ $(\mathrm{HeH}|\mathrm{HeHe})+(c_{\mathrm{He}a}^2c_{\mathrm{H}b}^2+2c_{\mathrm{He}a}c_{\mathrm{H}a}c_{\mathrm{He}b}c_{\mathrm{H}b}+c_{\mathrm{H}a}^2c_{\mathrm{He}b}^2)\ (\mathrm{HeH}|\mathrm{HeH})+$ $2c_{\mathrm{He}a}c_{\mathrm{H}a}c_{\mathrm{He}b}c_{\mathrm{H}b}\ (\mathrm{HH}|\mathrm{HeHe})+2(c_{\mathrm{H}a}^2c_{\mathrm{He}b}c_{\mathrm{H}b}+c_{\mathrm{He}a}c_{\mathrm{H}a}c_{\mathrm{H}b}^2)\ (\mathrm{HH}|\mathrm{HHe})$ $+\,c_{\mathrm{H}a}^2c_{\mathrm{H}b}^2(\mathrm{HH}|\mathrm{HH})$

E6.20a (b) $G[f]=\mathrm{d}(x^3)/\mathrm{d}x|_{x=1}=3$, (d) $G[f]=\int_1^3x^3\mathrm{d}x=20$

E6.21a $2\{|\psi_a(\boldsymbol{r})|^2+|\psi_b(\boldsymbol{r})|^2\}$

E6.23a (c) triple-zeta basis set, 6-311+G** basis set

Chapter 7

E7.1a identity E, C_3 axis, three vertical mirror planes σ_v

E7.3a (a) R_3, (b) C_{2v}, (c) D_{3h}, (d) $D_{\infty h}$

E7.4a (a) C_{2v}, (b) $C_{\infty v}$, (c) C_{3v}, (d) D_{2h}

E7.5a (a) C_{2v}, (b) C_{2h}

E7.6a (a) Pyridine, (b) Nitroethane

E7.8a $$D(\sigma_h)=\begin{pmatrix}-1&0&0&0\\0&-1&0&0\\0&0&-1&0\\0&0&0&-1\end{pmatrix}$$

E7.9a $$D(S_3)=\begin{pmatrix}-1&0&0&0\\0&0&-1&0\\0&0&0&-1\\0&-1&0&0\end{pmatrix}$$

E7.11a Integral must be zero

E7.12a Forbidden

E7.14a i, σ_h

E7.15a 3

E7.16a 2

E7.17a no orbits, d_{xy}

E7.19a $2A_1+B_1+E$

E7.20a (a) either E_{1u} or A_{2u}, (b) B_{3u} (x-polarized), B_{2u} (y-polarized), B_{1u} (z-polarized)

Chapter 8

E8.1a ClF_3, O_3, H_2O_2

E8.2a 1.4 D

E8.3a 37 D, $\theta=11.7°$

E8.4a 1.07×10^3 kJ mol^{-1}

E8.5a $$\frac{6l^4q_1^2}{\pi\varepsilon_0r^5}$$

E8.6a 4.9 μD

E8.7a 0.086 J mol^{-1}

E8.8a $28\bar{9}$ kJ mol^{-1}

E8.9a 1.24 bar, 1.24 bar

E8.10a 2.75 bar

E8.11a (a) 24 bar, (b) 22 bar

E8.12a 7.16×10^{-2} kg m^5 s^{-2} mol^{-2}, 2.26×10^{-5} m^3 mol^{-1}

E8.13a (a) 0.88, (b) 1.2 dm^3 mol^{-1}, attractive

E8.14a 140 atm

E8.15a 1.41×10^3 K

E8.16a 175 pm

E8.17a 0.46×10^{-4} m^3 mol^{-1}

E8.18a 0.66

Chapter 9

E9.1a centre of each edge

E9.2a $V=abc\sin\beta$

E9.4a 229 ppm, 397 ppm, 115 ppm

E9.5a 10.1°, 14.4°, 17.7°

E9.6a 1.07°

E9.7a $f_{\mathrm{Br}^-}=36$

E9.8a 4.01 g cm^{-3}

E9.9a 220 pm

E9.10a 8.16°, 4.82°, 11.75°

E9.11a face-centred cubic

E9.12a f

E9.13a $3f$ for $h+k$ even and $-f$ for $h+k$ odd

E9.17a 6.1 km s^{-1}

E9.18a 223 pm

E9.19a 0.9069

E9.20a (a) 0.5236, (b) 0.6902, (c) 0.7405

E9.21a (a) 74.9 pm, (b) 132 pm

E9.22a expansion

E9.23a 3500 kJ mol^{-1}

E9.24a 0.0039

E9.25a 1.7×10^{-3} cm^3

E9.27a 3.54 eV

E9.28a three unpaired spins

E9.29a -6.4×10^{-5} cm^3 mol^{-1}, -6.4×10^{-11} m^3 mol^{-1}

E9.30a 4.326, 5

E9.31a $+1.6\times10^{-8}$ m^3 mol^{-1}

E9.32a 6.0 K

Chapter 10

E10.1a 6.33×10^{-46} kg m^2

E10.3a (a) asymmetric, (b) oblate symmetric, (c) spherical, (d) asymmetric

E10.4a 3.07×10^{11} Hz

E10.5a (a) 2.642×10^{-47} kg m^2, (b) 127.4 pm

E10.6a 4.442×10^{-47} kg m^2, 165.9 pm

E10.7a 106.5 pm, 115.6 pm

E10.8a 20 475 cm^{-1}

E10.9a 198.9 pm

E10.10a (b) HCl, (d) CH_3Cl, (e) CH_2Cl_2, (b) $C_{\infty v}$, (d) C_{3v}, (e) C_{2h}(*trans*), C_{2v}(*cis*)

E10.11a (c) CH_4 is inactive

E10.12a $\frac{5}{3}$

E10.13a $16\ N\ m^{-1}$

E10.14a 1.089 per cent

E10.15a $327.8\ N\ m^{-1}$

E10.16a 967.0, 515.6, 411.8, 314.2

E10.17a $1580.38\ cm^{-1}$, 7.644×10^{-3}

E10.18a $2699.77\ cm^{-1}$

E10.19a (b) HCl, (c) CO_2, (d) H_2O

E10.20a (a) 3, (b) 6, (c) 12

E10.21a 37

E10.22a $G_q(v) = (v + \frac{1}{2})\tilde{v}_q \quad \tilde{v}_q = \frac{1}{2\pi c}\left(\frac{k_q}{m_q}\right)^{1/2}$ [10.46]

E10.23a (a) Nonlinear: all modes both infrared and Raman active, (b) Linear: the symmetric stretch is infrared inactive but Raman active

E10.24a Raman active

E10.25a $4A_1 + A_2 + 2B_1 + 2B_2$

E10.26a A_1, B_1 and B_2 are infrared active; all modes are Raman active.

E10.27a (a) $0.0469\ J\ m^{-3}\ s$, (b) $1.33 \times 10^3\ J\ m^{-3}\ s$, (c) $4.50 \times 10^{-16}\ J\ m^{-3}\ s$

E10.39a (b) 87.64 pm, 89.86 pm, 88.7 pm, $43.87\ cm^{-1}$, $21.93\ cm^{-1}$, $1783.0\ cm^{-1}$, $21.80\ cm^{-1}$, $10.37\ cm^{-1}$

Chapter 11

E11.1a 82.9 per cent

E11.2a $5.34 \times 10^3\ dm^3\ mol^{-1}\ cm^{-1}$

E11.3a $1.09\ mmol\ dm^{-3}$

E11.4a $1.3 \times 10^8\ dm^3\ mol^{-1}\ cm^{-2}$

E11.5a $450\ dm^3\ mol^{-1}\ cm^{-2}$

E11.6a $15\bar{9}\ dm^3\ mol^{-1}\ cm^{-1}$, 23 per cent

E11.7a (a) 0.87 m, (b) 2.9

E11.8a $1\sigma_g^1 1\pi_u^1$

E11.9a 3, u

E11.10a (a) allowed, (b) allowed, (c) forbidden, (d) forbidden, (e) allowed

E11.11a $\frac{2\sqrt{2}}{3} e^{-2ax_0^2/3}$

E11.12a $\frac{1}{32}\left(3 + \frac{4}{\pi}\right)^2$

E11.13a $\frac{1}{2}(\tilde{B}' + \tilde{B}) / |(\tilde{B}' - \tilde{B})|$

E11.14a $\frac{1}{2}(\tilde{B}' + \tilde{B}) / |(\tilde{B}' - \tilde{B})| - 1, 7$

E11.15a $30.4\ cm^{-1} < \tilde{B}' < 40.5\ cm^{-1}$, greater

E11.16a $\Delta_0 = P - \tilde{v}$, $14 \times 10^3\ cm^{-1}$

E11.17a $\frac{3}{8}\left(\frac{a^3}{b - \frac{1}{2}a}\right)^{1/2}$

E11.18a $\frac{1}{4} e^{-1/16} a$

E11.20a 280 nm has the $\pi^* \leftarrow n$ assignment, 189 nm has the $\pi^* \leftarrow \pi$ assignment

E11.22a lower, $\tilde{v} \approx 1800\ cm^{-1}$

E11.24a $\lambda = 60$ cm ($\nu = 500$ MHz)

E11.25a 20 ns, 70 MHz

Chapter 12

E12.1a $s^{-1}\ T^{-1}$

E12.2a 9.133×10^{-35} J s, $\pm 5.273 \times 10^{-35}$ J s, ± 0.9553 rad $= \pm 54.74°$

E12.3a 574 MHz

E12.4a $-1.473 \times 10^{-26}\ J \times m_I$

E12.5a 165 M Hz

E12.6a (a) 3.98×10^{-25} J, (b) 6.11×10^{-26} J, (a)

E12.7a 162.5 T

E12.9a (a) 1×10^{-6}, (b) 5.1×10^{-6}, (c) 3.4×10^{-5}

E12.10a 13

E12.11a 13, (a) independent, (b) 13

E12.12a (a) 11 μT, (b) 110 μT

E12.14a 0.39 ms, $2.6 \times 10^3\ s^{-1}$

E12.16a 753 MHz

E12.18a (a) If there is rapid rotation about the axis, the H nuclei are both chemically and magnetically equivalent, (b) Since $J_{cis} \neq J_{trans}$, the H nuclei are chemically but not magnetically equivalent

E12.19a 5.9×10^{-4} T, 20 μs

E12.20a (a) 2×10^2 T, (b) 10 mT

E12.21a 2.0022

E12.22a 2.3 mT, $2.002\bar{5}$

E12.23a equal intensity, 330.2 mT, 332.2 mT, 332.8 mT, 334.8 mT

E12.25a (a) 332.3 mT, (b) 1.206 T

E12.26a $I = \frac{3}{2}$

Chapter 13

E13.1a 21621600

E13.2a (a) 40320, (b) 5.63×10^3, (c) 3.99×10^4

E13.3a 1

E13.4a 524 K

E13.5a 7.43

E13.6a $35\bar{4}$ K

E13.7a (a) (i) 8.23×10^{-12} m, 8.23 pm, (ii) 2.60×10^{-12} m, 2.60 pm, (b) (i) 1.79×10^{27}, (ii) 5.67×10^{28}

E13.8a 0.3574

E13.9a 72.5

E13.10a (a) 7.97×10^3, (b) 1.12×10^4

E13.11a 18 K

E13.12a 37 K

E13.13a 4.5 K

E13.14a (a) 1, (b) 2, (c) 2, (d) 12, (e) 3

E13.15a 660.6

E13.16a 4500 K

E13.17a 2.571

E13.18a 42.3

E13.19a 4.292, 0.0353 to 0.0377 to 1

E13.20a 8.16×10^{-22} J

E13.21a 18.5 K

E13.22a 25 K

E13.23a 4.5 K

E13.24a 4600 K

E13.25a 10500 K

E13.26a 6500 K

E13.27a 4.033×10^{-21} J

E13.28a $1 + e^{-2\mu_B \beta \mathcal{B}}, \frac{2\mu_B \mathcal{B} e^{-2\mu_B \beta \mathcal{B}}}{1 + e^{-2\mu_B \beta \mathcal{B}}}, -\mu_B \mathcal{B} + \frac{2\mu_B \mathcal{B} e^{-2\mu_B \beta \mathcal{B}}}{1 + e^{-2\mu_B \beta \mathcal{B}}}$ (a) 0.71, (b) 0.996

E13.29a (a) He gas, (b) CO gas, (c) H_2O vapour

Chapter 14

E14.1a (a) $\frac{7}{2}R$, 8.671 kJ mol^{-1}, (b) $3R$, 7.436 kJ mol^{-1}, (c) $7R$, 17.35 kJ mol^{-1}

E14.3a 3.5×10^3 J needed, 5.7×10^2 J needed

E14.4a −75J

E14.5a (a) $\Delta U = \Delta H = 0$, 2.68 kJ, +2.68 kJ, (b) $\Delta U = \Delta H = 0$, −1.62 kJ, +1.62 kJ, (c) $\Delta U = \Delta H = 0$, $w = 0$, 0

E14.6a 30 J K^{-1} mol^{-1}, 22 J K^{-1} mol^{-1}

E14.7a 1.33 atm, +1.25 kJ, $w = 0$, +1.25 kJ

E14.8a 1.07×10^4 J, +10.7 kJ, -0.624×10^3 J, −0.624 kJ, +10.1 kJ

E14.9a +2.2 kJ, +2.2 kJ, +1.6 kJ

E14.10a (a) 0.236, (b) 0.193

E14.11a closer, closer

E14.12a 2.91×10^{-21} J

E14.13a $13\bar{1}$ K

E14.14a $0.0084\bar{6}$ m^3, $25\bar{7}$ K, -0.89×10^3 J

E14.15a −194 J

E14.16a 9.7 kPa

E14.17a 22.5 kJ, −1.6 kJ, 20.9 kJ

E14.18a −4564.7 kJ mol^{-1}

E14.19a +53 kJ mol^{-1}, −33 kJ mol^{-1}

E14.20a −167 kJ mol^{-1}

E14.21a −5152 kJ mol^{-1}, 1.58 kJ K^{-1}, +3.08 K

E14.22a (a) −114.40 kJ mol^{-1}, −111.92 kJ mol^{-1}, (b) −92.31 kJ mol^{-1}, −241.82 kJ mol^{-1}

E14.23a −1368 kJ mol^{-1}

E14.24a (a) +131.29 kJ mol^{-1}, +128.81 kJ mol^{-1}, (b) +134.14 kJ mol^{-1}, +130.17 kJ mol^{-1}

E14.26a −1892 kJ mol^{-1}

E14.27a 0.71 K atm^{-1}

E14.28a 50.6 mbar

E14.29a $+13\bar{1}$ K, $+7.52 \times 10^3$ J mol^{-1}, -7.39×10^3 J mol^{-1}

E14.30a 1.31×10^{-3} K^{-1}

E14.31a 2.0×10^3 atm

E14.32a −7.2 J atm^{-1} mol^{-1}, +8.1 kJ

Chapter 15

E15.1a not spontaneous

E15.2a I_2(g)

E15.3a (a) 126 J K^{-1} mol^{-1}, (b) 169 J K^{-1} mol^{-1}

E15.4a $T = 2.35 \times 10^3$ K

E15.5a 43.1, 22.36 K, 43.76 J K^{-1} mol^{-1}

E15.6a 11.5 J K^{-1} mol^{-1}

E15.7a 3.1 J K^{-1}

E15.8a (a) 34.72 J mol^{-1} K^{-1}, (b) 119.06 J mol^{-1} K^{-1}

E15.9a (a) 12.73 J K^{-1} mol^{-1}, (b) 66.94 J K^{-1} mol^{-1}

E15.10a $T_c = 191.2$ K

E15.11a (a) 366 J K^{-1}, (b) 309 J K^{-1}

E15.12a 30.0 kJ mol^{-1}

E15.13a 152.67 J K^{-1} mol^{-1}

E15.15a 0, 0

E15.16a $\Delta H = 0$, $\Delta H_{tot} = 0$, +2.7 J K^{-1}

E15.17a (a) +2.9 J K^{-1}, −2.9 J K^{-1}, 0, (b) +2.9 J K^{-1}, 0, +2.9 J K^{-1}, (c) 0, 0, 0

E15.18a +87.8 J K^{-1} mol^{-1}, −87.8 J K^{-1} mol^{-1}

E15.19a $\Delta S = 92.2$ J K^{-1}

E15.20a (a) 9.13 J K^{-1} mol^{-1}, (b) 13.4 J K^{-1} mol^{-1}, (c) 14.9 J K^{-1} mol^{-1}

E15.21a (a) −386.1 J K^{-1} mol^{-1}, (b) +92.6 J K^{-1} mol^{-1}, (c) −153.1 J K^{-1} mol^{-1}

E15.22a (a) −521.5 kJ mol^{-1}, (b) +25.8 kJ mol^{-1}, (c) −178.7 kJ mol^{-1}

E15.23a (a) −522.1 kJ mol^{-1}, (b) +25.78 kJ mol^{-1}, (c) −178.6 kJ mol^{-1}

E15.24a −503.05 kJ mol^{-1}

E15.25a −340 kJ mol^{-1}

E15.26a 817.90 kJ mol^{-1}

E15.27a −17 J

E15.28a −36.5 J K^{-1}

E15.29a 10 kJ

E15.30a +10 kJ

E15.31a +11 kJ mol^{-1}

E15.32a −13.8 kJ mol^{-1}, −0.20 kJ mol^{-1}

E15.33a (a) −6.42 kJ mol^{-1}, (b) −14.0 kJ mol^{-1}

Chapter 16

E16.3a $\mu_W(s) = \mu_W(l)$, $\mu_E(s) = \mu_E(l)$

E16.4a −0.38 J mol^{-1}

E16.5a -3×10^2 J mol^{-1}

E16.6a 1.81×10^5 J mol^{-1}

E16.7a 39°C

E16.8a +45 J K^{-1} mol^{-1}, $+1.59 \times 10^4$ J mol^{-1}, +15.9 kJ mol^{-1}

E16.9a +12487 J mol^{-1}, +12.487 kJ mol^{-1}

E16.10a 37°C

E16.11a (a) 1.1×10^3 g, 1.1 kg, (b) 2.0×10^4 g, 20 kg, (c) 1.4 g

E16.12a (a) $2.9\bar{7} \times 10^4$ J mol^{-1}, $29.\bar{7}$ kJ mol^{-1}, (b) 0.169 bar, 0.82 bar

E16.13a −0.35°C, 272.80 K

E16.14a $-1.0\bar{4}$ kJ, +3.5 J K^{-1}

E16.15a -1.37×10^3 J mol^{-1} = −1.37 kJ mol^{-1}, +4.6 J K^{-1} mol^{-1}, $\Delta_{mix}H = 0$

E16.16a (a) $\frac{1}{2}$, (b) 0.4624, 0.5376

E16.17a 6.4×10^3 kPa

E16.18a 0.268, 0.732, 58.6 kPa

E16.19a 2.0×10^2 kPa

E16.20a 85 g mol^{-1}

E16.21a (a) 3.4×10^{-3} mol kg^{-1}, (b) 3.4×10^{-2} mol kg^{-1}

E16.22a 88.2 kg mol^{-1}

E16.23a (a) 18 kg mol^{-1}, (b) 20 kg mol^{-1}, (c) 1.1

E16.24a 1.13, no phase separation

E16.25a 0.833, $0.9\bar{3}$, 0.125, $1.\overline{25}$, $2.\bar{8}$, $1.\overline{25}$

E16.26a 0.498, 0.667, 1.24, 1.11

E16.27a 0.694

E16.28a 1.55

E16.29a (a) 8.75 g, (b) 9.35 g

E16.30a 0.0190, 0.72, 0.85

E16.31a 2.01

Chapter 17

E17.2a $\nu_{Hg_2Cl_2} = -1$, $\nu_{H_2} = -1$, $\nu_{HCl} = 2$, $\nu_{Hg} = 2$ ($\Delta n_g = -1$)

E17.3a +12.3 kJ mol^{-1}

E17.4a −11.20 kJ mol^{-1}

E17.5a (a) −68.26 kJ mol^{-1}, 9.2×10^{11}, (b) −63.99 kJ mol^{-1}, 1.3×10^9, −69.7 kJ mol^{-1}

E17.6a −1108 kJ mol^{-1}

E17.7a (b) 0.33, (c) 0.33, (d) +2.8 kJ mol^{-1}

E17.8a -30 kJ mol^{-1}

E17.9a 3.70×10^{-3}

E17.10a K_x is reduced by 67%

E17.11a 0.045, $15\overline{00}$ K

E17.12a $+3.47$ kJ mol^{-1}, -14.8 J K^{-1} mol^{-1}

E17.13a (a) $+53$ kJ mol^{-1}, (b) -53 kJ mol^{-1}

E17.14a 1110 K

E17.17a (a) $Cd^{2+}(aq) + 2Br^-(aq) + 2\,Ag(s) \rightarrow Cd(s) + 2\,AgBr(s)$, (c) -0.62 V

E17.18a

	ν	$E^{\ominus}_{cell}/V$	$\Delta_r G^{\ominus}_{cell}/kJ\ mol^{-1}$
(a)	2	+1.56	−301
(b)	2	+0.40	−77
(c)	3	−1.10	+318
(d)	2	−0.62	$+12\bar{0}$

E17.19a (a) 6.4×10^{9}, (b) 8.9×10^{43}

E17.20a -1.46 V

E17.21a Elemental mercury cannot spontaneously displace the zinc(II) cation from solution under standard conditions.

E17.22a (a) 9.2×10^{-9} M, (b) 8.5×10^{-17}

E17.23a -52 kJ mol^{-1}, -58 J K^{-1} mol^{-1}, -69 kJ mol^{-1}

Chapter 18

E18.1a (a) 9.975, (b) 1

E18.2a 1904 m s^{-1}, 478 m s^{-1}

E18.3a 6.86×10^{-3}

E18.4a 333 m s^{-1}, 375 m s^{-1}, 530 m s^{-1}

E18.5a (a) 475 m s^{-1}, (b) 184 nm, (c) 2.58×10^{9} s^{-1}

E18.6a 0.276 Pa

E18.7a 9.7×10^{-7} m

E18.8a 397 m s^{-1}, (a) 5.0×10^{10} s^{-1}, (b) 5.0×10^{9} s^{-1}, (c) 5.0×10^{3} s^{-1}

E18.9a 1.2×10^{21}

E18.10a 537 s

E18.11a 16 mg

E18.12a $31\bar{9}$ g mol^{-1}

E18.13a 1.3 days

E18.14a 5.4×10^{-3} J K^{-1} m^{-1} s^{-1}

E18.15a

$D/m^2\ s^{-1}$	$J/mol\ m^{-2}\ s^{-1}$
1.04	43
1.04×10^{-5}	4.3×10^{-4}
1.04×10^{-7}	4.3×10^{-6}

E18.16a -0.055 J m^{-2} s^{-1}

E18.17a 0.0562 nm^2

E18.18a $10\bar{3}$ W

E18.19a 0.142 nm^2

E18.20a (a) 130 μP, (b) 130 μP, (c) 240 μP

E18.21a 16.8 J mol^{-1}

E18.22a 7.63×10^{-3} S m^2 mol^{-1}

E18.23a 283 μm s^{-1}

E18.24a 13.87 mS m^2 mol^{-1}

E18.25a 4.01×10^{-8} m^2 V^{-1} s^{-1}, 5.19×10^{-8} m^2 V^{-1} s^{-1}, 7.62×10^{-8} m^2 V^{-1} s^{-1}

E18.26a 1.05×10^{-9} m^2 s^{-1}

E18.27a 6.2×10^{3} s

E18.28a 420 pm

E18.29a 14.4 ps

E18.30a (a) 0.00 mol dm^{-3}, (b) 0.0121 mol dm^{-3}

E18.31a 381.5 nm

E18.32a 3.30 μm, 0.0308 μm

Chapter 19

E19.1a no change in pressure

E19.2a 8.1 mol dm^{-3} s^{-1}, 2.7 mol dm^{-3} s^{-1}, 2.7 mol dm^{-3} s^{-1}, 5.4 mol dm^{-3} s^{-1}

E19.3a $1.3\bar{5}$ mol dm^{-3} s^{-1}, $4.0\bar{5}$ mol dm^{-3} s^{-1}, 2.7 mol dm^{-3} s^{-1}, $1.3\bar{5}$ mol dm^{-3} s^{-1}

E19.4a dm^3 mol^{-1} s^{-1}, $k_r[A][B]$, $3k_r[A][B]$

E19.5a $\frac{1}{2}k_r[A][B][C]$, dm^6 mol^{-2} s^{-1}

E19.6a (a) $[k_r] = dm^3\ mol^{-1}\ s^{-1}$, $[k_r] = dm^6\ mol^{-2}\ s^{-1}$, (b) $[k_r] = kPa^{-1}\ s^{-1}$, $[k_r] = kPa^{-2}\ s^{-1}$

E19.7a second order

E19.8a $n = 2$

E19.9a 1.03×10^{4} s, (a) 498 Torr, (b) 461 Torr

E19.10a (a) $16.\bar{2}$ dm^3 mol^{-1} h^{-1}, 4.5×10^{-3} dm^3 mol^{-1} s^{-1}, (b) 5.1×10^{3} s, 2.1×10^{3} s

E19.11a (a) 0.098 mol dm^{-3}, (b) 0.050 mol dm^{-3}

E19.12a 1.11×10^{5} s, 1.28 days

E19.13a 4.0×10^{10} dm^{-3} mol s^{-1}, 7.1×10^{5} s^{-1}, 1.28×10^{4} dm^3 mol^{-1} s^{-1}

E19.14a (i) $k_2K^{1/2}[A_2]^{1/2}[B]$, (ii) $\dfrac{k_2^2[B]^2}{4k_1'}\left(\sqrt{1+\dfrac{16k_1'k_1[A_2]}{k_2^2[B]^2}}-1\right)$, $k_2K^{1/2}[A_2]^{1/2}[B]$ or $2k_1[A_2]$ with more approximations

E19.15a 1.9×10^{-6} Pa^{-1} s^{-1}, 1.9 MPa^{-1} s^{-1}

E19.16a 251, 0.996

E19.17a 0.125

E19.18a 3.3×10^{18}

E19.19a 0.52

E19.20a 0.56 mol dm^{-3}

E19.21a 7.1 nm

Chapter 20

E20.1a 79 kJ mol^{-1}, $1.8\bar{0} \times 10^{11}$ mol dm^{-3} s^{-1}

E20.2a 35 kJ mol^{-1}

E20.3a -3 kJ mol^{-1}

E20.4a 1.13×10^{10} s^{-1}, 1.62×10^{35} s^{-1} m^{-3}, 1.7 per cent

E20.5a (a)(i) 1.04×10^{-3}, (ii) 0.069, (b)(i) 1.19×10^{-15}, (ii) 1.57×10^{-6}

E20.6a (a)(i) 22%, (ii) 3%, (b)(i) 170%, (ii) 16%

E20.7a 1.03×10^{-5} m^3 mol^{-1} s^{-1}, 1.03×10^{-2} dm^3 mol^{-1} s^{-1}

E20.8a $4.\bar{5} \times 10^{7}$ m^3 mol^{-1} s^{-1}, $4.\bar{5} \times 10^{10}$ dm^3 mol^{-1} s^{-1}

E20.9a (a) 6.61×10^{6} m^3 mol^{-1} s^{-1}, 6.61×10^{9} dm^3 mol^{-1} s^{-1}, (b) 3.0×10^{7} m^3 mol^{-1} s^{-1}, 3.0×10^{10} dm^3 mol^{-1} s^{-1}

E20.10a 8.0×10^{6} m^3 mol^{-1} s^{-1} = 8.0×10^{9} dm^3 mol^{-1} s^{-1}, 4.2×10^{-8} s

E20.11a 0.79 nm^2, 1.16×10^{-3}

E20.12a 1.81×10^{8} mol dm^{-3} s^{-1}

E20.13a $+69.7$ kJ mol^{-1}, -25 J K^{-1} mol^{-1}

E20.14a $+73.4$ kJ mol^{-1}, $+71.9$ kJ mol^{-1}

E20.15a $-91\ \mathrm{J\,K^{-1}\,mol^{-1}}$
E20.16a $-74\ \mathrm{J\,K^{-1}\,mol^{-1}}$
E20.17a (a) $-46\ \mathrm{J\,K^{-1}\,mol^{-1}}$, (b) $+5.0\ \mathrm{kJ\,mol^{-1}}$, (c) $+18.7\ \mathrm{kJ\,mol^{-1}}$
E20.18a $7.1\ \mathrm{dm^6\,mol^{-2}\,min^{-1}}$
E20.19a 4×10^{-21} J, $2\ \mathrm{kJ\,mol^{-1}}$
E20.20a $12.\bar{5}\ \mathrm{nm^{-1}}$

Chapter 21

E21.1a $\dfrac{k_b K[\mathrm{AH}]^2[\mathrm{B}]}{[\mathrm{BH^+}]}$
E21.2a $1.50\ \mathrm{mmol\,dm^{-3}\,s^{-1}}$
E21.3a $1.18\times10^{5}\ \mathrm{s^{-1}}$, $7.9\times10^{6}\ \mathrm{dm^3\,mol^{-1}\,s^{-1}}$
E21.4a $2.0\times10^{-5}\ \mathrm{mol\,dm^{-3}}$
E21.5a (a)(i) $1.07\times10^{21}\ \mathrm{cm^2\,s^{-1}}$, (ii) $1.4\times10^{14}\ \mathrm{cm^2\,s^{-1}}$, (b)(i) $2.30\times10^{20}\ \mathrm{cm^2\,s^{-1}}$, (ii) $3.1\times10^{13}\ \mathrm{cm^2\,s^{-1}}$
E21.6a 0.13 bar
E21.8a $12\ \mathrm{cm^2}$
E21.9a $33.6\ \mathrm{cm^3}$
E21.10a chemisorption, 50 s
E21.11a 0.83, 0.36
E21.12a (a) 0.24 kPa, (b) 25 kPa
E21.13a 15 kPa
E21.14a $-12.\bar{4}\ \mathrm{kJ\,mol^{-1}}$
E21.15a $65\bar{1}\ \mathrm{kJ\,mol^{-1}}$, (a) 1.6×10^{97} min, (b) 2.8×10^{-6} min
E21.16a $61\bar{1}\ \mathrm{kJ\,mol^{-1}}$
E21.17a (a) 9.1 ps, 0.60 ps, (b) 4.1×10^{6} s, 6.6 μs
E21.18a (a) zeroth-order, (b) first-order

Solutions to odd-numbered problems

Chapter 1

P1.1 (a) 0.020, (b) 0.047, (c) 7×10^{-6}, (d) 0.5, (e) 0.61

P1.3 $a = \pi/2^{1/5}$

P1.5 π

P1.7 origin

P1.9 $i\hbar$

P1.11 (a) $\left(-\dfrac{\hbar^2}{2m_e}\dfrac{d^2}{dx^2} - \dfrac{e^2}{4\pi\varepsilon_0 x}\right)\psi = E\psi$, (b) $\left(-\dfrac{\hbar^2}{2m}\dfrac{d^2}{dx^2}\right)\psi = E\psi$,

(c) $\left(-\dfrac{\hbar^2}{2m}\dfrac{d^2}{dx^2} - cx\right)\psi = E\psi$

P1.17 (a) −1, (b) +1

P1.19 $\dfrac{\hbar^2 k^2}{2m_e}$

P1.21 (i) $3a_0/2$, $9a_0^2/2$, (ii) $5a_0$, $30a_0^2$

P1.27 496 nm, blue-green

Chapter 2

P2.1 1.24×10^{-39} J, 2.2×10^9, 1.8×10^{-30} J

P2.3 HI < HBr < HCl < NO < CO

P2.5 (a) $\frac{1}{2}$, (b) $\begin{cases} \frac{1}{4} & \text{for even } n \\ \frac{1}{4} - \frac{1}{2\pi n} & \text{for } n = 1, 5, 9, \text{etc.} \\ \frac{1}{4} + \frac{1}{2\pi n} & \text{for } n = 3, 7, 11, \text{etc.} \end{cases}$, (c) $\begin{cases} 0 & \text{for even } n \\ \dfrac{4\delta x}{L} & \text{for odd } n \end{cases}$

P2.7 $\frac{1}{2}mgL$

P2.9 (a) $\dfrac{3b}{4\alpha^4}$, (b) $\dfrac{a}{2\alpha^3\pi^{1/2}} + \dfrac{3b}{8\alpha^4}$, (c) $-\dfrac{a}{2\alpha^3\pi^{1/2}} + \dfrac{3b}{8\alpha^4}$

P2.11 $\left(\dfrac{1}{L}\right)^{1/2}$

P2.13 $\left(\dfrac{(e^{\kappa L} - e^{-\kappa L})^2}{16\varepsilon(1-\varepsilon)} + 1\right)^{-1}$, $16\varepsilon(1-\varepsilon)e^{-2\kappa L}$

P2.15 (a) $T = |A_3|^2 = A_3 \times A_3^* = \dfrac{4k_1^2k_2^2}{(a^2+b^2)\sinh^2(k_2L) + b^2}$ where

$a^2 + b^2 = (k_1^2 + k_2^2)(k_2^2 + k_3^2)$ and $b^2 = k_2^2(k_1 + k_3)^2$

P2.17 $\dfrac{1}{2}\left(\dfrac{mk}{\hbar^2}\right)^{1/2}$

P2.19 $0, \dfrac{3}{4}(2v^2 + 2v + 1)\alpha^4$

P2.21 (a) $L\left(\dfrac{1}{12} - \dfrac{1}{2\pi^2 n^2}\right)^{1/2}$, $\dfrac{nh}{2L}\dfrac{nh}{2\sqrt{3}}\left(1 - \dfrac{1}{24\pi^2 n^2}\right)^{1/2}$

(b) $\left\{\left(v + \dfrac{1}{2}\right)\dfrac{\hbar}{\omega m}\right\}^{1/2}$, $\left\{\left(v + \dfrac{1}{2}\right)\hbar\omega m\right\}^{1/2}$, $\left(v + \dfrac{1}{2}\right)\hbar$

P2.23 $\alpha\left(\dfrac{v+1}{2}\right)^{1/2}$, $\alpha\left(\dfrac{v}{2}\right)^{1/2}$

P2.27 (a) 3.30×10^{-19} J, (b) 4.98×10^{-14} s^{-1}, 6.02×10^{-7} m = 602 nm, (c) lower, increases

P2.29 2.68×10^{14} s^{-1}

Chapter 3

P3.1 8.6×10^{-31} J, 1.27×10^{-31} J

P3.3 1.6×10^{-6} m, 1.6 μm

P3.5 6.9×10^{-10} m, 0.69 nm

P3.7 (a) $N = \frac{1}{2}$, (b) $(-\hbar, 1/4)$, $(+\hbar, 3/4)$, (c) $\dfrac{\hbar^2}{2m_p r^2}$, (d) $\dfrac{\hbar}{2}$, $\dfrac{\hbar^2}{2m_p r^2}$

P3.9 1.53×10^{-3} m = 1.53 mm, $\pm\hbar$

P3.11 (a) $-\frac{3}{2}\hbar$, (b) $\dfrac{e^{3i\phi}}{2(2\pi)^{1/2}} + \left(\dfrac{3}{2\pi}\right)^{1/2}\dfrac{e^{-3i\phi}}{2}$, (c) 4.004×10^{-23} J

P3.13 $\theta = 0$ & π, $\theta = \pi/2$, $\theta = 0, \pi/2$, & π, $\theta = \pi/4$ & $3\pi/4$, $\theta = 0, \pi/2$, & π, $\theta = 0, \pi$, 63.4°, & 116.6°, $\theta = 0, \pi/2$, & π, $\theta = 0, \pi$, 54.7°, & 125.3°, $\theta = 0, \pi/2, 8\pi$,

P3.17 $\dfrac{h^2}{8m}\left(\dfrac{n_1^2}{L_1^2} + \dfrac{n_2^2}{L_2^2}\right) = E$

P3.19 (a) no uncertainty in Δl_z, $\dfrac{1}{2^{1/2}}$, (b) $\hbar$, $\dfrac{1}{2}$

P3.23 (a) 0, 0, 0, (b) $E = \dfrac{3\hbar^2}{I}$, $6^{1/2}\hbar$, (c) $E = \dfrac{6\hbar^2}{I}$, $2\sqrt{3}\hbar$

P3.25 1

P3.27 $\dfrac{\hbar}{i}\left(y\dfrac{\partial}{\partial z} - z\dfrac{\partial}{\partial y}\right)$, $\dfrac{\hbar}{i}\left(z\dfrac{\partial}{\partial x} - x\dfrac{\partial}{\partial z}\right)$, $\dfrac{\hbar}{i}\left(x\dfrac{\partial}{\partial y} - y\dfrac{\partial}{\partial x}\right)$, $-\dfrac{\hbar}{i}\hat{l}_z$.

P3.31 $\left(\dfrac{\hbar l}{2}\right)^{1/2} \times \left(\dfrac{1}{mkT}\right)^{1/4}$

P3.33 (c) $\dfrac{n^2h^2}{8ma^2}$

Chapter 4

P4.1 $n_2 \to 6$

P4.3 $R_{Li^{2+}} = 987\,663$ cm^{-1}, 137 175 cm^{-1}, 185 187 cm^{-1}, 122.5 eV

P4.5 $^2P_{1/2}$ and $^2P_{3/2}$, $^2D_{3/2}$ and $^2D_{5/2}$, $^2D_{3/2}$

P4.7 3.3429×10^{-27} kg, 1.000272

P4.9 (a) 0.9 cm^{-1}, (b) small

P4.11 (b) 4.115×10^5 cm^{-1}, 2.430×10^{-6} cm, 1.234×10^{14} s^{-1}, (c) $\dfrac{23}{2}a_0$, $\dfrac{3}{4}a_0$, $\dfrac{43}{4}a_0$

P4.13 (a) 0.323, (b) 141 pm

P4.15 (a) ±106 pm, (b) $r = \pm 1.76a_0/Z$

P4.17 (b) $\rho_{node} = 3 + \sqrt{3}$ and $\rho_{node} = 3 - \sqrt{3}$, $\rho_{node} = 0$ and $\rho_{node} = 4$, $\rho_{node} = 0$, (c) $\langle r\rangle_{3s} = \dfrac{27a_0}{2}$

P4.19 $r' = 2.66a_0$

P4.21 (a) $\dfrac{Z}{a_0}$, (b) $\dfrac{Z}{4a_0}$, (c) $\dfrac{Z}{4a_0}$

P4.23 $\Delta l = \pm 1$ and $\Delta m_l = 0$ or ± 1

P4.27 60957.4 cm^{-1}, 60954.7 cm^{-1}, 329170 cm^{-1}, 329155 cm^{-1}

P4.29 (a) receding, 1.128×10^{-3} c, 3.381×10^5 ms^{-1}

Chapter 5

P5.1 $1.87\times10^{6}\ \mathrm{J\,mol^{-1}} = 1.87\ \mathrm{MJ\,mol^{-1}}$

P5.9 1.9 eV, 130 pm

P5.11 (b) $2.5a_0 = 1.3\times10^{-10}$ m, $-0.555E_h = -15.1$ eV, $-0.565E_h = -15.4$ eV, $0.055E_h = 1.5$ eV, $0.065E_h = 1.8$ eV

P5.13 $\frac{1}{2}\left(\alpha_O + \alpha_C \pm (\alpha_O - \alpha_C)\sqrt{1+\frac{12\beta^2}{(\alpha_O-\alpha_C)^2}}\right)$,

$(\alpha_O - \alpha_C)\left(\sqrt{1+\frac{12\beta^2}{(\alpha_O-\alpha_C)^2}} - \sqrt{1+\frac{4\beta^2}{(\alpha_O-\alpha_C)^2}}\right), \frac{4\beta^2}{(\alpha_O-\alpha_C)}$

P5.15 (a)

Species	N	$\tilde{\nu}$/cm^{-1}	estimated β/eV
C_2H_4	2	61500	−3.813
C_4H_6	4	46080	−4.623
C_6H_8	6	39750	−5.538
C_8H_{10}	8	32900	−5.873

(b) 1.518β, 8.913 eV,

P5.21 all values of the internuclear distance

P5.23 $E_H - \frac{J+K}{1+S} + \frac{j_0}{R}, E_H - \frac{J-K}{1-S} + \frac{j_0}{R}$

P5.25 (a) $\frac{\alpha_A+\alpha_B}{2} \pm \frac{\alpha_A-\alpha_B}{2}\left(1+\frac{4\beta^2}{(\alpha_A-\alpha_B)^2}\right)^{1/2}$,

(b) $\alpha_A + \frac{\beta^2}{\alpha_A-\alpha_B}, \alpha_B - \frac{\beta^2}{\alpha_A-\alpha_B}$

P5.27 $P_N = xP_{N-1} - P_{N-2}$

P5.29 (a) $\alpha-\beta$, $\alpha-\beta$, and $\alpha+2\beta$, (b) $-413\ \mathrm{kJ\,mol^{-1}}$, (c) $-849\ \mathrm{kJ\,mol^{-1}}$, $3(\alpha/2) - 212\ \mathrm{kJ\,mol^{-1}}$, $3\alpha - 425\ \mathrm{kJ\,mol^{-1}}$

Chapter 6

P6.3 (a)

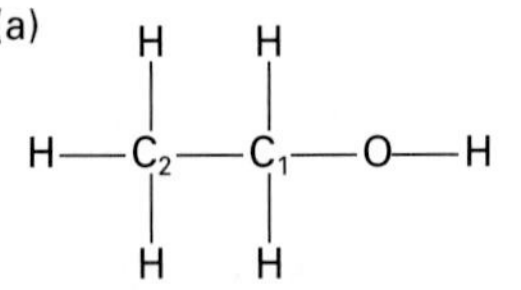

(b)

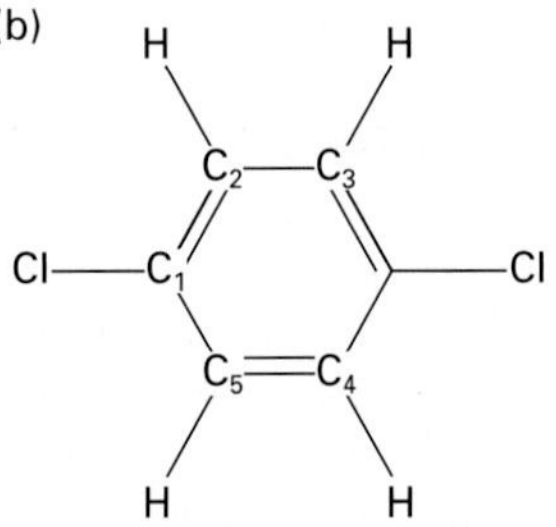

Ethanol	AM1	PM3	exp
$R(C_1C_2)$/pm	151.2	151.8	153.0
$R(C_1O)$/pm	142.0	141.0	142.5
$R(C_1H)$/pm	112.4	110.8	110
$R(C_2H)$/pm	111.6	109.7	109
R(OH)/pm	96.4	94.7	97.1
$\angle C_2C_1O$	107.34	107.81	107.8
$\Delta_f H^{\ominus}$ / kJ mol^{-1}	−262.180	−237.881	−235.10(g)

1,4-dichlorobenzene	AM1	PM3	exp
$R(C_1C_2)$/pm	139.9	139.4	138.8
$R(C_2C_3)$/pm	139.3	138.9	138.8
R(CCl)/pm	169.9	168.5	173.9
$\angle C_6C_1C_2$	120.58	121.09	
$\angle C_1C_2C_3$	119.71	119.450	
$\Delta_f H^{\ominus}$ / kJ mol^{-1}	33.363	42.306	24.6(g)

P6.5 (a) Enthalpies of formation in kJ mol^{-1}

Species	Computed	Experimental	% error
C_2H_4	69.580	52.46694	32.6
C_4H_6	129.834	108.8 ± 0.79	19.3
		111.9 ± 0.96	16.0
C_6H_8	188.523	168.0 ± 3	12.2
C_8H_{10}	246.848	295.9	16.6

P6.7 (b) The very small increase across the series in the experimental chemical shift of the para-^{13}C strongly correlates with decrease in the magnitude of the negative atomic charge across the series.

P6.9 (a), (b), (d), and (e)

P6.11 $\psi_0 = \psi_0^{(0)} + \psi_0^{(1)} = \psi_0^{(0)} + \sum_{n(\neq 0)}{}' \left\{\frac{\int \psi_n^{(0)} H^{(1)} \psi_0^{(0)} \mathrm{d}\tau}{E_0^{(0)} - E_n^0}\right\}\psi_n^{(0)}$

P6.15 (a) $\varepsilon_a = \frac{\alpha_{He} + \alpha_H - (\{\alpha_{He} - \alpha_H\}^2 + 4\beta^2)^{1/2}}{2}$,

$\varepsilon_b = \frac{\alpha_{He} + \alpha_H + (\{\alpha_{He} - \alpha_H\}^2 + 4\beta^2)^{1/2}}{2}$,

(b) $\varepsilon_a = \frac{\alpha_{Li} + \alpha_H - (\{\alpha_{Li} - \alpha_H\}^2 + 4\beta^2)^{1/2}}{2}$,

$\varepsilon_b = \frac{\alpha_{Li} + \alpha_H + (\{\alpha_{Li} - \alpha_H\}^2 + 4\beta^2)^{1/2}}{2}$

P6.19 $H_{22} = 2\int \psi_b(1) h_1 \psi_b(1)\mathrm{d}\tau_1 + \frac{1}{2}\{(AA|AA) - 4(AA|AB) + (AA|BB) + 2(AB|AB)\}$,
$H_{00} = 2E_H + \frac{1}{2}\{(AA|AA) + 4(AA|AB) + (AA|BB) + 2(AB|AB)\}$,
$H_{02} = \frac{1}{2}\{(AA|AA) - (AA|BB)\} = H_{20}$

P6.23 $V_{XC}(r) = \frac{5}{3}C\rho(r)^{2/3}$

P6.25 (a) 3.5, (b) slope = −1.49, intercept = −1.95, (c) 1.12×10^{-2}

Chapter 7

P7.1 (a) D_{3d}, (b) D_{3d}, C_{2v}, (c) D_{2h}, (d) D_3, (e) D_{4d}

P7.3 *trans*-CHCl=CHCl

P7.9 +1 or −1, 1, −1

P7.11 $A_1 + T_2$, s and p, (d_{xy}, d_{yz}, d_{zx})

P7.13 S_4, C_2, S_4

P7.15 (a) $2A_1 + A_2 + 2B_1 + 2B_2$, (b) $A_1 + 3E$, (c) $A_1 + T_1 + T_2$, (d) $A_{2u} + T_{1u} + T_{2u}$

P7.17 $4A_1 + 2B_1 + 3B_2 + A_2$

P7.19 (a) $7A_2 + 7B_1$, $\frac{1}{2}(a - a')$, $\frac{1}{2}(b - b')$, …, $\frac{1}{2}(g - g')$, $\frac{1}{2}(a + a')$, $\frac{1}{2}(b + b')$, …, $\frac{1}{2}(g + g')$

P7.21 there is no orbital of the central atom that forms a nonzero overlap with the p_1 combination so p_1 is nonbonding

P7.23 (a) 1.12×10^{-2}, (b) $11A_u + 11B_g$, $\frac{1}{2}(a + a')$, $\frac{1}{2}(b + b')$, … $\frac{1}{2}(k + k')$ $\frac{1}{2}(a - a')$, $\frac{1}{2}(b - b')$, … $\frac{1}{2}(k - k')$

Chapter 8

P8.1 (a) 1.1×10^8 V m^{-1}, (b) 4×10^9 V m^{-1}, (c) 4 kV m^{-1}

P8.5 2.4 nm

P8.7 196 pm

P8.9 0.123

P8.13 (a) 50.7 atm, (b) 35.1 atm, 0.693

P8.15 0.927, 0.208 dm^3

P8.17 (a) -1.32×10^{-2} dm^3 mol^{-1}, (b) -1.51×10^{-2} dm^3 mol^{-1}, 1.07×10^{-3} dm^6 mol^{-2}

P8.19 (a) α_{xy} must equal zero, (b) $\dfrac{e^2}{k}$

P8.21 $\dfrac{-2\pi N^2 C_6}{3Vd^3}, \dfrac{2\pi N_A^2 C_6}{3d^3}$

P8.29 $\left(\dfrac{2\pi}{3}\right) \times \left(\dfrac{N_A \rho}{M}\right)^2 \times \left(\dfrac{C_6}{d^3}\right)$

P8.31 -1.8×10^{-27} J $= -1.1 \times 10^{-3}$ J mol^{-1}

Chapter 9

P9.1 118 pm

P9.3 face-centred cubic, 408.55 pm, 10.507 g cm^{-3}

P9.5 4.8×10^{-5} K^{-1}, 1.6×10^{-5} K^{-1}

P9.7 834 pm, 606 pm, 870 pm

P9.9 4

P9.11 -146 kJ mol^{-1}

P9.13 0.127×10^{-6} m^3 mol^{-1}, 0.254×10^{-6} m^3 mol^{-1}, 0.423×10^{-6} m^3 mol^{-1}, 0.254 cm^3 mol^{-1}

P9.15 $\dfrac{1}{d^2} = \left(\dfrac{h}{a}\right)^2 + \left(\dfrac{k}{b}\right)^2 + \left(\dfrac{l}{c}\right)^2$

P9.21 0.907

P9.23 (a) no systematic absences, (b) alternation of intensity, whether $h+k+l$ is odd or even, (c) all $h+k+l$ odd lines are missing

P9.25 6.694

P9.27 (a) $\dfrac{2m^2|\varepsilon|}{9V_{\text{molecule}}}$ or $\dfrac{2m^2 N_A|\varepsilon|}{9V_m}$, (b) $\dfrac{1.3359|\varepsilon|}{V_{\text{molecule}}}$ or $\dfrac{1.3359 N_A|\varepsilon|}{V_m}$

P9.31 $\lim_{T\to 0} P(E) = 1$ when $E < \mu$, $\lim_{T\to 0} P(E) = 0$ when $E > \mu$, $(3N/8\pi)^{2/3}(h^2/2m_e)$, 3.1 eV

P9.37 $1 - \left(\dfrac{1}{4(p/K)+1}\right)^{1/2}$, decreases, increases

P9.39 3.61×10^5 g mol^{-1}

Chapter 10

P10.1 596 GHZ, 19.9 cm^{-1}, 0.503 mm, 9.941 cm^{-1}

P10.3 139.6 pm, $108.\bar{5}$ pm

P10.5 128.393 pm, 128.13 pm

P10.7 116.28 pm, 155.97 pm

P10.9 $\tilde{B}_0 = 10.433$ cm^{-1}, $\tilde{B}_1 = 10.126$ cm^{-1}

P10.11 14.35 m^{-1}, 26, 15

P10.13 $0.27877\bar{5}$ cm^{-1}, $0.27690\bar{5}$ cm^{-1}, 602.723 cm^{-1}, 608.392 cm^{-1}, $D_0 = 29\,032$ cm^{-1}

P10.15 $\tilde{\nu} = 2170.8$ cm^{-1}, $x_e\tilde{\nu} = 13.7$ cm^{-1}

P10.17 (a) 5.15 eV, (b) 5.20 eV

P10.19 (a) 152 m^{-1}, 2.72×10^{-4} kg s^{-2}, 2.93×10^{-46} kg m^2, 95.5 m^{-1}, (b) 293 m^{-1}, 0.96

P10.23 (a) 7, (b) C_{2h}, C_{2v}, C_2

P10.25 $m_A R^2 + m_C R'^2 - \dfrac{1}{m}(m_A R - m_C R')^2$

P10.27 $D_{JK} = 4.5 \times 10^2$ kHz, $D_J = 56$ kHz, $B = 25.5360$ GHz

P10.29 $\left(\dfrac{kT}{2hc\tilde{B}}\right)^{1/2} - \dfrac{1}{2}$, 30, $\left(\dfrac{kT}{hc\tilde{B}}\right)^{1/2} - \dfrac{1}{2}$, 6

P10.31 $2\tilde{D}_e/\tilde{\nu} - \frac{1}{2}$

P10.33 $\Delta_0 = \tilde{\nu}_O(J+2) - \tilde{\nu}_S(J-2) = 8\tilde{B}_0(J+\frac{1}{2})$, $\Delta_{1\,t} = \tilde{\nu}_O(J) - \tilde{\nu}_S(J) = 8\tilde{B}_1(J+\frac{1}{2})$

Chapter 11

P11.1 33 μg dm^{-3}

P11.3 $^2\Sigma_g^+ \leftarrow {}^2\Sigma_u^+$ is allowed, transition to $^2\Pi_g$ would be forbidden

P11.7 $\frac{1}{2}\Delta\tilde{\nu}_{1/2}\varepsilon_{\max}\sqrt{\pi/\ln(2)}$, $1.3\bar{0} \times 10^6$ dm^3 mol^{-1} cm^{-2}

P11.11 (a) 1.7×10^{-9} mol dm^{-3}, (b) 6.0×10^2

P11.13 498 kg mol^{-1}, 51.3 nm

P11.15 $A = \varepsilon'[\text{J}]_0(1 - e^{-l/\lambda})$, $A = \varepsilon l[\text{J}] \times (1 - l/2\lambda)$

P11.21 (a) allowed, (b) forbidden

P11.23 (a) $I_0 \times (1 - 10^{-\varepsilon[\text{J}]l})$, (b) $\phi_f I_0(\tilde{\nu})\varepsilon[\text{J}]l$

P11.25 4.4×10^3

P11.27 1.24×10^5 dm^3 mol^{-1} cm^{-2}

P11.29 3.1938 eV, 79.538 cm^{-1}, 2034.3 cm^{-1}, 2113.8 cm^{-1} $= \dfrac{4.1990 \times 10^{-20}\text{ J}}{hc}$, 1321 K, 10, 5.67×10^{28}

Chapter 12

P12.3 400×10^6 Hz $\pm$ 8 Hz, 0.29 s

P12.5 (a) $\Delta\omega_{1/2} = \dfrac{2}{T_2}$, (b) $\Delta\omega_{1/2} = \dfrac{2(\ln 2)^{1/2}}{T_2}$

P12.7 (b) $580 - 79\cos\phi + 395\cos 2\phi$

P12.9 2.8×10^{13} Hz

P12.11 6.9 mT, 2.1 mT

P12.13 0.10, 0.38, (a) 0.48, (b) 0.52, 3.8, 131°

P12.17 (a) Neither set of charges correlates well to the chemical shifts; however some correlation is apparent, particularly for the Mulliken Charges.

P12.19 0

P12.21 $\dfrac{1}{2}\dfrac{A\tau}{1 + (\omega_0 - \omega)^2\tau^2}$

P12.23 $\cos\phi = B/4C$

Chapter 13

P13.1 (a) 1, (b) most probable configurations are {2, 2, 0, 1, 0, 0} and {2, 1, 2, 0, 0, 0}

P13.5 0.030 K, 9.57

P13.7 not at equilibrium

P13.9 (a)(i) 5.000, (ii) 6.2622, (b)(i) 6.55×10^{-11}, (ii) 0.1221

P13.11 1.209, 3.004

P13.13 (a) 1.049, (b) 1.548, (a) 0.953, (b) 0.645, (a) 0.044, (b) 0.230, (a) 0.002, (b) 0.083

P13.15 $\ln q - \dfrac{j\varepsilon}{kT}$, 163 K, $\ln N_j = \ln q - \dfrac{j\varepsilon}{kT}$

P13.17 $\left(\frac{1}{q}\frac{d^2q}{d\beta^2}\right)^{1/2}, \left\{\frac{1}{q}\frac{d^2q}{d\beta^2} - \left(\frac{1}{q}\frac{dq}{d\beta}\right)^2\right\}^{1/2} = \frac{hc\tilde{\nu}e^{-\beta hc\tilde{\nu}/2}}{1-e^{-\beta hc\tilde{\nu}}}, \frac{hc\tilde{\nu}}{2\sinh(\beta hc\tilde{\nu}/2)}$

P13.19 $pV = NkT = nRT$

Chapter 14

P14.1 1.6 m, 0.80 m, 2.8 m

P14.3 alternate form for the heat capacity equation fits the data slightly better

P14.5 0.385, 0.0786, 0.0206

P14.7 4.2 J K^{-1} mol^{-1}, 15 J K^{-1} mol^{-1}

P14.9 28, 258 J mol^{-1} K^{-1}

P14.11 $-nRT\ln\left(\frac{V_2-nb}{V_1-nb}\right) - n^2a\left(\frac{1}{V_2}-\frac{1}{V_1}\right)$, (a) −1.7 kJ, (b) −1.8 kJ, (c) −1.5 kJ

P14.13 −1267 kJ mol^{-1}, +17.7 kJ mol^{-1}, +116.0 kJ mol^{-1}

P14.15 0.903, −73.7 kJ mol^{-1}

P14.17 −994.30 kJ mol^{-1}

P14.21 **1889 kJ mol^{-1}**

P14.23 $\frac{N\Delta\varepsilon^2}{kT^2}$

P14.25 $C_{V,\mathrm{m}} = R\left\{1 + \frac{1}{45}\left(\frac{\theta_\mathrm{R}}{T}\right)^2 + \frac{16}{945}\left(\frac{\theta_\mathrm{R}}{T}\right)^3 + \ldots\right\}$

P14.27 $C_{V,\mathrm{m}} = 3R\left\{\left(\frac{\theta_\mathrm{E}}{T}\right)^2 \frac{e^{\theta_\mathrm{E}/T}}{(e^{\theta_\mathrm{E}/T}-1)^2}\right\}$

P14.29 (a) 87.55 K, 6330 K, (b) and (c) $\alpha = (K/(K+4))^{1/2}$,
$2\alpha C_{V,\mathrm{m}}(\mathrm{H}) + (1-\alpha)C_{V,\mathrm{m}}(\mathrm{H_2})$, $K_{\mathrm{eq}i} = \frac{kT_i(\Lambda_{\mathrm{H_2},i})^3 e^{-(D/RT_i)}}{p^\ominus q_{\mathrm{V}i}q_{\mathrm{R}i}(\Lambda_{\mathrm{H}i})^6}$,
$1.5R,\ 2.5R + \left[\frac{\theta_\mathrm{V}}{T_i} \times \frac{e^{-(\theta_\mathrm{V}/2T_i)}}{1-e^{\theta_\mathrm{V}/T_i}}\right]^2 R$

P14.31 (a) $q^\mathrm{R} = 19.899$, $\dot{q}^\mathrm{R} = 19.558$, and $\ddot{q}^\mathrm{R} = 576.536$,
(b) $q^\mathrm{R} = 3.007$, $\dot{q}^\mathrm{R} = 2.979$, and $\ddot{q}^\mathrm{R} = 118.5$

P14.37 $\left(\frac{\partial C_V}{\partial V}\right)_T = \left(\frac{\partial}{\partial V}\left(\frac{\partial U}{\partial T}\right)_V\right)_T = \left(\frac{\partial}{\partial T}\left(\frac{\partial U}{\partial V}\right)_T\right)_V$,
$\left(\frac{\partial C_p}{\partial p}\right)_T = \left(\frac{\partial}{\partial p}\left(\frac{\partial H}{\partial T}\right)_p\right)_T = \left(\frac{\partial}{\partial T}\left(\frac{\partial H}{\partial p}\right)_T\right)_p$

P14.39 nR

P14.41 (a) $\left(\frac{\partial V}{\partial p}\right)_T dp + \left(\frac{\partial V}{\partial T}\right)_p dT, \left(\frac{\partial p}{\partial V}\right)_T dV + \left(\frac{\partial p}{\partial T}\right)_V dT$,
(b) $-\kappa_T\, dp + \alpha\, dT, \frac{1}{p\kappa_T}\left(\alpha\, dT - \frac{dV}{V}\right)$

P14.45 $\frac{(RV^2)\times(V-nb)}{(RTV^3)-(2na)\times(V-nb)^2}, \frac{V^2(V-nb)^2}{nRTV^3 - 2n^2a(V-nb)^2}$

P14.47 (a) $\frac{aT^2}{C_p}$, (b) $C_V = C_p - R\left(1+\frac{2apT}{R}\right)^2$

P14.49 (a) $\frac{1}{2}k_\mathrm{F}x_\mathrm{f}^2$, (b) (i) (ii) 9.1×10^{-16} N, (iii) (iv) $\frac{kNT}{2}[(1+\nu_\mathrm{f})\ln(1+\nu_\mathrm{f}) + (1-\nu_\mathrm{f})\ln(1-\nu_\mathrm{f})]$, (iv) (v) $kNT\ln 2$

P14.51 (a) −25 kJ, (b) 9.7 m, (c) 39 kJ, (d) 15 m

P14.53 (a) 29.9 K MPa^{-1}, (b) −2.99 K

Chapter 15

P15.3 $10.\bar{7}$ J K^{-1} mol^{-1}

P15.5 (a) 50.7 J K^{-1}, −11.5 J K^{-1}, (b) +3.46 kJ, is indeterminate, (c) 3.46×10^3 J, indeterminate, (d) +39.2 J K^{-1}, −39.2 J K^{-1}

P15.7 Path (a) −2.74 kJ, 0, −2.74 kJ, +9.13 J K^{-1}, 0, −9.13 J K^{-1},
Path (b) −1.66 kJ, 0, +1.66 kJ, +9.13 J K^{-1}, −5.53 J K^{-1}, +3.60 J K^{-1}

P15.9 +45.4 J K^{-1}, +51.2 J K^{-1}

P15.11 −21 K, +35.9 J K^{-1} mol^{-1}

P15.13 (a) 63.88 J K^{-1} mol^{-1}, (b) 66.08 J K^{-1} mol^{-1}

P15.17 46.60 J K^{-1} mol^{-1}, 46.60 J K^{-1} mol^{-1}

P15.19 (a) −7 kJ mol^{-1}, (b) +107 kJ mol^{-1}

P15.21 199.4 J mol^{-1} K^{-1}

P15.23 −21 kJ mol^{-1}

P15.25 7.79 km

P15.29 $\frac{dS}{dt} = -CA\ln(T_\mathrm{i} - T_\mathrm{S})$

P15.31 $\varepsilon = 1 - \left(\frac{V_\mathrm{B}}{V_\mathrm{A}}\right)^{1/c}$, 0.47, +33 J K^{-1}, +33 J K^{-1}, +33 J K^{-1}, +33 J K^{-1}

P15.35 $\left(\frac{\partial V}{\partial S}\right)_p = \left(\frac{\partial T}{\partial p}\right)_S, \left(\frac{\partial S}{\partial V}\right)_T = \left(\frac{\partial p}{\partial T}\right)_V$

P15.37 $C_p dT - \alpha TV\, dp, -\alpha TV\,\Delta p$, −0.50 kJ

P15.39 (a) $nRT\left(\frac{\dot{q}}{q}\right), nR\left\{\frac{\ddot{q}}{q} - \left(\frac{\dot{q}}{q}\right)^2\right\}, nR\left(\frac{\dot{q}}{q} + \ln\frac{eq}{N}\right)$, (b) 5.41 J K^{-1} mol^{-1}

P15.41 $\frac{N\hbar\omega}{e^x-1}, Nk\left(\frac{x}{e^x-1} - \ln(1-e^{-x})\right), NkT\ln(1-e^{-x})$

P15.45 9.57×10^{-15} J K^{-1}

P15.47 (a) +35 J K^{-1} mol^{-1}, (b) 12 W m^{-3}, 1.5×10^4 W m^{-3},
(c) $0.46\,\frac{\text{mol ATP}}{\text{mol glutamine}}$

P15.49 57.2 kJ mol^{-1}, 85.6 kJ mol^{-1}, 112.8 kJ mol^{-1}

Chapter 16

P16.1 (a) $y_\mathrm{M} = 0.36$, (b) $y_\mathrm{M} = 0.81$

P16.7 (a) $+1.84\times10^3$ Pa K^{-1}, (b) 0.7 per cent

P16.9 (a) 322 K, (b) $+7.32\times10^4$ J mol^{-1}, +73.2 kJ mol^{-1}

P16.11 1.25×10^5 g mol^{-1}, 1.23×10^4 dm^3 mol^{-1}

P16.17 (a) $V_1 = V_{\mathrm{m},1} + a_0x_2^2 + a_1(3x_1 - x_2)x_2^2$, $V_2 = V_{\mathrm{m},2} + a_0x_1^2 + a_1(x_1 - 3x_2)x_1^2$, (b) 75.63 cm^3 mol^{-1}, 99.06 cm^3 mol^{-1}

P16.23 363 K (90°C)

P16.27 G^E is reasonably consistent with the model regular solution

P16.29 72.53 cm^3 mol^{-1}

P16.35 56 μg N_2, 14 μg N_2, 1.7×10^2 μg N_2

P16.37 "molten globule" form is not stable

P16.39 (b) $Tb = 112$ K, (c) 8.07 kJ mol^{-1}

P16.41 (a) g cm K^{-1} mol^{-1}, (b) 84 784.0 g cm K^{-1} mol^{-1}, 1.1×10^5 g mol^{-1}, (d) $B' = 21.1$ cm^3 g^{-1}, $C' = 212$ cm^6 g^{-2}, (e) 28 cm^3 g^{-1}, 196 cm^6 g^{-2}

Chapter 17

P17.1 (a) +4.48 kJ mol^{-1}, (b) 0.101 atm

P17.3 0.02054

P17.5 3.89, 2.41

P17.7 (a) 1.24×10^{-9}, (b) 1.29×10^{-8}

P17.9 300 kJ mol^{-1}

P17.11 1.69×10^{-5}
P17.13 0.007 mol H_2, 0.107 mol I_2, 0.786 mol HI
P17.15 76.8 kJ mol^{-1}
P17.17 (a) and (b) perfect gas mixture: 156.5 bar, 81.8 bar, (a) and (b) van der Waals gas mixture: 132.5 bar, 73.7 bar, (c) confirm, $-T\left(\frac{\partial V}{\partial T}\right)_p + V$
P17.19 2.0
P17.21 +0.26843 V
P17.25 $pK_a = 6.736$, $B = 1.997$ $kg^{0.5}$ $mol^{-0.5}$, $k = -0.121$ kg mol^{-1}
P17.31 (a) $K_{Hill}(Mb) = 20.0$, $K_{Hill}(Hb) = 35.0$
P17.37 (b) +0.206 V
P17.39 (a) +1.23 V, (b) +1.11 V
P17.41 (iv) trihydrate

Chapter 18

P18.3 9.1
P18.5 18.9 s
P18.7 10.2 kJ mol^{-1}
P18.9 12.78 mS m^2 mol^{-1}, 2.57 mS m^2 (mol $dm^{-1})^{-3/2}$
P18.11 12.6 mS m^2 mol^{-1}, 7.30 mS m^2 (mol $dm^{-1})^{-3/2}$, (a) 11.96 mS m^2 mol^{-1}, (b) 119.6 mS m^{-1}, (c) 172.5 Ω
P18.13 830 pm
P18.15 (50 kN cm^{-1} mol^{-1})x, (8.2×10^{-23} kN cm^{-1} $molecule^{-1}$)x
P18.17 four, one to two
P18.19 (a) 0.00 mol dm^{-3}, (b) 4.25×10^{-5} mol dm^{-3}
P18.21 $v(\text{most probable}) = c^* = \left(\frac{2kT}{m}\right)^{1/2} = \left(\frac{2RT}{M}\right)^{1/2}$
P18.23 $f(v) = \left(\frac{m}{kT}\right) v e^{-mv^2/2kT}, \left(\frac{\pi kT}{2m}\right)^{1/2}$ or $\left(\frac{\pi RT}{2M}\right)^{1/2}$
P18.25 0.61, 61 per cent, 39 per cent, 0.533, 53 per cent, 47 per cent
P18.27 $\frac{2}{\pi^{1/2}}\left(\frac{n+1}{2}\right)!\left(\frac{2kT}{m}\right)^{n/2}, \frac{(n+1)!!}{2^{n/2}}\left(\frac{2kT}{m}\right)^{n/2}$
P18.29 $\eta = Ae^{E_a/RT}$ where $A = \eta_{ref}e^{-E_a/RT_{ref}}$, 17.5 kJ mol^{-1} at 20°C to 12.3 kJ mol^{-1} at 100°C
P18.33 (a) 0, (b) 0.016, (c) 0.054
P18.37 $N^{1/2}l$
P18.39 (a) 7.2×10^{13} Pa, (b) $\frac{2}{3}\rho_k$, (c) 1.1×10^2 TJ m^{-3}, (d) 2.0×10^{10} Pa, (e) 2.9×10^9 Pa
P18.41 0.25 J cm^{-3}

Chapter 19

P19.1 second order, $k_r = 0.059\bar{4}$ dm^3 mol^{-1} min^{-1}, 2.94 g
P19.3 7.0×10^{-5} s^{-1}, 7.3×10^{-5} dm^3 mol^{-1} s^{-1}
P19.5 55.4%
P19.7 first-order, 1.7×10^{-2} min^{-1}
P19.9 first-order kinetics, 7.2×10^{-4} s^{-1}
P19.11 steady-state approximation
P19.13 (a) $8k_ak_a'[A]_{tot} + (k_a')^2$, (c) $1.\bar{7} \times 10^7$ s^{-1}, $2.\bar{7} \times 10^9$ dm^3 mol^{-1} s^{-1}, $1.\bar{6} \times 10^2$
P19.15 (a) 6.9 ns, (b) $0.10\bar{1}$ ns^{-1}
P19.17 1.98×10^9 dm^3 mol^{-1} s^{-1}
P19.19 $\frac{3^{n-1}-1}{k_r(n-1)[A]_0^{n-1}}$
P19.21 $\left(\frac{1}{3A_0 - 2B_0}\right)\ln\left(\frac{(2x - A_0)B_0}{A_0(3x - B_0)}\right)$
P19.23 steady-state intermediate
P19.25 $\frac{k_1k_2k_3[A]_0}{k_1'k_2' + k_1'k_3 + k_2k_3}$
P19.27 $\left(\langle M^2\rangle_N - \langle M\rangle_N^2\right)^{1/2} = \frac{p^{1/2}M_1}{1-p}$, $M_1\{kt[A]_0(1 + kt[A]_0)\}^{1/2}$
P19.29 $k_r[\cdot M][I]^{-1/2}$
P19.31 $\left(\frac{I_a}{k_r}\right)^{1/2}$
P19.33 (a) first-order, (b) 0.00765 min^{-1} = 0.459 h^{-1}, (c) 1.51 h = 91 min
P19.35 6×10^{-14} mol dm^{-3} s^{-1}, 4.4×10^8 s = 14 yr
P19.37 5.9×10^{-13} mol dm^{-3} s^{-1}

Chapter 20

P20.1 9.70×10^4 J mol^{-1}, 97.0 kJ mol^{-1}
P20.3 +121.1 kJ mol^{-1}, +247.0 J K^{-1} mol^{-1}, 148.3 kJ mol^{-1}
P20.5 6.7×10^{-3}, 4.0×10^{-3} nm^2, 4.0×10^{-21} m^2
P20.7 $E_a = 86.0$ kJ mol^{-1}, 83.9 kJ mol^{-1}, +19.6 J K^{-1} mol^{-1}, +79.0 kJ mol^{-1}
P20.9 0.658 dm^2 mol^{-1} min^{-1}
P20.11 two univalent ions of the same sign
P20.13 +60.44 kJ mol^{-1}, +62.9 kJ mol^{-1}, −181 J K^{-1} mol^{-1}, +114.7 kJ mol^{-1}
P20.15 5.7×10^8 dm^3 mol^{-1} s^{-1}
P20.19 1.4×10^7, 900, 200, 1, 6.3×10^9 dm^3 mol^{-1} s^{-1}, 3.3×10^4 dm^3 mol^{-1} s^{-1}, $P = 2 \times 10^{-7}$
P20.21 $\frac{v^3}{(v^\ddagger)^2}e^{-\Delta E_0/RT}$, (a) 2.7×10^{-15} m^2 s^{-1}, (b) 1.1×10^{-15} m^2 s^{-1}
P20.29 (a) 6.23×10^6 m^3 mol^{-1} s^{-1}, 6.23×10^9 dm^3 mol^{-1} s^{-1}, (b) 4×10^{-10} m, 0.4 nm
P20.31 1.15 eV, 275 K
P20.35 (a) 2.1×10^{-16} mol dm^{-3} s^{-1}, (b) 4.3×10^{11} kg or 430 Tg

Chapter 21

P21.1 2.31 μmol dm^{-3} s^{-1}, 115 s^{-1}, 115 s^{-1}, 1.11 μmol dm^{-3}, 104 dm^3 μmol^{-1} s^{-1}
P21.3 (b) pH = 7.0
P21.5 (a) −0.11, (b) −0.51
P21.7 (a) 1.61×10^{15} cm^{-2}, (b) 1.14×10^{15} cm^{-2}, (c) 1.86×10^{15} cm^{-2}
P21.9 (a) 165, 13.1 cm^3, (b) 263, 12.5 cm^3
P21.11 2.4, 0.16
P21.13 0.138 mg g^{-1}, 0.58
P21.15 5.78 mol kg^{-1}, 7.02 Pa^{-1}
P21.17 (a) $\frac{[P]}{[P]_0} = (b+1)\frac{e^{at}}{1 + be^{at}}$, (c) $-\frac{1}{a}\ln(b)$,

(d) $\left(\frac{y}{1-y}\right) + \left(\frac{1}{1-p}\right)\ln\left(\frac{p+y}{p(1-y)}\right)$, $(A_0 + P_0)^2kt_{max} = \frac{1}{2} - p - \ln 2p$,

(e) $\left(\frac{y}{p(p+y)}\right) + \left(\frac{1}{1+p}\right)\ln\left(\frac{p+y}{p(1-y)}\right)$, $\frac{2-p}{2p} + \ln\frac{2}{p}$
P21.19 $v = \frac{v_{max}}{1 + \frac{1}{K[S]_0}}$ Rate law based on rapid pre-equilibrium approximation, $k_a' \gg k_b$
P21.23 $R = R_{eq}\{1 - e^{-k_rt}\}$ where $k_r = k_{on}a_0$, $R = R_{eq}e^{-k_rt}$ where $k_r = k_{off}$
P21.25 0.0247
P21.27 (a) $k = 0.2289$, $n = 0.6180$, (c) $k = 0.5227$, $n = 0.7273$

Index

(T) denotes a table in the Resource section.

A

B

D

E

F

G

H

I

J

K

N

O

P

Q

R

S

T